# Condensed Matter Physics

# Condensed Matter Physics

**Ajay Kumar Saxena**
*Professor and Head*
Department of Physics
A.P.S. University
Rewa (M.P.)

**I.K. International Pvt. Ltd.**
NEW DELHI

*Published by*
I.K. International Pvt. Ltd.
4435-36/7, Ansari Road, Daryaganj
New Delhi-110 002 (India)
E-mail: info@ikinternational.com
Website: www.ikbooks.com

ISBN: 978-93-90620-86-9



Published by Krishan Makhijani for I.K. International Pvt. Ltd., 4435-36/7, Ansari Road, Daryaganj, New Delhi-110 002.

# Preface

This book presents basic concepts of Condensed Matter Physics to the graduate and post-graduate students. Starting from the preliminary ideas of solid state physics, a systematic approach has been used in the subsequent chapters together with the inclusion of topics such as high temperature superconductivity, lasers and Mössbauer effect, Hartree and Hartree-Fock equations, nanophysics and nanostructures.

I am thankful to my students, Poonam, Indrawati, Anjali, Kanak, Rohit and Snehil for showing keen interest which inspired me in writing the book.

I am thankful to my Ph.D. supervisor (late) Professor O.N. Srivastava, who had been boosting my morale in every book written by me. I am also thankful to my family (Alka, Ankur and Akshat) for providing moral boost during manuscript preparation.

I also owe thanks to the authors and publishers of various books which have been helpful in making guidelines for preparing the manuscript.

Lastly, I thank the faculty at the Physics Department (Ex-HOD, Prof. A.P. Mishra and Dr. C. M. Tiwari and Dr. B.K. Tiwari) A.P.S. University, Rewa for providing moral encouragement.

Hoping to receive a favourable response from the students as well as teachers.

**Ajay Kumar Saxena**

# Contents

CHAPTER

# 1

# Anatomy of Matter and Basic Idea About Materials

## 1.1 THE GRANULAR NATURE OF MATTER

The Greek philosopher Democritus and the Roman poet Lucretius were the first to proclaim that the physical world was made up of empty space and indivisible particles called *atoms*. The first satisfactory atomic theory was advanced in 1808 by the English chemist John Dalton, with the atom still considered indivisible. Toward the end of the $19^{th}$ century, the phenomenon of radioactivity led to the belief that certain atoms were complex in structure and therefore capable of spontaneous sub-division. Still later, energy, too, was shown to be granular or discontinuous in character.

In dealing with gaseous and liquid states of matter, we shall be likely to encounter *molecules* rather than atoms. We can define a molecule as an electrically neutral grouping of two or more atoms of the same or different chemical elements bonded together in a definite proportion. Molecules vary considerably in shape and internal arrangement, as conjectured by the models shown in Fig. 1.1. Individual molecules escape detection even with the powerful *electron microscope* (Fig. 1.2) which magnifies, both visually and photographically, up to 1,000,000 diameters. On the other hand, huge molecular *clusters*, such as those of tobacco mosaic virus, may be readily observed with this instrument (Fig. 1.3). The recently perfected *field ion microscope* accomplishes the seemingly impossible task of revealing atoms themselves (Fig. 1.4). We know that on the basis of physical form, matter can be solid, liquid or a gas.

## 1.2 PROPERTIES OF GASES

We begin with the gaseous state since this is the state of matter best known to physics. Liquids are far more complex in character and we have only recently begun to investigate the solid state.

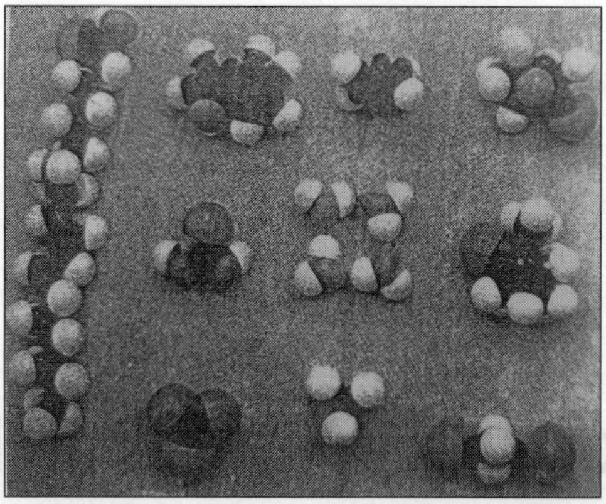

**Figure 1.1** Models of various molecules. *Lapine Scientific Co.*

Gases are aggregates of widely separated molecules, in a state of constant and random motion. *Ideal* gases may be pictured as composed of *dimensionless* or *point* molecules. In *real* gases, molecular diameters vary from about 2 to 4 one hundred-millionths of a centimetre [2 to 4 *angstrom* (Å) units]. The average spacing of gas molecules at standard atmospheric pressure is about 30 Å or 10 times the average molecular diameter. The negligible portion of the total gas volume occupied by the molecules themselves may justify calling gases empty, yet interstellar space is by far emptier. It is interesting to note that even with an ultra high laboratory vacuum, some $10^3$ (1000) gas molecules are left in every cubic centimetre of "evacuated" space.

**Figure 1.2** Electron microscope type JEM-6A; resolution to 1.8 Å; magnifies photographically up to 1,000,000 times. *Fisher Scientific Co.*

Because of large intermolecular spaces, gases may be readily compressed or reduced in volume. If relieved of external pressure, gases expand and spread into whatever space is offered to them. Marked changes in volume are brought about by changes in temperature as well as in pressure.

What evidence can we find for the ceaseless random motion of gas molecules? The *fine* movement of such small particles cannot be seen, but we can observe, even with an optical microscope, the *gross* movement of much larger *colloidal* particles, suspended in liquids or gases and presumably buffeted by the molecules themselves.

The so-called *Brownian movement* was discovered in 1827 by the English botanist Robert Brown, who examined with a conventional microscope a suspension of fine plant spores scattered over water and found the spores darting about in all directions. Figure 1.5 (a) shows a smoke filled type of Brownian movement apparatus and Fig. 1.5 (b) portrays the irregular course pursued by particles of smoke which vary in diameter from 500 to 10,000 Å. Note that the microscope points directly into the smoked filled cell, with no direct light reaching the observer. Actually, the light reaches the eye by scattered reflection from the individual smoke particles, much as it does with motes in a sunbeam.

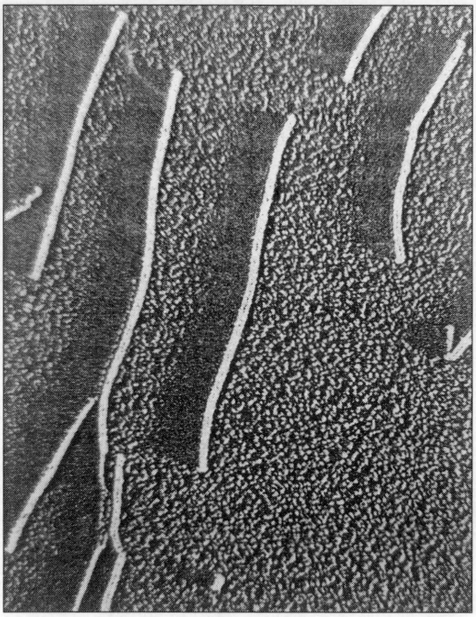

**Figure 1.3** Tobacco mosaic virus enlarged 30,000 times, photographically to 200,000 times. *Fisher Scientific Co.*

The zigzag paths followed by these particles suggest that frequent molecular collisions must

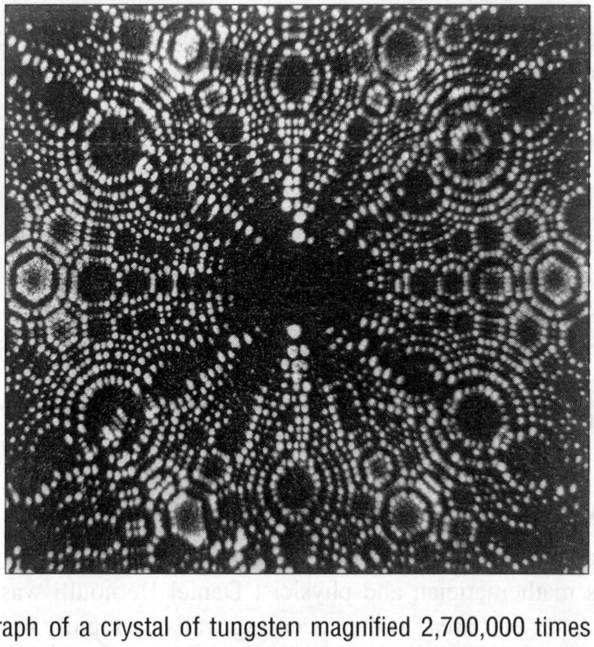

**Figure 1.4** Photograph of a crystal of tungsten magnified 2,700,000 times with the muller field ion microscope. Each luminous spot is an individual atom of matter. *Central Scientific Co.*

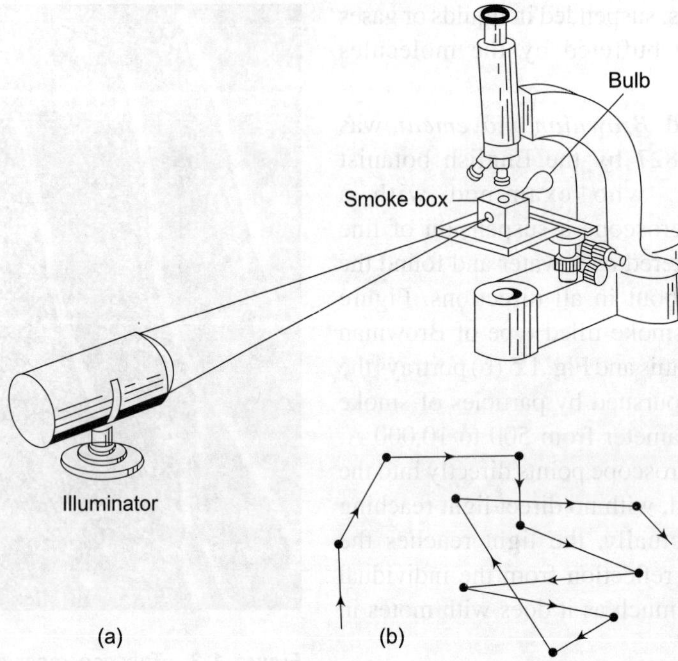


**Figure 1.5** (a) Brownian movement apparatus, and (b) the zig-zag path followed by a typical smoke particle

occur in a gas. When molecules come close to one another, the weak attractive forces between them are superseded by forces of repulsion which arise from the interaction of negatively charged electron "clouds" (Fig. 1.6) surrounding the nuclei of gas atoms. As a result, at a distance of about 3 Å units, gas molecules repel one another and change direction sharply. Calculations show that at 0°C and standard atmospheric pressure, each air molecule travels on an average of approximately 6.5 millionths of a centimetre before striking another molecule. This average length of path between two successive collisions, also known as the *mean free path*, rises to the fantastic figure of nearly one million miles with the highest attainable ultra high vacuum.

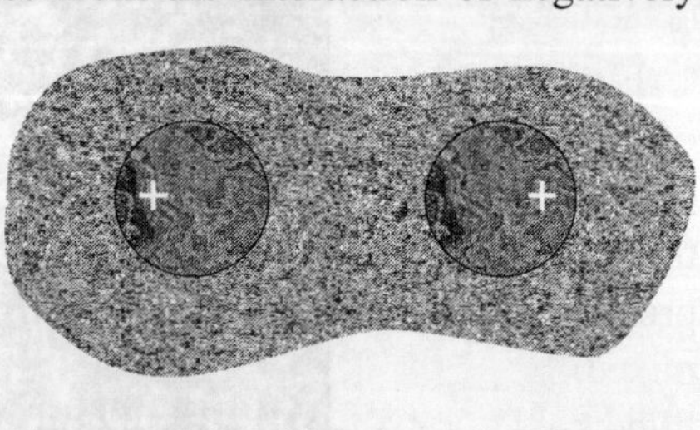


**Figure 1.6** Clouds of electrons surrounding two adjacent hydrogen atoms. Electrons are shared between two positively charged nuclei

## 1.3 KINETIC-MOLECULAR THEORY OF GASES

Early speculations about atoms were based on stationary or *static* configurations of these particles. The Swiss mathematician and physicist Daniel Bernoulli was the first to ascribe gas *pressure* (its push per unit area) to the unceasing walls by rapidly moving gas molecules. The *kinetic* aspect of gas structure was developed further by Mayer, Joule, and Clausius, who by the end of the $19^{th}$ century had erected a sound theoretical framework for explaining the behaviour of gases.

The resulting *kinetic-molecular* theory may be summarized as follows:

1. Gases are composed of very small particles (point-like or dimensionless in ideal gases). These particles, called *molecules*, occupy, under standard conditions, a volume approximately 1000 times as great as that occupied by the same particles in solid or liquid form.
2. Gas molecules move in a random fashion, the average number moving in any one direction being equal to the average number moving in any other direction.
3. Molecules of different gases have different masses but are roughly of the same diameter.
4. Molecules of a gas may be assumed to be perfectly elastic; accordingly they rebound from one another, conserving both momentum (product of mass and velocity) and kinetic energy.
5. The speed of molecular motion increases with a rise in temperature.
6. The impact of gas molecules against the walls of the containing vessel accounts for the pressure exerted by the gas.
7. The kinetic theory provides the basis for Boyle's, Gay-Lussac's, and Charles' laws of gases.

## 1.4 DIFFUSION AND ADSORPTION OF GASES

The random motion of gas molecules causes a gas to *spread* or *diffuse*. (Liquids and solids also diffuse, but more slowly.) We are familiar with the propagation of odours, but do not always realize that the smell of flowers, musk, incense, etc., is made possible only by the detachment and migration of small particles of the odoriferous materials.

Figure 1.7 (a) shows a bottle of oxygen and a bottle of hydrogen, placed mouth-to-mouth, with the bottle of oxygen uppermost. A lighted splint, introduced between the bottles (after a few moments to allow for diffusion), produces a slight explosion, proving that the gases have mixed. In Fig. 1.7 (b), we reverse the positions of the gas filled bottles, and again bring a lighted splint between them. The resulting explosion proves that *regardless of position*, molecules of the gas in one bottle intermingle with those in the other.

The rate of diffusion of a gas through small openings in a porous barrier (a process also called *effusion*), varies inversely with the square root of its density. This is known as *Graham's law*. More simply, it states that light gases escape through small openings more rapidly than heavy gases. (Can you deduce this from the kinetic theory of gases?)

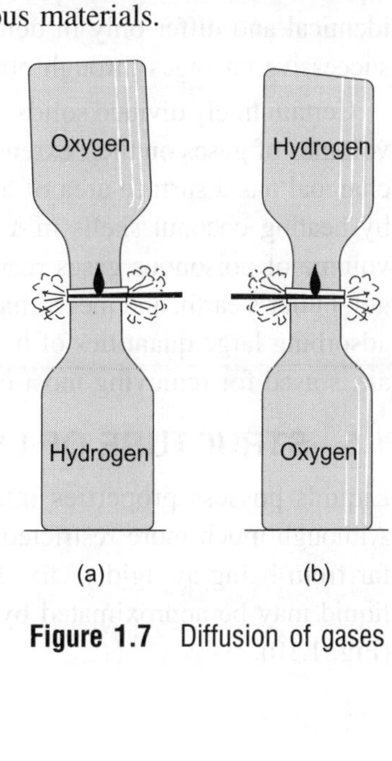


**Figure 1.7** Diffusion of gases

Figure 1.8 represents an unglazed or porous clay jar (A) containing air, over which a beaker of hydrogen (B) has been inverted. The clay jar is provided with a tightly fitting single-hole stopper through which a glass tube (C) is inserted. The tube extends below the surface of water in vessel (D).

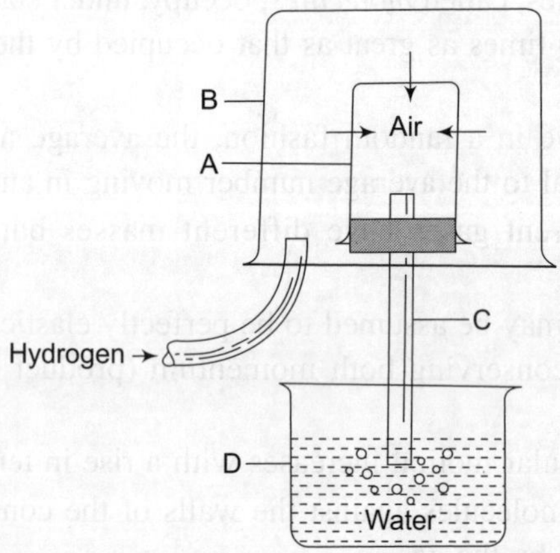


**Figure 1.8** Relative diffusion of hydrogen and air

Hydrogen, which is less dense than air, diffuses *into* the porous cup at a rate greater than that at which air diffuses *outward* into the beaker. As a result, the pressure of the accumulated gas in jar (A) increases, and bubbles are formed in vessel (D). This *differential* or unequal diffusion is utilized in the separation of certain gaseous *isotopes* (uranium hexa-fluorides), a process required for the production of the atomic fuel U-235. These gases are chemically identical and differ only in density. Since the differences in densities are only slight, many successive passages through porous barriers are required to produce an effective separation.

Certain finely divided solids, with a large *surface-to-volume* ratio, *adsorb* or *condense* large volumes of gases on their extensive surfaces. It is estimated that 1 g of thoroughly pulverized charcoal has a surface area of about 10,000,000 $cm^3$. Accordingly, gas mask charcoal, made by heating coconut shells in a limited supply of air, is able to adsorb many times its own volume of poisonous gases reaching the mask. Other excellent adsorbers of gases are silica gel, Fuller's earth, and meerschaum. Specially prepared platinum and palladium are capable of adsorbing large quantities of hydrogen and gaseous hydrocarbons. Charcoal cooled in liquid air is used for removing most of the traces of gas left after conventional exhaustion.

## 1.5 STRUCTURE OF LIQUIDS

Liquids possess properties intermediate in character between those of gases and solids. Although much more restricted in motion than gas molecules, molecules of liquids are still far from being as rigidly "fixed" in position as are the atoms of solids. The structures of a liquid may be approximated by a *ball-and-spring* model (Fig. 1.9) or a *space-filling* model (Fig. 1.10).

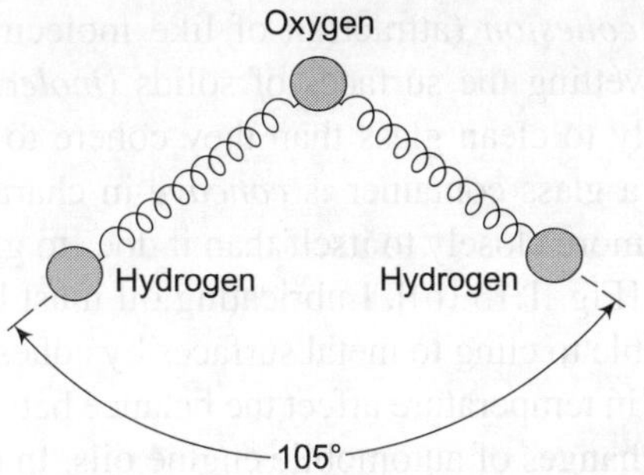


**Figure 1.9** Ball-and-spring model of a water molecule

**Figure 1.10** Space-filling model of four water molecules. *Fisher Scientific Co.*

Liquids exhibit a rough *short-range* order but lack the overall *long-range* pattern characteristic of true solids. In the more modern *tetrahedral* model of a water molecule (Fig. 1.11), two vertices of the pyramid are occupied by hydrogen atoms and the other two by pairs of electrons. Such pyramids may group themselves in a definite short-range pattern, but the existence of a number of unoccupied or vacant lattice sites throughout the full volume of the liquid makes possible the wandering of molecules from one site into another. The resulting "change of partners" accounts for the *mobility* or *diffusibility* of liquids (Fig. 1.12). The average intermolecular distance in liquids is approximately equal to the diameter of the molecules themselves, or about 3.5 Å.

Liquids have no definite shape of their own but exhibit a definite bulk or volume, subject to only a slight reduction under extremely high pressures. The rates of thermal expansion of liquids exceed those of solids but fall below those of gases. Unlike gases, different liquids do not always mix well together.

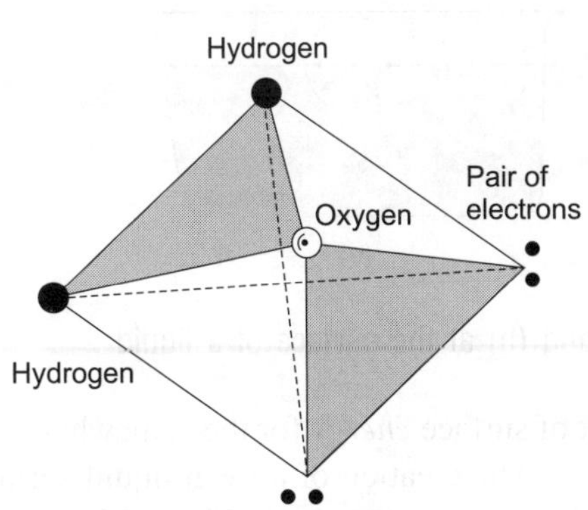


**Figure 1.11** Tetrahedral mode! of a water molecule

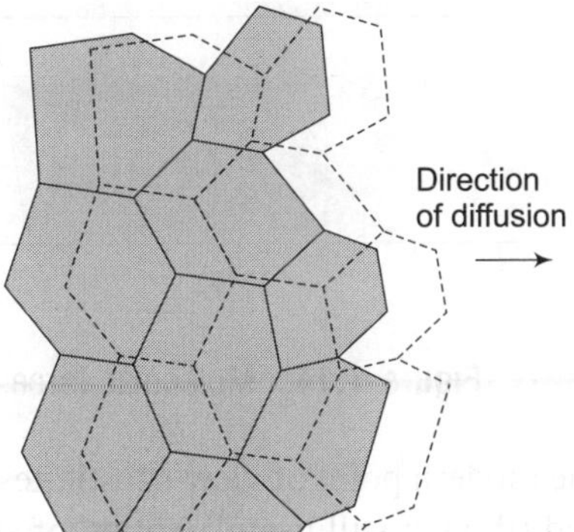


**Figure 1.12** Short range pattern in a liquid. Diagram shows a shift in the pattern and the accompanying diffusion of molecules

Liquids display varying degrees of *molecular cohesion* (attraction of like molecules), as well as different capacities for adhering to or wetting the surfaces of solids (*molecular adhesion*). Molecules of water adhere more strongly to clean glass than they cohere to one another. As a result, the *contact curve* of water in a glass container is *concave* in character [Fig. 1.13 (a)]. Mercury, on the other hand, adheres more closely to itself than it does to glass. Its contact curve in a glass dish is therefore *convex* [Fig. 1.13 (b)]. Lubricating oil must have sufficient "body" to hold together cohesion, yet be able to cling to metal surfaces by adhesion, thus, protecting them from excessive wear. Changes in temperature affect the balance between cohesive and adhesive forces and call for seasonal changes of automobile engine oils. In cold weather, for example, molecular cohesion of oil increases, with an accompanying decrease in its wetting or lubricating ability. As a result, *lighter* oils and greases are prescribed for winter use.

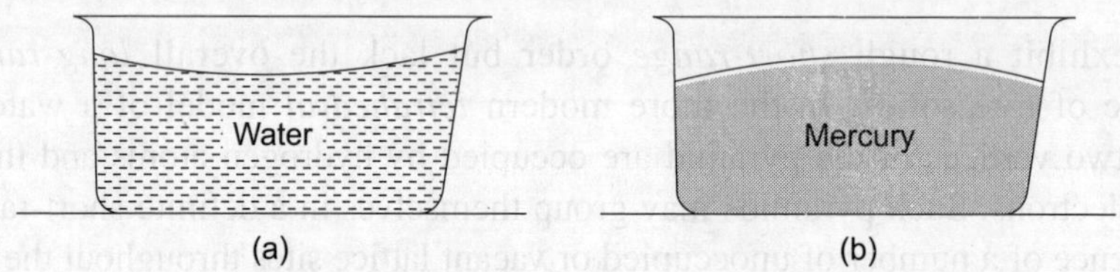


**Figure 1.13** Contact curves of water and mercury in glass dishes (exaggerated effect)

## 1.6 SURFACE TENSION OR SURFACE ENERGY

The free surface of a liquid is perhaps its most interesting feature. Below the surface, molecules are drawn to one another in all directions [Fig. 1.14 (a)]. At the surface, the attraction is only sidewise and inward [Fig. 1.14 (b)]. As a result, the molecules at the surface are pulled together and the entire surface takes on the character of an elastic membrane.

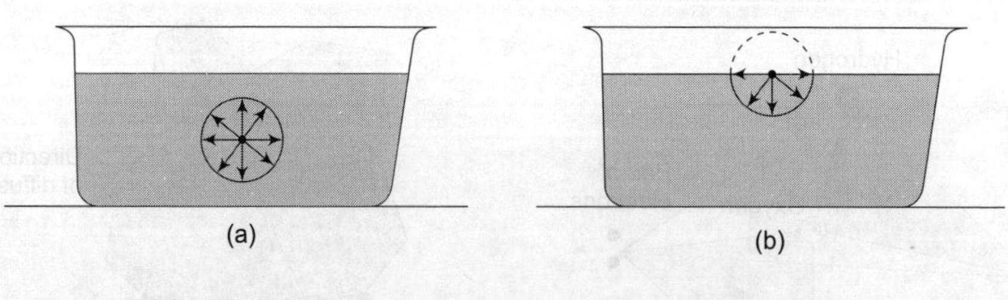


**Figure 1.14** Molecular force: (a) below, and (b) at the surface of a liquid

The modern point of view substitutes the concept of surface *energy* for the somewhat more limited (though still useful) notion of surface *tension*. The creation of a fresh liquid surface, as caused, for example, by the movement to the right of the cross-wire in Fig. 1.15, requires the performance of work and results in the storage of energy in the potential form. This is analogous to the work expended and potential energy accumulated when a sheet of rubber is stretched. Since, in any physical system, the amount of stored or potential energy *tends to a minimum value*, the surface of a liquid, left to itself, tends to *contract* rather than expand.

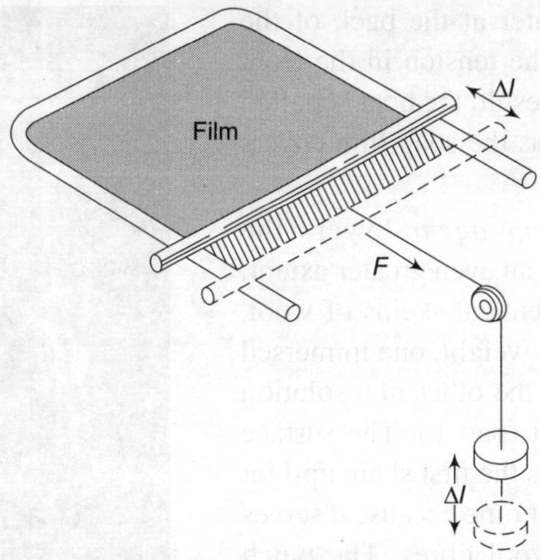


**Figure 1.15** Creation of an extra (cross-hatched) surface calls for an expenditure of work equal to the product of force *F* (weigh passing over pulley), and Δ*l* (distance through which force travels)

As an example of the rubber-like behaviour of surface films, consider drops of mercury deposited on a clean glass plate, or drops of mercury deposited on a clean glass plate, or drops of water scattered over an oiled or greasy sheet of glass. Such drops take on a spherical shape because a spherical surface is the smallest one that can enclose a given volume. Larger drops acquire a flattened appearance because of the effect of gravity. Note the shape of water drops scattered over a water-repellent material (Fig. 1.16). With forces of adhesion diminished, cohesive forces shape each drop, except the largest, into a spherical form.

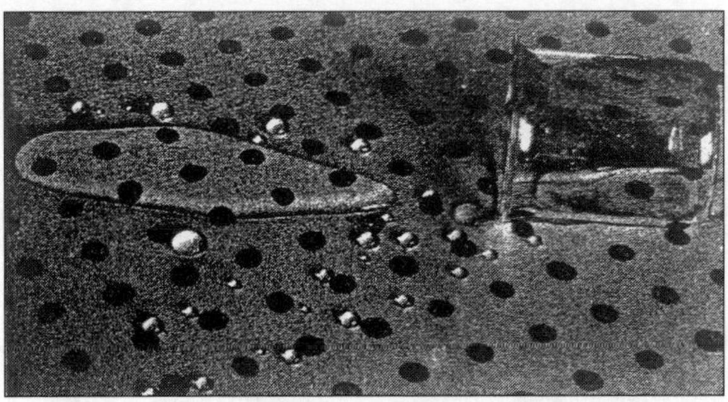

**Figure 1.16** Drops of water scattered over a water-repellent fabric.
*E. I. DuPont de Nemours* & Co., Inc.

Surface tension diminishes with a rise of temperature and is also lowered when certain substances are dissolved in pure liquids. A tiny bit of soap may be fastened to the stern of a small wooden boat which is then placed in a bathtub of water. As the soap dissolves, the surface tension of the water at the back of the boat is weakened, while the tension in the front remains unchanged. As a result, the boat is pulled forward. It stops as soon as the soap film covers the entire water surface.

A *detergent* or *wetting agent* lowers the surface tension of water to an even greater extent. Figure 1.17 shows two identical skeins of wool, each loaded with the same weight, one immersed in pure water (part a) and the other in a solution of a commercial detergent (part b). The surface tension of pure water keeps the first skein upright for a considerable period of time because it serves to trap air between the wool fibres. The much reduced surface tension of the detergent solution permits the escape of the air and the second skein, becoming less buoyant, sinks rapidly.

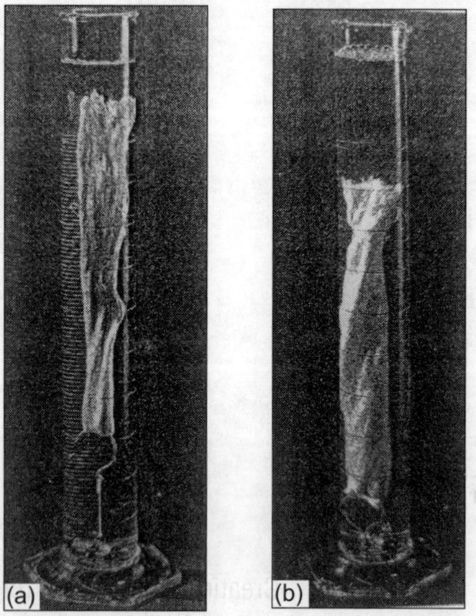


**Figure 1.17** (a) Skein of wool in water. (b) Skein of wool in a detergent solution. *E. I. DuPont de* Nemours & Co., Inc.

## 1.7 VISCOSITY OF LIQUIDS

Viscosity may be considered to be the *internal friction* exhibited by a fluid (liquid or gas). Because of viscosity, resistance is developed to the sliding of one layer of a fluid past another. Liquids are normally more viscous than gases.

Figure 1.18 shows a group of *bubble* type viscosity tubes, or *viscosimeters*. The greater the viscosity of a liquid, the slower the rise of the air bubble through the liquid. More elaborate viscosimeters permit a careful timing of the rate of discharge of a liquid through a narrow opening at the bottom of a metal container. Since viscosity depends on temperature, provision

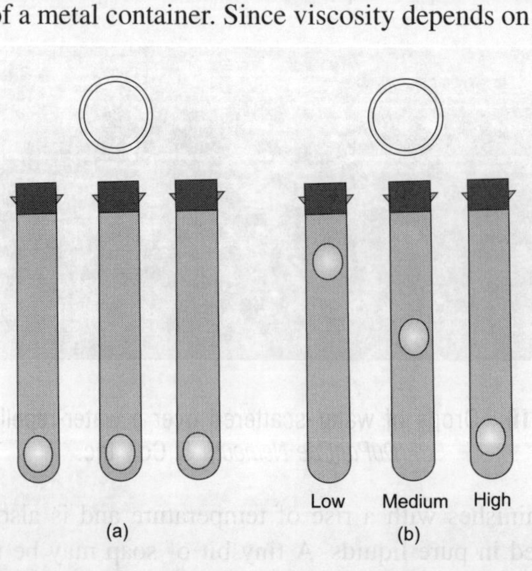


**Figure 1.18** Bubble viscosimeters measuring relative viscosity of three liquids

is made for maintaining the liquid in the container at some fixed prescribed temperature. New lubricants, known as *silicone oils*, show a nearly constant viscosity over a wide range of temperatures.

It is interesting to note in this connection the exceptional behaviour of so-called *helium-II*, a constituent of liquid helium obtained at a temperature of 2.2 degrees kelvin (absolute). Helium-II with a viscosity of less than 0.0001 that of hydrogen gas, flows readily through the narrowest opening [Fig. 1.19 (a)] or climbs up the sides of the containing vessel [Fig. 1.19 (b)], and overflows. This phenomenon, first discovered by Peter Kapitz, is known as *superfluidity*.

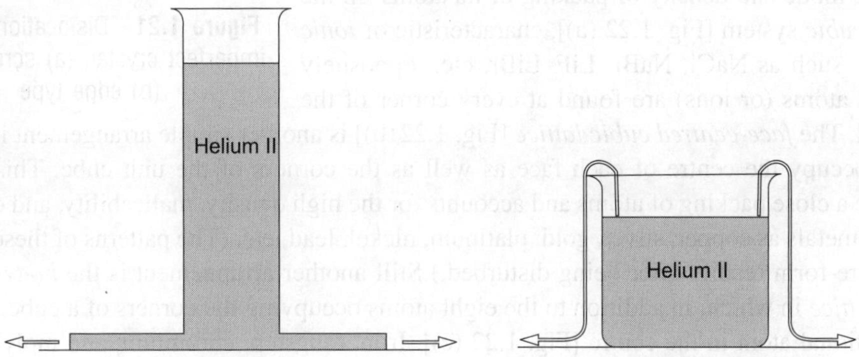


(a) Liquid flows readily through fine slit too narrow to permit ordinary liquid helium (He-I) to escape

(b) Helium-II climbs up the sides of a beaker and overflows

**Figure 1.19** Behaviour of Helium-II – A superfluid

## 1.8 ANATOMY OF SOLIDS

A true solid is marked by a relatively rigid crystalline structure, made up of a pattern of *unit cells* repeating itself at regular intervals throughout the body of a solid (Fig. 1.20). Unlike a gas, in which individual molecules move about in random fashion, or a liquid, in which molecular clusters display only a short-range order, the atoms of a solid are *geometrically oriented* about certain fixed positions which serve as centres for their limited vibrations. In metallic solids, clouds of electrons (so-called "electron gas") swarm between adjacent atoms and help hold together the positively charged metallic ions. Because of the presence of these

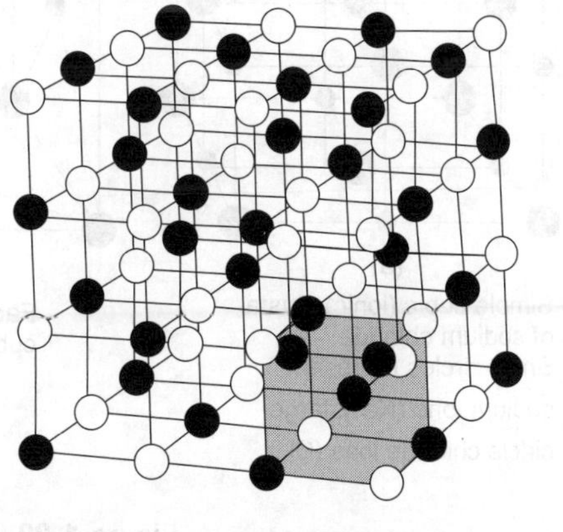

**Figure 1.20** A unit cell in the lattice of rock salt (NaCl). Black balls represent sodium ions, white balls chlorine ions

free electrons, *shared by* neighbouring atoms, metals make good electrical and thermal conductors.

*Single* perfect crystals (metallic "whiskers") show a surprising tensile strength. on the other hand, imperfect crystalline aggregates are easily deformed because of the presence of "dislocations" or "faults" in their overall structure (Fig. 1.21). This accounts for the *plastic flow* or progressive yield of imperfect solids.

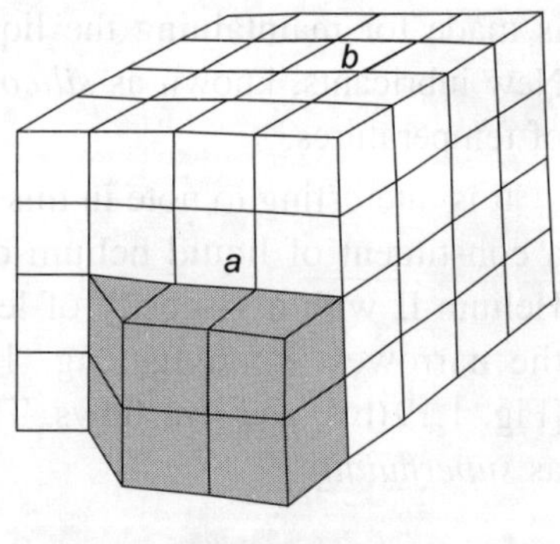


**Figure 1.21** Dislocations in an imperfect crystal: (a) screw type, (b) edge type

The structure of a crystalline solid is intimately connected with the mode and density of packing of its atoms. In the *simple cubic* system [Fig. 1.22 (a)], characteristic of *ionic crystals*, such as NaCl, NaBr, LiF, LiBr, etc., oppositely charged atoms (or ions) are found at every corner of the unit cell. The *face-centred cubic lattice* [Fig. 1.22 (b)] is another simple arrangement in which atoms occupy the centre of each face as well as the corners of the unit cube. This makes possible a close packing of atoms and accounts for the high density, malleability, and ductility of such metals as copper, silver, gold, platinum, nickel, lead, etc. (The patterns of these metals tend to re-form readily after being disturbed.) Still another arrangement is the *body-centred cubic lattice* in which, in addition to the eight atoms occupying the corners of a cube, there is an additional atom in the centre [Fig. 1.22 (c)]. Iron, tungsten, chromium, and molybdenum exhibit this system of crystallization. Such metals are strong and tough.

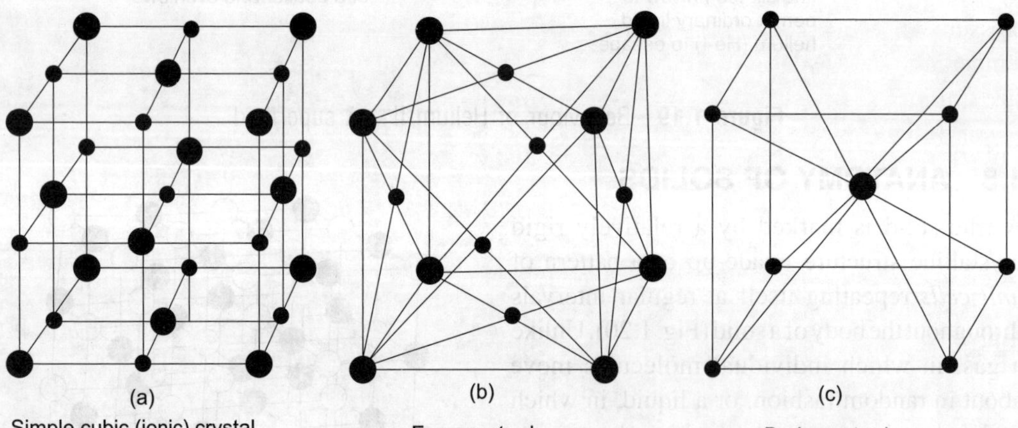

(a)
Simple cubic (ionic) crystal of sodium chloride. Small circles represent sodium ions ($Na^+$), large circircle chlorine ions ($Cl^-$)

(b)
Face-centred cubic lattice

(c)
Body-centred cubic lattice

**Figure 1.22** Typical crystal structures

Models of various crystalline solids are grouped together in Fig. 1.23. The second-from-the-left model is that of sodium chloride, or rock salt.

**Figure 1.23** Models of crystalline solids. *E. Leybold's Nachfolger*

Certain naturally occurring minerals, such as diamond (a form of carbon) exhibit an overall crystal pattern which is subsequently enhanced by expert cutting and polishing. A spatial (three-dimensional) model of a diamond crystal is shown in Fig. 1.24. Large perfect diamonds, such as the famous Cullinan (original weight 3106 carats), are carefully split along their crystal axes to form perfect smaller specimens. Recently, small industrial (as opposed to gem quality) diamonds, have been artificially prepared by subjecting mixtures of carbon and catalyst to a pressure of 1,500,000 pounds per square inch (psi) and a temperature of 5000°F. Figure 1.25 shows a group of carat-size artificial diamonds photographed along a 1" scale.

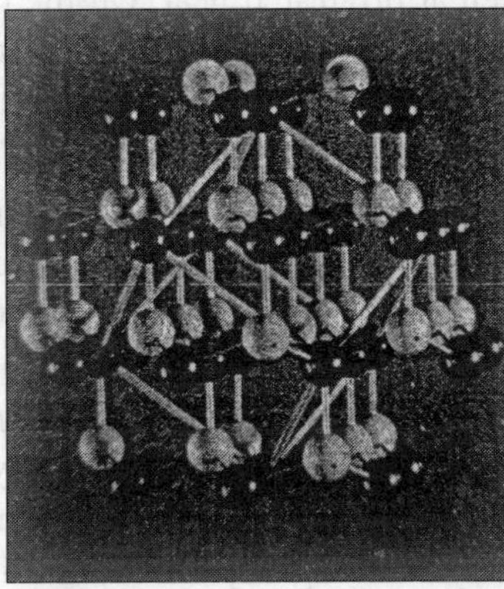

**Figure 1.24** Crystal pattern of diamond. *E. Leybold's Nachfolger*

**Figure 1.25** Group of carat-sized diamonds seen against a 1" scale. These diamonds are not yet of sufficient mechanical strength for industrial use

## 1.9 HARDNESS, MALLEABILITY, AND DUCTILITY

*Hardness,* or resistance to wear or abrasion, is directly related to the stability of crystal structure. Until 1957, diamond was the hardest substance known. It was the only substance capable of scratching or indenting all other substances. In that year, *cubic boron nitride* (borazon) was produced in the General Electric Research Laboratory. Borazon is made at a pressure of approximately 1,000,000 psi and a temperature of 3000°F. The structure of borazon is shown by X-ray diffraction tests to be made up of alternate layers of boron and nitrogen packed together in a cubic pattern almost as closely as the carbon atoms in diamond. The result is that the hitherto "unscratchable" diamond may be scratched by borazon and vice versa.

Next in hardness is *Carboloy*, an alloy of tungsten, tantalum, and titanium carbides, mixed with a cobalt binder. *Cemented carbides* keep their cutting edge 50 times as long as steel, even at bright-red heat. On the other hand, calcite gypsum, and talc are low in the scale of hardness. Graphite, another crystalline form of carbon, is even softer than talc and therefore useful as a lubricant. The question at once arises: "How can the same chemical element (carbon) be both hard and soft?"

*X-ray diffraction*, considered in more advanced texts, reveals that the atoms of diamond are arranged in a close-packed cubic system. Since each carbon atom is tightly bonded to four others, we can visualize the unit cell as a *tetrahedron* (a four-sided pyramid) with one carbon atom at the apex and four others at the corners of the base (Fig. 1.26). The pattern is extremely stable, resisting all attempts to upset it. The reader may recall from earlier study that a group of cannon balls, piled up in the form of a tetrahedral pyramid, constitutes a very stable arrangement, representing the closest possible packing of the individual cannon balls. By analogy, the close packing of carbon atoms in diamond ensures stability and hence makes diamond hard.

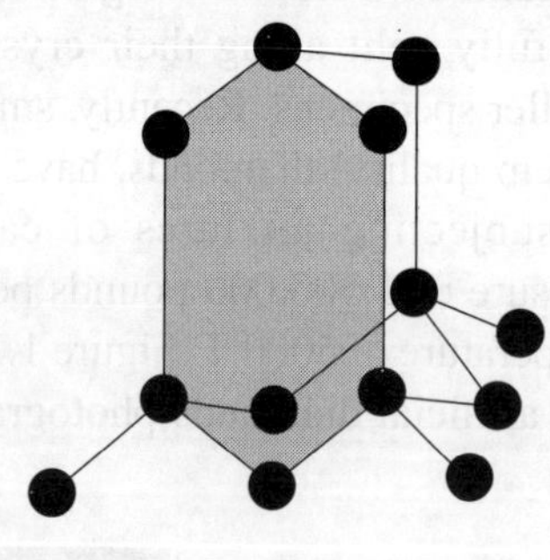

**Figure 1.26** Structure of diamond (Tetrahedral unit cells)

In the case of graphite, the tetrahedral pattern is flattened into a series of parallel *hexagonal* lattices or "layer" lattices approximately 2Å to 3Å apart (Fig. 1.27). The weakening of the bonds permits one layer of graphite to slip easily over another. As a result, graphite is cleavable, flaky, and soft.

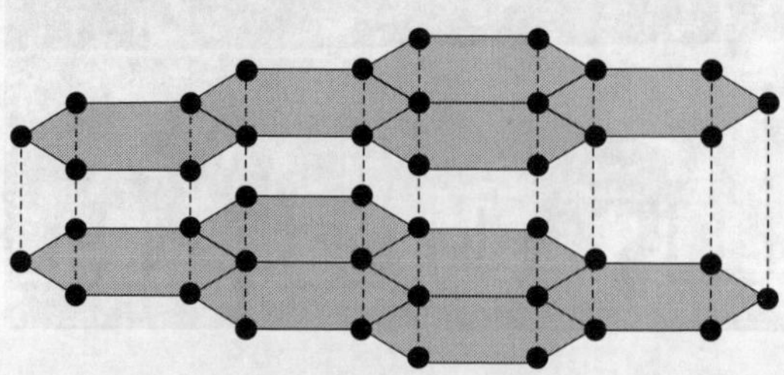

**Figure 1.27** Structure of graphite

Many properties of metals are today "made to order". High-speed tool steels can be made extremely hard by heating them to about 1300°C and then chilling quickly, sometimes in liquid nitrogen. The resulting hardness is, however, accompanied by *brittleness*, or susceptibility to fracture. If the steel is now reheated to about 200°C and then cooled very slowly, there is little loss of hardness but a marked gain in *toughness*, or resistance to impact and shock. Figure 1.28 attests to the success of modern metallurgical research as applied to the production of artillery shells, making them capable of penetrating thick slabs of steel.

**Figure 1.28** The result of metallurgical research applied to artillery shell design. *United States Steel Co.*

It is interesting to note that *glass* — an amorphous solid with some characteristics of the liquid state — may be tempered (heated and quenched slowly), so that it gains considerably in mechanical strength. The thick-walled "Bologna" bottle, produced in this fashion, survives being hammered or pounded against a table, but shatters violently if a small sharp-edged crystal of carborundum is dropped inside (see Fig. 1.29). The unequal stresses, originally "frozen in" and held in check, are suddenly released by this local "irritant" and cause a collapse of the entire structure.

**Figure 1.29** Bologna bottle. *Central Scientific Co.*

Metals may be hammered or rolled into thin sheets. We call this property, *malleability*. Sheets of gold less than 0.000004 inch thick may be obtained by gently hammering thicker sheets of metal placed between pieces of soft leather (gold-beater's skins). Such leaves are thin enough to be transparent. Other metals, such as copper and aluminum, are also malleable, though not to the same extent as gold. Malleability presupposes an actual *flow* of metal in the direction of any of the three axes of the face-centred cubic lattice of the metal.

Metals, as a rule, are *ductile* as well as malleable. Ductility permits plastic deformation in *one* direction under a state of tension, without accompanying rupture. The property of ductility is illustrated in Fig. 1.30 which shows how a rod is drawn through a die to make wire. The tungsten filament of a 5-W, 115-V incandescent lamp, for instance, only 0.0005". in diameter, and scarcely visible to the unaided eye, is produced by first heating sintered tungsten powder, and pounding it down to a small cross section, then its subjection to several successive drawings.

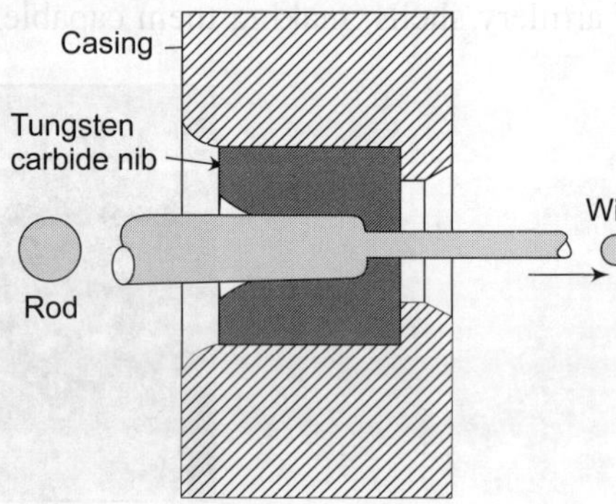


**Figure 1.30** Diagram of wire drawing

## 1.10 ELASTICITY, HOOKE'S LAW, AND THE ELASTIC LIMIT

We have already noted that the application of a force to a material may result in a plastic yield or deformation. When the force is removed, a *perfectly elastic* body regains its original shape, whereas a *perfectly inelastic* body fails to make any recovery. Nearly perfect elasticity may prevail up to a certain point called the elastic *limit*, beyond which full recovery is impossible.

The subject of elasticity is associated with *stress* and *strain*. We define stress as the ratio of a force, $F$ (applied to the ends of a test bar) to the cross sectional area of the bar, $A$ (stress $= F/A$). Strain is the ratio of the change in length ($\Delta l$) of the test bar to its original length, $l$ (strain $= \Delta l/l$). Stress may be expressed in pounds per square inch. Strain is a pure fraction.

Hooke's law, first stated in 1676, declares that, for an elastic body, *strain is directly proportional to stress*. The law is valid only up to the elastic limit (the *yield point*). At a certain point beyond the elastic limit, the specimen under test ruptures completely. Figure 1.31 represents a plastic bar breaking under a stress of 11,000 lb/in.$^2$. The apparatus used to find the breaking point is called a *tensile strength machine*.

## 1.11 METALS, SEMICONDUCTORS AND INSULATORS

Solids, in general, can be classified as metals, semiconductors and we know about metals, insulators (conducting materials such as silver, copper and aluminum, etc.) which are characterized by a large electrical conductivity and small resistivity. *Insulators* or insulating materials (e.g., porcelain, glass, quartz and rubber, etc.) are characterized by poor electrical conductivity and are used as to stop electrical current from flowing where it is not to be passed.

**Figure 1.31** A tensile strength test on a sample of plastic. The test pieces break under a stress of 11,000 ib./in.$^2$. *Standard Oil Co. (N.J.)*

Yet, there is another group of materials (e.g., germanium, silicon, selenium) which are neither good conductors nor good insulators. At room temperature, such materials have conductivities considerably lower than that of conductors and much higher than those of insulators. Such materials are called *semiconductors*. Their resistivities lie in a range from $10^{-4}$ ohm-m to about 0.5 ohm-m. The resistivities of conductors and insulators are ~$10^{-8}$ ohm-m and $10^{12}$ ohm-m respectively.

An advantageous property of semiconductors is that their resistance depends largely on various factors and so, can be controlled as per requirement in their device fabrication unlike conductors (metals), semiconductors have a negative temperature coefficient of resistance i.e., their resistivity decreases with the increase in temperature. As a result, semiconductors behave like insulators at very low temperatures but act as conductors at high temperatures.

The resistivity of semiconductors depends on the magnitudes of the electric field in the semiconductor, consequently, the current in it is not proportional to the voltage, as with the conductors that obey Ohm's law, but increases by far more than the voltage, i.e., semiconductors are *non-linear resistors*.

The semiconductors have another important property – the conductivity can be altered considerably by adding minute amounts of certain other elements called the *impurities*. (This process is known as *doping*. Introduction of certain impurities (see chapter 11) may even change the mechanism of electrical conduction in the semiconductor and certain physical phenomena begin to occur at the junction between sections of the semiconductor having different impurities. This enables us to fabricate important semiconductor devices such as diodes, transistors, SCR's, UJT's, varistors and thermistors.

## 1.12 THE ATOMIC IDEA FOR THE CONDENSED MATTER

Matter is made up from microscopic elementary particles interacting with each other to form composite particles; the latter in turn combine with each other and/or with other elementary particles to create the next level in the hierarchy of the structure of matter.

Thus, two up (*u*) quarks and one down (*d*) quark make up the *proton*, while two (*d*) quarks and one (*u*) quark make up the *neutron*. Protons and neutrons combine to create the various atomic nucleus. A nucleus attracts and traps electrons around it to form a neutral atom [when the number of trapped electrons equals the number (*z*) of protons in the nucleus]. Atoms (and / or ions) come together to form molecules; atoms and / or ions and and / or molecules in huge numbers are more or less close packed to form ordinary solids (and liquids) and other condensed macroscopic structures such as polymers.

The *materials science* refers to that branch of applied science concerned with investigation the relationship existing between the structure of materials and their properties, and it concerns with the inter-disciplinary study of materials for entirely practical purpose.

## 1.13 CLASSES OF MATERIALS

After defining the limits of materials that come within the scope of materials science, they may be classified in three broad groups according to their mode of occurrence:

1. Metals and alloys
2. Ceramics
3. Organic polymers

A *metal* is an elemental substance whilst an *alloy* is formed when two or more relatively pure metals are melted together to form a new metal, having in many instances, properties quite different from those of the two metals used in its manufacture. Among the solid materials, metals and alloys predominate because of their useful characteristics of hardness, strength, rigidity, formability, machinability, weldability, conductivity and dimensional stability.

*Ceramics* are materials consisting of phases. A phase is a physically separable and chemically homogeneous constituent of a material. These are themselves compounds of metallic and nonmetallic elements, and thus there is a multitude which falls into this group. For instance, all metallic compounds, rocks, minerals, glass, glass-fibre, abrasives and all fired clays are ceramics.

*Organic materials* are derived directly from carbon. They usually consist of carbon chemically combined with hydrogen, oxygen or other nonmetallic substances, and their structures are, in many instances, fairly complex. Plastics and synthetic rubbers are common organic engineering materials. Themed "polymers" because they are formed by polymerization reaction in which relatively simple molecules are chemically combined into massive long-chain molecules or "three-dimensional" structures.

The broad spectrum of engineering materials are listed in Table 1.1 which shows not only typical examples from each of these three groups but also gives a number of examples of materials which are composite made up of two groups. In general, in each and every engineering applications we find materials from all the three basic types of materials described above.

**Table 1.1** Some Important Grouping of Materials

| *Group of materials* | *Common examples of engineering use* |
|---|---|
| 1. Metals and alloys | Steels, aluminium, copper, brasses, bronze, invar, conductors, superalloys, etc. |
| 2. Ceramics | MgO, CdS, ZnO, Sic, $BaTio_2$, silica, soda-lime-glass, concrete, cement, ferrited and garnets, etc. |
| 3. Organic polymers | Plastics: PVC, PTEE, polyethylene<br>Fiber: terylene, nylon, cotton.<br>Natural and synthetic rubbers, leather, etc. |
| 4. Composites | |
| (i) Metals and alloys and ceramics: | Steel-reinforced concrete, dispersion-hardened alloys. |
| (ii) Metals and alloys and organic polymers: | Vinyl-coated steel, whisker-reinforced plastics. |
| (iii) Ceramics and organic polymers: | Fibre-reinforced plastics, carbon-reinforced rubber. |

Since the engineer must specify the materials for TV sets, computors, suspension bridges, oil refineries, rocket motors, nuclear reactors, or supersonic transports he must have sufficient knowledge to select the optimum material for each application. Although experience provides the engineer with a starting point for selection of materials, the skill of the engineer will be limited unless he understands the factors which contribute to the properties of materials.

## 1.14 ENGINEERING REQUIREMENTS OF MATERIALS

To know only the types of engineering materials is not all in materials science. What is very important is their use. The practical application of all types of engineering materials depends upon a thorough knowledge of their particular properties under a wide range of conditions. The range of properties found in different classes of materials is very large. Some of the most important properties of materials are:

| | |
|---|---|
| Mechanical | Strength, stiffness, ductility, elasticity, plasticity, toughness, brittleness, hardness, malleability. |
| Electrical | Conductivity, resistivity, dielectric permittivity, dielectric strength. |
| Magnetic | Permeability, coercive force, hysteresis. |
| Thermal | Specific heat, thermal expansion, conductivity. |
| Chemical | Corrosion resistance, acidity or alkalinity, composition. |
| Physical | Dimensions, density, porosity, structure. |
| Acoustical | Sound transmission, sound reflection. |
| Optical | Colour, light transmission, light reflection. |

A more meaningful classification for the purpose of this book is to distinguish between those properties that are determined by the actual crystal structure, the so-called *structure-sensitive* properties, and those that are independent of the crystal structure, namely, the *structure-insensitive* properties.

### *Level of Structure*

The internal structure of a material, simply called the structure, can be studied at various levels of observation. The degree of magnification required to study them by various methods is a measure of the level of observation. The details that are revealed or disclosed at a certain level of magnification are generally different from the details disclosed at some other level of magnification.

When considering structure at various levels it is useful to have some concept of relative sizes, and the so-called "logarithmic scale of the universe" (Table 1.2) provides one method of comparison.

**Table 1.2** Relative Size within the Universe: A Logarithmic Scale

| *Object* | *Average size* | *Logarithmic scale* |
|---|---|---|
| Distance to nearest galaxy | 1 million light year | $10^{21}$ |
| Distance to nearest star | 1 light year | $10^{16}$ |
| Diameter of the sun | 1 million kilometre | $10^{9}$ |
| Diameter of the earth | 10,000 kilometre | $10^{7}$ |
| Man (average dimension) | 1 metre | $10^{0}$ |
| Sand grain | 1 millimetre | $10^{-3}$ |
| Dust particle | 1 micrometre | $10^{-6}$ |
| Molecule | 1 nanometre | $10^{-9}$ |
| Atom | $10^{-1}$ nanometre | $10^{-10}$ |

Depending on the level, the structure of materials can be classified as:

1. Macrostructure
2. Microstructure
3. Substructure
4. Crystal structure
5. Electronic structure
6. Nuclear structure

*Macrostructure* is the structure of a material as seen by the naked eye or low-power magnification. It deals with the size, shape and atomic arrangements in a crystalline material. The individual crystals in a polycrystalline material such as a cast brass doorknob is visible when polished and etched by constant contact with human hand and sweat. Similarly, large, thin zinc crystals can sometimes be seen on the surface of heavily galvanized objects and they are called *macroscopic*. A macrostructure looks very similar to a sand grain and has more or less the same dimension (Table 1.2).

Macrostructure may be studied either directly on the surface of the work (a casting, for example) or on a fracture or, more frequently, on specimens or samples cut out of large billets (ingots, forgings, etc.) or articles. And there are standard procedures available for macro-examination to reveal flaws and segregation in a material.

*Microstructure* generally refers to the structure as observed under a microscope at magnifications from × 75 to × 1500. A microstructure looks very similar to a dust particle. Microscopic examination reveals greater detail of the structure of a material but in almost every instance some sort of special specimen preparation and mounting is necessary.

Optical microscopes are employed to investigate the structure of materials and they enable only those details of the structure to be observed that are larger than 0.15 to 0.2 microns (1,500 to 2,000 Å). Angstrom abbreviated Å is $10^{-10}$ m or $10^{-1}$ nm.

Macrostructure and microstructure are two words relating to the same general type of phenomena and differ only in the degree of magnification required to study them.

*Substructure* refers to structure obtained by using a microscope with a much higher magnification than optical microscope. And as such a wealth of new information on crystal imperfections such as dislocation and on very small particles of a few hundred angstroms in size is obtained and studied.

At the present time, the electron microscope is used to study the substructure of the material. An effective magnification of × 1,00,000 may be obtained. Details of the object may be distinguished which are from 30 to 50 Å in size. The field-ion microscope is now a very powerful and upcoming technique and it produces images of individual atom.

*Crystal structure* tells us the details of the atomic arrangement within a crystal, and it resembles a molecule.

## 1.15 VARIOUS KINDS OF CONDENSED MATTER

### 1.15.1 Monocrystalline and Polycrystalline Atomic Solids

This category includes elemental metals (such as copper), elemental semiconductors (such as silicon), elemental insulators (such as diamond), etc. A perfect periodic crystalline arrangement of the atoms is an idealization: there is always some kind of defects. Crystalline solids in the absence of two-dimensional defects along internal surfaces are called *monocrystals*. The preparation of monocrystals of any solid requires special sophisticated techniques and extra care. Most crystalline materials are *polycrystalline*, i.e., they consist of crystallites of various orientations and mesoscopic or macroscopic sizes coexisting in the solid; the interfaces of the crystallites constitute two-dimensional defects. Linear defects, the so-called *dislocations*, are always present and play a crucial role in determining the ductile vs. brittle behaviour exhibited by different solids, in influencing the crystal growth, etc. *Point defects*, such as *vacancies, interstitials, foreign atoms*, are also always present and are very important in determining the electrical and other properties of solids especially of semiconductors.

### 1.15.2 Atomic or Ionic Compounds and Alloys

These are usually polycrystalline solids such as ionic insulators (e.g., NaCl), composite semiconductors (e.g., GaAs), oxides (e.g., $SiO_2$, TiO, FeO, $V_2O_3$), perovskites (e.g., $SrTiO_3$, $Na_xWO_3$, $KFeF_3$), high temperature superconductors (e.g., $YBa_2Cu_3O_{7-x}$, $HgBa_2$ $Ca_2$ $Cu_3$ $O_8$), alloys (e.g., $Cu_xAu_{1-x}$, $Si_x$ $Ge_{1-x}$, SiC), etc. In certain alloys, e.g., $Al_{86}Mn_{14}$ fast cooled from the melt, the atoms arrange themselves in an ordered highly symmetric but *not periodic* manner, which usually consists of a nonperiodic assembly of clusters of 13 atoms forming regular icosahedra.

### 1.15.3 Molecular Solids

These solids are made up of molecules that retain their identity to a large degree in spite of the formation of solid. This is so because the binding of the atoms within each molecule is much stronger than the binding among neighbouring molecules. Typical examples are solids made up of nitrogen ($N_2$) or oxygen ($O_2$), or hydrogen ($H_2$) or $CO_2$ (dry ice), or $C_{60}$ (fulleride); these molecules are held together by weak van der Waals forces to form solids.

Water molecules in ice employ the relatively stronger hydrogen bond to keep themselves together in more or less fixed positions.

### 1.15.4 Glasses

Several materials, when they are solidified as their melt is cooled under usual conditions, do not form polycrystalline structures; instead, the atoms arrange themselves in a disordered manner. Order is present usually *only locally* (i.e., at the scale of a few angstroms), while for larger distances their relative positions are random. In other words, there is *no positional long-range order* as in crystalline materials; only *short-range order* may exist and even this is not perfect. Solids that exhibit this type of positional disorder are called *glasses*. Common examples of glasses are the ordinary window glasses (consisting of 70% $SiO_2$, 20% $Na_2O$, and 10% CaO), the chalcogenide glasses of the elements S, Se, and Te, such as $As_2S_3$, $As_2Se_3$, etc. Among the elements, S, P, and Se solidify to the glassy state. Very fast cooling from the melt drives many materials (e.g., transition metals) to a glassy state instead of their normal crystalline state, because the atoms have no time to place themselves in a lattice; instead, they are frozen more or less in the positions they had in the liquid phase. Molecular solids (e.g., polymers) consisting of medium- and/or large-size organic molecules form usually glasses. The transition from the liquid phase to the glassy state is mainly characterized by a huge change in the viscosity over a very narrow temperature range centred at the so-called *glass transition temperature*, $T_g$. Thus, glasses can also be characterized as "frozen liquids". However, neither, this characterization nor the one classifying them simply as "amorphous solids" would be sufficient. These two seemingly self-contradictory definitions are brought into agreement by noting that one is dealing with an intrinsically "non-ergodic" system. This tern simply means that the system does not pass through all possible states with equal probability according to the rules of thermodynamic equilibrium. In practice this term implies that on the sufficiently short time scales of usual experiments the system can be considered a solid, but on an extremely long time scale, e.g., that of decades or centuries, it may behave as a liquid.

### 1.15.5 Polymers

These are very important molecular materials both scientifically and technologically. They exhibit most of the characteristic properties of the so-called soft matter. Their structural length scales are intermediate between atomic and macroscopic; their time scales, $\tau$, may cover many orders of magnitude and depend strongly on the temperature; their interaction energies are comparable to $k_BT$ so that Brownian motion is important and the entropic term, $TS$, in competition with the energy (more correctly the enthalpy) determines the various equilibrium phases including self-assembled structures.

Polymeric materials are usually made of long linear molecular chains consisting of elementary molecular units called monomers. A polymer chain can be made by repeating the same monomer or by repeating several different monomers. Besides the linear polymers, other forms such as branched polymers or star polymers also exist. The so-called copolymers consisting of two (or more) different linear polymers joined together by a strong homopolar bond find many applications, including the possibility of forming self-assembled periodic structures.

In polymeric melts or glasses (containing typically large amorphous parts) the long flexible molecules are intertwined together "touching" each other at many points in a random orientation, a situation called *entanglement*. The latter plays a large role in creating a loose network. Under stress, slow flow can take place, during which polymers slide along their complicated intertwined path dictated by the entanglement; this slow sliding is known as *reptation*. There is a possibility of making chemical links at the points of "touching" between different polymers. This cross-linking, known as rubber vulcanization, inhibits the reptation and produces rubber-like response.

### 1.15.6 Colloids

Several systems consist of small (diameter of 10 μm or less) solid or liquid particles dispersed in a liquid medium. These systems are known as *colloidal dispersions*. The particles may arrange themselves in an ordered periodic structure, as their concentration increases beyond a certain critical value. The liquid medium may evaporate and the system may solidify in a periodic arrangement of the particles. In fact, the gemstone *opal* is a natural colloidal crystal consisting of silicon dioxide ($SiO_2$) spherical particles the diameter of which is less than 1 μm. Opal-like solids may be formed with polymeric particles instead of $SiO_2$ spheres. Colloidal structures, as every other equilibrium structure of matter, are determined by the interplay of the enthalpic and the entropic terms in the Gibbs free energy. The enthalpic term is, due mainly to various weak interactions among the particles of van der Waals origin, of screened coulombic nature (for charged particles in ionic solutions), and/or mediated, e.g., by polymers grafted on the surface of the particles or small polymer coils dispersed in the solution. For the ideal case of only hard sphere interaction, the structure is determined solely by the entropic term, which in an apparent paradox drives the system through a first-order phase transition to a periodic state for volume fractions between 0.494 and 0.545.

### 1.15.7 Gels

We met an example of gel in the case where polymers cross-link and undergo a rubber vulcanization. More generally, a gel is a system where its nano- or meso-scopic building units are bonded together to form a macroscopic connected network. Some examples, besides the vulcanized rubbers, are the thermosetting *resins* and the so-called *Sol–gel glasses*. The former consist of small polymers with up to four polymers; the Sol–gel glasses are organic derivatives of certain oxides that in the presence of water undergo hydrolysis and form substances similar to inorganic glasses.

### 1.15.8 Liquid Crystals

Liquid crystals are made of rigid polymers, or molecular aggregates with rigid rod-like shapes, or organic compounds with highly anisotropic molecular shapes. As a result of this almost rod-like structure of their constituents, liquid crystals may exhibit *orientational* long-range order, while they lack long-range positional order: a situation called *nematic* liquid crystallinity. In *smetic* liquid crystals, besides the orientational, there is also a one-dimensional long-range positional order, in the sense that the molecules arrange themselves in sheets; within each sheet, the molecules are aligned but their position is random.

Properties of material are directly related to the structures found within the materials and to the conditions under which the materials are used. To make it more clear, the structure-property relationship in metals, ceramics and polymers as examples are listed in Table 1.3.

**Table 1.3** Structure-Property Relationship in Metals, Ceramics, and Plastics

| *Material* | *Crystallinity* | *Bonding* | *Important properties* |
|---|---|---|---|
| Metals | Crystalline | Predominantly metallic | Excellent conductors, high to moderate strengths, usually malleable and ductile to some degree, opaque, lustrous, moderate to high melting points. |
| Ceramics | (a) Crystalline | Ionic and covalent | Nonconductors, high melting points, brittle, dull to lustrous, hard. |
| | (b) Non-crystalline (vitreous) | Ionic and covalent | Nonconductors, brittle, no definite melting points, transparent to translucent, hard. |
| | (c) Crystalline and non-crystalline phase | Ionic and covalent | Very poor conductors therefore insulators, brittle, hard, low strengths, opaque, high melting ranges. |
| Polymers (plastics and rubbers) | Predominantly non-crystalline | Covalent plus secondary | Very poor conductors therefore insulators, dull and usually translucent to opaque, low strength, deform plastically with little applied load, low softening temperature. |

This structure-property relationships give the materials engineer a basis for understanding the nature and behaviour of a wide variety of materials.

## 1.16 POLYMERS ('POLY': MANY; 'MER': PART)

Polymers are substances composed of molecules having a large number of repeating groups. The repeating group is called a *monomer unit*. A long chain molecule may contain as many as 10,000 repeat units or more. A few examples are polyethylene, polypropylene, polystyrene, polyvinyl chloride, polyvinyl acetate, etc. Besides synthetic polymers, which are produced in large quantities from petrochemicals, there are many natural polymers, e.g., wood, cotton and protein.

Polymers have rather loosely packed structures in the solid state because the primary bonds are completely satisfied within the long molecular chains and because the chains often contain bulky side groups of atoms which interfere with close packing. As a result, the polymers can develop non-crystalline structures relatively easily. This is because long and flexible molecular chains can become tangled easily and so, any atomic arrangement that contributes to loose packing of molecular chains will also favour the formation of non-crystalline structure.

The increasing use of polymers in engineering and other applications in the past few decades has been mainly due to some of the excellent properties they possess and the ease with which these materials can be mass-produced.

There exists a greater number of degrees of freedom for a molecule in a polymer. This helps in making the material appropriate for specific functions by suitable manipulations. Further, they can be moulded into any shape and size. *Polymerisation* is the process of linking monomers together to form polymeric molecules. The process of polymerization occurs through the grafting of monomers onto growing molecules.

### 1.16.1 Classification

Polymers are classified in a number of ways. Some of these are listed below.

(a) *Composition.* Polymers whose main chains comprise only carbon atoms with single bonds are called carbon-chain polymers or *homopolymers*, while those with backbones containing other atoms in addition to carbon (or carbon with multiple-bond) are called heterochain polymers. Examples of cabon-chain polymers are polyethylene, polypropylene, polystyrene, etc. Some of the heterochain polymers are bakelite, nylon, dacron. etc.

(b) *Formability.* On the basis of formability plastics are classified into three groups: thermoplastic, thermoset and elastomer. Some of the polymers can be repeatedly shaped and reshaped because they soften appreciably as temperature increases and liquify at sufficiently high temperatures. These are called *thermoplastics* (e.g., polyethylene, polystyrenc). On the other hand, there are polymers which once set cannot be reshaped at any temperature. These are called thermosets (e.g.,) polyesters, epoxies, bakelite, etc.). Their behaviour is due to the fact that they begin to decompose even before they soften. The behaviour of an elastomer is intermediate between that of a thermoplastic and a thermoset. *Elastomers* exhibit a large and reversible extensibility at room temperature. The structural criteria which they must satisfy, beyond being non-crystalline, are as given below:

1. The chains must be very long, with many 'bonds' in them (i.e., not straight).
2. Room temperature must provide enough thermal energy for chain segments to be in a state of constant motion.
3. The chains must be connected to one another every few hundred atoms by *cross links*, i.e., atoms or groups of atoms that form primary bonds between chains. The best known elastomer is natural rubber (polyisoprene) and its structure typifies those of elastomers in general. The chains of natural rubber are bent rather than straight because the methyl (—$CH_3$) group interferes with the hydrogen in the repeating unit:

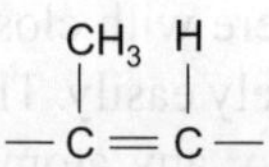

and causes the chain to bend at the double bond.

Polymerisation is of two types:

(i) *Addition Polymerisation.* When molecules just add on to form the polymer, the process is called 'addition polymerisation'. The monomer species in this case retains its structural identity when it gets converted into a identity when it gets converted into a polymer, e.g., the molecules of ethylene monomer can add on to form polyethylene in which, the structural identity of ethylene is retained.

(ii) *Condensation Polymerisation*. When molecules do not just add on but also undergo some reaction in forming the polymer, the process is called 'condensation polymerization'. Here, two molecules (of same or different monomers) condense to form a polymer. The condensation takes place between two reactive functional groups, like the carboxyl group (of an acid) and the hydroxyl group (of an alcohol). During the formation of polymer, water molecules also get eliminated.

Thus, in addition polymerisation, the molecular weight of polymer weight of polymer is roughly equal to that of all the molecules which combine to form the polymer, whereas in condensation polymerization, the molecular weight of polymer is less than the weight of the sample molecules eliminated during the condensation process.

(c) *Geometrical Structure.* Polymers are also classified as linear and three dimensional. A linear polymer is a substance in which each monomer unit is linked only to the two adjacent monomer units. The number of monomer units in a molecule is called the *degree of polymerisation.*

Linear molecules consist of long chains which are held together by the weaker van der Waals forces. Molecules in the chains are covalently bonded.

In three-dimensional network polymers, there are a number of branches which are so linked that a three-dimensional space network is formed. In case, the number of cross-links between linear polymer chains is small, the three-dimensional polymers is called cross-linked and the length between links is called network chains.

(d) *Molecular Structure.* When polymers are synthesized, the length of the chain as well as the arrangement of the molecules in space can vary. Depending on the regularity and arrangement of side groups, (in space) with respect to the main chain, there are three types of polymers. The effect of side group arrangements can be seen by considering the structure of *vinyl polymers*, i.e., those polymers with a repeating unit,

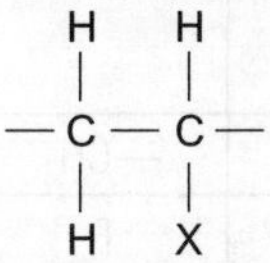

where X is some monovalent side group. The three possible arrangements of side groups are:

1. *Atactic*, or random.
2. *Isotactic*, or all on the same side of the chain.
3. *Syndiotactic*, or regularly alternating from one side to another.

These are illustrated as follows:

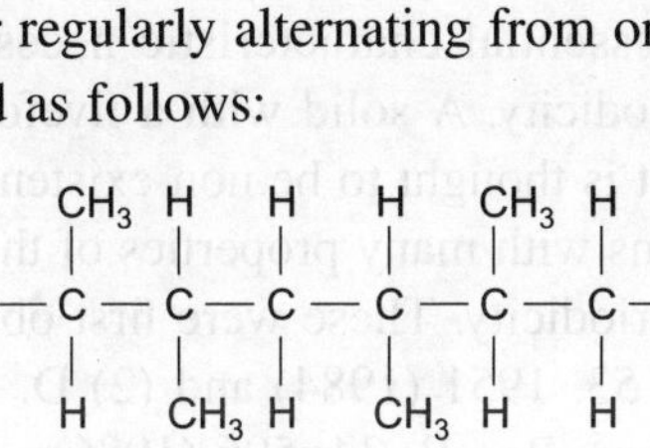

(Atactic polymer)

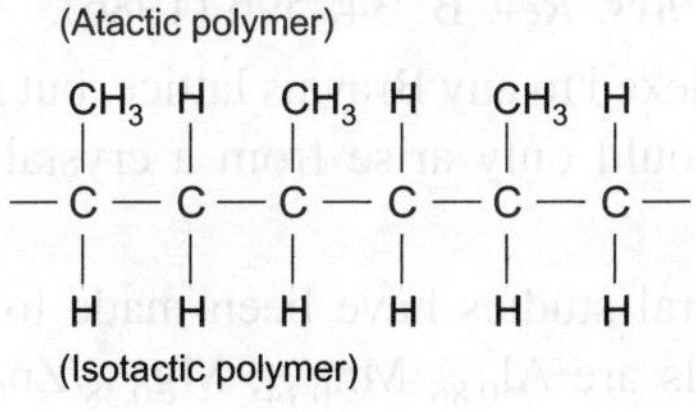

(Isotactic polymer)

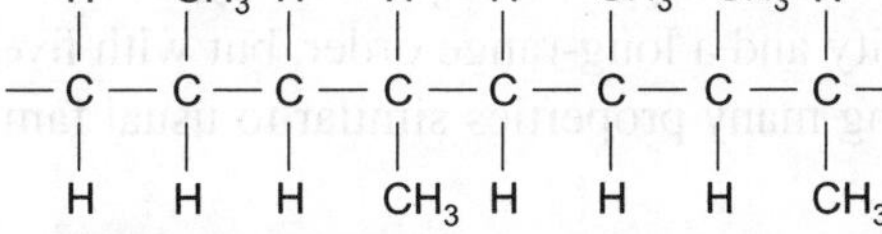

(Syndiotactic polymer)

Table 1.4 lists some selected vinyl polymers.

**Table 1.4** Structure of Selected Vinyl Polymers 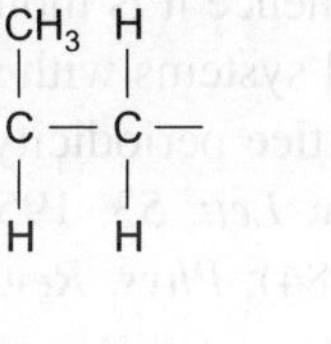

| *Polymers* | *1* | *2* | *3* | *4* | *International abbreviations* |
|---|---|---|---|---|---|
| Polyvinyls (general) | H | H | H | X | |
| Polyethylene | H | H | H | H | PE |

*Contd...*

*Table 1.4 Contd...*

| | | | | | |
|---|---|---|---|---|---|
| Polyvinyl chloride | H | H | H | —Cl | PVC |
| Polyvinyl alcohol | H | H | H | —OH | PVAl |
| Polypropylene | H | H | H | $—CH_3$ | PP |
| Polyvinyl acetate | H | H | H | O<br>‖<br>$—OCCH_3$ | PVAc |
| Polyisobutylene | H | H | $—CH_3$ | $—CH_3$ | PIB |
| Polytetra fluoroethylene | F | F | F | F | PTFE |

## 1.17 QUASI CRYSTALS

It is well known that crystals in nature cannot have lattice periodicity different from 1, 2, 3, 4, and 6-fold symmetries (i.e., 5-fold symmetry is non-existent). This is so because for having a crystalline phase, an essential characteristic necessary for crystal formation is the existence of translational periodicity. A solid with a fivefold symmetry will not have translational periodicity, and hence it is thought to be non-existent. However, it is interesting to note that there are physical systems with many properties of the usual crystalline state but without three-dimensional lattice periodicity. These were first observed in 1984 by (1) D.S. Schechtman et. al., *Phys, Rev. Lett.* 53, 1951 (1984) and (2) D. Levine and P.J. Stainhardt, *Phys. Rev. Lett.* 53, 2477 (1984); *Phys. Rev.* B. 34, 596 (1986).

These structures could not be indexed to any Bravais lattice, but revealed sharp Bragg peaks in the diffraction pattern, which could only arise from a crystal-like solid having fivefold symmetry.

During the last few years, several studies have been made to establish the existence of such crystals. A few such materials are $Al_{0.86}$ $Mn_{0.14}$, $Mg_{0.38}/Zn_{0.41}$ and $Cu_{0.2}$ $Al_{0.65}$ $Fe_{0.15}$. These systems do not have a strict crystal-like periodicity, but have only some subtle type of quasi/translational periodicity and a long-range order, but with fivefold symmetry axes. Such quasiperiodic crystals having many properties similar to usual familiar crystals are known as quasicrystals.

The full point symmetry group of the experimentally observed diffraction patterns of these fivefold symmetry crystals correspond to a platonic solid called an *icosahedron* (Fig. 1.32).

The icosahedron has six fivefold symmetry axes, 10 threefold and 15 twofold symmetry axes.

Quasiperiodicity was first observed in grains of size 2 μm in the alloy of Al with 14 atomic per cent Mn. The smaller Mn atoms are each surrounded by 12 Al atoms arranged at the corners of an icosahedron. The structure is made up of parallel icosahedra attached at their edges. As a result of tiling in two dimensions, we get the structure known as *Penrose patterns* (Fig. 1.33). There is no long-range translationally

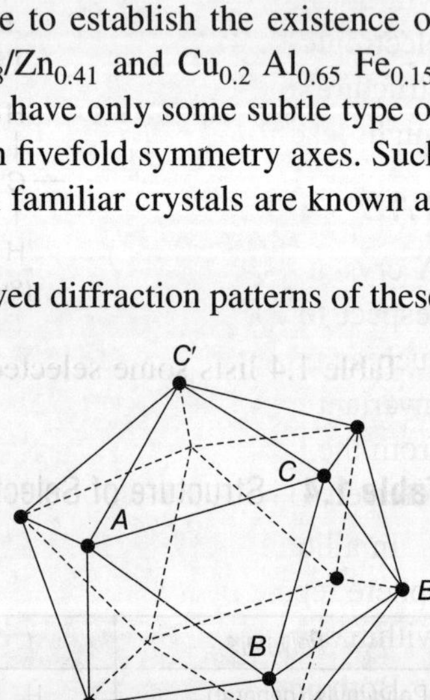

**Figure 1.32** An icosahedron

periodic order, but there is a long-range orientational order: all of the space cannot be filled by repeating the basic unit, but a crystal can be constructed by nucleation at a centre cell, followed by an outward growth from there; thus overall, the crystal could not exhibit the fivefold symmetry of an icosahedron but the X-ray diffraction pattern could give a fivefold symmetry.

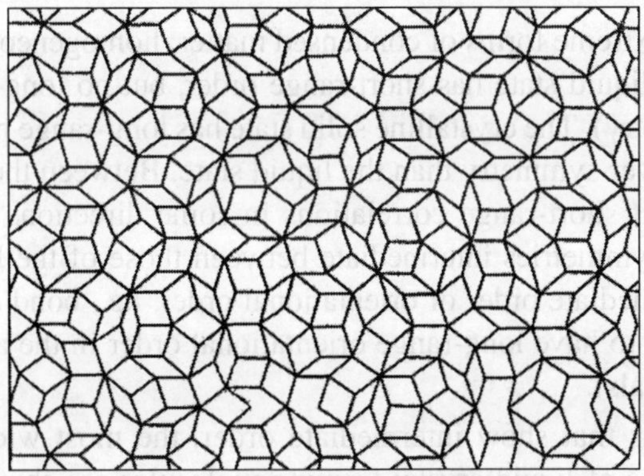

**Figure 1.33** Penrose pattern

The quasicrystalline materials have several curious characteristics and solid state properties, which can be investigated by working out afresh on these materials. For example, it is well known that the electronic band structure of solids is derived on the basis of Bloch theorem, which owes its origin to the existence of strict translational periodicity, the Bloch theorem will not be applicable and new methods would be necessary to be devised for calculation of electronic band structure of quasicrystals. It has been thought that in their electronic band structure, there would be a distribution of band gaps within the allowed energy bands. This might lead to new electronic properties.

## 1.18 LIQUID CRYSTALS

A crystal is not isotropic and homogeneous like a liquid. Rather than being invariant with respect to arbitrary rotations or displacements, it is invariant only under those operations, such as translation by a lattice spacing, that leave the periodic lattice unchanged. Since it is invariant under fewer operations, a crystal has a lower symmetry than a liquid. The transition from the liquid to the crystal breaks the symmetry of the liquid state, and the crystal is often referred to as a broken symmetry phase.

In a liquid, there are substantial local correlations in the position of the particles and hard sphere repulsion prevents two molecules from overlapping so that there will be no molecules within a molecular diameter of a given molecule.

Both liquids and gases are fluid. Fluids are spatially homogeneous and rotationally isotropic. This means that the average environment of any point in a fluid is identical to that of any other point and independent of direction.

The set of operations that leave a system unchanged form a group called the symmetry group. The group of arbitrary translations, rotations and reflections is the *Euclidean group*. Since a fluid is invariant under all these operations its symmetry group is the Euclidean group. Fluids have the highest possible symmetry and all other equilibrium phases of matter are invariant only under some subgroup of the Euclidean group and have lower symmetry than fluid phases.

Thus, we have two extreme forms of condensed matter: homogeneous isotropic liquids and crystalline solids. The liquid state has short-range order, but no long-range order (it has the highest possible symmetry). The crystalline solid state has long-range positional and rotational order, it has a much lower symmetry than the liquid state. Between these two extremes, there are systems that exhibit short-range correlations in some directions and long-range orders in others, and having symmetries intermediate between those of the liquids and the crystals (one form of this intermediate order of orientational order, i.e., bond angle order). It is to be noted that it is possible to have long-range orientational order in the absence of translational order (of periodic crystals).

Among the materials that show intermediate order, the most widely studied are liquid crystals. Some examples of liquid crystal forming molecules are shown in Fig. 1.34.

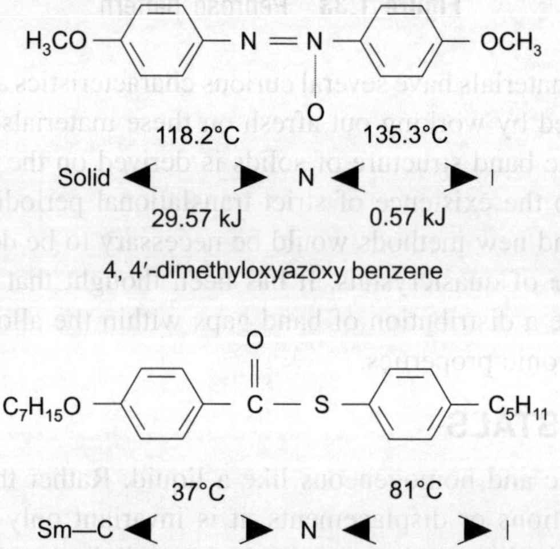


**Figure 1.34** Some molecules forming liquid crystal phases and their phase sequence as a function of temperatures (I-isotropic, N-nematic, C-cholesteric and Sm-smectic phases). (After: *Principles of Condensed Matter Physics*, P.M. Chaikin and T.C. Lubensky, Cambridge Universities Press, 1999)

The benzene rings give rigidity to the molecules. The molecules in a liquid crystal are in general, long. Liquid crystals are divided into three different phases: *nematic*, *cholesteric* and *smectic*.

The nematic phase has the simplest structure. The molecules are parallel to each other but their spatial distribution is random. There is an orientational order but the molecules are able to move around from one region to another. Molecules are rod like in shape. These have low viscosity.

In the cholesteric phase, the molecules are also aligned parallel to each other but the direction of alignment twists progressively resulting in a helical structure. As a result, the substance consists of parallel sheets or layers such that in each sheet, the molecules are aligned parallel to each other. the viscosity is somewhat higher than a nematic one.

A smectic phase consists of a series of layers in which the molecules are all parallel to each other and normal to the layer plane. The layers interact only weakly, and can slip readily past each other, or be made to rotate relative to each other. (The liquid crystalline phase is also called mesophase). The smectic is closest to a solid structure. It has lamellar structure and is essentially a two-dimensional solid.

## 1.19 ANISOTROPY OF THE PHYSICAL PROPERTIES OF SINGLE CRYSTALS

The physical properties of single crystals in general depend on the direction along which they are measured relative to the crystal axes; this phenomenon is called anisotropy. Some examples are the following: crystals do not grow in the form of spheres, but in polyhedral; certain types of atomic planes dissolve more readily than others; the coefficient of thermal expansion of Zn is $6.39 \times 10^{-5}$ along the hexagonal axis and $1.41 \times 10^{-5}$ per degree C perpendicular to it; the specific resistivities of Zn parallel to the hexagonal axis and perpendicular to it are, respectively, $\rho_{||} = 6.06 \times 10^{-6}$ and $\rho = 5.83 \times 10^{-6}$ ohm/cm. The reason for the anisotropy of the physical properties of crystals must be sought in the regular stacking of atoms. Thus, as one passes through a crystalline arrangement of atom or molecules along a given direction, one meets atoms or groups of atoms at different intervals and from different angles than one would along another direction. Single molecules are also anisotropic; however, in normal liquids or gases the orientation of the molecules is random and the physical properties become independent of the direction along which they are measured (isotropic) as long as a large number of molecules is involved or when a time average is taken for a single molecule. Polycrystalline material with a completely random distribution of the grain orientation are also isotropic.

As an example of anisotropy in a single crystal let us consider the electrical conductivity in which an electric field ($E$) gives rise to a current, $I$. In general, the current vector will not have the same direction as the electric vector. Thus, assuming a linear relationship between cause and effect, we may write for the current components relative to an arbitrarily chosen Cartesian coordinate system

$$
\begin{aligned}
I_x &= \sigma_{xx}E_x + \sigma_{xy}E_y + \sigma_{xz}E_z \\
I_y &= \sigma_{yx}E_x + \sigma_{yy}E_y + \sigma_{yz}E_z \qquad \text{...(1)} \\
I_z &= \sigma_{zx}E_x + \sigma_{zy}E_y + \sigma_{zz}E_z
\end{aligned}
$$

where the quantities ($\sigma_{ik}$) are components of the "conductivity tensor". It has been shown by Onsager that the tensor is symmetric, i.e., $\sigma_{ik} = \sigma_{ki}$.[1] Making use of this symmetry property and multiplying the Eqs. (1) respectively by $E_x$, $E_y$, and $E_z$, one obtains upon adding

$$I_xE_x + I_yE_y + I_zE_z = \sigma_{xx}E_x^2 + \sigma_{yy}E_x^2 + \sigma_{zz}E_z^2 + 2\sigma_{xy}E_xE_y + 2\sigma_{yz}E_yE_z + 2\sigma_{zx}E_zE_2 \quad ...(2)$$

The right-hand side represents a quadratic surface; by choosing our coordinates along the principal axes of this surface, the mixed terms disappear and one obtains in the new coordinate system

$$I_x = \sigma_1E_x; \quad I_y = \sigma_2E_y; \quad I_z = \sigma_3E_z \quad ...(3)$$

where $\sigma_1$, $\sigma_2$, and $\sigma_3$ are the principal conductivities. Thus, the electrical properties of any crystal, whatever low symmetry it may possess, may be characterized by three conductivities $\sigma_1, \sigma_2, \sigma_3$ or by three specific resistivities $\rho_1, \rho_2, \rho_3$. Note that $I$ and $E$ have the same direction only when the applied field falls along any one of the three principal axes of the crystal.

In cubic crystals the three quantities are equal and the specific resistivity does not vary with direction. In hexagonal, rhombohedral (trigonal), and tetragonal crystals the resistivity depends only on the angle $\phi$ between the direction in which $\rho$ is measured and the hexagonal, trigonal, or tetragonal axis, since in those crystals two of the three quantities $\rho_1$, $\rho_2$, $\rho_3$ are equal. One finds

$$\rho(\phi) = \rho_\perp \sin^2\phi + \rho_\parallel \cos^2\phi \quad ...(4)$$

where $\perp$ and $\parallel$ refer to directions perpendicular and parallel to the axis.

The effect referred to above may be called a "vector-vector" effect since an electric current (vector) is produced by an applied electric field (vector). The relations obtained may also be applied to other vector-vector effects such as thermal conductivity, where a thermal current vector is evoked by a thermal conductivity, where a thermal current vector is evoked by a thermal gradient; or diffusion under influence of a concentration gradient.

When one considers scalar-tensor effects, similar relationships are obtained. For example, the deformation (tensor) of a solid resulting from a change in temperature (scalar) may be characterized by three principal expansion coefficients, $\alpha_1$, $\alpha_2$, and $\alpha_3$. Here again, in cubic crystals $\alpha_1 = \alpha_2 = \alpha_3$ and such crystals are isotropic in this respect. The angular dependence of $\alpha$ for hexagonal, trigonal, and tetragonal crystals is given by an expression corresponding to Eq. (4).

Other effects such as vector-tensor effects and tensor-tensor effects may be treated along similar general lines. An example of a vector-tensor effect is piezoelectricity, in which an electric field (vector) gives rise to a deformation (tensor). The elastic deformation under influence of a stress tensor is an example of a tensor-tensor effect. These effects may require many more constants than appeared in the relatively simple case of a vector-vector effect outlined above.

[1] L. Onsager, *Phys. Rev.*, 37, 405 (1931); **38**, 2265 (1931); for the so-called "reciprocity relations" derived by Onsager on the basis of the principle of microscopic reversibility, see also C. Zwikker, *Physical Properties of Solid Materials*, Interscience, New York, 1954, Chap. 5; see also Chap. 4 for anisotropy.

## 1.20 THE ATOMIC IDEA: FROM ELEMENTARY PARTICLES TO SOLIDS

Matter is made up from microscopic elementary particles interacting with each other to form composite particles; the latter in turn combine with each other and/or with other elementary particles to create the next level in the hierarchy of the structures of matter, etc.

Thus, two *u* quarks and one *d* quark make up the proton, while two *d* quarks and one *u* quark make up the neutron; proton and neutrons combine to create the various atomic nuclei; a nucleus attracts and traps electrons around it to form a neutral atom (when the number of trapped electrons equals the number *Z* of protons in the nucleus).[2] Atoms (and/or ions) come together to form molecules; atoms and/or ions[3] and/or molecules in huge numbers are more or less close-packed to form ordinary solids and liquids and other condensed macroscopic structures such as polymers.

To go from the basic quantities characterizing the microscopic constituents of matter (essentially the atomic number *Z* of the participating atoms) to the various properties of a macroscopic system, such as a solid, it is a truly challenging task. It involves the employment of three essential ingredients (as shown in Fig. 1.35).

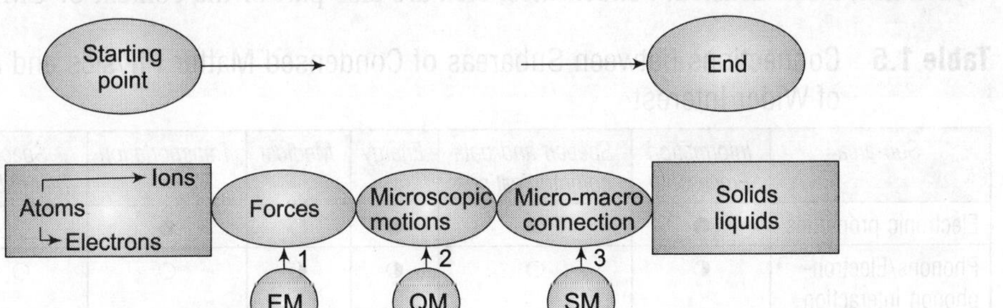


**Figure 1.35** To bridge the huge gap from atoms to solids (or liquids) three basic "Foundations" are needed: electromegnetism (EM), quantum mechanics (QM), and statistical mechanics (SM)

1. *The forces* (or more generally the interactions), which are responsible for binding the constituents together. From the scale of atoms all the way up to that of an asteroid, the only force that plays a non-negligible role is the *electromagnetic* (EM) one.[4]
2. The *microscopic motions* of the constituents under the action of the forces. The laws that govern these motions are those of *Quantum Mechanics* (QM) stemming from the fact that both interactions and particles have a dual wave–particle nature. We stress

[2] If the number of electrons is smaller than Z, a positively charged ion is formed called *cation*, while, if it is larger than Z, a negatively charged ion is formed called *anion*. Isolated ions are metastable structures, because they tend to combine with oppositely charged particles to form neutral structures.

[3] And possibly electrons (which are considered point particles).

[4] Living matter could be considered an indirect exception in the sense that its survival seems to require a planetary system with its central star; for the latter, gravitational and nuclear forces (as well as electromagnetic) are essential.

the fact that the world in all scales obeys quantum, not classical laws, although under some limited circumstances, it is possible that the conclusions of QM and Classical Mechanics practically coincide.

3. The *connection* between microscopic motions and macroscopic properties. This connection is provided by *Statistical Mechanics* and *Thermodynamics*.

## 1.21 SUBAREAS OF CONDENSED MATTER PHYSICS AND APPLICATIONS

As a matter fact, solids have been historically proved to be of paramount importance for technological and economic development (starting from the Neolithic Age and the Age of Metals (copper, iron) to the modern era of transistors, integrated circuits, chips, magnetic memories, computers, Magnetic Resonance Imaging (MRI), solid state lasers, etc.).

Table 1.5 (taken from a 1986 publication of the National Academy Press[5]) shows the impact of *Condensed Matter Physics (CMP)* on various aspects of a modern society. CMP includes that study of solids, liquids, and states of matter intermediate between solids and liquids such as polymers, liquid crystals, glasses, gels, etc. Special quantum states such as superfluids, Bose–Einstein condensates, etc., are also part of the content of CMP.

**Table 1.5** Connections between Subareas of Condensed Matter Physics and Applications of Wider Interest

| *Sub-area* | *Information Processing* | *Speech and data communications* | *Energy* | *Medical* | *Transportation* | *Space Technology* | *National Security* |
|---|---|---|---|---|---|---|---|
| Electronic properties | ● | ◐ | ● | ○ | ● | ○ | ● |
| Phonons/Electron-phonon interaction | ◐ | ○ | ◐ | ◐ | ○ | ○ | ○ |
| Phases transition | ◐ | ○ | ◐ | ○ | ◐ | ○ | ● |
| Magnetism | ● | ● | ● | ◐ | ● | ◐ | ◐ |
| Semiconductor | ● | ● | ● | ◐ | ● | ● | ● |
| Defects/diffusion | ● | ● | ● | | ● | ◐ | ● |
| Surfaces/interfaces | ● | ● | ● | ◐ | ● | ◐ | ● |
| Low-temperature physics | ◐ | ○ | ● | ◐ | ○ | | ○ |
| Liquids | ◐ | ○ | ◐ | ○ | ◐ | ◐ | ○ |
| Polymers | ● | ◐ | ◐ | ● | ● | ◐ | ◐ |
| Nonlinear dynamics, instabilities, chaos | ◐ | ○ | | ○ | ○ | ○ | ● |

A solid circle indicates a critical connection between the subarea of condensed-matter physics and the area of application: a half-filled circle indicates an important or emerging connection; an open circle denotes the possibility of a connection; a blank implies connection is not known.

5 "Physics through the 1990s: Condensed–Matter Physics," National Academy Press, 2101 Consitution Ave, NW Washington, DC 20418.

CHAPTER

# 2

# Solids and Crystal Structures

## 2.1 INTRODUCTION

In order to understand various solids, it is useful to consider their process of growth. The growth could be either from solution, melt or vapour, or from a suitable combination of these. If the growth process is slow, then the constituent particles have a tendency to settle down in positions for which the free energy of the configuration is minimum. This leads to a long range order in the solid. However, in certain extreme cases, when the growth process or the phase change is rather rapid, the constituent particles do not have sufficient time to achieve minimum energy configuration. Then, a long range order is not obtained in such cases. Nevertheless, there may be preferential ordering of neighbouring atoms around a particular atom and a short range order is still preserved.

In the subsequent section, we will classify solids on the basis of their degree and type of order.

## 2.2 SOLIDS: CRYSTALLINE AND AMORPHOUS

On the basis of order in structure, solids may be divided into two broad categories, viz., crystalline and amorphous. Crystalline solids may further be divided into single crystals and polycrystalline solids.

*Crystalline Solids.* A solid, in general, is said to be a crystal if the constituent particles (atoms or ions or molecules) are arranged in three dimensions in a periodic manner (i.e., stacked in a regular manner forming a 3-dimensional pattern by repetition of a pattern unit). The periodicity of pattern extends throughout the solid and there is a long range order, then such a solid is a (single) *crystal.*

The regularity in the appearance of crystals found in nature (Fig. 2.1) or grown in laboratory has led to believe that crystals are formed by a regular repetition of identical building blocks in three-dimensional space.

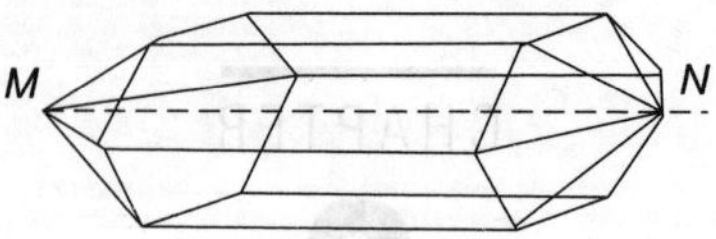


Quartz

**Figure 2.1** External form of crystal found in nature

When a crystal grows under constant environment, the external geometrical shape of the crystal often remains unchanged. The shape is a consequence of internal arrangement of constituent atoms or ions (or molecules) of which it is built up.

Observations show that crystals are bounded by optically plane faces, sharp straight edges and interfacial angles. A relationship among these can be expressed by the formula:

$$f + c = e + 2$$

where $f$ is the number of faces, $c$ is the number of angles and $e$ is the number of edges. Thus, if two are known, the third element can be easily obtained.

**Example 2.1** Determine the number of edges in a quartz crystal, if there are 18 faces and 14 angles in it.

***Solution*** Given $f = 18, c = 14$

$\therefore$ Number of edges $e = 18 + 14 - 2 = 30$

A crystal if more or less regular in shape is called a monocrystal or a single crystal. Quite common crystals such as rock salt, calcite, quartz, etc. (Fig. 2.2) are some examples of this class. A typical feature of a single crystal is its anisotropy, i.e., the difference in its physical properties in different directions. In addition, a crystalline solid has a sharp melting point.

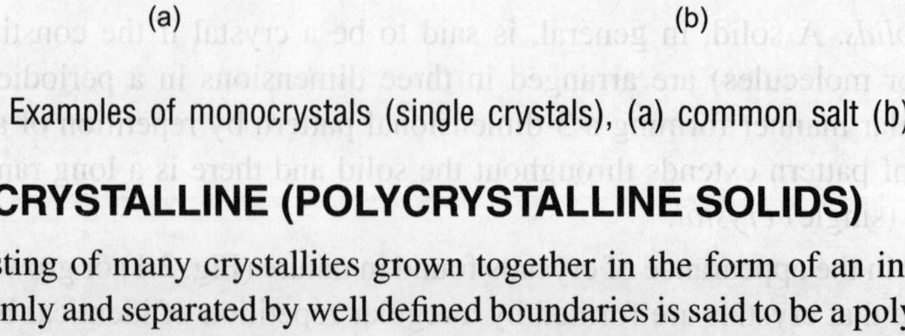


**Figure 2.2** Examples of monocrystals (single crystals), (a) common salt (b) quartz crystal

## 2.3 SEMICRYSTALLINE (POLYCRYSTALLINE SOLIDS)

A solid consisting of many crystallites grown together in the form of an interlocking mass, oriented randomly and separated by well defined boundaries is said to be a polycrystalline solid (Fig. 2.3). In general, the grains in such a solid are not related in shape to the crystal structure, the surface being random in shape rather than well defined crystal planes. A great majority of solids occurring in nature such as rock, sand, metals, salts, etc. are the polycrystalline structure. However, they can be grown as a single crystal under suitable conditions. Due to random distribution of crystallites, a polycrystalline solid is isotropic, i.e., its properties are same, on an average in all directions.

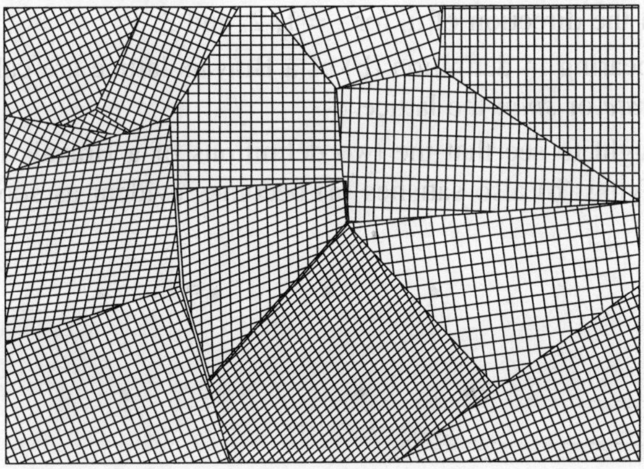

**Figure 2.3** Polycrystalline aggregates separated by well defined boundaries

## 2.4 BRAVAIS LATTICE

Any crystalline solid can be described in terms of a periodic array of discrete points, each of which may be single atom/group of atoms/molecule or ion. These are called *lattice points* and the repeat unit is called a *basis*. The regular periodic array of lattice points in space is called a *Bravais lattice* (Fig. 2.4) such that the Bravais lattice summarizes only the geometry of the underlying periodic structure, regardless of what the actual units (basis) may be. Thus,

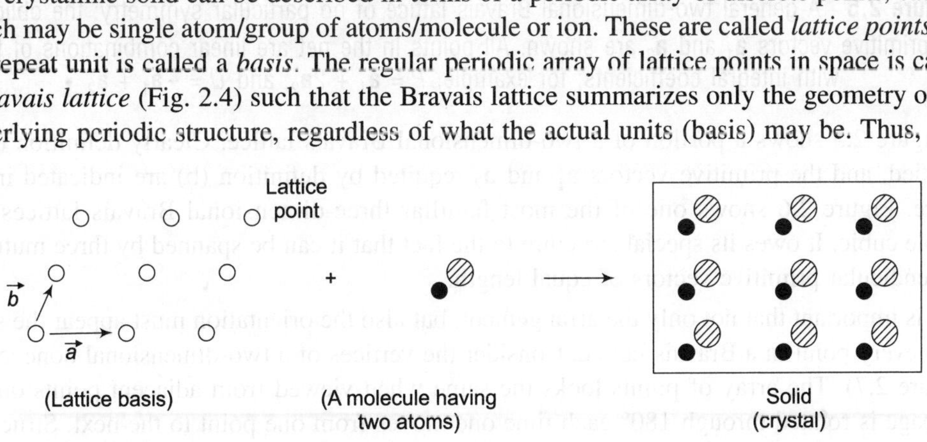


**Figure 2.4** A solid as a geometrical abstraction of lattice points

We have two equivalent definitions of a Bravais lattice:

(a) A Bravais lattice is an infinite array of discrete points with an arrangement and orientation that appears *exactly* the same, from whichever of the points the array is viewed.

(b) A (three-dimensional) Bravais lattice consists of all points with position vectors $\vec{R}$ of the form

$$\vec{R} = n_1 \vec{a}_1 + n_2 \vec{a}_2 + n_3 \vec{a}_3$$

where $a_1$, $a_2$, and $a_3$ are any three vectors not all in the same plane, and $n_1$, $n_2$ and $n_3$ range through all integral values. Thus, the point $\sum n_i a_i$ is reached by moving $n_i$ steps of length $a_i$ in the direction of $a_i$ for $i = 1, 2$, and 3.

The vectors $a_i$ appearing in definition (b) of a Bravais lattice are called *primitive vectors* and are said to *generate* or *span* the lattice.

It takes some thought to see that the two definitions of a Bravais lattice are equivalent. That any array satisfying (b) also satisfies (a) becomes evident as soon as both definitions are understood.

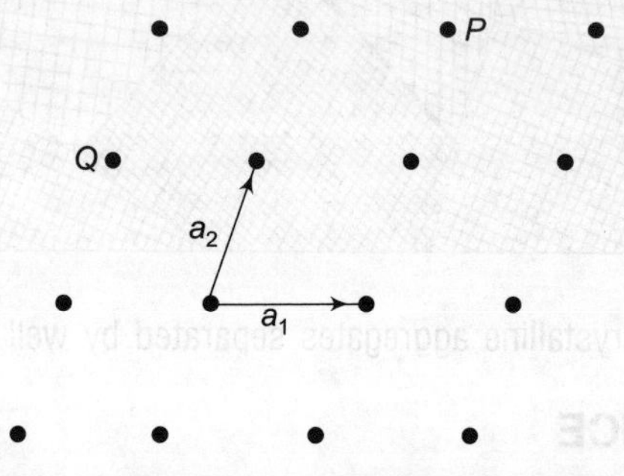


**Figure 2.5** A general two-dimensional Bravais lattice of no particular symmetry: the oblique net. primitive vectors $\mathbf{a}_1$ and $\mathbf{a}_2$ are shown. All points in the net are linear combinations of these with integral coefficients; for example, $P = \mathbf{a}_1 + 2\mathbf{a}_2$, and $Q = -\mathbf{a}_1 + \mathbf{a}_2$

Figure 2.5 shows a portion of a two-dimensional Bravais lattice. Clearly definition (a) is satisfied, and the primitive vectors $\mathbf{a}_1$ and $\mathbf{a}_2$ requited by definition (b) are indicated in the figure. Figure 2.6 shows one of the most familiar three-dimensional Bravais lattices, the simple cubic. It owes its special structure to the fact that it can be spanned by three mutually perpendicular primitive vectors of equal length.

It is important that not only the arrangement, but also the orientation must appear the same from every point in a Bravais lattice. Consider the vertices of a two-dimensional honeycomb (Figure 2.7). The array of points looks the same when viewed from adjacent points only if the page is rotated through 180° each time one moves from one point to the next. Structural relations are clearly identical, but *not* orientational relations, so the vertices of a honeycomb do not form a Bravais lattice.

## 2.5 IMPORTANT ILLUSTRATIONS

*Primitive vectors* of the two definitions of a Bravais lattice, definition (b) is mathematically more precise. It has, however, two minor shortcomings. First, for any given Bravais lattice the set of primitive vectors is not unique—indeed, there are infinitely many nonequivalent

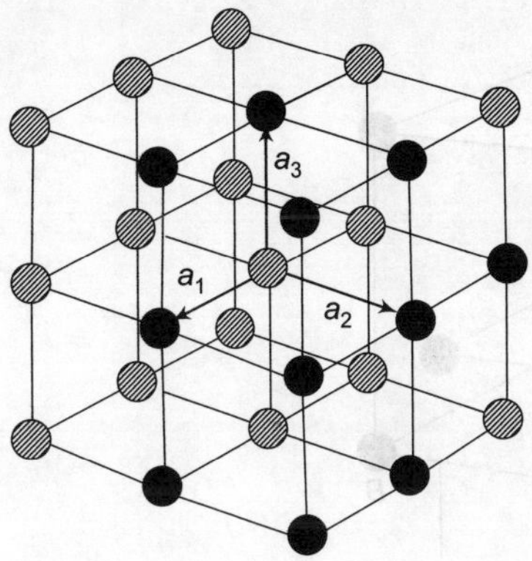


**Figure 2.6** A simple cubic three-dimensional Bravais lattice. the three primitive vectors can be taken to be mutually perpendicular, and with a common magnitude

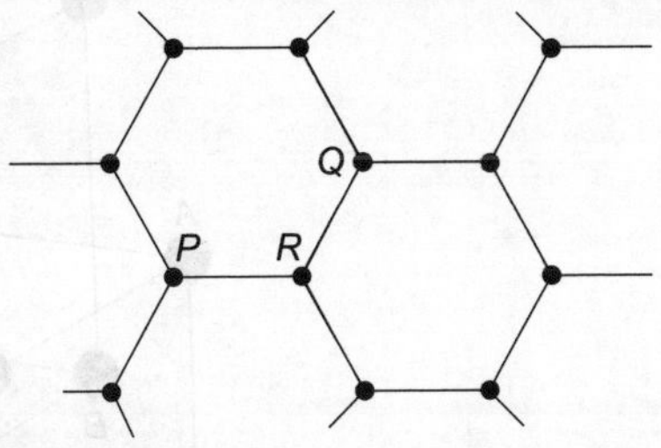


**Figure 2.7** The vertices of a two-dimensional honeycomb do not form a Bravais lattice. the array of points has the same appearance whether viewed from point *P* or point *Q*. However, the view from point *R* is rotated through 180°

choices (see Figure 2.8) and it is distasteful (and sometimes misleading) to rely too heavily on a definition that emphasizes a particular choice. Second, when presented with a particular array of points one usually can tell at a glance whether the first definition is satisfied, although the existence of a set of primitive vectors or a proof that there is no such set can be rather more difficult to perceive immediately.

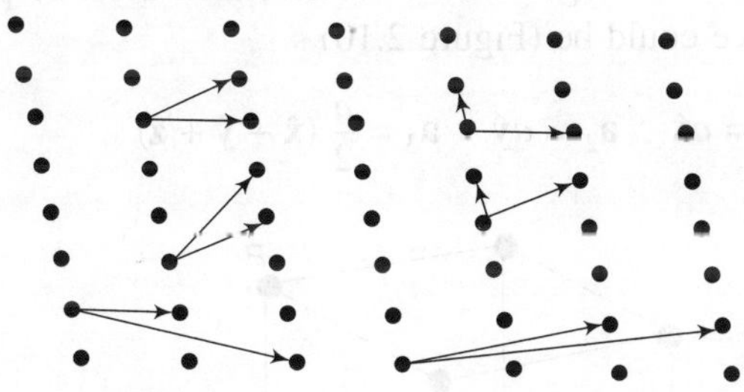

**Figure 2.8** Several possible choices of pairs of *primitive vectors* for a two-dimensional Bravais lattice. they are drawn, for clarity, from different origins

Consider, for example, the *body-centred cubic* (bcc) lattice, formed by adding to the simple cubic lattice of Figurc 2.6 (whose sites we now label *A*) an additional point, *B*, at the centre of each little cube (Figure 2.9). One might at first feel that the centre points *B* bear a different relation to the whole than the corner points, *A*. However, the centre point *B* can be thought of as corner points of a second simple cubic array.

In this new array the corner points *A* of the original cubic array are centre points. Thus, all points do have identical surroundings, and the body-centred cubic lattice is a Bravais lattice. If the original simple cubic lattice is generated by primitive vectors

$$a\hat{\mathbf{x}},\ a\hat{\mathbf{y}},\ a\hat{\mathbf{z}} \qquad ...(1)$$

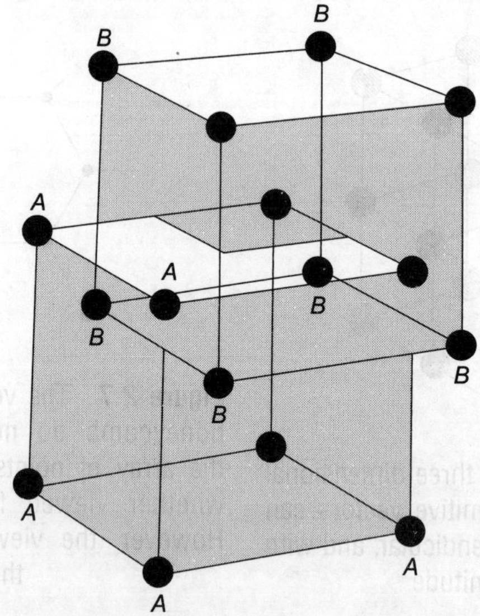


**Figure 2.9** A few sites from a body-centred cubic Bravais lattice. note that it can be regarded either as a simple cubic lattice formed from the points *A* with the points *B* at the cube centres, or as a simple cubic lattice formed from the points *B* with the points *A* at the cube centres. this observation establishes that it is indeed a Bravais lattice

where $\hat{\mathbf{x}}$, $\hat{\mathbf{y}}$, and $\hat{\mathbf{z}}$ are three orthogonal unit vectors, then a set of primitive vectors for the body-centred cubic lattice could be (Figure 2.10)

$$\mathbf{a}_1 = a\hat{\mathbf{x}}, \quad \mathbf{a}_2 = a\hat{\mathbf{y}}, \quad \mathbf{a}_3 = \frac{a}{2}(\hat{\mathbf{x}} + \hat{\mathbf{y}} + \hat{\mathbf{z}}) \qquad ...(2)$$

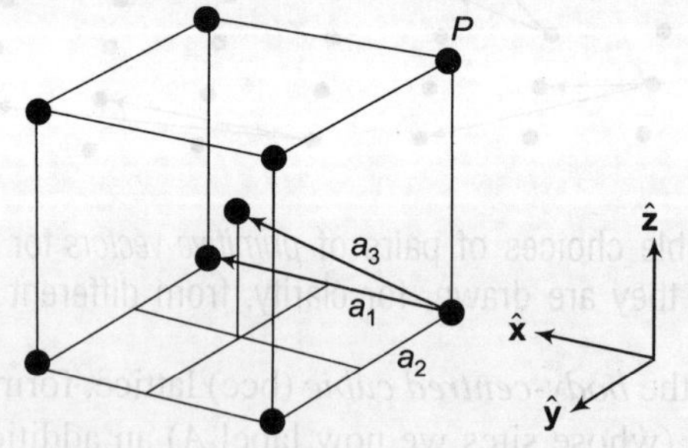


**Figure 2.10** Three primitive vectors, specified in Eq. (2) for the body-centred cubic Bravais lattice. the lattice is formed by taking all linear combinations of the primitive vectors with integral coefficients. The point *P*, for example, is $P = -\mathbf{a}_1 - \mathbf{a}_2 + 2\mathbf{a}_3$.

A more symmetric set (see Figure 2.11) is

$$\mathbf{a}_1 = \frac{a}{2}(\hat{\mathbf{y}} + \hat{\mathbf{z}} - \hat{\mathbf{x}}), \ \mathbf{a}_2 \ (\hat{\mathbf{z}} + \hat{\mathbf{x}} - \hat{\mathbf{y}}), \ \mathbf{a}_3 = \frac{a}{2}(\hat{\mathbf{x}} + \hat{\mathbf{y}} - \hat{\mathbf{z}}) \qquad ...(3)$$

It is important to convince oneself both geometrically and analytically that these sets do indeed generate the bcc Bravais lattice.

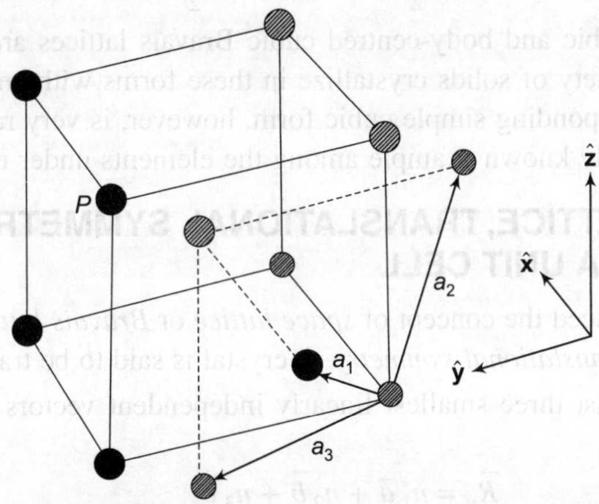


**Figure 2.11** A more symmetric set of primitive vectors, specified in Eq. (3), for the body-centred cubic Bravais lattice. The point $P$, for example, has the form $P = 2\mathbf{a}_1 + \mathbf{a}_2 + \mathbf{a}_3$

Another equally important example is the *face-centred cubic* (fcc) Bravais lattice. To construct the face-centred cubic Bravais lattice add to the simple cubic lattice of Figure 2.6 an additional point in the centre of each square face (Figure 2.12). For ease in description think of each cube in the simple cubic lattice as having horizontal bottom and top faces, and four vertical side faces facing north, south, east, and west. It may sound as if all points in this new array are not equivalent, but in fact they are.

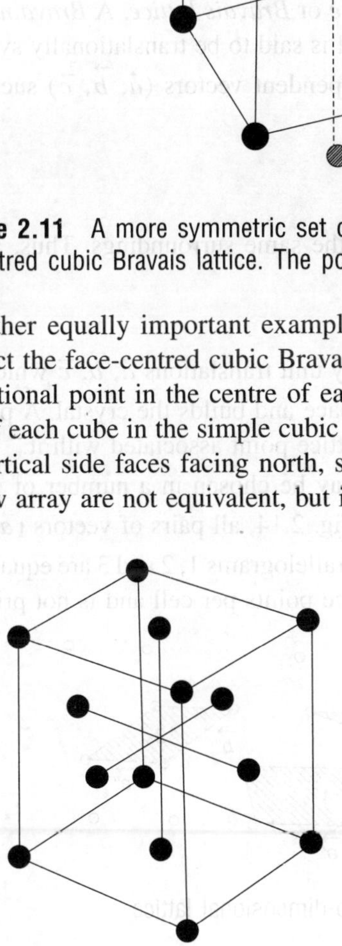

**Figure 2.12** Some points from a face-centred cubic Bravais lattice

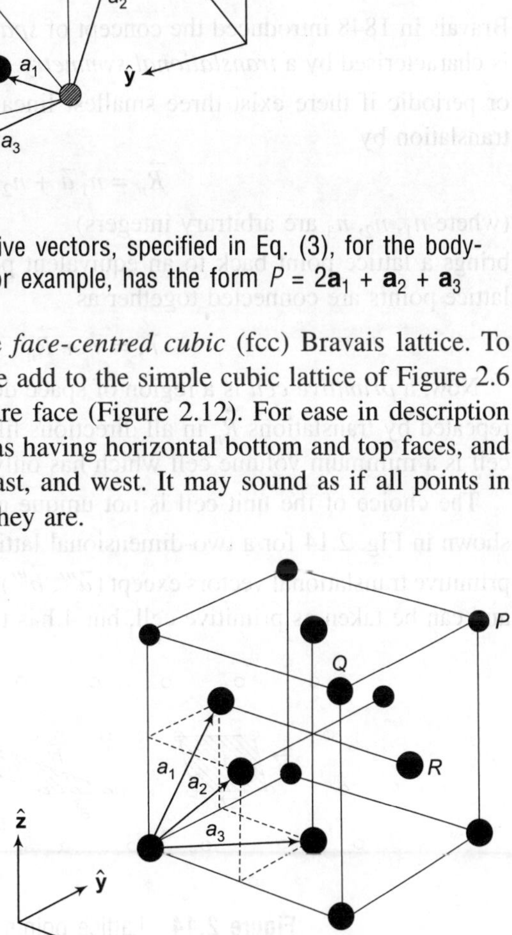


**Figure 2.13** A set of primitive vectors, as given in Eq. (4) for the face-centred cubic Bravais lattice. The labelled points are $P = \mathbf{a}_1 + \mathbf{a}_2 + \mathbf{a}_3$, $Q = 2\mathbf{a}_2$, $R = \mathbf{a}_2 + \mathbf{a}_3$, and $S = -\mathbf{a}_1 + \mathbf{a}_2 + \mathbf{a}_3$

A symmetric of primitive vector for the face-centred cubic lattice (see Figure 2.13) is

$$\mathbf{a}_1 = \frac{a}{2}(\hat{\mathbf{y}} + \hat{\mathbf{z}}), \quad \mathbf{a}_2 = \frac{a}{2}(\hat{\mathbf{z}} + \hat{\mathbf{x}}), \quad \mathbf{a}_3 = \frac{a}{2}(\hat{\mathbf{x}} + \hat{\mathbf{y}}) \qquad ...(4)$$

The face-centred cubic and body-centred cubic Bravais lattices are of great importance, since an enormous variety of solids crystallize in these forms with an atom (or ion) at each lattice site. (The corresponding simple cubic form, however, is very rare, the alpha phase of polonium being the only known example among the elements under normal conditions.)

## 2.6 BRAVAIS LATTICE, TRANSLATIONAL SYMMETRY AND CHOICE OF A UNIT CELL

Bravais in 1848 introduced the concept of *space lattice* or *Bravais lattice.* A *Bravains lattice* is characterised by a *translational symmetry.* A crystal is said to be translationally symmetry or periodic if there exist three smallest linearly independent vectors ($\vec{a}$, $\vec{b}$, $\vec{c}$) such that a translation by

$$\vec{R}_n = n_1 \vec{a} + n_2 \vec{b} + n_3 \vec{c}$$

(where $n_1$, $n_2$, $n_3$ are arbitrary integers)

brings a lattice point back to an equivalent point of the same surroundings. Thus, any two lattice points are connected together as

$$\vec{r}, = \vec{r} + \vec{R}_n$$

Now, a *primitive cell* is a region of space defined by unit translations $\vec{a}$, $\vec{b}$, $\vec{c}$ which when repeated by translations $\vec{R}_n$ in all directions fills all space and builds the crystal. A primitive cell is a minimum volume cell which has only one lattice point associated with it.

The choice of the unit cell is not unique and it may be chosen in a number of ways as shown in Fig. 2.14 for a two-dimensional lattice. In Fig. 2.14, all pairs of vectors ($\vec{a}$, $\vec{b}$) are primitive translational vectors except ($\vec{a}'''$, $\vec{b}'''$). The parallelograms 1, 2 and 3 are equal in area and can be taken as primitive cell, but 4 has two lattice points per cell and is not primitive.

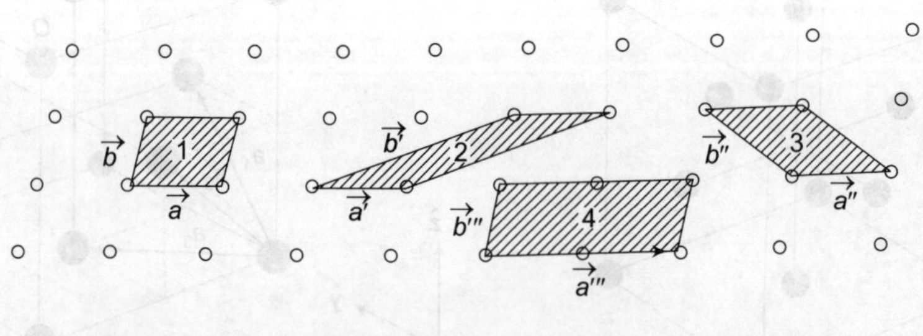


**Figure 2.14** Lattice points of a two-dimensional lattice

Thus, according to the requirement of the case either a primitive or a non-primitive cell can be selected as a unit cell.

The parallelopiped defined by translations $\vec{a}$, $\vec{b}$, $\vec{c}$ as the shortest possible sides along three crystallographic axes *x*, *y*, *z* is called a *primitive cell* (Fig. 2.15). The volume ($V_c$) of a primitive cell defined by primitive translations $\vec{a}$, $\vec{b}$, $\vec{c}$ is

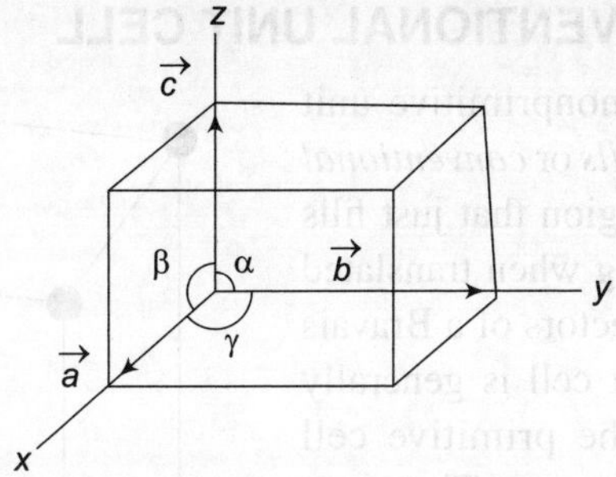


**Fig. 2.15** Crystal axes

$$V_c = |\vec{a} \times \vec{b} \cdot \vec{c}|; \quad (\vec{a} \equiv \vec{a}_1, \vec{b} \equiv \vec{a}_2, \vec{c} \equiv \vec{a}_3)$$

by elementary vector analysis.

## 2.7 COORDINATION NUMBER

The points in a Bravais lattice that are closest to a given point are called its *nearest neighbours*. Because of the periodic nature of a Bravais lattice, each point has the same number of nearest neighbours. This number is thus a property of the lattice, and is referred to as the *coordination number* of the lattice. A simple cubic lattice has coordination number 6; a body-centred cubic lattice, 8; and a face-centred cubic lattice, 12. The notion of a coordination number can be extended in the obvious way to some simple arrays of points that are not Bravais lattices, provided that each point in the array has the same number of nearest neighbours.

## 2.8 PRIMITIVE UNIT CELL

A volume of space that, when translated through all the vectors on a Bravais lattice, just fills all of space without either overlapping itself or leaving voids is called a *primitive cell* or *primitive unit cell* of the lattice. There is no unique way of choosing a primitive cells for a given Bravais lattice. Several possible choices of primitive cells for a two-dimensional Bravais lattice are illustrated in Fig. 2.16.

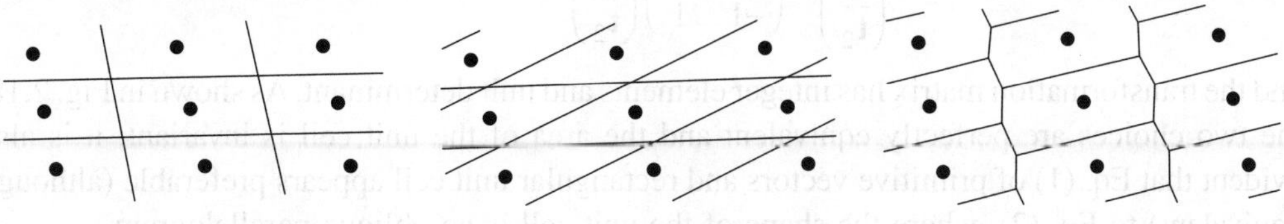

**Figure 2.16** Several possible choices of primitive cell for a single two-dimensional Bravais lattice

A primitive cell must contain precisely one lattice point (unless it is so positioned that there are points on its surface). It follows that if $n$ is the density of points in the lattice and $v$ is the volume of the primitive cell, then $nv = 1$. Thus, $v = 1/n$.

## 2.9 UNIT CELL, CONVENTIONAL UNIT CELL

One can fill space up with nonprimitive unit cells (known simply as *unit cells* or *conventional unit cells*). A unit cell is a region that just fills space without any overlapping when translated through some *subset* of the vectors of a Bravais lattice. The conventional unit cell is generally chosen to be bigger than the primitive cell and to have the required symmetry. Thus, one frequently describes the body-centred cubic lattice in terms of a cubic unit cell (Fig. 2.17) that is twice as large as a primitive bcc unit cell. Numbers specifying the size of a unit cell (such as the single number $a$ in cubic crystals) are called *lattice constants*.

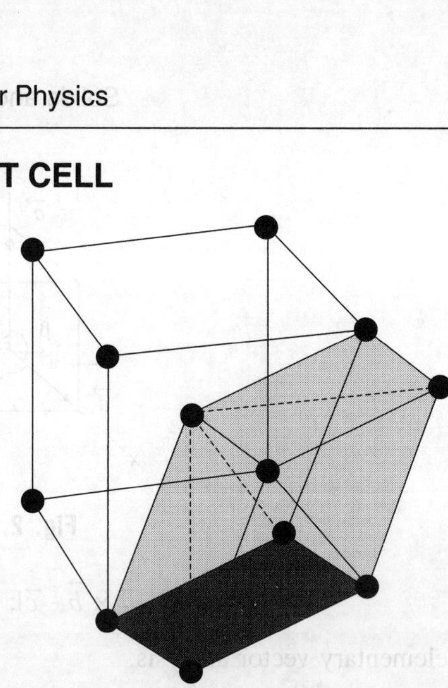

**Figure 2.17** Primitive and conventional unit cells for the body-centred cubic Bravais lattice. The primitive cell (shaded) has half the volume of the conventional cubic cell

### 2.9.1 Primitive and Conventional Unit Cells

An elementary example of the mentioned arbitrariness in the choice of the primitive translation vectors is illustrated in Fig. 2.18. We consider the two-dimensional rectangular Bravais lattice generated by the fundamental vectors $\mathbf{t}_1$ and $\mathbf{t}_2$, given by

$$\vec{\mathbf{t}}_1 = (a, 0, 0) \quad \vec{\mathbf{t}}_2 = (0, b, 0) \qquad \text{...(1)}$$

The same Bravais lattice can be described, for instance, by the primitive vectors

$$\vec{\tilde{\mathbf{t}}}_1 = (2a, -b, 0) \quad \vec{\tilde{\mathbf{t}}}_2 = (-a, b, 0) \qquad \text{...(2)}$$

In fact, the vectors $\tilde{\mathbf{t}}_1$, $\tilde{\mathbf{t}}_2$ are connected to $\mathbf{t}_1$, $\mathbf{t}_2$ by the relation

$$\begin{pmatrix} \tilde{\mathbf{t}}_1 \\ \tilde{\mathbf{t}}_2 \end{pmatrix} = \begin{pmatrix} 2 & -1 \\ -1 & 1 \end{pmatrix} \begin{pmatrix} \vec{\mathbf{t}}_1 \\ \vec{\mathbf{t}}_2 \end{pmatrix} \qquad \text{...(3)}$$

and the transformation matrix has integer elements and unit determinant. As shown in Fig. 2.18, the two choices are perfectly equivalent and the area of the unit cell is invariant; it is also evident that Eq. (1) of primitive vectors and rectangular unit cell appears preferable (although equivalent) to Eq. (2), where the shape of the unit cell is an oblique parallelogram.

From the definition of Bravais lattices, it is evidently possible to describe the whole set of lattice points (1) using non-primitive cells (called *conventional unit cells*), containing an integer number of primitive cells and a corresponding number of extra lattice with a conventional unit cell, is illustrated in Fig. 2.19, that reports the two-dimensional Bravais lattice, known as the centred rectangular lattice.

In Fig. 2.19 we choose as primitive lattice vectors

$$\vec{t}_1 = (a, 0, 0) \quad \text{and} \quad \vec{t}_2 = \left(\frac{a}{2}, \frac{b}{2}, 0\right) \qquad ...(4)$$

The primitive unit cell is a parallelogram of area $ab/2$. Instead of the primitive unit cell (which is a particular case of an oblique parallelogram), it is sometimes preferred (although equivalent) to describe the lattice by a *conventional unit cell of rectangular shape, double area and an extra lattice point at the centre of the rectangle*. The fundamental vectors of the conventional unit cell are

$$\vec{t}_1^{(c)} = (a, 0, 0) \quad \text{and} \quad \vec{t}_2^{(c)} = (0, b, 0) \qquad ...(5)$$

The matrix $M$ connecting $\mathbf{t}_1^{(c)}, \mathbf{t}_2^{(c)}$ with $\mathbf{t}_1, \mathbf{t}_2$ is given by

$$\begin{pmatrix} \vec{t}_1^{(c)} \\ \vec{t}_2^{(c)} \end{pmatrix} = \begin{pmatrix} 1 & 0 \\ -1 & 2 \end{pmatrix} \begin{pmatrix} \vec{t}_1 \\ \vec{t}_2 \end{pmatrix} \qquad ...(6)$$

and has determinant 2; the conventional cell defined by $\vec{t}_1^{(c)}$ and $\vec{t}_2^{(c)}$ contains two primitive cells.

Notice that the conventional cell is not the repeating block of smallest volume; thus, in the study of the effects related to the translational symmetry (for instance classification of states with wave vector quantum number, Brillouin zone concept, and in general in any problem directly or indirectly related to the *full* translational symmetry) we must *consider only the primitive translational vectors and the primitive unit cell.* Conventional cells are however of help for a pictorial description of a number of crystals.

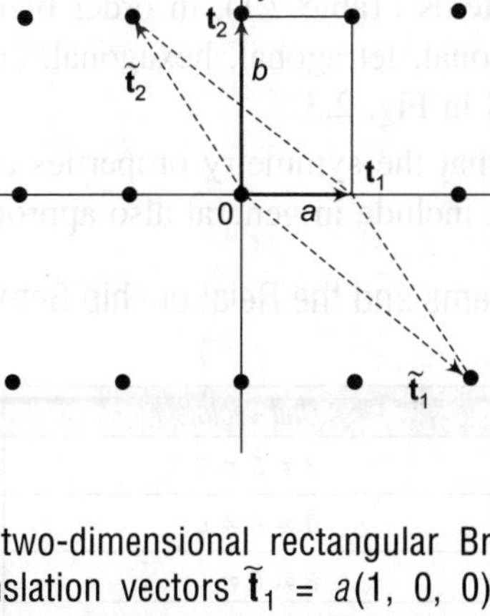


**Figure 2.18** Illustration of a two-dimensional rectangular Bravais lattice. the lattice can be described by the primitive translation vectors $\tilde{t}_1 = a(1, 0, 0)$ and $\tilde{t}_2 = b(0, 1, 0)$. Although this is the most natural choice, infinite other choices are possible; the figure illustrates the choice $\tilde{t}_1 = (2a, -b, 0)$ and $\tilde{t}_2 = (-a, b, 0)$. Whatever choice is done, the area of the unit cell is invariant and equals $ab$

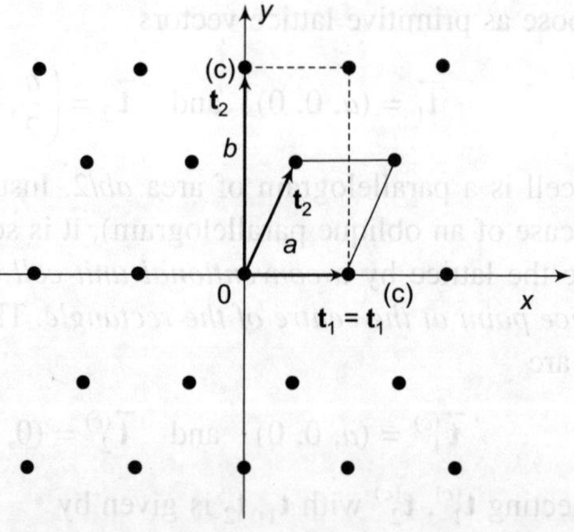


**Figure 2.19** Illustration of the primitive unit cell and conventional unit cell for a two-dimensional centred rectangular Bravais lattice

The number of different Bravais lattices is determined by symmetry considerations. In two dimensions, the different Bravais lattices are five: oblique, rectangular (primitive and centred), square, hexagonal.

In three dimensions, symmetry considerations lead to fourteen different Bravais lattices grouped into seven crystal systems: triclinic, monoclinic (primitive and base-centred), orthorhombic (primitive, base-centred, body-centred and face-centred), trigonal, tetragonal (primitive and base-centred), hexagonal, cubic (primitive, body-centred and face-centred). This classification is based on the equality or inequality of the lengths $a$, $b$, $c$ of the edges of the conventional cell, on the angles $\alpha$, $\beta$, $\gamma$ between each pair of them, and on possible occurrence of additional lattice points at the centre of opposite faces or at the centre of the cell. The seven crystal systems (Table 2.1), in order of increasing symmetry (triclinic, monoclinic, orthorhombic, trigonal, tetragonal, hexagonal, cubic) and the fourteen Bravais lattices (Table 2.2) are reported in Fig. 2.3.

Before closing, we remark that the symmetry properties of a crystal are not confined to the translational invariance, but include in general also appropriate point group operations.

**Table 2.1** Seven Crystal Systems and the Relationship Between the Primitive Unit Vectors and Angles

| *Crystal system* | *Unit vectors* | *Angles* |
|---|---|---|
| Triclinic | $a \neq b \neq c$ | $\alpha \neq \beta \neq \gamma$ |
| Monoclinic | $a \neq b \neq c$ | $\alpha = \gamma = 90 \neq \beta$ |
| Orthorhombic | $a \neq b \neq c$ | $\alpha = \gamma = \beta = 90$ |
| Tetragonal | $a = b \neq c$ | $\alpha = \beta = \gamma = 90$ |
| Rhombohedral | $a = b = c$ | $\alpha = \beta = \gamma \neq 90$ |
| Hexagonal | $a = b \neq c$ | $\alpha = \beta = 90$ $\gamma = 120$ |
| Cubic | $a = b = c$ | $\alpha = \beta = \gamma = 90$ |

**Table 2.2** Bravais Lattices and Crystal Types

| *Crystal type* | *Bravais lattices* |
|---|---|
| Triclinic | Simple triclinic |
| Monoclinic | Simple monoclinic; base-centred monoclinic |
| Orthorhombic | Simple orthorhombic; base-centred orthorhombic body-centred orthorhombic; face-centred orthorhombic |
| Rhombohedral | Rhombohedral |
| Tetragonal | Simple tetragonal; body-centred tetragonal |
| Hexagonal | Hexagonal |
| Cubic | Simple cubic; body-centred cubic; face-centred cubic |

## 2.10 THE SEVEN CRYSTAL SYSTEMS

Depending on the lengths of the primitive axes and angles between them there are seven crystal systems and 14 Bravais lattices in three dimensions. These are given in Table 2.3.

**Table 2.3** The Seven Crystal Systems. (14 Lattice Types in 3 Dimensions)

| *System* | *Number of lattices* | *Crystallographic axes (of conventional cell)* | *Essential symmetry* |
|---|---|---|---|
| Triclinic | 1 (P) | $a \neq b \neq c$<br>$\alpha \neq \beta \neq \gamma \neq 90°$ | No planes, no axes |
| Monoclinic | 2 (P, C) | $a \neq b \neq c$<br>$\alpha \neq \gamma \neq 90° \neq \beta$ | One 2-fold axis or one plane |
| Orthorhombic | 4 (P, C, F, I) | $a \neq b \neq c$<br>$\alpha = \beta = \gamma = 90°$ | Two mutually ⊥ 2-fold axes or two planes intersecting in a 2-fold axis |
| Tetragonal | 2 (P, I) | $a = b \neq c$<br>$\alpha = \beta = \gamma = 90°$ | One 4-fold axis or a 4-fold inversion axis |
| Rhombohedral (Trigonal) | 1 (R) | $a = b = c$<br>$\alpha = \beta = \gamma < 120° \neq 90°$ | One 3-fold axis |
| Hexagonal | 1 (P) | $a = b \neq c$<br>$\alpha = \beta = 90°$<br>$\gamma = 120°$ | One 6-fold axis |
| Cubic | 3 (P, F, I) | $a = b = c$<br>$\alpha = \beta = \gamma = 90°$ | Four 3-fold axes |

## 2.11 CRYSTAL STRUCTURE, LATTICE WITH A BASIS

A physical crystal can be described by giving its underlying Bravais lattice, together with a description of the arrangement of atoms, molecules, ions, etc., within a particular primitive

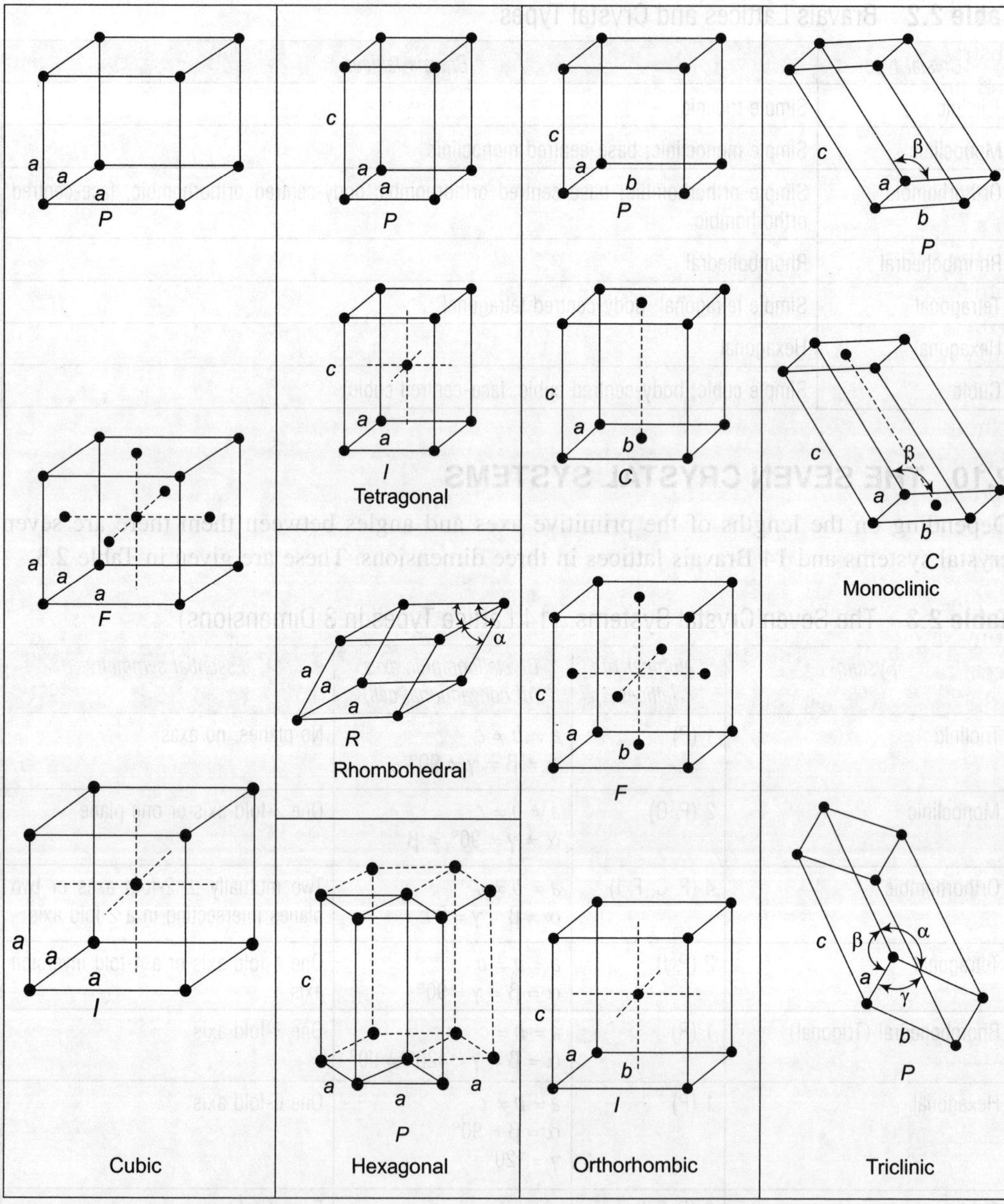


**Figure 2.20** Conventional unit cells of the 14 Bravais space lattices. The capital letters refer to the type of cell-*P*: primitive cell; *C*: cell with a lattice point in the centre of two parallel faces; *F*: cell with a lattice point in the centre of each face; *I*: cell with a lattice point in the centre of the interior; *R*: rhombohedral primitive cell. All points indicated are lattice points. There is no general agreement on the unit cell to use for the hexagonal Bravais lattice; some prefer the *P* cell shown with solid lines, and others prefer the *C* cell shown in dashed lines

cell. When emphasizing the difference between the abstract pattern of points composing the Bravais lattice and an actual physical crystal* embodying the lattice, the technical term "crystal structure" is used. *A crystal structure consists of idendical copies of the same physical unit, called the basis, located at all the points of a Bravais lattice* (or, equivalently, translated through all the vectors of a Bravais lattice). Sometimes the term lattice with a basis is used instead. However, "lattice with a basis" is also used in a more general sense to refer to what results even when the basic unit is *not* a physical object or objects, but another set of points. For example, the vertices of a two-dimensional honeycomb, though not a Bravais lattice, can be represented as a two-dimensional triangular Bravais lattice** with a two-point basis (Fig. 2.21). A crystal structure with a basis consisting of a single atom or ion is often called a monatomic Bravais lattice.

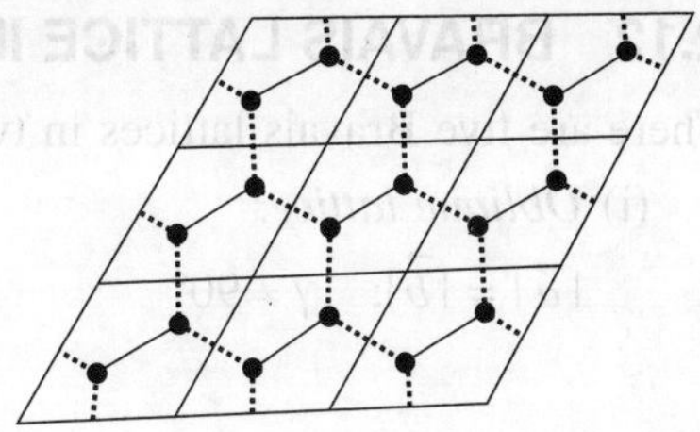

**Figure 2.21** The honeycomb net, drawn so as to emphasize that it is a Bravais lattice with a two-point basis. The pairs of points joined by heavy solid lines are identically placed in the primitive cells (parallelogram) of the underlying Bravais lattice

One can also describe a Bravais lattice as a lattice with a basis by choosing a non-primitive conventional unit cell. This is often done to emphasize the cubic symmetry of the bcc and fcc Bravais lattices, which are then described respectively, as simple cubic lattices spanned by $a\hat{\mathbf{x}}$, $a\hat{\mathbf{y}}$, and $a\hat{\mathbf{z}}$, with a two-point basis

$$\mathbf{0},\ \frac{a}{2}(\hat{\mathbf{x}}+\hat{\mathbf{y}}+\hat{\mathbf{z}}) \qquad \text{(bcc)} \qquad ...(1)$$

or a four-point basis

$$\mathbf{0},\ \frac{a}{2}(\hat{\mathbf{x}}+\hat{\mathbf{y}}),\ \frac{a}{2}(\hat{\mathbf{y}}+\hat{\mathbf{z}}),\ \frac{a}{2}(\hat{\mathbf{z}}+\hat{\mathbf{x}}) \qquad \text{(fcc)} \qquad ...(2)$$

The primitive unit cell volume of different lattice types are provided in Table 2.4.

**Table 2.4** Unit Cell Volume of Different Lattice Types

| *Lattice type* | *Volume* |
|---|---|
| Cubic | $a^3$ |
| Orthorhombic | $abc$ |
| Tetragonal | $a^2c$ |
| Hexagonal | $\frac{\sqrt{3}a^2c}{2}$ |
| Rhombohedral | $a^3\sqrt{1-3\cos^2\alpha+2\cos^3\alpha}$ |
| Monoclinic | $abc\sin\beta$ |
| Triclinic | $abc\sqrt{1-\cos\alpha-\cos\beta-\cos\gamma+2\cos\alpha\cos\beta\cos\gamma}$ |

* But still idealized in being infinite in extent.

** Spanned by two primitive vectors of equal length, making an angle of 60°.

## 2.12 BRAVAIS LATTICE IN TWO DIMENSIONS

There are five Bravais lattices in two dimensions.

(i) *Oblique lattice:*

$|\vec{a}| = |\vec{b}|$; $\gamma \neq 90°$

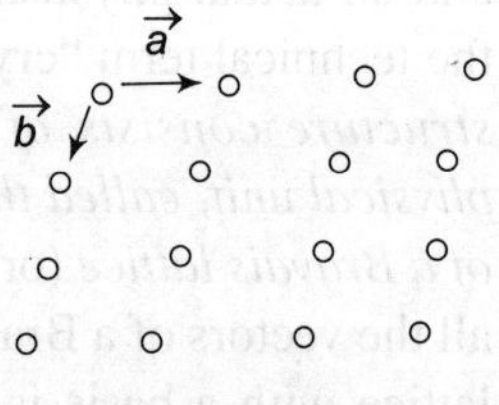


**Figure 2.22**

(ii) *Square lattice:*

$|\vec{a}| = |\vec{b}|$; $\gamma = 90°$

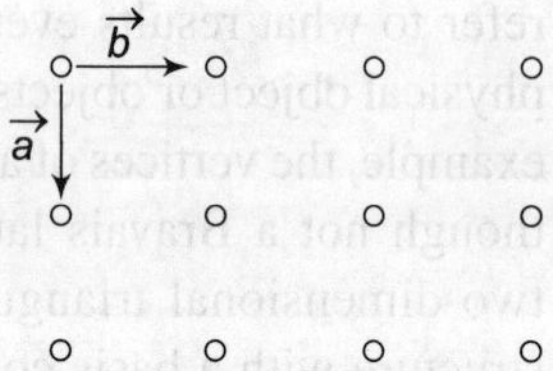


**Figure 2.23**

(iii) *Hexagonal lattice:*

$|\vec{a}| = |\vec{b}|$; $\gamma = 120°$

The conventional unit cell is a rhombus as shown (in adjacent Fig. 2.24).

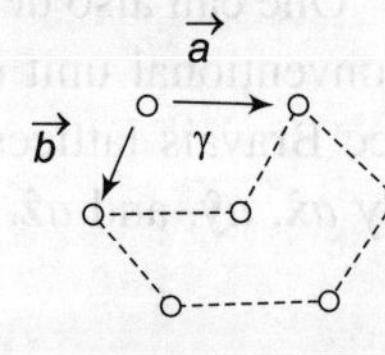


**Figure 2.24**

(iv) *Rectangular:*

$|\vec{a}| \neq |\vec{b}|$; $\gamma = 90°$

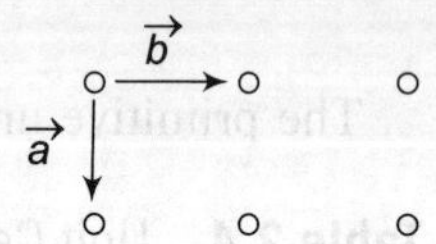


**Figure 2.25**

(v) *Centred-rectangular*

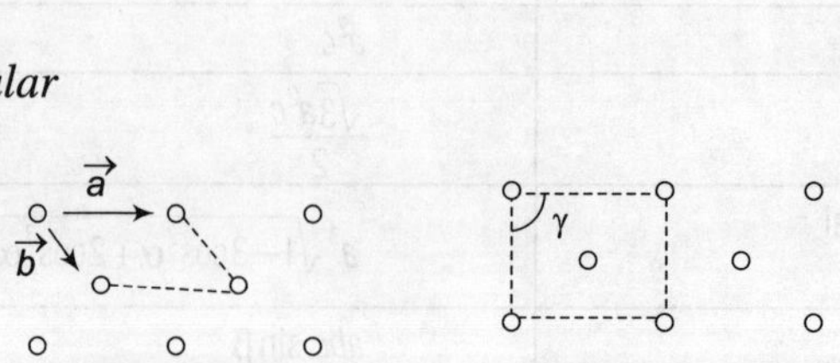


(a) (Primitive cell shown dotted) (b) Rectangular unit cell

**Figure 2.26** Both (a) and (b) are centred rectangular

$$\vec{a} = a_x\hat{i} \qquad |\vec{a}| \neq |\vec{b}|$$

$$\vec{b} = \frac{1}{2}a_x\hat{i} + b_y\hat{j}, \qquad \gamma = 90°$$

These above Bravais lattices satisfy the requirement that a lattice should be invariant under a rotation operation $2\pi/n$; ($n$ = 1, 2, 3, 4, 6) and/or restriction on the primitive translation vectors $\vec{a}$ and $\vec{b}$ under the mirror operation.

The various combinations of allowed rotation and reflection operations are found to give rise to 10 different two-dimensional point groups permissible in a crystal.

**Example 2.2** Find the operator associated with translational symmetry of the linear lattice.

***Solution*** Consider a finite translation by an operator $\hat{T}(\lambda a)$ in the form:

$$\hat{T}(\lambda a)\,\psi(x) = \psi\,(x - \lambda a)$$

where $a$ is the lattice constant and $\lambda$ is an integer. One may expand $\psi(x - \lambda a)$ as a power series.

$$\psi(x - \lambda a) = \psi(x) + \sum_{n=1}^{\infty}(-1)^n \frac{(\lambda a)^n}{n!}\frac{d^n\psi(x)}{dx^n}$$

which is dependent on the momentum operator since,

$$\frac{d}{dx}\psi(x) = \frac{i}{\hbar}\left(-i\hbar\frac{d}{dx}\right)\psi(x) = \frac{i}{\hbar}\,\hat{p}_x\psi(x)$$

We therefore obtain

$$\hat{T}(\lambda a)\,\psi(x) = \left[1 + \sum_{n=1}^{\infty}\frac{1}{n!}\left(-\frac{i}{\hbar}\lambda a p_x\right)^n\right]\psi(x)$$

$$= e^{-i\,\lambda a\,\hat{p}_x/\hbar}\psi(x)$$

or

$$\hat{T}(\lambda a) = e^{-i\,\lambda a\,\hat{p}_x/\hbar}.$$

If the translation operator is applied to a plane-wave state that represents a free particle, one obtains

$$\hat{T}(\lambda a)\,\psi(\vec{r}) = e^{-i\,\lambda a\,\hat{p}_x/\hbar}\,(Ae^{i\vec{k}\cdot\vec{r}})$$

$$= e^{-i\lambda a\,(-i\hbar\,\partial/\partial x)/\hbar}\cdot Ae^{i\vec{k}\cdot\vec{r}}$$

$$= e^{i\lambda a\,(i\,\partial/\partial x)}\,Ae^{i(k_x x + k_y y + k_z z)}$$

$$= e^{-\lambda a\,(i\,k_x x)}\,Ae^{i\vec{k}\cdot\vec{r}}$$

$$= e^{-i\lambda a k_x}\,\psi(\vec{r})$$

which means that the physical momentum of the free particle is given by the eigenvalue, $p_x = \hbar k_x$.

The wavenumber $k_x$ appears to be a quantum number characteristic of the translational symmetry of the linear lattice and $\vec{p} = \hbar k$ is known as the *crystal momentum*. Although for most purposes, crystal momentum behaves like a real momentum, it is not a true momentum.

## 2.13 SIMPLE AND COMPOSITE CRYSTAL STRUCTURES

The geometrical description of a crystal requires the specification of the *primitive translation vectors* $\vec{t}_1$, $\vec{t}_2$, $\vec{t}_3$ of the underlying Bravais lattice, as well as the specification of the atoms (or ions) in the primitive cell. The content of the unit cell is described by means of an appropriate set of *basis vectors* $\vec{d}_1, \vec{d}_2, \ldots \vec{d}_\nu$, which individuate the equilibrium positions of the nuclei of all the atoms (or ions) in the unit cell and, for consistency, the positions of the atoms in the basis cannot be related by translational vectors. The foregoing considerations are summarized as follows:

$$\text{Crystal structure} \Rightarrow \begin{cases} \vec{t}_1, \vec{t}_2, \vec{t}_3 \text{ (primitive translation vectors)} \\ \vec{d}_1, \vec{d}_2 \ldots \vec{d}_\nu \text{ (basis of equal or different atoms)} \end{cases}$$

The geometrical description of the crystal structure is thus specified by the triad of primitive translation vectors of the Bravais lattice and by the vectors forming the basis.

A crystal with a single atom in the primitive unit cell is called *simple crystal* (or *simple lattice*); in this case, the basis contains a single vector which can be taken to be zero (with appropriate choice of the origin in the crystal). In a simple lattice, the equilibrium atomic positions $\vec{R}_n$ coincide with the Bravais translation vectors $\vec{t}_n$, and we have

$$\vec{R}_n = n_1 \vec{t}_1 + n_2 \vec{t}_2 + n_3 \vec{t}_3$$

A crystal with two or more atoms (or ions) in the primitive unit cell is called a *composite crystal* (or a *composite lattice*). The equilibrium atomic positions are at the points:

$$\vec{R}_n^{(1)} = \vec{d}_1 + n_1 \vec{t}_1 + n_2 \vec{t}_2 + n_3 \vec{t}_3$$

$$\vec{R}_n^{(2)} = \vec{d}_2 + n_1 \vec{t}_1 + n_2 \vec{t}_2 + n_3 \vec{t}_3$$

$$\cdots\cdots\cdots\cdots\cdots\cdots$$

$$\cdots\cdots\cdots\cdots\cdots\cdots$$

$$\vec{R}_n^{(i)} = \vec{d}_i + n_1 \vec{t}_1 + n_2 \vec{t}_2 + n_3 \vec{t}_3$$

A composite lattice can be thought of as composed by a number of interpenetrating simple lattices (called *sublattices*) equal to the number of vectors of the basis. All the points of a given sublattice must thus be occupied by atoms of the same type; different sublattices are occupied by atoms of the same or different type.

## 2.14 WIGNER–SEITZ CELLS

We have seen that the smallest possible unit cell is a primitive cell. Frequently however, a larger non-primitive unit cell is used, which might have the advantage that the symmetry can be better recognized (bcc and fcc cells are examples of a conventional unit cell).

A convenient way of choosing the unit cell is to take the volume bounded by the planes that right bisect the nearest $R_n$. Such a unit cell is known as *Wigner-Seitz cell.*

It is a special type of primitive cell, which describes the symmetry of the cells. Figure 2.27 shows a Wigner-Seitz cell for a two-dimensional oblique lattice.

Figure 2.28 depicts Wigner-Seitz cell for the bcc lattice.

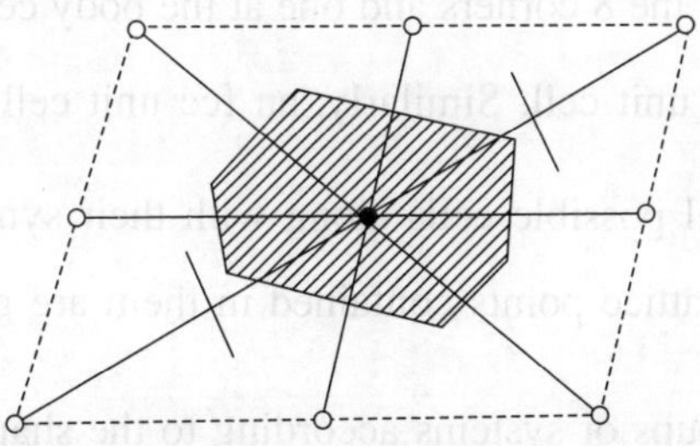

**Figure 2.27** A Wigner–Seitz cell for a two-dimensional oblique lattice

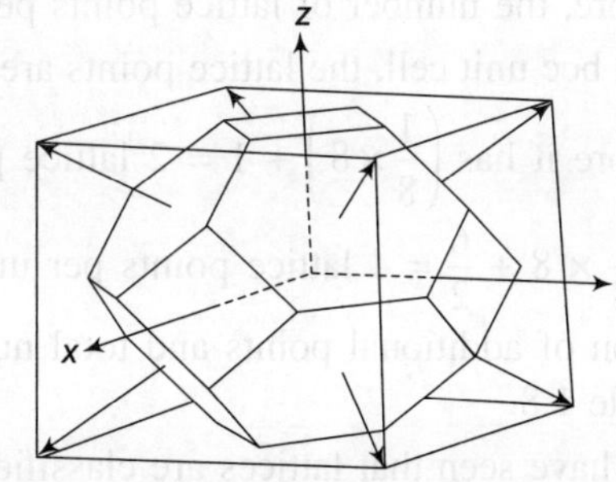


**Figure 2.28** Wigner–Seitz cell for the bcc structure

In the fcc lattice, the atoms are arranged on the corners and faces of a cube, which is equivalent to the centre points of the edges (Fig. 2.29). The Wigner-Seitz cell for fcc structure is shown in Fig. 2.30.

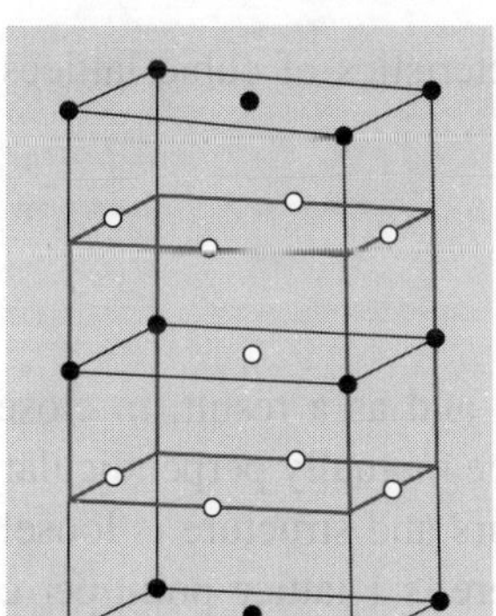

**Figure 2.29** Conventional unit cell of fcc structure. In the cell marked black, the atoms are situated on the corners and cube faces. In the white cell, the atoms are at the centres of the edges and the centre of the cell. (After: *Electronic Properties of Materials*, Rolf E. Hummel, Narosa.)

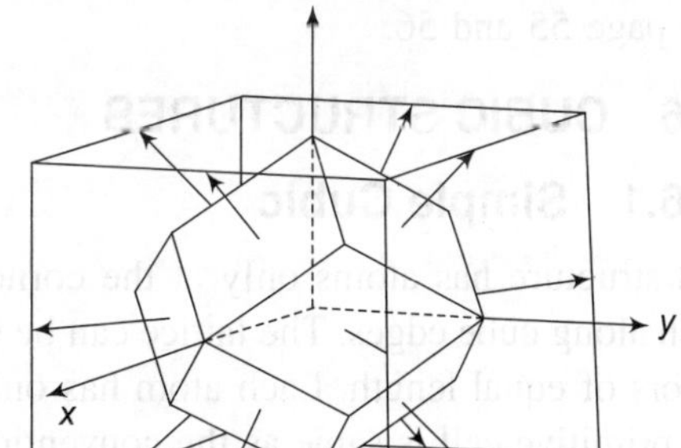


**Figure 2.30** Wigner-Seitz cell for fcc structure. It is constructed from white cell of Fig. 2.29

## 2.15 NUMBER OF LATTICE POINTS PER UNIT CELL

The number of lattice points in a unit cell depends on its nature and in a crystal, each lattice point is occupied by a group of atoms/molecules. In a primitive cell, lattice points are located only at the corners, while each conner of the cell is common to 8 neighbouring unit cells, so, the contributuion towards the unit cell per corner is only 1/8. Since the cell has 8 corners, therefore, the number of lattice points per unit cell is only one.

In a bcc unit cell, the lattice points are located at the 8 corners and one at the body centre, therefore it has $\left(\frac{1}{8}\times 8\right) + 1 = 2$ lattice points per unit cell. Similarly, an fcc unit cell will have $\frac{1}{8} \times 8 + \frac{6}{2} = 4$ lattice points per unit cell. All possible cells along with their symbol, location of additional points and total number of lattice points contained in them are given in Table 1.8.

We have seen that lattices are classified into groups or systems according to the shape of the unit cell which generates the entire lattice by translation. The shape of the unit cell is chosen for geometrical convenience and so may contain more than one lattice point. The unit cell is also a *primitive* ($p$) cell, if lattice points are located at the corners only. Otherwise it may be *body-centred* ($I$), with an extra point at the centre, *face-centred* ($F$), with an extra point at the centre of each face or *base-centred* ($c$), having a point at the centre of the base. Symmetry considerations restrict the number of different possible spatial arrangements, and this accounts for our hypothesis of treating the solid as a system of atoms with a periodic arrangement of discrete cells rather than a continuum.

Table 2.5 shows crystal systems and Table 2.6, characteristics of cubic lattices given on next page 55 and 56.

## 2.16 CUBIC STRUCTURES

### 2.16.1 Simple Cubic

This structure has atoms only at the corners of the cube and as a result, in close packing, touch along cube edges. The lattice can be spanned by three mutually perpendicular primitive vectors of equal length. Each atom has only six neighbours and structure is loosely packed. The primitive cell is same as the conventional cell, so there is 1 lattice point per cell.

Cell volume = $a^3$

nearest neighbour distance = $\frac{a}{2} + \frac{a}{2} = a$

(e.g., between $A$ and $B$)

$$AC^2 = AB^2 + BC^2 = a^2 + a^2$$

∴ Second neighbour distance

$$AC = \sqrt{2}\,a.$$

From Fig. 2.31 $R + R = a$ (considering $AB$)

(if $R$ = radius of sphere $A$ or $B$)

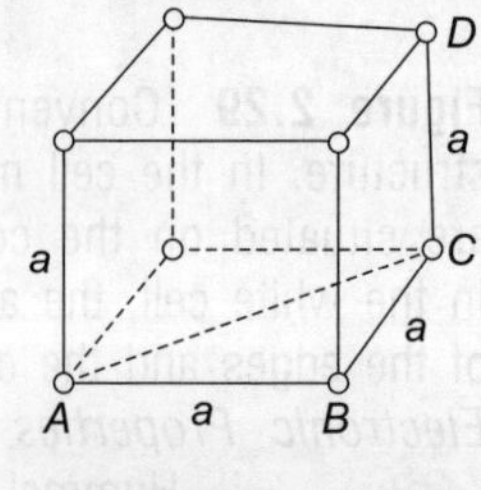


**Figure 2.31** Simple cubic structure

**Table 2.5** Specifications and Occurrences of Crystal Systems

| *Crystal system* | *No. in system* | *Unit cell* | *Schoenflies' symbol* | *International* | *Lattice symbol* | *Occurrence* |
|---|---|---|---|---|---|---|
| Triclinic | 1 | $a_1 \neq a_2 \neq a_3$<br>$\alpha \neq \beta \neq \gamma$ | $C_i$ | $\bar{1}$ | *P* | $FeSO_4 \cdot 5H_2O$,<br>$CaS_2O_3 \cdot 6H_2O$ |
| Monoclinic | 2 | $a_1 \neq a_2 \neq a_3$<br>$\alpha = \beta = 90° \neq \gamma$ | $C_{2h}$ | 2/m | *P*<br>*C* | $Na_2B_4O_7 \cdot 10H_2O$ (borex),<br>$Li_2SO_4 \cdot H_2O$ |
| Orthorhombic | 4 | $a_1 \neq a_2 \neq a_3$<br>$\alpha = \beta = \gamma = 90°$ | $D_{2h}$ | (2/m)(2/m)(2/m) | *P*<br>*C*<br>*I*<br>*F* | $PbSO_4$<br>$C_6H_2(NO_2)_3$ OH |
| Tetragonal | 2 | $a_1 = a_2 \neq a_3$<br>$\alpha = \beta = \gamma = 90°$ | $D_{4h}$ | 4/mmm | *P*<br>*I* | In, HgCl, Sn |
| Cubic | 3 | $a_1 = a_2 = a_3$<br>$\alpha = \beta = \gamma = 90°$ | $O_h$ | (4/m)($\bar{3}$)(2/m) | *P*<br>*I*<br>*F* | Most metals abd their alloys, NaCl |
| Trigonal | 1 | $a_1 = a_2 = a_3$<br>$\alpha = \beta = \gamma$<br>$<120°, \neq 90°$ | $C_{3v}$ | 3/m | *R* | $CaCO_3$, $MgSO_4 \cdot 6H_2O$ |
| Hexagonal | 1 | $a_1 = a_2 \neq a_3$<br>$\alpha = \beta = 90°, \gamma = 120°$ | $D_{6h}$ | 6/mmm | *P* | $Be_3Al_2(Si_6O_{18})$,<br>Be, Mg, Zn, Cd |

or $$a = 2R \qquad \therefore \quad a^3 = 8R^3$$

There is one lattice point per cell, so

$$\text{Packing fraction} = \frac{\text{Volume of a sphere}}{\text{Volume of cube}}$$

$$= \frac{\frac{4}{3}\Pi R^3}{a^3} = \frac{\frac{4}{3}\Pi R^3}{8R^3} = \frac{1}{6}\Pi.$$

### 2.16.2 Body-centred Cubic (bcc) Structure

The conventional unit cell is cubic with an atom at each corner and an atom at the centre of the cube. Each atom has eight neighbours and atoms are in contact along body diagonals (see Fig. 2.32). Thus,

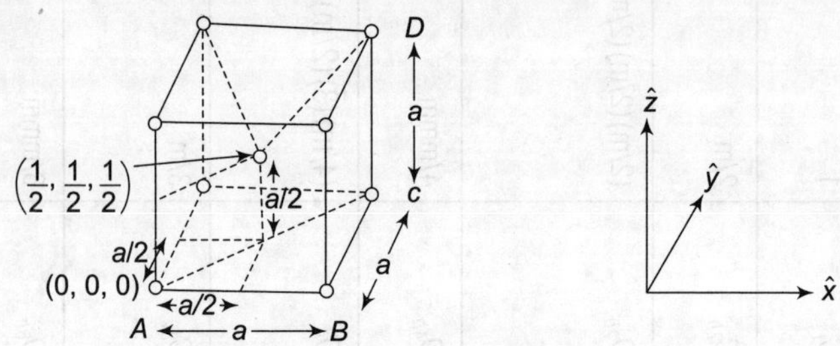


**Figure 2.32** bcc structure

There are $\left(8 \times \frac{1}{8}\right) + 1 = 2$ lattice points per cell. Let $a$ = cube edge.

From $$\Delta ABC,\ AC^2 = AB^2 + BC^2 = a^2 + a^2$$

$$\therefore \quad AC = \sqrt{2}\,a.$$

If $R$ = radius of an atom,

$$AD = R + 2R + R = 4R$$

For close packing $$AD^2 = AC^2 + CD^2$$

or $$(4R)^2 = (\sqrt{2}\,a)^2 + a^2$$

$$16\,R^2 = 2a^2 + a^2$$

$$\therefore \quad a = \frac{4R}{\sqrt{3}}$$

Volume of the cube $$a^3 = \left(\frac{4R}{\sqrt{3}}\right)^3$$

$$\text{Packing fraction} = \frac{\text{Volume of 2 spheres}}{\text{Volume of cube}} = \frac{2 \times \frac{4}{3}\Pi R^3}{\left(\frac{4R}{\sqrt{3}}\right)^3} = \left(\frac{\sqrt{3}}{8}\right)\Pi = 0.68.$$

**Table 2.6** Characteristics of Cubic Lattices

| *Characteristic* | *SC* | *bcc* | *fcc* |
|---|---|---|---|
| Volume of conventional cell | $a^3$ | $a^3$ | $a^3$ |
| Lattice points per cell | 1 | 2 | 4 |
| Volume, primitive cell | $a^3$ | $\frac{1}{2}a^3$ | $\frac{1}{4}a^3$ |
| Number of nearest neighbours | 6 | 8 | 12 |
| Nearest neighbour distance | $a$ | $\sqrt{3}\frac{a}{2}$ | $\frac{a}{\sqrt{2}}$ |
| Number of second neighbours | 12 | 6 | 6 |
| Second neighbour distance | $\sqrt{2}a$ | $a$ | $a$ |
| Packing Fraction | $\frac{\Pi}{6}$ | $\frac{\Pi\sqrt{3}}{8}$ | $\frac{\Pi\sqrt{2}}{6}$ |

Figures 2.33 and 2.34 show the primitive cell and primitive translational vectors for a bcc lattice.

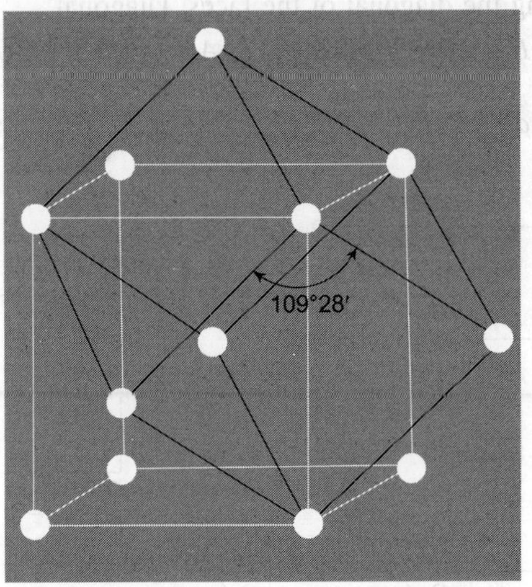


**Figure 2.33** bcc lattice showing a primitive cell (solid lines). It is a rhombohedron of edge $\sqrt{3}\,a/2$.

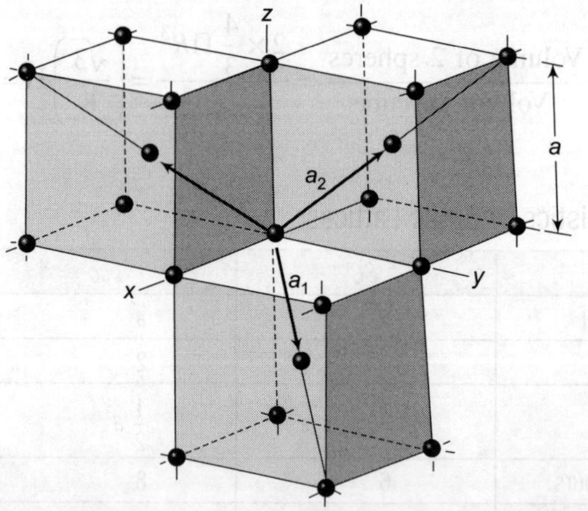


**Figure 2.34** Primitive translation vectors of a bcc lattice. These connect the lattice point at the origin to lattice points at the body centres, The primitive cell is obtained on completing the rhombohedron

### 2.16.3 Face-centred Cubic (fcc) Structure

The conventional unit cell is cubic with an atom at each of the eight corners and an atom at each of the six cube faces. So, there are

$$\left(8\times\frac{1}{8}\right)+\left(6\times\frac{1}{2}\right)=4$$

lattice points per unit cell, because each atom at a face centre is shared by two cells. The atoms are in contact along the diagonal of the faces. Diagonal

$$AD = R + 2R + R = 4R, \qquad AD^2 = AC^2 + CD^2$$

$$(4R)^2 = a^2 + a^2 \qquad \therefore \quad a = \frac{4R}{\sqrt{2}} \qquad \text{(see Fig. 2.35)}$$

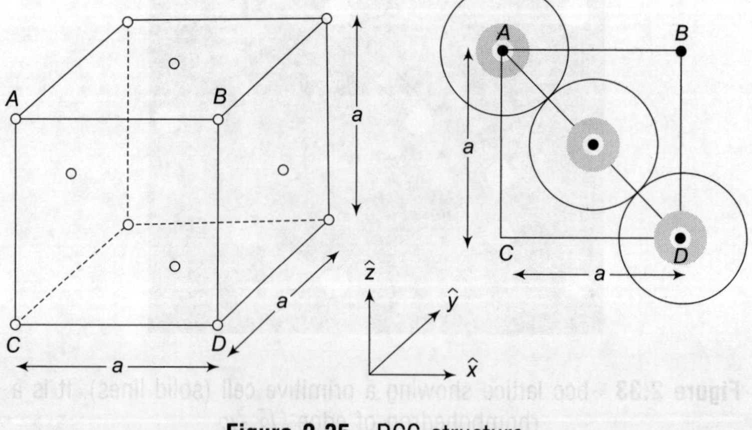


**Figure 2.35** RCC structure

$$\text{Volume of the cube} = \left(\frac{4R}{\sqrt{2}}\right)^3$$

$$\text{Packing fraction} = \frac{\text{Volume of 4 spheres}}{\text{Volume of cube}}$$

$$= \frac{4 \times \frac{4}{3}\pi R^3}{\left(\frac{4R}{\sqrt{2}}\right)^3} = \sqrt{2}\left(\frac{\pi}{6}\right) = 0.74.$$

Figure 2.36 shows the rhombohedral primitive cell of the fcc crystal. The primitive translational vectors connect the lattice point at the origin with lattice points at the face centres.

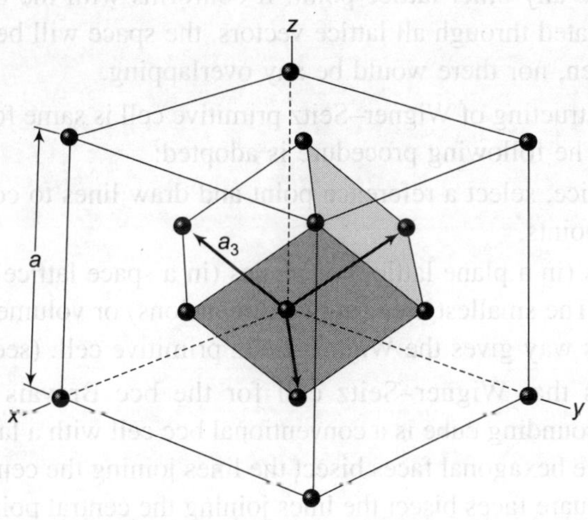


**Figure 2.36** The rhombohedral primitive cell of the fcc crystal

Figure 2.34 shows the primitive translation vectors $\vec{a}_1$, $\vec{a}_2$, $\vec{a}_3$, of the bcc lattice. As can be seen from the figure, these are

$$\vec{a}_1 = \frac{a}{2}(\hat{x} + \hat{y} + \hat{z})$$

$$\vec{a}_2 = \frac{a}{2}(-\hat{x} + \hat{y} + \hat{z})$$

$$\vec{a}_3 = \frac{a}{2}(\hat{x} - \hat{y} + \hat{z})$$

Figure 2.36 shows the primitive vectors of the fcc lattice. From the figure, these are

$$\vec{a}_1 = \frac{a}{2}(\hat{x} + \hat{y})$$

$$\vec{a}_2 = \frac{a}{2}(\hat{y} + \hat{z})$$

$$\vec{a}_3 = \frac{a}{2}(\hat{z} + \hat{x})$$

Here, $\hat{x}$, $\hat{y}$, $\hat{z}$ are the cartesian unit vectors. For Fig. 2.36 (i.e., fcc), the angles between the axes are 60°.

The primitive cell in each case contains one lattice point as per definition.

## 2.17 WIGNER-SEITZ UNIT CELL

Another way of choosing a primitive cell of equal volume as given by

$$V_c = |\vec{a} \times \vec{b} \cdot \vec{c}|$$

has been suggested by E.P. Wigner and F. Seitz and is known as Wigner–Seitz primitive unit cell.

The Wigner–Seitz unit cell about a lattice point is the region of space that is closer to that particular point than to any other lattice point. It conforms with the translational symmetry and hence when translated through all lattice vectors, the space will be just filled, leaving no empty space in between, nor there would be any overlapping.

The method of constructing of Wigner–Seitz primitive cell is same for both two- and three-dimensional lattices. The following procedure is adopted:

1. For a given lattice, select a reference point and draw lines to connect this point to all nearby lattice points.
2. Draw new lines (in a plane lattice) or planes (in a space lattice) bisecting each of the previous lines. The smallest area (in two dimensions) or volume (in three dimensions) enclosed in this way gives the Wigner–Seitz primitive cell. (see Fig. 2.37)

Figure 2.38 shows that Wigner–Seitz cell for the bcc Bravais lattice (a "truncated octahedron"). The surrounding cube is a conventional bcc cell with a lattice point at its centre and on each vertex. The hexagonal faces bisect the lines joining the central point to the points on the vertices. The square faces bisect the lines joining the central point to the central points in each of the six neighbouring cubic cells (not drawn). The hexagons are regular.

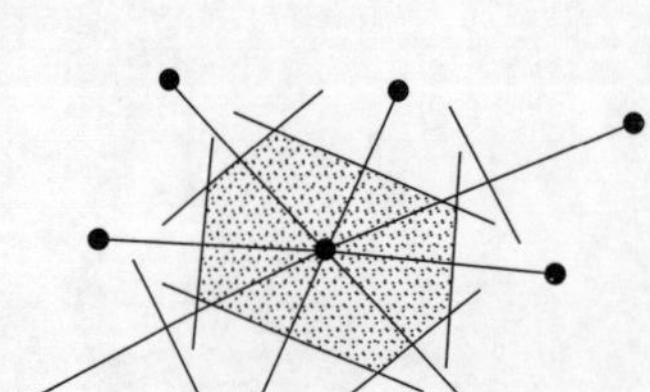

**Figure 2.37** A Wigner–Seitz cell for a two-dimensional oblique lattice

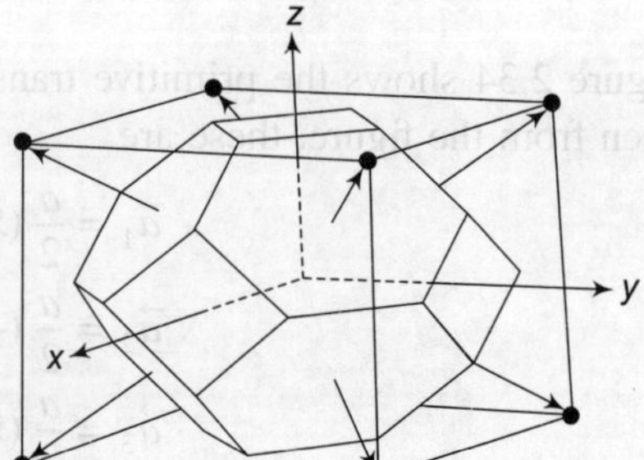


**Figure 2.38** Wigner–Seitz cell for the bcc structure

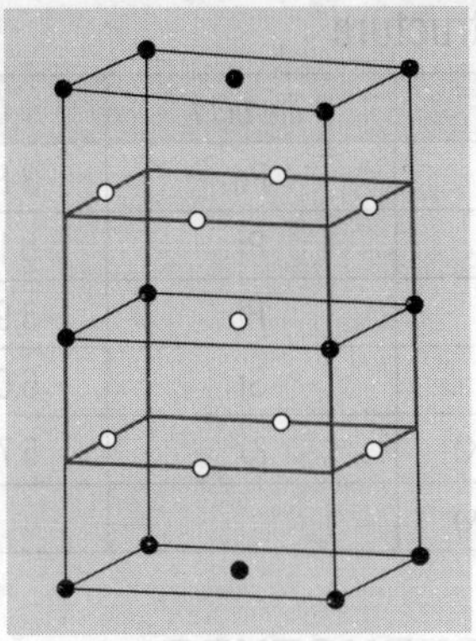

**Figure 2.39** Conventional unit cell of fcc structure. In the cell marked black, the atoms are situated on the corners and cube faces. In the white cell, the atoms are at the centres of the edges and the centre of the cell. (After: *Electronic Properties of Materials*, Rolf E. Hummel, Narosa)

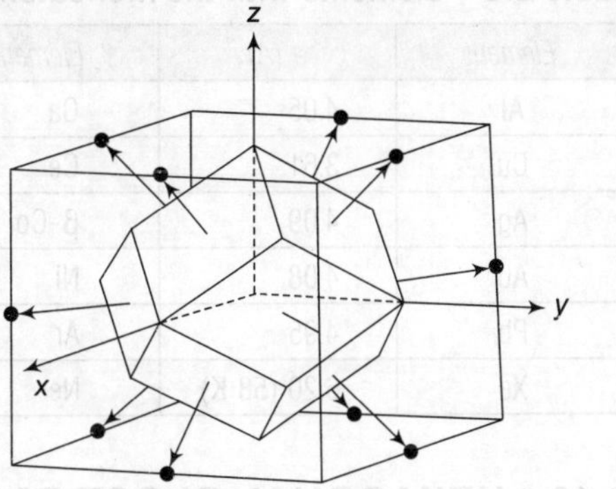


**Figure 2.40** Wigner–Seitz cell for fcc structure. It is constructed from white cell of Fig. 2.39

Figure 2.40 shows the Wigner–Seitz cell for the fcc Bravais lattice (a "rhombic dodecahedron"). The surrounding cube is not the conventional cubic cell but one in which lattice points are at the centre of the cube and at the centre of the 12 edges. Each of the 12 (congruent) faces is perpendicular to a line joining the central point to a point on the centre of an edge.

## 2.18 EXAMPLES OF bcc AND fcc STRUCTURES

All the alkali metals crystallize in the body centred cubic lattice. Table 2.7 shows lattice constants of some of the elements.

**Table 2.7** Elements with the Monatomic bcc Crystal Structure

| *Element* | *a (Å)* | *Element* | *a (Å)* |
|---|---|---|---|
| Ba | 5.02 | Fe | 2.87 |
| Li | 3.49 (78 K) | Mo | 3.15 |
| Na | 4.23 (5 K) | W | 3.16 |
| K | 5.23 (5 K) | Tl | 3.88 |
| Cs | 6.05 (78 K) | V | 3.02 |
| Nb | 3.30 | Ta | 3.31 |

Most other elements form face-centred cubic crystals (see Table 2.8).

**Table 2.8** Elements with the Monoatomic fcc Crystal Structure

| *Element* | *a (Å)* | *Element* | *a (Å)* | *Element* | *a (Å)* |
|---|---|---|---|---|---|
| Al | 4.05 | Ca | 5.58 | Pd | 3.89 |
| Cu | 3.61 | Ce | 5.16 | Pr | 5.16 |
| Ag | 4.09 | β-Co | 3.55 | Pt | 3.92 |
| Au | 4.08 | Ni | 3.52 | Sr | 6.08 |
| Pb | 4.95 | Ar | 5.26 (4.2 K) | Kr | 5.72 (58 K) |
| Xe | 6.20 (58 K) | Ne | 4.43 (4.2 K) | | |

## 2.19 HEXAGONAL CLOSE PACKED (hcp) STRUCTURE

Above, we have described bcc and fcc structures as Bravais lattice with a basis to emphasize their cubic symmetry. However, though not a Bravais lattice, the hcp structure ranks in importance with the bcc and fcc Bravais lattices. The conventional unit cell is shown in Fig. 2.41, where points represent atom positions. (These are not lattice points.)

The hcp structure is a simple hexagonal Bravais lattice, given by two-dimensional triangular nets directly above each other (Fig. 2.42). The direction of stacking is known as the *c*-axis. The nets are spanned by two primitive vectors $a_1$ and $a_2$ of equal lengths, making an angle of 60°. The three primitive vectors are

$$a_1 = a_x$$

$$a_2 = a/2_x + \sqrt{3}/2\ a_y$$

$$a_3 = c_z$$

The hcp structure consists of two interpenetrating simple hexagonal Bravais lattices, displaced from one another by $u_1/3 + u_2/3 + u_3/2$ (see Fig. 2.42).

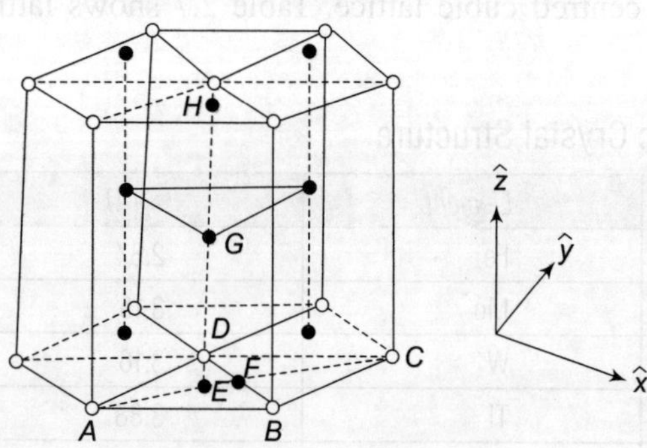


**Figure 2.41** hcp structure (conventional unit cell)

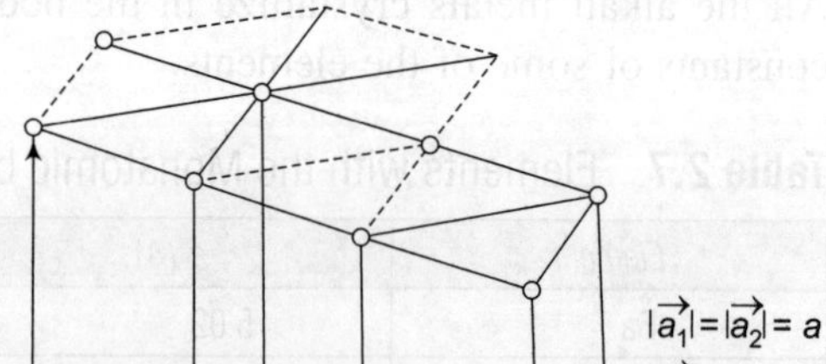


**Figure 2.42** The simple hexagonal Bravais lattice with two-dimensional triangular nets (shaded) stacked directly above one another, a distance *c* apart

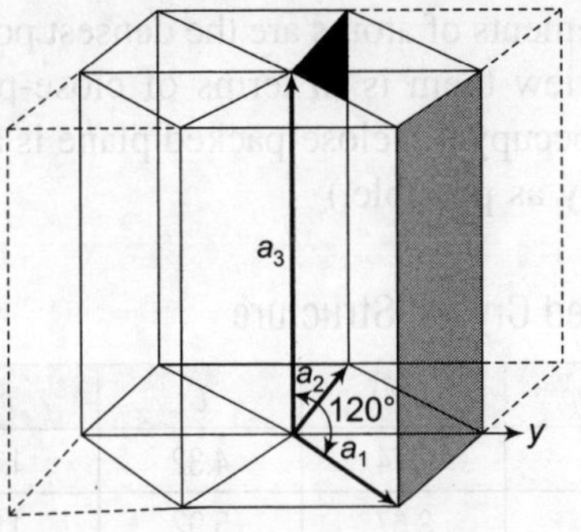


**Figure 2.43** Relation of the primitive cell in the hexagonal system (heavy lines) to a prism of hexagonal symmetry. Here $a_1 = a_2 \neq a_3$

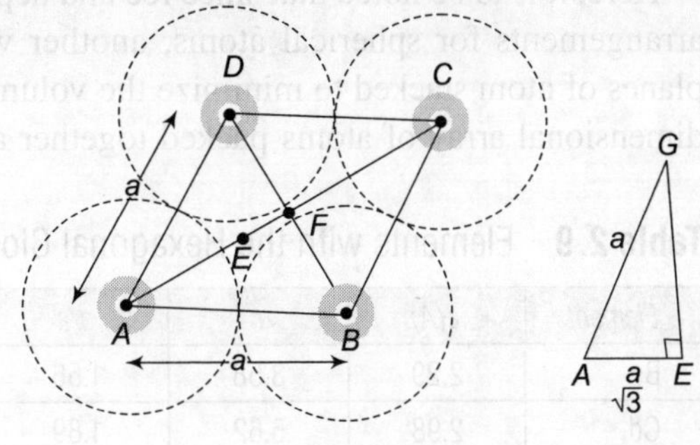


**Figure 2.44**

*ABCD* is the base of hexagonal unit cell, $AD = AB = a$. The sphere in the next upper layer of Fig. 2.41 has its centre ($G$) vertically above $E$ and it touches the three spheres whose centres are $A$, $B$ and $D$.

$AE$ is $\frac{2}{3}$ of median $AF$.

$$\therefore \quad AE = \frac{2}{3} \times \frac{\sqrt{3}}{2}\, a = a/\sqrt{3}$$

Hence,
$$EG = c/2 = \sqrt{a^2 - (a^2/3)} = a\sqrt{(2/3)}$$

$$\therefore \quad c/a = (2\sqrt{2})/\sqrt{3} = 1.63299$$

Spheres (lattice points) per conventional cell = (8 × 1/8) + 1 = 2

Therefore,

$$\text{Packing fraction} = \frac{\text{Volume occupied by atoms}}{\text{Volume of the unit cell}}$$

$$= \frac{2 \times (4/3)\,\pi R^3}{4 \times \text{area AFB} \times \text{height}} \quad \text{(from Fig. 2.44)}$$

$$= \frac{2 \times (4/3)\pi R^3}{4 \times (1/2.\sqrt{3}/2aR)c} = \frac{\left(\frac{8}{3}\right)\pi R^3}{\sqrt{3}a\,Rc} \quad \text{(from Fig. 2.44)}$$

Since $a = 2R$ (in ideal case),

∴ Packing fraction

$$= \frac{(8/3)\pi R^3}{2\sqrt{3}\,R^2 c} = \frac{(8/3)\pi R^3}{2\sqrt{3}\,R^2 \times 1.633\,(2R)} = \frac{(2/3)\pi}{\sqrt{3} \times 1.633} = 0.74$$

By convention, we refer to a crystal as hcp even if actual $c/a$ ratio departs somewhat from theoretical value of 1.633 as, for example, for zinc $c/a = 1.85$ and for magnesium $c/a = 1.62$.

Here, it is to be noted that since fcc and hcp arrangements of atoms are the densest possible arrangements for spherical atoms, another way to view them is in terms of close-packed planes of atom stacked to minimize the volume they occupy. (A close-packed plane is a two-dimensional array of atoms packed together as closely as possible.)

**Table 2.9** Elements with the Hexagonal Close-packed Crystal Structure

| *Element* | *a(Å)* | *c* | *c/a* | *Element* | *a(Å)* | *C* | *c/a* |
|---|---|---|---|---|---|---|---|
| Be | 2.29 | 3.58 | 1.56 | Os | 2.74 | 4.32 | 1.58 |
| Cd | 2.98 | 5.62 | 1.89 | Pr | 3.67 | 5.92 | 1.61 |
| Ce | 3.65 | 5.96 | 1.63 | Re | 2.76 | 4.46 | 1.62 |
| α-Co | 2.51 | 4.07 | 1.62 | Ru | 2.70 | 4.28 | 1.59 |
| Dy | 3.59 | 5.65 | 1.57 | Sc | 3.31 | 5.27 | 1.59 |
| Er | 3.56 | 5.59 | 1.57 | Tb | 3.60 | 5.69 | 1.58 |
| Gd | 3.64 | 5.78 | 1.59 | Ti | 2.95 | 4.69 | 1.59 |
| He (2K) | 3.57 | 5.83 | 1.63 | Tl | 3.46 | 5.53 | 1.60 |
| Hf | 3.20 | 5.06 | 1.58 | Tm | 3.54 | 5.55 | 1.57 |
| Ho | 3.58 | 5.62 | 1.57 | Y | 3.65 | 5.73 | 1.57 |
| La | 3.75 | 6.07 | 1.62 | Zn | 2.66 | 4.95 | 1.86 |
| Lu | 3.50 | 5.55 | 1.59 | Zr | 3.23 | 5.15 | 1.59 |
| Mg | 3.21 | 5.21 | 1.62 | | | | |
| Nd | 3.66 | 5.90 | 1.61 | 'Ideal' | | | 1.63 |

## 2.20 WURTZITE STRUCTURE

The Wurtzite structure consists of two interpenetrating hcp sublattices displaced along the *c*-axis by an amount $\vec{u}\cdot\vec{c}$; one of the two sublattices is occupied by one type of atoms, e.g., Zn, while the other by another type of atoms, e.g., *O*. Each atom has four nearest neighbours of the opposite type, when $u = 3/8$ and $c/a = \sqrt{8/3}$ each atom is at equal distance from its four neighbours. Important materials such as AlN, SiC, etc., crystallize in the Wurtzite structure.

## 2.21 DIAMOND/Si STRUCTURE

The crystal structure of silicon (or diamond) is of particular importance. The structure can be described as two interpenetrating face-centred cubic (fcc) lattices displaced by $(a/4)$ (1, 1, 1) (where (1, 1, 1) are the Miller indices), along the body diagonal of the conventional cube. The underlying Bravais lattice is fcc with two carbon atoms (for diamond) forming the basis (i.e., two lattice points per unit cell), the primitive translation vectors and the two vectors of the basis are

$$\vec{t}_1 = \frac{a}{2}(0, 1, 1); \qquad \vec{t}_2 = \frac{a}{2}(1, 0, 1); \vec{t}_3 = \frac{a}{2}(1, 1, 0)$$

$$\vec{d}_1 = 0, \qquad \vec{d}_2 = \frac{a}{4}(1, 1, 1)$$

The coordination number is 4, each atom in a given sublattice is surrounded by four atoms of the other sublattice, at distance $(a/4)\sqrt{3}$, in a tetrahedral configuration, there are 12 second neighbours at distance $(a/2)\sqrt{2}$. For instance, the carbon atoms at the origin ($\vec{t}_n = 0, \vec{d}_1 = 0$) is surrounded by four atoms in the positions

$$\vec{d}_2 = \frac{a}{4}(1, 1, 1); \qquad \vec{d}_2 - \vec{t}_1 = \frac{a}{4}(1, -1, -1)$$

$$\vec{d}_2 - \vec{t}_2 = \frac{a}{4}(-1, 1, -1); \qquad \vec{d}_2 - \vec{t}_3 = \frac{a}{4}(-1, -1, 1)$$

The diamond structure (Fig. 2.45) is also typical of Si, Ge and gray Sn. The lattice constants are $a$ = 3.57 Å for $C$ (diamond);

$a$ = 5.43 Å for Si; $\qquad a$ = 5.65 Å for Ge;

$a$ = 6.49 Å for $\alpha$ – Sn (grey tin)

The zinc blende (ZnS) structure is similar to that of diamond; now the two sublattices are occupied by two different types of atoms (or ions). Among crystals with this structure we mention: cubic ZnS ($a$ = 5.41 Å) AgI ($a$ = 6.47 Å), cubic BN ($a$ = 3.62 Å), CuCl ($a$ = 5.41 Å). This structure is also typical of several III-V and II-VI groups' semiconductor compounds; for example, GaAs ($a$ = 5.65 Å), Al As ($a$ = 5.66 Å), InAs ($a$ = 6.04 Å), GaSb ($a$ = 6.12 Å), CdTe ($a$ = 6.48 Å), HgTe ($a$ = 6.48 Å), ZnTe ($a$ = 6.09 Å).

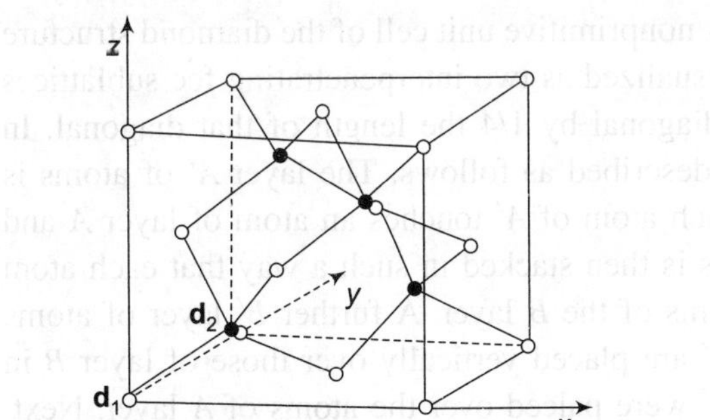


**Figure 2.45** Crystal structure of diamond and zinc blende; the end points of the vectors $\mathbf{d}_1$ and $\mathbf{d}_2$, of the basis are also indicated. in the diamond structure the two sublattices are occupied by atoms of the same type; in the zincblende structure the two sublattices are occupied by different types of atoms

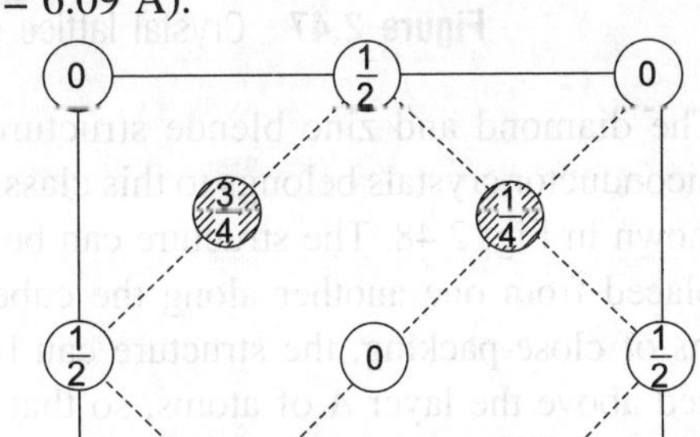


**Figure 2.46** Projection of atomic arrangement on a cube face in a diamond unit cell. The unit cell consists of two similar FCC lattices, displaced along the body diagonal of the cell by one-fourth of its length. The positions 0 and 1/2 refer to atoms on one lattice whole 1/4 and 3/4 give positions of atoms on the second lattice

## 2.22 SOME OTHER SIMPLE CRYSTAL STRUCTURES

The simplest crystal structures are those of NaCl and CsCl, shown in Fig. 2.47 (a) and (b), respectively. In the NaCl structure all the $Na^+$ ions are situated at the lattice points of the fcc sublattice, while all the $Cl^-$ ions are situated on the lattice points of another sublattice. The coordination number for this structure is 6. The CsCl structure is body centred with $Cl^-$ ions at the body centre position and the $Cs^+$ ions at the corner of the cube (or vice versa). The coordination number of this structure is 8.

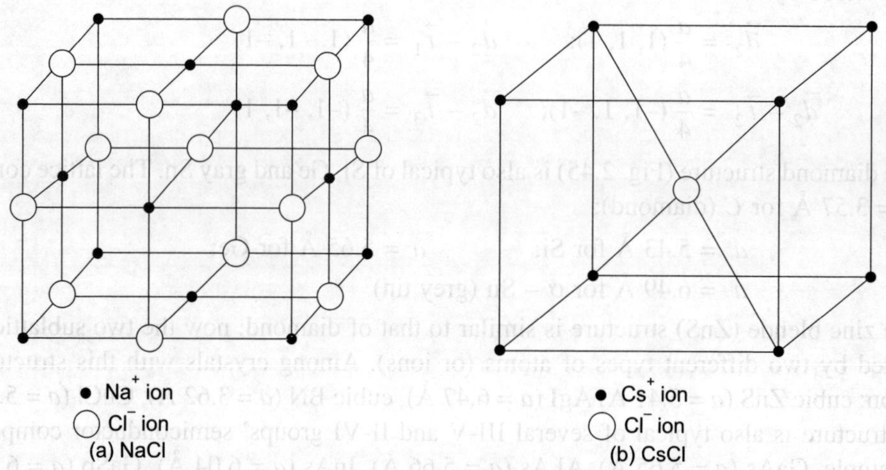


**Figure 2.47** Crystal lattice of (a) NaCl and (b) CsCl structures

The diamond and zinc blende structures are of interest to us because the majority of semiconductor crystals belongs to this class. A nonprimitive unit cell of the diamond structure is shown in Fig. 2.48. The structure can be visualized as two interpenetrating fcc sublattices displaced from one another along the cube diagonal by 1/4 the length of that diagonal. In terms of close-packing, the structure can be described as follows. The layer $A'$ of atoms is placed above the layer $A$ of atoms, so that each atom of $A'$ touches an atom of layer $A$ and rests vertically above it. The layer $B$ of atoms is then stacked in such a way that each atom of $A'$ rests in the hollow formed by three atoms of the $B$ layer. A further $B'$ layer of atoms is added in such a way that atoms of layer $B'$ are placed vertically over those of layer $B$ in exactly the same manner as atoms of layer $A'$ were palced over the atoms of $A$ layer. Next, a layer $C$ is added such that each atom of $B'$ is in the hollow formed by three atoms of layer $C$. Finally, a $C'$ layer of atoms is added to have atoms of this layer resting vertically above the atoms of $C$ layer. The resulting structure has two sublattices $ABC$ and $A'B'C'$, and the stacking follows the sequence $AA'BB'CC'AA'$.... The coordination number of this structure is 4.

The unit cell in Fig. 2.48 has 8 atoms at the corners, 6 face-centred atoms, and 4 atoms in the body of the cube. Thus, there are 8 atoms per unit cell. The position of various atoms can be conveniently specified in submultiples of the cube edge $a$. If we take the position of the

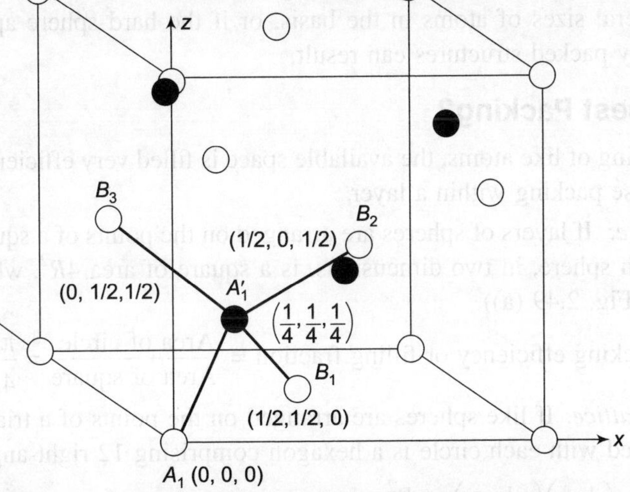


**Figure 2.48** A nonprimitive unit cell of the diamond lattice. The shaded atoms belong to the second sublattice which is displaced from the first along the major cube diagonal by one-fourth the length of the diagonal

corner atom $A_1$ as the origin, the coordinates of the three face-centred atoms $B_1$, $B_2$, and $B_3$, are as shown. The coordinates of the nearest atom $A_1'$ of the second sublattice are (1/4, 1/4, 1/4). If $A_1$ belongs to the $A$ layer, then $A_1'$ belongs to the $A'$ layer. Note that each of the atoms $A_1$, $B_1$, $B_2$, and $B_3$ is situated at a distance of $\sqrt{3}a/4$ from $A_1'$, and this is the nearest neighbour distance in the diamond structure. The diamond structure is relatively empty because only 34 per cent of the unit cell volume is occupied by atoms, compared to 74 per cent in close-packed fcc and hcp structures.

The zinc blende (ZnS) structure is similar to the diamond structure except that Zn atoms are placed on one fcc sublattice and S atoms on the other. Thus, each Zn atom is surrounded by four S atoms and vice versa. Elements Si and Ge crystallize in diamond lattice, and the compound semiconductors GaAs and GaP have the zinc blende structures.

Structures of crystalline solids are regular, three-dimensional patterens of atoms/constituents in space. The regularity of packing in solids arises from geometrical constraints which are imposed mainly by directional bonding and close packing.

## 2.23 SPHERES IN CLOSEST PACKINGS

Although we have pictured the crystal structure as an array of fairly separated lattice points (or bases), it would be more correct to think of the atoms to be in close contact or touching one another. Condensation into solid state is a result of some kind of attractive force between the constituent atoms, ions or molecules. As a first approximation, atoms may be regarded as hard spheres packed in such a way that energy is minimum in closest packings and that the spheres have maximum number of nearest neighbours. If the constituents are not spherical, or there are several sizes of atoms in the basis, or if the hard sphere approximation is not valid, less closely packed structures can result.

## What is Closest Packing?

In a closest packing of like atoms, the available space is filled very efficiently. Let us consider two cases of close packing within a layer.

*(a) Square Lattice:* If layers of spheres are arranged on the points of a square lattice, the area occupied by each sphere, in two dimensions, is a square of area $4R^2$, where $R$ is the radius of the atom (cf. Fig. 2.49 (a)).

$$\text{Therefore, packing efficiency or filling fraction} = \frac{\text{Area of circle}}{\text{Area of square}} = \frac{\pi R^2}{4R^2} = 0.785$$

*(b) Triangular Lattice:* If like spheres are arranged on the points of a triangular lattice, then the area associated with each circle is a hexagon comprising 12 right-angled triangles, each having an area of $\left(\frac{1}{2}R\right)\left(\frac{1}{\sqrt{3}}R\right) = \frac{R^2}{2\sqrt{3}}$ (cf. Fig. 2.49 (b)). The area of the hexagon is equal to 12 times the area of the triangle or $2\sqrt{3}\,R^2$.

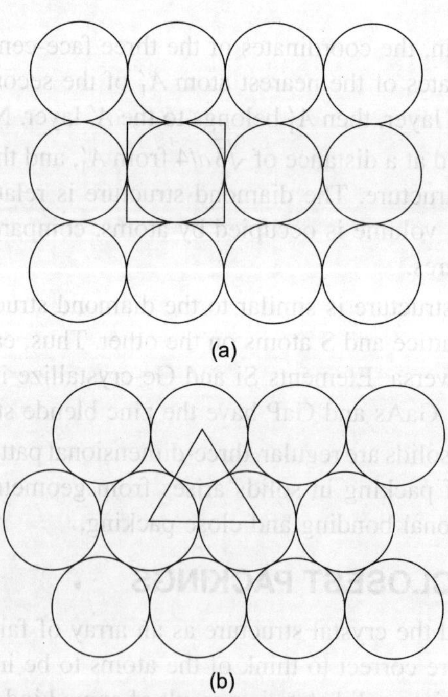

(a)

(b)

**Figure 2.49** (a) Square lattice. (b) Triangular lattice corresponding to hexagonal closest-packed layer

$$\text{Therefore, packing efficiency or filling fraction} = \frac{\text{Area of circle}}{\text{Area of hexagon}}$$

$$= \frac{\pi R^2}{2\sqrt{3}R^2} = 0.907$$

It should be obvious that a more efficient occupation of space in two dimensions is not possible, since more than six circles cannot be arranged such that each is in contact with one circle of the same size. Therefore, this is the closest packing possible in two dimensions. Incidentally, honeycombs found in nature have also this packing efficiency. Thus, we see that the closest packing possible in two dimensions is the hexagonal closest-packed layer (Fig. 2.49 (b)). This layer has the following properties: (a) each sphere is surrounded by six spheres; (b) each sphere is surrounded by six voids; and (c) each void is surrounded by three spheres.

On an average, 6 × 1/3 = 2 voids belong to each sphere. These two triangular voids differ from each other in that one has its apex pointing upwards and the other has its apex pointing downwards. In order to build up 3-dimensional structures, we proceed as follows.

Let us call the hexagonal closest packed layer as *A* layer. In order to achieve closest packing, the spheres in the next layer must occupy triangular voids either with their apices pointing up or down; if we place a sphere in a triangular void with the apex pointing up, then the other void with apex pointing down cannot be occupied due to geometrical or steric hindrance. Thus, in the second layer, spheres occupy the triangular voids with apices pointing up only. This layer may be called the B layer. The third layer may be packed in such a way that spheres occupy space directly above the triangular voids with apices pointing down (C layer). In 3 dimensions, such a type of packing, ABCABC ..., is indeed closest and called cubic closest packing (ccp) or face-centred cubic (fcc) packing. Alternative designations for this are abcabc, ccp and fcc. Maximum possible (74.04%) of the available space is occupied in this arrangement.

Alternatively, it may be possible to obtain another type of closest packing with the same efficiency by placing spheres in the third layer so that they are directly above those in the first (A) layer. Such ABAB ... or abab ... packing is called hexagonal closest packing (hcp or hcp). (The only difference between fcc and hcp is that in the hcp arrangement, the three spheres above the midplane are directly above the three spheres below the midplane and in the fcc arrangement they are in the alternate set of positions).

## 2.24 CLOSE PACKING OF SPHERES IN THREE DIMENSIONS

The close packed arrangement of equal spheres in a plane is shown in Fig. 2.50 (a) where each sphere is in contact with six other spheres. Since the symmetry of this layer is 6 mm, such a layer is called hexagonal close packed layer. Let this layer be called an A-layer. It contains two types of triangular voids or two sets of alternate valley BBB and CCC (Fig. 2.50 (b) and 2.50 (c) respectively).

In the three-dimensional packing, the next hexagonally close packed layer of spheres can occupy either the site *B* or *C*, but not both at the same time. Similarly, layer above the *B*-layer

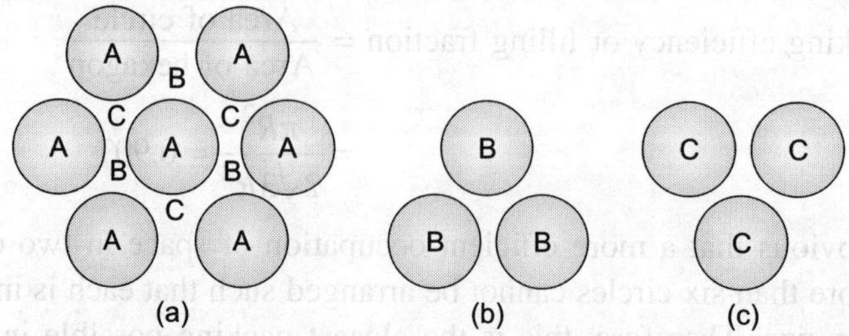


**Figure 2.50** (a) A central sphere surrounded by six spheres on a close packed place forms two sets of alternate valleys BBB and CCC (b) three spheres to fit in the set BBB (c) three spheres to fit in the set CCC. (Thus, there are three possible positions for a layer: A, B, and C)

can be either *C* or *A* and that above a C-layer, either *A* or *B*. No two successive layers can be alike in a close packed structure. Any sequence of letters *A*, *B* and *C* with no successive layers alike represents a possible manner of close packing of equal spheres. In such three-dimensional close packings, each sphere is in contact with 12 other spheres – 6 in its own layer-plane, three in the layer just above it and three in the layer just below it. Further, these three planes of the stacking are laterally shifted with respect to one another, i.e., the projections of the centres of the spheres on the base do not overlap. We have denoted these planes by *A*, *B* and *C*. Note that no other lateral position such as *D* is possible. By repeating this sequence of stacking, we obtain the (fcc) stacking:

$$\ldots ABC\ ABC\ ABCA\ldots$$

as this gives rise to the fcc monoatomic crystalline arrangement (Fig. 2.51).

In the fcc stacking, after choosing one set of valleys BBB for positioning the second plane, we used the other set CCC for the third plane. Instead, we could have positioned the spheres

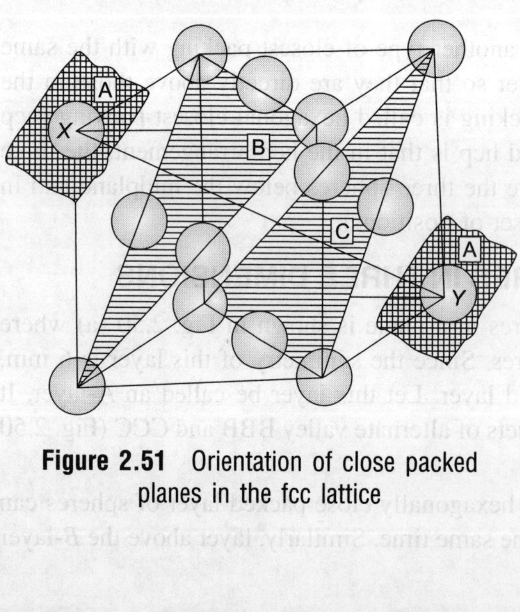


**Figure 2.51** Orientation of close packed planes in the fcc lattice

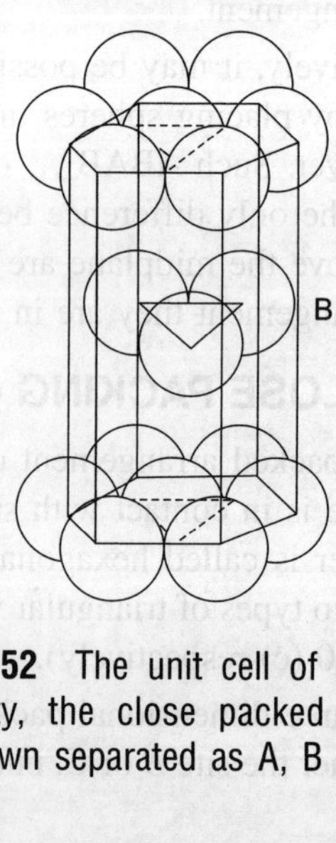


**Figure 2.52** The unit cell of hcp crystal. For clarity, the close packed planes are shown separated as A, B and A

of third plane such that their projections will coincide with those of the first plane AAA. Repetition of this sequence gives rise to the hcp stacking

...ABABABA...

as this sequence results in the hexagonal close packed crystal. Figure 2.52 shows the unit cell of the hcp crystal.

**Example 2.3** Assuming that the lattice points of a bcc crystal of lattice parameter $a$ are occupied by spherical atoms of radius $R$, determine the free volume per unit cell.

***Solution*** Volume of the unit cell = $a^3$.

Since there are two atoms per unit cell $\left(8 \times \frac{1}{8} \text{ for the corner atoms and 1 for the centre atom}\right)$.

Therefore, volume occupied by the atoms is

$$2 \times \frac{4}{3} \pi R^3$$

Since the body diagonal atoms touch one another, therefore

$$(4R)^2 = a^2 + \left(\sqrt{a^2 + a^2}\right)^2$$

or $$4R = \sqrt{3}a.$$

$\therefore$ the volume of atoms in terms of $a = \frac{8\pi}{3}\left(\frac{\sqrt{3}a}{4}\right)^2$

$$\text{Thus, free volume} = a^3 - \frac{8}{3} \times \frac{\pi a^3 3\sqrt{3}}{64}$$

$$= a^3 \left[1 - \frac{\pi\sqrt{3}}{8}\right].$$

## 2.25 CLOSE PACKING OF IDENTICAL SPHERES IN fcc AND hcp

fcc or hcp crystal arrays give 12 nearest neighbours per atom, the largest possible coordination number of identical atoms. The fcc and hcp structures are intimately related as one can see in Fig. 2.53. If one starts with a layer A of identical spheres having 6 touching neighbours and a second layer B is added so that the spheres fit into the depressions, each sphere in the second layer is in contact with 9 others, 3 below and 6 in the second layer. When a third layer is added, there is a choice which of the two sets of depressions the spheres occupy. If they lie at the A sites directly above those of the first layer and the crystal is built up of layers AB AB A ..., hexagonal close packing results; if they lie at the C sites and the crystal is built of layers ABC ABC ..., a face-centred cubic crystal is modelled. These two arrays are the closest possible packing for equal spheres; in each case the spheres occupy 74 per cent of the total volume ($\pi\sqrt{2} \times 100/6$). For metals like copper and silver, a fcc array of touching spheres is a fair model for the ion cores.

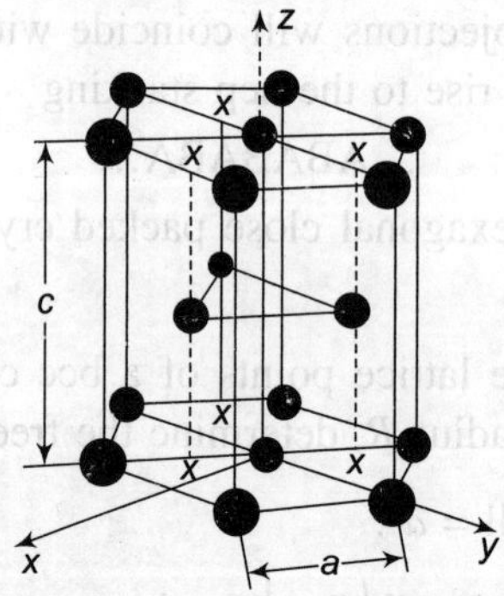


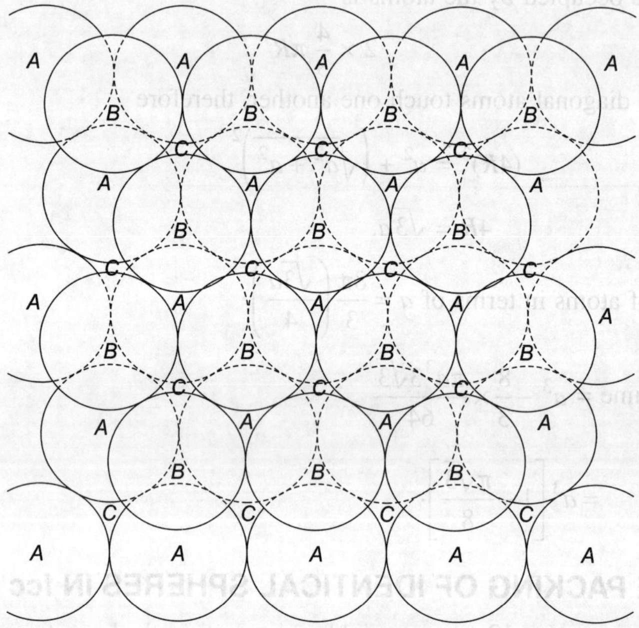


**Figure 2.53** (a) The hexagonal close-packed structure. (b) When close-packed layers are built up in the order ABABA ..., hexagonal close packing results; stacking in the order ABCABC ... leads to a face-centred cubic array

Table 2.10 depicts properties of sc, bcc, fcc and ideal hcp structures.

## 2.26 AXIAL RATIO AND LATTICE CONSTANTS

The extent to which a real crystal structure approximates to a close packing can be determined from its lattice constants. As seen above, any close packing can be conveniently referred to hexagonal axes. If $c$ denotes the height of the hexagonal unit cell, $n$ the identity period, and

**Table 2.10** Properties of Some Common Crystal Structures

| *Features* | *Crystal structures* | | | |
|---|---|---|---|---|
| | *sc* | *bcc* | *fcc* | *Ideal hcp* |
| Bravais lattice? | Yes | Yes | Yes | No |
| Similar atoms? | Yes | Yes | Yes | Yes or No |
| Lattice constant (*s*) | $a$ | $a$ | $a$ | $c = \sqrt{\frac{8}{3}}\, a^*$ |
| Primitive vectors | $a_1 = a_i$ | $a_1 = \frac{a}{2}(j + k - i)$ | $a_1 = \frac{a}{2}(j + k)$ | $a_1 = \frac{\sqrt{3}a}{2} i + \frac{a}{2} j$ |
| | $a_2 = a_j$ | $a_2 = \frac{a}{2}(k + i + j)$ | $a_2 = \frac{a}{2}(k + i)$ | $a_2 = \frac{\sqrt{3}a}{2} i + \frac{a}{2} j$ |
| | $a_3 = ak$ | $a_3 = \frac{a}{2}(i + j + k)$ | $a_3 = \frac{a}{2}(i + j)$ | $a_3 = ck$ |
| Basis of more than one atoms | – | – | – | $r_1 = 0$<br>$r_2 = \frac{2}{3} a_1 + \frac{1}{3} a_2 + \frac{1}{2} a_3$ |
| Wigner–Seitz primitive cell | Cube | Truncated octahedron (14 faces) (see Fig. 2.56) | Regular rhombic dodecahedron (see Fig. 2.57) | Regular rhombic prism |
| Unit cell | Cube | Cube | Cube | Regular hexagonal prism |
| $V_p$ | $a^3$ | $a^3/2$ | $a^3/4$ | $\frac{\sqrt{3}}{2} a^2 c = \sqrt{2} a^{3*}$ |
| $V_u/V_p$ | 1 | 2 | 4 | 3 |
| No. first nearst neighbours | 6 | 8 | 12 | 12 |
| No. of second nearest neighbours | 12 | 6 | 6 | 6 |
| Distance of first neighbours, $d$ | $a$ | $\frac{\sqrt{3}}{2} a$ | $\frac{\sqrt{2}}{2} a$ | $a^*$ |
| Distance of second neighbours, $d_2$ | $\sqrt{2}a$ | $a$ | $a$ | $\sqrt{2}a^*$ |
| Volume fraction of touching spheres | $\pi/6 \approx 0.52$ | $\sqrt{3}\,\pi/8 \approx 0.68$ | $\sqrt{2}\,\pi/8 \approx 0.74$ | $\sqrt{2}\,\pi/6 \approx 0.74$ |
| Radius $r_\alpha$, of a sphere of volume $V/N_\alpha$ | $(3/4\pi)^{1/3}$ d $\simeq 0.62\,d$ | $(1/\sqrt{3}\,\pi)^{1/3} d$ $\simeq 0.569\,d$ | $(3\sqrt{2}/8\pi)^{1/3} d$ $\simeq 0.553\,d$ | $(3\sqrt{2}/8\pi)^{1/3} d \simeq 0.553\,d$ |

* This equality may be violated in real hcp lattices

$V_p$ the volume of primitive cell, $V_u$, the volume of unit cell

$h$ the separation between successive close packed layers then $c = nh$. Also, we know that in a close packing, any sphere in a layer is in contact with three spheres in the layer below it (Fig. 2.54). The centres of these spheres lie at conners of a regular tetrahedron, as shown in Fig. 2.54b. The height of this tetrahedron is $h$, while the side equals the lattice constant "$a$" of the hexagonal unit cell (equals the diameter of the sphere). For a regular tetrahedron of side "$a$", a sin 60° is the median of any of the four bounding faces. Let $BP$ is the median of a side face, drawn from the apex $B$ of the tetrahedron and $BQ$ be the perpendicular from $B$ to the base of the tetrahedron. The triangle $BPQ$ is shown in Fig. 2.54. As $P$ is the mid point of one side of the base and $Q$ is its centroid,

$$PQ = \frac{1}{2} a \sin 60°$$

and

$$(BP)^2 = (BQ)^2 + (QP)^2$$

or

$$a^2 \sin^2 60° = h^2 + \frac{a^2}{9} \sin^2 60°$$

or

$$\frac{h}{a} = \frac{\sqrt{2}}{\sqrt{3}} = 0.8165$$

so that,

$$\frac{c}{a} = \frac{nh}{a} = 0.8165 \times n$$

The ratio $c/a$ is known as axial ratio, for an ideal close packed structure its value must be an integral multiple of 0.8165.

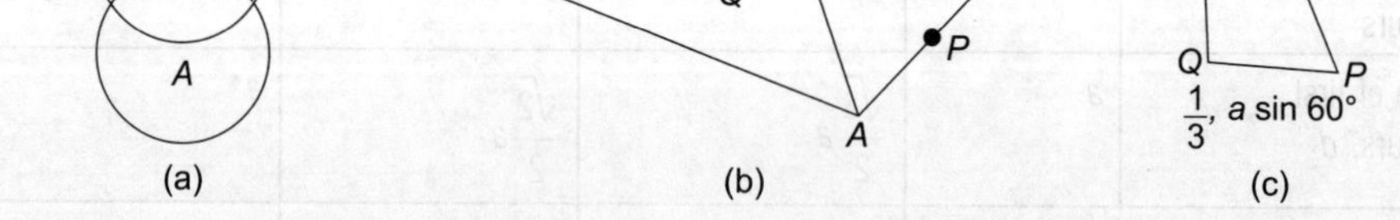


**Figure 2.54** (a) Tetrahedral voids in a close packing (b) Tetrahedron formed by the centres of the spheres (c) Section BPQ of Fig. (b)

In cubic close packing, we know the layers… $ABC$ … are stacked along [111] direction (body diagonal) of the unit cell. Therefore, the body diagonal of the cubic unit cell must be equal to $3h$. If the side of the cube is $a_c$ then the body diagonal will be $\sqrt{3}a_c$ (which is equal to $3h$). Therefore, $a_c = \sqrt{3}\,h = \sqrt{2}a$ from Eq. 9, where $a = 2R$ = the diameter of the sphere. Thus, the lattice parameter of an fcc unit cell is $\sqrt{2}$ times the diameter of the sphere.

## 2.27 VOIDS IN CLOSE PACKING

### 2.27.1 Classification of Voids

There are two kinds of voids that occur in close packing. If the triangular void in a close packed layer has a sphere directly above it, the resulting void will have four spheres around it as shown in Fig. 2.55 (a). These spheres are arranged on the corners of a regular tetrahedron (Fig. 2.55b) such a void is called tetrahedral void. On the other hand, if a triangular void pointing up in one close packed layer is covered by a triangular void pointing down in the next layer, the resulting void will be surrounded by six spheres (Fig. 2.56a). These spheres are arranged on the corners of an octahedron, such a void is known as an octahedral void and is shown in Fig. 2.56 (b). Thus, in three dimensions, where the possible close packed sequences are theoretically infinite, there are two kinds of possible voids.

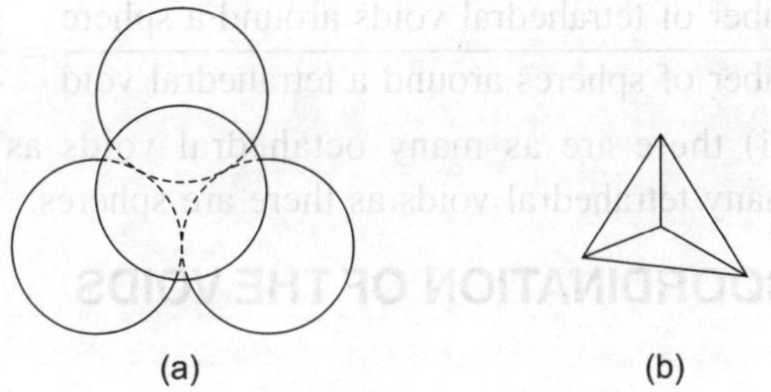

**Figure 2.55** (a) A tetrahedral void; (b) Projection of centre of spheres

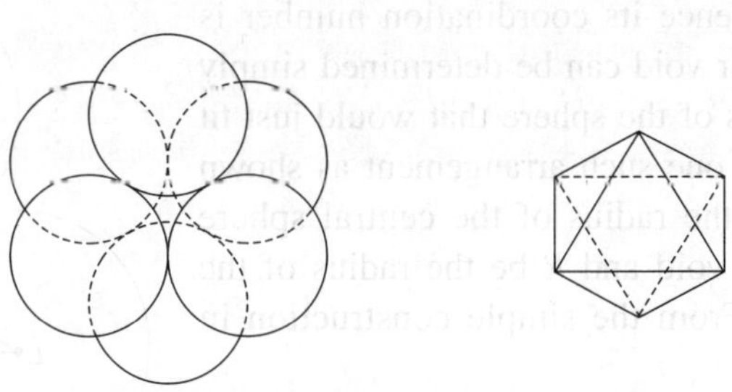

**Figure 2.56** (a) An octahedral void; (b) Projection of centre of spheres

The number of these voids in a three-dimensional close packing of spheres can readily be determined. In a single close packed layer *A*, each sphere is surrounded by three *B* voids and three *C* voids (Fig. 2.53). When the next layer is placed on the top of this, three voids of one kind are occupied and the other three are not. Thus, three become tetrahedral voids and the other three become octahedral voids. Similarly, the close packed layer below the *A* layer gives rise to three tetrahedral and three octahedral voids. Further, the reference sphere also covers a triangular void in layer above it and another in the layer below it. There are thus two more tetrahedral voids. Each sphere is therefore surrounded by 3 + 3 + 2 = 8 tetrahedral voids and 3 + 3 = 6 octahedral voids. Since the total number of spheres and voids in a close packing is very large and difficult to calculate. However, it is possible to find the average number of voids of each kind belonging to a sphere.

As we know that each octahedral void is surrounded by six spheres and each sphere, in turn, is surrounded by six voids, therefore, the number of octahedral voids belonging to a sphere is given by the ratio,

$$\frac{\text{Number of octahedral voids around a sphere}}{\text{Number of spheres around an octahedral void}} = \frac{6}{6} = 1$$

Similarly, each tetrahedral void is surrounded by four spheres and each sphere is surrounded by eight tetrahedral voids. Therefore, the number of tetrahedral voids belonging to a sphere is given by the ratio,

$$\frac{\text{Number of tetrahedral voids around a sphere}}{\text{Number of spheres around a tetrahedral void}} = \frac{8}{4} = 2$$

This follows that: (i) there are as many octahedral voids as there are spheres, and (ii) there are twice as many tetrahedral voids as there are spheres.

## 2.28 SIZE AND COORDINATION OF THE VOIDS

### *In Two Dimensions*

Referring to the close packed arrangement of equal spheres in a plane as shown in Fig. 2.53, we find that one particular void (*B* or *C* type) is surrounded by three spheres and hence its coordination number is 3. The size of the planar void can be determined simply by measuring the radius of the sphere that would just fit into the void. Consider one such arrangement as shown in Fig. 2.57. Let $r$ be the radius of the central sphere which just fits into the void and $R$ be the radius of the coordinating spheres. From the simple construction in Fig. 2.57, we have

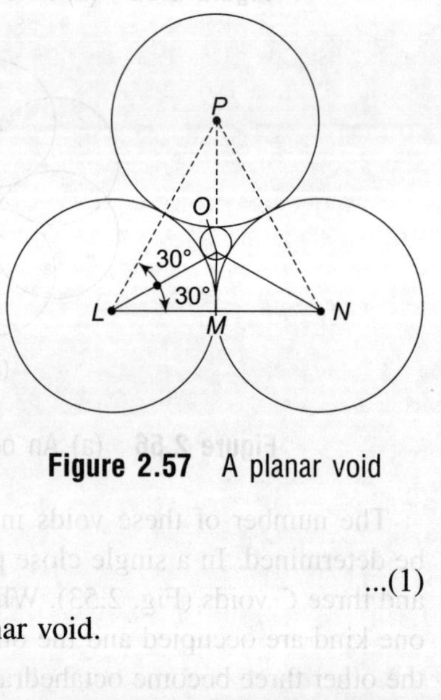

**Figure 2.57** A planar void

$$\frac{LM}{LO} = \frac{R}{R+r} = \cos 30°$$

or $$r = \frac{R}{\cos 30°} - R = \frac{2}{\sqrt{3}} R - R = 1.155R - R$$

or $$r = 0.155\, R \qquad ...(1)$$

This gives the radius of the sphere representing the planar void.

### *In Three Dimensions*

As we have seen earlier that there exist two kinds of voids, viz., tetrahedral and octahedral voids in three-dimensional close packings. The size of these voids can be found in a manner similar to the discussed above. Further, these voids in a close packing can be described in terms of the number of spheres that are arranged about each void. The number of spheres surrounding a void is called the coordination number of that void. Accordingly, the coordination number of a tetrahedral void is 4 and that of octahedral void is 6. A close packing can therefore be represented by the packing of coordination tetrahedral and octahedra (or in general coordination polyhedral) of voids as is done by the packing of spheres themselves. Size of these voids can be found as follows.

### *Tetrahedral Void*

Consider a tetrahedral void surrounded by four spheres of radius *R*. The centres of these spheres lie at the corners of a regular tetrahedron of side $a = 2R$. Let *r* be the radius of the sphere that just fits into this void whose centre is at *P* and is equidistant from all corners of the tetrahedron (Fig. 2.58). Let the bond length $p = r + R$. As shown earlier, the height $BQ = h$ is related to the side of the tetrahedron "*a*" as

$$h = \frac{\sqrt{2}}{\sqrt{3}}a$$

Hence, from the right angled triangle *PQR* shown in Fig. 2.58, we have

$$PQ = BQ - BP = h - p$$

Here, $$PR = p \text{ and } QR = \frac{2}{3} a \sin 60° = \frac{h}{\sqrt{2}}$$

Therefore, $$(PR)^2 = (h - p)^2 + (QR)^2$$

or $$p^2 = (h - p)^2 + \frac{h^2}{2}$$

so that $$p = \frac{3}{4} h = 0.75h \qquad \text{...(2)}$$

The bond length for an atom placed in a tetrahedral void in a close packing is thus equal to three-fourths the layer spacing. Substituting for *p* and *h* in terms of *r* and *R*, we have

$$r + R = \frac{3\sqrt{2}}{4\sqrt{3}} 2R = \frac{\sqrt{3}}{\sqrt{2}} R$$

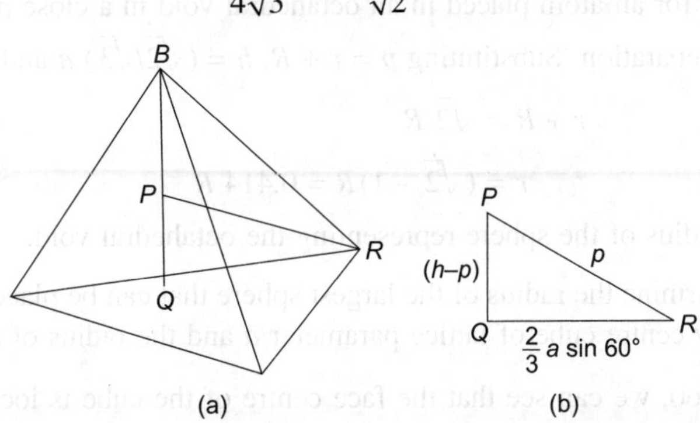


**Figure 2.58** (a) A tetrahedron; (b) section *PQR* of Fig. (a)

Therefore, $$r = \frac{\sqrt{3}-\sqrt{2}}{\sqrt{2}} R = 0.225R \qquad ...(3)$$

This gives the radius of the sphere representing the tetrahedral void.

### *Octahedral Void*

In an octahedral void the sphere that could just fit in would touch six spheres, three in the layer below and three in the layer above. The projection of the centres of the six spheres is shown in Fig. 2.59 (a). The point *O* represents the centre of the sphere fitted within the void. Again the bond length, $OR = p = R + r$. Now from the right-angled triangle $OO R$ shown in Fig. 2.59 (b) (where $O'$ lies in the plane of upper layer), we have

$$(OR)^2 = (OO')^2 + (O'R)^2$$

or $$p^2 = \left(\frac{h}{2}\right)^2 + \left(\frac{2}{3} a \sin 60°\right)^2$$

Therefore, $$p = \frac{\sqrt{3}h}{2} = 0.88h \qquad ...(4)$$

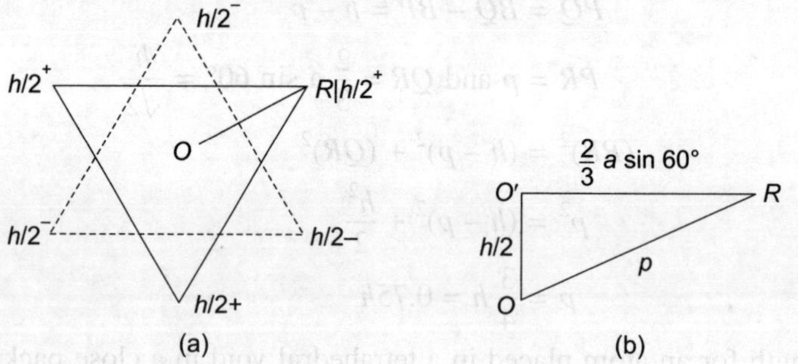


**Figure 2.59** Projection of centres of spheres with their position coordinates

The bond length for an atom placed in an octahedral void in a close packing is thus 0.88 times the layer of separation. Substituting $p = r + R$, $h = (\sqrt{2}/\sqrt{3})\ a$ and $a = 2R$, we have

$$r + R = \sqrt{2}\ R$$

Therefore, $$r = (\sqrt{2} - 1)R = 0.414\ R \qquad ...(5)$$

This gives the radius of the sphere representing the octahedral void.

**Example 2.4** Determine the radius of the largest sphere that can be placed at the centre of a cube face of a body centre cube of lattice parameter *a* and the radius of the lattice atom, *R*.

***Solution*** In Fig. 2.60, we can see that the face centre of the cube is located at $\frac{1}{2}\frac{1}{2}0$. This is an octahedral void as six lattice atoms are surrounding it. Let the largest sphere placed at this position just touching the two body centre atoms of the adjacent cube have a radius $R_x$. Then from simple geometry, the distance between the two body centre atoms is given by

$$2R + 2R_x = a$$

or

$$R + R_x = \frac{a}{2}$$

But for a bcc lattice,

$$4\,R = a\sqrt{3}$$

Therefore,

$$R + R_x = \frac{a}{2} = \frac{4}{2\sqrt{3}}R = 1.154R$$

So that,

$$R_x = 1.54R - R = 0.154R.$$

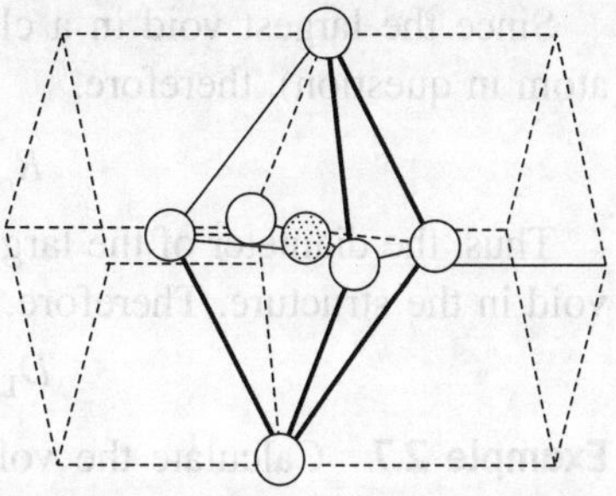

**Figure 2.60** An octahedral void, in bcc lattice

**Example 2.5** Determine the radius of the largest sphere that will fit into the void produced by the bcc packing of atoms of radius $R_L$.

***Solution*** The largest void in the body centred crystal is located at the position $\frac{1}{2}\,\frac{1}{4}\,0$ as shown in Fig. 2.61. This is a tetrahedral void as four lattice atoms are surrounding it. (Note that there are several other equivalent points in each cell.) If a new spherical atom of radius $R_x$ is introduced at this point, then from simple geometry we have

$$(R_L + R_x)^2 = \left(\frac{a}{4}\right)^2 + \left(\frac{a}{2}\right)^2 = \frac{5a^2}{16}$$

or

$$R_L + R_x = \frac{\sqrt{5}a}{4}$$

But

$$4\,R_L = a\sqrt{3}$$

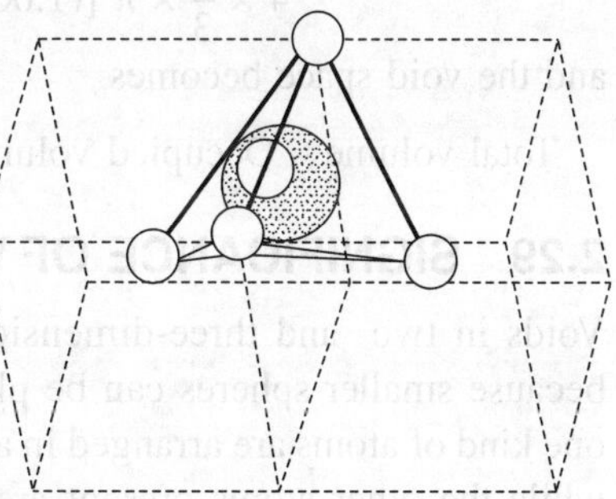

**Figure 2.61** A tetrahedral void in bcc lattice

Substituting the value of $a$ and rearranging the terms, we have

$$R_x = \frac{4R_L}{4\sqrt{3}}[\sqrt{5} - \sqrt{3}] = 0.291\,R_L$$

**Example 2.6** Find the diameter of the largest atom that would fit into the void in fcc nickel without distortion.

***Solution*** For Ni, the lattice parameter, $a$ = 3.53Å. Also for an fcc lattice, we have $\sqrt{2}a = 4R$. So that,

$$R = \frac{\sqrt{2}\times 3.52}{4} = 1.2445\text{Å}$$

Since the largest void in a close packing is an octahedral void (equivalent to the largest atom in question), therefore,

$$R_{oct} = 0.414R = 0.414 \times 1.2445 = 0.515\,\text{Å}$$

Thus, the diameter of the largest atom, $D_{LA}$ will be equal to the diameter of the octahedral void in the structure. Therefore,

$$D_{LA} = D_{oct} = 2R_{oct} = 1.03\,\text{Å}$$

**Example 2.7** Calculate the void space in a close packing of $n$ spheres of radius 1.000, $n$ spheres of radius 0.414 and $2n$ spheres of radium 0.225.

***Solution*** Let us consider the case of cubic close packing. In the case, side of the unit cell and the radius of the sphere is related as $\sqrt{2a} = 4R$, or $a = 2\sqrt{2}R$. Therefore, the volume of the unit cell

$$V = a^3 = 16\sqrt{2}R^3 = 16\sqrt{2} \text{ (for } R = 1)$$

Since an fcc unit cell contains 4 lattice points in it, therefore, $n = 4$. So that the occupied volume is

$$4 \times \frac{4}{3} \times \pi\ [(1.000)^3 + (0.414)^3 + 2 \times (0.225)^3] = 18.319$$

and the void space becomes

$$\text{Total volume} - \text{Occupied volume} = 16\sqrt{2} - 18.319 = 4.308 = 19\%$$

## 2.29 SIGNIFICANCE OF VOIDS

Voids in two- and three-dimensional close packings of spheres are of great significance, because smaller spheres can be placed within them. In fact, in many inorganic compounds, one kind of atoms are arranged in a close packed manner (not necessarily touching each other) while the other atoms present are distributed among the voids. For example, many oxides consist of oxygen atoms arranged in a hexagonal or cubic close packing with the smaller metal atoms occupying the voids. Also the voids have the capacity to hold interstitial atoms, which may be an impurity atom either added intentionally (or accidentally) or atoms of the parent structure itself accidentally displaced from the original positions. Such imperfections are known to play a very important role in influencing the properties of crystals. However, it is to be pointed out that in actual structures no voids can be occupied. In most inorganic close packed structures only certain voids related by symmetry are occupied.

## 2.30 PACKING OF UNEQUAL SPHERES IN 3 DIMENSIONS AND EFFECT OF RADIUS RATIO

In the preceding sections we have seen that the planar voids (in 2 dimensions), and tetrahedral and octahedral voids (in 3 dimensions) in close packing differ in size which increases in the order, octahedral > tetrahedral > planar voids. Therefore, in actual crystal structures a particular atom can best fit into one or the other kind of void depending on its size relative to that of the close packed atoms. Consequently, other packing arrangements of nondirectionally bonded atoms become possible when the sizes of the atoms are appreciably different (ionic solids provide the best example of such packings). The size of an atom depends on the size and the number of its coordinating atoms. In other words, the coordination number is a function of the difference in size of the central atom and the coordinating atoms. The greater the size difference, the smaller is the coordination number. The possible coordination numbers in a three-dimensional array are 1, 2, 3, 4, 6, 8 and 12. However, the coordination numbers of cations most frequently encountered in actual structures are 3, 4, 6, 8 and 12. Out of these, 4 and 6 are most common in structures having anions in close packing and either 4, 6, or 8 in almost all close packed structures. The coordination number 3 occurs in planar complexes, while 12 is the coordination number in close packing of like atoms. Each value of coordination number is found to be stable within a certain range of radius ratio (the ratio of the radius of the central atom to the atoms coordinating it).

As an example, let us consider the compound *AB* composed of equal numbers of *A* atoms and *B* atoms whose atomic radii are given by $r_A$ and $r_B$, respectively. Since the number of each type of atoms present in the structure is the same, they will have same coordinations, namely 1 : 1, 2 : 2, 3 : 3, 4 : 4, etc. The above calculated radius ratios for coordination numbers 3 (planar) 4 (tetrahedral) and 6 (octahedral) show that the coordination number imposes a limitation on the radius ratio and vice versa. Here, the coordination 1:1 and 2:2 imposes no limitations since the atoms can have any relative size and still have 1-fold or 2-fold coordination. Let us see the limitations imposed by other coordinations. According to Eq. 1 obtained in 2.28, $r_A/r_B$ must be at least 0.155 if the *A* atom is to touch the three *B* atoms coordinating it. Similarly, $r_B/r_A \geq 0.155$ if *B* atom is to touch the three *A* atoms coordinating it. Thus, a 3-fold coordination would require the radius ratio to lie between the limit 0.155 and 1/0.155. Let us analyse three different situations as illustrated in Fig. 2.62. Under the limiting situation when the central atom touches the three coordinating atom, the radius ratio ($r_A/r_B = 0.155$) is said to have a critical value. This is shown in Fig. 2.62(a). When the radius ratio varies in the range $0 < r_A/r_B < 0.155$, the central atom will "rattle" in the void and not touch the three coordinating atoms at the same time. This configuration is said to be unstable and is shown in Fig. 2.62 (a). Under this situation, the configuration can be made stable only if the coordination is reduced to 2. Finally, when the ratio $r_A/r_B > 0.155$, all the coordinating atoms touch the central atom but do not touch one another as shown in Fig. 2.62. This is a stable configuration. Identical arguments will be applicable to other coordination numbers also.

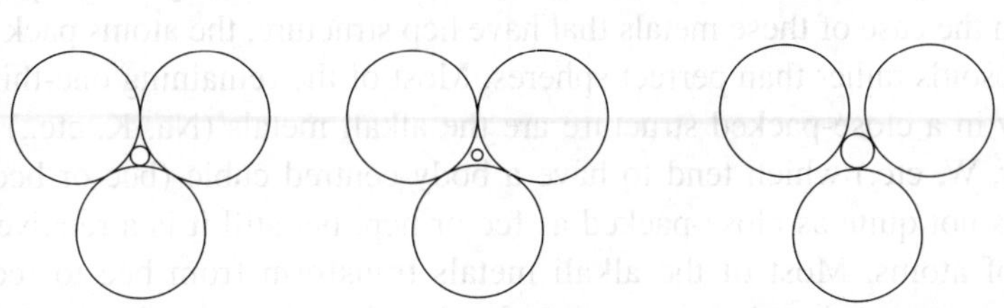


**Figure 2.62** Triangular coordination of anions around central cations (a) the critical configuration; (b) the unstable configuration; and (c) stable but not critical configuration

As far as the coordination is concerned, a given coordination number will be stable between that radius ratio at which the coordinating (close packed) atoms touch each other as well as the central atoms, and that the radius ratio at which the next higher coordination number is possible. For example, a coordination of 4 (or tetrahedral coordination), will be stable between the radius ratios of 0.225 and 0.414. Below the lower limit a triangular coordination is possible and above the upper limit an octahedral coordination results. The permissible range of radius ratio in *AB* compounds for coordination numbers most frequently encountered are given in Table 2.11.

**Table 2.11** Coordination Number as a Function of Radius Ratio in *AB* Compounds

| *Coordination number* | *Range of radius ratio* | *Coordination polyhedron* | *Packing* |
|---|---|---|---|
| 1:1 | 0–∞–0 | point | point |
| 2:2 | 0–∞–0 | line | linear |
| 3:3 | 0.155–1/0.155 | triangle | triangular |
| 4:4 | 0.225–1/0.225 | tetrahedron | tetrahedral |
| 6:6 | 0.414–1/0.414 | octahedron | octahedral |
| 8:8 | 0.732–1/0.732 | cube | cubic |
| 12:12 | 1.0 | twined cubo-octahedron (hcp)<br>cubo-octahedron (fcc) | hcp and fcc |

## 2.31 STRUCTURE OF CRYSTALS

fcc or hcp type of closest packings are exhibited by elements whose cohesive force is predominantly non-directional, i.e., by metals, ionic and molecular solids. Covalent solids are not expected to follow this rule since strongly covalent directional bonding usually gives rise to open structures. Further, polarization, presence of polarizing and polarizable ions, may also affect the structure. Several metals crystallize with the fcc or hcp structure. In this way, the number of bonds per unit volume is maximized and, hence, the bonding energy per unit volume is minimized.

All the solidified noble gases exhibit close-packed structure when they solidify at very low temperatures and about two-thirds of the metals exhibit either hcp or fcc packing at room temperature. In the case of these metals that have hcp structure, the atoms pack together more like slight ellipsoids rather than perfect spheres. Most of the remaining one-third metals, that do not solidify in a close-packed structure are the alkali metals (Na, K, etc.,) and transition metals (Fe, Cr, W, etc.) which tend to have a body-centred cubic (bcc or bcc) structure. A bcc structure is not quite as close-packed as fcc or hcp, but still it is a relatively low energy arrangement of atoms. Most of the alkali metals transform from bcc to fcc or HCPO at low temperatures suggesting that one reason for their less dense packing behaviour at room temperature is the effect of thermal energy. The reason that most of the transition metals solidify in a bcc structure is thought to be partly covalent and, therefore, directional nature of their bonds.

## 2.32 STRUCTURES OF SIMPLE COMPOUNDS

Table 2.12 describes how different crystal structures are generated by preferential occupation of tetrahedral and octahedral voids in closest packed array of larger atoms/ions by smaller atoms/ions of a different kind.

**Table 2.12** Some Simple Crystal Structures Described in Terms of Close-packed Array of Anions

| *Crystal structure* | *Anion packing* | *Fraction of cation sites occupied* |
|---|---|---|
| NaCl, MgO | fcc | All octahedral |
| $K_2O$ | fcc | All tetrahedral |
| ZnS (sphalerite) | fcc | All upright tetrahedral |
| ZnO (Wurtzite) | hcp | All upright tetrahedral |
| NiAs | hcp | All octahedral |
| $\alpha$-$Al_2O_3$ | hcp | Two-thirds octahedral |
| $CdCl_2$ | fcc | One-half octahedral (alternate layers) |
| $CdI_2$ | hcp | One-half octahedral (alternate layers) |
| Antifluorite ($Li_2O$) | ccp (fcc) | All tetrahedral |
| Fluorite ($CaF_2$) | ccp (fcc) (of cations) | All tetrahedral (by anions) |

### 2.32.1 Rock Salt (NaCl) Structure

The structure consists of two interpenetrating fcc sub-lattices*: lattice points on one of the sub-lattices are occupied by $Na^+$ ions and points on the other sub-lattice, by $Cl^-$ ions. Each $Na^+$ ions is surrounded by six $Cl^-$ ions and vice versa (cf. Fig. 2.63).

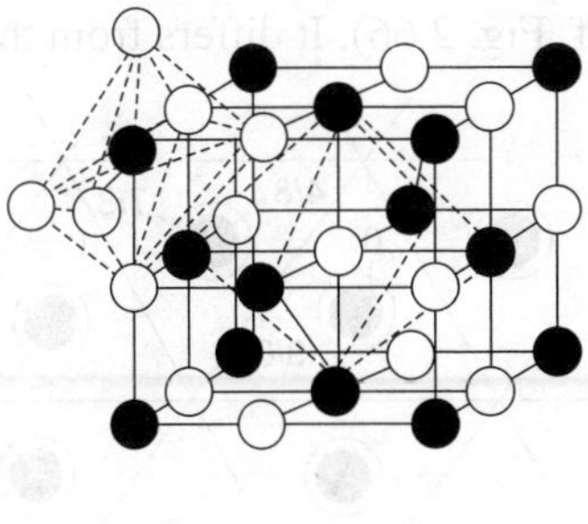

**Figure 2.63** Structure of sodium chloride

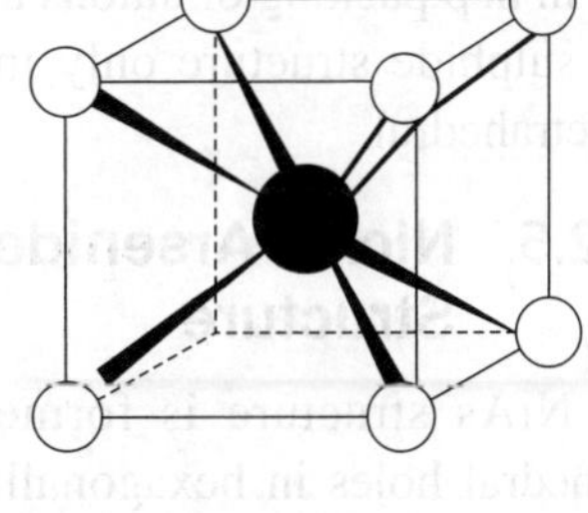

**Figure 2.64** Structure of caesium chloride

* According to some authors, it is erroneous to designate this structure as consisting of interpenetrating sub-lattices.

### 2.32.2 Caesium Chloride (CsCl) Structure

The CsCl structure consists of two interpenetrating simple cubic lattices of $Cs^+$ and $Cl^-$ ions (cf. Fig. 2.64). There is only one ion-pair per unit cell. Each caesium ion is surrounded by eight chloride ions and vice versa. Even when the radius ratio is favourable, ionic compounds do not prefer to adopt this structure.

### 2.32.3 Zinc Blende (or Sphalerite) Structure

The sphalerite or cubic zinc sulphide structure results when only half of the tetrahedral sites in fcc packed anions are occupied by cations (cf. Fig. 2.65).

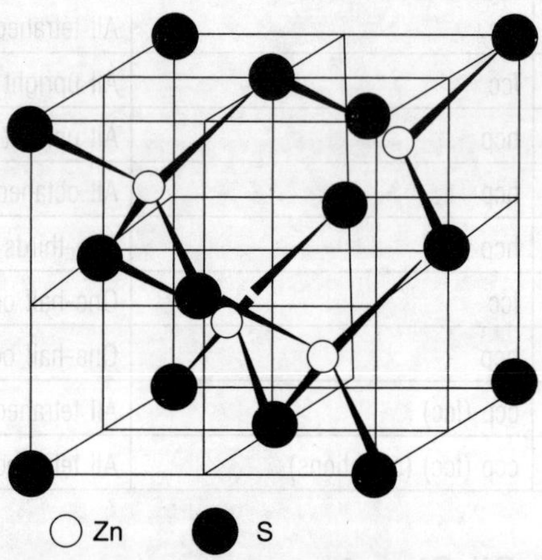


**Figure 2.65** Structure of zinc blende

### 2.32.4 Zinc Oxide (or Wurtzite) Structure

The wurtzite or hexagonal zinc sulphide structure results when only half of the tetrahedral sites in hcp packing of anions are occupied by cations (cf. Fig. 2.66). It differs from the cubic zinc sulphide structure only in the stacking of the tetrahedral.

### 2.32.5 Nickel Arsenide (NiAs) Structure

The NiAs structure is formed when all the octahedral holes in hexagonally closest packed arsenic atoms are occupied by nickel atoms (cf. Fig. 2.67). The nickel atoms have 8-fold coordination, being surrounded by six arsenic atoms and two additional nickel atoms, which are coplanar with four of the arsenic; on the other hand, the arsenic atoms are surrounded

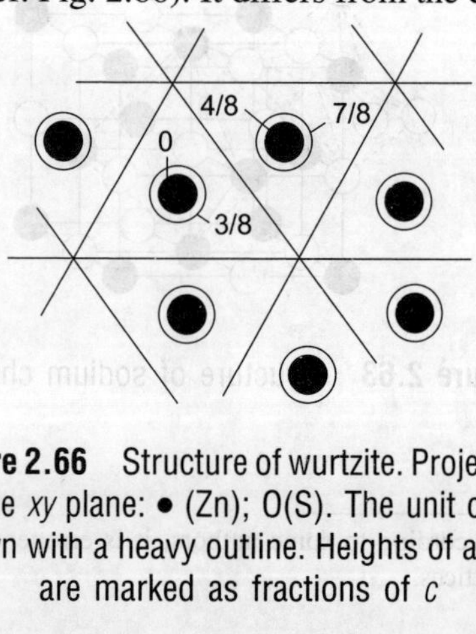


**Figure 2.66** Structure of wurtzite. Projection on the *xy* plane: • (Zn); O(S). The unit cell is shown with a heavy outline. Heights of atoms are marked as fractions of *c*

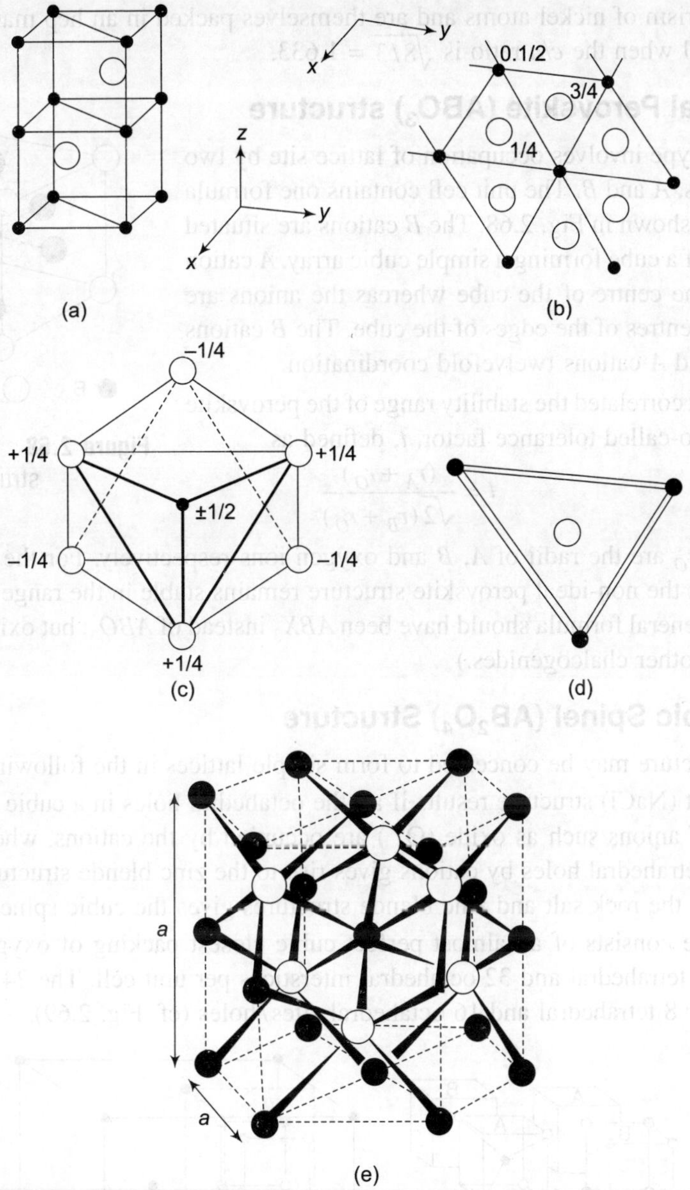


**Figure 2.67** Structure of nickel arsenide: (a) Perspective view of the unit cell (heavy outline: ● Ni; ○ As). (b) Projection on the *xy*-plane. (c) Projection on the *xy*-plane of the environment of the Ni atom, with heights marked. The As atoms at 1/4 form a flattened octahedron; addition of Ni atoms at 1/2 makes it into a rhombohedron. (d) Projection on the *xy*-plane of the environment of the As atom. Double lines indicate the edges of the trigonal prism which coincide in projection. (e) Overall view of the NiAs structure

by a trigonal prism of nickel atoms and are themselves packed in an hcp manner. The close packing is ideal when the $c/a$ ratio is $\sqrt{8/3} = 1.633$.

### 2.32.6 Ideal Perovskite ($ABO_3$) structure

This structure type involves occupation of lattice site by two kinds of cations, $A$ and $B$. The unit cell contains one formula unit ($ABO_3$) as shown in Fig. 2.68. The $B$ cations are situated at the corners of a cube forming a simple cubic array. $A$ cation is situated at the centre of the cube whereas the anions are located at the centres of the edges of the cube. The $B$ cations have sixfold and $A$ cations twelvefold coordination.

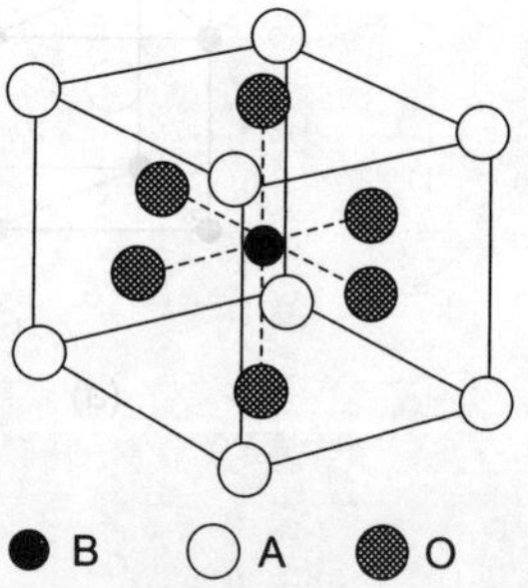


**Figure 2.68** Ideal perovskite structure

Goldschmidt correlated the stability range of the perovskite structure to a so-called tolerance factor, $t$, defined as

$$t = \frac{(r_A + r_O)}{\sqrt{2}\,(r_B + r_O)}$$

where, $r_A$, $r_B$, $r_O$ are the radii of $A$, $B$ and oxygen ions respectively. For the ideal structure $t = 1$. However, the non-ideal perovskite structure remains stable in the range $0.75 < t < 1.0$. (Actually the general formula should have been $ABX_3$ instead of $ABO_3$; but oxidic perovskites far outnumber other chalcogenides.)

### 2.32.7 Cubic Spinel ($AB_2O_4$) Structure

The spinel structure may be conceived to form simple lattices in the following way.

The rock salt (NaCl) structure results if all the octahedral holes in a cubic closest-packed arrangement of anions such as oxide ($O^{2-}$) are occupied by the cations, whereas filling up of half of the tetrahedral holes by cations gives rise to the zinc blende structure. An ordered combination of the rock salt and zinc blende structures gives the cubic spinel structure.

The structure consists of an almost perfect cubic closest packing of oxygen ions, 32 of which form 64 tetrahedral and 32 octahedral interstices per unit cell. The 24 metal ions are distributed over 8 tetrahedral and 16 octahedral sites/holes (cf. Fig. 2.69).

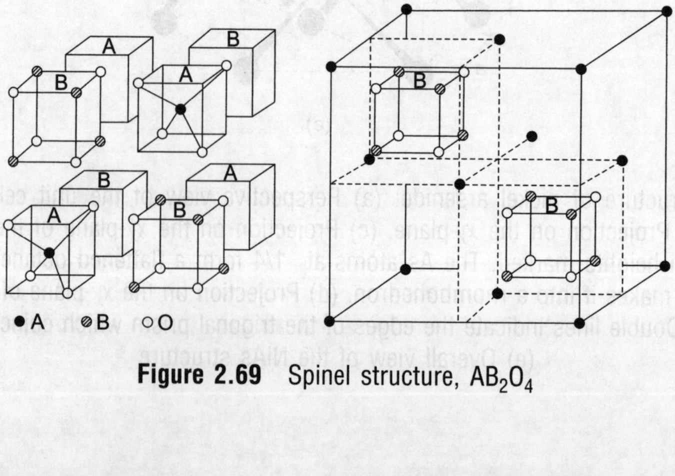


**Figure 2.69** Spinel structure, $AB_2O_4$

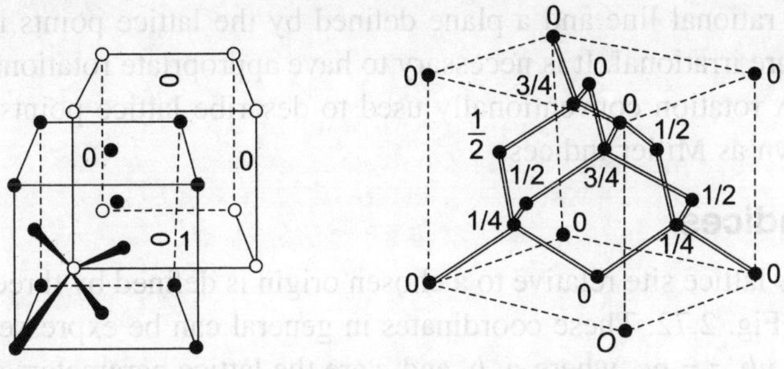


**Figure 2.70** The diamond lattice showing (a) How it is formed from two interpenetrating fcc lattice. (b) The tetrahedral arrangement of atoms

## 2.33 DIAMOND STRUCTURE

The crystal structure of diamond can be derived from the fcc lattice although the structure itself is not a closed pack one. Formally, it may be described as being built up from two interpenetrating fcc lattices which are displaced with respect to one another along the body diagonal (Fig. 2.70a). While this definition accurately describes the position of the atoms, it is not very useful when one actually visualizes the structure. A more useful model for all practical purposes is to consider each atom to be at the centre of a tetrahedron with its four nearest neighbours at the four corners of that tetrahedron (Fig. 2.70b). From (Fig. 2.70a) it can be seen that if the central atom of the tetrahedron is on one fcc sublattice, then the four corner atoms are all on the other sublattice. Further, since all atoms are equivalent, each of them can be treated as being the centre atom of the tetrahedron, although it would require a rather extended diagram to shown this.

As mentioned earlier, the space lattice is face-centred cubic (Fig. 2.70a).

The lattice can be considered to be formed by two interpenetrating fcc lattices (Fig. 2.70 (a)) displaced along the body diagonal by 1/4 of its length. The points at 0 and 1/2 are on one fcc lattice and those at 1/4 and 3/4 are on the other fcc lattice. In Fig. 2.71, fractions denote height above the base in units of a cube edge. Thus, with a fcc space lattice the basis consists of two identical atoms at 000 and 1/4 1/4 1/4. The atomic packing factor is 0.34 as compared to a value 0.74 for the close packed structure.

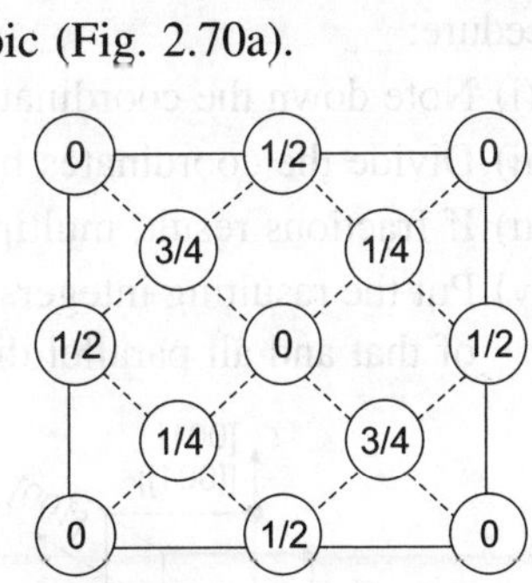


**Figure 2.71** Atomic positions in the cubic cell of diamond structure projected on a cube face

Apart from diamond (carbon), silicon, germanium and grey tin crystallize in the form of diamond structure.

## 2.34 RATIONAL FEATURES OF A CRYSTAL AND MILLER INDICES

The geometrical features of a crystal which are represented by lattice points are called rational, Thus, a lattice point (site) with respect to another lattice point (site) is a rational point, a row of lattice points is rational line and a plane defined by the lattice points is a rational plane. All other features are irrational. It is necessary to have appropriate rotations to represent their rational features. A rotation conventionally used to describe lattice points (sites), directions and planes is known as Miller indices.

### 2.34.1 Site Indices

The position of any lattice site relative to a chosen origin is defined by three of its coordinates *x*, *y*, *z* as shown it Fig. 2.72. These coordinates in general can be expressed in the following forms; $x = ma$, $y = nb$, $z = pc$, where *a*, *b*, and *c* are the lattice parameters and *m*, *n* and *p* are integers. However, if the lattice parameters are used as the units of length along the respective axes then the lattice coordinates will be obtained simply in the form of numbers *m*, *n* and *p*. These numbers are then termed site indices and are written in the form [*m n p*]. For a negative index the minus sign is written above the index. As an example, for a site with coordinates $x = -2a$, $y = 1b$, $z = -3c$, the site indices are written as $[[\bar{2}\ 1\ \bar{3}]]$.

### 2.34.2 Indices of Direction

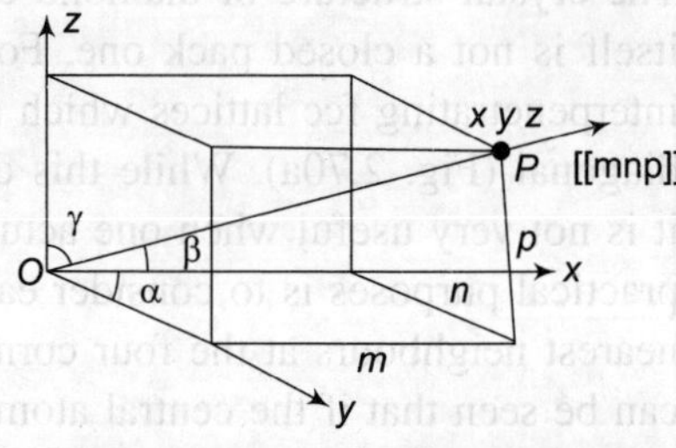


Figure 2.72 Indices of crystal lattice site and direction

To describe a direction in a crystal lattice, a straight line passing through the origin is chosen. It is defined by the indices of the first lattice site (point) through which the line passes (Fig. 2.72). In other words, it is defined by the three smallest integers that describe the position of the site nearest to the origin which lies on the given direction. Therefore, the site indices are at the same time the indices of direction (Fig. 2.73a). The indices of direction are calculated by adopting the following procedure:

(i) Note down the coordinates of the lattice site nearest to the origin in a given direction.
(ii) Divide the coordinates by appropriate unit translations.
(iii) If fractions result, multiply each of them by smallest common divisor.
(iv) Put the resulting integers in a set square brackets, i.e., [hkl] to get the required indices of that and all parallel directions.

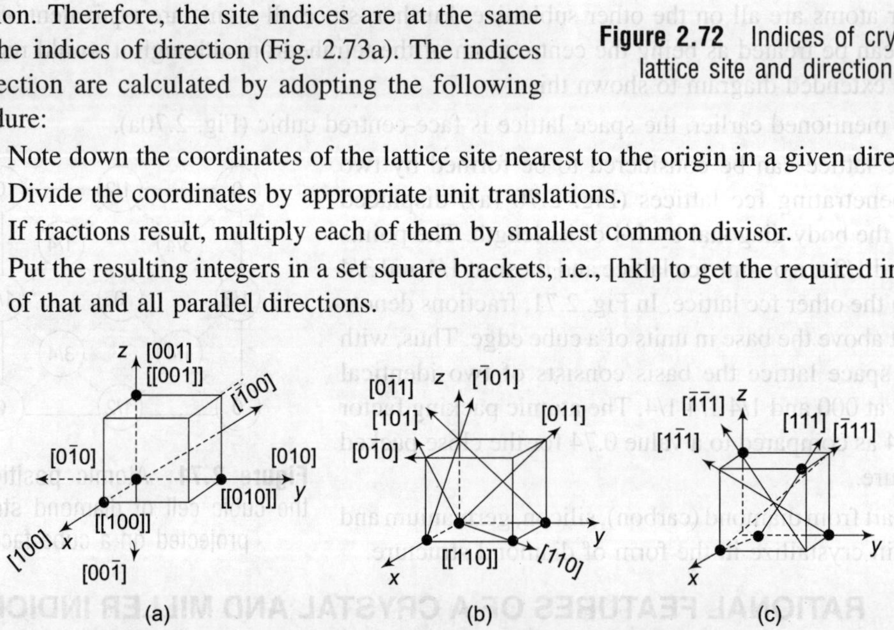


Figure 2.73 Indices of principal directions in a cubic crystal

It should be noted that the indices of direction are quite independent of the actual geometry of the unit cell and are determined solely by the ratios of the coordinates of any point on the given direction to the appropriate unit translations.

In a crystal, there exist an infinite number of families of directions. Out of these, there are sets of directions having identical spacing between the atoms (i.e., same unit translations) are called equivalent directions. Such sets of directions are always related by a symmetry operation. The shortened notation used to specify a family of directions is a set of angled brackets, i.e., <hkl>. For example (in the cubic system): the family <100> comprises the directions [100], [$\bar{1}$00], [010], [0$\bar{1}$0], [001] and [00$\bar{1}$]. Figure 2.73 shows all the principal directions (crystallorgraphic orientations) in a cubic crystal alongwith their notations. The unit translations for the low index directions in the three Bravais lattices of cubic system are given in Table 2.13.

**Table 2.13** The Unit Translation for Low Index Directions of a Cubic System

| *Family* | *Unit translation* | | |
|---|---|---|---|
| | *P* | *I* | *F* |
| <100> | $a$ | $a$ | $a$ |
| <110> | $\sqrt{2}a$ | $\sqrt{2}a$ | $\frac{a}{\sqrt{2}}$ |
| <111> | $\sqrt{3}a$ | $\frac{\sqrt{3}a}{2}$ | $\sqrt{3}a$ |

**Example 2.8** Write all directions from the family <110> and <111>.

***Solution*** The family <110> comprises the directions [110], [$\bar{1}$10], [1$\bar{1}$0], [$\bar{1}\bar{1}$0], [101], [$\bar{1}$01], [10$\bar{1}$], [$\bar{1}$0$\bar{1}$], [011], [011], and [011]. Similarly, the family <111> comprises the directions [111], [$\bar{1}$11], [1$\bar{1}$1], [11$\bar{1}$], [$\bar{1}\bar{1}$1], [$\bar{1}$1$\bar{1}$], [1$\bar{1}\bar{1}$] and [$\bar{1}\bar{1}\bar{1}$].

**Example 2.9** Classify the members of <100>, <110> and <111> families of direction in the tetragonal system.

***Solution*** For a tetragonal system, we know that $a = b \quad c$. Thus, the unit translation for the directions [100], [$\bar{1}$00], [010] and [0$\bar{1}$0] is the same as $a$, whereas for the directions [001] and [00$\bar{1}$] is $c$. Consequently, the first four directions are the members of <100> family and the last two are the members of <001> family.

Similarly, <110> family has four members, i.e., [110], [$\bar{1}$10], [1$\bar{1}$0] and [$\bar{1}\bar{1}$0], while the other eight directions [101], [$\bar{1}$01], [10$\bar{1}$], [$\bar{1}$0$\bar{1}$], [011], [0$\bar{1}$1], [01$\bar{1}$] and [0$\bar{1}\bar{1}$] are the members of <101> family or equivalently the <011> family.

Since all <111> directions have the same unit translation in the tetragonal system and therefore the family contains the same eight members as it does in the cubic system.

In certain calculations (e.g., in resolved shear stresses), it is necessary to determine the angle between two different crystal directions. The formula for a general case like triclinic is very complicated. Therefore, in Table 2.14, we enlist the formulae for some crystal systems which may be frequently used in calculations. The formula for cubic system is the simplest of all which can be determined by dot product.

**Table 2.14** Angle between Two Crystal Directions $[h_1k_1l_1]$ and $[h_2k_2l_2]$

| *Crystal System* | $\cos\phi$ |
|---|---|
| Cubic | $\dfrac{h_1 h_2 + k_1 k_2 + l_1 l_2}{(h_1^2 + k_1^2 + l_1^2)^{1/2}\,(h_2^2 + k_2^2 + l_2^2)^{1/2}}$ |
| Tetragonal | $\dfrac{a^2 (h_1h_2 + k_1k_2 + l_1l_2) + c^2 (l_1 l_2)}{\sqrt{\{a^2 (h_1^2 + k_1^2) + c^2 l_1^2\}\,\{a^2 (h_2^2 + k_2^2) + c^2 l_2\}}}$ |
| Orthorhombic | $\dfrac{a^2 h_1h_2 + b^2 k_1k_2 + c^2 l_1 l_2}{\sqrt{(a^2 h_1^2 + b^2k_1^2 + c^2 l_1^2)\,(a^2 h_2^2 + b^2k_2^2 + c^2 l_2^2)}}$ |
| Hexagonal | $\dfrac{h_1h_2 + k_1k_2 - \frac{1}{2}(h_1k_2 + k_1h_2) + \frac{c^2}{a^2} l_1l_2}{\sqrt{\left(h_1^2 + k_1^2\; h_1k_1 + \frac{c^2}{a^2} l_1^2\right)\left(h_2^2 + k_2^2 - h_2\, k_2 + \frac{c^2}{a^2} l_2^2\right)}}$ |

**Example 2.10** Calculate the angle between [111] and [001] directions in a cubic crystal.

***Solution*** From Table 2.14, we have

$$\cos\phi = \frac{h_1\, h_2 + k_1k_2 + l_1l_2}{\sqrt{(h_1^2 + k_1^2 + l_1^2)}\,\sqrt{(h_2^2 + k_2^2 + l_2^2)}} = \frac{0+0+1}{\sqrt{3}\times\sqrt{1}} = \frac{1}{\sqrt{3}}$$

so that $\phi = 54.75°$

**Example 2.11** Calculate the angle between [111] and $[\bar{1}\bar{1}1]$ directions in a cubic crystal.

***Solution*** From Fig. 2.74, we observe that [001] direction bisects the angle between [111] and $[\bar{1}\bar{1}1]$. Therefore, the angle $\phi$ between them is 2 × 54.75° = 109.5°. This can be verified from the dot product as mentioned above.

$$\cos\phi = \frac{-1-1+1}{\sqrt{3}\times\sqrt{3}} = -\frac{1}{3}$$

so that $\phi = 109.5°$

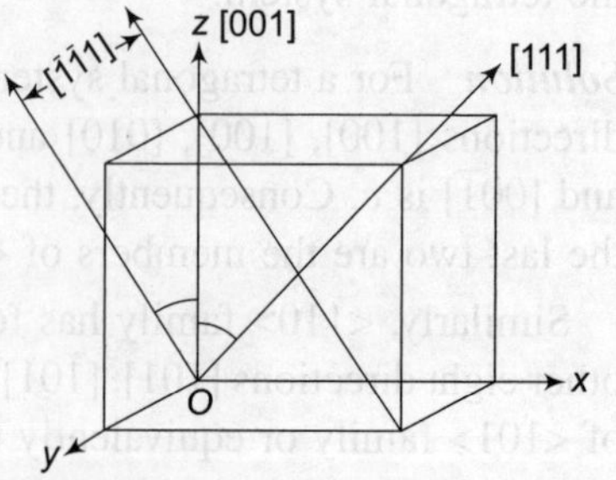


**Figure 2.74** Angles made by [111] and $[\bar{1}\bar{1}1]$ directions with respect to [100] in a cubic crystal

We will discuss indices of crystal planes (known as Miller indices) later, in the chapter on X-ray diffraction where we will consider Bragg's law and reciprocal lattice.

## 2.35 IMPORTANT FEATURES OF MILLER INDICES

Some important features of Miller Index notation are as follows:

1. The angle (θ) between two crystallographic directions $[u_1v_1w_1]$ and $[u_2v_2w_2]$ is given by the relation

$$\cos\theta = \frac{u_1u_2 + v_1v_2 + w_1w_2}{(u_1^2 + v_1^2 + w_1^2)^{1/2} \cdot (u_2^2 + v_2^2 + w_2^2)^{1/2}}$$

2. The normal to the plane with indices (hkl) is the direction [hkl]. This correspondence is of a tremendous aid in calculating the angle between the planes.
3. The distance ($d$) between adjacent planes for the set of (hkl) planes can be obtained for the seven crystal systems in terms of lattice constant $a$, $b$, $c$, $\alpha$, $\beta$, $\gamma$:

**For triclinic system:**

$$\frac{1}{d^2} = \frac{h^2}{a^2 \sin\beta} + \frac{k^2}{b^2} + \frac{l^2}{c^2 \sin^2\beta} - \frac{2hl\cos\beta}{ac\sin^2\beta}$$

**For orthorhombic system:**

$$\frac{1}{d^2} = \frac{h^2}{a^2} + \frac{k^2}{b^2} + \frac{l^2}{c^2}$$

**For cubic system:**

$$\frac{1}{d^2} = \frac{h^2 + k^2 + l^2}{a^2}$$

**For hexagonal system:**

$$\frac{1}{d^2} = \frac{4}{3}\left(\frac{h^2 + hk + k^2}{a^2}\right) + \frac{l^2}{c^2}$$

**For rhombohedral system:**

$$\frac{1}{d^2} = \frac{(h^2 + k^2 + l^2)\sin^2\alpha + 2(hk + kl + lh)\cos^2\alpha - \cos\alpha}{a^2(1 - 3\cos^2\alpha + 2\cos^3\alpha)}$$

**Example 2.12** Calculate the angle between two translation vectors $[p, q, s]$ and $[p', q', s']$ in a rectangular lattice.

***Solution*** We may use Cartesian coordinates, such that $\vec{a} = a\vec{e}_x$, $\vec{b} = b\,\vec{e}_y$, $\vec{c} = c\vec{e}_z$, and so:

$$\vec{R}_{pqs} = pa\vec{e}_x + qb\vec{e}_y + sc\vec{e}_z$$

$$\vec{R}_{p'q's'} = p'a\vec{e}_x + q'b\vec{e}_y + s'c\vec{e}_z$$

Assuming an orthorhombic lattice, where the axial ratio is *a*:*b*:*c*, we define the angle θ by

$$\cos\theta = \frac{\vec{R}\cdot\vec{R}'}{|\vec{R}||\vec{R}'|} = \frac{pp'a^2 + qq'b^2 + ss'c^2}{\sqrt{(p^2a^2 + q^2b^2 + s^2c^2)(p'^2a^2 + q'^2b^2 + s'^2c^2)}}$$

This result simplifies for tetragonal lattices, where the axial ratio is 1 : 1 : *c*, in the form:

$$\cos\theta = \frac{pp' + qq' + ss'c^2}{\sqrt{(p^2 + q^2 + s^2c^2)(p'^2 + q'^2 + s'^2c^2)}}$$

and then in cubic crystals, where the primitive vectors are in the ratio 1:1:1, reduces to

$$\cos\theta = \frac{pp' + qq' + ss'}{\sqrt{(p^2 + q^2 + s^2)(p'^2 + q'^2 + s'^2)}}$$

Figure 2.75 shows indices of important planes in a cubic crystal.

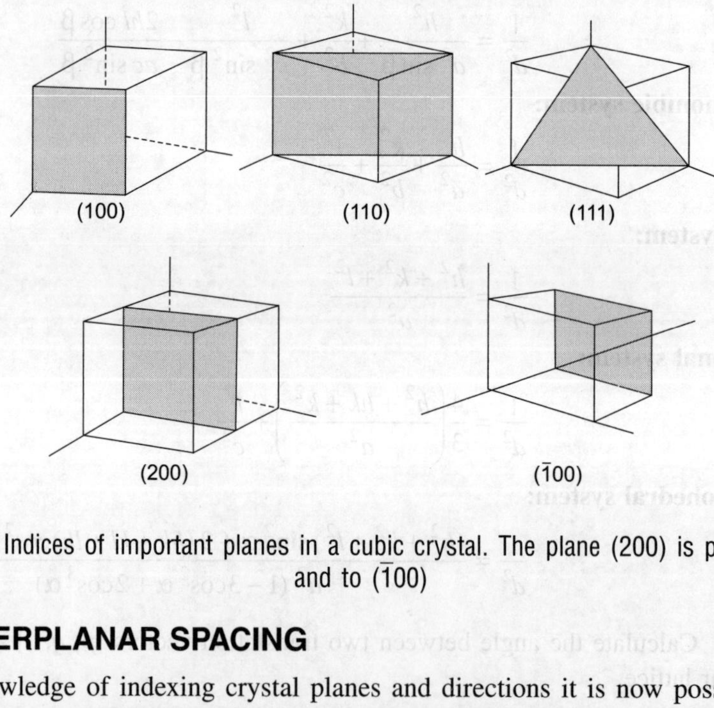

**Figure 2.75** Indices of important planes in a cubic crystal. The plane (200) is parallel to (100) and to $(\bar{1}00)$

## 2.36 INTERPLANAR SPACING

With the knowledge of indexing crystal planes and directions it is now possible to derive formula for the spacing between two consecutive parallel planes in a given unit cell. We shall limit our discussion to the unit cells which are expressed in terms of the orthogonal coordinate axes, so that simple Cartesian geometry is applicable. Thus, let us consider three mutually perpendicular coordinate axes, *ox*, *oy* and *oz*, and assume that a plane (hkl) parallel to the plane passing through the origin makes intercepts *a*/*h*, *b*/*k* and *c*/*l* on the three axes at *A*, *B* and *C* respectively as shown in Fig. 2.76. Furthur, let *OP* (= *d*, the interplanar spacing)

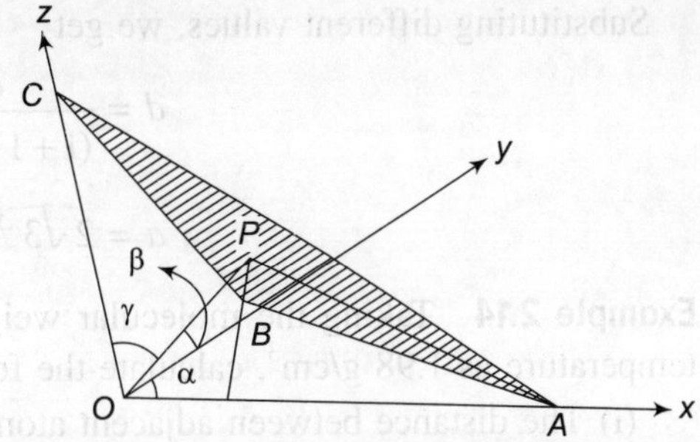


**Figure 2.76** (hkl) plane intercepting *x*, *y*, and *z* axes at *A*, *B* and *C*, respectively

be normal to the plane drawn from the origin and respectively make angles α, β and γ with the three axes. Therefore, $OA = a/h$, $OB = b/k$, $OC = c/l$ and $OP = d$.

From triangles *OPA* etc., we have

$$\cos\alpha = \frac{OP}{OA} = \frac{d}{(a/h)},$$

$$\cos\beta = \frac{OP}{OB} = \frac{d}{(b/k)},$$

$$\cos\gamma = \frac{OP}{OC} = \frac{d}{(c/l)},$$

Now, making use of the direction cosine, which states that

$$\cos^2\alpha + \cos^2\beta + \cos^2\gamma = 1 \qquad ...(1)$$

and substituting the values of cos α, and cos β, and cos γ in Eq. (1), we obtain

$$\frac{d^2}{(a/h)^2} + \frac{d^2}{(b/k)^2} + \frac{d^2}{(c/l)^2} = 1$$

or

$$d^2\left[\frac{h^2}{a^2} + \frac{k^2}{b^2} + \frac{l^2}{c^2}\right] = 1$$

So that

$$d = \left[\frac{h^2}{a^2} + \frac{k^2}{b^2} + \frac{l^2}{c^2}\right]^{-1/2} \qquad ...(2)$$

This is a general formula and is applicable to the primitive lattice of orthorhombic, tetragonal and cubic systems.

(i) Tetragonal system: $a = b \quad c$, Eq. (2) reduces to

$$d = \left[\frac{h^2 + k^2}{a^2} + \frac{l^2}{c^2}\right]^{-1/2} \qquad ...(3)$$

(ii) Cubic system: $a = b = c$, Eq. (2) reduces to

$$d = a\,(h^2 + k^2 + l^2)^{-1/2}$$

or

$$d = \frac{a}{(h^2 + k^2 + l^2)^{1/2}} \qquad ...(4)$$

**Example 2.13** The distance between consecutive (111) planes in a cubic crystal is 2 Å. Determine the lattice parameter.

***Solution*** For cubic crystals, we have

$$d = \frac{a}{(h^2 + k^2 + l^2)^{1/2}}$$

Substituting different values, we get

$$d = \frac{a}{(1+1+1)^{1/2}} = \frac{a}{\sqrt{3}}$$

$$a = 2\sqrt{3}\,\text{Å} = 3.46\,\text{Å}$$

**Example 2.14** Taking the molecular weight of KCl crystal as 74.45 and density at room temperature as 1.98 g/cm$^3$, calculate the following:

(i) The distance between adjacent atoms (lattice constant).

(ii) The spacing between

(a) {100} planes, (b) {110} planes, (c) {111} planes.

(iii) The distance between atoms of the same kind.

(iv) First nearest neighbour and second nearest neighbour distance.

***Solution*** (i) The mass of KCl molecule $= 74.45 \dfrac{g}{\text{molecule}} \times \dfrac{1 \text{ mole}}{N_A}$

$$= 74.45 \times \frac{1}{6.023 \times 10^{23}}\, g$$

$$= 12.37 \times 10^{-23}\, g$$

$$\frac{\text{Number of KCl molecules}}{\text{Unit volume}} = \frac{1.98\, g}{\text{cm}^3} \times \frac{1}{12.37 \times 10^{-23}\, g}$$

$$= 1.6 \times 10^{22} \text{ molecules/cm}^3$$

KCl molecule is diatomic, therefore,

$$\text{Number of atoms/volume} = 2 \times 1.6 \times 10^{22}/\text{cm}^3$$

$$= 3.2 \times 10^{22} \text{ atoms/cm}^3$$

Let $a$ = distance between adjacent atoms and let there be $n$ atoms along the edge of a 1 cm cube. Therefore, volume of unit cube = $(na)^3 = n^3a^3$, where $n^3$ is the number of atoms/cm$^3$.

$$n^3 = 3.2 \times 10^{22} \quad \Rightarrow \quad 3.2 \times 10^{22} \times a^3 = 1$$

or

$$a^3 = \frac{1 \times \text{cm}^3}{3.2 \times 10^{22}} \quad \Rightarrow \quad a = 3.15 \times 10^{-8} \text{ cm}$$

$\Rightarrow$ Lattice constant = 3.15 Å

(ii) $$d_{hkl} = \frac{1}{\sqrt{(h/a)^2 + (k/b)^2 + (l/c)^2}}$$

where $h, k, l$ are the Miller indices and $a, b, c$ are the lengths of the primitive translation vectors. For KCl, $a = b = c \equiv a$ (lattice constant). Hence,

$$d_{100} = \frac{1}{\sqrt{(1/a)^2 + 0 + 0}} = 3.15\,\text{Å}$$

$$d_{110} = \frac{1}{\sqrt{(1/a)^2 + (1/a)^2 + 0}} = \frac{a}{\sqrt{2}} = 2.23 \text{ Å}$$

$$d_{111} = \frac{1}{\sqrt{(1/a)^2 + (1/a)^2 + (1/a)^2}} = \frac{a}{\sqrt{3}} = 1.82 \text{ Å}$$

(iii) Distance between two atoms of the same kind,

$$2a = 2 \times 3.15 \text{ Å} = 6.3 \text{ Å}$$

(iv) The Bravais lattice of KCl crystal is fcc.

$\therefore$ The first nearest neighbour distance $= \sqrt{\left(\frac{a}{2}\right)^2 + \left(\frac{a}{2}\right)^2}$

$$= \frac{a}{2} = \frac{3.15}{2} = 1.575 \text{ Å}$$

and the second nearest neighbour distance is

$$a = 3.15 \text{ Å}.$$

## 2.37 SYMMETRY ELEMENTS

We know that the repetition of an object (or a lattice point) by a translation leaves the environment around that object (or lattice point) unchanged. In an extended array such as atomic pattern of a crystal, symmetries involving translation, rotation, reflection, inversion or their combinations are found to occur. The characteristic feature of these symmetries is that the translation operation repeats an object infinite number of times, while other operations (such as rotation, reflection, inversion or their combinations) repeat it only a finite number of times (Fig. 2.77). Further, the operations like translation or proper rotation leave the 'motif' (a fundamental group of atoms or molecules) unchanged whereas the operations like reflections, inversion or improper rotation change the character of the motif from a right handed to a left handed one and vice versa. The geometrical locus about which a group of finite operations act is known as *symmetry element*. Thus, in Fig. 2.77 the line normal to the page about which a rotation takes place, the plane about which the reflection occurs and the

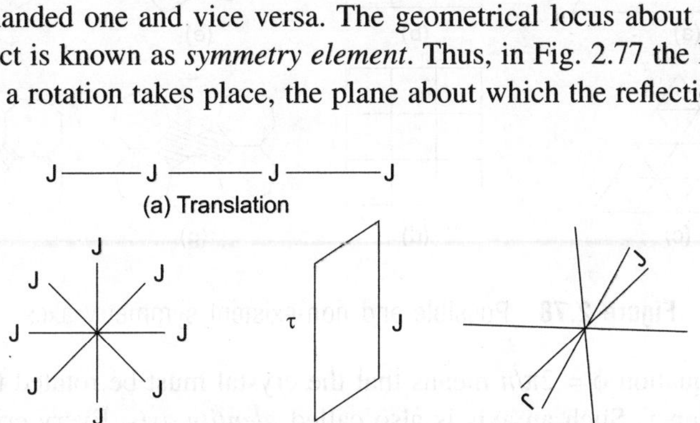


**Figure 2.77** Simple symmetry operations

centre about which the inversion takes place are simple examples of symmetry elements and are called, a rotation axis, a reflection or a mirror plane and an inversion centre, respectively. Crystals possess different symmetries or symmetry elements. The symmetry elements are described by certain operations.

A crystalline solid can have the following symmetry elements:

(i) Pure translation: $(\vec{R}_n)$ defined in Art. 2.6

(ii) Proper rotation: through an angle $\phi$

(iii) Reflection: across a line (in two dimensions)

(iv) Inversion: through a point

(v) Improper rotation: rotoreflection/rotoinversion.

However, it may be easily verified that the above symmetry elements are not all independent but are interconnected to each other. In the following, we shall discuss other symmetry elements.

### 2.37.1 Proper Rotation

The simplest way of representing the operation for proper rotation is shown in Fig. 2.77 (b). One can think of a line or axis passing through the centre and normal to the figure so that the J's (or in general any geometrical figure) are represented by a rotation through any angle $\phi = 2\pi/n$ about the axis of rotation, the axis is said to have $n$-fold symmetry. However, because of the reticular structure of crystals, only 1-, 2-, 3-, 4- and 6-fold symmetries (Fig. 2.78) are possible. They are known as symmetry elements corresponding to proper rotation, consequently, it means that a crystalline solid cannot possess either 5-fold or any other rotational symmetry higher than 6-folds. They are being demonstrated in Fig. 2.78 (e), 2.78 (g), 2.78 (h).

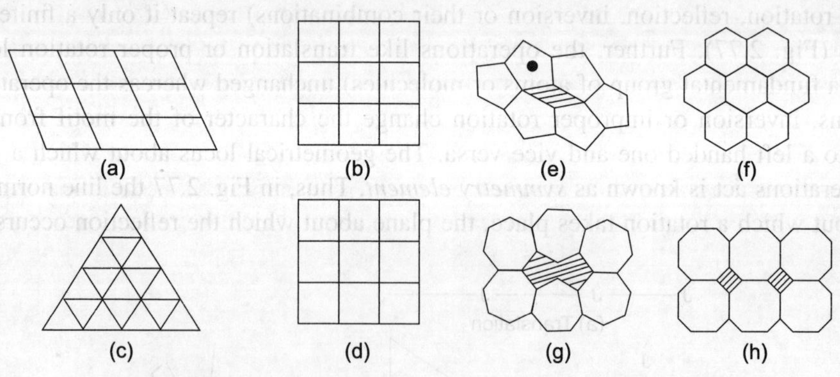


**Figure 2.78** Possible and non-existent symmetry axes

For $n = 1$ in equation $\phi = 2\pi/n$ means that the crystal must be rotated through $2\pi$ (360°) to achieve congruency. Such an axis is also called *identity axis*. Every crystal possesses an infinite number of such axes. Further, $n = 2$ meants the crystal is rotated through $\pi$ (180°) and the axis is said to have 2-fold symmetry. Such an axis is known as a *diad axis* and symbolically represented as (). In a similar way for $n = 3, 4$ and 6 the corresponding angles of rotation are 120°, 90° and 60°, respectively. They are said to have 3-fold, 4-fold and 6-fold symmetry, the axes are termed *triad, tetrad* and *hexad* and sumbolically represented as △, □ and ⬡ respectively.

In order to check the permissible rotational symmetry elements in a crystal, one may start with the fundamental requirement that all crystals must have periodic structure, that is, the symmetry elements present in the crystals must conform to their translational periodicities. This is the reason why the number of symmetry elements found in crystals is limited. Now, let us consider the combination of $n$-fold axis of rotation $A_n$ with a translation $t$ as shown Fig. 2.79. As we have already seen that a rotation axis repeats the translation $\phi°$ (= $2\pi/n$) away, and hence $n$ such rotations will bring back the object into its initial position (it is immaterial wether the rotation is clockwise or counterclockwise). Therefore, starting from the linear array as in Fig. 2.79, two such rotations will produce new lattice points $p$ and $q$ as shown in Fig. 2.80. Since they are equidistance from the original lattice row by construction, the two lattice points $p$ and $q$ must be connected by the same translation ($t$) or same integral multiple of it say $mt$, depending upon the magnitude of $\phi$. Consequently, the allowed values of $\phi$ can be determined from the construction shown in Fig. 2.80. From simple geometrical consideration, we have

$$mt = t + 2t \cos \phi, \text{ where } m = 0, \pm 1, \pm 2, \ldots \quad \ldots(1)$$

$\pm m$ is used depending whether the rotation is clockwise or counterclockwise. Dividing both sides by $t$, Eq. 1 can be written as

$$m = 1 + 2 \cos \phi \quad \text{or} \quad m - 1 = 2 \cos \phi \quad \ldots(2)$$

Since $m$ is an integer, $(m - 1)$ will also be an integer. Taking $(m - 1) = N$, we have

$$N = 2 \cos \phi \quad \text{or} \quad \cos \phi = N/2$$

where $$N = 0, \pm 1, \pm 2, \ldots \quad \ldots(3)$$

Since all possible values of $\cos \phi$ lie between $\pm 1$, corresponding possible values of $N$, $\phi$ and $n$ can be easily obtained and are given in Table 2.15. This Table clearly shows the non-existence of symmetry axes other than 1-, 2-, 3-, 4- and 6-fold. If $n = 2$, the crystal must be rotated through 180° to achieve congruence, and the axis is termed a *diad axis* (denoted by symbol ⬮). If $n = 3$, the corresponding angle of rotation is 120°, and the axis is called a *triad* and is denoted by the symbol ▲. If $n = 4$, the corresponding angle of rotation is 90°, and the axis is called a *tetrad* (denoted by the symbol ■). If $n = 6$, the corresponding angle of rotation is 60°, and the axis is called a *hexad* (denoted by the symbol ⬢).

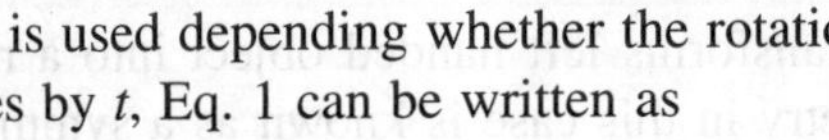


**Figure 2.79** $n$-fold axis of rotation with a translation $t$

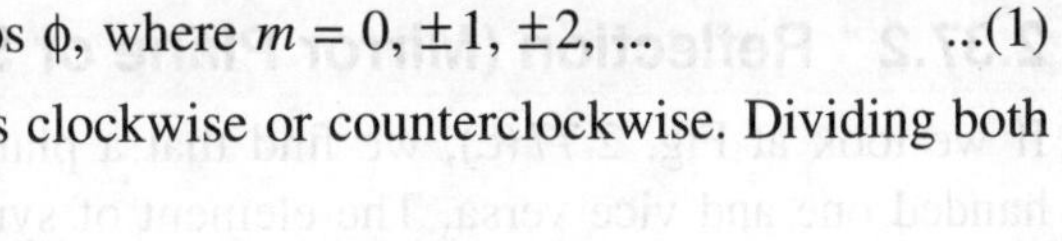


**Figure 2.80** Two-dimensional array of: (a) abjects, (b) points; a plane lattice

**Table 2.15** Allowed Rotational Axis in a Crystal Lattice

| *N* | *cos φ* | *φ* | *n allowed rotational axes* |
|---|---|---|---|
| −2 | −1 | 180 | 2 |
| −1 | −1/2 | 120 | 3 |
| 0 | 0 | 90 | 4 |
| +1 | +1/2 | 60 | 6 |
| +2 | +1 | 360 or 0 | 1 |

**Example 2.15** Construct a two-dimensional pentagonal and octagonal lattice.

***Solution*** Since each interior angle of a pentagon is 108°, which is not a quotient of 360°. So, if we try to construct a periodic array having 5-fold symmetry, we can see that the resulting pentagons do not fit together neatly and leave empty spaces in between as shown by the shaded region in Fig. 2.78 (e). Similarly, in an octagon each interior angle is 135°, which again is not a quotient of 360°. In this case too, the octagons cannot fill the whole space without leaving gaps (Fig. 2.78h). Both considerations indicate that neither 5-fold nor 8-fold symmetries can exist. In fact, all symmetries higher than 6-fold are not possible.

### 2.37.2 Reflection (Mirror Plane or Symmetry Plane)

If we look at Fig. 2.77 (c), we find that a plane transforms left handed object into a right handed one and vice versa. The element of symmetry in this case is known as a symmetry plane or a mirror plane and symbolically represented by the letter *m*. Mirror coincident figures of this type are known as enantiomorphous pairs and are related by reflection across a plane passing half way between them. If a crystal is cut along the symmetry plane and put on a mirror, the image will produce the other half of the crystal.

### 2.37.3 Inversion Centre (Centre of Symmetry)

This is a symmetry operation similar to reflection, with the difference that reflection occurs in a plane of mirror, while inversion is equivalent to reflection through a point called inversion centre or centre of symmetry. In other words, the enantiomorphous pair of figures in the case is related by inversion centre laying half way between the two (Fig. 2.77d). Each point of the figure is obtained from the other by reflection through centre and is symmetrically inverted. The complete figure thus obtained is inverted completely. Thus, the inversion centre or the centre of symmetry has the property of inverting all space (at present the latter *J*) through a point.

### 2.37.4 Improper Rotations

Symmetry operations encountered above, i.e., reflection and inversion, produce enantiomorphous sets of objects. However, it is also possible to have a rotational operation relating enantiomorphous objects. Such an operation is called improper rotation. The corresponding symmetry element is known as improper rotation axis. There exist two improper rotations, they are rotoreflection and rotoinversion.

### 2.37.5 Rotoreflection

If a rotation and a reflection is combined to form a single hybrid operation, the resulting operation is called rotoreflection. The corresponding symmetry element is called rotoreflection axis. For each proper rotation axis, there exists a corresponding rotoreflection axis. To distinguish the two kinds of axes, a tilde symbolically represented as (~) is placed over the numerical symbol of the corresponding rotational axis, e.g., $\tilde{1}$, $\tilde{2}$, $\tilde{3}$, etc. For a better understanding, let us see the following examples.

**Example 2.16** Show that $\tilde{1}$ (one "tilde") is equivalent to a mirror plane.

***Solution*** It is a combined operation of rotation and reflection. In this case, the proper 1-fold rotation will rotate the motif representing all space (here it is taken as *J*) through an angle of 0° or 360°, i.e., leaving it unchanged. Combining this with reflecting plane perpendicular to the rotation axis to produce a configuration as shown in Fig. 2.81 (a), which is identical to the configuration already encountered in Fig. 2.77 (c). Thus, the operation 1 is equivalent to a reflection through a plane (specifically a mirror plane).

**Example 2.17** Show that $\tilde{2}$ is equivalent to an inversion centre.

***Solution*** In the case, the proper 2-fold rotation will rotate the object (through an imaginary axis) through an angle 180° and then reflected across an imaginary plane placed at right angle to the imaginary rotation axis, the resulting configuration (Fig. 2.81b), is identical to the configuration shown in Fig. 2.77d. Thus, the operation $\tilde{2}$ is equivalent to an inversion centre.

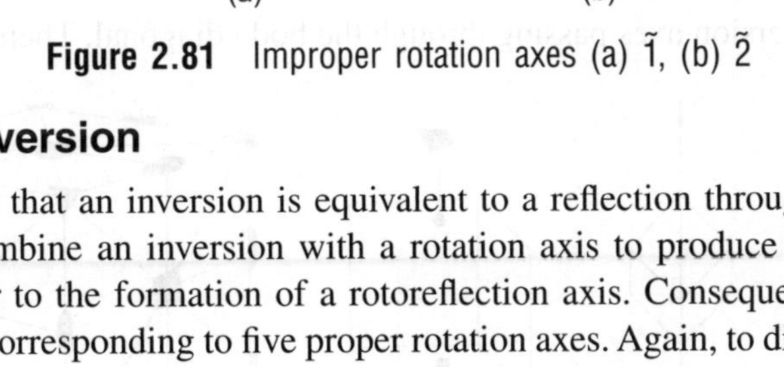


**Figure 2.81** Improper rotation axes (a) $\tilde{1}$, (b) $\tilde{2}$

### 2.37.6 Rotoinversion

As we stated above that an inversion is equivalent to a reflection through a point, therefore it is possible to combine an inversion with a rotation axis to produce a rotoinversion axis in a manner similar to the formation of a rotoreflection axis. Consequently, there exist five rotoinversion axes corresponding to five proper rotation axes. Again, to distinguish these axes, both from pure rotation and rotoreflection, a bar (–) is plaed over the numerical symbol of the corresponding rotational axis, e.g., $\bar{1}$, $\bar{2}$, $\bar{3}$, etc.

When these rotoinversion axes are compared with the corresponding rotoreflection axes, they are found to be equivalent in pairs and therefore, it is sufficient to adopt only one kind to represent these symmetry operations. Table 2.16 enlists the conventional symbols used to represent the improper axes.

**Table 2.16** Conventional Symbol for Improper Axes

| *Rotoinversion axes* | *Rotoreflection axes* | *Conventional symbol* | |
|---|---|---|---|
| $1 + \bar{1} \rightarrow \bar{1}$ | $\tilde{2}$ | Centre of symmetry | $\bar{1}$ |
| $2 + \bar{1} \rightarrow \bar{2}$ | $\tilde{1}$ | Mirror plane | $m$ |
| $3 + \bar{1} \rightarrow \bar{3}$ | $\tilde{6}$ | 3-fold rotoinversion | $\bar{3}$ |
| $4 + \bar{1} \rightarrow \bar{4}$ | $\tilde{4}$ | 4-fold rotoinversion | $\bar{4}$ |
| $6 + \bar{1} \rightarrow \bar{6}$ | $\tilde{3}$ | 6-fold rotoinversion | $\bar{6}$ |

With respect to these five rotary inversions it is necessary to note the following:

(i) The $\bar{1}$ is obviously identical with a centre of symmetry and is not a different symmetry element.

(ii) The $\bar{2}$ is also not a different symmetry element, it is exactly equivalent to mirror reflection across a plane passing through the centre at right angles to the axis.

(iii) The $\bar{3}$ is truly identical to the combination of a three-fold rotation axis and a centre, but for convenience it is usually considered to be a distinct element with the symbol ▲.

(iv) The $\bar{4}$ is irreducible, with symbol ◪ (a distinct symmetry element).

(v) The $\bar{6}$ is exactly equivalent to the combination of a three-fold rotation axis and mirror reflection across a plane at right angles to the axis, but for convenience it is usually considered to be a distinct element. Its symbol is ⬘.

It should be noted here that $\bar{2}$, i.e., a two-fold roto-reflection axis is neither a two-fold axis nor a reflection plane, rather it is a hybrid of the two operations – first a rotation by 180° and then reflection through a plane at right angles to the rotation axis.

**Example 2.18** Find the total number of symmetry elements that exist in a cube.

***Solution*** Looking at Fig. 2.82, we find that there exist

(i) 2-fold parallel to the face diagonal and passing through the centre of the cube. There are six such axes.

(ii) 3-fold rotoinversion axes passing through the body diagonal. There are four such axes.

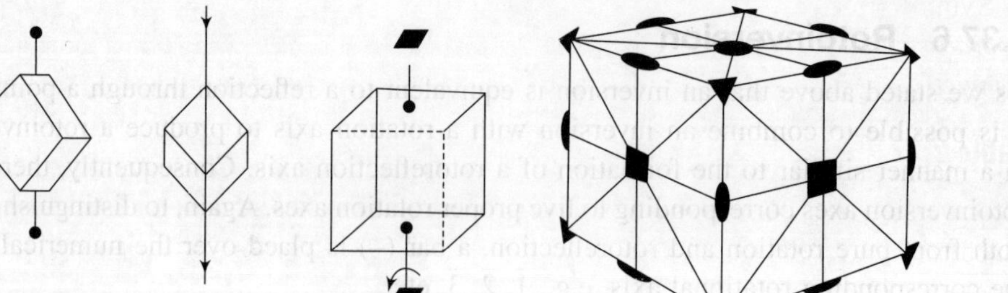

**Figure 2.82** Various symmetries of a cube

(iii) 4-fold symmetry about an axis passing through the centre of two opposite faces. There are three such symmetry axes.

(iv) Nine mirror planes bisecting the parallel faces, and connecting diagonal edges. All of them passing through the centre of the cube.

(v) One centre of symmetry at the centre of the cube.

Thus, there are: 6-diad + 4-triads + 3-tetrads = 13 axes, 3 surface planes + 6 diagonal planes = 9 planes, and 1 centre of symmetry. Therefore, the total symmetry elements in a cube = 13 + 9 + 1 = 23.

## 2.38 COMBINATION OF SYMMETRY ELEMENTS (POINT GROUP)

In the preceding section, we discussed various symmetry elements and their relationships. It was found that the symmetry elements are interlinked to each other and are not independent. In other words, a given symmetry element could be obtained from suitable combinations (if they are compatible) of other symmetry elements. Such a combination forms a group and gives rise to different symmetry points called *point groups*. Thus, a *point group* in the lattice is defined as the collection of symmetry operations which when applied about a lattice points, leave the lattice invariant. In other words, in point groups all possible symmetry elements must pass through a point. Suitable combinations of various symmetry elements give rise to thirty two (32) allowed point groups. Since crystals belonging to different crystal systems show different point group symmetries and therefore the classification of crystal systems can be easily made on the basis of point groups. A few examples illustrating the locations of equivalent positions forming groups are shown in Fig. 2.83.

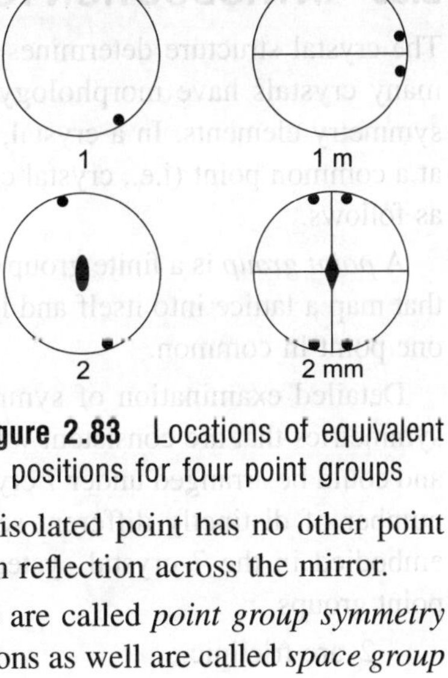


**Figure 2.83** Locations of equivalent positions for four point groups

Here, the dots represent equivalent points. The point group 1 has no symmetry elements and hence the single isolated point has no other point equivalent to it. For 1m, there is a mirror plane, one dot on reflection across the mirror.

Symmetry operations performed about a point or a line are called *point group symmetry operations* and *symmetry operations* performed by translations as well are called *space group symmetry operations*.

**Table 2.17** Permissible Point Group Symmetry Operations for Two Dimensional Crystals

| *Lattice* | *System* | *Point group symmetry about* |
|---|---|---|
| 1. Oblique | Oblique | 1, 2 |
| 2. Square | Square | 4, 4 mm |
| 3. Hexagonal | Hexagonal | 3, 3 m, 6 mm |
| 4. Primitive rectangular<br>5. Centred rectangular | Rectangle | 1 m, 2 mm |

The complete set of symmetry elements that may occur in space lattice is

$$\left.\begin{array}{ll} \text{the rotation axes} & : 1, 2, 3, 4, 6 \\ \text{the centre of symmetry} & : \bar{1} \\ \text{the roto-inversion axes} & : \bar{3}, \bar{4}, \bar{6} \\ \text{and the mirror plane} & : m \end{array}\right\} \quad ...(1)$$

The *macroscopic symmetry elements* are those symmetry elements which may influence the external shape of a freely formed crystal in such a way as to be detectable with the unaided eye and consist of elements described in set of Eq. (1).

There are other symmetry elements also which have no influence on the external shape at all, but only concern the structure at an atomic level. These are the *microscopic symmetry elements.* These are of two kinds: (i) the *glide plane* and (ii) the *screw axis.*

(We will not discuss here the microscopic symmetry elements and will consider point group symmetries and point groups.)

## 2.39 INTRODUCTION TO POINT GROUPS

The crystal structure determines the external form or morphology of the crystal. In nature, many crystals have morphology that displays the presence of one or more symmetry or symmetry elements. In a crystal, the different possible combinations of symmetry elements at a common point (i.e., crystal centre) is called a *point group*. We may define a point group as follows.

A *point group* is a finite group made up of the set of rotation and reflection transformations that map a lattice into itself and in which all rotation axed and reflection planes have at least one point in common.

Detailed examination of symmetry elements reveals that there are only 32 point group symmetries that are consistent with translational symmetry, and could be arranged under 7 crystal systems, i.e., the total number of distinctly different sets of symmetry elements embodied in the 7 crystal systems is only 32. Out of 32 point groups

2 are triclinic,
3 are monoclinic,
3 are orthorhombic,
7 are tetragonal,
5 are trigonal,
7 are hexagonal.
and 5 are cubic.

32 crystal classes in all.

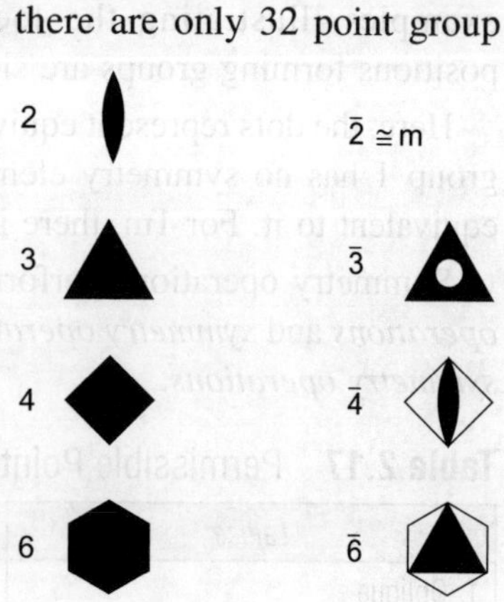


**Figure 2.84** The symbols for rotation, and rotoinversion axes (*m* is a mirror plane)

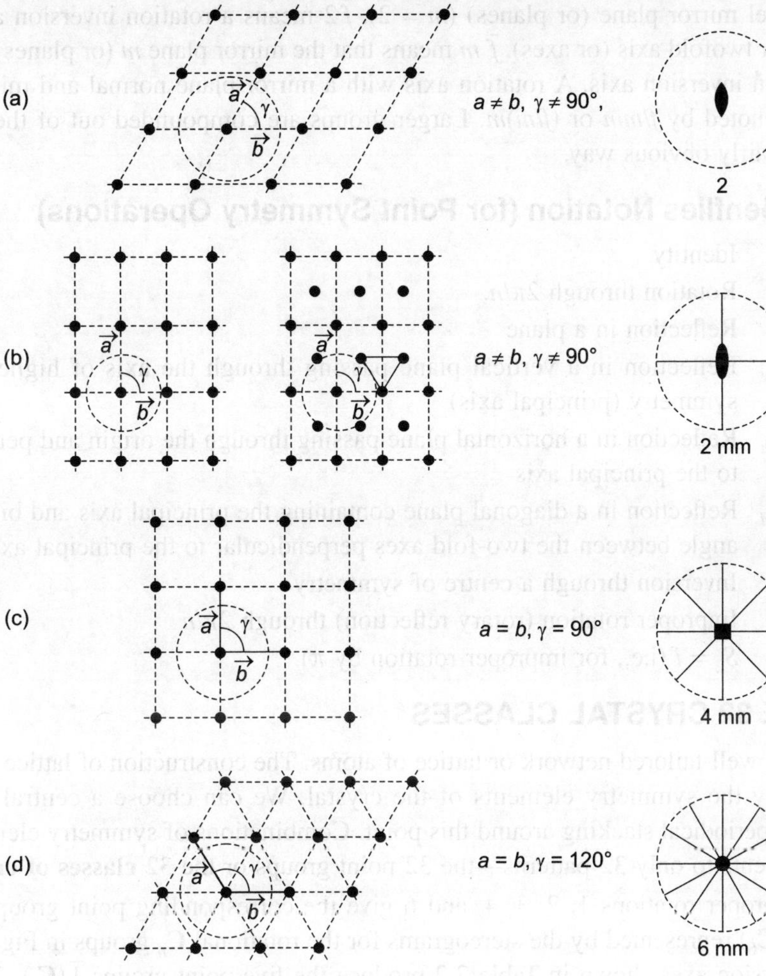


**Figure 2.85** Symmetry elements in the two-dimensional (a) oblique, (b) rectangular, (c) square and (d) hexagonal lattices

## 2.40 NOTATIONS

Prior to the enumeration of the 32 point groups, it would be worthwhile to consider the description of two classes of notation, viz., the *international* (sometimes called the *Hermann–Mauguin notation* and the *Schönflies notation.*)

### (A) International Notation

An *f*-fold axis of rotational symmetry will be specified by *f*. Also, *f* will stand for the group of *f*-fold rotations. For example, 2 means a two-fold axis of symmetry (previously called $C_2$), and it can also mean the group of twofold rotations. $\bar{f}$ will denote a rotation inversion of π followed by an inversion. *F*/*m* means a rotation axis with a perpendicular mirror plane. *f*2 means a rotation axis with a perpendicular twofold axis (or axes). *fm* means a rotation axis with a parallel mirror plane (or planes) ($m = \bar{2}$). $\bar{f}2$ means a rotation inversion axis with a perpendicular twofold axis (or axes). $\bar{f}\,m$ means that the mirror plane $m$ (or planes) is parallel to the rotation inversion axis. A rotation axis with a mirror plane normal and mirror planes parallel is denoted by *f/mm* or *(f/m)m*. Larger groups are compounded out of these smaller groups in a fairly obvious way.

### (B) Schöenflies Notation (for Point Symmetry Operations)

(1) $E$ Identity

(2) $C_n$ Rotation through $2\pi/n$.

(3) $\sigma$ Reflection in a plane

- (i) $\sigma_v$ Reflection in a vertical plane passing through the axis of highest rotation symmetry (principal axis)
- (ii) $\sigma_h$ Reflection in a horizontal plane passing through the origin and perpendicular to the principal axis
- (iii) $\sigma_d$ Reflection in a diagonal plane containing the principal axis and bisecting the angle between the two-fold axes perpendicular to the principal axis

(4) $i$ Inversion through a centre of symmetry

(5) $S_n$ Improper rotation (rotary reflection) through $2\pi/n$
$S_2 = i$ (i.e., for improper rotation by $\pi$).

## 2.41 THE 32 CRYSTAL CLASSES

A crystal is a well-tailored network or lattice of atoms. The construction of lattice is directed stringently by the symmetry elements of the crystal. We can choose a central point and consider the periodical stacking around this point. Combinations of symmetry elements limit the arrangements to only 32 patterns—the 32 point groups or the 32 **classes of crystals**.

The five proper rotations 1, 2, 3, 4, and 6 give the corresponding point groups ($\mathbf{C}_1$, $\mathbf{C}_2$, $\mathbf{C}_3$, $\mathbf{C}_4$, and $\mathbf{C}_6$) represented by the stereograms for the rotational $\mathbf{C}_n$ groups in Fig. 2.86. The rotation-inversion axes shown in Table 2.2 produce the five point groups $\bar{1}(\mathbf{C}_i)$, $\bar{2}(m = \mathbf{C}_s)$, $\bar{3}(\mathbf{S}_6)$, $\bar{4}(\mathbf{S}_4)$, and $\bar{6}(\mathbf{C}_{3h} = \mathbf{S}_3)$, where $\mathbf{C}_s = \mathbf{S}_1$ and $\mathbf{C}_i = \mathbf{S}_2$ shown in Fig. 2.86. Combinations of proper rotation axis produce six more rotational point groups (**D**, **T**, $\mathbf{D}_4$, $\mathbf{D}_3$, $\mathbf{D}_6$, and **O**), as diagrammed in Fig. 2.87. These are dihedral groups ($\mathbf{D}_n$) plus the rotational groups **T** and **O**. Combination of proper axes perpendicular to a mirror plane produces five point groups, $1/m = m = \bar{2}$, $2/m$, $3/m = \bar{6}$, $4/m$, and $6/m$. Two of these are identical with $\bar{2}$ and $\bar{6}$ shown in Fig. 2.86, leaving three additional point groups, $2/m$ ($\mathbf{C}_{2h}$), $4/m$ ($\mathbf{C}_{4h}$), and $6/m$ ($\mathbf{C}_{6h}$) represented in Fig. 2.87.

Combination of a proper rotation with a vertical mirror plane ($m$ or $\sigma_v$) produces the $\mathbf{C}_{nv}$ groups, ($1\,m = m$), $2\,mm$ ($= mm$), $3\,m$, $4\,mm$, and $6\,mm$. The point group $m$ ($\mathbf{C}_s$) is included in Fig. 2.86, leaving four $\mathbf{C}_{nv}$ groups in Fig. 2.87. Combination of a proper rotation with a horizontal mirror plane ($m$ or $\sigma_h$) and a vertical plane produces six more point groups: $2/m\ 2/m\ 2/m$ ($= mmm$) ($\mathbf{D}_{2h}$), $3/m\ m2$ ($= \bar{6}m2$)($\mathbf{D}_{3h}$), $4/m\ 2/m\ 2/m$ ($= 4/m\ mm$) ($\mathbf{D}_{4h}$), $6/m\ 2/m\ 2/m$ ($= 6/m\ mm$) ($\mathbf{D}_{6h}$), $2/m\ \bar{3}$ ($= m\bar{3}$) ($\mathbf{T}_h$), and $4/m\bar{3}2/m$ ($= m\bar{3}m$) ($\mathbf{O}_h$). These are shown in Fig. 2.88.

These proper rotations and improper rotations and combinations of proper rotations and mirror planes produce 29 point groups. The remaining three point groups are obtained by combinations of improper axes $\bar{3}$ and $\bar{4}$ with proper rotations. The products are point groups $\bar{4}2m$ ($\mathbf{D}_{2d}$), $\bar{3}m$ ($\mathbf{D}_{3d}$), and $\bar{4}3m$ ($\mathbf{T}_d$) as shown in Fig. 2.88. These complete the 32 crystallographic point groups or crystal classes.

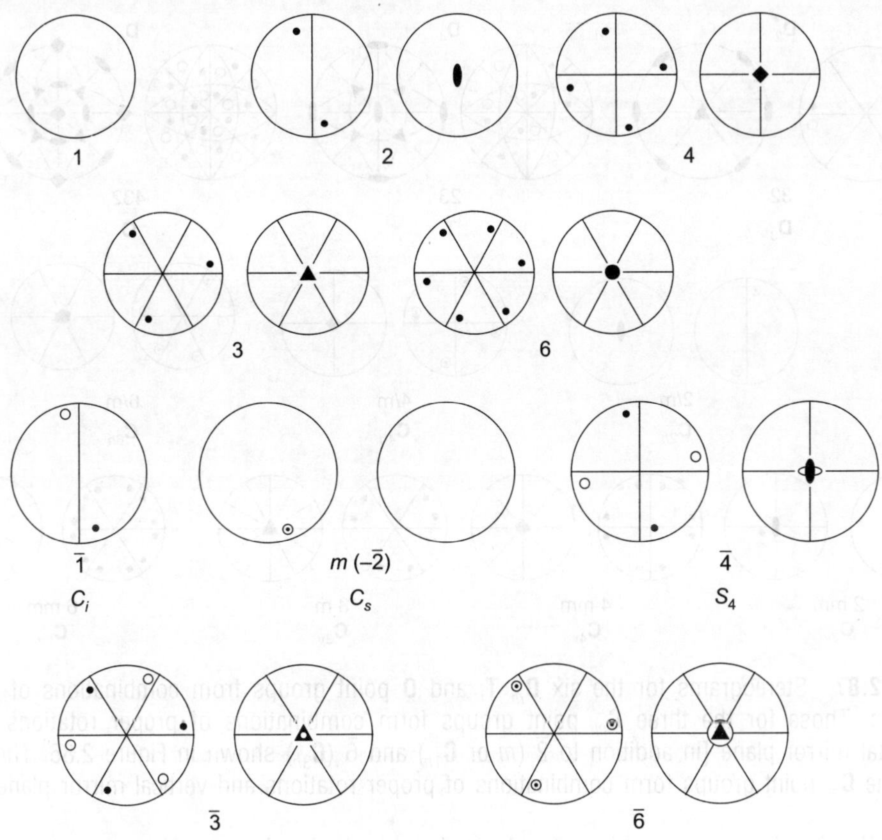


**Figure 2.86** Stereograms for the six $\mathbf{C}_n$ point groups produced by proper rotations and those for the point groups produced by improper rotations

## 2.42 ENUMERATION OF 32 POINT GROUPS

First we have rotational point groups $C_n$ (i.e., $C_1$, $C_2$, $C_3$, $C_4$ and $C_6$: five rotational point groups).

From each of the $C_n$ point groups, five new point groups can be obtained by adding a $\sigma_n$ plane. These are called $C_{nh}$ ($n$ = 1, 2, 3, 4, 6). Four new point groups can be generated by adding a vertical plane $\sigma_v$ to the rotational point groups $C_n$. These are termed $C_{nv}$ ($n$ = 2, 3, 4, 6). The case $n$ = 1 is the same as $C_{lh}$, since the vertical and the horizontal are not distinguishable for $C_1$.

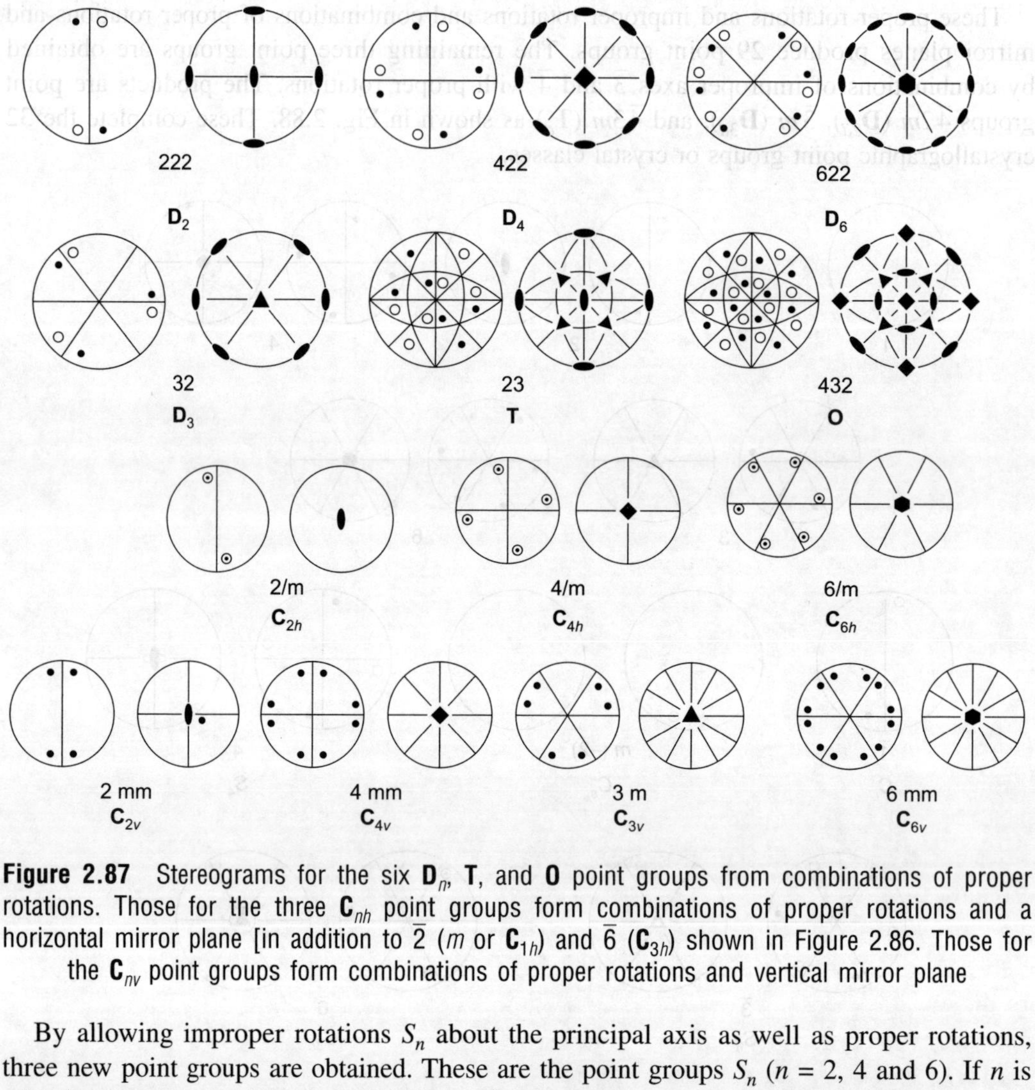


**Figure 2.87** Stereograms for the six $\mathbf{D}_n$, **T**, and **O** point groups from combinations of proper rotations. Those for the three $\mathbf{C}_{nh}$ point groups form combinations of proper rotations and a horizontal mirror plane [in addition to $\bar{2}$ ($m$ or $\mathbf{C}_{1h}$) and $\bar{6}$ ($\mathbf{C}_{3h}$) shown in Figure 2.86. Those for the $\mathbf{C}_{nv}$ point groups form combinations of proper rotations and vertical mirror plane

By allowing improper rotations $S_n$ about the principal axis as well as proper rotations, three new point groups are obtained. These are the point groups $S_n$ ($n = 2$, 4 and 6). If $n$ is odd, these groups are identical with $C_{nh}$ and hence need not be considered separately.

The 17 point groups mentioned above contain only one axis of $n$-fold symmetry. By introducing axes of symmetry in the horizontal plane in addition to the vertical axis, new points groups can be obtained. A horizontal axis of $m$-fold symmetry in conjunction with the vertical axes of $n$-fold symmetry implies the existence of $n - 1$ other horizontal axes of $m$-fold symmetry.

By taking the five $C_n$ point groups and adding a 2-fold symmetry ($m = 2$), one obtains four new point groups $D_n$ ($n = 2$, 3, 4, and 6).

Another four point groups $D_{nh}$ ($n = 2$, 3, 4, and 6) can be generated by adding a horizontal reflection plane $\sigma_h$ to the point groups $D_n$.

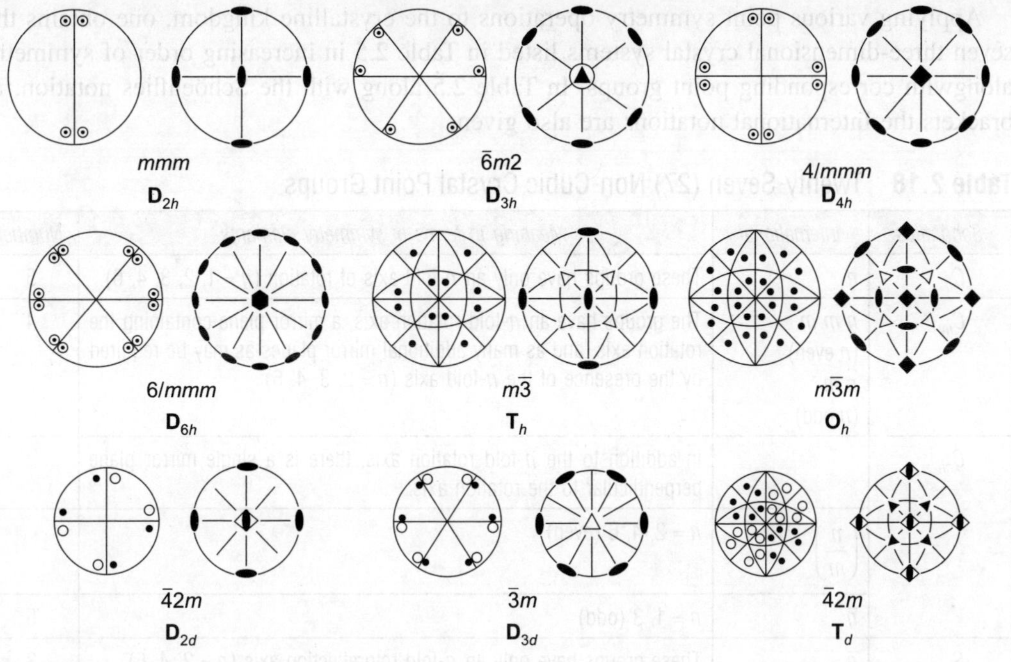


**Figure 2.88** Stereogram for point groups from proper rotations and a plane. Stereograms for the point groups from combinations of $\bar{4}$ of improper rotations $\bar{3}$ and $\bar{4}$ with proper rotations

Finally, two more groups $D_{nd}$ ($n$ = 2 and 3) can be obtained by adding diagonal reflection planes bisecting the angles between the horizontal twofold axes to the groups $D_n$.

The 27 point groups enumerated above have one principal n-fold axis (vertical) and may also have $n$ twofold axes ($m$ = 2) in the horizontal plane. These are also the point groups of two-dimensional crystals (Table 2.18).

In addition to these, there are five cubic point groups of higher symmetry which are distinguished by not possessing a unique axis but having four 3-fold axes. ($m > 2$). (Table 2.19).

$T$ is the smallest of these groups of higher symmetry, which takes a regular tetrahedron into itself. The point group $T_h$ is obtained from the point group $T$ by adding a mirror operation, whereas $T_d$ is the full tetrahedral group containing all the symmetry operations of a regular tetrahedron, including reflection $\sigma_d$.

$O_h$ is the highest of the groups of higher symmetry representing the full symmetry group of a cube or octahedron, including improper rotations and reflections. $O_h$ is obtained from the octahedral point group $O$, the group of proper rotations which takes a cube or an octahedron into itself.

It may be pointed out that if we add to these point group symmetry operations, the translation symmetry elements (screw operation ≡ coupling of a proper rotation with a nonprimitive translation and glide plane ≡ combination of an improper rotation with a nonprimitive translation), one gets the total 230 space groups of the crystal.

Applying various point symmetry operations to the crystalline kingdom, one obtains the seven three-dimensional crystal systems listed in Table 2.5 in increasing order of symmetry alongwith corresponding point groups. In Table 2.5 along with the Schöenflies notation, in brackets the International notations are also given.

**Table 2.18** Twenty-Seven (27) Non-Cubic Crystal Point Groups

| *Schönflies* | *International* | *Meaning in terms of symmetry elements* | *Number* |
|---|---|---|---|
| $C_n$ | $n$ | These groups have only an $n$-fold axis of rotation ($n$ = 1, 2, 3, 4, 6). | 5 |
| $C_{nv}$ | $n\,m\,m$ ($n$ even)<br>$n\,m$ ($n$ odd) | The groups have an $n$-fold rotation axis, a mirror plane containing the rotation axis, and as many additional mirror planes as may be required by the presence of the $n$-fold axis ($n$ = 2, 3, 4, 6) | 4 |
| $C_{nh}$ | | In addition to the $n$-fold rotation axis, there is a single mirror plane perpendicular to the rotation axis. | |
| | $\left(\frac{n}{m}\right)$ | $n$ = 2, 4, 6 (even) | |
| | $\bar{n}$ | $n$ = 1, 3 (odd) | 5 |
| $S_n$ | $\bar{n}$ | These groups have only an $n$-fold rotoreflection axis ($n$ = 2, 4, 6) | 3 |
| $D_n$ | $n\,2\,2$ ($n$ even)<br>$n\,2$ ($n$ odd) | These groups contain an $n$-fold axis of rotation, a 2-fold rotation axis perpendicular to the $n$-fold axis, and as many additional 2-fold axes as the presence of the $n$-fold axis demands ($n$ = 2, 3, 4, 6). | 4 |
| $D_{nh}$ | $\frac{n}{2}\frac{2}{m}\frac{2}{m}$ | In addition to all the elements of $D_n$, there is a mirror plane perpendicular to the $n$-fold rotation axis. | 4 |
| | $\left(\frac{n}{mmm}\right)$ | $n$ = 2, 4, 6 (even) | |
| | $\bar{n}\,2\,m$ | $n$ = 3 (odd) | |
| $D_{nd}$ | | In addition to all the elements of $D_n$, there are mirror planes containing the $n$-fold rotation axis such that the planes bisect the angles between the 2-fold axes of rotation. | 2 |
| | $\bar{n}\,2\,m$ | $n$ = 2 (even) | |
| | $\bar{n}\frac{2}{m}$ | $n$ = 3 (odd) | |

Both the nomenclature systems, the Schönflies and international, are used for the space group representation. Readers interested in more details may consult Buerger's *Elementary Crystallography* (Wiley, 1963).

**Table 2.19** The Five (5) Cubic Crystal Point Groups

| *Schönflies* | *International* | *Meaning in terms of symmetry elements* |
|---|---|---|
| $T$ | 2 3 | The tetrahedral point group (in a crystal the alternate corners of the cube are vacant). It contains three mutually perpendicular 2-fold axes, plus four 3-fold axes, angles between which are bisected by the 2-fold axes. |
| $T_d$ | $\bar{4}\ 3\ m$ | In addition to all the elements of $T$, there is a mirror plane through each pair of 3-fold axes (i.e., two mutually perpendicular mirror planes through each 2-fold axis. In all, there are thus six mirror planes. |
| $T_h$ | $\frac{2}{m}\bar{3}$ ($m\,3$) | The group has all symmetry elements of $T$, plus a centre of inversion $i$. |
| $O$ | 4 3 2 | The octahedral point group of $T$ contains three mutually perpendicular 4-fold axes and four 3-fold axes which have the same orientation with respect to one another as the 2-fold and 3-fold axes of the tetrahedral point group. |
| $O_h$ | $\frac{4}{m}\bar{3}\frac{2}{m}$ ($m\,3\,m$) | The group has all symmetry elements of $O$, plus a centre of inversion $i$. |

## 2.43 PICTORIAL REPRESENTATION OF POINT GROUPS (STEREOGRAMS)

A very useful pictorial way of thinking about point group symmetries is by the use of *stereogram* (*or stereographic projections*). Stereograms provide a way of representing the three-dimensional symmetry of the crystal in two dimensions. To construct a stereographic projection, a lattice point (or any other point about which one wishes to examine the point group symmetry) is surrounded by a sphere. Symmetry axes extending from the centre of the sphere intersect the sphere at points. These points are joined to the south pole (for points above the equator) by straight lines. Where the straight lines intersect a plane through the equator, a geometrical symbol may be placed to indicate the symmetry of the appropriate asymmetry axis. The stereogram is to be considered as viewed by someone at the north pole. Symmetry points below the equator can be characterized by turning the process upside down. Additional diagrams to show how typical points are mapped by the point group are often given with the stereogram. The idea is illustrated in Fig. 2.89. Wood and Brown have stereograms of the 32 point groups. Rather than going into great detail in describing stereograms, let us look at a stereogram for $D_3$ (or in the *international* notation 32).

The principal threefold axis is represented by the triangle in the centre of Fig. 2.90(b). The twofold symmetry axes perpendicular to the threefold axis are represented by the dark ovals at the end of the line through the centre of the circle.

In Fig. 2.90(a) the dot represents a point above the plane of the paper and the open circle represents a point below the plane of the paper. Starting from any given point, it is possible to get to any other point by using the appropriate symmetry operations. $D_3$ has no refection

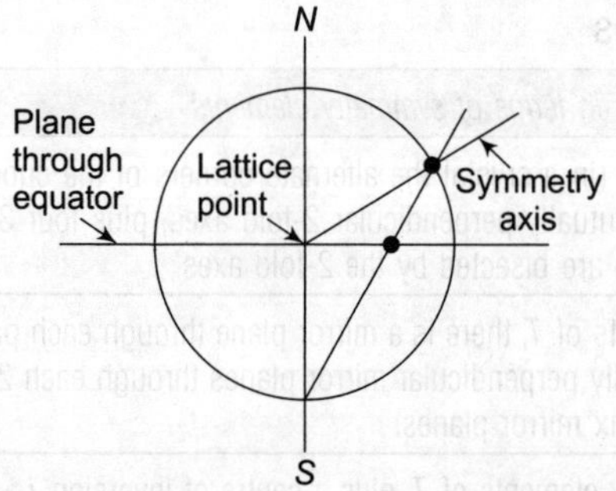


**Figure 2.89** Illustration of the way a stereogram is constructed

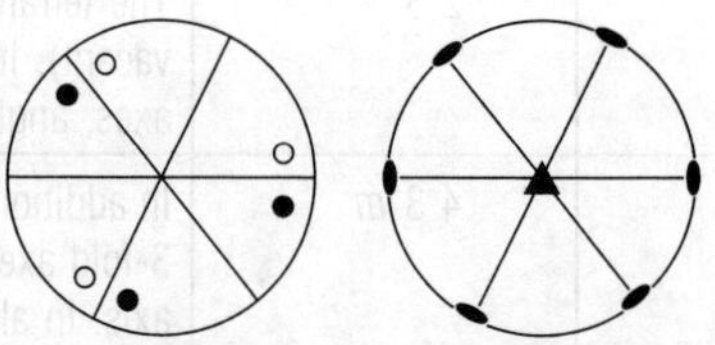


**Figure 2.90** Stereogram for $D_3$

planes. Reflection planes are represented by dark lines. If there had been a reflection plane in the plane of the paper, then the outer boundary of the circle in Fig. 2.90 (b) would have been dark.

The 32 point groups are shown in Table 2.20 crystal classes.

The stereographic projection of point group $D_{3d} \equiv \bar{3}\,\frac{2}{m}$ is shown in Fig. 2.91.

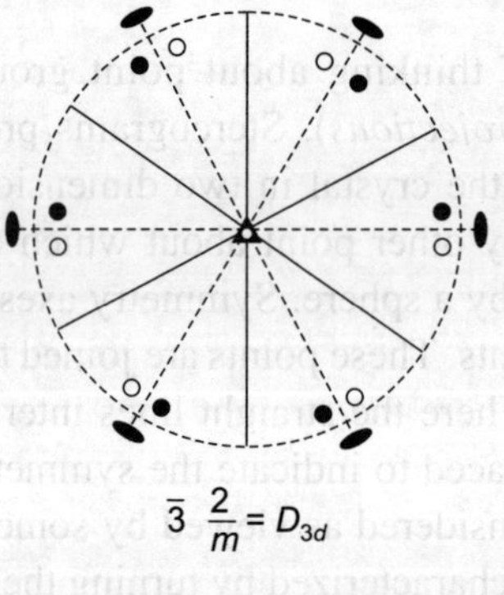


**Figure 2.91** Stereographic projection of the symmetry elements of group $D_{3d} \equiv \bar{3}\frac{2}{m}$. The symbol ⬮ and ▲ denote 2- and 3-fold axes of rotation, respectively. The full lines represent mirror planes

In $D_{3d}$, in addition to all the elements of $D_3$ there are mirror planes containing the 3-fold rotation axis such that the planes bisect the angles between the 2-fold axes are rotation.

For $D_3$, there is 3-fold axis of rotation and as many (i.e., 3) additional 2-fold axes as the presence of the 3-fold axis demands.

**Table 2.20** The Seven Crystal Systems and the 32 Crystal Classes

| *Crystal system* | *Schöenflies symbol* | *International symbol* | *Crystal system* | *Schöenflies symbol* | *International symbol* |
|---|---|---|---|---|---|
| Triclinic | $C_1$ | 1 | Rhombohedral | $C_3$ | 3 |
| | $C_i$ | $\bar{1}$ | | $S_6$ | $\bar{3}$ |
| | | | | $C_{3v}$ | $3m$ |
| Monoclinic | $C_s$ | $m$ | | $D_3$ | 32 |
| | $C_2$ | 2 | | $D_{3d}$ | $\bar{3}m$ |
| | $C_{2h}$ | $2/m$ | Hexagonal | $C_{3h}$ | $\bar{6}$ |
| | | | | $C_6$ | 6 |
| Orthorhombic | $C_{2v}$ | $2mm$ | | $C_{6h}$ | $6/m$ |
| | $D_2$ | 222 | | $D_{3h}$ | $\bar{6}m2$ |
| | $D_{2h}$ | $mmm$ | | $C_{6v}$ | $6mm$ |
| | | | | $D_6$ | 622 |
| Tetragonal | $C_4$ | 4 | | $D_{6h}$ | $6/m\ mm$ |
| | $S_4$ | $\bar{4}$ | | | |
| | $C_{4h}$ | $4/m$ | Cubic | $T$ | 23 |
| | $C_{4v}$ | $4mm$ | | $T_h$ | $m\bar{3}$ |
| | $D_{2d}$ | $\bar{4}2m$ | | $T_d$ | $\bar{4}3m$ |
| | $D_4$ | 422 | | $O$ | 432 |
| | $D_{4h}$ | $4/m\ mm$ | | $O_h$ | $m\bar{3}m$ |

Figure 2.92 shows 32 point groups.

## 2.44 ABOUT FIVEFOLD SYMMETRY IN A CRYSTAL

Molecules in a solid can have symmetry axes of arbitrary order. The requirement is translational symmetry in a direct lattice imposes the condition that there is the vector shorter than some minimum length vector and also places restrictions on possible rotational symmetries, as, subsequently we will see that it is impossible for a crystal to have fivefold symmetry, i.e., it would not be invariant with respect to rotation through $2\pi/5$. This is as follows:

Suppose, a crystal does have fivefold symmetry and let $\vec{a}_o = (1, 0)$ be the shortest vector in the lattice. Since the crystal is assumed to have fivefold symmetry, the vectors

$$\vec{a}_n = \left[\cos\left(\frac{2\pi n}{5}\right), \sin\left(\frac{2\pi n}{5}\right)\right]$$

with $n$ an integer, must also be in the direct lattice. But by the closure property of any lattice.

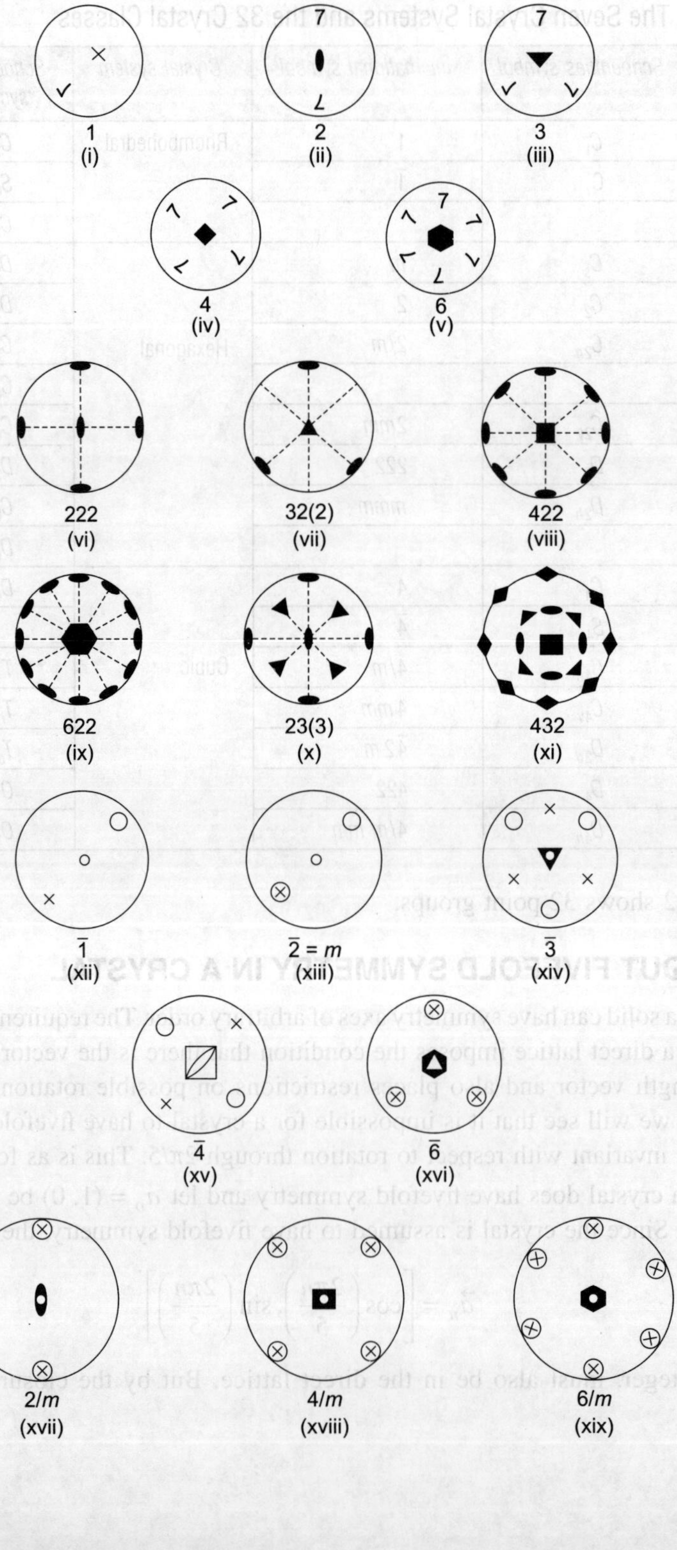

1
(i)
2
(ii)
3
(iii)
4
(iv)
6
(v)
222
(vi)
32(2)
(vii)
422
(viii)
622
(ix)
23(3)
(x)
432
(xi)
$\bar{1}$
(xii)
$\bar{2} = m$
(xiii)
$\bar{3}$
(xiv)
$\bar{4}$
(xv)
$\bar{6}$
(xvi)
2/m
(xvii)
4/m
(xviii)
6/m
(xix)

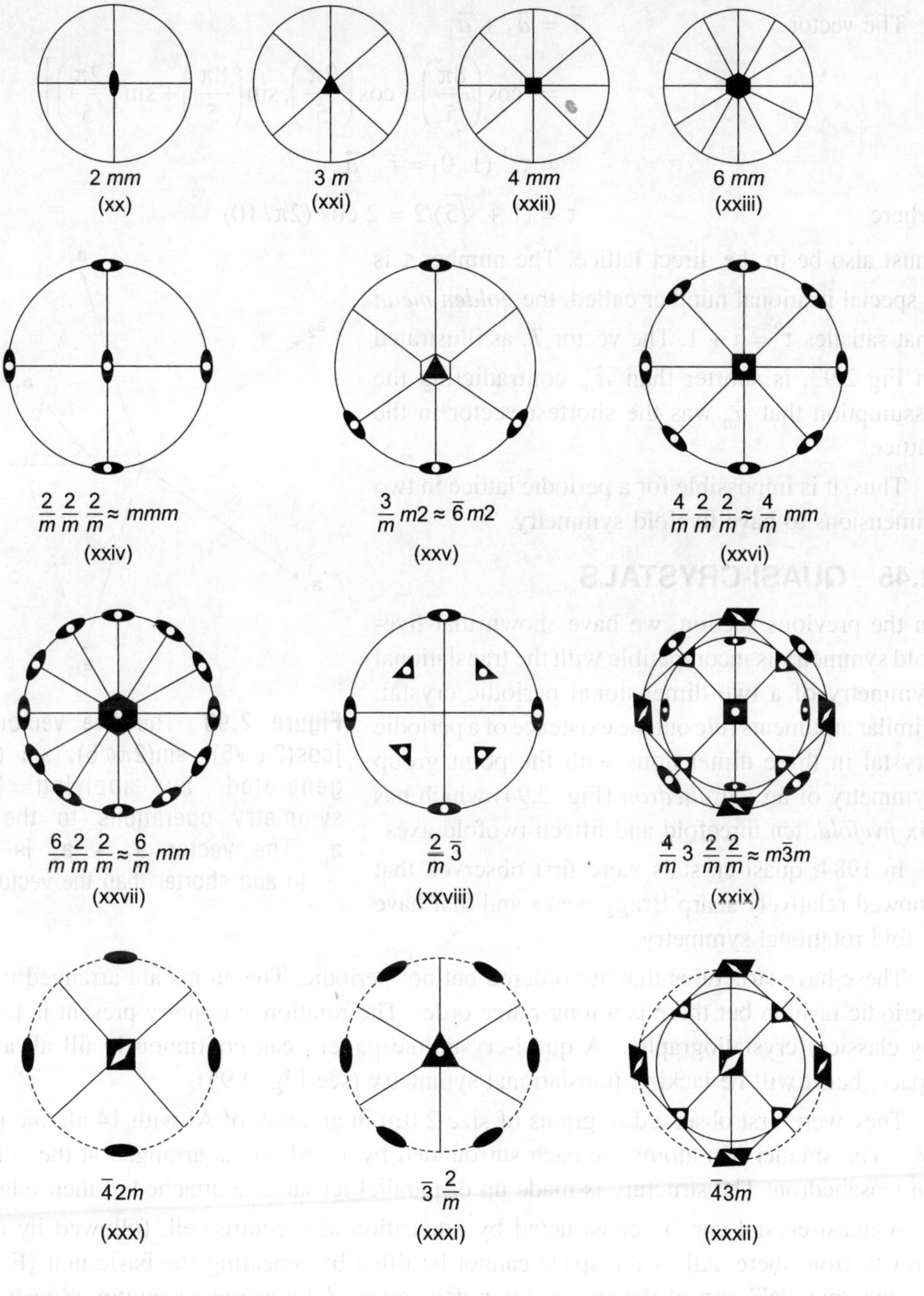


**Figure 2.92** The 32 point group [(i),(xii): Triclinic; (ii), (xiii), (xvii): monoclinic; (vi), (xx), (xxiv): orthorhombic; (iv), (xv), (xviii), (viii), (xxii), (xxx), (xxvi): tetragonal; (iii), (xiv), (vii), (xxi), (xxxi): trigonal; (vi), (xvi), (xix), (ix), (xxiii), (xxv), (xxvii): hexagonal; (x), (xxviii), (xi), (xxxii), (xxix): cubic]

The vector $$\vec{T} = \vec{a}_4 + \vec{a}_1$$

$$= \left[\cos\left(\frac{8\pi}{5}\right) + \cos\left(\frac{2\pi}{5}\right), \sin\left(\frac{8\pi}{5}\right) + \sin\left(\frac{2\pi}{5}\right)\right]$$

$$= \tau^{-1}\,(1, 0) = i^{-1}\,\vec{a}_o$$

where $$\tau = (1 + \sqrt{5})/2 = 2\cos(2\pi/10)$$

must also be in the direct lattice. The number $\tau$ is a special irrational number called, the *golden mean* that satisfies $\tau^2 = \tau + 1$. The vector $\vec{T}$, as illustrated in Fig 2.93, is shorter than $\vec{a}_o$, contradicting the assumption that $\vec{a}_o$ was the shortest vector in the lattice.

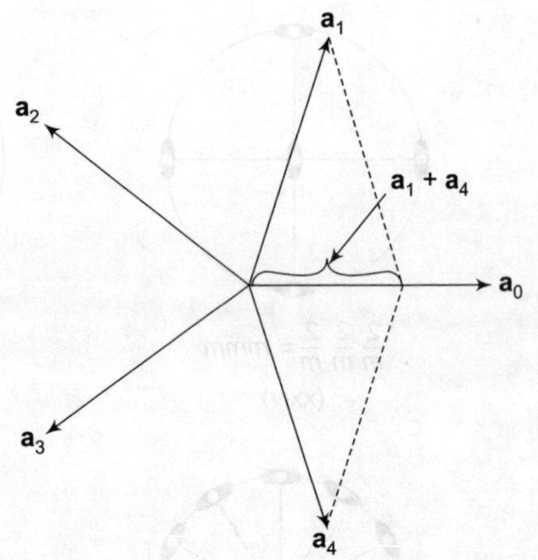


**Figure 2.93** The five vectors $\mathbf{a}_n$ = [cos(2π *n*/5), sin(2π*n*/5), sin (2π*n*/5)] generated by applying fivefold symmetry operations to the vector $\mathbf{a}_0$. The vector $\mathbf{a}_1 + \mathbf{a}_4$ is parallel to and shorter than the vector $\mathbf{a}_0$.

Thus, it is impossible for a periodic lattice in two dimensions to have fivefold symmetry.

## 2.45 QUASI-CRYSTALS

In the previous section, we have shown that five-fold symmetry is incompatible with the translational symmetry of a two-dimensional periodic crystal. Similar arguments rule out the existence of a periodic crystal in three dimensions with the point group symmetry of an *icosahedron* (Fig. 2.94) which has six *fivefold*, ten threefold and fifteen twofold axes.

In 1984, quasi-crystals were first observed that showed relatively sharp Bragg peaks and that have 5-fold rotational symmetry.

These have structures that are ordered but not periodic. The atoms are arranged in a non-periodic fashion but there is a long-range order. The rotation symmetry present is forbidden by classical crystallography.* A quasi-crystalline pattern can continuously fill all available space, but it will be lacking translational symmetry (see Fig. 2.95).

They were first observed in grains of size 2 μm in an alloy of Al with 14 atomic per cent Mn. The smaller Mn atoms are each surrounded by 12 Al atoms arranged at the corners of an icosahedron. The structure is made up of parallel icosahedra attached at their edges.

A quasi-crystal can be constructed by nucleation at a centre cell, followed by outward growth from there. All of the space cannot be filled by repeating the basic unit (Fig. 2.94) yet the "parallel" part of the specification does give a *long range orientational order*.

---

* (i) D. Levine and P.J. Steinhardt, Physical Review Letters **53**, 2477 (1984), Physical Review **34**, 596 (1986).
(ii) D.S. Schechtman and Others, Physical Review Letters **53**, 1951 (1984).

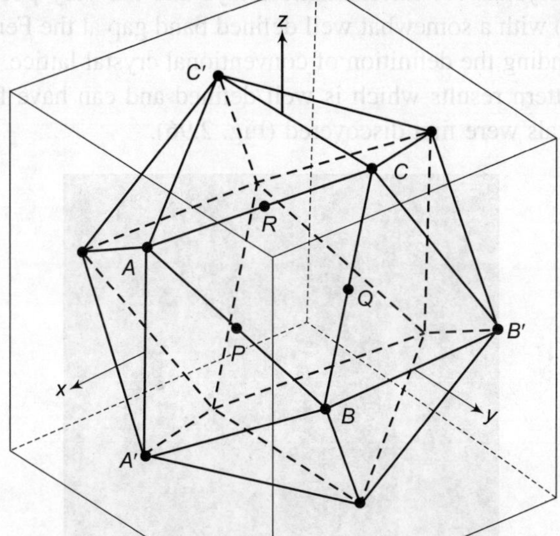


**Figure 2.94** An icosahedron showing its 12 vertices, 20 triangular faces, and 30 edges. The fivefold axes pass through the vertices, the threefold axes through the centres of the faces, and the twofold axes through the centres of the edges, The icosahedron with edges of unit length is inscribed in a cube. Its edges $AA'$, $BB'$, and $CC'$ are, respectively, parallel to the $z$-, $x$-, and, $y$-axes of the cube. The coordinates of $A$, $A'$, $B$, $B'$, $C$, and $C'$ are, respectively $(\tau, 0, \pm 1)$, $(\pm 1, \tau, 0)$, and $(0, \pm 1, \tau)$ where $\tau = (1 + \sqrt{5})/2$ is the golden mean. The coordinates of $P$, $Q$, and $R$ are, respectively, $(1 + \tau, \tau, 1)/2$, $(1, 1 + \tau, \tau)/2$, and $(\tau, 1, 1 + \tau)/2$.

[Adapted from P.M. Chaikin & T.C. Lubensky, Principles of Condensed Matter Physics, Cambridge University Press (1999)].

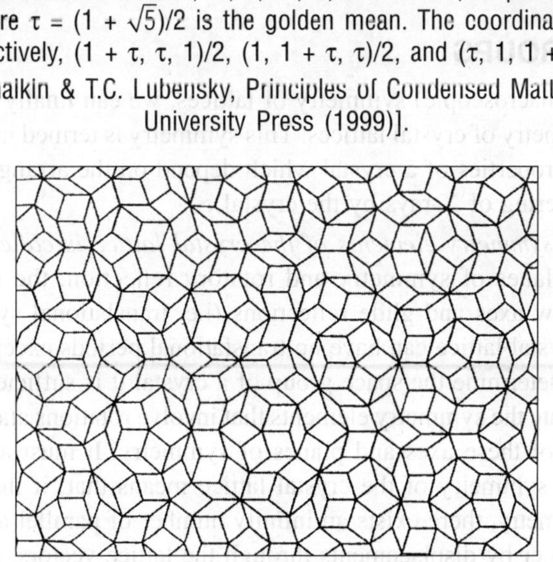

**Figure 2.95** A quasi-crystal tiling in two dimensions, after the work of Penrose. The long-range orientational order and the long-range nonperiodic order are shown

The known quasi-crystals are intermetallic alloys and are very poor electric conductors (are nearly insulators) with a somewhat well defined band gap at the Fermi-level. They are of great interest in expanding the definition of conventional crystal lattice. A distinctly different crystal diffraction pattern results which is well defined and can have fivefold symmetry by which the quasi-crystals were first discovered (Fig. 2.96).

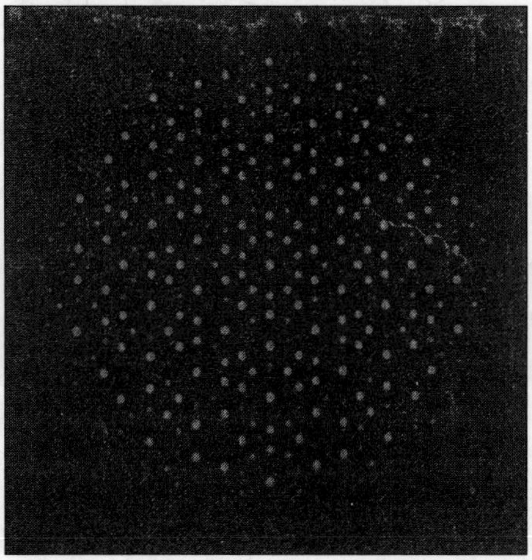

**Figure 2.96** Photograph of the calculated Fourier transform (diffraction pattern) of an icosahedral quasicrystal along one of the fivefold axes, illustrating the fivefold symmetry. The transform is calculated from a theoretical computer-generated model, by Micheal Jacob. [Adapted from C. Kittel, Introduction to Solid State Physics, John Wiley & Sons (1995)]

## 2.46 SPACE GROUPS

Having studied the (macroscopic) symmetry of lattices, we can finally go on to consider the complete actual symmetry of crystal lattices. This symmetry is termed microscopic symmetry. It determines those properties of a crystal which depend on the arrangement of atoms in its lattice (e.g., the scattering of X-rays by the crystal).

*The set of actual symmetry elements of the crystal lattice is called its space group.* In addition to simple planes of symmetry and rotatory reflection, the space group includes consideration of screw axes and glide reflections (i.e. translational symmetry). Now, since the definition of a crystal lattice can have no translational periods except those of its Bravais lattice, therefore, to determine the space group of a crystal it is sufficient to find the Bravais lattice and to enumerate the symmetry elements that involve rotations and reflections including the relative position of these axes and planes of symmetry. It must also be borne in mind that the translational symmetry of the crystal lattice means that, if the lattice possesses an axis or plane of symmetry, there exists an infinity number of parallel axes or planes that are carried into one another by displacements through the lattice vectors. Finally, in addition to these axes (or planes) of symmetry separated by lattice vectors, the simultaneous presence of translational symmetry and the axes (or planes) of symmetry results in the existence of other axes (or planes), which cannot be made to coincide with the former by a translation through any lattice vector (e.g., in a glide plane the reflection in any plane followed by translation through a distance ($d$) in a direction perpendicular to the plane is equivalent to reflection in a plane parallel to the first and at a distance $(1/2)d$ from it).

The possible space groups can be divided among the crystal classes, each space group being assigned to the class where the set of axes and planes of symmetry is the same as in the space group when no distinction is made between simple and screw axes and between simple and glide planes. Altogether there are 230 different space groups and they were first found by E.S. Fedorov in 1895. Their distribution among classes is shown in Table 2.21.

**Table 2.21** Space Groups in Different Crystal Classes (Total 230)

| *Class* | *Number of space groups* | *Class* | *Number of space groups* |
|---|---|---|---|
| $C_1$ | 1 | $S_6$ | 2 |
| $C_1$ | 1 | $C_{3v}$ | 6 |
| $C_5$ | 4 | $D_3$ | 7 |
| $C_2$ | 3 | $D_{3d}$ | 6 |
| $C_{2h}$ | 6 | $D_{3h}$ | 1 |
| $C_{2v}$ | 22 | $C_6$ | 6 |
| $D_2$ | 9 | $C_{6h}$ | 2 |
| $D_{2h}$ | 28 | $D_{3h}$ | 4 |
| $S_4$ | 2 | $C_{6v}$ | 4 |
| $C_4$ | 6 | $D_6$ | 6 |
| $C_{4h}$ | 6 | $D_{6h}$ | 4 |
| $D_{2d}$ | 12 | $T$ | 5 |
| $C_{uv}$ | 12 | $T_h$ | 7 |
| $D_4$ | 10 | $T_d$ | 6 |
| $D_{4h}$ | 20 | $O$ | 8 |
| $C_3$ | 4 | $O_h$ | 10 |

The symmetry elements of all these space groups may be found in manuals of crystallography (e.g., *International Table for X-ray Crystallography*, Vol. 1, Kynoch Press, Birmingham, 1952).

Table 2.22 shows the crystal systems, point groups, and the corresponding space groups. The numbers for space groups are those as derived and numbered by Schöenflies. 32 of the space groups are the same as the point groups. In the crystals belonging to the 32 point groups, each unit is reproduced without alteration. In the other space groups, each unit is reproduced without alteration. In the other space groups, each unit repeats with translation and rotation or reflection.

**Table 2.22** Space Groups, Crystal Systems, Points Groups and the Hermann-Mauguin Symbols

| *Space group number* | *Crystal system* | *Point group (International symbol)* | *Hermann-Manguin symbol* |
|---|---|---|---|
| 1 | Triclinic | 1 | *P*1 |
| 2 | | $\bar{1}$ | $P\bar{1}$ |
| 3-5 | Monoclinic | 2 (No. 3) | *P*2 |
| 6-9 | | m (No. 6) | *Pm* |
| 10-15 | | 2/*m* (No. 10) | *P*2/*m* |
| 16-24 | Orthorhombic | 222 (No. 16) | *P*22 |
| 25-46 | | *mm*2 | *Pmm*2 |
| 47-74 | | *mmm* (No. 74) | *Imma* |
| 75-80 | Tetragonal | 4 (No. 75) | *P*4 |
| 81-82 | | $\bar{4}$ | $P\bar{4}$ |
| 83-88 | | 4/*m* | *P*4/*m* |
| 89-98 | | 422 | *P*422 |
| 99-110 | | 4 *mm* | *P*4*mm* |
| 111-122 | | $\bar{4}$ 2*m* | $P\bar{4}2m$ |
| 123-142 | | 4/*m mm* (No. 123) | *P*4/*m mm* |
| 143-146 | Trigonal | 3 (No. 143) | *P*3 |
| 147-148 | | $\bar{3}$ | $P\bar{3}$ |
| 149-155 | | 312 | *P*312 |
| 156-161 | | 3 *m* | *P*3*m*1 |
| 162-167 | | $\bar{3}m$ (No. 167) | $R\bar{3}c$ |
| 168-173 | Hexagonal | 6 (No. 168) | *P*6 |
| 174 | | $\bar{6}$ | *P*6 |
| 175-176 | | 6/*m* | *P*6/*m* |
| 177-182 | | 622 | *P*622 |
| 183-186 | | 6mm | *P*6*mm* |
| 187-190 | | $\bar{6}m2$ | *P*6*m*2 |
| 191-194 | | 6/*m mm* (No. 194) | $P6_3/m\ mc$ |
| 195-199 | Cubic | 23 (No. 195) | *P*23 |
| 200-206 | | $m\bar{3}$ (No. 200) | $Pm\bar{3}$ |
| 207-214 | | 432 (No. 207) | *P*432 |
| 215-220 | | $\bar{4}3m$ (No. 215) | $P\bar{4}3m$ |
| 221-230 | | $m\bar{3}m$ (No. 221) | $Pm\bar{3}m$ |

## EXERCISE

2.1 Define the following terms in relation to crystal structure:

(i) Crystalline, polycrystalline and amorphous states of solids.

(ii) Lattice, basis and crystal structure.

(iii) Unit cell, primitive unit cell and Wigner–Seitz cell.

2.2 Define the terms primitive and conventional unit cells and lattice with a basis. What do you do you understand by a Bravais lattice?

2.3 Describe the possible Bravais lattices in two dimensions.

2.4 Describe the 7 crystal systems and conventional unit cells of the 14 Bravais space lattices.

2.5 Explain the structures of body-centred cubic (bcc) and face-centred cubic (fcc) structures and show that packing fractions are $(\sqrt{3}/8)\pi$ and $(\sqrt{2}/6)\,\pi$ respectively.

2.6 Show that an ideal hexagonal close packed (hcp) structure has a c/a ratio of $\sqrt{8/3} = 1.633$ and the density of atomic packing is equal to that of the face-centred cubic structure.

2.7 Explain the following:

(i) Wurtzite structure

(ii) Si/diamond structure

2.8 Explain the hexagonal close packed structure.

2.9 Explain voids in close packing and tetrahedral and octahedral voids. Give significance of the voids.

2.10 Describe caesium chloride (CsCl), zinc blende and ideal perovskite ($ABO_3$) structure.

2.11 Explain spinal structure.

2.12 (a) What do you understand by Miller indices?

Explain indices of crystal lattice site and indices of direction.

(b) Write all directions from the family <110> and <111>.

2.13 Show that interplanar spacing for a set of (h k l) planes is

$$d = \left(\frac{h^2}{a^2} + \frac{k^2}{b^2} + \frac{l^2}{c^2}\right)^{-1/2}$$

and hence deduce the spacing for (a) tetragonal system (b) cubic system.

2.14 If $|\vec{a}| = 4.50$ Å, $|\vec{b}| = 3.50$ Å, $|\vec{c}| = 2.50$ Å, calculate the lattice spacing for (110) planes in an orthorhombic lattice. [*Ans.* **0.36 Å**]

2.15 Determine the Miller indices of a plane that makes an intercept of 3 Å, 4 Å and 5 Å on the coordinate axes of an orthorhombic crystal with $a:b:c = 1:2:5$. [*Ans.* **(2 3 6)**]

2.16 NaCl crystallizes as a cubic crystal structure, Taking the molecular weight = 58.46 and the density at room temperature as 2.167 g/cc, calculate the lattice constant, i.e., the distance between adjacent atoms (Given 1amu = $1.66 \times 10^{-24}$ g). [*Ans.* **2.82 Å**]

2.17 Explain fivefold symmetry. What are quasi-crystals?

CHAPTER

# 3

# X-ray Diffraction Reciprocal Lattice, Electron Diffraction and Microscopy

## 3.1 COMPARATIVE ELECTRON, NEUTRON AND X-RAY DIFFRACTIONS: (PRELIMINARY IDEAS)

### 3.1.1 Diffraction of X-rays

X-rays are electromagnetic waves, (like ordinary light) which have wavelengths ~1 Å. Consequently, the ordinary (i.e., optical) diffraction gratings do not produce observable diffraction effects with X-rays. However, since the atoms in a crystal form a regular three-dimensional array, they can serve purpose of a grating (for X-rays), but with a difference that the crystal acts as a *space-grating* rather than a plane-grating as the diffraction centres are not a plane.

### 3.1.2 Diffraction of Other Kinds of Waves

Other types of waves, for example, de Broglie waves associated with electrons and with neutrons, can also interact with the atoms or ions in a solid. Just as with X-rays, if the wave vector (associated with the de Broglie wavelength (λ)) is more than half the basic reciprocal lattice dimension (i.e., $\sim 10^{10}$ per metre), then it is possible to observe diffraction. However, the practical details will be different for electrons and for neutrons. Now, de Broglie wavelength $\lambda = \frac{h}{p}$, where $p$ is the momentum of the particle.

and the wave-vector $\vec{k} = \frac{\vec{p}}{\hbar}$ (where $\hbar = h/2\pi$).

In terms of non-relativistic kinetic energy $E$,

$$k = \frac{\sqrt{2m\,E}}{\hbar} \quad \text{or} \quad E = \frac{\hbar^2 k^2}{2m}.$$

If $k \simeq 10^{10}$/m, then $E$ will be about $5 \times 10^{-19}$ J (i.e., 3.8 eV for electrons) and $E \sim 3.3 \times 10^{-22}$ J (or $2 \times 10^{-3}$ eV for neutrons).

Unfortunately, higher energies have to be used (greater than 3.8 eV) if we wish to observe *electron-diffraction* because the interaction of electrons with the atoms in a crystal is so strong that electrons with an energy of a few electron-volts would be completely absorbed by a specimen. If lower electron-energies are used (i.e., small penetration[1]) or if the specimen is very thin, then in either case, the crystal will act as a two-dimensional grating. Bragg's formula is then not applicable, since a diffracted beam will be obtained for any angle of incidence (as in an ordinary diffracted grating).

*Neutron Diffraction.* In this case, the situation is quite different. Their interaction with most atoms is very weak and may penetrate materials quite readily. Although the diffracted beams are not very intense, they (unlike electrons) escape easily from the sample and then can be detected.

If we equate the minimum energy for diffraction ($\sim 3.3 \times 10^{-22}$ J) to the thermal energy $\left(\frac{3}{2}\,kT\right)$, then this corresponds to the mean thermal energy at a temperature of about 25 K. Neutrons in thermal equilibrium with a system at room temperature (say 300 K), will have a wave-vector about three times the reciprocal lattice basic vector (since their energy varies as $k^2$) and so they will give several orders of diffraction which will be sufficient for structural analysis.

The only source of thermal neutrons are the slow neutrons in the moderator of a nuclear reactor. They are allowed to escape from the moderator through a hole in the wall of reactor and they are then collimated and are passed through a velocity-selector, so that they are mono-energetic, before being used as a beam in neutron diffraction investigations.

### The Choice of a Diffraction – Technique for Structure Analysis

The general form of the diffraction pattern will be the same whether X-rays, or electrons, or neutrons are used. In each case, the wavelength must lie in a region that allows Bragg-reflections to be produced, It is this condition that determines the respective energy-ranger of the beams, These are:

(i) 10 eV – 1 keV (electrons)

(ii) 10 MeV – 1 eV (neutrons/light atoms)

(iii) 1 keV – 100 keV (photons/X-rays)

At accelerating voltages of 50–100 kV (High Energy Electrons Diffraction: HEED), the equivalent wavelength is 0.05 – 0.04 Å, which is about $\frac{1}{30}$th to $\frac{1}{15}$th (for Cu and Mo respectively) of the range of X-ray wavelengths used. Insertion of this value into the Bragg equation ($2\,d \sin\theta = n\lambda$) shows diffraction angles $2\theta$ to be less than 1°, However, the diffracted beams are passed through an *electron-lens* which spreads them so that the patterns recorded

[1] In this case, the diffracted beam is reflected from the surface. This is a very useful technique for surface-structure studies.

on film has dimensions similar to those of X-ray patterns (electrons in this energy range penetrate to a depth about $\frac{1}{10}$th that of X-rays). When used in a reflection-geometry, electrons can thus be used to investigate thin surface layers that would be largely transparent to X-rays. Not only are materials much more opaque to electrons, but the depth of material in the path of the beam required to produce diffraction is very much less.

In the *transmission mode*, electron-diffraction can thus be used for films much too thin and for particles or crystallites much too small for X-rays work. This is the principal application of HEED. It is not preferable for specimens suitable for X-ray diffraction, because samples cannot be manipulated as well, the precision of *d*-spacings is less, and the interpretation of intensities is much more difficult.

At substantially lower voltages (Low Energy Electron Diffraction: LEED), the penetration depth is so reduced that essentially, only the top layer of atoms is involved ; the study of this layer is called two-dimensional or *surface-crystallography*. LEED is used to study absorbed monolayers on surfaces and the structure of surfaces themselves. The latter can be different from the bulk-structure because the outer layer of atoms is subject to less symmetrical force than are atoms in the interior.

*Thermal-neutrons* have wavelengths comparable to those of X-rays, used in diffraction and also yield diffraction patterns with crystalline materials. Neutrons are very penetrating and can be used only with large specimens. Beams of adequate intensity are available only from nuclear reactors. Despite these drawbacks, neutron diffraction is very important in crystallography because of two major differences from X-rays.

Neutrons are scattered by atomic nuclei rather than by electron shells. As a consequence, neutron scattering factors do not vary with atomic number in the same way that X-ray scattering factors do, and they do not vary as much. Therefore, the ratios of light atom to heavy atom scattering factors are generally more favourable, which make it possible to determine hydrogen positions in hydrides and in organic compounds and oxygen positions in the oxides of heavy metals with much better precision than with X-rays.

Because of the ready availability of X-rays sources, the majority of structure investigations, to date, have been carried out with X-ray beams. However, there are a number of questions that cannot be answered with X-ray studies. The atomic structure factors are proportional to the nuclear charge Z. Thus, the scattering intensities vary as $Z^2$. Therefore, when hydrogen (for example) occurs in combination with heavy elements, it is very difficult to detect it with X-rays, then one can use neutron beams. (The interaction cross sections for neutrons lie within a single order of magnitude for all elements for X-rays). On the other hand, the neutron cross sections for elements with adjacent atomic numbers which are difficult to distinguish with XRD, are quite different, Hence, it is possible to distinguish Fe, Co and Ni, e.g., in neutron scattering.

Some atoms have magnetic moments either from unpaired electrons or from orbital motions. X-rays do not respond to magnetic moments but the neutron, which has a magnetic moment of its own, will interact with them. Neutron diffraction with polarized neutrons, can thus be used to investigate the magnetic structure of materials. Symmetry elements beyond those observable by X-rays, come into play, and the number of magnetic space groups greatly exceeds that of the 230 conventional space groups. A difficulty with neutrons is, however, the requirement of nuclear reactors, and since, the cross sections are small, therefore, the detection of the beams is less straightforward. Thus, the necessary experimental outlay is far greater than for X-rays or electron scattering.

### 3.1.3 Practical Considerations

In general XRD is the cheapest and the most convenient method, therefore most widely used:

(a) X-rays can be detected photographically or with a counter and the latter enables the data to be recorded automatically in digital form for computer input.

(b) X-rays are not absorbed very much by air and so the specimen need not be in an evacuated chamber. However, X-rays do have disadvantage that they do not interact very strongly with lighter elements.

(i) Electrons are scattered strongly in air, so electron-diffraction experiments must be carried out in a high vacuum, since, nowadays, this is a standard routine, it is still an extra complication and expense.

(ii) Electrons being charged particles, interact strongly with all atoms, and this entails an electron beam of very high energy (50 keV to 1 MeV), as well as specimen which are tiresomely thin (100–1000 nm). The charge on the electron is however, by no means a nuisance (because, unlike X-rays and neutrons) electrons can be focussed into narrow beams by electrostatic or magnetic lenses. This enables the diffracted beams to be used to form a direct magnified image of the structure as in the electron microscope.

*Neutron diffraction* is a very special tool. A high flux nuclear reactor is necessary as a suitable intense source of thermal neutrons, and the number of such reactors available in the entire world is extremely limited. This implies that the investigations which involve neutron diffraction need to be carried out very carefully and selectively. Among the most straightforward are those in which light atoms positions are required. A second group of investigations rely on the fact that although uncharged, the neutron has an intrinsic magnetic moment and this therefore means that it will interact strongly with atoms or ions in a crystal which also have magnetic moments. Any special regularities in the alignment of the magnetic ions in a crystal can give rise to special 'magnetic' diffraction lines. For this reason, neutron diffraction has been a very important tool in the investigation of the magnetic ordering that occurs in some materials, particularly those that become antiferromagnetic.

## 3.2 CLASSICAL SCATTERING OF X-RAYS (BY ELECTRONS OF AN ELEMENT)

The phenomenon X-ray scattering has played an important part in the theory of the atom.

The first theory proposed by J.J. Thomson, considered a plane polarized electromagnetic wave moving along the z-axis with its electric vector in x-direction (Fig. 3.1).

If this wave interacts with a single free electron, the electron performs simple harmonic motion and in so doing reradiates. The acceleration of the electron is given by $a_x = -eE/m$, where $E$ is the electric intensity of the incident radiation. According to classical electromagnetic theory, an accelerated charge radiates electromagnetic wave with the electric vector at point $P$ of amplitude,

$$E_\phi = \frac{ea \sin \phi}{4\pi\varepsilon_0 c^2 r} \qquad ...(1)$$

where $r$ is the distance from the accelerated charge to $P$ and $\phi$ is the angle between **r** and **a**. When $a = -eE/m$ is substituted, we have the magnitude of $E_\phi$

$$E_\phi = \frac{e^2 E \sin \phi}{4\pi\varepsilon_0 mc^2 r} \qquad ...(2)$$

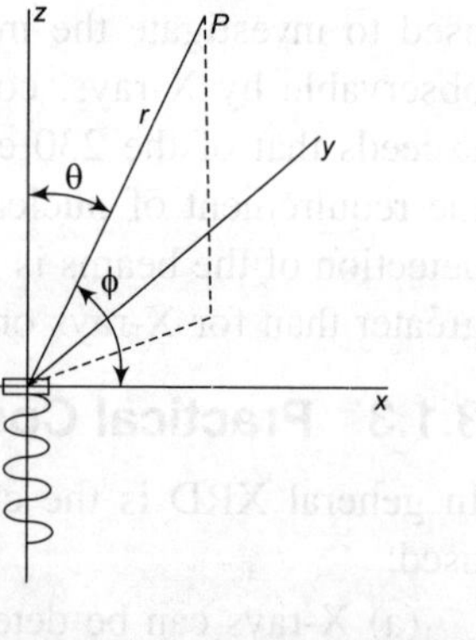


**Figure 3.1** An electron at the origin, set into harmonic oscillation by the electromagnetic wave from below, sectters a fraction of the incident intensity at angle θ

Since intensity is proportional to the square of the electric intensity, the ratio of the scattered intensity at $P$ to the incident intensity is

$$\frac{I_\phi}{I} = \frac{E_\phi^2}{E^2} = \frac{e^4 \sin^2 \phi}{16\pi^2 \varepsilon_0{}^2 m^2 c^4 r^2} \qquad ...(3)$$

If unpolarized radiation is incident on an electron, it is convenient to let **r** and the $z$-axis fix the $xz$ plane. The incident electric vector **E** can then be resolved into $x$ and $y$ components, which are equal on the average. For the $x$ components $\sin \phi$ in Eq. (3) is equal to $\cos \theta$, while for the $y$ components, $\sin \phi = 1$. The total intensity of the scattered wave at $P$ is given by

$$I_s = \frac{e^4}{16\pi^2 \varepsilon_0{}^2 m^2 c^4 r^2}\left[\left(\frac{I}{2} \times 1\right) + \left(\frac{I}{2} \times \cos^2 \theta\right)\right]$$

$$= \frac{Ie^4}{32\pi^2 \varepsilon_0{}^2 m^2 c^4 r^2}(1 + \cos^2 \theta) \qquad ...(4)$$

where $\theta$ is the angle between $r$ and the $z$-axis. Maxima in scattered intensity occur in the forward and backward direction; a minimum exists at $\theta = 90°$. By integrating Eq. (4) over a sphere of radius $r$ surrounding the electron, the scattered power $P_s$ is found to be

$$P_s = \int_0^\pi I_s \, 2\pi r^2 \sin \theta \, d\theta = \frac{8\pi}{3}\left(\frac{e^2}{4\pi\varepsilon_0 mc^2}\right)^2 I \qquad ...(5)$$

where $e^2/4\pi\varepsilon_0 mc^2$ is the classical radius of the electron.

The ratio of the power scattered to the primary intensity is called *scattering cross section* (or *scattering coefficient*) of the free electron, designated by $\sigma_e$.

$$\sigma_e = \frac{8\pi}{3}\left(\frac{e^2}{4\pi\varepsilon_0 mc^2}\right)^2 \qquad ...(6)$$

The scattering coefficient has the dimensions of area; of the radiation incident of a unit area, the electron scatters the amount which would fall on the area $\sigma_e$, which has the numerical value $0.666 \times 10^{-28}$ m$^2$, or 0.666 barn (1 barn = $10^{-24}$ cm$^2$ = $10^{-28}$ m$^2$). The Thomson (or classical) scattering coefficient is independent of wavelength. If each electron of an atom scatters independently, the atomic scattering cross section ($\sigma_a$) is the product of the atomic number of the atom and the electron cross section; $\sigma_a = Z\sigma_e$.

Thomson's theory predicts not only the intensity of the scattered beam but also its polarization. At $\theta = 0$ and $\theta = 180°$, the scattered beam from an unpolarized, but at $\theta = 90°$ the scattered beam is plane-polarized with electric vector in the y-direction.

When a beam of *x*-rays of initial intensity $I_0$ passes through a thin layer of thickness $\Delta x$ of any material, scattering by electrons in the layer changes the intensity by an amount

$$-\Delta I = \sigma I_0 \, \Delta x \qquad ...(7)$$

where $\sigma$ is called the *linear scattering coefficient* of the material. If each of the electrons of the material were to scatter independently of the others, the linear scattering coefficient would be

$$\sigma = n\sigma_e = \frac{8\pi}{3}\left(\frac{e^2}{4\pi\varepsilon_0 mc^2}\right)^2 n \qquad ...(8)$$

where *n* is the number of electrons per unit volume.

The ratio of $\sigma$ to the density $\rho$ of the scatterer is the *mass scattering coefficient* $\sigma_m$. For substances such as gases, the linear scattering coefficient depends, of course, on the density, but $\sigma_m = \sigma/\rho$ is constant. The first quantitative estimate of the number of electrons in an atom was made by Barkla and his coworkers in 1909 by applying Thomson's theory of data on the scattering of X-rays by carbon. They passed a beam of Mo *K*α X-rays through a graphite absorber and after correcting of photoelectric absorption found that the value of $\sigma_m$ is approximately 0.02 m$^2$/kg for carbon. If $\sigma = n\sigma_e$, then $\sigma_m = n\sigma_e/\rho$, and the number of electrons per kilogram is

$$\frac{n}{\rho} = \frac{\sigma_m}{\sigma_e} = \frac{0.02\ \text{m}^2/\text{kg}}{0.666\times10^{-28}\ \text{m}^2/\text{electron}} \qquad ...(9)$$

$$= 3 \times 10^{26}\ \text{electrons/kg}$$

The number of carbon atoms per kilogram is the ratio of Avogadro's number ($N_A$) to the atomic weight *A*, or

$$\frac{N_A}{A} = \frac{6.02 \times 10^{26}\ \text{atoms/kmol}}{12\ \text{kg/kmole}} \qquad ...(10)$$

$$= 3 \times 10^{25}\ \text{electrons/kg}$$

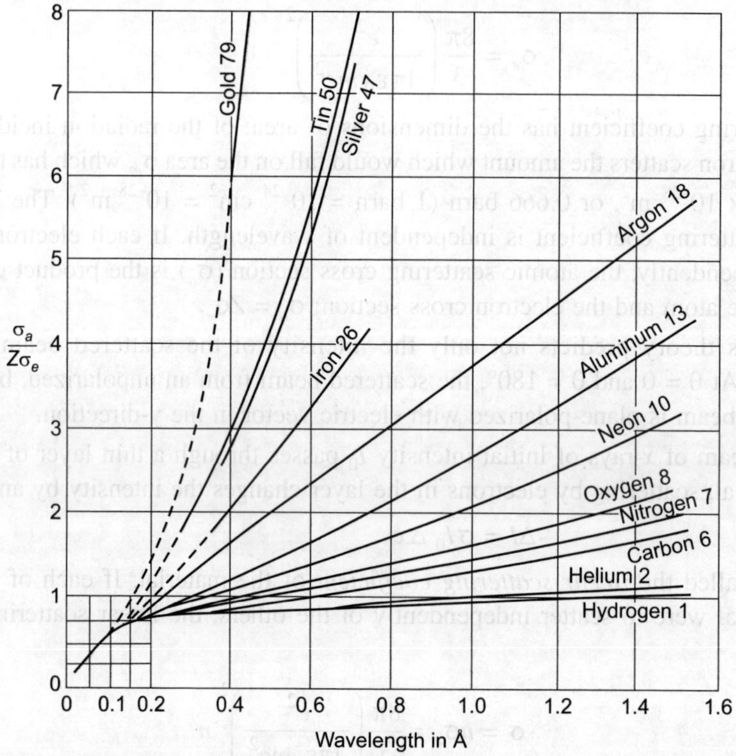


**Figure 3.2** Ratio of the scattering coefficient $\sigma_a$ of various atoms to $Z\sigma_e$, the value which would pertain if each electron scattered independently with cross section $\sigma_e$

$$\text{Number of electrons per atom} = \frac{\text{Number of electrons per kilogram}}{\text{Number of atoms per kilogram}} = \frac{30}{5} = 6.$$

Hence, the number of electrons per carbon atom must be 6. This determination was significant in the development of atomic theory, since it suggested the fact that the number of electrons per atom is less than the atomic weight and is perhaps about half of *A*.

Later measurements of scattering coefficients for $\lambda > 0.3$ Å showed that for heavier elements the scattering coefficients $\sigma$ exceeds $n\sigma_e$ and hence the atomic scattering cross section $\sigma_a$ is greater than $Z\sigma_e$ (Fig. 3.2). The excessive scattering is easily explained classically. When $\lambda$ becomes comparable with the distances between electrons in an atom, they no longer scatter independently; the waves scattered by neighbouring electrons are superposed more or less in the same phase, and so constructive interference occurs. As $\lambda$ increases from a very small value, at first the inner (*K*) electrons scatter coherently; for longer wavelengths more electrons scatter nearly in phase. When $\lambda$ is large compared to the atomic diameter, the relative phases of the waves from all *Z* electrons of an atom are small, and so the amplitude of the scattered wave from the atom is approximately proportional to *Z* and the scattered intensity to $Z^2$.

## 3.3 X-RAYS AND THEIR GENERATION

X-rays like a visible light are a form of electromagnetic radiation having shorter wavelengths, typically of the order of 1 Å, i.e., they represent a relatively high energy form of radiation. In the electromagnetic spectrum they lie between ultraviolet and γ-rays. X-rays are produced when a stream of high energy charged particles, e.g., a beam of electrons is allowed to strike a metal target. In the process, electrons lose certain amount of their energy which is converted into electromagnetic radiations, i.e., $\Delta E = h\nu$. Such processes give "white radiations" or continuous X-rays. Figure 3.3 shows continuous X-ray spectra corresponding to a tungsten target at several different accelerating potentials, which tails off gradually at higher wavelength side, but terminates abruptly at shorter wavelength limit $\lambda_{swl}$ (or $\lambda_{min}$). As the accelerating potential increases, both the maxima in the intensity distribution and the $\lambda_{min}$ shift towards the shorter wavelength side. At particular $V$, the $\lambda_{min}$ is fixed and are related to each other as

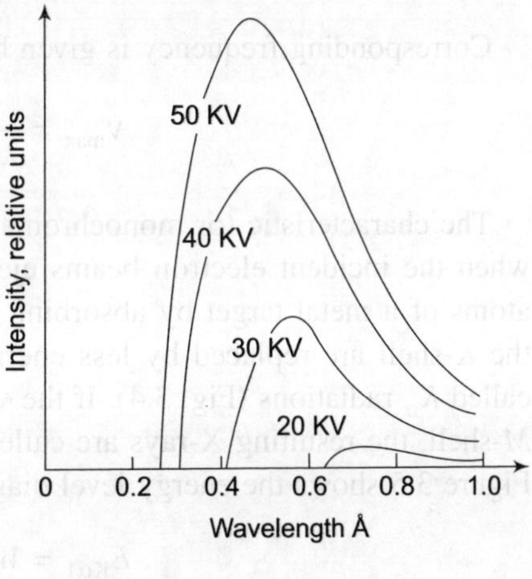


**Figure 3.3** X-rays spectra of tungsten at various accelerating potentials

$$\lambda_{min} = \frac{1.24 \times 10^{-6}}{V} \text{ V-m} \qquad ...(1)$$

$\lambda_{min}$ in Eq. 1 means a maximum frequency which consequently means a maximum photon energy, $h\nu_{max}$. The maximum energy that X-ray photon can have cannot exceed the maximum energy of the electron beam, which in turn, is determined by the voltage applied to the X-ray tube. Thus,

$$\Delta E_{max} = eV_{max} = h\nu_{max} = \frac{hc}{\lambda_{min}}$$

or

$$hc/\lambda_{min} = eV_{max} \quad \text{or} \quad \lambda_{min} = \frac{hc}{eV_{max}} \qquad ...(2)$$

Substituting the values of constants, we obtain

$$\lambda_{min} = \frac{6.63 \times 10^{-34} \text{ Js} \times 3 \times 10^{8} \text{ ms}^{-1}}{1.6 \times 10^{-10} \times C \times V}$$

$$= \frac{1.24 \times 10^{-6}}{V} \text{ V-m} \qquad ...(3)$$

which is just the same as the experimental relation given by Eq. 1.

**Example 3.1** Find the shortest wavelength that is present in the X-rays produced at an accelerating potential of 50 kV. Calculate the corresponding frequency also.

***Solution*** Given: Accelerating potential, $V = 50$ kV $= 5 \times 10^4$ V. Now using Eq. 3, we have

$$\lambda_{min} = \frac{1.24 \times 10^{-6}}{5 \times 10^4\ V} \text{ V-m} = 2.5 \times 10^{-11} \text{ m} = 0.25 \text{ Å}$$

Corresponding frequency is given by

$$\nu_{max} = \frac{c}{\lambda_{min}} = \frac{3 \times 10^8 \text{ ms}^{-1}}{2.5 \times 10^{-11} \text{ m}} = 1.2 \times 10^{10} \text{ Hz}$$

The characteristic (or monochromatic) X-rays result in the form of excess of energy $E_{K\alpha}$ when the incident electron beams ejects some of the most energetic $K$-electrons from the atoms of a metal target by absorbing an energy equal to $W_K$ and the resulting vacancies in the $K$-shell are replaced by less energetic electrons $W_L$ from the $L$-shell. Such X-rays are called $K_\alpha$ radiations (Fig. 3.4). If the vacancies in the $K$-shell are replaced by electrons from $M$-shell, the resulting X-rays are called $K_\beta$ radiations, however this less probable transition. Figure 3.5 shows the energy level diagram and possible transitions. They are:

$$E_{K\alpha 1} = W_K - W_{L111} \text{ due to } K \to L_{111} \text{ transition}$$

and

$$E_{K\alpha 2} = W_K - W_{L11} \text{ due to } K \to L_{11} \text{ transition}$$

where $K_{\alpha 1} = 1.54051$ Å and $K_{\alpha 2} = 1.54433$ Å respectively for copper target.

Complete X-ray spectrum ($I$ versus $\lambda$ curve) emitted the copper target is shown by solid line in Fig. 3.6. The $K$-radiations from elements of intermediate atomic weights have wavelengths comparable to interatomic dimensions and these are suitable for crystal diffraction experiments.

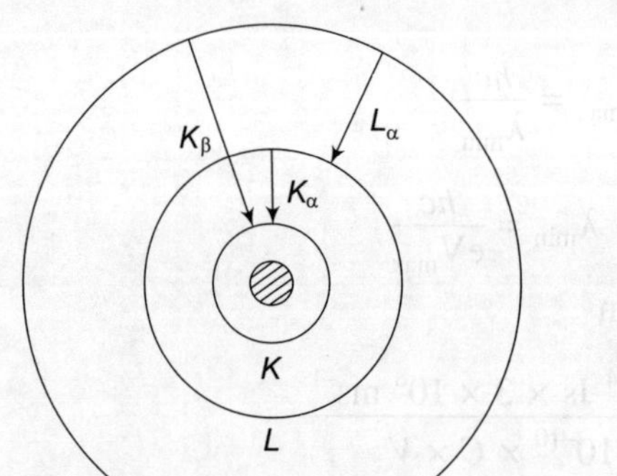


**Figure 3.4** Electron orbits in an atom, showing transitions associated with the production of characteristic X-rays

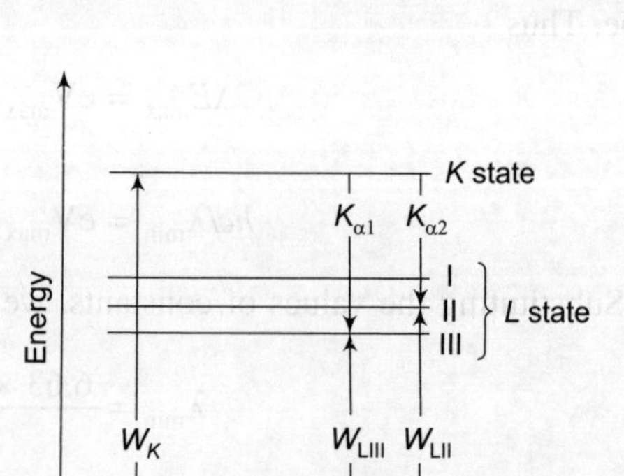


**Figure 3.5** Energy level diagram showing the allowed transitions

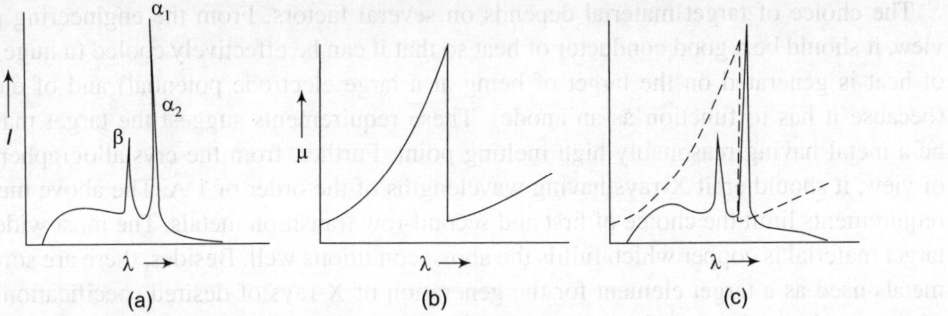


**Figure 3.6** Complete X-ray spectrum, I versus λ curve: solid line shows the characteristic lines superimposed on a background of white radiation. Dotted line shows *K* absorption edge suitable for use as a β-filter

The essential parts of an X-ray tube are shown in Fig. 3.7. A heated tungsten filament (acting as cathode) provides the electron beam, which under a large negative potential (~50 kV) is accelerated towards the target (acting as anode) to give reasonable amount of *K*-radiations. In order to prevent the electrons from losing their energy by colliding with gas molecules before they reach the target, the intervening space is evacuated. The negatively charged screen surrounding the filament serves to focus the electron beam into the target. The X-rays produced in this process leave the X-ray tube through windows ususlly made of beryllium, which when not in use are covered by shutters made of lead. Beryllium and lead are chosen because they have respectively low and high absorption coefficients for X-rays. In general absorption increases with the increase of atomic number of element.

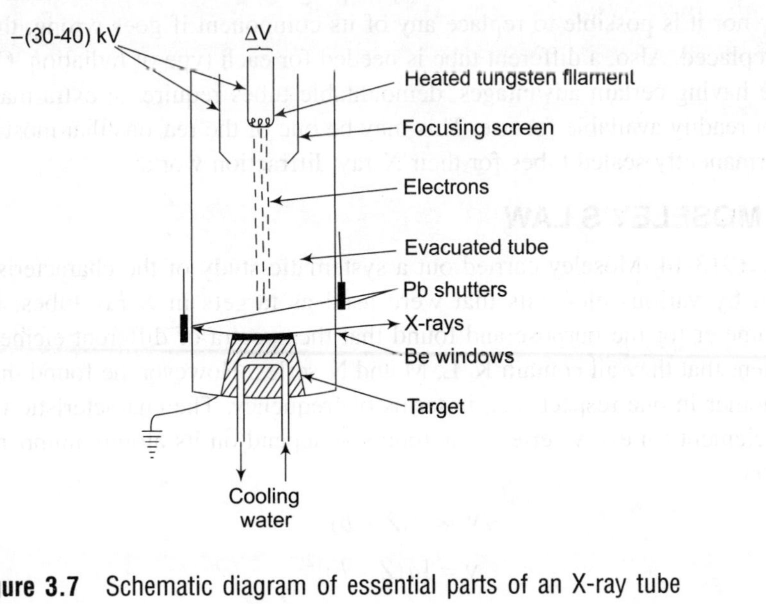


**Figure 3.7** Schematic diagram of essential parts of an X-ray tube

The choice of target material depends on several factors. From the engineering point of view, it should be a good conductor of heat so that it can be effectively cooled (a huge amount of heat is generated on the target of being at a large electrode potential) and of electricity (because it has to function as an anode). These requirements suggest the target material to be a metal having reasonably high melting point. Further, from the crystallographer's point of view, it should emit X-rays having wavelengths of the order of 1 Å. The above mentioned requirements limit the choice of first and second-row transition metals. The most widely used target material is copper which fulfils the above conditions well. Besides, there are some other metals used as a target element for the generation of X-rays of desired specifications. A list of commonly used target elements along with the wavelengths of their characteristic $K_\alpha$ lines is provided in Table 3.1.

**Table 3.1** Elements, Corresponding X-ray Wavelengths and Filter Elements

| *Elements* | *With at. no.* | *X-ray* | *Wavelengths* | *Density* | *Optimum thickness* |
|---|---|---|---|---|---|
| *Target* | *Filter* | $K_{\alpha 1}$ | $K_{\alpha 2}$ | | |
| Cr (24) | V (23) | 2.2896 | 2.2935 | 0.009 | 0.016 |
| Fe (26) | Mn (25) | 1.9360 | 1.9399 | 0.012 | 0.016 |
| Cu (19) | Ni (28) | 1.5405 | 1.5443 | 0.019 | 0.021 |
| Mo (42) | Nb (41) | 0.7093 | 0.7135 | 0.069 | 0.108 |
| Au (47) | Pd (46) | 0.5594 | 0.5638 | 0.030 | 0.030 |

X-ray tubes may be permanently sealed after evacuation, or may be demountable in which vacuum is maintained by continuous pumping. A sealed tube has neither rotating anode facility nor it is possible to replace any of its component if goes wrong, the whole tube has to be replaced. Also, a different tube is needed for each type of radiation. On the other hand, despite having certain advantages, demountable tubes require an extra maintenance facility and not readily available for use. This may be one of the reasons that most crystallographers use permanently sealed tubes for their X-ray diffraction work.

## 3.4 MOSELEY'S LAW

During 1913-14, Moseley carried out a systematic study of the characteristic X-ray spectra emitted by various elements that were used as targets in X-ray tubes. He used Bragg's spectrometer for the purpose and found that the spectra of different elements are similar to the extent that they all contain K, L, M and N-series. However, he found them to differ from one another in one respect, i.e., in terms of frequency. The characteristic frequency emitted by an element (in every series) was found to depend on its atomic number according to the equation

$$\sqrt{\nu} \propto a(Z - b)$$

or

$$\nu = [a(Z - b)]^2 \qquad ...(1)$$

where $a$ and $b$ are constants for particular series (their values are different for different series). The constant $a$ is related to Rydberg constant, $b$ is a small number called the nuclear screening constant and $(Z - b)$ is the effective mass number of the element. For $K$ spectrum; $b = 1$, and hence the Eq. (1) reduces to

$$\nu = [a(Z - 1)]^2 \qquad \text{...(2)}$$

Moseley plotted a graph between the atomic weight of the elements and the square root of the frequencies of characteristic X-rays and also between the square root of the frequencies and the atomic number of the elements emitting them. He found that in the first case, the points lie exactly on the straight line (Fig. 3.8). With the help of such a plot, missing elements like promethium (Pm) and technetium (Tc) were predicted. Further, Moseley's results made it possible to determine the exact atomic number of many elements which were not known correctly at that time. For example, he found that the atomic number of argon (Ar) should be less than potassium (K) and hence to be put first in the serial order. Earlier, they were wrongly placed in the order of increasing atomic weights. Moseley, thus came to an important conclusion that it is the atomic number rather than the atomic weight which determines the chemical and X-ray emitting properties of an atom. He identified the atomic number with the positive charge on the nucleus.

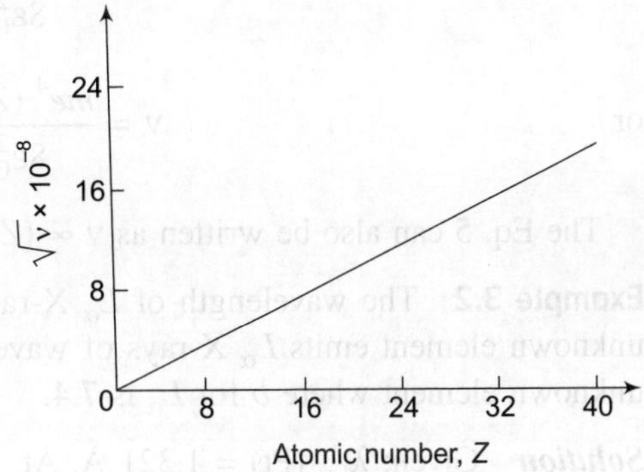


**Figure 3.8** $\sqrt{\nu}$ versus atomic number

Moseley's law can be deduced theoretically from Bhor's atomic model, according to which the energy of an electron moving in an orbit of principal quantum number $n_1$, is given by

$$E_1 = -\frac{me^4 Z^2}{8\varepsilon_0^2 n_1^2 h^2} \qquad \text{...(3)}$$

and for an electron moving in an orbit of principal quantum number $n_2$ is given by

$$E_2 = -\frac{me^4 Z^2}{8\varepsilon_0^2 n_2^2 h^2} \qquad \text{...(4)}$$

However, taking into account the screening effect of the nucleus, the effective atomic number of an element changes from Z to $(Z - b)$. Thus, the above expression can be written as

$$E_1 = -\frac{me^4 (Z - b_1)^2}{8\varepsilon_0^2 n_1^2 h^2}$$

and $$E_2 = -\frac{me^4 (Z - b_2)^2}{8\varepsilon_0^2 n_2^2 h^2}$$

so that, $$\Delta E = E_1 - E_2 = \frac{me^4}{8\varepsilon_0^2 h^2}\left(\frac{(Z-b_2)^2}{n_2^2} - \frac{(Z-b_1)^2}{n_1^2}\right)$$

For high $Z$ valume, $b_1 \simeq b_2 \simeq b$, therefore

$$\Delta E = \frac{me^4}{8\varepsilon_0^2 h^2}(Z-b)^2\left(\frac{1}{n_2^2} - \frac{1}{n_1^2}\right)$$

But, $\Delta E = h\nu$, therefore

$$h\nu = \frac{me^4}{8\varepsilon_0^2 h^2}(Z-b)^2\left(\frac{1}{n_2^2} - \frac{1}{n_1^2}\right)$$

or $$\nu = \frac{me^4}{8\varepsilon_0^2 h^3}(Z-b)^2\left(\frac{1}{n_2^2} - \frac{1}{n_1^2}\right) \quad ...(5)$$

The Eq. 5 can also be written as $\nu \propto (Z-b)^2$, which is the required Moseley's law.

**Example 3.2** The wavelength of $L_\alpha$ X-rays line of platinum (At. No. 78) is 1.321 Å. An unknown element emits $L_\alpha$ X-rays of wavelength 4.174 Å. Calculate the atomic number of unknown element where $b$ for $L_\alpha$ is 7.4.

***Solution*** Given: $\lambda_{L\alpha}$ (Pt) = 1.321 Å, At. No. (Pt) = 78, $b$ = 7.4, $\lambda_{L\alpha}$ of unknown element say, $x$ = 4.174 Å. At. No. ($x$) = ?

Making use of Eq. (1), we have

$$\nu = [a(Z-b)]^2$$

Therefore, $$\nu_{Pt} = \frac{c}{\lambda_{Pt}} = [a(78-7.4)]^2$$

Similarly, $$\nu_x = \frac{c}{\lambda_x} = [a(x-7.4)]^2$$

Dividing one from the other and solving for $x$, we have

$$x = 7.4 + \left(\frac{\lambda_{pt}}{\lambda_x}\right)^{1/2} \times 70.6 = 7.4 + 70.6\left(\frac{1321}{4174}\right)^{1/2} = 47.1$$

## 3.5 X-RAY DIFFRACTION (ACCORDING TO BRAGG), FROM A CRYSTAL

W.L. Bragg in 1913 pointed out that lattice planes in a crystal behave to X-rays as line in a diffraction grating, and incident wave are reflected specularly from lattice planes. The position of the spectral lines when diffracted by a line grating depends on the distance between the successive lines. Similarly, nature of the diffracted beam (by a crystal) is determined by the spacing $d$, between successive planes, i.e., crystal can be considered a space grating. X-rays of wavelength $\lambda$ falling on the crystal in a direction making a glancing angle $\theta$ (known as the Bragg angle) with a set of planes (of spacing $d$) will be diffracted, provided the condition

$$2d \sin \theta = n\lambda \ (n = 1, 2, 3 \ldots) \qquad \ldots(1)$$

is satisfied. This is Bragg's diffraction condition and is known as Bragg's law.

Figure 3.9 shows a monochromatic beam of parallel X-rays penetrating a stack of planes of spacing $d$ (in a crystal) at a glancing angle of incidence, θ. Each plane reflects a fraction of the incident beam energy specularly (i.e., as if by a mirror). The diffracted beams will be observed as reflected X-rays if these specularly reflected beams from the set of planes interfere constructively.

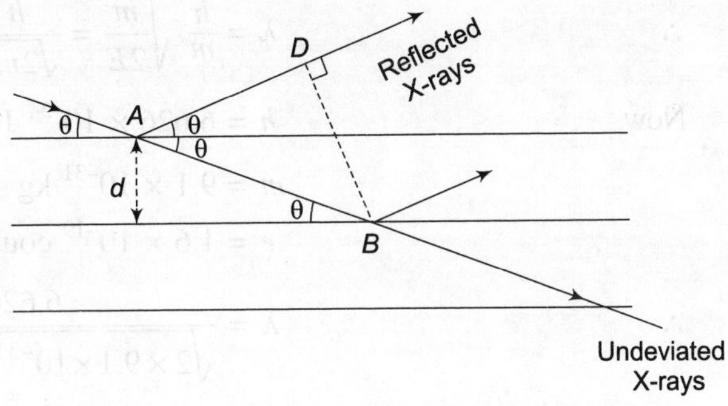


**Figure 3.9** Bragg's condition for reflection

If path difference is

$$AB - AD = n\lambda \ (n = 1, 2, 3, \ldots)$$

$$\frac{d}{AB} = \sin \theta \quad \text{or} \quad AB = \frac{d}{\sin \theta}$$

$$\frac{AD}{AB} = \cos 2\theta \quad \text{or} \quad AD = AB \cos 2\theta = \frac{d}{\sin \theta} \cos 2\theta$$

$$\therefore \quad n\lambda = AB - AD = \frac{d}{\sin \theta} - \frac{d \cos 2\theta}{\sin \theta}$$

$$= \frac{d}{\sin \theta}(1 - \cos 2\theta)$$

$$= \frac{d}{\sin \theta} 2 \sin^2 \theta$$

or $$n\lambda = 2d \sin \theta$$

This gives the angle θ at which a set of planes of spacing $d$ cooperatively reflect X-rays of wavelength λ in the $n$th order; i.e., for monoenergetic radiation used, the X-ray photons are monoenergetic and the equation can be satisfied for a particular set of planes having interplanar spacing $d$ provided

$$\theta_1 = \sin^{-1} (\lambda/2d)$$
$$\theta_2 = \sin^{-1} (2\lambda/2d)$$
$$\theta_3 = \sin^{-1} (3\lambda/2d) \ldots$$

These reflections are known as first, second third, ... order reflections.

**Example 3.3** Electrons are accelerated by 844 volts and are reflected from a crystal. The reflection maximum occurs when the glancing angle is 58°. Determine the interplanar spacing of the crystal.

***Solution*** $$\lambda = \frac{h}{mv} \text{ (de Broglie relation)}$$

$$E = \frac{1}{2} mv^2 \Rightarrow v = \sqrt{\frac{2E}{m}}$$

$$\therefore \quad \lambda = \frac{h}{m}\sqrt{\frac{m}{2E}} = \frac{h}{\sqrt{2mE}} = \frac{h}{\sqrt{2meV}} \qquad (\because \; E = eV)$$

Now,

$$h = 6.626 \times 10^{-34} \text{ J-s}$$

$$m = 9.1 \times 10^{-31} \text{ kg}$$

$$e = 1.6 \times 10^{-19} \text{ coulomb}$$

$$\therefore \quad \lambda = \frac{6.626 \times 10^{-34}}{\sqrt{2 \times 9.1 \times 10^{-31} \times 1.6 \times 10^{-19} \times 844}}$$

$$= 0.42 \times 10^{-10} \text{ m}$$

$$2d \sin\theta = n\lambda \quad \therefore \quad d = \frac{\lambda}{2 \sin\theta}$$

or

$$d = \frac{0.42 \times 10^{-10}}{2 \sin 58^\circ} = 0.0248 \times 10^{-9} \text{ m.}$$

## 3.6 DEFINITION OF RECIPROCAL LATTICE

Consider a set of point $\vec{R}$ constituting a Bravais lattice and a plane wave $e^{i\vec{k}\cdot\vec{r}}$. For general $\vec{k}$, such a plane wave will not have the periodicity of the Bravais lattice, rather for certain special choices of wave vector ($\vec{k}$), it will. *The set of all wave-vectors $\vec{K}$ that yield plane waves with the periodicity of a given Bravais lattice is known as its reciprocal lattice.*

Analytically, $\vec{K}$ belongs to the reciprocal lattice of a Bravais lattice of points $\vec{R}$, provided that

$$e^{i\vec{K}\cdot(\vec{r}+\vec{R})} = e^{i\vec{K}\cdot\vec{r}} \qquad \text{...(1)}$$

holds for any $\vec{r}$ and for all $\vec{R}$ in the Bravais lattice. Thus, we can characterize the reciprocal lattice as the set of wave-vectors $\vec{K}$ satisfying

$$e^{i\vec{K}\cdot\vec{R}} = 1 \qquad \text{...(2)}$$

for all $\vec{R}$ in the Bravais lattice.

That is, reciprocal lattice is the Fourier transform of its Bravais lattice. The Bravais lattice that determines a given reciprocal lattice is often referred to as the *direct lattice*, when viewed in relation to its reciprocal.

### 3.6.1 Von Laue Formulation and Bragg Planes

Here one regards the crystal as composed of identical microscopic objects (sets of ions or atoms) placed at the sites **R** of a Bravais lattice, each of which can reradiate the incident radiation in all directions. Sharp peaks will be observed only in directions and at wavelengths for which the rays scattered from all lattice points interfere constructively.

To find the condition for constructive interference, consider first two scatterers, separated by the displacement vector $\mathbf{d}$ (Fig. 3.10). Let an X-ray be incident from very far away, along a direction $\hat{\mathbf{n}}$, with wavelength $\lambda$, and wave vector $\mathbf{k} = 2\pi\hat{\mathbf{n}}/\lambda$. A scattered ray will be observed in a direction $\hat{\mathbf{n}}'$ with wavelengths[2] $\lambda$ and wave vector $\mathbf{k}' = 2\pi\hat{\mathbf{n}}'/\lambda$, provided that the path difference between the rays scattered by each of the two ions is an integral number of wavelengths. From Fig. 3.10 it can be seen that this path difference is just

$$d\cos\theta + d\cos\theta' = \mathbf{d}\cdot(\hat{\mathbf{n}} - \hat{\mathbf{n}}') \qquad ...(1)$$

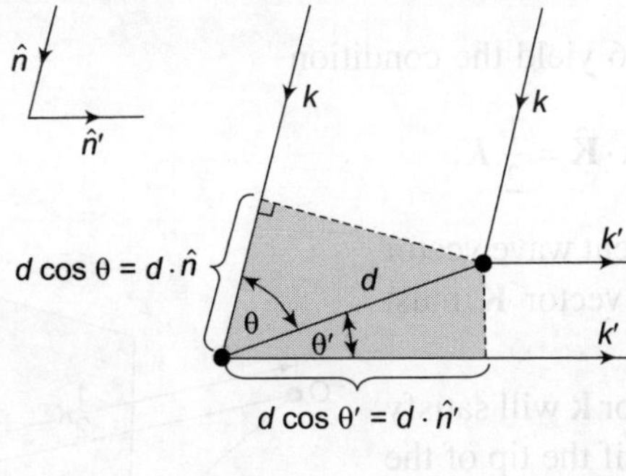


**Figure 3.10** Illustrating that the path difference for rays scattered from two point separated by **d** is given by Eq. (1) or (2)

The condition for constructive interference is thus

$$\mathbf{d}\cdot(\hat{\mathbf{n}} - \hat{\mathbf{n}}') = m\lambda \qquad ...(2)$$

for integral $m$. Multiplying both sides of Eq. (2) by $2\pi/\lambda$ yields a condition on the incident and scattered wave vectors:

$$\mathbf{d}\cdot(\mathbf{k} - \mathbf{k}') = 2\pi m \qquad ...(3)$$

for integral $m$.

Next, we consider not just two scatterers, but an array of scatterers, at the sites of a Bravais lattice. Since the lattice sites are displaced from one another by the Bravais lattice vectors $\mathbf{R}$, the condition that all scattered rays interfere constructively is that Eq. (3) hold simultaneously for all values of $\mathbf{d}$ that are Bravais lattice vectors:

$$\mathbf{R}\cdot(\mathbf{k} - \mathbf{k}') = 2\pi m, \quad \text{for integral } m \text{ and all Bravais lattice vectors } \mathbf{R}. \qquad ...(4)$$

This can be written in the equivalent form

$$e^{i(\mathbf{k}'-\mathbf{k})\cdot\mathbf{R}} = 1, \text{ for all Bravais lattice vectors } \mathbf{R}. \qquad ...(5)$$

[2] Here (and in the Bragg picture) we assume that the incident and scattered radiation has the same wavelength. In terms of photons this means that no energy has been lost in the scattering, i.e., that the scattering is elastic. To a good approximation the bulk of the scattered radiation is elastically scattered, though there is much to be learnt from the study of that small component of the radiation that is inelastically scattered.

Comparing this condition with the definition (section 3.6) of the reciprocal lattice, we arrive at the Laue condition that *constructive interference will occur provided that the change in wave vector,* **K** = **k**′ – **k**, *is a vector of the reciprocal lattice.*

It is sometimes convenient to have an alternative formulation of the Laue condition, stated entirely in terms of the incident wave vector, **k.** First note that because the reciprocal lattice is a Bravais lattice, if **k**′ – **k** is a reciprocal lattice vector, so is **k** – **k**′. Calling the latter vector **K,** the condition that **k** and **k**′ have the same magnitude is

$$k = |\mathbf{k} - \mathbf{K}| \qquad ...(6)$$

Squaring both sides of Eq. 6 yield the condition

$$\mathbf{k} \cdot \hat{\mathbf{K}} = \frac{1}{2} K; \qquad ...(7)$$

i.e., the component of the incident wave vector **k** along the reciprocal lattice vector **K** must be half the length of **K.**

Thus, an incident wave vector **k** will satisfy the Laue condition if and only if the tip of the vector lies in a plane that is the perpendicular bisector of a line joining the origin of *k*-space to a reciprocal lattice point **K** (Fig. 3.11). Such *k*-space planes are called *Bragg planes.*

It is a consequence of the equivalence of the Bragg and von Laue points of view, demonstrated in the following section, that the *k*-space Bragg plane associated with a particular diffraction peak in the Laue formulation is parallel to the family of direct lattice planes responsible for a peak in the Bragg formulation.

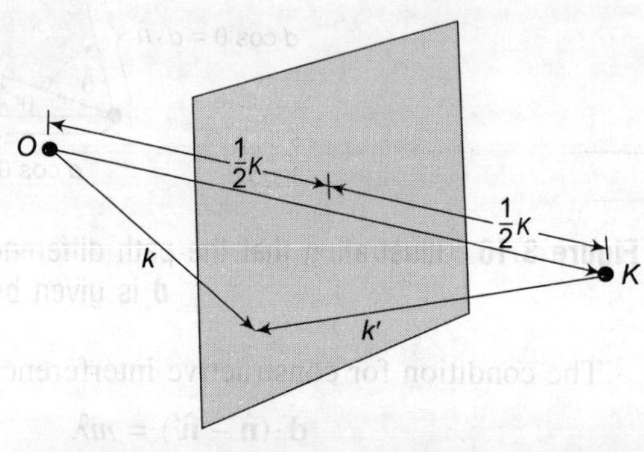


**Figure 3.11** The Laue condition. If the sum of **k** and –**k**′ is a vector **K,** and if **k** and **k**′ have the same length, then the tip of the vector **k** is equidistant from the origin *O* and the tip of the vector **K,** and therefore, it lies in the plane bisecting the line joining the origin to the tip of **K.**

## 3.7 EQUIVALENCE OF THE BRAGG AND VON LAUE FORMULATIONS

The equivalence of these two criteria for constructive interference of X-rays by a crystal follows from the relation between vectors of the reciprocal lattice and families of direct lattice planes (see section 3.8). Suppose the incident and the scattered wave sectors, **k** and **k**′, satisfy the Laue condition that **K** = **k**′ – **k** be a reciprocal lattice vector. Because the incident and scattered waves have the same wavelengths, **k**′ and **k** have the same magnitudes. It follows (see Fig. 3.12) that **k**′ and **k** make the same angle θ with the plane perpendicular to **K**. Therefore the scattering can be viewed as a Bragg reflection, with Bragg angle θ, from the family of direct lattice planes perpendicular to the reciprocal lattice vector **K.**

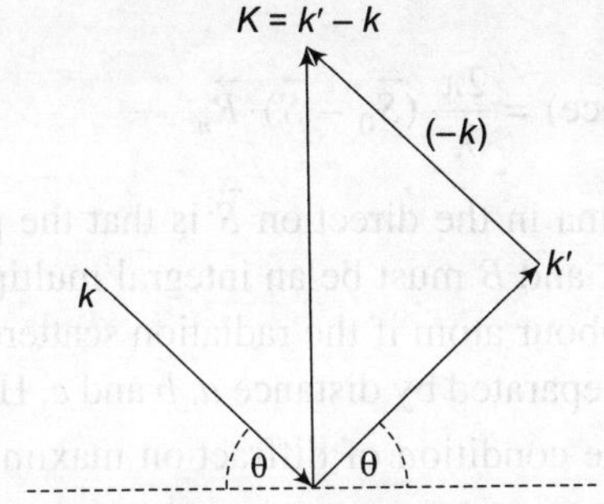


**Figure 3.12** The plane of the paper contains the incident wave vector **k**, the reflected wave vector **k′**, and their difference **K** satisfying the Laue condition. Since the scattering is elastic ($k' = k$), the direction of **K** bisects the angle between **k** and **k′**. The dashed line is the intersection of the plane perpendicular to **K** with the plane of the paper

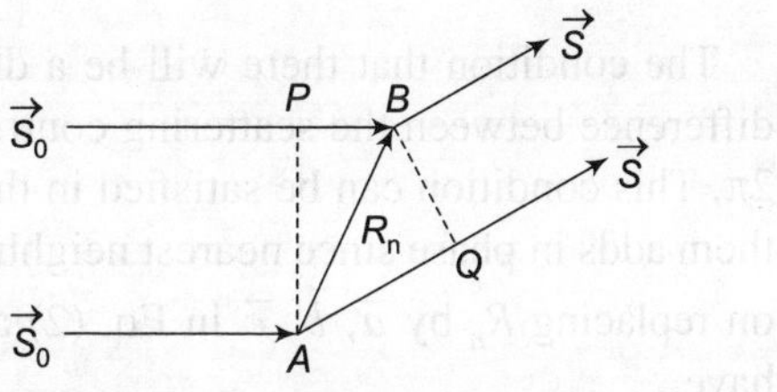


**Figure 3.13** Scattering of X-rays (Von Laue treatment) *A* and *B* are two lattice points

### 3.7.1 The Von Laue Treatment

A crystal may be considered as a series of scattering centres situated at the points of a space lattice. Thus, if a beam of X-rays of wavelength λ is incident on the lattice, each lattice point acts as a source of scattered X-rays, and diffracted beams appear in those directions in which the beams from the various lattice points are in phase. Consider, therefore, any two lattice points *A* and *B*, separated by $\vec{R}_n = n_1\,\vec{a} + n_2\,\vec{b} + n_3\,\vec{c}$ as shown in Fig. 3.13.

Let $\vec{S}_0$ be a unit vector along the incident beam and $\vec{S}$ a unit vector along the scattered beam.

The path difference between radiations scattered from centres *B* and *A* is

$$BP - AQ = \vec{S}_0 \cdot \vec{R}_n - \vec{S} \cdot \vec{R}_n = (\vec{S}_0 - \vec{S}) \cdot \vec{R}_n \qquad ...(1)$$

*BP* and *AQ* are projections of $\vec{R}_n$ on incident and scattered beams, respectively.

The vector along $(\vec{S}_0 - \vec{S})$ represents the direction normal to the plane that reflects the incident direction (Fig. 3.14). The plane may be regarded as the reflecting plane in the sense of Bragg treatment. If 2θ is the angle which $\vec{S}$ makes with $\vec{S}_0$ than $|\vec{S}_0 - \vec{S}| = 2\sin\theta$ because $\vec{S}_0$ and $\vec{S}$ are unit vectors along incident and scattered beams, respectively.

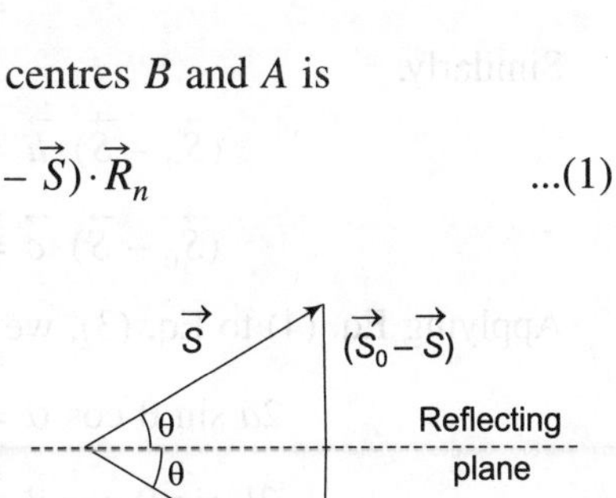


**Figure 3.14** Scattering (diffraction) in relation to reflecting plane

The phase difference

$$\phi_p = \frac{2\pi}{\lambda}\text{ (path difference)} = \frac{2\pi}{\lambda}(\vec{S}_0 - \vec{S})\cdot\vec{R}_n \qquad \text{...(2)}$$

The condition that there will be a diffraction maxima in the direction $\vec{S}$ is that the phase difference between the scattering contributions from *A* and *B* must be an integral multiple of $2\pi$. This condition can be satisfied in the nearest neighbour atom if the radiation scattered by them adds in phase since nearest neighbour atoms are separated by distance *a, b* and *c*. Hence, on replacing $R_n$ by $\vec{a}, \vec{b}, \vec{c}$ in Eq. (2) and applying the condition of diffraction maxima, we have

$$\left.\begin{aligned}\frac{2\pi}{\lambda}(\vec{S}_0 - \vec{S})\cdot\vec{a} &= 2\pi h' = 2\pi nh\\ \frac{2\pi}{\lambda}(\vec{S}_0 - \vec{S})\cdot\vec{b} &= 2\pi k' = 2\pi nk\\ \frac{2\pi}{\lambda}(\vec{S}_0 - \vec{S})\cdot\vec{c} &= 2\pi l' = 2\pi nl\end{aligned}\right\} \qquad \text{...(3)}$$

$h'$, $k'$ and $l'$ are three integers. These integers may form the smallest set of integers in such a way that they may be identical to Miller indices of a crystal plane or they may contain a common integer factor greater than unity. Let *n* be the common factor, so we can write $h' = nh$, $k' = nk$ and $l' = nl$. *h, k, l,* are now three smallest integers (Miller indices). In particular, *n* may be unity, then *h k l* and $h'\ k'\ l'$ have the same meaning.

Let $\alpha, \beta, \gamma$ be the angles between the scattering normal $(\vec{S}_0 - \vec{S})$ and $\vec{a}, \vec{b}$ and $\vec{c}$ the axes of the crystal, respectively. Now,

$$\begin{aligned}(\vec{S}_0 - \vec{S})\cdot\vec{a} &= 2\sin\theta\cdot a\cdot\cos\alpha\\ &= 2a\sin\theta\cos\alpha\end{aligned}$$

Similarly,

$$\left.\begin{aligned}(\vec{S}_0 - \vec{S})\cdot\vec{h} &= 2b\sin\theta\cos\beta\\ (\vec{S}_0 - \vec{S})\cdot\vec{c} &= 2c\sin\theta\cos\gamma\end{aligned}\right\} \qquad \text{...(4)}$$

Applying Eq. (4) to Eq. (3), we have

$$\left.\begin{aligned}2a\sin\theta\cos\alpha &= nh\lambda\\ 2b\sin\theta\cos\beta &= nk\lambda\\ 2c\sin\theta\cos\gamma &= nl\lambda\end{aligned}\right\} \qquad \text{...(5)}$$

These conditions are known as Laue equations. These serve to determine a unique value of $\theta$, thus defining a scattering direction. From these equations we note that direction cosines of the scattering normal $(\vec{S}_0 - \vec{S})$ are proportional to *h/a, k/d* and *l/c*. We also know that direction cosines of normal to *(hkl)* planes are proportional to *h/a, k/b,* and *l/c*, so we conclude that scattering normal $(\vec{S}_0 - \vec{S})$ is identical to the normal to (*hkl*) planes. Hence, the (*hkl*) planes may be regarded as the reflecting planes in the sense of Bragg treatment. If *d* is the spacing between two adjacent planes of a family (*hkl*), we have

$$d = (a/h)\cos\alpha = (b/k)\cos\beta = (c/l)\cos\gamma$$

Then, from either of Eq. (5), we get

$$2d \sin\theta = n\lambda \text{ (Bragg equation)}$$

## 3.8 RECIPROCAL LATTICE (CONSTRUCTION FROM A SPACE LATTICE)

We have earlier defined the reciprocal lattice (section 3.6). The concept of reciprocal lattice was first developed by Ewald for a better understanding of X-ray diffraction by crystals. According to Bragg the analysis of X-ray diffraction is in terms of the slopes of various lattice planes, which is rather tedious. However, it becomes easier if we consider normal to various sets of planes instead of planes themselves.

Thus, to obtain a reciprocal lattice to a given space lattice, we apply the following:

(i) From a common origin draw a normal to each crystal plane.

(ii) Make the length of each normal equal to (or $2\pi$ times) the reciprocal of interplanar spacing $d_{hkl}$.

(iii) Make the point at the end of each normal to represent the crystal plane.

A collection of point obtained in this way corresponding to various lattice planes forms an array, which is known as the *reciprocal lattice.*

Now let us consider a single family of parallel equidistant lattice planes, say (1 0 0), as shown in Fig. 3.15.

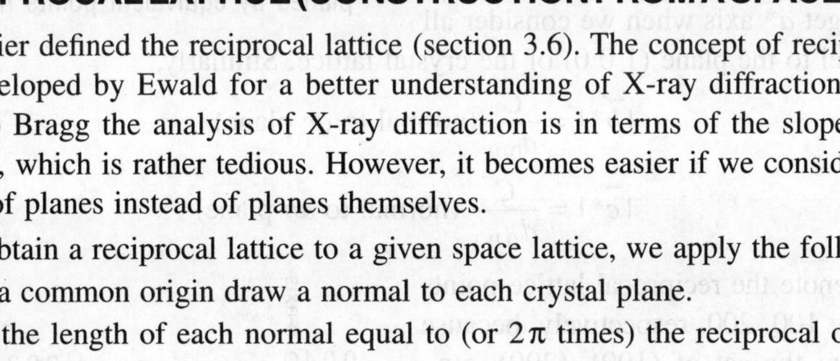


**Figure 3.15** A set of (1 0 0) planes

The reciprocal point for this set of planes is found by drawing a common normal *OP* to the planes from an origin *O* taken arbitrarily on any lattice point and then locating a point $R_1$ on this normal at a distance which is inversely proportional to the interplanar spacing $d_{100}$, i.e.,

$$OR_1 = \frac{C}{d_{100}} \qquad ...(1)$$

where *C* is constant of proportionality. It is to be noted that the entire set of planes is represented by just a single point, $R_1$. Similarly, we can obtain a reciprocal point for the set of (2 0 0) planes. We know that (2 0 0) planes, but have a spacing = $d_{100}/2$, so we get the reciprocal point $R_2$ along $OP$ at a distance $C/d_{200} = 2C/d_{100}$ as shown in Fig. 3.16.

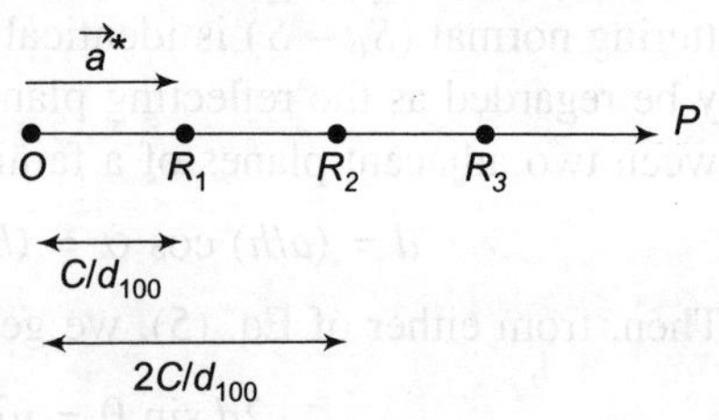


**Figure 3.16** Representation of set of lattice planes by equivalent points $R_1$, $R_2$ ... etc

The repeat period of this row of reciprocal lattice points is denoted by

$$a_i^* \; |\vec{a}^*| = |\overrightarrow{OR_1}| = c/d_{100} \qquad ...(2)$$

Thus, we get $a^*$ axis when we consider all planes parallel to the plane (1 0 0) of the crystal lattice. Similarly,

$$|\vec{b}^*| = \frac{C}{d_{010}} \text{ (normal to } ac \text{ plane)} \qquad ...(3)$$

and

$$|\vec{c}^*| = \frac{C}{d_{001}} \text{ (normal to } ab \text{ plane)} \qquad ...(4)$$

We can denote the reciprocal lattice points $R_1$, $R_2$ etc. by 100, 200, respectively, because they represent the set of (100), (200), etc., planes of the direct lattice. The array of lattice points $R_1$, $R_2$, $R_3$, etc., defined by the unit translations $|\vec{a}^*|$, $|\vec{b}^*|$, $|\vec{c}^*|$, forms the reciprocal lattice. Figure 3.17 shows the reciprocal lattice for a monoclinic crystal.

*b* is the unique axis and *a* and *c* are the unit cell edges (of the direct lattice). The array of point 1 0 0, 1 0 1, 1 0 2, ..., 2 0 0, 2 0 1, ..., 3 0 0, 3 0 1, ... and so on represent the reciprocal lattice for the monoclinic crystal. These points have been obtained by the procedure as outlined above, i.e., from a common origin *O*, normals are erected to each of the lattice planes *(h k l)*, etc., and points are placed on each of these normals at distances $1/d_{hkl}$.

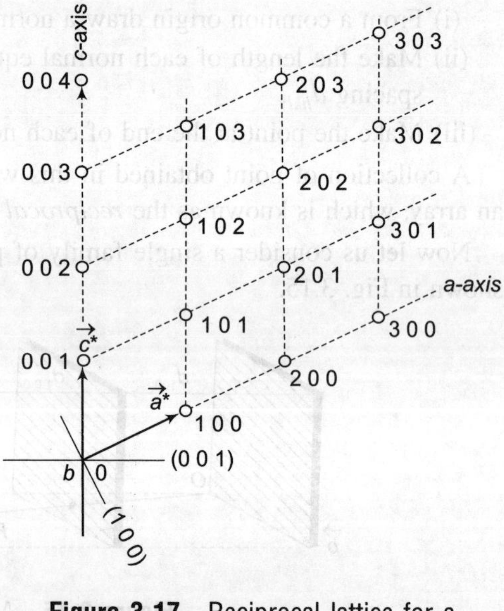


**Figure 3.17** Reciprocal lattice for a monoclinic crystal

Note that each of the reciprocal lattice points preserves all the important characteristics of the stack of planes it represents – the direction of the point from the origin preserves the orientation of the planes and the distance of the point from 0 preserves the interplanar spacing.

The vector distance of *h o o* reciprocal lattice point

$$\vec{r}^*_{hoo} = h\,\vec{a}^* \; (h = 1, 2, 3, ...) \qquad ...(5)$$

Similarly,

$$\vec{r}^*_{oko} = k\,\vec{b}^* \; (k = 1, 2, 3, ...) \qquad ...(6)$$

and $$\vec{r}^{\,*}_{ool} = l\,\vec{c}^{\,*} \; (l = 1, 2, 3, ...) \qquad ...(7)$$

In general, reciprocal lattice vector

$$\vec{r}^{\,*}_{hkl} = h\vec{a}^{\,*} + h\vec{b}^{\,*} + l\vec{c}^{\,*} \qquad ...(8)$$

Here, $\vec{a}^{\,*}$ and $\vec{b}^{\,*}$ are $\vec{c}^{\,*}$ basic unit translation vectors in the reciprocal lattice.

For $\vec{a}^{\,*}$ to be perpendicular to $\vec{b}$ and $\vec{c}$, $\vec{b}^{\,*}$ to be perpendicular to $\vec{c}$ and $\vec{a}$ and $\vec{c}^{\,*}$ to be perpendicular to $\vec{a}$ and $\vec{b}$, they should obey the following relationships.

$$\left.\begin{aligned} \vec{a}^{\,*}\cdot\vec{a} &= 2\pi \quad \vec{a}^{\,*}\cdot\vec{b} = 0 = \vec{a}^{\,*}\cdot\vec{c} \\ \vec{b}^{\,*}\cdot\vec{b} &= 2\pi \quad \vec{b}^{\,*}\cdot\vec{c} = 0 = \vec{b}^{\,*}\cdot\vec{a} \\ \vec{c}^{\,*}\cdot\vec{c} &= 2\pi \quad \vec{c}^{\,*}\cdot\vec{a} = 0 = \vec{c}^{\,*}\cdot\vec{b} \end{aligned}\right\} \qquad ...(9a)$$

$$|\vec{a}^{\,*}| = \frac{2\pi}{|\vec{a}|}; \quad |\vec{b}^{\,*}| = \frac{2\pi}{|\vec{b}|}; \quad |c^*| = \frac{2\pi}{|\vec{c}|} \qquad ...(9b)$$

It is apparent that Eq. (8) has the familiar form of the lattice generating equation, but with a different base set $\vec{a}^{\,*}$, $\vec{b}^{\,*}$ and $\vec{c}^{\,*}$. Thus, the vectors associated with arbitrary sets of parallel planes in the real lattice define a lattice with the reciprocal base set (Eq. 9) called the reciprocal lattice.

A geometrical procedure for constructing the two-dimensional reciprocal oblique lattice, using the normal to the unit cell sides of the direct (oblique lattice), is illustrated in Fig. 3.18.

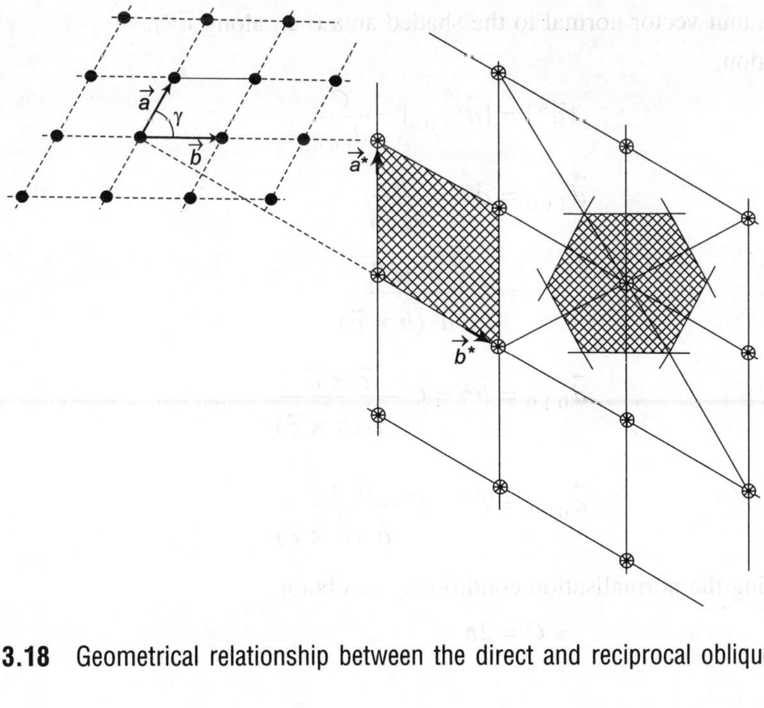


**Figure 3.18** Geometrical relationship between the direct and reciprocal oblique lattice

The Winger–Seitz cell in the reciprocal lattice shown, corresponding to the first Brillouiro zone (see Article 2.17).

### 3.8.1 Reciprocal Lattice Vectors in Terms of Direct Lattice Vectors

We can express the reciprocal unit vectors $\vec{a}*$, $\vec{b}*$ and $\vec{c}*$ in terms of $\vec{a}$, $\vec{b}$, and $\vec{c}$ of the direct lattice as follows.

Suppose, the direct crystal lattice has a unit cell defined by vectors $\vec{a}$, $\vec{b}$, $\vec{c}$.

Volume of the unit cell

$$V = (\text{area of base}) \cdot (\text{height})$$
$$= A \cdot (OP)$$

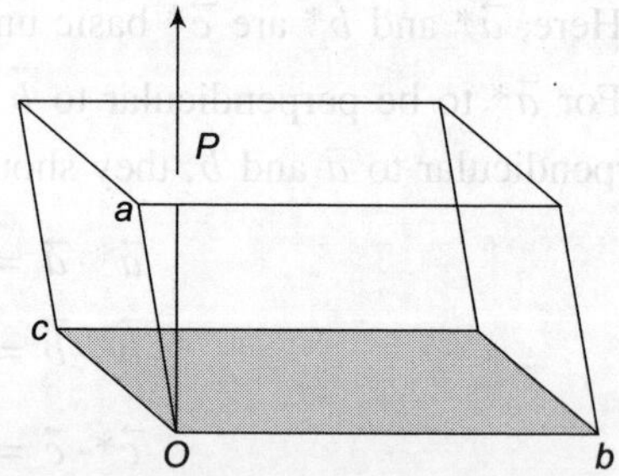


**Figure 3.19** For calculation of volume of unit cell

Since $A = \vec{b} \times \vec{c}$, therefore

$$V = \vec{a} \cdot (\vec{b} \times \vec{c}) \qquad \text{...(10)}$$

The height $OP$ is spacing of the $bc$ planes, i.e., $d_{100}$, therefore

$$V = A \cdot d_{100} \quad \text{or} \quad \frac{1}{d_{100}} = \frac{A}{V}$$

or

$$\frac{\vec{n}}{d_{100}} = \frac{\vec{b} \times \vec{c}}{\vec{a} \cdot (\vec{b} \times \vec{c})} \qquad \text{...(11)}$$

where $\vec{n}$ is a unit vector normal to the shaded area (i.e., along $\vec{a}*$).

By definition,

$$|\vec{a}*| = |\vec{d}*_{100}| = \frac{C}{d_{100}} \qquad \text{...(12)}$$

or

$$\vec{d}^*_{100} = \vec{a}* = \frac{\vec{n}}{d_{100}}$$

$$= C \frac{\vec{b} \times \vec{c}}{\vec{a} \cdot (\vec{b} \times \vec{c})} \qquad \text{...(13)}$$

Similarly,

$$\vec{d}^*_{010} = \vec{b}* = C \frac{\vec{c} \times \vec{a}}{\vec{a} \cdot (\vec{b} \times \vec{c})} \qquad \text{...(14)}$$

$$\vec{d}^*_{001} = \vec{c}* = C \frac{\vec{a} \times \vec{b}}{\vec{a} \cdot (\vec{b} \times \vec{c})} \qquad \text{...(15)}$$

On applying the normalisation conditions, we obtain

$$C = 2\pi$$

and

$$\left.\begin{aligned}\vec{a}^* &= 2\pi\,\frac{\vec{b}\times\vec{c}}{\vec{a}\cdot(\vec{b}\times\vec{c})}\\ \vec{b}^* &= 2\pi\,\frac{\vec{c}\times\vec{a}}{\vec{a}\cdot(\vec{b}\times\vec{c})}\\ \vec{c}^* &= 2\pi\,\frac{\vec{a}\times\vec{b}}{\vec{a}\cdot(\vec{b}\times\vec{c})}\end{aligned}\right\} \quad ...(16)$$

### 3.8.2 Properties of Reciprocal Lattice

(i) Every vector of the reciprocal lattice is normal to a set of lattice planes of the direct lattice.

To prove that $\vec{G}_{hkl}$ is normal to the crystal plane $(hkl)$,

We have to prove that the dot product of $\vec{G}_{hkl}$ and a vector lying in the $(h\ k\ l)$ plane vanishes.

The plane $(h\ k\ l)$ shown in Fig. 3.20 intercepts $\vec{a}$ at $\frac{\vec{a}}{h}$, $\vec{b}$ at $\frac{\vec{b}}{k}$ and $\vec{c}$ at $\frac{\vec{c}}{l}$.

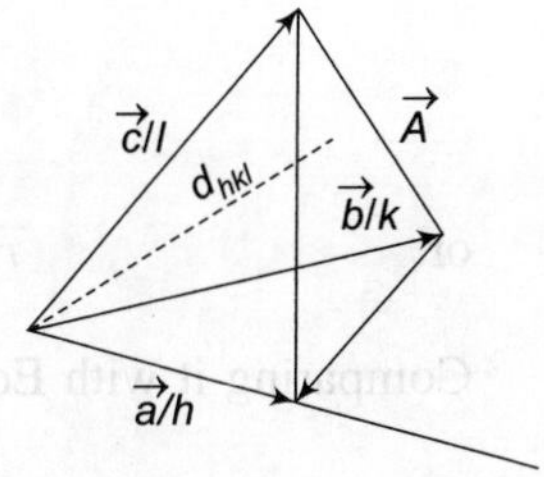


**Figure 3.20** A lattice plane containing the vectors $\vec{A}$ and $\vec{C}$

Now consider the vector

$$\vec{C} = \frac{\vec{a}}{h} - \frac{\vec{b}}{k} \quad ...(17)$$

which lies in this plane.

The dot product of $\vec{C}$ with $\vec{G}_{hkl}$ is given by

$$\vec{C}\cdot\vec{G}_{hkl} = \left(\frac{\vec{a}}{h} - \frac{\vec{b}}{k}\right)\cdot(h\vec{a}^* + k\vec{b}^* + l\vec{c}^*)$$

$$= \frac{\vec{a}}{h}\cdot(h\vec{a}^* + k\vec{b}^* + l\vec{c}^*) - \frac{\vec{b}}{k}\cdot(h\vec{a}^* + k\vec{b}^* + l\vec{c}^*)$$

$$= \left(\frac{h}{k} + 0 + 0\right) - \left(0 + \frac{k}{k} + 0\right) = 1 - 1 = 0 \quad ...(18)$$

Similarly, $$\vec{A}\cdot\vec{G}_{hkl} = \left(\frac{\vec{b}}{k} - \frac{\vec{c}}{l}\right)\cdot(h\vec{a}^* + k\vec{b}^* + l\vec{c}^*) = 0 \quad ...(19)$$

Since, $\vec{G}_{hkl}$ is normal to $\vec{A}$ and $\vec{C}$,

$\therefore$ It is normal to plane containing $\vec{A}$ and $\vec{C}$, i.e., to the plane $(h\ k\ l)$.

(ii) The magnitude of every reciprocal lattice vector is inversely proportional to the interplanar spacing of the corresponding set of lattice planes.

Let $\vec{r}$ be any vector from the origin of the crystal lattice to a plane $(h\ k\ l)$. Then the equation of the plane is

$$\vec{r}\cdot\hat{G}_{hkl} = d_{hkl} \qquad ...(20)$$

where $\hat{G}_{hkl}$ is the unit vector normal to the plane $(h\ k\ l)$.

Let us consider one particular $\vec{r}$, say $\vec{r} = \frac{\vec{a}}{h}$. Then

$$\vec{r}\cdot\vec{G}_{hkl} = \frac{\vec{a}}{h}\cdot(h\vec{a}^* + k\vec{b}^* + l\vec{c}^*)$$

$$= \frac{1}{h}(2\pi h) = 2\pi \qquad ...(21)$$

or

$$\vec{r}\cdot\hat{G}_{hkl} = \frac{2\pi}{G_{hkl}} \quad [\because \vec{G} = G\hat{G}] \qquad ...(22)$$

Comparing it with Eq. (22), we get

$$\frac{2\pi}{G_{hkl}} = d_{hkl} \qquad ...(23a)$$

or

$$G_{hkl} \cdot \frac{1}{(d_{hkl})} \qquad ...(23b)$$

(iii) The volume of a unit cell of the reciprocal lattice is inversely proportional to the volume of a unit cell of the crystal lattice.

The reciprocal lattice cell is described by $\vec{a}^*$, $\vec{b}^*$ and $\vec{c}^*$.

Its volume would be

$$\vec{a}^*\cdot\vec{b}^*\times\vec{c}^* = (2\pi)^3\ \frac{\vec{b}\times\vec{c}}{\vec{a}\cdot\vec{b}\times\vec{c}}\cdot\frac{(\vec{c}\times\vec{a})\times(\vec{a}\times\vec{b})}{(\vec{a}\cdot\vec{b}\times\vec{c})^2}$$

$$= \frac{(2\pi)^3}{(\vec{a}\cdot\vec{b}\times\vec{c})^3}\ (\vec{b}\times\vec{c})\cdot[\vec{c}\times\vec{a}\times(\vec{a}\times\vec{b})]$$

Using $\quad \vec{A}\times(\vec{B}\times\vec{C}) = \vec{B}(\vec{A}\cdot\vec{C}) - \vec{C}(\vec{A}\cdot\vec{B})$,

$$\vec{a}^*\cdot\vec{b}^*\times\vec{c}^* = \frac{(2\pi)^3}{(\vec{a}\cdot\vec{b}\times\vec{c})^3}\ (\vec{b}\times\vec{c})\cdot[\vec{a}(\vec{c}\times\vec{a}\cdot\vec{b}) - \vec{b}(\vec{c}\times\vec{a}\cdot\vec{a})]$$

(using the property of vector triple product that dot and cross are interchangeable)

Since $\quad \vec{c}\times\vec{a}\cdot\vec{a} = 0$,

$$\vec{a}^*\cdot\vec{b}^*\times\vec{c}^* = \frac{(2\pi)^3}{(\vec{a}\cdot\vec{b}\times\vec{c})^3}\ (\vec{b}\times\vec{c}\cdot\vec{c})\,(\vec{c}\times\vec{a}\cdot\vec{b})$$

$$= \frac{(2\pi)^3}{\vec{a} \cdot \vec{b} \times \vec{c}} \qquad ...(24)$$

where $\vec{a} \cdot \vec{b} \times \vec{c}$ is the volume of the unit cell of the crystal lattice.

(iv) The direct lattice is the reciprocal lattice to its own reciprocal lattice.

(v) Simple cubic lattice is self-reciprocal, whereas bcc and fcc lattices are reciprocal to each other.

### 3.8.3 Reciprocal Lattice to Simple Cubic (SC) Lattice

The primitive translation vectors of a simple cubic lattice may be written as

$$\left.\begin{aligned} \vec{a} &= a\hat{i} \\ \vec{b} &= a\hat{j} \\ \vec{c} &= a\hat{k} \end{aligned}\right\} \qquad ...(25)$$

Volume of the simple cubic unit cell

$$= \vec{a} \cdot \vec{b} \times \vec{c}$$

$$= a^3 (\hat{i} \cdot \hat{j} \times \hat{k}) = a^3 \qquad ...(26)$$

Using above equations the reciprocal lattice vectors to the simple cubic lattice are obtained as

$$\vec{a}^* = 2\pi \frac{\vec{b} \times \vec{c}}{\vec{a} \cdot \vec{b} \times \vec{c}} = 2\pi \frac{a\hat{j} \times a\hat{k}}{a^3} = \frac{2\pi}{a} \hat{i} \qquad ...(27a)$$

Similarly,

$$\vec{b}^* = 2\pi \frac{\vec{c} \times \vec{a}}{\vec{a} \cdot \vec{b} \times \vec{c}} = \frac{2\pi}{a} \hat{j} \qquad ...(27b)$$

and

$$\vec{c}^* = 2\pi \frac{\vec{a} \times \vec{b}}{\vec{a} \cdot \vec{b} \times \vec{c}} = \frac{2\pi}{a} \hat{k} \qquad ...(27c)$$

These equations indicate that all the three reciprocal lattice vectors are equal in magnitude. which means that the reciprocal lattice to simple cubic lattice is also simple cubic but with lattice constant equal to $2\pi/a$.

### 3.8.4 Reciprocal Lattice to bcc Lattice

The primitive translation vectors of a body-centred cubic lattice are

$$\left.\begin{aligned} \vec{a}' &= \frac{a}{2}(\hat{i} + \hat{j} - \hat{k}) \\ \vec{b}' &= \frac{a}{2}(-\hat{i} + \hat{j} - \hat{k}) \\ \vec{c}' &= \frac{a}{2}(\hat{i} - \hat{j} + \hat{k}) \end{aligned}\right\} \qquad ...(28)$$

where $a$ is the length of the cube edge and $\hat{i}, \hat{j}, \hat{k}$ are orthogonal unit vectors along the cube edges (Fig. 3.21). The volume of the primitive cell is given by

$$V = \vec{a}' \cdot \vec{b}' \times \vec{c}'$$

$$= \frac{a}{2}(\hat{i} + \hat{j} - \hat{k}) \cdot \left[\frac{a^2}{4}(-\hat{i} + \hat{j} - \hat{k}) \times (\hat{i} - \hat{j} + \hat{k})\right]$$

$$= \frac{a}{2}(\hat{i} + \hat{j} - \hat{k}) \cdot \frac{a^2}{4}(\hat{i} + \hat{j})$$

$$= \frac{a^3}{2} \quad ...(29)$$

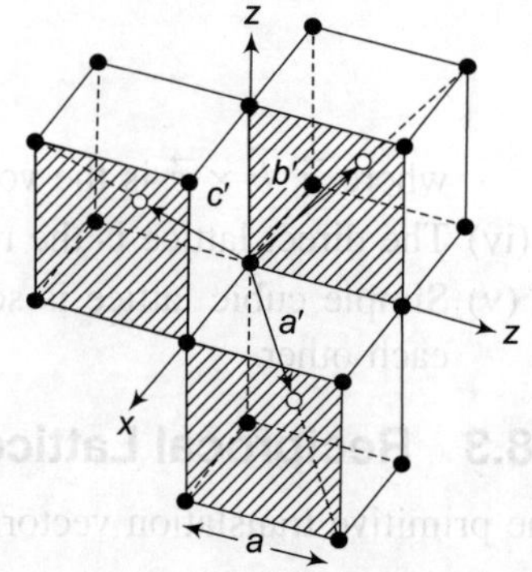


**Figure 3.21** Primitive translation vectors of an bcc lattice

Using Eq. (16), the reciprocal lattice vectors are obtained as

$$\vec{a}^* = 2\pi \frac{\vec{b}' \times \vec{c}'}{\vec{a}' \times \vec{b}' \times \vec{c}'} = \frac{2\pi(a^2/2)}{a^3/2}(\hat{i} + \hat{j}) = \frac{2\pi}{a}(\hat{i} + \hat{j}) \quad ...(30a)$$

Similarly,

$$\vec{b}^* = 2\pi \frac{\vec{c}' \times \vec{a}'}{\vec{a}' \cdot \vec{b}' \times \vec{c}'} = \frac{2\pi}{a}(\hat{j} + \hat{k}) \quad ...(30b)$$

and

$$\vec{c}^* = 2\pi \frac{\vec{a}' \times \vec{b}'}{\vec{a}' \cdot \vec{b}' \times \vec{c}'} = \frac{2\pi}{a}(\hat{k} + \hat{i}) \quad ...(30c)$$

These are the primitive translation vectors of an fcc lattice. Therefore, the reciprocal lattice to a bcc lattice is fcc lattice.

### 3.8.5 Reciprocal Lattice to fcc Lattice

The primitive translation vectors (Fig. 3.22) of an fcc lattice are

$$\left.\begin{aligned} \vec{a}' &= \frac{a}{2}(\hat{i} + \hat{j}) \\ \vec{b}' &= \frac{a}{2}(\hat{j} + \hat{k}) \\ \vec{c}' &= \frac{a}{2}(\hat{k} + \hat{i}) \end{aligned}\right\} \quad ...(31)$$

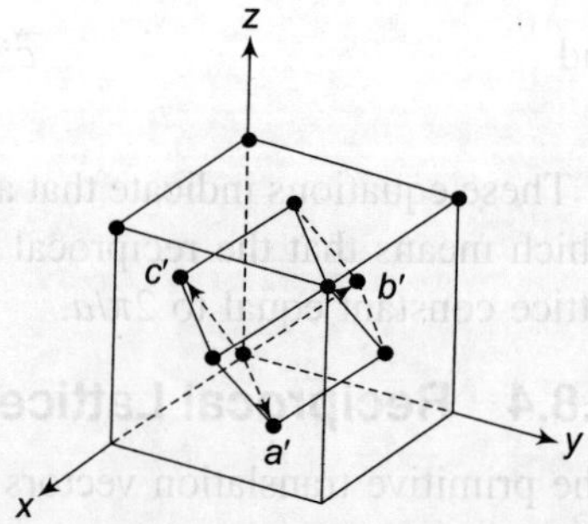


**Figure 3.22** Primitive translation vectors of an fcc lattice

Volume of the primitive cell is given by

$$V = \vec{a}' \cdot \vec{b}' \times \vec{c}'$$

$$= \frac{a}{2}(\hat{i} + \hat{j}) \cdot \frac{a^2}{4}[(\hat{j} + \hat{k}) \times (\hat{k} + \hat{i})]$$

$$= \frac{a}{2}(\hat{i} + \hat{j}) \cdot \frac{a^2}{4}(\hat{i} + \hat{j} - \hat{k})$$

$$= \frac{a^3}{4} \qquad ...(32)$$

Using Eq. (16), the primitive translation vectors of the reciprocal lattice are obtained as

$$\vec{a}^* = 2\pi \frac{\vec{b}' \times \vec{c}'}{\vec{a}' \cdot \vec{b}' \times \vec{c}'}$$

$$= 2\pi \frac{(a^2/4)(\hat{i} + \hat{j} - \hat{k})}{a^3/4} = \frac{2\pi}{a}(\hat{i} + \hat{j} - \hat{k}) \qquad ...(33)$$

Similarly,
$$\vec{b}^* = 2\pi \frac{\vec{c}' \times \vec{a}'}{\vec{a}' \cdot \vec{b}' \times \vec{c}'} = \frac{2\pi}{a}(-\hat{i} + \hat{j} + \hat{k}) \qquad ...(34)$$

and
$$\vec{c}^* = 2\pi \frac{\vec{a}' \times \vec{b}'}{\vec{a}' \cdot \vec{b}' \times \vec{c}'} = \frac{2\pi}{a}(\hat{i} - \hat{j} + \hat{k}) \qquad ...(35)$$

These are the primitive translation vectors of a bcc lattice having length of the cube edge as $2\pi/a$. Thus, the reciprocal lattice to an fcc lattice is a bcc lattice.

## 3.9 DEFINITION OF RECIPROCAL LATTICE IN TERMS OF WAVE VECTORS

Consider a set of point $\vec{R}$ constituting a Bravais lattice, and a plane wave $e^{i\vec{k}\cdot\vec{r}}$. For general $\vec{k}$ such a plane wave will not, of course, have the periodicity of the Bravais lattice, but for certain special choice of wave vector, it will. *The set of all wave vectors $\vec{K}$ that yield plane waves with the periodicity of a given Bravais lattice is known as its reciprocal lattice.*

Analytically, $\vec{K}$ belongs to the reciprocal lattice of a Bravais lattice of point $\vec{R}$, provided that the relation

$$e^{i\vec{K}\cdot(\vec{r} + \vec{R})} = e^{i\vec{K}\cdot\vec{r}} \qquad ...(1)$$

holds for any $\vec{r}$, and for all $\vec{R}$ in the Bravais lattice. Factoring out $e^{i\vec{K}\cdot\vec{r}}$, we can characterize. The reciprocal lattice as the set of wave-vectors $\vec{K}$ satisfying

$$e^{i\vec{K}\cdot\vec{R}} = 1 \qquad ...(2)$$

for all $\vec{R}$ in the (direct) Bravais lattice. Note that although one could define a set of vectors $\vec{K}$ satisfying Eq. (2) for an arbitrary set of vectors $\vec{R}$, such a set of $\vec{K}$ is called a reciprocal lattice only if the set of vectors $\vec{R}$ is a Bravais lattice.

### 3.9.1 The Reciprocal Lattice is also a Bravais Lattice

That the reciprocal lattice is itself a Bravais lattice follows most simply from the following definition of a Bravais lattice:

A Bravais lattice is a discrete set of vectors not all in a plane, closed under vector addition and subtraction (i.e., the sum and difference of any two vectors in the set are also in the set).

## 3.10 THE EWALD CONSTRUCTION

If one wishes to search experimentally for Bragg peak one must relax the constraint of fixed **k**. either varying the magnitude of **k** (i.e., varying the wavelength of the incident beam) or varying its direction (in practice, varying the orientation of the crystal with respect to the incident direction).

A simple geometric construction due to Ewald is of great help in visualizing the various methods and in deducing the crystal structure from the peaks so observed. We draw in *k*-space a sphere centred on the tip of the incident wave vector **k** of radius *k* (so that is passes through the origin). Evidently, (see Fig. 3.23) there will be *some* wave vector **k′** satisfying the Laue condition if and only if some reciprocal lattice point (in addition to the origin) lies on the surface of the sphere, in which case there will be a Bragg reflection from the family of direct lattice planes perpendicular to that reciprocal lattice vector.

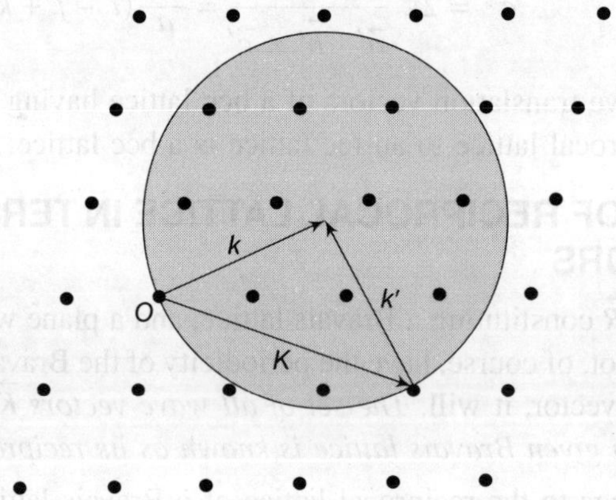


**Figure 3.23** The Ewald construction. Given the incident wave vector **k**, a sphere of radius **k** is drawn about the point **k.** Diffraction peaks corresponding to reciprocal lattice vectors **K** will be observed only if **K** gives a reciprocal lattice point on the surface of the sphere. Such a reciprocal lattice vector is indicated in the figure, together with the wave vector **k’** of the Bragg reflected ray

## 3.11 THE THREE EXPERIMENTAL METHODS OF X-RAY DIFFRACTION AND EWALD SPHERES

In general, a sphere in *k*-space with the origin on its surface will have no other reciprocal lattice points on its surface, and therefore the Ewald construction confirms our observation that for a general incident wave vector there will be no Bragg peaks. One can, however, ensure that some Bragg peaks will be produced by several techniques.

1. **The Laue Method.** One can continue to scatter from a single crystal of fixed orientation from a fixed incident direction ($\hat{\mathbf{n}}$) but can search for Bragg peaks by

using not a monochromatic X-ray beam, but one containing wavelengths from $\lambda_1$ up to $\lambda_0$. The Ewald sphere will then expand into the region contained between the two spheres determined by $\mathbf{k_0} = 2\pi\hat{\mathbf{n}}/\lambda_0$ and $\mathbf{k_1} = 2\pi\hat{\mathbf{n}}/\lambda_1$, and Bragg peaks will be observed corresponding to any reciprocal lattice vectors lying within this region (Fig. 3.24). By making the spread in wavelengths sufficiently large, one can be sure of finding some reciprocal lattice points within the region; whereas by keeping it from getting too large, one can avoid too many Bragg reflections, thereby keeping the picture fairly simple.

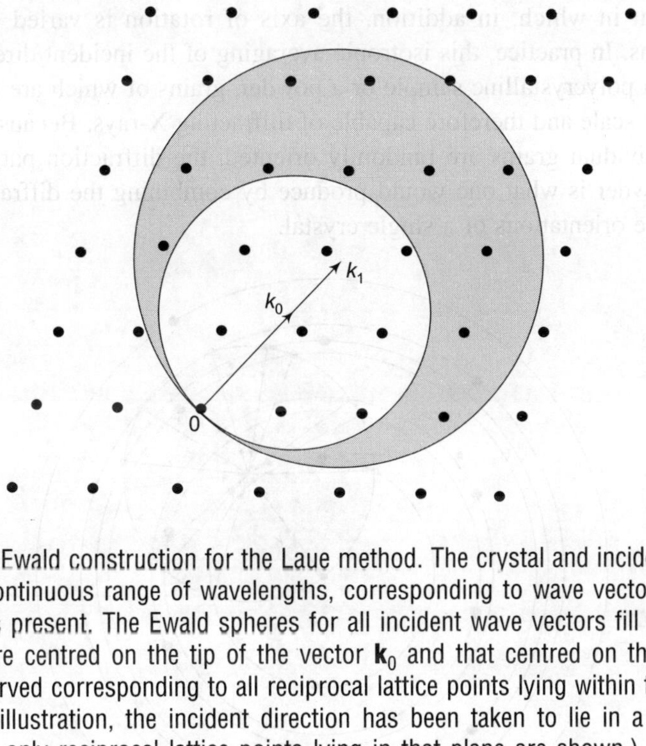


**Figure 3.24** The Ewald construction for the Laue method. The crystal and incident X-ray direction are fixed, and a continuous range of wavelengths, corresponding to wave vectors between $k_0$ and $k_1$ is magnitude, is present. The Ewald spheres for all incident wave vectors fill the shaded region between the sphere centred on the tip of the vector $\mathbf{k}_0$ and that centred on the tip of $\mathbf{k}_1$. Bragg peaks will be observed corresponding to all reciprocal lattice points lying within the shaded region. (For simplicity in illustration, the incident direction has been taken to lie in a lattice plane, and only reciprocal lattice points lying in that plane are shown.)

The Laue method is probably best suited for determining the orientation of a single crystal specimen whose structure is known, since, for example, if the incident direction lies along a symmetry axis of crystal, the pattern of spots produced by the Bragg reflected ray will have the same symmetry. Since solid state physicists generally do study substances of known crystal structure, the Laue method is probably the one of greatest practical interest.

2. **The Rotating Crystal Method.** This method uses monochromatic X-rays, but allows the angle of incidence to vary. In practice the direction of the X-ray beam is kept fixed, and the orientation of the crystal varied instead. In the rotating crystal method the crystal is rotated about some fixed axis, and all Bragg peaks that occur during the

rotation are recorded on a film. As the crystal rotates, the reciprocal lattice it determines will rotate by the same amount about the same axis. Thus, the Ewald sphere (which is determined by the fixed incident wave vector **k**) is fixed in *k*-space, while the entire reciprocal lattice rotates about the axis of rotation of the crystal. During this rotation each reciprocal lattice point traverses a circle about the rotation axis, and a Bragg reflection occurs whenever this circle intersects the Ewald sphere. This is illustrated in Fig. 3.25 for a particularly simple geometry.

3. **The Powder or Debye-Scherrer Method.** This is equivalent to a rotating crystal experiment in which, in addition, the axis of rotation is varied over all possible orientations. In practice, this isotropic averaging of the incident direction is achieved by using a polycrystalline sample or a powder, grains of which are still enormous on the atomic scale and therefore capable of diffracting X-rays. Because the crystal axes of the individual grains are randomly oriented, the diffraction pattern produced by such a powder is what one would produce by combining the diffraction patterns for all possible orientations of a single crystal.

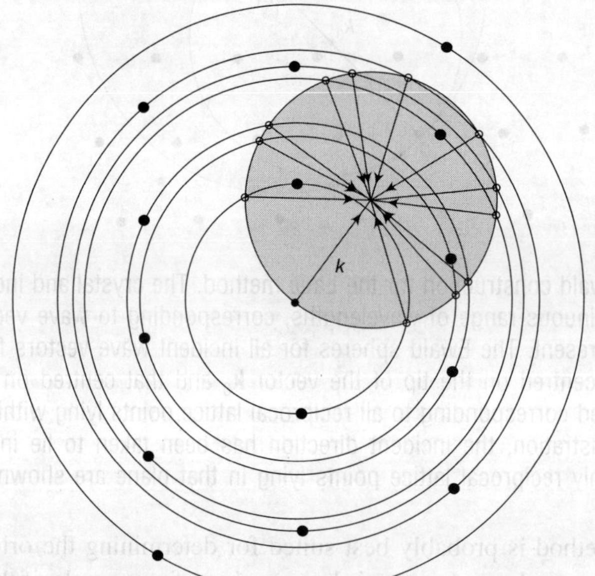


**Figure 3.25** The Ewald construction for the rotating-crystal method. For simplicity a case is shown in which the incident wave vector lies in a lattice plane, and the axis of rotation is perpendicular to that plane. The concentric circles are the orbits swept out under the rotation by the reciprocal lattice vectors lying in the plane perpendicular to the axis containing **k.** Each intersection of such a circle with the Ewald sphere gives the wave vector of a Bragg reflected ray. (Additional Bragg reflected wave vectors associated with reciprocal lattice vectors in other planes are not shown.)

The Bragg reflections are now determined by fixing the incident **k** vector, and with it the Ewald sphere, and allowing the reciprocal lattice to rotate through all possible angles about the origin, so that each reciprocal lattice vector **K** generates a sphere of radius $K$ about the origin. Such a sphere will intersect the Ewald sphere in a circle (Fig. 3.26a) provided that $K$ is less than $2k$. The vector joining any point on such a circle with the tip of the incident vector **k** is a wave vector **k′**, for which scattered radiation will be observed. Thus, each reciprocal lattice vector of length less than $2k$ generates a cone of scattered radiation at an angle $\phi$ to the forward direction, where (Fig. 3.26)

$$K = 2k \sin \frac{1}{2} \phi.$$

$$\sin \frac{1}{2} \phi$$

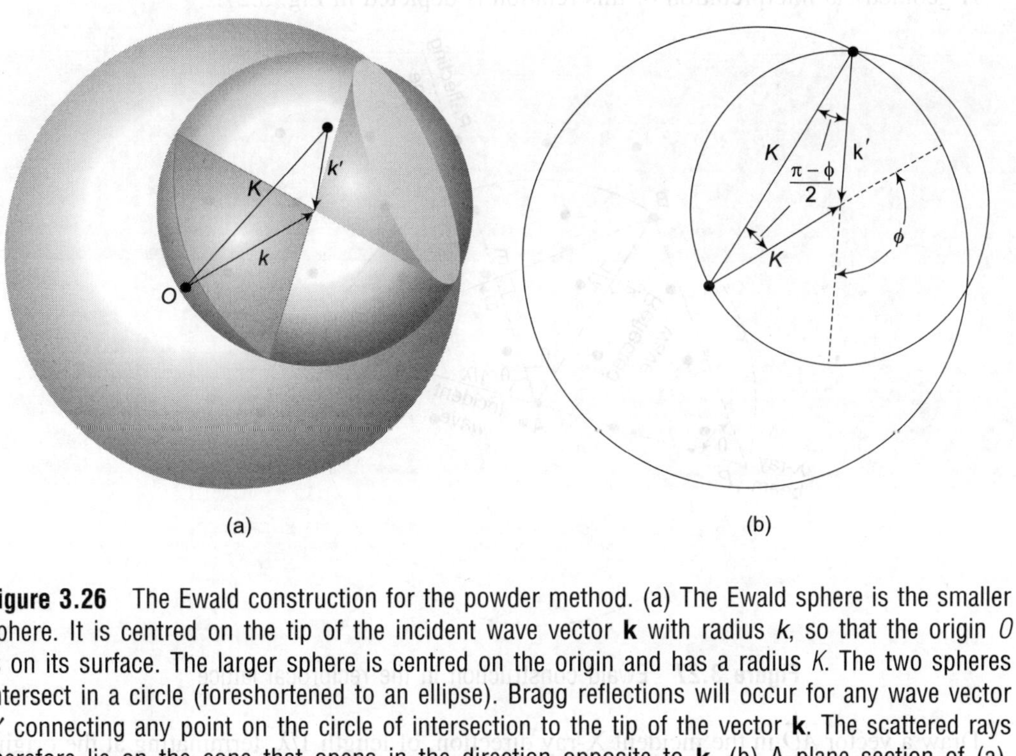


**Figure 3.26** The Ewald construction for the powder method. (a) The Ewald sphere is the smaller sphere. It is centred on the tip of the incident wave vector **k** with radius *k*, so that the origin *O* is on its surface. The larger sphere is centred on the origin and has a radius *K*. The two spheres intersect in a circle (foreshortened to an ellipse). Bragg reflections will occur for any wave vector **k′** connecting any point on the circle of intersection to the tip of the vector **k**. The scattered rays therefore lie on the cone that opens in the direction opposite to **k**. (b) A plane section of (a), containing the incident wave vector. The triangle is isosceles, and thus $K = 2k$

By measuring the angle $\phi$ at which Bragg reflections are observed, one therefore knows the lengths of all reciprocal lattice vectors shorter than $2k$. Armed with this information, some facts about the macroscopic crystal symmetry, and the fact that the reciprocal lattice is a Bravais lattice, one can usually construct the reciprocal lattice itself.

## 3.12 THE BRAGG'S DIFFRACTION CONDITION AND EWALD CONSTRUCTION

We shall subsequently see that the Bragg diffraction condition may be expressed as a relation between vector in the reciprocal lattice.

Bragg's equation is

$$\sin \theta_{hkl} = \frac{\lambda}{2\, d_{hkl}} \qquad ...(1)$$

(the order of reflection is already included). Rewriting,

$$2 \sin \theta_{hkl} = \frac{(1/d_{hkl})}{1/\lambda} \qquad ...(2)$$

A geometrical interpretation of this relation is depicted in Fig. 3.27.

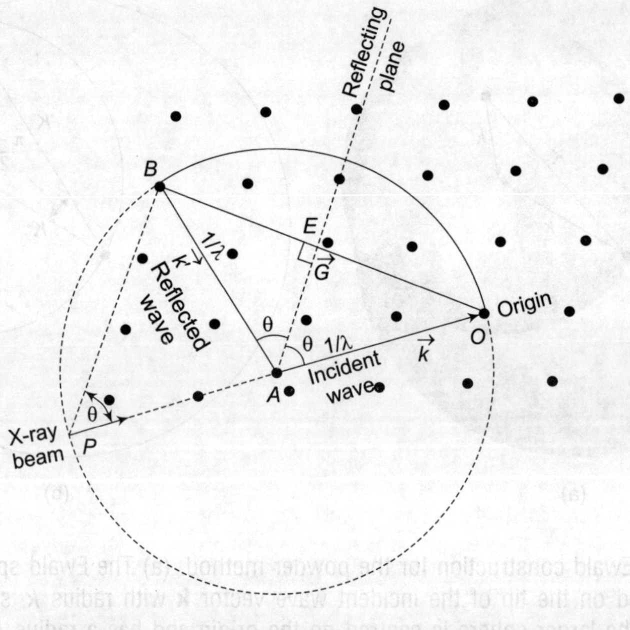


**Figure 3.27** Ewald construction in the reciprocal lattice

Draw a vector $\overrightarrow{AO}$ in the incident X-ray direction, of length $1/\lambda$, terminating at the origin, $O$. Construct a circle of radius $1/\lambda$ with centre at $A$.

Two possibilities arise:

(i) The circle does not pass through any reciprocal lattice point. This means that a particular wavelength in question would not be diffracted by that crystal in the orientation. Further, if $|\overrightarrow{OA}| < \left(\frac{1}{2a}\right)$, the circle would not pass through any point, showing that X-ray diffraction cannot occur if $\lambda > 2a$. It is to be noted that longer the vector $\overrightarrow{OA}$

(i.e., shorter the $\lambda$), the greater is the likelihood of circle intersecting a point and hence of diffraction.

(ii) The circle passes through any point $B$ of the reciprocal lattice. Join $A$ and $O$ to $B$. $\overrightarrow{OB}$ is a reciprocal lattice vector ($= \vec{G}$) and is normal to some set of lattice plane, e.g., $AE$. Hence, $OB = |\vec{G}| = \left(\frac{1}{d}\right)$, where $d$ is interplanar spacing.

Let $\vec{k} = \overrightarrow{AO}$

and $\vec{k}' = \overrightarrow{AB}$

be the incident wave vector and reflected wave vector, respectively. Thus,

$$\vec{k}' = \vec{k} + \vec{G} \quad \text{...(3)}$$

This shows that scattering changes only direction of $\vec{k}$ and scattered wave differs from the incident wave by a reciprocal lattice vector $\vec{G}$. For diffraction, it is necessary that

$$|\overrightarrow{AB}| = |\vec{k}'| = |\vec{k}| \quad \text{...(4)}$$

$$k'^2 = (k + G)^2 = k^2$$

or
$$2\,\vec{k}\cdot\vec{G} + G^2 = 0 \quad \text{...(5)}$$

$$\vec{k}\cdot\vec{G} + \frac{G^2}{2} = 0$$

or
$$\left(\vec{k} + \frac{\vec{G}}{2}\right).\,\vec{G} = 0$$

This is Bragg's law in vector form. The scalar from can be obtained by noting that

$$\overrightarrow{AE} = \left(\vec{k} + \frac{\vec{G}}{2}\right) \quad \text{...(6)}$$

is perpendicular to $OB$

$$\therefore \quad OB = 2OE = 2\sin\theta\left(\frac{1}{\lambda}\right) \quad \text{...(7)}$$

Also,
$$OB = \frac{1}{d}$$

$$\therefore \quad \frac{1}{d} = \left(\frac{2\sin\theta}{\lambda}\right) \quad \text{...(8)}$$

or
$$2d\sin\theta = \lambda$$

This shows that Bragg's equation has a simple geometrical significance in the reciprocal lattice space.

## 3.13 BRILLOUIN ZONES

The Wigner–Seitz cells of the reciprocal lattice are called *Brillouin zones*. These are constructed in the same way as the Wigner–Seitz cells in the real lattice, i.e., with origin of the reciprocal lattice (usually denoted by Γ) on a reciprocal lattice point, the Wigner–Seitz cell (i.e., Brillouin zone) is the polyhedron bounded by the set of planes

$$2\,\vec{k}\cdot\vec{g} + g^2 = 0$$

bisecting the joins of the origin ($\vec{k} = 0$) to the nearest reciprocal lattice points ($\vec{k} = \vec{g}$). Any wave whose wave vector $\vec{k}$ from the origin terminates on a surface of the Brillouin zone will be diffracted by the crystal. Energy distribution of electron in a solid is determined from its band structure. The function $E_n(\vec{k})$ is a many-valued function, periodic in $k$-space (i.e., reciprocal space). Then, Brillouin zone is the volume of periodicity of $E_n(\vec{k})$. From the symmetry properties of the Brillouin zone, certain points and lines within it and on its surface are particularly important. A knowledge of the band structure at such points and along such lines is often helpful for a discussion of the characteristics of a solid. We will now discuss Brillouin zones for one-dimensional monoatomic lattice, two-dimensional square lattice, a fcc lattice and a bcc lattice.

### 3.13.1 One-dimensional Monatomic Lattice

In case of a one-dimensional monatomic lattice, a line representing the value of wave vector $\vec{k}$ is divided up (by energy discontinuities) into segments of length $\pi/a$ (where $a$ is the lattice spacing).

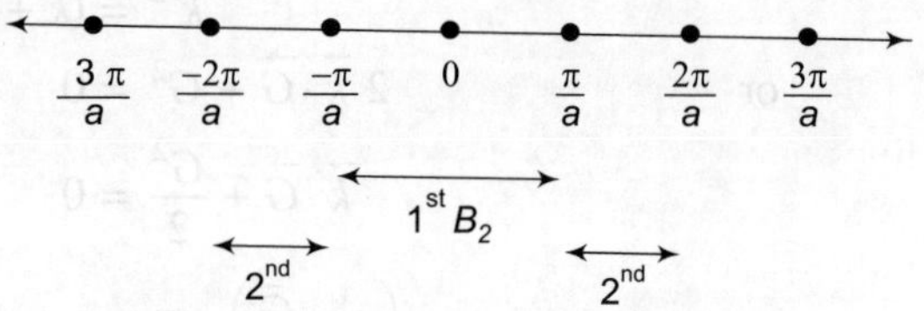


**Figure 3.28** Brillouin zones for a one-dimensional monatomic lattice

The segment $-\pi/a < k < \pi/a$ is the first Brillouin zone. The two segments $-\frac{2\pi}{a} < k < \frac{\pi}{a}$ and $\frac{\pi}{a} < k < \frac{2\pi}{a}$ from the second Brillouin zone (i.e., it consists, of two parts) (see Fig. 3.28)

$$2\,\vec{k}\cdot\vec{g} + g^2 = 0 \qquad ...(1)$$

In one dimension, the reciprocal lattice vector $\vec{g}$ and wave vector $\vec{k}$ may be represented by

$$\vec{g} = \left(\frac{2\pi}{a}\right)\hat{i}\,n_x; \quad \vec{k} = \hat{i}\,k_x \qquad ...(2)$$

From Eqs. (1) and (2), we get

$$|k_x| = \left(\frac{\pi}{a}\right) n_x \qquad ...(3)$$

For different values of $n_x = \pm 1, \pm 2, \ldots$

We obtain various zone boundaries.

### 3.13.2 Two-dimensional Square Lattice

In this case (Fig. 3.29), the reciprocal lattice vector and wave vector are given, respectively, by

$$\vec{g} = \frac{2\pi}{a}(\hat{i}\,n_x + \hat{j}\,n_y) \qquad ...(4)$$

and
$$\vec{k} = \hat{i}\,k_x + \hat{j}\,k_y \qquad ...(5)$$

Substituting these values in the equation

$$2\,\vec{k}\cdot\vec{g} + g^2 = 0$$

we get
$$2(\hat{i}\,k_x + \hat{j}\;k_y)\cdot\left[\frac{2\pi}{a}(\hat{i}\,n_x + \hat{j}\,n_y)\right] + \frac{4\pi^2}{a^2}(\hat{i}\,n_x + \hat{j}\,n_y)^2 = 0$$

or
$$k_x n_x + k_y n_y = -\frac{\pi}{a}(n_x^2 + n_y^2)$$

For $n_x = \pm 1$ and $n_y = 0$, we have

$$\pm k_x = -\frac{\pi}{a} \text{ or } k_x = \pm\frac{\pi}{a} \qquad ...(6)$$

These are boundaries of the first zone. Similary, for $n_x = 0$ and $n_y = \pm 1$ we have

$$k_y = \pm\frac{\pi}{a} \qquad ...(7)$$

These are also boundaries of the first zone. Thus, $k_x = \pm\pi/a$ and $k_y = \pm\pi/a$ are the four lines as boundary of the first zone. Now, for $n_x = \pm 1$ and $n_y = \pm 1$ we have (see Fig. 3.29)

$$\pm k_x \pm k_y = \frac{2\pi}{a} \qquad ...(8)$$

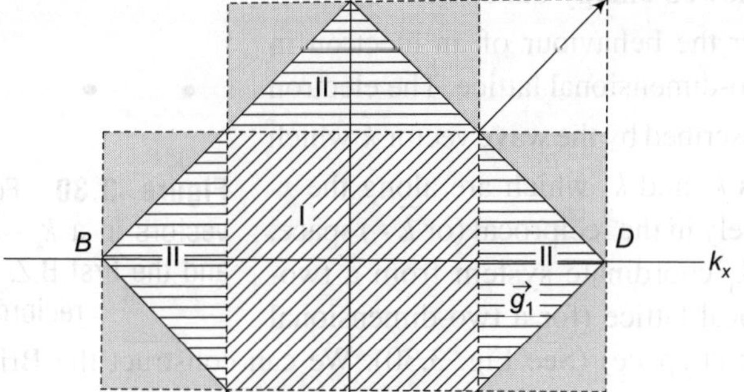


**Figure 3.29** First, second and third Brillouin zones (marked I, II and III, respectively) for a two-dimensional square lattice

These represent four lines, viz.,

$$\left.\begin{aligned} k_x + k_y &= \frac{2\pi}{a} \quad \text{(line } AD\text{)} \\ k_x + k_y &= -\frac{2\pi}{a} \quad \text{(line } BC\text{)} \\ -k_x + k_y &= \frac{2\pi}{a} \quad \text{(line } AB\text{)} \\ k_x - k_y &= \frac{2\pi}{a} \quad \text{(line } CD\text{)} \end{aligned}\right\} \quad ...(9)$$

The area enclosed by these four lines minus the first zone is the second Brillouin zone. Similar treatment may be done for third and other higher zones.

## 3.14 CONSTRUCTION OF BRILLOUIN ZONES FOR A TWO-DIMENSIONAL SQUARE LATTICE

We will see in the chapter on 'Band Theory' that when we consider motion of an electron in a periodic potential of the lattice and plot the energy $E$ vs wave vector $\vec{k}$, we obtain allowed and forbidden energy bands, i.e., allowed bands interrupted by Brillouin zones, that is, when we consider the reciprocal lattice (i.e., wave vector space or $\vec{k}$ - space), the region in $\vec{k}$-space between $-\pi/a$ and $+\pi/a$ is the first B.Z. The region between $\pi/a$ and $2\pi/a$ and also between $-\pi/a$ and $-2\pi/a$ is the second BZ the 1st BZ and 2nd BZ correspond respectively to the first allowed energy band and second allowed energy band.

We now consider the behaviour of an electron in the potential of a two-dimensional lattice. The electron movement can be described by the wave vector $\vec{k}$ which has the components $k_x$ and $k_y$ which are along the $x$- and $y$-axes respectively in the reciprocal (or $\vec{k}$ –) space. Points in the $k_x - k_y$ coordinate system from a two-dimensional reciprocal lattice (for a two-dimensional square lattice in direct space) (See Fig. 3.30). We can construct the Brillouin zones in this case as the following.

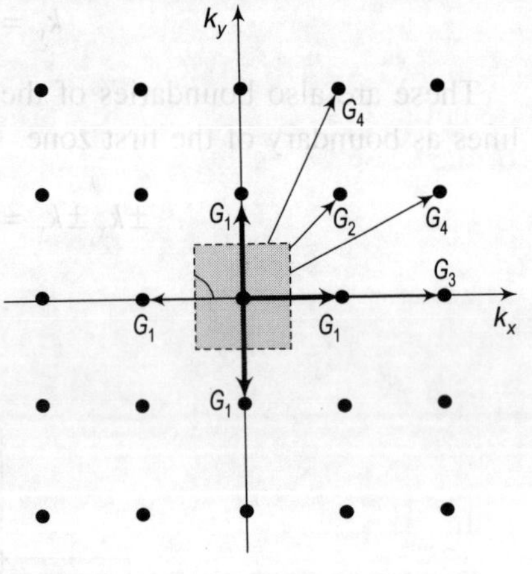


**Figure 3.30** Four shortest lattice vectors in a $k_x$ - $k_y$ coordinate system and the first B.Z. in a two-dimensional reciprocal lattice

For the first zone, one constructs the perpendicular bisectors on the shortest reciprocal lattice vectors, $G_1$. The area which is enclosed by these four "Bragg planes" is the first Brillouin zone. For other higher zones, the bisectors of the next shortest reciprocal lattice vectors are constructed. It is essential that for the zones of higher order, the extended limiting lines of the zones of lower order have to be used as additional limiting lines. The first four Brillouin zones we get are shown in Fig. 3.31. Note that all the zones have the same area.

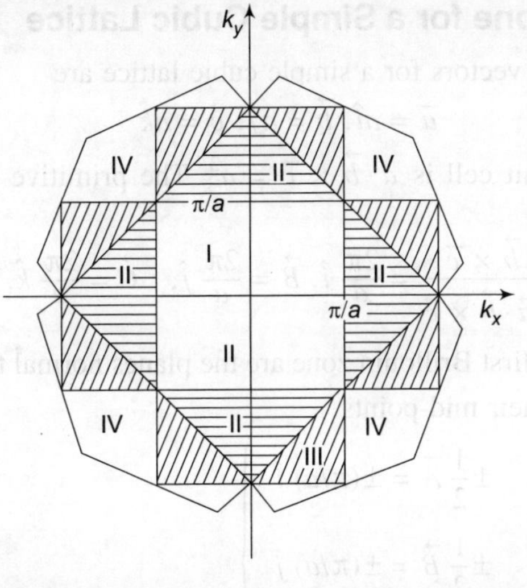


**Figure 3.31** The first four Brillouin zones of a two-dimensional reciprocal lattice

In the chapter on Band Theory, we will see (when we consider a reduced zone scheme), we do not need to plot $E$ vs $k$ curves for all Brillouin zones, the relevant information is, because of the $2\pi/a$ periodicity, already contained in the first Brillouin zone.

## 3.15 THE SIGNIFICANCE OF THE BRILLOUIN ZONES

The Brillouin zones are useful if one wants to calculate the behaviour of an electron which may travel in a specific direction in reciprocal space. For example, if in a two-dimensional lattice, an electron travels at 45° to the $k_x$-axis, the boundary of the Brillouin zone is reached for

$$k_{crit} = \frac{\pi}{a}\sqrt{2} \quad ...(1)$$

In contrast to this, the boundary of a BZ is reached at

$$k_{crit} = \frac{\pi}{a} \quad ...(2)$$

when an electron moves parallel to the $k_x$- or $k_y$-axes. The largest energies which electrons can assume in these two examples are therefore

$$E_{max} = \frac{\pi^2\hbar^2}{a^2 m} \left(\because \; E = \frac{\hbar^2 k^2}{2m}\right) \quad ...(3)$$

(using eqn (1)).

(at 45°, i.e., in the (110) direction) and

$$E_{max} = \frac{1}{2}\left(\frac{\pi^2\hbar^2}{a^2 m}\right) \quad ...(4)$$

along the $k_x$-axis in the (100) direction. Once this maximum energy has been reached, the electron waves form standing waves (or equivalently, the electrons are reflected back into the Brillounin zone.

### 3.15.1 Brillouin Zone for a Simple Cubic Lattice

The primitive translation vectors for a simple cubic lattice are

$$\vec{a} = a\hat{i},\ \vec{b} = a\hat{j},\ \vec{c} = a\hat{k} \qquad ...(1)$$

The volume of the unit cell is $\vec{a} \cdot \vec{b} \times \vec{c} = a^3$. The primitive translation vectors of the reciprocal lattice are

$$\vec{A} = 2\pi \frac{\vec{b} \times \vec{c}}{\vec{a} \cdot \vec{b} \times \vec{c}} = \frac{2\pi}{a}\hat{i},\ \vec{B} = \frac{2\pi}{a}\hat{j},\quad \vec{C} = \frac{2\pi}{a}\hat{k},\ (\because \hat{j} \times \hat{k} = i, \text{ etc.}) \qquad ...(2)$$

The boundaries of the first Brillouin zone are the planes normal to the six reciprocal lattice vectors $\pm\vec{A}$, $\pm\vec{B}$, $\pm\vec{C}$ at their mid-points

$$\left.\begin{aligned} \pm\frac{1}{2}\vec{A} &= \pm(\pi/a)\,\hat{i} \\ \pm\frac{1}{2}\vec{B} &= \pm(\pi/a)\,\hat{j} \\ \pm\frac{1}{2}\vec{C} &= \pm(\pi/a)\,\hat{k} \end{aligned}\right\} \qquad ...(3)$$

These six planes bound a cube of edge $2\pi/a$ and of volume $(2\pi/a)^3$ [see Fig. 3.32 (a)]. Figure 3.32 shows first Brillouin zones of the three important cubic lattices. The names of certain points of the Brillouin zone, such as Γ, W, L and X (e.g., for the fcc) are commonly used to refer to the origin (Γ), corners (W), centre of hexagonal faces (L) and centre of square faces (X) of the Brillouin zone. Such points are of high symmetry and play an important role in the computation of the electronic energy band structure and the vibrational spectrum of the solid (Fig. 3.32a).

### 3.15.2 Brillouin Zone of a fcc Lattice

The primitive translation vectors of a fcc lattice are

$$\vec{a} = \frac{a}{2}(\hat{i} + \hat{j}):\ b = \frac{a}{2}(\hat{j} + \hat{k}):\ \vec{c} = \frac{a}{2}(\hat{k} + \hat{i}) \qquad ...(4)$$

The primitive translation vectors of its reciprocal lattice are

$$\vec{a}^* = \frac{2\pi}{a}(\hat{i} + \hat{j} - \hat{k});\quad \vec{b}^* = \frac{2\pi}{a}(-\hat{i} + \hat{j} - \hat{k});\quad \vec{c}^* \frac{2\pi}{a}(\hat{i} - \hat{j} + \hat{k}) \qquad ...(5)$$

The *G*-type reciprocal lattice vector is

$$\begin{aligned} \vec{g} &= h\vec{a}^*\, k\vec{b}^* + l\vec{c}^* \\ &= \frac{2\pi}{a}[(h - k + l)\hat{i} + (h + k - l)\hat{j} + (-h + k + l)\hat{k}] \end{aligned} \qquad ...(6)$$

The shortest non-zero $\vec{g}'s$ are the following eight vectors

$$\frac{2\pi}{a}(\pm\hat{i} \pm\hat{j} \pm\hat{k}) \qquad ...(7)$$

The boundaries of the first Brillouin zone are determined by the normal bisector planes to the above eight vectors. However, the corners of the octahedron obtained in this manner are truncated by the planes that are normal bisectors to the following six reciprocal lattice vectors:

$$\frac{2\pi}{a}(\pm 2\hat{i});\quad \frac{2\pi}{a}(\pm 2\hat{j});\ \frac{2\pi}{a}(\pm 2\hat{k}) \qquad ...(8)$$

Hence, the first Brillouin zone has the shape of the *truncated octahedron* as shown in Fig. 3.32 (b). This is also one of the primitive cells of a bcc lattice (since it is in the reciprocal space and reciprocal of an fcc lattice is a bcc lattice).

### 3.15.3 Brillouin Zone of a bcc Lattice

The primitive translation vectors of a bcc lattice are

$$a = \frac{a}{2}(\hat{i} + \hat{j} - \hat{k});\ \vec{h} = \frac{a}{2}(-\hat{i} + \hat{j} + \hat{k});\ \vec{c} = \frac{a}{2}(\hat{i} - \hat{j} + \hat{k}) \qquad ...(9)$$

The primitive translation vectors of its reciprocal lattice are

$$\vec{a}^* = \frac{2\pi}{a}(\hat{i} + \hat{j});\quad \vec{b}^* = \frac{2\pi}{a}(\hat{j} + \hat{k});\quad \vec{c}^* = \frac{2\pi}{a}(\hat{k} - \hat{i}) \qquad ...(10)$$

The $\vec{G}$-type reciprocal lattice vector is

$$\vec{g} = h\vec{a}^* + k\vec{b}^* + l\vec{c}^*$$

$$= \frac{2\pi}{a}[(h + l)\,\hat{i} + (h + k)\,\hat{j} + (k + l)\,\hat{k}] \qquad ...(11)$$

The shortest non-zero $g$'s are the following 12 vectors:

$$\left(\frac{2\pi}{a}\right)(+\hat{i} + \hat{j});\quad \frac{2\pi}{u}(\pm\hat{j} \pm\hat{k});\quad \frac{2\pi}{a}(\pm\hat{k} \pm\hat{i}) \qquad ...(12)$$

The first Brillouin zone of the bcc lattice is then the region enclosed by the normal bisector planes to these 12 vectors. This zone has the shape as in Fig. 3.32 (c) and is called *rhombic dodecahedron* (a regular 12-faced solid).

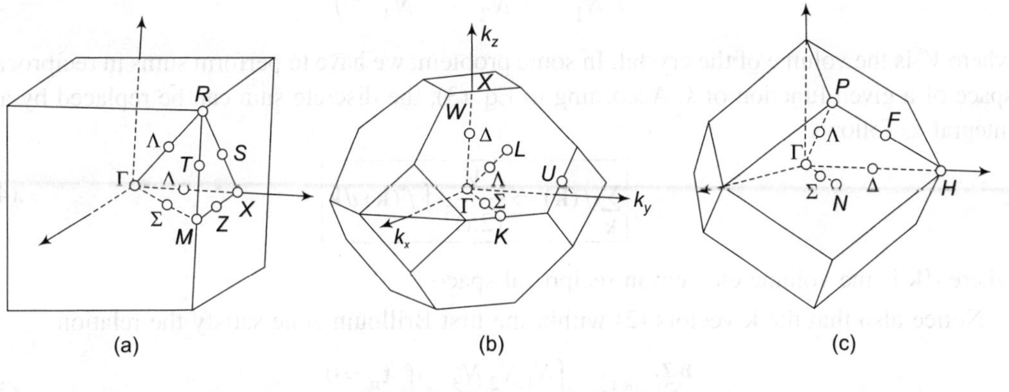


**Figure 3.32** Brillouin zones of: (a) simple cubic, (b) fcc and (c) bcc lattices. Also shown are the points of high symmetry

The names of certain points of the Brillouin zone, such as Γ, W, L And X for the fcc are commonly used by experts to refer to the origin (Γ), corners (W), centre of hexagonal faces (L) and centre of square faces (X) of the Brillouin zone. Such points are of high symmetry and play an important role in the computation of the electronic energy band structure and the vibration spectrum of the solid.

## 3.16 CYCLIC BOUNDARY CONDITIONS

In the previous sections we have considered ideal infinite crystal. Suppose now to have a microscopic (but finite) crystal, of volume $V$, in the form of a parallelepiped with edges $N_1\mathbf{t}_1$, $N_2\mathbf{t}_2$ and $N_3\mathbf{t}_3$, where $N_1$, $N_2$, $N_3$ are large (but finite) integer numbers. The Born-von Karman cyclic boundary conditions on the electronic wave functions require

$$\psi(\mathbf{r}) = \psi(\mathbf{r} + N_1\mathbf{t}_1) = \psi(r + N_2\mathbf{t}_2) = \psi(r + N_3\mathbf{t}_3)$$

Since $\psi(\mathbf{r})$ must be a Bloch function of vector $\mathbf{k}$, we have

$$e^{i\mathbf{k}\cdot N_1\mathbf{t}_1} = e^{i\mathbf{k}\cdot N_2\mathbf{t}_2} = e^{i\mathbf{k}\cdot N_3\mathbf{t}_3} = 1 \qquad ...(1)$$

The possible $\mathbf{k}$ vectors, compatible with cyclic boundary conditions (1), are thus

$$\mathbf{k} = \frac{m_1}{N_1}\mathbf{g}_1 + \frac{m_2}{N_2}\mathbf{g}_2 + \frac{m_3}{N_3}\mathbf{g}_3 \qquad ...(2)$$

with $m_i$ integer number.

The allowed $\mathbf{k}$ vectors (2) within the primitive unit cell of edges $\mathbf{g}_1$, $\mathbf{g}_2$, $\mathbf{g}_3$ are obtained choosing $0 \le m_i < N_i$ $(i = 1, 2, 3)$; their number is $N = N_1 \cdot N_2 \cdot N_3$. The first Brillouin zone has the same volume as the primitive unit cell in the reciprocal lattice. *Thus, the number of allowed* $\mathbf{k}$ *vectors in the first Brillouin zone equals the number of primitive unit cells of the crystal.*

The density of allowed $\mathbf{k}$ vectors (2) in the reciprocal space is uniform and is given by

$$W(k) = \frac{1}{\frac{1}{N_1}\mathbf{g}_1 \cdot \left(\frac{1}{N_2}\mathbf{g}_2 \times \frac{1}{N_3}\mathbf{g}_3\right)} = \frac{N_1N_2N_3}{\Omega_k} = \frac{V'}{(2\pi)^3}, \qquad ...(3)$$

where $V'$ is the volume of the crystal. In some problem, we have to perform sums in reciprocal space of a given function of $k$. According to Eq. (3), the discrete sum can be replaced by an integral as follows

$$\boxed{\sum_{\mathbf{k}} f(\mathbf{k}) \Rightarrow \frac{V}{(2\pi)^3}\int f(\mathbf{k})\,d\mathbf{k}} \qquad ...(4)$$

where $d\mathbf{k}$ is the volume element in reciprocal space.

Notice also that the $\mathbf{k}$ vectors (2) within the first Brillouin zone satisfy the relation

$$\sum_{\mathbf{k}}^{\text{B.Z.}} e^{i\mathbf{k}\cdot\mathbf{t}_n} = \begin{cases} N_1N_2N_3 & \text{if } \mathbf{t}_n = 0 \\ 0 & \text{if } \mathbf{t}_n \ne 0 \end{cases} \qquad ...(5)$$

where $\mathbf{t}_n$ is any translation of the crystal such that $0 \le n_i < N_i$ $(i = 1, 2, 3)$. The proof is simply based on the elementary algebraic property that the sum of the $N_i$- th roots of the unity equals zero. Similarly, we have

$$\sum_{\mathbf{t}_n}^{N_1 N_2 N_3} e^{i\mathbf{k}\cdot\mathbf{t}_n} = \begin{cases} N_1 N_2 N_3 & \text{if } k = \mathbf{g}_m \\ 0 & \text{if } k \ne \mathbf{g}_m \end{cases}, \quad \text{...(5a)}$$

where **k** is any vector (2) of the reciprocal space.

## 3.17 SPECIAL k POINTS FOR AVERAGING OVER THE BRILLOUIN ZONE

In several studies of crystal (for instance in the calculation of change density, total energies and other properties) we have to perform the average of appropriate **k**-dependent functions throughout the Brillouin zone. The average value of a function $F(\mathbf{k})$ is given by

$$F_0 = \frac{1}{N} \sum_{\mathbf{k}}^{\text{B.Z.}} F(\mathbf{k}), \quad \text{...(6)}$$

where $N$ is the (large) number of allowed **k** vectors in the Brillouin zone. The calculation of $F(\mathbf{k})$ at a given **k** vector may require laborious numerical work: then it would be highly desirable to find sets of "special **k** points" (composed by a reasonably modest number of **k** vectors) to determine efficiently the average, $F_0$.

To see the basic principle for this achievement we proceed as follows. Consider a function $F(\mathbf{k})$ with **k** defined in the first Brillouin zone, or, equivalently, consider a function $F(\mathbf{k})$ with the periodicity of the reciprocal lattice: its most general expansion is

$$F(\mathbf{k}) = F_0 + \sum_{\mathbf{t}_n \ne 0} F_n e^{i\mathbf{k}\cdot\mathbf{t}_n}, \quad \text{...(7)}$$

where $\mathbf{t}_n$ are translation vectors of the direct lattice, $F_n$ are the appropriate Fourier coefficients, and for convenience the average value $F_0$ has been separated from all the other terms. If we sum both members of Eq. (7) over all the allowed **k** vectors in the first Brillouin zone, use Eq. (5a) and divide by $N$, we recover expression (6). Thus, in general, only the sum over, a very large number of **k** vectors allows to obtain the exact average value of the function $F(\mathbf{k})$.

Consider now a function $G(\mathbf{k})$ represented in the fom

$$G(\mathbf{k}) = G_0 + \sum_{\mathbf{t}_i^{(I)}} G_i^{(I)} e^{i\mathbf{k}\cdot\mathbf{t}_i^{(I)}}, \quad \text{...(8a)}$$

where $\mathbf{t}_i^{(I)}$ denotes translation vectors belonging to the first shell of neighbours (all other Fourier coefficients are assumed to be zero). Let $\{\mathbf{k}_s\}$ indicate an appropriately selected set of $n_s$ points of the Brillouin zone, chosen in such a way that

$$\sum_{\mathbf{k}\in\{\mathbf{k}_s\}} e^{i\mathbf{k}\cdot\mathbf{t}_i^{(I)}} = 0 \quad \text{...(8b)}$$

for any translation vector of the first shell. Then the average value $G_0$ of the function $G(\mathbf{k})$ is rigorously expressed by

$$G_0 = \frac{1}{n_s} \sum_{\mathbf{k} \in \{\mathbf{k}_s\}} G(\mathbf{k}) \qquad \text{...(8c)}$$

As an example, consider a simple cubic lattice: the six translation vectors $\mathbf{t}_i^{(I)}$ of the first shell are: $(\pm a, 0{,}0)$, $(0, \pm a, 0)$, $(0, 0, \pm a)$. It is seen by inspection that the set of just two vectors $\mathbf{k}_1$ and $\mathbf{k}_2$, given by $\mathbf{k}_1 = -\mathbf{k}_2 = (\pi/a)$ (1/2, 1/2, 1/2), satisfies Eq. (8b) for any of the six vectors $\mathbf{t}_i^{(I)}$. The procedure can be extended to find systematically sets of wave vectors $\{\mathbf{k}_s\}$ for which $\Sigma_s \exp(i\mathbf{k}_s \cdot \mathbf{t}_n) = 0$ for all the translational vectors $\mathbf{t}_n$ belonging to the first shell, second shell, and so on, up to the outermost shell of interest, and we refer to the literature for details. When the number of shells to account for becomes very large, the set of special points approaches the uniform mesh of points (2) confined within the primitive reciprocal cell.

In several situations of physical interest, the Fourier coefficients in expression (7) depend only on the absolute value $|\mathbf{t}_n|$; in these situations, expansion (7) can be recast to the form

$$F(\mathbf{k}) = F_0 + F_1 \sum_{\mathbf{t}_i^{(I)}} e^{i\mathbf{k}\cdot\mathbf{t}_i^{(I)}} + F_2 \sum_{\mathbf{t}_i^{(II)}} e^{i\mathbf{k}\cdot\mathbf{t}_i^{(II)}} + \ldots \qquad \text{...(9)}$$

where $\mathbf{t}_i^{(I)}$ denotes the first shell of neighbour translation vectors, $\mathbf{t}_i^{(II)}$ denotes the second shell, etc. In this case, the determination of sets of special points is particularly instructive.

As an example let us consider a simple cubic lattice. The six translation vectors in the first shell are given by $(\pm a, 0, 0)$, $(0, \pm a, 0)$, $(0, 0, \pm a)$. The twelve vectors in the second shell are given by $(\pm a, \pm a, 0)$, $(\pm a, 0, \pm a)$, $(0, \pm a, \pm a)$. The eight vectors in the third shell are given by $(\pm a, \pm a, \pm a)$. For the simple cubic lattice, the expression (9) reads

$$\begin{aligned} F(k) &= F_0 + 2F_1(\cos k_x a + \cos k_y a + \cos k_z a) \\ &\quad + 4F_2(\cos k_x a \cos k_y a + \cos k_x a \cos k_z a + \cos k_y a \cos k_z a) \\ &\quad + 8F_3 \cos k_x a \cos k_y a \cos k_z a + \ldots . \end{aligned}$$

In the above expression, it is seen by inspection that a single point (the Baldereschi point) reduces to zero the contributions from the first, second and third shells: this mean value point is just $\mathbf{k} = (\pi/a)$ (1/2, 1/2, 1/2). Similar procedures can be performed with the other Bravais lattices. The mean value procedure, and the generated extensions, the particularly useful for instance in performing self-consistent calculations, keeping a relative manageable number of $k$ points into account.

## 3.18 SPECIAL POINT AND SPECIAL LINES IN THE FIRST BRILLOUIN ZONE OF A TWO-DIMENSIONAL SQUARE LATTICE

First, it is worthwhile to notice that the set of symmetry operations (e.g., mirror-reflection, rotary inversion axis. etc.). or in other words, the symmetry group in real space is also the set of symmetry operations in the reciprocal space, because

$$\vec{a}^* = 2\pi \frac{\vec{b} \times \vec{c}}{V_p}$$

$$\vec{b}^* = 2\pi \frac{\vec{c} \times \vec{a}}{V_p}$$

$$\vec{c}^* = 2\pi \frac{\vec{a} \times \vec{b}}{V_p}$$

where $V_p = \vec{a} \cdot (\vec{b} \times \vec{c})$, the volume of the primitive cell in direct space. Further, a direct lattice vector

$$\vec{r} = n_1 \vec{a} + n_2 \vec{b} + n_3 \vec{c}$$

and a reciprocal lattice vector

$$\vec{K} = h\vec{a}^* + k\vec{b}^* + l\vec{c}^*$$

of $\vec{k}$ space satisfy

$$e^{i\vec{K}\cdot\vec{r}} = 1$$

Any reciprocal lattice vector $\vec{K}$ and any vector in $k$-space can be written in terms of the vectors $\vec{a}^*$, $\vec{b}^*$ and $\vec{c}^*$ as

$$\vec{K} = h\vec{a}^* + k\vec{b}^* + l\vec{c}^*$$

$$\vec{K} = k_1\vec{a}^* + k_2\vec{b}^* + k_3\vec{c}^*$$

where $h$, $k$ and $l$ are integers and $k_i$ is any number. Thus, exp $(i\,\vec{k} \cdot \vec{r})$ has some complex value but when $\vec{k}$ is a reciprocal lattice vector, this expression equals unity.

Now, consider a two-dimensional square space lattice. The reciprocal lattice and Brillouin zones are shown in Fig. 3.33. For simplicity, assume that this lattice has point symmetry $C_{4v}$ (4 mm) with the eight symmetry operations $\{E, C_4, C_2, C_4^3, 2\sigma, 2\sigma_d\}$. The $2\sigma$ refers to mirror planes perpendicular to the $x$- and $y$-axes: $2\sigma_d$ are the two diagonal mirror planes and the first four symbols refer to rotations about the axis out of the paper by 0, $\pi/2$, $\pi$ and $3\pi/2$. These symmetry operations are applied to the crystal in real space. However, since the reciprocal lattice is mathematically "tied"to the real space lattice, when a symmetry operation $R$ operates on the real space lattice its effect is felt on the reciprocal lattice. Mathematically, it can be shown that $(R\,\vec{k})\cdot\vec{r} = \vec{k}\cdot(R^{-1}\,\vec{r})$, which means that if a symmetry operation acts on real space, its inverse operation acts on reciprocal space (in a group, the inverse of every symmetry operation is also a member of the group). In summary, the eight symmetry operations of this point group in real space are also the symmetry operations of the first Brillouin zone in reciprocal space.

Further, there are other related symmetry considerations. Consider $\vec{k}$ along $\vec{k}_x$ as shown in Fig. 3.32 (a) with the end point of the $k$-vector labeled $\Delta$. By applying all eight symmetry operations to this $k$-vector, we obtain the four $k$-values shown in this figure by the vectors with solid lines. Notice that under two of the symmetry operations, $E$ and $\sigma^y$, the original $k$-vector labelled $\Delta$ transforms into itself, i.e., $R\vec{k} = \vec{k}$ when $R = E, \sigma^y$. These two symmetry operations form the group of Δ. We now discuss other special points in the first Brillouin zone.

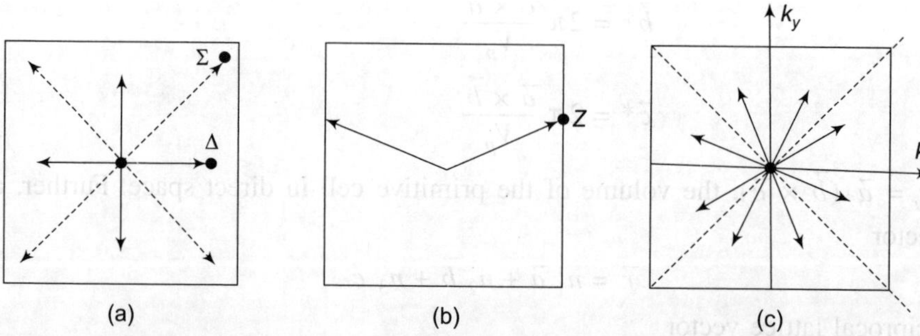


**Figure 3.33** The first Brillouin Zone of a two-dimensional square lattice showing special *k*-values: (a) the Δ- and Σ-lines, (b) the Z-line (c) a general point

Consider $\vec{k}$ along the diagonal $\vec{k}_x + \vec{k}_y$ with the end point of the vector labelled Σ in Fig. 3.33 (a). Again apply all right symmetry operations to the Σ-vector and the four *k*-values shown in the figure with dashed vectors are obtained. Again we notice that this *k*-value labeled Σ transforms into itself under the symmetry operations, *E* and $\sigma^d$, i.e., $R\vec{k} = k$ when $R = E$, $\sigma^d$. These two symmetry operations are called the *group of* Σ.

Finally, consider $\vec{k}$ to be along the line perpendicular to $\vec{k}_x$ at $|\vec{k}_x| = \pi/a$, labeled *Z* in Fig. 3.33 (b), i.e., the right boundary of the Brillouin zone. Under the symmetry operation $\sigma^x \vec{k}$ the right boundary becomes the left boundary. Modulo a reciprocal lattice vector $\vec{K} = \vec{a}^*$, both of these boundaries are equivalent in that any wave-function at these two $\vec{k}$ values are the same because the wave vector differs only by a reciprocal lattice vector. This equivalence is expressed mathematically as

$$R\vec{k} = \vec{k} + \vec{K} \quad \text{...(i)}$$

where *R* is a symmetry operation of the crystal $R\vec{k}$ and $\vec{k}$ are said to be equivalent (or invariant) if they differ by a reciprocal lattice vector. Or we say $R\vec{k} = \vec{k}$ modulo a reciprocal lattice vector. Thus, with this definition the *k*-value labelled *Z* transforms into itself (or its equivalent) under the symmetry operations *E* and $\sigma^x$. These two symmetry operations are called the *group of Z*.

All three of these lines, Δ, Σ and Z are called special lines because more than one symmetry operation of the group transforms them into an equivalent *k*-vector ($K \neq 0$ in Eq. (i)) or themselves ($K = 0$) Fig. 3.33 shows the first Brillouin zone with these three lines labelled Δ, Σ and *Z*.

Consider a $\vec{k}$ close to the Δ-line but with a small value of $\vec{k}_y$ so that this $\vec{k}$ is not on any special line. Then under all eight symmetry operations of this group, other non-equivalent *k*-values are generated as shown in Fig. 3.33 (c). Such a *k*-value is called a *general point.*

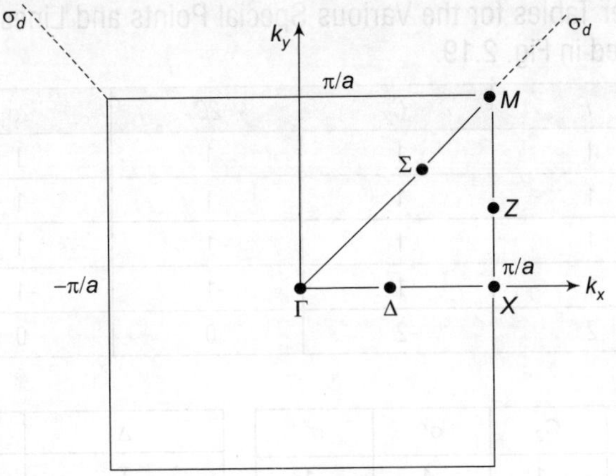


**Figure 3.34** The first Brillouin Zone of a two-dimensional square lattice showing the special points and lines. Writing the positions with a *z*-coordinate always equal to zero in this case, $\vec{a}^* = [1, 0, 0]\ 2\pi/a$ and $\vec{b}^* = [0, 1, 0]\ 2\pi/a$. Thus, $\vec{k}_x$ and $\vec{k}_y$ are in the [1, 0, 0] and [0, 1, 0] directions respectively. The *X* and *M* points are at [1, 0, 0] $\pi/a$ and [1, 1, 0] $\pi/a$ respectively

There are also special points (Fig. 3.34) labelled, Γ, *X* and *M*. For example, the X-point under $\sigma^y$ into itself ($\sigma^y X) = X$) and under $C_2$ and $\sigma^x$ transforms into itself modulo a reciprocal lattice vector, i.e., $C_2X = X + (2\pi/a)$. These four symmetry operations ($E$, $C_2$, $\sigma^y$, $\sigma^x$) are called the group of *X*. The Γ, $\vec{k} = 0$ and *M*, $\vec{k} = (a^*/2) + (\vec{b}^*/2)$ points transform into themselves (modulo a $\vec{K}$ for the *M*-point) under all symmetry operations of the group. Thus, the *group of* Γ and *group of M* consist of all eight symmetry operations.

*Importance of all the above consideration*: Along the special lines and at the special points there are some degeneracies that do not split no matter how strong the crystalline potential energy may be (an *essential degeneracy*). For the general points, there is no degeneracy. The general shape of the energy contours $E(\vec{k})$ are to a large extent determined by the splittings, degeneracies and shapes of the bands at the special points and lines. Also note (Fig. 3.33) that the bands need be calculated only in 1/8 of *k*-space, the rest can be determined by symmetry.

*The group of k and their irreducible representations:* There is a group, in the mathematical sense, of symmetry operations that transforms each of the special points or special line into themselves. This group, called the group of $\vec{k}$, is the group of Δ or the group of Σ and so on. Associated with every group is a set of irreducible representations that precisely describes how all functions (defined in ordinary space) transform under all of the symmetry operations of the group. (For example, there are eight symmetry operations in $C_{4v}$ and five irreducible representations.) The different irreducible representations are labelled (i.e., given a symbol to distinguish one from the other). The conventional notation for the irreducible representations in the group of Δ or group of Σ and so on, are subscripts, i.e., $\Delta_1$ and $\Delta_2$ or $\Sigma_1$ and $\Sigma_2$ and so on as in Table 3.2 which show the appropriate character tables.

**Table 3.2** Character Tables for the Various Special Points and Lines of the Square Lattice as Defined in Fig. 2.19.

| $\Gamma$, $M$ | $E$ | $C_2$ | $2C_4$ | $2\sigma_v$ | $2\sigma_d$ |
|---|---|---|---|---|---|
| $\Gamma_1$, $M_1$ | 1 | 1 | 1 | 1 | 1 |
| $\Gamma_2$, $M_2$ | 1 | 1 | 1 | –1 | –1 |
| $\Gamma_3$, $M_3$ | 1 | 1 | –1 | 1 | –1 |
| $\Gamma_4$, $M_4$ | 1 | 1 | –1 | –1 | 1 |
| $\Gamma_5$, $M_5$ | 2 | –2 | 0 | 0 | 0 |

| $X$ | $E$ | $C_2$ | $\sigma^y$ | $\sigma^x$ |
|---|---|---|---|---|
| $X_1$ | 1 | 1 | 1 | 1 |
| $X_2$ | 1 | 1 | –1 | –1 |
| $X_3$ | 1 | –1 | 1 | –1 |
| $X_4$ | 1 | –1 | –1 | 1 |

| $\Delta$ | $E$ | $\sigma^y$ |
|---|---|---|
| $\Sigma$ | $E$ | $\sigma_d$ |
| $Z$ | $E$ | $\sigma^x$ |
| $\Delta_1$, $\Sigma_1$, $Z_1$ | 1 | 1 |
| $\Delta_2$, $\Sigma_2$, $Z_2$ | 1 | –1 |

Figure 3.35 shows $\vec{k}$-dependence of the energy in empty lattice approximation, i.e., using only free-electron Bloch functions but considering the correct symmetry behaviour at special points and lines. For convenience, we define $E_0 \equiv (\hbar^2/2m)\,(2\pi/a)^2$ and plot reduced energy i.e., divided by this quantity. (For $X$ point, energy = $E_0/4$).

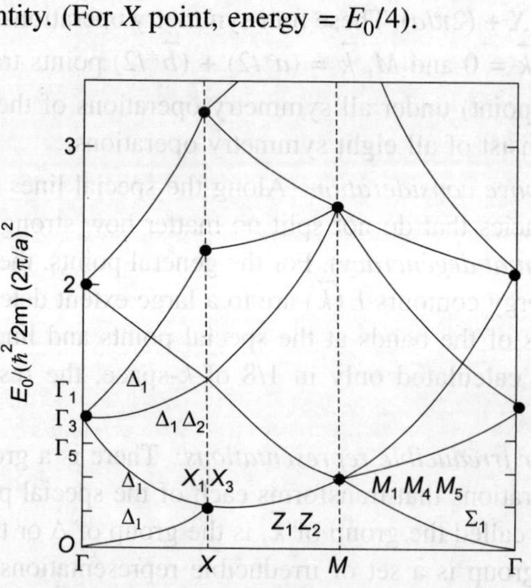


**Figure 3.35** $E$ vs $\vec{k}$ for the two-dimensional lattice. The energy is divided by $E_0 \equiv (\hbar^2/2m)\,(2\pi/a)^2$

## 3.19 THE CONCEPT OF BRILLOUIN ZONES FOR CUBIC AND HEXAGONAL LATTICES

The *first Brillouin zone* (or simply the *Brillouin zone*) of the reciprocal lattice has the same definition as the Wigner–Seitz cell in the direct lattice: it has *the property that any point of the cell is closer to the chosen lattice point* (say $\vec{g} \equiv 0$) *than to any other*. The first Brillouin zone can be obtained by bisecting with perpendicular planes nearest neighbours reciprocal lattice vectors, second nearest neighbours (and other orders of neighbours if necessary) and considering the smallest volume enclosed. Similarly, the second Brillouin zone is obtained continuing the bisecting operations and delimiting the second volume enclosed (with exclusion of the first zone), etc. The shape of the Brillouin zone is connected to the geometry of the direct Bravais lattice, irrespectively of the content of the basis. We illustrate the construction of the first Brillouin zone with some examples.

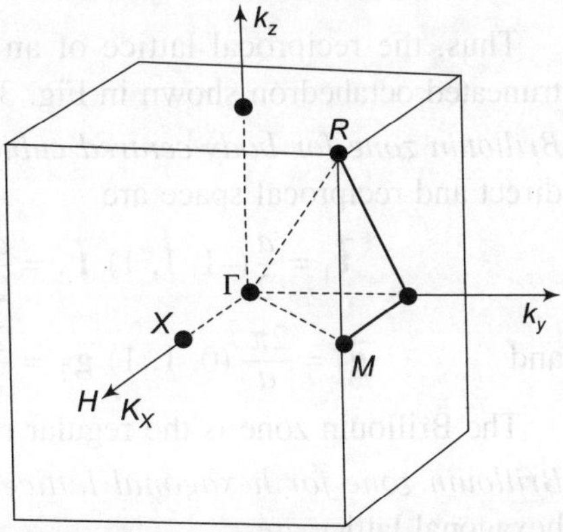


**Figure 3.36** Brillouin zone for the simple cubic lattice. Some high symmetry points are indicated: $\Gamma = 0$; $X = (2\pi/a)$ (1/2, 0, 0); $M = (2\pi/a)$ (1/2, 1/2, 0); $R = (2\pi/a)$ (1/2,1/2,1/2)

*Brillouin zone for the simple cubic lattice.* The fundamental vectors in direct space of a simple cubic lattice are

$$\mathbf{t}_1 = a(1, 0, 0)\ \mathbf{t}_2 = a(0, 1, 0)\ \mathbf{t}_3 = a(0, 0, 1). \quad ...(1a)$$

From Eq. (18), we have

$$\vec{g}_1 = (2\pi/a)(1, 0, 0)\ \vec{g}_2 = (2\pi/a)\ (0, 1, 0)\ \vec{g}_3 = (2\pi/a)\ (0, 0, 1). \quad ...(1b)$$

Thus, the reciprocal lattice is still a simple cube, with edge $2\pi/a$.

The first Brillouin zone is indicated in Fig. 3.36. Points of high symmetry in the cubic Brillouin zone are indicated by conventional letters: Γ denotes the origin of the Brillouin zone; *X* is the centre of a square face at the boundaries; *M* is the centre of a cube edge, and *R* is the vertex of the cube.

*Brillouin zone for the face-centred cubic lattice:* The fundamental vectors for an fcc Bravais lattice are

$$\vec{t}_1 = \frac{a}{2}(0, 1, 1)\ \vec{t}_2 = \frac{a}{2}(1, 0, 1)\ \vec{t}_3 = \frac{a}{2}(1, 1, 0). \quad ...(2a)$$

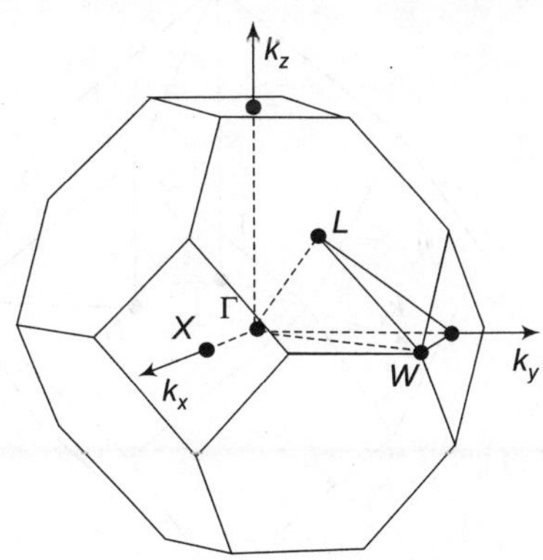


**Figure 3.37** Brillouin zone for the face-centred cubic lattice (truncated octahedron). Some high symmetry points are: $\Gamma = 0$; $X = (2\pi/a)$ (1, 0, 0); $L = (2\pi/a)$ (1/2, 1/2, 1/2); $W = (2\pi/a)$(1/2, 1, 0)

We have for the fundamental vectors of the reciprocal lattice

$$\vec{\mathbf{g}}_1 = \frac{2\pi}{a}(-1, 1, 1)\ \vec{\mathbf{g}}_2 = \frac{2\pi}{a}(1, -1, 1)\ \vec{\mathbf{g}}_3 = \frac{2\pi}{a}(1, 1, -1). \quad \text{...(2b)}$$

Thus, the reciprocal lattice of an fcc lattice is a bcc lattice. The Brillouin zone is the truncated octahedron shown in Fig. 3.37.

*Brillouin zone for body-centred cubic lattice:* The fundamental vectors for a bcc lattice in direct and reciprocal space are

$$\vec{\mathbf{t}}_1 = \frac{a}{2}(-1, 1, 1)\ \vec{\mathbf{t}}_2 = \frac{a}{2}(1, -1, 1)\ \vec{\mathbf{t}}_3 = \frac{a}{2}(1, 1, -1) \quad \text{...(3a)}$$

and

$$\vec{\mathbf{g}}_1 = \frac{2\pi}{a}(0, 1, 1)\ \vec{\mathbf{g}}_2 = \frac{2\pi}{a}(1, 0, 1)\ \vec{\mathbf{g}}_3 = \frac{2\pi}{a}(1, 1, 0) \quad \text{...(3b)}$$

The Brillouin zone is the regular rhombic dodecahedron shown in Fig. 3.38.

*Brillouin zone for hexagonal lattice*: The fundamental translation vectors of the Bravais hexagonal lattice are

$$\vec{\mathbf{t}}_1 = a\left(\frac{1}{2}, \frac{\sqrt{3}}{2}, 0\right)\ \vec{\mathbf{t}}_2 = a\left(\frac{1}{2}, \frac{\sqrt{3}}{2}, 0\right)\ \vec{\mathbf{t}}_3 = c(0, 0, 1), \quad \text{...(4a)}$$

and the fundamental vectors of the reciprocal lattice are

$$\vec{\mathbf{g}}_1 = \frac{2\pi}{a}\left(1, \frac{1}{\sqrt{3}}, 0\right)\ \vec{\mathbf{g}}_2 = \frac{2\pi}{a}\left(-1, \frac{1}{\sqrt{3}}, 0\right)\ \vec{\mathbf{g}}_3 = \frac{2\pi}{c}(0, 0, 1). \quad \text{...(4b)}$$

Thus, the reciprocal of the hexagonal lattice is still an hexagonal lattice. The Brillouin zone is shown in Fig. 3.39.

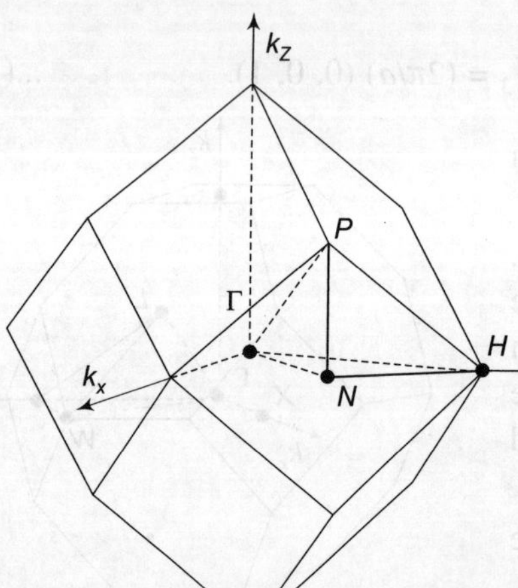


**Figure 3.38** Brillouin zone for the body-centred cubic lattice (rhombic dodecahedron). Some high symmetry points are also indicated: $\Gamma = 0$; $N = (2\pi/a)(1/2, 1/2, 0)$; $P = (2\pi/a)(1/2, 1/2, 1/2)$; $H = (2\pi/a)(0, 1, 0)$

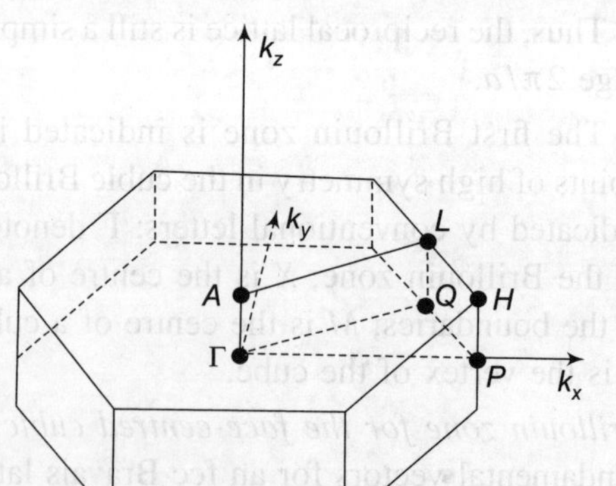


**Figure 3.39** Brillouin zone for the hexagonal Bravais lattice. Some high symmetry points are also indicated: $\Gamma = 0$; $P = (2\pi/a)(2/3, 0, 0)$; $Q = (\pi/a)(1, 1/\sqrt{3}, 0)$: $A = (\pi/c)(0, 0, 1)$

## 3.20 THE FIRST BRILLOUIN ZONE AND SYMMETRY POINT

The first Brillouin zone for the face-centred cubic, diamond, and zinc blende structures is the same (see Fig. 3.40). This is not surprising because all three structures can be formed from the same face-centred cubic lattice placing a different atomic basis into the lattice nodes. The electron energy is a function of $k_x$, $k_y$, and $k_z$ that varies within the first Brillouin zone. The four-dimensional plot, required to reproduce such a dependence, defies our imagination. The closest we can come to it is to reproduce three-dimensional surfaces of equal energy in the **k**-space. The alternative is to study different cross sections of this plot, varying only a certain component of $k$ (for example, $k_x$) and keeping other components ($k_y$ and $k_z$) constant. Fortunately, most important semiconductor properties are determined by regions in the first Brillouin zone corresponding to lines and points of symmetry. These symmetry points and lines of the Brillouin zone are marked on Fig. 3.40. A detailed study of the symmetry of electronic states in the zinc blende structure was done by Dresselhaus (1955) and Parmenter (1955). In silicon, germanium, diamond, and compound III-V and II-IV semiconductors, points Γ (in the centre of the Brillouin zone), $X$, and $L$, and direction Δ (see Table 3.3 and Fig. 3.40 are the most important regions. The dependencies of the electronic energy on the wave vector (energy bands) are usually calculated for the lines of symmetry in the first Brillouin zone.

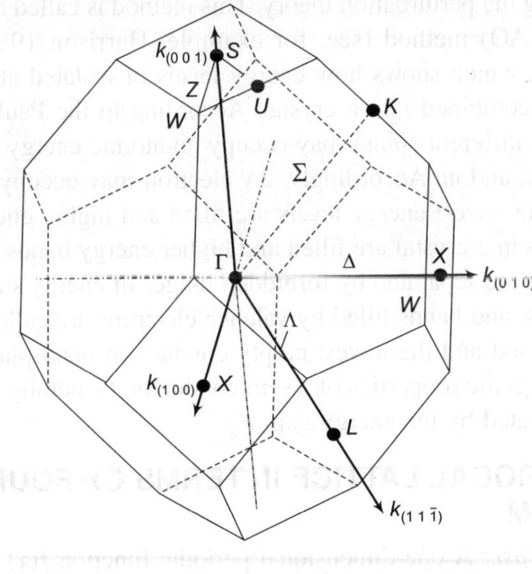


**Figure 3.40** The first Brillouin zone for the face-centred cubic, diamond, and zinc blende structures. Also shown are symmetry points and directions of the first Brillouin zone

Calculations of energy bands are based on the solution of the Schrödinger equation. Such calculations are difficult and computer-intensive. They are frequently done by starting from a known solution for a simpler system and using perturbation theory to introduce corrections for a given crystal. In particular, we can point out two complementary approaches. In the

**Table 3.3** Symmetry Points of the Brillouin Zone of the Zinc Blende Structure (See Fig. 3.25)

| Number of equiv. K vectors | Point | Symmetry operations | Point groups |
|---|---|---|---|
| 1 | $\Gamma(000)$ | $E, 8C_3, 3C_2, 6S_4, 6C_5$ | $T_d$ |
| 3 | $X\frac{2\pi}{a}(100)$ | $E, 3C_2, 2S_4, 2\sigma$ | $D_{3d}$ |
| 4 | $L\frac{2\pi}{a}\left(\frac{1}{2}, \frac{1}{2}, \frac{1}{2}\right)$ | $E, 2C_3, 3\sigma$ | $C_{3v}$ |
| 6 | $W\frac{2\pi}{a}\left(1\,0\,\frac{1}{2}\right)$ | $E, 2S_4, C_2$ | $S_4$ |
| 6 | $\Delta\,(K00)$ | $E, 2\sigma, C_2$ | $C_{2v}$ |
| 4 | $\Lambda\,(KKK)$ | $E, 2C_3, 3\sigma$ | $C_{3v}$ |
| 12 | $\Sigma(K0K)$ | $E, \sigma$ | $C_s$ |
| 12 | $K\frac{2\pi}{a}\left(\frac{3}{4}\,0\,\frac{3}{4}\right)$ | $E, C_2$ | $C_s$ |

first approach, an electron wave function of the crystal is represented as a linear combination of the atomic orbitals, and the crystal potential is considered a perturbation. The band structure is then determined using the perturbation theory. This method is called the linear combination of atomic orbitals (LCAO) method (see, for example, Harrison 1980). This approach is illustrated by Fig. 3.41, which shows how energy levels of isolated atoms split into energy bands when atoms are combined into a crystal. According to the Pauli exclusion principle, only two electron (with different spins) may occupy an atomic energy level with a given set of quantum number $n$, $l$, and $m$. Accordingly, $2N$ electron may occupy an energy band. Just as in an atom, where the lowest energy levels are filled and higher energy levels are empty, the lowest energy bands in a crystal are filled and higher energy bands are empty. The bands of allowed energy states are separated by forbidden ranges of energy states. Empty bands are called conduction bands, and bands filled by valence electrons are called valence bands. The highest filled valence band and the lowest empty conduction band play the most important role in determining electronic properties of a semiconductor. Generally, these conduction and valence bands are separated by the energy gap, $E_g$.

## 3.21 THE RECIPROCAL LATTICE IN TERMS OF FOURIER TRANSFORM

(a) *One-Dimensional Case:* A one-dimensional periodic function $f(x)$ with period $a$ so that $f(x + na) = f(x)$ for any integer $n$ can be written as a Fourier series

$$f(x) = \sum_{n=-\infty}^{\infty} F_n\, e^{i2\pi nx/a}. \qquad ...(1)$$

For sake of notational convenience, we define the set of points $G$ by $G_n = \frac{2\pi n}{a}$ for integers $n$. Then, we can write

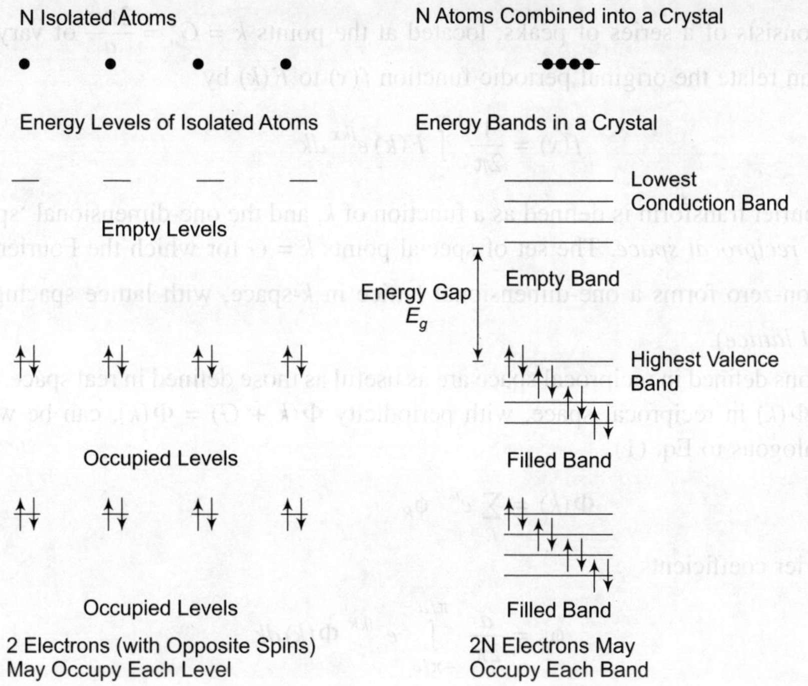


**Figure 3.41** Splitting of energy levels of isolated atoms into energy bands in a crystal. Arrows pointing up and down represent electrons with different values of spin ($S = \pm 1/2$)

$$f(x) = \sum_G F_G \, e^{iGx} \qquad ...(2)$$

The Fourier coefficients $F_G$ are given by

$$F_G = \frac{1}{a} \int_{-a/2}^{a/2} e^{-iGx} f(x)\, dx \qquad ...(3)$$

(This can be verified by multiplying both sides of Eq. 2 by $e^{-iGx}/a$ and integrating over one period of $f(x)$.

We can write Eq. (2) in integral form using the Dirac delta function $\delta(k)$, which is defined as

$$\delta(x) = 0 \quad \text{for} \quad x \neq 0$$

$$\int_{-\infty}^{\infty} \delta(x)\, dx = 1$$

Using a continuous variable $k$, we may define a new function $F(k)$, the Fourier transform of $f(x)$ by

$$F(k) = 2\pi \sum_G F_G \, \delta(k - G). \qquad ...(4)$$

$F(k)$ consists of a series of peaks, located at the points $k = G_n = \frac{2\pi n}{a}$ of varying height $F_G$. We can relate the original periodic function $f(x)$ to $F(k)$ by

$$f(x) = \frac{1}{2\pi} \int_{-\infty}^{\infty} F(k)\, e^{ikx}\, dk \qquad \text{...(5)}$$

The Fourier transform is defined as a function of $k$, and the one-dimensional 'space' $(\vec{k})$ is known as *reciprocal space*. The set of special points $k = G$ for which the Fourier transform $F(k)$ is non-zero forms a one-dimensional lattice in $k$-space, with lattice spacing $\frac{2\pi}{a}$, (the *reciprocal lattice*).

Functions defined in reciprocal space are as useful as those defined in real space. A periodic function $\Phi(k)$ in reciprocal space, with periodicity $\Phi(k + G) = \Phi(k)$, can be written as a series analogous to Eq. (1).

$$\Phi(k) = \sum_R e^{ikR}\, \phi_R \qquad \text{...(6)}$$

with Fourier coefficients

$$\phi_R = \frac{a}{2\pi} \int_{-\pi/a}^{\pi/a} e^{-ikR}\, \Phi(k)\, dk \qquad \text{...(7)}$$

where the points $R$ belong to the real space lattice $R_n = na$. Defining the real space function by

$$\phi(r) = \sum_R \phi_R\, \delta(r - R) \qquad \text{...(8)}$$

we can write the inverse Fourier transform as

$$\phi(k) = \int_{-\infty}^{\infty} \phi(r)\, e^{-ikr}\, dr \qquad \text{...(9)}$$

(b) *There-dimensional Case:* Two and three-dimensional Fourier transforms are simple extensions of the one-dimensional transform. In one dimension, we discussed a periodic function $f(x)$ with a period $a$, and found that the Fourier series involved a set of special points $G$ in reciprocal space given by $G_n = 2\pi n/a$. In two and three dimensions, for a function $f(\vec{r})$ that has the periodicity $\vec{R}$ of a real space lattice, the Fourier series involves a similar set of special points $\vec{G}$, which form a regular, periodic reaprocal lattice in $k$-space. In three dimensions, the reciprocal lattice is generated by three vectors $\vec{b}_1$, $\vec{b}_2$ and $\vec{b}_3$, so that any reciprocal lattice vector $\vec{G}$ can be written as a linear combination

$$\vec{G} = l\vec{b}_1 + m\vec{b}_2 + n\vec{b}_3 \qquad \text{...(10)}$$

for integers $l$, $m$ and $n$. The generating vectors $\vec{b}_1$, $\vec{b}_2$ and $\vec{b}_3$ are related to the real space generating vectors $\vec{a}_1$, $\vec{a}_2$, $\vec{a}_3$ through the relations

$$\vec{b}_1 = \frac{2\pi}{v_c}\, \vec{a}_2 \times \vec{a}_3;\ \vec{b}_2 = \frac{2\pi}{v_c}\, \vec{a}_3 \times \vec{a}_1;\ \vec{b}_3 = \frac{2\pi}{v_c}\, \vec{a}_1 \times a_2 \qquad ...(11)$$

where $v_c \cdot \vec{a}_1 \cdot (\vec{a}_2 \times \vec{a}_3)$ is the real space unit cell volume. As for the real space lattice, a unit cell for the reciprocal lattice is any volume which, when translated by the reciprocal lattice vector $\vec{G}$, fills reciprocal space without overlapping.

If now, we take a periodic real space function $f(\vec{r})$ with the periodicity of the lattice, the Fourier series can be written as

$$f(\vec{r}) = \sum_G F_G\, e^{i\vec{G}\cdot\vec{r}} \qquad ...(12)$$

with the vector $\vec{G}$ running over the reciprocal space lattice. The Fourier components $F_G$ are given by

$$F_G = \frac{1}{v_c} \int_{v_c} e^{-i\vec{G}\cdot\vec{r}} f(r)\, d\vec{r} \qquad ...(13)$$

where the integral is over a Wigner–Seitz unit cell volume $v_c$. We can then define the Fourier transform $F(\vec{k})$ as a sum of δ-functions

$$F(\vec{k}) = (2\pi)^3 \sum_{\vec{G}} F_{\vec{G}}\, \delta(\vec{k} - \vec{G}) \qquad ...(14)$$

The Fourier transform relation is

$$F(\vec{r}) = \frac{1}{(2\pi)^3} \int F(\vec{k})\, e^{i\vec{k}\cdot\vec{r}}\, d\vec{k} \qquad ...(15)$$

We see that just as for the one-dimensional case, the 3-dimensional Fourier transform of a periodic real space function is zero except for when the wave vector $\vec{k}$ is equal to a reciprocal space lattice vector, $\vec{G}$.

An analogous set of expressions exists for a function $\phi(\vec{k})$ that is periodic in $k$-space with the periodicity of the reciprocal lattice; we can write it as a Fourier series of real space terms as

$$\phi(\vec{k}) = \sum_{\vec{R}} \phi_{\vec{R}}\, e^{-i\vec{R}\cdot\vec{k}} \qquad ...(16)$$

with Fourier elements

$$\phi_{\vec{R}} = \frac{v_c}{(2\pi)^3} \int_{v_c} e^{i\vec{k}\cdot\vec{R}}\, \Phi(\vec{k})\, d\vec{k} \qquad ...(17)$$

where the integral is over the Brillouin zone defined about each reciprocal lattice point $\vec{G}$, in exactly the manner the Wigner–Seitz cell is defined in real space. The inverse Fourier transform is

$$F(\vec{k}) = \int f(\vec{r})\, e^{-i\vec{k}\cdot\vec{r}}\, d\vec{r} \qquad ...(18)$$

## 3.22 DIFFRACTION BY AN IDEAL CRYSTAL

If all atoms in a crystal were at their official lattice position, then all the measurable physical quantities pertaining to the crystal would be periodic functions. For example, the potential energy of an electron would satisfy the condition

$$V(r+1) = V(r) \qquad ...(1)$$

Let a fast electron beam be directed into the crystal, then due to scattering by this potential there will be transitions from one electron state to another. According to Born approximation of perturbation theory, the rate of transition between the initial state $\psi_k$ and the final state $\psi_k'$, is proportional to the square of the matrix element.

$$M_{k'k} = \int \psi_{k'}(r) \cdot V(r) \cdot \psi_k(r) \cdot dr \qquad ...(2)$$

Considering these to be plane wave sets

$$\psi_k = e^{ik\cdot r} \qquad ...(3)$$

Since the potential may be expanded in a Fourier series, therefore, we get

$$V(r) = \sum_G V_G \cdot e^{iG\cdot r} \qquad ...(4)$$

Thus, $$M_{k'k} = \int e^{-ikr} \sum_G \cdot V_G\, e^{iG\cdot r}\, e^{ik\cdot r} \cdot dr = \sum_G V_G \int e^{i(k+G-k')r} \cdot dr$$

$$= \begin{Bmatrix} V_G \text{ if } & k+G-k' = 0, \\ 0 & \text{otherwise} \end{Bmatrix} \qquad ...(5)$$

If the incident beam of electrons is monochromatic and well defined in direction, i.e., if $k$ is fixed, then the diffracted beams can be observed only in directions corresponding to wave vectors satisfying the condition

$$k' = k + G \qquad ...(6)$$

where $G$ is one of the reciprocal lattice vectors of the crystal. Considering that the energy of the diffracted beam is the same as that of the incident energy, i.e., the wave vectors have the same length, this will be the condition on the angle of scatter. If $2\theta$ is the angle between $k$ and $k'$, then we have

$$|G| = 2|k| \sin\theta \qquad ...(7)$$

In order to satisfy these geometrical conditions we construct the Ewald sphere, Fig. 3.42 (b), with radius $OP$ equal to the incident wave vector, $k$. If the origin of the reciprocal lattice is at $P$, then the wave vector $G$ must carry us to $Q$, where $OQ$ is the vector $k'$. Therefore, whenever the orientation of the crystal relative to the incident beam puts a point of the reciprocal lattice on the sphere, diffraction occurs. This can be expressed in another way. The length of reciprocal lattice vector is proportional to the reciprocal of the spacing, $d$, of the planes to which it is normal, i.e.,

$$|G| = \frac{2\pi n}{d} \qquad ...(8)$$

where $n$ is an integer. Thus, if $\lambda$ is the wavelength of the incident electrons, Eqs. (7) and (8) give

$$n\lambda = 2d \sin\theta \qquad ...(9)$$

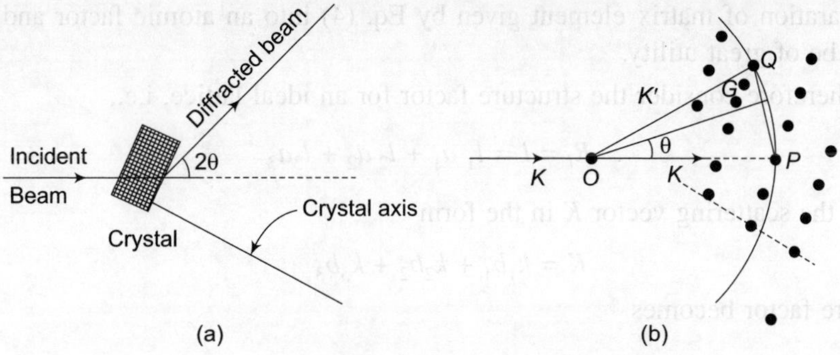


**Figure 3.42** (a) X-ray diffraction, (b) Ewald's construction

## 3.23 DIFFRACTION BY A CRYSTAL WITH LATTICE VIBRATIONS

Let us now consider a case, more general in nature, where the potential is not necessarily periodic. Suppose this results due to the superposition of atomic potentials

$$V(r) = \sum_{1} V_{\text{a}}(r - R_1) \quad ...(1)$$

where $R_1$ is the actual position of the atom at the lattice site 1, $V_a$ is the potential field of an individual atom. Let us assume a Bravais lattice, for the sake of simplicity of notation.

Matrix element for scattering from state $\psi_k$ to state $\psi_{k'}$, as in Eq. (1) may be calculated as

$$M_{k'k} = \int e^{-ik'\cdot r} \sum_{1} V_a(r - R_1)\, e^{ik\cdot r} \cdot dr = \sum_{1} \int e^{i(k-k')r}\ V_a(r - R_1) \cdot dr$$

$$= \sum_{1} e^{i(k-k')} \cdot R_1 \int e^{i(k-k')\cdot(r-R_1)} \cdot V_a(r - R_1) \cdot dr \quad ...(2)$$

Each of the integrals in the sum is nominally over all space or over the whole crystal. But $V_a(r)$ should be of very limited range in order that the result of integration can scarcely depend upon the centre, $R_1$.

Let us new introduce a **scattering vector, *K***,

$$K = k' - k \quad ...(3)$$

Then our matrix element becomes

$$M_{k'k} = V_a(K) \cdot \frac{1}{N} = \sum_{1} e^{ik\cdot R_1} \quad ...(4)$$

The Fourier transform of the atomic potential will be given by

$$V_a(K) \equiv \frac{1}{v_c} \int V_a(r)\, e^{-ik\cdot r} \cdot dr \quad ...(5)$$

The separation of matrix element given by Eq. (4) into an atomic factor and a structure factor will be of great utility.

Let us therefore consider the structure factor for an ideal lattice, i.e.,

$$R_l = l = l_1 a_1 + l_2 a_2 + l_3 a_3 \quad ...(6)$$

Writing the scattering vector $K$ in the form

$$K = k_1 b_1 + k_2 b_2 + k_3 b_3 \quad ...(7)$$

the structure factor becomes

$$\sum_l e^{-ik\cdot l} = \sum_{l_1, l_2, l_3} e^{-i(k_1 l_1 + k_2 l_2 + k_3 l_3)}$$

$$= \left(\sum_{0\le l_1 < L_1} e^{-ik_1 l_1}\right)\left(\sum_{0\le l_2 < L_2} e^{-ik_2 l_2}\right)\left(\sum_{0\le l_3 < L_3} e^{-ik_3 l_3}\right)$$

$$= \left(\frac{1-e^{-ik_1 L_1}}{1-e^{-ik_1}}\right)\left(\frac{1-e^{-ik_2 L_2}}{1-e^{-ik_2}}\right)\left(\frac{1-e^{-ik_3 L_3}}{1-e^{-ik_3}}\right) \quad ...(8)$$

where $L_1$, $L_2$, $L_3$ are the numbers of unit cells along the sides of the crystal block.

Each factor in Eq. (8) is only of the order of unity, and varies with $K$. The averaging process over a range of values of $K$ will give zero, except when it includes zero of all the denominators. In that case

$$e^{-ik_1} = e^{-ik_2} = e^{-ik_2} = 1 \quad ...(9)$$

this is true only if each of $k_1$, $k_2$, $k_3$ is an integral multiple of $2\pi$. But this is the condition that $K$ should coincide with one of the reciprocal lattice vectors, $G$.

When $K$ is actually equal to a reciprocal lattice vector, both numerator and denominator of each term in Eq. (8) vanish. We then return to the original sum, and note that each term being of the form $e^{-iG\cdot l}$ is exactly unity and the sum is just $N$. We then get an important rule given by

$$\frac{1}{N}\sum_l e^{ik\cdot l} = \delta_{K.G.} \quad ...(10)$$

where the $\delta$-symbol vanishes unless $k$ is one of the vectors of the set $G$.

Substituting Eq. (10) into Eq. (4), we get

$$M_{k'k} = V_a(G)\, d_{K'} - k\cdot G \quad ...(11)$$

Let us now suppose that the lattice modes of our crystal are excited. Each atom is displaced from its ideal lattice site.

$$R_1 = l + u_1 = l + \sum_{q>} (U_q e^{iql} + U_q^x e^{-iql}) \quad ...(12)$$

where $U_q$ is the vector amplitude of the mode of the mode of wave vectors $q$. The sum is over only half the range of $q$, so that we may display the complex conjugate term $U_q^x = U_{-q}$ making the displacements real.

Now substituting Eq. (12) in the structure factor of Eq. (4), we get

$$\frac{1}{N}\sum_l e^{-ik\cdot R_1} = \frac{1}{N}\sum_l \exp.\left[-ik\cdot\left\{1+\sum_{q>}(U_q e^{iql} + U_q^x e^{-iql})\right\}\right]$$

$$= \frac{1}{N}\sum_l \exp.\left[e^{-iql}\prod_{q>}\exp.\{-ik\cdot(U_q e^{iql} + U_q^x e^{-iql})\}\right] \quad ...(13)$$

Let us now expand the exponential function whose argument is proportional to $U_q$, supposedly a small displacement.

$$\exp.\{-ik\cdot(U_q e^{iql} + U_q^x e^{-iql})\} = 1 - ik\cdot(U_q e^{iql} + U_q^x e^{-iql}) = |K.u_q|^2 \quad ...(14)$$

If we substitute Eq. (14) in Eq. (13), it is obvious that, when the product is multiplied, there will be a term like (Eq. 8), giving the diffraction pattern of the ideal crystal structure, followed by a sum of all the terms that are linear in $K.\ U_q$, i.e., terms as given by Eq. (10).

$$\frac{1}{N}\sum_l e^{-ik\cdot l}\sum_q(-iK\cdot U_q)\cdot e^{iql} = \frac{1}{N}\sum_q(-iK\cdot U_q)\sum_l e^{i(q-K)}$$

$$= \sum_q(-iK\cdot U_q)\,\delta_{k-qG} \quad ...(15)$$

Let us now consider two cases in order to understand this result.

In Fig. (3.43 a), $K$ lies in the Brillouin zone, that is, we look for the scattered electron, or X-ray or neutron in the region of states where $k'$ does not differ by more than half a reciprocal lattice vector from the initial state $k$. Conventionally, the wave vectors $q$ are restricted to a Brillouin zone, so the only term left in Eq. (15) is the one for which $q = k$. The matrix element for the transition is

$$M_{k\cdot k} = i\{(k' - k)\cdot U_q\}\ V_a(k' - k) \quad ...(16)$$

where

$$q = k' - k \quad ...(17)$$

If $K$ does not lie in the first, Brillouin zone, Fig. (2.40. b), there is still one term left in the sum given by (Eq. 15). In other words, there is still a value of $q$ such that $k - q = G$, i.e.,

$$q = k' - k - G \quad ...(18)$$

Therefore, the Brillouin zone, being a unit cell of the reciprocal lattice, can be repeated so as to cover exactly the whole of the space of the vector $K$; any value of $K$ can be constructed uniquely one of a value of $G$, a transitional vector of the reciprocal lattice and a value of $q$, a vector restricted to lie in one unit cell of that lattice.

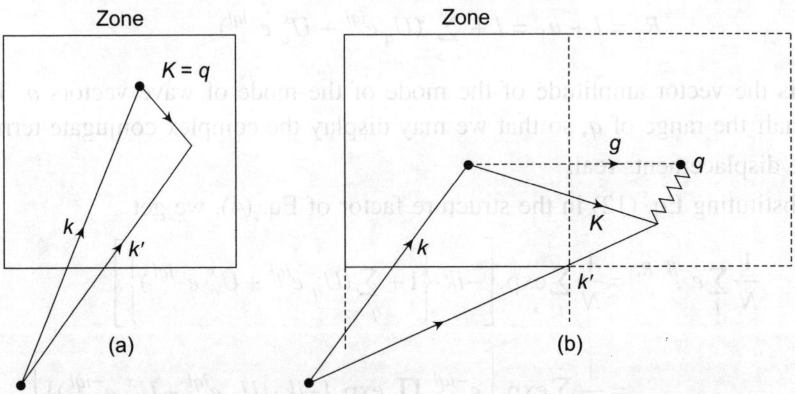


**Figure 3.43** (a) Normal process; (b) Umklapp process

Therefore, in both the cases, the expression (16) holds with the value of $q$ given by (17) or (18). If $G = O$ is included in the complete set of reciprocal lattice vectors, then the relation (18) is always valid. Some scattering will be there, irrespective of the direction we choose. Its amplitude will depend on the amplitude of lattice vibrations that happen to have the correct wave vector to satisfy these conditions.

## 3.24 THE DEBYE-WALLER FACTOR

The terms in Eq. (15), corresponding to scattering processes involving one phonon, are not all that may occur in the structure factor (13). The reason being that the product of factors like (14) and also higher terms in the expansion of exponential contribute to various products like $U_q \cdot e^{iql}$. For each such factor, the corresponding phonon has either been created or destroyed and hence these terms refer to multiphonon processes.

Generally, as we go to higher order, these processes fall off rapidly in rate and do not contribute heavily to the background of inelastic diffraction. However, there is an important class of terms, arising from the squared term $-|K.U_q|^2$ in (14), which do not average to a small contribution. If we look at Eq. (13) and Eq. (14), we find that the square of the matrix element for both elastic and inelastic scattering should have been multiplied by

$$e^{-2w} = \prod_q \{1 - K \cdot U_q^2\}$$

This is called the **Debye-Waller factor.**

### 3.24.1 The Debye-Waller Factor

If there is a plane wave of X-rays (or neutrons) impinging upon an electron or nucleus which is immobile, located at the origin, then this particle acts as an elastic scattering centre producing at large a spherical outgoing wave. The wave function of the incident wave is

$$\psi(\vec{r}) = e^{ik_i \cdot \vec{r}} + f(\theta, \phi)\,\frac{e^{ik_f r}}{r} \qquad ...(1)$$

where the initial wave vector $\vec{k}_i$, determines the direction of propagation and the energy quantum of the undisturbed plane wave, $f(\theta, \phi)$ is the so called *scattering-amplitude* coincident with the final wave vector $\vec{k}_f$ and $|\vec{k}_f| = |\vec{k}_i|$, because of the elastic character of the scattering. If the scattering centre is located at the point $\vec{R}$, Eq. (1) is modified as

$$\psi(\vec{r}) = e^{i\vec{k}_i \cdot \vec{r}} + e^{i\vec{k} \cdot \vec{R}} f(\theta, \phi) \frac{e^{ik_f r}}{r} \qquad ...(2)$$

where $\vec{k} = \vec{k}_i - \vec{k}_f$ is the momentum transfer (divided by $\hbar$) to the scattering centre.

If there are many identical particles (acting as scattering centres) located at different points $\vec{R}$, the total scattering amplitude is

$$f_t = f \sum_R e^{i\vec{k} \cdot \vec{R}} \qquad ...(3)$$

But in reality, phonons reduce the strength of the diffracted beam in a crystalline solid by a factor known as the Debye–Waller factor.

To simplify the derivation, we shall assume that there is only one atom per primitive cell.

In the presence of phonons, the position of an atom is not simply $\vec{R}$ (as in Eq. (3)), but $\vec{R} + \vec{U}(\vec{R}, t)$, where $\vec{u}(\vec{R}, t)$ implies

$$f_t = f \sum_{\vec{R}} e^{i\vec{k} \cdot \vec{R}} e^{ik \cdot \vec{u}} \qquad ...(4)$$

Expanding the last exponential factor in powers of the small quantity $\vec{k} \cdot \vec{u}$, we have

$$e^{ik \cdot \vec{u}} = 1 + i\,\vec{k} \cdot \vec{u} - \frac{1}{2}(k \cdot u)^2 + .... \qquad ...(5)$$

For simplicity if we assume that only one phonon, that is characterized by phonon momentum $\vec{q}$ and label $s$ is present, then quantum mechanically

$$\vec{u}(\vec{R}, t) = \vec{b}_{qs}\, e^{i(\vec{q} \cdot \vec{R} - w_{qs} t)}\, a_{qs} + b^*_{qs}\, e^{-i(\vec{q} \cdot \vec{R} - w_{qs} t)}\, a^\dagger_{qs} \qquad ...(6)$$

where $$\vec{b}_{qs} = \frac{1}{\sqrt{N_a}} \left( \frac{\hbar}{2m w_{qs}} \right)^{1/2} \vec{w}_{qs} \qquad ...(7)$$

$N_a$ is the number of atoms, and $w_q$ is the eigenfrequency of $s$ atoms.

$|w_{qs}| = 1$. Combining Eq. (4) to Eq. (6), we get

$$f_t = f \sum_{\vec{R}} e^{i\vec{k} \cdot \vec{R}} \left\{ \left[ 1 + i\vec{k} \cdot \vec{b}_{qs}\, e^{i\phi}\, a_{qs} + i^{\vec{k} \cdot b^*_{qs}}\, e^{-i\phi}\, a^\dagger_{qs} \right] \right.$$

$$\left. - \frac{1}{2}(\vec{k} \cdot \vec{b}_{qs}\, e^{i\phi}\, a_{qs} + \vec{k} \cdot b^*_{qs}\, e^{-i\phi}\, a^\dagger_{qs})^2 + .... \right\} \qquad ...(8)$$

where $$\phi = \vec{q} \cdot \vec{r} - w_{qs} t.$$

The term $(\vec{k}\cdot\vec{u})^2$ is equal to

$$(\vec{k}\cdot\vec{u})^2 = (\vec{k}\cdot\vec{b}_{qs})^2\, e^{2i\phi}\, a_{qs}\, a_{qs} + (\vec{k}\cdot\vec{b}_{qs}^{*})^2\, e^{-2i\phi}\, a_{qs}^{\dagger}\, a_{qs}^{\dagger} + |k\cdot b_{qs}|^2\,(a_{qs}\, a_{qs}^{\dagger} + a_{qs}^{\dagger}\, a_{qs}) \quad ...(9)$$

The average value of $a_{qs}\, a_{qs}$ and $a_{qs}^{\dagger}\, a_{qs}^{\dagger}$ for eigenstates of the number operator $a_{qs}^{\dagger}\, a_{qs}$ are zero and

$$\langle a_{qs}^{\dagger}\, a_{qs}\rangle \equiv n_{qs} \quad ...(10)$$

while

$$\langle a_{qs}\, a_{qs}^{\dagger}\rangle = 1 + \langle a_{qs}^{\dagger}\, a_{qs}\rangle \equiv 1 + n_{qs} \quad ...(11)$$

where $n_{qs}$ is the average number of phonons in the normal mode $q_s$. Hence

$$\langle|\vec{k}\cdot\vec{u}|^2\rangle = |\vec{k}\cdot\vec{b}_{qs}|^2\,(1 + \langle a_{qs}^{\dagger}\, a_{qs}\rangle + \langle a_{qs}^{\dagger}\, a_{qs}\rangle) = |\vec{k}\cdot\vec{b}_{qs}|^2\,(1 + 2\, n_{qs}) \quad ...(12)$$

Taking into account Eq. (5) and Eq. (12) and the relation $\phi = \vec{q}\cdot\vec{R} - w_{qs}t$, we can reexpress Eq. (8) as follows:

$$f_t = f\left[\sum_{\vec{R}} e^{i\vec{k}\cdot\vec{R}}\left(1-\frac{1}{2}(\vec{k}\cdot\vec{u})^2\right) + i\,\vec{k}\cdot\vec{b}_{qs}\, a_{qs}\, e^{-iw_{qs}t}\sum_{R} e^{i(\vec{q}+\vec{k})\cdot\vec{R}} + i\,\vec{k}\cdot\vec{b}_{qs}^{*}\, a_{qs}^{\dagger}\, e^{iw_{qs}t}\sum_{R} e^{i(\vec{k}-\vec{q})\cdot\vec{R}}\right]$$

since there are $N_a$ atoms, we get

$$f_t = f\,N_a\left[\delta_{\vec{k},\vec{G}}\left(1-\frac{1}{2}(\vec{k}\cdot\vec{u})^2\right) + i\,\vec{k}\cdot\vec{b}_{qs}\, a_{qs}\, e^{-iw_{qs}t}\,\delta_{\vec{k}+\vec{q},\vec{G}} + i\,\vec{k}\cdot\vec{b}_{qs}^{*}\, a_{qs}^{\dagger}\, e^{iw_{qs}t}\,\delta_{\vec{k}-\vec{q}\cdot\vec{G}}\right] \quad ...(13)$$

Thus, the average differential scattering cross section $\frac{d\sigma}{d\Omega} = \langle f_t^{*}\, f_t\rangle$, keeping terms up to second order in $|b_{qs}|$, is equal to

$$\langle f_t^{*}\, f_t\rangle = |f|^2\, N_a^2\left[\sum_{\vec{G}}\delta_{\vec{k},\vec{G}}\,(1 - <(\vec{k}\cdot\vec{u})^2>) + \left|\vec{k}\cdot\vec{b}_{qs}\right|^2 n_{qs}\sum_{\vec{G}}\delta_{\vec{q}+\vec{k},\vec{G}} + \left|\vec{k}\cdot\vec{b}_{qs}\right|^2 (1+n_{qs})\sum_{\vec{G}}\delta_{\vec{k}-\vec{q}\cdot\vec{G}}\right]. \quad ...(14)$$

The first term in brackets in Eq. (14) reduces the elastic differential cross section $|f|^2 N_a^2$ (in the absence of ionic displacements) by a factor (1 – 2 $W$), where 2 $W$ is given by

$$2W = \langle(\vec{k} \cdot \vec{u})^2\rangle \qquad ...(15)$$

Actually, if one keeps all orders in $|\vec{b}|$ the reduction factor is not (1–2 $W$) but exp (–2 $W$). This is the *Debye-Waller factor.*

The second term in brackets in Eq. (14) corresponds to the inelastic scattering associated with the absorption of a phonon, $\vec{q}^{\,s}$, while the last term is associated with the creation of a phonon, $\vec{q}^{\,s}$.

If the displacement $\vec{u}$ ($\vec{R}$, $t$) is due to an arbitrary phonon excitation, the second and the third term in the brackets of Eq. (14) will involve a summation over $\vec{q}\,s$ and

$$2W = \langle(\vec{k} \cdot \vec{u})^2\rangle = \sum_{qs} |\vec{k} \cdot \vec{b}_{qs}|^2 \,(1 + 2n_{qs}) \qquad ...(16)$$

or, taking into account Eq. (7), we have

$$2W = \frac{1}{N_a}\left(\frac{\hbar}{2m}\right)\sum_{qs} \frac{1}{w_{qs}} |\vec{w}_{qs} \cdot \vec{k}|^2 \,(1 + 2\, n_{qs})$$

or

$$\boxed{2W = \frac{\hbar k^2}{6mN_a}\sum_{q}\sum_{s}\left(\frac{1+2\,n_{qs}}{w_{qs}}\right)} \qquad ...(17)$$

In arriving at the last expression, we have taken into account $|\vec{w}_{qs}| = 1$ and $|\vec{w}_{qs} \cdot \vec{k}|^2 = k^2 \cos^2\theta$, finally, the average of $\cos^2\theta$ over the solid angle is 1/3. (These missing steps may be filled).

## 3.25 DIFFRACTION INTENSITY FOR A CRYSTAL

The Bragg's diffraction condition gives a simple statement of the condition for the constructive interference of waves scattered from the lattice. (Note that we have a basis associated with a lattice point) and a basis (i.e., a lattice point is a representative of a number of district atoms corresponding to a lattice point). So for the scattered intensity (which is the square of the scattered wave amplitude), we need to consider:

(1) scattering from electrons in the same atom (this corresponds to the atomic scattering factor or atomic form factor $f_i$ for the $j^{th}$ atom); and

(2) the summation of this scattering from all atoms in the unit cell (this is characterized by a geometrical structure factor $F$).

### 3.25.1 The Atomic Scattering Factor

The atomic scattering factor is a measure of an atom in scattering X-rays. It is defined as the ratio

$$f = \frac{\text{Amplitude scattered by the actual electrons distribution in an atom}}{\text{Amplitude scattered by one electron localized at a point}} \qquad ...(1)$$

To calculate $f$, we consider an atom containing electrons arranged in a spherically symmetric configuration around its centre, which is taken as origin. Let $r$ be the redius of the atom. For a small volume element $dV$ located at $\vec{r}$, the charge is $n(\vec{r})\,dV$ units (Fig. 3.44).

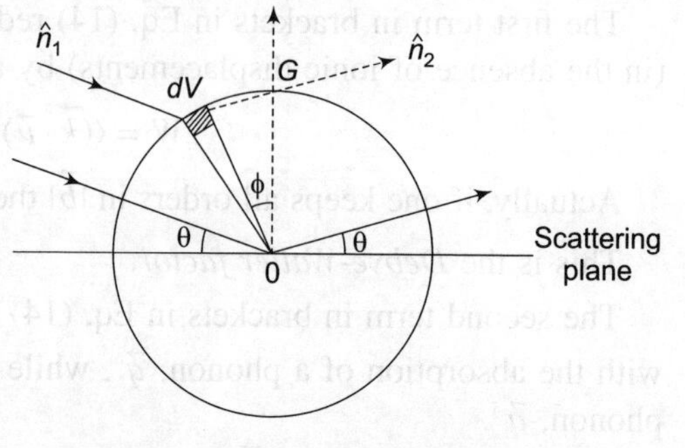


**Figure 3.44** Pertaining to calculation of atomic scattering factor

We first consider scattering from $dV$ and an electron located at the origin (Fig. 3.45). If $\hat{n}_1$ and $\hat{n}_2$ are directions of the incident and scattered beam, respectively, the phase difference between the waves scattered from $dV$ and from the centre is

$$\phi_r = \frac{2\pi}{\lambda}(\vec{r}\cdot\vec{N}) \qquad ...(2)$$

where $\vec{N} \equiv -\vec{G}.$

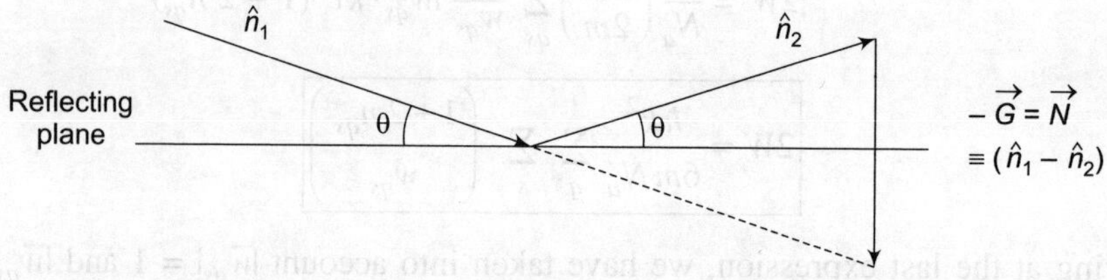


**Figure 3.45** Relationship between incident beam and scattered beam

Let the scattering amplitude from the point electron at $O$ in the direction $\hat{n}_2$ be $Ae^{i(kx-\omega t)}$. Then the scattering amplitude from the volume element $dV$ will be $Ae^{i(kx-\omega t)+i\phi r}\,n(\vec{r})\,dV$

$$\therefore \qquad df = \frac{Ae^{i(kx-\omega t)+i\phi r}\,n(\vec{r})\,dV}{Ae^{i(kx-\omega t)}} \qquad ...(3)$$

and

$$f = \int_v n(\vec{r})\,e^{i\phi r}\,dV \qquad ...(4)$$

where $V$ is the volume of the atom. Assuming spherical symmetry, we may use spherical polar coordinates. So,

$$dV = 2\pi\,r^2\sin\phi\,d\phi\,dr \qquad ...(5)$$

where $\phi$ is the polar angle.

Now, since

$$|N| = 2\sin\theta, \qquad ...(6)$$

$$\phi_r = \frac{2\pi}{\lambda}\,r\,N\cos\phi$$

$$= \frac{4\pi}{\lambda}\,r\sin\theta\cos\phi = \mu\,r\cos\phi \qquad ...(7)$$

where $$\mu = \frac{4\pi}{\lambda} \sin \theta$$

$$\therefore \quad f = \int_{r=0}^{R} \int_{\phi=0}^{2\pi} n(\vec{r})\, e^{i\mu r \cos\phi}\, 2\pi r^2 \sin\phi \, d\phi \, dr \quad ...(8)$$

$$= 2\pi \int_{r} \int_{-1}^{+1} dr\, r^2\, d(\cos\phi)\, n(\vec{r}) \times \exp(i\mu r \cos\phi)$$

$$= 2\pi \int_{r} dr\, r^2\, n(\vec{r}) \left( \frac{e^{i\mu r} - e^{-i\mu r}}{i\mu r} \right)$$

$$= 4\pi \int dr\, n(r)\, r^2 \frac{\sin \mu r}{\mu r} \quad ...(9)$$

Similarly, $$f_j = 4\pi \int dr\, n_j(r) r^2 \left( \frac{\sin Gr}{Gr} \right) \quad ...(10)$$

If the total electron density were concentrated at $r = 0$, then only $Gr = 0$ would contribute to the integrand. In this limit

$$\frac{\sin Gr}{Gr} = 1 \quad ...(11)$$

and $$f_j = 4\pi \int dr\, n_j(r) r^2 \equiv Z \quad ...(12)$$

which is the number of atomic electrons.

### 3.25.2 Geometrical Structure Factor

The intensity of an X-ray beam diffracted from a crystal not only depends upon the atomic scattering factors of the various atoms involved, but also on the number, type and distribution of atoms within the cell. It is taken into consideration by the structure factor $F(h', k' l')$. It is to be noted that the Bragg condition is only the minimum condition for efficient coherent reflection of X-rays by a set of parallel $(h' k' l')$ planes. How much intensity is actually reflected, when the angles of incidence and reflection are equal and the Bragg condition is satisfied for some integer $n$, depends on whether $F(h' k' l')$ is large, small or zero for the $(h' k' l')$ planes.

The total scattering amplitude or the structure factor $F(h' k' l')$ for the reflection $(h' k' l')$ is defined as the ratio of the amplitude of radiation scattered by the entire unit cell to the amplitude of radiation scattered by a single point electron placed at origin for the same wavelength. It is given by

$$F(h' k' l') = \sum_j f_j \, e^{i\phi j}$$

$$= \sum_j f_j \exp\left[ i \frac{2\pi}{\lambda} (\vec{r}_j \cdot \vec{N}) \right] \quad ...(13)$$

where $f_j$ is the atomic scattering factor for the $j$th atom, $\phi_j$ is the phase difference between the radiation scattered from the $j$th atom of the cell and that scattered from the electron placed at the origin and $N$ is the scattering normal. Also, the summation extends over all the atoms present in the unit cell. If $(u_j, v_j, w_j)$ represents the coordinates of the $j$th atom,we can write

$$r_j = u_j\, a + v_j\, b + w_j\, c$$

From the equations

$$\frac{2\pi}{\lambda}(a\cdot N) = 2\pi h'$$

$$\frac{2\pi}{\lambda}(b\cdot N) = 2\pi k'$$

$$\frac{2\pi}{\lambda}(c\cdot N) = 2\pi l'$$

we get
$$r_j\, N = \lambda(u_j h' + v_j k' + w_j\, l')$$

and
$$F(h'\, k'\, l') = \sum_j f_j\, e^{i2\pi(u_j h' + v_j k' + w_j r')} \qquad ...(14)$$

For identical atoms, all the $f_j'$ have the same value $f$. Therefore, Eq. (14) takes a simple form

$$F(h'\, k'\, l') = fS \qquad ...(15)$$

where
$$S = \sum_j e^{2\pi i(u_j h' + v_j k' + w_j l')} \qquad ...(16)$$

is called the geometrical structure factor as it depends upon the geometrical arrangement of atoms within the unit cell. Equation (15) defines the structure factor as the ratio of the total scattering amplitude to the atomic scattering factor. For dissimilar atoms, Eq. (14) should be used instead of (10).

Now, since the intensity of a radiation is proportional to the square of its amplitude, the intensity of diffracted beam may be written by using Eq. (14) as

$$I = |F|^2 = F^*\, F$$

$$= \left[\sum_j f_j \cos 2\pi\,(u_j h' + v_j k' + w_j l')\right]^2 + \left[\sum_j f_j \sin 2\pi\,(u_j h' + v_j k' + w_j l')\right]^2 \qquad ...(17)$$

where $F^*$ is the complex conjugate of $F$. The structure factors for some simple crystals are calculated and the intensity of various orders of reflections associated with these structures is discussed as follows.

### 3.25.2.1 Simple Cubic Crystals

The effective number of atoms in a unit cell of simple cubic structure is 1. Assuming that it lies at the origin, the structure factor given by Eq. (16) comes out to be unity. The diffraction amplitude from Eq. (15) becomes

$$F(h'\,k'\,l') = f \qquad ...(18)$$

which is independent of $h'$, $k'$ and $l'$.

Thus, all the diffraction lines predicted by the Bragg's law would appear in the diffraction pattern provided the value of $f$ is large enough to produce peaks of observable intensity.

### 3.25.2.2 Body-Centred Cubic Crystal

The effective number of atoms in a bcc unit cell is 2, one at a corner position and the other at the centre of the cube. If the coordinates of corner atom are arbitrarily taken as (0, 0, 0), then the coordinates of other atom become (1/2, 1/2, 1/2). Since both atoms are identical.

$$|F| = f\left[\left(\cos 2\pi(0) + \cos 2\pi\left(\frac{h'}{2}+\frac{k'}{2}+\frac{l'}{2}\right)\right)^2 + \left(0 + \sin 2\pi\left(\frac{h'}{2}+\frac{k'}{2}+\frac{l'}{2}\right)\right)\right]^{1/2}$$

$$= f\{[1 + \cos \pi\,(h' + k' + l')]^2 + \sin^2 \pi\,(h' + k' + l')\}^{1/2} \qquad ...(19)$$

We find from above expression that the structure factor becomes zero for odd values of $(h' + k' + l')$, and for even values of $(h' + k' + l')$. $F(h'\,k'\,l')$ equals $2f$ (and intensity proportional to $4f^2$). Thus, in a bcc structure, reflections like (1 0 0), (1 1 1), (2 1 0), etc., are missing, whereas the diffraction line corresponding to (1 1 0), (2 0 0 ), (2 2 2), etc., reflections are present.

### 3.25.2.3 Face-Centred Cubic Crystal

An fcc unit cell has four identical atoms. One of these atoms is contributed by corners and may arbitrarily assigned coordinates (0, 0, 0), whereas the other three are contributed by face centres and have the coordinates (1/2, 0, 1/2), (1/2, 1/2, 0) and (0, 1/2, 1/2). The diffraction amplitude becomes

$$|F| = \left\{\left(f(\cos 2\pi(0) + \cos 2\pi\left(\frac{h'}{2}+\frac{k'}{2}\right) + \cos 2\pi\left(\frac{h'}{2}+\frac{l'}{2}\right) + \cos 2\pi\left(\frac{k'}{2}+\frac{l'}{2}\right)\right)^2\right.$$

$$\left. + \left(\sin 2\pi\left(\frac{h'}{2}+\frac{k'}{2}\right) + \sin 2\pi\left(\frac{h'}{2}+\frac{l'}{2}\right) + \sin 2\pi\left(\frac{k'}{2}+\frac{l'}{2}\right)\right)^2\right\}^{1/2}$$

$$= f\{[1 + \cos \pi(h' + k') + \cos \pi(h' + l')]^2 + [\sin \pi(h' + k') + \sin \pi(h' + l') + \sin \pi(k' + l')]^2\}^{1/2}$$

$$= fS \qquad ...(20)$$

It is obvious that the structure factor is non-zero only if $h'$, $k'$ and $l'$ are all even or all odd and has a value equal to 4. Thus, the diffraction amplitude becomes $4f$ and the intensity becomes proportional to $16f^2$. The structure factor vanishes for all other odd-even combinations of $h'$ $k'$ and $l'$. Hence, reflections of the type (1 1 1), (2 0 0), (2 2 0) are present, whereas those of the type (1 0 0), (1 1 0), (2 1 1), etc., are absent for an fcc crystal.

The conclusions drawn above regarding allowed reflections for SC, bcc and fcc crystals are summarised in Table 3.4.

**Table 3.4** The Connection between Bravais Lattice of a Substance and its Diffraction Pattern

| *Bravais lattice* | *Reflections allowed* | *Reflections extinct* |
|---|---|---|
| SC | All | None |
| bcc | $(h + k + l)$ even | $(h + k + l)$ odd |
| fcc | $hkl$ unmixed (all odd or all even) | $h\ k\ l$ mixed |

**Example 3.4** The structure factor and form factor for sodium chloride (NaCl), lattices. It must be noted here that the diffraction maxima seen in diffractographs (obtained by X-ray diffraction technique) are indicative signatures of the structure of the crystal.

*Sodium Chloride*

The coordinates of the atoms in the unit cell (a cube of side $a = 2R$) are

$$Na^+ = 0\ 0\ 0;\ 1/2\ 1/2\ 0;\ 1/2\ 0\ 1/2;\ 0\ 1/2\ 1/2$$

Let us denote the atomic form factor of $Na^+$ by $(f_1)$ and that $Cl^-$ ion by $(f_2)$. These are obviously different because the number of electrons in an atom in either case is different (viz., 10 and 18). The structure factor for the $(h\ k\ l)$ reflection is given by

$$F(h\,k\,l) = f_1\{1 + \exp \pi i\,(h + k) + \exp \pi i\,(k + l) + \exp \pi i(l + h)\}$$
$$+ f_2\{\exp \pi i\,(h + k + l) + \exp(\pi i\,h) + \exp(\pi i\,k) + \exp(\pi i\,l)\}$$

In this case, we notice that

(i) When $h, k, l$ are all odd, $F_{hkl} = 4(f_1 - f_2)$

(ii) When $h, k, l$ are all even, $F_{hkl} = 4(f_1 + f_2)$

(iii) When $h, k, l$ are mixed (even + odd), $F_{hkl} = 0.$

Then

$F(1\,0\,0) = 0,$ but $(2\,0\,0) = 4(f_1 + f_2)$

$F(1\,1\,1) = 4(f_1 - f_2),$ but $F(2\,2\,2) = 4(f_1 + f_2)$

$F(1\,1\,0) = 0,$ but $F(2\,2\,0) = 4(f_1 + f_2)$

Thus, the first-order reflection from the cube face plane (1 0 0) will be absent, but the (2 0 0) reflection (which can be considered to be the second-order reflection) from (1 0 0) plane would be strong.

Similarly, for reflection from (1 1 0) and (2 2 0). The reflection from the (1 1 1) plane would be very weak, but that from (2 2 2); i.e., the second-order reflection would be strong. The intensities of the peaks observed by Bragg reflection for NaCl may then be explained satisfactorily. For the first peak obtained on reflection from the (1 0 0) plane at $\theta = 11.8°$ Bragg's relation would then be $2d_{2\,0\,0}$ sin (11.8°) = λ, and since $d = a(h_2 + k_2 + l_2)^{1/2}$, $d_{2\,0\,0} = a/2$. Hence, λ = 5.52 Å.

## 3.26 SCATTERING BY A UNIT CELL

It gives rise to geometrical structure factor and depends on the atomic arrangement in the unit cell. Consider a crystal structure consisting of three kinds of atoms *P, Q* and *R* (Fig. 3.46), reflecting in the first order from a set of $(h\,k\,l)$ planes of spacing $d_{h\,k\,l}$, (*P* being at the origin).

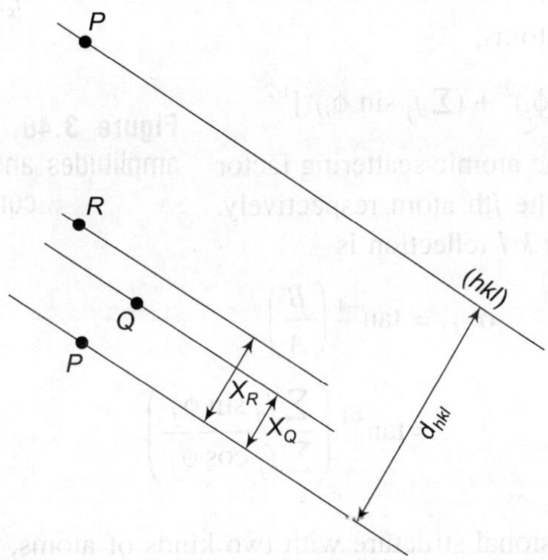


**Figure 3.46** Scattering by a unit cell having three kinds of atom (*P, Q* and *R*)

This means that the path difference for scattering from successive *P* plane is 1λ (or 2π phase change). The displacements $x_Q$, $x_R$ of the *Q*- and *R*-planes are equivalent to phase changes of,

$$\phi_Q = \frac{(x_Q)}{d_{hkl}} \times 2\pi \quad \text{and} \quad \phi_R = \frac{(x_R)}{d_{h\,k\,l}} \times 2\pi$$

respectively (Fig. 3.47).

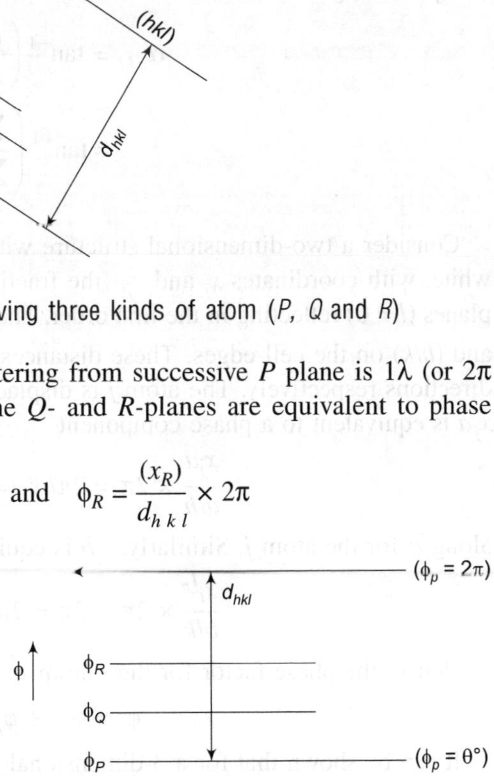


**Figure 3.47** Representation of phases

The amplitudes of the *hkl* waves scattered by the three atoms *P, Q, R* and their phase relations can be shown in a vector diagram in which each component wave is represented by a vector of length proportional to atomic scattering factor (*f*) and inclination to the horizontal equal to phase angle (ϕ) (Fig. 3.48).

The length of the resultant and its inclination gives the amplitude $F$ and the phase angle $\alpha$ of the composite (scattered) wave.

$$|F(hlk)| = \sqrt{A^2 + B^2}$$

$$= \sqrt{(F\cos\alpha)^2 + (F\sin\alpha)^2}$$

where,

$$(F\cos\alpha)^2 = (f_P\cos\phi_p + f_Q\cos\phi_Q + f_R\cos\phi_R)^2$$

and

$$(F\sin\alpha)^2 = (f_P\sin\phi_P + f_Q\sin\phi_Q + f_R\sin\phi_R)^2$$

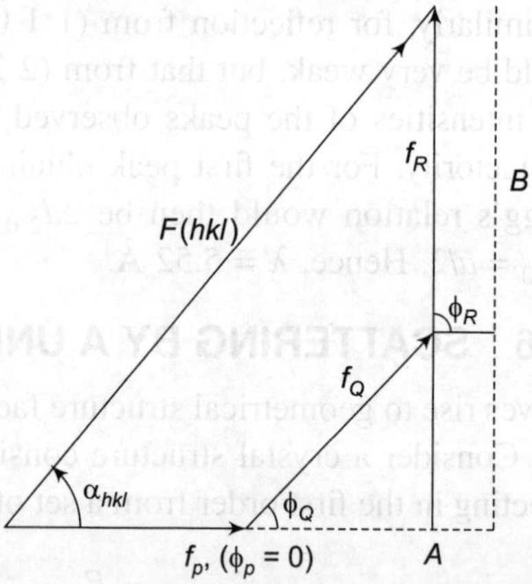


**Figure 3.48** Vector representation of amplitudes and phases of the scatterted component waves

For a unit cell with $j$ atoms,

$$|F(h\ k\ l)| = [(\Sigma f_j \cos\phi_j)^2 + (\Sigma f_j \sin\phi_j)^2]^{1/2}$$

where, $f_j$ and $\phi_j$ refer to the atomic scattering factor and the phase factor of the $j$th atom respectively. The phase angle for the $h\ k\ l$ reflection is

$$\alpha_{hkl} = \tan^{-1}\left(\frac{B}{A}\right)$$

$$= \tan^{-1}\left(\frac{\sum f_j \sin\phi_j}{\sum f_j \cos\phi_j}\right)$$

Consider a two-dimensional structure with two kinds of atoms, black at cell corners and white, with coordinates $x_j$ and $y_j$, (the fractional parts of cell edges). Assume a set of ($h\ k$) planes ($l = 0$) reflecting in the first order. The closest ($h\ k$) plane to origin has intercepts ($a/h$) and ($b/k$) on the cell edges. These distances are equivalent to $2\pi$ phase changes in $\vec{a}$ and $\vec{b}$ directions respectively. The atom $j$ is displaced from origin by $x_ja$ along $\vec{a}$ and $y_jb$ along $\vec{b}$. $x_ja$ is equivalent to a phase component

$$\frac{x_j a}{a/h} \times 2\pi = 2\pi h x_j\ (= \phi_{j,a}) \quad \text{(Fig. 3.49)}$$

along $\vec{a}$ for the atom $j$. Similarly, $y_jb$ is equivalent to phase component

$$\frac{y_j b}{b/k} \times 2\pi = 2\pi = 2\pi k y_j\ (= \phi_{j,b})$$

Hence, the phase factor for the $j$ atom is

$$\phi_j = \phi_{j,a} + \phi_{j,b} = 2p\ (hx_j + ky_j)$$

It can be shown that for a 3-dimensional structure

$$\phi_j = 2\,(hx_j + ky_j + lz_j)$$

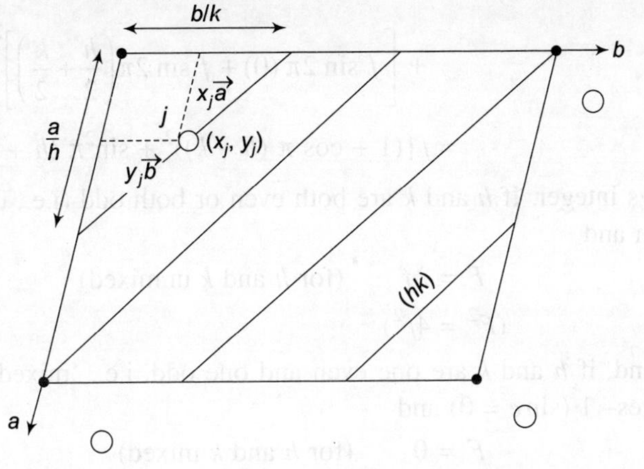


**Figure 3.49** A structure with two kinds of atoms

Therefore, the equation for the structure's amplitude becomes

$$|F\,(h\,k\,l)| = \{[\Sigma f_{oj} \cos 2\pi\,(hx_j + ky_j + lz_j)]^2 + [\Sigma f_{oj} \sin 2\pi\,(hx_j + ky_j + lz_j)]^2\}^{1/2}$$

The value of $|F|$ clearly depends on the kinds of atom (through $fo_j$) and on their positions (through cos and sine terms) in cell. This is often called the geometrical structure factor because it is solely a function of the coordinates of the atoms and therefore, characteristic of space group of the crystal. The phase angle for the $h\,k\,l$ reflection in terms of atomic coordinates becomes:

$$\alpha_{hkl} = \tan^{-1}\left[\frac{\Sigma f_{oj} \sin 2\pi(hx_j + ky_j + lz_j)}{\Sigma f_{oj} \cos 2\pi(hx_j + ky_j + lz_j)}\right]$$

The intensity of the beam diffracted by all atoms of the unit cell in a direction predicted by the Bragg's law is simply proportional to $|F|^2$.

**Example 3.5** Consider a unit cell containing only one atom at the origin, i.e., at (0, 0, 0) coordinates.

$$|F| = \{(fo \cos 2\pi(0))^2 + (fo \sin 2\pi(0))^2\}^{1/2}$$
$$= fo$$

and $$|F|^2 = fo^2$$

$F^2$ is thus independent of $h$, $k$ and $l$ and is the same for all reflections.

**Example 3.6** Base-centred unit cell.

It has two atoms of the same kind per unit cell located at 0 0 0 and $\frac{1}{2}\,\frac{1}{2}\,0$.

$$|F| = \left\{\left[f \cos 2\pi\,(0) + f \cos 2\pi\left(\frac{h}{2} + \frac{k}{2}\right)\right]^2\right.$$

$$+\left[f\sin 2\pi\,(0)+f\sin 2\pi\left(\frac{h}{2}+\frac{k}{2}\right)\right]^2\Bigg\}^{1/2}$$

$$= f\{(1+\cos\pi\,(h+k))^2+\sin^2\pi\,(h+k)\}^{1/2}$$

$(h + k)$ is always integer. If $h$ and $k$ are both even or both odd (i.e., unmixed), then their sum is always even and

$$F = 2f \qquad \text{(for } h \text{ and } k \text{ unmixed)}$$

$$(F^2 = 4f^2)$$

On the other hand, if $h$ and $k$ are one even and one odd, i.e., 'mixed', then their sum is odd and cosine gives– 1 ($\sin e = 0$) and

$$F = 0 \qquad \text{(for } h \text{ and } k \text{ mixed)}$$

$$(F^2 = 0)$$

Note that in either case the value of the $l$ index has no effect on the structure factor. Therefore, the reflections 1 1 1, 1 1 2 and 1 1 3 will have the same value of $F$, viz., $2f$.

It may be noted here that the structure factor is independent of the shape and size of the unit cell. For example, any body-centred cell will have missing reflections for those planes which have $(h + k + l)$ = odd irrespective of whether the cell is cubic, tetragonal or orthorhombic.

**Example 3.7** Close packed hexagonal cell.

$$F(h\,k\,l) = \sum_{j=1}^{N} f_j e^{2\pi i(hu_j + kv_j + lw_j)}$$

where, $N$ = number of atoms in the unit cell. The close packed hexagonal cell has two atoms of the same kind located at 0 0 0 and $\frac{1}{3}\frac{2}{3}\frac{1}{2}$

$$F = fe^{2\pi i(0)} + fe^{2\pi i\left(\frac{h}{3}+\frac{2k}{3}+\frac{l}{2}\right)}$$

$$= f\left[1 + e^{2\pi i\{(h+2k)/3 + l/2\}}\right]$$

For convenience put

$$\frac{h+2k}{3} + \frac{l}{2} = g$$

Then, $\qquad F = f[1 + e^{2\pi ig})$

Since $g$ may have fractional values, such as $\frac{1}{3}, \frac{2}{3}, \frac{5}{6}$, etc., this expression is still complex.

Multiplication by the complex conjugate, however, will give the square of the absolute value of the resultant wave amplitude

$$|F|^2 = f^2(1 + \exp(2\pi ig))\,(1 + \exp(-2\pi ig))$$

$$= f^2 [2 + e^{2\pi ig} + e^{-2\pi ig}]$$
$$= f^2 [2 + 2\cos 2\pi g]$$
$$= f^2 [2 + 2(2\cos^2 \pi g - 1)]$$
$$= 4f^2 \cos^2 \pi g$$
$$= 4f^2 \cos^2 \pi \left(\frac{h+2k}{3} + \frac{l}{2}\right)$$
$$= 0 \text{ when } (h + 2k) \text{ is multiple of 3 and } l \text{ is odd}$$

It is by these missing reflections such as 11.1, 11.3, 22.1, 22.3 that a hexagonal structure is recognized as being close packed, (*h c p*). Not all the reflections present have the same structure factor. For example, if $(h + 2k)$ is a multiple of 3 and $l$ even, then

$$\left(\frac{h+2k}{3} + \frac{l}{2}\right) = n \text{ (where, } n \text{ is an integer)}$$

and

$$\cos \pi n = \pm 1$$
$$\cos^2 \pi n = 1$$
$$|F|^2 = 4f^2$$

When all possible values of $h, k, l$ are considered, the results are:

| $(h + 2k)$ | $l$ | $\|F\|^2$ |
|---|---|---|
| $3n$ | odd | 0 |
| $3n$ | even | $4f^2$ |
| $3n \pm 1$ | odd | $3f^2$ |
| $3n \pm 1$ | even | $f^2$ |

In reflection by a typical crystal structure, it may be noted that,

1. The angular directions of the diffracted beams given by Bragg's law, are governed solely by the geometry of the lattice, i.e., the size and shape of the unit cell.
2. The arrangement of atoms within the unit cell affects the intensities of the diffracted beam, which involves two distinct contributions:
   (a) Scattering from electrons in the same atom (the atomic scattering factor).
   (b) Summation of this scattering from all the atoms of unit cell (the geometric structure factor).
3. Systematically, absent reflections are caused by some symmetry elements.

Thus, a critical study of reflection angles leads to determination of shape and size of the unit cell, and the unit cell, and the intensities of the reflections are used to deduce the atomic arrangements within the cell. Further, an analysis of the missing spectra gives the symmetry (space group) of the crystal structure.

**Example 3.8** A free electron of momentum $\hbar k$ is scattered through an angle $2\theta$ by a superposition of spherically symmetric atomic potentials of the form

$$v(r) = -u \quad r \le R$$
$$= 0 \quad r > R$$

within each atomic cell. Let us compute the atomic scattering factor.

The atomic scattering factor is given by the Fourier transform of the atomic potential, i.e.,

$$v(q) = \frac{1}{\Omega_0}\int dr \exp(-i\mathbf{q}\cdot\mathbf{r})v(\mathbf{r})$$
$$= -\frac{4\pi u}{\Omega_0}\int_0^R \frac{\sin qr}{qr} r^2\, dr$$
$$= \frac{4\pi u}{\Omega_0 q^3}(\sin qR - \cos qR)$$

where $$q = 2k\sin\theta \equiv \frac{4\pi\sin\theta}{\lambda}$$

from elementary geometry, as shown in Fig. 3.50.

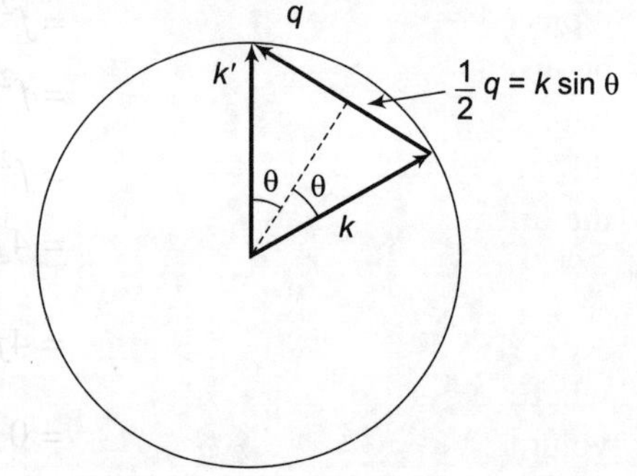


**Figure 3.50** Geometry for elastic electron scattering

## 3.27 FRESNEL'S THEORY OF X-RAY DIFFRACTION

The X-ray region of the electromagnetic spectrum has a wavelength which is suitable for the investigation of the structure of solids. Crystal structure is usually determined by the diffraction of X-rays of wavelength of about $10^{-10}$ m which is comparable to the distances between atoms in a crystal.

Diffraction can be described by carrying out a superposition of many individual waves, which originate in a distribution of point sources. The intuitive generalization of this procedure, used to describe interference phenomena, is called the *Fresnel theory*. Consider the superposition of *N* monochromatic spherical waves, in the arrangement shown in Figure 3.51 having the representation:

$$\Psi_j(r'_j, t) = \frac{A_j}{r_j} e^{i(kr'_j - \omega t)},\ A_j = C\frac{A_0}{r_j}e^{ikr_j}$$

where the complex amplitudes are expected to be proportional to the incident spherical waves, written in terms of the distances of every point scatterer from a primary source, $S_0$. We then obtain the resultant wave at a point *P* as

$$\Psi = CA_0 e^{-i\omega t}\sum_{j=1}^{N}\frac{e^{ik(r_j + r'_j)}}{r_j r'_j}$$

The diffraction patterns produced by a regular arrangement of atoms are described by introducing several simplifying approximations to this expression. The distances $r_j$ and $r'_j$ of the points $S_0$ and *P* from the crystal are large compared with its linear dimensions, and can be replaced by average values $r$, $r'$ so that the amplitude factors become all equal to $1/rr'$ and can be taken outside the sum. This means that different atoms are assumed to give contributions with equal amplitude. The change in $r_j + r'_j$, however, affects the phase contribution of different elements, due to the large value of the wave number, $k$. Under these assumptions the diffraction formula reduces to

$$\Psi = Ae^{-i\omega t} \sum_{j=1}^{N} e^{ik(r_j + r'_j)} \qquad \text{...(1)}$$

where $A$ includes the phase shift and all the slow varying parameters over the crystal volume.

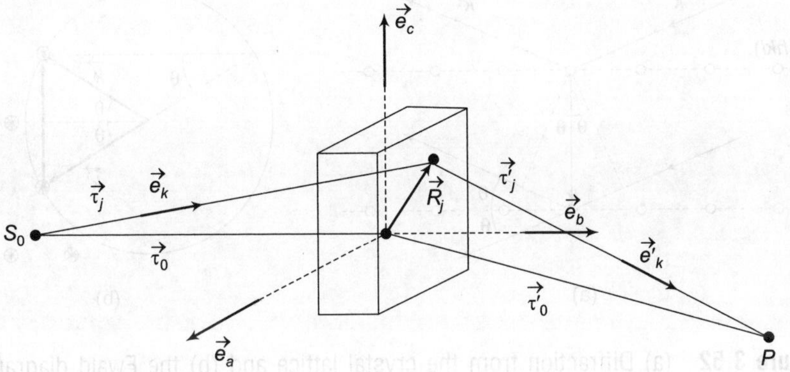


**Figure 3.51** Configuration of diffraction experiments

Approximate expressions for $r_j$ and $r'_j$ may be obtained in terms of position coordinates $\vec{R}_j$ of a given atomic scatterer. From Fig. 3.51 we have, to a first approximation:

$$r_j^2 = r_0^2 + 2\vec{R}_j \cdot \vec{r}_0 + R_j^2 = r_0^2 - 2r_0\vec{R}_j \cdot \vec{e}_k$$

$$= r_0^2\left(1 - \frac{2}{r_0}\vec{R}_j \cdot \vec{e}_k\right) = r_0^2(1 - \varepsilon)$$

which yields:
$$r_j = r_0(1-\varepsilon)^{\frac{1}{2}} = r_0\left(1 - \frac{\varepsilon}{2}\right) = r_0 - \vec{R}_j \cdot \vec{e}_k$$

In a similar manner, one obtains:

$$r'_j = r_0 + \vec{R}_j \cdot \vec{e}_k{}'$$

and therefore, the diffraction formula Eq. (1), can be written as

$$\Psi = Ae^{i\{k(r_0 + r'_0) - \omega t\}} \sum_i e^{ik\Delta(\vec{R}_j)} \qquad \text{...(2)}$$

The function $\Delta(\vec{R}_j)$ gives the path difference in terms of the lattice coordinates of the atomic scatterers only:

$$\Delta(\vec{R}_j) = \vec{R}_j \cdot (-\vec{e}_k) + \vec{R}_j \cdot (\vec{e}_{k'}) = \vec{R}_j \cdot (\vec{e}_{k'} - \vec{e}_k) \quad ...(3)$$

as illustrated at an atomic scale in Fig. 3.52 (a). Diffraction patterns are determined by the evaluation of the sum (2) over a scattering lattice. As long as we are only concerned with the relative amplitude and phase distribution, the phase term in front of the integral can be embedded in a single constant in a single constant and the diffraction formula simply reads

$$\Psi = \Psi_0 \sum_{j=1}^{N} e^{ik\Delta(\vec{R}_j)} \quad ...(4)$$

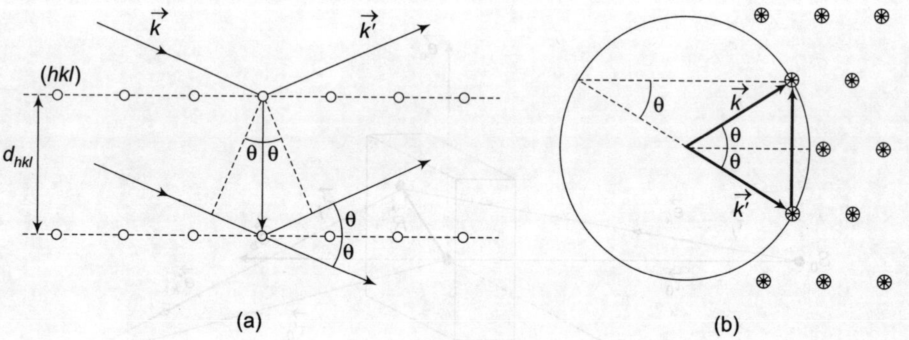


**Figure 3.52** (a) Diffraction from the crystal lattice and (b) the Ewald diagram in reciprocal space

Given the path difference $\Delta(\vec{R}_j)$, Eq. (3), for a point scatterer at lattice position $\vec{R}_j$, the corresponding phase shift can be written as

$$k\Delta(\vec{R}_j) = \vec{R}_j \cdot (\vec{k}' - \vec{k}) = \vec{R}_j \cdot \Delta\vec{k} \quad ...(5)$$

where $\Delta\vec{k}$ is called the *scattering vector*, which depends on the diffracted beam. Substituting Eq. (5), Eq. (4) becomes

$$\Psi = \Psi_0 \sum_{j=1}^{N} e^{i\vec{R}_j \cdot \Delta\vec{k}} \quad ...(6)$$

If we assume the crystal as a finite parallelepiped of dimensions $P\vec{a}, Q\vec{b}$ and $S\vec{c}$, and hence containing $N = PQS$ identical atoms, one at every lattice point, the position vector $\vec{R}_j$ has the form given by

$$\vec{R}_j = p_j\vec{a} + q_j\vec{b} + s_j\vec{c} \quad ...(7)$$

so that Eq. (6) takes the form

$$\Psi = \Psi_0 \sum_{p_j, q_j, s_j} e^{i(p_j\vec{a} + q_j\vec{b} + s_j\vec{c}) \cdot \Delta\vec{k}}$$

$$= \Psi_0 \sum_{p_j=0}^{P-1} e^{ip_j\vec{a}\cdot\Delta\vec{k}} \sum_{q_j=0}^{Q-1} e^{iq_j\vec{b}\cdot\Delta\vec{k}} \sum_{s_j=0}^{S-1} e^{is_j\vec{c}\cdot\Delta\vec{k}} \quad ...(8)$$

Each factor on the right hand side can be similarly transformed as follows

$$\sum_{p_j=0}^{P-1} (e^{i\vec{a}\cdot\Delta\vec{k}})^{p_j} = \frac{1-e^{iP\vec{a}\cdot\Delta\vec{k}}}{1-e^{i\vec{a}\cdot\Delta\vec{k}}} = \frac{e^{iP\vec{a}\cdot\Delta\vec{k}/2}}{e^{i\vec{a}\cdot\Delta\vec{k}/2}} \frac{\sin(P\vec{a}\cdot\Delta\vec{k}/2)}{\sin(\vec{a}\cdot\Delta\vec{k}/2)}$$

and the intensity distribution of the diffraction pattern reads

$$I = |\psi|^2 = I_0 \frac{\sin^2(P\vec{a}\cdot\Delta\vec{k}/2)}{\sin^2(\vec{a}\cdot\Delta\vec{k}/2)} \frac{\sin^2(Q\vec{b}\cdot\Delta\vec{k}/2)}{\sin^2(\vec{b}\cdot\Delta\vec{k}/2)} \frac{\sin^2(S\vec{c}\cdot\Delta\vec{k}/2)}{\sin^2(c\cdot\Delta\vec{k}/2)} \quad ...(9)$$

Maxima of intensity

$$I_{max} = P^2 Q^2 S^2 I_0 = N^2 I_0 \quad ...(10)$$

occur when the numerator and denominator both tend to zero, that is

$$\vec{a}\cdot\Delta\vec{k} = 2\pi h, \quad \vec{b}\cdot\Delta\vec{k} = 2\pi k, \quad \vec{c}\cdot\Delta\vec{k} = 2\pi l \quad ...(11)$$

where *h, k* and *l* are arbitrary integers. Multiplying each of these conditions by *p, q* and *s*, respectively and then, taking the sum, we obtain

$$(p\vec{a} + q\vec{b} + s\vec{c})\cdot\Delta\vec{k} = \vec{R}\cdot\Delta\vec{k} = 2\pi\,(ph + qk + sl) \quad ...(12)$$

By comparison with Eq. (3.13) it is seen that diffraction maxima occur if, and only if, the scattering vector $\varnothing\vec{k}$ is one of the reciprocal lattice vectors, that is,

$$\Delta\vec{k} = \vec{K}_{hkl} = h\vec{a}^* + k\vec{b}^* + l\vec{c}^* \quad ...(13)$$

The condition (13) on the scattering vector, which is equivalent to Eq. (12), is called the *Laue condition* for constructive interference in crystal diffraction. A simple geometrical interpretation of Eq. 13, known as the *Ewald diagram,* is illustrated in Fig. 3.52 (b). If we draw a sphere of redius $k = |\vec{k}|$ about the origin of the wave vector $\vec{k}$ of the incident radiation, where $\vec{k}$ is taken to pass through a point of the reciprocal lattice, then wherever the sphere intersects any other point of the reciprocal lattice constructive interference is observed. It is convenient to square both sides of Eq. (13), written as

$$k' = \vec{k} + \vec{K} \quad \text{or} \quad k^2 = k^2 + 2\vec{k}\cdot\vec{K} + \vec{K}^2$$

To obtain

$$2\vec{k}\cdot\vec{K} + \vec{K}^2 = 0 \quad ...(14)$$

Assuming that 2θ is the angle between the incident and scattered beams, Eq. (13) can be rewritten in the form

$$|\Delta\vec{k}| = 2k\sin\theta = |\vec{K}(hkl)| = \frac{2\pi}{d_{hkl}} \quad ...(15)$$

and reduces to the familiar *Bragg's law* for constructive interference by reflection planar arrays (*hkl*) in the real lattice

$$2d_{hkl}\sin\theta = \lambda \quad ...(16)$$

where $d_{hkl}$ is the interplanar spacing.

## 3.28 THEORY OF THE LAUE METHOD

The three-dimensional Laue conditions, Eq. (11 of Art 3.27), must be fulfilled for three directions simultaneously, in the form

$$\begin{aligned}\vec{a}\cdot(\vec{k}'-\vec{k}) &= 2\pi h \quad \text{or} \quad a(\cos\theta-\cos\theta_0) = h\lambda \\ \vec{b}\cdot(\vec{k}'-\vec{k}) &= 2\pi k \quad \text{or} \quad b(\cos\theta'-\cos\theta'_0) = k\lambda \quad ...(1) \\ \vec{c}\cdot(\vec{k}'-\vec{k}) &= 2\pi l \quad \text{or} \quad c(\cos\theta''-\cos\theta''_0) = l\lambda\end{aligned}$$

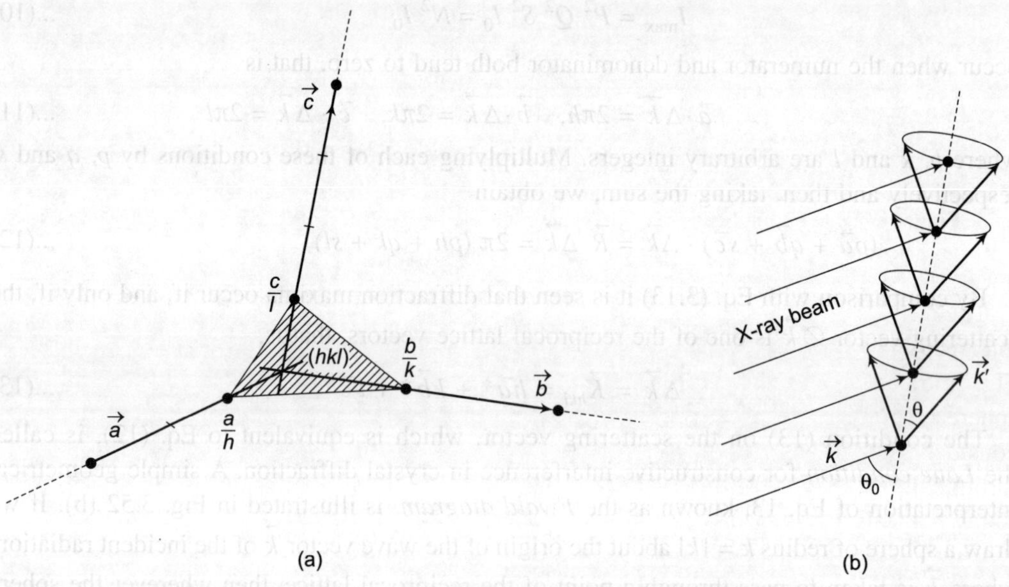


**Figure 3.53** (*hkl*) plane of constant phase shift (a) and cone diffraction from a linear atomic chain (b)

The path difference $h\lambda$ corresponds to a phase difference $h\lambda/2\pi$, and hence, X-ray scattered from an atom situated at a distance *a/h* from the origin, along the [100] direction, are dephased by $\lambda/2\pi$ with respect to those scattered from the origin. The same relative phase shift is exhibited by X-ray scattered from atoms situated at distances *b*/*k* and *c*/*l* along the [0 1 0] and [0 0 1] directions, respectively, and therefore from any atom which belongs to the (*hkl*) plane of the lattice, represented in Fig. 3.53 (a). For each of the linear atomic chains [1 0 0], [0 1 0] and [0 0 1], the directions of scattered rays form cones having the row of atoms as axis, as shown in Fig. 3.53 (b). Hence, the Laue conditions, Eq. (1), imply that only those reflections which correspond to the lines of intersection of all the three cones having axes $\vec{a}$, $\vec{b}$, and $\vec{c}$ are allowed. Using a single crystal of fixed orientation ($\vec{a}$, $\vec{b}$, $\vec{c}$) and an incident X-ray beam in a given direction, which is not monochromatic, but contains a continuous range of wavelengths, one obtains several possible reflections from the (*hlk*) planes of atoms. The experimental geometry of the Laue method, where the cones of diffracted rays give rise to elliptical or hyperbolic intersections with a photographic film placed perpendicular to the incident beam, is shown in Fig. 3.54.

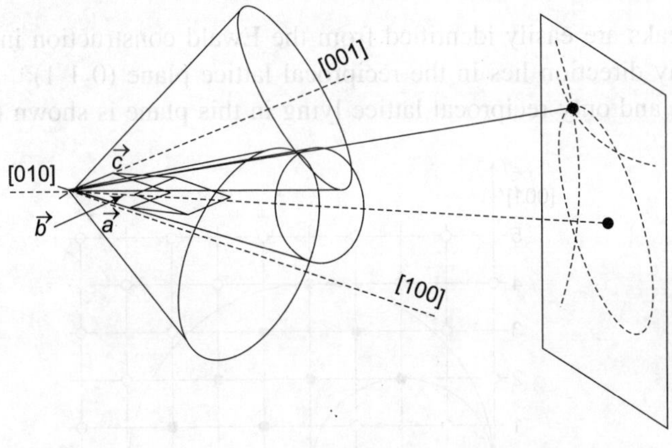


**Figure 3.54** Schematic representation of the Laue method for X-ray diffraction from single crystals

The diffraction pattern consists of a superposition of reflections recorded as points of triple intersection on the film, and can be interpreted in terms of the Ewald diagram, as shown in Fig. 3.55. As each component of wavelength $\lambda_i$ of the X-ray beam corresponds to the Ewald sphere of radius $2\pi/\lambda_j$ in the reciprocal lattice of the diffracting crystal, it is convenient to consider a series of spheres, of radius varying in the range $2\pi/\lambda_{max}$ to $2\pi/\lambda_{min}$, all passing through the origin of the reciprocal lattice such that their centres are collinear with the incident X-ray beam.

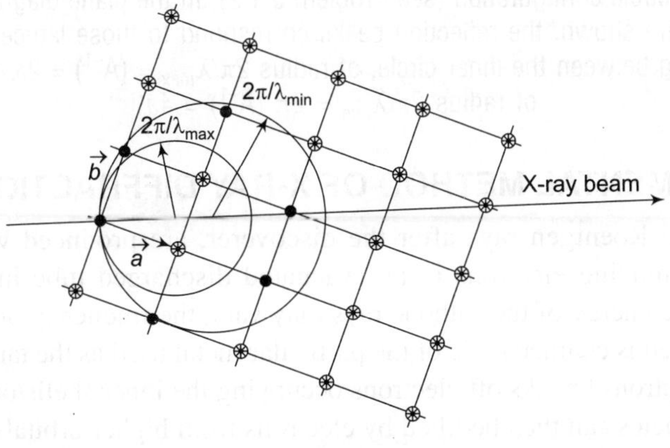


**Figure 3.55** The Ewald construction for the Laue method in the (*hk*0) plane

The observed reflections of the form (*hk*0) correspond to the lattice point contained in the (*hk*0) plane between the two Ewald spheres, are represented in Fig. 3.55 by solid circles. The Laue method is mainly used for the orientation of massive single crystals.

**Example 3.9** Determine the reflection peaks observed by the Laue method from a bcc crystal of lattice constant $a = 4$ Å, using an X-ray beam containing wavelengths ranging from 1 to 2 Å, which is normally incident on the (0 1 0) face.

***Solution*** The peaks are easily identified from the Ewald construction in reciprocal space. The incident X-ray direction lies in the reciprocal lattice plane (0 1 1)*, as it is parallel to [0 1 0][3] direction, and only reciprocal lattice lying in this plane is shown (Fig. 3.56).

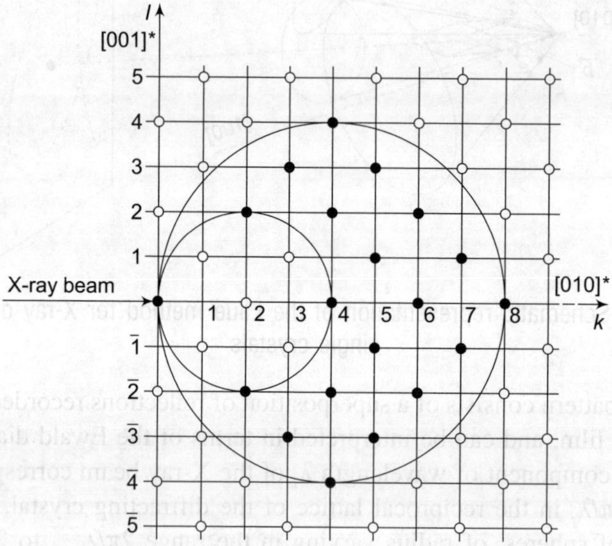


**Figure 3.56** If the reciprocal lattice constant is taken as $a^* = 2\pi/a = \pi/2$ (Å$^{-1}$), the lattice points have a face-centred cubic configuration (see Problem 3.1.2), In the plane diagram where the points of coordinates 0*kl* are shown, the reflection peaks correspond to those lattice points, represented by solid circles, lying between the inner circle, of radius $2\pi/\lambda_{max}$ = (Å$^{-1}$) = 2*a*, and the outer circle, of radius $2\pi/\lambda_{min} = 2\pi$ (Å$^{-1}$) = 4$a^*$

## 3.29 EXPERIMENTAL METHOD OF X-RAY DIFFRACTION[3]

X-rays, also called Röentgen rays after the discoverer, are produced when cathode rays consisting of fast-moving electrons in an evacuated discharged tube impinge on a metal target. Although the energy of the cathode rays may vary, the frequency or wavelength of the X-radiation produced is characteristic of the particular metal used as the target. Bombardment by fast-moving electrons knocks off electrons occupying the inner shells/orbitals of the target atoms. These vacancies can then be filled by electrons from higher orbitals. If electrons from

[3] L.V. Azaroff and M.J. Buerger, Powder Method in X-ray Crystallography, McGraw-Hill, New York, 1958.

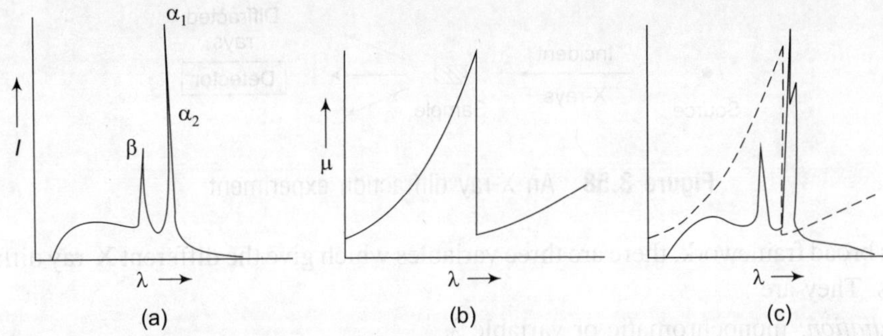


**Figure 3.57** (a) Intensity ($I$) of X-radiation as a function of wavelength ($\lambda$). (b) Absorption coefficient ($\mu$) of a filter as a function of wavelength ($\lambda$) of X-radiation (c) Superposition of (a) and (b) to indicate screening of β-radiation through the use of a filter

orbital with $n = 2$ or above jump to orbital with $n = 1$, where $n$ is the principal quantum number, the result is the $K$-series X-rays. If the electrons jump from $n = 3$ or above $n = 2$, we obtain the $L$-series X-rays and so on.

The X-ray powder diffraction technique can be conveniently employed for identification and characterisation of solids, especially inorganic solids (even in the powdered form). The fundamental equation employed is the Bragg's law, $n\lambda = 2d_{hkl} \sin \theta_{hkl}$; $n$ is the order of reflection and can be taken as unity, $\lambda$ is the wavelength of the incident X-radiation, $d_{hkl}$ is the distance between adjacent ($hkl$) planes, $hkl$ are the Miller indices and $\theta_{hkl}$ is the glancing angle corresponding to the respective Miller indices ($hkl$). For elements and compounds crystallizing in the simple crystal systems such as isometric (cubic), tetragonal, orthorhombic and rhombohedral/hexagonal, the Bragg's law enables determination of unit cell parameters with relative ease.

Figure 3.57 (a) depicts the X-ray intensity ($I$) as a function of wavelength ($\lambda$). Three characteristic peaks, $\beta$, $\alpha_1$, and $\alpha_2$ are observed, say, for $K$-radiation. Since the Bragg's law requires a unique value of $\lambda$, it is often convenient to screen the $\beta$ radiation through the use of filters and to lump $\alpha_1$, $\alpha_2$ together assuming $K_\alpha = \frac{1}{3}(K_{\alpha 2} + 2K_{\alpha 1})$. Usually, the best filters happen to be the preceding element in the Periodic Table. For example, nickel (Ni) is often used as a filter for a copper (Cu) target, vanadium (V) for chromium (Cr) target and manganese (Mn) for iron (Fe) target. For some purposes of identification, it may be advisable not to filter the $\beta$ radiation, so that additional data could be obtained from the same set of experimental measurements.

## 3.30 THE X-RAY DIFFRACTION EXPERIMENTS

Basically, the X-ray diffraction experiment necessitates an X-ray source, the sample under investigation and a photographic film (or a detector) to record the diffracted beam (Fig. 3.58).

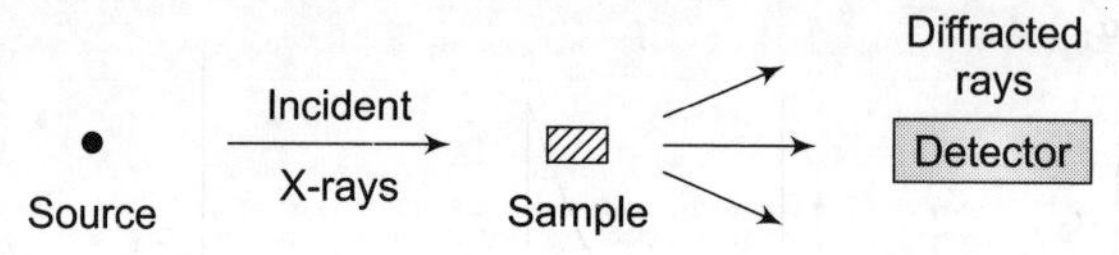


**Figure 3.58** An X-ray diffraction experiment

within this broad framework, there are three variables which give the different X-ray diffraction techniques. They are

(i) *Radiation:* monochromatic or variable λ.

(ii) *Sample:* single crystal, polycrystal (powder) or a solid (amorphous).

(iii) *Detector:* photographic film or radiation counter. These are summarized for the most important experimental techniques as below:

| *Wavelength* | *Sample* | *Detector* | *Method* |
|---|---|---|---|
| Fixed | Powder | Counter | Diffractometer |
| | | Film | Debye Scherrer |
| | | | Guinier (Focussing) |
| | Single crystal | Film | Rotation (Oscillation) |
| | | | Weissenberg |
| | | | Precession (Buerger) |
| | | Counter | Automatic diffractometer |
| Variable | Solid piece | Film | Laue |

Except, Laue method, monochromatic radiation is always used.

## 3.31 THE LAUE METHOD (EXPERIMENTAL)

As it has been pointed out in the beginning that in 1912 it was Max von Laue who first of all suggested the use of a crystal as three-dimensional diffraction grating to study the nature of X-rays. That way the Laue method is the oldest X-ray diffraction method. This is one of the most convenient and accurate methods of determining the orientation of a crystal or of individual grains in an aggregate and hence is widely used to orient crystals for solid state experiments. It quickly determines the crystal symmetry. However, since white radiation is used to get the Laue photograph (where different orders of reflection may superpose on a single spot), this method is hardly used for crystal structure determination.

A suitable crystal (as perfect as possible) is selected by examining the same under polarizing microscope. It is mounted on a glass fibre parallel to the principal axis using some adhesive (like shellac, glue, or even quick fix) material. The fibre along with the crystal is then mounted on a goniometer head with the help of modelling clay as shown in Fig. 3.59. The crystal is then aligned properly on the goniometer head with the help of four arcs (two mutually perpendicular arcs are for angular adjustment and the other two arcs are for linear adjustment).

The goniometer head alongwith the aligned crystal is then placed on a single crystal X-ray diffraction camera. The goniometer head is held fixed such that the incident X-ray beam (continuous radiation) is normal to one face of the crystal and a flat plate film cassette is placed either in between the source and the crystal for recording the back reflections on the X-rays film or behind the crystal to record the transmission Laue spots. A schematic

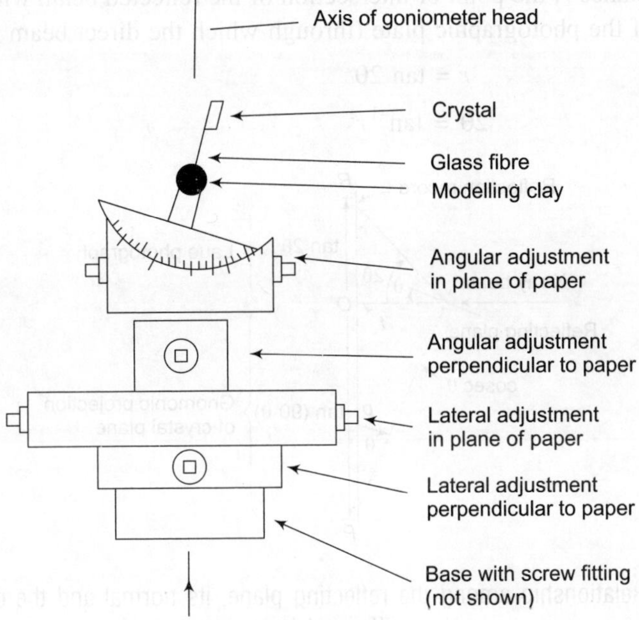


**Figure 3.59** Basic parts of goniometer head. A key is provided to fit the square pins for moving the arcs

representation of the Laue technique is shown in Fig. 3.60. The crystal-to-film distance in both the cases is normally kept small and is fixed between 3–5 cm. This increases the number of reflections intercepted by the film and reduces the exposure time. The two arrangements differ from one another in one respect. To intercept the direct incident beam a small lead cup is fixed at the centre of the cassette for recording the transmission Laue photograph, whereas in the back reflection a hole is made at the same point of the cassette to fit the collimator for the X-ray beam to pass.

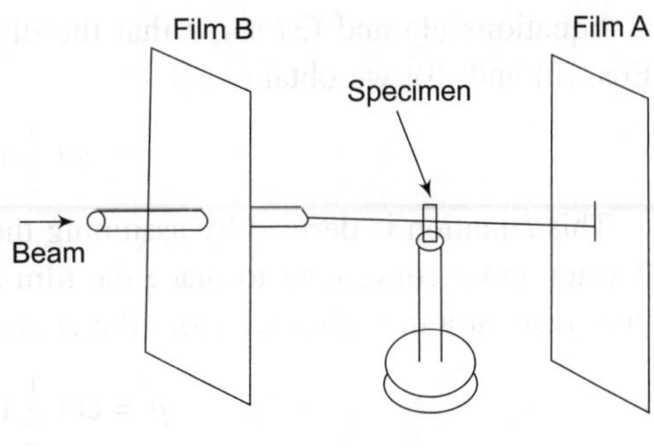


**Figure 3.60** Schematic representation of the Laue technique

### 3.31.1 Interpretation of Laue Photographs

Because of a close relationship between gnomonic projection and reciprocal lattice, the former is found to be extremely useful in the interpretation of Laue photographs. A relationship among the reflecting plane of a crystal, its normal, and the direction of the reflected beam is shown in Fig. 3.61. Here, it is assumed that the photographic film and the plane of projection are coincident and represent the vertical line, *PR*. If the crystal-to-film distance is taken to be unity, then the distance *r*, the point of intersection of the reflected beam with the vertical line from the centre of the photographic plate (through which the direct beam passes), is

$$r = \tan 2\theta \qquad ...(1)$$

or

$$2\theta = \tan^{-1} r \qquad ...(2)$$

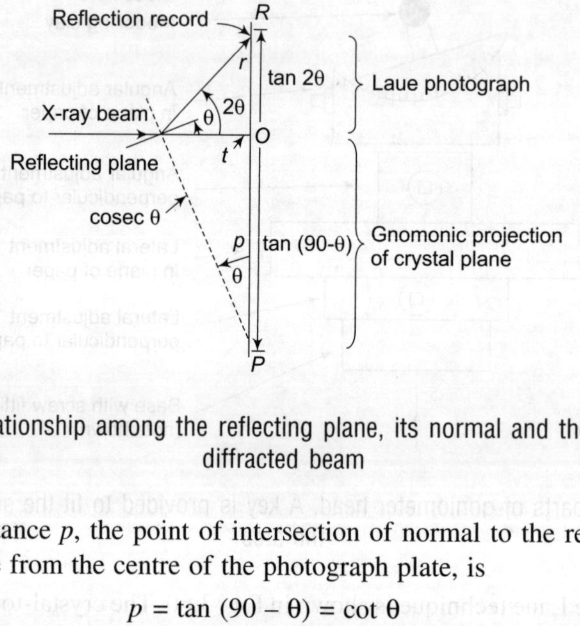


**Figure 3.61** Relationship among the reflecting plane, its normal and the direction of the diffracted beam

Similarly, the distance *p*, the point of intersection of normal to the reflecting plane (*hkl*) with the vertical line from the centre of the photograph plate, is

$$p = \tan (90 - \theta) = \cot \theta \qquad ...(3)$$

Equations (1) and (3) show that the distances *p* and *r* are related inversely. Combining Eqs. (2) and (3), we obtain

$$p = \cot \frac{1}{2} \tan^{-1} r \qquad ...(4)$$

This equation is derived by assuming the crystal-to-film distance to be unity. However, it is often more convenient to place the film at some other distance, say *D* and yet measure *p* in a plane at a unit distance from the crystal. In this case, the Eq. (4) is replaced by

$$p = \cot \frac{1}{2} \tan^{-1} \left( \frac{r}{D} \right) \qquad ...(5)$$

With the help of this equation, it is easy to transform the location of each spot of the Laue photograh to the gnomonic projection of the plane whose reflection produces it.

**Example 3.10** In a Laue photograph of an fcc crystal whose lattice parameter is 4.5Å. Determine the minimum distance from the centre of the pattern at which reflections can occur from the planes of maximum spacing, if the potential difference across the X-ray tube is 50 kV and the crystal to film distance is 5.0 cm.

***Solution*** Given: Crystal is fcc, $a$ = 4.5 Å, $V$ = 50 kV, crystal to film distance, $D$ = 5 cm = 0.05 m, $x$ = ?

In an fcc crystal we know that the planes of maximum spacing is (111).

Now, we know that in case of X-rays, the minimum wavelength is given by

$$\lambda_{min} = \frac{1.24 \times 10^{-6}}{V} = \frac{1.24 \times 10^{-6}}{5 \times 10^{4}} = 0.25 \text{ Å}$$

Similarly, for an fcc crystal, $d_{111} = \frac{4.5}{\sqrt{3}}$, so that

$$\theta_{111} = \sin^{-1}\left(\frac{\lambda}{2d}\right) = \sin^{-1}\left(\frac{0.25 \times \sqrt{3}}{2 \times 4.5}\right) = 2.76°$$

Now, from Fig. 3.62, we can obtain $\tan 2\theta = \frac{x}{5}$

or $$x = 5 \times \tan 2\theta = 5 \times \tan 5.52 = 0.48 \text{ cm}$$

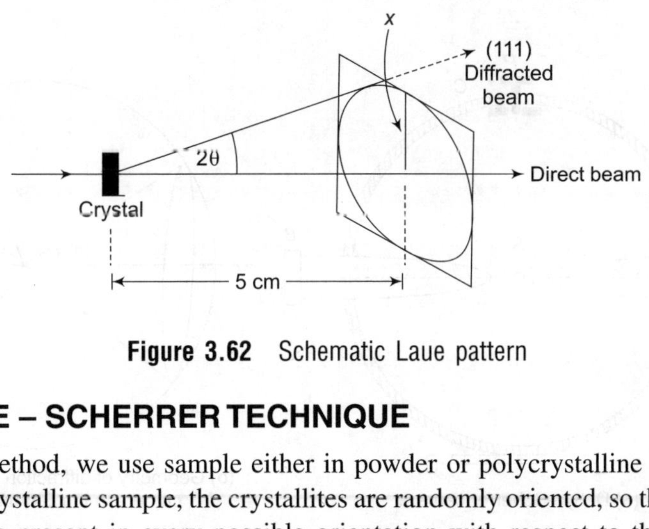


**Figure 3.62** Schematic Laue pattern

## 3.32 DEBYE – SCHERRER TECHNIQUE

In the powder method, we use sample either in powder or polycrystalline form. In an ideal powder or polycrystalline sample, the crystallites are randomly oriented, so that every possible lattice planes are present in every possible orientation with respect to the incident X-ray beam. Of all the available methods for recording X-ray diffraction pattern of powder of polycrystalline sample, the technique developed by Debye and Scherrer stands out to be the most useful and convenient because of the inherent simplicity and accuracy involved in the method.

A cylindrical camera (Fig. 3.63a) known as Debye-Scherrer camera is used for taking powder photographs of the sample. Cameras may be of different radii which are usually chosen to promote easy conversion of the distances on film into angles θ. (The cylindrical body *C* of the precision Debye-Scherrer camera of 114 mm diameter is about of 3 cm width and is mounted on a heavy metal base. The film is also cylindrical and is coaxial with the cylinder and is held tight against the inner surface of the cylinder by memo of pin system *P* at the top of the camera, Various methods of mounting the film in different types of Debye Scherrer camera are used. The one which is very common, known as Straumanis method and used in 11.4 cm precision camera, is shown in Fig. 3.63 (b).

The sample *S* is taken in the form of a uniformly cylindrical rod or wire of about 0.5 mm diameter which is mounted concentrically at the axis of the cylinder, on the sample holder. The axis of the sample is normal to the plane of two camera and coincident with the axis of the camera. *A* and *B* are two conical shaped slits placed at opposite ends at angular distance 180° apart. X-rays enter through a fine hole in *A* and being collimated by *A* fall on the sample *S*. The direct beam passes out through the other collimator *B* collinear with *A*. Both the collimators are fixed to the main body of the camera by punching holes through the films and some screw system. The primary X-ray beam enters through the entrance slit *A* and irradiates the samples at *S*. The diffraction cones with angles 4θ in front reflection and 4ϕ in the back reflection regions from different sets of planes intersect the cylindrical film forming pair of arcs as shown in Fig. 3.63(b). The film when laid flat appears as in Fig. 3.63 (c), with a large number of pairs of arcs around the two punched holes corresponding to *A* and *B*. After loading the camera with firm in the dark room, the cover lid is put back to make the camera light-light.

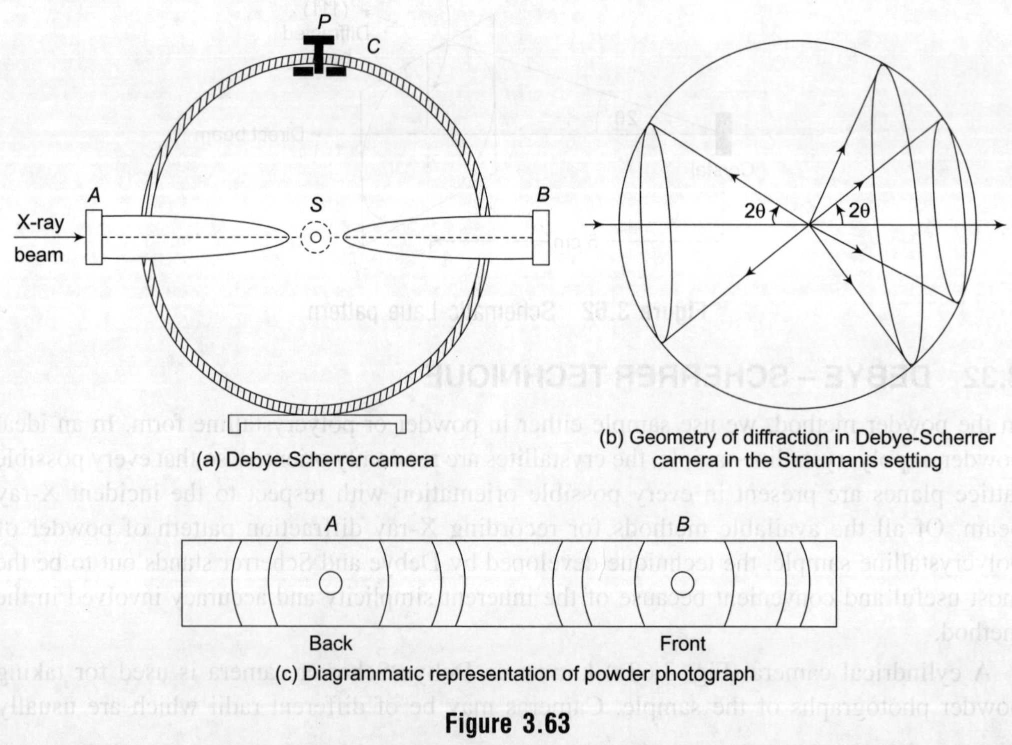


**Figure 3.63**

### 3.32.1 Recording and Evaluation of the Powder Pattern (Debye–Scherrer Method)

The sample, in the form of powder, is filled in a thin capillary glass tube which is sealed and then mounted in the X-ray camera. The sample eccentricity is ensured with the help of a simple microscope assembly. A narrow beam of monochromatic X-rays falls on the sample, which is continuously rotated. Since the crystallites of the powdered sample are randomly oriented, a large number of them corresponding to different Miller indices will be in a position to satisfy the diffraction conditions. As a result, the locus of directions making an angle $2\theta_{hkl}$ with a given direction forms a cone, which cuts the photographic film in a circle as shown in Fig. 3.64. Only a pair of arcs is actually recorded. (It may be remembered that the X-ray beam is directed normal to the photographic film.)

If $R$ is the radius of the camera, then

$$4\theta_{hkl} = \frac{S}{R} \text{ radians}$$

where $\theta_{hkl}$ is the glancing (Bragg) angle and $S$ the length of the arc.

Therefore, $$\theta_{hkl} = \left[\frac{180}{\pi} \cdot \frac{1}{4R}\right] S \text{ degrees}$$

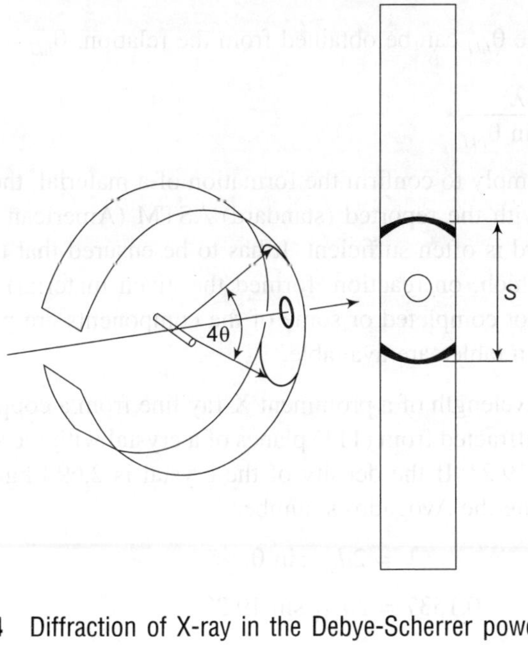


**Figure 3.64** Diffraction of X-ray in the Debye-Scherrer powder method

Now, $\frac{180}{\pi} \cong 57.3$, so that if $R = 57.3$ mm or its simple multiple, then the glancing angle can be directly obtained in degrees when the arc length $S$ is measured in millimetres (1 mm $= 10^{-3}$ m). This explains why commercial cameras often have a radius of 57.3 or 114.6 mm. Figure 3.65 shows a typical powder pattern. The determination of $\theta_{hkl}$ is made as follows:

(i) The coordinates $x_1$, $x_2$ corresponding to a particular pair of arcs belonging to a set of Miller (*hkl*) planes are measured.

(ii) The sum $x_1 + x_2$ should agree to within ±0.1 mm.

(iii) Then the arc length $S$ will be equal to $x_1 - x_2$ mm.

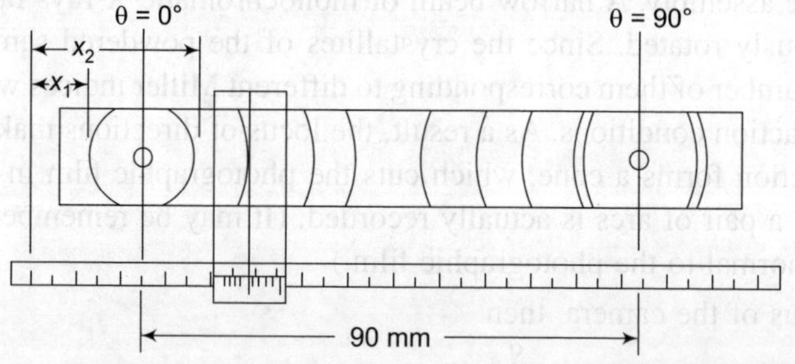


**Figure 3.65** A typical X-ray powder pattern

(iv) If the length between the two holes on the film differs from 90 mm by $p\%$, then the corrected arc length, $S' = S + \frac{PS}{100}$ mm.

(v) The Bragg angle $\theta_{hkl}$ can be obtained from the relation, $\theta_{hkl} = \frac{S'}{4}$

(vi) Then $d_{hkl} = \frac{\lambda}{2 \sin \theta_{hkl}}$

If the objective is simply to confirm the formation of a material, then a comparison of the observed $d_{hkl}$ values with the reported (standard) ASTM (American Society for Testing of Materials) data file/card is often sufficient. It has to be ensured that the lines corresponding to the components (which, on reaction, formed the given material) are absent; otherwise, either the reaction is not completed or some of the components are present in excess. Often $\theta$ versus $d_{hkl}$ conversion tables are available.

**Example 3.11** The wavelength of a prominent X-ray line from a copper target is 0.1537 nm. This radiation, when diffracted from (111) planes of a crystal with fcc structure, corresponded to a Bragg's angle of 19.2°. If the density of the crystal is 2,698 kg/m$^3$ and atomic weight 26.98 kg/kmol, calculate the Avogadro's number.

***Ans.***

$$\lambda = 2d_{hkl} \sin \theta_{hkl}$$

$$0.1537 = 2d_{111} \sin 19.2°$$

Therefore,

$$d_{111} = \frac{0.1537}{2 \times 0.3289} = 0.2336 \text{ nm}$$

Now the unit cell parameter,

$$a = d_{hkl} \sqrt{h^2 + k^2 + l^2}$$

$$= 0.2336 \times \sqrt{3}$$
$$= 0.4046 \text{ nm}$$

Volume of the unit cell, $a^3 = (0.4046)^3 \times 10^{-27} \text{ m}^3$

$$= 6.623 \times 10^{-29} \text{ m}^3$$

But, $$\text{Density} = \frac{\text{Mass}}{\text{Volume}}$$

or $$2698 = \frac{26.98 \times 4}{6.623 \times 10^{-29} \times N_A}$$

where $N_A$ is the Avogadro's number.

$$N_A = \frac{26.98 \times 4}{6.623 \times 10^{-29} \times 2698}$$
$$\cong 6.04 \times 10^{26} \text{ per k-mol}$$
$$\cong 6.04 \times 10^{23} \text{ per mol}$$

## 3.33 DETERMINATION OF UNKNOWN STRUCTURES

### 3.33.1 Isometric (Cubic) System

The experimentally observed '$d_{hkl}$' spacings are related to the lattice constant/unit cell edge $a$ by the relation,

$$d_{hkl} = \frac{a}{\sqrt{h^2 + k^2 + l^2}}$$

where $h$, $k$, $l$ are the Miller indices, which are integers that can have all values except $4^p$ $(8n + 7)$, where $p = 0, 1, 2, 3, \ldots$ and $n = 0, 1, 2, 3, \ldots$ Systematic absences/lattice extinctions place constraints on the permissible values of $h$, $k$ and $l$. Since the lattice constant should have the same value, it is relatively easy to identify the $h$, $k$, $l$ indices corresponding to the experimentally observed $d_{hkl}$ values.

It is also possible to make a graphical estimation of the lattice constant.[4] A chart (20" × 20") can be constructed, on which a plot of $a$ versus $d$ is made for various combinations of $h$, $k$, $l$ and for all possible values of $a$ say from 0 to 20 Å (0 to 2 nm). The indices are assigned by marking the measured $d$ value on a strip of paper, using the same scale as used in the chart and sliding the paper strip up and down, keeping the strip horizontal and keeping the point $d = 0$ on the y-axis, until a location is found for which all the lines on the strip correspond to lines on the graph.

### 3.33.2 Tetragonal System

Hull and Davey[5] proposed a graphical method, in which use is made of the relation,

---

[4] See L.V. Azaroff and M.J. Buerger, *The Powder Method* (McGraw-Hill, New York) 1958.

[5] A.W. Hull and W.P. Davey, *Phys. Rev.*, (1921), 549.

$$2 \log d_{hkl} = 2 \log a - \log \left\{ (h^2 + k^2) + \frac{l^2}{(c^2/a^2)} \right\}$$

If $a = l$ and $c/a = C$, then

$$2 \log d_{hkl} = -\log \left\{ \hbar^2 + k^2 + \frac{l^2}{C^2} \right\}$$

A chart is prepared by plotting $\log d_{hkl}$ versus $C$ on a semi-logarithmic paper. The procedure for indexing involves the following steps:

(a) Observed $d_{hkl}$ values are plotted on a strip of paper.

(b) $C$ (= $c/a$) value is read when a match is obtained.

(c) $a$ value is determined by substituting an $hk0$ value along with the $d_{hk0}$ value.

(d) $c$ value is determined by substituting a $d_{00l}$ value along with the $00l$ value.

Another chart, constructed by Bjurström,[6] may also be employed.

### 3.33.3 Sources of Error

The sources of error in the measured interplanar spacings may be listed as follows:

(a) The principal observational error lies in the location of the centre of a line on the film. This error can be minimized by repeating the reading several times or by preparing and measuring several films.

(b) The absorption of X-rays by the sample causes displacement of the centre from the true central position of the reflection. This effect decreases as the glancing angle $\theta_{hkl}$ increases, ultimately vanishing at $\theta_{hkl} = 90°$.

(c) The refraction of X-rays usually introduces a negligible error: however, unless this correction is applied, it may not be possible to compare unit cell dimensions measured with X-rays of different wavelengths. If this correction is not made, then the wavelength used should be specified.

(d) An uneven background causes a shifting of the peak position.

(e) If the sample axis is displaced from the cylinder axis during the rotation of the sample, a displacement of the reflection occurs. The factional error in $d_{hkl}$ caused by this displacement is proportional to $\cos^2 \theta$.

(f) The radius of the film rather than that of the camera is important. Further, the film may shrink during developing.

(g) It is virtually impossible to obtain a truly parallel X-ray beam. The fractional. error in '$d_{hkl}$' caused by the resulting displacement of the line is proportional to $\cot^2 \theta - 1$. Consequently, the lines are shifted to larger angles for $\theta_{hkl} < 45°$ and to smaller angles for $\theta_{hkl} > 45°$. However, the effect is small and can be eliminated by using small circular pinholes to collimate the X-ray beam.

---

6 T. Bjurström, *Z. Physik.*, **69** (1931), 346.

### 3.33.4 Accurate Determination of Unit Cell Dimensions

The following steps may lead to greater accuracy in the unit cell dimensions:

(a) A differentiation of the Bragg's law equation gives

$$\frac{\Delta d_{hkl}}{d_{hkl}} = \Delta\theta_{hkl} \cot \theta_{hkl}$$

Since $\cot \theta \to 0$ as $\theta \to 90°$, an error $\Delta\theta$ in $\theta$ will result in a relatively much smaller error $\Delta d$ in $d$. The recording of high angle reflections, therefore, will provide more precise results. This can be achieved by using back-reflection camera.

(b) Any procedure involving extrapolation to $\theta_{hkl} = 90°$ usually eliminates many of the errors.

(c) A simple and rapid procedure for obtaining reliable $d_{hkl}$ values consists of mixing the sample with another internal standard material, whose interplanar spacings are known to a desired accuracy.

## 3.34 ANALYSIS OF X-RAY POWDER DIFFRACTION PATTERNS

The following section outlines the methods used to analyze and characterize materials by X-ray diffraction, primarily using powder diffraction. The most common target used in X-ray sources is copper with a $K_\alpha$ wavelength defined as 1.5412Å. After the data has been recorded, the values of $\sin \theta$, $\sin^2\theta$, plane spacing $d$ calculated from $2d \sin \theta = \lambda$, and intensity are calculated for each line.

### 3.34.1 Powder Diffraction File

The powder diffraction pattern of every compound is distinctive in terms of both line positions and intensities. This information has been systematized by the JCPDS International Centre for Diffraction Data. All published diffraction spectra are included in a data file. This file is available as a card file, in book form as a search manual, and most recently on CD-ROM. Examples of the data available are shown below with courtesy of the publishers of the *Powder Diffraction File*.

The search manual gives a listing for the *d*-values and intensities of the 8 most intense lines in order of line intensity. The technique for using the file is to start with the most intense line, check the listing for the anticipated compound to see that the first three lines are present, and then compare the intensities and *d*-values for all other lines. Full data can then be obtained from the index card for the compound in the diffraction file. An example of the search procedure for a compound (sodium chloride) is shown below.

The diffraction pattern was recorded using a diffractometer with a copper target with a $K_\alpha$ wavelength of $\lambda = 1.5412$Å.

The *d*-values computed for lines of decreasing intensity are:

| Intensity Order | 1 | 2 | 3 | 4 | 5 | 6 |
|---|---|---|---|---|---|---|
| *d*-value: | 2.829 | 1.996 | 1.629 | 3.27 | 1.411 | 1.706 |
| Search manual: | 2.82 | 1.99 | 1.63 | 3.26 | 1.41 | 1.705 |
| Compound: | NaCl | | | | | |

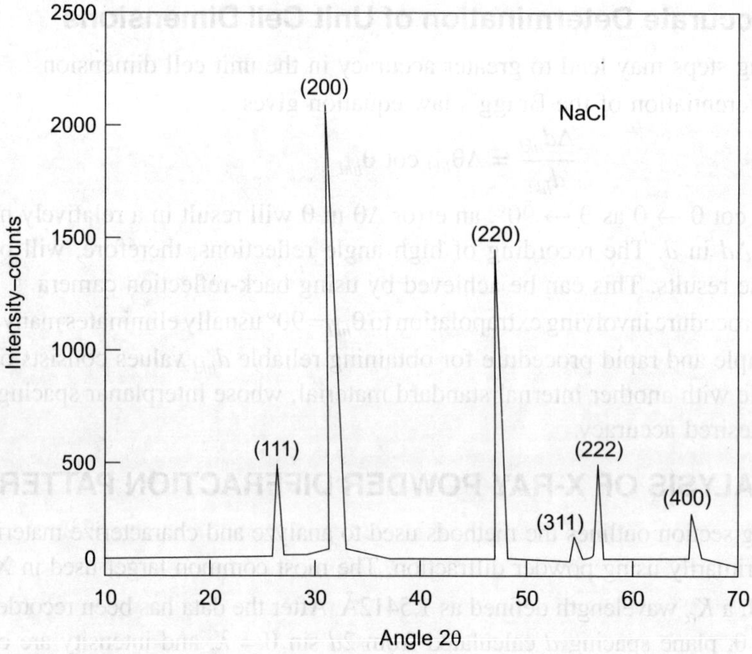


**Figure 3.66** Diffraction spectrum for NaCl using Cu $K_\alpha$ radiation

### 3.34.2 General Analysis for Cubic Structures

For a simple cubic lattice of spacing *a*, diffraction maxima occur for the (*hkl*) planes of the crystal at angles given by

$$\sin^2\theta = \frac{n^2 l^2}{4a^2}(h^2 + k^2 + l^2)$$

If the lattice is simple cubic, the maxima which occur are determined by the allowed values of $(h^2 + k^2 + l^2)$ as *h*, *k* and *l* take all values starting with 0. The allowed values are illustrated by the first line of Fig. 3.67. Mathematically, certain numbers are absent from the series (7, 15, …). These 'missing' numbers are given by $(8m - 1)$, where *m* is an integer.

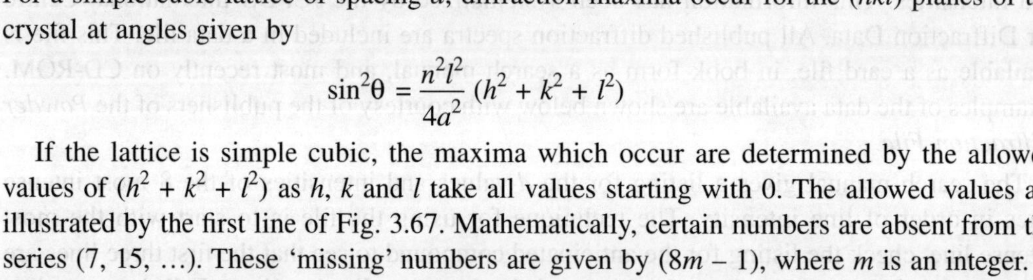


**Figure 3.67** Allowed planes for three cubic lattice

The effect of absent reflections due to additional planes of atoms in the face-centred cubic and body-centred cubic structures is to remove reflection from certain planes.

| | |
|---|---|
| Face-centred cubic: | $h$, $k$, and $l$ must be all even or all odd |
| Body centred cubic: | $h + k + l$ must be even. |

The experimental technique to determine the lattice spacing is to calculate $\sin^2\theta$ for all lines in the diffraction spectrum and to fit them into a series set by the allowed values of $h$, $k$, and $l$. This is shown for NaCl in Fig. 3.68.

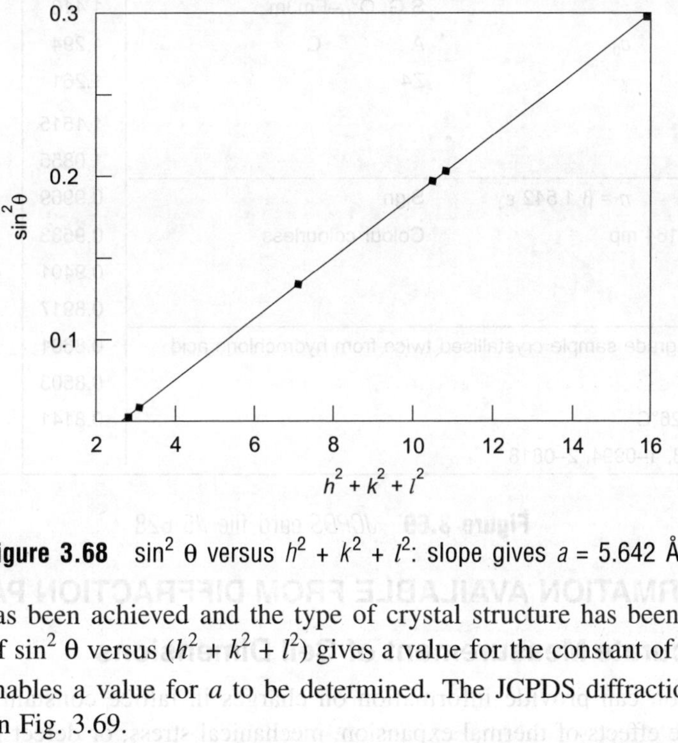


**Figure 3.68** $\sin^2\theta$ versus $h^2 + k^2 + l^2$: slope gives $a = 5.642$ Å

When a fit has been achieved and the type of crystal structure has been identified, the slope of a plot of $\sin^2\theta$ versus $(h^2 + k^2 + l^2)$ gives a value for the constant of proportionality $(\lambda^2/4a^2)$. This enables a value for $a$ to be determined. The JCPDS diffraction card file for NaCl is shown in Fig. 3.69.

Expressions for $\sin^2\theta$ for lattices of different crystal systems, and detailed procedures for fitting data for those lattices s are given in Chapter 5 of Lipson and Steeple.

### 3.34.3 X-ray Density

$$\text{Density} = \frac{\text{Mass of the unit cell}}{\text{Volume of the unit cell}}$$

$$= \frac{\Sigma Z}{6.023 \times 10^{23} \times V}$$

where $\Sigma Z$ is the sum of the atomic weights of the atoms in the unit cell, and $V$ is its volume.

| $d$ | 2.82 | 1.99 | 1.63 | 3.258 | NaCl | |
|---|---|---|---|---|---|---|
| $I/I_1$ | 100 | 55 | 15 | 13 | Sodium chloride | (Halite) |

| | | | $d$ Å | $I/I_1$ | $hkl$ |
|---|---|---|---|---|---|
| Rad. CuKα | λ 1.5404 | Filter Ni | | | |
| Dia. | Cut off | Coll. | 3.258 | 13 | 111 |
| $I/I_1$ | G.C. Diffractometer | D corr abs. | 2.821 | 100 | 200 |
| Ref. Swanson and Fuyat, NBS Circ. 539, II, 411953 | | | 1.994 | 55 | 220 |
| | | | 1.701 | 2 | 311 |
| | | | 1.629 | 15 | 222 |
| Sys. CUBIC | | S.G. $O^5_H$–Fm3m | 1.410 | 6 | 400 |
| $a_0$ 5.6402 $b_0$ | $c_0$ | A C | 1.294 | 1 | 331 |
| α β | γ | Z4 | 1.261 | 11 | 420 |
| Ref: ibid | | | 1.1515 | 7 | 422 |
| | | | 1.0855 | 1 | 511 |
| εα | $n$ = β 1.542 εγ | Sign | 0.9969 | 2 | 440 |
| 2V D 2.164 mp | | Colour colourless | 0.9533 | 1 | 531 |
| Ref: ibid | | | 0.9401 | 3 | 600 |
| | | | 0.8917 | 4 | 620 |
| An ACS reagent grade sample crystallised twice from hydrochloric acid | | | 0.8601 | 1 | 533 |
| | | | 0.8503 | 3 | 622 |
| X-ray pattern at 26°C | | | 0.8141 | 2 | 444 |
| Replaces 1–0993, 1–0994, 2–0818 | | | | | |

**Figure 3.69** *JCPDS* card file #5-628

## 3.35 INFORMATION AVAILABLE FROM DIFFRACTION PATTERNS

### 3.35.1 Accurate Measurement of Cell Dimensions

X-ray diffraction can provide information on charges in lattice constant in the study of alloys, or in the effects of thermal expansion, mechanical stress, or defect generation under radiation bombardment. The highest accuracy in lattice constant measurements is achieved at large θ.

The Bragg equation ($2d \sin\theta = \lambda$) may be written $d = \dfrac{\lambda \operatorname{cosec}\theta}{2}$

Differentiating:

$$\delta d = -\lambda \operatorname{cosec}\theta \cot\theta \, \delta\theta/2$$

$$\delta d = -d \cot\theta \, \delta\theta$$

Since $\cot\theta \to 0$ as $\theta \to 90°$, for a given $\delta\theta$ the relative error in the lattice parameter $d$ becomes very small when θ exceeds 80°.

In defect studies, comparison of the X-ray density with the density determined by, say, Archimedes method provides information on the relative volume of voids or defects.

### Crystallinity

A single crystal gives well resolved spots on a photographic plate taken by the Laue method. Polycrystalline material shows a pattern of full circles. Intermediate patterns (spotty circles) are good indicators of the crystalline state.

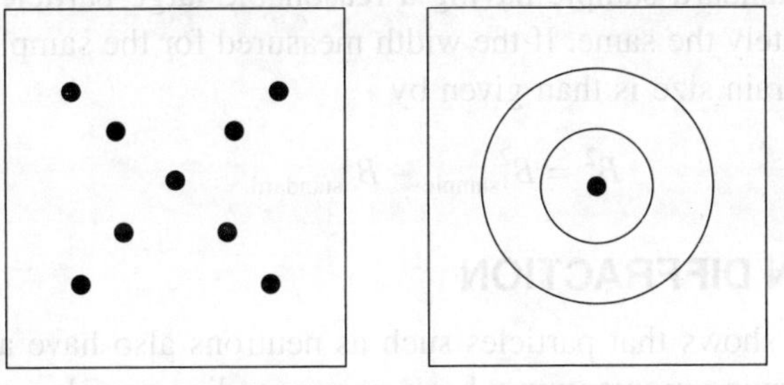

**Figure 3.70** X-ray Laue patterns of diffraction rings and spots characteristic of polycrystalline and single crystal material respectively

### Amorphous materials

These are characterized by diffuse rings corresponding to the average set of spacings present in the disordered structure.

This information is expressed in terms of the radial distribution function $\phi(r)$ which is a measure of the number of atoms at radius $r$ from any other atoms.

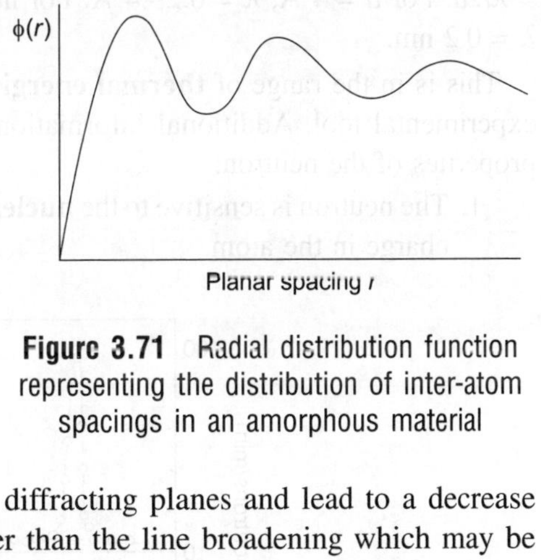


**Figure 3.71** Radial distribution function representing the distribution of inter-atom spacings in an amorphous material

### Temperature effects

The peak positions for a given pattern change as the lattice constants change under thermal expansion. The vibrations of the atoms about their mean positions in the lattice smear out the diffracting planes and lead to a decrease in the **intensity** of a given line. This occurs rather than the line broadening which may be expected. This decrease in intensity at angle θ is described by the Debye–Waller factor $B_T$.

$$I = I_o \exp\left(-2B_T\left(\frac{\sin\theta}{\lambda}\right)^2\right)$$

Expressions for $B_T$ are given in Cullity but may be approximated (within 20% for many metals) up to 900 K by $1.3T/A$, where $T$ is the temperature (K) and $A$ is the atomic weight.

### Grain size

As the particle size $t$ of a powder decreases, the line width of X-ray diffraction line increases. A quantitative estimate of the particle size $t$ in the nanometre size range may be obtained from

$$B = \frac{0.9\,\lambda}{t \cos \theta}$$

where $B$ is the broadening of a diffraction line (full width at half maximum) at angle θ.

The width of the line due to the spectrometer $B_{standard}$ must be evaluated by comparison with a line from a standard sample having a reasonable large particle size. The value of θ should be approximately the same. If the width measured for the sample is $B_{sample}$, the actual width $B$ due to the grain size is than given by

$$B^2 = B^2_{sample} - B^2_{standard}$$

## 3.36 NEUTRON DIFFRACTION

Quantum mechanics shows that particles such as neutrons also have a wave-like behaviour such that a particle having momentum $p$ has a corresponding wavelength λ such that $\lambda = h/p$. If the energy of the particle is $E = p^2/2m$, $\lambda = h/\sqrt{2}\,Em$.

Significant diffraction effects are found with *neutrons*. As noted earlier, the wavelength required to give appreciable diffraction angles can be estimated from Bragg's law as $\sin\theta = \lambda/2d$. For $d = 4$ Å, $\lambda = 0.2-4$ Å. For neutrons $m = 9.1 \times 10^{-31}$ kg, and $E = 0.021$ eV for $\lambda = 0.2$ nm.

This is in the range of **thermal energies,** and thermal neutron diffraction is a significant experimental tool. Additional information from neutron diffraction arises from the special properties of the neutron:

1. The neutron is sensitive to the **nuclear** position rather than to the distribution of electron charge in the atom.

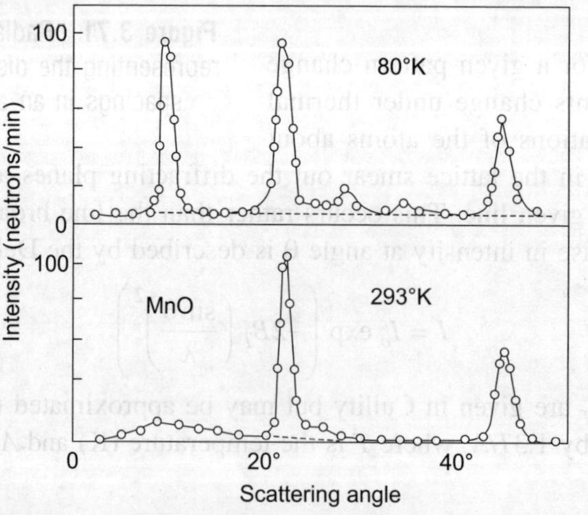


**Figure 3.72** Neutron diffraction spectra for MnO above and below a temperature range where magnetic ording takes place (from *C.G. Shull and J.S. Smart, Phys. Rev.,* **76**, 1256 (1949)).

2. The neutron has a magnetic moment and can therefore detect **magnetic structure** in a crystal, that is, the distribution of magnetic moments within a lager crystallographic structure.
3. The neutron is more sensitive to nuclear movement and therefore is a good probe of **vibrational spectra** in crystals.

An example of the value of neutron diffraction is seen in studies of the magnetic structure of manganese oxide MnO.

MnO is paramagnetic at room temperature, but develops an ordered antiferromagnetic structure at 80 K. This is shown by added peaks at low temperature in the neutron diffraction spectrum which can be indexed to a unit cell twice as large as that at 293 K.

## 3.37 THE POWDER DIFFRACTOMETRIC TECHNIQUES

X-ray diffrectometer has become a very popular and versatile instrument for routine X-ray analysis of powder and polycrystalline samples. In fact, the diffractometric technique is often preferred to Debye-Scherrer technique owing to its inherent merits. The most striking difference between the two methods is in the use of different intensity detection and measuring devices. In the diffractometer, the diffracted intensities from the various planes of the powder sample (which is often made in the form of rectangular slide or thin flat state) are measured directly by using some quantum counter and associated electronic circuit. The angular distribution of the diffracted intensities can be recorded directly on a chart recorder coupled to the counter through suitable circuitry. Thus, the entire diffraction pattern can be recorded in terms of a series of peaks corresponding to the different Bragg reflections from various planes. The heights of the peaks will be directly proportional to the intensities of the lines and the spread of the peaks to the defect state of the samples. The graphical record obtained in this way is called *diffractogram*, e.g., Fig. 3.73 shows diffraction pattern of silicon. Here, a Geiger counter measures the intensity of the diffracted X-rays as it moves through the 2θ circle. Each plane appears at a specific angle related to the interplanar spacing.

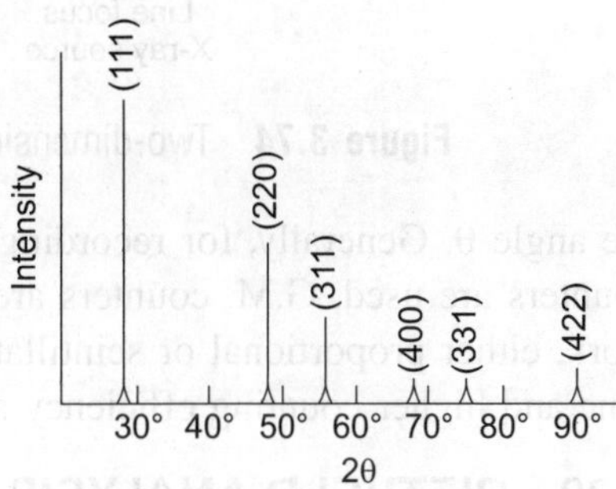


**Figure 3.73** Diffractogram of silicon (Geiger chart)

A simple geometry of the diffractometer showing the relative position of the X-ray line source, sample and the detector to achieve what is known as *parafocussing arrangement*, which ensures better intensity and resolution is shown in Fig. 3.74.

The figure depicts two-dimensional features of an ideal parafocussing arrangement employed in a standard powder diffractometer. The line focus $F$ of the X-ray tube, flat sample $S$ and the counter $G$ all lie on the circle of radius $R_F$ called focussing circle. The distance $FS = SG = r$ which is kept constant by coupling the goniometer with the counter in such a way that while the sample rotates through θ, the counter $G$ rotates through 2θ. This ensures that the $F$, $S$ and $G$ always lie on the circle and $r$ remains constant. However, $R_F$ depends on

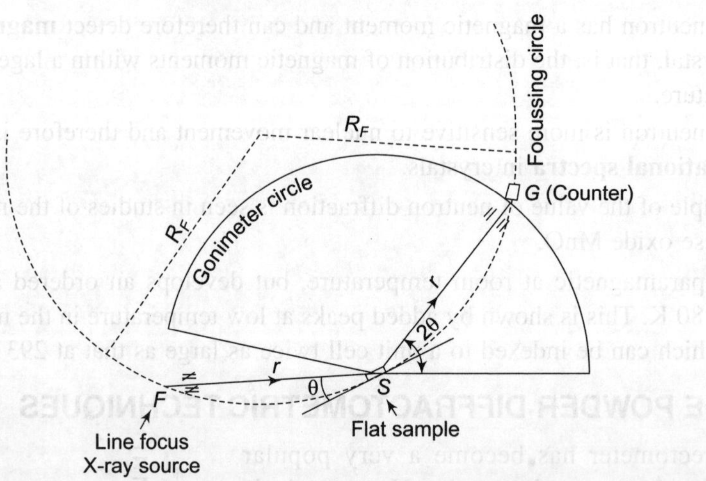


**Figure 3.74** Two-dimensional representation of parafocussing geometry

the angle θ. Generally, for recording intensities, Geiger Muller proportional or scintillation counters are used, G.M. counters are used for routine analysis whereas for good analytical work, either proportional or scintillation counters, both of which have very small resolving time and higher counting efficiency are used.

## 3.38 RIETVELD ANALYSIS

Rietveld method for structural refinement of powder diffraction data has been developed over the last four decades and has proved indispensable in solving crystal structures.

The process involves minimizing the difference between a crystallographic model and experimental data, via a least squares refinement; such intricate modelling of the height, width and position of Bragg reflections in an X-ray diffraction pattern can yield a lot of information about the crystal structure of a material. In 1964–1966 the need to refine crystal structures from powder where the peaks were too much overlapped and peak separation by least squares fitting (Gaussian profiles) cannot be used for severe overlapping. Therefore, in 1967 first refinement program by H.M. Rietveld, single reflections + overlapped, no other parameters than the atomic parameters was used. In 1969-first complete program with structures and profile parameters (ALGOL) was used. Then by 1972, FORTRAN version was distributed worldwide. By 1977 wide acceptance led it to extend to X-ray data.

Hugo Rietveld realized that if a pattern could be modelled, the fit between a computed pattern and observed data could be optimized. The Rietveld method involves recording the diffracted intensities at several thousand equal increments.

Developing a crystal structure model, i.e., symmetry, atomic positions, unit-cell size, site occupancies varying these and other parameters by least squares refinement to get best fit between calculated and observed data. Peak-shape and diffractometer setup are taken into account.

The Rietveld analysis is generally performed for solving an unknown crystal structure, calculating the amount of disorder or mixing on a Wyckoff site, quantitatively determining the percentages of different planes in the sample and determining the crystallite size in the sample.

## 3.39 TEMPERATURE DEPENDENCE OF THE X-RAY REFLECTION LINES AND DEBYE-WALLER FACTOR

As the temperature of the crystal is increased, the intensity of the Bragg-reflected beams decreases, but the angular width of the reflected line does not change. Experimental intensities for aluminum are shown in Fig. 3.75. It is surprising that we can get a sharp X-ray reflection from atoms undergoing large amplitude random thermal motion, with instantaneous nearest-neighbour spacings by 10 per cent at room temperature. Before the Laue experiment

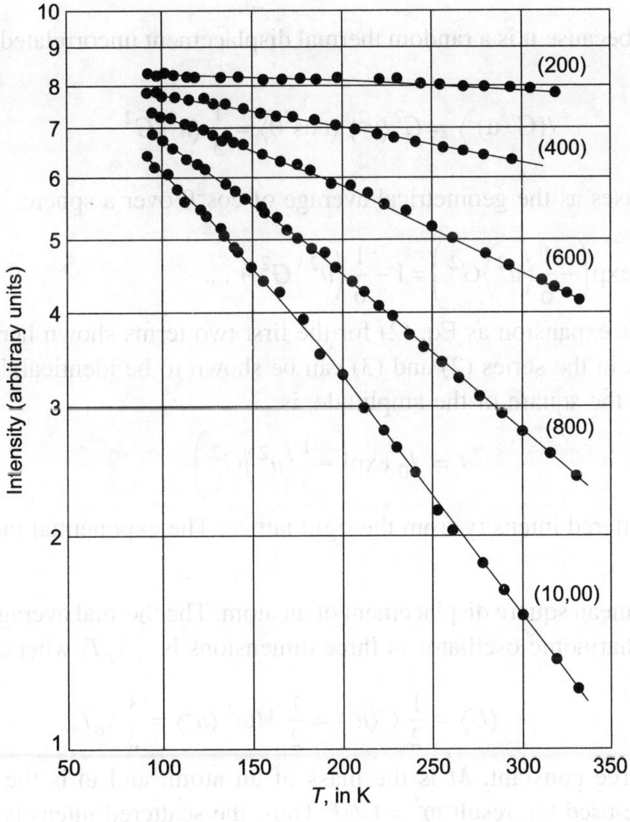


**Figure 3.75** The dependence of intensity on temperature for the ($h$00) X-ray reflections of aluminum. Reflections ($h$00) with $h$ odd are forbidden for an fcc structure. (After R.M. Nicklow and R.A. young)

was done, but when the proposal was discussed[7] in a coffee house in Munich, the objection was made that the instantaneous positions of the atoms in a crystal at room temperature are far from a regular periodic array, because of the large thermal fluctuation. Therefore, the argument went, one should not expect a wll-defined diffracted beam.

But such a beam is found. The reason was given by Debye. Consider the radiation amplitude scattered by a crystal: let the position of the atom nominally at $\mathbf{r}_j$ contain a term: $\mathbf{u}(t)$ fluctuating in time: $\mathbf{r}(t) = \mathbf{r}_j + \mathbf{u}(t)$. We suppose each atom fluctuates independently about its own equilibrium position.[8] Then the thermal average of the structure factor contains terms

$$f_j \exp(-i\mathbf{G}\cdot\mathbf{r}_j)\,\langle\exp(-i\mathbf{G}\cdot\mathbf{u})\rangle, \qquad ...(1)$$

where $\langle\cdots\rangle$ denotes thermal average. The series expansion of the exponential is

$$\langle\exp(-i\mathbf{G}\cdot\mathbf{u})\rangle = 1 - i\langle\mathbf{G}\cdot\mathbf{u}\rangle - \frac{1}{2}\langle(\mathbf{G}\cdot\mathbf{u})^2\rangle + .... \qquad ...(2)$$

But $\langle\mathbf{G}\cdot\mathbf{u}\rangle = 0$, because $\mathbf{u}$ is a random thermal displacement uncorrelated with the direction of $\mathbf{G}$. Further,

$$\langle(\mathbf{G}\cdot\mathbf{u})^2\rangle = G^2\langle u^2\rangle\,\langle\cos^2\theta\rangle = \frac{1}{3}\langle u^2\rangle G^2$$

The factor $\frac{1}{3}$ arises as the geometrical average of $\cos^2\theta$ over a sphere.

The function $\exp\left(-\frac{1}{6}\langle u^2\rangle G^2\right) = 1 - \frac{1}{6}\langle u^2\rangle G^2 + ....$ ...(3)

has the same series expansion as Eq. (2) for the first two terms shown here. For a harmonic oscillator all terms in the series (2) and (3) can be shown to be identical. Then the scattered intensity, which is the square of the amplitude, is

$$I = I_0 \exp\left(-\frac{1}{3}\langle u^2\rangle G^2\right) \qquad ...(4)$$

where $I_0$ is the scattered intensity from the rigid lattice. The exponential factor is the **Debye-Waller factor**.

Here $\langle u^2\rangle$ is the mean square displacement of an atom. The thermal average potential energy $\langle U\rangle$ of a classical harmonic oscillator in three dimensions is $\frac{3}{2}k_BT$, whence

$$\langle U\rangle = \frac{1}{2}C\langle u^2\rangle = \frac{1}{2}M\omega^2\langle u^2\rangle = \frac{3}{2}k_BT, \qquad ...(5)$$

where $C$ is the force constant, $M$ is the mass of an atom, and $\omega$ is the frequency of the oscillator. We have used the result $\omega^2 = C/M$. Thus, the scattered intensity is

$$I(hkl) = I_0 \exp(-k_BT\,G^2/M\omega^2), \qquad ...(6)$$

7 P. P. Ewald, private communication.

8 This is the Einstein model of a solid; it is not a very good model at low temperatures, but it works well at high temperatures. For a general treatment of scattering by thermal fluctuations, see *QTS*, Chapter 20.

where *hkl* are the indices of the reciprocal lattice vector **G**. This classical result is a good approximation at high temperatures.

For quantum oscillators $\langle u^2 \rangle$ does not vanish even at $T = 0$; there is zero-point motion. On the independent harmonic oscillator model the zero-point energy is $\frac{3}{2}\hbar\omega$: this is the energy of a three-dimensional quantum harmonic oscillator in its ground state referred to the classical energy of the same oscillator at rest. Half of the oscillator energy is potential energy, so that in the ground state

$$\langle U \rangle = \frac{1}{2} M\omega^2 \langle u^2 \rangle = \frac{3}{4}\hbar\omega; \quad \langle u^2 \rangle = 3\hbar/2M\omega, \qquad ...(7)$$

whence, by Eq. (4), $$I(hkl) = I_0 \exp(-\hbar G^2/2M\omega) \qquad ...(8)$$

at absolute zero. If $G = 10^9$ cm$^{-1}$, $\omega = 10^{i4}$ s$^{-1}$, and $M = 10^{-22}$ g, the argument of the exponential is approximately 0.1, so that $I/I_0 \simeq 0.9$. At absolute zero 90 per cent of the beam is elastically scattered and 10 per cent is inelastically scattered.

We see from Eq. (6) and from Fig. 3.75 that the intensity of the diffracted line decreases, but not catastrophically, as the temperature is increased. Reflections of low *G* are affected less than reflections of high *G*. The intensity we have calculated is that of the coherent diffraction or the elastic scattering in the well-defined Bragg directions. The intensity lost from these directions is the inelastic scattering and appears as a diffuse background. In inelastic scattering the X-ray photon causes the excitation or de-excitation of a lattice vibration, and the photon changes direction and energy.

At a given temperature the Debye–Waller factor of a diffraction line decreases with an increase in the magnitude of the reciprocal lattice vector **G** associated with the reflection. The larger |**G**| is the weaker the reflection at high temperatures. The temperature dependence of the reflected intensity for the (*h*00) reflections of aluminum is shown in Fig. 3.75. The theory we have worked out here for X-ray reflection applies equally well to neutron diffraction and to the **Mössbauer effect**, the recoilless emission of gamma rays by nuclei bound in crystals.

## 3.40 THE CRYSTALLOGRAPHIC ZONE-AXIS

In electron optical analysis, a property of considerable use in electron-diffraction is the zone-axis [*uvw*] of a crystal. Crystal planes (*h k l*) parallel to a zone-axis [*u v w*] are regarded as planes of the zone. Since $\vec{r}_{hkl}$, the reciprocal lattice vector, is normal to the plane (*h k l*), therefore

$$\vec{r}^{\,*} \cdot \vec{r} = 0 \qquad ...(1)$$

expresses the condition that (*h k l*) belongs to the zone [*u v w*] or

$$hu + kv + lw = 0 \qquad ...(2)$$

Then, the plane (*h k l*) will be perpendicular to the plane (*u v w*).

For the cubic system, we can specify the zone-axis by [*H K L*], which is normal to a crystallographic surface [*H K L*]. Consequently, for the incident electron beam parallell to

[*H K L*], or normal to (*H K L*), the corresponding reflections from allowed {*h k l*} planes are found from

$$Hh + Kk + Ll = 0] \quad ...(3)$$

*The Weiss zone law:*

It states that the plane (*h k l*) is in the zone [*U V W*] if and only if

$$hu + kv + lw = 0 \quad ...(4)$$

The numerical values of *U V W* for a given pair of planes in a zone can be evaluated by cross-multiplication rule.

Write each index twice, the second below the first. Omit the first and the last figures. Multiply the figures joined by the arms of each cross:

$$\begin{array}{c|ccccc|c} h_1 & k_1 & l_1 & h_1 & k_1 & & l_1 \\ & & \times & \times & \times & & \\ h_2 & k_2 & l_2 & h_2 & k_2 & & l_2 \end{array}$$

Then,

$$U = k_1 l_2 - k_2 l_1$$
$$V = l_1 h_2 - l_2 h_1$$
$$W = h_1 k_2 - h_2 k_1 \quad ...(5)$$

The indices of a plane (*h k l*) which lies in the zones $[U_1 V_1 W_1]$ and $[U_2 V_2 W_2]$ is given by the solution of the simultaneous equations

$$hU_1 + kV_1 + lW_1 = 0$$
$$hU_2 + kV_2 + lW_2 = 0$$

i.e.,

$$\frac{h}{\begin{vmatrix} V_1 & W_1 \\ V_2 & W_2 \end{vmatrix}} = \frac{k}{\begin{vmatrix} W_1 & U_1 \\ W_2 & U_2 \end{vmatrix}} = \frac{l}{\begin{vmatrix} U_1 & V_1 \\ U_2 & V_2 \end{vmatrix}}$$

or

$$(h\,k\,l) \equiv (V_1 W_2 - V_2 W_1,\ W_1 U_2 - W_2 U_1,\ U_1 V_2 - U_2 V_1)$$

Further, the condition that the three planes $(h_1 k_1 l_1)$, $(h_2 k_2 l_2)$ and $(h_3 k_3 l_3)$ belong to the same zone can be worked out to be from the determinant formed by then = 0, i.e.,

$$\begin{vmatrix} h_1 & k_1 & l_1 \\ h_2 & k_2 & l_2 \\ h_3 & k_3 & l_3 \end{vmatrix} = 0 \quad ...(6)$$

**Example 3.12** Verify that the planes $(\bar{1}10)$, $(\bar{3}10)$ and $(\bar{1}\bar{3}2)$ belong to the same zone. If so, what is the symbol for the zone-axis.

***Solution***

$$\begin{vmatrix} h_1 & k_1 & l_1 \\ h_2 & k_2 & l_2 \\ h_3 & k_3 & l_3 \end{vmatrix} = \begin{vmatrix} \bar{1} & 1 & 0 \\ \bar{3} & 1 & 1 \\ \bar{1} & \bar{3} & 2 \end{vmatrix}$$

$$= \overline{1}(2 + 3) - 1(\overline{6} + 1) + 0(9 + 1)$$

$$= -5 + 5 + 0 = 0$$

Therefore, they belong to the same zone.

$$U = (k_1 l_2 - k_2 l_1) = (1 \times 1 - 0 \times 1)$$

$$V = (l_1 h_2 - l_2 h_1) = (0 \times \overline{3} - \overline{1} \times 1)$$

$$W = (h_1 \times k_2 - h_2 \times k_1) = (\overline{1} \times 1 - \overline{3} \times 1)$$

$[U\ V\ W] \equiv [1\ 1\ 2]$ is the zone axis.

## 3.41 ELECTRON DIFFRACTION

In fact, a unit cell of a crystal defines the character of electron diffraction. In first approximation, diffraction from a crystal lattice can be considered a kinematic scattering process ideally described by a wave reflection from $\{h\ k\ l\}$ planes as expressed in Bragg equation. (X-ray diffraction results from photon reflection from nuclei, whereas electron diffraction results from diffraction due to lattice atoms).

Diffraction from a plane $(h\ k\ l)$ in a real crystal can be treated as a common vector in the reciprocal lattice feeding its intensity into a point. In other words, a diffraction intensity point in reciprocal space corresponds to a plane $(h\ k\ l)$ in the real crystal.

If the incident wave vector $\vec{K}_0$ is directed along a real crystal lattice direction $[H\ K\ L]$ normal to plane $(h\ k\ l)$ in the crystal, then the diffraction condition is satisfied in reciprocal space for intersection of Ewald sphere with reciprocal lattice points $h\ k\ l$ which are equivalent to reflecting planes $\{h\ k\ l\}$ in the real crystal. That is, with the incident electron beam normal to the Ewald sphere and directed along a zone-axis $[H\ K\ L]$ of the real crystal, diffraction spots are expected in the diffraction pattern satisfying the condition,

$$Hh + Kk + Ll = 0$$

These diffraction spots satisfy the diffraction condition,

$$\frac{P \sin \beta a_0}{\beta a_0} + \cos \beta a_0 = \cos |\vec{K}_0|\, a_0 \qquad \text{(see chapter 7, Eq. 7.49)}$$

The diffraction condition geometrically expressed in the Ewald sphere also characterises the Brillouin zone boundaries in the reciprocal lattice or $k$-space. In fact, the incident electron wave vector $\vec{k}_0$ (describing the Ewald sphere in reciprocal) lattice terminates on a Brillouin zone. The first Brillouin zone for the boundary condition

$$|\vec{K}_0|\, a_0 = \pm \pi \qquad ...(1)$$

is constructed in the reciprocal lattice by bisecting the shortest $\vec{g}_{hkl}$ on the Ewald sphere by a radius vector that represents the edge of a plane of the Brillouin zone boundary (Fig. 3.76).

The polyhedron generated by the intersection of these planes with the Ewald sphere forms the first Brillouin zone boundary.

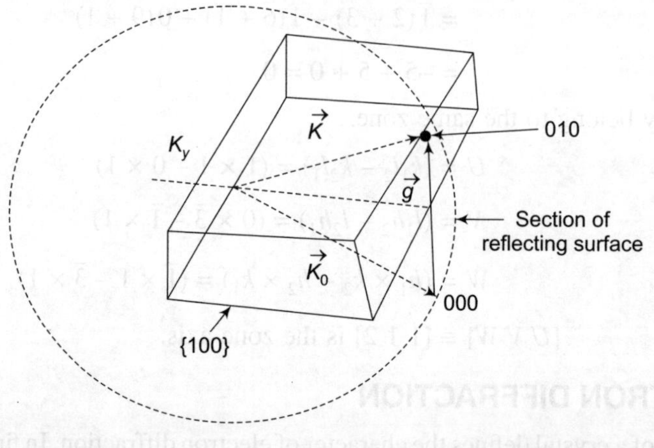


**Figure 3.76** Intersection of Ewald sphere with equivalent {1 0 0} reflection planes

## 3.42 GEOMETRICAL NATURE OF ELECTRON DIFFRACTION-PATTERNS

Since a crystal is a regular, repeating, geometric arrangement of atoms, it is expected that the resulting diffraction pattern would maintain a similar regularity, even if the diffraction zone is composed of random arrays of such crystals.

The diffraction condition is met when the Ewald sphere or sphere of reflection intersects a reciprocal lattice point. Further, the reciprocal lattice vectors are so much smaller than the corresponding reflecting sphere radius that to a good approximation, the Ewald sphere intersects a plane section of the reciprocal lattice. Considering this feature, the geometrical relationship between the reciprocal lattice and the diffraction pattern recorded on the fluorescent screen of an electron microscope (or a photographic plate) can be established by considering similar triangle in Fig. 3.77.

In Fig. 3.77, if $R$ is the distance from the centre spot 000 some reciprocal lattice point $h\ k\ l$, as measured from the photographic plane situated at an effective distance $L$ from the diffracting crystal or crystal normal to the direction of the incident beam vector $\vec{K}_0$, then

$$\frac{|\vec{g}_{hkl}|}{1/\lambda} = \frac{R}{L}$$

or

$$R = \lambda L\ |\vec{g}_{hkl}| \quad ...(2)$$

where, $\vec{g}_{hkl}$ is normal to $\vec{K}_0$ and $\vec{s}$, the deviation from the reflecting sphere is negligible.

Thus, the radii $R$ measured in the electron diffraction pattern are simply the magnitudes of the reciprocal lattice vectors multiplied by scale factor. The diffraction spots represent the concentrations of diffracted intensity as obtained due to intersection of the intensity profile by the Ewald sphere. The intensity profile at the reciprocal lattice points depends on crystal shape, crystallography and composition.

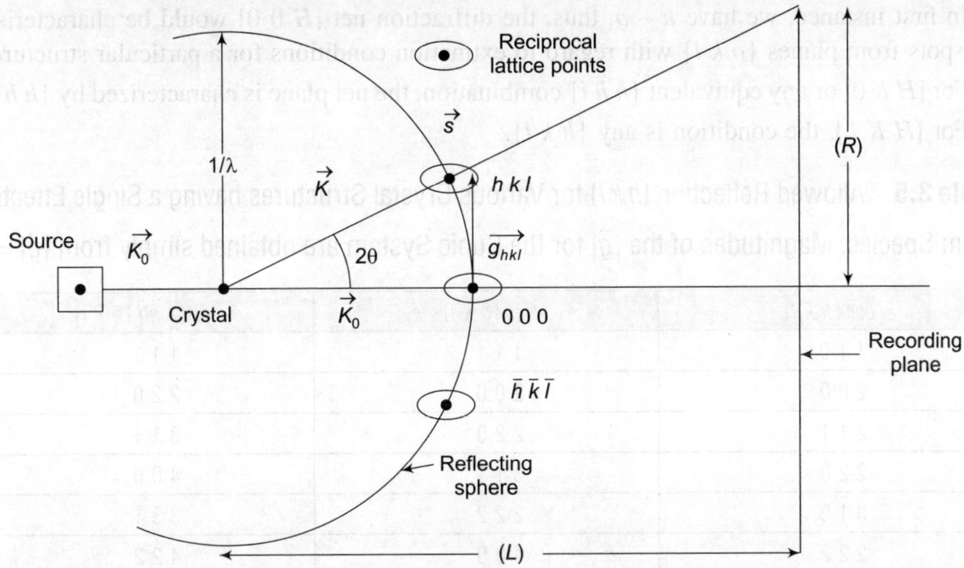


**Figure 3.77** We can think of the diffraction pattern simply as a projection of the reciprocal lattice points onto the recording plane (photographic plate)

The selected area electron diffraction (SAD) technique allow the various diffracted beams to be isolated, the image signal information specific to these beams is then observed in the dark field image. In order to characterise the crystallography of a selected grain area, the diffraction pattern must be recognizable as being unique to a particular reciprocal lattice projection. The basis of the method of identifying a spot diffraction pattern is indexing the spots.

## 3.43 INDEXING OF ELECTRON DIFFRACTION SPOT PATTERNS

If the incident electron beam is directed normal to a reciprocal lattice plane section, along zone axis $[H\,K\,L]$ of the real crystal, the resulting diffraction spots (intensities) due to projection of interference regions at the reciprocal points, is described by

$$Hh + Kk + Ll = 0 \qquad ...(1)$$

In indexing a spot pattern, we let the main beam spot represent the origin 0 0 0 and measure the distance to each unique diffraction spot symmetrically disposed with respect to main beam spot as reference. The distances (radii) for a known crystal structure represent the magnitude $|\vec{g}_{hkl}|$. Since the diffraction pattern is a reciprocal lattice and not a plane projection, the crystal surface plane (zone axis $[H\,K\,L]$ can be determined by considering the value of $|\vec{g}_{hkl}|$ measured from spots in the pattern.

We can construct the diffraction net from any zone-axis from Eq. (1). Let us consider the three possibilities

$$[H\,0\,0],\ [H\,K\,0] \text{ and } [H\,K\,L]$$

In first instance, we have $h - o$, thus, the diffraction net $[H\ 0\ 0]$ would be characterised by spots from planes $\{o\ k\ l\}$ with regard to extinction conditions for a particular structure.

For $[H\ K\ 0]$ or any equivalent $\{h\ \bar{h}\ l\}$ combination, the net plane is characterized by $\{h\ h\ l\}$.

For $[H\ K\ L]$, the condition is any $\{h\ k\ l\}$.

**Table 3.5** Allowed Reflection $\{h\ k\ l\}$ for Various Crystal Structures having a Single Effective Atom Species. Magnitudes of the $|\vec{g}|$ for the Cubic System are obtained simply from $|\vec{g}| = \frac{1}{d}$

| *bcc* $\{h\ k\ l\}$ | *fcc* $\{h\ k\ l\}$ | *diamond* $\{h\ k\ l\}$ |
|---|---|---|
| 1 1 0 | 1 1 1 | 1 1 1 |
| 2 0 0 | 2 0 0 | 2 2 0 |
| 2 1 1 | 2 2 0 | 3 1 1 |
| 2 2 0 | 3 1 1 | 4 0 0 |
| 3 1 0 | 2 2 2 | 3 3 1 |
| 2 2 2 | 4 0 0 | 4 2 2 |
| 3 2 1 | 3 3 1 | 5 1 1, 3 3 3* |
| 4 0 0 | 4 2 0 | 4 4 0 |
| 4 1 1, 3 3 0* | 4 2 2 | 5 3 1 |
| 4 2 0 | 5 1 1, 3 3 3* | 6 2 0 |
| 3 3 2 | 4 4 0 | 5 3 3 |
| 4 2 2 | 5 3 1 | 4 4 4 |

*equivalent $|\vec{g}|$

Suppose we wish to construct the diffraction net for the [1 0 0] zone-axis of an fcc crystal section. We observe from Table 3.5 that the first reflection $\{o\ k\ l\}$ will be $\{0\ 0\ 2\}$. Thus, we can have $(0\ 0\ 2)$, $(0\ 2\ 0)$, $(0\ 0\ \bar{2})$ and $(0\ \bar{2}\ 0)$ symmetrically disposed with respect to $(0\ 0\ 0)$. The next reflections are characterised by $\{0\ 2\ 2\}$ and we have $(0\ 2\ 2)$, $(0\ \bar{2}\ \bar{2})$, $(0\ \bar{2}\ 2)$ and $(0\ 2\ \bar{2})$. This scheme continues for higher order reflections with the resulting construction of a [1 0 0] diffraction net as shown in Fig. 3.78 (a). The [1 0 0] zone-axis is normal to the (1 0 0) plane. Consequently, the diffraction net can be characterised as the (1 0 0) diffraction pattern, i.e., the pattern that results for the (1 0 0) plane normal to the electron beam.

Let us consider now the (1 1 0) plane in fcc. That is, we construct the diffraction net for [1 1 0] in fcc. We observe again from table that the first expected reflections are $\{1\ \bar{1}\ 1\}$, i.e., $(1\ \bar{1}\ 1)$, $(\bar{1}\ 1\ \bar{1})$, $(\bar{1}\ 1\ 1)$ and $(1\ \bar{1}\ \bar{1})$. The next reflections are $\{0\ 0\ 2\}$ or $(0\ 0\ \bar{2})$ and $(0\ 0\ \bar{2})$. Then, $\{2\ \bar{2}\ 0\}$: $(2\ \bar{2}\ 0)$ and $(\bar{2}\ 2\ 0)$. [Note: $[H\ K\ O]$ for $\{h\ \bar{k}\ l\}$ satisfies $1h + 1(-h) + 0l = 0$ using [1 1 0]]. Figure 3.78 (b) illustrates the corresponding (1 1 0) diffraction net.

Thus, equation $Hh + Kk + Ll = 0$ allows all allowed expected reflections $\{h\ k\ l\}$ to be determined for the electron beam along $[H\ K\ L]$. These diffraction spot patterns correspond to a specific crystalline orientation plane $(H\ K\ L)$ of the lattice.

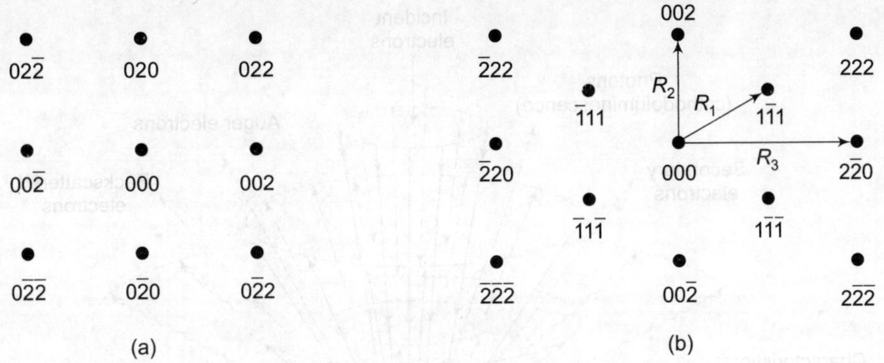


**Figure 3.78** Diffraction nets for (a) (1 0 0) plane normal to the electron beam and (b) (1 1 0) plane normal to the electron beam. The indices represent crystal directions with respect to the central beam or zero order reflection 0 0 0 or reflections from corresponding planes {*h k l*}

Having constructed such diffraction nets or equivalent surface plane projection (*H K L*) for a particular crystal system, the indexing of SAD patterns can be performed simply by comparing. For example, refer to Fig. 3.78 (b). Where the radii indicated as $R_1$, $R_2$ and $R_3$, etc., correspond to the magnitudes $|\vec{g}_{hkl}|$. Thus, by comparing $\frac{R_1}{R_2}, \frac{R_2}{R_3}, \frac{R_1}{R_3}$, etc., we can compare the indexed spots quite easily due to the fact that the camera constant cancels and only the relative dimensions (radii) must be measured. We observe from Fig. 3.78 (b) that

$$\frac{R_1}{R_2} = \frac{\sqrt{3}}{2}, \frac{R_2}{R_3} = \frac{2}{2\sqrt{2}}, \frac{R_1}{R_3} = \frac{\sqrt{3}}{2\sqrt{2}}$$

The ratio method of indexing is then expressed generally by

$$R_n/R_m = \left|\vec{g}_{h_n k_n l_n}\right| / \left|\vec{g}_{h_m k_m l_m}\right| \qquad ...(2)$$

## 3.44 ELECTRON BEAM – SPECIMEN INTERACTION

On the basis of the knowledge of interactions between the electron beam and specimen (as illustrated in Fig. 3.79) a number of different ways (analytical modes as signals of various kinds) are available.

The characteristics of an electron beam as it transmits through a specimen depend upon the thickness and chemical composition of the specimen, whereas, the signals generated on the surface of the specimen are independent of its thickness. Besides normal transmission and reflection of the incident electron beam by the specimen, a small portion excites characteristic X-rays. The characteristic X-rays so emitted can be analysed with regard to the area or volume of their origin within the specimen on the basis of measurement of either their wavelengths or energies, and precise determination or detection of the chemical elements present within the selected area or volume of the electron-specimen interaction can be made. These techniques are generally known as wavelength dispersive and energy dispersive X-ray

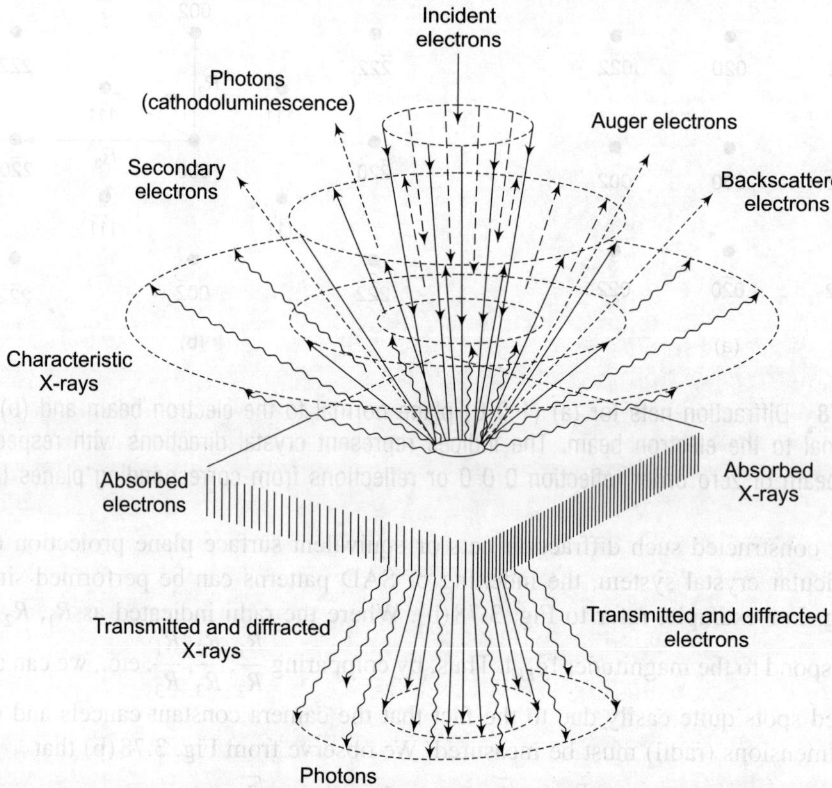


**Figure 3.79** Electron beam-specimen interaction and emission of various modes of signals

spectrometry respectively. Similarly, back-scattered electrons and secondary electrons emitted are monitored and used in the analysis of surface-composition, change and topography. The diffracted electrons on the other hand provide crystallographic and compositional information about the specimen.

A complimentary analytical tool is energy dispersive X-ray spectrometer with particularly high detection efficiency (Li through Cr) within biological and non-biological materials which has come up by the name *energy loss spectrometry* (ELS) in the domain of electron microscopy. This technique collects and analyses the inelastically scattered electrons through the specimen.

A modern versatille electron microscope incorporates built-in or optional facilities to qualitatively or quantitatively detect the emitted signals in various modes and to provide maximum possible information about a specimen under observation.

Increasing popularity of electron microscopy has triggered fresh researches and development work in several areas, e.g., development of ultrafine gram electron-sensitive photographic emulsions for recording electron micrographs of very high resolution, development and design of very clean and efficient, non-contaminating vacuum pumping systems and electronically controlled automatic functional systems, besides, considerable efforts have gone into the development of miniaturised sensitive detectors for various signals emitted from the specimen and their incorporation in instrument design without disturbing the general electron optics.

## 3.45 ELECTRON MICROSCOPES

The resolution of a conventional light microscope is limited by diffraction. Under ideal conditions, the smallest distance on the object that can be resolved is roughly half the wavelength of light. Using 400 nm as the shortest wavelength in the visible part of the spectrum, a light-microscope can resolve distances ~200 nm. This is a large distance on the scale of atoms or molecules: the distance between atoms in a solid is typically only about 0.2 nm.

For better resolution, one possibility is to use an ultraviolet microscope which use wavelengths down to about 200 nm. For wavelengths shorter than that, making effective lenses becomes too difficult.

A beam of electrons can easily be made to have a wavelength ~0.2 nm or smaller. To make electrons with a wavelength of 0.2 nm, we need to accelerate them through a potential difference of only 37.4 V. Typically, the electrons used in an electron microscope are more energetic than that, and so have shorter wavelengths. However, the resolution in this case is limited by imperfections in the electromagnetic "lenses" used to focus the electron beam and from the image.

The working of an electron microscope can be explained without talking explicitly about the wave nature of electrons. Similar to light microscopes using geometric optics for light rays, we have trajectories of electrons (in electron microscope) which are bent by magnetic lenses and scattered by the sample being studied. The advantage of the electron microscope over the light microscope is the smaller wavelength of the electrons, which extends "geometric electron optics" to much smaller objects.

Electron microscopes come in several forms. The one closest to the familiar light microscope is called the *transmission electron microscope* (TEM). When a beam of parallel electrons passes through the sample, electrons that are scattered by a point within the sample are focused back to a point on a screen by magnetic lenses, forming a real image of the sample on the screen. The electrons must pass through the sample without being slowed down appreciably, so, the TEM works only for thin samples—about 100 nm of maximum thickness. The TEM can resolve details as small as 0.2 nm—about 500 times better than an ultraviolet microscope with wavelength 200 nm.

Another kind of electron microscope, the *scanning electron microscope* (SEM), uses a magnetic lens to focus a beam of electrons onto one point on the sample at a time. These primary electrons knock secondary electrons out of the sample, an electron collector detects the number of secondary electrons produced. The electrical signal from the collector is fed to a monitor to be viewed directly or to a computer. The focused electron beam is swept across the sample by a beam deflector. The number of secondary electrons emitted at each spot on the sample is measured and the result displayed on a screen to give an image of the specimen.

The resolution of the SEM is not as good as TEM, ~about 10 nm at best. But the SEM does not require thin samples and since, it is sensitive to the surface contour of the specimen, it is much better at imaging thee-dimensional structures.

The *scanning transmission electron microscope* (STEM) scans the sample point by point, like the SEM, but it detects the electrons transmitted through the sample.

## 3.46 TRANSMISSION ELECTRON MICROSCOPY

In a transmission electron microscope (TEM), electrons from an electrons gun (i.e. electron source) enter the sample, are scattered as they pass through, are focused by an objective lens, are amplified by a magnifying (projector) lens and finally produce the desired image in the manner reading from left to right shown in Fig. 3.80.

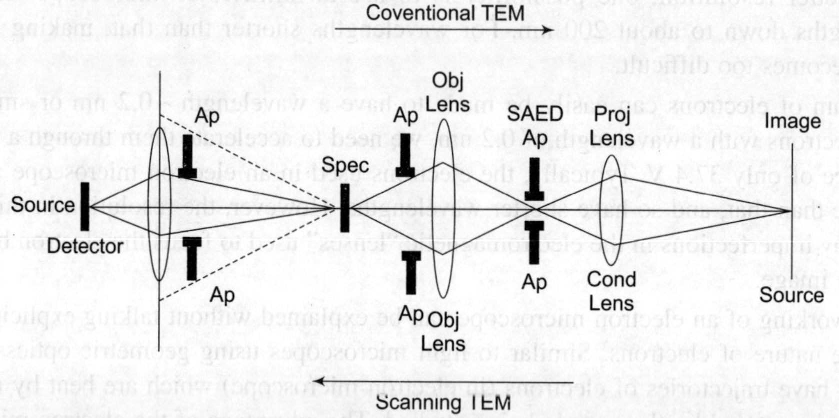


**Figure 3.80** Ray diagram of a conventional TEM (top path) and of a scanning transmission electron microscope (bottom path). The selected area electron diffraction (SAED) aperture (Ap) and the sample or specimen (spec) are indicated as well as the objective (obj) and projector (proj) or condenser (cond) lenses

In scanning electron microscopy, the electron optics given in Fig. 3.80 is similar to that of TEM except that in the (former) TEM case, the electrons travel from left to right, whereas in SEM, they move in the opposite direction, from right to left in the figure.

A transmission electron microscope can form images by the use of the selected area electron diffraction (SAED) aperture located between the objective and projector lenses shown in Fig. 3.80. The main part of the electron beam transmitted by the sample consists of electrons that have not undergone any scattering. The beam also contains electrons that have lost energy through inelastic scattering with no deviation of their paths, and electrons that have been reflected by various *hkl* crystallographic planes. To produce what is called a *bright field image*, the aperture is inserted so that it allows only the main undeviated transmitted electron beam to pass (Fig. 3.81). The bright field image is observed at the detector or viewing screen. If the aperture is positioned to select only one of the beams reflected from a particular *hkl* plane, the result is the generation of a *dark-field image* at the viewing screen.

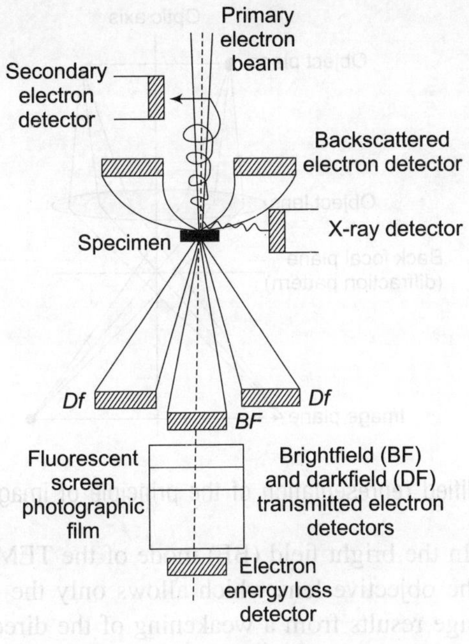


**Figure 3.81** Positioning of signal detectors in electron microscope column

The details of the dark field image that is formed can depend on the particular diffracted beam (particular *hkl* plane) that is selected for the imaging. Figure 3.80 shows the locations of the bright field (BF) and dark field (DF) aperture positions.

A TEM is based on electron diffraction by the specimen. Electron diffraction is of two types – *low energy electron diffraction* (LEED) and *high energy electron diffraction* (HEED). Elcctrons with cncrgy 10 to 200 eV are supposed to have low energy. Such electrons penetrate to depths of the order of tens of nanometers which renders the method of limited use for examination of the atomic layers close to the surface. High energy electron diffraction is generally carried out using electrons accelerated by potentials of 50–100 kV (or higher ~1000 kV). This method gives valuable direct information on the arrangement of atoms in crystallites of the sample film and their relative orientations.

## 3.47 IMAGE FORMATION IN TRANSMISSION ELECTRON MICROSCOPY

A simplified representation of the ray path leading to the first intermediate image in the image-plane of the object lens is depicted in Fig. 3.82. Electrons which come from the condenser system of the TEM are scattered by the sample, situated in the object plane of the object lens. Electrons scattered in the same direction are focussed in the back focal plane, and, as a result, a diffraction pattern is formed there. Electrons coming from the same point of the object are focused in the image plane. In the TEM, the first intermediate image is magnified by further lenses (projective system). In general, there are two image modes of TEM:

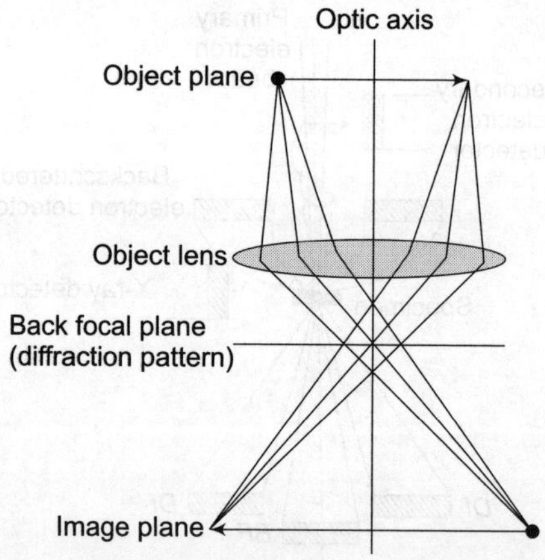


**Figure 3.82** Simplified representation of the principle of image formation in TEM

(i) *Bright field image.* In the bright field (BF) mode of the TEM, an aperture is placed in the back focal plane of the objective lens which allows only the direct beam to pass (Fig. 3.83). In this case the image results from a weakening of the direct beam by its interaction with the sample.

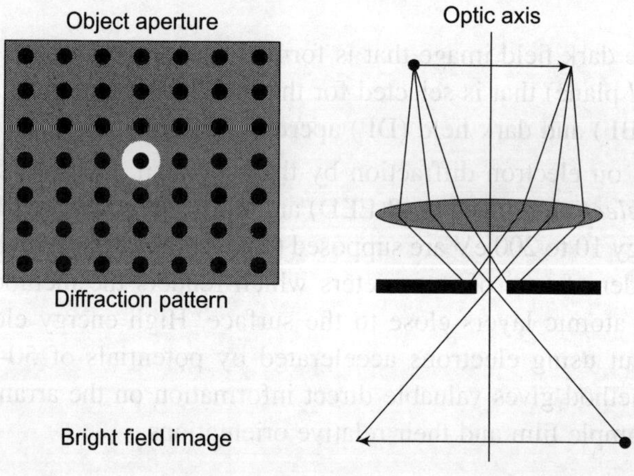


**Figure 3.83** Schematic illustration of bright filed (BF) image mode

Therefore, mass-thickness and diffraction contrast contribute to image formation: thick areas, areas in which heavy atoms are enriched, and crystalline areas appear with dark contrast.

(ii) *Dark field image mode.* In DF images, the direct beam is blocked by the aperture while one or more diffracted beams are allowed to pass the objective aperture (Fig. 3.84). Since diffracted beams have strongly interacted with the specimen, useful information is present in DF images, e.g., about planar defects, stacking faults or particle sizes.

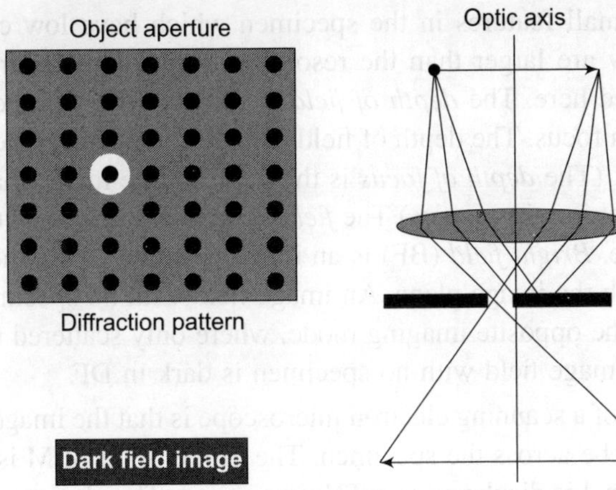


**Figure 3.84** Schematic illustration of dark field (DF) image mode

## 3.48 SCANNING ELECTRON MICROSCOPE

Microscopy is the study of structure and morphology of objects with the use of a microscope. Microscopes range from optical microscopes, which resolve details on the micrometre level, to transmission electron microscopes that can resolve details less than 1 nanometre across. Table 3.6 shows the basic properties of the different microscopes.

**Table 3.6** Properties of Various Microscopes

| *Technique* | *OM* | *SEM* | *TEM (incl, STEM)* |
|---|---|---|---|
| Resolution | 300 nm | 10 nm | 0.2 nm |
| Magnification range | 2–2000 | 20–1 × $10^5$ | 200–2 × $10^6$ |
| Can observe | surface, or bulk if transparent | surfaces only | 'bulk', but very thin films, less than 0.2 μm |
| Specimen environment | ambient | high vacuum | high vacuum |
| Radiation damage | none | little | severe |
| Specimen preparation | easy | easy | very difficult |
| Chemical analysis | no, unless connected to μ Raman | yes, x-ray | yes, x-ray and electron energy loss |
| Can detect molecular orientation | yes | no | yes |

In microscopy, resolution and contrast are key parameters. *Resolution* is the minimum distance between two object features at which they can still be seen as two features. The *contrast* of a feature is the fractional change in image brightness that it causes. If two features have high contrast but are separated by less than the resolution, they will appear as one object, visible but not resolved. If their contrast is very low (below about 0.05), they will not be visible at all. Small features in the specimen which have low contrast may not be observable, even if they are larger than the resolution limit of the instrument. Some more terms need to be defined here. The *depth of field* is the depth or thickness of the specimen that is simultaneously in focus. The depth of field is large for electron microscopes and small for optical microscopes. (The *depth of focus* is the depth of the image that is in focus at one time and is not important in microscopy.) The *field of view* is the area of the specimen which is included in the image. *Bright field* (BF) is an imaging mode where the direct unscattered beam is allowed to reach the image plane. An image field with no specimen in it is bright in BF. *Dark field* (DF) is the opposite imaging mode, where only scattered radiation is allowed to form the image. An image field with no specimen is dark in DF.

The essential feature of a scanning electron microscope is that the image is formed point by point, by scanning a probe across the specimen. The probe of an SEM is a focused electron beam and a detected signal is displayed as a TV type image. The electron beam in the display tube moves in synchrony with the probe on the specimen. In the SEM, there is *compositional contrast* and *topographical contrast*, the first depends on the composition of the material and the second on its form.

Figure 3.85 shows a block diagram of an SEM showing the connection between display scanning and probe scanning. The magnification is simply the ratio (linear size of the image)/(size of the region scanned on the specimen). The magnification can therefore be controlled without having any effect on other imaging conditions, much as focus. Other advantages of the SEM include higher resolution than in optical microscopy, with a much larger depth of field. The specimen chamber is large and samples from several millimetres to several inches in diameter can be accommodated, Specimen preparation is generally quite simple.

Sample must be fixed to a specimen stub. Non-conductive specimens require conductive coatings or the use of low accelerating voltages. SEM images are easy to interpret. They appear as though the specimen is viewed from the source of the scanning beam, and illuminated by a fight at the detector position.

Imaging by scanning allows any radiation from the specimen (or any result of its interaction with the beam) to be used to form the image. The appearance of the image will depend on the interaction involved and the detector and signal processing used. The optical resolution is determined by the size of the specimen region from which the signal is derived. This is related to the *interaction volume*, the region where the beam interacts with the specimen. Calculation of the interaction of the electron beam with a solid shows that the interaction volume increases at high accelerating voltage and for low atomic number of the specimen.

### 3.48.1 Imaging Signals

Three important signals from the specimen are backscattered electrons, secondary electrons and X-rays. Backscattered electrons are primary beam electrons which have been elastically scattered by nuclei in the sample and escape from the surface. The faction escaping varies from 0.06 for carbon to 0.5 for gold, so a backscattered electron image has compositional

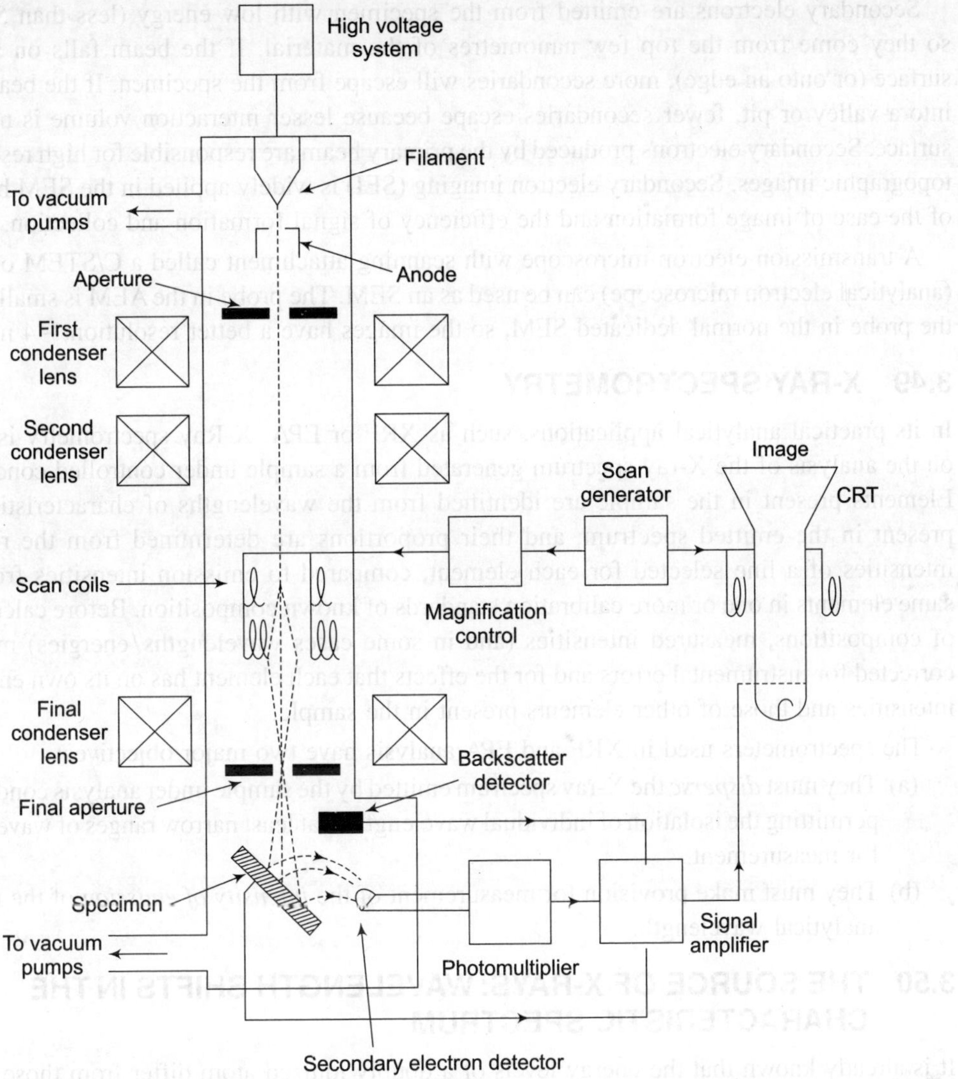


**Figure 3.85** Schematic diagram of a scanning electron microscope. There are two pairs of defection coils in the SEM column. Dobule deflection allows the scanning beam to pass through the final aperture. Four pairs are actually used for double defecation in *x* and *y* directions

contrast. Backscattered electrons have a high energy and they can come from depths of 1 μm or more within the specimen. They leave the surface from a wide area and this means that the resolution in backscattered electron image is low, only about 1 μm. Backscattered electron imaging combined with X-ray microanalysis is a powerful method of determination of the local chemical composition of a material. This is known as energy dispersive analysis of X-rays (EDAX).

Secondary electrons are emitted from the specimen with low energy (less than 50 eV), so they come from the top few nanometres of the material. If the beam falls on a tilted surface (or onto an edge), more secondaries will escape from the specimen. If the beam falls into a valley or pit, fewer secondaries escape because lesser interaction volume is near the surface. Secondary electrons produced by the primary beam are responsible for high resolution topographic images. Secondary electron imaging (SEI) is widely applied in the SEM because of the ease of image formation and the efficiency of signal formation and collection.

A transmission electron microscope with scanning attachment called a C/STEM or AEM (analytical electron microscope) can be used as an SEM. The probe in the AEM is smaller than the probe in the normal dedicated SEM, so the images have a better resolution, ~4 nm.

## 3.49 X-RAY SPECTROMETRY

In its practical analytical applications, such as XRF or EPA, X-Ray spectrometry is based on the analysis of the X-ray spectrum generated from a sample under controlled conditions. Elements present in the sample are identified from the wavelengths of characteristic lines present in the emitted spectrum, and their proportions are determined from the relative intensities of a line selected for each element, compared to emission intensities from the same elements in one or more calibration standards of known composition. Before calculation of compositions, measured intensities (and in some cases wavelengths/energies) must be corrected for instrumental errors and for the effects that each element has on its own emission intensities and those of other elements present in the sample.

The spectrometers used in XRF and EPA analysis have two major objectives:

(a) They must *disperse* the X-ray spectrum emitted by the sample under analysis conditions, permitting the isolation of individual wavelength or at least narrow ranges of wavelength for measurement.

(b) They must make provision for measurement of the *intensity of emission* of the related analytical wavelength.

## 3.50 THE SOURCE OF X-RAYS: WAVELENGTH SHIFTS IN THE CHARACTERISTIC SPECTRUM

It is already known that the energy levels of a doubly ionized atom differ from those of the same atom in the singly ionized state. Since elements in compounds have different electronic structures from those in the free state, particularly in so far as valance electrons are concerned, it is possible that the consequent shifts in energy levels might result in shifts in energy, or wavelength of characteristic spectral emissions. This would appear to be particularly likely for line emissions produced by transitions involving electrons from the valence or near-valance bands.

The lines commonly employed in X-ray spectrometry lie mostly between 1 and 10 Å, and most of them are produced by transitions from $L$ or $M$ levels to $K$ or $L$ levels (i.e., most analytical lines are selected from the $K$ or $L$ spectra of the various elements). For the heavier elements, valence electrons are not involved in these transitions, and the inner transition levels are "shielded" from the effects of small energy shifts in the valance band. Wavelength shifts in analytical lines of the *K* and *L* spectra are thus usually too small to be of practical consequence.

For light elements, however, the situation may be significantly different. For example, the elements between $Z = 13$ and $Z = 17$ have unfilled 3*p* orbitals, which are therefore involved in bonding with other atoms. Since the $K\beta$ line is produced by a transition from the 3*p* orbital, it is to be expected that $K\beta$ emission wavelengths from these elements (Al, Si, P, S, Cl) should show some dependence on chemical state. The same effect would be expected in the $K\alpha$ emissions of elements with unfilled 2*p* orbitals (*B*, *C*, *N*, *O*, *F*).

Ground state aluminium atoms, for example, have filled *K*, *L* and 3*s* levels, together with a single 3*p* electron. The $K\beta_{1.3}$ emission is thus weak and broad, and the profile of the peak depends very much on the nature of the chemical bonding of the Al atoms (Fig. 3.86). In $Al_2O_3$, for example, the Al $K\beta$ line has a band-like structure since it arises not from the discrete energy level characteristic of an aluminium atomic orbital but instead from the energy *band* of the molecular orbital of aluminium and oxygen.

These relationships (including corresponding shifts in the *L* spectra of some of the heavier elements) can therefore be used to derive similar information concerning chemical bonding, although for this purpose X-ray methods have been largely supplanted by more sensitive and more versatile electron spectroscopic techniques. Of more direct significance to spectrometry are the errors that can arise in XRF and EPA analysis if wavelength shifts are not recognized and properly compensated.

It will be recalled that XRF and EPA are both *relative* analytical methods, i.e., they depend on comparison of the intensities of identical lines, emitted under identical conditions, from analysis samples on one hand and calibration standards on the other.

It is often tacitly assumed that an X-ray spectrometer tuned, or 'peaked', to a particular emission from a calibration standard will be equally well peaked for the analysis samples (in particular, this assumption is often made when computer controlled systems are used). However, if there is a significant wavelength shift between the two, this will not be the case. To obtain accurate analysis, it will then be necessary:

(a) to repeak the spectrometer between sample and standard, or vice versa, or
(b) to deliberately sacrifice some spectrometer resolution (e.g., by using a coarse collimator, see below), to ensure that the maximum necessary *range* of wavelengths is measured, or
(c) to apply a mathematical correction to measured intensities, based on calibration of the peak shapes.

Of these, (a) is possibly the most widely used, since it can be easily adapted to computer controlled systems (which can rapidly scan the analytical peak on each sample or standard and ensure that the intensity measurement is made on the true peak centroid).

Whenever there is doubt concerning the possible magnitude of wavelength shifts, it is essential to determine accurate peak profiles on both samples and standards and derive an appropriate strategy before making the analysis.

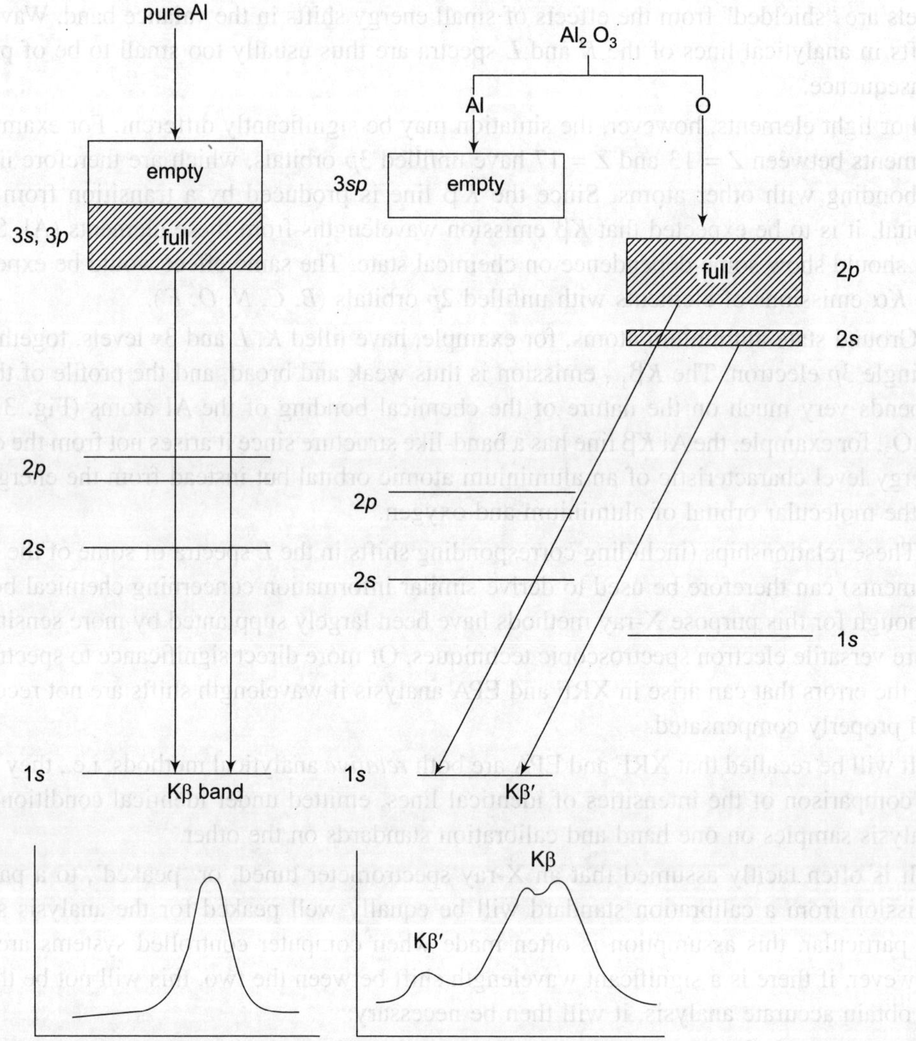


**Figure 3.86** Dependence of the profile of the Al *K*β peak on the involvement of the Al valence electrons in chemical bonding. The *K*β emission, or *peak*, is produced by transitions from the 3*s* and 3*p* levels to the 1*s* level. In Al metal, the 3*s* and 3*p* electrons are not involved in chemical bonding and the transitions produce a relatively narrow band of energies and hence a sharp, well defined peak. In $Al_2O_3$, the valence electrons combine with the oxygen 2*s* and 2*p* electrons to fill molecular orbitals which span a wider energy band. The *K*β emission becomes broader, asymmetric, and resolved into separate *K*β and satellite *K*β′ peaks

For electron excitation, either in a primary tube or in the electron microprobe, the intensity of any characteristic line (expressed as the rate of production of photons per unit time) is given by the empirically derived expression

$$I = C\, i\, (E_0 - E_c)^p$$

where $C$ is a constant, $i$ is the electron current, $E_0$ is the accelerating potential, $E_c$ is the critical excitation potential for the appropriate spectral series and the exponent $p$ has a value of about 1.7 when $E_0$ is less than about 1.7 $E_c$ and tends to smaller values at higher accelerating potentials. In other words, the efficiency of the characteristic spectrum excitation process depends on the extent to which the critical excitation potential is exceeded (but not indefinitely): in microprobe analysis this is called the *overvoltage effect*, and it partially explains why the microprobe is typically operated at voltage equal to or greater than $2E_c$ although other factors, such as background intensity or the magnitude of composition-dependent errors, may also have to be considered.

### 3.50.1 Dispersing Crystals

The first crystals to be used in X-ray spectrometry were those of natural materials, such as rock salt, mica, gypsum, calcite and quartz, whose structures had been among the first to be determined by diffraction methods. However, these have since been largely replaced by a group of synthetic crystals which give improved performance over a wider range of operating conditions.

The number of high-order reflections that can be diffracted by a particular crystal is limited by the relationship between $d$ and $\lambda$, $n\lambda$ cannot exceed $2d$, since this would require a value of sin $\theta$ greater than unity. Since we are concerned with X-ray spectrometry with analytical wavelengths between 1 and 100 Å, it would appear that the prime requirement is for a crystal of large $d$-spacing, necessary to diffract the long wavelengths. Such a crystal (e.g., lead stearate, $d$ = approximately 50 Å), however, can diffract the *short* wavelengths only in the very inefficient high orders (i.e., for relatively high values of $n$), and most of the elements emit their strongest characteristic radiations in the shorter wavelength portion of the X-ray band. Hence, it is apparent that a range of crystals is required for efficient cover of the complete energy range of interest, and most spectrometers are designed so that up to six or more dispersing crystals may be interchanged: most modern designs provide for interchange during the course of an analysis as successive elements are determined, albeit with varying degrees of convenience. Microprobes in particular, but also some XRF spectrographs, are also commonly equipped with three or more spectrometers so that several different wavelengths can be measured simultaneously. Complete wavelength coverage, and hence maximum analytical flexibility, is obtained by equipping each of these spectrometers with at least two or three interchangeable crystals.

The crystals most commonly used in Bragg spectrometers, together with some of their more important characteristics, are listed in Table 3.7. Some of these can be cleaved relatively easily parallel to the required reflecting planes (e.g., LiF, mica); others have to be carefully cut (e.g., quartz). For focusing spectrometers, the dispersing crystals have to be bent and ground to appropriate radii of curvature, for which some crystals are better suited than others. Mica, for example, can be easily bent and is often used in flat crystal of semi-focusing spectrometers: however, because of its perfect cleavage it cannot be as easily ground and it is not normally used in fully focusing spectrometers.

**Table 3.7** Characteristics of Commonly used Dispersion Crystals

| *Crystal* | *Abbrev.* | *(hkl)* | *2d (Å)* | *Lowest Z detectable* | | *(max Θ = 70°)* |
|---|---|---|---|---|---|---|
| | | | | *K* | *L* | *Efficiency* |
| topaz | — | (303) | 2.712 | 23(V) | 59(Pr) | average |
| lithium fluoride | LiF | (220) | 2.848 | 23(V) | 57(La) | high |
| lithium fluoride | LiF | (200) | 4.028 | 19(K) | 49(In) | intense |
| germanium | Ge | (111) | 6.532 | 16(S) | 40(Zr) | average |
| quartz | — | $(10\bar{1}1)$ | 6.686 | 15(P) | 40(Zr) | high |
| quartz | — | $(10\bar{1}0)$ | 8.50 | 14(Si) | 37(Rb) | average |
| pentaerythritol | PET | (002) | 8.742 | 14(Si) | 37(Rb) | high |
| ethylene diamine *d*-tartrate | EDDT | (020) | 8.804 | 14(Si) | 37(Rb) | average |
| ammonium hydrogen phosphate | ADP | (101) | 10.648 | 12(Mg) | 33(As) | low |
| mica (muscovite) | — | (002) | 19.8 | 9(F) | 26(Fe) | low |
| potassium acid phthalate | KAP | $(10\bar{1}0)$ | 26.6 | 8(O) | 23(V) | average |
| rubidium acid phthalate | RAP, RbAP | $(10\bar{1}0)$ | 26.1 | 8(O) | 23(V) | good |
| thallium acid phthalate | TAP, TIAP | $(10\bar{1}0)$ | 25.75 | 8(O) | 23(V) | high |
| lead stearate decanoate | PbSD | — | 100 | 5(B) | 20(Ca) | average |
| lead melissate | | | 160 | 4(Be) | 20(Ca) | average |

### 3.50.2 Selection of Dispersing Crystals

Selection of a crystal for a particular application must be based on several considerations (often conflicting and often requiring some measure of compromise), of which the following are the most important.

***d-spacing***

It is nearly always most desirable, other things being equal, to use first-order reflections of relatively strong spectral lines, since these yield the highest intensities and hence the best sensitivity and statistical precision. This requirement alone demands a range of analytical crystals whose first-order reflections cover the range of analytical interest.

For general purpose analysis, the $LiF_{200}$ crystal is probably the most widely used. With a 2*d*-spacing of 4.028 Å, however, its use is obviously restricted to wavelengths of less than 4 Å. Even this limit is mechanically impractical, since for a first-order emission of 4 Å

$$\sin\theta = n\lambda/2d = 4/4.028$$

whence $\sin\theta \approx 1$ and $2\theta \approx 180°$. This implies linear coincidence of sample and detector, which is mechanically impossible, so that the maximum wavelength that can be usefully dispersed by any crystal must be significantly less than 2*d*. The useful low wavelength limit of a crystal also depends partly on geometrical features of the spectrometer design (including such factors as the sample area from which radiation is excited and the size and shape of the crystal itself). Thus, the useful θ range of Bragg spectrometers is seldom more extensive than 10° – 70°, and may be appreciably less, particularly at the low-angle end. Performance in this respect varies appreciably from one design to another; small differences may be quite significant for certain applications and should be carefully examined when evaluating instrument specifications.

For $LiF_{200}$, the Bragg angle range of 10° – 70° corresponds to a wavelength range of 0.7 – 3.8 Å in the first order, which encompasses the $K\alpha$ emissions between of elements between potassium ($Z$ = 19) and molybdenum ($Z$ = 42) and the $L\alpha$ emissions of elements between indium ($Z$ = 49) and uranium ($Z$ = 92).

## 3.51 DETECTORS AND COUNTING EQUIPMENT

The basic function of an X-ray *detector* is to convert the energy of the X-ray photons into form which can be recognized, visually or electronically, and counted over a finite period of time to provide an estimate of the intensity of the detected radiation. A secondary, but extremely useful, characteristic of the detector is that its response should be in some way proportional to the energies of the detected X-ray photons: it then becomes possible to discriminate electronically against unwanted interfering responses (e.g., crystal fluorescence or scattered continuum) to ensure that only X-rays in a selected energy / wavelength band are measured.

There are many different types of X-ray detectors, all of them depending basically on the ability of X-rays to ionize atoms by photoelectric processes. For convenience, four major groups of detector can be distinguished, depending on the way in which the electron and ions produced by photoelectric interaction are used to produce a detectable signal.

(a) *Photographic film* types, in which the incident X-rays reduce silver halides to metallic silver by ionization processes. The integrated intensity of the X-rays detected over a finite time interval can be estimated from the degree of blackening of the film. This is the oldest and simplest technique of X-ray detection, and it is still widely used in X-ray crystallographic applications (e.g., Laue, Weissenberg, powder diffraction methods). However, it is inconvenient, lacks precision for intensity measurements (even if optical densitometers are used to measure the degree of film blackening) and does not have good proportional response (i.e., it is not easy to relate the strength of the detector 'signal'—the degree of blackening—to the energy of the X-ray photon that produced it). Hence, film measurement is no longer used in quantitative X-ray spectrometry and will not be further considered in this discussion.

(b) *Gas-filled detectors*, in which inert gases such as xenon or argon are ionized by the X-rays in a strong magnetic field produced by a high DC voltage applied across a cathode and anode. Electrons produced by the ionization process migrate to the anode, where they produce voltage pulses which can be amplified and fed to electronic counting circuits.

(c) *Scintillation detectors,* in which the X-ray photons interact with a phosphor material in such a way as to promote valence-band electron to higher energy levels. When these electrons revert to their stable valence-band state, energy is emitted in or close to the visible light range and is converted into voltage pulses by a photomultiplier. These pulses may be amplified and passed to counting or integration systems.

(d) *Semiconductor ('solid-state') detectors*, in which electrons produced by X-ray ionization of solid-state semiconductor atoms are promoted into conduction bands, thus causing the semiconductor (e.g., silicon) to become more conductive: a 'hole', equivalent to a free positive charge, is left in the valence band. Under the influence of an applied bias voltage, the charge carriers (electrons and holes) migrate to the bias electrodes where they produce small voltage pulses for subsequent amplification and counting. The proportional response of these detectors is better than that of either gas-filled or scintillation detectors, and in fact, they have sufficiently good inherent resolution to allow the electronic separation of output pulses produced by X-rays of different energy, with a resolution adequate for many analytical applications. Hence, detectors of this type form the basis of energy-dispersive spectrometers which need no dispersing crystals.

## 3.52 SOLID-STATE (SEMICONDUCTOR) DETECTORS

Solid-state detectors depend on the photoelectric processes initiated when an X-ray photon enters the detection system. However, these processes take place in a solid instead of a gaseous medium, and the detection process depends on the production of 'electron-hole' pairs rather than 'electron-ion' pairs.

The 'outer' orbital electrons of atoms in a crystalline material can only possess energies within discrete bands. Intrinsic semiconductors, such as pure crystalline silicon, have fully occupied valence bands separated by energy gaps from their higher-energy conduction bands. The energy of an X-ray photon absorbed by such materials is transferred by Auger and photoelectron interactions to the promotion of valence electron across the band gap to the conduction band. The conduction band electrons produced in this way are relatively mobile and can act as charge carriers, thus making the semiconductor material more highly conductive. In an intrinsic semiconductor, promotion of each valence electron also generates an electron 'hole' in the valence band: though less mobile, the 'hole' can also function as positive charge carriers.

If a suitable DC bias voltage is applied across a volume of such a semiconductor, the charge carriers can be 'swept out' and collected at the DC bias electrodes, in much the same way as the ion-pair charges are collected at the bias electrodes (anode and cathode) of a gas filled detector. Since the number of election-hole pairs produced will be proportional to the total energy of the incident X-ray photon, the magnitude of the charges collected at the bias electrodes will also be proportional to the X-ray photon energy, and a semiconductor detector of this type is also inherently a true proportional detector.

The most widely used solid-state detector material is crystalline silicon. When pure, silicon is an intrinsic semiconductor with a band gap of 1.1 eV. Under applied bias voltages of 500–1000 V its intrinsic conductivity (due to thermally excited electron-hole pairs) is relatively low, and it can be further reduced by maintaining the silicon at low temperatures (of the order of 100 K) with liquid nitrogen. For detector use, low intrinsic conductivity is essential; otherwise the applied DC bias produces 'leakage' currents which flow constantly, independently of the generation of charge carriers by photon absorption, and hence produce excessively high background signals.

Unfortunately, it is not technically feasible to produce silicon free of trace impurities, which significantly modify its electrical properties. The most common of these impurities is the Group III element boron, whose presence provides acceptor bands just above the silicon valence band and thus modifies the silicon to a *p*-type extrinsic semiconductor with increased conductivity. In practice, therefore, this undesirable effect has to be compensated by the deliberate, accurately controlled addition of another impurity chosen to provide donor bands to neutralize the boron acceptors. When properly compensated in this way silicon functions as an intrinsic semiconductor with high inherent resistivity.

The most widely used 'donor' element is lithium, which has a small atomic radius and hence can be easily diffused through the silicon, and which also has a low ionization potential (0.033 eV), making it an efficient electron donor.

A suitable solid-state detector medium can thus be prepared by adding the appropriate amount of lithium to silicon and diffusing it to a homogeneous distribution. Silicon with excess lithium will have *n*-type extrinsic properties, while lithium-deficient volumes will be *p*-type. The distribution of lithium must therefore be accurately controlled.

In practice, lithium is applied to the surface of a wafer of silicon, and diffused through it by heating briefly to about 350°C. This produces a gradient of diffused lithium in the silicon host (Fig. 3.87). The latter will be *n*-type in the high-Li region near the wafer surface, and

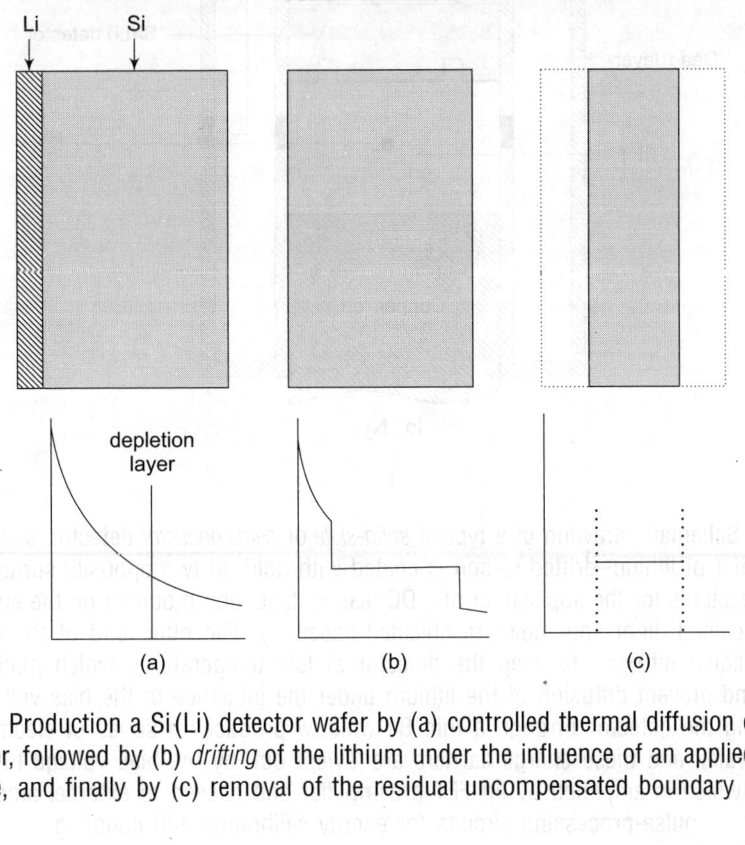


**Figure 3.87** Production a Si(Li) detector wafer by (a) controlled thermal diffusion of lithium into a silicon wafer, followed by (b) *drifting* of the lithium under the influence of an applied reverse bias voltage, and finally by (c) removal of the residual uncompensated boundary regions

*p*-type in the more remote low-Li region. There will be a narrow compensated zone, called a *depletion layer*, at the boundary between the *n*-type regions. This layer will lack free charge carriers and will thus have a high resistivity relative to the uncompensated regions on either side of it.

If a reverse bias voltage (*n*-type region positive with respect to the *p*-type region) is now applied – again under moderately increased temperatures to promote diffusion – a strong field in the resistive depletion layer will result in a migration of lithium ions from the *n*-region to the *p*-region. If properly controlled, the result of this drifting process will be an expansion of the compensated depletion layer to produce a wider region of high resistivity suitable for use as a detector (Fig. 3.87b). Excess uncompensated silicon (i.e., in the remaining *n*-type and *p*-type regions) is then removed, and the surfaces are coated with thin layers of gold to serve as electrodes. The result is a *lithium-drifted silicon detector*, or Si (Li) detector (Fig. 3.88).

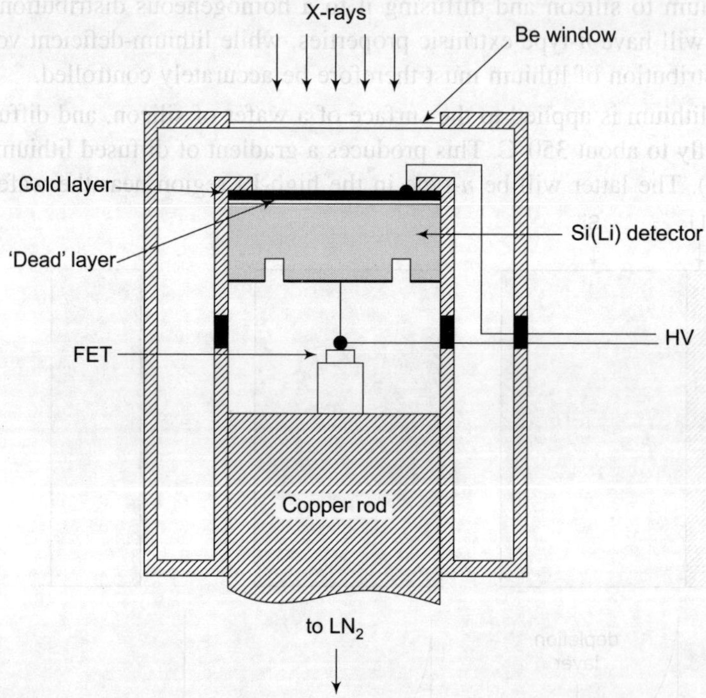


**Figure 3.88** Schematic drawing of a typical *solid-state* or *semiconductor* detector, such as an Si(Li) detector. A wafer of lithium-drifted silicon is coated with gold on two opposite surface, to provide electrical connections for the application of a DC bias voltage, and mounted on the end of a copper rod in an evacuated, light and moisture shielded assembly. The other end of the copper rod is immersed in liquid nitrogen to keep the detector at low temperatures which minimize thermal interference and prevent diffusion of the lithium under the influence of the bias voltage. An X-ray photon entering the detector through a thin Be window produces a series of electron-hole pairs in the Si(Li) wafer, and these charge carriers are 'swept out' by the bias voltage to yield a small voltage pulse which is amplified by an FET preamplifier and passed to external amplification and pulse-processing circuits for energy calibration and counting

Detectors of this type are operated at low temperatures, for two main reasons. First, as already noted, background noise due to thermally excited electron-hole pairs is minimized, and secondly, reverse diffusion of the lithium ions under the influence of the detector bias field is inhibited. At higher temperatures—even at room temperatures—an applied detector bias can otherwise produce a reverse drifting process and regenerate the *p*- and *n*-type regions (i.e., the detector can revert to a *p-n junction device*).

Effective cooling is achieved by mounting the detector on the end of a copper rod whose other end is kept immersed in liquid nitrogen ('$LN_2$' or 'LN2'). The crystal mounting is carefully designed to maintain the detector itself at a constant temperature of about 100 K. Some mountings include self-regulated heat source that act as thermostats to hold the detector at a constant temperature slightly above that of the LN2.

The basic principles of the Si (Li) detector are generally similar to those of the gas-filled detector. The electron-hole charge carriers produced by photoelectric absorption of the energy of an incoming X-ray photon are swept out by the bias field and collected at the bias electrodes to produce a pulse whose amplitude is determined by the number of charge carriers collected and hence is proportional to the energy of the incident photon (through again subject to statistical variation, since some of the photon energy is lost in interactions such as thermal lattice vibrations which do not produce charge carriers–just as some energy in the gas-filled detector is lost in non-ionizing collisions).

The pulse collected at the electrodes is then preamplified, amplified and passed to an electronic device called a *multi-channel analyser* (MCA), which fulfils the dispersive functions of a spectrometer by sorting the pulses electronically according to their amplitudes into a large number of different 'channels' (each corresponding to a different narrow range of X-ray photon energies) and counting the number of pulses collected in each channel.

Energy-dispersive spectrometers can function without need of a crystal to separate the different X-ray photon energies because their solid-state detectors have energy resolution characteristics which are superior to those of gas-filled or scintillation detectors. If an X-ray photon of energy $E$ enters a solid-state detector, it will produce an average of $N$ electron-hole pairs, such that

$$N = E/e$$

where $e$ is the *average* energy required to produce one electron-hole pair. For a Si (Li) detector operated under typical conditions, $e$ is approximately 3.8 eV: hence an Fe $K\alpha$ photon ($E$ = 6.398 keV) will produce an average of 6398/3.8, or 1683.7 electron-hole pairs. This is much higher than the equivalent charge-carrier yield produced by an Fe Kα photon in a gas-filled detector (Table 3.8). A series of Fe $K\alpha$ photon will still produce a statistical spread

**Table 3.8** Comparative Resolutions of Different Types of Detector

| *Detector* | Average ionization energy | *Average Fe Kα charge carrier yield* | *Theoretical resolution (%)** |
|---|---|---|---|
| Si(Li) | 3.8 | 1683.7 | 5.8 |
| Gas proportional (Ar) | 26.4 | 242.3 | 15.2 |
| Scintillation | 300.0 | 21.3 | 51.1 |

*W/V for Fe $K\alpha$

about this average value, but, because of the larger number of charge carriers generated and collected, the *proportional* spread about the mean will be less and the overall resolution will be considerably improved.

By contrast with the gas-filled detector, the Si (Li) detector has no internal gain, since it has no equivalent of the strong electron-accelerating field in the vicinity of the gas-filled detector's anode wire and thus it does not produce any 'avalanching' process. Hence, the swept electron pulse collected at the anode will consist only of the (average) 1683.7 electrons, or about $10^{-16}$ coulombs. This small signal must be electronically amplified before sorting and counting by a system which must obviously be characterized by high-gain, linear response (to maintain proportionality) and very low noise levels.

Again a preamplifier is usually attached as closely as possible to the detector to minimize capacitance effects, and a *field-effect transistor* (FET) is almost invariably used as the preamplifier because of its low-noise characteristics. The latter are further enhanced by LN2 cooling, so that the FET is also mounted within the detector cryostat assembly. The output signals from the FET are amplified by an external device similar to the main amplifier used by gas-filled detector systems, before processing in the multichannel analyser.

Electrons and holes can move freely only through a perfect crystal. In real crystals, impurities and crystal defects may produce intermediate electron energy levels between the silicon valence and conduction bands, and these can 'trap' the moving charge carriers, at least temporarily. Trapped carriers cannot be collected at the electrodes: the output pulse is then reduced in magnitude and the energy resolution characteristics of the detector are degraded. In severe cases an asymmetrical pulse amplitude distribution results, with a significant tail on the low-energy side, leading to extended overlap of the distributions produced by photons of different energies and thus to reduced ability of the system to *resolve* or distinguish between them.

Trapping losses can be reduced by using high bias voltages to ensure rapid collection of the charge carriers, or by operating the detector at higher temperatures to enhance detrapping by thermal excitation. However, high voltages and high temperatures both result in increased noise levels and thus degrade detector performance. Again some degree of compromise is always necessary. The actual extent of trapping problems can vary considerably from one detector to another, since it is largely a consequence of the quality of detector materials and manufacture.

As shown in Fig. 3.88 there is also a thin layer (up to about 0.1 mm thick) at the surface of the detector which does not function effectively as a detection medium. This 'dead layer' is transitional with the active region beneath it, and the transition zone typically contains a relatively high concentration of traps. Carrier loss effects due to trapping therefore tend to be most severe for low-energy photons which are more heavily absorbed in the surface layers.

Preamplifier noise cannot be entirely eliminated, and it always contributes a significant component to the statistical spread about the mean pulse amplitude distribution. A further term must be added in quadrature to the resolution equation to include the effect of random noise, so that

$$\Delta E = (= FWHM) = 2.36 \sqrt{(e\,F\,E)}$$

becomes

$$\Delta E^2 = (\varnothing E_{\text{noise}})^2 + (\varnothing E_{\text{ionization}})^2 \quad ...(1)$$

and hence

$$\Delta E = \sqrt{\{(\Delta E_{\text{noise}})^2 + [5.57\,(e\,F\,E)]\}} \quad ...(2)$$

(Remember that use of the term '$\Delta E$' in the above expressions as a measure of statistical spread about the mean is quite different from the use of the same expression to denote the window width in a pulse height discrimination system.)

$\Delta E_{\text{noise}}$ is typically about 100 eV and Fano factors are currently about 0.12 for good quality Si(Li) detectors. Thus, the *FWHM* for Fe $K\alpha$, for such a detector, is given by

$$\Delta E = \sqrt{\{(100^2 + 5.57 \times (3.8 \times 0.12 \times 6398)\}}$$

$$= 162 \text{ eV}$$

and

$$FWTM = 1.823 \times FWHM$$

$$= 295 \text{ eV}$$

This resolution is sufficient to separate most X-ray energies of analytical interest without the need for a dispersing crystal (except possible in the relatively 'crowded' *L* and *M* spectral regions of samples containing several heavy elements). It should be noted that for sales purposes absolute energy resolution is sometimes quoted for lower photon energies, often 1.0 keV which (because of window absorption etc.) marks the practical low-energy limit for the current generation of Si(Li) detectors. For 1.0 keV photons,

$$FWHM = \sqrt{\{(100^2 + 5.57 \times (3.8 \times 0.12 \times 1000)\}}$$

$$= 112 \text{ eV}$$

Most reputed manufacturers, however, quote their detector resolutions for a more realistic energy, usually that of Mn $K\alpha$ (5.894 keV). When these expressions are evaluated, it becomes clear that the noise term is actually the major component, and improved detectors with reduced noise characteristics can be expected to yield significantly better resolutions in the future (though it must also be noted that the rate of technical improvement in this respect has slowed considerably in recent years). It is also possible that new and superior semiconductor materials will ultimately replace lithium-drifted silicon, which at present is almost universally used in solid-state X-ray detection devices. Germanium detectors, for example, have already been used successfully for some applications.

Because of their superior energy resolution, solid-state detectors offer practical advantages over gas-filled or scintillation types for many analytical applications. Because they can be used without a dispersing crystal, they require none of the mechanical (goniometer) components needed to rotate the crystal and detector to appropriate Bragg angles: nor do they require the additional mechanical components necessary to exchange the crystal needed for different energy/wavelength ranges. In the absence of moving mechanical parts, there are no alignment problems resulting from backlash or wear.

A solid-state X-ray detector can be placed much closer to the emitting source (the sample) than the crystal in a conventional wavelength-dispersive spectrometer, and it does not need intensity-reducing collimator or slit systems to ensure Bragg relationships. Hence, it can sample a much wider cone of emitted X-rays, which in itself is equivalent to a substantial increase in effective X-ray intensity under constant excitation conditions. Furthermore, since it does not need a crystal to be set at different angles for different wavelengths, it can sample complete spectrum of emitted X-ray energies simultaneously, avoiding the time consuming necessity for sequential scans from one energy (wavelength) to the next.

These factors combine to give the solid-state, or energy-dispersive, system its most outstanding attribute – that of speed of analysis. Multi-element analyses can often be completed, with adequate levels of precision, in total measurement times of the order of seconds rather than the minutes or even hours required for wavelength-dispersive systems.

## 3.53 ENERGY-DISPERSIVE SPECTROMETERS

Because of their convenience and relative simplicity of operation and maintenance, energy-dispersive spectrometers have become widely used since the late 1960s, when advances in electronic and digital-processing technology first made them practicable for routine analysis of X-ray spectra. They are particularly well suited to applications in microprobe analysis, not the least because they provide a relatively economical way to add analytical capabilities to scanning electron microscopes. They are also useful adjuncts to wavelength-dispersive spectrometers, and they have found interesting applications in a number of compact, portable analytical system which use probes containing radio-isotope sources to excite fluorescent X-ray spectra directly from ‘field’ samples (rock outcrops, underground exposures, drill core samples, etc.). Energy-dispersive spectrometers are used to analyse the fluorescent spectra and provide rapid concentration estimates for one or more precalibrated elements.

An energy-dispersive spectrometer is basically an electronic system that processes the energy-proportional pulses produced by a solid-state detector, sorts them according to their amplitudes (and hence according to the energies of the X-ray photons that produced them), and counts the pulses produced in a series of amplitude ranges appropriate to the elements sought. Because of the superior energy resolution of the solid-state detector, EDS systems need no dispersing crystal and hence none of the mechanical components required to maintain Bragg relationships in wavelength-dispersive spectrometers. Since all elements of the X-ray spectrum are collected and processed simultaneously instead of sequentially, EDS systems can usually complete analysis much faster than their WDS equivalents.

Alternatively, since they work best with much lower total spectral intensities, they allow the use of lower excitation energy fluxes (such as the low electron beam current of a scanning electron microscope, or the low beam currents necessary to minimize thermal damage of some samples in the electron microprobe).

There is considerable variation in detail from one commercial EDS system to another, and EDS systems are still undergoing relatively rapid technological evolution. However, most of the units currently available commercially contain at least the following essential components (Fig. 3.89).

(a) A solid state *detector*, usually of the Si (Li) type, fitted with an FET-based preamplifier. The detector collects the charges produced when an X-ray photon interacts with the drifted silicon to produce electron-hole pairs, and the preamplifier integrates the collected charges and converts them to a low-voltage pulse whose amplitude is proportional to the number of charges collected and hence to the X-ray photon energy.

(b) A high-gain linear *amplifier*, which has two main functions:

  (i) It increases the preamplifier output pulse amplitudes to voltage levels suitable for sorting and counting.

  (ii) It shapes the pulses to a form appropriate for processing, while accurately maintaining the proportional relationship between pulse amplitude and X-ray photon energy on which the system depends.

(c) A *single-channel analyser* (SCA), which scans the amplifier output and rejects (filters) low-voltage electronic noise and other interfering signals.

(d) An *analog-to-digital converter* (ADC), which processes the filtered pulses from the SCA, converting each to a numerical (digital) value proportional to its original pulse amplitude and hence to the energy of the X-ray photon that produced it.

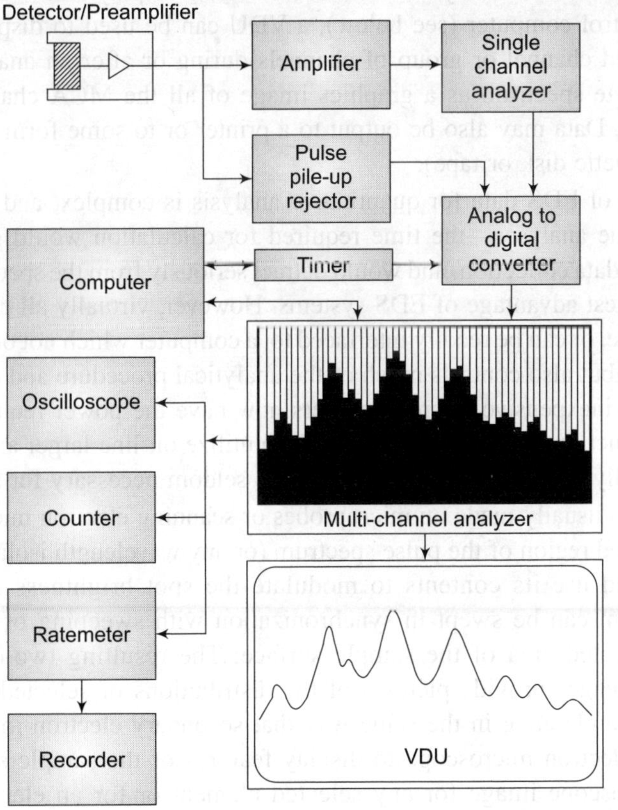


**Figure 3.89** Schematic diagram of the major components of a typical energy-dispersive X-ray spectrometer. The function of each component is described in the text

(e) A *multi-channel analyzer* (MCA), which sorts the digitized pluses into a series of computer memory registers, or 'channels'. Each channel is, in effect, a 'bin' used to store and count the pulses falling within narrow, accurately calibrated digitized energy ranges in the X-ray spectrum. A typical MCA has 1024 channels calibrated for consecutive 10 (or 20) eV increments, and thus spanning the range of X-ray energies from, say, 0 to 10.24 (or 20.48) keV. 10.24 keV corresponds to a wavelength of 1.2 Å, so that the range from 0 to 10.24 keV includes almost all of the characteristic X-ray lines routinely used in XRF or electron microprobe analysis.

(f) A *timer*, which is used to control the time intervals over which data are collected, so that the rates of pulse collection can be reduced to X-ray intensities. The timer is usually a 'live-time' clock – a device which provides automatic dead-time correction by stopping the clock during intervals when pulses are being processed and the system is temporarily insensitive to the arrival of further X-ray photons.

(g) Various types of output device which are used to display the collected data. These include scalers, ratemeters, chart recorders, etc., equivalent to those used in WDS systems, but virtually all except the most compact portable systems also utilize the convenience and flexibility of some kind of visual display unit (VDU). In conjunction with a control computer (see below), a VDU can be used to display the contents of any selected channel or group of channels during or after an analysis, or to display the complete spectrum as a graphics image of all the MCA channels, side by side (Fig. 3.89). Data may also be output to a printer or to some form of storage medium (e.g., magnetic disk or tape).

(h) Processing of EDS data for quantitative analysis is complex, and not practicable for most routine analyses—the time required for calculation would greatly exceed that needed for data collection, and would detract seriously from the speed of analysis which is the greatest advantage of EDS systems. However, virtually all commercial systems now include, or can be readily interfaced to, a computer which not only handles the data processing but also controls much of the analytical procedure and the system devices. Dedicated, inexpensive microcomputers now have the power required to perform all of these functions, but it is also possible to utilize on-line larger and faster computers for unusually complex applications. This is seldom necessary for routine analysis.

(i) Provision is usually made on microprobes or scanning electron microscopes to access any specified region of the pulse spectrum (or any wavelength isolated by a dispersing crystal) and use its contents to modulate the spot brightness of an oscilloscope whose beam can be swept in synchronization with sweeping of the electron beam over a selected area of the sample surface. The resulting two-dimensional X-ray intensity images provide pictures of the distributions of selected elements over the selected sample area, in the same way that secondary electron images are used in a scanning electron microscope to display features of the sample surface topography. The oscilloscope image for any selected element or for an electron image can be photographed for a permanent record, or can be computer-processed ('digitized') for enhancement of significant features eliminating many of the more tedious aspects of routine X-ray spectrographic analysis, these systems have done much to ensure widespread acceptance of XRF (and electron microprobe) analysis: in particular, they permit much of the work to be done on a routine basis by relatively unskilled operators. However, it should never be forgotten that the results that they produce are determined by such factors as the quality of the instrument to which they have been fitted (and of its maintenance), the standards of sample preparation, and the validity of the procedures that they have been programmed to follow. As noted elsewhere, the analyst using highly automated equipment must be particularly aware of the need for properly critical standards of appraisal, and of the dangers of excessive reliance on the computer for protection against errors over which it has, in fact, little or no control.

## 3.54 THE ELECTRON MICROPROBE

Since Castaing first demonstrated the practical feasibility of electron microprobe analysis in 1951, about a dozen instrument designs have been marketed commercially, and most of them have since passed through a very rapid process of technical evolution, particularly during the 1970s. Some aspects of instrumentation reflect differences in design philosophy or patent protection, and some are the result of technological advances over the past three decades. Nevertheless most of the basic features of Castaing's original designs are still found in current instruments. They include the essential components shown in Fig. 3.90.

### 3.54.1 Electron Gun

The function of the electron gun is to provide a source of electrons, of appropriate energy and current density, to form the electron beam used to excite X-rays from the analytical sample. In order to select optimum conditions for excitation of the sample X-ray spectra, provision must be made for variation of both energy and current, and each must be regulated within narrow tolerances. For maximum spatial resolution (i.e., production of an electron beam of minimum diameter at the sample surface), the electrons must be produced at an effective point source, or as near to it as practicably possible. The electron optics (see below) produce a demagnified image of this source at the sample surface.

In most microprobes these requirements are met by the use of a self-biasing triode type of electron gun, in which the electrons are produced by thermal emission from a pointed tungsten filament, held at high negative potential, and accelerated towards an anode plate at ground potential. A hole in the anode plate allows the accelerated electrons to leave the gun and enter the beam-forming electron optical system.

The filament emission current is controlled by a grid, usually in the form of a cylindrical cap ('Wehnelt cylinder') fitted over the filament. A small negative bias voltage on the grid is controlled with a variable resistor to determine the gun emission current: low bias voltages provide little opposition to the passage of the accelerated electrons and hence result is relatively high emission currents, and vice versa. The emitted electrons pass through a small hole in the centre of the grid cap, which itself has a focusing effect on the electron beam and produces a crossover whose demagnified image at the sample surface ultimately forms the electron 'probe'.

Other types of electron guns, including lanthanum hexaboride and field emission guns, have been investigated experimentally and found to offer some advantages. They have not yet come into general use in electron microprobes, but they are already widely used for certain types of electron microscope (e.g., scanning transmission electron microscopes, or 'STEMs'), some of which also include X-ray spectrometers to give them some degree of chemical analytical capability.

Many of the day-to-day problems encountered in routine electron microprobe analysis arise from instability in the electron beam, which can often be traced to the gun: filaments, for example, have limited lives (particularly if the vacuum system is not maintained at peak efficiency), and they almost invariably fail at the most inconvenient moment. This is one aspect of microprobe instrumentation which can be expected to improve considerably as the new types of electron gun come into wider use.

### 3.54.2 Electron Beam-forming Column (Electron Optics)

The gun crossover image is demagnified by a system of electron lenses (usually electromagnetic lenses, although there is no reason in principle for not using electrostatic types). Most beam columns contain two such lens systems, which are commonly described as 'condensing' and 'objective' lenses respectively because of a loose analogy with light microscope optics. The focal lengths of the lenses are controlled by means of variable, stabilized lens current supplies: adjustment of the condensing lens current serves effectively to control the probe current incident on the sample, and adjustment of the objective lens current controls the diameter of the electron beam at the sample surface (this is the basis of the analogy with light optics—a microscope condenser controls the intensity of light illuminating the object, and the objective aperture determines the resolution of the optical system; the analogy should not be pursued much further).

One or more apertures placed in the electron optical column (usually between the lenses) are used for further control of the probe current and also, indirectly, for its measurement. They also serve to minimize some of the aberrations inherent in electron optics, in a fashion analogous to the use of limiting apertures to control aberrations in light optical system. Other aberrations in the electron beam are controlled by auxiliary electromagnetic or electrostatic *stigmators,* which allow the operator deliberately to distort the beam cross section so as to oppose the aberration effects.

They very small effective beam diameter which can be attained in the electron microprobe means that X-ray spectrometric analyses can be made on extremely small sample volumes, and this, in turn, gives the microprobe its most powerful attribute – that of very high spatial resolution. To obtain maximum benefit from this capability, it must be possible to position the electron beam accurately on the sample surface. This is achieved partly by precise mechanical positioning of the sample under the beam (see below), and partly by the use of scanning coils or similar systems, similar in principle to those used in an ordinary television receiver to

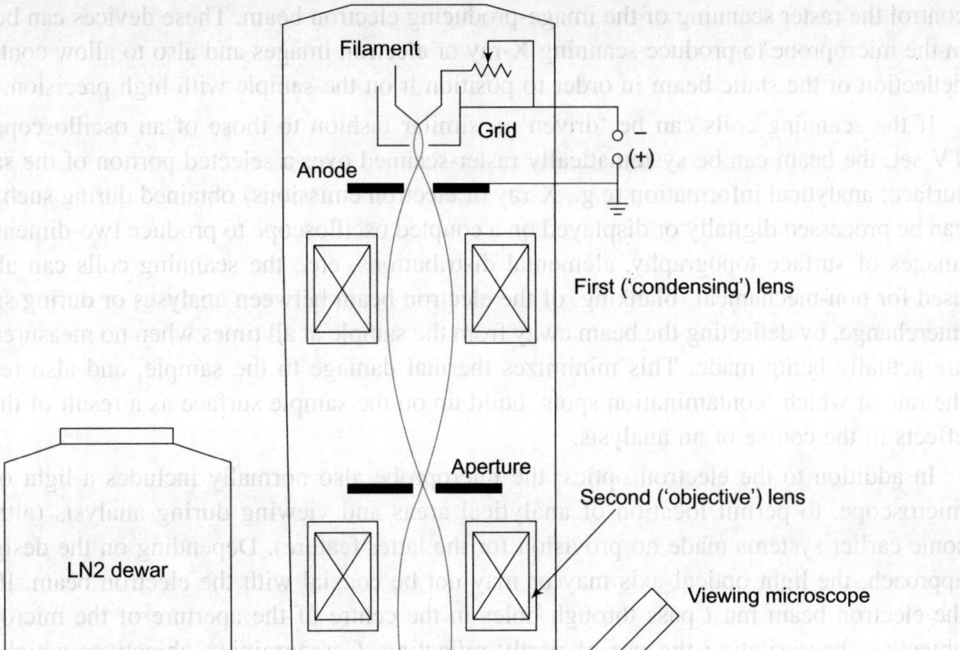


**Figure 3.90** Schematic diagram of the electron beam-forming column and spectrometer of an electron microprobe (only one wavelength-dispersive spectrometer is shown, but most microprobes have two or more, disposed radially around the beam column and sample chamber, in addition to an energy-dispersive spectrometer). Electronic components for processing the X-ray signals are generally similar to those shown in Figures 3.89. Electron microprobes also usually have additional facilities for processing signals from one or more electron detectors (e.g., secondary and/or backscattered electrons) and for the display of X-ray or electron images produces by raster-scanning of the electron beam over the sample surface

control the raster scanning of the image-producing electron beam. These devices can be used in the microprobe to produce scanning X-ray or electron images and also to allow controlled deflection of the static beam in order to position it on the sample with high precision.

If the scanning coils can be 'driven' in similar fashion to those of an oscilloscope or a TV set, the beam can be systematically raster-scanned over a selected portion of the sample surface: analytical information (e.g., X-ray or electron emissions) obtained during such scans can be processed digitally or displayed on a coupled oscilloscope to produce two-dimensional images of surface topography, elemental distributions, etc., the scanning coils can also be used for non-mechanical 'blanking' of the electron beam between analyses or during sample interchange, by deflecting the beam away from the sample at all times when no measurements are actually being made. This minimizes thermal damage to the sample, and also reduces the rate at which 'contamination spots' build up on the sample surface as a result of thermal effects in the course of an analysis.

In addition to the electron optics, the microprobe also normally includes a light optical microscope, to permit location of analytical areas and viewing during analysis (although some earlier systems made no provision for the latter feature). Depending on the designer's approach, the light optical axis may or may not be coaxial with the electron beam. If it is, the electron beam must pass through holes in the centre of the aperture of the microscope objective, necessitating the use of partly reflecting, Cassegrainian objectives which have poorer resolution than comparable fully refracting systems (though they are usually adequate for the positioning purpose for which they are primarily intended). In either case, resolution of the light optics at the high magnifications necessary for reasonably accurate location of the analytical area is limited by the need for long working distances and hence low numerical apertures—otherwise the light objective would interfere mechanically with the electron objective lens and with the emitted X-rays. In fact, the physical relationship between the electron lenses, the light optics, the sample and the X-ray spectrometers is one of the areas of major design differences between the various commercial microprobes, since it also necessarily represents an area of compromise.

### 3.54.3 Sample Chamber

The basic functions of the sample chamber are common to all microprobes, but the methods by which they are achieved vary considerably from one design to another. The principal requirements of a sample chamber include the following:

(a) It must provide for simple and rapid interchange of samples and standards, preferably by an airlock system so that it is not necessary to break the entire instrumental vacuum or to shut down the electron beam during interchange. The vacuum constraint is more critical than in XRF systems, whose sample changers must meet the same demands, because the operating pressures of the microprobe are several orders of magnitude lower than those of normal XRF systems and pumping times are correspondingly longer.

(b) In order to position the samples for analysis, provision must be made for orthogonal (*X*, *Y* and *Z*) and preferably also for rotational mechanical translations of the samples, with high standards of reproducibility – especially in automated systems in which sample movements are performed under computer control.
(c) To allow at least limited interchange of samples and/or standards without breaking any vacuum at all, the sample chamber should provide for simultaneous inclusion within its stage of at least three or four sample mounts. Some allow up to 12, but with improved vacuum technology this requirement is no longer as critical as it used to be and smaller, more compact stages have become more common.
(d) In conjunction with the light optics, the sample chamber must provide for simple, rapid and reproducible positioning of the analysis sample with respect to the electron beam and the X-ray spectrometers.
(e) On occasions when the vacuum must be broken to interchange samples, it must be possible to restore analytical conditions in a few minutes at most. Longer periods waste valuable instrument time and may introduce problems with time-dependent machine drift.
(f) The sample holder must be grounded ('earthed') to minimize or prevent charging of the sample during analysis by a build-up of absorbed electrons from the beam – which would result in erratic defection of the beam. The grounding circuit normally includes a sensitive microammeter to measure that portion of the beam current that has been absorbed by the sample and hence has contributed to X-ray excitation (variously called the 'absorbed current', 'sample current' or 'specimen current').

### 3.54.4 Vacuum System

The electron gun, beam column, sample chamber and X-ray spectrometers must all be maintained at low pressures during operation in order:

(a) to maintain satisfactory filament life by minimizing processes of corrosion (e.g., oxidation) and ionic bombardment;
(b) to minimize loss of electrons or of electron energy in the beam as a consequence of collisions with gas molecules; and
(c) to minimize absorption of the emitted X-rays, particularly those of relatively low energy.

The requirements are met by maintaining internal pressures of less than $10^{-4}$ torr (1 torr = 1 mm Hg); the figure of $10^{-5}$ torr is frequently quoted, but this usually refers to readings on a vacuum gauge whose sensing head is placed for convenience close to the vacuum diffusion pump. Actual pressures in more remote parts of the system, particularly in the electron gun and in the beam column (which is often pumped through the lens bores and even in some cases through the apertures), may well be as much as two orders of magnitude higher.

The vacuum must also be relatively 'clean' to minimize the deleterious effects of contamination deposits, produced on the surface of the sample during analysis by decomposition of hydrocarbon molecules derived from the vacuum pumps or from careless handling of the internal parts of the system (including samples and sample mounts). Once formed, these contamination deposits absorb both incident electrons and emitted X-rays and can have serious effects on analytical accuracy.

Mechanical rotary pumps such as those used in XRF systems are fast-pumping at relatively high pressures and are therefore effective in microprobe systems as 'roughing' pumps, i.e., pumps used to attain the first stages of a working vacuum. However, they are ineffective at pressures below $10^{-1} - 10^{-2}$ torr, so that microprobes require an additional pump, effective at lower pressures, to achieve a working vacuum. This is usually an *oil diffusion pump* – a type that has no mechanical components and is therefore simple, effective and (with proper care) very reliable.

Diffusion pumps are ineffective at high pressures (above $10^{-1}$ torr), and indeed will emit copious quantities of pump oil and produce very undesirable contamination effects if operated in a high-pressure regime. It is therefore necessary to *rough pump* the system down to about $10^{-1}$ torr with a rotary pump before the diffusion pump is switched in, and the rotary pump is then kept in operation to 'back' the diffusion pump, i.e., to maintain low pressures in the exhaust of the diffusion pump. The diffusion pump must *never* be operated at pressures higher than $10^{-1}$ torr, and *must never be operated without a backing pump.* Most microprobes include automatic valving to isolate the diffusion pump in the event of powder loss or of unexpected failure of the backing pump.

The same, or a second, rotary pump is also used to rough pump the sample chamber after sample interchange. The pumps therefore usually require a moderately complex system of valving to allow them to be switched in or out of various parts of the system as required. Early microprobes required manual operation of the valves and permitted some disastrous combinations to be selected by an inexperienced operator. On later models the vacuum system is usually automated to the push-button level, but it is still highly desirable for the operator to have a good understanding of the operating principles of the vacuum system.

Rotary and diffusion pumps both use oil as a pumping medium (though in very different ways), and the vacuum can therefore be contaminated by backstreaming of pump oil vapours. Backstreaming can be reduced (but not entirely eliminated) by water- or LN2-copled baffles above the diffusion pump and by adsorption traps in the backing line. Backstreaming can also be reduced by using narrow-bore pumping hoses, but at the expense of slower pumping speeds.

In some critical cases the diffusion pump may be replaced by a more powerful and more sophisticated device, such as a turbo-molecular pump or an ion pump. For most purpose, however, the simple diffusion pump will give satisfactory results if it is properly operated and maintained.

### 3.54.5 X-ray Spectrometers

X-rays emitted by the sample during analysis can be analysed with either wavelength- or energy-dispersive spectrometers. Each type has its advantages and limitations, as discussed in chapter 5. WDS types have better total resolution, but normally require much longer total counting times, even when fully focusing geometry is used (as is almost invariably the case). EDS spectrometers are better suited in many respects to the low count rates commonly encountered in microprobe analysis and, by contrast with XRF, are now probably more widely used in microprobes than their WDS equivalents. To provide maximum flexibility, a microprobe should incorporate at least two and preferably three WDS and one EDS spectrometers (some instruments have been built with as many as 12 spectrometers altogether).

The need for multiple WDS spectrometers arises from the relative difficulty of routine crystal interchange, imposed by the more rigid constraints of focusing geometry, and also from the difficulty of detector interchange. With longer counting times consequent upon lower X-ray intensities, multiple spectrometers also contribute usefully to lower overall analytical times by permitting the simultaneous measurement of several different elements.

There is some general advantage in keeping the *take-off angle* (the angle, measured at the sample surface, between the surface and those X-rays that enter the spectrometers) as high as possible – at least higher than 35°. This reduces the magnitude and hence the probable error of corrections that have to be applied for adsorption effects but it also imposes major design constraints.

The range of crystals available for microprobe WDS spectrometers is much the same as that for XRF spectrometers, except that some crystals are difficult or impossible to prepare in the bent and ground configuration required for focusing X-ray optics. Selection of crystals for specific analytical applications follows the same rules as in XRF analysis, but may be more constrained by limitations on the number of crystals available or needed for interchange and the number of spectrometers with which the microprobe is equipped. XRF spectrometers are commonly equipped with four or five interchangeable crystals, since most XRF systems have only one spectrometer, which has to handle the entire analytical wavelength range, and the geometrical constraints on the interchange mechanism are not severe. Microprobe spectrometers seldom provide for more than two crystals, or three at most: microprobes usually have at least three spectrometers, each of which can be set up to handle a portion of the complete wavelength range. Many microprobe analysts prefer to avoid crystal interchange altogether, if possible, because of the difficulty of maintaining the precise alignment demanded by focusing X-ray optics.

Few, if any, microprobe spectrometers use scintillation detectors, which are relatively large and unwieldy. Instead, sealed gas detectors are used for the shorter wavelength and thin-window gas flow detectors for the longer. A typical three-spectrometer microprobe configuration will have an LiF crystal with sealed detector on the first spectrometer, a PET (or ADP) crystal with a sealed or a gas flow detector on the second and a TlAP (or RbAP) crystal with a thin-window gas flow detector on the third. This configuration allows the detection of all elements heavier than oxygen, without need of any crystal or detector interchange.

### 3.54.6 Electron Detectors

The sample can be viewed during analysis of scanning by one or more electron detectors, which are used principally for secondary and/or backscattered electron current imaging.

The generation of secondary and backscattered electrons depends on factors such as sample composition and surface topography, so that the microprobe can be operated as a scanning electron microscope (SEM) by imaging the secondary electron or the backscattered electron distribution produced by a scanning probe beam. However, in the interests of improved beam stability, microprobes are usually designed to operate at lower electron gun brightnesses and higher beam currents, so that they cannot attain quite the same standards of resolution as a dedicated SEM unless they are equipped with versatile guns that provide for convenient control of gun brightness by simple methods such as adjustment of the filament-to-grid spacing. Similarly SEMs fitted with auxiliary spectrometers will often not yield microprobe standards of stability in the analytical mode.

### 3.54.7 Power Supplies and Support Electronics

These components include AC and DC amplifiers for X-ray and electron detector signals, discriminators used to reject harmonic and other unwanted signals, timing and counting systems for measurement of X-ray intensities, imaging systems (including display oscilloscopes), and an extensive range of power supplies. With the exception of the imaging systems and the electron detectors, these components are broadly similar to those used in XRF instruments.

## 3.55 AUGER ELECTRON SPECTROSCOPY

Auger electrons are emitted from a surface whenever incident radiation–photons, electrons, ions or neutral atoms—interacts with an atom with sufficient energy to remove an inner shell electron from the *K*, *L* or *M* shells. When higher level electrons fall into the empty state two things can happen—a characteristic X-ray is emitted, or the energy is transferred directly to a neighbouring electron which is released as an Auger electron. Figure 3.91 compares the mechanisms involved.

Since electron beams in the 1 to 30 keV range of energy can easily be produced and focused, this is the excitation of choice for Auger electron spectroscopy.

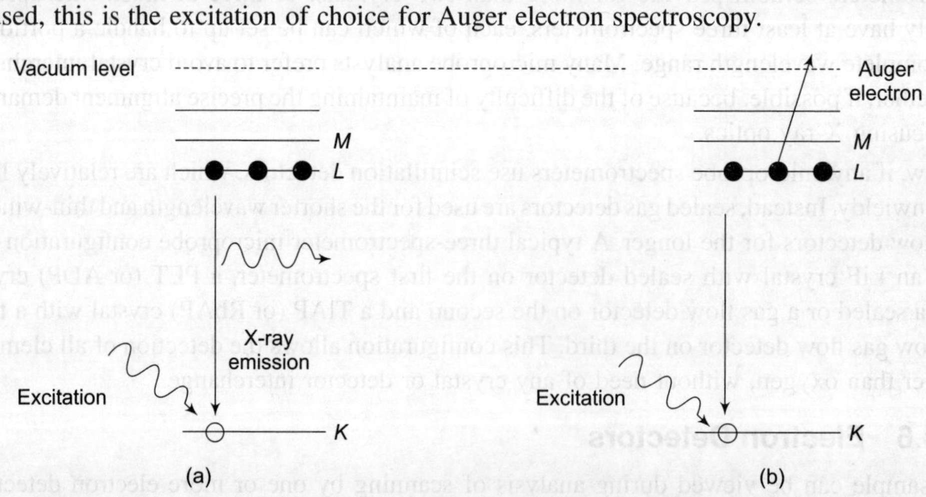


**Figure 3.91** The mechanism of: (a) X-ray emission. (b) Auger electron emission

The kinetic energy of an Auger electron resulting from a transition involving the $K$ and $L$ levels is

$$E_{KE} = E_K - 2E_L - \phi$$

where $\phi$, the work function for the surface, is the kinetic energy which is lost by the electron in escaping from the sample surface into vacuum. The low energy of the Auger electron means that only electrons in the top 0.5 to 3 nm (2–10 monolayers) escape from the surface. The energy of these electrons is analyzed by some form of electron energy analyzer.

Figure 3.92 (a) shows a chart of the principal Auger electron energies. Figure 3.92 (b) is a representative AES spectrum from silver. The sensitivity of the measurement is enhanced by calculating the differential of the simple count rate $N$ versus $E$ curve.

Information available from Auger electron spectroscopy includes:

(a) Electron energies characteristic of transitions within the source atoms.

(b) Shifts in the energies related to the chemical bonding in the surface layers.

## 3.56 X-RAY PHOTOELECTRON SPECTROSCOPY (XPS)

The excitation of core electrons by X-ray photons of frequency $\nu$ can release photoelectrons from the surface of a solid. Their kinetic energy upon release is

$$\text{KE} = h\nu - BE + \phi$$

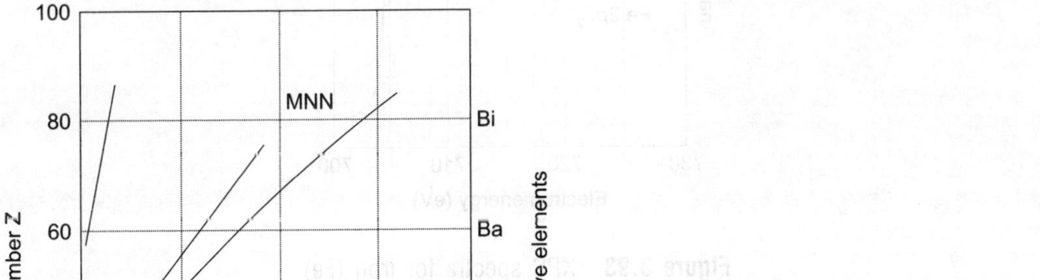


**Figure 3.92** (a) Chart of the principal Auger electron energies. (b) Experimental AES data from silver

Information is therefore available on the individual levels plus changes in work function ($\phi$) due to variations in chemical composition and environment at the surface. The photo-emitted electrons have low energy and therefore scattering ensures that only electrons within the top few atomic layers are emitted.

An XPS spectrum is shown in Fig. 3.93.

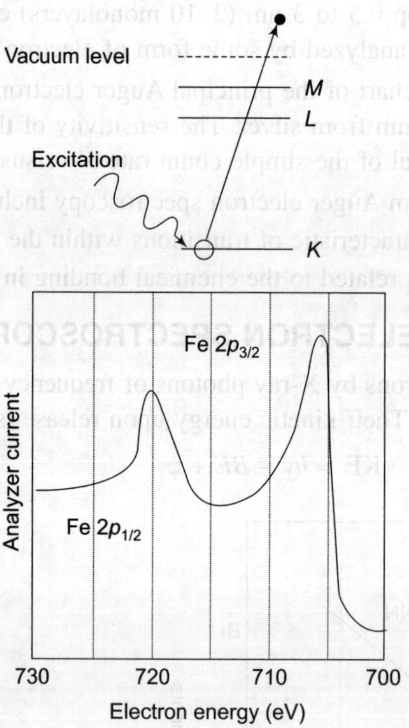


**Figure 3.93** XPS spectra for iron (Fe)

## 3.57 SECONDARY ION MASS SPECTROMETER (SIMS)

In secondary ion mass spectrometer (SIMS) an energetic beam of ions such as argon is directed at the sample surface. The bombarding ions sputter or dislodge atoms from the surface. The sample is held in a high or ultra-high vacuum environment. A sensitive mass spectrometer analyzes the nature of the sputtered atoms. Modern argon ion guns provide well controlled high sputtering rates for good sensitivity and a full surface analysis is achieved of the top 5 nm of the surface. Depth profiling is undertaken by use of a large area gun to sputter the sample uniformly over considerable distances. An analysis gun and ion analyzer then measures the local composition of the surface. Trace element analysis can be achieved with parts per million to parts per billion sensitivity. The technique is sensitive to light elements such as hydrogen.

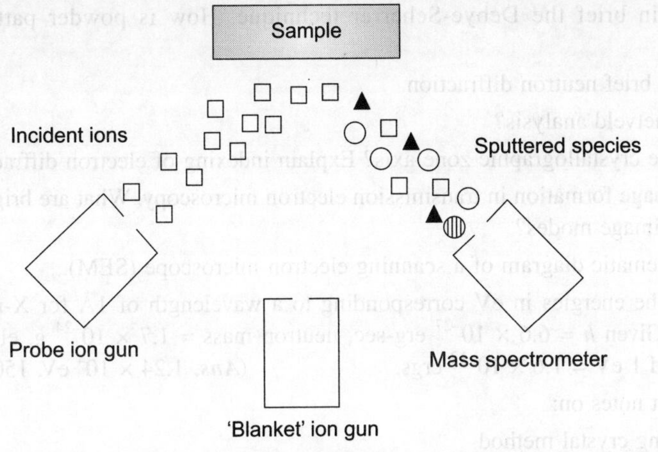


**Figure 3.94** Schematic representation of the principles of SIMS

## EXERCISE

3.1 What is a reciprocal lattice? Discuss some of its important properties.

3.2 Show that the fcc lattice is the reaprocal of the bcc lattice and vice versa.

3.3 Discuss Ewald construction and derive Braggs diffraction condition in terms of the reciprocal lattice vector.

3.4 How are Brillouin zones constructed? Describe and sketch the first Brillouin zones of bcc and fcc lattices. Mention their importance in crystal analysis.

3.5 Derive first, second and third Brillouin zones for a two-dimensional square lattice. Sketch the first four Brillouin zones of a two-dimensional square lattice. Give a sketch of special points and lines in the first Brillouin zone of a two-dimensional square lattice.

3.6 Describe the reciprocal lattice in terms of Fourier transform for one-dimensional case.

3.7 What is Debye-Waller factor. Derive an expression for the same (i.e., exp $(-2W)$).

3.8 Derive an expression for exp $(-2\,W)$ and prove that

$$2W = \frac{\hbar k^2}{6\text{m Na}} \sum_q \sum_s \left( \frac{1 + 2n_{qs}}{w_{qs}} \right)$$

where $k$ is the wave vector and

$$2W = \langle (\vec{k} \cdot \vec{u})^2 \rangle$$

3.9 Explain the terms atomic scattering factor and geometrical structure factor.

3.10 Explain scattering by a unit cell having three kinds of atoms.

3.11 Give theory of the Laue method for X-ray diffraction.

3.12 Describe experimental set-up used for the Laue method. How are Laue spots explained?

3.13 Describe in brief the Debye-Scherrer technique. How is powder pattern recorded and evaluated?

3.14 Explain in brief neutron diffraction.

3.15 What is Rietveld analysis?

3.16 What is the crystallographic zone axis? Explain indexing of electron diffraction spot patterns.

3.17 Explain image formation in transmission electron microscopy. What are bright field image and dark field image modes?

3.18 Give a schematic diagram of a scanning electron microscope (SEM).

3.19 Compute the energies in eV corresponding to a wavelength of 1Å for X-rays, electrons and neutrons. Given $h = 6.6 \times 10^{-27}$ erg-sec, neutron mass = $1.7 \times 10^{-24}$ g, electron mass = $9 \times 10^{-28}$ g and 1 eV = $1.6 \times 10^{-12}$ ergs. (***Ans.*** $1.24 \times 10^4$ eV, 150.33 eV, 0.082 eV)

3.20 Write short notes on:

(i) Rotating crystal method

(ii) Powder method.

3.21 What is Auger electron spectroscopy.

CHAPTER

# 4

# Bonding and Atoms Size

## 4.1 THE JELLIUM MODEL (JM)

Before discussing various types of bonding, we will first consider the Jellium model. The Jellium model (JM), the simplest possible model for a solid, assumes that *solid formation is associated with the splitting of each atom to ζ electrons and to a cation of charge ζ, where ζ is the bonding atomic valence; each cation is assumed to be mashed, as to cover uniformly the volume per atom, $v_a$, associated with it; as a result we end up with a uniform ionic background (a jelly) of mass density*

$$\rho_M = m_a/v_a; \qquad ...(1)$$

and charge density

$$\rho_e = \zeta e/v_a, \qquad ...(2)$$

*(instead of the actual discrete ions). This positive uniform ionic charge density is fully compensated by negative charge density of the valence electrons, which, being assumed* free of any force, *spread themselves uniformly over the whole volume V, of the solid.*

For every atom the atomic number Z, the mass[1] $m_a = A_w m_u$, and the bonding valence ζ are assumed known. It remains to find the volume per atom $v_a$, by minimizing the total energy. Having determined $v_a$ we can obtain immediately the concentration of atoms

$$n_a \equiv \frac{N_a}{V} \equiv \frac{1}{v_a}, \qquad ...(3)$$

the radius per atom

$$r_a \equiv (3v_a/4\pi)^{1/3}, \qquad ...(4)$$

and the density $\rho_M$ given by Eq. (1). In many books instead of $v_a$, $n_a$ and $r_s$, the quantities volume per valence electron, $v_e$, concentration of valence electrons, $n_e$, and the radius per valence electron, $r_s$, are introduced; they are directly related to $v_a$, $n_a$ and $r_a$ as follows

[1] The atomic weight $A_w$ is almost equal to A, the mass number, (if there is only one stable isotope); $m_u \simeq 1823$ $m_e$ is defined as the 1/12 of the mass of carbon-12.

$$v_e = \frac{v_a}{\zeta} \qquad ...(5)$$

$$n = \zeta n_a, \qquad ...(6)$$

$$r_\varepsilon = r_a/\zeta^{1/3}. \qquad ...(7)$$

The *dimensionless quantities* $\bar{r}_a \equiv r_a/a_B$ and $\bar{r}_s \equiv r_s/a_B$ are also widely used.

Is the JM capable of accounting approximately for some properties of solids? Or, is it so oversimplified as to be irrelevant? The answer is "yes" to the first question "no" to the second, in spite of the drastic omission of the force experienced by each (detached) electron due to the ions and the other electrons. The reason for this surprising answer (from a classical point of view which ignores the wave character of the electrons) is that the actual force on each electron is, to a first approximation, periodic and as such allows free-like propagation (for eigenenergies inside a band) and *in this respect resembles no force at all*. This is the reason that the JM is a responsible first approximation for metals, where the Fermi level is inside a band. On the other hand, for semiconductors and insulators, where the Fermi energy is inside a gap, the JM is expected to fail. The reason is that zero force produces no gaps, and hence, it is incapable of accounting for semiconducting or insulating properties.

## 4.2 GENERAL NATURE OF COHESION BETWEEN TWO ATOMS IN A SOLID

In a solid material, there are of two types of forces acting between constitutent atoms:

(i) attractive forces, which keep the atom together so as to form the solid, and

(ii) repulsive forces, which become noticeable while attempted to compress the solid.

Suppose two atoms $A$ and $B$ exert attractive or repulsive forces on each other such that potential energy of $B$ in the field of $A$ is given by

$$W(r) = -\frac{\alpha}{r^n} + \frac{\beta}{r^m} \qquad ...(1)$$

where $r$ is the separation between the two atoms ($n$ and $m$ are positive) $\alpha$ are $\beta$ positive constants which determine the strengths of the attractive and repulsive forces, respectively. It is assumed that for $r - \infty$, the potential energy of particles in each other's field is zero.

The condition under which particles form a stable molecule is that the function $W(r)$ exhibits a minimum for some finite value of $r$. If this minimum exists (say $r_0$) then the two atoms form a stable molecule, $r_0$ is termed the *equilibrium spacing* and $W(r_0)$ is called *bond energy*. The minimum energy is negative and hence the energy needed to dissolve the molecule equals the positive quantity $- W(r_0)$.

The condition for $W(r)$ to the extremum is $dW/dr = 0$ and for a minimum of $W(r)$, we must have $d^2W/dr^2 > 0$.

$$\left.\frac{dW}{dr}\right|_{r=r_0} = 0, \quad \frac{\alpha n}{r_0^{n+1}} - \frac{\beta m}{r_0^{m+1}} = 0$$

$$\Rightarrow \qquad r_0 = \left(\frac{\rho}{a}\cdot\frac{m}{n}\right)^{1/(m-n)}$$

$$\left.\frac{d^2W}{dr^2}\right|_{r=r_0} > 0, \frac{-\alpha n\,(n+1)}{r_0^{n+2}} + \frac{\beta m\,(m+1)}{r_0^{m+2}} = 0$$

or $$\beta m\,(m+1) > \alpha n\,(n+1)\, r_0^{m-n} \qquad ...(2)$$

Substituting for $r_0$ from Eq. (2), we get the condition for $W(r)$ to be minimum as

$$m > n \qquad ...(3)$$

This means that the attractive forces should vary more slowly with $r$ than the repulsive force. Figure 4.1 shows variation of the force $F$ and potential energy $W$ as a function of their separation, $r$.

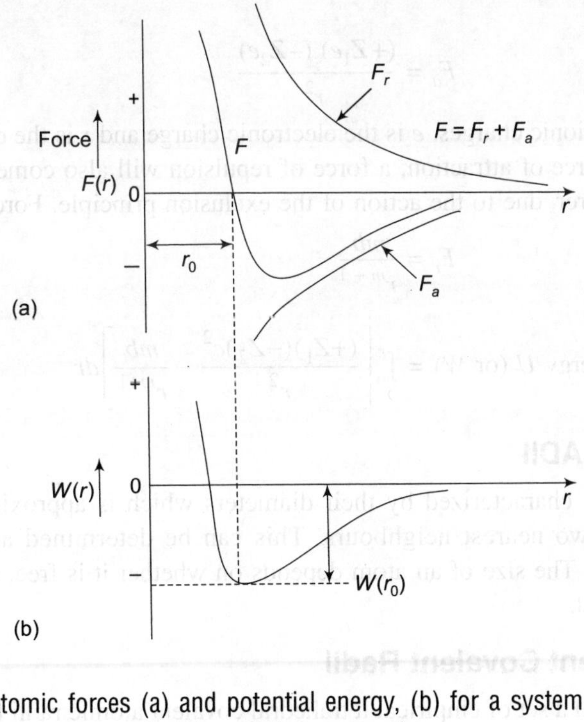


**Figure 4.1** Interatomic forces (a) and potential energy, (b) for a system of two atoms as a function of their separation $r$

## 4.3 IONIC BOND

An ionic bond is formed when atoms that have low ionization energies, and hence can lose electrons readily, interact with other atoms that have a high electron affinity. The former atom that loses electron (s) becomes positive ion, and the latter atom that gains electron(s) becomes negative ion. The coulomb attraction of the oppositely charged ions produces a stable union. The electrons that take part in the ionic formation are the outermost electrons of electropositive atoms (i.e., those when form metallic solids) and electronegative atoms are capable of picking up electrons, e.g.,

$$\text{Na}(11) = 1s^2 2s^2 2p^6 3s^1$$

$$\text{Cl}(17) = 1s^2 2s^2 2p^6 3s^2 3p^5$$

(Ne-structure)

$$\text{Na}^+ = 1s^2 2s^2 2p^6$$

$$\text{Cl}^- = 1s^2 2s^2 2p^6 3s^2 3p^6$$

(Ar-structure)

The sodium and chlorine ions have acquired the configuration of nearest inert gas. The bonding is a result of attraction between the positive and negative ions. In general, the force of attraction,

$$F_a = \frac{(+Z_1 e)\,(-Z_2 e)}{r^2} \qquad ...(1)$$

where $Z_1$ are $Z_2$ the ionic charges, $e$ is the electronic charge and $r$ is the distance between the ions. Besides the force of attraction, a force of repulsion will also come into play when the ions are bought nearer, due to the action of the exclusion principle. Force of repulsion

$$F_r = \frac{mb}{r^{m+1}} \qquad ...(2)$$

$$\text{The bond energy } U \text{ (or } W) = \int_{\infty}^{r} \left[ \frac{(+Z_1)(-Z_2)e^2}{r^2} - \frac{mb}{r^{m+1}} \right] dr \qquad ...(3)$$

## 4.4 ATOMIC RADII

The size of atom is characterized by their diameters which is approximately equal to the distance between two nearest neighbours. This can be determined accurately by X-ray diffraction methods. The size of an atom depends on whether it is free, or bound in a metal or in an ionic crystal.

### 4.4.1 Tetravalent Covalent Radii

Pauling has proposed a set of empirical tetrahedral covalent atomic radii for atoms in crystals with coordination number 4 such as diamond, cubic ZnS, and in crystals containing $PO_4$, $SO_4$, $SiO_4$, $MoO_4$, etc., in which the metal atom is at the centre of a tetrahedron and the oxygen atoms are at the corners. Table 4.1 gives the set of empirical tetrahedral covalent radii.

**Table 4.1** Atomic Radii (in Angstrom) when the Atoms Exist in Tetrahedral Covalent Bonds

| *Element* | *Radii* | *Element* | *Radii* | *Element* | *Radii* |
|---|---|---|---|---|---|
| Cu | 1.35 | Ag | 1.52 | Be | 1.06 |
| Mg | 1.40 | Zn | 1.31 | Cd | 1.48 |
| Hg | 1.48 | B | 0.88 | Al | 1.26 |
| Ga | 1.26 | In | 1.44 | C | 0.77 |
| Si | 1.17 | Ge | 1.22 | Sn | 1.40 |
| N | 0.70 | P | 1.10 | As | 1.18 |
| Sb | 1.36 | O | 0.66 | S | 1.04 |
| Se | 1.14 | Te | 1.32 | F | 0.64 |
| Ce | 0.99 | Br | 1.11 | I | 1.28 |

## 4.4.2 Ionic Crystal Radii

Bragg and Pauling have shown that the arrangement of ions in a crystal is largely determined by the respective sizes of the anion and cation. For ionic crystals, the ionic diameter ($d$) is given by

$$d = r_c + r_a + \Delta N \qquad ...(1)$$

where $N$ is the coordination number, $r_c$ and $r_a$ are respectively the radii of cation and anion and $\Delta N$ is the correction term for the coordination number. Table 4.2 gives the ionic crystal radii of certain metallic ions.

**Table 4.2** Ionic Crystal Radii (in Angstrom)

| *Ion* | *Radii* | *Ion* | *Radii* | *Ion* | *Radii* |
|---|---|---|---|---|---|
| $NH_4^+$ | 1.48 | $Ga^+$ | 1.13 | $In^+$ | 1.32 |
| $Te^-$ | 1.40 | $O^{--}$ | 1.45 | $Cl^-$ | 1.81 |
| $Br^-$ | 1.96 | $I^-$ | 2.19 | $S^{--}$ | 1.90 |
| $Te^{--}$ | 2.22 | $Li^+$ | 0.68 | $Na^+$ | 0.98 |
| $K^+$ | 1.33 | $Cs^+$ | 1.67 | $Cu^+$ | 0.95 |
| $Ba^{++}$ | 1.29 | $Sr^{++}$ | 1.10 | $Zn^{++}$ | 0.69 |
| $Cd^{++}$ | 0.92 | $Co^{++}$ | 0.70 | $Ni^{++}$ | 0.68 |
| $Ag^+$ | 1.13 | $Au^+$ | 1.37 | $Al^{+++}$ | 0.45 |

## 4.4.3 Ionic Core Crystal Radii

In an atom, the electrons in the valence orbits are called *conduction electrons*, whereas the electrons in all the other orbits are called *core electrons*. The nucleus and the core electrons in an atom form an ionic core. The total number of core electrons in an atom is given by

$$n_{\text{core}} = \sum 2(2l + 1) \qquad ...(2)$$

where $l$ is the angular momentum quantum number. The core electron orbits are divided into different shells which are represented by $s$, $p$, $d$, $f$ states. The number of core electrons in different states are as given below.

$s$ state ($l = 0$) : 2

$p$ state ($l = 1$) : 6

$d$ state ($l = 2$) : 10

$f$ state ($l = 3$) : 14

The pseudopotential[2] study gives us an idea about determining the ion core radii of the metals. The one parameter of pseudopotential model predicts that the condition which satisfies the cut-off region of the potential is given by

$$qr_c = \frac{\pi}{2}$$

where $r_c$ the ion core radii and $q$ is the momentum transfer vector.

If one know the correct point of the first node of pseudopotentials, the ionic core radii may be easily determined by the expression

$$r_c = \frac{\pi}{2q} \qquad ...(3)$$

Table 4.3 gives the value of ionic core radii for liquid metal.

**Table 4.3** Ionic Core Radii (in Angstrom)

| *Metal* | *Radii* | *Metal* | *Radii* | *Metal* | *Radii* |
|---|---|---|---|---|---|
| Li | 1.678 | Na | 1.758 | K | 2.133 |
| Rb | 2.256 | Cs | 2.425 | Be | 1.055 |
| Mg | 1.058 | Zn | 1.110 | Cd | 1.247 |
| Hg | 1.226 | Ca | 1.905 | Ba | 2.403 |
| Sr | 2.091 | Al | 1.131 | Ga | 1.116 |
| In | 1.194 | Te | 1.221 | Ge | 1.022 |
| Sn | 1.125 | Pb | 1.069 | As | 0.964 |
| Sb | 1.073 | Bi | 1.065 | Se | 0.947 |

## 4.5 IONIC BONDING

An ionic bonding can only be formed between two different atoms, one electropositive and the other electronegative. Electropositive elements readily give up electrons and are usually Group I or II elements, e.g., Na, K, and Ba, whereas electronegative elements readily take up electrons and are typically Group VI or VII elements, e.g., Cl, Br and O.

As an example of ionic bonding, consider a molecule of common salt, NaCl. When neutral atoms of Na and Cl are brought close together it is easy for the outer valence electron

[2] If the actual potential energy in the core region is replaced by an effective potential energy which gives the same wave functions outside the core as does the actual potential, then this effective potential energy is called *pseudopotential.*

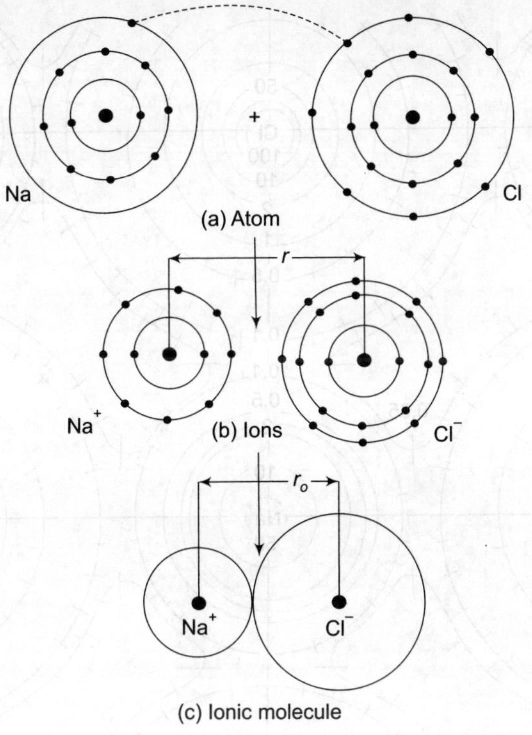


**Figure 4.2** Schematic representation of the formation of an ionic molecule of sodium chloride

of the sodium atom to be transferred to the chlorine atom so that both of them acquire a stable inert gas electronic configuration as shown in Fig. 4.2. Such configurations will have nearly spherically symmetric charge distribution around the ions with some distortion near the region of contact between neighbouring atoms. This is confirmed by X-ray studies of electron distribution (Fig. 4.3).

Since the sodium cation carries a positive charges and the chlorine anion a negative charge there exists an electrostatic attraction between the two ions. However, it is incorrect to expect that sodium ions ($Na^+$) and chlorine ions ($Cl^-$) would link up in pairs as shown, because then there would be a strong attractive force within the paired ions of a sodium chloride crystal while negligible attraction between the pairs. As a result, solid sodium chloride would not exist.

Actually, a negative charge attracts all positive charges in the neighbourhood, and vice versa. Consequently, in the crystalline solid $Na^+$ ions will be surrounded by $Cl^-$ ions and $Cl^-$ ions by $Na^+$ ions in such a way that the attraction between the neighbouring unlike charges exceeds the repulsion due to like charges. The resulting sodium chloride structure is shown in the Fig. 4.4.

**Figure 4.3** Electron density distribution in base plane of sodium chloride, after X-ray studies by *G. Schoknecht, Z. Naturforschg*, **12**, 983 (1957)

### 4.5.1 Bond Dissociation Energy of NaCl Molecule

NaCl is considered to be an ideal representative of ionic compound. In order to calculate the bond dissociation energy corresponding to a single NaCl molecule, let us begin with the consideration that the sodium and chlorine atoms are initially free when they are at infinite distance of separation. The energy required to remove the outer electron from the Na atom is equal to its ionization potential. When this energy is observed by the sodium atom, it gets ionized and a positively charged $Na^+$ ion is formed. The ionization process of a Na atom is represented by the reaction

$$\text{Na (g) + I.P. (energy)} \rightarrow \text{Na}^+ \text{ (g)} + \bar{e} \text{ (g)}$$

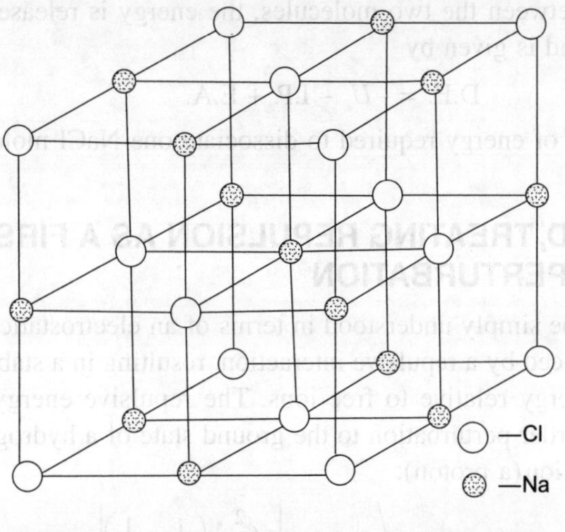


**Figure 4.4** Sodium chloride crystal structural

Similarly, when an electron is taken from infinity and added it to a neutral chlorine atom to form a negatively charged chlorine ion, an energy equivalent to its electron affinity is released. This formation of chlorine ion is given by the reaction

$$\text{Cl (g)} + \overline{e}\text{ (g)} \rightarrow \text{Cl}^- \text{ (g)} + \text{E.A. (energy)}$$

The net amount of energy absorbed (spent) in the process of formation of positive sodium ion and a negative chlorine ion at infinity is thus given by

$$\Delta E = \text{I.P.} - \text{E.A.}$$

Combining the above two reactions, we have

$$\text{Na (g)} + \text{Cl (g)} + \Delta \text{E} \rightarrow \text{Na}^+\text{(g)} + \text{Cl}^- \text{ (g)}$$

In the above reactions, the ions are infinitely separated, but now let the coulomb attraction between them bring the two ions closer together so that an ion-pair is formed. When the two ions are at an equilibrium separation, a stable NaCl molecule is formed. In this process, an energy equal to the potential energy is released at equilibrium separation, i.e.,

$$U_e = \frac{e^2}{4\pi\varepsilon_0 R_e}$$

where $\varepsilon_0$ is the absolute permittivity of free space and air. Its numerical value is given by

$$\varepsilon_0 = 8.854 \times 10^{-12} \text{ C}^2 \text{ N}^{-1} \text{ m}^{-2},$$

$$\text{and } \frac{1}{4\pi\varepsilon_0} = 9 \times 10^9 \text{ Nm}^2 \text{ C}^{-2}$$

Since the entire process (starting from neutral Na and Cl atoms till the formation of NaCl molecule) consists of three steps: (i) removal of an electron from Na atom (first ionization energy), (ii) addition of an electron to the Cl atom (electron affinity of Cl), and (iii) the

Coulomb attraction between the two molecules, the energy is released is equal to the bond dissociation energy and is given by

$$\text{D.E.} = -U_e - \text{I.P.} + \text{E.A.}$$

This is the amount of energy required to dissociate one NaCl molecule into $Na^+$ and $Cl^-$ ions.

## 4.6 IONIC BOND, TREATING REPULSION AS A FIRST ORDER QUANTUM PERTURBATION

The ionic bond may be simply understood in terms of an electrostatic attraction of spherical ions $H_p^-$ and $H_q^+$ balanced by a repulsive interaction, resulting in a stable configuration which has a net binding energy relative to free ions. The repulsive energy can be estimated by considering the first order perturbation to the ground state of a hydrogen atom $H_p$ caused by the approach of a $H_q^+$ ion (a proton):

$$E(R) = \left\langle \Psi_{100}(r) \left| \frac{e^2}{4\pi\varepsilon_0}\left(\frac{1}{R} - \frac{1}{d}\right) \right| \Psi_{100}(r) \right\rangle \quad ...(1)$$

where $\Psi_{100}$ is the ground state wave function of the hydrogen atom.

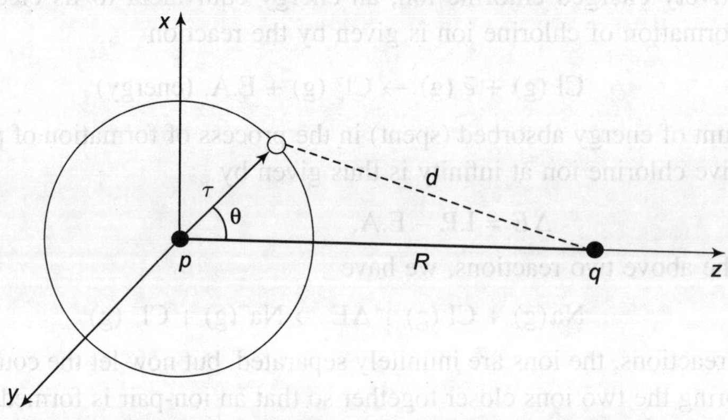


**Figure 4.5** Hydrogen-proton repulsive interaction

In the configuration illustrated in Fig. 4.5, where the distance $d$ is defined as $d = \sqrt{R^2 + r^2 - 2Rr\cos\theta}$, it is convenient to expand a root into a power series of either $r/R$ if $r < R$ or $R/r$ if $r > R$:

$$\frac{1}{d} = \frac{1}{\sqrt{R^2 + r^2 - 2Rr\cos\theta}}$$

$$= \frac{1}{R}\left(1 + \frac{r^2}{R^2} - 2\frac{r}{R}\cos\theta\right)^{\frac{1}{2}} = \frac{1}{R}\left\{1 + \frac{r}{R}\cos\theta + \frac{1}{2}(3\cos^2\theta - 1)\left(\frac{r}{R}\right)^2 \pm ...\right\}, r < R$$

$$=\frac{1}{r}\left(1+\frac{R^2}{r^2}-2\frac{R}{r}\cos\theta\right)^{-\frac{1}{2}}=\frac{1}{r}\left\{1+\frac{R}{r}\cos\theta+\frac{1}{2}(3\cos^2\theta-1)\left(\frac{R}{r}\right)^2\pm...\right\}, r>R$$

Substituting this result and the ground state wave function into Eq. (1) and remembering that upon integration all angular terms give a zero contribution, we obtain

$$E(R)=\frac{e^2}{4\pi\varepsilon_0}\frac{1}{\pi a_0^3}\int_0^R\int_0^\pi\int_0^{2\pi}e^{-2r/a_0}\left(-\frac{r}{R^2}\cos\theta-....\right)r^2\sin\theta d\theta d\varphi dr$$

$$+\frac{e^2}{4\pi\varepsilon_0}\frac{1}{\pi a_0^3}\int_R^\infty\int_0^\pi\int_0^{2\pi}e^{-2r/a_0}\left(\frac{1}{R}-\frac{1}{r}-\frac{R}{r^2}\cos\theta-....\right)r^2\sin\theta d\theta d\varphi dr$$

$$=\frac{e^2}{\pi\varepsilon_0 a_0^3}\int_R^\infty e^{-2r/a_0}\left(\frac{1}{R}-\frac{1}{r}\right)r^2\,dr=\frac{e^2}{4\pi\varepsilon_0}\left(1+\frac{a_0}{R}\right)e^{-2R/a_0} \quad ...(2)$$

Equation (2) represents a positive energy, which arises from a repulsive force, and varies rapidly with $R$. It is assumed that the overlap of the wave functions of the valence electron from two interacting adjacent ions can always be represented in an analytical form similar to Eq. (2), namely,

$$E(R) = \lambda e^{-R/\rho} \quad ...(3)$$

which is known as the *Born-Mayer potential*. The parameter $\lambda$ and $\rho$ are independent of the distance $R$ between the two ions. The attractive electrostatic energy in monovalent ionic solids can be written, if spherical ions are treated as point charges, in the familiar from $-e_0^2/R$. The electronic energy of a single ion pair can therefore be represented by the sum of the repulsive and attractive energies:

$$E(R) = \lambda e^{-R/\rho}-\frac{e_0^2}{R} \quad ...(4)$$

Plots of the expression (4) and its repulsive and attractive components as functions of the interionic spacing $R$ are given in Fig. 4.6. $R_0$ is the equilibrium interionic separation where the binding energy is a minimum.

For a system of $N$ ion pairs with alternating charges, the potential energy take the form:

$$U(R) = N\left(\sum_{j\neq i}\lambda_{ij}e^{-p_{ij}R/\rho}-\frac{e_0^2}{R}\sum_{j\neq i}\frac{\pm 1}{p_{ij}}\right) \quad ...(5)$$

where the distance $R_{ij} = p_{ij}R$ between the ion $i$ and $j$ is given in terms of the nearest neighbour separation $R$. Because of the exponential dependence of the repulsive energy, we may consider, without loss of

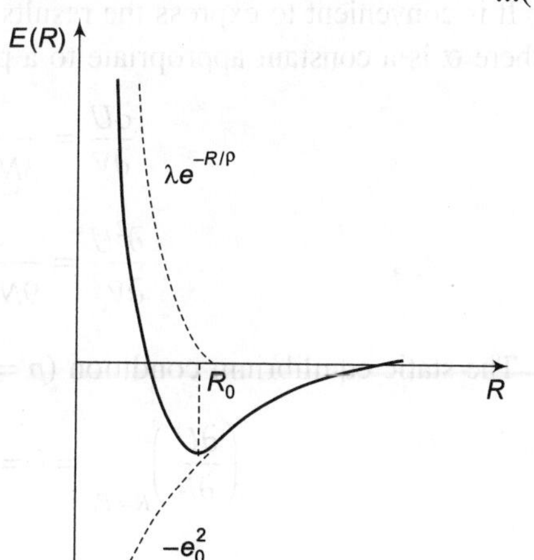


**Figure 4.6** Binding energy of a pair of ions

accuracy, only the nearest neighbour separation term in the first series on the right hand side. The second series in Eq. (5) is usually denoted by

$$A = \sum_{j \neq i} \frac{\pm 1}{p_{ij}} \quad ...(6)$$

and is known as the *Madelung constant*, which is clearly dependent on the geometry of the system. The evaluation of $A$ is straightforward for a regular array of ions with alternating unit charges. As a simple example, for a infinite linear chain of ions at separation $R$ we have

$$A = 2\left(\frac{1}{1} - \frac{1}{2} + \frac{1}{3} - \frac{1}{4} + ...\right) = 2 \ln 2$$

where use has been made of the series expansion.

$$\ln (1 + x) = x - \frac{x^2}{2} + \frac{x^3}{3} - \frac{x^4}{4} + .....$$

The potential energy Eq. 5 of the systems of ions then reduces to

$$U(R) = N\left(\lambda e^{-R/\rho} - \frac{Ae_0^2}{R}\right) \quad ...(7)$$

and we are left with two unknowns, $\lambda$ and $\rho$ to determine. They can be evaluated from thermodynamic consideration in terms of $R$, $A$ and the isothermal compressibility $k_T$ at low temperature, which is a measurable property of ionic structures, given by

$$\frac{1}{k_T} = -V\left(\frac{\partial p}{\partial V}\right)_T = V\left(\frac{\partial^2 U}{\partial V^2}\right)_T \quad ...(8)$$

It is convenient to express the results in terms of ionic separation $R$, by taking $V = N\alpha R^3$, where $\alpha$ is a constant appropriate to a particular ionic structure, so that we have

$$\frac{\partial U}{\partial V} = \frac{1}{3N\alpha R^2}\frac{\partial U}{\partial R} \quad ...(9)$$

$$\frac{\partial^2 U}{\partial V^2} = \frac{1}{9N^2\alpha^2 R^2}\left\{\frac{1}{R^2}\frac{\partial^2 U}{\partial R^2} - \frac{2}{R^3}\frac{\partial U}{\partial R}\right\} \quad ...(10)$$

The static equilibrium condition ($p = 0$) on the ionic structure at low temperature reads

$$\left(\frac{\partial U}{\partial R}\right)_{R=R_0} = 0 = N\left(\frac{Ae_0^2}{R_0^2} - \frac{\lambda}{\rho}e^{-R_0/\rho}\right) \quad ...(11)$$

i.e.,

$$\lambda = \frac{\rho A e_0^2}{R_0^2} e^{R_0/\rho} \quad ...(12)$$

In view of Eqs. (10) and (11). Eq. (8) reduces to

$$\frac{1}{k_T} = \frac{1}{9N\alpha R_0}\left(\frac{\partial^2 U}{\partial R^2}\right)_{R=R_0} = \frac{1}{9\alpha R_0}\left(\frac{\lambda}{\rho^2}e^{-R_0/\rho} - \frac{2Ae_0^2}{R_0^3}\right)$$

so that

$$\frac{R_0}{\rho} = 2 + \frac{9\alpha R_0^4}{Ae_0^2 k_T} \qquad ...(13)$$

Substituting Eq. (12), the binding energy is obtained from Eq. (7) in terms of the known ratio Eq. (13) as

$$U(R_0) = -\frac{NAe_0^2}{R_0}\left(1 - \frac{\rho}{R_0}\right) \qquad ...(14)$$

which is close to the total electrostatic energy at equilibrium separation, provided $\rho << R_0$. There is good agreement between the measured values and those calculated using Eq. (14) for typical ionic structures, such as the alkali halides.

## 4.7 THE BOND ENERGY AND MADELUNG CONSTANT FOR NaCl CRYSTAL

In a crystal, there are a large number of ion pairs arranged regularly in three dimensions. The bond energy of various ions from an ion then can be calculated by finding out the distances of various ions from an ion taken as the reference point. After taking into account the number of ions at the particular distance, we get the energy by summing over the expressions developed for the single ion pairs. Figure 4.7(a) shows the bonding in sodium chloride structure and Fig. 4.7(b) shows disposition of the atom of sodium and chlorine in the crystal.

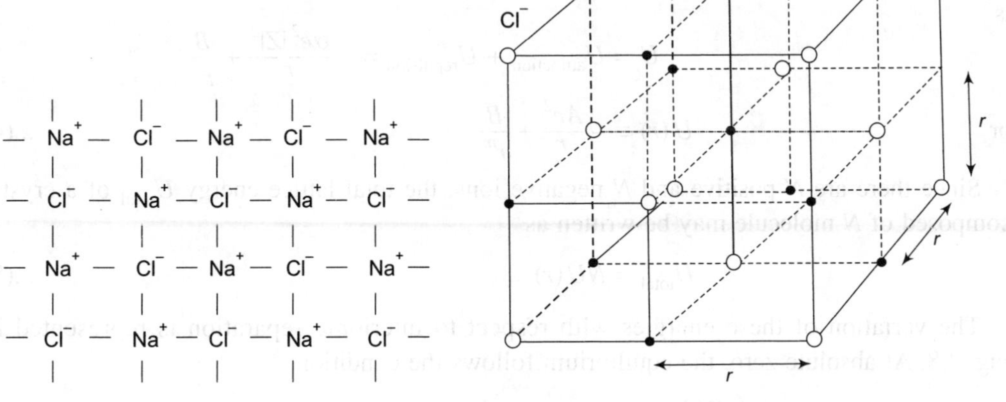


**Figure 4.7** (a) Two-dimensional representation of the bonding between sodium and chlorine ions forming the sodium chloride structure, (b) the sodium chloride structure

As can be seen from Fig. 4.2(b), each $Na^+$ ion is surrounded by

- 6 nearest neighbour $Cl^-$ ions at a distance $r$.
- 12 next nearest neighbour $Na^+$ ions at a distance $\sqrt{2}\,r$,
- 8 $Cl^-$ ions at the distance $\sqrt{3}\,r$,
- 6 $Na^+$ ions at the distance $2r$ and
- 24 $Cl^-$ ions at the distance $\sqrt{5}\,r$.

Now, the electrostatic energy of $Na^+$ ion with the surrounding ions can be written as

$$U_{\text{attraction}} = \frac{6e^2}{r}(+Z_1)(-Z_2) + \frac{12e^2}{\sqrt{2}r}(+Z_1)^2 + \frac{8e^2}{\sqrt{3}r}(+Z_1)(-Z_2) + \frac{6e^2}{\sqrt{2}r}(+Z_1)^2 + \ldots \quad \ldots(1)$$

In the present case, the two ions carry equal but opposite charges and hence

$$+(Z_1) = -(-Z_2)$$

or

$$Z_1 = Z_2 = |Z|$$

$$\text{Attraction} = \frac{e^2|Z|^2}{r}\left(6 - \frac{12}{\sqrt{2}} + \frac{8}{\sqrt{3}} - \frac{6}{\sqrt{2}} + \frac{24}{\sqrt{5}} - \ldots\right) \quad \ldots(2)$$

The terms within brackets is called the *Madelung constant* (represented by $\alpha$) of the crystal and represents the geometrical arrangements of ions (it has the same value for all crystals of the same structure). Its value for NaCl is 1.747558. A similar calculation for cesium chloride gives $\alpha = 1.763$ and for zinc blende $\alpha = 1.638$.

The potential energy contribution of the repulsive force has the form

$$U_{\text{repulsion}} = \frac{B}{r^m} \quad \ldots(3)$$

The $r^m$ dependence implies a short range force, it increases at the separation distance $r$ decreases. The total potential energy of each ion due to its interaction with all the other ions is

$$U = U_{\text{attraction}} + U_{\text{repulsion}} = -\frac{\alpha e^2|Z|^2}{r} + \frac{B}{r^m}$$

or

$$U(r) \equiv -\frac{Ae^2}{r} + \frac{B}{r^m} \quad \ldots(4)$$

Since there are $N$ positive and $N$ negative ions, the total lattice energy $U_{\text{total}}$ of a crystal composed of $N$ molecule may be written as

$$U_{\text{total}} = NU(r) \quad \ldots(5)$$

The variation of these energies with respect to interionic separation is represented in Fig. 4.8. At absolute zero, the equilibrium follows the condition

$$\left(\frac{dU}{dr}\right)_{r=r_0} = 0 \quad \ldots(6)$$

where $r_0$ is the smallest nearest neighbour distance at $T = 0$. Using the condition is Eq. (6), we have

$$\frac{Ae^2}{r_0} + B(-m)\, r_0^{-m-1} = 0$$

or $$B = \frac{Ae^2}{m} r_0(m^{-1}) \qquad ...(7)$$

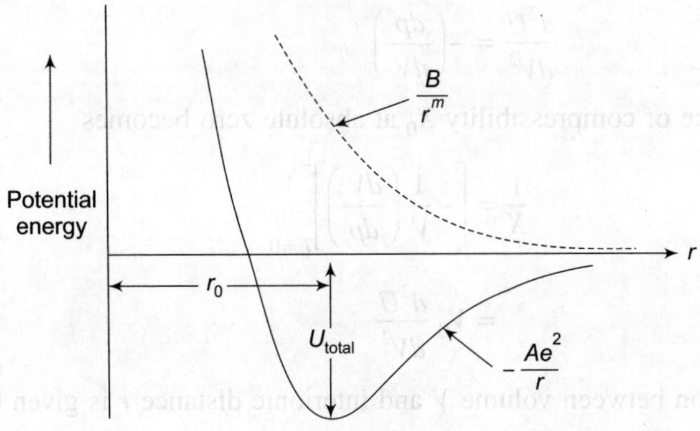


**Figure 4.8** Variation of potential energy with separation $r$

Substituting this value of $B$ in Eq. (4), and using Eq. (5), we obtain the following expression for lattice energy of ionic crystal

$$U_L = U_{\text{total}}(r_0) = NU(r_0)$$

$$= -\frac{NAe^2}{r_0}\left(1 - \frac{1}{m}\right) \qquad ...(8)$$

The *lattice energy* may be defined as the energy released when the constituent ions are placed in their respective position in the crystal lattice, i.e., the energy spent to separate the solid ionic crystal into its constituent ions.

*Evaluation of m*: Born determined the value of the factor ($m$) from the experimentally measured compressibility data of the crystals. The well-known expression for the compressibility $X$ of a solid may be expressed by

$$X = -\frac{1}{V}\frac{dV}{dp} \qquad ...(9)$$

where $V$ and $p$ are volume and pressure, respectively. From the first law of thermodynamics

$$dQ = dU + dW \qquad ...(10)$$

where $dQ$ is the heat supplied to the system which is consumed in performing work of the amount $dW$ band increasing the internal energy $dU$. We have

$$dQ = C_v dT \quad ...(11)$$

$$dW = pdV \quad ...(12)$$

An absolute zero, the value of $C_v$ is very small, hence on neglecting the quantity $dQ$, we obtain

$$dU = -pdV$$

$$\frac{d^2U}{dV^2} = -\left(\frac{dp}{dV}\right) \quad ...(13)$$

Hence, the value of compressibility $X_0$ at absolute zero becomes

$$\frac{1}{X} = \left[-\frac{1}{V}\left(\frac{dV}{dp}\right)\right]^{-1}_{T=0}$$

$$= V_0 \frac{d^2U}{dV^2} \quad ...(14)$$

Now, the relation between volume $V$ and interionic distance $r$ is given by

$$V = CNr^3 \quad ...(15)$$

where $C$ is a constant depending on the type of lattice (for NaCl, $C = 2$).

We represent the value of $\frac{d^2U}{dV^2}$ as follows

$$\frac{dU}{dV} = \frac{dU}{dr} \cdot \frac{dr}{dV}$$

$$\Rightarrow \quad \frac{d^2U}{dV^2} = \frac{dU}{dr} \cdot \frac{d^2r}{dV^2} + \frac{d^2U}{dr^2}\left(\frac{dr}{dV}\right)^2 \quad ...(16)$$

For $\quad r = r_0, \frac{dU}{dr} = 0$

hence

$$\frac{d^2U}{dV^2} = \left(\frac{d^2U}{dr^2}\right)_{r=r_0} \left(\frac{dr}{dV}\right)^2_{r=r_0} \quad ...(17)$$

From Eq. (15)

$$\frac{dV}{dr} = 3\,CNr^2$$

$$\left(\frac{dr}{dV}\right)^2_{r=r_0} = \frac{1}{9C^2N^2r_0^4} \quad ...(18)$$

From Eqs. (10), (13) and (14), we obtain

$$\frac{1}{X_0} = V_0 \left( \frac{d^2 U_{\text{total}}}{dr^2} \right)_{r=r_0} \frac{1}{9C^2 N^2 r_0^4} \qquad \text{...(19)}$$

From Eqs. (4) and (5), we have

$$U_{\text{total}} = -\frac{NAe^2}{r_0}\left[1 - \frac{1}{m}\left(\frac{r_0}{r}\right)^{m-1}\right]$$

$$\Rightarrow \qquad \frac{dU_{\text{total}}}{dr} = \frac{NAe^2}{r^2} - \frac{Ae^2 r_0^{m-1} N}{r^{m-1}}$$

and

$$\left(\frac{d^2 U_{\text{total}}}{dr^2}\right) = -\frac{2\,Ae^2 N}{r^3} + \frac{Ae^3\, r_0^{m-1}\,(m+1)N}{r^{m+2}} \qquad \text{...(20)}$$

Substituting Eqs. (15) and (20) in Eq. (19), we get

$$\frac{1}{X} = \frac{Ae^2\,(m-1)}{9Cr_0^4}$$

i.e.,

$$m = 1 + \frac{9\,C\,r_0^4}{X_0 e^2\,A} \qquad \text{...(21)}$$

Using the standard experimental data,[3] the different values of $m$ for various crystals are given in Table 4.4.

**Table 4.4** Values of *m* for Various Crystals

| *Crystal* | *m* |
|---|---|
| LiF | 5.9 |
| LiCl | 8.0 |
| LiBr | 8.7 |
| NaCl | 9.1 |
| NaBr | 9.5 |

Because of the large value of $m$, it appears that most of the lattice energy is due to the Coulomb interaction, repulsion gives only a relatively small contribution. Substituting the value of $m$ from Eq. (21) in Eq. (8), we obtain the final expression for lattice energy of an ionic crystal as

$$U_L = NU(r_0)$$

$$= -\frac{NAe^2}{r_0}\left[1 - \left\{\frac{9Cr_0^4}{X_0 e^2 A}\right\}^{-1}\right] \qquad \text{...(22)}$$

[3] J.C. Slater, *Phys. Rev.* **23** (1924) 488.

where $A$ has a value described by Eq. (4). The values obtained for lattice energy from the above relation have been presented in Table 4.5 along with values observed by the Born–Haber cycle.

**Table 4.5** Lattice Energies of Some Ionic Crystals

| *Crystal* | $r_0$ (Å) | $U_L$ (eV) | $U_L$ observed (eV) |
|---|---|---|---|
| LiF | 2.07 | 10.5 | – |
| LiCl | 2.57 | 8.4 | 8.6 |
| NaCl | 2.81 | 8.0 | 7.9 |
| Li Br | 2.74 | 7.9 | 8.2 |
| Na Br | 2.97 | 7.5 | 7.5 |

## 4.8 MADELUNG CONSTANT

Crystals that can be pictured as formed by positive and negative ions are referred to as *ionic crystals*. Some properties of these solids can be understood starting from the extreme approximation of a lattice of pointwise charge (*point ion lattice*); the influence of the actual internal constitution of ions can then be introduced in the model, and overall trends in electronic properties can often be understood at a reasonable elementary level.

An important quantity in the study of ionic crystals is the Coulomb interaction energy (Madelung energy) between ions of the crystals in the point ion approximation. As we shall see, this quantity accounts for most of (i) cohesive energy of the crystal, (ii) stability of the structure, and (iii) energy gap of the material. We thus begin the study of ionic crystals with the evaluation of the potential energy of a given reference ion (say, a negative ion taken at the origin) due to the Coulomb interaction with all the other positive and negative ions of the crystal. It is usual to express this potential energy in the Madelung form

$$V(0) = -\alpha_M \frac{e^2}{R}, \qquad \text{...(1)}$$

where $R$ is the nearest neighbour distance, $\alpha_M$ is the Madelung constant discussed below, $\pm e$ are the point charges at the lattice sites. The Madelung constant is a dimensionless constant, characteristic of the geometrical crystal structure under consideration.

The evaluation of the Madelung constant requires a special care, because of the long-range character of the Coulomb interaction. To give an idea of the slow convergence of the series that are encountered in the direct evaluation of the Madelung constant, let us consider the one-dimensional model of ionic crystal indicated in Fig. 4.9. In this case; we have for the potential of the ions at the origin

$$V(0) = -\frac{e^2}{R} 2\left[\frac{1}{1} - \frac{1}{2} + \frac{1}{3} - \frac{1}{4} + \frac{1}{5} - \ldots\right] = -\frac{e^2}{R} 2 \ln 2.$$

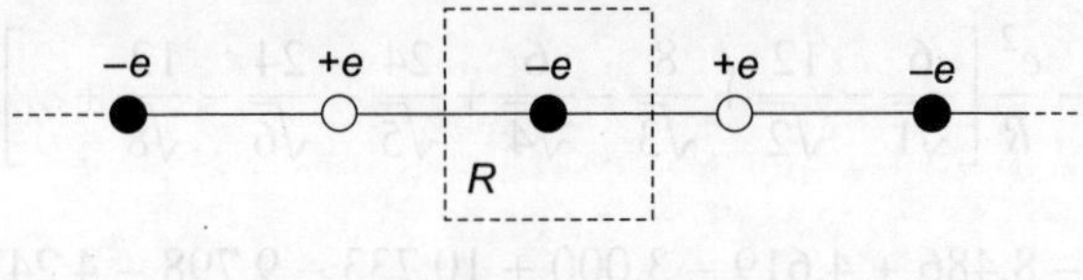


**Figure 4.9** Schematic representation of a one-dimensional ionic crystal in the point ion approximation for the calculation of the Madelung constant. The nearest neighbour distance is denoted by *R*. The ion at the origin has been surrounded by a box

**Table 4.6** List of Neighbours in NaCl Structure with $Cl^-$ at the Origin; All ions up to the Distance $R\sqrt{12}$ are Indicated. In the last Column, for each shell of Neighbours, We Indicate the Charge Inside the Cube of edge 4*R* Surrounding the Reference ion; Including the ion at the Origin, this Cube Contains 32 Positive and 32 Negative Electronic Charges

| *Reference atom at the origin* $Cl^-$ | | | |
|---|---|---|---|
| *Number and type of ions* | *Representative position* | *Distance from origin* | *Charge within the cube of edge 4R* |
| $6Na^+$ | *R*(1, 0, 0) | $R\sqrt{1}$ | +6*e* |
| $12Cl^-$ | *R*(1, 1, 0) | $R\sqrt{2}$ | –12*e* |
| $8Na^+$ | *R*(1, 1, 1) | $R\sqrt{3}$ | +8*e* |
| $6Cl^-$ | *R*(2, 0, 0) | $R\sqrt{4}$ | –3*e* |
| $24Na^+$ | *R*(2, 1, 0) | $R\sqrt{5}$ | +12*e* |
| $24Cl^-$ | *R*(2, 1, 1) | $R\sqrt{6}$ | –12*e* |
| $12Cl^-$ | *R*(2, 2, 0) | $R\sqrt{8}$ | –3*e* |
| $24Na^+$ | *R*(2, 2, 1) | $R\sqrt{9}$ | +6*e* |
| $6Na^+$ | *R*(3, 0, 0) | $R\sqrt{9}$ | 0 |
| $24Cl^-$ | *R*(3, 1, 0) | $R\sqrt{10}$ | 0 |
| $24Na^+$ | *R*(3, 1, 1) | $R\sqrt{11}$ | 0 |
| $8Cl^-$ | *R*(2, 2, 2) | $R\sqrt{12}$ | –*e* |

The series in the square brackets is a slowly, conditionally convergent, series; in this case, however, the sum can be performed analytically and gives ln $(1 + x)$ with $x = 1$.

The Madelung constant for a strictly one-dimensional model is thus $\alpha_M = 2 \ln 2$.

In actual three-dimensional crystals, the evaluation of the Madelung constant is faced with the problem of summing up numerically series, which are slowly and conditionally convergent. Let us consider for instance the NaCl structure. Suppose we take a $Cl^-$ anion at the origin; we find 6 ions $Na^+$ at distance *R* (where *R* denotes the nearest neighbour distance), 12 ions $Cl^-$ at $R\sqrt{2}$, 8 ions $Na^+$ at $R\sqrt{3}$, 6 ions $Cl^-$ at distance $R\sqrt{4}$ etc; the ions up to the distance $R\sqrt{12}$ are listed in Table 4.6. The Coulomb energy of this ion in the field of all the other ions is thus

$$V(0) = -\frac{e^2}{R}\left[\frac{6}{\sqrt{1}} - \frac{12}{\sqrt{2}} + \frac{8}{\sqrt{3}} - \frac{6}{\sqrt{4}} + \frac{24}{\sqrt{5}} - \frac{24}{\sqrt{6}} - \frac{12}{\sqrt{8}} + \dots\right]$$

$$= \frac{e^2}{R}\,[6.000 - 8.486 + 4.619 - 3.000 + 10.733 - 9.798 - 4.243 + \dots]. \qquad \dots(2)$$

By inspection of the terms indicated in the square brackets in Eq. (2), it is not possible to infer when the direct summation converges, even approximately, to the value $\alpha_M = 1.7476$ of the NaCl structure (see Table 4.7). The most used procedures to evaluate lattice sums of the type shown in Eq. (2) are the Evjen and the Ewald methods.

### 4.8.1 The Evjen Method for the Evaluation of the Madelung Constant

The Evjen method for the evaluation of the Madelung constant considers the crystal as formed by neutral (or approximately neutral) groups of ions (the Evjen cell). The electrostatic energy of a finite cell is evaluated by direct summation. As Evjen cells of increasing size are considered, one expects a rapidly convergent series, due to the short-range nature of the potential of groups of neutral atoms.

To give an exemplification of the Evjen method, we consider the NaCl structure, a reference ion (for instance $Cl^-$ taken at the origin), and the cube of edge $4R$ surrounding the reference ion. In the Evjen method, all the ions *inside* the cube are considered as they stand and all the ions *outside* the cube are neglected. Ions on the *border* of the cubes are assigned an appropriate fraction; this fraction is 1/2 for an ion at the surface, 1/4 for an ion at the edge and 1/8 for an ion at the corner of the cube. In the last column of Table 4.6, the charge within the cube of edge $4R$, contributed by each shell of neighbours, is reported; from it, we see that the potential energy of the point ion at the origin is given by

$$V(0) = -\frac{e^2}{R}\left[\frac{6}{1} - \frac{12}{\sqrt{2}} + \frac{8}{\sqrt{3}} - \frac{3}{\sqrt{4}} + \frac{12}{\sqrt{5}} - \frac{12}{\sqrt{6}} - \frac{3}{\sqrt{8}} + \frac{6}{\sqrt{9}} - \frac{1}{\sqrt{12}}\right] = -1.751\frac{e^2}{R},$$

a result already very near to the exact one $-\alpha_M\, e^2/R$ with $\alpha_M = 1.7476$.

### 4.8.2 The Ewald Method for the Evaluation of the Madelung Constant

The Ewald technique provides a very elegant procedure to transform one slow conditionally convergent lattice series into the sum of two fast absolutely convergent series in real and reciprocal space. The Ewald method essentially consists in using and balancing appropriately both real space and reciprocal space summations; this is achieved by splitting the function to be summed into a short-range part, to be computed as it stands in real space, and a long-range part to be represented and computed in reciprocal space. For the evaluation of the Madelung constant, the basic idea of Ewald is to replace point charges with Gaussian charge distributions, most conveniently handled in reciprocal space; the difference between the point charges and the Gaussian distribution is instead handled in real space. To implement the method, we need a few elementary facts concerning Gaussian functions, and the intimately related error function.

We remind the usual definition of the error function

$$\operatorname{erf}(x) = \frac{2}{\sqrt{\pi}} \int_0^x e^{-t^2}\, dt \qquad ...(3)$$

The error function increases monotonically from 0 to 1 as $x$ varies from zero to infinity. We noticed that

$$\operatorname{erf}(x) = \frac{2}{\sqrt{\pi}}\left( x - \frac{x^2}{3} + ....\right) \quad \text{for} \quad x << 1;$$

the derivative of the error function is the Gaussian function

$$\frac{d}{dx}\operatorname{erf}(x) = \frac{2}{\sqrt{\pi}} e^{-x^2}$$

It is also useful to consider the complimentary error function erfc $(x)$ defined from

$$\boxed{\operatorname{erf}(x) + \operatorname{erfc}(x) = 1} \qquad ...(4)$$

The reason for introducing the error function is the following. Let us consider a Gaussian charge distribution $\rho(r)$ (normalized to $+e$), centred at the origin in real space

$$\rho(\mathbf{r}) \equiv (+e)\left(\frac{\eta}{\pi}\right)^{3/2} e^{-\eta r^2} \qquad ...(5a)$$

The electrostatic potential $\phi(r)$, corresponding to the Gaussian charge density of Eq. (5a) is simply

$$\phi(r) = +\frac{e}{r}\operatorname{erf}(\sqrt{\eta}\, r) \qquad ...(5b)$$

Let us, apply the operator $-\nabla^2$ to the electrostatic potential (5b); we obtain

$$-\nabla^2\left[\frac{e}{r}\operatorname{erf}(\sqrt{\eta}\, r)\right] = -\frac{e}{r^2}\frac{\partial}{\partial r}\left[r^2 \frac{\partial}{\partial r}\frac{\operatorname{erf}(\sqrt{\eta}\, r)}{r}\right] = 4\pi e\left(\frac{\eta}{\pi}\right)^{3/2} e - \eta r^2 \qquad ...(5c)$$

The first passage in Eq. (5c) is just the action of the operator $\nabla^2$ on a spherically symmetric function; the second passage is obtained by explicitly performing the derivatives with respect to $r$. Equation (5c) shows that the charge density (5a) and the electrostatic potential (5b) indeed satisfy the Poisson equation $-\nabla^2 \phi(\mathbf{r}) = 4\pi\, \rho(\mathbf{r})$.

Notice that in the particular case of $\eta \to \infty$, the Gaussian charge density (5a) represents the point charge $(+e)\,\delta(\mathbf{r})$, and the potential (5b) gives just the Coulomb potential, $e/r$, which is singular at the origin. For a Gaussian charge density, the potential $\phi(r) = (e/r)\operatorname{erf}(\sqrt{\eta}\, r)$ is regular at the origin, where it takes the value of

$$\phi(0) = 2e\left(\frac{\eta}{\pi}\right)^{1/2} \qquad ...(6)$$

We now remember that the Fourier transform of a Gaussian function is still a Gaussian function. In particular, we have the relation

$$\left(\frac{\eta}{\pi}\right)^{3/2} \int_{-\infty}^{+\infty} e^{-\eta r^2} e^{i\mathbf{k}\cdot\mathbf{r}}\, d\mathbf{r} = e^{-k^2/4\eta} \qquad ...(7a)$$

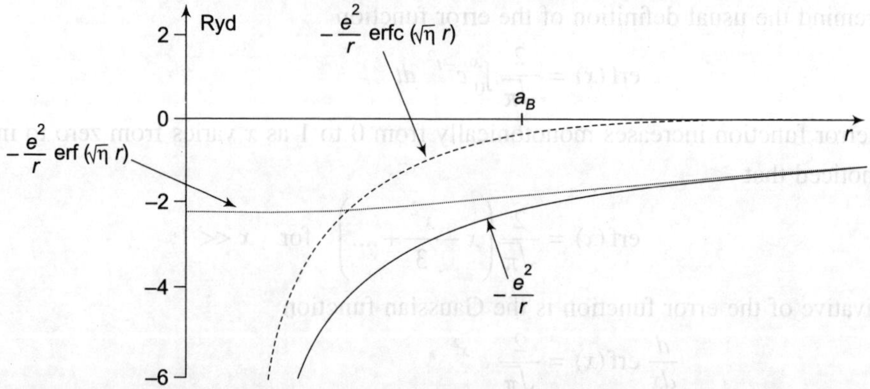


**Figure 4.10** Decomposition of the potential $-e^2/r$ (singular at the origin and of long-range nature) into a contribution $-(e^2/r)$ erf $(\sqrt{\eta}\,r)$ (regular at the origin and of long-range nature) and a contribution $-(e^2/r)$ erf $(\sqrt{\eta}\,r)$ (singular at the origin and of short-range nature). Energies are expressed in Rydberg, length in Bohr radius units; we have chosen $\sqrt{\eta} = 1a_B^{-1}$.

Another relation we need is

$$\int_{-\infty}^{+\infty} \frac{e^2}{r} \operatorname{erf}(\sqrt{\eta}\,r)\, e^{i\mathbf{k}\cdot\mathbf{r}}\, d\mathbf{r} = \frac{4\pi e^2}{k^2} e^{-k^2/4\eta} \quad \text{...(7b)}$$

It is seen by inspection that the derivation with respect to η of both members of Eq. (7b) gives back Eq. (7a)

In the following we use the simple trick (suggested by Eq. 14) to write a point charge potential $-e^2/r$ (or similarly, $+e^2/r$) in the form

$$-\frac{e^2}{r} \equiv -\frac{e^2}{r} \operatorname{erf}(\sqrt{\eta}\,r) - \frac{e^2}{r} \operatorname{erfc}(\sqrt{\eta}\,r) \quad \text{...(8)}$$

(see Fig 4.10). The first term on the right hand side of Eq. (18) is the potential generated by a Gaussian charge density; it is of *long-range nature and regular at the origin* (in lattice sums, it is conveniently handled in reciprocal space). The second term on the right hand side of Eq. (8) is the potential generated by a Gaussian charge density and a neutralizing point charge located at the origin; this contribution is of *short-range nature and singular at the origin* (in lattice sums with origin excluded, it is conveniently handled in real space).

### 4.8.3 Madelung Constant with the Ewald Methods for a Two-sublattice Structure

We consider now a crystal constituted by two sublattices, occupied by anions and cations of point charges $(-e)$ and $(+e)$ respectively; the two sublattices are described by the position $\mathbf{d}_1 + \mathbf{t}_n$ and $\mathbf{d}_2 + \mathbf{t}_n$ (with $\mathbf{d}_1 = 0$, $\mathbf{d}_2 \neq 0$ and $\mathbf{t}_n$ translational vectors). We take, for instance, a negative ion position as a reference site, as in Fig. 4.11. The potential energy $V(\mathbf{r})$ felt by a particle of charge $(-e)$ because of the field of all the ions of the crystal (except the reference one) is

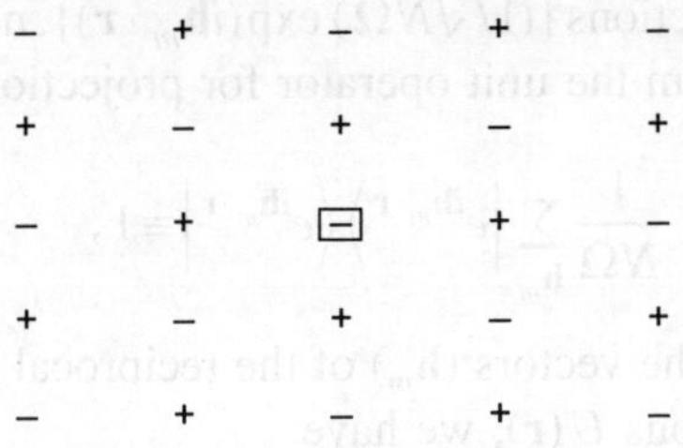

**Figure 4.11** Schematic representation of a two-sublattice ionic crystal in the point ion approximation, for the calculation of the Madelung constant. The ion at the origin has been surrounded by a box

$$V(\mathbf{r}) = \sum_{\mathbf{t}_n(\neq 0)} \frac{e^2}{|\mathbf{r}-\mathbf{t}_n|} - \sum_{\mathbf{t}_n} \frac{e^2}{|\mathbf{r}-\mathbf{t}_n-\mathbf{d}_2|} \qquad ...(9)$$

The first term in the right-hand side of Eq. (9) is the repulsive contribution due to the sublattice of negative ions, while the second term is the attractive contribution due to the sublattice of positive ions.

Taking into account Eq. (8), we can write Eq. (9) in the form

$$V(\mathbf{r}) = V_1(\mathbf{r}) + V_2(\mathbf{r}),$$

where $$V_1(\mathbf{r}) = \sum_{\mathbf{t}_n(\neq 0)} \frac{e^2}{|\mathbf{r}-\mathbf{t}_n|}\,\mathrm{erfc}(\sqrt{\eta}\,|\mathbf{r}-\mathbf{t}_n|) - \sum_{\mathbf{t}_n} \frac{e^2}{|\mathbf{r}-\mathbf{t}_n-\mathbf{d}_2|}\,\mathrm{erfc}(\sqrt{\eta}\,|\mathbf{r}-\mathbf{t}_n-\mathbf{d}_2|) \qquad ...(10a)$$

and $$V_2(\mathbf{r}) = \sum_{\mathbf{t}_n(\neq 0)} \frac{e^2}{|\mathbf{r}-\mathbf{t}_n|}\,\mathrm{erf}(\sqrt{\eta}\,|\mathbf{r}-\mathbf{t}_n|) - \sum_{\mathbf{t}_n} \frac{e^2}{|\mathbf{r}-\mathbf{t}_n-\mathbf{d}_2|}\,\mathrm{erf}(\sqrt{\eta}\,|\mathbf{r}-\mathbf{t}_n-\mathbf{d}_2|) \qquad ...(10b)$$

In particular, the term $V_1(0)$ is given by

$$V_1(0) = \sum_{\mathbf{t}_n(\neq 0)} \frac{e^2}{t_n}\,\mathrm{erfc}(\sqrt{\eta}\,t_n) - \sum_{\mathbf{t}_n} \frac{e^2}{|\mathbf{t}_n+\mathbf{d}_2|}\,\mathrm{erfc}(\sqrt{\eta}\,|\mathbf{t}_n+\mathbf{d}_2|) \qquad ...(11)$$

Because of the short-range nature of the complementary error function, the term $V_1(0)$ is conveniently calculated as it stands, i.e., by direct summation in real space.

We now elaborate the term $V_2(\mathbf{r})$ of Eq. (10b); this term is more conveniently transformed into a reciprocal space sum, with the following procedure. We write $V_2(\mathbf{r})$ in the form

$$V_2(\mathbf{r}) = U(\mathbf{r}) - \frac{e^2}{r}\,\mathrm{erf}(\sqrt{\eta}\,r), \qquad ...(12a)$$

where $U(\mathbf{r})$ is given by

$$U(\mathbf{r}) = \sum_{\mathbf{t}_n} \frac{e^2}{|\mathbf{r}-\mathbf{t}_n|}\,\mathrm{erf}(\sqrt{\eta}\,|\mathbf{r}-\mathbf{t}_n|) - \sum_{\mathbf{t}_n} \frac{e^2}{|\mathbf{r}-\mathbf{t}_n-\mathbf{d}_2|}\,\mathrm{erf}(\sqrt{\eta}\,|\mathbf{r}-\mathbf{t}_n-\mathbf{d}_2|); \qquad ...(12b)$$

Notice that $U(\mathbf{r})$ is a periodic function.

From the set of periodic functions $\{(1/\sqrt{N\Omega})\exp(i\mathbf{h}_m \cdot \mathbf{r})\}$, normalized to one in the volume $V = N\Omega$ of the crystal, we form the unit operator for projection of *periodic* functions

$$\frac{1}{N\Omega}\sum_{\mathbf{h}_m}\left|e^{i\mathbf{h}_m\cdot\mathbf{r}}\right\rangle\left\langle e^{i\mathbf{h}_m\cdot\mathbf{r}}\right| \equiv 1,$$

where the sum extends to all the vectors ($\mathbf{h}_m$) of the reciprocal lattice. By applying the above identity to the periodic functions $U(\mathbf{r})$, we have

$$U(\mathbf{r}) = \frac{1}{N\Omega}\sum_{\mathbf{h}_m} e^{i\mathbf{h}_m\cdot\mathbf{r}}\left\langle e^{i\mathbf{h}_m\cdot\mathbf{r}}|U(\mathbf{r})\right\rangle. \qquad \text{...(13a)}$$

The scalar product appearing in Eq. (13a) can be evaluated explicitly, and one obtains

$$\langle e^{i\mathbf{h}_m\cdot\mathbf{r}}|U(\mathbf{r})\rangle = \int_V e^{-i\mathbf{h}_m\cdot\mathbf{r}}\, U(\mathbf{r})\, d\mathbf{r}$$

$$= N(1 - e^{-i\mathbf{h}_m\cdot\mathbf{d}_2})\int_V e^{-i\mathbf{h}_m\cdot\mathbf{r}}\frac{e^2}{r}\operatorname{erf}(\sqrt{\eta}\, r)\, d\mathbf{r}$$

$$= N(1 - e^{-i\mathbf{h}_m\cdot\mathbf{d}_2})\frac{4\pi e^2}{h_m^2}e^{-h_m^2/4\eta} \quad (h_m \neq 0), \qquad \text{...(13b)}$$

where use has been made of Eq. (12b) and Eq. (7b). When $\mathbf{h}_m \equiv 0$, the scalar product (13b) vanishes; in fact the average of $U(\mathbf{r})$ on the crystal volume is evidently equal to zero, because of the overall neutrality of the system of positive and negative (identical) charge distributions.

From Eq. (13), we have

$$U(0) = \frac{4\pi e^2}{\Omega}\sum_{\mathbf{h}_m(\neq 0)}(1 - e^{-i\mathbf{h}_m\cdot\mathbf{d}_2})\frac{1}{h_m^2}e^{-h_m^2/4\eta}.$$

From the above expression and from Eq. (12a) we obtain

$$V_2(0) = \frac{4\pi e^2}{\Omega}\sum_{\mathbf{h}_m(\neq 0)}(1 - e^{-i\mathbf{h}_m\cdot\mathbf{d}_2})\frac{1}{h_m^2}e^{-h_m^2/4\eta} - 2e^2\sqrt{\frac{\eta}{\pi}}. \qquad \text{...(14)}$$

**Table 4.7** Madelung Constant for Some Cubic Structure

| *Crystal structure* | *Madelung constant* |
|---|---|
| Cesium chloride | $\alpha_M = 1.7627$ |
| Sodium chloride | $\alpha_M = 1.7476$ |
| Zinc blende | $\alpha_M = 1.6381$ |

Collecting Eqs. (11) and (14), we obtain for the Madelung energy the final expression

$$V(0) = \sum_{\mathbf{t}_n (\neq 0)} \frac{e^2}{t_n} \operatorname{erfc}(\sqrt{\eta}\, t_n) - \sum_{\mathbf{t}_n} \frac{e^2}{|\mathbf{t}_n + \mathbf{d}_2|} \operatorname{erfc}(\sqrt{\eta}\, |\mathbf{t}_n + \mathbf{d}_2|)$$

$$+ \frac{4\pi e^2}{\Omega} \sum_{\mathbf{h}_m (\neq 0)} (1 = e^{-i\mathbf{h}_m \cdot \mathbf{d}_2}) \frac{1}{h_m^2} e^{-h_m^2/4\eta} - 2e^2 \sqrt{\frac{\eta}{\pi}} \quad ...(15)$$

The arbitrary parameter (η) controls the convergence of the two summations, and is chosen to optimize simultaneously the convergence of the series in direct and reciprocal spaces.

Many ionic crystals occur in cubic structures: the cesium chloride structure (coordination number 8), sodium chloride structure (coordination number 6), zinc blende structure (coordination number 4). The Madelung constants is (slightly) higher for higher coordination number, as can be seen from Table 4.7.

## 4.9 THE ADIABATIC APPROXIMATION

A solid consists of a sufficiently large collection of atoms so that they exhibit the characteristic behaviour of the bulk material. In a solid the atom may lose some of its tightly bound electrons, leaving the *core* electrons. Such an entity is known as an *ion*. Hence, we will consider a system of ions with position vectors $\vec{R}_p$, masses $M_p$, charges $Z_p e$ and *valence* electrons $\vec{r}_j$, $m_e$, $-e$, described by the Hamiltonian operator:

$$\hat{H} = -\frac{\hbar^2}{2m_e}\sum_j \nabla_j^2 - \sum_p \frac{\hbar^2}{2M_p}\nabla_p^2 + \frac{1}{2}\sum_j \sum_{k \neq j} \frac{e_0^2}{r_{jk}}$$

$$-\sum_j \sum_p \frac{Z_p e_0^2}{|\vec{r}_j - \vec{R}_p|} + \frac{1}{2}\sum_p \sum_{q \neq p} \frac{Z_p Z_q e_0^2}{R_{pq}}$$

$$-\frac{\hbar^2}{2m_e}\sum_j \nabla_j^2 - \sum_p \frac{\hbar^2}{2M_p}\nabla_p^2 + U_{ee}(\vec{r}) + U_{ei}(\vec{r}, \vec{R}) + U_{ii}(\vec{R}) \quad ...(1)$$

The first two terms describe the kinetic energy, where $p$ is a running index over all the ions, the index ($j$) extends over all the electrons, whereas $U_{ee}(\vec{r})$, $U_{ei}(\vec{r}, \vec{R})$ and $U_{ii}(\vec{R})$ are the potential energies of the electron-electron, electron-ion and ion-ion interaction respectively, expressed in terms of $e_0 = e/\sqrt{4\pi\varepsilon_0}$, while $e$ is given in SI units. In the Schrödinger equation for the system of atoms, the motion of the valence electrons may be separated from that of the ions by introducing in Eq. (1) the Hamiltonian, $\hat{H}_e$, which describes the electron motion for a set of fixed ionic fixed ionic positions, denoted collectively by $\vec{R}$, in the form

$$\hat{H}\Phi(\vec{r}, \vec{R}) = \left\{\hat{H}_e - \sum_p \frac{\hbar^2}{2M_p}\nabla_p^2\right\}\Phi(\vec{r}, \vec{R}) = E\Phi(\vec{r}, \vec{R}) \quad ...(2)$$

The eigenvalue problem for the electronic motion reads:

$$\hat{H}_e \Psi(\vec{r}, \vec{R}) = \left\{ -\frac{\hbar^2}{2m_e} \sum_j \nabla_j^2 + U_{ee}(\vec{r}) + U_{ei}(\vec{r}, \vec{R}) + U_{ii}(\vec{R}) \right\} \Psi(\vec{r}, \vec{R}) = E(\vec{R}) \Psi(\vec{r}, \vec{R}) \quad ...(3)$$

where $\Psi(\vec{r}, \vec{R})$ and $E(\vec{R})$ are the eigenfunctions and eigenvalues of $\hat{H}_e$ with $\vec{R}$ as a fixed parameter. The electron coordinates, denoted collectively by $\vec{r}$, are the dynamical variables in Eq. (3). Assuming that the eigenfunctions $\Psi(\vec{r}, \vec{R})$ are known, the wave functions $\Phi(\vec{r}, \vec{R})$ which solve the Schrödinger Eq. (2) are usually approximated by the following expression

$$\Phi(\vec{r}, \vec{R}) = \Psi(\vec{r}, \vec{R})\, \varphi(\vec{R}) \quad ...(4)$$

which is known as the *adiabatic approximation*. Substituting Eq. (4), Eq. (2) becomes:

$$-\sum_p \frac{\hbar^2}{2M_p} \{\varphi(\vec{R}) \nabla_p^2 \Psi(\vec{r}, \vec{R}\} + 2\nabla_p \Psi(\vec{r}, \vec{R}) \cdot \nabla_p \varphi(\vec{R}) + \Psi(\vec{r}, \vec{R}) \nabla_p^2 \varphi(\vec{R})\}$$

$$+ E(\vec{R})\, \Psi(\vec{r}, \vec{R}\}\, \varphi(\vec{R}) = E\Psi(\vec{r}, \vec{R})\varphi(\vec{R}) \quad ...(5)$$

Then multiplying to the left of each term by $\Psi^*(\vec{r}, \vec{R})$ and integrating over all electron positions, we obtain an equation for $\varphi(\vec{R})$:

$$-\sum_p \frac{\hbar^2}{2M_p} \nabla_p^2 \varphi(\vec{R}) + E(\vec{R})\, \varphi(\vec{R}) = E\varphi(\vec{R}) + \sum_p \frac{\hbar^2}{2M_p} \varphi(\vec{R}) \int \Psi^*(\vec{r}, \vec{R}) \nabla_p^2 \Psi(\vec{r}, \vec{R}) d\vec{r}$$

$$+ \sum_p \frac{\hbar^2}{M_p} \nabla_p \varphi(\vec{R}) \cdot \int \Psi^*(\vec{r}, \vec{R}) \nabla_p\, \Psi(\vec{r}, \vec{R}) d\vec{r} \; ... \quad ...(6)$$

where the electron eigenfunctions are assumed to be normalized. If the valence electrons can be considered perfectly free, the last two terms on the right-hand side vanish because $\Psi(\vec{r}, \vec{R})$ becomes independent of the ion coordinates $\vec{R}$, so that $\nabla_p \Psi_{(\vec{r}, \vec{R})}$ is zero. Such an approximation is appropriate for metallic systems. For non-metals, it can be shown that the eigenfunctions $\Psi(\vec{r}, \vec{R})$ are slowly varying functions of the ionic coordinates. It is therefore always possible, to a first approximation, to reduce Eq. (6), for the ionic motion, to the form:

$$-\sum_p \frac{\hbar_2}{2M_p} \nabla_p^2 \varphi(\vec{R}) + E(\vec{R})\varphi(\vec{R}) = E_\varphi(\vec{R}) \quad ...(7)$$

As a result of the adiabatic approximation to the many-body eigenvalue problem of a system of atoms, it is possible either to suppress the ionic interactions while considering the motion of the valence electrons, described by Eq. (3) or to neglect the electronic interactions when solving the eigenvalue Eq. (7) for the motion of the ions. In other words, ionic motion can be ignored when describing the binding of the electrons to the nucleus, whereas the chemical bond may be regarded as a parameter when considering the ionic motion. The state functions introduced by Eq. (4) are obtained by first solving the electronic problem for fixed ionic positions and then using the total electronic energy $E(\vec{R})$ as a potential function for the ionic motion. The potential energy $E(\vec{R})$ exhibits a minimum, corresponding to the equilibrium positions $\vec{R}_0$ of the ions, which move in potential wells as illustrated in Figure 4.12 where the horizontal line at $E$ corresponds to the constant energy of a hypothetical bound state of the solid. Therefore the effect of the electrons appears as a coupling interaction of the ions, with forces which are dependent on the electronic states. Any change in the electronic states $\Psi(\vec{r}, \vec{R})$ modifies the potential energy $E(\vec{R})$ felt by the ions, and so modifies their coupling.

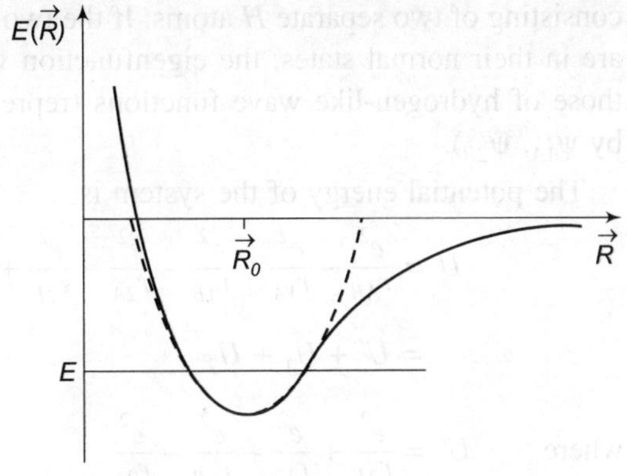


**Figure 4.12** Electronic potential energy for the ionic motion and its harmonic approximation (the broken line)

Because the interaction between ions is assumed to be conservative, the potential energy $E(\vec{R})$ for the ionic motion may be separated between the pairs of ions in the solid:

$$E(\vec{R}) = \frac{1}{2}\sum_{p}\sum_{q\neq p} E_{pq}(R_{pq}) \qquad \text{...(8)}$$

such that each component $E_{pq}$ reduces to a function of the interatomic separation $R_{pq} = |\vec{R}_p - \vec{R}_q|$. Then, $E_{pq}$ can be expanded in a Taylor series of ion displacements about the equilibrium positions $R_{0pq}$, such that the solution can be obtained to various degree of approximation. For many purposes it is sufficient to use the *harmonic approximation*, which limits the treatment to the harmonic motion of the ions because terms of higher than second order in displacement are neglected:

$$E_{pq}(R_{pq}) = E_{pq}(R_{0pq}) + \frac{1}{2}\left[\frac{\partial^2 E_{pq}(R_{pq})}{\partial R_{pq}^2}\right]_{R_{0pq}} (R_{pq} - R_{0pq})^2$$

$$= E_{pq}(R_{0pq}) + \frac{1}{2}\kappa_{pq}(R_{pq} - R_{0pq})^2 \qquad \text{...(9)}$$

where the first order derivatives in the displacements are zero, because of the equilibrium conditions, and $\kappa_{pq} = \kappa_{qp}$ are the *coupling constants*, having the dimensions of the harmonic force constants. The harmonic approximation to the potential field is represented by broken line in Fig. 4.12.

## 4.10 HEITLER AND LONDON THEORY OF COVALENT BONDING

The simplest example to understand the Heitler and London theory is the bonding of $H_2$ molecule.

In the molecule, $A$ and $B$ are two positively charged nuclei and 1 and 2 are two electrons. The various distances are as shown in Fig. 4.13. When $r_{AB}$ is large, the system may be considered

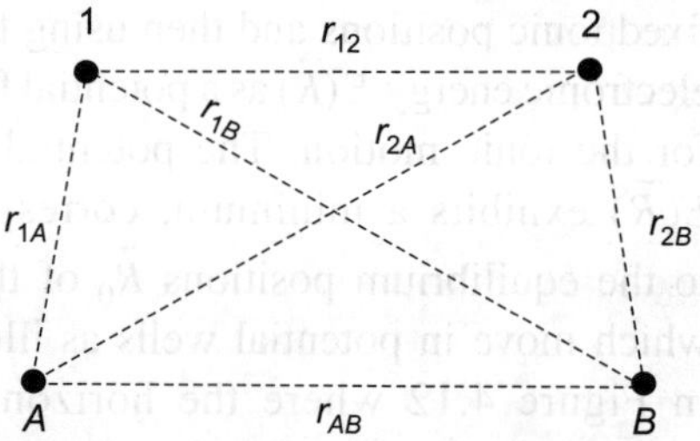


**Figure 4.13** $H_2$ molecule: *A* and *B* represent nuclei 1 and 2 represent the electrons of the two hydrogen atoms

consisting of two separate *H* atoms. If the two atoms are in their normal states, the eigenfunction will be those of hydrogen-like wave functions (represented by $\psi_{1A}$, $\psi_{2B}$).

The potential energy of the system is

$$U = \frac{e^2}{r_{AB}} - \frac{e^2}{r_{1A}} - \frac{e^2}{r_{1B}} - \frac{e^2}{r_{2A}} - \frac{e^2}{r_{2B}} + \frac{e^2}{r_{12}}$$

$$= U' + U_1 + U_2 \quad ...(1)$$

where $$U' = \frac{e^2}{r_{AB}} + \frac{e^2}{r_{12}} - \frac{e^2}{r_{1B}} - \frac{e^2}{r_{2A}} \quad ...(2)$$

and $$U_1 = \frac{e^2}{r_{1A}}, \quad U_2 = \frac{e^2}{r_{2B}}, \quad ...(3)$$

$U_1$ is the potential energyof electron 1 in the field of its own nucleus *A*, $U_2$ is potential energy of electron 2 in the field of its nucleus *B* and $U'$ is the potential energy due to perturbation term.

The Schrödinger equation for the two electrons having a wave function $\psi(x_1, y_1, z_1, s_1; x_2, y_2, z_2, s_2)$ depending on the position and the spin coordinates of the two electrons is given by

$$H\psi = E\psi$$

or $$-\frac{\hbar^2}{2m}(\nabla_1^2 + \nabla_2^2)\psi + U\psi = E\psi \quad ...(4)$$

where *E* is the energy eigenvalue. This equation cannot be solved analytically with the above mentioned potential energy expression. Heitler and London made some approximations by treating certain contribution to the potential energy as negligible. When $r_{AB}$ is large, one can treat the system as two free H-atoms and the Schrödinger equation reduces into two independent equations

$$\nabla_1^2 \psi_{1A} + \frac{2m}{\hbar^2}(E_1 - U_1)\psi_{1A} = 0 \quad ...(5a)$$

and $$\nabla_2^2 \psi_{2B} + \frac{2m}{\hbar^2}(E_2 - U_2)\psi_{2B} = 0 \quad ...(5b)$$

The solution of these equation is known from the H-atom problem and is given by

$$\psi_{1A} = \frac{1}{\sqrt{\pi a_0^3}} e^{-r_{2B}/a_0} : E_1 = \frac{e^2}{a_0} \quad ...(6a)$$

$$\psi_{2B} = \frac{1}{\sqrt{\pi a_0^3}} e^{-r_{2B}/a_0} : E_2 = \frac{e^2}{a_0} \quad ...(6b)$$

where $a_0$ is the Bohr radius. Since the density of electron cloud falls off very rapidly as the distance from the nucleus increases, so the probabilities to find electron 1 near nucleus *B* and electron 2 near nucleus $A$ are very small (for $r_{AB}$ large), hence, the total energy of the system

$$E = E_1 + E_2 = 2E_0$$

i.e., the sum of the two isolated atoms. Such a situation is described in Fig. 4.14 in terms of the electron density distribution. In this case wave function of the system

$$\psi = \psi_{1A} + \psi_{2B}$$

However, the electrons cannot be distinguished, therefore, by interchanging electrons 1 and 2 between the nuclei should not change the total energy of the system, but now the wave function will be $\psi = \psi_{2A}\psi_{1B}$ with same energy $E = 2E_0$. This is known as exchange degeneracy. The superposition of these two will also be a solution for the system. The superposition can be either symmetric or antisymmetric, accordingly.

$$\psi_s = P_S[\psi_{1A}\,\psi_{2B} + \psi_{2A}\,\psi_{1B}] \qquad \text{...(7)}$$

$$\psi_A = P_A[\psi_{1A}\,\psi_{2B} - \psi_{2A}\,\psi_{1B}] \qquad \text{...(8)}$$

Normalisation gives the value of $P_S$ and $P_A$ as

$$P_S = \frac{1}{\sqrt{2+2S}} \qquad \text{...(9)}$$

$$P_A = \frac{1}{\sqrt{2-2S}} \qquad \text{...(10)}$$

where $S = \iint \psi_{1A}\,\psi_{1B}\,\psi_{2A}\,\psi_{2B}\,dr_1\,dr_2$ is the overlap integral.

When the two atoms are brought closer, the electron density distribution is shown in Fig. 4.14(b). Now, $r_{AB}$ is not negligible and the potential energy.

$$U = U_0 + U' \qquad \text{...(11)}$$

(where $U_0 = U_1 + U_2$) and corresponding $H = H_0 + H'$.

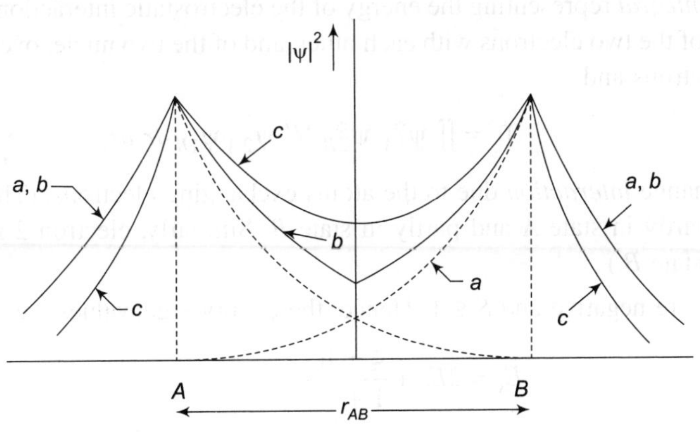


**Figure 4.14** $a \rightarrow$ Electron density distribution in independent H-atoms; $b \rightarrow$ electron density distribution when two H-atoms are at a distance $r_{AB}$; $c \rightarrow$ actual electron density distribution in an $H_2$ molecule

Writing $\psi = \psi_{1A}\psi_{2B} \pm \psi_{2A}\psi_{1B}$, and taking into account the perturbation effect ($E = 2E_0 + B$), the interaction energy $E'$ can be calculated as

$$E' = \frac{\int \psi^* H' \psi \, dv}{\int \psi^* \psi \, dv} \quad ...(12)$$

$$= \iint \psi_{1A}\psi_{2B} \pm \psi_{2A}\psi_{1B} \times \left( \frac{e^2}{r_{AB}} - \frac{e^2}{r_{12}} - \frac{e^2}{r_{1B}} - \frac{e^2}{r_{2A}} \right) \times \frac{(\psi_{1A}\psi_{2B} \pm \psi_{2A}\psi_{1B})\, dr_1\, dr_2}{\iint (\psi_{1A}\psi_{2B} \pm \psi_{2A}\psi_{1B})^2\, dr_1\, dr_2} \quad ...(13)$$

Denominator of this equation is equal to

$$\iint (\psi_{1A}\psi_{2B})^* (\psi_{1A}\psi_{2B})\, dr_1\, dr_2 + \iint (\psi_{2A}\psi_{1B})^* (\psi_{2A}\psi_{1B})\, dr_1\, dr_2$$

$$\pm 2 \iint \psi_{1A}\psi_{1B}\psi_{2A}\psi_{2B})\, dr_1\, dr_2 = 2 \pm 2S \quad ...(14)$$

where $S = \iint \psi_{1A}\psi_{1B}\psi_{2A}\psi_{2B}\, dr_1\, dr_2$ is the overlap integral. Considering numerator of Eq. (9), it is equal to

$$\iint \psi_{1A}^* \psi_{2B}^* H' \psi_{1A}\psi_{2B}\, dr_1\, dr_2 \pm \iint \psi_{1A}^* \psi_{2B}^* H' \psi_{2A}\psi_{1B}\, dr_1\, dr_2$$

$$\pm \iint (\psi_{2A}^* \psi_{1B}^* H' \psi_{1A}\psi_{2B})\, dr_1\, dr_2 + \iint (\psi_{2A}^* \psi_{1B}^* H' \psi_{2A}\psi_{1B})\, dr_1\, dr_2$$

$$= 2\xi \pm 2\eta \quad ...(15)$$

where

$$\xi = \iint \psi_{1A}^* \psi_{2B}^* H' \psi_{1A}\psi_{1B}\, dr_1\, dr_2 \quad ...(16)$$

is the *Coulomb integral* representing the energy of the electrostatic interaction of the electrons with the nuclei, of the two electrons with each other, and of the two nuclei over all the possible states of the electrons and

$$\eta = \iint \psi_{1A}^* \psi_{2B}^* H' \psi_{2A}\psi_{1B}\, dr_1\, dr_2 \quad ...(17)$$

denotes the exchange *interaction* due to the atoms exchanging electrons. (This integral show the electron is partly in state *A* and partly in state *B*. Similarly, electron 2 is partly in state *A* and partly in state *B*.)

Both $\xi$ and $\eta$ are negative and $S \leq 1$. Hence, the energy eigenvalues are

$$E_s = 2E_0 + \frac{\xi + \eta}{1 + S} \quad ...(18)$$

$$E_A = 2E_0 + \frac{\xi - \eta}{1 - S} \quad ...(19)$$

*The state with energy $E_s$ is termed symmetric with even parity (directions of spins antiparallel) that with energy $E_A$ is antisymmetric with odd parity (direction of spins parallel).*

The potential energy curves for both symmetric and antisymmetric modes are shown in Fig. 4.15. It shows that a molecule formation by an attractive interaction at an equilibrium distance is possible only when the eigenfunction is symmetric and an $H_2$ molecule cannot be formed in an antisymmetric state.

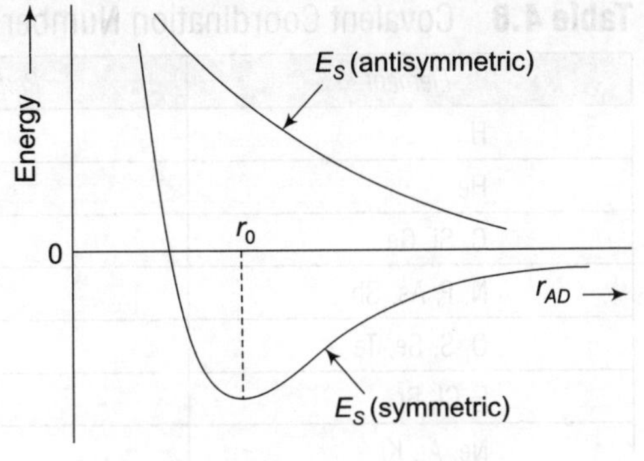


**Figure 4.15** Interaction energy of two hydrogen atoms for a symmetric and an antisymmetric state

Thus, in valence bond theory, the two participating electrons in a covalent bond pair up with parallel spins and the bonding is through an exchange interaction.

## 4.11 SATURATION IN COVALENT BONDS

A hydrogen molecule is formed with two hydrogen atoms and is stable. If a third hydrogen atom is brought near a hydrogen molecule, it experiences a repulsive force and, therefore, an $H_3$ molecule cannot be formed by covalent bonds. This is what we mean by the saturation of covalent bonds. In a hydrogen molecule, the two electrons align their spins in antiparallel manner (↑↓). An $H_3$ molecule can be formed only if the electron spin of an hydrogen atom is made to interact with a hydrogen molecule. This can be done in two different ways:

1. The hydrogen atom can exchange its spin (↑) with one electron of a hydrogen molecule having spin (↑). In this case, the two electrons will have their spins parallel. Consequently, the exchange force between the hydrogen atom and the hydrogen molecule will be repulsive and hence, an $H_3$ molecule cannot be formed.
2. The hydrogen atom can exchange its spin (↑) with the other electron of the hydrogen molecule having spin (↓). In this case, the electron spins of the hydrogen molecule will be aligned which is forbidden by Pauli's exclusion principle.

Thus, we consider that the only possible exchange between hydrogen atom and a hydrogen molecule is repulsive and, therefore, an $H_3$ molecule cannot be formed.

## 4.12 DIRECTIONAL NATURE OF A COVALENT BOND

According to quantum wave mechanical model of an atom. The electron distribution in different atoms is described in terms of orbitals of different shapes, e.g., *s, p, d, f* orbitals. A covalent bond between two atoms is formed by sharing of two electrons, one from each atom. Table 4.8 shows elements of the Periodic Table versus their covalent coordination number, i.e., the number of contacting neighbours in a covalent crystal. The covalent bond then lies along the direction of the overlapping atomic orbitals. For example, in a water ($H_2O$) molecule, the electronic structure of oxygen is $1s^2$, $2s^2$, $2p_x^{\ 2}\ 2p_y^{\ 1}\ 2p_z^{\ 1}$, so, two of the electrons ($p_y^{\ 1}$ and $p_z^{\ 1}$) form covalent bonds with two H-atoms, thus completing the octet

**Table 4.8** Covalent Coordination Number of Some Elements

| *Element* | *Periodic group* | *Covalent coordination number* |
|---|---|---|
| H | – | 1 |
| He | – | 0 |
| C, Si, Ge | 4 | 4 |
| N, P, As, Sb | 5 | 3 |
| O, S, Se, Te | 6 | 2 |
| F, Cl, Br | 7 | 1 |
| Ne, Ar, Kr | 8 | 0 |

(8 electrons in the 2nd orbit) (see Fig. 4.16 (a)). As a result, in a $H_2O$ molecule, the O atom has 8 outer electrons (four electron pairs), so, the structure is based on a tetrahedron. There are two bond pairs and two lone pairs (a lone pair is that which does not participate in bonding). The structure in Fig. 4.16 (a) is tetrahedral with two positions occupied by lone pairs (shown shaded). The two lone pairs distort the bond angle from 109°28' to 104°30', i.e., the electron density distribution $|\Psi|^2$ is large in certain directions, consequently the bonds are strong and concentrated along these directions. It is to be noted here that the directionality of bonds depends on the spatial symmetry of the orbitals involved in overlapping (pairing) (Fig. 4.16(b)).

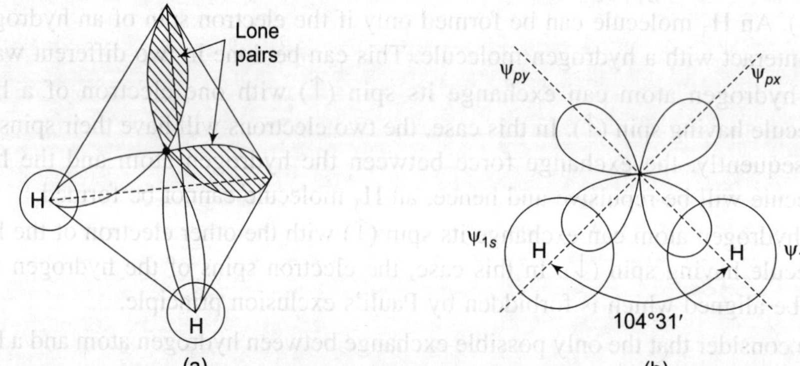


**Figure 4.16** (a) Tetrahedron based structure of $H_2O$ molecule and (b) directional nature of bonds in $H_2O$ in xy plane ($2s^2$ and $2p_z^2$ are lone pairs)

## 4.13 ISOELECTRONIC CONCENTRATION CONTOURS AND THE DIRECTIONAL NATURE IN A COVALENT COMPOUND

The covalent bond is the classical electron pair or homopolar bond of chemistry, i.e., bond formed by sharing two electrons (one from each atom) of the two atoms of the same type. It is a strong bond, e.g., two carbon atoms in diamond.

The covalent bond has strong directional properties (Fig. 4.17). Thus, carbon, silicon and germanium have the diamond structure with atoms joined to four nearest neighbours at tetrahedral angles, even though this arrangement gives a low filling of space 0.34 of the available space, compared with 0.74 for a close-packed structure. The tetrahedral bond allows only four nearest neighbours, whereas a close-packed structure has 12.

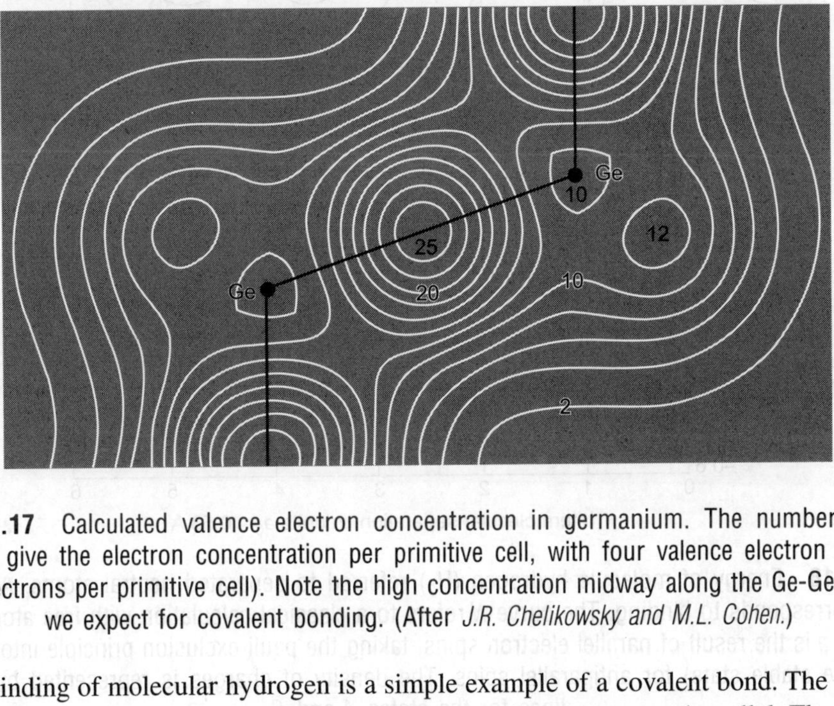


**Figure 4.17** Calculated valence electron concentration in germanium. The numbers on the contours give the electron concentration per primitive cell, with four valence electron per atom (eight electrons per primitive cell). Note the high concentration midway along the Ge-Ge bond, as we expect for covalent bonding. (After *J.R. Chelikowsky and M.L. Cohen.*)

The binding of molecular hydrogen is a simple example of a covalent bond. The strongest binding (Fig. 4.18) occurs when the spins of the two electrons are antiparallel. The bindings depend on the relative spin orientation not because there are strong magnetic dipole forces between the spins, but because the Pauli modifies the distribution of charge according to the spin orientation. This spin-dependent coulomb energy is called the exchange *interaction.*

The Pauli principle gives a strong repulsive interaction between atoms with filled shells. If the shells are not filled, electron overlap can be accommodated without excitation of electrons to high energy states and the bond will be shorter. Compare the bond length (2 Å) of $Cl_2$ with the interatomic distance (3.76 Å) of Ar in solid Ar; the difference between $Cl_2$ and $Ar_2$ is that the Cl atom has five electrons in the $3p$ shell and the Ar atom has six, filling the shell, so that the repulsive interaction is stronger in Ar than in Cl.

The elements C, Si, and Ge lack four electrons with respect to filled shells, and thus these elements (for example) can have an attractive interaction associated with charge overlapped. The electron configuration of carbon is $1s^2 2s^2 2p^2$. To form a tetrahedral system of covalent bonds the carbon atom must first be promoted to the electronic configuration $1s^2 2s^2 p^3$. This promotion from the ground state requires 4 eV, an amount more than regained when the bonds are formed.

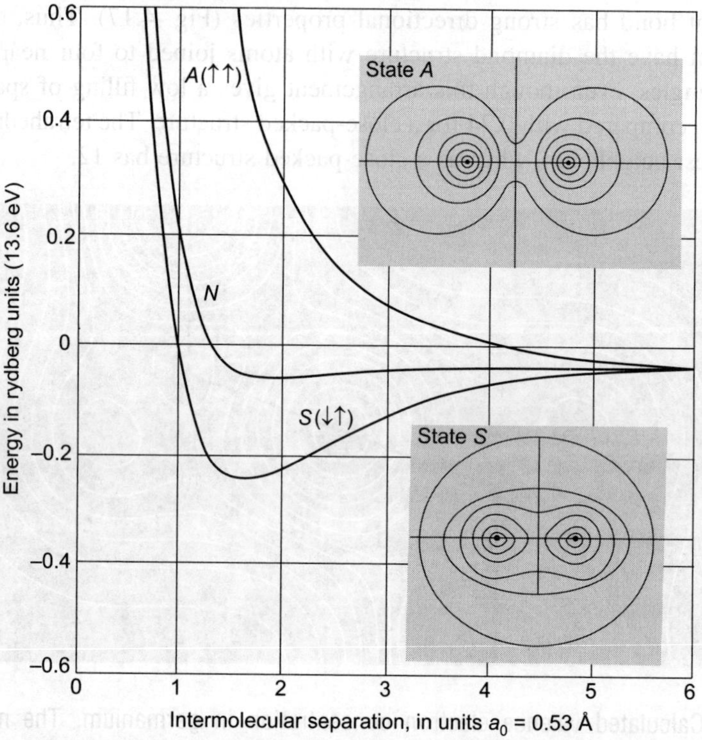


**Figure 4.18** Energy of molecule hydrogen ($H_2$) referred to separated neutral atoms. A negative energy corresponds to binding. The curve *N* refers to a classical calculation with free atom charge densities; *a* is the result of parallel electron spins, taking the pauli exclusion principle into account, and *S* (the stable state) for antiparallel spins. The density of charges is represented by contour lines for the states *A* and *S*

## 4.14 HYBRIDISATION

The simple approach of overlapping of *s* and *p* orbitals fails in case of covalent bonding in crystals formed by group IV elements, viz. C, Si, and Ge. Let us consider the diamond crystal formed by the bonding of carbon atoms. The ground state electronic configuration of carbon is $1s^2 2s^2 2p^2$. When carbon atoms are brought together, one of the two 2*s* electrons gets excited to the 2*p* subshell, giving four unpaired electrons in the $2s^1 2p^3$ hybrid state. The *hybridization* is the mixing of orbitals of valence shell of the same atom to form new orbitals with the same energy and similar shape. The resulting orbitals are called *hybrid orbitals.*

The four orbitals are the linear combination:

$$\psi_1 = \frac{1}{2}(\psi_s + \psi_{px} + \psi_{py} + \psi_{pz})$$

$$\psi_2 = \frac{1}{2}(\psi_s + \psi_{px} - \psi_{py} - \psi_{pz})$$

$$\psi_3 = \frac{1}{2}(\psi_s - \psi_{px} + \psi_{py} - \psi_{pz})$$

$$\psi_4 = \frac{1}{2}(\psi_s - \psi_{px} - \psi_{py} + \psi_{pz})$$

These four hybrid orbitals (or four unpaired electrons) then make covalent bonds with the four nearest atoms. The absolute squares of these wave functions have their maximum in the direction of the body diagonal of the octant formed by

$$x, y, z; \quad x, -y, -z; \quad -x, y, -z \quad \text{and} \quad -x, -y, z$$

axes, respectively, as shown in Fig. 4.19. The bonding directions are illustrated in Fig. 4.20.

Other hybrid bonds can also be formed, for example, in CuO, in which the four bonds are coplanar and directed towards the corners of a square.

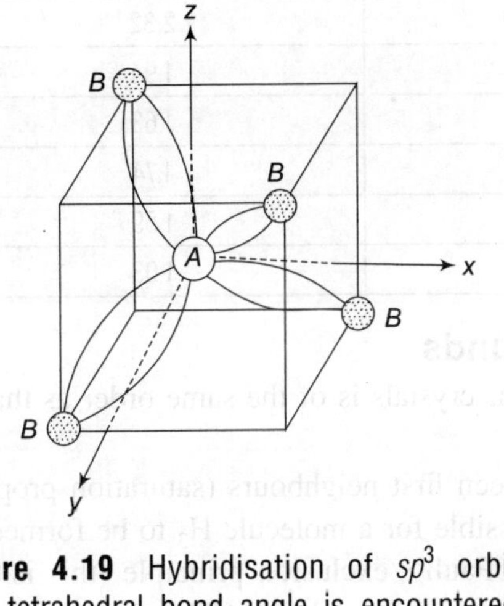


**Figure 4.19** Hybridisation of $sp^3$ orbitals (the tetrahedral bond angle is encountered in materials whenever four equal atoms are bonded to a central atom)

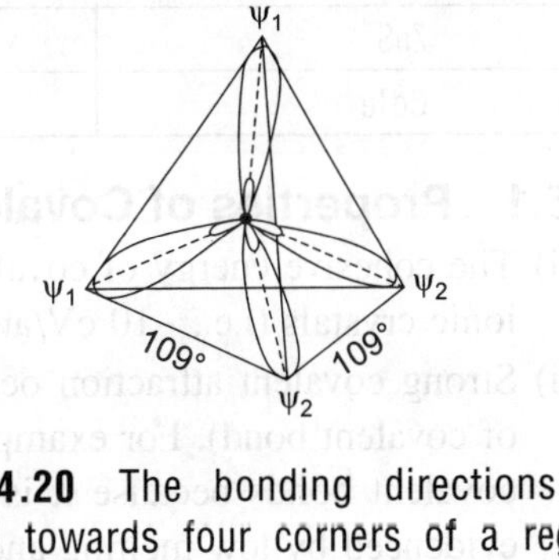


**Figure 4.20** The bonding directions are disposed towards four corners of a regular tetrahedron with bond angle ~109°

## 4.15 COVALENT COMPOUNDS

The covalent bond is a bond between the atoms of the same polarity, hence it is also known as a *homopolar bond*. Two (shared) electrons from adjacent atoms having the same energy and similar way of behaviour enter to form a covalent bond if they have opposite spins. The number of paired electrons in an atom (which can thus form pairs with electrons of other atoms) is given by the 8-$N$ rule, where $N$ is the number of columns of the Periodic Table containing the atom. For example, $O_2$ molecule will have 8 – 6 = 2 electron pair bonds and $N_2$ molecule will have 8 – 5 = 3 electron pair bonds.

Elements such as N, O, C, F and Cl form covalent bonds. Si, Ge, As and Se form bonds that are partly covalent (partly metallic). The transition metals (columns III-IV in the Periodic Table) are also thought to have a certain amount of covalent character to their bonds. A pre-requisite for a covalent bonding is that each atoms have at least one half-filled orbital. Only then the energy can be lowered substantially by having each bonding electron in the orbitals of the two atoms simultaneously. The more the orbitals' overlapping, more the energy is lowered and stronger the bond. The amount of overlap is limited by the electrostatic repulsion or by exclusion principle and the completely filled orbitals do not overlap.

Typical example of covalent compounds are homonuclear diatomic molecules like $H_2$, $N_2$, $O_2$, $F_2$, etc. Semiconductors like diamond Si, Ge, grey tin and organic solids are made up of carbon and hydrogen (e.g., $CH_4$, $CCl_4$, etc.) and compounds like $H_2O$, $NH_3$, SiC, etc. (see Table 4.9).

**Table 4.9** Crystal Data of Some Common Covalent Solids

| *Crystal* | *Interatomic Separation ($r_0$) ($10^{-10}$ m)* | *Cohesive energy ($-\phi$) eV/atom* |
|---|---|---|
| C | 1.54 | 3.68 |
| Si | 2.35 | 2.32 |
| Ge | 2.44 | 1.94 |
| GaAs | 2.45 | 1.63 |
| InP | 2.54 | 1.74 |
| ZnS | 2.34 | 1.59 |
| CdTe | 2.81 | 1.03 |

### 4.15.1 Properties of Covalent Compounds

(i) The cohesive energy of covalent or hompolar crystals is of the same order as that of ionic crystals (i.e., ~10 eV/atoms).

(ii) Strong covalent attraction occurs only between first neighbours (saturation property of covalent bond). For example, it is not possible for a molecule $H_3$ to be formed by covalent bonds because it is disallowed by Pauli's exclusion principle (this is also evidenced by low melting and boiling temperature of methane: $CH_4$).

(iii) (a) Multiple bonds have shorter bond lengths.

(b) Double bonds are stronger than single bonds (the latter situation is of importance in the formation of polymers and plastics) (see Table 4.10).

**Table 4.10** Covalent Bond Lengths and Associated Binding Energies

| *Bond* | *Bond length (Å)* | *Binding energy in (x $10^3$ kJ/k mol)* |
|---|---|---|
| H—H | 0.74 | 434 |
| C—H | 1.08 | 414 |
| C—C | 1.54 | 343 |
| C═C | 1.34 | 615 |
| C≡C | 1.20 | 841 |
| C—C | 1.20 | 355 |
| C═O | 1.22 | 745 |
| O—H | 0.97 | 463 |

(iv) Covalent crystals are usually hard with quite high bonding energies and they possess high melting and boiling points.

(v) Covalent crystals are typically semiconductors (e.g., Si and Ge), which are electrical insulators at 0 K. The electrical conductivity ($\sigma$) rises exponentially with increase in temperature ($T$) as:

$$\sigma \propto \exp\left(-\frac{E_g}{2\,k_B T}\right)$$

where $k_B$ is Boltzmann constant and $E_g$ is a constant having the dimensions of energy. The $E_g$ value is a measure of covalent bond strength; as the temperature is raised, the covalent bond gets successively broken up and valence electrons previously localised in covalent bonds become free to carry electric current. The electrical conductivity is quite sensitive to the presence of tiny amount of impurity atoms.

(vi) They are transparent to long wavelength radiation ($\lambda \geq$ infrared) but opaque to shorter wavelength, the transition being abrupt and occurring at a characteristic wavelength usually in the visible or near infrared.

(vii) Covalent substances are insoluble in polar solvents like water, but they are soluble in non-polar solvents like benzene, carbon disulphide, etc. The giant molecules are, however, not soluble in any solvent because of their large size.

(viii) A very interesting property of covalent crystals is the apparent lack of sensitivity of their physical properties to their bond type. For example, carbon in the diamond structure is the hardest substance and has a very high melting point of 3280 K. The hardness and melting point then decrease as we proceed to other elements in column IV of the Periodic Table from silicon to lead. Tin, for example, is very soft and has a low melting point. The variation in the electrical properties is also pronounced. Diamond is a very good insulator. Si and Ge are semiconductors while tin is a good conductor. This is due to variation in bond length and bond energy depending on the number of electrons shared in a covalent bond.

## 4.16 BONDS IN SILICON AND BONDING AND ANTIBONDING ORBITALS

The most stable electron configuration corresponds to complete $s$ and $p$ subshells of an outer (valence) shell. Indeed, in the periodic system, elements having completely filled $s$- and $p$-valence subshells correspond to inert gases. For example,

$$\text{Ne}\ \ 1s^2 2s^2 2p^6$$

$$\text{Ar}\ \ 1s^2 2s^2 2p^6 3s^2 3p^6$$

$$\text{Kr core} + 4s^2 4p^6$$

$$\text{Xe core} + 5s^2 5p^6$$

$$\text{Rn core} + 6s^2 6p^6$$

Bonds in a crystal are formed in such a way that neighbouring atoms share their valence electrons, satisfying (on the average) the valence requirements, i.e., having completed *s* and *p* subshells of the valence shell. For example, on Si or Ge there are four valence electrons per atom:

$$\text{Si core} + 3s^2 3p^2$$

$$\text{Ge core} + 4s^2 4p^2$$

In a silicon crystal, atom forms four bonds with four other silicon atoms (four nearest neighbours) and shares two valence electrons with each of them. Hence, it shares eight valence electrons with all four nearest neighbours. In other words, in a silicon or germanium crystals, each atom is tetrahedrally coordinated (see Fig. 4.21) in order to share eight electrons (two electrons per bond), corresponding to complete *p* and *s* valence subshells. These bonds can be presented as a linear combination of atomic wave functions (atomic orbitals)—*s*- and *p*-type atomic orbitals. One *s* orbital and three *p* orbitals, having two electrons with opposite spins each, from the outer subshell of a silicon atom. The *s*-type atomic orbital has a spherical symmetry. The angular dependence of the three *p*-type atomic orbitals may be visualized as linear combinations of three perpendicular lobes, as shown in Fig. 4.22. Four bonds can be formed using linear combinations of these atomic orbitals for each atom.

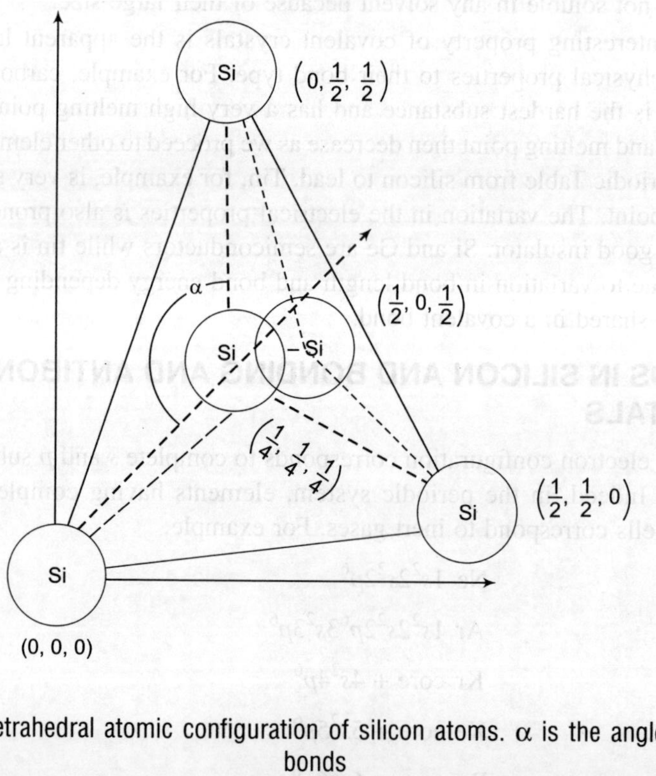


**Figure 4.21** Tetrahedral atomic configuration of silicon atoms. α is the angle between the bonds

$$\psi_1 = \frac{1}{2}(|s> + |p_x> + |p_y> + |p_z>) \quad ...(1)$$

$$\psi_2 = \frac{1}{2}(|s> + |p_x> - |p_y> - |p_z>) \quad ...(2)$$

$$\psi_3 = \frac{1}{2}(|s> - |p_x> + |p_y> - |p_z>) \quad ...(3)$$

$$\psi_4 = \frac{1}{2}(|s> - |p_x> - |p_y> + |p_z>) \quad ...(4)$$

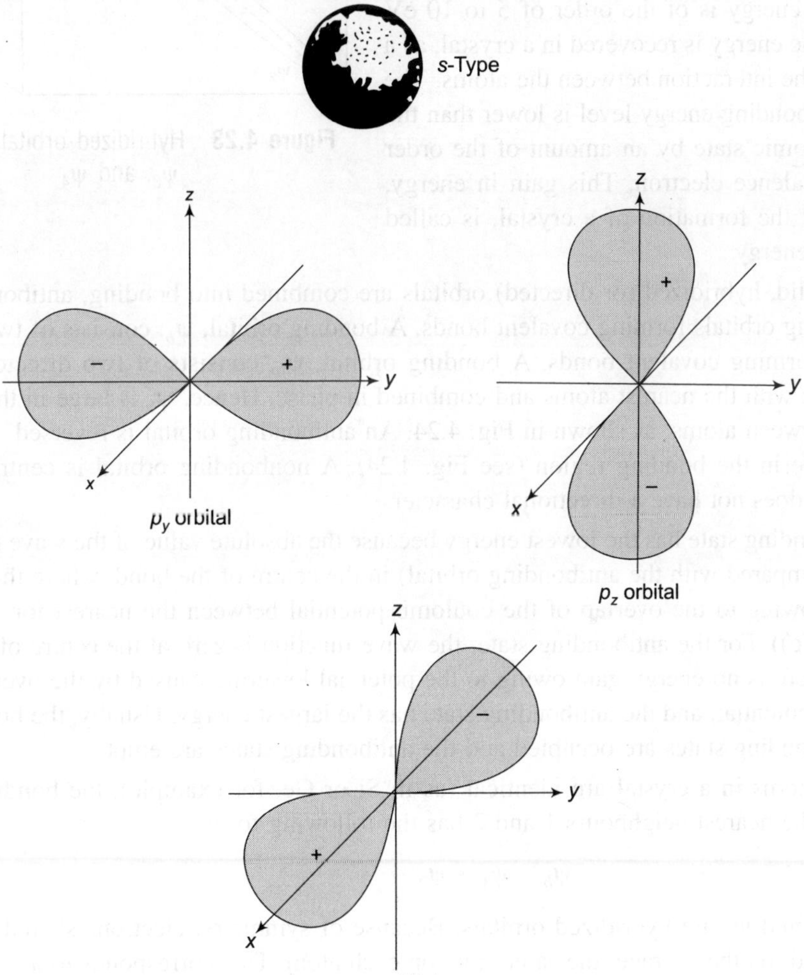


**Figure 4.22** Schematic representation of the angular dependence of *s*-type and *p*-type atomic orbitals

where |$s$>, |$p_x$>, |$p_y$>, and |$p_z$> denote the atomic orbitals. These functions are called *hybridized orbitals*. They have maximum values in the directions shown in Fig. 4.23. Comparison of Fig. 4.22 with Fig. 4.23 clearly illustrates the significance of the hybridized orbitals in the bond formation.

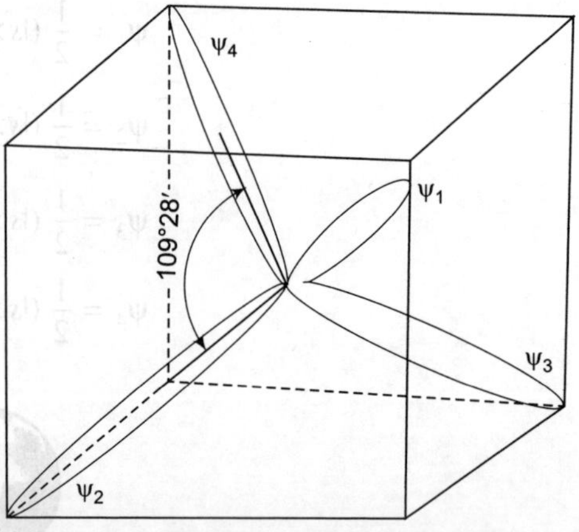


**Figure 4.23** Hybridized orbitals $\psi_1$, $\psi_2$, $\psi_3$, and $\psi_4$

Hybridized orbitals do not correspond to the ground state of an atom; atomic orbitals do. Their energy is of the order of 5 to 10 eV higher. The energy is recovered in a crystal, as a result of the interaction between the atoms. The resulting bonding energy level is lower than the ground atomic state by an amount of the order of 1 eV/valence electron. This gain in energy, caused by the formation of a crystal, is called cohesion energy.

In a solid, hybridized (or directed) orbitals are combined into bonding, antibonding, and nonbonding orbitals forming covalent bonds. A bonding orbital, $\psi_b$, consists of two directed orbitals forming covalent bonds. A bonding orbital, $\psi_b$, consists of two directed orbitals associated with the nearest atoms and combined in phase. Hence, $\psi_b$ is large in the bonding region between atoms, as shown in Fig. 4.24. An antibonding orbital is reversed. Hence, $\psi_a$ has a node in the bonding region (see Fig. 4.24). A nonbonding orbital is centred on one atom and does not have a directional character.

The bonding state has the lowest energy because the absolute value of the wave function is larger (compared with the antibonding orbital) in the centre of the bond, where the potential is lower owing to the overlap of the coulomb potential between the nearest ion cores (see Fig. 4.24(c)). For the antibonding state, the wave function is zero at the centre of the bond. Hence, there is no energy gain owing to the potential lowering caused by the overlap of the coulomb potential, and the antibonding state has the largest energy. Usually, the bonding and the non-bonding states are occupied and the antibonding states are empty.

If all atoms in a crystal are identical (as in Si or Ge, for example), the bonding orbital between the nearest neighbours 1 and 2 has the following form

$$\psi_b \sim \psi_1 + \psi_2 \qquad ...(5)$$

where $\psi_1$ and $\psi_2$ are hybridized orbitals. Because of symmetry, electrons shared in a bond must spend, on the average, the same time on each atom. This corresponds to a symmetrical distribution of the electron density and to a purely homopolar (covalent) bond. In a compound semiconductor, such as GaAs, the bonding orbital includes the weighting factor, $\lambda$,

$$\psi_b \sim \psi_1 + \lambda\psi_2 \qquad ...(6)$$

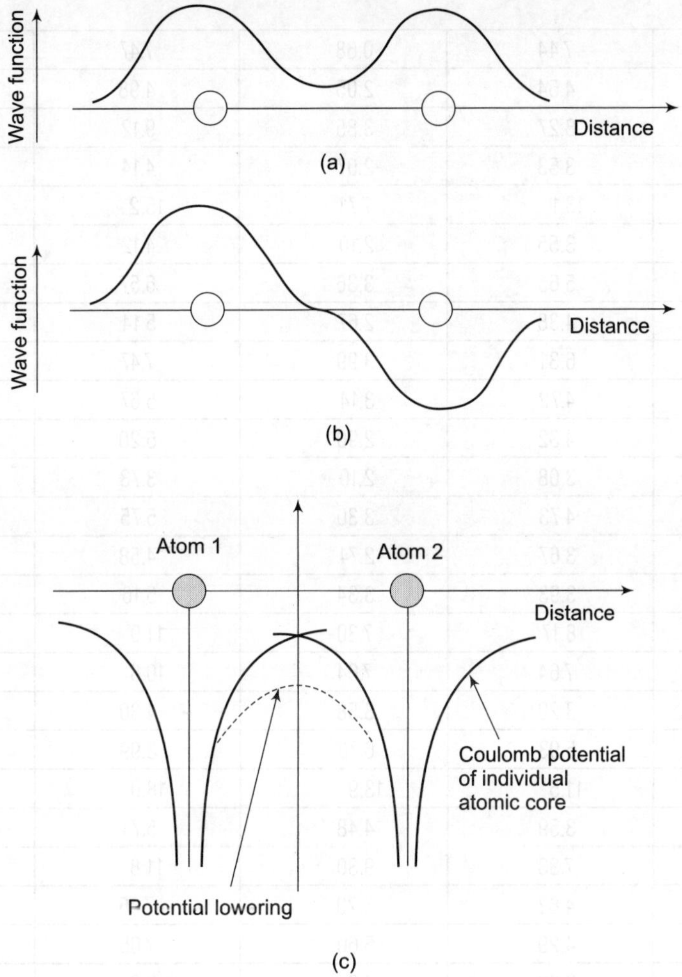


**Figure 4.24** (a) Bonding, (b) antibonding orbitals, and (c) decrease of potential in the middle of the bond due to overlap of the ion core coulombic potentials of the nearest atoms

**Table 4.11** Energies $E_h$, $C$, and $E_{ba}$ and Iconicity $f_i$ in Binary Tetrahedrally Coordinated Crystals (from Phillips 1973)

| *Crystal* | $E_h$ *(eV)* | *C (eV)* | $E_{ba}$ *(eV)* | $f_i$ |
|---|---|---|---|---|
| C | 13.5 | 0 | 13.5 | 0 |
| Si | 4.77 | 0 | 4.77 | 0 |
| Ge | 4.31 | 0 | 4.31 | 0 |
| Sn | 3.06 | 0 | 3.06 | 0 |
| BAs | 6.55 | 0.38 | 6.56 | 0.002 |

*Contd...*

*Table 4.11 Contd...*

| | | | | |
|---|---|---|---|---|
| BP | 7.44 | 0.68 | 7.47 | 0.006 |
| BeTe | 4.54 | 2.05 | 4.98 | 0.169 |
| SiC | 8.27 | 3.85 | 9.12 | 0.177 |
| AlSb | 3.53 | 2.07 | 4.14 | 0.250 |
| BN | 13.1 | 7.71 | 15.2 | 0.256 |
| GaSb | 3.55 | 2.10 | 4.12 | 0.261 |
| BeSe | 5.65 | 3.36 | 6.57 | 0.261 |
| AlAs | 4.38 | 2.67 | 5.14 | 0.274 |
| BeS | 6.31 | 3.99 | 7.47 | 0.286 |
| AlP | 4.72 | 3.14 | 5.67 | 0.307 |
| GaAs | 4.32 | 2.90 | 5.20 | 0.310 |
| InSb | 3.08 | 2.10 | 3.73 | 0.321 |
| GaP | 4.73 | 3.30 | 5.75 | 0.327 |
| InAs | 3.67 | 2.74 | 4.58 | 0.357 |
| InP | 3.93 | 3.34 | 5.16 | 0.421 |
| AlN | 8.17 | 7.30 | 11.0 | 0.449 |
| GaN | 7.64 | 7.64 | 10.8 | 0.500 |
| MgTe | 3.20 | 3.58 | 4.80 | 0.554 |
| InN | 5.93 | 6.78 | 8.99 | 0.578 |
| BeO | 11.5 | 13.9 | 18.0 | 0.602 |
| ZnTe | 3.59 | 4.48 | 5.74 | 0.609 |
| ZnO | 7.33 | 9.30 | 11.8 | 0.616 |
| ZnS | 4.82 | 6.20 | 7.85 | 0.623 |
| ZnSe | 4.29 | 5.60 | 7.05 | 0.630 |
| HgTe | 2.92 | 4.0 | 5.0 | 0.65 |
| HgSe | 3.43 | 5.0 | 6.1 | 0.68 |
| CdS | 3.97 | 5.90 | 7.11 | 0.685 |
| CuI | 3.66 | 5.50 | 6.61 | 0.692 |
| CdSe | 3.61 | 5.50 | 6.58 | 0.699 |
| CdTe | 3.08 | 4.90 | 5.79 | 0.717 |
| CuBr | 4.14 | 6.90 | 8.05 | 0.735 |
| CuCl | 4.83 | 8.30 | 9.60 | 0.746 |
| CuF | 8.73 | 15.8 | 18.1 | 0.766 |
| AgI | 3.09 | 5.70 | 6.48 | 0.770 |
| MgS | 3.71 | 7.10 | 8.01 | 0.786 |
| MgSe | 3.31 | 6.41 | 7.22 | 0.790 |
| HgS | 3.76 | 7.3 | 8.3 | 0.79 |

which may be interpreted in terms of functions of time $f_1$ and $f_2$, the bonding electrons spends on the first and the second atom, respectively

$$f_1 = \frac{1}{1+\lambda^2} \quad \text{and} \quad \frac{\lambda^2}{1+\lambda^2} \qquad ...(7)$$

Electrons in the bonding states spend a greater function of time on a anions (negatively charged atoms). This situation corresponds to a partially heteropolar (partially ionic) bond.

In covalent crystals the difference in the energies of bonding and antibonding states depends on the bond length. In ionic compounds there is also a contribution that depends on the difference of atomic potentials. According to Phillips (1973), this can be taken into account by presenting the crystal potential as the sum of the symmetric potential (as in a covalent crystal) and antisymmetric potential (as in a purely ionic crystal) and using the following empirical equation

$$E_{ba}^2 = E_h^2 + C^2 \qquad ...(8)$$

Here $E_{ba}$ is the energy gap between the bonding and the antibonding state, $E_h$ is the energy gap produced by the covalent (symmetrical) part of the potential, and $C$ is the magnitude of the energy gap produced by the antisymmetric (ionic) potential. This fractional ionicity and covalency of the bond are then defined as

$$f_i = C^2/E_{ba}^2 \qquad ...(9)$$

$$f_c = E_h^2/E_{ba}^2 \qquad ...(10)$$

Values of $E_h$, $C$, $E_{ba}$ and $f_i$ for binary tetrahedrally coordinated crystal are given in Table 4.11 (from Phillips 1973).

Many properties of semiconductors, such as their crystalline structure, cohesive energy, elastic constants and vibrational spectra, are strongly dependent on the degree of the ionicity of the bonds.

## 4.17 VAN DER WAALS—LONDON INTRACTION AND LENNARD JONES POTENTIAL

It is well known that atoms of the rare gases (He, Ar, and Ne) are chemically extremely inactive and form no compounds with other atoms. In other words, outer electron wave functions are not easily perturbed by the presence of other atoms. This implies a high degree of stability of the outer electron shells. This is also the reason why these materials remains in the gaseous state at normal temperatures. However, at very low temperatures, a gas like argon forms a solid and the question arises as to what keep these chemically inactive atoms together. In teams of classical picture, these weak attractive forces arise as a consequence of the fact that an electron revolving around a nucleus may be considered to represent a rotating electric dipole. Such a dipole will induce a dipole moment in a neighbouring atom such that dipole moment induced (in each other) will cause an attractive interaction between the atoms.

As a model, consider two identical linear harmonic oscillators 1 and 2 (representing two identical 'inert' gas atoms) separated by distance $R$ and places along the X-axis (Fig. 4.25).

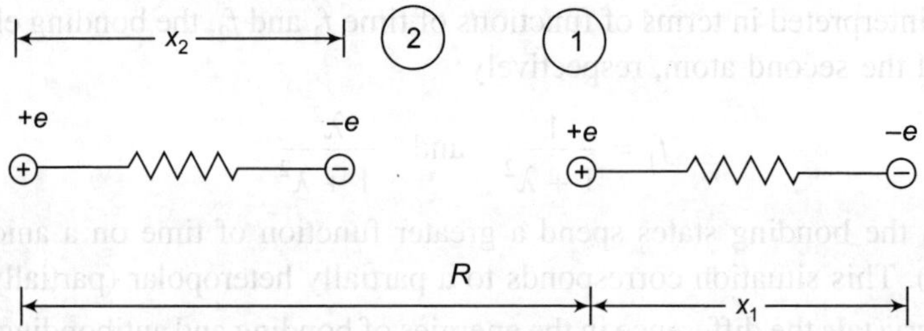


**Figure 4.25** Two identical linear harmonic oscillators 1 and 2 separated by a distance *R*

Let $p_1$ and $p_2$ denote their momenta. The force constant is *C*. then the Hamiltonian of the unperturbed systems is

$$H_0 = \frac{p_1^2}{2m} + \frac{1}{2}Cx_1^2 + \frac{p_2^2}{2m} + \frac{1}{2}Cx_2^2 \quad ...(1)$$

The uncoupled oscillators have the resonant frequency appropriate to a simple harmonic osscillators. Let $H_1$ be the coulomb interaction energy of the two oscillators. Then

$$H_1 = \frac{e^2}{R} + \frac{e^2}{R + x_1 - x_2} - \frac{e^2}{R + x_1} - \frac{e^2}{R - x_2} \quad ...(2)$$

$$= \frac{1}{R}\left[e^2 + e^2\left(1 + \frac{x_1 - x_2}{R}\right) - e^2\left(1 + \frac{x_1}{R}\right)^{-1} - e^2\left(1 - \frac{x_1}{R}\right)^{-1}\right]$$

or
$$H_1 \cong -\frac{2e^2 x_1 x_2}{R^3} \quad \text{(assuming } |x_1|, |x_2| << R) \quad ...(3)$$

Using the normal mode transformation

$$x_s = \frac{1}{2}(x_1 + x_2), \quad x_a = \frac{1}{2}(x_1 - x_2) \quad ...(4)$$

or
$$x_1 = \frac{1}{\sqrt{2}}(x_s + x_a), \quad x_2 = \frac{1}{\sqrt{2}}(x_s - x_a) \quad ...(5)$$

The subscript *s* and *a* denote symmetric and antisymmetric modes of motion. Furthermore, we have momenta

$$p_1 = \frac{1}{\sqrt{2}}(p_s + p_a) \quad p_2 = \frac{1}{\sqrt{2}}(p_s - p_a) \quad ...(6)$$

The total Hamiltonian $H_0 + H_1$ after the transformation from Eqs. (5) and (6) is

$$H = \left[\frac{-1}{2m}p_s^2 + \frac{1}{2}\left(C - \frac{2e^2}{R^3}\right)x_s^2\right]$$

$$+ \left[\frac{1}{2m}p_s^2 + \frac{1}{2}\left(C + \frac{2e^2}{R^3}\right)x_a^2\right] \quad ...(7)$$

The two frequencies of the coupled oscillation are found by inspection of this equation to be

$$\omega = \left[\left(C \pm \frac{2e^2}{R^3}\right)/m\right]^{1/2}$$

$$\cong \omega_0 \left[1 \pm \frac{1}{2}\left(\frac{2e^2}{CR^3}\right) - \frac{1}{8}\left(\frac{2e^2}{CR^3}\right)^2 + ....\right] \qquad ...(8)$$

with $\omega_0$ given by $(C/m)^{1/2}$. Therefore,

$$\hbar\omega_0 \cong \hbar\omega_0 \left[1 \pm \frac{1}{2}\left(\frac{2e^2}{CR^3}\right) - \frac{1}{8}\left(\frac{2e^2}{CR}\right)^2 + ....\right]$$

The zero-point energy of the system is $(1/2)\hbar(\omega_s + \omega_a)$; because of the interaction, this sum is lowered from $2 \times (1/2)\,\hbar\omega_0$ by

$$\Delta U = \frac{1}{2}\hbar(\Delta\omega_s + \Delta\omega_a)$$

$$= \hbar\omega_0 \frac{1}{8}\left(\frac{2e^2}{CR^3}\right)^2 = -\frac{A}{R^6} \qquad ...(9)$$

This is an attractive interaction which varies as the minus sixth power of the separation of the two oscillators (atoms). This is the van der Waals interaction, also known as the London interaction or the induced dipole interaction. It is the principal attractive interaction in crystals of inert gases and also crystals of many organic molecules. The interactive is a quantum effect in the sense that $\Delta U \to 0$ as $\hbar \to 0$. Thus, the zero point energy of the system is lowered by the dipole-dipole coupling of Eq. (3). The van der Waals interaction does not depend on its existence on any overlap of the charge densities of the two atoms.

There exists a finite 'zero-point energy', even in the lowest energy state because the atom has to be put in correspondence with a system of moving charges (electrons), consequently the atom carries a dipole oscillating with electronic frequency as a result, in many atoms or molecules without free valencies (spins), even before condensation, a sort of pairing occurs as a result of balancing of Heitler – London repulsion and van der Waals polarisation attracting force, since the latter sometimes fall off more slowly with increasing internuclear separation than the former. Certain examples are $Hg_2$, $Cd_2$, $Zn_2$ and mixed molecules that these elements form with alkali metals.

The repulsive forces provide an energy of the form $B/R^n$, where $n$ is an integer and $B$ is a constant. Experimental data on inert gases provide a value of $n$ as 12, then the interatomic potential energy is written in the form of Lennard–Jones potential as

$$\phi(R) = \phi_0 \times \left[\left(\frac{\sigma}{R}\right)^{12} - \left(\frac{\sigma}{R}\right)^6\right] \qquad ...(10)$$

Here $B = \phi_0\sigma^{12}$ and $A = \phi_0\sigma^6$

By differentiating with respect to $R$

$$\left.\frac{d\phi}{dR}\right|_{R=R_0} = 0 \text{ or } -12\sigma^{12}R^{-12-1} + 6\sigma^6 R^{-6-1} = 0$$

or $$-2\sigma^6 R^{-13} + R^{-7} = 0$$

or $$R^{-7} = 2\sigma^6 R^{-13}$$

or $$R_0^6 = 2\sigma^6$$

Therefore, $R_0 = 2^{1/6}\,\sigma = 1.12\,\sigma$ and so, value of $\phi(R)$ at the minimum $R$ (i.e., $R_0$) is $-\phi_0/4$, parameter $\sigma$ can be seen as the value of $R$ at which $\phi(R)$ equals zero. Bernardes in 1958 gave experimental Lennard Jones parameters, which are given in Table 4.12 for inert gases.

**Table 4.12** The Parameters $\sigma$ and Cohesive Energies of Some Inert Gas Solid Crystals

| *Gas* | *σ (Å)* | *$\phi_0$ (eV)* | *Cohesive energy (eV/atom)* | |
|---|---|---|---|---|
| | | | *Theory* | *Experimental* |
| Ne | 2.74 | 0.012 | −0.020 | −0.02 |
| Ar | 3.40 | 0.042 | −0.070 | −0.08 |
| Kr | 3.65 | 0.056 | −0.094 | −0.116 |
| Xe | 3.98 | 0.080 | −0.134 | −0.17 |

The cohesive energy of an inert gas crystal can be obtained by summing the Lennard Jones potential over all pairs of atoms in the crystal as follows:

$$\phi = 6\phi_0\left[\left(\frac{\sigma}{R_0}\right)^{12} - \left(\frac{\sigma}{R_0}\right)^6\right] + 3\phi_0\left[\left(\frac{\sigma}{a}\right)^{12} - \left(\frac{\sigma}{a}\right)^6\right] + .... \quad ...(11)$$

For the first term of the R.H.S., $R_0$ is the nearest neighbour distance and there are 12 nearest neighbours in an fcc inert gas crystal. A factor of 1/2 occurs to make it 6, to avoid counting twice each pair of atom.

Similarly, for the second term, $a$ is the next nearest negihbour distance and there are six such neighbours.

If we restrict our analysis to the second neighbour only, then the total cohesive energy is

$$\phi = 6\phi_0\left(\frac{1}{4} - \frac{1}{2}\right) + 3\phi_0\left(\frac{1}{256} - \frac{1}{16}\right) = -\frac{429}{256}\phi_0 \quad ...(12)$$

The value obtained from this expression is listed in Table 4.12 along with the measured value in case of inert gas crystal. The agreement obtained is quite satisfactory.

## 4.18 METALS

Metallic elements make up more than half of the total number of chemical elements in the Periodic Table. The physical properties of metals in the free atomic or solid state derive from the electronic structure of the metal-forming atom. Take sodium (Na) for example. The sodium atom consists of a nucleus of atomic number $Z = 11$ and 11 orbital electrons; 10 of these electrons associated with the $1s^2 2s^2 2p^6$ electronic shell of the atom are tightly bound to the nucleus to form the compact sodium ion, $Na^+$; the eleventh electron associated with the 3*s* electronic shell in the valence electron and is only loosely bound to the atomic nucleus. Consequently, when a collection of sodium atoms is brought together of form a solid, the compact ion cores provide the rigid framework (crystal lattice) through which the valence electrons move about freely like the particles of a gas, as shown in Fig. 4.26. There are nearly $10^{23}$ free (valence) electrons per cubic centimetre of sodium crystal.

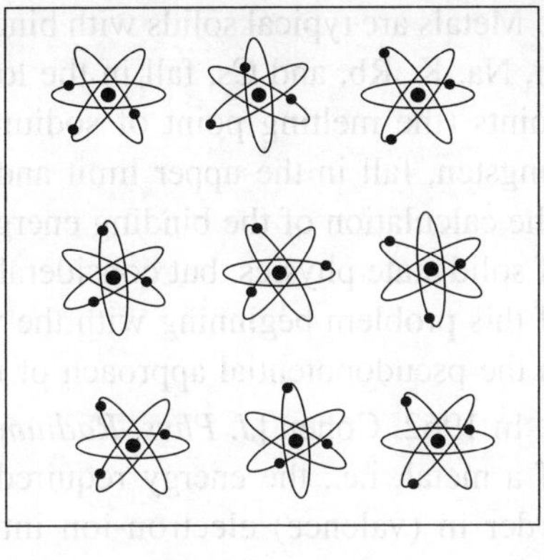

**Figure 4.26** Positive ions of a metal, such as sodium, embedded in a gas of the valence electrons

The presence of free electrons is responsible for the high electrical ($\sigma$) and thermal ($K$) conductivity of metal: For example, the electrical conductivity of sodium is $\sigma = 2 \times 10^5$ (ohm-cm)$^{-1}$ at room temperature. The high electrical conductivity is, in turn, responsible for the high optical reflection and absorption coefficients characteristic of the optical properties of metals. Well before the introduction of the quantum theory, Drude in 1900 explained the electrical and thermal conductivity of metals by treating a metal crystal as a container filled with free electron gas: For many metals, particularly the simple (nontransition) metals, the ion corcs occupy only about 10–20% of the value of the crystal, while the valence electrons gas occupies the remaining 80–91%. Consequently, the approximation of almost free electrons is quite good in simple metals, although a proper understanding of why this is possible (*vide* the strong electron-ion potential near each ion site) has only recently been clarified in terms of the concept of pseudopotentials.

Associated with the smallness of the metallic ion cores is the fact that most metallic structures tend to be close-packed. The most common metallic structures are the fcc, bcc, and hcp.

On the face of it, it might seem inconceivable that metals do exist as solids at all since one might expect that the free-electron gas would leak out from its crystal container. In practice this does not happen because the electrons are kept inside the crystal surface by some work function (of the order of 2 eV per atom for sodium): To get an electron out of the metal (via, for examples, the photoelectric effect) sufficient photon energy must be provided by the incident light to enable the electron to overcome the work function (or potential barrier). The existence of a sizable work function for photo ejection of electrons from metal surfaces is the reason why ordinary light does not cause deterioration of solid metals.

Metals are typical solids with binding energies between 1 and 5 eV per atom. Alkali metals, Li, Na, K, Rb, and Cs, fall in the lower limit, hence their relatively low melting and boiling points (the melting point of sodium is about 100°C); while the transition metals, such as tungsten, fall in the upper limit and hence have extremely high melting and boiling points. The calculation of the binding energy of a metal has been one of the most difficult problems of solid state physics, but considerable progress has been made toward the complete solution of this problem beginning with the works of Wigner and Seitz in the 1930s and culminating in the pseudopotential approach of the last decade.

In 1962, Cohen [*J. Phys. Radium* **23,** 643 (1962)] showed that if the total cohesive energy of a metal, i.e., the energy required to take the neutral atoms apart, is expanded to second order in (valence) electron-ion interaction potential, $V$ (which may be represented by a pseudopotential), then the resulting energy explicitly contains a term that represents a pair-wise interaction between atoms. The pair-wise interaction consists of two pieces, viz., the direct or "bare" Coulomb piece (of the type that we have associated with the Madelung energy of the crystal) and an indirect ion-electron-ion interaction that depends on $V$ and the dielectric constant of the electron gas. It is this effective pair-wise interaction between ions rather than the absolute value of the cohesive energy that one usually wishes to compute in order to account for the nature of metallic binding forces.

## 4.19 THE METALLIC BOND

The energy eigenvalue problem for a system of electrons is much simplified if formulated in the *one-electron approximation*, where each electron is assumed to move in an average field due partly to the net positive charge of the ion cores and partly to all the other electrons. The ions can be treated as a uniform positive background charge and the electrons, for which $k \neq j$, can be represented by the uniform charge distribution $-e\,|\psi_k(\vec{r}_k)|^2$). Hence,

$$\hat{H}_e \Psi(\vec{r}, R) = \sum_j \left[ -\frac{\hbar^2}{2m_e} \nabla_j^2 + U_{ee}(r_j) + U_{ei}(r_j, R) \right] \Psi(\vec{r}, R) = E(R)\Psi(\vec{r}, R) \qquad ...(1)$$

where the Coulomb potential energy of interaction $U_{ii}(R)$ between the ions has been omitted, since the electrons are allowed to move through an array of stationary ions, so that the interionic distance $R$ can be treated as a constant parameter. If $\Psi(\vec{r}, R)$ is assumed to be a product to be a product of one-electron functions $\psi_j(r_j)$, each corresponding to one or other of two possible eigenvalues $m_{si}\hbar = \pm\hbar/2$ of the spin component, we obtain the expectation value of energy as:

$$E(R) = \langle \psi | \hat{H}_e | \psi \rangle$$

$$= \sum_j \int \psi_j^*(\vec{r}_j) \left[ -\frac{\hbar^2}{2m_e} \nabla_j^2 + \sum_{k \neq j} e_0^2 \int \frac{|\psi_k(\vec{r}_k)|}{|\vec{r}_j - \vec{r}_k|} dV_k + U_{ei}(r_j, R) \right] \Psi_j(\vec{r}_j)\, dV_j$$

Assuming that an arbitrary variation $\delta\Psi(\vec{r}, R)$ is exclusively due to the variation $\delta\psi_j(\vec{r}_j)$, it follows that the energy eigenvalue $E_j$ and eigenfunction $\psi_j(\vec{r}_j)$ of the $j$th electron are determined by solving the single particle equation:

$$\left[-\frac{\hbar^2}{2m_e}\nabla_j^2 + e_0^2 \sum_{k\neq j} \int \frac{|\psi_k(\vec{r}_k)|^2}{|\vec{r}_j - \vec{r}_k|} dV_k + U_{ei}(r_j, R)\right] \psi_j(r_j) = E_j\, \psi_j(r_j) \qquad ...(2)$$

Since the second term on the left hand side of Eq. (2), $U_{ee}(r_j)$, depends on the functions of all the other electrons, the iteractive Hartree procedure must in general be used. An initial set of functions $\psi_j^{(0)}(\vec{r}_j)$ is assumed to calculate $U_{ee}^{(0)}(r_j)$ and the equation are numerically solved on this basis. The resulting set of single electron functions $\psi_j^{(1)}(\vec{r}_j)$ yields new potential energies $U_{ee}^{(1)}(r_j)$ and the prodecure is repeated until $\psi_j^{(k)}(\vec{r}_j)$ and $E_j^{(k)}$ have values no longer different from those introduced in the equations solved for them. Equation (2) might also be rewritten in the compact form

$$\left[-\frac{\hbar^2}{2m_e}\nabla^2 + U(\vec{r})\right]\psi(\vec{r}) = \varepsilon \quad \psi(\vec{r}) \qquad ...(3)$$

where $U(\vec{r})$ represents the last two terms on the left hand side of Eq. (2) which are assumed to depend on the position $\vec{r}$ of a given electron only. In other words, the $N$-electron system represented by Eq. (1) can be treated as $N$ one-electron systems described by Eq. (3).

Considering the metal to consist of positive ions located on lattice sites, bathed in a sea of conduction electrons, and taking the particular example where $N$ valence electrons move in a structure containing $N$ monovalent ion cores, it is common practice to drop $U(\vec{r})$ in Eq. (3), to a first approximation, in other words to treat the system of conduction electrons as a free electron gas, confined to a certain region in space. This provides a good description of many physical properties of simple metals. Assuming, for simplicity, $N$ free electrons inside a cubic box of volume $V = L^3$, the eigenvalue equation

$$-\frac{\hbar^2}{2m_e}\left(\frac{\partial^2}{\partial x^2} + \frac{\partial^2}{\partial y^2} + \frac{\partial^2}{\partial z^2}\right)\psi(\vec{r}) = \varepsilon\psi(\vec{r})$$

is separable in Cartesian coordinates, such that the normalized solutions vanishing at the boundaries are

$$\psi_{\vec{p}}(\vec{r}) = \left(\frac{2}{L}\right)^{\frac{3}{2}} \sin\frac{p_x \pi x}{L} \sin\frac{p_y \pi y}{L} \sin\frac{p_z \pi z}{L}$$

where $\vec{p}$ denotes a state with quantum numbers $(p_x, p_y, p_z)$ and energy

$$\varepsilon = \frac{\hbar^2\pi^2}{2m_e L^2} p^2$$

This shows that in $\vec{p}$-space, each set $(p_x, p_y, p_z)$ specifies an allowed energy value, and the surface of constant energy are spherical. As $p_x$, $p_y$ and $p_z$ are positive integers, and each $(p_x, p_y, p_z)$ value corresponds to an elementary cube of unit volume, as suggested in Fig. 4.27 (a), the counting is limited to the first octant of the sphere and the total number of energy levels, up to an energy $\varepsilon$, is

$$\frac{1}{8}\frac{4\pi}{3}p^3 = \frac{1}{6}\pi p^3 = \frac{V}{6\pi^2}\left(\frac{2m_e\,\varepsilon}{\hbar^2}\right)^{\frac{3}{2}}$$

According to the Pauli exclusion Principle, for each allowed energy there are two electron states $\phi(\vec{r}\, m_s) = \psi_{\vec{n}}(\vec{r})\chi_{m_s}$, with $m_s = \pm\frac{1}{2}$ and, because the energy spacing is small, it is convenient to introduce the density of electron states $N(\varepsilon)$, such that

$$2\frac{V}{6\pi^2}\left(\frac{2m_e\,\varepsilon}{\hbar^2}\right)^{\frac{3}{2}} = \int_0^{\varepsilon} N(\varepsilon)\,d\varepsilon$$

This yields: $$N(\varepsilon) = \frac{V}{3\pi^2}\frac{d}{d\varepsilon}\left(\frac{2m_e\varepsilon}{\hbar^2}\right)^{\frac{3}{2}} = \frac{V}{2\pi^2}\left(\frac{2m_e n}{\hbar^2}\right)^{\frac{3}{2}}\varepsilon^{\frac{1}{2}} \qquad ...(4)$$

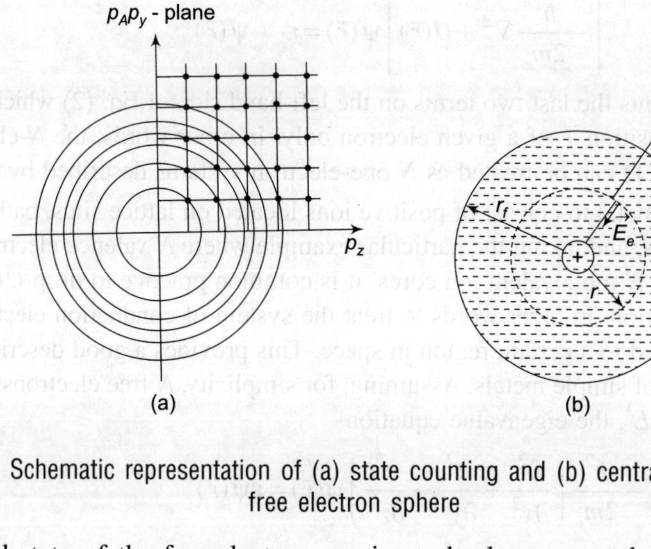


**Figure 4.27** Schematic representation of (a) state counting and (b) central interaction in the free electron sphere

The ground state of the free-electron gas is at absolute zero, where the $N$ electrons progressively fill the state up to an energy $\varepsilon_F$, called the Fermi energy, and the remaining states with $\varepsilon > \varepsilon_F$ are vacant

$$N = \int_0^{\varepsilon_F} N(\varepsilon)d\varepsilon = \frac{V}{3\pi^2}\left(\frac{2m_e}{\hbar^2}\right)^{\frac{3}{2}}\varepsilon_F^{\frac{3}{2}}$$

This gives $$\varepsilon_F = \frac{\hbar^2}{2m_e}\left(3\pi^2\frac{N}{V}\right)^{\frac{2}{3}} = \frac{e_0^2\,a_0}{2}(3\pi^2 n)^{\frac{2}{3}} \qquad ...(5)$$

where $n$ is the electron concentration and $a_0$ is the Bohr radius. The internal (kinetic) energy of the system reads

$$\int_0^{\varepsilon_F} \varepsilon N(\varepsilon)\,d\varepsilon = \frac{V}{5\pi^2}\left(\frac{2m_e}{\hbar^2}\right)^{\frac{3}{2}} = \varepsilon_F^{\frac{5}{2}} = \frac{3}{5}N\varepsilon_F$$

and hence the average kinetic energy per electron, at $T = 0$, is

$$E_k = \frac{3}{5}\varepsilon_F \quad ...(6)$$

In the frame of the one-electron approximation it is convenient to introduce the volume per conduction electron in terms of the radius $r_f$ of a spherical cell, called the *Wigner–Seitz sphere*

$$\frac{4\pi}{3}r_f^3 = \frac{V}{N} \quad \text{or} \quad r_f = \left(\frac{3}{4\pi n}\right)^{\frac{1}{3}} \quad ...(7)$$

The volume $V$ of the monovalent metal is divided this way into $N$ spherical cells, centred at the ion positions, where, in view of Eqs. (5) to (7), the free electron energy reads

$$E_k(r_f) = \frac{9e_0^2}{20}\frac{(3\pi^2/2)^{\frac{1}{3}}\,a_0}{r_f^2} \quad ...(8)$$

It is clear that $r_f$ decreases if we compress the electron gas, thus increasing the average kinetic energy of an electron. It follows that Eq. (8) can be interpreted as describing a repulsive energy inside the Wigner–Seitz sphere. We treat the conduction electron as a spherical charge distribution, with uniform density $\rho = -e/(4\pi r_f^3/3) = -3e/4\pi r_f^3$, which compensates the positive point charge of the ion, placed at the centre of the Wigner-Seitz sphere. The total potential is created by the central ion, in the form $V_i(r) = e/4\pi\varepsilon_0 r$, and by the electron distribution. The latter can be derived from the radial component of the electric field given by Gauss' law, in the configuration of Fig. 4.27 (b), as

$$4\pi r^2 E_e = \frac{1}{\varepsilon_0}\rho\frac{4\pi r^3}{3} \quad \text{or} \quad E_e = \frac{e}{4\pi\varepsilon_0 r_f^3}r$$

which leads to

$$V_e(r) - V_e(r_f) = -\int_{r_f}^{r}\vec{E}_e\cdot d\vec{r} = \frac{e}{4\pi r_f^3}\int_{r_f}^{r} r\,dr = -\frac{e}{4\pi\varepsilon_0 r_f^3}\frac{r_f^2 - r^2}{2}$$

Since outside the cell boundary (i.e., for $r > r_0$) this potential is expected to behave like that of a point charge $-e$ placed at the origin, the boundary condition gives $V_e(r_f) = -e/4\pi\varepsilon_0 r_f$, and hence

$$V_e(r) = -\frac{e}{4\pi\varepsilon_0 r_f}\left(\frac{3}{2} - \frac{r^2}{2r_f^2}\right)$$

The Coulomb potential energy inside the spherical cell follows as

$$E_C(r_f) = \int \rho V_i(r)dV + \frac{1}{2}\int \rho V_e(r)dV = -\frac{3}{2}\frac{e^2}{4\pi\varepsilon_0 r_f} + \frac{3}{5}\frac{e^2}{4\pi\varepsilon_0 r_f} = -\frac{9}{10}\frac{e_0^2}{r_f} \qquad ...(9)$$

By combining the two opposite contribution, Eqs. (8) and (9), the binding energy per cell, takes the form

$$U(r_f) = \frac{9e_0^2}{10}\left[\frac{1}{2}\frac{(3\pi^2/2)^{\frac{1}{3}}a_0}{r_f^2} - \frac{1}{r_f}\right]$$

such that, minimizing with respect to $r_f$, by writing $(\partial U/\partial r_f)_{r_f = r_{f0}} = 0$, one obtains the equilibrium radius of the free electron sphere as $r_{f0} = (3\pi^2/2)^{\frac{1}{3}} a_0 = 2.45\, a_0$, which is close to the range of the observed values for the monovalent metals. This yields

$$U(r_f) = \frac{9}{20}\frac{e_0^2}{r_{f0}}\left[\left(\frac{r_{f0}}{r_f}\right)^2 - 2\frac{r_{f0}}{r_f}\right] = U(r_{f0})\left[2\frac{r_{f0}}{r_f} - \left(\frac{r_{f0}}{r_f}\right)^2\right] \qquad ...(10)$$

Since, for $r_f = r_{f0}$, the magnitude of the attractive term is twice that of the kinetic energy, the binding energy is estimated in a straightforward way by Eq. (10) to be $U(r_{f0}) = -(3/5)\, \varepsilon_F\, (r_{f0}) = -4.9$ eV. This result provides an accurate order of magnitude, and also accounts for the empirical fact that in all alkali metals the binding energy is in a constant ratio to the Fermi energy.

## 4.20 HYDROGEN BONDING

We know that an $H_2$ molecule forms covalent bonds. However, the atom hydrogen in itself is unique in the following three important ways:

(i) The ion core of a hydrogen atom is a bare proton of ~ $10^{-13}$ cm in radius, a factor of $10^5$ smaller than any other ion core.

(ii) Hydrogen is only one electron less from the stable helium configuration.

(iii) The first ionisation potential of atomic hydrogen is unusually high (H: 13.59 eV; Li: 5.39 eV; Na: 5. 14 eV; K: 4.34 eV; Rb: 4.18 eV; Cs: 3.89 eV).

As a result of these properties, hydrogen can play a role unlike any other element in crystalline structures, e.g., bonding of an atom of hydrogen and an extremely electronegative atom (O, F, N, Cl, etc.). In this case, the two electronegative ions are drawn together by a positive hydrogen ion half-way between them as shown in Fig. 4.28 explaining the abnormally short F–F distance in $NaHF_2$. Hydrogen bonding plays a crucial role in water molecule, particularly in the crystal structure of ice.

Water molecules are exceptionally prone to form hydrogen bonds because the electrons around the O-atom in $H_2O$ are not symmetrically distributed but are more likely to be found in certain regions of high probability density. These regions project outward as through towards the vertices of a tetrahedron (Fig. 4.29).

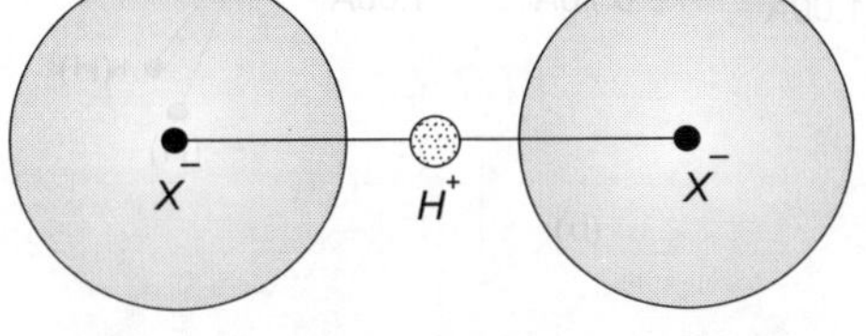


**Figure 4.28** Hydrogen bond in $NaHF_2$. *X* is an *F* atom in this case

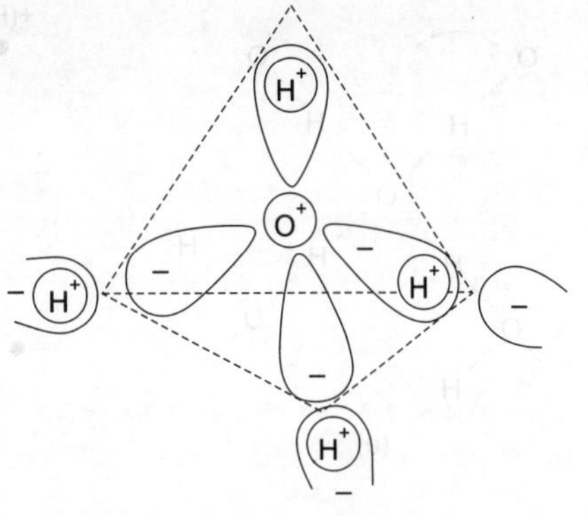


**Figure 4.29** In a $H_2O$ molecule, four pairs of valence electrons around the oxygen atom (six from O-atoms and one each by two H-atoms) preferentially occupy four regions forming a tetrahedron

Each $H_2O$ molecule can form hydrogen bonds with four other $H_2O$ molecules. Hydrogen atoms are at two of these vertices (which therefore exhibit localized positive charges) while the other two vertices exhibit somewhat more diffuse negative charges. Each $H_2O$ molecule can therefore, form hydrogen bonds with four other $H_2O$ molecules. In two of these bonds, the central molecule provides the bridging protons and in the other two, the attached molecules provide them. In the liquid states, the hydrogen bonds between adjacent $H_2O$ molecules are continually being broken and reformed due to thermal agitation, but at any instant, the molecules are combined in definite clusters. In the solid state, the clusters are large and stable and constitute ice. With only four nearest neighbours around each molecule (instead of 12 as in other solids), ice has extremely open structure, consequently ice has a low density. In ice the electron from the H-atom (like the proton) is fairly well localised near a single oxygen ion, along the line joining it to one of its neighbours thereby helping to bind the two oxygen ions together. The lack of regularity in the position of protons manifests itself in large (thermodynamic) residual entropy exhibited by ice at low temperatures, which corresponds to large number of ways of assigning a proton to either end of each bond (consistent with two protons being close to each oxygen atom).

The hydrogen bond is liquid water or ice has a binding energy of about 0.5 eV per molecule.

To form the hydrogen bond, a neutral (electropositive) *H* atom is attracted the two (electronegative) atoms such as two atoms of oxygen in ice, as shown in Fig. 4.30 (a): It is a purely configurational attribute. A single linear *H* bond holding two water molecules together is shown in Fig. 4.30 (b). In this case, each oxygen nucleus is surrounded symmetrically by a tetrad of four point charges ($\pm 0.19\ e$) the positive members of which represent partially shielded $H^+$ ions, i.e., protons.

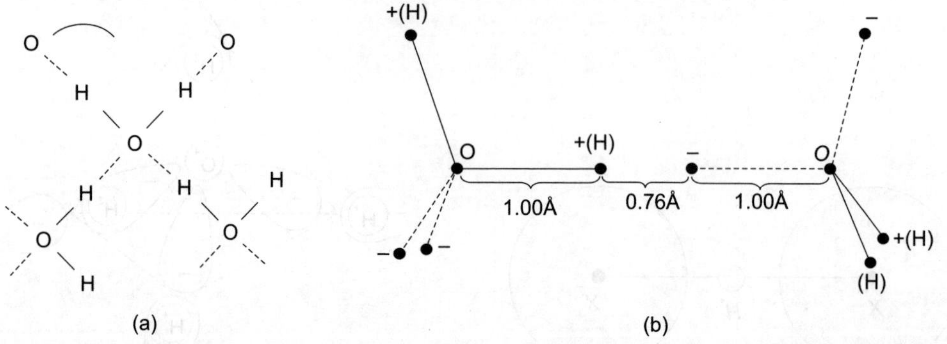


**Figure 4.30** (a) Hydrogen bond in ice (b) Hydrogen bond in water

The H bond is essentially ionic in character, and the binding energy may be discussed in terms of Coulomb interactions between charges and Lennard – Jones potentials. This shows a minimum energy configuration for two water molecules according to the potentials

$$V(1, 2) = V_{LJ}(r_{12}) + S(r_{12})\, V_{el}(1, 2)$$

where $r_{12}$ is the separation between the oxygen nuclei and

$$V_{L,J}(r) = 4e\left[\left(\frac{\sigma}{r}\right)^{12} - \left(\frac{\sigma}{r}\right)^{6}\right]$$

$$(\varepsilon = 5.01 \times 10^{-15} \text{ erg}, \sigma = 2.82 \text{ Å})$$

$V_{el}$ is the Coulomb interaction arising from the four charges $\pm q$ $(q = 0.10\, e)$ on each molecule and $S(r)$ is a correlation function. The configuration shown has a plane of symmetry and incorporates a single linear H bond.

## 4.20.1 Hydrogen Bonding in Ice

The hydrogen bonds have the directional property and are stronger than the van der Waals bonds.

In an effort to understand the mechanism of hydrogen bonding clearly we take up ice crystals which exhibit some unusual behaviour. The outer electron configuration of the oxygen atom in $H_2O$ is represented by four $sp^3$ hybrid orbitals. The two of these are involved in covalent bonds to the hydrogen atoms and the rest two are doubly occupied by oxygen electrons. Out of the six outer oxygen electrons, two are engaged in bonding and the other four find a place in spare orbitals. The bonding pattern gives a tetrahedral shape to bonded molecules in ice. Two vertices of a tetrahedron are occupied by hydrogen atoms which are essentially the localized protons. The other two vertices exhibit relatively diffused negative charges.

The negligible small size of the proton is responsible for the peculiar structure of ice crystal as it partically sits on the oxygen ions, something impossible for any other positive ion to do. One of the many phases of ice crystals is pictured in Fig. 4.31. Two oxygen atoms are bound by a proton that is localized close to one oxygen atom along the line joining it to one of its neighbouring oxygen atoms. There are two protons close to each oxygen atom, giving a large number of ways to attach a proton to either end of the bond. This is reflected in the irregular positions of protons which are very well accounted by the observed large residual entropy of ice crystals at low temperatures. Lastly, a word about the floating property of ice. The molecular clusters of water are smaller and less stable in liquid state than they are in ice because of which they are on the average more closely packed in this state. This increases the density of water and ice floats when immersed in it.

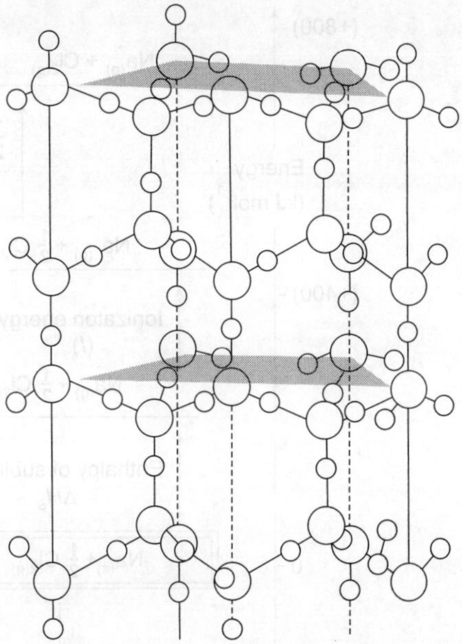

**Figure 4.31** The hydrogen bonding in ice. The large circles denote oxygen ions and the small circles are protons

## 4.21 BORN-HABER CYCLE

This cycle devised by Born and Haber in 1919 relates the lattice energy of a crystal to other thermodynamical data. The energy terms involved in building a crystal lattice such as sodium chloride may be taken in steps. The elements in their standard state are first converted to gaseous atoms, and then in ions, and finally packed into the crystal lattice. (Fig. 4.32).

The enthalpies of sublimation and dissociation and ionization energy are positive since energy is supplied to the system. The electron affinity and lattice energy are negative since energy is evolved in these processes.

According to Hess's law, the overall energy change in a process depends only on the energy of the initial and final states and not on the route taken. As can be seen from Fig. 4.32 the enthalpy of formation of $\Delta H_f$ is the algebraic sum of the terms going around the cycle (paying heed to the exothermic or endothermic nature of each step).

$$\Delta H_f = \Delta H_s + I + \frac{1}{2}\Delta H_d + E + U$$

All the terms except the lattice energy and electron affinity can be measured. Originally, the cycle was used to calculate electron affinities. By using known crystal structures, it was possible to calculate the lattice energy.

$$\Delta H_f = +\Delta H_s + I + \frac{1}{2}H_d + E + U$$

For NaCl $\quad 381.2 = +108.4 + 495.4 + 120.9 - 348.6 + U$

$\Rightarrow \quad U = -757.3 \text{ kJ mol}^{-2}$

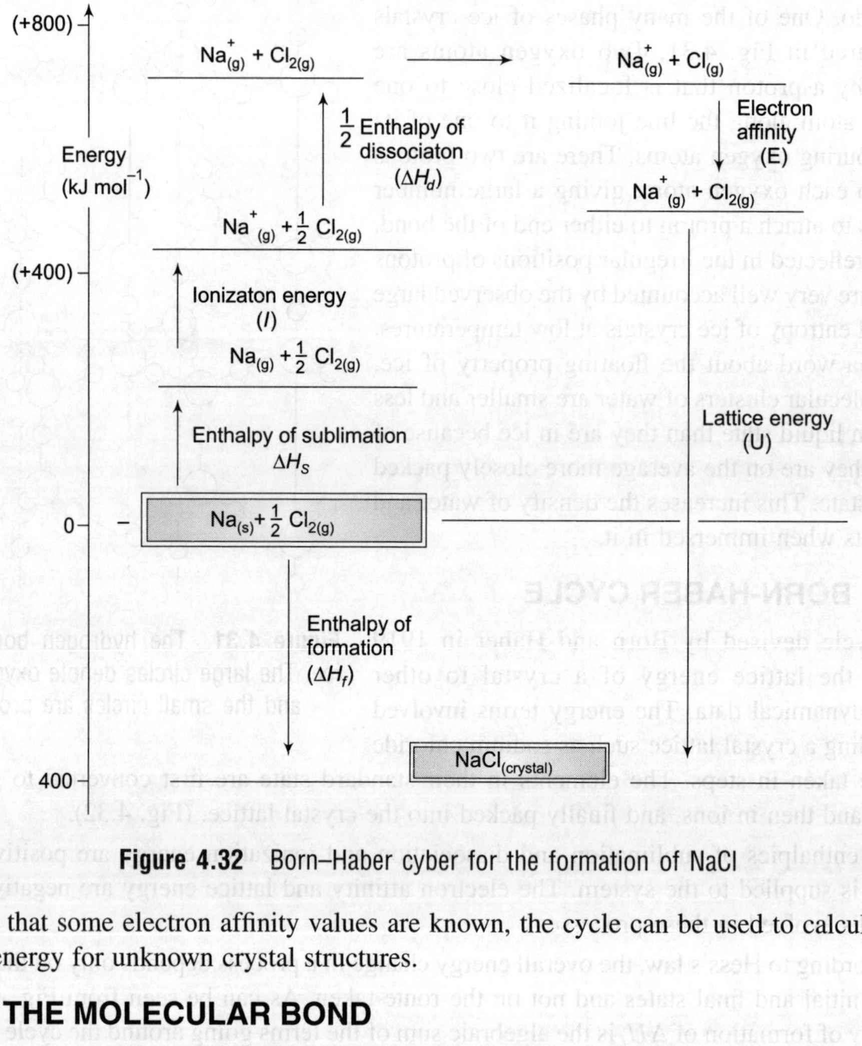


**Figure 4.32** Born–Haber cyber for the formation of NaCl

Now that some electron affinity values are known, the cycle can be used to calculate the lattice energy for unknown crystal structures.

## 4.22 THE MOLECULAR BOND

***Electric forces hold atoms together to form molecules.***

A molecule is an electrically neutral group of atoms held together strongly enough to behave as a single particle.

A molecule of a given kind always has a certain definite composition and structure. Hydrogen molecule, for instance, always consists of two hydrogen atoms each, and water molecules always consist of one oxygen atom and two hydrogen atoms each. If one of the atoms of a molecule is somehow removed or another atom becomes attached, the result is a molecule of a different kind with different properties.

A molecule exists because its energy is less than that of the system of separate noninteracting atoms. If the interactions among a certain group of atom reduce their total energy, a molecule can be formed. If the interaction increase their total energy, the atoms repel one another.

Let us see what happens when two atoms are brought closer and closer together, three extreme situations can occur:

1. *A covalent bond is format.* One or more pairs of electrons are shared by the two atoms. As these electrons circulate between the atoms, they spend more time between the atoms than elsewhere, which produces an attraction force. An example is $H_2$, the hydrogen molecule, whose electrons belong to both protons (Fig. 4.33). The attractive force the electrons exerts on the protons is more than enough to counterbalance the direct repulsion between them. If the protons are too closer together, however, there repulsion becomes dominant and the molecule is not stable.

   The balance between attraction and repulsion forces occurs at a separation of $7.42 \times 10^{-11}$ m, where the total energy of the $H_2$ molecule is –4.5 eV. Hence, 4.5 eV of work must be done to break a $H_2$ molecule into two H atoms

$$H_2 + 4.5 \text{ eV} \rightarrow H + H$$

   By comparison, the binding energy of the hydrogen atom is 13.6 eV

$$H + 13.6 \text{ eV} \rightarrow p^+ + e^-$$

   This is an example of the general rule that it is easier to break up a molecule than to break up an atom.

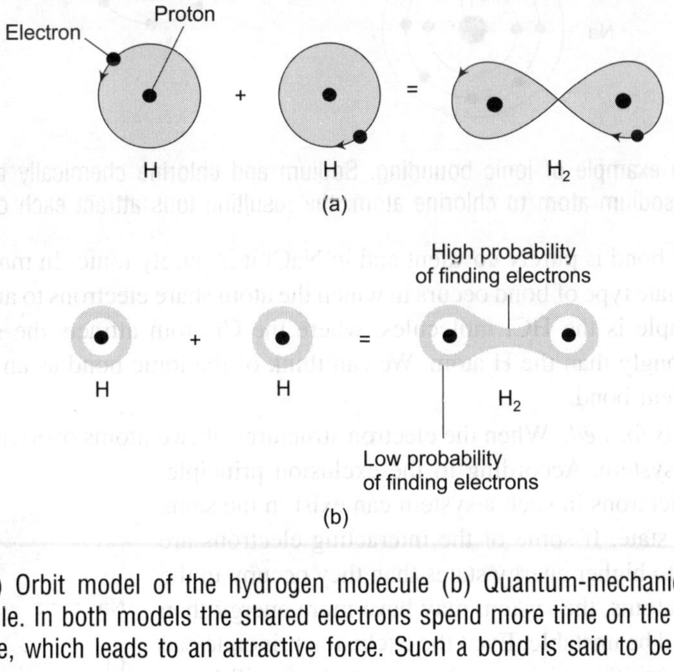


**Figure 4.33** (a) Orbit model of the hydrogen molecule (b) Quantum-mechanical model of the hydrogen molecule. In both models the shared electrons spend more time on the average between the nuclie, which leads to an attractive force. Such a bond is said to be covalent

2. *An ionic bond is formed.* One or more electrons from one atom may transfer to the other and the resulting positive and negative ions attract each other. An example as rock salt NaCl, where the bond exists between $Na^+$ and $Cl^-$ ions and not between

Na and Cl atoms (Fig. 4.34). Ionic bonds usually do not result in the formation of molecules. The crystals of the rock salt are aggregates of sodium and chlorine ions which, although always arranged in a certain definite structure (Fig. 4.35), do not pair off into molecules consisting of one $Na^+$ ion and one $Cl^-$ ion. Rock salt crystals may have any size and shape. There are always equal number of $Na^+$ and $Cl^-$ ions in rock salt, so that the formula NaCl correctly represents its composition. Molten NaCl also consists of $Na^+$ and $Cl^-$ ions. However, these ions form molecules rather than crystal only in the gaseous state.

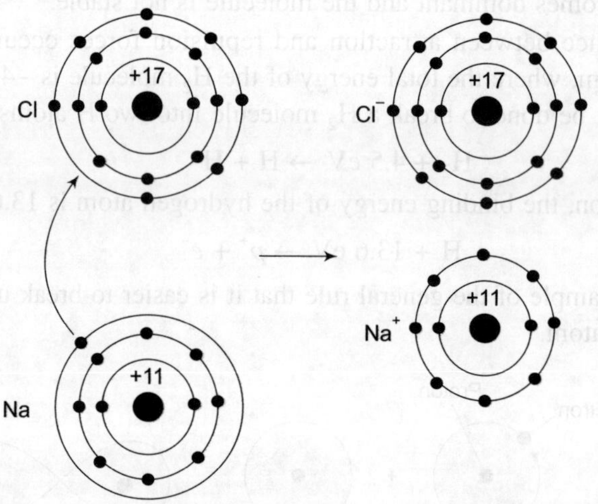


**Figure 4.34** An example of ionic bounding. Sodium and chlorine chemically by the transfer of electrons from sodium atom to chlorine atom; the resulting ions attract each other electrically.

In $H_2$ the bond is purely covalent and in NaCl it is purely ionic. In many molecules an intermediate type of bond occurs in which the atom share electrons to an unequal extent. An example is the HCl molecules, where the Cl atom attracts the shared electrons more strongly than the H atom. We can think of the ionic bond as an extreme case of the covalent bond.

3. *No bond is formed.* When the electron structures of two atoms overlap, they constitute a single system. According to the exclusion principle, no two electrons in such a system can exist in the same quantum state. If some of the interacting electrons are forced into higher energy states than they occupy in the separate atoms, the system may have more energy than before and be unstable. Even the exclusion principle can be obeyed with no increase in energy, there will be an electric repulsive force between the various electrons. This is a much less significant factor than the exclusion principle in influencing bond formation, however.

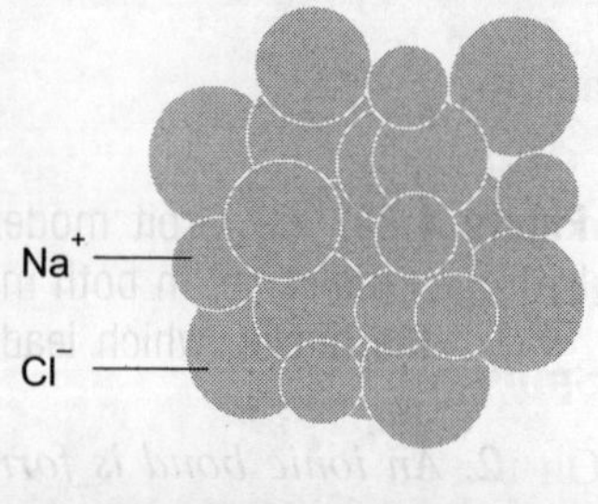


**Figure 4.35** Scale model of an NaCl crystal

### 4.22.1 The $H_2$ Molecule

***The spins of the electrons must be antiparallel.***

The $H_2$ molecule has two electrons instead of a single electron of $H_2^+$. According to the exclusion principle, both electrons can share the same orbital (that is, be described by the same wave function, $\psi_{nl\ ml}$) provided their spins are antiparallel.

With two electrons to contribute to the bond, $H_2$ ought to be more stable than $H_2^+$—at first glance, twice as stable, with a bond energy of 5.3 eV compared with 2.65 eV for $H_2^+$. However, the $H_2$ orbitals are not quite the same as those of $H_2^+$ because of the electric repulsion between the two electrons in $H_2$, a factor absent in the case of $H_2^+$. This repulsion weakens the bond in $H_2$, so that the actual energy is 4.5 eV instead of 5.3 eV. For the same reason, the bond length in $H_2$ is 0.074 nm, which is somewhat larger than the use of unmodified $H_2^+$ wave functions would indicate. The general conclusion in the case of $H_2^+$ that the symmetric wave function $\psi_S$ leads to a bound state and the antisymmetric wave function $\psi_A$ to an unbound one remains valid for $H_2$.

The exclusion principle was formulated in terms of the symmetry and antisymmetry of wave functions, and it was concluded that systems of electrons are always described by antisymmetric wave functions (that is, by wave functions that reverse sign upon the exchange of any pair of electrons). However, bound state in $H_2$ corresponds to both electrons being described by a symmetrical wave function $\psi_S$, which seems to contradict the above conclusion.

A closer look shows that there is really no contradiction. The *complete* wave function $\Psi$ (1, 2) of a system of two electrons is the product of a spatial wave function $\psi$ (1, 2) which describes the coordinates of the electrons and a spin function $s(1, 2)$ which describes the orientations of their spins. The exclusion principle requires that the complete wave function

$$\Psi(1, 2) = \psi(1, 2)\, s(1, 2)$$

be antisymmetric to an exchange of both coordinates and spin, not $\psi(1, 2)$ by itself. An antisymmetric complete wave function $\Psi_A$ can result from the combination of a symmetric coordinate wave function $\Psi_S$ and an antisymmetric spin function $s_A$ or from the combination of an antisymmetric coordinate wave function $\Psi_A$ and a symmetric spin function $s_S$. That is, only

$$\Psi(1, 2) = \psi_S S_A \quad \text{and} \quad \Psi(1, 2) = \psi_A s_S$$

are acceptable.

If the spins of the two electrons are parallel, their spin function is symmetric since it does not change sign when the electrons are exchanged. Hence, the coordinate wave function $\psi$ for two electrons whose spins are parallel must be antisymmetric

**Spins parallel** $\qquad \Psi(1, 2) = \psi_A s_S$

On the other hand, if the spins of the two electrons are antiparallel, their spin function is antisymmetric since it reverses sign when the electrons are exchanged. Hence, the coordinate wave function $\psi$ for two electrons whose spins are antiparallel must be symmetric:

**Spins antiparallel** $\Psi(1, 2) = \psi_S s_A$

Schrödinger's equation for the $H_2$ molecule has no exact solution. In fact, only for $H_2^+$ is an exact solution possible and all other molecular systems must be treated approximately. The results of a detailed analysis of the $H_2$ molecule are shown in Fig. 4.36 for the case when the electrons have their spin parallel and the case when their spins are antiparallel. The difference between the two curves is due to the exclusion principle, which leads to a dominating repulsion when the spins are parallel.

## 4.23 COMPLEX MOLECULES

***Their geometry depends on the wave function of the outer electrons of their atoms***

Covalent bonding in molecules other than $H_2$, diatomic as well as polyatomic, is usually a more complicated story. It would be yet more complicated but for the fact that any alteration in the electronic structure of an atom due to the proximity of another atom is confined to its outermost, or **valence**, electron shell. There are two reasons for this.

1. The inner electrons are much more tightly bound and hence less responsive to external influences, partly because they are closer to their parent nucleus and partly because they are shielded from the nuclear charge by fewer intervening electrons.
2. The repulsive interatomic force in a molecule becomes predominant while the inner shells of its atoms are still relatively far apart.

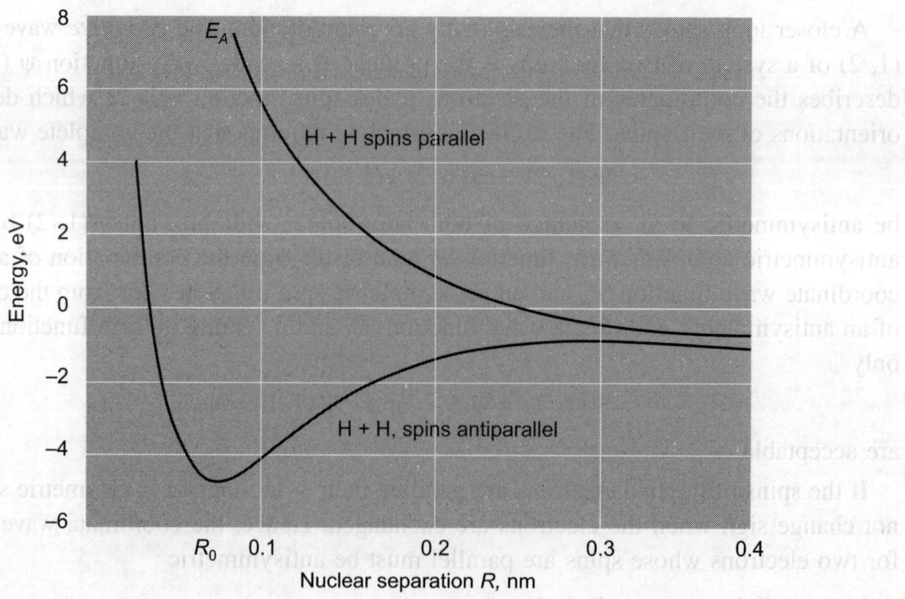


**Figure 4.36** The variation of the energy of the system h + H with their distance apart when the electron spains are parallel and antiparallel

The idea that only the valence electrons involved in chemical bonding supported by X-ray spectra that arise from transitions to inner-shell electron states. These spectra are virtually independent of how the atoms are combined in molecules or solids.

We have seen that H atoms can be combined to form an $H_2$ molecule; and, indeed, hydrogen molecules in nature always consist of two H atoms. The exclusion principle is what prevents molecules such as $He_2$ and $H_3$ from existing, while permitting such other molecules as $H_2O$ to be stable.

Every He atom in its ground state have a $1s$ electron of each spin. If it is to join with another He atom by exchanging electrons, each atom will have two electrons with the same spin for part of the time. That is, one atom will have both electron spins up (↑↑) and the other will have both spins down (↓↓). The exclusion principle, of course, prohibits two $1s$ electrons in an atom from having the same spins, which is manifested in a repulsion between He atoms. Hence, the $He_2$ molecules cannot exist.

A similar argument holds in the case of $H_3$. An $H_2$ molecule contains two $1s$ electrons whose spins are antiparallel (↑↓). Should another H atom approach whose electron spin is, say, up, the resulting molecules would have two spins parallel (↑↑↓), and this is impossible if all three electrons are to be in $1s$ states. Hence, the existing $H_2$ molecule repels the additional H atom. The exclusion principle argument does not apply if one of the three electrons in $H_3$ is in an excited state. All such states are of higher energy than the $1s$ state, however, the resulting configuration therefore has more energy than $H_2$ + H so will decay rapidly to $H_2$ + H.

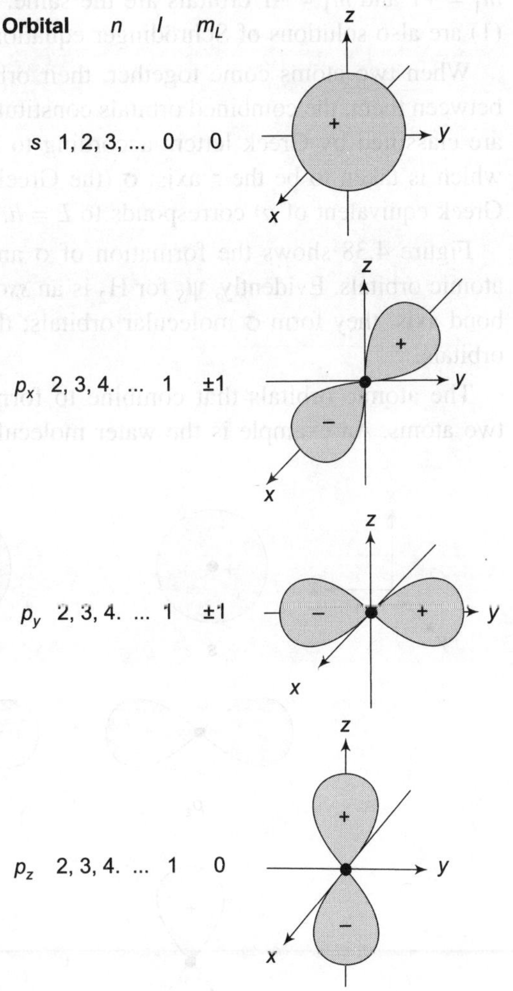


**Figure 4.37** Boundary surface diagrams for *s* and *p* atomic orbitals. Each orbital can "contain" two electrons. There is a high probability of finding an electron described by one of these orbitals in the shaded regions. The sign of the wave function in each lobe is indicated

### 4.23.1 Molecular Bonds

In Fig. 4.37 the *s* and $p_z$ orbitals are the same as the hydrogen-atom wave functions for *s* and *p* ($m_1 = 0$) states. The $p_x$ and $p_y$ orbitals are linear combinations of the *p* ($m_1 = +1$) and *p* ($m_1 = -1$) orbitals, where

$$\psi_{p_x} = \frac{1}{\sqrt{2}}(\psi_{+1} + \psi_{-1})$$

$$\psi_{p_y} = \frac{1}{\sqrt{2}}(\psi_{+1} + \psi_{-1}) \quad ...(1)$$

The $1/\sqrt{2}$ factors are needed to normalize the wave functions. Because the energies of the $m_1 = +1$ and $m_1 = -1$ orbitals are the same, the superposition of the wave functions in Eq. (1) are also solutions of Schrödinger equation.

When two atoms come together, their orbitals overlap. If the result is an increased $|\psi|^2$ between them, the combined orbitals constitute a bonding molecular orbital. Molecular bonds are classified by Greek letters according to their angular momenta $L$ about the bond axis, which is taken to be the $z$ axis: $\sigma$ (the Greek equivalent of $s$) corresponds to $L = 0$, $\pi$ (the Greek equivalent of $p$) corresponds to $L = h$, and so on in alphabetic order.

Figure 4.38 shows the formation of $\sigma$ and $\pi$ bonding molecular orbitals from $s$ and $p$ atomic orbitals. Evidently, $\psi_S$ for $H_2$ is an $ss\sigma$ bond. Since the lobes of $p_z$ orbitals are on the bond axis, they form $\sigma$ molecular orbitals; the $p_x$ and $p_y$ orbitals usually form $\pi$ molecular orbitals.

The atomic orbitals that combine to form a molecular orbital may be different in the two atoms. An example is the water molecule, $H_2O$. Although one $2p$ orbital in $O$ is fully

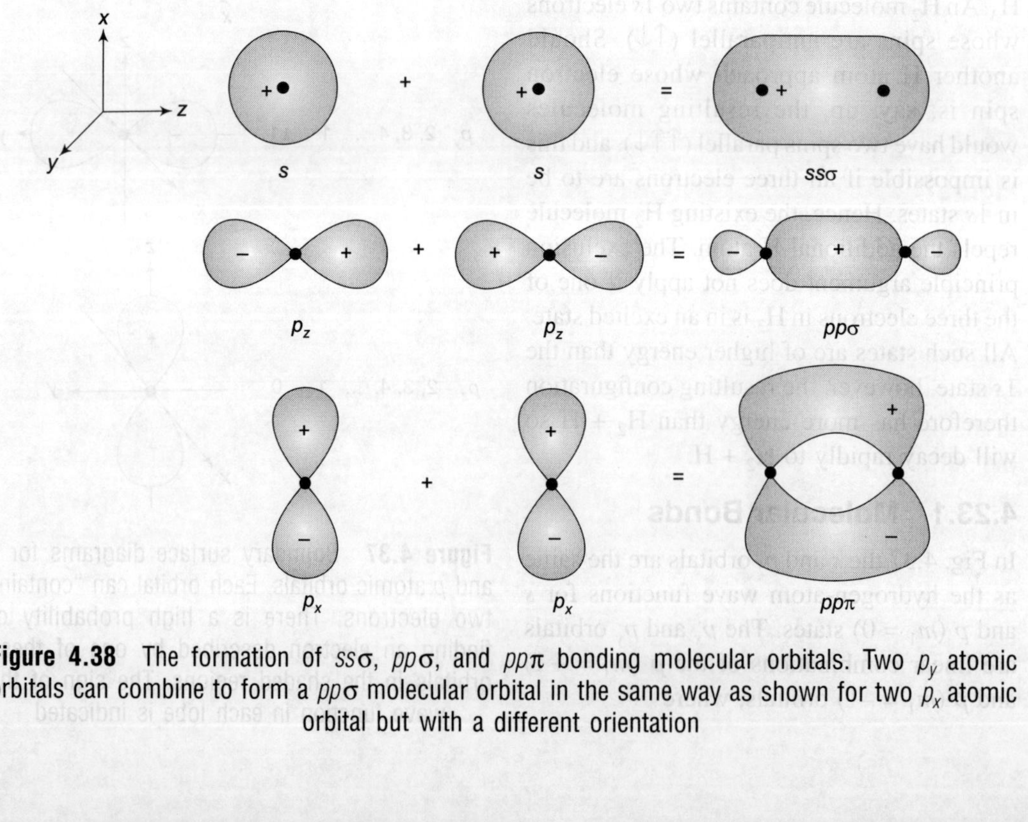


**Figure 4.38** The formation of $ss\sigma$, $pp\sigma$, and $pp\pi$ bonding molecular orbitals. Two $p_y$ atomic orbitals can combine to form a $pp\sigma$ molecular orbital in the same way as shown for two $p_x$ atomic orbital but with a different orientation

occupied by two electrons, the other two $2p$ orbitals are only singly occupied and so can join with the $1s$ orbitals of two H atoms to form $sp\sigma$ bonding orbitals (Fig. 4.39). The mutual repulsion between the H nuclei (which are protons) widens the angles between the bond axes from 90° to the observed 104.5°.

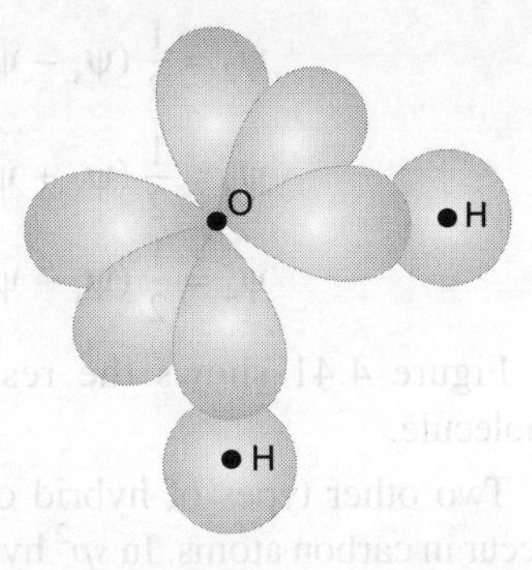


**Figure 4.39** Formation of an $H_2O$ molecule. Overlaps represent $sp\sigma$ covalent bonds. The angle between the bonds is 105.5°

### 4.23.2 Hybrid Orbitals

The straightforward way in which the shape of the $H_2O$ molecule is explained fails in the case of methane, $CH_4$. A carbon atom has two electrons in its $2s$ orbital and one electron in each of two $2p$ orbitals. Thus, we would expect the hydride of carbon to be $CH_2$, with two $sp\sigma$ bonding orbitals and a bond angle of a little over 90°. The $2s$ electrons should not participate in the bonding at all. Yet $CH_4$ exists and is perfectly symmetrical in structure with tetrahedral molecules whose C—H bonds are exactly equivalent to one another.

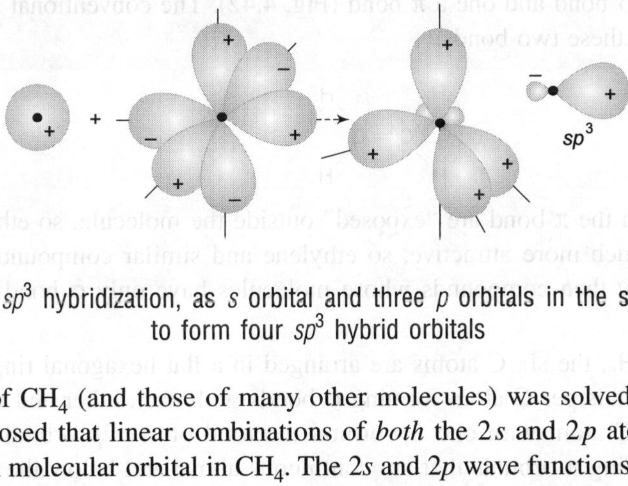


**Figure 4.40** In $sp^3$ hybridization, as $s$ orbital and three $p$ orbitals in the same atom combine to form four $sp^3$ hybrid orbitals

The problem of $CH_4$ (and those of many other molecules) was solved by Linus Pauling in 1928. He proposed that linear combinations of *both* the $2s$ and $2p$ atomic orbitals of $C$ contribute to each molecular orbital in $CH_4$. The $2s$ and $2p$ wave functions are both solutions of the same Schrödinger equation if the corresponding energies are the same, which is not true in the isolated C atom. However, in an actual $CH_4$ molecule the electric field experienced by the outer C electron is effected by the nearby H nuclei, and the energy difference between $2s$ and $2p$ state then can disappear. **Hybrid orbitals**, that consist of mixtures of $s$ and $p$ orbitals occur when the bonding energies they produced are greater than those which pure orbitals would produce. In $CH_4$ the four hybrid orbitals are mixtures of and $2s$ and three $2p$ orbitals, and accordingly are called $sp_3$ hybrids (Fig. 4.40). The wave functions of these hybrid orbitals are

$$\psi_1 = \frac{1}{2}(\psi_s + \psi_{p_x} + \psi_{p_y} + \psi_{p_z})$$

$$\psi_2 = \frac{1}{2}(\psi_s - \psi_{p_x} - \psi_{p_y} + \psi_{p_z})$$

$$\psi_3 = \frac{1}{2}(\psi_s + \psi_{p_x} - \psi_{p_y} - \psi_{p_z})$$

$$\psi_4 = \frac{1}{2}(\psi_s - \psi_{p_x} + \psi_{p_y} - \psi_{p_z})$$

Figure 4.41 shows the resulting structure of the $CH_4$ molecule.

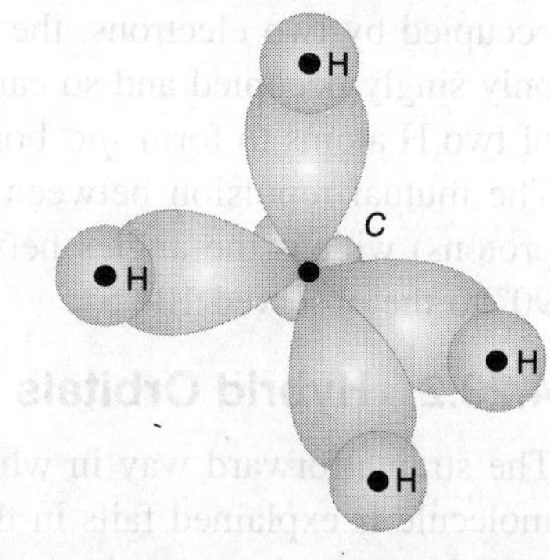


**Figure 4.41** The bonds in the $CH_4$ (methane) molecule involve $sp^3$ hybrid orbitals

Two other types of hybrid orbitals in addition to $sp^3$ can occur in carbon atoms. In $sp^2$ hybridization, one outer electron is in a pure $p$ orbital and the other three in hybrid orbitals that are $\frac{1}{3}s$ and $\frac{2}{3}p$ in character. In $sp$ hybridization, two outer electrons are in pure $p$ orbitals and the other two are in hybrid orbitals that are $\frac{1}{2}s$ and $\frac{1}{2}p$ in character.

Ethylene, $C_2H_4$, is an example of $sp^2$ hybridization in which the two C atoms are joined by two bonds, one a σ bond and one a π bond (Fig. 4.42). The conventional structural formula of ethylene shows these two bonds:

Ethylene

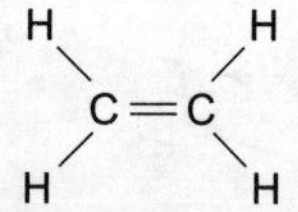

The electrons in the π bond are "exposed" outside the molecule, so ethylene and similar compounds are much more attractive, so ethylene and similar compounds are much more reactive chemically then compounds whose molecules have only σ bonds between their C atoms.

In benzene, $C_6H_6$, the six C atoms are arranged in a flat hexagonal ring, as in Fig. 4.43, with three $sp^2$ orbitals per C atom forming σ bonds with each other and with the H atoms. This leaves each C atom with one $2p$ orbital the total of six $2p$ orbitals in the molecule combine into bonding π orbitals that are continuous above and below the plane of the ring. The six electrons involved belong to the molecule as a whole and not bonding π orbitals that are continuous above and below the plane of the ring. The six electrons involved belong to the molecule as a whole and not to any particular pair of atoms, these electron are **delocalized**. An appropriate structural formula for benzene is therefore

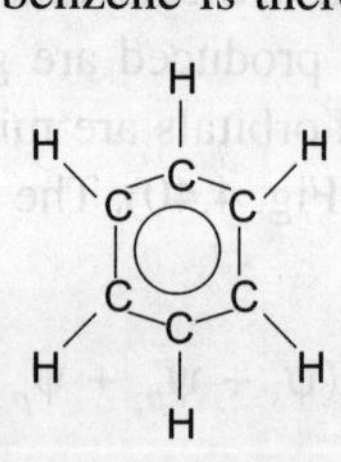

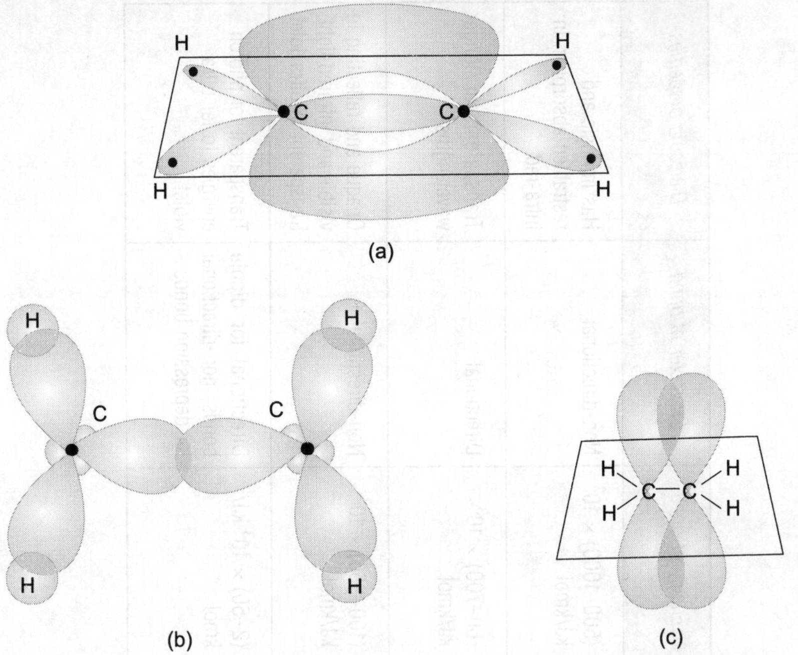


**Figure 4.42** (a) The ethylene ($C_2H_2$) molecule. All the atoms lie in a plane perpendicular to the plane of the paper. (b) Top view, showing the $sp^2$ hybrid orbitals that form σ bonds between the C atoms and between each C atom and two H atoms. (c) Side view, showing the pure $p_x$ orbitals that form a π bond between the C atoms

Table 4.13 given on page 326, compares various types of bonding in solids.

## 4.24 ATOMIC SIZE

The most stable state of an atom is that in which all the electrons occupy orbits of the lowest permitted energy, i.e., they are in the normal state or the ground state. This arrangement is slightly distributed if the atom is not electrically neutral. For example, an atom can lose one or more electrons and becomes positively charged. The electrons lying in the outermost orbit (called the valence electrons) of the atom can be removed most easily. The energy required to remove one valence electron from a neutral atom is called its first ionization potential, the energy required to remove the second valence electron is called the second ionization potential, and so on. The resulting positive ions have higher energies (they are said to be in the excited states) than the neutral atom and acquire a valence of +1, +2, etc. On the other hand , if one or more electrons are added to an atom, this becomes negatively charged and the resulting ions acquire a valence of –1, –2, etc. Most positive ions are stable, unless brought in the vicinity of other electrons with which they neutralize their negative charge, while most negative ions are unstable, the negatively charged ions repel the negatively charged electrons. However, there are some exceptions to this, notably the halogens, oxygen, and sulphur. These elements

**Table 4.13** Properties of Various Types of Bonding in Solids

| *Bond type* | *Bond formation condition* | *Bonding force* | *Electrical character* | *Energy of bond* | *Character of bond* | *Dielectric properties* |
|---|---|---|---|---|---|---|
| Ionic | One atom having smaller number of valence electorns | Electrostatic attraction of appositely charged ions | Insulator at low temperatures | $(500–1000) \times 10^3$ kJ/kmol | Non-directional | Has intrinsic and restrahlen absorption in infra-red. |
| Covalent | Atomic orbitals of two atoms overlap | Mutual sharing of valence electrons between atoms | Insulator or semicinductor | $(60–700) \times 10^3$ kJ/kmol | Directional | Transparent for longer wavelengths |
| Metallic | For elements having small number of valence electrons | Attraction between the ion cores and condition electron gas | Good conductor | $(100–350) \times 10^3$ kJ/kmol | Non-directional | Opaque and reflecting in visible and infrared light, transparent in ultraviolet |
| van der Waals | For molecules forming dipoles | Electrostatic force due to oscillating dipoles or permanent dipoles between molecules | Insulator | $(2–50) \times 10^3$ kJ/kmol | Directional for dipole bonds, non-directional for depression bonds | Transparent for photon energies to far ultra violet. |

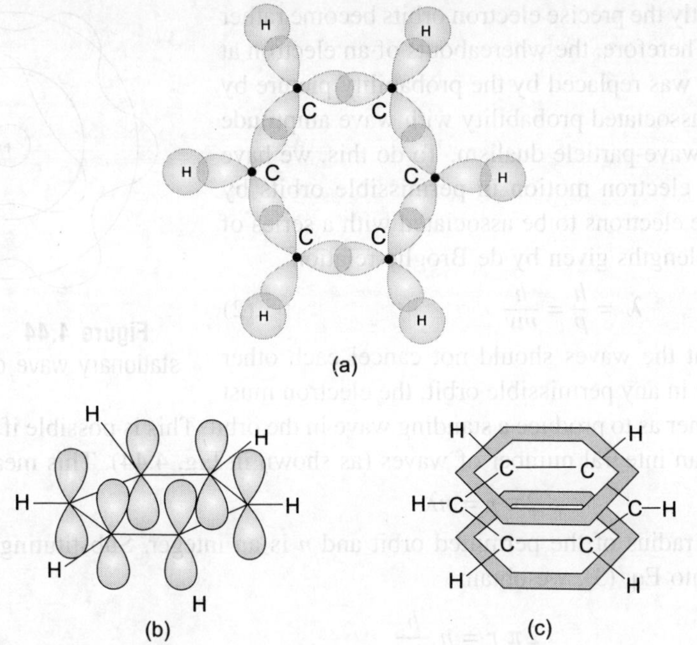


**Figure 4.43** The benzene molecule. (a) the overlaps between the $sp^2$ hybrids orbitals in the C atoms with each other and with the *s* orbitals of the H atoms lead to σ bonds. (b) Each C atom has a pure $p_x$ orbital occupied by one electron. (c) The bonding π molecular orbitals formed by the six $p_x$ atomic orbitals constitute a continuous electron probability distribution around the molecule that contain six delocalized electrons

actually lose energy on becoming negatively charged. This energy difference between the state of a neutral atom (ground state) and the state of a negative ion is called *electron affinity* and is used as a measure of the *electronegativity* of an atom. Similarly, the ease with which an atom gives up its valence electrons determines how much electropositive it is. The loss or gain of one or more electron does not greatly affect the spherical shape of an atom. Actually, if an atom loses electrons, the positively charged nucleus tends to draw in the remaining electrons more tightly. Conversely, if electrons are added to an atom, the repulsive forces tend to push the outer electrons farther away. This, however, generally affects the size and not the shape of an atom.

## 4.24.1 Wave Mechanical Concept of the Atom

From Hesenberg uncertainty principle

$$\Delta E\ \Delta t \geq \frac{h}{2\pi}\ (\equiv \hbar) \qquad ...(1)$$

where $\Delta E$ and $\Delta t$ are the uncertainties measuring energy and time respectively. If the energy is known exactly, then $\Delta t \to \infty$, i.e., error in measuring time will be extremely large. It follows that the motion of electron around the orbit (in an atom) will not be observable in terms of time and consequently the precise electron orbits become rather meaningless. Therefore, the whereabouts of an electron at a given instant was replaced by the probability picture by M. Born who associated probability with wave amptitude to resolve the wave-particle dualism. To do this, we have to replace the electron motion in permissible orbits by considering the electrons to be associated with a series of waves of wavelengths given by de Broglie relation

$$\lambda = \frac{h}{p} = \frac{h}{mv} \qquad ...(2)$$

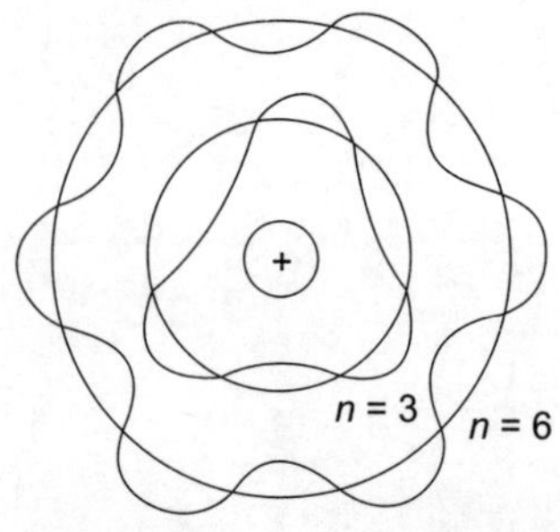


**Figure 4.44** De Broglie's stationary wave quantized orbits

In order that the waves should not cancel each other by interference in any permissible orbit, the electron must move in a manner as to produce a standing wave in the orbit. This is possible if the permissible orbits contain an integral number of waves (as shown in Fig. 4.44). This means that

$$2\pi\, r = n\lambda \qquad ...(3)$$

where $r$ is the radius of the permitted orbit and $n$ is an integer. Substituting the value of $\lambda$ from Eq. (2) into Eq. (3), we obtain

$$2\pi\, r = n\,\frac{h}{mv}$$

or

$$mv\,r = n\,\frac{h}{mv} \qquad ...(4)$$

This is just the Bohr condition for a stationary orbit which appeared so arbitrary from the old quantum theory, but now comes quite logically from the wave mechanical model.

According to the wave mechanical concept, the electron possesses a wave-like nature and can no longer be pictured as particle moving in an orbit of definite radius. Instead, we can speak of a probability distribution of electron density. Thus, in the case of hydrogen atom which has a single electron in its ground state, the circular orbit of the Bohr theory is replaced by a spherical probability distribution in which there is a finite probability of finding the electron at large distances from the nucleus, the most probable distance proves to be exactly equal to the radius of the Bohr orbit. Thus, the orbits representing excited states of the hydrogen atom are spheres of corresponding larger radii. The experimental proof that atoms are nearly spherical is given by the results of X-ray diffraction investigation of crystal structures. Because X-rays are scattered by the electrons in each atoms, it is possible to synthesize a picture of distribution of electron density directly from the experimentally observed intensities, after their relative phases have been determined. One such synthesis is shown Fig. 4.45, which is an electron density map of cubanite, $Cu_2FeS_3$, projected on the $xy$ plane. The contours of the figure join the points having same electron density. Such structural studies are useful in determining the effective radii of the spherical atom.

Actually, the size of an atom depends on many factors, the coordination number, radius ratio and the types of forces existing between the atoms, where these forces may differ in different compounds or even different structural arrangements of the same compound. It is therefore definite that the absolute value of atomic size are only approximately known.

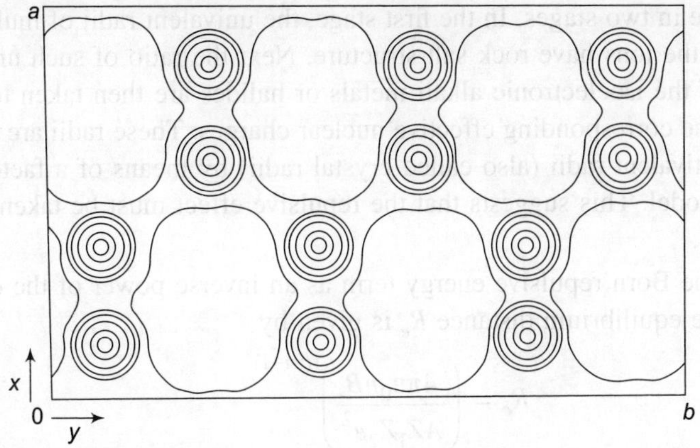


**Figure 4.45** The electron density map of cubanite

However, the knowledge of atomic size in a crystal is helpful in predicting the structure of other crystals or to ascertain whether an atom will fit into the voids of parent structure. To understand further, let us consider the following cases in somewhat details.

## 4.25 IONIC RADII

In an ionic crystal, the attractive Coulomb force is counterbalanced by a repulsive force due to interaction of the outer electron clouds of the ions, resulting in an equilibrium internuclear distance. As we have seen earlier, these outer electrons are responsible for the size of the ions, then it may be assumed that these are attracted towards the nucleus by the charge Ze except for the screening of this charge by the inner electrons. The effective charge may be written as $(Z - S)e$, where $S$ is a screening constant. The radius of an ion may then be written as

$$r_{\text{ion}} = \frac{C_n}{Z - S} = \frac{C_n}{Z_{\text{eff}}} \quad ...(1)$$

where $C_n$ is a constant determined by the quantum number and $Z_{\text{eff}}$ is the effective nuclear charge. Pauling obtained a set of values for $S$ by theoretical methods combined with other experimental data obtained from molar fraction and X-ray term values. $C_n$ takes the same value for all members of an isoelectronic series (ions having identical electronic structures). For ions with neon configuration, $S = 4.52$ per electron charge, and for argon, $S = 11.25$ per electron charge. Thus, for KCl, we have

$$r_{\text{K+}} = \frac{C_n}{19 - 11.25} \quad \text{and} \quad r_{\text{Cl}^-} = \frac{C_n}{17 - 11.25}$$

or

$$\frac{r_{\text{K+}}}{r_{\text{Cl}^-}} = \frac{17 - 11.25}{19 - 11.25} = \frac{5.75}{7.75} = 0.74 \quad ...(2)$$

From experiment $r_{\text{K+}} + r_{\text{Cl}^-} = 3.14$ Å. Solution of these equations yields $r_{\text{k+}} = 1.33$ Å, and $r_{\text{Cl}} = 1.81$ Å. Similar calculations can be made on different alkali halides. The radii obtained by this process are known as univalent radii. Pauling obtained the radii of multivalent ions having rock salt structure in two stages. In the first stage, the univalent radii of multivalent ions was determined as if the ions have rock salt structure. Next the ratio of such univalent radii and the ionic radii of the isoelectronic alkali metals or halides are then taken to be equal to the inverse ratio of the corresponding effective nuclear charges. These radii are then transformed into the real multivalent radii (also called crystal radii) by means of a factor which follows from the Born model. This suggests that the repulsive effect must be taken into account for such calculations.

Considering the Born repulsive energy term as an inverse power of the distance between the ions, $BR^{-n}$ the equilibrium distance $R_e$ is given by

$$R_e = \left(\frac{4\pi\varepsilon_0 nB}{AZ_1Z_2e^2}\right)^{1/(n-1)} \quad \text{...(3)}$$

This gives the interionic distance in the real crystal formed by multivalent ions ($Z_1 \not\equiv 1$, $Z_2 \not\equiv 1$). This distance can be represented in terms of the equilibrium interionic distance in the crystal formed by univalent ions ($Z_1 = Z_2 = 1$) having the same Born repulsive energy, by the equation

$$R_{\text{multivalent}} = R_{\text{univalent}} \times Z^{-2(n-1)} \quad \text{...(4)}$$

As an example of the Pauling procedure, the univalent radius of the $O^{2-}$ ion is equal to $(9 - 4.5)/(8 - 4.5)$ times the standard crystal radius of $F^-$ ion. To obtain the radius of $O^{2-}$ appropriate to an alkaline-earth oxide with the NaCl structure, strictly the radius of $O^{2-}$ appropriate to MgO crystals, the univalent radius is divided by $2^{1/3}$, since for MgO $Z$ is equal to 2 and $n$ is chosen by Pauling equal to 7. This yields the value 1.40 Å for the standard crystal radius of $O^{2-}$. A list of Pauling's crystal radii is given in Table 4.14.

**Table 4.14** Crystal Radii of Pauling (in Angstroms)

| −4 | −3 | −2 | −1 | 0 | +1 | +2 | +3 | +4 | +5 | +6 | +7 |
|---|---|---|---|---|---|---|---|---|---|---|---|
| | | | H | He | Li | Be | B | C | N | O | F |
| | | | 2.08 | 0.93 | 0.60 | 0.31 | 0.20 | 0.15 | 0.11 | 0.09 | 0.07 |
| C | N | O | F | Ne | Na | Mg | Al | Si | P | S | Cl |
| 2.60 | 1.71 | 1.40 | 1.36 | 1.12 | 0.95 | 0.65 | 0.50 | 0.41 | 0.34 | 0.29 | 0.26 |
| Si | P | S | Cl | Ar | K | Ca | Sc | Ti | V | Co | Mn |
| 2.71 | 2.12 | 1.84 | 1.81 | 1.54 | 1.33 | 0.99 | 0.81 | 0.68 | 0.59 | 0.52 | 0.46 |
| | | | | | Cu | Zn | Ga | Ge | As | Se | Br |
| | | | | | 0.96 | 0.74 | 0.62 | 0.53 | 0.47 | 0.42 | 0.39 |
| Ge | As | Se | Br | Kr | Rb | Sr | Y | Zr | Nb | Mo | |
| 2.72 | 2.22 | 1.98 | 1.95 | 1.69 | 1.48 | 1.13 | 0.93 | 0.80 | 0.70 | 0.62 | |
| | | | | | Ag | Cd | In | Sn | Sb | Te | I |
| | | | | | 1.26 | 0.97 | 0.81 | 0.71 | 0.62 | 0.56 | 0.50 |

*Contd...*

*Table 4.13 Contd...*

| Sn | Sb | Te | I | Xe | Cs | Ba | La | Ce | |
|---|---|---|---|---|---|---|---|---|---|
| 2.94 | 2.45 | 2.21 | 2.16 | 1.90 | 1.69 | 1.35 | 1.15 | 1.01 | |
| | | | | | Au | Hg | Tl | Pb | Bi |
| | | | | | 1.37 | 1.10 | 0.95 | 0.84 | 0.74 |

## 4.26 EMPIRICAL IONIC RADII

Experimentally determined maps (Fig. 4.46) for NaCl and LiF shows that the electron density rises towards each nucleus but falls approximately to zero in between the ions. A plot of this function in Fig. 4.47 shows this minimum very clearly for LiF. A few compounds like LiF, NaCl, KCl, $CaF_2$, MgO, etc., have been examined so far (Table 4.15) by this technique. However, their radii do not reproduce the internuclear separations for all the alkali halides to the desired accuracy. It is believed that there is some mutual interpenetration of $Li^+$ and $F^-$ ions, the radii necessary to reproduce the crystal distances for other alkali halides are actually larger, e.g., $Li^+$ 0.94, $F^-$ 1.16 Å. Taking these values together with a marginally different one for $Na^+$, Gourary and Adrian (1960) obtained the correct ionic radii for some alkali metal and halide ions (Table 4.16). They reproduced the intermolecular separation for all alkali halides (other than LiF) within about one per cent. The discrepancy in LiF is explained on the basis

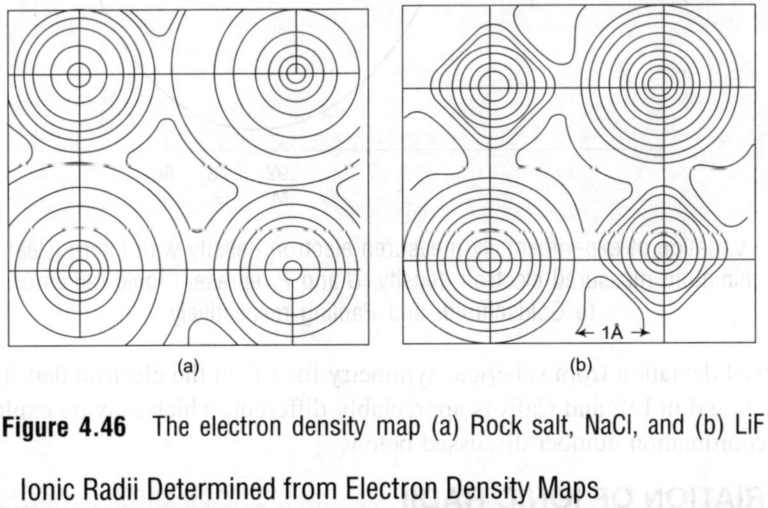


**Figure 4.46** The electron density map (a) Rock salt, NaCl, and (b) LiF

**Table 4.15** Ionic Radii Determined from Electron Density Maps

| | *Cation (Å)* | *Anion (Å)* |
|---|---|---|
| LiF | 0.92 | 1.09 |
| NaCl | 1.18 | 1.64 |
| KCl | 1.45 | 1.70 |
| $CaF_2$ | 1.26 | 1.10 |
| MgO | 1.02 | 1.09 |

**Table 4.16** Corrected Ionic Radii (in Å) from Electron Density Maps

| | *'Corrected' radius* | *Waddington radius* |
|---|---|---|
| $Li^+$ | 0.94 | 0.74 |
| $Na^+$ | 1.17 | 1.01 |
| $K^+$ | 1.49 | 1.32 |
| $Rb^+$ | 1.63 | 1.46 |
| $Cs^+$ | 1.86 | 1.72 |
| $F^-$ | 1.16 | 1.32 |
| $Cl^-$ | 1.64 | 1.82 |
| $Br^-$ | 1.80 | 1.98 |
| $I^-$ | 2.05 | 2.24 |

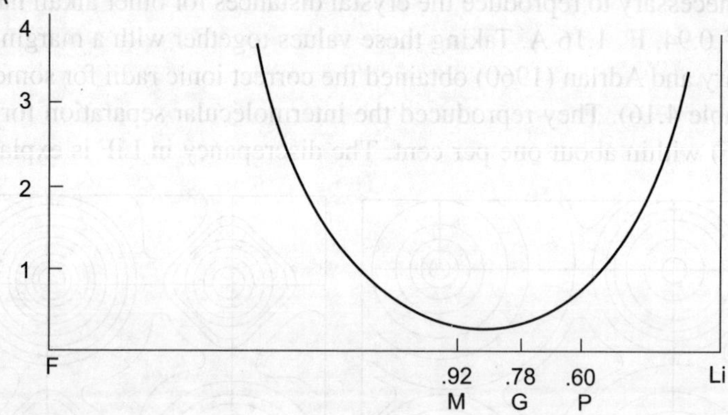


**Figure 4.47** Variation of experimentally measured electron density with internuclear positions for LiF. *M* is the minimum measured electron density. *G* and *P* represent ionic radius of $Li^+$ according to Goldschmidt and Pauling respectively

of the observed deviation from spherical symmetry for $Li^+$ in the electron density map. Also the value $F^-$ found in LiF and $CaF_2$ is appreciably different, which may be explained on the basis of its coordination number discussed below.

## 4.27 VARIATION OF IONIC RADII

The variation of ionic radius with coordination number may be represented as follows:

| Coordination number | 8 | 6 | 4 |
|---|---|---|---|
| Radius | 1.03 | 1.00 | 0.95 |

Thus, if the radii corresponding to sodium chloride structure is taken as standard, then the radii in the 8-coordinated cesium chloride structure are about 3% larger and in 4-coordinated zinc blende structure are about 5% smaller. On the basis of this, a simple conversion relation $1.052\ R_{CN} = 4$, $R_{CN} = 6$ and $0.97\ R_{CN} = 8$, may be used for calculation. It is therefore, desirable that the ionic radii should be expressed in a form appropriate to some particular coordination, and that of sodium chloride is conventionally chosen as standard. The crystal radii should be appropriately corrected on the basis of coordination before use in actual structure.

Another factor which influences ionic radii in the value of the radius ratio. At this stage, a natural question arises that why different MX-compounds should exist in different structural forms, such as rock salt, cesium chloride or zinc blende, in general, and why CsCl, CsBr and CsI should have a structure different from that of other alkali halides, in particular? In order to arrive at a satisfactory answer to this, we assume that:

1. the ions behave as hard spheres, the radii are fixed and independent of coordination number, and
2. the bonding between the ions is purely ionic and the cohesive energy is expressed in the form

$$U = -\frac{Ae^2N}{4\pi\varepsilon_0 R_e}\left(1 - \frac{\rho}{R_e}\right)$$

Now, the crystal will be as tightly bound as possible if $A$ is large and $R_e$ is small. The largest value of $A$ (for the three structures mentioned above) is that of cesium chloride structure, e.g., $A = 1.763$. Practically in all cases, anions are larger than cations. Accordingly, let us assume

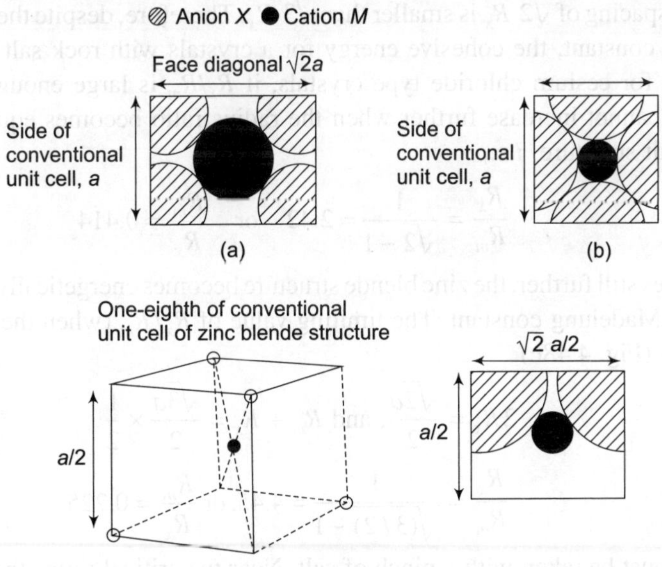


**Figure 4.48** Critical conditions in (a) cesium chloride structure (b) Rocksalt structure, and (c) Zincblende structure

that $R_m < R_x$ and that $R_x$ is fixed. As $R_m/R_x$ decreases from unity, $R_e = R_m + R_x$ decreases until a critical value is reached when all the anions are in contact with the cations and touch each other as shown in Fig. 4.48 (a) for the case of cesium chloride structure. This happens when

$$R_m + R_x = \text{half the body-diagonal of cube}$$

$$= \frac{\sqrt{3}}{2} \times \text{side of the cube}$$

$$= \frac{\sqrt{3}}{2} \times 2\,R_x = \sqrt{3}\,R_x$$

The interionic spacing $R_e = R_m + R_x$ is now fixed at $\sqrt{3}\,R_x$ and as $R_m$ is made smaller the binding energy will not be affected. The critical radius ratio becomes

$$\frac{R_x}{R_m} = \frac{1}{\sqrt{3}-1} = 1.37, \quad \text{or} \quad \frac{R_x}{R_m} = 0.732$$

No further increase in cohesive energy is possible with this structure. For the sodium chloride structure (Fig. 4.48 b) the limiting value of $R_e$ occurs when the ions touch (in this case the rock diagonal of the cube is equal to $4R$), i.e., when

$$R_m + R_x = \text{half the side of the cube}$$

$$= \frac{1}{2} \times \frac{4R_x}{\sqrt{2}} = \sqrt{2}\,R_x$$

An interionic spacing of $\sqrt{2}\,R_x$ is smaller than $\sqrt{3}\,R_x$. Therefore, despite the small difference in the Madelung constant, the cohesive energy for a crystals with rock salt structure can be greater than that for cesium chloride type crystals, if $R_x/R_m$ is large enough. However, the cohesive energy cannot increase further when the radius ratio becomes equal to the critical value for rock salt structure, i.e.,

$$\frac{R_x}{R_m} = \frac{1}{\sqrt{2}-1} = 2.42 \quad \text{or} \quad \frac{R_m}{R_x} = 0.414$$

As $R_m$ decreases still further, the zinc blende structure becomes energetically most favourable despite its small Madelung constant. The limiting value of $R_x/R_m$ (when the anions touch) is not reached until (Fig. 4.48c)

$$2R_x = \frac{\sqrt{2}a}{2}, \text{ and } R_m + R_x = \frac{\sqrt{3}a}{2} \times \frac{1}{2}$$

so that, $$\frac{R_x}{R_m} = \frac{1}{\sqrt{(3/2)}-1} = 4.45, \text{or } \frac{R_m}{R_x} = 0.225$$

These ratios must be taken with a pinch of salt. Near the critical ratios the stability of one structure as compared with another may be strongly influenced by the coordination number and its effect on the radius, which we have ignored for this calculation. Polarization effects may also be important enough to alter the above predictions. Some compounds can even change from one crystal structure to another as temperature and pressure are varied. However, it is found that generally when MX-compounds have ions with nearly equal radii, the cesium chloride structure is favoured; when the ionic radii differ by a factor of two or more the zinc blende structure is favoured: and in the intermediate cases the rock salt structure results.

Table 4.17 gives the actual radius ratio in cesium chloride, sodium chloride and zinc sulphide, the compounds after which the crystals types are named.

**Table 4.17** Actual Radius Ratio in Three Leading MX-compounds (after Pauling)

| *Radius of cation $r_m$ in Å* | *Radius of anion $r_x$ in Å* | *$r_x/r_m$* | *Approx. range expected from calculations* |
|---|---|---|---|
| $Cs^+$ 1.69 | $Cl^-$ 1.81 | 1.1 | 1.00 – 1.37 |
| $Na^+$ 0.95 | $Cl^-$ 1.81 | 1.9 | 1.37 – 2.42 |
| $Zn^{2+}$ 0.74 | $S^{2-}$ 1.84 | 2.5 | 2.42 – 4.45 |

**Example 4.1** Determine the radius of a cation occupying an octahedral position in an $M^+X^-$ ionic solid, when all anions are in contact with the cation and touch each other. The critical radius of anoin is 2.5 Å.

***Solution*** We know that for an octahedral case, the radius ratio is given by the relation

$$\frac{R_m}{R_x} = 0.414$$

Substituting the value of $R_x$, we can find

$$R_m = 2.5 \times 0.414 = 1.035 \text{ Å}$$

## 4.28 COVALENT RADII

The covalent radius in a covalent molecule is only half of the bond length (the distance between two bonded atoms) and is often called the atomic radius in the case of nonmetals. This suggests that covalent radii are simply additive, provided the bonding between atoms is a single electron pair bond. On the other hand, when atoms from multiple electron pair bonds with each other, the bond lengths depend on the number of electron pair bonds formed. Further, in case of hybrid bond formation, they also depend on the type of orbitals used. It should also be remembered that most atoms do not form purely covalent bonds in crystals. From these, it appears that the concept of characteristic covalent radii is subject to a number of qualifications:

(i) That the covalent radius of an atom (unlike its ionic radius) must not be interpreted as implying that the atom is the sphere of that size. The covalent radii are applicable only to the calculation of bond length between atom joined by purely covalent bonds, and tells nothing about the distance between atoms of the same kind when not so bonded. Thus, the single-bond radius of chlorine (i.e., 0.99 Å) is in perfect agreement with the observed Cl—Cl distance of 1.988 Å in the chlorine molecule. In the crystal structure of solid chlorine, however, the interatomic distance between contiguous chlorine atoms of different molecules, which are bound together only by van der Waals forces, is about 3.6 Å. the van der Waals radius of the chlorine atom is about 1.8 Å (nearly double the covalent radius) and consequently the $Cl_2$ molecule is pictured as having the form shown in Fig. 4.49.

(ii) That the bonds length depends upon the bond number. Single bonds are the largest one and the length decreases with the increase of number of bonds. The effect of multiple bonding can be produced fairly well by a simple observation. A double bond is about 0.86 as long as a single bond for first row elements, and triple bond is about 0.78 as long as a single bond. For second row and heavier elements (where multiple bonds are much rarer), the inner core electrons are less compressible and the factors are 0.91 for a double bond and 0.85 for a triple bond. For this reason separate radii are required for single, double and triple bonds, respectively. We must, however, also remember that a bond will not necessarily have an integral bond number (sometimes may have fractional) if the bond involves resonance. These intermediate bond lengths can be obtained by interpolation between the lengths corresponding to integral bond numbers, but with proper caution.

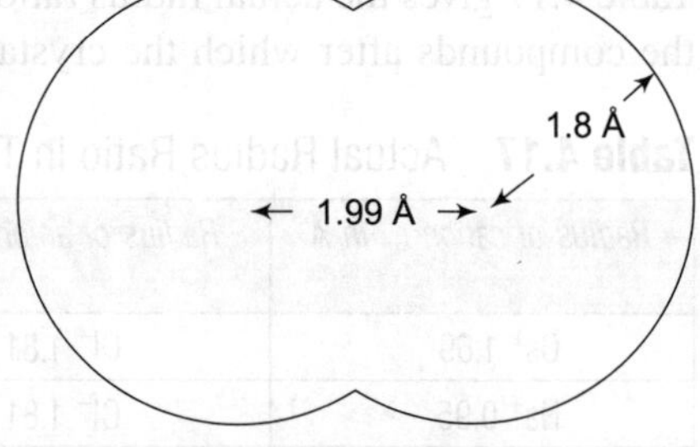


**Figure 4.49** The dimensions of chlorine molecule

(iii) That we must consider resonance not only between covalent bonds of different bond number but also between covalent and ionic bonds because many covalent bonds do, in fact, have considerable ionic character. The effect of this covalent/ionic resonance may be considerable, and in some cases reduces the length of a bond as much as 0.1 Å or more. Attempts have been made to apply necessary correction in terms of difference in electronegativity of the atoms involved. Schomaker and Stevenson observed that the degree of electron transfer (cause of ionic character) in a bond is roughly proportional to the electronegativity difference between the two atoms involved, and hence they proposed an empirical correction to the predicted bond length:

$$r_{AB} = r_A + r_B - 0.09\ (|x_A - x_B|) \qquad ...(1)$$

where $r_A$ and $r_B$ are covalent radii, $x_A$ and $x_B$ are electronegativity values, and the sign of electronegativity difference is disregarded. A somewhat better fit is provided by adjusting for the square of the electronegatively difference:

$$r_{AB} = r_A + r_B - 0.07\ (x_A - x_B)^2 \qquad ...(2)$$

It should be noted that the above correction predicts the bond length corresponding to a single bond only. When observed bond lengths are shorter than the predicted values in covalent molecules, it is usually assumed that some degree of multiple bonding is involved.

In some cases the "corrections" so applied increase the discrepancy between predicted and observed bond lengths. Therefore, the effect of covalent/ionic resonance may be better ignored while calculating bond lengths.

Based on similar argument, Pauling (1960) proposed an empirical formula for determining the percentage ionic character if the bond

$$p = 16\ |r_A - r_B| + 3.5\ |x_A - x_B|^2 \qquad ...(3)$$

The value of electronegativity calculated recently by Husain et al. (1989) along with the values of Pauling are provided in Table 4.18. Some important conclusions drawn from the data are summarized below:

**Table 4.18** The Electronegativities of the Elements

| Z | Element | $\chi_{cal.}$ | $\chi_p$ | Z | Element | $\chi_{cal.}$ | $\chi_p$ |
|---|---|---|---|---|---|---|---|
| 3 | Li | 0.88 | 1.0 | 39 | Y | 1.25 | 1.2 |
| 4 | Be | 1.84 | 1.5 | 40 | Zr | 1.53 | 1.4 |
| 5 | B | 2.02 | 2.0 | 41 | Nb | 1.76 | 1.6 |
| 6 | C | 2.60 | 2.5 | 42 | Mo | 1.88 | 1.9 |
| 7 | N | 3.06 | 3.0 | 43 | Te | 1.98 | 1.9 |
| 8 | O | 3.13 | 3.5 | 44 | Ru | 2.05 | 2.2 |
| 9 | F | 3.60 | 4.0 | 45 | Rh | 2.06 | 2.2 |
| 11 | Na | 0.83 | 0.9 | 46 | Pd | 1.71 | 2.2 |
| 12 | Mg | 1.17 | 1.2 | 47 | Ag | 1.84 | 1.9 |
| 13 | Al | 1.34 | 1.5 | 48 | Cd | 1.66 | 1.7 |
| 14 | Si | 1.69 | 1.6 | 49 | In | 1.45 | 1.7 |
| 15 | P | 2.04 | 2.1 | 50 | Sn | 1.68 | 1.8 |
| 16 | S | 2.17 | 2.5 | 51 | Sb | 1.80 | 2.0 |
| 17 | Cl | 2.54 | 3.0 | 52 | Te | 1.93 | 2.4 |
| 19 | K | 0.66 | 0.8 | 53 | I | 2.14 | 2.6 |
| 20 | Ca | 0.92 | 1.0 | 55 | Cs | 0.68 | 0.7 |
| 21 | Sc | 1.27 | 1.3 | 56 | Ba | 0.96 | 0.9 |
| 22 | Ti | 1.49 | 1.5 | 57 | La | 1.27 | 1.1 |
| 23 | V | 1.69 | 1.6 | 58 | Ce | 1.33 | 1.1 |
| 24 | Cr | 1.79 | 1.6 | 59 | Pr | 1.32 | 1.1 |
| 25 | Mn | 1.90 | 1.5 | 60 | Nd | 1.33 | 1.1 |
| 26 | Fe | 1.95 | 1.8 | 61 | Pm | 1.35 | 1.1 |
| 27 | Co | 1.89 | 1.8 | 62 | Sm | 1.37 | 1.1 |
| 28 | Ni | 1.98 | 1.8 | 63 | Eu | 1.11 | 1.1 |
| 29 | Cu | 1.93 | 1.9 | 64 | Gd | 1.42 | 1.1 |
| 30 | Zn | 1.87 | 1.6 | 65 | Tb | 1.40 | 1.1 |
| 31 | Ga | 1.52 | 1.6 | 66 | Dy | 1.43 | 1.1 |
| 32 | Ge | 1.81 | 1.8 | 67 | Ho | 1.47 | 1.1 |
| 33 | As | 2.05 | 2.0 | 68 | Er | 1.48 | 1.1 |
| 34 | Se | 2.14 | 2.4 | 69 | Tm | 1.89 | 1.1 |
| 35 | Br | 2.45 | 2.8 | 72 | Hf | 1.89 | 1.3 |
| 37 | Rb | 0.69 | 0.8 | 73 | Ta | 2.20 | 1.5 |
| 38 | Sr | 0.92 | 1.0 | 74 | W | 2.33 | 1.7 |

*Contd...*

*Table 4.18 Contd...*

| | | | | | | | |
|---|---|---|---|---|---|---|---|
| 75 | Re | 2.38 | 1.9 | 82 | Pb | 1.81 | 1.8 |
| 76 | Os | 2.54 | 2.2 | 83 | Bi | 1.82 | 1.9 |
| 77 | Ir | 2.58 | 2.2 | 84 | Po | 1.94 | 2.0 |
| 78 | Pt | 2.46 | 2.4 | 85 | At | 2.05 | 2.2 |
| 79 | Au | 2.35 | 2.4 | 90 | Th | 1.55 | 1.3 |
| 80 | Hg | 2.05 | 1.9 | 91 | U | 2.01 | 1.5 |
| 81 | Tl | 1.65 | 1.8 | | | | |

(a) An electronegativity difference of about 2.1 corresponds to a bond which is 50 per cent ionic. Bonds with a larger electronegativity difference are therefore primarily ionic, those with smaller difference are primarily covalent.

(b) Beryllium and aluminium have the same electronegativity. Bonds from these elements to fluorine are essentially ionic, to other elements essentially covalent.

(c) Bond between boron and hydrogen is almost purely covalent, between boron and other more electronegative elements is partially ionic with predominantly covalent character.

(d) Bonds between silicon and all other elements (except perhaps fluorine) are primarily covalent.

(e) Fluorine and oxygen are the most electronegative elements. Bonds between these elements and all the metals have considerable ionic character, and in most cases predominantly ionic.

(f) That even if a bond is purely covalent, its character and strength are not uniquely determined unless we know the orbitals involved in its formation. We must therefore use different effective radii in calculating the lengths of different types of hybrid bond. Tetrahedral radii are applicable when $sp^3$ hybridization takes place, octahedral radii correspond to $d^2sp^3$ and $sp^3d^2$ hybrids square radii are those which apply when a $dsp^2$ hybrid is formed and linear radii are those appropriate for the limited number of elements which form two *sp* hybrids bonds oppositely directly along a straight line. The square radii and the octahedral radii are numerically indistinguishable, however, they have been tabulated separately because in some cases the valency states corresponding to the types of coordination are not the same; for example, $Pt^{IV}$ forms a $d^2sp^3$ hybrid whereas $pt^{II}$ is $dsp^2$ hybridized. Values for the covalent radii of a number of atoms are given in Table 4.19.

**Example 4.2** Determine the percentage of ionic character in HF molecule, the electronegativities of the hydrogen and fluorine atoms are 2.1 and 4.0, respectively.

***Solution*** Using Eq. (3) and substituting the values of electronegativity given by Pauling, we obtain

$$P = 16\ (4.0 - 2.1) + 3.5\ (4.0 - 2.1)^2 = 43\%$$

**Table 4.19** Covalent Radii (in Angstrom Units)*

| *(a) Normal radii* | | | | | |
|---|---|---|---|---|---|
| | *H* | *C* | *N* | *O* | *F* |
| Single bond | 0.30 | 0.77 | 0.74 | 0.74 | 0.72 |
| Double bond | — | 0.67 | 0.62 | 0.62 | — |
| Triple bond | — | 0.60 | 0.55 | — | — |
| | | Si | P | S | Cl |
| Single bond | | 1.17 | 1.10 | 1.04 | 0.99 |
| Double bond | | 1.07 | 1.00 | 0.94 | — |
| Triple bond | | 1.00 | 0.93 | — | — |
| | | Ge | As | Se | Br |
| Single bond | | 1.22 | 1.21 | 1.17 | 1.14 |
| Double bond | | 1.12 | 1.11 | 1.07 | — |
| | | Sn | Sb | Te | I |
| Single bond | | 1.40 | 1.41 | 1.37 | 1.33 |
| Double bond | | 1.30 | 1.31 | 1.27 | — |

| *(b) Tetrahedral radii* | | | | | | |
|---|---|---|---|---|---|---|
| | Be | B | C | N | O | F |
| | 1.06 | 0.88 | 0.77 | 0.70 | 0.66 | 0.64 |
| | Mg | Al | Si | P | S | Cl |
| | 1.40 | 1.26 | 1.17 | 1.10 | 1.04 | 0.99 |
| Cu | Zn | Ga | Ge | As | Se | Br |
| 1.35 | 1.31 | 1.26 | 1.22 | 1.18 | 1.14 | 1.11 |
| Ag | Cd | In | Sn | Sb | Te | I |
| 1.52 | 1.48 | 1.44 | 1.40 | 1.36 | 1.32 | 1.28 |
| Au | Hg | Tl | Pb | Bi | | |
| 1.50 | 1.47 | 1.47 | 1.46 | 1.46 | | |

| *(c) Octahedral radii* | | | | | | |
|---|---|---|---|---|---|---|
| | $d^2sp^3$ hybrid | | | | | |
| $Fe^{II}$ | $Co^{II}$ | $Ni^{II}$ | $Ru^{II}$ | | $Os^{II}$ | |
| 1.23 | 1.32 | 1.39 | 1.33 | | 1.33 | |
| | $Cb^{III}$ | $Ni^{III}$ | | $Rh^{III}$ | | $Ir^{III}$ |
| | 1.22 | 1.30 | | 1.32 | | 1.32 |

*Contd...*

* Adapted from *Solid State Physics*, MA Wahab, Narosa (2005)

*Table 4.19 Contd...*

| $Fe^{IV}$ | | $Ni^{IV}$ | | | $Pd^{IV}$ | | $Pt^{IV}$ | $Au^{IV}$ |
|---|---|---|---|---|---|---|---|---|
| 1.20 | | 1.21 | | | 1.31 | | 1.31 | 1.40 |

$d^2sp^3$ hybrid

| $Ti^{IV}$ | $Zr^{IV}$ | $Sn^{IV}$ | $Te^{IV}$ | $Pb^{IV}$ |
|---|---|---|---|---|
| 1.36 | 1.48 | 1.45 | 1.52 | 1.50 |

(d) Square radii

| $Ni^{II}$ | $Pd^{II}$ | $Pt^{II}$ | $Au^{III}$ |
|---|---|---|---|
| 1.39 | 1.31 | 1.31 | 1.40 |

(e) Linear radii

| $Cu^{I}$ | $Ag^{I}$ | $Hg^{II}$ |
|---|---|---|
| 1.18 | 1.39 | 1.29 |

## 4.29 METALLIC RADII

To determine the metallic radii of metal elements is probably the simplest of all. Measurement of cell dimensions of the metallic crystals enables us to derive the characteristic atomic radii for given elements. In the case of metals with close-packed structures, the atomic radius is clearly half the distance of closest approach. In structure of lower coordination, a similar definition of atomic radius is applicable, but the values so deduced are not immediately comparable with those derived from closed-packed structures. Study of polymorphous metals having several structures of different coordination, and alloy systems in which metal atoms often occur in a state of different coordination from that in the pure element shows that a small but systematic decrease of atomic radius takes place with decreasing coordination. Goldschmidt summarized the observations on a series of elements and alloy systems by expressing the radius in 8-, 6-, and 4-coordination corresponding to unity in a close-packed structure (12-coordination) as follows:

| Coordination | 12 | 8 | 6 | 4 |
|---|---|---|---|---|
| Radius | 1.00 | 0.97 | 0.96 | 0.88 |

Problem is more difficult for B sub-group metals where the coordination is very irregular and so the concept of precise radius ceases to have definite meaning. However, in these cases, we can define the radius as half the distance of closest approach, but the values so obtained will be of significance only in the structure of the element and will have no immediate influence on its behaviour in alloy systems. Values of the radii derived from study of alloys are therefore often of a more general utility than those deduced from the elements themselves. For this reason two radii are, in general, given for each element as in Table 4.20. One of these corresponds to the distance of closest approach in the structure of the element and the other, if necessary is that appropriate to 12-coordination. For elements with closed-packed structures the two values are, of course, identical but in other cases they differ. For the B sub-group metals, in particular, the difference is often considerable.

**Table 4.20** The Atomic Radii of the Metals*

Representative and Transition Elements. All values are in Angstrom Units.

| *Metal* | *Atomic radius*[1] | *Atomic radius*[2] | *Metal* | *Atomic radius*[1] | *Atomic radius*[2] |
|---|---|---|---|---|---|
| Li | 1.52 | 1.56 | Tc | 1.35 | 1.35 |
| Be | 1.12 | 1.12 | Ru | 1.33 | 1.33 |
| Na | 1.85 | 1.91 | Rh | 1.34 | 1.34 |
| Mg | 1.60 | 1.60 | Pd | 1.37 | 1.37 |
| Al | 1.42 | 1.42 | Ag | 1.44 | 1.44 |
| K | 2.31 | 2.38 | Cd | 1.48 | 1.52 |
| Ca | 1.96 | 1.96 | In | 1.62 | 1.67 |
| Sc | 1.60 | 1.60 | Sn | 1.40 | 1.58 |
| Ti | 1.46 | 1.46 | Sb | 1.45 | 1.61 |
| V | 1.31 | 1.35 | Te | 1.43 | — |
| Cr | 1.25 | 1.28 | Cs | 2.62 | 2.70 |
| Mn | 1.12 | 1.36 | Ba | 2.17 | 2.24 |
| Fe | 1.23 | 1.27 | La | 1.87 | 1.87 |
| Co | 1.25 | 1.25 | Hf | 1.58 | 1.58 |
| Ni | 1.24 | 1.24 | Ta | 1.43 | 1.47 |
| Cu | 1.28 | 1.28 | W | 1.37 | 1.41 |
| Zn | 1.33 | 1.37 | Re | 1.37 | 1.37 |
| Ga | 1.21 | 1.35 | Os | 1.35 | 1.35 |
| Ge | 1.22 | 1.39 | Ir | 1.35 | 1.35 |
| As | 1.25 | — | Pt | 1.38 | 1.38 |
| Se | 1.16 | — | Au | 1.44 | 1.44 |
| Rb | 2.46 | 2.53 | Hg | 1.50 | 1.55 |
| Sr | 2.15 | 2.15 | Tl | 1.71 | 1.71 |
| Y | 1.81 | 1.81 | Pb | 1.74 | 1.74 |
| Zr | 1.60 | 1.60 | Bi | 1.55 | 1.82 |
| Nb | 1.43 | 1.47 | Po | 1.68 | — |
| Mo | 1.36 | 1.40 | | | |

The Lanthanide elements / The actinide elements

| *Metal* | *Atomic radius*[1] | *Atomic radius*[2] | *Metal* | *Atomic radius*[1] | *Atomic radius*[2] |
|---|---|---|---|---|---|
| Ce | 1.82 | 1.82 | Th | 1.80 | 1.80 |
| Pr | 1.82 | 1.82 | Pa | 1.60 | 1.63 |
| Nd | 1.81 | 1.81 | U | 1.38 | 1.54 |
| Pm | — | — | Np | 1.30 | 1.50 |

*Contd...*

* Adopted from *Solid State Physics*, MA Wahab, Narosa (2005).

*Table 4.19 Contd...*

| | | | | | |
|---|---|---|---|---|---|
| Sm | 1.98 | 2.04 | Pu | 1.64 | 1.64 |
| Eu | 1.98 | 2.04 | Am | — | — |
| Gd | 1.78 | 1.78 | Cm | — | — |
| Tb | 1.77 | 1.77 | Bk | — | — |
| Dy | 1.75 | 1.75 | Cf | — | — |
| Ho | 1.76 | 1.76 | E | — | — |
| Er | 1.73 | 1.73 | Em | — | — |
| Tm | 1.74 | 1.74 | Mv | — | — |
| Yb | 1.93 | 1.93 | No | — | — |
| Lu | 1.73 | 1.73 | | | |

## 4.30 VAN DER WAALS RADII

In crystals composed of molecules or noble gas atoms, non-bonded distances of closest approach may be divided to get the van der Waals radii, $r_w$, suitable for use in estimating various parameters of interest in crystal structure. A large number of research papers have been published by various investigators to find the most accurate values. A quantitative interpretation of van der Waals radii due to Morrison (1995) and Bondi (1964) shows that

$$r_w = C_{\lambda B} \qquad ...(1)$$

where $\lambda_B = h\sqrt{M_c I_0}$, the de Broglie wavelength of the outermost valence electron ($M_c$ is the rest mass of the electron; $I_0$ is the first ionization potential and $C$ varies from 0.48 to 0.61 according to the group). Collecting constants we may write

$$r_w = K\sqrt{I_0} \qquad ...(2)$$

**Table 4.21** Van der Waals radii (Å)

| | | | | |
|---|---|---|---|---|
| | | | H | He |
| | | | 1.20 | 1.40 |
| C | N | O | F | Ne |
| 1.70 | 1.55 | 1.52 | 1.47 | 1.54 |
| Si | P | S | Cl | Ar |
| 2.10 | 1.80 | 1.80 | 1.75 | 1.88 |
| | As | Se | Br | Kr |
| | 1.85 | 1.90 | 1.85 | 2.02 |
| | | Te | I | Xe |
| | | 2.06 | 1.98 | 2.16 |

This is a simple, logical and satisfying relation which can be used to obtain van der Waals radii of less commonly studied atoms in molecular crystal. Van der Waals radii of less commonly studied atoms in molecular crystal. Van der Walls radii of some atoms computed by Bondi is given in Table 4.21. These radii are used in computing the volumes of molecules and group which help in determining the preferred orientation of bulky constituents, are in estimating modes of packing. These radii do not have the same degree of consistency as the ionic, covalent and the metallic radii, and variation of as much as 0.1 Å may be found in different molecular radii structures. These radii are also larger than the corresponding covalent radii of the same element (Table 4.8), emphasizing once again that in molecular structures the effective radius of an atom within a molecule is quite different from its radius towards other molecules.

## EXERCISE

4.1 Given for NaCl crystal

$R_o = 2.81$ Å, $\alpha = 1.75$, $m = 9.4$

$e = 4.8 \times 10^{-10}$ e.s.u.

Calculate its binding energy/molecule.

$$\left(\text{Hint: } u = \frac{\alpha e^2}{4\pi e_o R_o}\left(1 - \frac{1}{m}\right)\right)$$ **[*Ans:*** 7.9 eV]

4.2 Discuss Heitler and London theory of covalent bonding. Explain the terms exchange integral and Coulomb integral.

4.3 Given a one-dimensional crystal consisting of alternate positive and negative ions with interionic distance *R*. show that the Madelung constant is given by $\alpha = 2 \ln 2$.

4.4 Rare gas atoms are chemically extremely inactive and interact via van der Waals - London interaction. As a model, two atoms can be represented by two identical linear harmonic oscillators 1 and 2 (each as a dipole), separated by a distance *R*. Prove that there is an attractive interaction potential which varies as the minus sixth power of the separation *R*.

4.5 Give in tabular form properties of various types of bonding in solids.

4.6 Discuss the Jellium model (JM) for a solid.

4.7 Write short notes on

(i) General nature of cohesion between two atoms in a solid.

(ii) Ionic bonding.

(iii) Bond dissociation energy of NaCl molecule.

(iv) Bond energy and Madelung constant for NaCl crystal.

(v) Directional nature of covalent bond.

(vi) Hydrogen bonding.

(vii) Molecular bonds.

CHAPTER

# 5

# Order and Disorder, Imperfections, Alloys and Deformation

## 5.1 ORDER AND DISORDER IN SOLIDS

The *ideal crystalline–solids* discussed in previous chapters, do not, and in fact, cannot actually exist in nature as single crystals in which the positions of atoms are perfectly ordered in space. *Perfect order* in a solid refers to the identical arrangement of atoms within each local atomic bonding unit and also relative to each lattice point. In real solids, at thermodynamic equilibrium, deviations from perfect order in the form of finite concentrations of *defects* such as vacancies and interstitials will always be present. They can exist at thermal equilibrium due to the resulting increase of entropy or *disorder*, which leads to a net decrease of the free energy of the solid.

Although only an idealization, the concept of a *perfect crystal* is nevertheless an extremely useful one. The perfect crystal serves as the natural starting point for discussion on real solids in which the order is imperfect (i.e., in which *structural, chemical* and *thermal disorder* are all present). Due to the presence of disorder, the physical properties of the solid will no longer be periodic in space, due to the absence of periodicity in the equilibrium positions of the atoms.

*Structural disorder* corresponds to the displacement of atoms from their equilibrium positions in a perfectly ordered crystal.

*Chemical disorder* corresponds to the occupation by atoms of sites in the structure which are normally occupied by other atoms. Specific examples of structural and chemical are discussed in detail later in this chapter.

The *degree of structural order* present in a solid is often specified in terms of the spatial extent of the order (i.e., by the size if the *ordered regions*). These ordered regions, often referred to as *crystallites*, can be thought of as the remnants of the original perfect order. The term *microstructure* is often used to describe the structure of a crystal on the scale on which deviations from perfect order become evident. The types of structural order known as polycrystalline, microcrystalline and nano-crystalline, as well as the structural disorder known as amorphous or noncrystalline are defined in Table 5.1, in terms of an average crystalline size.

**Table 5.1** Crystalline Size and Types of Structural Order in Solids

| *Crystallite size* | *Type of structural order* |
|---|---|
| ≈ 1 – 100 mm | Polycrystalline (or single crystal) |
| ≈ 1 – 1000 μm | Microcrystalline |
| ≈ 1 – 1000 nm | Nanocrystalline (nanophase) |
| ≈ 0.1 – 1 nm | Amorphous (non-crystalline) |

The term polycrystalline is often used to describe any crystalline solid which is not a single crystal but consists of grains, each bounded by a grain boundary. The amorphous state is a limit of maximum structural disorder in a solid, corresponding to the size of an ordered region or crystalline approaching that of an individual atom.

It is appropriate at this point to introduce and define some additional terms that are often used to describe the spatial extent of order in solids: short-range order (SRO), intermediate range (IRO), and long-range order (LRO).

1. *Short-range order* exists in a solid when the arrangement of NNs for each atom is the same as that found in the crystalline state. SRO is often retained in the amorphous state, where its presence is a strong indication that the type of bonding between atoms found in the crystal exists even in the presence of extreme structural disorder. The existence of SRO does not, however, necessarily imply that the bond lengths and angles found in the disordered solids are the same as those found in the crystal. For example, even though SRO is preserved to a large degree in amorphous covalent solids such as amorphous Si, since practically every Si atom remains at the centre of an $Si\text{-}Si_4$ bonding unit, the bond angles within the $Si\text{-}Si_4$ units are distributed continuously over a few degrees around the value of 109.47° found in crystalline Si.
2. Although difficult to define precisely, *intermediate range order* exists in a solid when the order extends beyond a given local atomic bonding unit to neighbouring units. An example of IRO can be found in amorphous carbon, a-C, where in addition to trigonal $sp^2$ coordination corresponding to the $C\text{–}C_3$ bonding units, there and tetrahedral to trigonal $sp^2$ coordination corresponding to $C\text{–}C_4$ bonding units, there can exist sixfold hexagonal rings of carbon atoms. These hexagonal rings in a-C are distorted versions of the regular planar rings found in crystalline graphite.
3. Perfect *long-range order* can exist only in perfectly ordered crystalline solids. Nevertheless, a type of LRO can also be said to exist within ordered regions of a disordered solid when crystalline sizes exceed 1 to 10 nm. With this definition, even the nanocrystalline solids as defines in Table 4.1 may have RO within small nanocrystallines. In some solids, LRO can exist in two dimensions can be modified or absent in the third. Examples include superlattice structures consisting of alternating

layers of materials, such as the dielectrics $PbTiO_3$ and $SrTiO_3$. Other examples include liquid crystals.

## 5.2 IMPERFECTIONS (DEFECTS)

Imperfections are mistakes in the crystallographic structure of crystals. The concept of an ideal crystal with a perfect arrangement of atom is, strictly speaking, valid only at the absolute zero temperature because then there is no entropy contribution. However, at a finite temperature, a certain native configurational disorder gets introduced into the structure (a direct consequence of laws of thermodynamics) and solid becomes structurally imperfect. Thus, infact, in almost all cases, a crystal just cannot be grown without an initial imperfection.

### 5.2.1 Necessity of Imperfections

Defects and imperfections affect especially the structure-sensitive properties of solids. Properties such as elasticity, density, specific heat, thermal expansions, compressibility and dia- and paramagnetism are only slightly affected by defects in crystals. On the other hand, colourability, diffusion and crystal growth (from vapour phase) are defect-sensitive properties. In the following, we will describe the experimental evidences which shows that actual crystals are not perfect unbroken arrays of atoms or ions but contain a variety of defects.

### 5.2.2 Imperfections (Statistical Viewpoint)

Deviations from the ideal crystal structure are known as imperfections. Imperfections could be of several types, as we shall see subsequently.

The presence of imperfections is not accounted for by the translational symmetry of the perfect crystal, though it forms the basis of the most of the interpretations in perfect crystals. One may be curious to know how the translational symmetry is consistent with laws of thermodynamics since these laws are applied to describe the growth of crystals which have one or the other imperfection.

We know that the Helmholtz free energy

$$F = U - TS \qquad ...(1)$$

must be minimum in the state of equilibrium at a certain temperature. In Eq. (1) $U$ is the internal energy and $S$, the entropy. We know that (from statistical mechanics), entropy

$$S = k \ln W \qquad ...(2)$$

when $W$ is the number of possible ways in which the elements of a system may be distributed. Now, we take the case of a defect that can be localized at a lattice point. In a perfect crystal, $W = 1$, and the entropy in zero. On the other hand, a defect at a site within a unit cell makes the unit cell look different from others. In this case, there can be as many ways of arranging the defect as the number of sites within the crystal, and entropy

$$S = k \ln N \qquad ...(3)$$

where $N$ is the number of sites.

Equation (3) expresses the contribution of the defect to the entropy of the crystal. Thus, any increase in the defect concentration raises the entropy, which, in turn, lowers the free energy at a finite temperature. In the equilibrium state, there is a finite concentration of imperfections of the crystal. The concentration of a particular type of imperfection depends on the type of the crystal lattice, the binding energy of the lattice and structure of the imperfection as well. The imperfections are critical to the interpretations of several properties that are not accounted by translational symmetry, e.g., colour of crystals, enhancement of conductivity of pure semiconductors, luminescence and diffusion of atoms in solids etc.

For a systematic description of imperfections, we classify them as follows:

1. Point imperfections
2. Line imperfections
3. Planar imperfections.

## 5.3 POINT IMPERFECTIONS

A point imperfection is a very localized interruption in the regularity of a lattice. A point imperfection comes about, as a rule, because of the absence of a matrix atom (an atom that would be present in a perfect crystal), the presence of an impurity atom, or a matrix atom in the "wrong" place (a site not occupied in the perfect crystal). The most common point imperfections in a crystal of a pure element are illustrated in Fig. 5.1; some of those that may occur in an ionic crystal are shown in Fig. 5.2. The absence of an atom from a normally occupied site is called a *vacancy*. A foreign atom which is in the site of a matrix atom is called a *substitutional impurity atom*, and a foreign atom in an interstice between matrix atoms is called an *interstitial impurity atom*. There are, in addition, point imperfections which are more complex, for example, condensed groups of vacancies such as *di-vacancies* and *tri-vacancies*. In a crystal, an atom may leave its site, creating a vacancy, and dissolve interstitially in the structure. The associated vacancy and interstitial atom is called a *Frankel imperfection*. When a cation vacancy is associated with an anion vacancy rather than an interstitial cation, the pair (the anion vacancy and the associated cation vacancy) is called a Schottky imperfection. In general, *Schottky imperfections* are more likely to form than Frankel imperfections, for a few sructures contain large enough interstices to dissolve cations without considerable strain.

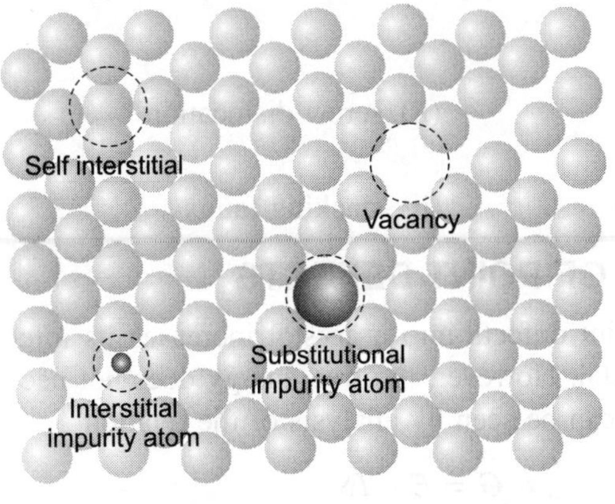


**Figure 5.1** A two-dimensional representation of a simple crystalline solid, illustrating some of the point imperfections that are possible

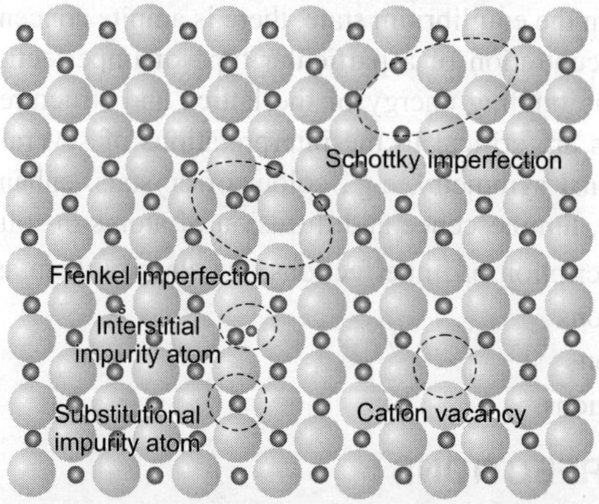


**Figure 5.2** A two-dimensional representation of an ionic solid, illustrating a number of possible imperfections. Cations are represented by small spheres, anion by large spheres

However, in a relatively open structure like that of fluorite ($CaF_2$), where the interstices are large, cations are quite likely to dissolve interstitially and produce Frankel imperfections.

In addition to structural imperfections, other point imperfections which are not structural in nature may be produced by raising the electrons in certain atoms to higher energy levels. This may be accomplished by bombarding the crystal with electrons (resulting in X-rays), microwave radiation (the principle of the maser), light (lasers and photoelectric materials) and other forms of radiation of suitable frequency.

Various point defects are:

(i) vacancy defects (ii) Schottky defects
(iii) self-interstitial defect (iv) Frenkel defects
(v) interstitial or substitutional impurity
(vi) colour centres (vii) polarons
(viii) excitons

Line defects are:

(i) edge dislocation (ii) screw dislocation

Surface defects are:

(i) grain boundaries (ii) stacking faults

## 5.4 FORMATION OF POINT DEFECTS

The introduction of point defects into a crystal increases its internal energy as compared to that of a perfect crystal. However, the free energy does not necessarily increase but may actually decrease. This can be seen by analysis of the relation

$$G = E - TS \qquad ...(1)$$

where $G$ is free energy, $E$, the total energy, $T$, the temperature and $S$, the entropy. Although the internal energy $E$ of the system may increase, the value of free energy $G$ may decrease because of the $TS$ terms. As the number of defects increases, the entropy ($S$) of the crystal increases (because of an increase in the number of possible arrangements). Consequently, the value of $T \Delta S$ will increase, thereby decreasing the value of $G$ (in spite of increase in $E$). When the addition of more defects to the system no longer lowers the value of $G$, the state of equilibrium has been reached at which free energy is the maximum.

The number of defects at equilibrium at a certain temperature can be determined from the equation

$$n = Ne^{-\varepsilon_v/kT} \quad \text{...(2)}$$

where $n$ is the number of defects, $N$, the total number of atomic sites/m$^3$ or per mole $\varepsilon_v$, the energy of activation necessary to form the defect $T$, the absolute temperature and $k$, Boltzmann constant. Figure 5.3 shows point imperfections in ionic solids.

### 5.4.1 Vacancy (Thermodynamic Solid)

It is to be noted that the defects under consideration are present as a result of thermodynamic conditions and do not include any accidental fault resulting from non-ideal growing conditions.

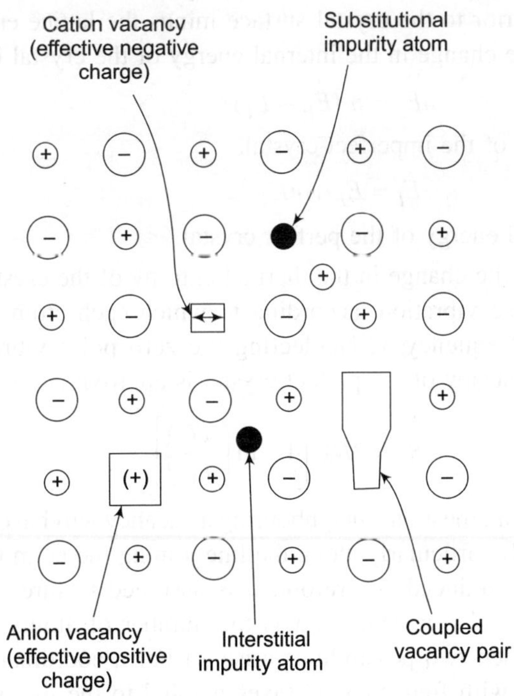


**Figure 5.3** Point imperfection in an ionic solid

As pointed out above that a vacancy is a missing atom from a regular atomic site in the structure. A vacancy can be created deliberately by transferring an interior atom from its regular site to a site on the surface of a two-dimensional crystal as shown in Fig. 5.4. The change in the internal energy as a result of the introduction of defects into the crystal occurs due to (i) the energy needed to produce defects in the static lattice, and (ii) the changed vibrational energy of the atoms coordinating the defects. A change in the entropy experienced by the crystal comes from two different sources: (i) due to change of thermal entropy as a result of the change in the frequency of atomic oscillators, and (ii) due to configurational entropy associated with the number of ways of selecting which atomic sites will be vacant. In the following calculations, we shall consider first the formation of vacancy in a mononatomic crystal.

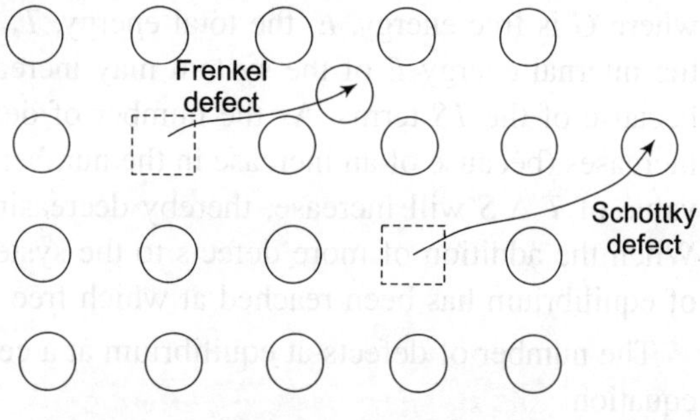


**Figure 5.4** The formation of Schottky and Frankel imperfections

The energy of formation of each imperfection $E_s$ taken in electron volts (also called the enthalpy of formation $\Delta H_f$, usually taken in joules per mole) is the energy, $E_H$, required to take an atom from interior to the crystal surface minus the lattice energy per atom $E_L$, i.e., $E_S = E_H - E_L$. Thus, the change in the internal energy of the crystal is

$$nE_s = n\,(E_H - E_L) \qquad ...(1)$$

and the internal energy of the imperfect crystal,

$$E_I = E_P + nE_S \qquad ...(2)$$

where $E_P$ is the internal energy of the perfect crystal.

In order to calculate the change in the thermal entropy of the crystal, we make use of the Einstein model of lattice vibration, according to which each atom in the crystal vibrates, independently with a frequency, $\nu$. Neglecting the zero point vibrational energy and for $h\nu << kT$ the thermal entropy of the perfect crystal is approximately given by

$$S_p = 3Nk\left[1 + \ln\left(\frac{kT}{h\nu}\right)\right] \qquad ...(3)$$

In the imperfect crystal, the atoms neighbouring a vacancy will have a frequency $\nu'$ (smaller than $\nu$ in perfect crystal) particularly along the line joining the atom to the vacancy, because the restoring forces are reduced. Therefore, it is assumed to three independent harmonic oscillators of frequency, $\nu' < \nu$. Thus, if $x$ is the number of atoms surrounding a vacancy ($x$-fold coordination), then two possibilities arise (i) For a coordination of 6-fold or below $nx$ atoms will oscillate with frequency, $\nu'$ (axes parallel to the line joining the atom to the vacancy are affected) and $(3N - nx)$ with $\nu$. (ii) For a coordination greater than 6-fold (when all the three axis are affected) $3nx$ atoms will oscillate with frequency $\nu'$ and $(3N - 3nx)$ with

$\nu$. The thermal entropy of the imperfect crystal is then obtained in analogy with Eq. 3, for the case (ii) as

$$S_I = 3nxk\left[1+\ln\left(\frac{kT}{h\nu'}\right)\right]+(3N-3nx)k\left[1+\ln\left(\frac{kT}{h\nu}\right)\right] \qquad ...(4)$$

Subtracting Eqs. 3 from 4, we obtain the change in the thermal entropy as

$$\Delta S_{th} = (S_I - S_P) = 3nxk\left[1+\ln\left(\frac{kT}{h\nu'}\right)\right]+(3N-3nx)k\left[1+\ln\left(\frac{kT}{h\nu}\right)\right]$$

$$- 3Nx\left[1+\ln\left(\frac{kT}{h\nu}\right)\right]$$

$$= 3nxk\left[1+\ln\left(\frac{kT}{h\nu'}\right)\right]-3nxk\left[1+\ln\left(\frac{kT}{h\nu}\right)\right]$$

$$= 3nxk\ \ln\left(\frac{\nu}{\nu'}\right) = nk\ \ln\left(\frac{\nu}{\nu'}\right)^{3x} \qquad ...(5)$$

In a similar way, the thermal entropy of the imperfect crystal for the case (i) can be obtained as

$$\Delta S_{th} = nk\ \ln\left(\frac{\nu}{\nu'}\right)^{x} \qquad ...(6)$$

As far as the configurational entropy of a crystal is concerned, it has nothing to do with the distribution of energy; it is solely determined by the number of different ways $E_{cf}$ in which the atoms may be arranged over the available number of lattice sites. For example, let us consider that there are $N$ lattice sites in the crystal and $n$ Schottky vacancies that are produced. Then the number of ways of arranging $n$ identical vacancies and $(N-n)$ identical atoms on $N$ sites is given by

$$W_{cf} = \frac{N!}{(N-n)!n!} \qquad ...(7)$$

The configurational entropy associated with $W_{cf}$ is obtained from the Boltzmann equation

$$S_{cf} = k\ \ln W_{cf} = k\ \ln\left[\frac{N!}{(N-n)!n!}\right] \qquad ...(8)$$

In the case of perfect crystal containing identical atoms, $W_{cf} = 1$ and $S_{cf} = 0$ because there is only one possible way of arranging the atoms. The total entropy occurring in the usual thermodynamic formulas is equal to the sum of the thermal and configurational entropies, i.e.,

$$S = S_{th} + S_{cf} \qquad ...(9)$$

Now using Eq. 5 (i.e., taking into account the coordination of vacancy as 6-fold or greater) we may write the free energy of the imperfect crystal as

$$F = E_p + nE_s - nkT \ln \left(\frac{v}{v'}\right)^{3x} - kT \ln \left[\frac{N!}{(N-n)!n!}\right] \qquad ...(10)$$

The equilibrium value of $n$ is obtained from this equation by setting $(\partial F/\partial n)_\tau = 0$ and making use of Stirling's approximation,

$$\ln x! \cong 8 \ln x \text{ for } x >> 1$$

so that $$\ln \left[\frac{N!}{(N-n)!n!}\right] = N \ln N - (N-n) \ln (N-n) - n \ln n$$

Consequently Eq. 10 become

$$F = E_P + nE_S - nkT \ln \left(\frac{v}{v'}\right)^{3x} - kT \left[N \ln (N) - (N-n) \ln (N-n) - n \ln (n)\right]$$

Differentiating with respect to $n$ partially, we get

$$\left(\frac{\partial F}{\partial n}\right)_T = E_S - kT \ln \left(\frac{v}{v'}\right)^{3x} - kT \ln \frac{(N-n)}{n} = 0$$

or $$E_s - kT \ln \left(\frac{v}{v'}\right)^{3x} - kT \ln \frac{(N-n)}{n} = 0$$

or $$kT \ln \frac{n}{N-n} = kT \ln \left(\frac{v}{v'}\right)^{3x} - E_S$$

or $$\ln \frac{n}{N-n} = \ln \left(\frac{v}{v'}\right)^{3x} - \frac{E_S}{kT}$$

or $$\frac{n}{N-n} = \left(\frac{v}{v'}\right)^{3x} \exp \left(-\frac{E_S}{kT}\right)$$

or $$\frac{n}{N} = \ln \left(\frac{v}{v'}\right)^{3x} \exp \left(-\frac{E_S}{kT}\right) \qquad ...(11)$$

where $n << N$. Use of Eq. 6 would result in

$$\frac{n}{N} = \left(\frac{v}{v'}\right)^{x} \exp \left(-\frac{E_S}{kT}\right) \qquad ...(12a)$$

For one mole of atoms, the internal energy $NkT = RT$, where $N$ is the Avogadro's number and $R$ is the gas constant. Therefore,

$$R = NK = 6.023 \times 10^{23} \times 1.38 \times 10^{-23} = 8.314 \text{ jmol}^{-1} \text{ K}^{-1}$$

Making use of this, Eq. 12(a) equivalently can be written as

$$\frac{n}{N} = \left(\frac{v}{v'}\right)^{x} \exp \left(\frac{\Delta H_f}{RT}\right) \qquad ...(12b)$$

where in Eq. 12 (a) the term $E_s$ is expressed in electron volts (eV) and in Eq. 12 (b) the $\Delta H_f$ is expressed in kjmol$^{-1}$ respectively. The energy required to form vacancies in some monoatomic crystals is listed in Table 5.2. Since $v > v'$, the change in thermal entropy favours the formation of vacancies as $v/v'$ is greater than 1. This factor may be large because $3x$ is rather a large number (24 for bcc and 36 for fcc).

**Table 5.2** Enthalpy of Formation of Vacancies in Some Crystals

| *Crystal* | *Kr* | *Cd* | *Pb* | *Zn* | *Mg* | *Al* | *Ag* | *Cu* | *Ni* |
|---|---|---|---|---|---|---|---|---|---|
| $\Delta H_f$ kJ mol$^{-1}$ | 7.7 | 38 | 48 | 49 | 56 | 68 | 106 | 120 | 168 |
| eV/vacancy | 0.08 | 0.39 | 0.50 | 0.51 | 0.58 | 0.70 | 1.10 | 1.24 | 1.74 |

The results obtained above may be used to explain qualitatively the reason for the existence of lattice defects at any temperature, $T > 0$. For example, suppose that in a perfect metallic crystal we produce certain number of vacancies by transferring the atoms from the interior of the crystal to the surface. This will require certain amount of energy so that the internal energy, $E$ of the crystal will increase. As a result of this, the free energy $F$ will also increase, which is unfavourable in the thermodynamic sense. On the other hand, creation of vacancies increases the disorder in the crystal which, in turn, increases the configurational entropy from zero to a certain value determined by the number of vacancies $n$ produced. Since the entropy appears in the free energy expression ($F = E - TS$) in the form $-TS$, an increase in the entropy reduces the free energy $F$, which is favourable in thermodynamic sense. As a result of the competition between internal energy and entropy, the stable state will always contain a certain number of vacancies.

**Example 5.1** Given that energy of vacancy formation for silver is 1.1 eV. Calculate the number of vacancies per cubic centimetre at 300°C. Also find out the fraction of lattice sites that are vacant at the melting temperature of silver (960°C). Given: atomic weight of silver $A = 107.9$ and density $\rho = 10{,}500$ kg/m$^3$.

***Solution***

$$N_v = Ne^{-\varepsilon_v/kT}$$

and the number of atomic sites per m$^3$ is

$$N = N_A \frac{\rho}{A} = 6.02 \times 10^{23} \text{ atoms/mol} \times \frac{10500}{1079 \times 10^{-3} \text{ kg/mol}}$$

$$= 5.86 \times 10^{28} \text{ atoms/m}^3$$

Boltzmann constant

$$k = 1.38 \times 10^{-23} \text{ J/K}$$

$$= 8.63 \times 10^{-5} \text{ eV/K}$$

Therefore, at 300°C,

$$nv = 5.86 \times 10^{28} \times \exp\left[-\frac{1.1\,\text{eV}}{8.63 \times p^{-5}\,\frac{\text{eV}}{\text{K}} \times 1233}\right]$$

$$= 1.096 \times 10^{19}\ \text{vacancies/m}^3.$$

The vacancy fraction at 960°C will be

$$\frac{nv}{N} = \frac{Ne^{-\varepsilon_v/kT}}{N} = \exp\left[-\frac{1.1\,\text{eV}}{8.63 \times 10^{-5}\frac{\text{eV}}{\text{K}} \times 1233}\right]$$

$$= e^{-10.34} \cong 3 \times 10^{-5}$$

That is, about three vacancies in every $10^5$ are vacant.

## 5.5 SCHOTTKY DEFECT IN IONIC CRYSTALS

Above, a vacancy is created in a metal by the migration of a metal atom (ion) to the surface. Similarly, in a binary ionic crystal (represented by $A^+ B^-$), if a cation-anion pair ($A^+$ and $B^-$) is missing from the respective lattice position, it is called a Schottky defect. The energy absorbed during the creation of this defect is more than compensated by the resultant disorder in the lattice/structure.

If $n$ is the number of Schottky defect produced by removing $nA^+$ cations and $nB^-$ anions, $N$ is the total number of cation-anion pairs present and $\bar{\varepsilon}_s$ is the average energy required to create a Schottky defect, then

$$S_{\text{config}} = k_B \ln\left[\frac{(N+n)!}{N!n!}\right]^2$$

$$= 2\,k_B \ln\left[\frac{(N+n)!}{N!n!}\right] \quad ...(1)$$

The exponent 2 appears in the above relation because then there are $n$ cation as well as $n$ anions, so that there will be $[(N + n)!/(N!n!)]^2$ number of ways of choosing the vacant sites, out of the $(N + n)$ ion pairs. So in this case, we would get

$$n = N \exp\left(\frac{\bar{\varepsilon}_s}{2k_BT}\right) \quad ...(2)$$

Equation (2) is an adequate expression for the determination of the number of Schottky defects present in binary ionic crystals such as NaCl and MgO at ordinary temperatures, at which the interaction between defects could be considered to be negligible since their number is relatively small. Thus, the number of Schottky defects depends on

(i) the total number of ion pairs $N$ (i.e., on mass of the ionic crystal)

(ii) the average energy required to produce a Schottky defect and temperature

(iii) The fraction of Schottky defect increases exponentially with increasing temperature.

## 5.6 SELF-INTERSTITIAL DEFECT IN ELEMENTAL SOLIDS

Interstitial site is the space normally unoccupied in between the normal crystal structure. Self-interstitial are the defects formed when atoms that initially occupy a normal lattice position migrate into interstitial lattice positions.

Let $\bar{\varepsilon}_i$ be the average energy required to displace an atom from its regular lattice position to an interstitial, $N_i$ be the number of interstitial sites and $N$ be the total number of atoms. Then the total number of ways in which an atom can be removed is given by

$$\frac{N(N-1)(N-2)\ldots(N-n+1)}{n!} = \frac{N!}{(N-n)!n!} \qquad \ldots(1)$$

where $n$ is the number of interstitial defects. Since $N_i$ is the total number of interstitial sites out of which $n$ are occupied, therefore, the number of different ways in which $n$ interstitial defects can be formed is given by

$$W_1 = \frac{N!}{(N-n)!n!} \times \frac{N_i!}{(N_i-n)!n!} \qquad \ldots(2)$$

Therefore, $$S_{\text{config}} = k_B \ln\left[\frac{N!}{(N-n)!n!} \times \frac{N_i!}{(N_i-n)!n!}\right]$$

$$= k_B\left[\ln\frac{N!}{(N-n)!n!} + \ln\frac{N_i!}{(N_i-n)!n!}\right]$$

$$\cong k_B[N\ln N - \cancel{N} - (N-n)\ln(N-n) + (\cancel{N} - \cancel{n}) - n\ln n + \cancel{n} + N_i\ln N_i - \cancel{N} - (N_i-n)\ln(N_i-n) - n\ln n]$$

or $$S_{\text{config}} \cong k_B\,[N\ln N + N_i\ln N_i - (N-n)\ln(N-n) - (N_i-n)\ln(N_i-n) - 2n\ln n] \qquad \ldots(3)$$

Now, since in this case the number of vacancies = $n$, the number of interstitial defects, therefore total volume = $(N+n)\,v_0$, where $v_0$ is the volume per atom

Therefore, $$G = U - T(S + S_{\text{config}}) + P(N+n)v_0$$

$\Rightarrow$ $$G = F_0 - TS_{\text{config}} + P(N+n)v_0$$

For $$\frac{\partial G}{\partial n} = \frac{\partial F_0}{\partial n} + Pv_0 - T\frac{\partial S_{\text{config}}}{\partial n} \qquad \ldots(4)$$

From Eq. (3) $$\frac{\partial S_{\text{config}}}{\partial n} = k_B\frac{\partial}{\partial n}\,[N\ln N - N_i\ln N_i - (N-n)\ln(N-n) - (N_i-n)\ln(N_i-n) - 2n\ln n]$$

$$= k_B\left[0 + 0 + \ln(N-n) - (-1) + \ln(N_i - n) - (-1) - 2\ln n - \frac{\partial n}{n}\right]$$

$$= k_B[\ln(N-n) + \ln(N_i - n) + 1 + 1 - 2\ln n - 2]$$

$$= k_B \ln\left[\frac{(N-n)(N_i-n)}{n^2}\right] \quad ...(5)$$

Now $\dfrac{\partial F_0}{\partial n} = \dfrac{\partial F_0}{n} = \varepsilon$ (thermal energy independent of $n$)

For $G$ to be minimum

$$\frac{\partial G}{\partial n} = 0 = \varepsilon = Pv_0 - k_B T \ln\left[\frac{(N-n)(N_i-n)}{n^2}\right]$$

If $N >> n$ and $N_i >> n$ then

$$0 = \varepsilon + Pv_0 - k_B T \ln\left(\frac{NN_i}{n^2}\right)$$

or
$$\varepsilon + Pv_0 = k_B T \ln\left(\frac{NN_i}{n^2}\right)$$

or
$$\frac{\varepsilon + Pv_0}{k_B T} = \ln\left(\frac{NN_i}{n^2}\right)$$

Therefore,
$$n^2 = NN_i \exp\left[-\frac{(\varepsilon + Pv_0)}{k_B T}\right]$$

or
$$n = (NN_i)^{1/2} \exp(\varepsilon/2k_B T) \exp\left(\frac{\bar{\varepsilon}_i}{2k_B T}\right) \quad ...(6)$$

where $\exp(\varepsilon/2k_BT)$ is a constant term = $\exp(\Delta S/2k_B)$ and $\bar{\varepsilon}_i = Pv_0$. $\Delta S$ is the change in thermal entropy per interstitial defect and $\bar{\varepsilon}_i$ is the energy of formation and self-interstitial.

The factor 2 appears in the exponentials because here, a defect has two components. In chemical language, the formation of a Frenkel defect may be written in the form of an equilibrium reaction as

Occupied lattice site + Unoccupied interstitial site – Vacancy + Interstitials occupied

From this, the law of mass action implies that $n$ should be proportional to $(NN_i)^{1/2}$.

## 5.7 FRENKEL DEFECT IN IONIC SOLID

A solid-interstitial defect in an ionic solid is known as Frenkel defect. It is present in open structure such as those of silver halides (Schottky defects are prevalent dominantly in closest packed structures, such as NaCl).

Let $\bar{\varepsilon}_F$ be the average energy required to displace a cation from its normal lattice position to an interstitial, $N_i$ be the number of interstitial sites, and $N$ be the total number of cations (in a binary ionic crystal $A^+ B^-$). Then the number of Frenkel defects

$$n = (NN_i)^{1/2} \exp\left(\frac{\Delta S}{2k_B}\right) \exp\left(-\frac{\varepsilon_F}{2k_B T}\right)$$

Usually, the number of Frenkel defects is small at low temperatures.

## 5.8 INTERSTITIAL OR SUBSTITUTIONAL IMPURITY

The introduction of impurity in an otherwise pure solid may cause various effects as follows. An impurity atom/ion may enter the lattice interstitially or substitutionally. In either case, physical and electrical properties will be greatly affected.

Whether an impurity atom/ion can enter interstitially or not is determined by the relative size of the host and the impurity atom should normally be close to that of the host atom for the formation of substitutional solid solutions, e.g., alloys. The physical properties such as malleability, ductility and electrical conductivity are drastically altered on alloying copper with zinc. Manganese metal becomes magnetic on alloying with copper. On the other hand, extra zinc atoms occupy interstitials in zinc oxide. Interstitial atoms/ions are relatively more mobile and reactive than those occupying the normal lattice positions. This is taken advantage of in the photographic process in which combination of free electrons with interstitial electrons with interstitial silver ions is believed to be one of the primary stages. Various point defects are illustrated in Fig. 5.5.

## 5.9 COLOUR (F) CENTRES

This type of defect imparts colour to an otherwise colourless crystal.

The alkali halides when perfectly pure are colourless. By heating a crystal of an alkali halide in an atmosphere containing vapours of an alkali metal or by subjecting the crystal to bombardment, the crystal becomes coloured. This colouration is due to the formation of F-centres (Farbzentrum in German means colour). By absorption and electron spin resonance studies, it has been established that the F-centre represents an electron trapped in a negative ion (i.e., anion) vacancy (Fig. 5.6).

It is a characteristic of the host crystal and is independent of the alkali metal used as vapour. It has been verified experimentally and it has been found that the same optical absorption spectra of the KCl single crystal is obtained whether the crystal is heated in potassium or sodium vapour. The optical absorption energies of the F-centres lie in the range 2–3 eV. It is possible to decolourise a coloured crystal by exposing it to an illumination with light within the absorption band. The bleaching occurs due to the ionization of the F-centre.

It is a worthwhile to note that in the experiment of heating alkali halide in a vapour of alkali metal, excess alkali atoms (number ranging from1 in $10^7$ to as high as 1 in $10^3$) are incorporated into the crystal. However, the mass density of the colour crystal diminishes in proportion to the concentration of excess alkali atoms, demonstrating that the atoms are not absorbed interstitially. Instead, the alkali metal atoms are ionized and assume positions in the sites of a perfect positively charged sublattice, and the excess electrons are bound to an equal number of negative ion vacancies.

We will discuss colour centres in detail in the subsequent sections.

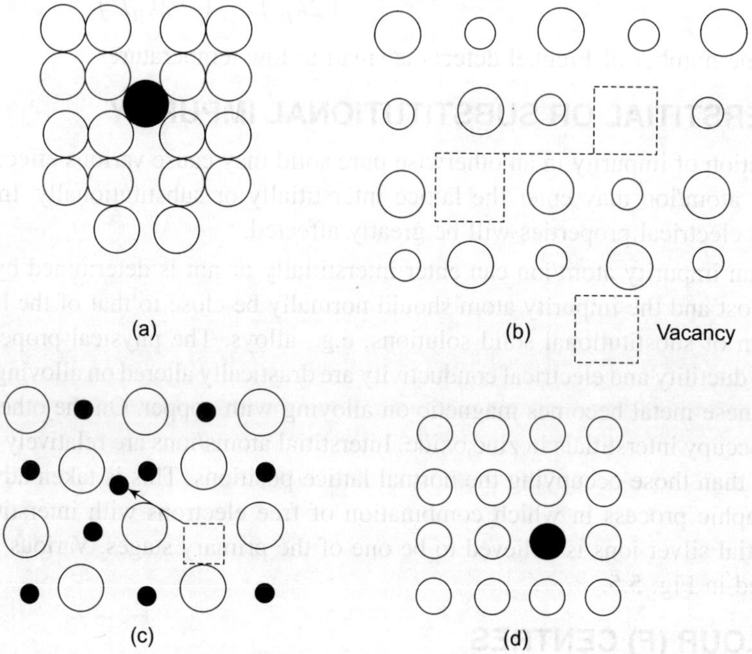


**Figure 5.5** Point defects: (a) interstitial impurity, (b) vacancy–Schottky, (c) vacancy–Frenkel and (d) substitutional impurity defects

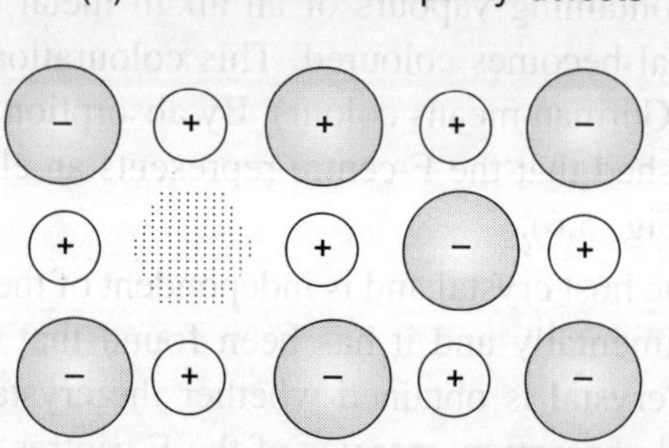

**Figure 5.6** An F-centre is a negative ion vacancy with one excess electron bound at the vacancy

*Point imperfections* are zero dimensional lattice defects which may be either chemical *impurities, vacant lattice sites* or *interstitial atoms*. The feature they have in common is the ability to bind and release electrons. They thus create *localized states* in the lattice.

In the subsequent description, we will see the effect of point imperfections on the band model. It will emerge that the delocalized band state and the localized states associated with defects can be treated together in the same energy diagram.

## 5.10 DESCRIPTION WITHIN THE FRAMEWORK OF THE BAND MODEL (LATTICE WITH A DEFECT)

Band theory assumes a strictly periodic lattice structure. This periodically is perturbed by defects. In this section we consider a single defect in an otherwise periodic potential.

Instead of the Schrödinger equation

$$H_0\psi_n(k, r) \equiv \left[-\frac{\hbar^2}{2m}\nabla^2 + V(r)\right]\psi_n(k,r) = E_n(k)\,\psi_n(k,r) \qquad ...(1)$$

$[V(r)$: periodic lattice potential, $\psi_n(k, r)$: Bloch function], we now have the equation

$$[H_0 + U(r)]\,\psi = E\psi. \qquad ...(2)$$

Here $U(r)$ is the additional potential introduced by the defect. For an impurity atom at a lattice position, for example, it is the potential of the impurity less the potential of the lattice atom which has been substituted. $U(r)$ may, however, also have to take account of a local lattice distortion caused by the impurity atom.

$U(r)$ can be negative or positive. Electrons can thus be bound to the defect or repelled by it. We consequently anticipate that (2) has solutions which are localized about the defect, and whose energy levels lie below or above the states of the energy band considered.

An examination of Eq. (2) for simple models reveals this behaviour. We reproduce one result in Fig. 5.7. The general result is that the presence of the defect in an otherwise periodic

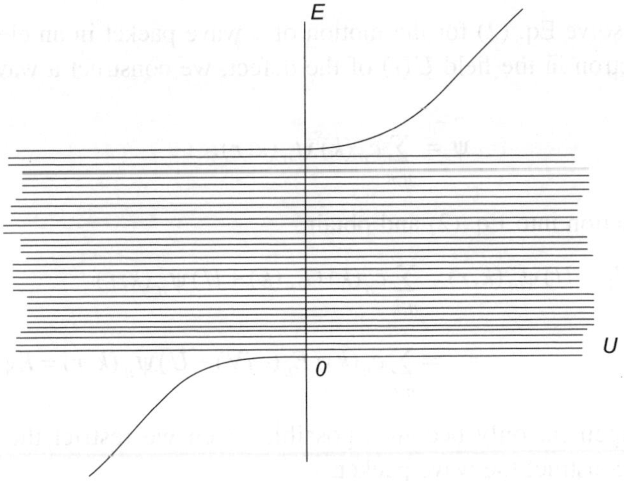


**Figure 5.7** Energy band for a simple three-dimensional potential model (periodic potential with an isolated defect) as a function of a deviation $U$ of the defect potential from the potential at an undisturbed lattice site [from *Koster* and *Slater*(*Slater*[106/xlxl])]. Depending on the sign of $U$, the state of highest or lowest energy splits out from the quasi-continuous band. The split-off state is spatially localized to the vicinity of the defect. The remaining delocalized band states are only displaced slightly in energy. This result justifies the retention of the band model and its extension with defect levels to describe crystals distorted by a small concentration of defects

potential leads to a splitting off of one state from the bend considered. If $U(r)$ is positive, the uppermost state splits off; if $U(r)$ is negative the lowest splits off. There are only minor displacements of the states within the band. While the wave function in the band remains approximately a delocalized Bloch function, the wave function associated with the split-off state is localized. We shall not carry out the calculations which lead to Fig. 5.7. The models to be discussed below show the same properties and are more valid for the problem of a defect in a crystal.

It is difficult to discuss Eq. (2) in general because of the unknown form of the function $U(r)$. We obtain the simplest form if we take the defect to be a donor atom in an elemental semiconductor, i.e., we take a $n$ + 1-valent atom in a covalent-bonded lattice of n-valent atoms. The donor atom then introduces one electron and one additional positive charge to the nucleus. This electron is not required for the covalent bonds to the nearest neighbours. U(r) in this case is the potential of the additional positive charge, in which the additional electron moves.

For large separation of the electron from the positive charge, the crystal lattice screens the Coulomb potential just like a homogenous medium of dielectric constant ε. If we use this continuum approximation, and thus put in Eq. (2) the potential

$$U(r) = -\frac{e^2}{4\pi\varepsilon\varepsilon_0 r} \qquad \text{...(3)}$$

We have reduced the perturbing potential to that of an H-atom in a medium of dielectric constant, ε.

We proceed to solve Eq. (2) for the motion of a wave packet in an electric field. For the motion of the electron in the field $U(r)$ of the defect, we construct a wave packet of Bloch functions

$$\psi = \sum_{n,k} c_n(k)\, \psi_n(k, r). \qquad \text{...(4)}$$

We put this function into Eq. (2) and obtain,

$$\sum_{n,k} c_n(k)\,(H_0 + U)\psi_n(k,r) = \sum_{n,k} c_n(k)\,(E_n(k) + U)\,\psi_n(k,r)$$

$$= \sum_{n,k} c_n(k)\,(E_n(-i\nabla) + U)\,\psi_n(k,r) = E\psi \qquad \text{...(5)}$$

Further rearrangement only becomes possible when we restrict the number of Bloch functions used to construct the wave packet.

As a first restriction we use only conduction band functions. This we can do if the energy with which the electron is bound to the defect is small compared to the energy with which the valence electrons are bound to the lattice (the energy gap, $E_G$). In the other case, Bloch functions from the valence band have also to be included. Defects to which these restrictions can be applied are called *shallow impurities*. The condition is fulfilled for most donors. Defects with binding energies comparable with $E_G$ act as traps and recombination centres. The binding energies for the most important donors in Ge and Si are less than 1% of the band gap.

Thus, in Eq. (5) we can omit the summation over $n$ and hence the index itself.

We find

$$[E(-i\nabla) + U]\psi = E\psi \quad \text{with} \quad \psi = \sum_k c(k)\,\psi(k, r) \qquad ...(6)$$

The second restriction follows from Eq. (3). Replacement of the lattice with a homogeneous medium of given dielectric constant contains the assumption that the orbit of the electron bound to the defect traverses many lattice cells. The extent of the wave packet in space is thus large compared with the lattice constant. Consequently, its extent in $k$-space is small compared with the dimensions of the Brillouin zone. Thus, only $k$-vectors from a narrow region around the band minimum contribute to Eq. (4). If we look first at the case of a simple, isotropic parabolic minimum at $k = 0$, the summation in Eq. (4) only runs over small value of $k$. Since the lattice periodic part $u$ in the Bloch function $\psi(k, r) = u(k, r)\cdot \exp(ik\cdot r)$ only changes slowly with $k$, we can replace $u(k, r)$ by $u(0, r)$. We thus obtain the wave packet

$$\psi = \sum_k c(k)\,u(0, r)\exp(ik\cdot r)$$

$$= \left[\sum_k c(k)\exp(ik\cdot r)\right]\psi(0, r) \equiv F(r)\,\psi(0, r) \qquad ...(7)$$

If one puts this into (6) then,[1] and

$$E(-i\nabla)F(r)\psi(0, r) = \sum_m E_m F(r + R_m)\,\psi(0, r + R_m)$$

$$= \psi(0, \mathrm{r})\sum_m E_m \exp(R_m\cdot\nabla)\,F(r) = \psi(0, r)\,E_n(-i\nabla)\,F(r) \qquad ...(8)$$

it follows that

$$[E(-i\nabla) + U]\,F(r) = EF(r) \qquad ...(9)$$

---

[1] The function $E_n(\vec{k})$ is periodic in $k$-space and can be expanded in a Fourier series

$$E_n(\vec{k}) = \sum_m E_{nm}\exp(i\vec{R}_m\cdot\vec{k}) \qquad ...(i)$$

We can construct $a_n^m$ operateor $E_n(-i\nabla)$ by replacing all $\vec{k}$ in $E_n(k)$ by $-i\nabla$:

$$E_n(-i\nabla)\,\psi_n(\vec{k}, \vec{r}) = \sum_m E_{nm}\exp(\vec{R}_m\cdot\nabla)\,\psi_n(\vec{k}, \vec{r})$$

$$= \sum_m E_{nm}\left[1 + \vec{R}_m\cdot\nabla + \frac{1}{2}(\vec{R}_m\cdot\nabla)^2 + ...\right]\psi_n(\vec{k}, \vec{r})$$

$$= \sum_m E_{nm}\,\psi_n(\vec{k}, \vec{r} + \vec{R}_m) = \sum_m E_{nm}\exp(i\vec{R}_m\cdot\vec{k})\,\psi_n(\vec{k}, \vec{r})$$

$$= E_n(\vec{k})\,\psi_n(\vec{k}, \vec{r}) \qquad ...(ii)$$

Equation (9) differs from Eq. (6) in that the rapidly changing function $\psi$ has been replaced by the slowly changing "envelope-functions" $F(r)$. One can then expand the operator $E(-i\nabla)$ and terminate the expansion after the quadratic term

$$E(-i\nabla) = E_C = \frac{\hbar^2}{2m^*}\nabla^2$$

(with $E_c$ = lower edge of the conduction band). ...(10)

This leads here to

$$\left(-\frac{\hbar^2}{2m^*}\nabla^2 - \frac{e^2}{4\pi\varepsilon\varepsilon_0 r}\right) F(r) = (E - E_C)\, F(r) \qquad ...(11)$$

where we have inserted the explicit from Eq. (3) for $U(r)$.

Equation (11) is the Schrödinger equation for an electron with effective mass $m^*$ in the field of a positive charge in a medium of dielectric constant, $\varepsilon$. We know the solution to this equation from the H-atom. The eigenvalues are

$$E_l = E_C - \frac{1}{l^2}\frac{e^4 m^*}{32\pi^2\varepsilon^2\varepsilon_0^2\hbar^2}, \quad l = 1, 2, .... \qquad ...(12)$$

The envelope function for the ground state is

$$F(r) = \frac{1}{\sqrt{\pi a_0^{*3}}}\exp\left(-\frac{r}{a_0^*}\right), \quad a_0^* = \frac{4\pi\varepsilon_0\hbar^2}{me^2}\frac{m}{m^*}\varepsilon \qquad ...(13)$$

The eigenfunctions for the excited state can readily be found in a similar way.

We note first of all that the bound states of the electron form a spectrum, similar to that of the H-atom, which lies below the lower edge of the conduction band. The Bohr radius for the orbit of the ground state is increased relative to that of the free H-atom (0.53 Å) by a factor $\varepsilon\,(m/m^*)$. For Si and Ge this leads to values between 20 and 50 Å. Further below we shall see that the binding energy for such orbits is reproduced fairly well by this approximation and that for the even larger orbits of the excited states there is almost exact agreement.

We obtain that the extent to the wave packet in $k$-space by evaluating the coefficients $c\,(k)$ of the expansion in Eq. (7) for known $F(r)$. We find for the coefficients

$$c\,(k) = \frac{8\pi^{1/2}}{V_g^{1/2}\, a_0^{*5/2}}\,(k^2 + a_0^{*-2})^{-2}. \qquad ...(14)$$

The wave packet is limited in $k$-space to a region having a radius of the order of $1/a_0^*$. In Si and Ge this is only a few per cent of the mean radius of the Brillouin zone.

The Eqs. (12) and (14) suggest that in the $E\,(k)$-diagram of the band model we should represent *donor levels* as discrete state below the minimum of the conduction band. The width of the level [given by Eq. (14)] can also be shown on the diagram, as a measure of the concentration of the wave packet Eq. (7) in $k$-space. This is shown in Fig. 5.8 for the case discussed so far of a donor atom in a semiconductor with isotropic, parabolic conduction band minimum.

We have used Si and Ge as our examples of the order of magnitude to be expected for the binding energies and Bohr radii, although thee semiconductors show marked deviations from the isotropic model. We can readily find the corrections needed when we note that these corrections concern assumptions we made in the context of the effective mass approximation. In Si and Ge we find, instead of an isotropic, parabolic minimum, a number of equivalent minima centred around the $k$-vectors of a star in the Brillouin zone. The energy surfaces

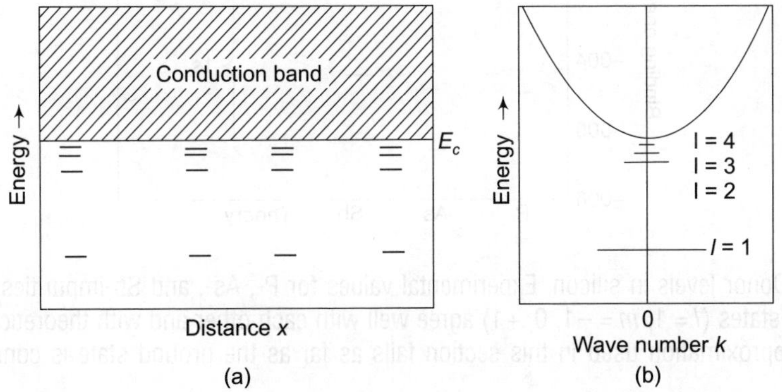


**Figure 5.8 (a) and (b)** Impurity levels in the band model. (a) in an energy space diagram, the localized impurity levels lie below the delocalized conduction band states. (b) in an $E$-$k$ diagram, the localized impurity levels lie below the minimum of the conduction band (here assumed isotropic and parabolic). The width of the impurity levels gives the extent of the state in $k$-space, according to Eq. (14)

around these minima are rotational ellipsoids, which are defined by two effective masses

$$E = E_C + \frac{\hbar^2}{2}\left(\frac{k_i^2}{m_1} + \frac{k_j^2 + k_k^2}{m_t}\right) \quad ...(15)$$

Here $m_1$ is the longitudinal and $m_t$ the transverse mass (reciprocal of the curvature parallel and perpendicular to the major axis of the ellipsoids). $k_i$, $k_j$, $k_k$ are the components of $k$ in a suitably oriented set of Cartesian coordinates, with centre at the respective $k$-vector of the minimum $k_c$. We handle this anomaly by setting up an effective mass Eq. (11) for each extremum, and write the kinetic energy operator in the form

$$-\frac{\hbar^2}{2m_1}\frac{\partial^2}{\partial x_i^2} - \frac{\hbar^2}{2m_t}\left(\frac{\partial^2}{\partial x_j^2} + \frac{\partial^2}{\partial x_k^2}\right) \quad ...(16)$$

The elliptical shape of the energy surfaces leads to a preferred direction. Eigenvalues of the Schrödinger Eq. (11) which were degenerate in the isotropic case then split. The split levels are classified by a further quantum number $m$, which corresponds to the magnetic quantum number in the free atom case. Deviations from the Coulomb spectrum thus appear.

In Fig. 5.9 we compare the results from a theory improved in this way with donor spectra in Si. One can see an almost quantitative agreement between theoretical predictions and

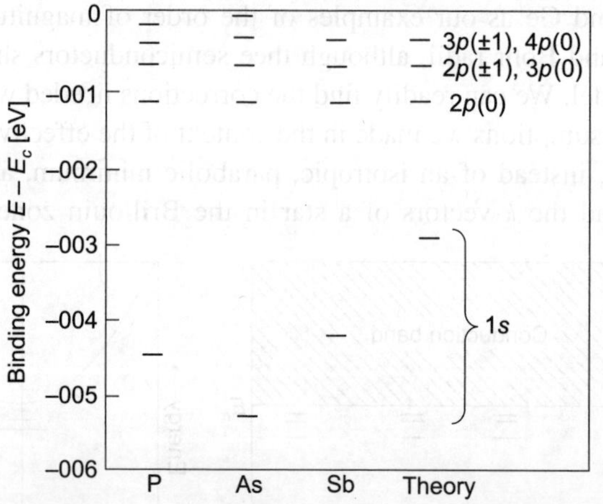


**Figure 5.9** Donor levels in silicon. Experimental values for P-, As-, and Sb-impurities. The energies of the excited states ($l = 1, m = -1, 0, +1$) agree well with each other and with theoretical predictions. The approximation used in this section fails as far as the ground state is connected

experiment for the excited states. The agreement is not so good for the ground state. Here the approximation (3) for the potential $U(r)$ is not adequate. It is apparent that the discrepancies stem from the failure of this approximation since the energies of the excited states are practically independent of the nature of the donor, while the energy of the ground state is different for P, AS, and Sb.

So far we have restricted our attention to *shallow donors*. A corresponding effective mass Eq. (11) can be set up for *shallow acceptors* (replacement of electrons in the conduction band by holes in the valence band). $F(r)$ is then the envelope function of a wave packet constructed out of states from the valence band. The Coulomb spectrum, Eq. (12) consists of a series of levels which lie above the valence band.

Defects with larger electron binding energies lead to levels which lie deeper in the energy gap. Electrons in these centres are strongly localized. An effective mass approximation is impossible. States from throughout the entire Brillouin zone and from several bands are needed to construct a wave packet. Defect such are these are called *deep defects*. (see Table 5.3)

**Table 5.3** Definition and Significance of Deep and Shallow Levels

| *Shallow levels are defect levels produced by the long-range Coulomb potential of defects. Deep Levels are defect levels produced by the central cell potential of defect.* | | |
|---|---|---|
| | *Deep level* | *Shallow level* |
| Energy | May or may not be near band edge. Spectrum is not hydrogen-like. | Near band edge. Spectrum is hydrogen-like. |
| Typical properties | Recombination centres. Compensators. Electron–hole generators. | Suppliers of carriers. |

There is a further obstacle to setting up a theory for deep impurities. Impurity atoms—whether at lattice sites or interstitial—do not fit in well to the host lattice. A displacement of the equilibrium positions of neighbouring atoms arises when they are inserted. Hence, $U(r)$ consists not only from contributions from the impurity atom itself, but also from the lattice atoms which surround it.

## 5.11 ELECTRON-PHONON INTERACTION IN IONIC CRYSTALS (POLARONS)

Consider an ionic crystal (e.g., KCl) and a conduction electron in a rigid lattice of KCl (Fig. 5.10). This 'extra' electron slightly shifts the ion from their equilibrium position due to its own electric field, i.e., it polarises the lattice (Fig. 5.10 (b)). In the absence of the electron such polarisation would be unstable, but the displacement of atoms in the lattice results in a potential well for the electron (due to its strain field). The electrons energy is reduced and it is trapped into the well which it has "dug-out" for itself. If this lowers the total energy (i.e., the energy of the electron plus that of the lattice), this complex identity will have the "right to exist". This combination of the electron and its associated strain field is called a *polaron.*

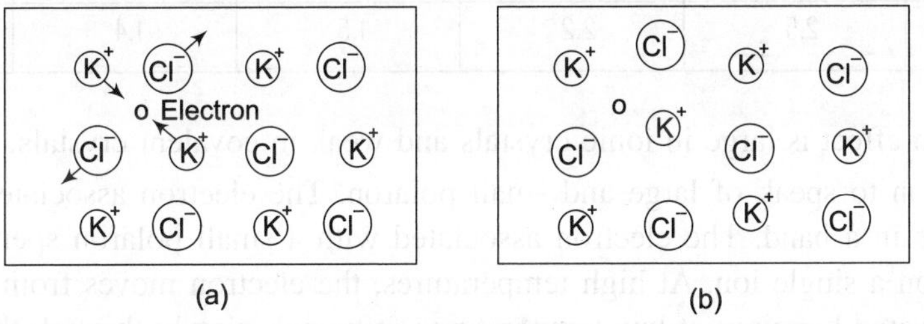


**Figure 5.10** The formation of a polaron due to polarisation of the lattice by a conduction electron

The force applied to the electron in the potential well is transferred to the atoms of the lattice, and therefore, the moving electron drags with itself, its potential well.

The formation of polarons is attributed to electron–phonon interaction. The strength of the electron–lattice interaction is measured by the dimensionless coupling constant $\alpha$ given by

$$\frac{1}{2}\alpha = \frac{\text{deformation energy}}{\hbar\omega_L} \quad ...(1)$$

where $\omega_L$ is the longitudinal optical phonon frequency near zero wave vector. $\alpha/2$ is viewed as the number of phonons that surround a slow moving electron in a crystal.

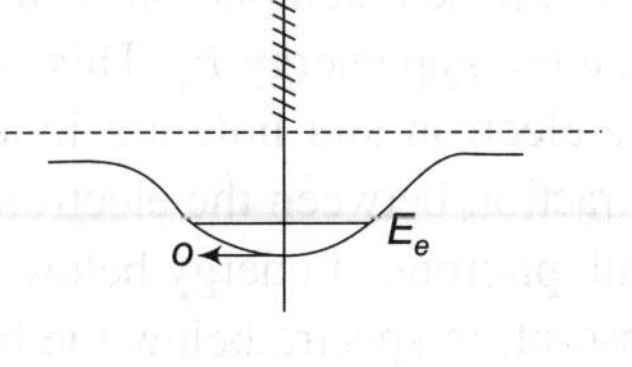


**Figure 5.11** An electrons energy $E_e$ diminishes if the electron is trapped in a potential well. The electron and potential well produced (due to lattice polarisation) can move together as an individual entity, the polaron

Inside a lattice, polaron's motion is more hindered than that of free electrons in a band, because the motion is accompanied by repolarisation of the surrounding. Therefore, the effective mass of a polaron is substantially more than that of an electron.

Values of α deduced from diverse experiments and theory are given in Table 5.4 after F.C. Brown. The values of the effective mass $m^*_{pol}$ of the polaron are from cyclotron resonance experiments. Theory relates $m^*_{pol}$ to the effective band mass $m^*$ of the electron in the undeformed lattice by the relation

$$m^*_{pol} \simeq m^* \left(1 + \frac{1}{6}\alpha\right) \quad (\text{for } \alpha << 1) \qquad ...(2)$$

Since α is always positive, $m^*_{pol}$ is always greater than the bare mass.

**Table 5.4** Coupling Constants α, Polaron Effective Masses $m^*_{pol}$ and Band Masses $m^*$ for Electrons in the Conduction Band

| *Crystal* | *KCl* | *KBr* | *AgCl* | *AgBr* | *InSb* |
|---|---|---|---|---|---|
| α | 3.97 | 3.52 | 2.00 | 1.69 | 0.014 |
| $m^*_{pol}/m^*$ | 2.5 | 2.2 | 1.5 | 1.4 | 1.0 |

The polaron effect is large in ionic crystals and weak in covalent crystals.

It is common to speak of large and small polaron. The electron associated with a large polaron moves in a band. The electron associated with a small polaron spends most of its time trapped on a single ion. At high temperatures, the electron moves from site to site by thermally activated hopping; at low temperatures, it tunnels slowly through the crystal.

## 5.12 EXCITONS

The optical effects are based largely on the independent-electron approximation and involves perfect solids. Now, effects that involve more than one electron or involve imperfect solids are considered. The study begins by looking at electron-hole pairs and the effect of the Coulomb interaction between them. This gives six to what are called *excitons.*

Suppose that a photon is incident on a semiconductor (or insulator) with an energy greater than the gap energy $E_g$. This is sufficient to create a free electron-hole pair (i.e. one in which the electron and hole are independent of each other). However, since there is a Coulomb attraction between the electron and hole, bound states also exist, and these may be produced with photons of energy below $E_g$. These states are excitons and give rise to sharp lines in the absorption spectra below the band edge. A description of excitons requires going beyond the independent electron approximation and is an example of a correlation effect.

Apart from photon-lattice interaction, phonon may also interact with the electrons of the crystal. Although the phonon energy is usually much smaller than electrons energy, but the collision may lead to a bound electron-hole pair (an exciton).

These are called *electronic imperfections*. An exciton can move through the crystal transporting excitation energy, but it is itself electrically neutral. There are two limiting cases:

(a) Frenkel exciton (small and tightly bound)

(b) Mott–Wannier exciton (weakly bound, with an electron hole separation large in comparison with lattice constant)

### 5.12.1 Frenkel Excitons

A Frenkel exciton is essentially an excited state of a single atom and the exciton is tightly bound, i.e., the hole is usually on the same atom as the electron but the pair may be anywhere in the crystal. The excitation, although localized but can hop from one atom to next by virtue of coupling between neighbours in a crystal (there is hopping-conduction) and the exciton travels as a wave through the crystal.

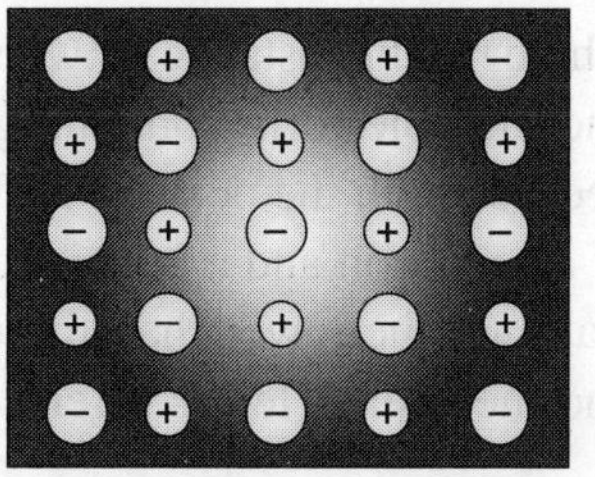

**Figure 5.12** A Frenkel exciton in an alkali-halide crystal (shown localized on one atom)

A Frenkel electron is an excited electronic state. Since, in an ionic crystal, the coupling between the ion's outer electron shells is strong, the excitation energy is actually transferred from ion to ion. Thus the Frenkel exciton can move through the crystal without causing the ions to shift. As a result (like polaron) it is far more mobile than vacancies, interstitials or substitutional impurities.

Frenkel excitons occur when the Coulomb interaction is not so strongly shielded by polarization charge. Consequently, $a_{ex}$ (i.e., size of an exciton) is comparable to the lattice constant, and the long range Coulomb potential no longer describes the interaction.

Suppose that there is a tightly bound electron–hole pair with a total energy $E_g - E_B$, on a given site ($E_B$ = exciton binding energy). This pair can tunnel to neighbouring sites and the system can be described by the tight-binding Hamiltonian:

$$H = \sum_{\vec{R}} (E_g - E_B)\left|R\right\rangle\left\langle\vec{R}\right| + \sum_{R,\vec{u}} (t\left|R+u\right\rangle\left\langle\vec{R}\right| + t^*\left|\vec{R}\right\rangle\left\langle\vec{R}+\vec{u}\right|) \quad ...(1)$$

where $\vec{u}$ denotes a set of vectors pointing to the NNs and $|\vec{R}\rangle$ denotes the state of a pair bound to site $\vec{R}$. The empirical parameter $t$ determines the hopping rate of excitons from site to site. Daigonalization of this Hamittonian occurs by writing the eigenstate as the phased sum:

$$|\psi\rangle = \sum_{\vec{R}} e^{i\vec{k}\cdot\vec{R}}\left|\vec{R}\right\rangle \quad ...(2)$$

The energy eigenvalues are distributed in a band:

$$E(k) = E_g - E_B + \sum_{\vec{u}} (te^{-i\vec{k}\cdot\vec{u}} + t^* e^{i\vec{k}\cdot\vec{u}}) \quad ...(3)$$

The Bloch-type states represent a set of excitons that are delocalized over the entire crystal. Thus, whereas the electron and hole remain tightly bound to each other, with a size comparable to a lattice constant, the correlated pair can have a wave-function that is spread out over the entire crystal.

## 5.12.2 Weakly Bound Excitons (Wannier–Mott Exciton)

Consider an electron in the conduction band and a hole in the valence band. The electron and hole attract each other by the Coulomb potential

$$U(r) = -\frac{e^2}{4\pi\in_o\in_r} \quad ...(4)$$

where $r$ is the distance between them and $\in$ is the dieletric constant of the medium. The electron and the hole form something similar to the hydrogen atom (or more precisely, positronium) because both the positive and the negative particles are light ones. This quasi-atom consisting of an electron and a hole is referred to as the Wannier–Mott exciton. The energy of the bound state will be lower than the energy $(\varepsilon_c - \varepsilon_v)$ of the non-interacting electron and hole by $E_{ex}$ and the energy levels referred to the top of the valence band are given by

$$E_n = E_g - \frac{\mu e^4}{2\hbar^2\in^2 n^2} \equiv E_g - E_{ex} \quad ...(5)$$

where $n$ is the principal quantum number and $\mu$ is the reduced mass

$$\frac{1}{\mu} = \frac{1}{m_e} + \frac{1}{m_\hbar} \quad ...(6)$$

formed from the effective masses $m_e$, $m_n$ of the electron and hole. The exciton ground state energy is obtained on setting $n = 1$ (this gives ionization energy of the exciton).

$$E_{ex} = \frac{\mu e^4}{2\hbar^2\in^2} = (\mu/m)\,(1/2)\,\frac{me^2}{\hbar^2}\,\frac{e^2}{\in^2} \quad ...(7)$$

Since $\dfrac{\hbar^2}{me^2} = a_o$ (Bohr radius)

$$\therefore \quad E_{ex} = \frac{\mu}{m}\frac{e^2}{2a_o\in^2}$$

or $$E_{ex} = \frac{\mu}{m}\frac{1}{\in^2}\,(13.6)\text{ eV} \quad ...(8)$$

where $\dfrac{e^2}{2a_o} = 13.6$ eV (° Rydberg)

Thus, the lowest bound state extends over a radius given by

$$a_{ex} = \frac{\hbar^2}{\mu(e^2/\in)} = \frac{\in m}{\mu}\frac{\hbar^2}{me^2} = \in\left(\frac{m}{\mu}\right)a_o \quad ...(9)$$

(which we get by replacing $m$ by $\mu$ and electronic charge $e$ by effective charge $e^2/\in$).

The validity of his model requires that $a_{ex}$ be large on the scale of the lattice (i.e., $a_{ex} >> a_o$) but insulators with small energy gaps tend to have small effective masses and large dieletric constants that is not difficult to achieve, particularly in semiconductors.

Thus, the system of exciton energy levels (Fig. 5.13) resembles that of hydrogen levels and this is just revealed by the reflectance and absorption spectra of some crystal which shows structure for photon energies just below the energy gap. The (Fig. 5.14) shows effect of an exciton level on the optical absorption of a semiconductor for photons of energy near the band gap $E_g$ in GaAs at 21 K (the gap $E_g$ = 1.5 21 eV and the exciton binding energy = 0.0034 eV).

**Table 5.5** Binding Energy of Exctions in meV

| Si | 14.7 | BaO | 56 | RbCl | 440 |
|---|---|---|---|---|---|
| Ge | 3.8–4.1 | InP | 4.0 | LiF | (1000) |
| Ga As | 4.2 | In Sb | (0.4) | Ag Br | 20 |
| GaP | 3.5 | KI | 480 | Ag Cl | 30 |
| CdS | 29 | KCl | 400 | TlCl | 11 |
| Cd Se | 15 | K Br | 400 | Tl Br | 6 |

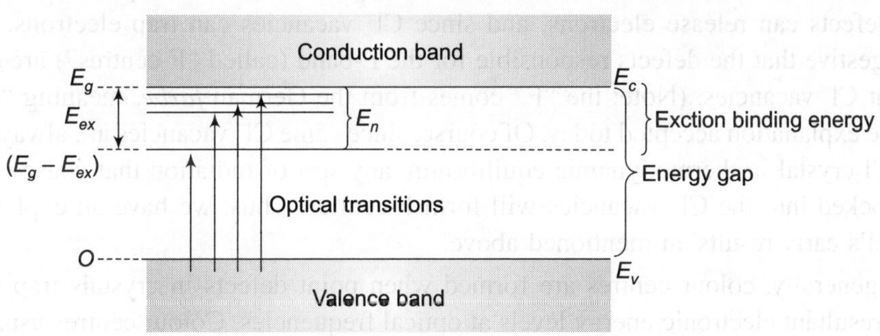


**Figure 5.13** Energy level $E_n$ of an exciton created in a direct process. The binding energy of the excition is $E_{ex}$, referred to a free electron and a free hole

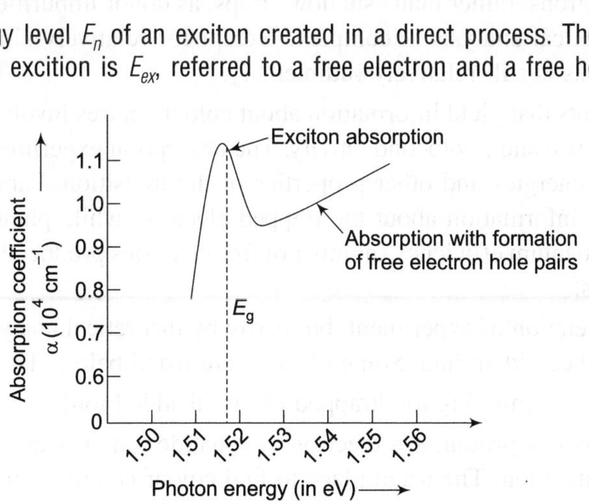


**Figure 5.14** Exciton absorption in gallium arsenide

## 5.13 COLOUR CENTRES

The study of colour centres arose out of the curiosity as to what caused the yellow colouration of rock salt (NaCl) and other colouration in similar crystals. This yellow colour was particularly noted in salt just removed from a mine. Becquerel found that NaCl could be coloured by placing the crystal near a discharge tube. From a fundamental point of view, NaCl should have an infrared absorption due to vibrations of its ions and an ultraviolent absorption due to excitation of the electrons. A perfect NaCl crystal should not absorb visible light, and should be uncoloured. The colouration of NaCl must be due, then, to defects in the crystal. The main absorption band in NaCl occurs at about 4650 Å (the "F-band"). This blue absorption is responsible for the yellow colour that the NaCl crystal can have. A further clue to the nature of the absorption is provided by the fact that exposure of a coloured crystal to white light can result in the bleaching out of the colour. Further experiments show that during the bleaching, the crystal becomes photoconductive, which means that electrons have been promoted to the conduction band. It has also been found that NaCl could be coloured by heating it in the presence of Na vapour. Some of the Na atoms become part of the NaCl crystal, resulting in a deficiency of Cl and, hence, $Cl^-$ vacancies. Since photoconductive experiments show that F-band defects can release electrons, and since $Cl^-$ vacancies can trap electrons, it seems very suggestive that the defects responsible for the F-band (called "F-centres") are electrons trapped at $Cl^-$ vacancies. (Note: the "F" comes from the German *farbe*, meaning "colour".) This is the explanation accepted today. Of course, since some $Cl^-$ vacancies are always present on a NaCl crystal in thermodynamic equilibrium, any sort of radiation that causes electrons to be knocked into the $Cl^-$ vacancies will form F-centres. Thus, we have an explanation of Becquerel's early results an mentioned above.

More generally, colour centres are formed when point defects in crystals trap electrons with the resultant electronic energy levels at optical frequencies. Colour centres usually form "deep" traps for electrons, rather than "shallow" traps, as donor impurities in semiconductors do, and, their theoretical analysis is complex. Except for relatively simple centres such as F-centres, the analysis is still relatively rudimentary.

Typical experiments that yield information about colour centres involve optical absorption, paramagnetic resonance and photoconductivity. The absorption experiments give information about the transition energies and other properties of the transition. Paramagnetic resonance gives wave function information about the trapped electron, while photoconductivity yields information on the quantum efficiency (number of free electrons produced per incident photon) of the colour centres.

Mostly by interpretation of experiment, but partly by theoretical analysis, several different colour centres have been identified. Some of these are listed below. The notation is

[missing ion |trapped electron| added ion],

where are notation is $p \equiv$ proton, $e \equiv$ electron, $- \equiv$ halide ion, $+ \equiv$ alkaline ion, and $M^{++} \equiv$ doubly charged positive ion. The usual place to find colour centres is in ionic crystals.

$[-|e|] = F$-centre

$$[-|2e|] = F'\text{-centre}$$

$$[\mp-|2e|] = M\text{-centre (?)}$$

$$[|e|p] = U_2\text{-centre}$$

$$[+|e|M^{++}] = Z_1\text{-centre (?)}$$

In Figs. 5.15 and 5.16 we give models for two of the less well-known colour centres. In these two figures, ions enclosed by boxes indicate missing ions, a dot means an added electron, and a circle includes a substitutionally added ion.

Colour centres turn out to be surprisingly difficult to treat theoretically with precision. But success has been obtained using modern techniques on, e.g., $F$ centres in LiCl. In recent years tunable solid-state lasers have been made using colour centres at low temperature.

**The F-centre**

We discuss here intrinsic defects of an ionic crystal for which there is just a single trapped carrier. The most famous and most studied of these centres in the F-centre, in which an

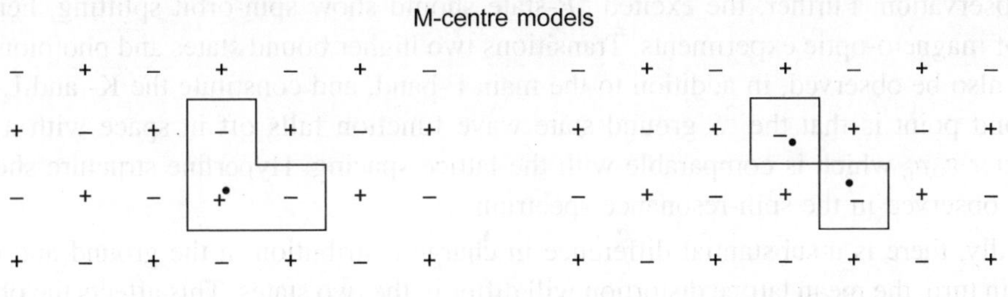


**Figure 5.15** Models of the M-centre: (a) Seitz, (b) Van Doorn and Haven

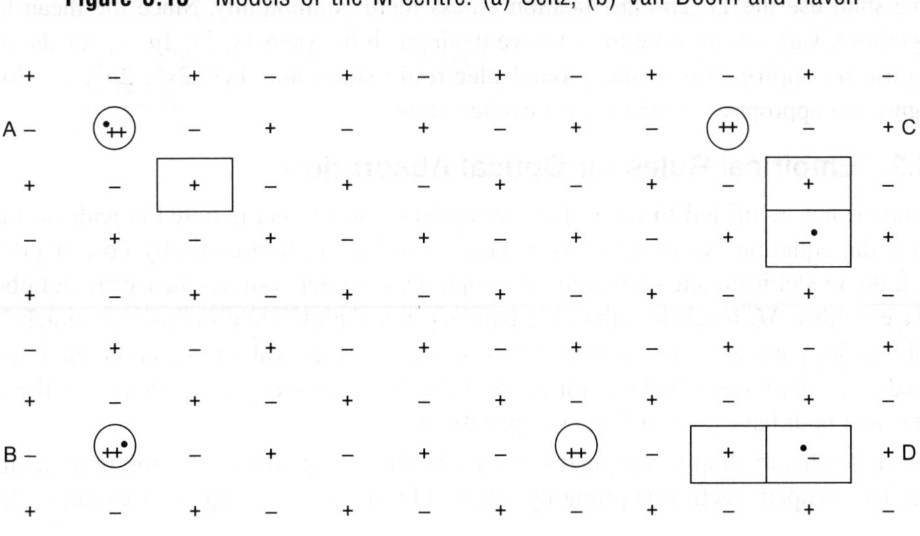


**Figure 5.16** Four proposed models for $Z_1$-centres

electron is trapped at an anion vacancy. Its importance comes from its high symmetry (cubic in the alkali halides, tetrahedral in the alkaline earth fluorides), from its ease of formation, and from the convenient values of its optical and spin-resonance parameters.

### 5.13.1 The Hydrogen Atom Model of the F-Centre

Several important aspects of the theory can be inferred from a very simple model. The *F*-centre is regarded as an electron of mass $m_0$ trapped in a Coulombic potential; screened by the optical dielectric constant, $\varepsilon_\infty$. Thus, the energy levels and wave functions are simply scaled from those of the hydrogen atom. The Rydberg is decreased by a factor $\varepsilon_\infty^{-2}$, and the Bohr radius is increased by $\varepsilon_\infty$. First, we see that the dominant optical transition should be the analogue of the $^2S \to {}^2P$ transition, with energy

$$\Delta E_H \simeq \left(\frac{3}{4}\varepsilon_\infty^{-2}\right) Ry \qquad ...(1)$$

These energies correspond to light in the visible range, and are in fairly good agreement with observation. Further, the excited $^2P$-state should show spin-orbit splitting, hence the value of magneto-optic experiments. Transitions two higher bound states and photoionization should also be observed, in addition to the main F-band, and constitute the K- and L-bands. A second point is that the $^2S$ ground state wave function falls off in space with a length parameter $\varepsilon_\infty a_0$ which is comparable with the lattice spacing. Hyperfine structure should be readily observed in the spin-resonance spectrum.

Thirdly, there is a substantial difference in charge distribution in the ground and excited states. In turn, the mean lattice distortion will differ in the two states. This affects the observed properties, notably spin-resonance and optical transition energies.

The hydrogenic model is particularly convenient for labelling the electronic eigenstates, and we shall use the 1*s*, 2*p* classification unless there is ambiguity. Since the mean lattice configuration varies from state to state, we distinguish between 1*s*, 2*s*, 2*p*, ... for the lattice configuration appropriate to the ground electronic state, and 1*s**, 2*s**, 2*p**, .... for the configuration appropriate to the lowest excited state.

### 5.13.2 Empirical Rules for Optical Absorption

One should not be misled to think that the model on which (1) is based is realistic merely because the equations work quite well. This point has been stressed by Fowler (1966b). Indeed, the model itself shown that the assumptions of effective-mass theory are not obeyed. We have neither $\Delta E$ much less than the bandgap nor the effective radius $\varepsilon_\infty a_0$ much larger than the lattice parameter. To emphasize this point, two empirical interpolation relations may be cited. They both described the variation of the F-band energy well throughout the alkali halides, and both have quite different implications.

The first scheme simply recognizes that $\varepsilon_\infty$ is the wrong sort of macroscopic parameter to use. The trapped electron is primarily affected by the nearest-neighbour cations, whereas

$\varepsilon_\infty$ is dominated by the anions: in 12 out of 17 alkali halides the anion polarizability is more than twice that of the cations. This feature is corrected by modifying (5.13.1) to give

$$\Delta E'_H = \Delta E_H/(A + BR_+/R_-) \quad ...(2)$$

The ionic polarizability increases with ionic radius, so that the correction can be written as a function of $R_+/R_-$. Equation (2) holds to about 20 per cent throughout the alkali halides. This is shown in Fig. 5.17. Analogous formulae hold for other electron-excess centres. The result suggests that atomic properties are important, but that these may be regarded as a small correction to the effective dielectric constant which depends on their detailed electronic structures. Levy (1956) has given a result similar to Eq. (2), but distinct from it.

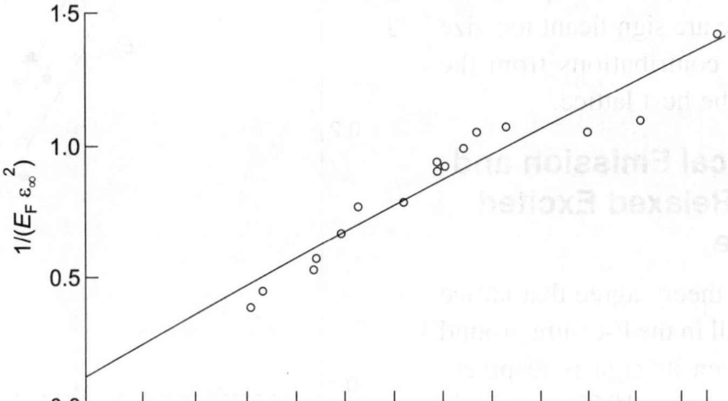


**Figure 5.17** F-centre transition energies. Ion-size effects within a continuum model are shown $R_+$ abd $R_-$ are Pauling ionic radii. $E_F$ is in units 0.75 *Ry* to show clearly the correspondence to the hydrogen atom

By contrast, there is another two-parameter interpolation relation. This is the Mollwo-Ivey rule (Ivey 1947) and is a generalization of an earlier results (Mollwo 1931, Fröehlich 1933)

$$\Delta E_{MI} = Ca^{-n}, \quad ...(3)$$

where $a$ is the nearest-neighbour spacing. It holds to 20 per cent to 30 per cent throughout the alkali halides. The relation is shown in Fig. 5.18. Theoretically, the transition energy will depend on lattice parameter alone only if the potential in which the defect electron moves scales with the lattice parameter, i.e., $V = V(r/a)$. The only simple potential of this sort is the point-ion potential (Wood 1965a; Hayes & Stoneham 1969). If, in addition, the electronic wavefuchtion scales with the lattice parameter $\psi = \psi(r/a)$, then Wood has shown that the coefficient $n$ lies in the range $1 < n < 2$, as observed. The Mollwo-Ivey rule suggests that details of atomic structure are almost irrelevant and that it is the Madelung-type terms, ignored in (5.13.3), which do, in fact, dominated.

These two models suggest very different interpretations of the trends in F-band transition energy throughout the alkali halides. We shall see later that certain features of both are valid. The successes of such different models stem from the fact that several factors (e.g., $\varepsilon_\infty$ and the ionic wave functions characterized by $R_+$, $R_-$) have similar trends. The major contribution to the transition energy can be attributed to the point-ion potential. In addition, there are significant ion-size corrections and contributions from the polarization of the host lattice.

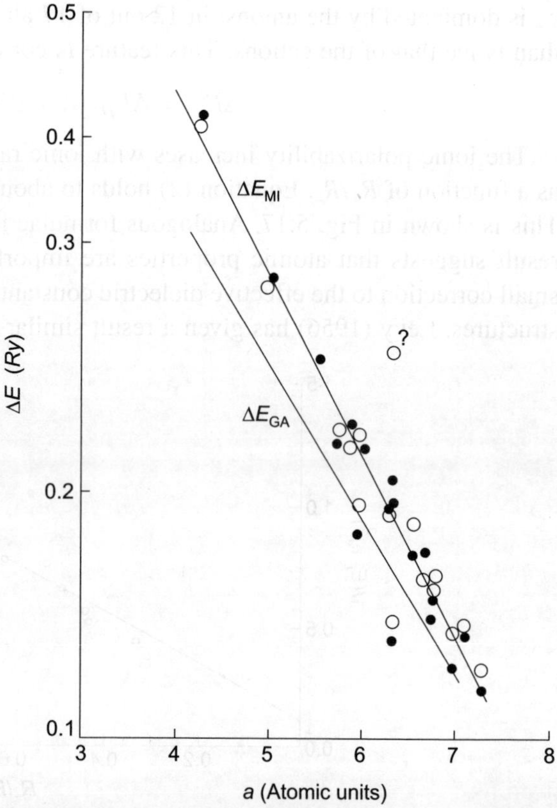


**Figure 5.18** F-centre transition energies. The energies in Rydbergs (*Ry*) are shown as a function of nearest neighbour distance. Open circles give experimental results, and full circles the Bartram-Stoneham-Gash results, including ion-size corrections with $\alpha = 0.53$. The Mollwo-Ivey rule given $\Delta E_{MI}$, and Gourary & Adrian's point-ion results are given as $\Delta E_{GA}$

## 5.13.3 Optical Emission and the Relaxed Excited State

Experiment and theory agree that lattice distortion is small in the F-centre ground state. Indeed, even its sign is disputed. Thus, Feuchwang's (1962) analysis of spin-resonance data indicated local contraction for LiF and KCl, but direct measurements of volume change indicate an expansion in KCl (Peisl, Balzer, & Waidelich 1966) and KBr (von Guerhard, Peisl, & Waidelich 1969).

Lattice distortion varies from state to state, and generally it is agreed that distortion and polarization are important in the relaxed excited state. This is partly because the charge distribution in the excited state differs greatly from that of the anion which has been removed, and partly because the 2*p*-state reached by optical absorption is degenerate and should show a Jahn–Teller effect. The significant distortion in the excited state is demonstrated by the large differences in absorption and emission energies and by other phenomena which will be discussed.

The simplest picture of the absorption-emission cycle is this. In optical absorption $1s \to 2p$, the host ions remain fixed (the Franck–Condon principle) although their electronic polarization changes. Once in the excited state, the lattice distorts rapidly. The lifetimes are long enough for equilibrium to be achieved. Radiative transitions $2p^* \to 1s^*$ to the ground electronic state usually dominate; the host ions remain fixed in optical emission. The lattice then distorts back to its ground-state configuration. Two types of different behaviours are observed. One example is Fulton & Fitchen's (1970) observation that, after emission in CsF, there is a delay before the characteristic F-centre absorption appears again. This suggests a bottleneck of some sort in the relaxation of the $1s^*$-state to the $1s$ ground state. The other cases are those (Li halides, NaBr, NaI, and alkaline earth flourides) where no emission at all is observed or (alkali flourides and NaCl) where the radiative efficiency is less than unity at the lowest temperatures. Competing non-radiative processes explain the observations. The absence of luminescence can be understood particularly easily since it appears to result from direct transitions from the excited electronic state to the ground state (Dexter, Klick, & Russell 1956; Bartram & Stoneham 1975). These transitions are primarily associated with vibronic states corresponding to the crossover of the configuration-coordinate curves. But, since the $2p$-state decays rapidly by phonon emission to the $2p^*$-state, this crossover configuration is only important if the unrelaxed $2p$-state lies higher in energy than the crossover. This leads to the rule: luminescence should be observed if the final state reached in absorption ($2p$) lies above the cross-over of the $1s$ and $2p$ configuration coordinate curves. Moreover, in the simplest model the rule is equivalent to:

$$\frac{1}{4}\text{ absorption energy} \geq \text{Lattice relaxation energy} \equiv S\hbar\omega$$

for luminescence. Since the Huang–Rhys factor $S$ and effective phonon energy $\hbar\omega$ can be found from the temperature dependence of the absorption line-shape, one can predict the observability of luminescence from absorption data alone. A fuller, quantitative discussion is given by Bartram & Stoneham (1975).

Whilst the form of the cycle is reasonably well established, the details are far from clear. Two pieces of evidence are particularly disturbing. First, the lifetime of the excited state is far longer than one would expect using the Einstein relations and transition matrix elements from absorption data (Fowler & Dexter 1962; Swank & Brown 1963; Fowler 1964). This argues that the electronic wave function is greatly modified by the lattice relaxation, although it gives no information about the symmetry of the distortion. Two main models have been suggested. The '*extended* $2p$-*state*' model argues that the distortion is such as to favour a $2p^*$ wave function of much larger radial extent than that which occurs in absorption. The overlap of the $1s$ and $2p$ functions is reduced, so that the transition matrix element and emission rate are small. On the other hand, the '$2s$-$2p$' *model* argues that the wave function of the relaxed state contains a substantial admixture of the $2s$-state. Since the $1s \leftrightarrow 2s$ transition is forbidden, such an admixture tends to increase the excited-state lifetime for radiative decay. The second piece of evidence shows an $s$-like state is involved in the relaxed excited state. Bogan & Fitchen's stark-effect data show an $s$-like state just below the $2p$-state in emission. Note that this order of energy levels is opposite to the $2s$-$2p$ order found in absorption.

Spin resonance results for the relaxed excited state have become available recently (Mollenauer, Pan & Yngvesson 1969; Ruedin, Schnegg, Jaccard & Aegerter 1970; Mollenauer, Pan, & Winnacker 1971; Mollenauer 1971; Mollenauer& Baldacchini 1972; Baldacchini &

Mollenauer 1973). The most recent results show that in KBr and KI the relaxed state is indeed widespread. In units of the nearest-neighbour distance, $\langle r \rangle$ is about 4.9 for KBr and 5.2 for KI. However, it is not possible to tell if it is the $2s^*$ state or $2p^*$ state observed. Nor is it simple to extrapolate to other systems; we shall see later that KCI can be understood without invoking a spread-out state. Even if the *s*-*p* model is appropriate, there is still ambiguity about the nature of the states. As Tomura (1967) observed, they may be vibronic states, rather than the electronic $2s$ or $2p$ state.

## 5.13.4 Empirical Rules for Optical Emission

As for absorption, there exist empirical rules for the transition energies. The usual Mollwo–Ivey law in the form $\Delta E \sim a^{-2.5}$ works adequately (Podini & Spinolo 1966). Better is the hydrogenic form for the optical emission energy.

$$\Delta E_{\text{emiss}} = \frac{3}{4}\frac{m^*/m_0}{\varepsilon_\infty^2}, \quad \text{...(4)}$$

where $m^* = 0.541\ m_0$ for the NaCl structure and 0.741 for the caesium halides (Gilbert 1968). However, agreement with experiment becomes worse if the measured effective masses ($m^*$) are used. Spinolo (1965) found a comparable expression is adequate for the thermal ionization energy of the excited state. This energy is significantly less than the value obtained by scaling $\Delta E_{\text{emiss}}$, suggesting that there is a substantial difference in lattice distortion between the empty vacancy and the relaxed excited state. Part of this may be a Jahn–Teller distortion.

## 5.13.5 Other One-carrier Centres

Studies of a number of simple centres related to the F-centre have given important results. We shall discuss three such centres in Sec. 5.14. They are:

1. The $F_A$-centre: an F-centre with a nearest-neighbour cation impurity, e.g., $Na^+$ replacing one of the $K^+$ ions in KCl.
2. The $Z_1$-centre: an F-centre with a divalent cation and a cation vacancy nearby; there is still some uncertainty as to the precise configuration.

**Table 5.6** Contributions to the Ground-state Energies of F-centres in Alkali Halides and Alkaline Earth Flourides

| Compound | $E_{pi}$ (Ry) | $E_{ls}$ (Ry) | $E_{pol}$ (Ry) | $E_{dist}$ (Ry) | $E$ (Ry) |
|---|---|---|---|---|---|
| CsF | −0.442 | −0.068 | −0.014 | −0.022 | −0.547 |
| RbF | −0.469 | −0.048 | −0.010 | −0.009 | −0.536 |
| KF | −0.490 | −0.017 | −0.007 | −0.003 | −0.516 |
| RbCl | −0.422 | −0.013 | −0.003 | −0.003 | −0.442 |
| RbBr | −0.407 | −0.010 | −0.003 | −0.002 | −0.422 |
| NaF | −0.543 | 0.036 | −0.003 | −0.000 | −0.510 |
| KCL | −0.437 | 0.001 | −0.002 | −0.002 | −0.440 |

*Contd...*

*Table 5.6 Contd...*

| | | | | | |
|---|---|---|---|---|---|
| KBr | −0.422 | 0.001 | −0.002 | −0.001 | −0.424 |
| RbI | −0.389 | −0.002 | −0.002 | −0.002 | −0.395 |
| KI | −0.402 | 0.007 | −0.001 | −0.002 | −0.398 |
| NaCl | −0.474 | 0.023 | −0.001 | −0.000 | −0.453 |
| LiF | −0.593 | 0.059 | −0.001 | −0.001 | −0.535 |
| NaBr | −0.455 | 0.017 | −0.001 | −0.000 | −0.439 |
| NaI | −0.429 | 0.016 | −0.000 | −0.001 | −0.414 |
| LiCl | −0.505 | 0.031 | −0.000 | −0.000 | −0.475 |
| LiBr | −0.482 | 0.022 | −0.000 | −0.000 | −0.460 |
| LiI | −0.453 | 0.020 | −0.000 | −0.001 | −0.434 |
| $CaF_2$ | −0.535 | −0.018 | −0.009 | −0.002 | −0.563 |
| $SrF_2$ | −0.511 | −0.057 | −0.011 | −0.007 | −0.586 |
| $BaF_2$ | −0.479 | −0.115 | −0.017 | −0.020 | −0.631 |

$E_{pl}$ includes the kinetic energy and point-ion potential. $E_{IS}$ is the ion-size correction. $E_{pol}$ and $E_{dist}$ are respectively the corrections for polarization and distortion, and $E$ is the total energy.

## 5.14 PERTURBED *F*-CENTRES

Several defects occur which can be regarded as an *F*-centre in an alkali halide with an adjacent impurity or other defect. Examples are the following.

1. *The $F_A$-centre.* Here one of the nearest-neighbour cations is replaced by a different alkali ion. In almost all observed cases the impurity is smaller than the replaced host ion. Thus, the *F*-centre in KCl might have one $Na^+$ neighbour. Most of the work in this chapter will be concerned with the $F_A$-centre, since it is the simplest and best understood of these defects.
2. *Photochromic F-centre—Rare-earth complexes.* Most of the theory treats centres consisting of an *F*-centre in the fluorite structure with an adjacent trivalent impurity ion (Alig 1970), although another charge state with an extra electron trapped is also observed.
3. *F-centres with vacancies at the nearest anion or cation sites.* The *F*-centre with an adjacent cation vacancy has only been identified in the alkaline earth oxides. It is a dipolar centre, with no net charge to bind the trapped electron. The *F*-centre with an adjacent anion vacancy is the $M^+$-centre, an analogue of the $H_2^+$molecular ion: the defective electron is trapped by two equivalent anion vacancies.
4. *The $Z_1$–centre.* This probably consists of an *F*-centre with a nearest-neighbour cation vacancy at, say, $(a, 0, 0)$ and a divalent metal ion at one of the sites $(a, \pm a, \pm a)$. The $Z_2$-centre, also associated with a divalent cation, appears to be a two-electron centre of uncertain structure.

## 5.15 CENTRES WITH TWO CARRIERS

We consider here the several centres in ionic crystals for which there are two trapped defect electrons or holes. One of the main features of interest is the electrostatic interaction between the two carriers.

Three types of centres will be considered. The first has two electrons trapped at an anion vacancy. The $F'$-centre.[2] The second involves two electrons trapped at close anion vacancies: the $M$-centre in alkali halides and centres like the $F_t$-centre in alkali earth oxides. Thirdly, there are defects with two holes trapped near cation vacancies in alkaline earth oxides. Obviously some of the methods of calculating the energy levels for these systems are virtually identical with those for one-carrier systems. In this chapter we concentrate mainly on the new features: the existence of bound state and the relative energies of states with parallel and anti-parallel spins.

## 5.16 THE $F'$-CENTRE

### 5.16.1 Bound States in the Alkali Halides

The $F$-centre is electrically neutral and gives no long-range Coulomb potential. The second electron is the $F'$-centre is bound by short-range interactions. Indeed, only a single bound state appears to exist, although there is some evidence of low-energy resonant states, The binding energy of the single state is quite large; as a rule the thermal ionization energy is about 0.5 eV, and the photoionization threshold is about 1 eV. (Thus, there is a strong resemblance between the $F'$-centre and the isoelectronic defects.

## 5.17 THE *M*-CENTRE

The M-centre consists of two electrons trapped at adjacent anion vacancies. It can be regarded as two adjacent $F$-centres and, as such, in the simplest of the $F$-aggregate centres proposed by Pick (1960) and reviewed by Compton & Rabin (1964). The R-centre (chapter 17) in another such aggregate.

The axis of the $M$-centre is $\langle 110 \rangle$ in the alkali halides, and is $\langle 100 \rangle$ in the alkaline earth fluorides and caesium halides. We shall concentrate on the alkali halides. There are three main optical transitions of the $M$-centre. One is induced by light with the electric-field vector along the axis. This $\sigma$ transition gives the main $M_1$ band. It is slightly lower in energy than the corresponding $F$-band, and the $M_1$ transition energy appear to obey Mollwo–Ivey rule. The other two transitions are included by light with a field vector normal to the axis. Their energies are very close to the $F$-band energy and probably obey a Mollwo-Ivey rule. The $M_2$ band is induced by light with a field vector in the [001] plane containing the [110] defect axis,

[2] There are two systems of nomenclature for two electrons trapped at an anion in the alkaline earth oxides. One, favoured by those interested in interaction between defects, calls the centre an $F$-centre. These authors argue that it is the net charge relative to the lattice which is most important, and so the neutral two-electron centre corresponds to the neutral one-electron F-centre in alkali halides. Others, including the original workers, have been more interested in electronic structure, and call the two electron centres $F'$-centres in both alkali halides and alkaline earth oxides. We adopt this second, original notation. It is hard to see that the slight benefits using 'F-centre' for the two-electron system have ever compensated for the confusion caused. On the rare occasions the net charge is needed, one is always free to use $F^0$ or $F^+$, as appropriate.

whereas the $M'_2$ band is induced by light with an [001] field vector. These are shown in Fig. 5.19. In addition, there are higher energy bands analogous to the *K* and *L* band of the *F*-vector centre. Some of these transitions may be to spin-triplet states (see Fowler 1968b, p. 112). In emission only the σ transition corresponding to the $M_1$ band is observed. The lifetime is short and the Stokes shift small, indicating a relaxed excited state very similar to the unrelaxed one (Fowler 1968, p. 114). The radiatives lifetimes can be obtained with reasonable accuracy using Einstein relations and scaling of $H_2$ properties (Bosi, Bussolati, & Cova 1972).

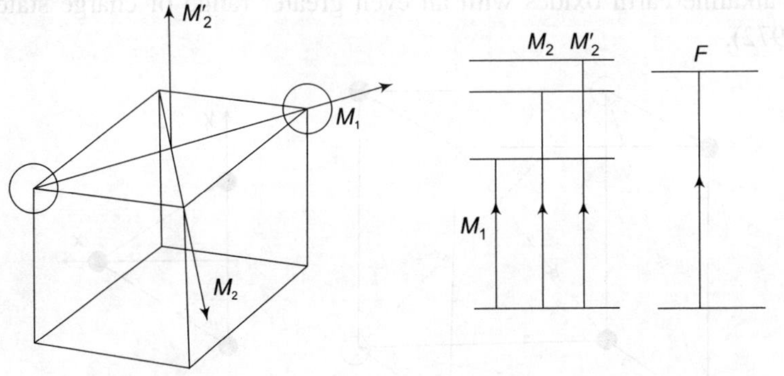


**Figure 5.19** M-centre transitions. The dipole directions for the three transitions are shown schematically for the (110) M-centre in alkali halides

The general magnitudes of the transition energies are predicted reasonably well by the types of theory used for the *F*-centre. These include continuum models (Nagajima & Tatsuma 1954; Herman, Wallis & Wallis 1956), modifications of the Heitler–London method (Nagajima, Kojima, & Kondoh 1954), variational calculations based on the point-ion model (Evarestov 1964), and extended-ion models (Meyer & Wood 1964; Wood & Meyer 1964). There are two special features which are very sensitive to the details of the calculation. One is the order and separation $M_2$ and $M'_2$ bands. The other is the separation of the lowest singlet and triplet states— a small separation of a few tenths of an electron volt.

**Table 5.7** Single-triplet Splitings for M-centres

| | *Splitting (eV)* | | | | | | |
|---|---|---|---|---|---|---|---|
| | *LiF* | *LiCl* | *NaF* | *NaCl* | *KCl* | *RbB* | $CaF_2$ |
| Berezin, semi-empirical | 0.249 | 0.142 | 0.201 | 0.115 | 0.086 | 0.082 | 0.049 |
| Berezin, variational | 0.569 | 0.320 | 0.411 | 0.259 | 0.199 | 0.161 | |
| Wood & Meyer | 0.180 | 0.180 | — | — | — | — | |
| Evarestov | — | — | — | 0.380 | --- | — | |
| Experiment | — | — | — | — | 0.064 | — | >0.1 |

## 5.18 THE *R*-CENTRE

The *R*-centre consists of three nearest-neighbour *F*-centres in an alkali-halide crystal. It is shown in Fig. 5.20. The *F*-centres are at the vertices of an equilateral triangle, and the defects has a [111] axis.

Centres of this type are observed in many alkali halides in one of three charge states: the neutral *R*-centre, the $R^+$-centre with one fewer trapped electron, and the R'-centre with one fewer trapped electron, and the R'centre with an extra defect electron. Similar centres are found in the alkaline earth oxides with an even greater range of charge states (Hughes & Henderson 1972).

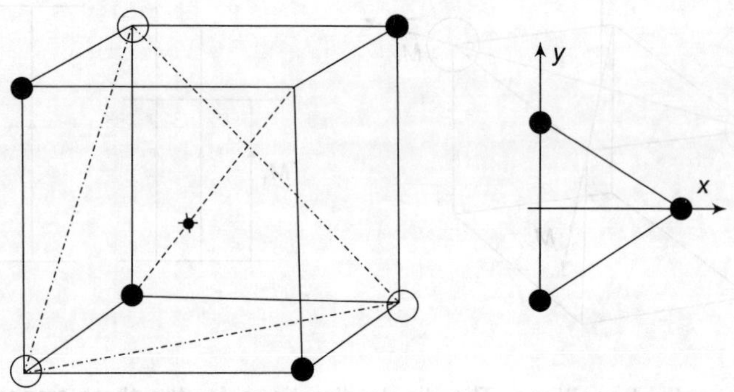


**Figure 5.20** R-centre geometry. (a) The trigonal arrangement of three F-centres, with their nearest cations. Displacement of the cation (full circle) on the (111) axis has a particularly profound effect. (b) Axes for defining mode symmetries

## 5.19 THE $V_K$-CENTRE

The $V_k$-centre is the self-trapped hole in an ionic crystal. It is not a defect in the usual sense: there are no atoms missing, no extra, atoms, nor any impurities. There is simply a hole in the valence band, which causes substantial local distortion of the host. This distortion renders the whole relatively in mobile, and the hole can be observed optically and in spin resonance, as if it were a conventional defect.

Self-trapped holes have been observed in the alkali halides (Kanzig 1955), in the alkaline earth fluorides (Beaumont, Hayes, Kirk, & Summers 1970), in more complicated crystals like $KMgF_3$ (Hall 1966), and in the silver halides (Ulrici 1970). In all but the silver halides the basic configuration can be regarded as an $X_2^-$ molecule replacing two neighbouring anions $(X^-)_2$ in the closed packed direction ([110] for alkali halides and [100] for alkaline earth fluorides). This description works very well: many of the properties of the $V_k$-centre are close to those of a free $X_2^-$ molecule at the appropriate internuclear distance. One of the main aims of this chapter is to assess the molecular approximation. The silver halides are different. Ulrici has proposed that the hole is trapped in the 4*d* cation shell. Obviously the optical and spin-resonance properties will be modified greatly.

This first question concerning self-trapped holes is, why are they self-trapped at all? In silicon, for instance, the holes are simply carriers like electrons, but described by different parameters. Gilbert (1966) has summarized the reason for self-trapping lucidly. He observes that one may go in three stages from a situation where there is a delocalized hole in an undistorted lattice to the self-trapped configuration. Self-trapping is favoured if the sum of the change in total energy in these stages in negative.

The first stage localizes the hole by constructing a wave packet from the band states of the undistorted lattice. The localization energy needed to do this is the difference between the mean energy in the band and the band extremum and is about half the bandwidth.

$$\Delta E_1 = E_{\text{loc}} > 0. \quad \text{...(1)}$$

The second stage allows polarization of the lattice by the localized hole whilst the separation of the two $V_k$ anions is kept fixed. This gives a reduction in energy,

$$\Delta E_2 = E_{\text{pol}} < 0. \quad \text{...(2)}$$

For holes in alkali halides the sum $(\Delta E_1 + \Delta E_2)$ is small, and its sign cannot be fixed without detailed calculation. The third stage allows the $X_2$ to relax to optimize its separation. Owing to changes in the positions of neighbouring atoms, there will be sundry small terms in the energy, but the main contribution will be the change in the bonding energy of the $X_2^-$ molecule.

$$\Delta E_3 \simeq E_{\text{bond}} < 0. \quad \text{...(3)}$$

Rough estimates by Gilbert indicate that the hole in alkali halides should be self-trapped, whereas the electron should not be.

| | Holc in KCl | Electron in KCl |
|---|---|---|
| $E_{\text{LOC}}$ | 0.3 eV (valence band) | 1.9 eV (conduction band) |
| $E_{\text{POL}}$ | – 0.5 eV | –0.3 eV |
| $E_{\text{BOND}}$ | – 1.5 eV ($Cl_2^-$ bond) | – 0.3 eV ($K_2^+$ bond) |
| Total | – 1.7 eV | + 1.3 eV |

The self-trapped hole is even more strongly stabilized in the alkaline earth fluorides (Norgette & Stoneham 1973). Note that electronic degeneracy is not involved at any stage. There is a no need to assume any particular form of band structure, nor to require any special symmetry. It is misleading to regard the self-trapping distortion as a Jahn–Teller effect. Whilst the valence band are usually degenerate, this is no direct influence.

The motion of the $V_k$-centre has been observed. The hopping from site to site occurs with a rate described well, over restricted temperature ranges, by the usual form for thermal activation

$$\tau^{-1} \sim A \exp(-E_a/gT). \quad \text{...(4)}$$

The activation energy is intimately related to the self-trapping distortion. We shall consider the theory of $\tau^{-1}$ when we discuss the $V_k$-centre as an example of a small polaron.

The difference between electrons and holes in ionic crystals is very strongly apparent for the exciton. In essence, the localized self-trapped hole binds an electron in a relatively diffuse orbital. The exciton is not a defect in the strict sense, but is usefully regarded as another charge state of the $V_k$-centre. We shall examine its properties in this light.

### 5.19.1 Atomic Configuration of the $V_k$-centre

The calculation of the atomic configuration of the $V_k$-centre is greatly simplified by the molecular model. There are three major terms in the energy: the interactions within the $X_2^-$ molecular ion, the interactions between the components of the $V_k$-centre and other crystal ions, and the interactions between ions outside the $V_k$-centre.

Two of these contributions can be estimated without difficulty. The interactions between ions outside the $V_k$-centre are assumed to be the same as those in the perfect crystal. Thus, they are a combination of Coulomb interactions, the various dipole-dipole interactions, and short-range repulsive forces. The interactions between the two components of the $V_k$-centre are readily related to the potential-energy curves for the free $X_2^-$ molecule. Gilbert & Wahl (1971) have given Hatree-Fock potential-energy curves for $F_2^-$ and $Cl_2^-$; results for $Br_2^-$ and $I_2^-$ are not available at present.

The third contribution, the interaction between the components of the $V_k$-centre and other ions, presents problems. What charge, polarizability, and repulsive interactions should be attributed to the $V_k$ ions? The charge distribution of $F_2^-$ and $Cl_2^-$ free molecular ions shows that they can be reasonably regarded as two ions $X^{-\frac{1}{2}}$. The repulsive interactions can be taken to be those of $X^-$ ions, or some form of scaling may be attempted to interpolate between values for $X^0$ and $X^-$. The polarizabilities can also be obtained by scaling, or by attempting to fit asymptotic parts of the potential energy curve. But there are ambiguities here which cannot be resolved without more extensive calculations. Jette, Gilbert, & Das (1969) use $X^-$ parameters for the $X_2^-$ components, whereas the analysis of the H-centre by Dienes *et al.*, attempts an interpolation. Norgett & Stoneham (1973) show that the choice of polarizability is not important in the alkaline earth fluorides; nor was the choice of repulsive parameters important, except in calculating certain activation energies for motion.

The most detailed calculations of $V_k$-centre properties are those of Norgett & Stoneham (1973) for the alkaline earth fluorides. These use a shell model for dielectric and lattice-dynamic properties of the host lattice and a very efficient procedure for calculating the equilibrium lattice configurations. The positions of many neighbours were adjusted explicitly, with the Mott-Littleton approximation used for distance ions. Figure 5.21 (a) labels the ions relaxed explicitly; no significant improvement was found by increasing this number. The predictions agree well with observation, notably for the separation of the $V_K$ ions and the corresponding optical and spin-resonance parameters. Remarkably, the $V_K$ separation is almost exactly that for the free $F^-$ ion; the crystalline environment has negligible effect on the molecular ion. The $V_K$ separation (about 1.9 Å) is much smaller than in an undistorted crystal, but is consistent with the similar values of the optical transition energies, hyperfine constants, and $g$-values observed in $CaF_2$, $SrF_2$ and $BaF_2$. The observed optical line-widths agree semiclassical estimates based on the energy surface near equilibrium.

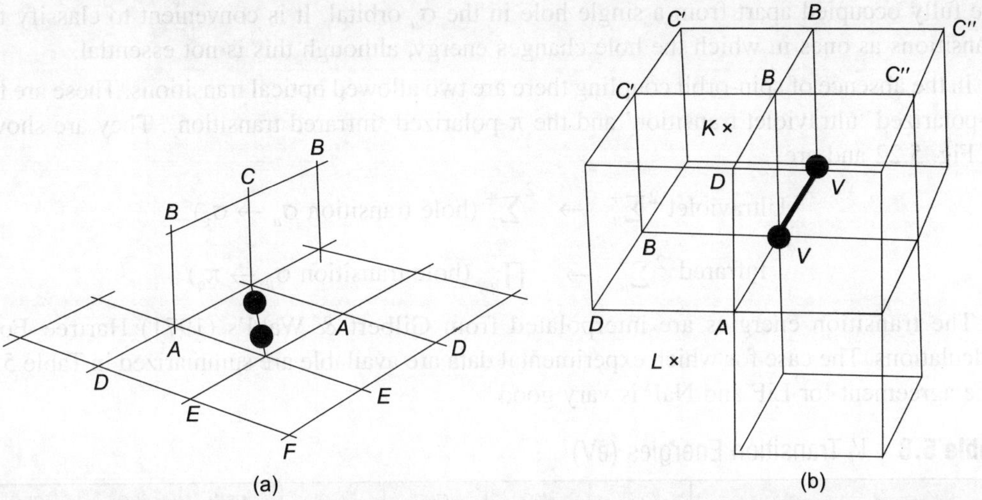


**Figure 5.21** Labelling of sites near a $V_k$-centre. The $V_k$ are shown as full circles. (a) Cation sites in alkali halides *A*, *C*, and *E*; anion sites are *B*, *D*, and *F*. (b) Cation sites in alkaline earth fluorides are *K* and *L*; anion sites are *A*, *B*, *C′*, *C′*, and *D*

Norgett & Stoneham's results were consistent with the earlier treatment of $CaF_2$ by Jette & Das (1969). Jette & Das allowed only a more restricted explicit relaxation (radial relaxation of ions *V*, *K*, *L* and *A* in Fig. 5.21 (a) inside the Mott–Littleton region, and used a much less reliable polarizable point-ion model of the host lattice. Further, only energy terms quadratic in the displacements of the ions were retained. This harmonic assumption is not adequate when there are large distortions; it is not too bad for alkaline earth and alkali fluorides, but does not work well for alkali chlorides.

### 5.19.2 Optical Transitions

In the 'molecule in a crystal' model the optical transitions of the $V_K$-centre are identified with those of the $X_2^-$ molecules ion at the appropriate internuclear separation. The crystal only enters the problem by determining the separation and by changing the line-width of the transition. Delocalization and distortion of the orbitals are ignored.

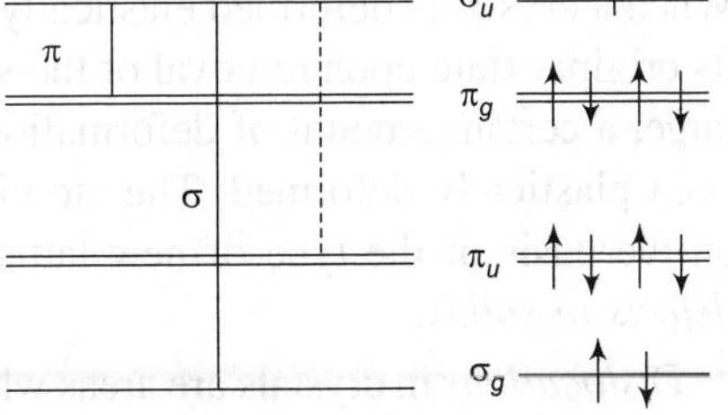


**Figure 5.22** Molecular orbitals of $X_2^-$ for the $V_K$-centre. The full lines donate optical transitions, and the broken lines is the energy separation involved in the *g*-factor. The occupancies of the orbitals in the ground state are also shown

Primarily, the outer *p* electrons of the halogen $X^-$ ions are involved. The molecular orbitals are shown in Fig. 5.22. Here the $\sigma_u$ orbital, for example, is a linear combination of axial *p* orbitals on the two atoms, such that the two atomic terms tend to cancel between the two sites. In the $X_2^-$ system the orbitals are fully occupied apart from a single hole in the $\sigma_u$ orbital. It is convenient to classify the transitions as ones in which the hole changes energy, although this is not essential.

In the absence of spin-orbit coupling there are two allowed optical transitions. These are the σ-polarized 'ultraviolet transition' and the π-polarized 'infrared transition'. They are shown in Fig. 5.22 and are:

$$\text{Ultraviolet } {}^2\Sigma_u^+ \rightarrow {}^2\Sigma_g^+ \text{ (hole transition } \sigma_u \rightarrow \sigma_g)$$

$$\text{Infrared } {}^2\Sigma_u^+ \rightarrow {}^2\Pi_g \text{ (hole transition } \sigma_u \rightarrow \pi_g)$$

The transition energies are interpolated from Gilbert & Wahl's (1971) Hartree–Fork calculations. The case for which experimental data are available are summarized in Table 5.8. The agreement for LiF and NaF is vary good

**Table 5.8** $V_k$ Transition Energies (eV)

| | $CaF_2$ | LiF | NaF | LiCl | NaCl | KCl | RbCl |
|---|---|---|---|---|---|---|---|
| Ultraviolat band | | | | | | | |
| Experiment | 4.04 | 3.48 | 3.38 | 3.15 | 3.28 | 3.39 | 3.40 |
| Jette *et al.* theory | 3.70 | 3.30 | 3.30 | 2.69 | 2.65 | 2.37 | 2.20 |
| Infrared band | | | | | | | |
| Experiment | 1.72 | 1.65 | | | | 1.65 | |
| Jette *et al.* theory | 1.87 | 1.60 | | | | 1.23 | |
| EPR band (mean of $\Delta E_x$ and $\Delta E_y$) | | | | | | | |
| Experiment ($f_p$ = 1) | 2.57 | | | | 2.63 | 2.75 | |
| Jette *et al.* theory | 2.55 | | | | 1.9 | 1.8 | |

## 5.20 DISLOCATION

When a crystal is deformed elastically under the influence of the applied stresses, it returns to its original state upon removal of the stresses. However, if the applied stresses are sufficiently large, a certain amount of deformation remains after the removal of stresses, the crystal has been plastically deformed. The atomic interpretation of plastic flow of crystal is explained on the basis of the type of new lattice defects, called *dislocations*. *Dislocations are linear defects in solids.*

*Dislocations* in crystals are areas where the atoms are out of position in the crystal structure. Dislocations, as evident, are generated and move when a stress is applied.

In many crystals, plastic flow results from sliding of *one* part of a crystal relative to another. The process in which a plane of atoms glides over another by an intergral multiple of an inter-atomic distance is called *slip*.

In contrast to elastic deformation where all atoms take part equally, a *plastic-deformation* is inhomogeneous and *only a relatively small number of atoms actually takes part in the slip-process, viz., only atoms that form layers on either side of a slip-plane.*

Besides being characterized by inhomogeneity, plastic flow is also anisotropic and slip usually takes place preferentially in planes of high atomic density, e.g., along {111} planes in an fcc lattice. Also the direction of slip commonly coincides with a direction along which the number of atomic per unit length is high.

## 5.21 DISLOCATION MODEL FOR SLIP

It may be introduced with reference to the crystal of Fig. 5.23 (a). Let the plane *PQR* be a slip plane. In the slip plane consider an arbitrary closed curve *ABC* (the region inside this curve is hatched [in Fig. 5.23 (b)]). Suppose now that somehow the material location over the hatched area in the upper half of the crystal is displaced by an atomic $\vec{b}$ relative to lower half of the crystal; at the same time, the material in the upper half lying outside area *ABC* is left undisplaced. In this way, we have a situation in which only a fraction of the upper half of the crystal has slipped relative to the lower half.

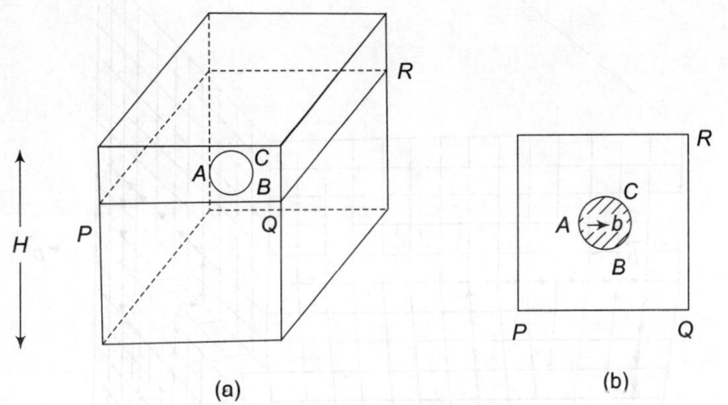


**Figure 5.23** Schematic representation of a dislocation *ABC* in a slip plane *PQR*. Slip has occurred only across the hatched area

The ratio of the area *ABC* to the total area *PQR* is referred to as the fraction of slip that has occurred in this plane. If in some way, the area *ABC* could be made to grow, *f* would increase and for $f = 1$ the whole upper half of the crystal would be displaced by an amount $\vec{b}$ relative to the lower half. For $f < 1$, the average displacement of the upper half relative to the lower half is $f\vec{b}$.

The line *ABC* in the above case marks the boundary in the slip-plane between slipped and unslipped material, this line is called a *dislocation line*. The vector $\vec{b}$, which defines the magnitude and direction of the slip, is called the *Burgers vector*. Since the atoms always seek positions of minimum energy, it implies that $\vec{b}$ must connect two atomic equilibrium positions when the displacement equals one lattice spacing, the dislocation is said to have unit strength.

## 5.22 BURGERS VECTOR

*The magnitude and direction of slip associated with a dislocation are represented by a vector called the Burgers vector. Its magnitude is a measure of the distortion caused by the dislocation*. This Burgers vector helps in identifying the type of dislocation; whether it is edge or screw. In the edge dislocation, the Burgers vector is perpendicular to the line vector of the dislocation, while in the screw dislocation it is parallel to the line vector. If we take a circuit around a dislocation, preferably in regions farther away from the dislocation such that each parallel pair of opposite sides has the same number of atoms of lattice sites, we shall find that the circuit can be closed by the Burgers vector (cf. Figs. 5.24 and 5.25). *Maintenance of lattice periodicity requires that the Burgers vector be equal to the shortest translation vector of the unit cell of the crystal structure.*

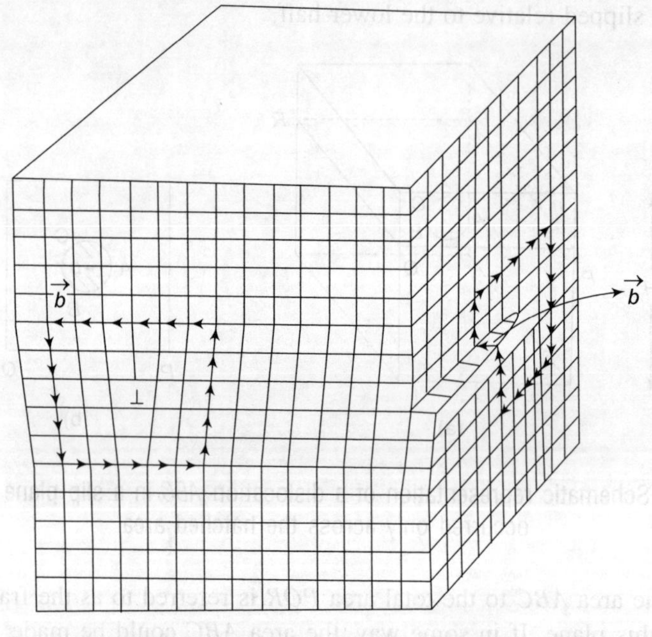


**Figure 5.24** Two Burgers circuits drawn around different segments of the same edge dislocation

## 5.23 THE INTERPRETATION OF SLIP: DISLOCATIONS

It has been observed that the theoretical critical shear stress (based in a perfect lattice) is much larger than the observed values for pure crystals. (In other words, crystals in practice show much smaller strength than expected.) To see this, we resort to a simplified model suggested by Frenkel (Fig. 5.26).

Figure 5.26 shows a closed packed structure and arrangement of atoms in successive layers.

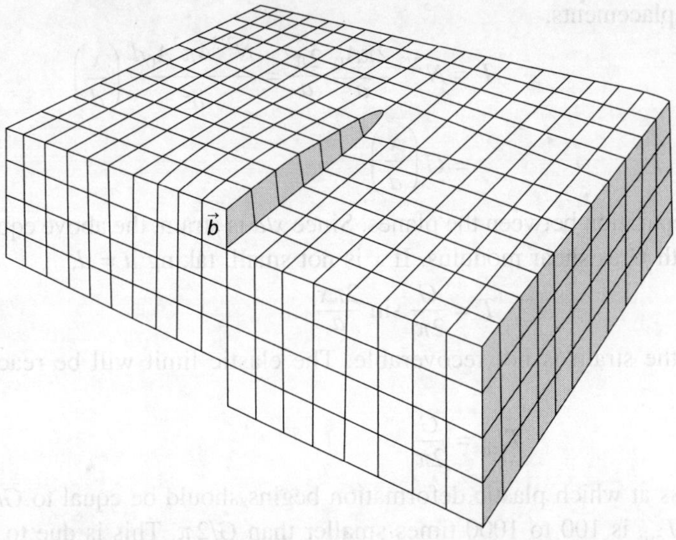


**Figure 5.25** The Burgers vector, $\vec{b}$, in a screw dislocation

Figure 5.26 also indicates variation of potential energy $U$ for the atoms $A$, $B$, $C$ on displacement from their equilibrium positions.

$$U(x) = -A_0 \cos \frac{2\pi x}{a} \quad ...(1)$$

The force on the atom,

$$F(x) = \frac{\partial V}{\partial x} = \frac{2\pi A_0}{a} \sin \frac{2\pi x}{a} \quad ...(2)$$

This must be balanced by the external force, if $T$ is the shear stress (i.e., force/area), then

$$T = N^{2/3} F(x) = N^{2/3} \frac{2\pi A_0}{a} \sin \frac{2\pi x}{a} \quad ...(3)$$

where $N$ is the number of atoms/volume.

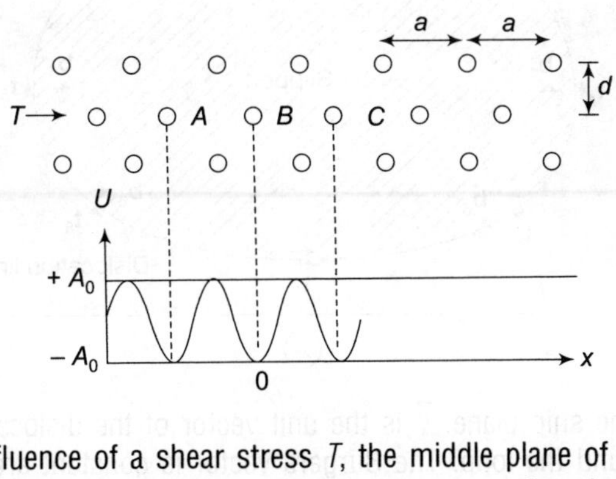


**Figure 5.26** Under influence of a shear stress $T$, the middle plane of atoms is displaced over a distance $x$. The lower curve shows potential energy as a function of displacement $x$

For small displacements,

$$T = N^{2/3}\frac{2\pi A_0}{a}\frac{2\pi}{a} = \frac{N^{2/3}4\pi^2 A_0 d}{a^2}\left(\frac{x}{d}\right)$$

$$= G\left(\frac{x}{d}\right) \qquad ...(4)$$

where $d$ is the separation between the planes. Since $x/d$ is strain, the above equation represents Hooke's law with $G$ as shear modulus. If $x$ is not small, taking $a = d$,

$$T = \frac{G}{2\pi}\sin\frac{2\pi x}{a}$$

For large $x$, the strain is not recoverable. The elastic limit will be reached at $x = a/4$. Therefore,

$$T_{crit} = \frac{G}{2\pi} \qquad ...(5)$$

i.e., critical stress at which plastic deformation begins should be equal to $G/2\pi$. Practically, it is found that $T_{crit}$ is 100 to 1000 times smaller than $G/2\pi$. This is due to the presence of dislocations.

## 5.24 TYPES OF LINE DISLOCATIONS (CAUSED BY SLIP)

Line dislocations are caused by slip and are of two types:

(i) Edge dislocation

(ii) Screw dislocation

Ether of these are described by two vectors. (see Fig. 5.27). The central portion has slipped and the outer portion has not. The plane (shown) across which slip has taken place, is called

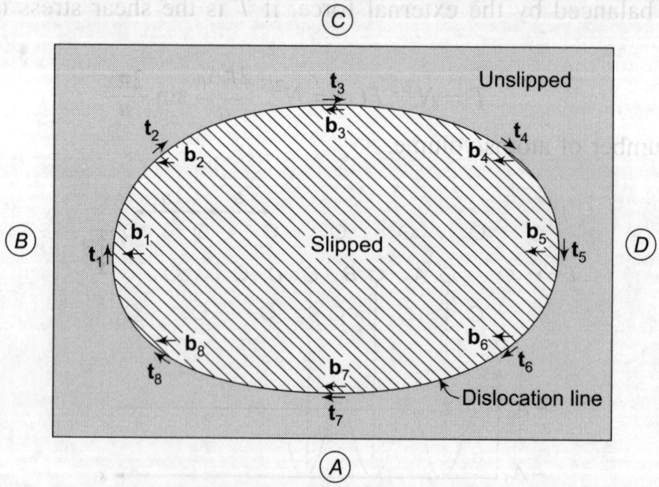


**Figure 5.27** *ABCD* is the ship plane. $\vec{t}$ is the unit vector of the dislocation line and is marked at various points around the loop. The Burgers vector is constant around the loop and is designation by the vector $\vec{b}$

the *slip plane*. The line imperfection which is the boundary between the slipped region and the unslipped region lies in the slip plane and is called a *dislocation*.

We are now in a position to define *edge* and *screw* dislocation segments. A dislocation segment is edge when $\vec{b}$ is perpendicular to $\vec{t}$ and a dislocation segment is screw when $\vec{b}$ is parallel to $\vec{t}$. When the angle between $\vec{b}$ and $\vec{t}$ is between 0° and 90°, the dislocation is called mixed.

The Burgers vector of any dislocation can be determined by using the feat that a series of lattice vectors that form a closed circuit in a perfect lattice will not form a closed circuit if they surround a dislocation, the vector necessary to close the circuit around the dislocation is the Burgers vector of the dislocation. Two such Burgers circuits are shown in Fig. 5.28.

We will consider now edge and screw dislocations separately.

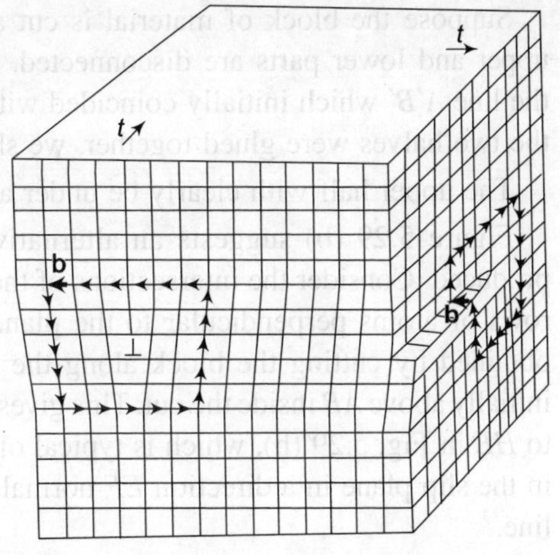


**Figure 5.28** Two Bugers circuits, drawn around different segments of the same dislocation. Each circuit is 5 × 5 and would close in a perfect lattice but here each circuit must be completed by a vector, **b**, the Burgers vector of the dislocation

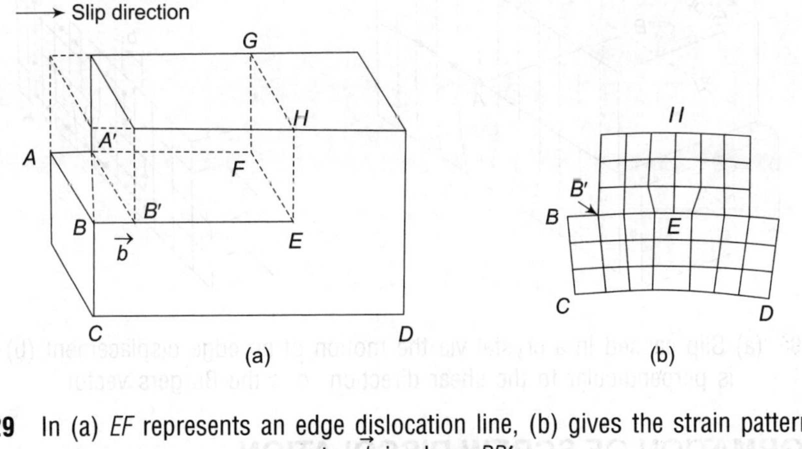


**Figure 5.29** In (a) *EF* represents an edge dislocation line, (b) gives the strain pattern. Burgers vector $\vec{b}$ is along *BB′*

## 5.25 FORMATION OF EDGE DISLOCATION

The simplest dislocation is one for which the dislocation line is straight. A pure edge dislocation is defined as a dislocation for which the Burgers vector $\vec{b}$ is everywhere perpendicular to the dislocation line.

Formation of an edge dislocation may be visualized in terms of a slip process with reference to Fig. 5.29.

Suppose the block of material is cut across the area *ABEF* so that across this area, the upper and lower parts are disconnected. The upper half is then pushed sideways such that the line *A′B′* which initially coincided with *AB* is shifted by an amount $\vec{b}$. If in this position, the two halves were glued together, we should have produced an edge dislocation.

The upper half with clearly be under a compression and the usual half under tension.

Figure 5.29 (b) suggests an alternative method by which an edge dislocation may be produced. Consider the intersections of the network of lines in Fig. 5.29 (b) as it representing rows of atoms perpendicular to the plane of the paper. The edge dislocation may then be obtained by cutting the block along the plane *EFGH* and putting the half plane of atoms initially above *AB* inside the cut. This gives rise to the 'extra' half plane of atoms corresponding to *HE* in Fig. 5.29 (b), which is typical of an edge dislocation. The edge dislocation extends in the slip plane in a direction *EF* normal to the slip direction. *EF* represents the dislocation line.

Figure 5.30 (a) illustrates slip in a crystal via the motion of an edge dislocation. *AB* is the *dislocation line.* Figure 5.30 (b) shows that the dislocation line is perpendicular to the shear direction.

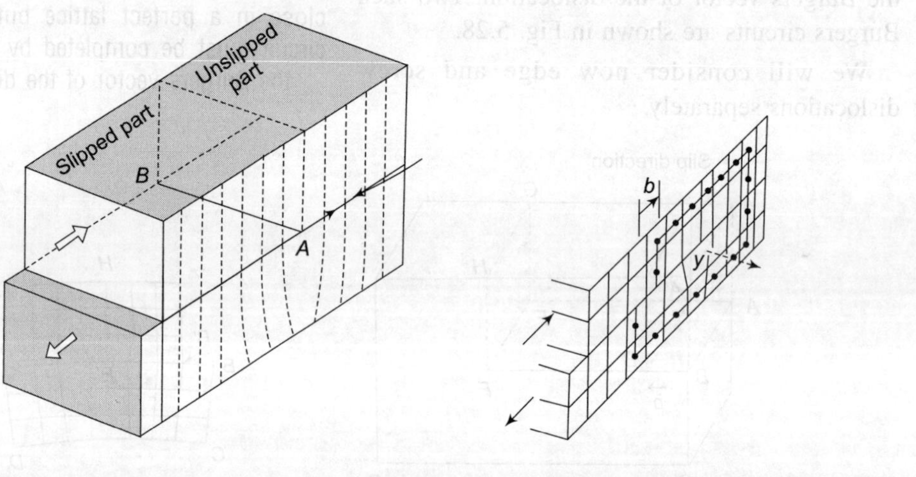


**Figure 5.30** (a) Slip caused in a crystal via the motion of an edge displacement (b) dislocation is perpendicular to the shear direction. $\vec{b}$ is the Burgers vector

## 5.26 FORMATION OF SCREW DISCOLATION

Screw dislocation may be thought of as having been produced by cutting a crystal along a straight line and then shearing the atoms by a Burgers vector $\vec{b}$ parallel to the line. Thus, for a screw dislocation, Burgers vector $\vec{b}$ is everywhere parallel to the dislocation line (Fig. 5.31)

Figure 5.32 represents the atomic configuration in the vicinity of a screw dislocation piercing the surface of a simple cubic lattice. This configuration may be obtained by cutting the block across the area *BFHM* and then pushing the upper part backward in the directon of the Burgers vector $\vec{b}$ as indicated.

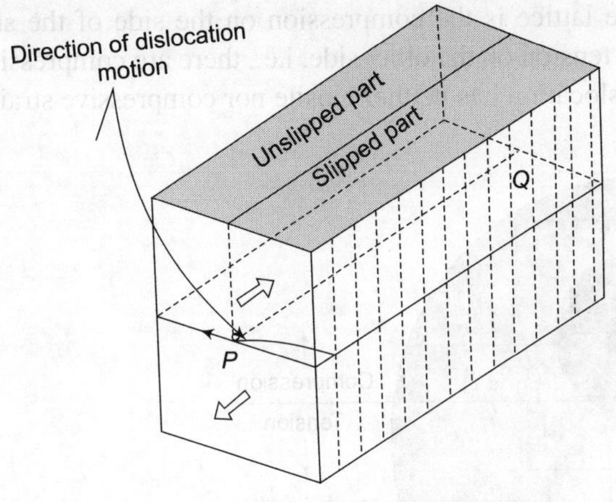


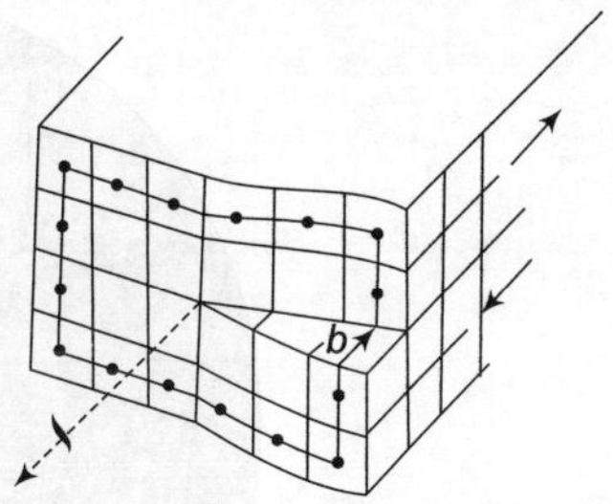

**Figure 5.31** (a) Screw dislocation caused by slip in the way as shown. (b) Dislocation line is parallel to the shear direction. $\vec{b}$ is the Burgers vector

The dislocation line *BM* is parallel to $\vec{b}$. As one moves around the dislocation line along a circuit such as *AKLCDE*, one advances in the direction of *BM* by an amount equal to $\vec{b}$ for every turn. Hence, the term screw dislocation.

The nature of the screw dislocation can be understood from Fig. 5.33. Figure 5.33 (a) is an array of four unit cells of an ideal simple cubic lattice. In Fig. 5.33 (a) have been displaced by the broad arrows resulting in a screw dislocation along the dashed line *PQ*. The thin arrows trace a helical path through the distorted lattice as we go around the dislocation line. Complete planes of atom normal to the dislocation no longer exist, rather atoms lie on a helical surface as onc spirals from one end of the crystal to the other. Note that slip direction is parallel to the dislocation line and so also is the Burgers vector.

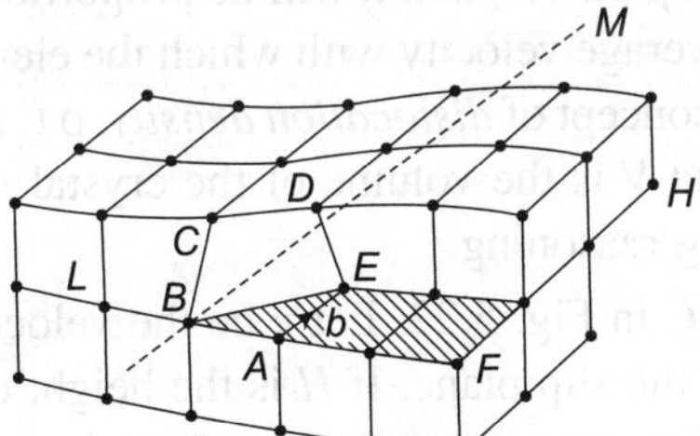


**Figure 5.32** Schematic representaion of a screw dislocationin a simple cubic lattice. *BM* is dislocation line. [After "Solid State Physics" by A.J. Dekker, Macmillan India Limited (1986).]

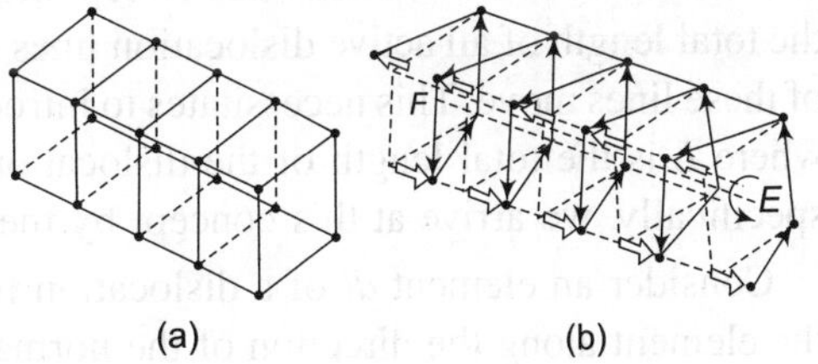


**Figure 5.33** The Nature of screw dislocations

## 5.27 STRESS FIELDS NEAR A DISLOCATION

A dislocation is a solid crystal produces a stress in the surrounding material. An edge dislocation differs from a screw dislocation in the type of stress the lattice experiences around it. Figure 5.34 illustrates the fact that the lattice is the compression on the side of the slip plane where the extra half *p*-lane is and in tension on the other side, i.e., there are compressive and tensite strains. In contrast, a screw dislocation has neither tensile nor compressive strains but rather shear strains associated with it.

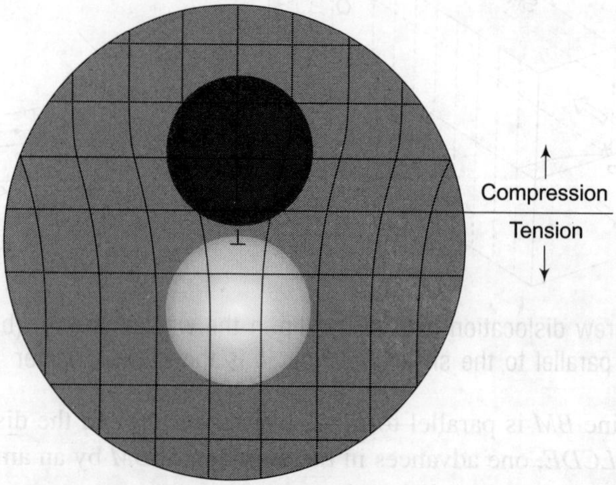


**Figure 5.34** The regions of tension (light) and compression (dark) around an edge dislocation in a simple cubic lattice

## 5.28 DISLOCATION DENSITY

A single dislocation line sweeping across a slip plane gives rise to a displacement in a few Angstroms. So, any appreciable plastic deformation would be the result of a large number of dislocations sweeping across many slip planes. The rate of plastic flow will be determined by the rate at which dislocation lines sweep through the slip planes, i.e., it will be proportional to the total length of all active dislocation lines and the average velocity with which the elements of these lines move. This necessitates to introduce the concept of *dislocation density*, $\rho$ ( $L/V$), where $L$ is the total length of the dislocation lines and $V$ is the volume of the crystal. More specifically, we arrive at this concept by the following reasoning.

Consider an element $dl$ of a dislocation line as $ABC$ in Fig. 5.22. Let $v$ be the velocity of the element along the direction of the normal to $dl$ in the slip plane. If $H$ is the height of the crystal and $A$, the area of the slip plane, the rate of increase in strain due to motion of the element $dl$ is

$$\frac{dy}{dt} = \frac{v\,dl\,b}{AH}\ (y = x/d) \qquad \text{...(1)}$$

If we now consider a rate of flow resulting from all dislocations in planes parallel to the plane $PQR$ in Fig. 5.22, we have to replace $dl$ by the total length $L$ of all these dislocations and $v$ by an average velocity $\langle v \rangle$, then

$$\frac{dy}{dt} = \frac{\langle v \rangle Lb}{V} = \rho b \langle v \rangle \qquad \text{...(2)}$$

where $\rho$ is the dislocation density.

## 5.29 ESTIMATION OF DISLOCATION DENSITY FROM X-RAY DIFFRACTION MEASUREMENTS

We know from Bragg's law that reflection occurs only if the Bragg condition is fulfilled. If a crystal has perfect periodicity, the angular spread about the Bragg angle should not be more than 5 seconds. However, many crystals show an angular spread of the order of several minutes. In order to explain such discrepancies, Darwin and Ewald introduced the notion of a mosaic crystal, i.e., an actual crystal, is assumed to be made up of a number of small blocks which themselves are perfect but slightly misoriented relative to each other. Now, it is believed that this mosaic structure is the result of the three-dimensional network of dislocations. Assuming this situation, an estimate of dislocation density in terms of the observed total angular speed θ of the X-ray pattern may be made as follows.

Suppose the surface area of the crystal involved in the X-ray measurements has a side $L$. Let the density of dislocations in the crystal be ρ, so that ρ dislocation lines pierce through a unit area of the surface of the crystal. We now define the edge λ of each block in the mosaic structure in such a way that $\lambda^2 = 1/\rho$, i.e., we associate one block with each dislocation line coming out of the surface. The average angular misfit between the blocks is then $\alpha = b/\lambda$ radians (these misfits may be positive or negative). Thus, as we pass across the crystal over a distance $L$, we pass $L/\lambda$ block. Consequently, the probable angle of misfit between the first and the last block is a $(L/\lambda)^{1/2}$. This angle equals the total observed spread (θ radians), i.e.,

$$\theta = \alpha\left(\frac{L}{\lambda}\right)^{1/2} \quad \text{or} \quad \theta = bL^{1/2}\rho^{3/4} \qquad \text{...(1)}$$

Thus, knowing θ, $L$ and $b$, ρ can be estimated.

## 5.30 DISLOCATION MOTION

Let us consider the simplest possible case of dislocation motion, i.e., an edge dislocation moving to the right under the influence of shear stress τ (Fig. 5.35). The shear stress will have to do some work in pulling atom 1, just on the right to the dislocation centre further away from its neighbour atom 2. The atom 3, however simultaneously moves closer to its equilibrium distance from its neighbour atom 4, and releases almost the same amount of energy as has been spent in moving the atom 1.

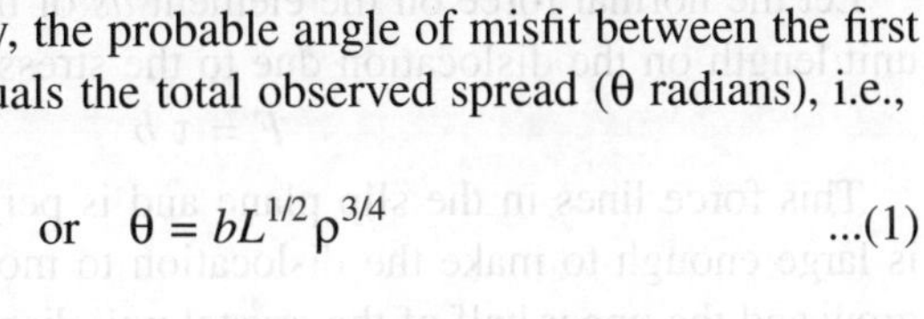


**Figure 5.35** Atomic arrangement in the vicinity of an edge dislocaion as it moves under stress

As the nature of interatomic forces near the dislocation line is somewhat different from the rest of the crystal, the atoms near and within the dislocation line almost completes the slip process when the next nearest neighbours are beginning to slip. Atoms away on either side of the dislocation line are not affected by the presence or motion of dislocation. To a first approximation, the increase in the attractive forces for the atoms on the left just equals the increase in the repulsive forces for the atom on the right, so that the forces on the dislocation line are balanced, and hence the dislocation should start moving under the influence of smallest shear stress.

Suppose a uniform shear shress $\tau$ is applied to the crystal along the direction of the Burgers vector, $b$. Mott and Nabarro have shown that this leads to a force on the dislocation line such that the slipped area tends to grow. In order to find the amount of force acting on the dislocation as a result of applied stress $\tau$, let us consider an element $ds$ of the dislocation line which has slipped outwardly by an amount $dl$ along a direction perpendicular to $ds$. The area swept out by this line element is then $ds \times dl$. This corresponds to an average displacement of the upper part of the crystal relative to the lower part by an amount $(ds \times dl \times b)/A$, where $A$ is the area of the slip plane. Thus, the work done by the shear stress is equal to the total shear force ($\tau A$) times the average shear displacement, i.e.,

$$\tau \times A \times \frac{ds \times dl \times b}{A} = \tau \times b \times dl \times ds$$

Let the normal force on the element $ds$ of the dislocation line be $F.\ ds$, then the force per unit length on the dislocation due to the stress $\tau$ is

$$F = \tau\, b \qquad \text{...(1)}$$

This force lines in the slip plane and is perpendicular to the dislocation line. If the force is large enough to make the dislocation to move in the direction of $F$, the slipped area will grow and the upper half of the crystal will displace over the lower half by an amount $b$ in the slip plane and thus creates a step (Fig. 5.36). Notice that the edge dislocation moves parallel to its Burgers vector while the screw dislocation moves perpendicular to its Burgers vector.

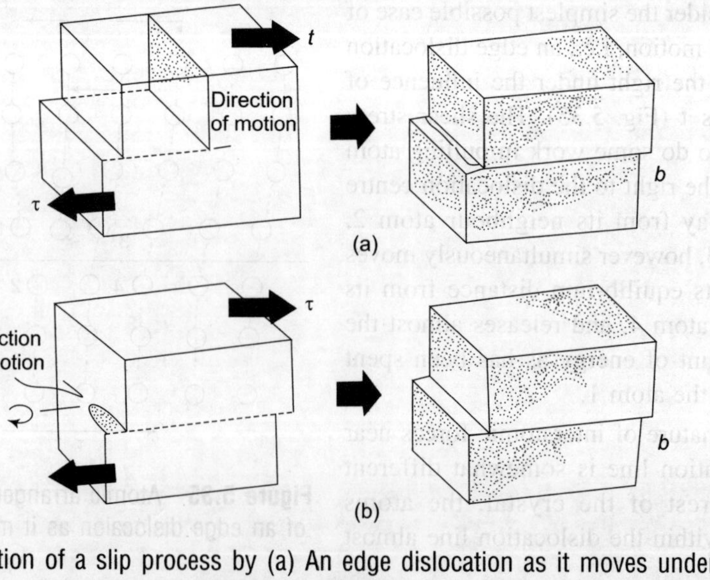


**Figure 5.36** Creation of a slip process by (a) An edge dislocation as it moves under stress (b) a screw dislocation

Now let us visualize the motion of dislocation in the presence of some other imperfections and impurities. Suppose that in gliding along its slip plane an edge dislocation meets obstacles in the form of a pair of precipitate particles at $B$ and $C$ as shown in the Fig. 5.37. As compared to the matrix material, these are not easily sheared. The applied stress $\tau$ gives a normal force, $\tau bl$ on the dislocation line, bowing it out between the points $B$ and $C$. This force is balanced by the parallel component of the dislocation line tension $T$, so that

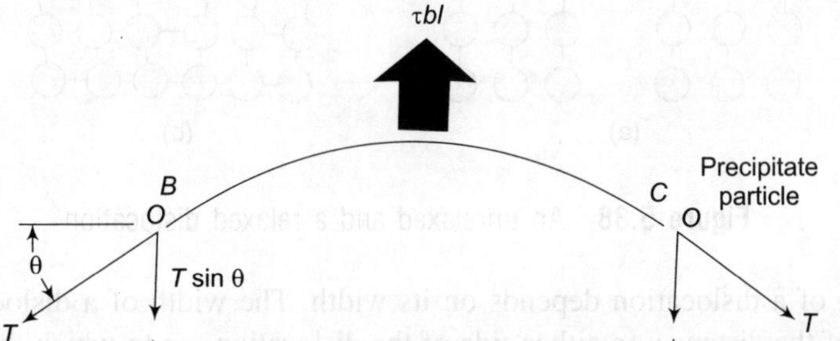


**Figure 5.37** Dislocation line tension when the dislocation encounters the obstacles $B$ and $C$

$$\tau bl = 2T \sin \theta \quad ...(2)$$

where $l$ is the distance, $B$ and $C$. The line tension, $T$ is a vector and is given by the expression

$$T = Gb^2 \quad ...(3)$$

where $G$ is the elastic shear modulus. From Eqs. (2) and (3), we have

$$\tau = \frac{2Gb \sin \theta}{l} \quad ...(4)$$

This tells us that the maximum stress required to cause maximum bowing in the line segment is when it becomes semicircular, i.e., at $\theta = 90°$, the stress assumes its maximum value:

$$\tau_{max} = \frac{2Gb}{l} \quad ...(5)$$

According to Eq. (4), a dislocation which does not meet any obstacle (i.e., $\theta = 0$) should be capable of moving at a vanishingly small stress; a higher stress is required for smaller $l$ when it does not meet any obstacle. Let us understand this further in a qualitative way by considering two edge dislocations as shown in Fig. 5.38. The dislocation in Fig. 5.38 (a) is said to be stiff, i.e., no relaxing displacements have taken place in the adjacent planes around dislocation region. The bond lengths in this region are normal except below the incomplete plane, where they are virtually broken. On the other hand, dislocation in Fig. 5.38 (b) is said to be relaxed, i.e., relaxing displacements have taken place in the adjacent planes around dislocation region. In this case, the strains are distributed more evenly among the bonds to give rise to compression above and tension below the slip plane.

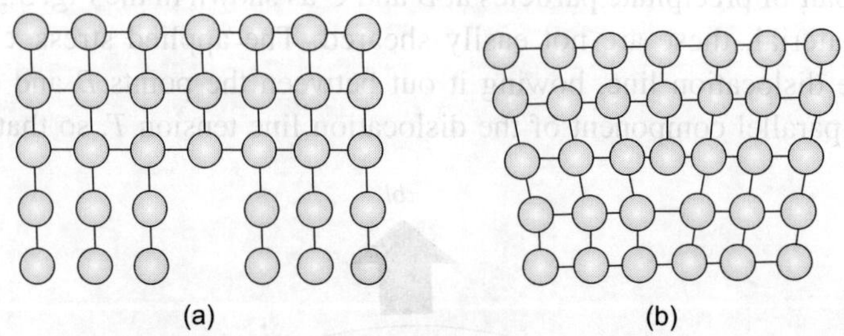


**Figure 5.38** An unrelaxed and a relaxed dislocation

The motion of a dislocation depends on its width. The width of a dislocation is defined as a measure of the distance on either side of the dislocation, up to which the stress-relaxing displacements are appreciable. Therefore, a dislocation is said to be wide if the displacements are distributed over an appreciable distance on either side from the core of dislocation. Accordingly, Fig. 5.38 (b) is treated as having a wide dislocation. When such a dislocation moves, realignment of a number of bonds takes place in the dislocation region, however, the change in any one bond length is very small. Contrary to this Fig. 5.37a is treated as having a narrow dislocation, where the bonds are normal (except below the slip plane) and atoms remain undisplaced. In order to move this type of dislocation, the rows of atoms below the slip plane will have to be moved through full interatomic distance. Therefore, the narrow dislocations are more difficult to move than the wide dislocation. Peierls and Nabarro calculated the stress necessary to move a dislocation in a crystal (in the absence of imperfections and impurities) in terms of its width $w$ and is given by

$$\tau_{PN} = \mu \exp\left(\frac{-2\pi w}{b}\right) \qquad ...(6)$$

where $\mu$ is the shear modulus of the crystal and $b$ the magnitude of the Burgers vector of the dislocation. Strong dependence of $\tau_{PN}$ on width $w$ is illustrated in Table 5.9. For an accurate calculation of $\tau_{PN}$, a very precise measurement of the dislocation width is required. Since such measurements are inherently difficult, therefore our discussion has been only qualitative.

**Table 5.9** Dependence of $\tau_{PN}$ on width $w$

| *Width* | *0* | *b* | *5b* | *10b* |
|---|---|---|---|---|
| $\tau_{PN}$ | $\mu$ | $\mu/400$ | $\mu/10^{14}$ | $\mu/10^{27}$ |

Dislocation motion can be assisted by thermal energy. The rate of plastic deformation, i.e., the strain rate $\varepsilon$ is proportional to Boltzmann probability factor:

$$\dot{\varepsilon} \propto \exp\left(-\frac{Q}{kT}\right) \quad ...(7)$$

where $Q$ is the activation energy for dislocation motion. This strain rate during plastic deformation can also be expressed in terms of the velocity of dislocation $v_d$ as

$$\dot{\varepsilon} = \rho b v_d \quad ...(8)$$

where ρ is the density of mobile dislocation and $b$ is their Burgers vector. Equation (7) indicates that thermal energy alone could move a dislocation from its initial position. However, such motions would be random, where a dislocation line will have equal probability of moving to the next minimum energy position on either side of the line. This kind of motion does not result in plastic deformation. Actually, a certain amount (threshold) stress is necessary to move the dislocation in a specified direction. Besides, the applied stress can lower the activation energy $Q$. Since Eq. 6 does not have a $kT$ term, therefore let us suppose that $\tau_{PN}$ is the stress in the absence of thermal energy (i.e., at 0K). Thus, at absolute zero $\tau_{PN}$ is equal to $\tau_{app}$, which is necessary to move a dislocation, as no assistance from thermal energy is available. At higher temperatures, the required external stress ($\tau_{app}$) is lower than $\tau_{PN}$. These two stress terms multiplied by a volume term called the activation volume $V$ is related to $Q$ as

$$Q = (\tau_{PN} - \tau_{app})\, V \quad ...(9)$$

From Eqs. (7) and (9), we have

$$\dot{\varepsilon} \propto \exp\left(\frac{(\tau_{PN} - \tau_{app})V}{kT}\right) \quad ...(10)$$

Equation 10 tells us that for a constant strain rate, the term within the exponential is constant. $\tau_{PN}$, as given by Eq. 6 is constant for a particular crystal. If the activation volume $V$ is taken to be constant then it follows that, with increasing temperature, less and less external stress is required to cause plastic deformation at an increasing strain rate at constant temperature, the external stress should be increased. Equation 10 can be written as

$$\ln \dot{\varepsilon} = -\frac{(\tau_{PN} - \tau_{app})V}{kT} + A' = A + B\tau_{app} \quad ...(11)$$

where $A$, $A'$ and $B$ are constants at constant temperature.

Dislocation motion is analogous to a worm-like motion. A worm moves forward by displacing its segment one after the other rather than simultaneously all the segments. In a similar way, dislocation moves in a sequential fashion, occupying successive positions during its motion.

When an edge dislocation moves in a direction perpendicular to its slip plane, the process is called climb. Such a motion must be accompanied by the generation or annihilation of vacancies or interstitial, depending upon the sense (positive or negative) of dislocation motion. Since the equilibrium concentration of vacancies (interstitials) depends on temperature, the dislocation climb is temperature sensitive. Climb of an edge dislocation is illustrated in Fig. 5.39 (a). The change of slip plane by moving a screw dislocation is called cross slip and is illustrated in Fig. 5.39 (b).

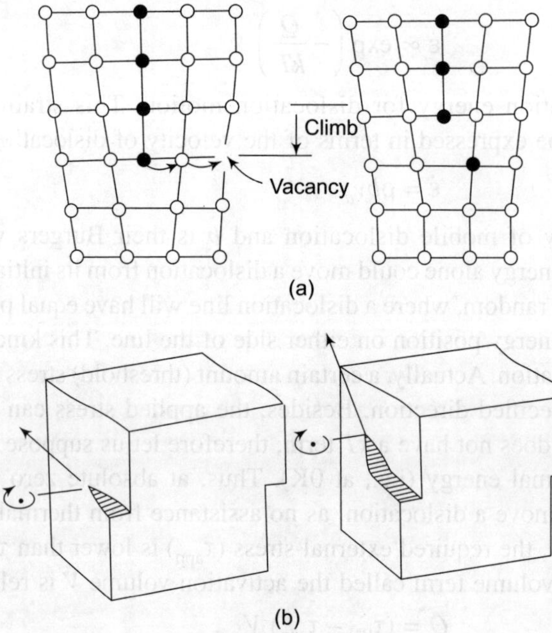


**Figure 5.39** (a) Climb of an edge dislocation (b) cross-slip of a screw dislocation

## 5.31 DISLOCATION MULTIPLICATION

Dislocations mainly arise due to:

1. Introduction during crystal growth.
2. Nucleation due to stress.
3. Regenerating multiplication.

We know large stresses are needed to nucleate dislocations in crystals. They may arise from impurity atoms or vacancies in a crystal in the immediate vicinities of which stresses may rise so high (due to applied stresses) as to induce nucleation of dislocations. In some cases, a single dislocation in a crystal may give rise to several dislocations. For this, explanation was given by Frank and Read, which is as follows.

The dislocations generally do not lie in a slip plane. In such a case, a segment of a dislocation of length $l$, which lies in a slip plane, may be considered to be pinned at two points. (The pinning may be caused by a pair of precipitate particles which are not easily sheared as the matrix.)

When a shear is applied, this segment attains equilibrium with a radius of curvature,

$$r_c = \frac{Gb}{2\sigma}$$

With increase in $\sigma$, $r_c$ decreases, becoming a semicircle at $\sigma = Gb/l$ (Fig. 5.40 (c) and (d)). The segment then curls upon itself leading ultimately to the formation of a large loop and a segment which, in turn, regenerates another loop (Fig. 5.40 (e)). This process is repeated resulting

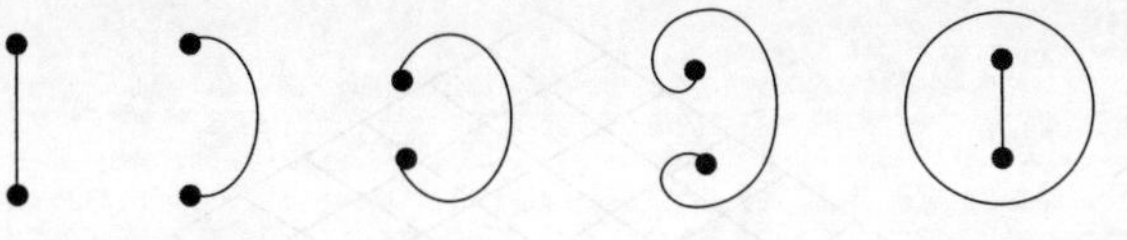


**Figure 5.40** Regeneration of dislocations from a Frank-Read source

in multiplication of dislocations. The two points at which the dislocations are pointed are pinned formed a Frank–Read source and the above mechanism of dislocation multiplication is known as Frank–Read mechanism.

Dislocations multiply in number in a variety of, more or less, complicated ways. Because dislocations multiply during plastic flow, the continued deformation of a solid increases its dislocation content. Eventually the dislocation become so numerous and so tangled together that they impede one another's motion, which decreases the plasticity of the material. This phenomenon is known as *work-hardening*. (This becomes prominent where there are about $10^8$ dislocations per $cm^2$ in a slip plane.)

If a work hardenered crystal is heated, its disordered atomic array tends to return to regularity; the rise in temperature (which increases the vibrational energy of atoms), provides a activation energy needed to initiate the release of energy stored in the dislocations. The process of heating a work-hardened crystal to restore its ductility is called *annealing*.

Annealing is a kind of heat treatment after which a metal or an alloy acquires a structure close to the equilibrium one.

For instance, in synthesis of high $T_c$ superconductors, pre-annealing heat treatment is followed by annealing, to improve the crystallinity of the material.

## 5.32 A SCREW-DISLOCATION MEETING A SURFACE

As atoms from a solution or the vapour state (during growth) strike the crystal surface and diffuse on it, they diffuse into the region of a step and become more tightly bound, since they can join to two neighbours rather than just one at a step. Once in place, successive atoms will join (i.e., bound by) three neighbours and growth occurs as the step builds up at the surface. By the nature of a screw-dislocation, there is always a step present, since a completed layer cannot be obtained.

Because the centre of a screw dislocation is fixed, the steps rotate about that point and a spiral develops, since the outer portions of the step have to go further to complete their filling in than do the inner portions (Fig. 5.41).

### 5.32.1 Interaction between Spirals

Two screw dislocations be situated at a critical distance apart from one another can combine, as do the edge-dislocation segment of a Frank–Read source, to give rise to growth rings rather than spirals.

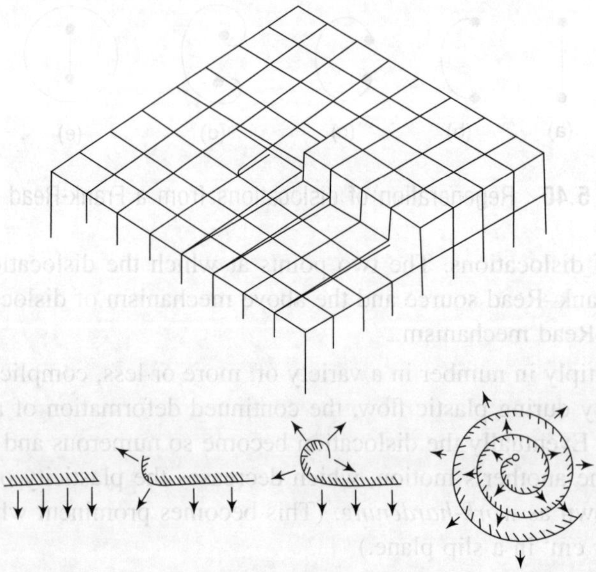

**Figure 5.41** The intersection of a screw dislocation with a free surface to produce a spiral step (After F.C. Frank, *Discussions Faraday Soc.* **5**,48 (1949))

Two dislocations of opposite sign give rise to closed growth fronts as shown in Fig. 5.42 (a). Mutual annihilation of the two dislocations just before cessation of growth leads to the features of Fig. 5.42 (b), i.e., to a sequence of closed growth-fronts without a visible centre.

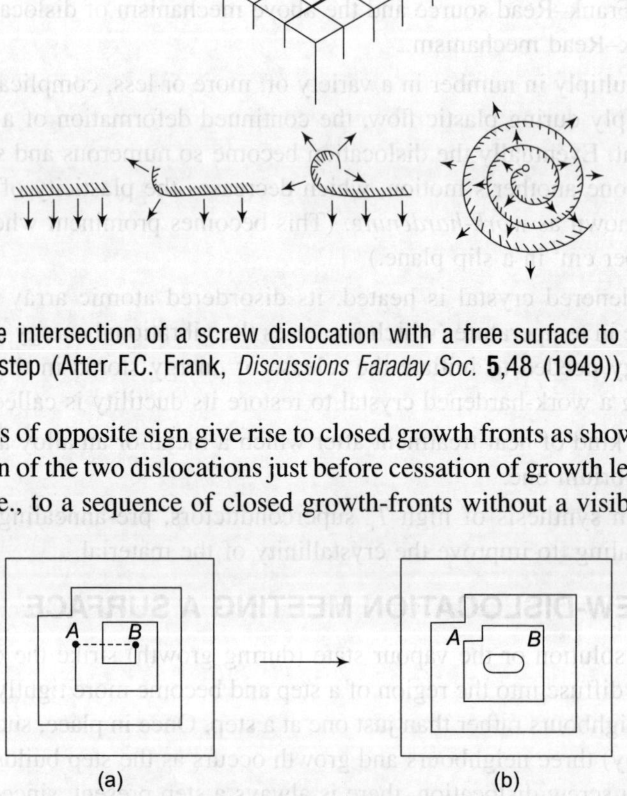


**Figure 5.42** The mutual annihilation of two dislocations of opposite side emergency in *A* and *B* as reflected in the pattern of surface steps: (a) During growth (b) after mutual annihilation of the dislocations in *A* and *B*

## 5.33 INTERACTION BETWEEN PARALLEL DISLOCATIONS

If the stress fields of two dislocations cancel each other, the dislocations attract each other and if, they reinforce, the dislocations will repel each other.

Dislocations of same sign and lying in the same slip plane attract each other and thus get annihilated to produce a perfect crystal.

Dislocations of opposite signs repel one another, leading to their segregation in particular planes.

By interaction, dislocations change their total free energy. The rate of change of energy with distance gives the force between them. The radial and angular components $(F/l)_r$ and $(F/l)_\theta$ of the force per unit length between two dislocations having a parallel line are given in polar coordinates by equation:

$$\left(\frac{F}{l}\right)_r = \frac{c\vec{b}_i\vec{b}_2}{r} \text{ (for both edge and screw dislocations)} \quad ...(1)$$

$$\left(\frac{F}{l}\right)_\theta = \begin{cases} \dfrac{cb_1b_2 \sin 2\theta}{r} & \text{(for edge dislocation)} \\ \text{zero} & \text{(for screw dislocation)} \end{cases} \quad ...(2)$$

Here force is proportional to the scalar product of the two Burgers vectors, $c$ is equal to $G/(2\pi)$ for screw dislocations and $G/\{2\pi\,(1-\mu)\}$ for edge dislocations. $\theta$ is the angle between the slip plane and the plane containing both the dislocation lines, and $r$ is the distance of separation of the dislocation lines, $\mu$ is Poisson's ratio.

Screw dislocations of opposite sign and the same Burgers vector will attract with a force per unit length of $Gb^2/(2\pi r)$. If unimpeded, they will come together, annihilate and leave a perfect lattice. Screw dislocations of like sign and Burgers vector will repel with a force per unit length of the same magnitude.

A number of different interactions can occur between parallel edge dislocations because of their stress fields. These are summarized as hereunder.

(1) Two parallel edge dislocations of the same sign on the same or nearby slip planes repel each other.

   Parallel edge dislocations of the same sign on widely separated planes attract or repel each other, depending on whether the angle between the slip direction and a line joining them is greater or less than 45°, respectively.

(2) Two parallel edge dislocations of opposite sign on the same or nearby slip planes attract each other.

   (i) If on the same slip plane, they may come together and be annihilated, leaning a perfect lattice.

   (ii) If on adjacent planes, they may attract and be annihilated, leaving only a row of vacancies or a row of interstitial atoms.

(3) Small interstitial solute atoms tend to migrate to the tension side of an edge dislocation, for there, they can contribute to lowering of the strain energy.

   Similarly, small substitution atoms, tend to migrate to the compression side of an edge dislocation, where, by replacing a larger matrix atom, they lower the strain energy.

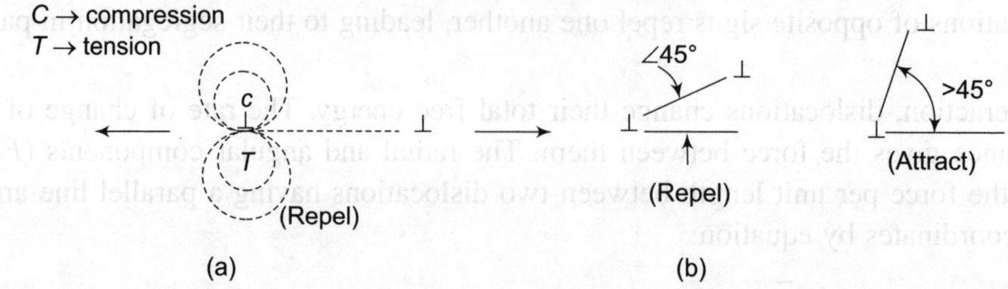


**Figure 5.43** (a) Like dislocations on same or nearby planes repel; (b) Like dislocations on widely separated planes

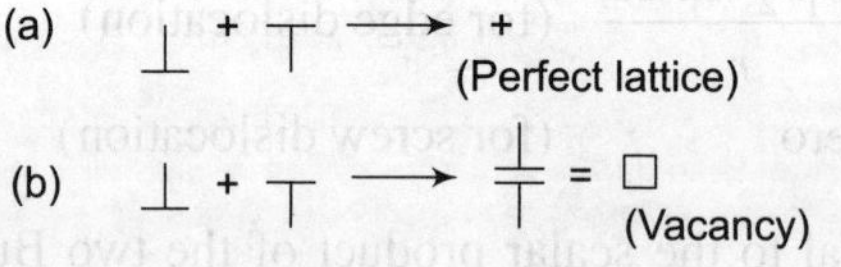


**Figure 5.44** Unlike dislocation on same or nearby planes attract. (a) If they are on the same plane; they annihilate and leave a perfect lattice. If on adjacent planes, they annihilate and leave vacancies (b) or interstitials (c) since no volume change is associated with the screw dislocation, however, there is no tendency for a preferential relocation of solute atoms of the kind described above

## 5.34 THE INTERACTION OF A DISLOCATION WITH AN OBSTACLE

If a dislocation line is subjected to a small shear stress, it will move until it reaches 'obstacles' (second phase particles). If the stress increases, it will begin to 'low-out' at the obstacles. The Figure 5.45 indicates that an advancing dislocation is held back in the vicinity of impurities or lattice defect, whereas other portions move through the lattice and the enfolding of the sections left back of the advancing from breakaway to from new dislocations.

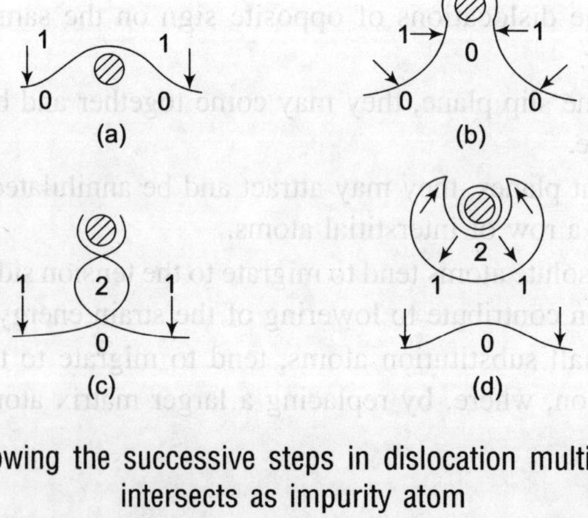


**Figure 5.45** Sketch showing the successive steps in dislocation multiplication as a dislocation intersects as impurity atom

**Examples 5.2** Calculate the line energies of dislocations in bcc iron. Given the Burgers vector in iron is of the 1/2 <111> type. Lattice parameter is $a$ = 2.87 Å and shear modulus of iron is 80.2 GN/m$^2$.

***Solution*** The strain energy per unit length of a dislocation ($E$) of Burgers vector $b$ is expressed approximately by

$$E = \frac{1}{2} Gb^2$$

where, $G$ is the shear modulus.

For bcc iron, $a = 2.87$ Å

$$\therefore \quad b = 2.87 \times \frac{\sqrt{3}}{2} = 2.49 \text{ Å}$$

∴ Line energy of dislocations

$$E = \frac{1}{2} Gb^2 = \frac{80.2 \times 10^9 \times (2.49)^2 \times 10^{-20}}{2}$$

$$= 2.49' \ 10^{-9} \text{ J/m.}$$

## 5.35 ENERGY OF A DISLOCATION

Energy added to the crystal by a dislocation = (elastic strain energy) + (energy of dislocation core)

i.e., $$W_{\text{disl}} = W_{\text{el}} + W_{\text{core}}$$

The elastic strain energy of volume $dV$ is

$$dW_{\text{el}} = \frac{1}{2} \sum_{i=x,y,z} \sum_{j=x,y,z} \sigma_{ij}\, e_{ij}\, dV$$

$$= \frac{1}{2} \sum_{i=r,\theta,z} \sum_{j=r,\theta,z} \sigma_{ij}\, e_{ij}\, dV$$

*For screw dislocation*

$$\sigma_{z\theta} = \sigma_{\theta z} = \frac{Gb^{(3)}}{2\pi r} \quad \text{and} \quad e_{z\theta} = e_{\theta z} = \frac{b}{4\pi r}$$

$$W_{\text{el}} = \int_0^L dz \int_0^{2\pi} d\theta \int_{r_0}^{R} (\sigma_{\theta z}\, e_{\theta z} + \sigma_{z\theta}\, e_{z\theta})\, r dr$$

$$= \frac{Gb^2 L}{4\pi} \int_{r_0}^{R} \frac{1}{r} dr = \frac{Gb^2 L}{4\pi} \ln\left(\frac{R}{r_0}\right)$$

Here integrator is from the surface of the core cylinder with radius $r_0$ to the external surface of the cylinder of radius $R$ of the specimen.

[3] $G$ is the shear modulus and $b$ is the Burgers vector.

Similarly *for edge dislocation*

$$W_{el} = \frac{Gb^2L}{4\pi (1 - \mu)} \ln \left( \frac{R}{r_0} \right)$$

In general,

$$W_{el} = \frac{Gb^2 L}{4\pi K} \ln \left( \frac{R}{r_0} \right) \quad ...(1)$$

where $K = 1$ for screw dislocation

and $K = 1 - \mu$ for edge dislocation.

where $\mu$ is Poisson's ratio.

From Eq. (1), we see that

(i) The total energy $W_{disl}$ is proportional to length $L \Rightarrow$ energy is minimized when dislocation segments between immobile points (e.g., modes) are straight.

(ii) The energy per unit length of an edge dislocation is larger than that of a screw dislocation:

Since $(1 = \mu) < 1$, for $\mu = 1/3$

$$W_{screw} = 0.66\, W_{edge}$$

$\Rightarrow$ Energy is minimized when dislocation has as large screw component as possible.

(iii) Relatively weak (logarithmic) dependence of the elastic energy on $R$:

(i) In polycrystalline materials, $R$ is limited by the size of the grains.

(ii) In crystals with many dislocations, the dislocations tend to form configurations in which superimposed elastic fields tend to cancel.

Table 5.10 describes point, line and planar imperfections.

**Table 5.10** Imperfections in Crystals

| *Type of imperfection* | *Description of imperfection* |
|---|---|
| **1. Point defects** | |
| (a) Interstitial | (a) Extra atom in an interstitial site. |
| (b) Schottky defect | (b) Atom missing from correct site. |
| (c) Frenkel defect | (c) Atom displaced to interstitial site creating nearby vacancy. |
| **2. Line defects** | |
| (a) Edge dislocation | (a) Row of atoms marking edge of a crystallographic plane extending only partway in a crystal. |
| (b) Screw dislocation | (b) Row of atoms about which a normal crystallographic plane appears spiral. |
| **3. Plane defects** | |
| (a) Tilt boundary | (a) Boundary between two adjacent perfect regions in the same crystal that are slightly tilted with respect to each other. |
| (b) Grain boundary | (b) Boundary between two crystallites in a polycrystalline solid. |
| (c) Stacking fault | (c) Boundary between two parts of in a closest packing having alternate stacking sequences. |

**Table 5.11** Edge and Screw Dislocations

| *Edge Dislocation* | *Screw Dislocation* |
|---|---|
| 1. Region of lattice disturbance extends along an edge inside a crystal. | 1. Region of lattice disturbance extends in two separate planes perpendicular to each other. |
| 2. It originates due to the presence of partial row of atoms along an edge. | 2. Originates due to a twist in stacking sequence of atoms bond forming a step around a line called screw axis. |
| 3. Burgers vector is perpendicular to the dislocation line. | 3. Burgers vector is parallel to the dislocation line. |
| 4. Both tensile and shear stress fields exist around the dislocation. | 4. Only shear stress field is present. Absence of tensile and compressional stresses is attributed to absence of extra half plane of atoms. |

## 5.36 STABILITY OF A DISLOCATION LOOP (CRITICAL RADIUS)

A dislocation loop is important in plastic deformation. Such a loop may consist of partly screw, partly edge and mostly of mixed dislocations. Strain energy increases with the length of dislocations. As a result, there is a tendency for the loop to shrink. This is opposed by a shear stress σ and an equilibrium thus exists.

For sake of simplicity assume that for all types of dislocations, energy per unit length is given by

$$E = \frac{Gb^2}{4\pi} \ln\left(\frac{R}{r_0}\right)$$

Consider a close loop of dislocation of radius $r$. Then the energy is

$$(2\pi r)\,\frac{Gb^2}{4\pi} \ln\left(\frac{R}{r_0}\right)$$

If the crystal is subjected to a shear σ, the work done by the internal stresses in expanding the loop to radius, $r$, is

$$\int_0^r 2\pi r\sigma b\,dr = \pi r^2\,\sigma b$$

since the force acting per unit length is $\sigma b$. Thus, the energy of the loop is

$$E = \frac{Gb^2}{4\pi} \ln\left(\frac{R}{r_0}\right) 2\pi r - \pi r^2\,\sigma b \qquad ...(1)$$

In equilibrium, $\left.\dfrac{dE}{dr}\right|_{r=r_0} = 0$

This gives $2\pi \times \dfrac{Gb^2}{4\pi} \ln\left(\dfrac{R}{r_0}\right) = 2\pi r_c \sigma b$

or critical radius $$r_c = \frac{Gb}{4\pi\sigma} \ln\left(\frac{R}{r_0}\right) \quad ...(2)$$

As $R$ of the same order as $r$, and $R/r_0$ lies in the range of $10^2$ to $10^5$, we may take

$$\ln\left(\frac{R}{r_0}\right) \cong 2\pi$$

So, $$r_c = \frac{Gb}{2\sigma} \quad ...(3)$$

## 5.37 DEFORMATION OF A SINGLE CRYSTAL

An understanding of the fundamental nature of plastic deformation that takes place in each crystallite of a metallic object can be simplified by singling out one single crystal and observing how it behaves under similar deformation. If a polycrystalline sample is used the result obtained is the average behaviour of all the differently oriented grains in the material. The behaviour of a single crystal under deformation will, therefore, reveal the study of an ordinary crystalline aggregate with a clearer view.

Single crystals may be produced by either of the following methods:

(1) the carefully controlled solidification of a melt, and

(2) the prolonged annealing of a piece of metal after it is suitably cold worked.

Using the latter method, pieces of wire of 1.5 mm in diameter and of considerable length can be readily used as single crystals, Plastic deformation of a single crystal occurs either by slip or by twining.

### 5.37.1 Deformation by Slip

*Slip* is a shear deformation that moves atoms by many interatomic distances relative to their initial positions. Figure 5.46 schematically illustrates the slip process in a single crystal under a shear load. As a result of plastic deformation, the crystal is devided into layers or slip blocks which are displaced in reference to each other and are separated by thin layers

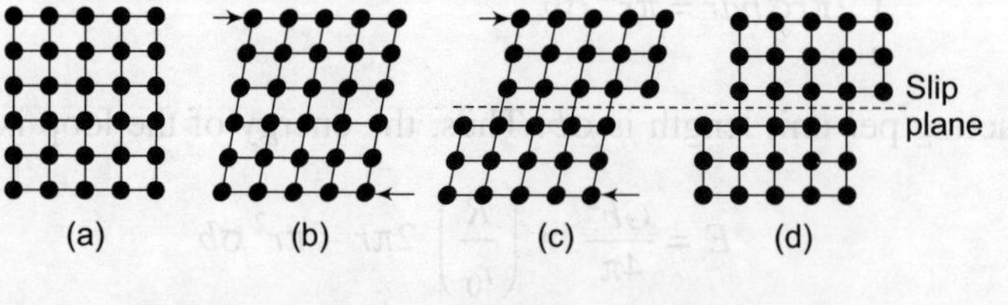


**Figure 5.46** Slip process in a single crystal

in which a considerable displacement of atoms has taken place. These intermediate layers with strongly distorted lattices are called *slip planes*. Due to the actual movement of the atomic blocks, no return to the original shape is possible after withdrawal of the straining force and the deformation is permanent. This is further illustrated in Fig. 5.47. As a very crude approximation, the slip or the glide, of a crystal can be considered analogous to the distortion produced in a pile of cards when it is pushed from one end.

Slip propagates gradually along the cross section of a crystal and, at each moment, only the atoms of a comparatively small portion of the cross section are displaced.

Slip occurs most readily in specific directions on certain crystallographic planes which are typical for each type of crystal lattice. Generally, the slip plane is the plane of greatest (linear) atomic density, and the slip direction is the closest packed direction within the slip plane. Since the planes of greatest atomic density are also the most widely spaced planes in the crystal structure, the resistance to slip is generally less for these planes than for any other set planes. The slip plane together with the slip direction establishes the *slip system*. In fcc and bcc crystals there are 12 slip systems, while in hcp there are only 3.

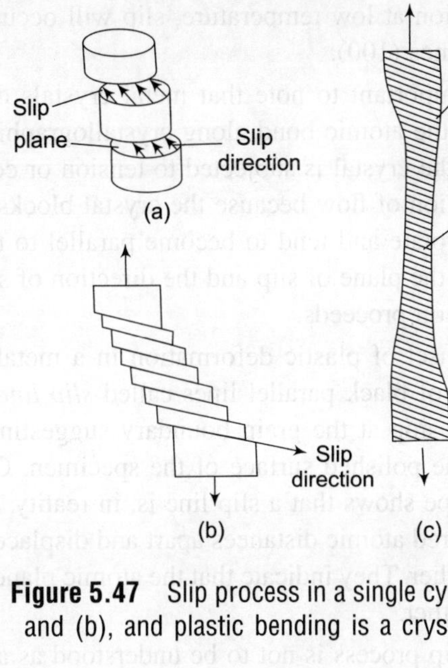


**Figure 5.47** Slip process in a single cyrstal in (a) and (b), and plastic bending is a crystal in (c)

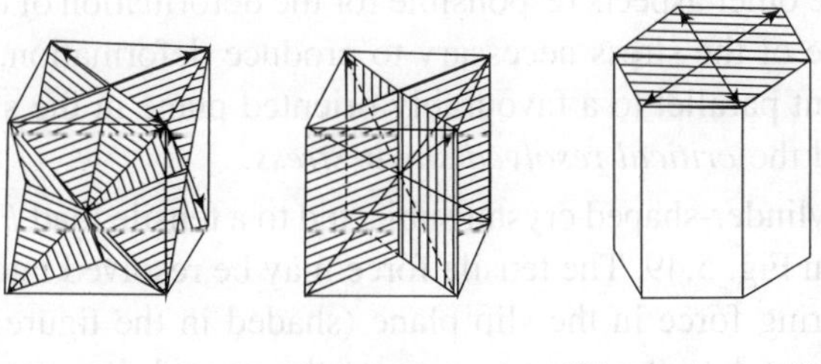

**Figure 5.48** Slip planes and directions for different crystallic systems

In fcc metals such as Al, Cu, Ag, Au, Ni, etc., the slip plane is usually a {111} plane, and the slip direction, i.e., the linear packing of atoms, are the <110> family (face diagonals). In bcc structures, several families of planes have almost the same density of packing. In *a*-Fe, the {110} or dodecahedral planes have the greatest density of packing but the {112} and {123} planes are almost as densely packed, and all three sets of planes can operate during slip. In all these planes, the <111> directions are the most densely packed. In hcp structures such as Mg, Zn, etc., they will be basal planes (0001) in closed-packed directions. Slip planes and directions for different crystallic systems are shown in Fig. 5.48. The more possible slip planes and directions a metal has the more plastic it is.

Slip planes and directions are subject to changes in accordance with the temperature, degree of preliminary deformation, and alloy composition. For example, if aluminium, subjected to deformation at low temperature, slip will occur along planes (111); at high temperatures—along planes (100).

It is important to note that metal crystals deform plastically, as already mentioned, by shearing the atomic bond along crystallographic planes of weakness, immaterial of the fact whether the crystal is subjected to tension or compression. The crystal become elongated in the direction of flow because the crystal blocks that are moving with respect to one another actually rotate and tend to become parallel to the axis of compression. Hence, we may say that both the plane of slip and the direction of slip tend to rotate into the direction of flow as deformation proceeds.

The onset of plastic deformation in a metal crystal is manifested by the appearance of a grating of black parallel lines called *slip lines* covering the polished face of a grain. The slip lines stop at the grain boundary suggesting that the grating is the trace of slip planes cutting the polished surface of the specimen. Careful examination by means of an electron microscope shows that a slip line is, in reality, a *slip band* made up of several slip planes, a few hundred atomic distances apart and displaced thousands of atomic distances with respect to each other. They indicate that the atomic planes within the crystal have sheared with respect to each other.

The slip process is not to be understood as a simultaneous displacement of the parts of a crystal. Such slip would require stresses considerably higher than those actually present.

### 5.37.2 Resolved Shear Stress

Before considering the other aspects responsible for the deformation of crystals, it is of interest to determine the value of the stress necessary to produce deformation. Slip occurs when the shear-stress component parallel to a favourably oriented plane in the slip direction reaches a threshold value called the *critical resolved shear stress*.

Let us consider a cylinder-shaped crystal subjected to a tensile load $F$ acting at the direction of the axis as shown in Fig. 5.49. The tensile force may be resolved into a force normal to the slip plane and a shearing force in the slip plane (shaded in the figure). Since the slip plane is inclined by an amount $\phi$ to the cross section of the crystal, its area is $A/\cos\phi$ where $A$ is the cross sectional (face) area of the crystal.

If the slip direction forms an angle $\lambda$ with the direction of the applied force, the component of the axial load acting in the slip plane in the slip direction is $F\cos\lambda$. Therefore, the critical resolved shear becomes

$$\tau_{cr} = \frac{\text{force}}{\text{area}}$$

$$= \frac{F\cos\lambda}{A/\cos\phi}$$

$$= \frac{F}{A}\cos\phi\cos\lambda \qquad \ldots(1)$$

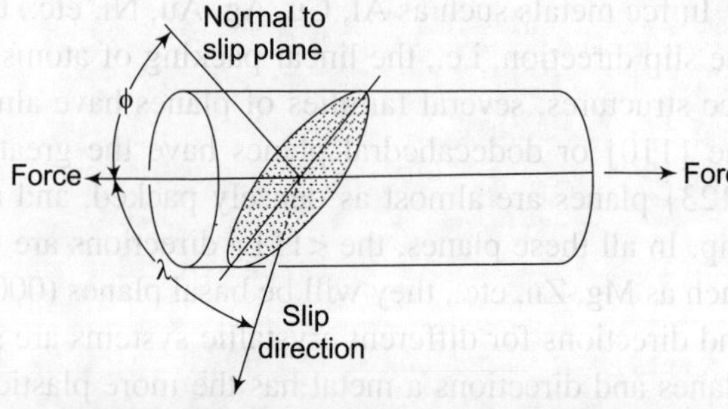


**Figure 5.49** Calculation for critical resolved shear stress

It is found that the shear in the slip plane will be maximum when $\phi$ is equal to 45°, but the same *resolved* shear stress $\tau_{cr}$ is necessary to produce the first slip in every case. Numerous experiments have shown that this is the only stress component that affects slip in most metal crystals. The stress component normal to the slip planes has no observable effect.

Critical shear stress is constant for each material and depends on the type of bonding forces. It is independent of the relative orientation of the applied stress and the slip direction but is affected by temperature, purity, extent of prior deformation, rate of deformation, and surface effects. Table 5.12 gives the critical shear for several pure metals.

**Table 5.12** Critical Shear Stresses of the Principal Slip Systems at 20°C

| *Metal* | *Total impurities, parts/thousand* | *Slip system* | | *Critical stress in $MN/m^2$* |
|---|---|---|---|---|
| | | *Plane* | *Direction* | |
| Cu | <1.0 | (111) | [1$\bar{1}$0] | 896 |
| Ag | 0.1 | | | 532 |
| Au | 0.1 | | | 812 |
| Ni | 2.0 | | | 4480 |
| Mg | 0.5 | (0001) | [11$\bar{2}$0] | 742 |
| Zn | 0.4 | | | 840 |
| Cd | 0.04 | | | 518 |
| β-Sn | 0.1 | (100) | (001] | 1680 |
| | | (110) | | 1176 |
| Bi | ~1 | (111) | [101] | 1960 |

## 5.38 DISLOCATION THEORY

Experiments show that slip in actual crystals of metal begins under a shear stress that is at least two orders of magnitude lower than the values of 1.5 per cent $E$ (Young's modulus) computed for a perfect crystal. This fact, coupled with the fact that plastic deformation is a local event, makes it plausible that slip should be particularly due to a lattice imperfection—

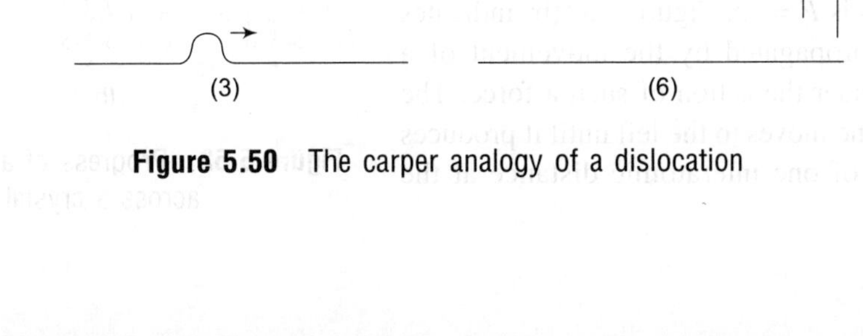


**Figure 5.50** The carper analogy of a dislocation

called dislocation. Dislocations in a crustal may occur during growth of crystals from a melt or they may be produced during slip.

This condition has been proved experimentally by means of filament-like crystals called "whiskers" which are free from any imperfection and dislocation. The ultimate strength of a whisker of iron, one micron (0.001 mm) in diameter and 2.5 mm long, reached $13 \times 10^2$ $MN/m^2$ compared to about $3 \times 10^2$ for a piece of pure iron and its elastic elongation was up to 5 per cent.

The analogy is moving a large carpet helps in understanding the role of dislocations. If the carpet is in contact with the floor all over its area, a large effort is required to move it. If, however, a wrinkle is present and the wrinkle is moved from one end of the carpet to the other, the carpet is shifted a distance of $L$ units to the right as illustrated in Fig. 5.50. The energy required to shift the wrinkle *progressively* is considerably less than that required to drag the carpet as a whole across the floor. Similarly en-block movement of the atoms is not needed in the process of slip.

Actually billions of dislocations ($10^8$ to $10^{10}$ per $m^2$) are known to exist in a crystal. Thus, they cause large amount of slip which can be seen visually. In addition, dislocations interact between themselves and generate new dislocations under excessive stress, which in turn, produce more plastic deformation.

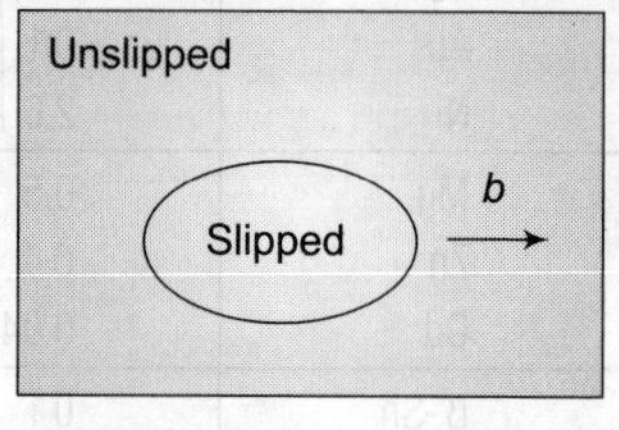


**Figure 5.51** Slipped and unslipped region of a crystal

A dislocation is considered to be a region of localised lattice disturbance separating the *slipped* and *unslipped* region of a crystal illustrated in Fig. 5.51. This is further explained by the schematic diagram of slip deformation given in Fig. 5.52. At first only the upper part of the crystalline is displaced to the left relative to the lower part. Plane *ABCD*, in this case, will be an extra half plane of atoms in the upper part of the crystalline. The edge *AD* of this plane is called, as mentioned earlier, an *edge dislocation* which is perpendicular to the direction of the slip. It is evident from Fig. 5.52(a) that the slip movement has not been transmitted to whole of the plane, but terminates at the dislocation line *AD*. Thus, the edge dislocation should be regarded as the frontal *boundary* of the part of the slip plane along which local displacement has occurred.

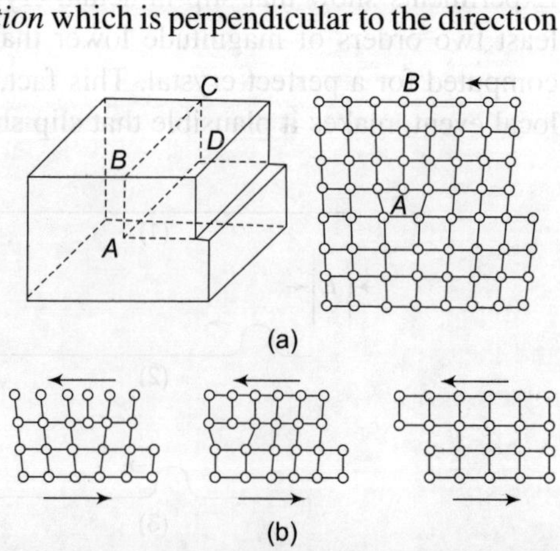


**Figure 5.52** Progress of a dislocation across a crystal

If the resolved shear stress on the slip plane is $\tau$ and the Burgers vector of the dislocation $b$, the force on the dislocation, i.e., force per unit length of dislocation is $F = \tau b$. Figure 5.52 (b) indicates how slip is propagated by the movement of a dislocation under the action of such a force. The extra half-plane moves to the left until it produces the slip step of one interatomic distance at the surface of the crystal. The same shear will be produced by a positive dislocation moving from left to the right as shown in Fig. 5.53. Here the extra half-plane of atoms is above the slip plane, and the dislocation is called a positive dislocation, denoted by the symbol . When the half plane is below the sip plane it is termed a negative dislocation denoted by the symbol T.

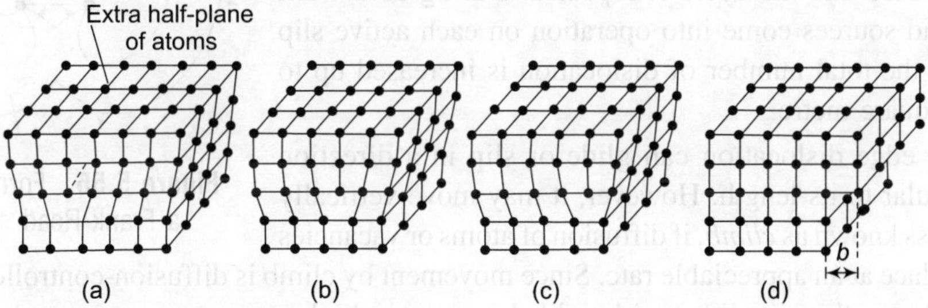


**Figure 5.53** Slip process in an edge dislocation

In addition to edge dislocation, there are also screw dislocations which occur when slip plane covers only a part and not the whole of crystalline cross section. It is seen that the slip plane in a screw dislocation is parallel and *not* perpendicular to the shear vector. The slip process as a result of a screw dislocation is shown in Fig. 5.54. Here also the magnitude of Burgers vector is equal to one atomic spacing. A screw dislocation maybe right-handed and left-handed similar to right-hand and left-hand screw convention and they may be denoted by +, and –, respectively.

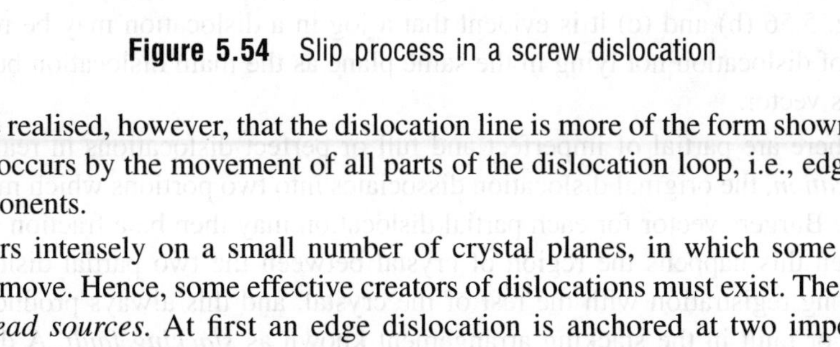


**Figure 5.54** Slip process in a screw dislocation

It must be realised, however, that the dislocation line is more of the form shown in Fig. 5.55 and the slip occurs by the movement of all parts of the dislocation loop, i.e., edge, screw and mixed components.

Slip occurs intensely on a small number of crystal planes, in which some hundreds of dislocations move. Hence, some effective creators of dislocations must exist. These are known as *Frank–Read sources.* At first an edge dislocation is anchored at two imperfections as shown in Fig. 5.55. The dislocation line then bulges under the influence of stress in the slip plane, and in a large stage the boundary of original dislocation becomes unstable and curves back around these two points as shown by the stages 1 to 3 in the diagram. Finally, in stage 4, the bulge further expands spontaneously and forms a closed loop that travels outwards to the grain boundary. As the loop disappears, the original edge dislocation is again created. If the stress persists another loop is formed and the cycle is repeated. During cold work Frank-Read sources come into operation on each active slip plane and the total number of dislocation is increased up to $10^{16}$ per square metre.

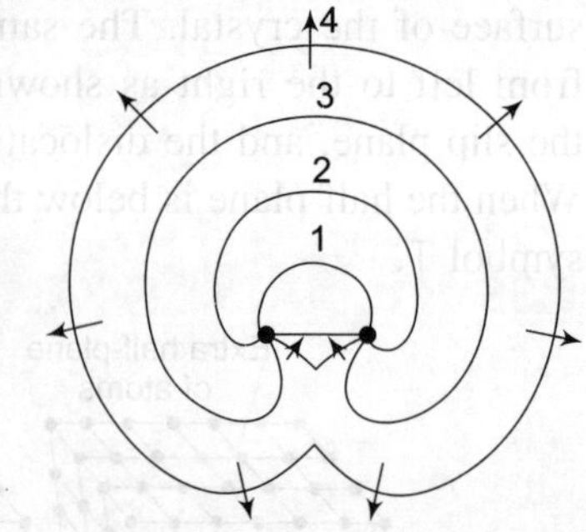


**Figure 5.55** Formation of a Frank-Read source

A pure edge dislocation can glide or slip in a direction perpendicular to its length. However, it may move vertically by a process known as *climb*, if diffusion of atoms or vacancies can take place at an appreciable rate. Since movement by climb is diffusion-controlled, motion is much slower than in glide and less likely except at high temperatures.

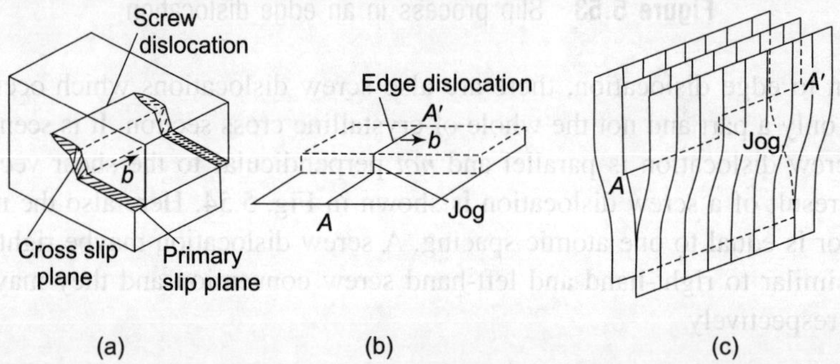


**Figure 5.56** Cross-slip and climb in a crystal (a) cross-slip, and (b) and (c) climb

The process whereby a screw dislocation glides into another slip plane having a slip direction in common with the original slip plane, as shown in Fig. 5.56 (a) is called a *cross-slip*. Usually, the cross-slip plane is a closed-packed plane, e.g., {111} in fcc crystals, and is suitably stressed. Where the dislocation jumps from one plane to another it is known as *jog*, and from Fig. 5.56 (b) and (c) it is evident that a jog in a dislocation may be regarded as a short length of dislocation not lying in the same plane as the main dislocation but having the same Burgers vector.

Besides, there are partial of imperfect and full or perfect dislocations in real crystals. In *partial dislocation*, the original dislocation dissociates into two portions which move together as a unit. The Burgers vector for each partial dislocation may then be a fraction of the lattice spacing. When this happens the region of crystal between the two partial dislocations will be in the wrong registration with the rest of the crystal, and this always produces a surface imperfection or fault in the stacking arrangement known as *stacking fault*. A dislocation in which the Burgers vector is an identity period in the lattice is spoken of as *full* or *perfect*. Passage of such a dislocation leaves the crystal unchanged in its atomic arrangement.

A combination of edge and screw dislocation, however, will give a composite dislocation. The mechanism for dislocation movement, described above, explains why the process of deformation may proceed at comparatively low stress at which simultaneous movement of all the atoms in a given slip plane would be impossible.

## 5.39 DEFORMATION BY TWINNING

The second important mechanism by which metals deform is the process known as *twinning*. This result when a portion of the crystal takes up an orientation that is related to the orientation of the rest of the untwinned lattice in a definite, symmetrical way. The twinned portion of the crystal is a mirror image of the parent crystal. The plane of symmetry between the two portions is called the twinning plane. Figure 5.57 illustrates the classical atomic picture of twinning.

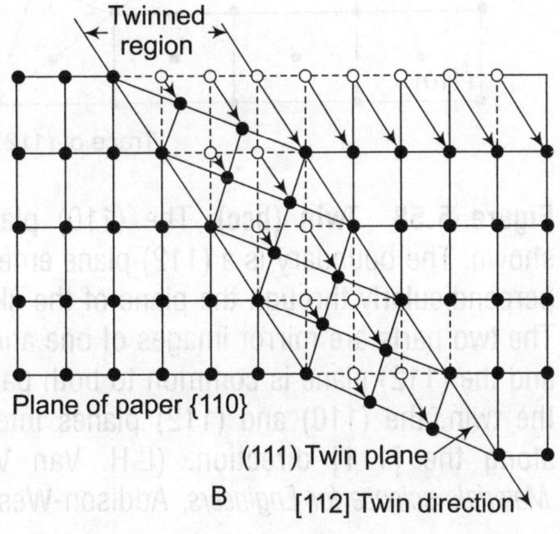


**Figure 5.57** Twinning in an fcc-crystal lattice

The twinning plane is perpendicular to the paper. If a shear stress is applied, the crystal will twin about the twinning plane. The region to the left of the twinning plane is undeformed. To the right of this plane, the planes of atoms have sheared in such a way as to make the lattice a mirror images across the twin plane. The mechanism of twinning may thus be described as the simple sliding of one plane of atoms over the next, the extent of the movement of each plane being proportional to its distance from the twining plane.

It should be noted that twinning differs from slip that every plane of atoms suffers some movement, and the crystallographic orientations of many unit cells are altered.

Twins maybe produced by mechanical deformation or as a result of annealing following plastic deformation. The first type is known as *mechanical twins*, while the latter are called *annealing twins*. Mechanical twins are produced in bcc or hcp metals under conditions of rapid rate of loading (shock loading) and decreased temperature.

The important role of twinning in plastic deformation comes not from the strain produced by the twinning process but from the fact that orientation changes resulting from twinning may place new slip systems in a favourable orientation with respect to the stress axis so that the additional slip can take place. Thus, twinning is important in the overall deformation of metals with a low number of slip systems, such as hcp metals. However, it should be understood that only a relatively small fraction of the total volume of a crystal is reoriented by twinning.

### 5.39.1 Examples of Deformation by Twinning

A crystal is *twinned* if it possesses a mirror boundary. In Fig. 5.58, the (110) plane of a bcc metal is sketched in two dimensions in the plane of the paper. Vertical to this plane is a twin boundary, across which the atoms are arranged in the same, but reflected pattern.

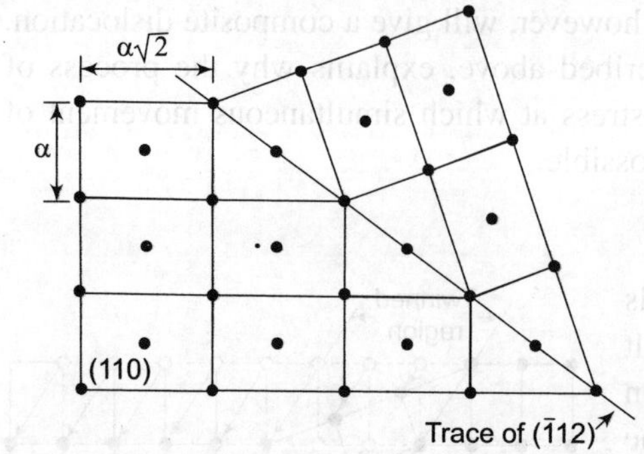


**Figure 5.58 Twin (bcc)**. The (110) plane is shown. The boundary is a (112) plane emerging perpendicularly through the plane of the sketch. The two parts are mirror images of one another, and the (112) plane is common to both parts of the twin. the (110) and (112) planes intersect along the [111] direction. (L.H. Van Vlack, *Materials science for Engineers*, Addison-Wessely)

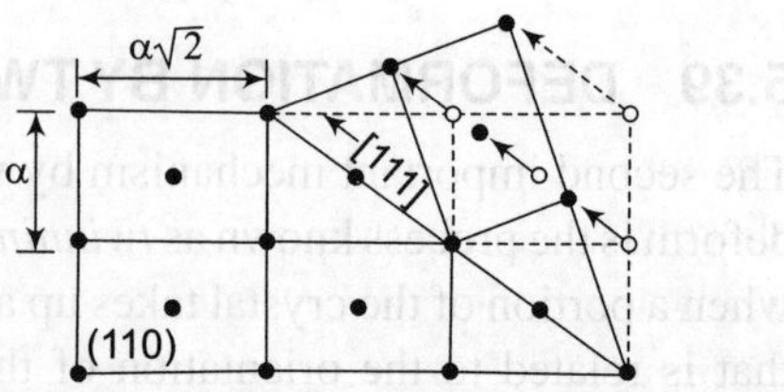


**Figure 5.59 Mechanical Twinning [bcc—(110) plane]**. A shear stress in the [111] direction Produces strain by twinning. Deformation by twinning is significant in hcp metals, because there are fewer hexagonal slip system

Twinning can be induced by shear stresses, as indicated in Fig. 5.59. This produces a permanent displacement, but is limited to relatively slow strains, because, once repositioned, the atom cannot move farther by twinning. For this reason, twinning deformation is overshadowed in most materials by plastic slip.

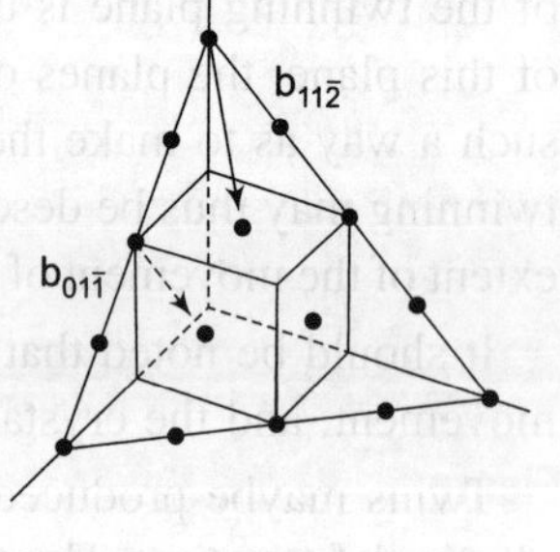


**Figure 5.60** Slip vector on (111) see example 5.3

**Example 5.3** Both $[01\bar{1}]$ and $(11\bar{2})$ lie in the (111) plane of fcc aluminium. Therefore, both $[01\bar{1}]$ (111) and $[11\bar{2}]$(111) slip are conceivable. (a) Make a sketch of the (111) plane and show the $[01\bar{1}]$ and [112] until vectors. (b) Compare the energies of the dislocation lines that have these two displacement vectors.

***Solution***

(a) See sketch (Fig. 5.59)

(b) $\dfrac{E_{01\bar{1}}}{E_{11\bar{2}}} = \dfrac{lGb^2_{01\bar{1}}}{lGb^2_{11\bar{2}}}$

Both involve the same slip plane (111); therefore, the same shear modulus applies. Based on unit length, *l*.

$$\frac{E_{01\bar{1}}}{E_{11\bar{2}}} = \left(\frac{b_{01\bar{1}}}{b_{11\bar{2}}}\right)^2 = \left(\frac{a/\sqrt{2}}{a\sqrt{6}/2}\right)^2 = \frac{1}{3}$$

$$E_{01\bar{1}} = \frac{1}{3} E_{11\bar{2}}$$

With this $\frac{1}{3}$ ratio, slip occurs appreciably more readily with the first of the two potential slip system for fcc metals.

## 5.40 STACKING FAULTS

A close-packed structure is generated by stacking close-packed layers on top of one structure is generally by stacking close-packed layers on top of another with the restriction that no two adjacent layers are in the same orientation. Given a layer *A*, a close packing can be extended by placing the next layer such that its atoms occupy either *B* or *C* sites (Fig. 5.61) Here *A*, *B* and *C* refer to the three possible layer positions in a projection normal to the close-packed layers. In the following, we shall discuss various ways of producing stacking faults in fcc and help crystals.

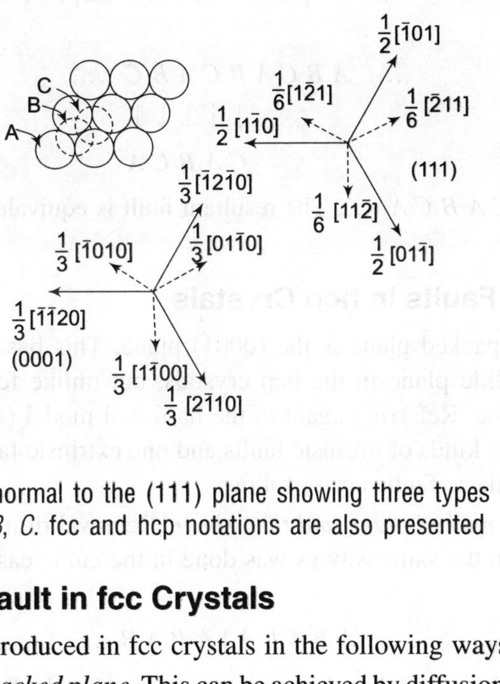


**Figure 5.61** Projection normal to the (111) plane showing three types of stacking positions *A*, *B*, *C*. fcc and hcp notations are also presented

### 5.40.1 Stacking Fault in fcc Crystals

Stacking faults may be produced in fcc crystals in the following ways.

(a) *By removing a close-packed plane*. This can be achieved by diffusion of a sheet of vacancies into the plane and by collapsing together the adjoining planes to eliminate the layer of void between them. For example, removing a *B* layer, the cubic-close packed sequence becomes

$$\ldots\ldots A\ B\ C\ A\ |\ C\ A\ B\ C \ldots\ldots$$

This is called an intrinsic fault. This can be regarded as a layer *CACA* of hcp or as two overlapping twin boundaries, *CAC* and *ACA*, across which next nearest layers are wrongly stacked. In such cases, the normal stacking sequence is maintained in the crystal on either side of the fault right up to the fault plane.

(b) *By inserting an extra plane.* This can be achieved by diffusion of interstitial atoms into the space between two close-packed planes. If say, a *B* layer is introduced as in the sequence

$$\downarrow$$
$$\ldots\ldots\ A\ B\ C\ B\ A\ B\ C\ \ldots\ldots$$

This is called an extrinsic fault. This fault is equivalent to two neighbouring twin boundaries, *BCB* and *BAB*.

(c) *By slip on a close-packed plane.* This can be produced by displacing all atoms above the reference plane by shearing operation on {111} planes. Let us consider an fcc crystal and suppose *A* as shown in Fig. 5.60 is a reference plane, and the next plane above it is a *B* plane. If the *B* plane and all other planes above it are displaced by the vector (1/6) $[\bar{2}11]$, the *B* plane moves into *C* positions, and the planes above it transforms in cyclic order, $A \to B \to C \to A$, relative to the reference plane. The shear displacement is represented by the arrows in the reaction

$$\ldots\ldots\ A\ B\ C\ A\ B\ C\ A\ B\ C\ \ldots\ldots$$
$$\downarrow\downarrow\downarrow\downarrow\downarrow$$
$$C\ A\ B\ C\ A$$

giving ...... *A B C A C A B C A*...... The resultant fault is equivalent to an intrinsic fault as obtained in (a).

### 5.40.2 Stacking Faults in hcp Crystals

In hcp crystals, close-packed plane is the (0001) plane. This basal plane is also the most frequently observed glide plane in the hcp crystals, but unlike fcc crystals, this does not correspond to twin plane. Referring again to the hard-ball model (Fig. 5.60), one can easily verify that there are two kinds of intrinsic faults and one extrinsic fault in hcp structures. The mode of formation of these faults are as follows.

(a) *By removing close packed plane and then shear.* Removal of a close-packed plane in an hcp structure is done in the same way as was done in the cubic case. The resulting stacking becomes

$$\ldots\ldots\ A\ B\ A\ B\ A \mid A\ B\ A\ B\ \ldots\ldots$$

Here, either side of boundary has the same plane (i.e., *A*) which violates the rule of close packing. Moreover, this is unstable also. In order to acquire the stable configuration, the layer above the boundary must shear.

Therefore, shearing them by the vector $\frac{1}{3}$ [1100], the *A* plane moves into *C* position, and the planes above it transform in anticyclic order, $A \to B \to C \to A$, relative to *A* plane, yielding the stacking sequence

$$\ldots\ldots\ A\ B\ A\ B\ A \mid C\ A\ C\ A\ \ldots\ldots$$

This fault is equivalent to cubic layer, *B A C*.

(b) *Simply by shear*. Another intrinsic fault can be produced in the hcp structure by directly shearing all the planes above a reference plane by the vector $\frac{1}{3}[\bar{1}100]$, i.e., the sequence

$$...... A\ B\ A\ B\ |\ A\ B\ A\ ......$$

after shearing yield a sequence

$$...... A\ B\ A\ B\ |\ C\ A\ C\ A\ .....$$

Similarly, this fault is equivalent to two overlapping cubic layers, *ABC* and *BCA*.

(c) *By inderting an extra plane*. This can be achieved by inserting a *C* plane in the normal hexagonal sequence *A B A B* ......

## 5.41 PLANAR OR SURFACE DEFECTS

Surface defects or imperfections arise from a change in the orientation or staking or atomic planes on or across a boundary. In geometric sense, a surface imperfection is a two-dimensional defect. These are of two types – external and internal. The external surface of the crystal marks the termination of crystal structure where the atoms have neighbours of one side only, whereas atoms inside the crystal have neighbours on their either side. Since the surface atoms are not entirely surrounded by others, they represent an external surface imperfection where the atoms possess higher energy than that of internal atoms.

Internal surface defects are basically of four types: grain boundaries, tilt boundaries, twin boundaries and stacking faults.

### 5.41.1 Grain Boundaries

Although it is described to prepare a single crystal for some specific purposes, yet some solids are polycrystalline, i.e., they contain a large number of crystallites or grains. Gain boundaries are those planar imperfections in polycrystalline materials that separate grains of different orientation, i.e., adjacent grains are inclined towards one another. When the included angle is large (~10° – 15°) it is called a *wide-angle grain boundary*, and when small (a few degrees). it is called a *low-angle tilt boundary*.

The grain between two crystals, which have different crystalline arrangements or differ in composition, is known as inter-phase boundary or simply an interface.

### 5.41.2 Tilt and Twist Boundary

Tilt boundary may be defined as a boundary between two adjacent perfect regions in the same crystal that are slightly tilted, with respect to one another. In other words, when the angle between the crystalline orientations of the two grains is small (less than 10°), the boundary is said to be a tilt boundary. They are also called low angle boundaries (Fig. 5.62). The structure of such boundaries can be described by means of arrays of dislocations. It consists of a series of equally space edge dislocations of the same sign located one above the other.

The angle of tilt θ is related to the Burgers vector *b* of the edge dislocations by

$$\frac{b}{h} = \tan\theta \qquad ...(1)$$

where $h$ is the vertical spacing between two neighbouring edge dislocations. For small angle of tilt,

$$\frac{b}{h} = \theta \qquad ...(2)$$

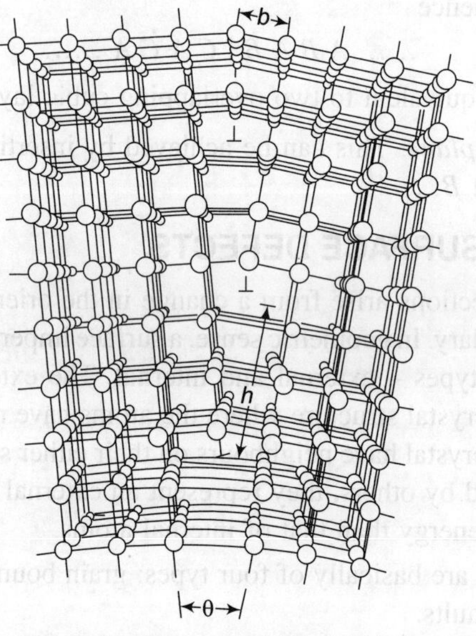


**Figure 5.62** A tilt boundary consists of equally spaced edge dislocations of the same sign one above the other

If the two parts of a crystal are rotated through a small angle, about an axis which is perpendicular to the grain boundary (rather than about one that lies in the boundary as in the case of tilt boundary), a twist boundary is formed. Twist boundary consists of a set of screw dislocations, in contrast to the tilt boundary which consists of a set of edge dislocations.

If the misprinted single crystal sections are identical but are joined together in such a way that boundary acts as a reflecting plane, the pair of crystals constitute a twin and the resulting boundary is said to be a twin boundary. In such cases, the atomic arrangement on one side of the boundary is a mirror reflection of the arrangement on the other side (Fig. 5.63).

There are several ways in which twinning can be produced. Growth twnis are formed during the growth of a crystal. The transformation between high and low temperature modifications of a crystal may take place via twinning. Twinning produced during plastic deformation of a crystal is called deformation twinning. The simplest example of twinning in an fcc structure can be represented by the sequence

$$\downarrow$$
$$\ldots\ldots A\,B\,C\,A\,B\,C\,B\,A\,C\,B\,A\ldots\ldots$$

where the arrow indicates the twin plane and the centre of the fault. The occurrence of twinning is common especially in metals with bcc or hcp structure.

**Example 5.4** Calculate the spacing between dislocations in a tilt boundary in fcc nickel, when the angle of tilt is 2°.

***Solution*** Given: θ = 2°. Crystal is fcc nickel, so that $a$ = 352 Å and hence

$h = a/\sqrt{2}$ = 2.489 Å. Now, we have

$$h = \frac{b}{\tan\theta} = \frac{2.489}{\tan 2^\circ} = 71.3 \text{ Å}$$

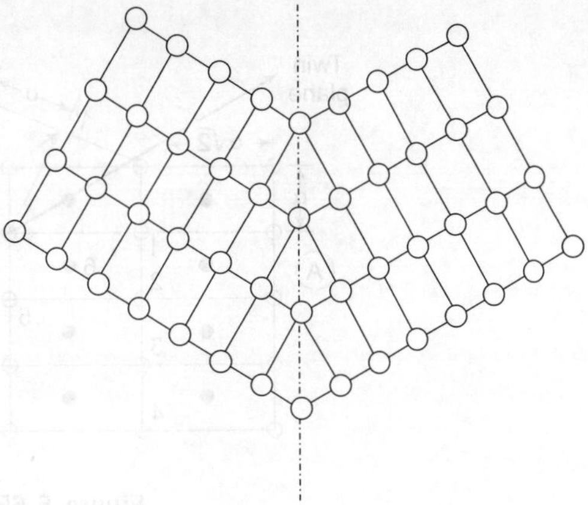

**Figure 5.63** A twin plane

### *Grain Boundary Region*

These are the regions formed by the differently oriented adjacent grains. As such in these regions lattice of different orientations will overlap each other and as a consequence the distortion in the lattice will be higher energy regions as compared to the central part of the grains.

In the case of fine-grained structure the higher proportion per unit volume will be occupied by the grain boundaries and hence their mechanical properties are dominated by the properties of grain boundaries or higher energy regions. In the case of coarse grained structure the higher proportion will be occupied by the grains and hence the properties are governed by the lattices of the grains, Fig. 5.64.

Grain boundary regions will have high distortion of the lattices or higher energy regions. Such regions will bc two-dimcnsional on a lattice plane.

Hence, grain boundary regions are called *surface defects*.

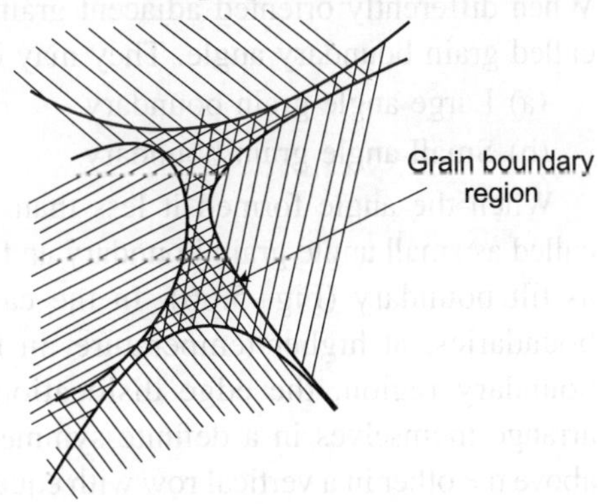


**Figure 5.64** Grain boundary region

## 5.41.3 Twin Boundaries

Another type of surface defects is known as twin boundaries defect. In this case also the distortions of lattice take place about the twin plane.

Two identical objects are generally known as twins. If you see your image in a mirror, it is an exactly replica of yours and we can say that twin is formed.

Twin is a process of deformation in which the atom planes can move adjacent to each other in a specified direction. In this process, a part of the crystal can move against the other in a definite pattern on a plane called *twin* plane such that one part of the crystal becomes the mirror image of the other part Fig. (5.65).

Such distortion produced due to twinning is known as surface defect is called *twin boundaries*.

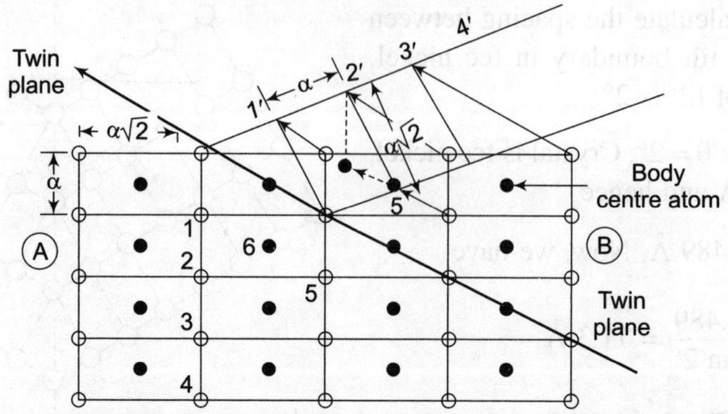


**Figure 5.65** Twin process

A bcc structure (110) plane is shown in Fig. (5.64) to illustrate twin. 1′, 2′, 3′, 4′, 5′ and 6′ are the images of the atoms 1, 2, 3, 4, 5 and 6. Hence, the movement of atom will be fixed for the part B of the crystal so as to form the twin about the twin plane. Such distortions as rotation of the (110) plane as shown in Fig. (5.65) produces a suface defects known as twin boundaries.

## 5.42 GRAIN BOUNDARY ANGLE

When differently oriented adjacent grains meet, an angle is formed between them which is called grain boundary angle. They may be classified as

(a) Large angle grain boundary
(b) Small angle grain boundary.

When the angle formed it less than 12°; it is called as small angle grain boundary and it known as tilt boundary (Fig. 5.66). In the case of tilt boundaries, at higher temperature, in the grain boundary region, the edge dislocation present arrange themselves in a definite symmentry, one above the other in a vertical row with equal vertical distance between them.

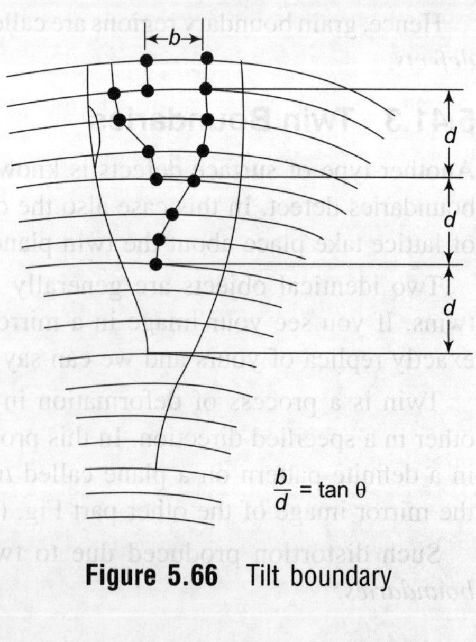


**Figure 5.66** Tilt boundary

Such an arrangement provides a parallel slip planes (plane perpendicular to the dislocation line is called slip plane) which promotes deformation process through slip. Hence, at higher temperatures fine grained structure show more ductility as compared to course grained structure, since a larger proportion of the region is occupied by the grain boundaries. The tilt angle θ is given by $\tan^{-1} b/d$ which leads to $\theta = b/d$ for small values of θ.

## 5.43 SOLID SOLUTION

If we take a glass of water and add sugar into it and stir, the water becomes uniformly sweet and its sweetness is homogeneous throughout. The water is solvent and sugar is called solute. In other words we can say the property is homogeneous throughout. However, if we mix oil into water, it is not soluble and the properties become different and different regions, that is, it is heterogeneous in properties. Similar is the case when atoms of different types are mixed together to form an alloy. At least two types of atoms are required to form an alloy.

In certain cases the solidification of an alloy results in the formation of one kind of crystal in which both types of atoms are present, but they cannot be detected even by a high power microscope, although the properties of the crystal are greatly altered. In such a case we have a solid metal in which the inter-atomic state which existed in the liquid solution is preserved after solidification and it is known as *solid solution*. A single phase alloy is called solid solution. In an A-B alloy, A atoms are called solvent and B atoms are called solute atoms. In contrast to liquid solution, A-B alloy is different than B-A alloy.

A phase is physically distinct, chemically homogeneous and mechanically separable region of a system. A homogeneous region of a system means properties are same in that region.

An alloy cannot be perfectly homogeneous, because it will contain at least two types of atoms and no two atoms of different type will have the same size. There will always be a distortion surrounding the solute atoms as they may be smaller or bigger than the solvent atoms. However, in practice certain reasonable distortions are tolerated to consider that the alloy is homogeneous or it is having a single phase. Hence, in a solid solution the atoms occur in a definite geometrical pattern, which is usually a slightly distorted form of that one of the constituent metals. In general, a solid solution demands a more special solvent-solute relationship than a liquid solution because the structure is less adaptable. When a solute atom is much smaller than the solvent atoms, it may dissolve interstitially occupying a void space in the parent or the solvent structure.

If zinc is added into copper to produce brass having a single phase, it may happen that only up to a certain percentage of zinc the property remains homogeneous giving rise to limit of solid solubility. This gives the solid solution range or extended solid solubility. Solid solutions are of two types:

(1) Substitutional solid solution

(2) Interstitial solid solution.

Substitutional solid solution means that the solute atoms are substituted at some lattice points in place of solvent atoms.

## 5.44 HUME ROTHERY'S EMPIRICAL RULES FOR THE FORMATION OF SUBSTITUTIONAL SOLID SOLUTION

1. **Size Factor.** This factor arises due to the strains which are induced in the crystal lattice by a misfitting solute atom. The greater is the difference in sizes of solvent and solute atoms, the greater is the distortion of atomic pattern and the more likely it is that the atoms can adopt some other arrangement (intermetallic compound) which may be more stable.

[The difference in the atomic diameter of the solvent and solute atoms should not be more than 15%].

Larger difference than this is unfavourable and the range of the solid solution is restricted.

2. **Chemical affinity factor.** The electronegativity difference between the solvent and solute atoms should be small.

   The more electronegative the solute atom and the more electropositive the solvent atom, or vice versa, the greater is the tendency to restrict the solid solution range and to form intermetallic compounds.
3. **Relative valency factor.** The valencey of solvent atoms and solute atoms should be same. A metal of lower valency tends to dissolve a metal of higher valency more readily than vice versa. For example, silicon with a valency four dissolves little copper which has a valency of one; but copper dissolves an appreciable amount of silicon.
4. **Structure factor.** Both solvent and solute elements forming solid solution should have similar structure, that is, same lattice pattern so as to facilitate the substitution. Thus, two fcc metals like copper and nickel will form a continuous solid solution assisted by the fact the atom sizes of these two metals are within the prescribed limit.

   A high degree of substitutional solid solubility between two metals will be obtained when

   (a) The difference in atomic radii is less than 15%

   (b) Electrochemical properties are similar

   (c) The two atoms have similar lattice patterns.

In a substitutional solid solution the atoms of the solvent metals are replaced by the atoms of the solute metal. Such replacements take place at the lattice points. This substitution can be either *ordered* or *disordered* or *random* as illustrated in Figs. 5.67 and 5.68.

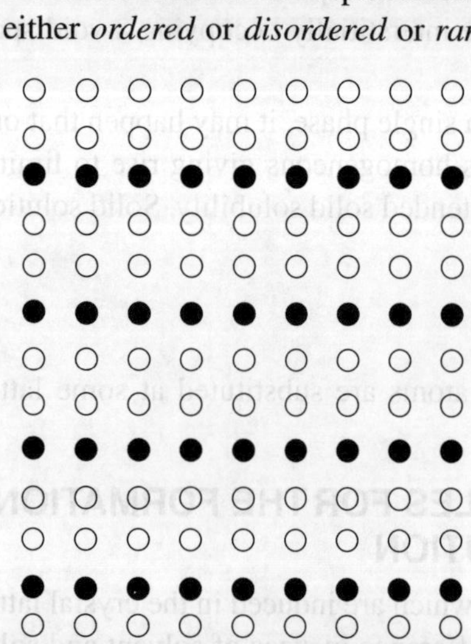

Figure 5.67 Ordered substitutional solid solution

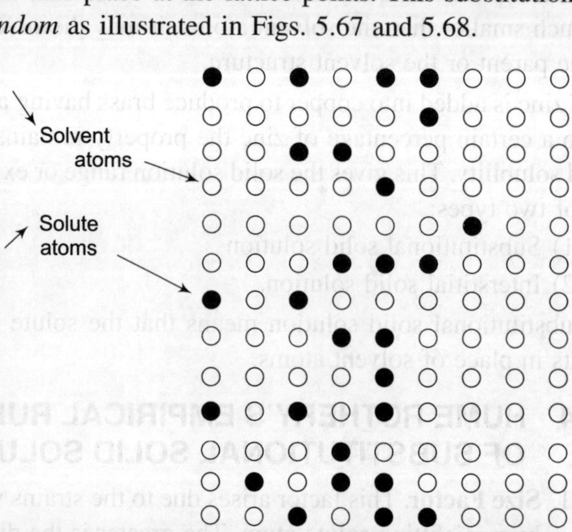


Figure 5.68 Disordered or random substitutional solid solution

## 5.45 INTERSTITIAL SOLID SOLUTION

Interstitial solid solution can be formed only when the atoms of the solute atoms or added atom are very small compared with the solvent or parent atoms, thus enabling then to fit into the interstitial spaces in the crystal lattice of the solvent metal. The carbon atoms can form interstitial solid solution with fcc iron during the solidification of iron-carbon alloy called steel. This can also happen when the solvent metal is already solid, that is, carbon atom can be absorbed by the solid iron provided that solvent metal is heated to a temperature which is suitable for such diffusions.

An ordered solid solution on heating will become disordered. A disordered solid solution is more in existence due to the presence of defects and the thermal energy due to temperature.

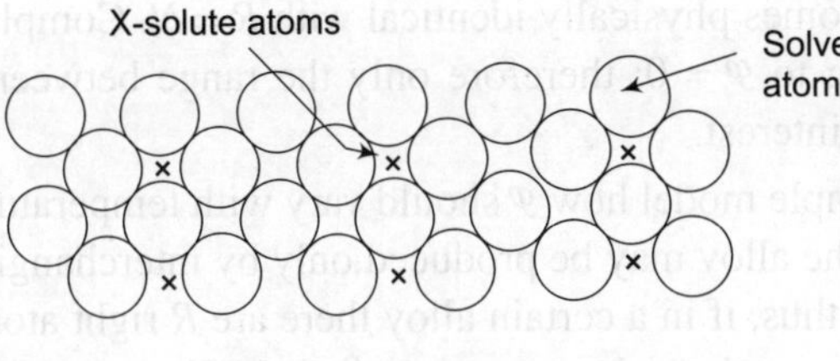


**Figure 5.69** Ordered interstitial solid solution

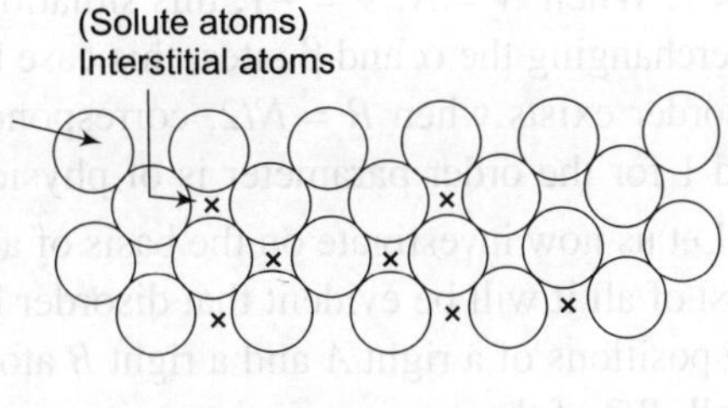


**Figure 5.70** Disordered interstitial solid solution

## 5.46 INTER-METALLIC COMPOUNDS

When the conditions of extended solid solubility are not satisfied either partly or wholly, the two metals or one metal and the other non-metal form an alloy, they are called intermetallic compounds. Formation of carbides, nitrides, etc., fall in this category. There are two types of intermatallic compounds.

1. **Valency Compounds:** Mixing of electropositive elements like Na, K, Al, and Mg with electronegative elements like Sn, Pb, Sb, As, etc., obey the normal valency rules of solid solubility. The bondings are either ionic or covalent. They are usually hard and brittle. They behave as insulators. They have melting point above the melting points of the either element. Examples are $Mg_2Sn$, $Mg_2Pb$.
2. **Electron Compounds:** These compounds are formed in those metals which have similar electrochemical properties and a favourable size factor, but they have different number of valence electrons. They do not obey the normal valency rules. The crystal structure depends on the ratio of the total number of valence electrons to the total number of atoms that make the compound. These ratios may be 3 : 2, 7 : 4, 21 : 13 etc.

**Examples** (i) $CuZn_3$ — 7/4 (ii) $Cu_3Al$ — 6/4 = 3/2

(iii) $Cu_5Zn_8$ — 21/13 (iv) $Cu_9Al_4$ — 21/13

## 5.47 THE LONG DISTANCE ORDER THEORY OF BRAGG AND WILLIAMS[4]

We shall consider an alloy of the simple composition *AB* which in the completely ordered structure may be represented by two interoperating lattice of *A* and *B* atoms. If the lattices run through the whole crystal without discontinuities, one may define the sites corresponding to one lattice as α sites and those of the other lattice as β sites. In the completely ordered structure all α sites are occupied by *A* atoms, and all β sites by *B* atoms. In an incompletely ordered alloy one may then define right (*A* on α, *B* on β) and wrong atoms (*A* on β, *B* on α). When *R* and *W* distance order parameter is defined by

$$\varphi = (R - W)/N = (2R - N)/N \qquad ...(1)$$

where $N = R + W$ is the total number of atoms. When $R = N$ there is complete order and $\varphi = 1$. When $W = N$, $\varphi = -1$; this situation also describes a state of complete order, since by interchanging the α and β sites, this case becomes physically identical with $R = N$. Complete disorder exists when $R = N/2$, corresponding to $\varphi = 0$; therefore only the range between 0 and 1 for the order parameter is of physical interest.

Let us now investigate on the basis of a simple model how $\varphi$ should vary with temperature. First of all it will be evident that disorder in the alloy may be produced only by interchanging the positions of a right *A* and a right *B* atom; thus, if in a certain alloy there are *R* right atoms in all, *R*/2 of these occur on the α sites and an equal number occurs on β sites. Since the alloy at absolute zero tends to be completely ordered, the energy required to produce disorder must be positive. Now suppose that the alloy in thermal equilibrium contains *R* right atoms and *W* wrong atoms. If in this state we were to interchange the positions of a right *A* and a right *B* atom, the change in *W* would be $\Delta W = -\Delta R = 2$. Since in thermal equilibrium the change $\Delta F$ in the free energy associated with $\Delta W$ must vanish ($\Delta W \ll W$), one can readily find the equilibrium values for *R* and *W* in the following manner : suppose that the energy required to produce a pair of wrong atoms in the state *R*, *W* is $\phi(R, W)$. The configurational entropy associated with the state *R*, *W* is

$$S_{cf} = k(N \log N - R \log R - W \log W) \qquad ...(2)$$

Thus, the change $\Delta S_{ef}$ associated with $\Delta W$ is

$$\Delta S_{cf} = -k(\Delta R \log R + \Delta W \log W) = 2k \log (R/W) \qquad ...(3)$$

Neglecting thermal entropy changes for simplicity, we may write

$$\Delta F = 0 = \phi(R, W) - 2kT \log (R/W)$$

or

$$R/W = \exp [\phi(R, W)/2kT] \qquad ...(4)$$

Let us now inquire about the nature of the function $\phi(R, W)$. Consider a given right *A* atom; the probability for an arbitrarily chosen nearest neighbour of this atom to be a *B* atom (right) is *R*/*N*; the probability that a nearest neighbour atom is an *A* atom is *W*/*N*. The potential energy of the *A* atom in the field of its nearest neighbours is then

[4] W.L. Bragg and E. J. Williams, *Proc. Roy. Soc.* (*London*), **A145,** 699 (1934).

$$-z(\phi_{AB} R/N + \phi_{AA} W/N)$$

Similarly, the potential energy of a right $B$ atom in the field of its nearest neighbours is

$$-z(\phi_{AB} R/N + \phi_{BB} W/N)$$

This model of the nearest neighbour interactions leads, for the energy required to interchange the position of the right $A$ and $B$ atoms, to the expression

$$\phi(R, W) = \phi_0 (R - W)/N = \phi_0 \varphi \quad ...(5)$$

where $\phi_0 = z(2\phi_{AB} - \phi_{AA} - \phi_{BB})$ is a positive quantity since the dissociation energy $\phi_{AB}$ for an unlike pair is larger than that for a pair of similar atoms. The physical meaning of $\phi_0$ is that it represents the energy required to produce 2 wrong atoms in the completely ordered lattice ($\varphi = 1$). It is observed that according to that energy required produce 2 wrong atoms decreases as the amount of order decreases. Qualitatively, this can readily be understood; for example, if two atoms in the completely disordered state ($\varphi = 0$) are interchanged the energy is, on an average, zero because in the long-distance theory the distribution of $A$ and $B$ atoms around a given $A$ and $B$ atom is then random. Actually, the simple linear relationship (5) between $\phi(R, W)$ and $\varphi$ was introduced by Bragg and Williams as an assumption. We see that this assumption is equivalent to the model employed above in which the interaction between the atom simplified to nearest neighbour interact with constant $\phi_{AA}$, $\phi_{AB}$, and $\phi_{BB}$ values.

When Eqs. (4) and (5) are substituted into Eq. (1) we obtain the following implicit equation for the long-distance order parameter.

$$\varphi = \tanh (\phi_0 \varphi / 4kT) \quad ...(6)$$

This equation may be solved graphically by introducing the variable

$$x = \phi_0 \varphi / 4kT \quad \text{or} \quad \varphi = 4kTx/\phi_0 \quad ...(7)$$

where $\varphi$ must satisfy the equation $\varphi = \tanh x$ as well as Eq. (7). The function $\tanh x$ is represented in Fig. 5.71. The $\varphi(x)$ curves corresponding to Eq. (7) produce a set of straight line,

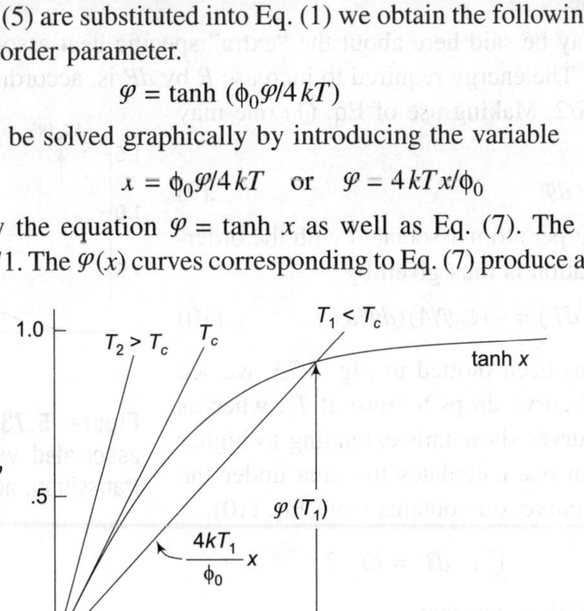


**Figure 5.71** Graphical solution of equation (6) as explained in text

of slope equal to $4kT/\phi_0$, i.e., proportional to $T$. The intersection of the tanh $x$ curve with one of the straight lines then gives the value of $\varphi$ for the corresponding temperature. Since the slope of the tanh $x$ curve for very small values of $x$ is equal to unity, there exists a critical temperature $T_c$ above which $\varphi = 0$, viz.,

$$T_c = \phi_0/4k \qquad \text{...(8)}$$

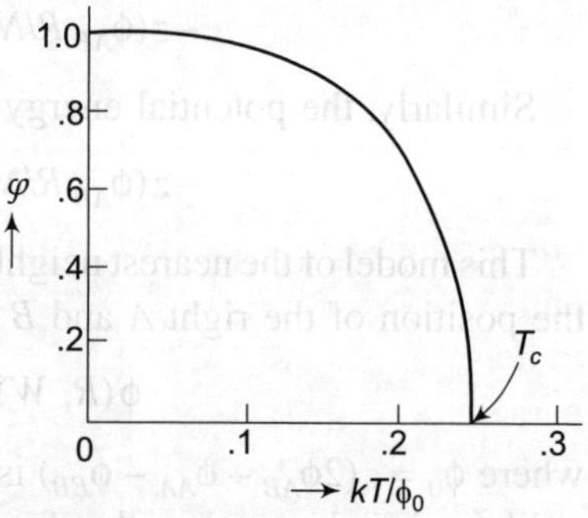


**Figure 5.72** The long distance order parameter as function of temperature, according to the Bragg-Williams theory

Thus, in this theory the order disappears altogether at $T_c$. The temperature dependence of $\varphi$ according to the Bragg–William theory is given in Fig. 5.72. Note the rapid drop in the vicinity of $T_c$. The reason for this lies in the fact that once a certain amount of disorder is present, it becomes easier for the thermal motions to produce more disorder. One therefore speaks of a cooperative phenomenon. Other cooperative phenomena are ferromagnetism and ferroelectricity, it is instructive to compare the theoretical treatment of those phenomena with the order-disorder treatment given above.

For β brass, $T_c \simeq 740°\text{K}$, so that according to Eq. (8) the quantity $\phi_0$ is approximately 0.25 eV in this case.

A few words may be said here about the "extra" specific heat associated with the order-disorder transition. The energy required to increase $R$ by $dR$ is, according to the definition of $\phi$, equal to $-f_0\varphi dR/2$. Making use of Eq. (1) one may thus write

$$dE = -(N\phi_0/4)\varphi\, d\varphi \qquad \text{...(9)}$$

The specific heat per atom associated with the order-disorder transformation is thus given by

$$\Delta c_v = (1/\text{N})(dE/dT) = -(\phi_0\varphi/4)(d\varphi/dT) \qquad \text{...(10)}$$

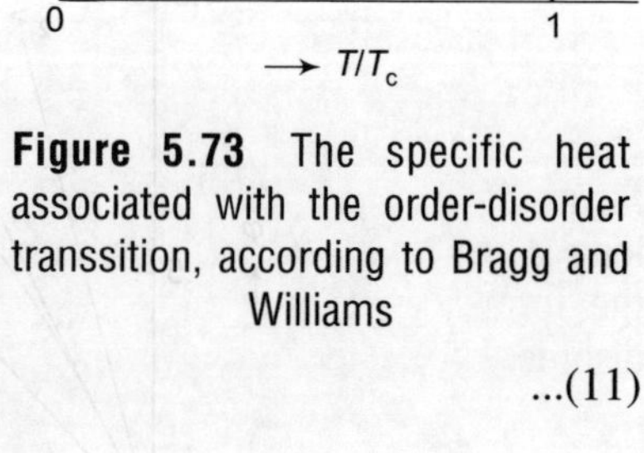


**Figure 5.73** The specific heat associated with the order-disorder transsition, according to Bragg and Williams

This function has been plotted in Fig. 5.73, we see that the theoretical curve drops to zero at $T_c$, whereas the experimental curves show tails extending to higher temperatures. When one calculates the area under the extra specific heat curve, one obtains from Eq. (10):

$$\int_0^{\infty} c_v\, dT = kT_c/2 \qquad \text{...(11)}$$

in fair agreement with experiment.

## 5.48 SHORT-DISTANCE ORDER THEORIES

The essential difference between the long-distance and short-distance theories of order may be illustrated with reference to Fig. 5.73. From the long-distance order point of view, this lattice would be highly disordered and yet we observed that nearly all atoms have unlike atoms as nearest neighbours. In other words, if one were to employ the relative number of unlike nearest neighbours as a criterion for order, the lattice of Fig. 5.74 has a high degree of order. Theories based on this concept have been worked out by many investigators.[5] Short-range order may be defined in terms of the number of right pair (*AB*) and the number of wrong pairs (*AA*, *BB*). Thus, consider an *A* atom and let a probability for a given nearest neighbour to be a *B* atom be $(1 + \sigma)/2$ and to be an *A* atom $(1 - \sigma)/2$; $\sigma$ is then the short distance order parameter. For complete order, $\sigma = 1$, for a random distribution of atoms, $\sigma = 0$. Suppose that the dissociation energies $\phi_{AA}$ and $\phi_{BB}$ are equal, and, of course, smaller than $\phi_{AB}$, in that case we should have, according to the Boltzmann distribution,

$$(1 + \sigma)/(1 - \sigma) = \exp\,[(\phi_{AB} - \phi_{AA})/kT] = \exp\,(\phi/kT) \qquad ...(1)$$

Thus, if $\phi$ were a constant independent of the degree of order, $\sigma$ would decrease slowly to zero at high temperature and there would be no critical temperature. In Bethe's theory $\phi$ is calculated in terms of the long-range order which exists in the crystal; since $\phi$ decreases with decreasing long-range order, $\sigma$ decreases more rapidly to zero in the vicinity of the critical point than for constant, $\phi$. As an example we give in Fig. 5.75 the long-range and short-range order parameters $\varphi$ and $\sigma$ for a superlattice of the $AB_3$ structure. For the details of short-distance order theory we refer the reader to the literature. We may mention here an approximate theory developed by Cowley in which the order parameter is expressed in terms of the coefficients of the Fourier series, which determines the intensity of X-ray scattering. This makes a direct comparison between theory and experiment possible; the agreement is very good.

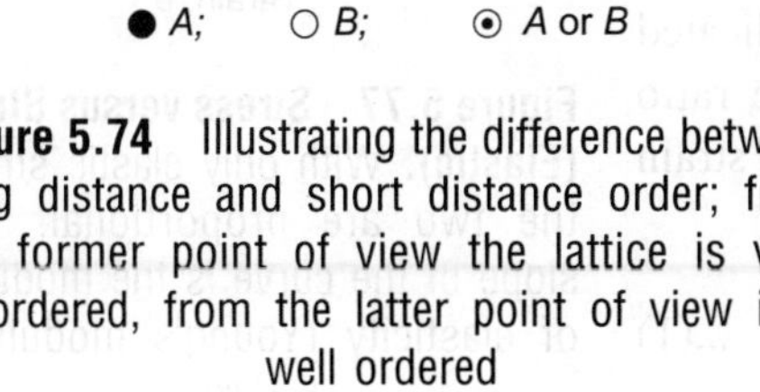


**Figure 5.74** Illustrating the difference between long distance and short distance order; from the former point of view the lattice is very disordered, from the latter point of view it is well ordered

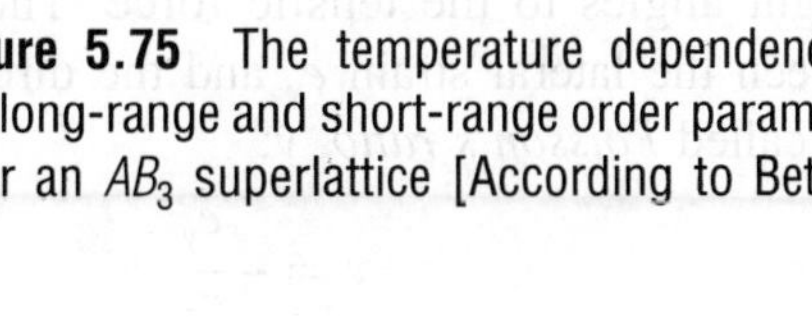


**Figure 5.75** The temperature dependence of the long-range and short-range order parameters for an $AB_3$ superlattice [According to Bethe]

5 H.A. Bethe, *Proc. Roy. Soc.* (*London*), **A150,** 552 (1935); E. J. Williams, *Proc. Roy. Soc.* (*London*), **A152,** 231 (1935); R. Peierls, *Proc. Roy. Soc.* (*London*), **A154,** 207 (1936); J. G. Kirkwood, *J. Chem. Phys.*, **6,** 70 (1938); C. N. Yang, *J. Chem. Phys.*, **13,** 66 (1949); Y. Y. Li, *J. Chem. Phys.*, **17,** 447 (1949); H. A. Kramers and G. H. Wannier, *Phys. Rev.*, **60,** 252, 263 (1941); L. Onsager, *Phys. Rev.*, **65,** 117 (1944).

## 5.49 ELASTIC DEFORMATION

Elastic deformation is a reversible strain. If a stress is applied in tension, the material becomes slightly longer; removal of the load permits the material to return to its original dimension. Conversely, when it is under compression, the material becomes slightly shorter. The dimension of the unit cells change when the material undergoes elastic strain (Fig. 5.76).

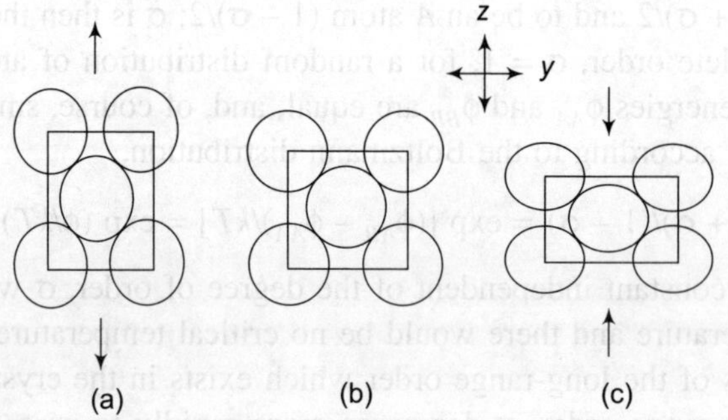


**Figure 5.76 Elastic Normal Strain (Greatly Exaggerated).** Atoms are not permanently displaced from their original neighbours. (a) Tension (+). (B) No strain. (c) Compression (–)

### 5.49.1 Elastic Moduli

When only elastic deformation exists, the strain is proportional to the applied stress (Fig. 5.77). The ratio of stress to strain is the *modulus of elasticity* (Young's modulus), and is a property of the material. The greater the forces of attraction between atoms in a material, the higher is the modulus of elasticity.

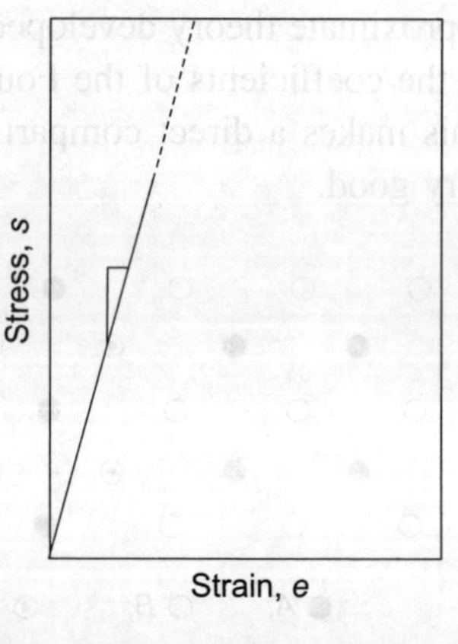


**Figure 5.77 Stress versus Strain (Elastic).** With only elastic strain, the two are proportional. The slope of the curve is the modulus of elasticity (Young's modulus), $E = s/e$.

Any lengthening or compression of the crystal structure in one direction, due to a uniaxial force, produces an adjustment in the dimensions at right angles to the force. In Fig. 5.76 (a), for example, a small contraction is indicated at right angles to the tensile force. The negative ratio between the lateral strain $e_y$ and the direct tensile strain $e_z$ is called *Poisson's ratio*, $v$:

$$v = -\frac{e_y}{e_z} \quad \text{...(1)}$$

Engineering materials may be loaded in shear as well as in tension (and compression). In shear loading, the two forces are parallel but are not aligned (Fig. 5.78 b). As a result, the *shear stress*, τ, is the shear force $F_s$ divided by the sheared area $A_3$:

$$\tau = F_s / A_s \quad \text{...(2)}$$

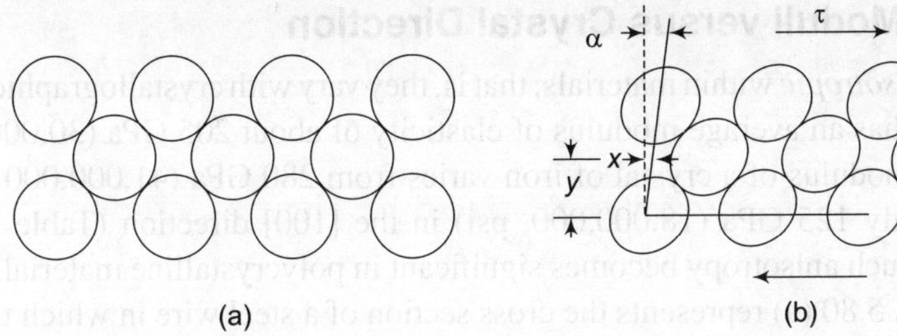


**Figure 5.78 Elastic Shear Strain.** Shear couple produce a relative displacement of one plane of atoms past the next. This strain is elastic as long as atoms keep their original neighbours. (a) No strain. (b) Shear strain

A shear stress produces an angular displacement, α. We define shear strain γ as the tangent of that angle; that is, as $x/y$ in Fig. 5.78 (b). The recoverable or elastic shear strain is proportional to the shear stress:

$$G = \tau/\overline{\gamma} \qquad ...(3)$$

where $G$ is the *shear modulus*, also called the modulus of rigidity, the shear modulus is different from the modulus of elasticity, $E$; however, the two are related at small strains by

$$E = 2G(1 + \nu) \qquad ...(4)$$

Since Poisson's ratio $\nu$ is normally between 0.25 and 0.5, the value of $G$ is approximately 35 per cent of $E$.

A third elastic modulus is the *bulk modulus*, $K$. It is the reciprocal of the compressibility β of the material and is equal to the hydrostatic pressure $P_h$ per unit of volume compression, $\Delta V/V$:

$$K = \frac{P_h V}{\Delta V} = \frac{1}{\beta} \qquad ...(5)$$

The bulk modulus is related to the modulus of elasticity as follows:

$$K = \frac{E}{3(1 - 2\nu)} \qquad ...(6)$$

## 5.49.2 Elastic Moduli versus Temperature

Elastic moduli decrease as temperature increases, as shown in Fig. 5.78 for four common metals. A thermal expansion reduces the value of $dF/da$, and therefore decreases the modulus of elasticity. The discontinuity in the curve for iron in Fig. 5.79 is due to the change from bcc to fcc at 912°C (1673°F). Not surprisingly, more densely packed fcc polymorph requires greater stresses for a given strain; that is, the elastic modulus is greater for fcc, Also note from Fig. 5.79 that higher-melting temperature metals have greater elastic moduli.

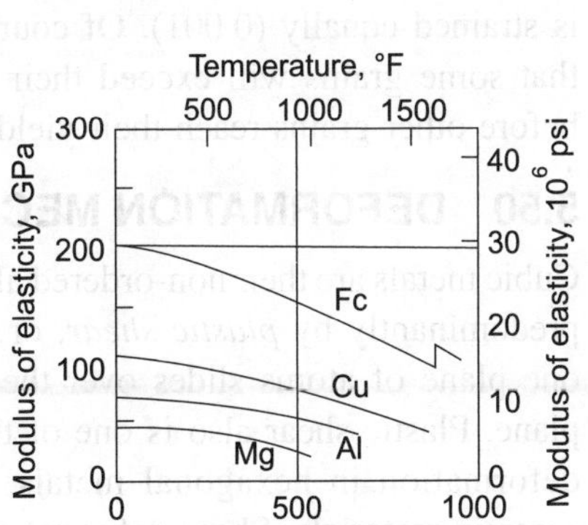


**Figure 5.79 Modulus of elasticity versus temperature.** (Adapted from A.G. Guay and J.J. Hren, *Elements of Physical Metallurgy*. Addison-Wesley.)

### 5.49.3 Elastic Moduli versus Crystal Direction

Elastic moduli are *anisotropic* within materials; that is, they vary with crystallographic direction. As an example, iron has an average modulus of elasticity of about 205 GPa (30,000,000 psi); however, the actual modulus of a crystal of iron varies from 280 GPa (41,000,000 psi) in the [111] direction to only 125 GPa (18,000,000, psi) in the [100] direction (Table 5.13). The consequence of any such anisotropy becomes significant in polycrystalline materials. Assume, for example, that Fig. 5.80 (a) represents the cross section of a steel wire in which the average

**Table 5.13** Moduli of Elasticity (Young's Modulus)*

| *Metal* | *Maximum* | | *Minimum* | | *Random* | |
|---|---|---|---|---|---|---|
| | *GPa* | *$10^6$ psi* | *GPa* | *$10^6$ psi* | *GPa* | *$10^6$ psi* |
| Aluminium | 75 | 11 | 60 | 9 | 70 | 10 |
| Gold | 110 | 16 | 40 | 6 | 80 | 12 |
| Copper | 195 | 28 | 70 | 10 | 110 | 16 |
| Iron (bcc) | 280 | 41 | 125 | 18 | 205 | 30 |
| Tungsten | 345 | 50 | 345 | 50 | 345 | 50 |

* Adapted from E. Schmid and W. Boas, *Plasticity in Crystals.* English translation, London: Chapman Hall

stress is 205 MPa (30,000 psi). If the grains are randomly oriented the elastic strain is 0.001, because the average modulus of elasticity is 205 GPa (30,000,000 psi). However, in reality, the stress varies from 125 MPa (18,000 psi) to 280 MPa (41,000 psi) as shown in Fig. 5.80 (b), bacause grains have different orientations, but each is strained equally (0.001). Of course, this means that some grains will exceed their yield strength before other grains reach their yield strength.

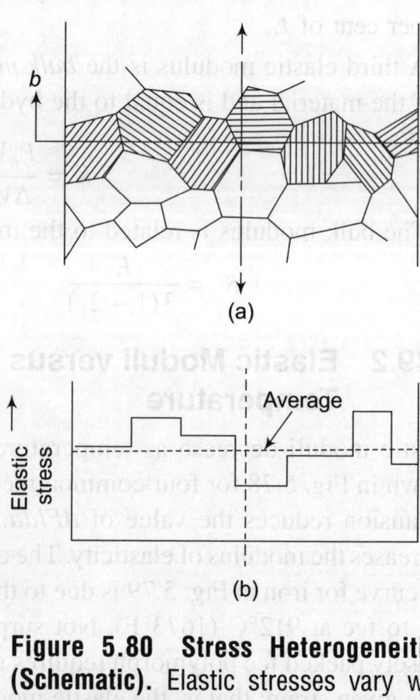


**Figure 5.80 Stress Heterogeneities (Schematic).** Elastic stresses vary with grain orientation, because the moduli of elasticity are not isotropic

## 5.50 DEFORMATION MECHANISMS

Cubic metals are their non-ordered alloys deformed predominantly by *plastic shear*, or *slip*, in which one plane of atoms slides over the next adjacent plane. Plastic shear also is one of the methods of deformation in hexagonal metals, and in a few ceramic materials. Shear deformation occurs even when compression or tension forces are applied, because the stresses may be resolved into shear stresses.

### 5.50.1 Slip Systems

Slip occurs more readily along certain crystal directions and planes than along others. This is illustrated in Fig. 5.81, where a single crystal of an hcp metal was deformed plastically. The shear stress required to produce slip on a crystal plane is called the *critical shear stress*, $\tau_c$.

The predominant sets of *slip systems* in several familiar metals are summarized in Table 5.14. A slip system includes the *slip plane* (*hkl*), and a *slip direction* [*uvw*]. There are a number of slip systems, because of the multiple planes in a family of planes and the multiple directions in a family of directions. Two facts stand out in Table 5.14.

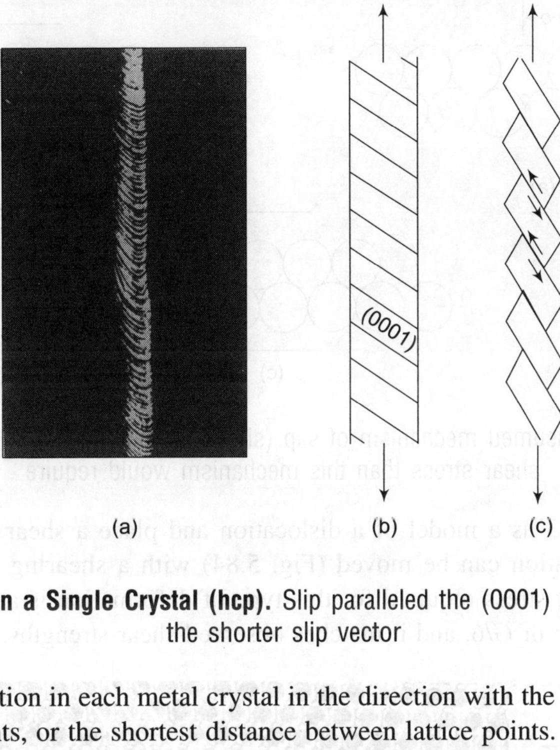


**Figure 5.81 Slips in a Single Crystal (hcp).** Slip paralleled the (0001) plane, which contains the shorter slip vector

1. The slip direction in each metal crystal in the direction with the highest linear density of lattice points, or the shortest distance between lattice points.
2. The slip planes are planes that have wide interplanar spacings and, therefore, high planar densities.

### 5.50.2 Mechanism of Slip

Figure 5.82 shows a simplified mechanism of slip. If we were to calculate the strength of metals on this basis, the result would indicate that the strength of metals should be approximately $G/6$, where $G$ is the shear modulus. Since metals have only a fraction of that strength, a different slip mechanism must be operative. All experimental evidence support a mechanism involving dislocation movements.

**Table 5.14** Predominant Slip System in Metals

| *Structure* | *Examples* | *Slip Directions* | *Slip Planes* | *Number of Independent Slip System* |
|---|---|---|---|---|
| bcc | α-Fe,Mo,Na,W | $\langle\bar{1}11\rangle$ | {101} | 12 |
| bcc | α-Fe,Mo,Na,W | $\langle\bar{1}11\rangle$ | {211} | 12 |
| fcc | Ag,Al,Cu,*-Fe,Ni,Pb | $\langle\bar{1}10\rangle$ | {111} | 12 |
| hcp | Cd,Mg,α-Ti,Zn | $\langle 11\bar{2}0\rangle$ | {0001} | 3 |
| hcp | α-Ti | $\langle 11\bar{2}0\rangle$ | $\{10\bar{1}0\}$ | 3 |

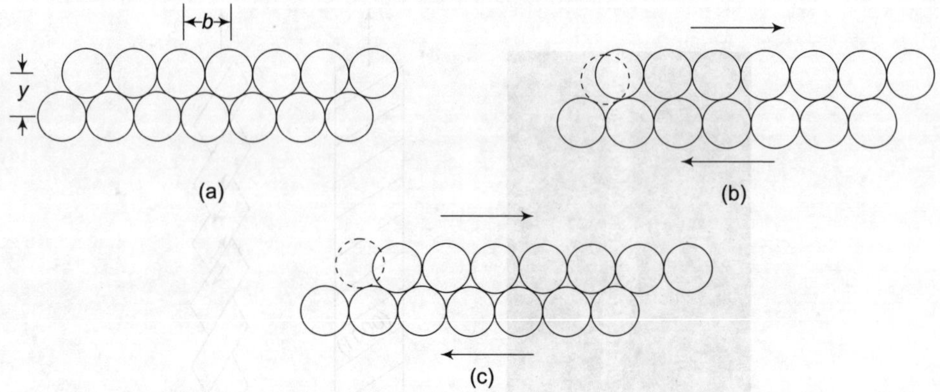


**Figure 5.82** An assumed mechanism of slip (simplified). Metals actually deform with less shear stress than this mechanism would require

If we use Fig. 5.83 as a model of a dislocation and place a shear along the horizontal direction, the dislocation can be moved (Fig. 5.84) with a shearing displacement within the crystal. The shear stress required for this type of deformation is a small fraction of the previously cited value of *G*/6, and it matches observed shear strengths.

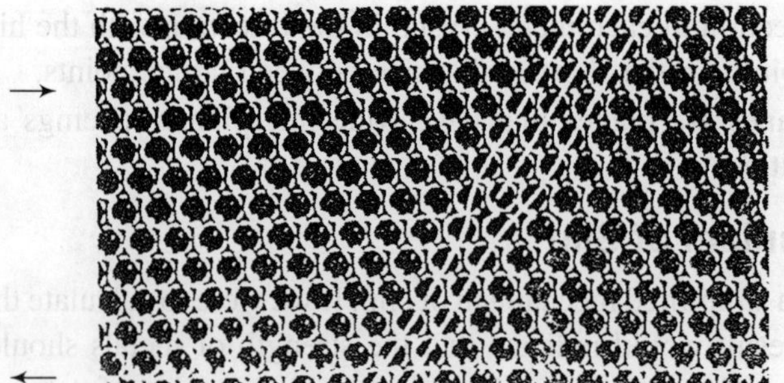

**Figure 5.83 Edge Dislocation.** "Bubble-raft" model of an imperfection in a crystal structure. Note the extra row of atoms. (Bragg and Nye, *Proc. Roy. Soc.* (*London*)).

The mechanism of slip requires the growth and movement of a dislocation line; therefore, energy is required. The energy, *E*, of a dislocation line is proportional to the length of the dislocation line, *l*, the product of the shear modulus, *G*, and the square of the unit slip vector **b**.

$$E \propto lG\mathbf{b}^2 \qquad ...(1)$$

Thus, the easiest dislocations to generate and expand for plastic deformation are those with the shortest unit slip vector, **b**, particularly since the **b** term.

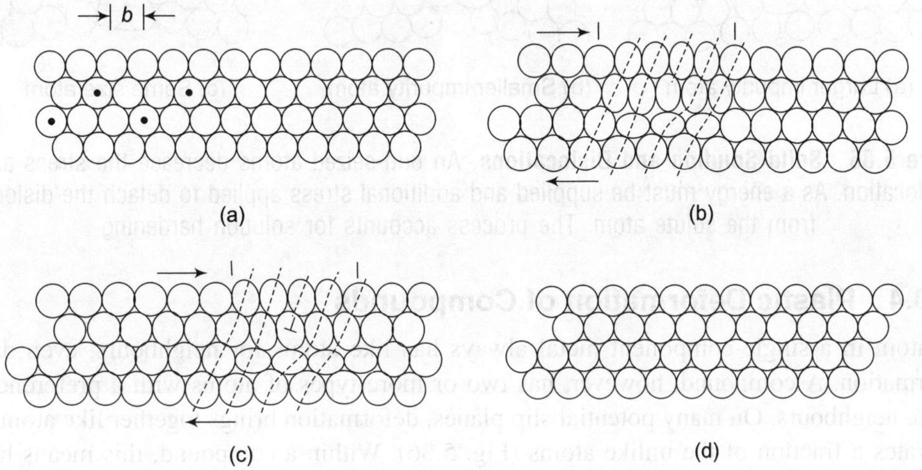

**Figure 5.84 Slip by Dislocation.** In this model, only a few atoms at a time are moved from their low-energy positions. Less stress is therefore required to produce slip than would be needed if all the atom moved at once as proposed in Fig. 5.8

is squared. The distance in a metal with a shortest slip vector will be the direction with the greater linear density of atoms. The lowest value for the shear modulus *G*, accompanies the planes that are farthest apart and hence have the greatest planar density of atoms. Thus, we can develop the rule of thumb that predicts that the lowest critical shear stresses will occur on the most densely packed planes and in the most densely packed direction.

### 5.50.3 Dislocation Movement in Solid Solutions

The energy associated with an edge dislocation is the same whether the dislocation is located at (b) or at (c) in Fig. 5.84. Therefore, no net energy is required for the movement between the two,[6] such is not the case when solute atoms are present as shown in Fig. 5.85, when an impurity atom is present, the energy associated with a dislocation is less than it is in a pure metal. Thus, when a dislocations encounter foreign atoms, their movement retrained

[6] The statement does not apply if (1) the movement includes an increase in the length of the dislocation, or (2) there is a pile-up of dislocations.

because energy always has higher strengths than do pure metals. We call this process *solution hardening*.

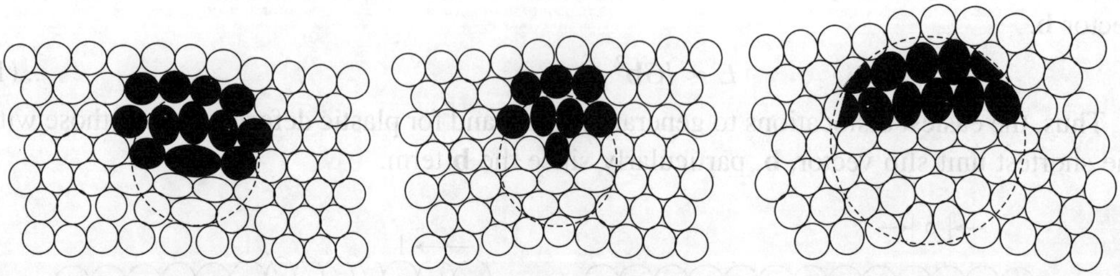

(a) Larger impurity atom (b) Smaller impurity atom (c) Same size atom

**Figure 5.85 Solid Solution and Dislocations.** An odd-seized atoms decrease the stress around a dislocation. As a energy must be supplied and additional stress applied to detach the dislocation from the solute atom. The process accounts for solution hardening

## 5.50.4 Plastic Deformation of Compounds

An atom in a single-component metal always has like atoms for neighbours, even during deformation. A compound, however, has two or more types of atoms with a preference for unlike neighbours. On many potential slip planes, deformation brings together like atoms and separates a fraction of the unlike atoms (Fig. 5.86). Within a compound, this means higher energy, which shows up as a resistance to shear. Consequently, the critical shear stresses on some planes are sufficiently high that slip is essentially impossible. In effect, the number of possible slip systems is reduced, and ductility decreases. The result is revealed in Table 5.15, which sites the slip systems for several relatively simple compounds, For example, only six sets are operative in MgO. In more complex compounds—such as in $Ni_8Fe_{16}O_{24}$, used in magnet ceramics, and in $PbZrO_3$, used in piezoelectric transducers—the possibility of slip is negligible at normal temperatures; hence, the material behaves in a brittle manner.

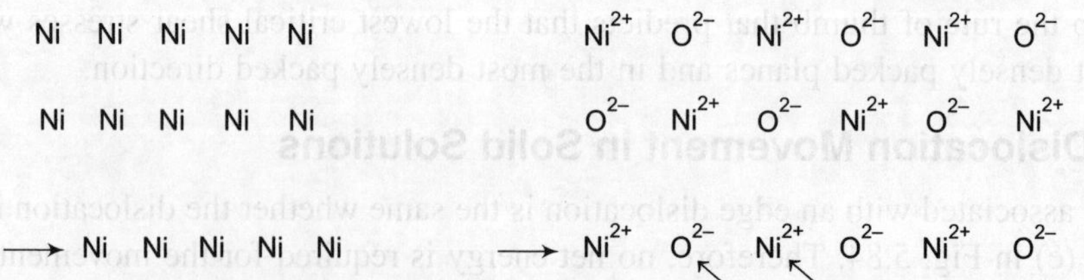


(a) (b)

**Figure 5.86** Comparison of slip process (Metallic Nickel and Nickel Oxide). More force is required to displace the ions in NiO then is needed for the atom is nickel. The strong repulsive forces between like ions become significant. Nickel also has more slip system than does nickel oxide

The shortest repeating distance (i.e., the slip vector) in an NaCl type crystal is along the several ⟨110⟩ directions. Quite expectedly therefore, the ⟨110⟩ directions have been found by experiment to be the slip directions. The most common slip plane for NaCl-type crystals is one of those in the {110} form. This is particularly true in those compounds, such as LiF and MgO that have small, "non-deformable" ions. Other NaCl-type compounds with large ions, such as PbS and MnSe, possess other slip planes, but still have the ⟨110⟩ slip directions (Table 5.15).

**Table 5.15** Slip Systems in Simple Compounds

| *Structure* | *Examples* | *Slip Directions* | *Slip Planes* | *Number of Combinations* |
|---|---|---|---|---|
| NaCl | LiF, MgO, MnS, TiC | ⟨110⟩ | $\{1\bar{1}0\}$ | 6 |
| NaCl | PbS | ⟨110⟩ | {001} | 6 |
| NaCl | MnSe | ⟨110⟩ | $\{\bar{1}11\}$ | 12 |
| CaCl | CaCl | ⟨100⟩ | {001} | 6 |
| $Al_2O_3$ | $Al_2O_3$ | $\langle 11\bar{2}0\rangle$ | {0001} | 3 |

We conclude that both intermetallic and ceramic compounds are inherently less deformable than pure metals and their solid solution phases, because critical shear stresses are high and the number of slip planes is small. Consequently, fracture stresses are commonly exceeded before plastic deformation is initiated.

## EXERCISE

5.1 Discuss order and disorder in solids.

5.2 What are the imperfections? Discuss necessity of imperfections in crystals.

5.3 Discuss imperfection from statistical viewpoint.

5.4 Discuss the formation of point defects, viz., vacancy, Schottky and Frankel imperfections.

5.5 Discuss self-interstitial defects in elemental solids for statistical viewpoints.

5.6 Discuss briefly the formation of polarons in ionic crystals.

5.7 What are the excitons? How are these formed?

5.8 Write short notes on:

(i) Frenkel excitons
(ii) Wannier-Mott excitons
(iii) Colour centres
(iv) The hydrogen atom model of $F$-centre
(v) The $F$ -centre
(vi) The $M$-centre
(vii) The $R$-centre
(viii) The $V_k$-centre

5.9 What are the dislocations? Give dislocation model of slip.

5.10 Explain the concept of Burgers vector in case of (a) edge dislocation and (b) screw dislocation.

5.11 Explain the formation of (a) edge dislocation (b) screw dislocation. Explain the concept of dislocation density and stability of a dislocation loop.

5.12 Explain how dislocation density may be estimated from X-ray diffraction measurements.

5.13 What is a brain boundary? Explain the Burgers model of low angle grain boundaries.

5.14 Explain the terms: tilt boundaries, twin boundaries, and stacking faults.

5.15 Discuss energies of dislocations. Derive an expression in case of screw dislocation.

CHAPTER

# 6

# Lattice Vibrations and Lattice Dynamics

## 6.1 INTRODUCTION

An important aspect of the study of solid state physics is the lattice dynamics, which concerns itself with the vibrations of atoms about their equilibrium sites in a solid. These vibrations occur at any temperature, even at absolute zero. They are almost entirely responsible for the thermal properties—heat capacity, thermal conductivity, etc., of insulator and contribute the greater part of the heat capacity of metals.

A convenient way to visualize lattice vibrations is to consider a large number of tiny spheres, each connected to its neighbours by springs (Fig. 6.1).

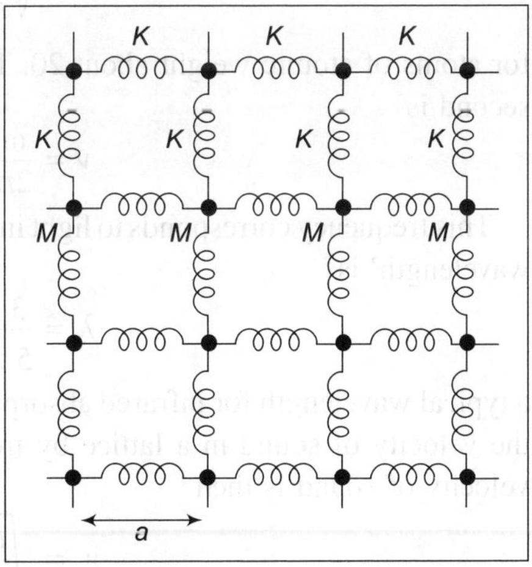


**Figure 6.1** Elastic coupling of vibrating atoms in a simple cubic lattice. ($K$ = force constant).

These springs provide restoring forces to the spheres when they are displaced from their equilibrium positions. Springs connecting only nearest neighbours characterise short range forces. If each of the particles represented by the spheres is imagined to be vibrating about its equilibrium position, a fairly accurate picture of lattice vibrations emerges.

The problem of finding the normal modes and characteristic frequencies of a crystal is a classic one. The high degree of symmetry is very useful in determining these modes. The crystal will have independent oscillations in these modes.

## 6.2 ELASTICITY AND ATOMIC FORCE CONSTANTS

Consider Fig. 6.1, we may guess an order of magnitude for $K$ by remembering that the potential energy of a pair of atoms is raised by about 1 eV when their separation is changed from the equilibrium value by about 1 Å. If $\delta a$ is the change in separation from the equilibrium value we can write the increase in potential energy as

$$\Delta V = \frac{1}{2} K (\delta a)^2$$

or
$$1.6 \times 10^{-19} = \frac{1}{2} K \times 10^{-20} \text{ joule}$$

∴
$$K = 30 \text{ Nm}^{-1}.$$

This is readily related to the macroscopic elastic modulus $E$ for extension parallel to one of the principal directions of the cubic lattice. We have

$$E = \frac{\text{Froce/Area}}{\text{Extension/Length}} = \frac{K\delta a/a^2}{\delta a/a} = \frac{K}{a}$$

since a typical interatomic separation is 3Å, we have

$$E \approx \frac{30}{3 \times 10^{-10}} = 10^{11} \text{ Nm}^{-2} \ (= 10^{12} \text{ dyne cm}^{-2})$$

which is indeed of the order of magnitude of Young's modulus.

Although each atom is coupled to several others, the vibrations of the atoms are not independent of each other. Collectively, we note here that a typical order of magnitude of vibration frequency is that of a single mass on a single spring

$$\omega \sim \sqrt{\frac{K}{M}} \approx \left(\frac{30}{3\times 10^{-26}}\right)^{1/2} \approx 3\times 10^{13} \ s^{-1}$$

for atoms of atomic weight about 20. This is angular frequency. The frequency in cycle per second is

$$\nu = \frac{\omega}{2\pi} \approx 5 \times 10^{12} \text{ Hz}$$

This frequency corresponds to light in the infra-red region of the spectrum; the corresponding wavelength' is

$$\lambda = \frac{3 \times 10^8}{5 \times 10^{12}} \text{ m} = 60 \ \mu\text{m},$$

a typical wavelength for infrared absorption due to molecular vibrations. We can also calculate the velocity of sound in a lattice by treating it, for long wavelengths, as a continuum. The velocity of sound is then

$$v_s = \sqrt{\frac{E}{\rho}}$$

where ρ, the density is $M/a^3$, therefore

$$v_s = \left(\frac{K}{a} \times \frac{a^3}{M}\right)^{1/2} = a\sqrt{\frac{K}{M}}$$

$$\approx 3 \times 10^{-10} \times 3 \times 10^{13} \cong 10^4 \text{ ms}^{-1}.$$

This is also correct in order of magnitude; most solids have sound velocities in the range $1 - 10 \text{ km s}^{-1}$.

## 6.3 DYNAMIC MOTION OF A TWO-ATOM CHAIN

We now consider the dynamic behaviour of two atoms connected together by a spring of spring constant $k$ and equilibrium separation, $r_o$. The atoms wave so that they have a kinetic energy $T$ and potential energy, $U = \phi(r)$.

We assume that the centre of mass of our system remains at rest. The two atoms have masses $M_1$ and $M_2$ which are at distance $r_1$ and $r_2$ from the origin, respectively and the location of the centre of mass is

$$r_{cm} = \frac{M_1 r_1 + M_2 r_2}{M_1 + M_2} \qquad ...(1)$$

The atomic positions can be written in terms of $r_{cm}$ as

$$\left.\begin{aligned} r_1 &= r_{cm} - \frac{M_2}{M_1 + M_2} r \\ r_2 &= r_{cm} + \frac{M_1}{M_1 + M_2} r \end{aligned}\right\} \qquad ...(2)$$

If the centre of mass is at rest, $\dot{r}_{cm} = 0$ $\left(\text{where } \dot{r} = \frac{dr}{dt}\right)$ then the atom velocities must satisfy

$$M_1\dot{r}_1 + M_2\dot{r}_2 = 0 \qquad \text{(from Eq. (1))}.$$

or

$$M_1\dot{r}_1 = -M_2\dot{r}_2 \qquad ...(3)$$

The kinetic energy is

$$T = \frac{1}{2} M_1\dot{r}_1^2 + \frac{1}{2} M_2\dot{r}_2^2 = \frac{1}{2} \mu\dot{r}^2 \qquad ...(4)$$

using the reduced mass $\mu$ defined by

$$\frac{1}{\mu} = \frac{1}{M_1} + \frac{1}{M_2} \qquad ...(5)$$

with the momentum $p = \mu\dot{r}$, the kinetic energy can be written

$$T = \frac{p^2}{2\mu} \qquad ...(6)$$

The Hamiltonian for the system $H = T + U$ is then

$$H = \frac{p^2}{2\mu} + \phi(r) \qquad ...(7)$$

and Hamiltonian equation of motion, then gives

$$\mu\ddot{r} = -\frac{d\phi(r)}{dr} = f(r) \qquad ...(8)$$

Let us consider only small displacements

$$u = r - r_0$$

from the equilibrium spacing $r_0$, using the Taylor expansion for the interaction potential, the equation of motion for $u(t)$ would be

$$\mu \ddot{u} = -\left.\frac{d^2\phi}{dr^2}\right|_{r_0} u. \qquad ...(9)$$

This is the equation of motion of a simple harmonic oscillator, and has a harmonic solution of the form

$$u(t) = u_0 \cos(\omega_0 t + \varphi) \qquad ...(10)$$

where the resonant frequency $\omega_0$ is given by

$$\omega_0 = \sqrt{\frac{1}{\mu}\frac{d^2\phi}{dr^2}}, \qquad ...(11)$$

The amplitude $u_0$ and phase $\varphi$ are determined by the initial conditions. Using complex notations the harmonic solutions can be written as

$$u(t) = u_0 e^{-i\omega_0 t}, \qquad ...(12)$$

where the amplitude $u_0$ can be complex, to allow for the plane factor $\varphi$, and the actual solution is obtained by taking the real part of (12).

We see that the natural resonance frequency $\omega_0$ of our system is determined by the reduced mass $\mu$ and by the curvature of the interaction potential, $d^2\phi/dr^2$. This close relation is due to our use of the harmonic approximation for the potential; another way to represent the interaction in this approximation is to think of the atoms as being linked by a simple linear spring, with spring constant $k$ given by

$$k = \left.\frac{d^2\phi}{dr^2}\right|_{r_0} \qquad ...(13)$$

The equilibrium length of the spring is $r_0$, and the potential energy for separation $r$ is thus $U = k(r - r_0)^2/2 = ku^2/2$, the same as for the harmonic approximation, with the trivial change that our zero of energy is now at the equilibrium point $r_0$ rather than at infinite separation. Both the response to weak external forces, and the natural dynamic resonant, are captured by this simple model, shown schematically in Fig. 6.2.

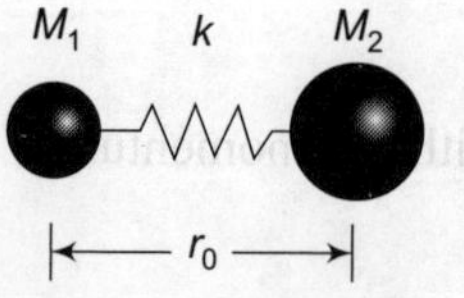


**Figure 6.2** Spring model for the interaction potential between two atoms

For the Lennard–Jones potential, and two argon atoms, with masses $M_1 = M_2 = 2\mu = 6.6 \times 10^{-23}$ g, we have a spring constant $k = 52$ meV/Å$^2$ = 0.83 N/m; we now find a natural resonance frequency $\omega_0/2\pi = 0.8$ THz. This is somewhat low for a mechanical atomic resonance frequency.

## 6.4 THE THREE-ATOM CHAIN

We now add a third atom to our system, and for simplicity assume all the atoms are identical, with mass $M$. We arrange the atoms in a line, and again restrict motion to one dimension, along the line of atoms. We further assume that the force which holds the three atoms together is a *two-particle* interaction, depending only on the positions $r_1$ and $r_2$ of, say, atoms 1 and 2. Hence, we write the interaction potential energy as $\phi(r_1, r_2)$. Furthermore, we assume the potential is a function only of the distance between the atoms, so that $\phi(r_1, r_2) = \phi(r_2 - r_1)$. For three identical atoms, the total potential energy $U$ in the absence of any external forces is then given by

$$U(r_1, r_2, r_3) = \phi(r_2 - r_1) + \phi(r_3 - r_2) + (r_1 - r_3) \qquad ...(1)$$

We will further assume that the potential interaction is very short-range, so that we only need to include the interactions between atoms adjacent to one another; for a chain with the atoms in the sequence (1, 2, 3), as shown in Fig. 6.3, this means we can drop the third term in Eq. (1).

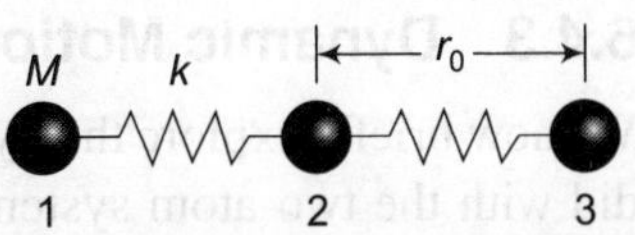


**Figure 6.3** Spring model for the three atom chain

Our final simplification will be to replace each potential interaction by its harmonic approximation, so that for atoms 2 and 3,

$$\phi(r = r_3 - r_2) = \frac{1}{2}\left.\frac{d^2\phi}{dr^2}\right|_{r_0}(r - r_0)^2, \qquad ...(2)$$

choosing the zero of energy at the equilibrium separation, $r_3 - r_2 = r_0$.

As discussed above, the response of the two-atom potential to both weak external forces and for small dynamic motion is entirely equivalent to that when the atoms are connected by a linear spring with spring constant $k$, whose value is given by the curvature of $\phi(r)$. We therefore find that our simplified, one-dimensional three-atom system now consists of three masses $M$ connected by linear springs $k$, as shown in Fig. 6.3.

### 6.4.1 External Force on a Trimer

We first consider what happens when we pull on either end of the chain with equal and opposite external forces $f_{ext}$, as shown in Fig. 6.4. By symmetry, atom 2, in the middle of the chain, does not move. The two end atoms 1 and 3 will be displaced until the restoring force provided by their interaction with atom 2 matches that of the external force. The displacement $u_1$ of atom 1 from its rest point is therefore identical to that obtained when pulling on the two-atom chain,

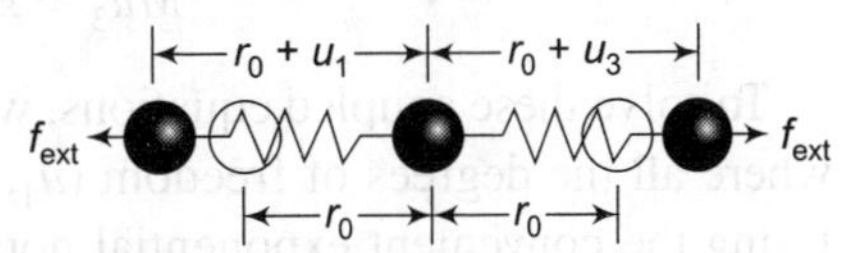


**Figure 6.4** Three-atom chain stretched by an external force $f_{ext}$

$$\begin{aligned} u_1 &= r_2 - r_1 - r_0 = f_{ext}/k \\ &= \frac{1}{d^2\phi/dr^2} f_{ext}, \end{aligned} \qquad ...(3)$$

with an identical expression for $u_3 = r_3 - r_2 - r_0$.

### 6.4.2 Strain in the Trimer

We are now in a position to define the *strain*, a very useful quantity when describing distortions of solid objects. The strain, which we will write as $e$ (we will be careful to distinguish this from the electron charge), is defined as the fractional displacement of a point in the solid from its rest point. For our three-atom chain, the strain $e$ between atoms 1 and 2 is

$$e = \frac{u_1}{r_0} = \frac{f_{\text{ext}}}{kr_0} \qquad ...(4)$$

The strain between atoms 3 and 2 is the same, $e = u_3/r_0 = f_{\text{ext}}/kr_0$. The strain is a dimensionless quantity, and for a uniform external force applied to a uniform object, will be constant through the object.

### 6.4.3 Dynamic Motion

We now briefly explore the dynamic behaviour of our three-atom chain; we will assume, as we did with the two-atom system, that the centre of mass of the system is at rest. The dynamic motion can be described by the relative displacements, $u_1$, $u_2$ and $u_3$, of the atoms from their rest points; the centre of mass condition is then, for equal mass atoms,

$$\dot{u} + \dot{u}_2 + \dot{u}_3 = 0. \qquad ...(5)$$

The momentum associated with each atom is $p_n = M\dot{u}_n$.

The kinetic energy of the three atoms, and their (nearest neighbour) spring interactions, yield the Hamiltonian

$$H = T + U = \frac{1}{2M}(p_1^2 + p_2^2 + p_3^2) + \frac{k}{2}(u_1 - u_2)^2 + \frac{k}{2}(u_2 - u_3)^2 \qquad ...(6)$$

The corresponding equations of motion are

$$\left.\begin{aligned} M\ddot{u}_1 &= k(u_2 - u_1) \\ M\ddot{u}_2 &= k(u_1 - 2u_2 + u_3), \\ M\ddot{u}_3 &= k(u_2 - u_3) \end{aligned}\right\} \qquad ...(7)$$

To solve these coupled equations, we look for the *normal modes* of the system, the solutions where all the degrees of freedom ($u_1$, $u_2$ and $u_3$) have the same, harmonic, time dependence. Using the convenient exponential notation, this means we look for solutions of the form

$$u_n = A_n e^{-i\omega t}, \quad (n = 1 \text{ to } 3) \qquad ...(8)$$

where the common frequency $\omega$ and the (complex) amplitudes $A_n$ are still to be determined. Inserting these solutions in Eq. (7), we find the linear system of equations

$$\left.\begin{aligned} -M\omega^2 A_1 &= k(A_2 - A_1), \\ -M\omega^2 A_2 &= k(A_1 - 2A_2 + A_3), \\ -M\omega^2 A_3 &= k(A_2 - A_3), \end{aligned}\right\} \qquad ...(9)$$

where we have cancelled out the common time dependence. Defining the frequency $\omega_0 = (k/M)^{1/2}$, the system of Eq. (9) can be written as an *eigenvalue-eigenvector equation*,

$$\left(\frac{\omega}{\omega_0}\right)^2 \begin{bmatrix} A_1 \\ A_2 \\ A_3 \end{bmatrix} = \begin{bmatrix} 1 & -1 & 0 \\ -1 & 2 & -1 \\ 0 & -1 & 1 \end{bmatrix} \begin{bmatrix} A_1 \\ A_2 \\ A_3 \end{bmatrix} \qquad ...(10)$$

The eigenvalues are the three values of $\omega/\omega_0$ that yield non-trivial solutions to this equation, and the eigenvectors the set of amplitudes ($A_1$, $A_2$, $A_3$) that correspond to each eigenvalue.

With $z = (\omega/\omega_0)^2$, we subtract the term on the left from both sides, leaving the equation

$$\begin{bmatrix} 1-z & -1 & 0 \\ -1 & 2-z & -1 \\ 0 & -1 & 1-z \end{bmatrix} \begin{bmatrix} A_1 \\ A_2 \\ A_3 \end{bmatrix} = 0. \qquad ...(11)$$

This has a non-trivial solution only if the determinant of the matrix is zero

$$\det \begin{vmatrix} 1-z & -1 & 0 \\ -1 & 2-z & -1 \\ 0 & -1 & 1-z \end{vmatrix} = -z^3 + 4z^2 - 3z = 0. \qquad ...(12)$$

This is known as the characteristic equation. The solutions to this cubic equation are the set $z = (\omega/\omega_0)^2 = 0$, 1 and 3.

Let us deal with the non-zero solutions. If we take the solution $z = 1$, i.e., $\omega = \omega_0$, the corresponding eigenvector ($A_1$, $A_2$, $A_3$) is obtained from Eq. (11). We find that the amplitudes must satisfy $A_1 = -A_3$ and $A_2 = 0$. The overall amplitude of the eigenvector is arbitrary; the eigenvector Eq. (11) only tells us the *relative* amplitudes of motion for the three atoms. It is convenient to make the length of the eigenvector ($A_1$, $A_2$, $A_3$) equal to unity, that is, to normalize the solution, so we find the solution pair

$$\omega = \omega_0 \leftrightarrow \begin{bmatrix} A_1 \\ A_2 \\ A_3 \end{bmatrix} = \begin{bmatrix} 1/\sqrt{2} \\ 0 \\ -1/\sqrt{2} \end{bmatrix}. \qquad ...(13)$$

The second non-zero frequency is $z = 3$ or $\omega = \sqrt{3}\,\omega_0$; we find the corresponding eigenvector must satisfy $A_2 = -2A_1 = -2A_3$, so the normalized solution is

$$\omega = \sqrt{3}\,\omega_0 \leftrightarrow \begin{bmatrix} A_1 \\ A_2 \\ A_3 \end{bmatrix} = \begin{bmatrix} -1/\sqrt{6} \\ 2/\sqrt{6} \\ -1/\sqrt{6} \end{bmatrix}. \qquad ...(14)$$

The final frequency is $\omega = 0$. Inserting the solution in Eq. (11) we find the corresponding eigenvector ($A_1$, $A_2$, $A_3$) = ($1/\sqrt{3}$, $1/\sqrt{3}$, $1/\sqrt{3}$). This last solution, with equal values for all the displacement amplitudes $A_n$, is actually somewhat special: the zero frequency solution corresponds to an overall motion of the centre of mass, where the actual time dependence is not an exponential $e^{-i\omega t}$, but instead has the form $u_n = at + b$, corresponding to the uniform translation of the chain at constant velocity. This can be seen by inserting this solution into the original system of Eq. (7); however, it violates our condition on the centre of mass, Eq. (5), unless the velocity $a$ is zero. We therefore ignore this now trivial solution; we have, after all, found two independent solutions to a system with two independent degrees of freedom, the third degree of freedom having been removed by our condition on the centre of mass motion.

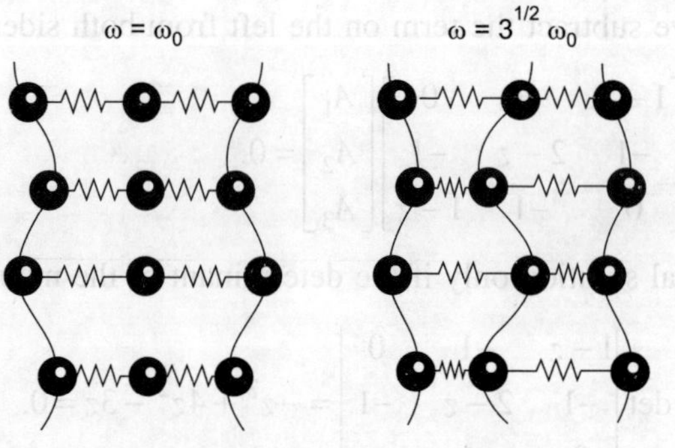


**Figure 6.5** The two normal modes for the three-atom chain. the time axis runs vertically; note that if the time axis were proportional, the oscillations for the higher frequency mode on the right would be three time faster than for that on the left. The relative amplitudes and phase of motion for the three atoms are however correct

We note that if we had used the centre of mass equation to eliminate one degree of freedom, say $u_2$, in terms of the other two, and then written the corresponding Hamiltonian and two equations of motion, we would have obtained the same two (nontrivial) eigenfrequencies and eigenvectors found above.

In Fig. 6.5 we illustrate the two normal modes, Eqs. (13) and (14).

## 6.5 PHONONS

As a result of their thermal motion, the lattice ions in a solid vibrate about their equilibrium positions. All the elastic properties, compressibility and propagation of acoustic waves in solids are then described in terms of a continuum theory disregarding the atomic structure of the lattice.

When we supply thermal energy to a lattice ion, it will be rapidly distributed throughout the entire lattice by the mutual interaction between the ions. As a result, the ions vibrate about their mean positions at all temperatures above absolute zero, the restoring force being provided by the force between the ions due to the chemical bonds. Thus, local excitations lead to *collective vibrations* of the whole ion system. Collective coordinates (normal coordinates) are generally used for the mathematical description for describing oscillations in continuum theory. Similar to light quanta (photons), the lattice vibrations can be quantized and the associated quanta are elementary excitations termed *phonons*. Phonons are bosons and are, therefore, described by statistics different from that of electrons. Electrons follow Femi–Dirac statistics whereas phonon follow the Bose–Einstein Statistics. Some important characteristics of phonons are given below.

(i) Like photons, phonons have dual nature (wave or particle).

(ii) These are indistinguishable particles obeying Bose–Einstein statistics.

(iii) The energy of a photon is $h\nu = h\omega$, where $\nu$ is the frequency of lattice vibration and phonon momentum = $\hbar k = h/\lambda$.

(iv) At finite temperatures, the crystals are full of phonons. Like photons, they are not conserved. At a given temperature, the total energy of phonons is the thermal energy of the solid. By increasing the temperature, the number of phonons increases. A decrease in temperature results in destruction of phonons.

(v) The phonon interaction may be viewed as a scattering collision.

(vi) When a lattice vibration is involved in an interaction, not only is its energy always an integral multiple of $\hbar\omega$, but also, at a time, only a single phonon is observed or excited. This is confirmed by neutron scattering experiments.

(vii) There are two types of phonons—acoustical and optical, which are associated with acoustical and optical modes of vibration, respectively. At low temperatures, acoustic modes are more populated than the optical modes.

We know that the photoelectric effect is a direct evidence of quantization of electromagnetic waves. So far, no direct analogous experiment has been carried out for phonons. However, there are some indirect evidences as follows:

(i) It is practically observed that the zero temperature, the lattice contribution to heat capacity is zero. This is only possible when lattice waves are quantized.

(ii) X-rays and neutrons are scattered inelastically by crystals with energy and momentum changes corresponding to creation or absorption of one or more phonons. By measuring the recoil of the scattered neutron, it is possible to determine the properties of individual phonons. Such experiments provide a means for determining the dispersion relations [a relation between frequency, energy and phonon wave vector i.e., $\omega(q)$ which provide the strongest evidence for existence of phonons].

### 6.5.1 Phonons (Waves or Particles)

A phonon has energy $\hbar\omega$ as quantum of excitation of the lattice, and consequently the phonons are not localized particles. The uncertainty principle demands that the position cannot be determined because the momentum $\hbar k$ is exact However, just as with photons (or electrons), one can construct a fairly localized wave packet by combining modes of slightly different frequency and wavelength. Thus, if we take waves with a spread in $\vec{k}$ of order $\left(\frac{\pi}{10a}\right)$, we can make a wave packet localized to within about 10 atomic spacings. Such a wave packet represents a fairly localized phonon moving with group velocity $(d\omega/d\vec{k})$. We can therefore treat phonons as localized particles within the limits of the uncertainty principle. The $\omega(\vec{k})$ curve for lattice vibrations can be interpreted, if both axes are multiplied by $\hbar$, as a relation between energy and crystal momentum for phonons ($E = \hbar\omega$, $\vec{p} = \hbar\vec{k}$).

## 6.5.2 Energy of a Harmonic Oscillator

We know that the energy levels of a quantum harmonic oscillator are

$$E_n = \left(n + \frac{1}{2}\right)\hbar\omega \qquad \text{...(1)}$$

so that the average energy $\overline{E}$ in thermal equilibrium at temperature $T$ is given by

$$\overline{E} = \sum_n p_n E_n \qquad \text{...(2)}$$

Here $p_n$ is the probability of the oscillator being in level $E_n$; this is given by a Boltzmann factor so that

$$\overline{E} = \frac{\sum_{n=0}^{\infty}\left(n + \frac{1}{2}\right)\hbar\omega \exp[-E_n / k_B T]}{\sum_{n=0}^{\infty} \exp[-E_n / k_B T]}$$

$$= \frac{\sum_{n=0}^{\infty}\left(n + \frac{1}{2}\right)\hbar\omega \cdot \exp\left[-\left(n + \frac{1}{2}\right)\hbar\omega / k_B T\right]}{\sum_{n=0}^{\infty} \exp\left[-\left(n + \frac{1}{2}\right)\hbar\omega / k_B T\right]}$$

$$= \frac{1}{2}\hbar\omega + \frac{\sum_{n=0}^{\infty} n\hbar\omega \exp(-n\hbar\omega / k_B T)}{\sum_{n=0}^{\infty} \exp(-n\hbar\omega / k_B T)} \qquad \text{...(3)}$$

$$= \frac{1}{2}\hbar\omega + k_B T^2 \frac{\partial}{\partial T}\left[\ln \sum_{n=0}^{\infty} \exp(-n\hbar\omega / k_B T)\right]$$

But $\quad \sum_{n=0}^{\infty} e^{-n\hbar\omega / k_B T} = 1 + e^{-\hbar\omega/k_B T} + e^{-2\hbar\omega/k_B T} + \ldots$

$$= (1 - e^{-\hbar\omega}/k_B T)^{-1}$$

so that

$$\overline{E} = \frac{1}{2}\hbar\omega + \frac{\hbar\omega}{e^{\hbar\omega/k_B T} - 1} \qquad \text{...(4)}$$

This mean energy is readily interpreted in terms of phonons. For bosons of energy $\hbar\omega$, which are not conserved, the average number present in thermal equilibrium at temperature $T$ is

$$n(\omega) = \frac{1}{e^{\hbar\omega/kBT} - 1} \qquad ...(5)$$

as in the case of photons $C_{\text{in}}$ black body radiation.

Multiplication of Eq. (5) by $\hbar\omega$ gives the second term in Eq. (4) as the contribution of phonons to the energy. The first term in Eq. (4) viz. $\frac{1}{2}\hbar\omega$ is the zero point energy, which cannot be frozen out because of the uncertainty principle. The mean energy tends to this value in the low temperature limit $k_BT \ll \hbar\omega$. At high temperatures $k_BT \gg \hbar\omega$ we can expand the exponential as

$$\bar{E} = \frac{1}{2}\hbar\omega + \hbar\omega / \left[\frac{\hbar\omega}{k_BT} + \frac{1}{2}\left(\frac{\hbar\omega}{k_BT}\right)^2 + ...\right]$$

$$= \frac{1}{2}\hbar\omega + k_BT\left[1 - \frac{1}{2}(\hbar\omega / k_BT) + ...\right]$$

$$= \frac{1}{2}\hbar\omega + k_BT - \frac{1}{2}\hbar\omega + ....$$

$$\cong k_BT\left[1 + o\left(\frac{\hbar\omega}{k_BT}\right)^2\right]$$

so that a classical equipartition value is obtained in this limit; note that a one-dimensional oscillator gives $k_BT$, not $\frac{1}{2}k_BT$, because the potential energy as well as the kinetic energy, is a quadratic contribution to the total energy.

## 6.6 COUPLED OSCILLATORS AND LATTICE VIBRATION QUANTIZATION

We now consider the linear chain of coupled oscillators in the form of a whole assembly of vibrating masses interacting with each other. We will consider quantum mechanical description and find that the energies of the harmonic oscillators are quantized showing thereby the *quantization of lattice vibrations*.

Our model is characterized by the following:

(i) A large number of identical masses are arranged in regular array, i.e., a *chain* of "atoms" having interatomic spacing '*a*'.

(ii) The atoms are coupled by short range forces. There is a spring of force constant $K$ between nearest neighbours; the second and higher order interactions are negligible.

(iii) The *linear chain* of atoms is closed so that the cyclic boundary conditions are valid and end effects are eliminated.

The Hamiltonian is

$$H = \frac{1}{2M} \sum_l p_l^2 + \frac{1}{2} K \sum_i (u_i - u_{i+a})^2 \quad \text{...(1)}$$

where $i$ is the atomic index. The fact that displacement operator $u_i$ does not interfere with the momentum operator $p_i$ is expressed by the commutation relation

$$[u_l, p_{l'}] = i\hbar\, \delta_{ll'} \quad \text{...(2)}$$

To diagonalize the Hamiltonian, we make a Fourier transformation to a new set of operators:

$$\left.\begin{aligned} U_k &= \frac{1}{\sqrt{N}} \sum_l \exp(ikl)\, u_l \\ P_k &= \frac{1}{\sqrt{N}} \sum_l \exp(-ikl)\, p_l \end{aligned}\right| \quad \text{...(3)}$$

Now, $$\sum_k U_k \exp(-ikl') = \frac{1}{\sqrt{N}} \sum_l u_l \exp[ik(l-l')]$$

$$= \frac{1}{\sqrt{N}} \sum_l N u_l\, \delta_{ll'} = \sqrt{N}\, u_{l'}$$

$$\left.\begin{aligned} \therefore\ u_l &= \frac{1}{\sqrt{N}} \sum_k \exp(-ikl)\, U_k \\ \text{Similarly,}\quad p_l &= \frac{1}{\sqrt{N}} \sum_k \exp(ikl)\, P_k \end{aligned}\right| \quad \text{...(4)}$$

It is to be noted that the above definitions preserve the commutation relations Eq. (2). It can be seen as follows:

$$[U_k, P_{k'}] = \frac{1}{N} \sum_{l,l'} \exp\{i(kl - k'l')\}\, [u_l, p_{l'}]$$

$$= \frac{1}{N} \sum_{l,l'} \exp\{i(kl - k'l')\}\, i\hbar\, \delta_{ll'}$$

$$= i\hbar \frac{1}{N} \sum_l \exp\{i(k - k')\}\, l$$

$$= i\hbar\, \delta_{kk'} \quad \text{...(5)}$$

To get the final form of the kinetic energy term of the Hamiltonian Eqn. (1), we use Eq. (4):

$$\frac{1}{2M} \sum_l p_l^2 = \frac{1}{2M} = \sum_l p_l^* p_l = \frac{1}{2MN} \sum_{k,k'} P_k^* P_{k'} \left[ \sum_l \exp\{-i(k - k')l\} \right]$$

$$= \frac{1}{2M} \sum_{k,k'} P_k^* P_{k'} \delta_{kk'} = \frac{1}{2M} \sum_k P_k^* P_k \qquad ...(6)$$

Now $$\sum_l (u_l - u_{l+a})^2 = \frac{1}{N} \sum_{l,k} [\exp(-ikl) U_k - \exp\{-ik(l+a)\} U_k]^2$$

$$= \frac{1}{N} \sum_{l,k,k'} [\exp(ikl) - \exp\{ik(l+a)\}] U_k^*$$

$$\times [\exp(-ik'l) - \exp\{ik'(l+a)\}] U_{k'}$$

$$= \frac{1}{N} \sum_{k,k'} \left[ \sum_l \exp\{i(k-k')l\} \right] [1 - \exp(ik'a) - \exp(ika)$$

$$+ \exp\{i(k-k')a\}] U_k^* U_{k'}$$

$$= \frac{1}{N} \sum_{k,k'} N \delta_{kk'} [1 - \exp(ik'a) - \exp(ika)$$

$$+ \exp\{i(k-k')a\}] U_k^* U_{k'}$$

$$= \sum_k [1 - \exp(-ika) - \exp(ika) + 1] U_k^* U_k$$

$$= \sum_k 2(1 - \cos ka) U_k^* U_k$$

$$\therefore \quad \frac{1}{2} K \sum_l (u_l - u_{l+a})^2 = \frac{1}{2} \sum_k K'(k) U_k^* U_k \qquad ...(7)$$

where $$K'(k) = 2 K (1 - \cos k \alpha) \qquad ...(8)$$

Thus, using Eqs. (6) and (7) in Eq. (1)

$$H = \frac{1}{2} \sum_k \left[ \frac{1}{M} P_k^* P_k + K'(k) U_k^* U_k \right] \qquad ...(9)$$

We recognise this as the reduction to a sum of squares of moments and displacements of the normal coordinates. The formula for the frequency of the mode of wave vector $k$ is

$$\omega_k = \left( \frac{K'(k)}{M} \right)^{1/2} = 2\sqrt{\frac{K}{M}} \sin\left( \left| \frac{ka}{2} \right| \right) \qquad ...(10)$$

which is identical to the expression for one-dimensional monatomic lattice. Now, by convention, we make all frequencies positive so that $\omega_k$ is identically equal to $\omega_k$.

We now introduce annihilation and creation operators by demanding Hermitian conjugation

$$U_k^* = \frac{1}{\sqrt{N}} \sum_l \exp(i(-k)l) u_l = U_{-k}$$

$$P_k^* = p_{-k} \qquad ...(11)$$

Using the general form of the Hamiltonian

$$\frac{H}{\hbar\omega} = \frac{1}{2\hbar\omega}\left[\frac{1}{M}p^2 - i^2 Ku^2\right] \qquad ...(12)$$

We can factorise the sum of squares into the product of two special operators defined as

$$a_k = (1/\sqrt{2\hbar M\omega_k})\,[P_k - i\,M\,\omega_k\,U_k^*] \qquad ...(13)$$

$$a_k^+ = (1/\sqrt{2\hbar M\omega_k})\,[P_k^* + i\,M\,\omega_k\,U_k] \qquad ...(14)$$

It can be seen that

$$[a_k, a_{k'}] = [a_k^+, a_{k'}^+] = 0$$

and

$$[a_k, a_{k'}^+] = \delta_{kk'} \qquad ...(15)$$

Remembering $P_k^* = P_{-k}$ and $U_k^* = U_{-k}$, $\omega_{-k} = \omega_k$,

We can solve Eqns. (13) and (14) as simultaneous equation to get

$$P_k = \frac{1}{2}(2\hbar\,M\omega_k)^{1/2}\,(a_k + a_{-k}^+) \qquad ...(16)$$

$$U_k = \frac{1}{2iM\omega_k}(2\hbar\,M\omega_k)^{1/2}\,(a_k^+ - a_{-k}) \qquad ...(17)$$

Then

$$P_k^*\,P_k = \frac{1}{2}\hbar M\omega_k\left\{a_k^+\,a_k + a_k^+\,a_{-k}^+ + a_{-k}\,a_k + a_{-k}\,a_{-k}^+\right\}$$

$$U_k^*\,U_k = \frac{\hbar}{2M\omega_k}\left\{a_k a_k^+ - a_k\,a_{-k} - a_{-k}^+\,a_k^+ + a_{-k}^+\,a_{-k}\right\}$$

Substituting these in Eq. (9) and remembering that $K'(k) = M\omega_k^2$ we get

$$H = \frac{1}{2}\sum_k\left(\frac{1}{2}\hbar\omega_k\right)\{(a_k^+\,a_k + a_k^+\,a_{-k}^+ + a_{-k}\,a_k + a_{-k}\,a_{-k}^+)$$

$$+ (a_k\,a_k^+ - a_k\,a_{-k} - a_{-k}^+\,a_k^+ + a_{-k}^+\,a_{-k})\}$$

Then using $\quad a_k^2\,a_{-k}^+ - a_{-k}^+\,a_k^+ = 0 = a_{-k}\,a_k - a_k\,a_{-k}$

we get

$$H = \frac{1}{4}\sum_k \hbar\omega_k\{(a_k^+\,a_k + a_k\,a_k^+) + (a_{-k}^+\,a_{-k} + a_{-k}\,a_{-k}^+)\}$$

or

$$H = \frac{1}{2}\sum_k \hbar\omega_k(a_k^+\,a_k + a_k\,a_k^+) \qquad ...(18)$$

Since a summation over 'allowed values of $-k$' merely duplicates the sum over $k$.

We now give matrix representation for $a_k$ and $a_k^+$ that satisfy the commutation relations (15) and in which $H$ is diagonalized. We define a set of ket vectors by

$$a_k|n_k\rangle = \sqrt{n_k}\,|n_k - 1\rangle$$
$$a_k^+|n_k\rangle = \sqrt{n_k + 1}\,|n_k + 1\rangle \quad ...(19)$$

Now
$$[a_k, a_k^+]\,|n_k\rangle = a_k\,a_k^+\,|n_k\rangle - a_k^+\,a_k\,|n_k\rangle$$
$$= (n_k + 1)\,|n_k\rangle - n_k\,|n_k\rangle$$

$\therefore$
$$[a_k, a_k^+]\,|n_k\rangle = |n_k\rangle \text{ or } [a_k\, a_k^+] = 1.$$

In general
$$[a_k, a_{k'}^+] = \delta_{kk'}$$
$$[a_k, a_{k'}] = [a_k^+, a_{k'}^+] = 0$$

Thus, we regain Eq. (15). Then from Eq. (18)

$$H\,|n_k\rangle = \left(\frac{1}{2}\hbar\omega_k\right)\{a_k^+\,a_k\,|n_k\rangle + a_k\,a_k^+\,|n_k\rangle\}$$
$$= \left(\frac{1}{2}\hbar\omega_k\right)[n_k\,|n_k\rangle + (n_k + 1)\,|n_k]$$

or
$$E_k\,|n_k\rangle = \left(n_k + \frac{1}{2}\right)\hbar\omega_k\,|n_k\rangle$$

$\therefore$ The energy of the $k^{\text{th}}$ oscillator is quantized according to

$$E_k = \left(n_k + \frac{1}{2}\right)\hbar\omega_k.$$

## 6.7 GREEN'S FUNCTIONS AND ITS APPLICATIONS TO SIMPLE HARMONIC OSCILLATOR

Historically Green's functions were first used by Green in solution of potential problems. In those days the concepts of functional analysis, vector spaces and operators, etc., were not very much in vogue. Green's work, which was primarily a study of the methods of solving Laplace's and Poisson's equations with various boundary conditions contained the germs of a much wider formulation for solving a variety of eigenvalue equations and the corresponding inhomogeneous equations. This formulation has been possible because of the development of functional analysis and has been exploited during the recent decades in various branches of physical sciences.

Perhaps the simplest way of understanding the technique of Green's functions is to solve the classical simple one-dimensional harmonic oscillator problem using Green's functions. The equation of motion is

$$m\frac{d^2x}{dt^2} + mw_0{}^2x = f(t) \quad ...(1)$$

where $f(t)$ is a time-dependent force and the other symbols have their usual meanings. The formal solution of Eq. (1) can be written as

$$x(t) = \int_{-\infty}^{\infty} G(t-t')\, f(t')\, dt' \qquad ...(2)$$

where $G(t - t')$ is the Green's functions satisfying the equation

$$\left( m\frac{d^2}{dt^2} + m\omega_0^2 \right) G(t-t') = \delta(t-t') \qquad ...(3)$$

If we use the Fourier integral representation of the δ-function,

$$\delta(t-t') = \frac{1}{2\pi} \int_{-\infty}^{\infty} e^{i\omega(t-t')}\, d\omega \qquad ...(4)$$

we obtain from Eq. (3)

$$G(t-t') = \left( m\frac{d^2}{dt^2} + m\omega_0^2 \right)^{-1} \frac{1}{2\pi} \int_{-\infty}^{\infty} e^{i\omega(t-t')}\, d\omega$$

$$= \frac{1}{2\pi} \int_{-\infty}^{\infty} \frac{e^{i\omega(t-t')}\, d\omega}{m(\omega_0^2 - \omega^2)}$$

$$= \frac{1}{\sqrt{2\pi}} \int_{-\infty}^{\infty} G(\omega)\, e^{i\omega(t-t')}\, d\omega \qquad ...(5)$$

where $G(\omega)$ which is the Fourier transform of $G(t)$ is given by

$$G(\omega) = \frac{1}{\sqrt{2\pi}\, m(\omega_0^2 - \omega^2)} \qquad ...(6)$$

Substituting (1.5) in (1.2) we obtain

$$x(t) = \frac{1}{\sqrt{2\pi}} \int_{-\infty}^{\infty} dt' \left[ \frac{1}{\sqrt{2\pi}} \int_{-\infty}^{\infty} d\omega\, G(\omega)\, e^{i\omega t} \right] e^{-i\omega t'} f(t')$$

$$= \frac{1}{\sqrt{2\pi}} \int_{-\infty}^{\infty} d\omega\, G(\omega)\, F(\omega)\, e^{i\omega t} \qquad ...(7)$$

where $$F(\omega) = \frac{1}{\sqrt{2\pi}} \int_{-\infty}^{\infty} e^{i\omega t} f(t)\, dt \qquad ...(8)$$

Evaluation of the integrals in Eq. (5) or Eq. (7) pose problems because of the occurrence of poles in $G(\omega)$ at $-\omega = \pm\, \omega_0$ on the contour of integration, i.e., the real axis.

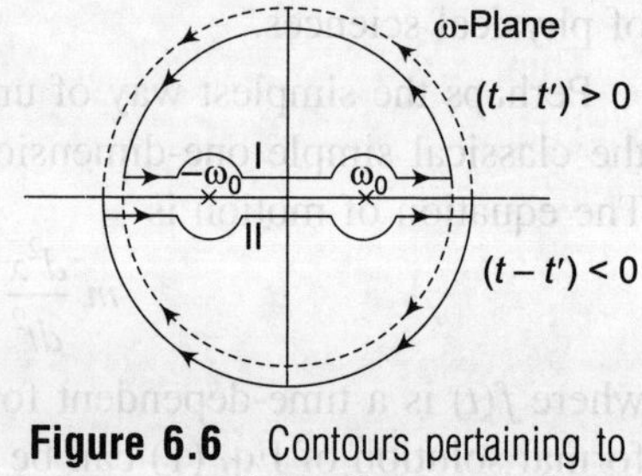

**Figure 6.6** Contours pertaining to integration

The contour can be indented about these singularities in many ways, and each choice leads to a particular Green's functions. The contours I and II shown in Fig. 6.6 just above and below the real axis and closed by semicircle of infinite radius in the upper and lower half-planes depending on whether $(t - t')$ $> 0$ or $(t - t') < 0$ lead to two well known Green's functions,

$$G_A\,(t - t') = \frac{1}{2\pi} \oint_I \frac{e^{i\omega}\,(t - t')}{m(\omega_0^2 - \omega^2)}\, d\omega$$

$$= -\,\frac{\sin\,\omega_0 (t - t')}{m\,\omega_0}\,\theta\,(t' - t) \qquad ...(9)$$

and

$$G_R\,(t - t') = -\,\frac{\sin\,\omega_0 (t - t')}{m\,\omega_0}\,\theta\,(t - t') \qquad ...(10)$$

Here $\theta\,(t)$ is the Heaviside step function having the property

$$\theta\,(t) = 0; \quad t < 0$$

$$\theta\,(t) = 1; \quad t > 0 \qquad ...(11)$$

In this classical mechanical problem $G_R\,(t - t')$ is the useful Green's functions, since use of it in Eq. (2) gives the solution for $x(t)$ with the desirable feature that its value at $t$ is not influenced by $f(t')$ for $t' > t$, i.e., the present displacement of the oscillator is not influenced by the future form of the force. In other words, use of the retarded Green's functions gives a solution that satisfies the "Principle of Causality", according to which the cause (force) must precede the effect (displacement).

Equation (7) has an important significance. The Fourier integral can be inverted to give us

$$x(\omega) \equiv \frac{1}{\sqrt{2\pi}} \int_{-\infty}^{\infty} x(t)\, e^{-i\omega t}\, dt = G(\omega)\, F(\omega) \qquad ...(12)$$

From this it follows that the Fourier transform of the response (displacement) is proportional to the Fourier transform of the stimulus (force), and the constant of proportionality is the Green's functions, $G(\omega)$. Thus, $G(\omega)$ has the character of a generalised susceptibility, i.e., the capacity of the system to respond to stimuli. For instance, if $f(t)$ is a sinusoidal force of the form

$$f(t) = f_0 e^{i\omega_1 t}$$

then $\quad F(\omega) \sim f_0 \delta\,(\omega - \omega_1)$

and $\quad x(\omega) \sim f_0 G\,(\omega_1)\,\delta\,(\omega - \omega_1)$

so that $\quad x(t) \sim f_0 G(\omega_1)\, e^{i\omega_1 t}$

Thus, $G\,(\omega_1)$ is the ratio of the displacement to the sinusoidal force of frequency $\omega_1$. If $\chi_D^{(\omega)}$ denotes the suaceptibility of the displacement to a sinusoidal force of frequency $\omega$, then clearly,

$$\chi_D(\omega) \equiv G(\omega) \qquad ...(13)$$

One can similarly define other types of susceptibility, for instance, $\chi_v(\omega)$ defined as

$$\chi_v(\omega) = i\omega\, G(\omega) \qquad ...(14)$$

which will connect the velocity of the particle to the sinusoidal force of frequency $\omega$.

## 6.8 QUANTUM MECHANICAL TREATMENT OF THE OSCILLATOR IN HEISENBERG REPRESENTATION FOR A TIME DEPENDENT PERTURBATION

The quantum mechanical treatment of the oscillator can be introduced through the Heisenberg picture.

The Hamiltonion (in unperturbed form) is

$$H_0 = \frac{p^2}{2m} + \frac{1}{2} m\omega_0^2 x^2 \qquad ...(1)$$

The equation of motion for the position operator is given by

$$i\hbar \frac{\partial x}{\partial t} = [x, H_0] \quad \text{or} \quad \dot{x} = \frac{1}{i\hbar}[x, H_0] \qquad ...(2)$$

where the commutator, $[x, H_0] \equiv xH_0 - H_0x$

Using the commutation rules

$$[x, x] = [p, p] = 0 \quad \text{and} \quad [x, p] = i\hbar \qquad ...(3)$$

We obtain

$$\dot{x} = p/m \qquad ...(4)$$

and

$$\ddot{x} = \frac{1}{i\hbar}[\dot{x}, H_0] = -\omega_0^2 x \qquad ...(5)$$

i.e., the classical equation of motion for the displacement $x$ of a harmonic oscillator is the same as the quantum mechanical one, except that in the latter case, $x$ has to be interpreted as an operator.

The susceptibility of such a system to an external stimulus can be determined by studying the expectation value of a given observable under the influence of the stimulus, since only expectation values have physical significance in quantum mechanics For this, we shall make use of time-dependent perturbation theory: Hamiltonian

$$H = H_0 + H_I(t) \qquad ...(6)$$

where the external stimulus has been taken into account through the addition time-dependent/ perturbation term $HI(t)$. If $HI(t)$ is small, a linear relationship can be established between the expectation value of a physical quantity and the perturbation by using the first order time dependent wave function.

The eigenfunctions $\psi_n$ (corresponding to the eigenvalues $E_n$) of $H_0$ enable us to write for the wave function

$$\psi(t) = \sum_n a_n(t)\psi_n \exp\left(-\frac{iE_n t}{\hbar}\right) \qquad ...(7)$$

Substituting this in the Schrödinger equation, we get

$$i\hbar \frac{\partial \psi}{\partial t} = \sum_n \{i\hbar\, \dot{a}_n(t) + E_n\, a_n(t)\}\, \psi_n \exp\left(-\frac{iE_n t}{\hbar}\right) \qquad ...(8)$$

$$= H\,\psi = \sum_n a_n(t)\,\{H_0 + H_I\}\,\psi_n \exp(-iE_n t/\hbar)$$

$$= \sum_n a_n(t)\,[E_n + H_I]\,\psi_n \exp(-iE_n t/\hbar) \quad \text{...(9)}$$

Multiplying both sides by $\psi_m^*$ and integrating, we get

$$\dot{a}_m(t) = \frac{1}{i\hbar}\sum_n a_n(t)\,H_I(m, n; t)\exp[-i(E_n - E_m)t/\hbar] \quad \text{...(10)}$$

or
$$a_m(t) = \frac{1}{i\hbar}\sum_n \int_0^t dt'\,a_n(t').\,H_I(m, n; t')\exp[-i(E_n - E_m)t'/\hbar] \quad \text{...(11)}$$

where
$$H_I(m, n, t) = \int \psi_m^*\,H_I(t)\,\psi_n\,d\tau \equiv \langle m|H_I(t)|n\rangle \quad \text{...(12)}$$

and it his been presumed that $H_I(t)$ is switched on at $t = 0$.

If we assume that intially, the system was in the state $\psi_{n'}$, the perturbation (switched on at $t = 0$) will alter $\psi(t)$ as follows

$$\psi_n(t) = \psi_n \exp(-i\,E_n t/\hbar)$$

$$+ \frac{1}{i\hbar}\sum_{n'}\left\{\int_0^t H_I(n', n; t')\exp[i(E_{n'} - E_n)t'/\hbar]\,dt'\right\}$$

$$\times\,\psi_{n'} \exp(-i\,E_{n'}\,t/\hbar) \quad \text{...(13)}$$

If we are interested in the expectation value of an operator $A$, initially it will be

$$A_{nn} = \int \psi_n^*\,A\,\psi_n\,d\tau \quad \text{...(14)}$$

and after switching on the perturbation it becomes

$$A_{nn}(t) = A_{nn} + \frac{1}{i\hbar}\left[\sum_{n'}\int_0^t \{H_I(n', n; t')\,A_{nn'}\exp[i\omega_{n'n}(t' - t)]\right.$$

$$\left. - H_I(n, n; t')\,A_{n'n}\exp[-i\omega_{n'n}(t' - t)]\}\,dt'\right] \quad \text{...(15)}$$

where
$$A_{n'n} = \int \psi_{n'}^*\,A\,\psi_n\,d\tau \quad \text{...(16)}$$

and
$$\omega_{n'n} = (E_{n'} - E_n)/\hbar \quad \text{...(17)}$$

From Eq. (26) it can be noted that

$$A_{n'n}\exp(i\omega_{n'n}t) = \left\langle n\left|\exp\left(-\frac{iH_0 t}{\hbar}\right)A\exp\left(+\frac{iH_0 t}{\hbar}\right)\right|n'\right\rangle \quad \text{...(18)}$$

But here, the operator within the bracket on the r.h.s. is the Heisenberg representation of the operator $A(-t)$, if the time evolution of the operator were to occur in the unperturbed system. Hence, the expression for $A_{nn}(t)$ becomes (using the closure property of the set $\psi_n$) as

$$A_{nn}(t) = A_{nn} + \frac{1}{i\hbar}\int_0^t [A(t-t'), H_I(t')]_{nn}\, dt' \qquad ...(19)$$

or

$$\Delta A_{nn}(t) = \frac{1}{i\hbar}\int_0^t [A(t-t'), H_I(t')]_{nn}\, dt' \qquad ...(20)$$

This equation is the fundamental equation of the *theory of linear response* in which *the change in the physical quantity A is connected linearly to the strength of the perturbation $H_I(t)$.*

From a practical viewpoint, $\Delta A_{nn}(t)$ is not of much significance since the system can rarely be expected to be in the state $\psi_n$. What is more likely is that *at any given time determined by statistical mechanical considerations*. A particularly simple result ensues if we determine the weightage of each such expectation value using the canonical distribution function, Thus, the statistical average (also called the thermal average) of $\Delta A_{nn}(t)$ is given by

$$\Delta A(t) = \sum_n \Delta A_{nn}(t)\exp(-\beta E_n)/\sum_{n'}\exp(-\beta E_{n'})$$

$$= \frac{1}{Z(\beta)}\, Tr\left[\frac{1}{i\hbar}\exp(-\beta H_0)\,[A(t-t')],\, H_I(t')\right] dt' \qquad ...(21)$$

where $Tr[\ldots]$ is the trace of the operator within the bracket, $\beta \equiv \frac{1}{k_B T}$ ($k_B$ being the Boltzmann constant) and $Z(\beta)$ is the canonical partition function defined by

$$Z(\beta) = \sum_n \exp(-\beta E_n) = Tr[\exp(-\beta H_0)] \qquad ...(22)$$

Equation (21) is called the *Kubo formula.*

It would be impressive at this point to analyze in some detail the structure of the commutator in the integrand of Eq. (21). The form of $H_I(t)$ is usually such that it can be expressed as the product of a function of time and an operator which does not depend explicitly on time, i.e.,

$$H_I(t) = B\, f(t). \qquad ...(23)$$

where $f(t)$ is a force. With $H_I(t)$ in this form, Eq. (21) can be written as

$$A(t) = \int_{-\infty}^{+\infty} G_{AB}(t-t')\, f(t')\, dt' \qquad ...(24i)$$

where

$$G_{AB}(t-t') = -\frac{i}{h}\,\theta(t-t')\,\frac{1}{Z(\beta)}$$

$$\times\, T_r\,\{\exp(-\beta H_0)\,[A(t-t'),\, \beta]\} \qquad ...(24ii)$$

The lower limit in Eq. (24) is effectively the time at which $f(t)$ is switched on. The upper limit is $t$, because of the $O$-function.

The term "Green's functions" is used for quantities such as are defined in Eq. (24). Using this Green's functions, the change in a physical property of a system due to a perturbation can be written in a form analogous to what we would get if the physical property satisfies a linear inhomogeneous differential equation, the inhomogeneity arising out of the perturbation.

In statistical physics, the term 'Green's functions' or the term thermal Green's functions is used to denote expressions of the sort given as

$$G_{AB}(t) = -\frac{i}{\hbar Z(\beta)}\theta(t)\, Tr\{\exp(-\beta H_0)\,[A(t), B]\}$$
$$\equiv \langle\langle A(t), B(0)\rangle\rangle \qquad ...(25)$$

where $A$ and $B$ are the operators for any two physical observables. From the preceding discussion, it is clear that such a Green's functions will appear in a natural manner in linear response theory when the system is perturbed by $H_I(t)$ of the form given by Eq. (23), since the response is indicated in Eq. (24) (i).

The operator $A(t)$ and the operator $B$ which is explicitly time independent (23) as connected by the relation (23) will be characterized by a Green's function of the sort given by Eq. (25). The time dependence of the operator $A(t)$ will depend on the Green's function for $A(t)$ and $B$ through the integral given by Eq. (24) (i).

## 6.9 A MANY-PARTICLE SYSTEM: (THE SYSTEM OF COUPLED OSCILLATORS)

A system of coupled oscillators is essentially a mass-spring system with $N$ particles and $N(N-1)/2$ springs connecting the various pairs. This is a special case of many-particle system where $V(\vec{r}_{jj'})$ is a quadratic form in $r_{j,\sigma}$ when $r_{j,\sigma}$ is measured from the equilibrium position of the $j^{\text{th}}$ particle at which point

$$\frac{\partial}{\partial r_{j,\sigma}} \sum_{j<j'} V(\vec{r}_{jj'}) = 0 \qquad ...(1)$$

The Hamiltonian of the system is given by

$$H_0 = \sum_{j,\sigma} (p_{j,\sigma}^2 / 2m_j) + \frac{1}{2} \sum_{\substack{j,\sigma \\ j',\sigma'}} A_{\sigma\sigma'}(j, j')\, r_{j,\sigma}\, r_{j',\sigma'} \qquad ...(2)$$

where

$r_{j,\sigma}$ the σ-component, i.e., $x$, $y$, or $z$ component of the displacement of the $j^{\text{th}}$ particle and

$$A_{\sigma\sigma'}(j, j') = \frac{\partial^2}{\partial r_{j,\sigma}\,\partial r_{j',\sigma'}} \sum_{j<j'} V(\vec{r}_{jj'})\Bigg|_{r_{j,\sigma} = r_{j',\sigma'} = 0} \qquad ...(3)$$

$A_{\sigma\sigma}(j, j')$ is the spring constant of the spring connecting the σ-displacement of the $j$-th particle with the σ′-displacement of the $j'$-th particle.

Equality of action and reaction implies

$$A_{\sigma\sigma'}(j, j') = A_{\sigma'\sigma}(j', j) \qquad ...(4)$$

The classical equations of motion of the system are

$$m_j\, \ddot{r}_{j,\sigma} + \sum_{j',\sigma'} A_{\sigma\sigma'}(j, j')\, r_{j',\sigma'} = F_{j,\sigma}(t) \qquad ...(5)$$

($A_{\sigma\sigma'}$ being the spring constant).

$F_{j,\sigma}(t)$ is the force on the $j$-th particle in the σ-direction. The perturbing Hamiltonian corresponding to such a force is of the form

$$H_I(t) = -\sum_j \vec{r}_j \cdot \vec{F}_j \qquad ...(6)$$

Equation (5) can be written in a compact matrix form as

$$\left[\hat{M}\frac{d^2}{dt^2} + \hat{A}\right]\vec{R} = \vec{F} \qquad ...(7)$$

where $\hat{M}$ is a diagonal $3N \times 3N$ matrix with elements

$$M_{\sigma\sigma'}(j, j') = m_j\, \delta_{\sigma\sigma'}\, \delta_{jj'} \qquad ...(8)$$

$\hat{A}$ is a $3N \times 3N$ symmetric matrix having the elements $A_{\sigma\sigma'}(j, j')$ and $\vec{R}$ and $\vec{F}$ are 3N-dimensionel vectors having the components $r_{j,\sigma}$ and $F_{j,\sigma}$. In analogy with Eq. (2), Art. 6.7, we can define a Green's functions matrix $\hat{G}(t - t')$ satisfying the equations:

$$\vec{R}(t) = \int_{-\infty}^{+\infty} \hat{G}(t - t')\, \vec{F}(t')\, dt' \qquad ...(9)$$

and
$$\left[\hat{M}\frac{d^2}{dt^2} + \hat{A}\right]\hat{G}(t - t') = \hat{I}\,\delta(t - t') \qquad ...(10)$$

where $\hat{I}$ is the $3N \times 3N$ unit matrix.

$$G(t - t') = \frac{1}{\sqrt{2\pi}}\int_{-\infty}^{+\infty} \hat{G}(\omega)\exp[i(t - t')]\, d\omega \qquad ...(11)$$

where
$$\hat{G}(\omega) = \frac{1}{\sqrt{2\pi}}[\hat{A} - \hat{M}\omega^2]^{-1} \qquad ...(12)$$

The detailed structure of the Green's functions matrix in Eq. (12) can be studied using some properties of the $\hat{A}$ and $\hat{M}$ matrices which are well known in the classical theory of small oscillations [Goldstein, H., Classical Mechanics, Chapter 10, Addison Wesley (1950)]. There exist eigenvectors (corresponding to normal modes] $\Phi_n$ such that

$$\hat{A}\,\vec{\Phi}_n = \omega_n^2\, \hat{M}\vec{\Phi}_n \qquad (13)$$

where $\omega_n^2$ is the square of the $n$-th normal mode frequency and $\Phi_n$ satisfies the relation.

$$\sum_{j,\sigma} \Phi_n^*(j, \sigma)\, m_j\, \Phi_{n'}(j, \sigma) = \delta_{nn'} \qquad ...(14)$$

$$\sum_{n=1}^{3N} \Phi_n(j, \sigma)\, \Phi_n^*(j', \sigma') = \frac{\delta_{\sigma\sigma'}\,\delta_{jj'}}{\sqrt{m_j\, m_{j'}}} \qquad ...(15)$$

Here $\Phi_n(j, \sigma)$ is the $(j, \sigma)$-component of $\vec{\Phi}_n$. Using these equations, we obtain for the matrix elements of $\hat{G}(\omega)$ the expression,

$$G_{\sigma\sigma'}(j, j'; \omega) = \frac{1}{\sqrt{2\pi}} \sum_{n=1}^{3N} \frac{\phi_n(j,\sigma)\,\phi_n^*(j',\sigma')}{\omega_n^2 - \omega^2} \quad \text{...(16)}$$

This has poles at the normal mode frequencies, i.e., at $\omega = \pm\omega_n$, $n = 1, 2 \ldots 3N$. Its Fourier transform in Eq. (11) can be taken by choosing a suitable contour. For instance, the retarded Green's functions is given by

$$G_{\sigma\sigma'}(j, g', t - t') = \theta(t - t') \sum_{n=1}^{3N} \Phi_n^*(j',\sigma')\,\Phi_n(j,\sigma) \frac{\sin \omega_n (t - t')}{\omega_n} \quad \text{...(17)}$$

An interesting property of $\hat{G}(\omega)$ is that

$$Tr\,\hat{M}\hat{G}(\omega) = \sum_{j,\sigma} m_j\, G_{\sigma\sigma}(j, j; \omega)$$

$$= \frac{1}{\sqrt{2\pi}} \sum_{n=1}^{3N} \frac{1}{(\omega_n^2 - \omega^2)} \quad \text{...(18)}$$

If we give a small positive imaginary part to $\omega^2$ and use equation.

$$\frac{1}{x \pm i\varepsilon} = \frac{P}{x} \mp i\pi\,\delta(x) \quad \text{(Dirac identity)} \quad \text{...(19)}$$

We obtain,

$$\lim_{\omega^2 + i0^+} \operatorname{Im}[Tr\,\hat{M}\,\hat{G}(\omega)] = \sqrt{\frac{\pi}{2}} \sum_{n=1}^{3N} \delta(\omega_n^2 - \omega^2) \quad \text{...(20)}$$

The right hand side is proportional to the distribution function $\rho(\omega^2)$ which is defined by $\rho(\omega^2)\,d(\omega^2) = \frac{1}{3N}$ [i.e., the number of normal mode frequencies (squared) that lie in the interval between $\omega^2$ and $\omega^2 + d(\omega^2)$]. ...(21)

We thus get the important result

$$\frac{1}{3N}\sqrt{\frac{2}{\pi}}\left[\lim_{\omega^2 + i0^+} \operatorname{Im}\{Tr\,\hat{M}\,\hat{G}(\omega)\}\right] = \rho(\omega^2) \quad \text{...(22)}$$

The frequency distribution function $g(\omega)$ is related to $\rho(\omega^2)$ by the equation

$$g(\omega) = 2\omega \cdot \rho(\omega^2). \quad \text{...(23)}$$

To develop a quantum mechanical description of the coupled oscillator system, we start with the canonical commutation relations:

$$[r_{j,\sigma}, p_{j',\sigma'}] = i\hbar\,\delta_{\sigma\sigma'}\,\delta_{jj'}, \text{ with } p_{j,\sigma} = m_j\, r_{j,\sigma} \quad \text{...(24)}$$

and all other commutators of the type $[r_{j,\sigma}, r_{j',\sigma'}]$ and $[p_{j,\sigma}, p_{j',\sigma'}]$ vanish (= 0).

Now, the Green's functions which given the expectation value of the displacement, $r_{j,\sigma}(t)$, when the force is applied on the $j$-th particle in the $\sigma'$-direction is

$$G_{\sigma\sigma'}(j, j; t - t') = -\ll r_{j,\sigma}(t), r_{j',\sigma'}(t')\gg \quad ...(25)$$

If we adopt the equation of motion method for evaluating the Green's functions, using Eq. (4) we note that

$$[r_{j,\sigma}, H_0] = \frac{i\hbar}{m_j} p_{j,\sigma} \quad ...(26)$$

and
$$[[r_{j,\sigma}, H_0], H_0] = \frac{i\hbar}{m_j}\left[p_{j,\sigma}, \frac{1}{2}\sum_{\substack{j',\sigma' \\ j'',\sigma''}} A_{\sigma'\sigma''}(j', j'')\, r_{j',\sigma'}\, r_{j'',\sigma''}\right]$$

$$= \frac{\hbar^2}{2m_j}\left[\sum_{j'',\sigma''} A_{\sigma\sigma''}(j', j'')\, r_{j'',\sigma''} + \sum_{j',\sigma'} A_{\sigma'\sigma}(j', j)\, r_{j',\sigma'}\right]$$

$$= \frac{\hbar^2}{m_j}\sum_{j',\sigma'} A_{\sigma\sigma'}(j, j')\, r_{j',\sigma'} \quad ...(27)$$

We thus get

$$i\hbar\frac{d}{dt}G_{\sigma\sigma'}(j, j'; t - t') = -\frac{i\hbar}{m_j}\ll p_{j,\sigma}(t), r_{j',\sigma'}(t')\gg \quad ...(28)$$

and
$$(i\hbar)^2\frac{d^2}{dt^2}G_{\sigma\sigma'}(j, j'; t - t') = -\frac{\hbar^2}{m_j}\delta(t - t')\,\delta_{\sigma\sigma'}\,\delta_{jj'}$$

$$-\frac{\hbar^2}{m_j}\sum_{j'',\sigma''} A_{\sigma\sigma'}(j', j')\ll r_{j',\sigma'}(t), \varepsilon_{j',\sigma'}(t')\gg \quad ...(29)$$

or
$$m_j\frac{d^2}{dt^2}G_{\sigma\sigma'}(j; j'; t - t') + \sum_{j'',\sigma''} A_{\sigma\sigma''}(j, j'')\, G_{\sigma''\sigma'}(j'', j'; t - t')$$

$$= \delta(t - t')\,\delta_{\sigma\sigma'}\, d_{jj'} \quad ...(30)$$

In matrix form, Eq. (30) can be written as

$$\left[\hat{M}\frac{d^2}{dt^2} + \hat{A}\right]\hat{G}(t - t') = \hat{I}\,\delta(t - t') \quad ...(31)$$

which is the same as Eq. (10) follows, therefore, that the analytic properties of the Fourier transform of the Green's functions are the same as those of the one in Eq. (12) and whose explicit form is given in Eq. (17).

## 6.10 VIBRATIONAL EXCITATIONS AND THE CONCEPT OF PHONONS

The orthogonality and closure properties of the eigenvectors of $\hat{A}$ (given in Eqs. (14) and (15) are helpful in forming useful operators by taking linear combinations of $r_{j,\sigma}$, which have very interesting properties. For instance, we define two operators

$$b_n^+ = \frac{-i}{\sqrt{2\hbar\omega_n}} \sum_{j,\sigma} [m_j \dot{r}_{j,\sigma} + i\omega_n m_j r_{j,\sigma}] \Phi_n (j, \sigma) \qquad ...(32)$$

and
$$b_n = \frac{i}{\sqrt{2\hbar\omega_n}} \sum_{j,\sigma} [m_j \dot{r}_{j,\sigma} - i\omega_n m_j r_{j,\sigma}] \Phi_n^* (j, \sigma) \qquad ...(33)$$

Then, it can be seen that

$$[b_n, b_{n'}^+] = \delta_{nn'}; \; [b_n, b_{n'}] = 0 = [b_n^+, b_{n'}^+] \qquad ...(34)$$

Further
$$b_{n'}^+ b_n = \frac{1}{2\hbar\omega_n} \sum_{\substack{j,\sigma \\ j',\sigma'}} [\{m_j m_{j'} \dot{r}_{j,\sigma} \dot{r}_{j',\sigma'} + \omega_n^2 m_j m_{j'} r_{j,\sigma} r_{j',\sigma'}\} \Phi_n (j, \sigma) \Phi_n^* (j', \sigma') + i\omega_n \phi_n (j, \sigma) \phi_n^* (j', \sigma') m_j m_{j'} \{t_{j,\sigma} \dot{r}_{j',\sigma'} - \dot{r}_{j,\sigma'} r_{j'\sigma'}\}] \qquad ...(35)$$

or
$$\sum_n \hbar\omega_n \left(b_n^+ b_n + \frac{1}{2}\right) = \sum_{j,\sigma} p_{j,\sigma}^2 / 2m_j + \frac{1}{2} \sum_{\substack{j,\sigma \\ j',\sigma'}} A_{\sigma\sigma'} (j, j') r_{j,\sigma} r_{j,\sigma'} = H_0. \qquad ...(36)$$

In obtaining the Eq. (36), we have used Eq. (13) to write

$$m_j \omega_n^2 \phi_n (j, \sigma) = \sum_{j'',\sigma''} A_{\sigma\sigma''} (j, j'') \phi_n (j'', \sigma'') \qquad ...(37)$$

and subsequently the sum over $n$ is taken and the closure relation (15) utilized.

The operators $b_n$ and $b_n^+$ have the properties

$$\left.\begin{aligned} [b_n, II_0] &= \hbar \omega_n b_n \\ [b_n^+, H_0] &= -\hbar \omega_n b_n^+ \end{aligned}\right| \qquad ...(38)$$

The equations of motion for the operators $b_n$ and $b_n^+$ are given by

$$\dot{b}_n = \frac{1}{i\hbar} [b_n, H_0] = -i\omega_n b_n \quad \text{or} \quad b_n(t) = b_n(0) \exp(-i\omega_n t) \qquad ...(39)$$

$$\dot{b}_n^+ = \frac{1}{i\hbar} [b_n, H_0] = i\omega_n b_n^+ \quad \text{or} \quad b_n^+(t) = b_n^+(0) \exp(i\omega_n t) \qquad ...(40)$$

If we start with a state $|E\rangle$ such that

$$H_0 |E\rangle = E |E\rangle \qquad ...(41)$$

the state $b_n^+ |E\rangle$ has the property

$$H_0 \{b_n^+ |E\rangle\} = b_n^+ H_0 |E\rangle - [b_n^+, H_0] |E\rangle \qquad ...(42)$$

$(\because \; [b_n^+, H_0] = b_n^+ H_0 - H_0 b_n^+)$

or
$$H_0 \{b_n^+ |E\rangle\} = (E + \hbar\omega) \{b_n^+ |E\rangle\} \qquad ...(43)$$

Similarly, $$H_0\{b_n|E\rangle\} = (E - \hbar\omega)\{b_n|E\rangle\} \quad ...(44)$$

Thus, $b_n$ operating on a state creates a state with energy less than the previous energy and $b_n^+$ creates one with energy more than the previous energy, in each case by amount $\hbar\omega_n$, which is the quantum of energy associated with an oscillator of frequency, $\omega_n$. The lowest energy of this oscillator is $\frac{1}{2}\hbar\omega_n$. The Hamiltonian of the system becomes expressible as seen in Eq. (36)

$$H_0 = \sum_n \hbar\omega_n\left(b_n^+ b_n + \frac{1}{2}\right) \quad ...(45)$$

as the sum of 3$N$ independent harmonic oscillators with frequencies $\omega_n$; ($n$ = 1, 2 ... 3$N$). The lowest energy state (ground state $|0\rangle$) satisfies

$$b_n|0\rangle = 0 \text{ for all } n.$$

Then, $$H_0|0\rangle = \left[\sum_n \frac{1}{2}\hbar\omega_n\right]|0\rangle \quad ...(46)$$

Since $$[b_n^+ b_n, b_n^+] = b_n^+$$

$$[b_n^+ b_n, (b_n^+)^2] = \quad ...(47)$$

$$[b_n^+ b_n, (b_n^+)^2] = 2(b_n^+)^2$$

= = = = = = = = = = = = =

$$[b_n^+ b_n, (b_n^+)^s] = s(b_n^+)^s \quad ...(48)$$

it follows that starting from $|0\rangle$, if we generate a state $\{(b_n^+)^s|0\rangle\}$, this state has the property that

$$\begin{aligned} b_n^+ b_n\{(b_n^+)^s|0\rangle\} &= [b_n^+ b_n, (b_n^+)^s]|0\rangle \\ &\quad + (b_n^+)^s b_n^+ b_n|0\rangle \\ &= [b_n^+ b_n, (b_n^+)^s]|0\rangle + 0. \\ &\equiv s\{(b_n^+)^s|0\rangle\}. \end{aligned} \quad ...(49)$$

Thus, we get the important result that $b_n^+ b_n$ operating on an eigenstate of $H_0$ gives the number of quanta ($s$), each of energy $\hbar\omega_n$, in that state $b_n^+$ and $b_n$ are respectively called creation and annihilation operators and $b_n^+ b_n$ is called the number operator.

What are these oscillators whose quanta are generated by $b_n^+$? Since $b_n^+$ is a linear combination of $r_{j\sigma}$ and $\dot{r}_{j,\sigma}$, the oscillators associated with a quantum $\hbar\omega_n$ for any $n$ involves the displacements of all the particles. These quanta associated with the collective oscillations of all the particles are called phonons. We can specify an excited state of the system by the symbol $|s_1, s_2, ... s_{3N}\rangle$ where $s_n$ is the number of phonons present with frequency $\omega_n$.

The energy of this state is

$$E(\{s_n\}) = \sum_n \hbar\omega_n\left(s_n + \frac{1}{2}\right) \quad ...(50)$$

The representation of states by $|s_1, s_2, \dots s_{3N}\rangle$ is an example of the use of the method of second quantization in which the states of a many-particle system are prescribed by giving the number of particles in each quantum state of a single particle (e.g., $s_n$ phonons in $n$-th state). For an assembly of Bosons, the creation and annihilation operators satisfy commutation relation give in Eq. (34). For Fermions, the corresponding operators (denoted by $a_n^+$ and $a_n$, $n$ being the quantum number of a single particle state) satisfy anti-commutation relations

$$[a_n, a_{n'}^+]_+ = \delta_{nn'} \quad \text{...(51)}$$

$$[a_n, a_{n'}]_+ = 0 = [a_n^+, a_{n'}^+]_+ \quad \text{...(52)}$$

(where $[a_n, a_{n'}^+]_+ = a_n\, a_{n'}^+ + a_{n'}^+\, a_n$ etc.).

It is to be noted here that phonons behave as Bosons, and since they are not exactly like particles which have mass, position etc., they are called '*quasi-particles*' or elementary excitations, It can be further noted here that many particle systems have excitations which can be thought of as arising out of creation of quasi-particles of the appropriate type (e.g., cooper-pairs in a superconducting state).

## 6.11 PHONONS AS PARTICLES

Having discussed quantization of lattice vibrations (in the previous article), we will now consider, how phonons may be treated as wave packets, i.e., particle-like quanta.

We develop this picture first with a one-dimensional model, which can be extrapolated to the three-dimensional case.

Consider a one-dimensional system with $N$ atoms each of mass $m$, connected by springs (of spring constant $k$) in a chain of length $L = Nr_0$, where $r_0$ is the equilibrium spacing. Using periodic boundary conditions, the dispersion relation is given by

$$\omega(q) = 2\sqrt{\frac{k}{M}}\left|\sin\frac{qr_0}{2}\right| \quad \text{...(1)}$$

where the values of $q$ are given by

$$q_m = \frac{2\pi m}{L}$$

with $\quad -N/2 < m \le N/2.$

The displacement associated with a mode is given by

$$u_q\,(x, t) = \frac{1}{\sqrt{N}}\, e^{iqx} \cdot e^{-i\omega(q)t}. \quad \text{...(2)}$$

where we have the displacement $u\,(x, t)$ a continuous function of position $x$, giving the normal mode solution $u_n(t)$ at the equilibrium atomic positions $x = r_n = nr_0$. These solutions represent waves of wavelength $\frac{2\pi}{q}$, spread throughout the one-dimensional crystal, that travel with the phase velocity $c_\phi = \omega(q)/q$.

Consider now building a superposition of the modes, known as a *wave packet*, as

$$u(x, t) = \frac{L}{2\pi} \sum_q A(q)\, u_q(x, t)$$

$$= \frac{L}{2\pi} \frac{1}{\sqrt{N}} \sum_q A(q)\, e^{iqx}\, e^{-i\omega(q)t} \quad ...(3)$$

The amplitude $A(q)$ is an envelope function that picks out the modes in a range $\Delta q$ about a central value, $q_0$. Assume that the chain is very long, and that the range $\Delta q$ is much larger than the discrete spacing in $q$ of $2\pi/L$, but smaller than the Brillouin zone width $2\pi/r_0$. Then the summation in the above equation can be replaced by the integration, i.e.,

$$u(x, t) = \frac{1}{\sqrt{N}} \int_{-\infty}^{+\infty} A(q)\, e^{iqx}\, e^{-i\omega(q)t}\, dq. \quad ...(4)$$

At any fixed time say $t = 0$, Eq. (4) has the form of an inverse Fourier transform with $u(x, 0)$ the spatial transform of $A(q)$. A general property of such transform is that if $A$ has a spread $\Delta q$ in wave vector space, then the *wave packet* $u(x, 0)$ will have a spread $\Delta x$ in real space, such that

$$\Delta x\, \Delta q \geq \frac{1}{2}$$

If we take a spread in $q$ of 1% of the Brillouin zone $\Delta q \approx 0.01\ (2\pi/r_0)$, then the spread of the wave packet in physical space will be $\Delta x \approx 8\, r_0$. Hence, by making the packet energy, or frequency, spread out by 1%, we have reduced the real space size of the mode from the length $L$ of the chain to that of a few atomic spacings. (see Fig. 6.7).

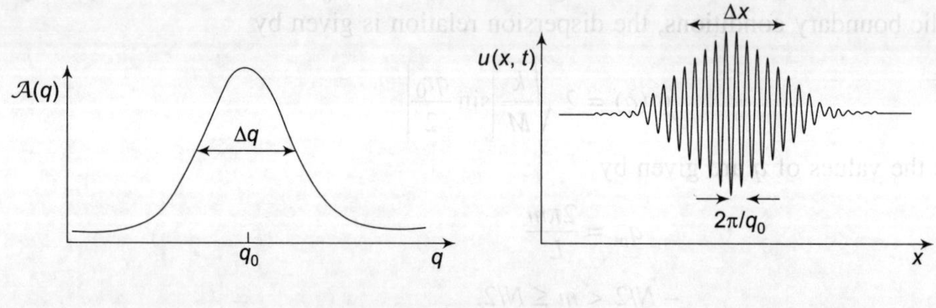


**Figure 6.7** Example of a wave packet envelope function $A(q)$ and the corresponding spatial wave packet $u(x, t)$.

If the dispersion relation $\omega(q)$ is linear, with $\omega = c_\phi q$, the wave packet will travel in $x$ at the phase velocity $c_\phi$. For the dispersion relation (1), this holds for small $q$, but at large $q$ the relation is strongly nonlinear, In that case, any individual Fourier component with wave vector $q$ will travel at the local phase velocity $c_\phi = \omega/q$, but the packet as a whole travels at the *group velocity* $c_g = d\omega/dq$.

This can be seen as follows: Assume $A(q)$ is sharply peaked about $q_0$; we then expand the dispersion relation $\omega(q)$ in a Taylor series about this value, so that

$$\omega(q) \approx \omega(q_0) + \frac{d\omega}{dq}(q - q_0) = \omega_0 + c_g(q - q_0). \qquad ...(5)$$

In the wave packet expression Eq. (3), we then make the approximation

$$u(x, t) \approx \frac{1}{\sqrt{N}} e^{it(c_g q_0 - \omega_0)} \int \mathcal{A}(q)\, e^{iq(x - c_g t)}\, dq. \qquad ...(6)$$

The integral part of this expression is $u(x - c_q t, 0)$, the wave packet at $t = 0$, displaced along $x$ by $c_g t$; the remaining exponential is a simple phase factor.

Hence, we see that, to this order of approximation, the wave packet travels undistorted at speed $c_g$, as promised. Note however that the approximation in Eq. (6) is not exact, and that at long times the wave packet will spread out in real space; the spreading is more severe for packets that initially contain a small range $\Delta q$, and much slower for those with larger spreads. A specific example of building such a wave packet is left to the exercises.

In three dimensions, the construction of the wave packet is the same. We build the packet using a three-dimensional envelope $\mathcal{A}(q)$, superposing modes in a volume $v_q$ in $q$-space about a central wave vector $q_0$. Typically, the superposed modes would all have the same polarization $\varepsilon_{q0n}$, so the construction is given by

$$u(r, t) = \varepsilon_{q0n} \int \mathcal{A}(q)\, e^{iq\cdot r - i\omega(q)t}\, dq \qquad ...(7)$$

The spread in real space of the wave packet $u$ follows the same rules as for the one-dimensional packets; the spread $\Delta x_\alpha$ along each axis $x_\alpha$ is related to the spread $\Delta q_\alpha$ by $\Delta x_\alpha \Delta q_\alpha \geq 1/2$.

## 6.12 THE COUPLED OSCILLATOR PROBLEM (USING GREEN'S FUNCTION) PHONONS IN CRYSTALS

A system of coupled oscillators is essentially a mass-spring system with $N$ masses and $\frac{N(N-1)}{2}$ springs connecting various pairs. The classical equation of motion of the system is given by (assuming that each mass has one degree of freedom, for simplicity),

$$m_j \frac{d^2 x_j}{dt^2} + \sum_{j'} A_{jj'}\, x_{j'} = f_j(t);\ j, j' = 1, 2, ..., N. \qquad ...(1)$$

In a compact matrix notation this becomes

$$\left[\mathbf{M}\frac{d^2}{dt^2} + \mathbf{A}\right]\mathbf{X} = \mathbf{F} \qquad ...(2)$$

where $\mathbf{M}$ is a $N \times N$ diagonal mass matrix, $\mathbf{A}$ is a $N \times N$ matrix, having the elements $A_{jj}'$, $\mathbf{X}$ and $\mathbf{F}$ are $N$-dimensional vectors having the components $x_j$ and $f_j(t)$. We could define a Green's functions matrix $G(t - t')$ satisfying the equations

$$\mathbf{X} = \int_{-\infty}^{\infty} \mathbf{G}(t - t')\, \mathbf{F}(t')\, dt' \qquad ...(3a)$$

and $$\left[\mathbf{M}\frac{d^2}{dt^2}+\mathbf{A}\right]G(t-t')=\mathbf{I}\,\delta(t-t') \qquad ...(3b)$$

$$\mathbf{G}\,(t-t')=\frac{1}{\sqrt{2\pi}}\int_{-\infty}^{\infty} e^{i\omega(t-t')}\,\mathbf{G}(\omega)\,d\omega \qquad ...(4)$$

where $$\mathbf{G}(\omega)=\frac{1}{\sqrt{2\pi}}[\mathbf{A}-\mathbf{M}\,\omega^2]^{-1} \qquad ...(5)$$

The detailed structure of the Green's functions matrix can be studied using some of the properties of the **A** and **M** matrices which are well explained in the classical theory of small oscillations. There exist eigenvectors $\Phi_n$ such that

$$\mathbf{A}\,\Phi_n=\omega_n^2\,\mathbf{M}\,\Phi_n \qquad ...(6)$$

where $\omega_n^2$ is the square of the $n$th normal mode frequency and $\Phi_n$ satisfies the relations

$$\sum_{p=1}^{N}\Phi_n^*(p)\,m_p\,\Phi_{n'}(p)=\delta_{nn'} \qquad ...(7a)$$

$$\sum_{n=1}^{N}\Phi_n^*(p)\,\Phi_n(p')=\frac{\delta_{pp'}}{\sqrt{m_p\,m_{p'}}} \qquad ...(7b)$$

Using these equations we get for the matrix elements $G(\omega)$,

$$\mathbf{G}\,(p,p';\,\omega)=\frac{1}{\sqrt{2\pi}}\sum_{n=1}^{N}\frac{\Phi_n^*(p')\,\Phi_n(p)}{\omega_n^2-\omega^2} \qquad ...(8)$$

This has poles at the normal mode frequencies, i.e., at $\omega=\pm\omega_n$; $n=1, 2, \ldots, N$, and the various Green's functions can be evaluated by choosing different contours. For instance,

$$G_R\,(p,p';\,t-t')=\theta(t-t')\sum_{n=1}^{N}\Phi_n^*(p')\,\Phi_n(p)\,\frac{\sin\omega_n\,(t-t')}{\omega_n} \qquad ...(9)$$

An interesting property of this Green's functions matrix is that,

$$\mathbf{T_r\,M\,G}\,(\omega)=\sum_p m_p\,G(p,p;\omega)=\frac{1}{\sqrt{2\pi}}\sum_{n=1}^{N}\frac{1}{\omega_n^2-\omega^2}. \qquad ...(10)$$

If we give a small positive imaginary part to $\omega^2$ and use the identity

$$\lim_{\varepsilon\to 0}\frac{1}{x\pm i\varepsilon}=\frac{P}{x}\mp i\pi\,\delta(x) \qquad ...(11)$$

we get $$\lim_{\omega^2+io^+}\left[\mathrm{I_m}\,T_r\,\mathbf{M}\,\mathbf{G}\,(\omega)\right]=\sqrt{\frac{\pi}{2}}\sum_{n=1}^{N}\delta(\omega_n^2-\omega^2)$$

The right side is proportional to the distribution function, $\rho(\omega^2)$, where $\rho(\omega^2)\, d(\omega^2)$ = the number of normal mode frequencies (squared) that lie in the interval between $\omega^2$ and $\omega^2 + d(\omega^2)$.

We thus get the important result,

$$\left[\frac{1}{N}\sqrt{\frac{2}{\pi}}\right] \lim_{\omega^2 + io^+} [\text{Im}\, T_r\, \mathbf{MG}(\omega)] = \rho(\omega^2) \qquad ...(12)$$

The quantum mechanical description of the coupled oscillator system can be developed in the same way as in the case of the single oscillator. Here, the Hamiltonian is

$$H = \sum_j \frac{p_j^2}{2m_j} + \frac{1}{2}\sum_{j,j'} A_{jj'}\, x_j\, x_{j'} \qquad ...(13)$$

Using the canonical commutation relations

$$[x_j, p_k] = i\hbar\, \delta_{jk},\ p_k = m_k\, \dot{x}_k \qquad ...(14)$$

we obtain

$$\ddot{x}_j = \frac{1}{i\hbar}[\dot{x}_j, H] = -\frac{1}{\hbar^2}[[x_j, H], H]$$

$$= -\frac{1}{m_j}\sum_{j'} A_{jj'}\, x_{j'} \qquad ...(15)$$

which is the same as Eq. (1) without any external force. The method of solution of Eq. (1) as discussed earlier is applicable in this case also, regarding $x_j$ as an operator.

The Green's functions $G(j, j'; t - t')$ of Eq. (9) or its Fourier transform $G(j, j'; \omega)$ of Eq. (8) assume a new significance, which we shall proceed to examine taking the example of a linear lattice with all masses equal and nearest neighbour interactions. For this case, Eq. (7) becomes

$$\sum_{p=1}^{N} \Phi_j^*(p)\, \Phi_{j'}(p) = \frac{\delta_{jj'}}{m} \qquad ...(16a)$$

$$\sum_{n=1}^{N} \Phi_n^*(p)\, \Phi_n(p') = \frac{\delta_{pp'}}{m} \qquad ...(16b)$$

Here a possible form of $\Phi_n(p)$,

$$\Phi_n(p) = \frac{1}{\sqrt{Nm}}\, e^{2\pi i\, np/N}. \qquad ...(17)$$

A general expression for $x_j$ can be written as

$$x_j = \sum_n a_n\, \Phi_n(j) \qquad ...(18)$$

where the coefficient $a_n$ is an operator defined as

$$a_n = \sum_n m\, \Phi_n^*(j)\, x_j = \sum_j m\, \Phi_{(-n)}(j)\, x_j \qquad ...(19)$$

Since $x_j$ is Hermitian it follows that

$$a_n^\dagger = \sum_j m\Phi_{(-n)}^*(j)\, x_j = a_{(-n)} \qquad ...(20)$$

From Eq. (14) for this case of equal masses, we have

$$[x_j, \dot{x}_{j'}] = \frac{i\hbar}{m}\,\delta_{jj'} \qquad ...(21)$$

Using the above equation, we get

$$[a_n, \dot{a}_{n'}] = \sum_{j,j'} m^2\, \Phi_n^*(j)\, \Phi_{(-n')}(j')\,[x_j, x_{j'}]$$

$$= i\hbar \sum_j m\, \Phi_n^*(j)\, \Phi_{(-n')}(j) = i\hbar\, \delta_{n,(-n')} \qquad ...(22)$$

and all other commutators such as $[a_n, a_{n'}]$ and $[\dot{a}_n, \dot{a}_{n'}]$ vanish. Making $n' \equiv -n'$ and using Eq. (20), we get

$$[a_n, \dot{a}_{n'}^\dagger] = i\hbar\, \delta_{nn'} \qquad ...(23)$$

The Hamiltonian can be re-expressed in terms of these operators

$$H = \frac{1}{2}\sum_n \dot{a}_n^\dagger \dot{a}_n + \frac{1}{2}\sum_n \omega_n^2\, a_n^\dagger a_n \qquad ...(24)$$

If we define two operators

$$b_n^\dagger = \frac{1}{\sqrt{2\hbar\omega_n}}\, \dot{a}_n^\dagger + i\sqrt{\frac{\omega_n}{2\hbar}}\, a_n^\dagger \qquad ...(25a)$$

$$b_n = \frac{1}{\sqrt{2\hbar\,\omega_n}}\, \dot{a}_n - i\sqrt{\frac{\omega_n}{2\hbar}}\, a_n \qquad ...(25b)$$

Equation (24) can be written as

$$H = \sum_n \hbar\omega_n \left(b_n^\dagger b_n + \frac{1}{2}\right) \qquad ...(26)$$

We also have $$[b_n, b_{n'}^\dagger] = \delta_{nn'} \qquad ...(27)$$

The equations of motion for the operator $b_n$ and $b_n^+$ are given by

$$\dot{b}_n = \frac{1}{i\hbar}[b_n, H] = -i\omega_n b_n, \quad \text{or} \quad b_n(t) = e^{-i\omega_n t}\, b_n(0); \qquad ...(28a)$$

$$\dot{b}_n^\dagger = \frac{1}{i\hbar}[b_n^\dagger, H] = i\omega_n b_n^\dagger \quad \text{or} \quad b_n^\dagger(t) = e^{i\omega_n t}\, b_n^\dagger(0) \qquad ...(28b)$$

From Eq. (28) one notes that

$$[b_n, H] = \hbar\,\omega_n\, b_n \qquad ...(29a)$$

$$[b_n^\dagger, H] = -\hbar\,\omega_n\, b_n^\dagger \qquad ...(29b)$$

Thus, if we start with a state $|E\rangle$ such that

$$H\,|E\rangle = E\,|E\rangle$$

the state $b_n^\dagger\,|E\rangle$ has the property,

$$H\{b_n^\dagger\,|E\rangle\} = (E + \hbar\omega_n)\,\{b_n^\dagger\,|E\rangle\}$$

and the state $\hbar\,|E\rangle$ has the property

$$H\{b_n\,|E\rangle\} = (E - \hbar\omega_n)\,\{b_n\,|E\rangle\}$$

$b_n$ operating on a state creates a state with energy less than the previous energy and $b_n^\dagger$ creates a state with energy more than the previous energy, in each case by an amount $\hbar\omega_n$, which is the quantum of energy associated with an oscillator of frequency, $\omega_n$. The lowest energy of this oscillator is $\frac{1}{2}\hbar\omega_n$. The Hamiltonian of the linear chain in this form (called the second quantised form) becomes expressible as the sum of the Hamiltonians of $N$ independent oscillators with frequencies $\omega_n$, $n = 1, 2, \ldots, N$.

Since $$[b_n^\dagger\, b_n, b_n^\dagger] = b_n^\dagger;\ [b_n^\dagger\, b_n, (b_n^\dagger)^2] = 2\,(b_n^\dagger)^2;\ \ldots\ [b_n^\dagger\, b_n, (b_n^\dagger)^s] = s\,(b_n^\dagger),\ \text{etc.}$$

it follows that starting from the lowest energy state $|0\rangle$ if one generates the state $(b_n^\dagger)^s\,|0\rangle$ containing $s$ quanta of energy $\hbar\omega_n$, we have

$$b_n^\dagger\, b_n\,\{(b_n^\dagger)^s\,|0\rangle\} = [b_n^\dagger\, b_n, (b_n^\dagger)^s]\,|0\rangle + (b_n^\dagger)^s\, b_n^\dagger\, b_n\,|0\rangle = s\,\{(b_n^\dagger)^s\,|0\rangle\}$$

since $b_n\,|0\rangle = 0$, $|0\rangle$ is the lowest energy state by hypothesis. Thus, we get the important result that $b_n^\dagger\, b_n$ operating on an eigen state of the Hamiltonian gives the number of quanta of energy in that state. $b\dagger$ and $b_n$ are called creation and annihilation operators and $(b_n^\dagger\, b_n)$ is often called the number operator.

What are these oscillators? From the way we have proceeded it is obvious that such oscillations involve displacements of all the masses in the system; $a_n$ and hence $b_n$ is a linear combination of $X_j$'s. These collective oscillations are called phonons. In this case we can characterise each phonon by the number $n$, $n = 1, 2, \ldots N$. The ground state energy of the system is given by $\left[\sum_n \frac{1}{2}\hbar\omega_n\right]$, and an excited state can be specified by the symbol $|S_1, S_2 \ldots, \ldots, S_N\rangle$, where $S_n$ is the number of phonons present each having energy $\hbar\omega_n$. The energy of this state is

$$E_{\{S_n\}} = \sum_n \hbar\omega_n\left(S_n + \frac{1}{2}\right) \qquad \ldots(30)$$

Going back to Eq. (18), we note that

$$x_j = \sum_n \frac{1}{i}\sqrt{\frac{\hbar}{2\omega_n}}\,[b_{-n}^{\dagger} - b_n]\,\phi_n(j)$$

$$= \sum_n -\frac{1}{i}\sqrt{\frac{\hbar}{2\omega_n}}\,[b_{-n} - b_n^{\dagger}] \qquad ...(31)$$

so that,

$$[x_j(t), x_{j'}(o)] = i\hbar \sum_n \phi_n(j)\,\phi_n^*(j')\,\frac{\sin \omega_n t}{\omega_n} \qquad ...(32)$$

Comparing Eq. (32) with Eq. (9) we obtain

$$G_R(j, j'; t) = -\frac{i}{\hbar}[x_j(t), x_{j'}(o)]\,\theta(t) \qquad ...(33)$$

and its Fourier transform is given by

$$G_R(j, j'; \omega) = \frac{1}{\sqrt{2\pi}} \sum_n \frac{\phi_n(j)\,\phi_n^*(j')}{\omega_n^2 - \omega^2 + io^+} \qquad ...(34)$$

An interesting point to note here is that in this coupled harmonic oscillator system, since $[x_i(t), x_{j'}(o)]$ is a C-number, statistical averaging dose not bring in a temperature dependent term. When trace of $e^{-}H_o\,[x, (t), x_j, (o)]\,f(t)$ is taken and the result divided out by $Z(\beta)$ the result is independent of temperature. One can thus state that the displacement susceptibility of a mass in this system for a time-dependent force at another mass is independent of temperature. This is not necessarily so for other susceptibilities, because then the commutators may not be C-numbers.

Explicit evaluation of $G(j, j', \omega)$ is possible for the linear chain case, since here

$$\omega_n^2 = \frac{2\gamma}{m}\left(1 - \cos\frac{2\pi n}{N}\right) \qquad ...(35)$$

with $\gamma$ being the nearest neighbour spring constant. Then, using Eqs. (17) and (35) in Eq. (34) we obtain for large $N$ and a value of $\omega^2$ which is complex,

$$G(j, j', \omega) = \frac{1}{\sqrt{2\pi}}\frac{1}{mN}\sum_{n=1}^{N} \frac{e^{2\pi i n(j-j')/N}}{\frac{2\gamma}{m}\left(1 - \cos\frac{2\pi n}{N}\right) - \omega^2}$$

$$= \frac{1}{(2\pi)^{3/2}\, m}\int_0^{2\pi} \frac{e^{i\varphi(j-j')}\,d\phi}{\frac{\omega_L^2}{2}(1-\cos\varphi) - \omega^2}$$

$$= \frac{[\sqrt{(\omega^2 - \omega_L^2)\omega^2} - (\omega^2 - \omega_L^2/2)]^{1j-j'1}}{\sqrt{2\pi}\, m\sqrt{(\omega^2 - \omega_L^2)\,\omega^2}\;[\omega_L^2/2]^{1j-j'1}} \qquad ...(36)$$

where $\omega_L^2 = 4\gamma/m$.

The applications of this function, for instance in computation of $\rho(\omega^2)$ using Eq. (12) is discussed in literature.

It may be stressed that the equation of motion of a Green's functions $G_{AB}(t)$ is essentially the same as that of the operator $A(t)$, apart from a δ-function occurring in the right hand side. This property of the Green's functions has been explicitly demonstrated in the above example, because $\ll x_j(t), x_{j'}(0) \gg$ obtained quantum mechanically satisfies the classical equation for $x_j(t)$. For displacements of masses in coupled oscillators (in the harmonic approximation) the Green's functions $\ll x_j(t), x_{j'}(0) \gg$ can be evaluated purely classically as in Eq. (34). But other Green's functon need not have this simple property. We shall discuss these details later.

## 6.13 APPLICATION OF THE GREEN'S FUNCTIONS METHOD FOR PHONONS

We shall examine in this section a application of the Green's functions method in lattice dynamics.

### 6.13.1 Local Dynamics of Atoms in a Lattice

In a crystal in thermal equilibrium if one of the atoms is acted upon by an impulsive force, it is of interest to know how the dynamical disturbance spreads, i.e., how the other atoms respond to this impulsive force. Such an impulsive force may arise, for instance, when an atom is struck by a neutron, or when it receives a recoil due to emission of a γ-ray. Let the impulsive force be on the atom at the origin, then, we know that[1]

$$x_l(t) = \sum_{l'} \int_{-\infty}^{\infty} \ll x_l(t-t'), x_l'(0) \gg F_l'(t')\,dt' \qquad ...(1)$$

where, by assumption

$$F_l(t) = f_o\,\delta_{l,o}\,\delta(t) \qquad ...(2)$$

Thus, we obtain

$$x_l(t) = f_o \ll \underline{x}_l(t), x_o(o) \gg$$

$$= f_o \sum_n \Phi_n(l)\,\Phi_n^*(o)\frac{\sin \omega_n t}{\omega_n};\ t > o \qquad (3)$$

Equation (3) is a complete solution to the problem. If we apply this equation to the example of the linear chain, we get the result

$$x_l(t) = f_o \times \frac{1}{Nm}\sum_{n=1}^{N} e^{2\pi i n l/N}\,\frac{\sin\left[\sqrt{\omega_L^2\left(1-\cos\dfrac{2\pi n}{N}\right)}\,t\right]}{\sqrt{\dfrac{\omega_L^2}{2}\left(1-\cos\dfrac{2\pi n}{N}\right)}}$$

[1] See Article 6.7.

$$= \frac{f_o}{m} \times \frac{1}{2\pi} \int_o^{2\pi} e^{il\varphi} \frac{\sin(\omega_L t \sin\varphi/2)}{\omega_L \sin\varphi/2} d\varphi$$

$$= \frac{f_g}{m} \int_o^t J_{2l}(\omega_L t')\, dt' \qquad ...(4)$$

where $J_{2l}$ is the Bessel function of order $2l$.

The velocity of the $l^{th}$ mass will be given by

$$u_l(t) \equiv \dot{x}_l = \frac{f_o}{m} J_{2l}(\omega_L t) \qquad ...(5)$$

The appropriate Green's functions for a three-dimensional crystal can be worked out to give the answer in the case of more realistic crystals.

## 6.14 BORN-VON KARMAN BOUNDARY CONDITION FOR DISPERSION RELATION, $\omega(k)$

In case of vibration of a linear monatomic chain, we will see that $\omega = 0$ at $k = 0$ and angular frequency $\omega = \omega_m$ at $k = \pm\pi/a$, i.e., all the normal mode frequencies lie in the range of $k$-values from $-\pi/a$ to $+\pi/a$ defining the extent of the first Brillonin zone of the linear lattice.

The complete spectrum of normal modes is usually determined by invoking the periodic boundary condition originally given by Born and von Karman for the electron gas. The condition requires the two ends of the crystal to be joined. It is visualized most easily in case of the electronic conduction. The electrical conductivity depends on the details of electronic motion within the crystal. It implies as if an electron enters the crystal at one end as soon as it leaves the other end. This is equivalent to stating that the two ends of the crystal behave as being in contact. The linear chain is transformed into a ring.

Thus, if our metal were one dimensional, we would simply replace the line from $O$ to $L$ of which the electrons were confined, by a circle of circumference, $L$. In one dimension, the circular model of a metal results in the boundary condition $\psi(x + L) = \psi(x)$ (for the time independent Schrödinger equation) and the generalization to a three-dimensional cube is

$$\begin{aligned} \psi(x, y, z + L) &= \psi(x, y, z) \\ \psi(x, y + L, z) &= \psi(x, y, z) \\ \psi(x + L, y, z) &= \psi(x, y, z). \end{aligned} \qquad ...(1)$$

The above equations are known as the Born-von Karman or periodic boundary conditions.

By imposing an appropriate boundary condition on the wave functions we can find that the wave vector $\vec{k}$ must be real, and arrive at a condition restricting the allowed values of $\vec{k}$ in case of a lattice. The conditions chosen will be a natural generalization of the conditions used in the Sommerfeld theory of free electrons in a cubical box.

Initially, we may consider a one-dimensional monatomic chain, and we know that its vibration corresponds to a dispersion relation

$$\omega(k) = \omega_m \left|\sin\left(\frac{ka}{2}\right)\right|$$

which shows that $\omega = 0$ at $k = 0$ and $\omega = \omega_m$ at $k = \pm\pi/a$, corresponding, all the normal mode frequencies lie in the range $-\pi/a < k < \pi/a$, i.e., the first Brillouin zone. The complete spectrum of normal modes is usually determined by invoking the boundary conditions. For electronic conduction in a lattice, the conditions require that the two ends of the crystal are joined. It implies as if an electron enters the crystal at one end as soon as it leaves the other end. The fact is equivalent to stating that length of the chain $L = Na$.

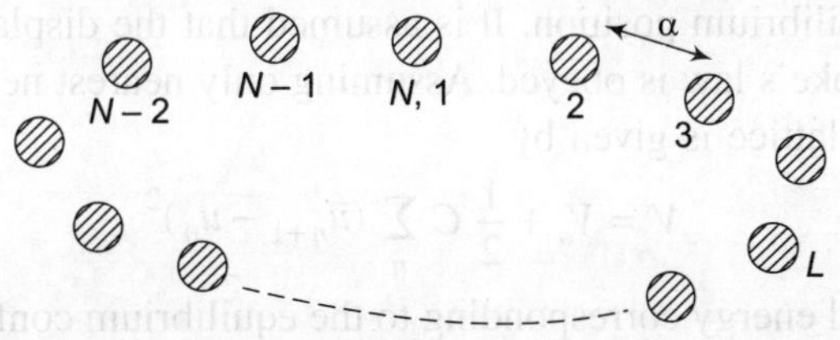


**Figure 6.8** The Born-von karman boundary condition for a linear chain of $N$ identical atoms. If $s_e$ denotes the magnitude of the displacement of an ion at position $l$, then $s_e = s_{L+Na}$ where $a$ is the interatomic spacing at rest position

The two ends of the crystal behave as in contact. In the present case, it has been assumed that the first atom and the last atom are joined by an additional spring similar to that joining two successive atoms in the chain. Thus, linear chain becomes a ring as shown in Fig. 6.8. We require

$$s_l = s_{l+Na} \qquad \text{...(2)}$$

where $l = na$, $n$ is an integer

$Na = L$ (the length of the chain).

We know that $$s_l = s_o \exp\,[i\,(\vec{k}\cdot\vec{l} - \omega t)] \qquad \text{...(3)}$$

$$s_{l+Na} = s_o \exp\,[i\,(\{\vec{k}\cdot(\vec{l} + Na)\} - \omega t)] \qquad \text{...(4)}$$

Using Eqs. (2), (3) and (4), we have

$$\exp\,(i\,k\,Na) = 1 = \exp\,[i\,(2\pi n)]$$

$$\Rightarrow \qquad k = \frac{2\pi}{a}\frac{n}{N} \quad \text{or} \quad k = n\cdot\frac{2\pi}{L}$$

where $n$ is an integer. So the allowed values of $k$ are

$$k = 0, \pm\frac{2\pi}{L}, \pm 2\frac{2\pi}{L}, \ldots.$$

## 6.15 VIBRATION IN ONE-DIMENSIONAL MONATOMIC LATTICE

We now consider the classical description of dynamics of a linear chain of identical atoms. A convenient way to visualize lattice vibration in this case is to consider the system of identical atoms as tiny spherical balls elastically coupled to each other by identical springs.

Consider a lattice consisting of an infinite number of equally spaced identical particles (lattice spacing = $a$) each of mass $M$, along a straight line, say $X$-axis. Let us assume the case of longitudinal vibrations (Fig. 6.9). Since the chain is infinite, any sphere may be chosen as origin. '$a$' denotes the equilibrium spacing between any two nearest neighbours. Let $C$ be the force-constant of any of the connecting springs. Let $u_n$ denote the displacement of the $n^{\text{th}}$ particle from its equilibrium position. It is assumed that the displacements are within the elastic limit so that Hooke's law is obeyed. Assuming only nearest neighbour interaction, the potential energy of the lattice is given by

$$V = V_o + \frac{1}{2} C \sum_n (u_{n+1} - u_n)^2 \qquad \text{...(1)}$$

where $V_0$ is the potential energy corresponding to the equilibrium configuration. If $F_p$ denotes the force acting on the $p^{\text{th}}$ particle

$$F_p = \frac{\partial V}{\partial u_p} = -\frac{1}{2} C \frac{\partial}{\partial u_p} [(u_p - u_{p-1})^2 + (u_{p+1} - u_p)^2] \qquad \text{...(2)}$$

Thus,

$$M \frac{d^2 u_p}{dt^2} = F_p = -\frac{1}{2} C [2(u_p - u_{p-1}) + 2(u_{p+1} - u_p)(-1)]$$

$$= C[u_{p+1} - 2u_p + u_{p-1}]$$

$$\left( \because \frac{\partial u_{p-1}}{\partial u_p} = 0 = \frac{\partial u_{p+1}}{\partial u_p} \right)$$

or

$$\frac{M d^2 u_n}{dt^2} = C(u_{n+1} - 2u_n + u_{n-1}) \qquad \text{...(3)}$$

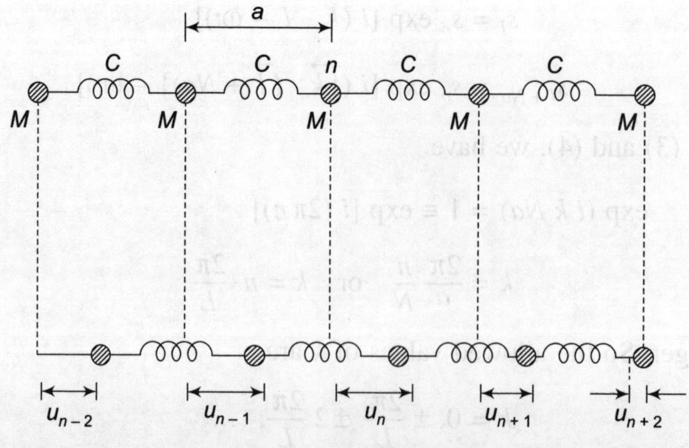


**Figure 6.9** Longitudinal vibrations of atom of one-dimensional monatomic chain

Assuming simple harmonic motion (since the displacements are within the elastic limit), the solution of Eq. (3) would be of the form

$$u_n = A \exp [i(kx - \omega t)] \quad ...(4i)$$

where $k = \omega/c$ is the wave vector and $c$ is the wave velocity $x = na$ is the position of the $n^{th}$ particle from the origin. So,

$$u_n = A \exp [i(kna - \omega t)] \quad ...(4ii)$$

This represents a travelling wave in which all atoms vibrate with the same frequency $\omega$, and have the same amplitude, $A$. Substituting this in Eq. (3), we get

$$-\omega^2 Mu_n = C(e^{ika} - 2 + e^{-ika})\, u_n = 2C(\cos ka - 1)\, u_n$$

or

$$\omega^2 M = 2C(1 - \cos ka)$$

$$= 4C \sin^2 \frac{ka}{2}$$

$\Rightarrow$

$$\omega = \pm\left(\frac{4C}{M}\right)^{1/2} \left\{\sin\frac{ka}{2}\right\} \quad ...(5)(i)$$

or

$$\omega = \pm\omega_m \left\{\sin\frac{ka}{2}\right\} \quad ...(5)(ii)$$

where $\omega_m = \sqrt{4C/M}$ is the maximum (natural cut-off) frequency and is a meascre of the spring constant of the lattice.

It may be noted here that the one-dimensional monatomic chain thus may be considered a low-pass filter which allows the propagation of frequencies lying between $o$ and $\omega_m$ only (In contrast a continuons string has no limit.) The plus and minus signs in Eq. (5) denote that waves travel either to the right or to the left, the motion at any point is obviously periodic in time.

## 6.15.1 Dispersion Relation

A medium is said to display *dispersion* if the velocity of propagation $v$ is a function of $k$, i.e., $\omega$ is not a linear function of $k$. In non-dispersive medium like a continuous string, $v$ is constant and independent of $k$ (or $\omega \propto k$) because for a string $\omega = 2\pi\nu = 2\pi\, v/\lambda = vk$.

The functional dependence of $\omega$ on $k$ shown by Eq. (5) is called a *depersion relation*. This is shown in Fig. 6.9 and is referred to as a dispersion curve.

In the limiting case, when $k \to 0$ (or $\lambda \to \infty$), we have from Eq. (5)

$$\left.\begin{aligned} \omega &= vk \\ v_o &= a\left(\frac{C}{M}\right)^{1/2} \quad ka \ll 1. \end{aligned}\right\} \quad ...(6)$$

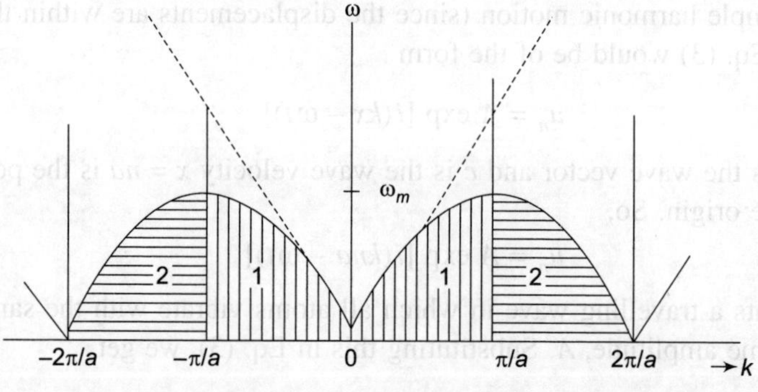


**Figure 6.10** Dispersion curve of a one-dimensional monatomic lattice; 1 and 2 respectively represent the region of first and second BZ

Thus, $\omega \propto k$ when $\lambda \gg a$. Hence, an array of identical mass points and a continuous string yield identical results when wavelength is large. This situation is represented by broken lines in Fig. 6.10.

The region of $k$-values defined by

$$-\frac{\pi}{a} \le k \le +\frac{\pi}{a}$$

is called the first *Brillouin zone* (BZ). Because of the periodic neture of $\omega$ vs. $k$ relation, it is sufficient to confine to the first BZ to completely describe the wave propagation. The first BZ contains the entire spectrum.

It may be noted here that the dispersion curve flattens gredually as $k \to \pm\pi/a$. We may think of the flattening as being due to a component of the reflected wave, which gradually grows until at $k = \pm\pi/a$, the amplitude is equal to the incident wave. Thus, a wave corresponding, to the BZ boundary is a standing wave. It has a wavelength appropriate for Bragg reflection through an angle 180°.

### 6.15.2 Phase and Group Velocity

Using Eq. (5).

Phase velocity $$v_p = \frac{\omega}{k} = v\left[\frac{\sin(ka/2)}{ka/2}\right] \quad ...(7)$$

and group velocity $$v_g = \frac{d\omega}{dk} = v\cos\left(\frac{ka}{2}\right) \quad ...(8)$$

where $v = a\sqrt{C/M}$, is the ordinary speed of sound in the lattice.

When $$k \to 0, \frac{\sin(ka/2)}{ka/2} \to 0$$

and $$v_p = v_g = v.$$

This is to be expected as waves of long wavelength would not be sensitive to the discreteness of the chain, i.e., large number of atoms would participate in all displacements, as a homogeneous line.

From Eq. (7) we see that $v_p = 2v/\pi$ at the zone boundary, since, there, $k = \pm\pi/a$. This means that there $\lambda = 2a$ or the nearest neighbour atoms are moving in antiphase and $\cos(ka/2) = 0$, i.e., $v_g = 0$, as seen from Eq. (8). Since, physically $v_g$ signifies the transfer of energy, this means that at $k = \pm\pi/a$, the wave is a standing wave. The situation is analogous to Bragg reflection and X-ray diffraction.

### 6.15.3 Density of States

Although any vibrational pattern is possible for an elastic continuum, there is a limit to the number of distinguishable modes of vibration for a lattice of finite size consisting of discrete atoms. For a linear chain of $N + 1$ atoms and length $L = Na$, if the end atoms are coincident, we have $x_1 = x_{N+1} = 0$. Then either longitudinal or transverse vibrations can exist for which there are 1, 2, 3 … or $N$ half wavelengths in the distance $Na$. Corresponding to these allowed vibrational modes, there are $N$ linearly spaced values of wave vector:

$$k = \frac{\pi}{Na}, \frac{2\pi}{Na}, \frac{3\pi}{Na}, \ldots \text{ or } \frac{\pi}{a}.$$

We refer $D(k)$ as the *density of states* with respect to $k$ and define it as the number of states per unit length per unit interval of $k$

$$D(k) = \frac{N}{(Na)(\pi/a)} = \frac{1}{\pi}.$$

Then the number of states in the range $k$ to $k + dk$ is given by

$$D(k)\, dk = \frac{dk}{\pi}. \qquad \text{...(9)}$$

In using DOS for calculation of physical characteristics such as specific heat (by summing over all normal modes), we require the DOS per unit frequency range $D(\omega)$, if $dn$ is the number of states covering a range $dk$, centred at $k$ and a range $d\omega$ of $\omega$, we have

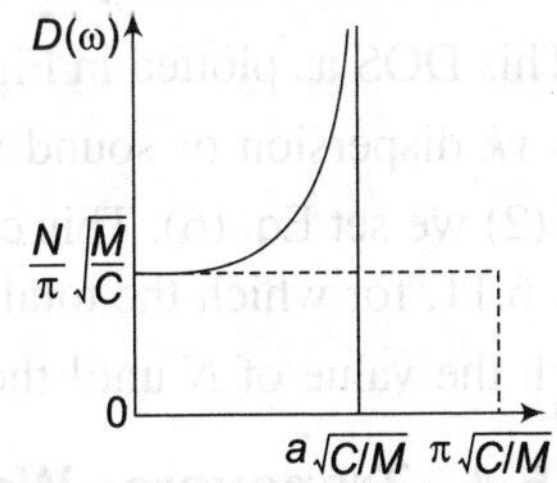


**Figure 6.11** Density of states for a monatomic chain

$$dn = D(k)\, dk = D(\omega)\, d\omega$$

$$\therefore \qquad D(\omega) = D(k)\frac{dk}{d\omega}$$

or

$$D(\omega) = \frac{D(k)}{(d\omega/dk)} = \frac{1}{\pi(d\omega/dk)} \qquad \text{...(10)}$$

Using Eq. (5)

$$D(\omega) = \frac{1}{\pi a}\left(\frac{M}{C}\right)^{1/2}\frac{1}{\cos(ka/2)} \qquad \text{...(11)}$$

$$= \frac{1}{\pi a}\left(\frac{M}{C}\right)^{1/2}\left[\frac{1}{\left(1-\sin^2\frac{ka}{2}\right)^{1/2}}\right]$$

$$= \frac{1}{\pi a}\left(\frac{M}{C}\right)^{1/2}\left[\frac{1}{\left(1-\frac{\omega^2 M}{4C}\right)}\right]^{1/2}$$

$$= \frac{1}{\pi a}\left(\frac{M}{C}\right)^{1/2}\left[\frac{4C}{4C-\omega^2 M}\right]^{1/2}$$

$$= \frac{2}{\pi a}\left(\frac{4C}{M}-\omega^2\right)^{-1/2} \quad ...(12)$$

$$= \frac{2}{\pi a}\frac{1}{\sqrt{\omega_m^2-\omega^2}} \quad ...(13)$$

If $\omega = 0,\ D(\omega) = \dfrac{2}{\pi a}\dfrac{1}{\sqrt{4C/M}}$

or $D(\omega) = \dfrac{1}{\pi a}\sqrt{M/C}$ ...(14)

As the zone boundary is approached $\omega \to \omega_m$ and $D(\omega) \to \infty$.

This DOS as plotted in Fig. 6.11. If we ignore the dispersion of round waves, we can write $\omega = vk$ dispersion of sound waves, we can write $\omega = vk$ or $d\omega/dk = v = a\sqrt{C/M}$ and from Eq. (2) we set Eq. (6). This corresponds to the constant DOS shown by the broken like in the Fig. 6.11, for which the total number of states does not reach the value of $N$ until the frequency $\omega = \pi\sqrt{C/M}$.

## 6.15.4 Transverse Waves

If the atoms can vibrate in a plane perpendicular to their line, then transverse modes are possible. For small displacements, it can readily be shown that the equations of motion are very similar to those for longitudinal motion (Eq. (3)). There is, of course, a new force constant. All the analysis of the longitudinal motion will apply and for a linear lattice with cylindrical symmetry, there will be two dispersion curves (Fig. 6.12).

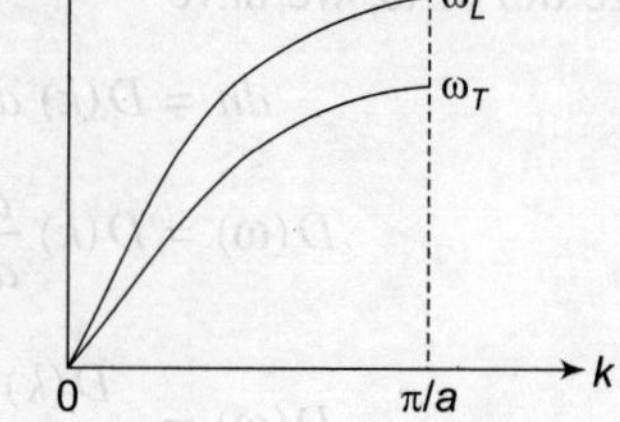


**Figure 6.12** Transverse and longitudinal waves in a one-dimensional monatomic lattice with $\omega_T$ and $\omega_L$ are the corresponding cut-off frequencies

## 6.16 VIBRATIONS OF A LINEAR DIATOMIC CHAIN OF ATOMS

Earlier we considered vibration of a linear monatomic chain of atoms. This was an example of a simple Bravais lattice which could be generated by translation of a unit cell containing two particles.

However, now we consider a Bravais lattice with a basis of two particles, i.e., a linear diatomic chain of atoms (Fig. 6.13). In this case, it is assumed that spacing of the atoms is uniform and alternate atoms have different masses, $M$ and $m$.

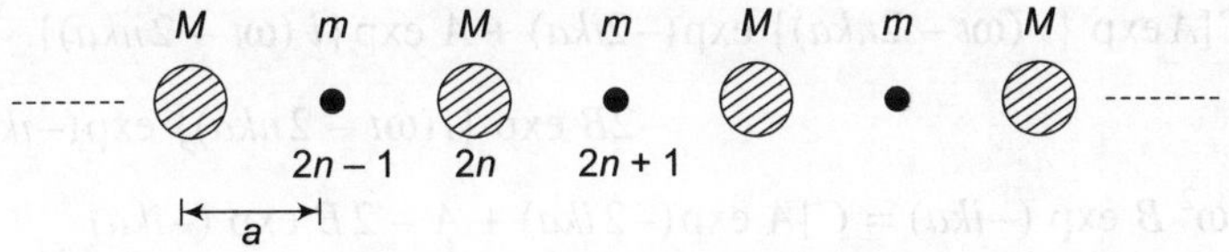


**Figure 6.13** A linear diatomic chain with a uniform spacing and alternate atoms with different masses

In the present case, the nearest neighbour distance (which is constant) is $a$ and the particles are numbered in such a way that even numbers correspond to mass $M$ and odd ones to $m$. Without loss of generality, we assume $M > m$. The equations of motion of two neighbouring atoms can be written as

$$m \frac{d^2 x_{2n+1}}{dt^2} = C[x_{2n+2} - x_{2n+1}] - C[x_{2n+1} - x_{2n}]$$

$$= C[x_{2n+2} + x_{2n} - 2x_{2n+1}] \quad \text{...(1a)}$$

$$M \frac{d^2 x_{2n}}{dt^2} = C[x_{2n+1} - x_{2n}] - C[x_{2n} - x_{2n-1}]$$

$$= C[x_{2n+1} + x_{2n-1} - 2x_{2n}] \quad \text{...(1b)}$$

where $C$ is the force constant

$$x_{2n+1} = B \exp [i\{\omega t - (2n + 1)ka\}] \quad \text{...(2a)}$$

$$x_{2n} = A \exp [i(\omega t - 2nka)] \quad \text{...(2b)}$$

$A$ and $B$ are the amplitudes

$$\therefore \quad x_{2n+2} = A \exp[i\{\omega t - (2n + 2)ka\}]$$

$$= A \exp[i(\omega t - 2nka)] \exp(-2ika)$$

$$= x_{2n} \exp(-2ika) \quad \text{...(2c)}$$

$$x_{2n-1} = B \exp[i\{\omega t - (2n - 1)ka\}]$$

$$= B \exp[i\{\omega t - (2n + 1)ka\}] \exp(2ika) \quad \text{...(2d)}$$

Further

$$\left.\begin{aligned}\frac{d^2x_{2n+1}}{dt^2} &= -\omega^2 x_{2n+1}\\ \frac{d^2x_{2n}}{dt^2} &= -\omega^2 x_{2n}\end{aligned}\right\} \qquad \text{...(3a)}$$

Substituting for respective values in Eq. (1), we get $-m\omega^2 B \exp[i\{\omega t - (2n+1)ka\}]$

$$= C[A\exp\{i(\omega t - 2nka)\}\exp(-2ika) + A\exp\{i(\omega t - 2nka)\}$$

$$-2B\exp\{i(\omega t - 2nka)\}\exp(-ika)]$$

or $\quad -m\omega^2 B\exp(-ika) = C[A\exp(-2ika) + A - 2B\exp(-ika)$

(because $\exp[i(\omega t - 2nka)]$ cancels out on both sides.)

or $\quad -m\omega^2 B = C[A\exp(-ika) + A\exp(ika) - 2B]$

or $\quad -m\omega^2 B = CA(2\cos ka) - 2BC] \qquad$ ...(3b)

$$\left(\because \cos ka = \frac{e^{-ika} + e^{ika}}{2}\right)$$

or $$(2C - m\omega^2)B - (2C\cos ka)A = 0 \qquad \text{...(4a)}$$

Similarly, from Eq. (1b)

$$(-2C\cos ka)B + (2C - M\omega^2)A = 0 \qquad \text{...(4b)}$$

Equations (4a) and (4b) have a non-trivial silution (for $A$ and $B$), only if the determinant of the coefficients vanishes, i.e.,

$$\begin{vmatrix}(2C - M\omega^2) & -2C\cos ka\\ -2C\cos ka & (2C - m\omega^2)\end{vmatrix} = 0 \qquad \text{...(5)}$$

$\therefore \quad (2C - M\omega^2)(2C - m\omega^2) - 4C^2\cos^2 ka = 0$

$Mm\omega^4 - 2C(M+m)\omega^2 + 4C^2(1 - \cos^2 ka) = 0$

or $\quad Mm\omega^4 - 2C(M+m)\omega^2 + 4C^2(\sin^2 ka) = 0$

or $$\omega^4 - \left[\frac{2C(M+m)}{Mm}\right]\omega^2 + \frac{4C^2\sin^2 ka}{Mm} = 0 \qquad \text{...(6)}$$

This is an equation quadratic in $\omega^2$. The roots are given by

$$\omega^2 = \frac{1}{2}\left\{\frac{2C(M+m)}{Mm} \pm \sqrt{\frac{4C^2(M+m)^2}{(Mm)^2} - \frac{4\times 4C^2\sin^2 ka}{Mm}}\right\}$$

or the two possible solutions are

$$\omega^2 = C\left\{\frac{M+m}{Mm}\right\} \pm C\left[\left(\frac{M+m}{Mm}\right)^2 - \frac{4\sin^2 ka}{Mm}\right]^{1/2} \qquad ...(7)$$

This is the dispersion relation for the linear chain of diatomic lattice.

### 6.16.1 Acoustic and Optic Branches

From Eq. (7) for a single value of *k*, we have two angular frequencies $\omega_+$ and $\omega_-$. A plot of $\omega$ vs. *k* is given in Fig. 6.14. Thus, we observe two branches, one corresponding to $\omega_+$ and other for $\omega_-$. The curve corresponding to $\omega_L$ is called the *acoustical branch* and that corresponding to $\omega_+$ is called the *optical branch.*

For small *k* (i.e., $\to 0$), the sine term can be neglected. For the root $\omega_+$ we have

$$\omega_+^2 = 2C\left(\frac{M+m}{Mm}\right)$$

$$\omega_+ = \left[2C\left(\frac{1}{M}+\frac{1}{m}\right)\right]^{1/2} \equiv \omega \text{ (optical branch, for } k = 0) \qquad ...(8)$$

Thus, optical branch corresponds to a non-zero value of $\omega (\equiv \omega_1)$ for $k = 0$. It is called *optical branch* because the frequencies in this branch are of the order of infrared frequencies.

For $\qquad k = 0, \omega_- = 0.$

Thus, the other branch that corresponds to $\omega = 0$ ($\omega = \omega_- = 0$) for $k = 0$ is called the acoustic branch, because the corresponding frequencies are of the same order as acoustical or supersonic vibrations.

The *acoustic branch* is mathematically defined as

$$\lim_{k\to 0} (\omega_-) \to 0 \qquad ...(9)$$

Now

$$\omega_-^2 = C\left(\frac{1}{M}+\frac{1}{m}\right) - C\left[\left(\frac{1}{M}+\frac{1}{m}\right)^2 - \frac{4k^2a^2}{Mm}\right]^{1/2}.$$

$$= C\left(\frac{1}{M}+\frac{1}{m}\right) - C\left(\frac{1}{M}+\frac{1}{m}\right)\left[1 - \frac{4k^2a^2}{Mm}\frac{1}{\left(\frac{1}{M}+\frac{1}{m}\right)^2}\right]^{1/2}$$

$$= \left(\frac{1}{M}+\frac{1}{m}\right) - C\left(\frac{1}{M}+\frac{1}{m}\right)\left[1 - \frac{4k^2a^2\, Mm}{(M+m)^2}\right]^{1/2}$$

$$\cong C\left(\frac{1}{M}+\frac{1}{m}\right)\left\{1-\left[1-\frac{2k^2a^2\,Mm}{(M+m)^2}\right]\right\}$$

$$= C\left(\frac{1}{M}+\frac{1}{m}\right)\frac{2k^2a^2\,Mm}{(M+m)^2} \quad ...(10)$$

$$\therefore \qquad \omega_- \approx \sqrt{\frac{2Ca^2}{(M+m)}}\,k \quad \text{for the acoustic branch} \quad ...(11)$$

A simple limiting case corresponds to the vibration of the lattice for $ka \simeq \pi/2$ (or $|\vec{k}| \simeq \pi/2a$), i.e., vibration for wave vectors close to the Brillouin zone.

$$\omega^2 = C\left[\left(\frac{1}{M}+\frac{1}{m}\right) \pm \sqrt{\left(\frac{1}{M}+\frac{1}{m}\right)^2 - \frac{4}{Mm}\sin^2\left(\frac{\pi}{2}\right)}\right]$$

$$= C\left[\left(\frac{1}{M}+\frac{1}{m}\right) \pm \left(\frac{1}{m}-\frac{1}{M}\right)\right] \quad ...(12)$$

Thus, the optical branch terminates at $\omega = \sqrt{2C/M} \equiv \omega_2$ and the acoustical branch terminates at $\omega = \sqrt{2C/M} \equiv \omega_3$ (Fig. 6.14). Since $M > m$, waves corresponding to frequencies $\omega_3 < \omega < \omega_2$ and

$$\omega > \omega_1 \equiv \sqrt{2C\left(\frac{M+m}{Mm}\right)} \quad ...(13)$$

are not propagated through the lattice.

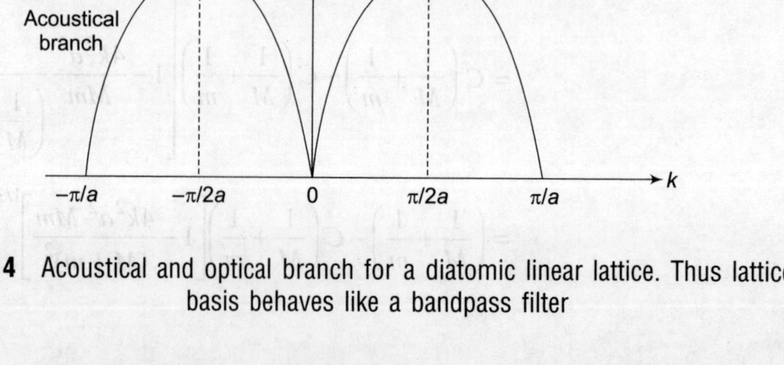


**Figure 6.14** Acoustical and optical branch for a diatomic linear lattice. Thus lattice with a basis behaves like a bandpass filter

## 6.16.2 Particle Displacements in the Two Branches

There is an important physical difference between the two branches, which further becomes clear by calculating the ratio of the amplitudes in the two branches.

For $k = 0$ ($\cos ka = 1$) from Eq. (4a) we have for the optical branch,

$$-M\omega^2 A = +2CB - 2CA \quad \text{or} \quad \frac{\omega^2}{2C} = \frac{A-B}{MA}$$

$$-\omega^2 mB = 2CA - 2CB - \frac{\omega^2}{2C} = \frac{A-B}{mB}$$

$$\frac{A-B}{mB} = -\left(\frac{A-B}{MA}\right)\frac{A}{B} = -\frac{m}{M} \qquad \text{...(14)}$$

For the acoustic branch,

$\cos(ka) \simeq 1 - \dfrac{k^2a^2}{2}$ (the second-order term in the expansion cannot be neglected)

Then, from Eq. (14), we have

$$-\omega^2 MA = 2CB\left(1 - \frac{k^2a^2}{2}\right) - 2CA$$

$$-\omega^2 mB = 2CA\left(1 - \frac{k^2a^2}{2}\right) - 2CB$$

Adding these two equations, we have

$$\omega^2 (MA + mB) = -k^2a^2C\,(A + B)$$

Substituting the value of $\omega^2$ as $2Ck^2a^2/(M + m)$ for this acoustical branch, we get

$$\frac{2\,Ck^2a^2}{(M + m)}(MA + mB) = -k^2\,a^2\,C\,(A + B)$$

$$\Rightarrow \qquad \frac{A}{B} = +1 \qquad \text{...(15)}$$

Thus, we see that in optical branch the atoms vibrate in opposite and their amplitudes are inversely in the ratio of their masses, in such a way that their centres of mass remain fixed during the period of motion. For other values of $k$, the ratio $A/B$ can be calculated and the results are shown in Fig. 6.15. The existence of optical type vibrations does not necessarily depend on mass difference or even on oppositely charged ion. It rather depends only on two atoms per primitive cell so that they can vibrate in and out from the centre of mass. It is also seen from Eq. (15) that in the acoustical branch,

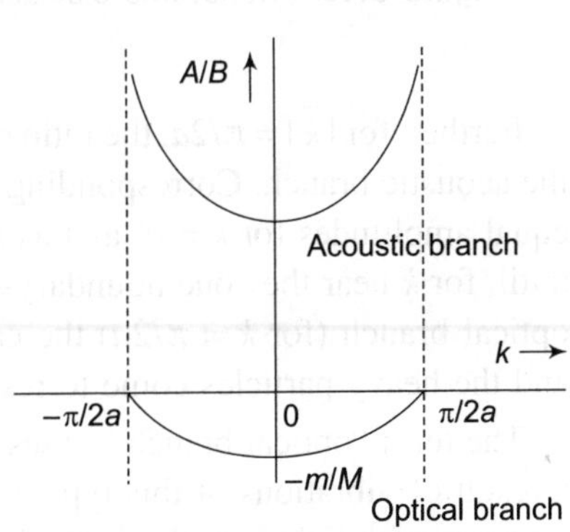


**Figure 6.15** The ratio of the amplitudes $A/B$ for the optical and acoustic branches as a function of $k$

the atoms and their centres of mass move in the same direction with the same amplitude as in long wavelength acoustic vibrations.

We have seen that for $k = 0$, the amplitude ratios are $-m/M$ and $+1$ respectively for the optical and acoustic branches. Let us first consider the acoustic branch. Here, in the limit $k \to 0$, the particles $m$ and $M$ vibrate with equal amplitude and zero phase difference, i.e., all the particles are displaced by the same amount in the same direction. Consequently, the lattice is displaced as a whole (infinite wavelength and zero frequency [Fig. 6.16 (a)]. Corresponding to the optical branch, for $k \to 0$, the particles $m$ and $M$ vibrate out of phase by 180° and the displacements of all the particles of a particular type are the same at all times.

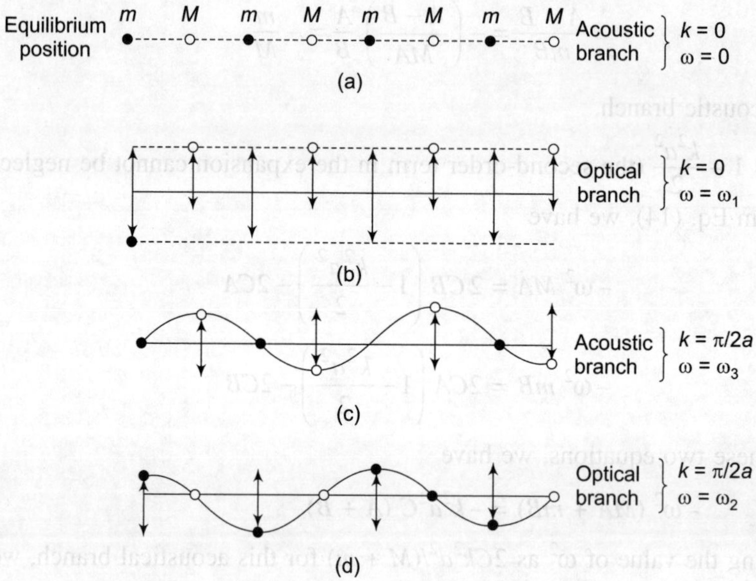


**Figure 6.16** Transverse displacements of the diatomic chain for the acoustic and optic branches

Further, for $|\vec{k}| = \pi/2a$, the ratio of $A/B$ goes to zero for the optical branch and to infinity for the acoustic branch. Corresponding to the acoustic branch, the particles vibrate in phase with equal amplitudes for $k = 0$; and as $k$ increases, the amplitude of the light particles decreases until, for $k$ near the zone boundary, it goes to zero and only heavy particles oscillate. For the optical branch (for $k = \pi/2a$) the case is just reversed, i.e., only the light particles oscillate and the heavy particles come to rest.

The term "optical branch" arises because an oscillating electric dipole moment appears as a result of vibrations of this type in ionic crystals. The dipole moment of the lattice actively interacts with light (usually in the infrared region of the spectrum and hence the term "optical branch". The energy of optical lattice vibrations is of the same order as the vibrational energy of nuclei in a diatomic molecule, i.e., ~0.1 eV. Accordingly, the frequency gap for exciting optical lattice vibrations is $\Delta v = 2 \times 10^{13}$ Hz (infrared region).

The diatomic linear chain model is a popular tool for studying vibrations in ionic crystals. If an optical mode is present at $k = 0$, then the positive ions within the unit cell move in the opposite direction to the negative ones, i.e., an oscillating dipole moment is produced. This can couple to an external oscillating electric field (e.g., infrared radiations). Thus, such vibrations are "infrared active", i.e., they cause strong infrared light absorption and resonance takes place when the frequency of radiation matches the frequency of transverse optical mode of the crystal. This happens for a range of frequencies called *"rest strah len band"*.

## 6.17 LOCAL PHONON MODES

When certain impurities are introduced in a crystal, then due to the loss of translational periodicity various types of modes are associated with imperfections.

Consider, for example, a KCl crystal in which one heavy $Cl^-$ ion is replaced by a lighter $H^-$ ion, as shown in Fig. 6.17.

This impurity atom is called a U-centre. This hydrogen atom is associated with high frequency local phonon modes and vibrates between the heavy $K^+$ ion Since the impurity atom ($H^-$ ion) is lighter than the other surrounding atoms of the crystal, it has a higher natural frequency. As a result of this vibration, the crystal lattice near $H^-$ ion is slightly deformed during motion, but the amplitude of deformation decreases rapidly with increasing distance from the $H^-$ ion. Such type of vibration is known as *localised phonon.* Schaefer calculated the optical absorption due to localised phonons in case of KCl crystal and the results obtained have been shown is Fig. 6.18.

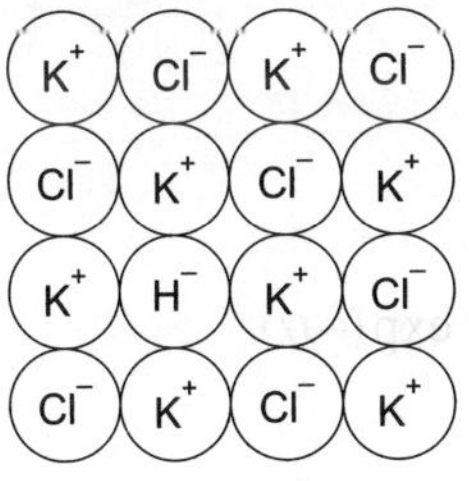


**Figure 6.17** Lighter $H^-$ ions in place of heavier $Cl^-$ ions in KCl crystal

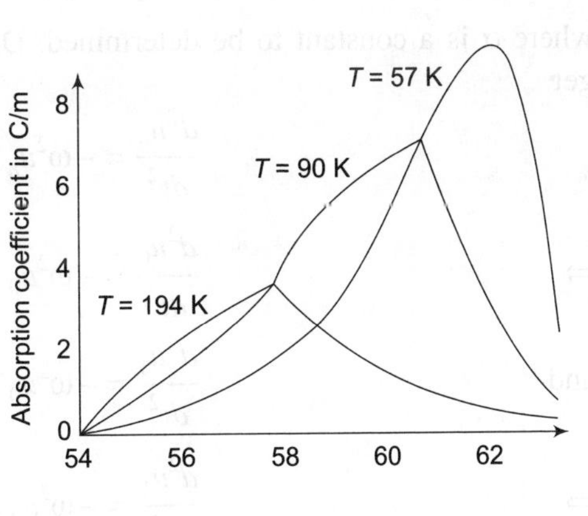


**Figure 6.18** Optical absorption for KCl at different temperatures

The problem of local phonons may be analyzed mathematically by considering a linear lattice of atoms, all of equal mass $M$ except the one impurity atom of mass $m$ such that $m < M$, as shown in Fig. 6.19. We are now going to show that one of the normal modes of the lattice is localized around the impurity atom of mass $m$ and that the corresponding frequency exceeds the maximum frequency, $\omega_{max}$, of the unperturbed lattice.

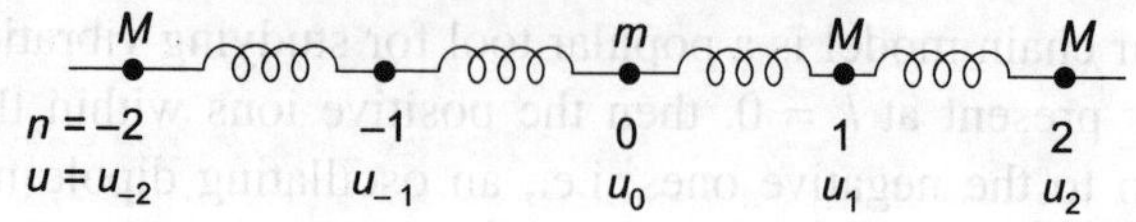


**Figure 6.19** KCl lattice with one impurity atom of mass $m$

Assuming only the nearest neighbour interactions and that the light, impurity atom is at $n = 0$, the equations of motions for the lattice may be written as

$$m\frac{d^2u_0}{dt^2} = \beta\,[u_1 + u_{-1} - 2u_0] \quad ...(1)$$

and
$$M\frac{d^2u_1}{dt^2} = \beta\,[u_2 + u_0 - 2u_1] \quad ...(2)$$

where $\beta$ is the force constant.

Let the trial solution of the above equations be of the exponentially damped form as we move away from the impurity atom at $n = 0$ and which in the limit $m \to M$, approaches the form of the normal mode of the perfect lattice corresponding to the highest frequency $\omega_{max}$.

Let the trial solution of the perturbed lattice be represented by

$$u_n = u_0\,(-1)^n \exp\,(-i\omega t)\exp\,(-|n|\,\alpha) \quad ...(3)$$

where $\alpha$ is a constant to be determined. Differentiating Eq. (3) twice with respect to $t$, we get

$$\frac{d^2u_n}{dt^2} = -\omega^2 u_n = -\omega^2 u_0\,(-1)^n \exp(-i\omega t)\exp(-|n|\,\alpha)$$

$$\Rightarrow \quad \frac{d^2u_0}{dt^2} = -\omega^2 u_0 \exp\,(-i\omega t) \quad ...(4)$$

and
$$\frac{d^2u_1}{dt^2} = -\omega^2 u_0\,(-1)^1 \exp(-i\omega t)\exp(-\alpha)$$

$$\Rightarrow \quad \frac{d^2u_1}{dt^2} = -\omega^2 u_0 \exp(-i\omega t)\exp(-\alpha) \quad ...(5)$$

Also, from Eq. (3)
$$u_1 = u_0\,(-1)\exp(-i\omega t)\exp(-\alpha)$$

$$\Rightarrow \quad u_1 = -u_0 \exp(-i\omega t)\exp(-\alpha) \quad ...(6)$$

$$u_2 = u_0\,(-1)^2 \exp(-i\omega t)\exp(-2\alpha)$$

$$\Rightarrow \quad u_2 = -u_0 \exp(-i\omega t)\exp(-2\alpha) \quad ...(7)$$

$$u_{-1} = u_0\,(-1)^{-1} \exp(-i\omega t)\exp(-\alpha)$$

$$\Rightarrow \qquad u_{-1} = -u_0 \exp(-i\omega t) \exp(-\alpha) \qquad ...(8)$$

$$u_0 = u_0 (-1)^0 \exp(-i\omega t) \exp(0)$$

$$u_0 = u_0 \exp(-i\omega t) \qquad ...(9)$$

Substituting the respective values in Eq. (1), we get

$$m(-\omega)^2 u_0 \exp(-i\omega t) = \beta[-\exp(-\alpha) - \exp(-\alpha) - 2]\, u_0 \exp(-i\omega t)$$

$$\Rightarrow \qquad m\omega^2 = 2\beta\,(1 + \exp(-\alpha)]$$

$$\Rightarrow \qquad \omega^2 = \frac{2\beta}{m}\,[1 + \exp(-\alpha)] \qquad ...(10)$$

Similarly, substituting the respective values in Eq. (2), we get

$$M\omega^2 u_0 \exp(-i\omega t) \exp(-\alpha) = \beta[\exp(-2\alpha) + 1 + 2\exp(-\alpha)]\, u_0 \exp(i\omega t)$$

$$\Rightarrow \qquad M\omega^2 \exp(-\alpha) = \beta[\exp(-2\alpha) + 1 + 2\exp(-\alpha)]$$

$$\Rightarrow \qquad M\omega^2 = \beta[\exp(-\alpha) + \exp(\alpha) + 2]$$

$$\Rightarrow \qquad \omega^2 = \frac{\beta}{M}\,[\exp(-\alpha) + \exp(\alpha) + 2] \qquad ...(11)$$

Equations (10) and (11) will be consistent only if

$$\frac{2\beta}{m}\,[1 + \exp(-\alpha)] = \frac{\beta}{M}\,[\exp(-a) + \exp(a) + 2]$$

$$\Rightarrow \qquad \frac{2M}{m} = \frac{[\exp(-\alpha) + \exp(\alpha) + 2]}{[1 + \exp(-\alpha)]}$$

Subtracting 1 from both sides, the above equation gives

$$\frac{2M - m}{m} = \frac{1 + \exp\alpha}{1 + \exp(-\alpha)}$$

$$\Rightarrow \qquad (2M - m)\,[1 + \exp(-\alpha)] = [1 + \exp\alpha]m$$

$$\Rightarrow \qquad \frac{2M - m}{\exp\alpha}\,[1 + \exp\alpha] = m[1 + \exp\alpha]$$

$$\Rightarrow \qquad \frac{1}{\exp\alpha} = \frac{m}{2M - m}$$

$$\Rightarrow \qquad \exp(-\alpha) = \frac{m}{2M - m} \qquad ...(12)$$

Substituting for $\exp(-\alpha)$ from Eq. (12) in Eq. (10), we get

$$\omega^2 = \frac{2\beta}{m}\left[1 + \frac{m}{2M - m}\right]$$

$$\Rightarrow \qquad \omega^2 = \frac{2\beta}{m}\left[\frac{2M}{2M-m}\right] = \frac{4\beta}{m}\left[\frac{M}{2M-m}\right]$$

$$\Rightarrow \qquad \omega^2 = \frac{4\beta}{M}\left[\frac{M^2}{2Mm-m^2}\right]$$

$$\Rightarrow \qquad \omega^2 = \omega_{max}^2\left(\frac{M^2}{2Mm-m^2}\right) \qquad ...(13)$$

where

$$\omega_{max}^2 = 2\sqrt{\frac{\beta}{M}} \qquad ...(14)$$

represents the cut-off frequency of the unperturbed lattice for which $M = m$. For $m \ll M$, Eq. (13) gives

$$\omega^2 = \omega_{max}^2\left(\frac{M}{2m}\right) \qquad ...(15)$$

Equation (15) shows that the upper frequency limit of the perturbed lattice is greater than that of the unperturbed lattice.

It has been observed that the presence of substitutional impurity atoms brings about the following effects:

1. The frequencies of all the normal vibrational modes vary.
2. A few vibrational states appear within the forbidden frequency range, which lie above the acoustical branch and below the optical branch.

If the mass of the impurity atoms is equal to that of the original atom at the lattice, localized vibrational states are observed. In general, we may say that the impurities will differ from the host atom both in mass and in charge.

## 6.18 PHONON MOMENTUM

A phonon on a lattice does not have momentum but in real practice, when a phonon of wave-vector $\vec{K}$ interacts with other particles and fields it appears that it has a momentum $\hbar\vec{K}$. In crystals, wave vector and momenta follow the conservation laws.

Suppose $\vec{k}$ is a wave vector of a photon at a place. Now, if after scattering with a phonon, $\vec{k}$ changes to $\vec{k}'$ and one phonon of wave vector $\vec{k}$ is absorbed. Then, from the law of conservation of momentum

$$\hbar\vec{k}' = \hbar\vec{k} + \hbar\vec{K}. \qquad ...(1)$$

(see Fig. 6.20).

or

$$\hbar(\vec{k}' - \vec{k}) = \hbar\vec{K} \qquad ...(1a)$$

when a X-ray photon is scattered with a crystal then the Bragg's diffraction follows the following conservation law of momentum:

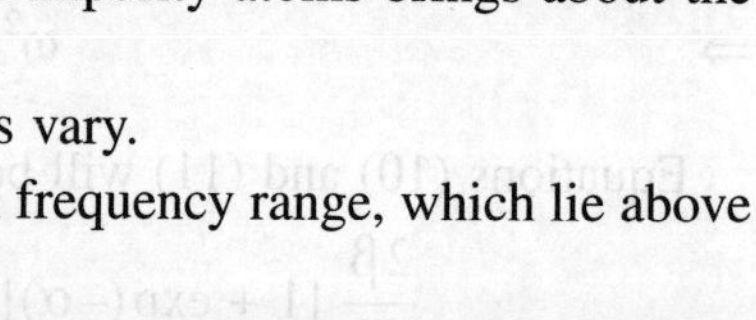


**Figure 6.20** Wave vectors

$$\hbar \vec{k}' = \hbar \vec{k} + \hbar \vec{G}$$

i.e., $$\vec{k}' = \vec{k} + \vec{G} \qquad ...(2)$$

where $\vec{G}$ is the reciprocal lattice vector. In the reflection process, the whole crystal recoils with momentum $-\hbar G$ and the above expression becomes

$$\vec{k}' = \vec{k} - \vec{G} \qquad ...(3)$$

But this happens rarely.

If the scattering of photon is inelastic and a phonon of wave vector $\vec{K}$ is absorbed in the process, the conservation law becomes

$$\vec{k}' = \vec{k} + \vec{G} + \vec{K} \qquad ...(4)$$

But if a phonon of wave vector $\vec{K}$ is created in the process, then the conservation law is

$$\vec{k}' + \vec{K} = \vec{k} + \vec{G} \qquad ...(5)$$

### 6.18.1 Phonon Processes

In elastic collision, the frequency of the radiation in unchanged and scattering is governed by the wave vector conservation law

$$\vec{k}' = \vec{k} + \vec{G}$$

In inelastic collision, there is change of energy and momentum corresponding to the absorption or creation of phonons and the scattering is governed by different conservation laws (Eqs. (4) and (5)).

In inelastic scatterings, the phonon collisions with different scatters depend on the number of phonons created or absorbed during the process.

That process in which one, two or three phonons take part or destroy in an interaction is called one-, two- or three-phonon process, i.e., depending on the number of phonons, which take part or destroy during the process.

If many phonons take part, it is called a multi-phonon process.

For various transport properties, it necessitates to establish a true equilibrium distribution of phonons. The collisions of a phonon with static imperfection or a crystal boundary do not establish thermal equilibrium themselves due to the reason of no change in the energy of individual phonons. In such case

$$\omega_1 = \omega_2 \qquad ...(6)$$

where $\omega_1$ is the frequency of incident phonon and $\omega_2$, that of the scattered phonon.

Let us consider a three-phonon collision process

$$\vec{K}_1 + \vec{K}_2 = \vec{K}_3 \qquad ...(7)$$

The above process does not establish thermal equilibrium and the total momentum of the phonon gas is not changed by such a collision. As a result, the whole distribution of phonons can move down the crystal with a drift velocity. For such collisions, the phonon momentum $\vec{J}$ is conserved and it may be described by

$$\vec{J} = \sum_n \hbar \vec{K}_n \qquad ...(8)$$

where $n$ is the number of phonons.

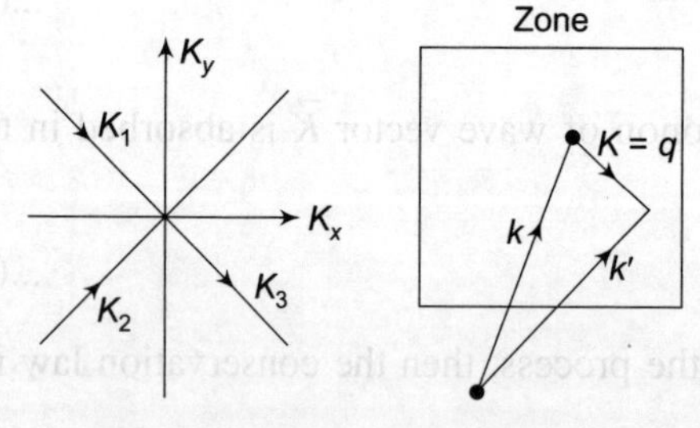


**Figure 6.21** Normal process

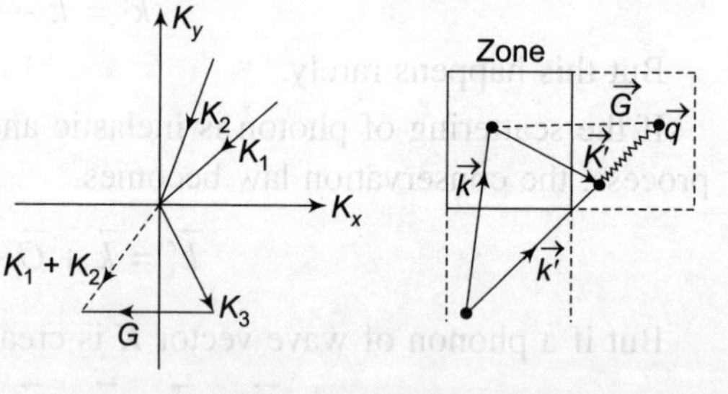


**Figure 6.22** Umklapp process

During collision, the change in $\vec{J}$ is

$$\vec{K}_3 - \vec{K}_2 - \vec{K}_1 = 0$$

But in three-phonon collision process

$$\vec{J} \neq 0$$

i.e., there is no change in phonon momentum, The role of phonon momentum during scattering gives rise to two types of processes:

(i) Normal process (N-process)

or (ii) Umklapp process (U-process).

In N-process, there is no change of phonon momentum and during scattering

$$\vec{K}_1 + \vec{K}_2 = \vec{K}_3$$

In this process no reciprocal lattice vector is involved. This process corresponds to repeated zone scheme.

In case of U-process, there is change of phonon momentum and

$$\vec{K}_1 + \vec{K}_2 = \vec{K}_3 + \vec{G}.$$

i.e., a reciprocal lattice vector is involved. This process corresponds to reduced zone scheme. Figure 6.22 describes this process:

$$\vec{k}' - \vec{k} = \vec{K}$$

$$\vec{q} = \vec{K} - \vec{G} = \vec{k}' - \vec{k} - \vec{G}$$

where $\vec{q}$ is phonon wave vector which is constrained to lie in a unit cell of the lattice. It can be said that $\vec{q}$ is the value of $\vec{K}$, reduced to the first Brillouin zone.

For both the processes, the energy is conserved, i.e.,

$$\omega_1 + \omega_2 = \omega_3$$

where $\omega_1$, $\omega_2$, $\omega_3$ are the corresponding frequencies of the wave vectors $\vec{K}_1$, $\vec{K}_2$, $\vec{K}_3$ respectively.

## 6.19 NEUTRON-PHONON INTERACTION (INELASTIC SCATTERING CASE)

In case of neutron scattering, the neutrons are scattered by nuclear forces. This type of scattering may be coherent (elastic) or incoherent (inelastic). The scattering of neutron beam with a crystal lattice may be described by the conservation of wave vector relation

$$\mathbf{k}' = \mathbf{k} + \mathbf{G} \pm \mathbf{q}, \qquad ...(1)$$

where $\mathbf{k}$ is the wave vector of incident neutron and $\mathbf{k}'$ of the scattered neutron. $+\mathbf{q}$ is the wave vector of the phonon created or $-\mathbf{q}$ of the phonon absorbed. The law of conservation of energy may be described by

$$\frac{\hbar^2 k^2}{2M_n} = \frac{\hbar^2 k'^2}{2M_n} \pm \hbar\omega_q$$

or

$$\mathbf{k}^2 - \mathbf{k}'^2 = \pm \frac{2M_n}{\hbar}\,\omega_q\,(\mathbf{k}' - \mathbf{k}), \qquad ...(2)$$

where $+\omega_q$ is the frequency of created phonon and $-\omega_q$ of the absorbed phonon.

When cold or thermal neutrons are scattered by a crystal lattice, the de Broglie wavelength becomes large and the same neutron can interact with many phonons. The scattered waves interfere with each other and the interference is maximum for the nuclei with isotopes.

The neutron scattering technique is an ideal method for determining the phonon spectrum. If the absorption of neutrons by the crystal nuclei is high, this technique cannot be used. The neutron scattering experiments have an advantage of measuring directly the energy shift.

## 6.20 INELASTIC SCATTERING OF PHOTONS BY PHONONS

A crystal can be considered to be a continuum of refractive index, $\mu$. If a photon of frequency $\upsilon = \omega/2\pi$ propagates in a crystal then its wave-vector $k$ is given by

$$\left.\begin{aligned} \omega &= \frac{ck}{\mu} \\ \text{or} \quad \lambda\upsilon &= v_p = \frac{c}{\mu} \end{aligned}\right\} \qquad ...(1)$$

where $c$ is the velocity of light.

Now, let us suppose that such a photon interacts with a phonon beam i.e. a sound wave in the crystal, and gets scattered by the wave. This can take place because the refractive index of the crystal changes due to a change in the concentration of atoms brought about by the elastic strain field of the sound wave. In this way, the optical properties of the medium are also changed by the sound waves. The converse is also true, i.e., the elastic field of light waves sets up a periodic mechanical strain in the medium thereby changing the elastic properties of the medium.

In an inelastic scattering by a crystal, a photon can either create or absorb a phonon. The photon is scattered in the process and let the photon wave vector change from $\vec{k}$ to $\vec{k}'$ and its frequency change from ω to ω′. Let a phonon be created having wave vector $\vec{K}$ and angular frequency Ω. Then energy conservation principle gives

$$\hbar\omega = \hbar\omega' + \hbar\Omega \qquad ...(2)$$

or from momentum conservation

$$\hbar\vec{k} = \hbar\vec{k}' + \hbar\vec{K} \qquad ...(3)$$

This scattering is shown in Fig. 6.23.

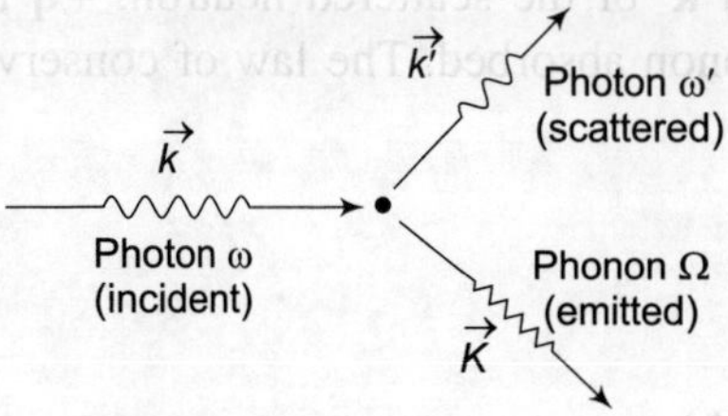


**Figure 6.23** Inelastic scattering

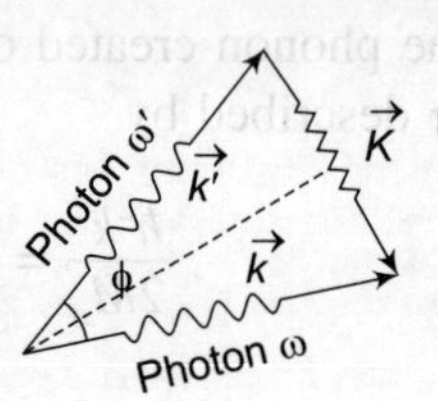


**Figure 6.24** Momentum conservation (Eq. 3)

If the velocity of sound $v_s$ is constant and because

$$v_s = \frac{\lambda\Omega}{2\pi} \qquad ...(4)$$

∴ We have $\Omega = v_k K$. The emitted phonon can have only a fraction of the energy of the incident photon and since $v_s \ll c$, if $\vec{K}$ is comparable in magnitude with $\vec{k}$, we have $v_s K \ll ck$ or, in other words $\Omega \ll \omega$. Showing that we have long wavelength phonons. From Eq. (2), it follows that $\omega' = \omega$ and $k' \approx k$ the vector triangle becomes isosceles with a base

$$K = 2k \sin \phi/2. \qquad ...(5)$$

Putting the value of $k = \omega\mu/c$ from Eq. (1) in Eq. (4) and using Eq. (5), we get

$$v_s K \approx 2\, v_s \frac{\omega\mu}{c} \cdot \sin(\phi/2).$$

or $\Omega = \dfrac{2v_s\omega\mu}{c} \sin(\phi/2)$. This gives the frequency of the created phonon when photons are scattered.) inelastically at an angle φ from the incident direction.

## 6.21 PHOTONIC CRYSTALS AND PHONONIC CRYSTALS

Photonic crystals are artificial structures, i.e., periodic structures which can be prepared technologically to manipulate the flow of electromagnetic waves.

Similarly, phononic crystals are periodic structures which can be technologically prepared to control the flow of acoustic waves.

A photonic crystal is a periodic optical nanostructure that affects the motion of photons in much the same way that ionic lattices affect electrons in solids. Photonic crystals are periodic dielectric structures that have a band-gap that forbids propagation of a certain frequency range of light. This property enables one to control light with amazing facility and produce effects that are impossible with conventional optics.

The control and manipulation of acoustic/elastic waves is a fundamental problem with many potential applications, especially in the field of information and communication technologies. For instance, confinement, guiding, and filtering phenomena at the scale of the wavelength are useful for signal processing, advanced nanoscale sensors and acoustic optic on-chip devices.

## 6.22 DISPERSION CURVES [ω (k)] FOR A TWO-DIMENSIONAL SQUARE LATTICE

In the previous chapter, we have discussed special points and special lines for the first Brillouin zone of a two-dimensional square lattice. We will now see their importance in relation to the dispersion curves.

### *What is the importance of all such considerations?*

Along the special lines and at the special points there are some degeneracies that do not split no matter how strong the crystalline potential energy may be (an *essential degeneracy*). For the general points, there is no degeneracy. The general shape of the energy contours $E(\vec{k})$ are to a large extent determined by the splittings, degeneracies and shapes of the bands at the special points and lines. Also note (Fig. 6.25) that the bands need be calculated only in 1/8 of $k$-space, the rest can be determined by symmetry.

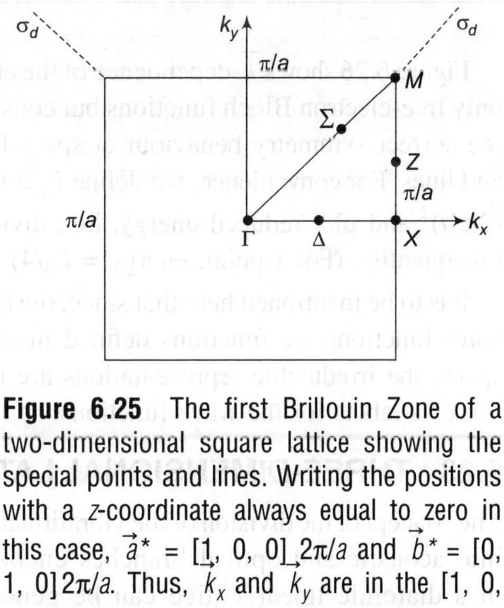


**Figure 6.25** The first Brillouin Zone of a two-dimensional square lattice showing the special points and lines. Writing the positions with a $z$-coordinate always equal to zero in this case, $\vec{a}^* = [1, 0, 0]\, 2\pi/a$ and $\vec{b}^* = [0, 1, 0]\, 2\pi/a$. Thus, $k_x$ and $k_y$ are in the [1, 0, 0] and [0, 1, 0] directions respectively. The $X$ and $M$ points are at [1, 0, 0] $\pi/a$ and [1, 1, 0] $\pi/a$ respectively

### *The group of k and their irreducible representations*

There is a group, in the mathematical sense, of symmetry operations that transforms each of thc special points or special lines into themselves. This group is called the group of $\vec{k}$, i.e., the group of Δ or the group of ∑ and so on. Associated with every group is a set of irreducible representations that precisely describes how all functions (defined in ordinary space) transform under all of the symmetry operations of the group (for example, there are eight symmetry operations in $C_{4v}$ and five irreducible representations). The different irreducible representations are labelled (i.e., given a symbol to distinguish one from the other). The conventional notation for the irreducible representations in the group of $\Delta$ or group of $\Sigma$ and so on, are subscripts, i.e., $\Delta_1$ and $\Delta_2$ or $\Sigma_1$ and $\Sigma_2$ and so on as in Table 6.1 which shows the appropriate character tables.

**Table 6.1** Character Tables for the Various Special Points and Lines of the Square Lattice as Defined in Fig. 6.25

| $\Gamma, M$ | $E$ | $e_2$ | $2C_4$ | $2\sigma_v$ | $2\sigma_d$ |
|---|---|---|---|---|---|
| $\Gamma_1, M_1$ | 1 | 1 | 1 | 1 | 1 |
| $\Gamma_2, M_2$ | 1 | 1 | 1 | –1 | –1 |
| $\Gamma_3, M_3$ | 1 | 1 | –1 | 1 | –1 |
| $\Gamma_4, M_4$ | 1 | 1 | –1 | –1 | 1 |
| $\Gamma_5, M_5$ | 2 | –2 | 0 | 0 | 0 |

| $X$ | $E$ | $C_2$ | $\sigma_y$ | $\sigma_x$ |
|---|---|---|---|---|
| $X_1$ | 1 | 1 | 1 | 1 |
| $X_2$ | 1 | 1 | –1 | –1 |
| $X_3$ | 1 | –1 | 1 | –1 |
| $X_4$ | 1 | –1 | –1 | 1 |

| $\Delta$ | $E$ | $\sigma_y$ |
|---|---|---|
| $\Sigma$ | $E$ | $\sigma_d$ |
| $Z$ | $E$ | $\sigma_x$ |
| $\Delta_1, \Sigma_1, Z_1$ | 1 | 1 |
| $\Delta_2, \Sigma_2, Z_2$ | 1 | –1 |

Figure 6.26 shows $\vec{k}$-dependence of the energy in the empty lattice approximation, i.e., using only free-electron Bloch functions but considering the correct symmetry behaviour at special points and lines. For convenience, we define $E_0 \equiv (\hbar^2/2m)$ $(2\pi/a)^2$ and plot reduced energy, i.e., divided by this quantity. (For $x$ point, energy = $E_0/4$).

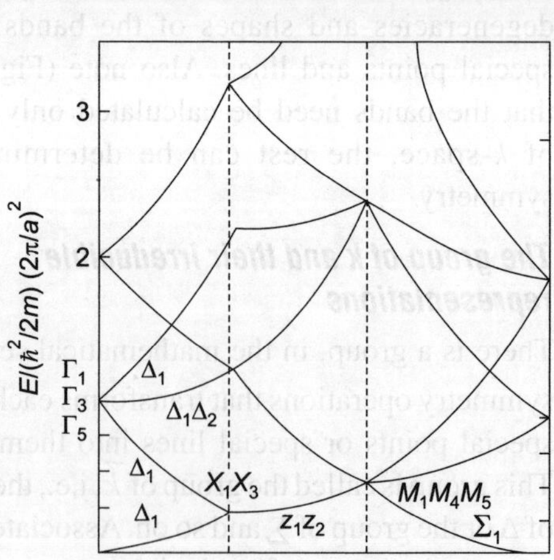


**Figure 6.26** $E$ vs $\vec{k}$ for the two-dimensional lattice. The energy is divided by $E_0 \equiv (\hbar^2/2m)\ (2\pi/a)^2$

It is to be mentioned here that since, the electron wave functions are functions defined in ordinary space, the irreducible representations are used to serve as labels for the wave functions.

## 6.23 THREE-DIMENSIONAL LATTICE

The concept of the division of the vibrational modes into acoustic and optical branches encountered in a diatomic linear lattice can be generalized to be applicable to three-dimensional structures. However, then the number of transverse modes is double that of the longitudinal modes. Thus, for

$sN$ atoms in a solid with a polyatomic basis (where $s$ = number of atoms in the primitive basis and $N$ = number of basis groups in the entire crystals), the number of modes in various classes will be as follows:

| | |
|---|---|
| Longitudinal acoustic modes | $= N$ |
| Transverse acoustic modes | $= 2N$ |
| Longitudinal optical modes | $= (s - 1) N$ |
| Transverse optical modes | $= 2(s - 1) N$ |
| Total modes | $= N + 2N + (s - 1) + N + 2(s - 1) N$ |
| | $= 3sN$ |

This shows that the number of longitudinal and transverse acoustic modes are always 1 and 2 per atom, irrespective of the basis structure. Also, the optical modes are more numerous than acoustic modes. For a crystal with $s = 2$, there will be one longitudinal optical mode and two transverse modes, as shown in Fig. 6.27. However, for a general wave vector, there is no degeneracy and there is no definite distinction between longitudinal and transverse modes. For each branch of the spectrum

$$\omega^2(\mathbf{k}) = \omega^2(-\mathbf{k})$$

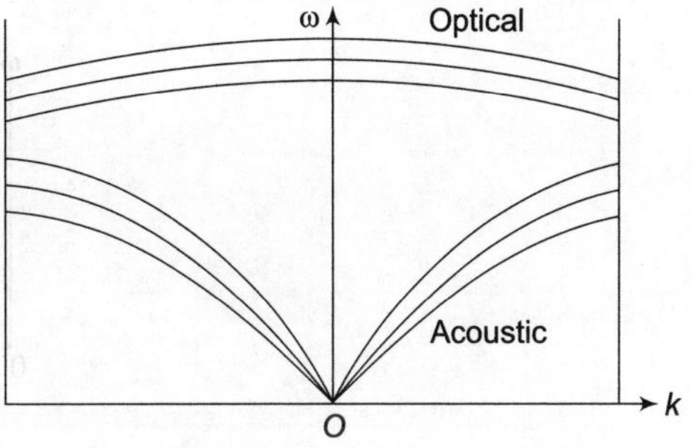


**Figure 6.27** Dispersion relation for a three-dimensional crystal with two atoms per unit cell

## 6.23.1 Three-dimensional Lattice

Let us now extend our discussion to the three-dimensional lattice. To avoid mathematical details, which become quite involved here, we shall present an essentially qualitative discussion. However, the results follow smoothly and logically from the one-dimensional case treated previously.

Consider first the monatomic Bravais lattice, in which each unit cell has a single atom. The equation of motion of each atom can be written. Here also the atoms are coupled together because of their mutual interactions. In attempting a normal mode solution, we write

$$\mathbf{u}_n = \mathbf{A}e^{i(\mathbf{q}\cdot\mathbf{r} - \omega t)}. \qquad ...(1)$$

where the wave vector $\mathbf{q}$ specifies both the wavelength and direction of propagation. A vector is necessary here because propagation takes place in three dimensions. The vector $\mathbf{A}$ specifies the amplitude as well as the direction of vibration of the atoms. Thus, this vector specifies the *polarization* of the wave, i.e., whether the wave is longitudinal ($\mathbf{A}$ parallel to $\mathbf{q}$) or transverse ($\mathbf{A} \perp \mathbf{q}$). (In general, the wave in a lattice is neither purely longitudinal nor purely transverse, but a mixture of both.)

When we substitute Eq. (1) into the equation of motion, we obtain three simultaneous equations involving $A_x$, $A_y$, $A_z$, the components of $\mathbf{A}$. These equations are coupled together and are equivalent to a 3 × 3 matrix equation. Writing the secular equation for this matrix, we arrive at a 3 × 3 detrimental equation, which is cubic in $\omega^2$. The roots of this equation lead to three different dispersion relations, or three dispersion curves, as shown in Fig. 6.28 (a). All three branches pass through the origin, which means that in this lattice all the branches are acoustic. This is, of course, to be expected, since we are dealing with a monatomic Bravais lattice.

Note that, in this three-dimensional situation, the dispersion relations are not necessarily isotropic in **q**-space, and the dispersion curve in Fig. 6.28 represents only the "profile" of the dispersion in a certain **q**-direction. If the dispersion relations are plotted in another direction, a new profile will result which may look quite different from the previous one. In the three-dimensional case, therefore, a complete representation of the dispersion relations requires giving the frequencies for points throughout the three-dimensional **q**-space. This is often accomplished by plotting the frequency contours in this space.

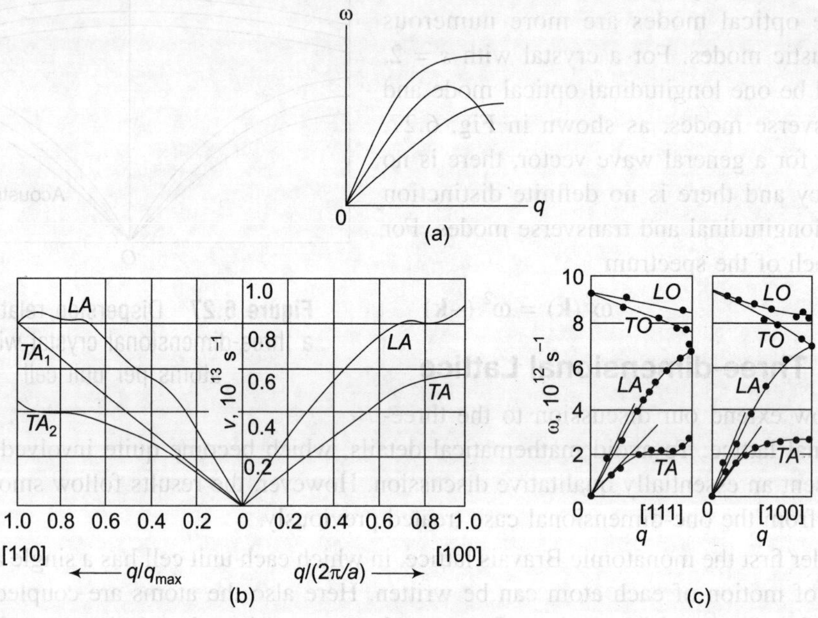


**Figure 6.28** (a) The three acoustic branches in a three-dimensional Bravais lattice. (b) Dispersion curves for Al in [100] direction (right portion) and in [110] direction, left portion. The TA branch in the [100] direction actually represents two coincident, or degenerate, branches. (Note that because each branch is individually symmetric relative to the origin, only half of each branch is plotted.) (c) Dispersion curves of Ge in the [100] and [111] directions

The three branches in Fig. 6.25 differ in their polarization. When **q** lies along a direction of high symmetry—for example, the [100] or [110] directions these waves may be classified as either pure longitudinal or pure transverse waves. In that case, two of the branches are transverse and one is longitudinal.[2] We usually refer to these as the TA (*transverse acoustic*) and LA (*longitudinal acoustic*) branches, respectively. However, along nonsymmetry directions the wave may not be pure longitudinal of pure transverse, but has a mixed character. One may still refer to the branches as TA or LA on the basis of their polarization along the directions of high symmetry. Figure 6.25 (b) shows the dispersion curves for Al in the [100] and [111] directions. Note that in certain high-symmetry directions, such as the [100] in Al, the two transverse branches coincide. The branches are then said to be *degenerate*.

As we have seen, the polarization and degeneracy of dispersion curves are intimately related to the crystal symmetry relative to the direction of propagation. We could carry the subject much further by using the elegant mathematical methods of group theory, but that lies beyond the scope of this book.

We turn our attention now to the non-Bravais three-dimensional lattice. Here the unit cell contains two or more atoms. If there are $r$ atoms per cell, then on the basis of our previous experience we conclude that there are $3r$ dispersion curves. Of these, three branches are acoustic, and the remaining $(3r - 3)$ are optical. The mathematical justification for this assertion is as follows: We write the equation of motion for each atom in the cell, which results in $r$ equations. Since these are vector equations, they are equivalent to $3r$ scalar equations, or to a single matrix equation of the order of $3r \times 3r$. Therefore, the secular equation is of degree $3r$ in $\omega^2$, and has three roots, leading to $3r$ branches. It can be shown that three of the roots always vanish at $q = 0$, which results in three acoustic branches. The remaining $(3r - 3)$ roots, therefore, belong to the optical branches, as stated above.

The acoustic branches may be classified, as before, by their polarizations as $TA_1$, $TA_2$, and $LA$. The *optical* branches can also be classified as longitudinal or transverse when $q$ lies along a high-symmetry direction, and one speaks of $LO$ and $TO$ branches. As in the one-dimensional case, one can also show that, for an optical branch, the atoms in the unit cell vibrate out of phase relative to each other. As an example of a non-Bravais lattice, the dispersion curves for Ge are shown in Fig. 6.25 (c). Since there are two atoms per unit cell in germanium, there are six branches: three acoustic and three optical. Note that the two transverse branches are degenerate along the [100] direction, as indicated earlier.

Lattice dispersion curves are measured by inelastic X-ray or neutron scattering methods.

## 6.24 CLASSICAL THEORY OF THE LATTICE

Here we will develop the formalism for finding the normal modes for the vector field displacement $\vec{u}(\vec{r}, \vec{t})$ in two and three dimensions. The formalism gives an accurate description of sound waves in crystals (we will find the dispersion relations $\omega(\vec{q})$ and we will discover that the dispersion relation depends on the direction of the wave vector $\vec{q}$ in the crystal. (The theory is classical because the atoms are treated as point masses moving according to the laws of classical mechanics.)

[2] Usually the longitudinal branch is higher than the transverse branches because the restoring forces associated with longitudinal oscillations are greater.

We take a crystal with one atom per unit cell, with atomic mass, $M$. We locate the origin of our coordinate system so that when the crystal is at equilibrium, the atom in the $j$th unit cell is located at the lattice point, $\vec{R}_j$. In order to probe the dynamic behaviour of the lattice, we add a continuous lattice displacement $\vec{u}\,(\vec{r},\ \vec{t})$, with the displacement $\vec{u}$ nominally defined at every point $\vec{r}$ in space (i.e., not just the lattice points $\vec{R}_j$). In the presence of this displacement, the atom in the $j$th cell will move to $\vec{r}_j\,(t) = \vec{R}_j + u(\vec{r}_j,\ t)$, where the displacement $\vec{u}$ is evaluated at the instantaneous location of the atom. We further assume that the displacements $\vec{u}$ are very small, and approximate by evaluating the displacement field at the rest position $\vec{R}_j$ rather than at the instantaneous position $\vec{r}_j$, so $\vec{u}\,(\vec{r}_j,\ t) \simeq \vec{u}\,(\vec{R}_j,\ t)$ and for convenience we write

$$\vec{u}_j(t) \equiv \vec{u}\,(\vec{R}_j,\ t) \qquad \text{...(1)}$$

Hence, at time $t$, the $j$th atom is located at

$$\vec{r}_j(t) = \vec{R}_j + \vec{u}_j(t).$$

The total kinetic energy $T$ of the lattice is given by

$$T = \frac{1}{2} M \sum_{j=1}^{N} \left| \frac{\partial \vec{r}_j}{\partial t} \right|^2$$

$$= \frac{1}{2} M \sum_{j=1}^{N} \left| \frac{d\vec{u}_j}{dt} \right|^2 \qquad \text{...(2)}$$

The sum over $j$ runs through the $N$ unit cells in the crystal (so for an infinite crystal we take $N \to \infty$).

The total interaction potential energy is, in general, a function of all the atoms coordinates

$$U = U(\vec{r}_1,\ \vec{r}_2 \ldots\ \vec{r}_N) \qquad \text{...(3)}$$

For small displacements $\vec{u}_j$ from the equilibrium position, the interaction can be expanded in a Taylor series about the static positions, $\vec{R}_j$:

$$U(\vec{r}_1, \vec{r}_2 \ldots \vec{r}_N) = U(\vec{R}_1, \vec{R}_2 \ldots \vec{R}_N) + \sum_{j=1}^{N} \sum_{\alpha=1}^{3} \frac{\partial U}{\partial r_{j\alpha}} u_{j\alpha}$$

$$+ \sum_{j,\,k=1}^{N} \sum_{\alpha,\,\beta=1}^{3} \frac{1}{2} \frac{\partial^2 U}{\partial r_{j\alpha}\, \partial r_{k\beta}} u_{j\alpha}\, u_{k\beta} + \ldots \qquad \text{...(4)}$$

The derivatives are with respect to the components $\alpha$ of each atom $i$'s coordinates $r_{i\alpha}$, and the derivatives are all evaluated with the atoms at their equilibrium positions $\vec{R}_j$.

The first term sets the zero of energy and may be dropped. The second term must vanish, as the positions $\vec{R}_j$ are assumed to be the equilibrium positions. We thus retain only the third term, dropping the higher order terms involving the third and higher order derivatives This is known as the harmonic approximation (the higher order terms generate anharmonic, non-linear responses). Hence, we can write

$$U_{har}(\vec{r}_1, \vec{r}_2 \dots \vec{r}_N) = \frac{1}{2} \sum_{j,k=1}^{N} \sum_{\alpha,\beta=1}^{3} \frac{\partial^2 U}{\partial r_{j\alpha}\, \partial r_{k\beta}} u_{j\alpha}\, u_{k\beta} \qquad \text{...(5)}$$

We now define a tensor (a 3 × 3 matrix) whose entries are functions of position, viz., $\phi(\vec{R}_i, \vec{R}_j)$ with tensor elements $\phi_{\alpha\beta}$, in terms of the curvature of the interaction energy $U(\vec{r}_1 \dots \vec{r}_N)$:

$$\Phi_{\alpha\beta}(\vec{R}_i, \vec{R}_j) = \frac{\partial^2 U}{\partial r_{i\alpha}\, \partial r_{j\beta}} \qquad \text{...(6)}$$

where the derivatives are evaluated with the atoms at their equilibrium positions, $\vec{R}_i$. Note that the tensor is a function of the atom positions $\vec{R}_i$ and $\vec{R}_j$, as the derivatives are taken with respect to the components of the displacements of these two atoms.

Our crystal does not change under lattice translation, so the entries in the tensor $\Phi(\vec{R}_j, \vec{R}_k)$ also should not change. In other words, the tensor depends only on the difference of the atom positions $(\vec{R}_j - \vec{R}_k)$, so we can write

$$\Phi_{\alpha\beta}(\vec{R}_i, \vec{R}_j) = \Phi_{\alpha\beta}(\vec{R}_i - \vec{R}_j) \qquad \text{...(7)}$$

Then potential energy

$$U_{\text{har.}}(\vec{r}_1 \dots \vec{r}_N) = \frac{1}{2} \sum_{j,k=1}^{N} \sum_{\alpha,\beta=1}^{3} u_{j\alpha}\, \Phi_{\alpha\beta}(\vec{R}_j - \vec{R}_k)\, u_{k\beta}$$

$$= \frac{1}{2} \sum_{jk} \vec{u}_j \cdot \Phi(\vec{R}_j - \vec{R}_k) \cdot \vec{u}_k \qquad \text{...(8a)}$$

where we have used the tensor summation convention (i.e., inner product) to represent the double sum.

## 6.25 DIATOMIC INTERACTION POTENTIALS

The expression (8a) is rather complicated. Here we consider somewhat a simpler expression. We consider the form of the interaction potential energy when the atoms interact with one another only through a diatomic potential $\phi(\vec{r})$ i.e., one that involves interactions only between pairs of atoms. i.e. $\phi = \phi(\vec{r}_i - \vec{r}_j)$. Then, total potential energy is the sum over all atom pairs:

$$U(\vec{r}_1 \dots \vec{r}_N) = \frac{1}{2} \sum_{i,j=1}^{N} \phi(\vec{r}_i - \vec{r}_j) \qquad \text{...(8b)}$$

The factor $\frac{1}{2}$ comes to avoid counting a pair twice Taylor expansion for a function $\phi$ is

$$\phi(\vec{r}_i - \vec{r}_j) = \phi(\vec{R}_i - \vec{R}_j) + \sum_{\alpha=1}^{3} \left.\frac{\partial \phi}{\partial r_\alpha}\right|_{(\vec{R}_i - \vec{R}_j)} (u_{i\alpha} - u_{j\alpha})$$

$$+ \frac{1}{2} \sum_{\alpha,\beta=1}^{3} \left.\frac{\partial^2 \phi}{\partial r_\alpha \, \partial r_\beta}\right|_{(\vec{R}_i - \vec{R}_j)} (u_{i\alpha} - u_{j\alpha})(u_{i\beta} - u_{j\beta}) + \ldots \quad \ldots(9)$$

Substituting this in Eq. (8b) and keeping only the second order derivatives (neglecting higher order derivatives), we are left with the equivalent of the harmonic approximation

$$U_{\text{har.}} = \frac{1}{4} \sum_{j,k=1}^{N} \sum_{\alpha,\beta=1}^{3} (u_{j\alpha} - u_{k\alpha}) \left.\frac{\partial^2 \phi}{\partial r_\alpha \, \partial r_\beta}\right|_{(\vec{R}_j - \vec{R}_k)} (u_{j\beta} - u_{k\beta})$$

$$= \frac{1}{4} \sum_{j,k}^{N} \sum_{\alpha,\beta}^{3} (u_{j\alpha} u_{j\beta} - u_{j\alpha} u_{k\beta} - u_{k\alpha} u_{j\beta} + u_{k\alpha} u_{k\beta})$$

$$\times \left.\frac{\partial^2 \phi}{\partial r_\alpha \, \partial r_\beta}\right|_{(\vec{R}_j - \vec{R}_k)}$$

$$= \frac{1}{4} \sum_{j,k}^{N} \sum_{\alpha,\beta}^{3} [(u_{j\alpha} u_{j\beta} + u_{k\alpha} u_{k\beta}) - (u_{j\alpha} u_{k\beta} + u_{k\alpha} u_{j\beta})]$$

$$\times \left.\frac{\partial^2 \phi}{\partial r_\alpha \, \partial r_\beta}\right|_{(\vec{R}_j - \vec{R}_k)}$$

$$= \frac{1}{2} \sum_{j,k} \sum_{\alpha,\beta} (u_{j\alpha} u_{j\beta} - u_{j\alpha} u_{k\beta}) \left.\frac{\partial^2 \phi}{\partial r_\alpha \, \partial r_\beta}\right|_{(\vec{R}_j - \vec{R}_k)} \quad \ldots(10)$$

Substituting this into the general expression (8a), we can work out the equivalent expressions for the tensor $\Phi(\vec{R}_j - \vec{R}_k)$ in terms of the diatomic potential. We then find

$$\left.\begin{aligned} \Phi_{\alpha\beta}(\vec{R}_k = 0) &= \sum_{l=1}^{N} \left.\frac{\partial^2 \phi}{\partial r_\alpha \, \partial r_\beta}\right|_{\vec{R}_l} \\ \Phi_{\alpha\beta}(\vec{R}_k \neq 0) &= -\left.\frac{\partial^2 \phi}{\partial r_\alpha \, \partial r_\beta}\right|_{\vec{R}_k} \end{aligned}\right\} \quad \ldots(11)$$

Using this tensor, we can write

$$U_{\text{har}} = \frac{1}{2} \sum_{j,k} \sum_{\alpha,\beta} u_{j\alpha} \, \Phi_{\alpha\beta}(\vec{R}_j - \vec{R}_k) \, u_{k\beta} \quad \ldots(12)$$

## 6.26 SYMMETRY OF THE INTERACTION TENSOR

The function $\Phi$ is inversion symmetric in $\vec{R}$:

$$\Phi_{\alpha\beta}(\vec{R}) = \Phi_{\alpha\beta}(-\vec{R}) \qquad ...(13)$$

This can be seen by inverting the order of differentiation. Further, for any given $\vec{R}$, the tensor is symmetric in its indices:

$$\Phi_{\alpha\beta}(\vec{R}) = \Phi_{\beta\alpha}(\vec{R}). \qquad ...(14)$$

This symmetry follows from the inversion symmetry of the Bravais lattice. Finally, if we take the $\alpha\beta$ tensor element $\Phi_{\alpha\beta}(\vec{R})$ and sum over all $\vec{R}$ in the Bravais lattice, the result is zero:

$$\sum_{j=1}^{N} \Phi_{\alpha\beta}(\vec{R}_j) = 0. \qquad ...(15)$$

This can be seen by examining Eq. (11). Physically it means that if all the ions in the lattice are given the same displacement $\vec{u}$, shifting the crystal as a unit, the potential energy $U$ should not change.

Now, the Hamiltonian $H$ for the system can be formed from the sum of the total kinetic and potential energies ($H = T + U$):

$$H = \frac{1}{2} M \sum_{j=1}^{N} \left|\frac{\partial \vec{u}_j}{\partial t}\right|^2 + \frac{1}{2} \sum_{j,k} \sum_{\alpha,\beta} u_{j\alpha}\, \Phi_{\alpha\beta}(\vec{R}_j - \vec{R}_k)\, u_{k\beta} \qquad ...(16)$$

The Hamiltonian equations of motion then generate a second order differential equation of motion for each index $\alpha$ of the atomic displacement vector $\vec{u}_j$

$$M \frac{\partial^2 u_{j\alpha}}{\partial t^2} = -\sum_{k=1}^{N} \sum_{\beta=1}^{3} \Phi_{\alpha\beta}(\vec{R}_j - \vec{R}_k)\, u_{k\beta} \qquad ...(17)$$

where the component index $\alpha$ runs from 1 to 3 (in 3 dimensions) and the atom index $j$, from 1 to $N$, the number of atoms in the crystal.

## 6.27 NORMAL MODES OF THE LATTICE

The set of Eq. (17) is solved by finding the normal modes of the coupled equations, i.e., those solutions where every atom has the same harmonic time dependence. These solutions form a complete set, so that any arbitrary motion of the crystal can be written as a linear superposition of the normal modes. We find the normal modes by taking the same harmonic time dependence for all the atoms:

$$\vec{u}_j(t) = \vec{A}_j e^{i\omega t}. \qquad ...(1)$$

where $\vec{A}_j$ is the vector displacement (which may be complex) for the $j$th atom. Substituting this into Eq. (17) yields a set of $3N$ coupled linear equations

$$\omega^2 MA_{j\alpha} = \sum_{k=1}^{N} \sum_{\beta=1}^{3} \Phi_{\alpha\beta}(\vec{R}_j - \vec{R}_k) A_{k\beta} \quad \text{...(2)}$$

We solve this set of equation by taking the Bloch from for the amplitude $\vec{A}_j$

$$\vec{A}_j = \vec{A} e^{i\vec{q}\cdot\vec{R}_j} \quad \text{...(3)}$$

where $q$ is the wave vector for the mode with frequency $\omega$ (we write $\omega_{\vec{q}}$ for frequency to indicate this relationship). The amplitude and polarization of the mode is given by $\vec{A}$, with components, $A_\alpha$.

With the periodic boundary conditions, the values of $\vec{q}$ are restricted to a discrete finite set. For a two-dimensional case, (length $L$ and $N$ atoms along each axis), the values of $\vec{q}$ are restricted to the set $\vec{q} = (q_1, q_2) = 2\pi\left(\frac{m}{L}, \frac{n}{L}\right)$ with $-N/2 < m < N/2$ and $-N/2 < n < N/2$. Similarly for a 3-dimensional solid, we have $\vec{q} = (q_1, q_2, q_3)$. However, here, we ignore the discretized values, and assume $\vec{q}$ is a continuous vector.

Substituting Eq. (3) in Eq. (2), the equation for the $j$th atom takes on the form

$$\omega_{\vec{q}}^2 M A_\alpha = \sum_{k=1}^{N} \sum_{\beta=1}^{3} \Phi_{\alpha\beta}(\vec{R}_j - \vec{R}_k) e^{i\vec{q}\cdot(\vec{R}_k - \vec{R}_j)} A_\beta \quad \text{...(4)}$$

This expression cannot depend on the index $j$ of the atom we started with. Taking $\vec{R}_l = 0$ for atom $l$, we can write the equation for $j = l$ leaving the equivalent relation

$$\omega_{\vec{q}}^2 M A_\alpha = \sum_{k=1}^{N} \sum_{\beta=1}^{3} \Phi_{\alpha\beta}(\vec{R}_k) e^{i\vec{q}\cdot\vec{R}_k} A_\beta \quad \text{...(5)}$$

where we have used the inversion symmetry. We now define a new tensor $D(\vec{q})$, as the Fourier transform of the original tensor $\Phi(\vec{R}_j)$. The components $D_{\alpha\beta}(\vec{q})$ of the new tensor are given by

$$D_{\alpha\beta}(\vec{q}) = \sum_{k=1}^{N} \Phi_{\alpha\beta}(\vec{R}_k) e^{i\vec{q}\cdot\vec{R}_k} \quad \text{...(6)}$$

Substituting this form in Eq. (5) generates the set of equations

$$\omega_{\vec{q}}^2 M A_\alpha = \sum_{\beta=1}^{3} D_{\alpha\beta}(\vec{q}) A_\beta \quad \text{...(7)}$$

The tensor $D(\vec{q})$ is known as the *dynamical matrix*. The set of three Eqs. (7) is now easier to solve than the original set of $3N$ Eq. (17). They form an *eigenvector-eigenvalue* problem, which can we written

$$M\,\omega_{\vec{q}}^2 = \begin{bmatrix} A_1 \\ A_2 \\ A_3 \end{bmatrix} = \begin{bmatrix} D_{11} & D_{12} & D_{13} \\ D_{21} & D_{22} & D_{23} \\ D_{31} & D_{32} & D_{33} \end{bmatrix} \begin{bmatrix} A_1 \\ A_2 \\ A_3 \end{bmatrix} \quad \text{...(8)}$$

where $\vec{A}$ and $D$ are functions of $\vec{q}$. For each wave vector $\vec{q}$, there are three distinct eigenfrequencies $\omega_{\vec{q}}$ and corresponding to each eigenfrequency, an eigenvector $\vec{A}$ that determines the *polarization* of the mode.

The eigenfrequencies are found by solving the characteristic equation resulting from setting the determinant of the matrix in Eq. (8) equal to zero. Once these frequencies have been found, the three corresponding polarizations $\vec{A}(\vec{q})$ are found by inserting each eigenfrequency $\omega_{\vec{q}}$ in Eq. (8). If there are $N$ atoms in the lattice, there will be $N$ distinct values of $\vec{q}$ and therefore a total of $3N$ modes for the $3N$ degrees of freedom. In an infinite crystal, there are infinite number of distinct wave vectors, so $\vec{q}$ becomes a continuous variable.

We note that the expression (6) for the tensor $D(\vec{q})$ can be manipulated. Using the symmetries (13) to (15) yielding.[3]

$$D_{\alpha\beta}(\vec{q}) = -2\sum_{\vec{R}} \Phi_{\alpha\beta}(\vec{R})\sin^2\left(\frac{\vec{q}\cdot\vec{R}}{2}\right) \quad \text{...(9)}$$

## 6.28 DISPERSION RELATIONS FOR FCC LATTICE

We now calculate the dispersion relations for a three-dimensional face-centred cubic crystal where we assume the atoms interact only with their nearest neighbours through the spherically-symmetric interaction potential $\phi(r)$, where $r$ is the radial coordinate. The twelve atoms nearest the atom at the cube corner (which we place at the origin $\vec{R} = 0$) are located at

$$\vec{R}_{1-4} = \frac{d}{2}(\pm\hat{x} \pm\hat{y}), \quad \vec{R}_{5-8} = \frac{d}{2}(\pm\hat{x} \pm\hat{z}), \quad \text{...(1)}$$

$$\vec{R}_{9-12} = \frac{d}{2}(\pm\hat{y} \pm\hat{z})$$

where $d$ is the length of the cube edge; the nearest neighbour distance is $a = d/\sqrt{2}$.

The interaction tensor $\psi(\vec{r})$ has the elements

$$\psi_{\alpha\beta}(\vec{r}) = \frac{\partial^2\phi}{\partial x_\alpha \partial x_\beta}. \text{ With } r = (x^2 + y^2 + z^2)^{1/2}, \text{ we have}$$

$$\phi(r) = \phi(\sqrt{x^2 + y^2 + z^2}) \quad \text{...(2)}$$

$$\frac{\partial\phi}{\partial x} = \frac{\partial r}{\partial x}\cdot\frac{\partial\phi}{\partial r} = \frac{1}{2}\cdot 2x\cdot(x^2 + y^2 + z^2)^{-1/2}\cdot\frac{\partial\phi}{\partial r} \quad \text{...(3)}$$

$$= \frac{x}{r}\frac{\partial\phi}{\partial r}$$

$$\frac{\partial^2\phi}{\partial x^2} = \frac{\partial}{\partial x}\left(\frac{x}{r}\cdot\frac{\partial\phi}{\partial r}\right)$$

[3] N.W. Ashcroft and N.D. *Mermin, Solid State Physics*. Saunders College, Philadelphia P.A. (1976).

$$= \frac{1}{r}\frac{\partial\phi}{\partial r} + \frac{x}{r}\frac{\partial}{\partial x}\left(\frac{\partial\phi}{\partial r}\right)$$

$$= \frac{1}{r}\frac{\partial\phi}{\partial r} + \frac{x}{r}\cdot\frac{\partial}{\partial r}\left(\frac{\partial\phi}{\partial x}\right)$$

$$= \frac{1}{r}\cdot\frac{\partial\phi}{\partial r} + \frac{x}{r}\frac{\partial}{\partial r}\left(\frac{\partial\phi}{\partial r}\cdot\frac{\partial r}{\partial x}\right)$$

$$= \frac{1}{r}\frac{\partial\phi}{\partial r} + \frac{x}{r}\frac{\partial}{\partial r}\left(\frac{x}{r}\cdot\frac{\partial\phi}{\partial r}\right)$$

$$= \frac{1}{r}\frac{\partial\phi}{\partial r} + \frac{x}{r}\left[\frac{\partial\phi}{\partial r}\left(-\frac{x}{r^2}\right) + \frac{x}{r}\frac{\partial^2\phi}{\partial r^2}\right]$$

$$= \frac{1}{r}\frac{\partial\phi}{\partial r} - \frac{x^2}{r^3}\frac{\partial\phi}{\partial r} + \frac{x^2}{r^2}\frac{\partial^2\phi}{\partial r^2}$$

Thus,
$$\Phi_{xx}(\vec{r}) \equiv \frac{\partial^2\phi}{\partial x^2} = \left(\frac{x}{r}\right)^2\frac{\partial^2\phi}{\partial r^2} + \frac{1}{r}\frac{\partial\phi}{\partial r}\left(1 - \frac{x^2}{r^2}\right) \quad ...(4)$$

Similarly,
$$\frac{\partial^2\phi}{\partial x\partial y} = \frac{\partial^2\phi}{\partial y\partial x} = \frac{\partial}{\partial y}\left(\frac{\partial\phi}{\partial x}\right)$$

$$= \frac{\partial}{\partial y}\left(\frac{x}{r}\cdot\frac{\partial\phi}{\partial r}\right)$$

$$= \frac{\partial}{\partial r}\left(\frac{x}{r}\frac{\partial\phi}{\partial r}\right)\cdot\frac{\partial r}{\partial y}$$

$$= \left[\frac{\partial\phi}{\partial r}\left(-\frac{x}{r^2}\right) + \frac{x}{r}\frac{\partial^2\phi}{\partial r^2}\right]\cdot\frac{y}{r} \quad ...(5)$$

or
$$\Phi_{xy}(\vec{r}) = \frac{xy}{r^2}\frac{\partial^2\phi}{\partial r^2} - \frac{xy}{r^3}\frac{\partial\phi}{\partial r} \quad ...(6)$$

with $d_1 = \frac{1}{r}\cdot\frac{\partial\phi}{\partial r}$ and $d_2 = \frac{\partial^2\phi}{\partial r^2}$, we can write

$$\Phi(\vec{r}) = \begin{bmatrix} 1 & 0 & 0 \\ 0 & 1 & 0 \\ 0 & 0 & 1 \end{bmatrix} d_1 + \begin{bmatrix} x^2 & xy & xz \\ yx & y^2 & yz \\ zx & zy & z^2 \end{bmatrix} \times \left(\frac{d_2 - d_1}{r^2}\right) \quad ...(7)$$

The dynamical matrix $D(\vec{q})$ is given by the sum Eq. (9), Art. 6.27 as

$$D_{\alpha\beta}(\vec{q}) = -2\sum_{\vec{R}}\Phi_{\alpha\beta}(\vec{R})\sin^2\left(\frac{\vec{q}\cdot\vec{R}}{2}\right) \quad ...(8)$$

where the sum is over the nearest neighbour vectors given in Eq. (1).

The parameters $d_1$ and $d_2$ determine the detailed shape and scale of the interaction matrix elements; the eigenfrequencies $\omega_n(\vec{q})$ can then be calculated by solving the eigenvalue equation

$$M\,\omega_n^2(\vec{q})\,A_\alpha = \sum_{\beta=1}^{3} D_{\alpha\beta}(\vec{q})\,A_\beta \qquad ...(9)$$

or written in vector form

$$M\,\omega_n^2(\vec{q})\,\vec{A} = D(\vec{q})\,\vec{A}. \qquad ...(10)$$

Figure 6.29 shows the calculated dispersion curves $\omega_n(\vec{q})$ for $d_1/M = 3TH_z^2$ and $d_2/M = 30\,TH_z^2$, choices which yield dispersion relations similar to those for lead, an fcc metal.

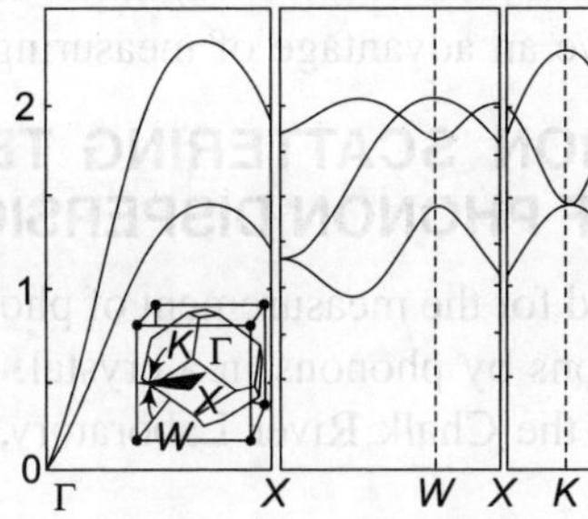


**Figure 6.29** Phonon dispersion relations for an fcc lattice, using a spherically symmetric potential $\phi(r)$. The potential parameters are chosen to give a dispersion relation similar to that of *Pb*.[4]

The calculations are along different directions in the reciprocal lattice; the first segment from the origin Γ to the Brillouin zone edge $X = \left(0, \frac{2\pi}{a}, 0\right)$, the second from $X$ to *a* point outside the zone at $\left(\frac{2\pi}{a}, \frac{2\pi}{a}, 0\right)$, and the third from that point back to the zone edge point $K = (3\pi a/2\ ,\ 3\pi a/2,\ 0)$ and then returning to the origin Γ. (Note that the symmetry of the crystal yields degenerate values for the two lower, transverse acoustic branches along the Γ–$X$ line.)

(v) *Neutron–Phonon Interaction* (*Inelastic scattering case*). In case of neutron scattering, the neutrons are scattered by nuclear forces. This type of scattering may be coherent (elastic) or incoherent (inelastic). The scattering of neutron beam with a crystal lattice may be described by the conservation of wave vector relation

$$\mathbf{K'} = \mathbf{k} + \mathbf{G} \pm \mathbf{q} \qquad ...(1)$$

where $\mathbf{k}$ is the wave vector of incident neutron and $\mathbf{k'}$ of the scattered neutron. $+\mathbf{q}$ is the wave vector of the phonon created or $-\mathbf{q}$ of the phonon absorbed. The law of conservation of energy may be described by

4 After: Andrew N. Cleland, *Foundations of Nanomechanics*, Springer-Verlag (2003).

$$\frac{\hbar^2 k^2}{2M_n} = \frac{\hbar^2 k'^2}{2M_n} \pm \hbar\,\omega_q$$

or
$$\mathbf{k}^2 - \mathbf{k}'^2 = \pm\frac{2M_n}{\hbar}\,\omega_q\,(\mathbf{k}' - \mathbf{k}) \qquad ...(2)$$

where $+\omega_q$ is the frequency of created phonon and $-\omega_q$ of the absorbed phonon.

When cold or thermal neutrons are scattered by a crystal lattice, the de Broglie wavelength becomes large and the same neutron can interact with many phonons. The scattered waves interfere with each other and the interference is maximum for the nuclei with isotopes.

The neutron scattering technique is an ideal method for determining the phonon spectrum. If the absorption of neutrons by the crystal nuclei is high, this technique cannot be used. The neutron scattering experiments have an advantage of measuring directly the energy shift.

## 6.29 INELASTIC NEUTRON SCATTERING TECHNIQUE FOR THE DETERMINATION OF PHONON DISPERSION

A widely used experimental method for the measurement of phonon dispersion relation is due to the inelastic scattering of neutrons by phonons in a crystal. Brockhouse had developed a triple axes crystal spectrometer at the Chalk River Laboratory, Ontario in Canada which is schematically shown in Fig. 6.30.

In this spectrometer, a neutron beam from a nuclear reactor is Bragg reflected by a single crystal monochromator. This Bragg reflected neutron beam passes through a collimator and falls on the specimen in the form of a single crystal. Here the inelastic scattering takes place and the scattered neutron beam is analysed by an analysing crystal and a detector. The detector measures the magnitude and the direction of the scattered neutron beam. Since rotation about three axes is required to adjust the three scattering angles $\theta_M$, $\phi$ and $\theta_A$, the whole arrangement is known as triple axes spectrometer.

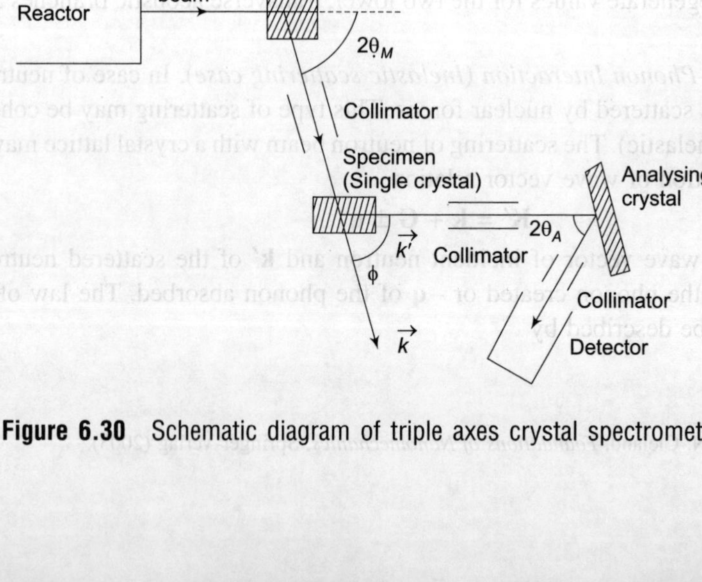


**Figure 6.30** Schematic diagram of triple axes crystal spectrometer

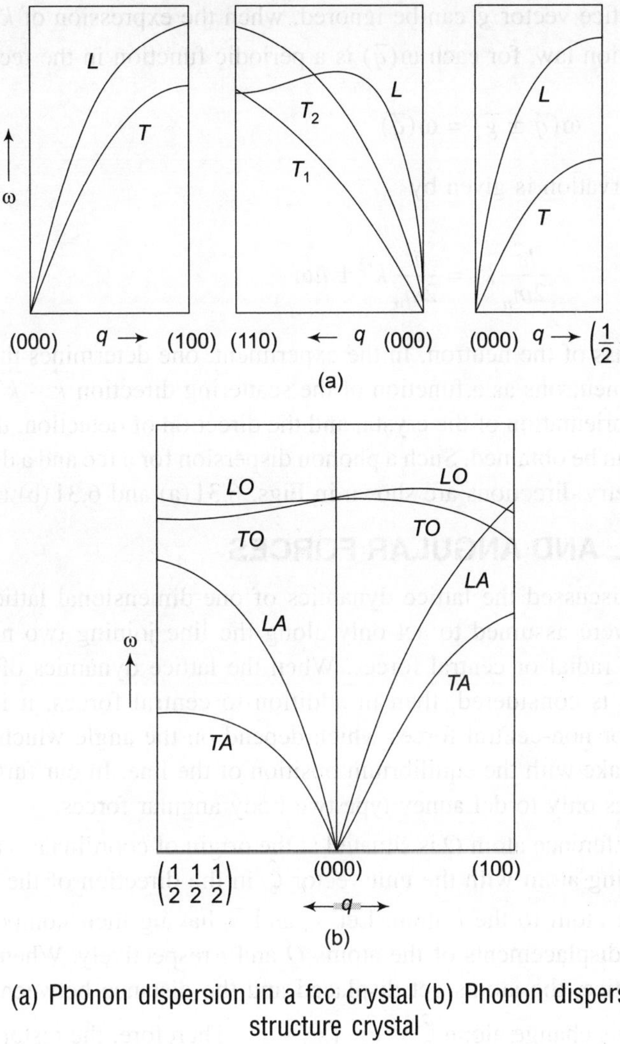


**Figure 6.31** (a) Phonon dispersion in a fcc crystal (b) Phonon dispersion in a diamond structure crystal

Let the incident and the scattered neutron wave vectors be $\vec{k}$ and $\vec{k}'$. Similar to Bragg diffraction of X-rays, the elastic scattering of neutrons will follow the selection rule

$$\vec{k}' = \vec{k} + \vec{g} \quad ...(1)$$

where $\vec{g}$ is a reciprocal lattice vector. If the scattering of the neutron is inelastic, with the creation of a phonon of wave vector $\vec{q}$, then the selection rule will be

$$\vec{k}' + \vec{q} = \vec{k} + \vec{g} \quad ...(2)$$

If instead the phonon is absorbed in the process, then the selection rule will be

$$\vec{k}' - \vec{q} = \vec{k} + \vec{g} \quad ...(3)$$

The reciprocal lattice vector $\vec{g}$ can be ignored, when the expression of $k$ is substituted in the energy conservation law, for each $\omega(\vec{q})$ is a periodic function in the reciprocal lattice

$$\omega(\vec{q} \pm \vec{g}) = \omega(\vec{q}) \qquad ...(4)$$

The energy conservation is given by

$$\frac{\hbar^2}{2m_n} k^2 = \frac{\hbar^2}{2m_n} k'^2 \pm \hbar\omega \qquad ...(5)$$

Here $m_n$ is the mass of the neutron. In the experiment, one determines the energy gain or loss of the scattered neutrons as a function of the scattering direction $\vec{k} - \vec{k}'$. By varying the incident energy, the orientation of the crystal and the direction of detection, the entire phonon dispersion relation can be obtained. Such a phonon dispersion for a fcc and a diamond structure crystal in the symmetry directions are shown in Figs. 6.31 (a) and 6.31 (b) respectively.

## 6.30 CENTRAL AND ANGULAR FORCES

Till now, we have discussed the lattice dynamics of one-dimensional lattice, in which the interatomic forces were assumed to act only along the line joining two neighbours. Such forces are known as radial or central forces. When the lattice dynamics of more than one-dimensional lattices is considered, then in addition to central forces, it is worthwhile to include the angular or non-central forces which depend on the angle which the line joining the moving atoms make with the equilibrium position of the line. In our further analysis, we will confine ourselves only to deLauney type two body angular forces.

In Fig. 6.32, the reference atom $O$ is situated at the origin of coordinates and let $i$ represent any other neighbouring atom with the unit vector $\hat{\zeta}_i$ in the direction of the equilibrium line joining the reference atom to the $i$ atom. Let $\vec{s}_o$ and $\vec{s}_i$ having their components $u_o$, $v_o$, $w_o$ and $u_i$, $v_i$, $w_i$ be the displacements of the atoms $O$ and $i$ respectively. When an elastic wave travels through a lattice, the atoms get displaced and the distance between atoms changes. The component of this change along $\hat{\zeta}_i$ is $\hat{\zeta}_i \cdot (\vec{s}_o - \vec{s}_i)$. Therefore, the restoring central force is

$$\vec{F}_{ci} = -\alpha[\hat{\zeta}_i \cdot (\vec{s}_o - \vec{s}_i)]\, \hat{\zeta}_i \qquad ...(1)$$

where $\alpha$ is central force constant, different for different neighbours. For angular forces, one needs the component of the change perpendicular to the unit vector $\hat{\zeta}_i$. The magnitude of this component is $\hat{\zeta}_i \times (\vec{s}_o - \vec{s}_i)$. In order to get the required direction, another cross-product will provide the restoring angular force as

$$\vec{F}_{ai} = -a' \; [\hat{\zeta}_i \times (\vec{s}_o - \vec{s}_i)] \times \hat{\zeta}_i \qquad ...(2)$$

where $\alpha'$ is the angular force constant. Using the vector identity

$$\vec{A} \times (\vec{B} \times \vec{C}) = \vec{B}(\vec{A} \cdot \vec{C}) - \vec{C}(\vec{A} \cdot \vec{B}) \qquad ...(3)$$

One obtains

$$\vec{F}_{ai} = -\alpha' (\vec{s}_o - \vec{s}_i) + \alpha' [\hat{\zeta}_i \cdot (\vec{s}_o - \vec{s}_i)] \hat{\zeta}_i \qquad ...(4)$$

Hence, the resultant restoring force acting on $O$ due to $i$ will be given by

$$F_i = -\alpha' (\vec{s}_o - \vec{s}_i) - (\alpha - \alpha') [\hat{\zeta}_i \cdot (\vec{s}_o - \vec{s}_i)] \hat{\zeta}_i \qquad ...(5)$$

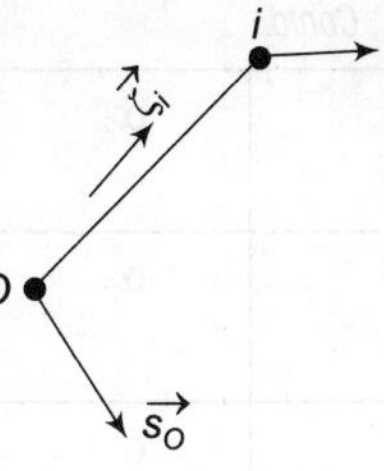


**Figure 6.32** The displacement vector $\vec{s}_o$ and $\vec{s}_i$ of atoms $O$ and $i$.[5]

In component form,

$$X_i = -\alpha' (u_o - u_i) - (\alpha - \alpha)\, \zeta_{i1} [\zeta_i (u_o - u_i) + \zeta_{i2} (v_o - v_i) + \zeta_{i3} (w_o - w_i)]$$

$$Y_i = -\alpha' (v_o - v_i) - (\alpha - \alpha')\, \zeta_{i2} [\zeta_i (u_o - u_i) + \zeta_{i2} (v_o - v_i) + \zeta_{i3} (w_o - w_i)]$$

$$Z_i = -\alpha' (w_o - w_i) - (\alpha - \alpha')\, \zeta_{i3} [\zeta_i (u_o - u_i) + \zeta_{i2} (v_o - v_i) + \zeta_{i3} (w_o - w_i)] \qquad ...(6)$$

Here $X_i$, $Y_i$, $Z_i$ are the components of the force $\vec{F}_i$ along $x$, $y$, $z$ axes respectively and $\zeta_{i1}$, $\zeta_{i2}$, $\zeta_{i3}$ represent the direction cosines of the unit vector $\zeta_i$. These angular forces are rotationally variant. It should be mentioned, however, that three body angular forces due to Born are rotationally invariant but these are not discussed here.

## 6.31 LATTICE DYNAMICS OF MONATOMIC TWO-DIMENSIONAL LATTICES

Let us consider the simplest two-dimensional example of a square lattice of identical masses $M$ as shown in Fig. 6.33. We will consider the interatomic interactions up to second nearest neighbour and both the central and the angular forces. In Table 6.2, the lattice details for the dynamical calculations are given.

**Table 6.2** Lattice Details Related to Lattice Dynamics of Square Lattice.[6]

| Atom number | Force constant | | $x_i$ | $y_i$ | $\zeta_{i1}$ | $\zeta_{i2}$ |
|---|---|---|---|---|---|---|
| | Central | Angular | | | | |
| 1 | $\alpha_1$ | $\alpha'_1$ | $a$ | 0 | 1 | 0 |
| 2 | $\alpha_1$ | $\alpha'_1$ | $-a$ | 0 | –1 | 0 |
| 3 | $\alpha_1$ | $\alpha'_1$ | 0 | $a$ | 0 | 1 |
| 4 | $\alpha_1$ | $\alpha'_1$ | 0 | $-a$ | 0 | –1 |

*Contd...*

5 [After: H.C. Gupta, *Solid State Physics*, Vikas Publishing, New Delhi (1997)].

6 [Adapted from: H.C. Gupta, *Solid State Physics,* Vikas Publishing, New Delhi (1997)]

*Table 6.2 Contd...*

| | | | | | | |
|---|---|---|---|---|---|---|
| 5 | $\alpha_2$ | $\alpha'_2$ | $a$ | $a$ | $\frac{1}{\sqrt{2}}$ | $\frac{1}{\sqrt{2}}$ |
| 6 | $\alpha_2$ | $\alpha'_2$ | $-a$ | $-a$ | $-\frac{1}{\sqrt{2}}$ | $-\frac{1}{\sqrt{2}}$ |
| 7 | $\alpha_2$ | $\alpha'_2$ | $-a$ | $a$ | $-\frac{1}{\sqrt{2}}$ | $\frac{1}{\sqrt{2}}$ |
| 8 | $\alpha_2$ | $\alpha'_2$ | $a$ | $-a$ | $\frac{1}{\sqrt{2}}$ | $-\frac{1}{\sqrt{2}}$ |

When this data is substituted in Eq. (6) and remembering that it is a two-dimensional lattice, the force equations in $x$ direction will be

$$X_1 = -\alpha'_1 (u_o - u_1) - (\alpha_1 - \alpha_1)\, 1\, [1\, (u_o - u_1)] = -\alpha_1 (u_o - u_1)$$

$$X_2 = -\alpha'_1 (u_o - u_2) - (\alpha_1 - \alpha'_1)\, (-1)\, [(-1)\, (u_o - u_2)] = -\alpha_1 (u_o - u_2)$$

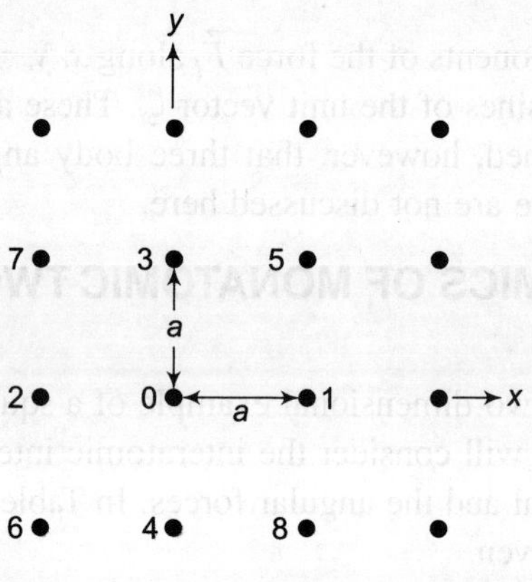


**Figure 6.33** A two-dimensional square lattice

$$X_3 = -\alpha'_1 (u_o - u_3)$$

$$X_4 = -\alpha'_1 (u_o - u_4)$$

$$X_5 = -\alpha'_2 (u_o - u_5) - (\alpha_2 - \alpha'_2)\left(\frac{1}{\sqrt{2}}\right)\left[\frac{1}{\sqrt{2}}(u_o - u_5) + \frac{1}{\sqrt{2}}(v_o - v_5)\right]$$

$$= -\frac{(\alpha^2 + \alpha'_2)}{2}(u_o - u_5) - \frac{(\alpha_2 - \alpha'_2)}{2}(v_o - v_5)$$

$$X_6 = -\alpha'_2 (u_o - u_6) - (\alpha_2 - \alpha'_2)\left(\frac{-1}{\sqrt{2}}\right)\left[\frac{-1}{\sqrt{2}}(u_o - u_6) - \frac{1}{\sqrt{2}}(v_o - v_6)\right]$$

$$= -\frac{(\alpha_2 + \alpha'_2)}{2}(u_o - u_6) - \frac{(\alpha_2 + \alpha'_2)}{2}(v_o - v_6)$$

$$X_7 = -\alpha'_2 (u_o - u_7) - (\alpha_2 - \alpha'_2)\left(\frac{-1}{\sqrt{2}}\right)\left[\frac{-1}{\sqrt{2}}(u_o - u_7) + \frac{1}{\sqrt{2}}(v_o - v_7)\right]$$

$$= -\frac{(\alpha_2 + \alpha'_2)}{2}(u_o - u_7) - \frac{(\alpha_2 + \alpha'_2)}{2}(v_o - v_7)$$

$$X_8 = -\alpha'_2 (u_o - u_8) - (\alpha_2 - \alpha'_2)\left(\frac{1}{\sqrt{2}}\right)\left[\frac{1}{\sqrt{2}}(u_o - u_8) - \frac{1}{\sqrt{2}}(v_o - v_8)\right]$$

$$= -\frac{(\alpha_2 + \alpha'_2)}{2}(u_o - u_8) - \frac{(\alpha_2 + \alpha'_2)}{2}(v_o - v_8) \qquad ...(7)$$

Therefore, the equation of motion in $x$ direction will be

$$M\frac{d^2 u_o}{dt^2} = -\alpha_1 (2u_o - u_1 - u_2) - \alpha'_1 (2u_o - u_3 - u_4) - \frac{(\alpha_2 + \alpha'_2)}{2}$$

$$\left(4u_o - \sum_5^8 u_i\right) - \frac{(\alpha_2 - \alpha'_2)}{2}(-v_5 - v_6 + v_7 + v_8) \qquad ...(8)$$

In a similar manner, the equation of motion in $y$ direction will be

$$\frac{d^2 v_o}{dt^2} = -\alpha_1 (2v_o - v_3 - v_4) - \alpha'_1(2v_o - v_1 - v_2) - \frac{(\alpha_2 + \alpha'_2)}{2}$$

$$\left(4v_o - \sum_5^8 v_i\right) - \frac{(\alpha_2 - \alpha'_2)}{2}(-u_5 - u_6 + u_7 + u_8) \qquad ...(9)$$

To solve these equations, we take the plane wave solution as

$$u_i = \xi_1 \exp i\,(q_1 x_i + q_2 y_i - \omega t)$$

$$v_i = \xi_2 \exp i\,(q_1 x_i + q_2 y_i - \omega t) \qquad ...(10)$$

Substituting these solutions and converting the exponentials into trigonometric functions using $C_j = \cos aq_j$, $S_j = \sin aq_j$, one obtains

$$\xi_1 [\alpha_1 (2 - 2C_1) + \alpha'_1 (2 - 2C_2) + 2(\alpha_2 + \alpha'_2)(1 - C_1C_2) - M\omega^2] + 2(\alpha_2 - \alpha'_2) S_1S_2 \xi_2 = 0$$

$$2(\alpha_2 - \alpha'_2) S_1S_2 \xi_1 + \xi_2 [2\alpha_1 (1 - C_2) + 2\alpha'_1 (1 - C_1) + 2(\alpha_2 + \alpha'_2)(1 - C_1C_2) - M\omega^2] = 0 \qquad ...(11)$$

The secular determinant is given by

$$\begin{vmatrix} 2\alpha_1(1-C_1)+2\alpha'_1(1-C_2)+2(\alpha_2+\alpha'_2)(1-C_1C_2)-M\omega^2 & 2S_1S_2(\alpha_2-\alpha'_2) \\ 2S_1S_2(\alpha_2-\alpha'_2) & -M\omega^2+2\alpha_1(1-C_2)+2\alpha'_1(1-C_1)+2(\alpha_2+\alpha'_2)(1-C_1C_2) \end{vmatrix} = 0 \quad ...(12)$$

This will give the phonon dispersion of a two-dimensional square lattice in [10] and [11] symmetry directions as shown in Fig. 6.34.

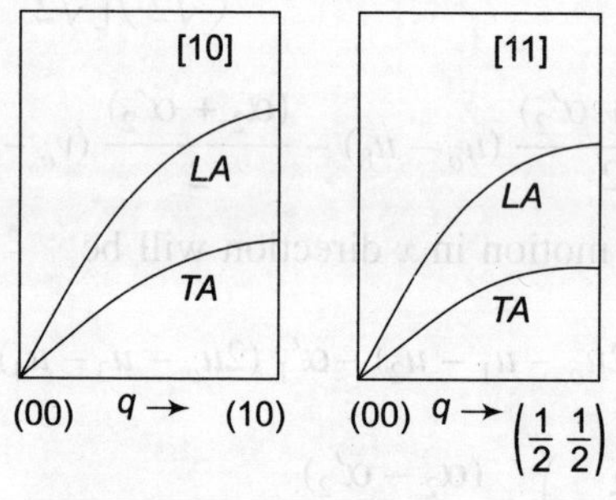


**Figure 6.34** Phonon dispersion for a square lattice in symmetry directions

In [10] direction, at the first Brillouin zone boundary

$$aq_1 = \pi,\ aq_2 = 0,\ C_1 = -1,\ C_2 = 1,\ S_1 = 0,\ S_2 = 0 \quad ...(13)$$

giving

$$M\omega_L^{\,2} = 4\alpha'_1 + 4(\alpha_2 + \alpha'_2)$$

$$M\omega_T^2 = 4\alpha'_1 + 4(\alpha_2 + \alpha'_2) \quad ...(14)$$

Similarly, in [11] direction, at the first Brillouin zone boundary,

$$aq_1 = \frac{\pi}{2},\ aq_2 = \frac{\pi}{2},\ C_1 = 0,\ C_2 = 0,\ S_1 = 1,\ S_2 = 1 \quad ...(15)$$

giving

$$M\omega_L^2 = 2\alpha_1 + 2\alpha'_1 + 4\alpha'_2$$

$$M\omega_T^2 = 2\alpha_1 + 2\alpha'_1 + 4\alpha'_2 \quad ...(16)$$

One should note the important feature of polarization of lattice waves in two-dimensional lattice, which was not existing in case of a linear chain. In case of longitudinal branches the polarization vector is parallel to wave vector $\vec{q}$ and for transverse branches, the polarization vector is perpendicular to wave vector $\vec{q}$.

## 6.32 MODIFICATION OF LATTICE—DIFFRACTION BY LATTICE WAVES

Now let us see what the diffraction pattern of a vibrating lattice looks like, by examining the behaviour of the structure factor

$$S(q) = \frac{1}{N}\sum_j \exp(-i\mathbf{q}\cdot\mathbf{R}_j) \qquad ...(1)$$

when it is supposed that the atomic coordinates $R_j$ are functions of the time. First of all, we recall that if $\mathbf{R}_j = \mathbf{R}_j^0$, the equilibrium lattice configuration, then the summation here simply yields $\delta_{\mathbf{q},\mathbf{g}}$, where $\mathbf{g}$ is a reciprocal lattice vector. To see how this *rigid lattice* diffraction pattern is modified by lattice vibrations, we set

$$\mathbf{R}_j = \mathbf{R}_j^0 + \mathbf{u}_j(t) \qquad ...(2)$$

where $\mathbf{u}_j(t)$ represents the displacement of the $j$th atom from its equilibrium lattice position $\mathbf{R}_j^0$. We suppose further that the state of vibration of the crystal may be characterized by a particular value $\mathbf{Q}$, say, of the lattice wave vector, so that we may write

$$\mathbf{u}_j(t) = \mathbf{A}\exp[i(\mathbf{Q}\cdot\mathbf{R}_j^0 - \omega t)] + \text{c.c.}$$

where c.c. stands for complex conjugate of the preceding term.

The structure factor now becomes

$$S(q) = \frac{1}{N}\sum_j \exp\{-i\mathbf{q}\cdot[\mathbf{R}_j^0 + \mathbf{A}\exp(i\mathbf{Q}\cdot\mathbf{R}_j^0) + \text{c.c.}]\}$$

in which the time dependence is absorbed in the amplitude $A$. For lattice waves of small amplitudes, $|\mathbf{A}\cdot\mathbf{u}_j| \ll 1$, we may make the expansion

$$\exp(i\mathbf{q}\cdot\mathbf{u}_j) = 1 + i\mathbf{q}\cdot\mathbf{u} - \frac{1}{2}(\mathbf{q}\cdot\mathbf{u})^2 + \ldots \qquad ...(3)$$

Thus, if we drop terms of order $\mathbf{u}_j^2$ or higher in this expansion, then we find

$$S(q) = \frac{1}{N}\sum_j \exp(-i\mathbf{q}\cdot\mathbf{R}_j^0) + (i\mathbf{q}\cdot\mathbf{A})\frac{1}{N}\sum_j \exp[-i(\mathbf{q}-\mathbf{Q})\cdot\mathbf{R}_j^0]$$

$$+ (i\mathbf{q}\cdot\mathbf{A}^*)\frac{1}{N}\sum_j \exp[-i(\mathbf{q}+\mathbf{Q})\cdot\mathbf{R}_j^0] \qquad ...(4)$$

$$= \delta_{\mathbf{q},\mathbf{g}} + (i\mathbf{q}\cdot\mathbf{A})\,\delta_{\mathbf{q},\mathbf{g}-\mathbf{Q}} + (i\mathbf{q}\cdot\mathbf{A}^*)\,\delta_{\mathbf{q},\mathbf{g}+\mathbf{Q}}.$$

This result may be interpreted physically as follows. The first term is the elastic scattering part that gives rise to the Laue pattern at reciprocal lattice points characterized by $\mathbf{g}$; the remaining two terms are the coherent inelastic scattering parts that appear as a pair of satellites associated with each reciprocal lattice vector in the Laue pattern as shown in Fig. 6.35.

We may accordingly write the selection rule for inelastic scattering in the form

$$\mathbf{k}_i - \mathbf{k}_f = \mathbf{g} \pm \mathbf{Q}. \qquad ...(5)$$

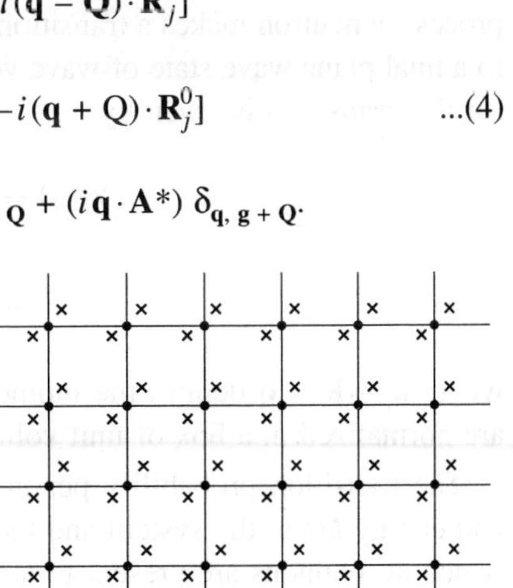

**Figure 6.35** Laue pattern for neutron diffraction from a lattice with lattice vibrations

Since the phonon frequency $\omega_Q$ is periodic in **q**-space with the periodicity of the reciprocal lattice, the value of **Q** in this relation may be restricted to the Brillouin zone. With this convention, there are two possible ways in which neutron scattering with quantized lattice waves (phonons) can take place. In one process (called normal or $N$ process), $\mathbf{g} = 0$, so that $\mathbf{k}_i - \mathbf{k}_f$ lies inside the first Brillouin zone. In the other process (called Umklapp or $U$ process), $\mathbf{k}_i - \mathbf{k}_f$ does not lie inside the first Brillouin zone; i.e., $\mathbf{g} \neq 0$.

This distinction leads to important physical effects, such as finiteness of the lattice thermal conductivity.

## 6.33 DYNAMIC FORM FACTOR

In order to gain additional insight into the nature of inelastic neutron scattering, let us now develop the concept of dynamic form factor, that is, the analog of atomic scattering factor. Neutron scattering in solids is of two types, viz., magnetic and nonmagnetic scattering. In the former, which does not concern us here, the scattering is between neutrons and the electrons: the neutron has associated with it a magnetic moment that gets coupled to the electron spin density. In the latter, the scattering is mainly between the neutrons and the atomic nuclei. This interaction may be represented by a Fermi pseudopotential:

$$V_I(\mathbf{R}) = \frac{\hbar^2}{2m_n} \sum_j 4\pi a_j \, \delta(\mathbf{R} - \mathbf{R}_j) \qquad ...(1)$$

where $\mathbf{R}_j$ is the position of the $j$th nucleus whose scattering length is $a_j$ and **R** is the position of an incoming neutron, $m_n$ being the neutron mass.

We proceed to determine the neutron scattering cross section on the basis of the assumption that $V_I$ is sufficiently weak that the Born approximation may ne used. In a typical scattering process, a neutron makes a transition from a plane wave state of wave vector $\mathbf{k}_i$ and energy $E_i$ to a final plane wave state of wave vector $\mathbf{k}_f$ and energy $E_f = (\hbar^2/2m_n)\, k_f^2$: The matrix element for this transition is given by

$$V_I(q) \equiv \langle \mathbf{k}_f | V_I | \mathbf{k}_i \rangle = \int d\mathbf{R} \exp(-i\mathbf{k}_f \cdot \mathbf{R}) \, V_I(R) \exp(+i\mathbf{k}_i \cdot \mathbf{R})$$

$$= \frac{2\pi\hbar^2}{m_n} \sum_j a_j \exp(-i\mathbf{q} \cdot \mathbf{R}_j) \qquad ...(2)$$

where $\mathbf{k}_f - \mathbf{k}_i = \mathbf{q}$ defines the momentum $\hbar q$ transferred to the system and the plane waves are normalized in a box of unit volume.

The transition probability per unit time that the neutron beam transfers momentum $\hbar q$ and energy $\hbar\omega$ to the system and thereby causes the system to go from an initial state $|0\rangle$ in which no phonons are present to a final state $|s\rangle$ in which one or more phonons are created (or annihilated) is, by virtue of Fermi's golden rule.

$$R(q, \omega) = \frac{2\pi}{\hbar} \sum_s |\langle s| V_I(q) |0\rangle|^2 \, \delta[\hbar\omega - (E_s - E_o)]$$

in which $E_0$ and $E_s$ designate the energies of the initial and the final states. In practice, we do not have one single initial state but rather a statistical distribution over initial states given by

$$\rho_{s'} = \frac{\exp(-\beta E_{s'})}{\sum_n \exp(-\beta E_n)}$$

where $\beta = 1/k_\beta T$. Then the most general formula for the transition rate becomes

$$R(q, \omega) = \frac{2\pi}{\hbar} \sum_{ss'} \rho_{s'} |\langle s| V_1(q)| s'\rangle|^2 \, \delta[\hbar\omega - (E_s - E_{s'})]. \quad ...(3)$$

To calculate the differential scattering cross section, we need the transition rate per unit time into an interval $d\mathbf{p}_f = \hbar^3 d\mathbf{k}_f$ of the final momentum state given by

$$R' = \frac{R(q, \omega)\hbar^3 \, d\mathbf{k}_f}{(2\pi)^3}.$$

We further express $d\mathbf{k}_f$ in the form

$$d\mathbf{k}_f = \mathbf{k}_f^2 \, dk_f \, d\gamma = \frac{m_n k_f \, d\omega \, d\gamma}{\hbar}$$

in terms of an element of solid angle $d\gamma$ and an energy transfer interval $\hbar \, d\omega$. The differential cross section $d\sigma$ is defined as the ratio of $R'$ to the flux of incident particles,

$$d\sigma = \frac{R'}{|\mathbf{v}_i|}$$

where $\mathbf{v}_i = \hbar\mathbf{k}_i/m_n$ is the velocity of particles in the incident (monochromatic) beam. Thus, on gathering results, we find

$$\frac{d^2\sigma}{d\omega \, d\gamma} = \left(\frac{m_n}{2\pi}\right)^2 \frac{|\mathbf{k}_f|}{|\mathbf{k}_i|} \sum_{ss'} |\langle s| \, V_I(q)|s'\rangle|^2 \, \delta[\hbar\omega - (E_s - E_{s'})]. \quad ...(4)$$

For the case of *coherent* scattering, characterized by having all $a_j = a$ (constant, i.e., independent of $j$), the differential cross section collapses to a simple expression of the form

$$\frac{d^2\sigma}{d\omega d\gamma} = AF(q, \omega)$$

where, by virtue of Eq. (2), $A = (a\hbar^2)^2 \, |\mathbf{k}_f|/|\mathbf{k}_i|$, and

$$F(q, \omega) = \sum_{ss'} |\langle s| \sum_j \exp(-i\mathbf{q} \cdot \mathbf{R}_j) \, |s'\rangle|^2 \, \delta(\omega - \omega_{ss'}), \quad ...(5)$$

$\hbar\omega_{ss'} = (E_s - E_{s'})$ being the excitation energy of the system. The function $F(q, \omega)$ is called the *dynamic form factor*: It depends only on the system itself rather than the nature of the probe that can be neutron beam, X-rays, or electrons; and it contains all possible information that can be extracted from a scattering experiment about a system.

The dynamic form factor has a simple physical significance that can be exhibited as follows. If we make use of the standard substitution

$$\delta(\omega) = \frac{1}{2\pi}\int_{-\infty}^{+\infty} dt \exp(i\omega t) \quad \text{...(6)}$$

and the completeness relation $\Sigma s\, |s\rangle\langle s| = 1$, as well as set $s' = 0$, then we may re-express the dynamic form factor in the form

$$F(q, \omega) = \frac{1}{2\pi}\int dt \exp(i\omega t) \sum_s \langle 0| \sum_j \exp(-i\mathbf{q}\cdot\mathbf{R}_j)\, |s\rangle$$

$$\langle s| \sum_k \exp(+i\mathbf{q}\cdot\mathbf{R}_k)\, |0\rangle$$

$$= \frac{1}{2\pi}\int dt \exp(iwt)\, \langle 0| \sum_{j,k} \exp[i\mathbf{q}\cdot\mathbf{R}_k - \mathbf{R}_j)]\, |0\rangle$$

$$= \frac{1}{2\pi}\int dt\, d\mathbf{R} \exp[i(\omega t - \mathbf{q}\cdot\mathbf{R})]\, \langle n(R, t)\, n(0, 0)\rangle$$

$$= \frac{N}{2\pi}\int dt\, d\mathbf{R} \exp[i(\omega t - \mathbf{q}\cdot\mathbf{R})]\, P(\mathrm{R}, t). \quad \text{...(7)}$$

Here, $$n(\mathbf{R}, t) = \sum_j \delta[\mathbf{R} - \mathbf{R}_j(t)] \quad \text{...(8)}$$

is the atomic density function, and

$$P(\mathbf{R}, t) = \frac{1}{N}\int d\mathbf{R}'\, \langle n(\mathbf{R} + \mathbf{R}', t)\, n(R', 0)\rangle \quad \text{...(9)}$$

is called the density-density correlation function.

The Fourier transform of $n(\mathbf{R}, t)$ is simply

$$S(q) = \frac{1}{N}\int d\,\mathbf{R} \exp(i\mathbf{q}\cdot\mathbf{R})\, n(\mathbf{R}, t)$$

$$= \frac{1}{N}\sum_j \exp[-i\mathbf{q}\cdot\mathbf{R}_j(t)],$$

which is the structure factor introduced in Eq. (6). It depends on the instantaneous configuration of the atoms at any time. Equation (7) is now seen to represent the space and time Fourier transform of the density-density correlation function. This is a very general result.

## 6.33.1 The Debye-Waller Factor

Another way of analyzing the effect of lattice vibrations on neutron scattering is provided by the overall reduction of the scattered neutron intensity by the Debye–Waller factor, exp $(-2W)$, defined as follows. Since the intensity or, equivalently, the scattering cross section in elastic or inelastic coherent scattering is proportional to $S^*(\mathbf{q})\, S(\mathbf{q})$ or $|S(\mathbf{q})|^2$, we write

$$\langle |S(\mathbf{q})|^2\rangle = \left|\left\langle \frac{1}{N}\sum_j \exp(i\mathbf{q}\cdot\mathbf{R}_j^0)\exp(i\mathbf{q}\cdot\mathbf{u}_j)\right\rangle\right|^2 = \frac{(2\pi)^3}{\Omega}\sum_g \delta_{\mathbf{q},\mathbf{g}} \exp(-2W) \quad \text{...(10)}$$

where $$\exp(-2W) = \langle \exp(i\mathbf{q}\cdot\mathbf{u}_j)\rangle. \quad \text{...(11)}$$

The factorization leading to the separation of the Debye–Waller factor from the sum has been made possible by the fact that the thermal average in Eq. (12) does not depend on the particular displacement, $\mathbf{u}_j$, under consideration.

In order to appreciate the significance of the Debye–Waller factor, we may take another look at the expansion in Eq. (3), Art. 6.32. Then we see that $\langle \mathbf{q} \cdot \mathbf{u}_j \rangle = 0$ because $\mathbf{u}_j$ is a random thermal motion uncorrelated with the direction of $\mathbf{q}$. Thus, the important contribution (due to thermal motion) come from the quadratic term, $-\frac{1}{2} \langle \mathbf{q} \cdot \mathbf{u}_j \rangle^2$, which may be expressed in the form $-\frac{1}{6} \langle \mathbf{u}_j^2 \rangle q^2$. The term $\langle \mathbf{u}_j^2 \rangle$ is the *mean square displacement* of the atom.

In the Einstein model, the mean square displacement in a given direction $\langle \mathbf{u}_j^2 \rangle$ is readily calculated by making use of the equation of motion,

$$M \frac{d^2 \mathbf{u}_j}{dt^2} + b\mathbf{u}_j = 0 \left( \frac{b}{M} = \omega_E^2 \right),$$

and the fact that at high temperatures ($T > \Theta_E \equiv \hbar\omega_E / k_B$),

$$\frac{1}{2} b \langle \mathbf{u}_j^2 \rangle = \text{mean potential energy} = \frac{1}{2} k_E T.$$

The result is

$$\langle \mathbf{u}_j^2 \rangle = \frac{k_B T}{b} = \frac{\hbar^2 T}{M k_B \Theta_E^2}, \qquad ...(12)$$

from which it follows that $W$ is of order $T/\Theta_E^2$.

## EXERCISE

6.1 How lattice vibrations are quantized? Give evidence.

6.2 What are phonons? What information can be obtained by studying the phonon spectrum of solids?

6.3 Discuss the dynamic motion of three-atom chain.

6.4 Give some important characteristics of phonons.

6.5 Discuss how phonons may be treated as wave packets, i.e., particle like quanta.

6.6 Discuss Born–von Karman boundary conditions for a linear chain of $N$ identical atoms.

6.7 Describe and discuss vibrations in a one-dimensional monatomic lattice, i.e., linear chain of identical atoms.

6.8 Obtain the various vibrational modes of a linear monatomic lattice. Show that the frequencies for mode of propagation vectors $\vec{k}$ and $\vec{k} + \vec{G}$ are the same, where $\vec{G}$ is a reciprocal lattice vector.

6.9 From the expression or dispersion relation of a monatomic lattice of $N$ atoms with nearest neighbour interactions, show that the density of modes is

$$D(\omega) = \frac{2N}{(\pi)} \frac{1}{(\omega_m^2 - \omega^2)^{1/2}}$$

where $\omega_m$ is the maximum frequency.

6.10 Obtain the vibrational spectrum of a linear diatomic lattice and show that the spectrum cousists of two branches. Discuss the main features of thee branches.

6.11 (a) Discuss briefly the factors influencing the phonon mean free path in insulators.

(b) Explain the terms: U-process, N-process.

6.12 Describe the inelastic scattering of neutrons for the experimental determination of phonon spectra.

6.13 Discuss inelastic scattering of photons by phonons.

6.14 Calculate the dispersion relations for a three-dimensional fcc crystal. Also give a plot of phonon dispersion relations.

CHAPTER

# 7

# Basics of Thermodynamics, Statistical Physics and Boltzmann Transport Equation

## 7.1 INTRODUCTION

The laws of thermodynamics are generalizations of common experience. Simple measurements may be made of pressure, volume, temperature, chemical composition, and other appropriate quantities; such data specify the *state* of the object or region of interest (called the *system*) and all its properties. If a system is left undisturbed, it will eventually come to equilibrium, and all its properties will no longer vary with time. The first law of thermodynamics is the statement of the conversation of energy, which leads to the conclusion that the internal energy of any system is a function only of the thermodynamic state. Although heat and work are equivalent forms of energy, they are not completely interchangeable. According to the second law, heat may never be entirely converted to work, and heat never changes spontaneously to work. Such behaviour is described by the maximization of a new state function, the *entropy* which increases as a system approaches equilibrium and is maximized at equilibrium. Such quantities are relatively easy to measure and permit the prediction of the equilibrium state of any thermodynamic system. The spontaneous occurrence of a natural event can then be described in terms of a decrease in the appropriate free energy function.

The basis of thermodynamics is common, everyday experience. The laws of thermodynamics are, it is true, exact and mathematical in nature; however, each law follows from simple and direct observation of natural events. By way of example, consider what we must know in order to specify the properties of such complicated system as a solid object or a body of gas. If the position and velocity of every atom therein were precisely known, and all inter-atomic and other forces specified, and all of the properties of individual atoms known, it might be possible, by tremendous labour, to calculate or predict all of the physical properties of whatever portion of the universe happened to be of interest. This portion of the universe is, incidentally, generally referred to as a *system*. Thermodynamics is not concerned with such atomic or microscopic detail but deals instead with macroscopic, directly measurable quantities. Pressure, volume, temperature, and chemical composition are typical of the microscopic quantities. If necessary, fields, whether gravitational or electromagnetic, may also be measured. In solids, the mechanics of elasticity and plasticity are important.

It is always possible, if the set of measurements is complete enough, to determine all other physical properties of the system. Such a set of numbers is said to fix the thermodynamic state of the system. With each measurement, or set of measurements, an equilibrium is associated. If, for instance, no motion occurs, owing to complete balancing of forces, mechanical equilibrium has been attained. If the chemical composition does not change with time anywhere, chemical equilibrium has occurred. Thermal equilibrium is associated with the absence of heat flow because temperature is everywhere equalized. When all of the variables which specify the thermodynamic state are not changing, the system is in thermodynamic equilibrium.

If our measurement of the state of any system is to be free of embarrassing uncertainties, the state involved should be an equilibrium state. Observations made "in transit" are likely to depend on where we have been, rather than where we are. This dependence upon past history is common in non-equilibrium situations, which are by far the most frequently encountered. After all equilibrium is by definition the last state arrived at by an unperturbed system; it is a universal rarity. It is important only as a guide to behaviour, that is, as an indication of the direction in which natural events move. This direction is really the fundamental problem of thermodynamics. Given a certain set of initial conditions and the knowledge that the system will spontaneously move towards equilibrium, one must predict the equilibrium state. The experimental and analytical determination of quantities that define the state and changes in states provide the principal tools of thermodynamics.

## 7.2 ENERGY CONSERVATION, STATE FUNCTIONS, AND THE FIRST LAW

We will assume a certain familiarity with the nature of energy and the fact that it may exist in the equivalent forms of heat and work. The most familiar principle involving energy is that energy must be conserved. Such a statement, however, leads immediately to the following conclusion: the internal energy content of any system is a single valued function only of the thermodynamic state.

If this conclusion were not so, contradiction of the energy conservation principle would be a simple matter. A system is chosen which is relatively isolated from the rest of universe; a change of state is effected and the change in energy content noted. The system is then returned to its initial state by a completely different method. The net change of the internal energy on completion of the "cycle" must be zero; otherwise energy would have appeared or disappeared, in contradiction to a principle previously agreed upon. The methods by which the changes in state were accomplished are completely arbitrary and, therefore, the energy content change which accompanies any change of state must not depend on the path, but only on the initial and final states, whether they are equilibrium states or non-equilibrium states. Similar reasoning applies to any conserved quantity, which is then referred to as a *state function*.

For any system which undergoes a change in state, we may then write

$$\Delta E = q - w \qquad ...(1)$$

where $E$ is the internal energy, $q$ the total heat input to the system, and $w$ the total work output during the change. The first law states that the change in internal energy corresponding to a change of state between stipulated initial and final states does not depend on the method used to achieve the transformation. The quantities $q$ and $w$ are not properties of the system and will depend on methods; $\Delta E$, however, is a function only of the states, and is therefore independent of the path between them.

A particularly convenient change of state to observe is the rise in temperature due to input of heat. Such observations made under common laboratory conditions lead to the definition.

$$C_p \equiv \left(\frac{\partial q}{\partial T}\right)_P \qquad ...(2)$$

where $C_p$ is called the heat capacity at constant pressure. No linear relationship is implied between $q$ and $T$ by Eq. (2); $C_p$ is merely the instantaneous slope of a more complicated curve. It is useful to connect such experimental quantities with the state functions. If only the atmosphere does work on our system, the differential of $E$ is then

$$dE = dq - P\,dV \qquad ...(3)$$

If the volume is held constant, then

$$\left[\frac{\partial E}{\partial T}\right]_v = \left[\frac{dq}{dT}\right]_v = C_V \qquad ...(4)$$

where $C_v$ is called the heat capacity at constant volume. It is usually difficult, however, to hold the volume constant. Consequently, it is convenient to define a new state function, the *enthalpy*:

$$H \equiv E + PV \qquad ...(5)$$

Because the thermodynamic state includes values of $P$ and $V$, and $E$ is a state function, $H$ is also a single valued function of state. The differential of $H$ at constant pressure leads immediately to

$$\left[\frac{\partial H}{\partial T}\right]_P = \left[\frac{dq}{dT}\right]_P = C_P \qquad ...(6)$$

The enthalpy, which is also conserved, is therefore a more useful state function in some ways than the energy, as it is more directly related to our measurements.

## 7.3 THE SECOND LAW

The previous section refers briefly to the equivalence of heat and work as forms of energy. Experience shows, however, that such a statement is not completely and unconditionally acceptable. If heat and work were always completely interchangeable, truly remarkable events might occur. Stones, cooling themselves by converting their heat contents to potential energy (i.e., work) could roll uphill unaided; one could never trust hot water to "stay put" as it might at any time convert its heat to kinetic or potential energy, with uncomfortable results. In general then, nature does not allow heat to be spontaneously converted to work without the accompaniment of other changes. This last statement is actually Kelvin's formulation of the second law. Steam, gasoline, and other engines all convert heat to work; what "other changes" occur during such conversion?

For the sake of simplicity, friction should be eliminated from consideration. As a limiting case, the idea of a *reversible process*, or a reversible changes, is devised. The condition of reversibility is exactly what the term implies. A reversible process may be exactly retraced, or reversed, to its beginning. The process may be carried out backwards and everything restored to its original state. *Dissipation*, or friction, would then be out of the question, as the energy dissipated is irrecoverable. A pendulum, for instance, can retrace its path exactly only if friction is absent reversible processes operate with maximum efficiency. If a number of different reversible methods accomplish the same change of state, they must all be equally efficient. Even in the absence of obvious sources of friction, a process can be reversible only if the system is, at all times, very close to equilibrium. The situation must be one of *virtual equilibrium*; the process is a series of very minute (ideally infinitesimal) perturbations from equilibrium. Two examples, comparisons of reversible and irreversible means of accomplishing the same process, are seen in Figs. 7.1 and 7.2.

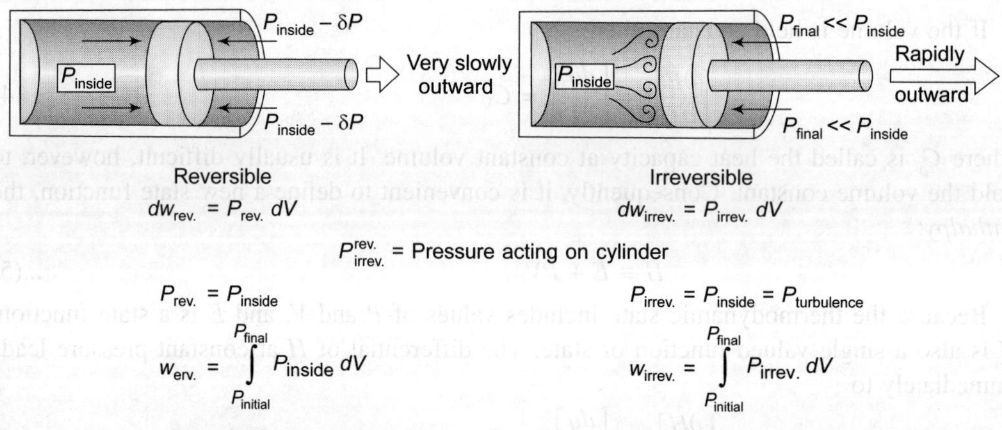


$$dw_{rev.} = P_{rev.}\, dV \qquad\qquad dw_{irrev.} = P_{irrev.}\, dV$$

$$P^{rev.}_{irrev.} = \text{Pressure acting on cylinder}$$

$$P_{rev.} = P_{inside} \qquad\qquad P_{irrev.} = P_{inside} = P_{turbulence}$$

$$W_{erv.} = \int_{P_{initial}}^{P_{final}} P_{inside}\, dV \qquad\qquad W_{irrev.} = \int_{P_{initial}}^{P_{final}} P_{irrev.}\, dV$$

$$PV = NRT \qquad\qquad P_{irrev.} < P_{rev.}$$

$$w_{rev.} = NRT\ln\left(\frac{P_{initial}}{P_{final}}\right) \qquad\qquad w_{irrev.} < w_{rev.}$$

**Figure 7.1** Reversible and irreversible expansion

Temporarily then, friction may be dropped from the discussion, and the reversible conversion of heat to work analyzed. It is simplest to utilize a hypothetical engine, which absorbs heat, does work, and is then restored to its initial state, all in a reversible manner. The changes occurring over the entire cycle may be regarded as that attendant on the conversion of heat to work in the most efficient manner possible. Detailed analyses of reversible heat engines of this sort will be found in many of the references at the end of this chapter. We will use in many of the references at the end of this chapter. We will use only the result of primary interest, i.e., the theorem of Clausius:

$$\oint \frac{dq}{T} \text{"} 0 \qquad \text{...(7)}$$

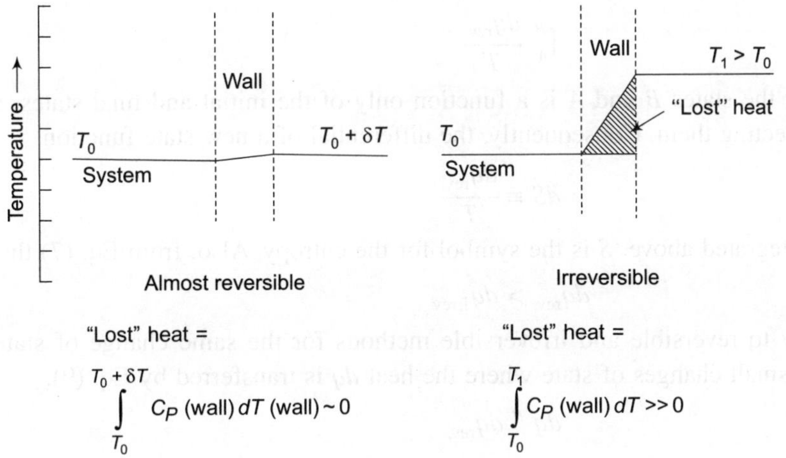


**Figure 7.2** Reversible and irreversible heat flow

The integral is taken around one complete cycle. The inequality applies to irreversible cycles, and the equality to reversible cycles. $T$ is an "absolute" temperature. We could not, according to Clausius' theorem, just extract heat and convert it to work. During the cycle it would be necessary to reject heat as well as in order to satisfy Eq. (7). The rejected heat, of course, represents that portion of the heat intake which cannot be converted to work. Also, rejection of heat must take place at lower temperatures than the heat intake, if work is to be done. This is easily seen. To satisfy the first law (Eq. 1), the heat input minus the heat output must be equal to the work done. If the work is positive, the heat output is always less than the original heat input. Consequently, the range of temperature over which heat rejection occurs is, in general, lower than the temperature range where heat is absorbed in order for Eq. (7) to be satisfied. If heat is absorbed at a single temperature, and no lower temperature is available, Eq. (7) requires that the same heat be rejected to that temperature, and no work done. Thus, one cannot convert to work the large amounts of heat energy available in the oceans for instance, as no longer, low temperature *'reservoirs"* exist for the dumping of the amounts of heat required by Eq. (7). The inequality in the above example corresponds to the lower efficiency of irreversible methods; more of the heat taken in must be rejected, and consequently less converted to work.

We should not lose sight of our intention. Equation (7) is not intended to be an "explanation" of the natural phenomena we have discussed, but rather a description. The scope of this description may now be expanded greatly by the introduction of a new thermodynamic variable, the *entropy*. Thus far, the purposeful conversion of heat to work, by various cyclic devices, has been analyzed; now, the second law must be extended to natural, "spontaneous" events mentioned at the beginning of this section. The simplest way to accomplish the above, is to use Clausius' theorem to define a new state function. For a reversible cycle, the integral in Eq. (7) is zero. Two thermodynamic states connected by *any* two reversible paths constitute such a cycle, regardless of path. Therefore, the integral

$$\int_A^B \frac{dq_{rev.}}{T}$$

connecting the states *B* and *A* is a function only of the initial and final states, and not the paths connecting them. Consequently, the differential of a new state function

$$dS \equiv \frac{dq_{rev.}}{T} \qquad ...(8)$$

is being integrated above. *S* is the symbol for the entropy. Also, from Eq. (7) the relation

$$dq_{rev.} > dq_{irrev.} \qquad ...(9)$$

must apply to reversible and irreversible methods for the same change of state. Consider arbitrarily small changes of state where the heat *dq* is transferred by Eq. (9).

$$dq\ ''\ dq_{rev.} \qquad ...(9a)$$

The change in entropy of the system, regardless of reversibility, is

$$dS_{sys.} = \frac{dq_{rev.}}{T}$$

as entropy is determined by the initial and final states only. Because *dq* is so small, the surroundings may be returned to their initial state by restoration of *dq*; all other changes are presumed to be second order. The entropy changes of the surroundings may be obtained by restoring *dq* reversibly:

$$dS_{surr.} = -\frac{dq}{T}$$

The *net* entropy change during any such differential state change is then

$$dS_{net} = dS_{sys.} + dS_{surr.} = \frac{dq_{rev.} - dq}{T} \geq 0 \qquad ...(10)$$

Once again, the equality refers to the reversibility. A spontaneously occurring process is by nature irreversible, otherwise it would not have occurred. Reversibility is attainable only when equilibrium is so near that no observable changes will occur. According to Eq. (10) then, *any spontaneous (natural) event implies an increase in total entropy*. The equality in Eq. (10) simply means that *entropy* is *maximized at equilibrium*.

These statements or the equations, are the most general form of the second law. *It is our experience,* then, that no event occurs spontaneously which decreases entropy, including those of the hot water and the rolling stone, which were discussed at the beginning of this section.

## 7.4 ENTROPY, TEMPERATURE, AND THE THIRD LAW

There is an important omission in the preceding sections; the temperature is undefined. Pressure, volume, work, heat, and the other variables have definitions which can be carried over from classical mechanics. To say, "temperature is what is measured with a thermometer," is unsatisfactory as no ordinary thermometer will measure the absolute temperature, which is the only useful thermodynamic scale. The only thermodynamically consistent temperature scale is obtained by using the second law. To do this, we consider a reversible, cyclic engine, as in Section 7.3. This engine, however, has the special ability to absorb heat isothermally at the upper temperature $T_2$ and to reject it isothermally at the lower temperature, $T_1$. The Clausius' theorem, (Eq. 7) applied to such a cycle gives

$$\frac{Q_2}{T_2} + \frac{Q_1}{T_1} = 0 \qquad ...(11)$$

where $Q_2$ is the heat absorbed at temperature $T_2$ and so on. The temperature scale is now established by the ratio of heat input to outflow

$$\frac{Q_2}{Q_1} = -\frac{T_2}{T_1} \qquad ...(11a)$$

This relation is independent of the manner of construction of the engine. We may now arbitrarily assign a number to some easily recognized temperature (such as the freezing point of pure water at one atmosphere pressure), and all other such temperatures follow from Eq. (11a).

Having used the absolute temperature to define the second law (entropy), we now find the second law is the only consistent way to define temperature. There is only one way out; entropy must be accepted as a fundamental variable, as basic in nature as temperature.

In mechanics, for instance, we have a consistent system: length, time, mass, force, momentum, energy. Only one of the last four need can be chosen, and the rest can be defined in terms of it, and length and time. Such a choice is arbitrary and is usually made according to the way in which one wishes to see the laws of mechanics formulated. So it is with entropy and temperature. We may choose either to define the other, but neither is to be regarded as "more fundamental".

One more requirement on the entropy must be made. Integration of Eq. (8) will give the entropy, up to some arbitrary constant of integration. In 1906, Nernst, following certain suggestions of Planck, found that at absolute zero the entropies of all systems must be equal. This universal constant has been set equal to zero:

$$[S]_{T=0} = 0 \qquad ...(12)$$

The above relation, sometimes called the third law of thermodynamics, allows the calculation of *absolute entropy* by setting the lower limit of integration at the absolute zero of temperature. The integral itself is defined by combining Eqs. (2) and (8)

$$dS = \frac{dq_{\text{rev.}}}{T} = \frac{C_p dT}{T} \qquad ...(13)$$

and therefore $$S(T) = \int_0^T \frac{C_p}{T'}\,dT' \qquad ...(13a)$$

## 7.5 FREE ENERGIES: THE THERMODYNAMIC POTENTIALS

We have now almost done what we set forth to do in the introductory section of this chapter. A new variable, the entropy, has been defined. This variable increases as a system moves toward equilibrium, and it is maximised at equilibrium. Consequently, if we can calculate the entropy of any system, we can also predict what changes will occur and where equilibrium will be reached. As a final step, a criterion for equilibrium will be derived which is in a more familiar form and involves quantities which are more directly measurable. Sometimes, when we seek to compute the net change in entropy accompanying an event of interest, we find the coupling of the system to the surroundings so complicated that such computations become laborious, if not possible. The equilibrium criteria which follow circumvent this problem.

In mechanics, the significance of a minimum in the potential energy function is well understood. Such minima represent equilibrium positions. Similarly, a good "thermodynamic potential" will be minimized at thermodynamic equilibrium. The problem is one of the choice. The internal energy is not suitable. We have shown, in the previous sections, that all of the internal energy may not be utilized to do work. Work is really the basis of a well-defined potential energy. We do have, however, an accurate way to calculate how much of the internal energy may not be utilized thus: the entropy. According to the second law, if the heat $Q_0 = T_0\,\Delta S$ is taken on at temperature $T_0$, with the intention of converting it to work, with all processes carried out in the most efficient (i.e., reversible) manner, we shall still have to reckon with the necessity of dropping off the heat $Q_1 = -T_1\,\Delta S$ at some lower temperature $T_1$. Otherwise, the system itself cannot be returned to its original state. As the heat $Q_1$ represents the portion of $Q_0$ which cannot be converted to work, and we can obtain $Q_1$ for any temperature provided the entropy change is known, the entropy is commonly associated with *unavailable* energy. Consequently, as a first try, we consider the function

$$A \equiv E - TS \qquad ...(14)$$

*A* is referred to as the *Helmholtz free energy*. If Eq. (14) is differentiated and the first law, $dE = dq - dw$, and the definition of entropy Eq. (8) are both used, the relation

$$\begin{aligned} dA &= dE - T\,dS - S\,dT \\ &= dq - dq_{\text{rev.}} - dw - S\,dT \end{aligned} \qquad ...(14a)$$

results. Now, if we stimulate that the temperature be constant, and no work be done on or by the system,

$$[dA]_{w,T} = dq - dq_{\text{rev.}}$$

From the second law, Eq. (1.9a) states that

$$dq\,''\,dq_{\text{rev.}}$$

Consequently, $$[dA]_{W,T} \le 0 \qquad ...(15)$$

where the equality applies to equilibrium, and the inequality to spontaneous changes. In an isothermal system on which no work is done, the Helmholtz free energy decreases for all natural changes and is minimized at equilibrium. Maintaining constant temperature is a reasonable condition; no work, however is another story. How, for instance, do we prevent, expansion or contraction which together with atmospheric pressure, do the work

$$w = \int P\, dV$$

Similar problems exist for changes in magnetic and electric polarization and for many other properties. In section 7.2 the problem of volume changes was dealt with successfully for the case of heat capacities by considering enthalpy ($H$) rather than energy. We shall do the same here. The Gibbs free energy is defined

$$F \equiv H - TS = E + PV - TS \qquad ...(16)$$

where the last equality comes from Eq. (5) the definition of the enthalpy. We now subject the Gibbs free energy to the same routine that was performed on the Helmholtz free energy

$$\begin{aligned} F &= dE + P\, dV + V\, dP - T\, dS \;\; S\, dT \\ &= dq - dw + P\, dV + V\, dP - dq_{\text{rev.}} - S\, dT \qquad ...(17) \end{aligned}$$

If only the atmosphere does not work on the system, $dw = P\, dV$, and

$$dF = dq - dq_{\text{rev.}} + V\, dP - S\, dT \qquad ...(18)$$

Now we need only hold pressure and temperature constant, and

$$[dF]_{P,T} = dq - dq_{\text{rev.}} \qquad ...(19)$$

and therefore $$[dF]_{P,T} \,''\, 0 \qquad ...(20)$$

The Gibbs free energy is minimized at equilibrium if temperature and pressure are held constant. Such condition is relatively easy to attain. If magnetic, electric, or other fields are present, and cannot be neglected when the work is considered, the situation becomes more complicated. For our purposes, such necessities will not occur. The goal is now attained. From statements (i.e., thermodynamic laws) of universal experience, a state function is derived which decreases for all spontaneous changes and is minimized at thermodynamic equilibrium. Furthermore, this state function, the Gibbs free energy, is minimized under simple condition and is easy to measure.

## 7.6 RELATION BETWEEN THERMODYNAMIC AND EQUILIBRIUM STRUCTURE OF MATTER

It is well known from mechanics that *the stable equilibrium of a system corresponds to the absolute minimum of its total energy and vice versa.* What makes this principle very important for determining the properties and the structure of matter is that physical system do possess a finite minimum of their total energy; this minimum is realized because of the presence of both the potential energy (which as we shall see, is negative and tends to squeeze the system) and the quantum kinetic energy (which is always positive and tends to blow the system apart).

In other words, the total energy $E$ of a system is the sum of two[1] terms opposing each other; the total potential energy $E_p$ and the total kinetic energy $E_k$:

$$E = E_K + E_P. \qquad ...(1)$$

Both $E_P$ and $E_K$ depend on many parameters: for example, the volume $V$ of the system, the electronic concentration $n_e$ ($r$), the placement $R_\alpha$ of the atoms or ions of a solid in space, the orientation of the spins of the atoms or the ions, etc.

If one succeeds in minimizing the total energy $E$ with respect to all of these free parameters, one has obtained a full determination of the *ground state* of the system (assuming that no external pressure is exercised on the system). As an example, let us minimize $E$ with respect to the volume (assuming zero absolute temperature)

$$\frac{\partial E}{\partial V} = \frac{\partial E_K}{\partial V} + \frac{\partial E_P}{\partial V} = 0 \qquad ...(2)$$

However, $-[\partial E/\partial V]_{T=0}$ is the total pressure[2] $-(\partial E_K/\partial V)_{T=0}$ is the positive pressure $P_K$ due to the finite kinetic energy and $-(\partial E_P/\partial V)$ $T = 0$ is the negative pressure $P_P$ due to the attractive potential energy. Thus, we conclude that the minimization of the total energy with respect to the volume implies that the total pressure must be equal to zero, or equivalently, that $P_K = |P_P|$; hence, it follows from Eq. (2) that the squeezing pressure of the potential energy is balanced exactly at equilibrium by the repulsive pressure of the kinetic energy. If there is a non-zero external squeezing pressure of the kinetic energy. If there is a non-zero *eternal* squeezing pressure $P_0$, equilibrium demands that the difference $P_K - |P_P|$ must be equal to $P_0$

$$P_0 = P_K - |P_P|, \qquad ...(3)$$

This is equivalent to the minimization (with respect to the volume) of the quantity $E + P_0V$, which is called *enthalpy.*

Finally, if the system is placed in an environment of pressure $P_0$ and absolute temperature $T_0$, the stable equilibrium corresponds to the minimization of the quantity $E + P_0V - T_0S$, where $S$ is the *entropy* of the system.[3] Since $T$ at equilibrium $P = P_0$ and $T = T_0$, the quantity $E + P_0V - TS$ reduces to the so called *Gibbs free energy*, $G \equiv E + PV - TS$.

Under conditions of constant pressure and temperature (and no exchange of matter), the stable equilibrium of a system corresponds to the minimization of its Gibbs free energy

---

1 A third term in the total energy is the so called rest energy, $\Sigma m_{io}c^2$ where $m_{io}$ are the rest masses of the nuclei and the electrons making up the solid. However, the formation, transformations and other changes in solids do not involve measurable changes in $m_{io}$, thus, this third term is a constant which can be eliminated by a redefinition of the zero of the energy.

2 Remember the First Law of Thermodynamic, $dE = -PdV + TdS$ and set $T = 0$.

3 This is very important principle in an immediate consequence of the First Law ($dE = dQ - PdV$) and the Second Law ($dQ \le TdS$) of Thermodynamics. Indeed, by combining the two laws we obtain $dE \le TdS - PdV$, or $dE + PdV - TdS \le 0$, or (taking into account $T$ and $P$ are constant) $dG \le 0$, where $G \le E + PV - TS$. The last inequality under the said conditions, means that the Gibbs free energy $G$ decreases with time (or, in extreme cases remains constant), until its minimum value is reached upon which no further spontaneous change can occur, and consequently, stable equilibrium is established.

$$G \equiv E + PV - TS \qquad ...(4)$$

Notice that when $P = T = 0$, the Gibbs free energy ($G$) reduces to the energy $E$, and thus, recaptures the initial statement of this section. The important message to be remembered is that the process recipe to predict (or to justify a posteriori). The equilibrium structures of matter and their properties. Here is the recipe:

*For the system under study, obtained the Gibbs free energy G (under condition of constant P and T), or the energy E (under conditions P = T = 0), as a function of various free parameters (such as the volume, the positions of the atom and ions, the electronic concentration* $n_e(r)$*, etc.); minimize G or E with respect to all these free parameters; the resulting state is one of stable equilibria to be observed in nature.*

## 7.7 STATISTICAL MECHANICS

In 1906, the kinetic theory of gases emerged as a real mathematical theory of the gases. However, a real contact with the thermodynamics was made when Boltzmann developed his H-theorem and thereby established a direct connection between entropy on one hand and the dynamics of molecules on the other. Almost simultaneously, the kinetic theory began giving way to its more sophisticated successor—the ensemble theory. The power of the technique that finally emerged, reduced thermodynamics to the state of an essential consequence of the get-together of the statistics and the mechanics of the molecules constituting a given physical system and the resulting formalism was the *statistical mechanics.*

## 7.8 THE MACROSCOPIC AND THE MICROSCOPIC STATES

A system in thermodynamics is defined by the variables ($E$, $V$, $N$), where $N$ is the number of atoms measured by the mass. These are measurable and define the *microscopic state* of the system. A typical *macrostate* is a small volume of the gas. It contains about $10^{23}$ interacting atoms.

While the macroscopic variables are few, the microscopic description requires the knowledge of the positions and the velocities (or momenta) of all the $10^{23}$ atoms at any instant. As time lapses, these will change due to mutual collisions. An enormous number of variables and their associated equations of motion will have to be solved which would be tedious and impossible.

However, in kinetic theory, the reduction in the number of variables can be achieved by averaging over the velocities of the molecules. On the other hand, quite a different approach is provided by the method of statistical mechanics. It aims to study most probable behaviour of the system. To calculate it, according to Gibbs, an *ensemble* is constructed.

A large collection of macroscopically identical systems used for studying probability characteristics is called an *ensemble*. All the elements (identical imagined systems) of the ensemble are in the same macroscopic state or *macrostate* e.g. ($T$, $V$, $N$). They may however differ at any given instant in their microscopic state or *microstates*, specifying the maximum possible information about the gas molecules (i.e., their instantaneous positions and velocity of each molecule).

## 7.9 THE EXTENSIVE AND INTENSIVE PROPERTIES

Consider a physical system composed of $N$ identical particles confined to a space of volume $V$. In the thermodynamic limit $N \to \infty$ and $V \to \infty$ such that the ratio $N/V$ called the particle density remains at a preassigned fixed value then the extensive properties of the system become indirectly proportional to the size of the system (i.e., to $N$ or $V$) while the intensive properties become independent thereof. The particle density ($N/V$) remains, however, an important parameter for all physical properties of the system. Those variables which are proportional to the amount of matter are called *extensive variables*, e.g., volume $V$ and heat capacity. Those variables which are independent of the amount of matter are called *intensive variables*, e.g., temperature $T$, pressure $P$ and velosity, $\eta$.

## 7.10 GENERALIZED COORDINATES AND THE LAGRANGIAN

The number of *degrees of freedom* a given system has is equal to the minimum number of independent parameters necessary to uniquely determine the location and orientation of the system in physical space. Such independent parameters are called *generalized coordinates.* Thus, for example, a rigid dumbbell molecule free to move in three-space has five degrees of freedom: three to locate the centre of mass of the molecule and two (angles) to determine the orientation of the molecule. A fluid of $N$ such molecules has $5N$ degrees of freedom. A rigid triangular molecule has six degrees of freedom: five to locate an edge of the triangle and an additional angle to fix the orientation of the plane of the molecule about this axis.

For a system with $N$ degrees of freedom, generalized coordinates are labelled $q_1, q_2, \ldots, q_N$. These parameters may be taken to comprise the components of an $N$-dimensional vector,

$$q = (q_1, q_2, \ldots, q_N) \qquad \text{...(1)}$$

The corresponding *generalized velocities* also comprise components of $N$-dimensional vector, and we write

$$\dot{q} = (\dot{q}_1, \dot{q}_2, \ldots, \dot{q}_N) \qquad \text{...(2)}$$

The composite vector $(q, \dot{q})$ determines the *state* of the system.

Let the system at hand exist in a conservative force field with kinetic energy $T(q, \dot{q})$ and potential, $V(q)$. The *Lagarangian* of the system is then given by

$$L(q, \dot{q}) = T(q, \dot{q}) - V(q) \qquad \text{...(3)}$$

The dynamics of the system may be formulated in terms of *Hamilton's principle*. This principle states that the motion of the system between two fixed points, $(q, t)_1$ and $(q, t)_2$ renders the action integral

$$S = \int_1^2 L(q, \dot{q}, t)\, dt \qquad \text{...(4)}$$

an extremum.[4] That is,

$$\delta \int_1^2 L(q, \dot{q}, t)\, dt = 0 \qquad \text{...(5)}$$

[4] More precisely, a minimum.

where $\delta$ denotes a variation about the motion of the system.[5] See Fig. 7.3.

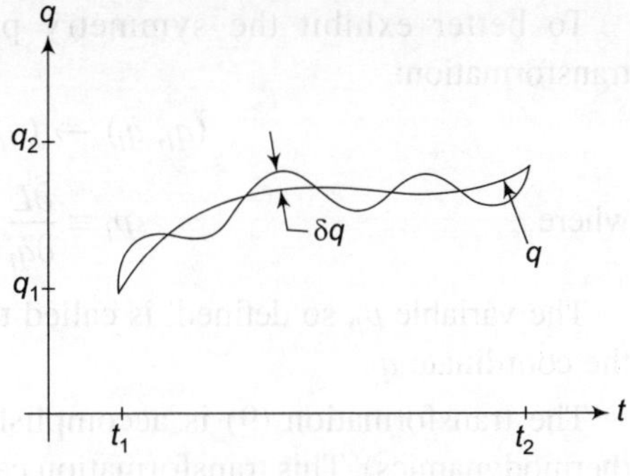


**Figure 7.3** End points $q_1$ and $q_2$ are held fixed in the variation (5)

## Lagrange's Equations

The dynamical principle (5) may be employed to obtain a differential equation for $L(q, \dot{q}, t)$. That is let $q(t)$ be the motion that renders $S$ an extremum.

Effecting the variation (5) and Taylor series expanding the integrand gives

$$\delta S = \int_1^2 L(q+\delta q, \dot{q}+\delta\dot{q}, t)\,dt - \int_1^2 L(q, \dot{q}, t)\,dt$$

$$= \int_1^2 \left( L + \frac{\delta L}{\delta q}\cdot\delta q + \frac{\delta L}{\delta \dot{q}}\cdot\delta\dot{q} \right) dt - \int_1^2 L(q, \dot{q}, t)\,dt \quad ...(6)$$

Note that $$\frac{\partial L}{\partial q}\frac{d}{dt}(\delta q) = \frac{d}{dt}\left(\frac{\partial L}{\partial \dot{q}}\cdot\delta q\right) - \frac{d}{dt}\left(\frac{\partial L}{\partial \dot{q}}\right)\cdot\delta q$$

Inserting this expression into the preceding equation gives

$$\left[\frac{\partial L}{\partial \dot{q}}\cdot\delta q\right]_1^2 + \int_1^2 \left[\frac{\partial L}{\partial q} - \frac{d}{dt}\left(\frac{\partial L}{\partial \dot{q}}\right)\right]\cdot\delta q\,dt = 0$$

Since the end points of the trajectory are held fixed in the variation, the first bracketed "surface terms" vanish. Insofar as the remaining integral must vanish for any arbitrary, infinitesimal variation $\delta q$, it is necessary that

$$\boxed{\frac{d}{dt}\left(\frac{\partial L}{\partial \dot{q}_l}\right) - \frac{\partial L}{\partial q_l} = 0 \quad (l = 1, 2, ..., N)} \quad ...(7)$$

Here we have returned to the component representation of $(q, \dot{q})$. Equation (12) is called Lagrange's equations.

## 7.11 HAMILTON'S EQUATIONS

When written in the form

$$\frac{d}{dt}\left(\frac{\partial L}{\partial \dot{q}_l}\right) = \frac{\partial L}{\partial q_l} \quad ...(8)$$

Lagrange's Eq. (7) reveals the following important observation: If the coordinate $q_l$ is missing in the Lagrangian, the quantity $\partial L/\partial \dot{q}_l$ is constant in time. In this event we say that the coordinate $q_1$ is *cyclic* or *ignorable.*

[5] Furthermore, the variation is a *virtual* displacement, that is, one that occurs in zero time.

To better exhibit the symmetry principle of mechanics, we effect the following transformation:

$$(q_l, \dot{q}_l) \rightarrow (q_l, p_l) \quad ...(9)$$

where
$$p_l \equiv \frac{\partial L}{\partial \dot{q}_l} \quad ...(10)$$

The variable $p_l$, so defined, is called the *canonical momentum* or momentum *conjugate* to the coordinate $q_l$.

The transformation (9) is accomplished through a Legendre transformation (familiar to thermodynamics). This transformation carries the Lagrangian $L(q, \dot{q})$ to the *Hamiltonian* $H(q, p)$ by the following recipe

$$H = \sum_l \frac{\partial L}{\partial \dot{q}_l} \dot{q}_l - L \quad ...(11)$$

Equations of motion in terms of the Hamiltonian are obtained by forming the differential of the latter equation.

$$dH(q, p, t) = \sum_l \left[ d\left(\frac{\partial L}{\partial \dot{q}_l}\right) \dot{q}_l + \frac{\partial L}{\partial \dot{q}_l} d\dot{q}_l - \frac{\partial L}{\partial \dot{q}_l} d\dot{q}_l - \frac{\partial L}{\partial \dot{q}_l} dq_l - \frac{\partial L}{\partial t} dt \right]$$

With Lagrange's Eq. (7) and the definition Eq. (10), we obtain

$$dH(q, p, t) = \sum_l (\dot{q}_l \, dp_l - \dot{p}_l \, dq_l) - \frac{\partial L}{\partial t} dt$$

Expanding the left side of this equation and identifying terms gives

$$\boxed{\dot{p}_l = -\frac{\partial H}{\partial q_l}, \qquad \dot{q}_l = \frac{\partial H}{\partial p_l}} \quad ...(12)$$

together with $\partial H/\partial t = -\partial L/\partial t$. Equation (12) is called Hamilton's equation.

## 7.12 CONSTANTS OF THE MOTION

If a dynamical function

$$W = W(q_1, ..., q_N, p_1, ..., p_N, t)$$

remains constant as the motion of the system unfolds in time, $W$ is a *constant of the motion or an integral of the system.*

How many such constants are there? For a system with $N$ degrees of freedom, integration of Hamilton's equation gives

$$\left.\begin{aligned} \dot{p}_l &= -\frac{\partial H}{\partial q_l} \rightarrow p_l(t) = p_l(0) - \int_0^t \frac{\partial H}{\partial q_1} dt \\ \dot{q}_l &= -\frac{\partial H}{\partial p_l} \rightarrow q_l(t) = q_l(0) + \int_0^t \frac{\partial H}{\partial p_l} dt \end{aligned}\right\} \quad ...(1)$$

Thus, we find that the state of the system is specified by $2N$ constants, $\{q_1(0), p_l(0)\}$; This conclusion may also be inferred geometrically.

In $2N + 1$-dimensional space, $2N$ surfaces define a curve. For example, in three- space, the two surfaces

$$f(x, y, z) = C_1$$

$$g(x, y, z) = C_2$$

define a curve. Only one of three variables $x, y, z$ is independent. A curve is a one-dimensional. Thus, the intersection of the two surfaces written above may be written in the parametric form as

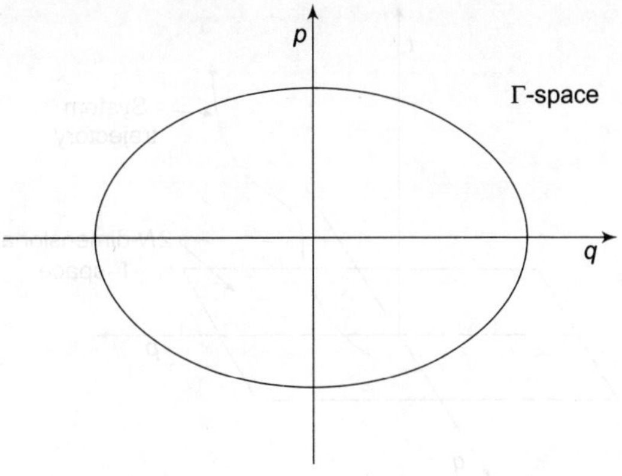


**Figure 7.4** For a harmonic oscillator, the state of the system lies on an ellipse in Γ-space

$$x = x(t), \quad y = y(t), \quad z = z(t)$$

Before proceeding further with this geometric construction, we define Γ-space.

## 7.13 Γ-SPACE

For a system with $N$ degrees of freedom. Γ-space is a $2N$ dimensional Cartesian space whose axes are the $\{q_1, p_t\}$ variables. Thus, the state of the system at any given instant $(q_1, ..., q_n, p_1, ..., p_N)$ is a single point in $2N$ dimensional Γ-space.

Let the system be a single harmonic oscillator for which Γ-space is two dimensional. The orbit of the oscillator may be written

$$q = a \cos \omega t, \quad p = b \sin \omega t$$

from which we find

$$\left(\frac{q}{a}\right)^2 + \left(\frac{p}{b}\right)^2 = 1$$

Thus, the state of the system $(q, p)$ lies on an ellipse Γ-space. See Fig. 7.4.

This example illustrates that time is suppressed in Γ-space. All we know from the curve shown in Fig. 7.4 is that the particle has $q, p$ values that lie somewhere on the ellipse. We do not know what these values are at specific time.

Time is exhibited explicitly in $2N + 1$ $\tilde{\Gamma}$-Space. This Cartesian space is comprised of Γ-space and an additional orthogonal time axis. The *system trajectory* or *dysfunctional orbit,*

$$[q_1(t), ..., q_N(t); P_1(t), ..., p_N(t)]$$

is a curve in $\tilde{\Gamma}$-space. See Fig. 7.5. The system trajectory for the harmonic oscillator is shown in Fig. 7.6.

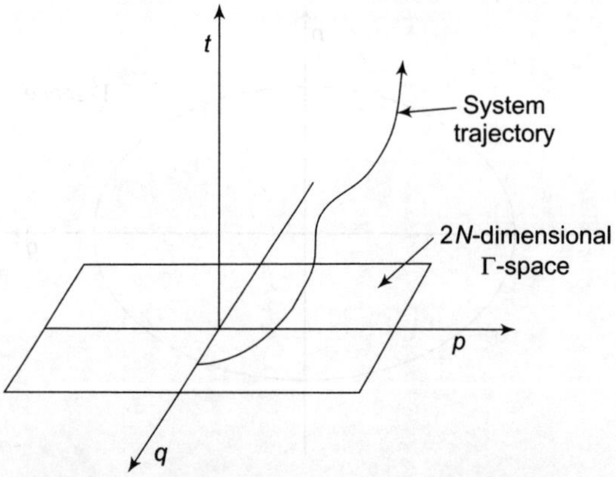


**Figure 7.5** The system trajectory is a curve in $\tilde{\Gamma}$-space. At any instant of time, the curve intersects Γ-space at a single point (*q*, *p*)

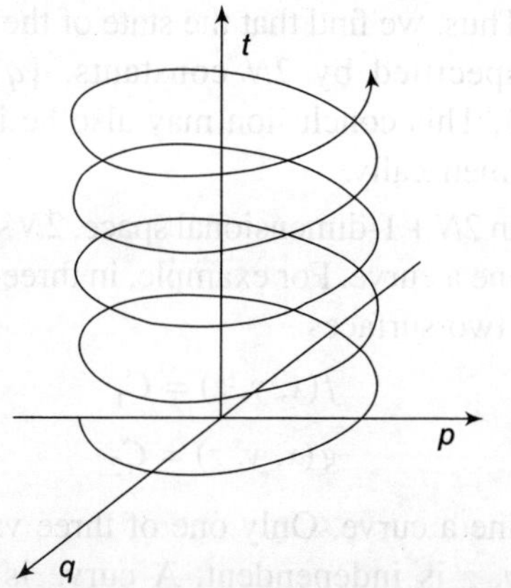


**Figure 7.6** Dynamical trajectory of a harmonic oscillator in three-dimensional (2 + 1) $\tilde{\Gamma}$-space. Projection onto Γ-space is an ellipse

## 7.14 EQUATION OF MOTION FOR DYNAMICAL VARIABLES

Let $u(q, p, t)$ denote a dynamical variable. Then we may write

$$\frac{du}{dt} = \sum_l \left( \frac{\partial u}{\partial q_l} \dot{q}_l + \frac{\partial u}{\partial p_l} \dot{p}_l \right) + \frac{\partial u}{\partial t} \quad ...(1)$$

With Hamilton's equation, the preceding equation may be rewritten

$$\frac{du}{dt} = \sum_l \left( \frac{\partial u}{\partial q_l} \frac{\partial H}{\partial p_l} - \frac{\partial H}{\partial q_l} \frac{\partial u}{\partial p_l} \right) + \frac{\partial u}{\partial t} \quad ...(2)$$

Let us introduce the following *Poisson bracket* notation relevant to two dynamical variables *A* and *B*

$$[A, B] = \sum \left( \frac{\partial A}{\partial q_l} \frac{\partial B}{\partial p_l} - \frac{\partial B}{\partial q_l} \frac{\partial A}{\partial p_l} \right) \quad ...(3)$$

Then Eq. (2) may be written

$$\boxed{\frac{du}{dt} = [u, H] + \frac{\partial u}{\partial t}} \quad ...(4)$$

This is the *equation of motion* for the dynamical variable $u(q, p, t)$. The following properties of Poisson brackets are easily verified:

$$[A, B] = -[B, A] \quad ...(5a)$$

$$[A, c] = 0 \ (c \text{ is constant}) \quad ...(5b)$$

$$[A + B, F] = A[B, F] + [B, F] \quad ...(5c)$$

$$[AB, F] = A[B, F] + [A, F]B \quad ...(5d)$$

$$[A, [B, F]] + [B[F, A]] + [F, [A, B]] = 0 \quad ...(5e)$$

$$\frac{\partial}{\partial t}[A, B] = \left[\frac{\partial A}{\partial t}, B\right] + \left[A, \frac{\partial B}{\partial t}\right] \quad ...(5f)$$

$$[A, B(A)] = 0 \quad ...(5g)$$

The relation (5e) is called *Jacob's identity*. In (5g), the function $B(A)$ is assumed to have a Taylor series expansion.

In the event that a dynamical variable $u$ is not explicitly dependent on time, then $\partial u/\partial t = 0$ and Eq. (4) becomes

$$\frac{du}{dt} = [u, H] \quad ...(6)$$

If, further, $u = u(H)$, then with the property (5e) we may write

$$\frac{du}{dt} = 0 \quad ...(7)$$

So we may conclude that any dynamical variable that is function only of the Hamiltonian is a constant of the motion.

With Eq. (4), Hamilton's equation may be rewritten

$$\dot{q} = [q, H] \quad ...(8a)$$

$$\dot{p} = [q, H] \quad ...(8b)$$

Note also the *fundamental* Poisson bracket relations

$$[q_l, q_{l'}] = [p_l, p_{l'}] = 0, \quad [q_l, p_{l'}] - d_{ll'} \quad ...(9)$$

## 7.15 CANONICAL TRANSFORMATIONS

Consider the transformation of variables

$$q, p \rightarrow q', p' \quad ...(1)$$

Under an arbitrary transformation $(q, p) \rightarrow (q', p')$, a volume integral in Γ-space transforms as follows:

$$\iint_{\Omega} dq\, dp = \iint_{\Omega'} J\left(\frac{qp}{q'\,p'}\right) dq'\, dp' \quad ...(2)$$

In this expression, Ω denotes volume in phase space.

If the transformation is canonical, then Eq. (2) becomes

$$\iint_{\Omega} dq\, dp = \iint_{\Omega'} dq'\, dp' \quad ...(3)$$

We may conclude that volume elements in Γ-space are canonical invariants. See Fig. 7.7.

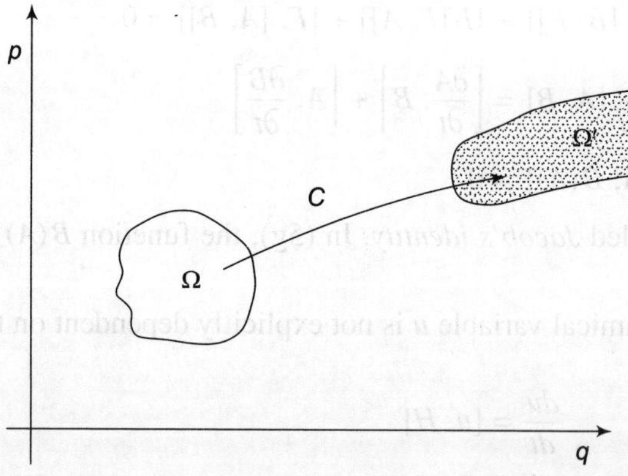


**Figure 7.7** Volume elements in Γ-space are canonical invariants $\Omega = \Omega'$

### 7.15.1 Action Generates the Motion

We now come to an important relation that is a central point in connecting dynamics to kinetic theory.

Consider the action integral corresponding to motion in the interval from $t$ to $t + T$.

$$S(t, T) = \int_t^{t+T} L(q, \dot{q})d\tau$$

$$= \int_t^{t+T} [\sum p\dot{q} - H]d\tau - \int_0^t [\sum p\dot{q} - H]\,d\tau \qquad ...(4)$$

Differentiating, we obtain

$$\frac{dS}{dt} = \sum p'\dot{q}' - \sum p\dot{q} + (H - H') \qquad ...(5)$$

Here we have labelled

$$p' \equiv p(t + T), \quad q' \equiv q(t + T) \qquad ...(6)$$

Assuming that $H$ is time independent, Eq. (5) gives

$$dS = \sum p'dq' - \sum p\,dq \qquad ...(7)$$

Three important conclusions are evident from this relation:

1. The action $S$ is a generating function for the actual physics motion in time.
2. The differential motion in time is a canonical transformation.
3. Because of the group property of canonical transformation, (2.23), the extended motion in time is a canonical transformation in time.

These properties will come into play in our first derivation of the Liouville equation to follow.

## 7.16 LIOUVILLE EQUATION

### 7.16.1 The Ensemble: Density in Phase Space

As mentioned previously, the state of a system is a single point in Γ-space. As time evolves, the system point moves on the system trajectory:

$$[q_1(t), ..., q_N(t); p_1(t), ..., p_N(t)].$$

Now imagine a large number of independent replicas of the same system. Such an abstract collection of identical system is called an *ensemble*. Suppose there are $\mathcal{N}$ systems in the ensemble. Then at $t = 0$ the state of the ensemble is $\mathcal{N}$ points in Γ-space. See Fig. 7.8.

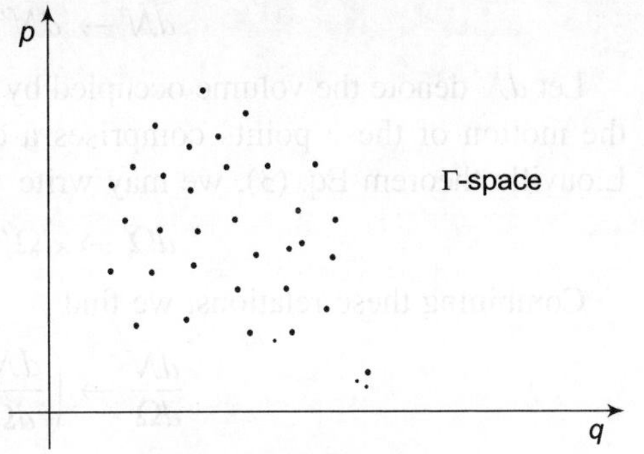


**Figure 7.8** The ensemble comprises $\mathcal{N}$ points in Γ-space

### 7.16.2 First Interpretation of *D* (*q*, *p*, *t*)

Let us introduce the function $D(q, p, t)$, which is defined as follows. The product

$$D(q, p, t)\, dq\, dp \equiv D(q, p, t)\, d\Omega \qquad \text{...(8a)}$$

represents the number of system points in the phase volume $d\Omega$ about the point $(q, p)$ at the time $t$. We may write

$$D(q, p, t) = \frac{d\mathcal{N}}{d\Omega} \qquad \text{...(8b)}$$

An important property of the ensemble is that trajectories of the *ensemble* never cross in Γ-space.

Consider at a given instant of time, ensemble points in a differential volume of Γ-space are contained within a continuously closed surface. As time evolves, with the property that trajectories do not cross, we conclude that point interior to the surface remains interior and that surface points remain surface points. See Figure 7.9.

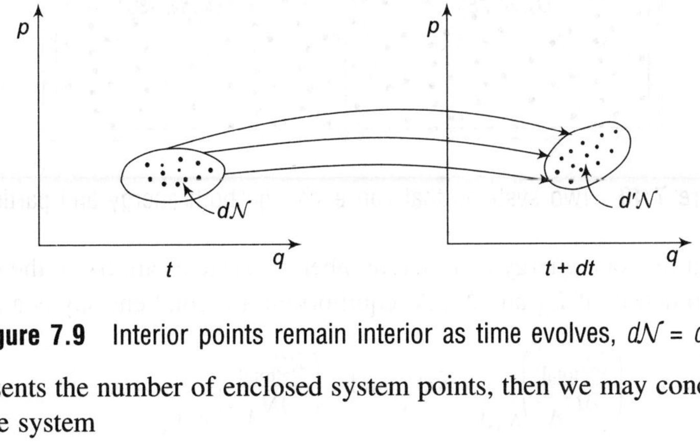


**Figure 7.9** Interior points remain interior as time evolves, $d\mathcal{N} = d\mathcal{N}'$

If $d\mathcal{N}$ represents the number of enclosed system points, then we may conclude that in the evolution of the system

$$d\mathcal{N} \to d\mathcal{N}' = d\mathcal{N} \quad ...(9)$$

Let $d\mathcal{N}$ denote the volume occupied by these ensemble points. As established by Eq. (10), the motion of these points comprises a canonical transformation. Consequently, with the Liouville theorem Eq. (3), we may write

$$d\Omega \to d\Omega' = d\Omega \quad ...(10)$$

Combining these relations, we find

$$\frac{d\mathcal{N}}{d\Omega} \to \left(\frac{d\mathcal{N}}{d\Omega}\right) = \frac{d\mathcal{N}}{d\Omega} \quad ...(11)$$

Recalling the definition (8)

$$D = \frac{d\mathcal{N}}{d\Omega}$$

## 7.17 DIFFUSIVE EQUILIBRIUM AND CHEMICAL POTENTIAL

When two systems are in *thermal* equilibrium, their temperatures are the same. When they are in *mechanical* equilibrium, their pressures are the same. What quantity is the same when they're in the *diffusive* equilibrium?

We can find out by applying the same logic as in the previous section. Consider two systems, *A* and *B*, that are free to exchange both energy and particles, as shown in Fig. 7.10. (The volumes of the systems could also vary, but I'll take these to be fixed for simplicity.) We have drawn a system of two interesting gases, but it could just as well be a gas interacting with a liquid or solid, or even two solids in which atoms gradually migrate around. We are assuming, though, that both systems are made of the same *species* of particles, for instance, $H_2O$ molecules.

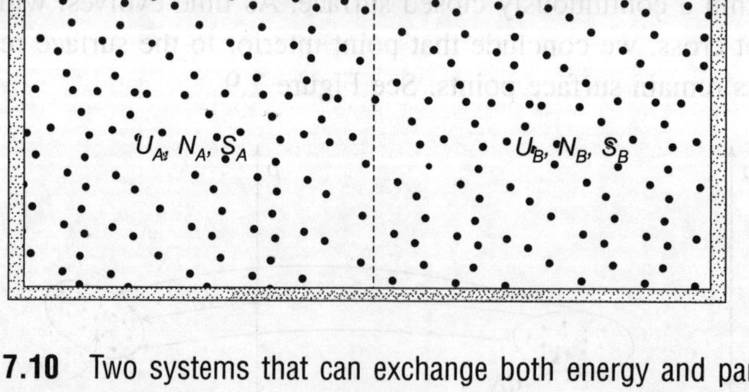


**Figure 7.10** Two systems that can exchange both energy and particles

Assuming that the total energy and total number of particles are fixed, the total entropy of the system is a function of $U_A$ and $N_A$. At equilibrium, the total entropy is a maximum, so

$$\left(\frac{\partial S_{\text{total}}}{\partial U_A}\right)_{N_A, V_A} = 0 \text{ and } \left(\frac{\partial S_{\text{total}}}{\partial N_A}\right)_{U_A, V_A} = 0. \quad ...(1)$$

(If the volumes of the systems are allowed to vary, then $\partial S_{\text{total}}/\partial V_A = 0$ as well.)

Again, the first condition says that the two systems must be at the same temperature. The second condition is new, but is entirely analogous to condition on volume from the previous section. Following the same reasoning as these, we can conclude

$$\frac{\partial S_A}{\partial N_A} = \frac{\partial S_B}{\partial N_B} \quad \text{at equilibrium,} \qquad ...(2)$$

where the partial derivatives are taken at fixed energy and volume. We're free to multiply this equation through by a factor of $T$, the temperature, since the systems are also in thermal equilibrium. By convention; we also multiply by $-1$:

$$-T\frac{\partial S_A}{\partial N_A} = -T\frac{\partial S_B}{\partial N_B} \quad \text{at equilibrium.} \qquad ...(3)$$

The quantity $-T(\partial S/\partial N)$ is much less familiar to most of us than temperature or pressure, but it's still extremely important. It is called the **chemical potential**, denoted $\mu$:

$$\mu \equiv -T\left(\frac{\partial S}{\partial N}\right)_{U,V} \qquad ...(4)$$

This is the quantity that's the same for both systems when they are in diffusive equilibrium:

$$\mu_A = \mu_B \quad \text{at equilibrium.} \qquad ...(5)$$

If the two systems are *not* in equilibrium, then the one with the larger value of $\partial S/\partial N$ will tend to gain particles, since it will thereby gain more entropy than the other loses. IIowever, becausc of thc minus sign in definition (4), this system has the *smaller* value of $\mu$. Conclusion: For particles to flow from the system with higher $\mu$ into the system with lower $\mu$ (see Fig. 7.11).

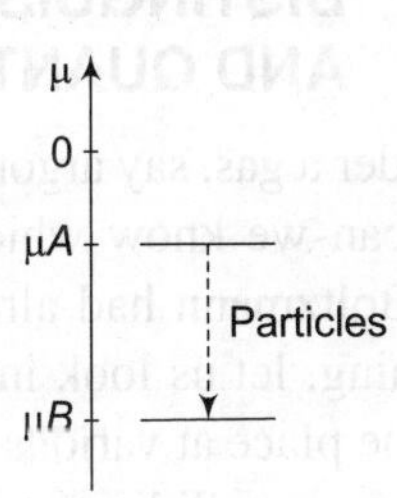


**Figure 7.11** Particles tend to flow toward lower values of the chemical potential, even if both values are negative

It's not hard to generalize the thermodynamic identity to include processes in which $N$ changes'. If we imagine changing $U$ by $dU$, $V$ by $dV$, and $N$ by $dN$, then, by the same logic as in the previous section, the total change in the entropy is

$$dS = \left(\frac{\partial S}{\partial U}\right)_{N,V} dU + \left(\frac{\partial S}{\partial V}\right)_{N,U} dV + \left(\frac{\partial S}{\partial N}\right)_{U,V} dN \qquad ...(6)$$

$$= \frac{1}{T}\,dU + \frac{P}{T}\,dV - \frac{\mu}{T}\,dN.$$

Solving for $dU$ as before, we obtain

$$dU = T\,dS - P\,dV + \mu\,dN. \qquad ...(7)$$

Just as the $-P\,dV$ term is usually associated state with mechanical work, the $\mu\,dN$ the term is sometimes referred to as "chemical work."

This genetalized thermodynamic identity is a great way to remember the various partial-derivative formulas for $T$, $P$, and $\mu$ to generate other similar formulas. Notice that four qualities are changing in this equation $U$, $S$, $V$, and $N$. Now just imagine a process in which any two of these are fixed. For instance, in a process which fixed $U$ and $V$,

$$0 = T\,dS + \mu\,dN, \quad \text{that is,} \quad \mu = -T\left(\frac{\partial S}{\partial N}\right)_{U,V} \qquad ...(8)$$

Similarly, in a process with fixed $S$ and $V$,

$$dU = \mu\,dN, \quad \text{that is,} \quad \mu = \left(\frac{\partial U}{\partial N}\right)_{S,V} \qquad ...(9)$$

This last result is another useful formula for the chemical potential. It tells us directly that $\mu$ has units of energy; specifically, $\mu$ is the amount by which a system's energy changes, when you add one particle and keep the entropy and volume fixed. Normally to hold the entropy (to multiplicity) fixed, you must *remove* some energy as you add a particle, so $\mu$ is negative. However, if you have to give the particle some potential energy (gravitational, if the system lives on a mountain top, or chemical, if the system is a solid crystal) to get it into the system, this energy also contributes to $\mu$.

## 7.18 DISTINGUISHABILITY OF PARTICLES AND CLASSICAL AND QUANTUM STATISTICS

Consider a gas, say argon, all the atoms look alike. Where is the question of distinguishability? How can we know which atom is which? After all, they are not pointed or marked. Quite true. Boltzmann had already thought about it and had no answer. To understand his line of reasoning, let us look into the gas and see what is going on there. We notice atom flying all over the place at various speeds in various directions. Occasionally, there would be collisions, whereupon colliding atoms would be deflected or scattered rather like billiard balls.

In classical mechanics, it is possible to follow the motion of particles via their trajectories. And so, unless the atoms come infinitely close to each other (which, of course, is not possible), one can always keep track of what the different atoms are doing which is as good as giving them individual names. This is (in sense) what happens when we watch billiards and see two white balls collide. The balls are not marked but be know which balls is which because we are able to observe their individual trajectories. Boltzmann assumed this was possible with atoms also, which effectively made them distinguishable. In physics particles are broadly classified as follows:

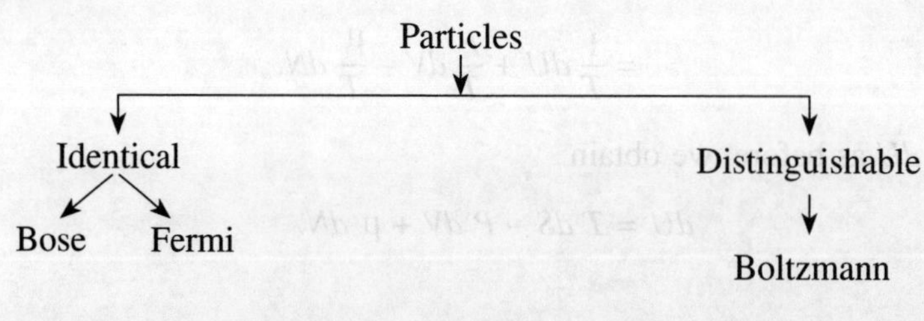

Particles inside the atom are quantum particles. Bose and Fermi particles are also termed Bosons and Fermions, respectively (quantum particles).

What is the difference in the behaviour of these particles? The answer is the way in which they can be packed into available states/levels (see Fig. 7.12). The result is that any number of identical Bosons can occupy a given quantum state but not more than *one* Fermion can occupy a given state.

What about the Boltzmann (i.e., classical) particles? In nature, all elementary particles are either Bosons or Fermions. These are not the particles of the Boltzmann type. Why then did Boltzmann invent something which today we say has nothing to do with reality?

There is one good reason. One should notice that both Fermions and Bosons behave like Boltzmann particles in the *classical limit.*

Bosons and Fermions are the only types encountered so far in the quantum world. They differ in their *distributions*, i.e., their statistics. Particle physicists have to deal with particles in ones and twos.

Elementary particles (of an atom) follow quantum (i.e., either Bose or Fermi) statistics. Solid state physicists on the other hand have to deal with particles in large numbers (~ $10^{23}$). Under such circumstances, one may want to know, what the total energy $E$ of the system is at a given temperature $T$. Then $E = \sum_i n_i E_i$ where $i$ is a given energy level, $n_i$ is the number of particles occupying the level $i$.

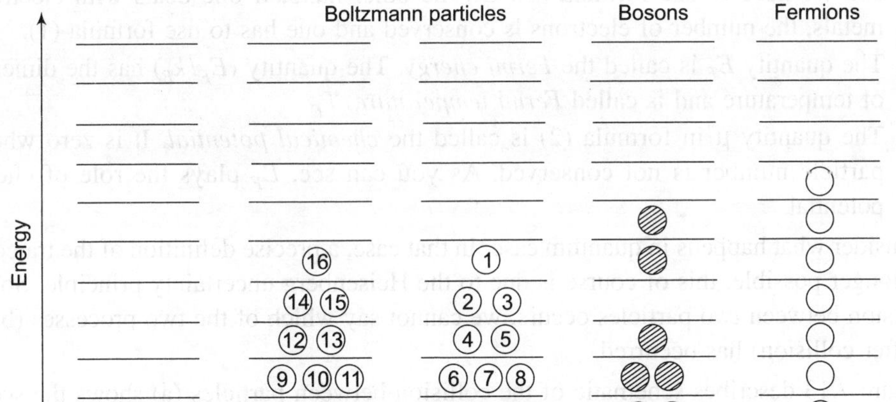


**Figure 7.12** Particles of distribution. This figure illustrates that given the same choice of energy levels say (as in (a)), How the Boltzmann, the Bose and the Fermi particles can distribute themselves

Let $\langle n_i(T)\rangle$ denote the *average* number of particles in energy level $i$ of a system kept at a temperature, $T$. If $E_i$ denotes the energy of level $i$, then we have the following:

**Fermions** (number conserved)

$$\langle n_i(T)\rangle = \frac{1}{e^{(E_i - E_F)/k_BT} + 1} \quad ...(1)$$

**Bosons** (number conserved)

$$\langle n_i(T)\rangle = \frac{1}{e^{(E_i - \mu)/k_BT} - 1} \quad ...(2)$$

**Bosons** (number not conserved)

$$\langle n_i(T)\rangle = \frac{1}{e^{E_i/k_BT} - 1} \quad ...(3)$$

**Comments:**

1. $k_B$ is the Boltzmann constant. You will hear more about it in the next chapter. For the present you may take it as

$$k_B = R/N_A$$

where $R$ is the gas constant (remember the law $PV = RT$?) and $N_A$ is the Avogadro number.

2. The number of particles in the system need not always remain the same. If it varies, we say the particle number is not conserved; otherwise it is *conserved*. When one deals with the photon problem (as we shall in later chapters), the number is not conserved and we have to use formula (3). On the other hand, if one deals with electrons in metals, the number of electrons is conserved and one has to use formula (1).
3. The quantity $E_F$ is called the *Fermi energy*. The quantity $(E_F/k_B)$ has the dimensions of temperature and is called *Fermi temperature*, $T_F$.
4. The quantity $\mu$ in formula (2) is called the *chemical potential*. It is zero when the particle number is not conserved. As you can see, $E_F$ plays the role of chemical potential.

Consider what happens in quantum case. In that case, a precise definition of the trajectories is no longer possible, this of course is due to the Heisenberg uncertainty principle. So when a collision between two particles occurs, we cannot say which of the two processes (b(ii) or (iii) after collision) has occurred.

Figure 7.13 describes schematic of the collision between particles (a) shows the scenario in classical mechanics (it is like collision of two billiard balls). Since individual trajectories could be clearly and precisely determined (at least in principle), the atoms were effectively distinguishable according to Boltzmann. Trouble arises when Heisenberg's principle comes into picture. What then happens then in quantum mechanics, one could have only fuzzy trajectories as in Fig. 7.13 b(i). One may say that is OK, but after the collision we should certainly be able to say which particle went where, since they are now so far apart that fuzziness should not matter any longer. In other words, we must at least be able to say that the collision occurred as per Fig. 7.13 b(ii) or Fig. 7.13 b(iii). However, in quantum mechanics, this kind of argument does not work, we must allow for both Fig. 7.13 b(ii) and Fig. 7.13 b(iii) each having its own probability amplitude.

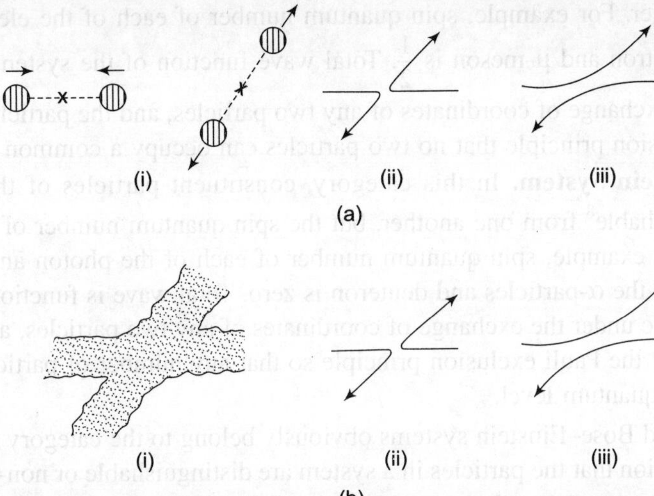


**Figure 7.13** Collision between 2 particles (a) in classical mechanics, (b) in quantum mechanics

We have no means of knowing either Fig. 7.13 b(ii) or Fig. 7.13 b(iii). In fact, both are possible, each with its own probability amplitude, and when we perform scattering calculations we must allow for both–if the particles are Bosons, then the allowance is made in one way while if they are Fermions, the allowance is made somewhat in different way. The main point is that there is a certain amount of indistinguishability. One may ask that if quantum mechanics is the reality, then how is that the Boltzmann formula works at high temperatures? This is explained as follows.

Quantum fuzziness depends on a quantity $\lambda_{th}$ called the *thermal wavelength*. At high temperatures $\lambda_{th}$ is small which also means that fuzziness is small. So it is no wonder that use of the Boltzmann statistics does not lead to atrocious results at high temperatures.

## 7.19 STATISTICAL DISTRIBUTION SYSTEMS

When constituent particles of a gas do not interact with one another, it is known as an *ideal gas*. Potential energy of the gas is zero, and the total energy is the sum of kinetic energies of its constituent particles. In true sense, there is no ideal gas, but a gas of weakly interacting particles may be considered an ideal one. A system of gas may belong to one of the following categories:

**(i) Maxwell–Boltzmann System.** In this category, constituent particles of the gas are distinguishable[6] from one another. This system is also known as the chemical system, and there is no restriction on the number of particles which can occupy any energy level.

**(ii) Fermi-Dirac System.** In this category, constituent particles of the gas are indistinguishable[7] from one another and the spin quantum number of the particles is

[6] Mean separation between the particles is more than the thermal length.

[7] Mean separation between the particles is smaller than the thermal length.

a half integer. For example, spin quantum number of each of the electron, positron, proton, neutron and μ-meson is $\frac{1}{2}$. Total wave function of the system is asymmetric under the exchange of coordinates of any two particles, and the particles obey and the Pauli exclusion principle that no two particles can occupy a common quantum level.

**(iii) Bose-Einstein System.** In this category, constituent particles of the gas are also indistinguishable[8] from one another, but the spin quantum number of a particle is an integer. For example, spin quantum number of each of the photon and π-meson is 1 and each of the α-particles and deuteron is zero. Total wave is function of the system is symmetric under the exchange of coordinates of any two particles, and the particles do not obey the Pauli exclusion principle so that any number of particles can occupy a common quantum level.

Fermi–Dirac and Bose–Einstein systems obviously belong to the category of the quantum systems. The criterion that the particles in a system are distinguishable or non-distinguishable decides whether the system is to be treated in a classical or in a quantum one. When the mean distance between the particles is larger than the de Broglie wavelength, we apply classical distribution, and when the de Broglie wavelength is larger than the mean distribution. Further, in the quantum scenario, spin quantum number of particles is the basis of the two different treatments.

In the present chapter, we shall discuss the distribution laws for these three categories of the systems.

## 7.20 MAXWELL–BOLTZMANN DISTRIBUTION

Let us consider an isolated system of $N$ identical, distinguishable and weakly interacting particles contained in a volume $V$ at an equilibrium temperature $T$ with total energy $U$. Here, we do not account for spin quantum number of the particles. Since the particles are weakly interacting with one another, the potential energy of the system is zero and its total energy is the sum of the kinetic energies of the particles. Example of such system is an ideal monatomic gas enclosed in a container. Mean separation between the particles is larger than the thermal length and therefore they are distinguishable and any number of particles can occupy any energy level in the system.

Suppose there are $l$ states with energies, $E_1, E_2, \dots E_L$, in which the particles are distributed. If the number of particles in these states are $n_1, n_2, \dots n_l$, respectively, we have

$$n_1 + n_2 + \dots + n_l = N \qquad \sum_{i=1}^{l} n_i = N \qquad \text{...(1)}$$

$$n_1 E_1 + n_2 E_2 + \dots + n_l E_l = U \qquad \sum_{i=1}^{l} n_i E_i = U \qquad \text{...(2)}$$

[8] Mean separations between the particles is smaller than the thermal length.

The total energy $U$ and the total number of particles $N$ of the system remain conserved. Further, the energies of the level $E_i$'s remain unchanged. Thus, differentiation of Eqs. (1) and (2) gives

$$\sum_{i=1}^{l} dn_i = dN = 0 \quad \text{and} \quad \sum_{i=1}^{l} E_i \, dn_i = dU = 0 \qquad \text{...(3)}$$

The particles in the system can be distributed among the energy states in a number of ways, keeping in view that there is no restriction to the number of particles to occupy any energy state and there is equal probability for a particle to occupy any of the energy states. The set of the numbers $n_1$, $n_2$, ... $n_l$ is one of the probable ways of distribution of particles among the states. It shows that in the equilibrium situation, distribution of particles has maximum probability of occurrence. Let us first count the total number $W$ of independent ways to distribute particles among the states corresponding to the set of the numbers $n_1$, $n_2$, ... $n_l$, and then find the set of the numbers for which $W$ is maximum.

For the first state with energy $E_1$, let us find out the number of ways in which $n_1$ particles are chosen from the $N$ distinguishable particles. The number of ways of choosing the first particle, out of $N$ particles is $N$. After choosing the first particle, we are left with $(N - 1)$ particles. The number of ways of choosing the second particle is $(N - 1)$. Then we are left with $(N - 2)$ particles. The number of ways of choosing the third, fourth, ..., $n_1$-th particles are $(N - 2)$, $(N - 3)$, ..., $(N - n_1 + 1)$. Thus, the number of ways of selecting $n_1$ particles out of $N$ particles is

$$= N(N-1)(N-2)(N-3) \ldots (N-n_1+1)$$

$$= \frac{N(N-1)(N-2)(N-3) \ldots (N-n_1+1)(N-n_1)!}{(N-n_1)!} = \frac{N!}{(N-n_1)!}$$

This number includes the $n_1!$ arrangements $n_1$ particles among themselves. Any sequence of the particles in the state gives the same distribution of $n_1$ particles. Thus, number of independent ways of selecting $n_1$ particles for the first energy state with energy $E_1$ is

$$W_1 = \frac{N!}{n_1!\,(N-n_1)!}$$

After selecting $n_1$ particles for the first energy state, we are left with $(N - n_1)$ particles. In the similar manner as discussed for the first state, the number of independent ways of selecting $n_2$ particles out of $(N - n_1)$ particles for the second energy state with energy $E_2$ is

$$W_2 = \frac{(N-n_1)!}{n_2!\,(N-n_1-n_2)!}$$

Similarly, the number of independent ways of selecting $n_3$, $n_4$, ..., $n_l$ particles for the third, fourth, ..., $l$-th states are

$$W_3 = \frac{(N-n_1-n_2)!}{n_3!\,(N-n_1-n_2-n_3)!}$$

$$W_4 = \frac{(N - n_1 - n_2 - n_3)!}{n_4!\,(N - n_1 - n_2 - n_3 - n_4)!}$$

$$\dots \quad \dots \quad \dots \quad \dots \quad \dots$$

$$\dots \quad \dots \quad \dots \quad \dots \quad \dots$$

$$W_l = \frac{(N - n_1 - n_2 \dots - n_{l-1})!}{n_l!\,(N - n_1 - n_2 \dots - n_l)!}$$

For these numbers $W_1$, $W_2$, $W_3$ ..., $W_l$, the total number $W$ of distinct ways of obtaining the distribution of the $n_1$, $n_2$, ... $n_l$ particles among the states with energies, $E_1$, $E_2$, ... $E_l$, is the product of them

$$W = W_1\, W_2\, W_3 \dots W_l$$

$$= \frac{N!}{n_1!\,(N - n_1)!}\,\frac{(N - n_1)!}{n_2!\,(N - n_1 - n_2)!}\,\frac{(N - n_1 - n_2)!}{n_3!\,(N - n_1 - n_2 - n_3)!}\dots\frac{(N - n_1 - n_2 \dots - n_{l-1})!}{n_l!\,(N - n_1 - n_2 \dots - n_l)!}$$

$$= \frac{N!}{n_1!\,n_2! \dots n_l!\,(N - n_1 - n_2 \dots - n_l)!}$$

$$= \frac{N!}{n_1!\,n_2! \dots n_l!\,0!} = \frac{N!}{n_1!\,n_2! \dots n_l!}$$

So far, we have assumed that the energy states are non-degenerate. Suppose, $g_1$, $g_2$, ..., $g_l$ are degeneracies of the states with energies $E_1$, $E_2$, $E_l$ respectively. Then, for example, for the $i$-th energy state with energy $E_i$ containing $n_i$ particles there are $g_i$ levels, and each of the $n_i$ particles can be in any of the $g_i$ levels without any preference. Since we cannot distinguish between $g_i$ levels, there are $g_i$ choices for each of the $n_i$ particles to occupy these $g_i$ levels. Thus, there are $g_i \cdot g_i \dots \{n_i \text{ times}\}$ $(= g_i^{n_i})$ ways of arranging $n_i$ particles among the $g_i$ levels. Hence, states with degeneracies $g_1$, $g_2$, ... $g_l$ and $n_1$, $n_2$, ... $n_l$ particles, respectively, have $g_1^{n_1}$, $g_2^{n_2}$, ..., $g_l^{n_l}$, respectively, ways of arranging the particles. After accounting for the degeneracies of the states, the total number of arrangements of the particles in the given distribution is

$$W = \frac{N!}{n_1!\,n_2! \dots n_l!}\,g_1^{n_1}\,g_2^{n_2} \dots g_l^{n_l} = N! \prod_{i=1}^{l}\left[\frac{g_i^{n_i}}{n_i!}\right] \qquad \dots(4)$$

The most probable distribution of $N$ particles among $l$ energy states is that for which the $W$ is the maximum. For mathematical convenience, we take logarithm of $W$ and consider the maximum of ln ($W$).

In Eq. (4), $n$'s are the only variables. Thus, we have

$$d \ln W = 0$$

$$\frac{\partial}{\partial n_1}(\ln W)\,dn_1 + \frac{\partial}{\partial n_2}(\ln W)\,dn_2 + \dots + \frac{\partial}{\partial n_l}(\ln W)\,dn_l = 0$$

$$\sum_{i=1}^{l} \frac{\partial}{\partial n_i}(\ln W)\, dn_i = 0$$

This equation along with Eq. (3) can be written as

$$\sum_{i=1}^{l} \left[ \frac{\partial}{\partial n_i}(\ln W) - \alpha - \beta E_i \right] dn_i = 0 \quad \text{...(5)}$$

where $\alpha$ and $\beta$ are constants of the system. Equation (5) must be true for the individual increments, $dn_i$. Hence, for each level, term in the bracket of Eq. (5) must be zero.

$$\frac{\partial(\ln W)}{\partial n_i} - \alpha - \beta E_i = 0 \quad \text{...(6)}$$

Taking logarithm of Eq. (4), we have

$$\ln W = \ln N! + \ln\left( \prod_{i=1}^{l} \left[ \frac{g_i^{n_i}}{n_i!} \right] \right)$$

$$= \ln N! + \sum_{i=1}^{l} (n_i \ln g_i - \ln n_i!)$$

Using Stirling formula, we have

$$\ln W = N \ln N - N + \sum_{i=1}^{l} (n_i \ln g_i - n_i \ln n_i + n_i) \quad \text{...(7)}$$

Differentiating Eq. (7) with respect to $n_1$ and remembering that except the $i$-th term, differentiation of all other terms would be zero, we get

$$\frac{\partial}{\partial n_i}(\ln W) = 0 + \ln g_i - n_i \frac{1}{n_i} - \ln n_i + 1$$

$$= \ln g_i - \ln n_i = -\ln\left(\frac{n_i}{g_i}\right) \quad \text{...(8)}$$

Using Eq. (8) in Eq. (6), we have

$$\ln\left(\frac{n_i}{g_i}\right) = -\alpha - \beta E_i$$

$$n_i = g_i\, e^{(-\alpha - \beta E_i)} = g_i\, e^{-\alpha}\, e^{-\beta Ei} \quad \text{...(9)}$$

This is the Maxwell–Boltzmann distribution of the particles among various states.

**Example 7.1** A Maxwell–Boltzmann gas has two particles in the $i$-th state whose degeneracy is three. Find out the number of independent ways of selecting the particles in the state.

***Solution*** Here, we have $n_i = 2$, $N = 2$ and $g_i = 3$. The number of independent ways $W_i$ of selecting the particles in the $i$-th state is

$$W_i = N! \frac{g_i^{n_i}}{n_i!} = 2! \frac{3^2}{2!} = 9$$

Denoting the two distinguishable particles by the alphabets $A$ and $B$, 9 independent ways of arranging the particles in 3 levels of the state are given in the Table below:

| *Ways* | *Levels* | | |
|---|---|---|---|
| | *1* | *2* | *3* |
| 1 | A B | - | - |
| 2 | - | A B | - |
| 3 | - | - | A B |
| 4 | A | B | - |
| 5 | B | A | - |
| 6 | A | - | B |
| 7 | B | - | A |
| 8 | - | A | B |
| 9 | - | B | A |

## 7.21 FERMI-DIRAC DISTRIBUTION

Let us consider an isolated system of $N$ identical, indistinguishable and weakly interacting particles contained in a volume $V$ at an equilibrium temperature $T$ with total energy $U$. Here, we account for the particles whose spin quantum number is a half integer. These particles are known as Fermi particles of fermions. Examples of the fermions are electron, positron, proton, neutron, μ-meson; each of them with the spin quantum number $\frac{1}{2}$. Since the particles are weakly interacting, the potential energy of the system is zero, and its total energy is the sum of the kinetic energy of the particles. Mean separation between the particles is smaller than the thermal length and therefore they are indistinguishable. Further, these particles with a half-integer spin follow the Pauli exclusion principle, which states that two or more particles can never be in a common quantum level. Hence, in any one level we can have only one particle or no particles.

Suppose there are $l$ states with energies, $E_1$, $E_2$, ... $E_l$, and degeneracies $g_1$, $g_2$, ... $g_l$, respectively, in which the particles are distributed. If the number of particles in these states are, respectively, $n_1$, $n_2$, ..., $n_l$, we have

$$n_1 + n_2 + ... + n_l = N \qquad \sum_{i=1}^{l} n_i = N \qquad ...(10)$$

$$n_1 E_1 + n_2 E_2 + ... + n_l E_l = U \qquad \sum_{i=1}^{l} n_i E_i = U \qquad ...(11)$$

Total energy $U$ and total number of particles $N$ of the system remain conserved. Further, the energies of the levels $E_i$'s remain unchanged. Thus, differentiation of Eqs. (10) and (11) gives

$$\sum_{i=1}^{l} dn_i = dN = 0 \quad \text{and} \quad \sum_{i=1}^{l} E_i\, dn_i = dU = 0 \qquad ...(12)$$

Following Pauli's exclusion principle, for the $i$-th state, the number of $n_i$ particles should not be larger than the degeneracy $g_i$ of the state. Figure 7.14 shows $g_i$ levels of the $i$-th state in which we want to accommodate $n_i$ particles.

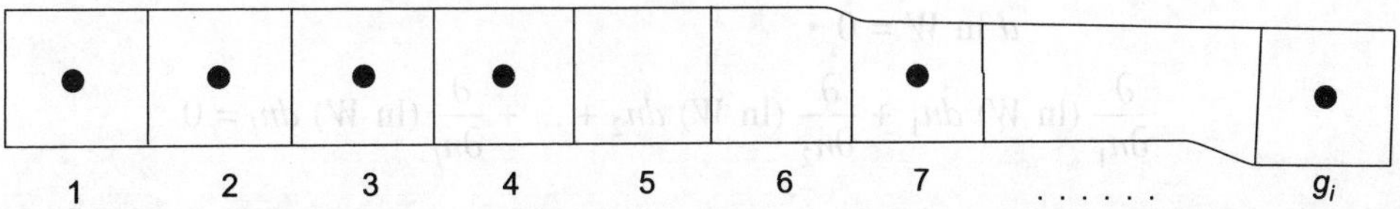


**Figure 7.14** A possible distribution of $n_i$ particles among the levels in the $i$-th state with degeneracy $g_i$

Now, we want to have possible permutations of $n_i$ indistinguishable particles in place in $g_i$ in levels with the condition that only one particle or no particle can be placed in each level. The number of ways in which the first of the $n_i$ particles can be placed in one of the $g_i$ the levels is $g_i$. After replacing the first particle in one of the $g_i$ levels, the remaining number of level is $(g_i - 1)$. Thus, number of ways in which the second particle can be placed in one of the $(g_i - 1)$ levels is $(g_i - 1)$. Similarly, the number of ways in which the third, fourth, ..., $n_i$-th particles can be placed in one of the remaining levels are $(g_i - 2)$, $(g_i - 3)$, ..., $(g_i - n_i + 1)$.

Thus, the total number of ways of putting $n_i$ particles in $g_i$ levels in the $i$-th state is

$$= g_i(g_i - 1)(g_i - 2)(g_i - 3) \ldots (g_i - n_i + 1)$$

$$= \frac{g_i(g_i - 1)(g_i - 2)(g_i - 3) \ldots (g_i - n_i + 1)(g_i - n_i)!}{(g_i - n_i)!} = \frac{g_i!}{(g_i - n_i)!}$$

The number of permutations of $n_i$ particles among themselves is $n_i!$. Because of indistinguishable nature of the particles, these permutations do not give independent arrangements. Thus, actual number of ways of putting $n_i$ particles in $g_i$ levels in the $g_i$ levels in the $i$-th state is

$$W_i = \frac{g_i!}{n_i!\,(g_i - n_1)!}$$

For these numbers $W_1$, $W_2$, $W_3$, ..., $W_l$, for $l$ states, total number $W$ of distinct ways of obtaining the distribution of the particles among the states is their product.

$$W = W_1\, W_2\, W_3 \ldots W_l$$

$$= \frac{g_1!}{n_1!(g_1-n_1)!}\,\frac{g_2!}{n_2!(g_2-n_2)!}\,\frac{g_3!}{n_3!(g_3-n_3)!}\cdots\frac{g_l!}{n_l!(g_l-n_l)!}$$

$$= \prod_{i=l}^{l} = \frac{g_i!}{n_i!(g_i-n_i)!} \qquad \text{...(13)}$$

The most probable distribution of $N$ articles among $l$ energy states in that for which the $W$ is the maximum. For mathematical convenience, we take logarithm of $W$ and consider the maximum of $\ln(W)$.

In Eq. (13) $n$'s are the only variables, Thus, we have

$$d \ln W = 0$$

$$\frac{\partial}{\partial n_1}(\ln W)\,dn_1 + \frac{\partial}{\partial n_2}(\ln W)\,dn_2 + \ldots + \frac{\partial}{\partial n_l}(\ln W)\,dn_l = 0$$

$$\sum_{i=1}^{l} \frac{\partial}{\partial n_i}(\ln W)\,dn_i = 0$$

This equation along with Eq. (12) can be written as

$$\sum_{i=1}^{l}\left[\frac{\partial}{\partial n_i}(\ln W) - \alpha - \beta E_i\right] dn_i = 0 \qquad \text{...(14)}$$

where $\alpha$ and $\beta$ are constants of the system. Equation (14) must be true for the individual increments $dn_i$. Hence, for each level, term in the bracket of Eq. (14) must be zero.

$$\frac{\partial}{\partial n_i}(\ln W) - \alpha - \beta E_i = 0 \qquad \text{...(15)}$$

Taking logarithm of Eq. (13), we have

$$\ln W = \ln\left[\prod_{i=l}^{l} \frac{g_i!}{n_i!(g_i-n_i)!}\right] = \sum_{i=l}^{l} \ln\left[\frac{g_i!}{n_i!(g_i-n_i)!}\right]$$

$$= \sum_{i=l}^{l} \left[\ln g_i! - \ln n_i! - \ln(g_i - n_i)!\right]$$

Using Sterling formula, we have

$$\ln W = \sum_{i=l}^{l} [g_i \ln g_i - g_i - n_i \ln n_i + n_i - (g_i - n_i)\ln(g_i - n_i) + (g_i - n_i)]$$

$$= \sum_{i=l}^{l} [g_i \ln g_i - n_i \ln n_i - (g_i - n_i)\ln(g_i - n_i)] \qquad \text{...(16)}$$

Differentiating Eq. (16) with respect to $n_i$ and remembering that except the $i$-th term, differentiation of all other terms would be zero, we get

$$\frac{\partial}{\partial n_i}(\ln W) = 0 - \frac{n_i}{n_i} - \ln n_i + \frac{(g_i - n_i)}{(g_i - n_i)} + \ln (g_i - n_i)$$

$$= -\ln n_i - \ln (g_i - n_i) = \ln \left[\frac{g_i - n_i}{n_i}\right] \qquad ...(17)$$

Using Eqs. (5.17) in (5.15), we have

$$\ln \left[\frac{g_i - n_i}{n_i}\right] - \alpha - \beta\, E_i = 0$$

$$\ln \left[\frac{g_i - n_i}{n_i}\right] - \alpha - \beta\, E_i$$

$$\frac{g_i - n_i}{n_i} = e^{(\alpha + \beta E_i)}$$

$$\frac{g_i}{n_i} = e^{(\alpha + \beta E_i)} + 1$$

$$n_i = \frac{g_i}{e^{(\alpha + \beta E_i)} + 1} \qquad ...(18)$$

This is the Fermi–Dirac distribution of the particles among various states.

**Example 7.2** A Fermi–Dirac gas has two particles in the $i$-th state whose degeneracy is three. Find out the number of independent ways of selecting the particles in the state.

***Solution*** Here, we have $n_i = 2$ and $g_i = 3$. The number of independent ways $W_i$ of selecting the particles in the level is

$$W_i = \frac{g_i!}{n_i!\,(g_i - n_i)!} = \frac{3!}{2!\,(3-2)!} = \frac{3!}{2!\,1!} = 3$$

Denoting the two indistinguishable particles by the alphabets $A$ and $A$, 3 independent ways of arranging the particles, obeying Pauli's exclusion principle, in 3 levels of the state are given in the Table below:

| Ways | Levels | | |
|---|---|---|---|
| | 1 | 2 | 3 |
| 1 | A | A | - |
| 2 | A | - | A |
| 3 | - | A | A |

## 7.22 BOSE–EINSTEIN DISTRIBUTION

Let us consider an isolated system of $N$ identical, indistinguishable and weakly interacting particles contained in a volume $V$ at an equilibrium temperature $T$ with total energy, $U$. Here, we account for the particles whose spin quantum number is an integer. These particles are known as Bose particles or boson. Examples of bosons are photon, $\pi$-meson each of them with the spin quantum number 1; $\alpha$-particle, deuteron, each of them with the spin quantum number 0. Since the particles are weakly interacting, the potential energy of the system is zero, and its total energy is the sum of the kinetic energies of the particles. Mean separation between the particles is smaller than the thermal length and therefore they are indistinguishable. Further, these particles with an integer spin do not follow the Pauli exclusion principle and any number of particles can be in a common level.

Suppose there are $l$ states with energies, $E_1$, $E_2$, ... $E_l$ and degeneracy's $g_1$, $g_2$, ..., $g_l$, respectively, in which the particles are distributed. If the number of particles in these levels are, respectively, $n_1$, $n_2$, ... $n_l$, we have

$$n_1 + n_2 + ... + n_l = N \qquad \sum_{i=1}^{l} n_i = N \qquad ...(19)$$

$$n_1 E_1 + n_2 E_2 + ... + n_l E_l = U \qquad \sum_{i=1}^{l} n_i E_i = U \qquad ...(20)$$

Total energy $U$ and the total number of particles $N$ of the system remain conserved. Further, the energies of the levels $E_i$'s remain unchanged. Thus, differentiation of Eqs. (19) and (20) gives

$$\sum_{i=1}^{l} dn_i = dN = 0 \text{ and } \sum_{i=1}^{l} E_i\, dn_i = dU = 0 \qquad ...(21)$$

Figure 7.15 shows $g_i$ quantum levels of the $i$-th state in which we want to accommodate $n_i$ particles.

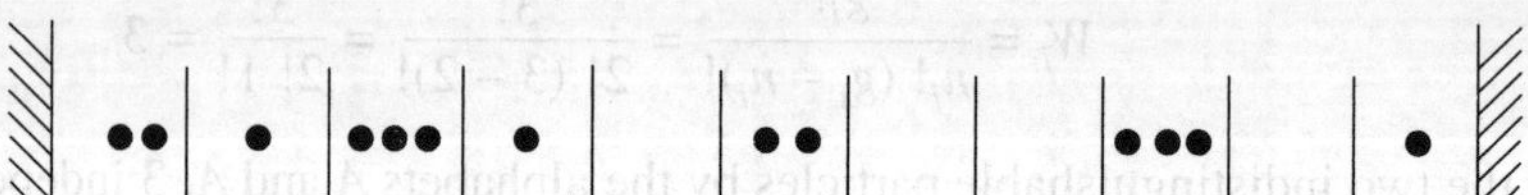

**Figure 7.15** One of the possible distributions of 13 particles among 11 levels with 10 partitions

Now, we want to have possible permutations of $n_i$ indistinguishable particles to place in $g_i$ levels with the freedom that any number of particles can be placed in any level. For arranging $n_i$ particles in $g_i$ in levels, we require $(g_i - 1)$ partitions as shown in Fig. 7.15. The number of permutations of $n_i$ particles among $g_i$ levels is equal to the number of permutations of $n_i$ particles and $(g_i - 1)$ partitions. Obviously, we shall obtain all possible permutations of the particles plus the partitions, i.e., of $n_i + (g_i - 1)$ objects.

Now, the number of ways of arranging $(n_i + g_i - 1)$ objects is $(n_i + g_i - 1)!$. Here, the number of permutations of $n_i$ particles among themselves is $n_i!$, and that of $(g_i - 1)$ partitions among themselves is $(g_i - 1)!$. Since the particles among themselves and the patitions among themselves are indistinguishable, the number of independent permutations of $n_i$ particles in $g_i$ levels is

$$W_i = \frac{(n_i + g_i - 1)!}{n_i!\,(g_i - 1)!}$$

For these numbers $W_1, W_2, W_3, \ldots W_l$, for $l$ states, total number $W$ of distinct ways of obtaining the distribution of the particles among the states is their product

$$W = W_1\, W_2\, W_3 \ldots W_l$$

$$= \frac{(n_1 + g_1 - 1)!}{n_1!(g_1 - 1)!} \frac{(n_2 + g_2 - 1)!}{n_2!(g_2 - 1)!} \frac{(n_3 + g_3 - 1)!}{n_3!(g_3 - 1)!} \cdots \frac{(n_l + g_l - 1)!}{n_l!(g_l - 1)!}$$

$$= \prod_{i=l}^{l} \frac{(n_i + g_i - 1)!}{n_i!(g_i - 1)!}$$

Since $n_i$ and $g_i$ are large numbers, we can omit 1 in comparison to them, and we have

$$W = \prod_{i=l}^{l} \frac{(n_i + g_i)!}{n_i!\,g_i!} \qquad \text{...(22)}$$

The most probable distribution of $N$ particles among $l$ states is that for which the $W$ is the maximum. For mathematical convenience, we take logarithm of $W$ and consider the maximum of $\ln(W)$. In Eq. (22), $n$'s are the only variables. Thus, we have

$$d \ln W = 0$$

$$\frac{\partial}{\partial n_1}(\ln W)\, dn_1 + \frac{\partial}{\partial n_2}(\ln W)\, dn_2 + \ldots + \frac{\partial}{\partial n_l}(\ln W)\, dn_l = 0$$

$$\sum_{i=1}^{l} \frac{\partial}{\partial n_i}(\ln W)\, dn_i = 0$$

This equation along with Eq. (21) can be written as

$$\sum_{i=1}^{l}\left[\frac{\partial}{\partial n_i}(\ln W) - \alpha - \beta E_i\right] dn_i = 0 \qquad \text{...(23)}$$

where $\alpha$ and $\beta$ are the constants of the system. Equation (23) must be true for the individual increments, $dn_i$. Hence, for each level, term in the bracket of Eq. (23) must be zero.

$$\frac{\partial}{\partial n_i}(\ln W) - \alpha - \beta E_i = 0 \qquad \text{...(24)}$$

Taking logarithm of Eq. (22), we have

$$\ln W = \ln\left[\prod_{i=l}^{l} \frac{(n_i + g_i)!}{n_i!\,g_i!}\right] = \sum_{i=l}^{l} \ln\left[\frac{(n_i + g_i)!}{n_i!\,g_i!}\right]$$

$$= \sum_{i=l}^{l} \left[ \ln (n_i + g_i)! - \ln n_i! - \ln g_i! \right]$$

Using Sterling formula, we have

$$\ln W = \sum_{i=l}^{l} [(n_i + g_i) \ln (n_i + g_i) - (n_i + g_i) - n_i \ln n_i + n_i - g_i \ln g_i - g_i]$$

$$= \sum_{i=l}^{l} [(n_i + g_i) \ln (n_i - g_i) - n_i \ln n_i - g_i \ln g_i] \qquad ...(25)$$

Differentiating Eq. (25) with respect to $n_i$ and remembering that except the $i$-th, term differentiation of all other terms would be zero, we get

$$\frac{\partial}{\partial n_i} (\ln W) = \frac{n_i + g_i}{n_i + g_i} - \ln (n_i + g_i) \frac{n_i}{n_i} - \ln n_i - 0$$

$$= \ln n_i - \ln (g_i - n_i) = \ln \left[ \frac{n_i + g_i}{n_i} \right] \qquad ...(26)$$

Using Eq. (26) in Eq. (24), we have

$$\ln \left[ \frac{n_i + g_i}{n_i} \right] - \alpha - \beta E_i = 0$$

$$\ln \left[ \frac{n_i + g_i}{n_i} \right] - \alpha + \beta E_i$$

$$\frac{n_i + g_i}{n_i} = e^{(\alpha + \beta E_i)}$$

$$\frac{g_i}{n_i} = e^{(\alpha + \beta E_i)} - 1$$

$$n_i = \frac{g_i}{e^{(\alpha + \beta E_i)} - 1} \qquad ...(27)$$

This is the Bose–Einstein distribution of the particles among various states.

**Example 7.3** A Bose–Einstein gas has two particles in the $i$-th state whose degeneracy is three. Find out the number of dependent ways of selecting the particles in the state.

***Solution*** Here, we have $n_i = 2$, $g_i = 3$. The number of independent ways $W_i$ of selecting the particles in the state is

$$W_i = \frac{(n_i + g_i - 1)!}{n_i! \, (g_i - 1)!} = \frac{(2 + 3 - 1)!}{2! \, (3 - 2)!} = \frac{4!}{2! \, 2!} = 6$$

Denoting the two indistinguishable particles by the alphabets $A$ and $A$, 6 independent ways of arranging the particles in 3 levels of state are given in the Table below.

| Ways | Levels | | |
|---|---|---|---|
| | 1 | 2 | 3 |
| 1 | A A | – | – |
| 2 | – | A A | – |
| 3 | – | – | A A |
| 4 | A | A | – |
| 5 | A | – | A |
| 6 | – | A | A |

**Example 7.4** The energy ε of an oscillator with frequency $v$ is

$$\varepsilon = \left(n + \frac{1}{2}\right) hv, \qquad n = 0, 1, 2, \ldots$$

When a system consisting of $N$ independent oscillators has a total energy

$$E = \frac{1}{2} Nhv + Mhv$$

find out (i) statistical weight, $W_i$, and (ii) relation between the temperature $T$ and energy $E$ of this system.

***Solution*** (i) When we denote quantum number of the $i$-th state by $n_i$, the statement of the total energy of the system is equal to $\frac{1}{2} Nh_v + Mh_v$ shows that

$$n_1 + n_2 = \ldots + n_N = M$$

and the number of quantum level is $N$. Thus, from section 3 (Bose–Einstein distribution), we have

$$W_i = \frac{(M + N - 1)!}{M!\,(N - 1)!}$$

(ii) The entropy of the system is

$$S = k \ln W_i$$

Substituting $W_i$ and using Sterling formula under the assumption that $N \gg 1$ and $M \gg 1$, we get

$$S = k[M + N) \ln (M + N) - N \ln M - N \ln N]$$

The temperature $T$ of the system is

$$\frac{1}{T} = \left(\frac{\partial S}{\partial E}\right)_{N,V} = \frac{\partial S}{\partial M}\frac{\partial M}{\partial E} = k \ln\left(\frac{M+N}{M}\right)\frac{\partial M}{\partial E}$$

$$= \frac{k}{hv} \ln\left(\frac{M + N/2 + N/2}{M + N/2 - N/2}\right) = \frac{k}{hv} \ln\left(\frac{E/N + hv/2}{E/N - hv/2}\right)$$

or $$\frac{E + Nh\nu}{2/E - Nh\nu/2} = \exp\,(h\nu/kT)$$

On solving this for $E$, one gets

$$E = N\left(\frac{1}{2}h\nu + \frac{h\nu}{\exp(h\nu/kT) - 1}\right)$$

This is the required relation between the temperature $T$ and energy $E$ of the system.

**Example 7.5** Show that the Maxwell–Boltzmann distribution is the limiting case of the Fermi–Dirac and Bose–Einstein distributions, and the Maxwell–Boltzmann distribution is applicable to the rarefied gases.

***Solution*** The three distributions can be expressed as

$$n_i = \frac{g_i}{e^{(\alpha + \beta E_i)} + J}$$

where $$J = \begin{cases} 0 & \text{for Maxwell–Boltzmann distribution} \\ 0 & \text{for Fermi–Dirac distribution} \\ -1 & \text{for Bose–Einstein distribution} \end{cases}$$

Obviously, when $$e^{(\alpha + \beta E_i)} >> 1 \qquad \text{...(28)}$$

the Fermi–Dirac as well as the Bose–Einstein distributions reduce to the Maxwell–Boltzmann distribution. Hence, the Maxwell–Boltzmann distribution is the limiting case and the Fermi–Dirac and Bose–Einstein distributions.

For the Maxwell–Boltzmann distribution, we have

$$\frac{n_i}{g_i} = \frac{1}{e^{(\alpha + \beta E_i)}} \qquad \text{...(29)}$$

Equations (28) and (29) give

$$\frac{n_i}{g_i} << 1$$

It shows that in the Maxwell–Boltzmann gas the number of particles per level is very less. Hence, the Maxwell–Boltzmann distribution is applicable to a rarefied gas.

### 7.22.1 Distributions – Combined

For the three categories, discussed in the last three sections, the total number $W$ of distinct ways of obtaining the distribution the $n_1$, $n_2$, ... $n_l$ particles among the states with energies, $E_1$, $E_2$, ... $E_l$ with degeneracies $g_1$, $g_2$, ... $g_l$, respectively, is

$$W = N!\prod_{i=1}^{l}\left[\frac{g_i^{n_i}}{n_i!}\right] \qquad \text{for Maxwell–Boltzmann distribution}$$

$$W = \prod_{i=1}^{l} \frac{g_i!}{n_i!(g_i - n_i)!} \quad \text{for Fermi–Dirac distribution}$$

$$W = \prod_{i=1}^{l} \frac{(n_i + g_i)!}{n_i!\, g_i!} \quad \text{for Bose–Einstein distribution}$$

## 7.23 BOSE–EINSTEIN GAS

Consider a gas consisting of $N$ identical, indistinguishable particles contained in a volume $V$ at an equilibrium temperature $T$ total energy, $U$. The thermal length ($\lambda = h/\sqrt{2\pi mkT}$) of the system is larger than the mean interparticle distance. For such gas, we do account for spin quantum number of the particles and the gas with an integral spin quantum number is termed the Bose–Einstein gas. For example, spin quantum number of each of the photon and $\pi$-meson is 1, of each of the $\alpha$-particle, deuteron is 0. Total wave function of the system is symmetric under the exchange of coordinates of any two particles and the particles do not obey the Pauli exclusion principle. So, any number of particles can occupy a quantum state. The particles are not interacting with each other, the potential energy of the gas is zero, and the total energy of the gas is equal to the sum of the kinetic energies of the constituent particles.

The Bose–Einstein energy distribution is

$$f(E) = \frac{1}{e^{\alpha+\beta E} - 1} = \frac{1}{Ae^{\beta E} - 1} \qquad ...(1)$$

where $b = 1/(kT)$

and $$A \equiv e^{\alpha} = e^{-\mu/kT} = \frac{V}{N}\left(\frac{2\pi mkT}{h^2}\right)^{3/2} \qquad ...(2)$$

where, $\mu$ is chemical potential which is in general, a function of temperature.

When $A >> 1$, Bose–Einstein gas reduces to the Maxwell–Boltzmann gas.

At a high temperature, most of the particles are in the excited states. The number of particles in the ground state (with $E = 0$) is negligibly small. For a Bose–Einstein gas, total number $N$ of particles and total energy $U$ are

$$N = \int_0^{\infty} f(E)\, g(E)\, dE, \qquad U = \int_0^{\infty} E(f)\,(E) g\, dE$$

where, $$f(E) - \frac{1}{Ae^{\beta E} - 1} \qquad ...(3)$$

and the number of energy states $g(E)dE$ in the energy range from $E$ to $E + dE$ is

$$g(E)\, dE = 2\pi\, gV \left(\frac{2m}{h^2}\right)^{3/2} E^{1/2}\, dE \qquad ...(4)$$

where $g$ ($\equiv 2s + 1$) is the spin degeneracy for the spin quantum number, $s$. For spinless particles, $g = 1$. Thus, we have

$$N = 2\pi\, gV \left(\frac{2m}{h^2}\right)^{3/2} \int_0^\infty \frac{E^{1/2}}{Ae^{\beta E} - 1} dE \quad ...(5)$$

$$U = 2\pi\, gV \left(\frac{2m}{h^2}\right)^{3/2} \int_0^\infty \frac{E^{3/2}}{Ae^{\beta E} - 1} dE \quad ...(6)$$

### 7.23.1 Bose–Temperature

At high temperature, most of the particles in a Bose–Einstein are in the excited states and the population of the ground state is negligible. As the temperature decreases, the particles start to reach the ground state ($E = 0$). For the Bose particles, there is no restriction on the number of particles to occupy any level in the system, and any number of particles can be in any of the levels.

Let $N_0$ be the number of particles in the ground state and $N_e$, the number in the excited state, so that $N = N_0 + N_e$. (In the FD gas, population of the ground state does not increase even at $T = 0\,K$). At lower temperatures population of the ground state for the Bose–Einstein gas increases. Thus, in a Bose–Einstein gas, at low temperature, the number of particles $N_e$ in the excited states and total energy $U$ of the system are

$$N_e = \int_0^\infty f(E)\, g(E)\, dE; \qquad U = \int_0^\infty Ef\ (E) g\ (E)\ dE \quad ...(7)$$

where $f(E)$ is Bose–Einstein distribution function, and $g(E)$ is the number of energy states in the energy states in the energy range from $E$ to $E + dE$ (i.e., density of states)

Thus, we have

$$N_e = 2\pi\, gV \left(\frac{2m}{h^2}\right)^{3/2} \int_0^\infty \frac{E^{1/2}}{Ae^{\beta E} - 1} dE \quad ...(8)$$

$$U = 2\pi\, gV \left(\frac{2m}{h^2}\right)^{3/2} \int_0^\infty \frac{E^{3/2}}{Ae^{\beta E} - 1} dE \quad ...(9)$$

Let $\beta E = u$ then $\beta dE = du$

and $$N_e = 2\pi\, gV \left(\frac{2mkT}{h^2}\right)^{3/2} \int_0^\infty \frac{u^{1/2}}{Ae^{u} - 1} du = \frac{gV}{\lambda^3} g_{3/2}/(A) \quad ...(10)$$

$$U = 2\pi\, gV\, kT \left(\frac{2mkT}{h^2}\right)^{3/2} \int_0^\infty \frac{u^{3/2}}{Ae^{u} - 1} du = \frac{3}{2} kT \frac{gV}{\lambda^3} g_{5/2}/(A) \quad ...(11)$$

where $$\lambda = \left(\frac{h^2}{2\pi m\, kT}\right)^{1/2} \quad ...(12)$$

is the thermal length and

$$g_n(A) = \frac{1}{\Gamma(n)} \int_0^\infty \frac{u^{n-1}}{Ae^{u}-1} du \qquad ...(13)$$

Here $\Gamma(n)$ is a gamma function. When $A$ is large ($A > 1$), we can express

$$\frac{1}{Ae^u - 1} = \frac{1}{Ae^u}\left(1 - \frac{1}{Ae^u}\right)^{-1} = \frac{1}{Ae^u}\sum_{r=1}^{\infty}\left(\frac{1}{Ae^u}\right)^{r-1} = \sum_{r=1}^{\infty}\left(\frac{1}{Ae^u}\right)^{r} \qquad ...(14)$$

Hence, the integral $g_n(A)$ can be evaluated as follows:

$$g_n(A) = \frac{1}{\Gamma(n)}\int_0^\infty \sum_{r=1}^{\infty}\left(\frac{1}{Ae^u}\right)^{r} u^{n-1} du$$

$$= \frac{1}{\Gamma(n)}\sum_{r=1}^{\infty}\frac{1}{A^r}\left(\int_0^\infty e^{-ru}\, u^{n-1}\, du\right) \qquad ...(15)$$

Let $ru = x$, then $rdu = dx$

and
$$g_n(A) = \frac{1}{\Gamma(n)}\sum_{r=1}^{\infty}\frac{1}{A^r}\frac{1}{r^n}\left(\int_0^\infty e^{-x}\, x^{n-1}\, dx\right) = \sum_{r=1}^{\infty}\frac{1}{A^r}\frac{1}{r^n}$$

$$= \frac{1}{A} + \frac{1}{A^2\, 2^n} + \frac{1}{A^3\, 3^n} + ... \qquad ...(16)$$

For a Bose–Einstein gas, the chemical potential at $\mu$ is always negative and therefore $A \geq 1$. Further, let us consider spinless particles, so that $g = 1$, with the decrease of temperature, $A$ decreases and can attain the minimum value 1. The value of $A$ is 1 at a temperature $T_B$ called the *Bose temperature*. This Bose temperature can be determined from Eq. (10) by putting $A = 1$. At this temperature $T_B$, $N$ is practically equal to $N_e$, showing that up to the Bose temperature, the number of particles in the ground state is negligible. Hence,

$$N = V\left(\frac{2\pi m\, kT_B}{h^2}\right)^{3/2} g_{3/2}(A), A = 1. \qquad ...(17)$$

where
$$g_{3/2}(1) = 1 + \frac{1}{2^{3/2}} + \frac{1}{3^{3/2}} + \frac{1}{4^{3/2}} + ... \equiv \zeta(3/2) \qquad ...(18)$$

is the Riemann zeta function whose value is 2.612. Hence, the Bose temperature is given by

$$T_B = \left(\frac{h^2}{2\pi mk}\right)\left(\frac{N}{2.612\, V}\right)^{2/3} \qquad ...(19)$$

($k$ is Boltzmann constant)

Thus, the Bose temperature is the minimum temperature at which all the particles are in excited states. For all known Bose–Einstein gases value of $T_B$ is very low. For example, for $He^4$, it is about 3.14 K. Though this is very low, yet there is a temperature $0 \leq T \leq T_B$ where Bose–Einstein gases show a specific behaviour.

### 7.23.2 Bose–Einstein Gas at $T < T_B$ (Bose–Einstein Condensation)

In this temperature range $A = 1$ and number of particles in the excited state is

$$N_e = N - N_o = 2.612\ V\left(\frac{2\pi mkT}{h^2}\right)^{3/2} \quad ...(20)$$

It shows that for the given values of $V$ and $T$,

when $$N > 2.612\ V\left(\frac{2\pi mkT}{h^2}\right)^{3/2}$$

the excited states can occupy the number of particles given by Eq. (17) and therefore rest ($N_o$) of them would be in the ground state ($E = 0$).

Hence, as long as $$N \leq 2.612\ V\left(\frac{2\pi mkT}{h^2}\right)^{3/2}$$

all the particles are in the excited states ($E \neq 0$). From Eqs. (17) and (20), we get the number of particles in the excited state

$$N_e = N\left(\frac{T}{T_B}\right)^{3/2} \quad ...(21)$$

and the number of particles in the ground state

$$N_o = N - N_e = N\left[1 - \left(\frac{T}{T_B}\right)^{3/2}\right] \quad ...(22)$$

showing that a large number of particles accumulate in the ground state when $T < T_B$.

This phenomenon is known as *Bose–Einstein condensation*. At $T = 0$, all the particles occupy the ground state and at $T = T_B$, all the particles occupy the excited state. For $T > T_B$, the particles occupy higher excited states only Fig. 7.16.

It is to be noted that the fraction of particles that are condensing into the ground state are *all in the same quantum state*. The *condensate*, as it is called, is a macroscopic quantum system—a quantum system that makes its quantum properties known through large scale observable effects. It is interesting that although Einstein proposed the condensation phenomenon for a gas of non-interacting Bose-particles in 1924, it took until 1995 actually to observe the phenomenon visually.

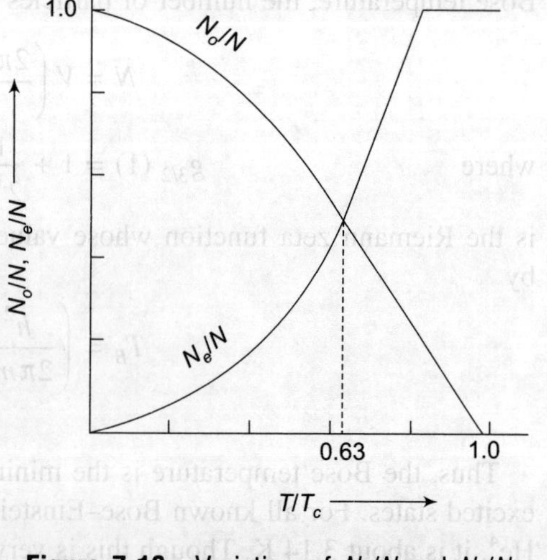


**Figure 7.16** Variation of $N_o/N$ and $N_e/N$ with temperature

In 1995, a group of physicists headed by E. Cornell and C. Weiman in US performed experiments that showed the Bose–Einstein condensation in the most direct way possible. A gas of $^{87}$Rb atoms, which are bosons, was cooled to a temperature about 300 times lower than any material had ever been cooled – about 20 billionth of a degree above absolute zero. (The atoms at this temperature have average speeds of only about one metre per hour.) The experiments used a set of ingenuous schemes in which laser light played an important role. At a critical temperature $T_c = 190 \times 10^{-9}$ K, a Bose–Einstein condensation occurred. A study of the speed distribution showed a transition from the Maxwell–Boltzmann distribution above $T_c$ to one in which pronounced peak around velocity $V = 0$ as the temperature was lowered beyond $T$.

The beautiful thing about this experiment was that the condensate and rest of the atoms could be separated spatially, and one could study the pure condensate for extended periods of time (up to fifteen seconds in the initial experiments). A new state of matter explicable only by the quantum theory has been created.

## 7.24 THE BOLTZMANN EQUATION

Let us define a distribution functions $f(x, k, t)$ that represents the probability of finding a particle at position $x$ with $k$-vector at the time $t$. Clearly this is a classical distribution function since it simultaneously specifies both the position $x$ and the momentum $k$ (remember that $k$ is simply related to $\vec{p}$ for free particles).

The question we think to answer is how does $f(x, k, t)$ evolve in space, time and momentum space? The distribution $f(x, k, t)$ governs the behaviour of the system in what is called phase space. Phase space is defined as the collective space consisting of three dimensions in real space and three dimensions in $k_0$ or momentum space. Therefore six coordinates and time are needed to describe the system.

We can derive the basic equation that governs the behaviour of a non-equilibrium system by recognizing that the probability distribution function $f(x, k, t)$ must be conserved.

In other words, $f(x, k, t)$ must obey a continuity equation. In the special case in which there are no scatterings, the total derivative of $f(x, k, t)$ must be zero, i.e.,

$$\frac{df}{dt} = 0 \qquad ...(1)$$

Equation (1) states that the total time rate of change of the probability distribution function must be zero. There is no loss of probability and hence, no loss of particles; the probability flux is conserved. However, in semiconductors, carriers undergo scattering between themselves and their surroundings. Carriers can be scattered by the ions within the lattice forming the host crystal or by impurities. In addition, carriers can transfer between bands through generations and recombination events.

Therefore, in general, the distribution function changes because of scatterings and generations and recombination events, when all these effects are lumped together into one term, the general equation for the time rate of change of the non-equilibrium distribution function becomes

$$\frac{df}{dt} = \left(\frac{\partial f}{\partial t}\right)_{\text{scatterings}} \qquad ...(2)$$

Equation (2) is called the Boltzmann equation. We can recast it into a more convenient form by expanding the total time derivative on the left-hand side. Since the non-equilibrium distribution function $f$ is a function $x$, $k$ and $t$, the total time derivative can be expanded as

$$\frac{df}{dt} = \frac{\partial f}{\partial t} + \frac{\partial f}{\partial x}\frac{dx}{dt} + \frac{\partial f}{\partial y}\frac{dy}{dt} + \frac{\partial f}{\partial z}\frac{dz}{dt} + \frac{\partial f}{\partial k_x}\frac{dk_x}{dt} + \frac{\partial f}{\partial k_y}\frac{dk_y}{dt} + \frac{\partial f}{\partial k_z}\frac{dk_z}{dt} \quad ...(3)$$

We can simplify Eq. (3) by recognizing the following relationships

$$\left.\begin{aligned} \frac{dx}{dt}\hat{i} + \frac{dy}{dt}\hat{j} + \frac{dz}{dt}\hat{k} &= \vec{u} \\ \frac{dk_x}{dt}\hat{i} + \frac{dk_y}{dt}\hat{j} + \frac{dk_z}{dt}\hat{k} &= \frac{e\vec{F}}{\hbar} \end{aligned}\right\} \quad ...(4)$$

where $\vec{F}$ represents all the external forces acting on the system. Recognizing that,

$$\left.\begin{aligned} \frac{\partial f}{\partial x}\hat{i} + \frac{\partial f}{\partial y}\hat{j} + \frac{\partial f}{\partial z}\hat{k} &= \nabla_r f \\ \frac{\partial f}{\partial k_x}\hat{i} + \frac{\partial f}{\partial k_y}\hat{j} + \frac{\partial f}{\partial k_z}\hat{k} &= \nabla_k f \end{aligned}\right\} \quad ...(5)$$

We find that the Boltzmann equation becomes

$$0 = \frac{\partial f}{\partial t} + (\nabla_r f)\cdot\vec{v} + (\nabla_r f)\cdot\vec{F}/\hbar - \left(\frac{\partial f}{\partial t}\right)_{\text{scatt}} \quad ...(6)$$

where the last term on the right hand side above includes the change in the distribution function due to scattering between particles and their surroundings.

The scattering term is defined in the following way:

$$\left(\frac{\partial f}{\partial t}\right)_{\text{scatt}} \equiv -\int\{f(x,k,t)[1-f(r,k',t)]\,s(k,k') - f(r,k',t)[1-f(r,k,t)]\,s(k',k)\}dk' \quad ...(7)$$

where $[1 - f(x, k, t)]$ is the probability of a vacancy existing in the state denoted by $(x, k)$ at the time $t$. $S(k, k')$ is the rate at which an electron makes a transition from the initial state $k$ to the final state $k'$. Equivalently, $S(k', k)$ is the rate at which an electron makes a transition from the initial state $k'$ to the final state $k$. The meaning of the terms in Eq. (7) can then be understood as follows. For a transition to occur from an initial state $k$ to a final state $k'$, a particle must first exist within the initial state. The probability that the initial state is occupied is given by $f(x, k, t)$. In addition, the final state $k'$ must be unoccupied. This probability is simply $1 - f(x, k', t)$. Finally, there is some transition probability rate itself that is expressed by $S(k, k')$. The sum of the product of all these terms together gives the rate at which a transition occurs from $k$ to $k'$. Hence, the scattering time is related to the scattering rate of the particles from $k$ to $k'$ and from $k'$ back to $k$.

Substituting for each term in Eq. (2) gives

$$\frac{\partial f}{\partial t} + \frac{\vec{F}_{\text{ext}}}{\hbar} \cdot \vec{\nabla}_k f + \vec{v} \cdot \vec{\nabla}_x f$$

$$= -\int\{f(x,k,t)[1 - f(x,k',t)]S(k,k')$$

$$- f(x,k',t)[1 - f(x,k,t)]S(k',k)\}\,dk' \equiv \left(\frac{df}{dt}\right)_{\text{coll}} \qquad \text{...(8)}$$

which is known as the *Boltzmann transport equation,* or BTE. We will often refer to the BTE as simply the Boltzmann equation as well.

Before we consider some approximate methods of solving the Boltzmann equation, it is important to consider the assumptions made in its derivation. The principal assumption is that the distribution function can be specified in terms of position and momentum simultaneously. In other words, the distribution function, by design, is a strictly classical distribution function since it is specified as $f(x, k, t)$. The simultaneous description of the real space position and momentum of the particle implicit in the definition of the distribution function is in direct contradiction to the Heisenberg uncertainty principle. Therefore, the Boltzmann equation is not a valid description of transport when quantum mechanical effects area into play.

### 7.24.1 Approximate Solutions of the Boltzmann Equation

How can a solution of the Boltzmann equation be obtained? The basis of the relaxation-time approximation is that it is assumed that a scattering term, which involves an integral, can be replaced by a constant relaxation term, thus reducing the original integral differential equation into only differential equation. The partial derivative with respect to time due to collisions is assumed to be inversely proportional to a lifetime that characterizes the mean free time between collisions. To see how this works in practice, let us first make a few definitions. Let $f_0$ represent the equilibrium distribution function and $f$ represent the non-equilibrium distribution function for which the Boltzmann equation is solved. The left-hand side of the Boltzmann equation can then rewritten as

$$\left(\frac{\partial f}{\partial t}\right)_{\text{collisions}} = -\frac{f - f_0}{\tau} \qquad \text{...(1)}$$

Equation (1) implies that the system will relax to the equilibrium distribution $f_0$ after a characteristic time, $\tau$, which is called the relaxation time. The time $\tau$ physically represents some average time in which, through the action of collisions, the system relaxes from an initial non-equilibrium state back into a final equilibrium state.

The meaning of $\tau$ can be further understood from the following. If an electric field $F_0$ is applied to the system at $t < 0$ and no diffusion gradients exist, $\nabla_x f = 0$, then the Boltzmann equation reduces to

$$\frac{\partial f}{\partial t} = -\frac{q\vec{F}_0}{m} \cdot \vec{\nabla}_w f(x, v, t) + \left(\frac{\partial f}{\partial t}\right)_{\text{collisions}} \qquad ...(2)$$

If at $t = 0$ the field is turned off, then the Boltzmann equation at $t > 0$ further reduces to

$$\frac{\partial f}{\partial t} = \left(\frac{\partial f}{\partial t}\right)_{\text{collisions}} \qquad ...(3)$$

with the relaxation time approximation, the time rate of change of $f$ due to collisions is given by Eq. (1). Subsequently Eq. (3) becomes

$$\frac{\partial f}{\partial t} = -\frac{f - f_0}{\tau}. \qquad ...(4)$$

The time of rate of change of the equilibrium distribution function $f_0$ is zero, so it can be added to the left-hand side of Eq. (4) to give

$$\frac{\partial}{\partial t}(f - f_0) = -\frac{f - f_0}{\tau}, \qquad ...(5)$$

which can be simply solved as

$$y \equiv (f - f_0), \quad \frac{dy}{dt} = -\frac{y}{\tau},$$

$$y = y_0\, e^{-\frac{t}{\tau}}. \qquad ...(6)$$

Therefore, resubstituting for $y(f - f_0)$ in the last of Eq. (6) yields

$$(f - f_0) = (f - f_0)\, e^{-\frac{t}{\tau}}. \qquad ...(7)$$

The nonequilibrium distribution $f$ is equal to

$$f(t) = f_0 + [f(t = 0) - f_0]\, e^{\frac{t}{\tau}}. \qquad ...(8)$$

From the above result it is clear that as it approaches infinity, the non-equilibrium distribution function $f$ relaxes to $f_0$, the equilibrium distribution function. The system relaxes to equilibrium at a time $\tau$, called the relaxation time. It is important to realize that the system relaxes through scattering events. In the absence of scattering the time rate of change of the distribution due to scattering would be zero. Therefore Eq. (2) would reduce to

$$\frac{\partial f}{\partial t} = 0, \qquad ...(9)$$

which clearly implies that the system never relaxes. Once driven out of equilibrium by the action of an external field the system never returns to equilibrium. Such a condition exist in a superconducting system.

## 7.25 BOLTZMANN EQUATION IN RELATION TIME APPROXIMATION

We have

$$\frac{\partial f}{\partial t} + \frac{1}{\hbar}\mathbf{F} \cdot \nabla_k f + v \cdot \nabla_r f = \left(\frac{df}{dt}\right)_{\text{coll}} \qquad ...(1a)$$

This is the **Boltzmann transport equation.**

### 7.25.1 Collision Term and Relaxation Time

Let us consider a uniform crystal and assume the distribution function to be independent of position $\boldsymbol{r}$. Then the distribution function is written as $f(\boldsymbol{k})$. The rate of change in the distribution function $f(\boldsymbol{k})$ includes two terms. The rate of increase in $f(\boldsymbol{k})$ due to transitions from all possible $\boldsymbol{k}'$ states (excluding $\boldsymbol{k}$) to the $\boldsymbol{k}$ state and the rate of decrease in $f(\boldsymbol{k})$ due to transitions from the $\boldsymbol{k}$ state to other possible $\boldsymbol{k}'$ states. Defining the respective transition rates per unit time by $P(\boldsymbol{k}', \boldsymbol{k})$ and $P(\boldsymbol{k}, \boldsymbol{k}')$, the collision term is written as

$$\left(\frac{df}{dt}\right)_{\text{coll}} = \sum_{\boldsymbol{k}'} \{P(\boldsymbol{k}', \boldsymbol{k})\, f(\boldsymbol{k}')[1 - f(\boldsymbol{k}')] - P(\boldsymbol{k}, \boldsymbol{k}')\, f(\boldsymbol{k})[1 - f(\boldsymbol{k}')]\} \qquad ...(1)$$

where the prefactor $f(\boldsymbol{k}')\,[1 - f(\boldsymbol{k})]$ of $P(\boldsymbol{k}', \boldsymbol{k})$ represents the probability of electron occupation in the initial state $\boldsymbol{k}'$ and of electron vacancy in the final state $\boldsymbol{k}$.

For simplicity we consider the case where the Fermi level lies below the bottom of the conduction band in the band gap; then, the conditions $f(\boldsymbol{k}) \ll 1$ and $f(\boldsymbol{k}') \ll 1$ hold. Defining the distribution function for thermal equilibrium by $f_0(\boldsymbol{k})$, we have $(df/dt)_{\text{coll}} = 0$ in Eq. (1), and thus we obtain the relation in:

$$P(\boldsymbol{k}', \boldsymbol{k})\, f_0(\boldsymbol{k}')] = P(\boldsymbol{k}, \boldsymbol{k}')\, f_0(\boldsymbol{k}). \qquad ...(2)$$

This relation is called the principle of detailed balance. Using this Eq. (1) is written as

$$\left(\frac{df}{dt}\right)_{\text{coll}} = \sum_{\boldsymbol{k}'} P(\boldsymbol{k}, \boldsymbol{k}')\left[f(\boldsymbol{k}) - f(\boldsymbol{k}')\frac{f_0(\boldsymbol{k})}{f_0(\boldsymbol{k}')}\right] \qquad ...(3)$$

When we replace the summation of $\mathbf{k}'$ by an integral, we have

$$\left(\frac{df}{dt}\right)_{\text{coll}} = -\frac{V}{(2\pi)^3}\int d^3\boldsymbol{k}'\, P(\boldsymbol{k}, \boldsymbol{k}')\left[f(\boldsymbol{k}) - f(\boldsymbol{k}')\frac{f_0(\boldsymbol{k})}{f_0(\boldsymbol{k}')}\right], \qquad ...(4)$$

where $V = L^3$ is the crystal volume. When the external force is very weak and the displacement of the distribution function from the thermal equilibrium value is small, we may write (or expand in Taylor's series)

$$f(\boldsymbol{k}) = f_0(\boldsymbol{k}) + f_1(\boldsymbol{k}), \quad (f_1(\boldsymbol{k}) \ll f_0(\boldsymbol{k})). \qquad ...(5)$$

Assuming that the energy change due to scattering is small and that the energy $\varepsilon(\boldsymbol{k})$ of the initial state $\boldsymbol{k}$ is very close to the energy $\varepsilon(\boldsymbol{k}')$ of the final state $\boldsymbol{k}'$ (this condition is called an **elastic collision**), we may put $f_0(\boldsymbol{k}) \cong f_0(\boldsymbol{k}')$. In this case we obtain the following relation from

$$\left(\frac{df}{dt}\right)_{\text{coll}} = -f_1(\boldsymbol{k}) - \frac{V}{(2\pi)^3}\int d^3\boldsymbol{k}'\, P(\boldsymbol{k}, \boldsymbol{k}')\left[1 - \frac{f_1(\boldsymbol{k}')}{f_1(\boldsymbol{k})}\right],$$

$$\equiv -\frac{f_1(\boldsymbol{k})}{\tau(\boldsymbol{k})} \equiv -\frac{f(\boldsymbol{k})-f_0(\boldsymbol{k})}{\tau(\boldsymbol{k})}, \qquad ...(6)$$

where $\tau(\boldsymbol{k})$ is called the **relaxation time** of the collision and is given by

$$\frac{1}{\tau(\boldsymbol{k})} = \frac{V}{(2\pi)^3}\int d^3\boldsymbol{k}'\, P(\boldsymbol{k},\boldsymbol{k}')\left[1-\frac{f_1(\boldsymbol{k}')}{f_1(\boldsymbol{k})}\right], \qquad ...(7)$$

which is a function of the electron wave vector $\boldsymbol{k}$ and thus of the energy. This approximation is called the **relaxation approximation**.

After the electron system reaches the steady state under an external field, the external field is removed at time $t = 0$ and we have $\partial f/\partial t = (\partial f/\partial t)_{\text{coll}}$ from Boltzmann Eq. (1a). When we assume that relaxation approximation and we use Eq. (6), it may be expressed as

$$\frac{\partial f}{\partial t} = \left(\frac{df}{dt}\right)_{\text{coll}} = -\frac{f-f_0}{\tau}. \qquad ...(8)$$

For simplicity we assume that the relaxation time $\tau$ is constant, and then we obtain

$$f - f_0 = (f-f_0)_{t=0}\exp\left(-\frac{t}{\tau}\right), \qquad ...(9)$$

where $(f-f_0)_{t=0}$ is the shift of the distribution function from the thermal equilibrium at time $t = 0$. The above result indicates that the distribution function recovers exponentially with time constant $\tau$ towards the thermal equilibrium value after removal of the external field at $t = 0$.

When the relaxation time approximation is valid, the Boltzmann transport equation is written as

$$\frac{\partial f}{\partial t} + \frac{1}{\hbar}\boldsymbol{F}\cdot\nabla_k f + v\cdot\nabla_r f = -\frac{f-f_0}{\tau}. \qquad ...(10)$$

Spatial uniformity results in $\nabla_r f = 0$ and the steady state gives rise to $\partial f/\partial t = 0$. Therefore, we obtain

$$\frac{1}{\hbar}\boldsymbol{F}\cdot\nabla_k f = -\frac{f_1}{\tau} \qquad ...(11)$$

and when the external field is applied in the $x$ direction it reduces to

$$f_1 = -\frac{\tau}{\hbar}F_x\frac{\partial f}{\partial k_x}. \qquad ...(12)$$

The energy of the electron is a function of the wave vector and is written as $\varepsilon = \hbar^2k^2/2m^*y$ for an electron with isotropic effective mass $m^*$, which leads to the relation $\hbar k_x = m^*v_x$. Therefore, we obtain

$$f_1 = -\frac{\tau}{\hbar}F_x\frac{\partial f}{\partial \varepsilon}\frac{\partial \varepsilon}{\partial k_x} = -\tau v_x F_x\frac{\partial f}{\partial \varepsilon}. \qquad ...(13)$$

When we use the relations $f = f_0 + f_1$ and $f_0 \gg f_1$ in the above equation we may approximate it as

$$f_1 = -\tau v_x F_x \frac{\partial f_0}{\partial \varepsilon}. \quad ...(14)$$

When we assume elastic scattering and that the relaxation time $\tau$ is a function of the energy $\varepsilon$, the magnitude of the relaxation time $\tau$ is not changed after the scattering event and the relaxation time $\tau(\boldsymbol{k})$ is expressed as follows by inserting Eq. (14) into Eq. (7)

$$1/\tau(\boldsymbol{k}) = \frac{V}{(2\pi)^3} \int d^3\boldsymbol{k}'\, P(\boldsymbol{k}, \boldsymbol{k}') \left(1 - \frac{k'_x}{k_x}\right)$$

$$= \frac{V}{(2\pi)^3} \int d^3\boldsymbol{k}'\, P(\boldsymbol{k}, \boldsymbol{k}')\,(1 - \cos\theta), \quad ...(15)$$

where $k_x$ and $k'_x$ are the components of the electron wave vectors $\boldsymbol{k}$ and $\boldsymbol{k}'$ along the direction of the external force ($x$ component) before and after scattering, respectively, and $\theta$ the angle between $\boldsymbol{k}$ and $\boldsymbol{k}'$.

In the case of a degenerate semiconductor, the principle of detailed balance is obtained from Eq. (1) as

$$P(\boldsymbol{k}', \boldsymbol{k})\, f_0(\boldsymbol{k}')\, [1 - f_0(\boldsymbol{k})] = P(\boldsymbol{k}, \boldsymbol{k}')\, f_0(\boldsymbol{k})\, [1 - f_0(\boldsymbol{k}')] \quad ...(16)$$

Using this relation, the relaxation time of a degenerate semiconductor is given by

$$\frac{1}{\tau(\boldsymbol{k})} = \frac{1}{1 - f_0(\boldsymbol{k})} \sum_{\boldsymbol{k}'} P(\boldsymbol{k}, \boldsymbol{k}')[1 - f_0(\boldsymbol{k}')] \left(1 - \frac{k'_x}{k_x}\right)$$

$$= \frac{1}{1 = f_0(\boldsymbol{k})} \frac{V}{(2\pi)^3} \int d\boldsymbol{k}' P(\boldsymbol{k}, \boldsymbol{k}')[1 - f_0(\boldsymbol{k}')] \left(1 - \frac{k'_x}{k_x}\right). \quad ...(17)$$

Equation (17) reduces to (15) for the non-degenerate case when we let $f_0(\boldsymbol{k}) \ll 1$ and $f_0(\boldsymbol{k}') \ll 1$. It may be pointed out that the final state is occupied by an electron, $1 - f_0(\boldsymbol{k}') = 0$ and the transition (scattering) is not allowed.

### 7.25.2 Mobility and Electrical Conductivity

When the relaxation time is given by a function of the electron energy or electron wave vector, the electron mobility and electrical conductivity (referred to as mobility and conductivity hereafter, respectively, unless otherwise specified) are calculated as follows. In the presence of an electric field applied in the $x$ direction, we may write $F_x = -eE_x$; (5), and (14), then give the following relation:

$$f(\boldsymbol{k}) = f_0(\boldsymbol{k}) + eE_x \tau v_x \frac{\partial f_0}{\partial \varepsilon}. \quad ...(18)$$

Therefore, the current density in the $x$ direction is given by

$$J_x = \frac{2}{(2\pi)^3}\int(-e)v_x f(\boldsymbol{k})d^3\boldsymbol{k}$$

$$\frac{e}{4\pi^3}\int v_x f_0(\boldsymbol{k})d^3\boldsymbol{k}\ \frac{e^2E_x}{4\pi^3}\int \tau v_x^2\frac{\partial f_0}{\partial\varepsilon}d^3\boldsymbol{k}, \qquad ...(19)$$

where the factor 2 arises from the spin degeneracy and $v_x = \hbar k_x/m^*$. A function $f_0(\boldsymbol{k})$ is given by Fermi (or Fermi–Dirac) distribution function or the Boltzmann distribution function, which is a function of the electron energy $\varepsilon(\boldsymbol{k})$. Since $\varepsilon$ is an even function of $\boldsymbol{k}$, $v_x$, $f_0(k)$ is an odd function $v_x$. Since integration with respect to $dk_x$ ranges from $-\infty$ to $+\infty$, the first term on the right hand side becomes zero, and only the second term remains, resulting in

$$J_x = -\frac{e^2E_x}{4\pi^3}\int\tau v_x^2\frac{\partial f}{\partial\varepsilon}d^3\boldsymbol{k}. \qquad ...(20)$$

This result is interrupted with the help of the Fig. 7.17. In the absence of the electric field, the distribution function of the thermal equilibrium state is isotropic in $\boldsymbol{k}$ space and thus the current in any direction has the same magnitude, resulting in mutual cancellation and hence in zero current. In the presence of an applied electric field, the distribution function is given by $f = f_0 + f_1$ and is replaced by $f_1$ in the electric field direction, resulting in the current proportional to the displacement along the electric field direction. Using the electron field density, $n$ is given by

$$n = \frac{2}{(2\pi)^3}\int f_0 d^3\boldsymbol{k}, \qquad ...(21)$$

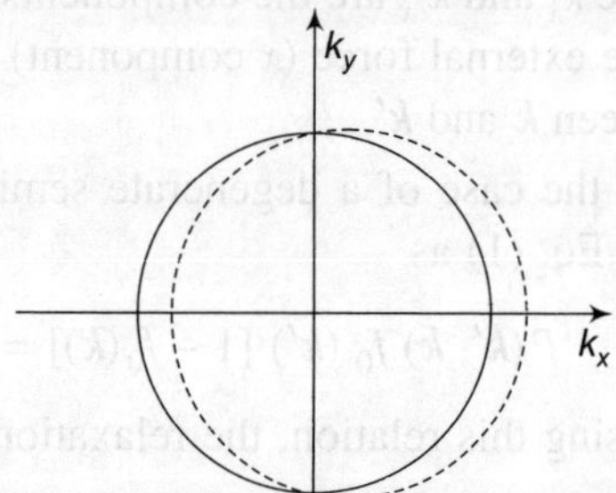


**Figure 7.17** Change in the distribution function under an applied electric field. The distribution function in $\boldsymbol{k}$ space is slightly displaced along the field direction

Equation (20) is written as

$$J_x = -e^2nE_x\frac{\int\tau v_x^2\frac{df_0}{d\varepsilon}d^3\boldsymbol{k}}{\int f_0 d^3\boldsymbol{k}} = \frac{e^2nE_x}{k_BT}\frac{\int\tau v_x^2 f_0(1-f_0)d^3\boldsymbol{k}}{\int f_0 d^3\boldsymbol{k}}, \qquad ...(22)$$

where the final results obtained by using the following relations:

$$f_0(\varepsilon) = \frac{1}{e(\varepsilon-\varepsilon_F)/k_BT+1} \qquad ...(23)$$

$$\frac{\partial f_0}{\partial\varepsilon} = -\frac{1}{k_BT}f_0(1-f_0). \qquad ...(24)$$

Since the term $\tau f_0(1-f_0)$ in the integral of (22) is a function of the energy $\varepsilon$, we express it as $\phi(\varepsilon)$. Noting $d^3\boldsymbol{k} = dk_x dk_y dk_z$, we have

$$\int v_x^2\,\phi(\varepsilon)d^3\boldsymbol{k} = \int v_y^2\,\phi(\varepsilon)d^3\boldsymbol{k} = \int v_x^2\,\phi(\varepsilon)d^3\boldsymbol{k} = \frac{1}{3}\int v^2\phi(\varepsilon)d^3\boldsymbol{k}. \quad ...(25)$$

The integral with respect to $d^3\boldsymbol{k}$ is changed into an integral with respect to ε by using the relation $\varepsilon = \hbar^2k^2/2m^*$. We then obtain

$$d^3\boldsymbol{k} = 4\pi k^2dk = \frac{8\pi m^{*3/2}}{\sqrt{2}\hbar^3}\varepsilon^{1/2}\,d\varepsilon \equiv A\varepsilon^{1/2}\,d\varepsilon. \quad ...(26)$$

In Eq. (26) the effective mass $m^*$ is the density-of-states mass for isotropic (scalar) effective mass and is often notified by $m^*_d$. In semiconductors such as Ge and Si, we know that the conduction bands consist of the many-valley structures and their constant energy surface is expressed by an ellipsoid with transverse effective mass $m_t$ and longitudinal effective mass, $m_1$. In such a case the density-of-state mass $m_d^*$ is given by $(m_t^2\, m_1)^{1/3} \equiv m_d^*$. Inserting this relation, Figs. (25), and (26) into Eq. (22), we obtain

$$J_x = \frac{2e^2nE_x}{3k_BTm^*}\frac{\int_0^\infty \tau\varepsilon^{3/2}f_0(1-f_0)d\varepsilon}{\int_0^\infty \varepsilon^{1/2}f_0\,d\varepsilon}. \quad ...(27)$$

Although the upper limit of the integral should be the upper edge of the band, it is replaced by infinity, Taking account of the exponential decay of the distribution function in the higher energy region. As stated above, the density-of-states mass appears in the denominator and numerator and is cancelled, which is evident from Eq. (22). The effective mass $m^*$ in Eq. (27) results from the term $v_x = \hbar k_x/m^*$, assuming, an isotropic effective mass. The effective mass, usually defined by $m_c^*$, is called the **conductivity effective mass** because it comes from the term of the velocity component. In the case of many-valley semiconductors such as Ge and Si, however, the energy is expressed by an ellipsoid, and the conductivity effective mass is given by $1/m^*_c \equiv (1/3)\,(2/mt + 1/m_1)$. Next, we carry out the calculation of Eq. (27) for the following two cases separately.

**(a) Metal and Degenerate Semiconductor.** In the case the Fermi energy $\varepsilon_F$ is located in the conduction band, and the term $-df_0/d\varepsilon = f_0(1-f_0)/k_BT$ has a large value near the Fermi energy only. Therefore, the term $-df_0/d\varepsilon$ may be approximated by the Dirac δ-function. In addition, the following relations,

$$f_0 \cong 1 \qquad (\varepsilon \le \varepsilon_F),$$

$$f_0 \cong 0 \qquad (\varepsilon \le \varepsilon_F),$$

hold for the Fermi distribution function, and therefore we obtain

$$\frac{1}{k_BT}\int_0^\infty \tau\varepsilon^{3/2}f_0(1-f_0)d\varepsilon$$

$$= -\int_0^\infty \tau(\varepsilon)\varepsilon^{3/2}\frac{df_0}{d\varepsilon} \cong \tau(\varepsilon_F)\varepsilon_F^{3/2} \quad ...(28)$$

$$\int_0^\infty \varepsilon^{1/2}f_0d\varepsilon = \int_0^\infty \varepsilon^{1/2}d\varepsilon = \frac{2}{3}\varepsilon_F^{3/2}. \quad ...(29)$$

Inserting these relations into Eq. (27), we obtain

$$J_x = \frac{ne^2\tau(\varepsilon_F)}{m^*}E_x, \quad ...(30)$$

where $t(\varepsilon_F)$ is the relaxation time of an electron at $\varepsilon = \varepsilon_F$ (or at the Fermi surface). The relation between the Fermi energy $\varepsilon_F$ and the electron density $n$ is obtained from Eq. (21), and Eq. (28) and is given by

$$\varepsilon_F = \frac{\hbar^2}{2m^*}(3\pi^2 n)^{2/3}. \quad ...(31)$$

**(b) Non-degenerate Semiconductor.** For non-degenerate semiconductors, we have $f_0 \ll 1$ and $1 - f_0 \cong 1$ and we may approximate as

$$f_0 = \exp\left(\frac{\varepsilon_F}{k_BT}\right)\exp\left(-\frac{\varepsilon}{k_BT}\right) \equiv A_F \exp\left(-\frac{\varepsilon}{k_BT}\right), \quad ...(32)$$

and Eq. (27) reduces to

$$J_x = \frac{2e^2nE_x}{3k_BTm^*}\frac{\int_0^\infty \tau\varepsilon^{3/2} f_0\, d\varepsilon}{\int_0^\infty \varepsilon^{1/2} f_0\, d\varepsilon}. \quad ...(33)$$

Integration by parts gives rise to

$$\int_0^\infty \varepsilon^{3/2}\, f_0 d\varepsilon = \frac{3}{2}k_BT\int_0^\infty \varepsilon^{1/2}\, f_0 d\varepsilon. \quad ...(34)$$

Using this relation we obtain the following relation

$$J_x = \frac{e^2nEx}{m^*}\frac{\int_0^\infty \tau\varepsilon^{3/2} f_0\, d\varepsilon}{\int_0^\infty \varepsilon^{3/2} f_0\, d\varepsilon}. \quad ...(35)$$

When we write Eq. (35) as

$$J_x = \frac{ne^2\langle\tau\rangle}{m^*}E_x, \quad ...(36)$$

The relaxation time $\langle\tau\rangle$ is given by

$$\langle\tau\rangle = \frac{\int_0^\infty \tau\varepsilon^{3/2} f_0\, d\varepsilon}{\int_0^\infty \varepsilon^{3/2} f_0\, d\varepsilon} \quad ...(37)$$

and $\langle\tau\rangle$ represents the average of $\tau(\varepsilon)$.

When we define the average velocity of the electron by $\langle v_x\rangle$ in the presence of an electric field in the $x$ direction, it is also calculated from Eq. (37). The average velocity is often referred to as the **drift velocity**. This is because the electron contributes to the current via **drift motion** along the electric field direction suffering from collision. The average time between the scattering events (collisions) or the scattering rate (collision rate) is given by the average value of the relaxation time $\langle\tau\rangle$, where the scattering rate is not equivalent to relaxation time in the strict sense. The difference will be discussed later. Using the electron density $n$, the current density is written as

$$J_x = n(-e)\langle v_x \rangle \quad ...(38)$$

where the drift velocity $\langle v_x \rangle$ is given by Eq. (36) as

$$\langle v_x \rangle = -\frac{e\langle \tau \rangle}{m^*} E_x \equiv -\mu E_x. \quad ...(39)$$

Here μ is the average electron velocity (drift velocity) under a unit electric field and gives a measure of the mobility; it is called the electron mobility. This mobility is often referred to as the drift velocity to distinguish it from the Hall mobility described in section 5.5. Using $\langle \tau \rangle$ the electron mobility is expressed as

$$\mu = \frac{e\langle \tau \rangle}{m^*}, \quad ...(40)$$

and when we define the **conductivity** (electrical conductivity) σ by

$$J_x = \sigma E_x, \quad ...(41)$$

the conductivity is given by

$$\sigma = \frac{ne^2\langle \tau \rangle}{m^*} = ne\mu. \quad ...(42)$$

These relations hold for degeneration semiconductors by replacing $\langle \tau(\varepsilon) \rangle$ with $\tau(\varepsilon_F)$.

As a simple example we consider the case of the relaxation time τ given by a function of energy ε such as

$$\tau = a\varepsilon^{-s}. \quad ...(43)$$

Then from Eq. (37), we obtain

$$\langle \tau \rangle = \frac{a\int_0^\infty \varepsilon^{3/2-s} \exp(-\varepsilon/k_BT)d\varepsilon}{\int_0^\infty \varepsilon^{3/2} \exp(-\varepsilon/k_BT)d\varepsilon} = a(k_BT)^{-s} \frac{\Gamma\left(\frac{5}{2}-s\right)}{G\left(\frac{5}{2}\right)}, \quad ...(44)$$

where $\Gamma(s)$ is the Gamma function and has properties such as

$$\Gamma(s) = \int_0^\infty x^{x-1} e^{-x} dx, \quad ...(45)$$

$$\Gamma(s+1) = s\Gamma(s), \quad \Gamma(1) = 1, \quad \Gamma\left(\frac{1}{2}\right) = \sqrt{\pi}. \quad ...(46)$$

## EXERCISE

7.1 Distinguish between classical and quantum particles. Differentiate between bosons and fermions.

7.2 Derive Maxwell–Boltzmann distribution of the particles among various states.

7.3 Derive the expression for Fermi-Dirac distribution of particles.

7.4 Derive the Bose – Einstein distribution of particles.

7.5 Explain Bose – Einstein condensation.

7.6 Derive the Boltzmann transport equations.

7.7 Derive an approximate solution of the Boltzmann transport equation. What is relaxation time?

7.8 Derive the Boltzmann transport equation in relaxation time approximation.

7.9 Derive an expression for current density $J_x$ using Boltzmann transport equation.

CHAPTER

# 8

# Elastic Properties of Solids

## 8.1 INTRODUCTION (SOLIDS SUBJECTED TO EXTERNAL FORCE)

Since solids are not perfectly rigid, apart from translation in space or rotation, other types of response to external force may be considered. In particular, solids can undergo mechanical formations or changes in dimensions or shapes, depending on their strength, i.e., ability to withstand mechanical deformation.

The mechanical properties of materials include elastic, anelastic or inelastic behaviour.

A solid is said to be *elastic* when any deformations disappears "quickly" once the external forces are removed (i.e., the solid returns to its initial state). In the proportional elastic region (i.e., within the elastic limit), the deformations are directly proportional to the external forces (i.e., the mechanical strain tensor components $\varepsilon_{ij}$ are directly proportional to the stress tensor or components, $\sigma_{ij}$).

The *anelastic* behaviour of materials is a type of time-dependent mechanical behaviour in which the applied stresses and the resulting strains are not uniquely related to each other due to relaxation effects.

Other types of response, including permanent deformations, plasticity, viscoelasticity, fracture and so on correspond to inelastic behaviour.

The definitions of stress and strain in materials are introduced first, along with the material properties known as elastic constants.

First, we shall discuss the elasticity of single crystal and find their elastic constants. While dealing with this, we treat the crystal as a homogeneous continuous medium rather than a periodic array of atoms. This continuous approximation is usually valid for elastic waves of wavelength longer than $10^{-6}$ cm, i.e., for frequencies below $10^{12}$ Hz.

## 8.2 THE STRESS

Under the equilibrium condition, stresses are the forces applied per unit area to a body.

Now, to analyse stress in crystals, it is convenient to imagine that one removes from the solid at a point in question, a small cube of sides *dx*, *dy*, *dz* and then apply to it the same stresses to which it was subjected in the solid. Now, on each of the cube faces, there will in general be forces which can be resolved into three mutually perpendicular components: the one that is normal to the plane it is called the normal stress, while the two lying in the plane are called the shear stress.

When the cube is in dynamic equilibrium, the forces on the opposite faces must be equal in magnitude and opposite in sign. Thus, the stress condition of the cube may be described by nine components – three stresses normal to the cube faces and six shear stresses acting across the cube faces. These can be expressed in the matrix form as

$$\begin{bmatrix} X_x & X_y & X_z \\ Y_x & Y_y & Y_z \\ Z_x & Z_y & Z_z \end{bmatrix}$$

A capital letter indicates the direction of the force and its subscript, the direction of the outward normal to the plane on which the force is acting. Thus, in accordance with Fig. 8.1, the stress component $X_z$ denotes a force acting in the *x*-direction per unit area of a face whose normal lies in the *x*-direction; the stress component $x_y$ represents a force acting in the *x*-direction per unit area of a face whose normal lies in the *y*-direction. The number of independent stress component is reduced further to 6 if the cube does not rotate. For this, the condition is

$$Y_z = Z_y, \quad Z_x = X_z, \quad X_y = Y_x \text{ (Fig. 8.2)}$$

Then, the six independent stress components may be taken as $X_x$, $Y_y$, $Z_z$, $Y_z$, $Z_x$, $X_y$. These six components are sufficient to describe the state of stress at any point in the crystal. If these stresses are the same at every point in the crystal, it is said to be homogeneously stressed.

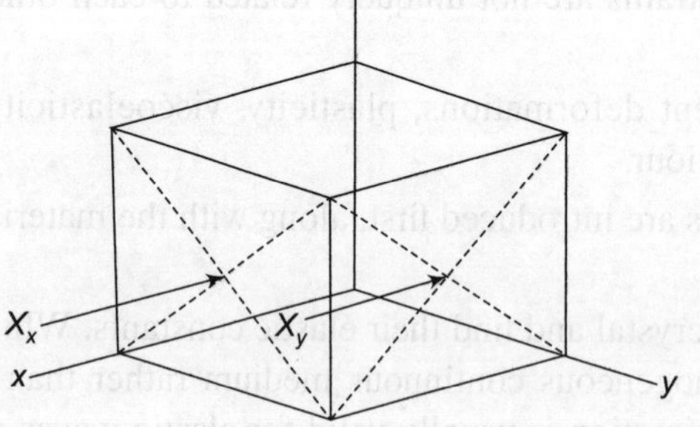


**Figure 8.1** Meaning of the stress components $X_x$ and $X_y$

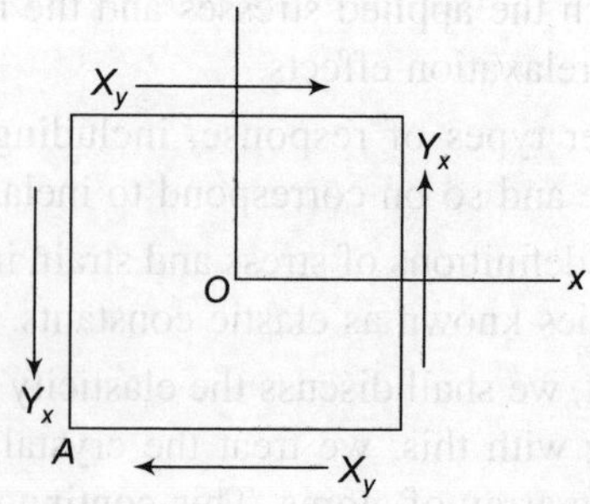


**Figure 8.2** The condition for no rotation, demonstrated for $Y_x = X_y$ (static equilibrium)

## 8.3 THE STRAIN

The elastic strain of a body from a deformation is usually specified in terms of six components $e_{xx}, e_{yy}, e_{zz}, e_{xy}, e_{yz}, e_{zx}$ which are defined below.

Imagine that three orthogonal unit vectors $\vec{i}, \vec{g}, \vec{k}$ are embedded firmly in the unstrained solid. Next suppose that a small uniform deformation (i.e., deformation in which each primitive cell of the crystal is deformed in the same way) of the solid has taken place, and the axes are distorted in orientation and length so that new axes $i', j', k'$ may be written in terms of the old axes as

$$\left.\begin{aligned} \vec{i}' &= (1+\varepsilon_{xx})\,\vec{i} + \varepsilon_{xy}\,\vec{j} + \varepsilon_{xz}\,\vec{k} \\ \vec{j}' &= \varepsilon_{yx}\,\vec{i} + (1+\varepsilon_{yy})\,\vec{j} + \varepsilon_{yz}\,\vec{k} \\ \vec{k}' &= \varepsilon_{zx}\,\vec{i} + \varepsilon_{zy}\,\vec{j} + (1+\varepsilon_{zz})\,\vec{k} \end{aligned}\right\} \quad ...(1)$$

The coefficients, $\varepsilon_{xx}, \varepsilon_{xy}, \varepsilon_{yz}$, etc., define the deformation. They are dimensionless quantities and have values $\ll 1$ if the the strain is small. The above expressions express the fact that the new axes will not necessarily be of unit length, though the original axes were. For example,

$$\vec{i}' = 1 + \varepsilon_{xx} + ....$$

as is seen by taking the square root of

$$\vec{i}' \cdot \vec{i}' = i'^2 = 1 + 2\varepsilon_{xx} + \varepsilon_{xx}^2 + \varepsilon_{xy}^2 + \varepsilon_{xz}^2$$

We also observe that the fractional changes in lengths of $\vec{i}, \vec{j}, \vec{k}$ axes are then $\varepsilon_{xx}, \varepsilon_{yy}, \varepsilon_{zz}$ respectively to the first order and thus $\varepsilon_{xx}, \varepsilon_{yy}, \varepsilon_{zz}$ represent the linear strain components. We now define

$$e_{xx} \equiv \varepsilon_{xx}, \quad e_{yy} \equiv \varepsilon_{yy}, \quad e_{zz} \equiv \varepsilon_{zz} \quad ...(2)$$

Note also that the original axes were mutually orthogonal but the new axes are not necessarily so, since

$$\vec{i}' \cdot \vec{j}' \cong \varepsilon_{yx} + \varepsilon_{xy}\ (\neq 0)$$

Further, since $\vec{i} \cdot \vec{j} = 0$, $\therefore$ $\vec{i}' \cdot \vec{j}'$ as such gives a measure of the change in orientation between $\vec{i}'$ and $\vec{j}$ as a result of deformation. Thus, if we define $e_{xy}$ in terms of the change in angle between the axes, we see that $(\varepsilon_{yx} + \varepsilon_{xy})$ represents $e_{xy}$. Similarly,

$$e_{yz} = \vec{j}' \cdot k' \simeq \varepsilon_{zy} + \varepsilon_{yz}$$

and $$e_{zx} = \vec{k}' \cdot \vec{i}' \simeq \varepsilon_{zx} + \varepsilon_{xz}. \quad ...(3)$$

We now note that the fact that merely rotating the axes does not change the angle between them. Using Eq. (3), we see that for pure rotation $\varepsilon_{yx} = -\varepsilon_{xy}$, $\varepsilon_{zy} = -\varepsilon_{yz}$, $\varepsilon_{zx} = -\varepsilon_{xz}$. If we exclude pure rotations (as not of interest here), we may take $\varepsilon_{yx} = \varepsilon_{xy}$, $\varepsilon_{zy} = \varepsilon_{yz}$, $\varepsilon_{zx} = \varepsilon_{xz}$ and so we may have

$$e_{xy} = 2\varepsilon_{xx}; \quad e_{yz} = 2\varepsilon_{yz}; \; e_{zx} = 2\varepsilon_{zx} \quad ...(4)$$

Now, let us consider the effect of the deformation (1) on a point (i.e., an atom) originally at $\vec{r} = x\hat{i} + y\hat{j} + z\hat{k}$. The origin is taken at some other point. If the deformation is uniform, then, after the deformation the above point will be at the position

$$\vec{r} = x\hat{i}' + y\hat{j}' + z\hat{k}'$$

and the displacement $\vec{R}$ of the deformation is defined by

$$\vec{R} \simeq \vec{r}' - \vec{r} = x(\vec{i}' - \vec{i}) + y(\vec{j}' - \vec{j}) + z(\vec{k}' - \vec{k}) \qquad ...(5)$$

or using Eq. (1)

$$\left.\begin{aligned} \vec{R}(\vec{r}) = \vec{i}\,(x\varepsilon_{xx} + y\varepsilon_{yx} + z\varepsilon_{zx}) \\ + \vec{j}\,(x\varepsilon_{xy} + y\varepsilon_{yy} + z\varepsilon_{zy}) \\ + \vec{k}\,(x\varepsilon_{xz} + y\varepsilon^{yz} + z\varepsilon_{zz}) \end{aligned}\right\} \qquad ...(6)$$

Considering that *u*,*v* and *w* are the components of the displacement $\vec{R}$ along the three coordinate axes, Eq. (6) may be written in a more general form

$$\vec{R}(\vec{r}) = u(r)\,\vec{i} + v(r)\,\vec{j} + w(r)\,\vec{k} \qquad ...(7)$$

If the deformation is non-uniform, we must relate *u*, *v*, *w* to the local strain, i.e., we must take the origin of $\vec{r}$ close to the region of interest comparing now Eqs. (6) and (7), after using Taylor's series expansion of $\vec{R}$ and using $\vec{R}(0) = 0$, we get

$$x\varepsilon_{xx} \simeq x\frac{\partial u}{\partial x}; \quad y\varepsilon_{yx} \simeq y\frac{\partial u}{\partial y} \text{ etc.}$$

(Note that the derivatives are independent of the origin chosen for $\vec{R}$). Therefore, we can alternatively define the strain components

$$e_{xx} \equiv \varepsilon_{xx} = \frac{\partial u}{\partial x}, \quad e_{yy} \equiv \varepsilon_{yy} = \frac{\partial u}{\partial y}, \quad e_{zz} \equiv \varepsilon_{zz} = \frac{\partial w}{\partial z} \qquad ...(8)$$

The other strain components are as hereunder:

$$\left.\begin{aligned} e_{xy} \equiv \vec{i}'\cdot\vec{j}' \simeq \varepsilon_{yx} + \varepsilon_{xy} = \frac{\partial u}{\partial y} + \frac{\partial v}{\partial x} \\ e_{yz} \equiv \vec{j}'\cdot\vec{k}' \simeq \varepsilon_{zy} + \varepsilon_{yz} = \frac{\partial u}{\partial z} + \frac{\partial w}{\partial y} \\ e_{zx} \equiv \vec{k}'\cdot\vec{i}' \simeq \varepsilon_{zx} + \varepsilon_{xz} = \frac{\partial u}{\partial z} + \frac{\partial w}{\partial x} \end{aligned}\right\} \qquad ...(9)$$

In these expressions, we have neglected the terms of the order of $\varepsilon^2$. These six components completely define the strain. The strains thus defined are dimensionless.

## 8.4 INTERNAL FORCES AND STRESS

When external forces are applied to a solid material, stresses can be created within the solid. For a solid in mechanical equilibrium, the stresses result from the internal forces that are generated within the solid to counterbalance the externally applied forces. These internal forces are exerted by adjacent elements of the solid for each other and ultimately arise from the forces between atoms. The response of the solid to the stresses σ are the deformations known as strains, ε.

To define the stresses σ in terms of the internal forces in a solid, situation where an external force $F_{ext}$ is applied to the end of a solid bar is examined. This is shown schematically in Fig. 8.3. Consider the forces acting on a small element of volume

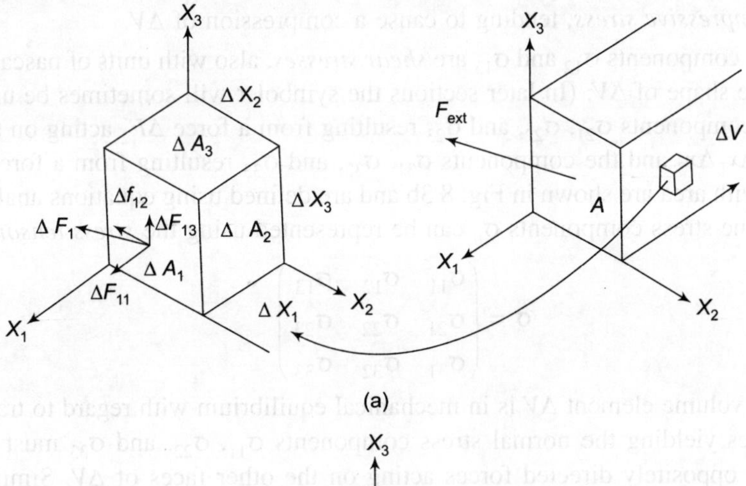


**Figure 8.3** Internal forces and stresses in a solid by resulting from the application of an external force $F_{ext}$: (a) internal force $F_1$ acting on one face of a small element of volume $\Delta V = \Delta x_1\ \Delta x_2\ \Delta x_3$ (b) the three components $\sigma_{11}$, $\sigma_{12}$ and $\sigma_{13}$ of the stress acting on this face

$\Delta V = \Delta x_1\ \Delta x_2\ \Delta x_3$ within the solid. The internal force $\Delta F_1$ acting on the face with area $\Delta A_1 = \Delta x_2\ \Delta x_3$ is exerted by the different material in the solid. The volume $\Delta V$ is chosen to be small enough so that the force $\Delta F_1$ acts uniformly over the area $\Delta A_1$. The components of the force $\Delta F_1$ acting along the three coordinate axes $x_1$, $x_2$, and $x_3$ are $\Delta F_{11}$, $\Delta F_{12}$, and $\Delta F_{13}$, respectively. as shown in Fig. 8.3a.

The three components $\sigma_{ij}$ off the *stress* σ which act on the face shown have magnitude given by

$$\sigma_{11} = \frac{\Delta F_{11}}{\Delta A_1}, \quad \sigma_{12} = \frac{\Delta F_{12}}{\Delta A_1}, \quad \sigma_{13} = \frac{\Delta F_{13}}{\Delta A_1}. \qquad ...(1)$$

The direction of $\sigma_{11}$, $\sigma_{12}$, and $\sigma_{13}$ are shown in Fig. 8.3b and are the same as the directions of the corresponding force components (i.e., the first subscript indicates direction of the normal of the face to which the stress is applied, and the second subscript indicates the direction in which the free component acts). For example, the stress component $\sigma_{11}$ is a *normal (tensile) stress*, with units of N/m$^2$ or pascal (1 Pa = 1 N/m$^2$), which tends to cause an elongation of the volume element in the $x_1$ direction. If the direction of $\Delta F_{11}$ were reversed, $\sigma_{11}$ would be a *normal compressive stress*, tending to cause a compression of $\Delta V$.

The stress components $\sigma_{12}$ and $\sigma_{13}$ are *shear stresses*, also with units of pascal, which tend to deform the shape of $\Delta V$. (In later sections the symbol $\tau$ will sometimes be used for shear stress). The components $\sigma_{21}$, $\sigma_{22}$, and $\sigma_{23}$ resulting from a force $\Delta F_2$ acting on the face with area $\Delta A_2 = \Delta x_1 \Delta x_3$ and the components $\sigma_{31}$, $\sigma_{32}$, and $\sigma_{33}$ resulting from a force $\Delta F_3$ acting on the face with area are shown in Fig. 8.3b and are defined using equations analogous to Eq. (1). These nine stress components $\sigma_{ij}$ can be represented using the *stress tensor* $\sigma$, given by

$$\sigma = \begin{pmatrix} \sigma_{11} & \sigma_{12} & \sigma_{13} \\ \sigma_{21} & \sigma_{22} & \sigma_{23} \\ \sigma_{31} & \sigma_{32} & \sigma_{33} \end{pmatrix} \quad ...(2)$$

When the volume element $\Delta V$ is in mechanical equilibrium with regard to translation, the internal forces yielding the normal stress components $\sigma_{11}$, $\sigma_{22}$, and $\sigma_{33}$ must be balanced by equal but oppositely directed forces acting on the other faces of $\Delta V$. Similarly, for $\Delta V$ to be in rotational equilibrium, the couple $\Delta F_{12}\Delta x_1$ due to the two shear stress components corresponding to $\sigma_{12}$ acting on the two opposite faces with areas $\Delta A_1$ must be balanced by an opposing couple acting on the two opposite faces with areas $\Delta A_2$. This opposing torque is given by $\Delta F_{21}\Delta x_2$ and results from the shear stress components corresponding to $\sigma_{21}$ (see Figs. 8.3b and 8.4). It therefore follows that $\sigma_{12} = \sigma_{21}$ and similarly, that $\sigma_{13} = \sigma_{31}$ and $\sigma_{23} = \sigma_{32}$ for rotational equilibrium of the volume element $\Delta V$.

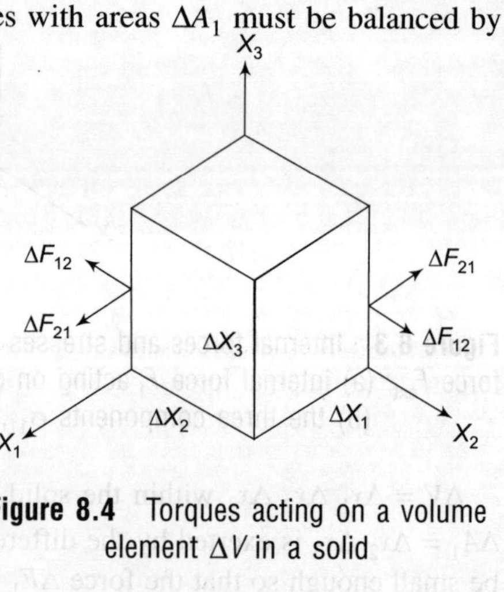


**Figure 8.4** Torques acting on a volume element $\Delta V$ in a solid

In the limit $\Delta V \to 0$, the state of stress at a given point in a solid can therefore be specified by just six independent stress components instead of the nine possible $\sigma_{ij}$. These six stress components are often labelled using a single subscript, as follows:

$$\sigma_1 = \sigma_{11}, \quad \sigma_2 = \sigma_{22}, \quad \sigma_3 = \sigma_{33},$$
$$\sigma_4 = \sigma_{23} = \sigma_{32} \quad \sigma_5 = \sigma_{31} = \sigma_{13} \quad \sigma_6 = \sigma_{12} = \sigma_{21}. \quad ...(3)$$

The stress tensor $\sigma$ defined in Eq. (2) is therefore symmetric, with the form

$$\sigma = \begin{pmatrix} \sigma_{11} & \sigma_{12} & \sigma_{13} \\ \sigma_{12} & \sigma_{22} & \sigma_{23} \\ \sigma_{13} & \sigma_{23} & \sigma_{33} \end{pmatrix} = \begin{pmatrix} \sigma_1 & \sigma_6 & \sigma_5 \\ \sigma_6 & \sigma_2 & \sigma_4 \\ \sigma_5 & \sigma_4 & \sigma_3 \end{pmatrix} \quad ...(4)$$

The relationship between the external forces and the resulting internal forces and stresses σ are simply only for special cases. For example, if external forces $F_{ext}$ act on the axis of a long bar of original cross-sectional area $A_o$, as is usually the case in the tensile testing of materials, the internal forces $F_{int}$ and normal tensile stresses also act along the axes of the bar. It follows that $\sigma = F_{ext}/A_o$ as long as changes in $A_o$ are small. When changes in $A_o$ can not be neglected, the *nominal stress* $\sigma = F_{ext}/A_o$ no longer represents the true tensile stress in the material. The *true stress* is defined by $\sigma_{true} = F_{ext}/A$ where $A$ is the actual cross-sectional area, which can vary along the length of the bar.

Within the elastic limit of a material, the stresses can often be calculated analytically from the forces applied. In practice, however, the actual of stress in a solid cannot be calculated easily or measured directly. The only general way to determine the stresses in a solid through experimental study of the resulting strains.

## 8.5 DEFORMATION AND STRAIN

Having defined the stresses σ that can exist within a solid due to the application of external forces, the next step is to describe of the response of the solid (i.e., its deformation) in terms of the resulting *strains* ε created within the solid. There are in general six strain components $\varepsilon_i$, $i$ = 1 to 6, analogous to the 6 stress components $\sigma_i$ define in Eq. (3). The $\varepsilon_i$ are illustrated using Fig. 8.5, where a solid with initial volume $V_0 = x_{10}x_{20}x_{30}$ and with edges lying initially along the $x_1$, $x_2$, $x_3$ coordinate axes is shown both before and after a general deformation. This general *deformation* can be described as follows:

$$\mathbf{x}_1 = \mathbf{x}_{10} + \mathbf{u}_1, \quad \mathbf{x}_2 = \mathbf{x}_{20} + \mathbf{u}_2, \quad \mathbf{x}_3 = \mathbf{x}_{30} + \mathbf{u}_3, \qquad ...(5)$$

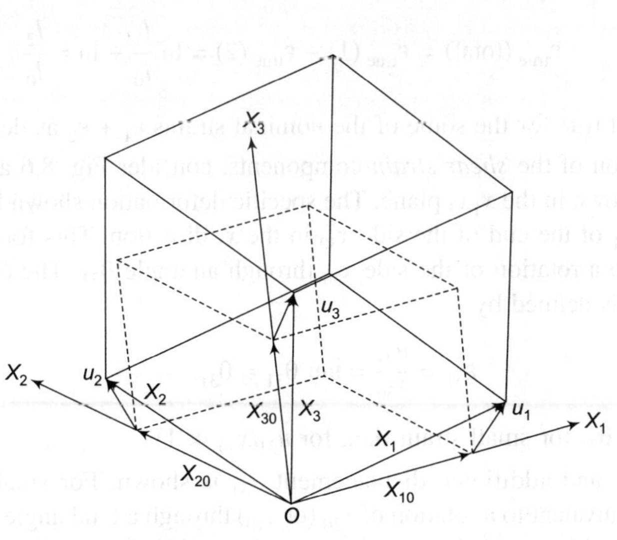


**Figure 8.5** Solid with an initial volume $V_0 = x_{10}, x_{20}, x_{30}$ both before and after undergoing a general deformation. Also shown are the vector displacements $\mathbf{u}_1$, $\mathbf{u}_2$, $\mathbf{u}_3$, of the three edges that are used to define the strains, ε

where the vectors $\mathbf{x}_{10}$, $\mathbf{x}_{20}$, $\mathbf{x}_{30}$ and $\mathbf{x}_1$, $\mathbf{x}_2$ and $\mathbf{x}_3$, define the edges of the solid before and after the deformation, respectively. The vectors $\mathbf{u}_1$, $\mathbf{u}_2$, $\mathbf{u}_3$ are therefore the displacements corresponding to the deformation, as shown. The components $\mathbf{u}_1$ along the three cooordinate axes are $u_{11}$, $u_{12}$, $u_{13}$.

The three *normal strain* components $\varepsilon_i$ (= $\varepsilon_{ii}$) are defined in terms of the displacements $u_{ii}$ by

$$\varepsilon_1 (= \varepsilon_{11}) = \frac{u_{11}}{x_{10}},\ \varepsilon_2 (= \varepsilon_{22}) = \frac{u_{22}}{x_{20}},\ \varepsilon_3 (= \varepsilon_{33}) = \frac{u_{33}}{x_{30}}. \qquad ...(6)$$

The strains $\varepsilon_i$ as defined are fractional changes in length and so are dimensionless quantities. Deformations for which $\varepsilon_i > 0$ are *elongations*, while those for which even smaller than zero are *compressions*.

An alternative definition for normal strain involves the changes in length relative to the instantaneous length $l$ instead of the relative to the initial length $l = 0$. Following the definition given in Eq. (6), the *nominal strain* ε can be defined by $\Delta l/l_0 = (l - l_0)/l_0$. In terms of changes relative to the instantaneous length $l$, the *true strain* $\varepsilon_{\text{true}}$ is defined as follows:

$$\varepsilon_{\text{true}} = \int_{l0}^{l} \frac{dl}{l_0} = \ln\frac{l}{l_0} = \ln\frac{l_0 + \Delta l}{l_0} = \ln\left(1 + \frac{\Delta l}{l_0}\right) = \ln(1+\varepsilon) \qquad ...(7)$$

In the limit of very small strain (i.e., $\varepsilon \ll 1$) it can be seen that the nominal strain ε and $\varepsilon_{\text{true}}$ are equivalent. This definition of true strain has the advantage that consecutive true strain $\varepsilon_{\text{true}}$ (1) and $\varepsilon_{\text{true}}$ (2) are additive that is,

$$\varepsilon_{\text{true}}\ (\text{total}) = \varepsilon_{\text{true}}\ (1) + \varepsilon_{\text{true}}\ (2) = \ln\frac{l_1}{l_0} + \ln = \frac{l_2}{l_0}. \qquad ...(8)$$

The same is not true for the some of the nominal strains $\varepsilon_1 + \varepsilon_2$ as defined in Eq. (6).

For the definition of the *shear strain* components, consider Fig. 8.6 a, in which the side view of solid is shown in the $x_1x_3$ plane. The specific deformation shown here corresponds to a displacement $u_{31}$ of the end of the side $x_{30}$ in the $x_1$ direction. This formation can be seen to be equivalent to a rotation of the side $x_{30}$ through an angle $\theta_{31}$. The corresponding shear strain component is defined by

$$\varepsilon_{31} = \frac{u_{31}}{x_{30}} = \tan\theta_{31} \approx \theta_{31}. \qquad ...(9)$$

Note that $\varepsilon_{31} \approx \theta_{31}$ for small strain (i.e., for $u_{31}/x_{30} \ll 1$).

In Fig. 8.6 (b), and additional displacement $u_{31}$ is shown. For small strains, the total displacement is equivalent to a rotation of $x_{30}$ (or $x_{10}$) through a total angle $\theta_{31} + \theta_{13}$, as shown in Fig. 8.6. The resulting total shear strain components therefore can be defined as follows:

$$\varepsilon_5\ (= \varepsilon_{31} = \varepsilon_{13}) = \frac{u_{31}}{x_{30}} + \frac{u_{13}}{x_{10}} \qquad ...(10)$$

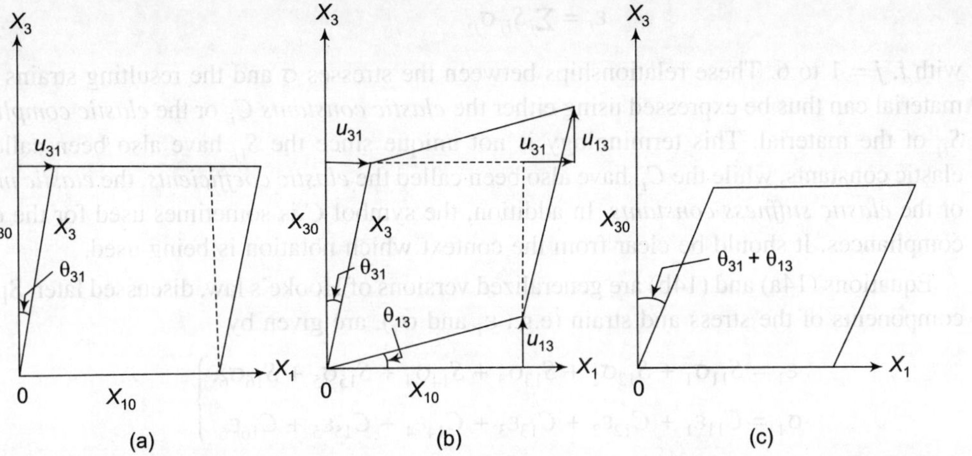


**Figure 8.6** Components of shear strain in a solid defined using symmetric diagram. The deformation shown in (a) corresponds to the displacement $u_{31}$ and is equivalent to a rotation of the side $x_{30}$ through an angle $\theta_{31}$. In (b) an additional displacement $u_{13}$ is shown. For small strain, the total displacement (b) is equivalent to a rotation of $u_{30}$ (or $x_{10}$) Through a total angle $\theta_{31} + \theta_{13}$, as shown in (c)

The two remaining shear strain components $\varepsilon_4$ and $\varepsilon_6$ are defined similarly:

$$\left.\begin{aligned} \varepsilon_4\ (= \varepsilon_{32} = \varepsilon_{23}) &= \frac{u_{32}}{x_{30}} + \frac{u_{23}}{x_{20}}, \\ \varepsilon_6\ (= \varepsilon_{21} = \varepsilon_{12}) &= \frac{u_{21}}{x_{20}} + \frac{u_{12}}{x_{10}}. \end{aligned}\right\} \qquad ...(11)$$

It follows from these definitions that the *strain tensor* $\varepsilon$ is also symmetric with the form.

$$\varepsilon = \begin{pmatrix} \varepsilon_{11} & \varepsilon_{12} & \varepsilon_{13} \\ \varepsilon_{12} & \varepsilon_{22} & \varepsilon_{23} \\ \varepsilon_{13} & \varepsilon_{23} & \varepsilon_{33} \end{pmatrix} = \begin{pmatrix} \varepsilon_1 & \varepsilon_6 & \varepsilon_5 \\ \varepsilon_6 & \varepsilon_2 & \varepsilon_4 \\ \varepsilon_5 & \varepsilon_4 & \varepsilon_3 \end{pmatrix}. \qquad ...(12)$$

## 8.6 RELATIONSHIP BETWEEN STRESS AND STRAIN (ELASTIC CONSTANT)

Linear elasticity refers in general to a relationship between two second-rank tensors, the stress tensor $\sigma_{ij}$ and the strain tensor $\varepsilon_{mn}$, through the fourth-rank *elastic coefficient tensor* $C_{ijmn}$; that is,

$$\sigma_{ij} = \Sigma_{m,n}\, C_{ijmn}\, \varepsilon_{mn}. \qquad ...(13)$$

Due to symmetry mentioned earlier of $\sigma_{ij}$ and of $\varepsilon_{mn}$, the number of independent elements of $C_{ijmn}$ is reduced from 81 to 36 in general. Hence, the following contracted notations are used:

$$\sigma_i = \Sigma_j C_{ij} \varepsilon_j, \qquad ...(14a)$$

$$\varepsilon_i = \sum_j S_{ij}\sigma_j, \quad \text{...(14b)}$$

with $i, j = 1$ to 6. These relationships between the stresses $\sigma$ and the resulting strains $\varepsilon$ in a material can thus be expressed using either the *elastic constants* $C_{ij}$ or the *elastic compliances* $S_{ij}$ of the material. This terminology is not unique since the $S_{ij}$ have also been called the elastic constants, while the $C_{ij}$ have also been called the *elastic coefficients*, the *elastic moduli*, or the *elastic stiffness constants*. In addition, the symbol $C$ is sometimes used for the elastic compliances. It should be clear from the context which notation is being used.

Equations (14a) and (14b) are generalized versions of Hooke's law, discussed later. Specific components of the stress and strain (e.g., $\varepsilon_1$ and $\sigma_1$), are given by

$$\left.\begin{aligned} \varepsilon_1 &= S_{11}\sigma_1 + S_{12}\sigma_2 + S_{13}\sigma_3 + S_{14}\sigma_4 + S_{15}\sigma_5 + S_{16}\sigma_6, \\ \sigma_1 &= C_{11}\varepsilon_1 + C_{12}\varepsilon_2 + C_{13}\varepsilon_3 + C_{14}\varepsilon_4 + C_{15}\varepsilon_5 + C_{16}\varepsilon_6. \end{aligned}\right\} \quad \text{...(15)}$$

In Eq. (14a) the stresses are determined from the strains, whereas in Eq. (14b) the reverse is true. Although the second approach is more logical, the mechanical properties of materials are more often expressed using the elastic constant $C_{if}$. It is clear that the two approaches are closely related since the matrices $S_{ij}$ and $C_{ij}$ are inverse of each other.

The region in which the strains $\sigma$ and the stresses $\varepsilon$ directly proportional to each other [i.e., in which the $S_{ij}$ and independent of the stresses $\sigma_j$ and the $C_{ij}$ are independent of the strains $\varepsilon_j$ (and both $S_{ij}$ and $C_{ij}$ are independent of time)], is known as the *proportional elastic limit*. The mechanical response of the material in this limit obeys *Hooks's law*, which states that the response of the material (i.e., the deformation or strain) is directly proportional to the force or stress applied. When the $S_{ij}$ depend on the $\sigma_j$ or when the $C_{ij}$ depend on the $\varepsilon_j$, the material is said to be *inelastic*. When the $S_{ij}$ or $C_{ij}$ is time dependent the material is said to be *anelastic*.

Four different types of mechanical behaviour observed for materials are illustrated in Fig. 8.7 in the usual form of curves of nominal stress versus nominal stress. These curves are determined using a procedure known as *tensile testing*, taken with the material in the form of a bar and under tension. In Fig. 8.7 the response of alumina, $Al_2O_3$, can be seen to be elastic initially, followed immediately by brittle fracture. In Fig. 8.7 the elastic region observed for a low-carbon steel is followed by an extensive region of plastic deformation and then finally by ductile fracture. In Fig. 8.7 the type of elastomeric response found for a cross-linked natural rubber is shown, while in Fig. 8.7 d the viscoelastic behaviour of a plastic polymer, polymethylmethacrylate (PMMA), is shown.

It should be noted that within the elastic limit, typical strains in material such as metals and ceramics extend up to about $10^{-3}$, beyond which point inelasticity involving permanent deformations and ultimately, fracture can occur. In other materials, such as polymers elastic strains up to unity or beyond are not unusual. Clearly, as illustrated for the materials shown here, the mechanical response of materials can be quite complicated.

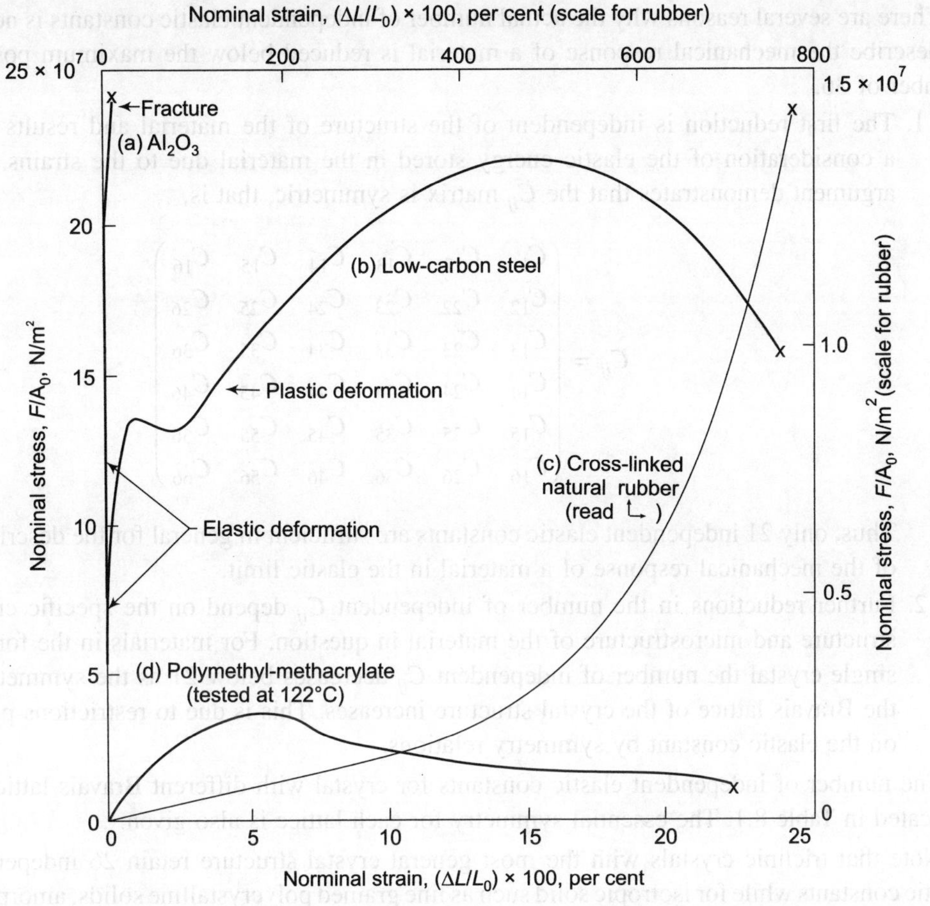


**Figure 8.7** Four examples of the mechanical response of material corresponding to (a) elastic, (b) plastic, (c) elastomeric and (d) viscoelastic behaviour are illustrated using nominal stress strain curves. (From A. G. Guy, *Introduction to Materials Science*, (1972) McGraw-Hill.)

While a solid material in principle has 36 independent elastic compliances $S_{ij}$ and also 36 independent elastic constants $C_{ij}$, in practice the actual number of independent $S_{ij}$ and $C_{ij}$ is always less than 36 (and can even be equal to 2 for isotropic solids). The 36 elastic constant $C_{ij}$ of a single crystal of a uniform, homogeneous material conveniently displayed in a 6 × 6 matrix as follows:

$$C_{ij} = \begin{pmatrix} C_{11} & C_{12} & C_{13} & C_{14} & C_{15} & C_{16} \\ C_{21} & C_{22} & C_{23} & C_{24} & C_{25} & C_{26} \\ C_{31} & C_{32} & C_{33} & C_{34} & C_{35} & C_{36} \\ C_{41} & C_{42} & C_{43} & C_{44} & C_{45} & C_{46} \\ C_{51} & C_{52} & C_{53} & C_{54} & C_{55} & C_{56} \\ C_{61} & C_{62} & C_{63} & C_{64} & C_{65} & C_{66} \end{pmatrix} \quad ...(16)$$

There are several reasons why the actual number of independent elastic constants is needed to describe the mechanical response of a material is reduced below the maximum possible number of 36.

1. The first reduction is independent of the structure of the material and results from a consideration of the elastic energy stored in the material due to the strains. This argument demonstrates that the $C_{ij}$ matrix is symmetric, that is,

$$C_{ij} = \begin{pmatrix} C_{11} & C_{12} & C_{13} & C_{14} & C_{15} & C_{16} \\ C_{12} & C_{22} & C_{23} & C_{24} & C_{25} & C_{26} \\ C_{13} & C_{23} & C_{33} & C_{34} & C_{35} & C_{36} \\ C_{14} & C_{24} & C_{34} & C_{44} & C_{45} & C_{46} \\ C_{15} & C_{25} & C_{35} & C_{45} & C_{55} & C_{56} \\ C_{16} & C_{26} & C_{36} & C_{46} & C_{56} & C_{66} \end{pmatrix} \quad ...(17)$$

   Thus, only 21 independent elastic constants are sufficient in general for the description of the mechanical response of a material in the elastic limit.

2. Further reductions in the number of independent $C_{ij}$ depend on the specific crystal structure and microstructure of the material in question. For materials in the form of single crystal the number of independent $C_{ij}$ decreases below 21 as the symmetry of the Bravais lattice of the crystal structure increases. This is due to restrictions placed on the elastic constant by symmetry relations.

The number of independent elastic constants for crystal with different Bravais lattices is indicated in Table 8.1. The essential symmetry for each lattice is also given.

Note that triclinic crystals with the most general crystal structure retain 26 independent elastic constants while for isotropic solid such as fine grained polycrystalline solids, amorphous solids, or glasses, only two elastic constants are sufficient.

**Table 8.1** Number of Independent Elastic Constants $C_{ij}$ for Crystals with Different Bravais Lattices

| *Bravais Lattice* | *Essential Symmetry* | *Number of Independent $C_{ij}$* |
|---|---|---|
| Triclinic | None | 21 |
| Monoclinic | One twofold axis | 13 |
| Orthorhombic | Two orthogonal twofold axes | 9 |
| Tetragonal | One fourfold axis | 6 or 7 |
| Trigonal (rhombohedral) | One threefold axis | 6 or 7 |
| Hexagonal | One sixfold axis | 5 |
| Cubic (Isotropic) | Four threefold axes | 3<br>2 |

## 8.7 STRESS, STRAIN AND ELASTIC COMPLIANCE CONSTANTS

The stress at a given point $P$ in the crystal will depend on the orientation of the element of area at that point. We can draw three mutually perpendicular axes $x$, $y$ and $z$ through $P$. We, then can have three elements of area $A_x$, $A_y$ and $A_z$ in the $yz$, $xz$ and $xy$ planes respectively each containing the point $P$. A general force can be resolved into three components. By taking proper account of the area on which the force acts, one can talk of stress instead of restoring force.

We write $\tau_{xx}$ to represent the component of stress working in the $x$-direction and on the area element $A_x$. Similarly, $\tau_{yz}$ and $\tau_{zx}$, define the stress component in the $y$-direction and $z$-direction respectively on the area $A_x$. Similarly $\tau_{xy}$, $\tau_{yy}$, $\tau_{zy}$ on the area $A_y$ and $\tau_{xz}$, $\tau_{yz}$, $\tau_{zz}$ on the area $A_z$. Out of the nine components of the second order stress tensor, the three $\tau_{xx}$, $\tau_{yy}$, $\tau_{zz}$ are called the *normal stresses* and the rest six components $\tau_{xy}$, $\tau_{yz}$, $\tau_{xz}$, $\tau_{zx}$, $\tau_{yz}$, $\tau_{zy}$ are known as shear stresses. If the shear stresses are in equilibrium,

$$\tau_{xy} = \tau_{yx},\ \tau_{yz} = \tau_{zy},\ \tau_{xz} = \tau_{zx}. \qquad ...(1)$$

Therefore, only six stress components

$$\tau_{xx},\ \tau_{xy},\ \tau_{yy},\ \tau_{yz},\ \tau_{xz},\ \tau_{zz}$$

are sufficient to describe the state of stress at any point in the crystal. Corresponding to every stress there is a strain and the strain components also form a second order tensor similar to the stress tensor. The six strain components sufficient to define the strain components are denoted by $\gamma_{xx}$, $\gamma_{xy}$, $\gamma_{yy}$, $\gamma_{yz}$, $\gamma_{xz}$, $\gamma_{zz}$.

Within the elastic limits, Hooke's law states that stress is directly proportional to strain. In tensor notation

$$\tau_{ik} = C_{ikmn}\,\gamma_{mn} \qquad ...(2)$$

($i$, $k$, $m$, $n$ run over $x$, $y$, $z$).

The proportionality constants $C_{ikmn}$ are known as elastic compliance constants (or simply elastic constants), a tensor of fourth order having in general $3^4 = 81$ components. Because of the symmetry of the stress and the strain components, i.e., $\tau_{ik} = \tau_{ki}$ and $\gamma_{mn} = \gamma_{nm}$, these $9 \times 9$ components reduce to $6 \times 6$ components. For practical purposes, it has been conventional to replace the two suffixes $i$, $k$ or $m$, $n$ by a single digit according to the following scheme

| *ik* | *xx* | *yy* | *zz* | *yz*, *zy* | *zx*, *xz* | *xy*, *yx* |
|---|---|---|---|---|---|---|
| *i* | 1 | 2 | 3 | 4 | 5 | 6 |

with these substitutes, the expansion of Eq. (2) can be given as

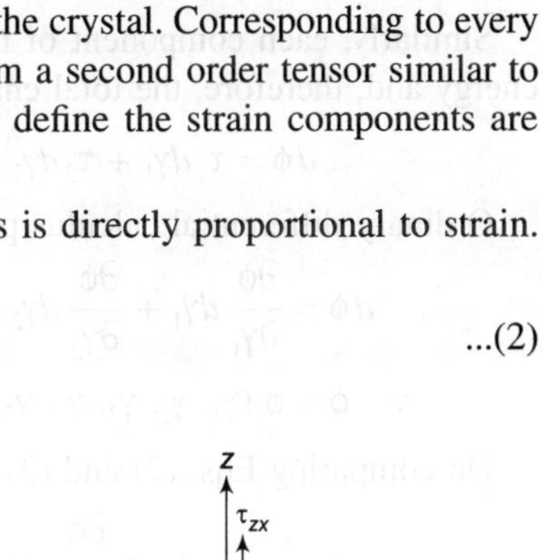


**Figure 8.8** The three stress components $\tau_{xx}$, $\tau_{yx}$, $\tau_{zx}$ acting on the element of area $A_x$

$$\left.\begin{aligned}
\tau_1 &= C_{11}\gamma_1 + C_{12}\gamma_2 + C_{13}\gamma_3 + 2C_{16}\gamma_6 + 2C_{14}\gamma_4 + 2C_{15}\gamma_5 \\
\tau_2 &= C_{21}\gamma_1 + C_{22}\gamma_2 + C_{23}\gamma_3 + 2C_{26}\gamma_6 + 2C_{24}\gamma_4 + 2C_{25}\gamma_5 \\
\tau_3 &= C_{31}\gamma_1 + C_{32}\gamma_2 + C_{33}\gamma_3 + 2C_{36}\gamma_6 + 2C_{34}\gamma_4 + 2C_{35}\gamma_5 \\
\tau_4 &= C_{41}\gamma_1 + C_{42}\gamma_2 + C_{43}\gamma_3 + 2C_{46}\gamma_6 + 2C_{44}\gamma_4 + 2C_{45}\gamma_5 \\
\tau_5 &= C_{51}\gamma_1 + C_{52}\gamma_2 + C_{53}\gamma_3 + 2C_{56}\gamma_6 + 2C_{54}\gamma_4 + 2C_{55}\gamma_5 \\
\tau_6 &= C_{61}\gamma_1 + C_{62}\gamma_2 + C_{63}\gamma_3 + 2C_{66}\gamma_6 + 2C_{64}\gamma_4 + 2C_{65}\gamma_5
\end{aligned}\right\} \quad ...(3)$$

The reason for the introduction of factor $z$ before terms involving $\gamma$'s with unequal suffixes is that $\gamma_{mn} = \gamma_{nm}$.

## 8.8 ELASTIC ENERGY

The elastic potential energy is determined by the forces applied and the deformation they produce. We have termed $\tau_1$ as the force per unit area (stress) acting on the crystal in the $x$-direction which has contributed to change in length per unit $\gamma_1$ (strain) in the same direction. Let the crystal be further deformed by an amount $d\gamma_1$ in the $x$-direction. The change in potential energy is equal to the work done as given by

$$d\phi = \tau_1\, d\gamma_1 \quad ...(1)$$

Similarly, each component of the strain tensor will contribute to the change in potential energy and, therefore, the total change of potential energy is given by

$$d\phi = \tau_1 d\gamma_1 + \tau_2 d\gamma_2 + \tau_3\, d\gamma_3 + 2\tau_4\, d\gamma_4 + 2\tau_5\, d\gamma_5 + 2\gamma_6\, d\gamma_6 \quad ...(2)$$

Ordinary differential calculus provides

$$d\phi = \frac{\partial\phi}{\partial\gamma_1} d\gamma_1 + \frac{\partial\phi}{\partial\gamma_2} d\gamma_2 + \frac{\partial\phi}{\partial\gamma_3} d\gamma_3 + \frac{\partial\phi}{\partial\gamma_4} d\gamma_4 + \frac{\partial\phi}{\partial\gamma_5} d\gamma_5 + \frac{\partial\phi}{\partial\gamma_6} d\gamma_6$$

$$\because \quad \phi = \phi\,(\gamma_1, \gamma_2, \gamma_3, \gamma_4, \gamma_5, \gamma_6) \quad ...(3)$$

On comparing Eqs. (2) and (3), we get

$$\tau_1 = \frac{\partial\phi}{\partial\gamma_1}, \tau_2 + \frac{\partial\phi}{\partial\gamma_2}, ..., 2\tau_6 + \frac{\partial\phi}{\partial\gamma_6} \quad ...(4)$$

Also, the differentiation of Eq. (3), section 8.7 partially with respect to $\gamma$s gives

$$\frac{\partial\tau_1}{\partial\gamma_1} = C_{11}, \frac{\partial\tau_1}{\partial\gamma_2} = C_{12}, ... \frac{\partial\tau_1}{\partial\gamma_6} = 2C_{16} \text{ etc.}$$

$$\frac{\partial\tau_2}{\partial\gamma_1} = C_{21}, \frac{\partial\tau_2}{\partial\gamma_2} = C_{22}, ... \frac{\partial\tau_2}{\partial\gamma_1} = 2C_{26} \text{ etc.} \quad ...(5)$$

Substituting Eq. (4) into Eq. (5), one obtains

$$C_{12} = \frac{\partial\tau_1}{\partial\gamma_1} = \frac{\partial^2\phi}{\partial\gamma_1\partial\gamma_2} = C_{21} = \frac{\partial\tau_2}{\partial\gamma_1} = \frac{\partial^2\phi}{\partial\gamma_2\partial\gamma_1} \text{ etc.}$$

In general, $C_{ik} = C_{ki}$ ...(6)

This will reduce further the independent elastic constant to 21 from 36, represented in the following manner.

$$\begin{matrix} C_{11} & C_{12} & C_{13} & 2C_{14} & 2C_{15} & 2C_{16} \\ & C_{22} & C_{23} & 2C_{24} & 2C_{25} & 2C_{26} \\ & & C_{33} & 2C_{34} & 2C_{35} & 2C_{36} \\ & & & 4C_{44} & 4C_{45} & 4C_{46} \\ & & & & 4C_{55} & 4C_{56} \\ & & & & & 4C_{66} \end{matrix} \quad ...(7)$$

## 8.9 EFFECT OF CRYSTAL SYMMETRY ON ELASTIC CONSTANTS

The presence of crystal symmetry imposes certain conditions on these elastic constants. The transformation of the fourth order tensor $C_{ikmn}$ is given by

$$C'_{ikmn} = \alpha_{ip}\,\alpha_{kq}\,\alpha_{mr}\,\alpha_{ns}\,C_{pqrs} \quad ...(8)$$

If $\alpha = (\alpha_{uv})$ belongs to the symmetry group of the crystal, $C'_{ikmn}$ must coincide with $C_{ikmn}$, so we get

$$C_{ikmn} = \alpha_{ip}\,\alpha_{kq}\,\alpha_{mr}\,\alpha_{ns}\,C_{pqrs} \quad ...(9)$$

which should be obeyed for all *i*, *k*, *m* and *n* as well as for all α belonging to the symmetry group of the crystal. The set of Eq. (9) constitute the conditions imposed on the elastic constants by the crystal symmetry. Consider now belonging to the basic symmetry elements of all crystals.

### *Triclinic system*

Triclinic system possesses only inversion centre symmetry, giving each α value to –1. The product of the α's in Eq. (9) is always +1 and the transferred *C* is equal to the original *C*. Hence, in the scheme of elastic constants Eq. (7), there are no zero terms and a triclinic has 21 elastic constants.

### *Monoclinic system*

A monoclinic system has a twofold axis corresponding to the rotation through π around, say, the *z*-axis. This will change the sign of the third component of any vector, while the sign of other components will not be affected. In terms of the matrix for the transformation of the axes.

$$\begin{matrix} & x & y & z \\ x' & -1 & 0 & 0 \\ y' & 0 & -1 & 0 \\ z' & 0 & 0 & 1 \end{matrix} \quad ...(10)$$

Taking the elastic constants for the triclinic crystal in the order and applying in the above transformations, we obtain

$$
\begin{aligned}
C_{11} &= \alpha_{11}\,\alpha_{11}\,\alpha_{11}\,\alpha_{11}\,C_{1111} = C_{11}\\
C_{12} &= \alpha_{11}\,\alpha_{11}\,\alpha_{22}\,\alpha_{22}\,C_{1122} = C_{12}\\
C_{13} &= \alpha_{11}\,\alpha_{11}\,\alpha_{33}\,\alpha_{33}\,C_{1133} = C_{13}\\
C_{14} &= \alpha_{11}\,\alpha_{11}\,\alpha_{22}\,\alpha_{33}\,C_{1123} = (-1)(-1)(-1)(1)\,C_{14} = -C_{14} \text{ or } C_{14} = 0\\
C_{15} &= \alpha_{11}\,\alpha_{11}\,\alpha_{11}\,\alpha_{33}\,C_{1113} = (-1)(-1)(-1)(1)\,C_{15} = -C_{15} \text{ or } C_{15} = 0\\
C_{16} &= \alpha_{11}\,\alpha_{11}\,\alpha_{11}\,\alpha_{22}\,C_{1112} = (-1)(-1)(-1)(-1)\,C_{16} = C_{16}\\
C_{22} &= \alpha_{22}\,\alpha_{22}\,\alpha_{22}\,\alpha_{22}\,C_{2222} = C_{22}\\
C_{23} &= \alpha_{22}\,\alpha_{22}\,\alpha_{33}\,\alpha_{33}\,C_{2233} = C_{23}\\
C_{24} &= \alpha_{22}\,\alpha_{22}\,\alpha_{22}\,\alpha_{33}\,C_{2223} = -C_{24} \text{ or } C_{24} = 0\\
C_{25} &= \alpha_{22}\,\alpha_{22}\,\alpha_{11}\,\alpha_{33}\,C_{2213} = -C_{25} \text{ or } C_{25} = 0\\
C_{26} &= \alpha_{22}\,\alpha_{22}\,\alpha_{11}\,\alpha_{22}\,C_{2212} = C_{26}\\
C_{33} &= \alpha_{33}\,\alpha_{33}\,\alpha_{33}\,\alpha_{33}\,C_{3333} = C_{33}\\
C_{34} &= \alpha_{33}\,\alpha_{33}\,\alpha_{22}\,\alpha_{33}\,C_{3323} = -C_{34} \text{ or } C_{34} = 0 \qquad \text{...(11)}\\
C_{35} &= \alpha_{33}\,\alpha_{33}\,\alpha_{11}\,\alpha_{33}\,C_{3313} = -C_{35} \text{ or } C_{35} = 0\\
C_{36} &= \alpha_{33}\,\alpha_{33}\,\alpha_{11}\,\alpha_{22}\,C_{3312} = C_{36}\\
C_{44} &= \alpha_{22}\,\alpha_{33}\,\alpha_{22}\,\alpha_{33}\,C_{2323} = (-1)(1)(-1)(1)\,C_{44} = C_{44}\\
C_{45} &= \alpha_{22}\,\alpha_{33}\,\alpha_{11}\,\alpha_{33}\,C_{2313} = (-1)(1)(-1)(1)\,C_{45} = C_{45}\\
C_{46} &= \alpha_{22}\,\alpha_{33}\,\alpha_{11}\,\alpha_{22}\,C_{2312} = -C_{46} \text{ or } C_{46} = 0\\
C_{55} &= \alpha_{11}\,\alpha_{33}\,\alpha_{11}\,\alpha_{33}\,C_{1313} = C_{55}\\
C_{56} &= \alpha_{11}\,\alpha_{33}\,\alpha_{11}\,\alpha_{22}\,C_{1312} = -C_{56} \text{ or } C_{56} = 0\\
C_{66} &= \alpha_{11}\,\alpha_{22}\,\alpha_{11}\,\alpha_{22}\,C_{1212} = C_{66}
\end{aligned}
$$

The scheme of the elastic constants is therefore as follows

$$
\begin{matrix}
C_{11} & C_{12} & C_{13} & 0 & 0 & 2C_{16}\\
 & C_{22} & C_{23} & 0 & 0 & 2C_{26}\\
 & & C_{33} & 0 & 0 & 2C_{36}\\
 & & & 4C_{44} & 4C_{45} & 0\\
 & & & & 4C_{55} & 0\\
 & & & & & 4C_{66}
\end{matrix} \qquad \text{...(12)}
$$

### Orthorhombic system

An orthorhombic crystal has three mutually perpendicular twofold axes which implies that the only $C_{ikmn}$ different from zero are those in which each suffix appears an even number of times. This provides the elastic constants scheme as

$$\begin{matrix} C_{11} & C_{12} & C_{13} & 0 & 0 & 0 \\ & C_{22} & C_{23} & 0 & 0 & 0 \\ & & C_{33} & 0 & 0 & 0 \\ & & & 4C_{44} & 0 & 0 \\ & & & & 4C_{55} & 0 \\ & & & & & 4C_{66} \end{matrix} \qquad ...(13)$$

### Tetragonal system

A tetragonal crystal has fourfold axis and, therefore, a rotation through 90° around, say, $z$-axis change components of a vector according to the following transformation.

$$\begin{matrix} & x & y & z \\ x' & 0 & 1 & 0 \\ y' & -1 & 0 & 0 \\ z' & 0 & 0 & 1 \end{matrix} \qquad ...(14)$$

A fourfold axis is simultaneously a twofold axis also, as two rotations of 90° are equal to one rotation of 180°, the non-zero elastic constants cannot include any that are zero in the monoclinic system. Applying the transformation, to the monoclinic crystal scheme of the elastic constants, one obtains

$$C_{11} = \alpha_{12}\, \alpha_{12}\, \alpha_{12}\, \alpha_{12}\, C_{2222} = C_{22}$$

$$C_{12} = \alpha_{12}\, \alpha_{12}\, \alpha_{21}\, \alpha_{21}\, C_{2211} = C_{21}$$

$$C_{13} = \alpha_{12}\, \alpha_{12}\, \alpha_{33}\, \alpha_{33}\, C_{2233} = C_{23}$$

$$C_{16} = \alpha_{12}\, \alpha_{12}\, \alpha_{12}\, \alpha_{21}\, C_{2221} = -C_{26}$$

$$C_{22} = \alpha_{21}\, \alpha_{21}\, \alpha_{21}\, \alpha_{21}\, C_{1111} = C_{11}$$

$$C_{23} = \alpha_{21}\, \alpha_{21}\, \alpha_{33}\, \alpha_{33}\, C_{1133} = C_{13}$$

$$C_{26} = \alpha_{21}\, \alpha_{21}\, \alpha_{12}\, \alpha_{21}\, C_{1121} = -C_{16}$$

$$C_{33} = \alpha_{33}\, \alpha_{33}\, \alpha_{33}\, \alpha_{33}\, C_{3333} = C_{33}$$

$$C_{36} = \alpha_{33}\, \alpha_{33}\, \alpha_{12}\, \alpha_{21}\, C_{3321} = -C_{36} \text{ or } C_{36} = 0$$

$$C_{44} = \alpha_{21}\, \alpha_{33}\, \alpha_{21}\, \alpha_{33}\, C_{1313} = C_{55}$$

$$C_{45} = \alpha_{21}\,\alpha_{33}\,\alpha_{12}\,\alpha_{33}\,C_{1323} = -C_{54} = -C_{45} \text{ or } C_{45} = 0$$

$$C_{55} = \alpha_{12}\,\alpha_{33}\,\alpha_{12}\,\alpha_{33}\,C_{2323} = C_{44}$$

$$C_{66} = \alpha_{12}\,\alpha_{21}\,\alpha_{12}\,\alpha_{21}\,C_{2121} = C_{66} \qquad ...(15)$$

The scheme for a tetragonal crystal, therefore, will be

$$\begin{matrix} C_{11} & C_{12} & C_{13} & 0 & 0 & C_{16} \\ & C_{11} & C_{13} & 0 & 0 & -C_{16} \\ & & C_{33} & 0 & 0 & 0 \\ & & & C_{44} & 0 & 0 \\ & & & & C_{44} & 0 \\ & & & & & C_{66} \end{matrix} \qquad ...(16)$$

However, it has been found[1] that on the appropriate choice of the coordinate axes, the elastic constant $C_{16}$ is not an independent elastic constant in a case of tetragonal crystals. Therefore, one has only six independent elastic constants for a tetragonal crystal.

### Cubic System

In this system, there are three mutually perpendicular twofold axes and also a threefold axis lying along a body diagonal of the cube. The axes $x$, $y$, $z$ cannot be distinguished from one another and thus a transformation of 1 into 2, 2 into 3, 3 into 1 will not change the result. Starting from orthorhombic scheme and applying this, one obtains

$$C_{12} = C_{23} = C_{13} \quad \text{and} \quad C_{44} = C_{55} = C_{66}.$$

This gives in a cubic crystal only three elastic constants $C_{11}$, $C_{12}$, and $C_{44}$ with the scheme

$$\begin{matrix} C_{11} & C_{12} & C_{12} & 0 & 0 & 0 \\ & C_{11} & C_{12} & 0 & 0 & 0 \\ & & C_{11} & 0 & 0 & 0 \\ & & & C_{44} & 0 & 0 \\ & & & & C_{44} & 0 \\ & & & & & C_{44} \end{matrix} \qquad ...(17)$$

### Trigonal system

A trigonal crystal has threefold axis and a similar derivation as done above in other class of crystals leads to the following matrix for the elastic constants having six independent parameters.

1 F.I. Fedorov, *Theory of Elastic Waves in Crystals*. Plenum Press, New York, 1968.

$$\begin{matrix} C_{11} & C_{12} & C_{13} & C_{14} & 0 & 0 \\ & C_{11} & C_{13} & -C_{14} & 0 & 0 \\ & & C_{33} & 0 & 0 & 0 \\ & & & C_{44} & 0 & 0 \\ & & & & C_{44} & C_{14} \\ & & & & & \frac{1}{2}(C_{11}-C_{12}) \end{matrix} \qquad ...(18)$$

### *Hexagonal System*

In this system, simultaneous presence of identically direct twofold and threefold axis exists. The twofold axis will give the elastic constants scheme of a monoclinic crystal and a threefold axis will give the elastic constant scheme of a trigonal crystal. Therefore, for hexagonal crystal, which has the property of both the schemes, the elastic constants matrix will be

$$\begin{matrix} C_{11} & C_{12} & C_{13} & 0 & 0 & 0 \\ & C_{11} & C_{13} & 0 & 0 & 0 \\ & & C_{33} & 0 & 0 & 0 \\ & & & C_{44} & 0 & 0 \\ & & & & C_{44} & 0 \\ & & & & & \frac{C_{11}-C_{12}}{2} \end{matrix} \qquad ...(19)$$

As is obvious, it has 5 independent elastic constants.

## 8.10 DILATION

The fractional increase in volume associated with deformation is called the dilation. To calculate it, consider a unit cube of edges $\vec{i}$, $\vec{j}$, $\vec{k}$ and volume $V = 1$. Its volume of the deformation is

$$V' = \vec{i}' \cdot \vec{j}' \cdot \vec{k}'$$

Using now Eq. (1), we have

$$\vec{i}' \cdot \vec{j}' \times \vec{k}' = \begin{vmatrix} 1+\varepsilon_{xx} & \varepsilon_{xy} & \varepsilon_{xz} \\ \varepsilon_{yx} & 1+\varepsilon_{yy} & \varepsilon_{yx} \\ \varepsilon_{zx} & \varepsilon_{zy} & 1+\varepsilon_{zx} \end{vmatrix}$$

$$\cong 1 + e_{xx} + e_{yy} + e_{zz}$$

where squares and product of the two strain components have been neglected. Then, dilation,

$$\delta = \left(\frac{V'-V}{V}\right) \cong e_{xx} + e_{yy} + e_{zz} \qquad ...(1)$$

It is to be noted that the dilation is negative for hydrostatic pressure and positive for tension.

## 8.11 ELASTIC COMPLIANCE AND STIFFNESS CONSTANTS

Hooke's law states for sufficiently small deformations, the strain is directly proportional to the stress. Consequently, strain components are linear functions of the stress components.

$$\begin{aligned}
e_{xx} &= S_{11} X_x + S_{12} Y_y + S_{13} Z_z + S_{14} Y_z + S_{15} Z_x + S_{16} X_y \\
e_{yy} &= S_{21} X_x + S_{22} Y_y + S_{23} Z_z + S_{24} Y_z + S_{25} Z_x + S_{26} X_y \\
e_{zz} &= S_{31} X_x + S_{32} Y_y + S_{33} Z_z + S_{34} Y_z + S_{35} Z_x + S_{36} X_y \\
e_{yz} &= S_{41} X_x + S_{42} Y_y + S_{43} Z_z + S_{44} Y_z + S_{45} Z_x + S_{46} X_y \\
e_{zx} &= S_{51} X_x + S_{52} Y_y + S_{53} Z_z + S_{54} Y_z + S_{55} Z_x + S_{56} X_y \\
e_{xy} &= S_{61} X_x + S_{62} Y_y + S_{63} Z_z + S_{64} Y_z + S_{65} Z_x + S_{66} X_y
\end{aligned} \quad ...(2)$$

Conversely, the stress components are

$$\begin{aligned}
X_x &= C_{11} e_{xx} + C_{12} e_{yy} + C_{13} e_{zz} + C_{14} e_{yz} + C_{15} e_{zx} + C_{16} e_{xy} \\
Y_y &= C_{21} e_{xx} + C_{22} e_{yy} + C_{23} e_{zz} + C_{24} e_{yz} + C_{25} e_{zx} + C_{26} e_{xy} \\
Z_z &= C_{31} e_{xx} + C_{32} e_{yy} + C_{33} e_{zz} + C_{34} e_{yz} + C_{35} e_{zx} + C_{36} e_{xy} \\
Y_z &= C_{41} e_{xx} + C_{42} e_{yy} + C_{43} e_{zz} + C_{44} e_{yz} + C_{45} e_{zx} + C_{46} e_{xy} \\
Z_x &= C_{51} e_{xx} + C_{52} e_{yy} + C_{53} e_{zz} + C_{54} e_{yz} + C_{55} e_{zx} + C_{56} e_{xy} \\
X_y &= C_{61} e_{xx} + C_{62} e_{yy} + C_{63} e_{zz} + C_{64} e_{yz} + C_{65} e_{zx} + C_{66} e_{xy}
\end{aligned} \quad ...(3)$$

The coefficients $S_{11}$, $S_{12}$, .... etc., are called *elastic compliance constants* or simply elastic constants and the coefficients $C_{11}$, $C_{12}$, ... are called the *elastic stiffness constants* or moduli of elasticity. These constants differ from material to material. The $\delta_s$ have the dimensions of (area/force) or (volume/energy) and the $C_s$ have the dimensions of (force/area) or (energy/volume). Note that Eqs. (2) or (3) describe a large number of elastic constant (36). Fortunately, these may be reduced to a number by several considerations which we shall consider in the subsequent sections.

## 8.12 ELASTIC ENERGY DENSITY

Work is required to deform a solid elastically. This work is stored in the solid in the form of potential energy as long as the deformation persists. The elastic energy density is defined as the limit of the ratio of energy to the volume of an elementary cube of the solid. Thus, we first calculate the incremental work as $\delta W$ done by the stress-system in deforming the small cube of side, $L$. With the origin take at one of the corners and the coordinates axes parallel to the cube edges, $\delta W$ can be written

$$\delta W = \vec{F} \cdot \delta \vec{R} \quad ...(4)$$

where $\vec{F}$ is the applied force and $\delta\vec{R}$ is the increment of the displacement $\vec{R}$, which (using Eq. 7) can be written as

$$\delta\vec{R} = \vec{i}\,\delta u + \vec{j}\,\delta v + \vec{k}\,\delta w \qquad ...(5)$$

Now, if $X, Y, Z$ denote the components of $\vec{F}$ per unit area, then Eq. (13) becomes

$$\delta W = L^2\,(X\delta u + Y\delta v + Z\delta w) \qquad ...(6)$$

Now, by the definition of the strain components used in connection with Eqs. (6) and (7) (for $\vec{R}$), we have

$$\delta u = x\,\delta\varepsilon_{xx} + y\,\delta\varepsilon_{yx} + z\,\delta\varepsilon_{zx}$$

Using Eqs. (2) and (4), we get

$$\delta u = x\,\delta e_{xx} + \frac{1}{2}\,y\,\delta e_{yx} + \frac{1}{2}\,z\,\delta e_{zx}$$

Since, the displacement of the three cube faces containing the origin is zero, we must concern ourselves with the $\vec{R}$ of the point $(x = L, y = L, z = L)$ of the cube; it will represent the total displacement of the cube. Thus, we must have

$$\delta u = L\left(\delta e_{xx} + \frac{1}{2}\,\delta e_{yx} + \frac{1}{2}\,\delta e_{zx}\right)$$

$$\delta v = L\left(\frac{1}{2}\,\delta e_{xy} + \delta e_{yy} + \frac{1}{2}\,\delta e_{yz}\right) \qquad ...(7)$$

$$\delta w = L\left(\frac{1}{2}\,\delta e_{zx} + \frac{1}{2}\,\delta e_{yz} + \delta e_{zz}\right)$$

Substituting Eq. (7) in Eq. (6) and using proper suffixes with $X, Y, Z$ in Eq. (6), we get

$$\delta W = L^3\,(X_x\,\delta e_{xx} + Y_y\,\delta e_{yy} + Z_z\,\delta e_{zz} + Y_z\,\delta e_{yz} + Z_x\,\delta e_{zx} + X_y\,\delta e_{xy}).$$

The increment $\delta U$ of elastic energy per unit volume, i.e., the elastic energy density is then

$$\delta U = X_x\,\delta e_{xx} + Y_y\,\delta e_{yy} + Z_z\,\delta e_{zz} + Y_z\,\delta e_{yz} + Z_x\,\delta e_{zx} + X_y\,\delta e_{xy} \qquad ...(8)$$

From Eq. (8), it can be seen that

$$\frac{\partial U}{\partial e_{xx}} = X_x \quad \text{and} \quad \frac{\partial U}{\partial e_{yy}} = Y_y$$

Differentiating again

$$\frac{\partial X_x}{\partial e_{yy}} = \frac{\partial^2 U}{\partial e_{xx}\partial e_{yy}} = \frac{\partial Y_y}{\partial e_{xx}}$$

Thus, $C_{12} = C_{21}$

and in general $$C_{ij} = C_{ji} \qquad ...(9)$$

which represents 15 relations among the 30 non-diagonal terms of the matrix of the *C*s? Thus, due to Eq. (9), the number of independent elastic coefficient is reduced to 21. These are represented as

$$\begin{bmatrix} C_{11} & C_{12} & C_{13} & C_{14} & C_{15} & C_{16} \\ C_{12} & C_{22} & C_{23} & C_{24} & C_{25} & C_{26} \\ C_{13} & C_{32} & C_{33} & C_{34} & C_{35} & C_{36} \\ C_{14} & C_{24} & C_{34} & C_{44} & C_{45} & C_{46} \\ C_{15} & C_{25} & C_{35} & C_{45} & C_{55} & C_{56} \\ C_{16} & C_{26} & C_{36} & C_{46} & C_{56} & C_{66} \end{bmatrix} \qquad ...(10)$$

The matrix of *C*s is now symmetrical; the symmetry of the matrix of the $\delta_s$, viz., $S_{ij} = S_{ji}$ is a corollary.

Fortunately, for some solids, several of these may become equal to each other or vanish if the crystal possesses certain symmetry elements. Then, the number of independent constants is reduced further. We now concern ourselves only with the elastic constants of the cubic crystals.

## 8.13 ELASTIC STIFFNESS CONSTANTS OF CUBIC CRYSTALS

We shall show that in a cubic crystal, there are only three independent stiffness constants. We start with the assertion that the elastic energy density of a cubic crystal is given as

$$U = \frac{1}{2} C_{11} (e_{xx}^2 + e_{yy}^2 + e_{zz}^2) + \frac{1}{4} C_{44} (e_{yz}^2 + e_{zx}^2 + e_{xy}^2)$$

$$+ C_{12} (e_{yy} e_{zz} + e_{zz} e_{xx} + e_{xx} e_{yy}) \qquad ...(11)$$

and that no other quadratic terms

$$(e_{xx} e_{xy} + ...); (e_{yz} e_{zx} + ...); (e_{xx} e_{yz} + ...) \qquad ...(12)$$

occur. This result follows essentially from the minimum symmetry requirement for a cube crystal, which is the existence of four threefold rotation axes running along the body diagonal direction. We first confirm it. For this, consider the effect of rotation of $2\pi/3$ about one of these four axes on the expressions (11) and (12). Thus, with reference to Fig. (8.9), such a rotation changes $x \to y$, $y \to z$ and $z \to x$ and the expression (11) is thus changed into

$$U = \frac{1}{2} C_{11} (e_{yy}^2 + e_{zz}^2 + e_{xx}^2) + \frac{1}{2} C_{44} (e_{zx}^2 + e_{xy}^2 + e_{yz}^2)$$

$$+ C_{12} (e_{zz} e_{xx} + e_{xx} e_{yy} + e_{yy} e_{zz}) \qquad ...(13)$$

Clearly, the expression (11) is invariant under the operation considered. Similarly, it will be invariant with the other rotational operations too. But, Since each term in Eq. (12) is odd in one or more indices, a rotation can always be found which will change the sign of the terms because $e_{xy} = -e_{x(-y)}$, for example. Thus, the term (12) is not invariant under the required rotation operations, and such will not occur in the elastic energy density expression.

Here, it remains to verify that the numerical factors in Eq. (11) are correct. This will be implicit in the coming discussion and there would be no need to consider separately.

Returning now to the problem of finding the elastic constants of cubic crystals, we would now compare the value of $X_x$ obtained from Eq. (11) (as $X_x = \partial U/\partial e_{xx}$) with that expressed in Eq. (3), i.e., we compare

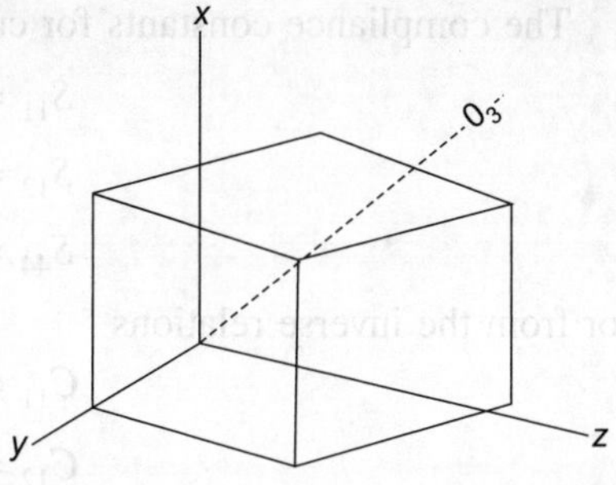


**Figure 8.9** Rotation by $2\pi/3$ about the axis marked 3 changes $x \to y$, $y \to z$ and $z \to x$

$$X_x = \frac{\partial U}{\partial e_{xx}} = C_{11}\, e_{xx} + C_{12}\,(e_{yy} + e_{zz})$$

with $$X_x = C_{11}\, e_{xx} + C_{12}\, e_{yy} + C_{13}\, e_{zz}\, C_{14}\, e_{yz} + C_{15}\, e_{zx} + C_{16}\, e_{xy}$$

We find, $$C_{12} = C_{13} \quad \text{and} \quad C_{14} = C_{15} = C_{16} = 0 \qquad ...(14)$$

Comparing further

$$X_x = \frac{\partial U}{\partial e_{xy}} = C_{44}\, e_{xy}$$

with $$X_y = C_{61}\, e_{xx} + C_{62}\, e_{yy} + C_{63}\, e_{zz} + C_{64}\, e_{yz} + C_{65}\, e_{zx} + C_{66}\, e_{xy} \qquad ...(15)$$

Carrying out other similar comparisons along with Eqs. (14) and (15), we find that the array of values of the elastic stiffness constants is reduced for a cubic crystal to the matrix showing that there are only three independent elastic stiffness constants for cubic crystal: $C_{11}$, $C_{12}$, $C_{44}$.

| | $e_{xx}$ | $e_{yy}$ | $e_{zz}$ | $e_{yz}$ | $e_{zx}$ | $e_{xy}$ |
|---|---|---|---|---|---|---|
| $X_x$ | $C_{11}$ | $C_{12}$ | $C_{12}$ | 0 | 0 | 0 |
| $X_y$ | $C_{12}$ | $C_{11}$ | $C_{12}$ | 0 | 0 | 0 |
| $Z_z$ | $C_{12}$ | $C_{12}$ | $C_{11}$ | 0 | 0 | 0 |
| $Y_z$ | 0 | 0 | 0 | $C_{44}$ | 0 | 0 |
| $Z_x$ | 0 | 0 | 0 | 0 | $C_{44}$ | 0 |
| $X_y$ | 0 | 0 | 0 | 0 | 0 | $C_{44}$ |

...(16)

It may be noted here that constant $C_{11}$ relates the compression stress and strain along the *X*, *Y*, or *Z* axis while $C_{44}$ relates the shear stress and strain in the same direction. The constant $C_{12}$ relates the compression stress in one direction to strain in another. These may, for example, be the *x*- and *y*-directions.

The compliance constants for cubic crystal may be obtained from the relations:

$$S_{11} = (C_{11} + C_{12})/(C_{11} - C_{12})\,(C_{11} + 2C_{12}) \qquad ...(17)$$

$$S_{12} = -C_{12}/(C_{11} - C_{12})\,(C_{11} + 2C_{12})$$

$$S_{44} = 1/C_{44}.$$

or from the inverse relations

$$C_{11} = (S_{11} + S_{12})/(S_{11} - S_{12})\,(S_{11} + 2S_{12}) \qquad ...(18)$$

$$C_{12} = -S_{12}/(S_{11} - S_{12})\,(S_{11} + 2S_{12})$$

$$C_{44} = 1/S_{44}$$

These relations follow an evaluating inverse matrix:

|  | $X_x$ | $Y_y$ | $Z_z$ | $Y_z$ | $Z_x$ | $X_y$ |
|---|---|---|---|---|---|---|
| $e_{xx}$ | $S_{11}$ | $S_{12}$ | $S_{12}$ | 0 | 0 | 0 |
| $e_{yy}$ | $S_{12}$ | $S_{11}$ | $S_{12}$ | 0 | 0 | 0 |
| $e_{zz}$ | $S_{12}$ | $S_{12}$ | $S_{11}$ | 0 | 0 | 0 |
| $e_{yz}$ | 0 | 0 | 0 | $S_{44}$ | 0 | 0 |
| $e_{zx}$ | 0 | 0 | 0 | 0 | $S_{44}$ | 0 |
| $e_{xy}$ | 0 | 0 | 0 | 0 | 0 | $S_{44}$ |

...(19)

From Eqs. (16) and (19), we conclude that no shear stress is introduced by tensile stresses $X_x$, $Y_y$, and $Z_z$, all the terms indicating such connections are zero. Also, pure shear stresses $Y_z$, $Z_x$ and $X_y$ produce the pure shear strain only, whatever be the shear axis. In the height of this conclusion, the term $S_{11}$ measures the amount of tensile strain (in the *x*-direction) produced by an *x*-axis tensile stress of unit magnitude, and the term $S_{12}$ measures the amount of *y*-axis (or *z*-axis) strain which accompanies *x*-axis tensile stress. Clearly $S_{12}$ will be the opposite kind, since a tensile strength is always accompanied by a compressive strain of opposite kind in all directions at right angles to the direction of the applied tensile stress. Since, there is no distinction between *x*, *y*, and *z* axes in the actual materials, we may also interpret $S_{11}$ as measuring the *y*-axis strain produced by a unit *y*-axis tensile stress and $S_{12}$ as measuring the *z*-axis (or *x*-axis) strain, etc. With this view, it can be understood that $S_{11}^{-1}$ is completely equivalent to the Young's modulus *E*, $(-S_{12}/S_{11})$ equivalent to the Poisson's ratio $\nu$ and $S_{44}^{-1}$ equivalent to shear modulus *G*. In terms of *C*s, we have

$$E = \frac{1}{S_{11}} = \frac{(C_{11} - C_{12})\,(C_{11} + 2C_{12})}{C_{11} + C_{12}}$$

$$\nu = -\frac{S_{12}}{S_{11}} = \frac{C_{12}}{C_{11} + C_{12}}$$

$$G = \frac{1}{S_{44}} = C_{44} \qquad ...(20)$$

## 8.14 BULK MODULUS AND COMPRESSIBILITY

Consider a uniform dilation of the crystal

$$e_{xx} = e_{yy} = e_{zz} = \frac{1}{3}\,\delta$$

The energy density Eq. (11) for a cubic crystal for this deformation becomes

$$U = \frac{1}{6}\,(C_{11} + 2\,C_{12})\,\delta^2$$

The bulk modulus $B$ is defined by the relation

$$U = \frac{1}{2}\,B\,\delta^2$$

Thus, for cubic crystals,

$$B = \frac{1}{3}\,(C_{11} + 2\,C_{12}) \qquad \text{...(21)}$$

The compressibility $K$ is then defined as

$$K = \frac{1}{B} = \left(\frac{3}{C_{11} + 2C_{12}}\right) \qquad \text{...(22)}$$

If $C_{12} = 0$, we have from Eqs. (20), (21), and (22)

$$E = C_{11};\ \nu = 0;\quad B = \frac{1}{3}\,C_{11};\ K = 3/C_{11} \qquad \text{...(23)}$$

This corresponds to the case of no transverse contraction accompanying longitudinal expansion. Thus, approximation is applicable to a one-dimensional lattice. Further, if the crystal is isotropic,

$$C_{12} + 2\,C_{44} = C_{11} \quad \text{or} \quad S_{44} = 2\,(S_{11} - S_{12})$$

leaving only two independent modulii. The mechanical engineering tables usually list only two modulii for the most materials used for structural purposes. This is because even though the materials may have isotropic crystalline structures, they are usually polycrystalline and hence, of isotropic forms. The modulii listed are usually the shear modulus $G = C_{44}$ and Young's modulus, $E$.

## 8.15 ELASTIC WAVES IN CUBIC CRYSTAL

When a system of stresses (to which a body is subjected), is suddenly altered, we have (i) a corresponding motion of the body itself and (ii) the propagation of changes in stress through it, the two occurring simultaneously and constituting a stress wave or an 'elastic wave' propagating through it. These are small amplitude waves identical with sound waves and their properties may be studied by using Newtonian mechanics.

To treat these waves, consider the forces acting on an elementary cube of a solid having a volume $\Delta x\ \Delta y\ \Delta z$. It is shown in Fig. 8.10, with a stress $X_x(x)$ acting on the face at $x$ and

$$X_x\,(x + \Delta x) \simeq X_x\,(x) + \frac{\partial X_x}{\partial x}\,\Delta x$$

on the parallel face at $x + \Delta x$. The net force due to the variation of $X_x$ is then $\left(\dfrac{\partial X_x}{\partial x}\Delta x\right)$ $\Delta y\ \Delta z$.

The other force contributions in the $x$-direction arises from the variation, across the cube, of the stresses $X_y$ and $X_z$ which are not shown. The net $x$-component of the force on the cube is thus

$$F_x = \left(\frac{\partial X_x}{\partial x} + \frac{\partial X_y}{\partial y} + \frac{\partial X_z}{\partial z}\right)\Delta x\ \Delta y\ \Delta z.$$

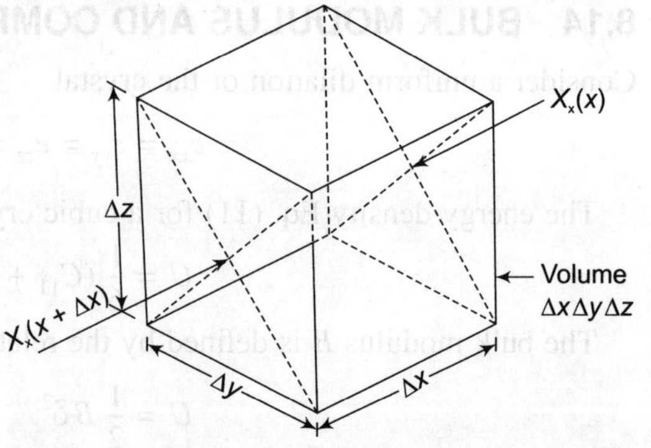


**Figure 8.10** Elementary cube

This force is equal to the mass of the cube times the component of the acceleration in the $x$-direction. The mass is $\Delta x\ \Delta y\ \Delta z$ where $\rho$ is the density of the solid and the acceleration is $\partial^2 u/\partial t^2$ where $u$ is the displacement in the $x$-direction. Thus, equation of motion for $x$-direction is

$$\rho\frac{\partial^2 u}{\partial t^2} = \frac{\partial X_x}{\partial x} + \frac{\partial X_y}{\partial y} + \frac{\partial X_z}{\partial z} \quad ...(24)$$

These are similar equations for $y$- and $z$-directions. Substituting now the values of the stress components $X_x$, $X_y$ and $X_z$ from Eqs. (3) and (16), Eq. (24) becomes

$$\rho\frac{\partial^2 u}{\partial t^2} = C_{11}\frac{\partial e_{xx}}{\partial x} + C_{12}\left(\frac{\partial e_{yy}}{\partial x} + \frac{\partial e_{zz}}{\partial x}\right) + C_{44}\left(\frac{\partial e_{xy}}{\partial y} + \frac{\partial e_{zx}}{\partial z}\right) \quad ...(25)$$

Here $x$-, $y$-, $z$-directions are parallel to the cube edges. Using then the definitions (Eqs. (8)) and (9) section 8.3) of the strain components, we have

$$\rho\frac{\partial^2 u}{\partial t^2} = C_{11}\frac{\partial^2 u}{\partial x^2} + C_{44}\left(\frac{\partial^2 u}{\partial y^2} + \frac{\partial^2 u}{\partial z^2}\right) + (C_{12} + C_{44})\left(\frac{\partial^2 v}{\partial x\partial y} + \frac{\partial^2 w}{\partial x\partial z}\right) \quad ...(26a)$$

where $u$, $v$ and $w$ are the components of the displacement $R$ is defined by Eq. (7) (section 8.3) along the $x$-, $y$-, $z$-axes respectively. The corresponding equations of motion for $\dfrac{\partial^2 v}{\partial t^2}$ and $\dfrac{\partial^2 w}{\partial t^2}$ can be written directly from Eq. (26a) by symmetry:

$$\rho\frac{\partial^2 v}{\partial t^2} = C_{11}\frac{\partial^2 v}{\partial y^2} + C_{44}\left(\frac{\partial^2 v}{\partial x^2} + \frac{\partial^2 v}{\partial z^2}\right) + (C_{12} + C_{44})\left(\frac{\partial^2 u}{\partial x\partial y} + \frac{\partial^2 w}{\partial y\partial z}\right) \quad ...(26b)$$

$$\rho\frac{\partial^2 v}{\partial t^2} = C_{11}\frac{\partial^2 w}{\partial z^2} + C_{44}\left(\frac{\partial^2 w}{\partial x^2} + \frac{\partial^2 w}{\partial y^2}\right) + (C_{12} + C_{44})\left(\frac{\partial^2 u}{\partial x\partial z} + \frac{\partial^2 v}{\partial y\partial z}\right) \quad ...(26c)$$

The complexity of Eq. (26) does not suggest the forms of the possible vibrational waves occurring in the solids. This may be inferred however, by considering waves travelling in a given direction. After the direction of propagation is chosen, account is also taken for the direction of the displacement of the material by the wave. In simple isotropic solids, there can be three basic directions of the displacement—one along the axis of propagation and the two transverse to the axis of propagation. In former case, the wave is termed longitudinal or compression wave and in the later cases, the waves are termed transverse or their shear waves, the two shear waves themselves have orthogonal displacements. It is always possible in cubic solids to single out three non-interacting waves, all in general having different wave velocities. We shall now discuss the propagation of these waves along some simple directions in cubic crystals.

## 8.16 WAVES IN THE [100] DIRECTION

First consider a transverse or shear wave with the wave vector $\vec{k}$ along the $x$-cube edge and the particle displacement $v$ in the $y$-direction

$$v = v_o \exp [i(kx - \omega t)] \quad ...(27)$$

Here, the wave vector $k$ is $= 2\pi/\lambda$ and $\omega$ is the the angular frequency $= 2\pi\ \nu$ ($\nu$ being the frequency). Equation (27) will represent the solution of Eq. (26b),

if $$\omega^2\rho = C_{44}\, k^2$$

as found on substituting Eq. (27) into Eq. (26b).

or $$\frac{\omega}{k} = \left(\frac{C_{44}}{\rho}\right)^{1/2}$$

But $\frac{\omega}{k}$ is the velocity of the wave (Eq. (27)). Thus, if the velocity $v_s$ of a transverse wave in the [100] direction is given by

$$v_s = \left(\frac{C_{44}}{\rho}\right)^{1/2} \quad ...(28)$$

then it represents the solution of Eq. (26b). Similarly, Eq. (26c) also represents another transverse wave (with the wave vector along the $x$-cube edge and the particle displacement $w$ (along the $z$-direction) propagating with the identical velocity.

Thus, for $\vec{k}$ parallel to [100], the two independent shear waves have equal velocities. This is not true for $\vec{k}$ in a general direction in the crystal.

Next, consider a longitudinal or compression wave with both the wave vector and the particle displacement along the $x$ cube edge:

$$u = u_o \exp [i(kx - \omega t)] \quad ...(29)$$

This will represent the solution of Eq. (26a) if

$$\omega^2\rho = C_{11}\, k^2 \quad \text{or} \quad \omega/k = (C_{11}/\rho)^{1/2}$$

Thus, the velocity of a longitudinal wave in the [100] direction is

$$v_s = \omega/k = (C_{11}/\rho)^{1/2}. \quad ...(30)$$

In arriving at this expression we have tacitly assumed that the wave propagation is through a crystal which extends to infinity in the $y$- and $z$-directions; this in effect, means that there is no appreciable associated transverse movement of the crystal.

## 8.17 WAVES IN THE [110] DIRECTION

The waves which propagate in the [100] direction (i.e., the face diagonal direction) are of special interest, because their propagation velocities enable us to determine all the three elastic constants of cubic crystal: $C_{11}$, $C_{12}$, $C_{44}$.

First consider a shear which propagates in the $xy$-plane with particle displacement $w$ in the $z$-direction.

$$w = w_a \exp [i(k_x x + k_y y - \omega t)] \quad ...(31)$$

On substituting in Eq. (26c), this gives

$$\omega^2 \rho = C_{44}(k_x^2 + k_y^2) = C_{44} k^2$$

or

$$\frac{\omega}{k} = \left(\frac{C_{44}}{\rho}\right)^{1/2} \quad ...(32)$$

Note that the velocity of this shear wave is independent of the propagation direction in the $xy$-plane, i.e., the result (32) does not depend whether the wave propagates in [110] direction or any other direction in the $xy$-plane. It involves $C_{44}$.

Next consider other waves which propagate in the $xy$-plane.

Let

$$\left.\begin{aligned} u &= u_o \exp [i(k_x x + k_y y - \omega t)] \\ v &= v_o \exp [i(k_x x + k_y y - \omega t)] \end{aligned}\right| \quad ...(33)$$

represent the component along the $x$- and $y$-directions respectively. On substitution in Eq. (26a) and (26b), we get

$$\omega^2 \rho u = (C_{11} k_x^2 + C_{44} k_y^2) u + (C_{12} + C_{44}) k_x k_y v$$

$$\omega^2 \rho v = (C_{11} k_y^2 + C_{44} k_x^2) v + (C_{12} + C_{44}) k_x k_y u$$

This pair of equations have a relatively simple solution for a wave travelling in the [110] direction for which $k_x = k_y = k/\sqrt{2}$. For this case, these equations reduce to

$$\left.\begin{aligned} \omega^2 \rho u &= (C_{11} + C_{44}) \frac{k^2}{2} u + (C_{12} + C_{44}) \frac{k^2}{2} v \\ \omega^2 \rho v &= (C_{11} + C_{44}) \frac{k^2}{2} v + (C_{12} + C_{44}) \frac{k^2}{2} u \end{aligned}\right| \quad ...(34)$$

Now, we are interested in the non-zero solution of these equations, the condition for which is that the determinant of the coefficients of $u$ and $v$ should be equal to zero, i.e.,

$$\begin{vmatrix} -\omega^2\rho + \frac{1}{2}(C_{11}+C_{44})k^2 & \frac{1}{2}(C_{12}+C_{44})k^2 \\ \frac{1}{2}(C_{12}+C_{44})k^2 & -\omega^2\rho + \frac{1}{2}(C_{11}+C_{44})k^2 \end{vmatrix} = 0$$

or $\quad \omega^4\rho^2 - \omega^2\rho\,(C_{11}+C_{44})\,k^2 + \frac{1}{4}k^2\,(C_{11}^2 - C_{12}^2 + 2C_{11}C_{44} - 2C_{12}C_{44}) = 0$

or $$\omega^2\rho = \frac{(C_{11}+C_{44})\,k^2 \pm k^2\,(C_{12}+C_{44})}{2}$$

∴ Either $$\omega^2\rho = \frac{1}{2}(C_{11}+C_{12}+2C_{44})k^2$$

or $$\omega^2\rho = \frac{1}{2}(C_{11}+C_{12})\,k^2 \qquad ...(35)$$

The upper root describes a longitudinal wave; for, if it is substituted into the upper of Eq. (34), we get

$$\frac{1}{2}(C_{11}+C_{12}+2C_{44})\,k^2u = \frac{1}{2}(C_{11}+C_{44})\,k^2u + \frac{1}{2}(C_{12}+C_{44})\,k^2v$$

consequently, $u = v$ indicating that the particle displacement is along [110] and thus parallel to the $\vec{k}$ vector. Similarly, the lower root in Eq. (35) describes a shear wave: this may be seen by substituting it into upper of Eq. (34), whereby

$$\frac{1}{2}(C_{11}-C_{12})\,k^2u = \frac{1}{2}(C_{11}+C_{44})\,k^2u + \frac{1}{2}(C_{12}+C_{44})\,k^2v$$

⇒ $u = -v$, indicating that the particle displacement in this case is along $[1\bar{1}0]$ and thus perpendicular to the $k$-vector. The velocity of these two waves are then

$$v_S = \frac{\omega}{k} = \left(\frac{C_{11}+C_{12}+2C_{44}}{2\rho}\right)^{1/2} \quad \text{(longitudinal wave)}$$

or $$v_s = \frac{\omega}{k} = \left(\frac{C_{11}+C_{12}}{2\rho}\right)^{1/2} \quad \text{(shear wave)} \qquad ...(36)$$

## 8.18 WAVES IN THE [111] DIRECTION

Now, let us consider the waves propagating in the body diagonal direction [111]. For this direction, the wave vector $\vec{k}$ satisfy $k_x = k_y - k_z = k/\sqrt{3}$.

Let $$u = u_o \exp\left[i(x+y+z)\frac{k}{\sqrt{3}} - i\omega t\right]$$

$$\left.\begin{aligned} v &= v_o \exp\left[i(x+y+z)\frac{k}{\sqrt{3}} - i\omega t\right] \\ w &= w_o \exp\left[i(x+y+z)\frac{k}{\sqrt{3}} - i\omega t\right] \end{aligned}\right\} \quad \text{...(37)}$$

represent component waves in the $x$-, $y$- and $z$-directions respectively. Substituting Eq. (37) in Eq. (26) in proper order, we get

$$\left.\begin{aligned} \rho\omega^2 u &= (C_{11} + 2C_{44})\frac{k^2}{3}u + (C_{12} + C_{44})\frac{k^2}{3}v + (C_{12} + C_{44})\frac{k^2}{3}w \\ \rho\omega^2 v &= (C_{11} + 2C_{44})\frac{k^2}{3}v + (C_{12} + C_{44})\frac{k^2}{3}u + (C_{12} + C_{44})\frac{k^2}{3}w \\ \rho\omega^2 w &= (C_{11} + 2C_{44})\frac{k^2}{3}w + (C_{12} + C_{44})\frac{k^2}{3}u + (C_{12} + C_{44})\frac{k^2}{3}v \end{aligned}\right\} \quad \text{...(38)}$$

We are interested in the non-trivial solutions of these equations (i.e., different from $u = v = w = 0$). The condition for this is that the determinant of the coefficients of $u$, $v$, and $w$ should be equal to zero, i.e.,

$$\begin{vmatrix} -\omega^2\rho + (C_{11} + 2C_{44})\frac{k^2}{3} & (C_{12} + C_{44})\frac{k^2}{3} & (C_{12} + C_{44})\frac{k^2}{3} \\ (C_{12} + C_{44})\frac{k^2}{3} & -\rho\omega^2 + (C_{11} + 2C_{44})\frac{k^2}{3} & (C_{12} + C_{44})\frac{k^2}{3} \\ (C_{12} + C_{44})\frac{k^2}{3} & (C_{12} + C_{44})\frac{k^2}{3} & -\rho\omega^2 + (C_{11} + 2C_{44})\frac{k^2}{3} \end{vmatrix} = 0 \quad \text{...(39)}$$

After a tedious calculation, we get an equation of the form $(X + 2Y)(X - Y)^2 = 0$, i.e., equation in $(\omega^2)^3$ cubic equation with two identical roots. The identical roots are given by

$$-\rho\,\omega^2 + (C_{11} + 2C_{44})\frac{k^2}{3} = (C_{12} + C_{44})\frac{k^2}{3}$$

or
$$\rho\,\omega^2 = (C_{11} - C_{12} + C_{44})\frac{k^2}{3} \quad \text{...(40)}$$

The two roots describe two shear waves. Substituting it into upper of Eq. (38), we get

$$(C_{11} - C_{12} + C_{44})\frac{k^2}{3}u = (C_{11} + 2C_{44})\frac{k^2}{3}u + (C_{12} + C_{44})\frac{k^2}{3}v + (C_{12} + C_{44})\frac{k^2}{3}w$$

or
$$u + v + w = 0$$

indicating that the particle displacement is in a direction perpendicular to [111], and thus perpendicular to $\vec{k}$. The velocity of propagation of these shear waves is

$$v_s = \frac{\omega}{k} = \left[\frac{C_{11} - C_{12} + C_{44}}{3\rho}\right]^{1/2}. \qquad ...(41)$$

The other root is given by

$$-\rho w^2 + (C_{11} + 2\,C_{44})\frac{k^2}{3} + 2(C_{12} + C_{44})\frac{k^2}{3} = 0$$

or

$$\rho w^2 = (C_{11} + 2\,C_{12} + 4\,C_{44})\frac{k^2}{3} \qquad ...(42)$$

It describes a longitudinal wave. Substituting it into upper of Eq. (38), we get

$$(C_{11} + 2\,C_{12} + 4\,C_{44})\frac{k^2}{3}u = (C_{11} + 2C_{44})\frac{k^2}{3}u + (C_{12} + C_{44})\frac{k^2}{3}(v + w)$$

so that $2u = v + w$. The possible solution of this is $u = v = w$, indicating that the particle displacement is this case is along [111], and thus parallel to the $\vec{k}$ vector. The velocity of propagation of this wave is

$$v_s = \frac{\omega}{k} = \left(\frac{C_{11} + 2C_{12} + 4C_{44}}{3\rho}\right)^{1/2} \qquad ...(43)$$

This treats the wave propagation in some simple directions in cubic crystals. In principle, wave propagation can be in any direction, but then, the treatment may be complex.

Solution of the wave equation for elastic waves in the ⟨100⟩, ⟨110⟩ and ⟨111⟩ direction in cubic crystals correspond to the propagation of a single longitudinal wave and two transverse waves, with velocities listed in Table 8.2. These wave velocities for elastic waves correspond to the speed of sound in the elastic medium.

**Table 8.2** Velocities of Propagation of Elastic Waves in Cubic Crystals

| *Direction* | *Modes* | *Wave Velocity u* |
|---|---|---|
| ⟨100⟩ | Longitudinal | $\sqrt{\frac{C_{11}}{\rho}}$ |
| | Transverse (2) | $\sqrt{\frac{C_{44}}{\rho}}$ |
| ⟨110⟩ | Longitudinal | $\sqrt{\frac{C_{11} + C_{12} + 2C_{44}}{2\rho}}$ |
| | Transverse (polarized along [001]) | $\sqrt{\frac{C_{44}}{\rho}}$ |
| | Transverse (polarized along [110]) | $\sqrt{\frac{C_{11} - C_{12}}{2\rho}}$ |

*Contd...*

*Table 8.2 Contd...*

| Direction | Modes | Wave Velocity u |
|---|---|---|
| ⟨111⟩ | Longitudinal | $\sqrt{\dfrac{C_{11}-2C_{12}+4C_{44}}{3\rho}}$ |
| | Transverse (2) | $\sqrt{\dfrac{C_{11}-C_{12}+C_{44}}{3\rho}}$ |

## 8.19 DETERMINATION OF ELASTIC CONSTANTS

### 8.19.1 Experimental Methods

There are two types of experimental methods for measuring elastic constants:

(a) Static methods

(b) Dynamic methods.

Generally, the static methods were used in the past. The static methods presented the isothermal moduli whereas the dynamic methods give the adiabatic moduli. The dynamic methods have advantage over static methods in the viewpoint that in the former, even smaller size of specimen can be used with maximum accuracy. The difference between isothermal and adiabatic moduli is clear from the following analysis.

Let us assume that the elastic energy is a function of stress and temperature. In tensor notation

$$[S_{ij}]_{\text{adiab}} = \left[\frac{\partial e^i(T_i, T)}{\partial T_j}\right]_S$$

$$= \left[\frac{\partial e_i}{\partial T_j}\right]_T + \left[\frac{\partial e_i}{\partial T}\right]_{T_i}\left[\frac{\partial T}{\partial T_j}\right]_S$$

$$\equiv [S_{ij}]_T - \alpha_i \left[\frac{\partial T}{\partial S}\right]_{T_j}\left[\frac{\partial S}{\partial T_j}\right]_T$$

where $S$ is the entropy, $\alpha_i \equiv -\left(\frac{\partial e_i}{\partial T}\right)_{T_i}$ is the coefficient of thermal expansion for the compressive strains, and $e_i$ the change of strain component with respect to temperature. We have

$$\left[\frac{\partial T}{\partial S}\right]_{T_j} = \frac{T}{C_p} \qquad \left(\because\ C_p = \left(\frac{\partial S}{\partial T}\right)T\right)$$

$$\left[\frac{\partial S}{\partial T_j}\right]_T = \frac{-\partial^2 F}{\partial T_j \partial T} \qquad \left(\because\ S = -\left(\frac{\partial F}{\partial T}\right)\right)$$

$$= \frac{\partial^2 F}{\partial T \partial T_j} = \frac{\partial}{\partial T}\left(\frac{\partial F}{\partial T_j}\right)$$

$$= -\frac{V \partial e_j}{\partial T} \qquad \left(\because \frac{1}{V}\left(\frac{\partial F}{\partial T_j}\right) = -ej\right).$$

$C_p$ is the specific heat at constant pressure, i.e., at constant stress and $F$ is the Helmholtz free energy.

Hence, $$[S_{ip}]_S - [S_{ij}]_T = -\alpha_i \left[\frac{\partial T}{\partial S}\right]_{T_j} \frac{V \partial e_j}{\partial T}$$

$$= -\alpha_i V \left[\frac{\partial T}{\partial S}\right]_{T_j} \alpha_{j'}$$

$$= -\alpha_i \alpha_j \frac{T}{C_p/V}$$

$$= \frac{-\alpha_i \alpha_j T}{\rho C_p}$$

This difference was found to be small.

### (a) Static methods

First of all, Voigt used the static methods for measuring the crystal constants. Young's modulus, shear molecules, Poisson's ratio were measured directly for single crystals from which the elastic constants were derived. Bridgman and Slater, simultaneously gathered information on elastic properties from the measurements of linear and volume comprehensibility.

There are other techniques like optical and electrical methods which are also static methods. The Young's modulus of class can be determined with the help of Newton's rings method. X-ray diffraction technique determines the lattice parameters but one can also measure the small strains by it. The accuracy of static methods in measurement of small strains.

### (b) Dynamic method

In these methods we measure the frequency of response rather than the small displacement as in static methods. Hence, this technique is more advanced as compared to static methods. Generally, in this technique the whole system is affected by forced vibrations from which the maximum response is determined as a function of frequency. A critical review on this method is given elsewhere.

Among the modern methods the pulse ultrasonic method has been widely used. The experiment arrangement is presented in Fig. 8.11. A piezoelectric material like α-quartz is fixed to one plane face of the material specimen. A short pulse of high frequency alternating voltage is applied to the quartz crystal, so that mechanical vibrations are set up in the specimen.

The total time $t$ taken by the pulse to travel top the opposite plane face of the specimen and back is determined and this givers the ultrasonic velocity, $V$.

$$V = \frac{d_{st}}{t} \quad ...(1)$$

where $d_{s \cdot t}$ is sound trip distance.

The pulse voltage is generated in an oscillator and fed to the crystal. The reflected waves are picked up and amplified and fed to an oscilloscope. The time base signal is obtained through a trigger oscillator.

The first order elastic constants $C_{11}$, $C_{12}$, $C_{44}$ of a cubic crystal can be determined from the velocity of different waves along symmetry direction as discussed earlier.

Bhagavantam developed an optical method useful for opaque media. In this method, the specimen is kept in contact with a wedge material shaped crystal of quartz or tourmaline. The back-face of the material is immersed in a transparent liquid.

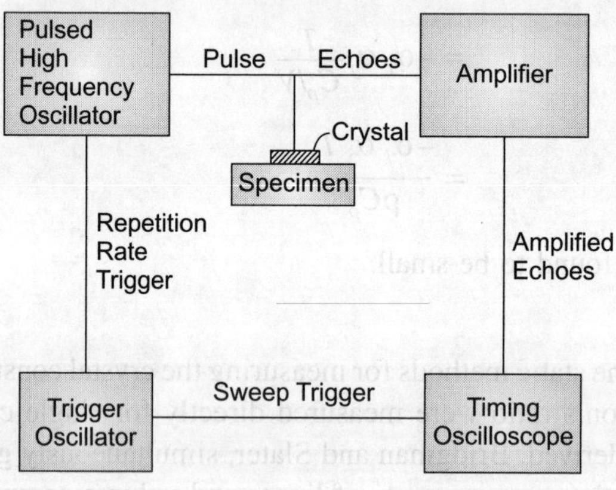


**Figure 8.11** The block diagram of pulse ultrasonic method for the measurement of elastic constants

With the help of the intensity of diffracted light through liquid the resonance condition is detected. The velocity in the specimen is determined by measuring the frequency difference between the resonances.

The experimental values of first order elastic constants $C_{11}$, $C_{12}$ and $C_{44}$ of certain solids are summarized in Table (8.3).

### 8.19.2 Theoretical Methods

Theoretically, one can easily determine elastic constants with the help of lattice dynamics of solids. It provides a relation between elastic constants and force constants which may be obtained by setting up equations of motion of the ions in terms of the force constants and analyzing it through plane waves which gives the usual secular equation. Now applying the method of long waves of Born and Huang and De Launay, on equations of motion which shows the following relationship between interatomic force constant and elastic constants

$$C_{11} = \frac{4}{a}\sum_{S}\frac{n^{S}}{4^{8}}\sum_{v}(ls_{v})^{2}\,Ks_{vv}$$

$$C_{44} = \frac{2}{a}\sum_{S}\frac{n^{S}}{4^{8}}\left[\sum_{v}(ls_{v+1})^{2}\,(ls_{v+2})^{2}\right]Ks_{vv}$$

$$C_{44} + C_{12} = \frac{8}{a}\sum_{S}\frac{n^{S}}{4^{8}}\left[\sum_{v}(ls_{v+1}\,ls_{v+2})\,K_{v+1,\,v+2}\right] \quad ...(2)$$

Where $n^S$ denotes the number of atoms in the $s$ shell and the sum of $s$ means a sum over small shells. The sum over $v$ means a sum over the $x$, $y$, $z$, components with $K_{vv}$ being the corresponding element of the interatomic force constant matrix element, $a$ is the lattice parameter. The interatomic force constants may be determined with the help of tangential and radial force constants, which are associated with the first and second derivations of the interatomic potential.

**Table 8.3** First Order Elastic Constant of the Solids in Units of $10^{11}$ dynes/cm$^2$

| *S. No.* | *Solid* | $C_{11}$ | $C_{12}$ | $C_{44}$ | *Temp. (°K)* |
|---|---|---|---|---|---|
| 1. | Li | 1.481 | 1.248 | 1.077 | 78 |
| 2. | Na | 0.741 | 0.624 | 0.419 | 300 |
| 3. | K | 0.372 | 0.315 | 0.188 | 295 |
| 4. | Rb | 0.296 | 0.244 | 0.160 | 80 |
| 5. | Cs | 0.245 | 0.208 | 0.159 | 0 |
| 6. | Cu | 16.839 | 12.142 | 7.539 | 300 |
| 7. | Ag | 12.399 | 9.367 | 4.612 | 300 |
| 8. | Au | 19.234 | 16.314 | 4.195 | 300 |
| 9. | Al | 11.430 | 6.190 | 3.160 | 0 |
| | | 10.680 | 6.070 | 2.820 | 300 |
| 10. | Pb | 5.550 | 4.540 | 1.940 | 0 |
| | | 4.950 | 4.230 | 1.490 | 300 |

## EXERCISE

8.1 How are stress and strain in a crystal analyzed? What do you understand by the dilation in crystals?

8.2 Define the elastic constant for crystals. Prove that the elastic stiffness constants are symmetrical, i.e., $C_{ij} = C_{ji}$.

8.3 Consider the effect of symmetry on the elastic stiffness constants of crystals and show that the cubic crystals have only three independent elastic constants.

8.4 Describe the characteristics of stress-strain relations in cubic crystals.

8.5 Write short notes on:

(a) Elastic energy density

(b) Elastic waves in the [100] direction in cubic crystals

(c) Waves in the [110] direction

8.6 Describe static and dynamic methods for determination of elastic constants of crystals.

8.7 Solve the equation governing the propagation of elastic waves for a solid possessing cubic symmetry when the propagation direction is [110]. What is the importance of waves propagating in this direction?

CHAPTER

# 9

# Thermal Properties of Solids

## 9.1 INTRODUCTION

The thermal properties of solids are mainly concerned with the flow of heat through the solid. The two main carriers of heat are the quanta of lattice vibrations (i.e., the phonons) and the electrons. The total heat transferred is shared by these two. Depending upon the temperature, one may dominate the other. Usually, the former predominates at the lower temperatures while the latter at the higher temperatures.

Thus, heat transfer in solids may be provided by free electrons as well as by phonons. In insulators (which do not contain any free electrons) the heat must be conducted exclusively by phonons. In metals and alloys, on the other hand, the heat conduction is dominated by electrons, because of the presence of a large number of free electrons. Thus, in this case, the phonon contribution is neglected.

## 9.2 PHONON PROCESS

For various transport properties, it is essential to establish a true equilibrium distribution of phonons. Phonon collisions with a static imperfection or a crystal boundary do not by themselves establish thermal equilibrium because such collisions do not change the energy of individual phonons: the frequency, $\omega_2$, of the scattered phonon is equal to the frequency, $\omega_1$, of the incident phonon. Depending on the number of photons taking part or destroyed during a process, the process is known as a one-, two- or three-phonon process.

It is rather remarkable that a three-phonon collision process

$$\vec{k}_1 + \vec{k}_2 = \vec{k}_3 \qquad ...(1)$$

does not establish equilibrium because total momentum of the phonon gas is not changed by a such a collision. Then, an equilibrium distribution of photons at a temperature $T$ can move down the crystal with a drift velocity that is not disturbed by three-phonon collision of the form Eq. (1). For such collisions, the phonon momentum

$$\vec{J} = \sum_k n_k \hbar_k \quad \text{...(2)}$$

is conserved, because on collision the change in $\vec{J}$ is $\vec{k}_3 - \vec{k}_2 - \vec{k}_1 = 0$ ($n_k$ is the number of phonons having wave vector $\vec{k}$).

For distribution with $\vec{J} \neq 0$, collisions such as Eq. (1) are also incapable of establishing complete thermal equilibrium because $\vec{J}$ is unchanged. As a result, there is no thermal resistance. The situation is analogous to collisions between molecules of a gas in a straight open tube with frictionless walls. Peierls has pointed out that the important three-phonon processes for thermal resistivity are not of the form $\vec{k}_1 + \vec{k}_2 = \vec{k}_3$, but are of the form $\vec{k}_1 + \vec{k}_2 = \vec{k}_3 + \vec{g}$, where $\vec{g}$ is reciprocal lattice vector.

Thus, there are two types of processes as follows:

(i) N-process (normal process) and

(ii) U-process (umklapp process).

In N-process, there is no change of phonon on momentum during scattering (Fig. 9.1)

$$\vec{k}_1 + \vec{k}_2 = \vec{k}_3 \quad \text{...(3)}$$

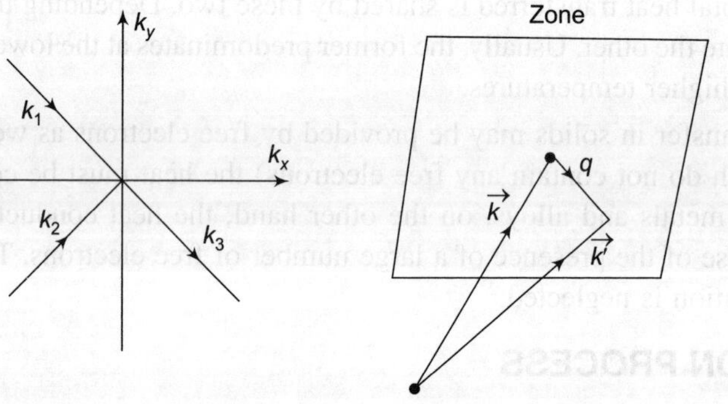


**Figure 9.1** Normal process in a two-dimensional square lattice

In this process $\vec{g} = 0$. This process corresponds to repeated zone scheme.

In U-process, $\vec{k}_1 + \vec{k}_2 = \vec{k}_3 + \vec{g}$, i.e., $\vec{g} \neq 0$ and there is change of momentum. This process corresponds to reduced zone scheme. In this case, the Brillouin zone (i.e., the unit cell of the reciprocal lattice) can be repeated so as to cover exactly the whole space occupied by the vactor $\vec{k}$ (Fig. 9.2). In German, Umklapp stands for 'flipping over'.

It may be noted from the figure that is the umklapp process, the direction of the $x$-component of the phonon flux has been reserved. Further, Fig. 9.2 (b) describes

$$q = (\vec{k}' - \vec{k} - \vec{g}) \equiv \vec{K} - \vec{g} \quad \text{...(4)}$$

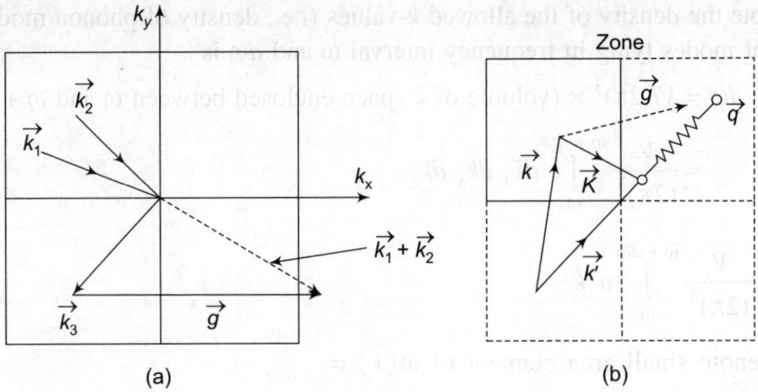


**Figure 9.2** Umklapp process in a two-dimensional square lattice. The square in figure represents the first Brillouin zone in the phonon $k$-space. Vectors $\vec{K}$ with the arrowheads at the centre of the zone represent phonons absorbed in the collision process; those source with arrowheads away from the centre represent phonon emitted

where $q$ is phonon wave vector, which is constrained to lie in a unit cell of that lattice. We can say that $\vec{q}$ is the value of $k' - \vec{k} \equiv \vec{K}$ reduced to the Forst Brillouin zone.

For both the processes ($N$ or $U$), the energy is conserved, i.e.,

$$\omega_1 + \omega_2 = \omega_3 \quad \text{...(5)}$$

where $\omega_1$, $\omega_2$ and $\omega_3$ are the frequencies of the wave vectors $\vec{k}_1$, $\vec{k}_2$ and $\vec{k}_3$ respectively.

## 9.3 PHONON DENSITY OF STATES

One of the important problems of the theory of lattice vibrations is to determine the thermal energy of a solid by determining the frequency distribution of normal modes.

Let us consider a cube of a periodic solid of edge $L$. Let there be $N^3$ primitive cells within the cube. The periodic boundary condition over these cells for a three-dimensional case yields

$$e^{i(k_x x + k_y y + k_z z)} = \exp i\,[k_x(x + L) + k_y\,(y + L) + k_z\,(z + L)] \quad \text{...(1)}$$

where $(x, y, z)$ are the coordinates of any lattice site. The above equation is valid provided

$$k_x, k_y, k_z = 0; \pm\frac{2\pi}{L}, \pm\frac{4\pi}{L}, \ldots, \pm\frac{N\pi}{L} \quad \text{...(2)}$$

The above set shows that there is one allowed value of $k$ within a volume of $(2\pi/L)^3$ in the wave vector space. Hence, the number of allowed $k$-values per unit volume of $k$-space is

$$\left(\frac{L}{2\pi}\right)^3 = \frac{V}{(2\pi)^3} \quad \text{...(3)}$$

where $V$ is the volume of the crystal.

If we denote the density of the allowed $k$-values (i.e., density of phonon modes) by $Z(\omega)$, the number of modes lying in frequency interval $\omega$ and $d\omega$ is

$$Z(\omega)\, d\omega = V/(2\pi)^3 \times (\text{volume of } k \text{ space enclosed between } \omega \text{ and } \omega + d\omega)$$

$$= \frac{V}{(2\pi)^3} \int_{\omega}^{\omega + d\omega} dk_x\, dk_y\, dk_z$$

or

$$\frac{V}{(2\pi)^3} \int_{\omega}^{\omega + d\omega} d^3k \qquad ...(4)$$

Let $ds_\omega$ denote small area element of $\omega(\vec{k})$ = constant and $d\vec{k}_\perp$, a vector normal to it, with its tip touching the surface $\omega + d\omega$ = constant (Fig. 9.3). Then volume element,

$$d^3\vec{k} = ds_\omega \cdot \overrightarrow{dk_\perp}$$

With $d\omega = |\text{grad}_k\, \omega|\, dk_\perp$, we obtain

$$Z(\omega)\, d\omega = \frac{V}{(2\pi)^3} d\omega \int_{\omega=\text{const}} \frac{ds_\omega}{|\text{grad}_k \omega|} \qquad ...(5)$$

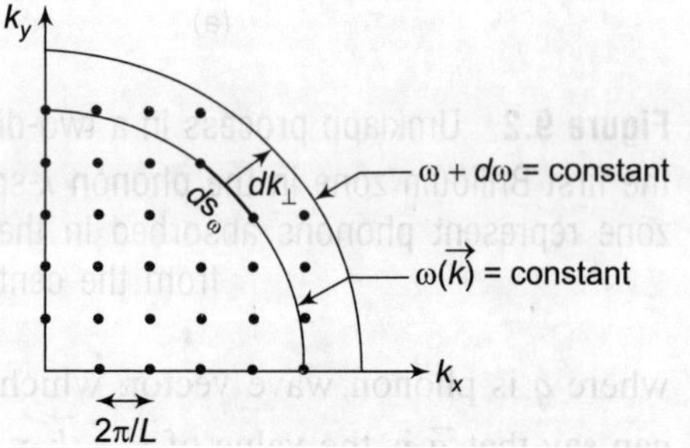


**Figure 9.3** Allowed values of $k$ in the wave vector space for a square lattice. $L$ is the repeat distance (periodicity) in real space

The density of states $z(\omega)\, d\omega$ is high in region where the dispersion curve is flat. The frequency derivative of the density of states has singularities (known as van Hove singularities) point where the tangents on the dispersion curve ($\omega(k)$) are horizontal. One-dimensional linear monoatomic chain is a special case for which the density of states itself is singular.

## 9.4 THE DEBYE MODEL

The dispersion relation can be used to calculate the specific heat of a solid if one makes, according to Debye, two hypotheses.

The first hypothesis is that the crystal is considered an isotropic elastic medium. All atoms are elastically coupled to one another, so unlike Einstein's model (atoms vibrating at a constant frequency), there is a spectrum of frequencies. In a solid for each wave vector $\vec{k}$ there are three modes of vibration: one longitudinal mode and two degenerate transfer modes.

Thus, effectively we have three acoustical branches and the velocity of sound in the crystal is a constant ($v = \omega/k$). The speeds of propagation in the longitudinal and transverse directions (denoted by $v_L$ and $v_T$, respectively) are assumed to be independent of the direction and magnitude of the wave vector, since for each branch the surface $\omega(k)$ = constant is a sphere. Then, $|\text{grad}_k\, \omega|$ is equal to the sound velocity $v_i$ for a launch $i$ (and is independent of $\vec{k}$). Consequently, $|\text{grad}\, \omega|$ can be taken outside the integral sign and

$$\int_{\omega(\vec{k})=\text{const}} dS\omega = 4\pi\, k^2 \qquad ...(1)$$

Hence, for each branch, we have

$$Z_i(\omega)\, d\omega = \frac{V}{2\pi^2}\frac{k^2}{v_i}\, d\omega = \frac{V}{2\pi^2}\frac{k^2}{v_i}\, d\omega \qquad ...(2)$$

The total density of phonon states is

$$Z(\omega)\, d\omega = \frac{V}{2\pi^2}\left(\frac{1}{v_L^3}+\frac{2}{v_T^3}\right)\omega^2\, d\omega \qquad ...(3)$$

that is, density of states increases quadratically with frequency. However, for every solid there is a maximum possible frequency known as Debye frequency, ($\omega_D$).

The second hypothesis of Debye is that the number of modes of vibration in a solid is equal to $3N$. This hypothesis is quite reasonable since this is precisely the number of vibrational modes of a system which has $3N$ decrease of freedom

$$\int_0^{\omega_D} Z(\omega)\, d\omega = 3N \qquad ...(4)$$

where $N$ is the total number of atoms

$$\therefore \quad \frac{V}{2\pi^2}\left(\frac{1}{v_L^3}+\frac{2}{v_T^3}\right)\int_0^{\omega^2}\omega^2\, d\omega = 3N \qquad ...(5i)$$

$$\Rightarrow \quad \frac{V}{2\pi^2}\left(\frac{1}{v_L^3}+\frac{2}{v_T^3}\right) = \frac{9N}{\omega_D^3} \qquad ...(5ii)$$

Putting $\omega_D = 2\pi\, \nu_D$, we have

$$\nu_D^3 = \frac{9N}{4\pi V}\left(\frac{1}{v_L^3}+\frac{2}{v_T^3}\right)^{-1} \qquad ...(6)$$

Further, Eq. (3) gives

$$Z(\nu)\, d\nu = 4\pi V\left[\frac{1}{v_L^3}+\frac{2}{v_T^3}\right]\nu^2 d\nu \qquad ...(7)$$

Substituting Eq. (6) in Eq. (7), we obtain

$$Z(\nu) = \frac{9N}{\nu_D^3}\,\nu^2 \qquad ...(8)$$

This gives

$$Z(\nu) n = K\nu^2 \quad (\text{for } \nu \le \nu_D)$$
$$= 0 \quad (\text{for } \nu \ge \nu_D)$$

where, $K = 9N/\nu_D^3$ is a constant.

The distribution function is compared with the Einstein's function in Fig. 9.4.

While we discuss contribution of lattice vibrations to thermal properties of solids, we may classify the properties in two groups

(i) Properties whose broad features can be explained under harmonic approximation.

(ii) Properties whose broad features cannot be explained without considering anharmonic effects.

The first group accounts for heat capacity and the second group is responsible for properties such as thermal conductivity and thermal expansion.

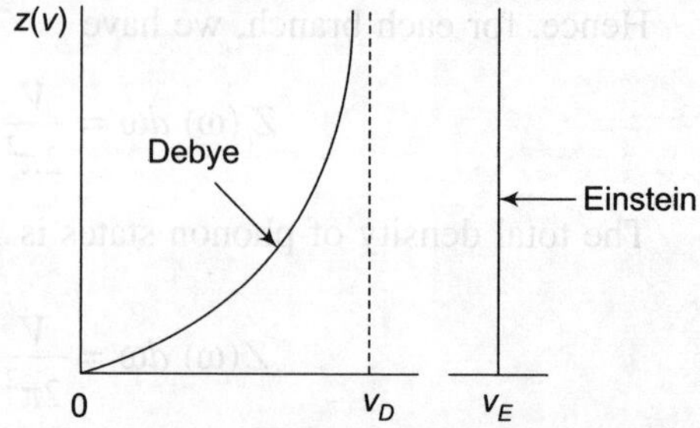


**Figure 9.4** Frequency distribution function according to debye model and Einstein model

## 9.5 THE SPECIFIC HEAT AT CONSTANT VOLUME AND AT CONSTANT PRESSURE

According to the first law of thermodynamics, the amount of heat $dQ$ added to a system must be equal to the increase in energy $dE$ of the system plus the amount of work done by the system. In case the work done by the system is of a mechanical nature only, one may thus write

$$dQ = dE + p\,dV \qquad ...(1)$$

Now, $E$ is, except for an arbitrary constant, determined uniquely by the temperature and volume of the system. Hence

$$dE = \left(\frac{\partial E}{\partial T}\right)_V dT + \left(\frac{\partial E}{\partial V}\right)_T dV$$

and Eq. (1) may be rewritten in the form

$$dQ = \left(\frac{\partial E}{\partial T}\right)_V dT + \left[\left(\frac{\partial E}{\partial V}\right)_T + p\right] dV \qquad ...(2)$$

The specific heat, in general, is defined by $dQ/dT$, and unless stated otherwise, will be assumed to refer to 1 gram molecule of the solid. However, unless one specifies in which way the increase in temperature takes place, the specific heat is undetermined; in particular one must specify the corresponding change in volume, as is evident from Eq. (2). Thus, there exist an infinite number of specific heats, but in general one is interested in only two: the specific heat at constant volume $C_V$ and the specific heat at constant pressure $C_p$. According to Eq. (2), to the former is given by

$$C_V = \left(\frac{dQ}{dT}\right)_V = \left(\frac{\partial E}{\partial T}\right)_V \qquad ...(3)$$

Theoretically speaking, this is the most interesting quantity, as it is obtained immediately from the energy of the system; most of the following discussion will therefore refer to $C_V$.

From the experimental point of view, however, it is much more convenient to measure the specific heat of a solid at constant pressure than at constant volume. As shown in textbooks on thermodynamics, the second law leads to the following relationship between $C_p$ and $C_V$:[1]

$$C_p - C_V = -T\left(\frac{\partial V}{\partial T}\right)_p^2 = \left(\frac{\partial p}{\partial V}\right)_T \qquad ...(4)$$

This may be rewritten in terms of the volume expansion coefficient $\alpha_V$ and the compressibility $K$, defined by

$$\alpha_V = (1/V)(\partial V/\partial T)_p \quad \text{and} \quad K = -(1/V)(\partial V/\partial p)_T \qquad ...(5)$$

Expression (4) then takes the form

$$C_p - C_V = \alpha_V^2\, TV/K \qquad ...(6)$$

Thus, $C_V$ may be calculated from $C_p$ measurements if at the same time $\alpha_V$ and $K$ are known at the temperature of interest. Since both $\alpha_V$ and $K$ are positive qualities, $C_p - C_V \geq 0$.

By way of illustration, we have given in Fig. 9.5 $C_p$ and $C_V$ as functions of temperature for copper; note that at low temperature as their difference becomes very small and that both go to zero at $T = 0$. It is essentially the temperature variation of the specific heat at constant volume which will be discussed in the present chapter.

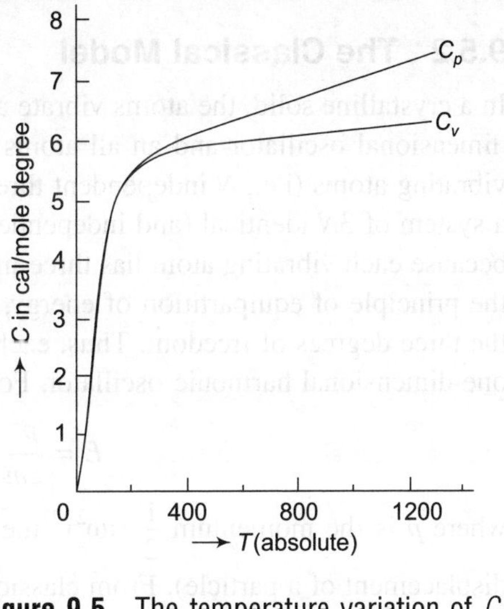


**Figure 9.5** The temperature variation of $C_p$ and $C_V$ for copper. [By permission from M. W. Zemansky, *Heat and Thermodynomic*, 2d ed., McGraw-Hill, New York, 1943, p. 237]

It may be noted that if no direct compressibility data are available for the temperature range of interest, one frequently employs the relationship

$$C_p = C_V(1 + \gamma\alpha_V T)$$

with $\qquad \gamma = \alpha_V V/KC_V \qquad ...(7)$

The quantity $\gamma$ is called the constant and is practically independent of temperature.[2] Thus, by calculating $\gamma$ at some arbitrary temperature from available data, one may obtain an approximation for $C_V$ at other temperatures from a knowledge of the coefficient of volume expansion.

## 9.5.1 Specific Heat of Solids

We know from elementary knowledge that the specific heat is the amount of heat energy required to be supplied to one mole of a solid to raise its temperature by one degree. Specific heat of a substance is given at constant volume and constant pressure by

1 See, for example, M. W. Zemansky, *Heat and Thermodynamics*, 2d ed., McGraw Hill, New York, 1943, p. 227.

2 E. Grüneisen, *Handbuch der Physik,* **10,** 1–59 (1926); see also J. C. Slater, *Phys. Rev.*, 57, 744 (1940).

$$C_v = \left.\frac{dQ}{dT}\right|_{v=\text{const.}} = \left.\frac{\partial E}{\partial T}\right|_{v=\text{const}} \quad ...(8)$$

$$C_p = \left.\frac{dQ}{dT}\right|_{p=\text{const}} \quad ...(9)$$

respectively, where $Q$ is heat supplied, $E$ the internal energy and $T$ the temperature. $C_p$ is different from $C_v$ for gases. However, in case of solids, $C_p$ is almost same as $C_v$. In solids when heat is supplied, it raises its internal energy. This increase in internal energy is manifested mainly as

(i) brisk vibrations of atoms about their mean position; and

(ii) an increase in kinetic energy of free electrons.

Thus, in general, the specific heat of a solid is expressed as

$$C_{\text{solid}} = C_{\text{lattice}} + C_{\text{electronic}} \quad ...(10)$$

First, we will restrict ourselves to only the first term (on the r.h.s.), i.e., thermal vibrations of the lattice.

## 9.5.2 The Classical Model

In a crystalline solid, the atoms vibrate about their mean positions, each behaving as a three-dimensional oscillator and an all atoms vibrate independent of one another. A system of $N$ vibrating atoms (i.e., $N$ independent three-dimensional harmonic oscillators) is equivalent to a system of $3N$ identical (and independent) one-dimensional harmonic oscillators. This is so because each vibrating atom has three independent vibrational degrees of freedom, and from the principle of equipartition of energy, the vibrational energy is distributed equally among the three degrees of freedom. Thus, each vibrational degree of freedom can be regarded as a one-dimensional harmonic oscillator. For a linear oscillator, energy

$$E = \frac{p^2}{2m} + \frac{1}{2}m\omega^2x^2 \quad ...(11)$$

where $p$ is the momentum $\frac{1}{2}m\omega^2x^2$ the potential energy ($\omega$ is angular frequency and $x$ the displacement of a particle). From classical statistical mechanics, the mean energy of such an oscillator is $k_BT\left(\frac{1}{2}k_BT \text{ for each degree of freedom}\right)$. For three-dimensional motion

$$\langle E\rangle = 3\,k_BT \quad ...(12)$$

where $k_B$ is the Boltzmann constant. For $N$ oscillators, the total energy is $3N\,k_BT$. If we have one mole of substance, $N = N_A$ (the Avogadro number), then

$$C_v = \frac{\partial}{\partial T}(3N_Ak_BT) = 3\,N_Ak_B \equiv 3R$$

$$= 24.9 \text{ J mole}^{-1}\text{ K}^{-1} \quad ...(13)$$

Further, $$C_v = 5.96 \text{ cal mole}^{-1}\text{ K}^{-1} \quad ...(14)$$

that is, the molar heat capacity of all the solids is constant (independent of temperature and frequency). This is known as the *Dulong and Petits law*. This result is in good agreement with the observed heat capacity for a number of solids. However, the following experimental facts indicate that the law fails in certain cases.

(i) At low temperatures, the specific heat of all elemental solids is found to approach zero as $T^3$ is insulators and as $T$ in metals.

(ii) In magnetic solid, there is a large contribution to the specific heat near the temperature at which the magnetic moments become ordered.

(iii) A number of low atomic number and high melting points element (e.g., B, Be and C) show lower specific heat as predicted by Dulong and Petits law.

## 9.6 EINSTEIN THEORY OF SPECIFIC HEAT

Einstein in 1907 applied the Planck's quantum theory to account for variation of specific heat of solids with temperatures.[3] Einstein's model has the following points.

(i) A crystal consists of atoms which may be regarded as identical and independent harmonic oscillations.

(ii) A solid having $N$ atoms is equivalent to $3N$ one-dimensional harmonic oscillators.

(iii) All atomic oscillators vibrate with the same frequency.

(iv) The oscillators are quantum oscillators having discrete energies rather than a continuous spectrum is for classical case.

(v) The atomic oscillators obey Maxwell–Boltzmann distribution law of energies.

(vi) The atomic oscillators for an assembly of systems which are distinguishable due to their location at distinct and different lattice sites and hence obey Maxwell–Boltzmann distribution law of energies.

The energy levels of the harmonic oscillator are

$$E_n = (n + 1/2)h\nu$$
$$= (n + 1/2)\hbar\omega \; (n = 0, 1, 2, ...) \qquad ...(1)$$

where $\frac{1}{2}\hbar\omega$ is the temperature independent zero-point energy contribution to the internal energy of the oscillator.[4] Using Maxwell–Boltzmann distribution of energy

$$\langle E \rangle = \frac{\int_0^\infty E_n \exp(-E_n / k_B T)\, dE}{\int_0^\infty \exp(-E_n / k_B T)\, dE} \qquad ...(2)$$

[3] A Einstein, *Ann. Physik* 22, 180 (1907).

[4] The zero-point energy $\frac{1}{2}\hbar\omega$ exists at all temperatures (including absolute zero). (This is a consequence of 'modern' quantum mechanics.) Since it is independent of temperature, inclusion of this term in the energy does not affect the specific heat.

or
$$\langle E\rangle = \frac{\sum_n \left(n+\frac{1}{2}\right)\hbar\omega \exp\left\{-\left(n+\frac{1}{2}\right)\hbar\omega/k_BT\right\}}{\sum_n \exp\left\{-\left(n+\frac{1}{2}\right)\hbar\omega/k_BT\right\}}$$

$$= \frac{\hbar\omega \sum_{n=0}^{\infty}\left(n+\frac{1}{2}\right)\exp\left\{\left(n+\frac{1}{2}\right)x\right\}}{\sum_{n=0}^{\infty}\exp\left[\left(n+\frac{1}{2}\right)x\right]}$$

where, $$x = -\frac{\hbar\omega}{k_BT}$$

$$\therefore \quad \langle E\rangle = \frac{\hbar\omega(1/2\, e^{x/2} + 3/2\, e^{3x/2} + 5/2\, e^{5x/2} + \ldots)}{(e^{x/2} + e^{3x/2} + e^{5x/2} + \ldots)}$$

$$= \hbar\omega \frac{d}{dx} \ln\,[e^{x/2} + e^{3x/2} + e^{5x/2} + \ldots]$$

$$= \hbar\omega \frac{d}{dx} \ln\,[e^{x/2}(1 + e^{x} + e^{2x} + \ldots)]$$

Now, $\ln(1 + e^x + e^{2x} + \ldots) = -\ln(1 - e^x)$

$$\therefore \quad \langle E\rangle = \hbar\omega \frac{d}{dx}[x/2 - (1 - e^x)]$$

$$= \hbar\omega\left(\frac{1}{2} + \frac{e^x}{1-e^x}\right) = \hbar\omega\left(\frac{1}{2} + \frac{e^x}{e^{-x}-1}\right)$$

$$= \hbar\omega\left(\frac{1}{2} + \frac{1}{e^{\hbar\omega/k_BT}-1}\right)$$

$$= \frac{1}{2}\hbar\omega + \frac{\hbar\omega}{\exp\left(\frac{\hbar\omega}{k_BT}\right)-1} \quad \ldots(3)$$

Now, the first term ($\hbar\omega/2$) is the zero point energy of the oscillator (and corresponding to $T = 0$). Thus, according to quantum mechanics, the atoms are not at rest even at 0 K and each atom possesses the vibrational energy of ($\hbar\omega/2$). This energy, however, has no contribution to $C_v$. It is to be noted that expression (3) also contains the frequency of the oscillator in contrast to the corresponding expression of the classical theory. For one mole of the solid, the total energy of the crystal

$$U = 3N_A\langle E\rangle = \frac{3N_A\hbar\omega}{2} + \frac{3N_A\hbar\omega}{\exp(\hbar\omega/k_BT)-1}$$

The molar specific heat is

$$C_v = \left(\frac{\partial U}{\partial T}\right)_v = 3N_A \frac{\hbar\omega}{[\exp(\hbar\omega / k_B T) - 1]^2} \frac{\hbar\omega}{k_B T^2} \exp\left(\frac{\hbar\omega}{k_B T}\right)$$

$$= 3R\left(\frac{\hbar\omega}{k_B T}\right)^2 \frac{\exp(\hbar\omega/k_B T)}{[\exp(\hbar\omega/k_B T) - 1]^2} \quad ...(4)$$

Let $\theta_E = \hbar\omega/k_B$, ...(5)

where $\theta_E$ is the characteristic temperature, known as *Einstein temperature*. Then,

$$C_v = 3N_A k_B \left(\frac{\theta_E}{T}\right)^2 \frac{e^{\theta_E/T}}{(e^{\theta\varepsilon/T} - 1)^2} \quad ...(6)$$

We now consider the following two cases.

(i) **High Temperatures:** For temperatures such that $k_B T \gg \hbar\omega$ or $T \gg \theta_E$, we can write

$$e^{\hbar\omega/kBT} - 1 \equiv 1 + \frac{\hbar\omega}{wk_B T} + ... - 1 = \frac{\hbar\omega}{k_B T} \quad ...(7i)$$

and from Eq. (7), we get

$$\langle E\rangle = \frac{1}{2}\hbar\omega + k_B T \cong k_B T$$

Thus, at high temperatures, the average vibrational energy is the same as that obtained from the classical theory. Also, Eq. (4) becomes

$$C_v = 3N_A k_B \left(\frac{\hbar\omega}{k_B T}\right)^2 \frac{(1 + \hbar\omega / k_B T)}{(\hbar\omega / k_B T)^2} \quad ...(7ii)$$

$$= 3N_A k_B (1 + \hbar\omega/k_B T) \quad ...(8)$$

For large $T$, $\hbar\omega/k_B T \to 0$ and then

$$C_v = 3N_A k_B = 3R$$

which is the Dulong and Petit's law as obtained from classical theory.

(ii) **Low Temperatures:** For temperature such that $k_B T \gg \hbar\omega$ or $T \gg \theta_E$, we can write

$$e^{\hbar\omega/k_B T} = -1 \cong e^{\hbar\omega/k_B T}$$

then Eq. (3) becomes

$$\langle E\rangle = \frac{1}{2}\hbar\omega + \hbar\omega\, e^{-\hbar\omega/k_B T} \quad ...(9)$$

It shows that at low temperatures, the average vibrational energy decreases exponentially with decrease in temperature. The expression (4) for $C_v$ becomes

$$C_v = 3N_A k_B \left(\frac{\hbar\omega}{k_B T}\right)^2 e^{-\hbar\omega/k_B T}$$

or in terms of $\theta_E$,

$$C_u = 3N_A k_B \left(\frac{\theta_E}{T}\right)^2 e^{-\theta_E/T} \qquad ...(10)$$

Thus, for $T \ll \theta_E$, the heat capacity is proportional to $e^{-\theta_E/T}$ which is the dominating factor, i.e., $C_u$ falls off almost exponentially with decreasing $T$. However, experimentally, $C_u$ is found to decrease much more slowly. Thus, though the decrease of $C_u$ with decrease of $T$ can be explained quantitatively by Einstein's theory, there is no quantitative agreement between theory and experiment at low temperatures. The discrepancy is due to the oversimplified assumption of the Einstein's model in which the atomic oscillators have been considered to vibrate independently at the same frequency.

## 9.7 DEBYE'S THEORY OF LATTICE HEAT CAPACITY

In 1912, Debye put forward his model of lattice heat capacity.[5] In this model, the vibrational motion of the crystal as a whole was considered to be equivalent to the vibrational motion of a system of coupled harmonic oscillators (because in a solid, the atoms are bound together by strong interatomic forces). The coupled harmonic oscillators can propagate a range of frequencies rather than a single frequency. Debye proposed that a crystal can propagate elastic wave of wavelengths ranging from low frequencies (sound waves) to high frequencies corresponding to infrared absorption. Therefore, there can be a number of modes of vibration. The number of vibrational modes per unit frequency range is called *density of modes, g (v)*.

The number of possible modes of elastic standing waves (of any kind) with frequencies between $v$ and $v + dv$ in a solid is given by

$$g(v)\,dv = \left(\frac{4\pi V_0}{v_s^3}\right) v^2\,dv \qquad ...(1)$$

where $V_0$ is the volume of a kilomole of the solid cube and $v_s$ is wave speed. There are two kinds of elastic waves that can propagate in solid, longitudinal waves whose speed is $v_1$ and transverse waves whose speed is $u_r$. Also, there are two perpendicular directions of polarisation for a transverse wave. Hence, the total number of possible standing elastic waves with differences between $v$ and $v + dv$ is

$$g(v)\,dv = 4\pi V_0 \left(\frac{1}{v_l^3} + \frac{2}{v_t^3}\right) v^2\,dv \qquad ...(2)$$

By analogy with black body radiation, the average energy $\overline{E}$ per standing wave is assumed to be

$$\overline{E} = \frac{hv}{(e^{\hbar\omega/k_BT} - 1)} \qquad ...(3)$$

(In other words, it has been assumed that a phonon gas behaves statistically similar to a photon gas.) The total internal energy

[5] P. Debye, *Ann. Physik* 39, 789 (1912).

$$U = V_0 \int_0^{\nu_m} g(\nu)\,\bar{E}\,d\nu \qquad ...(4)$$

$$= 4\pi V_0 \left(\frac{1}{v_l^3} + \frac{2}{v_t^3}\right) \int_0^{\nu_m} \frac{h\nu^3\,d\nu}{(e^{h\nu/k_BT} - 1)} \qquad ...(5)$$

Here, the upper limit of integration in some maximum frequency $\nu_m$, which indicates that there cannot be an infinite number of standing waves. Debye assumed if there are $N_A$ number of atoms in a solid then total number of vibrational modes is limited to $3N_A$, so that

$$3N_A = V_0 \int_0^{\nu_m} g(\nu)\,d\nu \qquad ...(6)$$

$$= 4\pi V_0 \left(\frac{1}{v_l^3} + \frac{2}{v_t^3}\right) \int_0^{\nu_m} \nu^2\,d\nu$$

$$= \frac{4}{3}\pi V_0 \left(\frac{1}{v_l^3} + \frac{2}{v_t^3}\right) \nu_m^3$$

$$\therefore \qquad \nu_m = \left[\frac{9N_A}{4\pi V_0 \left(\frac{1}{v_l^3} + \frac{2}{v_t^3}\right)}\right]^{1/3} \qquad ...(7)$$

Equation (5) can be rewritten in terms of $\nu_m$ as

$$U = \frac{9N_A}{\nu_m^3} \int_0^{\nu_m} \frac{h\nu^3\,d\nu}{(e^{h\nu/k_BT} - 1)}$$

$$U = 9N_A \left[\frac{1}{\omega_m^3} \int_0^{\nu_m} \frac{\hbar\omega^3\,d\omega}{\exp(\hbar\omega/k_BT) - 1}\right] \qquad ...(8)$$

neglecting the contribution from zero-point energy. It would be independent of $T$ and would not contribute to $C_v$, so that

$$C_v = \left(\frac{\partial U}{\partial T}\right)_v = \frac{9N_A}{\omega_m^3} \int_0^{\omega_m} \frac{\hbar\omega^3 (\hbar\omega/k_BT^2) \exp\left(\frac{\hbar\omega}{k_BT}\right) d\omega}{[\exp(\hbar\omega/k_BT) - 1]^2}$$

Putting $x = \hbar\omega/k_BT$ and $x_m = \hbar\omega/k_BT$, we have

$$C_v = \frac{9N_A}{\omega_m^3} \int_0^{x_m} \frac{\hbar^2}{k_BT^2} \left(\frac{k_BT}{\hbar}\right)^5 \frac{x^4\,(e^x)dx}{(e^x - 1)^2}$$

$$= 3\,R.3\left(\frac{T}{\theta_D}\right)^2 \int_0^{x_m} \frac{x^4\,(e^x)dx}{(e^x-1)^2} \qquad ...(9)$$

Here $\theta_D$ is the *Debye temperature*

$$\theta_D = \hbar\omega_m/k_B \qquad ...(10)$$

The inregral in Eq. (9) is a function of $x_m = \hbar\omega\; m/k_BT = \theta_D/T$. Hence, we have

$$C_u = 3RF_D\,(\theta_D/T) \qquad ...(11)$$

where the Debye function $F_D$ is given by

$$F_D\,(\theta_D/T) = 3R.3\left(\frac{T}{\theta_D}\right)^3 \int_0^{\theta_D/T} \frac{x^4\,(e^x)\,dx}{(e^x-1)^2} \qquad ...(12)$$

From the equation, integrating by parts, we get

$$\int_0^{x_m} \frac{x^4\,(e^x)\,dx}{(e^x-1)^2} = -\left(\frac{x^4}{e^x-1}\right)_0^{x_m} + 4\int_0^{x_m} \frac{x^3 dx}{e^x-1}$$

$$= 4\int_0^{x_m} \frac{x^3 dx}{e^x-1} - \frac{x_m^4}{e^{x_m}-1} \qquad ...(13)$$

Hence, we get from Eq. (9)

$$C_v = 3R\,\frac{3}{x_m^3}\left[4\int_0^{x_m} \frac{x^3\,dx}{e^x-1} - \frac{x_m^4}{e^{x_m}-1}\right]$$

$$= 3R\left[\frac{12}{x_m^3}\int_0^{x_m} \frac{x^3\,dx}{e^x-1} - \frac{3\,x_m}{e^x m-1}\right] \qquad ...(14)$$

We now consider this equation for two limiting cases.

(i) For $T \gg \theta_D$: If $\omega_m$ is such that $\hbar\omega_m \ll k_BT$, then $x_m = \hbar\omega_m/k_BT \ll 1$. That is, in this case $x = \hbar\omega_m k_BT \ll 1$, since $\omega < \omega_m$. So we have $e^x - 1 \approx x$ and $e^{x_m} - 1 \approx x_m$.

Hence, $$C_v = 3R\left[\frac{12}{x_m^3}\int_0^{x_m} \frac{x^3\,dx}{e^x-1} - \frac{3\,x_m}{x_m}\right]$$

$$= 3R\left[\frac{12}{x_m^3}\frac{x_m^3}{3} - 3\right] = 3R \qquad ...(15)$$

This is in agreement with the Dulong and Petit's law.

(ii) For $T \ll \theta_D$: $x_m \gg 1$, so that for $x_m \to \infty$, we have

$$\frac{3\,x_m}{e^{x_m}-1} \approx 3\,x_m\,e^{(-x_m)} \to 0$$

and
$$\int_0^{x_m} \frac{x^3\,dx}{e^x - 1} = \int_0^{x_m} x^3 e^{(-x)} [1 - e^{(-x)}]^{-1}\,dx$$

$$= \int_0^{x_m} x^3 e^{(-x)} [1 + e^{-x} + e^{-2x} + \ldots]\,dx$$

$\therefore$
$$\lim_{x_m \to \infty} \int_0^{\infty} \frac{x^3\,dx}{e^x - 1} = \int_0^{x_m} x^3 [e - x + e^{-2x} + \ldots]\,dx$$

$$= \sum_{n=1}^{m} \int_0^{\infty} x^3 \exp(-nx)\,dx$$

But, we know that for an integer $r$,

$$\int_0^{\infty} x^r \exp(-nx)\,dx = \frac{r!}{n^{r+1}} \quad \ldots(16)$$

Hence,
$$\int_0^{\infty} x^3 \exp(-nx)\,dx = \frac{6}{n^4} \quad \ldots(17)$$

Also,
$$\sum_{n=1}^{\infty} \frac{1}{n^4} = \frac{\pi^4}{90} \quad \ldots(18)$$

So, we get
$$\sum_{n=1}^{\infty} \int_0^{\infty} x^3 e^{-x}\,dx = \frac{6\pi^4}{90} = \frac{\pi^4}{15} \quad \ldots(19)$$

Therefore, at low temperatures

$$C_v = 3R \frac{12}{x_m^2} \int_0^{x_m} \frac{x^3 dx}{e^x - 1} = 3R \frac{12}{x_m^2} \frac{\pi^4}{15}$$

$$= 3R \frac{4\pi^4}{5} \left(\frac{T}{\theta_D}\right)^3 \quad \ldots(20)$$

Thus, at very low temperature, $C_v \propto T^3$. This is known as Debye's $T^3$ law. According to this, $C_v \to 0$, as $T \to 0$ (the law holds for $T \leq \theta_D/10$). There is an excellent agreement between theory and experiment.

### 9.7.1 Limitations of the Debye Model

(i) The model is valid for long wavelengths (low frequencies) only.

(ii) The total number of vibrational modes are assumed to be $3N$. But the solid is an elastic continuum and should possess infinite frequencies.

(iii) The cut-off frequency is assumed to be the same for both longitudinal and transverse waves. But, different velocities of transverse and longitudinal waves should imply different values of cut-off frequency for these waves.

(iv) According to the Debye's theory, $\theta_D$ is independent of temperature, whereas actually it has been found to vary by 10% or more.

(v) The theory cannot be applied to crystals comprised of more than one type of atoms.

(vi) The theory does not take into account the interaction among the atoms and the contribution of electrons to the specific heat.

**Example 9.1** The Einstein frequency for copper is $\nu_E = 4.8 \times 10^{12}$ Hz. Calculate its Einstein temperature.

***Solution***

$$\theta_E = \frac{h\nu_E}{k_B} = \frac{6.6 \times 10^{-34} \times 4.8 \times 10^{12}}{1.38 \times 10^{-23}} = 229.6 \text{ K}$$

## 9.8 THERMAL CONDUCTIVITY OF SOLIDS

A theory of thermal conductivity of solids can be developed by knowing the rate at which the energy is transferred across the solid by electrons and photons which get scattered by collisions among themselves as well as by impurities and imperfections in the solid. In crystals the phonons move with the velocity of sound and, therefore, heat conduction also takes place with the velocity of sound. In the present case we limit our discussion to conduction by phonons, leaving the conduction by electrons to be discussed later. This theory will, therefore, be applicable only to insulators.

The amount of heat $Q$ that flows through unit cross-sectional area of a solid per second where the temperature gradient $\frac{d\theta}{dx}$ is given by

$$Q = -K\frac{d\theta}{dx} \qquad ...(1)$$

where $K$ is the thermal conductivity. The negative sign in Eq. (1) is used to indicate that as the distance $x$ increases, the temperature $\theta$ decreases.

The thermal conductivity of solids can most easily be explained if we regard the solid to be a container enclosing a phonon gas which is found to behave like an ideal monatomic gas. We shall, therefore, first derive an expression for the thermal conductivity of a monatomic gas and shall then apply it to the lattice of a crystalline solid.

Let us find the number of particles of an ideal gas having velocity line between $v$ and $v + dv$ and crossing and elementary area $ds$ per second from a direction line between angle $\theta$ and $\theta + d\theta$. Referring to Fig. 9.6, we see that this number will evidently be just half

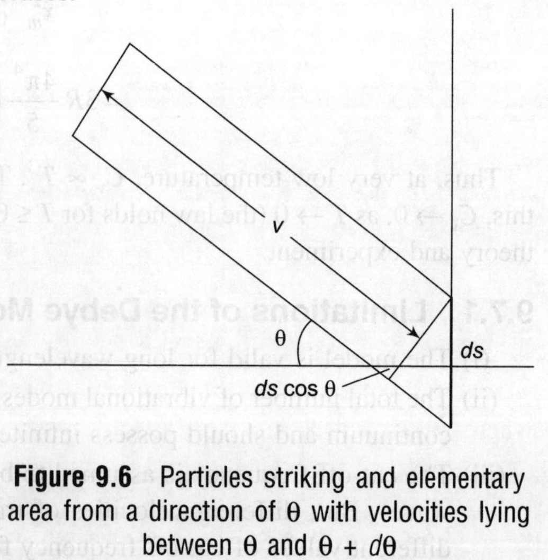


**Figure 9.6** Particles striking and elementary area from a direction of θ with velocities lying between θ and θ + dθ

the number of particles within this velocity range inside of column of cross sectional area $ds \cos\theta$ and of length $v$, because the other half may go in the opposite direction.

If $n(v)\,dx$ be the number of molecules per unit volume having velocity lying between $v$ and $v + dv$, then the number of molecules crossing the area $ds \cos\theta$ in a direction lying between angle $\theta$ and $\theta + d\theta$ is

$$dn = \frac{1}{2}\,v\cdot n(v)\,dv\cdot\frac{4\pi\sin\theta\,d\theta}{4\pi}\cdot ds\cos\theta$$

$$\Rightarrow \quad dn = \frac{1}{2}\,v\cdot n(v)\cdot\sin\theta\,\cos\theta\,d\theta\,dv\,ds \qquad ...(2)$$

where $\left(\dfrac{4\pi\sin\theta\,d\theta}{4\pi}\right)$ represents the solid angle subtended by the element $d\theta$ about $\theta$ to the solid angle $4\pi$ subtended by the entire space around it.

From Maxwell–Boltzmann distribution law, we have

$$n(v)\,dv = 4\pi n\left(\frac{m}{2\pi k_B T}\right)^{3/2} v^2 \exp\left(-\frac{mv^2}{2k_B T}\right)dv$$

Substituting this value of $n(v)\,dv$ in Eq. (2) and integrating over $v$ (remembering that the velocity distribution is independent of angle $\theta$) we get the number of molecules striking the elementary area $ds$ per second from a direction line between angle $\theta$ and $\theta + d\theta$ as

$$n(\theta)\,d\theta = \frac{1}{2}\cdot 4\pi n\left(\frac{m}{2\pi k_B T}\right)^{3/2}\cdot\frac{1}{2}\left(\frac{2k_B T}{m}\right)^2 \sin\theta\cos\theta\,d\theta\,ds$$

$$\Rightarrow \quad n(\theta)\,d\theta = \frac{\pi n}{\pi^{3/2}}\cdot\left(\frac{m}{2k_B T}\right)^{3/2}\cdot\left(\frac{2k_B T}{m}\right)^2 \sin\theta\cos\theta\,d\theta\,ds$$

$$\Rightarrow \quad n(\theta)\,d\theta = n\cdot\sqrt{\frac{2k_B T}{\pi m}}\,\sin\theta\cos\theta\,d\theta\,ds$$

$$\Rightarrow \quad n(\theta)\,d\theta = \frac{1}{2}\cdot 2n\cdot\sqrt{\frac{2k_B T}{\pi m}}\,\sin\theta\cos\theta\,d\theta\,ds$$

$$\Rightarrow \quad n(\theta)\,d\theta = \frac{1}{2}\,n\sqrt{\frac{8k_B T}{\pi m}}\,\sin\theta\cos\theta\,d\theta\,ds$$

$$\Rightarrow \quad n(\theta)\,d\theta = \frac{1}{2}\,n\bar{c}\,\sin\theta\cos\theta\,d\theta\,ds \qquad ...(3)$$

where

$$\bar{c} = \sqrt{\frac{8k_B T}{\pi m}} \qquad ...(4)$$

is the average thermal speed of the particle.

It may be noted that the statistical distribution applies two systems in thermal equilibrium. The thermal conduction involves a flow of energy due to temperature gradient and so, we are dealing here with a system which is not in thermal equilibrium. The use of statistical distribution in this case is, therefore, only an approximation.

## 9.9 LATTICE THERMAL CONDUCTIVITY

Consider a system made up of ideal harmonic oscillators. The various normal modes of vibration of the system are then completely independent of one another and the energy of vibration is propagated directly by phonons which move with the velocity of sound, instead of the much slower process associated with thermal conduction. The energy is thus not randomized through the solid. Such a transport of energy suffers no thermal resistance and the thermal conductivity of the solid is infinity. This result, which contradicts the reality, arises because we have assumed the harmonic oscillators to be ideal, an assumption that is far from reality.

In fact, since almost all solids expand on heating, the presence of a small amount of anharmonicity is indicated. The different modes of vibration are, therefore, not independent, but coupled with one another. In terms of phonons, we may say that the phonons corresponding to different normal modes interact (or collide) with one another and in this process of collision, the momentum and the direction of propagation of a group of phonons representing a single normal mode is altered in a random fashion. The energy in this case, thus, randomized. This thermal transport, therefore, suffers thermal resistance and the conductivity is limited to a finite value.

Since the magnitude of anharmonicity increases with the amplitude of vibration which, in turn, increases with the rise in temperature, we may expect the anharmonicity to increase with increase in temperature and a consequent decrease in the lattice thermal conductivity. This effect is quite generally observed experimentally at sufficiently high temperatures.

Let us consider a situation where a temperature gradient exists along the $x$-axis. The articles at the origin are assumed to have energy $E$ and that at a short distance away to the left, $E + \Delta E$, as shown in Fig. 9.7.

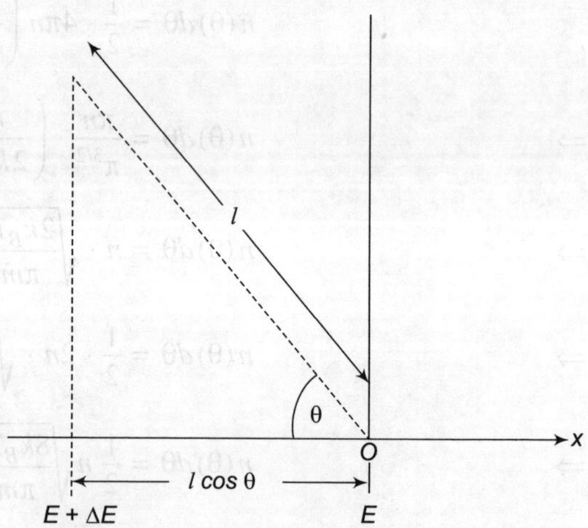


**Figure 9.7** Figure showing the path of a particle in a system having the temperature gradient along $x$-axis

Using Taylor's theorem, we may write

$$E = \Delta E = E + \left(\frac{\partial E}{\partial x}\right)\Delta x \qquad ...(1)$$

The energy of a particle at a distance $l$ away from the origin in a direction $\theta$ may then be expressed as

$$E = \Delta E = E + (l \cos \theta) \frac{\partial E}{\partial x} \quad ...(2)$$

If this particle gives up all its energy (= $E + dE$) to the distribution at the origin through collision at origin, then there will be a particle leaving the origin and carrying energy[6] $E$ so as to reach the starting point of the first particle.[7] The net transport of energy per particle to the right is, therefore,

$$E = \Delta E - E = (l \cos \theta) \frac{\partial E}{\partial x}$$

$$\Rightarrow \quad \Delta E = (l \cos \theta) \frac{\partial E}{\partial x} \quad ...(3)$$

Since the average value of $l$ is simply the mean free path $\lambda$ between randomizing collisions, the net transport of energy per particle to the right is

$$\Delta E = (\lambda \cos \theta) \frac{\partial E}{\partial x} \quad ...(4)$$

Therefore, the net energy flux for particles coming from a distance $\theta$ is[8]

$$\Delta F = \frac{n(\theta)\, d\theta \cdot \Delta E}{ds}$$

$$\Rightarrow \quad \Delta F = \frac{1}{2} n\bar{c} \sin\theta \cos\theta \, d\theta \cdot \lambda \cos\theta \frac{\partial E}{\partial x}$$

$$\Rightarrow \quad \Delta F = \frac{1}{2} n\bar{c}\lambda \frac{\partial E}{\partial x} \cos^2\theta \sin\theta \, d\theta \quad ...(5)$$

where we have used Eq. (3) section 9.8.

Integrating Eq. (5) over $\theta$ between the limits $\theta$ and $\pi$ we get the total energy flux, $F$.

Thus,
$$F = \int \Delta F = \frac{1}{2} n\bar{c}\lambda \frac{\partial E}{\partial x} \int_0^\pi \cos^2\theta \sin\theta \, d\theta$$

$$\Rightarrow \quad F = \frac{1}{2} n\bar{c}\lambda \left[ -\frac{\cos^3\theta}{3} \right]_0^\pi \frac{\partial E}{\partial x}$$

$$\Rightarrow \quad F = \frac{1}{6} n\bar{c}\lambda \, [-\cos\pi - (-\cos 0)] \frac{\partial E}{\partial x}$$

$$\Rightarrow \quad F = \frac{1}{6} n\bar{c}\lambda \, [1 + 1] \frac{\partial E}{\partial x}$$

$$\Rightarrow \quad F = \frac{1}{3} n\bar{c}\lambda \frac{\partial E}{\partial x} \quad ...(6)$$

---

[6] The energy of the particle at the origin is $E$.

[7] This is necessary for the particle density to be the same throughout.

[8] $n(\theta)\, d\theta$ is the number of particles striking area $ds$ per second from a direction $\theta$ and $\Delta E$ denotes the energy imparted by with particle. Therefore, the total energy imparted per second is $n(\theta)\, d\theta \;\; \Delta E$.

According to the definition of thermal conductivity, this energy flux is equal to $K\dfrac{\partial\theta}{\partial x}$.

Hence, $$K\frac{\partial\theta}{\partial x} = \frac{1}{3}n\bar{c}\lambda\frac{\partial E}{\partial x}$$

$$\Rightarrow \qquad K = \frac{1}{3}n\bar{c}\lambda\left(\frac{\partial E}{\partial\theta}\right) = \frac{1}{3}n\bar{c}\lambda C_v \qquad ...(7)$$

where $$C_v = \left(\frac{\partial E}{\partial\theta}\right)$$

Since the velocity of sound $\bar{c}$ and the specific heat could be easily determined, it follows from Eq. (7) that the lattice thermal conductivity can be determined if the mean free path for the phonon–phonon interaction in the crystal is known.

## 9.10 PHONON MEAN FREE PATH

Two main processes which determine the mean free path of phonons in solids are the phonon–phonon scattering and the geometrical scattering. If the forces between the atoms were purely harmonic, there would have been no mechanism for collision between different phonons and the mean free path would be determined solely by collision of phonon with the crystal boundary and lattice imperfections which constitutes the geometrical scattering. But, if the forces between atoms are anharmonic, there is collision between different phonons and the phonon–phonon scattering limits the value of mean free path. We discuss these two processes separately.

### 9.10.1 Phonon-Phonon Scattering

The anharmonic interaction between atoms leading to phonon–phonon scattering is the only mechanism in solids which may bring about the distribution of phonons locally into thermal equilibrium. Under the condition of thermal equilibrium, we can speak of phonons at one end of the crystal as being in thermal equilibrium at temperature $T_1$ and those at the other end in thermal equilibrium at temperature $T_2$. In thermal conductivity, we do not need only a way of limiting the mean free path $\lambda$, but also a way of establishing and equilibrium distribution of phonons because phonon collision with a lattice imperfection or crystal boundary will not, by themselves, establish thermal equilibrium. This is so because such collisions do not change the energy of individual phonons.

Although the theory of the effect of anharmonic coupling on the mean free path $\lambda$ is complicated, we can get a qualitative picture. We know that the mean free path of a particle is inversely proportional to the particle density, that is,

$$\lambda \propto \frac{1}{N_p} \qquad ...(8)$$

where $N_p$ is the phonon density.

Since at high temperature, the average energy of a vibrational mode of any frequency is the number of photons of frequency $v$ is $\frac{k_B T}{hv}$. Thus,

$$N_p \propto T \qquad ...(9)$$

as $k_B$, $h$ and $v$ are constants. Moreover, at temperatures well above the Debye temperature, the specific heat at constant volume, $C_v$, is independent of $T$.

Combining Eqs. (8) and (9), we conclude that

$$\lambda \propto \frac{1}{T} \quad \text{for} \quad T \gg \theta_D \qquad ...(10)$$

This result is in agreement with the experiment. At low temperature, however,

$$N_p \propto \left(\frac{T}{\theta_D}\right)^3 \qquad ...(11)$$

and

$$C_v \propto \left(\frac{T}{\theta_D}\right)^3 \qquad ...(12)$$

Hence, from Eq. (7),

$$K \propto \bar{c}\lambda C_v$$

$$\Rightarrow \quad K \propto \bar{c}\cdot\frac{1}{N_p}\cdot C_v \quad \text{[Refer to Eq. (8)]}$$

$$\Rightarrow \quad K \cdot \bar{c}\left(\frac{\theta_D}{T}\right)^3 \cdot \left(\frac{T}{\theta_D}\right)^3 \quad \text{[Refer to Eqs. (11) and (12)]}$$

$$\Rightarrow \quad K = \text{a constant} \qquad ...(13)$$

But, in reality, the thermal conductivity increases as temperature decreases. At lower temperatures, the amplitude of vibration and hence, the anharmonicity decreases, resulting in a rise in the value of thermal conductivity. But, the simple reasoning we are employing here cannot account for it.

Peierls considered that this problem in great detail and was able to account for the correct variation of thermal conductivity with temperature. Let us consider a solid in which a temperature gradient exists and assume that the phonon distribution in it is different from that exciting in thermal equilibrium. Then, the phonon–phonon collisions would tend to restore the equilibrium and the rate of this restoring process would be determined the thermal resistance. The kinematics of collisions between two phonons is described by the following selection rules:

$$\omega_1 + \omega_2 = \omega_3 \qquad ...(14)$$

and

$$\vec{k}_1 + \vec{k}_2 = \vec{k}_3 + \vec{G} \qquad ...(15)$$

Equation (14) expresses the principle of conservation of energy of phonons.[9] $\vec{G}$ in Eq. (15) is the reciprocal lattice vector which occurs in all momentum conservation laws in crystal lattice.

The processes of collisions in which $\vec{G} \neq 0$ are called *Umklapp processes* or *U* processes after the German 'flipping over'. Collisions in which $\vec{G} = 0$ are called *normal processes* or *N* processes. At temperature $T > \theta_D$, all phonons are excited because $k_B T > h\nu_{max}$.

### 9.10.2 Phonon–Mean Free Path

A substantial proportion of all phonon collisions will then be *U* processes with high momentum change in collisions.

We now proceed to estimate the contribution made by Umklapp process to the thermal conductivity at low temperatures, much less than the Debye temperature. At low temperatures, energy is low and hence, only the small *k*'s are important.

The average number of phonons in the low temperature range ($T \ll \theta_D$) is given by

$$N_p = \frac{1}{\exp.\left(\dfrac{h\nu}{k_B T}\right) - 1} \approx \exp.\left(-\frac{h\nu}{k_B T}\right) \approx \exp.\left(-\frac{\theta_D}{T}\right)$$

According to Eq. (8), $\lambda \propto (1/N_p)$ and, therefore,

$$\lambda \propto \exp.\left(\frac{\theta_D}{T}\right)$$

and $$K \propto \exp.\left(\frac{\theta_D}{T}\right), \text{ as } K \propto \lambda, \text{ from Eq. (7)}$$

Thus, the Umklapp process leads to the conclusion that thermal conductivity decreases exponentially with temperature in the low temperature region. In the high temperature region, it varies as $T^{-1}$. These results are in agreement with the experiments. *We can now conclude that the phonon mean free path which enters into the expression for thermal conductivity in Eq. (7) is the mean free path for Umklapp collisions between phonons and not for all collisions between phonons.*

## 9.11 THERMAL EXPANSION

The thermal expansion of solids is the most common effect of heat on solids. All solids increase in size when heated. Different solids expand to different extents for the same rise of temperature. In order to specify this expansion of solids, we often determine the quantity $\alpha$, called the coefficient of linear expansion, is defined by

$$\alpha = \frac{l_2 - l_1}{l_1(t_2 - t_1)} \qquad ...(1)$$

9 Two phonons of energy $\hbar\omega_1$ and $\hbar\omega_2$ may give rise to a third phonon of energy $\hbar\omega_3$.

where $l_1$ and $l_2$ are the lengths of a rod at temperatures $t_1$ and $t_2$ respectively, or

$$\alpha = \frac{l_t - l_0}{l_0 t} \equiv \frac{1}{l}\frac{dl}{dt} \qquad ...(2)$$

if $l_0$ is the length at 0°C and $l_t$ at a temperature t°C. This equation is the form in which the coefficient of linear expansion is often defined. One might ask why we do not choose to work with the other two coefficients, namely, the coefficient of superficial expansion β and the coefficient of volume expansion, ν. The answer lies in that β and ν are related to α through the relations:

$$\beta = 2\alpha \quad \text{and} \quad \nu = 3\alpha \quad \text{(for isotropic solids)} \qquad ...(3i)$$

$$\nu = \alpha_1 + \alpha_2 + \alpha_3 \text{ (for anisotropic solids)} \qquad ...(3ii)$$

(crystal are generally anisotropic solids)

where $\alpha_1$, $\alpha_2$, $\alpha_3$ represent three coefficients of linear expansion (also called principal coefficients of expansion) measured along the three mutually perpendicular directions, viz., the principal axes of direction. The quantity α is more fundamental in nature.

### 9.11.1 Origin of Thermal Expansion

A solid may be considered an assembly of oscillators, oscillating about a periodic array of equilibrium positions. So, as the temperature increases, the amplitude of their oscillation increases, becoming finally ~ 12% of the interatomic distance at the melting point of the solid. However, this fact alone is not sufficient to account for the thermal expansion, as this may or may not lead to an increase in the average interatomic distance corresponding to thermal extension.

If the oscillators are purely harmonic (i.e., potential energy, a parabolic function of position of an oscillator), the average positions of different oscillators would not be the functions of temperature (or energy) despite increase in amplitude, as is illustrated in Fig. 9.8 (a). No change in average interatomic distance would take place with energy increase, and no thermal expansion. On the other hand, if the oscillators are anharmonic, i.e., if the potential energy is an asymmetric function of position, the average positions of different oscillators would be the functions of temperature as illustrated in Fig. 9.8 (b). Here, the average interatomic distance would be greater than the zero temperature lattice constant, and thus, the thermal expansion is observed. In conclusion, we may say that the thermal expansion of solids is due mainly to the anharmonicity of the lattice oscillations.

We shall now find an expression for the coefficient of the thermal expansion, α. To start with, we shall first calculate the average value of $x$ as appearing in the expression:

$$U(x) = \underset{\text{(harmonic term)}}{ax^2} - \underset{\text{(anharmonic terms)}}{\delta x^2 - fx^4} \qquad ...(4)$$

This expression proximates the true potential energy of an atomic of oscillator as a function of position with the origin of position taken at $R_0$. The lattice constant at $T = 0$, $x$ is therefore a displacement representing the increase in lattice constant on the thermal expansion. Using the relation $\alpha = \frac{\partial \bar{x}}{\partial T}$, we calculate α, where $\bar{x}$ is the average value of $x$.

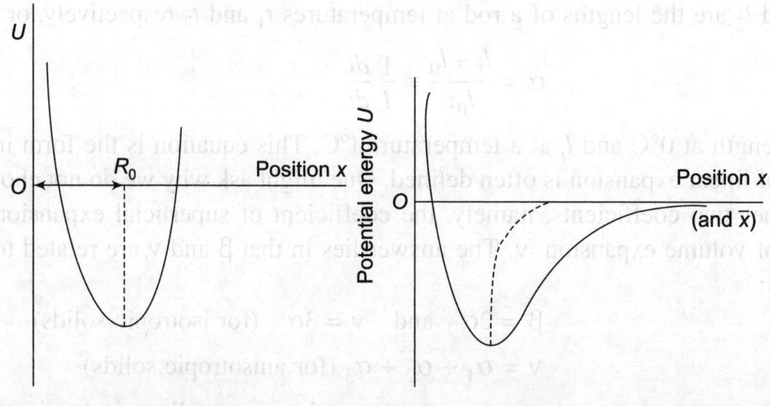


**Figure 9.8** Potential energy $U$ vs. $x$ curves (continuous) and potential energy vs. average position $\bar{x}$ curves (discontinuous) for a harmonic oscillator in (a) and for an anharmonic oscillator in (b). Note that in the average position is at the zero temperature lattice constant $R_0$ at all energies (a) (temperature) whereas in (b), it moves away from $R_0$ as the (a) temperature is increased. (b) leads to the thermal expansion while does not

It is to be noted here that the term involving $x^3$ steepens the left side and the flattens the right side of the potential energy curve; it therefore accentuates the effect of the repulsive forces of the atoms. The term involving $x^4$ flattens the bottom of the curve and represents softening of the vibrations at large amplitudes since it reduces the energy necessary to produce a displacement.

Using the Boltzmann distribution function, we have

$$\bar{x} = \frac{\int_{-\infty}^{+\infty} x \exp[-(ax^2 - gx^3 - fx^4)/kT]dx}{\int_{-\infty}^{+\infty} \exp[-(ax^2 - gx^3 - fx^4)/kT]dx} \quad ...(5)$$

where $k$ is Boltzmann constant.

If the anharmonic terms are small in comparison with $kT$, we may expand the exponential function in Eq. (5) as

$$\exp\left[-\frac{(ax^2 - gx^3 - fx^4)}{kT}\right] \to \exp\left(-\frac{ax^2}{kT}\right)\exp\left[\frac{gx^3 + fx^4}{kT}\right]$$

$$= \exp\left(-\frac{ax^2}{kT}\right)\left[1 + \frac{gx^3}{kT} + \frac{fx^4}{kT}\right]$$

This gives

$$\bar{x} = \frac{\int_{-\infty}^{+\infty} x \exp\left(-\frac{ax^2}{kT}\right)\left(x - \frac{gx^4}{kT} + \frac{fx^5}{kT}\right)dx}{\int_{-\infty}^{+\infty} \exp(-ax^2/(kT))\,dx} \quad ...(6)$$

In writing this, we have neglected the anharmonic terms of denominator. Using a table of standard integrals, we can write

$$\bar{x} = \frac{0 + \frac{3}{4}\frac{9}{kT}\sqrt{\frac{\pi k^2 T^5}{a^5}} + 0}{\sqrt{\frac{\pi kT}{a}}} = \frac{3}{4}\frac{8}{a^2} \cdot kT \qquad ...(7)$$

Thus, $$\alpha = \frac{\partial \bar{x}}{\partial T} = \frac{3}{4}\frac{8k}{a^2} \qquad ...(8)$$

According to the above calculations, we find that the thermal expansion is directly proportional to the temperature and α is independent of temperature. This is what actually has been observed experimentally for most solid but only at room temperatures and above. However, at low temperatures, α for most solids is observed to become much smaller than the above predicted value and indeed approaches zero as $T \to 0$. Thus, the above expression is not a perfect one. It is therefore apparent that the classical picture of atomic oscillators used above breaks down and we should consider atoms of the crystal as quantum oscillators. Then, we have

$$\bar{x} = \frac{3g}{4a^2} \frac{\hbar\omega_0}{\exp\left[\frac{\hbar\omega_0}{kT}\right] - 1} \qquad ...(9)$$

instead of Eq. (7). At high temperatures, this reduces to Eq. (7) but for $T \to 0$, we have

$$\bar{x} = \frac{3g}{4a^2} \hbar\omega_0 \exp\left(-\frac{\hbar\omega_0}{kT}\right) \qquad ...(10)$$

Then $$\alpha = \frac{\partial \bar{x}}{\partial T} = \frac{3gk}{4a^2}\left(\frac{\hbar\omega_0}{kT}\right)^2 \exp\left(\frac{-\hbar\omega_0}{kT}\right) \qquad ...(11)$$

This gives results which are in reasonably good agreement with experiments over a wide range of low temperatures. Thus, the result Eq. (11) combined with Eq. (7) gives an expression for α which may be taken as satisfactory for describing the experiments on thermal expansion of crystals.

### 9.11.2 Gruneisen Relation

On the basis of oscillator model of the solid, Gruneisen predicted that the three important physical properties of a solid, viz., the linear coefficient of thermal expansion α, the specific heat at constant volume $C_V$ and the compressibility $K$ are linked together through the relation

$$\gamma = \frac{3\alpha V}{C_V K} \qquad ...(12)$$

where γ is a constant independent of temperature, called the Gruneisen constants. $V$ is the volume of the solid under consideration. In other words, the relation predicts that at a constant pressure, the coefficient of thermal expansion $\alpha$ must be proportional to $C_v$ at all temperatures and thus $\alpha$ and $C_v$ should have the same temperature dependence. The relation has been found experimentally valid for a number of solids and the value of $\gamma$ generally lies between 1 and 2.

To understand the physical basis of the Gruneisen relation, we examine the consequences of the assumption that the frequency of lattice vibration of specified wave vector change with the volume of the solid; this is an anharmonic effect. For simplicity, we assume that a given change of volume gives rise to the same relative change of frequency of every mode.

Now, since any change in volume of a solid must cause a change in its energy, we therefore apply energy, considerations. The free energy may be written approximately,

$$F = E_0(V) + F_D(T, V) \qquad ...(13)$$

where $E_0$ $(V)$ is the internal energy at 0°K and $F_D$ is the contribution in the Debye approximation of the lattice vibrations to the free energy. This latter contribution is an effect of the anharmonic interactions between different normal mode vibrations of the lattice. (Note that in the harmonic model of lattice vibrations, $F_D$ is the independent of the volume but depends on the number of atoms in the solid.)

Now, in the spirit of the Debye approximation, we may suppose that the dependence of $F_D$ on the volume $V$ is adequately described by specifying the dependence of $\theta_D$ (the Debye temperature) on $V$.

Then,
$$\left(\frac{\partial F}{\partial V}\right)_T = \left(\frac{\partial E_0}{\partial V}\right)_T + \left(\frac{\partial F_D}{\partial \theta_D}\frac{\partial \theta_D}{\partial V}\right)_T \qquad ...(14)$$

Employing the thermodynamic relation

$$P = -\left(\frac{\partial F}{\partial V}\right)_T \qquad ...(15)$$

we get
$$P = \left(\frac{\partial F}{\partial V}\right)_T = \left(\frac{\partial E_0}{\partial V}\right)_T + \left(\frac{\partial F_D}{\partial \theta_D}\frac{\partial \theta_D}{\partial V}\right)_T \qquad ...(16)$$

Now, we know the Debye internal energy $E_D$ due to lattice vibrations involves $T$ times a function of $(\theta_D/T)$. Employing the thermodynamics relation

$$E = F - T\left(\frac{\partial F}{\partial T}\right)_V \qquad ...(17)$$

it is then clear that $F_D$ must also be of the same form:

$$F_D = Tf\left(\frac{\theta_D}{T}\right) \qquad ...(18)$$

so that
$$\frac{\partial F_D}{\partial \theta_D} \equiv f'\left(\frac{\theta_D}{T}\right) = \frac{E_D}{\theta_D} \qquad ...(19)$$

Substituting this in Eq. (16) gives

$$P = -\left(\frac{\partial E_0}{\partial V}\right) - \frac{E_D}{\theta_D}\left(\frac{\partial \theta_D}{\partial V}\right)$$

$$= -\left(\frac{\partial E_0}{\partial V}\right) + \gamma\left(\frac{E_D}{V}\right) \quad ...(20)$$

where

$$\gamma = -\left(\frac{V}{\theta_D}\right)\left(\frac{\partial \theta_D}{\partial V}\right)$$

$$= -\frac{d(\log \theta_D)}{d(\log V)} \quad \left(\because \frac{d(\log x)}{dx} = \frac{1}{x}\right) \quad ...(21)$$

Equation (20) referred to as the Debye equation of state of solids. To identify the parameter $\gamma$, we first differentiate Eq. (20) w.r.t. $T$, and we obtain

$$\left(\frac{\partial P}{\partial T}\right)_V = \gamma\left(\frac{\partial E_D}{\partial T}\right)\frac{1}{V} = \gamma\frac{C_v}{V} \quad ...(22)$$

Next, we introduce the linear expansion coefficient $\alpha$ recalling that it is one-third of the volume expansion coefficient, i.e.,

$$\alpha = \frac{1}{3V}\left(\frac{\partial V}{\partial T}\right) = \frac{1}{3V}\frac{(\partial P/\partial T)_V}{(\partial P/\partial T)_T} = \frac{K}{3}\left(\frac{\partial P}{\partial T}\right)_V = \frac{\gamma C_v K}{3V} \quad ...(23)$$

where $K$ is the compressibility. The relation $\alpha = \frac{\gamma C_v K}{3V}$ is the same as the *Gruneisen relation.* Thus, constant $\gamma$ defined by Eq. (22) is therefore the Gruneisen constant.

It is thus clear that if $w_k$ is an eigenfrequency of the solid, $-d(\log w_k)/d(\log V) = \gamma_k$ for this mode. But in view of our earlier assumption, $\gamma$ is independent of $K$ and therefore has the same value for every mode. Thus, the constant $\gamma$ offers a convenient way to discuss the effects of anharmonicity in solids. It is a dimensionless constant.

### 9.11.3 Experimental Evidence for Anharmonicity

(i) To demonstrate the anharmonic effect, Shiren described an experiment in which a beam of longitudinal phonons of frequency 9.20 GHz interacts in a mgO crystal with a parallel beam of longitudinal phonons at 9.18 GHz. The interaction of the two beams produced are third beam of longitudinal phonons at 9.20 + 9.18 = 18.38 GHz.

(ii) Rollins carried out an experiment in which to narrow ultrasonic beams are allowed to interact and observe that a third beam is generated at the volume of interaction only when energy and wave vector are conserved between the three beams.

## 9.12 COMPREHENSIBILITY OF THE OSCILLATOR MODEL

In case of a solid, the compressibility is defined as

$$\chi = -\left(\frac{V - V_0}{V_0}\right)\frac{1}{p} \quad ...(1)$$

where $V - V_0$ is the difference of volume due to the application of pressure $p$. At absolute zero, the work done to compress the solid is

$$W = \int_{V_0}^{V} p\,dv = \frac{1}{\chi}\int_{V_0}^{V}\left(\frac{V - V_0}{V_0}\right)dv = \frac{1}{2}\frac{(V - V_0)^2}{\chi V_0} \quad ...(2)$$

The work done can also be considered in terms of a increase in energy. Let $(\varepsilon - \varepsilon_0)$ be the increase in energy of each atom above equilibrium, then work done for $N$ atom is $N(\varepsilon - \varepsilon_0)$ using Taylor's theorem, we have

$$N(\varepsilon - \varepsilon_0) = N(V - V_0)\left(\frac{\partial \varepsilon}{\partial V}\right)_{V=V_0} + \frac{1}{2}N(V - V_0)^2\left(\frac{\partial^2 \varepsilon}{\partial V^2}\right)_{V=V_0} + ... \quad ...(3)$$

where $V_0$ is the equilibrium volume, i.e., the volume at which the crystal energy is minimum. Hence, at equilibrium volume

$$N(V - V_0)\left(\frac{\partial \varepsilon}{\partial V}\right)_{V=V_0} = 0$$

$$\therefore \quad N(\varepsilon - \varepsilon_0) = \frac{1}{2}N(V - V_0)^2\left(\frac{\partial^2 \varepsilon}{\partial V^2}\right)_{V=V_0} \quad ...(4)$$

Comparing Eqs. (2) and (4), we get

$$\frac{1}{2}\frac{1}{\chi}\frac{(V - V_0)^2}{V_0} = \frac{1}{2}N(V - V_0)^2\left(\frac{\partial^2 \varepsilon}{\partial V^2}\right)_{V=V_0}$$

or

$$\frac{1}{\chi} = NV_0\left(\frac{\partial^2 \varepsilon}{\partial V^2}\right)_{V=V_0} \quad ...(5)$$

## 9.13 ANHARMONIC TURN IN THE POTENTIAL/THE GRUNEISEN PARAMETER

We wish to address the topic of thermal expansion, which would not exist without anharmonic terms in the potential (for then the average position of the atoms would be independent of their amplitude of vibration). Other effects of the an harmonic terms are the existence of finite thermal conductivity and the increase of the specific heat beyond the classical Dulong and Petit value at high temperature. We wish to obtain an approximate expression for the coefficient of thermal expansion (which would vanish if there were no anharmonic terms).

We first derive an expression for the free energy of the latest due to the thermal vibrations. The free energy is given by

$$F_L = -k_B T \ln Z, \qquad ...(1)$$

where $Z$ is the partition function. The partition function is given by

$$Z = \sum_{\{n\}} \exp(-\beta E_{\{n\}}),\ \beta = \frac{1}{k_B T}, \qquad ...(2)$$

where $$E_{\{n\}} = \sum_{k,j}\left(n_k + \frac{1}{2}\right)\hbar\omega_j(\boldsymbol{k}) \qquad ...(3)$$

in the harmonic approximation and $\omega_j(\boldsymbol{k})$ labels the frequency of the different modes at wave vector $\boldsymbol{k}$. Each $n_k$ can vary from 0 to . The partition function can be written as

$$Z = \sum_{n_1}\sum_{n_2} ... \exp(-\beta E_{\{n_k\}})$$

$$= \prod_{k,j}\prod_{n_k}\exp\left[-\beta\left(n_k + \frac{1}{2}\right)\hbar\omega_j(\boldsymbol{k})\right]$$

$$= \prod_{k,j}\exp[-\hbar\omega_j(\boldsymbol{k})/2]\prod_{n_k}\exp[-\beta_{n_k}\hbar\omega_j(\boldsymbol{k})],$$

which leads to

$$F_L = k_B T \sum_{k,j} \ln\left[2\sinh\left(\frac{\hbar\omega_j(\boldsymbol{k})}{2k_B T}\right)\right]. \qquad ...(4)$$

We must add to this the free energy at absolute zero due to the increase in elastic energy if the crystal changes its volume by $\Delta V$. We call this term $U_0$.[10]

$$F = k_B T \sum_{k,j} \ln\left[2\sinh\left(\frac{\hbar\omega_j(\boldsymbol{k})}{2k_B T}\right)\right] + U_0. \qquad ...(5)$$

We calculate the volume coefficient of thermal expansion $\alpha$

$$\alpha = \frac{1}{V}\left(\frac{\partial V}{\partial T}\right)_P \qquad ...(6)$$

But, $$\left(\frac{\partial V}{\partial T}\right)_P\left(\frac{\partial P}{\partial V}\right)_T\left(\frac{\partial T}{\partial P}\right)_V = -1.$$

The isothermal compressibility is defined as

$$\kappa = -\frac{1}{V}\left(\frac{\partial V}{\partial P}\right)_T, \qquad ...(7)$$

then we have $$\alpha = \kappa\left(\frac{\partial P}{\partial T}\right)_V. \qquad ...(8)$$

---

[10] $U_0$ is included for completeness, but we end up only using a vanishing temperature derivative so it could be left out.

But $$P = -\left(\frac{\partial F}{\partial V}\right)_T,$$

so $$P = -\frac{\partial U_0}{\partial V} = k_B T \sum_{k,j} \coth\left(\frac{\hbar\omega_j(\boldsymbol{k})}{2k_B T}\right)\frac{\hbar}{2k_B T}\frac{\partial\omega_j(\boldsymbol{k})}{\partial V}. \qquad ...(9)$$

The anharmonic terms come into play by assuming the $\omega_j(k)$ depend on volume. Since the average number of photons in the mode $k$, $j$ is

$$\bar{n}_j(\boldsymbol{k}) = \frac{1}{\exp\left(\frac{\hbar\omega_j(\boldsymbol{k})}{k_B T}\right) - 1} = \frac{1}{2}\left[\coth\left(\frac{\hbar\omega_j(\boldsymbol{k})}{k_B T}\right) - 1\right]. \qquad ...(10)$$

Thus, $$P = \frac{\partial U_0}{\partial V} - \sum_{k,j}\left(\bar{n}_j(\boldsymbol{k}) + \frac{1}{2}\right)\hbar\frac{\partial\omega_j(\boldsymbol{k})}{\partial V}. \qquad ...(11)$$

We define the Gruneisen parameter for the mode $\boldsymbol{k}$, $j$ as

$$\gamma_j(\boldsymbol{k}) = \frac{V}{\omega_j(\boldsymbol{q})}\frac{\partial\omega_j(\boldsymbol{k})}{\partial V} = \frac{\partial \ln \omega_j(\boldsymbol{k})}{\partial \ln V}. \qquad ...(12)$$

Thus, $$P = -\frac{\partial}{\partial V}\left[U_0 + \sum_{k,j}\frac{1}{2}\hbar\omega_j(\boldsymbol{k})\right] + \sum_{k,j}\bar{n}_j\frac{\hbar\omega_j(\boldsymbol{k})\gamma_j}{V}. \qquad ...(13)$$

However, the latest internal energy is in the harmonic approximation)

$$U = \sum_{k,j}\left(\bar{n}_j(\boldsymbol{k}) + \frac{1}{2}\right)\hbar\omega_j(\boldsymbol{k}). \qquad ...(14)$$

So, $$\frac{\partial U}{\partial V} = \sum_{k,j}\hbar\omega_j(\boldsymbol{k})\frac{\partial\bar{n}_j(\boldsymbol{k})}{\partial T}, \qquad ...(15)$$

$$c_v = \frac{1}{V}\frac{\partial U}{\partial V} = \frac{1}{V}\sum_{k,j}\hbar\omega_j(\boldsymbol{k})\frac{\partial\bar{n}_j(\boldsymbol{k})}{\partial T} = \sum c_{v_j}(\boldsymbol{k}), \qquad ...(16)$$

which defines a specific heat for each mode. Since the first term of $P$ in Eq. (13) is independent of $T$ at constant $V$, and using

$$\alpha = \kappa\left.\frac{\partial P}{\partial T}\right)_V$$

we have $$\alpha = \kappa\frac{1}{V}\sum_{k,j}\hbar\omega_j(\boldsymbol{k})\gamma_j(\boldsymbol{k})\frac{\partial\bar{n}_j(\boldsymbol{k})}{\partial T}. \qquad ...(17)$$

Thus, $$\alpha = \kappa\sum_{k,j}\gamma_j(\boldsymbol{k})c_{v_j}(\boldsymbol{k}). \qquad ...(18)$$

Let us define the overall Gruneisen parameter $\gamma_T$ as the average Gruneisen parameter for mode $\boldsymbol{k}$, $j$ weighted by the specific heat for that mode. Then by Eqs. (12) and (16) we have

$$c_v\gamma_T = \sum_{k,j}\gamma_j(\boldsymbol{k})c_{v_j}(\boldsymbol{k}). \qquad ...(19)$$

We then find $$\alpha = \kappa\gamma_T c_v. \quad ...(20)$$

If $\gamma_T$ (the Gruneisen parameter) were actually a constant $\alpha$ would tend to follow the changes of $c_v$, which happens for some materials.

From thermodynamics

$$c_P = c_V + \frac{\alpha^2 T}{\kappa}, \quad ...(21)$$

so $c_p = c_v\,(1 + \gamma\alpha T)$ and $\gamma$ is often between 1 and 2.

**Table 9.1** Gruneisen Constants

| *Temperature* | *LiF* | *NaCl* | *KBr* | *KI* |
|---|---|---|---|---|
| 0 K | 1.7 0.05 | 0.9 0.03 | 0.29 0.03 | 0.28 0.02 |
| 283 K | 1.58 | 1.57 | 1.49 | 1.47 |

## EXERCISE

9.1 Explain briefly the Dulong and Petit's law at low temperatures.

9.2 Derive an expression for heat capacity of solids on the basis of Einstein's model. What are the chief merits and demerits of this model?

9.3 State the assumptions made in the Debye's model of the heat capacity of solids.

9.4 Describe the salient features of Debye's theory of heat capacity of solids. How it explains the decrease in heat capacity of solids at low temperatures.

9.5 At what temperature does the electron heat capacity become larger than the phonon heat capacity? Express this temperature in terms of the Debye temperature and the electron density.

9.6 Derive an expression for phonon density of states.

9.7 Explain thermal expansion in solids and origin of thermal expansion. Derive the Gruneisen relation.

9.8 Derive an expression for lattice thermal conductivity in terms of specific heat $C_v$ and mean free path, $\lambda$.

CHAPTER

# 10

# Theory of Metals and Fermi Surfaces

## 10.1 FREE ELECTRONS IN A METAL AS AN ELECTRON GAS

In a metal, the atoms are packed together with their centres only a few angstroms apart and atoms are large compared to their spacing. Consequently, mean free path between electron–atom collisions is very small. As a result, if a lattice is perfect, the electrons are free to move through the crystals smoothly (see Drude's theory). The electron can pump into one atom or another almost immediately, as if the motion is in vacuum. (This allows metals to conduct electricity easily and has also permitted the development of many practical devices.)

## 10.2 THE FREE ELECTRON MODEL

An electron is a metal, or for the matter in any solid, finds itself in the field of all nuclei and all other electrons. The potential energy for such an electron may therefore be expected to be periodic, the periodicity being that of the lattice. In the model employed by Sommerfeld, however, it is assumed that the "free" electrons, i.e., those giving rise to conductivity, find themselves in a potential which is constant everywhere inside the metal.[1] Since one does not observe electron emission from metals at room temperature, it seems evident that the potential energy of an electron at rest inside the metal must be lower than that of an electron at rest outside the metal. This is confirmed by relatively simple theoretical arguments.[2] The change in potential energy of an electron $E_s$ as one crosses the metal-vacuum boundary may, for a number of problems, be considered abrupt (see Fig. 10.1). For some problems, however, it is necessary to consider the variation of potential at the surface in some more detail. Then one thus arrives at a physical model in which the interior of the metal is represented by a potential energy box of depth $E_s$ as indicated in Fig. 10.1; the energy of an electron at rest outside the

[1] A. Sommerfeld, *Z. Physik*, **47,** 1 (1928); see also the article by A. Sommerfeld, and H. Bethe, *Handbuch der Physik*, Vol 24/2.

[2] See, for example, H. Fröhlich, *Elektronen Theorie der Metalle*, Springer, Berlin, 1936, p. 11.

metal is used as a reference and is commonly referred to as the vacuum level.

It may be of interest to note that $E_s$ may be determined experimentally from electron diffraction experiments with slow electrons (a few hundred eV). An electron impinging on the metal from the outside with an initial energy $E_0$ gains an amount $E_s$ upon entering the metal. Thus, from a knowledge of the lattice structure and $E_0$, it is possible to determine $E_s$. For nickel one has found in this way $E_s$ = 14.8 eV. In general, $E_s$ is of the order of 10 eV.

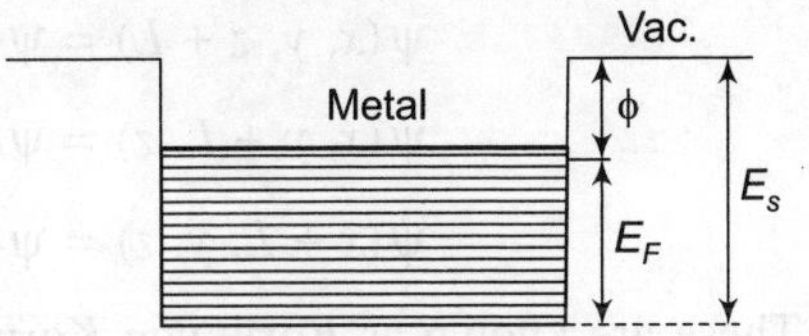


**Figure 10.1** The Sommerfeld model. $E_s$ is the energy difference between an electron at rest inside the metal and one at rest is vacuum. At $T$ = 0, all energy levels upto $E$ are filled, all higher ones are empty, the work function $\phi = E_s - E_F$

In the Sommerfeld model, the free electrons are assumed to be the valence electrons of the composing atoms. Thus, the alkali metals are assumed to contain one free electron per atom; aluminum supposedly has three free electrons per atom.

## 10.3 GROUND STATE OF THE ELECTRON GAS (THE ALLOTTED WAVE VECTORS IN $\vec{k}$ - SPACE)

We are interested in the ground state properties of the $N$ electrons confined to a volume $V$, in the independent electron approximation (i.e., no mutual interaction).

A single electron can be described by a wave function $\psi(\vec{r})$ and associated with a level of energy $E$ satisfies the time independent Schrödinger equation

$$-\frac{\hbar^2}{2m}\left(\frac{\partial^2}{\partial x^2}+\frac{\partial^2}{\partial y^2}+\frac{\partial^2}{\partial z^2}\right)\psi(\vec{r}) - -\frac{\hbar^2}{2m}\nabla^2\psi(\vec{r})$$

$$= E\,\psi(\vec{r}) \qquad \text{...(1)}$$

We may consider the electrons in a cubic metal of side $L = V^{1/3}$. The electron is so confined to the volume $V$ (by the attraction of the ions) that the bulk properties are unaffected by the detailed configuration of the surface.

Next, we must have to append a boundary condition to Eq. (1). One may think $\psi(\vec{r})$ to vanish whenever $\vec{r}$ is on the surface of the cube. But it is not sought for, as it would lead to standing wave solutions of Eq. (1). For transport of charge and energy, we should have running waves. We can do this by imagining each face of the cube to be joined to the face opposite to it, so that an electron coming to the surface is not reflected back in, but leaves the metal, simultaneously re-entering at a corresponding point on the opposite surface. Thus, in one-dimensional case, we should simply replace the line from $O$ to $L$, by the circle of circumference, $L$. This results in the boundary condition

$$\psi(x + L) = \psi(x)$$

and the generation to a three-dimensional cube gives

$$\left.\begin{aligned}\psi(x, y, z+L) &= \psi(x, y, z)\\ \psi(x, y+L, z) &= \psi(x, y, z)\\ \psi(x+L, y, z) &= \psi(x, y, z).\end{aligned}\right| \qquad ...(2)$$

These are known as *Born–von Karman* (or *periodic*) *boundary conditions*.

We can solve Eq. (1) subject to the boundary conditions (2). It can be verified (by differentiation) that a solution, neglecting the boundary condition is

$$\psi_{\vec{k}}(\vec{r}) = \frac{1}{\sqrt{V}} e^{i\vec{k}\cdot\vec{r}} \qquad ...(3)$$

with energy $$E(\vec{k}) = \frac{\hbar^2 k^2}{2m} \qquad ...(4)$$

where $\vec{k}$ is any position independent vector. We have chosen the normalization constant in Eq. (3) so that the probability of finding the electron *somewhere* in the whole volume $V$ is unity

$$\int d\vec{r}\, |\psi(\vec{r})|^2 = 1 \qquad ...(5)$$

To see the significance of the vector $\vec{k}$, note that the level $\psi_{\vec{k}}(\vec{r})$ is an eigenstate of the momentum operator

$$\hat{\vec{p}} = \frac{\hbar}{i}\frac{\partial}{\partial \vec{r}} = \frac{\hbar}{i}\nabla, \qquad ...(6)$$

$$\left(\because \hat{p}_x = \frac{\hbar}{i}\frac{\partial}{\partial x}\right).$$

with eigenvalue $\vec{p} = \hbar\,\vec{k}$

$$\because \quad \frac{\hbar}{i}\frac{\partial}{\partial \vec{r}}\, e^{i\vec{k}\cdot\vec{r}} = \hbar\, k\, e^{i\vec{k}\cdot\vec{r}} \qquad ...(7)$$

Since a particle in an eigenstate of an operator has a definite value of the corresponding observable given by an eigenvalue, an electron in the level $\psi_{\vec{k}}(\vec{r})$ has a definite momentum $\hbar\vec{k}$

$$\vec{p} = \hbar\,\vec{k} \qquad ...(8)$$

and a velocity $$\vec{v} = \frac{\hbar\,\vec{k}}{m} \qquad ...(9)$$

In view of this the energy (Eq. (4)) can be written in classical form

$$E = \frac{p^2}{2m} = \frac{1}{2}\,\mathrm{mV}^2 \qquad ...(10)$$

We can also interpret $\vec{k}$ as a wave vector. The plane wave $e^{i\vec{k}\cdot\vec{r}}$ is constant in any plane perpendicular to $\vec{k}$ (since such planes are defined by the equation $\vec{k}\cdot\vec{r}$ = constant) and it is periodic along lines parallel to $\vec{k}$, with wavelength

$$\lambda = \frac{2\pi}{k} \qquad ...(11)$$

known as the de Broglie wavelength. We now invoke the boundary condition (Eq. (2)).

This permits only certain discrete values of $\vec{k}$, since Eq. (2) will be satisfied by the general wave function (Eq. 3) only if

$$e^{ik_xL} = e^{ik_yL} = e^{ik_zL} = 1 \qquad ...(12)$$

Since $e^z = 1$ only if $z = 2\pi$ in where $n$ is an integer, the components of the wave vector $\vec{k}$ must be of the form

$$k_x = \frac{2\pi n_x}{L}, k_y = \frac{2\pi n_y}{L}, k_z = \frac{2\pi n_z}{L} \qquad ...(13)$$

$$(n_x, n_y, n_z \text{ integers}).$$

Thus, in a 3-dimensional space with Cartesian axes $k_x$, $k_y$, $k_z$ (known as $k$-space) the allowed wave vectors are those whose coordinates along the three axes are given by integral multiples of $2\pi/L$. This is illustrated (in two dimensions) in Fig. 10.2.

A region on $k$-space of volume $\Omega$ will contain

$$\frac{\Omega}{(2\pi/L)^3} = \frac{\Omega V}{8\pi^3} \qquad ...(14)$$

allowed values of **k**, or, equivalently, that the number of allowed $k$-values per unit volume of $k$-space (also known as the $k$-space density of levels) is just

$$\frac{V}{8\pi^3}. \qquad ...(15)$$

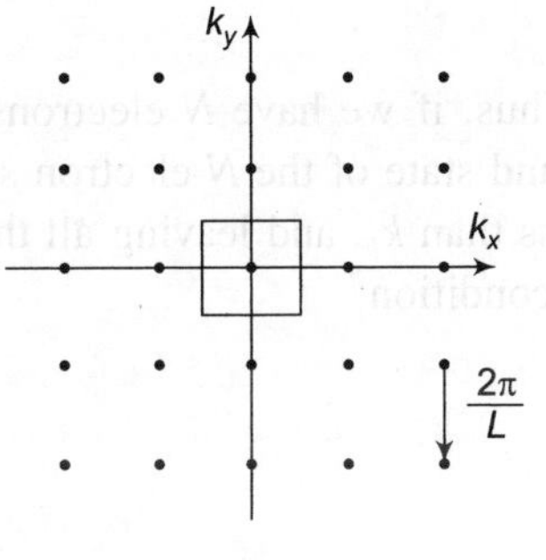


**Figure 10.2** Points in a two-dimensional $k$-space of the form $k_x = 2\pi n_x/L$, $k_y = 2\pi n_y/L$. Note that the area per point is just $(2\pi/L)^2$. In $d$ dimensions the volume per point is $(2\pi/L)^d$

In practice we shall deal with $k$-space regions so large (~ $10^{22}$ points) and so regular (typically spheres) that to all intense and purposes, Eqs. (14) and (15) can be regarded as exact. We shall begin to apply these important counting formulae shortly.

Because we assume the electrons are non-interacting, we can build the $N$-electron ground state by placing electrons into the allowed one-electron levels we have just found. The Pauli exclusion principle plays a vital role in this construction (as it does in building the states of many electron atoms): we may place at least one electron in each single electron level. The one-electron levels are specified by the wave vectors **k** and by the projection of the electron's spin along an arbitrary axis which can take either of the two values $\hbar/2$ or $-\hbar/2$. Therefore, associated with each allowed wave vector **k** are *two* electronic levels, one for each direction of the electron's spin.

Thus, in building up the $N$-electron ground state we begin by placing two electrons in the one-electron level, $\mathbf{k} = 0$, which has the lowest possible one-electron energy, $\varepsilon = 0$. We then continue to add electrons, successively filling the one-electron levels of lowest energy that are not already occupied. Since the energy of a one-electron level is directly proportional to the square of its wave vector (see Eq. (4)), when $N$ is enormous the occupied region will be indistinguishable from a sphere.[3] The radius of this sphere is called $k_F$ ($F$ for Fermi), and its volume $\Omega$ is $4\pi k_F^3/3$. According to Eq. (14) the number of allowed values of **k** within the sphere is

$$\left(\frac{4\pi k_F^3}{3}\right)\left(\frac{V}{8\pi^3}\right) = \frac{k_F^3}{6\pi^2}V. \qquad ...(16)$$

Since each allowed $k$-value leads to two one-electron levels (one for each spin value), in order to accommodate $N$ electrons, we must have

$$N = 2\cdot\frac{k_F^3}{6\pi^2}V = \frac{k_F^3}{3\pi^2}V. \qquad ...(17)$$

Thus, if we have $N$ electrons in a volume $V$ (i.e., electronic density $n = N/V$), then the ground state of the $N$-electron system is formed by occupying all single-particle levels with $k$ less than $k_F$, and leaving all those with $k$ greater than $k_F$ unoccupied, where $k_F$ is given by the condition:

$$\boxed{n = \frac{k_F^3}{3\pi^2}.} \qquad ...(18)$$

This free and independent electron ground state is described by some rather unimaginative nomenclature:

The sphere of radius $k_F$ (the *Fermi wave vector*) containing the occupied one-electron levels is called the *Fermi sphere*.

The surface of the Fermi sphere, which separates the occupied from the unoccupied levels is called the *Fermi surface*. (We shall see that the Fermi surface is one of the fundamental constructions in the modern theory of metals, in general, is not spherical.)

The momentum $\hbar k_F = p_F$ of the occupied one-electron levels of highest energy is known as the *Fermi momentum*; their energy, $\varepsilon_F = \hbar^2 k_F^2/2m$ is the *Fermi energy*, and their velocity, $v_F = p_F/m$, is the *Fermi velocity*. The Fermi velocity plays a role in the theory of metals comparable to the thermal velocity, $v = (3k_BT/m)^{1/2}$, in a classical gas.

## 10.4 ELECTRICAL CONDUCTIVITY AND Ohm's LAW

$$E = \frac{\hbar^2 k^2}{2m_e} = \frac{p^2}{2m_e} \qquad ...(1)$$

where $m_e$ is the electronic mass and the momentum $\vec{p} = \hbar\vec{k}$.

[3] If it were not spherical it would not be the ground state, for we could then construct a state of lower energy by moving the electrons in levels farthest away from $\mathbf{k} = 0$ into unoccupied levels closer to the origin.

In the presence of an external electric field $\vec{\varepsilon}$, the force exerted on the electron is $-e\vec{\varepsilon}$. But the force is also the rate of change of momentum.

Therefore, $$-e\vec{\varepsilon} = \frac{dp}{dt} = \frac{d(\hbar k)}{dt} = \hbar\left(\frac{dk}{dt}\right) \qquad \text{...(2)}$$

Suppose the field were applied at time $t = 0$ to the electron gas filling the *Fermi sphere*,[4] whose centre lies at the origin of $k$-space; after and infinitesimal time $dt$ the sphere will be displaced so as to have a new origin/centre at $\delta k$ such that,

$$\delta\vec{k} = \frac{-e\vec{\varepsilon}\,dt}{\hbar} \qquad \text{...(3)}$$

Because of the collisions with imprefections and lattice waves (called phonons), the displaced sphere may be maintained in a steady state in the presence of the electric field (Fig. 10.3). If the average collision time is $\tau$, the centre of the Fermi sphere will be displaced to $\Delta k$ such that,

$$\Delta k = -\frac{e\varepsilon\tau}{\hbar}$$

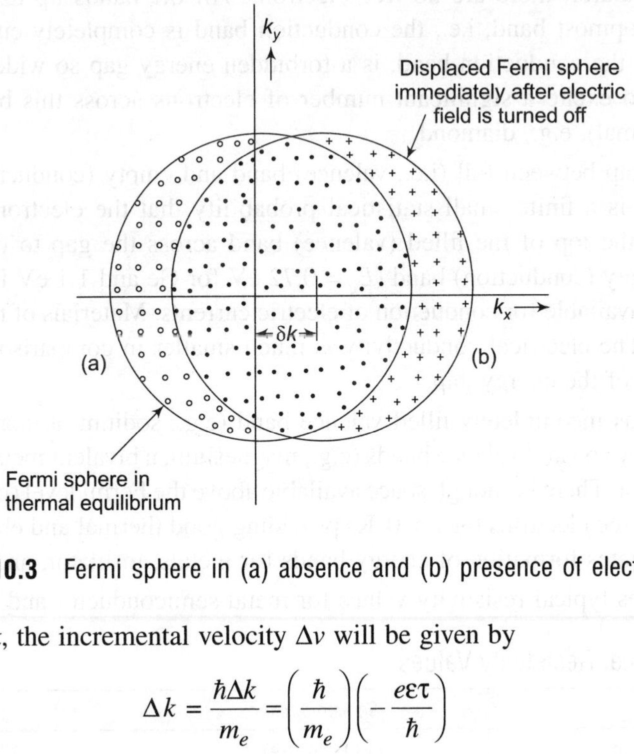


**Figure 10.3** Fermi sphere in (a) absence and (b) presence of electric field

Since $m_e v = \hbar k$, the incremental velocity $\Delta v$ will be given by

$$\Delta k = \frac{\hbar\Delta k}{m_e} = \left(\frac{\hbar}{m_e}\right)\left(-\frac{e\varepsilon\tau}{\hbar}\right)$$

4 In the ground state of a system containing free electrons, the occupied states can be represented as points inside a sphere in $k$-space. (see C. Kittel, *Introduction to Solid State Physics*, Wiley Eastern Limited, New Delhi 1971).

$$= \frac{e\varepsilon\tau}{m_e} \qquad ...(4)$$

If the density of electrons, i.e., the number of electrons per unit volume, is $n$, the current density $J$ will be given by

$$J = n(-e)\Delta v$$

$$= \frac{ne^2\varepsilon\tau}{m_e}$$

$$= \sigma\varepsilon \text{ (Ohm's law)} \qquad ...(5)$$

where $\sigma$ is the electrical conductivity.

The free electron theory (of metals) gives us a great deal of insight into several of the physical properties of metals; yet, there are some other properties which cannot be explained by this theory.

## 10.5 DISTINCTION BETWEEN INSULATOR, SEMICONDUCTOR AND METAL

In case of an insulator, there are no free electrons. All the bands up to the valance band are full and the topmost band, i.e., the conduction band is completely empty. Between the valance band and the conduction band, is a forbidden energy gap so wide ($E_g > 5$ eV) that it is impossible to excite a significant number of electrons across this by ordinary means (electrical or thermal), e.g., diamond.

If the energy gap between full (i.e., valence) band and empty (conduction) band is quite small, then there is a finite small statistical probability that the electrons will be excited from states near the top of the filled (valence) band across the gap to the states near the bottom of the empty (conduction) band ($E_g \simeq 0.72$ eV for Ge and 1.1 eV for Si), thus, a few electrons will be available for conduction of electric currents. Materials of this type are called semiconductors. The electrical conductivity is much smaller in comparison to metals and it is also a function of the energy gap.

If a material has incompletely filled valence band (e.g., sodium, a monovalent metal) or overlapping conduction and valence bands (e.g., magnesium, a bivalent metal), then it behaves as a good conductor. There is enough space available above the Fermi level (partially full states) for movement of free electrons (at $T > 0$°K) providing good thermal and electrical behaviour. Figure 10.4 illustrates formation of energy bands for metals, insulator, and semiconductor.

Table 10.1 gives typical resistivity values for metal semiconductor and insulator.

**Table 10.1** Typical Resistivity Values

| *Metal* | *Semiconductor* | *Insulator* |
|---|---|---|
| $10^{-6}$ Ω-cm | 50 Ω-cm (Ge) | $10^{12}$ Ω-cm |
| (copper) | $50 \times 10^3$ Ω-cm (Si) | (mica) |

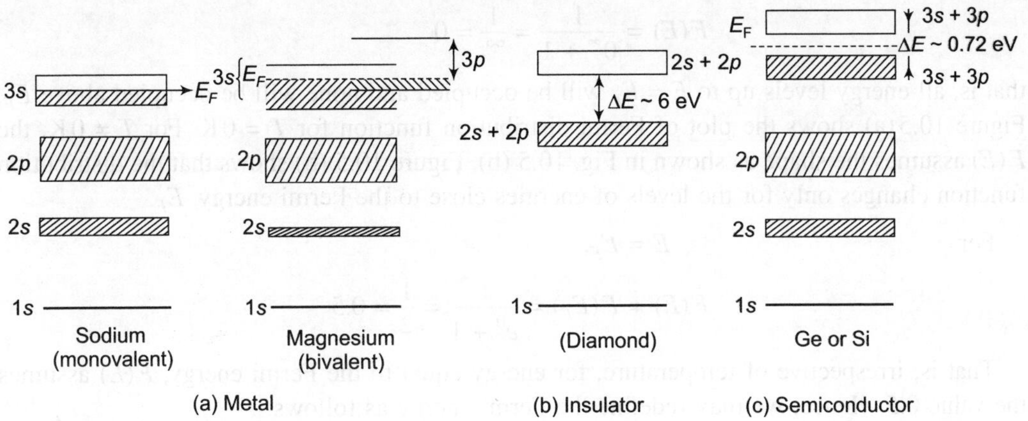


**Figure 10.4** Schematic illustration of formation of energy bands for (a) metal (b) insulator and (c) semiconductor

## 10.6 CONCEPT OF FERMI ENERGY AND FERMI LEVEL

The Fermi level is defined as the highest energy level which the electrons assume (may assume) in a metal (a semiconductor) at 0 K temperature.

In case of a metal, *the bottom of the partially filled band corresponds to zero of the free-electron energies* (i.e., $E \equiv 0$). At 0 K, the highest occupied level in this partly full band is the Fermi level ($E_F$). The levels above $E_F$ are empty at 0 K. For copper $E_F$ = 7.05 eV. The electron velocity corresponding to Fermi energy is called the *Fermi velocity*. For copper $v_F \simeq 1.6 \times 10^6$ m/s. This fact indicates that electrons are not motionless at zero $k$ temperature. The electrons ordinarily available for conduction at room temperatures are only stacked up to the Fermi energy at absolute zero temperature giving zero thermal conductivity; however electrical conduction is possible because there are electrons close to the Fermi level and empty states are available in the (partially full) conduction band just above the Fermi energy. As the temperature is raised, in a metal, a few of the conduction (band) electrons near the Fermi energy acquire energies slightly above $E_F$ and become available for conduction. We have seen in Sommerfeld's theory that conduction electrons in a metal behave like a free-electron gas (fermions), therefore they obey FD statistics. Hence, the probability of occupancy of an energy level at any temperature $T$ is given by the FD distribution function.

$$F(E) = \frac{1}{e^{(E - E_F)/k_BT} + T} \quad ...(1)$$

where $E$ is the energy, $k_B$ is the Boltzmann constant and $T$ is the absolute temperature.

At $T = 0$ K for $E < E_F$;

$$F(E) = \frac{1}{e^{-\infty} + 1} = \frac{1}{0 + 1} = 1$$

and for $E > E_F$;

$$F(E) = \frac{1}{0^{\infty}+1} = \frac{1}{\infty} = 0$$

that is, all energy levels up to $E = E_F$ will be occupied and none will be occupied above $E_F$. Figure 10.5 (a) shows the plot of Fermi distribution function for $T = 0\,\text{K}$. For $T \neq 0\,\text{K}$, the $F(E)$ assumes the shape as shown in Fig. 10.5 (b). Figure 10.5 (b) shows that the distribution function changes only for the levels of energies close to the Fermi energy, $E_F$.

For $$E = E_F,$$

$$F(E) = F(E_F) = \frac{1}{e^0+1} = \frac{1}{2} = 0.5$$

That is, irrespective of temperature, for energy equal to the Fermi energy, $F(E)$ assumes the value 0.5. Hence, we may redefine the Fermi energy as follows.

*The Fermi energy of a given material is the energy of that quantum state which has the probability equal to 1/2 of being occupied by the conduction electrons.*

Now, for temperatures of the order of 1000 K, we have $k_BT \simeq 0.086$ eV, i.e., high temperatures alone are insufficient to make (conduction) electron energies close to $E_F$. (This however becomes feasible by applying an electric field). Still, at sufficiently high temperatures, fermions from even the lowest energy states will begin to be excited to energies $\geq E_F$, then $F(E)$ near $E = 0$ will drop below the value 1. There are more filled states just above the Fermi energy in the partially filled band, consequently lesser number of unoccupied levels, so resistivity increases with temperature. $F(E)$ will then assume the shape as shown in Fig. 10.5 (c).

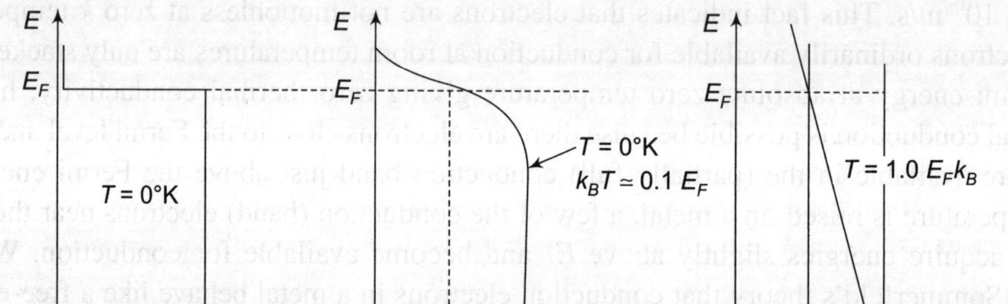


**Figure 10.5** The Fermi distribution function $F(E)$ vs. energy $E$ of fermions for different temperatures ($E_F \sim 7.05$ eV for copper)

The Fermi energy is an important part of an electron's energy band diagram of a material because many electronic properties such as optical, electrical and magnetic properties depend upon the location of Fermi level within a band.

**Table 10.2** Fermi Energy for Some Metals

| *Metal* | $E_F$ *(eV)* |
|---|---|
| Li | 4.87 |
| Na | 3.24 |
| K | 2.12 |
| Rb | 1.86 |
| Cs | 1.58 |
| Cu | 7.05 |
| Ag | 5.49 |
| Au | 5.52 |
| Ca | 4.68 |
| Ba | 3.65 |
| Al | 11.65 |

## 10.6.1 The Fermi Sphere and Free Electron Density

The number of allowed $k$-values per unit volume of $k$-space is just $1/(2\pi/L)^3$

$$\left(\frac{V}{8\pi^3}\right)$$

In $k$-space, the radius of Fermi sphere is $k_F$ and its volume is

$$\Omega = \frac{4\pi\, k_F^3}{3}$$

Accordingly, the number of allowed values of $\vec{k}$ within the sphere is

$$\left(\frac{4\pi\, k_F^3}{3}\right) \times \frac{V}{8\pi^3} = \frac{k_F^3}{6\pi^2} V.$$

Since each allowed $k$-value leads to two one-electron levels (one for each spin value), in order to accommodate $N$ electrons, we must have

$$N = 2 \cdot \frac{k_F^3}{6\pi^2} V = \frac{k_F^3}{3\pi^2} V$$

i.e., if we have $N$ electrons in a volume $V$ then, electron density

$$n = \left(\frac{k_F^3}{3\pi^2}\right) \quad \text{...(1)}$$

$$\therefore \quad \left(\frac{n}{\varepsilon_F}\right) = \frac{k_F^3}{3\pi^2} \times \frac{1}{\varepsilon_F} = \frac{k_F^3}{3\pi^2}\left(\frac{2m}{\hbar^2 k_F^2}\right)$$

or
$$\left(\frac{n}{\varepsilon_F}\right)=\left(\frac{2m\,k_F}{3\pi^2\hbar^2}\right) \qquad ...(2)$$

### 10.6.2 Application of the Fermi Dirac Distribution to the Free Electron Gas (Chemical Potential)

We know that the F.D. function

$$f(\varepsilon) = \frac{1}{e^{(\varepsilon-\mu)/k_BT}+1} \qquad ...(3)$$

In a free (and independent) electron gas, the one electron levels are specified by the wave vector $\vec{k}$ and spin quantum number $s$, with energies that are independent of $s$ (in the absence of a magnetic field) and are given by

$$\varepsilon(\vec{k}) = \frac{\hbar^2k^2}{2m} \qquad ...(4)$$

In the ground state ($T = 0°K$), those and only those levels are occupied with $\varepsilon(\vec{k}) \leq \varepsilon_F$, so the ground state distribution function must be

$$\begin{aligned} f_{\vec{k}s} &= 1,\ \varepsilon(\vec{k}) < \varepsilon_F \\ &= 0,\ \varepsilon(\vec{k}) > \varepsilon_F \end{aligned} \qquad ...(5)$$

On the other hand, as $T \to 0$, the limiting form of the F.D. distribution is

$$\begin{aligned} \lim_{T\to 0} f_{\vec{k}s} &= 1,\ \ \varepsilon(\vec{k})<\mu \\ &= 0,\ \ \varepsilon(\vec{k})>\mu \end{aligned} \qquad ...(6)$$

$\mu$ is the chemical potential.

For these to be consistent, it is necessary that

$$\lim_{T\to 0} \mu = \varepsilon_F \qquad ...(7)$$

We shall see subsequently that for metals the chemical potential remains equal to the Fermi energy to a high degree of precision, all the way up to room temperature. As a result, one frequently fails to make any distinction between the two when dealing with metals. This, however can be very misleading. In precise calculations, it is essential to keep track of the extent to which $\mu$, the chemical potential, differs from its zero temperature value, $\varepsilon_F$.

Now, energy density $\qquad u = U/V. \qquad ...(8)$

In the independent electron approximation, the internal energy $U$ is just the sum over one-electron levels of $\varepsilon(\vec{k})$ times the mean number of electrons in the level

$$U = 2\sum_{\vec{k}} \varepsilon(\vec{k})\, f(\varepsilon(\vec{k})) \qquad ...(9)$$

where
$$f(\varepsilon) = \frac{1}{e^{(\varepsilon-\mu)/k_BT}+1} \qquad ...(10)$$

(The factor of 2 in Eq. (9) reflects the fact that each $k$-level can contain two electrons of opposite spin orientations.) From Eqs. (8) and (9)

$$u = 2\int \frac{d\vec{k}}{8\pi^3}\, \varepsilon(\vec{k})\, f(\varepsilon(\vec{k}))$$

or

$$u = \int \frac{d\vec{k}}{4\pi^3}\, \varepsilon(\vec{k})\, f(\varepsilon(\vec{k})) \qquad ...(11)$$

Also, total number of electrons

$$N = \sum_i f_i$$

where the sum is over all levels, therefore

$$N = \sum_v \frac{1}{e^{(\varepsilon_i - H)/k_B T} + 1}$$

$$n \equiv \frac{N}{V} = \frac{1}{V}\sum_i f_i$$

or

$$n = \int \frac{d\vec{k}}{4\pi^3}\, f(\varepsilon(\vec{k})) \qquad ...(12)$$

In evaluating integrals like Eq. (11) Eq. (12) of the form

$$\int \frac{d\vec{k}}{4\pi^3}\, F(\varepsilon(\vec{k})) \qquad ...(13)$$

We employ the fact that the integrand depends on $\vec{k}$ only through the electronic energy $\varepsilon = \hbar^2 k^2/2m$, by evaluating the integral in spherical coordinates and changing variables from $k$ to $\varepsilon$ (using $d\vec{k} = 4\pi k^2\, dk$)

$$\int_0^\infty \frac{d\vec{k}}{4\pi^3} F(\varepsilon) = \int_0^\infty \frac{k^2 dk}{\pi^2} F = \frac{\hbar^2 k^2 dk}{(\hbar^2 \pi^2)} F$$

$$= \int_0^\infty \frac{m}{(\hbar^2 \pi^2)} \sqrt{\frac{2m\varepsilon}{\hbar^2}} \cdot \frac{kdk}{m} \times \hbar^2 F \qquad \left(\because k = \sqrt{\frac{2m\varepsilon}{\hbar^2}}\right)$$

$$= \int_0^\infty \frac{m}{\hbar^2 \pi^2} \sqrt{\frac{2m\varepsilon}{\hbar^2}} \left[\hbar^2 \frac{2kdk}{2m}\right] F$$

$$= \int_0^\infty \frac{m}{\hbar^2 \pi^2} \sqrt{\frac{2m\varepsilon}{\hbar^2}}\, d\varepsilon\, F$$

$$= \int_{-\infty}^{\infty} g(\varepsilon)\, d\varepsilon \cdot F(\varepsilon). \qquad ...(14)$$

where
$$g(\varepsilon) = \frac{m}{\hbar^2\pi^2}\sqrt{\frac{2m\varepsilon}{\hbar^2}}, \quad \varepsilon > 0$$
$$= 0 \qquad , \quad \varepsilon < 0 \qquad \text{...(15)}$$

$g(\varepsilon)$ is known as the density of the levels per unit volume or simply the density of levels:

$g(\varepsilon)\, d\varepsilon = \frac{1}{V} \times$ (the number of one-electron levels in the energy range $\varepsilon$ to $\varepsilon + d\varepsilon$).

$$g(\varepsilon) = \frac{m}{\hbar^2\pi^2}\sqrt{\frac{2m\varepsilon}{\hbar^2}} = \frac{mk}{\hbar^2\pi^2}$$
$$= \frac{mk_F}{\pi^2\hbar^2}\left(\frac{k}{k_F}\right)$$
$$= \frac{mk_F}{\pi^2\hbar^2}\left(\frac{k^2}{k_F^2}\right)^{1/2}$$
$$= \frac{mk_F}{\pi^2\hbar^2}\left[\frac{\hbar^2k^2}{2m}\cdot\frac{2m}{\hbar^2k_F^2}\right]^{1/2}$$
$$= \frac{mk_F}{\pi^2\hbar^2}\left(\frac{\varepsilon}{\varepsilon_F}\right)^{1/2}$$
$$= \frac{3}{2}\times\frac{2mk_F}{3\pi^2\hbar^2}\left(\frac{\varepsilon}{\varepsilon_F}\right)^{1/2}$$

or
$$g(\varepsilon) = \frac{3}{2}\frac{n}{\varepsilon_F}\left(\frac{\varepsilon}{\varepsilon_F}\right)^{1/2}, \quad \varepsilon > 0$$
$$= 0 \qquad , \quad \varepsilon < 0 \qquad \text{...(16)}$$

(using Eq. (2)).

A quantity of importance is the density of levels at the Fermi energy:

$$g(\varepsilon_F) = \frac{m\,k_F}{\hbar^2\pi^2} \qquad \text{...(17)}$$

or
$$g(\varepsilon_F) = \frac{3}{2}\frac{n}{\varepsilon_F} \qquad \text{...(18)}$$

In general, the integrals Eqs. (13) and (14) have a rather complex structure. There is, however, a simple systematic expansion that exploits the fact that at almost all temperatures of interest in metals, $T$ is very much smaller than the Fermi temperature. In Fig. 10.6, the Fermi function $f(\varepsilon)$ is plotted at $T = 0$ and at room temperature for typical metallic densities

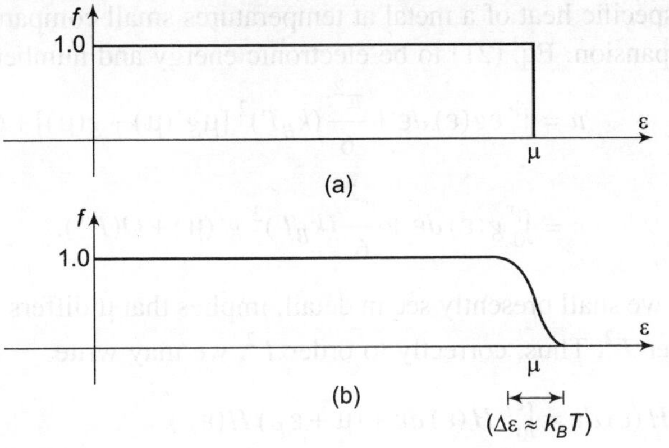


**Figure 10.6** The Fermi function, $f(\varepsilon) = 1/(e^{\beta(\varepsilon - \mu)} + 1]$ versus ε for given μ, at (a) $T = 0$ and (b) $T \approx 0.01\,\mu$ (of order room temperature, at typical metallic densities). The two curves differ only in a region of order $k_BT$ about μ

($k_BT/\mu \approx 0.01$). Evidently $f$ differs from its zero temperature form only in a small region about μ of width a few $k_BT$. Thus, the way in which integrals of the form $\int_{-\infty}^{\infty} H(\varepsilon)\, f(\varepsilon)\, d\varepsilon$ differ from their zero temperature values, $\int_{-\infty}^{\varepsilon_F} H(\varepsilon)\, d\varepsilon$, will be entirely determined by the form of $H(\varepsilon)$ near $\varepsilon = \mu$. If $H(\varepsilon)$ does not vary rapidly in the energy range of the order of $k_BT$ about μ, the temperature dependence of the integral should be given quite accurately by replacing $H(\varepsilon)$ by the first few terms in its Taylor expansion about $\varepsilon = \mu$:

$$H(\varepsilon) = \sum_{n=0}^{\infty} \frac{d^n}{d\varepsilon^n} H(\varepsilon)\bigg|_{\varepsilon=\mu} \frac{(\varepsilon - \mu)^n}{n!}. \qquad \text{...(19)}$$

This procedure is carried out in Appendix C. The result is a series of the form:

$$\int_{-\infty}^{\infty} H(\varepsilon)\, f(\varepsilon)\, d\varepsilon = \int_{-\infty}^{\infty} H(\varepsilon)\, d\varepsilon + \sum_{n=1}^{\infty} (k_BT)^{2n}\, a_n \frac{d^{2n-1}}{d\varepsilon^{2n-1}} H(\varepsilon)\bigg|_{\varepsilon=\mu} \qquad \text{...(20)}$$

which is known as the Sommerfeld expansion. The $a_n$ are dimensionless constants of the order of unity. The functions $H$ one typically encounters have major variations on an energy scale of the order of μ, and generally $(d/d\varepsilon)^n\, H(\varepsilon)|_{\varepsilon=\mu}$ is of the order of $H(\mu)/\mu^n$. When this is the case, successive terms in the Sommerfeld expansion are smaller by $O(k_BT/\mu)^2$ which is $O(10^{-4})$ at room temperature. Consequently, in actual calculations only the first and (*very* occasionally) the second terms are retained in the sum in Eq. (20). The explicit form for these is:

$$\int_{-\infty}^{\infty} H(\varepsilon)\, f(\varepsilon) d\varepsilon$$
$$= \int_{-\infty}^{\mu} H(\varepsilon)\, d\varepsilon + \frac{\pi^2}{6}(k_BT)^2\, H'(\mu) + \frac{7\pi^4}{360}(k_BT)^4\, H'''(\mu) + O\left(\frac{k_BT}{\mu}\right)^6. \qquad \text{...(21)}$$

To evaluate the specific heat of a metal at temperatures small compared with $T_F$ we apply the Sommerfeld expansion, Eq. (21) to be electronic energy and number densities.

$$u = \int_0^{\mu} \varepsilon g(\varepsilon)\, d\varepsilon + \frac{\pi^2}{6}(k_B T)^2 [\mu g'(\mu) + g(\mu)] + O(T^4), \qquad ...(22)$$

$$n = \int_0^{\mu} g(\varepsilon)\, d\varepsilon + \frac{\pi^2}{6}(k_B T)^2\, g'(\mu) + O(T^4). \qquad ...(23)$$

Equation (23) as we shall presently see in detail, implies that $\mu$ differs from its $T = 0$ value, $\varepsilon_F$, by terms of order $T^2$. Thus, correctly to order $T^2$, we may write

$$\int_0^{\mu} H(\varepsilon)\, d\varepsilon = \int_0^{\varepsilon_F} H(\varepsilon)\, d\varepsilon + (\mu + \varepsilon_F)\, H(\varepsilon_F). \qquad ...(24)$$

If we apply this expansion to the integrals in (22) and (23) and replace $\mu$ by $\varepsilon_F$ in terms already of order $T^2$ in these equations, we find

$$u = \int_0^{\varepsilon_F} \varepsilon_g(\varepsilon)\, d\varepsilon + \varepsilon_F \left\{ (\mu + \varepsilon_F) g(\varepsilon_F) + \frac{\pi^2}{6}(k_B T)^2\, g'(\varepsilon_F) \right\}$$

$$+ \frac{\pi^2}{6}(k_B T)^2 g\,(\varepsilon_F) + O(T^4), \qquad ...(25)$$

$$n = \int_0^{\varepsilon_F} g(\varepsilon)\, d\varepsilon + \left\{ (\mu + \varepsilon_F) g\,(\varepsilon_F) + \frac{\pi^2}{6}(k_B T)^2\, g'(\varepsilon_F) \right\}. \qquad ...(26)$$

The temperature-independent first terms on the right sides of (25) and (26) are just the values of $u$ and $n$ in the ground state. Since we are calculating the specific heat at constant density, $n$ is independent of temperature, and (26) reduces to

$$0 = (\mu - \varepsilon_F)\, g\,(\varepsilon_F) + \frac{\pi^2}{6}\,(k_B T)^2\, g'\,(\varepsilon_F), \qquad ...(27)$$

which determines the deviation of the chemical potential from $\varepsilon_F$:

$$\mu = \varepsilon_F - \frac{\pi^2}{6}(k_B T)^2 \frac{g'(\varepsilon_F)}{g\,(\varepsilon_F)}. \qquad ...(28)$$

The free electron theory satisfactorily explained the high electrical and thermal conductivity of metals. Free electrons readily move under the influence of an electrical field; when a temperature gradient is applied, free electrons quickly carry away heat from the hotter to the colder end. The theory also satisfactorily explains the characteristic opacity and lustre of metals. When an electromagnetic radiation strikes a metal surface, the free electrons can absorb energy of almost any wavelength and the metal appears opaque; when the radiation is re-emitted, only the rays directed towards the surface pass through and the metal appears lustrous.

## 10.7 ELECTRICAL CONDUCTIVITY OF METALS AND RELAXATION TIME

According to the free electron theory, electrons move freely in a solid. If an external electric field ($\varepsilon$) is applied, movement of electrons is accelerated under the action of a force $e\varepsilon$, where $e$ is the electronic charge. Thus,

$$a = \frac{e\varepsilon}{m_e}$$

where $a$ is the acceleration and $m_e$ the electron mass. However, the electrons do not get accelerated indefinitely; through collisions with scattering centres, their velocity momentarily falls to zero. Figure 10.7 describes the behaviour of the free electron (s) as a function of time. The average electronic velocity is given by

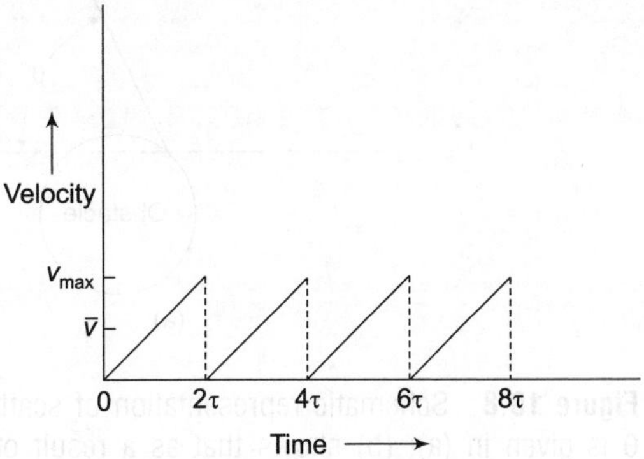


**Figure 10.7** Plot of velocity ($v$) versus time for free electrons in solids

$$\bar{v} = \frac{\tau_e\varepsilon}{m_e},\ v_{\max} = \frac{2\tau_e\varepsilon}{m_e}$$

where $2\tau$ represents the time between two collisions and $\tau$ is called the relaxation time. If the density of electrons is $n$, then the current density is given by

$$J = ne\bar{v} = \frac{ne^2\tau\varepsilon}{m_e}$$

Since $$J = \sigma\varepsilon$$

Therefore, $$\sigma = \frac{ne^2\tau}{m_e} = ne\mu \qquad ...(1)$$

where $\mu \equiv e\tau/m_e$ is the average drift velocity acquired per unit electric field (mobility). Eq. (1) is the free electron expression for the electrical conductivity.

## 10.8 RELAXATION TIME, COLLISION TIME, AND MEAN FREE PATH

In the preceding section we introduced the concept of a relaxation time on phenomenological grounds. In this section we shall see that such a relaxation time may be obtained from a picture in which one assumes the electrons to collide with "obstacles"; the nature of these obstacles will be discussed in subsequent sections.

Suppose an electron moves with a velocity $v_x$ along the $x$-direction of a Cartesian coordinate system. Let us assume that the electron collides with an obstacle in such a way that the probability for it to be scattered over an angle between $\theta$ and $\theta + d\theta$ is given by $P(\theta)2\pi \sin\theta\, d\theta$, where $P(\theta)$ is an arbitrary function of the scattering angle $\theta$ [see Fig. 10.8 (a)]. Evidently then, we should have

$$2\pi \int_{\theta=0}^{\pi} P(\theta)\sin\theta\, d\theta = 1 \qquad ...(1)$$

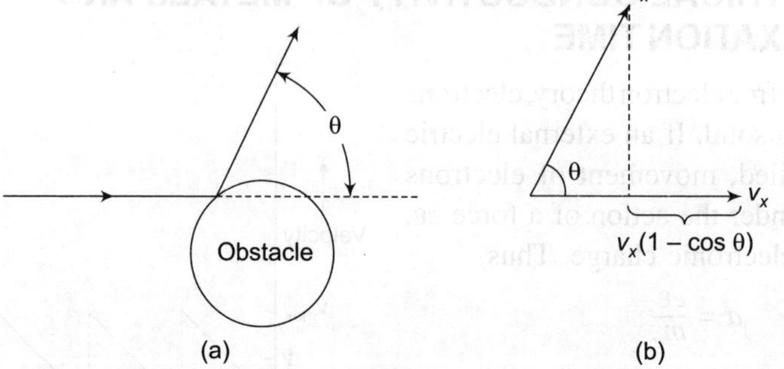


**Figure 10.8** Schematic representation of scattering of an electron by an obstacle over an angle θ is given in (a); (b) shows that as a result of this scattering process, the *x*-component of the velocity has been reduced by the amount $v_x(1 - \cos\theta)$, assuming that the magnitude of the velocity is the same after scattering as before

We shall assume that the magnitudes of the velocities before and after scattering are the same. In connection with this assumption we should comment that actually some energy is transferred in the conduction process from the electrons to the lattice (represented in the present picture by the obstacles); this is, after all, the reason for the dissipation of Joule heat. However, one can show that the percentage change in velocity resulting from the collisions is very small indeed. In discussing the collision process, therefore, we may assume equal velocities before and after the collision. Under these circumstances, an electron scattered over an angle θ has a velocity component along the *x*-direction equal to $v_x \cos\theta$. Hence, it has lost along the *x*-direction a velocity component amounting to (see Fig. 10.8 b).

$$v_x\,(1 - \cos\theta) \qquad ...(2)$$

Consider now a large group of electrons, all moving with the same velocity $v_x$ along the *x*-direction. Assuming all these particles carry out one collision with an obstacle, what is the average velocity component lost by the particles? The answer is evidently

$$v_x\,(1 - \langle\cos\theta\rangle) \qquad ...(3)$$

where $\langle\cos\theta\rangle$ is the average value of the cosine of the scattering angle. Since the probability for scattering over a given angle is described by $P(\theta)$, we must have

$$\langle\cos\theta\rangle = 2\pi \int_{\theta=0}^{\pi} P(\theta)\cos\theta\sin\theta\, d\theta \qquad ...(4)$$

Note that if the scattering is spherically symmetric, i.e., if $P(\theta)$ is constant, $\langle\cos\theta\rangle = 0$.

Let us now consider the collisions from the point of view of time. Assume that the probability for an electron to collide with an obstacle during a short time interval $dt$ is given by $dt/\tau_c$, where $\tau_c$ is a constant with the dimensions of time. Let $F(t)$ represent the probability that an electron moves for a time $t$ without suffering a collision, and let $F(t + dt)$ represent the same quantity for a period $t + dt$. We may then write

$$F(t + dt) = F(t) + \frac{dF}{dt}\,dt \qquad ...(5)$$

on the other hand, we may also write

$$F(t + dt) = F(t)\,F(dt) = F(t)\,[1 - dt/\tau_c] \qquad ...(6)$$

in view of the definition of $\tau_c$. Hence, the last two equations give

$$dF/dt = -F/\tau_c \quad \text{or} \quad F(t) = e^{-t/\tau_c} \qquad ...(7)$$

since $F(t) = 1$ for $t = 0$, which fixes the integration constant. What is the average time which elapses between collisions? The answer is

$$\langle t\rangle = \int_0^\infty t\,\frac{dF}{dt}\,dt \qquad ...(8)$$

because $-(dF/dt)\,dt$ gives the probability that the electron collides between $t$ and $t + dt$. Substituting $F(t)$ from Eq. (7) into Eq. (8) we find

$$\langle t\rangle = \tau_c \qquad ...(9)$$

Hence, the physical meaning of $\tau_c$ introduced above is that it represents the *average time between collisions.*

We are now prepared to answer the question: what is the rate of change of $\langle v_x\rangle$ for a group of electrons resulting from collisions with the obstacles described earlier? In accordance with (3) an electron loses per collision on the average

$$\langle v_x\rangle\,(1 - \langle\cos\theta\rangle)$$

Since the number of collisions carried out per second is on an average equal to $1/\tau_c$, we have

$$\left[\frac{\partial}{\partial t}\langle v_x\rangle\right]_{\text{coll}} = -\frac{\langle v_x\rangle}{\tau_c}(1 - \langle\cos\theta\rangle) \qquad ...(10)$$

Thus, we see that the relaxation time is related to the mean time between collisions as follows

$$\tau = \frac{\tau_c}{1 - \langle\cos\theta\rangle} \qquad ...(11)$$

When the scattering is isotropic, $\langle\cos\theta\rangle = 0$ and $\tau = \tau_c$. If the scattering takes place predominantly in the direction of incidence, $\langle\cos\theta\rangle$ will have a positive value, and the relaxation time $\tau > \tau_c$. This is understandable qualitatively, because the electrons then will "remember" longer what their initial direction of motion was, and consequently the time required to make the velocity distribution random increases.

The mean free path of an electron, $\lambda$, will be defined as

$$\lambda = v\tau_c \qquad ...(12)$$

where $v$ is the total velocity of an electron. It will be evident that if the obstacle are hard spheres, the quantity $\lambda$ would be determined by the concentration of obstacles. Thus, in such a case the collision time $\tau_c$ would be inversely proportional to the velocity $v$ of the electrons for a given concentration of obstacles. Consequently, when $\langle \cos \theta \rangle$ is independent of the velocity of the electrons, the relaxation time will be inversely proportional to $v$. This raises an important question, because in the preceding section we introduced a single relaxation time $\tau$, without any reference to the distribution of electron velocities. In the case of metals, however, it can be shown that the relaxation time $\tau$ which determines the conductivity always refers to a single total velocity, viz., the one corresponding to the *Fermi energy*. In order to explain this properly, one would have to carry through a procedure involving Fermi statistics, a subject which lies outside the scope of the present treatment. A few qualitative remarks may suffice here. The free electrons in a metal can accept a series of discrete energy levels. At absolute zero, all energy levels below a certain value $W_F$ are filled, all those above $W_F$ being empty; $W_F$ is called the Fermi level of the electrons. The magnitude of $W_F$ is of the order of several electron volts. At any temperature $T$, the probability of a state corresponding to the energy $W$ being occupied by an electron is given by

$$f(W) = \frac{1}{\exp\,[(W - W_F)/kT] + 1} \qquad \text{...(13)}$$

For $T = 0$, this function is a step function, as indicated in Fig. 10.9; for $T > 0$, one obtains a curve which differs slightly from the one corresponding to $T = 0$, as indicated in an exaggerated fashion in Fig. 10.9. Since $kT$ at room temperature is only 0.025 eV, $kT$ is very small compared to $W_F$, and, consequently, the energy region over which temperature influences the energy distribution of the electrons is very narrow. The velocity of the electrons with the Fermi energy, $v_F$, is given by

$$\frac{1}{2}mv_F^2 = W_F \qquad \text{...(14)}$$

For $E_F = 5$ eV, which is a representative figure, one finds $v_F \cong 4 \times 10^6$ metres per second (1 eV = $1.6 \times 10^{-19}$ joules). The relaxation time appearing in the theory of conductivity of metals refers to electrons with a velocity $v_F$.

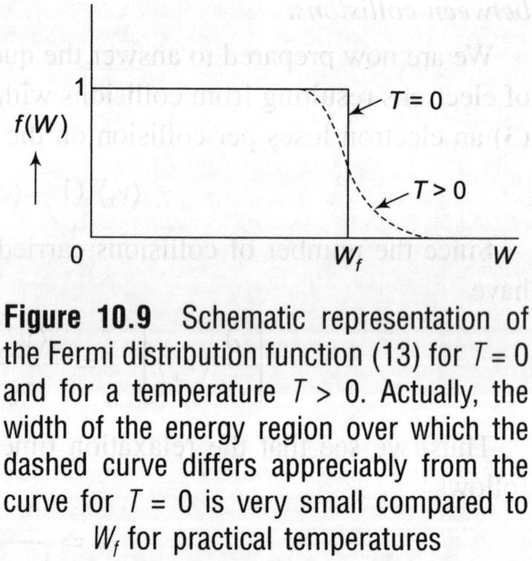


**Figure 10.9** Schematic representation of the Fermi distribution function (13) for $T = 0$ and for a temperature $T > 0$. Actually, the width of the energy region over which the dashed curve differs appreciably from the curve for $T = 0$ is very small compared to $W_f$ for practical temperatures

We have seen that it is possible to interpret the relaxation time in terms of collisions of the electrons with obstacles. Out next task will be to determine what these obstacles are, and if the model so obtained agrees with the observed dependence of the conductivity on temperature, impurity concentration, etc.

## 10.9 ELECTRON SCATTERING AND THE RESISTIVITY OF METALS

If one assumes for monovalent metals that the number of free electrons per m$^3$ is equal to the number of atoms per m$^3$, one obtains at 0°C values for $\tau$ as given in Table 10.3. It is observed that the relaxation times are of the order of $10^{-14}$ seconds. Since this number

**Table 10.3** Conductivity ($\sigma$) Fermi Level ($W_F$), Mean Free Path ($\lambda$), and Relaxation Time ($\tau$) at 0°C for Some Monovalent Metals

| *Metal* | $\sigma$ ($10^8$ ohm$^{-1}$ m$^{-1}$) | $W_F$ (eV) | $\lambda$ (angstroms) | $\tau$ ($10^{-14}$ sec) |
|---|---|---|---|---|
| Li | 0.12 | 4.7 | 110 | 0.9 |
| Na | 0.23 | 3.1 | 350 | 3.1 |
| K | 0.19 | 2.1 | 370 | 4.4 |
| Cu | 0.64 | 7.0 | 420 | 2.7 |
| Ag | 0.68 | 5.5 | 570 | 4.1 |

may not convey to the reader any more than the fact that it is small, it is useful to consider the results in terms of the mean free path defined by Eq. (12). It was mentioned in the preceding section that $\tau$ in metals refers to electrons with the Fermi velocity $v_F$. Thus, if we assume that scattering in these metals is isotropic (it can be shown that this is a good approximation for the temperatures of interest), we have $\tau = \tau_c$ and $\lambda = v_F T$. The values of $\lambda$ so obtained are also indicated in Table 10.3. It is noted that at 0°C, the mean free path for scattering is several hundred Angstroms. We may mention further that the resistivity of metals at low temperatures decreases to very small values, as may be seen from the example in Fig. 10.10 for sodium. Although some of the assumptions made above are not applicable for liquid helium temperatures, we may state that the *resistivity of a perfect, pure single crystal of a metal goes to zero as T approaches zero.* In other words, at extremely low temperature, $\lambda$ reaches macroscopic values. From these arguments one can draw an important, though somewhat negative conclusion, viz., that the *obstacles* introduced earlier *are not the individual*

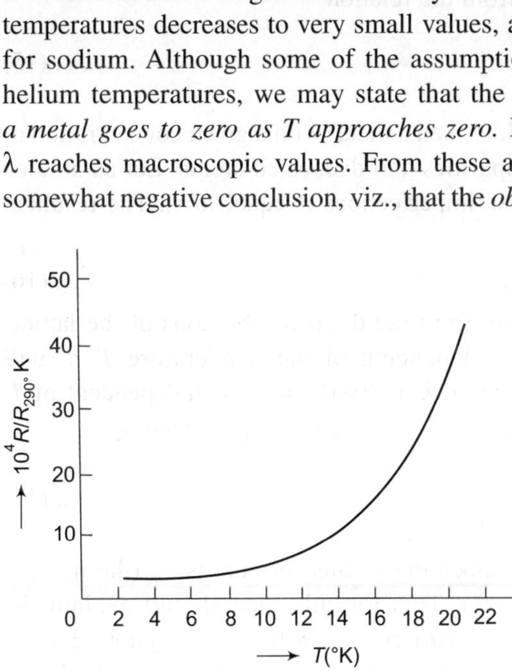


**Figure 10.10** The resistivity of sodium as a function of temperature in the low temperature region. the constant value approached at low temperatures is due to lattice imperfections and impurities, and corresponds to $\rho_i$ in the text

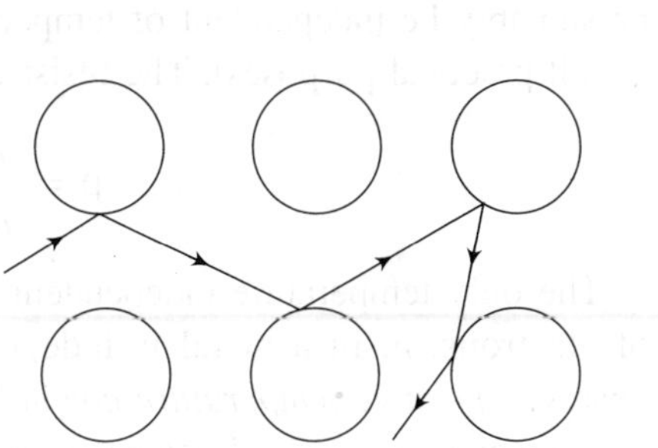

**Figure 10.11** Schematic representation of the classical picture of scattering of an electron by the ion cores; this picture predicts a mean free path of a few angstroms, in contradiction with experimental information

*ion cores* in the metallic lattice. In fact, if we were to picture the conduction electrons as moving between the positive ion cores, one would expect, in terms of classical physics, mean free paths of the order of a few Angstroms, as indicated in Fig. 10.11. This classical picture thus breaks down, and we have to look for another one. The reason for the breakdown of the classical picture lies in the wave aspect of elementary particles. In fact, it can be shown theoretically that when the wave nature of the electrons is taken into account properly, *an electron moving in a perfectly periodic lattice would not be scattered at all.* In other words, the waves representing a moving electron in a perfect crystal resemble the electromagnetic waves corresponding to visible light which passes undeviated through a crystal such as sodium chloride. Thus, the present interpretation of the electrical resistivity of metals is based on the notion that *scattering of the electrons is due to deviations from perfect periodicity of the lattice*. In pure metals, such deviations may be caused, for example, by the vibrations of the atoms in the lattice; they may also be caused by atoms which occupy "wrong" positions in the lattice, or by lattice sites which are unoccupied, or by impurity atoms. Thus, there are various kinds of obstacles which we have to consider in the scattering of electrons in metals, each of them giving rise separately to a certain relaxation time $\tau_j$. Since the probability for scattering by a particular kind of obstacle per unit time is equal to $1/\tau_j$ (for isotropic scattering) the resultant relaxation time may be obtained from the relation

$$\frac{1}{\tau} = \sum_j \frac{1}{\tau_j} \qquad ...(15)$$

since scattering probabilities are additive. Consider then a metal with a given concentration of imperfections, i.e., a given concentration of impurities and dislocated atoms. Let us assume that the relaxation time associated with all these imperfections is equal to $\tau_i$. The resultant relaxation time may then be written as

$$1/\tau = 1/\tau_i + 1/\tau_T \qquad ...(16)$$

where $\tau_T$ represents the relaxation time resulting from the thermal vibrations of the lattice. When the concentration of imperfections is independent of the temperature $T$, $\tau_i$ will presumably be independent of temperature (because $\lambda_i$ is fixed and $v_F$ is independent of $T$, for all practical purposes). The resistivity of the metal may then be represented by

$$\rho = \frac{m}{ne^2}\left(\frac{1}{\tau_i} + \frac{1}{\tau_T}\right) \qquad ...(17)$$

The only temperature-independent quantity appearing in this formula is $\tau_T$ (the number of electrons, $n$, in a metal is independent of $T$!). This formula then should explain the observed *positive temperature coefficient of the resistivity of metals* (we are ignoring some very special cases). As $T$ increases, the amplitude of the atomic vibrations increases; hence the deviations from periodicity become more severe and the scattering by lattice vibrations increases. Consequently, as $T$ increases, $\lambda$ decreases and since $v_F$ is constant, $\tau_T$ decreases. Thus, as the temperature of a metal (or alloy) is raised, the electrical resistivity increases. At temperatures above the so-called Debye temperature of the material, the energy of the lattice vibrations increases linearly with $T$, and consequently, the resistivity increases linearly with $T$ in this region, obeying the formula

$$\rho = \rho_i + aT \qquad ...(18a)$$

where $\rho_i$ and $a$ are constants. The separation of the resistivity into a temperature-independent part and a temperature-dependent part is referred to as Matthiessen's rule. The Debye temperature may be considered a measure for the maximum frequency of the lattice vibrations; hard materials have higher Debye temperatures than soft materials, as may be seen from the values given in Table 10.4. The Debye temperature plays an important role in the specific heat of a material in that above the Debye temperature the specific heat is approximately constant, whereas below it, the specific heat drops rapidly, approaching zero as $T \to 0$.

**Table 10.4** Debye Temperature ($\theta_D$) for Some Metals, in Degrees Absolute

| *Metal* | *Ag* | *Al* | *Au* | *Cu* | *Fe* | *Pb* |
|---|---|---|---|---|---|---|
| $\theta_D$ | 215 | 398 | 180 | 315 | 420 | 88 |

As temperatures fall below the Debye temperature, $\tau_T$ increases rapidly and as $T$ goes to zero, $\tau_T$ approaches infinity. There exists a close *relationship between the resistivity and the specific heat* of the lattice vibrations; in fact, the Debye temperature may be calculated from the resistivity versus temperature curve of a metal. Since the electrical resistivity at very low temperatures resulting from scattering of electrons by lattice vibrations becomes negligible, the total resistivity is essentially determined by the imperfections and impurities. Thus, with reference to Fig. 10.10, the constant resistivity near 4°K is a measure for the content of imperfections and impurities in the sodium sample.

The arguments given also make the following observation intelligible: when one adds to a metal like nickel a small amount of copper, the conductivity decreases, notwithstanding the fact that copper has a higher conductivity than nickel. The reason is, of course, that in replacing some of the nickel atoms in the lattice, the copper atoms destroy the lattice periodicity in their immediate neighbourhood and thus act as new scattering centres for the conduction electrons.

In Fig. 10.12 we give an example of the influence of temperature and of impurity content on the resistivity of copper. It is noted that $\rho$ varies linearly with $T$ in accordance with (18 a) and that the resistivity of the alloys increases with increasing nickel content.

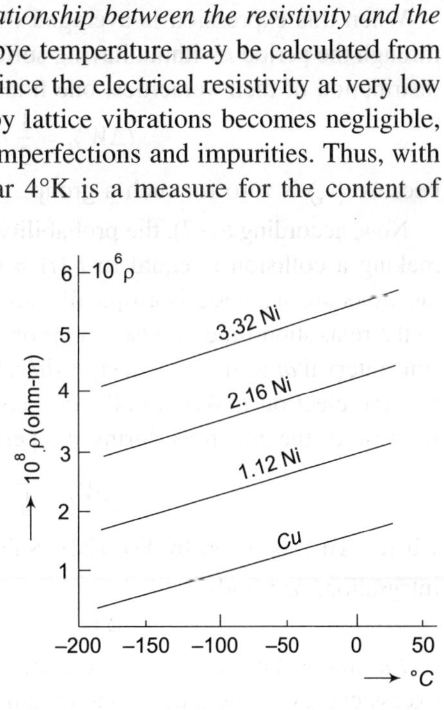


**Figure 10.12** The resistivity of alloys of nickel in copper as a function of temperature; the numbers indicate the atomic percentage of nickel in the alloy

## 10.10 THE HEAT DEVELOPED IN A CURRENT CARRYING CONDUCTOR

In this section, we shall give an elementary derivation of the well known experimental fact that the heat developed per m$^3$ per second in a conductor carrying a current density $J$ as a result of an applied field $E$ is given by

$$W = JE = \sigma E^2 \qquad ...(19)$$

If $J$ is in amperes per m$^2$ and $E$ is in volts per m, $W$ is found in watts per m$^3$. Consider a particular electron which at the instant $t = 0$ has carried out a collision with the lattice; at that instant let the velocity components of the electron be $v_x$, $v_y$, $v_z$. Assuming that at the instant $t (> 0)$ the electron has not yet collided again, and assuming the field $E$ is applied along the *negative* $x$-direction, the velocity components of the electron at the time $t$ will be given by

$$v_x + (e/m)\, Et,\ v_y, \text{ and } v_x \qquad ...(20)$$

The increment $(e/m)Et$ results from the acceleration along the $x$-direction produced by the field on the electron of charge $-\varepsilon$. The increase in energy of the electron over the period $t$ is therefore

$$(\Delta W)_t = \frac{1}{2}\, m\,[2e/m)\, Ev_x t + (e/m)^2\, E^2 t^2] \qquad ...(21)$$

When this expression is averaged over a large number of electrons which have all lived through the period $t$ without having suffered a collision, but which presumably have a random distribution of their velocities, one finds

$$\langle \Delta W \rangle_t = \frac{1}{2}\, m\ [(e/m)^2\, E^2 t^2] = e^2 E^2 t^2 / 2m \qquad ...(22)$$

because $\langle v_x \rangle$ is zero for such a group.

Now, according to (7), the probability that an electron will live through the period $t$ without making a collision is equal to $F(t) = \exp[-t/\tau_c]$. For simplicity we shall assume that the electrons are scattered isotropically, so that the average time between collisions, $\tau_c$, is equal to the relaxation time $\tau$. Also, the probability that an electron will suffer a collision during a time interval $dt$ is given by $dt/\tau_c = dt/\tau$. Hence, the product $e^{-t/\tau}\,(dt/\tau)$ gives us the probability that the electron makes a collision between $t$ and $t + dt$. Consequently, the average energy increase of the electrons during the period between two collisions is equal to

$$\langle \Delta W \rangle = \int_{t=0}^{\infty} \langle \Delta W \rangle_t\, e^{-t/\tau}\, \frac{dt}{\tau} \qquad ...(23)$$

where $\langle \Delta W \rangle_t$ is given by Eq. (22). Substituting Eq. (22) into Eq. (23) and carrying out the integration, one finds

$$\langle \Delta W \rangle = \tau^2 e^2 E^2 / m \qquad ...(24)$$

Assuming there are $n$ electrons per m$^3$, and assuming that these electrons transfer their excess energy to the lattice, one obtains for the total energy dissipated per m$^3$ per second

$$W = \frac{n}{\tau} \langle \Delta W \rangle = \left( \frac{ne^2 \tau}{m} \right) E^2 = \sigma E^2 \qquad ...(25)$$

in agreement with the experimental relation (19).

## 10.11 THE THERMAL CONDUCTIVITY OF METALS

When a homogeneous isotropic material is subjected to a temperature gradient, a flow of heat results in the direction opposite to the gradient. Thus, when $dT/dx$ represents the temperature gradient and $Q$ represents the heat flow density, one defines the thermal conductivity $\boldsymbol{K}$ by means of the equation

$$Q = -K(dT/dx) \qquad ...(26)$$

When $\boldsymbol{Q}$ is expressed in watts per $m^2$, and $\boldsymbol{dT/dx}$ in degrees **C** per metre, one finds $\boldsymbol{K}$ in watt $m^{-1}$ degree$^{-1}$. It is well known that, at least at normal temperatures, metals conduct heat much better than insulators. In Table 10.5 we gives values for $\boldsymbol{K}$ at room temperature for various materials to illustrate the range of thermal conductivities encountered. In insulating solids, the heat current is carried by the lattice vibrations. This, in part, is also the case in metals, but the thermal conductivity due to the conduction electrons predominates in these materials. We shall consider only the *thermal conductivity arising from the conduction electrons.*

**Table 10.5** Thermal Conductivity $K$ at Room Temperature in Watt Meters$^{-1}$ Degrees$^{-1}$, for Various Materials

| *Material* | *K* | *Material* | *K* |
|---|---|---|---|
| Al | 209 | Mica (⊥ cleavage plane) | 0.75 |
| Constantan (60 Cu, 40 Ni) | 22.5 | Quartz (‖ axis ) | 12.6 |
| Cu | 385 | (⊥ axis) | 6.7 |
| Fe | 67 | Carbon | 4.2 |
| Ni | 54 | Glass (window) | 1.0 |

Assuming a temperature gradient $dT/dx$ along the $x$-direction, the average energy of the electrons will also be a function of $x$. Thus, let $W(0)$ be the average energy per electron in the plane $x = 0$. At a distance $\Delta x$ from this plane, the energy of the electrons is then given by

$$W(\Delta x) = W(0) + \frac{dW}{dx}\Delta x = W(0) + \frac{dW}{dT}\frac{dT}{dx}\Delta x \qquad ...(27)$$

The atomic picture of heat conduction by electrons is the following: One considers a unit area of the plane at $x = 0$, and calculates the net transport of energy through that plane resulting from electrons coming from the two regions $x > 0$ and $x < 0$. It is assumed that the energy of an electron which passes through the plane at $x = 0$ is equal to the average electron energy at $\Delta x$, where $\Delta x$ gives the location along the $x$-direction where the electron last collided with the lattice. Thus, let $vx_i$ represent the velocity component of a particular electron along the $x$-direction, and let $-\Delta x_i$ represent the location of the plane where it last collided. If there were $n_i$ such electrons per $m^3$, the number passing the plane $x = 0$ per second would be $n_i vx_i$. Since the energy carried by electron $i$ is obtained by replacing $\Delta x$ by $-\Delta x_i$ in Eq. (27), the total heat current density through the plane $x = 0$ in the positive $x$-direction may be written as

$$Q = \sum_i \left[ W(0) - \frac{dW}{dT}\frac{dT}{dx}\Delta x_i \right] v_{xi} \qquad ...(28)$$

where the summation extends over all electrons per m$^3$. Since the average velocity of the electrons per unit volume is zero, the sum containing the quantity $W(0)$ vanishes. We are thus left with

$$Q = \sum_i \frac{dW}{dT}\frac{dT}{dx} v_{xi}\,\Delta x_i = -n\frac{dW}{dT}\frac{dT}{dx}\langle v_x\,\Delta x\rangle \qquad ...(29)$$

where $n$ stands for the total number of conduction electrons per m$^3$, and where $\langle v_x \Delta x\rangle$ is the average value of $v_x \Delta x$. Now, an electron moving in a direction making an angle $\theta$ with the $x$-axis has a velocity component $v_x = v\cos\theta$, where $v$ is the total velocity. Similarly, if $\Delta x$ measures the distance from the plane $x = 0$ to the plane where it collided last, the electron has actually covered a path length $\Delta r = \Delta x/\cos\theta$ between the instant it collided and the instant it arrives at the plane $x = 0$, as indicated in Fig. 10.13. We may thus write Eq. (29) in the form

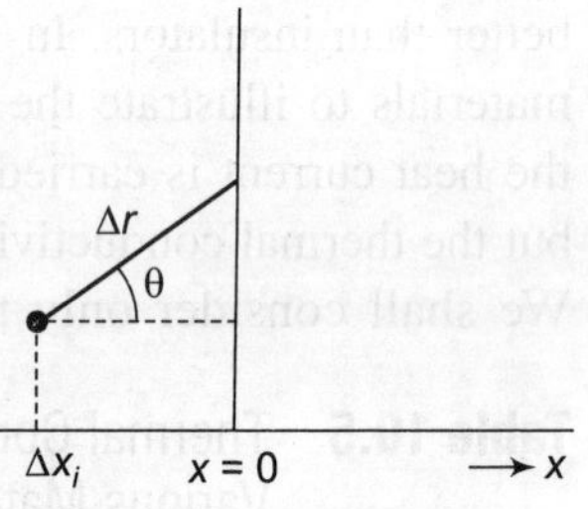


**Figure 10.13** Illustrating the geometry used in the derivation of the heat conductivity

$$Q = -n\frac{dW}{dT}\frac{dT}{dx}\langle v\,\Delta r\cos^2\theta\rangle = -\frac{1}{3}n\frac{dW}{dT}\frac{dT}{dx}\langle v\,\Delta r\rangle \qquad ...(30)$$

where the last equality follows from the fact that

$$\langle\cos^2\theta\rangle = \frac{\int_0^\pi \cos^2\theta\sin\theta\,d\theta}{\int_0^\pi \sin\theta\,d\theta} = \frac{1}{3} \qquad ...(31)$$

At this point we shall make a statement that will not be proved here, but which bears relation to an earlier remark in connection with the electrical conductivity of metals: from a more rigorous treatment it can be shown that only the mean free path or the relaxation time corresponding to electrons with the Fermi energy is of importance. Thus, $v$ in Eq. (30) may be replaced by the velocity $v_F$ (Fermi velocity), and $\Delta r$ by $\lambda$; here $\lambda$ is the mean free path for scattering corresponding to electrons with the velocity $v_F$. Hence,

$$Q = -\frac{1}{3}n\frac{dW}{dT}\frac{dT}{dx}v_F\lambda \qquad ...(32)$$

The quantity $n(dW/dT)$ represents the energy increase of the electron gas per m$^3$ per degree; i.e., this quantity is simply the *specific heat of the electron gas at constant volume, C*. Since $\lambda = v_F\tau$, where $\tau$ if the relaxation time, we have

$$Q = \frac{1}{3}Cv_F^2\,\tau\frac{dT}{dx} \qquad ...(33)$$

From Fermi–Dirac statistics it can be shown that the specific heat of the electron gas is equal to

$$C = \frac{n\pi^2 k^2 T}{m v_F^2} \qquad ...(34)$$

The last two equations thus give

$$Q = -\frac{1}{3}\frac{n\pi^2 k^2 T\tau}{m}\frac{dT}{dx} \qquad ...(35)$$

Comparing this expression with Eq. (26) we find for the thermal conductivity

$$K = \frac{1}{3}\frac{n\pi^2 k^2 T\tau}{m} \qquad ...(36)$$

We can draw several conclusions from this result. First, we may estimate $K$ from (36) and compare the result with the values given in Table 10.4. For $T$ = 300°K, $n \cong 5 \times 10^{28}$ m$^{-3}$, $\tau \cong 10^{-14}$ seconds, $k$ = 1.38 × $10^{-23}$ joule degree$^{-1}$, $m = 9 \times 10^{-31}$ kg, we find $k \cong$ 100 watt m$^{-1}$ degree$^{-1}$, which is of the correct order of magnitude. How does the thermal conductivity depend on temperature? All quantities appearing in expression (36) for $K$ are independent of temperature, except $T$ and $\tau$. As mentioned in section 10.9, $\tau$ varies as $T^{-1}$ for temperatures above the Debye temperature. Hence, in this temperature region the *thermal conductivity is independent of temperature*. At low temperatures the behaviour is more complicated, as illustrated in Fig. 10.14 for copper.

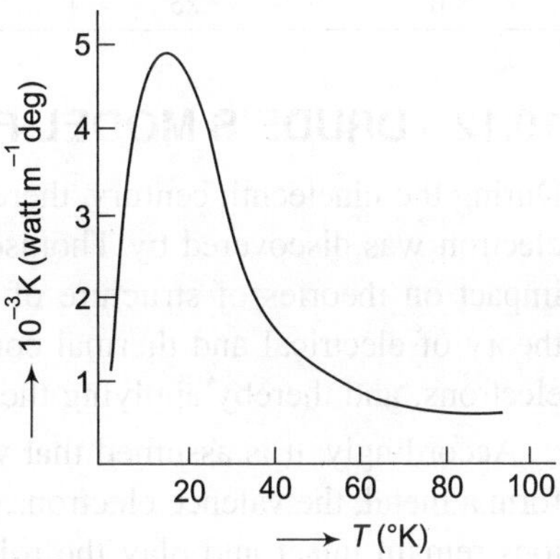


**Figure 10.14** The thermal conductivity of copper at low temperatures. [After R. Berman and D. K. C. MacDonald, *Proc. Roy. Soc.* (London) **A211**, 122 (1952)]

There exists an interesting relationship between the electrical conductivity and the thermal conductivity of a metal.

We know that $\sigma = \dfrac{ne^2 i}{m}$

So, we may write

$$\frac{K}{\sigma} = \frac{\pi^2}{3}\frac{k^2}{e^2}T \quad \text{or} \quad \frac{K}{\sigma T} = \frac{\pi^2}{3}\frac{k^2}{e^2} \equiv L \qquad ...(37)$$

The theory thus predicts that $K/\sigma T$ for all metals should be equal to a universal constant; this law is known as the *Wiedemann–Franz law*. The constant represented by $L = 2.45 \times 10^{-8}$ watt ohm degree$^2$ is known as the *Lorenz number*. In Table 10.6, we give experimental values of the *Lorenz number* for various metals, at 0°C and at 100°C. It is observed that the agreement between theory and experiment is satisfactory. It should be remarked here that the contribution to $K$ from the lattice vibrations has been neglected in our considerations. At low temperatures, the theory given does not apply, and a relaxation time for the thermal conductivity and electrical conductivity *cannot* be defined uniquely. Thus, the Lorenz number for copper near

15°K is an order of magnitude smaller than at room temperature. The Wiedemann–Franz law holds only above the Debye temperature.

**Table 10.6** Experimental Lorenz Number for Various Metals at 0°C and at 100°C, Expressed In $10^{-8}$ Watt Ohm Degree$^2$

| *Metal* | *0°C* | *100°C* | *Metal* | *0°C* | *100°C* |
|---|---|---|---|---|---|
| Ag | 2.31 | 2.37 | Pb | 2.47 | 2.56 |
| Au | 2.35 | 2.40 | Pt | 2.51 | 2.60 |
| Cu | 2.23 | 2.33 | Mo | 2.61 | 2.79 |

## 10.12 DRUDE'S MODEL FOR CONDUCTION

During the nineteenth century, there was no precise concept known for atomic structure. The electron was discovered by Thomson in 1897, and this discovery has a vast and immediate impact on theories of structure of matter. Three years later (i.e., in 1900) Drude gave his theory of electrical and thermal conduction by considering the metals to be containing free electrons, and thereby applying the kinetic theory of gases to metals.

Accordingly, it is assumed that when atoms of a metallic element are brought together to form a metal, the valence electrons get detached and wander freely within the metal, whereas ions remain intact and play the role of immobile particles. Surrounding the nucleus are $Z_a$ electrons of total charge $-eZ_a$. Out of these, there are $Z$ relatively weakly bound valence electrons and remaining $(Z_a - Z)$ are relatively tightly bound and are known as core electrons. When atoms condense to form the material, core electrons remain bound to the nucleus, but valence electrons detach from their parent atoms and are called conduction electrons. They are akin to atoms in a gas moving against a background of heavy immobile ions (see Fig. 10.15).

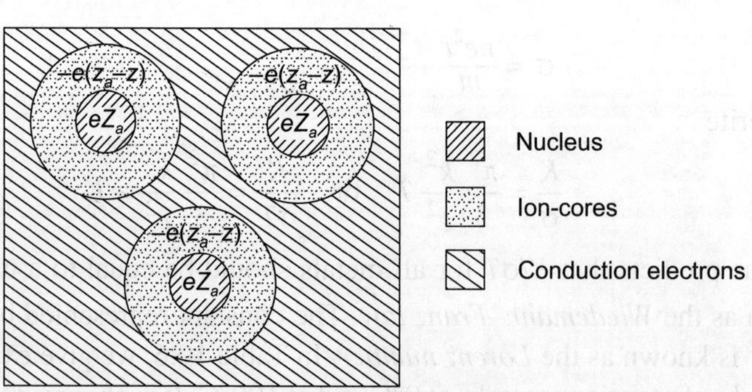


**Figure 10.15** Schematic picture of a metal. Valence electrons leave the atom to form the electron gas

Further, since at temperature close to room temperature, there is no electron emission from metal surface, potential outside the metal is taken very high ($\simeq \infty$), whereas inside, the potential energy is neglected and the total energy of the electrons is assumed to be totally kinetic.

The free-electron density may range from ~2 × $10^{28}$ $m^{-3}$ to upwards depending on the location of the metallic elements in the periodic table.

In this, above classical free-electron model, conduction electrons are constantly colliding with the ion cores. The mean free path of a conduction electron is controlled primarily by these elastic collisions. In a metal containing $N$ ion cores per unit volume, each of effective scattering radius $R$, the mean free path of an electron would be

$$\lambda \simeq \frac{1}{\pi R^2 N} \quad ...(1)$$

from purely geometric considerations. The mean free time between two successive collisions would be

$$\tau_m = \frac{\lambda}{v_{th}} \quad ...(2)$$

where $v_{th}$ denotes the thermal speed with which an electron moves from one scattering centre to another. Drift velocity exerted by external field would be small but systematic velocities superimposed on a thermal motion of much larger amplitude, which cancels out as an ensemble average.

The successes of the Drude's model were considerable and failures were resolved by using quantum theory in place of classical theory. Despite its limitations, the model did introduce some ideas that are still incorporated in the more elaborate treatments.

### 10.12.1 Lorentz Modification of Drude's Model

Lorentz modified the oversimplified Drude model and added the following points

(i) The assumption that all electrons move with the same velocity was dropped. (There is a distribution of velocities.)

(ii) The classical Maxwell-Boltzmann velocity distribution is perturbed by the presence of an electric field or a thermal gradient. They tend to displace the equilibrium velocity distribution and distort its symmetry.

(iii) Boltzmann transport equation was employed to describe the transport of charge and kinetic energy of electrons which constitute the electron gas.

### 10.12.2 Some Features of the Electrical Conductivity of Metals

Apart from deviations in special cases, the general features of the electrical conductivity of metals are as follows:

1. In accordance with Ohm's law, the current density in the steady state is proportional to the field strength.
2. The specific resistivity of metals at room temperature is of the order of $10^{-5}$ ohm-cm (1 ohm cm ≅ 1.1 × $10^{-12}$ cgs units).
3. Above the Debye temperature, the resistivity of metals increases linearly with temperature.

4. At low temperatures (~ >20°K) the resistivity of many metals is proportional to $T^5$; at liquid helium temperatures, some metals exhibit a minimum in the resistivity versus temperature curve.
5. For most metals, the resistivity decreases with increasing pressure.
6. According to Matthiessen's rule, the resistivity $P$ of a metal containing small amounts of impurities may be written
$$p = p_0 + p(T)$$
where $p_0$ is a constant which increases with increasing impurity content and $P(T)$ is the temperature dependent part of the resistivity.
7. Above the Debye temperature, the ratio of thermal to electrical conductivity is proportional to $T$, the constant of proportionality being approximately the same for all metals (Wiedemann-Franz Law).
8. A number of metals exhibit the phenomenon of superconductivity, i.e., their resistivity disappears at temperatures above absolute zero.[5]

## 10.13 A SIMPLE MODEL LEADING TO A STEADY STATE, DRIFT VELOCITY AND RELAXATION TIME

In order to appreciate the essential problem in the theory of conductivity it is useful to consider a simple model which shows the features of the more sophisticated theory. From the macroscopic point of view, the electrical conductivity of a metal is defined by

$$I_x = \sigma E_x \qquad ...(1)$$

where $I_x$ is the current density resulting from an applied electric field $E_x$ in the $x$-direction. In the case of an anisotropic solid, the conductivity depends on direction, and σ becomes a tensor; we shall assume an isotropic solid. From an atomic viewpoint, we may ascribe the current to a flow of electrons, i.e.,

$$I_x = -ne\,\langle v_x \rangle \qquad ...(2)$$

where $n$ is the number of electrons per unit volume, $-e$ is the electronic charge, and $\langle v_x \rangle$ is the average velocity of the electrons in the $x$-direction (the average being taken over the electrons per unit volume). In the absence of an external field, the velocity distribution is isotropic and $\langle v_x \rangle$ vanishes. Now, a free electron under influence of an external field $E_x$ obtains an acceleration $a_x = -eE_x/m$, and thus its velocity would continue to increase with time. It is evident that the influence of the electric field alone would not lead to a steady state as required by Ohm's law; it is therefore necessary to assume the occurrence of some kind of "frictional" process. This process together with the influence of the external field should then lead to an average velocity $\langle v_x \rangle$ which, according to Eqs. (1) and (2), should be proportional to $E_x$. The origin of the "frictional" process must obviously be sought in a possible interaction of the conduction electrons with the atomic lattice, since collisions between the electrons themselves cannot provide the required result (the latter would not destroy momentum in the field direction).

[5] This topic will not be discussed here; for an introductory survey and refrences to the literature, see C. Kittel, *Introduction to Solid State Physics*, Wiley, New York, 1953 Chap. 20.

In its simplest phenomenological form the interaction of the electrons with the lattice may be described in the following manner: suppose the probability for an electron to collide with the lattice during a small time interval $dt$ is $dt/\tau$. For the moment we shall assume for simplicity that $\tau$ is constant, independent of the energy of the electron and of the direction of motion. Furthermore, let it be assumed that in a collision with the lattice, the electron loses all the energy it has gained from the external field and that its velocity after the collision is random (independent of the direction of motion before the collision). In other words, the collisions are assumed to be so designed that immediately after the collision the electron has no memory of what happened before the collision. Under the terms of the model specified above, we may argue in the following way: The rate of change of the average velocity in the $x$-direction due to the field alone is

$$(\partial\langle v_x\rangle/\partial t)_{\text{field}} = -eE_x/m \qquad ...(3)$$

Also, the rate of change of $\langle v_x\rangle$ due to collisions with the lattice alone is

$$(\partial\langle v_x\rangle/\partial t)_{\text{coll}} = -\langle v_x\rangle/\tau \qquad ...(4)$$

since $1/\tau$ is the probability for a collision per second and after the collisions the velocities are random. In the steady state we must have

$$d\langle v_x\rangle/dt = 0 = (\partial\langle v_x\rangle/\partial t)_{\text{field}} + (\partial\langle v_x\rangle/\partial t)_{\text{coll}} \qquad ...(5)$$

From the last three equations it then follows that the average drift velocity in the field direction is given by

$$\langle v_x\rangle = (-e\tau/m)\,E_x \qquad ...(6)$$

From Eqs. (2) and (6) it then follows that the conductivity is given by

$$\sigma = ne^2\tau/m \qquad ...(7)$$

Suppose that under the influence of an electric field $E_x$ the electrons have a certain average drift velocity and that at the instant $t = 0$ the field is suddenly switched off. As a result of the collisions with the lattice the average drift velocity will gradually approach zero; since the rate of change of $\langle v_x\rangle$ by collisions alone is given by Eq. (4), the decay will follow the expression

$$\langle v_x(t)\rangle = \langle v_x(0)\rangle\, e^{-t/\tau} \qquad ...(8)$$

where $\langle v_x(0)\rangle$ is the average drift velocity at $t = 0$. Because of the exponential form of Eq. (8), the quantity $\tau$ is called the relaxation time.

We may note here already that with $n \simeq 10^{22}$ cm$^{-3}$, expression (7) requires $\tau \simeq 10^{-14}$ second in order to obtain agreement with experimental room temperature data.

For the particular type of collisions postulated above, $\tau$ also represents the mean free time between collisions. This may be shown as follows: Let $P(t)$ be the probability that $t$ seconds after a certain collision has occurred, an electron has not yet collided again; $P(t + dt)$ represents the same quantity after $(t + dt)$ seconds. Then

$$P(t + dt) = P(t) + (dP/dt)\,dt$$

On the other hand, we may also write

$$P(t + dt) = P(t)\, P(dt) = P(t)\,(1 - dt/\tau)$$

where $(1 - dt/\tau)$ represents the probability for an electron not to collide during the interval $dt$. From the last two equations one finds

$$P(t) = e^{-t/\tau}$$

since $P = 1$ for $t = 0$. Hence, the mean free time between collisions is

$$\langle t \rangle = \int_0^\infty t(dP/dt)\,dt = \tau \qquad ...(9)$$

It must be emphasized, however, that the relaxation time and the mean free time between collisions are identical only if the velocity after collision is random. For example, if the scattering is not isotropic and $\tau_c$ is the mean free time between collisions, the relaxation time can readily be shown to be

$$\tau = \tau_c/(1 - \langle \cos \beta \rangle) \qquad ...(10)$$

where $\langle \cos \beta \rangle$ is the average of the cosine of the scattering angle. Thus, when nearly all collisions involve small angles, the electron has a rather strong "memory" and it takes a relatively large number of collisions to erase this memory, i.e., $\tau \gg \tau_c$ in that case.

## 10.14 THE BOLTZMANN TRANSPORT EQUATION

Suppose we have a metal or a semiconductor solid. It will be evident that in a state of steady flow of heat or electricity, the distribution function $f$ for the velocity components and spatial coordinates of the electrons will be different from that in thermal equilibrium in the absence of flow. Thus, the theory of transport phenomenon is concerned with determining this distribution function for given external fields. We shall see in this section that the determination of the distribution function requires solving an integrodifferential equation, viz., the Boltzmann transport equation.[6]

Let $p_x$, $p_y$, $p_z$ represent the components of the momentum of an electron and let

$$f(p_x p_y p_z;\, xyz;\, t)\; dp_x dp_y dp_z\; dx\, dy\, dz \qquad ...(11)$$

represent the number of electrons in the volume element $dx\,dy\,dz$ which at the instant $t$ have momenta in the range $dp_x dp_y dp_z$. The steady state is then defined by

$$df/dt = 0 \qquad ...(12)$$

In order to obtain information about the function $f$, it is necessary to consider the causes which, when operative by themselves, would tend to produce a change of $f$ with time. First of all, we must consider the rate of change of $f$ resulting from the velocities of the electrons and from the components $X$, $Y$, $Z$ of the external forces which are assumed to act on the electrons. Consider the group of particles defined by (11) at an instant $t + \delta t$, where $\delta t$ is a very small time interval. The moments and spatial coordinates of this group of electrons at $t + \delta t$ are then to be found about the point

$$p_x + X\,\delta t; \quad p_y + Y\delta t; \quad p_z + Z\,\delta t$$

[6] L. Boltzmann, *Vorlesungen über Gastheorie*, Barth, Leipzig, 1923.

$$x + p_x\,\delta t/m; \quad y + p_y\,\delta t/m \quad z + p_z\,\delta t/m \qquad ...(13)$$

However, according to the definition of the distribution function (11) the number of electrons which at the instant $t + \delta t$ have their representative points in an element $dp_x\, dp_y\, dp_z\, dx\,dy\,dz$ around the point defined by Eq. (13) must be equal to

$$f(p_x + X\,dt, ...; x + p_x\,\delta t/m ...; t + \delta t)\; dp_x\, dp_y\, dp_z\, dx\,dy\,dz \qquad ...(14)$$

Since Eqs. (11) and (14) must be equal, one is led to the following result, obtained by expanding (14),

$$(\partial f/\partial t)_{\text{fields}} = -\frac{\partial f}{\partial p_x}X - \frac{\partial f}{\partial p_y}Y - \frac{\partial f}{\partial p_z}Z - \frac{\partial f}{\partial x}v_x - \frac{\partial f}{\partial y}v_y - \frac{\partial f}{\partial z}v_z \qquad ...(15)$$

where $v_x$, $v_y$, $v_z$ represent the velocity components. In the steady state there must be other processes which just balance the rate of change (15) produced by fields and gradients. As we have noted in the preceding section, such processes are provided by electron-lattice interactions. Thus, condition (12) may be written in the form

$$(\partial f/\partial t)_{\text{fields}} + (\partial f/\partial t)_{\text{coll}} = 0 \qquad ...(16)$$

where the first term is given by Eq. (15) and where the second term refers to electron-lattice scattering (compare (5)). Since the force exerted on an electron by a combined electric and magnetic field is given by the Lorentz expression

$$\mathbf{F} = -e\left(\mathbf{E} + \frac{1}{c}\mathbf{v} \times \mathbf{H}\right)$$

where $\mathbf{v}$ is the velocity vector, we may write Eq. (14) combined with Eq. (15) in the general form

$$(\partial f/\partial t)_{\text{coll}} = -e\left(E + \frac{1}{c}v \times H\right)\cdot \text{grad}_p f + v\cdot\text{grad}_r f \qquad ...(17)$$

which is the Boltzmann transport equation for electrons.

The left hand side of this equation involves an integral operator, making the equation an integrodifferential equation; this may be seen as follows: The number of electrons per unit volume which, due to collisions with the lattice, change their momenta per unit time from the range $dp_x\, dp_y\, dp_z$ to another range $dp'_x\, dp'_y\, dp'_z$ can be represented by

$$f(\boldsymbol{p}, \boldsymbol{r}, t)\; dp_x dp_y dp_z\; P(\boldsymbol{p}, \boldsymbol{p}', \boldsymbol{r})\; dp'_x\, dp'_y\, dp'_z$$

where the transition probability $P(\boldsymbol{p}, \boldsymbol{p}', \boldsymbol{r})$ is determined by the type of electron-lattice interaction. Similarly, the corresponding number of electrons thrown from the range $dp'_x\, dp'_y\, dp'_z$ into $dp_x\, dp_y\, dp_z$ per unit time is

$$f(\boldsymbol{p}', \boldsymbol{r}, t)\; dp'_x\, dp'_y\, dp'_z\, P(\boldsymbol{p}', \boldsymbol{p}, \boldsymbol{r})\; dp_x dp_y dp_z$$

The net difference between the above quantities integrated over $dp'_x\, dp'_y\, dp'_z$ determines $(\partial f/\partial t)_{\text{coll}}$ i.e.,

$$(\partial f/\partial t)_{\text{coll}} = \iiint [f(\boldsymbol{p}', \boldsymbol{r}, t)\, P(\boldsymbol{p}', \boldsymbol{p}, \boldsymbol{r}) - f(\boldsymbol{p}, \boldsymbol{r}, t) P(\boldsymbol{p}, \boldsymbol{p}', \boldsymbol{r})]\, dp'_x\, dp'_y\, dp'_z \quad ...(18)$$

It is evident that since the left-hand side of Eq. (17) is given by expression (18), containing the transition probabilities $P$, the distribution function in the state of steady flow depends explicitly on the mechanism of interaction between the electrons and the lattice. From the atomic theory of electron scattering it can be shown that under certain circumstances it is possible to define a relaxation time such that $(\partial f/\partial t)_{\text{coll}}$ takes the form

$$(\partial f/\partial t)_{\text{coll}} = -\frac{f(\boldsymbol{p}, \boldsymbol{r}) - f_0}{\tau(\boldsymbol{p}, \boldsymbol{r})} \quad ...(19)$$

(compare (4). Here $f_0$ represents the distribution function in thermal equilibrium in the absence of external fields. The physical meaning of the relaxation time $\tau(\boldsymbol{p}, \boldsymbol{r})$ is analogous to that of $\tau$ introduced in the preceding section: when the external fields are suddenly removed, $(f - f_0)$ decays to zero in the fashion

$$(f - f_0)_t = (f - f_0)_{t=0}\, e^{-1/\tau} \quad ...(20)$$

When for a certain problem a relaxation time exists, the treatment is strongly simplified, since the integrodifferential equation then becomes an ordinary equation. An example of this type will be discussed in the next section.

Special cases for which a relaxation time can be defined consistently may be mentioned here.[7]

(i) In processes whereby the electrons may be considered to be scattered by elastic spheres; this is of importance for that part of the resistivity which is due to impurity scattering.

(ii) From the (approximate) theory of the interaction between electrons and lattice vibrations it follows that a relaxation time can be defined when $(\theta/T)^2 \ll 1$, where $\theta$ is the Debye temperature. This simplifies the theory of electrical and thermal conductivities at high temperatures.

## 10.15 THE SOMMERFELD THEORY OF ELECTRICAL CONDUCTIVITY BASED ON FREE ELECTRON APPROXIMATION

A theory of metallic conductivity based on average velocities, as employed in section 10.13, was developed by Drude in 1900. Lorentz in 1905 reinvestigated the problem, using the Boltzmann transport equation and a simplified model for the collisions between the electrons and atoms in the lattice. However, the use of classical statistics led to serious difficulties; for a review of these theories, we refer to the literature.[8] In 1928 Sommerfeld recalculated the conductivities along the lines of Lorentz, theory, but replacing classical statistics by Fermi–Dirac statistics. Sommerfeld did not investigate the actual mechanism of interaction between the electrons and the lattice any further but assumed that a relaxation time can be defined which is a function of the energy of the electrons only. As an application to the Boltzmann

---

[7] See, for example, A. H. Wilson, *The Theory of Metals*, 2nd ed., Cambridge, London, 1953, pp. 8, 264.

[8] Boltzmann Transport Equation.

transport equation, we shall discuss below Sommerfeld's theory of the electrical conductivity based on the free electron approximation; the thermal conductivity will be discussed later.

The number electronic states per unit volume associated with an element $dp_x\ dy_y\ dp_z$ in momentum space is (including the spin) $(2/h^3)\ dp_x\ dp_y\ dp_z$. In thermal equilibrium and in the absence of fields, let the average number of occupied states be

$$(2/h^3)\ F_0(p)\ dp_x\ dp_y\ dp_z$$

where $F_0$ is the Fermi distribution function in terms of the total momentum $p$. Suppose we apply an electric field $E_x$ along the $x$-direction, other fields or gradients being absent. When in the state of steady current, the average number of electrons per unit volume in the range $dp_x\ dp_y\ dp_z$ is represented by

$$(2/h^3)\ F(\boldsymbol{p})\ dp_x\ dp_y\ dp_z \qquad ...(21)$$

we may write immediately for the current density,

$$I_x = -\frac{2e}{h^3} \iiint v_x\,(F - F_0)\ dp_x\ dp_y\ dp_z \qquad ...(22)$$

which is a generalization of Eq. (2). The term $F_0$ does not contribute anything to the current, since it is spherically symmetric with respect to $\boldsymbol{p}$; it has been added, however, to emphasize the fact that the current is essentially determined by the deviation $(F - F_0)$ from the Fermi distribution. Thus, if one can calculate $(F - F_0)$, $I_x$ may be obtained.

The Boltzmann transport Eq. (17) reduces for the case under consideration to

$$(\partial F/\partial t)_{\text{coll}} = -e\,E_x(\partial F/\partial p_x) \qquad ...(23)$$

We shall now assume that there exists a relaxation time $\tau$ such that

$$(\partial F/\partial t)_{\text{coll}} = -(F - F_0)/\tau \qquad ...(24)$$

(compare Eq. (19)). Thus, according to the last two equations,

$$(F - F_0)/\tau = eE_x\,(\partial F/\partial p_x) \simeq eE_x\,(\partial F_0/\partial p_x) \qquad ...(25)$$

where the last approximation is valid for small fields so that $(F - F_0)$ is relatively small (physically speaking this assumption is equivalent with a linear dependence of $I$ and $E$). Making use of the fact that the energy of the electrons is given by $\varepsilon = (p_x^2 + p_y^2 + p_z^2)/2m$, one may write instead of Eq. (25),

$$(F - F_0)/\tau = ev_xE_x\,(\partial F_0/\partial \varepsilon) \qquad ...(26)$$

Substituting Eq. (26) into Eq. (22), one obtains for the current density,

$$I_x = -\,(2e^2/h^3)\ E_x \iiint v_x^2\,\tau\,(\partial F_0/\partial \varepsilon)\ dp_x\ dp_y\ dp_z \qquad ...(27)$$

We shall assume that $\tau$ is a function only of the energy and not of the direction of motion (compare Eq. 19). Since $\partial F_0/\partial\varepsilon$ is also a function of $\varepsilon$ alone, one may transform Eq. (27) into a single integral by replacing $v_x^2$ by $v^2/3$ and $dp_x\ dp_y\ dp_z$ by $4\pi p^2\ dp$. Expressing the integrand in terms of $\varepsilon$, one obtains

$$I_x = -\frac{16\pi e^2 (2m)^{1/2}}{3h^2} E_x \int_0^\infty \varepsilon^{3/2} \tau(\varepsilon) \left(\frac{\partial F_0}{\partial \epsilon}\right) d\varepsilon \quad ...(28)$$

Now we know that $\partial F_0/\partial \varepsilon$ has an appreciable value only in an energy range of a few $kT$ about the Fermi level, $\varepsilon_F$. To a good approximation $\varepsilon^{3/2} \tau(\varepsilon)$ under the integral sign may thus be replaced by the quantity $\varepsilon_F^{3/2} \tau_F$ in front of the integral. Furthermore,

$$\int_0^\infty (\partial F/\partial \varepsilon)\, d\varepsilon = -1$$

and if one substitutes for $\varepsilon_F$, one finally obtains the simple result,

$$I_x/E_x = \sigma = ne^2 \tau_F/m \quad ...(29)$$

where $n$ is the number of electrons per unit volume. It is interesting to note that although all electrons take part in the conduction mechanism only the relaxation time of the electrons at the Fermi level occurs in the conductivity. The reason for this may be explained with reference to Fig. 10.16. The full circle represents the Fermi distribution for a two-dimensional case in the absence of an external field. In the presence of a field along the $x$-direction, the velocity of all electrons is shifted by an amount $\Delta v$ (the average drift velocity), leading to the dashed circle. It is evident that the distribution is changed only in the vicinity of the Fermi level, so that only the relaxation time of electrons near $\varepsilon_F$ is of importance.

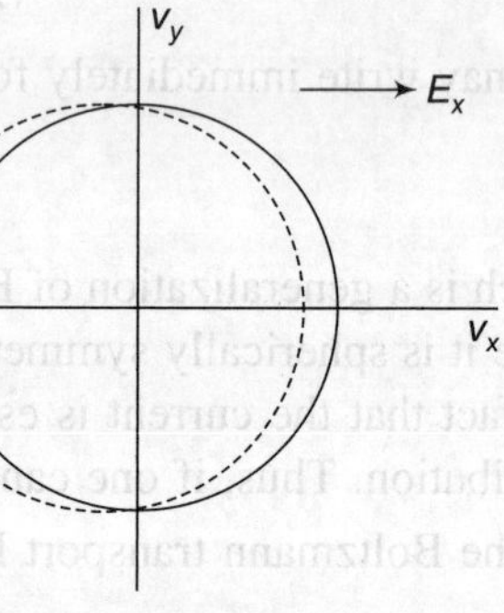


**Figure 10.16** Exaggerated representation of the influence of an electric field on the velocity distribution for a two-dimensional crystal. The fully drawn circle corresponds to the Fermi distribution in the absence of a field; the field $E_x$ produces a shift $\Delta_v$ opposite to the field direction (dashed curve)

Note that (29) is essentially the same as Eq. (7), except that $\tau$ has been replaced by $\tau_F$. Although the treatment given here was based on the free electron approximation, a similar treatment may be given for the band approximation.[9] The result of such a calculation is

$$\sigma = n_{\text{eff}}\, e^2\, \tau_F/m \quad ...(30)$$

i.e., $n$ is replaced by the effective number of free electrons $n_{\text{eff}}$. It must be noted that Eq. (30) is based on the assumption that the energy of the electron as well as $\tau$ are functions of the absolute value of the wave vector only.

## 10.16 CLASSICAL THEORY OF ELECTRICAL CONDUCTION

When a constant electric field is established in some materials, current flows. Assuming a local relationship, one writes $\mathbf{J}(\mathbf{r}) = \sigma \mathbf{E}(\mathbf{r})$, where $\mathbf{J}$ is the current per unit area (current density),

[9] See, for example, N. F. Mott and H. Jones, *Theory of the Properties of Metals and Alloys*, Oxford, New York, 1936, p. 258.

σ the electrical conductivity and **E** the electric field. For an isotropic medium, σ is a scalar and the vectors **J** and **E** are parallel. For an anisotropic material σ becomes a tensor and these vectors need no longer be parallel. The microscopic form of Ohm's law states that σ is independent of the electric field. Its magnitude determines whether the material is a conductor (high σ), an insulator (low σ), or a semiconductor (intermediate σ). The difference in the value of σ among materials can be enormous, varying by more than 20 orders of magnitude in going from a good conductor to a good insulator. At first sight it is a mystery how, simply by changing the atomic number a few units in the periodic table, one can observe such a huge variation in a physical parameter. The answer to the mystery will involve understanding how tightly (relative to $k_BT$) the valence electrons are bound to the atoms and ions. We begin this section by studying metals and then proceed to semiconductors and insulators.

### 10.16.1 Drude Theory

An early attempt at a theory of Ohm's law was made by Drude, who applied Newtonian mechanics to study the motion of electrons through a metal. The electrons that are not tightly bound to atoms or ions (i.e., the "conduction" electrons), are accelerated by an applied electric field and are assumed to collide with "scatterers," which deflect them and randomize their velocities to a thermal distribution. Consider a sequence of collisions labelled by the index $j$. Let the velocity of an electron just after the $j$th collisions be $\mathbf{v}_j$. Between collisions $j$ and $j+1$ the velocity is

$$\mathbf{v}(t) = \mathbf{v}_j - \frac{e\mathbf{E}}{m}(t - t_j), \qquad \text{...(1)}$$

since the acceleration is given by $-e\mathbf{E}/m$. The probability of surviving to time $t$ without making a collision and then making a collision between times $t$ and $t + dt$ is

$$dP = e^{-(t-tj)/\tau}\frac{dt}{\tau}, \qquad \text{...(2)}$$

where τ is the mean time between collisions. The mean velocity between times $t_j$ and $t_{j+1}$ is

$$\langle v_j(t)\rangle = \int \mathbf{v}(t)\,dP = \int_{t_i}^{\infty}\left[\mathbf{v}_j - \frac{e\mathbf{E}}{m}(t-t_j)\right]e^{-(t-t_j)/\tau}\frac{dt}{\tau}. \qquad \text{...(3)}$$

The direction of $\mathbf{v}_j$ is random so will average to zero for all the conduction electrons. The term $-eE\tau/m$ is the drift velocity $\langle\mathbf{v}\rangle$. The average current density is

$$\mathbf{J} = -ne\langle\mathbf{v}(t)\rangle = \frac{ne^2\tau}{m}\mathbf{E} = ne\mu\mathbf{E} \qquad \text{...(4)}$$

where $n$ is the number of conduction electrons per unit volume. The parameter $\mu = \langle v\rangle/E = e\tau/m$ is called the mobility of the electron. Thus, the *Drude conductivity formula is obtained*:

$$\sigma = \frac{ne^2\tau}{m}. \qquad \text{...(5)}$$

The *Joule heating formula* may be derived similarly by examining the kinetic energy loss in a typical collision:

$$\Delta K = \frac{m}{2}\mathbf{v}_{j+1}^2 - \frac{m}{2}\left[\mathbf{v}_j - \frac{e}{m}\mathbf{E}(t_{j+1} - t_j)\right]^2. \quad ...(6)$$

On an average $\langle \mathbf{v}_{j+1}^2 \rangle = \langle \mathbf{v}_j^2 \rangle$ and $\langle \mathbf{v}_j \rangle = 0$, so

$$\langle \Delta K \rangle = -\frac{e^2\tau^2 E^2}{m}. \quad ...(7)$$

Since the mean time between collisions is $\tau$, the power produced per unit volume is

$$\mathscr{P} = -n\frac{\langle \Delta K \rangle}{\tau} = \sigma E^2. \quad ...(8)$$

Use has been made of the integral $\int t^2\, dP = 2\tau^2$, with $dP$ given by Eq. (2). This power is dissipated as heat through the production of phonons or other excitations of the metal.

Representative values for the electrical conductivity of metals are given in Table 10.7. The electron density, $n$, is computed from the relation $n = N_A \nu\rho_m/A$, where $N_A$ Avogadro's number, $z$ the valence, $\rho_m$ the mass density, and $A$ the atomic weight. Note that the SI unit for conductivity is $(\Omega \cdot \text{m})^{-1}$. The values of the collision times computed from Eq. (5) are displayed.

Without collisions $\sigma$ and $\tau$ are infinite and the metal is perfect conductor. This state may actually be achieved in superconductors.

**Table 10.7** Parameters of Some Metals at $T = 295$ K

| *Metel* | *Atomic Number A* | *Valence z* | *Mass Density $\rho_m$ ($10^3$ kg/m$^3$)* | *Electron Densityn n ($10^{29}$ m$^{-3}$)* | *Conductivity $\sigma$ [$10^6$ ($\Omega$ m)$^{-1}$]* | *Collision Time $\tau$ ($10^{-15}$ s)* |
|---|---|---|---|---|---|---|
| Ag | 107.9 | 1 | 10.5 | 0.585 | 62.1 | 37.6 |
| Al | 26.98 | 3 | 2.70 | 1.81 | 36.5 | 7.17 |
| Ba | 137.3 | 2 | 3.59 | 0.315 | 2.6 | 2.93 |
| Be | 9.012 | 2 | 1.82 | 2.43 | 30.8 | 4.50 |
| Ca | 40.08 | 2 | 1.53 | 0.460 | 27.8 | 21.5 |
| Cd | 112.4 | 2 | 8.65 | 0.927 | 13.8 | 5.29 |
| Cs | 132.3 | 1 | 2.00 | 0.091 | 5.0 | 19.5 |
| Cu | 63.55 | 1 | 8.93 | 0.846 | 58.8 | 24.7 |
| In | 114.8 | 3 | 7.29 | 1.15 | 11.4 | 3.53 |
| K | 39.10 | 1 | 0.91 | 0.140 | 13.9 | 35.2 |
| Li | 6.939 | 1 | 0.54 | 0.469 | 10.7 | 8.11 |
| Mg | 24.31 | 2 | 1.74 | 0.862 | 22.3 | 9.18 |

## 10.17 VARIATION OF ELECTRICAL RESISTIVITY OF METALS (MATTHIESEN RULE)

For metals, the electrical conductivity σ and mobility μ are given by

$$\sigma \simeq \frac{ne^2\tau(E_F)}{m^*} \quad ...(1)$$

$$\mu = \frac{e\tau(E_F)}{m^*} \quad ...(2)$$

where $\tau(E_F)$ is the relaxation time and $m^*$ is the effective mass of the electron.

To understand the temperature dependence of resistance, we need to consider the temperature dependence of $\tau(E_F)$ or μ because the electron concentration is independent of temperature. There are basically two scattering mechanisms: *phonon* and *defect* scattering. Assuming that the two mechanisms are independent of each other, the total scattering probability is the sum of the individual scattering probabilities. The scattering probability is inversely proportional to the average free time of flight of a carrier and hence, is inversely proportional to the relaxation time consequently,

$$\frac{1}{\tau} = \frac{1}{\tau_{ph}} + \frac{1}{\tau_{def}} \quad ...(3)$$

where $\tau_{ph}$ and $\tau_{def}$ are the average times between scattering from phonons and from defects respectively.

Now, the number of collisions per unit time is proportional to Ω (scattering cross section) and $v$ (the velocity of the particle). Therefore,

$$\frac{1}{\tau} \propto \Omega v \quad ...(4)$$

For metals, $v$ is simply the velocity at the Fermi surface and is therefore temperature independent. For defects, the cross section Ω is also temperature independent. Defect scattering therefore leads to a temperature-independent component of resistivity, $\rho_{def}$.

For phonon scattering, the scattering cross section can be set proportional to $\langle A^2(\vec{q})\rangle$, the mean square vibrational amplitude of the appropriate phonon (wave vector $\vec{q}$ and frequency $\omega_q$). In the limiting case of high temperature, it follows from the equipartition theorem,

$$M\omega_q^2 \langle A^2(\vec{q})\rangle = kT, \ (T \gg \theta_D) \quad ...(5)$$

where $M$ is the mass of the heavy ion core, and $k$, the Boltzmann constant. Thus,

$$\frac{1}{\tau_{ph}} \propto \langle A^2(\vec{q})\rangle \propto \frac{kT}{M\omega_q^2} \quad ...(6)$$

A rough estimate is obtained by replacing $\omega_q$ by the Debye frequency $\omega_D = k\theta_D/\hbar$ where $\theta_D$ is the Debye temperature. This yields

$$\tau_{ph} \propto \frac{M\theta_D^2}{T} \text{ for } T \gg \theta_D. \quad ...(7)$$

For $T < \theta_D$, the phonon excitation probability decreases rapidly and due to decreasing phonon energy, this leads to predominantly small angle scattering. In this case ($T < \theta_D$), an exact theory has been given by Gruneisen (1933) and according to him

$$\rho_{ph}(T) \propto T^5, \ (T < \theta_D) \qquad ...(8)$$

According to Eq. (3) we can write the resistivity $\rho = \frac{1}{\sigma} \propto \frac{1}{\tau}$ of a metal as the sum of a temperature-independent residual resistivity $\rho_{def}$ (due to defects) and a part due to phonon scattering $\rho_{ph}(T)$ which is linear in temperature at high temperature, i.e.,

$$\rho = \rho_{ph}(T) + \rho_{def} \qquad ...(9)$$

This behaviour is known as *Matthiesen rule*.

## 10.18 QUANTUM THEORY: SOMMERFELD MODEL

To explain the discrepancies between the theoretical and experimental values of different physical quantities, Sommerfeld put forward the quantum mechanical model of the electron. The following are the main features of this theory:

- The classical statistics is replaced by the quantum statistics, viz., Fermi–Dirac statistics which leads to the conclusion that the energy possessed by an electron is quantized.
- The crystal is assumed to consist of a large number of arrays of electrons which are placed in a potential energy box with its wall at a very high potential so that the electrons are restricted to remain within the crystal, but free to move within its confines.
- Inside the crystal, the electrons move in a constant potential field ($V = 0$).

### 10.18.1 Energy Levels of Electron in a One-dimensional Potential Box

Consider an electron confined in a one-dimensional crystal of length $L$ whose ends are at very high potential $V_0$ whereas the potential everywhere within the crystal is constant and equal to zero, as shown in Fig. 10.17. The high potential at the two ends of crystal prevents the electrons from leaving the crystal.

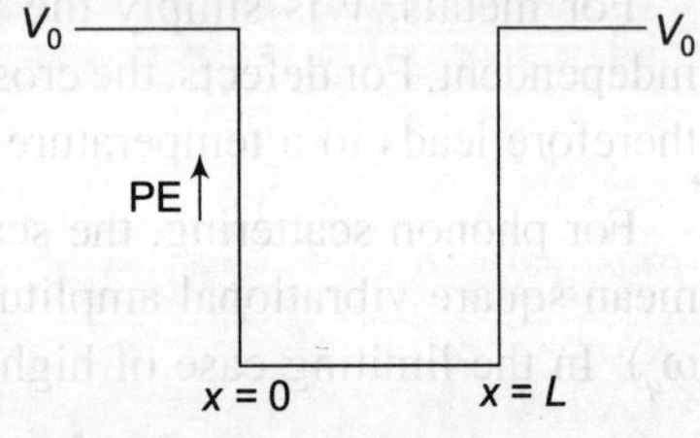

**Figure 10.17** Potential energy of electrons confined in one-dimensional potential box

In one dimension, the Schrödinger equation takes the form

$$\frac{d^2\psi}{dx^2} + \frac{2mE}{\hbar^2}\psi = 0$$

$$\Rightarrow \quad \frac{d^2\psi}{dx^2} + \frac{8\pi^2 mE}{h^2}\psi = 0 \qquad ...(1)$$

The general solution of Eq. (1) is

$$\psi = A\sin\frac{2\pi}{h}\sqrt{2mEx} + B\cos\frac{2\pi}{h}\sqrt{2mEx} \qquad ...(2)$$

where $A$ and $B$ are constants whose values can be determined by imposing the boundary conditions existing at $x = 0$ and $x = L$.

On either side of the crystal where the potential is very high, Schrödinger equation has the form

$$\frac{d^2\psi}{dx^2} - \frac{2m(V_0 - E)}{\hbar^2}\psi = 0$$

$$\Rightarrow \qquad \frac{d^2\psi}{dx^2} - \frac{8\pi^2(V_0 - E)}{h^2}\psi = 0 \qquad ...(3)$$

The general solution of Eq. (3) is

$$\psi = C\exp\left(\frac{2\pi}{h}\sqrt{2m(V_0 - E)}\,x\right) + D\exp\left(-\frac{2\pi}{h}\sqrt{2m(V_0 - E)}\,x\right) \qquad ...(4)$$

An inspection of Eq. (4) shows that as $x$ approaches $\pm\infty$, the wave function also approaches infinity which means that the probability of finding an electron outside the crystal, increases with increasing distance from the crystal. This is simply absurd.

Consequently, $C = 0$ for $x > L$

$D = 0$ for $x < 0$

We thus get two different solutions, one each for either side of the crystal. Thus,

$$\psi = C\exp\left(\frac{2\pi}{h}\sqrt{2m(V_0 - E)}\,x\right) \text{ for } x \le 0 \qquad ...(5)$$

and
$$\psi = D\exp\left(-\frac{2\pi}{h}\sqrt{2m(V_0 - E)}\,x\right) \text{ for } x \ge 0 \qquad ...(6)$$

Since there is finite probability of finding the electron anywhere in space, the wave function must satisfy the following two conditions:

1. It must be everywhere continuous.
2. Its derivative, $d\psi/dx$, must be everywhere continuous.

We now apply the condition that $d\psi/dx$ is continuous at $x = L$. Differentiating Eq. (6) and substituting $x = L$, we get

$$\left(\frac{d\psi}{dx}\right)_{x=L} = -\frac{2\pi}{h}\sqrt{2m(V_0 - E)}\; D\exp\left(-\frac{2\pi}{h}\sqrt{2m(V_0 - E)}\,x\right)$$

$$\Rightarrow \qquad \left(\frac{d\psi}{dx}\right)_{x=L} = -\frac{2\pi}{h}\sqrt{2m(V_0 - E)}\;\psi_{x=L} \qquad ...(7)$$

It follows from Eq. (7) that as the potential energy at the crystal surface approaches infinity[10], $d\psi/dx$ becomes infinite[11] unless $\psi_{x=L}$ tends to zero. Thus, the wave function $\psi$ must vanish at $x = L$. Similarly, by differentiating Eq. (5) and by imposing the condition that $d\psi/dx$ is continuous at $x = 0$, we can show that the wave function must vanish at $x = 0$.

We thus arrive at the conclusion that the wave function must vanish at the boundaries of the crystal where the potential is infinite.

Putting $\psi = 0$ at $x = 0$ in Eq. (2), we get

$$0 = A \sin 0 + B \cos 0$$

$$\Rightarrow \quad B = 0 \qquad \text{...(8)}$$

Again, putting $\psi = 0$ at $x = L$ in Eq. (2), we get

$$0 = A \sin \frac{2\pi}{h} \sqrt{2mE} \cdot L$$

As $A$ cannot be zero[12],

$$\sin \frac{2\pi}{h} \sqrt{2mE} \cdot L = 0 = \sin n\pi, \text{ where } n = 1, 2, 3, \ldots$$

$$\Rightarrow \quad \frac{2\pi}{h} \sqrt{2mE} \cdot L = n\pi$$

$$\Rightarrow \quad \sqrt{2mE} = \frac{nh}{2L} \qquad \text{...(9)}$$

$$\Rightarrow \quad E_n = \frac{n^2 h^2}{8mL^2} \qquad \text{...(10)}$$

Thus, an electron can possess only discrete values of energy given by Eq. (10), which are closely spaced.

The general solution of Eq. (1), as given by Eq. (2), now reduces to

$$\psi = A \sin \frac{2\pi}{h} \sqrt{2mE} x$$

$$\Rightarrow \quad \psi = A \sin \frac{2\pi}{h} \cdot \frac{nh}{2L} \cdot x, \text{ using Eq. (9)}$$

$$\Rightarrow \quad \psi_n = A \sin \frac{n\pi x}{L} \qquad \text{...(11)}$$

---

10 Although infinitely high potentials are not encountered in practice, this asumption greatly simplifies the derivation.

11 This is impossible.

12 If both $A$ and $B$ are zero, the wave function vanishes for all values of $x$. The solution of Eq. (1), therefore. becomes trivial.

The value of the constant $A$ has to be chosen in such a way that the total probability of finding the electron within the crystal is unity. Since $\psi_n(x)\,\psi_n^*(x)$ represents the probability of finding the electron in the line segment $dx$ at $x$, we must have

$$\int_0^L \psi_n(x)\psi_n^*(x)\,dx = 1$$

$$\Rightarrow \quad \int_0^L A^2 \sin^2\left(\frac{n\pi x}{L}\right) dx = 1$$

$$\Rightarrow \quad A^2 \cdot \frac{L}{2} = 1$$

$$\Rightarrow \quad A = \sqrt{\frac{2}{L}}$$

Thus,
$$\psi_n = \sqrt{\frac{2}{L}} \sin\left(\frac{n\pi x}{L}\right) \qquad ...(12)$$

### 10.18.2 Filling Up of Energy Levels

We now consider the distribution of conduction electrons around various allowed energy levels. The Pauli exclusion principle states that no two electrons in an atom can have all their quantum numbers identical. Each orbital can thus be occupied by only one electron. This applies to electrons in atoms, molecules or solids.

A conduction electron in a linear solid is characterized by two quantum numbers $n$ and $m_s$, where $n$ can take any positive integral values and $m_s = \pm 1/2$, according to its spin orientation. Thus, two electrons can be accommodated in a pair of orbitals characterized by quantum number $n$, one with spin up and the other with spin down.

Consider now the building of a free electron gas by dropping electrons one by one in a box. The first electron when dropped into the box will take the position of minimum kinetic energy. The second electron when dropped into the box can also take the position of minimum kinetic energy, provided it orients its spin anti-parallel to the spin orientation of the first electron. The third electron when dropped into the box cannot take the minimum kinetic energy, as two of the states belonging to this energy level have already been occupied and the Pauli exclusion principle forbids multiple occupancy. The third electron will, therefore, go to the next slightly higher energy level. The fourth electron when dropped will also go to the same energy level as the third electron provided it orients itself anti-parallel to the spin orientation of the third electron. Assuming that there are $N$ conduction electrons to be accommodated (where $N$ is even), if this process is repeated, the first $N/2$ energy levels will be occupied and all the higher energy levels will be completely empty. The highest filled energy level is known as the Fermi level and the energy corresponding to the Fermi level (or Fermi surface), as the *Fermi energy*.

If $n_F$ denotes the value of $n$ for the uppermost filled level then

$$n_F = \frac{N}{2} \qquad ...(13)$$

Substituting $n = n_F$ in Eq. (10), we get the expression for the Fermi energy $E_F$ as

$$E_F = \frac{n_F^2 h^2}{8mL^2} = \frac{h^2}{8mL^2}\left(\frac{N}{2}\right)^2$$

$$\Rightarrow \qquad E_F = \frac{\hbar^2}{8m}\left(\frac{N\pi}{2L}\right)^2 \qquad ...(14)$$

## 10.18.3 Average Kinetic Energy of Conduction Electron in Ground State

This can be determined by dividing the total energy by the total number of electrons. The total energy $E_0$ possessed by all the electrons in the ground state of the system can be obtained by taking summation of energies of all the levels from $n = 1$ to $n = N/2$. Using Eq. (10) this gives[13]

$$E_0 = 2\sum_{n=1}^{N/2} E_n = 2\times\frac{h^2}{8mL^2}\sum_{n=1}^{N/2} n^2$$

$$\Rightarrow \qquad E_0 = 2\times\frac{h^2}{8mL^2}(1^2 + 2^2 + 3^2 + ... + n^2)$$

$$\Rightarrow \qquad E_0 = 2\times\frac{h^2}{8mL^2}\left[\frac{n(n+1)(2n+1)}{6}\right]$$

$$\Rightarrow \qquad E_0 = \frac{h^2}{8mL^2}\times\frac{n(n+1)(2n+1)}{3}$$

$$\Rightarrow \qquad E_0 = \frac{h^2}{8mL^2}\times\frac{n(2n^2 + 3n + 1)}{3}$$

$$\Rightarrow \qquad E_0 \approx \frac{h^2}{8mL^2}\times\frac{2n^3}{3}, \text{ as } n = \frac{N}{2} >> 1$$

$$\Rightarrow \qquad E_0 \approx \frac{h^2}{8mL^2}\times\frac{2}{3}\left(\frac{N}{2}\right)^3 = \frac{h^2}{8mL^2}\times\frac{2N}{3\times 2}\times\left(\frac{N}{2}\right)^2$$

$$\Rightarrow \qquad E_0 = \frac{N}{3}\left[\frac{h^2}{8mL^2}\left(\frac{N}{2}\right)^2\right] = \frac{N}{3}\left[\frac{\hbar^2}{2m}\left(\frac{N\pi}{2L}\right)^2\right] = \frac{N}{3}\times E_F, \text{ using Eq. (14)}$$

---

[13] $\sum_{n=1}^{N/2} E_n$ has been multiplied by 2 as there are two electrons in each level.

$$\Rightarrow \qquad \frac{E_0}{N} = \frac{E_F}{3} \qquad ...(15)$$

Thus, in one-dimensional problems, the average kinetic energy possessed by a conduction electron is about one-third of the Fermi energy.

## 10.19 THE FREE ELECTRON GAS IN THREE DIMENSIONS

Suppose the object in question is a rectangular solid, with dimensions $l_x$, $l_y$, $l_z$, and imagine that an electron inside experiences no forces at all, except at the impenetrable walls:

$$V(x, y, z) = \begin{cases} 0, & \text{if } 0 < x < l_x, \quad 0 < y < l_y, \text{and } 0 < z < l_z; \\ \infty, & \text{otherwise.} \end{cases} \qquad ...(1)$$

The Schrödinger equation,

$$-\frac{\hbar^2}{2m}\nabla^2\psi = E\psi,$$

separates, in Cartesian coordinates: $\psi(x, y, z) = X(x)\,Y(y)\,Z(z)$, with

$$-\frac{\hbar^2}{2m}\frac{d^2X}{dx^2} = E_xX; \; -\frac{\hbar^2}{2m}\frac{d^2Y}{dy^2} = E_yY; \; -\frac{\hbar^2}{2m}\frac{d^2Z}{dz^2} = E_zZ,$$

and $E = E_x + E_y + E_z$.

Let $$k_x \equiv \frac{\sqrt{2mE_x}}{\hbar}, \quad k_y \equiv \frac{\sqrt{2mE_y}}{\hbar} \quad k_z \equiv \frac{\sqrt{2mE_z}}{\hbar},$$

we obtain the general solutions

$$X(x) = A_x \sin(k_x x) + B_x \cos(k_x x),$$
$$Y(y) = A_y \sin(k_y y) + B_y \cos(k_y y),$$
$$Z(z) - A_z \sin(k_z z) + B_z \cos(k_z z).$$

The boundary conditions require that $X(0) = Y(0) = Z(0) = 0$, so $B_x = B_y = B_z = 0$, and $X(l_x) = Y(l_y) = Z(l_z) = 0$, so that

$$k_x l_x = n_x\pi, \quad k_y l_y = n_y\pi, \quad k_z l_z = n_z\pi, \qquad ...(2)$$

where each $n$ is a positive integer:

$$n_x = 1, 2, 3 ..., \quad n_y = 1, 2, 3, ..., \quad n_z = 1, 2, 3, .... \qquad ...(3)$$

The (normalized) wave functions are

$$\psi_{n_x n_y n_z} = \sqrt{\frac{8}{l_x l_y l_z}} \sin\left(\frac{n_x\pi}{l_x}x\right) \sin\left(\frac{n_y\pi}{l_y}y\right) \sin\left(\frac{n_z\pi}{l_z}z\right). \qquad ...(4)$$

and the allowed energies are

$$E_{n_x n_y n_z} = \frac{\hbar^2\pi^2}{2m}\left(\frac{n_x^2}{l_x^2} + \frac{n_y^2}{l_y^2} + \frac{n_z^2}{l_z^2}\right) = \frac{\hbar^2 k^2}{2m}, \qquad ...(5)$$

where $k$ is the magnitude of the **wave vector,** $\mathbf{k} \equiv (\mathbf{k_x}, \mathbf{k_y}, \mathbf{k_z})$.

If you imagine a three-dimensional space, with axes $k_x$, $k_y$, $k_z$ and planes drawn in at $k_x = (\pi/l_x), (2\pi/l_x), (3\pi/l_x), ...,$ at $k_y = (\pi/l_y), (2\pi/l_y), (3\pi/l_y), ...$ and at $k_z = (\pi/l_z), (2\pi/l_z), (3\pi/l_z), ...,$ each interaction point represents a distinct (one-particle) stationary state (Fig. 10.18).

Each block in this grid, and hence also each state, occupies a volume

$$\frac{\pi^3}{l_x l_y l_z} = \frac{\pi^3}{V} \qquad ...(6)$$

of "$k$-space", where $V \equiv l_x l_y l_z$ is the volume of the object itself. Suppose our sample contains $N$ atoms, and each atom contributes $q$ free electrons. (In practice, $N$ will be enormous — of the order of Avogadro's number, for an object of macroscopic size— whereas $q$ is a small number—1 or 2, typically.) If electrons were bosons (or distinguishable particles), they would all settle down to the ground state, $\psi_{111}$. But electrons are, in fact, identical *fermions*, subject to the Pauli exclusion principle, so only two of them can occupy any given state. They will fill up one octant of a *sphere* in $k$-space,[14] whose radius, $k_F$, is determined by the fact that each pair of electrons requires a volume $\pi^3/V$ (Eq. (6)).

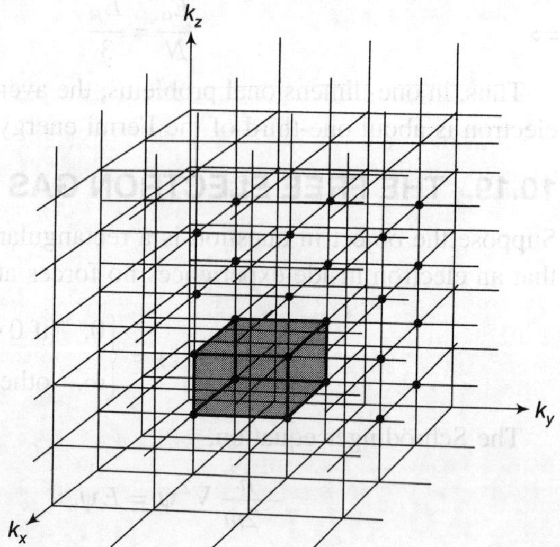


**Figure 10.18** Free electron gas. Each intersection on the grid represents a stationary state. shading indicates one "block"; there is one state for every block

$$\frac{1}{8}\left(\frac{4}{3}\pi k_F^3\right) = \frac{Nq}{2}\left(\frac{\pi^3}{V}\right).$$

Thus, $$k_F = (3\rho\pi^2)^{1/3}, \qquad ...(7)$$

where $$\rho \equiv \frac{Nq}{V} \qquad ...(8)$$

is the *free electron density* (the number of free electrons per unit volume).

The boundary separating occupied and unoccupied states, in $k$-space, is called the **Fermi surface** (hence the subscript **F**). The corresponding energy is called the **Fermi energy,** $E_F$; for a free electron gas,

$$E_F = \frac{\hbar^2}{2m}(3\,\rho\pi^2)^{2/3}. \qquad ...(9)$$

[14] Because $N$ is such a huge number, we need not worry about the distinction between the actual jagged edge of the grid and the smooth spherical surface that approximates it.

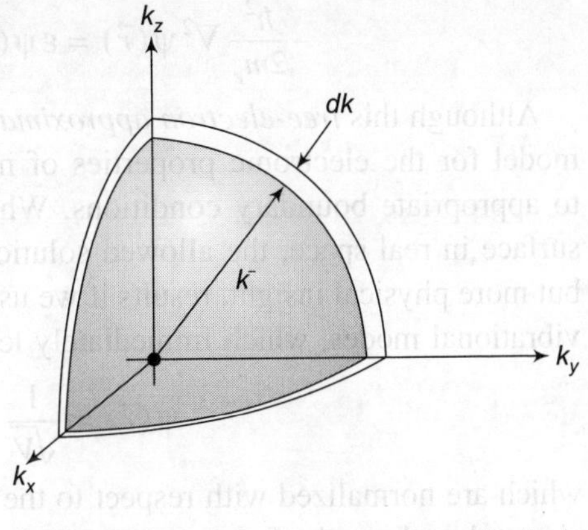

**Figure 10.19** One octant of a spherical shell in *k*-space

The *total* energy of the electron gas can be calculated as follows: A shell of thickness $dk$ (Fig. 10.19) contains a volume

$$\frac{1}{8}(4\pi k^2)\,dk,$$

so the number of electron states in the shell is

$$\frac{2[1/2\pi k^2 dk]}{(\pi^3/V)} = \frac{V}{\pi^2}k^2 dk.$$

Each of these states carries an energy $\hbar^2 k^2/2m$ (Eq. (5)), so the energy of the shell is

$$dE = \frac{\hbar^2 k^2}{2m}\frac{V}{\pi^2}k^2\,dk, \qquad ...(10)$$

and hence the total energy is

$$E_{\text{tot}} = \frac{\hbar^2 V}{2\pi^2 m}\int_0^{k_F} k^4\,dk = \frac{\hbar^2 k_F^5 V}{10\pi^2 m} = \frac{\hbar^2 (3\pi^2 Nq)^{5/3}}{10\pi^2 m}V^{-2/3}. \qquad ...(11)$$

This quantum mechanical energy plays a role rather analogous to the internal *thermal* energy ($U$) of an ordinary gas. In particular, it exerts a *pressure* on the walls, for if the box expands by an amount $dV$, the total energy decreases:

$$d\,E_{\text{tot}} = -\frac{2}{3}\frac{\hbar^2 (3\pi^2 Nq)^{5/3}}{10\pi^2 m}V^{-5/3}\,dV = -\frac{2}{3}E_{\text{tot}}\frac{dV}{V},$$

and this shows up as work done on the outside ($dW = P\,dV$) by the quantum pressure $P$. Evidently,

$$P = \frac{2}{3}\frac{E_{\text{tot}}}{V} = \frac{2}{3}\frac{\hbar^2 k_F^5}{10\pi^2 m} = \frac{(3\pi^2)^{2/3}\hbar^2}{5m}\rho^{5/3} \qquad ...(12)$$

Here, then, is a partial answer to the question of why a cold solid object doesn't simply *collapse*: There is a stabilizing internal pressure, having nothing to do with electron-electron repulsion (which we have ignored) or thermal motion (which we have excluded), but is strictly quantum mechanical, and derives ultimately from the antisymmetrization requirement for the wave functions of identical fermions. It is sometimes called **degeneracy pressure**, though "exclusion pressure" might be a better term.

## 10.20 FREE–ELECTRON APPROXIMATION

A convenient approach to the electronic properties of a solid consists of using the one-electron model, where the core electrons are treated as almost completely localized, and the energy eigenvalue problem for the valence electron system, problem reduces to solving the Schrödinger wave equation for each valence electron as it moves in the spatially dependent potential energy $U(\vec{r})$. If forces on the valence electrons can be ignored, $U(\vec{r})$ does not depend on $\vec{r}$ such that, if we drop $U(\vec{r})$, it reduces to the free-electron problem:

$$-\frac{\hbar^2}{2m_e}\nabla^2\psi(\vec{r}) = \varepsilon\psi(\vec{r})$$

Although this *free-electron approximation* is a drastic assumption, it provides an excellent model for the electronic properties of metals. Solutions of this wave equation are subject to appropriate boundary conditions. When assuming that $\psi(\vec{t})$ must vanish over a closed surface in real space, the allowed solutions are standing waves. The same density of states, but more physical insight, results if we use the periodic boundary conditions, as for the lattice vibrational modes, which immediately lead to allowed plane wave solutions of the form:

$$\psi(\vec{r}) = \frac{1}{\sqrt{V}}e^{i\vec{k}\cdot\vec{r}} \qquad ...(1)$$

which are normalized with respect to the volume $V$ of the crystal. The corresponding energy eigenvalues have the form:

$$\varepsilon = \frac{\hbar^2 k^2}{2m_e} \qquad ...(2)$$

and the momentum eigenvalues of free electrons are given by:

$$\hat{\vec{p}}\,\psi(\vec{r}) = \frac{\hbar}{i}\nabla\left\{\frac{1}{\sqrt{V}}e^{i\vec{k}\cdot\vec{r}}\right\} = \hbar\vec{k}\psi(\vec{r}) \quad \text{or} \quad \vec{p} = \hbar\vec{k} \qquad ...(3)$$

If we assume the volume $V$ to be a finite parallelepiped of sides $P\vec{a}$, $Q\vec{b}$ and $S\vec{c}$, the *periodic boundary conditions* require that the solutions are periodic over the distances $Pa$, $Qb$ and $Sc$ respectively, namely:

$$\psi(\vec{r} + P\vec{a}) = \psi(\vec{r} + Q\vec{b}) = \psi(\vec{r} + S\vec{c}) = \psi(\vec{r})$$

Substituting Eq. (1) yields:

$$P\vec{k}\cdot\vec{a} = 2\pi p, \quad Q\vec{k}\cdot b = 2\pi q, \; S\vec{k}\cdot\vec{c} = 2\pi s$$

where $p$, $q$ and $s$ are arbitrary integers. Comparing these equations with the definitions of the reciprocal base set, it is clear that $\vec{k}$ can be considered to be a vector in $\vec{k}$-space:

$$k = \frac{p}{P}\vec{a}* + \frac{q}{Q}\vec{b}* + \frac{s}{S}\vec{c}*$$

Note that, for a general vector in reciprocal space, the components $p/P$, $q/Q$, and $s/S$ can be fractions of an integer. If they are integers $h = p/P$, $k = q/Q$ and $l = s/S$, then $\vec{k}$ becomes a reciprocal lattice vector, denoted by $\vec{K}$. In view of the definition of the first Brillouin zone, we can write $k^2 = (\vec{k} + \vec{K})^2$, it follows that the propagation vector may always be decomposed into the sum of a reciprocal lattice vector $\vec{K}$ and a vector $\vec{k}$ lying in the first Brillouin zone. We note that the lattice vibration properties have been discussed in terms of $\vec{k}$ values in the first zone only, but there is no such restriction for electrons, where two or more zones will often be considered. Hence, Eq. (1) can be rewritten as:

$$\psi_{\vec{k}}(\vec{r}) = \frac{1}{\sqrt{V}}e^{i\vec{k}\cdot\vec{r}}e^{i\vec{K}\cdot\vec{r}} \qquad ...(4)$$

where $e^{i\vec{K}\cdot\vec{r}}$ is a function with the periodicity of the lattice, and the corresponding eigenvalue, Eq. (2), becomes a function of the *reduced wave vector* $\vec{k}$:

$$\varepsilon_{\vec{k}} = \frac{\hbar^2}{2m_e}(\vec{k} + \vec{K})^2 \quad ...(5)$$

In this representation, each possible $\vec{k}$-state outside the first zone is translated by a specific $\vec{K}_{hkl}$ and remapped into the zone. The electronic energy calculations are usually presented for wave numbers along symmetry lines of the crystal structure.

Consider the specific example of the simple cubic structure with lattice parameter $a$ and reciprocal base set $(2\pi/a)\vec{e}_x$, $(2\pi/a)\vec{e}_y$, $(2\pi/a)\vec{e}_z$, such that the reciprocal lattice vector, equation reads

$$\vec{K}_{hkl} = \frac{2\pi}{a}(he_x + ke_y + le_z)$$

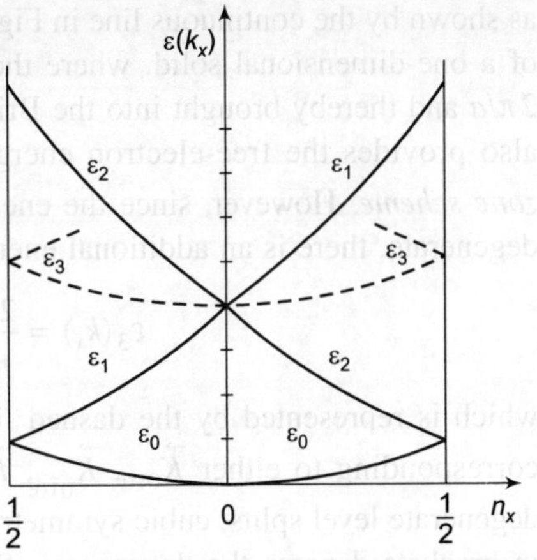


**Figure 10.20** Free-electron energy bands along [100] in the simple cubic structure

The first Brillouin zone is a cube with edge $2\pi/a$, thus each of $k_x$, $k_y$ and $k_z$ is in the range from $-\pi/a$ to $+\pi/a$, and the propagation vector can be written as:

$$\vec{k} = k_x\vec{e}_x + k_y\vec{e}_y + k_z\vec{e}_z = \frac{2\pi}{a}(q_x\vec{e}_y + q_y\vec{e}_y + q_z\vec{e}_z)$$

where $-\frac{1}{2} \leq q_x, q_y, q_z \leq +\frac{1}{2}$ are quasicontinuous variables, because the allowed wave numbers are very closely spaced in a large system. It is then sensible to represent the eigenvalues, Eq. (5), as continuous functions of the reduced wave number, called *energy brands*:

$$\varepsilon(\vec{k}) = \frac{\hbar^2}{2m_e}(\vec{k} + \vec{K}) = \frac{\hbar^2}{2m_e}\frac{4\pi^2}{a^2}[(q_x + h)^2 + (q_y + k)^2 + (q_z + k)^2] \quad ...(6)$$

If we consider, for simplicity, the [100] direction in the Brillouin zone, the energy becomes a function of $k_x$ only:

$$\varepsilon(k_x) = \frac{2\pi^2\hbar^2}{m_e a^2}[(q_x + h)^2 + k^2 + l^2] \quad ...(7)$$

The lowest energy band is provided by Eq. (7), for $\vec{K}_{hkl} = 0$ or $(h = k = l = 0)$ as:

$$\varepsilon_0(k_x) = \frac{2\pi^2\hbar^2}{m_e a^2} q_x^2$$

which is a nondegenerate parabolic band. The minimum of this band lies at $\vec{k} = 0$. The next energy bands are obtained by translations $\vec{K}_{100}$ or $\vec{K}_{\bar{1}00}$ along [100] in the form:

$$\varepsilon_1(k_x) = \frac{2\pi^2\hbar^2}{m_e a^2}(q_x + 1)^2, \; \varepsilon_2(k_x) = \frac{2\pi^2\hbar^2}{m_e a^2}(q_x - 1)^2$$

as shown by the continuous line in Fig. 10.20. In this special case the problem reduces to that of a one-dimensional solid, where the energy bands are simply translated by a multiple of $2\pi/a$ and thereby brought into the Brillouin zone. Hence, the continuous line representation also provides the free-electron energy bands in one dimension, in the so-called *reduced zone scheme*. However, since the energy level $\varepsilon_{1,2}$ $(k_x = 0) = 2\pi^2\hbar^2/m_e a^2$ should be sixfold degenerate, there is an additional energy band in simple cubic solids:

$$\varepsilon_3(k_x) = \frac{2\pi^2\hbar^2}{m_e a^2}(q_x^2 + 1)$$

which is represented by the dashed line in Fig. 10.19. This is a fourfold degenerate band, corresponding to either $\vec{K}_{010}$, $\vec{K}_{0\bar{1}0}$, $\vec{K}_{001}$ or $\vec{K}_{00\bar{1}}$ translations. In other words, the sixfold degenerate level splits, cubic symmetry, into three levels $\varepsilon_1(1)$ $\varepsilon_2(1)$ $\varepsilon_3(4)$, where the number in brackets denotes the degeneracy. Similar energy bands can be further derived from Eq. (7) for higher $\vec{K}_{hkl}$ values, which are determined by all the allowed combinations of $h$, $k$ and $l$ integer values.

In the unit cell volume $V_0^* = \vec{a}^* \cdot (\vec{b}^* \times \vec{c}^*)$ of reciprocal space, defined the same way as $V_0$ for the direct lattice, there are $PQS$ $\vec{k}$-states of the free electron, so that the number of states per unit volume of reciprocal space is

$$\frac{PQS}{V_0^*} = \frac{PQS}{\vec{a}^* \cdot (\vec{b}^* \times \vec{c}^*)} = \frac{(PQS)\,V_0}{(2\pi)^3} = \frac{V}{(2\pi)^3} \quad \text{...(8)}$$

where $V$ is the crystal volume. Since the states in $\vec{k}$-space are uniformly distributed, the electron density of states is the product of this number of states per unit volume of reciprocal space and the volume between the surfaces of constant energies $\varepsilon$ and $\varepsilon + d\varepsilon$:

$$N(\varepsilon)\,d\varepsilon = \frac{V}{(2\pi)^3}\int_{S_\varepsilon} d^3k = \frac{V}{(2\pi)^3}\int_{S_\varepsilon} dS_\varepsilon\, dk = \frac{V}{(2\pi)^3}\int_{S_\varepsilon} \frac{dS_\varepsilon d\varepsilon}{d\varepsilon/dk}$$

where the integral is over the surface $S_\varepsilon$ of constant energy. In reciprocal space, we can write:

$$d\varepsilon = \frac{\partial\varepsilon}{\partial k_x}dk_x + \frac{\partial\varepsilon}{\partial k_y} + \frac{\partial\varepsilon}{\partial k_z}dk_z = (\nabla_{\vec{k}}\varepsilon)\cdot dk = |\nabla_{\vec{k}}\varepsilon|\,dk$$

and substituting $d\varepsilon/dk$ yields:

$$N(\varepsilon)\,d\varepsilon = \frac{V}{(2\pi)^3}\int_{S_\varepsilon}\frac{dS_\varepsilon}{|\nabla_{\vec{k}}\varepsilon|}\,d\varepsilon$$

Because each $\vec{k}$-state corresponds to one of the two possible eigenvalues $\pm\hbar/2$ of the $z$-component of spin, a factor of 2 should be included, to give:

$$N(\varepsilon) = \frac{2V}{(2\pi)^3}\int_{S_\varepsilon}\frac{dS_\varepsilon}{|\nabla_{\vec{k}}\varepsilon|} \quad \text{...(9)}$$

For free electrons of energy given by Eq. (2), one obtains:

$$|\nabla_{\bar{k}}\varepsilon| = \frac{\hbar^2}{2m_e}\left|\frac{\partial k^2}{\partial k_x}\vec{e}_x + \frac{\partial k^2}{\partial k_y}\vec{e}_y + \frac{\partial k^2}{\partial k_z}\vec{e}_z\right| = \frac{\hbar^2}{m_e}|k_x\vec{e}_x + k_y\vec{e}_y + k_z\vec{e}_z| = \frac{\hbar^2 k}{m_e}$$

and this is a constant on the surface $S_\varepsilon$, which is a sphere with radius $k = (2m_e\varepsilon/\hbar^2)^{\frac{1}{2}}$. It follows that:

$$N(\varepsilon) = \frac{2V}{(2\pi)^3}\frac{4\pi\ k^2}{|\nabla_{\bar{k}}\varepsilon|} = \frac{V}{2\pi^2}\left(\frac{2m_e}{\hbar^2}\right)k = \frac{V}{2\pi^2}\left(\frac{2m_e}{\hbar^2}\right)^{\frac{3}{2}}\varepsilon^{\frac{1}{2}} = C\varepsilon^{\frac{1}{2}} \qquad ...(10)$$

If we wish to accommodate a total of $N$ free electrons, one by one, into the volume $V$ of real space, *at absolute zero*, only one electron will be permitted by the Pauli exclusion principle, to be in a given state of the system, at a given time. In other words, occupation numbers $n_k$ of the electron states can assume only the values 0 or 1.

The lowest energy states will be progressively filled up to a maximum value $\varepsilon_F(0)$, called the *Fermi energy*, obtained by equating with $N$ the number of states with energy less than $\varepsilon = \varepsilon_F(0)$, which reads:

$$N = \frac{2V}{(2\pi)^3}\left(\frac{4\pi}{3}k^3\right) = \frac{Vk^3}{3\pi^2} = \frac{V}{3\pi^2}\left(\frac{2m_e\varepsilon}{\hbar^2}\right)^{\frac{3}{2}} \qquad ...(11)$$

Hence, we have

$$\varepsilon_F(0) = \frac{\hbar^2}{2m_e}\left(3\pi^2\frac{N}{V}\right)^{\frac{2}{3}} = \frac{\hbar^2 k_F^2}{2m_e} \qquad ...(12)$$

where the Fermi wave vector was introduced as

$$k_F = \left(3\pi^2\frac{N}{V}\right)^{\frac{1}{3}} \qquad ...(13)$$

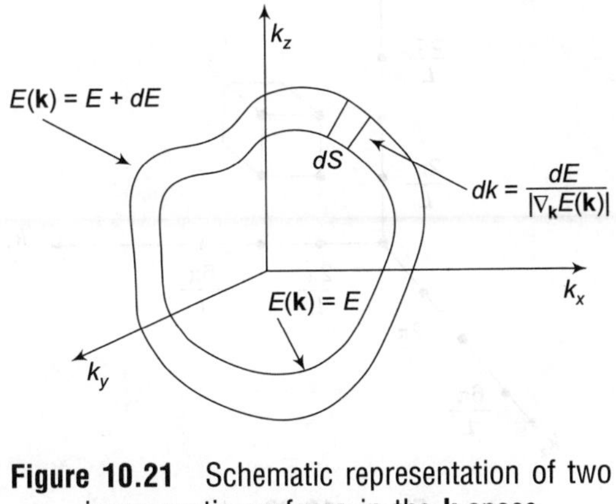


**Figure 10.21** Schematic representation of two isoenergetic surfaces in the **k**-space

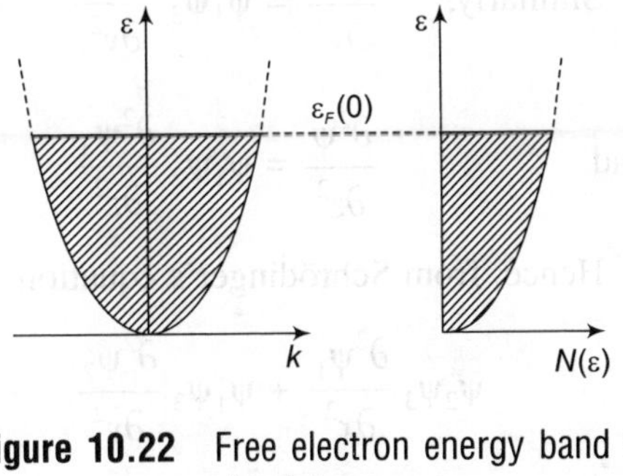


**Figure 10.22** Free electron energy band at absolute zero

There is one electron in each state below $\varepsilon_F(0)$ and none in the states situated above $\varepsilon_F(0)$, as illustrated in Fig. 10.22. The result is a parabolic energy band for free electrons at absolute zero, which is plotted in Fig. 10.22 as a function of both $k$, Eq. (3) and $N(\varepsilon)$, Eq. (10).

## 10.21 THE DENSITY OF STATES

A function that gives the number of available electron states per unit volume in an energy interval between $E$ and $E + dE$ is designated by $D(E)\,dE$, where $D(E)$ is known as the density of states function and represents the number of states per unit volume per electron volt.

We will now derive an expression for $D(E)$. For this, let us consider a crystal of finite volume. We assume that the effective mass is a scalar quantity and the $E(k)$ curve is parabolic.

Assume that we have a free carrier (electron or hole) of mass $m^*$ in the solid. Further, let potential $V = 0$. The time-independent Schrödinger equation is

$$\left.\begin{aligned} &\nabla^2\psi + \frac{2m^*E}{\hbar^2}\psi = 0 \\ \text{or} \quad &\nabla^2\psi + k^2\psi = 0 \end{aligned}\right\} \qquad ...(1)$$

where $$k^2 = 2m^*E/\hbar^2$$

Consider a volume in $k$-space with components $k_x$, $k_y$ and $k_z$ as shown in Fig. 10.23.

Since $$k^2 = k_x^2 + k_y^2 + k_z^2 \qquad ...(2)$$

$$E = \frac{\hbar^2}{2m^*}(k_x^2 + k_y^2 + k_z^2) \qquad ...(3)$$

Using the method of separation of variables, we can write

$$\psi(x, y, z) = \psi_1(x)\,\psi_2(y)\,\psi_3(z) \qquad ...(4)$$

Since $\psi_2$ and $\psi_3$ are independent of $x$,

$$\frac{\partial\psi}{\partial x} = \psi_2\psi_2\frac{\partial\psi_1}{\partial x} \quad \text{and} \quad \frac{\partial^2\psi}{\partial x^2}\psi_2\psi_3\frac{\partial^2\psi_1}{\partial x^2}$$

Similarly, $$\frac{\partial^2\psi}{\partial y^2} = \psi_1\psi_3\frac{\partial^2\psi_2}{\partial y^2}$$

and $$\frac{\partial^2\psi}{\partial z^2} = \psi_1\psi_2\frac{\partial^2\psi_3}{\partial z^2}$$

Hence, from Schrödinger's equation

$$\psi_2\psi_3\frac{\partial^2\psi_1}{\partial x^2} + \psi_1\psi_3\frac{\partial^2\psi_2}{\partial y^2}$$

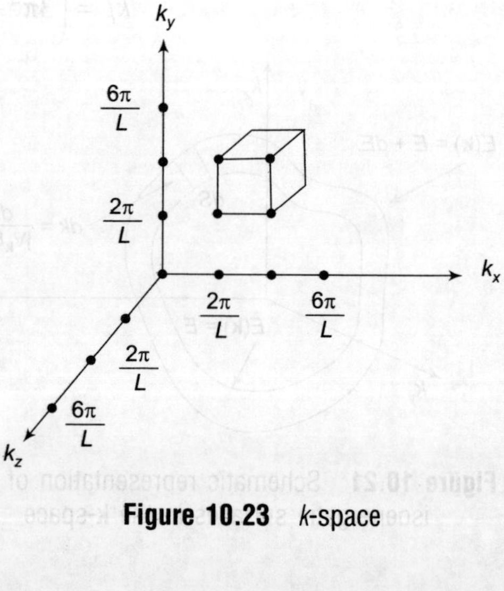


**Figure 10.23** $k$-space

$$+ \psi_1\psi_2 \frac{\partial\psi_3}{\partial z^2} = -k^2 \psi_1\psi_2\psi_3$$

or $$\frac{1}{\psi_1}(\partial^2\psi_1/\partial x^2) + \frac{1}{\psi_2}(\partial^2\psi_2/\partial y^2) + \frac{1}{\psi_3}(\partial^2\psi_3/\partial z^2) = -k^2 \quad ...(5)$$

This can only be true if each term on L.H.S. is a constant.

$$1/\Psi_1 \frac{\partial^2\psi_1}{\partial x^2} = -k_x^2; \; + 1/\Psi_2 \frac{\partial^2\psi_2}{\partial y^2} = -k_y^2; \quad 1/\Psi_3 \frac{\partial^2\psi_3}{\partial y^2} = -k_z^2$$

or $$\frac{\partial^2\psi_1}{\partial x^2} + k_x^2\,\psi_1 = 0; \quad \frac{\partial\psi_2}{\partial y^2} + k_y^2\,\psi_2 = 0; \quad \frac{\partial^2\psi_3}{\partial y^2} = k_z^2\,\psi_3 = 0 \quad ...(6)$$

The general solutions of these equations are of the form

$$\psi_1(x) = A \sin k_x x + B \cos k_x x \quad ...(7)$$

Assume $\psi(x)$ to be periodic with periodicity $L$, so when $x = 0$; $\psi_1 = 0$; hence $B = 0$.

If $x = L$ and $\psi_1 = 0$, then $A \sin k_x x = 0$,

or $k_x L = 2n_x\pi$

or $$k_x = \frac{2\pi}{L} n_x$$

where $n_x = 1, 2, 3$ ... (a positive integer)

Thus, the vector $\vec{k}$ has the components

$$k_x = \frac{2\pi}{L} n_x; \quad k_y = \frac{2\pi}{L} n_y; \quad k_z = \frac{2\pi}{L} n_z \quad ...(8)$$

($n_x$, $n_y$ and $n_z$ are positive integers).

From Fig. 10.23, the volume of a unit cell in $k$-space is $(2\pi/L)^3 = (8\pi^3/L^3)$.

Therefore, the volume of a unit cell in $k$-space for a crystal of unit volume is $(8\pi^3)$. For a crystal of unit volume, the volume of a spherical shell in $k$-space having radius between $k$ and $k + dk$ is given by

$$4\pi k^2 \, dk \quad ...(9)$$

Thus, the number of unit cells in the spherical shell is

$$\frac{4\pi k^2 \, dk}{8\pi^3} = \frac{k^2 \, dk}{2\pi^2} \quad ...(10)$$

The density of available electron states is twice of this because each cell can contain two electrons with different (i.e., opposite) spins,

$$\therefore \quad D(k)\,dk = \frac{k^2}{\pi^2}\,dk \quad ...(11)$$

is the number of electron states per unit volume within an infinitesimal range $dk$ of $k$. Now $k^2 = 2m^*E/\hbar^2$, therefore

$$2kdk = 2m^*dE/\hbar^2 \qquad ...(12)$$

and

$$D(E)\,dE = D(k)\,dk\,\frac{k^2}{\pi^2}\,dk \qquad ...(13i)$$

or

$$D(E)\,dE = \left(\frac{2m^*E}{\hbar^2}\right)\frac{1}{\pi^2}\frac{m^*\,dE}{\hbar^2 k} \qquad ...(13ii)$$

$$= \frac{2m^{*2}}{\hbar^2}\frac{E}{\pi^2\hbar^2}\left(\frac{\hbar}{\sqrt{2m^*}}\right)\frac{1}{\sqrt{E}}\,dE \text{ (putting the value of } k)$$

$$= \frac{2m^{*2}}{\pi^2\hbar^2}\frac{E^{112}}{(2m^*)^{1/2}}\,dE$$

$$= \frac{2m^{*2}}{\pi^2(h^3/8\pi^3)}\frac{E^{1/2}}{\sqrt{2m^{*1/2}}}\,dE$$

$$= \frac{4\pi}{h^3}\frac{(2m^*)^2}{(2m^*)^{1/2}}\,E^{1/2}\,dE$$

$$= \frac{4\pi}{h^3}(2m^*)^{3/2}\,E^{1/2}\,dE \qquad ...(14)$$

This gives the density of states $D(E)$ for a particle of mass $m^*$ with energy between $E$ and $E + dE$.

### 10.21.1 Density of States and Critical Points

In several problems the primary interest is not in the detailed wave vector dependence of the crystal band structure but only on the density of states in a given energy range. Consider for simplicity a non-degenerate band $E(k)$; the corresponding density of states for this band is given by

$$D(E) = 2\sum_k \delta(E(\mathbf{k}) - E) = 2\int_{B.Z.}\frac{V}{(2\pi)^3}\delta(E(\mathbf{k}) - E) - E)\,d\mathbf{k}, \qquad ...(1)$$

where $d\mathbf{k}$ is the volume element of the reciprocal space and the factor 2 accounts for the spin degeneracy (we suppose this degeneracy is not removed). Equation (1) shows that contributions to the density of states $D(E)$ at energy $E$ occur from the band states (if any) such that $E(\mathbf{k}) = E$; the factor $V/(2\pi)^3$ gives the uniform density of allowed $\mathbf{k}$ vectors in $\mathbf{k}$-space.

In Fig. 10.21 we indicate schematically the two isoenergetic surfaces $E(\mathbf{k}) = E$ and $E(\mathbf{k}) = E + dE$. The distance $dk$ between the two isoenergetic surfaces is obtained by observing that $dE = \nabla_{\mathbf{k}}E(\mathbf{k})\cdot d\mathbf{k} = |\nabla_{\mathbf{k}}E(\mathbf{k})|\,dk$. Counting the volume in $\mathbf{k}$-space enclosed between the two surfaces, the expression (1) for the density of states can be transformed into the integral on the constant energy surface $E(\mathbf{k}) = E$ in the form

$$D(E) = \int_{E(\mathbf{k})=E} \frac{V}{(2\pi)^3} \frac{dS}{|\nabla_{\mathbf{k}} E(\mathbf{k})|} \qquad ...(2)$$

Expression (2) clearly shows that singularities in the density of states are expected at the *critical points*, defined as those points in $k$-space for which

$$\nabla_{\mathbf{k}} E(\mathbf{k}) = 0; \qquad ...(3)$$

for these points the density of states is expected to exhibit anomalies as a function of energy.

At the critical points, Eq. (1) or Eq. (2) can be integrated analytically. In more general situations the density of states can be obtained numerically by appropriate sampling of the **k** vectors in the Brillouin zone.

Near a critical point, where Eq. (3) holds, we can expand the energy as a function of the wave vector, in quadratic form. Indicating with $k_x$, $k_y$, $k_z$ the principal axes of the quadratic form and taking the origin at the critical point itself, we have

$$E(\mathbf{k}) = E_c \pm \frac{\hbar^2}{2m_x} k_x^2 \pm \frac{\hbar^2}{2m_y} k_y^2 \pm \frac{\hbar^2}{2m_z} k_z^2 \qquad ...(4)$$

(where we choose $m_x$, $m_y$, $m_z > 0$, while the occurrence of plus or minus sign specifies the kind of critical point). The critical point $M_0$ denotes zero negative signs in expression (4) and is thus a minimum of $E(\mathbf{k})$; the critical point $M_3$ denotes three negative signs in expression (4) and is thus a maximum; $M_1$ and $M_2$ denote one and two negative signs, respectively, and are thus two saddle points. Thus, in three dimensions we have four types of critical points. Similarly, in two-dimensional crystals we have three types of critical points $M_0$, $M_1$ and $M_2$: a minimum, a saddle point and a maximum, respectively. In one-dimensional crystals we have two types of critical points: a minimum $M_0$ and a maximum $M_1$.

Near a three-dimensional critical point, the expression (1) for the density of states becomes

$$D(E) = \frac{2V}{(2\pi)^3} \int \delta\left( E_c \pm \frac{\hbar^2 k_x^2}{2m_x} \pm \frac{\hbar^2 k_y^2}{2m_y} \pm \frac{\hbar^2 k_z^2}{2m_z} - E \right) dk_x\, dk_y\, dk_z. \qquad ...(5a)$$

For the two-dimensional case, we have

$$D(E) = \frac{2L_x L_y}{(2\pi)^2} \int \delta\left( E_c \pm \frac{\hbar^2 k_x^2}{2m_x} \pm \frac{\hbar^2 k_y^2}{2m_y} - E \right) dk_x\, dk_y. \qquad ...(5b)$$

For the one-dimensional case, we have

$$D(E) = \frac{2L_x}{2\pi} \int \delta\left( E_c \pm \frac{\hbar^2 k_x^2}{2m_x} - E \right) dk_x. \qquad ...(5c)$$

All types of integrals (5) can be easily calculated analytically with the help of the following property of the delta function

$$\delta[f(x)] = \sum_n \frac{\delta(x - x_n)}{|f'(x_n)|}, \qquad ...(6)$$

where $x_n$ are the simple zeroes of the function $f(x)$.

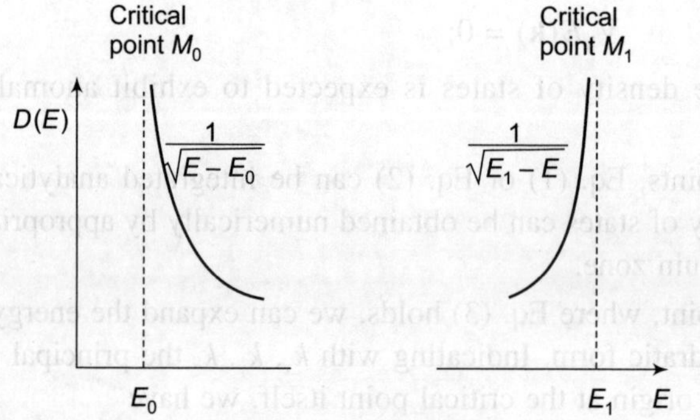


**Figure 10.24** Density of states at the critical points $M_0$ (minimum) and $M_1$ (maximum) of a one-dimensional crystal

### *Critical points in one-dimensional crystals*

In one-dimensional crystals, for the critical point of type $M_0$ at energy $E_0$, we consider the integral

$$D(E) = \frac{2L_x}{2\pi} \int \delta\left(E_0 + \frac{\hbar^2 k_x^2}{2m_x} - E\right) dk_x = L_x \frac{\sqrt{2m_x}}{\pi\hbar} \int \delta(E_0 + q_x^2 - E)\, dq_x \qquad ...(7)$$

where the change of variable $q_x = (\hbar/\sqrt{2m_x})\, k_x$ has been performed. Consider the function $f(q_x) = E_0 + q_x^2 - E$; the zeroes of this function occur for $q_{x_0} = \pm\sqrt{E - E_0}$ and $f'(q_{x_0}) = 2q_{x_0} = \pm 2\sqrt{E - E_0}$. The integral, Eq. (7) then becomes

$$D(E) = L_x \frac{\sqrt{2m_x}}{\pi\hbar} \frac{1}{\sqrt{E - E_0}} \qquad E > E_0. \qquad ...(8)$$

Similarly, for a critical point of type $M_1$ at energy $E_1$, we have

$$D(E) = L_x \frac{\sqrt{2m_x}}{\pi\hbar} \frac{1}{\sqrt{E_1 - E}} \qquad E > E_1. \qquad ...(9)$$

In Fig. 10.24 we give the behaviour of the density of states for the two possible critical points in one-dimensional crystals.

### *Critical points in two-dimensional crystals*

In two-dimensional crystals, for a critical point of type $M_0$ at energy $E_0$, we have to calculate the integral

$$D(E) = \frac{2L_xL_y}{(2\pi)^2}\int\delta\left(E_0 + \frac{\hbar^2k_x^2}{2m_x} + \frac{\hbar^2k_y^2}{2m_y} - E\right)dk_x\,dk_y. \quad ...(10)$$

It is convenient to introduce the variable $q_x = (\hbar/\sqrt{2m_x})\,k_x$ and $q_y = (\hbar/\sqrt{2m_y})\,k_y$ in Eq. (10), and denote by $S = L_xL_y$ the surface of the two-dimensional crystal; we have

$$D(E) = S\frac{\sqrt{m_xm_y}}{\pi^2\hbar^2}\int\delta(E_0 + q_x^2 - q_y^2 - E)\,dq_x\,dq_y.$$

We pass to cylindrical coordinates and easily obtain

$$D(E) = S\frac{\sqrt{m_xm_y}}{\pi\hbar^2} \quad \text{for} \quad E > E_0. \quad ...(11)$$

A similar expression holds for the critical point which is a maximum, for $E < E_2$. Thus, the density of states is step like at the points $M_0$ and $M_2$ of a two-dimensional crystal.

For the saddle critical point $M_1$ we have to evaluate the expression

$$D(E) = S\frac{\sqrt{m_xm_y}}{\pi^2\hbar^2}\int\delta(E_1 + q_x^2 - q_y^2 - E)\,dq_x\,dq_y. \quad ...(12)$$

We suppose momentarily $E > E_1$. Consider the function $f(q_x) = E_1 + q_x^2 - q_y^2 - E$; the zeroes of this function occur for $q_{x_0} = \pm\sqrt{E - E_1 + q_y^2}$ and $f'(q_{x_0})$. The integral (12) thus becomes

$$D(E) = S\frac{\sqrt{m_xm_y}}{\pi^2\hbar^2}\int_{-q_c}^{q_c}\frac{1}{\sqrt{E - E_1 + q_y^2}}\,dq_y,$$

where the integral has been confined to a cutoff where the series expansion is supposed to hold. With the help of the indefinite integral

$$\int\frac{dx}{\sqrt{a^2 + x^2}} = \ln(x + \sqrt{a^2 + x^2}),$$

we obtain
$$D(E) = S\frac{\sqrt{m_xm_y}}{\pi^2\hbar^2}\ln\frac{q_c + \sqrt{E - E_1 + q_c^2}}{-q_c + \sqrt{E - E_1 + q_c^2}} \quad ...(13)$$

For $E \approx E_1$, with appropriate series development in the second member of Eq. (13), we obtain

$$D(E) = S\frac{\sqrt{m_x m_y}}{\pi^2 \hbar^2} \ln \frac{4q_c^2}{|E - E_1|} \qquad ...(14)$$

and we see that a logarithmic divergence occurs at the critical point (with the absolute value $|E - E_1|$, Eq. (14) holds also for $E < E_1$). In Fig. 10.25 we show the behaviour of the density of states at the two-dimensional critical points.

***Critical points in three-dimensional crystals***

For three-dimensional crystals we have four types of critical points: $M_0$ (minimum), $M_1$ and $M_2$ (saddle points), $M_3$ (maximum). The analytic behaviour at the three-dimensional critical points can be obtained with procedures similar to the previous, reported calculations; we have:

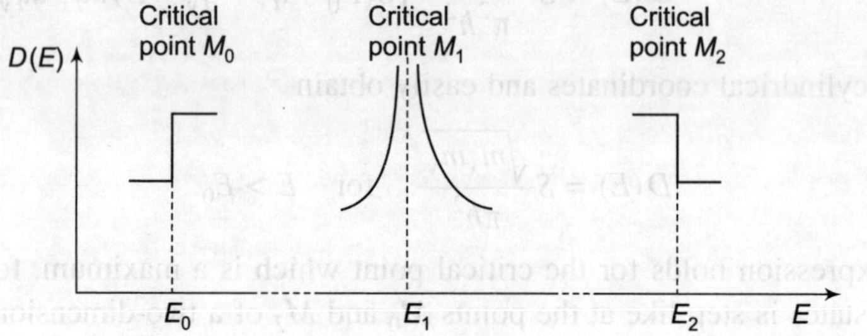


**Figure 10.25** Density of states at the critical points of type $M_0$ (minimum), $M_1$ (saddle point), and $M_2$ (maximum) of a two-dimensional crystal

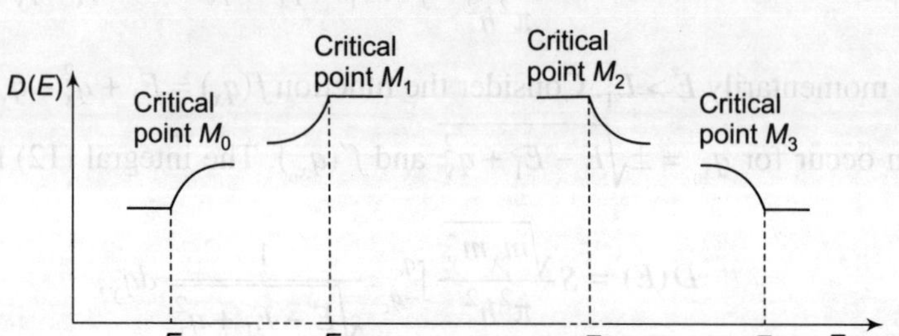


**Figure 10.26** Density of states at the critical points of type $M_0$ (minimum), $M_1$ and $M_2$ (saddle points), and $M_3$ (maximum) of a three-dimensional crystal

Critical point $M_0$ $$D(E) = C_0 + V\frac{\sqrt{2m_x m_y m_z}}{\pi^2 \hbar^3}\sqrt{E - E_0} \quad \text{for } E > E_0 \qquad ...(15a)$$

Critical point $M_1$ $$D(E) = C_1 - V\frac{\sqrt{2m_x m_y m_z}}{\pi^2 \hbar^3}\sqrt{E_1 - E} \quad \text{for } E > E_1 \qquad ...(15b)$$

Critical point $M_2$ $$D(E) = C_2 - V\frac{\sqrt{2m_x m_y m_z}}{\pi^2\hbar^3}\sqrt{E - E_2} \text{ for } E > E_2 \quad ...(15c)$$

Critical point $M_3$ $$D(E) = C_3 - V\frac{\sqrt{2m_x m_y m_z}}{\pi^2\hbar^3}\sqrt{E_3 - E} \text{ for } E > E_3. \quad ...(15d)$$

In the above expressions $C_i$ indicates either a constant (including zero) or a smoothly varying energy dependent quantity, while the terms with the square root are present only when the argument is positive. The results (15) are schematically indicated in Fig. 10.26.

## 10.22 THE NUMBER OF ELECTRONS/VOLUME, THE POPULATION DENSITY AND $E_{F_0}$

The number of electrons per unit volume having an energy $E$ within a band is given by[15]

$$N(E) = D(E), F(E) \quad ...(1)$$

where $$D(E) = \frac{4\pi}{h^3}(2m^*)^{3/2} E^{1/2} \quad ...(2)$$

and $$F(E) = \frac{1}{\exp\left(\frac{E - E_F}{k_B T}\right) + 1} \quad ...(3)$$

$N(E)$ is called the (electron) population density.

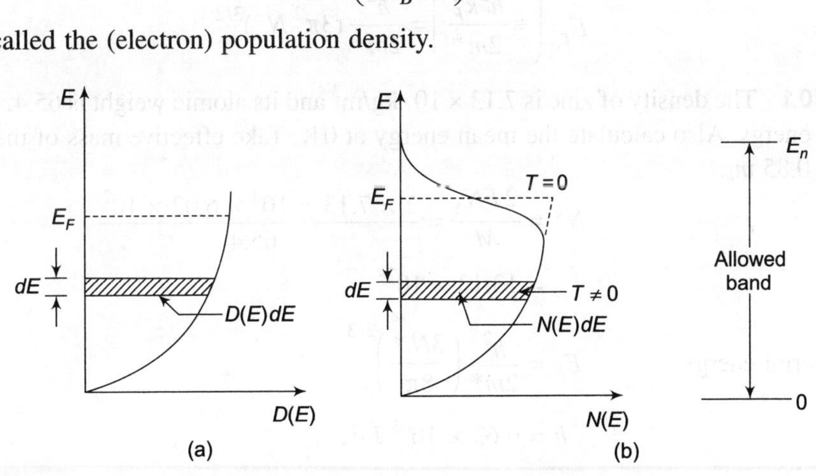


**Figure 10.27** (a) Density of states, (b) Population density within a band

The area under the curve in Fig. 10.27 (b) represents the number density $N^*$ of electrons that have an energy equal to or smaller than $E_n$. For an energy interval $dE$

$$dN^* = N(E)\, dE \quad ...(4)$$

[15] See Eq. (14), section 10.21.

We can now calculate the Fermi energy in the simple case of $T \rightarrow 0$ and $E < E_F$. In this case, $F(E) = 1$

$$N^* = \int_0^{E_F} N(E)dE = \int_0^{E_F} D(E)dE \qquad ...(5)$$

$$= \int_0^{E_F} \frac{4\pi}{h^3}(2m^*)^{3/2} E^{1/2} dE$$

$$= \frac{4\pi}{h^3}(2m^*)^{3/2} \frac{E_F^{3/2}}{3/2}$$

$$= \frac{8\pi}{3h^3}(2m^*)^{3/2} E_F^{3/2}$$

$$\therefore \quad E_F^{3/2} = \frac{3h^3}{8\pi} \frac{N^*}{(2m^*)^{3/2}}$$

or

$$\boxed{E_{F_0} = \left(\frac{3N^*}{8\pi}\right)^{2/3} \frac{h^2}{2m^*}} \qquad ...(6)$$

where $N^*$ is the number of electrons per unit volume.

We can rewrite, Fermi energy

$$E_{F_0}\left(\equiv \frac{\hbar^2 k_F^2}{2m^*}\right) = \frac{\hbar^2}{2m^*}(3\pi^2 N^*)^{3/2} \qquad ...(7)$$

**Example 10.1** The density of zinc is $7.13 \times 10^3$ kg/m$^3$ and its atomic weight is 65.4. Calculate the Fermi energy. Also calculate the mean energy at 0 K. Take effective mass of the electron in zinc as 0.85 $m_e$.

$$N^* = \frac{2FN_A}{M} = \frac{2 \times 7.13 \times 10^3 \times 6.02 \times 10^{26}}{65.4}$$

$$= 13.13 \times 10^{28}$$

Now, Fermi energy

$$E_F = \frac{h^2}{2m^*}\left(\frac{3N^*}{8\pi}\right)^{2/3}$$

$$h \simeq 6.63 \times 10^{-3} \text{ J–s},$$

$$E_F = \frac{(6.63)^2 \times 10^{-68}}{2 \times 9.1 \times 0.85 \times 10^{-31}}\left(\frac{3}{8\pi}\right)^{2/3}(N^*)^{2/3}$$

$$= \frac{43.95}{15.47} \times 10^{-37} \times (0.1193)^{2/3} (N^*)^{2/3}$$

$$= 0.688 \times 10^{-37} \times (N^*)^{2/3}$$

$$= \frac{0.688}{1.6 \times 10^{-19}} \times 10^{-37} \times (N^*)^{2/3} \text{ eV}.$$

$$= 4.2 \times 10^{-19} (N^*)^{2/3} \text{ eV}$$

$$= 4.2 \times 10^{-19} \times (13.13 \times 10^{28})^{2/3} \text{ eV}.$$

$$= 9.3 \text{ eV}$$

$$\overline{E}_0 = \frac{3}{5} E_F = 5.58 \text{ eV}.$$

## 10.23 FREE ELECTRONS IN A METAL AT 0°K

The state of the conduction electrons at the 0°K temperature forms the basis for studying the properties of metals. Free electrons in a metal at 0°K behave like a free electron gas which obeys Fermi-Dirac statistics. At 0°K, all the levels upto $E_F$ are full and all above $E_F$ are empty. Using this fact, we can calculate the total number of electrons $N$ in terms of Fermi energy, $E_F$.

$$N = \int_0^{E_F} N(E)\, F(E)\, dE \qquad ...(1)$$

Since $$N(E)\,dE = \frac{V}{2\pi^2} \frac{(2m)^3}{^2/\hbar^3} E^{1/2}\, dE \qquad ...(2)$$

Therefore, $$N = \frac{V}{2\pi^2} = \frac{(2m)^{3/2}}{\hbar^3} \int_0^{E_F} E^{1/2}\, dE$$

$$(\because \text{ for } E < E_F,\ F(E) = 1).$$

or $$N = \frac{V}{3\pi^2} \frac{(2m\, E_F)^{3/2}}{\hbar^3} \qquad ...(3)$$

$\therefore$ $$\ln N = \frac{3}{2} \ln E_F + a \text{ constant}$$

Differentiating, we have $$\frac{dN}{dE_F} = \frac{3N}{2E_F} = N(E_F)$$

Denoting $N/V \equiv n$, we have

$$n = \frac{N}{V} = \frac{1}{3\pi^2} \left( \frac{2mE_F}{\hbar^2} \right)^{3/2} \qquad ...(4)$$

This gives Fermi energy

$$\boxed{E_F = \frac{\hbar^2}{2m} (3\pi^2 n)^{2/3}} \qquad ...(5)$$

i.e., knowing the electron density $n$, we can calculate the Fermi energy $E_F$. $E_F$ is also a measure of the Fermi wave vector $k_F$ which is radius of the spherical Fermi energy surface in the $k$-space (i.e., Fermi momentum $p_F = \hbar k_F$).

so that $$k_F = (3\pi^2 n)^{1/3} \quad (\because\ E_F = P_F^2/2m) \qquad ...(6)$$

The electron moves on the Fermi surface with a constant speed given by

$$v_F = \frac{\hbar k_F}{m} = \frac{\hbar}{m}(3\pi^2 n)^{1/3} \qquad ...(7)$$

We can estimate the energy of the free electron at 0°K per unit volume by

$$E_0 = \frac{1}{V}\int_0^{E_F} EN(E)\,dE \qquad ...(8)$$

Using the expression for $N(E)$, we get

$$E_0 = \frac{E_F^{5/2}}{5\pi^2}\left(\frac{2m}{\hbar^2}\right)^{3/2} \qquad ...(9)$$

Using Eq. (5), we have from Eq. (9)

$$E_0 = \frac{E_F^{3/2}}{5\pi^2}\cdot E_F\cdot\left(\frac{2m}{\hbar^2}\right)^{3/2}$$

$$= \left(\frac{\hbar^2}{2m}\right)^{3/2}\frac{3(\pi^2)n}{5\pi^2}\left(\frac{2m}{\hbar^2}\right)^{3/2}\cdot E_F$$

or $$\boxed{\frac{E_0}{n} = \frac{3}{5}E_F} \qquad ...(10)$$

## 10.24 ELECTRICAL CONDUCTIVITY (QUANTUM TREATMENT)

The maximum velocity which the electrons are able to assume is the Fermi velocity, $v_F$ (i.e., the velocity of the electrons at the Fermi energy). The sphere having $v_F$ as a radius represents, therefore, the Fermi surface. All points inside the Fermi sphere are occupied. As a consequence the velocity vectors cancel each other pairwise at equilibrium and no net velocity of the electrons results.

If an electric field is applied, the Fermi sphere is displaced opposite to the field direction, i.e., towards the positive end of the electric field, due to the net velocity gain of the electrons (Fig. 10.27 (b) dashed circle). The great majority of the electron velocities still cancel each other pairwise (shaded area). However, some electrons remain uncompensated; their velocities $v + \Delta v$ are shown cross hatched in Fig. 10.27 (b). These electrons cause the observed current. The Drude description of conduction thus needs a modification. In the *classical* picture one would assume that *all* electrons drift, under the influence of an electric field, with a modest velocity. Quantum mechanics, instead, teaches us that *only specific electrons* participate in conduction and that these electrons drift with a high velocity which is approximately the Fermi velocity, $v_F$.

An additional point needs to be discussed and leads to an even deeper understanding. The largest energy which the electrons can assume in a metal at $T = 0$ is the Fermi energy, $E_F$. A large number of electrons actually possess this very energy since the density of states and thus the population density is highest around $E_F$ (Fig. 10.28). Thus, only a little extra energy $\Delta E$ is needed to raise a substantial number of electrons from the Fermi level into slightly higher states. As a consequence, the energy (or the velocity) of electrons accelerated by the electric field $\mathscr{E}$ is only slightly larger than the Fermi energy $E_F$ (or the Fermi velocity, $v_F$) so that for all practical purposes the mean velocity can be approximated by the Fermi velocity, $v_F$. We have implied this fact already in our previous discussions.

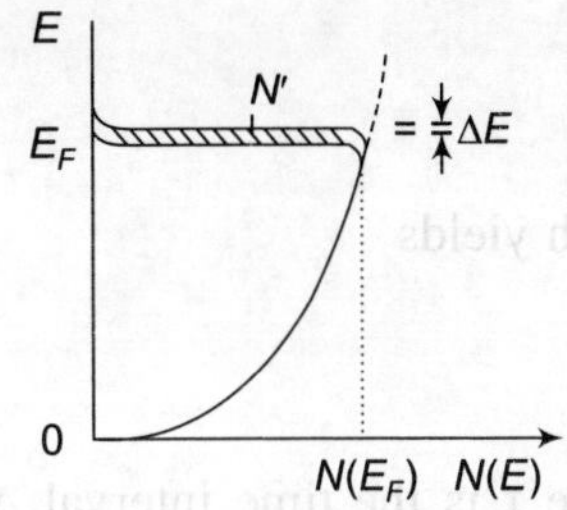


**Figure 10.28** Population density $n(e)$ versus energy for free electrons (see Fig. 10.26) and displacement by an electric field. $N'$ is the number of displaced electrons per unit volume in the energy interval $\Delta E$. $N(E)$ is, in the present case, also defined per unit volume

We now calculate the conductivity by quantum mechanical means and apply, as before, Ohm's law, $j = \sigma\mathscr{E}$. The current density $j$ is, as stated in present case, we know that the velocity of the electrons which are responsible for the electron conduction is essentially the Fermi velocity $v_F$. Further, the number of electrons which need to be considered here is $N'$, i.e., the number of displaced electrons per unit volume, as shown in Fig. 10.28. Thus,

$$j = v_F e N'. \quad ...(1)$$

The number of electrons displaced by the electric field $\mathscr{E}$ is

$$N' - N(E_F)\Delta E \quad ...(2)$$

(see Fig. 7.4), which yields for the current density

$$j = v_F e N(E_F)\Delta E = v_F e N(E_F)\frac{dE}{dk}\Delta k. \quad ...(3)$$

The factor $dE/dk$ is calculated by using $E$ versus $|\mathbf{k}|$ relationship known for free electrons, i.e.,

$$E = \frac{h^2}{2m}k^2. \quad ...(4)$$

Taking the first derivative of Eq. (4) yields, with $k = p/\hbar$ Eq. (3) of section 10.4,

$$\frac{dE}{dk} = \frac{\hbar^2}{m}k = \frac{\hbar^2 p}{m\hbar} = \frac{\hbar m v_F}{m} = \hbar v_F. \quad ...(5)$$

Inserting Eq. (5) into Eq. (3) yields

$$j = v_F^2\, e N(E_F)\, \hbar\Delta k. \quad ...(6)$$

The displacement $\Delta k$ of the Fermi sphere in $k$-space under the influence of an electric field can be calculated using $p = \hbar k$

$$F = m\frac{dv}{dt} = \frac{d(mv)}{dt} = \frac{dp}{dt} = \hbar\frac{dk}{dt} = e\mathscr{E}, \qquad ...(7)$$

which yields
$$dk = \frac{e\mathscr{E}}{\hbar}dt$$

or
$$\Delta k = \frac{e\mathscr{E}}{\hbar}\Delta t = \frac{e\mathscr{E}}{\hbar}\tau \qquad ...(8)$$

where $\tau$ is the time interval $\Delta t$ between two "collisions" or the relaxation time. Inserting Eq. (8) into Eq. (6) yields

$$j = v_F^2 e^2 N(E_F)\mathscr{E}\tau. \qquad ...(9)$$

One more consideration needs to be made. If the electric field vector points in the negative $v(k)_x$ direction, then only the components of those velocities which are parallel to the positive $v(k)_x$ direction contribute to the electric current (Fig. 10.29). The $v(k)_x$ components cancel each other pairwise. In other words, only the projections of the velocities $v_F$ on the positive $v(k)_x$-axis ($v_{Fx} = v_F \cos\theta$) contribute to the current. Thus, we have to sum up all contributions of the velocities in the first and fourth quadrants in Fig. 10.29 which yields

$$j = e^2 N(E_F)\mathscr{E}\tau \int_{-\pi/2}^{+\pi/2} (v_F \cos\theta)^2 \frac{d\theta}{\pi}$$

$$= e^2 N(E_F)\mathscr{E}\tau \frac{v_F^2}{\pi}\int_{-\pi/2}^{+\pi/2} \cos^2\theta\, d\theta$$

$$= e^2 N(E_F)\mathscr{E}\tau \frac{v_F^2}{\pi}\left[\frac{1}{4}\sin 2\theta + \frac{\theta}{2}\right]_{-\pi/2}^{+\pi/2},$$

$$j = e^2 N(E_F)\mathscr{E}\tau\, v_F^2.$$

Similar calculation for a spherical Fermi surface yields

$$j = \frac{1}{3}e^2 N(E_F)\mathscr{E}\tau\, v_F^2. \qquad ...(10)$$

Thus, the conductivity finally becomes, with $\sigma = j/\mathscr{E}$,

$$\boxed{\sigma = \frac{1}{3}e^2 v_F^2 \tau N(E_F)} \qquad ...(11)$$

This quantum mechanical equation reveals that the conductivity depends on the Fermi velocity, the relaxation time, and the population density (per unit volume). The latter is, as we know, proportional to the density of states. Equation (11) is more meaningful than the expression derived from the classical electron theory. Specifically, (11) contains the information that not all free electrons $N_f$ are responsible for conduction, i.e., the conductivity in metals depends to a large extent on the population density of the electrons near the Fermi surface. Figure 10.29 shows the schematic representation of the density of states.

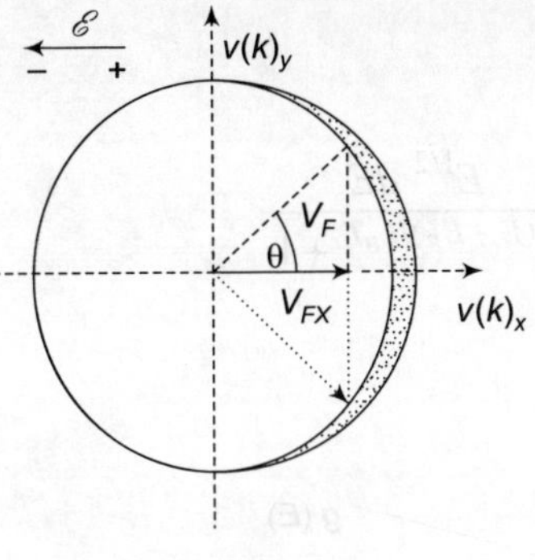


**Figure 10.29** Two-dimensional velocity space

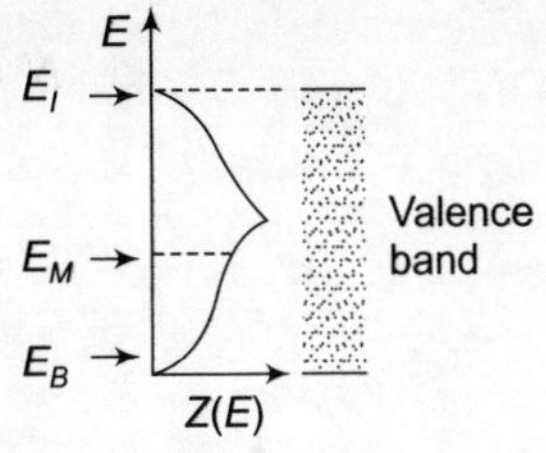


**Figure 10.30** Schematic representation of the density of states and thus, with minor modifications, also the population density. examples for highest electron energies for a monovalent metal ($E_M$), for a bivalent metal ($E_B$), and for an insulator ($E_1$) are indicated

The representation of the DOS for monovalent metal, bivalent metal and insulator.

## 10.25 FERMI ENERGY FOR FREE ELECTRONS IN A METAL AS A FUNCTION OF TEMPERATURE

The Fermi–Dirac distribution function is given by

$$f(E) = \frac{1}{e(E - E_F)/k_BT + 1} \quad ...(1)$$

As absolute zero, for $E > E_F$, the Eq. (1) gives $f(E) = 0$ and for $E < E_F$, $f(E) = 1$. This is what we have seen in last section that below Fermi level, all states are occupied and the probability of finding an electron on a level with energy greater than Fermi energy is zero. At temperatures above absolute zero, for $E << E_F$, $f(E) = 1$ in view of the Pauli exclusion principle that limits state occupancy to only one electron and for $E >> E_F$, $f(E)$ must approach zero. At high temperatures, when $k_BT >> E_F$ or $T \to \infty$, electrons lose their quantum mechanical character and Fermi–Dirac distribution function reduces to classical Boltzmann distribution of $e^{-E/k_BT}$. These details of the Fermi–Dirac distribution function are shown in Fig. 10.31.

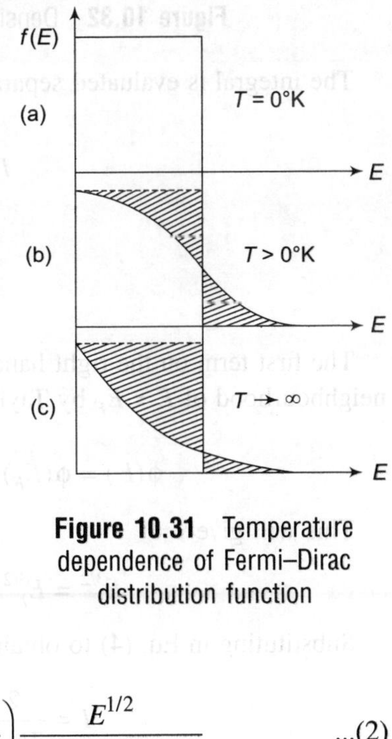


**Figure 10.31** Temperature dependence of Fermi–Dirac distribution function

The equation for the occupied density of states can now be written as

$$N(E) = f(E)\, g(E) = \frac{V}{2\pi^2}\left(\frac{2m}{\hbar^2}\right)\frac{E^{1/2}}{e^{(E - EF)/k_BT} + 1} \quad ...(2)$$

This is shown in Fig. 10.32 by the shaded area. The total number of free electrons will now be

$$N = \int_0^\infty N(E)\,dE$$

$$= \frac{V}{2\pi^2}\left(\frac{2m}{\hbar^2}\right)^{3/2} \int_0^\infty \frac{E^{1/2}\,dE}{e^{(E-E_F)/k_BT}+1} \qquad ...(3)$$

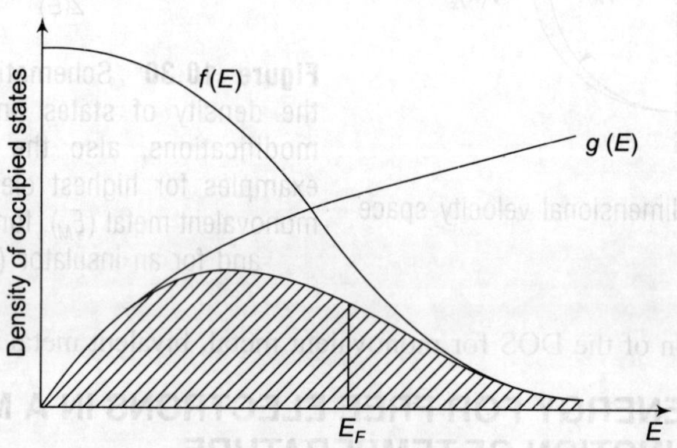


**Figure 10.32** Density of occupied states shown by shaded portion

The integral is evaluated separately. Integrating by parts,

$$I = \int_0^\infty \frac{E^{1/2}\,dE}{e^{(E-E_F)/k_BT}+1} = \frac{2}{3}E^{3/2}\left.\frac{1}{\{e^{(E-EF)/k_BT}+1\}}\right|_0^\infty$$

$$+\frac{2}{3}\int_0^\infty \frac{E^{3/2}\,e^{(E-E_F)/k_BT}}{(e^{(E-E_F)/k_BT}+1)^2}\cdot\frac{dE}{k_BT} \qquad ...(4)$$

The first term on the right hand side is zero for both the limits. Any function $\phi(E)$ in the neighbourhood of $E = E_F$ by Taylor's series in powers of $(E - E_F)$ can be written as

$$\phi(E) = \phi(E_F) + (E - E_F)\,\phi'(E_F) + \frac{(E-E_F)^2}{2!}\,\phi'(E_F) + ... \qquad ...(5)$$

This will give

$$E^{3/2} = E_F^{3/2} + (E - E_F)\,\frac{3}{2} + E_F^{1/2}\,\frac{(E-E_F)^2}{2}\,\frac{3}{4}\,E_F^{-1/2} + ... \qquad ...(6)$$

Substituting in Eq. (4) to obtain

$$I = \frac{2}{3k_BT}\int_0^\infty \frac{E^{(E-E_F)/k_BT}}{(e^{(E-E_F)/k_BT}+1)^2}$$

$$\left[E_F^{3/2} + \frac{3}{2}(E-E_F)\,E_F^{1/2} + \frac{3}{8}(E-E_F)^2\,E_F^{-1/2} + ...\right] dE \qquad ...(7)$$

Put $(E - E_F)/k_BT = x$ to get $dE = k_BT\, dx$. Also the lower limit is taken as $(-\infty)$ instead of $(-E_F/k_BT)$, because of the $E_F$ in the neighbourhood of $E$,

$$I = \frac{2k_BT}{3k_BT}\int_{-\infty}^{\infty}\frac{e^x\,dx}{(e^x+1)^2}E_F^{3/2} + \frac{2k_BT}{3k_BT}\int_{-\infty}^{\infty}\frac{e^x\,dx}{(e^x+1)^2}\frac{3}{2}k_BT\,xE_F^{1/2}$$

$$+\frac{2k_BT}{3k_BT}\int_{-\infty}^{\infty}\frac{e^x\,dx}{(e^x+1)^2}\frac{3}{8}k_B^2T^2x^2E_F^{1/2}+\dots$$

$$= \frac{2}{3}E_F^{3/2}\,2\cdot\frac{1}{2} + 0 + \frac{2}{3}E_F^{-1/2}\cdot\frac{3}{8}k_B^2T^2\,2\cdot\frac{\pi^2}{6}+\dots$$

$$= \frac{2}{3}E_F^{3/2}\left[1+\frac{\pi^2}{8}\left(\frac{k_BT}{E_F}\right)^2+\dots\right] \qquad \dots(8)$$

Using the standard integrals $\int_{-\infty}^{\infty}\frac{e^x\,dx}{(e^x+1)^2} = 1$, $\int_{-\infty}^{\infty}\frac{xe^x\,dx}{(e^x+1)^2} = 0$

and $$\int_{-\infty}^{\infty}\frac{x^2e^x}{(e^x+1)^2}dx = \frac{\pi^2}{3}.$$

Substituting the value of this integral in Eq. (3) and retaining only up to second term in the integral, we have

$$N = \frac{V}{3\pi^2}\left(\frac{2mE_F}{\hbar^2}\right)^{3/2}\left[1+\frac{\pi^2}{8}\left(\frac{k_BT}{E_F}\right)^2\right] \qquad \dots(9)$$

At absolute zero, $T = 0$. Therefore,

$$N = \frac{V}{3\pi^2}\left(\frac{2mE_{F0}}{\hbar^2}\right)^{3/2} \qquad \dots(10)$$

or $$E_{F0} = \frac{\hbar^2}{2m}\left(\frac{3\pi^2N}{V}\right)^{2/3} \qquad \dots(11)$$

Substituting Eq. (10) in Eq. (9) will give

$$E_{F0}^{3/2} = E_F^{3/2}\left[1+\frac{\pi^2}{8}\left(\frac{k_BT}{E_{F0}}\right)^2\right]$$

or $$E_F \cong E_{F0}\left[1-\frac{\pi^2}{8}\left(\frac{k_BT}{E_{F0}}\right)^2\right] \qquad \dots(12)$$

Because of the small value of the factor $\left(\frac{k_B T}{E_{F0}}\right)$ at room temperature, the Fermi energy is considered to be a constant with respect to temperature and therefore '0' subscript is dropped from Eqs. (10) and (11).

Now we shall discuss some examples where these results describe various phenomena in terms of free-electron model.

## 10.26 ELECTRONIC SPECIFIC HEAT

Sommerfeld was able to calculate the specific heat of the electron gas from the analysis of the previous section and thereby shows that the specific heat of the electrons in a metal is very much small compared to the lattice specific heat. In order to calculate the electronic specific heat we need to know the average energy of an electron in the metal at temperature $T$, given by

$$\langle E_n \rangle_T = \frac{1}{N} \int_0^\infty E N(E) dE$$

$$= \frac{V}{2\pi^2 N} \left(\frac{2m}{\hbar^2}\right)^{3/2} \int_0^\infty \frac{E^{3/2} \, dE}{e^{(E - E_F)/kBT} + 1} \quad ...(13)$$

The integral is evaluated by parts to give

$$I_E = \frac{2}{5} E^{5/2} \left. \frac{1}{\{e^{(E - E_F)/k_BT} + 1\}} \right|_0^\infty + \frac{2}{5} \int_0^\infty \frac{E^{5/2} \, e^{(E - E_F)/k_BT}}{(e^{(E - E_F)/k_BT} + 1)^2} \frac{dE}{k_BT} \quad ...(14)$$

The first term on the right-hand side is zero and in the second term use of Eq. (5) for $E^{5/2}$ as

$$E^{5/2} = E_F^{5/2} + (E - E_F) \frac{5}{2} E_F^{3/2} + \frac{(E - E_F)^2}{2} \cdot \frac{15}{4} E_F^{1/2} + ... \quad ...(15)$$

will give

$$I_E = \frac{2}{5k_BT} \int_0^\infty \frac{e^{(E - E_F)/k_BT}}{(e^{(E - E_F)/k_BT} + 1)^2}$$

$$\left[ E_F^{5/2} + \frac{5}{2} (E - E_F) E_F^{3/2} + \frac{15}{8} (E - E_F)^2 E_F^{1/2} + ... \right] dE \quad ...(16)$$

Again the substitution $\frac{E - E_F}{k_BT} = x$ will provide

$$I_E = \frac{2}{5} E_F^{5/2} + 0 + \frac{2}{5} \cdot \frac{15}{8} k_B^T T^2 E_F^{1/2} \frac{\pi^2}{3} + ... \quad ...(17)$$

Retaining up to second term only and then substituting this integral value in Eq. (13), we have

$$\langle E_n \rangle_T = \frac{V}{2\pi^2 N} \left(\frac{2m}{\hbar^2}\right)^{3/2} \frac{2}{5} E_F^{5/2} \left[ 1 + \frac{5\pi^2}{8} \left(\frac{k_BT}{E_F}\right)^2 \right] \quad ...(18)$$

At absolute zero,

$$\langle E_n \rangle_0 = \frac{V}{5\pi^2 N}\left(\frac{2m}{\hbar^2}\right)^{3/2} E_{F0}^{5/2}$$

$$= \frac{3}{5}\frac{V}{3\pi^2}\left(\frac{2m}{\hbar^2}\right)^{3/2} E_{F0}^{3/2}\,\frac{1}{N}\,E_{F0} = \frac{3}{5}E_{F0} \quad ...(19)$$

Equation (19) in Eq. (18) will provide

$$\langle E_n \rangle_T \simeq \langle E_n \rangle_0 \frac{E_F^{5/2}}{E_{F0}^{5/2}}\left[1 + \frac{5\pi^2}{8}\left(\frac{k_B T}{E_{F0}}\right)^2\right]$$

$$\simeq \langle E_n \rangle_0 \left[1 - \frac{5\pi^2}{24}\left(\frac{k_B T}{E_{F0}}\right)^2\right]\left[1 + \frac{5\pi^2}{8}\left(\frac{k_B T}{E_{F0}}\right)^2\right]$$

{from Eq. (12)}

$$\simeq \langle E_n \rangle_0 \left[1 + \frac{5\pi^2}{12}\left(\frac{k_B T}{E_{F0}}\right)^2\right] \quad ...(20)$$

Differentiating this equation with respect to $T$ will give the specific heat at constant volume per electron as

$$C_v \cong \langle E_n \rangle_0 \frac{5}{6}\pi^2 \frac{k_B^2 T}{(E_{F0})^2} = \frac{3}{5}E_{F0}\cdot\frac{5}{6}\frac{\pi^2 k_B^2 T}{(E_{F0})^2} = \frac{\pi^2 k_B^2 T}{2\,E_{F0}} \quad ...(21)$$

$$= \frac{\pi^2 k_B T}{2T_F} \qquad (\because E_{F0} = k_B T_F) \quad ...(22)$$

which as $T << T_F$ at normal temperatures, is much less than $\frac{3}{2}k_B$ suggested by a classical free electron model and agrees much better with experiment. It should be noted that the electronic specific heat varies linearly with $T$ and at low temperatures, lattice specific heat follows Debye $T^3$ law, so that the total specific heat of a metal can be represented by

$$C_v = \gamma T + AT^3 \quad ...(23)$$

where $\gamma$ and $A$ are constants and characteristics of the material. This is the behaviour observed at low temperatures for alkali metals, where a plot of $\frac{C_v}{T}$ vs $T^2$ yields a straight line with slope $A$ and intercept $\gamma$, as shown in Fig. 10.33.

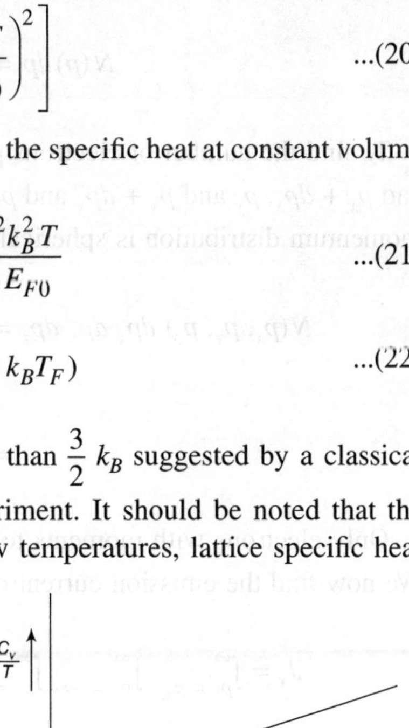


**Figure 10.33** Observed specific heat as a function of $T^2$ for alkali metals

## 10.27 RICHARDSON'S THERMIONIC EMISSION

As the temperature of a metal is increased, electrons near the Fermi surface are excited to higher energy states; eventually some have enough energy to surmount the surface barrier and escape. The energy required is $E_F + \phi$, where $\phi$ is the work function, but for an electron to escape it must not only have at least this energy but must also arrive at the surface with its momentum **p** suitably directed. If we consider escape through a surface defined by the pane $x = 0$, then $x$ component of the momentum must be equal to or greater than the critical value

$$p_{xc} = +\sqrt{2m(E_F + \phi)} \qquad \text{...(1)}$$

since the kinetic energy associated with the $y$ and $z$ components of the momentum does not aid in the escape.

Let us now assume that all electrons which reach the surface with $p_x > p_{xc}$ are thermionically emitted. Then the emission current density $J_x$ has the magnitude $ev_x$ times the number of electrons per unit volume with $p_x > p_{xc}$. To calculate $J_x$ we convert the energy distribution to a momentum distribution by using $E = p^2/2m$, thereby obtaining for the number of electrons per unit volume with momentum magnitude between $p$ and $p + dp$,

$$N(p)\,dp = \frac{8\pi p^2\,dp}{h^3\left(e^{(p^2/2m - E_F)/kT} + 1\right)} \qquad \text{...(2)}$$

To find the number of electrons per unit volume with momentum components between $p_x$ and $p_x + dp_x$, $p_y$ and $p_y + dp_y$ and $p_z$ and $p_z + dp_z$, we note from Fig. 10.34 that, because the momentum distribution is spherically symmetrical,

$$N(p_x, p_y, p_z)\,dp_x\,dp_y\,dp_z = \frac{dp_x\,dp_y\,dp_z}{4\pi p^2\,dp} N(p)\,dp$$

$$= \frac{2\,dp_x\,dp_y\,dp_z}{h^3\left(e^{(p^2/2m - E_F)/kT} + 1\right)} \qquad \text{...(3)}$$

Only electrons with moments to the right of the plane $p_x = p_{xc}$ in Fig. 10.33 can escape. We now find the emission current density by evaluating

$$J_x = \int_{p_x = p_{xc}}^{\infty} \int_{p_y = -\infty}^{\infty} \int_{p_z = -\infty}^{\infty} e\frac{p_x}{m} \frac{2dp_x\,dp_y\,dp_z}{h^3\left(e^{[(p_x^2 + p_y^2 + P_z^2)/2m - E_F]/kT} + 1\right)} \qquad \text{...(4a)}$$

Since only electrons with energy greater than $E_F + \phi$ can be omitted and $\phi$ is many times $kT$, we may neglect the unity in the Fermi factor and write

$$J_x = \frac{2ee^{EF/kT}}{h^3 m} \int_{p_{xc}}^{\infty} p_x e^{-p_x^2/2mkT}\,dp_x \int_{-\infty}^{\infty} e^{-p_y^2/2mkT}\,dp_y \int_{-\infty}^{\infty} e^{-P_z^2/2mkT}\,dp_z \qquad \text{...(4b)}$$

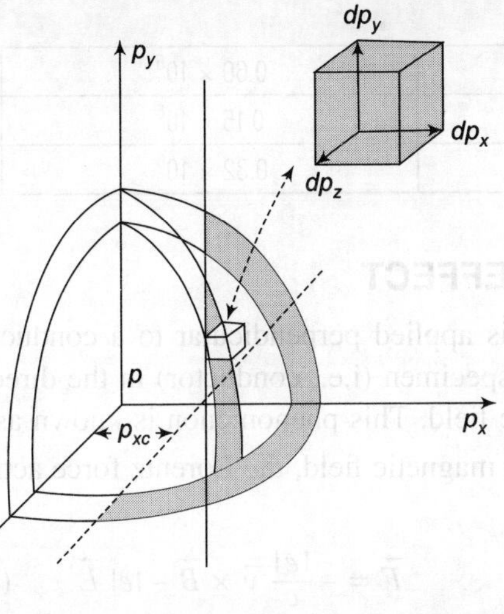


**Figure 10.34** Only electrons with $p_x > p_{xc}$ can escape from the metal. because the momentum distribution is spherically symmetric, the number of electrons with a momentum terminating in $dp_x\, dp_y\, dp_z$ is $N(p)\, dp$ multiplied by the ratio of the volume $dp_x\, dp_y\, dp_z$ to the volume of the spherical shell $4\pi p^2\, dp$

These three integrals are standard ones; upon substituting the appropriate values, using $p_{xc} = \sqrt{2m(E_F + \phi)}$, one obtains

$$J = \frac{4\pi m e k^2}{h^3} T^2 e^{-\phi/kT}$$

$$= A_0 T^2 e^{-\phi/kT} \qquad ...(5)$$

This is the *Richardson-Dushman equation* for thermionic emission, in which $A_0$ is a universal constant with value $1.2 \times 10^6$ A/m$^2$-°K$^2$.

Some representative thermionic-emission data are given in Table 10.8.

**Table 10.8** Thermionic-emission Constants

| *Emitter* | *$A_0$, A/m$^2$-°K$^2$* | *$\phi$, eV* |
|---|---|---|
| Cr | $0.48 \times 10^6$ | 4.60 |
| Cs | $1.6 \times 10^6$ | 1.6 |
| Mo | $0.55 \times 10^6$ | 4.3 |
| Ni | $0.30 \times 10^6$ | 4.61 |
| Ta | $0.55 \times 10^6$ | 4.19 |

*Contd...*

*Table 10.8 Contd...*

| | | |
|---|---|---|
| W | $0.60 \times 10^6$ | 4.52 |
| Ba on W | $0.15 \times 10^6$ | 1.56 |
| Cs on W | $0.32 \times 10^6$ | 1.36 |

## 10.28 THE HALL EFFECT

When a magnetic field is applied perpendicular to a conductor carrying current, a voltage is developed across the specimen (i.e., conductor) in the direction perpendicular to both the current and the magnetic field. This phenomenon is known as *Hall effect.*

In a metal placed in a magnetic field, the Lorentz force acting on a conduction electron is given by

$$\vec{F} = -\frac{|e|}{c}\vec{v} \times \vec{B} - |e|\,\vec{E} \qquad \text{(in CGS units)} \qquad ...(1)$$

where $\vec{B}$ is the magnetic induction and $\vec{E}$ is the external electric field.

If $\vec{E}$ is zero, then the electron would process under the action of the magnetic force, i.e., the path will be a circle around the $\vec{B}$ direction with a precession frequency known as cyclotron frequency, $\omega = \frac{|e|B}{mc}$, $m$, is the mass of the electron and $c$, the speed of light.

Let us now discuss the modification of electrical conductivity of free electrons in a metal due to an applied magnetic field in the classical relaxation-time approximation. The equation of motion in this case becomes

$$m\left(\frac{d^2\mathbf{r}}{dt^2} + \frac{1}{\tau}\frac{d\mathbf{r}}{dt}\right) = -e\left(\mathbf{E} + \frac{1}{c}\frac{d\mathbf{r}}{dt} \times \mathbf{B}\right) \qquad \text{(in CGS units)} \qquad ...(2a)$$

where $d^2\mathbf{r}/dt^2$ is the acceleration and $d\mathbf{r}/dt$ is velocity, $\tau$ is the relaxation time, $e$ is the magnitude of the electron charge

$$m\left(\frac{d^2\vec{r}}{dt^2} + \frac{1}{c}\frac{d\vec{r}}{dt}\right) = -e\left(\vec{E} + \frac{1}{c}\frac{d\vec{r}}{dt} \times \vec{B}\right); \qquad \text{(in MKS units)} \qquad ...(2b)$$

$$|\vec{B}| = B(0, 0, \hat{z}) \qquad ...(3)$$

### 10.28.1 Classical Hall Effect, Hall Coefficient and Hall Voltage

We first reconsider the classical Hall effect for electrons only.

Let electrons move in the $(x, y)$-plane with a magnetic field in the $z$-direction and an electric field also in the $(x, y)$-plane. In MKS units and standard notation $(e > 0)$

$$F_x = -eE_x - eV_yB - \frac{mV_x}{\tau}, \qquad ...(4)$$

$$F_y = -eE_y + eV_xB - \frac{mV_y}{\tau}, \quad ...(5)$$

where the term involving the relaxation time $\tau$ is due to scattering. The current density is given by

$$J_x = -neV_x, \quad ...(6)$$

$$J_y = -neV_x, \quad ...(7)$$

where $n$ is the number of electrons per unit volume. Letting the dc conductivity be

$$\sigma_0 = \frac{ne^2\tau}{m}, \quad ...(8)$$

We can write (in the steady state when $F_x$, $F_y = 0$ using Eqs. (4) and (8))

$$\begin{pmatrix} E_x \\ E_y \end{pmatrix} = \frac{1}{\sigma_0}\begin{pmatrix} 1 & \omega_c\tau \\ -\omega_c\tau & 1 \end{pmatrix}\begin{pmatrix} J_x \\ J_y \end{pmatrix} \quad ...(9)$$

where $n$ is the cyclotron frequency and we can show (by Eq. (8))

$$\frac{B}{ne} = \frac{\omega_c\tau}{\sigma_0}. \quad ...(10)$$

The inverse to Eq. (9) can be written

$$\begin{pmatrix} J_x \\ J_y \end{pmatrix} = \frac{\sigma_0}{1+(\omega_c\tau)^2}\begin{pmatrix} 1 & -\omega_c\tau \\ \omega_c\tau & 1 \end{pmatrix}\begin{pmatrix} E_x \\ E_y \end{pmatrix}. \quad ...(11)$$

We will use the geometry as shown in Fig. (4). We rederive the Hall coefficient. Setting $J_y = 0$, then

$$E_y = |\boldsymbol{V} \times \boldsymbol{B}| = -\frac{\omega_c\tau}{\sigma_0} J_x = -\frac{B}{ne} J_x, \quad ...(12)$$

where $\boldsymbol{V} = V_x = J_x/ne$ from Eq. (6). The Hall coefficient is defined as usual.

$$R_{\mathrm{H}} = \frac{E_y}{J_xB_z} = -\frac{1}{ne} \quad ...(13)$$

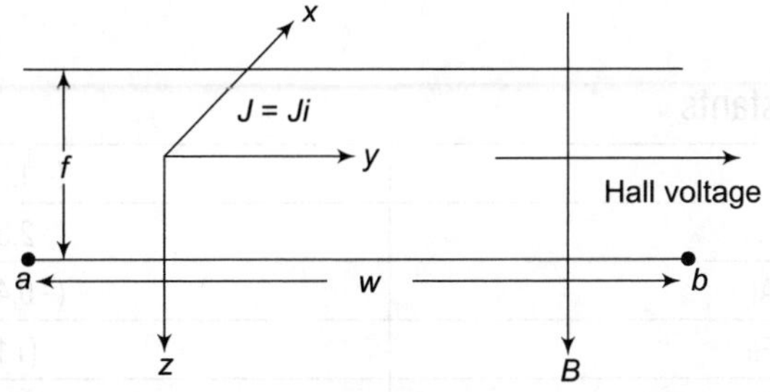


**Figure 10.35** Schematic diagram of classical Hall effect

The Hall voltage over the length $w$ would then be

$$V_{\mathrm{H}} = -E_y w = \frac{BJ_x w}{ne}. \quad ...(14)$$

The current through the segment of area $tw$ is

$$I_x = J_x tw, \quad ...(15)$$

so

$$V_{\mathrm{H}} = \frac{BI_x}{nte}. \quad ...(16)$$

Define $n_a$ as the number of electrons per unit area (projected into the $(x, y)$-plane) so the Hall voltage can be written

$$V_{\mathrm{H}} = \frac{I_x B}{n_a e}. \quad ...(17)$$

The Hall conductance, $1/R_{xy}$ is

$$\frac{1}{R_{xy}} = \frac{I_x}{V_{\mathrm{H}}} = \frac{n_a e}{B}. \quad ...(18)$$

Longitudinally over a length $L$, the voltage change is

$$V_L = E_x L = \frac{J_x L}{\sigma_0} = \frac{I_x L}{tw\sigma_0}, \quad ...(19)$$

which we find to be independent of $B$. This is the usual Drude result. However, this result is based on the assumption that all electrons are moving with the same velocity. If we allow the electrons to have a distribution of velocities by doing a proper Boltzmann equation calculation, we find there is a magnetoresistance effect. The result is (Blakemore),

$$\sigma = \frac{\sigma_0}{1 + (\sigma_0 R_{\mathrm{H}})^2 \dfrac{|\boldsymbol{J} \times \boldsymbol{B}|^2}{|\boldsymbol{J}|^2}}. \quad ...(20)$$

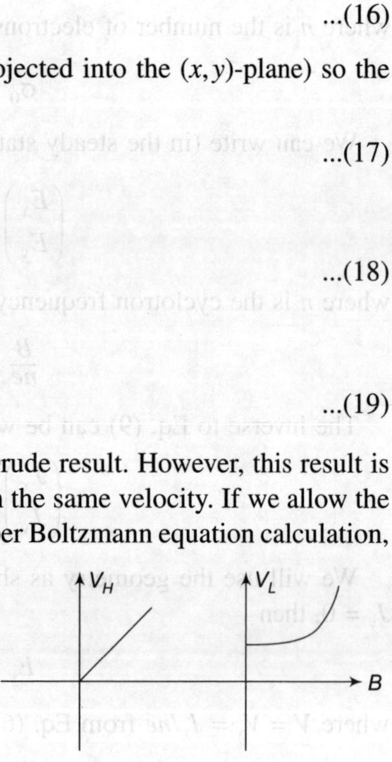


**Figure 10.36** Schematic diagram of classical Hall effect behaviour. see Eq. (17) for $V_{\mathrm{H}}$ and (29) for $V_L$

In addition, when band structure effects are taken into account one finds there also may be a magnetoresistance even when $\boldsymbol{J} \times \boldsymbol{B} = 0$. Classically then we predict behaviour for the Hall effect (with $I_x$ = constant) as shown in Fig. 10.36.

**Table 10.9** Hall Constants

| | |
|---|---|
| Li | (−1.89) |
| Na | (−2.30) |
| Al | (−0.43) |
| Fe | (+14) |
| Cu | (−0.8) |
| Zn | (+0.3) |
| Ag | (−0.8) |

| | |
|---|---|
| Au | (−0.7) |
| Cd | (+0.6) |
| Hg | (−0.6) |

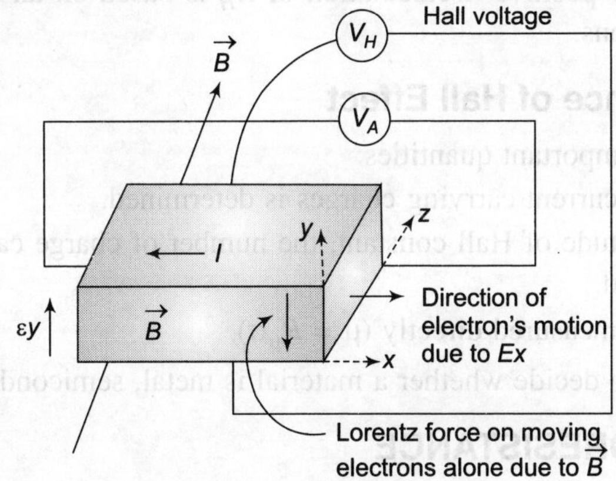


**Figure 10.37** The Hall effect

## 10.28.2 Mobility and Hall Angle

Mobility $\quad \mu = \dfrac{v_x}{E_x} \quad \therefore \; v_x = \mu\, E_x$

$$0 = -e\, E_y + v_x H_z \text{ or } E_y = (v_x H_z)$$

$$\therefore \quad E_y = \mu\, E_x H_z \qquad ...(21)$$

Using Eqs. (13) and (21)

$$R_H j_x\, H_z = \mu\, E_x\, H_z \text{ or } \mu = \frac{R_H j_x}{E_x}$$

or

$$\mu = R_H \sigma$$

In emu

$$\mu = R_H\, \sigma = \frac{E_y}{H_z j_x}\, \sigma = \frac{\theta}{H_z}\; (\because j_x = \sigma E_x)$$

where

$$\theta = \left(\frac{E_y}{E_x}\right) \text{ is called } \textit{Hall angle}$$

$$\theta = \mu \text{Hz}.$$

$\mu$ is the mobility.

The factor $R_H$ is called the *Hall constant*. Since $e$ is taken positive, the Hall constant is negative for electrons and is inversely proportional to the carrier concentration. For a real crystal, $m$ has to be replaced by $m^*$, i.e., the effective mass. The ratio $t/m^*$ provides a measure of the deviation of a band picture from the free electron model.

Since actual Fermi surface of metals have both electron surfaces and hole surfaces, the two being differentiated by positive sign of $m^*$ for electrons and negative sign of $m^*$ for holes, accordingly, $R_H$ will be positive if calculation of $R_H$ is based on an isotropic hole surface, and negative for electrons.

### 10.28.3 Importance of Hall Effect

It gives the following important quantities:

1. The sign of the current carrying charges is determined.
2. From the magnitude of Hall constant, the number of charge carriers per unit volume can be calculated.
3. The mobility is measured directly ($\mu = R_\mu \sigma$).
4. It can be used to decide whether a material is metal, semiconductor on insulator.

## 10.29 MAGNETORESISTANCE

During the experiment on Hall effect, it is noticed that the resistance of the metal sample along the direction of current $j_x$ generally increases (or the conductance decreases) after the magnetic field is switched on. The fractional increase in the zero-field resistivity $\rho(0)$ is defined as the magnetoresistance, $\Delta P/\rho(0)$. This effect of $j_x$ is due to the fact that when the magnetic field is applied, the paths of the electrons become curved, and so the electrons do not go exactly in the direction of the superimposed electric field. This effect was discovered by Thomson in 1856.

The effect of great interest is the transverse magnetoresistance, which is usually studied in the following geometrical arrangement.

A long thin wire is aligned along $x$-axis and a dc electric field is established in the wire by means of an external power supply. A uniform magnetic field $\vec{H}$ is applied along the $z$-axis. In this geometry, the effect of a weak magnetic field ($\omega_c\tau \ll 1$) is to increase the resistance by an additive term proportional to $H^2$.

We shall derive, subsequently, an expression for this transverse magnetoresistance.

### 10.29.1 Transverse Magnetoresistance

Consider a wire subjected to an electric field $\varepsilon_x$ and a magnetic field $H_z$ (i.e., normal to the axis of the wire). The Lorentz force is given by

$$F = m\frac{d^2x}{dt^2} = e\left(\varepsilon_x + \frac{1}{c}\vec{v} \times \vec{H}\right)$$

$$= e\left[\varepsilon_x + \frac{1}{c}(v_y H_z = v_z H_y)\right]$$

$$= e\left(\varepsilon_x + \frac{1}{c} v_y H_z\right) \qquad (\because \;\; H_y = 0)$$

or
$$\frac{d^2x}{dt^2} = \frac{e}{m}\left(\varepsilon_x + \frac{1}{c}\frac{dy}{dt} H_z\right) \qquad \text{...(1)}$$

Similarly,
$$m\,\frac{d^2y}{dt^2} = e\left[\varepsilon_y + \frac{1}{c}(v_2\, H_x - v_x\, H_z)\right]$$

$$= -\frac{1}{c} H_x \frac{dx}{dt} \qquad (\because \;\; \varepsilon_y = 0 \text{ and } H_x = 0) \qquad \text{...(2)}$$

For $z$-direction
$$m\,\frac{d^2z}{dt^2} = 0 \qquad \text{...(3)}$$

Integrating Eqs. (1) and (2), we get

$$m\,\frac{dx}{dt} = e\,\varepsilon_x t + \frac{eH_y}{c}\,y + C_1$$

$$m\,\frac{dy}{dt} = -\frac{e}{c} H_z x + C_2$$

where $C_1$ and $C_2$ are constants of integration.

These are determined as follows

At
$$t = 0, \quad x = 0, \quad \frac{dx}{dt} = u_x, \quad \therefore \quad C_1 = mu_x$$

$$t = 0, \quad y = 0, \quad \frac{dy}{dt} = u_y, \qquad C_2 = mu_y$$

$\Rightarrow$
$$m\,\frac{dx}{dt} = e\,\varepsilon_x t + \frac{e}{c} II_z y + mu_x$$

or
$$\frac{dx}{dt} = \frac{e\varepsilon_x}{m}\,t + \frac{e}{mc} H_z y + u_x \qquad \text{...(4)}$$

and
$$\frac{dy}{dt} = \frac{-e}{mc} H_z x + u_y \qquad \text{...(5)}$$

Integrating Eq. (4), we have

$$x = \frac{e\varepsilon_x}{m}\frac{t^2}{2} + \frac{e}{m}\frac{H_z}{c}\,yt + u_x t + C_3$$

where $C_3$ is a constant of integration.

At
$$t = 0 \text{ and } x = 0,\ C_3 = 0,$$

So,
$$x = \frac{e\varepsilon_x}{m}\frac{t^2}{2} + \frac{e}{mc} H_z yt + u_x t \qquad \text{...(6)}$$

Substituting the value of $x$ from Eq. (6) into Eq. (5), we get

$$\frac{dy}{dt} = -\frac{eH_z}{mc}\left[\frac{e\varepsilon_x}{m}\frac{1}{2} + \frac{eH_z}{mc}yt + u_x t\right] + u_y \quad ...(7)$$

Substituting this value of $\frac{dy}{dt}$ in Eq. (1), we have

$$\frac{d^2x}{dt^2} = \frac{e}{m}\left[\varepsilon_x + \frac{H_z}{c}\left\{-\frac{eH_z}{mc}\left(\frac{e\varepsilon_x t^2}{2m} + \frac{eH_z}{mc}yt + u_x t\right) + u_y\right\}\right] \quad ...(8)$$

Integrating it

$$\frac{dx}{dt} = \frac{e}{m}\left[\varepsilon_x t + \frac{H_z}{e}\left\{-\frac{eH_z}{mc}\left(\frac{e\varepsilon_x}{2m}t^3 + \frac{eH_z}{mc}\frac{yt^2}{2} + \frac{u_x t^2}{2}\right) + u_y t\right\}\right] \quad ...(9)$$

If $\tau$ is the time between two successive collisions, then

$$\left\langle\frac{dx}{dt}\right\rangle = \frac{1}{\tau}\int_0^\tau \frac{dx}{dt}\,dt$$

$$= \frac{e}{m\tau}\left[\int_0^\tau \varepsilon_x t\,dt + \frac{H_z}{c}\left\{-\frac{eH_z}{mc}\left(\int_0^\tau \frac{e\varepsilon_x}{6m}t^3\,dt + \int_0^\tau \frac{eH_z}{mc}\frac{yt^2}{2}dt + \int_0^\tau \frac{u_x t^2}{2}dt\right) + \int_0^\tau u_y t\,dt\right\}\right]$$

$$= \frac{e}{m\tau}\left[\varepsilon_x\frac{\tau^2}{2} + \frac{H_z}{c}\left\{-\frac{eH_z}{mc}\left(\frac{e\varepsilon_x}{24}\frac{\tau^4}{m} + \frac{eH_z y\tau^3}{6mc} + \frac{4x}{6}\tau^2 + \frac{u_y\tau^2}{2}\right)\right\}\right]$$

$$\simeq \frac{e}{m\tau}\left[\frac{\varepsilon_x\tau^2}{2} - \frac{e^2H_z^2\,\varepsilon_x\tau^4}{24\,m^2\,c^2}\right] \quad \left(\because\ H_z y \ll \frac{\tau^3}{6}\text{ and }\langle u_x\rangle = \langle u_y\rangle = 0\right) \quad ...(10)$$

Now, current density

$$J_x = N_e\left\langle\frac{dx}{dt}\right\rangle$$

$$= \frac{Ne^2}{\tau m}\left(\varepsilon_x\frac{\tau^2}{2} - \frac{e^2H_z^2\,\varepsilon_x\tau^4}{24\,m^2c^2}\right)$$

$$= \frac{Ne^2\varepsilon_x}{2m}\left(\tau - \frac{e^2H_z^2\,\tau^3}{12\,m^2c^2}\right) \quad ...(11)$$

Now, conductivity

$$\sigma = \frac{J_x}{\varepsilon_x} = \frac{Ne^2\tau}{2m}\left(1 - \frac{e^2H_z^2\,\tau^2}{12\,m^2c^2}\right) \quad ...(12)$$

When $H_z = 0$, the conductivity $\sigma_0$ is given by

$$\sigma_0 = \left(\frac{Ne^2}{2m}\right)\tau_0 \qquad \text{...(13)}$$

$$\Rightarrow \qquad \frac{\sigma - \sigma_0}{\sigma_0} = \frac{\tau - \tau_0}{\tau_0} - \frac{e^2 H_z^2}{12 m^2 c^2}\frac{\tau^3}{\tau_0}$$

When $\tau \approx \tau_0$, we have

$$\frac{\Delta\sigma}{\sigma_0} = \frac{\Delta\tau}{\tau_0} - \frac{e^2 H_z^2}{12\, m^2 c^2}\tau_0^2$$

$$= \frac{\Delta\tau}{\tau_0} - \frac{e^2 H_z^2}{12\, m^2 c^2} = \left(\frac{2\, m\sigma_0}{Ne^2}\right)^2 = \frac{\Delta\tau}{\tau_0} - \frac{1}{3}\left(\frac{H_z}{c}\right)^2 \frac{\sigma_0^2}{N^2 e^2}$$

So, change of resistivity

$$\frac{\Delta\rho}{\rho_0} = -\frac{\Delta\sigma}{\sigma} = -\frac{\Delta\tau}{\tau_0} + \frac{1}{3}\left(\frac{\sigma_0}{ecN}\right)^2 H_z^2 \qquad \text{...(14i)}$$

or $\qquad (\Delta\rho/\rho_0) \simeq (\sigma_0/(ecN)^2)\, H_z^2$, i.e., $(\Delta\rho/\rho_0) \propto H_z^2$. ...(14ii)

This is for one type of carriers only.

## 10.29.2 Experimental Observations of Magnetoresistance

When the magnetic field is applied normal to the current flow, the effect is termed transverse magnetoresistance and when the field is applied parallel to current flow, the effect is termed longitudinal magnetoresistance.

In case of a magnetic metal, the external magnetic field increases the alignment of magnetic moments opposite to thermal vibrations, which tend to decrease the alignment. If the temperature is below the Curie point, magnetoresistance is decreased and if the temperature is above the Curi point, there will be no change in it. At low magnetic fields, there are changes in resistivity which are basically associated with *magnetostriction*, i.e., elongation or contraction of metal depending on the direction of magnetic moments (these changes are assumed to be due to changes in Fermi surface). Figure 10.38 shows the longitudinal and transverse magnetoresistances of Ni at room temperature. The resistivity for low-field region is governed by magnetostrictive effects while at higher fields, gradual decrease due to moments alignment is predominant.

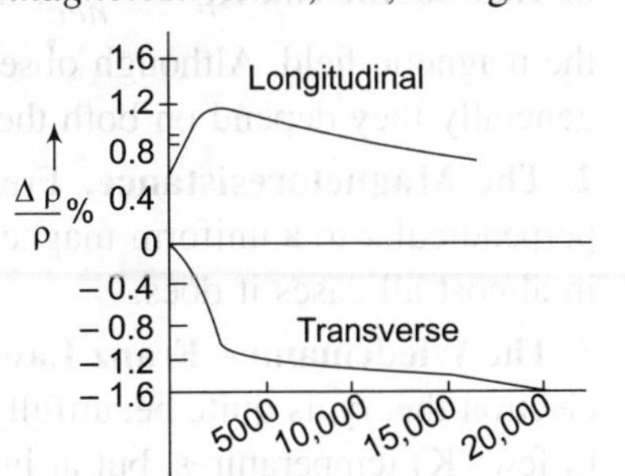


**Figure 10.38** Longitudinal and transverse resistance of Ni at room temperature

The transverse magnetoresistance measurements on a single crystal predict pronounced dependence on magnetic field direction relative to the crystal axes, especially in pure metals.

Figure 10.39 shows a polar plot of magnetoresistance of copper (at 4.2) as a function of current direction relative to crystal axes in presence of magnetic field of strength 18000 gauss applied in [100] direction.

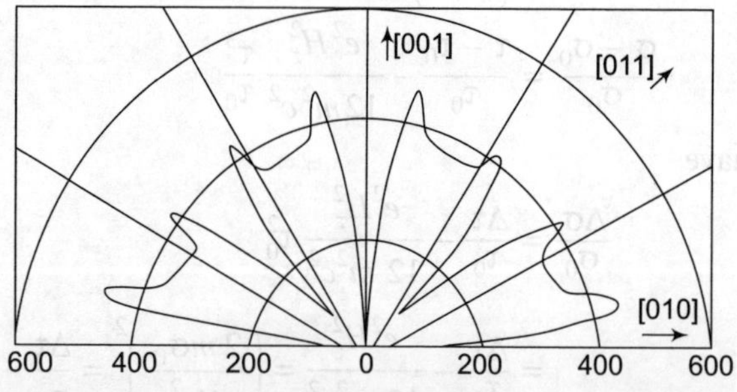


**Figure 10.39** Variation of transverse magnetoresistance in copper at 4.2 K at 18000 gauss, as a function of the orientation of the current direction relative to the crystal axes, the magnetic field is in [100] direction

It shows that variation of magnetoresistance ($\Delta R/Ro$) with the magnetic field is different for crystallographic directions. In some directions it increases and saturates at some constant value whereas in some other directions, it falls off and approaches near zero values. Further, a slight variation in orientation of the magnetic field is found to lead to a significant change in variation of the magnetoresistance with the current direction, in presence of such high fields (~ 18000 gauss).

## 10.30 DIFFICULTIES WITH THE SOMMERFELD'S FREE ELECTRON MODEL: (INADEQUACIES IN THE FREE ELECTRON TRANSPORT COEFFICIENTS)

**1. The Hall Coefficient.** Only the Hall coefficient of the alkali metals satisfies the predictions of free electron theory. Free electron theory predicts (at metallic densities) a constant value of Hall coefficient, $R_H = -\frac{1}{nec}$, independent of temperature, relaxation time or strength of the magnetic field. Although observed Hall coefficient values have this order of magnitude, generally they depend on both the magnetic field strength and the temperature.

**2. The Magnetoresistance.** Free electron theory predicts that the resistance of a wire perpendicular to a uniform magnetic field should not depend on the strength of the field, but in almost all cases it does.

**3. The Wiedemann – Franz Law.** The Wiedemann – Franz law, a great triumph of the free electron theory, is quite beautifully obeyed at high (room) temperatures and also at very low (a few °K) temperatures, but at intermediate temperatures it fails.

**4. Temperature Dependence of DC Electrical Conductivity.** Nothing in free electron theory can account for the temperature dependence of the dc conductivity (it has to be mechanically added into the theory as an ad hoc temperarure dependence in the relaxation time, $\tau$.

In the subsequent section, we will describe the metal work function and the Schottky effect at a metal-vacuum interface. In the next section we will discuss thermionic emission which is useful for carrier transport in a metal-semiconductor contact.

## 10.31 METAL WORK FUNCTION AND SCHOTTKY EFFECT (FIELD EMISSION)

The schematic energy band diagram under equilibrium conditions for a metal in free space is shown in Fig. 10.40. The energy difference between the vacuum level and the Fermi level is known as the work function of a metal. The work function, $\phi_m$, is defined as the minimum kinetic energy required for an electron to escape from the metal surface (or the Fermi level) into free space at $T = 0$ K. The probability for an electron to escape from the metal surface into the vacuum depends on the velocity of electrons perpendicular to the metal surface. The minimum kinetic energy required for an electron to escape from the metal

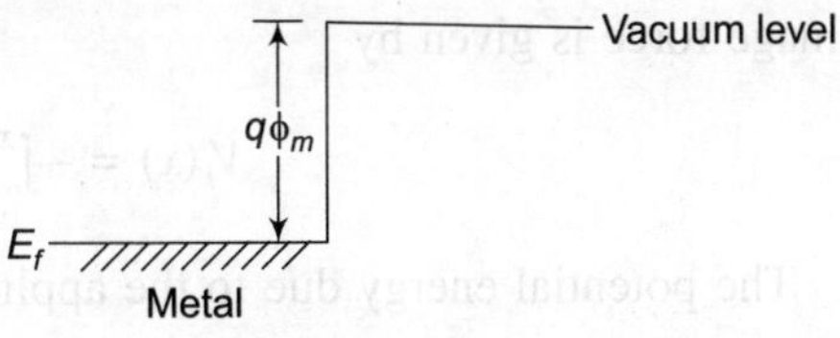


**Figure 10.40** Energy band diagram at a metal-vacuum interface: $\phi_m$ is the metal work function and $E_f$ Fermi level

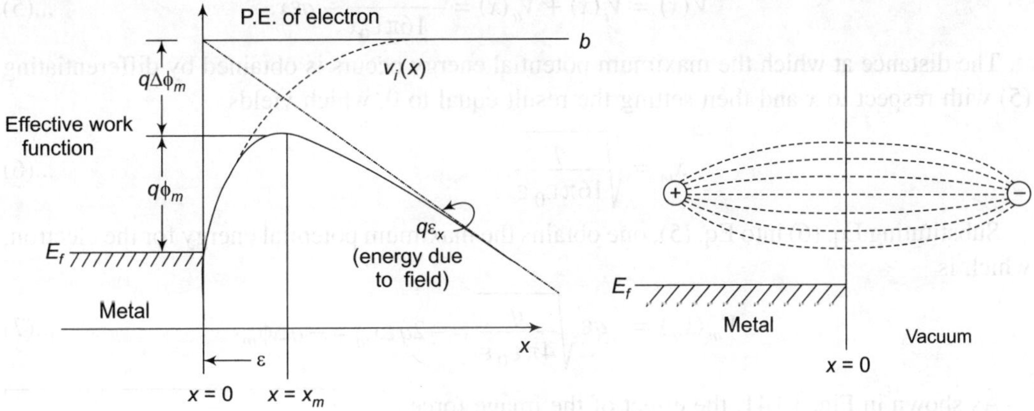


**Figure 10.41** Schottky (or image lowering) effect at the metal-vacuum interface in the presence of an applied electric field: (a) energy band diagram, showing the applied field ($g_x$), the Image potential $V_i(x)$ and the image-lowering potential $\Delta\phi_m$; (b) the induced image charge (positive) inside the metal

surface into vacuum is given by

$$\frac{1}{2}\,(m_0 v_1^2) \geq q\phi_m \qquad ...(1)$$

where $v_1$ is the electron velocity normal to the metal surface, and $m_0$ is the free electron mass. The Schottky effect, or image-lowering effect, occurs when an external electric field is applied to the metal surface. To understand the Schottky effect, consider the energy band diagram shown in Fig. 10.41. When an electric field is applied to the metal surface, electrons that escape from the metal surface will experience two external forces: the image force that arises from the Coulomb attractive force as a result of the positive image charges induced inside the metal by the escaping electrons, and the Lorentz force due to the applied electric field. The positive image charges create a Coulomb attractive force, which tends to pull the escaping electrons back into the metal. The image force can be expressed by

$$F_i = \frac{q^2}{16\pi\varepsilon_0 x^2}, \quad ...(2)$$

where $x$ is the distance from the metal surface. The potential energy associated with this image force is given by

$$V_i(x) = -\int_{\infty}^{x} F_i\, dx = -\frac{q^2}{16\pi\,\varepsilon_0 x}. \quad ...(3)$$

The potential energy due to the applied electric field can be written as

$$V_a(x) = -q\varepsilon x. \quad ...(4)$$

The total potential energy of the electron is equal to the sum of Eqs. (3) and (4), namely,

$$V(x) = V_i(x) + V_a(x) = -\frac{q^2}{16\pi\varepsilon_0 x} - q\varepsilon x. \quad ...(5)$$

The distance at which the maximum potential energy occurs is obtained by differentiating (5) with respect to $x$ and then setting the result equal to 0, which yields

$$x_m = \sqrt{\frac{q}{16\pi\varepsilon_0\,\varepsilon}}. \quad ...(6)$$

Substituting Eq. (6) into Eq. (5), one obtains the maximum potential energy for the electron, which is

$$V_m(x_m) = -q\varepsilon\sqrt{\frac{q}{4\pi\varepsilon_0\,\varepsilon}} = -2q\varepsilon x_m = -q\Delta\phi_m. \quad ...(7)$$

As shown in Fig. 10.41, the effect of the image force and the applied electric field is to lower the work function of a metal. Therefore, the effective metal work function under the applied electric field can be obtained from Fig. 10.41, and the result is

$$q\phi'_m = q\phi_m + V_m = q\phi_m - q\Delta\phi_m$$

$$= q\phi_m - q\sqrt{\frac{q\varepsilon}{4\pi\varepsilon_0}} \quad ...(8)$$

where $q\Delta\phi_m = -V_m$ is the image lowering potential energy. To see the effect of the image lowering effect, one can consider two electric field strengths. If the applied electric field $\varepsilon$ is equal to $10^5$ V/cm, then $x_m$ is equal to

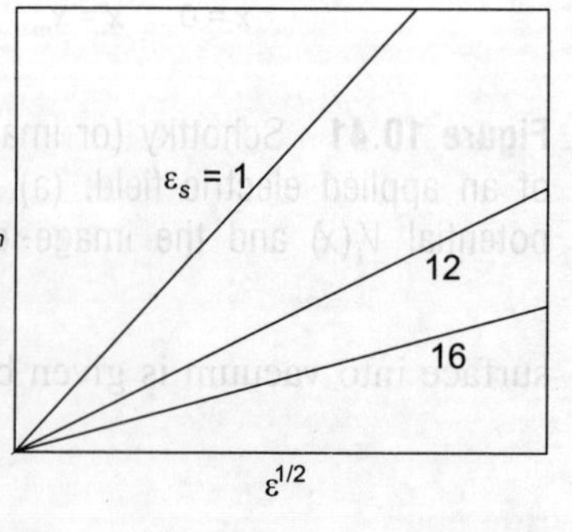


**Figure 10.42** Image lowering potential versus square root of the applied electric field with dielectric constant $\varepsilon_s$ as parameter

60 Å and $q\Delta\phi_m$ = 0.12 eV. On the other hand, if $\varepsilon = 10^7$ V/cm, then $x_m$ = 6 Å and $q\Delta\phi_m$ = 1.2 eV. Therefore, it is obvious that the effective metal work function is greatly reduced in high electric fields as a result of the image lowering effect. Figure 10.42 shows the image lowering potential versus square root of the applied electric field with dielectric constant $\varepsilon_s$ as a parameter. Table 10.10 lists the work function data for some metals.

**Table 10.10** Metal Work Functions for a Clean Metal Surface in Vacuum

| *Metals (s)* | *Work function (eV)* | *Metal (s)* | *Work function (eV)* | *Metal* | *Work Function (eV)* |
|---|---|---|---|---|---|
| Ti, Al, Ta, Ag | 4.33 | Sn | 4.4 | Ir | 5.3 |
| Au, Pd | 5.10 | W, Mo, Sb | 4.63 | Tl | 3.9 |
| Pt | 5.65 | Ga, Cd | 4.28 | In | 4.2 |
| Cr, Hg | 4.5 | Ni | 5.15 | Zn | 4.4 |
| Mg | 3.65 | Rh | 5.05 | Fe | 4.45 |
| Cu | 4.65 | In | 4.2 | Mn | 4.15 |
| Si | 4.85 | Se | 5.9 | Co | 5.0 |

## 10.32 FERMI SURFACE AND FREE ELECTRON FERMI SURFACE

A metal may be defined as a solid with a Fermi surface. The *Fermi surface* is defined as the surface of constant energy $E_F$ in *k*-space (inside which all the states are occupied by the valence electrons, while all the states lying outside are empty. This becomes further clear when we consider band structure of solids. From Bloch's, theorem,

$$U(\vec{r} + \vec{R}) = U(\vec{r}) \quad \text{...(1)}$$

$$H\psi = \left(-\frac{\hbar^2}{2m}\nabla^2 + U(\vec{r})\right)\psi = E\psi \quad \text{...(2)}$$

$$\psi_{n\vec{k}}(\vec{r}) = e^{i\vec{k}\cdot\vec{r}}\, v_{n\vec{k}}(\vec{r}) \quad \text{...(3)}$$

$\psi_{n\vec{k}}$ are known as the Bloch wave functions, $\vec{R}$ is a reciprocal lattice vector.

$$u_{n\vec{k}}(\vec{r} + \vec{R}) = u_{n\vec{k}}(\vec{r}) \quad \text{...(4)}$$

and

$$\psi(\vec{r} + \vec{R}) = e^{i\vec{k}\cdot\vec{R}}\,\psi(\vec{r}) \quad \text{...(5)}$$

From Bloch's theorem, the index *n* indicates that for each $\vec{k}$ there are many solutions $E_n(\vec{k})$ (the energy levels).

Now, instead of holding $\vec{k}$ fixed, we may allow it to vary. Eigenstates and eigenfunctions for $\vec{k}$, that differ by a reciprocal lattice vector are identical due to periodicity:

$$\psi_{n,\vec{k}+\vec{K}}(\vec{r}) = \psi_{n\vec{k}}(\vec{r}) \qquad ...(6)$$

$$E_{n,\vec{k}+\vec{K}} = E_{n\vec{k}} \qquad ...(7)$$

For band structure, we can construct a ground state of $N$ Bloch electrons (for $N$ unit cells) which may result in two distinct configurations.

1. Configuration involving a number of bands completely filled with all other higher one's empty.
2. A number of bands being partially filled.

For each partially filled band, there is a surface in $\vec{k}$-space separating the filled and unfilled levels in the partially filled band. This is the *Fermi-surface* (FS). This definition holds actually at absolute zero. The effect of temperature on the FS is very slight and the surface remains sharp even at room temperature. *The shape of the FS is determined by the geometry of the energy contours in a Brillouin zone* (BZ). For one free electron per atom, the FS is a sphere of radius $\vec{k}_F$ and it lies well within the first BZ. However, non-spherical and complicated shapes are observed when the FS and BZ are close to or touch one another under the effect of pseudopotential. A study of the shape of the FS and its proximity to the BZ is necessary for a better understanding of the intricate details of various properties of solids such as heat capacity, Pauli paramagnetism, electrical conductivity, etc.

*Free-Electron Fermi Surface*. In case of one electron model, we may ignore the interaction between the valence electrons and the Hartree or Hartree-Fock potential $V(\vec{r})$ seen by each valence electron, and may be taken as constant as the electron moves through the crystal.

The electrons are then described by plane wave states

$$|\vec{k}\rangle \equiv \psi_{\vec{k}} = e^{i\vec{k}\cdot\vec{r}} \qquad ...(8)$$

having energies

$$E(\vec{k}) = \frac{\hbar^2 k^2}{2m_e} \qquad ...(9)$$

We know that these states satisfy the Bloch theorem. The surfaces of constant energy are spheres in $\vec{k}$-space. The allowed values of $\vec{k}$ are distributed with density $V/(8\pi^3)$ in this $k$-space $\left(V = \frac{4}{3}\pi k_F^3\right)$. But for each value of $\vec{k}$, these are two electron states of opposite spin. If there are $z$ electrons per atom, 1 atom per cell and $N$ unit cells per unit volume, then there will be $ZN$ electrons per unit volume in real space and to satisfy the Pauli's principle, we need to fill the states up to a wave vector number $k_F$ given by

$$2 \times \frac{4}{3}\pi k_F^3/(8\pi^3) = ZN$$

i.e.,

$$k_F = (3\pi^2 ZN)^{1/3} \qquad ...(10)$$

It is then said that the *Fermi-surface* is a sphere of radius, $k_F$. The *Fermi-energy* $E_F$ (or *Fermi-level* is the energy corresponding to top of the distribution:

$$E_F = \frac{\hbar^2 k_F^2}{2m_e} \quad ...(11)$$

The Fermi surface separates filled from the unfilled states. For most of the materials, the electron energy as a function of the wave vector $\vec{k}$ is usually not so simple as Eq. (9) and the Fermi surface of most solids deviate from a spherical shape depending on the interaction of electrons with the lattice (Fig. 10.43).

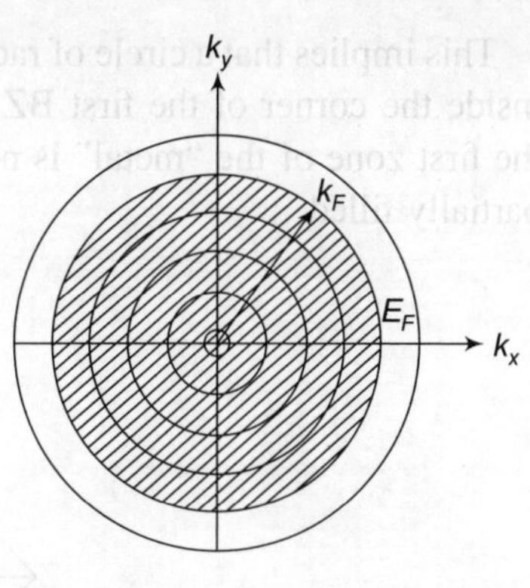


**Figure 10.43** The Fermi surface for free electrons

## 10.33 FERMI SURFACES IN TWO DIMENSIONS

For simplicity, we consider a square lattice of lattice periodicity $a$ to illustrate the construction of Fermi surfaces. The Brillouin zones are also constructed. In order to know the size of the Harrison's (Fermi) circle, let us consider the following cases of increasing electron concentration.

*Monovalent Metal.* We know that the area of a BZ corresponding to a square lattice of periodicity $a$ is given by

$$A = \frac{4\pi^2}{a^2} \quad ...(1)$$

Since, for a monovalent case only half the area of the BZ is occupied, therefore,

or $$\pi k_F^2 = \frac{1}{2}\frac{4\pi^2}{a^2} \quad \text{or} \quad k_F = \sqrt{\frac{2}{\pi}}\left(\frac{\pi}{a}\right) = 0.798\frac{\pi}{a} \quad ...(2)$$

where $\pi/a$ is the distance of the zone boundary from the centre of the zone, since the value of $k_F$ lies between $0 < k_F < \pi/a$. This indicates that the Fermi circle lies well within the first BZ as shown in Fig. 10.44 Hence, the FS remains undistorted.

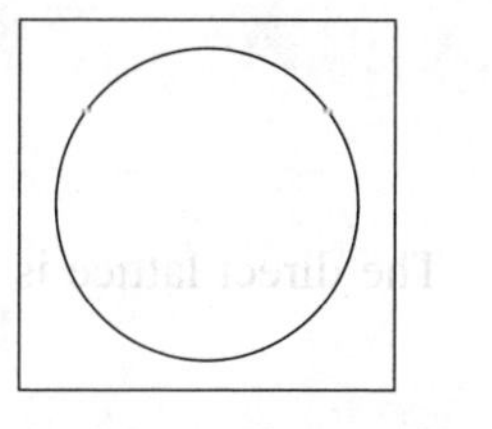

**Figure 10.44** Free electron FS within the first BZ

*Divalent Metal.* For a divalent metal with the same type of the crystalline lattice as above, the area of the Fermi circle is

$$\pi k_F^2 = \frac{2}{2}\frac{4\pi^2}{a^2}$$

or $$k_F = \frac{2}{\sqrt{\pi}}\left(\frac{\pi}{a}\right) = 1.228\frac{\pi}{a} \quad ...(3)$$

Further, the distance of a corner of the first zone from the centre is

$$\sqrt{2}\,\frac{\pi}{a} = 1.414\,\frac{\pi}{a} \quad ...(4)$$

From Eqs. (3) and (4), we have

$$\frac{\pi}{a} < k_F < 1.414\,\frac{\pi}{a}$$

This implies that a circle of radius $k_F$ will go beyond the first zone boundary but will remain inside the corner of the first BZ as shown in Fig. 10.45. In other words, it can be said that the first zone of the "metal" is not completely filled with electrons while the second zone is partially filled.

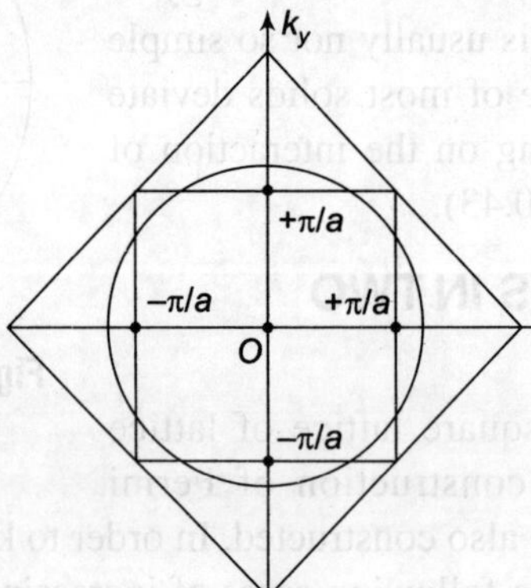


**Figure 10.45** FS partially on first and second BZ

**Example 10.2** In 2*D*, find the reciprocal lattice for the lattice defined by the unit cell, given next.

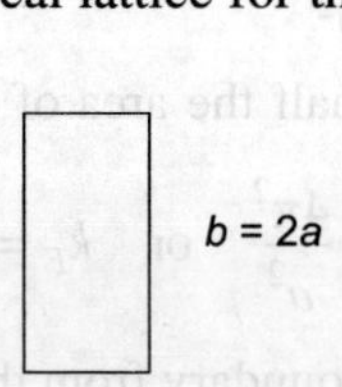


**Figure 10.46**

The direct lattice is defined by

$$\boldsymbol{a} = a\boldsymbol{i} \quad \text{and} \quad \boldsymbol{b} = b\boldsymbol{j} = 2a\boldsymbol{j}. \qquad \text{...(1)}$$

The reciprocal lattice is defined by vectors

$$\boldsymbol{A} = A_x\boldsymbol{i} + A_y\boldsymbol{j} \quad \text{and} \quad \boldsymbol{B} = B_x\boldsymbol{i} + B_y\boldsymbol{j}, \qquad \text{...(2)}$$

with

$$\boldsymbol{A}\cdot\boldsymbol{a} = \boldsymbol{B}\cdot\boldsymbol{b} = 2\pi \quad \text{and} \quad \boldsymbol{A}\cdot\boldsymbol{b} = \boldsymbol{B}\cdot\boldsymbol{a} = 0.$$

Thus,

$$\boldsymbol{A} = \frac{2\pi}{a}\boldsymbol{i}, \qquad \text{...(3)}$$

$$\boldsymbol{B} = \frac{2\pi}{b}\boldsymbol{j} = \frac{\pi}{a}\boldsymbol{j}, \qquad \text{...(4)}$$

where the $2\pi$ now inserted in an alternative convention for reciprocal lattice vectors. The unit cell of the reciprocal lattice looks like:

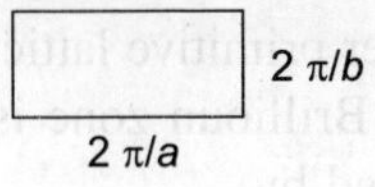


**Figure 10.47**

Now we suppose there is one electron per atom and one atom per unit cell. We want to calculate (a) the radius of the Fermi surface and (b) the radius of an energy surface that just manages to touch the first Brillouin zone boundary. The area of the first Brillouin zone is

$$A_{BZ} = \frac{(2\pi)^2}{ab} = \frac{4\pi^2}{a^2}. \quad ...(5)$$

The radius of the Fermi surface is determined by the fact that its area is just ½ of the full Brillouin zone area

$$\pi k_2^F = \frac{1}{2} A_{BZ} \text{ or } k_F = \frac{\sqrt{\pi}}{a}. \quad ...(6)$$

The radius to touch the Brillouin zone boundary is

$$k_T = \frac{1}{2}\frac{2\pi}{b} = \frac{\pi}{2a}. \quad ...(7)$$

Thus, $$\frac{k_T}{k_F} = \frac{\sqrt{\pi}}{2} = 0.89,$$

and the circular Fermi surface extends into the second Brillouin zone. The first two zones are sketched in Fig. 10.48.

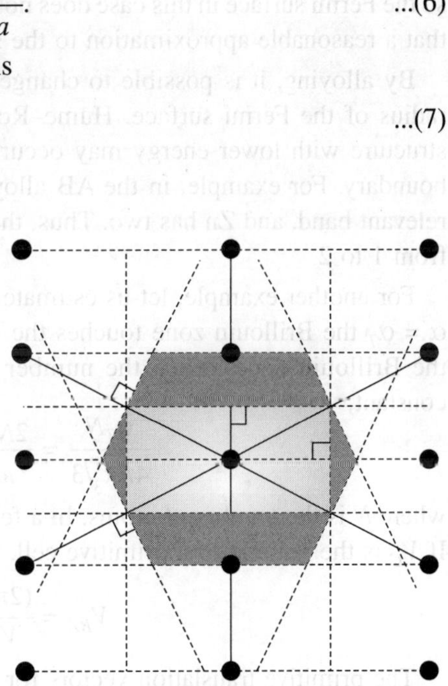

**Figure 10.48** First (light-shaded area) and second (dark-shaded area) Brillouin zones

As another example, let us consider a body-centred cubic lattice (bcc) with a standard, nonprimitive, cubic unit cell containing two atoms. The reciprocal lattice is fcc. Starting from a set of primitive vectors, one can show that the first Brillouin zone is a dodecahedron with twelve faces that are bounded by planes with perpendicular vector from the origin at

$$\frac{\pi}{a}\left\{(\pm 1, \pm 1, 0), (\pm 1, 0, \pm 1), (0, \pm 1, \pm 1)\right\}.$$

Since there are two atoms per unit cell, the volume of a primitive unit cell in the bcc lattice is

$$V_C = \frac{a^3}{2}. \quad ...(8)$$

The Brillouin zone, therefore, has volume

$$V_{BZ} = \frac{(2\pi)^3}{V_C} = \frac{16\pi^3}{a^3}. \quad ...(9)$$

Let us assume we have one atom per primitive lattice point and each atom contributes one electron to the band. Then, since the Brillioun zone is half-filled, if we assume a spherical energy surface, the radius is determined by

$$\frac{4\pi k_F^3}{3} = \frac{1}{2} \cdot \frac{16\pi^3}{a^3} \quad \text{or} \quad k_F = \frac{\sqrt[3]{6\pi^2}}{a}. \qquad ...(10)$$

From Eq. (11), a sphere of maximum radius $k_T$, as given below, can just be inscribed within the first Brillouin zone

$$k_T = \frac{\pi}{a}\sqrt{2}. \qquad ...(11)$$

Direct computation yield

$$\frac{k_T}{k_F} = 1.14,$$

so the Fermi surface in this case does not touch the Brillouin zone. We might expect, therefore, that a reasonable approximation to the shape of the Fermi surface would be spherical.

By alloying, it is possible to change the effective electron concentration and, hence, the radius of the Fermi surface. Hume–Rothery has predicted that phase changes to a crystal structure with lower energy may occur when the Fermi surface touches the Brillouin zone boundary. For example, in the AB alloy $Cu_{1-x}Zn_x$, Cu has one electron to contribute to the relevant band, and Zn has two. Thus, the number of electrons on average per atom, $\alpha$, varies from 1 to 2.

For another example, let us estimate for a fcc structure (bcc in reciprocal lattice) at what $\alpha = \alpha_T$ the Brillouin zone touches the Fermi surface. Let $k_T$ be the radius that just touches the Brillouin zone. Since the number of states per unit volume of reciprocal space is a constant,

$$\frac{\alpha_T N}{4\pi k_3^T/3} = \frac{2N}{V_{BZ}} \qquad ...(12)$$

where $N$ is the number of atoms. In a fcc lattice, there are 4 atoms per nonprimitive unit cell. If $V_C$ is the volume of a primitive cell, then

$$V_{BZ} = \frac{(2\pi)^3}{V_C} = \frac{4}{a^3}(2\pi)^3. \qquad ...(13)$$

The primitive translation vectors for a bcc unit cell are

$$\boldsymbol{A} = \frac{2\pi}{a}(\boldsymbol{i} + \boldsymbol{j} - \boldsymbol{k}), \qquad ...(14)$$

$$\boldsymbol{B} = \frac{2\pi}{a}(-\boldsymbol{i} + \boldsymbol{j} + \boldsymbol{k}), \qquad ...(15)$$

$$\boldsymbol{C} = \frac{2\pi}{a}(\boldsymbol{i} + \boldsymbol{j} - \boldsymbol{k}), \qquad ...(16)$$

From this we easily conclude

$$k_T = \left(\frac{2\pi}{a}\right)\left(\frac{1}{2}\right)\sqrt{3}.$$

So we find

$$a_T = 2\left(\frac{a^3}{4}\right)\frac{1}{8\pi^3}\cdot\frac{4}{3}\pi\left[\frac{(2\pi)^3}{a^3}\right]\frac{1}{8}3^{3/2} \quad \text{or} \quad \alpha_T = 1.36.$$

## 10.34 FERMI SURFACE FOR A SQUARE LATTICE

We begin our discussion with a square lattice. Figure 10.49 shows Brillouin zones of a square lattice in two dimensions.

The circle shown is a surface of constant energy for free electrons, it will be Fermi surface for some particular value of the electron concentration. The total area of the field region in $\vec{k}$-space depends only on the electron concentration, independent of interaction of electrons with the lattice. The slope of the Fermi surface depends on the lattice interactions and it will not be an exact circle in actual lattice.

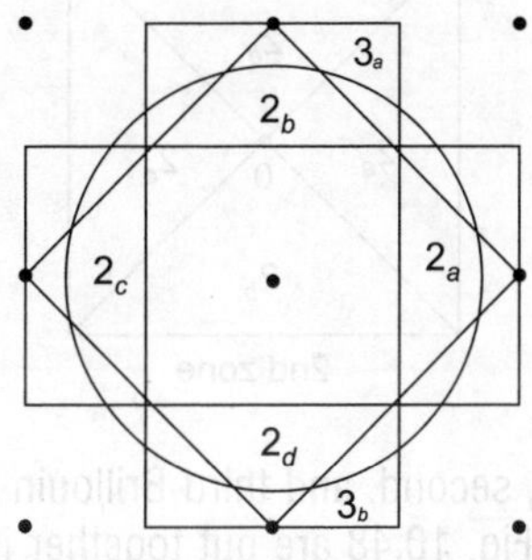


**Figure 10.49** Brillouin zones of a square lattice in two dimensions. The circle shown is a surface of constant energy for free electrons. It will be the Fermi surface for some particular value of the electron concentration. The total area of the filled region in **k** space depends only on the electron concentration and is independent of the interaction of the electrons with the lattice. The shape of the fermi surface depends on the lattice interaction, and the shape will not be an exact circle in an actual lattice. The labels within the sections of the second and third zones refer to Fig. 7. [After Charles Kittel, *Introduction to Solid State Physics,* John Wiley & Sons, Inc. (1996)]

We know that the equation for the zone boundary is

$$2\vec{k}\cdot\vec{G} + \vec{G}^2 = 0 \qquad ...(1)$$

It is satisfied if $\vec{k}$ terminates on the plane perpendicular to $\vec{G}$ at the mid-point of $|\vec{G}|$. The first Brillouin zone is the area enclosed by perpendicular bisectors of $|\vec{G}_1|$ (i.e., mid-points of four reciprocal lattice vectors $\pm$ $(2\pi/a)\,\hat{k}_x$ and $\pm$ $(2\pi/a)\,\hat{k}_y$. The second zone is constructed from $\vec{G}_2s$ and similarly for the third zone.

To determine the boundaries of some zones, we have to consider sets of several (non-equivalent) reciprocal vectors, e.g., boundaries of section $3a$ of the third zone are formed from the perpendicular bisectors of three $G$s, namely, $(2\pi/a)\hat{k}_x$, $(4\pi/a)\hat{k}_y$ and $(2\pi/a)$ $(\hat{k}_x + \hat{k}_y)$.

The energy $E_x'$ for $k'$, outside the first Brillouin zone, is identical to the energy $E_x$ within the first zone if a suitable translation vector $\vec{G}$ can be found so that a wave vector $\vec{k}'$ becomes

$$k' = \vec{k} + \vec{G} \qquad \text{...(2)}$$

This feature is used in plotting energy bands in the form of reduced zone scheme, for example, the triangle labelled $2a$ can be moved by a reciprocal lattice vector $\vec{G} = -(2\pi/a)\,k_x$ such that the triangle reappears in the area of the first Brillouin zone. Other reciprocal lattice vectors will shift the triangles $2b$, $2c$, $2d$ to the other parts of the first zone completing the mapping on the second zone into the reduced zone scheme (Fig. 10.50). The parts of the Fermi surface falling in the second zone are now connected (Fig. 10.51), whereas for the third zone, the parts of the Fermi surface still appear disconnected. When one looks at it in the periodic zone scheme, the FS forms a lattice of rosetts (Fig. 10.52).

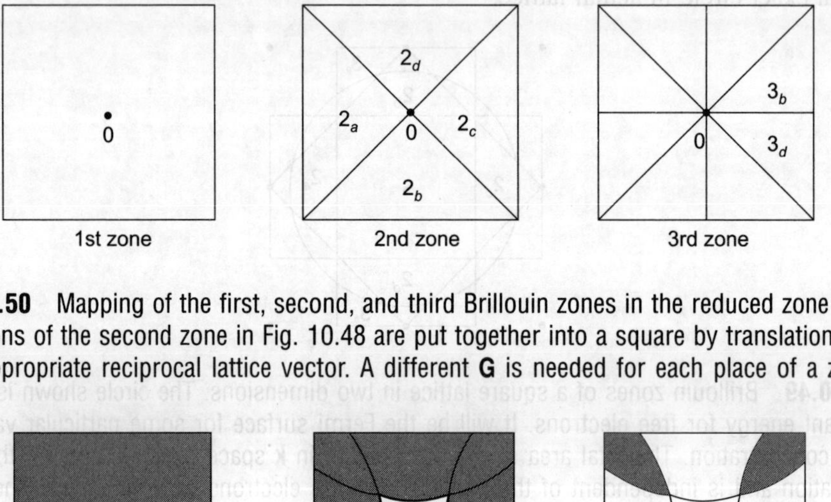


**Figure 10.50** Mapping of the first, second, and third Brillouin zones in the reduced zone scheme. The sections of the second zone in Fig. 10.48 are put together into a square by translation through an appropriate reciprocal lattice vector. A different **G** is needed for each place of a zone

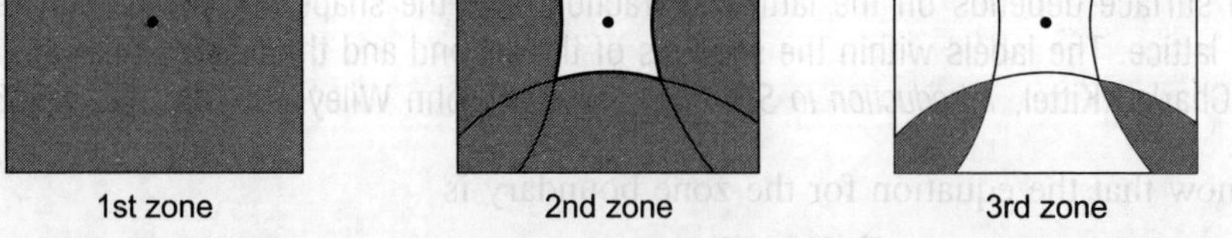


**Figure 10.51** The free-electron Fermi surface of Fig. 10.48, as viewed in the reduced zone scheme. The shaded areas represent occupied electron states. Parts of the Fermi surface fall in the second, third, and fourth zones. The fourth zone is not shown. The first zone is entirely occupied

Note that the Fermi surface enables us to visualize the relative fullness or occupation of the allowed empty lattice geometrically in $k$-space.

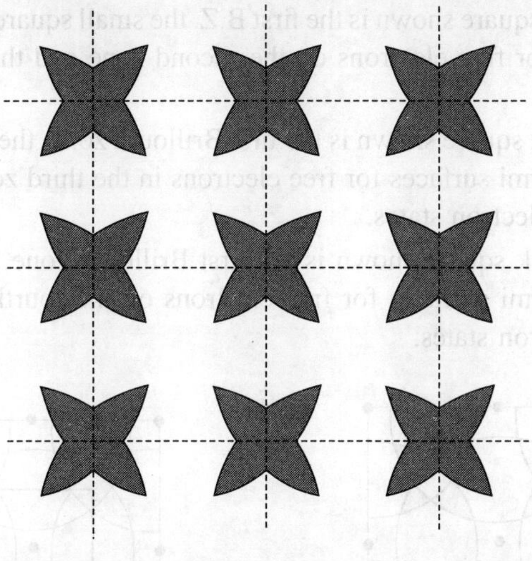

**Figure 10.52** The Fermi surface in the third zone as drawn in the periodic zone scheme. The figure was constructed by repeating the third zone of Fig. 10.50

## 10.35 HARRISON CONSTRUCTIONS

Fermi surfaces for free electrons are constructed by Harrison's method as follows.

The reciprocal lattice points of a square lattice are determined, and free-electron sphere of radius appropriate to the electron concentration is drawn around each reciprocal lattice point. Any point in $\vec{k}$-space that lies within at least one sphere corresponds to an occupied state in the first zone. Points within at least two spheres correspond to occupied states in the second zone, and similarly for points in three or more spheres. In Fig. 10.53 the black (dark) square shown is the first B.Z. This first B.Z. is completely occupied.

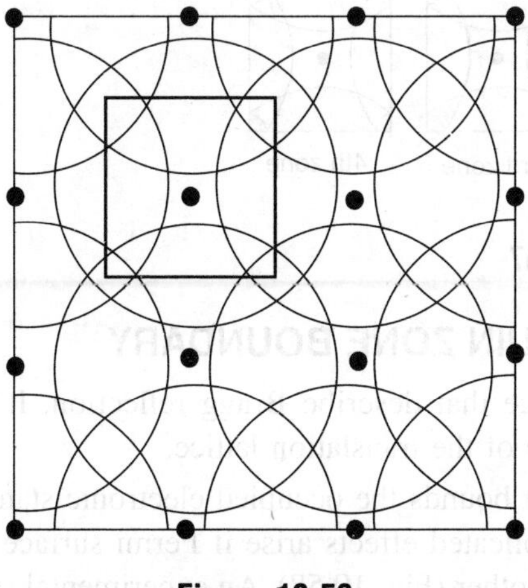

**Figure 10.53**

**Figure 10.54**

In Fig. 10.54 the dark square shown is the first B.Z. the small square shaped lines (unshaded) are the Fermi surfaces for free electrons on the second zone and the shaded areas represent occupied electron states.

In Fig. 10.55 the black square shown is the first Brillouin zone, the dark curved lines (about shaded areas) are the Fermi surfaces for free electrons in the third zone and the shaded areas represent the occupied electron states.

In Fig. 10.56 the black square shown is the first Brillouin zone, the dark lines about the shaded areas are the Fermi surfaces for free electrons on the fourth zone and shaded areas represent occupied electron states.

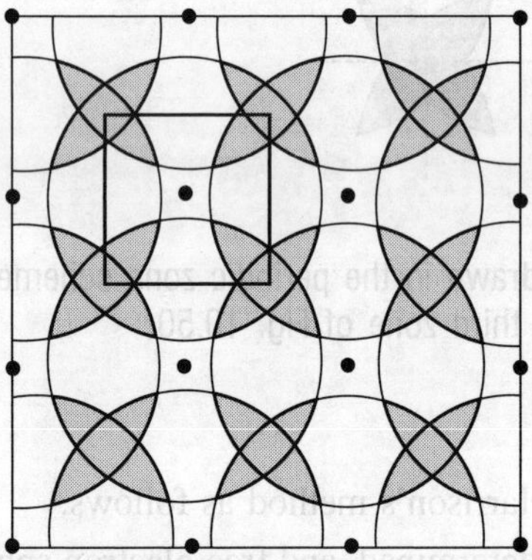

**Figure 10.55**

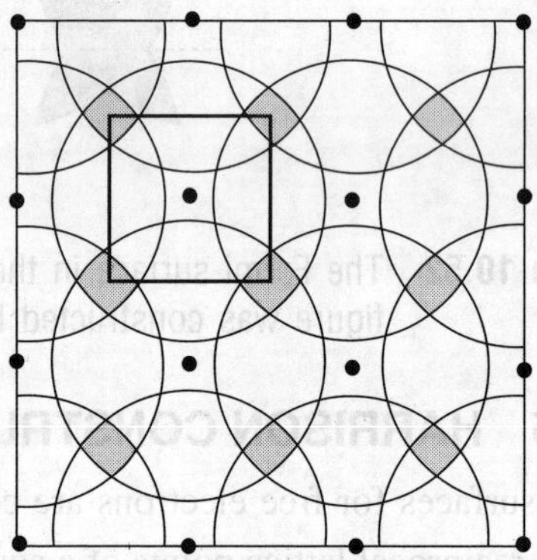

**Figure 10.56**

Thus, in Fig. 10.57 we show the free electron, FS, as viewed in the reduced zone scheme. The shaded areas represent occupied electron states. Parts of FS (dark lines about shaded areas) fall in the second, third and fourth zones. The first zone is entirely occupied.

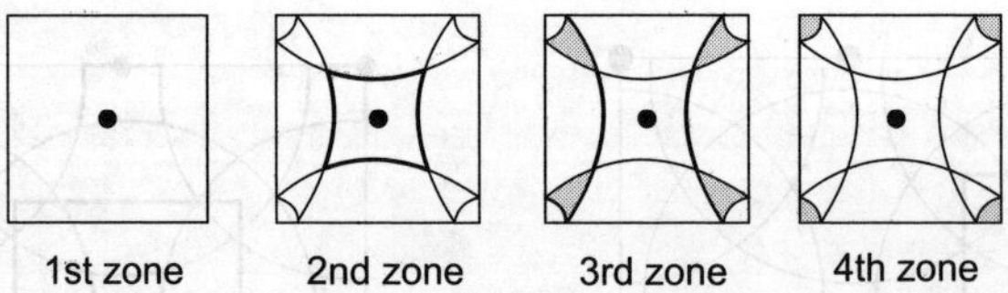


**Figure 10.57**

## 10.36 FERMI SURFACE AND BRILLOUIN ZONE BOUNDARY

A Brillouin zone is the locus of points in $\vec{k}$-space that describe Bragg reflection. It is a geometric construction determined by the structure of the translation lattice.

The Fermi surface is the surface in $\vec{k}$ space that bounds the occupied electronic states at $T = 0\,\text{K}$. Experiments have shown that more complicated effects arise if Fermi surface and Brillouin zone boundary are close to or touch each other (Fig. 10.58). An experimental study of the shape of the Fermi surface and its proximity to the Brillouin zone is fundamental to a detailed understanding of the physics of electrons in metals. Such investigations are usually made by investigating the magnetic or electrical behaviour as a function of an applied magnetic field. In connection with the intersection of the Fermi surface with the Brillouin zone boundaries, we may observe the following.

(i) If the Fermi surface lies entirely within the first Brillouin zone, it is exactly spherical in shape.

(ii) The gradient of the energy function (if non-degenerate) must vanish normal to the zone boundary, which makes the constant energy surface to always come out at right angles to the boundary.

This renders the surfaces quite different from spheres near the corners of the zone.

(iii) The free electron energy contours ($E(k) = \hbar^2k^2/2m$) are perturbed by the lattice potential and it can be noted from the free electron energy parabola that (a) depression of energy just inside a zone boundary moves the energy contours out towards the boundary; (b) and the increased energy just outside a boundary moves a constant energy contour in towards the boundary, i.e., zone boundaries attract the energy contours.

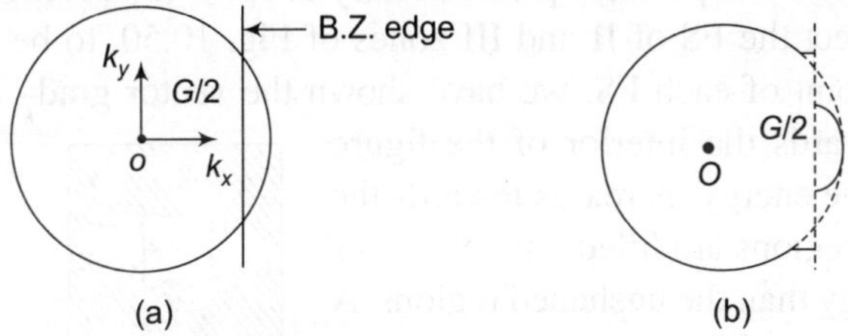


**Figure 10.58** (a) Fermi circle cut by a Brillouin zone edge. (b) The Fermi surface for nearly free electron case, showing the interactions which are present in the free electron approximation

Very briefly, for a crystal with a mirror plane perpendicular to $\vec{k}$, we show that $\partial E/\partial k = 0$ at the zone boundary.

Since $\vec{k}$ is assumed to be perpendicular to a mirror plane $E(\vec{k})$, must be symmetric along this direction. Also $E(\vec{k})$ is a periodic function of the reciprocal lattice vectors. These two conditions lead to

$$E(\vec{k}) = E(-\vec{k}), \quad (\partial E/\partial k)_{\vec{k}} = -(\partial E/\partial k)_{-\vec{k}}$$

$$E(\vec{k}) = E(\vec{k} + \vec{G}), \quad (\partial E/\partial k)_{\vec{k}} = (\partial E/\partial k)_{\vec{k}+\vec{G}}$$

However, at the zone boundary, $\vec{k} = \pm \vec{G}/2$ and the above equations lead to different results.

$$(\partial E/\partial k)_{\vec{G}/2} = -(\partial E/\partial k)_{-\vec{G}/2}$$

$$(\partial E/\partial k)_{\vec{G}/2} = +(\partial E/\partial k)_{-\vec{G}/2}$$

Both of these equations can be true only if

$$\partial E/\partial k = 0$$

at the zone boundary.

## 10.37 HARRISON'S METHOD OF CONSTRUCTION OF FERMI SURFACES

Harrisons method is based on a weak pseudopotential, $V_{\text{eff}}(r)$. The effect of this potential is to introduce a small perturbation into the motion of free electrons in the whole $k$-space except the region near the zone boundaries. It is this potential which causes discontinuities as well as distortions in the FS at the zone boundaries. Therefore, $V_{\text{eff}}(r)$ is when arbitrarily low, the energy discontinuities and the distortions in the FS can be removed.

The effectiveness of a weak periodic potential on the surfaces constructed from the free electron sphere is simply to round off sharp edges and corners. If however, a branch of the FS consists of very small pieces of surface (surrounding either occupied or unoccupied levels, known respectively as *pockets of electrons* or *pocket of holes*), then a weak periodic potential may cause these to disappear. In addition, if the free electron, FS, has parts with a very narrow cross section, a weak periodic potential may cause it to become disconnected at such points. Thus, we expect the FS of II and III zones of Fig. 10.50, to be changed as shown in Fig. 10.59. At one point of each FS, we have shown the vector $\text{grad}_{\vec{k}} E$. In the II zone, the energy increases towards the interior of the figure and in the III zone, the energy increases towards the exterior. The shaded regions are filled with electrons and are lower in energy than the unshaded regions. A FS like that of the III zone is *electron-like* (the FS is convex to the region of occupied electronic states), whereas FS like that of II zone is *hole-like* (the FS is convex to the region of unoccupied states).

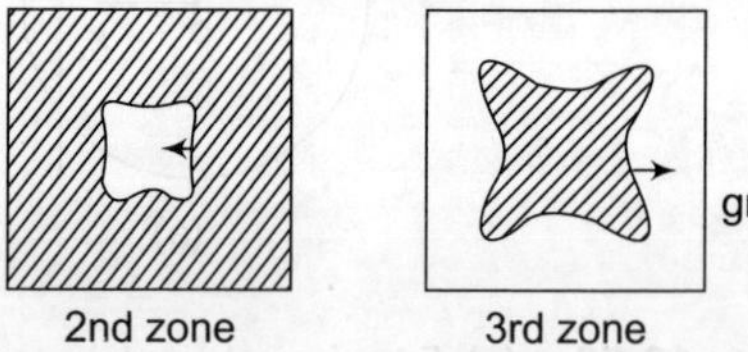


**Figure 10.59** Qualitative impression of the effect of a weak periodic crystal potential on the FS of Fig. 10.57

### 10.37.1 Divalent Metal

For $V_{\text{eff}}(r) = 0$, Fig. 10.60= shows the FS partially on first and second BZ, i.e., a circle of radius $k_F$ goes beyond the first zone boundary but remains inside the corner of the first BZ. It can therefore be said that first zone of the metal is not completely filled with electrons while the second zone is partially filled.

However, if we assume that $V_{\text{eff}}(r) > 0$ but small (i.e., a weak pseudopotential), then it will have the following effects.

(i) Energy discontinuities will appear at the zone boundaries.

(ii) Constant energy curves will be a periodic function in $k$-space with a period $\left(\frac{2\pi}{a}\right)$ along the $k_x$ and $k_y$ axes.

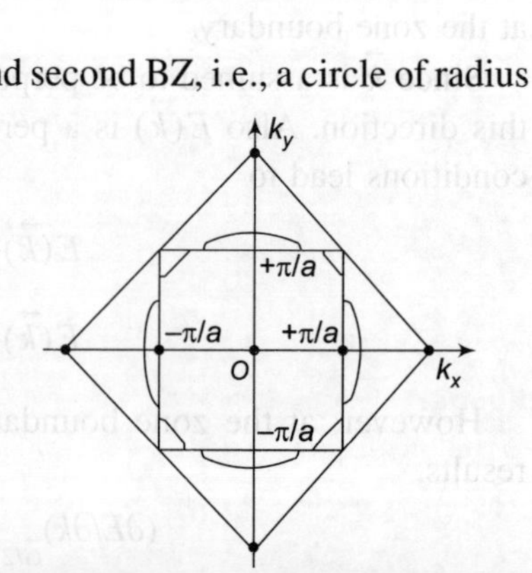


**Figure 10.60** Separate pieces of Harrison's circle at zone boundaries

As a result of energy discontinuities, Fermi circle breaks up into a number of separate 'pieces' of constant energy curves at the zone boundaries as shown in Fig. 10.60. If we translate these pieces through a distance of $2\pi/a$ along the $k_x$ and $k_y$ axes, we can construct closed curves around the corners and boundaries of the zone as shown in Fig. 10.61 in a periodic zone scheme. In the first case, the constant energy curves (surfaces in three dimensions) surround empty areas (spaces in three dimensions) and are known as first zone holes while in the second case these curves which surround the areas filled with electrons are known as second zone electrons within the first Brillouin zone, beyond the first Brillouin zone is the extended zone scheme, the occupied regions are shown as *ABC, DEF, GHI* and *JKL* in Fig. 10.62 (a).

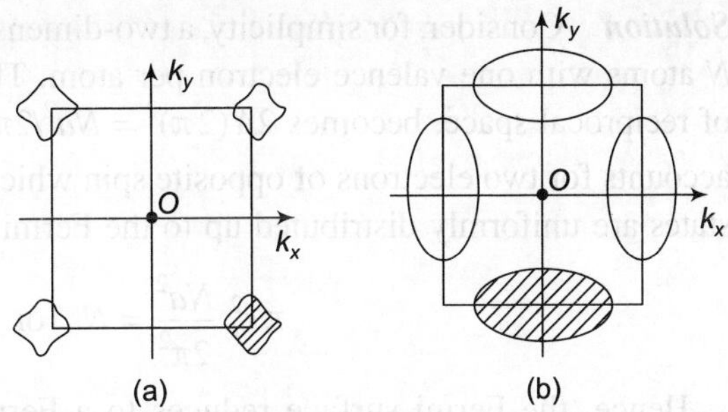


**Figure 10.61** Representation in periodic zone scheme (a) First zone holes, (b) Second zone electrons

Figure 10.62 (b) shows band 1 and unoccupied regions are not shaded. The occupied regions *ABC*, *DEF*, *GHI* and *JKL* in Fig. 10.62 (a) are brought inside the first Brillouin zone by translation through appropriate reciprocal lattice vectors. Then they are shown as *AMC*, *DNF*, *GOI* and *JPL* (see Fig. 10.62 (c)).

If the sharp corners are rounded off due to periodic lattice potential, then the unoccupied regions of Fig. 10.62 (b) will give a cross section centred at $x$ representing the holes in band 1 (Fig. 10.62 (d)). The occupied states take the shape of elliptical orbits representing electrons in band 2, shown as *ABCM*, *DEFN*, *GHIO* and *JKLP* [see Fig. 10.62(d)].

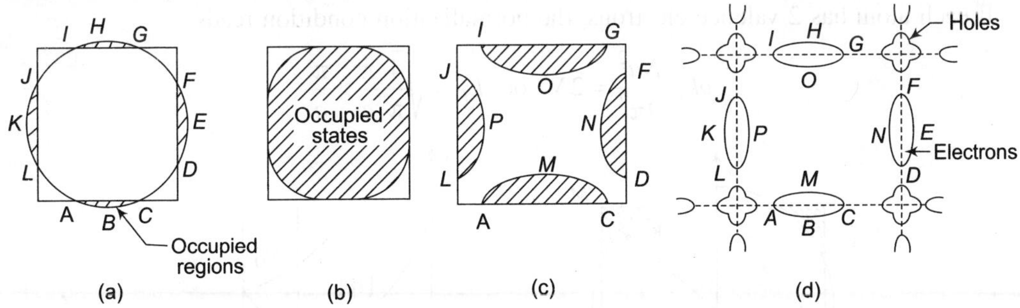


**Figure 10.62** Valence 2 metal: (a) Extended zone scheme; (b) Reduced zone scheme; (c) Valence 2 (reduced zone scheme); (d) Holes in band 1 centred around $x$ and electrons in band 2 in elliptical orbits in a Fermi surface

**Example 10.3** Find for the Fermi surface in a two-dimensional metal is modified, as the electron concentration increases from one to two valence electrons per atom.

***Solution*** Consider, for simplicity, a two-dimensional square lattice of parameter $a$, containing $N$ atoms with one valence electron per atom. The density of electron states per unit volume of reciprocal space, becomes $2A\,(2\pi)^2 = Na^2/2\pi^2$, for a crystal of area $A$, where the factor 2 accounts for two electrons of opposite spin which can be accommodated by each $\vec{k}$-state. The states are uniformly distributed up to the Fermi wave vector $k_F$, such that

$$\pi k_F^2 \frac{Na^2}{2\pi^2} = N \quad \text{or} \quad k_F = \sqrt{\frac{2}{\pi}}\,\frac{\pi}{a} < \frac{\pi}{a}$$

Hence, the Fermi surface reduces to a Fermi circle, which is entirely within the first Brillouin zone. The Fermi level $\varepsilon_F = \pi\hbar^2/m_e a^2$ is shown as a dashed line in the free-electron band diagrams along [10] and [11] in the square lattice, which are described by

$$\varepsilon_{[100]} = \frac{\hbar^2}{2m_e}k_x^2, \quad \varepsilon_{[110]} = \frac{\hbar^2}{2m_e}(k_x^2 + k_y^2)$$

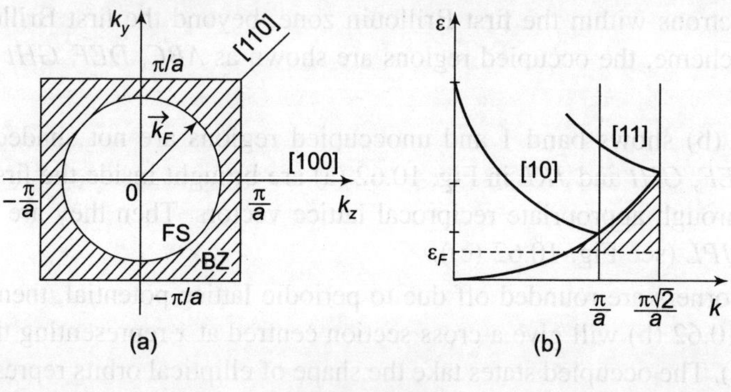


**Figure 10.63**

If each atom has 2 valence electrons, the normalization condition reads

$$pk_F^2 \frac{Na^2}{2\pi^2} = 2N \quad \text{or} \quad k_F = \sqrt{\frac{4}{\pi}}\,\frac{\pi}{a} > \frac{\pi}{a}$$

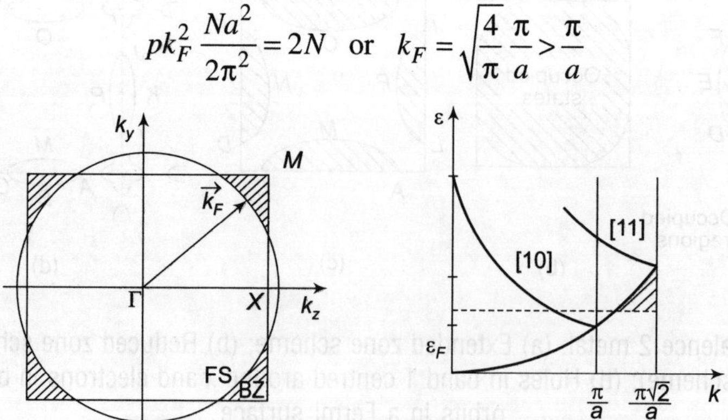


**Figure 10.64**

and hence, the Fermi surface has two sheets, one in the first energy band and one in the second. The new position of the Fermi level in the energy band diagram indicates that the occurrence of different symmetry directions may cause band overlap when the allowed energy states are integrated over all possible $\vec{k}$-values.

## 10.38 FERMI LINES

A periodic pseudopotential, if strong enough, can, under certain additional conditions, transform a free-electron metal into a semiconductor or insulator. This transition can be studied in a more illuminating way by following how the Fermi line (a circle in the unperturbed case) is deformed and eventually may disappear in the semiconducting state, as the strength of the periodic potential increases.

To determine the Fermi wave number, $k_F$, for the free-electron 2D case, we shall use the relation

$$2\frac{SS_{k_F}}{(2\pi)^2} = N_e = N_{pc}N_{ep}, \quad \text{...(1)}$$

where $S$ is the total area in real space of our $2D$ system, $S_{k_F} = \pi k_F^2$ is the area in $k$-space inside the free-electron Fermi circle, and $N_e$ is the total number of electrons, which is equal the number of primitive cells, $N_{pc}$, times the number of electrons, $N_{ep}$, per primitive cell. The L.H.S. of (1) gives the total number of states (including the two orientations of spin) inside the area $S_{k_F}$ of $\vec{k}$-space. For a square lattice of lattice constant $a$, we have $S = N_{pc}a^2$. Hence, the unperturbed Fermi wave vector is given by

$$k_F = \frac{\sqrt{2\pi N_{ep}}}{a}. \quad \text{...(2)}$$

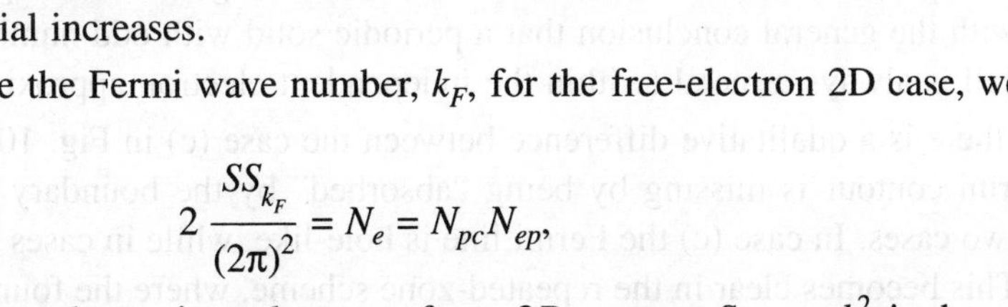


**Figure 10.65** The 1st BZ for a 2D square lattice and the Fermi line for one electron per primitive cell. The periodic potential is zero (a), or weak (b), or very strong (c) [After E.N. *Economor, the Physics of Solids*, Springer (2010)]

In Fig. 10.65 we plot the 1st BZ of a square lattice together with the Fermi line, when $N_{ep} = 1$, for the free-electron case, Fig. 10.65; for a weak periodic pseudopotential, Fig. 10.65b; and for a strong periodic pseudopotential, Fig. 10.65c. It is clear that the unperturbed Fermi circle lies entirely within the 1st BZ, since $k_F/(\Gamma X) = \sqrt{2\pi}/\pi = 0.798$, for $N_{ep} = 1$. The introduction of a weak periodic potential deforms the Fermi line, which develops bulges in the $\Gamma X$ direction (and its symmetric ones) and depressions in the $\Gamma M$ direction (and its symmetric ones), while keeping the area inside the Fermi line constant (according to (1)). *Question*: Can you explain why the deformation has this shape?

As the periodic potential becomes stronger, the bulges become more pronounced, until eventually, they touch the boundary of the 1st BZ, Fig. 10.65c, which is not part of the Fermi line (except the eight points of intersection). Since the area inside the Fermi line is conserved, there is no way for the latter to disappear entirely no matter how strong the periodic potential is. Hence, the 2D periodic "solid" with one electron per primitive cell will always have a nonzero DOS at $E_F$ and would never[16] become semiconducting or insulating. This result is in agreement with the general conclusion that a periodic solid with odd number of electrons per primitive cell is always a metal (within the independent-electron approximation).

Notice that there is a qualitative difference between the case (c) in Fig. 10.65 c (in which part of the Fermi contour is missing by being "absorbed" by the boundary of the 1st BZ) and the other two cases: In case (c) the Fermi line is hole-like, while in cases (a) and (b) it is electron-like, This becomes clear in the repeated-zone scheme, where the four disjoint pieces of the Fermi line in Fig. 10.65c would give a closed contour, the interior of which has only empty states.

## 10.39 CONSTANT ENERGY CONTOURS

Suppose, we have an empty crystal and pour electrons in it. These will fill one constant energy surface after another until all the electrons are filled up. We have now reached the Fermi level. In reciprocal space, we shall find compact volumes inside which all the states are occupied, bounded by a constant energy surface or surfaces. The FS is this boundary between filled and empty states in *k*-space, in the ground state of the crystal.

We already know that the Fermi surface for a free electron system is a simple sphere of radius given by ($k_F = 3\pi^2 n)^{1/3}$. This ideal shape is not usually maintained for most metals, although it may be a useful first approximation. In fact, as we approach the zone boundaries, the constant energy surfaces may be strongly distorted. There is a rule that the gradient of the energy function (if non-degenerate) must vanish normal to the zone-boundary. This makes the constant energy surfaces always come out at right angles to the

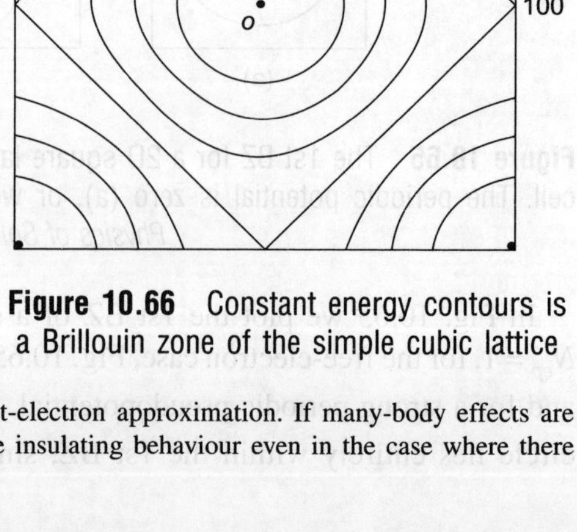


**Figure 10.66** Constant energy contours is a Brillouin zone of the simple cubic lattice

[16] This conclusion is valid within the periodic independent-electron approximation. If many-body effects are included, it is possible, under certain conditions, to have insulating behaviour even in the case where there were an odd number of electrons per primitive cell.

boundary. As shown in Fig. 10.65 this makes the surfaces quite different from spheres near the zone-corners.

It is obviously important to know the topology of the Fermi surface (FS),—for example, whether it actually intersects the zone boundaries. If part of the filled volume is held in by a portion of the boundary of the first zone, then this does not count, since it may correspond to states deep down in energy. The area of the FS may then be quite small, even through we have a large number of electrons (Fig. 10.67). The geometry of such situations is then clarified by repeated zone scheme (repetition of unit cell of the reciprocal lattice), with its full and empty regions, trusting to the irrelevance of the translation of any wave vector by a reciprocal lattice vector. The FS may then simply enclose one or other empty region and may be treated as a whole (Fig. 10.68).

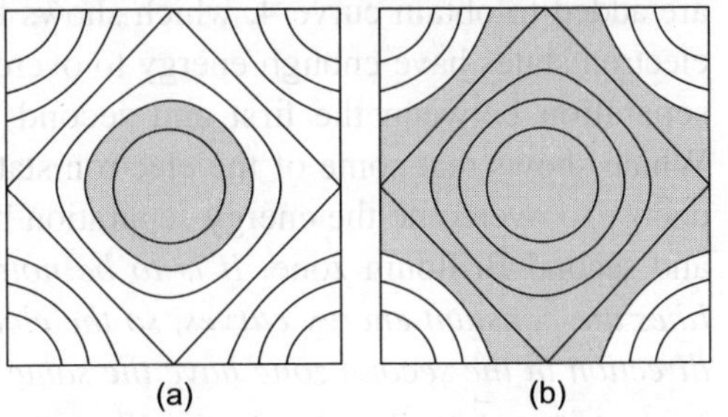

(a) (b)

**Figure 10.67** Fermi surfaces for (a) Few electrons (b) Many electrons

**Figure 10.68** Repeated zone scheme, showing original zone of Fig. 10.66 (b)

## 10.39.1 Constant Energy Contours in First Brillouin Zone

As more and more electrons are added, the electron lattice interaction (perturbation) increases and there is larger variation in the shape of the Fermi surface (deviation from a spherical shape).

The only real metals that have spherical Fermi surface are the monovalent alkali metals (Rb, Cs). For these elements, all the inner shells are filled, but the highest band (different of each element) is only half full of electrons. Consequently, at the top of the electron distribution of the first Brillouin zone boundary as more electrons are added, the electrons occupy states of successively increasing energy. The square lattice is indicated in Fig. 10.69.

The curve labelled 1 represents a low concentration of electrons where $k_F$ is much smaller than the distance of any Brillouin zone face.

For curve 2, there is a bigger perturbation of some of the electron states nearest the Brillouin zone boundary.

Curve 3 represents electron states for which there is a large perturbation for some states, yet all the filled electron states remain within the first Brillouin zone. More electrons are added to obtain curve 4, which shows that some of the electron states have enough energy to overcome the energy separation between the first and second Brillouin zone. Which shows that some of the electron states have enough energy to overcome the energy separation between the first and second Brillouin zone. *It is to be noted here that the lines are constant energy curves, so the electrons in the* $\vec{a}^*$ *direction in the second zone have the same energy as those in the* $(\vec{a}^* + \vec{b}^*)$ *direction in the* $1^{st}$ *zone.*

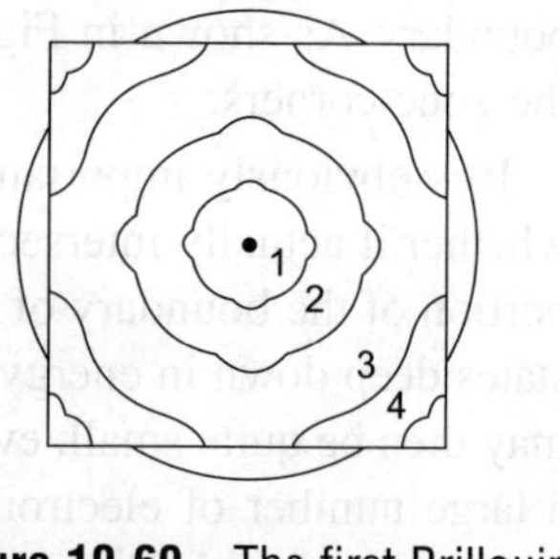


**Figure 10.69** The first Brillouin zone for a two-dimensional square lattice with various concentrations of electrons, increasing from 1 to 4

Thus, we can make approximate (free hand) constructions for the Fermi surfaces in nearly free electron case, by using the following four facts.

(i) The interaction of the electron with periodic lattice potential causes energy gaps at the zone boundaries.

(ii) The Fermi surface (almost) always intersects the zone boundaries perpendicularly.

(iii) The crystal potential will smoothen (round-out) sharp corners in the Fermi surface.

(iv) The total volume (i.e., number of states) enclosed by the Fermi surface depends only on the electron concentration and is independent of the details of the lattice interaction.

### 10.39.2 Constant Energy Contour in a Divalent Metal

Figure 10.70 shows perturbation of a free electron energy contour by a lattice potential in case of a divalent metal. Since, energy at a zone corner is twice of the centre of zone edge $(\alpha(\sqrt{2})^2)$, therefore, the lowest states in the second zone are filled first and are below the highest states in the first zone and we have overlapping bands. One or more of the zones are always partly empty whenever the zones overlap. This explains *why divalent atoms are* good

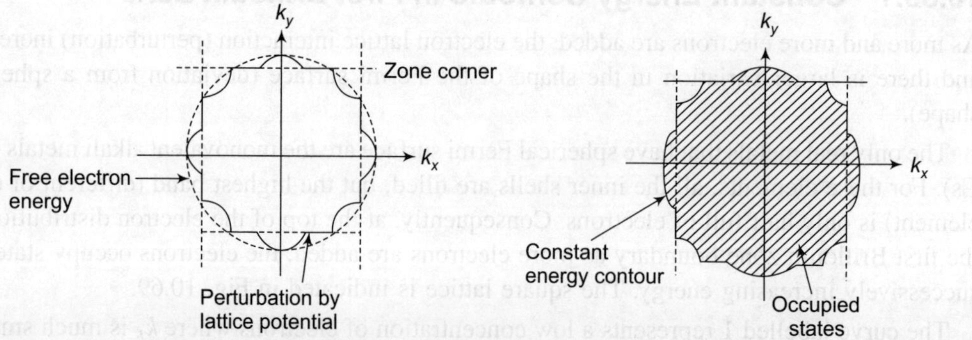


**Figure 10.70** Constant energy contour in a divalent metal

conductors. Although the FS is very different from a free electron, it does have same free area, which can give current carrying states when FS is displaced (see section 10.4). If the energy gap increases we reach a situation when the bands no longer overlap, the first zone is entirely full and the second zone entirely empty. Then occupied space is entirely bounded by BZ boundary (there is no free area FS) and we have an insulator.

The Fermi surface in Fig. 10.69 consists of a number of disconnected pieces, broken up boundaries. These can be brought together by translation of $2\pi/a$ as in Fig. 10.61 (d). Then, the roughly circular surface of holes is the first zone and the almost elliptical surface contains electrons in the second zone. The number of electrons and holes are equal. This explains the metallic behaviour exhibited by divalent metals (viz., alkaline earths: Ca, Sr, Ba, etc.) The wide variety of Hall observed in Group II metals, e.g., Br$(1/(R_H\text{Ne})) = +0.2$, Mg $(1/(R_H\text{Ne}))$ $= -1.4$, is accounted differences in electron and hole mobilities.

It is to be noted that the divalent metals 9, (e.g., Be or Mg) have weak lattice interactions and spherical Fermi surface (like alkali metals). However, since they have two valence electrons each, the surface encloses twice the volume (in $\vec{k}$-space) than for alkalis, i.e., the volume enclosed by surface is exactly equal to that of a zone, but because the surface is spherical, it gets extended the first zone (into the second zone).

## 10.40 ELECTRON ENERGY CONTOURS FOR METALS AND INSULATORS

The electrical behaviour of a solid depends upon the degree of occupancy of its energy bands and also upon the nature of the band structure. A solid will be an insulator if two conditions are satisfied: (1) There are even number of valence electrons per structural unit (by structural unit, we mean an atom in a metal or covalent elemental solid such as diamond, a molecule in a molecular solid and an ion pair in an ionic solid), and (2) the band that contains the highest energy electrons must be separated from the (next) allowed band (i.e., above it) by an energy gap larger compared with $kT$. The reason for condition (1) is that it ensures that the highest band be completely filled and the reason for (2) is that none of the electrons be able to cross the gap to reach unfilled states. A conductor is characteirzed by its violation of either (or both) of the above two conditions.

Figure 10.71 (a) shows the energy contours of a hypothetical insulator. In case of alkali metals, with an odd number of valence electrons per structural unit (i.e., one per atom), we have a (conducting) metal, as also are divalent metals, e.g., Mg and Zn which have overlapping energy bands. The corresponding energy contours and shown in Fig. 10.71 (b) and (c) respectively.

Hall effect studies indicate that the conductivity of divalent metals (e.g., Zn or Cd) is largely due to positive charge carriers and not due to electrons. This is readily accounted for on the basis of the band picture by assuming that the overlap of the Fermi surface into the highest band is small, leaving vacant states, which are holes in the band below it. Thus, holes in the lower band carry the bulk of the current while the electrons in the upper band play a minor role.

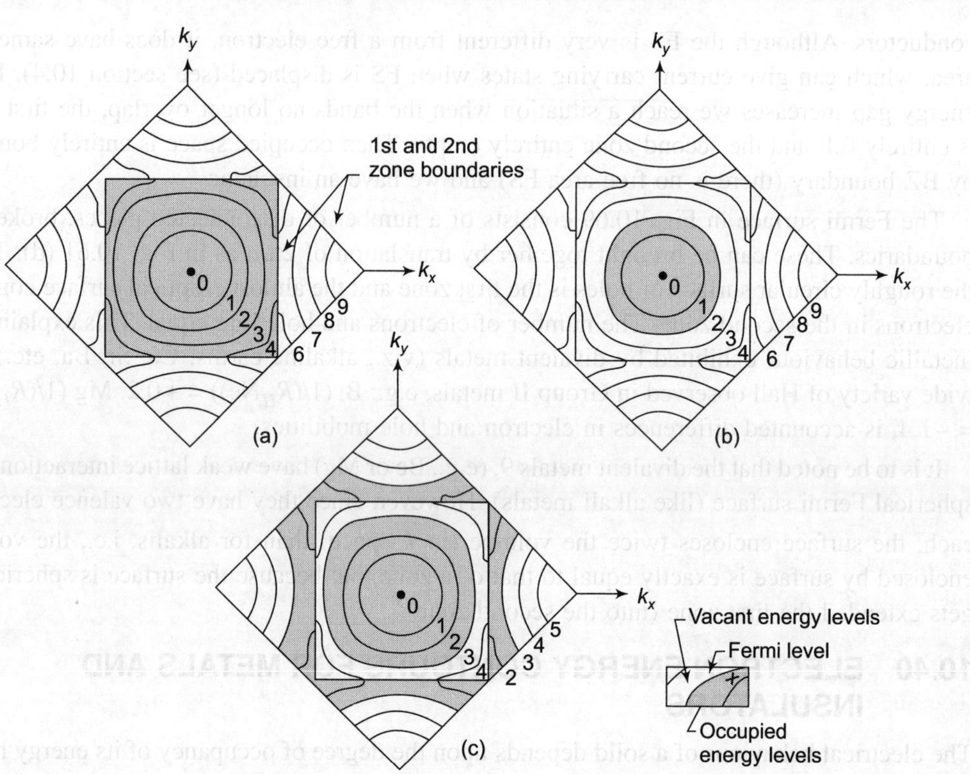


**Figure 10.71** Electron energy contours and Fermi levels in (a) Insulator (b) Monovalent metal and (c) Divalent metal

## 10.41 CHARACTERISTICS OF FERMI SURFACES

From the preceding discussions, we find that the FS has the following characteristic features:

1. The FS represents the dynamic and inertial properties of conduction electron in *k*-space.
2. The volume of the FS represents the number of conduction electrons.
3. The FS has spherical shape within the first BZ non-spherical in higher zones.
4. The FS always meets the zone boundary at right angles along the line of intersection.
5. For a spherical FS (i.e., for free electron case), the velocity of an electron is

$$V = \frac{\hbar k}{m_0} \quad ...(1)$$

where *v* is proportional to and parallel to the wave vector *k* as shown in Fig. 10.72 (a). On the other hand, for non-spherical Fermi surfaces (i.e., for electrons under periodic potential), the velocity is a non-linear function of *k*, which is given by

$$v = \frac{1}{\hbar} \nabla_k E(k) \qquad ...(2)$$

where the velocity is proportional to the gradient of the energy in $k$-space. Since the gradient vector is perpendicular to the contour lines (i.e., the constant energy curves), a well known fact from vector analysis, it follows that the velocity $v$ at every point in $k$-space is normal to the energy curve passing through that point and hence the velocity may or may not be parallel to $k$. This is shown in Fig. 10.72 (b).

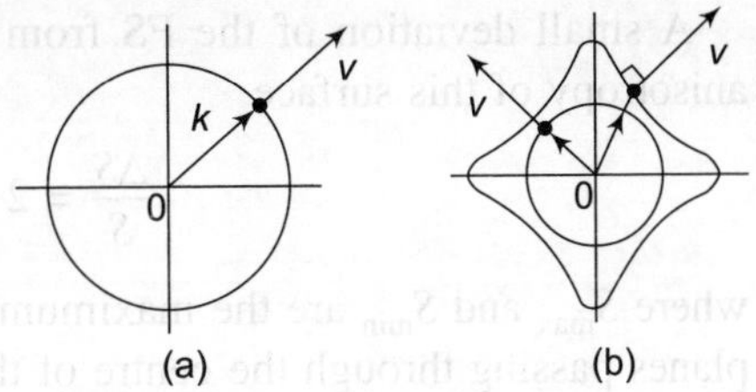


**Figure 10.72** The velocity of (a) a free-electron, and (b) a Bloch electron

6. Study of FS provides us to know many important properties of solids, such as heat capacity, Pauli'sparamagnetism, electrical conductivity, etc. Thus, FS is important for characterising the thermal, electrical, magnetic and optical properties of crystalline metals and semiconductors.

FSs are closely related to the atomic lattice, which is the underlying feature of all crystalline solids, and to the energy band theory which describes how electrons are distributed in such materials.

## 10.42 FERMI SURFACES OF ALKALI AND NOBLE METALS

The alkali metals have body-centred lattice, whose first BZ is the *rhombo-dodecahedron* in which all the twelve faces are identical and are located at the same distance $(\sqrt{2}\,\pi/a)$ from the centre of the zone. $k_F$ is defined by

$$\frac{k_3^F}{3\pi^2} = n = \frac{2}{a^3}$$

$$k_F = \left(\frac{3}{4\pi}\right)^{1/3} \frac{2\pi}{a} = 0.620\left(\frac{2\pi}{a}\right)$$

The shortest distance from the centre of the zone to zone face is

$$d = \frac{2\pi}{a}\sqrt{\left(\frac{1}{2}\right)^2 + \left(\frac{1}{2}\right)^2 + 0^2} = 0.707\left(\frac{2\pi}{a}\right)$$

Thus,
$$\frac{k_F}{d} = \frac{0.620}{0.707} = 0.88$$

Since the magnitude of the effective potential, generally speaking, is not great, we can expect that with such a relatively large distance from the Fermi sphere to the faces of the BZ their distorting effect on the shape of the FS will also not be great.

However, $k_F$ is not all that small so to make completely unnoticeable, the distortion of the real constant energy surface in alkalis at close distances to the zone boundary. The expected small deviations of the FS from the spherical shape evidently have the form of twelve projections located in the directions of shortest distances from the centre of the zone to its faces.

A small deviation of the FS from the sphere can be characterised qualitatively by the anisotropy of this surface

$$\frac{\Delta S}{S} = 2\,\frac{(S_{max} - S_{min})}{(S_{max} + S_{min})}$$

where $S_{max}$ and $S_{min}$ are the maximum and minimum areas of the section of the FS with the planes passing through the centre of the BZ. For a sphere, evidently, $\Delta S/S = 0$.

The percentage anisotropy of Fermi surfaces for alkali metals is given in Table 10.11 which shows that the anisotropy of Fermi surfaces does not exceed 1.5 per cent, so that replacement of a real FS with a sphere gives only an insignificant error. It is relevant to remark that the FS of Li is not well-known because it has a so-called *martensitic transformation* to a mixture

**Table 10.11** Anisotropy of FS of Metals

| *Metal* | *Na* | *K* | *Rb* | *Cs* |
|---|---|---|---|---|
| $\frac{\Delta S}{S} \times 100$ | 0.2 | 0.6 | 0.7 | 1.4 |

of crystalline phases at 77 K. Thus, bcc phase only exists at temperatures too high to observe the de Hass van Alphen effect, and the low-temperature phase lacks the crystallinity necessary for a de, Haas van Alphen study.

The noble metals have face-centred lattice, whose first BZ is a *cubo-octahedron* (which is less symmetrical than the rhombododecahedron) having eight hexagonal faces and six square faces. The distances from the centre of the zone to these are respectively $d_h = \sqrt{3},\ \pi/a$ and $d_s = 2\pi/a$. On the other hand,

$$k_F^3 = 3\pi^2 n = 3\pi^2\left(\frac{4}{a^3}\right)$$

$$\therefore \qquad k_F = 0.78\left(\frac{2\pi}{a}\right)$$

Thus, $$\frac{k_F}{d_x} = \frac{2 \times 0.78}{\sqrt{3}} = 0.90$$

and $$\frac{k_F}{d_x} = 0.78$$

Thus, the FS near the centres of the hexagonal faces must be substantially distorted. The FS near these points opens and adheres to the zone boundaries. This results in a qualitative change in the topology of FS, i.e., the closed single bonded surface becomes an open one.

Figure 10.73 shows the FS in the first BZ for Cu, Ag and Au and its section with a vertical plane of symmetry passing through $k_z$-axis and the bisector of the angle $k_x$ and $k_y$-axes. The shape of the FS is shown in Fig. 10.74.

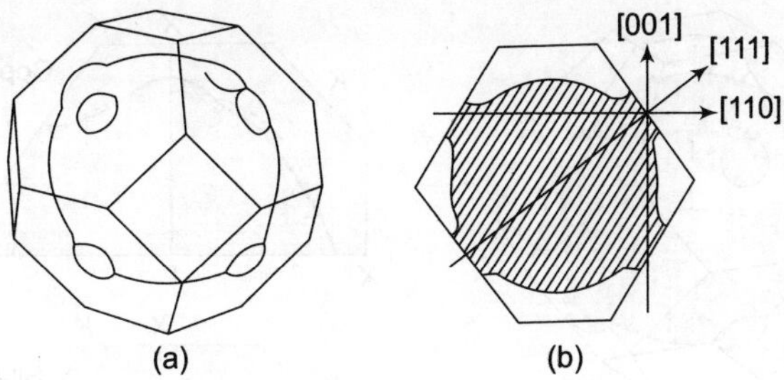


**Figure 10.73** The FS in 1 Brillouin zone for noble metals

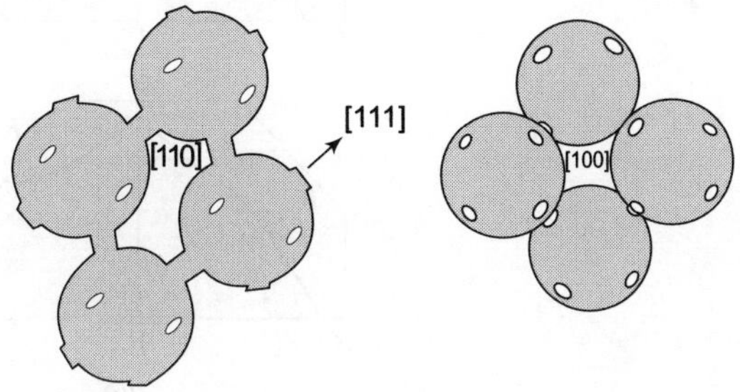


**Figure 10.74** The shape of fs in noble metals

The de Haas van Alphen effect in all three noble metals reveals that their FSs are closely related to the free electron sphere, however, in the $\langle 111 \rangle$ directions contact is actually made with the zone faces and the measured FSs have the shapes shown in Fig. 10.75.

Thus, the FS noble metals form a system of spheres connected by thin *necks* in the direction of body diagonals. Eight *necks* each out to touch the eight hexagonal faces of the zone, but otherwise the surface is not grossly distorted from spherical nature. The *necks* between the spheres strongly depend on the crystal potential. With the growth of the atomic number in the Cu, Ag, Au series, the ionic structure of the metal becomes more complicated, which results in the increase of the potential and diameter of the *neck* of the FS.

From the two periods existing in de Haas van Alphen oscillations, we can experimentally determine the extremal *belly* (maximum), which is spherical, and *neck* (minimum) orbits of FS, when a magnetic field is applied along a [111] direction (Table 10.12).

**Table 10.12** FS Areas of Noble Metals

| *Metal* | $A_{111}$ *(belly)*/$A_{111}$ *(neck)* |
|---|---|
| Cu | 27 |
| Ag | 51 |
| Au | 29 |

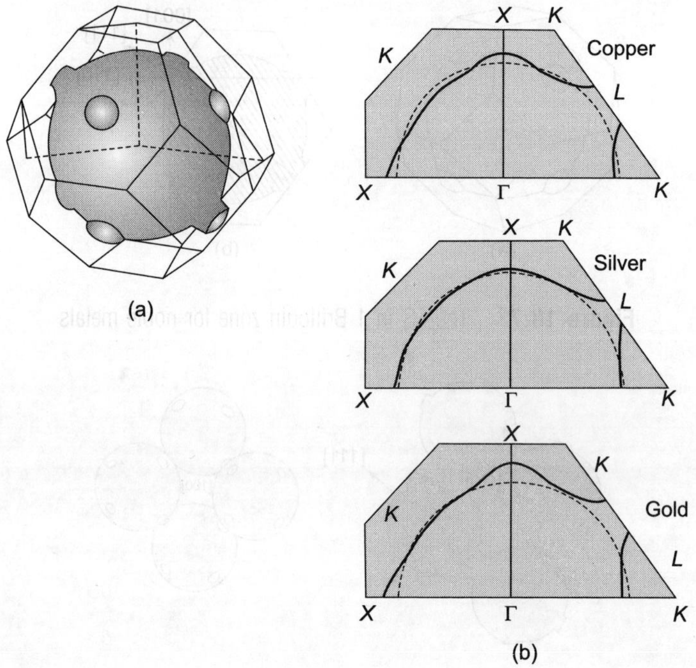


**Figure 10.75** (a) In the three noble metals the free electron sphere bulges out in the ⟨111⟩ directions to make contact with the hexagonal zone faces. (b) Detailed cross sections of the surface for the separate metals. The cross sections may be identified by a comparison with (a)

The ratio of the two periods directly determines the ratio of the maximal to the minimal ⟨111⟩ cross sections.

Although a distorted sphere, bulging out to make contact with the hexagonal zone faces, is still a fairly simply structure, when viewed in the repeated zone scheme, the noble metal FS reveals a variety of exceedingly complex orbits. Some of the simplest are shown in Fig. 10.76. The open orbits are responsible for the very dramatic behaviour of the magnetoresistance of the noble metals (Fig. 10.77).

Figure 10.76 indicates only a few of the surprisingly many types of orbits an electron can pursue in $k$-space when a uniform magnetic field is applied to a noble metal. (Recall that the orbits are given by slicing the Fermi surface with planes perpendicular to the field). The figure displays (a) a closed particle orbit; (b) a closed hole orbit; (c) an open orbit, which continues in the same general direction indefinitely in the repeated zone scheme.

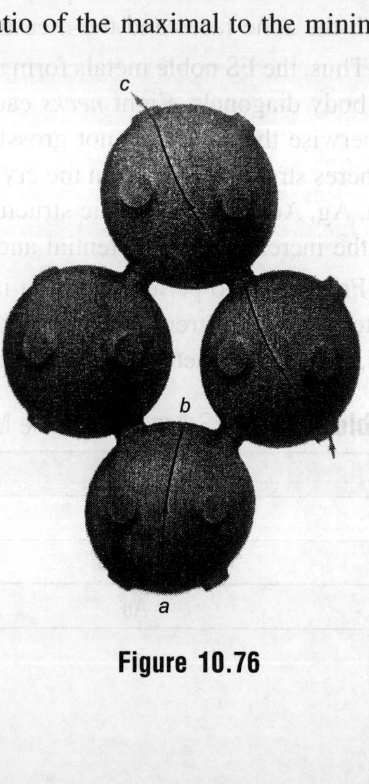


**Figure 10.76**

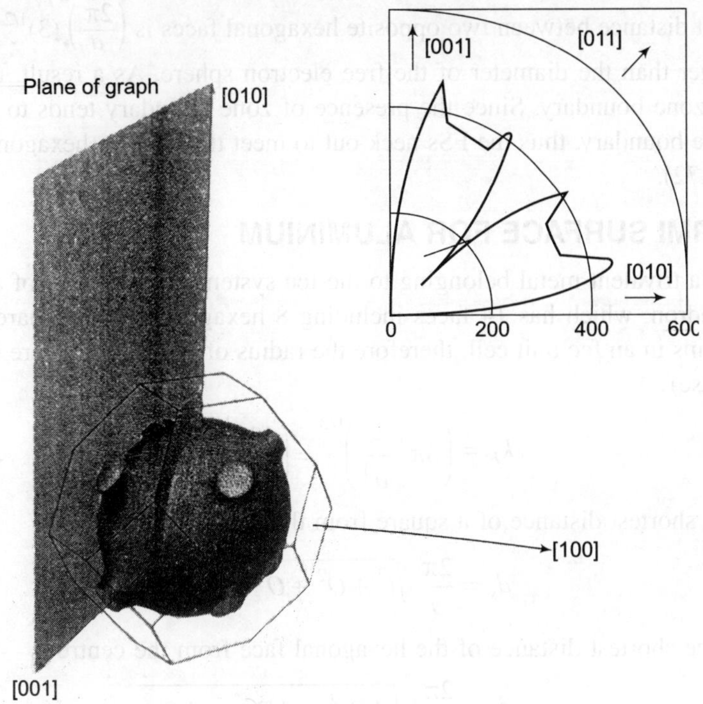


**Figure 10.77**

Figure 10.76 shows the spectacular direction dependence of the high-field magnetoresistance in copper that is characteristic of a Fermi surface supporting open orbits. The [001] and [010] directions of the copper crystal are as indicated in Fig. 10.76, and the current flows in the [100] direction perpendicular to the graph. The magnetic filed is in the plane of the graph. Its magnitude is fixed at 18 kilo-gauss, and its direction varied continuously from [001] to [010]. The graph is a polar plot of

$$\frac{\rho(H) - \rho(0)}{\rho(0)}$$

vs. orientation of the field. The sample is very pure and the temperature very low (4.2 K— the temperature of liquid helium) to ensure the highest possible value for $\omega_c\tau$.

## 10.43 FERMI SURFACE OF COPPER

Copper has an fcc structure and the 1st Brillouin zone is a truncated octahedron. The FS is distinctly non-spherical with 8 necks making contact with the hexagonal faces of the BZ. For monovalent copper, the electron concentration is $n = 4/a^3$, i.e., there are 4 electrons in a cube of volume $a^3$. The radius of a free electron Fermi sphere is

$$k_F = (3\pi^2 n)^{1/3} = (12\pi^2/a^3)^{1/3} = \left(\frac{4.9}{a}\right),$$

and the diameter is $9.8/a$.

The shortest distance between two opposite hexagonal faces is $\left(\frac{2\pi}{a}\right)(3)^{1/2} = \left(\frac{10.88}{a}\right)$, i.e., somewhat larger than the diameter of the free electron sphere. As a result, the sphere does not touch the zone boundary. Since the presence of zone boundary tends to lower the band energy near the boundary, thus the FSs neck out to meet the closest (hexagonal) faces of the zone. (Fig. 10.72).

## 10.44 FERMI SURFACE FOR ALUMINIUM

Aluminium is a trivalent metal belonging to the fcc system. The first BZ of an fcc structure is cubo-octahedron, which has 14 faces including 8 hexagonal and 6 square we know that there are 4 atoms in an fcc unit cell, therefore the radius of the Fermi sphere is given by (for monovalent case)

$$k_F = \left(3\pi^2 \frac{4}{a^3}\right)^{1/3} = \left(\frac{12}{\pi}\right)^{1/3} \frac{\pi}{a} = 1.563\left(\frac{\pi}{a}\right) \qquad \text{...(1)}$$

Further, the shortest distance of a square from the centre of the zone is

$$d_s = \frac{2\pi}{a}\sqrt{l^2 + O^2 + O^2} = \frac{2\pi}{a} \qquad \text{...(2)}$$

Similarly, the shortest distance of the hexagonal face from the centre is

$$d_h = \frac{2\pi}{a}\sqrt{(1/2)^2 + (1/2)^2 + (1/2)^2}$$

$$= 3\left(\frac{\pi}{a}\right) = 1.732\left(\frac{\pi}{a}\right) \qquad \text{...(3)}$$

Since Al is trivalent, the radius of the Fermi sphere for Al is given by

$$k_F = (3)^{1/3} \times 1.563\left(\frac{\pi}{a}\right) = 2.225\left(\frac{\pi}{a}\right) \qquad \text{...(4)}$$

From Eqs. (2), (3) and (4), we have

$$d_h < k_F < \frac{3\pi}{a},\ d_s < k_F < \frac{3\pi}{a}$$

$$\frac{k_F}{d_h} = 1.30 \quad \text{and} \quad \frac{k_F}{d_s} = 1.128$$

These imply that the Fermi sphere completely encloses the first BZ (indicating that the first BZ is completely filled with electrons) and passes through the second, third and fourth zones. In order to construct the constant energy surfaces in these zones, let us consider the first BZ in periodic zone scheme (Fig. 10.78). We observe that the two zones share the faces, three share the edges and four share the corners.

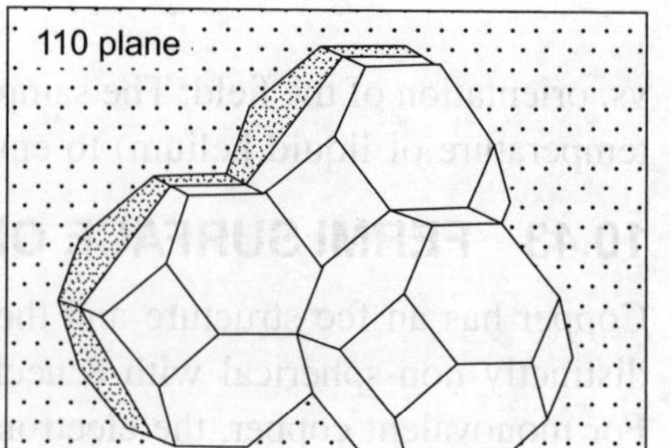


**Figure 10.78** Periodic zones of an fcc lattice sectioned by (110) plane

Accordingly, two Fermi spheres (which completely enclose the first BZ) overlap near zone faces, three near zone edges and four near zone corners. Next, consider the system of BZ and the corresponding Fermi sphere in a periodic zone scheme and draw two vertical sections of the system, one along the coordinate plane and the other along the diagonal plane as shown in Fig. 10.79. The empty space left inside the Fermi sphere (Fig. 10.79a) resembles a cubo-octahedron with concave faces and edges and hence according to the rule 1 (as pointed out earlier), it belongs to the second zone holes as shown in Fig. 10.80(a) in a reduced zone scheme. On the other hand, in Fig. 10.79(b), the shaded volumes which belong simultaneously to three spheres and bounded by convex surfaces belong to third zone electrons as shown in Fig. 10.80(b). In the absence of effective potential, they have triangular shape "tubes" of variable cross sections, i.e., wider at the middle and narrower towards the end of the edges. They give rise to the third zone "monster". In the fourth zone, only small electron pockets exist. However, when $V_{eff}(r) \neq 0$ these electron pockets vanish.

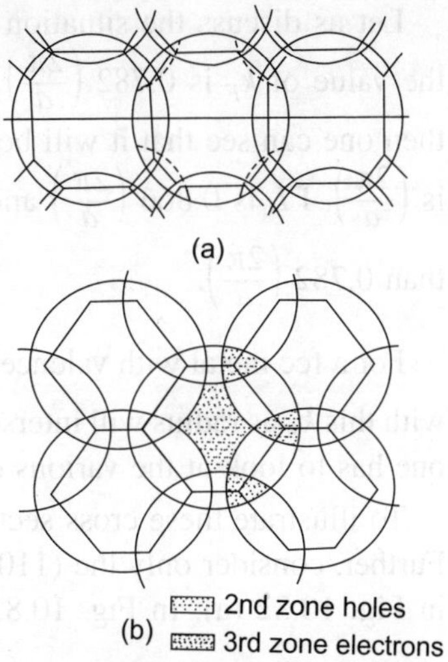


**Figure 10.79** (a) (111) and (b) (110) sections of an fcc lattice in periodic zone scheme

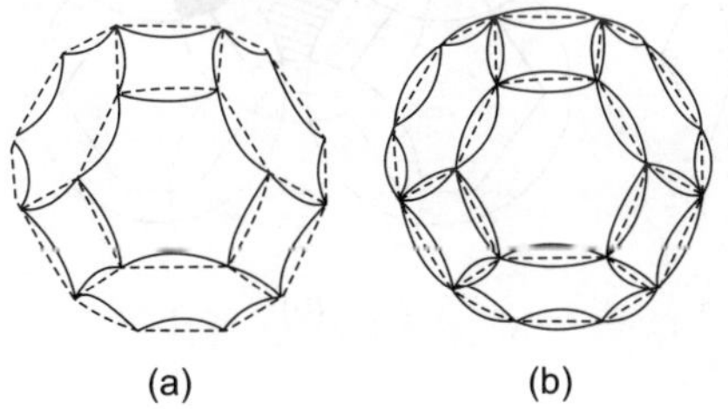


**Figure 10.80** FS of (a) Second zone hole (b) Third zone monster

For tetravalent solids such as lead with an fcc structure, the size of the second zone hole decreases, while the cross section of third zone monster and fourth zone electron pockets increases because of greater size of Fermi sphere. The fourth zone electron pockets are shown in Fig. 10.81.

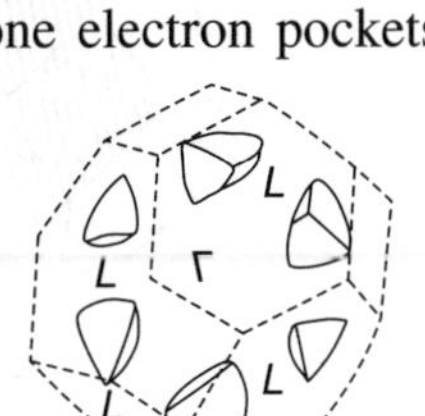


**Figure 10.81** Fourth zone electron pockets in reduced zone scheme

## 10.45 FERMI SURFACE FOR LEAD

In case of three-dimensional lattices, the volume of the unit cell is given by $\vec{a}\cdot(\vec{b}\times\vec{c})$, as explained in chapter 2. This value, for example, in fcc, b.c.c. and hcp structures is $\frac{a^3}{4}, \frac{a^3}{2}$ and $\frac{\sqrt{3}}{2}a^2c$ respectively. If we substitute this value in the expression of Fermi wave vector $k_F$, then one can determine the radius of the free electron Fermi surface.

Let us discuss the situation in case of a fcc metal. If the valency of the metal is 1, then the value of $k_F$ is 0.782 $\left(\frac{2\pi}{a}\right)$. Now, if we draw a Fermi surface with this value as radius, then one can see that it will be completely within the first Brillouin zone as the distance $\Gamma X$ is $\left(\frac{2\pi}{a}\right)$, $\Gamma L$ is 0.866 $\left(\frac{2\pi}{a}\right)$ and similarly other distances from the centre $\Gamma$ are also longer than 0.782 $\left(\frac{2\pi}{a}\right)$.

For a fcc metal with valence 2, the $k_F$ value is 0.985 $\left(\frac{2\pi}{a}\right)$ and, therefore, the Fermi sphere with this $k_F$ as radius will intersect the first Brillouin zone. In order to understand more clearly, one has to look at the various cross sections of the Brillouin zone and the Fermi surface.

To illustrate these cross sections, let us consider the case of lead, a metal with valence 4. Further, consider only the (110) section through the centre of the Billouin zone $\Gamma$, as shown in Fig. 10.82 (a). In Fig. 10.82 (b), we exhibit the hexagonal shaped zone boundaries and

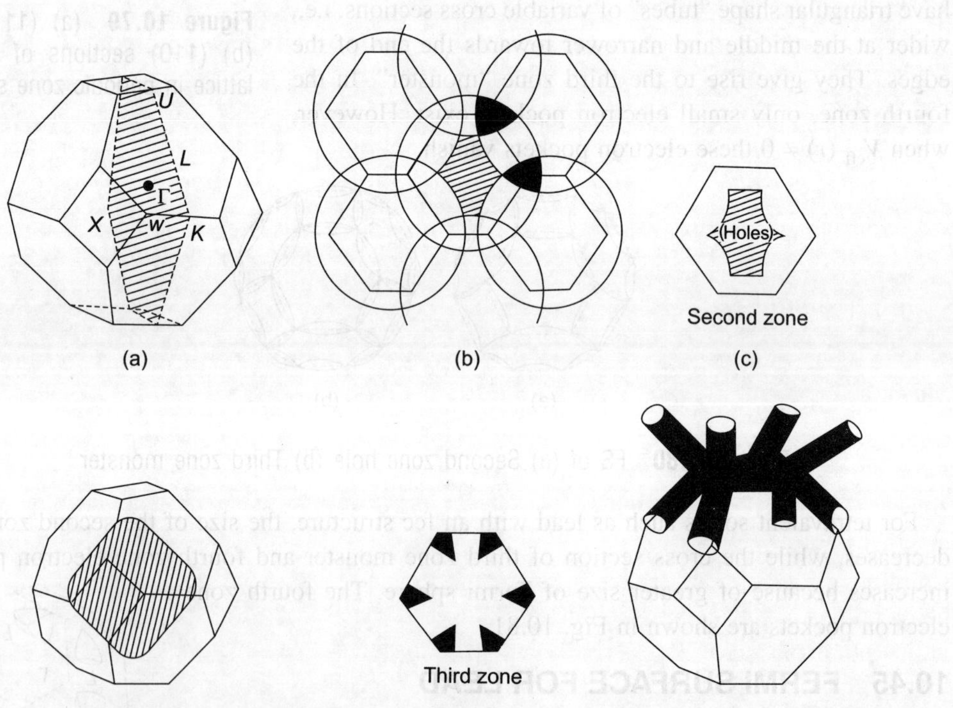


**Figure 10.82** (a) (110) section of the first Brillouin zone of a fcc metal, say lead (b) Construction of Fermi surfaces (c) and (d) Fermi surface redrawn in reduced zone scheme for second zone in free electron and periodic potential approximation (e) and (f) Fermi surface redrawn in reduced zone scheme for third zone in free electron and periodic potential approximation

the corresponding sections of the Fermi sphere drawn about neighbouring reciprocal lattice points. One will observe that the first zone is fully occupied, but in the second zone there is a hexagonal shaped contour between the occupied and the unoccupied points as shown in Fig. 10.82 (c). If the sharp corners are rounded off by the periodic lattice potential, then the cross section of the second zone surface representing holes is like Fig. 10.82 (d).

This also defines the orbit of a hole in a magnetic field applied perpendicular to (110) plane, known as α orbit enclosing an area about $1.2\left(\frac{2\pi}{a}\right)^2$, in the free electron approximation. In the third zone, there are occupied states corresponding to electron pockets, defining a ζ orbit of area about $0.16\left(\frac{2\pi}{a}\right)^2$ in the free electron approximation as shown in Fig. 10.82 (e). In fact, one can see the (110) section as in Fig. 10.82 (b), the dark portions indicate that this cross section is multiply connected. The fourth and the higher bands are empty in this section of the Brillouin zone. Similarly, one has to see various cross sections along various orientations and then combine them to get the exact shape of the Fermi surface. This is not a simple job as Harrison described this as the 'Pippard's Puzzle'. An example can be of the third band of the present case itself. A rather interesting feature of the third zone surface as 'monster' in Fig. 10.82 (f) is obtained by assembling various pieces.

### 10.45.1 Harrison's Construction for Lead (Pb)

Harrison's procedure for Pb reads something like the following. Pb is a fcc metal with valence $z = 4$. Thus, the radius of the Fermi sphere containing the $z$ electrons is given by

$$k_F = \left(\frac{3\pi^2 z}{\Omega_0}\right)^{1/3} = 1.24\left(\frac{2\pi}{a}\right)$$

where $\Omega_0 = \frac{1}{4}a^3$ is the atomic volume appropriate for fcc lattice, $a$ being the lattice constant. Since the radius of a sphere inscribed in the Brillouin zone is $\frac{1}{2}\sqrt{3}\,(2\pi/a) = 0.87(2\pi/a)$, it is clear that the Fermi sphere will have portions in the second and higher zones. Accordingly the reduced zone scheme may be used to construct the portion of the Fermi surface in each zone.

Such a construction can be bewildering in three dimensions, but Harrison's procedure simplifies the construction in the following way. He noted that a sphere with centre at $\mathbf{k} = 0$, on being translated by a reciprocal lattice vector, becomes a sphere with centre at that reciprocal lattice point. Thus, the translation of each segment of the Fermi sphere to bring it into the first Brillouin zone (in the reduced zone scheme) may be achieved by constructing spheres centred on all reciprocal lattice points, as indicated for the (100) and (110) sections of the first Brillouin zone (Figs. 10.83 and 10.84).

The result of such a construction is a number of intersecting spheres [or circles in the (100) and (110) sections shown]. To assign these segments to portions of the energy band to which they belong, Harrison again noted that each segment of the sphere divides some band into a region of occupied and unoccupied states. Thus, a point within the first zone that lies within

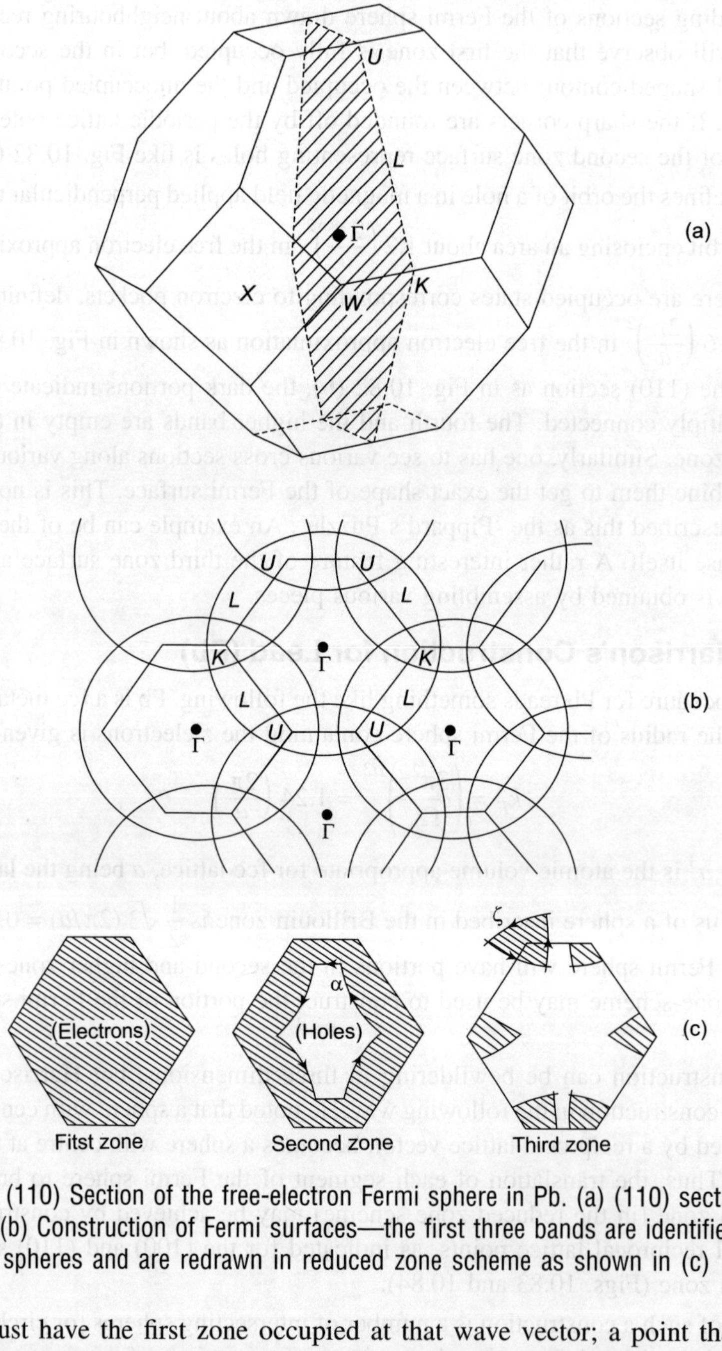


**Figure 10.83** (110) Section of the free-electron Fermi sphere in Pb. (a) (110) section of the first Brillouin zone; (b) Construction of Fermi surfaces—the first three bands are identified by counting spheres and are redrawn in reduced zone scheme as shown in (c)

one sphere must have the first zone occupied at that wave vector; a point that lies within two spheres must have both the first and the second bands occupied, and so on. Therefore, by counting the spheres one can readily assign each point of the segment of the sphere to its band.

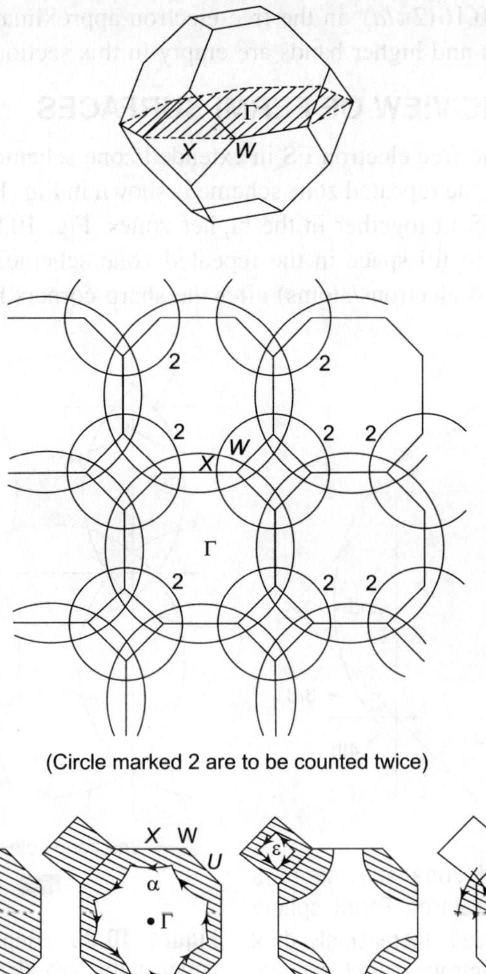


(Circle marked 2 are to be counted twice)

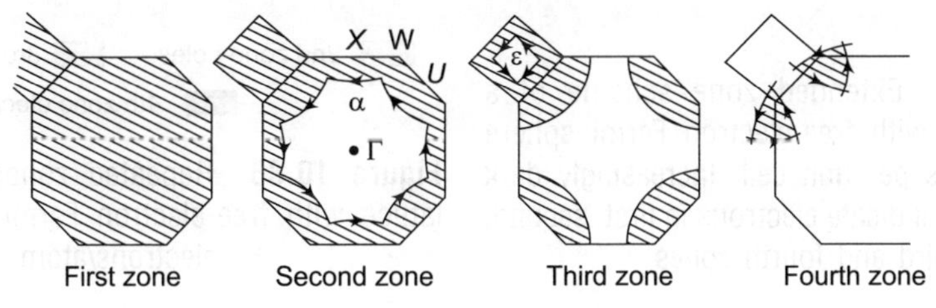


**Figure 10.84** (100) Section of the free-electron Fermi sphere in Pb, trough *W*

To exemplify, consider the (110) section through the centre of the Brillouin zone, Γ, shown in Fig. 10.83 (a); the hexagonal-shaped zone boundaries so obtained and the corresponding sections of the Fermi sphere drawn about neighbouring reciprocal lattice points are as shown in Fig. 10.83 (b); and the remapped contours in the reduced zone scheme as shown in Fig. 10.83 (c). The first zone is fully occupied; there is a hexagonal-shaped contour separating the occupied and the unoccupied points in the second zone, which defines the orbit of a hole in a magnetic field applied perpendicular to the (110) plane. This is called the α orbit (Anderson and Gold, *Phys. Rev.* **139**, A1459, 1965). The orbit encloses an area $A_\alpha \approx 1.2\,(2\pi/a)^2$ in the free-electron approximation; The actual area deduced from experiment is $1.0(2\pi/a)^2$, which is about 20% smaller. In the third band, there are electron pockets that define the ζ orbit enclosing an area $A_\zeta \approx 0.16\,(2\pi/a)^2$ in the free-electron approximation; the observed area is $0.11\,(2\pi/a)^2$. The fourth and higher bands are empty in this section of the Brillouin zone.

## 10.46 SCHEMATIC VIEW OF FERMI SURFACES

Figure 10.85 presents the free electron FS in extended zone scheme for a square length. The Harrison construction in the repeated zone scheme is shown in Fig. 10.86 which indicates how the separate pieces of FS fit together in the higher zones, Fig. 10.86 shows a few Brillouin zones stacked together to fill space in the repeated zone scheme. Figure 10.87 shows the schematic (for fcc with 4 electrons/atoms) after the sharp corners have been rounded off by

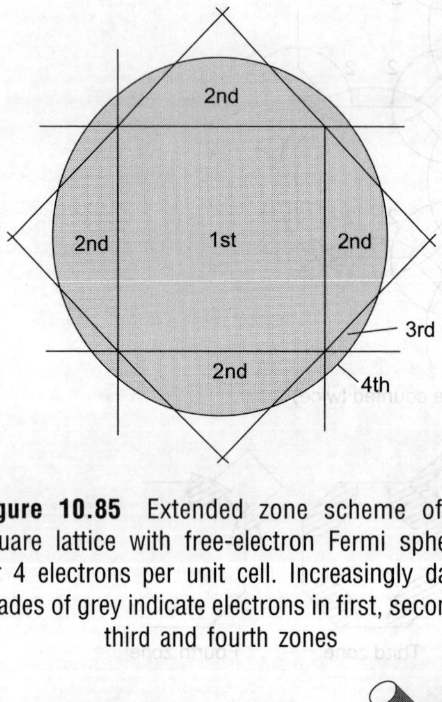


**Figure 10.85** Extended zone scheme of a square lattice with free-electron Fermi sphere for 4 electrons per unit cell. Increasingly dark shades of grey indicate electrons in first, second, third and fourth zones

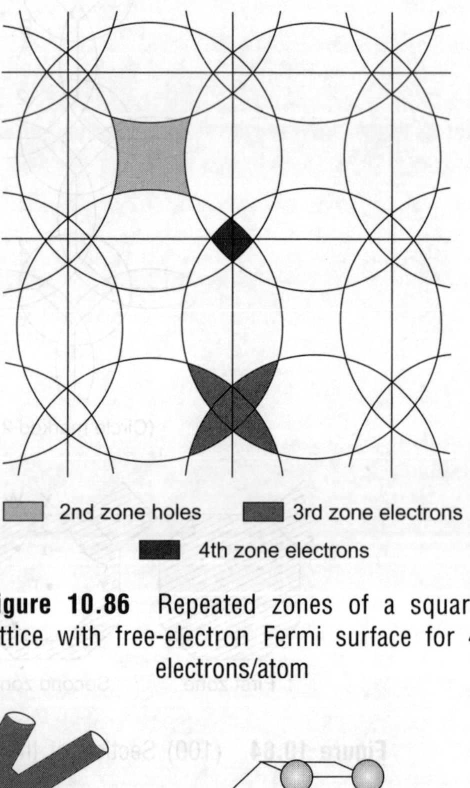


**Figure 10.86** Repeated zones of a square lattice with free-electron Fermi surface for 4 electrons/atom

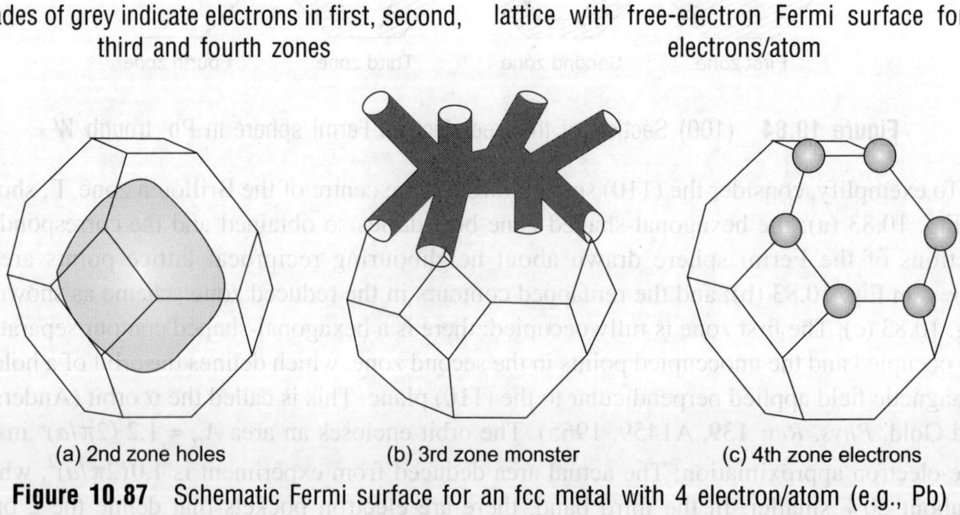


**Figure 10.87** Schematic Fermi surface for an fcc metal with 4 electron/atom (e.g., Pb)

the periodic lattice potential. The most interesting feature is the 3rd zone surface known as the 'monster'. The three-dimensional view makes it clear that this cannot unambiguously be. called 'electrons' or 'holes' because it is multiply connected cross sections.

## 10.47 EFFECT OF ELECTRIC FIELD ON FERMI SURFACE

If an electric field $E$ is applied, the external force $eE$ equals the rate of change of momentum of the electron, i.e.,

$$\frac{d}{dt}(\hbar \vec{k}) = -e\vec{E} \quad \text{or} \quad \frac{d\vec{k}}{dt} = -\frac{e\vec{E}}{\hbar} \qquad ...(1)$$

i.e., the electric field causes the wave vector of an electron to change. This implies that the electron moves to new states. A wave vector $\vec{k}$ specifies some *position* of the electron in $k$-space, so $d\vec{k}/dt$ is the *velocity* of the electron in $\vec{k}$-space. In a constant electric field Eq. (1) states that the velocity of each electron in $\vec{k}$-space is a constant. Thus, all the electrons move with the same velocity through $\vec{k}$-space. Consequently, the entire Fermi surface moves through $\vec{k}$-space, retaining its original shape.

However, there are also collisions taking place, which try to bring the Fermi surface back to original position, i.e., collisions oppose the action of the electric field, which tries to move the Fermi surface. At equilibrium, both effects cancel each other and the Fermi surface suffers a net displacement $\Delta\vec{k}$ in $\vec{k}$-space.

Consider, for example, the case of copper. A cross section of its Fermi surface is shown in Fig. 10.88. Under an applied electric filled pointing to the left, the Fermi surface gets displaced to the right. (Fig. 10.88 (b)).

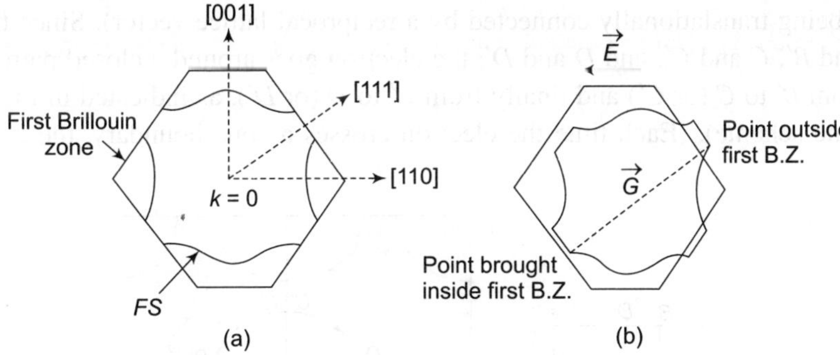


**Figure 10.88** (a) Cross section of the Fermi surface of copper, (b) Displacement of its Fermi surface in an electric field

Since some of the states at the first Brillouin zone boundary are occupied, a displacement of the Fermi surface causes some of the electrons to go outside the first Brillouin zone. We can relabel the wave vectors of these electrons so that they are inside the first Brillouin zone (by adding some reciprocal lattice vector), as shown in Fig. 10.88 (b). Actually, these electrons did not really leave the first Brillouins zone, instead they simply jumped to the opposite side (when they crossed the boundary). If we now redraw Fig. 10.88 relabelling all electron states (so that they are inside the first Brillouin zone), we obtain a Fermi surface which looks like Fig. 10.89 (a).

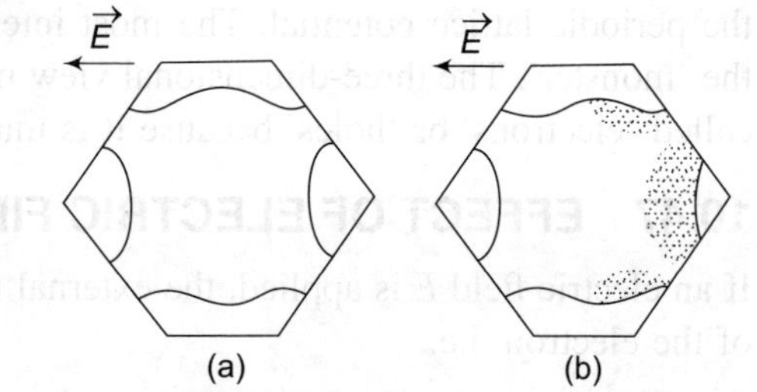


**Figure 10.89** (a) Displaced Fermi surface in the first Brillouin zone, (b) Uncompensated electrons

With the Fermi surface displaced, some electrons do not come in $+\vec{k}$ and $-\vec{k}$ pairs and so, are uncompensated. These uncompensated electrons are shown by the shaded area in Fig. 10.89 (b). As a result, the average wave vector $(\vec{k})$ of the electrons points to the right, giving rise to an electric current towards left.

The size of electric current depends on the number of uncompensated electrons, which in turn, depends on the displacement of the Fermi surface as well as the shape and area of the Fermi surface. For this reason, the electrical conductivity of aluminium is approximately the same as that of copper, although aluminium has three times as many conduction electrons as copper. The area of the FS is approximately the same in both cases, therefore they have approximately the same number of uncompensated electrons to carry current.

## 10.48 CLOSED ORBITS AND OPEN ORBITS

Consider a Fermi surface shown in Fig. 10.90 (a). The Fermi surface touches the zone boundary (reduced zone scheme). An electron in this plane section starts moving from point $D'$ towards point $A$. Its arrival at point $A$ is the same as reaching point $A'$ (which is equivalent to point $A$ in $\vec{k}$-space, being translationally connected by a reciprocal lattice vector). Since the same is true for $B$ and $B'$, $C$ and $C'$, and $D$ and $D'$, the electron goes around a closed path from $A'$ to $B$ (or $B'$), from $B'$ to $C$ (or $C'$) and finally from $C'$ to $D$ (or $D'$), as indicated in Fig. 10.90 (b) (periodic zone scheme). (Each time the electron crosses a zone boundary, the continuity in

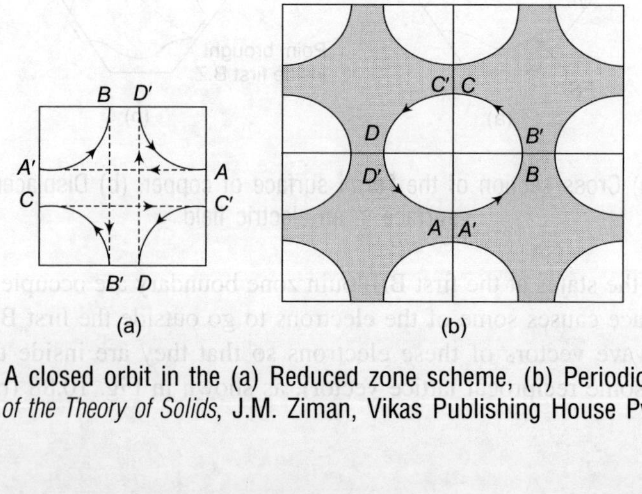


**Figure 10.90** A closed orbit in the (a) Reduced zone scheme, (b) Periodic zone scheme. (After *Principle of the Theory of Solids*, J.M. Ziman, Vikas Publishing House Pvt. Ltd (1979))

path is preserved by providing another cell of the reciprocal lattice on the other side.) Since the orbit encloses an empty region of $\vec{k}$-space, it is termed a *hole orbit*. A three-dimensional multiply-connected Fermi surface is depicted in Fig. 10.91. It shows electron orbits and hole orbits (closed orbits).

The possibility of an open orbit is shown in Fig. 10.92. It is clear that orbits in a plane; whose section with the Fermi surface is not a closed curve, are open. In this case, the electron continues to move with wave vector increasing towards infinitely large values in the periodic zone scheme.

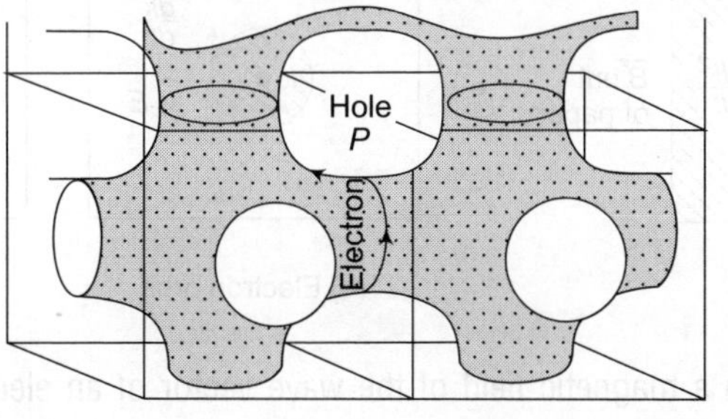


**Figure 10.91** Electron and hole orbits on a multiply-connected Fermi surface (Closed Orbits). (After *Principles of the Theory of Solids*, J.M. Ziman, Vikas Publishing House Pvt. Ltd. (1979))

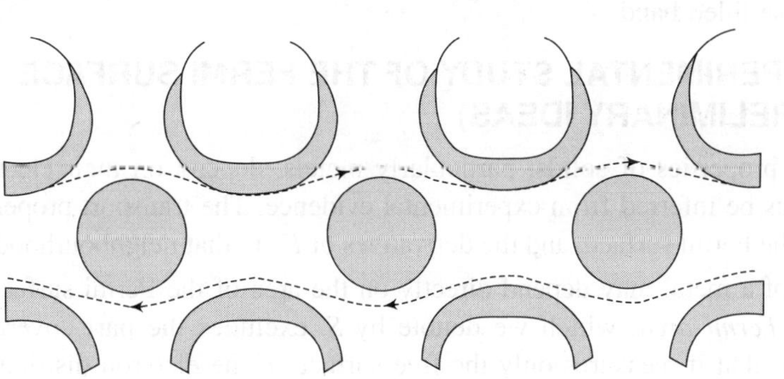

**Figure 10.92** An open orbit. (After *Principles of the Theory of Solids*, J.M. Ziman, Vikas Publishing House Pvt. Ltd. (1979))

## 10.49 TOPOLOGY OF ORBITS IN MAGNETIC FIELD

An electron describing an orbit in $\vec{k}$-space also describes an orbit in real space.

An electron in a static magnetic field $\vec{B}$ moves on a curve of constant energy on a plane normal to $\vec{B}$. An electron on the Fermi surface will, therefore, move in a curve on the Fermi surface, as this is a surface of constant energy. Three types of orbits are shown in Fig. 10.93. The closed orbits of (a) and (b) are traversed in opposite senses: we say that one orbit is electron-like and the other orbit is hole-like; electrons in a hole-like orbit move in a magnetic field as if endowed with a positive charge. Also, a hole moves like a particle of charge $+e$.

An electron orbit enclosed states of lower energy and the velocity vector $d\vec{k}/dt$ points outward. A hole orbit encloses states of high energy and the velocity vector points inward.

In Fig. 10.93 (c) the orbit is not closed; the particle on reaching the zone boundary at $A$ is instantly umklapped back to $B$, where $B'$ and $B$ are equivalent points, connected by a reciprocal lattice vector. This is an open orbit. Open orbits have an important effect on the magnetoresistance.

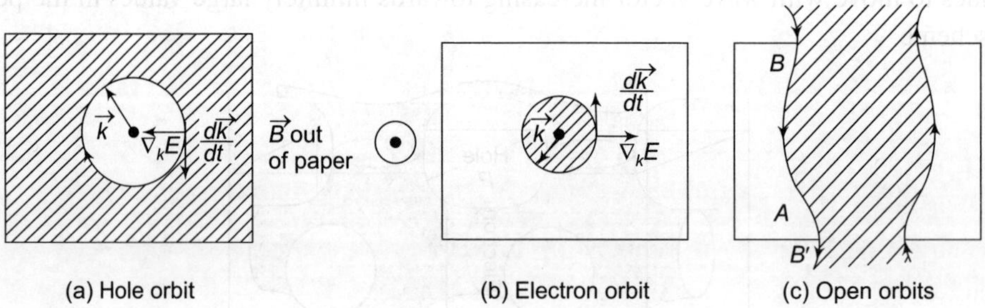


**Figure 10.93** Motion in a magnetic field of the wave vector of an electron on the Fermi surface

Notice that orbits that enclose filled states are electron orbits. Orbits that enclose empty states are hole orbits. Hole-like orbits are formed when there are vacant orbitals near the top of an otherwise filled band.

## 10.50 EXPERIMENTAL STUDY OF THE FERMI SURFACE (PRELIMINARY IDEAS)

The physical properties of solids, particularly metals, depend on their electronic structure which can thus be inferred from experimental evidence. The transport properties depend on the shape of the Fermi surface, and the derivatives of $E_{\vec{k}}$ in that neighbourhood. The electrical conductivity of a metal may depend directly on the area of the Fermi surface, among other factors. This *Fermi area*, which we denote by $S$, excludes the part covered by the zone boundaries, so that it measures only the free surface of the electron distribution. When the Fermi surface is spherical, with radius $R$ (as for free electrons), we have

$$S = 4\pi k_F^2 = 4\pi(3\pi^2 n)^{2/3} \quad ...(1)$$

which is a function only of the electron density. Except for monovalent metals, the true Fermi area must be rather less than this, because of the effect of zone boundaries.

Another property giving similar information is the thermoelectric power of a metal. This depends, in part, on the rate of change of the Fermi area as the Fermi level is changed, i.e., on $dS/dE_F$.

A scalar parameter of great importance is the value of the level density at the Fermi level, $N(E_F)$. This is measured directly by the electronic contribution to the specific heat, given by

$$C_{el} = \frac{1}{3}\pi^2 k^2 T N(E_F) + O(T^3) = \gamma T \qquad ...(2)$$

This linear term can be observed at low temperatures (when the lattice specific heat falls off more rapidly, as $T^3$) and hence provides $N(E_F)$.

Another measure of the density of states comes from the Pauli spin paramagnetism of the electrons. It can be shown to contribute a term

$$\chi_s = \mu_B^2 N(E_F) \qquad ...(3)$$

to the susceptibility (where $\mu_B$ is the Bohrmagneton). However, this term is partially masked by the diamagnetic susceptibility of the electrons, so cannot be measured directly. Moreover, due to exchange and other effects (which have not been taken into account), the value of $N(E_F)$ derived from this measurement would not agree with the corresponding one occurring in the formula for the specific heat.

Assuming a constant DOS per unit volume of $k$-space, the level density, i.e., the number of levels between the surfaces $E$ and $E + dE$ (distant $dk$ apart) is given by

$$N(E)\,dE = \frac{2}{V}\frac{V}{8\pi^3}\int dk\, d\delta = \frac{1}{4\pi^3}\int \frac{ds}{\partial E_k/\partial k}\,dE \qquad ...(4)$$

$$\left(\because \frac{dE}{dk} = \frac{\partial E_k}{\partial k}\right)$$

$$\Rightarrow \quad N(E_F) = \frac{1}{4\pi^3}\int \frac{ds}{\hbar v_k}\left(\because v_k = \frac{1}{\hbar}\frac{\partial E_k}{\partial k}\right) \qquad ...(5)$$

If we define the Fermi velocity ($v_F$) as the harmonic mean velocity of the electrons on the Fermi surface then Eq. (5) is equivalent to

$$N(E_F) = \frac{S}{4\pi^3 \hbar v_F} \qquad ...(6)$$

It must be noted here that a high density of states would correspond to a low electron velocity since $S$ is usually limited in magnitude.

Now, due to the advent of the free electron theory, it is usual to quote the value of $m^*$, the effective mass, instead of the measured quantity $N(E_F)$. The argument is that the band may be represented by the approximate relation.

$$E_k = \frac{\hbar^2 k^2}{2m^*} \qquad ...(7)$$

as if, it were isotropic and parabolic. For this model, one finds

$$v_F = \frac{\hbar k_F}{m^*} \qquad ...(8)$$

and DOS function must be

$$N(E) = \frac{3}{2} n E^{1/2} / (E_F)^{3/2} \qquad ...(9)$$

since $$N(E_F) = \frac{3}{2}\left(\frac{n}{E_F}\right) \quad ...(10\text{i})$$

$$\therefore \quad N(E_F) = n\frac{3}{2}\left(\frac{2}{m^* v_F^2}\right) = n\frac{3}{2}\left(\frac{2m^*}{m^{*2} v_F^2}\right)$$

$$= n\frac{3m^*}{\hbar^{*2}k_F^2} = n\frac{3m^*}{\hbar^2} \times \frac{1}{k_F^2}$$

or $$N(E_F) = n\frac{3m^*}{\hbar^2}\left(\frac{1}{3\pi^2 n}\right)^{2/3} \quad ...(10\text{ii})$$

Thus, values of $m^*$ are quoted, from measurements of $N(E_F)$, using the formula 10 (i). Measurements of the overall width of a band, derived from observation of the X-ray emission spectrum, have even been used to define $E_F$ and hence a value of $m^*$.

The effective mass ($m^*$) is actually a tensor, the properties of which show up in a series of more complex phenomena especially in presence of a magnetic field. These properties are studied in relation to the Hall effect and magnetoresistance, where the anisotropy of the Fermi surface can appear directly.

There are two more important phenomena which need to be mentioned here. These are the static and dynamic components of the diamagnetic susceptibility—the de Haas van Alphen effect and cyclotron resonance—which depend upon the fact that the magnetic field causes the wave vector of an electron to follow a path of constant energy in the reciprocal (i.e., $\vec{k}$ –) space, and hence to go through a periodic motion exactly equivalent to cyclotron rotation. These two phenomena we will discuss in the subsequent sections.

The Fermi surface can be determined experimentally by the following methods:

(i) Magnetoresistance
(ii) Anomalous skin effect
(iii) Cyclotron resonance
(iv) Magnetoacoustic effect
(v) Shubnikov-de Haas effect
(vi) De Haas-van Alphen effect

We propose to study first the de Haas-van Alphan effect because it exhibits very well the characteristic periodicity in $1/B$ of the properties of a metal in a uniform magnetic field. Before going into details of this method, we will like to discuss the effect of magnetic field on the Fermi surface.

## 10.51 QUANTISATION OF ORBITS IN A MAGNETIC FIELD

The momentum $\vec{p}$ of a particle in a magnetic field is the sum of two parts, the kinetic momentum ($hk$) and the field momentum, $e_{\vec{A}}$, where, $e$ is the value of charge on the particle and $\vec{A}$, the vector potential ($B = \nabla \times \vec{A}$), i.e., $$\vec{p} = \hbar \vec{k} + e_{\vec{A}} \quad \text{(SI units)} \qquad ...(1)$$

It is assumed that the orbits in a magnetic field are quantised by the Bohr–Ommerfeld relation

$$\oint \vec{p} \cdot d\vec{r} = (n + \gamma) h \qquad ...(2)$$

where $h$ is Planck's constant, $n$ is an integer and $\gamma$ is a phase correction factor (= 1/2 for free electrons).

$$\oint \vec{p} \cdot d\vec{r} = \oint \hbar \vec{k} \cdot d\vec{r} + e \int \vec{A} \cdot d\vec{r}$$

The equation of motion of a charge in a magnetic field is

$$\hbar \frac{d\vec{k}}{dt} = e\vec{V} \times \vec{B}$$

or $$\hbar \frac{d\vec{k}}{dt} = e \frac{d\vec{r}}{dt} \times \vec{B} \qquad ...(3i)$$

Integrating this, we obtain

$$\hbar \vec{k} = e\vec{r} \times \vec{B} \qquad ...(3ii)$$

or $$\hbar \vec{k} \cdot d\vec{r} = e\vec{r} \times \vec{B} \cdot d\vec{r}$$

$$\oint \hbar \vec{k} \cdot d\vec{r} = e \oint \vec{r} \times \vec{B} \cdot d\vec{r}$$

$$= e\vec{B} \cdot \oint \vec{r} \times d\vec{r} \qquad ...(4)$$

Using $\oint \vec{r} \times d\vec{r}$ = area of the orbit, we have

$$\vec{e} \oint \vec{B} \cdot d\vec{a} = e \iint (\nabla \times \vec{A}) \cdot d\vec{a}$$

$$= e \iint \vec{B} \cdot d\vec{a} = e\Phi \qquad ...(5)$$

where $\Phi$ is the magnetic flux enclosed within the orbit in real space.

Using Stockes theorem and $d\vec{a}$ is the area element in real space.

Combining Eqs. (4) and (5), we get

$$\oint \vec{p} \cdot d\vec{r} = e\Phi = \left(n + \frac{1}{2}\right) h$$

It follows that the orbit is quantised in such a way that the flux through it is

$$\Phi_n = \left(n + \frac{1}{2}\right) \frac{2\pi\hbar}{e} \qquad ...(6)$$

(where $\hbar = h/2\pi$). The flux unit $2\pi\hbar/e = 4.14 \times 10^{-7}$ tesla–m$^2$. The above expression gives the flux through the orbit in real space. However, in the de Haas-van Alphen effect (discussed in the next section), we need the area of the orbit in wave vector space.

Now, from Eqn. (6):

$$\left.\begin{aligned}\vec{k} &= \frac{e}{\hbar}(\vec{r}\times\vec{B})\\ |\Delta\vec{r}| &= \left(\frac{\hbar}{eB}\right)\Delta k\end{aligned}\right\} \qquad \text{...(7)}$$

This implies that the orbits in $\vec{k}$-space and real space are identical, the former being obtained from the latter by rotating through $\pi/2$ about the axis of $\vec{B}$ and magnifying by a numerical factor $(eB/\hbar)$ as shown in Fig. 10.94.

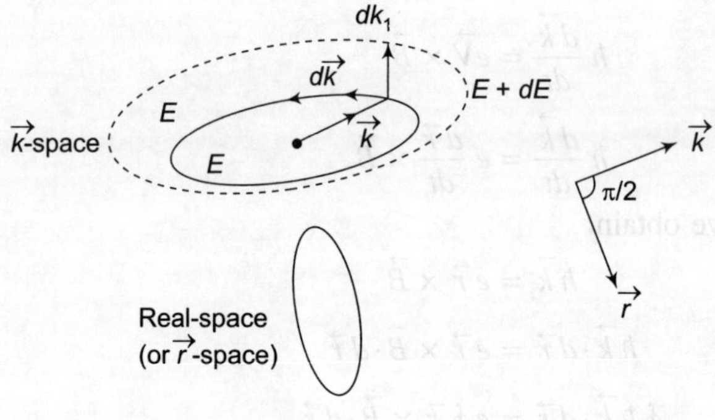


**Figure 10.94** Electron orbit in a magnetic field

The orbit may be closed or open, but here we consider only the closed orbits.

So, the area $S_n$ in $\vec{k}$-space is related to the area $A_n$ of the orbit in $\vec{r}$-space by

$$A_n = \left(\frac{\hbar}{eB}\right)^2 S_n \qquad \text{...(8)}$$

and hence $\Phi_n = A_n B$

$$\Phi_n = \left(\frac{\hbar}{e}\right)^2 \frac{1}{B} S_n = \left(n+\frac{1}{2}\right)\frac{2\pi\hbar}{e} \qquad \text{...(9)}$$

This means that the area of an orbit in $\vec{k}$-space will satisfy

$$S_n = \left(n+\frac{1}{2}\right)\frac{2\pi e}{\hbar} B$$

or

$$\boxed{S_n = \frac{2\pi eB}{\hbar}\left(n+\frac{1}{2}\right)} \qquad \text{...(10)}$$

This is known as *Onsagar–Lifshitz quantization condition* and forms on the basis of de Haas-van Alphen effect.

## 10.52 de HAAS-VAN ALPHEN EFFECT AND LANDAU TUBES

We know that when a metal is placed in a magnetic field, the electrons do not move in a straight path but rotate in orbits around the Fermi surface in a plane perpendicular to the field. Based on this concept, it was discovered by de Haas and van Alphen in 1931 that at low temperatures the diamagnetic susceptibility of pure bismuth shows periodic oscillations when plotted against the reciprocal of the applied (static) magnetic field at high fields. This is known as the de Haas-van Alphen effect (Fig. 10.95). It has developed into a powerful method for measuring the cross sectional areas of the Fermi surface of a metal.

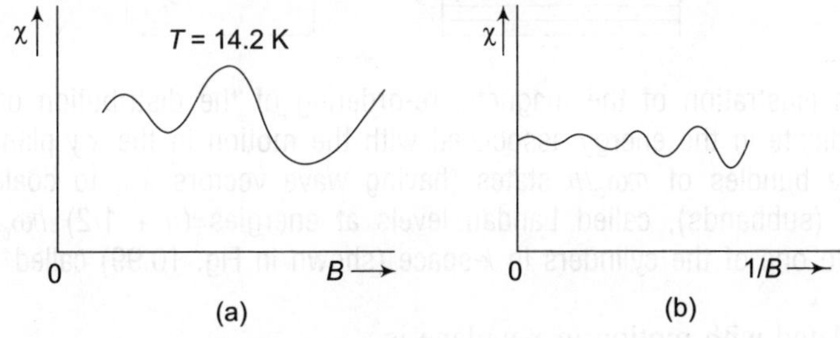


**Figure 10.95** de Haas-van Alphen effect in (a) Bismuth, (b) Copper
χ is diamagnetic susceptibility

In 1930, Landau had shown theoretically that conduction electrons in a metal should have a diamagnetic susceptibility due to the orbital motion of electrons, although classically it is predicted that a gas of free electrons should have zero susceptibility. This non-zero diamagnetism appears because the energy levels of the electrons (which have a quasi-continuous character in the absence of a magnetic field) are grouped into highly degenerate discrete levels in the presence of a magnetic field. (These are known as *Landau levels*). The energies of electron in the *n*th Landau level is given by

$$E_n = \left(n + \frac{1}{2}\right)\hbar\omega_c + \frac{\hbar^2 k_z^2}{2m^*},\ n = 0,\ 1,\ 2.... \qquad ...(11)$$

when the magnetic field is applied parallel to the $z$-axis and effective mass ($m^*$) takes into account the band structure.

Expression shows that the motion of the electron in the $z$-direction is that of a free electron but in the $k_{(x-y)}$ plane, the motion is quantised, and electron states on the orbital path are degenerate. (Electrons of a particular energy are located on one of concentric set of cylinders, so called *Landau tubes*.) Each Landau tube corresponds to a Landau level, depicted in Fig. 10.96.

The formation of the Landau tubes can be understood as follows.

The magnetic field does not change the longtitudinal component of motion, so the energy related with this is

$$E_g = \frac{\hbar^2 k_z^2}{2m^*} \qquad ...(12a)$$

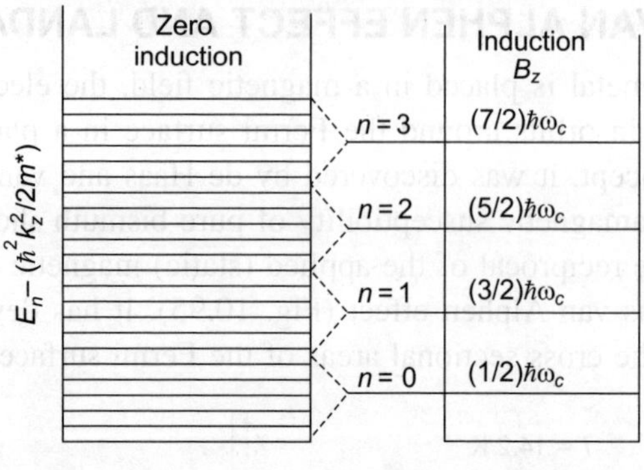


**Figure 10.96** An illustration of the magnetic re-ordering of the distribution of electron states in energy. The ordinate in the energy associated with the motion in the *x-y* plane. The magnetic induction $B_z$ cause bundles of $m\omega_c/\hbar$ states (having wave vectors $k_n$), to coalesec into highly degenerate states (subbands), called Landau levels at energies $(n + 1/2)\ \hbar\omega_c$. Each of these corresponds to one of the cylinders in *k*-space (shown in Fig. 10.99) called *Landau tubes*

The energy related with motion in *x-y* plane is

$$E_{\perp} = \left(n + \frac{1}{2}\right)\hbar\omega_c \qquad ...(12b)$$

$$E_n = E_g + E_{\perp} = \left(n + \frac{1}{2}\right)\hbar\omega_c + \frac{\hbar^2 k_z^2}{2m^*} \qquad ...(13)$$

For considering the nature of distribution of electrons in $\vec{k}$-space under the action of magnetic field, let us consider the case of a one-zone metal with spherical Fermi surface without spin. For $B = 0$, the allowed states are distributed uniformly inside the Fermi sphere.

Application of a magnetic field does not change the total number of electrons in the metal, instead it causes redistribution between the bands. When the magnetic field is applied (along *z*-axis), the discrete energy levels are given by

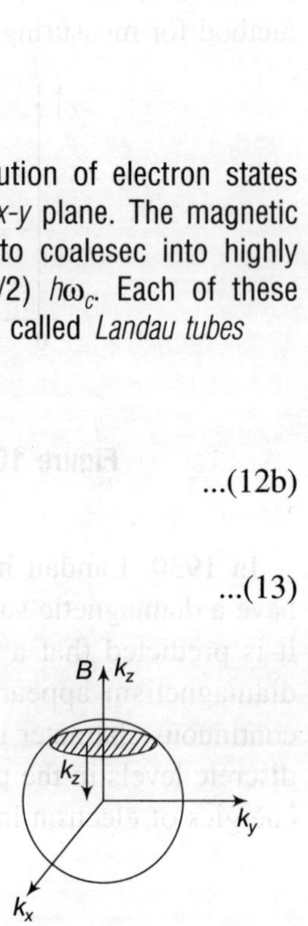


**Figure 10.97** Motion of an electron in a magnetic field

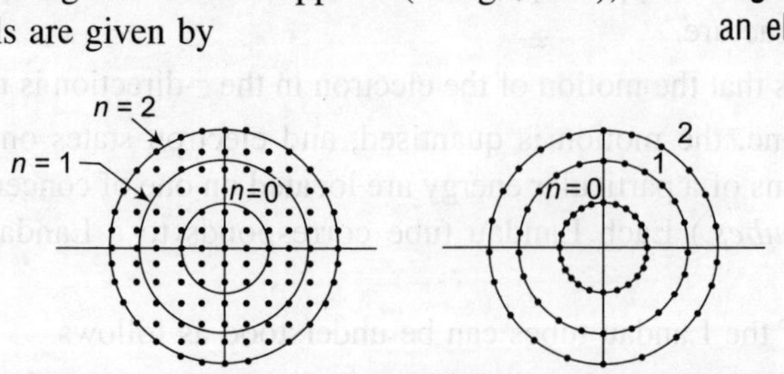


**Figure 10.98** Quantisation (in $\vec{k}$-space) for electron: (a) without magnetic field, (b) in a magnetic field

$$E_{\perp} = \left(n + \frac{1}{2}\right)\hbar\omega_c$$

determine the allowed electron orbits in the planes $k_z$ = constant in $\vec{k}$-space. The radius $k_n$ of the $n$th orbit is found by

$$\frac{\hbar^2 k_n^2}{2m^*} = \left(n + \frac{1}{2}\right)\hbar\omega_c \quad \text{or} \quad k_n = \left[\frac{2m^*\omega_c}{\hbar}\left(n + \frac{1}{2}\right)\right]^{1/2} \qquad ...(14)$$

This shows that the states lying between the orbits of radii $k_n$ ($n = 0, 1, 2, \ldots$) are forbidden. Hence, when we consider three-dimensional case this means that in presence of magnetic field, all allowed states within the Fermi surface are Onsager–Lifshitz quantisation condition

$$S_{n+1} - S_n = \frac{2\pi eB}{\hbar} \qquad ...(15)$$

i.e., *the difference between the areas is constant (for large n) and independent* of $k_z$ and in magnetic field the areas of electron path are quantised in $\vec{k}$-space. The allowed states in $\vec{k}$-space 'condense' onto the Landau tubes as soon as a minimal field is switched on.

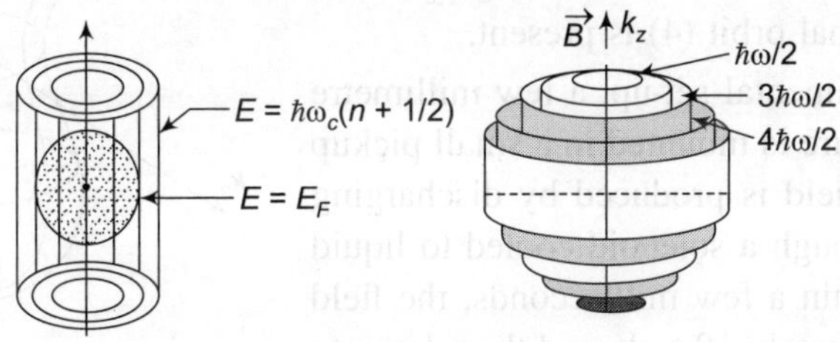


**Figure 10.99** In a magnetic field, the allowed states of the electrons lie on the Landau tubes

So far, we have considered a constant magnetic field. However, when the magnetic field $\vec{B}$ is slowly increased, Eq. (15) tells us that Landau tubes slowly expand and successively have the Fermi surface, $E(\vec{k}) = E_F$, passing through a contour of contact. If the contact contour of the Fermi level lies midway between two Landau tubes, the highest occupied states lie below the Fermi level. The average energy is then lower than when the Landau tube is just leaving the Fermi surface at the contact contour. The average energy therefore oscillates.

The period of oscillation can be easily specified by nothing that *orbits belonging to different quantum numbers can be manipulated to have a common value of the area by adjusting the magnetic field*. The areas of two successive orbits are

$$\left.\begin{aligned} S_n &= \left(n + \frac{1}{2}\right) \frac{2\pi eB_1}{\hbar} \\ S_{n+1} &= \left(n + 1 + \frac{1}{2}\right) \frac{2\pi eB_2}{\hbar} \end{aligned}\right\} \qquad ...(16)$$

If $B_1$ and $B_2$ are such that

$$S_n = S_{n+1} = S$$

then

$$\left(\frac{1}{B_2} - \frac{1}{B_1}\right) = \frac{2\pi e}{\hbar S}$$

or

$$\delta\left(\frac{1}{B}\right) = \frac{2\pi e}{\hbar S} \qquad \text{...(17a)}$$

Since $n \propto 1/B$ (from Eq. (16), $B_2 < B_1$, and by decreasing $B$ slowly, we can proceed to orbits of higher quantum number. Here $\delta$ is any extremal (maximal or minimal) cross sectional area of the Fermi surface in a plane perpendicular to the magnetic field (see Fig. 10.100). If the field is along $z$-axis, the cross sectional area of the Fermi surface at a height $k_z$ is $S(k_z)$ and the extremal areas $S$ are the values of $S(k_z)$ at the $k_z$ where $dS/dk_z = 0$. Thus, for $B$ along $k_z$ axis, (1) and (2) are maximal orbits and (3) is a minimal orbit. When the field is along $k_x$ (or $k_z$) axis, only one extremal orbit (4) is present.

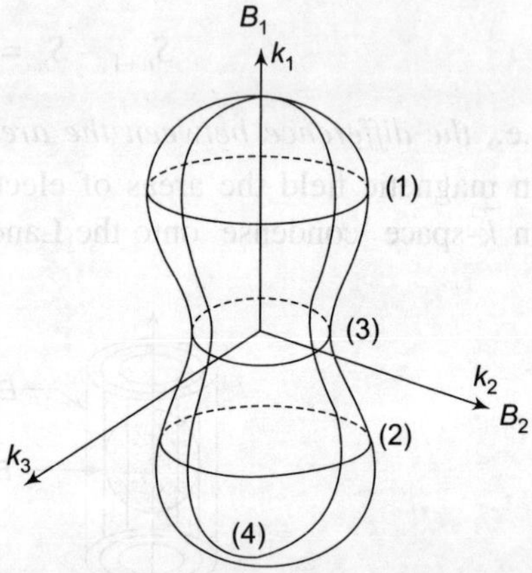

**Figure 10.100** Various extremal orbit on the Fermi surface

In a typical experimental set-up, a few millimetre long single crystal wire is mounted in a small pickup coil. The magnetic field is produced by discharging a large capacitor through a solenoid cooled to liquid $N_2$ temperature. Within a few milliseconds, the field increases to approximately 10 tesla and then drops to zero. With charging magnetic field, an emf is induced which is proportional to

$$\frac{dM}{dt} = \frac{dM}{dt} = \frac{dB}{dt} \qquad \text{...(17b)}$$

where $M$ is the magnetic moment.

Accordingly, the induced emf produces an oscillatory signal, which is proportional to the differential susceptibility, $dM/dB$.

Another method uses superconducting coils and persistent currents. When the solenoids get energized, a persistent current is set up with no external source of emf (because of no resistance) coils provide highly stable steady field ~ 5 tesla. This field can be modulated at a fixed frequency which provides phase-sensitive detections with improved signal-to-noise ratio.

## 10.53 EXPERIMENTAL DETERMINATION OF FERMI SURFACES OF COPPER

Electronic Fermi surfaces have been measured through observation of the oscillation of transport properties in magnetic field $\vec{H}$, viz., The de Haas-van Alphen effect and the

Shubnibov-de Haas effect. The former is an oscillation in *magnetic susceptibility* and the latter in *resistivity*. The oscillations are periodic versus $1/\vec{H}$ and occur because of the quantization of energy levels in the plane perpendicular to a magnetic field, a phenomenon first predicted by Lev Landau. The new states are called *Landau levels*. The determination of the periods of oscillation for various applied field directions allows mapping of the Fermi surface.

Observation of the de Haas-van Alphen and Shubnikov de-Haas oscillations requires magnetic fields large enough that the circumference of the cyclotron orbits is smaller than a mean free path. Therefore, experiments are usually performed at high field facilities like High Field Magnet Laboratory (Netherlands), Grenoble High Magnetic Field Laboratory (France), Tsukuba Magnet Laboratory (Japan) or National High Magnetic Field Laboratory (United States).

We will now indicate how we deal with the experimental determination of FS of a real metal, taking copper (which is fcc) as an example.

Figure 10.101 (a) shows the first BZ of the fcc lattice, marked with points of high symmetry. In Fig. 10.101 (b), a (110) section of the BZ is shown. The dashed circle represents a sphere whose volume is one half that of the BZ. It is close to the *L*-point, but farther from the *x*-point. Since copper metal has one free electron per atom, so the first BZ is half filled.

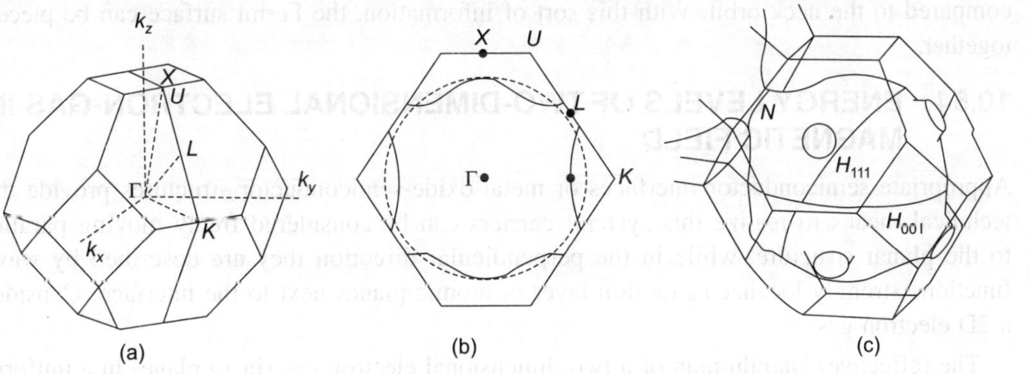


**Figure 10.101** Various aspects of FS of copper. (a) BZ of an fcc lattice (b) A (110) section of the BZ (c) the proposed FS of copper

However, the half-filled Fermi sphere is strongly perturbed near the *L*-point due to its proximity to the face of the BZ for copper. Figure 10.101 (c) shows the expected results. "Necks" appear near the *L*-point.

For determination of the FS, de Haas van Alphen technique is generally applied in which we measure the oscillations of the (dia-) magnetic moment of the metal sample as a function of the magnetic filed, *H*. These oscillations occur as the Landau levels pass through the FS causing slight changes in the total energy (and related properties) of the conduction electrons. We know that these changes in the total energy will have a constant period in $H^{-1}$ given by

$$\Delta(H^{-1}) = \frac{e\hbar}{mcE_F} \qquad ...(18)$$

using
$$E_F = \left(\frac{\hbar^2}{2m}\right) k_F^2 \qquad ...(19)$$

we obtain
$$\Delta(H^{-1}) = \frac{2\pi e}{c\hbar S_k} \qquad ...(20)$$

where $S_k = \pi k_F^2$ is the area in $k$-space of a cut in the FS perpendicular to the applied magnetic field. Thus, by measuring the periods of oscillation for $\vec{H}$ applied in different directions, $S_k$ can be determined. This enables one to piece together what the FS looks like.

For example, if a magnetic field is applied in [001] direction, the Landau levels cut the Fermi surface in the circle called $H_{001}$ in Fig. 10.100 (c). From the oscillation frequency, the area of this orbit in $\vec{k}$-space can be determined. (For subtle phase relationships regions, orbits that are parallel but have smaller areas tend to cancel. The dominant response comes from orbits of the type $H_{001}$. Such orbits are called *extremal orbits.*) By applying the magnetic field in the [111] direction, two extremal orbits are obtained and are labelled $N$ and $H_{111}$ in Fig. 10.100 (c). In the de Haas van Alphen measurements, oscillations corresponding to the areas of these two extremal orbits are found and we must separate the two. With such results, we can appreciate how the areas of the $H_{111}$ and $H_{001}$ orbits can be separated and then compared to the neck orbit. With this sort of information, the Fermi surface can be pieced together.

## 10.54 ENERGY LEVELS OF TWO-DIMENSIONAL ELECTRON-GAS IN MAGNETIC FIELD

Appropriate semiconductor interfaces or metal-oxide-semiconductor structures provide the technical means to realize this system: carriers can be considered freely moving parallel to the planar structure, while in the perpendicular direction they are described by wave functions strongly localized in a thin layer of atomic planes next to the interface. Consider a 2D electron gas.

The (effective) Hamiltonian of a two-dimensional electron gas (in $xy$ plane) in a uniform magnetic field (parallel to $z$) can be written as

$$H_0 = \frac{1}{2m}\left(p_x + \frac{e}{c}A_x\right)^2 + \frac{1}{2m}\left(p_y + \frac{e}{c}A_y\right)^2, \qquad ...(1)$$

where $e$ is the absolute value of the electron charge, the effective mass ($m$) is taken equal to the electron mass (for simplicity), and $\mathbf{A}(\mathbf{r})$ is the vector potential of the applied magnetic field. The Hamiltonian of Eq. (1) focuses on the modification of the *orbital motion* of carriers due to the magnetic field, and neglects the interaction of the spin magnetic moment with the applied field, spin degeneracy is taken into account introducing a factor two, whenever appropriate.

A possible choice of the vector potential, corresponding to a uniform magnetic field of intensity $H$ in the $z$-direction, is given by the so-called first Landau gauge

$$\mathbf{A}(\mathbf{r}) = (-H_y, 0, 0); \qquad ...(2)$$

it is seen by inspection that curl $\mathbf{A}(\mathbf{r}) = H(0, 0, 1)$. The vector potential is not uniquely determined; other frequently used gauges are the second Landau gauge $\mathbf{A}(\mathbf{r}) = (0, H_x, 0)$ and the symmetric gauge $\mathbf{A}(\mathbf{r}) = (1/2)(-H_y, H_x, 0)$.

The energies of the Hamiltonian, $H_0$, in the absence of the external magnetic field, are given by

$$E(k_x, k_y) = \frac{\hbar^2}{2m}(k_x^2 + k_y^2).$$

The corresponding two-dimensional density-of-states (excluding spin degeneracy) for a sample in a rectangular box of surface $S$ is given by

$$\boxed{D(E, H=0) = \frac{2\pi m}{\hbar^2} S \quad \text{for} \quad E > 0} \qquad \text{...(3)}$$

In the presence of the magnetic field, and adopting the gauge specified by Eq. (2), the Hamiltonian of Eq. (1) becomes

$$H_0 = \frac{1}{2m}\left(p_x - \frac{eH}{c}y\right)^2 + \frac{1}{2m}p_y^2. \qquad \text{...(4)}$$

The above Hamiltonian does not contain $x$, and then $p_x$ is a constant of motion with values $hk_x$; the Hamiltonian of Eq. (4) can be rewritten in the form

$$H_0 = \frac{1}{2m}p_y^2 + \frac{1}{2m}\left(\hbar k_x - \frac{eH}{c}y\right)^2$$

$$= \frac{1}{2m}p_y^2 + \frac{1}{2}m\omega_c^2(y - y_0)^2, \qquad \text{...(5a)}$$

where

$$\omega_c = \frac{eH}{mc} \qquad \text{...(5b)}$$

is the *cyclotron resonance frequency*, and

$$y_0 = \frac{\hbar c}{eH}k_x \qquad \text{...(5c)}$$

is the centre of oscillations. Eq. (5a) is the familiar Hamiltonian of a displaced linear oscillator of frequency $\omega_c$; its eigenvalues are

$$E_{nkz} = \left(n + \frac{1}{2}\right)\hbar\omega_c, \qquad \text{...(6)}$$

where $n = 0, 1, 2, \ldots$ indicates the *Landau level*. Thus, the presence of a magnetic field modifies the continuous kinetic energy $E(k_x, k_y) = (\hbar^2/2m)(k_x^2 + k_y^2)$ into discretized Landau levels, as indicated schematically in Fig. 10.102. The energy separation between adjacent Landau levels is $\hbar\omega_c = (e\hbar/mc)H = 2\mu_B H$; the value of the Bohr magneton is $\mu_B = (e\hbar/2mc) = 0.05788$ meV/tesla, and then $\hbar\omega_c/H = 0.1158$ meV/tesla (1 tesla = $10^4$ gauss).

The eigenfunctions corresponding to the eigenvalues of Eq. (6) are

$$\phi_{nk_x}(\mathbf{r}) = \frac{1}{\sqrt{L_x}} e^{ik_x x} H_n(y - y_0),$$

where $L_x$ is the length, in the $x$-direction, of the rectangular box where the electron is confined and $H_n$ are the Hermite functions for the harmotic oscillator. The wave functions, $\phi nk_x(\mathbf{r})$, in the presence of a magnetic field (in the chosen Landau gauge) are plane waves in the $x$-direction, multiplied by harmonic oscillator wave functions in the $y$-direction, with typical extension $l_n = \sqrt{2n+1}\, l_0$, where $l_0 = (\hbar c/eH)^{1/2}$ is the classical cyclotron radius of an electron with kinetic energy $\hbar\omega_c/2$. For $H = 10^4$ gauss, with $hc/e = 4.136 \times 10^{-7}$ gauss·cm$^2$, the magnetic length $l_0$ becomes $l_0 = 2.57 \times 10^{-6}$ cm = 257 Å.

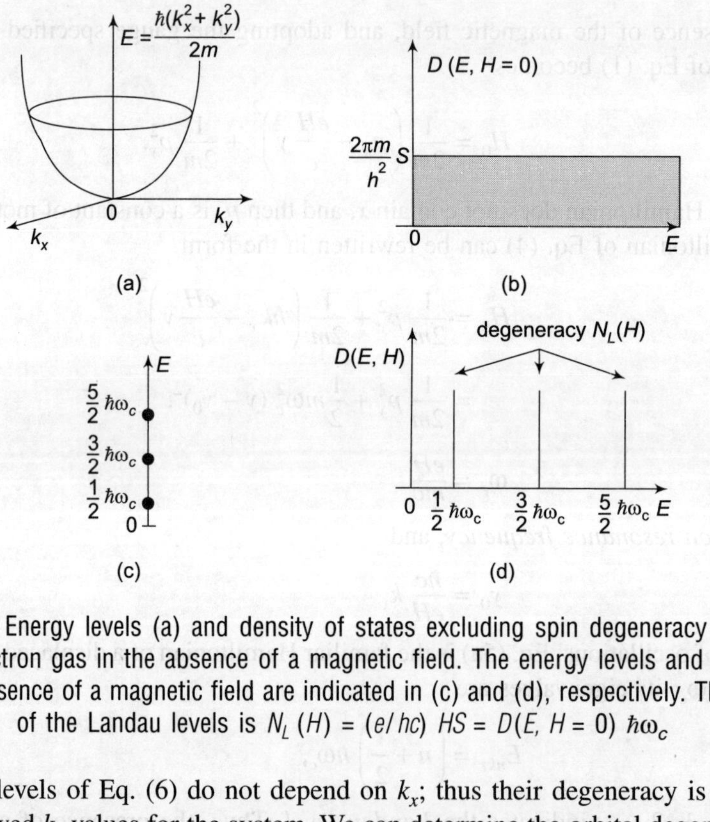


**Figure 10.102** Energy levels (a) and density of states excluding spin degeneracy (b) of a two-dimensional electron gas in the absence of a magnetic field. The energy levels and the density of states in the presence of a magnetic field are indicated in (c) and (d), respectively. The degeneracy of the Landau levels is $N_L(H) = (e/hc)\, HS = D(E, H = 0)\, \hbar\omega_c$

The energy levels of Eq. (6) do not depend on $k_x$; thus their degeneracy is given by the number of allowed $k_x$ values for the system. We can determine the orbital degeneracy $N_L(H)$ of a Landau level by imposing the condition

$$0 \le y_0 = \frac{\hbar c}{eH} k_x < L_y \quad \text{or equivalently} \quad 0 \le k_x < \frac{eH}{\hbar c} L_y.$$

The number $N_L$ of allowed $k_x$ values in the above interval gives the number of harmonic oscillators, whose origin is confined within the sample; the degeneracy $N_L$ of each Landau level is then

$$\boxed{N_L(H) = \frac{L_x}{2\pi}\frac{eH}{\hbar c}L_y \equiv \frac{e}{hc}H\,S.} \qquad ...(7)$$

The orbital degeneracy $N_L(H)$ is *thus proportional to the magnetic field, whose effect is that of piling up a large number of states into discrete Landau levels.*

The orbital degeneracy $N_L$ can be written also in the effective form

$$N_L(H) = \frac{\Phi(H)}{\Phi_0},$$

where $\Phi(H) = H\,S$ is the flux of $H$ through the sample, and

$$\Phi_0 = \frac{hc}{e} = 4.136 \times 10^{-7}\ \text{gauss}\cdot\text{cm}^2$$

provides the natural unit of flux.

The degeneracy $N_L(H)$ of each Landau level can be written also in the following alternative form

$$N_L(H) = D(E, H = 0)\,\hbar\omega_c, \qquad ...(8)$$

where $D(E, H = 0) = 2\pi m\,S/h^2$ indicates the constant two-dimensional density of states in the absence of the magnetic field (see Eq. 3). The quantity $D(E, H = 0)\,\hbar\omega_c$ represents the number of states in the energy interval $\hbar\omega_c$ between two successive Landau levels. Eq. (8) explicitly shows that the overall number of states is unaffected by the magnetic field.

There is another instructive way to express the Landau degeneracy of Eq. (7). Let us consider a sample with $N$ free electrons (of both spin directions) in the surface $S$, and surface density $n_s = N/S$. In the presence of a magnetic field $H$ perpendicular to the sample, the energy levels are quantized in Landau levels, whose degeneracy increases with $H$; thus, it is always possible to find a *threshold value* $(H_t)$ *of the magnetic field*, at which all the available electrons fully occupy the ground Landau level (above $H_t$, all the available electrons are accommodated in the ground Landau level without filling it). The *threshold field* $H_t$, *for which all the available electrons fully occupy the ground Landau level, assuming no removal of spin degeneracy*, is determined by the relation

$$2\,N_L(H_t) = 2\,\frac{e}{hc}\,H_t S \equiv N.$$

Thus, the characteristic field $H_t$ is given by

$$H_t = \frac{1}{2}\frac{N}{S}\frac{hc}{e} = \frac{1}{2}\,n_s\,\frac{hc}{e}. \qquad ...(9)$$

For $n_s$ in the range $10^{10} - 10^{12}$ electrons/cm$^2$, and $hc/e = 4.136 \times 10^{-7}$ gauss$\cdot$cm$^2$, we have $H_t$ in the range $10^3 - 10^5$ gauss. The orbital degeneracy of the Landau levels can be cast in the form

$$N_L(H) = \frac{1}{2}\,N\,\frac{H}{H_t}. \qquad ...(10)$$

Expressions (7), (8) and (10) for the degeneracy of the Landau levels are all equivalent and particularly significant in discussing magnetic field effects.

In a quite similar way, we can define the characteristic field ($H_t$*) as the field for which *all the available electrons fully occupy the ground Landau level with definite spin orientation*; $H_t^*$ is given by the expression

$$H_t^* = n_s \frac{hc}{e} \qquad ...(11)$$

and we evidently have $H_t^* = 2\, H_t$.

## EXERCISE

10.1 Distinguish between metals, insulators and semiconductors.

10.2 Define Fermi energy. Write down and explain the expression for Fermi–Dirac distribution law.

10.3 Based on Fermi–Dirac statistics, state the nature of the Fermi distribution function. How does it vary with temperature?

10.4 On the basis of the Fermi–Dirac distribution function, derive an expression for chemical potential of a free electron gas.

10.5 Derive an expression for thermal conductivity ($K$) of metals and prove the Wiedemann–Franz law $\left(\frac{K}{\sigma T} = \frac{\pi^2}{3}\frac{k^2}{e^2} \equiv L\right)$, where $L$ is the Lorenz number.

10.6 Explain Drude's model for conduction in metals.

10.7 Discuss Sommerfeld's free electron theory of metals and explain filling of electrons in a one-dimensional potential box.

10.8 Write short notes on:
(a) Hall effect
(b) Magnetoresistance

10.9 Derive Richardson's formula for thermiomic emission.

10.10 What is Hall effect? Find an expression for the Hall coefficient of a metal and describe an experimental set-up to measure it.

10.11 Explain Matleisen rule for metals.

10.12 What is a Fermi surface? Give its importance. How it is experimentally determined for metals?

10.13 Prove that Fermi energy for free electrons in a metal as a function of temperature is given by

$$E_{F(T)} = E_{\varepsilon(0)}\left[1 - \frac{\pi^2}{l^2}\left(\frac{k_B T}{E_{F(0)}}\right)^2\right]$$

where $k_B$ is Boltzmann constant.

10.14 Distinguish between Fermi surface and Brillouin zone. How is the FS of copper determined experimentally?

10.15 Explain Harrison's construction of Fermi surfaces. Give examples.

10.16 Explain the quantization of orbits in a magnetic field and derive the Onsager–Lifshitz quantization condition.

10.17 Write short notes on

(a) Characteristics of Fermi surfaces

(b) Open orbits and closed orbits

(c) Electron orbits and hole orbits.

10.18 Discuss effect of an electric field on Fermi surface.

10.19 Discuss de Haas-van Alphen effect and explain the formation of Landau tubes.

10.20 Discuss Harrison construction of FS in case of a divalent metal.

# CHAPTER 11

# Physics of Semiconductors

## 11.1 SEMICONDUCTORS

Semiconductors are materials that have resistivity ranging from about $10^{-3}$ to $10^8$ $\Omega$-cm. Semiconductor materials at 0K have basically the same energy-band structure as insulators – a filled valence band separated from an empty conduction band by a forbidden gap which is generally less than 2.5 eV. So that a significant number of electrons can be thermally excited from the valence band to the conduction band at room temperature. Conduction occurs in semiconductors—the result of a net movement under the influence of an electric field of electrons in the conduction band and holes in the valence band, i.e., charge carriers are electrons and holes.

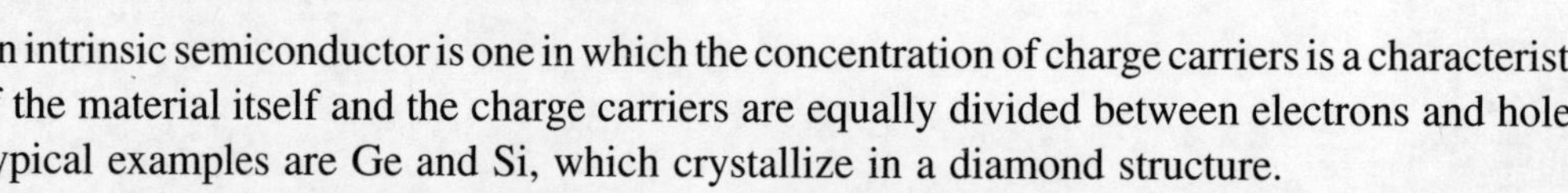


**Figure 11.1** Resistivity scale of materials

### 11.1.1 Intrinsic Semiconductor

An intrinsic semiconductor is one in which the concentration of charge carriers is a characteristic of the material itself and the charge carriers are equally divided between electrons and holes. Typical examples are Ge and Si, which crystallize in a diamond structure.

Consider a pure crystal of Si containing no impurities and no lattice defects. Figure (11.2) shows two-dimensional representation of such a crystal.

Here the circles represent the atom—cores with four positive charges and dots represent valence electrons. At 0K, all the electrons are tightly bound in the bonds and the crystal is a perfect insulator. At higher temperatures, vibrations of lattice atoms occasionally impart

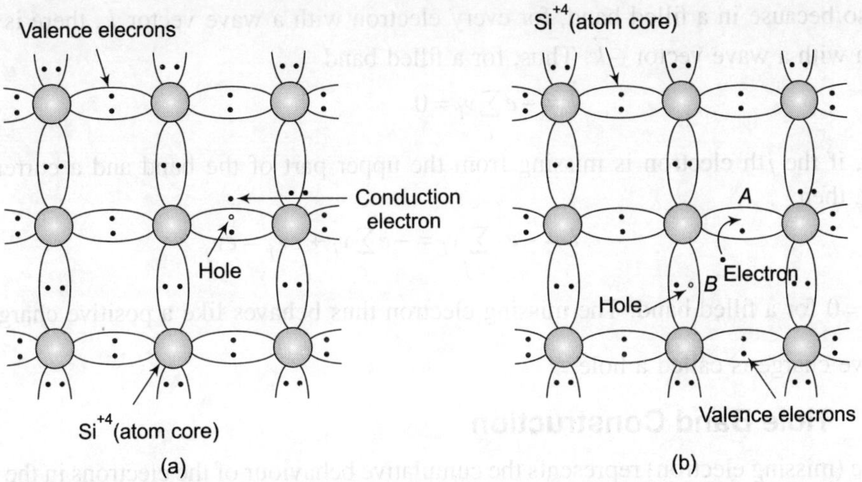


**Figure 11.2** A two-dimensional schematic representation of silicon crystal showing the atom cores and valence electrons (a) thermal generation of an electron-hole pair and (b) motion of a hole from location *A* to *B* (After S. M. Sze, *Semiconductor Devices: Physics & Technology*, Wiley Eastern Ltd. 1991)

sufficient energy to some of the bound electrons enabling them to leave the bond and move freely in the crystal. Figure 11.2 (a) illustrates the formation of an electron-hole pair by thermal generation.

In Fig. 11.2 (b), as a result of thermal motion, an electron in a covalent bond at *B* adjacent to the hole jumps into the original vacant site *A*, thus transferring the location of the hole to the site (*B*) from where the electron came (*A*). In this way, the motion of the hole can be regarded as the transfer of ionization from one atom to next by the jumping of the bound electron from a covalent bond to a neighbouring vacant site. This process does not affect the motion of a free electron. Thus, conduction electrons and holes in the valence band can move freely independent of each other.

If $p$ and $n$ represent the hole and electron concentrations respectively, then

$$p = n = n_i \qquad ...(1)$$

For an intrinsic semiconductor, the pair-concentration ($n_i$) varies with temperature and is different for different semiconductors. (There is a recombination of electron-hole at the same rate at which they are generated. The generation rate increases with the rise in temperature, and a new carrier concentration $n_i$ is established such that the higher recombination rate just balances generation.)

## 11.2 CONCEPT OF HOLES

If we have a number of electrons in the band and the electron is moving with a velocity $v_i$ then $I = -e\ å,\ v_i$. If a band is completely filled, the current is zero, since for every electron moving with a velocity ($v_i$) there is one moving in the opposite direction with the same speed;

this is so because in a filled band, for every electron with a wave vector $\vec{k}$, there is another electron with a wave vector $-\vec{k}$. Thus, for a filled band

$$I = -e\sum_i v_i = 0 \qquad ...(1)$$

Now, if the $j$th electron is missing from the upper part of the band and a current ($I$) is flowing, then

$$I = -e \sum_{i \neq j} v_i = -e\sum_i v_i + ev_j = ev_j \qquad ...(2)$$

as $\sum_i v_i = 0$ for a filled band. The missing electron thus behaves like a positive charge. Such a positive charge is called a hole.

### 11.2.1 Hole Band Construction

The hole (missing electron) represents the cumulative behaviour of the electrons in the valence band. For a completely filled band, the momentum is zero, i.e.,

$$\hbar\sum \vec{k} = 0 \quad \text{or} \quad \sum \vec{k} = 0 \qquad ...(3)$$

If an electron is excited to the conduction band (e.g., by absorption of a photon), the momentum of valence band is

$$\hbar(\sum \vec{k} - \vec{k}_e) = -\hbar \vec{k}_e$$

i.e., an electron of wave vector $\vec{k}_e$ becomes a missing electron and the total wave vector of the system is $-\vec{k}_e$, which is attributed to the hole.

$$\vec{k}_h = -\vec{k}_e \qquad ...(4)$$

According to Eq. (4), the wave vector of the hole is equal to that of the unpaired electron at point $C$ in the valence band whose counterpart (the missing electron at $A$) has been transferred to the conduction band (Fig. 11.3), since for a symmetric band

$$E_e(\vec{k}_e) = E_e(-\vec{k}_c) \qquad ...(5)$$

Now, let the zero of energy of the valence band be at the top of the band. The lower the missing electron lies in this band, the higher the energy of the system. The energy of the hole is opposite in sign to the energy of the missing electron, because it takes more work to remove an electron from a deeper orbital. Thus, for a symmetric band

$$E_e(k_e) = E_e(-\vec{k}_e) = -E_h(-\vec{k}_e) = -E_h(\vec{k}_h) \qquad ...(6)$$

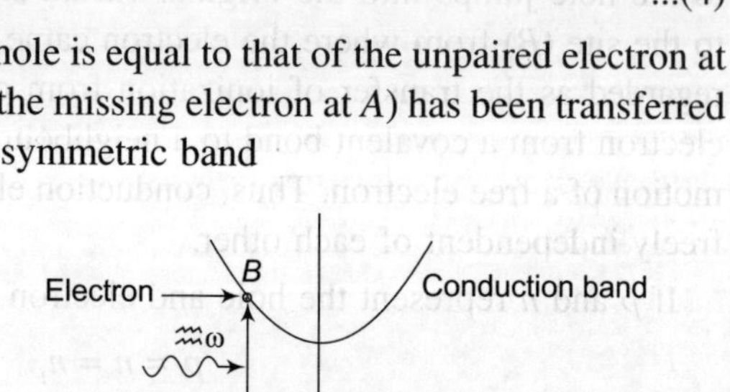


**Figure 11.3** Excitation of an electron with wave vector $\vec{k}_e$ from state $A$ in valence band to state $B$ in the conduction band (by absorption of a photon). This creates a hole with wave vector $\vec{k}_h = -\vec{k}_e$ in the state C of the valence hand

## 11.3 CRYSTAL GROWTH

Consider a system consisting of $N$ atoms (or molecules) of the same type in a container of volume $V$ with concentration $n = N/V$. The atoms could be in a vapour, about to form a droplet of condensate. Alternatively, the atoms could be in a melt, about to nucleate a solid-state particle; or they may be atoms in solution, which nucleate to form a small particle. For the sake of definiteness, the focus of attention will be on the vapour-liquid transition. If the pressure of the gas exceeds the vapour pressure (for that temperature), the system is thermodynamically unstable and droplets will begin to nucleate.

Let $\mu_V$ be the chemical potential of a vapour atom and $\mu_D$ be the chemical potential of an atom in the droplet. The number of atoms in vapour is $N_V$ and the number in the droplet is $N_D$. The droplet is assumed to be a sphere with radius $r$ and surface energy (assumed to be uniform) $\sigma$. In forming a droplet, one begins with a Gibbs free energy $G_i = N_V\mu_V$ and changes it to $G_f = (N_V - N_D)\mu_V + N_D\mu_D + 4\pi\sigma r^2$.

The change in $G$ is

$$\Delta G = N_D(\mu_D - \mu_V) + 4\pi\sigma r^2. \qquad ...(1)$$

If $\Omega$ denotes the volume of an atom in the droplet,

$$N_D = \frac{4\pi r^3}{3\Omega}. \qquad ...(2)$$

Therefore,

$$\Delta G(N_D) = -N_D\delta\mu + 4\mu\sigma\left(\frac{3\Omega N_D}{4\pi}\right)^{2/3} \qquad ...(3)$$

where $\delta\mu = \mu_V - \mu_D$. Assuming that $\delta\mu > 0$, the function rises and then falls as $N_D$ is increased from zero. The function $\Delta G(N_D)$ will peak at

$$N_D^* = \frac{32\pi}{3}\frac{\Omega^2\sigma^3}{(\delta\mu)^3} \qquad ...(4)$$

and have the maximum value

$$\Delta G^* = \frac{16\pi}{3}\frac{\Omega^2\sigma^3}{(\delta\mu)^2} = \frac{1}{2}N_D^*\,\delta\mu. \qquad ...(5)$$

The number $N_D^*$ represents the critical size that a cluster must reach before it becomes established. When this is large enough to have physical significance, the corresponding critical radius is given by $r_c = 2\sigma\Omega/\delta\mu$. If $N_D < N_D^*$ (or $r < r_c$), the cluster will shrink, whereas if $N_D > N_D^*$, it will continue to grow in size. The probability that a given thermal fluctuation will nucleate a particle of a given size is proportional to a factor $P = \exp[-\beta\Delta G(N_D)]$ multiplied by the probability of having $N_D$ particles found together within some critical distance of each other.

A volume $V$ contains $nV$ atoms, and any of these atoms could serve as the site at which nucleation could begin. The number of nuclei of a given size that will form will be given by $nVP$. The number of nuclei per unit volume of size $N_D$ is therefore

$$n(N_D) = ne^{-\beta\Delta G(N_D)}. \qquad ...(6)$$

The same analysis given for droplet formation in a vapour applies to crystals nucleating in the melt. Crystals, however, possess a lattice and there is an anisotropy associated with the solid. After nucleation the crystals will continue to grow as atoms adsorb on the nucleated crystallite. As the crystals grow in size, their shapes will also change. Facets will form as the crystals attempt to find the state of thermodynamic equilibrium (i.e., their state of lowest Gibbs free energy). The stable crystals assume polyhedral shapes and adjust themselves so as to minimize their total surface energies, subject to the constraint that the volume is fixed. Consider one such crystallite. Let $h_i$ denote the distance from a face of area $A_i$ to the centre of mass of the polyhedron and $\sigma_i$ denote the corresponding surface energy. The higher the surface energy is, the less stable a given surface is. The net surface energy is $E_s = \sum \sigma_i A_i$ and the volume is the sum of a collection of pyramidal volumes, $V = \sum A_i h_i/3$. Minimizing the net surface energy subject to the constraint of fixed volume yields the equation $\delta(E_s - \lambda V) = 0$, with $\lambda$ being a Lagrange multiplier. Variation with respect to $A_i$ gives

$$h_i = \frac{3}{\lambda}\sigma_i \qquad \text{...(7)}$$

One finds that $\lambda = E_s/V$. Those surfaces with the higher surface energies will have higher growth velocities $v_i$, will be further away from the centre of mass, and will generally have smaller areas. As one constructs the resulting polyhedral shape, it may turn out that the geometry is such that some surface faces will be completely absent. In describing the approach to equilibrium, one may assign growth velocities to the various faces to describe the kinetics.

In the early phases of crystal growth, the crystals draw their material from the melt (or solution or vapour) and the crystallites grow independently. Later, as the melt becomes depleted, the crystallites compete with each other for growth. In the phenomenon of Ostwald ripening, small crystallites shrink and supply atoms to larger crystallites, which grow. To see how this comes about, consider two spheres with numbers of atoms $N_1$ and $N_2$, respectively. The total number of atoms is $N = N_1 + N_2$ and is fixed. The Gibbs free energy for the pair is

$$G(N_1, N - N_1) = N\mu_D + 4\pi\sigma\left(\frac{3\Omega}{4\pi}\right)^{2/3}[N_1^{2/3} + (N - N_1)^{2/3}] \qquad \text{...(8)}$$

This function has a maximum at $N_1 = N_2 = N/2$. If $N_1 > N_2$ crystal 1 will grow at the expense of crystal 2 until crystal 2 disappears, and vice versa. The final energy drops to

$$G(N, 0) = N\mu_D + 4\pi\sigma\left(\frac{3\Omega}{4\pi}\right)^{2/3} N^{2/3}. \qquad \text{...(9)}$$

For example, the difference in energy between having one combined sphere and two separate spheres is

$$\Delta G = -4\pi\sigma\left(\frac{3\Omega}{4\pi}\right)^{2/3}(2^{1/3} - 1)N^{2/3}. \qquad \text{...(10)}$$

The degree of crystal growth depends to a large extent on the time spent at high temperatures when the atoms are free to move around. If the material is cooled prematurely, the growth of the crystals is stunted and the microstructure existing at that time is frozen into the solid. The number of grains, grain sizes, and grain boundary area depend to a large extent on how far one permits this crystallization to proceed before quenching the system.

## 11.4 SINGLE CRYSTAL GROWTH OF SILICON AND REFINING

Silicon is obtained by the chemical reduction of $SiO_2$ commonly produced from sand. The $SiO_2$ is reduced by carbon in an arc furnace equipped with graphite electrodes, according to the following reaction: $SiO_2 + C \rightarrow Si + CO_2$. The silicon produced by this method has a low purity and is called "metallurgical-quality silicon". The silicon then reacts with hydrochloric acid to produce trichlorosilane, according to the reaction $Si + 3\ HCl \rightarrow SiHCl_3 + H_2$. The obtained trichlorosilane is then filtered and purified by distillation. Finally, trichlorosilane is decomposed at high temperature into HCl and silicon. Once pure silicon has been obtained it must be converted into a single crystal.

This is achieved using a technique called Czochralski growth, (CZ growth) which allows one to grow single-crystal rods up to 40 cm in diameter and over a metre long. During CZ growth the silicon is melted in a quartz container. The seed crystal is spun and slowly pulled out of the molten silicon. As the crystal is pulled upwards the temperature differential between the molten silicon bath and the gas ambient above it causes the silicon to crystallize. A rod-like silicon ingot is thus produced, as illustrated in Fig. 11.4. Impurities such as boron or phosphorus can be added to the molten silicon to dope the silicon P- or N-type and give it the desired resistivity. The pulling of a silicon crystal by the Czochralski technique typically takes 24 hours. Once cooled, the ingot is cut like a salami into 0.5 to 2 mm-thick slices called silicon wafers. The wafers are then mirror polished using a combination of mechanical and chemical polishing agents.

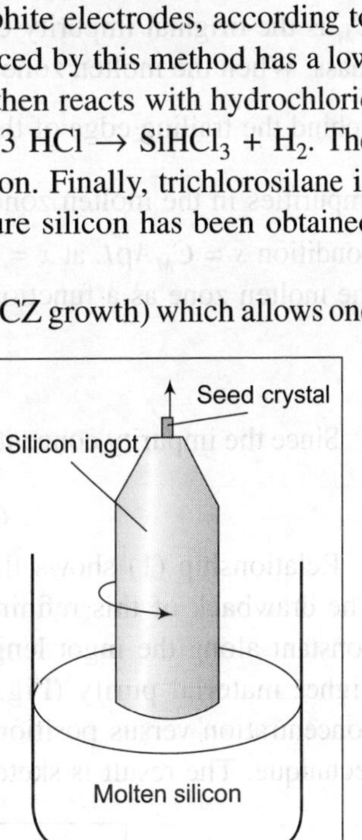


**Figure 11.4** Czochralski growth

The silicon ingot can be further refined using the float-zone technique, in which a section of the ingot is melted by induction using an RF coil. The molten zone is then swept from one side of the ingot to the other side (Fig. 11.5). Several passes can be applied to further improve crystal purity.

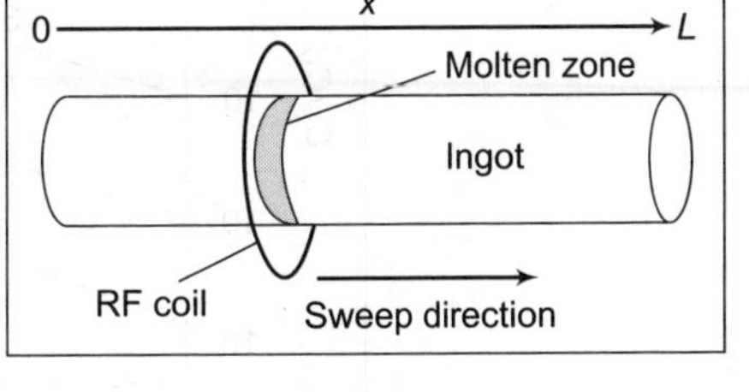


**Figure 11.5** Float-zone purification technique

The principle behind float-zone (FZ) refining is the following. The segregation coefficient for an impurity at the solid-liquid interface of the molten zone, $k$, is defined as the ratio of the concentrations of the impurity in the solid and the liquid: $k = \frac{C_S}{C_L}$. As an illustration, here are some segregation coefficients of some impurities in silicon: P: 0.32, As: 0.27, Sb: 0.02, B: 0.72, Ga: 0.0072, Au: 0.0000225. Since $k$ is smaller than unity impurities are extracted from the solid and trapped in the molten zone during the float-zone refining process. To analyze the physics of the float-zone refining process let us define the following parameters: $L$ is the length of the ingot, $x$ is the position along the ingot, $s$ is the amount of impurities dissolved in the molten zone, $A$ is the cross-sectional area of the ingot, $C_M$ is the original impurity concentration (per gram of silicon), and $\rho$ is the silicon volumic mass. When the molten zone moves a distance $dx$ the quantity of impurities introduced in the behind the trailing edge of the molten zone is equal to $ks\frac{dx}{L}$. The variation of the quantity of impurities in the molten zone is, therefore, equal to $ds = C_M A\rho dx - ks\frac{dx}{L}$. Using the initial condition $s = C_M A\rho L$ at $x = 0$ we can calculate the variation of the quantity of impurities in the molten zone as a function of position:

$$s = \frac{C_M A\rho L}{k}(1 - (1 - k)\,e^{-kx/L})$$

Since the impurity concentration left behind the molten zone, $C_S$, is equal to $\frac{ks}{A\rho L}$ we obtain:

$$C_S(x) = C_M\,(1 - (1 - k)\,e^{-kx/L}) \qquad \text{...(1)}$$

Relationship (1) shows that the smaller the value of $x$, the more purified the crystal is. The drawback of this refining method is, of course, that the impurity concentration is not constant along the ingot length. However, the use of several float-zone passes can produce higher material purity (Fig. 11.6). Although Eq. (1) is valid for the first pass only, the concentration versus position in the crystal can be calculated using a numerical calculation technique. The result is sketched in Fig. 11.6.

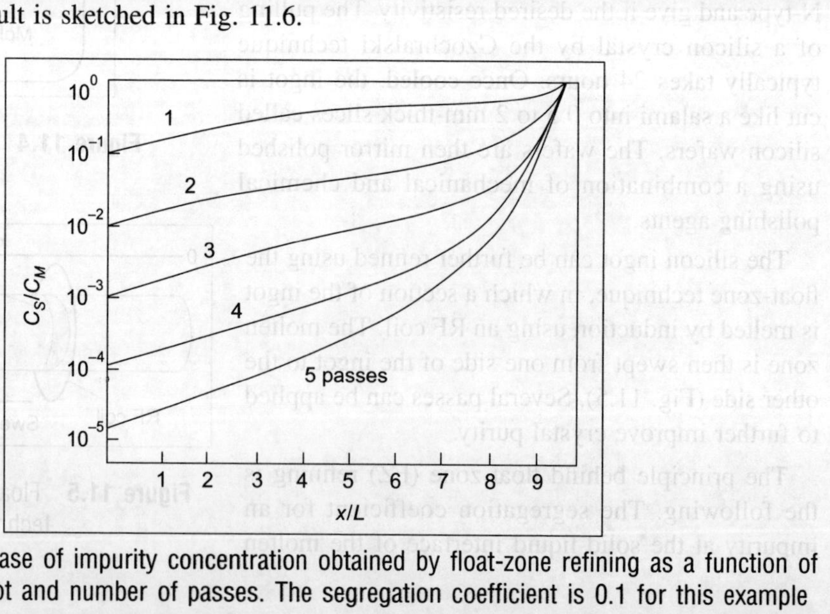


**Figure 11.6** Decrease of impurity concentration obtained by float-zone refining as a function of position in the ingot and number of passes. The segregation coefficient is 0.1 for this example

## 11.5 DOPING AN INTRINSIC SEMICONDUCTOR BY IMPURITIES

In an *intrinsic* (i.e., very pure) semiconductor, condition electrons and holes always occur in equal numbers. Such a semiconductor is not useful for practical applications, because the conductivity is low as there are no free lattice sites available for carrier motion and low carrier density in conduction or valance bands. In fact, the development of practical semiconductor devices would not be possible without the freedom of adding minute, controlled amounts (~ 1 part per million) of appropriate foreign atoms (impurities) to otherwise pure material. In this way, we get semiconductors, which exhibit extrinsic conductivity over a reasonably wide temperature range, called *extrinsic semiconductors*. These semiconductors have either more conduction electrons than holes, or vice versa. Such a result cannot, generally, be accomplished on a large scale by simply adding electrons to, or removing them from, pure material, since this would upset the electrical neutrality of the whole crystal and bring into play tremendous electrical restoring forcers. We can, however, substitute for a few of the original semiconductor atoms by some different ones which have either one more or one less valence electron.

## 11.6 EXTRINSIC SEMICONDUCTORS

Apart from thermal generation, charge carriers in semiconductors can also be created by adding a substitution impurity, the process is called doping and doped semiconductors are extrinsic semiconductors because the type of conduction (electron- or hole-type) that predominates depends on the valence (and number) of impurity atoms added, e.g., if impurity atoms with a valence of five, such as P or As or Sb is added to the lattice, there will be an extra electron per atom available for conduction, i.e., one that is not required to pair up with the four valance electrons of the Ge or Si (Fig. 11.7).

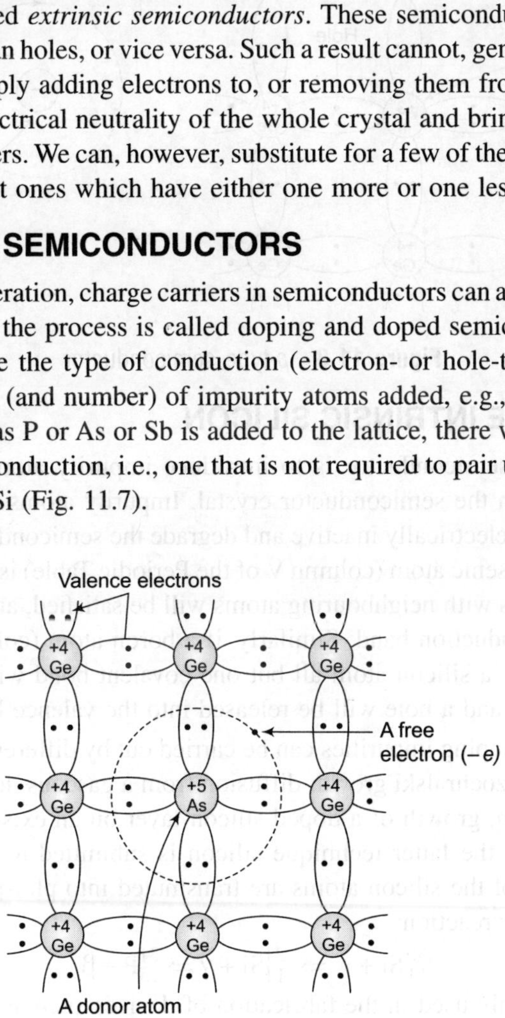


**Figure 11.7** *n*-type semiconductor

Thus, extrinsic semiconductors doped with atoms of valence 5 give rise to crystals with electrons as majority carriers, the so called *n-type semiconductors*. Similarly, if the impurity atoms have a valance of three (such as B, Al, In, Ga), one hole per atoms is created by an unsatisfied bond and we get *p-type semiconductors* (with holes as majority carriers) (see Fig. 11.8).

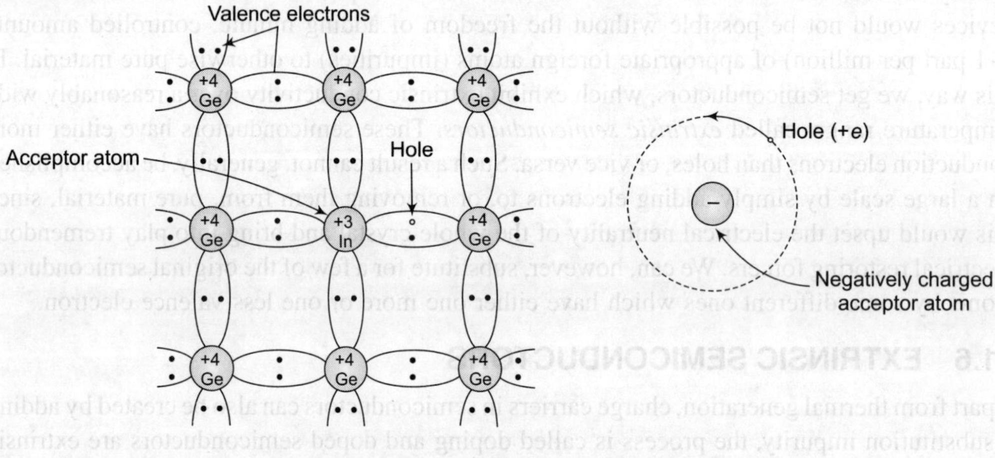


**Figure 11.8** *p*-type semiconductor

## 11.7 DOPING THE INTRINSIC SILICON

The goal of doping a semiconductor is to introduce impurity atoms of a desired species into substitution sites in the semiconductor crystal. Impurity atoms in interstitial (i.e., non-substitutional) sites are electrically inactive and degrade the semiconductor properties such as carrier mobility. If an arsenic atom (column V of the Periodic Table) is substituted for a silicon atom the covalent bonds with neighbouring atoms will be satisfied, and an extra electron will be released into the conduction band. Similarly, if a boron atom (column III of the Periodic Table) is substituted for a silicon atom all but one covalent bond with neighbouring silicon atoms will be satisfied, and a hole will be released into the valence band.

The introduction of doping impurities can be carried out by different techniques: doping of molten silicon before Czochralski growth diffusion from a gaseous doping substance into the silicon, ion implantation, growth of a doped silicon layer on an existing substrate (epitaxy), and neutron doping. In the latter technique silicon is submitted to a neutron ($v$) flux in a nuclear reactor. Some of the silicon atoms are transmuted into phosphorus atoms according to the following nuclear reaction:

$$^{30}_{14}\text{Si} + v \Rightarrow \ ^{31}_{14}\text{Si} + \gamma \Rightarrow \ ^{31}_{15}\text{P} + \beta$$

Neutron doping is only used in the fabrication of devices such as power thyristors where a low doping concentration and a high uniformity of doping concentration are needed.

### 11.7.1 Doping: Impurity Diffusion

Impurity diffusion is generally carried out at high temperature (800° – 1100°C) in a furnace. At those temperatures, the impurity atoms can diffuse throughout the crystal lattice through interactions with point defects (interstitials and vacancies). The equation governing the diffusion of impurities can be derived using Flick's first law of diffusion. Accordingly, the variation of impurity concentration in an elementary volume in the crystal is given by

$$\frac{\partial C}{\partial t}dx = F(x) - F(x + dx)$$

where $C$ is the average impurity concentration in the volume under consideration and $F$ is the flux of impurity atoms (Fig. 11.9). Letting $dx$ approach 0 we obtain $\frac{\partial C}{\partial t} = -\frac{dF}{dx}$.

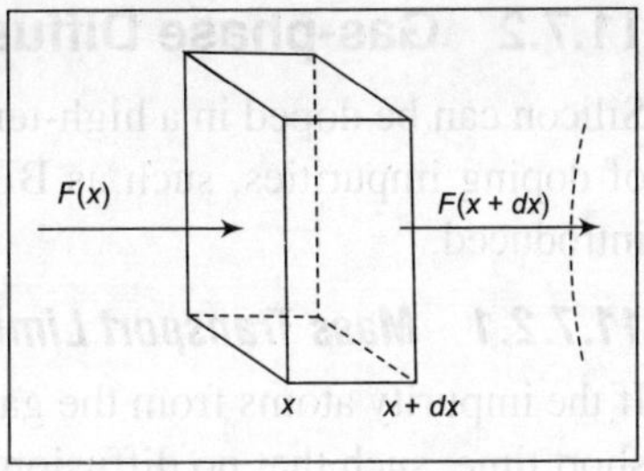


**Figure 11.9** Elementary volume and flux of impurity atoms

The flux is proportional to the impurity concentration gradient: $F = -D\frac{dC}{dx}$ where $D$ is the diffusion constant of the impurity in silicon. Combining the previous equations we obtain the following relationship, known as Flick's second law of diffusion:

$$\frac{\partial C}{\partial t} = D\frac{d^2C}{dx^2} \quad ...(1)$$

Strictly speaking, the diffusion constant is not a real constant since it depends on both the temperature and the concentration of impurities in the silicon (the diffusing impurity or other impurities). The solution of Eq. (1) yields a value called the "characteristic diffusion length", $L$, which is a function of temperature and time of diffusion: $L(T, t) = 2\sqrt{Dt}$. If the impurity concentration distribution before diffusion is Gaussian, it remains Gaussian after diffusion. The depth of the peak concentration, $R_p$, remains unchanged, but the peak concentration, $C_p$, decreases, and the distribution spreads out, such that its standard deviation increases. The new standard deviation, noted $L'$, is a function of the pre-diffusion standard deviation, $\Delta R_p$ and the diffusion length, $2\sqrt{Dt}$. It is equal to:

$$L' = \sqrt{2\Delta R_p^2 + 4Dt} \quad ...(2)$$

and the impurity concentration profile after diffusion is given by:

$$C(x, t) = \frac{N'}{L'\sqrt{\pi}}\exp\left(-\left(\frac{x - R_p}{L'}\right)^2\right) \quad ...(3)$$

It is worthwhile to note that the concentration profile of implanted impurities before and after diffusion are described by the same equation, $C(x) = \frac{N'}{\sqrt{\pi}L}\exp\left(-\left(\frac{x - R_p}{L}\right)^2\right)$, in which $L = \sqrt{2}\,\Delta R_p$ before diffusion and $L = \sqrt{2\Delta R_p^2 + 4Dt}$ after diffusion.

### 11.7.2 Gas-phase Diffusion

Silicon can be doped in a high-temperature furnace in which an ambient gas containing atoms of doping impurities, such as $BBr_3$ (for boron doping) or $POCl_3$ (for phosphorus doping) is introduced.

#### *11.7.2.1 Mass Transport Limited Case*

If the impurity atoms from the gas phase are present at the surface of the silicon in a relatively short time, such that no diffusion into the crystal takes place, the doping concentration profile can be approximated by a delta function. The impurity concentration is, therefore, given by $C(x = 0) = N'$ ($cm^{-2}$) and $C(x > 0) = 0$. If the sample is then submitted to a thermal annealing step called "drive-in" the impurity profile can be found by solving the diffusion equation, (Eq. 1). The impurity concentration follows a Gaussian distribution and the peak concentration is located at the silicon surface ($x = 0$)

$$C(x, t) = \frac{N'}{L'\sqrt{\pi}} \exp\left(-\left(\frac{x - R_p}{L'}\right)^2\right) \quad ...(4)$$

The standard deviation of the Gaussian profile is equal to $L' = 2\sqrt{Dt}$ and depends on temperature (through the diffusion coefficient $D$) and the annealing time. If several annealing steps are carried out (the number of annealing steps being $n$) the final standard deviation is equal to

$$L' = \sqrt{\sum_{i=1}^{n} Li^2} = \sqrt{\sum_{i=1}^{n} 4D_i t_i} \quad ...(5)$$

#### *11.7.2.2 Diffusion Limited Case*

If the doping impurity gas is fed into the furnace continually the impurity concentration at the silicon surface, $C_s$, is maintained constant. In such a case the impurity concentation profile is no longer Gaussian, and is rather given by a complementary error function, *erfc* ($x$):

$$C(x, t) = C_s \, erfc\left(\frac{x}{2\sqrt{Dt}}\right) = \frac{2Cs}{\sqrt{\pi}} \int_{x/2\sqrt{Dt}}^{\infty} e^{-v^2} dv \quad ...(6)$$

The complementary error function is defined by

$$erfc(\eta) = 1 - erf(\eta) = 1 - \frac{2}{\sqrt{\pi}} \int_0^{\eta} \exp(-v^2) dv$$

One of the properties of Eq. (6) is $C(0, t) = C_s$

The total concentration of impurities in the silicon is equal to:

$$N'(t) = \int_0^{\infty} C(x, t) dx = 2\sqrt{\frac{Dt}{\pi}} C_S \quad ...(7)$$

The maximum impurity concentration that can be introduced in silicon is fixed by the solid solubility of each doping species in silicon. Solid solubility is a function of temperature and is equal to $3.2 \times 10^{20}$, $2.1 \times 10^{20}$, and $10^{20}$ $cm^{-3}$ for P, As and B in silicon at 1000°C, respectively. Any attempt to introduce more doping atoms would result in the formation of impurity precipitates in the silicon crystal.

Modern silicon technology requires the formation of very shallow junctions. For that reason annealing techniques have been developed to activate implanted impurities with negligible diffusion. The *rapid thermal annealing* (RTA) technique achieves this goal: the silicon wafer is rapidly heated up to a high temperature (e.g., 1100°C) for a few seconds, and then is quickly cooled down. Such an annealing step allows for the restoring of the silicon crystal and the placement of doping atoms into the substitutional sites without causing diffusion over an appreciable distance.

## 11.8 OXIDATION

Silicon is oxidized by oxygen or steam at high temperature according to the following chemical reactions:

$$\text{Si(Solid)} + O_2\text{(gas)} \rightarrow SiO_2\text{(solid) (dry oxidation)}$$

or

$$\text{Si(Solid)} + 2H_2O\text{ (gas)} \rightarrow SiO_2\text{(solid)} + 2H_2\text{(gas) (wet oxidation)}$$

Two mechanisms influence the growth rate of the oxide. The first one is the actual chemical reaction rate between silicon and oxygen. The second one is the diffusion rate of the oxidizing species through an already grown oxide layer. When there is no or little oxide on the silicon the oxidizing agent easily reaches the silicon surface and the factor determining the growth rate is the kinetics of the silicon-oxide chemical reaction. In that case the oxidation process is reaction limited and the oxide thickness increases linearly as a function of time. If, on the other hand, the silicon is already covered by a sufficiently thick layer of oxide the oxidation process is mass-transport limited and the factor limiting the growth rate is the diffusion rate of $O_2$ or $H_2O$ through the oxide, in which case oxide growth increases as a square root function of time. A steam ambient is usually preferred to a dry oxygen ambient for the growth of thick oxides: $H_2O$ molecules are smaller than $O_2$ molecules, and as a result, they can diffuse more readily through $SiO_2$, which gives rise to higher oxidation rate.

To derive the equation describing silicon oxidation we will consider the mass transport of oxygen molecules from the gas ambient towards the silicon through a layer of already grown oxide (Fig. 11.10). The flux of oxygen flux towards the oxide, $F_1$, is thus differential in oxygen molecules is proportional to differential in oxygen concentration between the ambient, $C^*$, and the oxide surface, $C_0$. The oxygen flux towards the oxide, $F_1$, is thus given by the following equation. $F_1 = h(C^* - C_O)$ where $h$ is the mass transport coefficient for oxygen in the gas phase.

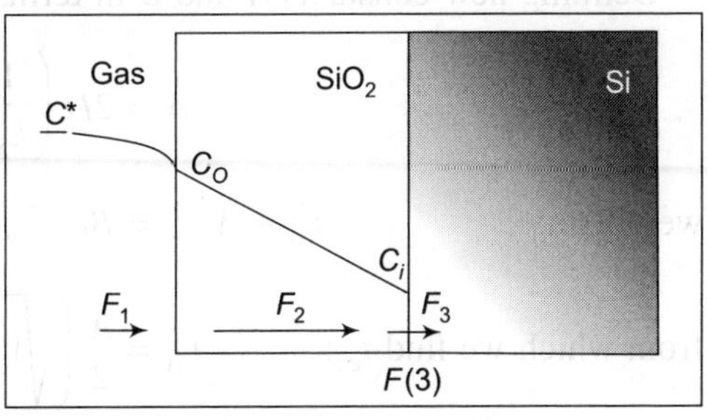


**Figure 11.10** Oxygen concentration profile during oxidation

The diffusion of oxygen through the oxide is proportional to the difference of oxygen concentration between the oxide surface and the silicon/$SiO_2$ interface. The flux of oxygen through the oxide, $F_2$, is given by $F_2 = D\dfrac{C_o - C_i}{t_{ox}}$ where $C_i$ is the oxygen concentration at the silicon/$SiO_2$ interface, $D$ is the diffusion coefficient of either $O_2$ or $H_2O$ in oxide, and $t_{ox}$ is the oxide thickness. Finally, the kinetics of the chemical reaction $k$, such that we have: $F_3 = k_s C_i$.

In steady state all flux terms are equal: $F_1 = F_2 = F_3 \equiv F$. Eliminating $C_o$ from the flux equation we obtain

$$C_i = \frac{C^*}{1 + \dfrac{k_S}{h} + \dfrac{k_S t_{ox}}{D}} \qquad ...(1)$$

If $N_{ox}$ is a constant representing the number of oxidizing gas molecules necessary to grow a unit thickness of oxide one can write

$$\frac{dt_{ox}}{dt} = N_{ox}\, F = N_{ox} k_s\, C_i = \frac{N_{ox} k_S C^*}{1 + \dfrac{k_S}{h} + \dfrac{k_S t_{ox}}{D}} \qquad ...(2)$$

The solution to this differential equation is

$$\int_0^{tox} \frac{1 + \dfrac{k_S}{h} + \dfrac{k_S t_{ox}}{D}}{N_{ox} k_S C^*}\, dt_{ox} = \int_0^t dt$$

If $t_{ox} = 0$ when $t = 0$, the integration yields

$$\frac{t_{ox}^2}{2} + \left(\frac{D}{k_S} + \frac{D}{h}\right) t_{ox} - N_{ox}\, C^*\, Dt = 0$$

or:

$$\frac{t_{ox}^2}{2} + 2D\left(\frac{1}{k_S} + \frac{1}{h}\right) t_{ox} = 2DN_{ox} C^* t$$

Defining new constants $A$ and $B$ in terms of $D$, $k_s$, $N_{ox}$ and $C^*$:

$$A = 2D\left(\frac{1}{k_S} + \frac{1}{h}\right) \quad \text{and} \quad B = 2DC^* N_{ox}$$

we obtain:

$$t_{ox}^2 + A\, t_{ox} = Bt$$

from which we find $t_{ox}$:

$$t_{ox} = \frac{A}{2}\left(\sqrt{1 + \frac{(t+\tau)}{A^2/4B}} - 1\right) \qquad ...(3)$$

Parameter $\tau$ is introduced to account for the possible presence of an oxide layer on the silicon before thermal oxide growth is performed. This pre-existing oxide layer can either be a native oxide layer due to the oxidation of bare silicon by ambient air or a thermally grown oxide produced during a prior oxidation step. $\tau = 0$ if the thickness of the initial oxide is equal to zero. Equation (3) is referred to as the Deal–Grove model of oxidation.

When thin oxides are formed the growth rate is limited by the kinetics of the chemical reaction between silicon and oxygen. In that case Eq. (3) can be approximated by: $t_{ox} = \frac{B}{A}(t + \tau)$ which is linear with time. The $\frac{B}{A}$ ratio is called the "linear growth coefficient" and is dependent on the crystal orientation of silicon. When thick oxides are formed the growth rate is limited by the diffusion rate of the oxygen through the oxide. In that case relationship (3) can be approximated by: $t_{ox} = \sqrt{B(t+\tau)} \cong \sqrt{Bt}$. The coefficient $B$ is called the "parabolic growth coefficient" and is independent of the crystal orientation of silicon. The parabolic growth coefficient can be increased by increasing the pressure of the ambient oxygen up to 10 to 20 atmospheres (high-pressure oxidation, HIPOX).[10] The linear growth coefficient can be increased if the silicon contains a high concentration of impurities such as phosphorus. These impurities increase the concentration of point defects in the crystal which increase the oxidation reaction rate at the silicon/silicon dioxide interface. Similarly, the oxidation process generates point defects in silicon, which accelerates the diffusion of doping impurities (oxidation enhanced diffusion, OED). Therefore, some doping impurities diffuse faster when annealing is performed in an oxidizing ambient than when it is carried out in a neutral gas such as nitrogen.

Oxide growth consumes silicon. As a rule of thumb one considers that the thickness of silicon consumed is 44% of the grown oxide. When an oxide is grown the doping atoms in the silicon redistribute between the silicon and the oxide and there exists a constant ratio between the impurity concentration on both sides of the $Si/SiO_2$ interface called the "segregation coefficient" defined by: $m = \frac{\text{CSi}}{\text{CSiO}_2}$. The segregation coefficient of arsenic and phosphorus in silicon is larger than unity, while that of boron is smaller than unity. In other words, the concentration of arsenic and phosphorus in the oxide is less than in silicon, and the concentration of boron is larger. This effect results in a "pile-up" of arsenic or phosphorus in the silicon at a $Si/SiO_2$ interface or a depletion of boron in the silicon at a $Si/SiO_2$ interface.

## 11.9 IMPURITY STATES AND BAND MODEL FOR *n*-TYPE AND *p*-TYPE SEMICONDUCTORS

Suppose an impurity atom from group V of the Periodic Table (e.g., P or As) is added to pure Ge or Si crystal. Such an atom in the crystal will behave as a fixed ion of charge $+e$ to which an electron is loosely bound. The orbit of the donated electron will be quite large compared to the interatomic spacing so that many atoms will be encompassed by the path of the electron about the ion. The electrostatic force between the electron and the impurity ion will essentially be modified by the polarization force between the electron and the impurity ion will essentially be modified by the polarization of the intervening atoms. Under these circumstances, we may consider the electron as if it has been immersed in a uniform polarizable medium of the semiconductor crystal and, the electron plus impurity ion system forms a hydrogen like atom in the uniform continuous medium of dielectric constant, $\in$.

The problem of calculating the energy levels of the impurity electron in the crystal may be taken up; these levels are called the impurity levels or *impurity states*. These may be worked out by modifying the Bohr theory of the hydrogen atom so as to take into account both the dielectric constant of the crystal and the effective mass of the electron in the periodic potential of the crystal. The necessary modification is introduced by replacing $m$ by $m^*$ and using $[e^2/(\in r_n^2)]$ as the electrostatic force between the impurity ion and the associated electron instead of just $[e^2/r_n]$. Then, the quantization condition for orbit angular momentum becomes

$$m^* r_n^2 \omega_n = n\hbar \qquad ...(1)$$

and the force equation is

$$\frac{e^2}{\in r_n^2} = \frac{m^* v_n^2}{r_n} = m^* r_n \omega_n^2 \qquad ...(2)$$

These two equations may be combined to give

$$r_n = \frac{n^2 \hbar^2 \in}{m^* e^2} \qquad ...(3)$$

the total energy

$E_n$ = (Potential energy) + (Kinetic energy)

$$= \frac{e^2}{\in r_n} + \frac{e^2}{2\in r_n} = -\frac{e^2}{2\in r_n}$$

or

$$E_n = \frac{m^* e^4}{2\in^2 n^2 \hbar^2} \qquad ...(4)$$

Application of Eqs. (3) and (4) to Ge and Si is, however, complicated because of the anisotropy of the effective mass of the conduction electrons. However, it enters only as the first power whereas, dielectric constant, as the second in the energy value of $m_e^* \approx 0.1\, m_e$ for electrons in Ge and $m_e^* \approx 0.2\, m_e$ for electrons in Si. The dielectric constant $\in$ is 156.8 for Ge and 11.7 for Si.

Now let the ionization energy come into picture. It is the amount of energy required to detach the electron completely from the normal atom. In other words, the ionization energy of the impurity atom in Ge is the energy of the ground state of electron, i.e.,

$$E_d = \frac{m_e{}^* e^4}{2\in^2 \hbar^2} = \left(\frac{m_e e^4}{2\hbar^2}\right)\frac{0.1}{\in^2} \qquad ...(5)$$

(using $m_e{}^* = 0.1\, m_e$)

$$= (\text{ionisation energy of the } H \text{ atom}) \times 0.1/\in^2$$

$$= \frac{13.6 \text{ eV} \times 0.1}{(15.8)^2} \approx 0.0055 \text{ eV} \qquad ...(6)$$

i.e., the ionisation energy of the impurity atom is smaller than that of the hydrogen atom by a factor $(2 \times 10^{-4})$. The corresponding result for Si is 0.04 eV. Calculation using the correct anisotropic mass tensor predict 0.0091 eV for Ge and 0.0298 eV for Si.

Thus, the energy states of the donor impurity atom lie within the forbidden energy gap specifically a few hundredth of an electron volt below the conduction band as shown in Fig. (11.11a) which depicts the band model for an $n$-type semiconductor.

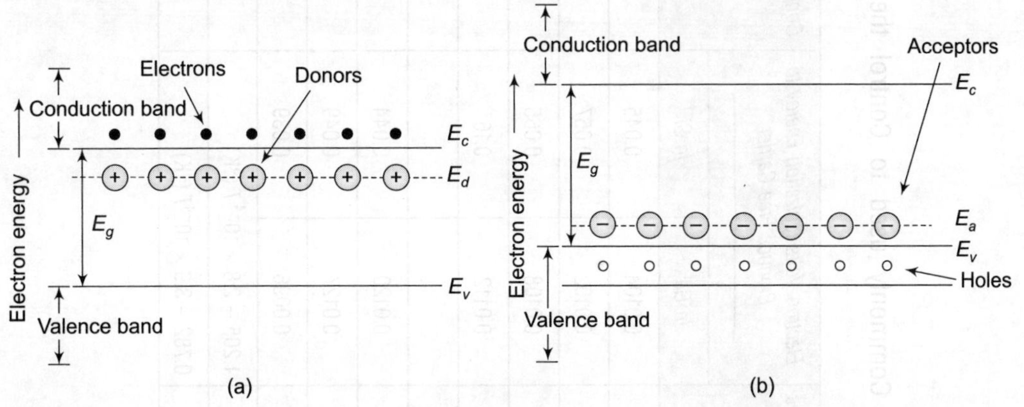


**Figure 11.11** Energy band diagram of extrinsic semiconductor (a) $n$-type (b) $p$-type. The impurity levels (or states) are as shown (i.e., $E_d$ and $E_a$ respectively)

The ionization energy for $n$-type semiconductors $E_1 = E_c - E_d$. Although there could be many excited states between $E_d$ and $E_c$, but these states are of little importance because at ordinary temperatures, almost all the donors are ionized and their electrons are set free into the conduction band.

Similarly, for a group III impurity addition to pure Ge or Si crystal, the impurity is shown to consist of a fixed ion of charge $(-e)$ to which a hole is loosely bound. Again the hole plus impurity ion constitute a H-like atom in a uniform dielectric medium and having associated impurity states. These acceptor states are located in the forbidden energy gap (shown by $Ea$), just above the valence band as shown in Fig. (11.11b). At finite temperatures, the holes acquire thermal energy sufficient enough to cause conduction.

*Hopes* are the *majority carriers* in $p$-type material, and the *minority carriers* in $n$-type material. Similarly, *conduction electrons* are the *majority carriers* in $n$-type material and the minority carriers in $p$-type material. These matters are summarized in Table 11.1 for several donor and acceptor impurities that are commonly added to germanium and silicon.

Consider valence-5 arsenic added a few ppm to a Si crystal. Each As-atom occupies one of the Si sites, bringing to it 5 electrons when only 4 are required for covalent bonds with nearest neighbours. The Si-crystal is a medium of high ($\approx 12$) dielectric constant; partly as a result, the $5^{th}$ electron of the As atom has a binding energy of only 0.049 eV. By virtue of thermal excitation, this electron is rarely held by the impurity atom and is in the conduction band most of the time, making Si $n$-type and the conduction is primarily by electron from the donors (As). The donor atoms introduce isolated donor levels near the bottom of the conduction band.

**Table 11.1** Some Properties of Silicon and Germanium, and Impurities Commonly used to Control their Electron Conductivities

| *Element* | *Number of Valence Electrons* | *Atomic Number* | *Radius in Crystal (Angstroms)* | *Function* | *Majority Free Carrier* | *Electron Volts Ionization Energy to Produce Free Carrier* | | *Conductivity Type Created* |
|---|---|---|---|---|---|---|---|---|
| | | | | | | *In Ge*[†] | *In Si*[††] | |
| Boron (B) | 3 | 5 | 0.81 | Acceptor | Hole | 0.0104 | 0.045 | *p*-type |
| Aluminum (Al) | 3 | 13 | 1.26 | Acceptor | Hole | 0.0102 | 0.057 | *p*-type |
| Gallium (Ga) | 3 | 31 | 1.26 | Acceptor | Hole | 0.0108 | 0.065 | *p*-type |
| Indium (In) | 3 | 49 | 1.44 | Acceptor | Hole | 0.0112 | 0.16 | *p*-type |
| Tin (Sn) | 4 | 50 | 1.40 | Neutral Solder | ... | ... | ... | ... |
| Phosphorus (P) | 5 | 15 | 1.13 | Donor | Electron | 0.0120 | 0.044 | *n*-type |
| Arsenic (As) | 5 | 33 | 1.18 | Donor | Electron | 0.0127 | 0.049 | *n*-type |
| Antimoney (Sb) | 5 | 51 | 1.36 | Donor | Electron | 0.0096 | 0.039 | *n*-type |
| Silicon (Si) | 4 | 14 | 1.17 | Host | Equal | $1.205 - 2.8 \times 10^{-4} T$ (°K) | | Intrinsic |
| Germanium (Ge) | 4 | 32 | 1.22 | Host | Equal | $0.782 - 3.9 \times 10^{-4} T$ (°K) | | Intrinsic |

[†] R. A. Smith, *Semiconductors*, Cambridge University Press, 1959.
[††] *E. M. Conwell, Proc. IRE,* **46** (June 1958), pp 1281–1300.

When a semiconductor is doped by adding a few ppm of a valence-3 element such as indium (In), the impurity atom brings 3 electrons but occupied a site calling for four covalent bonds. Such an atom can readily capture an electron, thereby establishing the fourth band. The captured electron is not available for conduction but there remains a hole in valance band from which the electron was excited, making the semiconductor *p*-type. Such acceptor (In) introduce isolated energy levels in the forbidden gap about 0.01 eV (in Ge) above the top of the valence band. Figure 11.1 shows properties of Si, Ge and impurities commonly used for doping.

Table 11.2 shows the energy of levels introduced in Ge and Si by light doping of common donor and acceptor atoms.

**Table 11.2** Levels Introduced by Light Doping of a Few Parts per Million (ppm)

| *Host crystal* | *Donor ionization energy ($E_c - E_d$) in eV* | | | *Acceptor ionization energy ($E_a - E_v$) in eV* | | |
|---|---|---|---|---|---|---|
| | *P* | *As* | *Sb* | *B* | *Ge* | *In* |
| Ge | 0.012 | 0.0127 | 0.0096 | 0.0104 | 0.0108 | 0.0112 |
| Si | 0.045 | 0.050 | 0.039 | 0.045 | 0.065 | 0.160 |

## 11.10 IMPURITY LEVELS (SHALLOW AND DEEP LEVELS)

In an energy band scheme, dopants are characterized by the discrete energy levels which they introduce. The energy levels of *shallow donors* are located below end close to the bottom of a conduction band. The energy levels of *shallow acceptors* are above and close to the top of a valence band. This means that a very small energy (in many cases less than a thermal energy $k_BT$) is needed to ionize a shallow impurity and supply an electron from a shallow donor into a conduction band, or a hole from a shallow acceptor into a valence band.

Table 11.2 shows the energy of levels introduced in Ge and Si by light doping of common donor and acceptor atoms.

It is to be noted that only a few impurities behave as shallow donors or acceptors even at room temperature ($k_BT/q \cong 0.02584$ eV). Some impurity atoms, *deep donors* and *deep acceptors*, create "deep" impurity levels, which are separated from the conduction or valence band edges by more than several $k_BT/q$ even at room temperature. As previously described, certain impurity atoms (such as Si in GaAs) may behave either as donors or acceptors, depending on which site (Ga or As) they occupy. Such impurity are said to be amphoteric and their behaviour is quite complicated.

## 11.11 COMPOUND SEMICONDUCTORS

For Si, Ge, Ga and As, – elements that are particularly important for semiconductor electronics, the valence configurations are given by

$$\text{Si } (z = 14) : \text{core} + 3s^2\, 3p^2$$

$$\text{Ge } (z = 32) : \text{core} + 4s^2\, 4p^2$$

$$\text{Ga } (z = 31) : \text{core} + 4s^2\ 4p^1$$
$$\text{As } (z = 33) : \text{core} + 4s^2\ 4p^3$$

Here "core" means all inner shells. The first digit in the second term is the principal quantum number of the outer (valence) shell–3 for Si, 4 for Ge, Ga and As. Superscripts denote the number of electrons in the subshells.

We can form semiconductor compounds by combining elements of Column III of the Periodic Table (having 3 valence electrons: two *s* electrons and one *p* electron) with electrons from Column V (having 5 valence electrons, two *s* electrons and three *p* electrons). Some of the examples of III-V compounds are GaAs, InAs, InP, AeAs, GaP, InSb, AlN, GaSb and GaN. Elements from the second and sixth columns of the Periodic Table (having two and 6 valence electrons respectively) can also be combined to from II-VI compound semiconductors. CdS, ZnS, CdTe and CdSe are the examples. Moreover, many III-V and II-VI compound semiconductors may form solid-solutions, such as $Al_x$, $Ga_{1-x}$ As, where $x$ is a molar fraction of Al. Such materials are called *ternary compounds.* By varying $x$ from 0 to 1, one can change the properties of the compound (from those of GaAs to those of AlAs). This particular compound is especially important because the lattice constants of GaAs and AlAs are very close (5.6533 Å and 5.6605 Å at 300 K for GaAs and AlAs respectively). That is why AlGaAs can easily be grown on GaAs (substantiate), forming a *heterostructure*. Heterostructures have numerous important applications in novel semiconductor devices and circuits.

In recent years a number of compound semiconductors have found applications in various devices. The most important among them are the intermetallic III-V and II-VI compounds. Some IV-VI compounds also exhibit semiconducting properties, but most of them have a small band gap that limits their use to infrared detectors and lasers.

**III-V Compound Semiconductors** These semiconductors are compounds formed from the elements in group IIIA (all metals) with those in group VA. Ignoring the compounds of thallium and bismuth, we present the remaining sixteen:

| *Group III A* ↓ | *Group V A* → | | | |
|---|---|---|---|---|
| | N | P | As | Sb |
| B | BN | BP | Bas | BSb |
| Al | AlN | AlP | AlAs | AlSb |
| Ga | GaN | **GaP** | **GaAs** | **GaSb** |
| In | InN | **InAs** | **InAs** | **InSb** |

Not all these compounds are potential semiconductors. Boron and nitrogen compounds, for example, have been included only for the sake of completeness. Aluminum compounds are not very stable and usually disintegrate with time. The six most important semiconductors are printed in bold type.

Most of the III-V semiconductors have the zinc blende crystal structure. Eight valence electrons between a pair of nearest atoms, so that, on an average, each atom has four

**Table 11.3** Some Properites of Group IV Elements

| *Element* | *Atomic Number* | *Band Gap $E_g$ at 300 K (eV)* |
|---|---|---|
| Carbon (C) | 6 | 5.30 |
| Silicon (Si) | 14 | 1.12 |
| Germanium (Ge) | 32 | 0.67 |
| Tin ($\alpha$-Sn) | 50 | 0.08 |

valence electrons. This suggests that the bonding has a covalent character, and to the first approximation, the cohesion between the atoms is homopolar. Therefore, we would expect the properties of these compounds to be similar to those of corresponding group IV elements. However, since the elements of group III are more electropositive, and those of group V are more electronegative than the group IV elements, the bonding in III-V compounds has a partial ionic character as well.

"Because of their predominantly character, the III-V semiconductors form a link between group IV elements and II-VI compounds. Therefore, most of their characteristics can be deduced from those of neighbouring materials. To apply these considerations, Table 11.4 is drawn by assigning isoelectronic III-V compounds to their group IV elements. Here, the average total number of electrons (core plus valence electrons) per atom in a row is nearly the same, and this is what is meant by the term *isoelectronic*. It is seen that certain material properties change gradually as one moves from one end to the other in horizontal and vertical directions. These characteristics were predicted by H.J. Welker even before many of the compounds were obtained in the laboratory. They can be summarized as follows:

"**Vertical Direction** The cohesive energy and the energy gap ($E_g$) decrease with increasing atomic number. This is because in an atom with a large number of electrons, the core electrons repel each other with a greater force. This results in weak cohesion of atoms and a loose binding of valence electrons. Decreasing binding energy results in wider allowed bands and narrower band gaps."

**Table 11.4** Variation in the Properties of Isoelectronic Group IV and III-V Compounds

| *Group IV semiconductors* | *III-V Compounds* |
|---|---|
| C | BN |
| SiC | BP AlN |
| Si | BAs AlP GaN |
| SiGe | Bsb AlAs GaP InN |
| Ge | AlSb GaAs InP |
| GeSn | GaSb InAs |
| $\alpha$-Sn | InSb |
| Cohesive energy | |
| Band gap | |

"**Horizontal Direction** Both the cohesive energy and the energy gap increase in horizontal transition from elemental to compound semiconductors. This happens because of increased cohesion due to the ionic contribution.

The technical importance of III-V semiconductors is that they provide a wider choice of band gap than do elemental semiconductors. Some of these compounds have energy gaps corresponding to the visible region of spectrum and have found applications in light-emitting diodes. Others like InSb and InAs, are used in galvanomagnetic devices."

"The III-V compounds have a general formula $A^{III}B^{V}$. However, several of these compounds have identical structures and similar lattice constants. Such compounds exhibit compounds soild solubility in each other and structures of the from $A_x^{III}\ B_{1-x}^{III}\ C^{V}$ are possible, with $x$ ranging continuously from 0 to 1. The properties of such compounds vary smoothly with composition. Thus, it is possible to grow crystal of a compound like $Al_xGa_{1-x}As$. Similarly, the growth of crystal having two different group V elements like $GaP_yAs_{l-y}$ is also possible."

**II-VI Compounds Semiconductors** These semiconductors are formed by combining the elements in groups IIB and VIA in the Periodic Table. A list of all the II-VI compounds is given below:

| *Group III B* ↓ | *Group VI A* → | | | |
|---|---|---|---|---|
| | O | S | Se | Te |
| Zn | ZnO | **ZnS** | **ZnSe** | **ZnTe** |
| Cd | CdO | **CdS** | **CdSe** | **CdTe** |
| Hg | HgO | HgS | HgSe | HgTe |

Among the oxides, ZnO has found use in both electrophotography as a photosensitive material and in paints and varnishes. Little information is available about the semiconducting properties of CdO and HgO. The remaining three group VI elements are called *chalcogenides*. Among the chalcogenides of mercury both HgSe and HgTe are semimetals, whereas HgS is a semiconductor. The remaining six compounds printed in bold letters form a group of potential II-VI semiconductors.

The crystal structure of these compounds is rather complex. CdS and CdSe crystallize in the hcp Wurtzite structure, and ZnTe and CdTe in the zinc blende structure, whereas ZnS and ZnSe can exist in both these forms. The bonding in these compounds is a mixture of covalent and ionic types. This is because the average number of electrons per atom is still 4, but group VI atoms are considerably more electronegative than group II atoms, and this introduces iconicity. The ionic character varies considerably over the whole range and increases as the atomic size decreases. Thus, while ZnS is predominantly ionic, the bonding is nearly covalent in HgTe. The ionic character has the effect of binding the valence electrons rather tightly to the lattice atoms. Thus, the band gaps of these compounds are larger than those of the covalent semiconductors of comparable atomic weights, although their hardness does not match that of covalent semiconductors. Because of their large band gaps, these materials are considered potential source of electroluminescent devices.

Like III-V compounds, ternary compounds of the form $A^{II}B_x^{VI}C_{1-x}^{VI}$ can also be obtained from many of these semiconductors. Thus, it has been possible to grow crystal of $ZnS_xSe_{1-x}$ with $x$ ranging from 0 to 1. Here again the properties of the resulting compounds vary gradually with the fraction $x_{lt}$.

## 11.12 THE CONNOTATION PROCESS

The most important thing about semiconductors is their ability to conduct electric current. The current will depend on both the number of charges free to move and the speeds at which they move. The current is a result of the motions of a great many individual charges and not the "component" current produced by the motion of any one.

We have to think of the current carriers (electrons and holes) as particles with a certain effective mass and to describe the electronic behaviours in these terms. There will be collisions (scattering) of the carriers with crystal atoms sitting in their equilibrium positions. It is to be recalled that all the effects of the *periodic lattice* potentials in the crystal are buried entirely in the effective mass parameter, $m^*$, so that only departures from periodicity act as deflecting forces, provided the irregularities should occur over distances large compared to a lattice spacing. There are *four types* of such disturbances which act to deflect the conduction electrons or holes in their motion within the lattice.

(i) *Lattice Scattering*. At normal temperatures, by far, the most important deflecting or "scattering" events arise from distortions of the crystal associated with thermal vibrations. Atoms are either squeezed together over a region, or pulled apart over a region, as compared to their normal spacing in the ideal periodic lattice. Electric potential alterations caused thereby correspond to "applied" electric fields which alter the direction and magnitudes of the carrier momentum (called *thermal* or lattice *scattering*).

(ii) *Scattering from Charged Impurity Atoms*. This is caused by an ionized donor or acceptor. We know that even for an electron *bound to* (i.e., orbiting about) such an impurity, the radius is quite a few lattice spacings; so it is clear that appreciable deflection of a moving charge carrier will occur at distance much greater than this value. The extent depends on the temperature and the impurity concentration. The charged impurities of either sign will deflect carriers of either sign by attraction or repulsion as appropriate.

(iii) *Scattering from an Electrically Neutral Impurity Atom*: An electrically neutral atom can also scatter a carrier, when the carrier gets close enough to the impurity to detect its positive core separately from its orbiting outer electrons, or when the impurity atom produces a local lattice strain and a corresponding bump of electrical potential. This kind of scattering is weak and usually obscured by the other mechanisms.

(iv) Finally, we must consider the possibility of conduction electrons and holes deflecting each other in passing. The situation is similar to that of scattering by an oppositely charged impurity, except that both particles deflect about their common centre of mass. This scattering is rarely of major importance in electrical conduction.

We should have also concerned ourselves with possible deflections produced by encounters between two conduction electrons or two holes.

The current is

$$\vec{J} = \sum_i q\,\vec{v}_i = \frac{q}{m^*}\sum_i p_i = \frac{q}{m^*}\,\vec{p}$$

i.e., the current is proportional to the momentum of the *entire group* for carriers of a single type. Whereas the path of any individual carrier is much altered by collisions with others of its own type, the net effect upon the measured current from all the carriers of that type is zero for their scattering process.

The foregoing discussion is for the behaviour of carriers in the absence of any field.

## 11.13 DRIFT AND MOBILITY

The faster the carrier moves, the more collisions it makes *per unit time*. With the nature of each collision substantially *unaltered*, the resulting (opposing) force increases with speed which leads to *terminal velocities*. Thus, the field produces a *drift speed* rather than a steady acceleration (if the field is not too large).

It is important to realize that it is not primarily the loss of *energy* in each collision that limits the current, rather the random changes in momentum *direction*. The energy loss *per collision* is very small, essentially because the carrier mass is smaller compared to the mass of the crystal atoms or the energy loss per event is very small as compared to random thermal energy.

To carry out the analysis explicitly, denoted by $\mathbf{v_i}$ the *average incremental* velocity imparted by an applied force $\mathbf{F}$ *per particle* to the particles in a group on ($E$) all of whose energies are in the small range between $E$ and $E + dE$. This group is small enough to permit $dE$ to be small, *but large anough so that random effects cancel out*. As a result of the increased number of collisions per unit time, the incremental velocity gives rise to an effective "resisting force" on the group proportional to $\mathbf{v_i}$; say $\alpha(E)\,\mathbf{v_i}\,\delta n(E)$, where $\alpha$ may depend on the kinetic energy $E$ for the reasons discussed above. Thus, the dynamical equation of motion relating the applied force and the incremental velocity *of the whole group* is

$$\mathbf{F}\,\delta n(E) = m^*\,\delta n(E)\,\frac{d\mathbf{v_i}}{dt} + \alpha(E)\,\mathbf{v_i}\,\delta n(E) \qquad \text{...(1a)}$$

Note that $\mathbf{F}$ includes only forces over and above the periodic forces of the crystal lattice, the latter being included already in the effective mass, $m^*$. Also, all random velocities cancel out over the group $\delta n(E)$; that is why the equation of motion relates only to the incremental velocity ($\mathbf{v_i}$) of the group.

Dividing through Eq. (1a) by the factor $m^*\,\delta n(E)$ gives

$$\frac{\mathbf{F}}{m^*} = \frac{d\mathbf{v_i}}{dt} + \frac{\mathbf{v_i}}{\tau_c(E)} \qquad \text{...(1b)}$$

In which $\alpha(E)/m^* \equiv 1/\tau_c(E)$. We can show that $\tau_c(E)$ has the significance of a "relaxation time" for energy group $\delta n(E)$. That is, suppose no force is applied ($\mathbf{F} = 0$), but at $t = 0$ there was an initial disturbance $\mathbf{v_i}(0)$ of the group from its random equilibrium state. Then Eq. (1b) can be solved to yield

$$\mathbf{v_i} = \mathbf{v_i}(0)\, e^{-t/\tau_c(E)}; \quad t \geq 0 \qquad ...(2a)$$

showing that the group of $\delta n(E)$ carriers "decays" or "relaxes" back to its thermal equilibrium state ($\mathbf{v_i} = 0$) with a time constant, $\tau_c(E)$. What is occurring here is that all the members of the group $\delta n(E)$ with kinetic energy near $E$ had an additional small velocity $\mathbf{v_i}(0)$ at $t = 0$; but with the passage of time an increasing number of these have their energy changed by collisions to a value outside that of the group, and they are replaced in the group by new members that have kinetic energy near $E$, but no incremental velocity component at all. Thus, the amount of *organized* velocity *of the group* decreases with time. In another way of saying it, the number of particles in the group stays the same, so the *current* carried by the group in the direction of $\mathbf{v}_i$ simply decays with time. It is replaced by random current. The same reasoning applies to the net momentum of the group as to the current.

Next, for our introductory discussion of the effect of applied forces, let us consider the simple case in which the relaxation time ($\tau_c$) is the *same* for all kinetic energy groups; that is, $\tau_c$ is *independent* of $E$. Then, if the force $\mathbf{F}$ on the negatively charged conduction electrons is produced by an electric field $\mathbf{E}$, we know $\mathbf{F} = -q\mathbf{E}$. In the steady state, $d\mathbf{v}_i/dt$ and the solution to Eq. (1b) is just a steady *drift velocity* $\mathbf{v}_d$, independent of $E$

$$\mathbf{v}_d = \frac{-q\tau_c}{m_e^*}\mathbf{E} = -\mu_e\mathbf{E} \qquad ...(2b)$$

If the velocity is given in cm/sec, and field in volts/cm, the proportionality constant ($\mu_e$) has dimensions of cm$^2$/volt sec. The quantity ($\mu_e$) is called the *mobility* (for conduction electrons) and is by convention always defined to be a positive number.

$$\mu_e = \frac{q\tau_c}{m_e^*} \qquad ...(3)$$

Clearly the mobility is the incremental average speed per particle, per unit electric field.

In this simple case, the relaxation time for *all* the electrons is the same as that of any group, namely, $\tau_c$. Sometimes the time ($\tau_c$) *in the mobility expression* Eq. (3) is called the *mean free time*. Such an idea comes from thinking about the collisions as if they were sudden discrete events, separated by absolutely "free times" between them. If the collisions occur properly at random in this kind of a model, we find a result like Eq. (3), in which the time that appears is just the average time between the idealized collisions. Inasmuch as we have seen that collisions do not really occur this way, all we shall use from such models is the name: mean free time.

Collisions of electrons with lattice atoms which are out of their equilibrium positions, because of thermal vibration, would be expected to happen more often at higher temperatures as the atoms develop larger amplitudes of vibration. Thus, the mobility would be expected to *decrease* with increasing temperature when thermal vibrations dominate the scattering process.

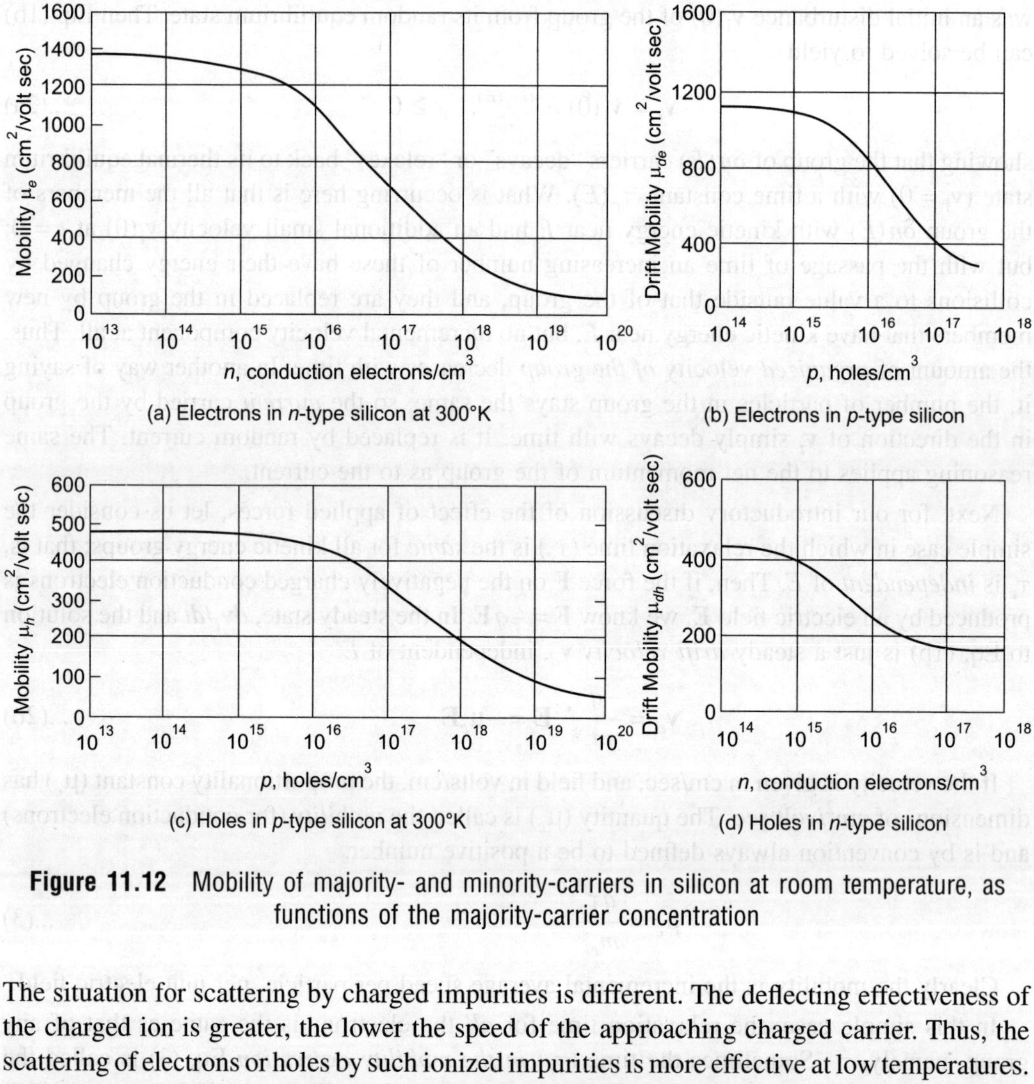


**Figure 11.12** Mobility of majority- and minority-carriers in silicon at room temperature, as functions of the majority-carrier concentration

The situation for scattering by charged impurities is different. The deflecting effectiveness of the charged ion is greater, the lower the speed of the approaching charged carrier. Thus, the scattering of electrons or holes by such ionized impurities is more effective at low temperatures. This means that mobility tends to have a broad maximum at intermediate temperatures. On the other hand, at low *enough* temperatures the donors and acceptors may not be ionized, so their scattering effect may eventually drop again. Some data confirming the aforementioned trends of the dependence of mobility upon temperature appear in Table 11.5. The effect of charged impurities on mobility at normal temperatures is shown more clearly, for both majority and minority carriers, in Fig. 11.12. At room temperature and above, collisions with the vibrating atoms usually dominate, although the data shows that the impurities do interfere more as their concentration is increased. We shall soon understand better why the majority-carrier concentration ($n$ and $p$ for $n$- and $p$-type material, respectively) is a relatively direct measure of the impurity concentration, except at low values of $n$ and $p$.

**Table 11.5** Mobility of Carriers in Very Pure Germanium and Silicon.[†]
*(Thermal vibration of the lattice should be by far the dominant scattering process)*

| *Material* | *Mobility at 300°K ($cm_2$/volt-sec)* | *Approximate temperature dependence of mobility* |
|---|---|---|
| Ge | | |
| Free electron | 3900 | $4.9 \times 10^7\, T^{-1.66}$ (100–300°K) |
| Holes | 1900 | $1.05 \times 10^9\, T^{-2.33}$ (125–300°K) |
| Si | | |
| Free electrons | 1350 | $2.1 \times 10^9\, T^{-2.5}$ (160–400°K) |
| Holes | 480 | $4.9 \times 10^9\, T^{-2.7}$ (150–400°K) |

[†] E. M. Conwell, *Proc.* IRE, **46** (June 1958), pp. 1281–1300.

### 11.13.1 Effect of Temperature on Mobility

The two basic types of scattering mechanisms that influence electron and hole mobility are lattice scattering and impurity scattering. Lattice scattering is due to vibration of the lattice (i.e., phonons) and impurity scattering is due to scattering from ionized impurity atoms (donors and acceptors). Other imperfections such as dislocation also contribute to scattering but to a lesser degree in most useful materials.

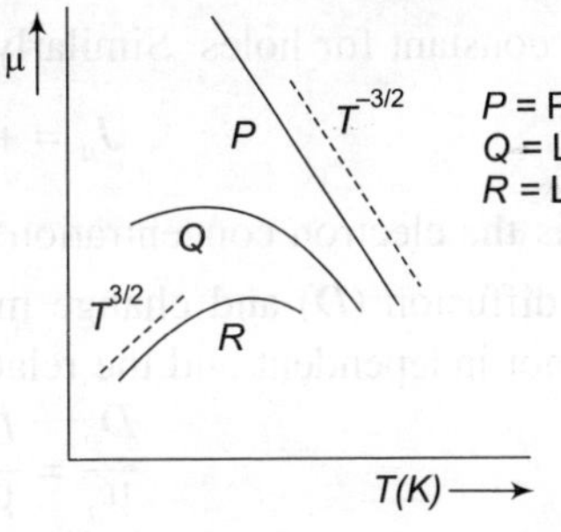


**Figure 11.13** Temperature dependence of mobility for pure and impure semiconductors

The frequency of lattice scattering events increases as the temperature increases (due to increased thermal agitations). Therefore, we expect the mobility to decrease as the sample is heated. On the other hand, at low temperatures, lattice atoms are less agitated, lattice scattering is less important. However, at low temperature, the thermal motion of carriers is also slower.

Since a slowly moving carrier is likely to be scattered more strongly by an interaction with an impurity atom than is a carrier with increased momentum, impurity scattering events cause a decrease in mobility with decreasing temperature. The lattice mobility ($\mu_L$) and ion mobility ($\mu_I$) are theoretically found to be

$$\mu_L = AT^{-3/2} \text{ and } \mu_I = BT^{3/2} \qquad ...(1)$$

where $A$ and $B$ are constants for given materials. Resultant resistivity

$$\rho = \frac{1}{\sigma} = \rho_L + \rho_I \qquad (\sigma = \text{conductivity})$$

or

$$\frac{1}{\sigma} = \frac{1}{\sigma_L} + \frac{1}{\sigma_I}$$

or

$$\frac{1}{ne\mu} = \frac{1}{ne\mu_L} + \frac{1}{ne\mu_I}$$

$$\frac{1}{\mu} = \frac{1}{\mu_L} + \frac{1}{\mu_I} = \frac{1}{AT^{-3/2}} + \frac{1}{BT^{3/2}} \qquad ...(2)$$

$$= \frac{T^{3/2}}{A} + \frac{T^{-3/2}}{B} \quad ...(3)$$

## 11.14 DIFFUSION OF CARRIERS IN A SEMICONDUCTOR

When an electric field is applied across a semiconductor, there is a drift of carriers (electrons as well as holes) within the semiconductor. In addition to this *drift* – motion, there is one more mechanism of charge transport, known as *diffusion* which is due to the concentration gradient. (This diffusion is not ordinarily encountered in metals due to absence of concentration gradient.) We will now consider the diffusion process.

There are two types of carriers in a semiconductor, viz., electrons and holes. Suppose, we have a semiconductor having a non-uniform concentration of holes, and the concentration $p$ of holes varies with distance $x$ due to the presence of a concentration gradient $dp/dx$ in the hole density. Then, in a given time interval, more holes will cross a given cross section from the side of greater concentration to the side of smaller concentration. The diffusion hole current density $J_p$ A/m$^2$ is given by $J_p = -e D_p \dfrac{dp}{dx}$, (hole diffusion current) where $D_p$ (m$^2$/s) is the diffusion constant for holes. Similarly, for electrons

$$J_n = +e D_n \frac{dn}{dx} \quad \text{(electron diffusion current)}$$

where $n$ is the electron concentration and $D_n$, the diffusion constant for the electrons.

Since diffusion ($D$) and change mobility ($\mu$) are statistical thermodynamic phenomena, they are not independent and the relation is

$$\frac{D_p}{\mu_p} = \frac{D_n}{\mu_n} \equiv V_T.$$

This is known as the Einstein relation.

Here $$V_T = \frac{k_B T}{e} = \frac{T}{11{,}600}$$

known as the *thermal voltage*.

At room temperature (300° K):

$$\mu = 39\, D$$

and for silicon (Si):

$$D_n = 34 \text{ cm}^2/\text{s}$$

$$D_p = 13 \text{ cm}^2/\text{s}.$$

## 11.15 DRIFT CURRENT AND HOT CARRIER EFFECT (SATURATION OF DRIFT VELOCITY)

Current density due to drift in presence of an electric field (ε) is given as

$$J = \sigma E \quad \text{(Ohm's law)} \quad ...(1)$$

where the conductivity, $\sigma = en\mu_e$

$e$ is the electronic charge, $n$ the electron density and $\mu_e$, the electron mobility, given by

$$\frac{ne^2\tau}{m_e} \equiv \sigma, \quad \mu_e = \frac{e\tau}{m_e} \qquad ...(2)$$

$\tau$ represents the mean time between two consecutive scattering events and is called the *mean free time*.

In the above equations, it has been assumed that the draft current is proportional to the electric field and that means Ohm's law is valid in the drift process, i.e., $\sigma$ is not a function of the field. This is valid over a wide range of $\varepsilon$ values, but large electric fields (> $10^3$ V/cm) can cause the drift velocity and therefore the current

$$J = -env_d \qquad ...(3)$$

to exhibit a sub-linear dependence on the field $\varepsilon$. This dependence of $\sigma$ upon the field $\varepsilon$ is an example of a *hot carrier effect*, which implies that the carrier drift velocity ($v_d$) is comparable to the thermal velocity, $v_{th}$. In many cases, an upper limit is reached for the carrier drift velocity in a high field. This limit occurs near the mean thermal velocity $\simeq 10^7$ cm/s, and represents the point at which added energy imparted by one field is transferred to the lattice rather than increasing the carrier velocity. The result of this *scattering limited velocity* is a fairly constant current at high field. This behaviour is typical of Si, Ge and some other semiconductors. However, there are other important effects in some other semiconductors, for example, there is a *decrease* in electron velocity at high field for GaAs (and certain other materials) which results in *negative conductivity*.

## 11.16 A RELATION BETWEEN THE EQUILIBRIUM CARRIER CONCENTRATIONS

Imagine a perfectly pure, intrinsic semiconductor without any crystalline faults conduction electrons and hole are produced in pair at a rate (number per unit volume, per unit time) that depends in part on the semiconductor material under consideration, i.e., the energy and momentum to produce a pair and on temperature $T$ at which the material is in thermal equilibrium.

In the state of thermal equilibrium, the concentration of carriers in any sample are not changing with time on an average. So, on the average, the pair generation rate must balance the pair recombination rate everywhere,

$$G_{\text{ith}}(T) = R_i\,(n_o, p_o, T) \qquad ...(1)$$

But regarded as a function of $n_o$ and $p_o$, at fixed $T$ (i.e., for different samples of the same basic ideal semiconductor, characterized by different small amounts of donor and/or acceptor impurities, but all at the same temperature), we can make a Taylor expansion of $R_i$, in two variables

$$R_i\,(n_o, p_o, T) = a_i(T) + b_i(T)n_o + c_i(T)p_o + r_i(T)\,n_o\,p_o + e_i(T)n_o^2\,p_o + f_i(T)\,n_o p_o{}^2 + ... \qquad ...(2)$$

Here, however, it is clear that

$$a_i(T) = b_i(T) = c_i(T) \equiv 0,$$

because there can be no *pair-recombination* rate in a sample in which only *one* carrier type is present; and certainly none in a sample in which *neither* is present. So, for the small carrier concentration which are of greater interest in electronic devices, the leading non-zero term of Eq. (2) is dominant and Eq. (1) becomes

$$G_{\text{ith}}(T) = r_i(T)\, p_o n_o \qquad ...(3)$$

or

$$p_o n_o = \frac{G_{\text{ith}}(T)}{r_i(T)} \qquad ...(4)$$

of course, Eq. (4) holds for intrinsic material as a special case, and there $p_o = n_o$. If we denote by $n_i$ the concentration of either conduction electrons or holes in intrinsic material, it therefore follows from Eq. (4) that

$$\frac{G_{\text{ith}}(T)}{r_i(T)} = n_i^2 \qquad ...(5)$$

from which we draw the relatively obvious conclusion that, for a given ideal semiconductor, $n_i$ is a function of $T$ only, $n_i(T)$, and that in intrinsic material both carrier concentrations are identical functions of temperature

$$p_o = n_o = n_i(T).$$

Accordingly, Eq. (5) permits us to write Eq. (4) in the form

$$p_o n_o = n_i^2(T) \qquad ...(6)$$

The result is that an effort to "sprinkle in" conduction electrons by adding donors to a semiconductor, thus making $n_o > n_i$, has to result in *fewer* holes, $p_o < n_i$, in order to prevent the additional conduction electrons from recombining with holes faster than the unchanged thermal generation rate of pains will permit.

It is dear that addition of acceptors alone can be discussed in the same manner and will again lead to Eq. (6).

## 11.17 EQUILIBRIUM CARRIER CONCENTRATIONS IN 'DOPED' HOMOGENEOUS SEMICONDUCTOR

Given the intrinsic concentration ($n_i$) of conduction electrons or holes in a parent semiconductor (like Ge), at a temperature high enough so that any excess donors or acceptors in the material will be completely ionized, Eq. (6) furnishes *one* of the relations needed to determine the new hole and conduction-electron concentration in the presence of "doping" impurities.

For the special case of *homogeneous* material containing *both* donors and acceptors, an additional relation comes from the fact that in every volume of such material large enough to contain many atoms of all the types involved, every conduction electron must have come from either a broken bond or an ionized donor, and every hole must be accounted for by either an ionized acceptor or the broken bond left by one of the conduction electrons.

In presence of Na ionized donors and Na ionized acceptors, if we subtract $N_d$ from $n_o$, and also $N_a$ from $p_o$, we would be left with only electron hole pairs and the resulting number of conduction electrons and holes will be equal, i.e.,

$$\begin{aligned} &\text{(a) } n_o - N_a = p_o - N_a. \\ \text{or} \quad &\text{(b) } N_a - N_d = p_o - n_o \\ \text{or} \quad &\text{(c) } N_o - N_a = p_o + N_d \end{aligned} \qquad ...(7)$$

$$\Rightarrow \qquad q(p_o - n_s) = q(N_a - N_d) \qquad ...(8)$$

where $q$ is the electronic charge.

i.e., *net positive mobile charge* equals the *not negative immobile charge.* (Recall that ionized acceptors are negatively charged and ionized donors are positively charged.)

We now donote $N_d - N_a \equiv N$, the "excess" donor concentration. Therefore

$$N_o - p_o = N$$

or

$$N = N_d - N_a \qquad ...(9)$$

is, in effect, the "excess" carriers "sprinkled in" by the impurities to affect the equilibrium balance between pair generation and pair recombination. If $N$ is positive, it represents the *effective* concentration of donors sprinkling in conduction electrons. If $N$ is negative, acceptors are more than donors, and $N$ represents in magnitude the *effective* acceptor concentration capable of furnishing holes at ordinary temperatures.

Provided all the impurities are ionized, we can determine $p_o$ and $n_o$ if $n_i(T)$ is known, for the parent semiconductor.

It is clear from Eq. (9) that if $N \gg n_i$, $n_o \gg n_i$. Accordingly, from Eq. (6), we see that $p_o << n_i$. It follows from Eq. (9) again that $p_o$ is negligible there, and

$$n_o \approx N.$$

As long as the excess impurity is ionized and $N$ remains large compared to $n_i$, the conduction, electron concentration $n_o$ remains approximately constant with temperature. The hole concentration, however, is given approximately by Eq. (6) as

$$p_o = \left(\frac{n_i}{n_o}\right) n_i \approx \left(\frac{n_i}{N}\right) n_i \ll n_i \text{ if } N \gg n_i$$

which is small.

The condition $N \gg n_i(T)$ may at low enough temperature in a given $n$-type material, or for a large enough donor concentration at a given temperature.

Similarly in $p$-type material, for which $N < o$, the temperature and acceptor concentration may be such that $|N| >> n_i$. Then from Eq. (9) we find $p_o >> n_i$ and from Eq. (6) $n_o << n_d$. It then follows from Eq. (9) again that $n_o$ is negligible, and

$$p_o \approx |N|$$

virtually independent of temperature, and from Eq. (6) that

$$n_o \approx \frac{n_i^2}{N} = \left(\frac{n_i}{N}\right) n_i \ll n_i$$

which varies with temperature like $n_i^2(T)$.

Under the two circumstances descried above, in which $|N| >> n_i$, the semiconductor in question is said to be extrinsic in question is said to be *extrinsic* $n$-type or $p$-type (as the case may be). Its *majority*-carrier concentration is large and nearly equal to the excess impurity concentration $|N|$, virtually independent of temperature. The *minority* carrier concentration is small and varies [like $n_i^2(T)$] almost exponentially with temperature.

If we consider a given sample at various temperatures, the temperature range in which all the impurities are ionized and $|N| >> n_i$ is called the *extrinsic temperature range* (sometimes) also called the "saturation" or 'exhaustion' range.

Eventually, however, at high temperatures $n_i$ is bound to become comparable to $|N|$. The number of hole-electron pairs generated from broken bonds compete with the carriers supplied by the impurities. Both $p_o$ and $n_o$ become large compared to $|N|$ when $n_i(T) >> N$ and therefore $p_o \approx n_o \approx n_i(T)$. We often say that the semiconductor has entered the *intrinsic temperature range* under these conductions, although it is certainly no "purer" than it was before.

Finally, there is a very low temperature range in which the impurities can no longer remain ionized. The material tends to become an insulator.

## 11.18 CONDUCTIVITY

To understand the conduction process, consider a portion of a one carrier conductor (Fig. 11.14). First, let there be *no* applied field and assume that the conductor is in thermal equilibrium. Under these circumstances let there be $n$ conduction electrons per m$^3$ in the conductor, on the average. This number is fixed by the atomic and crystal structure, along with the neutrality requirement. A given macroscopic volume of the conductor (e.g., that shown between planes $A$ and $B$) contains many electrons at any instant, some of which are leaving it and all of which are moving about rapidly at random. A moment later, some electrons have left; but on the average just as many new ones have entered. Moreover, in the absence of any applied field, not only is the *net* time-average number of electrons entering or leaving the volume per unit time zero, but the *net* time-average number crossing any plane (such as $A$ or $B$) is *also* zero. Even further, if we observe only those electrons with thermal energy in the range between $E$ and $E + dE$, then for *these alone* there would be no net rate of increase in the number within the volume considered, and no net flow across any plane (principle of detailed valence).

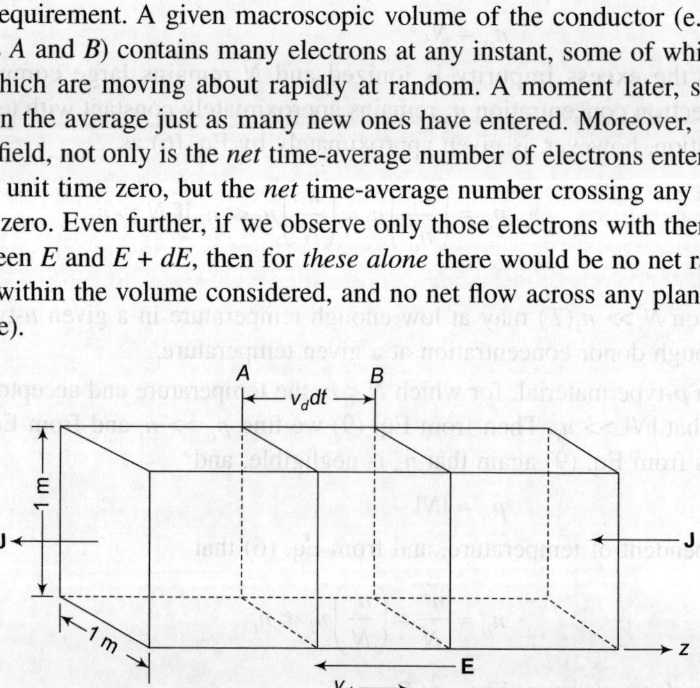


**Figure 11.14** The electronic conduction process

Next, let an electric field **E** be applied in the $z$ direction. We know that this field will cause a small common drift velocity $\mathbf{v}_d(E)$ for *all* the electrons with energy between $E$ and $E + dE$. Suppose there are $\delta n(E)$ such electrons per unit volume, so that

$$n = \int_{\text{all } E} \delta n(E) \qquad \text{...(1)}$$

Then if we again consider a macroscopic volume of the conductor, like that between planes $A$ and $B$ of Fig. 11.14, the net number of electrons in it which have energies near $E$ will still not change with time, on an average. There *is* now, however, an average drift of such electrons *into* the volume across plane $A$, and equally *out* of it across plane $B$, because of the organized drift velocity, $\mathbf{v}_d$.

To determine the amount of charge crossing a plane per unit time, let the distance between planes $A$ and $B$ be $dz = v_d\, dt$. Then the average number of electrons which have energy near $E$ and are between the planes is just $\delta n(E) v_d dt$. Because all of these have the same drift component, $\mathbf{v}_d$, all of them on an average cross plane $B$ from left to right in time $dt$, all other random motions across plane $B$ being cancelled out on the average by the thermal velocities of particles on both sides of the plane.

Noting that each electron carries a charge $-q$, and that the cross-sectional area in Fig. 11.14 is $1\text{m}^2$, we recognize that the charge per unit time crossing plane $B$ has the dimensions of amperes per square metre. It is referred to as a *current density*, with symbol **J**. (Actually, in the general case, plane $B$ is defined as that one which is oriented to intercept the greatest flow; the current density **J** is then a vector normal to that plane, of magnitude equal to the flow rate.) Thus, the group $\delta n(E)$ of conduction electrons considered here contributes to the current density an amount

$$\delta \mathbf{J} = \frac{-q\,\mathbf{v}_d(E)\,\delta n(E)\,dt}{dt} = -q\,\mathbf{v}_d(E)\;\delta n(E) \qquad \text{...(2)}$$

The total current density produced by all energy groups is obtained by adding (integrating) all the separate contributions

$$\mathbf{J} = -\int_{\text{all } E} q\mathbf{v}_d(E)\,\delta n(E) \qquad \text{...(3)}$$

in which we know that in the steady state $[(d\mathbf{v}_i/dt) = 0)]$ there is a steady drift velocity $\mathbf{v}_d(E)$:

$$\mathbf{v}_d(E) = -\frac{q\tau_c(E)}{m_e^*}\mathbf{E} \qquad \text{...(4)}$$

So

$$\mathbf{J} = \frac{q^2}{m_e^*}\left[\int_{\text{all } E} \tau_c(E)\delta n(E)\right]\mathbf{E} \qquad \text{...(5)}$$

Let us now define and use the average relaxation time $<\tau_c>$ for all the conduction electrons, of which we have said there are $n$ per unit volume (Eq. 1).

$$<\tau_c> \equiv \left[ \frac{\int_{\text{all } E} \tau_c(E)\,\delta n(E)}{\int_{\text{all } E} \delta n(E)} \right] = \frac{1}{n} \int_{\text{all } E} \tau_c(E)\,\delta n(E) \qquad \text{...(6)}$$

On this basis Eq. (5) becomes

$$\mathbf{J} = q\left(\frac{q<\tau_c>}{m_e^*}\right) n\mathbf{E} \equiv q\mu_e n\mathbf{E} \qquad \text{...(7)}$$

where now we have put

$$\mu_e \equiv \frac{q<\tau_c>}{m_e^*} \qquad \text{...(8)}$$

In other words, the mobility is related to the *mean* relaxation time $<\tau_c>$.

Notice that Eq. (7) may be written

$$\mathbf{J} = (q\mu_e n)\,\mathbf{E} = \sigma_e \mathbf{E} \qquad \text{...(9)}$$

which is precisely Ohm's law, with $\sigma_e$ the conductivity of the material (mhos per m) and $E$ the voltage drop per meter!

In a semiconductor there are two kinds of mobile carriers present at the same time. Because they are oppositely charged, they move in opposite directions in an electric field. But, again because they are oppositely charged, they carry current in the *same* direction. The conduction electrons and holes have different effective masses, and they may be scattered differently. Using subscripts $e$ and $h$ for conduction electrons and holes, respectively, the addition of currents for the two carriers yields

$$\mathbf{J} = q(\mu_e n + \mu_h p)\,\mathbf{E} = (\sigma_e + \sigma_h)\,\mathbf{E} = \sigma\mathbf{E} \qquad \text{...(10)}$$

where evidently

$$\sigma = \sigma_e + \sigma_h = q(\mu_e n + \mu_h p) \qquad \text{...(11)}$$

## 11.19 CARRIER FLOW BY DRIFT AND DIFFUSION

Diffusion is a manifestation of random thermal motion of particles and shows up as a net current of particles from regions of higher concentration to neighbouring regions of lower concentration.

Suppose, there is a decreasing concentration ($C$) of particles along $x$-direction. Then, there will be a net flow of particles from left to right. Since the net flow is along the positive $x$-direction while the concentration gradient is along the negative $x$-direction, therefore the particle flow ($F$) is given by

$$F = -D\,\frac{dC}{dx} \equiv -D\,\nabla_x C \qquad \text{...(1)}$$

The constant of proportionality ($D$) is known as the diffusion coefficient.

Thus, for holes, the diffusion current density

$$J_{pD} = -q\, D_p\, \nabla_p \qquad ...(2)$$

where $D_p$ and $\nabla_p$ are the diffusion coefficient for holes and the hole concentration gradient respectively.

Similarly for electrons $\quad J_{nD} = -(-q)\, D_n \nabla_n = q\, D_n \nabla_n \qquad ...(3)$

**Net carrier flow.** When both drift and diffusion are present, current density for holes

$$J_p = q\mu^p p\varepsilon - q\, D_p\, \nabla_p \qquad ...(4)$$

and for electrons $\quad J_n = q\, \mu_n\, n\varepsilon + q\, D_n\, \nabla_n \qquad ...(5)$

## 11.20 DRIFT, DIFFUSION AND THE EINSTEIN RELATION

The mobile carriers produce current by both the conduction (on drift) mechanism and the diffusion mechanism. For holes, the total hole current density (amperes/m$^2$) would be the sum of the drift and diffusion components:

$$\vec{J}_h = q\, \mu_h\, p\, \vec{\varepsilon} - q\, D_h\, \nabla_p \qquad ...(1)$$

and for conduction electrons

$$\vec{J}_e = q\, \mu_e n\, \vec{\varepsilon} + q\, D_e\, \nabla_n \qquad ...(2)$$

Note that the drift terms in both equations above have the same algebraic sign because the electric field produces current in the *same* direction for appositely charged particles; but the diffusion terms have *opposite* algebraic signs when written as amperes (rather than as particle flows), because a given concentration gradient would cause both kinds of particles to diffuse in the *same* direction, and therefore to carry electric current in *opposite* directions on account of the opposite charges.

Equations (1) and (2) are very important in non-equilibrium situations when we take up the internal electronics of devices, but at the moment, we wish only to apply there under thermal equilibrium conditions. Then $\vec{J}_h = 0$. So, from Eq. (1), we have $\vec{J}_h = 0$

$$\Rightarrow \qquad q\mu_h\, p_o\, \vec{\varepsilon} - q\, D_n\, \nabla p_o = 0 \qquad ...(3)$$

But, the electrostatic potential $V$ is related to $\vec{\varepsilon}$ as

$$\vec{\varepsilon} = -\nabla V \qquad ...(4)$$

which may be used in Eq. (3) to obtain

$$\frac{\nabla p_o}{p_o} = -\left(\frac{\mu_h}{D_h}\right)\nabla V$$

or

$$\nabla\left(\ln p_o + \frac{\mu_h}{D_h} V\right) = 0$$

or

$$\ln p_o + \frac{\mu_h}{D_h} V = \text{constant}$$

or finally, if $C$ is a constant independent of position

$$p_o = C e^{-(\mu_h/D_h)V} \qquad \text{...(5)}$$

Similarly, for the conduction electrons

$$h_o = C' e^{-(\mu_e/D_e)V} \qquad \text{...(6)}$$

where $C'$ is independent of position.

Now, we know that[1]

$$p_o(x) = n_i e^{-qV/kT} \qquad \text{...(7)}$$

for non-interacting holes in thermal equilibrium and similarly for free electrons

$$n_o(x) = n_i e^{+qV/kT} \qquad \text{...(8)}$$

Comparing Eqs. (5) and (7), and Eqs. (6) and (8), we get

$$\boxed{\frac{D_n}{\mu_n} = \frac{D_e}{\mu_e} = \frac{kT}{q}} \qquad \text{...(9)}$$

This is known as the *Einstein's relation.*

## 11.21 THE EQUATIONS OF CONTINUITY

Consider a region of semiconductor occupying volume $V$ and bounded closed surface $S$, as shown in Fig. 11.15. Focus attention on one type of carrier, in this case holes. The number of holes within $V$ is given by the volume intergral

$$\int_V p\, dV = \text{number of holes in } V \qquad \text{...(1)}$$

The electric current density of holes we denote by $\mathbf{J}_h$ amperes per m$^2$, and, because holes positive charge $q$, the *flux* density of holes is $\mathbf{J}_h/q$ holes per m$^2$ per sec. Therefore, the total flux of holes (number per unit time) *leaving* the region $V$ via its bounding closed surface $S$ is

$$\int_S \frac{1}{q} \mathbf{J}h \cdot \mathbf{n}\, ds = \text{number of holes leaving } V \text{ per unit time} \qquad \text{...(2)}$$

where $\mathbf{n}$ is the *outward* normal to $S$ (Fig. 11.15).

But holes are not lost from $V$ only because they leave; some hole-electron pairs are generated, and some holes recombine with conduction electrons inside $V$. If the number of recombinations per unit volume, per unit, is a function $R(n, p)$ of the two carrier concentrations, and if the number of pair generations is $G$ per unit volume, per unit time, the net excess of recombinations over generation taking place within volume $V$ per unit time is

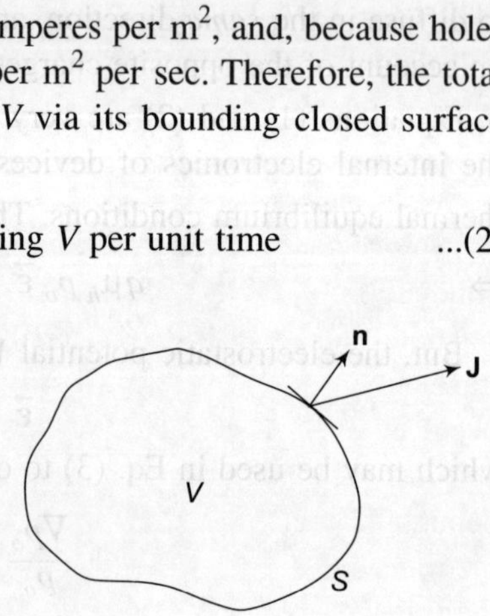

**Figure 11.15** For the development of the continuity equation

[1] See Appendix A.

$$\int_V [R(n,p) - G]\, dV = \textit{\textbf{net}} \text{ number of recombinations in } V \text{ per unit time} \quad ...(3)$$

Now the number of holes that disappear from the fixed region $V$ per unit time is

$$-\frac{d}{dt}\int_V p\,dV = -\int_V \left(\frac{\partial p}{\partial t}\right) dV \quad ...(4)$$

That is, the total number disappearing from $V$ per unit time is the sum of all those disappearing in each elemental volume of $V$ per unit time. And this total number must be accounted for by the holes that leave $V$ via its boundary $S$ and those that leave it by recombination. In other words,

$$-\int_V \left(\frac{\partial p}{\partial t}\right) dV = \int_V [R(n,p) - G] dV + \int_S \frac{1}{q}\mathbf{J}_h \cdot \mathbf{n}\, ds \quad ...(5)$$

But the definition of the divergence operation $(\nabla\cdot)$ on a vector as the outward flux of the vector per unit volume (or the vector relation known as the divergence theorem) tells us that

$$\int_S \mathbf{J}_h \cdot \mathbf{n}\, ds = \int_V \nabla \cdot \mathbf{J}_h\, dV \quad ...(6)$$

as long as the derivatives involved in $\nabla\cdot$ are continuous within $V$. Accordingly, Eq. (5) because

$$\int_V \left(\frac{\partial p}{\partial t} + R(n,p) - G + \frac{1}{q}\nabla \cdot \mathbf{J}_h\right) dV = 0 \quad ...(7)$$

The important point now is that Eq. (7) must hold for *every* volume $V$ of the semiconductor, however small (provided the necessary derivatives are continuous in $V$ of course). The only way the integral involved can vanish for *every* such choice of $V$ is to require the integrand to vanish. Therefore, the equation of continuity becomes

$$\frac{1}{q}\nabla \cdot \mathbf{J}_h + R(n,p) - G = -\frac{\partial p}{\partial t} \quad ...(8)$$

We have tacitly been discussing Eqs. (5) to (8) as if the only process which adds holes to or removes them from circulation within $V$ is that of pair generation or recombination with conduction electrons. This is indeed the usual case. But it does leave out the possibility that separate mobile carriers may be "trapped" temporarily at irregularities or impurities. Such transient "trapping" and "releasing" of the *separate* free carriers is in fact possible, and does occur in some materials at some temperature. Moreover, Eq. (8) can include such cases if $G$ and $R$ include the emptying and filling rates of the hole traps. But the effect is not important enough in the usual electronic devices for us here to add the complication of keeping track of the transient state of occupancy of such traps, as we should have to do, in adding to following the transient concentration of holes and conduction electrons.

Thus, on the basis of the foregoing comments, we shall only consider the situation in which every hole that appears or recombines brings or takes with it a conduction electron; so $R(n, p)$ and $G$ are also the number of such electrons recombining or being generated per unit volume, per unit time.

Noting that, on account of the negative charge $-q$ of the electron, the *flux* density of electrons is related to the *electrical* current density $\mathbf{J}_e$ by relationship

$$\mathbf{J}_{\text{particle}, e} = -\frac{1}{q}\mathbf{J}_e$$

we find by methods analogous to those we used for holes the following continuity equation for conduction electrons

$$-\frac{1}{q}\nabla \cdot \mathbf{J}_e + R(n, p) - G = -\frac{\partial n}{\partial t} \quad ...(9)$$

### 11.21.1 Continuity Equation

If we disturb the equilibrium concentration of carriers in a semiconductor, the concentration of holes or electrons will vary with time. (This we have seen in the preceding section.) In the general case, however, in the body of a semiconductor, the carrier concentration is a function of both time and distance. We now derive the differential equation which governs this functional relationship. This equation is based upon the fact that charge can neither be created nor destroyed.

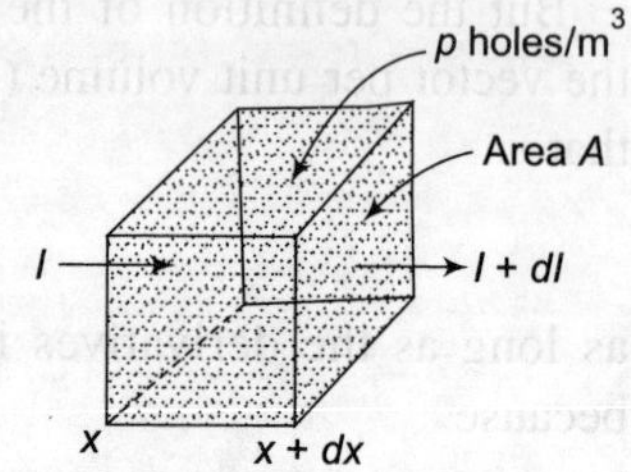


**Figure 11.16** Relating to the conservation of charge

Consider the infinitesimal element of volume of area $A$ and length $dx$ (Fig. 11.16) within which the average hole concentration is $p$. If $\tau_p$ is the mean lifetime of the holes, then $p/\tau_p$ equals the holes per second lost by recombination per unit volume. If $e$ is the electronic charge, then, because of recombination, the number of coulombs per second.

*Decreases* within the volume $= eAdx\,\dfrac{p}{\tau_p}$ ...(1)

If $g$ is the thermal rate of generation of hole-electron pairs per unit volume, the number of coulombs per second.

*Increases* within the volume $= eA\,dx\,g$ ...(2)

In general, the current will vary with distance within the semiconductor. If as indicated in Fig. 11.16, the current entering the volume at $x$ is $I$ and leaving at $x + dx$ is $I + dI$, the number of coulombs per second.

*Decreases* within the volume $= dI$ ...(3)

Because of the three effects described above, the hole density must change with time, and the total number of coulombs per second.

*Increases* within the volume $= eA\,dx\,\dfrac{dp}{dt}$ ...(4)

Since charge must be conserved, we have

$$A\,dx\,\frac{dp}{dt} = -eA\,dx\,\frac{p}{\tau_p} + eA\,dx\,g - dI \quad ...(5)$$

The hole current is the sum of the diffusion current

$$I_{\text{diff.}} = -AeD_p \frac{dp}{dx}$$

and the drift current

$$I_{\text{drif.}} = Ape\, \mu_p\, \varepsilon$$

i.e.,
$$I = -AeD_p \frac{dp}{dx} + Ape\, \mu_p\, \varepsilon \qquad ...(6)$$

where $\varepsilon$ is the electric field intensity within the volume. If the semiconductor is in thermal equilibrium with its surroundings and is subjected to no applied fields ($\varepsilon = 0$), the hole density will attain a constant value $p_o$. Under these conditions $I = 0$ and $dp/dt = 0$, so that from Eq. (5)

$$g = \frac{p_o}{\tau_p} \qquad ...(7)$$

This equation indicates that the rate at which holes are generated thermally just equals the rate at which the hole are lost because of recombination under equilibrium conditions. Combining Eqs. (5), (6) and (7), yields the equation of conservation of charge, or the *continuity equation*[2]

$$\frac{dp}{dt} = -\frac{p - p_o}{\tau_p} + D_p \frac{d^2p}{dx^2} \qquad ...(8)$$

If we are considering holes in the $n$-type material, the subscript $n$ is added to $p$ and $p_o$. Also, since $p$ is a function of both $t$ and $x$, partial derivatives should be used. Then, we have finally

$$\frac{\partial p_n}{\partial t} = -\frac{p_n - p_{no}}{\tau_p} + D_p \frac{\partial^2 p_n}{\partial x^2} - \mu_p \frac{\partial (p_n \varepsilon)}{\partial x} \qquad ...(9)$$

***This is the basic continuity equation for holes minority carriers in an n-type semiconductor.*** In the case of a homogeneous semiconductor in the absence of electric field and carrier gradients, we see that the majority carrier concentration (and hence also the majority carrier concentration) relaxes towards the equilibrium value with time constant $\tau_p$ (typically $\tau_p \simeq 10^{-5}$ s). Similarly,

$$\frac{\partial n_p}{\partial t} = -\frac{n_p - n_{po}}{\tau_n} + D_n \frac{\partial^2 n_p}{\partial x^2} - \mu_n \frac{\partial (n_p \varepsilon)}{\partial x} \qquad ...(10)$$

*This is the basic continuity equation for electrons minority carriers in a p-type semiconductor.*

---

[2] $$eAdx \frac{dp}{dt} = -eA\, dx \frac{p}{\tau_p} + eA\, dx \frac{p_o}{\tau_p} - dI$$

$$= -eA\, dx \frac{(p - p_o)}{\tau_p} + AeD_p\, d\left(\frac{dp}{dx}\right) - Ae\mu_p\, d(p\varepsilon)$$

or $$\frac{dp}{dt} = -\frac{(p - p_o)}{\tau_p} + D_p \frac{d^2p}{dx^2} - \mu_p \frac{d(p\varepsilon)}{dx}$$

## 11.22 SOLUTIONS OF TYPICAL TRANSPORT EQUATIONS IN UNIFORMLY DOPED SEMICONDUCTORS

### 11.22.1 Injection of Minority Carriers at One End of an *n*-type Semiconductor in Steady Conditions

Consider a bar of homogeneously doped *n*-type semiconductor and suppose that an external source (for instance light irradiation or particle bombardment) maintains an excess of holes $\Delta p(x) = \Delta p(0)$, and an excess of electrons $\Delta n(x) = \Delta n(0) = \Delta p(0)$ for any $x \leq 0$. Carriers diffuse in the $x > 0$ part of the semiconductor, and we determine the excess concentration $\Delta p(x)$ and $\Delta n(x)$ in stationary conditions; we also determine the internal electric field, that automatically sets in and accompanies the diffusion process (even in the open circuit arrangement of Fig. 11.17).

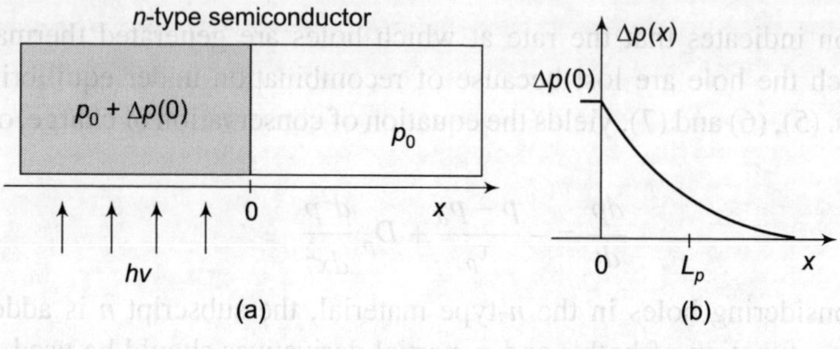


**Figure 11.17** (a) Schematic representation of an *n*-type sample subject to irradiation (b) Distribution of excess hole (and excess electrons) for $x > 0$

The basic Eq. (9) controls the *dynamics of the minority carriers* in the region $x > 0$ (where only the thermal electron-hole generation-recombination mechanism is present). We ignore (momentarily) the internal electric field, which may accompany the diffusion processes, and also consider steady conditions; Eq. (9) simplifies in the form

$$D_p \frac{d^2p}{dx^2} - \frac{p - p_0}{t_p} = 0 \qquad \text{...(11a)}$$

The boundary conditions for the carrier concentration $p(x)$ at $x = 0$ and at $x = \infty$ are given by

$$p(0) \equiv p_0 + \Delta p(0) \quad \text{and} \quad p(\infty) \equiv p_0 \qquad \text{...(11b)}$$

The solution of Eq. (11a), which fulfills the boundary conditions (11b), is

$$p(x) - p_0 = D_p(0)\, e^{-x/L_p}. \qquad \text{...(12a)}$$

where

$$L_p^2 = D_p t_p \qquad \text{...(12b)}$$

The length $L_p$ is called the *minority carrier diffusion length*. With $D_p \approx 10$ cm$^2$/sec, $t_p \approx 10^{-5}$ sec the diffusion length of minority carriers is $L_p \approx 0.01$ cm. We plot in Fig. 11.17b the behaviour of the excess of minority carriers concentration $\Delta p(x) = p(x) - p_0$ for $x > 0$. Notice that the mean square displacement of the excess particle distribution (12a) is

$$\langle x^2 \rangle = \frac{\int_0^\infty x^2 \exp(-x/L_p)\, dx}{\int_0^\infty \exp(-x/L_p)\, dx} = 2L_p^2 = 2D_p t_p. \quad ...(12c)$$

The hole diffusion current in the $x$-direction becomes

$$J_p^{(\mathrm{diff})}(x) = -e D_p \frac{\partial p}{\partial x} = e D_p \frac{\Delta p(0)}{L_p} e^{-x/L_p}; \quad ...(13a)$$

In particular for $x = 0$, we have

$$J_p^{(\mathrm{diff})}(0) = -e D_p \frac{\Delta p(0)}{L_p} e L_p \frac{\Delta p(0)}{t_p}. \quad ...(13b)$$

The expression of the hole diffusion current at $x = 0$ is particularly meaningful; in fact $L_p \Delta p(0)/t_p$ represents the number of holes that must be supplied per unit time through the plane $x = 0$ to keep approximately an excess of carriers $\Delta p(0)$ (with lifetime, $t_p$) in the hostile region of width $L_p$ in the $n$-type semiconductor.

Due to the local quasi-neutrality conditions alredy discussed, the excess of majority carriers is given with good approximation

$$\Delta n(x) \approx \Delta p(x) \equiv \Delta p(0)\, e^{-x/L_p}.$$

The electron diffusion current, $J_n^{(\mathrm{diff})}(x) = e D_n\, \partial n/\partial x$ in the $x$-direction becomes

$$J_n^{(\mathrm{diff})}(x) = e D_n \frac{\Delta p(0)}{L_p} e^{-x/L_p} = -b\, J_p^{(\mathrm{diff})}(x) \quad ...(14a)$$

where $b = D_n/D_p = \mu_n/\mu_p$.

From Eq. (13a) and Eq.(14a) we see that the sum of the diffusion currents due to electrons and holes is not zero. However, *the total current* (diffusion current plus drift current of electron and holes) obeys Eq. (6), and so it is a constant in steady situation, and actually vanishes in the open circuit situation we are investigating. We thus must have at any $x$ point

$$J_{\mathrm{tot}}(x) = J_p^{(\mathrm{diff})}(x) + J_n^{(\mathrm{diff})}(x) + \sigma_n E(x) \equiv 0, \quad ...(14b)$$

where $\sigma_n = e n \mu_n$ is the conductivity of the electrons in the $n$-type material, and the much smaller conductivity of the minority carriers is neglected. From Eq. (14b) we recover for the electric filed

$$E(x) = \frac{1}{\sigma_n}(b - 1)\, e D_p \frac{\Delta p(0)}{L_p} e^{-x/L_p}. \quad ...(14c)$$

This shows that also the electric field decreases exponentially in the region $x > 0$; it is also evident that for $b = 1$ the electron and hole diffusion currents are equal and opposite, and the electric field is zero. In most ordinary conductivity situations, the electric field Eq. (14c) is anyway so small that corrections to the field-free Eq. (11a) are hardly of any significance.

### 11.22.2 Injection (or extraction) of Carriers at Both Ends of an *n*-type Semiconductor in Steady Conditions

Consider a homogeneously doped *n*-type semiconductor and suppose to maintain for $x \leq 0$ an excess (or a deficit) of holes $\Delta p(0)$ and for $x \geq w$ an excess (or deficit) of holes $\Delta p(w)$ (see Fig. 11.18).

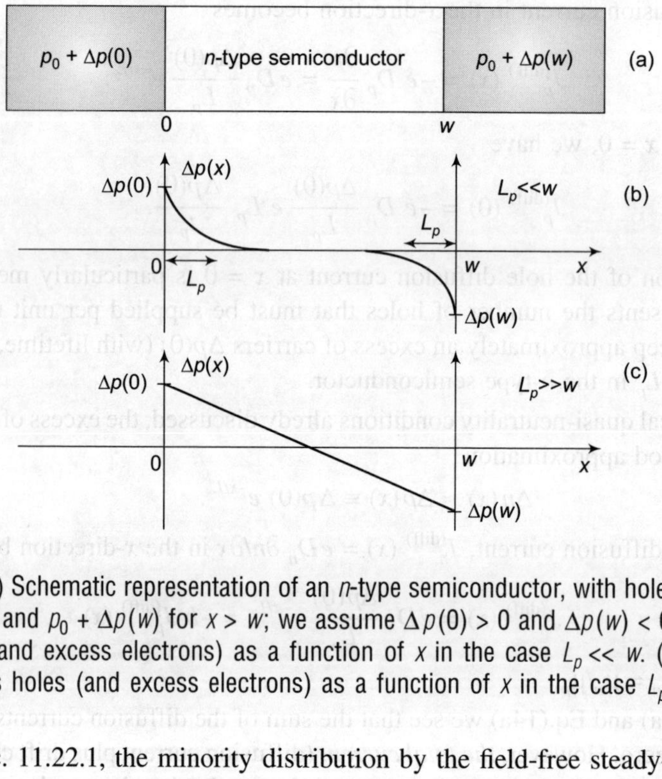


**Figure 11.18** (a) Schematic representation of an *n*-type semiconductor, with hole concentration $p_0 + \Delta p(0)$ for $x < 0$ and $p_0 + \Delta p(w)$ for $x > w$; we assume $\Delta p(0) > 0$ and $\Delta p(w) < 0$. (b) distribution of excess holes (and excess electrons) as a function of *x* in the case $L_p << w$. (c) distribution of excess holes (and excess electrons) as a function of *x* in the case $L_p >> w$

Similar to sec. 11.22.1, the minority distribution by the field-free steady-state continuity equation

$$D_p \frac{d^2 p}{dx^2} - \frac{p - p_0}{\tau_p} = 0. \qquad \text{...(15a)}$$

The boundary conditions for the carrier concentration $p(x)$ at $x = 0$ and $x = w$ are

$$p(0) \equiv p_0 + \Delta p(0) \quad \text{and} \quad p(w) \equiv p_0 + \Delta p(w). \qquad \text{...(15b)}$$

We consider the two linearly independent solutions $\exp(\pm x/L_p)$ of the second order differential Eq. (15a), and determine the linear combination that fulfills the boundary condition (15b). We obtain

$$\Delta p(x) = \Delta p(0)\,\frac{\sinh (w - x)/L_p}{\sinh w/L_p} + \Delta p(w)\,\frac{\sinh x/L_p}{\sinh w/L_p} \quad 0 \leq x \leq w \qquad \text{...(16a)}$$

The hole diffusion current $J_p^{(\text{diff})}(x) = -e\, D_p\, \partial \Delta p(x)/\partial x$ associated to the above carrier distribution is

$$J_p^{(\text{diff})}(x) = -e\, D_p \frac{\Delta p(0)}{L_p} \frac{\cosh[(w-x)/L_p]}{\sinh(w/L_p)} - e\, D_p \frac{\Delta p(w)}{L_p} \frac{\cosh(x/L_p)}{\sinh(w/L_p)}. \qquad ...(16b)$$

In particular, at the points $x = 0$ and $x = w$ we have

$$J_p^{(\text{diff})}(0) = e\, D_p \frac{\Delta p(0)}{L_p} \frac{1}{\tanh(w/L_p)} - e\, D_p \frac{\Delta p(w)}{L_p} \frac{1}{\sinh(w/L_p)}. \qquad ...(16c)$$

$$J_p^{(\text{diff})}(w) = e\, D_p \frac{\Delta p(0)}{L_p} \frac{1}{\sinh(w/L_p)} - e\, D_p \frac{\Delta p(w)}{L_p} \frac{1}{\tanh(w/L_p)}. \qquad ...(16d)$$

We discuss the concentration of minority carriers and diffusion currents, given by Eq. (16), in the two limiting cases $w >> L_p$ and $w << L_p$. In the case the width ($w$) is much larger than the diffusion length, Eq. (16) simplify, and we have in particular

$$\Delta p(x) = \Delta p(0)\, e^{-x/L_p} + \Delta p(w)\, e^{-(w-x)/L_p} \quad 0 \le x \le w \quad w >> L_p$$

and $$J_p^{(\text{diff})}(0) = e\, D_p \frac{\Delta p(0)}{L_p}, \quad J_p^{(\text{diff})}(w) = -e\, D_p \frac{\Delta p(w)}{L_p}. \qquad ...(17)$$

We thus see that the concentration of minority carriers, as well as diffusion current penetrating in the semiconductors from the two ends, are completely decoupled (as shown in Fig. 11.18b).

A much more interesting situation occurs in the case $L_p >> w$. With appropriate series expansion of the terms at right member of Eq. (16), we obtain

$$\Delta p(x) = \Delta p(0) \frac{w-x}{w} + \Delta p(w) \frac{x}{w} + \ldots = \Delta p(0) + [\Delta p(w) - \Delta p(0)] \frac{x}{w} + \ldots$$

and $$J_p^{(\text{diff})}(0) = J_p^{(\text{diff})}(w) = e\, D_p \frac{\Delta p(0) - \Delta p(w)}{w}. \qquad ...(18)$$

From Eqs. (18) and (17), we see that diffusion current is greatly increased when $w << L_p$.

Due to the local quasi-neutrality condition the excess majority carrier concentration is given approximately by $\Delta n(x) \approx \Delta p(x)$; from the requirement that the total current must be zero, the (small) electric field in the sample can be obtained, with considerations similar to those worked out in sec. 11.22.1.

### 11.22.3 Injection of a Narrow Pulse of Minority Carriers and Measurement of Drift Mobility (Haynes–Shockley Experiment)

Consider an $n$-type semiconductor and suppose that a narrow pulse of holes (and electrons) is produced by a source in the plane $x = 0$ at the time ($t = 0$) (for instance by irradiation of light, or in practice by a voltage-pluse applied to an emitting contact). We wish to determine the excess hole (and electron) concentration $\Delta p(x, t)$ at successive instants $t > 0$ in the semiconductor.

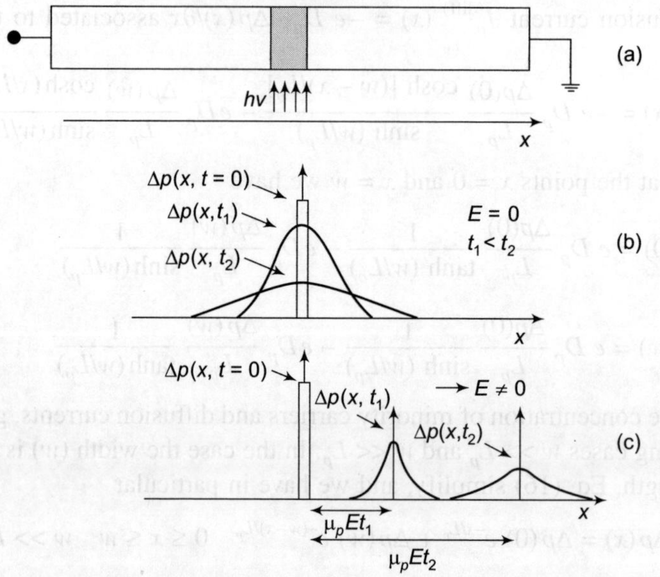


**Figure 11.19** (a) Schematic representation of an *n*-type semiconductor with a narrow pulse of holes (and electrons) optically injected at $x = 0$ and $t = 0$. (b) distribution of excess holes (and electrons) in presence of a constant and uniform electric field

The equation that governs the dynamics of the holes minority carriers in the injection experiment is the basic Eq. (9); neglecting momentarily the terms containing the electric field, we have

$$\frac{\partial p}{\partial t} = D_p \frac{\partial^2 p}{\partial x^2} - \frac{p - p_0}{t_p}. \qquad ...(19)$$

The solution of this equation with boundary condition $p(0,0) = p_o + N\,\delta(t)\,\delta(x)$ is

$$\Delta p(x, t) = \frac{N}{\sqrt{4\pi D_p t}} \exp\left(-\frac{x^2}{4 D_p T}\right) \exp\left(-\frac{t}{t_p}\right); \qquad ...(20)$$

in fact, it is easily seen by inspection that the above expression of $\Delta p(x, t)$ satisfies the differential Eq. (19), as well as the boundary conditions $\Delta p(0, 0) = N\,\delta(t)\,\delta(x)$. The behaviour of $\Delta p(x, t)$ at different times in the semiconductor is illustrated in Fig. (11.19b).

It is of interest to evaluate the average square distance $\langle x^2(t)\rangle$ of the carriers at time $t$; we have

$$\langle x^2(t)\rangle = \int_0^\infty x^2\, \Delta p(x,t)\,dx \,/ \int_0^\infty x^2\, \Delta p(x,t)\,dx = 2\,D_p t. \qquad ...(21)$$

The average distance $(x^2)$, travelled by the carriers during their lifetime is then $\langle x^2\rangle = 2 D_p t_p$, in agreement with the previous expression in Eq. (12c). The calculation of $\langle x^2(t)\rangle$ has been carried out using the following elementary integrals of the Gaussian function

$$\int_0^{\infty} e^{-ax^2}\,dx = \frac{1}{2}\sqrt{\frac{\pi}{a}} \qquad \int_0^{\infty} x^2 e^{-ax^2}\,dx = \frac{1}{4a}\sqrt{\frac{\pi}{a}} \qquad a > 0.$$

We discuss now the injection of a narrow pulse of holes in the presence of a uniform and constant electric field **E** (in the *x*-direction); instead of Eq. (19), we have to consider the partial differential equation of the from

$$\frac{\partial p}{\partial t} = D_p \frac{\partial^2 p}{\partial x^2} - \frac{p - p_0}{t_p} - \mu_p E \frac{\partial p}{\partial x};$$

the solution of the above equation with boundary condition $\Delta p(0, 0) = N\,\delta(t)\,\delta(x)$ is

$$\Delta p(x, t; E) = \frac{N}{\sqrt{4\pi D_p t}} \exp\left[-\frac{(x - \mu_p Et)^2}{4D_p t}\right] \exp\left(-\frac{t}{t_p}\right).$$

The expression $\Delta p(x, t; E)$ has the same form as $\Delta p(x, t; E \equiv 0)$ of Eq. (20), except that $x$ is now replaced by $x - \mu_p E$; thus the excess carriers (both holes and electrons) follow the electric field with drift velocity, $\mu_p E$, as shown in Fig. (11.19c).

The above result are the basis understanding the Haynes – Shockley experiment for measuring the drift mobility of minority carriers. The method consists in the injection of a narrow pulse of minority carriers into a filament and sweeping the pulse along the filament by means of an electric field; if $L$ is the distance between the emitter and collector contacts, from the measurement of the transit time, $t = L/\mu_p E$ the mobility of minority carriers can be obtained.

## 11.23 TIME DEPENDENT DIFFUSE EQUATION

We have

$$\vec{J}_p = q\,\mu_p p\,\vec{\varepsilon} - q\,D_p\,\nabla p \qquad \text{...(1)}$$

$$\vec{J}_n = q\,\mu_n n\,\vec{\varepsilon} - q\,D_n\,\nabla n$$

These equations describe the flow of mobile carriers in semiconductors the influence of an electric field and concentration gradients. Considering only one-dimensional current flow, i.e., only along $x$-direction, we have

$$\nabla p = \frac{\partial p}{\partial x} \text{ and } \nabla n = \frac{\partial n}{\partial x}. \qquad \text{...(2)}$$

The hole and electron current densities are given by

$$J_p = e\mu_p p\varepsilon - eD_p \frac{\partial p}{\partial x}$$

$$J_n = e\mu_n n\varepsilon - eD_n \frac{\partial n}{\partial x} \qquad \text{...(3)}$$

If we divide the hole current density by $(+e)$ and the electron current density by $(-e)$, we obtain each particle flux and equation become

$$\frac{J_p}{(+e)} \equiv F_p^+ = \mu_p p\varepsilon - D_p \frac{\partial p}{\partial x} \qquad \text{...(4)}$$

$$\frac{J_n}{(-e)} \equiv F_n^- = \mu_n n\varepsilon - D_n \frac{\partial n}{\partial x} \qquad ...(5)$$

Taking the divergence of Eqs. (4) and (5) and substituting into the continuity equations

$$\frac{\partial p}{\partial t} = -\frac{\partial F_p^+}{\partial x} + gp - \frac{p}{\tau_{pt}} \qquad ...(6)$$

$$\frac{\partial n}{\partial t} = -\frac{\partial F_n^-}{\partial x} + gn - \frac{n}{\tau_{nt}} \qquad ...(7)$$

(for one dimensional flow), we get

$$\frac{\partial p}{\partial t} = -\mu_p \frac{\partial (p\varepsilon)}{\partial x} + D_p \frac{\partial^2 p}{\partial x^2} + gp - \frac{p}{\tau_{pt}} \qquad ...(8)$$

$$\frac{\partial n}{\partial t} = -\mu_n \frac{\partial (n\varepsilon)}{\partial x} + D_n \frac{\partial^2 n}{\partial x^2} + gn - \frac{n}{\tau_{nt}} \qquad ...(9)$$

where $\tau_{pt}$ on $\tau_{nt}$ includes the thermal carrier lifetime and the excess carrier lifetime. Keeping in view dimensional case, we have

$$\frac{\partial (p\varepsilon)}{\partial x} = \varepsilon \frac{\partial p}{\partial x} + p \frac{\partial \varepsilon}{\partial x} \qquad ...(10)$$

Then, Eqs. (8) and (9) can be written in the from

$$D_p \frac{\partial^2 p}{\partial x^2} - \mu_p \left( \varepsilon \frac{\partial p}{\partial x} - p \frac{\partial \varepsilon}{\partial x} \right) + g_p - \frac{p}{\tau_{pt}} = \frac{\partial p}{\partial t} \qquad ...(11)$$

$$D_n \frac{\partial^2 n}{\partial x^2} - \mu_n \left( \varepsilon \frac{\partial n}{\partial x} - n \frac{\partial \varepsilon}{\partial x} \right) + g_n - \frac{n}{\tau_{nt}} = \frac{\partial n}{\partial t} \qquad ...(12)$$

These equations are the time-dependent diffusion equations for holes and electrons respectively to describe the space and time behaviour of the excess carriers.

Now, the hole and electron concentrations are functions of the thermal equilibrium and the excess values as given by

$$n = n_0 + \delta_n \qquad ...(13)$$

and

$$p = p_0 + \delta p$$

The thermal equilibrium concentrations $n_0$ and $p_0$ are functions of the time. For the case of a homogeneous semiconductor, $n_0$ and $p_0$ are also independent of the space coordinates. Equations (11) and (12) may then be written as

$$D_p \frac{\partial^2 (\delta p)}{\partial x^2} - \mu_p \left( \varepsilon \frac{\partial (\delta p)}{\partial x} + p \frac{\partial \varepsilon}{\partial x} \right) + g_p - \frac{p}{\tau_{pt}} = \frac{\partial (\delta p)}{\partial t} \qquad ...(14)$$

$$D_n \frac{\partial^2(\delta n)}{\partial x^2} - \mu_n \left( \varepsilon \frac{\partial(\delta n)}{\partial x} + n \frac{\partial \varepsilon}{\partial x} \right) + g_n - \frac{p}{\tau_{nt}} = \frac{\partial(\delta n)}{\partial t} \quad ...(15)$$

Here $p$ and $n$ are total concentrations and $\delta p$ and $\delta n$ are the excess concentrations.

## 11.24 AMBIPOLAR TRANSPORT

Initially, it was assumed that the electric field in Eq. (3) was applied electric field. This electric field term appears in the time dependent diffusion equations given by Eqs. (4) and (5). If a pulse of excess electron and a pulse of excess holes are created at a particular point in a semiconductor with an applied electric field, the excess holes and electrons will tend to drift in opposite direction. However, because the electrons and charged particles, any separation will induce an internal electric field ($\varepsilon_{int}$) between the two sets of particles. This $\varepsilon_{int}$ will create a force attracting the electrons and holes back towards each other (Fig. 11.20). Then net ε-field may be written as

$$\varepsilon = \varepsilon_{app} + \varepsilon_{int} \quad ...(16)$$

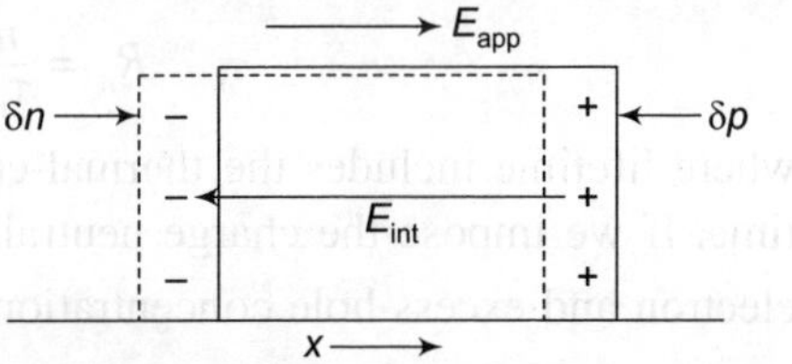

**Figure 11.20** The creation of an internal electric field as excess electrons and holes tend to separate

where $\varepsilon_{app}$ is the applied field,

Since the field $E_{int}$ creates a force attracting the electrons and holes, this field will hold pulses of excess electrons and excess holes together. The negatively charged electrons and positively charged holes phenomenon is called ambipolar transport.

### 11.24.1 Ambipolar Transport Equation

The time-dependent diffusion Eqs. (14) and (15) describe the behaviour of the excess carriers. However, a third equation is required to relate the excess electrons and hole-condensation to the internal electric field. This is Poisson's equation which may be written as

$$\nabla \cdot \varepsilon_{int} = \frac{e(\delta p - s_n)}{\varepsilon_s} = \frac{\partial \varepsilon_{int}}{\partial x} \quad ...(17)$$

where $\varepsilon_s$ is the permittivity of the semiconductor material. For solutions of Eqs. (14), (15) and (17) we assume that $|\varepsilon_{int}| << |\varepsilon_{app}|$, however $\nabla \cdot \varepsilon_{int}$ may not be negligible. Further, we impose the condition of charge neutrality: It is assumed that the excess electron concentration is just balanced by an excess hole concentration at any point in space and time, only a very small difference in the excess electron concentration and excess hole concentration will set up an internal ε-field sufficient to keep the particles diffusing and drafting together (i.e., we assume quasi-neutrality condition).

Now, thermal generation rates of electrons and hole are equal, i.e.,

$$G_{no} = G_{po} \quad ...(18)$$

Also, generation rates of excess electrons and holes

$$g'_n = g'_p \quad ...(19)$$

Considering Eqs. (18) and (19), we can define

$$g_n = g_p \equiv g$$

Recombination rates of electrons and holes

$$R_{n0} = R_{p0} \qquad ...(20)$$

The excess electron and holes and electrons recombine in pairs so, their recombination rates

$$R'_n = R'_p \qquad ...(21)$$

Considering Eqs. (20) and (21), we can define

$$R_n = \frac{n}{\tau_{nt}} = R_p = \frac{p}{\tau_{pt}} \equiv R \qquad ...(22)$$

where lifetime includes the thermal-equilibrium carrier lifetime and the excess carrier lifetime. If we impose the charge neutrality condition then $\delta_n \approx \delta_p$, we denote both the excess electron and excess hole concentrations in Eqs. (14) and (15) by $\delta n$. Then

$$D_p \frac{\partial^2(\delta_n)}{\partial x^2} - \mu_p\left(\varepsilon \frac{\partial(\delta n)}{\partial x} + p\frac{\partial \varepsilon}{\partial \varepsilon}\right) + g - R = \frac{\partial(\delta n)}{\partial t} \qquad ...(23)$$

and

$$D_p \frac{\partial^2(\delta_n)}{\partial x^2} - \mu_n\left(\varepsilon \frac{\partial(\delta n)}{\partial x} + p\frac{\partial \varepsilon}{\partial x}\right) + g - R = \frac{\partial(\delta n)}{\partial t} \qquad ...(24)$$

Multiply Eqs. (23) by $\mu_n n$ and (24) by $\mu_n p$ and add the two equations (the $\nabla \cdot \varepsilon = \partial\varepsilon/\partial x$ term will be eliminated). This gives

$$(\mu_n n D_p + \mu_p p D_n)\frac{\partial^2(\delta_n)}{\partial x^2} + (\mu_n \mu_p)(p - n)\, e \frac{\partial(\delta n)}{\partial x}$$

$$+ (\mu_n n + \mu_p p)(g - R) = (\mu_n n + \mu_p p)\frac{\partial(\delta n)}{\partial t} \qquad ...(25)$$

If we divide this equation by $(\mu_n n + \mu_p p)$, we get

$$D' \frac{\partial^2(\delta_n)}{\partial x^2} + \mu' \varepsilon \frac{\partial^2(\delta n)}{\partial x^2} + g \;\; R = \frac{\partial(\delta n)}{\partial t} \qquad ...(26\text{i})$$

where

$$D' = \frac{\mu_n n\, D_p + \mu_p p\, D_n}{\mu_n n + \mu_p p} \qquad ...(26\text{ii})$$

and

$$\mu' = \frac{\mu_n \mu_p\,(p - n)}{\mu_n n + \mu_p p} \qquad ...(26\text{iii})$$

Equation (26) is called the *ambipolar transport equation* and describes the behaviour of excess electrons and holes in the time and space. $D'$ is called the *ambipolar diffusion coefficient* and $\mu'$ is called the *ambipolar mobillity*.

Using Einstein relation

$$\frac{\mu_n}{D_n} = \frac{\mu_p}{D_p} = \frac{e}{kT} \quad ...(27)$$

We can write

$$D' = \frac{D_n D_p (n + p)}{D_n n + D_p p} \quad ...(28)$$

## 11.25 HALL EFFECT IN SEMICONDUCTORS

We have $$m^* \left( \frac{d\bar{v}}{dt} + \frac{\bar{v}}{\tau} \right) = e\,(\varepsilon + \vec{v} \times \vec{B}) \quad \text{(in MKS units)} \quad ...(1)$$

In case of an intrinsic semiconductor both electrons and holes are present, and the Hall field developed by deflection of electrons would be in the opposite direction to Hall field developed by deflection of holes. At some proper ratio between electrons and hole densities, there exists a possibility of a zero Hall field. In that case, current in the $y$-direction must be zero. In such a steady state,

$$0 + \frac{v_y}{t} = \frac{e}{m^*} (\varepsilon_y - v_x H_z) \quad ...(2)$$

$$\therefore \quad 0 = J_y = J_{ye} + J_{yh} = n_e e v_{ey} + n_h e v_{hy} \quad ...(3)$$

Since $v_{hv}$ is opposite to $v_{ey}$ and $q_e = -q_h = e$, so

$$0 = n_e \mu_e e (\varepsilon_y - v_{ex} H_z) + n_h \mu_h e (\varepsilon_y - v_{hx} H_z) \quad ...(4)$$

and

$$J_x = e\, \varepsilon_x (n_e \mu_e + n_h \mu_h) \quad ...(5)$$

Here $e$ is a positive quantity and we have to take minus sign for electrons explicitly.

$$\therefore \quad 0 = e\, \varepsilon_y (n_e \mu_e + n_h \mu_h) - e H_z (n_e \mu e v_{ex} + n_h\, \mu_n v_{hx}) \quad ...(6)$$

The average drift velocities in the $x$-direction in the absence of magnetic field are

$$v_{ex} = \mu_e \varepsilon_x \text{ and } v_{hx} = \mu_h \varepsilon_x \quad ...(7)$$

$\therefore$ From Eqs. (6) and (7), we have

$$0 = e\, \varepsilon_y (n_e \mu_e + n_h\, \mu_h) + e H_z e_x (n_e \mu_e^2 - n_h\, \mu_h^2)$$

Eliminating $\varepsilon_x$ between Eq. (5) and Eq. (7) gives

$$\varepsilon_y = \frac{-J_x H_z (n_e \mu_e^2 + n_e \mu_n^2)}{e (n_e \mu_e + n_h \mu_h)^2} \quad ...(8)$$

Hence, we get two carrier Hall constants

$$R_H = \frac{\varepsilon_y}{J_x H_z} \frac{(n_h \mu_h^2 + n_e \mu_e^2)}{e (n_e \mu_e + n_h \mu_h)^2}$$

$$= \frac{(n_h \mu_n^2 + n_e \mu_e^2)}{(e n_e \mu_e + e n_h \mu_h)(n_e \mu_e + n_h \mu_h)} \quad ...(9)$$

and two carrier Hall mobilities

$$\mu_H = \sigma R_H = \frac{(n_h \mu_h^2 + n_e \mu_e^2)}{(n_e \mu_e + n_h \mu_h)^2} \quad ...(10)$$

This gives the Hall mobility of an intrinsic semi-conductor ($n_e = n_h$) as ($\mu_h - \mu_e$). Usually, $\mu_e - \mu_h$, so Hall mobility for an intrinsic semiconductor is negative (Fig. 11.21).

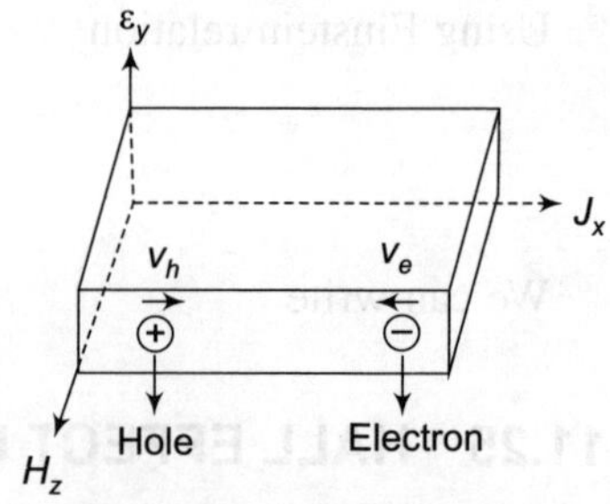


**Figure 11.21** Hall effect in a semiconductor

## 11.26 HALL EFFECT MEASUREMENTS IN A SEMICONDUCTOR

If a current is passed through a semiconductor and a magnetic field $\vec{B}$ is applied flow, at right angles to the direction of current flow, an electric field is induced in a direction mutually perpendicular to $\vec{B}$ and the direction of current flow. This is known as the *Hall-effect.*

A Hall measurement experimentally confines the validity of the concept that it is possible for two independent types of charge carrier, electrons and holes to exist in a semiconductor. We shall subsequently see that a measurement of the Hall coefficient determines not only the sign of the charge of the carriers but also their density.

Now, let us for simplicity as shown in Fig. 11.22 and assume the carriers to be positive holes of charge $e$ only.

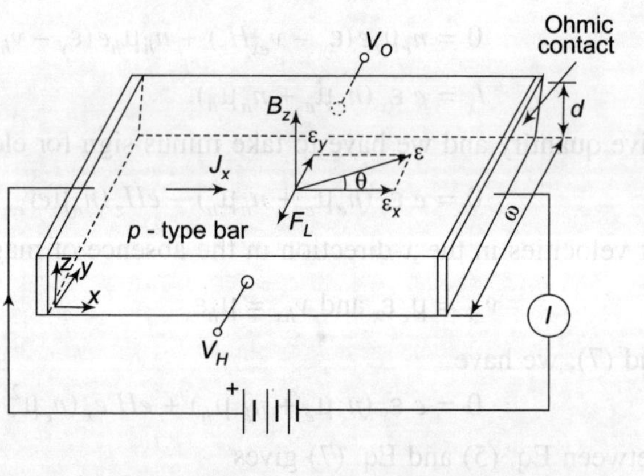


**Figure 11.22** Hall effect in a *p*-type semiconductor

assume that a current density $J_x$ is produced in the bar by application of an electric field $E_x$ and that a magnetic field of flux density $B_z$ is applied in the $z$-direction. Since the holes are flowing with some drift velocity ($v_{Dx}$) under the influence of the applied field, they experience a Lorentz force:

$$F_L = ev \times B \quad ...(1)$$

of magnitude

$$|F_L| = ev_{Dx} B_z \quad ...(2)$$

and directed in the negative $y$-direction. This force tends to drive holes towards the front face of the block (see the figure); there is an excess of holes there and a deficiency of holes at the back face. Since there can be no net flow in the $y$-direction, the movement of holes to front and back creates an electric field in the positive $y$-direction, $\mathscr{E}_y$. This produces a force on the holes which exactly compensates for the Lorentz force field and prevents transverse current flowing. Hence, in equilibrium

$$e\mathscr{E}_y = F_L - ev_{Dx}B_z \qquad \text{...(3)}$$

Also, the current density $J_x$ is given by

$$J_x \simeq pev_{Dx}$$

which can be substituted in Eq. (3) to give

$$\mathscr{E}_y = \frac{J_x B_z}{pe} \qquad \text{...(4)}$$

Thus, if voltage probes are attached to front and back faces, assumed separated by a distance $w$, a voltage

$$V_{\text{H}} = \mathscr{E}_y w$$

may be measured.

The Hall coefficient, $R_{\text{H}}$, defined from Eq. (4), is

$$R_{\text{H}} = \frac{\mathscr{E}_y}{J_x B_z} = \frac{1}{pe}$$

Alternatively, if the current through the bar of thickness $d$ is $I$

$$R_{\text{H}} = \frac{V_{\text{H}}}{wIB_z} \cdot wd = \frac{V_{\text{H}} d}{IB_z} = \frac{1}{pe} \qquad \text{...(5)}$$

It will be easily verified that for $n$-type material where the current is carried by majority, electrons, the polarity of the Hall voltage, $V_{\text{H}}$, is reversed and

$$R_{\text{He}} = -\frac{1}{ne} \qquad \text{...(6)}$$

We see that the measurement of current and magnitude and sign of the Hall voltage for any given magnetic flux density gives the sign of the charge carriers, that is, it shows whether the doped semiconductor is $n$- or $p$-type, together with the density of the majority carriers.

It will be noticed that the net electric field in the semiconductor, which is the vector sum of $\mathscr{E}_x$ and $\mathscr{E}_y$ is not directed along the axis but is at some angle $\theta$ to it. From Fig. 11.22 it will be seen that $\theta$, called the *Hall angle*, is given by

$$\tan\theta = \frac{\mathscr{E}_y}{\mathscr{E}_x} \qquad \text{...(7)}$$

which, using Eq. (4) and remembering that $J_x = \sigma\mathscr{E}_x$, gives

$$\tan\theta = \frac{J_x B_z}{pe} \cdot \frac{\sigma}{J_x} = \mu_h B_z \qquad \text{...(8)}$$

or alternatively $\mu_h = R_H\sigma$ ...(9)

Thus, provided the assumptions made in this simple model are correct, a simultaneous measurement of $R_H$ and $\sigma$ can lead to an experimental value for the carrier drift mobility.

We have tacitly assumed that all the carriers move with an average drift velocity, $v_D$, and hence have the same collision time. If the problem is treated on a more rigorous statistical basis, it can be shown that, if the principal scattering mechanism is thermal scattering, then an additional numerical factor is introduced into the expression for the Hall coefficient and

$$R_H = \frac{3\pi}{8}\cdot\frac{1}{pe} \qquad ...(10)$$

If other types of scattering predominate, the numerical factor has a different value. We see that, in this case, Eq. (11) becomes

$$\tan\theta = \frac{3\pi}{8}\cdot\mu_h B_z = \mu_H B_z \qquad ...(11)$$

where $\mu_h$ is known as the *Hall mobility*. It is related to drift velocity via Eq. (11)

$$\mu_H = \frac{\tan\theta}{B_z} = \frac{3\pi}{8}\mu_h \qquad ...(12)$$

and is equal to the product of Hall coefficient and conductivity:

$$\mu_H = R_H\sigma \qquad ...(13)$$

We have further assumed that the conduction process is by means of one type of carrier only. In some semiconductors, for example, in extrinsic or nearly fully compensated material, the assumption is not valid and the Hall coefficient must be modified to account for the presence of two types of charge carrier, as follows.

Both types of carrier will drift under the influence of the applied $\mathscr{E}_x$ field with drift velocities:

$$v_{Dh} = \mu_h\mathscr{E}_x \text{ in the positive } x\text{-direction}$$

and $$v_{De} = \mu_h\mathscr{E}_x \text{ in the negative } x\text{-direction}$$

using the same coordinate system as in Fig. 11.22. As a consequence, the holes and electrons each experience a Lorentz force given by

$$F_h = e(v_{Dh}\times B) = -ev_{Dh}B_z$$

and $$F_e = e(v_{De}\times B) = -ev_{De}B_z$$

Both deflecting forces are in the same direction, deflecting electrons and holes to the front face, as before. The carriers recombine at the surface and the net charge there produces an electron field in the $y$-direction, $\mathscr{E}_y$. In equilibrium, the current due to the deflected electrons and holes must be exactly cancelled out by current flowing in the opposite direction due to the Hall field. If the transverse velocities of the deflected carriers are $v_{yh}$ and $v_{ye}$, the current in the transverse direction due to deflection of electrons and holes by the magnetic field is $(epv_{yh} - env_{ye})$ and, therefore, in equilibrium:

$$\sigma\mathscr{E}_y = e(pv_{yh} - mv_{ye}) \qquad ...(14)$$

Expressions for the transverse velocities can be obtained from the Hall angle Eqs. (7) and (8), which give

$$v_{yh} = \mu_h(\mathscr{E}_x \tan\theta) = \mu_h(\mathscr{E}_x \mu_h B_z)$$

and, similarly, $$v_{ye} = \mu_e^2 \mathscr{E}_x B_z$$

These are substituted in Eq. (14) to give

$$\sigma\mathscr{E}_y = e(p\mu_h^2 - n\mu_e^2)\,\mathscr{E}_x B_z \qquad \text{...(15)}$$

The generalized version of Ohm's law is used to eliminate $\varepsilon_x$ from this equation, which can then be rearranged to give the Hall coefficient:

$$R_{\text{H}} = \frac{\mathscr{E}_y}{J_x B_z} = \frac{e(p\mu_h^2 - n\mu_e^2)}{\sigma^2}$$

Finally, we include the general expression for the conductivity[3] to give:

$$R_{\text{H}} = \frac{3\pi}{8}\frac{p\mu_h^2 - n\mu_e^2}{e(p\mu_h + n\mu_e)^2} \qquad \text{...(16)}$$

Again, a more rigorous treatment, which does not assume that time between scattering events is the same for all carriers, result in the introduction of an additional numerical factor in the Hall coefficient expression. For example, if lattice scattering is the principle loss mechanism

$$R_{\text{H}} = \frac{3\pi}{8}\frac{p\mu_h^2 - n\mu_e^2}{e(p\mu_h + n\mu_e)^2} \qquad \text{...(17)}$$

Comparison of this expression with Eqs. (5) and (6) show that the Hall coefficient, and hence the Hall voltage, are generally smaller for near-intrinsic materials than for the more highly doped extrinsic materials, Notice, too, that measurement of the Hall coefficient of the intrinsic material does not lead immediately to values for the carrier concentrations, as it does in the extrinsic material.

## 11.27 FOUR-POINT PROBE METHOD FOR CONDUCTIVITY MEASUREMENTS

In principle, the conductivity of a semiconductor may be found by measuring the current drawn when a voltage is applied between two contacts formed at the ends of a bar of the material. In practice, such a measurement is not straightforward because of the difficulty of making uniform ohmic contacts to the sample. Further, it is not always convenient to produce a specimen of known dimensions for a conductivity measurement. The four-point probe method eliminates accurately, the conductivity of ingots or slices, both thick and thin, of semiconductor crystals.

The probe head basically employs four spring-loaded, equispaced needles which make contact with a plane lapped surface on the specimen, as shown in Fig. 11.23. A stabilized current is passed through the outer pair of probes, *A* and *D*, and the voltage, *V*, between

[3] See section 11.18 Eq. (11).

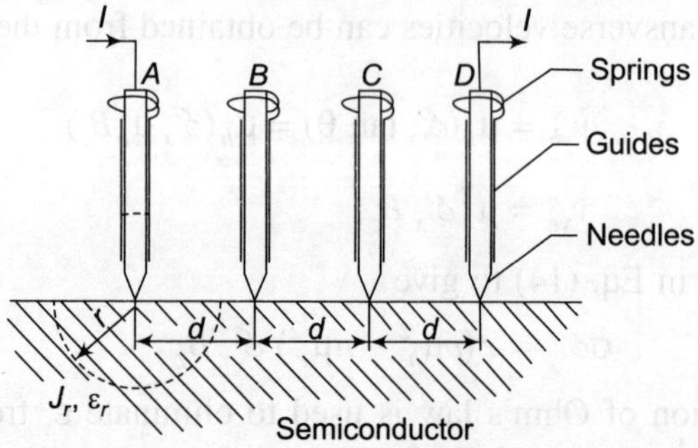


**Figure 11.23** Four-point conductivity measurement

the other two, $B$ and $C$, is measured by, say, a digital voltmeter which draws negligible current.

Let us first consider the sample dimensions to be large compared with the probe spacing, $d$. We can treat the source current at $A$ and the sink at $D$ independently, using the principle of superposition to find their combined effect later. Thus, the current density, $J_r$, at radius $r$ from $A$, due to the current $I$ entering at $A$ is

$$J_r = \frac{I}{2\pi r^2}$$

since current only flows in the bottom half-plane. We then use Ohm's law to find the corresponding electric field at $r$, $\mathscr{E}_r$

$$\mathscr{E}_r = \frac{J}{\sigma} = \frac{I}{2\pi\sigma r^2}$$

where $\sigma$ is the conductivity of the material.

By definition, the floating potential at some radius $a$ is then

$$V_a = -\int_{-\infty}^{a} \mathscr{E}_r\, dr = -\frac{I}{2\pi\sigma}\int_{-\infty}^{a} \frac{1}{r^2}\, dr = \frac{I}{2\pi\sigma a} \qquad \text{...(18)}$$

The potential difference between probes $B$ and $C$ due to the current source at $A$ is then

$$V_{B-C} = \frac{I}{2\pi\sigma d} - \frac{I}{2\pi\sigma(2d)} = \frac{I}{4\pi\sigma d} \qquad \text{...(19)}$$

If, now, we consider the current leaving at $D$ separately, there exists a potential difference between $B$ and $C$ which, because of symmetry, is the same as that due to the current source above, given in Eq. (19). By superposition, the total voltage measured between $B$ and $C$ is thus twice that for current source or sink alone or

$$V = \frac{I}{2\pi\sigma d}$$

from which the conductivity is obtained:

$$\sigma = \frac{1}{2\pi d}\cdot\frac{I}{V} \qquad \text{...(20)}$$

This equation is valid only for materials whose dimensions are large compared with the probe spacing, *d*. If this is not the case, for example, when measurements are made on thin slices or small dice of material, then a correction factor, *F*, is introduced and the conductivity is given by

$$\sigma = \sigma_0 F$$

where $\sigma_0$ is the uncorrected value of σ. Factor *F* is a function of slice thickness, breadth, and width, normalized to the probe spacing; the reader is referred to the more specialized literature for the derivation and tabulation of values of *F* for various slice geometries.[4]

## 11.27.1 Hall Effect Sensors (Geometric Magnetoresistance)

A *Hall-effect sensor generates an electrical signal (the Hall voltage) in response to a changing magnetic field.*

### *Magnetoresistance*

During the experiment on the Hall effect, it is noticed that the resistance of the metal sample along the direction of the current $j_x$ generally increases, (i.e., the conductance decreases) after the magnetic field is switched on.

The fractional increase in the zero field resistivity ($\rho(0)$) is defined by $\frac{\Delta\rho}{\rho(o)}$. This effect was discovered by Thomson in 1856. The amount of $\Delta\rho/\rho(o)$ is found to be dependent on the geometry of the sample, consequently, termed the *geometric-magnetoresistance*. Thus, a Hall effect sensor is based on the geometric magnetoresistance. Hall sensors are used in computer keyboards, where pushing a key moves a permanent magnet close to a Hall-effect sensor. Such a sensor is simply a piece of an *n*-type semiconductor with four contacts, as shown in Fig. 11.23. (In practical devices, all contacts are made on top of a silicon wafer to make the fabrication process compatible with integrated circuit technology.)

The movement of the magnet creates a magnetic field of the order of several hundred gausses. A typical sensitivity of a Hall sensor (defined as a ratio of the Hall voltage over the magnetic field) is of the order of tens of microvolts per gauss. Hence, pushing a computer keyboard key produces a Hall voltage of the order of several tens of millivolts which is then amplified to a desired level.

The Hall electric field develops as a result of the carrier flux in the direction perpendicular to the current flow. This perpendicular motion is caused by the magnetic field when it is turned on. At first, the carriers move under the angle called the *Hall angle*.

$$\theta = \tan^{-1}\left(\frac{v_y}{v_x}\right) = \tan^{-1}\frac{\mu v_x B}{v_x} = \tan^{-1}(\mu B) \qquad ...(1)$$

Here $v_y$ and $v_x$ are components of the carrier velocity in the *x* and *y* directions (see Fig. 11.24). Then carriers accumulate at one of the sample side surfaces depleting the other side and building up the Hall electric field. The Hall electric field compensates the Lorentz force and stops the further motion of carriers in the direction perpendicular to the current flow.

4 L.B. Valdes 'Resistitrity Measurements in Ge for Transistors', Proco *I.R.E.*, **42**, 420 (1954).

This implies, however, that the sample dimensions should satisfy the requirement $L >> d$ for the Hall electric field to appear. In the opposite limiting case, $d << L$, most of the carriers moving under the Hall angle arrive at the contact, not at sample side, as illustrated by Fig. 11.24b, and hence, the Hall electric field is almost totally shorted by the contact. As a result, carriers travel a longer distance,

$$L_H = L(1 + \tan^2 \theta)^{1/2} \quad ...(2)$$

and the component of their velocity in the direction of the applied electric field decreases owing to the component of the Lorentz force related to the component of the electron velocity parallel to the contacts. Assuming a zero Hall field we find

$$\mathbf{v} = \mu\,(\mathbf{F} + \mathbf{vxB}) \quad ...(3)$$

where $\mu$ is a low field mobility. Hence,

$$v_x = \mu\,(F_x + v_y B) \quad ...(4)$$

$$v_y = -\mu v_x B \quad ...(5)$$

(because $F_y = 0$). Substituting Eq. (5) into Eq. (4), we find

$$v_x = \frac{\mu F_x}{1 + \mu^2 B^2} \quad ...(6)$$

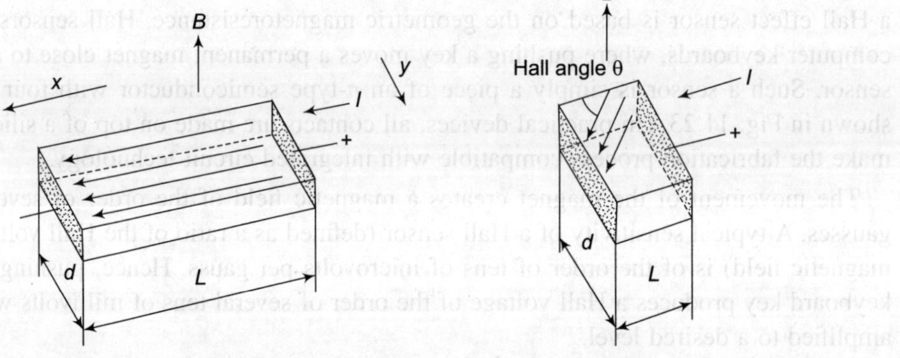


**Figure 11.24** Streamlines of electric current in semiconductor sample with $d << L$ (a) and $d >> L$ (b)

Equation (6) means that the effective low-mobility, $\mu_B$, for a sample with $L/d << 1$ in the magnetic field $B$ is given by

$$\mu_B = \frac{\mu}{1 + \mu^2 B^2} \quad ...(7)$$

The related increase in the low-field resistivity, $\Delta\rho$, in the magnetic field (geometric magnetoresistance) is given by

$$\frac{\Delta\rho}{\rho_o} = (\mu B)^2 \qquad ...(8)$$

Here $$\rho_o = \frac{1}{qn\mu} \qquad ...(9)$$

is the low-field resistivity, and $n$ is the carrier concentration.

Measuring the geometric magnetoresistance allows one to measure the low-field mobility, μ, (see for example, Jervis and Johnson 1970). Equation (8) for the geometric magnetoresistance is valid when the Hall field is completely shorted by the contacts. This may only be achieved when the ratio $L/d$ tends to infinity.

## 11.28 CYCLOTRON RESONANCE

The electrons in conduction band and holes in valence band describe spiral orbits about the axis of a constant magnetic field $B$. The angular frequency of rotation is given by

$$m^* \frac{v^2}{r} = \pm evB \qquad ...(1a)$$

or $$\omega_c = \pm \frac{eB}{m^*} \qquad ...(1b)$$

where $r$ is radius of orbit. The plus or minus sign indicates the opposite senses of rotation of electrons and holes. Now, if a radio frequency electric field is applied perpendicular to static magnetic field, the absorption of energy may place under resonance condition.

$$\frac{d\vec{v}}{dt} = \frac{e}{m^*} \vec{v} \times \vec{B} \qquad ...(2)$$

period of an orbit $$T = \oint dt = \oint \frac{dr}{v} \qquad ...(3)$$

$$\vec{v} = \frac{\partial E}{\partial p} = \frac{1}{\hbar}\frac{\partial E}{\partial k} = \frac{1}{\hbar}\frac{dE}{dk_\perp} \qquad ...(4)$$

where $dk_\perp$ is normal distance in $\vec{k}$-space projected a plane perpendicular to $\vec{B}$.

Now $$\hbar \frac{d\vec{k}}{dt} = e\vec{v} \times B \Rightarrow \vec{k} = \frac{e}{\hbar} \vec{r} \times \vec{B} \Rightarrow d\vec{r} = \frac{\hbar}{eB} d\vec{k} \qquad ...(5)$$

Hence, $$T = \oint \frac{d\vec{r}}{u} = \hbar \oint\int d\vec{r} \cdot \frac{d\vec{k}_\perp}{dE} = \int\oint \frac{\hbar^2}{eB} d\vec{k} \cdot \frac{d\vec{k}_\perp}{dE} = \frac{\hbar^2}{eB} \frac{\oint \vec{k}_\perp dk}{dE}$$

(Using Eq. (4) for $\vec{v}$ and Eq. (5) for $d\vec{r}$)

or $$T = \frac{\hbar^2}{eB}\frac{dA}{dE} \quad (A = \text{Area of orbit in } k\text{-space}) \qquad ...(6)$$

$$\omega_c = \frac{eB}{m^*} = \frac{2\pi}{T} = \frac{2\pi eB}{\hbar^2}\frac{dE}{dA} \qquad ...(7)$$

$$\Rightarrow \qquad m^* = \frac{\hbar^2}{2\pi}\frac{dA}{dE} \qquad ...(8)$$

i.e., the time period $T$ is correlated with effective mass and hence with Fermi-surface. By measuring $\omega_c$ for different directions of $\vec{B}$. One measures the effective mass in different directions.

### 11.28.1 Effective Mass Tensor from Cyclotron Resonance

For electron at the bottom of the conduction band or for holes at the top of the valence band one may assume that $E(\vec{k})$ has a minimum or a maximum respectively at the edge of the band. If the edge of the band occurs at $\vec{k} = \vec{k}_0$, then using Taylor's expansion

$$E(\vec{k}) = E(\vec{k}_0) + \frac{1}{2}\sum_i \sum_j \left(\frac{\partial^2 E}{\partial k_i \partial k_j}\right)_{k_0} \Delta k_i \Delta k_j \quad ...(9)$$

where $\Delta\vec{k} = (\vec{k} - \vec{k}_0)$. But since this is a position-definite quantity for electrons in the conduction band or a negative quantity for electron in the valance band; one can choose the axes for the $\vec{k}$'s such that only terms with $\Delta k_x^2$, $\Delta k_y^2$ and $\Delta k_z^2$ remain. In such cases there are only three effective masses, and the constant energy surfaces are ellipsoids around $\vec{k} = \vec{k}_0$.

It may be that the ellipsoids has rotational symmetry around one of its axes. The constant energy surfaces are then spheroids and there are only different terms in the mass tensor:

$$\begin{pmatrix} m_t & 0 & 0 \\ 0 & m_t & 0 \\ 0 & 0 & m_l \end{pmatrix} \quad ...(10)$$

a transverse mass $m_t$ and a longitudinal mass $m_l$. (e.g., this occurs in the conduction bands of germanium and silicon). Their constant energy surfaces and the associated effective masses can be explored by cyclotron resonance experiments.

The material under study is kept at liquid helium temperature and a static magnetic field is applied, then the carriers rotate in circles with an angular frequency

$$\omega_c = \frac{eB}{m^*} \quad ...(11)$$

In addition, a microwave electric field is applied at right angles to the magnetic field. As a result, energy is absorbed which is maximum at the frequency $\omega_c$. By measuring microwave absorption as a function of the magnetic field $B$ and noting at what field absorption is maximum, we can determine the effective mass $m^*$.

Let the $z$-direction point in the direction of the axis of the spheroid and let the applied field $\vec{B}$ make an angle $\theta$ with the $z$-axis in the $xz$-plane. The equation of motion of the electron is

$$\frac{d\vec{p}}{dt} = e\vec{v} \times \vec{B} \quad ...(12)$$

where $\vec{v}$ has components $p_x/m_t$, $p_y/m_t$ and $p_z/m_t$ and the field has copponents $B_x = B \sin\theta$, $B_y = 0$ and $B_z = B\cos\theta$.

$$\vec{p} = (p_x\hat{i} + p_y\hat{j} + p_z\hat{k}) \quad ...(13)$$

and $$p \sim e^{j\omega t}$$

$$\vec{v} \times \vec{B} = \begin{vmatrix} \hat{i} & \hat{j} & \hat{k} \\ v_x & v_y & v_z \\ B_x & B_y & B_z \end{vmatrix} \qquad ...(14)$$

so that $$\left\{ \begin{aligned} j\omega p_x &= e(v_y B_z - v_z B_y) \\ j\omega p_y &= e(v_z B_x = v_x B_z) \\ j\omega p_z &= e(v_x B_y - v_y B_x) \end{aligned} \right. \qquad ...(15)$$

Putting $$\omega_t = \frac{eB}{m_t} \text{ and } \omega_l = \frac{eB}{m_l} \qquad ...(16)$$

we get $$\left\{ \begin{aligned} j\omega p_x - \omega_t \cos\theta\, p_y + 0 &= 0 \\ \omega_t \cos\theta\, p_x + j\omega p_y = \omega_l \sin\theta\, p_z &= 0 \\ 0 + \omega_t \sin\theta\, p_y + j\omega p_z &= 0 \end{aligned} \right. \qquad ...(17)$$

It has a solution if the determinant of coefficient vanishes, i.e.,

$$\begin{vmatrix} j\omega & -\omega_t \cos\theta & 0 \\ \omega_t \cos\theta & j\omega & -\omega_l \sin\theta\, p_z \\ 0 & \omega_t \sin\theta & j\omega \end{vmatrix} = 0 \qquad ...(18)$$

This gives $$\omega^2 = \omega_c^2 = \omega_t^2 \cos^2\theta + \omega_t\, \omega_l \sin^2\theta$$

$$= \frac{e^2B^2}{m_t} \cos^2\theta + \frac{e^2B^2}{m_t m_l} \sin^2\theta \qquad ...(19)$$

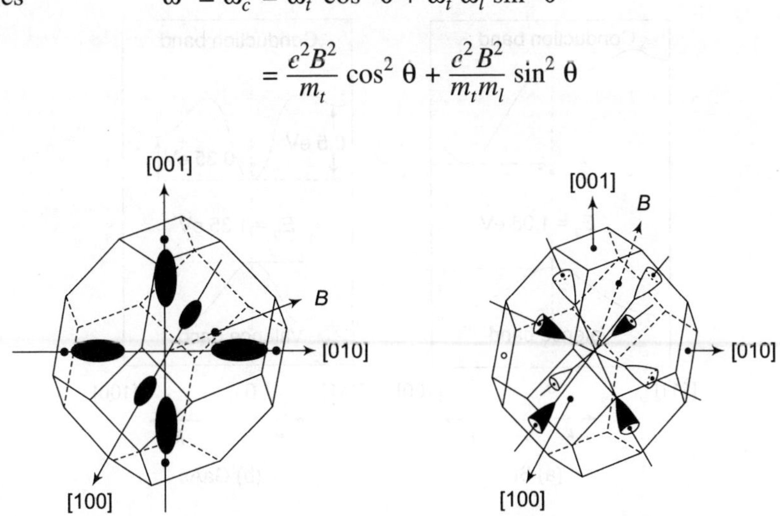


**Figure 11.25** Constant energy ellipsoids near the conduction band minimum in: (a) Silicon, (b) Germanium

The effective mass is then given by

$$\left(\frac{1}{m^*}\right)^2 = \frac{\cos^2\theta}{m_t^2} + \frac{\sin^2\theta}{m_t m_l} \quad \left(\because w_c^2 = \frac{e^2B^2}{m^{*2}}\right) \qquad ...(20)$$

By measuring $m^*$ as a function of θ, one can determine $m_t$ and $m_l$. It is found for silicon that $m_t = 0.18\ m_e$ and $m_t = 0.98\ m_e$ where $m_e$ is the mass of a free electron. The energy minima in Si lies in the [100], [010] and [001] axes.

In germanium, the minima lie in the [111], $[\bar{1}11]$, $[1\bar{1}1]$ and $[11\bar{1}]$ directions.

## 11.29 DIRECT AND INDIRECT SEMICONDUCTORS

The wave function of an electron in a lattice is assumed to be in the form of a plane wave, e.g., for motion in $x$-direction

$$\psi_{\vec{k}}(x) = V(\vec{k}_x, x)\ e^{ik_x x}$$

This is the space-dependent wave function for the electron where the function $V(\vec{k}_x, x)$ modulates the wave function according to the periodicity of the lattice. In such a calculation, allowed values of energy can be plotted against the propagation constant vector. Since the periodicity of most lattices is different in various directions, the $E(\vec{k})$ diagram is plotted for the various crystal directions, e.g., as in Fig. 11.26 for silicon (Si) and gallium arsenide (GaAs) respectively. The full relationship between $E$ and $\vec{k}$ is visualized as a complex surface in three dimensions.

It is clear from Fig. 11.26 that maxima of the dispersion curves for the valence bands of these crystals lie in the middle of the Brillion zone and curves resemble a parabola with its

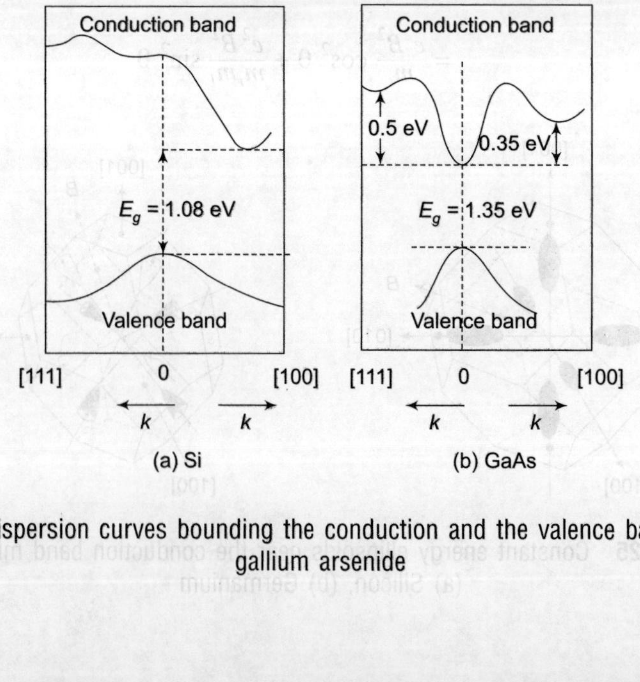


**Figure 11.26** Dispersion curves bounding the conduction and the valence band in silicon and gallium arsenide

vertex facing upwards. The minima of the dispersion curve for the conduction band of silicon are found to lie at a distance $k < \pi/a$ from the middle of the zone in the [1 0 0] directions. GaAs has two minima of this type; one in the middle of the Brillion's zone and the other at a distance 0.75 $\pi/a$ from the middle of the zone in the [1 0 0] directions. Customarily, the minimum of the dispersion curves is referred to as the *bottom* of the energy band and a maximum as the *top* of the energy band. GaAs has a minimum in the conduction band and a maximum in the valence band for the same $\vec{k}$ value ($\vec{k} = 0$). On the other hand, Si has its valence band maximum at a different value of $\vec{k}$ than its conduction band minimum. Thus, an electron making a smaller energy-transition (in GaAs) from conduction band to valence band can do so without a change in $\vec{k}$-value. On the other hand, a transition from conduction band minimum to valence band maximum in Si requires some change in $\vec{k}$. Thus, GaAs and Si are termed respectively *direct* and *indirect band gap (B.G.) semiconductors* Fig. (11.27).

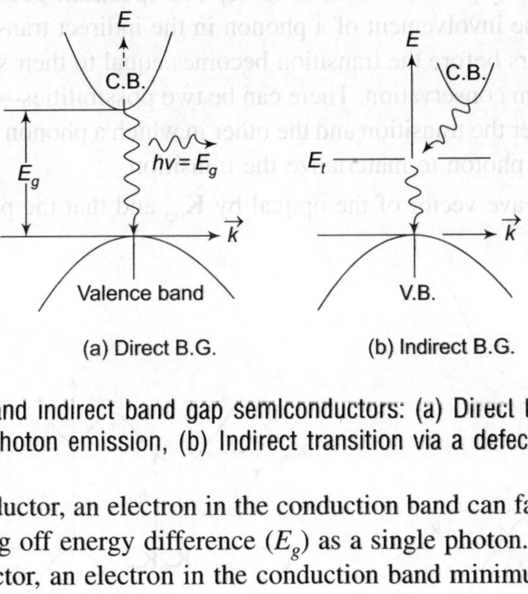


**Figure 11.27** Direct and indirect band gap semiconductors: (a) Direct transition accompanied by a photon emission, (b) Indirect transition via a defect level

In a direct semiconductor, an electron in the conduction band can fall to an empty state in the valence band giving off energy difference ($E_g$) as a single photon. On the other hand, in an indirect semiconductor, an electron in the conduction band minimum cannot fall directly to the valence band maximum, but must undergo a change in momentum change as well as change in energy, e.g., it involvers a defect state or phonons. Then for indirect transition

$$\hbar u = E_g + \hbar\omega \qquad ...(1)$$

where $\hbar\omega$ is the energy of phonons.

The energy difference is given up as heat to the lattice. This concept is very important for deciding which semiconductors can be used for device requiring light output, e.g., LED's and lasers. These usually make use of materials capable of direct transition or indirect materials with transition between defect states.

## 11.30 METHODS OF DETERMINATION OF BAND GAP

A number of methods are used to determine the band gap of semiconductors. The technique of continuous optical absorption is more often used on account of its accuracy and the important information gained about the band structure. An abrupt increase in the absorption of optical radiation takes place as soon as the energy of the incident radiation $\hbar\omega$ becomes large enough to exceed the energy gap.

In direct-gap semiconductors (e.g., GaAs, InSb), where the conduction band minimum and the valence band maximum occur at the same *k*-value in the *k*-space, the optical threshold at $\omega$ (= $E_g / h$) directly gives the band gap (Fig. 11.28 (a)). But in indirect-gap materials (e.g., Si, Ge, GaP), the direct photon absorption accompanied by the transfer of an electron from the top of the valence band to the bottom of the conduction band would not conserve the crystal momentum because the initial and final points of the transition in the *k*-space do not have the same *k*-value. Hence, such a direct transition is not allowed. The transition process will have to be indirect in which the absorption of an optical photon must be accompanied by some other process with whose involvement the condition of momentum conservation may be satisfied. The intensity of a continuous absorption spectrum gets sufficient contribution from phonons. With the involvement of a phonon in the indirect transition under discussion, the sum of wave vectors before the transition becomes equal to their sum after the transition, showing the momentum conservation. There can be two possibilities—one in which a phonon is emitted (created) after the transition and the other in which a phonon is absorbed (destroyed) along with the optical photon to materialize the transition.

Let us denote the wave vector of the optical by $\mathbf{K}_{op}$ and that the phonon by $\mathbf{K}_{ph}$. If the

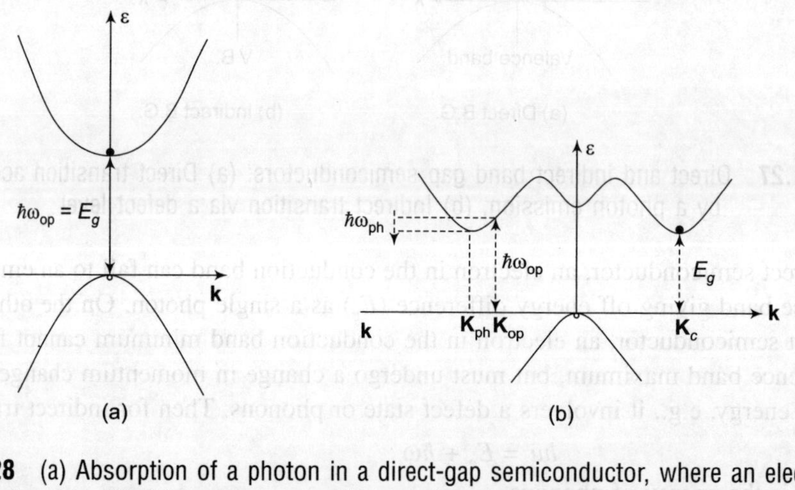


**Figure 11.28** (a) Absorption of a photon in a direct-gap semiconductor, where an electron from the valence band is lifted into the conduction band and the crystal momentum is conserved, where it is shown that a phonon is emitted on the phonon absorption to satisfy the momentum conservation. The $\mathbf{K}_{op}$ and $\mathbf{K}_{ph}$ repectively, denote the wave vectors of the absorbed optical photon and the emitted phonon

wave vector at the conduction band minimum as measured from the valence band maximum be represented by $\mathbf{K}_c$, then in the case of phonon emission [Fig. 11.28 (b)], we have

$$\underset{\text{(before the transition)}}{\mathbf{K}_{op}} = \underset{\text{(after the transition)}}{\mathbf{K}_c + \mathbf{K}_{ph}} \qquad \text{...(1)}$$

But the photon wave vectors $\mathbf{K}_{op}$ are negligibly small in the range of energies concerned. Therefore, from Eq. (1),

$$\mathbf{K}_c \simeq -\mathbf{K}_{ph} \qquad ...(2)$$

Relation (2) states that the increase of crystal momentum by $\hbar\mathbf{K}_c$ during the transition is offset by an equal amount owing to the emission of a phonon with wave vector $(-\mathbf{K}_c)$. Since the phonon takes away a part of the energy of the incident photon (though very small), the optical threshold energy is greater than $E_g$ [see Fig. 11.28(b)]. In this case,

$$\hbar\omega_{op} = E_g + \hbar\omega_{ph} \qquad ...(3)$$

At high temperatures a good number of phonons are present in the crystal. With the absorption of a phonon during the transition, the momentum conservation is given by

$$\underset{\text{(before the transition)}}{\mathbf{K}_{op} + \mathbf{K}_{ph}} = \underset{\text{(after the transition)}}{\mathbf{K}_c} \qquad ...(4)$$

This gives optical threshold as

$$\hbar\omega_{op} = E_g - \hbar\omega_{ph} \qquad ...(5)$$

So, the optical threshold is lower than $E_g$. The change in optical threshold in these processes is generally of little consequence because the phonon energies are characteristically small (~ a few hundredths of an eV) as compared to the size of the band gap. This change, however, is of considerable value in semiconductors having a small band gap.

Since the indirect transitions generate heat when phonons are created, the direct-gap materials are preferred in order to have efficient devices. A mixture of two types of materials is a desired manipulation in some devices such as injection lasers where it is practised on account of the demand for higher power.

Another method commonly used for determining the band gap is based on the temperature dependent study of the electrical conductivity $\sigma$. We will see later in this chapter that electrical conductivity of semiconductors depends on temperature according to the following proportionality:

$$\sigma \propto \exp(-E_g/2k_BT) \qquad ...(6)$$

An approximate value of $E_g$ is given by the $\ln(\sigma)$ versus $1/T$ graph. The gap is also determined from the intrinsic carrier concentration derived from the experimental values of Hall coefficient.

## 11.31 THE EFFECTIVE MASSES FOR ELECTRONS AND HOLES

As described earlier, the most generalized solution of the electron wave functions in a periodic crystal is a plane wave modulated by the Bloch function, $u_k(r)$. For the time-dependent electron wave functions, this can be written as

$$\phi_k(r, t) = u_k(r)\, e^{i(k\cdot r - \omega t)}. \qquad ...(1)$$

Since the wave function for a Bloch-type wave packet extends over the entire crystal lattice, the group velocity for such a wave packet is given by

$$\upsilon_g = \frac{d\omega}{dk} = \left(\frac{1}{\hbar}\right)\nabla_k E(k). \qquad \text{...(2)}$$

Note that the electron energy $E(k) = \hbar\omega$ is used in Eq. (2) to define the group velocity, $\upsilon_g$. According to Eq. (2), the group velocity of an electron wave packet is in the direction perpendicular to the constant-energy surface at a given wave vector $k$ in $k$-space. The group velocity can be determined by the gradient of energy with respect to the wave vector $k$.

If a Lorentz force $\boldsymbol{F}$, which may be due to either an electric field or a magnetic field, is applied to the electron inside a crystal, then the wave vector of electrons will change with the applied Lorentz force according to the following relation

$$\boldsymbol{F} = -q\,(\varepsilon + \boldsymbol{v}_g \times \boldsymbol{B})\,\hbar\left(\frac{dk}{dt}\right) = \hbar k, \qquad \text{...(3)}$$

where $\varepsilon$ is the electric field, and $\boldsymbol{B}$ is the magnetic flux density. The product $\hbar k$ is referred to as the change of crystal momentum. Equation (3) shows that the external applied force acting on an electron tends to change the crystal momentum or the electron wave vector in a crystal lattice. The electron effective mass in a crystal lattice can be defined by

$$\boldsymbol{F} = m_n^* \mathbf{a} = m_n^*\left(\frac{d\boldsymbol{v}_g}{dt}\right). \qquad \text{...(4)}$$

Solving Eq. (2) through Eq. (4) one obtains an expression of acceleration for electrons due to the applied Lorentz force, which is given by

$$\boldsymbol{a} = \frac{d\boldsymbol{v}_g}{dt} = \left(\frac{1}{\hbar}\right)\left(\frac{d\nabla_k E}{dk}\right)\left(\frac{dk}{dt}\right) = \left(\frac{1}{\hbar^2}\right)\left(\frac{d^2E}{dk^2}\right)\cdot \boldsymbol{F}. \qquad \text{...(5)}$$

Solving Eqs. (4) and (5), one obtains an expression for the reciprocal effective mass tensor for electron, whose component is given by

$$(m_n^*)_{ij}^{-1} = \left(\frac{1}{\hbar^2}\right)\left(\frac{\partial^2 E(k)}{\partial k_i\,\partial k_j}\right), \qquad \text{...(6)}$$

where $i, j = 1, 2, 3...$ are the indices used to define the crystal orientations. From Eq. (6), it is noted that the reciprocal effective mass is directly proportional to the curvature of the energy band structure in the $E$ versus $k$ plot. A large curvature near the conduction band minimum implies a small effective mass of electrons, and vice versa. For example, a comparison of the curvatures of the energy band diagrams near the bottom of the conduction band for silicon and GaAs shows that silicon has a smaller curvature than GaAs near the conduction band minimum, and hence has a larger electron effective mass than that of a GaAs crystal.

Another important concept to be discussed in this section is concerned with holes in the valence bands of a semiconductor. A hole in the valence band marks the absence of a valence electron or the creation of an empty state in the valence band. Furthermore, the motion of a hole can be regarded as the motion of a missing electron in the valence band. Since most of the holes reside near the top of the valence band maximum in which the curvature of the $E$ versus $k$ diagram is always negative, which implies a negative electrons effective mass, it is appropriate to replace the missing electrons by the positively charged holes. This arrangement greatly simplifies the treatment of electronic conduction in the valence band of a semiconductor. By using the concept of holes, which have a positive effective mass and a positive charge, the inverse hole effective mass can be derived from the expression

$$\frac{d\boldsymbol{v}_g}{dt} = -\left(\frac{1}{m_n^*}\right)\boldsymbol{F} = \left(\frac{1}{\hbar^2}\right)\nabla_{k'}^2 E_{k'} \cdot \boldsymbol{F} = \left(\frac{1}{m_h^*}\right)\boldsymbol{F}, \quad \text{...(7)}$$

which yields $$\frac{1}{m_h^*} = \left(\frac{1}{\hbar^2}\right)\nabla_{k'}^2 E_{k'}, \quad \text{...(8)}$$

where $\boldsymbol{F}$ is the Lorentz force experienced by a hole. Thus, a hole in the valence band may be considered a charged particle with a positive charge $q$ and a positive effective mass, $m_h^*$. Figure 11.29 shows the electrons near the bottom of the conduction band and holes near the top of the valence band.

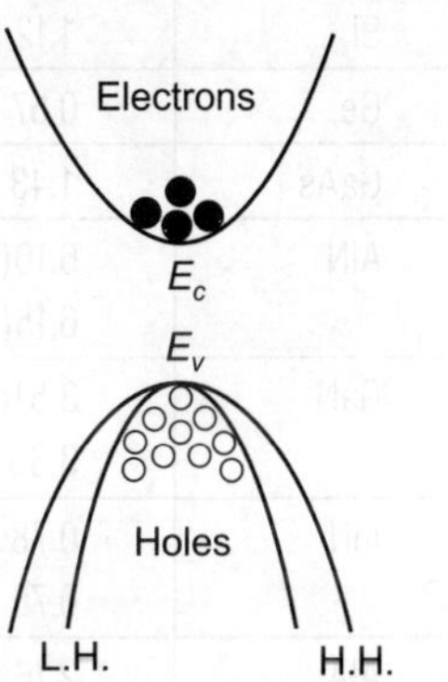


**Figure 11.29** Electrons near the bottom of the conduction band and holes near the top of the valence band

The effective mass concept presented above is particularly useful for describing the transport properties of a semiconductor. In a semiconductor, most of the electrons reside near the bottom of the conduction band, and holes are located near the top of the valence bands. If the energy band structures near the bottom of the conduction band and the top of the valence bands have spherical constant-energy surfaces, then the effective masses for both electrons and holes are given by a scalar quantity. If one assumes that both the conduction band minimum and the valence band maximum are located at $k = 0$ (i.e., at the zone centre ($\Gamma$-point)), then the $E - k$ relation can be expressed by

$$E_k = E_c + \frac{\hbar^2 k^2}{2m_n^*} \quad \text{...(9)}$$

for electrons in the conduction band, and

$$E_{k'} = E_v - \frac{\hbar^2 k'^2}{2m_h^*} \quad \text{...(10)}$$

for holes in the valence bands, both the heavy- and light-hole bands degenerate into a single band at the top of valence band edge.

Equations (9) and (10) may be used to describe the $E - k$ relation for electrons near the bottom of the conduction bands and holes near the top of the valence bands with parabolic bands structure. These relations are valid for direct band gap semiconductors such as GaAs, InP, and InAs, in which the constant-energy surfaces near the conduction band minimum and the valence band maximum are assumed parabolic. If the constant-energy surface near the band edge is nonparabolic, then an effective mass tensor given by Eq. (8) should be used instead. For silicon and germanium, the constant-energy surface near the bottom of the conduction band is ellipsoidal, and the electron effective mass may be expressed in terms of its *transverse* and *longitudinal effective masses* (i.e., $m_t^*$ and $m_l^*$). Both these masses can be determined using the cyclotron resonance experiment performed at very low temperature. The effective masses of electrons and holes for some semiconductors are given in Table 11.6. Using the effective mass concept for electrons in the conduction band and holes in the valence bands, one can treat both the electrons and holes as quasifree particles, which in turn, greatly simplify the mathematics of solving the carrier transport problem in a semiconductor.

**Table 11.6** Energy Band Gaps and Effective Masses for Elemental and Compound Semiconductors at 300 K.

| Element | $E_g$ (eV) | Electron mass ($m_e^*/m_0$) | Hole mass ($m_e^*/m_0$) |
|---|---|---|---|
| Si | 1.12 | $m_t^* = 0.19$, $m_l^* = 0.97$ | $m_{lh}^* = 0.16$, $m_{hh}^* = 0.50$ |
| Ge | 0.67 | $m_t^* = 0.082$, $m_l^* = 1.6$ | $m_{lh}^* = 0.04$, $m_{hh}^* = 0.30$ |
| GaAs | 1.43 | 0.068 | $m_{lh}^* = 0.074$, $m_{hh}^* = 0.62$ |
| AlN | 6.10(WZ)<br>6.15(ZB) | $m_t^* = 0.33$, $m_l^* = 0.32$ | $m_{hht} = 0.73$, $m_{hhl} = 3.52$;<br>$m_{lh} = 0.471$ |
| GaN | 3.51(WZ)*<br>3.35 (ZB) | $m_t^* = 0.22$, $m_l^* = 0.20$ | $m_{hht} = 0.39$, $m_{hhl} = 2.04$<br>$m_{lht} = 0.39$, $m_{lhl} = 0.74$ |
| InN | 0.78 (WZ)<br>0.70 (ZB)* | $m_t^* = 0.07$, $m_l^* = 0.06$ | $m_{hht} = 0.14$, $m_{hhl} = 2.09$<br>$m_{hht} = 0.13$, $m_{hhl} = 0.50$ |
| AlAs | 2.16 | $m_l^* = 2.0$ | $m_{lh}^* = 0.15$, $m_{hh}^* = 0.76$ |
| GaP | 2.26 | $m_l = 1.12$, $m_t^* = 0.22$ | $m_{hh}^* = 0.79$, $m_{lh}^* = 0.14$ |
| GaSb | 0.72 | 0.045 | $m_{hh}^* = 0.62$, $m_{hl}^* = 0.074$ |
| InP | 1.29 | 0.08 | $m_{hh}^* = 0.85$, $m_{hl}^* = 0.089$ |
| InAs | 0.33 | 0.023 | $m_{hh}^* = 0.60$, $m_{hl}^* = 0.027$ |
| InSb | 0.16 | 0.014 | $m_{hh}^* = 0.60$, $m_{hl}^* = 0.027$ |
| CdS | 2.42 | 0.17 | 0.60 |
| CdSe | 1.70 | 0.13 | 0.45 |
| CdTe | 1.50 | 0.096 | 0.37 |
| ZnSe | 2.67 | 0.14 | 0.60 |
| ZnTe | 2.35 | 0.18 | 0.65 |
| ZnS | 3.68 | 0.28 | — |
| PbTe | 0.32 | 0.22 | 0.29 |

*Note:* $m_t^*$ denotes transverse effective mass $m_l^*$ longitudinal effective mass $m_{lh}^*$ light-hole mass, $m_{hh}^*$ heavey-hole mass, and $m_0$ free-electron mass ($9.1 \times 10^{-31}$ kg).

* WZ: Wurtzite structure, ZB: Zinblende structure.

## 11.32 THE POSITION OF THE FERMI LEVEL IN SEMICONDUCTOR (QUALITATIVE IDEA)

Electrons in solids obey Fermi–Dirac statistics. The distribution of electron over a range of allowed energy levels at thermal equilibrium is given by the Fermi–Dirac distribution function

$$f(E) = \frac{1}{1 + \exp\left\{\frac{E - E_F}{kT}\right\}} \quad ...(1)$$

where $k$ is Boltzmann constant.

$$k = 8.62 \times 10^{-5} \text{ eV/K} = 1.38 \times 10^{-23} \text{ J/K}.$$

$f(E)$ is the probability that an available energy state at $E$ will be occupied by an electron at absolute temperature $T$. (Fig. 11.30).

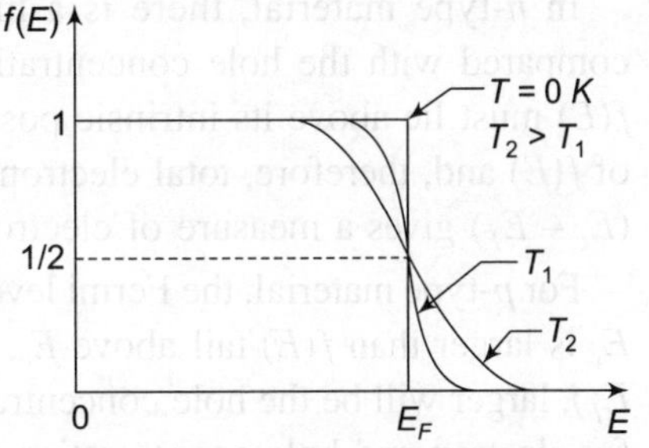


**Figure 11.30** The Fermi distribution function

At temperature 0K, every available energy state up to $E_F$ is full and all states above $E_F$ are empty. At temperatures higher than 0K, some probability exists for states above the Fermi level to be filled, e.g., at $T = T_1$ there is some probability $f(E)$ that states above $E_F$ are filled and there is a corresponding probability $[1 - f(E)]$ that states below $E_F$ are empty. (Note that function $f(E)$ is symmetrical about $E_F$ for all temperatures.) In applying the Fermi–Dirac distribution to semiconductors, we can best visualize the relation between $f(E)$ and the band structure by turning the $f(E)$ vs $E$ diagram on its side so that the $E$ scale corresponds to the energies of the band diagram (Fig. 11.31).

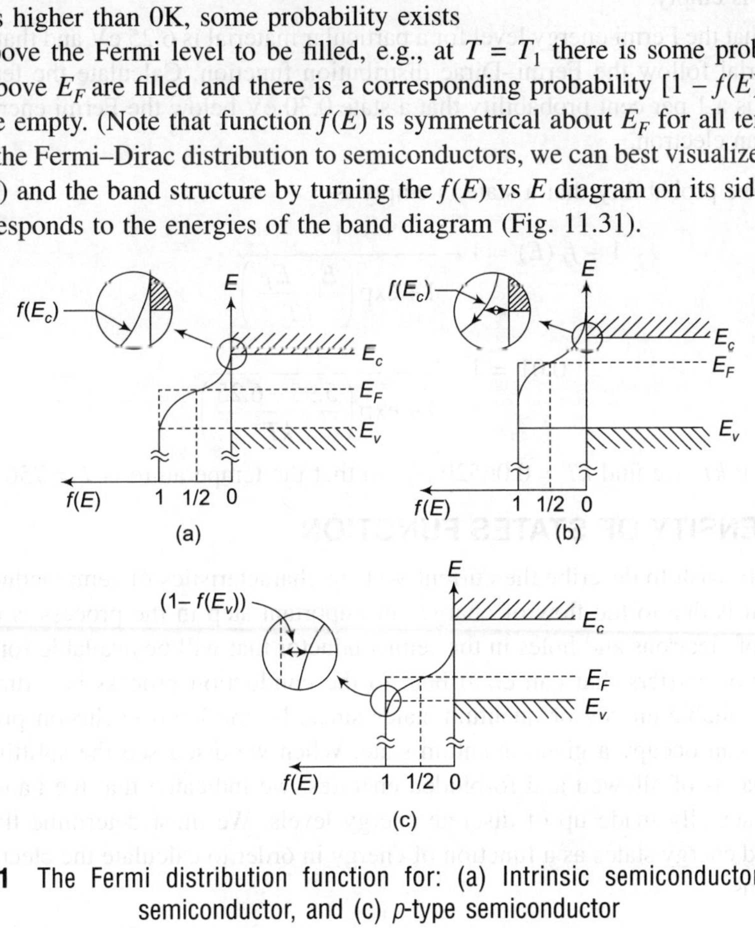


**Figure 11.31** The Fermi distribution function for: (a) Intrinsic semiconductor, (b) *n*-type semiconductor, and (c) *p*-type semiconductor

For intrinsic material, we know that there is an equal concentration of electrons (in conduction band) and holes (in valence band) therefore the Fermi level $E_F$ must lie at the middle of the band gap. Though distribution function has values within the band gap (between $E_v$ and $E_c$), there are no energy states available, consequently no electron occupancy from $f(E)$ in this range [Fig. 11.31 (a)].

In *n*-type material, there is a high concentration of electrons in the conduction band compared with the hole concentration in the valence band. Thus, the distribution function $f(E)$ must lie above its intrinsic position on the energy scale (in *n*-type material). The value of $f(E)$ and, therefore, total electron concentration increases as $E_F$ moves closer to $E_c$. Thus, $(E_c - E_F)$ gives a measure of electron concentration [Fig. 11.31(b)].

For *p*-type material, the Fermi level lies near the valence band such that $[1 - f(E)]$ tail below $E_v$ is larger than $f(E)$ tail above $E_c$. Larger the value of $(E_F - E_v)$ (or smaller the value $E_v - E_F$), larger will be the hole concentration [Fig. 11.31(c)]. We subsequently derive expressions for electron and hole concentrations.

**Example 11.1** To determine the temperature at which is a 1 per cent probability that an energy state is empty.

Assume that the Fermi energy level for a particular material is 6.25 eV and that the electron in this material follow the Fermi–Dirac distribution function. Calculate the temperature at which there is a 1 per cent probability that a state 0.30 eV below the Fermi energy level will not contain an electron.

***Solution*** The probability that a state is empty is

$$1 - f_F(E) = 1 - \frac{1}{1 + \exp\left(\frac{E - E_F}{kT}\right)}$$

Then

$$0.01 = 1 - \frac{1}{1 + \exp\left(\frac{5.95 - 6.25}{kT}\right)}$$

Solving for $kT$, we find $kT = 0.06529$ eV, so that the temperature is $T = 756$ K.

## 11.33 DENSITY OF STATES FUNCTION

We eventually wish to describe the current-voltage characteristics of semiconductor devices. Since current is due to the flow of charge, an important step in the process is to determine the number of electrons and holes in the semiconductor that will be available for conduction. The number of carriers that can contribute to the conduction process is a function of the number of available energy or quantum states since, by the Pauli exclusion principle, only one electron can occupy a given quantum state. When we discussed the splitting of energy levels into bands of allowed and forbidden energies, we indicated that the band of allowed energy was actually made up of discrete energy levels. We must determine the density of these allowed energy states as a function of energy in order to calculate the electron and hole concentrations.

### 11.33.1 Mathematical Derivation

To determine the density of allowed quantum states as a function of energy, we need to consider an appropriate mathematical model. Electrons are allowed to move relatively freely in the conduction band of a semiconductor, but are confined to the crystal. As a first step, we will consider a free electron confined to a three-dimensional infinite potential well, where the potential well represents the crystal. The potential of the infinite potential well is defined as

$$V(x, y, z) = 0 \quad \text{for} \quad \begin{matrix} 0 < x < a \\ 0 < y < a \\ 0 < z < a \end{matrix} \qquad ...(1)$$

$$V(x, y, z) = \infty \quad \text{elsewhere}$$

where the crystal is assumed to be a cube with length $a$. Schrödinger's wave equation in three dimensions can be solved using the separation of variables technique. Extrapolating the results from the one-dimensional infinite potential well, we can show that

$$\frac{2mE}{\hbar^2} = k^2 = k_x^2 + k_y^2 + k_z^2 = (n_x^2 + n_y^2 + n_z^2)\left(\frac{\pi^2}{a^2}\right) \qquad ...(2)$$

where $n_x$, $n_y$ and $n_z$ are positive integers. (Negative values of $n_x$, $n_y$ and $n_z$ yield the same wave function, except for the sign, as the positive integer values, resulting in the same probability function and energy, so the negative integers do not represent a different quantum state.)

We can schematically plot the allowed quantum states in $k$ space. Figure. 11.32a shows a two-dimensional plot as a function of $k_x$ and $k_y$. Each point represents an allowed quantum state corresponding to various integral values of $n_x$ and $n_y$. Positive and negative values of $k_x$, $k_y$ or $k_z$ have the same energy and represent the same energy state. Since negative values of

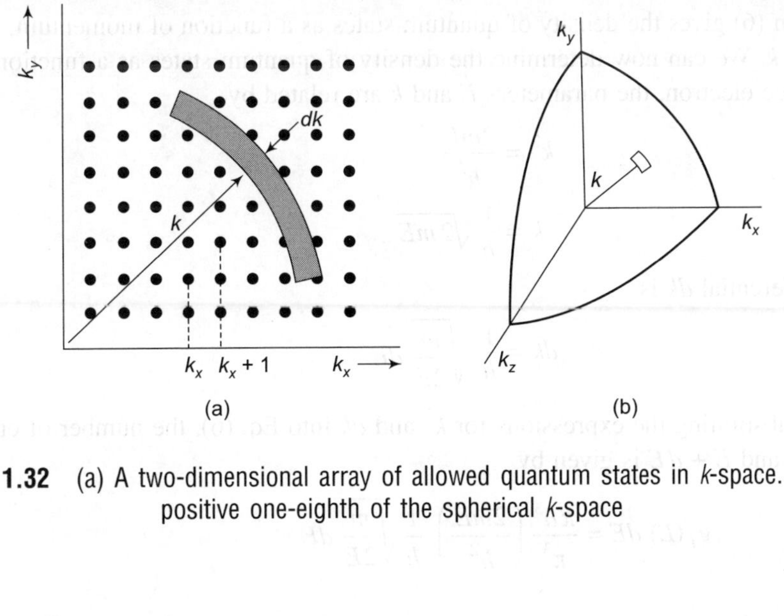


**Figure 11.32** (a) A two-dimensional array of allowed quantum states in $k$-space. (b) The positive one-eighth of the spherical $k$-space

$k_x$, $k_y$ or $k_z$ do not represent additional quantum states, the density of quantum states will be determined by considering only the positive one-eighth of the spherical $k$ space as shown in Fig. 11.32b.

The distance between two quantum states in the $k_x$ direction, for example, is given by

$$k_{x+1} - k_x = (n_x + 1)\left(\frac{\pi}{a}\right) - n_x\left(\frac{\pi}{a}\right) = \frac{\pi}{a} \quad ...(3)$$

Generalizing this result to three dimensions, the volume $V_k$ of a single quantum state is

$$V_k = \left(\frac{\pi}{a}\right)^3 \quad ...(4)$$

We can now determine the density of quantum states in $k$-space. A differential volume in $k$-space is shown in Figure 11.32b and is given by $4\pi k^2 dk$, so the differential density of quantum states in $k$ space can be written as

$$g_T(k)\, dk = 2\left(\frac{1}{8}\right)\frac{4\pi k^2\, dk}{\left(\frac{\pi}{a}\right)^3} \quad ...(5)$$

The first factor, 2, takes into account the two spin states for each quantum states; the next factor, $\frac{1}{8}$, take into account that we are considering only the quantum states for positive values of $k_x$, $k_y$, and $k_x$. The factor $4\pi k^2\, dk$ is again the differential volume and the factor $(\pi/a)^3$ is the volume of one quantum state. Equation (5) may be simplified to

$$g_T(k)\, dk = \frac{\pi\, k^2\, dk}{\pi^3} \cdot a^3 \quad ...(6)$$

Equation (6) gives the density of quantum states as a function of momentum, through the parameter, $k$. We can now determine the density of quantum states as a function of energy, $E$. For a free electron, the parameters $E$ and $k$ are related by

$$k^2 = \frac{2mE}{\hbar^2} \quad ...(7a)$$

or

$$k = \frac{1}{\hbar}\sqrt{2mE} \quad ...(7b)$$

The differential $dk$ is

$$dk = \frac{1}{\hbar}\sqrt{\frac{m}{2E}}\, dE \quad ...(8)$$

Then, substituting the expressions for $k^2$ and $dk$ into Eq. (6), the number of energy states between $E$ and $E + dE$ is given by

$$g_T(E)\, dE = \frac{\pi a^3}{\pi^3}\left(\frac{2mE}{\hbar^2}\right) \cdot \frac{1}{\hbar}\sqrt{\frac{m}{2E}}\, dE \quad ...(9)$$

Since $\hbar = h/2\pi$, Eq. (9) becomes

$$g_T(E)\,dE = \frac{4\pi a^3}{h^3}\cdot(2m)^{3/2}\cdot\sqrt{E}\,dE \qquad ...(10)$$

Equation (10) gives the total number of quantum states between the energy $E$ and $E + dE$ in the crystal space volume of $a^3$. If we divide by the volume $a^3$, then we will obtain the density of quantum states per unit volume of the crystal. Equation (10) then becomes

$$g(E) = \frac{4\pi(2m)^{3/2}}{h^3}\sqrt{E} \qquad ...(11)$$

The density of quantum states is a function of energy, $E$. As the energy of this free electron becomes small, the number of available quantum states decreases. This density function is really a double density, in that the units are given in terms of states per unit energy per unit volume.

**Example 11.2** To calculate the density of states per unit volume over a particular energy range.

Consider the density of states for a free electron given by Eq. (9). Calculate the density of states per unit volume between 0 and 1 eV.

***Solution*** The volume density of quantum states, from Eq. (9) is

$$N = \int_0^{1\text{eV}} g(E)\,dE = \frac{4\pi(2m)^{3/2}}{h^3}\cdot\int_0^{1\text{eV}}\sqrt{E}\,dE$$

or

$$N = \frac{4\pi(2m)^{3/2}}{h^3}\cdot\frac{2}{3}\cdot E^{3/2}$$

The density of states is now

$$N = \frac{4\pi[2\,(9.11\times10^{-31})]^{3/2}}{(6.625\times10^{-34})^3}\cdot\frac{2}{3}\cdot(1.6\times10^{-19})^{3/2}$$

$$= 4.5\times10^{27}\ \text{m}^{-3}$$

or

$$N = 4.5\times10^{21}\ \text{states/cm}^3$$

The density of quantum states is typically a large number. An effective density of states in a semiconductor, as we will see in the following sections and in the next chapter, is also a large number, but is usually less than the density of atoms in the semiconductor crystal.

## 11.33.2 Extension to Semiconductors

In the last section, we derived a general expression for the density of allowed electron quantum states using the model of a free electron with mass $m$ bounded in a three-dimensional infinite potential well. We can extend this same general model to a semiconductor to determine the density of quantum states in the conduction band and the density of quantum states in the valence band. Electrons and holes are confined within the semiconductor crystal so we will again use the basic model of the infinite potential well.

The parabolic relationship between energy and momentum of a free electron is given as $E = p^2/2m = \hbar^2k^2/2m$. Figure 11.33a shows the conduction energy band in the reduced $k$ space. The $E$ versus $k$ curve near $k = 0$ at the bottom of the conduction band can be approximated as a parabola, so we may write

$$E = E_c + \frac{\hbar^2k^2}{2m_n^*} \qquad ...(12)$$

where $E_c$ is the bottom edge of the conduction band and $m_n^*$ is the electron effective mass. Equation (12) may be rewritten to give

$$E - E_c = \frac{\hbar^2k^2}{2m_n^*} \qquad ...(13)$$

The general form of the $E$ versus $k$ relation for an electron in the bottom of a conduction band is the same as the free electron, except the mass is replaced by the effective mass. We can then think of the electron in the bottom of the conduction band as being a "free" electron with its own particular mass. The right side of Eq. (13) is of the same form as the right side of equation, $E = \frac{\hbar^2k^2}{2m}$, which was used in the derivation of the density of states function. Because of this similarity, which yields the "free" conduction electron model, we may generalize the free electron results of Eq. (11) and write the density of allowed electronic energy states in the conduction band as

$$\boxed{g_c(E) = \frac{4\pi(2m_n^*)^{3/2}}{h^3}\sqrt{E - E_c}} \qquad ...(14)$$

Equation (14) is valid for $E \geq E_c$. As the energy of the electron in the conduction band decreases, the number of available quantum states also decreases.

The density of quantum states in the valence band can be obtained by using the same infinite potential well model, since the hole is also confined in the semiconductor crystal and can be treated as a "free" particle. The effective mass of the hole is $m_p^*$. Figure 11.33b shows the valence energy band in the reduced $k$ space. We may also approximate the $E$ versus $k$ curve near $k = 0$ by a parabola for a "free" hole, so that

$$E = E_u - \frac{\hbar^2k^2}{2m_p^*} \qquad ...(15)$$

Equation (15) may be rewritten to give

$$E_u - E = \frac{\hbar^2k^2}{2m_p^*} \qquad ...(16)$$

Again, the right side of Eq. (16) is of the same form used in the general derivation of the density of states function. We may then generalize the density of states function from Eq. (11) to apply to the valence band, so that

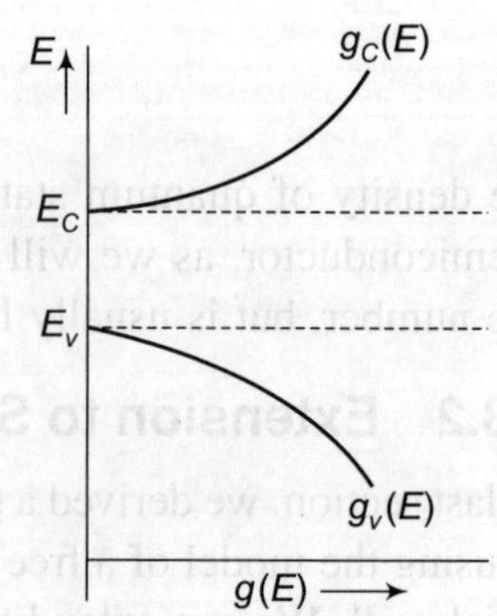


**Figure 11.33** The density of energy states in the conduction band and the density of energy states in the valence band as a function of energy

$$g_u(E) = \frac{4\pi(2m_p^*)^{3/2}}{h^3}\sqrt{E_u - E} \qquad ...(17)$$

Equation (17) is valid for $E \leq E_u$.

We have argued that quantum states do not exist within the forbidden energy band, so $g(E) = 0$ for $E_u < E < E_c$. Figure 11.33 shows the plot of the density of quantum states as a function of energy. If the electron and hole effective masses were equal, then the functions $g_c(E)$ and $g_u(E)$ would be symmetrical about the energy midway between $E_c$ and $E_u$, or the midgap energy, $E_{\text{midgap}}$.

Figure 11.34 shows the band diagram, DOS, FD distribution and the carrier concentrations.

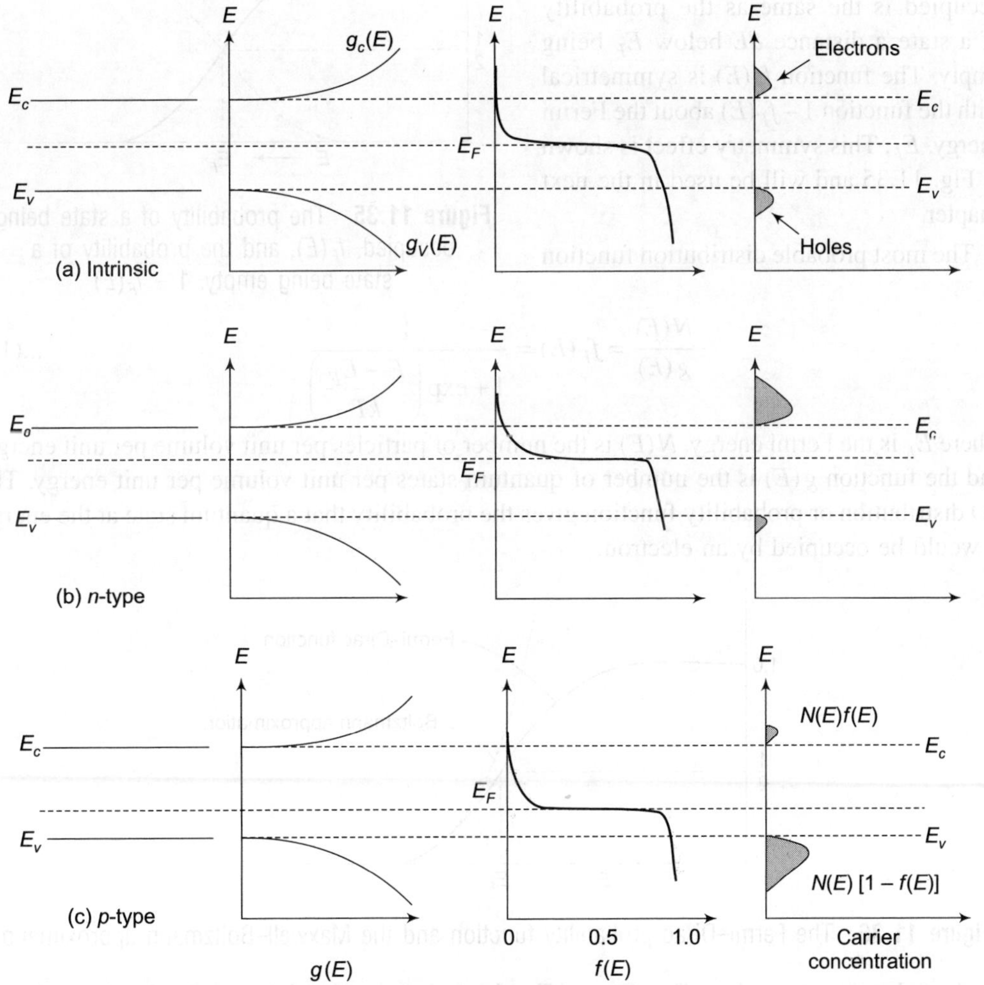


**Figure 11.34** Density of states, Fermi-Dirac distribution, and the carrier concentrations for (a) intrinsic, (b) *n*-type, and (c) *p*-type semiconductors at thermal equilibrium

### 11.33.3 The FD Probability of a State being Occupied or Empty and Maxwell Boltzmann Approximation

We see from Fig. 11.30 that for temperatures above absolute zero, there is a nonzero probability that some energy states above $E_F$ will be occupied by electrons and some energy states below $E_F$ will be empty. This means that some electrons have jumped to higher energy levels with increasing thermal energy. It can also be seen from Fig. 11.30 that the probability of an energy above $E_F$ being occupied increases as the temperature increases and the probability of a state below $E_F$ being empty increases as the temperature increases.

We may note that the probability of a state at distance $dE$ above $E_f$ being occupied is the same as the probability of a state a distance $dE$ below $E_F$ being empty. The function $f_F(E)$ is symmetrical with the function $1 - f_F(E)$ about the Fermi energy, $E_F$. This symmetry effect is shown in Fig. 11.35 and will be used in the next chapter.

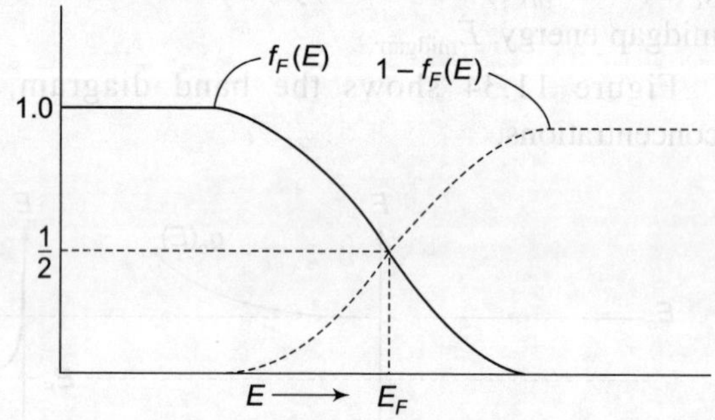


**Figure 11.35** The probability of a state being occupied, $f_F(E)$, and the probability of a state being empty, $1 - f_F(E)$

The most probable distribution function is

$$\frac{N(E)}{g(E)} = f_F(E) = \frac{1}{1 + \exp\left(\frac{E - E_F}{kT}\right)} \quad ...(18)$$

where $E_F$ is the Fermi energy, $N(E)$ is the number of particles per unit volume per unit energy, and the function $g(E)$ is the number of quantum states per unit volume per unit energy. The FD distribution or probability function gives the probability that a quantum state at the energy $E$ would be occupied by an electron.

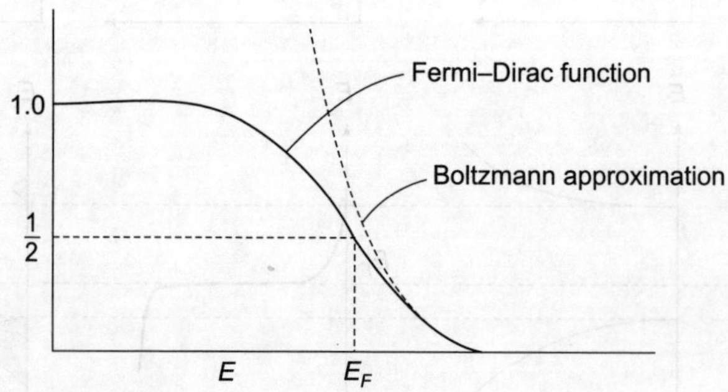


**Figure 11.36** The Fermi–Dirac probability function and the Maxwell–Boltzmann approximation

Consider the case when $E - E_F \gg kT$, where the exponential term in the denominator of Eq. (10) is much greater than unity. We may neglect the 1 in the denominator, so the Fermi–Dirac distribution function becomes

$$\boxed{f_F(E) \approx \exp\left[\frac{-(E-E_F)}{kT}\right]} \quad ...(19)$$

Equation (19) is known as the Maxwell–Boltzmann approximation, or simply the Boltzmann approximation, to the Fermi–Dirac distribution function. Figure 11.36 shows the Fermi–Dirac probability function and the Boltzmann approximation. This figure gives an indication of the range of energies over which the approximation is valid.

**Example 11.3** To determine the energy at which the Boltzmann approximation may be considered valid.

Calculate the energy, in terms of $kT$ and $E_F$, at which the difference between the Boltzmann approximation and the Fermi–Dirac function is 5 per cent of the Fermi function.

***Solution*** We can write

$$\frac{\exp\left[\frac{-(E-E_F)}{kT}\right] - \frac{1}{1+\exp\left(\frac{E-E_F}{kT}\right)}}{\frac{1}{1+\exp\left(\frac{E-E_F}{kT}\right)}} = 0.05$$

If we multiply both numerical and denominator by the 1 + exp ( ) function, we have

$$\exp\left[\frac{-(E-E_F)}{kT}\right]\cdot\left\{1+\exp\left[\frac{E-E_F}{kT}\right]\right\} - 1 = 0.05$$

which becomes

$$\exp\left[\frac{-(E-E_F)}{kT}\right] = 0.05$$

or

$$(E - E_F) = kT \ln\left(\frac{1}{0.05}\right) \approx 3kT$$

As seen in this example and in Figure 11.36 the $E - E_F \gg kT$ notation is somewhat misleading. The Maxwell–Boltzmann and Fermi–Dirac functions are within 5 per cent of each other when $E - E_F \approx 3kT$.

## 11.34 CHARGE CARRIERS IN SEMICONDUCTORS

Current is the rate at which charge flows. In a semiconductor, two types of charge carrier, the electron and the hole, can contribute to a current. Since the current in a semiconductor is determined largely by the number of electrons in the conduction band and the number of holes in the valence band, an important characteristic of the semiconductor is the density of these charge carriers. The density of electrons and holes is related to the density of states function and the Fermi distribution function, both of which we have considered. A qualitative discussion of these relationships will followed by a more rigorous mathematical derivation of the thermal-equilibrium concentration of electrons and holes.

### 11.34.1 Equilibrium Distribution of Electrons and Holes

The distribution (with respect to energy) of electrons in the conduction band is given by the density of allowed quantum states multiplied by the probability that a state is occupied by an electron. This statement is written in equation form as

$$n(E) = g_c(E)\, f_F(E) \qquad \text{...(1)}$$

where $f_F(E)$ is the Fermi–Dirac probability function and $g_c(E)$ is the density of quantum states in the conduction band. The total electron concentration per unit volume in the conduction band is then found by integrating Eq. (1) over the entire conduction-band energy.

Similarly the distribution (with respect to energy) of holes in the valence band is the density of allowed quantum states in the valence band multiplied by the probability that a state is not occupied by an electron. We may express this as

$$p(E) = g_v(E)\,[1 - f_F(E)] \qquad \text{...(2)}$$

$$g_c(E) = \frac{4\pi}{h^3}(2m_n)^{3/2}(E - E_c)^{1/2}\, dt \qquad \text{...(2a)}$$

and

$$g_v(E) = \frac{4\pi}{h^3}(2m_p)^{3/2}(E_V - E)^{1/2}\, dt \qquad \text{...(2b)}$$

where the energy $E$ is measured from the conduction band edge.

The total hole concentration per unit volume is found by integrating this function over the entire valence band energy.

To find the thermal equilibrium electron and hole concentrations, we need to determine the position of the Fermi energy ($E_F$) with respect to the bottom of the conduction-band energy $E_c$, and the top of the valence band energy, $E_v$. To address this question, we initially consider an intrinsic semiconductor. An ideal intrinsic semiconductor is a pure semiconductor with no impurity atoms and no lattice defects in the crystal (e.g., pure silicon). We have argued in the previous chapter that, for an intrinsic semiconductor at $T = 0\,\text{K}$, all energy states in the valence band are filled with electrons and all energy states in the conduction band are empty of electrons. The Fermi energy must therefore, be somewhere between $E_c$ and $E_u$. (The Fermi energy does not need to correspond to an allowed energy.)

As the temperature begins to increase above 0K, the valence electrons will gain thermal energy. A few electrons in the valence band may gain sufficient energy to jump to the conduction band. As an electron jumps from the valence band to the conduction band, an empty state, or hole, is created in the valence band. In an intrinsic semiconductor, then, electrons and holes are created in pairs by the thermal energy so that the number of electrons in the conduction band is equal to the number of holes in the valence band.

Figure 11.37 (a) shows a plot of the density of states function in the conduction band, $g_c(E)$, the density of states function in the valence band $g_u(E)$, and the Fermi–Dirac probability function for $T > 0\,\text{K}$ when $E_F$ is approximately halfway between $E_c$ and $E_u$. If we assume, for the moment, that the electron and hole effective masses are equal, then $g_c(E)$ and $g_u(E)$ $g_v(E)$ are symmetrical functions about the midgap energy (the energy midway between $F_c$ and $E_u$). We noted earlier that the function $f_F(F)$ for $E > E_F$ is symmetrical to the function $1 - f_F(E)$

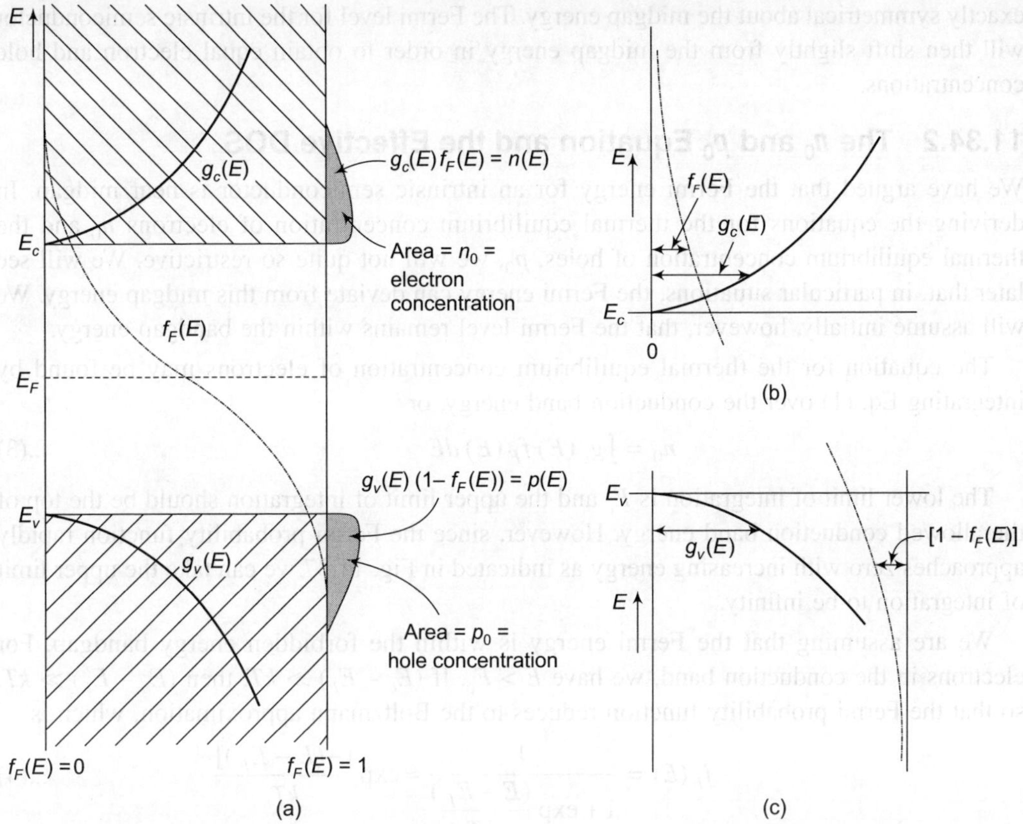


**Figure 11.37** (a) Density of states functions Fermi–Dirac probability function, and areas representing electron and hole concentrations for the case when $E_F$ is near the midgap energy; (b) expanded view near the conduction band energy; and (c) expanded view near the valence band energy

for $E < E_F$ about the energy $E = E_F$. This also means that the faction $f_F(E)$ for $E = E_F + dE$ is equal to the function $1 - f_F(E)$ for $E = E_F - dE$.

Figure 11.37 is an expanded view of the plot in Fig. 11.37 (a) showing $f_F(E)$ and $g_c(E)$ above the conduction band energy, $E_c$. The product of $g_c(E)$ and $f_F(E)$ is the distribution of electrons $n(E)$ in the conduction band given by Eq. 1. This product is plotted in Fig. 11.37 (a). Figure 3.1 (c) is an expanded view of the plot in Fig. 11.37 (a) showing $[1 - f_F(E)]$ and $g_v(E)$ below the valance band energy, $E_v$. The product of $g_v(E)$ and $[1 - f_F(E)]$ is the distribution of holes $p(E)$ in the valence band given by Eq. (2). This product is also plotted in Fig. 11.37 (a). The areas under these curves are then the total density of electrons in the conductors in the conduction band and total density of holes in the valence band. From this we see that if $g_c(E)$ and $g_v(E)$ are symmetrical, the Fermi energy must be at the midgap energy in order to obtain equal electron and hole concentrations. If the effective masses of the electron and hole are not exactly equal, then the effective density of states functions, $g_c(E)$ and $g_v(E)$, will not be exactly symmetrical about the midgap energy. The Fermi level for the intrinsic semiconductor will then shift slightly from the midgap energy in order to obtain equal electron and hole concentrations.

### 11.34.2 The $n_0$ and $p_0$ Equation and the Effective DOS

We have argued that the Fermi energy for an intrinsic semiconductor is near midgap. In deriving the equations for the thermal equilibrium concentration of electrons $n_0$ and the thermal equilibrium concentration of holes, $p_0$, we will not quite so restrictive. We will see later that, in particular situations, the Fermi energy can deviate from this midgap energy. We will assume initially, however, that the Fermi level remains within the bandgap energy.

The equation for the thermal equilibrium concentration of electrons may be found by integrating Eq. (1) over the conduction band energy, or

$$n_0 = \int g_c(E) f_F(E)\, dE \qquad ...(3)$$

The lower limit of integration is $E_c$ and the upper limit of integration should be the top of the allowed conduction band energy. However, since the Fermi probability function rapidly approaches zero with increasing energy as indicated in Fig. 11.37 we can take the upper limit of integration to be infinity.

We are assuming that the Fermi energy is within the forbidden energy bandgap. For electrons in the conduction band, we have $E > E_c$. If $(E_c - E_F) \gg kT$, then $(E_c - E_F) \gg kT$, so that the Fermi probability function reduces to the Boltzmann approximation, which is

$$f_F(E) = \frac{1}{1+\exp\dfrac{(E-E_F)}{kT}} \approx \exp\frac{[-(E-E_F)]}{kT} \qquad ...(4)$$

Applying the Boltzmann approximation to Eq. (3), the thermal equilibrium density of electrons in the conduction band is found from

$$n_0 = \int_{E_c}^{\infty} \frac{4\pi\,(2m_n^*)^{3/2}}{h^3} \sqrt{E-E_c}\, \exp\left[\frac{-(E-E_F)}{kT}\right] dE \qquad ...(5)$$

The integral of Eq. (5) may be solved more easily by making a change of variable. If we let

$$n = \frac{E-E_c}{kT} \qquad ...(6)$$

then Eq. (5) becomes

$$n_0 = \frac{4\pi\,(2m_n^* kT)^{3/2}}{h^3} \exp\left[\frac{-(E_c-E_F)}{kT}\right] \int_0^{\infty} \eta^{1/2} \exp(-\eta)\, d\eta \qquad ...(7)$$

The integral is the gamma function, with a value of

$$\int_0^{\infty} \eta^{1/2} \exp(-\eta)\, d\eta = \frac{1}{2}\sqrt{\pi} \qquad ...(8)$$

Then Eq. (7) becomes

$$n_0 = 2\left(\frac{2\pi m_n^* kT}{h^2}\right)^{3/2} \exp\left[\frac{-(E_c - E_F)}{kT}\right] \quad ...(9)$$

We may define a parameter $N_c$ as

$$N_c = 2\left(\frac{2\pi m_n^* kT}{h^2}\right)^{3/2} \quad ...(10)$$

so that thermal equilibrium electron concentration in the conduction band can be written as

$$\boxed{n_0 = N_c \exp\left[\frac{-(E_c - E_F)}{kT}\right]} \quad ...(11)$$

The parameter $N_c$ is called the *effective density of states function in the conduction band.* If we were to assume that $m_n^* = m_0$, then the value of the effective density of states function at $T$ = 300 K is $N_c$ = 2.5 × $10^{19}$ $cm^{-3}$, which is the order of magnitude of $N_c$ for most semiconductors. If the effective mass of the electron is larger or smaller than $m_0$, then the value of the effective density of states function changes accordingly, but is still of the same order of magnitude.

**Example 11.4** Calculate the probability that a state in the conduction band is occupied by electron and calculate the thermal equilibrium electron concentration in silicon at $T$ = 300 K.

Assume the Fermi energy is 0.25 eV below the conduction band. The value of $N_c$ for silicon at $T$ = 300 K is $K_c$ = 2.8 × $10^{19}$ $cm^{-3}$.

***Solution*** The probability that an energy at $E = E_c$ is occupied by an electron is given by

$$f_F(E_c) = \frac{1}{1+\exp\left(\dfrac{E_c - E_F}{kT}\right)} \simeq \exp\left[\frac{-(E_c - E_F)}{kT}\right]$$

or

$$f_F(E_c) = \exp\left(\frac{-0.25}{0.0259}\right) = 6.43\times10^{-5}$$

The electron concentration is given by

$$n_0 = N_c \exp\left[\frac{-(E_c - E_F)}{kT}\right] = (2.8\times10^{19})\exp\left(\frac{-0.25}{0.0259}\right)$$

or

$$n_0 = 1.8 \times 10^{15} \text{ cm}^{-3}$$

The probability of a state being occupied can be quite small, but the fact that there is a large number of states means that the electron concentration is a reasonable value. The thermal equilibrium concentration of holes in the valence band is found by integrating Eq. (2) over the valence band energy, or

$$p_0 = \int g_v(E)[1 - f_F(E)]\, dE \quad ...(12)$$

We may note that

$$1 - f_F(E) = \frac{1}{1 + \exp\left(\frac{E_F - E}{kT}\right)} \quad ...(13a)$$

For energy states in the valence band, $E < E_v$. If $(E_F - E_v) \gg kT$ (the Fermi function is still assumed to be within the bandgap), then we have a slightly different form of the Boltzmann approximation. Equation (13a) may be written as

$$1 - f_F(E) = \frac{1}{1 + \exp\left(\frac{E_F - E}{kT}\right)} \approx \exp\left[\frac{-(E_F - E)}{kT}\right] \quad ...(13b)$$

Applying the Boltzmann approximation of Eq. (13b) to Eq. (12), we find the thermal equilibrium concentration of holes in the valence band is

$$p_0 = \int_{-\infty}^{E_v} \frac{4\pi(2m_p^*)^{3/2}}{h^3} \sqrt{E_v - E} \exp\left[\frac{-(E_F - E)}{kT}\right] dE \quad ...(14)$$

where the lower limit of integration is taken as minus infinity instead of the bottom of the valence band. The exponential term decays fast enough so that this approximation is valid.

Equation (14) may be solved more easily by again making a change of variable. If we get

$$\eta = \frac{E_v - E}{kT} \quad ...(15)$$

then Eq. (14) becomes

$$p_0 = \frac{-4\pi(2m_p^* kT)^{3/2}}{h^3} \exp\left[\frac{-(E_F - E_v)}{kT}\right] \int_{+\infty}^{0} (\eta')^{1/2} \exp(-\eta')\, d\eta' \quad ...(16)$$

where the negative sign comes from the differential $dE = -kT\, d\eta'$. Note that the lower limit of $\eta'$ becomes $+\infty$ when $E = -\infty$. If we change the order of integration, we introduce another minus sign. From Eq. (8), Eq. (6) becomes

$$p_0 = 2\left(\frac{2\pi m_p^* kT}{h^2}\right)^{3/2} \exp\left[\frac{-(E_F - E_v)}{kT}\right] \quad ...(17)$$

We may define a parameter $N_v$ as

$$N_v = 2\left(\frac{2\pi m_p^* kT}{h^2}\right)^{3/2} \quad ...(18)$$

which is called the *effective density of states function in the valence band*. The thermal equilibrium concentration of holes in the valence band may be written as

$$\boxed{p_0 = N_v \exp\left[\frac{-(E_F - E_v)}{kT}\right]} \quad ...(19)$$

The magnitude of $N_v$ is also in the order of $10^{19}$ cm$^{-3}$ at $T$ = 300K for most semiconductors.

The effective density of states functions, $N_c$ and $N_v$, are constant for a given semiconductor material at a fixed temperature. Table 11.7 gives the values of the density of states function and of the effective masses for silicon, gallium arsenide, and germanium. Note that the value of $N_c$ for gallium arsenide is smaller than the typical $10^{19}$cm$^{-3}$ value. This difference is due to the small electron effective mass in gallium arsenide.

**Table 11.7** Effective Density of States Function Mass Values

| | $N_c$ (cm$^{-3}$) | $N_v$ (cm$^{-3}$) | $m_n^*/m_0$ | $m_p^*/m_0$ |
|---|---|---|---|---|
| Silicon | $2.8 \times 10^{19}$ | $1.04 \times 10^{19}$ | 1.08 | 0.56 |
| Gallium arsenide | $4.7 \times 10^{17}$ | $7.0 \times 10^{18}$ | 0.067 | 0.48 |
| Germanium | $1.04 \times 10^{19}$ | $6.0 \times 10^{18}$ | 0.55 | 0.37 |

The thermal equilibrium concentrations of electrons in the conduction band of holes in the valence band are directly related to the effective density of states constants and to the Fermi energy level.

### 11.34.3 The Intrinsic Carrier Concentration and Law of Action

For an intrinsic semiconductor, the concentration of electrons in the conduction band is equal to the concentration of holes in the valence band. We may denote $n_i$ and $p_i$ as the intrinsic electron and hole concentrations in an intrinsic semiconductor. We have

$$\left.\begin{aligned} n_o &= N_c\, e^{-(E_c - E_F)/kT} \\ p_o &= N_v\, e^{-(E_F - E_v)/kT} \end{aligned}\right\} \quad ...(20)$$

where
$$\left.\begin{aligned} N_c &= 2\left(\frac{2\pi m_n^* kT}{h^2}\right)^{3/2} \\ N_u &= 2\left(\frac{2\pi m_p^* kT}{h^2}\right)^{3/2} \end{aligned}\right\} \quad ...(21)$$

Equation (21) predicts that the hole concentration increases as $E_F$ moves closer to the valence band. The electron and hole concentrations predicted by Eqs. (20) and (21) are valid whether the material is intrinsic or doped, provided thermal equilibrium is maintained. For intrinsic material, thus, $E_F$ lies at some intrinsic level $E_i$ near the middle of the band gap and the intrinsic concentrations are

$$n_i = N_c\, e^{-(E_c - E_i)/kT}, \quad p_i = N_v\, e^{-(E_i - E_v)/kT} \quad ...(22)$$

The product of $n_o$ and $p_o$ at equilibrium is a constant for a particular material and temperature *even if the doping is varied*:

$$n_o p_o = N_c\, e^{-(E_c - E_F)/kT} \cdot N_v\, e^{-(E_F - E_v)/kT} = N_c N_v\, e^{-(E_c - E_v)/kT} \quad ...(23a)$$

$$n_i p_i = N_c\, e^{-(E_c - E_i)/kT} \cdot N_v\, e^{-(E_i - E_v)/kT} = N_c N_v\, e^{-(E_c - E_v)/kT} \qquad ...(23b)$$

Thus, $$n_o p_o = n_i p_i \cdot N_c N_v\, e^{-E_g/kT} \qquad ...(24)$$

Since $n_i = p_i$ (carriers are reacted in pairs),

Therefore $$\boxed{n_o p_o = n_i^2} \qquad ...(25)$$

This is an important relation, known as *law of mass action*.

and $$n_i = \sqrt{N_c N_v}\, e^{-E_g/2kT} \qquad ...(26)$$

Comparing Eq. (22) and Eq. (26), we note that $E_c - E_i = E_g/2$ (if $N_c = N_v$), i.e., the intrinsic level $E_i$ is the middle of the band gap.

Further, using Eq. (22) and Eq. (20).

$$\boxed{\begin{aligned} n_o &= n_i\, e^{(E_F - E_i)/kT} \\ p_o &= n_i\, e^{(E_i - E_F)/kT} \end{aligned}} \qquad ...(27)$$

Table 11.8 lists the commonly accepted values of $n_i$ for silicon, gallium arsenide, and germanium at $T = 300$ K.

The intrinsic carrier concentration is a very strong function temperature.

**Table 11.8** Commonly Accepted Values of $n_i$ at $T = 300$ K

| | |
|---|---|
| Silicon | $n_i = 1.5 \times 10^{10}$ cm$^{-3}$ |
| Gallium arsenide | $n_i = 1.8 \times 10^{6}$ cm$^{-3}$ |
| Germanium | $n_i = 2.4 \times 10^{13}$ cm$^{-3}$ |

**Example 11.5** To calculate the intrinsic carrier concentration in gallium arsenide at $T$ = 300 K and at $T$ = 450 K.

The values of $N_c$ and $N_v$ at 300 K for gallium arsenide are $4.7 \times 10^{17}$ cm$^{-3}$ and $7.0 \times 10^{18}$ cm$^{-3}$, respectively. Both $N_c$ and $N_v$ vary as $T^{3/2}$. Assume the band gap energy of gallium arsenide is 1.42 eV and does not vary with temperature over this range. The value of $kT$ at 450 K is

$$kT = (0.0259)\left(\frac{450}{300}\right) = 0.03885 \text{ eV}$$

***Solution*** Using Eq. (24), we find for $T$ = 300 K

$$n_i^2 = (4.7 \times 10^{17})(7.0 \times 10^{18}) \exp\left(\frac{-1.42}{0.0259}\right) = 5.09 \times 10^{12}$$

so that $$n_i = 2.26 \times 10^{6} \text{ cm}^{-3}$$

At $T$ = 450 K, we find

$$n_i^2 = (4.7 \times 10^{17})\,(7.0 \times 10^{18})\left(\frac{450}{300}\right)^3 \exp\left(\frac{-1.42}{0.03885}\right) = 1.48 \times 10^{21}$$

so that $$n_i = 3.85 \times 10^{10}\ \text{cm}^{-3}$$

We may note from this example that the intrinsic carrier concentration increased by over 4 orders of magnitude as the temperature increased by 150°C.

Figure 11.38 is a plot of $n_i$ from Eq. (23) for silicon, gallium arsenide, and germanium as a function of temperature. As seen in Fig. (11.38), the value of $n_i$ for these semiconductors may easily vary over several orders of magnitude as the temperature changes over a reasonable range.

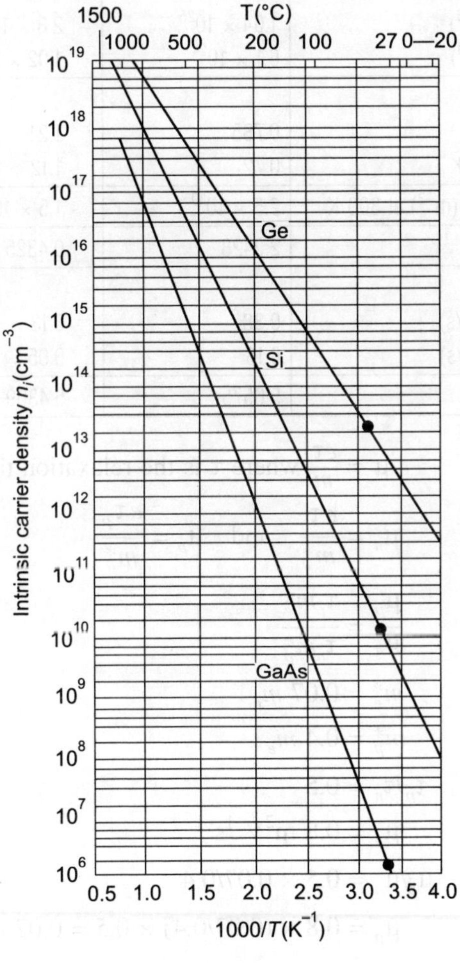


**Figure 11.38** The intrinsic carrier concentration of Ge, Si, and GaAs as a function of temperature

**Example 11.6** In a semiconductor the effective mass of an electron is 0.07 $m_e$ and that of a hole is 0.4 $m_e$ ($m_e$ = free electron mass). Assuming that the average relaxation time for the holes is half that for electrons calculate the mobility of the hole when the mobility of the electron is 0.8 m$^2$ V$^{-1}$ s$^{-1}$.

**Table 11.9** Properties of Ge, Si and GaAs

| *Property* | *Ge* | *Si* | *GaAs* |
|---|---|---|---|
| Atoms/m$^3$ | $4.42 \times 10^{28}$ | $5.0 \times 10^{28}$ | $2.21 \times 10^{28}$ |
| Breakdown Field (V/m) | $\sim 10^7$ | $\sim 3 \times 10^7$ | $\sim 4 \times 10^5$ |
| Crystal structure | Diamond | Diamond | Zinc blende |
| Density (kg/m$^3$) | $5.32 \times 10^3$ | $2.33 \times 10^3$ | $5.32 \times 10^3$ |
| Effective density of states: | | | |
| $N_c$(m$^{-3}$) | $1.04 \times 10^{25}$ | $2.8 \times 10^{25}$ | $4.7 \times 10^{22}$ |
| $N_v$(m$^{-3}$) | $6.1 \times 10^{24}$ | $1.02 \times 10^{25}$ | $7.0 \times 10^{24}$ |
| Forbidden gap $E_g$(eV): | | | |
| at 0 K | 0.785 | 1.21 | 1.43 |
| at 300 K | 0.72 | 1.12 | 1.32 |
| Intrinsic carrier concentration $n_i$(m$^{-3}$) at 300 K | $2.5 \times 10^{19}$ | $1.5 \times 10^{16}$ | $9.0 \times 10^{12}$ |
| Intrinsic conductivity $\sigma$ (mho m$^{-1}$) | 2.2428 | $0.4325 \times 10^{-3}$ | $1.2832 \times 10^{-6}$ |
| (Drift) mobility at 300 K | | | |
| $\mu_n$ (m$^3$/Vs) | 0.38 | 0.13 | 0.85 |
| $\mu_h$ (m$^2$/Vs) | 0.18 | 0.05 | 0.04 |
| Lattice constant a (Å) | 5.6575 | 5.4309 | 5.6534 |

***Solution*** Mobility $\mu = \dfrac{e\tau}{m}$ where $\tau$ is the relaxation time.

$$\mu_n = \frac{e\tau_n}{m_e^*} \quad \text{and} \quad \mu_p = \frac{e\tau_p}{m_e^*}$$

$$\frac{\mu_p}{\mu_n} = \frac{\tau_p m_e^*}{\tau_n m_p^*}$$

Now

$$m_e^* = 0.07\ m_e$$

$$m_p^* = 0.4\ m_e$$

$$\tau_p/\tau_n = 0.5$$

and

$$\mu_n = 0.8\ \text{m}^2\text{V}^{-1}\text{s}^{-1}$$

Thus

$$\mu_p/\mu_n = 0.5 \times 0.07/0.4$$

$$\mu_p = 0.8 \times (0.07/0.4) \times 0.5 = 0.07\ \text{m}^2\text{V}^{-1}\text{s}^{-1}$$

## 11.35 DETERMINATION OF BAND GAP OF INTRINSIC SEMICONDUCTORS

For an intrinsic semiconductor

$$n = p = n_i$$

and intrinsic conductivity

$$\sigma_i = n_i e(\mu_e + \mu_h) \qquad \text{...(1)}$$

Substituting the value of $n_i$ from Eq. (26) we get

$$\sigma_i = 2e\left(\frac{2\pi kT}{h^2}\right)^{3/2}(m_e^* m_h^*)^{3/4}\exp\left(-\frac{E_g}{2\,kT}\right)(\mu_e+\mu_h) \quad ...(2)$$

At higher temperatures, the scattering is dominated by phonons and mobility is proportional to $T^{3/2}$.

Then,
$$\sigma_i = \sigma_i(0)\exp\left(-\frac{E_g}{2\,kT}\right) \quad ...(3)$$

where
$$\sigma_i(0) \equiv 2e\left(\frac{2\pi kT}{h^2}\right)^{3/2}(m_e^* m_h^*)^{3/4}(\mu_e+\mu_h) \quad ...(4)$$

is practically independent of temperature. Taking the logarithm of Eq. (3), we obtain

$$\ln\sigma_i = \ln\sigma_i(0) - \frac{E_g}{2\,kT} \quad ...(5)$$

A plot of ln $\sigma_i$ versus $1/T$ will be a straight line with a slope of $E_g/(2k)$. Thus, knowing the slope of the line, the band gap $E_g$ can be determined.

## 11.36 KINETICS OF EXTRINSIC CASE

If the temperature is quite low, then electrons are excited only from donor levels and no electrons are coming from valence band. Then it can be assumed that donor level is playing the role of valence band.

The density of ionised donors can be written as

$$n_d = N_d\,[1 - f(E_d)] \simeq N_d \exp\left[\frac{E_d - E_F}{kT}\right] \quad ...(6)$$

The density of electron in the conduction band

$$n_e = 2\left[\frac{2\pi m^* kT}{h^2}\right]^{3/2}\exp\left[\frac{(E_F - E_c)}{kT}\right] \quad ...(7)$$

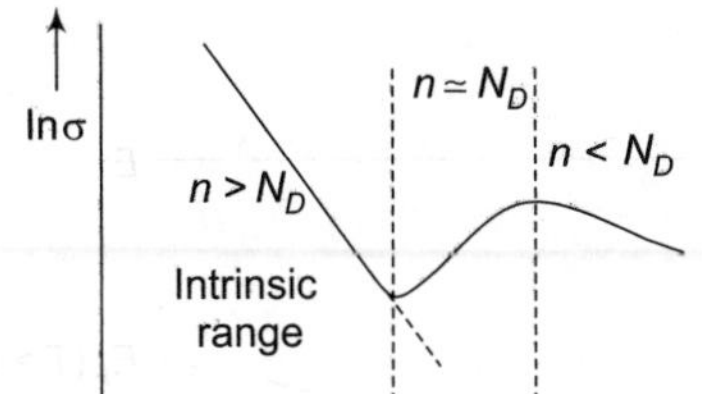


**Figure 11.39** A plot of ln σ as a function of 1/*T* for an *n*-type semiconductor

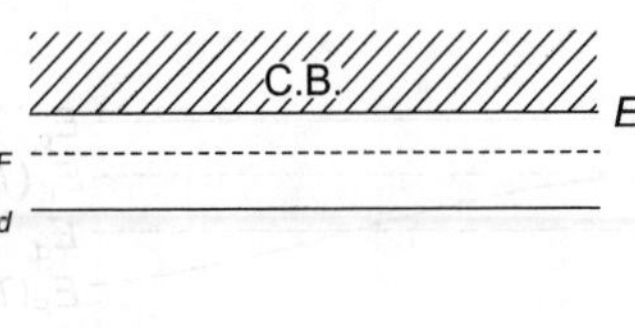


**Figure 11.40** Location of Fermi levels with respect to donor states in *n*-type semiconductor at 0 K

Now, since all the electrons are coming from the donor level, therefore, $n_e = n_d$

Solving, we get

$$E_F = \frac{E_d + E_c}{2} + \frac{kT}{2} \log \left[ \frac{N_d}{2(2\pi m^* kT/h^2)^{3/2}} \right] \qquad ...(8)$$

i.e., at absolute zero, the Fermi level will be exactly midway between $E_d$ and $E_c$. Putting the value of $E_F$ from above Eq. (8) into $\exp\left[\frac{(E_F - E_c)}{kT}\right]$, we get

$$\exp\left[\frac{(E_F - E_c)}{kT}\right] = \exp\left[\frac{E_F - E_c}{2\,kT} + \frac{1}{2}\log\left\{\frac{N_d}{2\,(2\pi m^* kT/h^2)^{3/2}}\right\}\right]$$

$$= \exp\left(\frac{E_d - E_c}{2\,kT}\right)\frac{(N_d)^{1/2}}{\left\{2\left(\frac{2\pi m^* kT}{h^2}\right)^{3/2}\right\}^{1/2}}$$

Substituting this in Eq. (7), we get

$$n_e = 2\left[\frac{2\pi m^* kT}{h^2}\right]^{3/2} \frac{N_d^{1/2}}{\sqrt{2}\left(\frac{2\pi m^* kT}{h^2}\right)^{3/4}} \exp\left(\frac{E_d - E_c}{2\,kT}\right)$$

$$n_e = \sqrt{2 N_d}\left(\frac{2\pi m^* kT}{h^2}\right)^{3/4} \exp\,(-\Delta E/2\,kT) \qquad ...(9)$$

where $\Delta E = E_c - E_d$ represents the ionisation energy of donors.

As the temperature increases, the Fermi level will drop down. However, at higher temperatures, there will be contributions from valence band also, as a result, the Fermi level will not fall below the centre of band gap.

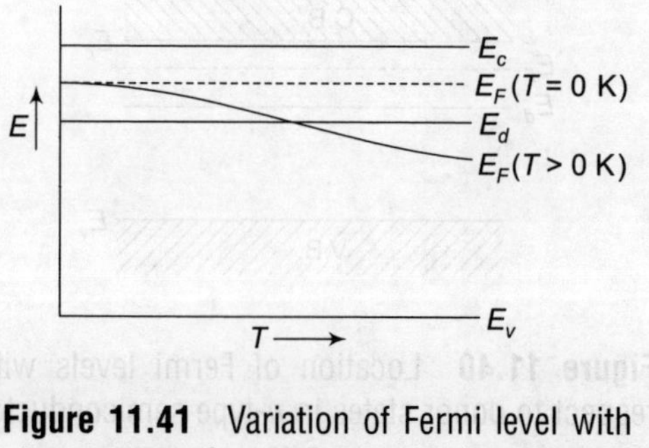


**Figure 11.41** Variation of Fermi level with temperature for an *n*-type semiconductor

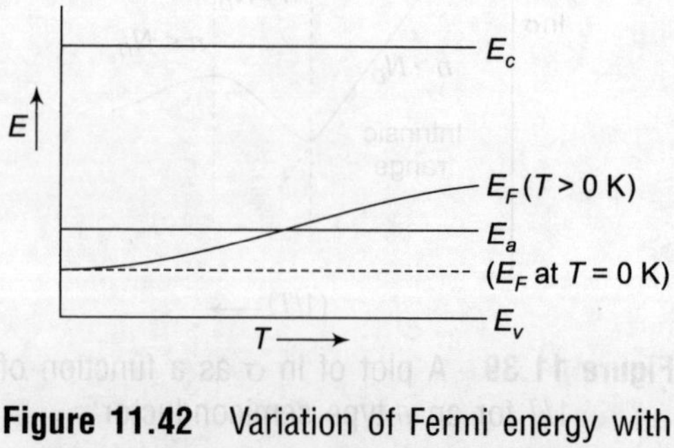


**Figure 11.42** Variation of Fermi energy with temperature for a *p*-type semiconductor

Similarly, we can discuss a $p$-type semiconductor having an acceptor level $E_a$ and acceptor atoms, density $N_a$:

The number of holes $$n_p = \sqrt{2N_a}\left(\frac{2\pi m^* kT}{h^2}\right)^{3/4} \exp\left(-\frac{\Delta E}{2kT}\right) \quad ...(10)$$

where $$\Delta E = E_a - E_v$$

If both types of impurities are present simultaneously, the material will behave as $n$-type or $p$-type depending on which of the impurities dominates. When all the impurities are ionised (Fig. 11.43), the donor atoms will contain an mobile +ve charge and acceptor atoms immobile – ve charge. Balancing the charges, we get

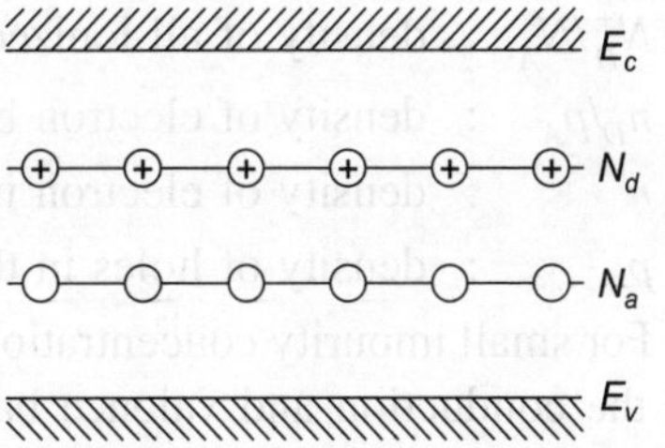


**Figure 11.43** A mixed semiconductor

$$(n_e + N_a)\, e = (n_p + N_d)\, e \quad ...(11)$$

or $$n_e + N_a = N_p + N_d$$

$$\frac{n_i^2}{n_p} + N_a = n_p + N_d \quad (\because n_e n_p = n_i^2)$$

or $$n_p^2 + (N_d - N_a)\, n_p - n_i^2 = 0 \quad ...(12)$$

Solving, we get $$n_p = \frac{-(N_d - N_a) + \sqrt{(N_d - N_a)^2 + 4n_i^2}}{2} \quad ...(13)$$

and $$n_e = \frac{(N_d - N_a) + \sqrt{(N_d - N_a)^2 + 4n_i^2}}{2} \quad ...(14)$$

For $n$-type material, assuming $N_d - N_a \gg 2n_i$

we get $$n_e \cong N_d - N_a \simeq N_d \quad (\text{for } N_d \gg N_a) \quad ...(15)$$

and $$n_p = \frac{n_i^2}{N_d - N_a} \simeq \frac{n_i^2}{N_d} \quad ...(16)$$

Similarly, for a $p$-type material, assuming $N_a - N_d >> 2n_p$ we get

$$n_p = N_a - N_d \simeq N_a \quad (\text{for } N_a \gg N_d) \quad ...(17)$$

and $$n_e = \frac{n_i^2}{N_a - N_d} \simeq \frac{n_i^2}{N_a} \quad ...(18)$$

(For $N_o = N_d$, the material becomes intrinsic in nature)

## 11.37 POPULATION OF DONOR AND ACCEPTOR LEVELS IN THE STATE OF THERMAL EQUILIBRIUM

Let us first define the symbols that will be used in our future descriptions

$N_D/N_A$ : density of all available donors/acceptors
$N_D^0/N_A^0$ : density of all neutral donors/acceptors
$N_D^+/N_A^-$ : density of all ionized donors/acceptors
$n_D/p_A$ : density of electron bound to donors/density of holes bound to acceptors
$n$ : density of electron in the conduction band
$p$ : density of holes in the valence band

For small impurity concentrations (i.e., for non-degenerate semiconductors) the occupancy of the conduction and valence bands is as usual described by the Boltzmann distribution function. Therefore, at these concentrations the law of mass action which was derived on the basis of the above conditions must apply even to extrinsic semiconductors. But in the extrinsic case the value neutrality condition. For homogeneous doping, the neutrality condition can be expressed as

$$n + N_A^- = p + N_D^+ \quad ...(1)$$

with

$$N_D = N_D^0 + N_D^+ \quad ...(2a)$$

$$N_A = N_A^0 + N_A^- \quad ...(2b)$$

The above terminology may be appreciated better with the help of Fig. 11.44. Let us first consider the calculation of population of the donor levels. At low concentrations (up to $10^{17}$ cm$^{-3}$), the interaction between electron bound to separate donors may be neglected. We further assume for simplicity of calculations that the impurity only a single one-electron level.

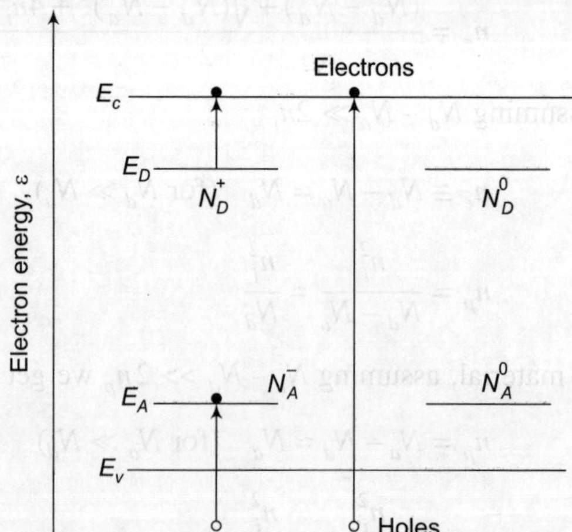


**Figure 11.44** Electrons in the conduction band and holes in the valence band made available either by inter-band excitation or by impurity ionization

Under these approximations the level could either be empty, or contain one electron of either spin, or contain two electrons with opposite spins. The double occupation is not favoured on account of its high energy arising from the column repulsion between two localized electrons. In the state of thermal equilibrium, the mean number of electron in a system is expressed as

$$\langle n \rangle = \frac{\sum_j N_j \exp[-(\varepsilon_j - \varepsilon_F N_j)/k_B T]}{\sum_j \exp[-(\varepsilon_j - \varepsilon_F N_j)/k_B T]} \quad ...(3)$$

where the sum is over all states of the system; the $\varepsilon_j$ and $N_j$ denote the energy and the number of electron in states, $j$. With a single impurity we have just three states: one with no electron, thereby making no contribution to the energy, and two distinguishable spins with a single electron of energy, $E_d$. Hence, from Eq. (3), we have

$$\langle n \rangle = \frac{2\exp[-(E_D - \varepsilon_F)/k_B T)}{1 + 2\exp[-(E_D - \varepsilon_F)/k_B T]}$$

$$= \frac{1}{1 + \frac{1}{2}[E_D - \varepsilon_F)/k_B T]} \quad ...(4)$$

The factor of 1/2 serves to modify the Fermi–Dirac distribution in this case. However, this factor is generally ignored which amounts to treating two single electron spins of equal energy as one state. The population density of donor levels in the states of thermal equilibrium is given by

$$n_D = N_D \langle n \rangle$$

or $$n_D = \frac{N_D}{1 + \exp[E_D - \varepsilon_F)/k_B T]} \quad \text{(ignoring the factor of 1/2)} \quad ...(5)$$

Also note that, $$n_D = N_D^0 \quad ...(6)$$

Similarly, it can be shown that the population density of the acceptor level in thermal equilibrium as given by

$$p_A = \frac{N_A}{1 + \exp[(\varepsilon_F - E_A)/k_B T]} \quad ...(7)$$

and also, $$p_A = N_A^0 \quad ...(8)$$

### 11.37.1 Extrinsic Carrier Densities

The case of an extrinsic semiconductor in which both donors and acceptors are present is difficult to handle. Because of this reason we consider a pure $n$-type semiconductor and calculate its carrier concentration that is contributed by donors alone. Generally, every $n$-type material has a few accepted and every $p$-type material has a few donors because of practical limitations on growing 100 per cent pure crystals. But the concentration of naturally present impurities in a pure crystal is expected to be negligibly small compared to the concentration of the impurity doped. Therefore, the relation (1) in the present case is rewritten as

$$n = N_D^+ + p_i \qquad ...(9)$$

The net electron density $n$ has contributions of electrons from donors and from the valence band (see Fig. 11.14). The density of the later type ($n_i$, being intrinsic) is equal to the density of holes ($p_i$). Normally, $N_D^+ \gg n_i$ (or $p_i$). For example, in Si this condition is satisfied even at low levels of the donor concentration since the intrinsic concentration is mere $1.5 \times 10^{10}$ cm$^{-3}$ around room temperature. Therefore for our $n$-type material, $p_i$ in Eq. (9) may be dropped giving,

$$n \simeq N_D^+ = N_D - N_D^0 \quad \text{[using Eq. (1)]} \qquad ...(10)$$

or

$$n = N_D - n_D \quad (\text{since } n_D = N_D^0)$$

$$= N_D\left(1 - \frac{1}{1 + \exp[(E_D - \varepsilon_F)/k_BT]}\right) \quad \text{[using Eq. (5)]}$$

$$= \left(\frac{N_D}{1 + \exp[E_D - \varepsilon_F/k_BT]}\right) \qquad ...(11)$$

As already mentioned in the beginning of Section 11.37, the relations for $n$, $p$ and $np$ (the law of the mass action) as derived in section 11.35 are applicable to extrinsic materials with low doping levels. We write the relation (see section 11.34.2):

$$n = N(c)\exp\left[\frac{-(E_c - \varepsilon_F)}{k_BT}\right] \qquad ...(12a)$$

with

$$N(c) = 2\left(\frac{m_e^* k_BT}{2\pi\hbar^2}\right)^{3/2} \qquad ...(12b)$$

and Eq. (19) (Art. 11.34.1)

$$p = N(v)\exp\left(\frac{-(\varepsilon_F - E_v)}{k_BT}\right) \qquad ...(13a)$$

with

$$N(v) = 2\left(\frac{m_h^* k_BT}{2\pi\hbar^2}\right)^{3/2} \qquad ...(13b)$$

From (12a), we have

$$\exp\left(\frac{\varepsilon_F}{k_BT}\right) = \frac{n\exp(E_c/k_BT)}{N(c)} \qquad ...(14)$$

Eliminating $\varepsilon_F$ in Eq. (11) with the help of Eq. (14), we get

$$n = \frac{N_D}{1 + \dfrac{n\exp(E_d/k_BT)}{N(c)}} \quad (\text{with } E_c - E_D = E_d)$$

or $$n^2\left(\frac{\exp(E_d/k_BT)}{N(c)}\right) + n - N_D = 0 \qquad ...(15)$$

Using the substitution,

$$\frac{\exp(E_d k_BT)}{N(c)} = X \qquad ...(16)$$

we get $$Xn^2 + n - N_D = 0 \qquad ...(17)$$

The physically meaningful solution to the above equation is

$$n = \frac{-1+\sqrt{1+4N_DX}}{2X} \qquad ...(18)$$

Rationalizing Eq. (18), we get

$$n = 2\,N_D[1+\sqrt{1+4\,N_DX}]^{-1}$$

Using Eq. (16), we have $$n = 2\,N_D\left[1+\sqrt{1+4N_D\frac{\exp(E_d/k_BT)}{N(c)}}\right]^{-1} \qquad ...(19)$$

There are three limiting cases of the above expression as discussed below:

***Case I*** At low temperatures,

$$4\left(\frac{N_D}{N(c)}\right)\exp\left(\frac{E_d}{k_BT}\right) \gg 1 \qquad ...(20)$$

This transforms Eq. (19) to the following form,

$$n \simeq \sqrt{N_DN(c)}\,\exp\left(\frac{-E_d}{2\,k_BT}\right) \qquad ...(21)$$

At these low temperatures a large number of donors stay in the unionized state. The range of temperatures over which this condition exists is known as the *freeze-out range*. $n$ depends exponentially on temperature and the exponential dependence dominates the other dependence entering through $N(c)$. In the donors' case a much smaller quantity $E_d$ appears in the exponential as against $E_g$ in the intrinsic case. This accounts for a larger carrier concentration in the $n$-type material.

***Case II*** For temperatures at which

$$4\left(\frac{N_D}{N(c)}\right)\exp\left(\frac{E_d}{k_BT}\right) \ll 1 \qquad ...(22)$$

the relation (19) reduces to

$$n \simeq N_D \qquad ...(23)$$

i.e., a constant.

This means that at these temperatures all donors are ionized and the electron density reaches its maximum with the excitations of electron from the valence band considered negligible in the first approximation. This is referred to as the *saturation range* or *exhaustion range* of carriers.

***Case III*** When temperatures are still higher, the concentration of conduction electrons contributed by the valence band becomes appreciable. Since the concentration of donor electrons in the conduction band no more increases on account of saturation, the intrinsic electron density overtakes the density of donor electrons at some stage. In this condition an *n*-type semiconductor is characterized by intrinsic behaviour and we speak of the *intrinsic range* of carriers. The variations of electron density and energy as functions of temperature are sketched in Fig. 11.45, identifying the three ranges.

The calculation of electron density in the conduction band as carried out above is based on the assumption that there are no acceptor impurities in the *n*-type semiconductor under consideration. Experiments, however, confirm that traces of acceptors are always present. Therefore, quantitative estimate made on the basis of the above theory may slightly from the experimental values.

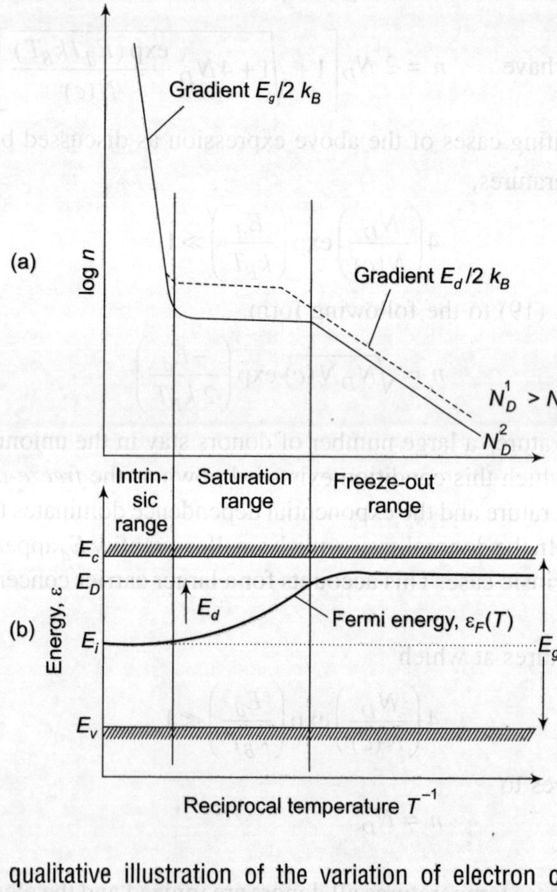


**Figure 11.45** (a) A qualitative illustration of the variation of electron concentration ($n$) in the conduction band with change in temperature in two samples of different donor concentrations $N_D^1$ and $N_D^2$. (b) A qualitative illustration of the temperature dependence of the Fermi energy $\varepsilon_F$ in the same semiconductor. The $E_i$ represents the Fermi energy in the material. All other symbols have their usual meaning

There are several applications in which there is a need to monitor the electron and hole concentrations. Since at low doping levels the law of mass action, $n_o p_o = n_i^2$, is valid, it is exploited to manipulate the carrier concentration in non-degenerate semiconductors. When the electron concentration $n$ is increased by adding the trace of a suitable donor impurity, the hole concentration $p$ goes down so that the product $(np)$ remains constant at a temperature in accordance with the law of mass action. This way it is possible to reduce the sum $(n + p)$ considerably. This method of reduction is in wide practice and known as *compensation*. The sign of majority carriers is quickly determined by detecting the sign of Hall voltage.

## 11.38 SHALLOW AND DEEP IMPURITY LEVELS IN SEMICONDUCTORS (EF)

We start by considering a simple chemical model of shallow donor and acceptor defects. We will give a better definition later, but for now, by "shallow", we mean energy levels near the bottom of the conduction band for donor level and near the top of the valence band for acceptors.

Consider $Si^{14}$ as the prototype semiconductor. In the usual one-electron shell notation, its electron structure is denoted by

$$1s^2\, 2s^2\, 2p^2\, 3s^2\, 3p^2.$$

There are four valence electrons in the $3s^2\, 3p^2$ shell, which require eight to be filled. We think of neighbouring Si atoms sharing electrons to fill the shells. This sharing lowers energy and binds the electrons. We speak of covalent bonds. Schematically, in two dimensions, we picture this occurring as in Fig. 11.46. Each line represents a shared electron. By sharing, each Si in outer shell has eight electrons. This is, of course, like the bonding of *C* to form diamond.

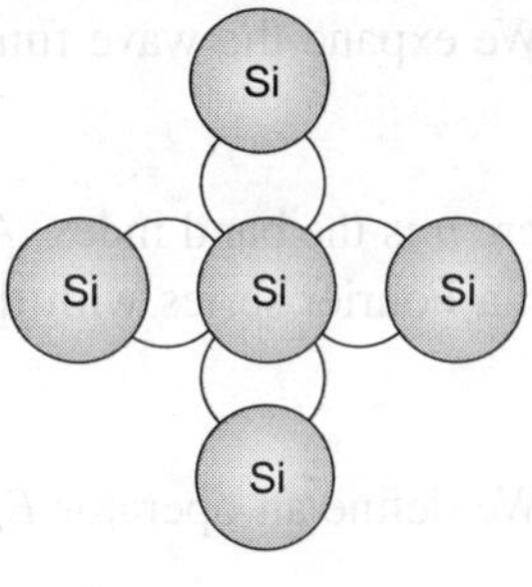

**Figure 11.46** Chemical model of covalent bond in Si

Now, suppose we have an atom, say As, which substitutionally replaces a Si. The *sp* shell of As has five electrons $(4s^2 4p^3)$ and only four are needed to "fill the shell". Thus, As acts as a donor with an additional loosely bound electron (with a large orbit encompassing many atoms), which can be easily excited into the conduction bond at room temperature.

An acceptor like In (with three electrons in its outer *sp* shell $(5s^2 5p)$) needs from electron to complete its covalent bonds. Thus, In can accept an electron from the valence band, leaving behind a hole. The combined effects of effective mass and dielectric constant cause the carrier to be bound much less tightly than in an analogous hydrogen atom. The result is that donors introduce energy levels just below the conduction-band minimum and acceptors introduce levels just above the top of the valence band. We discuss this in more detail as the following.

In brief, it turns out that ground-state donor energy level is given by (atomic units),

$$E_n = -\frac{m^*/m}{2n^2 \varepsilon^2} \qquad \text{...(1)}$$

where $m^*/m$ is the effective mass ratio typically about 0.25 for Si and $\varepsilon$ is the dielectric constant (about 11.7 in Si). $E_n$ in Eq. (1) is measured from the bottom of the conduction band. Except for the use of the dielectric constant and the effective mass, this is the same result as obtained simple result holes for acceptor states. These results arise from pioneering work by Kohn and Luttinger and we develop the basics below.

## 11.38.1 Effective Mass Theory, Shallow Defects

### *(a) Envelope Functions*

The basic model we will use here is called the envelope approximation. It will allow us to justify our treatments of effective mass theory and shallow defects in semiconductors. With a few more comments, we will then be able to relate it to a simple approach to superlattices.

Let
$$\mathcal{H}_0 = \frac{\hbar^2}{2m}\nabla^2 + V(r) \qquad ...(2)$$

where $V(r)$ is the periodic potential. Let $\mathcal{H} = \mathcal{H}_0 + U$ where $U = V_D(r)$ is the extra defect potential. Now $\mathcal{H}_0 \psi_n(k, r) = E_n \psi_n(\boldsymbol{k}, \boldsymbol{r})$ and $\mathcal{H}\psi = E\psi$.

We expand the wave function in Bloch functions

$$\psi = \sum\nolimits_{n,k} a_n(k)\,\psi_n(\boldsymbol{k}, \boldsymbol{r}), \qquad ...(3)$$

where $n$ is the band index. Also, since $E_n(\boldsymbol{k})$ is a periodic function in $\boldsymbol{k}$-space, we can expand it in a Fourier series with the sum restricted to lattice points

$$E_n(\boldsymbol{k}) = \sum\nolimits_m F_{nm}\, e^{ik\cdot R_m} \qquad ...(4)$$

We define an operator $E_n(-i\nabla)$ by substituting $-i\nabla$ for $k$:

$$E_n(-i\nabla)\,\psi_n(\boldsymbol{k}, \boldsymbol{r}) = \sum\nolimits_m F_{nm}\, e^{R_m\cdot\nabla}\,\psi_n(\boldsymbol{k}, \boldsymbol{r})$$

$$= \sum\nolimits_m F_{nm}\left[1 + \boldsymbol{R}_m\cdot\nabla + \frac{1}{2}(\boldsymbol{R}_m\cdot\nabla)^2 + ...\right]\psi_n(\boldsymbol{k}, \boldsymbol{r}) \qquad ...(5)$$

$$= \sum\nolimits_m F_{nm}\,\psi_n(\boldsymbol{k}, \boldsymbol{r} + \boldsymbol{R}_m),$$

by the properties of Taylor's series. Then using Bloch's theorem

$$E_n(-i\nabla)\psi_n(\boldsymbol{k}, \boldsymbol{r}) = \sum\nolimits_m F_{nm}\, e^{ik\cdot R_m}\,\psi_n(\boldsymbol{k}, \boldsymbol{r}) \qquad ...(6)$$

and by Eq. (4) $E_n(-i\nabla)\,\psi_n(\boldsymbol{k}, \boldsymbol{r}) = E_n(k)\,\psi_n(\boldsymbol{k}, \boldsymbol{r})$. ...(7)

Substituting Eq. (3) into $\mathcal{H}\psi = E\psi$ we have (using the fact that $\psi_n$ is an eigenfunction of $\mathcal{H}_0$ with eigenvalue $E_n(\boldsymbol{k})$)

$$\sum\nolimits_{n,k} E_n(\boldsymbol{k})\, a_n(\boldsymbol{k})\psi_n(\boldsymbol{k}, \boldsymbol{r}) + \sum\nolimits_{n,k} V_D a_n(\mathbf{k})\psi_n(\boldsymbol{k}, \boldsymbol{r})$$
$$= \sum\nolimits_{n,k} E a_n(\boldsymbol{k})\,\psi_n(\boldsymbol{k}, \boldsymbol{r}). \qquad ...(8)$$

If we use Eqs. (4) and (6), this becomes

$$\sum\nolimits_{n,k} a_n(\boldsymbol{k})\,[E_n(-i\nabla) + V_D]\,\psi_n(\boldsymbol{k}, \boldsymbol{r}) = E\psi. \qquad (9)$$

### (b) First Approximation

We neglect band-to-band interactions and hence, neglect the summation over $n$. Dropping $n$ entirely from Eq. (9), we have

$$\psi = \sum_k a(\boldsymbol{k})\ \psi(\boldsymbol{k}, \boldsymbol{r}), \qquad ...(10)$$

and

$$[E(-i\nabla) + V_D]\ \psi(\boldsymbol{k}, \boldsymbol{r}) = E\psi. \qquad ...(11)$$

### (c) Second Approximation (A)

We assume a large extension in real space that means that only a small range of $k$ values are important – say the ones near a parabolic (assumed for simplicity) minimum at $k = 0$.

We assume, then

$$\psi(\boldsymbol{k}, \boldsymbol{r}) = e^{ik\cdot r}\ u(k, r) \cong e^{ik\cdot r}\ u(0, r) = e^{ik\cdot r}\ \psi(0, r) \qquad ...(12)$$

so using Eqs. (10) and (12),

$$\psi = F(\boldsymbol{r})\ \psi(0, \boldsymbol{r}), \qquad ...(13)$$

where

$$F(r) = \sum_k a(\boldsymbol{k})\ e^{ik\cdot r}. \qquad ...(14)$$

So, we have by Eq. (11)

$$[E(-i\nabla) + V_D]F(\boldsymbol{r})\ \psi(0, r) = EF(\boldsymbol{r})\ \psi(0, r). \qquad ...(15)$$

Using the definition of $E(-i\nabla)$ as in Eq. (5) we have with $n$ suppressed,

$$\sum_m F_m F(\boldsymbol{r} + \boldsymbol{R}_m)\ \psi\ (0, \boldsymbol{r} + \boldsymbol{R}_m) = (E - V_D)\ F(\boldsymbol{r})\ \psi(0, \boldsymbol{r}). \qquad ...(16)$$

But, $\psi(0, r + R_m) = \psi(0, r)$, so it can be cancelled. Thus, retracing our steps, we have

$$[E\ (-i\nabla) + V_D]\ F(r) = EF(\boldsymbol{r}). \qquad ...(17)$$

This simply means that a rapidly varying function has been replaced by a slowly varying function $F(r)$ called the "envelope" function. This immediately leads to the concept of shallow donors. Consider the bottom of a parabolic conductor band near $k = 0$ and expand about $k = 0$;

$$E(-i\nabla) = E_c + \left.\frac{dE}{dk}\right|_{k=0} (-i\nabla) + \frac{1}{2}\frac{d^2E}{dk^2}(-i\nabla)^2. \qquad ...(18)$$

Also,

$$\left.\frac{dE}{dk}\right|_{k=0} = 0, \qquad ...(19)$$

and

$$\left.\frac{d^2E}{dk^2}\right|_{k=0} = \frac{\hbar^2}{m^*}, \qquad ...(20)$$

where $m^*$ is the effective mass. Thus, we find

$$E(-i\nabla) = E_c - \frac{\hbar^2}{2m^*}\nabla^2. \qquad ...(21)$$

And, if $V_D = e^2/4\pi\varepsilon r$, our resulting equation is

$$\left(-\frac{\hbar^2}{2m^*}\nabla^2 - \frac{e^2}{4\pi\varepsilon r}\right) F(\mathbf{r}) = (E - E_c)\, F(r) \qquad ...(22)$$

Except for the use of ε and $m^*$, these solutions are just hydrogenic wave functions and energies, and so use of the hydrogenic solution (1) is justified.

Now let us discuss briefly electron and hole motion in a perfect crystal. If $U = 0$, we simply write

$$\left(\frac{\hbar^2}{2m_e^*}\nabla^2 + E_c\right) F = EF. \qquad (23)$$

On the other hand, suppose $U$ is still 0, but consider a valence band with a maximum at $k = 0$. We then can expand about that point with the following result:

$$E(-i\nabla) = E_v + \frac{1}{2}\frac{d^2E}{dk^2}(-i\nabla^2). \qquad ...(24)$$

Using the hole mass, which has the opposite sign for the electron mass ($m_h = -m_e$), we can write

$$E(-i\nabla) = E_v + \frac{\hbar^2}{2m_h^*}(\nabla^2), \qquad ...(25)$$

so the relevant Schrödinger equation becomes

$$\left[-\frac{\hbar^2}{2m_h^*}\nabla^2 - E_v\right] F = -EF. \qquad ...(26)$$

Looking at Eqs. (23) and (26), we see how discontinuities in band energies can result in effective changes in the potential for the carriers, and we see why the hole energies are inverted from the electron energies.

## 11.39 IMPURITY STATES AND THE EFFECTIVE HAMILTONIAN METHOD

Hitherto we have consider energy band formation and the methods of calculating the band structure for pure ideal crystal for which the assumption of lattice periodicity is valid. In real crystals, a large number of lattice defects are usually present. The most common types of lattice defects include atoms, may also be introduced by diffusion or implantation as a means of controlling their electronic properties.

For this reason the determination of the energy states of an impurity atom in an otherwise perfect crystal lattice is an important aspect of energy band theory. If the impurity potential is week, one may hope to obtain approximate solution of the Schrödinger equation by perturbation theory. Usually, the impurity potential is numerically large but may for all practical purposes be considered localized. In this situation, it is possible to describe the electronic states associated with the localized impurity by the "effective Hamiltonain" method, which we now proceed to describe.

### 11.39.1 Wannier Functions

In order to derive the effective Hamiltonian, it is necessary to introduce basic kinematical concepts like momentum and the position canonically conjugates to it within a band of index, $n$. The Wannier localized description of energy bands provides a natural framework for doing this. If the $n$th band is no degenerate, then both the energy band, $E_n(\mathbf{k})$, and the Bloch function, $\chi_{k,n}$ associated with it may be expanded in Fourier series as follows:

$$E_n(\mathbf{k}) = \sum_l \exp(i\mathbf{k}\cdot\mathbf{l})\, E_n(\mathbf{l}) \qquad ...(1)$$

$$\chi_{\mathbf{k},n}(\mathbf{r}) = \frac{1}{\sqrt{N}} \sum_l \exp(i\mathbf{k}\cdot\mathbf{l})\, a_n(\mathbf{r}, \mathbf{l}) \qquad ...(2)$$

$N$ being the number of primitive cells in the crystal. These expansions relate $E_n(\mathbf{k})$ in Bloch representation to $E_n(\mathbf{l})$ in a localized (Wannier) representation.

The functions, $a_n(\mathbf{r}, \mathbf{l})$, have the property that they depend only on the difference of the two arguments. This may be proved as follows. First we modify the expansion of $\chi_{\mathbf{k},n}(\mathbf{r})$ to read

$$\chi_{\mathbf{k},n}(\mathbf{r}-\mathbf{l}') = \frac{1}{\sqrt{N}} \sum_l \exp(i\mathbf{k}\cdot\mathbf{l})\, a_n(\mathbf{r}-\mathbf{l}', \mathbf{l}).$$

Also by Bloch's theorem, we may write

$$\chi_{\mathbf{k},n}(\mathbf{r}-\mathbf{l}') = \frac{1}{\sqrt{N}} \sum_l \exp[i\mathbf{k}\cdot(\mathbf{l}-\mathbf{l}')]\, a_n(\mathbf{r}, \mathbf{l})$$

or, by shifting the summation index,

$$\chi_{\mathbf{k},n}(\mathbf{r}-\mathbf{l}') = \sum_l \exp(i\mathbf{k}\cdot\mathbf{l}) a_n(\mathbf{r}, \mathbf{l}+\mathbf{l}').$$

By identifying the coefficients in the two series, and setting $\mathbf{l} = 0$, we finally obtain

$$a_n(\mathbf{r}-\mathbf{l}, 0) = a_n(\mathbf{r}, \mathbf{l}').$$

We drop the zero argument and write $a_n(\mathbf{r}, \mathbf{l}) \equiv a_n(\mathbf{r}-\mathbf{l})$ subsequently.

Then
$$\chi_{\mathbf{k},n}(\mathbf{r}) = \frac{1}{\sqrt{N}} \sum_l \exp(i\mathbf{k}\cdot\mathbf{l}) a_n(\mathbf{r}-\mathbf{l}). \qquad ...(3)$$

The function, $a_n(\mathbf{r}-\mathbf{l})$, is called the *Wannier* function associated with the band of index $n$; it is analogous to a tight-binding function except for the important fact that Wannier functions form orthogonal and complete sets, whereas tight-binding functions do not.

The expansions may be Fourier-inverted to yield

$$a_n(\mathbf{r}-\mathbf{l}) = \frac{1}{\sqrt{N}} \sum_k \exp(-i\mathbf{k}\cdot\mathbf{l}) \chi_{\mathbf{k},n}(\mathbf{r}) \qquad ...(4)$$

and
$$E_n(\mathbf{l}) = \frac{1}{\sqrt{N}} \sum_k \exp(-i\mathbf{k}\cdot\mathbf{l}) E_n(\mathbf{k}). \qquad ...(5)$$

[Remember that in the reduced zone scheme **k** is restricted to the primitive zone, so that

$$\sum_{\mathbf{k}} \equiv \frac{\Omega}{(2\pi)^3} \int_{\Omega_0} d\mathbf{k},$$

$\Omega_0^*$ being the volume of the primitive zone **k**-space.] It follows that the Bloch representation and the Wannier representation are related to each other by a unitary transformation.

Now, if the phase factor in $\chi_{\mathbf{k},n}(\mathbf{r})$ is chosen appropriately, the probability density of $a_n(\mathbf{r})$ can be concentrated near the orgin. The probability density of $a_n(\mathbf{r}-\mathbf{l})$ is then concentrated at the lattice point, **l**. For example $a_n(\mathbf{r}-\mathbf{l}) = \delta(\mathbf{r}-\mathbf{l})$ for free electrons. Thus, Wannier functions provide a definition of localization compatible with the existence of a band index $n$. This fact is usually expressed more formally in operator language by defining a *lattice translation operator*, **R**, according to the relation

$$\mathbf{R}a_n(\mathbf{r}-\mathbf{l}) = \mathbf{l}a_n(\mathbf{r}-\mathbf{l}) \quad ...(6)$$

From Eq. (6) it then follows that the result of operating on $\chi_{\mathbf{k},n}(\mathbf{r})$ is

$$\mathbf{R}\chi_{\mathbf{k},n}(\mathbf{r}) = -i\frac{\partial}{\partial k}\chi_{\mathbf{k},n}(\mathbf{r}) \quad ...(7)$$

The proper relation between **R** and **k** results from this by treating an arbitrary wave function, $\psi$ as a superposition of such Bloch functions; i.e.,

$$\psi(r) = \sum_n \int d\mathbf{k} C_{k,n}\, \chi_{k,n}(\mathbf{r}).$$

The effect of the operator $R$ on $\psi$ then takes the form

$$\mathbf{R}\psi = \sum_n \int d\mathbf{k} C_{\mathbf{k},n}\left[-i\frac{\partial}{\partial \mathbf{k}}\chi_{k,n}(\mathbf{r})\right].$$

Since both $C_{\mathbf{k},n}$ and $\chi_{\mathbf{k},n}$ must be periodic in **k**, we can integrate with respect to **k** by parts and set the integrated out term equal to zero. This yields

$$\mathbf{R}\psi = \sum_n \int d\mathbf{k}\, i\frac{\partial C_{\mathbf{k},n}}{\partial k}\chi_{k,n}(\mathbf{r})$$

or

$$\mathbf{R}C_{\mathbf{k},n} = i\frac{\partial C_{\mathbf{k},n}}{\partial \mathbf{k}}. \quad ...(8)$$

It follows from this that $\hbar\mathbf{k}$ *and* **R** *are canonically canjugate within the band of index n*, just as **p** and **r** are conjugate in free-electron theory.

### 11.39.2 The Effective Hamiltonian

The time-dependent equation satisfied by $C_{\mathbf{k},n}$ is easily deduced from the wave equation

$$[H_0 + U(\mathbf{r})]\psi = i\hbar\frac{\partial \psi}{\partial t} \quad ...(9)$$

where $H_0\,\chi_{\mathbf{k},n} = E_n(\mathbf{k})\chi_{\mathbf{k},n}$ and $U$ is the potential due to an impurity. We readily obtain

$$\sum_n [E_n(\mathbf{k})\,\delta_{n'n} + U_{k,nn'}]C_{k,n'} = i\hbar \frac{\partial C_{k,n}}{\partial t} \qquad \text{...(10)}$$

where $$U_{\mathbf{k},n'n} \equiv \langle \chi_{\mathbf{k},n} | U | \chi_{\mathbf{k},n} \rangle$$

$$= \frac{1}{N}\sum_{\mathbf{ll}} \exp[i\mathbf{k}\cdot(\mathbf{l}-\mathbf{l}')]\, U_{n'n}(\mathbf{l},\mathbf{l}')$$

and $$U_{n'n}(\mathbf{l},\mathbf{l}') \equiv \langle a_{n'}(\mathbf{r}-\mathbf{l}) | U | a_n(\mathbf{r}-\mathbf{l}) \rangle$$

$$\equiv \int d\mathbf{r} a_{n'}^{*}(\mathbf{r}-\mathbf{l}')U(\mathbf{r})\,a_n(\mathbf{r}-\mathbf{l}). \qquad \text{...(11)}$$

Evidently Eq. (10) is exact and its solution is as complicated as the original problem. We therefore make two approximations. First, we assume that the impurity potential does not have interband matrix elements so that $U_{n'n}(\mathbf{l},\mathbf{l}')$ is small if $n' \neq n$. Second, we assume that $U(r)$ is slowly varying compared with the lattice positions, so that $U_{nn}(\mathbf{l},\mathbf{l}')$ vanishes for $\mathbf{l} \neq \mathbf{l}'$ by virtue of the orthogonality of Wannier functions, and we may write $U_{nn}(\mathbf{l},\mathbf{l}')|_{\mathbf{l}'=\mathbf{l}} = U(\mathbf{R})$. Then Eq. (10) simplifies to

$$[E_n(\mathbf{k}) + U(\mathbf{R})]\, C_{k,n} = i\hbar \frac{\partial C_{\mathbf{k},n}}{\partial t}. \qquad \text{...(12a)}$$

The operator $$H_{\text{eff}} = E_n(\mathbf{k}) + U(\mathbf{R}) \qquad \text{...(12b)}$$

is called the "effective Hamiltonian." Since $\mathbf{R}$ and $i(\partial/\partial\mathbf{k})$ are equivalent and vice versa, we see that

$$H_{\text{eff}} = E_n(\mathbf{k}) + U\left(i\frac{\partial}{\partial\mathbf{k}}\right) \quad \text{or} \quad E_n\left(i\frac{\partial}{\partial\mathbf{k}}\right) + U(\mathbf{R}). \qquad \text{...(13)}$$

### 11.39.3 The Effective Mass Approximation and the k·p Method

Obviously, for the effective Hamiltonian problem in Eq. (12a) to be tractable when we make the replacement $\mathbf{k} \to i\partial/\partial\mathbf{k}$ in $E_n(\mathbf{k})$, it is necessary to take a simple form of $E_n(\mathbf{k})$. This is achieved by using the effective mass approximation in conjunction with the effective Hamiltonian method. A systematic way of introducing this approximation is provided by a special perturbation technique called $\mathbf{k}\cdot\mathbf{p}$ method, which works quite well if the impurity levels in silicon illustrated in Fig. 11.47.

The "$\mathbf{k}\cdot\mathbf{p}$ perturbation" is introduced in the wave equation for $E_n(\mathbf{k})$ by writing the equation for the periodic part, $u_{n\mathbf{k}}(\mathbf{r})$, of the Bloch function for the $n$th band, $\psi_{n\mathbf{k}}(\mathbf{r}) = \exp(i\,\mathbf{k}\cdot\mathbf{p})\,u_{n\mathbf{k}}(\mathbf{r})$:

$$\left[-\frac{\hbar^2}{2m}\nabla^2 + V(\mathbf{r})\right]\psi_{n\mathbf{k}}(\mathbf{r}) = E_n(\mathbf{k})\,\psi_{n\mathbf{k}}(\mathbf{r}).$$

We readily obtain

$$\left[-\frac{\hbar^2}{2m}\nabla^2 + V(\mathbf{r}) + \frac{\hbar^2}{2m}\mathbf{k}\cdot\mathbf{p}\right]u_{n\mathbf{k}} = \left[E_n(\mathbf{k}) - \frac{\hbar^2}{2m}k^2\right]u_{n\mathbf{k}}$$

or $$\frac{\hbar^2}{2m}(\mathbf{p}+\mathbf{k})^2 u_{n\mathbf{k}} + V(\mathbf{r})u_{n\mathbf{k}} = E_n(\mathbf{k})u_{n\mathbf{k}} \qquad ...(14)$$

where $\mathbf{p} \equiv -i\nabla$. For any given $\mathbf{k}$, the set of functions $u_{n\mathbf{k}}$, obtained by letting $n$ run over all bands, forms a complete set of functions having the same periodicity as the lattice.

If we now suppose that we know all solutions $u_{n\mathbf{k}0}$ and $E_n(\mathbf{k}_0)$ at some point $\mathbf{k}_0$ in the Brillouin zone, so that

$$H_{\mathbf{k}0}u_{n\mathbf{k}0} \equiv \left[\frac{\hbar^2}{2m}(\mathbf{p}+\mathbf{k}_0)^2 + V(\mathbf{r})\right]u_{n\mathbf{k}0} = E_n(\mathbf{k}_0)u_{n\mathbf{k}0}, \qquad ...(15)$$

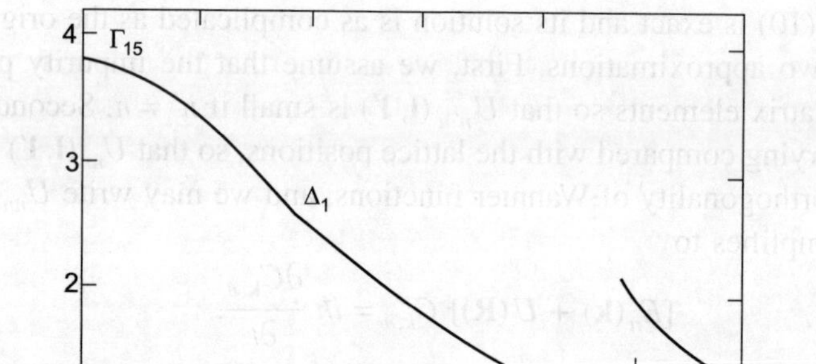


**Figure 11.47** Sulphur level in silicon. (After Pantelides)

Then we may rewrite Eq. (14) in the form

$$(H_{\mathbf{k}0} + H_{\mathbf{k}})u_{n\mathbf{k}} = E_n(\mathbf{k})u_{n\mathbf{k}} \qquad ...(16)$$

where $$H_{\mathbf{k}} = \frac{\hbar^2}{m}(\mathbf{k}-\mathbf{k}_0)\,.\,\mathbf{p} + \frac{\hbar^2}{2m}(\mathbf{k}^2 - \mathbf{k}_0^2). \qquad ...(17)$$

This extra term can be treated as a perturbation if $\mathbf{k} - \mathbf{k}_0$ is small, hence the name "**k·p** perturbation".

Two cases of the **k·p** perturbation usually arise according to the degeneracy of the bond at $\mathbf{k}_0$. For example, at the top of the valence band that occurs at $\mathbf{k}_0 = 0$, we have a triply degenerate band. We need therefore a degenerate perturbation theory, which we shall not discuss here. Near the bottom of the conduction band, however, that occurs here. Near the bottom of the conduction band, however, that occurs along the (100) axis in silicon ($\Delta_1$ point) as shown in Fig. 11.47, $\mathbf{k}_0 \neq 0$, and nondegenerate perturbation theory can be used. The result is

$$E_c(\mathbf{k}) = E_c(\mathbf{k}_0) + \langle u_{c\mathbf{k}0} | H'_{\mathbf{k}} | u_{c\mathbf{k}0} \rangle + \sum_{m \neq c} \frac{\langle u_{c\mathbf{k}0} | H'_{\mathbf{k}} | u_{c\mathbf{k}0} \rangle \langle u_{m\mathbf{k}0} | H'_{\mathbf{k}} | u_{c\mathbf{k}0} \rangle}{E_c(\mathbf{k}_0) - E_m(\mathbf{k}_0)} \quad ...(18)$$

On using Eq. (17), $E_c(k)$ can be rewritten in the form

$$E_c(\mathbf{k}) = E_c(\mathbf{k}_0) + \frac{\hbar^2}{2m} \sum_{i,j} \left(\frac{1}{m^*}\right)_{ij} (k_i - k_{0i})(k_j - k_{0j}) \quad ...(19)$$

where $$\left(\frac{1}{m^*}\right)_{ij} = \frac{1}{\hbar^2} \left(\frac{\partial^2 E_c}{\partial k_i \, \partial k_j}\right)_{\mathbf{k}=\mathbf{k}_0}$$

$$= \frac{1}{m} \delta_{ij} + \frac{2}{m^2} \sum_{n \neq c} \frac{\langle u_{c\mathbf{k}0} | p_i | u_{n\mathbf{k}0} \rangle \langle u_{n\mathbf{k}0} | p_j | u_{c\mathbf{k}0} \rangle}{E_c(\mathbf{k}_0) - E_n(\mathbf{k}_0)} \quad ...(20)$$

defines the effective mass tensor $m^*_{ij}$. The relation in Eq. (20) is called the *f*-sum rule.

Now, if we denote by κ the wave vector measured from $\mathbf{k}_0$ and make a proper choice of orthogonal axes, then Eq. (19) may be diagonalized in the form

$$E_c(\mathbf{k}) = E_c(\mathbf{k}_0) + \frac{\hbar^2}{2m} \sum_i \left(\frac{m}{m^*}\right)_{ii} \kappa_i^2 \quad ...(21)$$

In silicon and most semiconductor at least two of the coefficients are equal and we may write

$$E_c(\mathbf{k}) = E_c(\mathbf{k}_0) + \frac{\hbar^2}{2m_t} \kappa_t^2 + \frac{\hbar^2}{2m_t} \kappa_l^2 \quad ...(22)$$

where $m_t$ and $m_l$ are, respectively, called the **transverse** and **longitudinal effective masses** and $\kappa_l$ and $\kappa_t$ are, respectively, the components of κ along and perpendicular to the symmetry axis.

In silicon, there are six conduction band minima, all occurring at the same energy. In the vicinity of each minimum a surface of constant, $E_c(k)$ = constant, is an ellipsoid of revolution about the [100] axis, characterized by a longitudinal effective mass $m_l$ = 0.98 m and a transverse effective mass $m_t$ = 0.19 m.

We should observe that the conduction band masses are associated with parabolic bands that are concave upwards: These define average electron masses, $m_e^*$. A similar analysis for the valence band would give (apart from complications due to the triple degeneracy of the valence band edge) effective characteristic of parabolic bands that are concave downwards. Such masses are negative (because they represent the curvature of the band at the valence band maximum). The dynamics of an electron in energy band states characterized by negative effective mass is much easier to visualize if the sign of the mass is chosen positive while the sign of the electric charge is reversed: The result is a positively charged *hole* with masses, positive effective mass. In general, therefore, we speak of electron and hole masses, and because of the splitting of the triply degenerate valence band edge by the crystal field, there are usually two (light and heavy) average hole masses, $m^*_{h1}$ and $m^*_{h2}$, as illustrated in Fig. 11.48.

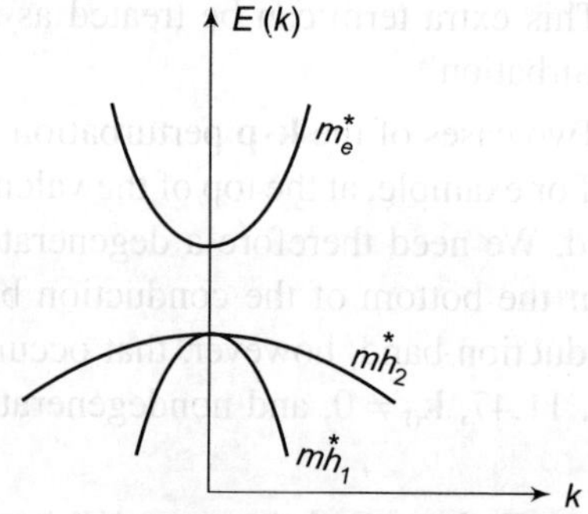


**Figure 11.48** Typical semiconductor energy bands near $k = 0$ showing the electron effective mass ($m^*_e$) and the hole effective masses ($m^*_{h1}$ and $m^*_{h2}$)

On gathering result, the effective mass approximation leads to an explicit wave equation for the $C_n$'s of Eq. (12a) of the form

$$\left[-\frac{\hbar^2}{2m_t}\left(\frac{\partial^2}{\partial x^2}+\frac{\partial^2}{\partial y^2}\right)-\frac{\hbar^2}{2m_t}\frac{\partial^2}{\partial z^2}+U(R)\right]C = EC, \qquad ...(23)$$

which follows from Eq. (22) upon writing $k_x = \partial/\partial x$ etc., and setting $\mathbf{R} = (x, y, z)$. For a "hydrogenic" impurity, the potential $U(R)$ takes the form

$$U(R) = -\frac{e^2}{\varepsilon R} \qquad ...(24)$$

where ε is the dielectric constant of the semiconductor. Then Eq. (23) differs from the Schrödinger equation only in the anisotropic effective mass and the dielectric constant. The rather large dielectric constant of silicon ($\varepsilon \simeq 12$) together with the small effective mass, reduces the binding energy of the electron from its value of 13.6 in H atom to about 30 meV for shallow donor impurity. Central cell corrections (deviations from effective mass approximation near the impurity atom) increases the latter value to 50–80 meV, depending on the impurity. Basically, it is this small binding energy that allows one to obtain *n*-type and *p*-type silicon easily, the latter being due to shallow acceptors.

## 11.40 THE *pn*-JUNCTION (FORMATION OF A DEPLETION REGION)

In *n*-type semiconductor, the concentration of free electrons is higher than that of holes whereas in *p*-type semiconductor, the concentration of holes is higher than that of electrons. So, when a *pn*-junction is formed, some electrons from *n*-side diffuse into the *p*-region and some holes from *p*-side diffuse into *n*-region near the metallurgical junction.

The diffusion equations are

$$I_{nd} = q\,AD_n\,dn/dx \quad \text{and} \quad I_{pd} = -qAD_p\,dp/dx$$

Subscript $d$ stands for diffusion, $n$ for electrons and $p$ for holes. The negative sign appears for holes due to difference between the direction of carrier flow.

This leads to free carrier recombination (electrons combining with holes) near the junction: electrons on $n$ side neutralized by holes diffused from $p$-side and holes on $p$-side neutralized by electrons diffused from $n$-side. As a result, we are left with ionized acceptor atoms on $p$-side and ionized donor atoms on the $n$-side near the junction. These ions are known as uncovered charges. This region near the junction is known as *depletion region*, since it is depleted of free carriers. Further, due to ionic density gradient in the vicinity of junction there develops a barrier voltage $V_B$ across the junction which prevents further diffusion of free carriers.

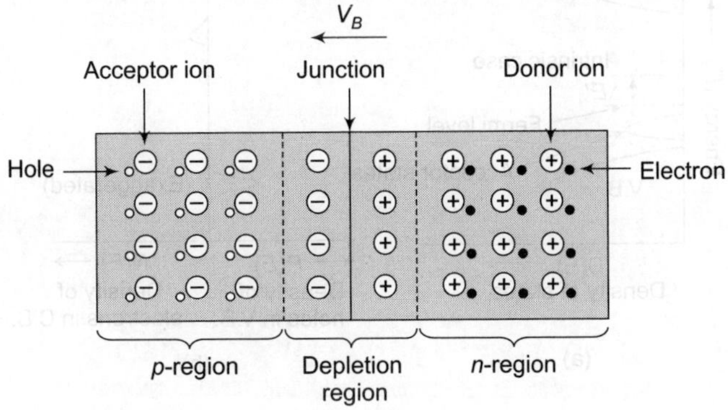


**Figure 11.49** Formation of depletion region in a *pn*-junction

## 11.41 ENERGY BAND DIAGRAM FOR *pn*-JUNCTION AND JUNCTION BARRIER VOLTAGE

We first consider $n$- and $p$-type semiconductors separately.

Figure 11.50 (a) shows the position of Fermi level and (b) the density of carriers in an $n$-type semiconductors.

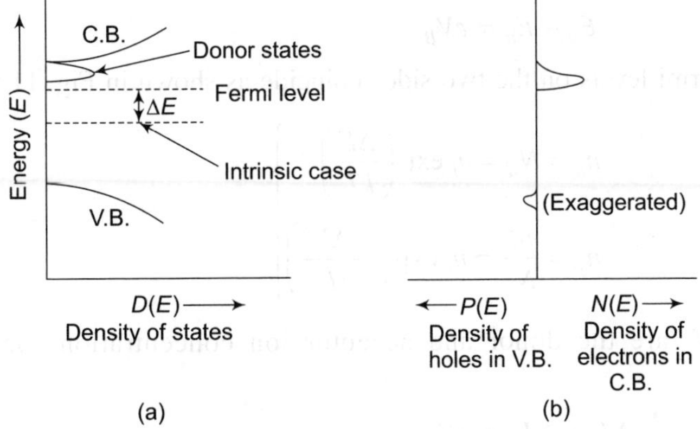


**Figure 11.50** Fermi level and carrier density in *n*-type semiconductor

The Fermi level is at an energy $\Delta E'$ above the centre of the gap. The total number of electrons in the conduction band is given by

$$n = n_i \exp\left(\frac{\Delta E}{kT}\right) \qquad ...(1)$$

where $n_i$ is the carrier density for the intrinsic case.

Figure 11.51 (a) shows the position of Fermi level and (b) carrier density for *p*-type semiconductor.

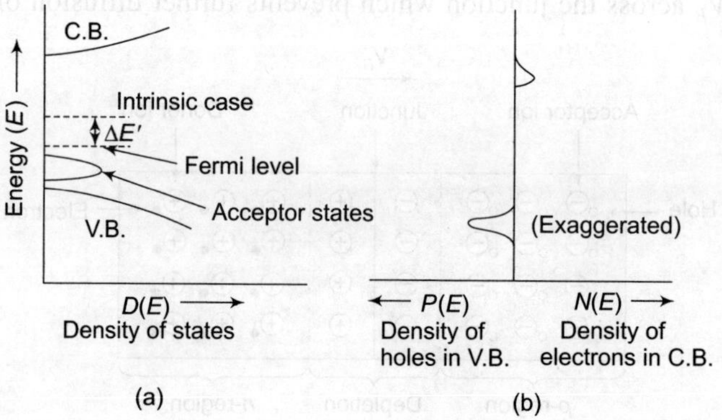


**Figure 11.51** Fermi level and carrier concentration in *p*-type semiconductor

The Fermi level is at an energy $\Delta E'$ below the centre of the gap. The total number of holes in the valence band is given by

$$p = n_i \exp\left(\frac{\Delta E'}{kT}\right) \qquad ...(2)$$

When the junction is formed, due to the barrier voltage, conduction band on *p*-side is higher by an amount $eV_B$ from the conduction band on *n*-side, i.e., if $E_p$ and $E_n$ are the energy levels of the bottom of the conduction bands on *p*-side and *n*-side respectively, then

$$E_p - E_n = eV_B \qquad ...(3)$$

Further, the Fermi levels on the two sides coincide as shown in Fig. 11.52.

$$\left.\begin{aligned} n_n &= N_d = n_i \exp\left(\frac{\Delta E}{kT}\right) \\ \text{and} \quad n_p &= \frac{n_i^2}{N_a} = n_i \exp\left(\frac{-\Delta E'}{kT}\right) \end{aligned}\right\} \qquad ...(4)$$

where $N_d$ and $N_a$ are the donor and acceptor ion concentrations on *n*- and *p*-sides respectively.

$$\Delta E + \Delta E = eV_B \qquad ...(5)$$

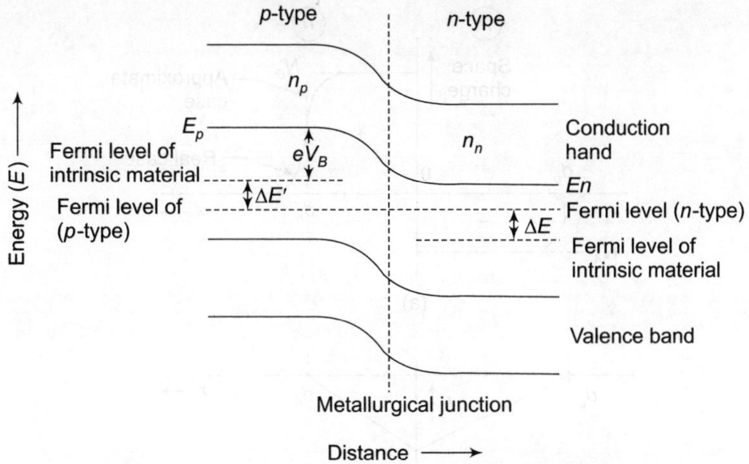


**Figure 11.52** Energy band diagram for *pn*-junction

Thus
$$\frac{n_n}{n_p} = \exp\left(\frac{\Delta E + \Delta E'}{kT}\right) = \exp\left(\frac{eV_B}{kT}\right) \quad ...(6)$$

or
$$V_B = \frac{kT}{e}\log_e\left(\frac{n_n}{n_p}\right) = \frac{kT}{e}\log_e\left(\frac{N_d N_a}{n_i^2}\right) \quad ...(7)$$

The quantity $(kT/e)$ is called thermal voltage.

**Example 11.7** A sample of Ge is made *p*-type by acceptor atoms at a rate of one atom per $4 \times 10^8$ Ge atoms. $n_i = 2.5 \times 10^{19}/\text{m}^3$ at 300 K and all the acceptor atoms are ionized at 300 K. The density of germanium is $4.4 \times 10^{28}$ atoms/m$^3$. Compare the density of electrons with intrinsic charge carriers.

***Solution*** For $4 \times 10^8$ Ge atoms, one acceptor atom is added. For $4.4 \times 10^{28}$ Ge atoms, the acceptor atoms added will be $1.1 \times 10^{20}$.

Now
$$n_i^2 = np, \; n = n_p \quad \text{and} \quad p = N_a$$

$$\therefore \quad n_p = \frac{n_i^2}{N_a} = \frac{(2.5 \times 10^{19})^2}{1.1 \times 10^{20}} = 5.7 \times 10^{18}$$

$$\Rightarrow \quad \frac{n_p}{n_i} = \frac{5.7 \times 10^{18}}{2.5 \times 10^{19}} = 0.22$$

## 11.42 WIDTH OF THE DEPLETION REGION

The width of the depletion region tells us how wide the region is where there is an upset in the bulk equilibrium conditions. We will now derive an expression for the depletion region width $d \equiv (d_p + d_n)$. This width is determined by the impurity concentration and the voltage.

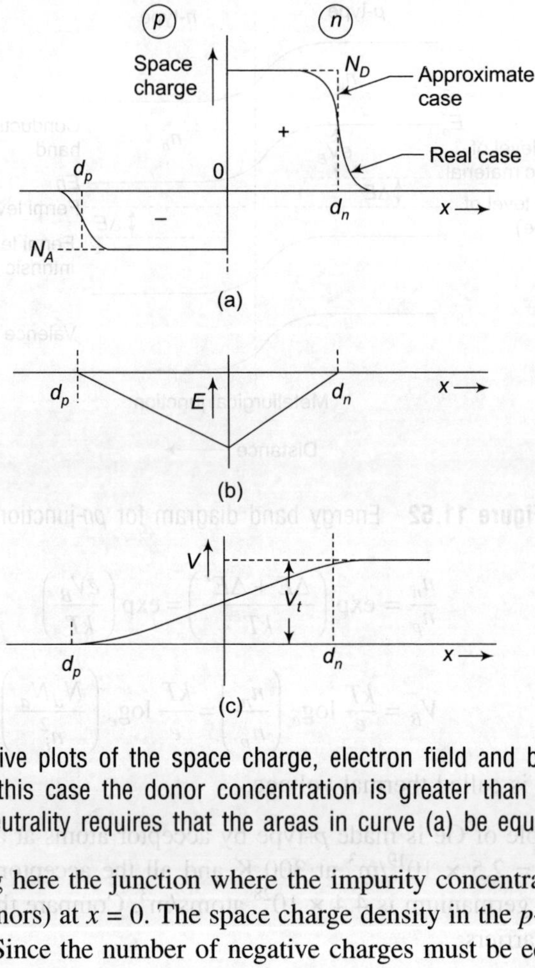


**Figure 11.53** Qualitative plots of the space charge, electron field and barrier potential near an abrupt *pn*-junction. In this case the donor concentration is greater than acceptor concentration, but charge neutrality requires that the areas in curve (a) be equal about $x = 0$

We are considering here the junction where the impurity concentration changes abruptly (from acceptors to donors) at $x = 0$. The space charge density in the *p*-region is $-qN_A$ and in the *n*-region is $qN_D$. Since the number of negative charges must be equal to the sum of the positive charges, we have

$$(qA)N_D d_n = (qA)N_A d_p \quad \text{...(8a)}$$

or

$$N_D d_n = N_A d_p \quad \text{...(8b)}$$

(*A* is the junction area.)

We carry the calculations in two columns, one for the negative and the other for the positive distances. We will obtain two voltages at about $x = 0$, which will be added to obtain the desired result.

*Negative x* ; *Positive x*

$$\frac{d^2V}{dx^2} = \frac{qN_A}{\varepsilon_r\varepsilon_o} \quad ; \quad \frac{d^2V}{dx^2} = \frac{-qN_D}{\varepsilon_r\varepsilon_o} \quad \text{...(9)}$$

$$\left(\text{because, from Poisson's equation } -\frac{d^2V}{dx^2} = \frac{\text{space charge density}}{\varepsilon_r\varepsilon_o}\right)$$

Integrating Eq. (9), we get

$$\frac{dV}{dx} = \frac{qN_A}{\varepsilon_r\varepsilon_o} x + C_1 \quad ; \quad \frac{dV}{dx} = \frac{-qN_D}{\varepsilon_r\varepsilon_o} + C_2 \qquad ...(10)$$

Now, refer to Fig. 11.53 (b), the electron field ($E = -dV/dx$), must be zero at $d_p$ on the left and at $d_n$ on the right. This demands

$$\frac{dV}{dx} = \frac{qN_A}{\varepsilon_r\varepsilon_o}(x - d_p) \quad ; \quad \frac{dV}{dx} = \frac{-qN_D}{\varepsilon_r\varepsilon_o}(x - d_n) \qquad ...(11)$$

Integrating again, we obtain the voltage distribution as

$$V = \frac{qN_A}{\varepsilon_r\varepsilon_o}\left(\frac{x^2}{2} - d_x x\right) + C \quad ; \quad V = \frac{-qN_D}{\varepsilon_r\varepsilon_o}\left(\frac{x^2}{2} - d_n x\right) + C \qquad ...(12)$$

The voltage at $x = 0$ must be the same coming from either side, this requires that the integration constant must be the same ($\equiv C$). The voltage at $x \le d_p$ and at $x \ge d_n$ are respectively

$$V_p = \frac{qN_A}{2\varepsilon_r\varepsilon_o} d_p^2 + C \quad ; \quad V_n = \frac{qN_D}{2\varepsilon_r\varepsilon_o} d_n^2 + C \qquad ...(13)$$

The total voltage across the junction is $V_n - V_p$. Let us denote this as

$$V_t = V_B + V_a \qquad ...(14)$$

where $V_B$ is the built in (or barrier) voltage and $V_a$ is some externally applied voltage. Thus,

$$V_t = V_n - V_p = \frac{qN_A}{2\varepsilon_r\varepsilon_o}(N_A d_p^2 + N_D d_n^2) \qquad ...(15)$$

Now, we evaluate $d_p$ and $d_n$ with the help of Eqs. (8) and (15) as

$$d_n = \left[\frac{2\varepsilon_r\varepsilon_o V_t}{q} \cdot \frac{N_A}{N_A N_D + N_D^2}\right]^{1/2} \qquad ...(16a)$$

and

$$d_p = \left[\frac{2\varepsilon_r\varepsilon_o V_t}{q} \cdot \frac{N_A}{N_A N_D + N_A^2}\right]^{1/2} \qquad ...(16b)$$

Adding these two expressions, we obtain the *width of the depletion region*

$$d \equiv d_n + d_p = \left(\frac{2\varepsilon_r\varepsilon_o V_t}{q(N_A + N_D)}\right)^{1/2} \left[\sqrt{\frac{N_A}{N_D}} + \sqrt{\frac{N_D}{N_A}}\right] \qquad ...(17)$$

## 11.43 BARRIER RECTIFICATION (AND DEVELOPMENT OF CONTACT-POTENTIAL) AT A *pn*-JUNCTION

When a *pn*-junction is formed, the *Fermi level on the p-side is lower than that on the n-side*, because electron concentration is lower on *p*-side then on the *n*-side (see Fig. 11.54) before transfer of carriers (charges) while the junction is just formed. As a result of the junction formation, electrons move across the boundary from the *n*-side to the *p*-side by diffusion due to the concentration difference till the Fermi levels equalize on the two sides of the junction. This gives rise to the contact potential difference $V_0$ (also denoted as $V_B$, the barrier potential) (the holes are the majority carriers on the *p*-side and electron, that on the *n*-side). Due to the movement of these carriers across the junction (holes from *p* to *n*-side and electrons from *n*- to *p*-side), (see Fig. 11.49). The region across the junction on either side becomes devoid of mobile charges (electron or holes) and so, called as the *depletion region*. Figure such as (11.54) is an energy-level diagram.

Electrons (free carriers) on *n*-side have higher energy than holes on the other (i.e., *p*-side). Electrons can lower their energy by sinking and holes lower their energies by rising.

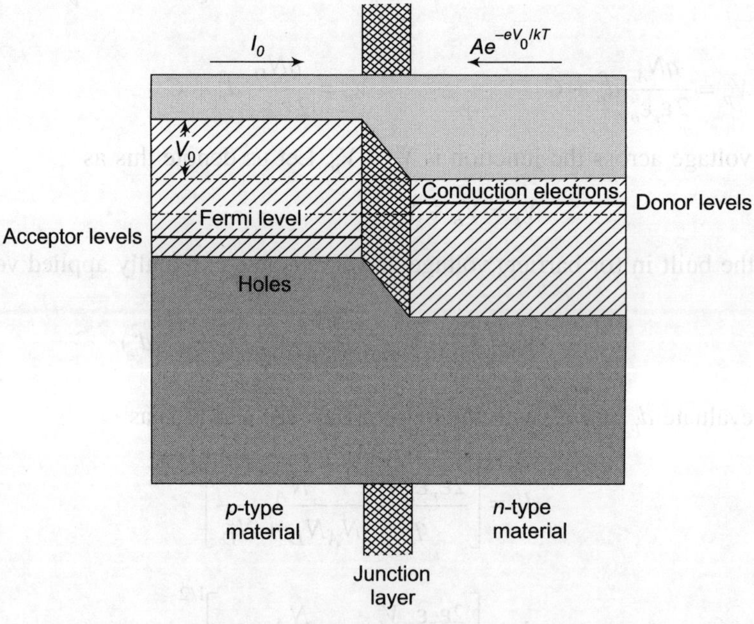


**Figure 11.54** Energy band diagram for a *pn*-junction rectifier. The contact potential $V_0$ is shown. The filling of the states is designated by the shading. The flow of minority electrons from the *p*-region is equal to the number of majority electrons from the *n*-regions surmounting the barrier; thus, the net current is zero

### 11.43.1 Junction Rectification

Apart from the majority carrier movement across the junction in the depletion region formation, we also have minority carrier movement also. The flow of electrons (minority carriers) from the $p$-side to the right, must be balanced at equilibrium by an equal flow to the left. This means that the flow of minority carriers from the $p$-type material is balanced by a flow of majority carriers (electrons again) from the $n$-type material, i.e.,

$$I_0 = Ae^{-eV_0/kT} \quad ...(1)$$

[Because, conduction electrons on the $n$-side have to face the contact potential barrier ($V_0 \equiv V_B$) and the crossing probability is proportional to ($e^{-eV_0/kT}$)]. If the flow to the left is $Ae^{-eV_0/kT}$, the current at equilibrium is given by Eq. (1). A similar expression can be written for flow of holes. The band scheme shown in Fig. 11.55 for forward and reverse bias can be used to visualize $p$-$n$ junction rectification.

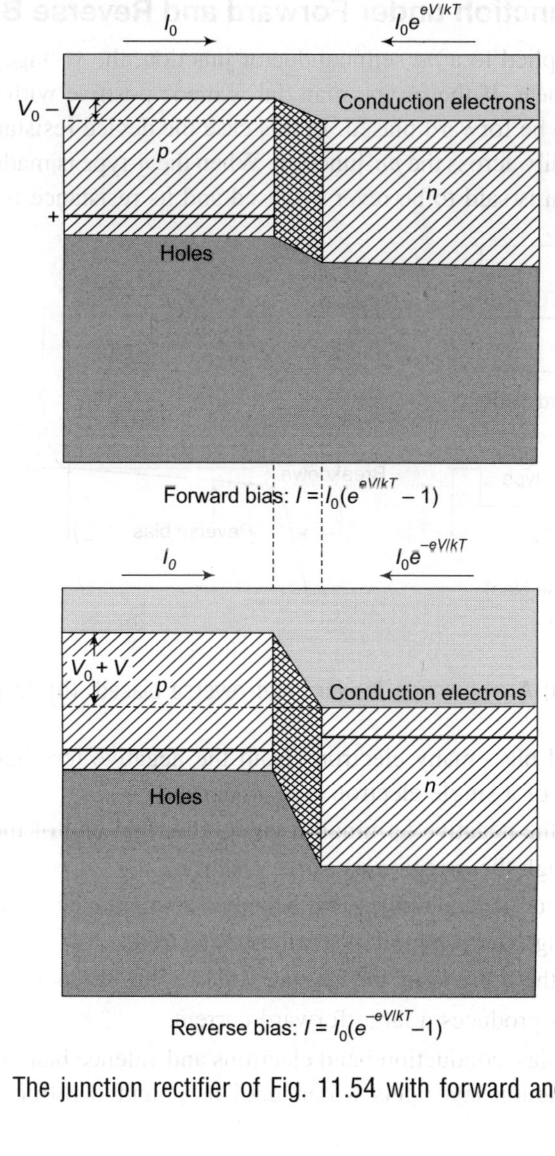


**Figure 11.55** The junction rectifier of Fig. 11.54 with forward and reverse bias

The electrons in the $n$-type material face a barrier which may be lowered or raised by the applied voltage. Thus, for reverse bias, the majority flow is $Ae^{-e(V_0 + V)/kT}$, or $I_0 e^{-eV_0/kT}$, where $V$ is the applied voltage. The net current is the sum of two currents: (1) the minority flow from the left, and (2) the majority flow from the right. It is small and equal to $I_0(e^{eV_0/kT} - 1)$. For forward bias, the majority flow is $Ae^{-(V_0 - V)/kT}$, or $I_0 e^{eV/kT}$. The net flow in the opposite direction is $I_0(e^{-eV/kT} - 1)$ which increases exponentially with voltage.

The diode consists of a crystal, usually silicon or germanium, with acceptor impurities on one side of the junction and donor impurities on the other side. The barrier and rectification effect arise from the difference in the Fermi levels.

### 11.43.2 A *pn*-Junction under Forward and Reverse Bias

When a voltage is applied to a $pn$ semiconductor junction, the voltage current characteristic is found to be non-linear. If the $p$-type material is made positive with respect to the $n$-type, the junction is said to be forward biased. In this case the device resistance is low and a large (forward) current readily across the $pn$-junction. When the $n$-type is made positive with respect to $p$-type, the junction is said to be reverse-biased and its resistance is high (Fig. 11.56).

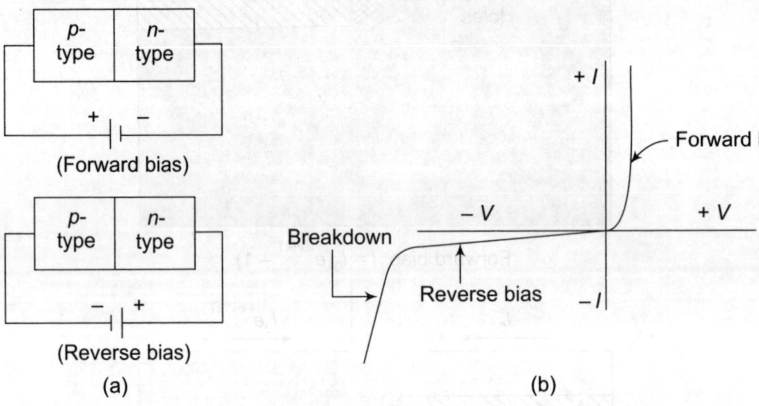


**Figure 11.56** (a) A *pn*-junction forward and reverse biased, (b) its I-V characteristics

In case of forward bias, many electrons from the negative terminal of battery enter the $n$-side. The life story of a single electron is as under:

1. After leaving the negative terminal, it enters the right end of the crystal.
2. It travels through the $n$-region as a free electron.
3. Near the junction it recombines and becomes a valence electron.
4. It travels through the $p$-region as a valence electron.
5. After leaving the left end of the crystal, it flows into the positive source terminal.

Thus, forward bias produces a large forward current.

In case of reverse bias, conduction band electrons and valence band holes move away from the junction, the depletion layer gets wider, until its potential difference equals the supply voltage. Under this condition, free electrons and holes stop moving. However, because of thermal generation, a small current due to minority carriers exist in the circuit. This reverse current caused by minority carriers is called the saturation current (because we cannot get more of this current by increasing the reverse voltage unless we increase the temperature).

### 11.43.3 The Energy Band Diagrams

When only a small current is flowing in a typical *pn*-junction, there is negligible voltage drop in the bulk material itself, and all the applied voltage is dropped across the junction. The presence of an applied voltage across the junction adds to (for reverse bias) or subtracts from (for forward bias) the barrier voltage, $V_B$.

Under the forward bias *V*, when an electron passes across the junction from the *n*- to the *p*-type material, it is moving in the same direction as the force due to the applied electric field. As a result, the potential energy in the *p*-type material is less than what would have been in an unbiased junction by an amount *V* electron volts. Thus, the conduction and valence bands on the *p*-side are lowered by this value.

Figure 11.58 illustrates the case of the *pn*-junction with reverse bias.

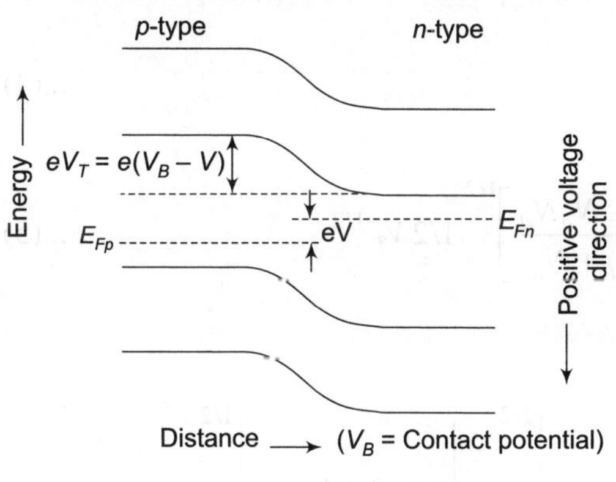


**Figure 11.57** Energy-band diagram for a forward biased *pn*-junction

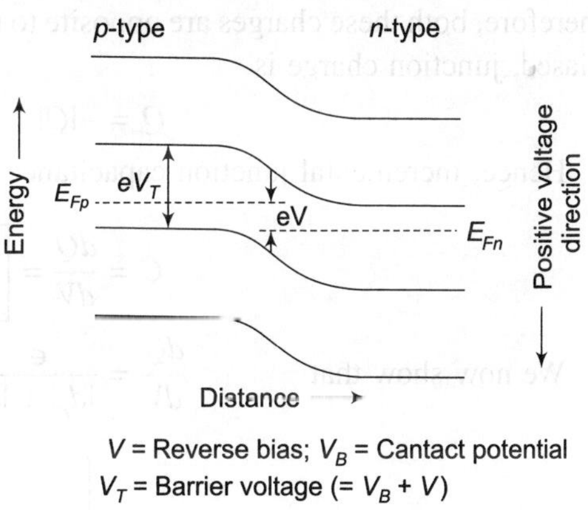


**Figure 11.58** Energy-band reverse for a forward biased *pn*-junction

### 11.43.4 Junction Capacitance

A reverse bias voltage causes majority carriers to move away from the junction, thereby uncovering more immobile charges, this may be considered a capacitive effect. We define an incremental capacitance by $C_T = dQ/dV$, where $dQ$ is an increase in charge caused by a change $dV$ in voltage (see Fig. 11.57).

The conduction band, the valance band and the Fermi level in *p*-type side are lowered by a value (eV) with respect to counterparts on *n*-side,

$$V_T = V_B - V \quad ...(1)$$

Now, each space charge region of the junction contains a charge which is numerically

$$|Q| = eN_a |d_p| = eN_d |d_n| \qquad ...(2)$$

where $$|d_p| = \left\{ \frac{2\in V_T}{eN_a\left(1+\frac{N_a}{N_d}\right)} \right\}^{1/2} ; \quad |d_n| = \left\{ \frac{2\in V_T}{eN_d\left(1+\frac{N_d}{N_a}\right)} \right\}^{1/2}$$

$$\therefore \qquad Q = eN_a |d_p|$$

$$= \left[ \frac{e^2 N_a^2 \, 2\in V_T N_d}{eN_a (N_d + N_a)} \right]^{1/2}$$

$$= \left\{ \frac{eN_a \, N_d (2\in V_T)}{N_d + N_a} \right\}^{1/2} \qquad ...(3)$$

Since *n*-type region contains positive charges and *p*-type region contains negative charges, therefore, both these charges are opposite to that on an equivalent capacitor, i.e., when forward biased, junction charge is

$$Q = -|Q| \qquad ...(4)$$

Hence, incremental junction capacitance

$$C = \frac{dQ}{dV} = \left[ \frac{2e\in N_a N_d}{N_d + N_a} \right]^{1/2} 1/2 \, V_T^{-1/2} \qquad ...(5)$$

We now show that $$\frac{dQ}{dV} = \frac{\in}{|d_p| + |d_n|}$$

$$|d_p| + |d_n| = \left\{ \frac{2\in V_T}{eN_a\left(1+\frac{N_a}{N_d}\right)} \right\}^{1/2} + \left\{ \frac{2\in V_T}{eN_d\left(1+\frac{N_d}{N_a}\right)} \right\}^{1/2}$$

$$= \sqrt{\frac{2\in V_T}{e(N_a + N_d)}} \left\{ \left(\frac{N_d}{N_a}\right)^{1/2} + \left(\frac{N_a}{N_d}\right)^{1/2} \right\}$$

$$= \sqrt{\frac{2\in V_T}{e(N_a + N_d)}} \, \frac{(N_a + N_d)}{\sqrt{N_a \, N_d}}$$

$$= \left| \frac{2\in V_T \, (N_a + N_d)}{eN_a \, N_d)} \right|^{1/2} ;$$

We have, therefore,

$$C = \frac{\in}{|d_p| + |d_n|} = \in \left[ \frac{e N_a N_d}{2 \in V_T (N_a + N_d)} \right]^{1/2} \qquad ...(6)$$

Thus, across an abrupt semiconductor junction, there is a non-linear capacitance. (This is of more importance when the junction is reverse biased, because then its reactance may be smaller than the resistance.)

## 11.44 CURRENT ACROSS A *pn*-JUNCTION: THE RECTIFIER EQUATION FOR A FORWARD BIASED *pn*-JUNCTION

Consider Fig. 11.59, which shows the energy bands of a *pn*-junction, In the diagram, a forward bias (*V* volts) has been applied. As a result, the Fermi level on *n*-side $E_{Fn}$ is higher than that on the *p*-side by the value eV.

There are four possible components of current flowing across the junction Fig. 11.59. (In Fig. 11.59, suffixes 1 and 2 represent the majority and minority carriers respectively.) In the valence band, holes can flow to the right or to the left (indicated by $J_1^+$ and $J_2^+$). In the conduction band, electrons can flow to the right or to the left. Due to the negative charge, a flow of electrons to the left produces a conventional current to the right therefore arrows in conduction band indicate the direction of flow of electrons and a negative sign is placed before the current density.

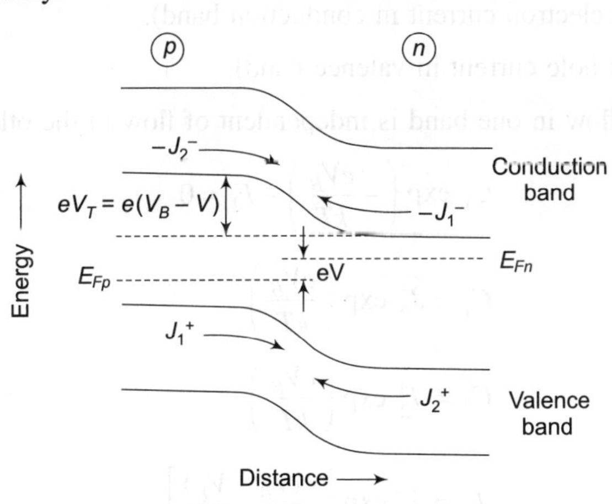


**Figure 11.59** Energy band diagram of a *pn*-junction, forward biased by V volts. Superscript + refers to hole current and –, to electron current

Thus, current density $J_1^-$ is flowing to the right and current density $J_2^-$ is flowing to the left. This means that the current densities and $J_1^-$ and $J_1^+$ add together in the right hand direction and current densities $J_2^-$ and $J_1^+$ add together in the left hand direction. The net current flowing to the right is

$$J = (J_1^- + J_1^+) - (J_2^- + J_2^+) \qquad ...(7)$$

Here $J_2^-$ is the minority carrier electron current drifting from $p$-side to $n$-side. It is independent of the applied voltage $V$(provided $V < V_B$). It is known as *saturation current density*. Similarly, $J_2^+$ is a minority carrier hole current drifting from $n$-side to $p$-side (also saturation current).

In the conduction band, $J_1^-$ is the majority carrier electron current diffusing from $n$-side to $p$-side. Provided the current is not excessive, the only restriction to the flow of this current is the barrier potential, $V_T = V_B - V$. For potential of height $V_T$, the number of electrons which are available sources of current is proportional to exp $(-eV/kT)$.

Thus, $$J_1^- = C_1 \exp\left(-\frac{eV_T}{kT}\right) \quad ...(8)$$

where $C_1$ is a constant to be determined. Since holes carry a positive sign, holes flowing downwards in the figure are effectively flowing up the barrier. The majority carrier hole current diffusing from $p$- to $n$-side is given by

$$J_1^+ = C_2 \exp\left(-\frac{eV_T}{kT}\right) \quad ...(9)$$

where $C_2$ is another constant.

Since there is no net current flow across the junction when applied voltage $V = 0$ (or $V_T = V_B$) then,

$J_1^- - J_2^- = 0$ (net electron current in conduction band).

and $J_1^+ - J_2^+ = 0$ (net hole current in valence band). ...(10)

(because the current flow in one band is independent of flow in the other).

$$\Rightarrow \quad C_1 \exp\left(-\frac{eV_B}{kT}\right) - J_2^- = 0$$

or $$C_1 = J_2^- \exp\left(\frac{eV_B}{kT}\right) \quad ...(11a)$$

Similarly, $$C_2 = J_2^+ \exp\left(\frac{eV_B}{kT}\right) \quad ...(11b)$$

$$\therefore \quad J_1^- = J_2^- \exp\left[\frac{e(V_B - V_T)}{kT}\right] \quad ...(12)$$

and $$J_1^+ = J_2^+ \exp\left[\frac{e(V_B - V_T)}{kT}\right] \quad ...(13)$$

$$\Rightarrow \quad J = (J_1^+ + J_1^-) - (J_2^+ + J_2^-)$$

or $$J = (J_2^- + J_2^+)\left[\exp\left(\frac{\text{eV}}{kT}\right) - 1\right]$$

or $$J = J_2\left\{\exp\left(\frac{eV}{kT}\right) - 1\right\} \quad ...(14)$$

where $J_2 = J_2^- + J_2^+$ is the sum of two saturation current densities. This equation is known as the rectifier equation and is plotted in Fig. 11.60.

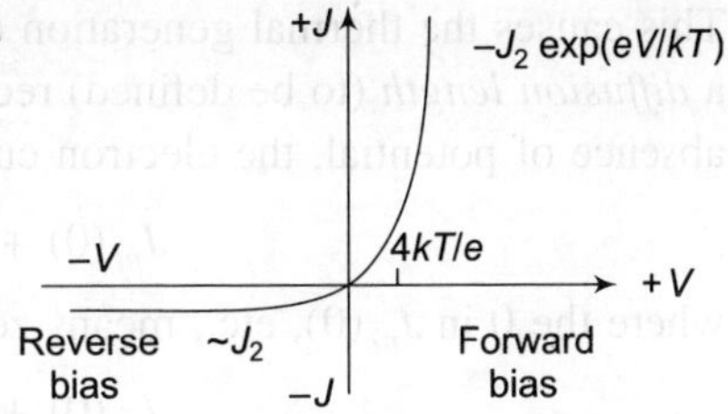


**Figure 11.60** *V-J* characteristic curve of a *pn*-diode

If forward bias is $V \gg \frac{4\,kT}{e}$ ($kT/e \simeq 0.0259$ V at room temperature), then

$$\exp\left(\frac{eV}{kT}\right) \gg 1 \quad \text{and} \quad J \simeq J_2 \exp\left(\frac{eV}{kT}\right) \quad ...(15)$$

When a reverse bias is applied such that $-V \geq \frac{4\,kT}{e}$, then exp $(eV/kT) \ll 1$ (exponential term approaches zero), and

$$J \simeq -J_2, \quad ...(16)$$

that is, current at reverse bias remains constant at the saturation current.

## 11.45 *pn*-JUNCTION UNDER BIAS: (Inhomogeneous Semiconductors not in Equilibrium)

Here we will discuss *pn*-junction under bias and how this leads to electron and hole injection. We will start with a qualitative treatment and then do a more quantitative analysis. The study of *pn*-junctions is fundamental for the study transistors.

We start by looking at a *pn*-junction in equilibrium where there are two types of electron flow that balance in equilibrium (as well as two types of hole flow which also balance in equilibrium).

From the *n*-side to the *p*-side, there is an electron recombination ($r$) or diffusion current ($J_{nr}$) where $n$ denotes electrons. This is due to the majority carrier electrons, which have enough energy to surmount the potential barrier. This current is very sensitive to a bias field that would change the potential barrier. On the *p*-side, there are thermally generated electrons, which in the space-charge region may be swiftly swept downhill into the *n*-region.

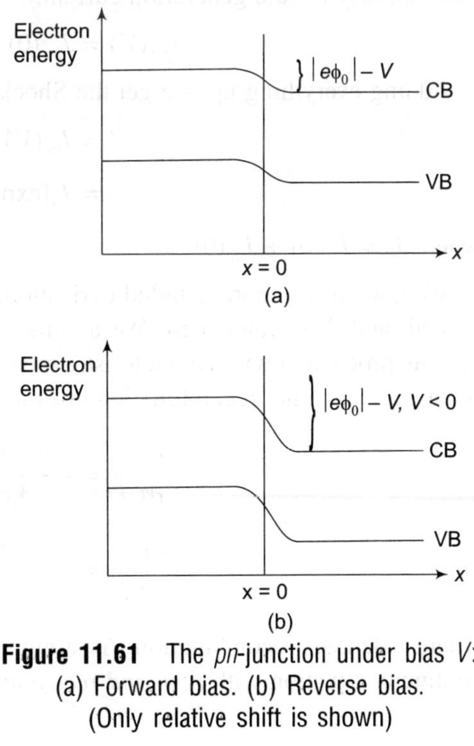


**Figure 11.61** The *pn*-junction under bias *V*: (a) Forward bias. (b) Reverse bias. (Only relative shift is shown)

See also, e.g., Kittel [6.17, p. 572] or Ashcroft and Mermin [6.2, p. 600].

This causes the thermal generation ($g$) or drift current ($J_{ng}$). Electrons produced farther than a *diffusion length* (to be defined) recombine before being swept across. As mentioned, in the absence of potential, the electron currents balance and we have

$$J_{nr}(0) + J_{ng}(0) = 0, \quad \text{...(1)}$$

where the 0 in $J_{nr}(0)$, etc., means zero bias voltage. Similarly, for holes, denoted by $p$,

$$J_{pr}(0) + J_{pg}(0) = 0. \quad \text{...(2)}$$

We set the notation that forward bias ($V > 0$) is when the $p$-side is higher in potential than the $n$-side. (See Fig. 11.61) Since the barrier responds exponentially to the bias voltage, we might expect the electron *injection current*, from $n$ to $p$ to be given by

$$J_{nr}(V) = J_{ng}(0) = -J_{nr}(0) \exp\left(\frac{eV}{kT}\right). \quad \text{...(3)}$$

The thermal generation current is essentially independent of voltage so

$$J_{ng}(V) = J_{ng}(0) = -J_{nr}(0) \quad \text{...(4)}$$

Similarly, for injection of holes from $p$ to $n$, we expect

$$J_{pr}(V) = J_{pr}(0) \exp\left(\frac{eV}{kT}\right), \quad \text{...(5)}$$

and similarly for the generation current,

$$J_{pg}(V) = J_{pg}(0) = -J_{pr}(0). \quad \text{...(6)}$$

Adding everything up, we get the Shockley diode equation for a $pn$-junction under bias

$$J = J_{nr}(V) + J_{ng}(V) + J_{pr}(V) + J_{pg}(V) \quad \text{...(7)}$$

$$= J_0[\exp(eV/kT) - 1]$$

where $J_0 = J_{nr}(0) + J_{pr}(0)$.

We now give a more detailed derivation, in which the exponential term is more carefully argued, and $J_0$ is calculated. We assume that both electrons and holes recombine (due to various processes) with characteristic recombination times, $\tau_n$ and $\tau_p$. The usual assumption is, that as far as net recombination goes *with no flow*,

$$\left.\frac{\partial p}{\partial \tau}\right)_r = -\frac{p - p_0}{\tau_p}, \quad \text{...(8)}$$

and

$$\left.\frac{\partial n}{\partial \tau}\right)_r = -\frac{n - n_0}{\tau_n}, \quad \text{...(9)}$$

where $r$ denotes recombination. Assuming no external generation of electrons or holes, the continuity equation with *flow and recombination* can be written (in one dimension):

$$\frac{\partial J_p}{\partial x} + e\frac{\partial p}{\partial t} = -e\left(\frac{p - p_0}{\tau_p}\right), \qquad ...(10)$$

$$\frac{\partial J_n}{\partial x} - e\frac{\partial n}{\partial t} = +e\left(\frac{n - n_0}{\tau_n}\right) \qquad ...(11)$$

The electron and hole current densities are given by

$$J_p = -eD_p\frac{\partial p}{\partial x} + ep\mu_p E, \qquad ...(12)$$

$$J_n = eD_n\frac{\partial n}{\partial x} + en\mu_n E. \qquad ...(13)$$

And, as always, we assume Gauss' law, where ρ is the total charge density

$$\frac{\partial E}{\partial x} = \frac{\rho}{\varepsilon}. \qquad ...(14)$$

We will also assume a steady state, so

$$\frac{\partial p}{\partial t} = \frac{\partial n}{\partial t} = 0. \qquad ...(15)$$

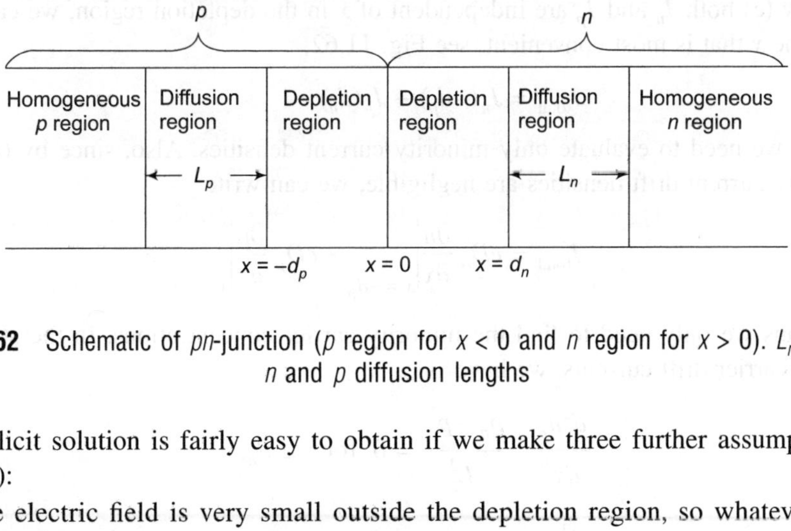


**Figure 11.62** Schematic of *pn*-junction (*p* region for $x < 0$ and *n* region for $x > 0$). $L_n$ and $L_p$ are *n* and *p* diffusion lengths

An explicit solution is fairly easy to obtain if we make three further assumptions (see Fig. 11.62):

(a) The electric field is very small outside the depletion region, so whatever drop in potential there is occurs across the depletion region.

(b) The concentrations of injected minority carriers in the region outside the depletion region is negligible compared to the majority carrier concentration. Also, the majority carrier concentration is essentially constant beyond the depletion and diffusion regions.

(c) Finally, we assume negligible generation or recombination of carriers in the depletion region. We can argue that this ought to be a good approximation if the depletion layer is sufficiently thin. Under this approximation, the electron and hole currents are constant across the depletion region.

A few further comments are necessary before we analyze the *pn*-junction. In the depletion region there are both drift and diffusion currents that are large. In the nonequilibrium case they do not quite cancel. Consistent with this the electric fields, gradient of carrier densities and space charge are all large. Electric fields can be so large here as to lead to the validity of the semiclassical model being open to question. However, we are only trying to develop approximate device equations so our approximations are probably OK.

The diffusion region only exists under applied voltage. The minority drift current is negligible here but the gradient of carrier densities can still be appreciable as can the drift current even though electric fields and space charge are small. The majority drift current is not small as the majority density is large.

In the homogeneous region the whole current is carried by drift and both diffusion currents are negligible. The carrier densities are nearly the same as in equilibrium, but the electric field, space charge, and gradient of carrier densities are all small.

For any $x$ (the direction along the *pn*-junction, see Fig. 11.62.), the total current should be given by

$$J_{\text{total}} = J_n(x) + J_p(x) \quad \text{...(16)}$$

Since by (c) both $J_n$ and $J_p$ are independent of $x$ in the depletion region, we can evaluate them for the $x$ that is most convenient, see Fig. 11.62,

$$J_{\text{total}} = J_n(-d_p) + J_p(d_n). \quad \text{...(17)}$$

That is, we need to evaluate only minority current densities. Also, since by (a) and (b). the minority current drift densities are negligible, we can write

$$J_{\text{total}} = eD_n \left.\frac{\partial n}{\partial x}\right|_{x=-d_p} - eD_p \left.\frac{\partial p}{\partial x}\right|_{x=-d_n}, \quad \text{...(18)}$$

which means we only need to find the minority carrier concentrations. In the steady state, neglecting carrier drift currents, we have

$$\frac{d^2 n_p}{dx^2} - \frac{p_n - p_{n_0}}{L_p^2} = 0, \text{ for } x \geq d_n, \quad \text{...(19)}$$

and
$$\frac{d^2 n_p}{dx^2} - \frac{n_p - n_{p_0}}{L_n^2} = 0, \text{ for } x \leq -d_p, \quad \text{...(20)}$$

where the diffusion lengths are defined by

$$L_p^2 = D_p \tau_p, \quad \text{...(21)}$$

and
$$L_n^2 = D_n \tau_n. \quad \text{...(22)}$$

Diffusion lengths measure the distance a carrier goes before recombining. The solutions obeying appropriate boundary conditions can be written

$$p_n(x) - p_{n_0} = [p_n(d_n) - p_{n_0}] \exp\left(-\frac{(x-d_n)}{L_p}\right), \qquad ...(23)$$

and $$n_p(x) - n_{p_0} = [n_p(-dp) - n_{p_0}] \exp\left(+\frac{(x+d_p)}{L_n}\right). \qquad ...(24)$$

Thus, $$-\left.\frac{\partial p_n}{\partial x}\right|_{x=-d_n} = \frac{[p_n(d_n) - p_{n_0}]}{L_p}, \qquad ...(25)$$

and $$+\left.\frac{\partial n_p}{\partial x}\right|_{x=-d_p} = \frac{[n_p(-d_p) - n_{p_0}]}{L_n}. \qquad ...(26)$$

Thus, $$J_{\text{total}} = \left(\frac{eD_n}{L_n}\right)[n_p(-d_p) - n_{p_0}] + \left(\frac{eD_p}{L_p}\right)[p_n(d_n) - p_{n_0}] \qquad ...(27)$$

To finish the calculation, we need expressions for $n_p(-d_p) - n_{p_0}$ and $p_n(-d_n) - p_{n_0}$, which are determined by the injected minority carrier densities.

Across the depletion region, even with applied bias, $J_n$ and $J_p$ are very small compared to individual drift and diffusion currents of electron and holes (which nearly cancel). Therefore, we can assume $J_n \cong 0$ and $J_p \cong 0$ across the depletion regions. Using the Einstein relations, as well as the definition of drift and diffusion currents, we have

$$kT\frac{\partial n}{\partial x} = en\frac{\partial \varphi}{\partial x}, \qquad ...(28)$$

and $$kT\frac{\partial p}{\partial x} = -ep\frac{\partial \varphi}{\partial x}. \qquad ...(29)$$

Integrating across the depletion region

$$\frac{n(d_n)}{n(-d_p)} = \exp\left(+\frac{e}{kT}[\varphi(d_n) - \varphi(-d_p)]\right), \qquad ...(30)$$

and $$\frac{p(d_n)}{p(-d_p)} = \exp\left(-\frac{e}{kT}[\varphi(d_n) - \varphi(-d_p)]\right). \qquad ...(31)$$

If $\Delta\varphi$ is the built-in potential and $\varphi_h$ is the bias voltage with the conventional sign

$$\varphi(d_n) - \varphi(-d_p) = \Delta\varphi - \varphi_b. \qquad ...(32)$$

Thus, $$\frac{n(d_n)}{n(-d_p)} = \exp\left(\frac{e\Delta\varphi}{kT}\right)\exp\left(-\frac{e\varphi_b}{kT}\right) = \frac{n_{n_0}}{n_{p_0}}\exp\left(-\frac{e\varphi_b}{kT}\right), \qquad ...(33)$$

and $$\frac{p(d_n)}{p(-d_p)} = \exp\left(-\frac{e\Delta\varphi}{kT}\right)\exp\left(\frac{e\varphi_b}{kT}\right) = \frac{p_{n_0}}{p_{p_0}}\exp\left(-\frac{e\varphi_b}{kT}\right). \quad ...(34)$$

By assumption (b) $$n(d_n) \cong n_{n_0} \quad ...(35)$$

and $$p(-d_p) \cong p_{p_0} \quad ...(36)$$

So, we find $$n_p(-d_p) = n_{p_0}\exp\left(\frac{e\varphi_b}{kT}\right), \quad ..(37)$$

and $$p_n(d_n) = p_{n_0}\exp\left(\frac{e\varphi_b}{kT}\right). \quad ...(38)$$

Substituting, we can find the total current, as given by the Shockley diode equation

$$J_{\text{total}} = e\left(\frac{D_n}{L_n}n_{p_0} + \frac{D_p}{L_p}p_{n_0}\right)\left[\exp\left(\frac{e\varphi_b}{kT}\right) - 1\right]. \quad ...(39)$$

### 11.45.1 Reverse Bias Breakdown

The Shockley diode equation indicates that the current attains a constant value of $-J_0$ when the reverse bias is sufficiently strong. Actually, under large reverse bias, the Shockley diode equation is no longer valid and the current becomes arbitrarily large and negative. There are two mechanisms for this reverse current breakdown, as we discuss below (which may not destroy the device).

One is called the Zenger breakdown. This is due to quantum-mechanical interband tunnelling and involves a breakdown of the quasiclassical approximation. It can occur at lower voltages in narrow junctions with high doping. At higher voltages, another mechanism for reverse bias breakdown is dominant. This is the avalanche mechanism. The electric field is the junction accelerates electron in the electric field. When the electron gains kinetic energy equal to the gap energy, then the electron can create an electron-hole pair ($e^- \to e^- + e^- + h$). If the sample is wide enough to allow further accelerations and/or if the electrons themselves retain sufficient energy, then further electron-hole pairs can form, etc. Since a very narrow junction is required for tunnelling, avalanching is usually the mode by which reverse bias breakdown occurs.

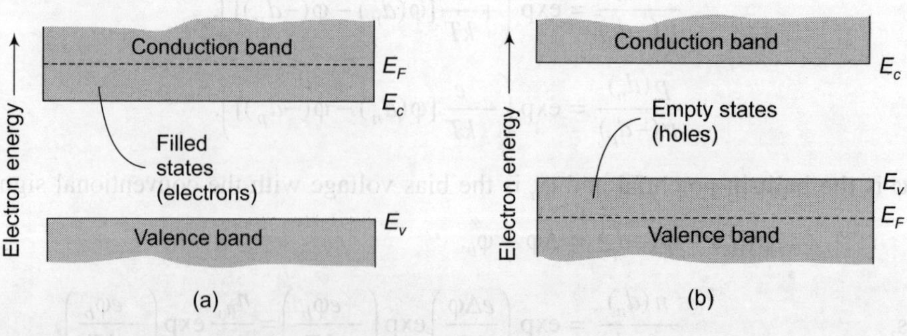


**Figure 11.63** Simplified energy band diagram for degenerately doped (a) *n*-type and (b) *p*-type semiconductor

## 11.46 DEGENERATE SEMICONDUCIORS

If we continue to dope a semiconductor (*p*-type or *n*-type) with impurities of either type, a point is reached at which the impurities become so closely packed within the lattice that interactions between them cannot be ignored. For example, donors present in high concentrations (e.g., $10^{20}$ donors/cm$^3$) are so close together that the donor states instead of discrete non-interacting states form a band which may overlap the bottom of the conduction band. If the conduction band electron concentration *n* exceeds the effective density of states $N_c$, the Fermi level is no longer within the band gap but lies within the conduction band. Under this condition, the material is called *degenerate n*-type. The analogous case of degenerate *p*-type material occurs when the acceptor concentration is very high and the Fermi level lies in the valence band. In a degenerate *n*-type material, the region between $E_c$ and $E_F$ (for the most part) filled with electrons and in degenerate *p*-type material, the region between $E_v$ and $E_F$ is almost completely filled with holes.

### 11.46.1 Tunnel Diode

The tunnel diode is a *pn*-junction device that operates in certain regions of its *I*–*V* characteristic by the quantum mechanical tunnelling of electrons through the potential barrier of the junction. The tunnelling process for reverse current is known as the Zener effect but negligible reverse bias is needed to initiate this process in tunnel diodes. It was invented by Esaki in 1973 and he received the Nobel prize for discovery of this effect. The tunnel diode exhibits the important feature of *negative resistance* over a portion of its *I*–*V* characteristic.

#### *Operation*

A *pn*-junction between two degenerate semiconductors (in terms of energy bands) is illustrated in Fig. 11.64 (a). In this equilibrium condition, the Fermi level is constant throughout the junction. It is to be noted that $E_{Fp}$ lies below the valence band edge on the *p*-side and $E_{Fn}$ above the conduction band edge on the *n*-side, i.e., the bands overlap on the energy scale in order for $E_F$ to be constant [Fig. 11.64 (a)].

Suppose, now a small reverse bias is applied [Fig. 11.64 (b)]. It allows electron tunnelling from the filled valence band states below $E_{F_p}$ to the empty conduction band states above $E_{F_n}$.

Increase in reverse bias increases the electron-tunnelling from *p* to *n*.

When a small forward bias *V* is applied, $E_{F_n}$ moves up in energy with respect to $E_{F_p}$ by the amount eV. Consequently, electrons below $E_{F_n}$ (on the *n*-side) are placed opposite empty states above $E_{F_p}$ on the *p*-side [Fig. 11.64 (c)], and electrons tunnelling occurs from *n* to *p* with the resulting conventional current from *p* to *n*. [Part *OA* of the curve shown in [Fig. 11.65 (a).]

This forward tunnelling current continues to increase with increased forward bias as more filled states are placed opposite empty states [Fig. 11.64 (d)]. However as $E_{F_n}$ continues to move up with respect to $E_{F_p}$, a point is reached at which the bands begin to pass by each other.

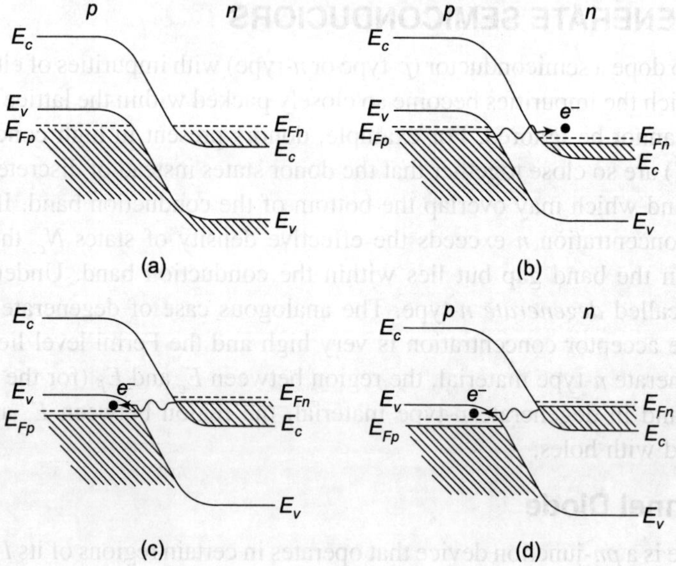


**Figure 11.64** The Tunnel diode energy band diagrams: (a) zero-bias condition (no tunnelling), (b) small reverse bias (electron tunnelling from *p* to *n*). (c) small forward bias (electron tunnelling from *n* to *p*). (d) forward biased ($e^-$ tunnelling from *n* to *p* decreases as bands pass by each other)

Then, the number of filled states opposite empty states decreases. The resulting decrease in tunnelling current is illustrated in Fig. 11.65 (a) by *AB*. This region of the *I–V* characteristic corresponds to the *decrease* of tunnelling current with *increased* bias, i.e., it produces a region of negative slope (negative dynamic resistance *dV*/*dI*). This negative resistance region is useful in a number of applications.

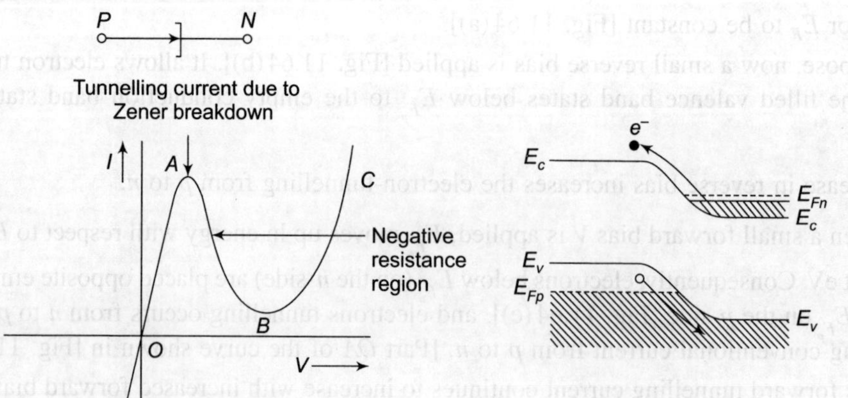


**Figure 11.65** (a) Tunnel diode symbol and characteristic. (b) energy band diagram for the diode beyond the tunnelling condition (corresponding to part BC of the curve in Fig. (a)

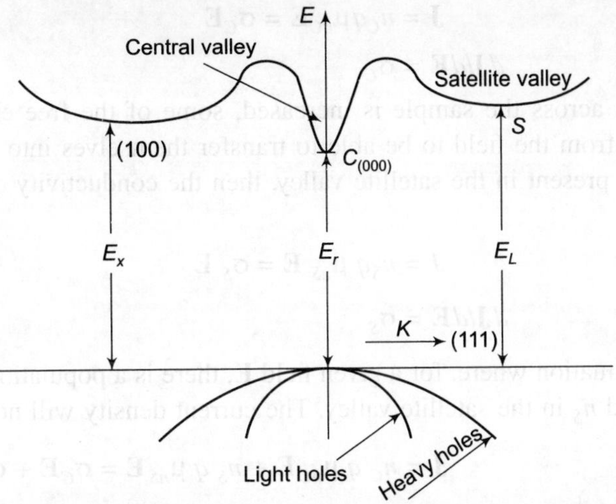


**Figure 11.66** Sketch of *E*-**k** diagram of direct band gap material e.g., GaAs

## 11.47 THE GUNN EFFECT: IT IS EXHIBITED BY DEGENERATED S.C.

To understand the Gunn effect, consider, for example, the direct band gap semiconductor GaAs. A typical *E*-**k** diagram of such a direct band semiconductor is shown in Fig. 11.66. The band gap $E_g$ is the difference between $E_C$, the lowest minima in the conduction band and $E_V$, the highest maximum in the valence band. For GaAs, it is 1.43 eV. However, in GaAs, there exists another minimum *S* in the conduction band only 0.036 eV above the other minima as shown in Fig. 11.66.

Recall that the effective $m_e^*$ of electrons is positive and inversely proportional to the curvature of the band, i.e.,

$$m_e^* = \hbar^2 \cdot \left[\frac{\partial^2 E}{\partial \mathbf{k}^2}\right]^{-1} \qquad ...(1a)$$

As shown in the diagram, the curvature is much larger in the Central Valley than in the satellite. So, for GaAs, the effective mass of electrons in the Central Valley is only 0.07 $m_0$ as opposed to 1.2 $m_0$ for electrons in the satellite valley, where $m_0$ is the free electron mass. So we should expect that the electron mobility, which depends inversely on effective mass, would be much greater for electrons in the Central Valley than in the satellite valley.

i.e., $$\mu_{nC} \gg \mu_{nS}$$

The conductivity of electrons in a band is given by $\sigma_n$:

$$\sigma_n = nq\mu_n \qquad ...(2a)$$

and by Ohm's law: $$\mathbf{J} = \sigma \mathbf{E} \qquad ...(3)$$

Let us consider the **J** vs. **E** curve for a sample of GaAs. For low fields, the free electrons $n_C$ present would be occupying states near *C* in the Central Valley. Hence, here

$$\mathbf{J} = n_C q \mu_{nC} \mathbf{E} = \sigma_C \mathbf{E} \quad ...(4)$$

And so, $$d\mathbf{J}/d\mathbf{E} = \sigma_C \quad ...(5)$$

As the field **E** across the sample is increased, some of the free electrons may acquire sufficient energy from the field to be able to transfer themselves into the satellite valley. if $n_s$ electrons were present in the satellite valley, then the conductivity due to these electrons would be

$$J = n_S q\, \mu_{nS}\, \mathbf{E} = \sigma_S\, \mathbf{E} \quad ...(6)$$

And so, $$d\mathbf{J}/d\mathbf{E} = \sigma_S \quad ...(7)$$

Consider the situation where, for a given field **E**, there is a population density of $n_C$ in the Central Valley and $n_S$ in the satellite valley. The current density will now become

$$\mathbf{J} = n_C\, q\, \mu_{nC} \mathbf{E} + n_S\, q\, \mu_{nS} \mathbf{E} = \sigma_C \mathbf{E} + \sigma_S \mathbf{E} \quad ...(8)$$

If the field **E** is now increased by $\Delta\mathbf{E}$, some additional electrons $\Delta n$ will acquire sufficient energy to transfer from the Central Valley to the satellite valley. Hence now,

$$\mathbf{J} + \Delta\mathbf{J} = (n_C - \Delta n)\, q\, \mu_{nC}\, (\mathbf{E} + \Delta\mathbf{E}) + (n_S + \Delta n)\, q\, \mu_{nS}\, (\mathbf{E} + \Delta\mathbf{E})$$

$$\mathbf{J} + \Delta\mathbf{J} = \sigma_C(\mathbf{E} + \Delta\mathbf{E}) + \sigma_S(\mathbf{E} + \Delta\mathbf{E}) + \Delta n \cdot q \cdot \Delta\mu.\ (\mathbf{E} + \Delta\mathbf{E}) \quad ...(9)$$

where $$\Delta\mu = \mu_{nS} - \mu_{nC}$$

$$\therefore \quad \Delta\mathbf{J} = \sigma_C\, \Delta\mathbf{E} + \sigma_S\, \Delta\mathbf{E} + \Delta n \cdot q \cdot \Delta\mu.\ (\mathbf{E} + \Delta\mathbf{E}) \quad ...(10)$$

The ratio $\Delta\mathbf{J}/\Delta\mathbf{E}$ is referred to as the *differential conductivity* $\Delta\sigma$ and represents the slope of the **J**-**E** characteristic of the sample.

$$\therefore \quad \Delta\sigma = \sigma_C + \sigma_S + \Delta_n\, q\, \Delta\mu\, (\mathbf{E}/\Delta\mathbf{E} + 1) \quad ...(11)$$

$$\therefore \quad \Delta\sigma = T_1 + T_2 + T_3$$

In the above equation, the terms $T_1$, and $T_2$ are always positive, while $T_3$ is always negative since $\Delta\mu$ is always negative. Hence, for certain fields, $\Delta\sigma$ could be negative, depending on the actual values of the material parameters and the magnitude of **E**.

A material like GaAs in fact does show such a negative differential conductivity as sketched in Fig. 11.67. When the field is low, the conductivity is $\sigma_C$. When the applied field exceeds $\mathbf{E}_C$, the transfer of electrons to the satellite becomes significant and the conductivity reduces from its low field values. As **E** increases further, the term $T_3$ predominates and a negative resistance region is observed. The field $\mathbf{E}_{th}$

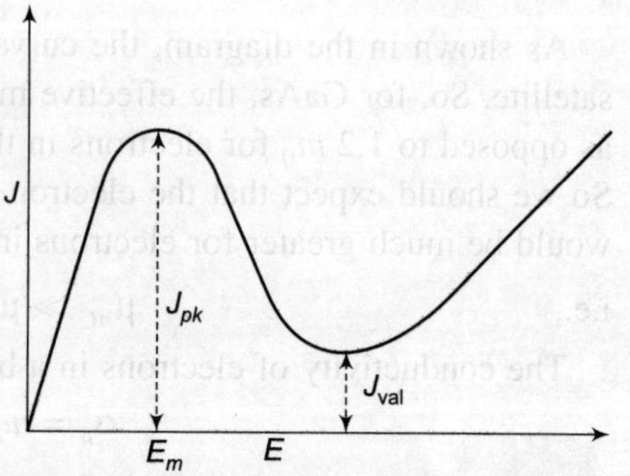


**Figure 11.67 J vs E characteristics of a Gunn diode**

beyond which a negative resistance is observed is the threshold field, and the corresponding current density is the peak current density, $\mathbf{J}_{pk}$. When the field **E** is sufficiently large, virtually all electrons get transferred to the satellite valley and the conductivity becomes positive determined solely by the mobility of electrons in the satellite valley. The **J** versus **E** curve thus goes through a minimum. The current density at the minima is referred to as the valley current, $\mathbf{J}_{val}$.

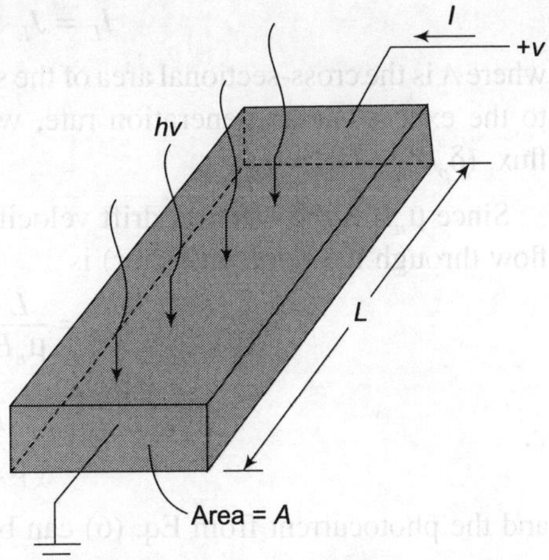


**Figure 11.68** A photoconductor

## 11.48 PHOTOCONDUCTOR

There are several semiconductor devices that can be used to detect the presence of photons. These devices are known as *photodetectors*, they convert optical signals into electrical signals. Due to optical excitation, excess electrons and holes are generated in a semiconductor which increases the conductivity of the material. This change in conductivity is the basis of the photoconductor, the simplest type of photodetector.

Figure 11.68 shows a bar of semiconductor material with ohmic contacts at each end and a voltage applied between the terminals. The initial thermal equilibrium conductivity is

$$\sigma_0 = e(\mu_n n_0 + \mu_p p_0) \qquad ...(1)$$

If excess carriers are generated in the semiconductor. the conductivity becomes

$$\sigma = e[\mu_n(n_0 + \delta n) + \mu_p(p_0 + \delta p)] \qquad ...(2)$$

where $\delta n$ and $\delta p$ are the excess electron and hole concentrations respectively. If we consider a $n$-type semiconductor, then from charge neutrality, we can assume that $\delta n = \delta p \equiv \delta p$. We will use $\delta p$ as the concentration of excess carriers. In steady state $\delta p = G_L \tau_p$ where $G_L$ is the generation rate of excess carriers ($cm^{-3}.s^{-1}$) and $\tau_p$ is the excess minority carrier lifetime. The conductivity can be written as

$$\sigma = e(\mu_n n_0 + \mu_p p_0) + e\delta\, p\,(\mu_n + \mu_p) \qquad ...(3)$$

The change in conductivity due to optical excitations known as the *photoconductivity*, is

$$\Delta\sigma = e \cdot \delta\, p \cdot (\mu_n + \mu_p) \qquad ...(4)$$

An electric field is induced in the semiconductor by the applied voltage, which produces a current. The current density can be written as

$$J = (J_0 + J_L) = (\sigma_0 + \Delta\sigma)E \qquad ...(5)$$

where $J_0$ is the current density in the semiconductor prior to optical excitation and $J_L$ is the photo current density $J_L = \Delta\sigma \cdot E$. If the excess electrons and holes are generated uniformly in the semiconductor, then photo current

$$I_L = J_L \cdot A = \Delta\sigma AE = eG_L\tau_p(\mu_n + \mu_p)AE \qquad ...(6)$$

where $A$ is the cross-sectional area of the semiconductor. The photocurrent directly proportional to the excess carrier generation rate, which in turn, is proportional to the incident photon flux, $(\delta_p/\tau_p = G_L)$.

Since $\mu_n E$ is the electron drift velocity, the electron transit time (i.e., the time required to flow through the semiconductor) is

$$t_n = \frac{L}{\mu_n E} \qquad ...(7)$$

$$\therefore \qquad E = \left(\frac{L}{\mu_n t_n}\right) \qquad ...(8)$$

and the photocurrent from Eq. (6) can be rewritten as

$$I_L = eG_L\left(\frac{\tau_p}{t_n}\right)\left(1 + \frac{\mu_p}{\mu_n}\right)AL \qquad ...(9)$$

We may define a photoconductor gain ($\Gamma_{ph}$) as the ratio of the rate at which charge is collected by the contacts to the rate of which charge is generated within the photoconductor

i.e.,
$$\Gamma_{ph} = \frac{I_L}{e \cdot G_L AL} \qquad ...(10a)$$

which using Eq. (9) can be written as

$$\Gamma_{ph} = \frac{\tau_p}{t_n}\left(1 + \frac{\mu_p}{\mu_n}\right) \qquad ...(10b)$$

## 11.49 PHOTODIODE

A photodiode is a *pn*-junction diode operated with an applied reverse bias voltage. We will consider a long diode in which excess carriers are generated uniformly throughout the entire device. Figure 11.69 (a) shows the reverse biased diode and Fig. 11.69 (b) shows the minority carrier distribution in the reverse biased junction prior to photon illumination. Let $G_L$, be the generation rate of excess carriers. The excess carriers generated within the space charge region are swept out of the depletion region very quickly by the electric field, the electrons are swept into the $n$-region and the holes into the $p$-region. The photon-generated current density from the space charge region is given by

$$J_{L1} = e\int G_L dx \qquad ...(11)$$

where the integral is over the space charge region width. If $G_L$, is constant throughout the space charge volume then

$$J_{L1} = e\,G_L W \qquad ...(12)$$

where $W$ is the space charge width.

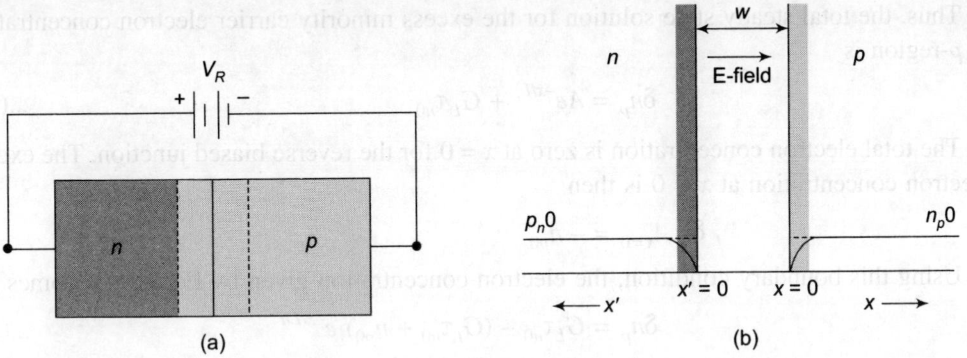


**Figure 11.69** (a) A reverse-biased *pn*-junction (b) Minority carrier concentration in the reverse biased *pn*-junction

It is to be noted here that $J_L$, is in the reverse-bias direction through the *pn*-junction.

Comparing Eqs. (10a) and (12), we note that the photodiode gain is unity.

Excess carriers are also generated within the neutral *n* and *p* regions. The excess minority carrier electron-distribution in the *p*-region is given by the ambipolar transport equation

$$D_n \frac{\partial^2(\delta n_p)}{\partial x^2} + G_L - \frac{\delta n_p}{\tau_{n0}} = \frac{\partial(\delta n_p)}{\partial t} \quad ...(13)$$

We assume that the *E*-field is zero in the neutral regions. In steady state $\partial(\delta n_p)/\partial t = 0$ so that

$$\frac{d^2(\delta n_p)}{dx^2} - \frac{\delta n_p}{L_n^2} = -\frac{G_L}{D_n} \quad ...(14)$$

where $L_n^2 = D_n \tau_{n0}$

The solution to Eq. (14) can be found as the sum of the homogeneous and particular solutions. The homogeneous solution is the solution of

$$\frac{d^2(\delta n_{\text{ph}})}{dx^2} - \frac{\delta n_{\text{ph}}}{L_n^2} = 0 \quad ...(15)$$

where $\delta n_{\text{ph}}$ is the homogeneous solution, given by

$$\delta n_{\text{ph}} = Ae^{-x/L_n} + Be^{+x/L_n} \; (x \geq 0) \quad ...(16)$$

One boundary condition is that $\delta n_{\text{ph}}$ must remain finite; this implies $B \equiv 0$ for the long diode particular solution is found from

$$-\frac{\delta n_{\text{ph}}}{L_n^2} = -\frac{G_L}{D_n} \quad ...(17)$$

which gives

$$\delta n_{pp} = \frac{G_L L_n^2}{D_n} = \frac{G_L(D_n \tau_{n0})}{D_n} = G_L \tau_{n0} \quad ...(18)$$

Thus, the total steady state solution for the excess minority carrier electron concentration in $p$-region is

$$\delta n_p = Ae^{-x/L_n} + G_L \tau_{n0} \quad ...(19)$$

The total electron concentration is zero at $x = 0$ for the reverse biased junction. The excess electron concentration at $x = 0$ is then

$$\delta n_p |_{x=0} = -n_{p0}$$

Using this boundary condition, the electron concentration given by Eq. (19) becomes

$$\delta n_p = G_L \tau_{n0} - (G_L \tau_{n0} + n_{p0})\, e^{-x/Ln} \quad ...(20)$$

We can find the excess minority carrier hole concentration in the $n$-region using similar analysis. Using $x'$ notation (as in Fig. 11.69 b), we can write

$$\delta p_n = G_L \tau_{p0} - (G_L \tau_{p0} + p_{n0})\, e^{-x/L_p} \quad ...(21)$$

Equations (20) and (21) are plotted in Fig. 11.70.

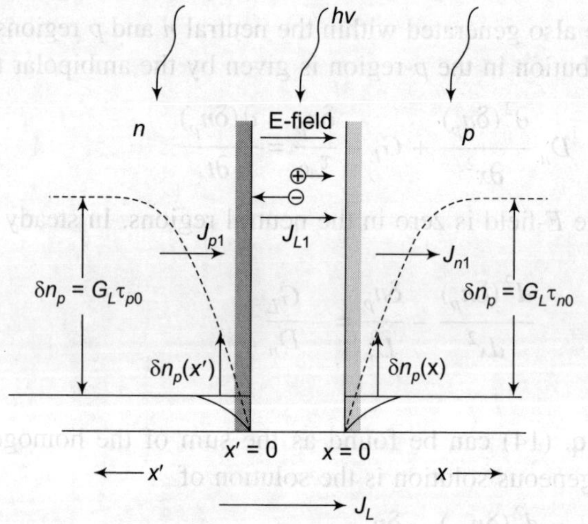


**Figure 11.70** Steady-state, photoinduced minority carrier concentrations and photocurrents in a "long" reverse-biased $pn$-junction

The gradient in the minority carrier concentrations will produce diffusion currents in the $pn$-junction. The diffusion current density at $x = 0$ due to minority carrier electrons is

$$J_{n1} = eD_n \left.\frac{d(\delta n_p)}{dx}\right|_{x=0} = eD_n \frac{d}{dx}\left[G_L t_{n0} - (G_L \tau_{n0} + n_{p0})e^{-x/L_n}\right]_{x=0}$$

$$= \frac{eD_n}{L_n}(G_L \tau_{n0} + n_{p0})$$

or $$J_{n1} = eG_L L_n + \frac{eD_n n_{p0}}{L_n} \quad ...(22)$$

The first term in this equation is the steady state photocurrent density and the second term is the ideal reverse saturation current density due to the minority electrons.

The diffusion current density (in the $x$-direction) at $x' = 0$ due to minority carrier holes is

$$J_{p1} = eG_L L_p + \frac{eD_p p_{n0}}{L_p} \quad ...(23)$$

The first term is the steady state photocurrent density and the second term is the ideal reverse saturation current density.

The total steady-state diode photocurrent density for the long diode is

$$J_L = eG_L W + eG_L L_n + eG_L L_p$$

$$= e(W + L_n + L_p)G_L \quad ...(24)$$

It is to be noted that the photocurrent is in the reverse bias direction through the diode. The resulting expression (24) is due to assuming uniform generation of excess carriers throughout the semiconductor, a long diode and steady state.

## 11.49.1 Expressions for Current

When a $pn$-junction is formed, a region devoid of mobile charge carriers is formed at the junction known as depletion region which has a high internal field across it given by $V_B$. An electron and a hole (i.e., a pair) formed by a photon (absorption gets separated (Fig. 11.71). We may detect this charge separation in two ways. If the device is left on open circuit, an externally measurable potential will appear between the $p$ and $n$ regions.

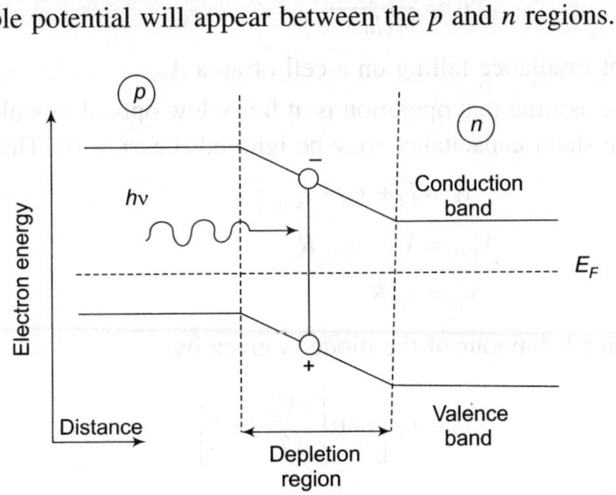


**Figure 11.71** Generation and subsequent separation of an electron-hole pair by photon absorption within the depletion region of a $pn$-junction

This is known as the photovoltaic mode of operation (i.e. a voltage develops because of the incident light). On the other hand, we may short circuit the device externally (or operate it under reverse bias) in which case an external current flow (from $p$ to $n$): this is known as the photoconductive mode of operation.

In operation, we may represent the photodiode as an ideal diode and a constant current generator $i_\lambda$ (in parallel) as shown in Fig. 11.72. The internal characteristics of the photodiode may be better modelled by the introduction of a shunt resistor $R_{sh}$, a shunt capacitor ($C_j$) and a series resistor, $R_s$.

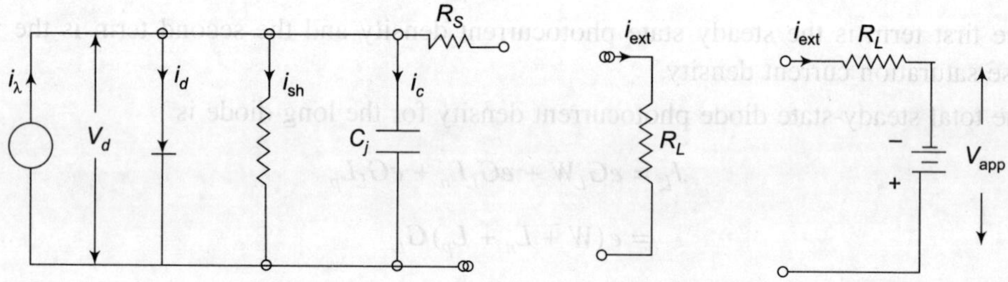


**Figure 11.72** Photodiode equivalent circuit. In the photovoltaic mode, an external high resistor $R_L$ is connected across the output and the voltage across this measured. In the photoconductive mode, an external voltage $V_{app}$ is applied (along with $R_L$) and current through $R_L$ is monitored by measuring the voltage across it

Let $\eta$ denote the quantum efficiency for photon absorption, then, assuming that all the incident radiation is absorbed within the diode, we have

$$i_\lambda = \frac{\eta I_0 A e \lambda}{(hc)} \quad \text{...(25)}$$

where, $I_0$ is the light irradiance falling on a cell of area $A$.

For simplicity, we assume that operation is at fairly low optical modulation frequencies so that the effect of the shunt capacitance may be ignored (i.e., $i_c = 0$). Then

$$i_\lambda = i_d + i_{sh} + i_{ext} \quad \text{...(26)}$$

Also,
$$V_{ext} = V_d - i_{ext} R_s \quad \text{...(27)}$$

and
$$V_d = i_{sh} R_{sh} \quad \text{...(28)}$$

The current-voltage behaviour of the diode is given by

$$i_d = i_0 \left[\exp\left(\frac{eV_d}{kT}\right) - 1\right] \quad \text{...(29)}$$

(i) *Photovoltaic-mode*

The external current flow is very small ($i_{ext} = 0$): Equations (26) and (27) become

$$i_\lambda = i_d + i_{sh} \quad \text{...(30)}$$

$$V_{ext} = V_d \quad ...(31)$$

respectively, substituting for $i_d$ and $i_{sh}$ in Eq. (30), we obtain

$$i_\lambda = \tau_0 \left[\exp\left(\frac{eV_d}{kT}\right) - 1\right] + \frac{V_d}{R_{sh}}$$

or $$\exp\left(\frac{eV_d}{kT}\right) = 1 + \frac{i_\lambda}{i_B} - \frac{V_d}{i_0 R_{sh}}$$

Under normal operating conditions $i \gg i_0$ and $V_d = i_0 R_{sh}$. Thus,

$$\exp\left(\frac{eV_d}{kT}\right) = \frac{i_\lambda}{i_0}$$

$$\Rightarrow \quad V_d (= V_{ext}) = \frac{kT}{e} \ln\left(\frac{i_\lambda}{i_0}\right)$$

Substituting for $i_\lambda$ from Eq. (25), we get

$$V_{ext} = \frac{kT}{e} \ln\left(\frac{\eta I_0 e \lambda A}{i_0 hc}\right)$$

i.e., the external voltage is a logarithmic function of the incident light irradiance.

(ii) *Photo conductive mode*

A relatively large reverse bias (~ 10V) is applied across the diode. Since the diode current saturates at $i_0$ for relatively small values of reverse bias (i.e., for a few tenths of a volt), Eq. (26) can be written as

$$i_\lambda = i_0 + i_{sh} + i_{ext}$$

and $i_0$ and $i_{sh}$ are negligible in comparison with $i_\lambda$ (which is of the order of μ Amp.) Then, $i_{ext} = i_\lambda$. Consequently

$$i_{ext} = \frac{\eta I_0 A e \lambda}{hc} \quad ...(32)$$

i.e., in the photo conductive mode. the external current is directly proportional to the incident, light irradiance.

The silicon photodiode is one of the most popular of all-purpose radiation detectors in the wavelength range 0.4 to 0.9 μm. It has the virtues of high quantum efficiency (~ 80%), small size, good linearity of response, large band width, simple biasing requirements and relatively low cost.

## 11.49.2 Operation in Current Mode and Current-Voltage Characteristics

In a photodiode we know that electron-hole pairs are produced and this is referred to as the inner photoelectric effect (due to incident light). The incident light flux controls the reverse current when photons of energy greater than the band gap of the device material are incident, electron-hole pairs are generated, the conductivity of the diode increases and the reverse current builds up (Fig. 11.73). This is the *current mode* of operation. The current voltage characteristics ($I = f(v)$) at light flux $\phi$ = constant for the current mode of operation are not unlike the output characteristics of a bipolar transistor in a common base circuit. When there is no light flux incident, the usual reverse leakage current $I_0$ (called the dark current) is flowing through the device. When, however there is a light flux incident, the current through the photodiode increases and the characteristics run higher, the stronger the incident radiation, the greater the current is (Fig. 11.74). An increase in the reverse voltage across a photodiode will raise the current only slightly.

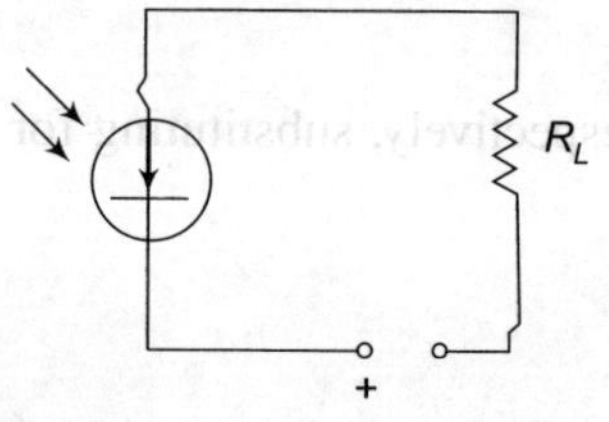


**Figure 11.73** Connection of a photodiode for operation in the current mode

At some applied voltage, however, there occurs a breakdown (the dashed portions of the characteristics. The current-flux characteristics ($I = f(\phi)$) photodiodes at $V$ = constant are linear and almost independent of the applied voltage (Fig. 11.75).

The operating voltage of photodiodes is usually 10 to 30 volts. The dark current does not exceed 10–20 μA for germanium devices and 1.2 μA for silicon devices. The light current is several hundred micro-amperes.

Photodiodes generally come in several basic types. Improved performance is displayed by *heterojunction photodiodes* (*p*- and *n*-regions of different materials). In avalanche photodiodes, the carriers are generated at the *pn*-junction by the avalanche mechanism which substantially raises the sensitivity. In *Schottky barrier photodiodes*, the semiconductor is brought in contact with a metal owing to which they show a high speed of response.

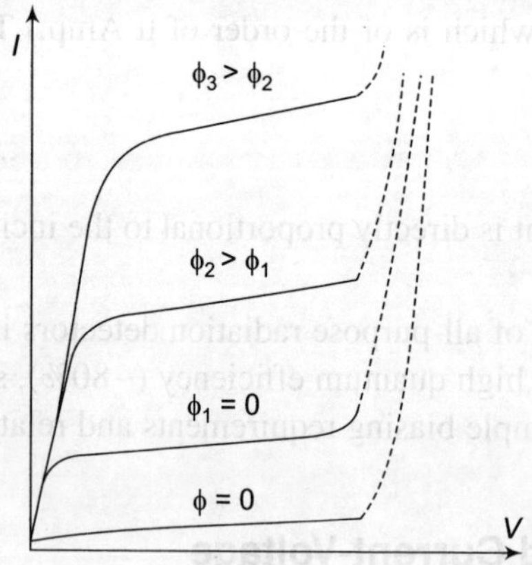


**Figure 11.74** Current-voltage characteristics of a photodiode in the currrent mode

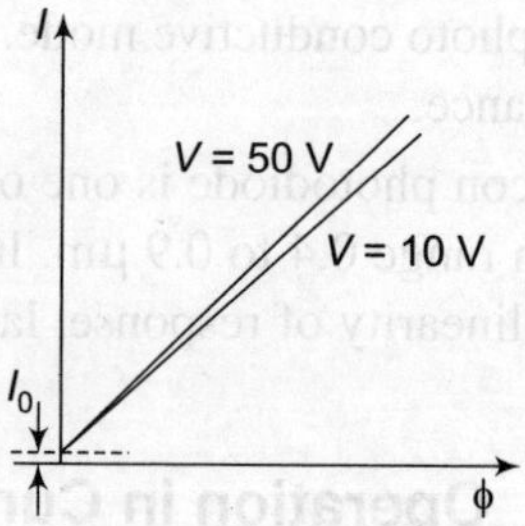


**Figure 11.75** Current vs flux characteristics of a photodiode

## 11.50 PHOTOTRANSISTOR

A bipolar phototransistor is similar in construction to a conventional bipolar junction transistor, except that, the top surface has a window (or lens) so that its base region can be exposed to incident radiation.

The phototransistor is an optical device that provides internal gain and is therefore attractive for many applications. It is a bipolar junction transistor that is excited by an optical signal, and the latter is converted into a photocurrentt, which is the collector current of the device. Usually, the base contact is not made and there is no base current. This is therefore called the floating base mode of operation: the base region is exposed to incident radiation. Then, in a sense, the optical signal acts as the base current in this device and the device is analogous to a junction photodiode followed by an amplifier [Fig. 11.76(a)].

To understand the operation of the device, we consider the external currents to be as shown in Fig. 11.76(b). The base current $i_b$ is supplied by the photogenerated current.

$$i_c = i_e - i_b$$

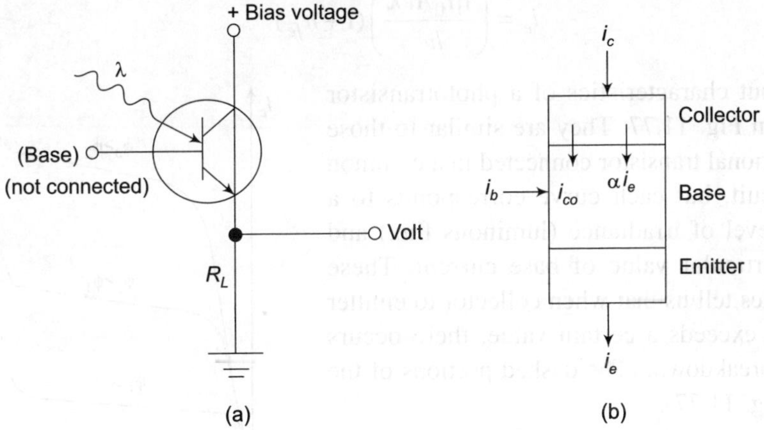


**Figure 11.76** External connections made to *npn* phototransistor (a) Due to light absorbed in base, an emitter current flows through load resistance $R_L$ (a signal voltage appears across it) (b) Currents flowing in the phototransistor

The collector current $i_c$ has two components (i) the normal diode reverse saturation current $i_{co}$, and (ii) that part of emitter current that manages to cross the collector (i.e., that carried by minority carrier diffusion across the base and not all minority carriers leaving the emitter will reach the collector). We write this latter current as $\alpha i_e$, where α is slightly less than 1 (α is known as the common base current gain).

$$\therefore \qquad i_{co} + \alpha i_e = i_e - l_b \qquad \text{...(1a)}$$

$$i_e = \frac{i_b + i_{co}}{1 - \alpha}$$

$$= (i_b + i_{co})\left(\frac{\alpha}{1-\alpha}+1\right)$$

$$= (i_b + i_{co})\,(h_{fe} + 1) \qquad ...(1b)$$

where, $h_{fe}\left(=\frac{\alpha}{1-\alpha}\right)$ is the common emitter current gain of the transistor. With no incident radiation $i_{co} = 0$ and the current flowing is $i_{co}\,(h_{fe} + 1)$ (i.e., the dark current of the device). ($h_{fe} \simeq 100$)

When illuminated, there will be a base current of magnitude, $i_\lambda$

where
$$i_\lambda = \left(\frac{\eta I_0 A e\lambda}{hc}\right). \qquad ...(2)$$

The external current flowing now is $(i_\lambda + i_{co})\,(1 + h_{fe})$, which if $i_\lambda \gg i_{co}$, is equal to $i_\lambda\,(1 + h_{je})$. The device, thus gives us internal gain

$$i_e = \left(\frac{\eta I_0 A e\lambda}{hc}\right)(1+h_{fe}) \qquad ...(3)$$

The output characteristics of a phototransistor are shown in Fig. 11.77. They are similar to those of a conventional transistor connected in a common emitter circuit, but each curve corresponds to a particular level of irradiance (luminous flux) and not to a particular value of base current. These characteristics tell us that when collector to emitter voltage $v_{CE}$ exceeds a certain value, there occurs an electric breakdown. (The dashed portions of the curves in Fig. 11.77).

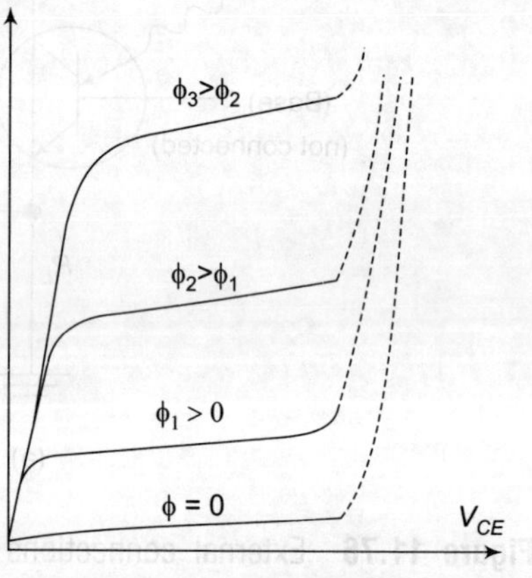


**Figure 11.77** Output characteristics of a phototransistors

The parameters of phototransistor are sensitivity, the operating voltage (which usually is ~ 10–15 V), the dark current (which may run into several hundred microamperes), the operating current (up to tens of microamperes), the maximum allowable power dissipation (up to tens of milliwatts), and the upper frequency limit of several kilohertz, and those manufactured by the diffusion method (planar phototransistors) are able to operate at frequencies up to several megahertz. However, a limitation of phototransistors is a relatively high inherent noise level.

## 11.51 SOLAR CELL

A solar cell is based on the photovoltaic effect, i.e., direct conversion of sunlight into electric current. When radiations are incident upon a semiconductor, the absorption of photons of energy (hυ) greater than or equal to the land gap energy ($E_g$) breaks the bonds and creates electron-hole pairs (e.g., for silicon hυ ≥ 1.1 eV is required). For photovoltaic effect a mechanism is necessary to separate the electron-hole pairs created to ensure a preferred direction of charge flow in an external circuit without the application of an external field. In a photodiode which is basically a *pn*-junction (homojunction) of desired $E_g$, this charge-separating mechanism (separation of electrons and holes and their subsequent opposite movement) is provided by the depletion region potential barrier which is due to immobile ionized impurities. This potential barrier opposes the flow of majority carriers and favours the flow of minority carriers across the junction. The minority carriers generated by the absorption of photons (energy $h\upsilon \geq E_g$) cross over the junction by diffusion in the negative direction of the concentration gradient (Fig. 11.78). For photons of energy $\geq E_g$ incident on the junction excess electron-hole pairs are generated above the equilibrium value, holes cannot diffuse from *p*- to *n*-side and electrons from *n*-side to *p*-side because of the barrier voltage at the junction. Thus, electron generated on *p*-side and holes generated on *n*-side (i.e., minority carriers) are separated by the junction field (produced by immobile ions), electrons roll down into the *n*-region and holes and swept into the *p*-region. These charge carriers which cross-over the junction result in useful charge carriers and flow through external resistance $R_L$. The current forward biased and results in the flow of current opposite in direction to the diode current inside (i.e., $J_D$).

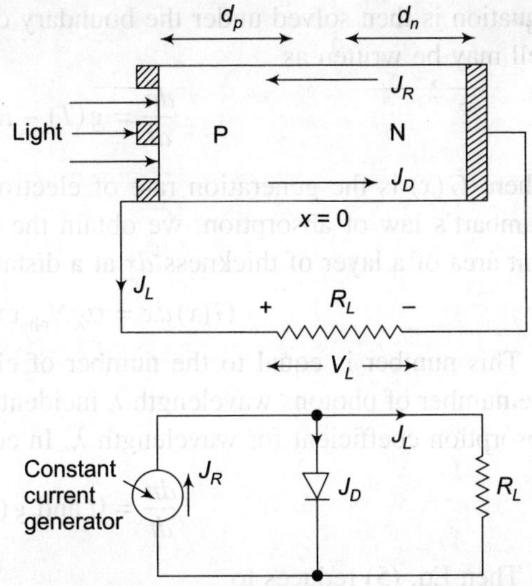


**Figure 11.78** *pn*-junction Solar-cell (top) with its equivalent circuit (bottom) hatched squares on the left on *p* side are metal – "Fingers" to allow absorption of solar radiation and collection of charges

Figure 11.78 shows a schematic diagram of a *pn*-junction with radiation being absorbed and current flowing through an external resistance and an equivalent circuit of the solar cell.

Taking the direction of generated current to be positive, the net current in an ideal Shockely *pn*-junction solar cell is given by

$$J_L = J_R - J_0\left[\exp\left(\frac{qV_L}{kT}\right) - 1\right] \qquad \text{...(4)}$$

where $V_L$ is the voltage across the load

$J_L$ = load current density

and $J_R$ = photocurrent density

Equation (4) can be derived with the help of continuity equation for the minority carriers in a *p*-type semiconductor in the presence of light radiation falling on it. This continuity equation is then solved under the boundary conditions The continuity equation for the solar cell may be written as

$$\frac{dn}{dt} = g(T) - R + G(x) + D_n \frac{d^2n}{dx^2} \quad ...(5)$$

where $G(x)$ is the generation rate of electrons due to incident light at a distance $x$. From Lambart's law of absorption, we obtain the number of photon absorbed per unit time in a unit area of a layer of thickness $dx$ at a distance $x$ below the surface.

$$G(x)\,dx = \alpha_\lambda N_{\text{ph}} \exp(-\alpha_\lambda x)\,dx \quad ...(6)$$

This number is equal to the number of electron-hole pairs generated. In Eq. (6), $N_{\text{ph}}$ is the number of photons wavelength $\lambda$ incident per unit area on $p$-semiconductor and $\alpha_\lambda$ is the absorption coefficient for wavelength $\lambda$. In equilibrium

$$\frac{dn}{dt} = 0 \text{ and } g(T) - R = \frac{n_p - n}{\tau_n} \quad ...(7)$$

Then Eq. (5) reduces to

$$\frac{n_p - n}{\tau_n} + G(x) + D_n \frac{d^2n}{dx^2} = 0 \quad ...(8)$$

For simplification, unit quantum efficiency of photons to create electron hole pairs has been assumed. In equilibrium, the number of electron hole pairs generated by incident radiation in the entire $p$-semiconductor is assumed to be constant = $N_e$, i.e.,

$$G(x) = \alpha_\lambda N_{\text{ph}} \exp(-\alpha_\lambda x) \equiv N_e \quad ...(9)$$

Substituting this into Eq. (8), we get

$$\frac{n_p - n}{\tau_n} + N_e + D_n \frac{d^2n}{dx^2} = 0 \quad ...(10)$$

The solution of the Eq. (10) is given by

$$n = C_1 \exp\left(\frac{x}{L_n}\right) + D_1 \exp\left(\frac{x}{L_n}\right) + n_p + N_e\tau_n \quad ...(11)$$

Constants $C_1$ and $D_1$ are evaluated with the help of boundary conditions

$$\left.\begin{aligned} &\frac{dn}{dx} = 0 \text{ at } x = -d_p \\ \text{and} \quad &n = n_p \exp\left(\frac{qV_L}{kT}\right) \text{ at } x = 0 \end{aligned}\right\} \quad ...(12)$$

Using these equations in Eq. (11) and substituting the values of $C_1$ and $D_1$ into

$$\frac{dn}{dx} = \frac{C_1}{L_n} \exp\left(\frac{x}{L_n}\right) + \frac{D_1}{L_n} \exp\left(\frac{x}{L_n}\right) \quad ...(13)$$

we get

$$\left.\frac{dn}{dx}\right|_{x=0} = \frac{N_e \tau_n}{L_n}\left\{\tanh\left(\frac{d_p}{L_n}\right)\right\} - \frac{n_p}{L_n}\left\{\tanh\left(\frac{d_n}{L_n}\right)\right\} \times \left\{\exp\left(\frac{qV_L}{kT}\right) - 1\right\} \quad ...(14)$$

Substituting for $\frac{dn}{dx}$ from Eq. (14) into

$$J_n = qD_n \left.\frac{dn}{dx}\right|_{x=0}$$

we get

$$J_n = \frac{qN_e \tau_n D_n}{L_n}\left\{\tanh\frac{d_p}{L_n}\right\} - \frac{qD_n n_p}{L_n}\left\{\tanh\frac{d_p}{L_n}\right\} \times \left\{\exp\left(\frac{qV_L}{kT}\right) - 1\right\} \quad ...(15)$$

If we assume electron-hole pairs generated in *n*-region of the *pn*-junction to be negligible, the current density due to holes from *n*-region can be written as

$$J_p = \frac{qD_p p_n}{L_p}\left\{\tanh\frac{d_n}{L_p}\right\}\left\{\exp\left(\frac{qV_L}{kT}\right) - 1\right\} \quad ...(16)$$

It is similar to the unexposed diode current contribution.

The total current density (for an ideal solar cell), is obtained by summing up Eqs. (15) and (16) and is given by

$$J_L = J_R - J_0\left[\exp\left(\frac{qV_L}{kT}\right) - 1\right] \quad ...(17)$$

where

$$J_R = \frac{qN_e \tau_n D_n}{L_n}\left\{\tanh\frac{d_p}{L_n}\right\}$$

and

$$J_0 = \frac{qD_n n_p}{L_n}\left\{\tanh\frac{d_p}{L_n}\right\} + \frac{qD_p p_n}{L_p}\left\{\tanh\frac{d_n}{L_p}\right\}$$

Equation (17) can be expressed as

$$J_L = J_R - J_D \quad ...(18)$$

When the solar cell delivers power to the laod, its *pn*-junction is forward biased and $J_D$ is the dark current density. Figure 11.79 shows the characteristics of a solar cell. (Equations (17) or (18) hold in case of power delivery mode).

## 11.51.1 Solar Cell Parameters

We have

$$J_L = J_R - J_0\left[\exp\left(\frac{qV_L}{kT}\right) - 1\right] \quad ...(19)$$

where $J_R$ is the photocurrent density.

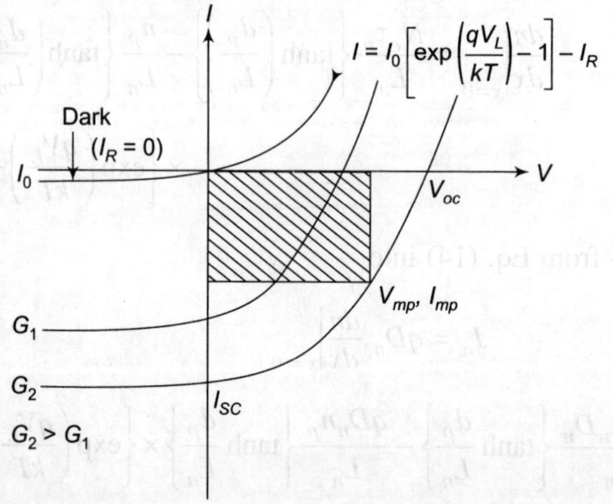


**Figure 11.79** Dark and illuminated characteristics of a solar cell

This equation under condition of open-circuit voltage, $V_{oc}$, i.e., voltage output for infinite load resistance (i.e., $J_L = 0$) gives

$$V_{0c} = \frac{kT}{q} \ln\left(\frac{J_R}{J_0} + 1\right) \equiv V_T \ln\left(1 + \frac{J_R}{J_0}\right) \quad ...(20)$$

and under short circuit condition $V_L = 0$ gives

$$J_L = J_{sc} = J_R \quad ...(21)$$

The three most important parameters of a solar $PV$ cell are the short-circuit current $I_{sc}$, the open-circuit voltage and the fill factor (FF).

$$I_{sc} = I_R \text{ (for } V_L = 0) \quad ...(22)$$

and

$$V_{oc} = V_T \ln\left(1 + \frac{I_R}{I_0}\right) \text{ (for } I_L \text{ or } J_L = 0) \quad ...(23)$$

When a resistance $R_L$ is connected across the cell terminals; $V_L$ is less than $V_{oc}$ and $I_L$ is less than $I_R$. Then, the power $P$ delivered to the load is

$$P = I_L V_L = I_R V_L - I_0 V_L \left[\exp\left(\frac{V_L}{V_T}\right) - 1\right] \quad ...(24)$$

The condition for maximum power is obtained by making $\partial P/\partial V_L = 0$. The voltage, $V_{mp}$, corresponding to the maximum value of $P$ is then obtained from the solution of the equation

$$\frac{\partial P}{\partial V_L} = 0 = I_R - \frac{\partial}{\partial V_L}\left[I_0 V_L \exp\left(\frac{V_L}{V_T}\right) - I_0 V_L\right]$$

$$I_R - I_0 \exp\left(\frac{V_L}{V_T}\right) - I_0 \frac{V_L}{V_T} \exp\left(\frac{V_L}{V_T}\right) + I_0 = 0$$

or $$\left(1 + \frac{V_{mp}}{V_T}\right) \exp\left(\frac{V_{mp}}{V_T}\right) = \left(1 + \frac{I_R}{I_0}\right) \quad ...(25)$$

or $$\exp\left(\frac{V_{mp}}{V_T}\right) = \frac{V_T}{(V_{mp} + V_T)}\left(\frac{I_0 + I_R}{I_0}\right) \quad ...(26)$$

The maximum power current, $I_{mp}$, is given by

$$I_{mp} = I_R - I_0 \left[\exp\left(\frac{V_{mp}}{V_T}\right) - 1\right]$$

$$= I_R - I_0 \left[\frac{V_T}{(V_{mp} + V_T)}\left(\frac{I_0 + I_R}{I_0}\right) - 1\right]$$

(using Eq. 26)

$$= (I_R + I_0) - \left[\frac{V_T}{(V_{mp} + V_T)}\right](I_R + I_0)$$

or $$I_{mp} = (I_0 + I_R)\left[\frac{V_{mp}}{V_{mp} + V_T}\right] \quad ...(27)$$

The maximum obtainable power $P_m$ is then given by

$$P_m = I_{mp} V_{mp} = (I_0 + I_R)\left[\frac{V_{mp}^2}{V_{mp} + V_T}\right] \quad ...(28)$$

and the optimum load resistance $R_{op}$ which achieves the maximum power is

$$R_{op} = \frac{V_{mp}}{I_{mp}} = \frac{(V_{mp} + V_T)}{(I_0 + I_R)} \quad ...(29)$$

The fill factor is defined by the relation

$$FF = \frac{V_{mp} \times I_{mp}}{V_{oc} \times I_{sc}} \quad ...(30)$$

and is a measure of the squarenes of the $I - V$ curve in Fig. 11.79. The ideal power conversion efficiency is given by

$$\eta = \frac{V_{mp} \times I_{mp}}{P_{in}} = F.F.\ \frac{V_{oc}\, I_{sc}}{P_{in}} \quad ...(31)$$

where $P_{in}$ is the incident solar power.

Typical values for a single cell are

$$V_{oc} = 0.5 \text{ to } 0.8 \text{ V}$$

$$I_{oc} = 10 \text{ to } 40 \text{ mA/cm}^2 \text{ of cell area}$$

$$FF = 0.6 \text{ to } 0.8$$

and $$\eta = 6 \text{ to } 20\%$$

### 11.51.2 Effect of Illumination on *pn*-Junction

Diode current without illumination is given by

$$I = qA\left(\frac{D_p p_n}{L_p} + \frac{D_n n_p}{L_n}\right)(e^{V/\gamma V_T} - 1)$$

$$\equiv I_0\,(e^{V/\gamma V_T} - 1) \qquad ...(32)$$

where $I_0$ is the minority carrier generation current also called reverse saturation current which is opposite to the diode current, $I$. $\gamma$ is the diode factor which depends on the diode mechanism ($\gamma = 1$ for injection dominated current and $\gamma = 2$ for recombination in the transition region).

When the junction is illuminated, the minority carrier concentration on either side of the transition region increases above the previous thermal equilibrium value. The majority carrier concentration is essentially constant. Because of increase in concentration of minority carriers on either side carriers which are produced within a diffusion length ($L_n$ and $L_p$) on either side of the depletion region, diffuse into the depletion region and due to the barrier voltage of the depletion region $\left(V_B = \frac{kT}{e}\ln\left[\frac{N_d N_a}{n_i^2}\right]\right)$, drift to the opposite side where they become the majority type. Because of this carrier movement , there results a lowering of the Fermi level on the side and raising on the *n*-side. (Fig. 11.80).

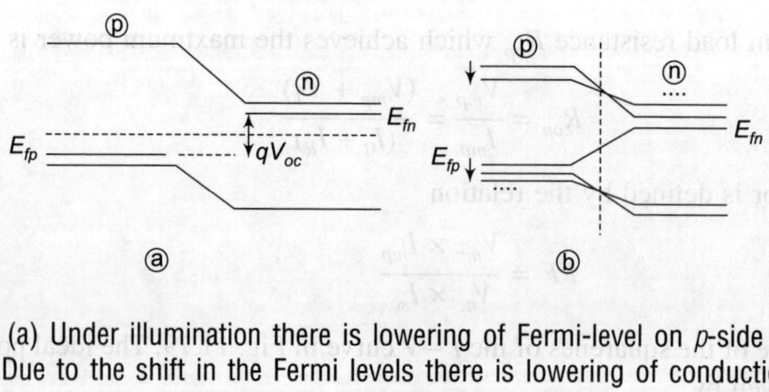


**Figure 11.80** (a) Under illumination there is lowering of Fermi-level on *p*-side and raising on the *n*-side. (b) Due to the shift in the Fermi levels there is lowering of conduction and valence bands on *p*-side than on *n*-side

Hence, illumination results in an additional generation current, $I_{op}$ ($\equiv I_R$) which is opposite to the normal diode current.

$$I = qA\left(\frac{D_p p_n}{L_p} + \frac{D_n n_p}{L_n}\right)(e^{V/V_T} - 1) - I_{op} \equiv I_0(e^{V/V_T} - 1) - I_{op} \qquad ...(33)$$

where $I_{op} = e\,AG\,(L_p + L_n)$ ...(34)

$G$ is the optical generation rate.

Here it should be noted that the diffusion lengths ($L$) should be small compared to the absorption ($1/\alpha$), because only then carrier generated farther from the transition region will be able to diffuse to transition region.

For a symmetrical junction

$$V_{oc} = \frac{kT}{q}\ln\left(\frac{G}{g(T)}\right) \qquad ...(35)$$

where $g(T)$ is thermal generation rate for minority carriers. Due to optical generation of carriers, the normal $I - V$ curve is lowered by an amount proportional to the optical generation rate $G$ (Fig. 11.79). Such a diode operated in the third or the fourth quadrant of the $I - V$ characteristics. For

$$I \equiv -(I_0 + I_{op}) \qquad ...(36)$$

the operation is in the third quadrant and the diode is reverse biased by a few volts and carries a reverse current $I$. The diode operated in this mode acts as a photodiode and can be used to measure the light intensity. Operation in the fourth quadrant is achieved by connecting a load resistance across the diode terminals. The diode operated in this manner is a *photovoltaic cell* and this appearance of a forward voltage ($V_{oc}$) across an illuminated cell is termed the *photovoltaic effect.*

It should also be noted here that the Fermi-level shift $qV_{oc}$ cannot exceed the barrier voltage $qV_B$, because it is the maximum forward voltage that can appear across the junction.

Further, junction cells have a shunt resistance $R_{sh}$ as a result of processing and a series resistance $R_1$, due to body materials and contacts. As a result the "effective" equivalent circuit of a solar cell is as shown in Fig. 11.81.

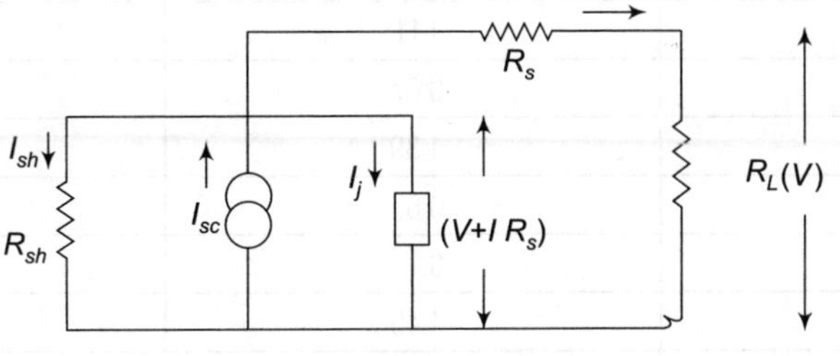


**Figure 11.81** Equivalence circuit of a solar cell with $R_s$ and $R_{sh}$

Thus, idealised junction current in dark

$$I = I_0\left[e^{q(V-IR_s)/\gamma kT} - 1\right] \qquad ...(37)$$

where for $\gamma = 1$ (i.e., injection dominated current)

$$I_0 \propto \exp\left(-\frac{E_g}{kT}\right)$$

and for $\gamma = 2$, (i.e., for recombination in the transition region)

$$I_0 \propto \exp\left(-\frac{E_g}{2kT}\right)$$

Here $I_0$ and $\gamma$ (diode factor) depend only on diode mechanism and are independent of illumination.

∴ Under illumination

$$I = I_0\left\{\exp\left[\frac{q(V = IR_s)}{\gamma kT}\right] - 1\right\} - I_{0p} \qquad ...(38)$$

As, $E_g$ is increased, the diode parameter $I_0$ decreases strongly implying increase in voltage obtainable from the cell. However, an increase in $E_g$ also decreases $I_{op}$ since, less and less of solar spectrum is absorbed. A compromise between these two effects produces a maximum in the solar efficiency at band gap energies of ~1.4 eV to 1.5 eV (Fig. 11.82).

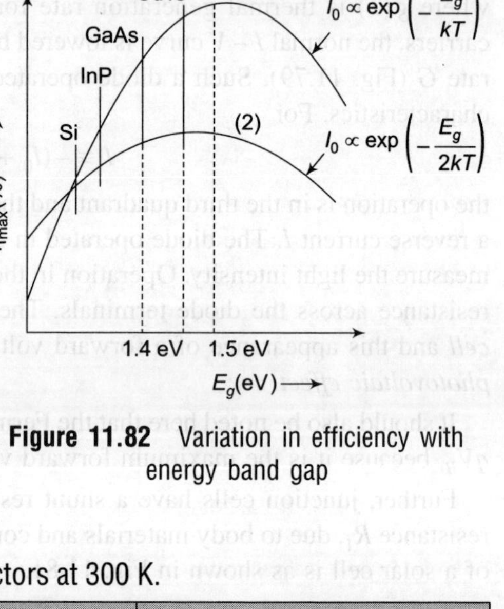


**Figure 11.82** Variation in efficiency with energy band gap

Table 11.10 depicts the band gap energies of some semiconductors.

**Table 11.10** Energy Gaps of Some Semiconductors at 300 K.

| *Material* | $E_g$ *(eV)* | *Structure* |
|---|---|---|
| Si | 1.11 | Diamond |
| Ge | 0.67 | Diamond |
| Ga As | 1.39 | Zinc blende |
| Ga Sb | 0.67 | Zinc blende |
| In Sb | 0.17 | Zinc blende |
| Cd Te | 1.45 | Zinc blende |
| Cd S | 2.45 | Wurtzite, Z.B. |
| Cu In $Se_2$ | 1.02 | – |
| Cu In $S_2$ | 1.53 | – |

### 11.51.3 Solar Spectrum and Its Matching to the Solar Cell Material

If the photon energy is > $E_g$, the band gap of the semiconductor material (λ very small), the absorption coefficient α is large. As the photon energy becomes smaller than $E_g$ (λ larger), α falls off drastically. This can be seen from Fig. 11.83 which shows α vs. wavelength for four semiconductors of interest for solar cell work.

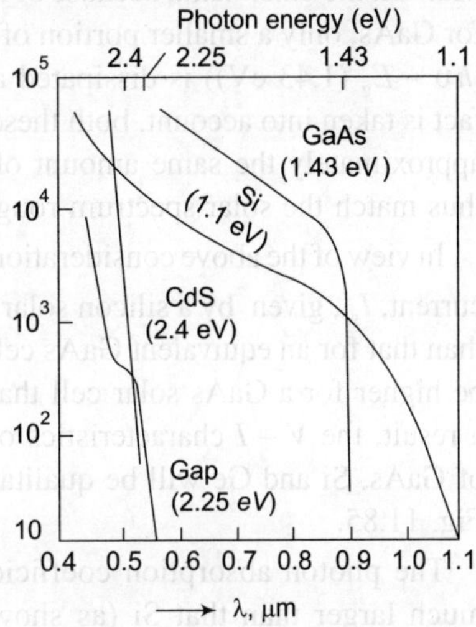


**Figure 11.83** Absorption coefficient α vs. photon wavelength for four typical semiconductors

Figure 11.84 gives the solar energy spectrum in space (AMO—air mass zero) and on the surface of earth under clear sky condition where maximum sun occurs (AM1). The total energy content per $m^2$ per second is given by total area under each of these curves. Under AM1 condition, the total solar power incident is 100 mW/cm$^2$. A major portion of this power is contained in the visible spectrum. It is also to be noted that because of smaller band gap of the silicon compared to GaAs, silicon can convert energy contained in a larger part of the solar spectrum into electrical energy. Thus, it may be apparent that Si could be better than GaAs in this result. However, Si converts only a smaller portion of the energy contained in each photon (an amount equal to $h\upsilon \equiv 1.1$ eV $(= E_g)$ into

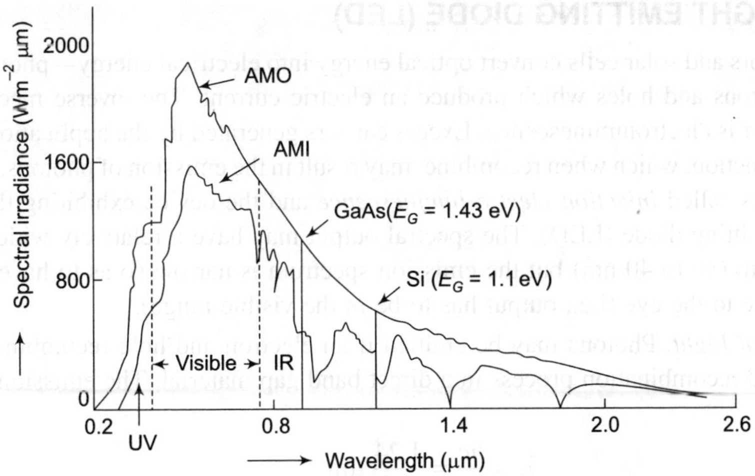


**Figure 11.84** Solar spectra under AMO and AM1 conditions showing the cut-off wavelengths for photon absorption in Si and GaAs. AM1 is the solar spectrum at ground level under a clear sky for vertical incidence on earth's surface. For oblique incidence defined by AM 1.5, AM2 etc. The spectral irradiance curves will be below this AMO refers to vertical incidence outside earth's atmosphere

electrical energy and a higher portion is dissipated as heat. On the other hand, because of the larger band gap, for GaAs, only a smaller portion of the photon energy ($h\upsilon - E_g$ (1.43 eV)) is dissipated as heat. When this fact is taken into account, both these materials convert approximately the same amount of solar energy and thus match the solar spectrum roughly equally.

In view of the above considerations, the short-circuit current, $I_{sc}$, given by a silicon solar cell will be higher than that for an equivalent GaAs cell, whereas $V_{oc}$ will be higher for a GaAs solar cell than for a Si-cell. As a result, the $V - I$ characteristics of solar cells made of GaAs, Si and Ge will be qualitatively as shown in Fig. 11.85.

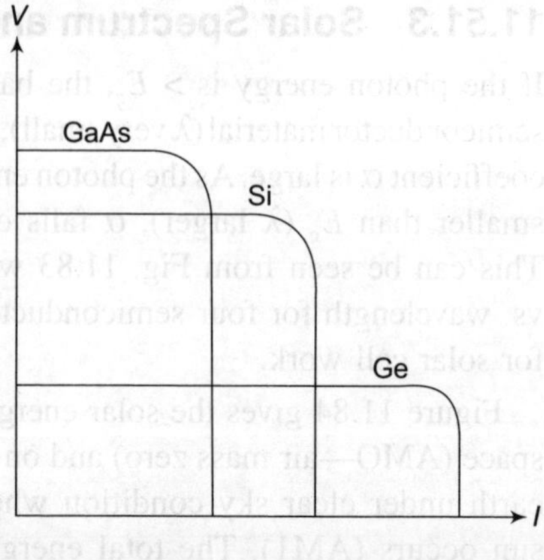


**Figure 11.85** Qualitative plots (not to scale) of V-I characteristics of GaAs, Si and Ge solar cells under the same illumination

The photon absorption coefficient $\alpha$, in GaAs is much larger than that Si (as shown in Fig. 11.83), hence most of the photon energy is absorbed in a thin layer close to the surface in GaAs. Therefore, the material thickness to be used can be made very small in GaAs which would lead to thin solar cells and higher power/weight ratio. In case of Si, photons penetrate much deeper and wafer thickness has to be larger to ensure that all the photons up to the energy of 1.1 eV are absorbed. Thus, the material requirement is larger for silicon cells leading to lower power/weight ratio, which is a disadvantage in space application.

## 11.52 LIGHT EMITTING DIODE (LED)

Photodetectors and solar cells convert optical energy into electrical energy—photons generate excess electrons and holes which produce an electric current. The inverse mechanism of a photodetector is electroluminescence. Excess carriers generated by the application of forward bias to *pn*-junction, which when recombine, may result in the emission of photons. This inverse mechanism is called *injection electro-luminescence* and the device exhibiting this is known as a light-emitting diode (LED). The spectral output may have a relatively wide band width of wavelength (30 to 40 nm) but the emission spectrum is narrow so as to have a particular colour visible to the eye (i.e., output has to be in the visible range).

*Generation of Light*. Photons may be emitted if an electron and hole recombine by a direct band-to-band recombination process in a direct band gap material. The emission wavelength is given by

$$\lambda = \frac{hc}{E_g} = \frac{1.24}{E_g} \mu\text{m}$$

where $E_g$ is the energy gap measured in electron volts.

Suppose, a voltage is applied across a *pn*-junction. Then, electron and holes are injected across the space charge region where they become excess minority carriers. These excess minority carriers diffuse into the neutral semiconductor regions where they recombine with majority carriers. *If the recombination process is a direct band-to-band process, photons are emitted.*

The diode diffusion current is directly proportional to the recombination rate, so the output photon intensity will also be proportional to the recombination rate, so the output photon intensity will also be proportional to the ideal diode diffusion current. In GaAs, electroluminiscence originates primarily on the *p*-side of the junction because the efficiency for electron injection is higher than that for hole injection.

### 11.52.1 LED Materials

Most commercially available LEDs are made from III-V compound semiconductors, while some LEDs are fabricated from ZnSe and SiC materials. An LED, being an electroluminescent device, The emission of light is accomplished by applying a sufficiently large forward bias voltage across the *pn*-junction, followed by the minority carrier injection and radiative recombination taking place in the quasi-neutral *n*- and *p*-regions of the diode.

**Table 11.11** Energy Band Gap, Emitting Wavelength, and Colour for Some LED Materials

| *Material* | *Energy band gap (eV)* | *Wavelength (μm)* | *Colour of LEDs* |
|---|---|---|---|
| GaN | 3.3 | 0.375 | UV, blue |
| InGaN | 0.67–3.3 | 0.375–1.85 | Blue, gree, red |
| AlGaN | 3.3–6.2 | 0.2–0.375 | UV, blue |
| SiC | 2.86 | 0.435 | Blue |
| ZnSe | 2.70 | 0.46 | Blue |
| GaP | 1.98 | 0.63 | Green, red |
| AlGaInP | 1.35–1.98 | 0.63–0.92 | Yellow, orange, amber, red |
| AlGaAs | 1.43–2.19 | 0.57 0.83 | Orange, red |
| GaAsP | 1.41–1.95 | 0.63–0.88 | Yellow, orange, red |
| GaAs | 1.43 | 0.84 | IR emitter |
| InGaAs | 0.34–1.43 | 1.3 | IR |

Table 11.11 lists the most widely used LED material systems. These include a majority of the III-V compound semiconductors as well as some II-VI and IV-IV compound semiconductors such as ZnSe and SiC.

### 11.52.2 Injection Mechanisms

The basic requirement for radiative recombination to take place in a semiconductor is the injection of minority carriers into the bulk semiconductor. To explain the injection mechanism in an LED, Fig. 1.86 shows the schematic drawing and Fig. 11.86 (b) the energy band diagram of AlGaAs/GaAs *pn*-junction IR emitter under thermal equilibrium. The wide-band gap $p^+$–$Al_{0.3}Ga_{0.7}As$ window layer is employed to reduce the surface recombination velocity and

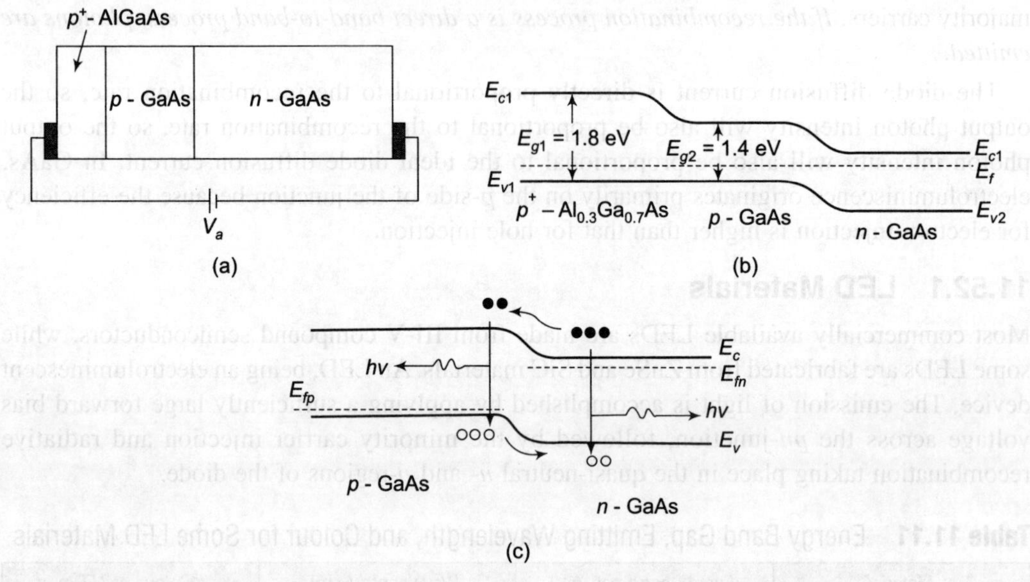


**Figure 11.86** (a) Schematic drawing of GaAs IR emitter with an AlGaAs window layer, (b) the energy band diagram in equilibrium, and (c) the energy band diagram under forward bias condition showing light emission from both *p*- and *n*-regions

to increase the luminescent efficiency of the GaAs IR emitter. Figure 11.86 shows the energy band diagram of the same LED under forward-bias conditions. It is seen that under forward-bias conditions, electrons are injected from the *n*-region into the *p*-region while holes are injected from the *p* into the *n*-region of the junction. For a direct band material such as GaAs, if the band-to-band radiative recombination is dominated on both sides of the junction, then the emission of optical radiation can be readily achieved.

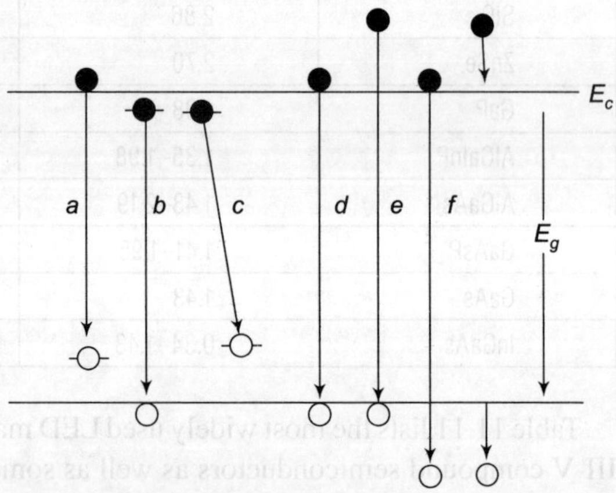


**Figure 11.87** Possible electronics transitions that lead to radiative recombination in a semiconductor. (a) conduction band to acceptor states, (b) donor states to valence band, (c) donor to acceptor states (pair emission), (d) conduction band to valence band (intrinsic emission), (e) hot carrier or avalanche emission, and (f) intraband transition.

### 11.52.3 Electronic Transitions

Figure 11.87 shows the possible electronic transition in a semiconductor due to external excitations. These transitions may lead to either radiative or nonradiative recombination processes, which include (a) conduction band to acceptor states, (b) donor states to valence band, (c) donor to acceptor states (pair emission), (d) conduction band to valence band (intrinsic emission), (e) hot carrier or avalanche emission, and (f) intraband transition. For an efficient luminescent material, the radiative transition usually dominates the nonradiative process. In a direct-band-gap semiconductor such as GaAs, the emission of optical radiation is due mainly to the band-to-band radiative recombination, as shown by process (d) in Fig. 11.87. The emission spectrum for such a transition is given by

$$I(h\nu) = \nu^2 (h\nu - E_g)^{1/2} \exp[-(h\nu - E_g)/k_B T] \qquad ...(1)$$

which shows that the peak intensity occurs near the band gap energy of the semiconductor, and the theoretical full width at half maximum (FWHM) line width for the emission spectrum of an LED is $\Delta E = 1.8\ k_B T$, as shown in Fig. 11.88.

If the electronic transition is form the band edge of one energy band to the impurity level near the opposite band (see process (a) and (b) in Fig. 11.87) then the energy of the emitted photons will be slightly smaller than the band gap energy of the material. The emission spectrum for the electronic transition from the conduction band edge to the acceptor level near the valence band is given by

$$I(h\nu) = \nu^2 (h\nu - E_g + E_a)^{1/2} \{\exp[h\nu - E - E_g + E_{fn})/k_B T + 1\}^{-1} \qquad ...(2)$$

where $E_a$ is the ionization energy of the acceptor level. The peak intensity occurs near $(E_g - E_a)$, and the width of the emission spectrum is also proportional to $k_B T$.

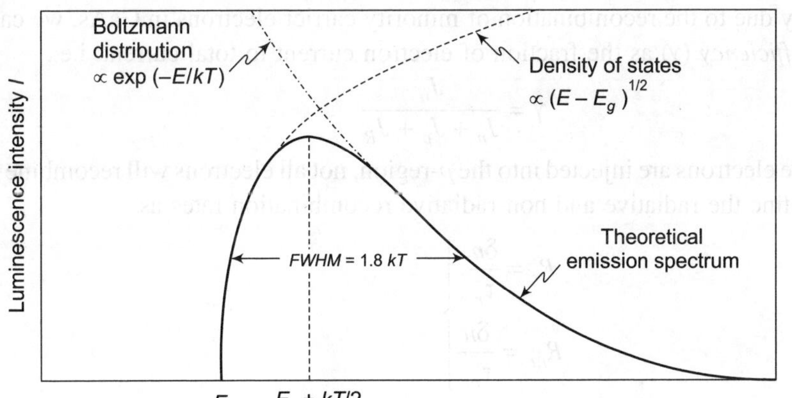


**Figure 11.88** The emission spectrum of an LED due to band-to-band radiative recombination in a direct band gap semiconductor (see process "*d*" in Figure 13.2). After E. F. Schubert.[21]

The peak emission wavelength of an LED for band-to-band radiative recombination can be calculated using the formula

$$\lambda_p = \frac{hc}{E_g} = \frac{1.24}{E_g}\ \mu\text{m}, \qquad ...(3)$$

where $E_g$ is the band gap energy of the semiconductor. For a GaAs IR emitter, with $E_g$ = 1.43 eV, the peak emission wavelength is $\lambda_p$ = 0.873 μm at 300 K.

## 11.52.4 Internal Quantum Efficiency

The internal quantum efficiency of an LED is the fraction of diode current that will produce luminescence. It is a function of the injection efficiency and the percentage of radiative recombination events compared with the total number of recombination events.

The three current components in a forward-biased diode are the minority carrier electron diffusion current, the minority carrier hole diffusion current and the space charge recombination current. These current densities can be written, respectively as

$$J_n = \frac{eD_n n_{po}}{L_n}\left[\exp\left(\frac{\text{eV}}{kT}\right) - 1\right] \quad ...(1)$$

$$J_p = \frac{eD_p p_{no}}{L_p}\left[\exp\left(\frac{\text{eV}}{kT}\right) - 1\right] \quad ...(2)$$

and
$$J_R = \frac{en_i W}{2\tau_0}\left[\exp\left(\frac{\text{eV}}{2kT}\right) - 1\right] \quad ...(3)$$

The recombination of electrons and holes within the space charge region is, in general. through traps (in band gap) near midgap and is a non-radiative process. Since luminescence is primarily due to the recombination of minority carrier electrons in GaAs, we can define an *injection efficiency* (γ) as the fraction of electron current to total current, i.e.,

$$\gamma = \frac{J_n}{J_n + J_p + J_R} \quad ...(4)$$

Once the electrons are injected into the *p*-region, not all electrons will recombine radiatively. We can define the radiative and non-radiative recombination rates as

$$\left.\begin{aligned} R_r &= \frac{\delta n}{\tau_r} \\ \text{and} \quad R_{nr} &= \frac{\delta n}{\tau_{nr}} \end{aligned}\right\} \quad ...(5)$$

where $\tau_r$, and $\tau_{nr}$ are the radiative and non-radiative recombination lifetimes respectively and $\delta n$ is the excess carrier concentration. The total recombination rate is

$$R = R_r + R_{nr} = \frac{\delta n}{\tau} = \frac{\delta n}{\tau_r} + \frac{\delta n}{\tau_{nr}} \quad ...(6)$$

where τ is the next excess carrier lifetime.

The radiative efficiency η is defined as the fraction

$$\eta = \frac{R_r}{R_r + R_{nr}} = \frac{\frac{1}{\tau_r}}{1/\tau_r + 1/\tau_{nr}} = \tau/\tau_r \quad ...(7)$$

The nonradiative recombination rate is proportional to the density of nonradiative trapping sites within the forbidden band ($\equiv N_i$). Obviously, the radiative efficiency increases as $N_i$, is reduced.

The internal quantum efficiency is

$$\eta_i = \gamma\eta \qquad ...(8)$$

The radiative recombination rate is proportional to the $p$-type doping. As the $p$-type doping increases, the radiative recombination rate increases. However, the injection efficiency decreases, therefore, there is an optimum doping will maximize the internal quantum efficiency $\eta_i$.

### 11.52.5 External Quantum Efficiency

The important parameter of the LED is the *external quantum efficiency*, the fraction of generated photons that are actually emitted from the semiconductor. It is normally a much smaller number than the internal quantum efficiency. Once a photon has been produced in the semiconductor, there are three loss mechanisms the photon to encounter: photon absorption within the semiconductor, Fresnel loss and critical angle loss.

Figure 11.89 shows a *pn*-junction LED. Photons can be emitted in any direction. Since the emitted photon energy $h\upsilon$ must be $\geq E_g$, these emitted photons can be reabsorbed within the semiconductor material. Majority of photons, in fact, will be emitted away from the surface and reabsorbed.

Photons must be emitted from the semiconductor into the air; thus photons must be transmitted across a dielectric interface. Figure 11.90 shows the incident, reflected and transmitted waves.

The parameter $\overline{n_2}$ is the index refraction for the semiconductor and $\overline{n_1}$, that for air. The reflection coefficient is

$$\Gamma = \left[\frac{\overline{n_2} - \overline{n_1}}{\overline{n_2} + \overline{n_1}}\right]^2 \qquad ...(9)$$

This effect is called Fresnel loss. $\Gamma$ is the fraction of incident photons that are reflected back into the semiconductor.

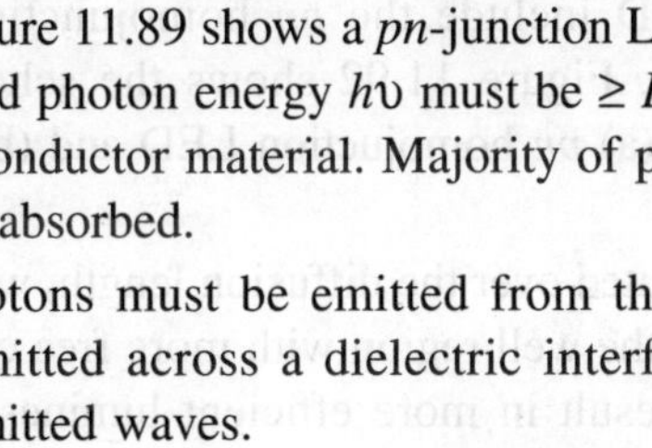


**Figure 11.89** Schematic of photon emission in a *pn*-junction of an LED

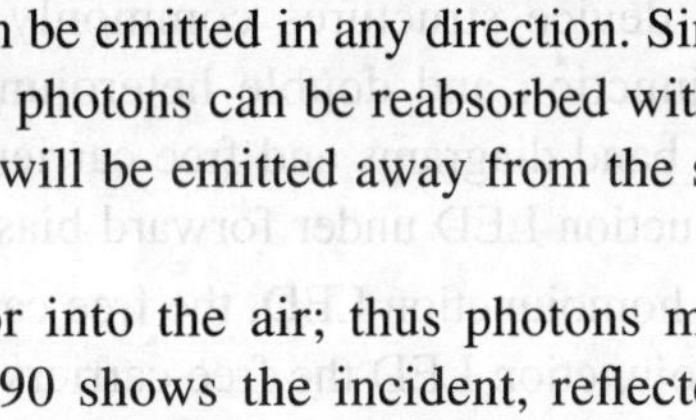


**Figure 11.90** Schematic of a incident reflected and transmitted photons at a dielectric interface

Photons incident on the semiconductor-air interface at an angle refracted as shown in Fig. 11.91. If the photons are incident at an angle greater than the critical angle $\theta_c$, they will experience total internal reflection. The critical angle is given by Snell's law as

$$\theta_c = \sin^{-1}\left(\frac{n_1}{n_2}\right) \quad \text{...(10)}$$

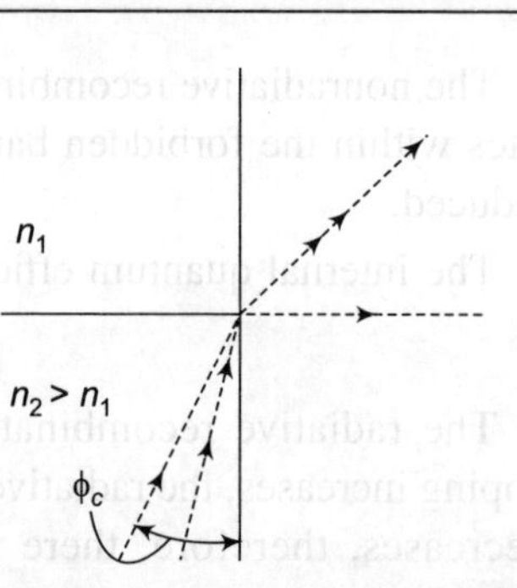


**Figure 11.91** Schematic showing refraction and total internal reflection at a semiconductor – air interface

## 11.52.6 LED Devices and Electrical Characteristics

The wavelength of the output signal of an LED is determined by the energy band gap of the semiconductor. GaAs, a direct band gap material, has a band gap energy of 1.42 eV which yields a wavelength of $\lambda = 0.873$ μm. Comparing this wavelength to the visible spectrum, we find the output of a GaAs LED is not in the visible range. For a visible output, the emitted wavelength should be in the range 0.4 to 0.72 μm. This range of $\lambda$ corresponds to band gap energies between ~1.7 eV and 3.1 eV.

The device structures commonly used in an LED include the *pn* homojunction, *pn* heterojunction and double heterojunction structures. Figure 11.92 shows the schematic energy band diagrams and free carrier distribution in (a) *pn* homojuction LED and (b) a *pn* heterojuction LED under forward bias conditions.

In a homojunction LED, the free carriers are distributed over the diffusion length, while in a heterojunction LED the free carriers are confined to the well region with more free carriers available for radioactive recombination, and hence result in more efficient luminescence. Heterojunction are widely used in the fabrication of high performance LEDs and LDs. Figure 11.93 shows the schematic drawing of the most commonly used DH LED, consisting of a bulk or a multiquantum well (MQW) active region and two confinement layers. The confinement layers are often called the cladding layers, which are formed using a wider gap material than that in the active layer.

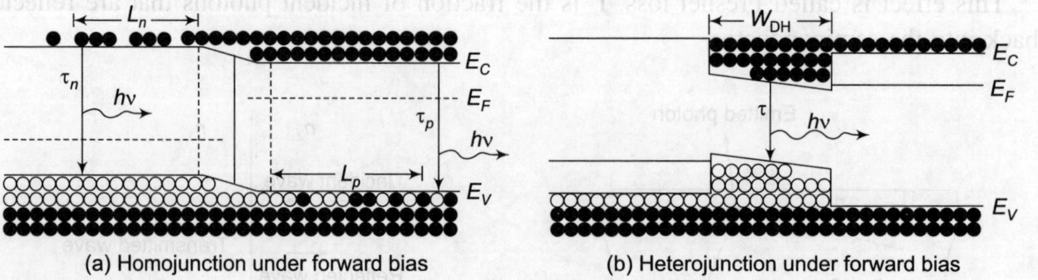


**Figure 11.92** Schematic energy band diagrams for (a) a *pn* homjunction LED and (b) a *pn* heterojunction LED under forward-bias conditions, showing free-carrier distribution and radioactive recombination emission. In homojunctions carriers are distributed over the diffusion length, while in hetrojunctions more carriers are confined to the well region leading to more efficient iuminescence. After E. F. Schubert

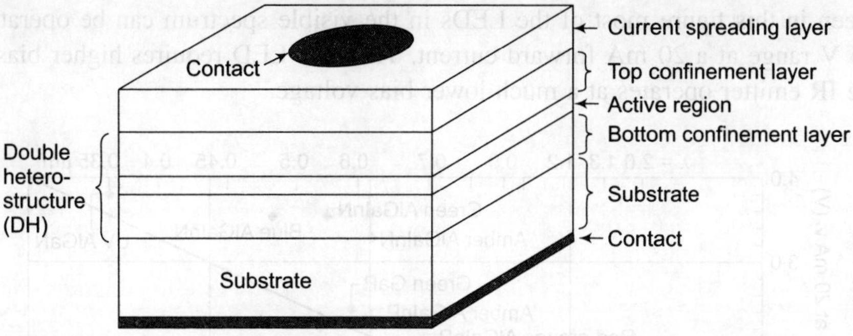


**Figure 11.93** A Schematic drawing of a double-heterostructure (DH) LED consisting of a bulk or multiquantum well (MQW) active region and two confinement layers. The confinement layers are those in the active layer After E. F. Schubert

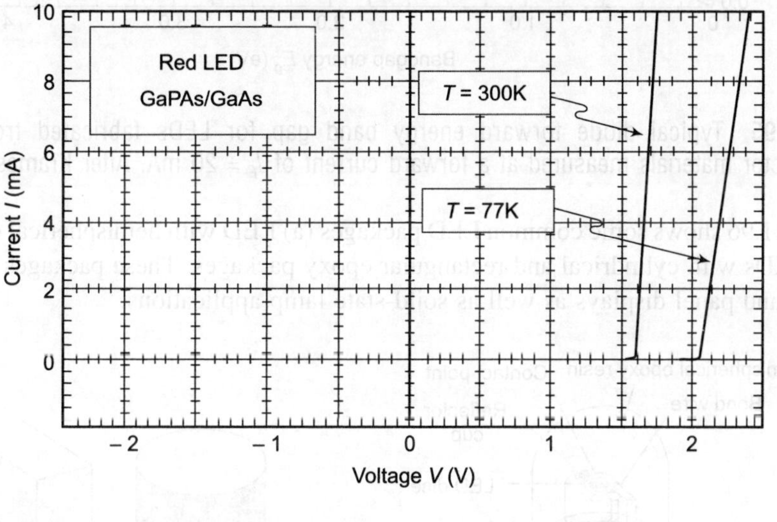


**Figure 11.94** Current–voltage (*I*– *V*) characteristics of a GaAsP/GaAs LED emitting in the red part of the visible spectrum, measured at 77 K and 300 K. The threshold voltages are 2 V and 1.6 V at 77 and 300 K, respectively. After E. F. Schubert

The electrical characteristics of LEDs can be described by the forward current-voltage ($I_F - V_F$) characteristics of a *p-n* junction diode. Figure 11.94 shows the *I–V* characteristic of a GaAsP/GaAs LED emitting in the red part of the visible spectrum, measured at 77 and 300 K, and the threshold voltages for this LED are 2 and 1.6 V at 77 and 300 K, respectively. The forward-bias voltage required to drive an LED at a constant forward current (e.g., $I_F$ = 20 mA) depends on the band gap energy of the semiconductor; typically, the larger-band gap LED (e.g., GaN-LEDs) needs higher bias voltage. Figure 11.95 shows the forward voltage versus energy band gap for different LED materials, operating at 20 mA forward current. As can be seen in this figure most of the LEDs in the visible spectrum can be operated in the 1.5 – 3.5 V range at a 20 mA forward current. The blue LED requires higher bias voltage, while the IR emitter operates at a much lower bias voltage.

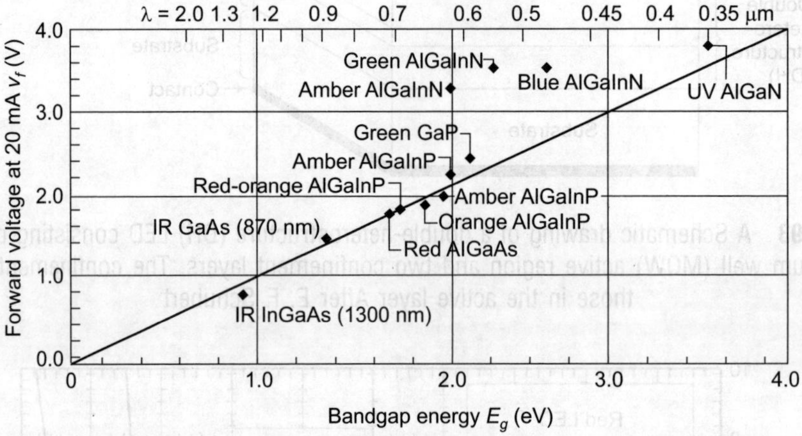


**Figure 11.95** Typical diode forward energy band gap for LEDs fabricated from different semiconductor materials measured at a forward current of $I_F$ = 20 mA. After Krames *et al.*, 2000

Figure 11.96 shows some common LED packages (a) LED with hemispherical epoxy dome and (b) LEDs with cylindrical and rectangular epoxy packages. These packages are suitable for signal and panel displays as well as soild-state lamp applications.

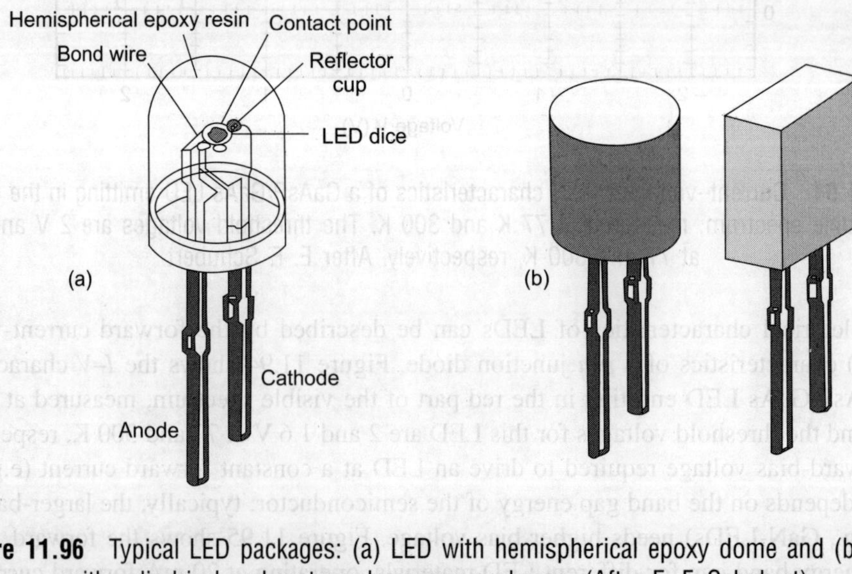


**Figure 11.96** Typical LED packages: (a) LED with hemispherical epoxy dome and (b) LEDs with cylindrical and rectangular epoxy packages. (After E. F. Schubert)

## 11.53 THE HAYNES–SHOCKLEY EXPERIMENT

In this section an example is given to illustrate how the basic semiconductor equations described in section 11.52 can be applied to solve the space- and time-dependent excess carrier phenomena in a semiconductor. First consider a uniformly doped *n*-type semiconductor bar in which *N* electron-hole pairs are generated instantaneously at $x = 0$ and $t = 0$. If one assumes that the semiconductor bar is infinitely long in the *x*-direction (see Fig. 11.98), then the continuity equation given by (Eq. (14), section 11.23) for the excess holes under a constant applied electron field can be reduced to a 1-D equation, which is given by

$$\frac{\partial \Delta p}{\partial t} = D_p \frac{\partial^2 \Delta p}{\partial x^2} - \mu_p \varepsilon \frac{\partial \Delta p}{\partial t} - \frac{\partial p}{\tau_p}. \quad ...(1)$$

assuming that the external generation rate $g_E$ is zero. The solution of Eq. (1) is given by

$$\Delta p(x, t) = \left[\frac{Ne^{-t/\tau_p}}{(4\pi D_p t)^{1/2}}\right] \exp[-(x - \mu_p \varepsilon t)^2 / 4D_p t]. \quad ...(2)$$

From Eq. (2), it is seen that the initial value of $\Delta p(x, 0)$ is zero except at $x = 0$, where $\Delta p(x, 0)$ approaches infinity. Thus, the initial hole concentration distribution corresponds to a Dirac delta function. For $t > 0$, the distribution of $\Delta p(x, t)$ has a Gaussian shape. The half-width of $\Delta p(x, t)$ will increase with time, and its maximum amplitude will decrease with distance along the direction of the applied electric field, with a drift velocity $v_d = \mu_p \varepsilon$. The total excess carrier density injected at time *t* into the semiconductor is obtained by integrating Eq. (2) with respect to *x* from $-\infty$ to $+\infty$, which yields

$$\Delta p(t) = \int_{-\infty}^{+\infty} \Delta p(x, t)\, dx = Ne^{-t/\tau_p}. \quad ...(3)$$

Equation (3) shows that $\Delta p(t)$ decays exponentially with time and with a time constant equal to the hole lifetime $\tau_p$. Figure 11.97 shows the space and time dependence of the excess carrier density in an *n*-type extrinsic semiconductor under a constant applied electric field. As shown in this figure, in order to maintain the original injection hole density profile, a large hole lifetime ($\tau_p$) is needed. This implies that the semiconductor specimen should be of high quality with a very low defect density. Figure 11.98 shows the schematic diagram of the Haynes–Shockley experiment for measuring both the diffusivity and drift mobility of minority carriers in a semiconductor. In this experiment, $P_1$ and $P_2$ denote the injection and collector for the minority carriers (i.e., holes in the

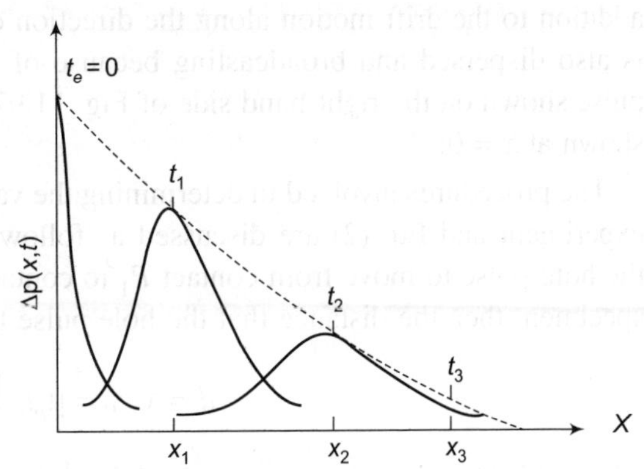

**Figure 11.97** Space and time dependence of the excess hole density in an *n*-type extrinsic semiconductor bar under a constant applied field

present case), and $V_1$ and $V_2$ are the voltage applied to the respective contacts in order to create a uniformed electric field along the specimen and to provide a reverse bias voltage to the collector contact. The injection of minority carriers at contact $P_1$ can be achieved by using either an electron pulse generator or a pulsed laser. An oscilloscope is used to display the pulse shape at contacts $P_1$ and $P_2$ and to measure the time delay of minority carrier travelling between the injecting and collecting contacts.

The Haynes-Shockley experiment is described as follows. At $t = 0$, holes are injected at point $P_1$ of the sample in the form of a pulse of very short duration (on the order of a few microseconds or less.) After this initial hole injection, the excess holes will move along the direction of the applied electron field (i.e., the $x$-direction) and are collected at contact $P_2$. This collection results in a current flow and a voltage drop across the load resistor $R$. The time elapsed between the initial injection pulse at $P_1$ and the arrival of the collection pulse

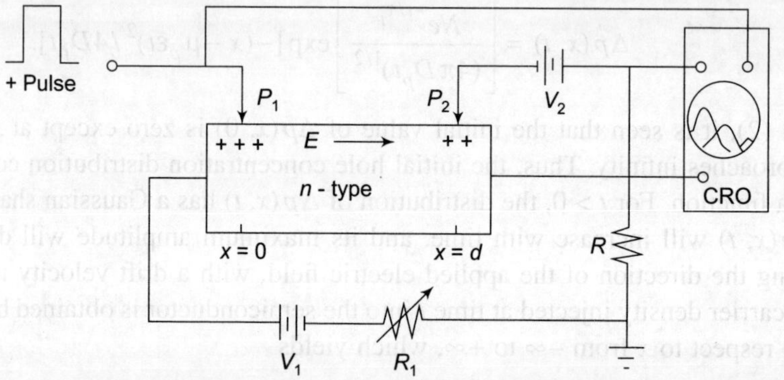


**Figure 11.98** Schematic diagram for the Haynes-Shokley experiment

at $P_2$ is a measure of the **drift velocity** of the holes in the $n$-type semiconductor bar. In the addition to the drift motion along the direction of the applied electric field, the hole density is also dispersed and broadcasting because of the diffusion effect. This explains why the pulse shown on the right-hand side of Fig. 11.97 is not as sharp as the initial injection pulse shown at $x = 0$.

The procedures involved in determining the values of $\mu_p$ and $D_p$ from the Haynes-Shockley experiment and Eq. (2) are discussed as follows: If $t_0$ is the time required for the peak of the hole pulse to move from contact $P_1$ to contact $P_2$ when an electric field is applied to the specimen, then the distance that the hole pulse travelled is given by

$$d = \upsilon_d t_0 = \mu_p t_0 \left(\frac{V_a}{l}\right), \qquad ...(4)$$

where $d$ is the distance between the injection and collecting contacts of the specimen, and $V_a$ is the applied voltage across the sample of length $l$. If values of $d$ and $t_0$ are known, then the hole drift mobility $\mu_p$ can be easily calculated from Eq. (4).

The hole diffusion constant $D_P$ can be determined from width of the Gaussian distribution function $\Delta p(x, t)$. The output voltage ($V_R$) of the hole pulse will drop to 0.367 of its peak value when the second exponential factor on the right-hand side of Eq. (2) is equal to unity. Thus, one obtains

$$(d - \mu_p \varepsilon \Delta t)^2 = 4\, D_p \Delta t. \qquad ...(5)$$

If $t_1$ and $t_2$ denote the delay time constants that satisfy Eq. (5) and $\Delta t = t_2 - t_1$, then $D_p$ can be determined from the expression given by

$$D_p = (\mu_p \varepsilon)^2 (\Delta t)^2 / 16 t_0. \qquad ...(6)$$

The approximation given above is valid as long as the exponential factor $(t/\tau_p)$ given by Eq. (5) does not change appreciably over the measured time interval $\Delta t$. In practice, the diffusion constants for electrons and holes are determined from the electron and hole mobilities using the Einstein relations. Values of the electron and hole drift mobilities for silicon and germanium determined by the Haynes-Shockley experiment at room temperature are listed in Table 11.12.

**Table 11.12** Drift Mobilities (cm²/V · s) for Si and Ge Measured at 300K Using the Haynes-Shockley Experiment

| *Silicon* | *Germanium* |
|---|---|
| $\mu_n = 1350 \pm 100$ | $\mu_p = 3900 \pm 100$ |
| $\mu_p = 480 \pm 15$ | $\mu_p = 1900 \pm 50$ |

## 11.54 BIPOLAR JUNCTION TRANSISTOR

During the period 1904 to 1947, the vacuum tubes were the devices of interest and development. In December 1947, the electronics industry experienced the advent of a completely new direction of interest and development: John Bardeen and W. Brattain demonstrated the amplifying action of the first transistor at the Bell Telephone Laboratories. The first junction transistor was invented by Shockley in 1951. The transistor's impact on electronics has been enormous. The discovery of transistor led to all kinds of related inventions like integrated circuits, up to electronic devices and modern computers.

### 11.54.1 Transistor Construction

The transistor is a three-layer semiconductor device consisting of either two *n*- and one *p*-type layers of material or two *p*- and one *n*-type layers of material. The former is called an *npn-transistor* while the latter is called a *pnp-transistors* (Fig. 11.9). In the *npn-transistor*, the emitter is heavily doped; its job is to emit or inject electrons into the base. The base is lightly doped and very thin, it passes most of the emitter injected electrons onto the collector. The doping level of collector is intermediate (between the heavy doping of emitter and light doping of base). The collector is so named because it collects electrons from the base. The collector is the largest of the three regions, as it has to dissipate more heat than the emitter or the base.

The *pnp*-transistor is the complement of *npn*-transistor and the (majority) carriers in the emitter are holes (instead of free electrons).

The arrow in the circuit symbol points in the direction of conventional current flow.

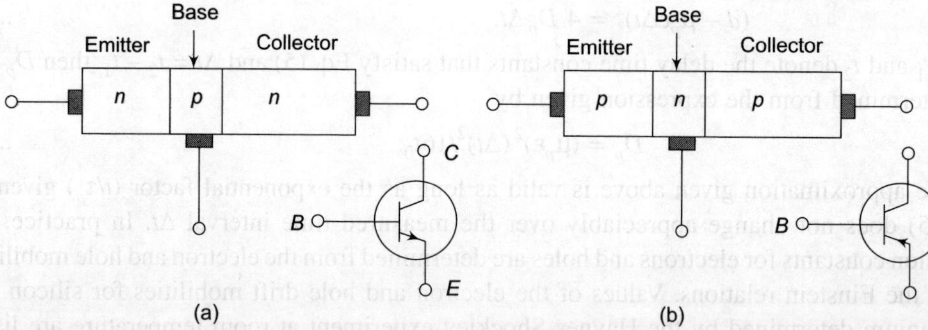


**Figure 11.99** The two types of bipolar junction transistor and their circuit symbol (a) *npn*-transistor, (b) *pnp*-transistor

The diffusion of free electrons across the junction produces two depletion layers. For each of these depletion layers, the barrier potential is approximately 0.7 V (at 25°C) for a silicon transistor and 0.3 V for a germanium transistor. Due to different doping levels of the three regions, the two depletion layers do not have the same width. The emitter depletion layer is small and the collector depletion layer is large. (Fig. 11.100).

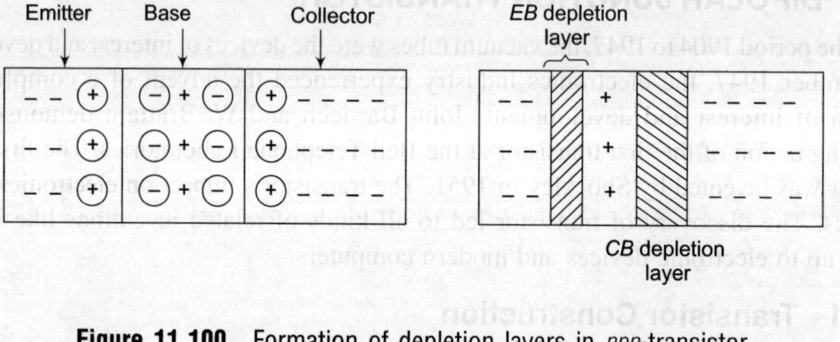


**Figure 11.100** Formation of depletion layers in *npn*-transistor

## 11.54.2 Transistor Biasing

For basic operation of the transistor, the *npn*- and *pnp*-transistors are biased as shown in the Fig. 11.101. The emitter base diode is forward biased and collector-base diode is reverse biased. The abbreviation BJT is used for the junction transistor. The term *bipolar* reflects the fact that both holes and electron participate in the injection process into the oppositely polarized crystal.

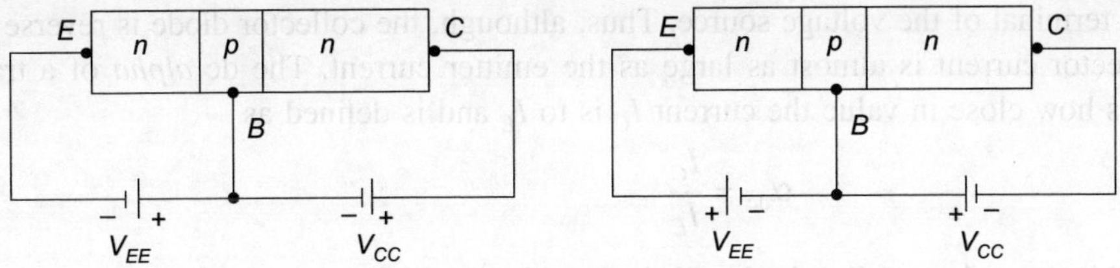


**Figure 11.101**

## 11.54.3 Transistor Operation

1. *npn-transistor.* Figure 11.102 shows the energy band diagram of a properly biased *npn*-transistor.

Forward biasing the emitter diode allows some of the free electrons to move from emitter into the base. If the width of base region is made much smaller than the electron diffusion length only little recombination occurs and due to higher density of minority carriers (i.e., electrons), almost 95 per cent of them dffuse into the collector depletion layer and consequently fall the collector energy hill. Only about 5 per cent of the emitter injected electrons fall along the recombination path shown in Fig. 11.102, those that recombine become valence electrons and flow through base holes into the external base lead.

The larger the forward bias $V_{BE}$, the greater is the number of electrons injected from emitter into the base. The reverse bias $V_{CB}$ has a little influence on the number of electrons entering the collector.

The depletion layer field of the collector-base junction pushes steady stream of electron into the collector region. These electrons enter the external collector lead and flow into the

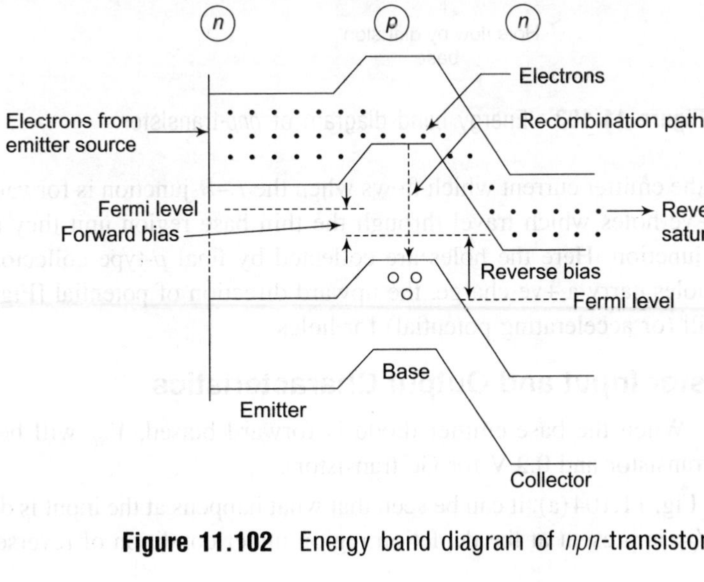


**Figure 11.102** Energy band diagram of *npn*-transistor

positive terminal of the voltage source. Thus, although, the collector diode is reverse biased, the collector current is almost as large as the emitter current. The dc *alpha* of a transistor indicates how close in value the current $I_C$ is to $I_E$ and is defined as

$$\alpha_{dc} = \frac{I_C}{I_E} \quad \text{...(1)}$$

The thinner and more lightly doped the base is, the higher the $\alpha_{dc}$.

2. *pnp-transistor.* Figure 11.103 shows the energy band diagram of a biased *pnp*-transistor. The emitter is made of high conductivity *p*-type material and base is low conductivity *n*-type

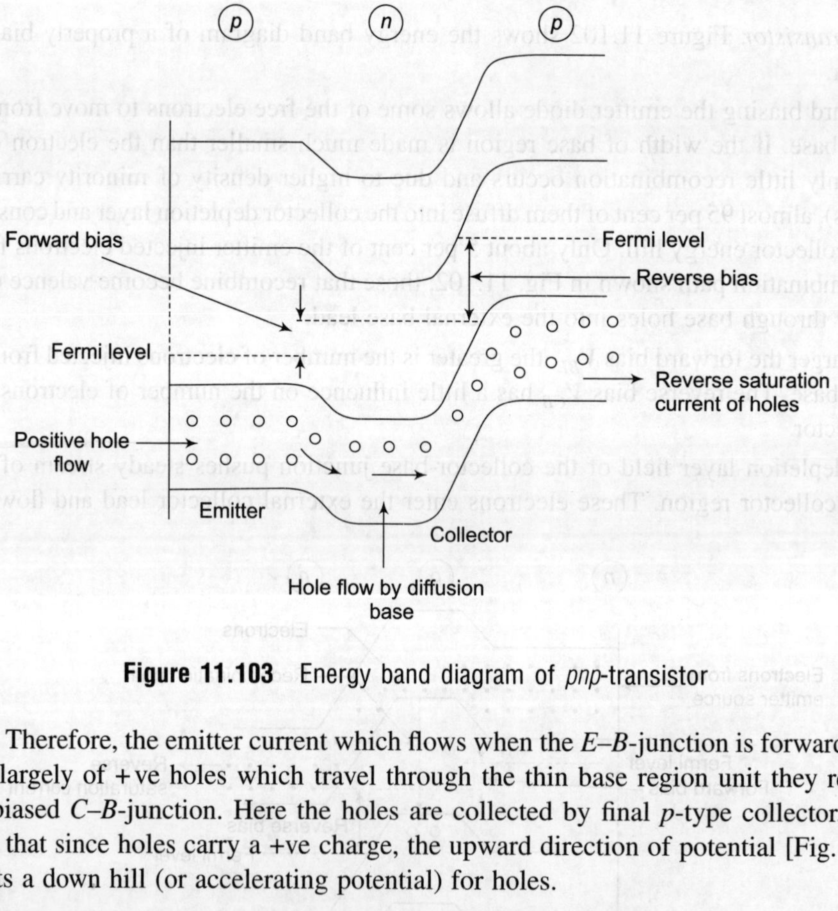


**Figure 11.103** Energy band diagram of *pnp*-transistor

material. Therefore, the emitter current which flows when the *E–B*-junction is forward biased, consists largely of +ve holes which travel through the thin base region unit they reach the reverse biased *C–B*-junction. Here the holes are collected by final *p*-type collector. It is to be noted that since holes carry a +ve charge, the upward direction of potential [Fig. 11.103] represents a down hill (or accelerating potential) for holes.

## 11.54.4 Transistor Input and Output Characteristics

*Input characteristic.* When the base-emitter diode is forward biased, $V_{BE}$ will be initially small (0.6 V for Si transistor and 0.2 V for Ge transistor).

From the curve in Fig. 11.104 (a), it can be seen that what happens at the input is dependent upon what voltage $V_{CE}$ is present at the depletion region under condition of reverse bias.

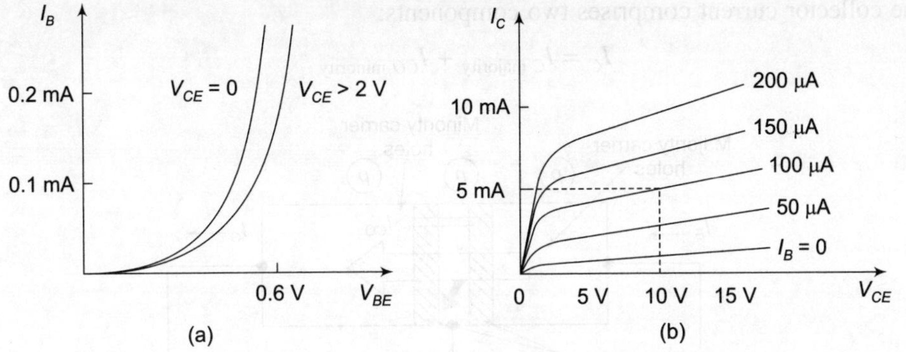


**Figure 11.104** Typical (a) input characteristic, (b) output characteristic for a silicon transistor in the common emitter connection (see Fig. 11.107)

*Output (or collector) characteristic.* In the output circuit, one may think that although collector base diode is reverse biased yet, a (large) current flows in the collector circuit. The answer lies in the heavy doping of collector (with respect to base) and the narrow width of the base. This provides an effect of a *'leaky diode'* so that a large reverse current can flow if the base emitter junction is forward biased. The *pn*-junction from base to emitter acts like a control valve. The larger the forward bias $V_{BE}$ and the base current $I_B$, the larger will be $I_C$ for a given $V_{CC}$.

### 11.54.5 Collector Cut-Off Current $I_{CO}$

If no forward bias is applied ($I_B = 0$), then the only current that will flow in the collector circuit will be the leakage current that can flow. Through the *B–E*-junction with the base open. This is denoted by $I_{CO}$ (i.e., the leakage current or cut-off current from collector to emitter with the base open).

### 11.54.6 Majority and Minority Carrier Flow in a *pnp*-Transistor

A large number of majority carriers diffuse across the forward biased emitter-base junction into the *n*-type base. Due to smallness of the thickness of base and low conductivity, a very small number of these carriers will take the path to base lead (due to high resistance). The magnitude of minority carrier base current is typically of the order of microamperes as compared to milliamperes for the emitter and collector currents. From Kirchhoff's law

$$I_E = I_C + I_B$$

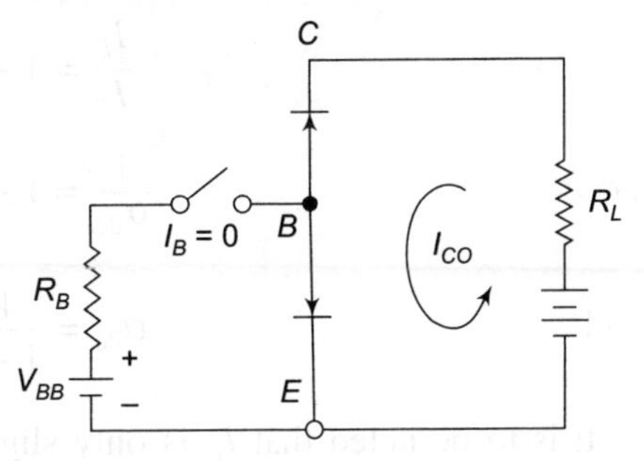


**Figure 11.105** Using two-diode analogy to explain the collector cut-off current $I_{CO}$

The collector current comprises two components:

$$I_C = I_{C \text{ majority}} + I_{CO \text{ minority}}$$

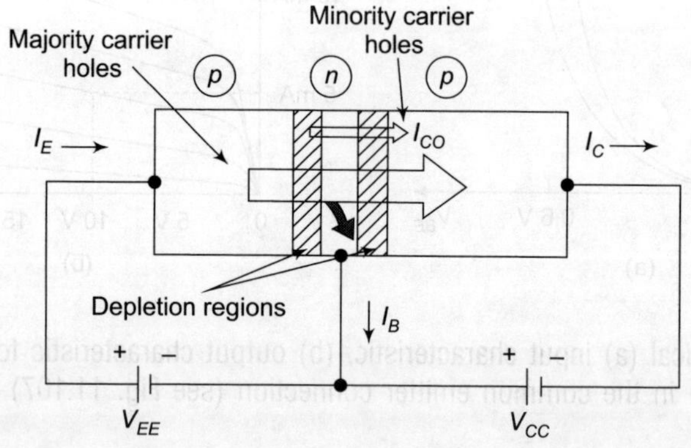


**Figure 11.106** Majority and minority carrier flow in a *pnp*-transistor

## 11.54.7 DC Beta ($\beta_{dc}$) and Relation between $\alpha_{dc}$ and $\beta_{dc}$

We have earlier related the collector current to the emitter current by using $\alpha_{dc}$. However, we can also relate collector current to base current by defining dc beta of a transistor as

$$\beta_{dc} = \frac{I_C}{I_B} \qquad ...(2)$$

where $I_C$ and $I_B$ are determined at a particular operating point on the characteristics. For practical devices, the value of β typically ranges from about 50 to 400 (with most in the mid-range).

We know that $$I_E = I_C + I_B \qquad ...(3)$$

Dividing both sides by $I_C$, we get

$$\frac{I_E}{I_C} = 1 + \frac{I_B}{I_C}$$

or $$\frac{1}{\alpha_{dc}} = 1 + \frac{1}{\beta_{dc}} \quad \text{or} \quad \beta_{dc} = \frac{\alpha_{dc}}{1 - \alpha_{dc}}$$

and $$\alpha_{dc} = \frac{\beta_{dc}}{1 + \beta_{dc}} \qquad ...(4)$$

It is to be noted that $I_C$ is only slightly less than $I_E$ and $I_B$ is much smaller than $I_C$ (or $I_E$). Thus, $\alpha_{dc}$ is slightly less than unity (typically $\alpha_{dc} \sim 0.98$) and $\beta_{dc}$ is large (50 or more). Further, the definitions of α and β are valid only when the emitter junction is forward biased and the collector junction is reverse biased.

### 11.54.8 Common-Emitter Action

Consider a *pnp*-transistor in the common-emitter configuration (Fig. 11.107). The emitter is full of holes. When the base-emitter forward bias $V_{BE}$ is greather than 0.7 V, the emitter injects these holes into the base. The thin, lightly doped base gives almost all (–97%). These holes have enough lifetime to diffuse into the collector, constituting a large reverse current (giving the effect of a leaky diode). The remaining 3 per cent of holes recombine (in the *n*-type base) with electrons. $V_{BB}$ serves to replenish these electrons (constituting a base current $I_B$). With reverse biased collector diode the collector-base depletion layer field pushes the holes into the collector region where they flow out to the external voltage source.

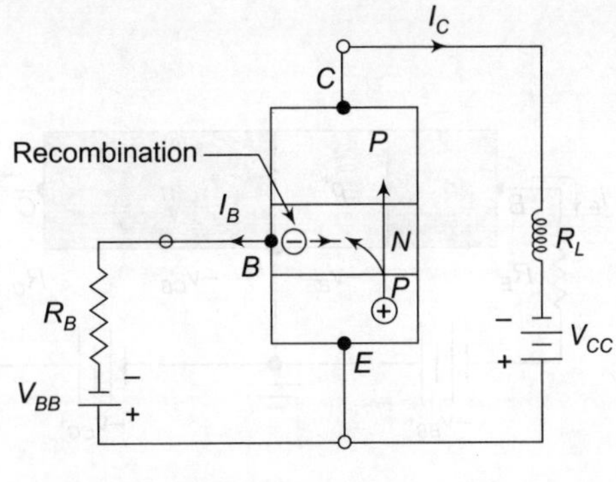


**Figure 11.107** Movement of holes and electron in a forward biased *pnp*-transistor in *CE* configuration

### 11.54.9 Doping Profile and Current Relations in a Transistor

We now consider the *npn*-transistor. Figure 11.108 shows the idealized impurity doping profile in an *npn*-bipolar transistor for the case when each region is uniformly doped. Typical impurity doping concentration in the emitter, base and collector may be on the order of $10^9$, $10^{17}$, and $10^{15}$ $cm^{-3}$ respectively.

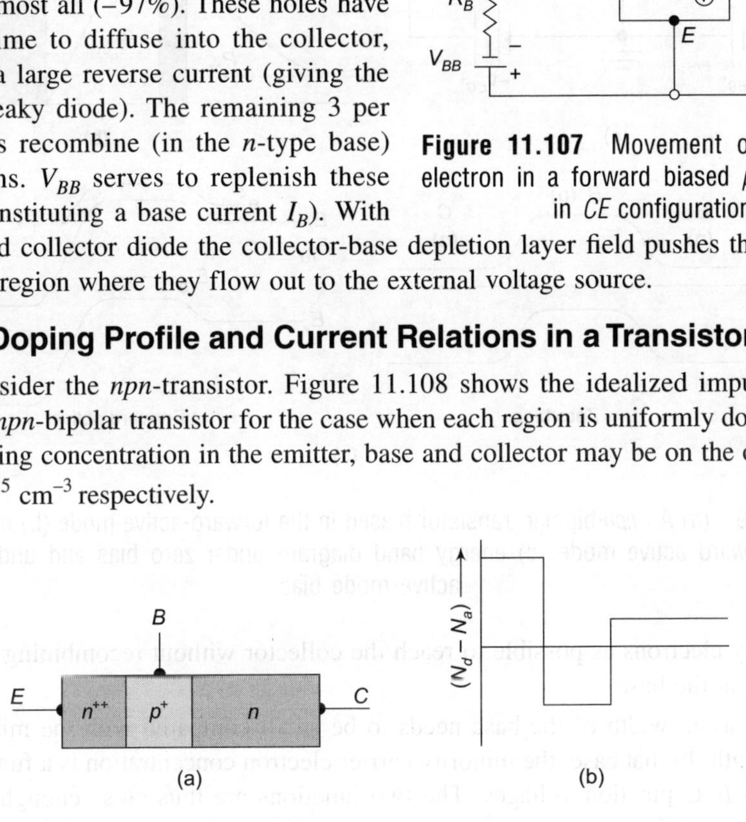


**Figure 11.108** Idealized doping profile of uniformly doped *npn* bipolar transistor

In normal bias configuration *B–E pn*-junction is forward biased and *B–C pn*-junction is reverse biased. (11.109). This is called the *forward-active* operation mode of the transistor. Due to forward bias, electrons from the emitter are injected across the base-emitter junction into the base. These injected electrons create an excess concentration of minority carriers in the base. Since base-collector junction is reverse biased, so the minority carrier electron concentration at the edge of the *B–C* junction is ideally zero. We expect the electron concentration in the base to be like that shown in Fig. 11.109 (b). The large gradient in electron concentration means that electrons injected from the emitter will diffuse across the base region into the *BC* space charge region, where the electric field will sweep these electrons into the collector. We

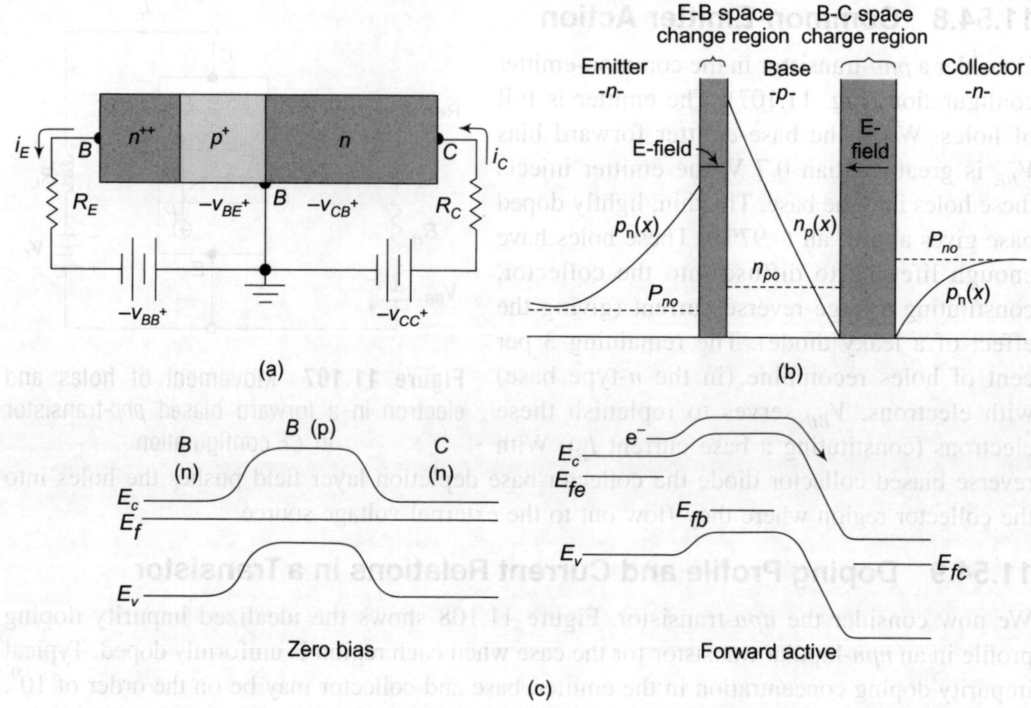


**Figure 11.109** (a) An *npn*-bipolar transistor biased in the forward-active mode (b) minority carrier profile in forward active mode (c) energy band diagram under zero bias and under a forward-active mode bias

want as many electrons as possible to reach the collector without recombining any majority carrier holes in the base.

For this reason, width of the base needs to be small compared with the minority carrier diffusion length. In that case, the minority carrier electron concentration is a function of both the *B–E* and *B–C* junction voltages. The two junctions are thus close enough to be called *interacting pn*-junctions.

The minority carrier concentrations are again shown in Fig. 11.110 for an *npn*-bipolar transistor biased in the forward active mode. Ideally, the minority carrier electron concentration in the base is a linear function of distance, so there is no recombination. The electrons diffuse across the base and are swept into the collector by the electron field in the *B–C* space-charge region.

*Collector current.* Assuming the ideal linear electron distribution in the base, the collector current can be written as a diffusion current given by

$$i_c = e\, D_n A_{BE} \frac{dn(x)}{dx} = e D_n\, A_{BE} \left[ \frac{n_B(0) - 0}{0 - x_B} \right]$$

$$= \frac{-eD_n A_{BE}}{x_B} n_{B_o} \exp\left(\frac{v_{BE}}{V_T}\right) \quad ...(5)$$

where $A_{BE}$ is the cross-sectional area of the *B-E* junction, $n_{B_o}$ is the thermal equilibrium electron concentration in the base and $V_T$ is the thermal voltage (= *kt/e*). The diffusion of electron is in the $+x$ direction so that the conventional current is the $-x$ direction. Considering magnitudes only, Eq. (5) can be written as

$$i_c = I_s \exp\left(\frac{v_{BE}}{V_T}\right) \quad ...(6)$$

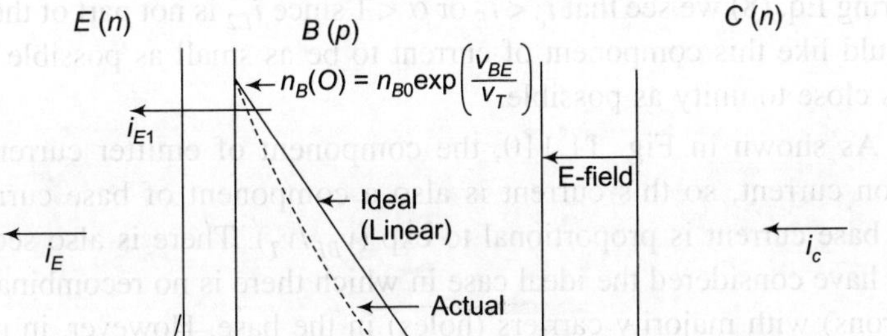


**Figure 11.110** Minority carrier distribution and basic currents in a forward biased *npn*-transistor

The collector is controlled by the base-emitter voltage, i.e., the current at one terminal is controlled by the voltage applied to the other two terminals. This is the basic transistor action.

*Emitter current.* One component of emitter current, $i_{E1}$, shown in Fig. 11.110 is due to the flow of electron injected from the emitter into the base. This current then, is equal to the collector current given by Eq. (6).

Since the base-emitter junction is forward biased, majority carrier holes in the base are injected across the *B-E* junction into the emitter. These injected holes produce a *pn*-junction current $i_{E2}$ as indicated in Fig. 11.110. This current is only a *B–E* junction current so this component of emitter current is not part of the collector current. Since $i_{E2}$ is a forward-biased *pn*-junction current, we can write

$$i_{E2} = I_{s2} \exp\left(\frac{v_{BE}}{V_T}\right) \quad ...(7)$$

where $I_{s2}$ involves the minority carrier hole parameters in the emitter. The total emitter current is

$$i_E = i_{E1} + i_{E2} = i_c + i_{E2} = I_{SE} \exp\left(\frac{v_{BE}}{V_T}\right) \quad ...(8)$$

since all current components in above equation are function of exp $(v_{BE}/v_T)$, the ratio of collector current to emitter current is a constant.

$$\frac{i_c}{i_E} = \alpha, \quad ...(9)$$

the *common-base current gain.*

By considering Eq. (8) we see that $i_c < i_E$ or $\alpha < 1$ since $i_{E2}$ is not part of the basic transistor action, we would like this component of current to be as small as possible. We would then like $\alpha$ to be as close to unity as possible.

*Bass current.* As shown in Fig. 11.110, the component of emitter current $i_{E2}$ is a base-emitter junction current, so this current is also a component of base current as $i_{Ba}$. This component of base current is proportional to exp $(v_{BE}/v_T)$. There is also second component of current. We have considered the ideal case in which there is no recombination of minority carriers (electrons) with majority carriers (holes) in the base. However, in reality there will be some recombination. Since majority carrier holes in the base are disappearing, they must be re-supplied by a flow of positive charge into the base material. This flow of charge is indicated as a current $i_{Bb}$ in Fig. 11.110. The number of holes per unit time recombining in the base is directly related to the number of minority carrier electrons in the base. Therefore, the current $i_{Bb}$ is also proportional to exp $(v_{BE}/V_T)$. The total base current is the sum of $i_{Ba}$ and $i_{Bb}$ and is proportional to exp $(v_{BE}/V_T)$.

The ratio of collector current to base current is a constant since both are directly proportional to exp $(v_{BE}/V_T)$, i.e.,

$$\frac{i_c}{i_B} = \beta, \quad ...(10)$$

the common-emitter current gain.

Normally, the base current will be relatively small, so, in general, the common-emitter current gain *b* is much larger than unity (on the order of 100 or more).

## 11.54.10 Basic Common-Emitter Transistor Amplifier

Figure 11.111 shows basic common emitter amplifier using *NPN* transistor. A load resistance $R_L$ is required to develop the output voltage due to the collector current $I_C$ flowing through it. $V_{CC}$ is the supply voltage (usually 12 to 20 V). The collector to emitter voltage is referred to as $V_{CE}$ and base to emitter voltage as $V_{BE}$. (It is to be noted that double subscript notation is very

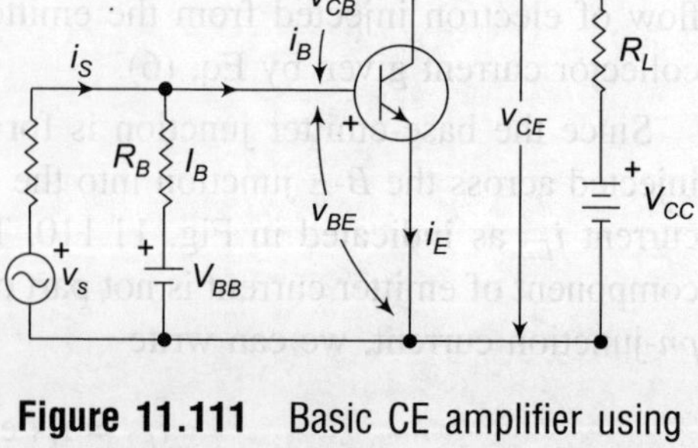


**Figure 11.111** Basic CE amplifier using *npn*-transistor showing various voltages

common in transistor work and the small letter $v$ designates instantaneous value. $V_{BB}$ is used to bias the transistor in the centre of its characteristics (i.e., active region (see Fig. 11.113). With no signal $V_s$ applied $V_{BB}$ supplies a dc current into the base $I_B$. Correspondingly, a dc collector current $I_C$ flows in the circuit, so that

$$I_E = I_B + I_C$$

The application of the $v_s$ will cause a current $i_s$ to be added to $I_B$ (see Fig 11.112), resulting in a current, $i_B$. This, is turn, causes an ac current ($i_C$) to flow in the collector circuit which added to $I_C$ is called $i_C$. Here, a small base current ($i_B$) controls a large collector current, $i_C$. Hence, the transistor has a current gain, $A_i$. Also, since the voltage ($v_{BE}$) is small and the voltage in the collector circuit is large, there will be a voltage gain, $A_v$. This will, therefore, give a power gain, $A_P = A_v A_i$.

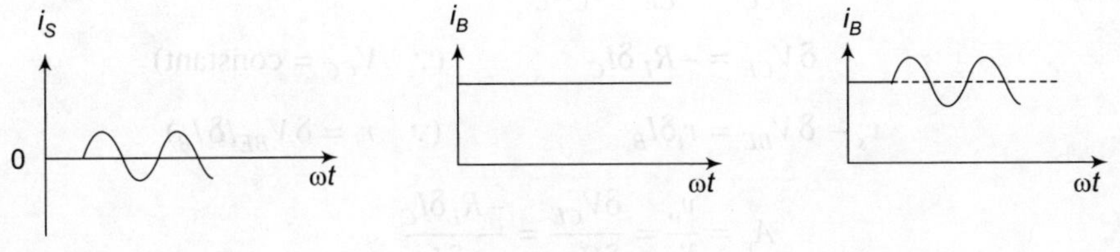


**Figure 11.112** Figure showing that the total base current $i_n$ consists of an ac (signal) component $i_s$ and a dc component $I_B$

In order to get a faithful amplification, the quiescent point must be well within the active region of the transistor characteristic (Fig. 11.113).

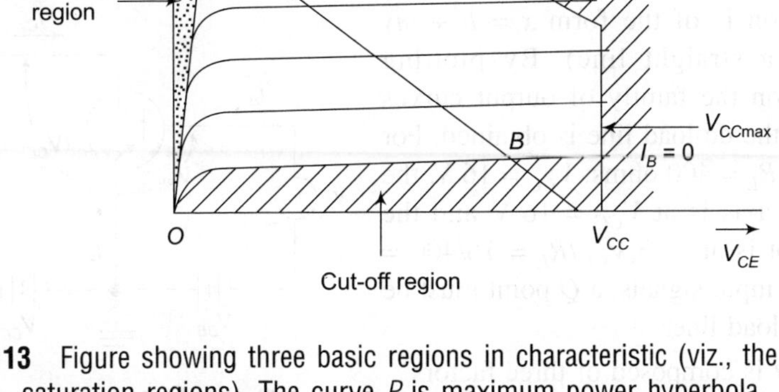


**Figure 11.113** Figure showing three basic regions in characteristic (viz., the cut-off and saturation regions). The curve $P$ is maximum power hyperbola

The curve $P$ indicates the maximum power the transistor can handle.

$$\alpha_{ac} = \frac{\Delta I_C}{\Delta I_E}\ \beta_{ac} = \frac{\Delta I_C}{\Delta I_B} \qquad ...(11)$$

*Saturation state.* When both the base and the collector are forward biased, the transistor is said to be in the saturation state. In this state, the normal transistor action is lost and

$$I_{C(\text{sat})} = \frac{V_{CC}}{R_L} \quad \text{and} \quad I_{B(\text{sat})} = \frac{I_{C(\text{sat})}}{\beta_{dc}} \qquad ...(12)$$

*Cut-off state.* When the emitter base junction becomes reverse biased, (the normal transistor action is again lost) and the transistor is said to be in the cut-off state. In this state both $I_B$ and $I_C$ are almost zero and the output $V_{CE} = V_{CC}$.

*Amplification:*

$$V_{CC} = V_{CE} + I_C R_L \qquad ...(13a)$$

$$\Rightarrow \quad \delta V_{CE} = -R_L \delta I_C \qquad (\because \ V_{CC} = \text{constant}) \qquad ...(13b)$$

Now,

$$v_s - \delta V_{BE} = r_i \delta I_B \qquad (\because \ r_i = \delta V_{BE}/\delta I_B) \qquad ...(14)$$

$$\therefore \quad A_v = \frac{v_o}{v_s} = \frac{\delta V_{CE}}{\delta V_{BE}} = \frac{-R_L \delta I_C}{r_i \delta I_B}$$

$$= -R_L \frac{\beta_{ac}}{r_i} \qquad (\beta_{ac} \text{ is large and } r_i \text{ is small}) \qquad ...(15)$$

Further,

$$\beta_{ac} = g_m \times r_i \qquad ...(16)$$

$$\therefore \quad A_v = -g_m R_L \qquad ...(17)$$

$$\text{Current gain } A_i = \beta_{ac}$$

$$\therefore \quad \text{Power gain} = A_v \times A_i = -\beta_{ac}^2 \left(\frac{R_L}{r_i}\right) \qquad ...(18)$$

*The load-line.* Writing a voltage equation around the output or collector loop gives

$$V_{CE} = V_{CC} - i_C R_L \qquad ...(19)$$

This equation is of the form $x = b + my$ (equation of a straight line). By plotting this equation on the family of output curves (Fig. 11.115), the dc load line is obtained, For example, with $R_L = 400$ ohms, $V_{CC} = 16$ V, the intercept on $x$-axis is at $V_{CC} = 16$ V and the $y$-axis intercept is at $i_C = V_{CC}/R_L = 16/400 = 40$ mA. For ac input signals, a $Q$ point must be chosen on the load line.

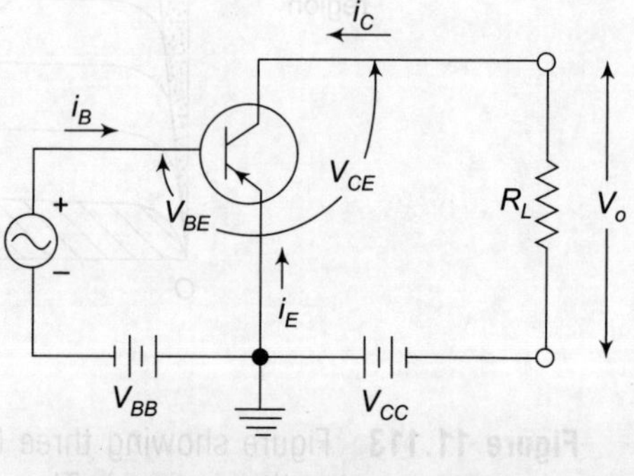


**Figure 11.114** Common emitter amplifier using *pnp*-transistor

In general $\alpha$ is composed of three factors

$$\alpha = \gamma b\ \delta;\ \gamma b\ \delta \le I \qquad ...(20)$$

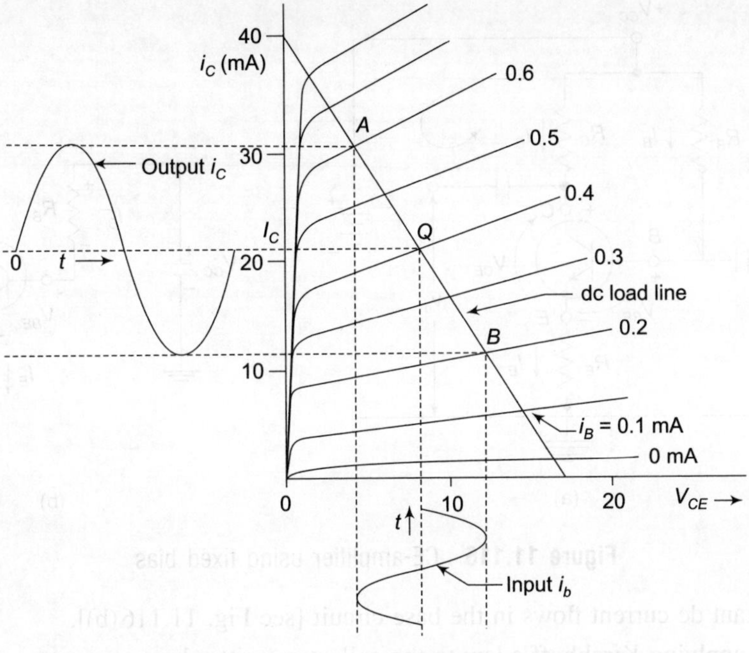


**Figure 11.115** Output curve and dc load line for a *CE*-transistor

$\gamma$ is the *emitter injection efficiency*, *b the base transport factor* and $\delta$ is a surface and depletion layer *recombination factor*. $\gamma$ expresses the portion of the total emitter current injected into the base. *b* tells what current portion of the injected current reaches the collector. The surface recombination factor is explained as follows: Just as recombination goes on in the base of a transistor, it occurs in emitter depletion layer as well, expressed by $\delta$.

## 11.54.11 Fixed Bias Circuit

For normal operation of a transistor, particularly as an amplifier, we have to apply the bias voltages. The use of the fixed bias circuit eliminates the battery voltage $V_{BB}$ by employing a resistance $R_B$ between $V_{CC}$ and the base of the transistor. The resistor $R_E$ provides the emitter bias. This makes the base-emitter junction forward biased (as necessary). The resulting circuit is shown in Fig. 11.116.

Applying Kirchhoff's voltage law to the base emitter loop in clockwise direction [Fig. 11.116(b)], we get

$$-V_{CC} + I_B R_B + V_{BE} + I_E R_E = 0 \qquad ...(21)$$

Also $$I_E = I_C + I_B$$

$$= \beta_{IB} + I_B = (\beta + 1) I_B \qquad ...(22)$$

$\therefore$ From Eqs. (21) and (22)

$$I_B = \frac{V_{CC} - V_{BE}}{R_B + (\beta + 1) R_E} \qquad ...(23a)$$

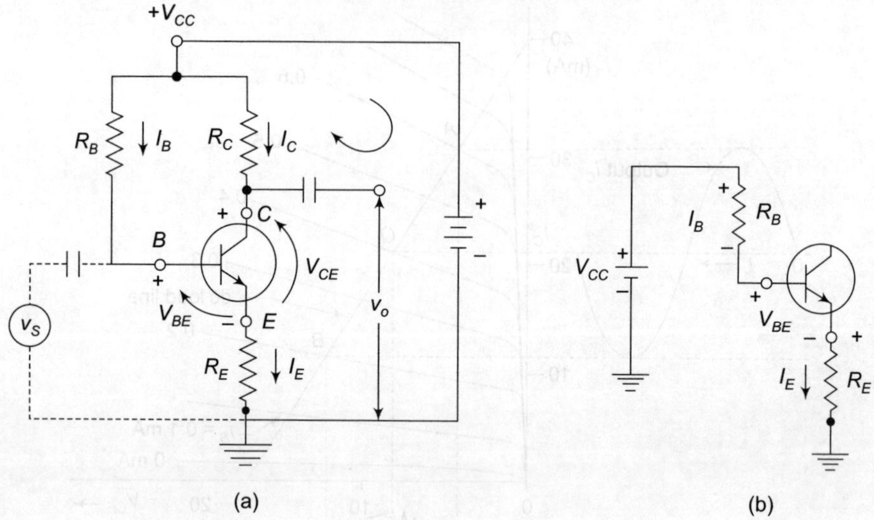


**Figure 11.116** CE-amplifier using fixed bias

This constant dc current flows in the base circuit [see Fig. 11.116(b)].

Similarly, applying Kirchhoff's law to the collector-emitter loop, we get

$$-I_E R_E - V_{CE} - I_C R_C + V_{CC} = 0$$

or
$$V_{CE} \simeq V_{CC} - I_C (R_C + R_E) \qquad (\because \quad I_E \simeq I_C)$$

## 11.55 BASE SPREADING RESISTANCE AND EMITTER CURRENT CROWDING IN BIPOLAR JUNCTION TRANSISTORS

As can be seen from the schematic structure of a bipolar junction transistor (shown in Fig. 11.117), a base current flows in the direction parallel to the *E-B* junction. This distribution current flows leads to a voltage drop, $V_{bb'}$, between the central portion of the emitter-base junction and the periphery of the junction. This voltage drop, $V_{bb'}$, is proportional to the base current:

$$V_{bb'} = r_{bb'} I_b$$

Resistance $r_{bb'}$ is called the base spreading resistance (an estimate of the base spreading resistance could be obtained by calculating the ohmic resistance of the base region between the *E-B* junction and the base contact). This is a non-uniformity and is directly related to the voltage drop along the base caused by the base current. Thus, if the voltage drop across the periphery of the emitter base junction is $V_{eb'}$, + $I_b\, r_{bb'}$. As the emitter current density is exponentially dependent on the voltage across the emitter-base junction, the emitter current density at the periphery of the junction may be much larger than in the centre of the junction. This effect is called emitter current crowding. This is illustrated schematically in Fig. 11.117, which shows the streamlines of the base current. In the extreme case, in which the voltage drop $V_{bb}$, becomes much larger than the thermal voltage $k_BT/q$, most of the emitter-current

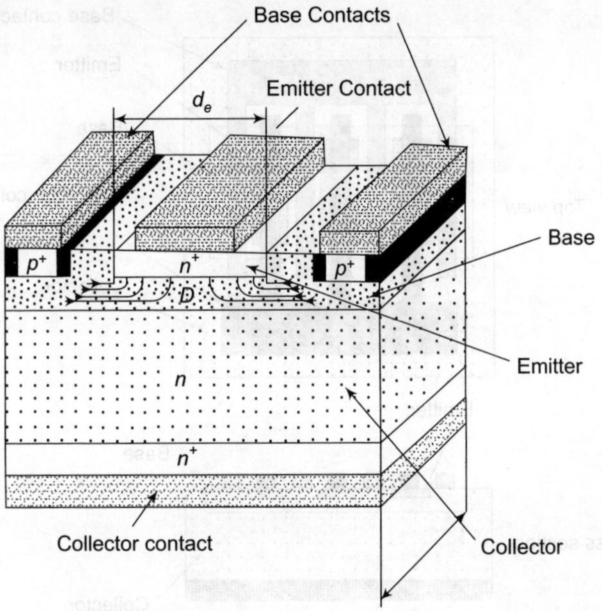


**Figure 11.117** Schematic structure or a bipolar junction transistor. Also shown are the streamlines of the base current

is concentrated near the edges of the emitter-base (E–B) junction. The length of the edge region, $\Delta de$, where most of the emitter current flows is roughly given as

$$\Delta de \approx V_{\text{th}}/F_{\text{base}}$$

where

$$F_{\text{base}} \approx I_b/(q\, N_{ab}\, \mu_p\, t_e\, W)$$

$$= \frac{I_C}{(q\, N_{ab}\, \mu_p\, t_e\, W_B)}$$

Here $N_{ab}$ is the concentration of acceptors in the base region, $t_e$ is the width of the emitter stripe (see Fig. 11.117), $H_p$ is the hole mobility (we considered the transistor to be an *npn* type). Assuming $I_C = 2\mu\text{A}$, $\beta = 50$, $W = 0.5\ \mu\text{m}$, $t_e = 5\ \mu\text{m}$, $\mu_p = 400\ \text{cm}^2/V_s$, $N_{ab} = 10^{17}\ \text{cm}^{-3}$, and $T = 300$ K ($V_{\text{th}} = 0.0258\ V$), we obtain $\Delta de \cong 1\ \mu\text{m}$. Hence, the emitter current in this case is primarily flowing in two 1-μm - wide regions near the edges of the emitter stripe. To, make the emitter-current distribution at high injection levels more uniform, the emitter stripes should be made narrow, i.e., we should have $d_e < 2\Delta d_e$. This leads to the transistor design with *inter-digitated emitter and base fingers* (see Fig. 11.118). Such a design considerably improves the uniformity of the emitter-current distribution.

## 11.56 EBERS-MOLL MODEL

In order to analyse a transistor circuit, one needs a mathematical model or equivalent circuit of the transistor. There are several possible models, each one having certain advantages and disadvantages, e.g., the Ebers-Moll model, the hybrid-pi model, etc.

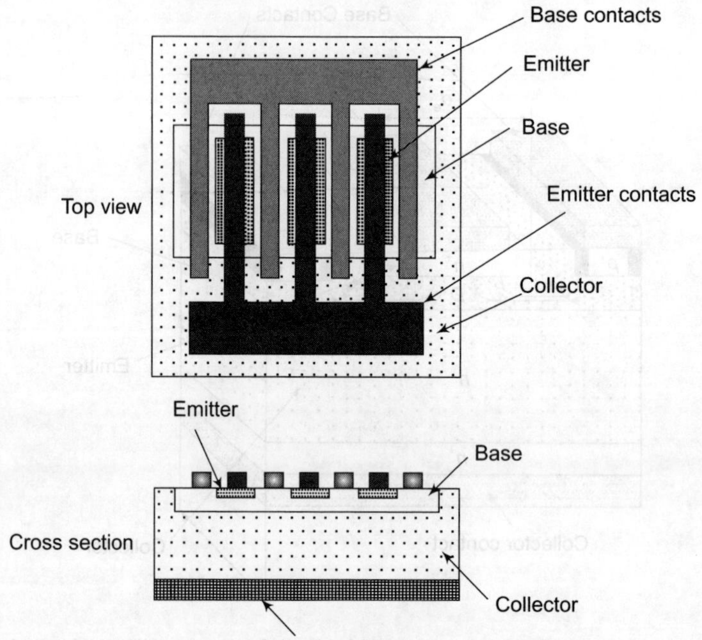


**Figure 11.118** Transistor design with interdigitated emitter and base fingers

The bipolar transistors can be divided into two categories from application point of view, viz., switching and amplification. Switching usually involves turning a transistor from its "off' state (or cut-off) to its "on" state, either forward active or saturation, and then back to its "off" state. Amplification usually involves superimposing sinusoidal signals on dc values so that amplified voltages or currents are obtained. The Ebers-Moll model is used in switching applications whereas the hybrid-pi model is used in amplification applications. We will discuss here Ebers-Moll model only.

The Ebers-Moll model is based on the interacting diode junctions and is applicable in any of the transistor operating modes.

Figure 11.119 shows the current directions and voltage polarities used in this equivalent circuit current model. The currents are defined as all entering the terminals so that

$$I_E + I_B + I_C = 0 \qquad ...(1)$$

The direction of the emitter current is opposite to that usually considered, but as long as we are consistent in the analysis, the defined direction does not matter.

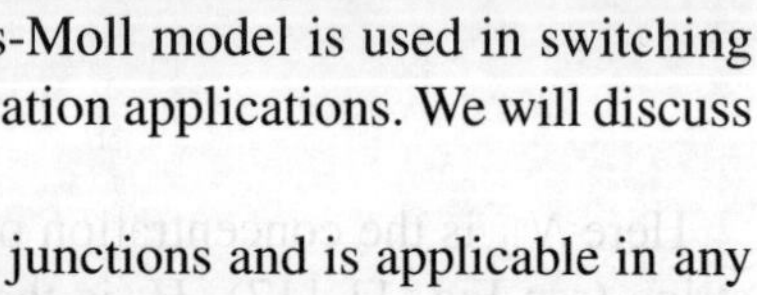


**Figure 11.119** Current direction and voltage polarity definitions for the Ebers-Moll model

The collector current can be written in general as

$$I_c = \alpha_F I_F - I_R \qquad ...(2)$$

where $\alpha_F$ is the common base current gain in the forward active mode. In this mode, Eq. (2) becomes

$$I_c = \alpha_F I_F + I_{cs} \quad ...(3)$$

where the current $I_{CS}$ is the reverse-bias *B–C* junction current. The current $I_F$ is given by

$$I_F = I_{ES}\left[\exp\left(\frac{eV_{BE}}{kT}\right) - 1\right] \quad ...(4)$$

The current $I_{ES}$ is the reverse-bias *B–E* junction. If the *B–C* junction becomes forward biased (such as in saturation), then, we can write the current $I_R$ as

$$I_R = I_{cs}\left[\exp\left(\frac{eV_{BC}}{kT}\right) - 1\right] \quad ...(5)$$

using Eqs. (4) and (5), the collector current from Eq. (2) can be written as

$$I_c = \alpha_f I_{ES}\left[\exp\left(\frac{eV_{BE}}{kT}\right) - 1\right] - I_{cs}\left[\exp\left(\frac{eV_{Bc}}{kT}\right) - 1\right] \quad ...(6)$$

We can also write the emitter current as

$$I_E = \alpha_R I_R - I_F \quad ...(7)$$

or

$$I_E = \alpha_R I_{cs}\left[\exp\left(\frac{eV_{BC}}{kT}\right) - 1\right] - I_{ES}\left[\exp\left(\frac{eV_{BE}}{kT}\right) - 1\right] \quad ...(8)$$

The current $I_{ES}$ is the reverse-bias *B–E* junction current and $\alpha_R$ is the common base current gain for the inverse-active mode. Equations (6) and (8) are known as the Ebers-Moll equations. Figure 11.120 shows the equivalent circuit corresponding to Eqs. (6) and (8). The current sources in the equivalent circuit represent current components that depend on voltages across other junctions. The Ebers-Moll model has four parameters: $\alpha_F$, $\alpha_R$, $I_{ES}$ and $I_{cs}$. However, only three parameters are independent. Also

$$\alpha_F I_{ES} = \alpha_R I_{cs} \quad ...(9)$$

Since the Ebers-Moll model is valid in each of the four operating modes (cut-off, forward active, saturation and inverse active[5]), we can, here, use the model for the transistor in saturation. In the saturation mode, both B-E and B-C junctions are forward biased so that $V_{BE} > 0$ and $V_{BC} > 0$. The $V_{BE}$ is a known parameter, as this is the voltage applied across the junction. The B–C voltage is the result of driving the transistor into saturation and is the unknown to the determined from the Ebers-Moll equations.

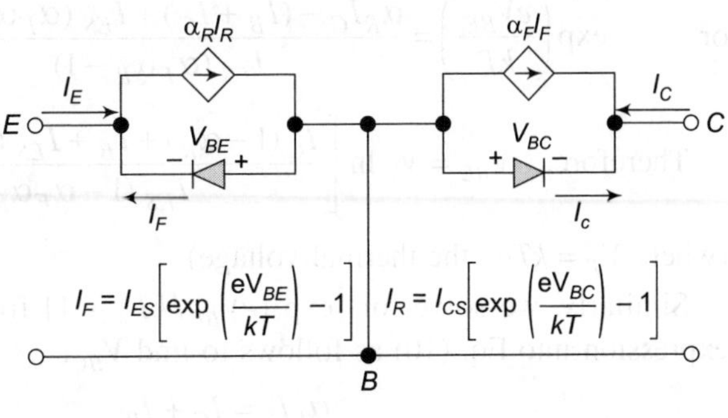


**Figure 11.120** Basic Ebers-Moll equivalent circuit

Combining Eqs. (1) and (8), we have

[5] This occurs when B-E junction is reverse-biased and B-C junction is forward-biased.

$$-(I_B + I_C) = \alpha_R I_{cs}\left[\exp\left(\frac{eV_{BC}}{kT}\right)-1\right] - I_{ES}\left[\exp\left(\frac{eV_{BE}}{kT}\right)-1\right] \quad ...(10)$$

Now, we solve for $\{\exp(eV_{Bc}/kT) - 1\}$ from Eq. (10) and substitute the resulting expression into Eq. (6) as follows:

$$\alpha_R I_{cs}\left[\exp\left(\frac{eV_{BC}}{kT}\right)-1\right] = -(I_B + I_C) + I_{ES}\left[\exp\left(\frac{eV_{BE}}{kT}\right)-1\right]$$

or $$\left[\exp\left(\frac{eV_{BC}}{kT}\right)-1\right] = \left\{\frac{-(I_B+I_C)+I_{ES}\left[\exp\left(\frac{eV_{BE}}{kT}\right)-1\right]}{\alpha_F I_{ES}}\right\} \quad ...(11)$$

Substituting Eq. (11) into Eq. (6), we get

$$I_C = \alpha_F I_{ES}\left[\exp\left(\frac{eV_{BE}}{kT}\right)-1\right] - \frac{I_{CS}}{\alpha_F I_{ES}}\left\{-(I_B+I_C)+I_{ES}\left[\exp\left(\frac{eV_{BE}}{kT}\right)-1\right]\right\}$$

or $$I_C = \alpha_F I_{ES}\exp\left(\frac{eV_{BE}}{kT}\right) - \alpha_F I_{ES} - \frac{1}{\alpha_R}\left\{-(I_B+I_C)+I_{ES}\exp\left(\frac{eV_{BE}}{kT}\right) - I_{ES}\right\}$$

(using Eq. (9))

or $$I_C = \left[\alpha_F I_{ES} - \frac{I_{ES}}{\alpha_R}\right]\exp\left(\frac{eV_{BE}}{kT}\right) - \alpha_F I_{ES} + \frac{(I_B+I_C)}{\alpha_R} + \frac{I_{ES}}{\alpha_R}$$

or $$I_C = \frac{(\alpha_F\alpha_R - 1)}{\alpha_R} I_{ES}\exp\left(\frac{eV_{BE}}{kT}\right) + I_{ES}\left(\frac{1}{\alpha_R} - \alpha_F\right) + \frac{I_B + I_C}{\alpha_R}$$

or $$I_{ES}\frac{(\alpha_F\alpha_R - 1)}{\alpha_R}\exp\left(\frac{eV_{BE}}{kT}\right) = I_C - \frac{(I_B+I_C)}{\alpha_R} - I_{ES}\frac{(1-\alpha_F\alpha_R)}{\alpha_R}$$

or $$\exp\left(\frac{eV_{BE}}{kT}\right) = \frac{\alpha_R I_C - (I_B+I_C) + I_{ES}(\alpha_F\alpha_R - 1)}{I_{ES}(\alpha_F\alpha_R - 1)}$$

Therefore, $$V_{BE} = V_T \ln\left[\frac{I_C(1-\alpha_R) + I_B + I_{ES}(1-\alpha_F\alpha_R)}{I_{ES}(1-\alpha_F\alpha_R)}\right] \quad ...(12)$$

(where $V_T = kT/e$, the thermal voltage)

Similarly, we solve for $[\exp(eV_{BE}/kT) - 1]$ from Eqs. (2) and (5) and substitute this expression into Eq. (10) as follows to find $V_{BC}$:

$$\alpha_F I_F = I_C + I_R$$

or $$\alpha_F I_{ES}\left[\exp\left(\frac{eV_{BE}}{kT}\right)-1\right] = I_C + I_{CS}\left[\exp\left(\frac{eV_{BC}}{kT}\right)-1\right]$$

$$\left[\exp\left(\frac{eV_{BE}}{kT}\right)-1\right]=I_C+I_{CS}\left[\exp\left(\frac{eV_{BC}}{kT}\right)-1\right] \quad ...(13)$$

From Eqs. (13) and (10), we get

$$-(I_B+I_C)=\alpha_R I_{CS}\left[\exp\left(\frac{eV_{BC}}{kT}\right)-1\right]-I_{ES}\left\{\frac{I_C+I_{CS}\left[\exp\left(\frac{eV_{BC}}{kT}\right)-1\right]}{\alpha_F I_{ES}}\right\}$$

$$=\alpha_R I_{CS}\exp\left(\frac{eV_{BC}}{kT}\right)-\alpha_R I_{CS}-\frac{1}{\alpha_F}\left\{I_C+I_{CS}\left[\exp\left(\frac{eV_{BC}}{kT}\right)-1\right]\right\}$$

$$=\alpha_R I_{CS}\exp\left(\frac{eV_{BC}}{kT}\right)-\frac{I_C}{\alpha_F}-\frac{I_{CS}}{\alpha_F}\exp\left(\frac{eV_{BC}}{kT}\right)+\frac{I_{CS}}{\alpha_F}-\alpha_R I_{CS}$$

$$=\left(\alpha_R-\frac{1}{\alpha_F}\right)I_{CS}\exp\left(\frac{eV_{BC}}{kT}\right)-\frac{I_C}{\alpha_F}+\frac{I_{CS}}{\alpha_F}(1-\alpha_F\alpha_R)$$

or $$\exp\left(\frac{V_{BC}}{V_T}\right)=\left[\frac{I_C}{\alpha_F}-(A_B+I_C)-\frac{(1-\alpha_F\alpha_R)}{\alpha_F}I_{CS}\right]\Big/\frac{(\alpha_F\alpha_R-1)}{\alpha_F}I_{CS}$$

$$=\frac{-I_C+\alpha_F(I_B+I_C)+I_{CS}(1-\alpha_F\alpha_R)}{(1-\alpha_F\alpha_R)I_{CS}}$$

$\therefore$ $$V_{BC}=V_T\ln\left[\frac{\alpha_F I_B-(1-\alpha_F)I_C+I_{CS}(1-\alpha_F\alpha_R)}{I_{CS}(1-\alpha_F\alpha_R)}\right] \quad ...(14)$$

Normally, in electronic circuit applications, the collector-emitter voltage at saturation can be found from

$$V_{CE}^{(\text{sat})}=V_{BE}-V_{BC}.$$

## 11.57 FIELD EFFECT TRANSISTOR (SYMBOLS)

The field effect transistor is a unipolar device depending only upon majority current flow (either electrons in the *N*-channel type or holes in the *P*-channel type). A cross-sectional view of an *N*-channel FET is shown in Fig. 11.121 together with the supply voltages and circuit symbol. The circuit shown indicates normal application of bias voltages to obtain its characteristic curves. Note that *pn*-junction is reverse biased, so that gate draws negligible current. The arrow points in the direction in which the conventional current would flow if the junction were forward biased.

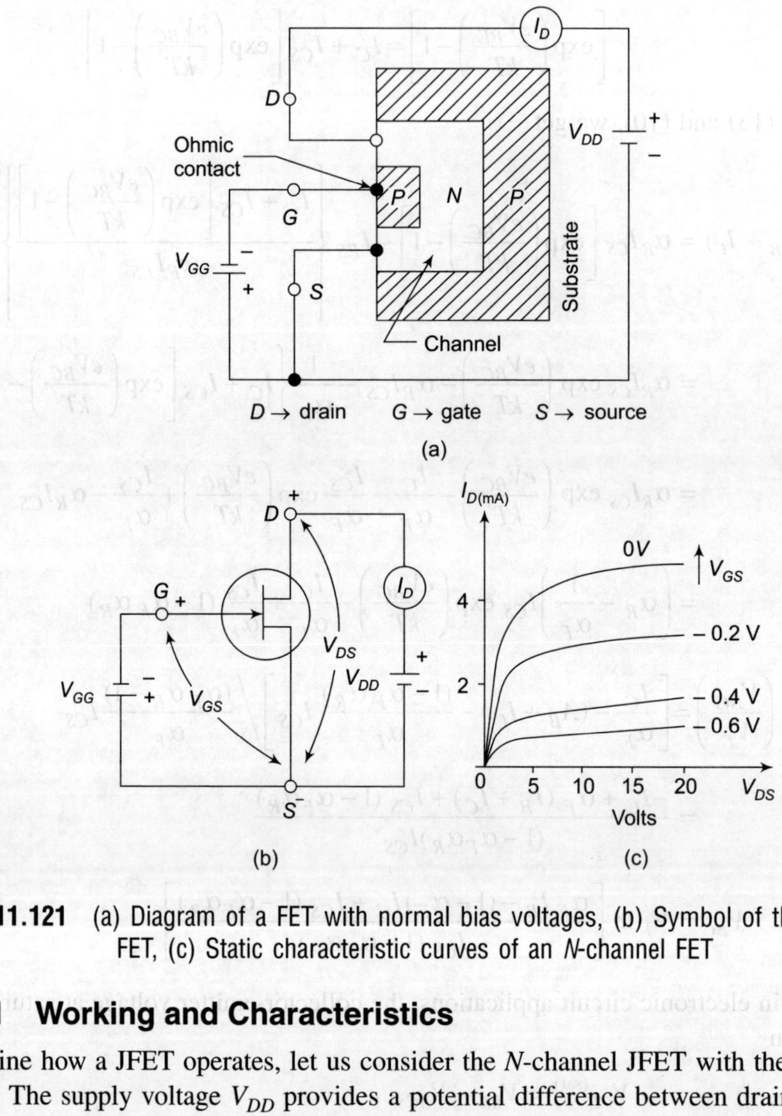


**Figure 11.121** (a) Diagram of a FET with normal bias voltages, (b) Symbol of the *N*-channel FET, (c) Static characteristic curves of an *N*-channel FET

## 11.57.1 Working and Characteristics

To examine how a JFET operates, let us consider the *N*-channel JFET with the applied bias voltages. The supply voltage $V_{DD}$ provides a potential difference between drain and source,

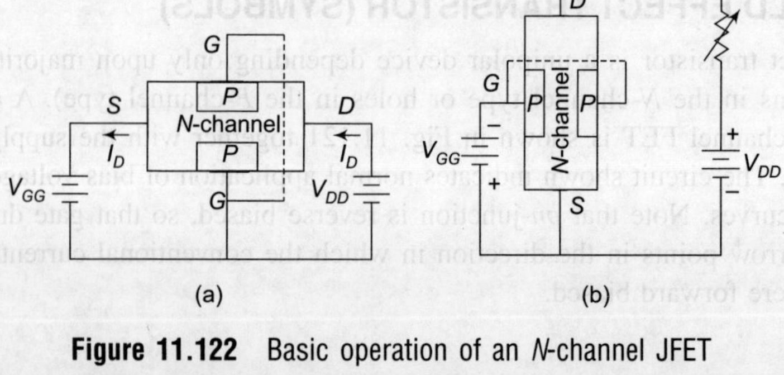


**Figure 11.122** Basic operation of an *N*-channel JFET

$V_{DS}$, which results in a current $I_D$, from drain to source (electrons in an N-channel actually move from the source, hence the name 'to drain'). This drain current passes through the channel surrounded by the *P*-type gate. A voltage between gate and source, $V_{GS}$ is set up by the supply voltage, $V_{GG}$. Since this voltage reverse biases the gate source junction, no gate current will result. The effect of the gate voltage will be to create a depletion region in the channel and thereby reduce the channel width to increase the drain-source resistance resulting in less drain current.

Let us first consider the device operating with $V_{GS} = 0$, i.e., with gate and source shorted, and then by increasing the reverse bias voltage, $V_{GS}$. Figure 11.123 shows that $I_D$ through *N*-type material of the drain-source junction provides a voltage drop along the channel, which is more positive at the gate-drain junction than at the gate-source junction.

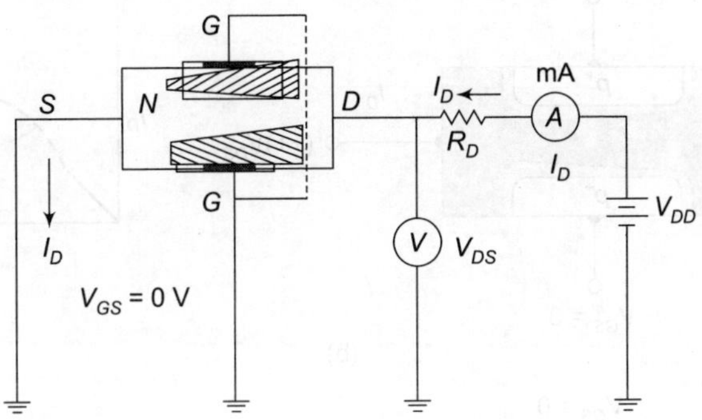


**Figure 11.123** A JFET circuit with $V_{GS} = 0$ V. The device is always operated with the gate-to-source junction reverse biased. The depletion layer is wider towards the drain end because reverse bias potential between the gate and drain is greater than that between the gate and source

As $V_{DD}$ (and therefore $V_{DS}$) is increased gradually from 0 V, $I_D$ increases proportionately (Fig. 11.125 between *A* and *B*) in this region the channel resistance is essentially constant because the depletion layer is not large enough to have any significant effect. This is called ohmic region, because $V_{DS}$ and $I_D$ are related by Ohm's law.

At *B*, the curve levels off and $I_D$ is relatively constant called $I_{DSS}$ the drain-to-source saturated current or the drain-to-source current under shorted gate condition. At this point, the reverse bias voltage between the gate and drain junction ($V_{GD}$) produces a depletion region which narrows down the channel width such that its resistance begins to increase sufficiently. The value of $V_{GD}$ at this point is called the *pinch-off voltage*, $V_P$. As $V_{DSP} = V_{DB} + V_{GS}$, we may write

$$V_{GD} = V_{GS} - V_{DSP}$$

Therefore with $V_{GS} = 0$,

$$V_{GD} = -V_{DSP}.$$

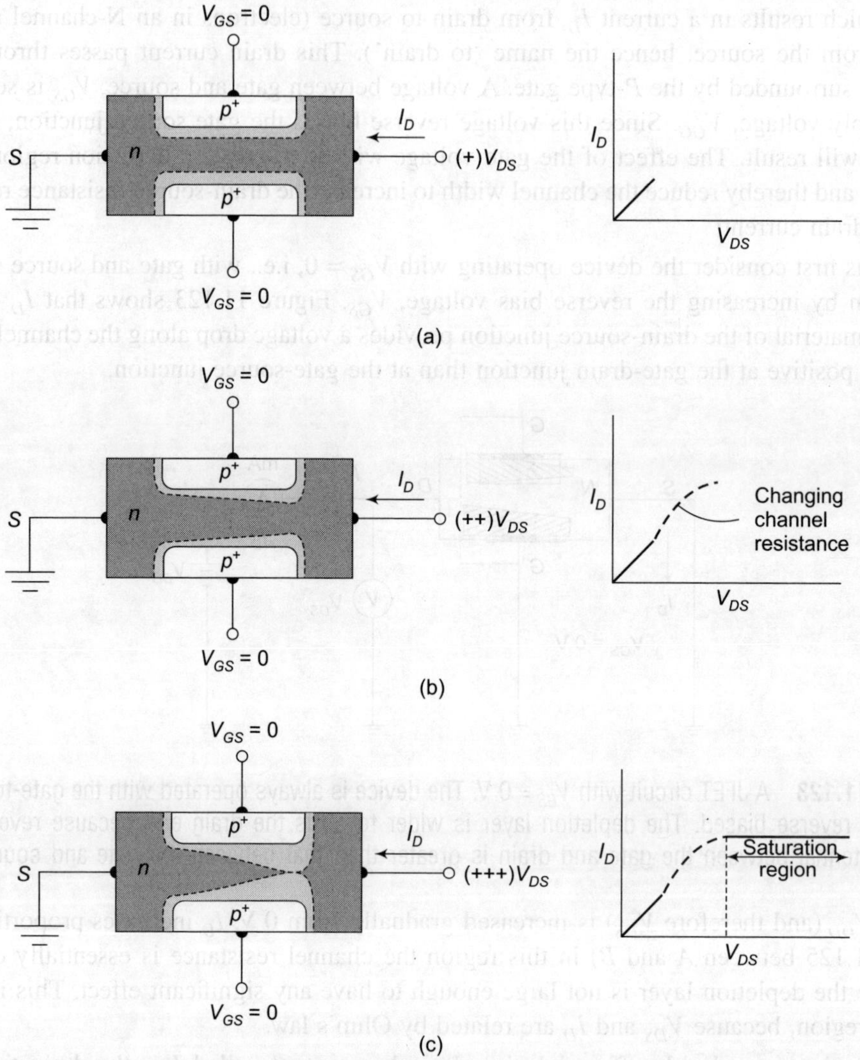


**Figure 11.124** Gate-to-channel space charge regions and I-V characteristics for zero gate voltage and for (a) a small drain voltage, (b) a larger drain voltage, and (c) a drain voltage to achieve pinch-off at the drain terminal

It is noted that $V_P$ is constant for a given JFET and represents a fixed parameter. The value of $V_{DS}$ at pinch-off ($V_{DSP}$) is a variable depending upon $V_{GS}$. This means that as $|V_{GS}|$ is increased (more negative for *N*-channel), the point corresponding to $V_{DSP}$ of Fig. 11.125 shifts towards the origin. If $V_{DS}$ is further increased beyond the point *B*, $I_D$ remains constant to a specific value $I_{DSS}$, the maximum value of $I_D$ when $V_{GS} = 0$. At the point *C*, breakdown occurs and $I_D$ increases very rapidly. The JFETs are always operated below the breaksown point and within the active region.

The mechanism of current control clearly depends upon the degree to which the electric field due to the charge in the depletion region extends into the channel and provides the effect of decreasing the conductivity through the transistor, and hence the name 'field effect transistor'.

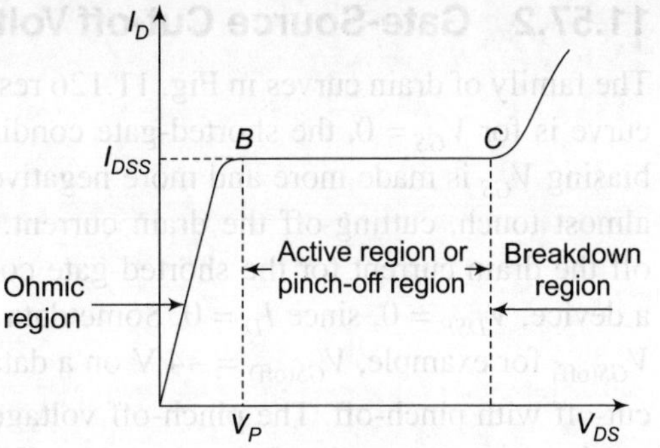


**Figure 11.125** Drain characteristic curve for $V_{GS}$ = 0 V

Now let us consider a negative gate bias, for example, $V_{GS} = -1$ V, as shown in Fig. 11.126. As $V_{DD}$ is increased from 0V, pinch-off occurs at a lower value of $V_{DS}$ as shown. The reason for this is that the pinch-off voltage $V_P$ is constant and therefore for a certain negative gate voltage, the drain voltage must only reach a value sufficient to make the sufficient gate-to-drain voltage equal to $V_P$ in order to produce pinch-off. As $V_{GS}$ is set to increasingly negative values, a family of curves is produced.

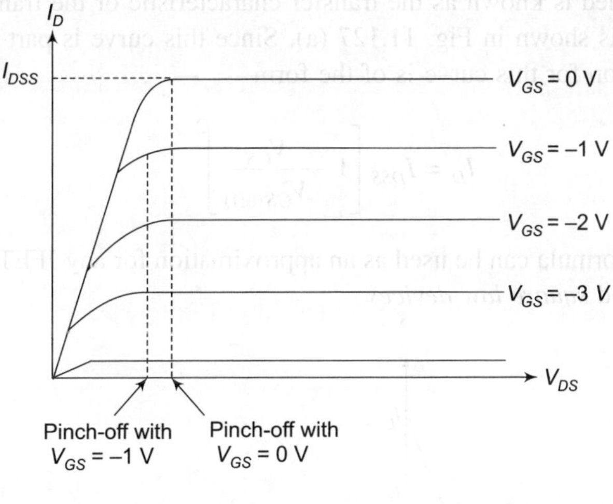


**Figure 11.126** Family of drain curves for N-channel JFET

In the ohmic region ($r_{DS} = V_{DS}/I_D$), the JFET is useful as a voltage variable resistor (VVR) or voltage dependent resistor (VDR) as $r_{DS}$ is related to $V_{GS}$ in almost a linear manner. The pinch-off region is used for high impedance voltage amplifier.

The pinch-off voltage $V_P$ is the drain voltage above which $I_D$ becomes almost constant for the shorted gate condition. When $V_{DS}$ equals $V_P$, the conducting channel becomes extremely narrow and the depletion layers almost touch. Because of the small passage between the depletion layers, any further increase in $V_{DS}$ produces only a very small increase in $I_D$.

## 11.57.2 Gate-Source Cut-off Voltage

The family of drain curves in Fig. 11.126 resembles the collector curves of a BJT. The highest curve is for $V_{GS} = 0$, the shorted-gate condition. The pinch-off voltage is $V_P$. As the reverse biasing $V_{GS}$ is made more and more negative, for a value, $V_{GS} = V_{GS(\text{off})}$, the depletion layers almost touch, cutting-off the drain current. Since $V_P$ is the drain voltage that also pinches off the drain current for the shorted-gate condition, $V_P = +V_{GS(\text{off})}$, Since $V_P$ is constant for a device, $V_{DSP} = 0$, since $I_D = 0$. Some data sheets do not list $V_P$, but they almost always list $V_{GS(\text{off})}$ for example, $V_{GS(\text{off})} = -4$ V on a data sheet means that $V_P = +4$ V. Let us not confuse cut-off with pinch-off. The pinch-off voltage $V_P$ is the gate-to-drain voltage $V_{GD}$ at which $I_D$ reaches a constant value for a given $V_{GS}$. Whereas the cut-off voltage $V_{GS(\text{off})}$ is that value of $V_{GS}$ at which $I_D = 0$, and $I_D$ will remain zero when $|V_{GS}| \geq V_P$ and will be non-zero for less negative values of $V_{GS}$ ($I_D = 0$ corresponds to off condition).

## 11.57.3 Transconductance Curve

From Fig. 11.126 for a given value of $V_{DS}$, we can read off the values of $I_D$ for different values of $V_{GS}$. Then a curve for $I_D$ versus $V_{GS}$ can be plotted for a JFET with $V_{DS}$ as a parameter. The curve so obtained is known as the transfer characteristic or the transconductance curve for the device and is shown in Fig. 11.127 (a). Since this curve is part of a parabola, from calculus, the equation for this curve is of the form

$$I_D = I_{DSS}\left[1 - \frac{V_{GS}}{V_{GS(\text{off})}}\right]^2 \qquad \text{...(1)}$$

This is an ideal formula can be used as an approximation for any JFET. This is why JFETs are sometimes called *square law devices*.

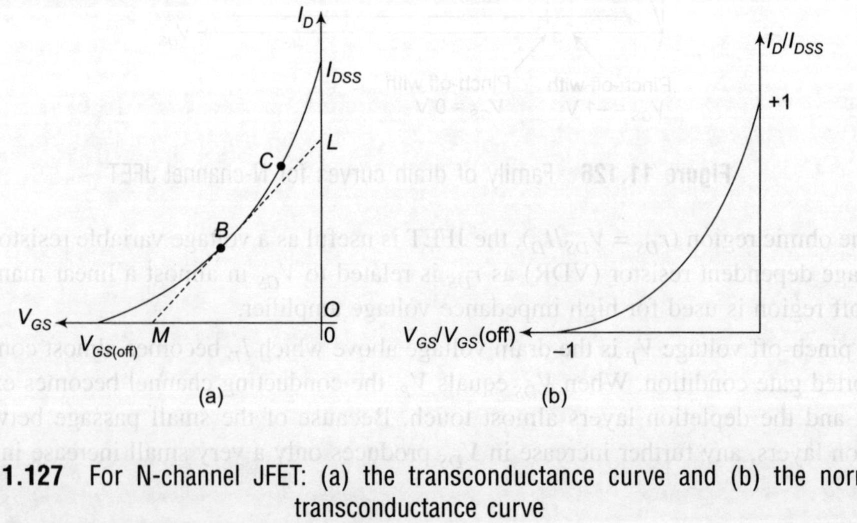


**Figure 11.127** For N-channel JFET: (a) the transconductance curve and (b) the normalized transconductance curve

### 11.57.4 Normalized Transconductance Curve

Equation (1) can be written as

$$\frac{I_D}{I_{DSS}} = \left[1 - \frac{V_{GS}}{V_{GS(\text{off})}}\right]^2 \quad ...(2)$$

From the curve of Fig. 11.127 (a), we can find out the values of $V_{GS(\text{off})}$ and $I = D_{SS}$, and then compute $V_{GS}/V_{GS(\text{off})}$ and $I_D/I_{DSS}$ and plot the curve as in the Fig. 11.127 (b) which is known as the normalized transconductance curve. This can also be done by taking the value of $V_{GS(\text{off})}$ from the data sheet and using different values of $V_{GS}$ in Eq. (2).

### 11.57.5 The Pinch-Off Effect in FET

The control of drain current by $V_{GS}$, through the channel is not a repelling effect but a 'narrowing of the channel' effect. Consider what happens to the *PN*-junction between gate and source under reverse bias (Fig. 11.128).

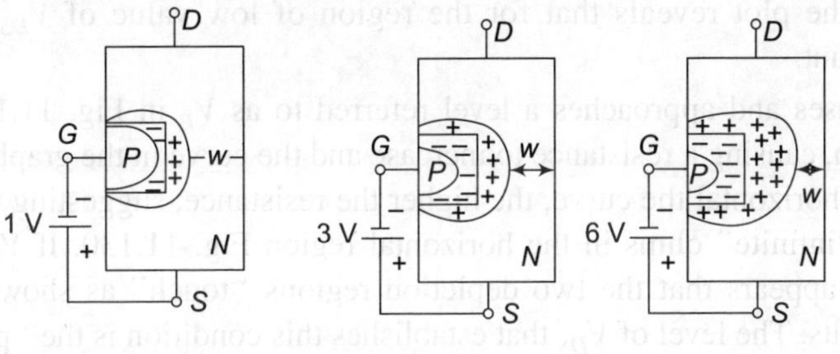


**Figure 11.128** Effect of reverse bias on the width of the channel

Assume, initially that no voltage is applied between *D* and *S*. With a small reverse bias, say $V_{GS} = 1$ V, the depletion region at the junction is quite narrow. Electrons have been drawn from the junction in the *N*-type material, leaving bound, immobile positive ions in that region of the channel, effectively reducing the width and conductivity of the channel. Similarly, negatively charged ions are left in the *p*-type material near the junction. As the reverse bias is increased to –3 V and –6 V, the depletion region or space charge region widens reaching even further into the channel and restricting the passage of electrons from *S* to *D* when a voltage is applied. Finally, a value of voltage from gate to source can be applied that *pinches-off* channel conduction altogether and cuts-off drain current. This value of gate voltage is called the pinch-off voltage, $V_P$, and is referred to in the specification as $V_{GS(\text{off})}$.

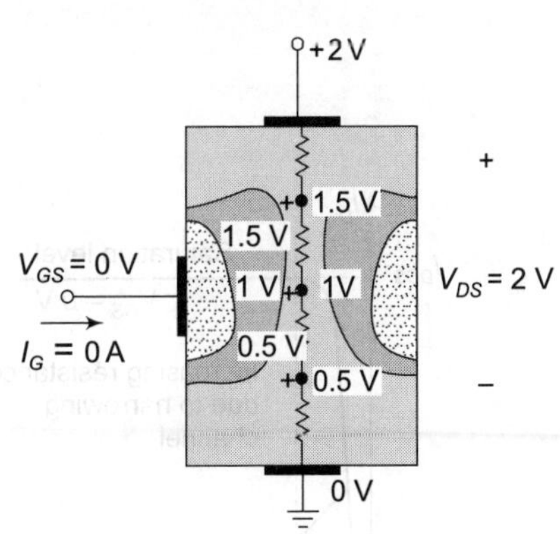


**Figure 11.129** JFET with $V_{GS} = 0$ V and $V_{DS} = +2$ V. As a result, there is varying reverse bias potentials across the PN-junction of this *n*-channel JFET

### 11.57.6 Pinch-Off for $V_{GS}$ = 0 V

In Fig. 11.129, a positive voltage $V_{DS}$ (= 2 V) has been applied across the (*n*) channel and the gate has been connected directly to the source, to establish the condition $V_{GS}$ = 0 V. Electrons will flow from source to the drain and ($I_D$ will be equal to $I_S$). It is important to note that the depletion region is wider near the top of both *p*-type materials (i.e., nearer to the drain). The reason for the change in width of the depletion region can be understood as follows.

Assuming a uniform-resistance in the *N*-channel, the resistance of the channel can be divided into the division shown. Thus, the midpoint will be at 1 V and the upper region of the *P*-type material will be reverse biased by ~1.5 V and the lower region, reverse biased by ~0.5 V. Since the greater the applied reverse bias, the wider the depletion region—hence the distribution of the depletion region will be as shown in figure.

### 11.57.7 Pinch-Off Voltage for $V_{GS}$ = 0 V

As the voltage $V_{DS}$ is increased from 0 to a few volts, the current will increase as determined by Ohm's law and the plot of $I_D$ versus $V_{DS}$ will appear as shown in Fig. 11.130. The relative straightness of the plot reveals that for the region of low value of $V_{DS}$ the resistance is essentially constant.

As $V_{DS}$ increases and approaches a level referred to as $V_P$ in Fig. 11.130, the depletion region will widen, causing a resistance to increase and the curve in the graph of Fig. 11.130 to occur. The more horizontal the curve, the higher the resistance, suggesting that the resistance is approaching "infinite" ohms in the horizontal region Fig. 11.130. If $V_{DS}$ is increased to a level where it appears that the two depletion regions "touch" as shown in Fig. 11.131, "pinch-off" results. The level of $V_{DS}$ that establishes this condition is the "pinch-off" voltage and is denoted by $V_P$ as shown in Fig. 11.131.

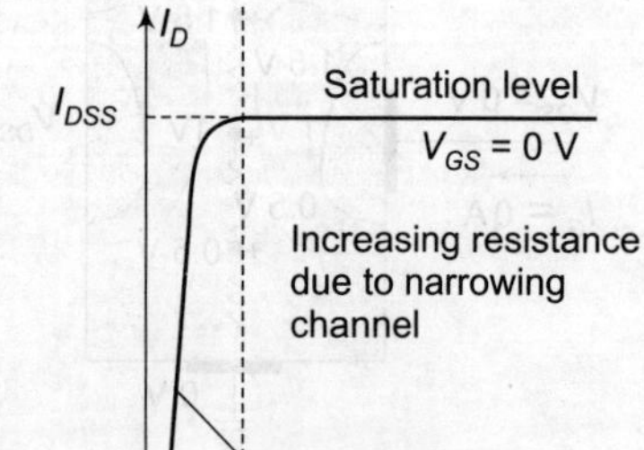


**Figure 11.130** $I_D$ versus $V_{DS}$ for $V_{GS}$ = 0 V

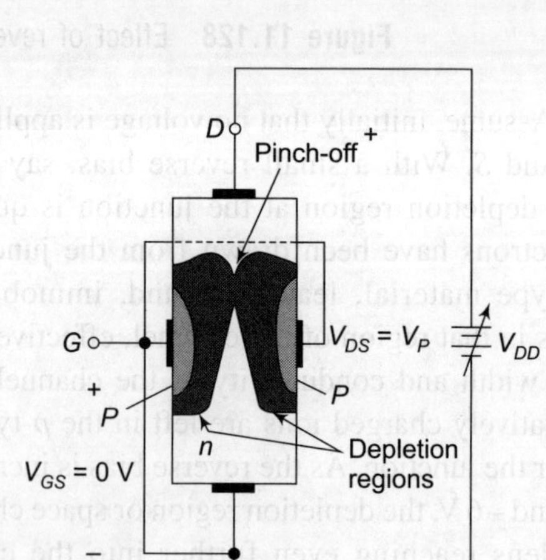


**Figure 11.131** Pinch-off ($V_{GS}$ = 0 V, $V_{DS}$ = $V_P$)

The term "pinch-off" is a misnomer in that it suggests that the current $I_D$ is pinched-off and drops to zero amperes. As shown in Fig. 11.130 however, this is not the case – $I_D$ maintains a saturation level defined as $I_{DSS}$ in Fig. 11.133.

Figure 11.132 shows an expanded view of the pinch-off region in the channel. The $n$-channel and the drain terminal are now separated by a space charge region which has a length, $\Delta L$. The electrons move through the $n$-channel from the source and are injected into the space charge region where, subjected to the $E$-field force, they are swept through into the drain contact area. If we assume that $\Delta L \ll L$, then the electric field in the $n$-channel region remains unchanged from the $V_{DS}$ (sat) case; the drain current will remain constant $V_{DS}$ changes. Once the carriers are in the drain region, the drain current will be independent of $V_{DS}$; thus, the device looks like a constant current source.

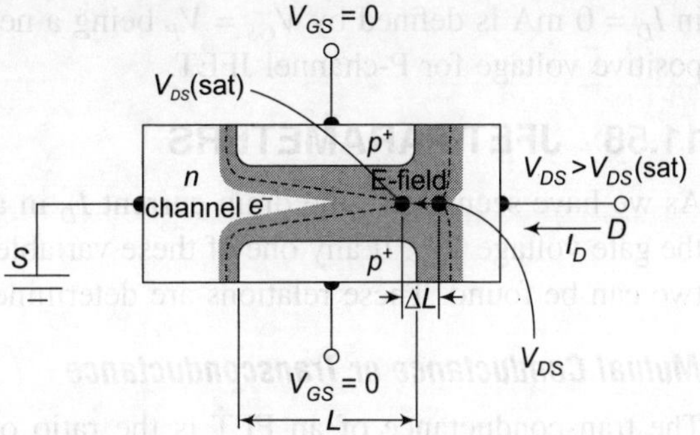


**Figure 11.132** Expanded view of the space charge region in the channel for $V_{DS} > V_{DS}$ (sat)

## 11.57.8 $V_{GS} < 0$ V

The voltage from gate to source, i.e., $V_{GS}$ is the controlling voltage of the JFET. Thus, curves of $I_D$ versus $V_{DS}$ for various levels of $V_{GS}$ can be obtained. For the $N$-channel device, the controlling voltage $V_{GS}$ is made more and more negative from its $V_{GS} = 0$V level. Consequently, the resulting saturation level for $I_D$ reduces and will continue to decrease as $V_{GS}$ is made more and more negative, as shown in Fig. 11.133 and ultimately, the level of $V_{GS}$ that results

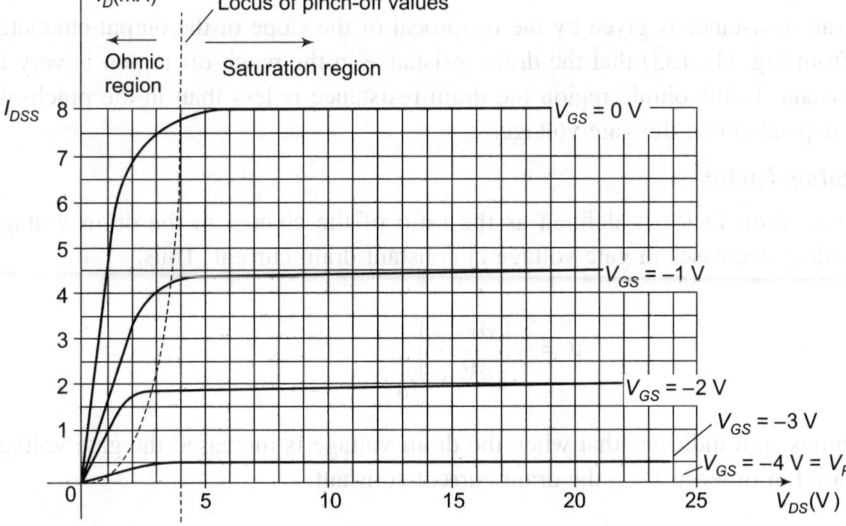


**Figure 11.133** $N$-channel JFET characteristics with $I_{DSS}$ = 8 mA and $V_P$ = −4 V

in $I_D = 0$ mA is defined by $V_{GS} = V_P$ being a negatively voltage for N-channel device, and a positive voltage for P-channel JFET.

## 11.58 JFET PARAMETERS

As we have seen above the drain current $I_D$ in a FET depends on the drain voltage $V_{DS}$ and the gate voltage $V_{GS}$. If any one of these variables is kept fixed the relation between the other two can be found. These relations are determined by three parameters of the FET.

### *Mutual Conductance or Transconductance*

The transconductance of an FET is the ratio of a small charge in the drain current to the corresponding change in the gate source voltage at constant drain-source voltage. The changes are to be considered in the straight portion of the characteristics. Thus,

$$\text{Transconductance, } g_m = \left(\frac{\partial I_D}{\partial V_{GS}}\right)_{V_{DS}} \quad \text{...(1)}$$

The transconductance is measured in the unit mho or siemens (s).

The transconductance is given by the slope of the transfer characteristics. For example, in Fig. 11.127 the value of $g_m$ at the point $B$ is given by $OL$ in ampere/$OM$ in volt.

### *Drain Resistance*

The drain resistance of an FET is the ratio of a small change in drain voltage to the resulting change in the drain current at constant gate voltage. Its unit is then ohm. Thus,

$$r_d = \left(\frac{\partial V_{DS}}{\partial I_D}\right)_{V_{GS}} \quad \text{...(2)}$$

The drain resistance is given by the resiprocal of the slope of the output characteristic. It is clear from Fig. (11.133) that the drain resistance in the pinch-off region is very high and fairly constant. In the ohmic region the drain resistance is less than in the pinch-off region and also depends upon the gate voltage.

### *Amplification Factor*

The amplification factor is defined as the ratio of the change in the drain voltage to the corresponding decreases in gate voltage at constant drain current. Thus,

$$\mu = -\left(\frac{\partial V_{DS}}{\partial V_{GS}}\right)_{I_D} \quad \text{...(3)}$$

The minus sign indicates that when the drain voltage is increased the gate voltage has to be decreased in order to keep the drain current constant.

## 11.58.1 Relation between the FET Parameters

As the drain current $I_D$ depends on the drain voltage $V_{DS}$ and the gate voltage $V_{GS}$, we can write

$$I_D = f(V_{DS}, V_{GS}) \quad \text{...(4)}$$

Hence, we get

$$dI_D = \left(\frac{\partial I_D}{\partial V_{DS}}\right)_{V_{GS}} dV_{DS} + \left(\frac{\partial I_D}{\partial V_{GS}}\right)_{V_{DS}} dV_{GS}$$

Dividing both sides by $dV_{GS}$ we get

$$\frac{dI_D}{dV_{GS}} = \left(\frac{\partial I_D}{\partial V_{DS}}\right)_{V_{GS}} \left(\frac{dV_{DS}}{dV_{GS}}\right) + \left(\frac{\partial I_D}{\partial V_{GS}}\right)_{V_{DS}}$$

If $I_D$ is constant then $\frac{dI_D}{dV_{GS}} = 0$.

$$\therefore \quad 0 = \left(\frac{\partial I_D}{\partial V_{DS}}\right)_{V_{GS}} \left(\frac{\partial V_{DS}}{\partial V_{GS}}\right) + \left(\frac{\partial I_D}{\partial V_{GS}}\right)_{V_{DS}}$$

Substituting for the partial derivatives from Eqs. (2), (3) and (4)

we get

$$0 = \left(\frac{1}{r_d}\right)(-\mu) + g_m$$

$$\therefore \quad m = g_m r_d \quad \text{...(5)}$$

Thus, the amplification factor is equal to the product of the transconductance and the drain resistance.

## 11.58.2 Variation of Transconductance with Drain Current and Gate Voltage

Diffrentiating $I_{DS}$ in Shockley's equation with respect to $V_{GS}$, we get

$$\frac{\partial I_D}{\partial V_{GS}} = I_{DSS} \times 2\left(1 - \frac{V_{GS}}{V_P}\right)\left(-\frac{1}{V_P}\right)$$

But

$$\left(\frac{\partial I_D}{\partial V_{DS}}\right) = g_m$$

$$\therefore \quad g_m = -\frac{2I_{DSS}}{V_P}\left(1 - \frac{V_{GS}}{V_P}\right) \quad \text{...(6)}$$

Now from Eq. (2) of section 11.57.4, we have

$$\left(1 - \frac{V_{GS}}{V_P}\right) = \sqrt{\frac{I_{DS}}{I_{DSS}}}$$

Substituting this value in Eq. (6)

$$g_m = -2\frac{\sqrt{I_{DS}I_{DSS}}}{V_P} \quad ...(7)$$

Let $g_m = g_{mo}$ when $V_{GS} = 0$. Then from Eq. (7), we get

$$g_{mo} = -\frac{2I_{DSS}}{V_P} \quad ...(8)$$

Combining Eqs. (7) and (8), we get

$$g_m = g_{mo}\left(1 - \frac{V_{GS}}{V_P}\right) \quad ...(9)$$

Equation (7) shows that the transconductance varies as the root of the saturation drain current, $I_{DS}$. Equation (9) shows that the transconductance decreases linearly with increase of gate voltage, $V_{GS}$.

### 11.58.3 Channel JFET

We have so far confined our discussion of FETs to the *N*-channel type, *P*-channel JFETs are also available. The above discussion can be applied to the *P*-channel JFETs by interchanging *N* and *P* and reversing the voltages and currents. The Table 11.13 gives the comparison of the *N*-channel and *P*-channel JFETs.

**Table 11.13**

| *N-channel JFET* | *P-channel JFET* |
|---|---|
| 1. Current carriers are electrons | Current carriers are holes |
| 2. Mobility of electrons large | Mobility of holes poor |
| 3. Input noise low | Input noise large |
| 4. Transconductance large | Transconductance small |

The above facts show that *N*-channel FETs serve better than *P*-channel FETs.

## 11.59 GENERAL DESCRIPTION OF THE JFET

The voltage-variable depletion region width of the junction is used to control the effective cross-sectional area of the conductivity channel. In the device of Fig. 11.134, the current $I_D$ flows an *n*-type channel between two $p^+$ region. A reverse bias between these $p^+$ regions and the channel causes the depletion regions to intrude into the *n* material, and therefore the effective width of the channel can be restricted. Since the resistivity of the channel region is fixed by its doping, the channel resistance varies with changes in the effective cross-sectional area. By analogy, the variable depletion regions serve as the two doors of a gate, which open and close on the conducting channel.

In Fig. 11.134 electrons in the *n*-type channel drift from right to left, opposite to current flow. The end of the channel from which electrons flow is called the *source*, and the end toward which they flow is called the *drain*. The $p^+$ regions are called *gates*. If the channel were $p$-type, holes would flow from the source to the drain, in the same direction as the current flow, and the gate regions would be $n^+$.[6] It is common practice to connect the two gate regions electrically, therefore, the voltage $V_G$ refers to the potential from each gate regions $G$ to the source $S$. Since the conductivity of the heavily doped $p^+$ regions is high, we can assume that the potential is uniform throughout each gate. In the lightly doped channel material, however, the potential varies with position (Fig. 11.134). If the channel of Fig. 11.134 is considered a distributed resistor carrying a current $I_D$ it is clear that the voltage from the drain end of

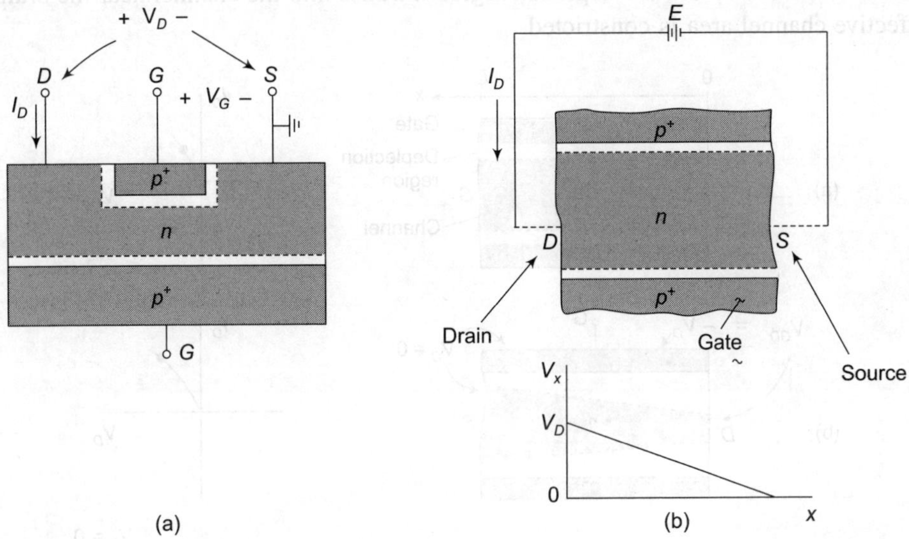


**Figure 11.134** Simplified cross-sectional view of a junction FET: (a) Transistor geometry; (b) detail of the channel and voltage variation along the channel with $V_G = 0$ and small $I_D$.

the channel $D$ to the source electrodes $S$ must be greater than the voltage from a point near the source end, $S$. For low values of current we can assume a linear variation of voltage $V_x$ in the channel, varying from $V_D$ at the drain end to zero at the source end (Fig. 11.134).

### 11.59.1 Pinch-Off and Saturation

In Fig. 11.135 we consider the channel in a simplified way by neglecting voltage drops between the source and drain electrodes and the respective ends of the channel. For example, we assume that the potential at the drain end of the channel is the same as the potential at the electrode, $D$. This is a good approximation if the source and drain regions are relatively large,

6 In early treatments of field-effect transistors $p$-channel devices were dominant, and it was most convenient to draw the source on the left and drain on the right, with hole current flowing from left to right on the diagram. Since $n$-channel devices are now more common, we choose in this chapter to draw FET devices with the current flowing left to right, from drain to source. In comparing figures in this chapter with other books, the student will find that many use the convention of placing the source on the left.

so that there is little resistance between the ends of the channel and the electrodes. In Fig. 11.135, the gates are short circuited to the source ($V_G = 0$), such that the potential at $x = L$ is the same as the potential everywhere in the gate regions. For very small currents, the width of the depletion regions is close to the equilibrium values (Fig. 11.135). As the current $I_D$ is increased, however, it becomes important that $V_x$ is large near the drain and small near the channel. Since the reverse bias across each point in the gate-to-gate channel junction is simply $V_x$ when $V_G$ is zero, we can estimate the shape of the depletion regions as in (Fig. 11.135). The reverse bias is relatively large near the drain ($V_{GD} = -V_D$) and decreases towards zero near the source. As a result, the depletion region intrudes into the channel near the drain, and the effective channel area is constricted.

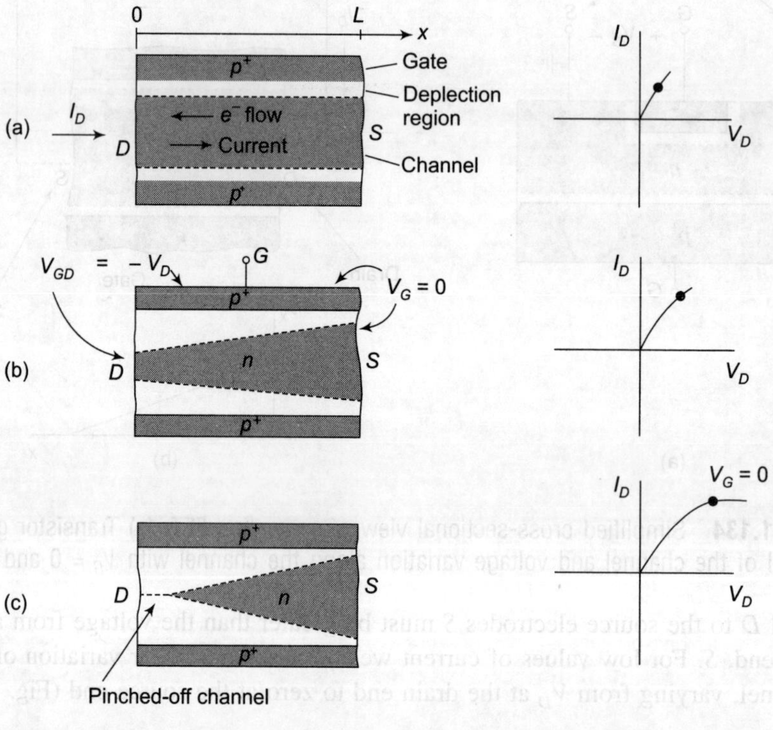


**Figure 11.135** Depletion regions in the channel of a JFET with zero gate bias for several values of $V_D$; (a) linear range (b) near pinch-off; (c) beyond pinch-off

Since the resistance of the constricted channel is higher, the *I–V* plot for the channel begins to depart from the straight line that was valid at low current levels. As the voltage $V_D$ and current $I_D$ are increased still further, the channel region near the drain becomes more constricted by the depletion regions and the channel resistance continues to increase. As $V_D$ is increased, there must be some bias voltage at which the depletion regions meet near the drain and essentially *pinch-off* the channel (Fig. 11.135). When this happens, the current $I_D$ cannot increase significantly with further increase in $V_D$. Beyond pinch-off the current is *saturated* approximately at its value at pinch-off.[7] Once electrons from the channel enter the electric field of the depletion region, they are swept through the ultimate flow to the positive drain contact. After the current saturates beyond pinch-off, the differential channel resistance $dV_d/dI_D$ becomes very high. To a good approximation, we can calculate the current at the critical pinch-off voltage and assume there is no further increase in current as $V_D$ is increased.

### 11.59.2 Gate Control

The effect of a negative gate bias $-V_G$ is to increase the resistance of the channel and induce pinch-off at a lower value of current (Fig. 11.136). Since the depletion regions are larger with $V_G$ negative, the effective channel width is smaller and its resistance is higher in the low-current range of the characteristics. Therefore, the slopes of the $I_D$ vs. $V_D$ curves below pinch-off become smaller as the gate voltage is made more negative (Fig. 11.136 b). The pinch-off condition is reached at a lower drain-to-source voltage, and the saturation current is lower than for the case of zero gate bias. As $V_G$ is varied, a family of curves is obtained for the $I-V$ characteristics of the channel, as in Fig. 11.136 (b).

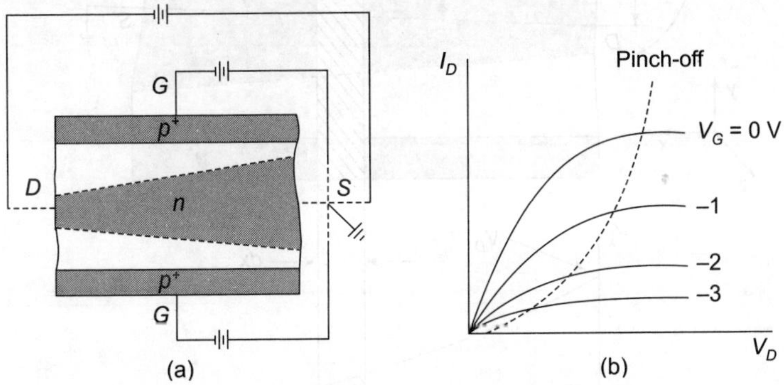


**Figure 11.136** Effects of a negative gate bias: Increase of depletion region width with $V_C$ negative; Family of current-voltage curves for the channel as $V_C$ is varied

Beyond the pinch-off voltage the drain current $I_D$ is controlled by $V_G$. By varying the gate bias we can obtain amplification of an ac signal. Since the input control voltage $V_G$ appears across the reverse-biased gate junctions, the input impedance of the device is high.

We can calculate the pinch-off voltage rather simply by representing the channel in the approximate from Fig. 11.137. If the channel is symmetrical and the effects of the gates are the same in each half of the channel region, we can restrict our attention to the channel half-width $h(x)$, measured from the centre line ($y = 0$). The metallurgical half-width of the channel (i.e., neglecting the depletion region) is $a$. We can find the pinch-off voltage by calculating

[7] *Saturation* is used by device engineers in more different contexts than any other word. We have discussed velocity saturation, reverse saturation current of junction, saturation of a BJT, and here the saturation of FET characteristics.

the reverse bias between the $n$ channel and the $p^+$ gate at the drain end of the channel ($x = 0$). For simplicity we shall assume that the channel width at the drain decreases uniformly as the reverse bias increases the pinch-off. If the reverse bias between the gate and the drain is $-V_{GD}$, the width of the depletion region at $x = 0$ can be found as

$$W(x=0) = \left[\frac{2\varepsilon(-V_{GD})}{qN_d}\right]^{1/2} \qquad (V_{GD} \text{ negative}) \qquad \text{...(1)}$$

In this expression we assume the equilibrium contact potential $V_0$ is negligible compared with $V_{GD}$ and the depletion region extends primarily into the channel for the $p^+$-$n$ junction.

Pinch-off occurs at the drain end of the channel when

$$h(x=0) = a - W(x=0) = 0 \qquad \text{...(2)}$$

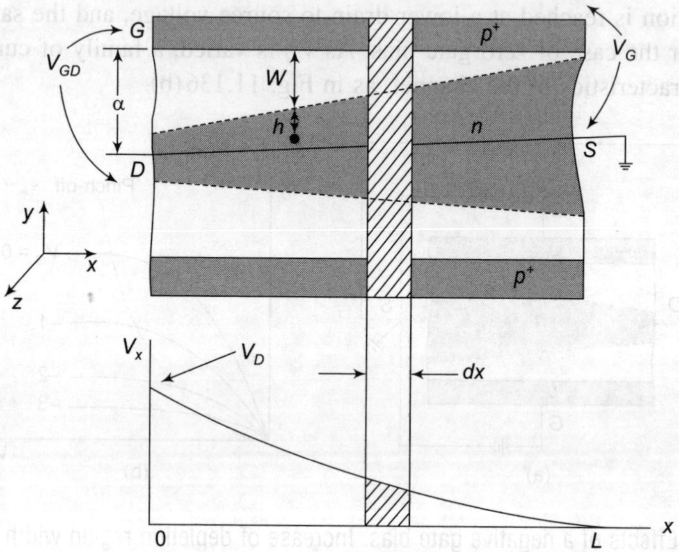


**Figure 11.137** Simplified diagram of the channel with definitions of dimensions and differential volume for calculations

that is, when $W(x = 0) = a$. If we define the value of $-V_{GD}$ at pinch-off as $V_P$, we have

$$\left[\frac{2\varepsilon V_P}{qN_d}\right]^{1/2} = a$$

$$\boxed{V_P = \frac{qa^2N_d}{2\varepsilon}} \qquad \text{...(3)}$$

The pinch-off voltage $V_P$ is a positive number; its relation to $V_D$ and $V_G$ is

$$V_P = -V_{GD} \text{ (pinch-off)} = -V_G + V_D \qquad \text{...(4)}$$

where $V_G$ is zero or negative for proper device operation. A forward bias on the gate would cause hole injection from the $p^+$ regions into the channel, eliminating the field-effect control of the device. From Eq. (4) it is clear that pinch-off results from a combination of gate-to-source voltage and drain-to-source voltage. Pinch-off is reached at lower value of $V_D$ (and therefore a lower $I_D$) when a negative gate bias is applied, in agreement with Fig. 11.136.

### 11.59.3 Current-Voltage Characteristics

Calculation of the exact channel current is complicated, although the mathematics is relatively straightforward below pinch-off. The approach we shall take is to find the expression for $I_D$ just at pinch-off, and then assume the saturation current beyond pinch-off remains fairly constant at thus value.

In the coordination system defined in Fig. 11.137, the centre of the channel at the drain end is taken as the origin. The length of the channel in the $x$-direction $L$, and the depth of the channel in the $z$-direction is $Z$. We shall call the resistivity of the $n$-type channel material $\rho$ (valid only in the neutral $n$ material, outside the depletion regions). If we consider the differential volume of neutral channel material $Z2h(x)\,dx$, the resistance of the volume element is $\rho dx/Z2h(x)$. Since the current does not change with the distance along the channel, $I_D$ is related to the differential voltage drop in the element $-dV_x$ by the conductance of the element

$$I_D = -\frac{Z2h(x)}{\rho}\frac{dV_x}{dx} \qquad ...(5)$$

The minus sign associated with $dV_x$ simply indicates that $V_x$ decreases as $x$ increases along the channel. The term $2h(x)$ is the channel width at $x$.

The half-width of the channel at point $x$ depends on the local reverse bias between gate and channel $-V_{GS}$

$$h(x) = a - W(x) = a - \left[\frac{2\varepsilon(-V_{Gx})}{qN_d}\right]^{1/2} = a\left[1-\left(\frac{V_x - V_G}{V_P}\right)^{1/2}\right] \qquad ...(6)$$

since $V_{Gx} = V_G - V_S$ and $V_P = qa^2 N_d/2\varepsilon$. Implicit in Eq. (6) is the assumption that the expression for $W(x)$ can be obtained by a simple extension of Eq. (1) to point $x$ in the channel. This is called the gradual approximation; it is valid if $h(x)$ does not vary abruptly in any element $dx$.

The voltage $V_{Gx}$ will be negative since the gate voltage $V_G$ is chosen zero or negative for proper operation. Substituting Eq. (6) into Eq. (5), we have

$$\frac{2Za}{\rho}\left[1-\left(\frac{V_x - V_G}{V_P}\right)^{1/2}\right] dV_x = -I_D dx \qquad ...(7)$$

We can solve this equation to obtain

$$\boxed{I_D = G_0 V_P\left[\frac{V_D}{V_P} + \frac{2}{3}\left(-\frac{V_G}{V_P}\right)^{3/2} - \frac{2}{3}\left(\frac{V_D - V_G}{V_P}\right)^{3/2}\right]} \qquad ...(8)$$

where $V_G$ is negative and $G_0 \equiv 2aZ/\rho L$ is the conductance of the channel for negligible $W(x)$, i.e., with no gate voltage and low values of $I_D$. This equation is valid only up to pinch-off, where $V_D - V_G = V_P$. If we assume the saturation current remains essentially constant at its value at pinch-off, we have

$$I_D\,(\text{sat.}) = G_0 V_P\left[\frac{V_D}{V_P} + \frac{2}{3}\left(-\frac{V_G}{V_P}\right)^{3/2} - \frac{2}{3}\right]$$

$$= G_0 V_P\left[\frac{V_G}{V_P} + \frac{2}{3}\left(-\frac{V_G}{V_P}\right)^{3/2} + \frac{1}{3}\right] \quad \text{...(9)}$$

where $\dfrac{V_D}{V_P} = 1 + \dfrac{V_G}{V_P}$

The resulting family of $I-V$ curves for the channel agrees with the results we predicted qualitatively (Fig. 11.136). The saturation current is greatest when $V$ is zero and becomes smaller as $V_G$ is made negative.

We can represent the device biased in the saturation region by an equivalent circuit where changes in drain current are related to gate voltage change by

$$g_m\,(\text{sat.}) = \frac{\partial I_D\,(\text{sat.})}{\partial V_G} = G_0\left[1 - \left(-\frac{V_G}{V_P}\right)^{1/2}\right] \quad \text{...(10)}$$

The quantity $g_m$ is the *mutual transconductance*, with units (A/V) called siements (s), sometimes called mhos. As a figure of merit for FET devices it common to describe the transconductance per unit channel width $Z$. This quantity $g_m/Z$ is usually given in units of millisiemens per millimetre.

It is found experimentally that a square-law characteristic closely approximates the drain current in saturation

$$I_D(\text{sat.}) \simeq I_{DSS}\left(1 + \frac{V_G}{V_P}\right)^2, \qquad (V_G \text{ negative}) \quad \text{...(11)}$$

where $I_{DSS}$ is the saturated drain current with $V_G = 0$.

## 11.60 METAL-OXIDE SEMICONDUCTOR FIELD EFFECT TRANSISTOR

These are of two types: (a) depletion type, and (b) enhancement type.

### 11.60.1 The Depletion-Mode MOSFET

The metal-oxide semiconductor FET or MOSFET, has a source, gate and drain. Unlike a JFET, however, the gate is electrically insulated from the channel. Because of this, the gate current is extremely small whether the gate is positive or negative. It is also referred to as an insulated gate FET (IGFET).

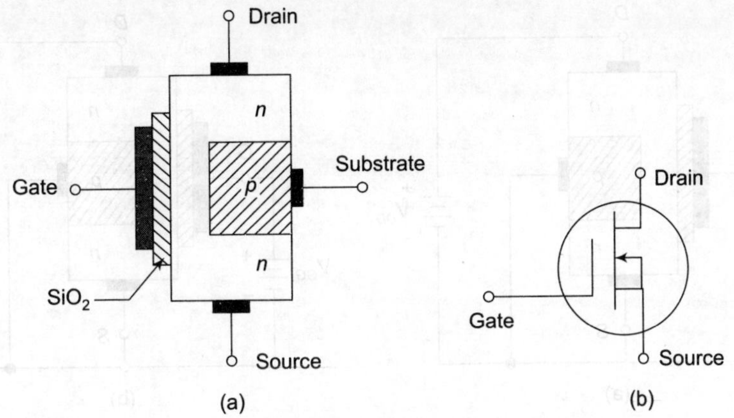


**Figure 11.138** (a) *n*-channel depletion mode MOSFET. (b) Schematic symbol

### 11.60.1.1 Construction of an N-Channel Depletion Mode MOSFET

Figure 11.138 (a) shows an *n*-channel depletion mode MOSFET.

It consists of a piece of *n*-material with a *p*-region on the right and a gate (insulated by ($SiO_2$) on the left. Free electrons can flow from the source to the drain through the narrow *n*-channel. The *p*-region is called the substrate.

A thin layer of insulator ($SiO_2$) is deposited on the left side of the channel. The gate is metallic and insulated from the channel ( by $SiO_2$ layer). Since the gate is insulated, negligible gate current flows even when the gate voltage is positive.

Figure 11.138 (b) shows the schematic symbol, and just to the right of the gate is the thin vertical line respresenting the channel. The arrow on the *p* substrate points to the *n*-material. Usually the manufacturer internally connects the substrate to the source.

So, the device generally has three leads: drain, gate and source.

### 11.60.1.2 Working

Figure 11.139 (a) shows a depletion mode MOSFET with a negative gate. The $V_{DD}$ supply forces free electron to flow from the source to drain (through the narrow *n*-channel) like JFET, the gate voltage controls the width of the channel. (The more negative gate voltage, the smaller the drain current). So when $V_{GS}$ is negative, the operation is similar to that of a JFET.

Since the gate is insulated from the channel, we can apply a positive voltage to the gate as shown in Fig. 11.139 (b). This positive voltage increases the number of free electrons flowing through the channel. The more positive the gate voltage, the greater is the current from source to drain. Being able to use a positive gate voltage is what distinguishes the depletion mode MOSFET from the JFET.

### 11.60.1.3 Static and Transfer Characteristics

Figure 11.140 (a) shows typical drain curves for *n*-channel MOSFET. Note that the upper curves have a positive $V_{GS}$ and the lower curves, a negative $V_{GS} = V_{GS(\text{off})}$. Along this cut-off

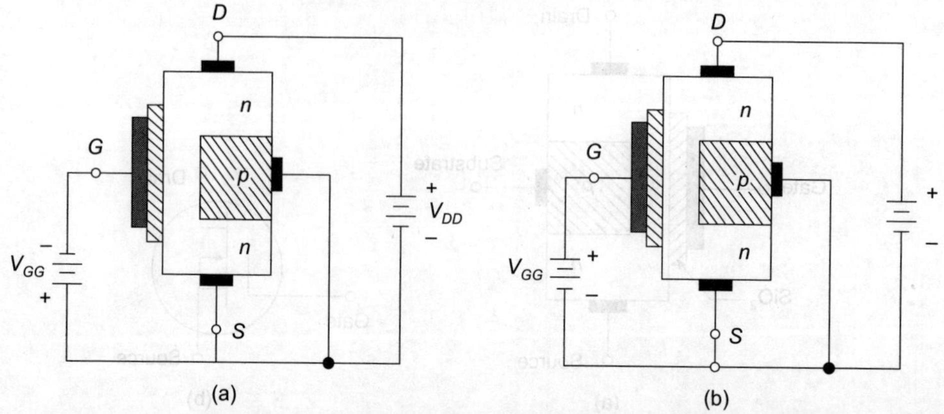


**Figure 11.139** (a) Negative gate voltage, (b) Positive gate voltage

curve, the drain current is approximately zero. When $V_{GS}$ is between $V_{GS(\text{off})}$ and zero, we get a depletion mode operation. $V_{GS} > 0$ gives an enhancement mode operation.

Figure 11.140 (b) is the transconductance curve of a depletion mode MOSFET and $I_{DSS}$ is the drain current with a shorted gate. Since the curve extends to the right of the origin, $I_{DSS}$ is not the maximum possible drain current.

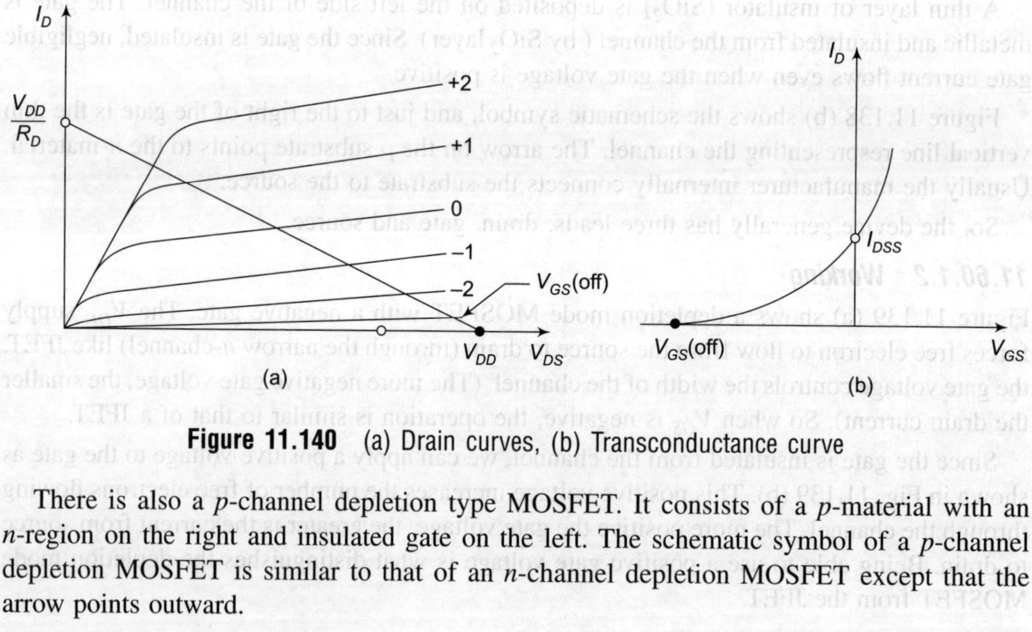


**Figure 11.140** (a) Drain curves, (b) Transconductance curve

There is also a *p*-channel depletion type MOSFET. It consists of a *p*-material with an *n*-region on the right and insulated gate on the left. The schematic symbol of a *p*-channel depletion MOSFET is similar to that of an *n*-channel depletion MOSFET except that the arrow points outward.

## 11.60.2 The Enhancement Mode MOSFET

The enhancement mode MOSFET device has revolutionized the electronic industry and has become enormously important in digital electronics and computers.

### 11.60.2.1 Construction and Working

Figure 11.141 shows an $n$-channel enhancement type MOSFET. The substrate extends all the way to the insulator $SiO_2$. Note that there is no channel between the source and the drain.

Figure 11.141 (b) shows normal biasing of $n$-channel enhancement type MOSFET. When the gate voltage is zero, the $V_{DD}$ supply tries to force free electrons from the source to the drain, but the $p$ substrate has only a few thermally produced free electrons. Besides these minority carriers (and some surface leakage), the current between the source and drain is zero. For this reason, the MOSFET is normally off when the gate voltage is zero.

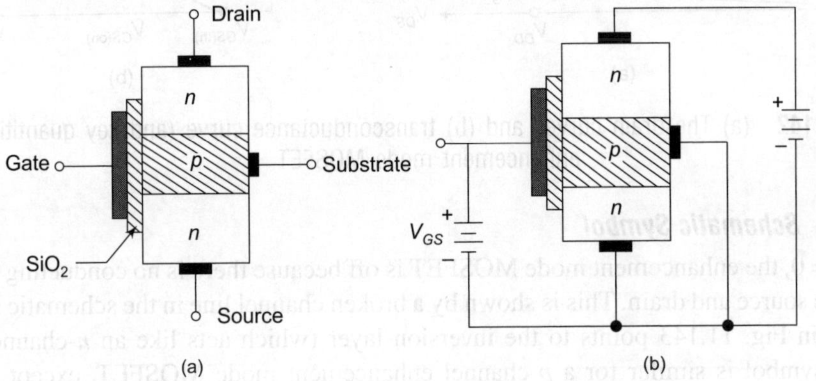


**Figure 11.141** Enhancement mode MOSFET

When the gate is positive enough, it attracts free electrons into the $p$-region. The free electrons recombine with holes next the $SiO_2$. Thus, when the gate is positive enough, all the holes touching the $SiO_2$ are filled and free electrons begin to flow from the source to the drain. Thus, effectively, a thin layer of $n$-type material is created next to $SiO_2$. This conducting layer is called the $n$-type inversion layer.

The minimum $V_{GS}$ that creates the $n$-type inversion layer is called the threshold voltage, denoted by $V_{GS(\text{th})}$. When $V_{GS}$ is less than $V_{GS(\text{th})}$, the drain current is zero, but when $V_{GS}$ is $> V_{GS(\text{th})}$ an inversion layer connects to the source to drain and a large drain current flows in the circuit.

### 11.60.2.2 Depletion Mode MOSFET

Depletion mode MOSFET's (and JFET's) are classified as depletion mode devices because their conductivity depends on the action of depletion layers. The enhancement mode MOSFET's are so called because their conductivity depends on the action of ($n$-type) inversion layer. Depletion mode devices are normally on when the gate voltage is zero, whereas enhancement mode devices are normally off when the gate voltage is zero.

Figure 11.142 shows (a) the drain curves and (b) transconductance curve for enhancement mode, MOSFETs.

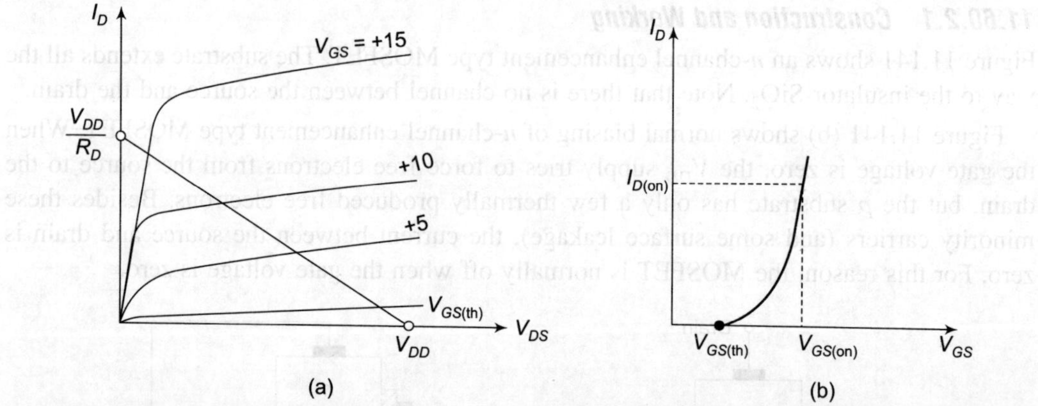


**Figure 11.142** (a) The drain curves and (b) transconductance curve (and key quantities) for enhancement mode MOSFET

### *11.60.2.3 Schematic Symbol*

When $V_{GS} = 0$, the enhancement mode MOSFET is off because there is no conducting channel between the source and drain. This is shown by a broken channel line in the schematic symbol. The arrow in Fig. 11.143 points to the inversion layer (which acts like an *n*-channel). The schematic symbol is similar for a *p*-channel enhancement mode MOSFET, except that the arrow points outwards.

## 11.61 SILICON CONTROLLED RECTIFIER

It is a four layered device shown in Fig. 11.144, having a cathode, an anode and a control gate. It provides current only in one direction and hence it can be described as unidirectional, reverse-blocking thyristor. [Thyristors are those solid state devices which have two or more junctions and can be switched from ON to OFF or from OFF to ON, i.e., between two

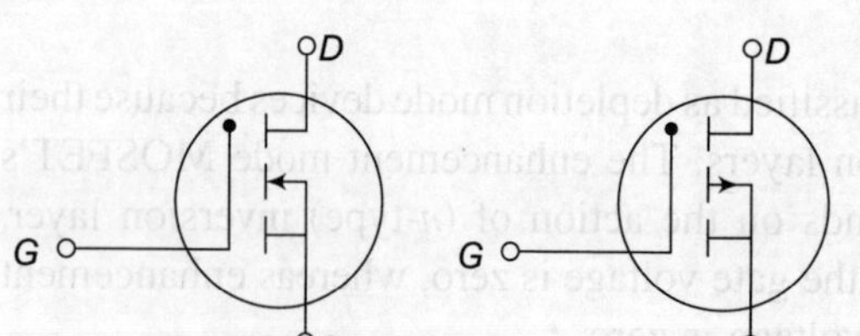


**Figure 11.143** Schematic symbols for (a) *n*-channel, (b) *p*-channel enhancement mode MOSFET

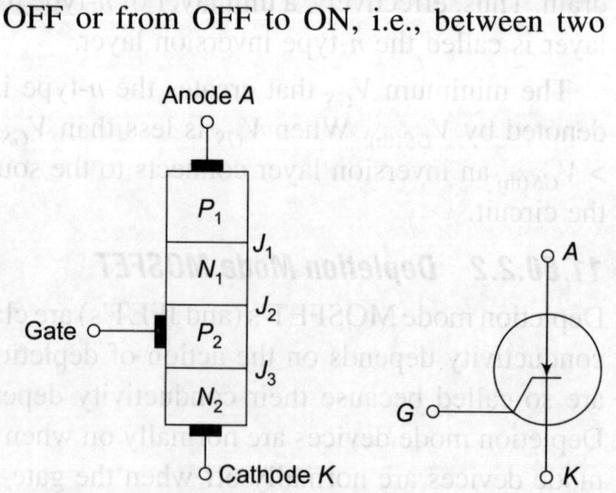


**Figure 11.144** Silicon controlled rectifier SCR: (a) structure; (b) symbol

conducting states]. The SCR can be switched from OFF to ON states by a positive trigger plus of right character applied to the gate.

The operation of the SCR may be explained by considering it to be divided into two parts each being a transistors: a PNP transistor $T_1$ and another NPN transistor $T_2$ as shown in Fig. 11.145 (a). Figure 11.145 (b) shows its electrical equivalent circuit.

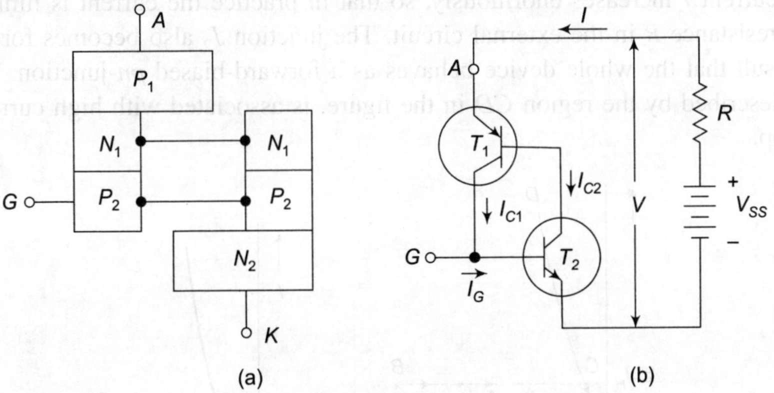


**Figure 11.145** SCR equivalent circuit

First, let it be assumed that the gate terminal $G$ is not connected anywhere and the gate current $I_G$ is zero. A voltage supply $V_{SS}$ is connected in series with a protecting resistor $R$ between the anode and the cathode, as shown in Fig. 11.145 (b). The junctions $J_1$ and $J_3$ are forward biased and the junction $J_2$ is reverse biased by the applied voltage, $V_{SS}$. This biasing of the junctions is just correct for the operation of the two transistors, $T_1$ and $T_2$. The collector-currents of the two transistors are then given by

$$\left.\begin{aligned} I_{C1} &= \alpha_1 I + I_{CO1} \\ \text{and} \quad I_{C2} &= \alpha_2 I + I_{CO2} \end{aligned}\right\} \quad ...(1)$$

where $\alpha$ and $I_{CO}$ are the current gain and collector reverse saturation current respectively, pertaining to the two transistors.

The total current $I$ in the anode terminal is given by

$$I = I_{C1} + I_{C2} = (\alpha_1 + \alpha_2)I + I_{CO1} + I_{CO2}$$

$$I = \frac{I_{CO1} + I_{CO2}}{1-(\alpha_1+\alpha_2)} \quad ...(2)$$

Though for an ordinary transistor, the value of $\alpha$ is very near to but less than unity but the SCR is designed so that $(\alpha_1 + \alpha_2)$ is maintained less than unity for normal voltages between anode and cathode, and gate open circuited; the values of $\alpha_1$ and $\alpha_2$ being individually less than 0.5. It is the characteristics property of silicon junction that the value of $\alpha$ in such junctions is quite small (less than 0.5) at very small current levels, and it increases as the current level is increased. Such property is not exhibited by germanium and, hence, the controlled rectifiers are always made of silicon.

The volt-ampere characteristics of an SCR for $I_G = 0$ is shown in Fig. 11.146 (a).

At low currents in the region *OA* of the curves, the value of $(\alpha_1 + \alpha_2) < 1$ and junction $J_2$ behaves as a reverse bias *pn*-junction. As the applied voltage *V* is increased, the current slowly increases and the value of $(\alpha_1 + \alpha_2)$ also goes on increasing till it becomes unity at point *B* corresponding to the voltage $V_{BO}$, called the *breakover* or *triggering voltage* and current is $I_H$. The current *I* increases enormously, so that in practice the current is limited only by protecting resistance *R* in the external circuit. The junction $J_2$ also becomes forward biased with the result that the whole device behaves as a forward-biased *pn*-junction. This part of operation described by the region *CD* in the figure, is associated with high current and low voltage drop.

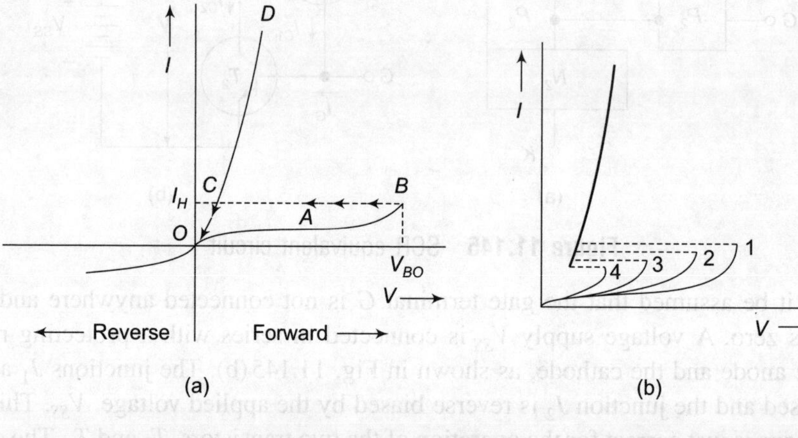


**Figure 11.146** Volt-ampere characteristic of a SCR for (a) $I_G = 0$, (b) Varying values of $I_G$; $I_{G1} < I_{G2} < I_{G3} < I_{G4}$

Once the SCR is fired to bring it in its 'ON' condition, the conduction can be stopped only by reducing the current below the value $I_H$ called the *holding current*. As soon as the current falls below this value, conduction ceases and the operating point goes directly from *C* to *O* as indicated.

In the above discussion, the effect of the gate current $I_G$ was not taken into account. Now, suppose that the supply voltage $V_{SS}$ is less than that needed to fire the SCR. If a gate current $I_G$ is now supplied, the junction $J_3$ becomes more forward biased, and it is possible that the value of $\alpha_2$ increases to such an extent so as to satisfy the condition, $\alpha_1 + \alpha_2 = 1$. The SCR would then fire even with much lower anode voltage be merely introducing small amount of the gate current, $I_G$. This is the principle on which the operation of SCR is biased upon. The additional base current $I_G$ is effective in raising $\alpha_2$ independently of the applied voltage *V*. As shown in Fig. 11.146 (b) higher the value of the triggering gate current $I_G$, the lower is the value of applied voltage *V* at which breakover occurs.

Once the SCR has fired, the gate loses its controlling action and the SCR cannot act and the SCR cannot be turned off by gate signals. The only way to bring the SCR back to its normal OFF condition is to reduce the anode voltage sufficiently for anode current to drop below the holding current, $I_H$.

The four-layer devices, which have a control mechanism, are commonly referred to as *thyristors*. However, the term is most frequently applied to SCRs. (Note that SCR has a gate terminal for control purposes.)

SCRs find application in regulated power supplies, time-delay circuits, motor-controls, inverters, battery chargers, heater controls and phase controls.

## 11.62 UNIJUNCTION TRANSISTOR (UJT)

Like SCR, UJT is also a triggering device which may be used to fire thyristors. It also has a negative resistance region which is used to advantage in a relaxation oscillator (whose frequency may be varied) to provide motor speed control. Other areas of application include trigger circuits, sawtooth generators, phase control, timing circuits and voltage—or current regulated power supplies. It is a low power absorbing device.

### 11.62.1 Construction

The basic construction of a unijunction transistor (UJT) is shown in Fig. 11.147 (a).

It is a three-dimensional device. There is a slab of high resistivity (i.e., lightly doped) silicon (*n*-type); having two base contacts attached to both ends of its one surface. An aluminium rod is alloyed to the other surface of the slab. The alloyed-end serves as a *pn*-junction with Al-rod acting as *p*-type material. The single *pn*-junction accounts for the *terminology unijunction*.

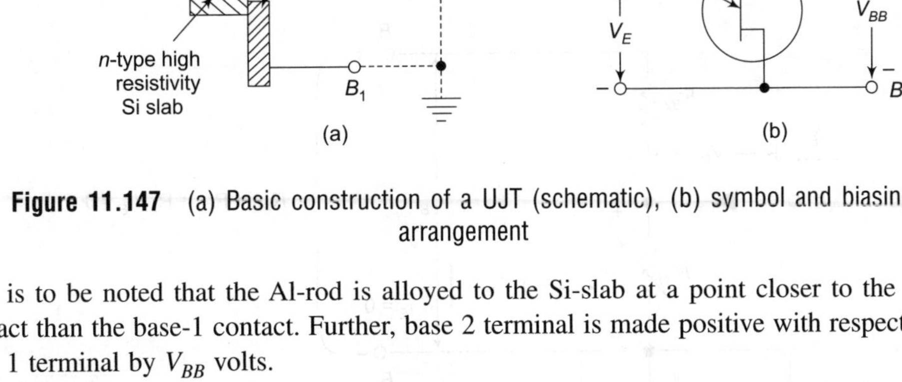


**Figure 11.147** (a) Basic construction of a UJT (schematic), (b) symbol and biasing arrangement

It is to be noted that the Al-rod is alloyed to the Si-slab at a point closer to the base-2 contact than the base-1 contact. Further, base 2 terminal is made positive with respect to the base 1 terminal by $V_{BB}$ volts.

The symbol and basic biasing is shown in Fig. 11.147 (b). The arrowhead points in the direction of conventional current (hole) flow when the device is in the forward biased (active) or conducting state.

### 11.62.2 Static Characteristics Curve

The characteristics of a representative unijunction transistor are shown in Fig. 11.148. If $V_{BB}$ is made zero, and a voltage is applied to $V_E$, the resulting current $I_E$ that flows gives the emitter to the base-1 diode characteristics. ($I_{B2} = 0$). With $V_{BB} = 10$ V, there is a leakage current $I_{EO}$ from base-2 to the emitter. It takes approximately 7 volts from emitter to base-1 to reduce this current to zero and then cause current to flow in the opposite direction, reaching a peak point $V_P$. After $V_P$, an attempt to increase $V_E$ is followed by a sudden increase in emitter current $I_E$ with a corresponding decrease in $V_E$. This corresponds to unstable negative-resistance portion of the characteristics which is followed by valley voltage and then, the device saturates.

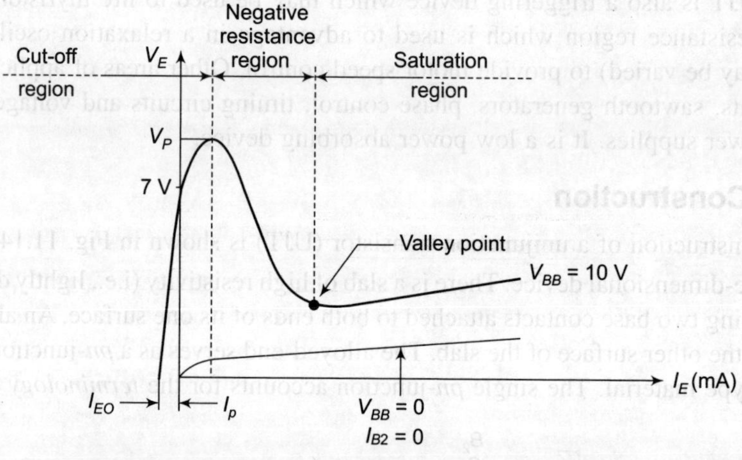


**Figure 11.148** Staticemitter characteristic curve of a UJT

### 11.62.3 Equivalent Circuit and Working

The equivalent circuit is simple and straightforward (Fig. 11.149). The resistance of the silicon bar is represented by two series resistors: $R_{B2}$ is the resistance of the base-2 portion while $R_{B1}$

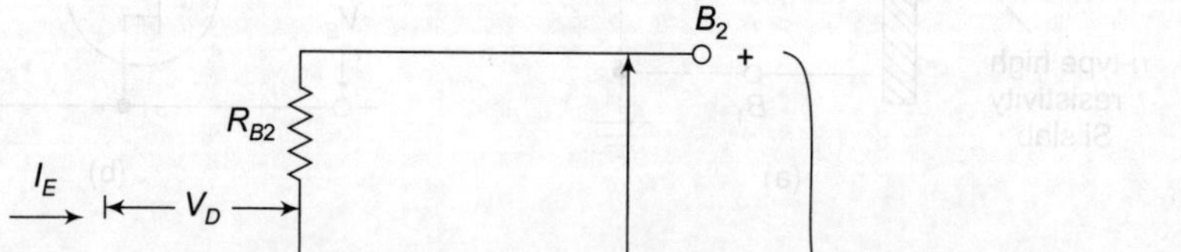


**Figure 11.149** Equivalent circuit of a UJT

is shown as a variable resistance, its value depending on the current, $I_E$. For a representative UJT, $R_{B1}$ may vary from 5 kΩ to 50 kΩ for a corresponding change of $I_E$ from 0 to 50 μA.

The interbase resistance $R_{BB}$ is the resistance between terminals $B_1$ and $B_2$ when $I_E = 0$.

i.e., $$R_{BB} = (R_{B_1} + R_{B_2})| \, I_E = 0 \quad ...(1)$$

Typically, $R_{BB}$ ranges from 4 to 10 kΩ.

The magnitude of $V_{RB_1}$ (with $I_E = 0$, i.e., emitter open) is determined as

$$V_{RB_1} = \frac{R_{B_1} V_{BB}}{R_{B_1} + R_{B_2}} \equiv \eta V_{BB} \Bigg|_{I_E=0} \quad ...(2)$$

where $$\eta \equiv \frac{R_{B_1}}{R_{B_1} + R_{B_2}}\Bigg|_{I_E=0} = \frac{R_{B_1}}{R_{BB}} \quad ...(3)$$

is called the *intrinsic stand-off ratio* of the device. It is typically between 0.51 and 0.82.

Thus, with $E$ open $V_{BB}$ applied, the diode is reverse biased. Now, with a fixed voltage at $V_{BB} = 10$ V (say), consider increasing $V_E$ until $V_E$ just exceed $\eta\ V_{BB}$. The diode will become forward biased and holes will be injected into the silicon bar, with a resulting decrease in the resistance from $E$ to $B_1$. Therefore, for an applied $I_E$ (with the resistance decreasing) the voltage $V_E$ must decrease, giving a negative resistance effect in which current increases with decreasing voltage. (At the peak voltage $V_p = \eta\ V_{BB} + V_D$.)

After the negative resistance region, the valley point is reached and further increase in $I_E$ places the diode in the saturation region (with characteristic approaching that of a semiconductor diode).

## 11.63 UNIJUNCTION OSCILLATOR

A particular device, the unijunction transistor can be used in a single-stage oscillator circuit to provide a pulse signal suitable for digital-circuit applications. The unijunction transistor can be used in what is called a *relaxation oscillator* as shown in the basic circuit of Fig. 11.150. Resistor $R_T$ and capacitor $C_T$ are the timing components that set the circuit oscillating rate. The oscillating frequency may be calculated using Eq. (1) which includes the unijunction transistor *intrinsic stand-off ratio* η, as a factor (in addition to $R_T$ and $C_T$) in the operating frequency.

$$f_o \cong \frac{1}{R_T C_T \ln\,[1/(1-9)]} \quad ...(1)$$

Typically, a unijunction transistor has a stand-off ratio from 0.4 to 0.6. Using a value of η = 0.5, we get

$$f_o \cong \frac{1}{R_T C_T \ln\,[1/(1-0.5)]} = \frac{1}{R_T C_T} \ln 2 = \frac{1.44}{R_T C_T}$$

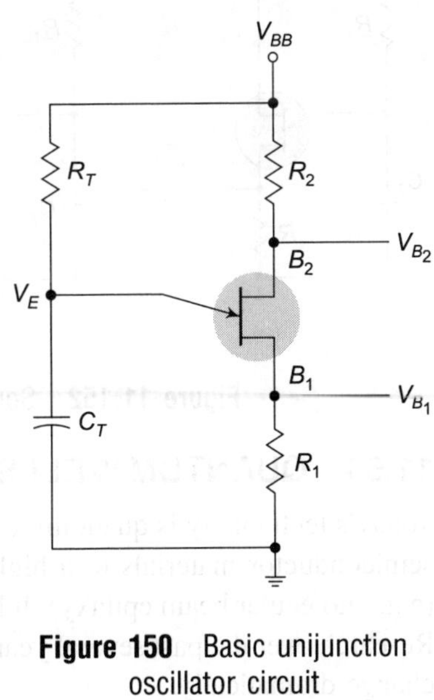


**Figure 150** Basic unijunction oscillator circuit

$$\cong \frac{1.5}{R_T C_T} \qquad ...(2)$$

Capacitor $C_T$ is charged through resistor $R_T$ toward supply voltage $V_{BB}$. As long as the capacitor voltage $V_E$ is below a stand-off voltage ($V_P$) set by the voltage across $B_1 - B_2$, and the transistor stand-off ratio η

$$V_P = \eta\, V_{B_1} V_{B_2} - V_D \qquad ...(3)$$

the unijunction emitter lead appears as an open circuit. When the emitter voltage across capacitor $C_T$ exceeds this value ($V_P$), the unijunction circuit fires, discharging the capacitor, after which a new charge cycle begins. When the unijunction fires, a voltage rise is developed across $R_1$ and a voltage drop is developed across $R_2$ as shown in Fig. 11.151. The signal at the emitter is a sawtooth voltage waveform that at base 1 is a positive-going pulse, and at base 2 is a negative-going pulse. A few circuit variations of the unijunction oscillator are provided in Fig. 11.152.

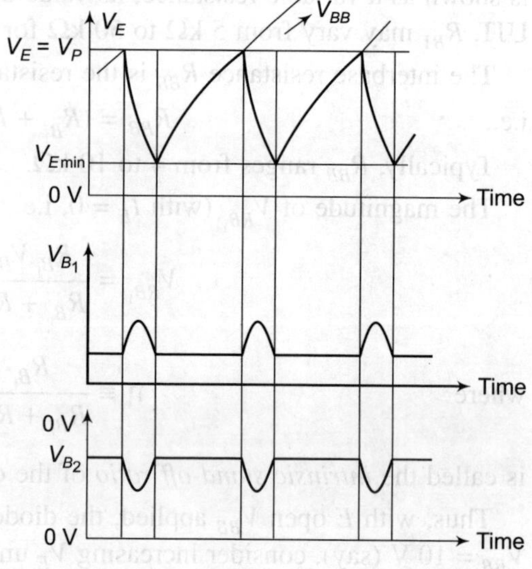


**Figure 11.151** Unijunction oscillator waveforms

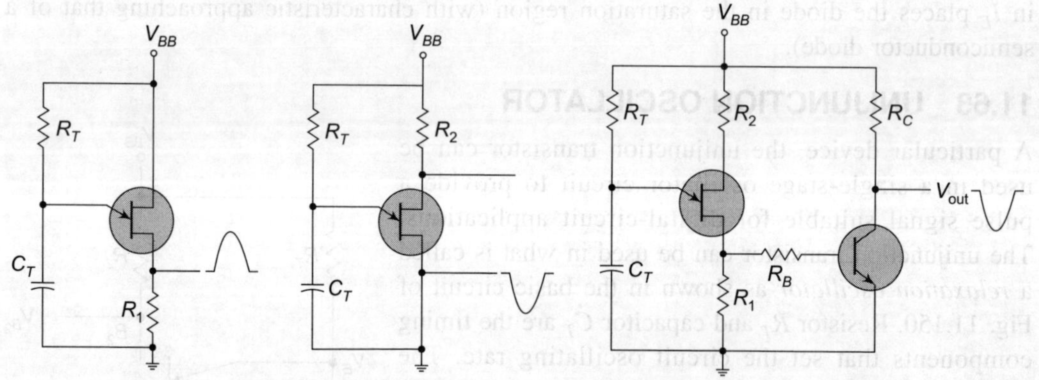


**Figure 11.152** Some unijunction oscillator circuit configurations

## 11.64 QUANTUM WELLS AND SUPERLATTICES

Today's technology is quantum technology. With the development of techniques to purify the semiconductor materials to a high degree and various high vacuum evaporation techniques (e.g., molecular beam epitaxy). It has become possible to tailor materials as per requirements. Research over the past several years has revealed that for smaller sizes, properties of materials change dramatically.

In most cases, we deal with the electronic properties in bulk material where electron mean free path or de Broglie wavelength (~ 200 – 400 Å) is much smaller than sample dimensions. However, when the dimensions become comparable to the mean free path of carriers, *quantum sized effect* become important and dominate the electronic properties of materials.

The first step in this direction was taken by Esaki and Tsu who fabricated the materials wherein ultra thin layers of GaAs and $Ga_xAl_{1-x}$ As were deposited alternately, producing one-dimensional periodic potential with the period shorter than electron mean free path. This structure is named superlattice. Further efforts in this direction have opened up a new area of interdisciplinary investigations in the field of device physics.

## 11.64.1 Heterojunction

There are certain quite different materials which crystallize in the similar manner and have exactly same lattice constants.

As a consequence, they may be grown together as a single crystal. At the boundary between two such materials, there is a change in the band gap energy, in the electron affinity, in the permittivity and other material properties. Such a junction in single crystal, between two materials whose lattices match but which are otherwise different, is known as *heterojunction*. These are commonly grown with semiconductors which may be undoped or doped (*n*- or *p*-type). Figure 11.153 (a) illustrates two semiconductors *A* and *B* with conduction and valence band edges $E_C$ and $E_V$ having different magnitudes of forbidden gap.

An electron at conduction band edge in *A* will have lower energy than in material *B*. When junction is made, a potential-barrier is formed at the interface [Fig. 11.153 (b)]. This restricts the motion of electrons incident from *A* side. This situation is prevalent in case of GaAs and $Al_x Ga_{1-x}$ As junctions.

## 11.64.2 Double Heterostructure

It is possible to have more than one heterojunction in a structure. When there are two heterojunctions, it is called double heterostructure. In this case, two heterojunctions are used to sandwich a layer of narrow band gap material between (two) wide gap materials

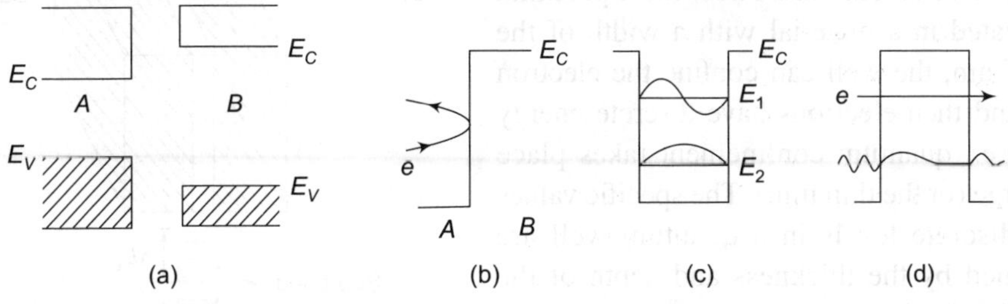


**Figure 11.153** (a) Conduction and valence-band edges of two semiconductors in the two semiconductors lined up together (b) Heterostructure-interface formed due to line up of two semiconductors of Fig. 11.153 (a). (c) Two heterostructure interface forming a quantum well. (d) Two heterostructure interface forming a potential barrier

[Fig. 11.153(c)]. Such double heterostructures are becoming common for light emitting diodes and laser for optical communication.

### 11.64.3 Quantum Well

The use of hetroepitaxial crystal growth techniques such as molecular beam epitaxy and metallorganic chemical vapour deposition enables heterostructures to be formed with very thin layers. In such structures, the motion of electrons is controlled by enforcing restrictions on them in terms of potential barriers [Fig. 11.153 (c)]. The behaviour of an electron in an infinite well has discrete allowed energy levels given by

$$E_n = \frac{\hbar^2}{2\,m^*}\left(\frac{n\pi}{L}\right)^2 \qquad ...(1)$$

$$\psi_n = \sqrt{\frac{2}{L}}\,\sin\left(\frac{n\pi}{L}\right) \qquad ...(2)$$

The wave functions correspond to a standing wave pattern of the wave associated with the particle which is reflected from the walls of the well. For electron wave to be confined within the well, the width of potential well must have dimensions comparable to the de Broglie wavelength of electron. The energy of electron

$$E = \frac{\hbar^2 k^2}{2\,m^*} \quad k_B T \qquad ...(3)$$

where $k$ is wave vector and $k_B$, the Boltzmann constant. ($T$ is absolute temperature). The associated de Broglie wavelength

$$\lambda = \frac{2\pi}{k} = 2\pi\left(\frac{\hbar^2}{2m^* k_B T}\right)^{1/2} = \frac{2\pi\hbar}{\sqrt{2m^* k_B T}} \qquad ...(4)$$

Taking effective mass $m^* = 4\,m_e$ (where $m_e$ is mass of free electron) and $T = 4$K, $L$ comes out to be 1000 Å or 100 nm. Thus, for a potential well created in a material with a width of the order of nm, the well can confine the electron waves and then electrons have discrete energy levels, i.e., quantum confinement takes place and charge (or the thin film). The specific values of the discrete levels in a quantum well are determined by the thickness and depth of the well. (The depth of quantum well for electrons is the discontinuity $\Delta E_c$ of conduction band and for the holes it is $\Delta E_v$).

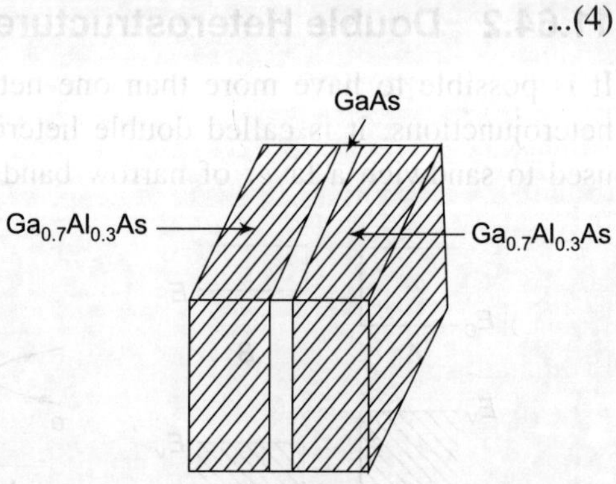


**Figure 11.154** A GaAs/A/GaAs carriers are confined to, move only in a plane quantum well structure

### 11.64.4 Superlattice

When a succession of very thin layers of two or more semiconductors with same lattice constants are grown as a single crystal, it forms a number of quantum wells on top of the another. The band structure is as shown in Fig. 11.155.

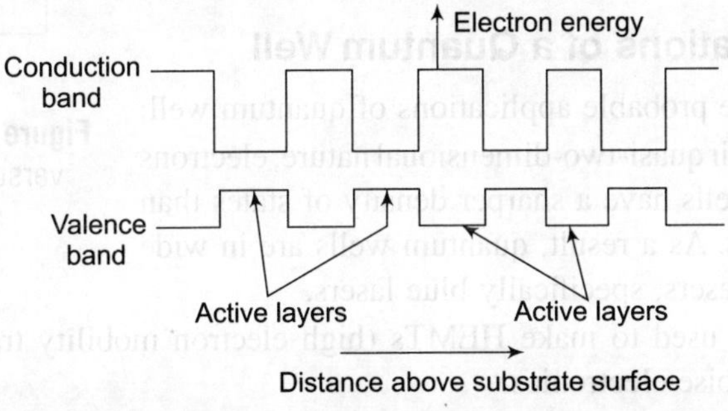


**Figure 11.155** The energy band diagram of a multiple quantum well structure

It has periodicity in one direction with the lattice constant ~ a few nm as compared to a few Å in a normal crystal. Therefore, it is called a superlattice due to large lattice constant. There are alternating active and confining layers. When the confining barrier layers are also sufficiently thin (< 5 nm). tunnelling between active layers is significant. The resulting structure is a superlattice with properties controlled by the artificial periodicity of the multilayered structure. The properties of superlattice can be varied by choices of materials involved, the thicknesses of the layers used and the spacing between them.

## 11.65 A TYPICAL QUANTUM-WELL

A quantum well, shown in Figure 11.156, is a potential well that confines particles, which were orginally free to move in three dimensions, to two dimensions, forcing them to occupy a planar region. The effects of quantum confinement take place when the quantum well thickness becomes comparable at the de Broglie wavelength of the carriers, with the result that the carriers can have only discrete energy values.

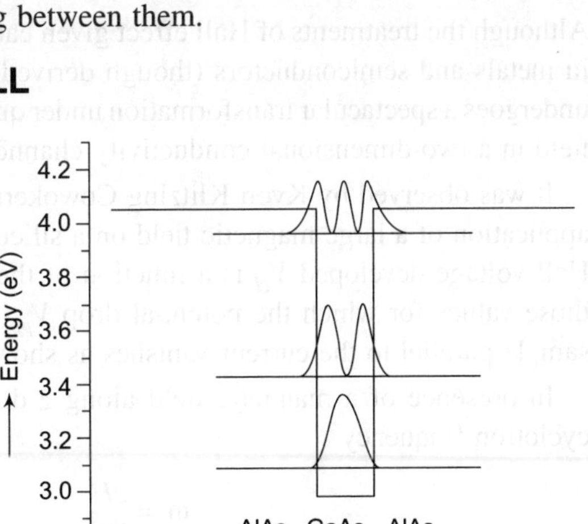


**Figure 11.156** A typical 5 nm GaAs quantum well

A quantum well confines the motion of electrons or holes in one dimension and allows free propagation in two dimensions. Quantum wells are formed in semiconductors by having a material, like gallium arsenide sandwiched between two layers of a material with a wider bandgap, like aluminium arsenide. These structures can be grown by molecular beam epitaxy or chemical deposition with control of the layer thickness down to monolayers. This is now common in industry, in research, and even for academic students.

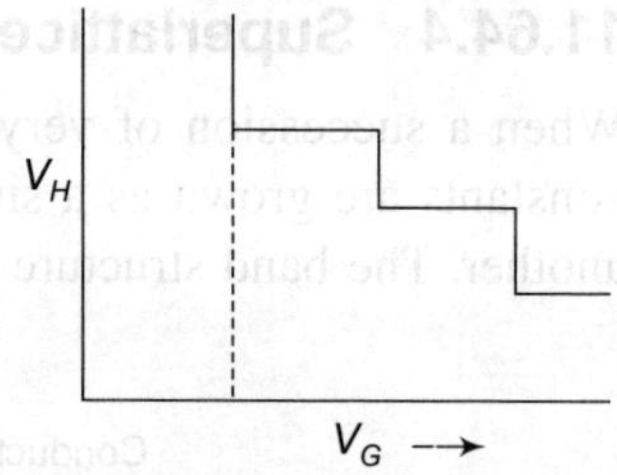


**Figure 11.157** Hall voltage versus Gate voltage in a silicon MOSFET

### 11.65.1 Applications of a Quantum Well

The following are the probable applications of quantum well:

1. Because of their quasi-two-dimensional nature, electrons in quantum wells have a sharper density of states than bulk materials. As a result, quantum wells are in wide use in diode lasers, specifically blue lasers.
2. They are also used to make HEMTs (high electron mobility transistors), which are used in low-noise electronics.
3. Quantum well infrared photodetectors are also based on quantum wells, and are used for infrared imaging.
4. By doping either the well itself, or preferably, the barrier of a quantum well with donor impurities, a two-dimensional electron gas (2DEG) may be formed. This quasi-two-dimensional system has interesting properties at low temperature. One such property is the *quantum Hall effect*, seen at high magnetic fields. Acceptor dopants can lead to a two-dimensional hole gas (2DHG).

## 11.66 QUANTUM HALL EFFECT

Although the treatments of Hall effect given earlier give a good account of electrical transport in metals and semiconductors (though derived on classical considerations) yet, the scenario undergoes a spectacular transformation under quantum conditions of temperature and magnetic field in a two-dimensional conductivity channel.

It was observed by Kvon Klitzing Cowokers [*Phys. Rev. Lett.* 45, 434 (1980)] that on the application of a large magnetic field on a silicon MOSFET at liquid helium temperature, the Hall voltage developed $V_H$ is a function of the gate voltage $V_G$ and exhibits plateau around those values for which the potential drop $V_{pp}$ (the integral quantum hall voltage) along the sample parallel to the current vanishes as shown in Fig. 11.158.

In presence of a magnetic field along $z$ direction, electrons acquire circular orbits and cyclotron frequency

$$\omega = \frac{eH}{mc}$$

For very large $H$, orbits are small and the associated energy is quantised as

$$E_{\perp} = \left(n + \frac{1}{2}\right)\hbar\omega \text{ (in } xy \text{ plane)}$$

where, $n$ is an integer.

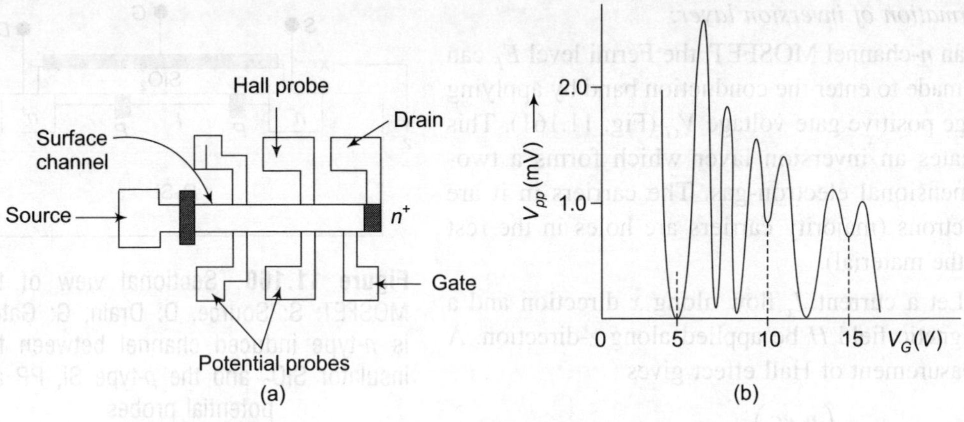


**Figure 11.158** (a) The MOS transistor setup for quantum Hall effect meauserments (top view), and (b) the IQHE voltage $V_{PP}$ as a function of the gate voltage, $V_G$

Due to formation of the Landau levels $E_{nl}$ (at low temperature, in presence of $H$)

$$E_{nl} = E_{\perp} + E_z = \left(n + \frac{1}{2}\right)\hbar\omega + \frac{\hbar^2 (k_z)_l^2}{2m} \qquad \text{...(1a)}$$

For a thin film of thickness $d_z$, the $k_z$ (perpendicular to film thickness) also gets quantised in units of $2\pi/d_z$, i.e.,

$$(k_z)_l = l\left(\frac{2\pi}{d_z}\right), l = 0, \pm 1, \pm 2. \ldots \text{ along } z$$

For small $d_z$, the energy levels are separated enough to stratify the electron states (Landau levels). For a given $k_z$ or $l$

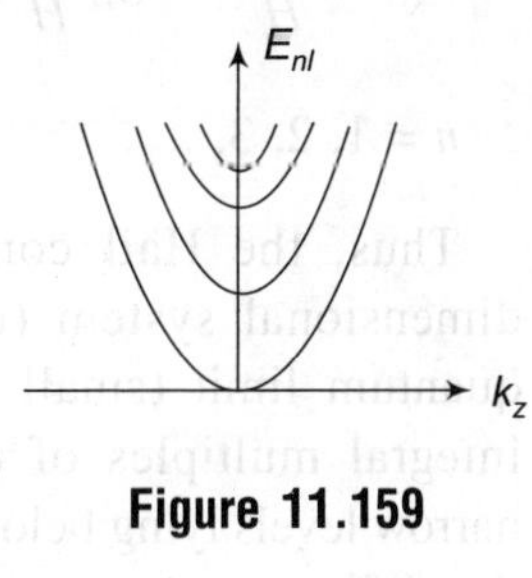


**Figure 11.159**

$$E_{\perp} = \left(n + \frac{1}{2}\right)\left(\frac{e\hbar}{mc}\right)H$$

i.e.,

$$\text{Separation} = \left(\frac{e\hbar}{mc}\right)H$$

If $\frac{p^2}{2m} = \left(\frac{e\hbar}{mc}\right)H$ defines $p$, then the degeneracy (i.e., number of states within each Landau level) is (area) $\times\ \pi p^2/h^2$ or per unit area,

$$g_L = \pi p^2/h^2 = 2\pi\,\frac{e\hbar}{c}\,H = \left(\frac{e}{hc}\right)H \qquad \text{...(1b)}$$

(neglecting spin)

The degeneracy of a Landau level $E_{nl}$ for a given $l$ is same for all $n$.

*Formation of inversion layer:*

In an $n$-channel MOSFET, the Fermi level $E_F$ can be made to enter the conduction band by applying large positive gate voltage $V_G$ (Fig. 11.161). This creates an inversion layer which forms a two-dimensional electron-gas. The carriers in it are electrons (majority carriers are holes in the rest of the material).

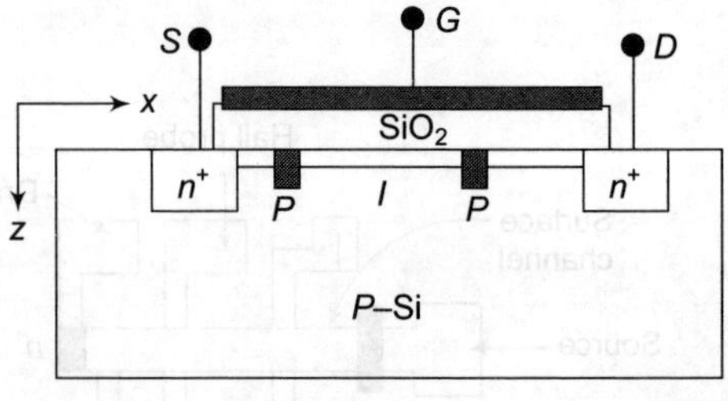


**Figure 11.160** Sectional view of the MOSFET: S: Source, D: Drain, G: Gate $I$ is $n$-type induced channel between the insulator $SiO_2$ and the $p$-type Si, PP are potential probes

Let a current $J_x$ flow along $x$ direction and a magnetic field $H$ be applied along $z$-direction. A measurement of Hall effect gives

$$J_x = \left(\frac{n_e ec}{H}\right) E_y = \sigma_{xy} E_y \quad \text{...(1c)}$$

where, $n_e$ is the density of charge carriers and $\sigma_{xy} = n_e ec/H$, relating $J_x$ and $E_y$, is the Hall conductivity. Its reciprocal is the Hall resistivity $\rho_{xy}$, and

$$R_H = \frac{\rho_{xy}}{H} = \frac{1}{n_e ec} \quad \text{...(1d)}$$

is the Hall coefficient

Writing the carrier concentration as $n_e = ng_L$, $n = 1, 2, 3, \ldots$

We get from Eqs. (1b) and (1c)

$$\sigma_{xy} = \frac{n_e ec}{H} = ng_L \frac{ec}{H} = n\left(\frac{e^2}{h}\right),$$

$n = 1, 2, 3, \ldots$

Thus, the Hall conductivity of the two-dimensional system (inversion layer) in the quantum limit (small orbits) is quantised in integral multiples of $e^2/h$. Here there are $n$ narrow levels lying below $E_F$. As a consequence, the difference between the Hall voltages for two successive plateaus is found to be exactly 1/25813 ohms, and it serves as the definition of standard ohm in terms of Planck's constant and electronic charge.

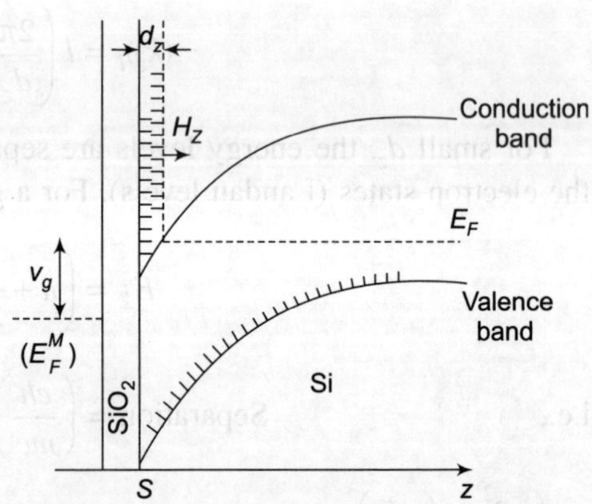


**Figure 11.161** Formation of the inversion layer, i.e., $n$-type induced channel in $p$-silicon at the Interface with the insulator $SiO_2$ in a MOSFET. $V_G$ = Gate voltage, A: acceptor states (negatively charged), $z$ is distance from interface into the crystal. $E_F$ = Semiconductor Fermi level. $E_F^M$ is metal Fermi level

The electrons in the channel of the MOSFET form a two-dimensional electron gas (in $XY$ plans) and when a strong magnetic field is applied (say along $z$-axis), the electron energies in the electron gas are quantised into Landau levels of energy $\left(n+\frac{1}{2}\right)\hbar\omega_c$ where $\hbar = h/2\pi$ and $\omega_c$ is the cyclotron frequency. The potential drop $V_{pp}$ tends to zero when a Landau level coincides with the Fermi level. Since the gate voltage $V_G$ controls the number of electrons/area of the channel and, therefore, the Fermi energy, hence minima in $V_{pp}$ (and plateau in $V_H$) occur at intervals $\hbar\omega_c$.

## 11.67 MOLECULAR BEAM EPITAXY

Molecular beam epitaxy (MBE) was developed in the 1970s and is by now a common technology for use in making low-dimensional solid-state structure. MBE is on ultrathin film vacuum technique in which several atomic and/or molecular beams collide with and stick to the substrate. Exitaxy means that at the interface of two materials, there is a common crystal

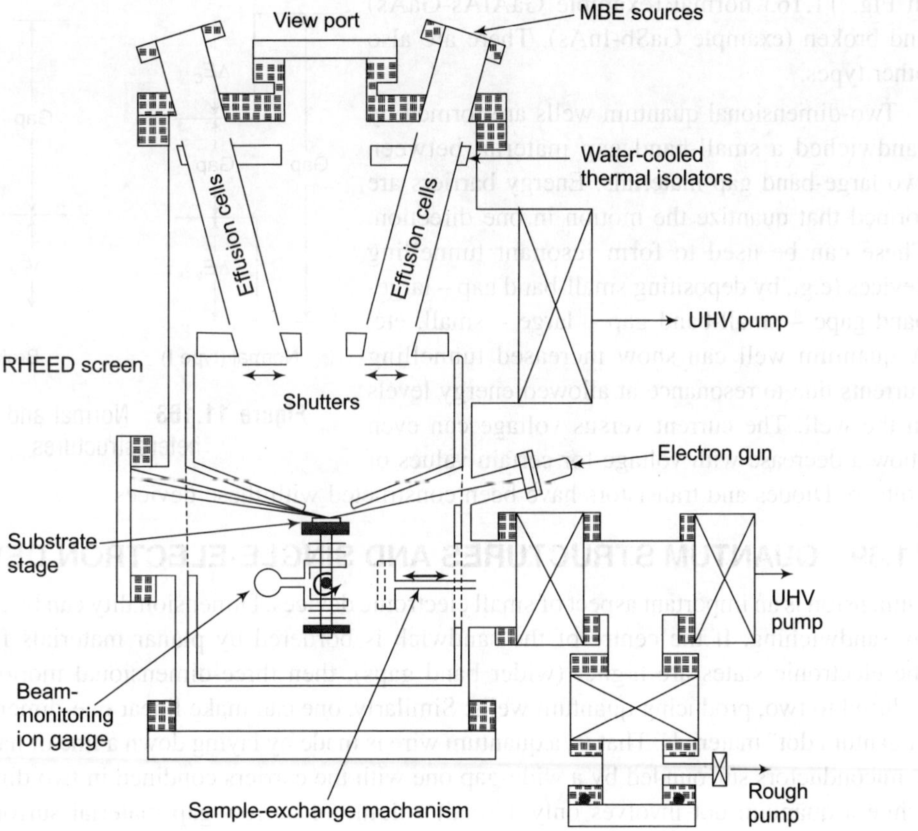


**Figure 11.162** Schematic diagram of an ultrahigh vacuum, molecular beam growth system (adapted from James D. Patterson and Bernard C. Bailey, *Solid State Physics*, Springer Verlag (2007)). Reflection High-energy Electron Diffraction (RHEED) is used for monitoring the growth

orientation and registry of atoms. The substrates are heated to temperature $T$ and mounted suitably. Each effusion cell, from which the molecular beams originate, are held at appropriate temperatures to maintain a suitable flux. The effusion cells also have shutters so that the growth of layers due to the molecular beams can be controlled (see Fig. 11.162). MBE produces high-purity layers in ultrahigh vacuum. Abrupt transitions on an atomic scale can be grown at a rate of a few (tens of) nanometers per second. Other techniques for producing layered structures include chemical vapour deposition and electrochemical deposition.

## 11.68 HETEROSTRUCTURES AND QUANTUM WELLS (USING MBE)

By use of MBE or other related techniques, heterostructures and quantum wells can be formed. Heterostructures are layers of semiconductors with crystal structure, grown coherently, but with different band gaps. Their properties depend heavily on their type. Two types are shown in Fig. 11.163 normal (example GaAlAs-GaAs) and broken (example GaSb-InAs). There are also other types.

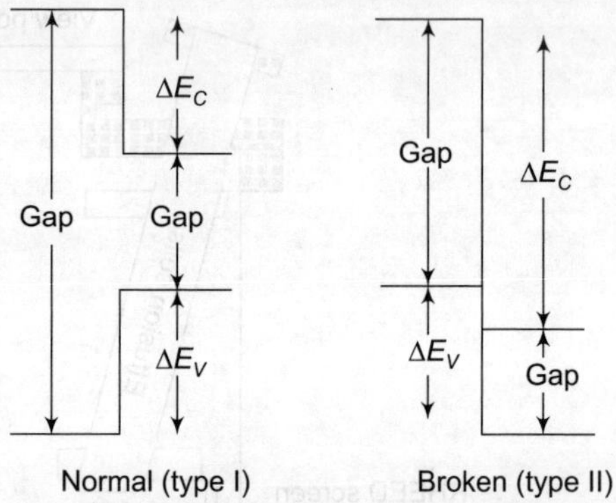


**Figure 11.163** Normal and broken heterostructures

Two-dimensional quantum wells are formed by sandwiched a small-band gap material between two large-band gap materials. Energy barriers are formed that quantize the motion in one direction. These can be used to form resonant tunnelling devices (e.g., by depositing small-band gap – large-band gape – small-band gap – large – small, etc. A quantum well can show increased tunnelling currents due to resonance at allowed energy levels in the well. The current versus voltage can even show a decrease with voltage for certain values of voltage. Diodes and transistors have been constructed with these devices.

## 11.69 QUANTUM STRUCTURES AND SINGLE-ELECTRON DEVICES

Dimension is an important aspect of small electronic devices. Dimensionality can be controlled by sandwiching. If the centre of the sandwich is bordered by planar materials for which the electronic states are higher (wider band gaps), then three-dimensional motion can be reduced to two, producing quantum wells. Similarly, one can make linear one-dimensional or "quantum dot" materials. That is, a quantum wire is made by laying down a line of narrow-gap semiconductors surrounded by a wide-gap one with the carriers condined in two dimensions, while a quantum dot involves only a small volume of narrow-gap material surrounded by wide-gap material and the carriers are confined in all three dimensions. With the quantum-dot structure, one may confine or exchange one electron at a time and develop single-electron transistors that would be fast, low power, and have essentially error-free signals. These three types of quantum structures are summarized in Table 11.14.

**Table 11.14** Summary of Three Types of Quantum Structurs

| *Nanostructures* | *Comments* |
|---|---|
| Quantum wells | Superlattices can be regarded as quantum well layers – alternating layers of different crystals (when the wells are not too far apart) |
| Quantum wires | A crystal enclosed on two sides by another crystalline material, with appropriate wider band gaps |
| Quantum dots | A crystal enclosed on three sides by another crystalline material—sometimes descriptively called a quantum box |

*Note:* Nanostructures have a least one dimension of a size between approximately one to ten nanometers.

### 11.69.1 Coulomb Blackade

The Coulomb blockade model shows how electron-electron interactions can give rise to effects that in certain circumstances are very easy to understand. It relates to the ideas of single-electron transistors, quantum dots, charge quantization leading to an energy gap in the density of states for tunnelling, and is sometimes even qualitatively likened to a dripping of states for tunnelling, and is sometimes even qualitatively likened to dripping faucet. For purpose of illustration, we consider a simple model of an artificial atom represented by the metal particle shown in Fig. 11.164.

Experimentally, the conductance (current per voltage bias) from source to drain shows large changes with gate voltage. We wish to analyze this with the Coulomb blockade model. Let $C$ be the total capacitance between the metal particle and the rest of the system, which we will assume is approximately the capacitance between the metal particle and the gate. Let $V_g$ be the gate voltage, relative to source, and assume the source, particle, and drain voltages are close (but sufficiently different to have the possibility of drawing current from source to drain). If there is a charge $Q$ on the metal particle, then its electrostatic energy is

$$U = QV_g + \frac{Q^2}{2C}. \quad ...(1)$$

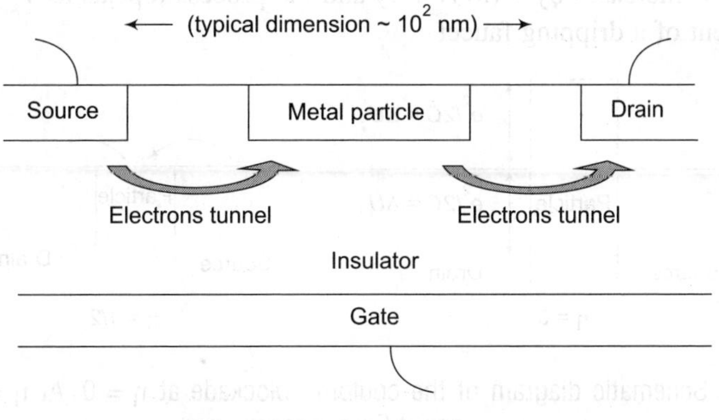


**Figure 11.164** Model of a single-electron transistor

Setting $\partial U/\partial Q = 0$, we find $U$ has a minimum at

$$Q = Q_0 = -CV_g. \quad \text{...(2)}$$

If $N$ is an integer, let $Q_0 = -(N + \eta)e$, where $e > 0$, so

$$CV_g = (N + \eta)e. \quad \text{...(3)}$$

Note that while $Q_0$ can be any value, the actual physical situation will be determined by the integer number of electrons on the artificial atom (metal particle) that makes $U$ the smallest. This will only be at a mathematical minimum if $V_g$ is an integral multiple of $e/C$.

For $-1/2 < \eta < 1/2$, and $V_g = (N + h)e/C$, the minimum energy is obtained for $N$ electrons on the metal particle. The Coulomb blockade arises because of the energy required to transfer an electron to (or from) the metal particle (you can't transfer less than one electron). We can easily calculate this as follows. Let us consider $\eta$ between zero and one half. Combining Eqs. (1) and (3).

$$U = \frac{1}{C}\left[Q(N+\eta)e + \frac{Q^2}{2}\right]. \quad \text{...(4)}$$

Let the initial charge on the particle be $Q_1 = -Ne$ and the final charge $Q_f = -(N \pm 1)e$. Then for the energy difference,

$$\Delta U^+ = U_f^{\pm} - U_i,$$

we find $\Delta U^+ = \dfrac{e^2}{C}\left(\dfrac{1}{2} - \eta\right)$, $\Delta U^- = \dfrac{e^2}{C}\left(\dfrac{1}{2} + \eta\right)$.

We see that for $\eta < 1/2$ there is an energy gap for tunnelling: the Coulomb blockade. For $\eta = 1/2$ the energies for the metal particle having $N$ and $N + 1$ electrons are the same and the gap disappears. Since the source and the drain have approximately the same Fermi energy, one can understand this result from Fig. 11.165. Note $\Delta U^+$ is the energy to add an electron and $\Delta U^-$ is the energy to take away an electron (or to add a hole). Thus, the gap in the allowed states of the particle is $e^2/C$, just above $\eta = 1/2$, the number of electrons on the artificial atoms increases by 1 (to $N + 1$) and the process repeats as $V_g$ is increased. It is indeed reminiscent of a dripping faucet.

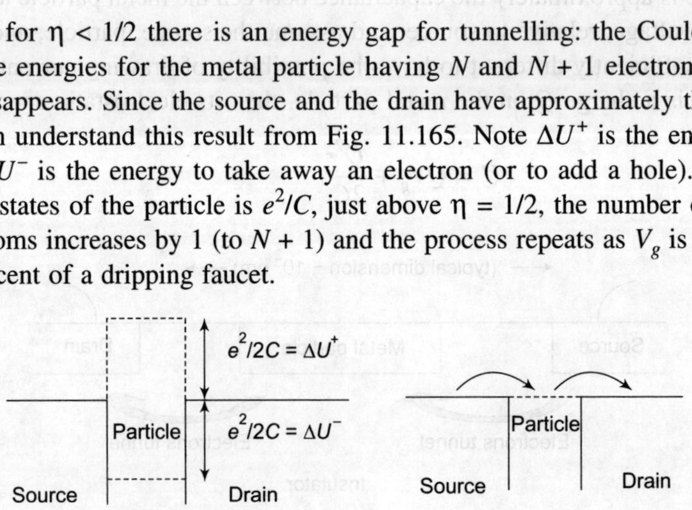


**Figure 11.165** Schematic diagram of the coulomb blockade at $\eta = 0$. At $\eta = 1/2$ the energy gap $\Delta E$ disappears

The total voltage change from one turn on to the next turn on occurs, e.g., when $\eta$ goes from 1/2 to 3/2 or

$$\Delta V_g = \frac{e}{C}\left[N + \frac{3}{2} - \left(N + \frac{1}{2}\right)\right] = \frac{e}{C}.$$

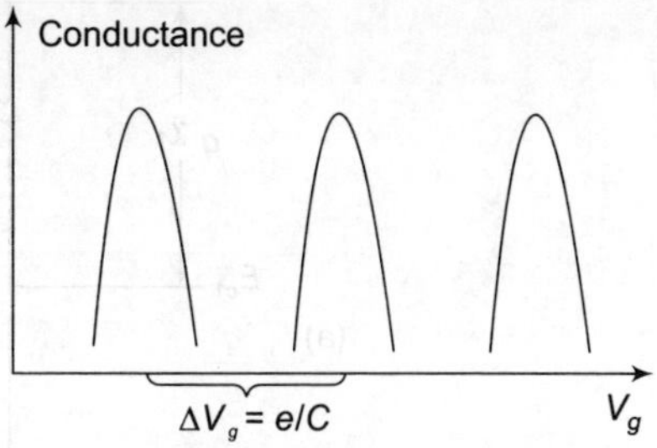


**Figure 11.166** Periodic conductance peaks

A sketch of the conductance versus gate voltage in Fig. 11.166 shows periodic peaks. In order to conduct an electron must go from source to particle, and then from particle to drain (or a hole from drain to particle, etc.).

Low temperature are required to see this effect, as one must have

$$kT < \frac{e^2}{2C},$$

so that the thermal effects do not wash out the gap. This condition requires small temperatures and small capacitances, such as encountered in nanodevices. In addition the metal particle-artificial atom has discrete energy levels that may be observed in tunnelling experiments by fixing $V_g$ and varying the drain-to-source voltage. See Kasner *op cit.*

## 11.70 HETEROJUNCTIONS

Thus, far we have discussed *p-n* junction formed within a single semiconductor (*homojunctions*) and junctions between a metal and a semiconductor. The third important class of junctions consists of those between two lattice matched semiconductors with different band gaps (*heterojunctions*). The interface between two such semiconductors may be virtually free of defects, and continuous crystals containing single or multiple heterojunctions can be formed such that there is good lattice matching at the interface. The availability of heterojunctions and multilayer structures in compound semiconductors opens a broad range of possibilities for device development. We will discuss many of these applications in later chapters, including heterojunction bipolar transistors, field-effect transistors, and semiconductor lasers.

When semiconductors of different band gaps and electron affinities are brought together to form a junction, we expect discontinuities in the energy bands as the Fermi levels line up at equilibrium (Fig. 11.167). The discontinuities in the conduction band $\Delta E_c$ and the valence band $\Delta E_v$ accommodate the difference in band gap between the two semiconductors, $\Delta E_g$. In an ideal case, $\Delta_{Ec}$ would be the difference in electron affinities $q(\chi_2 - \chi_1)$, and $\Delta_{Ev}$ would be found from $\Delta E_g - \Delta E_c$. In practice, the band discontinuities are found experimentally for particular semiconductor pairs. For example, in the commonly used system GaAs-AlGaAs, the direct band gap difference $\Delta E_g^{\Gamma}$ between the wider band gap AlGaAs and the narrower band gap GaAs is apportioned approximately $\frac{2}{3}$ in the conduction band and $\frac{1}{3}$ in the valence band for heterojunction. The built-in contact potential is divided between the two semiconductors as required to align the Fermi levels at equilibrium. The resulting depletion region on each side of the heterojunction and the amount of built-in potential on each side (making up the contact potential, $V_0$) are found by solving Poission's equation with the boundary condition

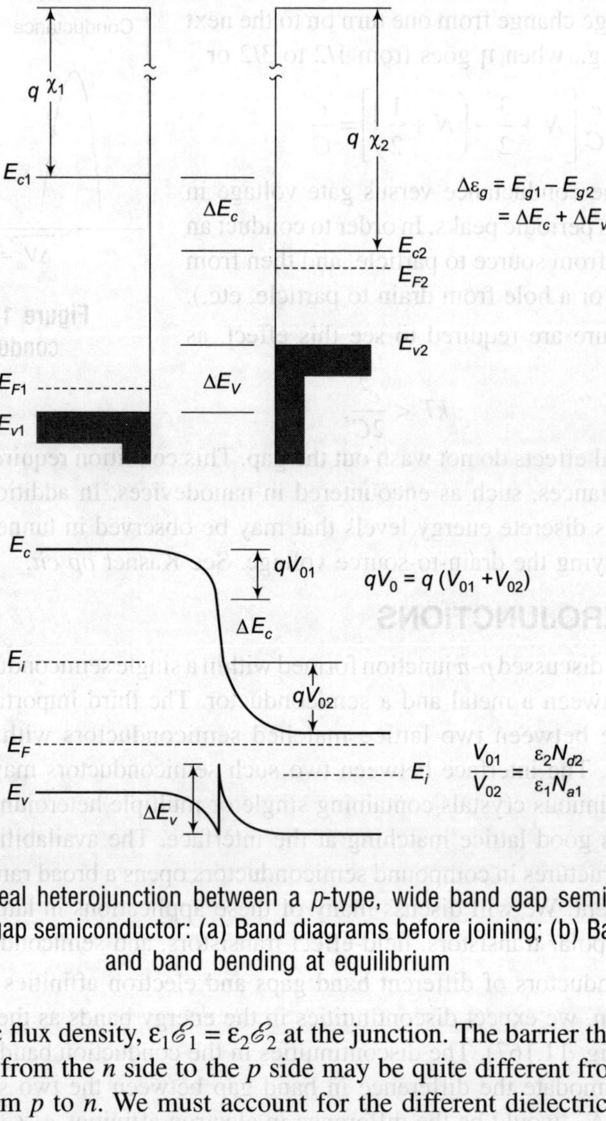


**Figure 11.167** An ideal heterojunction between a *p*-type, wide band gap semiconductor and an *n*-type narrower band gap semiconductor: (a) Band diagrams before joining; (b) Band discontinuities and band bending at equilibrium

of continuous electric flux density, $\varepsilon_1\mathscr{E}_1 = \varepsilon_2\mathscr{E}_2$ at the junction. The barrier that electrons must overcome in moving from the *n* side to the *p* side may be quite different from the barrier for the holes moving from *p* to *n*. We must account for the different dielectric constants in the two semiconductors.

To draw the band diagram for a heterojunction accurately, we must not only use the proper values for the band discontinuities but also account for the band bending in the junction. To do this, we must solve Poission's equation across the heterojunction, taking into account the details of the doping and space charge, which generally requires a computer solution. We can, however, sketch an approximate diagram without a detailed calculation. Given the experimental band offsets $\Delta E_v$ and $\Delta E_c$, we can proceed as follows:

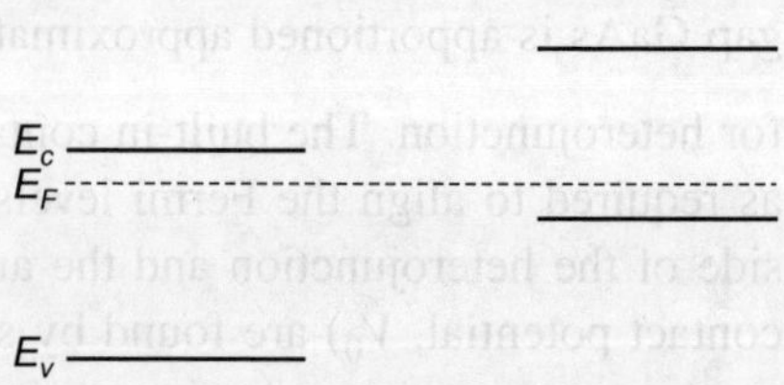

1. Align the Fermi level with the two semiconductor bands separated. Leave space for the transition region.
2. The metallurgical junction ($x = 0$) is located near the more heavily doped side. At $x = 0$ put $\Delta E_v$ and $\Delta E_c$, separated by the appropriate band gaps.
3. Connect the conduction band and the valence band regions, keeping the band gap constant in each material.

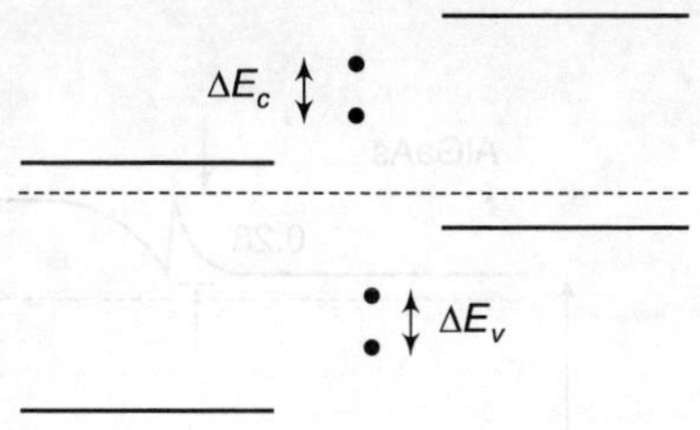


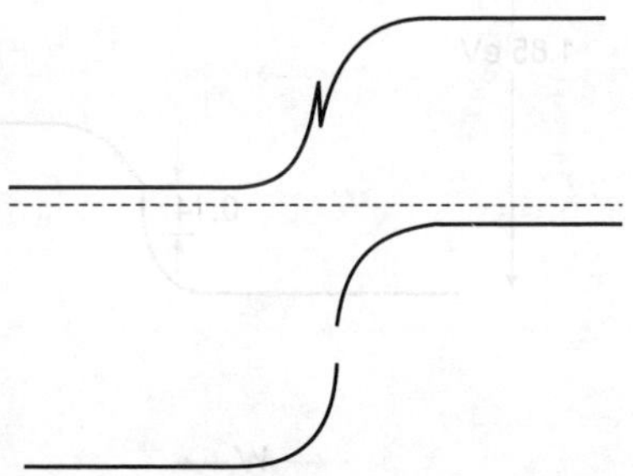

Steps 2 and 3 of procedure are where the exact band bending is important and must be obtained by solving Poisson's equation. In step 2 we must use the band offset values $\Delta E_c$ and $\Delta E_v$ for the specific pair of semiconductors in the heterojunction.

**Example 11.8** For heterojunctions in the GaAs-AlGaAs system, the direct ($\Gamma$) band gap difference $\Delta E_g^{\Gamma}$ is accommodated approximately $\frac{2}{3}$ in the conduction band and $\frac{1}{3}$ in the valence band. For an Al composition of 0.3, the AlGaAs is direct with $E_g^{\Gamma} = 1.85$ eV. Sketch the band diagrams for two heterojunction cases: $N^+$-$Al_{0.3}Ga_{0.7}As$ on *n*-type GaAs, and $N^+$-$Al_{0.3}Ga_{0.7}As$ on $p^+$-GaAs.[8]

***Solution*** Taking $\Delta E_g = 1.85 - 1.43 = 0.42$ eV, the band offsets are $\Delta E_c = 0.28$ eV and $\Delta E_v = 0.14$ eV. In each case we draw the equilibrium Fermi level, add the appropriate bands far from the junction, add the band offsets while estimating the relative amounts of band bending and position of $x = 0$ for the particular doping on the two sides, and finally sketch the band edges so that $E_g$ is maintained in each separate semiconductor right up to the heterojunction at $x = 0$.

A particularly important example of a hetrojunction is shown in Fig. 11.168 in which heavily *n*-type AlGaAs is grown on lightly doped GaAs. In this example the discontinuity in the conduction band allows electrons to spill over from the $N^+$-AlGaAs into the GaAs, where they become trapped in the potential well. As a result, electrons collect on the GaAs side of the hetrojunction and move the Fermi level above the conduction band in the GaAs near the interface. These electrons are confined in a narrow potential well in the GaAs conduction band. If we construct a device in which conduction occurs parallel to the interface, the electrons in such a potential well form a *two-dimensional electron gas* with very interesting device properties. Electron conduction in such a potential well can result in very high mobility electrons. This high mobility is due to the fact that the electrons in this well come from the AlGaAs, and not from the doping in the GaAs. As a result, there is negligible impurity scattering in the GaAs well, and the mobility is controlled almost entirely by lattice scattering (phonons). At low temperatures, where phonon scattering is low, the mobility in this region

[8] In discussing heterojunctions, it is common to use capital *N* or *P* designate the wide band gap material.

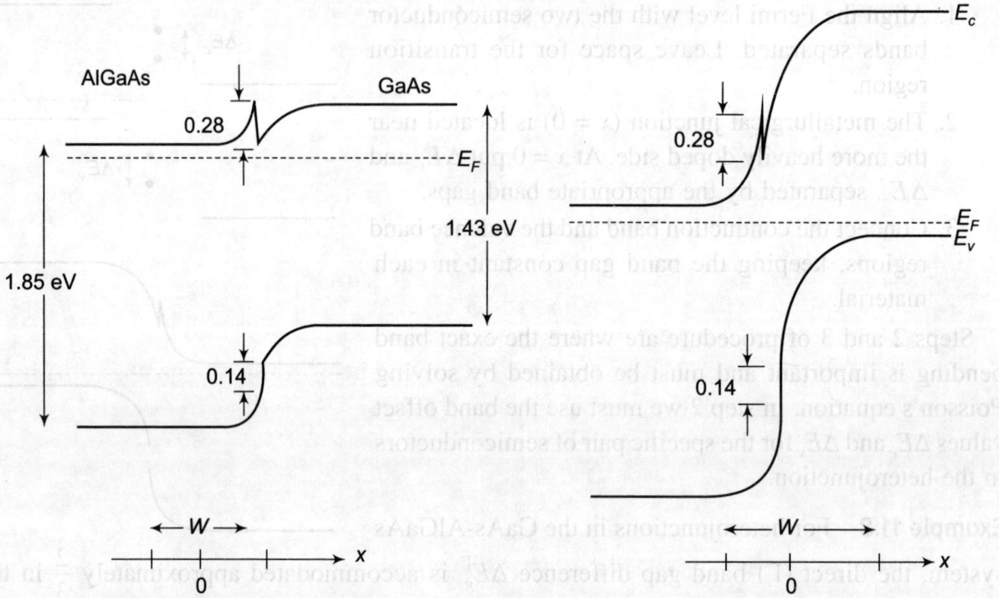


can be very high. If the band-bending in the GaAs conduction band is strong enough, the potential well may be extremely narrow, so that discrete states such as $E_1$ and $E_2$ in Fig. 11.168 are formed.

Another obvious feature of Fig. 11.168 is that the concept of a contact potential barrier $qV_0$ for both electrons and holes in a homojunction is no longer valid for the heterojunction. In Fig. 11.168 the barrier for the electrons $qV_n$ is smaller than the barrier for holes $qV_p$. This property of a heterojunction can be used to alter the relative injection of electrons and holes.

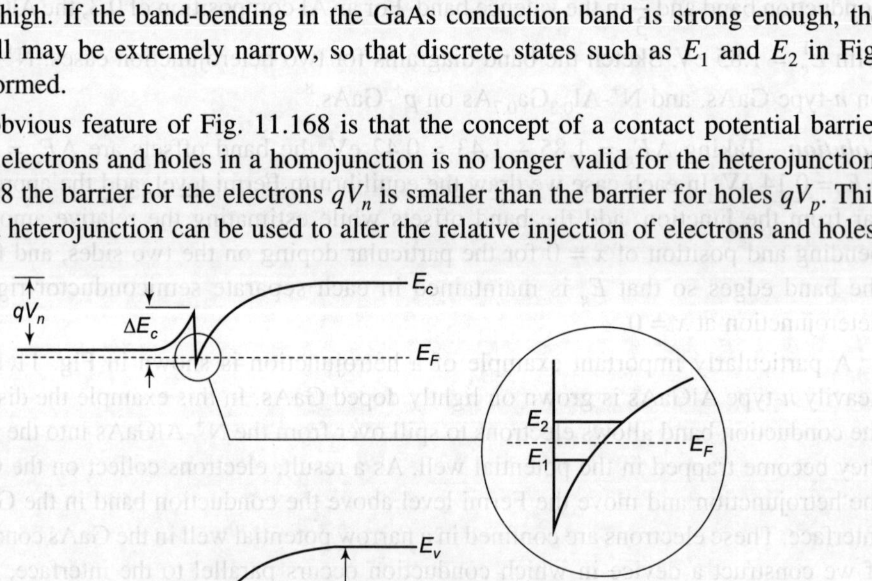


**Figure 11.168** A Heterojunction between $N^+$-AlGaAs and lightly doped GaAs, illustrating the potential well for electrons formed in the GaAs conduction band. If this well is sufficiently thin, discrete states (such as $E_1$ and $E_2$) are formed

## 11.71 CYCLOTRON RESONANCE EXPERIMENTS FOR SEMICONDUCTORS

We know that cyclotron resonance provides direct means of determining the effective mass of current carriers in semiconductors and since

$$\frac{1}{m_i g} = \frac{1}{\hbar^2}\left(\frac{\partial^2 E}{\partial k_i \, \partial k_j}\right)$$

therefore it enables us to determine *E-k* relation and shape of the constant energy surfaces. In the experiment the free carriers are caused to rotate in helical orbits about the axis of a static magnetic field. These orbits are just the orbits imposed by Lorentz force and have nothing to do with the atomic orbits. The frequency of rotation (known as the cyclotron frequency) is

$$\left(\frac{2\pi}{t}\right)\omega_c = \frac{eH}{m^* c} \quad (e \text{ is in esu and } H \text{ in oersteds}) \qquad ...(1)$$

(*t* is the time taken by a carrier to travel once along the orbit)

Further, a circularly polarized *rf* electric field is applied in the plane of the orbits. The force from this field (being parallel to the direction of the carriers) accelerates them to higher energies and enlarges the orbits, and this exactly occurs in a cyclotron. For resonance, frequency of the electric field

$$\omega = \omega_c = \frac{eH}{m^* c} \quad \text{(cyclotron resonance)} \qquad ...(2)$$

Evidently, knowing $\omega$ and $H$, $m^*$ can be determined. There is a problem however, the carriers constantly suffer collisions as they move about. And if they collide too often, during a period, they will be unable to absorb much energy from the field. That is, sharpness of resonance depends on the energy broadening or

$$\Delta E = \frac{\hbar}{\tau} \quad \text{(from uncertainty principle)} \qquad ...(3)$$

where $\tau$ is the relaxation time. Therefore, if

$$\frac{\hbar}{\tau} > \hbar\omega \qquad ...(4)$$

no, resonance features will be observed or, for getting observable resonance absorption, the basic condition is

$$\omega\tau \gg 1 \qquad ...(5)$$

*Analysis.* Let $v_0$ = drift velocity of carriers then,

$$m^*\left(\frac{dv_0}{dt} + \frac{v_0}{\tau}\right) = e\left(\vec{\varepsilon} + \frac{\vec{v}_0 \times \vec{H}}{c}\right) \qquad ...(6)$$

where, $\vec{\varepsilon}$ is the electric field

$$\vec{\varepsilon} = \vec{\varepsilon}_x = \vec{\varepsilon}_0 \, e^{iwt} \qquad ...(7)$$

along the $x$-axis and

$$\vec{H} = \vec{H}_z \qquad ...(8)$$

along the $z$-axis.

For simplicity, it is assumed that $m^*$ and $\tau$ are independent of velocity and direction.

For $x$- and $y$-components of Eq. (6), we have

$$m^*\left(i\omega + \frac{1}{\tau}\right)v_x = e\varepsilon_x + \frac{e}{c}v_y H \qquad ...(9)$$

$$m^*\left(i\omega + \frac{1}{\tau}\right)v_y = -\frac{ev_x H}{c} \qquad ...(10)$$

Letting
$$1 + i\omega\tau = A \qquad ...(11)$$

we have
$$v_x = \frac{e\varepsilon_x\tau + (e/c)H\tau v_y}{Am^*} \qquad ...(12)$$

$$v_y = \frac{-(e/c)\,H\tau v_x}{Am^*} \qquad ...(13)$$

Now substituting $v_y$ into $v_x$, we get

$$v_x = \frac{e\varepsilon_x\tau\, Am^*}{A^2 m^{*2} + (e^2/c^2)\,H^2\tau^2} = \frac{e\varepsilon_x\tau(A/m^*)}{A^2 + (e^2H^2/m^{*2}c^2)\tau^2} \qquad ...(14)$$

Since $\frac{eH}{m^{*2}} = \omega_c$, we have

$$v_x = \frac{e\varepsilon_x\tau(A/m^*)}{A^2 + \omega_c^2\,\tau^2} = \frac{(e\varepsilon_x\tau/m^*)\,(1 + i\omega\tau)}{-\omega^2\tau^2 + 1 + 2\,i\omega\tau + \omega_c^2\,\tau^2} \qquad ...(15)$$

$$= \frac{(e\varepsilon_x\tau/m^*)\,(1 + i\omega\tau)}{1 + (\omega_c^2 - \omega^2)\,\tau^2 + 2\,i\omega\tau} \qquad ...(16)$$

Thus, the conductivity at angular frequency $\omega$ in the $x$-direction is

$$\sigma = \frac{nev_x}{\varepsilon_x} = \left(\frac{ne^2\tau}{m^*}\right)(1 + i\omega\tau)/[1 + (\omega_c^2 - \omega^2)\tau^2 + 2i\omega\tau] \qquad ...(17)$$

Letting $\frac{ne^2t}{m^*} \equiv \sigma_0$, which is ordinary (or dc) conductivity of the solid at zero frequency in the absence of the magnetic field with $n$ as the number of current carriers per unit volume, we have

$$\sigma = \frac{\sigma_0(1 + i\omega\tau)}{1 + (\omega_c^2 - \omega^2)\,\tau^2 + 2\,i\omega\tau} \qquad ...(18)$$

This equation shows that the high frequency conductivity is complex. On solving for the real part $\sigma_R$ of the conductivity, we find

$$\frac{\sigma_R}{\sigma_0} = \frac{1 + \omega^2\tau^2 + \omega_c^2\tau^2}{(1 + \omega_c^2\tau^2 - \omega^2\tau^2)^2 + 4\omega^2\tau^2} \quad ...(19)$$

A cyclotron resonance experiment is designed to detect a change in the power absorbed from an oscillating electric field, as the magnetic field strength is varied. Then, remembering that the power absorption is proportional to $\sigma_R$, and since it is usual to work at a fixed frequency, $\omega$, and measure the power absorption as $H$ is varied, it is useful to plot the quantity $(\sigma_R/\sigma_0)$ as a function of $(\omega_c/\omega)$ for various values of the parameter, $\omega\tau$. This is shown in Fig. 11.169. This shows that the when $\omega\tau \ll 1$, there is no significant change in the absorption for different frequencies $\omega_c$ and a little change occurs as $\omega\tau \to 1$. However, when $\omega\tau > 1$, a distinct resonance effect is observed with maximum power absorption at a point closer to the cyclotron resonance frequency.

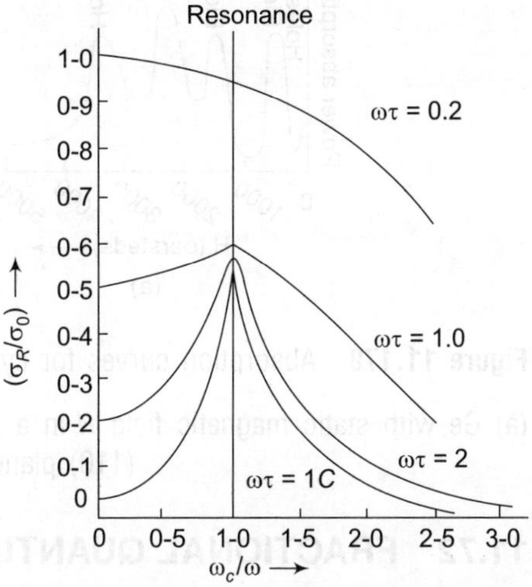


**Figure 11.169** Theoretical cyclotron resonance curves showing the development of resonance as $\omega\tau > 1$

There is one thing missing in our above analysis: the sign of $\omega_c$, which distinguishes cyclotron resonance of electrons from that of holes. The reason is that the absorption (or $\sigma_R/\sigma_0$) is proportional to $\omega_c^2$. If instead of linearly polarized electric field, the applied fields were circularly polarised (as is usually the case), we would obtain

$$\frac{\sigma_R}{\sigma_0} = \frac{1}{1 + (\omega + \omega_c)^2\tau^2} \quad ...(20)$$

in place of Eq. (19). This is easily seen by writing the equations of motion for $\varepsilon$ in the $xy$-plane as

$$m^*\left(i\omega + \frac{1}{\tau}\right)v_x = e\varepsilon_x + \frac{e}{c}v_y H \quad ...(21a)$$

$$m^*\left(i\omega + \frac{1}{\tau}\right)v_y = e\varepsilon_y - \frac{ev_x H}{c} \quad ...(21b)$$

and proceeding, as in the above analysis. We note that Eq. (20) is indeed sensitive to the sign of $\omega_c$.

The above theory predicts only two resonances: one for electrons and one for holes. However, cyclotron resonance effects studied for Si and Ge reveal that for both substances, there are more than two resonances (Fig. 11.170). These have been explained on the basis of the model in which the constant energy surfaces for electrons in $k$-space are far from being spherical. (i.e., $m^*$ not independent of direction of motion) and in which the presence of three separate valence bands (two of which are degenerate at $k = 0$) are considered.

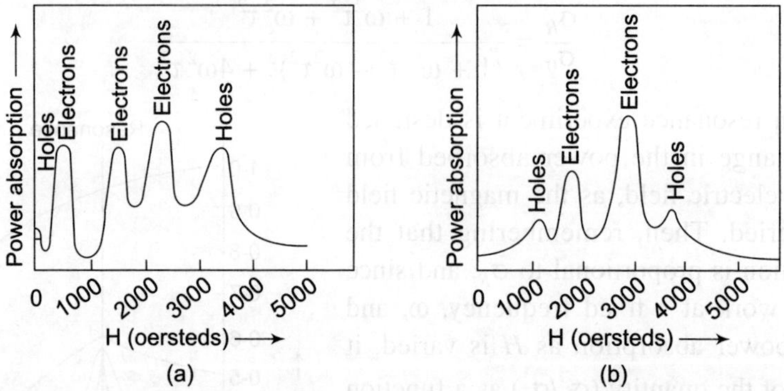


**Figure 11.170** Absorption curves for cyclotron resonance at $\frac{\omega}{2\pi}$ = 2.4 × $10^4$ MHz and $T$ = 4°K (a) Ge with static magnetic field $H$ in a (110) plane at 60° from [100] axis (b) Si with H in a (110) plane at 30° from [100] Axis

## 11.72 FRACTIONAL QUANTUM HALL EFFECT

*Hall effect.* In Hall effect, a planar sample of conductive material is placed in a magnetic field $\vec{B}$ perpendicular to its surface, An electric current $I$ is made to pass from one end to other ($\perp$ to $\vec{B}$) by means of a potential, $V_L$. The longitudinal resistivity is the ratio $R_L = V_L/I$. Because of the Lorentz force ($e\vec{v} \times \vec{B}$) where $\vec{v}$ is velocity of electrons, more electrons accumulate on one side of the sample than the other. Thus, a measurable voltage $V_H$ develops across the sample (perpendicular to both $\vec{v}$ and $\vec{B}$), called Hall voltage.

The ratio $\rho_H = V_H/I$ is called *Hall resistivity* $\left(\rho_H = \frac{h}{ne^2}\right)$.

*Integral quantum effect.* It was observed by Klaus von Klitzing (1980) that for samples cooled to within 1°K and placed in strong magnetic fields, the Hall resistivity exhibits a series of plateaus, i.e., intervals in which the Hall resistivity appears not to vary at all with the magnetic field. (Fig. 11.171).

**Figure 11.171** The integral quantum Hall effect appears as plateaus in the Hall resistivity of a sample which coincide with the disappearance of the sample's electrical resistivity, as the magnetic field strength is varied

*Fractional quantum Hall effect.* In 1982, Daniel Tsui, H. Strömer and A. Gossard discovered other plateaus at which $n$ has specific fractional values (1/3, 2/5 and 3/7). This is the fractional quantum Hall effect.

The fractional quantum Hall effect is seen only when a Landau level is partially filled.

For instance, a plateau is seen when the lowest Landau level is approximately 1/3 full. In this case the Hall resistivity is equal to 1/3 of the square of the electron charge ($e^2$) divided by the Planck's constant.

To explain the fractional quantum Hall effect, we must consider the interaction between electrons. It is helpful to write the total wave function of particles moving in a Landau level as a Slater determinant

$$\psi = \left(\frac{\sqrt{2}}{x_c}\right)^{M_l+1} \frac{1}{\sqrt{(M_l+1)!}}\left(\prod_{n=0}^{n=M_l}\frac{1}{\sqrt{n!}}\right)\Phi(1,2,...,M_1+1) \quad ...(1)$$

where
$$\Phi = \exp\left(-\frac{1}{2}\sum_{i=1}^{M_l+1}|z_i|^2\right)\begin{vmatrix} 1 & 1 & -- & z_{M_l+1} \\ z_1 & z_2 & & z_{M_l+1} \\ z_1^2 & z_2^2 & -- & z_{M_l+1}^2 \\ | & | & | & | \\ z_1^{M_l} & z_2^{M_l} & -- & z_{M_l+1}^{M_l} \end{vmatrix} \quad ...(2)$$

or
$$\Phi = \exp\left(-\frac{1}{2}\sum_{i=1}^{M_l+1}|z_i|^2\right)\prod_{i>j=0}^{i>j=M_l-1}(z_i - z_j) \quad ...(3)$$

$$z \equiv u\exp(-i\,\phi),\ \phi = \tan^{-1}(y/x),\ u = r/x_c$$

$$x_c = \sqrt{\hbar/(m\omega)},\ r = \sqrt{x^2+y^2} \quad ...(4)$$

for planar circular motion in $x$-$y$ plane. $|M_l|$ is the maximum angular momentum for the first Landau level. In 1983, Robert Laughlin modified the wave–function, $\Phi$ as

$$\Phi_\upsilon = \exp\left(-\frac{1}{2}\sum_{i=1}^{M_l+1}|z_i|^2\right)\prod_{i>j=0}^{i>j=M_l-1}(z_i - z_j)^\upsilon$$

These states are exact ground states in the limit when electron – electron repulsions become infinitely short-ranged. The exponent $\upsilon$ measures the fraction of filled states – the wave functions have the required stability when $\upsilon$ equals 1/3, 1/5, 1/7, 2/3, 4/5, 6/7. (The denominator in the fraction $\upsilon$ must be an odd number in order to satisfy the Pauli principle). The fractional quantum Hall effect is explained by the concept of *anyone*, which are intermediate between bosons and fermions.

## 11.73 SEMICONDUCTOR QUANTUM WELLS, QUANTUM WIRES AND QUANTUM DOTS

In a QW semiconductor, a very thin layer ($L_z \simeq 5 \div 20$ nm) of a smaller band gap material $Eg_1$ is sandwiched between two layers of a large band gap material, $Eg_2$. Technically, this involves sophisticated techniques like molecular beam epitaxy (MBE) or metallo-organic chemical vapour deposition (MOCVD). Since $Eg_1 < Eg_2$, potential wells are established for electrons at the top of the valence band and for the holes, at the bottom of the conduction band (Fig. 9.100). Due to the electron and hole confinement in these potential wells, and since the semiconductor dimension is now comparable to the electron and hole De Broglie wavelength, energy levels of electrons and holes show very marked quantum – size effects. Further, due to small thickness of the layer, we can allow lattice constants for the two materials to differ significantly, as a result strain develops within the thin quantum layer. The strain changes the quantum properties of the QW semiconductor considerably and in particular, it changes the effective masses. The quantum size effects (and for a strained QW), the change in effective masses causes optical properties of the semiconductors' QW to differ markedly from those of the corresponding bulk material. The semiconductor QWs (either strained or unstrained type) have become most widely used semiconductor – laser materials.

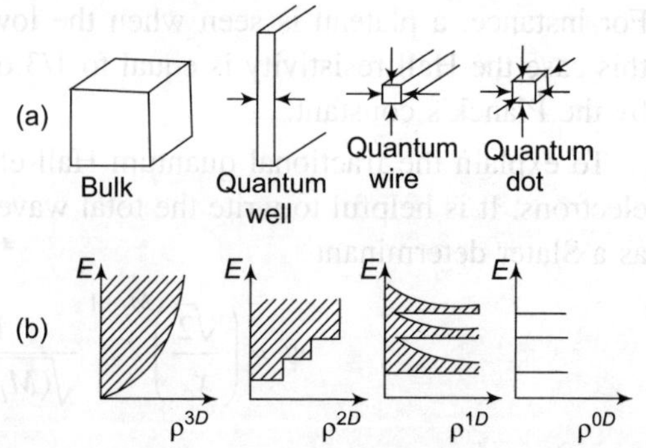


**Figure 11.172** (a) Different configurations and (b) and (b) corresponding forms of the density of states for bulk, quantum well, quantum wire and quantum dot semiconductors

According to the foregoing discussion, the improvement in optical properties obtained on going from bulk materials to the corresponding QW material is essentially due to a quantum confinement effect arising from the fact one dimension of the semiconductor has become comparable to the De Broglie wavelength. This idea can be extended to the other two possible cases of quantum confinement, viz., *quantum wires and quantum dots*, where two or all three dimensions become comparable to the De Broglie wavelength (Fig. 11.172).

As for a QW, the fundamental difference between these quantum confined structures and bulk materials arises from the different forms of the density of states. Fig. 11.172 (b) shows the qualitative behaviour of the density of states for quantum wires and quantum dots as compared to bulk and QW materials. Using these different forms of density of states (DOS), one can calculate the expected laser gain. However, we do not pursue this here and limit ourselves to showing in Fig. 11.173 (as

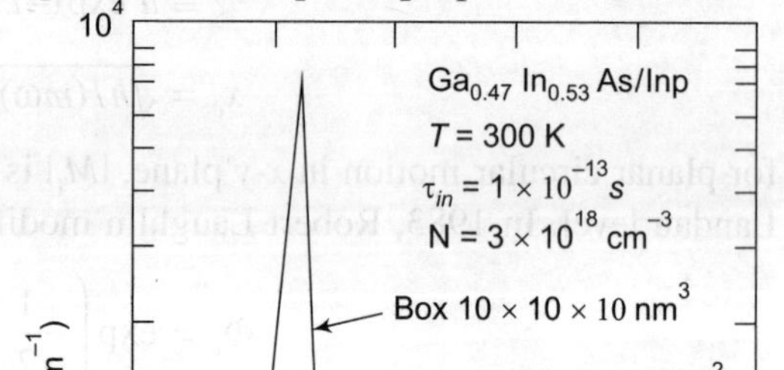


**Figure 11.173** Plot of calculated gain coefficient vs emission wavelength at $N = 3 \times 10^{18}$ $cm^{-3}$ electron injection for a $Ga_{0.47}In_{0.51}As$ [After *Principal of Lasers* by O. Svello, Springer, Inc. (1998)]

a representative example) the expected material gain vs photon energy for a $Ga_{0.47}In_{0.51}As/InP$ system. (Here, $Ga_{0.47}\ In_{0.51}$. As constitutes the quantum confinement material). In the figure, curves of predicted material gain vs. photon energy in the bulk case for a 10 nm quantum well, 10 nm × 10 nm quantum wire and 10 nm × 10 nm × 10 nm quantum dot are plotted for the same electron injection ($N = 3 \times 10^{18} cm^{-3}$). The peak gain increases on going from bulk to QW, from QW to quantum wire and from quantum wire to quantum dot. The gain bandwidth on the other hand decreases from QW to quantum wire to quantum dot.

As a laser media quantum wires and dot will perhaps be used in the form of an array, such as the planar array in Fig. 11.174. These are considerable technological difficulties due to mainly to the need of high packing density, low size fluctuation and low defect density.

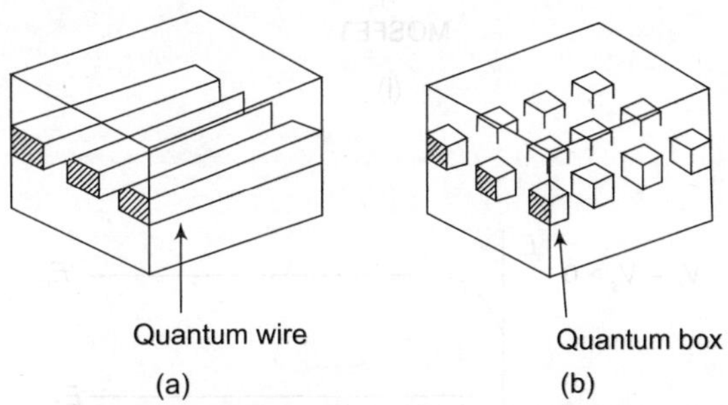


**Figure 11.174** Planar array of (a) Multiple quantum wires and (b) Multiple quantum dots

## 11.74 APPLICATION OF THE MOSFET: THE CHARGE–COUPLED DEVICES

We know that a MOSFET is a unipolar device where the resistance of semiconducting channel is controlled electrostatically. A simplified structure is as shown in Fig. 11.175 (i) when a voltage is applied across such a system, a charge appears on the two ends, i.e., we have a sort of capacitor whose one plate is metallic and the other, a semiconductor. The carriers distribute themselves so as to create a total charge opposed to that at the metal surface. As a result, there appears a space charge near the semiconductor surface. Depending on the sign and size of the charge, we may have accumulation or depletion of free carriers, or even inversion (i.e., the appearance of a layer of minority carriers at the surface). In this system, the presence of the insulator prevents charge transport perpendicular to the surface, thus there is no current in the semiconductor, and the semiconductor carriers are in thermodynamic equilibrium.

When the metal is negative with respect to the semiconductor ($V_m < V_s$), the negative potential of the metal attracts holes. The energy levels of the semiconductor increase near to the insulator. (Fig. 11.175 (ii) (a) shows the band scheme for this case for *p*-type semiconductor).

The Fermi level is constant since we are in equilibrium and is closer to the valence band at the oxide-semiconductor interface and there is accumulation of hole concentration at the interface.

When $V_m > V_s$ the holes are repelled and the band scheme is shown in Fig. 11.175 (ii) (b). There is depletion of mobile carriers and there remain only ionized acceptors near the interface with concentration, $N_a$.

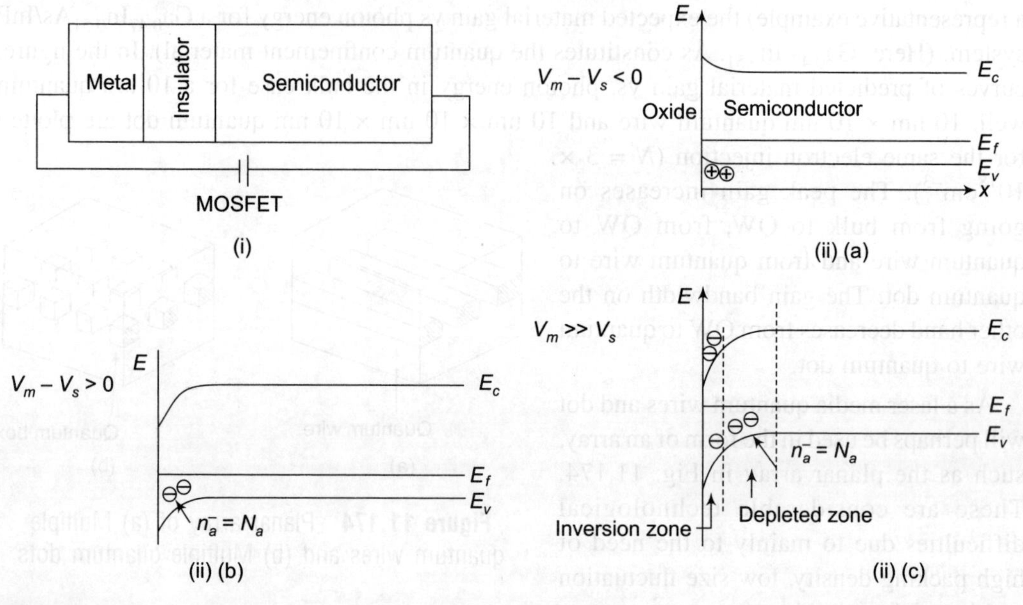


**Figure 11.175** (i) MOSFET, (ii) Energy levels in the semiconductors of a MOSFET as a function of distance from the insulator. (a) Negative voltage applied to the insulator (b) Small positive voltage (c) Large positive voltage

Above a certain threshold, for $V_m \gg V_s$ we obtain the band scheme shown in Fig. 11.175 (ii) (c). There are three zones at the interface, the Fermi level is close to the conduction band, and electrons appear. This is called an inversion layer. The electron number in this inversion layer depends directly on the voltage applied to the metal. Deeper in the semiconductor, we encounter a depletion region where the charges are only ionized acceptors, before reaching the neutral semiconductor.

### *The Charge Coupled Device (CCD)*

We have seen above the MOSFET made of a *p*-type semiconductor, we can create a potential well for the electrons in the inversion layer if we apply a strong positive voltage to the metal.

A CCD structure involves a *p*-type semiconductor covered by an insulating layer and a series of very close metal gates. When we apply to the system (Fig. 11.176 (a)) positive potentials $V_1 = V_3 = V_4$ and $V_2 > V_1$, a potential well for the electrons appears under electrode 2.

Now, assume that some electrons are introduced under electrode 2 where they are stored. As there are no holes in the inversion layer, there is no recombination. If, now the potentials are changed so that $V_1 = V_4$, $V_1 < V_2 < V_3$, the charges move towards the deeper potential well, ending up under electrode 3. (Fig. 11.176 (b)).

By applying an appropriate sequence of voltage on the various gates. We can transfer the charges from one region of surface to another and scan it point by point and line by line. We can identify which charges come from which region by analyzing the resulting sequence of electrical signals.

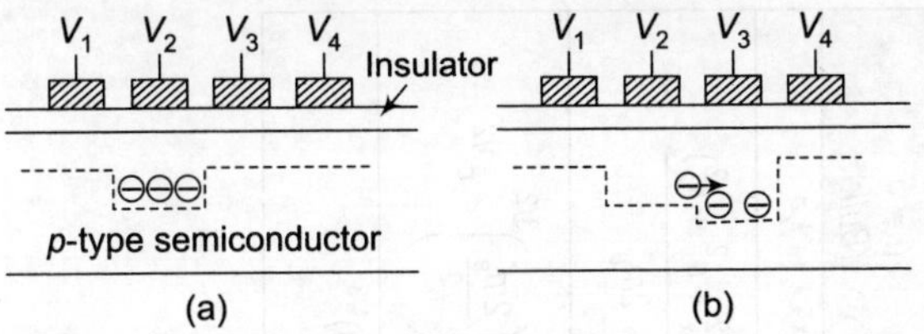


**Figure 11.176** Schematic of a charge coupled device. The metallic gates are shows hatched. The dashed line shows the potential profile of electrons in Fig. (a) $V_2 > V_1 = V_3 = V_4$ and potential well is under electrode 2. In Fig. (b), $V_3 > V_2 > V_1$ (= $V_4$) and the electrons transfer from electrode 2 to 3

The minority carriers (electron) can be produced by luminous excitation. Each surface element (pixel) provides a charge and consequently, we have the conversion of an image consisting of a set of light and dark points into an electro-optical signal. This provides an electrical signal whose amplitude varies with the light intensity at various points of the original image.

Charge-coupled devices are very sensitive light detectors. They have now replaced Vidicon tubes in cameras and TV and they are used in spectroscopy and astronomical detection.

## 11.75 DIMENSIONALITY AND QUANTUM CONFINEMENT

The size and effective dimensionality of a solid can have important effects on its physical properties. For example, in a semiconductor, the electronic wave functions, energy levels, and densities of states are all dependent on whether the semiconductor is a bulk, macroscopic solid (i.e., 3-dimensional) or a microscopic solid (i.e., 2-, 1- or 0-dimensional). Here, the terms two-, one-, and zero-dimensional refer respectively, to structures in which one, two, or all three physical dimensions are comparable to some characteristic length $L$, with the remaining dimensions much greater than $L$. (see Fig. 11.177).

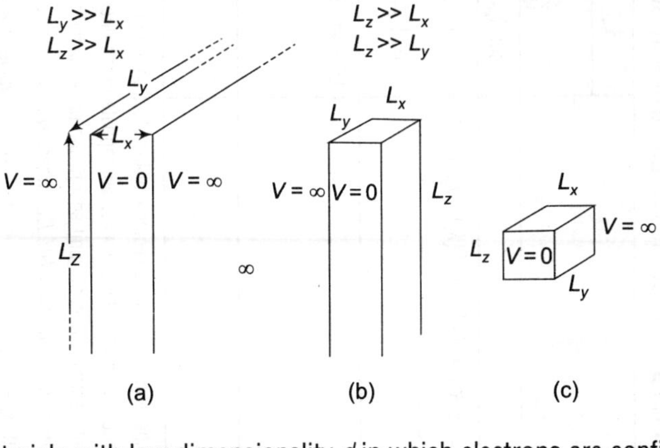


**Figure 11.177** Materials with low dimensionality $d$ in which electrons are confined: (a) A quantum well ($d$ = 2); (b) A quantum wire ($d$ = 1); (c) A quantum dot ($d$ = 0). The boundaries of these regions are infinite potential barriers

**Table 11.15** Properties of Electrons in Solids of Reduced Dimensionality

| | Dimensionality | | | |
|---|---|---|---|---|
| | $d = 0$ (Quantum dot) | $d = 1$ (Quantum wire) | $d = 2$ (Quantum wire) | $d = 3$ (Bulk) |
| $\psi_k(\mathbf{r})^a$ | $A \sin k_x x (\sin k_y y)(\sin k_z z)$ | $A \sin k_x x (\sin k_y y)\, e^{ik_z z}$ | $A(\sin k_x x)\, e^{i(k_y y + k_z z)}$ | $Ae^{i(k_x x + k_y y + k_z z)}$ |
| $E(\mathbf{k})^b = E(k_x) + E(k_y) + E(k_z)$; $n_2, n_y, n_2 = 1, 2, 3. \ldots$ | $\frac{h^2}{8m_e^*}\left(\frac{n_x^2}{L_x^2}+\frac{n_y^2}{L_y^2}\right)+\frac{n_z^2}{L_z^2}$ | $\frac{h^2}{8m_e^*}\left(\frac{n_x^2}{L_x^2}+\frac{n_y^2}{L_y^2}\right)+\frac{\hbar^2 k_z^2}{2m_e^*}$ | $\frac{h^2 n_x^2}{8m_e^* L_x^2}+\frac{\hbar^2(k_y^2+k_z^2)}{2m_e^*}$ | $\frac{\hbar^2(k_x^2+k_y^2+k_z^2)}{2m_e^*}$ |
| $\rho_e(E)^{b,c}$ | Discrete states | $\frac{\sqrt{2m_e^*}}{\pi\hbar L_x L_y} E^{-1/2}$ | $\frac{m_e^*}{\pi\hbar^2 L_x}$ | $\frac{1}{2\pi^2}\left(\frac{2m_e^*}{\hbar^2}\right)^{3/2} E^{1/2}$ |
| $k_F{}^d$ | $-^e$ | $\frac{\pi n L_x L_y}{2}$ | $\sqrt{2\pi n L_x}$ | $(3\pi^2 n)^{1/3}$ |

[a]The components of the wave vector $k$ of the electron are given by $k_x = n_x \pi/L_x$, $n_x = 1, 2, \ldots$.

[b]The electron (or hole) effective mass appropriate to the direction or the plane of motion should be used in $E(\mathbf{k})$ and $\rho_e(E)$.

[c]Density of electron states per unit energy and unit volume.

[d]$n$ = Electron concentration = $N/L_xL_yL_z$, where $N$ is the number of electrons confined in the region. Note that $n_{3d} = n$, $n_{2d} = nL_x$ and $n_{1d} = nL_xL_y$.

[e]For $d = 0$, $k_F = k_{max}$, where $k_{max}$ is the maximum value of $k = \pi(n_x^2 + n_y^2/n_y^2 + n_z^2/n_z^2)^{\frac{1}{2}}$ for any electron in the quantum dot.

Note that when an electron is spatially confined within a region of width $L_x$, its ground state energy corresponding to $n_x = 1$ is increased above the unconfined value by an amount equal to $h^2/8m_e^*$ due to the quantum confinement. The amount of this increase grows rapidly as $L_x$ decreases, increasing from $1.5 \times 10^{-12}$ eV to $1.5 \times 10^{-6}$ eV as $L_x$ decreases from 1 nm to 1 μm, and then to 1 nm. This increase in energy is a result of *Heisenberg uncertainty principle*, which states that a particle confined within a region of size $L_x$ will have a resulting minimum uncertainty in its momentum given by $\Delta p_x \approx \hbar/L_x$. The corresponding minimum in its uncertainty in its energy, the *zero-point energy*, will therefore be $\Delta E \approx (\Delta p_x)^2/2m_e^*$, which corresponds to the ground-state energy given above. This energy enhancement can also be expected to occur for the binding energies of donors and acceptors atom in low-dimensional doped semiconductors.

## EXERCISE

11.1 What is an intrinsic semiconductor? Obtain an expression for the electrical conductivity of an intrinsic semiconductor. Also derive the law of mass action for semiconductors.

11.2 What is an extrinsic semiconductor? What are impurity states? What role do they play in determining the electrical conductivity of a doped semiconductor?

11.3 Discuss qualitatively the variation of the Fermi level and carrier concentration in an $n$-type semiconductor with temperature.

11.4 Locate $E_F$ at room temperature in a Ge crystal containing $10^{23}$ donors/m$^3$. Assume $E_g = 0.72$ eV, and the ionization energy of a donor = 0.012 eV. [*Ans.* 0.52 eV]

11.5 Find the expression for the Hall coefficient of a semiconductor in which both electrons and holes are present in appreciable concentrations.

11.6 Discuss the phenomenon of cyclotron resonance in semiconductors and explain its applications.

11.7 What is a unijunction transistor? Give its equivalent circuit. Describe its static characteristics and explain its working.

11.8 Assume concentration $n$, $p$, mobilties $\mu_e$ and $\mu_n$, and masses $m_e$ and $m_{nn}$ that the Hall coefficient in the drift velocity approximation is

$$R_H = \frac{1}{e}\frac{p - nb^2}{(p + nb)^2}$$

where $b = \mu_e/\mu_n$ is the mobility ratio.

11.9 Describe the construction of a $N$-channel junction field effect transistors (JFET) and explain the principle of its operation.

11.10 What is a tunnel diode? Discuss its operation and explain its characteristics.

11.11 Describe and discuss quantum Hall effect.

11.12 What is a photoconductor? Explain its working.

11.13 What is a photodiode? Derive an expression for current-voltage behaviour in (a) photo voltaic mode, (b) photoconductive mode.

11.14 Explain the working of phototransistor.

11.15 What is a solar cell? Discuss qualitatively its working principle.

11.16 What is a charge coupled device (CCD)? Explain its working.

11.17 Explain the working of a light emitting diode (LED). What do you understand by its internal quantum efficiency and external quantum efficiency?

CHAPTER

# 12

# Band Theory and Energy – Band Structures

Prior to discussion on the formation of energy bands and various band theories, it would be encouraging to discuss the motion of an electron quantum mechanically in a one-dimensional lattice and a three-dimensional lattice.

## 12.1 AN ELECTRON MOVING IN A ONE-DIMENSIONAL LATTICE

We want to analyze quantitatively the quantum mechanisation. Imagine a one-dimensional system made of a long line of atoms as shown in Fig. 12.1 (a). (A crystal is, of course, threc dimensional but the physics is very much the same: once you understand the one-dimensional case you will be able to understand what happens in three dimensions.) Next, we want to see what happens if we put a single electron on this line of atoms. Of course, in a real crystal there are already millions of electrons. But most of them (nearly all for an insulating crystal) take up positions in some pattern of motion each around its own atom—and everything is quite stationary. However, we now want to think about what happens if we put an *extra* electron in. We will not consider what the other ones are doing because we suppose that to change their motion involves a lot of excitation energy. We are going to add an electron as if to produce one slightly bound negative ion. In watching what the *one* extra electron does we are making an

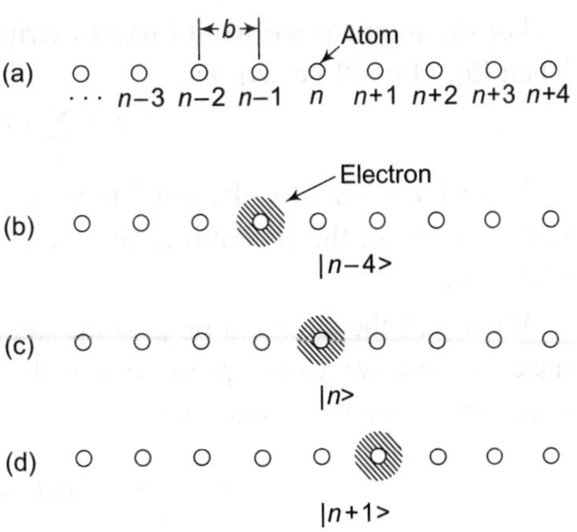


**Figure 12.1** The base status at an electron in a one-dimensional crystal

approximation which disregards the mechanics of the inside working of the inside workings of the atoms.

Of course the electron could then move to another atom, transferring the negative ion to another place. We will suppose that just as in the case of an electron jumping between two protons, the electron can jump from one atom to the neighbour on either side with a certain amplitude.

Now, how do we describe such a system? What will be reasonable base states? If you remember what we did when we had only two possible positions, you can guess how it will go. Suppose that in our line of atoms the spacings are all equal; and that we number the atoms in sequence, as shown in Fig. 12.1 (a). One of the base states is that the electron is at atom number 6, another base state is that the electron is at atom number 7, or at atom number 8, and so on. We can describe the *n*th base state by saying that the electron is at atom number *n*. Let's say that this is the base state $|n\rangle$. Figure 12.1 shows what we mean by the three base states

$$|n-1\rangle, |n\rangle, \quad \text{and} \quad |n+1\rangle.$$

Using these base states, any state $|\phi\rangle$ of our one-dimensional crystal can be described by giving all the amplitudes $\langle n|\phi\rangle$ that the state $|\phi\rangle$ is in one of the base states—which means the amplitude that it is located at one particular atom. Then we can write the state $|\phi\rangle$ as a superposition of the base states

$$|\phi\rangle = \sum_n |n\rangle \langle n|\phi\rangle. \qquad ...(1)$$

Next, we are going to suppose that when the electron is at one atom, there is a certain amplitude that it will leak to the atom on either side. And we'll take the simplest case for which it can only leak to the nearest neighbour—to get to the next-nearest neighbour, it has to go in two steps. We'll take that the amplitudes for the electron jump from one atom to the next is $iA/\hbar$ (per unit time).

For the moment we would like to write the amplitude $\langle n|\phi\rangle$ to be on the *n*th atom as $C_n$. Then Eq. (1) will be written

$$|\phi\rangle = \sum_n |n\rangle C_n. \qquad ...(2)$$

If we knew each of the amplitudes $C_n$ at a given moment, we could take their absolute squares and get the probability that you would find the electron if you looked at atom *n* at that time.

What will the situation be at some later time? By analogy with the two-state systems we have studied, we would propose that the Hamiltonian equations for this system should be made up of equations like this

$$i\hbar \frac{dC_n(t)}{dt} = E_0 C_n(t) - AC_{n+1}(t) - AC_{n-1}(t). \qquad ...(3)$$

The first coefficient on the right, $E_0$, is, physically, the energy the electron would have if it couldn't leak away from one of the atoms. (It doesn't matter what we call $E_0$; as we have seen many times, it represents really nothing but our choice of the zero of energy.) The next term represents the amplitude per unit time that the electron is leaking into the $n$th pit from the $(n + 1)$st pit; and the last term is the amplitude for leakage from the $(n - 1)$st pit. As usual, we'll assume that $A$ is a constant (independent of $t$).

For a full description of the behaviour of any state $|\phi\rangle$, we would have one equation like Eq. (3) for every one of the amplitudes, $C_n$. Since we want to consider a crystal with a very large number of atoms, we'll assume that there are an indefinitely large number of states—that the atoms go on forever in both directions. (To do the finite case, we will have to pay special attention to what happens at the ends.) If the number $N$ of our base states is indefinitely large, then also our full Hamiltonian equations are infinite in number! We'll write down just a sample

$$\left.\begin{array}{l} \vdots \qquad\qquad \vdots \\ i\hbar \dfrac{dC_{n-1}}{dt} = E_0 C_{n-1} - AC_{n-2} - AC_n, \\ i\hbar \dfrac{dC_n}{dt} = E_0 C_n - AC_{n-1} - AC_{n+1}, \\ i\hbar \dfrac{dC_{n+1}}{dt} = E_0 C_{n+1} - AC_n - AC_{n+2}, \\ \vdots \qquad\qquad \vdots \end{array}\right\} \qquad ...(4)$$

## 12.2 STATES OF DEFINITE ENERGY

We could study many things about an electron in a lattice, but first let's try to find the states of definite energy. As we have seen in earlier chapters this means that we have to find a situation in which the amplitudes all change at the same frequency if they change with time at all. We look for solutions of the form

$$C_n = a_n\, e^{-iEt/\hbar}. \qquad ...(5)$$

The complex number $a_n$ tells us about the non-time-varying part of the amplitude to find the electron at the $n$th atom. If we put this trial solution into the equations of Eq. (4) to test them out, we get the result

$$Ea_n = E_0 a_n - Aa_{n+1} - Aa_{n-1}. \qquad ...(6)$$

We have an infinite number of such equations for the infinite number of unknowns $a_n$—which is rather petrifying.

All we have to do is take the determinant ... but wait! Determinants are fine when there are 2, 3, or 4 equations. But if there are a large number—or an infinite number—of equations, the determinants are not very convenient. We'd better just try to solve the equations directly. First, let's label the atoms by their *positions*; we'll say that the atom $n$ is at $x_n$ and the atom $(n + 1)$ is at $x_{n+1}$. If the atomic spacing is $b$—as in Fig. 12.1—we will have that $x_{n+1} = x_n + b$. By choosing our origin at atom zero, we can even have it that $x_n = nb$. We can rewrite Eq. (5) as

$$C_n = a(x_n)e^{-iEt/\hbar}, \quad ...(7)$$

and Eq. (6) would become

$$Ea(x_n) = E_0 a(x_{n+1}) - Aa(x_{n+1}) - Aa(x_{n-1}). \quad ...(8)$$

or, using the fact that $x_{n+1} = x_n + b$, we could also write

$$Ea(x_n) = E_0 a(x_n) - Aa(x_n + b) - Aa(x_n - b). \quad ...(9)$$

This equation is somewhat similar to a differential equation. It tells us that a quantity, $a(x)$, at one point, $(x_n)$, is related to the same physical quantity at some neighbouring points, $(x_n \pm b)$. (A differential equation relates the value of a function at a point to the values at infinitesimally nearby points.) Perhaps the methods we usually use for solving differential equations will also work here; let's try.

Linear differential equations with constant coefficients can always be solved in terms of exponential functions. We can try the same thing here; let's take as a trial solution

$$a(x_n) = e^{ikx_n}. \quad ...(10)$$

Then Eq. (9) becomes

$$Ee^{ikx_n} = E_0 e^{ikx_n} - Ae^{ik(x_n+b)} - Ae^{ik(x_n-b)}. \quad ...(11)$$

We can now divide the common factor $e^{ikx_n}$; we get

$$E = E_0 - Ae^{ikb} - Ae^{-ikb}. \quad ...(12)$$

The last two terms are just equal to ($2A\cos kb$), so

$$E = E_0 - 2A\cos kb. \quad ...(13)$$

We have found that for *any* choice at all for the constant $k$ there is a solution whose energy is given by this equation. There are various possible energies depending on $k$, and each $k$ corresponds to a different solution. There are infinite number of solutions—which is not surprising, since we started out with an infinite number of base states.

Let's see what these solutions mean. For each $k$, the $a$'s are given by Eq. (10). The amplitudes $C_n$ are then given by

$$C_n = e^{ikx_n} e^{-(i/\hbar)Et}, \quad ...(14)$$

where you should remember that the energy $E$ also depends on $k$ as given in Eq. (13). The *space dependence* of the amplitudes is $e^{ikx_n}$. The amplitudes oscillate as we go along from one atom to the next.

We mean that, in space, the amplitude goes as a *complex* oscillation—the *magnitude* is the same at every atom, but the phase at a given time advances by the amount ($ikb$) from one atom to the next. We can visualize what is going on by plotting a vertical line to show just the real part at each atom as we have done in Fig. 12.2. The envelope of these vertical lines (as shown by the broken-line curve) is, of course, a cosine curve. The imaginary part of $C_n$ is also an oscillating function, but is shifted 90° in phase so that the absolute square (which is the sum of the squares of the real and imaginary parts) is the same for all the $C$'s.

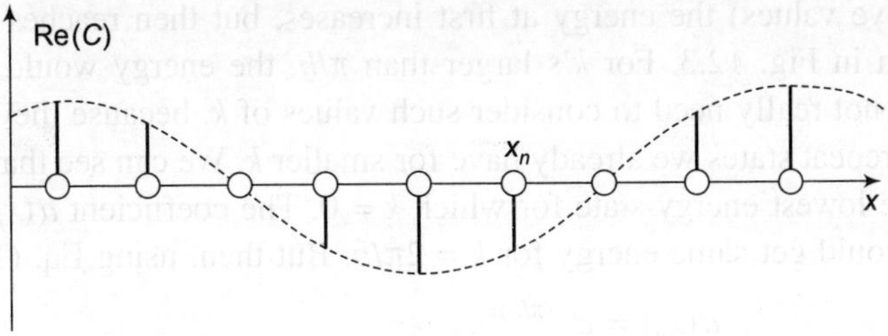


**Figure 12.2** Variation of the real part of $C_n$ with $x_n$

Thus, if we pick a $k$, we get a stationary state of a particular energy $E$. And for any such state, the electron is equally likely to be found at every atom—there is no preference for one atom or the other. Only the phase is different for different atoms. Also, as time goes on the phases vary. From Eq. (14) the real and imaginary parts propagate along the crystal as waves—namely, as the real or imaginary parts of

$$e^{i[kx_n - (E/\hbar)t]}. \quad ...(15)$$

The wave can travel towards positive or negative $x$ depending on the sign we have picked for $k$.

Notice that we have been assuming that the number $k$ that we put in our trial solution, Eq. (10), was a real number. We can see now why that must be so if we have an infinite line of atoms. Suppose that $k$ were an imaginary number, say $ik'$. Then the amplitudes $a_n$ would go as $e^{k'x_n}$, which means that the amplitude would get larger and larger as we go towards large $x$'s-or towards large negative $x$'s if $k'$ is a negative number. This kind of solution would be O.K., if we were dealing with line of atoms that ended, but cannot be a physical solution for an infinite chain of atoms. It would give infinite amplitudes—and, therefore, infinite probabilities—which can't represent a real situation. Later on we will see an example in which an imaginary $k$ does make sense.

The relation between the energy $E$ and the wave number $k$ as given in Eq. (13) is plotted in Fig. 12.3. As you can see from the figure, the energy can go from $(E_0 - 2A)$ at $k = 0$ to $(E_0 + 2A)$ at $k = \pm\pi/b$. The graph is plotted for positive $A$; if $A$ were negative, the curve would simply be inverted, but the range would be the same. The significant result is that any energy is possible within a certain range or "band" of energies, but no others. According to our assumptions, if an electron in a crystal is in a stationary state, it can have no energy other than values in this band.

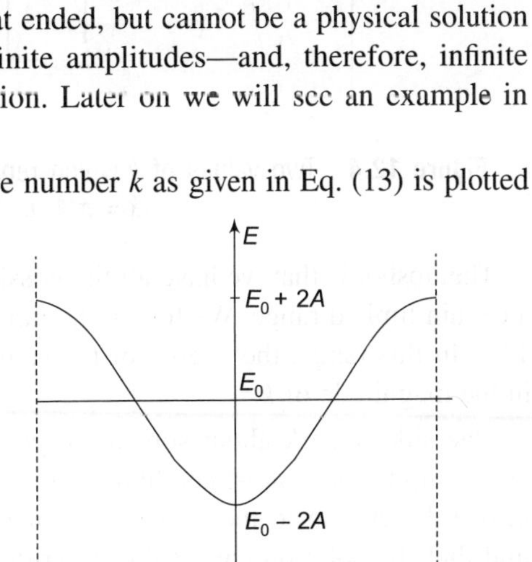


**Figure 12.3** The energy of the stationary states as a function of the parameter $k$.

According to Eq. (10), the smallest $k$'s correspond to low-energy states—$E \approx (E_0 - 2A)$. As $k$ increases in magnitude (towards either positive or negative values) the energy at first increases, but then reaches a maximum at $k = \pm\pi/b$, as shown in Fig. 12.3. For $k$'s larger than $\pi/b$, the energy would start to decrease again. But we do not really need to consider such values of $k$, because they do not give new states—they just repeat states we already have for smaller $k$. We can see that in the following way. Consider the lowest energy state for which $k = 0$. The coefficient $a(x_n)$ is the same for all $x_n$. Now we would get same energy for $k = 2\pi/b$. But then, using Eq. (10), we have that

$$a(x_n) = e^{i(2\pi/b)x_n}.$$

However, taking $x_0$ to be at the origin, we can set $x_n = nb$; then $a(x_n)$ becomes

$$a(x_n) = e^{i2\pi n} = 1.$$

The state described by these $a(x_n)$ is physically the same state we got for $k = 0$. It does not represent a different solution.

As another example, suppose that $k$ were $\pi/4b$. The real part of $a(x_n)$ would vary as shown by curve 1 in Fig. 12.4. If $k$ were seven times larger ($k = 7\pi/4$), the real part of $a(x_n)$ would vary as shown by curve 2 in the figure. (The complete cosine curves don't mean anything, of course; all that matters is their values *at the points* $x_n$. The curves are just to help you see how things are going.) You see that both values of $k$ give the same amplitudes at all of the $x_n$'s.

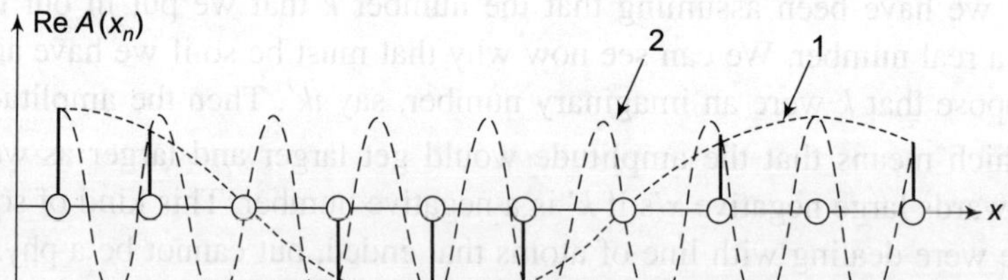


**Figure 12.4** Two values of $k$ which represent the same physical situation; curve 1 is for $k = \pi/4$, curve 2 is for $k = 7\pi/4$

The upshot is that we have all the possible solutions of our problem if we take only $k$'s in a certain limited range. We'll pick the range between $-\pi/b$ and $+\pi/b$—the one shown in Fig. 12.3. In this range, the energy of the stationary states increases uniformly with an increase in the magnitude of $k$.

One side remark about something you can play with. Suppose that the electron cannot only jump to the nearest neighbour with amplitude $iA/\hbar$, but also has the possibility to jump in one direct leap to the *next nearest* neighbour with some other amplitude $iB/h$. You will find that the solution can again be written in the form $a_n = e^{ikx_n}$—this type of solution is universal. You will also find that the stationary states with wave number $k$ have an energy equal to $(E_0 - 2A \cos kb - 2B \cos 2kb)$. This shows that the shape of the curve of $E$ against $k$ is not universal, but depends upon the particular assumptions of the problem. It is not always a cosine wave—it's not even necessarily symmetrical about some horizontal line. It is true, however, that the curve always repeats itself outside of the interval from $-\pi/b$ to $\pi/b$, so you never need to worry about other values of $k$.

Let's look a little more closely at what happens for small $k$—that is, when the variations of the amplitudes from one $x_n$ to the next are quite slow. Suppose we choose our zero of energy by defining $E_0 = 2A$; then the minimum of the curve in Fig. 13.3 is at the zero of energy. For small enough $k$, we can write that

$$\cos kb \simeq 1 - k^2b^2/2,$$

and the energy of Eq. (13) becomes

$$E = Ak^2b^2. \quad \text{...(16)}$$

We have that the energy of the state is proportional to the square of the wave number which describes the spatial variations of the amplitudes, $C_n$.

## 12.3 TIME-DEPENDENT STATES

In this section we would like to discuss the behaviour of states in the one-dimensional lattice in more detail. If the amplitude for an electron to be at $x_n$ is $C_n$, the probability of finding it there is $|C_n|^2$. For the *stationary* states described by Eq. (12), this probability is the same for all $x_n$ and does not change with time. How can we represent a situation which we would describe roughly by saying an electron of a certain energy is localized in a certain region—so that it is more likely to be found at one place than at some other place? We can do that by making a superposition of several solutions like Eq. (12) with slightly different values of $k$—and, therefore, slightly different energies. Then at $t = 0$, at least, the amplitude $C_n$ will vary with position because of the interference between the various terms, just as one gets beats when there is a mixture of waves of different wavelengths. So we can make up a "wave packet" with a predominant wave number $k_0$, but with various other wave numbers near $k_0$.[1]

In our superposition of stationary states, the amplitudes with different $k$'s will represent states of slightly different energies, and, therefore, of slightly different frequencies; the interference pattern of the total $C_n$ will, therefore, also vary with time—there will be a pattern of "beats" and the peaks of the beats [the place where $|C(x_n)|^2$ is large] will move along in $x$ as time goes on; they move with the speed we have called the "group velocity." We found that this group velocity was related to the variation of $k$ with frequency by

$$v_{\text{group}} = \frac{d\omega}{dk}; \quad \text{...(17)}$$

the same derivation would apply equally well here. An electron state which is a "clump"—namely, one for which the $C_n$ vary in space like the wave packet of Fig. 12.5—will move along our one-dimensional "crystal" with the speed $v$ equal to $d\omega/dk$, where $\omega = E/\hbar$. Using (16) for $E$, we get that

1 Provided we do not try to make the packet too narrow.

$$v = \frac{2Ab^2}{\hbar}\, k. \qquad ...(18)$$

In other words, the electrons move along with a speed proportional to the typing $k$. Equation (16) then says that the energy of such an electron is proportional to the square of its velocity—*it acts a classical particle.* So long as we look at things on a scale gross enough that we don't see the fine structure, our quantum mechanical picture begins to give results like classical physics. In fact, if we solve Eq. (18) for $k$ and substitute into (Eq. 16), we can write

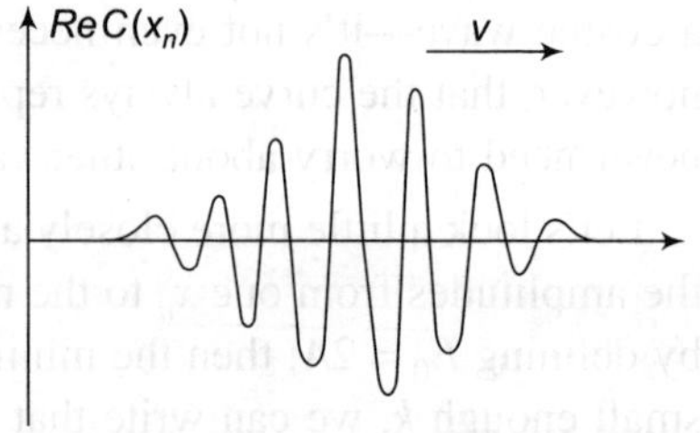


**Figure 12.5** The real part of $C(x_n)$ as a function of $x$ for a superposition of several states of similar energy. (The spacing $b$ is very small on the scale of $x$ shown.)

$$E = \frac{1}{2}\, m_{\text{eff}}\, v^2, \qquad ...(19)$$

where $m_{\text{eff}}$ is a constant. The extra "energy of motion" of the electron in a packet depends on the velocity just as for a classical particle. The constant $m_{\text{eff}}$ —called the "effective mass"—is given by

$$m_{\text{eff}} = \frac{\hbar^2}{2Ab^2}. \qquad ...(20)$$

Also notice that we can write

$$m_{\text{eff}}\, v = \hbar k. \qquad ...(21)$$

If we choose to call $m_{\text{eff}}\, v$ the "momentum," it is related to the wave number $k$ in the way we have described earlier for a free particle.

Don't forget that $m_{\text{eff}}$ has nothing to do with the real mass of an electron. It may be quite different—although in real crystal it often happens to turn out to be the same general order of magnitude, about 2 to 20 times the free-space mass of the electron.

We have now explained a remarkable mystery—how an electron in a crystal (like an extra electron put into germanium) can ride right through the crystal and flow perfectly freely even through it has to hit all the atoms. It does so by having its amplitudes going pip-pip-pip from one atom to the next, working its way through the crystal. That is how a solid can conduct electricity.

## 12.4 AN ELECTRON IN A THREE-DIMENSIONAL LATTICE

Let's look for a moment at how we could apply the same ideas to see what happens to an electron in three dimensions. The results turn out to be very similar. Suppose we have a rectangular lattice of atoms with lattice spacings of $a$, $b$, $c$ in the three directions. (If you want a cubic lattice, take the three spacings all equal.) Also suppose that the amplitude to leap in the $x$-direction to a neighbour is $(iA_x/\hbar)$, to leap in the $y$-direction is $(iA_y/\hbar)$, and to leap in the $z$-direction is $(iA_z/\hbar)$. Now how should we describe the base states? As in the one-dimensional case, one base state is that the electron is at the atom whose locations are $x$, $y$, $z$, where $(x, y, z)$ is one of the lattice points. Choosing our origin at one atom, these points are all at

$$x = n_x a, \quad y = n_y b, \quad \text{and} \quad z = n_z c,$$

where $n_x$, $n_y$, $n_z$ are any three integers. Instead of using subscripts to indicate such points, we will now just use $x$, $y$, and $z$, understanding that they take on only their values at the lattice points. Thus, the base state is represented by the symbol $|\text{electron at } x, y, z\rangle$, and the amplitude for an electron in some state $|\psi\rangle$ to be in this base state is $C(x, y, z) = \langle\text{electron at } x, y, z\,|\,\psi\rangle$.

As before, the amplitudes $C(x, y, z)$ may vary with time. With our assumptions, the Hamiltonian equations should be like this

$$i\hbar\frac{dC(x, y, z)}{dt} = E_0 C(x, y, z) - A_z C(x + a, y, z) - A_x C(x - a, y, z)$$
$$- A_y C(x, y + b, z) - A_y C(x, y - b, z)$$
$$- A_z C(x, y, z + c) - A_z C(x, y, z - c). \quad ...(22)$$

It looks rather long, but you can see where each term comes from.

Again we can try to find a stationary state in which all the $C$'s vary with time in the same way. Again the solution is an exponential

$$C(x, y, z) = e^{-iEt/\hbar}\, e^{i(k_x x + k_y y + k_z z)}. \quad ...(23)$$

If you substitute this into Eq. (22) you see that it works, provided that the energy $E$ is related to $k_x$, $k_y$, and $k_z$ in the following way

$$E = E_0 - 2A_x \cos k_x a - 2A_y \cos k_y b - 2A_z \cos k_z c. \quad ...(24)$$

The energy now depends on the *three* wave numbers $k_x$, $k_y$, $k_z$, which, incidentally, are the components of a three-dimensional vector, $k$. In fact, we can write Eq. (23) in vector notation as

$$C(x, y, z) = e^{-iFt/\hbar}\, e^{-ik\cdot r} \quad ...(25)$$

The amplitude varies as a complex *plane wave* in three dimensions, moving in the direction of $k$, and with the wave number $k = (k_x^2 + k_y^2 + k_z^2)^{1/2}$.

The energy associated with these stationary states depends on the three components of $k$ in the complicated way given in Eq. (24). The nature, of the variation of $E$ with $k$ depends on relative signs and magnitudes of $A_x$, $A_y$, and $A_z$. If these three numbers are all positive, and if we are interested in small values of $k$, the dependence is relatively simple.

Expanding the cosines as we did before to get Eq. (16), we can now get that

$$E = E_{\min} + A_x a^2 k_z^2 + A_y b k_y^2 + A_z c k_z^2. \quad ...(26)$$

For a simple cubic lattice with lattice spacing $a$ we expect that $A_x$ and $A_y$ and $A_z$ would be equal—say all are just $A$—and we would have just

$$E = E_{\min} + Aa^2(k_x^2 + k_y^2 + k_z^2),$$

or

$$E = E_{\min} + Aa^2k^2. \quad ...(27)$$

This is just like Eq. (16). Following the arguments used there, we would conclude that an electron packet in *three* dimensions (made up by superposing many states with nearly equal energies) also moves like a classical particle with some effective mass.

In a crystal with a lower symmetry than cubic (or even in a cubic crystal in which the state of the electron at each atom is not symmetrical) the three coefficients $A_x$, $A_y$, and $A_z$ are different. Then the "effective mass" of an electron localized in a small region *depends on its direction of motion*. It could, for instance, have a different inertia for motion in the *x*-direction than for motion in the *y*-direction. (The details of such a situation are sometimes described in terms of an "effective mass tensor.")

## 12.5 FORMATION OF ENERGY BANDS

There are two approaches to discuss formation of energy bonds in solids: (i) atomistic approach and (ii) one-electron approach.

In the atomistic approach, electrons are assumed to be tightly bound to individual atoms. As atoms are brought together to form a crystal, interaction between the neighbouring atoms causes the electron energy levels of individual atoms to spread into bands of energies.

In the one-electron approximation, we study the behaviour of a single electron in the potential field established by the lattice atom cores and modified by the presence of all the other free electrons. The various penaissible energy values (levels) obtained for this electron represent the allowed energy levels of all the electrons.

We have seen that when two H atoms are brought together, the original 1*s* wave function of the two (isolated) atoms combine to form two (symmetric and antisymmetric) joint wave functions and as a result there is a splitting of the 1*s* energy level (of isolated H atom) into two levels $E_A$ and $E_S$. Now, suppose two atoms (other than H) are brought together (to form a diatomic molecule), the levels begin to split. Producing an energy difference that increases for each level (*n*) as the distance between the two atoms decreases.

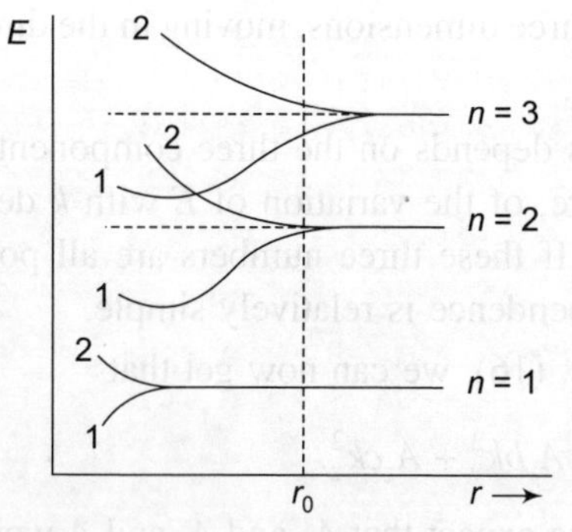


**Figure 12.6** Energy states for electron in diatomic molecule (1 and 2 refer to two states created when stabilising the molecule)

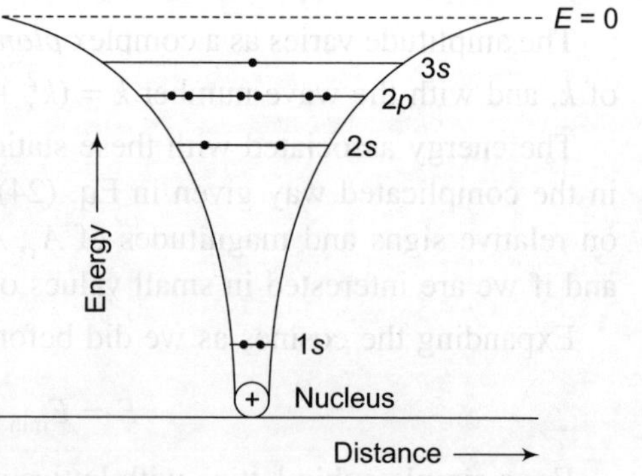


**Figure 12.7** Electronic energy levels in a single Na atom. Note that the highest energy is zero and the levels 1*s*, 2*s*, 2*p*, 3*s*, etc., have negative energies so that the electrons are bound to the nucleus

The greater the number of interacting atoms, the greater is the number of levels produced by the mixing of their respective valence wave functions. As an example, consider a single Na atom. Its energy levels are depicted in Fig. 12.7.

## 12.6 THE ENERGY BANDS IN SODIUM ATOM

As an illustration, we first consider the case of *sodium* crystal (i.e., a metal) when the atoms are far apart, there is no mutual interaction and the allowed states of the system are just the states of a single atom repeated $N$ times in space where $N$ is the number of atoms in the crystal.

As we decrease the interatomic spacing, the valence orbitals of the $N$ atoms begin to overlap and due to degeneracy, splitting of the original levels begins to take place, i.e., the 3$s$ level splits into $N$-levels which are quite close and form a *quasi-continuous band*. Similarly, the 3$p$ level also spreads into an energy band. Further reduction in the interatomic spacing causes broadening of the higher bands and overlapping, there, takes place (see Fig. 12.8).

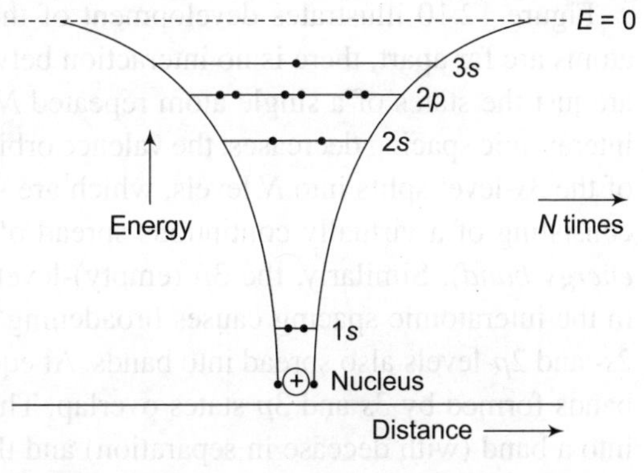


**Figure 12.8** Energy levels in a single sodium atom. Note that the highest energy is zero and the energy levels 1$s$, 2$s$, 2$p$, 3$s$, etc., have negative energies so that the electrons are bound to the nucleus

### 12.6.1 Energy Bands in Sodium Crystal

In order to understand the energy band development of a sodium crystal, consider a crystal having atoms arranged in a linear (periodic) array. Figure 12.9 shows the potential energy profile for the linear array of equilibrium spacing, $r_0$.

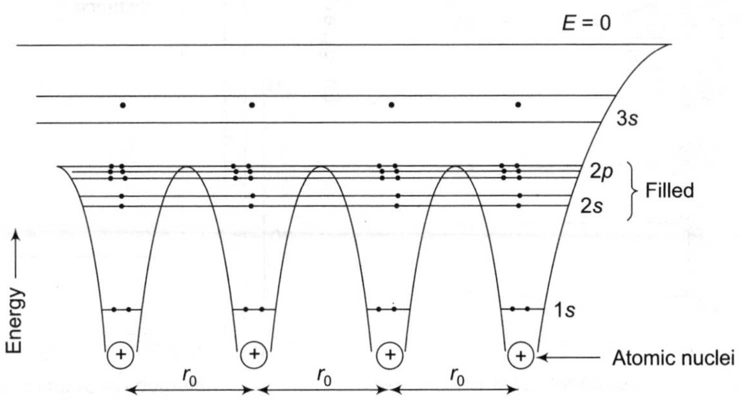


**Figure 12.9** Potential profile and electron energy levels along an array of atoms in a sodium crystal

Outside the crystal (represented in the right-hand side of the figure), the potential energy rises towards $E = 0$ asymptotically. However, between neighbouring atoms, because of the mutual interaction, the energy lies considerably below $E = 0$ value. The 2*s*- and 2*p*-levels of atoms are separated by potential humps, whereas the band formed by 3*s* electrons is only half filled and electrons in this band have their energies lying above the potential humps; they behave as free electrons, making the sodium metal a good conductor.

Figure 12.10 illustrates development of the energy bands in a sodium crystal. When the atoms are far apart, there is no interaction between them, and the allowed states of the system are just the states of a single atom repeated $N$ times in space ($N$ = number of atoms). As the interatomic spacing decreases, the valence orbitals of the atoms begin to overlap and the energy of the 3*s*-level splits into $N$ levels, which are so close that together they form an energy band consisting of a virtually continuous spread of permitted energies (called a *quasi-continuous energy band*). Similarly, the 3*p* (empty)-level also spreads into a band. A further reduction in the interatomic spacing causes broadening of the bands formed by 3*s*- and 3*p*-levels. The 2*s*- and 2*p*-levels also spread into bands. At equilibrium separation $r_0$ (= 0.367 nm for Na), the bands formed by 3*s* and 3*p* states overlap. The 3*s*-level is the first occupied level to broaden into a band (with decease in separation) and the average energy in the 3*s*- band drops at first, signifying the attractive forces between the atoms. (The actual internuclear distance in solid sodium corresponds to the minimum average 3*s* electron energy.)

Figure 12.11 shows the band structure of sodium.

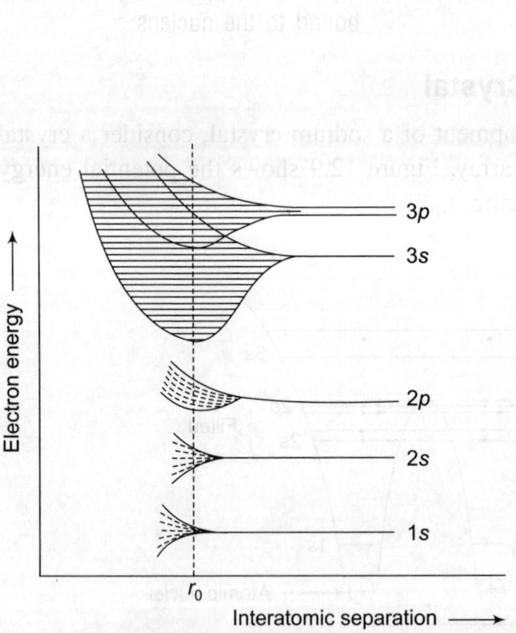


**Figure 12.10** Energy bands in sodium crystal

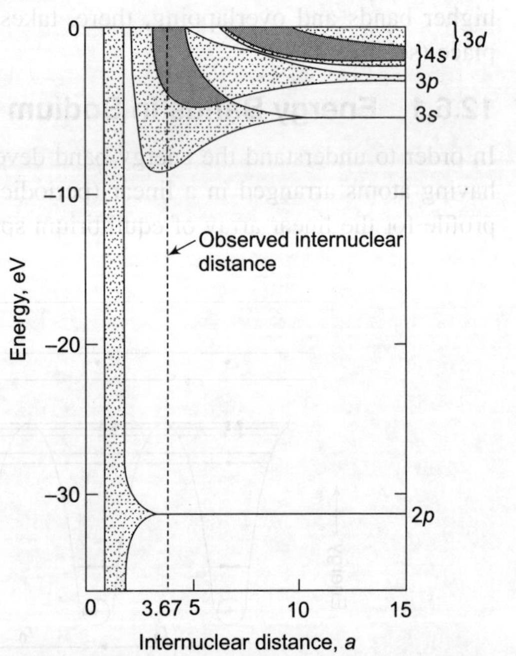


**Figure 12.11** The band structure of sodium as a function of internuclear spacing. Note the overlapping of bands with (extreme) increase in density

Thus, we see that in real solid, a large number of atoms are involved, their interaction with one another leads to a splitting of each of their energy levels. Then, the energy levels lie on a quasi-continuous scale and energy bands are formed (Fig. 12.12).

The bands formed may overlap in some materials whereas they may not overlap in other materials. The space between the bands are called forbidden energy gaps, as electrons in the solid may not have these energies. The energy of the highest filled level is called the *Femi energy* ($E_F$). At 0 $K$ all states up to $E_F$ are full and all above $E_F$ are empty. Because of the Pauli's exclusion principle, each $s$-band of a crystal, consisting of $N$ atoms, has space for $2N$ electrons (i.e., for two electrons per atom). If the highest filled $s$-band of a crystal is occupied by two electrons per atom (i.e., completely filled), then the electrons cannot drift within the crystal on application of an external electric field. Solids in which the highest filled band is completely occupied by electrons are therefore *insulators*.

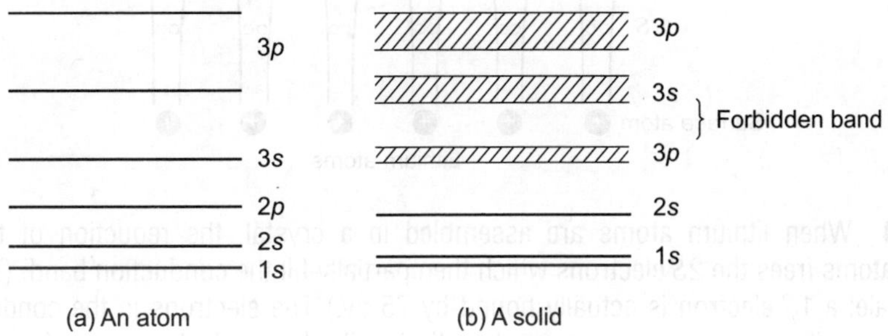


**Figure 12.12** Allowed energy levels for: (a) Bound electrons in an atom, (b) Electrons in a solid

## 12.7 ENERGY BANDS FOR LITHIUM

When lithium atoms are brought together to form a crystal, the net electrostatic field at any point in the crystal is the sum of all of the individual fields. Consequently, the potential between the atoms never rises to zero potential. In fact, the potential at a particular point between atoms in a crystal is reduced substantially below the potential at the same distance from an isolated atom (Fig. 12.13). The reduction is so pronounced in the crystal that the 2$S$ electron, which was *bound* in the isolated atom, no longer encounters a potential barrier that is sufficient to constrain it to the vicinity of any particular atom. The 2$S$ electrons in a lithium crystal are *free* electrons or *conduction* electrons.

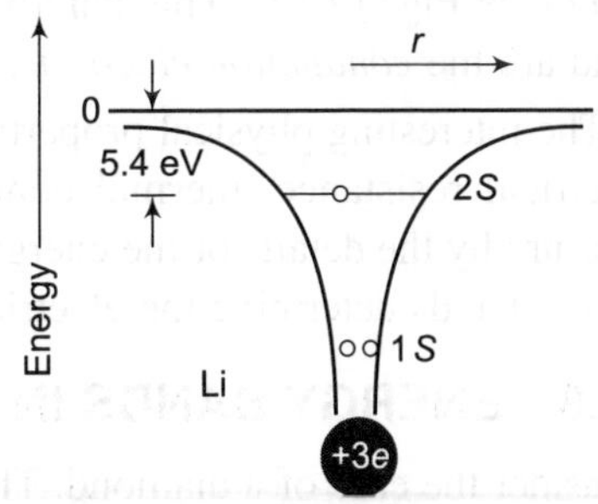


**Figure 12.13** Schematic potential energy diagram for an isolated lithium atom. The outermost electron (i.e., the 2$S$ electron) is bound by 5.4 eV; it requires ~75 eV to remove one of the 1$S$ electrons and an additional ~120 eV to remove the second 1$S$ electron

The conduction electrons "belong," not to individual atoms, but to the crystal as a whole. That is, the 2$S$

electron wave functions are not localized but extend throughout the crystal. The 2*S* energy state of the isolated lithium atom becomes an energy "state" of the crystal. If there are *N* atoms in the crystal, there are *N* electrons in this "state." But we know that the exclusion principle does not permit more than two electrons to exist in any single energy state. Therefore, the atomic 2*S* state must expand into a series of closely spaced crystal states, each of which can accommodate two electrons.

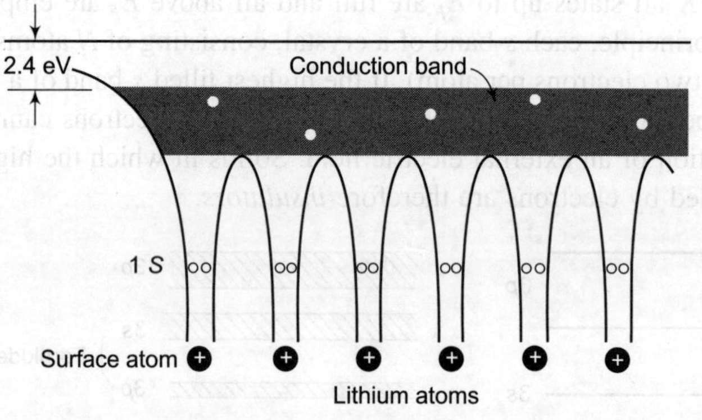


**Figure 12.14** When lithium atoms are assembled in a crystal, the reduction of the potential between the atoms frees the 2*S* electrons which then partially fill the conduction band. (The energies are not to scale; a 1*S* electron is actually bound by 75 eV.) The electrons in the conduction band, shown schematically in the diagram, are not really localized; the electron wave functions extend throughout the crystal

The spacing between these crystal states is so small that the distribution of states is essentially continuous. As a result, the discrete atomic state becomes, in the crystal an energy *band* (see Fig. 12.14). This band is the *conduction band* and the electrons that exist in this band are the *conduction electrons*.

The interesting physical properties of an element or compound in bulk form (for example, electrical resistance, thermal conductivity, magnetic properties) are determined in large measure by the details of the energy band structure. We shall investigate next how the crystal energy bands determine the electrical conductivity of the material.

## 12.8 ENERGY BANDS IN DIAMOND

Consider the case of a diamond. The isolated carbon atoms have the electronic configuration $1s^2$, $2s^2$, $2p^6$. However, in the diamond crystal, as a result of the formation of $sp^3$ hybrid (a mixture of 2*s*- and 2*p*-wave functions with covalent tetrahedral bounding), there is modification of the *s*- and *p*-levels (Fig. 12.15).

As the interatomic spacing is decreased, the 6*N* states of the upper band and the 2*N* states of the lower band merge to give a wider energy band with 8*N* states and 4*N* electrons. With still further reduction in interatomic separation (i.e., at the equilibrium separation), the 2*N* states in the 2*s* subshell and the filled 2*N* states in the 2*p* subshell come together and spread

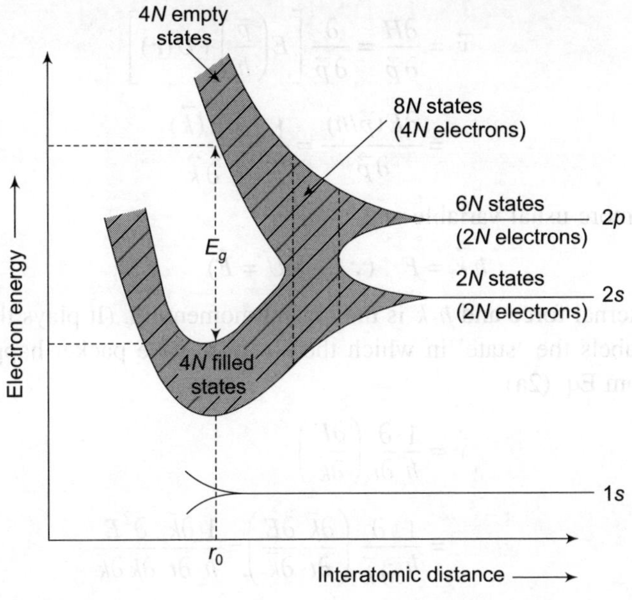


**Figure 12.15** Development of energy bands in a diamond crystal

into a band of 4*N* states, all of which are occupied by electrons. The remaining 4*N* empty states in the 2*p* subshell spread into another band, which is separated from the completely filled lower band by a forbidden energy region (*Energy gap* = $E_g$). The completely filled lower band contains the valence electrons and is known as a *valence band.* The upper band is termed the conduction band because the excitation of electrons from the lower valence band into this band may cause electrical conduction.

For diamond, the $sp^3$ hybrid stems from the 2*s* and $2p^3$ atomic states, for silicon from the 3*s* and $3p^3$ states, whereas for Germanium, from the 4*s* and $4p^3$ states. $E_g$ = 5.5 eV for diamond, 0.78 eV for Ge (at 0°K) and 1.2 eV for Si (at 0°K). Diamond is an insulator whereas Ge and Si are *semiconductors.*

## 12.9 THE CONCEPT OF EFFECTIVE MASS

We consider the case in which electrons moving in a lattice follow quasi-classical dynamics, i.e., our electron wavepackets behave like classical point particles following Hamilton's equations of motion

$$\dot{\vec{r}} = \frac{\partial H}{\partial p} \quad \text{and} \quad \dot{\vec{p}} = \frac{\partial H}{\partial p} \qquad ...(1)$$

where
$$H(\vec{r}, \vec{p}) = E\left(\frac{\vec{p}}{\hbar}\right) + U(\vec{r})$$

is the Hamiltonian $E$ is energy, a function of $\vec{p}/\hbar$ and $U(\vec{r})$ is a perturbing potential. The velocity of the electron is

$$\vec{v} = \frac{\partial H}{\partial \vec{p}} = \frac{\partial}{\partial \vec{p}}\left[E\left(\frac{\vec{p}}{\hbar}\right) + U(\vec{r})\right]$$

$$= \frac{\partial E(\vec{p}/\hbar)}{\partial \vec{p}} = \frac{1}{\hbar} = \frac{\partial E(\vec{k})}{\partial \vec{k}} \qquad ...(2a)$$

Now, $\vec{k}$ is our more usual variable and

$$\hbar \dot{\vec{k}} = F \quad (\because \quad -\nabla U = \vec{F}) \qquad ...(2b)$$

where $\vec{F}$ is the external force and $\hbar\,\vec{k}$ is the crystal momentum. (It plays the role of classical momentum and labels the 'state' in which the electron wave packet happens to be mainly concentrated). From Eq. (2a)

$$\dot{\vec{v}} = \frac{1}{\hbar}\frac{\partial}{\partial t}\left(\frac{\partial E}{\partial \vec{k}}\right)$$

$$= \frac{1}{\hbar}\frac{\partial}{\partial \vec{k}}\left(\frac{\partial \vec{k}}{\partial t}\frac{\partial E}{\partial \vec{k}}\right) = \frac{1}{\hbar}\frac{\partial \vec{k}}{\partial t}\frac{\partial^2 E}{\partial \vec{k}\,\partial \vec{k}}$$

$$= \frac{1}{\hbar^2}(\hbar \dot{\vec{k}})\frac{\partial^2 E}{\partial \vec{k}\,\partial \vec{k}}$$

$$= \left(\frac{1}{\hbar^2}\frac{\partial^2 E}{\partial \vec{k}\,\partial \vec{k}}\right)\vec{F} \qquad ...(3)$$

If we compare this with the Newton's force equation

$$m\dot{\vec{V}} = \vec{F} \qquad ...(4)$$

we see that electron in the lattice behaves as if it possesses a mass represented by $\hbar^2/(\partial^2 E/\partial \vec{k}\,\partial \vec{k})$. It is, in general, a tensor and is known as effective mass ($m^*$). Physically, it represents an exchange of momentum with the lattice, as the electron gets retarded near the Brillouin zone boundary. Thus

$$\frac{1}{m^*} = \frac{\partial^2 E}{\hbar^2\,\partial k\,\partial k} \qquad ...(5)$$

$$m^* = \frac{1}{\partial^2 E/(\hbar^2\,\partial k^2)} = \left(\frac{\partial^2 E}{\partial k^2}\right)^{-1} \qquad ...(6)$$

where $\vec{p} = \hbar\,\vec{k}$ is the crystal momentum. The effective mass $m^*$ is usually different from the electron mass $m$ because of the crystal potential and the deviation of $m^*$ from its nominal value $m$ is a measure of crystal potential.

Figure 12.16 (a) shows $E$ vs. $k$ curve for a periodic potential. Figures 12.16 (b), (c) and (d) represent $dE/dk$, $d^2E/dp^2$ and $m^*$, respectively, for electron in the range $0 \le k \le \pi/a$.

For small $k$, the curve (a) corresponds to the free electron curve and $m^* \sim m$. At this point (i.e., the point of inflection), the $v_g$ reaches its maximum value; $d^2E/dk^2$ or $d^2E/dp^2$ is zero and

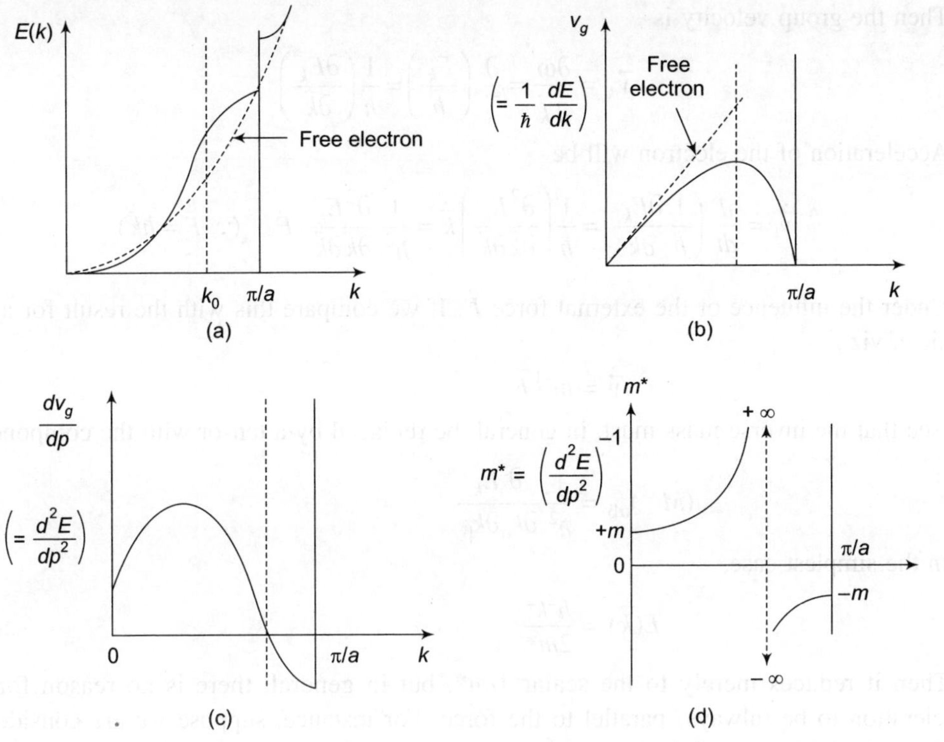


**Figure 12.16** Variation of effective mass $m^*$ and other related parameters with wave vector $k$

$m^*$ becomes infinite. Between $k = k_0$ and $\pi/a$, $d^2E/dk^2$ is negative, corresponding to negative values of the effective mass $m^*$ (as $k \to \pi/a$, $v_g \to 0$ and $m^* \to -m$). The negative effective mass is interpreted as follows.

When $k$ is essentially $\pi/a$, on applying an accelerating electric field to the electron, the lattice exerts such a large retarding force that the force produces a retardation (negative acceleration) leading to a Bragg reflection.

The momentum transfer to the lattice from the electron is larger than the momentum transfer from the applied force to the electron, i.e., force on the electron in one direction leads to electron gaining momentum in the opposite direction. In other words, electron behaves as though it had a negative effective mass at the top of a band.

## 12.10 EFFECTIVE MASS TENSOR

In case of crystalline solid, there is periodic field of the ion-cores and for electron-motion, we can no longer use simple plane waves (even in case of free-particle gas). We must construct *wave packets* whose motion corresponds to particle-motion as in quantum wave mechanics. But we know that wave packets in a dispersive medium travel with the group velocity. If $\vec{k}$ represents the wave vector of an electron-wave, its 'frequency' from the time dependent Schrödinger equation is $\omega = \frac{E_k}{\hbar}$.

Then the group velocity is

$$\vec{v}_k = \frac{\partial \omega}{\partial \vec{k}} = \frac{\partial}{\vec{k}}\left(\frac{E_k}{\hbar}\right) = \frac{1}{\hbar}\left(\frac{\partial E_k}{\partial \vec{k}}\right) \qquad ...(7)$$

Acceleration of the electron will be

$$\dot{\vec{v}}_k = \frac{d}{dt}\left(\frac{1}{\hbar}\frac{\partial E_k}{\partial \vec{k}}\right) = \frac{1}{\hbar}\left(\frac{\partial^2 E_k}{\partial \vec{k}\partial \vec{k}}\right)\dot{\vec{k}} = \frac{1}{\hbar^2}\frac{\partial^2 E_k}{\partial \vec{k}\partial \vec{k}} \cdot \vec{F} \quad (\because \vec{F} = \hbar\dot{\vec{k}}) \qquad ...(8)$$

Under the influence of the external force $\vec{F}$. If we compare this with the result for a free particle, viz.,

$$\dot{\vec{v}} = m^{-1}\vec{F} \qquad ...(9)$$

we see that the inverse mass must, in general, be replaced by a tensor with the components:

$$(M^{-1})_{\alpha\beta} = \frac{1}{\hbar^2}\frac{\partial^2 E_k}{\partial k_\alpha \partial k_\beta} \qquad ...(10)$$

In the simplest case,

$$E(\vec{k}) = \frac{\hbar^2 k^2}{2m^*} \qquad ...(11)$$

Then it reduces merely to the scalar $1/m^*$, but in general, there is no reason for the acceleration to be (always) parallel to the force. For instance, suppose we are considering electrons at the bottom of the conduction band of a semiconductor such as germanium. The lowest energy states are then often not at the centre of the zone, and there are energy minima at a number of different points in the zone. Near any one of these, the energy may be expanded as a quadratic function of $\vec{k}$, i.e.,

$$E_{\vec{k}} = E_0 + \frac{1}{2}\sum_{\alpha\beta} \hbar^2 \,(M^{-1})_{\alpha\beta}\, \Delta k_\alpha \Delta k_\beta \qquad ...(12)$$

where $\Delta\vec{k}$ is the vector from $\vec{k}$ to the centre, $k_0$, of the minimum. The constant energy surfaces are thus similar ellipsoids, which may well be extremely anisotropic (Fig. 12.17) $\dot{\vec{v}}_{\vec{k}}$ and $\vec{k}$ will then be by no means parallel, nor will $\vec{v}_{\vec{k}}$ and $\vec{k}$.

In this case, we should choose as special local axes in $k$-space the principal axes of these ellipsoids and write

$$E_{\vec{k}} = E_0 + \frac{\hbar^2}{2}\left[\frac{(\Delta k_1)^2}{m_1^*} + \frac{(\Delta k_2)^2}{m_2^*} + \frac{(\Delta k_3)^2}{m_3^*}\right] \qquad ...(13)$$

so that $m_1^*$, $m_2^*$ and $m_3^*$ are three *principal effective masses* of the inverse mass tensor $M^{-1}$. An electron can thus be accelerated much more rapidly in directions where the effective mass is small (e.g., across the length of the ellipsoid in the figure) than in directions where $m^*$ is large.

An interesting effect is found near the top of a band or near the boundary of a zone. The derivative of $E_k$ normal to the boundary must vanish there so that the shape of $E_k$, going out from the origin in a suitable direction must be as shown in Fig. 12.17. Since $E_k$ is here drawn with a minimum at the origin, we must have $\vec{v}_k$ zero at both the origin and the boundary of the zone. In between, the slope of $\vec{v}$ will go through a maximum. Above this point $\vec{v}$ is decreasing as $\vec{k}$ increases, i.e.,

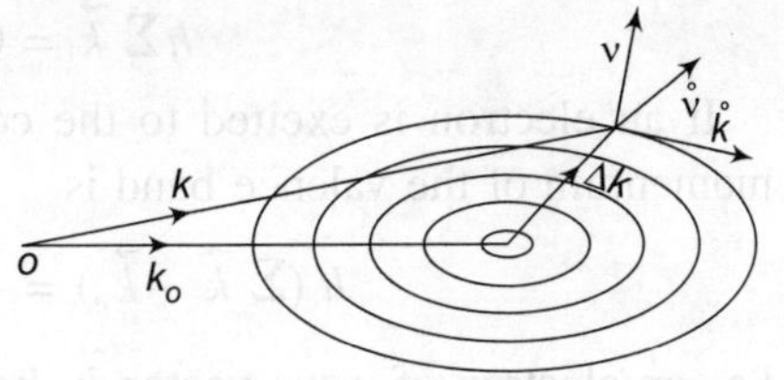


**Figure 12.17** Velocity and acceleration on an anisotropic energy surface (After *Electrons and Phonons* by J.M. Ziman, Clarendon Press, Oxford (1972))

$$\frac{1}{\hbar^2}\frac{\partial^2 E_k}{\partial k^2} = \frac{1}{\hbar}\frac{\partial v_k}{\partial k} \quad ...(14)$$

is negative and we find ourselves in a region of *negative effective mass*. Here, the application of a force *decelerates* the electron wave packet. The wave vector is increasing steadily, but this brings the state nearer to the zone boundary, where the interaction with the lattice is *stronger*. This necessitates the concept of hole in this case (and the situation is like an Everest climber, the nearer he approaches the peak, the slower he goes) (See Fig. 12.16 (b). Negative acceleration means that close to the top of the band (beyond the inflection point), the acceleration is opposite to the applied force, implying a negative mass (the lattice exerts such a large retarding force on the particle that it overcomes the applied force and decelerates).

## 12.11 CONCEPT OF HOLES

If we have a number of electrons in the band and the electron is moving with a velocity $v_i$ then $I = -e\,\mathring{a}_i v_i$. If a band is completely filled, the current is zero, since for every electron moving with a velocity $v_i$ there is one moving in the opposite direction with the same speed; this is so because in a filled band, for every electron with a wave vector $\vec{k}$, there is another electron with a wave vector $-\vec{k}$. Thus, for a filled band

$$I = -e\sum_i v_i = 0 \quad ...(15)$$

Now, if the $j$th electron is missing from the upper part of the band and a current $I$ is flowing, then

$$I = -e\sum_{i \neq j} v_i = -e\sum_i v_i + ev_j = ev_j \quad ...(16)$$

as $\sum_i v_i = 0$ for a filled band. The missing electron thus behaves like a positive charge. Such a positive charge is called a hole.

### 12.11.1 Hole Band Construction

The hole (missing electron) represents the cumulative behaviour of the electrons in the valence band. For a completely filled band, the momentum is zero, i.e.,

$$\hbar\Sigma \vec{k} = 0 \quad \text{or} \quad \Sigma\vec{k} = 0 \qquad \text{...(17)}$$

If an electron is excited to the conduction band (e.g., by absorption of a photon), the momentum of the valence band is

$$h\,(\Sigma\,\vec{k} - \vec{k}_e) = -\hbar\vec{k}_e$$

i.e., an electron of wave vector $\vec{k}_e$ becomes a missing electron and the total wave vector of the system is $-\vec{k}_e$, which is attributed to the hole.

$$\vec{k}_h = -\vec{k}_e \qquad \text{...(18)}$$

According to Eq. (18), the wave vector of the hole is equal to that of the unpaired electron at point $C$ in the valence band whose counterpart (the missing electron at $A$) has been transferred to the conduction band (Fig. 12.18), since for a symmetric band

$$E_e(\vec{k}_e) = E_e(-\vec{k}_e) \qquad \text{...(19)}$$

Now, let the zero of energy of the valence band be at the top of the band. The lower the missing electron lies in this band, the higher the energy of the system. The energy of the hole is opposite in sign to the energy of the missing electron, because it takes more work to remove an electron from a deeper orbital. Thus, for a symmetric band

$$E_e(\vec{k}_e) = E_e(-\vec{k}_e) = -E_h(-\vec{k}_e) = -E_h\,(k_h)$$

or

$$E_h\,(\vec{k}_h) = -E_e\,(\vec{k}_e) \qquad \text{...(20)}$$

This enables us to construct a band scheme (Fig. 12.19) to represent the properties of a hole.

The wave vector and energy of the hole are equal but opposite in sign to the wave vector and energy of the empty electron orbital in the valence band. (In the hole band, we do not

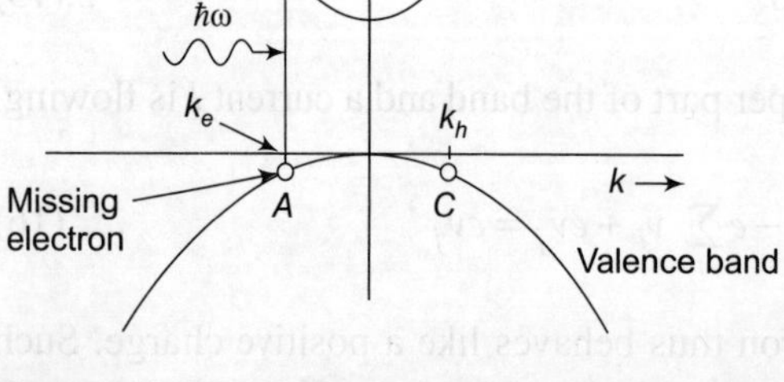


**Figure 12.18** Excitation of an electron with wave vector $\vec{k}_e$ from state $A$ in valence band to state $B$ in the conduction band (by absorption of a photon). This creates a hole with wave vector $\vec{k}_h = -\vec{k}_e$ in the state $C$ of the valence hand

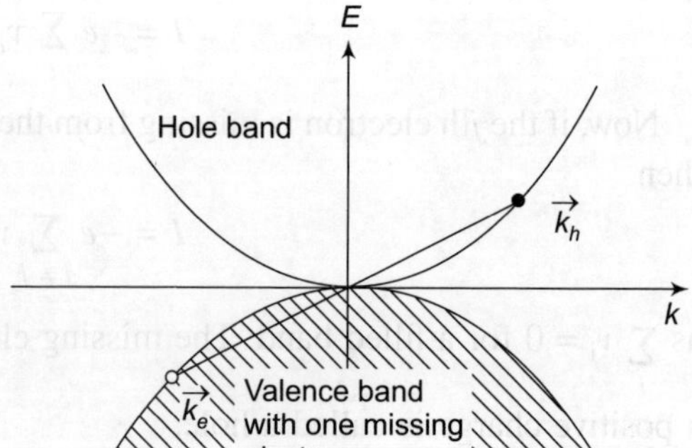


**Figure 12.19** The upper half of the figure shows the hole band that simulates the dynamics of a hole

show the disposition of the electron removed from the valence band at $\vec{k}_e$.) In the presence of an electric field, the current contribution by the unpaired electron at $C$ is

$$-e\vec{v}(-\vec{k}_e) = [-e\vec{v}(\vec{k}_e)] \quad (\because \; v(-\vec{k}_e) = -\vec{v}(\vec{k}_e))$$

$$= -(\text{Current of the missing electron, had it been there})$$

$$= \text{Current contributed by the hole with charge } +e$$

Thus, $$-e\,\vec{v}(\vec{k}_e) = +\vec{e}\,\vec{v}(-\vec{k}_e)$$

$$= e\vec{v}(\vec{k}_h) \qquad \text{...(21)}$$

The velocity of the hole is equal in magnitude and direction to that of the missing electron.

We define the hole mass as

$$\frac{1}{m_h^*} = \frac{1}{\hbar^2}\frac{\partial^2 E_h}{\partial k_h^2}$$

Then $$\frac{1}{m_h^*} = -\frac{1}{\hbar^2}\frac{\partial^2 E_e}{\partial(-k_e^*)^2} = -\frac{1}{\hbar^2}\frac{\partial^2 E_e}{\partial k_e^2} = -\frac{1}{m_e^*}$$

Thus, if $E_e = Ak^2$ (convex down band), $A$ being a constant then $m_e > 0$ (i.e., positive) whereas if $E_e = -Ak^2$ then $m_h^* = -m_e^*$, i.e., concave down bands have negative electron masses but positive hole masses.

## 12.12 METALS, INSULATORS AND SEMICONDUCTORS

The existence of energy bands (allowed and forbidden bands) provides a simple explanation for the general properties of metals, insulators and semiconductors (Fig. 12.20).

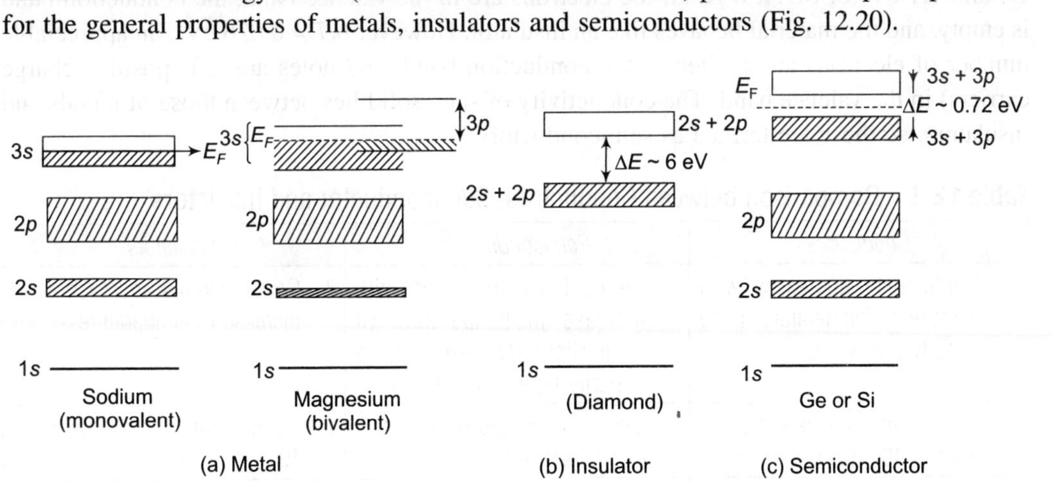


**Figure 12.20** Schematic illustration of energy bands for: (a) Metal (b) Insulator, (c) Semiconductor

Let us consider the order in which the energy levels of the energy bands are filled. Electrons in the inner shells (i.e., in narrow energy bands) are filled first. These are not influential in determining the general properties. The relevant electrons are the valence electrons and they occupy the valence band. At 0 K, the valence electrons occupy the lower levels of the valence band. The occupancy of higher levels determines the conductivity properties of the solids. There arise three cases. In the first case, the electrons fill only the lower half of the valence band at 0 K. For example, in sodium there are $N$ electrons and $2N$ levels in the $3s$ band. This band is therefore half full. This partially full band is responsible for good electrical (and thermal) conductivity of sodium. Therefore, it is metallic in behaviour. (The partially full $3s$ band is also the conduction band in this case as it provides space for movement of carriers (free electrons); similar is the case if the energy bands overlap (e.g., in Mg or Zn). For instance, in magnesium, $3p$ and $3s$ bands overlap. A $p$ atomic level can hold six electrons, so a $p$-band (formed by $N$ atoms) can hold $6N$ electrons. Similarly, the $3s$ band can hold $2N$ electrons. Consequently, $(3s + 3p)$ in Mg can hold $8N$ electrons in all. With only $2N$ electrons (since it is bivalent) in the band, it is only 1/4th filled, therefore, it is a conductor (metal).

In the second case, the valence electrons completely fill the valence band, which is separated by a large energy gap $\Delta E$ from the conduction band. For example, in diamond the $2s$ and $2p$ levels hybridise, forming two bands (each of which is a mixture of $2s$ and $2p$ states), which are separated by an energy gap $\Delta E \sim 6$ eV.

The lower band is completely full and the upper band is completely empty. The energy equal to energy gap (~6 eV) cannot be provided by ordinary means (electrical or thermal), so diamond exhibits insulator behaviour ($\Delta E \gg kT$).

The third case, viz. semiconductor, is qualitatively similar to that of insulator except that the energy gap between the conduction band and the valence band is much smaller, 0.72 eV for Ge and 1.1 eV for Si. At 0 K, all the electrons are in the valence band, the conduction band is empty, and the material behaves like an insulator. However ($kT = 0.026$ eV), an appreciable number of electrons are excited to the conduction band, and holes are left (positive charge carriers) in the valence band. The conductivity of such solid lies between those of metals and insulators, so these are termed as semiconductors.

**Table 12.1** Comparison between Conductors, Semiconductor and Insulators

| *Conductors* | *Semiconductors* | *Insulators* |
|---|---|---|
| 1. Conductivity decreases with increase in temperature up to a nearly zero value. | 1. Conductivity increases with increase in temperature (it is particularly sensitive to impurity type and content). | 1. Conductivities increase with increase in temperature. |
| 2. Conductivity of metals is of the order of $10^7$ $(\Omega\text{–m})^{-1}$. | 2. Conductivity ranges from $10^{-6}$ to $10^4$ $(\Omega\text{–m})^{-1}$. | 2. Conductivity ranges between $10^{-10}$ and $10^{-20}$ $(\Omega\text{–m})^{-1}$ |
| 3. Electrical resistivity is very low and ranges from $10^{-9}$ to $10^{-4}$ $\Omega$–m. | 3. Resistivity is normally high and ranges between $10^{-3}$ to $10^3$ $\Omega$–m. | 3. Resistivity is very high and ranges between $10^4$ and $10^{17}$ $(\Omega\text{–m})$ |

*Contd...*

*Table 12.1 Contd...*

| | | |
|---|---|---|
| 4. Resistivity increases with rise in temperature for metals. | 4. Resistivity decreases with rise in temperature. | 4. Negative resistance temperature coefficient. |
| 5. At low temperature, they exhibit semiconductivity. At every low temperatures, some elements and their alloys exhibit infinite (i.e., super) conductivity. | 5. At low temperatures, semiconductors become dielectrics. | 5. No charge in properties is observed with decrease in temperature. |
| 6. Current carriers are electrons. | 6. Electrons and holes, both serve as current carriers. | 6. No carriers. |
| 7. Have unfilled overlapping energy bands. | 7. Have filled energy bands and small forbidden zones. | 7. There is a large energy gap between valence and conduction bands. |

We will discuss typical energy bands in semiconductors in detail in the chapter on "Semiconductor Physics".

### 12.12.1 Comparison of Semiconductors, Metals and Insulators

1. A *semiconductor* has an electrical conductivity $\sigma$ (at 300 K), typically in the range $10^{-8}$ to $10^{3}$ $(\Omega\text{-m})^{-1}$.

   Typical *metals* such as Na and Cu have $\sigma$ (300 K) > $10^{6}$ $(\Omega\text{-m})^{-1}$, typical *semimetals* such as As, Bi, have $\sigma$(300 K) in the range $10^{3}$ to $10^{6}$ $(\Omega\text{-m})^{-1}$ and typical *insulators* such as diamond, NaCl and $S_iO_2$ have $\sigma$(300 K) < $10^{-8}$ $(\Omega\text{-m})^{-1}$.
2. Semiconductors have an electrical conductivity that is zero at $T = 0$ K and *increases exponentially with increasing T*, according to $\sigma(T) \propto \exp\left(-\frac{Eg}{k_B T}\right)$, due to increasing concentration of charge carriers (electrons in the conduction band and holes in the valence band) resulting from the thermal activation of electrons across the energy gap.
3. A semiconductor has a concentration of electrons in the conduction band at T = 300 K in the range $10^{10}$ $m^{-3} < n < 10^{20}$ $m^{-3}$. Typical *metals* have $n \approx 10^{28}$ to $10^{29}$ $m^{-3}$, typical *semimetals* have $n \approx 10^{23}$ to $10^{26}$ $m^{-3}$, and typical *insulators* have $n \lesssim 10^{10}$ $m^{-3}$.

## 12.13 ELECTRONS IN A PERIODIC POTENTIAL

In fact, electrons in a metal move not in a constant potential but in a perfectly periodic potential whose periodicity is equal to that of the lattice. If we consider the motion of electrons in a periodic potential, we come to the conclusion that there are allowed energy bands in a crystal which are separated by forbidden regions. This we shall see when we will discuss the Kronig–Penney model.

Consider an electron in a crystal moving with velocity v in a region of periodically varying potential (rather than a constant potential) as shown in Fig. 12.21.

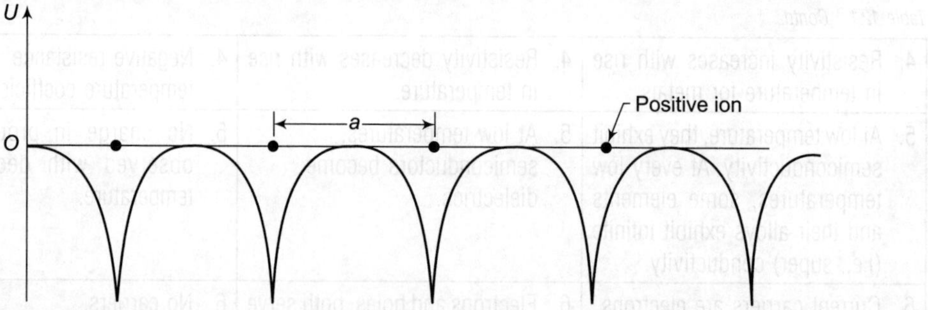


**Figure 12.21** Variation of potential energy of a conduction electron in a periodic array of positive ions

The de Broglie wavelength of a free electron moving with velocity ν is

$$\lambda = \frac{h}{m\nu} = \frac{h}{p} \qquad ...(1)$$

where *m* is the mass of the electron. Low energetic free electrons travel freely through the crystal since their wavelengths are larger compared to the periodicity of the lattice, *a*. More energetic electrons, with wavelength comparable with *a* are diffracted the same way as X-rays or electrons in a beam directed at the crystal from outside. An electron of wavelength λ undergoes Bragg's reflection from one of the atomic planes in the crystal when it approaches the plane at an angle θ satisfying

$$n\lambda = 2a \sin\theta. \qquad ...(2)$$

In terms of wave number, $k = 2\pi/\lambda$, the Eq. (2) can be written as

$$n \cdot \frac{2\pi}{k} = 2a \sin\theta$$

therefore, $$k = \frac{n\pi}{a \sin\theta} \qquad ...(3)$$

Figure 12.22 shows the Bragg's reflection in a two-dimensional square lattice. Referring to this figure, we can express the Bragg's condition by saying that reflection from the vertical rows of ions occurs when the component of wave vector $\vec{k}$ in the *x*-direction, $k_x$, is equal to $\frac{n\pi}{a}$. That is, Bragg's reflection from vertical rows of ions occurs when

$$k_x = \frac{n\pi}{a} \qquad ...(4)$$

Similarly, Bragg's reflection from horizontal rows of ions occurs when

$$k_y = \frac{n\pi}{a} \qquad ...(5)$$

Let us first consider the electrons whose wave numbers are sufficiently small. If $k < \pi/a$, the electrons can move freely through the lattice in any direction. When the wave number *k* satisfies the relation

$$k = \pm\frac{\pi}{a} \quad ...(6)$$

they are Bragg-reflected and cannot propagate through the lattice. In fact, an electron with $k = \pm\frac{\pi}{a}$, instead of being described by a travelling wave, is described by standing waves which are linear combination of the former.

Let the travelling wave in the X-direction be represented by

$$\psi = e^{ikx} \quad ...(7)$$

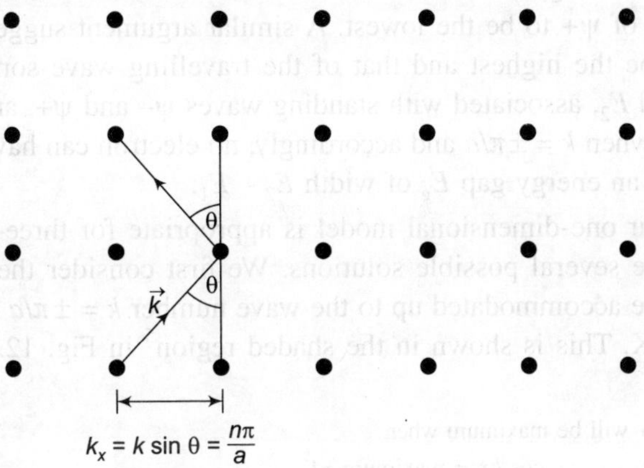


**Figure 12.22** Bragg's reflection from vertical rows of ions

Then, the reflected wave will be given by

$$\psi = e^{-ikx} \quad ...(8)$$

We can form two different standing waves from the travelling waves, given by Eqs. (7) and (8)

$$\psi+ = e^{ikx} + e^{-ikx} = 2\cos kx \quad ...(9)$$

$$\psi- = e^{ikx} - e^{-kx} = 2i\sin kx \quad ...(10)$$

The standing waves are labelled ψ+ and ψ–, according to whether they do or do not change sign when $x$ is replaced by $-x$. The two standing waves, ψ+ and ψ–, pile up electrons at two different regions, and therefore, two waves have two different values of potential energy. This is the origin of energy gap.

From quantum mechanics, the probability density of a particle is

$$\rho = \psi\psi^* = |\psi|^2 \quad ...(11)$$

For a pure travelling wave,

$$\rho = e^{ikx} \cdot e^{-ikx} = 1 \quad ...(12)$$

That is, for a pure travelling wave, the charge density is uniform.

For the stationary wave $\psi+$,

$$\rho+ = |\psi+|^2 \propto \cos^2 kx \qquad ...(13)$$

which piles up electrons on or near the lattice points.[2]

For the stationary wave $\psi-$,

$$\rho- = |\psi-|^2 \propto \sin^2 kx \qquad ...(14)$$

which distributes the electrons midway between the ion cores.[3]

Since the potential energy of an electron will tend to be $-\infty$ at the ion core,[4] we expect the potential energy of $\psi+$ to be the lowest. A similar argument suggests that the potential energy of $\psi-$ will be the highest and that of the travelling wave somewhere in-between. The energies, $E_1$ and $E_2$, associated with standing waves $\psi-$ and $\psi+$, are different. No other solution is possible when $k = \pm\pi/a$ and accordingly, no electron can have energy between $E_1$ and $E_2$ and we have an energy gap $E_g$ of width $E_2 - E_1$.

Assuming that our one-dimensional model is appropriate for three-dimensional system, we now examine the several possible solutions. We first consider the situation where the electrons may first be accommodated up to the wave number $k = \pm\pi/a$ (or the first Brillouin zone is filled) at 0 K. This is shown in the shaded region[5] in Fig. 12.23 (a). There are no

---

2 The charge density $\rho+$ will be maximum when

$$\cos^2 kx = \text{maximum} = 1$$

$$\Rightarrow \quad \cos kx = \pm 1 = \cos n\pi$$

$$\Rightarrow \quad x = \frac{n\pi}{k} = \frac{n\pi}{\pi/a} = na$$

$$\Rightarrow \quad x = 0, a, 2a, 3a, \ldots$$

Electron will be piled up near the lattice points which are occupied by positive ions.

3 Charge density $\rho-$ will be maximum when

$$\sin^2 kx = 1$$

$$\Rightarrow \quad \sin kx = \pm 1 = \sin \frac{(2n+1)\pi}{2}$$

$$\Rightarrow \quad kx = \pm \frac{(2n+1)\pi}{2}$$

$$\Rightarrow \quad x = \pm \frac{(2n+1)\pi}{2k} = \pm \frac{(2n+1)\pi}{2\pi/a} = \pm \frac{(2n+1)a}{2}$$

$$\Rightarrow \quad x = \pm\frac{a}{2}, \pm\frac{3a}{2}, \pm\frac{5a}{2}, \ldots$$

4 The potential energy of an electron carrying a charge $-q$ in the field of charge $+Q$ is

$$U = \frac{1}{4\pi\varepsilon_0} \cdot \frac{Q(-q)}{r}$$

when $r \to 0$, $U \to \infty$.

5 We have,

$$E_k = E - V = \frac{p^2}{2m} = \left(\frac{hk}{2\pi}\right)^2 \cdot \frac{1}{2m} = \frac{h^2k^2}{8\pi^2 m}$$

$$\Rightarrow \quad E_k \propto k^2$$

Thus, $E_k - k$ curve is a parabola.

empty states available just above the filled levels, the application of an electric field cannot increase the kinetic energy of the electrons and the material will not conduct electricity. If the temperature of the material is raised, it might be possible for one of the electrons to cross the band gap and hence, to go over to the next Brillouin zone. This will, however, depend upon the magnitude of the band gap. In case of NaCl, it is of the order of 10 eV, whereas the thermal energy available to the electrons at room temperature is only of the order of 0.025 eV. Hence, NaCl is an insulator at all temperature. In germanium, it is about 0.2 eV, so that we have the situation at a finite temperature as shown in Fig. 12.23 (b). A semiconductor has the energy band structure of an insulator with a narrow band gap which allows a considerable amount of conduction even at room temperature. The energy band structure of a conductor is shown in Fig. 12.23 (c) which is typically a half-filled band. Since there is no energy gap just above the field levels, electrons are always free to move throughout the crystal under the influence of an electric field.

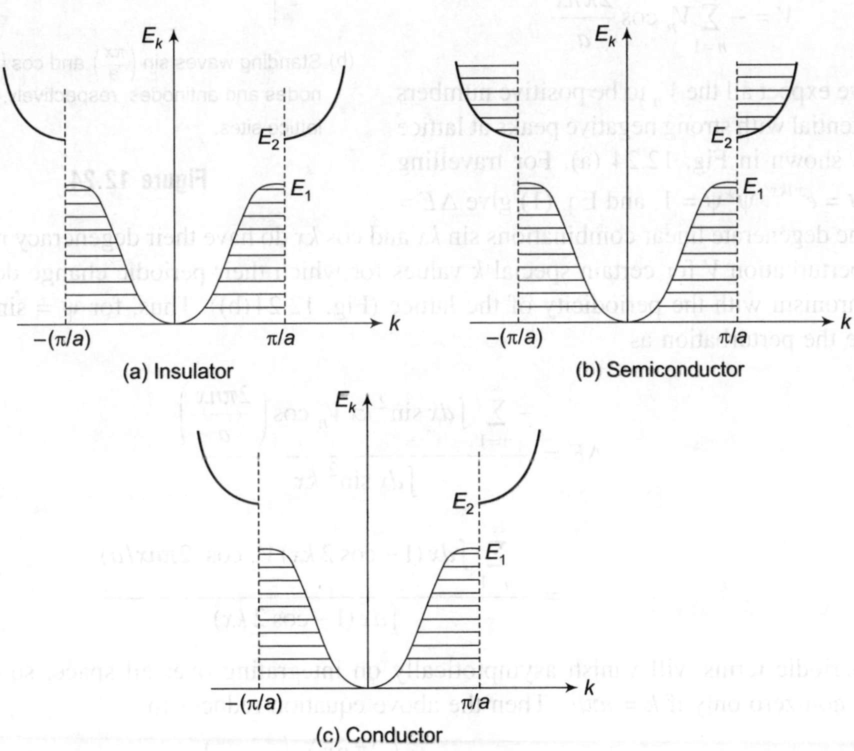


**Figure 12.23** Distinction between insulator, semiconductor and conductor

## 12.14 THE EFFECT OF A PERIODIC LATTICE POTENTIAL

In this section, we will now make an attempt to improve the free electron model by taking into account the fact that the positive ions produce not a uniform attractive potential, but one which has strong negative peaks at the lattice sites. We consider the one-dimensional periodic potential such as shown in Fig. 12.24. The period is equal to the lattice spacing $a$. We shall estimate the correction to the free electron energy $E = \hbar^2 k^2/(2m)$ by using the following standard formula of first order perturbation theory

$$\Delta E = \frac{\int \psi^* V \psi \, dx}{\int \psi^* \psi \, dx}$$

For convenience, we take our zero of energy as the mean value of the potential, then in our present case, the potential can be written a Fourier series in the form

$$V = -\sum_{n=1}^{\infty} V_n \cos \frac{2\pi n x}{a}$$

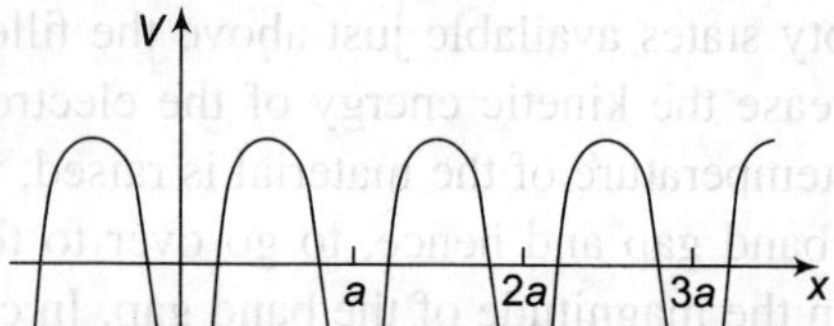


(a) A periodic potential in one dimention

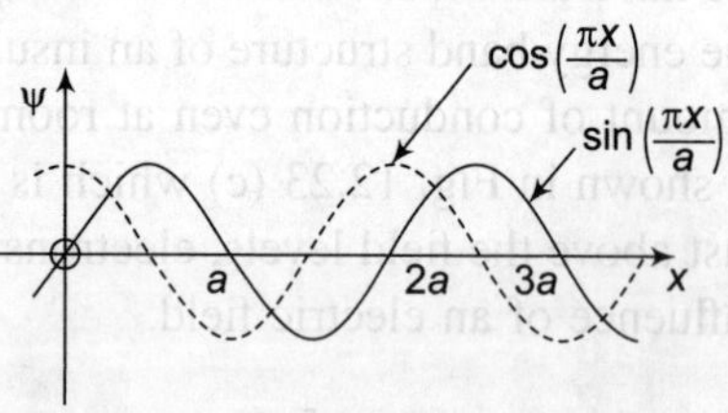


(b) Standing waves $\sin\left(\frac{\pi x}{a}\right)$ and $\cos\left(\frac{\pi x}{a}\right)$, with nodes and antinodes, respectively, at the lattice sites.

**Figure 12.24**

where we expect all the $V_n$ to be positive numbers for a potential with strong negative peaks at lattice sites, as shown in Fig. 12.24 (a). For travelling waves $\psi = e^{\pm ikx}$, $\psi^*\psi = 1$, and Eq. (1) give $\Delta E = 0$. But the degenerate linear combinations sin $kx$ and cos $kx$ do have their degeneracy removed by the perturbation $V$ for certain special $k$ values for which their periodic change density is in synchronism with the periodicity of the lattice (Fig. 12.24 (b)). Thus, for $\psi = \sin kx$, we calculate the perturbation as

$$\Delta E = \frac{-\sum_{n=1}^{\infty} \int dx \sin^2 kx \, V_n \cos\left(\frac{2\pi n x}{a}\right)}{\int dx \sin^2 kx}$$

$$= -\frac{\sum_{n=1}^{\infty} \int dx \, (1 - \cos 2\, kx) \, V_n \cos (2\pi n x/a)}{\int dx \, (1 - \cos 2\, kx)}$$

All periodic terms will vanish asymptotically on integrating over all space, so that the result is non-zero only if $k = n\pi/a$. Then the above equation reduces to

$$\Delta E = \int dx \, V_n \cos^2\left(\frac{2\pi n x}{a}\right) \Big/ \int dx = \frac{1}{2} V_n$$

all other terms in the series having integrated to zero.

Similarly, for $\psi = \cos kx$, by using $\cos^2 kx = \frac{1}{2}(1 + \cos^2 k)$

We find $\qquad \Delta E = -\frac{1}{2} V_n$

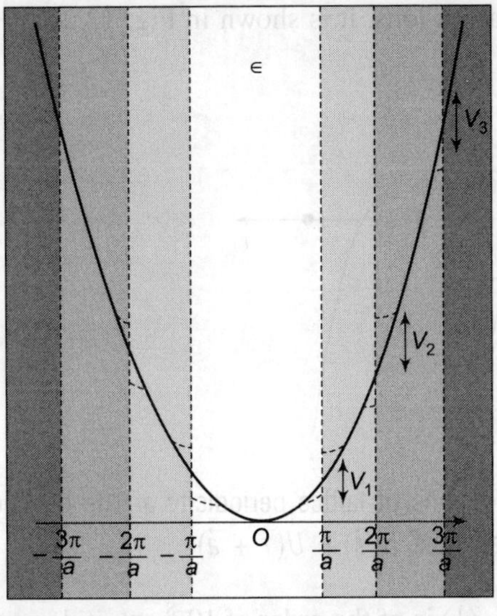


**Figure 12.25** Free electron parabola $\varepsilon = \hbar^2k^2/2m$, with perturbed energies for standing waves at $|k| = n\pi/a$; continuation to other $k$ values shown by broken lines

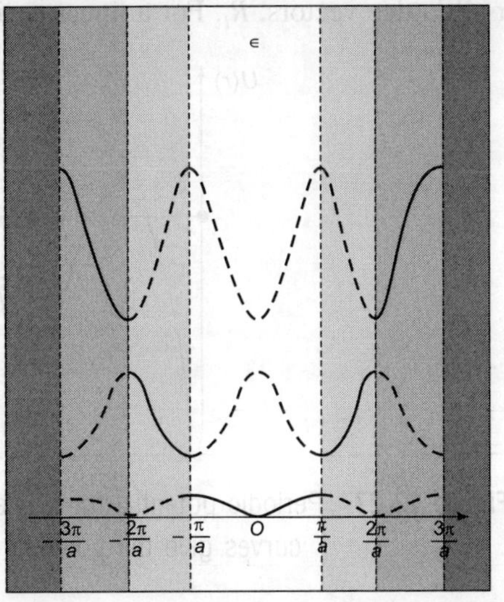


**Figure 12.26** Perturbed energies of Fig. 12.25 continued to physically equivalent $k$ values, to show the correspondence to a series of energy bands

The physical reasons for these results, for the case $n = 1$, become obvious on inspection of Fig. 12.24. $\psi = \sin(\pi x/a)$ has antinodes where the potential is repulsive, so its energy is raised; and $\psi = \cos(\pi x/a)$ has antinodes where the potential is attractive, so its energy is lowered.

In Fig. 12.25, we show these perturbed energies in relation to the free electron parabola $E = h^2k^2/2m$. Since the stationary states for $k = (n\pi/a)$ are standing waves, it follows that the group velocity, and hence $d\varepsilon/dk$ must be zero at these points. To calculate other points on the $\varepsilon(k)$ cure requires second order perturbation theory, since the first order result is zero. But knowing $\varepsilon$ and $d\varepsilon/dk$ at $k = (n\pi/a)$, we can easily sketch in the rest of the curve by continuity, as shown by the dotted curves in Fig. 12.25. We see that the continuous parabola is broken into a number of separate branches by energy gaps (Fig. 12.26). The lowest branch is derived from an atomic ground state and the higher energy branches can be regarded as derived from atomic excited states.

## 12.15 ELECTRON MOTION IN A CRYSTAL (ONE-ELECTRON APPROACH)

In case of a crystal, electron's motion in it is not free. For example, in one-dimensional case, it may be considered to move in a potential $U(x)$ having the periodicity of the underlying Bravais lattice

$$U(x + R_l) = U(x) \qquad ...(1)$$

for all lattice vectors, $R_l$. For a linear periodic array of ions, it is shown in Fig. 12.27.

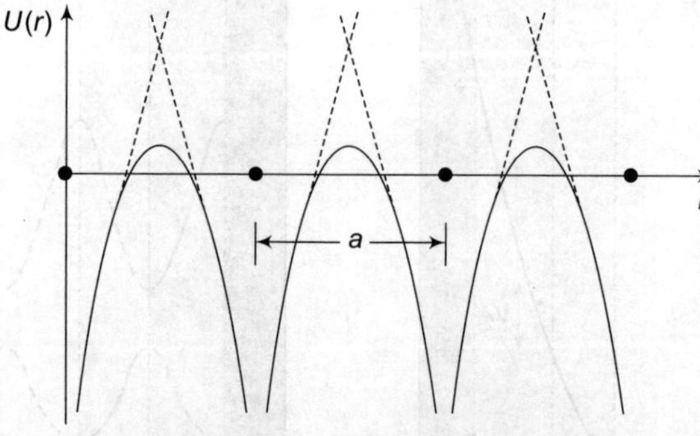


**Figure 12.27** Periodic potential for a linear array of ions of lattice periodicity *a*. The dashed curves give the potential of isolated Ions $U(\vec{r}) = U(\vec{r} + \vec{a})$

Since the scales of periodicity of potential of Eq. (1) is of the order of $10^{-8}$ cm, it demands the use of Schrödinger wave equation.

$$H\psi(\vec{k}, \vec{r}) = \left[-\frac{\hbar^2}{2m}\nabla^2 + V(\vec{r})\right]\psi(\vec{k}, \vec{r}) = E\psi(\vec{k}, \vec{r}) \qquad ...(2)$$

where $V(\vec{r})$, in general, corresponds to the averaged interaction term of Hartree–Fock approximation.

## 12.16 BLOCH WAVES AND BLOCH THEOREM

A crystal lattice possesses a translational symmetry which requires that the potential energy $V(\vec{r})$ of the one electron problem, is a periodic function which can be written as

$$V(\vec{r} + \vec{R}) = V(\vec{r}) \qquad ...(1)$$

where $\vec{R}$ is any lattice vector that connects equivalent points in two unit cells. We will see subsequently in next articles how the periodic potential splits the allowed electron energies into bands separated by forbidden regions called energy gaps. We start by substituting $\vec{r} + \vec{R}$ for $\vec{r}$ in the Schrödinger equation giving

$$\left[-\frac{\hbar^2}{2m_e}\nabla^2 + V(\vec{r} + \vec{R})\right]\psi(\vec{r} + \vec{R}) = E\,\psi(\vec{r} + \vec{R})$$

or in view of Eq. (1)

$$\left[-\frac{\hbar^2}{2m_e}\nabla^2 + V(\vec{r})\right]\psi(\vec{r} + \vec{R}) = E\,\psi(\vec{r} + \vec{R}) \qquad ...(2)$$

Comparing the Eq. (2) with the Schrödinger equation

$$\left[-\frac{\hbar^2}{2m_e}\nabla^2 + V(\vec{r})\right] \psi(\vec{r}) = E\,\psi(\vec{r})$$

allows us to infer that $\psi(\vec{r} + \vec{R})$ and $\psi(\vec{r})$ correspond to the same eigenvalue $E$, that is

$$\psi(\vec{r} + \vec{R}) = C(\vec{R})\,\psi(\vec{r}) \qquad ...(3)$$

where $|C(R)|^2 = 1.$

In other words, under any translation $\vec{R}$, which leaves the lattice invariant, the wave function is multiplied by a phase factor $C(\vec{R})$. Substituting $\vec{R} + \vec{R}'$ for $R$ in Eq. (3) yields

$$\begin{aligned}\psi(\vec{r} + \vec{R} + \vec{R}') &= C(\vec{R} + \vec{R}')\,\psi(\vec{r}) \\ &= C(\vec{R})\,\psi(\vec{r} + \vec{R}') \\ &= C(R)\,C(R')\,\psi(\vec{r})\end{aligned}$$

that is $C(\vec{R} + \vec{R}') = C(\vec{R})\,C(\vec{R}').$

This shows that we can choose

$$C(\vec{R}) = e^{i\vec{k}\cdot\vec{R}} \qquad ...(4)$$

so that Eq. (3) reduces to the *Bloch condition*

$$\psi(\vec{r} + \vec{R}) = e^{i\vec{k}\cdot\vec{R}}\,\psi(\vec{r}) \qquad ...(5)$$

The significance of $\vec{k}$ may be obtained from the periodic boundary conditions on the wave function (Eq. 5), which read

$$\psi(\vec{r} + P\vec{a}) = \psi(\vec{r} + Q\vec{b}) = \psi(\vec{r} + S\vec{c}) = \psi(\vec{r})$$

and require that

$$P\,\vec{k}\cdot\vec{a} = 2\pi p;\; Q\,\vec{k}\cdot\vec{b} = 2\pi q;\; S\,\vec{k}\cdot\vec{c} = 2\pi s$$

where $p$, $q$ and $s$ are arbitrary integers.

It means that $\vec{k}$ is a vector of reciprocal space:

$$\vec{k} = \frac{p}{P}\vec{a}^* + \frac{q}{Q}\vec{b}^* + \frac{s}{S}\vec{c}^* \qquad ...(6)$$

where the components $p/P$, $q/Q$ and $s/S$ are in general fractions of an integer.

With each $\vec{k}$ value, we may associate a one-electron wave function of the form

$$\psi_{\vec{k}}(\vec{r}) = e^{i\vec{k}\cdot\vec{R}}\,U_{\vec{k}}(\vec{r}) \qquad ...(7)$$

called a *Bloch wave*.

Substituting the above expression into the Bloch-condition (Eq. 5) leads to

$$e^{i\vec{k}\cdot(\vec{r}+\vec{R})}\,u_{\vec{k}}(\vec{r} + \vec{R}) = e^{i\vec{k}\cdot\vec{R}}\,e^{i\vec{k}\cdot\vec{r}}\,u_k(\vec{r})$$

or $$u_{\vec{k}}(\vec{r}+\vec{R}) = u_{\vec{k}}(\vec{r}) \qquad ...(8)$$

Equations (7) and (8) are known as the *Bloch theorem*, which states that in a periodic potential (Eq. (1)) we can always choose the one electron wave functions to be Bloch waves of the from (Eq. 7) where $u_{\vec{k}}(\vec{r})$ has the periodicity of the lattice.

Substituting Eq. (7) into the one electron Schrödinger equation, we find an equation for $u_{\vec{k}}(\vec{r})$

$$-\frac{\hbar^2}{2m_e}[\nabla^2 + 2i\vec{k}\cdot\nabla - k^2]\,u_{\vec{k}}(\vec{r}) + V(\vec{r})\,u_{\vec{k}}(\vec{r}) = E(\vec{k})\,u_{\vec{k}}(\vec{r})$$

($\because$ $e^{i\vec{k}\cdot\vec{r}}$ cancels out on both sides) ...(9)

or $$\left[\frac{(-i\hbar\nabla)^2}{2m_e} + \frac{2(-i\hbar\nabla)}{2m_e}\cdot\hbar\vec{k} + \frac{\hbar^2k^2}{2m_e} + V(\vec{r})\right]u_{\vec{k}}(\vec{r})$$

$$= E(\vec{k})\,u_{\vec{k}}(\vec{r})$$

or $$\left[\frac{(\vec{p}+\hbar\vec{k})^2}{2m_e} + V(\vec{r})\right]u_{\vec{k}}(\vec{r}) = E(\vec{k})\,u_{\vec{k}}(\vec{r}) \qquad ...(10)$$

where $\hbar k$ is called the *crystal momentum.* It is apparent from Eq. (10) that, if the electrons are not free, the crystal momentum is no longer their true momentum. Considering the complex conjugate equation, we obtain $u^*_{\vec{k}}(\vec{r}) = u_{-\vec{k}}(\vec{r})$, which leads to

$$\psi^*_{\vec{k}}(\vec{r}) = \psi_{-\vec{k}}(\vec{r}) \text{ and } E(-\vec{k}) = E(\vec{k})$$

This means that energy eigenvalues are symmetric about the origin of reciprocal space.

### 12.16.1 Proof of the Bloch Theorem

We know that any function $\psi(r)$ obeying the Born-von Karman boundary conditions can be expanded in the set of all plane waves that satisfy the boundary condition:

$$\psi(\vec{r}) = \sum_k C_k\, e^{i\vec{k}\cdot\vec{r}} \qquad ...(1)$$

Because the potential $V(\vec{r})$ is periodic in the lattice, its plane wave expansion will only contain plane waves with the periodicity of the lattice end therefore, with wave vectors that are vectors $\vec{K}$ of the reciprocal lattice

$$U(\vec{r}) = \sum_k U_{\vec{K}}\, e^{iK\cdot\vec{r}} \qquad ...(2)$$

The Fourier coefficients $U_{\vec{k}}$ are

$$U_{\vec{k}} = \frac{1}{v}\int_{\text{cell}} d\vec{r}\, e^{-\vec{K}\cdot\vec{r}}\, U(\vec{r}) \qquad ...(3)$$

where $v \equiv V/N$ is the volume of a direct lattice primitive cell. We are at liberty to change the potential energy by an additive constant for a primitive cell as

$$U_0 = \frac{1}{v} \int_{\text{cell}} d\vec{r}\, U(\vec{r}) \equiv 0. \qquad ...(4)$$

Note that because the potential $U(\vec{r})$ is real, it follows from Eq. (3) that the Fourier coefficients satisfy

$$U_{-\vec{K}} = U_K^* \qquad ...(5)$$

If we assume that the crystal has inversion symmetry so that, for a suitable choice of origin, $U(\vec{r}) = U(-\vec{r})$, then Eq. (3) implies that $U_{\vec{K}}$ is real and so,

$$U_{-\vec{K}} = U_{\vec{K}} = U_K^* \text{ (for inversion symmetry).} \qquad ...(6)$$

We now place the expansions (Eqs. (1) and (2) into the Schrödinger equation. The kinetic energy term gives

$$\frac{p^2}{2m}\psi = -\frac{\hbar^2}{2m}\nabla^2\psi = \sum_k \frac{\hbar^2}{2m} k^2 c_k e^{i\vec{k}.\vec{r}}. \qquad ...(7)$$

and

$$U\psi = \left(\sum_{\vec{K}} U_{\vec{K}}\, e^{i\vec{K}.\vec{r}}\right)\left(\sum_k c_k\, e^{i\vec{k}\cdot\vec{r}}\right)$$

or

$$U\psi = \sum_{\vec{K}\vec{k}} U_{\vec{k}}\, c_{\vec{k}}\, e^{i(\vec{K}+\vec{k})\cdot\vec{r}}$$

$$= \sum_{\vec{K}\vec{k}'} U_{\vec{K}}\, c_{\vec{k}'-K}\, e^{i\vec{k}'.\vec{r}} \qquad ...(8)$$

where $\qquad \vec{k}' = (\vec{K} + \vec{k})$

and the Schrödinger equation becomes

$$\sum_k e^{i\vec{k}\cdot\vec{r}} \left\{ \left(\frac{\hbar^2}{2m} k^2 - E\right) c_k + \sum_{\vec{K}'} U_{\vec{K}}\, c_{\vec{k}-\vec{K}'} \right\} = 0$$

(on replacing $\vec{k}' \to \vec{k}$ and $\vec{K} \to K'$) ...(9)

Since the plane waves satisfying the Born-von Karman boundary conditions are an orthogonal set, the coefficient of each separate term in Eq. (9) must vanish, i.e., for all allowed wave vectors $\vec{k}$,

$$\left(\frac{\hbar^2}{2m} k^2 - E\right) c_k + \sum_{\vec{K}'} U_{\vec{K}'}\, c_{k-\vec{K}'} = 0. \qquad ...(10)$$

Replacing $\vec{k} \to (\vec{k} - \vec{K})$

$$\left(\frac{\hbar^2}{2m} (\vec{k} - \vec{K})^2 - E\right) c_{\vec{k}-\vec{K}} + \sum_{\vec{K}'} U_{\vec{K}'}\, c_{\vec{k}-\vec{K}-\vec{K}'} = 0. \qquad ...(11)$$

and change $(\vec{K}') \to \vec{K}' - \vec{K}$

$$\left[\frac{\hbar^2}{2m}(\vec{k}-\vec{K})^2 - E\right] c_{\vec{k}-\vec{K}} + \sum_{\vec{K}'} U_{(\vec{K}'-\vec{k})}\, c_{\vec{k}-K'} \quad ...(12)$$

We emphasize here that Eqs. (10) and (12) are nothing but restatements of the original. Schrödinger equation in momentum space, with the fact that because of the periodicity of the potential, $U_{\vec{k}}$ is non-vanishing only when $\vec{k}$ is a vector of the reciprocal lattice.

Putting the above information back into Eq. (1) of the wave function ψ, we see that if the wave vector $\vec{k}$ only assumes the values $\vec{k}$, $\vec{k} - \vec{K}'$, $\vec{k} - \vec{K}''$ ... then the wave-function will be of the form

$$\psi_{\vec{k}} = \sum_{\vec{k}} c_{\vec{k}-\vec{K}}\, e^{i(\vec{k}-\vec{K})\cdot\vec{r}}$$

(where $\vec{k}$ is a reciprocal lattice vector in the first Brillouin zone), we can rewrite

$$\psi_{\vec{k}}(\vec{r}) = e^{i\vec{k}\cdot\vec{r}} + \left(\sum_{\vec{k}} c_{\vec{k}-\vec{K}}\, e^{-i\vec{K}\cdot\vec{r}}\right)$$

which is of the Bloch form with

$$\psi(\vec{r}) \equiv \sum_{\vec{k}} c_{\vec{k}-\vec{K}}\, e^{-i\vec{K}\cdot\vec{r}}$$

### 12.16.2 Second Proof of Bloch Theorem

Before proceeding to prove Bloch's theorem, we may note that

$$\psi(x+a) = e^{\pm ik(x+a)}\, u_k(x+a)$$

$$= e^{\pm ikx} \cdot e^{\pm ika} \cdot u_k(x), \text{ as } u_k(x+a) = u_k(x)$$

$$= e^{\pm ika}\, \psi(x), \text{ using Eq. (11)}$$

$$\Rightarrow \qquad \psi(x+a) = Q\psi(x) \quad ...(13)$$

where $$Q = e^{\pm ika} = \text{a constant} \quad ...(14)$$

If we can prove that Schrödinger equation satisfies the property given by Eqs. (13) and (14) Bloch theorem will be proved.

Schrödinger equation is a differential equation of second degree and so, it has two independent linear solutions, say $f(x)$ and $g(x)$. The general solution will, therefore, be given by the linear combination of $f(x)$ and $g(x)$. Let

$$\psi(x) = Af(x) + Bg(x) \quad ...(15)$$

where $A$ and $B$ are constants.

Since $f(x)$ and $g(x)$ are solutions of Schrödinger equation, $f(x+a)$ and $g(x+a)$ are also the solutions of Schrödinger equation. It follows directly from Eq. (13). Let

$$f(x+a) = \alpha_1 f(x) + \alpha_2 g(x) \qquad ...(16)$$

and
$$g(x+a) = \beta_1 f(x) + \beta_2 g(x) \qquad ...(17)$$

where $\alpha_1$, $\alpha_2$, $\beta_1$ and $\beta_2$ are real functions of energy.

From Eq. (15),

$$\begin{aligned} \psi(x+a) &= Af(x+a) + Bg(x+a) \\ &= A[\alpha_1 f(x) + a_2 g(x)] + B[\beta_1 f(x) + \beta_2 g(x)] \\ &= (A\alpha_1 + B\beta_1) f(x) + (A\alpha_2 + B\beta_2) g(x) \\ &= AQf(x) + BQg(x) \\ &= [Af(x) + Bg(x)]Q \end{aligned}$$

$$\Rightarrow \qquad \psi(x+a) = Q\psi(x)$$

where
$$A\alpha_1 + B\beta_1 = AQ \qquad ...(18)$$

$$\Rightarrow \qquad A(\alpha_1 - Q) + B\beta_1 = 0 \qquad ...(19)$$

and
$$A\alpha_2 + B\beta_2 = BQ$$

$$\Rightarrow \qquad A\alpha_2 + B(\beta_2 - Q) = 0 \qquad ...(20)$$

This proves the theorem.

Equations (19) and (20) have non-vanishing solution for $A$ and $B$ only if the determinant of their coefficients vanishes. That is,

$$\begin{vmatrix} \alpha_1 - Q & \beta_1 \\ \alpha_2 & \beta_2 - Q \end{vmatrix} = 0$$

$$\Rightarrow \qquad (\alpha_1 - Q)(\beta_2 - Q) - \alpha_2\beta_1 = 0$$

$$\Rightarrow \qquad Q^2 - (\alpha_1 + \beta_2) Q + (\alpha_1\beta_2 - \alpha_2\beta_1) = 0 \qquad ...(21)$$

Since $f(x)$ and $g(x)$ are two independent solutions of Schrödinger equation, they must satisfy Schrödinger equation. Hence,

$$\frac{d^2 f(x)}{dx^2} + \frac{2m}{\hbar^2}[E - V] f(x) = 0$$

$$\Rightarrow \qquad \frac{d^2 f(x)}{dx^2} - \frac{2m}{\hbar^2} V f(x) = -\frac{2m}{\hbar^2} Ef(x)$$

Multiplying both sides by $g(x)$, the above equation gives

$$g(x)\frac{d^2 f(x)}{dx^2} - \frac{2m}{\hbar^2} V f(x) g(x) = -\frac{2m}{\hbar^2} Ef(x) g(x) \qquad ...(22)$$

and
$$\frac{d^2g(x)}{dx^2} + \frac{2m}{\hbar^2}[E - V]\,g(x) = 0$$

$$\Rightarrow \quad \frac{d^2g(x)}{dx^2} - \frac{2m}{\hbar^2}V\,g(x) = -\frac{2m}{\hbar^2}Eg(x)$$

Multiplying both sides by $f(x)$, the above equation gives

$$f(x)\frac{d^2g(x)}{dx^2} - \frac{2m}{\hbar^2}V f(x)\,g(x) = -\frac{2m}{\hbar^2}Ef(x)\,g(x) \qquad ...(23)$$

Subtracting Eq. (22) from Eq. (23), we get

$$f(x)\frac{d^2g(x)}{dx^2} - g(x)\frac{d^2f(x)}{dx^2} = 0$$

$$fg'' - gf'' = 0$$

$$\Rightarrow \quad \frac{d}{dx}[fg' - gf'] = 0$$

$$\Rightarrow \quad \begin{vmatrix} f & g \\ f' & g' \end{vmatrix} = \text{constant} = W(x), \text{ the Wronskian} \qquad ...(24)$$

Now,
$$W(x+a) = f(x+a)\cdot\frac{dg(x+a)}{dx} - g(x+a)\cdot\frac{df(x+a)}{dx}$$

$$= [\alpha_1 f(x) + \alpha_2 g(x)]\frac{d}{dx}[\beta_1 f(x) + \beta_2 g(x)]$$

$$- [\beta_1 f(x) + \beta_2 g(x)]\frac{d}{dx}[\alpha_1 f(x) + \alpha_2 g(x)]$$

$$\Rightarrow \quad W(x+a) = [\alpha_1 f(x) + \alpha_2 g(x)]\,[\beta_1 f'(x) + \beta_2 g'(x)]$$

$$- [\beta_1 f(x) + \beta_2 g(x)]\,[\alpha_1 f'(x) + \alpha_2 g'(x)]$$

$$\Rightarrow \quad W(x+a) = (\alpha_1\beta_2 - \alpha_2\beta_1)\,\{f(x)g'(x) - g(x)f'(x)\}$$

$$\Rightarrow \quad W(x+a) = (\alpha_1\beta_2 - \alpha_2\beta_1)\,W(x)$$

Since the wave function is periodic,

$$W(x+a) = W(x)$$

$$\Rightarrow \quad \alpha_1\beta_2 - \alpha_2\beta_1 = 1$$

Equation (21), therefore, reduces to

$$Q^2 - (\alpha_1 + \beta_2)\,Q + 1 = 0 \qquad ...(25)$$

Equation (25) is a quadratic equation in $Q$. It has, therefore, two roots $Q_1$ and $Q_2$ which are given by

$$Q_1 = \frac{(\alpha_1 + \beta_2) + \sqrt{(\alpha_1 + \beta_2)^2 - 4}}{2} \quad ...(26)$$

and

$$Q_2 = \frac{(\alpha_1 + \beta_2) - \sqrt{(\alpha_1 + \beta_2)^2 - 4}}{2} \quad ...(27)$$

Two different cases may arise:

1. For certain range of energy $E$, for which $(\alpha_1 + \beta_2)^2 < 4$, the roots will be complex and since $Q_1 Q_2 = 1$, they will be conjugate. In these regions, we may, therefore, write

$$Q_1 = e^{ika} \text{ and } Q_2 = e^{-ika}$$

and
$$\psi(x + a) = e^{ika}\,\psi_1(x) \text{ and } \psi_2(x + a) = e^{-ika}\,\psi_2(x)$$

2. In other regions of energy $E$, for which $(\alpha_1 + \beta_2)^2 > 4$, the two roots $Q_1$ and $Q_2$ are real and reciprocal of each other (as $Q_1 Q_2 = 1$). These roots correspond to the solution of the Schrödinger equation of the type

$$\psi_1 = e^{\mu x} u_k(x) \text{ and } \psi_2 = e^{-\mu x} u_k(x)$$

where $\mu$ is real.

Both these solutions are divergent for an infinite crystal and do not represent allowed eigen states. Thus, there are no electronic states in the region corresponding to the real roots $Q_1$ and $Q_2$. The above discussion, therefore, leads to the conclusion that the energy spectrum of an electron moving in a periodic potential consists of allowed and forbidden energy regions or bands.

### 12.16.3 The Property of Translation

All the solutions of a wave equation in a *homogeneous medium* can have the form of an exponential *plane wave* of spatial part,

$$\psi_{\vec{k}}(\vec{r}) = A \exp(i\,\vec{k} \cdot \vec{r}) \quad ...(1a)$$

$A$ is a normalization constant, $\vec{k}$ is the wave vector, and $\hbar\vec{k}$ is the momentum of the quantum of this wave. For such a solution, we have the relation

$$\psi_{\vec{k}}(\vec{r} + \vec{R}) = e^{i\,\vec{k}\cdot\vec{R}}\,\psi_{\vec{k}}(\vec{r}) \quad ...(1b)$$

where $\vec{R}$ is any vector in real space. In words, Eq. (1) says that in homogeneous media, there is a complete set of solutions having the following property:

*Any* translation in real space from $\vec{r}$ to $\vec{r}' = \vec{r} + \vec{R}$, where $\vec{R}$ is *arbitrary*, multiply the solution by a factor exp $(i\vec{k}.R)$. This property stems from the fact that in a homogeneous medium, *any* translation from $\vec{r}$ to $\vec{r}' = \vec{r} + \vec{R}$ leaves the wave equation invariant.

This feature can be generalized to waves propagating in periodic media (the periodicity of the latter is always characterized by a Bravais lattice): In periodic media, there is a complete set of wave equation solutions having the following property: *Any* translation connecting *equivalent points* in space from $r$ to $r' = r + R_n$ (where, now $\vec{R}_n$ is not completely arbitrary, but it is restricted to any of the Bravais lattice vectors characterizing the periodicity of the medium) multiplies the solution by $\exp(i\vec{k}\cdot\vec{R}_n)$:

$$\psi_k(\vec{r}+\vec{R}_n) = e^{i\vec{k}\cdot\vec{R}n}\,\psi_k(\vec{r}), \qquad \text{...(2)}$$

where $k$ is called *crystal wave vector* and $\hbar k$ is called *crystal momentum*. Again, this property stems from the fact that any translation by a vector of the Bravais lattice leaves the wave equation invariant. Equation (2) is called Bloch's theorem; it can be recast in an equivalent form

$$\psi_k(\vec{r}) = w_{\vec{k}}(\vec{r})\,e^{i\vec{k}\cdot r}, \qquad \text{...(3)}$$

where $w_k(r)$ is a periodic function with the same periodicity as that of the medium, i.e.,

$$w_{\vec{k}}(\vec{r}+\vec{R}_n) = w_k(\vec{r}). \qquad \text{...(4)}$$

Equation (4) follows from Eqs. (2) and (3).

In Table 12.2, we compare the Bloch waves with plane waves and we point out the similarities and the differences. It must be stressed that propagation in a perfect periodic medium is free-like (since the average velocity of a Bloch wave is in general non-zero), and as a result, the mobility is in general infinite as in the case of zero force.

## 12.17 DESCRIPTION OF THE SEMICLASSICAL MODEL OF ELECTRONS IN A SOLID

In the Bloch description, we have two quantum numbers $\vec{k}$ and $n$ for an electron. $\hbar\vec{k}$ is the crystal momentum and $n$ is the band index. For each $n$, $\vec{k}$ runs through all wave vectors in a single primitive cell of the reciprocal lattice consistent with the Born-von Karman periodic boundary condition; $n$ runs through an infinite set of discrete values. For a given band index $n$, $E_n(k)$ has no simple explicit form, and is periodic in the reciprocal lattice

$$E_n(\vec{k}+\vec{K}) = E_n(\vec{k}).$$

where $\vec{K}$ is a reciprocal lattice vector.

Given the function $E_n(\vec{k})$, the semiclassical model associates with each electron a position $\vec{r}$, a wave vector $\vec{k}$ and a band index $n$. In the course of time, and in the presence of external electric and magnetic fields $\vec{\varepsilon}(\vec{r}, t)$ and $\vec{H}(\vec{r}, t)$, the position, wave vector, and band index are taken to evolve according to the following rules:

1. The band index $n$ is a constant of the motion. (The semiclassical model ignores the possibility of "inter band transitions".
2. The time evolution of the position and wave vector of an electron with band index $n$ are determined by the equations of motion

$$\dot{\vec{r}} = u_n(\vec{k}) = \frac{1}{\hbar}\frac{\partial E_n(\vec{k})}{\partial \vec{k}} \qquad \text{...(1a)}$$

**Table 12.2** Comparison of Plane and Bloch Electronic Waves

| *Property* | *Plane wave* | *Bloch wave* |
|---|---|---|
| Force | Zero | Non-zero; periodic function of $r$ |
| Potential $V(r)$ | Constant, independent of $r$ | Periodic function of $r$ |
| Wave form | $A \exp(i\mathbf{k}\cdot r)$ | $w_{n,k}(r) \exp(i\mathbf{k}\cdot r)$ |
| Amplitude | $A$ is constant, independent of $r$ | $w_{n,k}(r)$ is a periodic function of $r$ with the same periodicity as $V(r)$ |
| $k$ varies over | The whole $\mathbf{k}$ space | A Brillouin zone, usually the first one |
| Other parameter characterizing the wave | None | The natural number $n$ = 1, 2, 3, … called band index. It compensates for the restriction of $\mathbf{k}$ in a single BZ. |
| Relation of $k$ to the momentum $p$ and the velocity $v$ | $p = \hbar k$, $v = \hbar k/m_e$ | The quantity, $\hbar k$, in spite of being called crystal momentum, is not the momentum and $\hbar \mathbf{k}/m_e$ is not the velocity. Bloch wave are not eigen functions of either $\hat{p}$ or $\hat{v}$. |
| Translation property | $\psi_k(r+R) = \exp(i\mathbf{k}\cdot R)\ \psi_k(r)$ for any $R$ | $\psi_{n,k}(r + R_l) = \exp(i\mathbf{k}\cdot R_l)\ \psi_{n\cdot k}(r)$ only for vectors $R_1$ of the direct Bravais lattice |
| Dispersion relation connecting energy and $k$ | $\varepsilon = \dfrac{\hbar^2 k^2}{2m_e}$ | $\varepsilon_n(k)$ is, in general, a complicated function of $\mathbf{k}$ for each $n$, to be determined by solving Schrödinger's equation |
| Average velocity | $\langle u\rangle = v = \hbar \mathbf{k}/m_e$ | $\langle v_n\rangle = \dfrac{1}{\hbar}\dfrac{\partial \varepsilon_n(k)}{\partial(k)}$ |
| Density of states | $\rho(\varepsilon) \sim \varepsilon^{1/2}$ | $\rho(\varepsilon)$ is, in general, a complicated, nonmonotonic function of $\varepsilon$, with gaps and van Hove singularities different for each solid. |
| Finding the amplitude of the wave | $A = \dfrac{1}{\sqrt{V}}$ | To find $w_{n\cdot k}(r)$ one has to solve the PDE resulting from Schrödinger's equation by substituting $\psi_{n\cdot k}(r) = w_{n\cdot k}(r) \exp(i\mathbf{k}\cdot r)$. |
| Metal or insulator? | Always metal | Metal when the number of valence electrons per primitive cell is odd: otherwise either metal, or insulator or semiconductor depending on the particular solid. |
| Number of states (per spin) in the $k$-space volume element $\Delta^3 k$ | $\dfrac{V\Delta^3 k}{(2\pi)^3}$ | $\dfrac{V\Delta^3 k}{(2\pi)^3}$ |
| Equation of motion | $\hbar \dfrac{dk}{dt} = F_{tot}$ | $\hbar \dfrac{dk}{dt} = F_{\mathbf{addit}}$ |
| Mobility | Infinite | Infinite |

$$\hbar \vec{k} = -e\left[\vec{\varepsilon}\,(\vec{r}, t) + \frac{1}{c}\,\vec{v}_n\,(\vec{k}) \times \vec{H}\,(\vec{r}, t)\right] \qquad ...(1b)$$

3. The wave vector of an electron is only defined to within an additive reciprocal lattice vector $\vec{K}$. One cannot have two *distinct* electrons with the same band index $n$ and position $\vec{r}$, whose wave vectors $\vec{k}$ and $\vec{k}'$ differ by a reciprocal lattice vector $\vec{K}$, i.e., the labels $n$, $\vec{r}$, $\vec{k}$ and $n$, $\vec{k} + \vec{K}$ are completely equivalent ways of describing the *same* electron.

All distinct wave vectors for a single band therefore lie in a single primitive cell of the reciprocal lattice. In thermal equilibrium, the contribution to the electronic density from those electrons in the $n^{\text{th}}$ band with wave vectors in the volume element $d\vec{k}$ of $k$-space is given by F.D. statistics as

$$f\,(E_n(\vec{k}))\,\frac{d^3\vec{k}}{4\pi^3} = \frac{d^3\vec{k}/4\pi^3}{[E_n(\vec{k}) - \mu]/k_B T},$$

The semiclassical model predicts how, in the absence of collisions, the position $\vec{r}$ and wave vector $\vec{k}$ of each electron evolve in the presence of externally applied electric and magnetic fields. This prediction is based entirely upon a knowledge of the band structure of the metal. i.e., upon the forms of the functions $E_n\,(\vec{k})$ and upon no other explicit information about the periodic potential of the ions.

### 12.17.1 Filled Bands

A filled band is one in which all the energies lie below $E_F$. Electrons in a filled band with wave vectors in a region of $k$-space of volume $d^3\vec{k}$ contribute $d^3\vec{k}/4\pi^3$ to the total electronic density. Thus, the number of such electrons in a region of position space of volume $d^3\vec{r}$ will be $d^3\vec{r}\;d^3\vec{k}/4\pi^3$. One can therefore characterize a filled band semiclassically by the fact that the density of electrons in a six-dimensional $k$-space (called phase space, in analogy to the $rp$-space of ordinary classical mechanics) is $1/4\pi^3$. *The filled bands are inert*. The semiclassical Eq. (1) implies that a filled band remains a filled band at all times, even in the presence of space- and time-dependent electric and magnetic fields.

## 12.18 KRONIG–PENNEY MODEL

Kronig and Penney in 1931 solved the Schrödinger equation for electrons in a simple idealized periodic field. The model is of considerable importance because it interprets the main features of the band structure of metals. The periodic field assumed was in the form of potential wells of zero potential energy of width $a$ separated by rectangular barriers of width $b$ and height $\phi_0$, as shown in Fig. 12.28. For the potential of Fig. 12.28, the Schrödinger equation has to be solved in two regions (a) Region I, $0 < x < a$, potential well region in which $\phi = 0$ and (b) Region II, $b < x < 0$, barrier region in which $\phi = \phi_0$. In order to find the allowed energies, we solve the Schrödinger equation in these two regions and apply the appropriate boundary conditions. In region I, the Schrödinger equation is

$$\frac{d^2\psi}{dx^2} + \frac{2mE}{\hbar^2}\psi = 0 \text{ or } \frac{d^2\psi}{dx^2} + \beta^2\psi = 0, \left[\beta^2 = \frac{2mE}{\hbar^2}\right] \qquad ...(1)$$

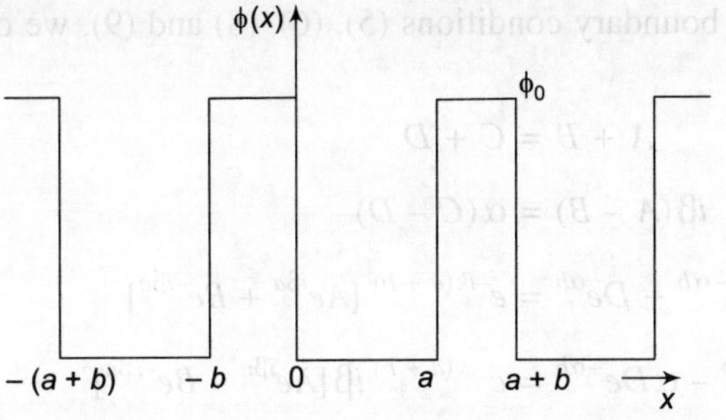


**Figure 12.28** One-dimensional chain of potential wells

Solution of this equation is

$$\psi_1 = Ae^{i\beta x} + Be^{-i\beta x} \qquad \text{...(2)}$$

where $A$ and $B$ are constants. In Region II, the Schrödinger equation is

$$\frac{d^2\psi}{dx^2} + \frac{2m}{\hbar^2}(E - \phi_0)\,\psi = 0 \text{ or } \frac{d^2\psi}{dx^2} - \alpha^2\psi = 0$$

$$\left[\alpha^2 = \frac{2m(\phi_0 - E)}{\hbar^2}\right] \qquad \text{...(3)}$$

Solution of this equation, for $E < \phi_0$ is

$$\psi_2 = Ce^{\alpha x} + De^{-\alpha x} \qquad \text{...(4)}$$

where $C$ and $D$ are constants. For continuity at $x = 0$,

$$\psi_1(0) = \psi_2(0) \qquad \text{...(5)}$$

$$\left.\frac{\partial\psi_1}{\partial x}\right|_{x=0} = \left.\frac{\partial\psi_2}{\partial x}\right|_{x=0} \qquad \text{...(6)}$$

In a periodic lattice with $\phi(x + a) = \phi(x)$ it is expected that the wave function will also exhibit this periodicity. Since the wave function must be a Bloch wave function, therefore Bloch theorem gives

$$\psi_k(x + a + b) = e^{ik(a+b)}\,\psi_k(x) \qquad \text{...(7)}$$

We incorporate this requirement at $x = -b$ and $x = a$, then we obtain the continuity boundary conditions as

$$\psi_2(-b) = e^{-ik(a+b)}\,\psi_1(a) \qquad \text{...(8)}$$

$$\left.\frac{\partial\psi_2}{\partial x}\right|_{x=-b} = e^{-ik(a+b)}\left.\frac{\partial\psi_1}{\partial x}\right|_{x=a} \qquad \text{...(9)}$$

When we apply the boundary conditions (5), (6), (8) and (9), we obtain four equations in four unknown

$$A + B = C + D \qquad ...(10)$$

$$i\beta(A - B) = \alpha(C - D) \qquad ...(11)$$

$$Ce^{-\alpha b} + De^{\alpha b} = e^{-ik(a+b)}[Ae^{i\beta a} + Be^{-i\beta a}] \qquad ...(12)$$

$$\alpha Ce^{-\alpha b} - \alpha De^{+\alpha b} = e^{-ik(a+b)}\, i\beta[Ae^{i\beta a} - Be^{-i\beta a}] \qquad ...(13)$$

By solving these equations simultaneously, e.g., by requiring that the determinant of the coefficients of *A, B, C* and *D* vanish, we get

$$\begin{vmatrix} 1 & 1 & -1 & -1 \\ i\beta & -i\beta & -\alpha & \alpha \\ -e^{-ik(a+b)+i\beta a} & -e^{-ik(a+b)-i\beta a} & e^{-\alpha b} & e^{\alpha b} \\ -i\beta e^{-ik(a+b)+i\beta a} & i\beta e^{-ik(a+b)-i\beta a} & \alpha e^{-\alpha b} & -ae^{\alpha b} \end{vmatrix} = 0 \qquad ...(14)$$

On expanding this determinant, one obtains

$$\cos k(a+b) = \left(\frac{\alpha^2 - \beta^2}{2\alpha\beta}\right) \sinh \alpha b \sin \beta a + \cosh \alpha b \cos \beta a \qquad ...(15)$$

Since the wave vector *k* is real, the left-hand side of Eq. (15) lies only between –1 and +1. Since the right hand side of Eq. (15) is an oscillating function with increasing energy, there will be certain values of *E* for which the value of right-hand side lies between –1 and 1. These are the allowed energy levels separated by, ranges of *E* for which the value of the right-hand side lies outside –1 and 1, the forbidden energy levels. In order to understand this important conclusion, Kronig and Penney considered a simplified problem by assuming that $b - 0$ and $\phi_0 \to \infty$ such that the product $b\phi_0$ is finite. This assumption is equivalent to the consideration of Dirac δ-function type potential energies $\phi_0$ separated by a distance *a*, the potential energy being zero in between the δ-function spikes. With this assumption,

$$\frac{\alpha^2 - \beta^2}{2}(\sinh \alpha b) \simeq \frac{2m\,\phi_0}{2\hbar^2}(\alpha b) = \left(\frac{ma}{\hbar^2}\right)\frac{\alpha}{a}\, b\phi_0 \qquad ...(16)$$

$$= P\frac{\alpha}{a}\left[P = \frac{ma}{\hbar^2} b\phi_0 = \text{const.}\right]$$

Therefore, Eq. (15) is

$$\cos ka = \frac{P}{a\beta}\sin \beta a + \cos \beta a \qquad ...(17)$$

A plot of the right-hand side of Eq. (17) is given in Fig. 12.29 which leads to following inferences:

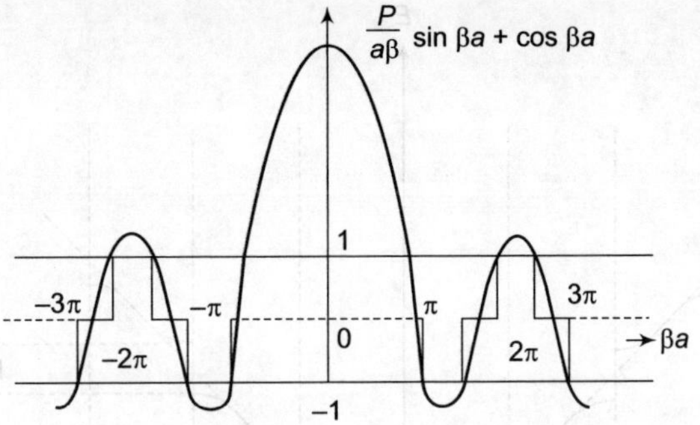


**Figure 12.29** Plot of $\frac{P}{a\beta}$ sin $\beta\alpha$ + cos $\beta a$ versus $\beta a$ to show allowed and forbidden bands for the Kroning–Penney approximation

1. Since cos $ka$ can have values between –1 and 1, only those values of $\beta a$ are allowed for which the right-hand side of Eq. (17) falls in this range. Thus, the boundaries of allowed ranges of $\beta a$ are given by cos $ka = \pm 1$, i.e.,

$$k = \pm \frac{n\pi}{a} \qquad ...(18)$$

   These values actually define the boundaries of Brillouin zones of one-dimensional lattice.
2. When $P = 0$, i.e., there is no potential barrier, we obtain cos $ka$ = cos $\beta a$, i.e., $\beta a = 2n\pi \pm ka$ ($n$ = 0, 1, 2, ...). Then all values of energy are allowed and there are no forbidden energy regions.
3. The energy $E$ is related to $\beta$ in terms of Eq. (1)

   i.e.,
$$E = \frac{\hbar^2}{2m}\beta^2 \qquad ...(19)$$

   This exhibits that the allowed values of $E$ follow the allowed ranges of $\beta a$. The $E$-$k$ curve for a one-dimensional lattice is then of the nature sketched in Fig. 12.30. It shows the periodic dependence of $E$ on $k$ and demonstrates the existence of forbidden energy gaps.
4. This model also indicates that allowed energy bands increase in width with increasing energy and forbidden bands decrease in width with increasing energy. Thus, for very high energies, all energies are allowed and the quantum mechanical description tends to the classical description.

## 12.19 NUMBER OF POSSIBLE WAVE FUNCTIONS PER BAND

For a linear crystal of length $L$, the boundary condition may be written as

$$\psi(x + L) = \psi(x) \qquad ...(1)$$

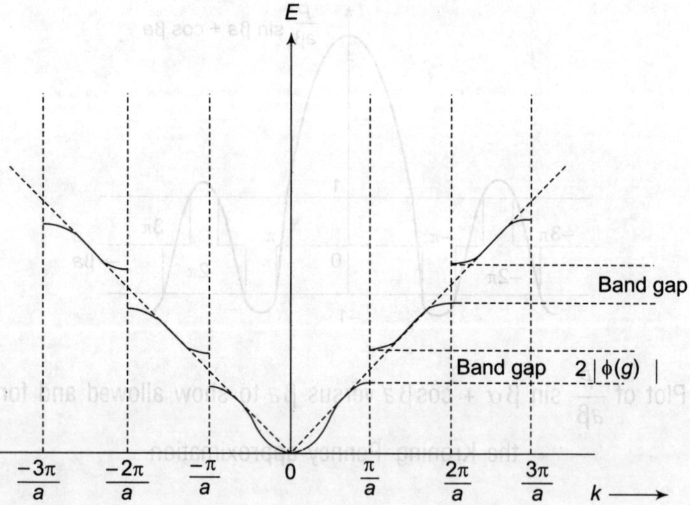


**Figure 12.30** *E-k* curve for a one-dimensional lattice

As we are dealing with Bloch functions, we must have

$$e^{ik(x+L)}u_k(x+L) = e^{ikx}u_k(x) \quad ...(2)$$

Because of the periodicity of $u_k(x)$, we have

$$u_k(x+L) = u_k(x)$$

and hence, Eq. (2) reduces to

$$e^{ikx(x+L)} = e^{ikx}$$

$$\Rightarrow \quad e^{ikx}\cdot e^{ikL} = e^{ikx}$$

$$\Rightarrow \quad e^{ikL} = 1$$

$$\Rightarrow \quad \cos kL + i\sin kL = 1 = \cos 2n\pi + i\sin 2n\pi \quad ...(3)$$

Equating real and imaginary parts, Eq. (3) gives

$$kL = 2n\pi$$

$$\Rightarrow \quad k = \frac{2n\pi}{L}, \text{ where } n = \pm 1, \pm 2, \ldots \quad ...(4)$$

$$\Rightarrow \quad n = \frac{kL}{2\pi}$$

The number of possible wave functions in the interval $dk$ is

$$dn = \frac{L}{2\pi}dk \quad ...(5)$$

Hence, the total number of possible wave functions in a band is

$$n = \int dn = \frac{L}{2\pi}\int_{-\pi/a}^{\pi/a} dk = \frac{L}{a}$$

$$\Rightarrow \qquad L = na \qquad ...(6)$$

With $b \to 0$, $(a + b) \to a$. Also, the length of the crystal $L = N(a + b) = Na$. Comparing this result with Eq. (6), we get $n = N$, where $N$ is the number of unit cells. This leads to the following very important conclusion:

*The total number of possible states in a band is equal to the number of unit cells in it.*

According to Pauli exclusion principle, each energy state can be occupied by two electrons having opposite spins. Thus, each energy band provides place for a maximum number of electrons equal to twice the number of unit cells. Stated otherwise, if there are $N$ unit cells, then the band is completely filled if the number of electrons in it is $2N$.

Since every band can accept up to two electrons per primitive cell, it follows that, when the number of electrons per primitive cell in even, i.e., equal to $2p$, then, for $T = 0$, the $p$ lower bands are fully occupied and the rest (above) are completely empty. Hence, in this case, the one-dimensional "solid" would be an insulator or semiconductor. On the other hand, if the number of electrons per primitive cell is odd, then the highest occupied band would be exactly half full, the $1D$ "solid" would exhibit conducting behaviour with the Fermi wavenumber $k_F$ being equal to $\pi/(2a)$ in the reduced zone scheme.

## 12.20 ORIGIN OF FORBIDDEN BANDS

The significance of the Brillouin zones becomes apparent when we examine the energies of the electrons in each zone. The energy of a free electron is related to its momentum $p$ by

$$E = \frac{p^2}{2m} \qquad ...(1)$$

and hence to its propagation constant $k$ by

$$E = \frac{\hbar^2 k^2}{2m} \qquad ...(2)$$

In the case of an electron in a crystal for which $k \ll \pi/a$, there is practically no interaction with the lattice, and Eq. (2) is valid. Since the energy of such an electron depends upon $k^2$, the contour lines of constant energy in a two-dimensional $k$-space are simply circles of constant $k$, as in Fig. 12.31. With increasing $k$ the constant-energy contour lines become progressively closer together and also more and more distorted. The reason for the first effect is merely that $E$ varies with $k^2$. The reason for the second is almost equally straightforward. The closer an electron is to the boundary of a Brillouin zone in $k$-space, the closer it is to being diffracted by the actual crystal lattice. But in particle terms the diffraction occurs by virtue of the interaction of the electron with the periodic array of positive ions that occupy the lattice points, and the stronger the interaction, the more the electron's energy is affected. Figure 12.32 shows how $E$ varies with $k$ in the $x$ direction. As $k$ approaches $\pi/a$, $E$ increases more slowly than $\hbar^2k^2/2m$, the free particle figure. At $k = \pi/a$, $E$ has two values, the lower belonging to the first Brillouin zone and the higher to the second zone. There is a definite gap between the possible energies in the first and second Brillouin zones which corresponds to the forbidden band spoken of earlier. The same pattern continues as successively higher Brillouin zones are reached.

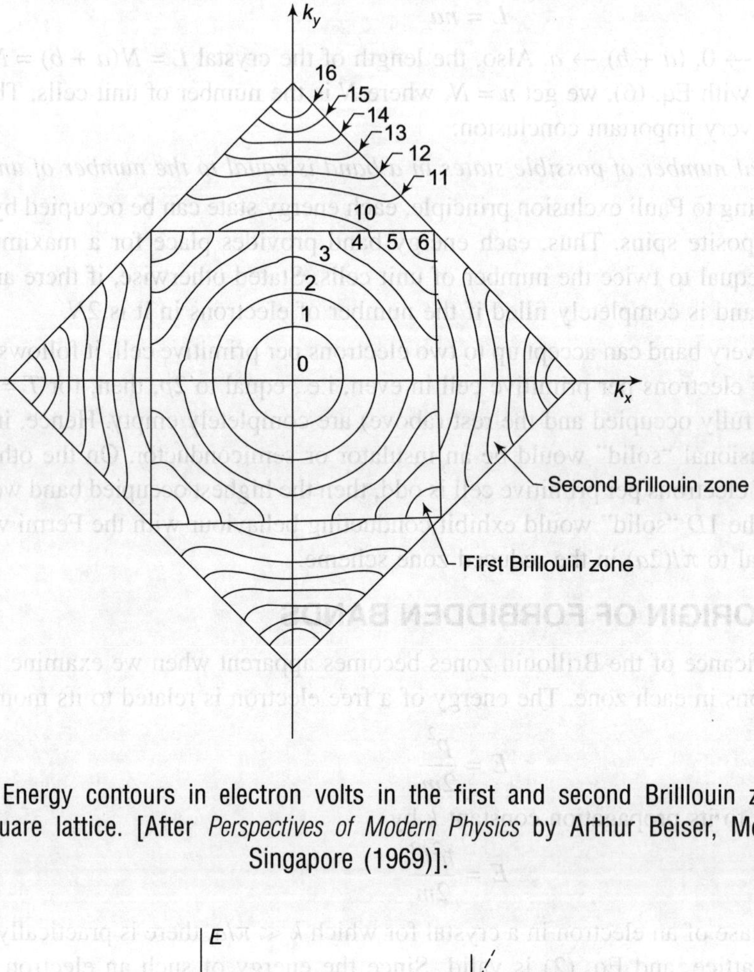


**Figure 12.31** Energy contours in electron volts in the first and second Brillouin zones of a hypothetical square lattice. [After *Perspectives of Modern Physics* by Arthur Beiser, McGraw-Hill, Singapore (1969)].

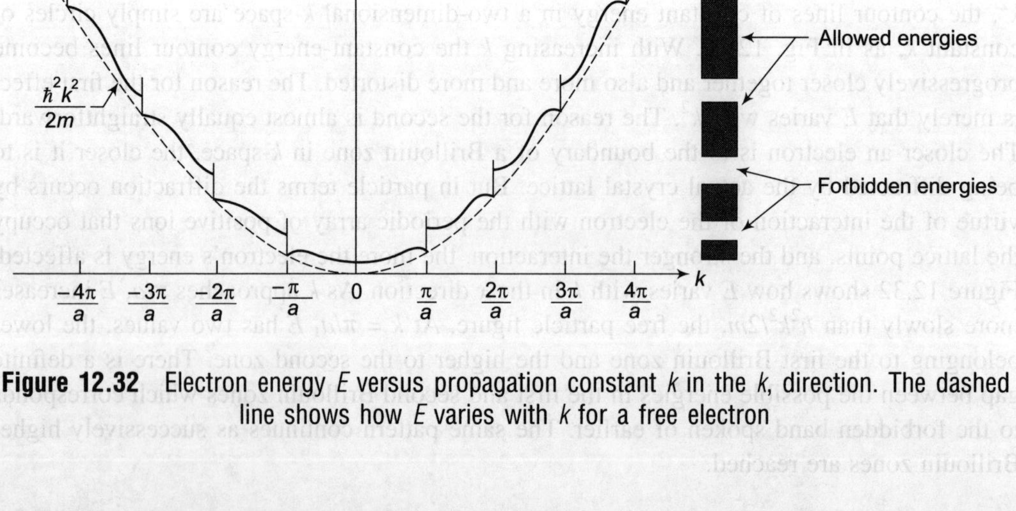


**Figure 12.32** Electron energy $E$ versus propagation constant $k$ in the $k$, direction. The dashed line shows how $E$ varies with $k$ for a free electron

The energy discontinuity at the boundary of a Brillouin zone follows from the fact that the limiting values of $k$ correspond to standing waves rather than travelling waves. For clarity we shall consider electrons moving in the $x$ direction; the extension of the argument to any other direction is straightforward. When $k = \pm\pi/a$, as we have seen, the waves are Bragg-reflected back and forth, so the only solutions of Schrödinger's equation consist of standing waves whose wavelength is equal to the periodicity of the lattice. There are two possibilities for these standing waves for $n = 1$, namely

$$\psi_1 = A \sin \frac{\pi x}{a} \qquad \text{...(3)}$$

$$\psi_2 = A \cos \frac{\pi x}{a} \qquad \text{...(4)}$$

The probability densities $|\psi_1|^2$ and $|\psi_2|^2$ are plotted in Fig. 12.33. Evidently, $|\psi_1|^2$ has its minima at the lattice points occupied by the positive ions, while $|\psi_2|^2$ has its maxima at the lattice points. Since the charge density corresponding to an electron wave function $\psi$ is $e|\psi|^2$, the charge density in the case of $\psi_1$ is concentrated *between* the positive ions while in the case of $\psi_2$ it is concentrated at the positive ions. The potential energy of an electron in a lattice of positive ions is greatest midway between each pair of ions and least at the ions themselves, so the electron energies $E_1$ and $E_2$ associated with the standing waves $\psi_1$ and $\psi_2$ are different. No other solutions are possible when $k = \pm\pi/a$, and accordingly no electron can have an energy between $E_1$ and $E_2$.

Figure 12.34 shows the distribution of electron energies that corresponds to the Brillouin zones pictured in Fig. 12.31. At low energies (in this hypothetical situation for $E <$ ~2 eV)

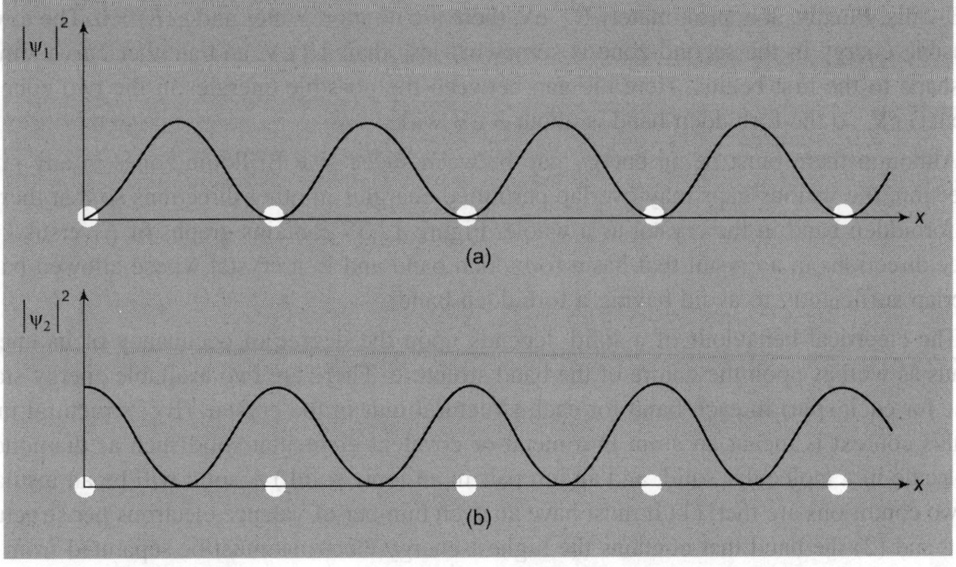


**Figure 12.33** Distributions of the probability densities, $|\psi_1|^2$ and $|\psi_2|^2$

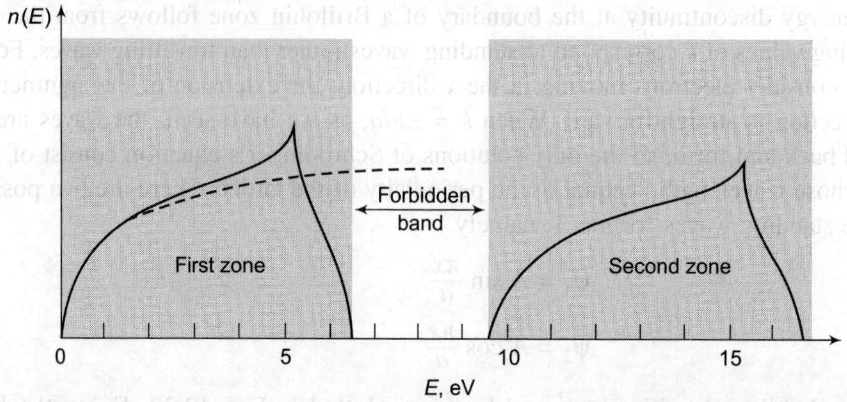


**Figure 12.34** The Distributions of electron energies in the Brillouin zones of Fig. 20.16. The dashed line is the distribution predicted by the free-electron theory

the curve is almost exactly the same as that (Fig. 12.31) based on the free-electron theory, which is not surprising since at low energies $k$ is small and the electrons in a periodic lattice then do behave like free electrons. With increasing energy, however, the number of available energy states goes beyond that of the free-electron theory owing to the distortion of the energy contours by the lattice: there are more different $k$ values for each energy. Then, when $k = \pm\pi/a$, the energy contours reach the boundaries of the first zone, and energies higher than about 4 eV (in this particular model) are forbidden for electrons in the $k_x$ and $k_y$ directions although permitted in other directions. As the energy goes farther and farther beyond 4 eV, the available energy states become restricted more and more to the corners of the zone, and $n(E)$ falls. Finally, at approximately $6^{1/2}$ eV, there are no more states and $n(E) = 0$. The lowest possible energy in the second zone is somewhat less than 10 eV, and another curve similar in shape to the first begins. Here the gap between the possible energies in the two zones is about 3 eV, so the forbidden band is about 3 eV wide.

Although there must be an energy gap between successive Brillouin zones in any given direction, the various gaps may overlap permitted energies in other directions so that there is no forbidden band in the crystal as a whole. Figure 12.35 contains graphs of $E$ versus $k$ for three directions in a crystal that has a forbidden band and in a crystal whose allowed bands overlap sufficiently to avoid having a forbidden band.

The electrical behaviour of a solid depends upon the degree of occupancy of its energy bands as well as upon the nature of the band structure. There are two available energy states (one for each spin) in each band for each structural unit in the crystal. (By "structural unit" in this context is meant an atom in a metal or covalent elemental solid such as diamond, a molecule in a molecular solid, and an ion pair in an ionic solid.) A solid will be an insulator if two conditions are met: (1) It must have an even number of valence electrons per structural unit, and (2) the band that contains the highest-energy electrons must be separated from the allowed band above it by an energy gap large compared with $kT$. The reason for condition

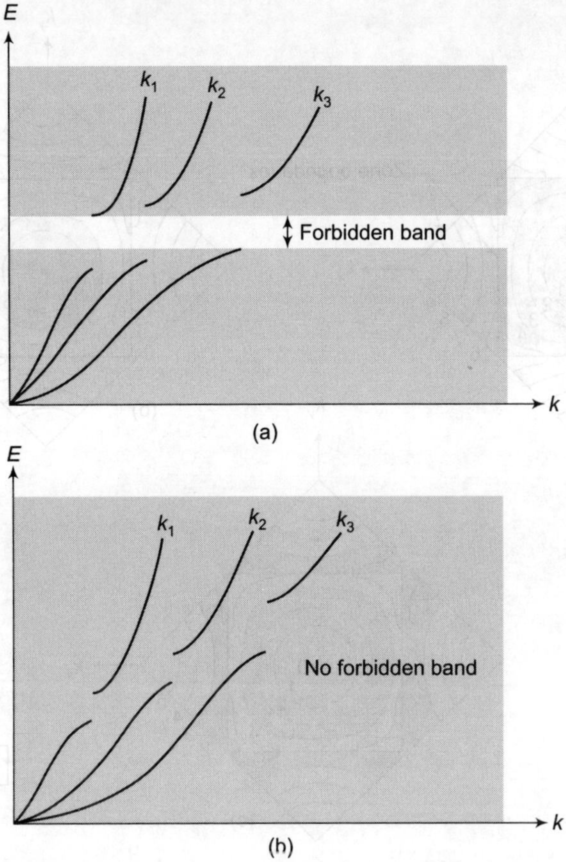


**Figure 12.35** *E* versus *k* curves for three directions in two crystals. In (a) There is a forbidden band, while in (b) The allowed energy bands overlap and there is no forbidden band

(1) is that it ensures that the highest-energy band be completely filled, and the reason for (2) is that none of the electrons be able to cross the gap to reach unfilled states. Thus, diamond, with four electrons per atom, solid hydrogen, with two electrons per $H_2$ molecule, and NaCl, with eight electrons per $Na^+$–Cl-ion pair, all have wide forbidden bands in addition and are insulators. Figure 12.36 a shows the energy contours of a hypothetical insulator.

A conductor is characterized by its violation of either (or both) of the above conditions.

Thus, the alkali metals, with an odd number of valence electrons per structural unit (namely one per atom), are conductors, as are such divalent metals as magnesium and zinc which have overlapping energy bands. Figure 12.36 (b) and (c) respectively show the energy contours of these two types of metal. When the forbidden band in an insulator is narrow or the amount of overlap in a metal is small, the electrical conductivity falls in the semiconductor region, and it is not really correct to speak of the substance as either a metal or a nonmetal. The

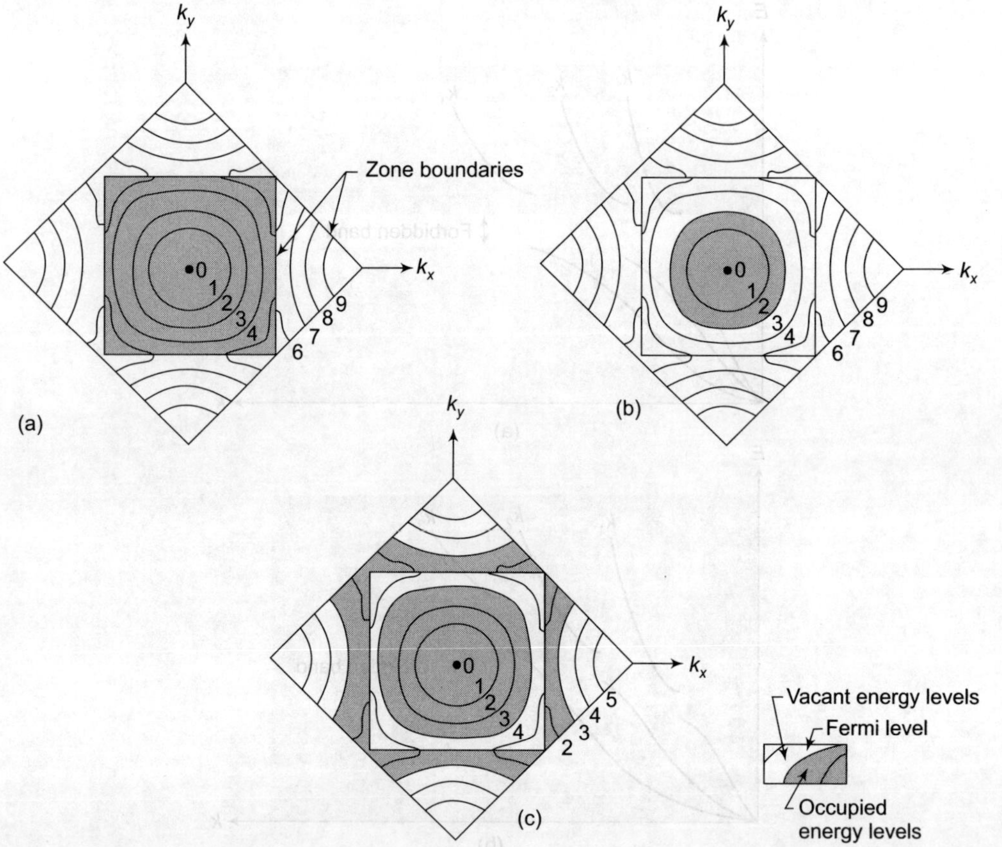


**Figure 12.36** Electron energy contours and Ferml levels in three types of solid: (a) Insulator, (b) Monovalent metal; (c) Divalent metal. Energies are in electron volts

boundary between filled and empty electron energy states in three-dimensional $k$-space is called the *Fermi surface.*

## 12.21 ZONE BOUNDARY EFFECTS

When the crystal potential $V$ is small compared with the total kinetic energy of the electrons, it is more meaningful to consider nearly free-electron Bloch waves in order to gain a better understanding of the behaviour of electrons in a periodic potential. As per pseudo-potential theory, this case has gained considerable importance in the description of real metals.

In the presence of a position dependent crystal potential $V(\vec{r})$ (no matter how small), Bloch's theorem tells us that the wave function of an electron moving in such a potential has the form

$$\psi(\vec{r}) = \psi_{\vec{k}}(\vec{r}) = u_k(\vec{r}) \exp(i\,\vec{k}\cdot\vec{r}) \qquad ...(1)$$

where $u_k(\vec{r} + 1) = u_k(\vec{r})$

i.e., $u_k(\vec{r})$ is periodic and has the periodicity of the crystal potential. This implies that we can expand $u_{\vec{k}}(\vec{r})$ in Fourier series as

$$u_{\vec{k}}(\vec{r}) = \sum_{\vec{g}} a_k(g) \exp(i\,\vec{g}\cdot\vec{r}) \qquad ...(2)$$

where the $a_k(g)$'s are the Fourier components of $u_{\mathbf{k}}(\mathbf{r})$. To determine the Fourier components, we must substitute the resulting expression for $\psi_{\mathbf{k}}$, i.e.,

$$\psi_{\mathbf{k}}(\mathbf{r}) = \sum_{g} a_k(g) \exp[i(\mathbf{k} + g)\cdot\mathbf{r}] = 0 \qquad ...(3)$$

into the Schrödinger equation. The result is

$$\sum_{g} [E^0_{k+g} - E(k) + V(\mathbf{r})]\, a_k(g) \exp[i(\mathbf{k} + \mathbf{g})\cdot\mathbf{r}] = 0$$

where $E^0_{k+g} = \hbar^2 (k+g)^2/2m$. If we now multiply this on the left by $\exp[-i(\mathbf{k} + \mathbf{g}')\cdot\mathbf{r}]$ and integrate with respect to $\mathbf{r}$ over all space, we find the relation

$$[E^0_{k+g} - E(k)]\, a_k(g) + \sum_{g' \neq g} V(\mathbf{g}' - \mathbf{g})\, a_k(g') = 0 \qquad ...(4)$$

where $V(\mathbf{g}' - \mathbf{g})$ is again the Fourier transform of $V(\mathbf{r})$, and we have chosen the zero of energy so that $V(\mathbf{g} = 0)$ is zero. This result holds for all $\mathbf{g}$ and $\mathbf{g}'$ (including $\mathbf{g} = 0$ and $\mathbf{g}' = 0$).

In order to solve for the $a_{\mathbf{k}}(g)$'s, we may proceed by making a system of approximations, the simplest being to set all $a_{\mathbf{k}}(g) = 0$ except when $\mathbf{g} = 0$. This simply gives the free-electron result $E(\mathbf{k}) = E^0_k$. A better approximation may be obtained by supposing that all $a_{\mathbf{k}}(g)$'s are small, except when $g = 0$. Then we find

$$a_{\mathbf{k}}(g) \approx \frac{V(g)}{E^0_{\mathbf{k}} - E^0_{\mathbf{k}+\mathbf{g}}}. \qquad ...(5)$$

The corresponding expression for the energy in this approximation is

$$E(\mathbf{k}) = E^0_{\mathbf{k}} + \sum_{\mathbf{g}} \frac{|V(\mathrm{g})|^2}{E^0_{\mathbf{k}} - E^0_{\mathbf{k}+\mathbf{g}}}, \qquad ...(6)$$

which is the standard result of second-order perturbation theory.

The assumption that $a_{\mathbf{k}}(g) \ll a_{\mathbf{k}}(0)$ for all values of $\mathbf{g}$ other than zero is see from Eq. (5) to be justified if $V$ is small, unless for any $\mathbf{g} \neq 0$, the denominator is small or zero. If, however, $\mathbf{k}$ lies on or near the Bragg plane defined by $E^0_{\mathbf{k}+\mathbf{g}} = E^0_{\mathbf{k}}$, for some value of $\mathbf{g}$, then the corresponding $a_{\mathbf{k}}(g)$ will *not* be small compared with $a_{\mathbf{k}}(0)$. Accordingly, in order to obtain approximate values of the energy and approximate form of the Bloch wave in this case, we set for the wave function $\psi_{\mathbf{k}}$,

$$\psi_{\mathbf{k}}(\mathbf{r}) = \exp(i\mathbf{k}\cdot\mathbf{r})\, [a_{\mathbf{k}}(0) + a_{\mathbf{k}}(\mathbf{g}) \exp(i\mathbf{g}\cdot\mathbf{r})] \qquad ...(7)$$

thereby neglecting all other terms that are really small. Substituting this into the Schrödinger equation, we again find

$$[E_{\mathbf{k}}^0 - E + V(r)]\, a_{\mathbf{k}}(0) \exp(i\mathbf{k}\cdot\mathbf{r}) + [E_{\mathbf{k+g}}^0 - E + V(r)]\, a_{\mathbf{k}}(g) \exp[i(\mathbf{k}+\mathbf{g})\cdot\mathbf{r}] = 0.$$

If we now multiply this equation either by

$$\exp(-i\mathbf{k}\cdot\mathbf{r}) \text{ or by } \exp[-i(\mathbf{k}+\mathbf{g})\cdot\mathbf{r}]$$

and integrate over all space, we obtain in the two cases

$$(E_k^0 - E)a_0 + V(g)\,a_g = 0,$$

$$V^*(g)a_0 + (E_{\mathbf{k+g}}^0 - E)\,a_g = 0 \qquad \text{...(8)}$$

where we have defined $a_0 = a_{\mathbf{k}}(0)$ and $a_g = a_{\mathbf{k}}(g)$ (fixed $g$). Eliminating $a_0$ and $a_g$ from these equations, we find

$$(E - E_{\mathbf{k}}^0)\,(E - E_{k+g}^0) - |V(g)|^2 = 0,$$

which is a quadratic equation for $E$ with solutions

$$E = \frac{1}{2}(E_{\mathbf{k}}^0 + E_{\mathbf{k+g}}^0) \pm \frac{1}{2}\sqrt{[(E_{\mathbf{k}}^0 - E_{\mathbf{k+g}}^0)^2 + 4|V(g)|^2]}. \qquad \text{...(9)}$$

In order to calculate $a_0$ and $a_g$, it is convenient to introduce the parameter

$$x = \left(\frac{\hbar^2}{2m}\right) g \cdot \frac{\frac{1}{2}\mathbf{g}+\mathbf{k}}{|V(g)|} \qquad \text{...(10)}$$

so that $x = 0$ defines a Brillouin zone face corresponding to the reciprocal lattice vector **g** under consideration and so that distances in **k**-space along the normal to the Brillouin zone face are proportional to $x$. Then, after a straightforward algebra, we find

$$E = E_{\mathbf{k}}^0 + |V(g)|\,[x \pm (1 + x^2)^{1/2}]. \qquad \text{...(11)}$$

And from the equations for $a_0$ and $a_g$, we determine the ratio

$$\frac{a_0}{a_g} = \frac{E - E_{\mathbf{k}}^0}{V(g)} = x \pm (1 + x^2)^{1/2}, \qquad \text{...(12a)}$$

which, on being taken together with the normalization condition

$$|a_0|^2 + |a_g|^2 = 1, \qquad \text{...(12b)}$$

yields the following results

$$|a_0|^2 = \frac{1}{2} \mp \frac{x}{(1 + x^2)^{1/2}},$$

$$|a_g|^2 = \frac{1}{2} \pm \frac{x}{(1 + x^2)^{1/2}}. \qquad \text{...(13)}$$

We are now in a position to discuss the mathematical schemes for representing $E$ and the $a$'s in **k**-space. We observe first that if there were no crystal potential at all, then $V(g)$ is zero and the energy $E$, considered a function of **k**, reduces simply of $E = \hbar^2\mathbf{k}^2/2m$ (parabola) and a surface in **k**-space with $k = k_F$, say, is a sphere corresponding to constant energy $E = E_F$.

Now, if $V(g) \neq 0$, then $E$ no longer is a simple parabola, although it resembles a parabola rather closely except near the Brillouin zone face. In particular it has two branches, corresponding to the respective choices of plus or minus signs in the expression (9) or (11). In an *extended zone scheme* of representation, it is quite possible to think of $E$, in spite of its two roots, as a *single-valued* function, as depicted in Fig. 12.37 (a) and (c). To arrive at this figure, we proceed as follows: We start along the normal $\Gamma A$ from the centre of a free-electron sphere in **k**-space and plot $E$ versus $k$ with $x < 0$ for one of the two branches and with $x > 0$ for the other branch. A similar plot of the cross section of the constant energy

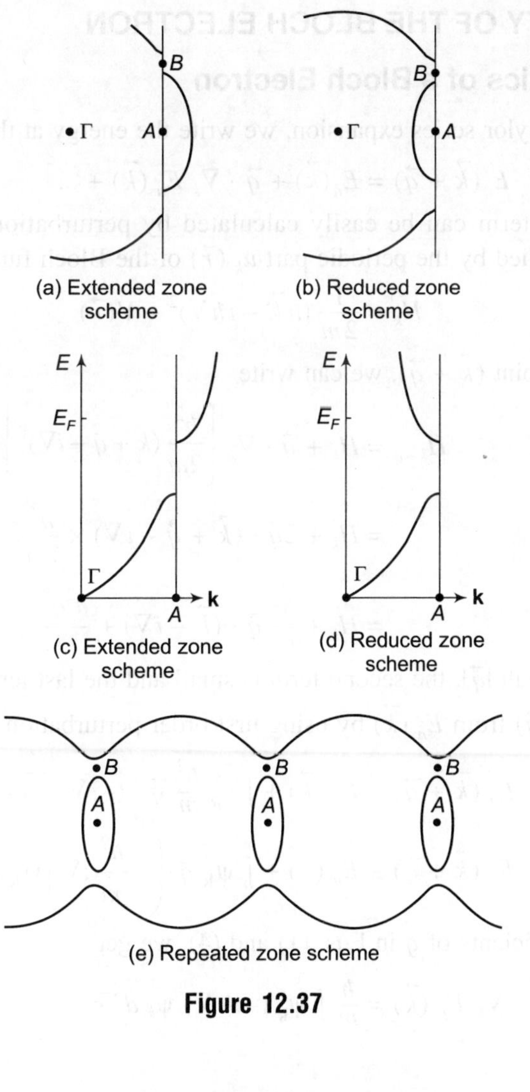


**Figure 12.37**

surface $E = E_F$ is shown in Fig. 12.37 (a). In this representation, $E$ is evidently a single-valued function of $k$ with discontinuity appearing at the Brillouin zone boundary: The magnitude of the discontinuity is $2|V(g)|$ and is called a *forbidden gap* or simply an *energy gap*. The gap disappears as $V(g)$ is allowed to tend to zero.

Another way of representing the same $E$ versus $k$ relation is as a *multiple-valued* function of $k$, where $k$ is restricted to the Brillouin zone. This is called the *reduced zone scheme* and is illustrated in Fig. 12.37 (b) and (d). To arrive at this representation, both branches of $E$ versus $k$ are plotted for only negative values of $x$ (so that $k$ lies in the range $\Gamma A$ only). The lower curve in Fig. 12.37 (d) for $E$ versus $k$ is called the *first band,* the next branch above it is called the *second band*; etc.

A combination of the two schemes, as in Fig. 12.37 (e), called *repeated zone scheme,* is sometimes useful.

## 12.22 VELOCITY OF THE BLOCH ELECTRON

### 12.22.1 Dynamics of a Bloch Electron

Using a first-order Taylor series expansion, we write the energy at the point $\vec{k} + \vec{q}$

$$E_n(\vec{k} + \vec{q}) = E_n(\vec{k}) + \vec{q} \cdot \nabla_k E_n(\vec{k}) + \ldots \qquad \ldots(1)$$

Now, the second term can be easily calculated by perturbation theory. We write the Hamiltonian $H_k$ satisfied by the periodic part $u_k(\vec{r})$ of the Bloch function:

$$H_k = \frac{1}{2m}(\hbar \vec{k} - i\hbar\nabla)^2 + V(\vec{r}) \qquad \ldots(2)$$

Similarly, at the point $(\vec{k} + \vec{q})$, we can write

$$H_{k+q} = H_k + \vec{q} \cdot \nabla_k \left\{ \frac{\hbar^2}{2m}(\vec{k} + \vec{q} - i\nabla)^2 \right\}$$

$$= H_k + 2\vec{q} \cdot (\vec{k} + \vec{q} - i\nabla) \times \frac{\hbar^2}{2m}$$

$$= H_k + \frac{\hbar^2}{m}\vec{q} \cdot (\vec{k} - i\nabla) + \frac{\hbar^2 q^2}{m} \qquad \ldots(3)$$

We see that for small $|\vec{q}|$, the second term is small and the last term is negligible. We can thus obtain $E_n(\vec{k} + \vec{q})$ from $E_n(\vec{k})$ by using first order perturbation theory which gives

$$E_n(\vec{k} + \vec{q}) = E_n(\vec{k}) + \int u_R^* \frac{\hbar^2}{m}\vec{q} \cdot (-i\nabla + \vec{k})\, u_k\, d^3\vec{r} + \ldots$$

or

$$E_n(\vec{k} + \vec{q}) \cong E_n(\vec{k}) + \int \psi_{\mathbf{k}}^* \vec{q} \cdot \left( -\frac{\hbar^2}{m} i\nabla \right) \psi_{\mathbf{k}}\, d^3\vec{r} + \ldots \qquad \ldots(4)$$

Equating the coefficients of $q$ in Eqs. (1) and (4), we get

$$\nabla_{\mathbf{k}} E_n(\vec{k}) = \frac{\hbar}{m} \int \psi_{\mathbf{k}}^* (-i\hbar\nabla)\, \psi_k\, d^3\vec{r}$$

or $$\nabla_{\mathbf{k}} E_n(\vec{k}) = \frac{\hbar}{m}\langle \vec{p}\rangle \quad ...(5)$$

Ehrenfest theorem then gives the group velocity of the wave group representing the electron as

$$\vec{v} = \frac{d\langle \vec{r}\rangle_{n,\vec{k}}}{dt} = \frac{1}{m}\langle \vec{p}\rangle_{n,\vec{k}} \quad ...(6)$$

from which we deduce the velocity of an electron in the Bloch state $\psi_{n,\vec{k}}$ as

$$\vec{v} = \frac{1}{\hbar}\nabla_{\vec{k}} E_n(\vec{\mathbf{k}}) \quad ...(7)$$

## 12.22.2 Acceleration in the Reciprocal Space

$\vec{k}$-space is the reciprocal space. The work $dW$ done on an electron of speed $\vec{v}$ and charge-$e$ over the time interval $dt$ changes its energy $E_n(\vec{k})$ (in an eternal electric field $\vec{\varepsilon}$) by modifying the value of $\vec{k}$ and thus the crystal momentum. Hence,

$$dW = -e\,\vec{\varepsilon}\cdot\vec{v}\,dt = \frac{d\,E_n(\vec{k})}{dt}\,dt \quad ...(8)$$

Using the expression (7) for $\vec{v}$ we get

$$-\varepsilon\,\vec{\varepsilon}\cdot\frac{1}{\hbar}\nabla_k E_n(\vec{k}) = \frac{dE_n(\vec{k})}{dt}$$

$$= \frac{d\,E_n(\vec{k})}{d\vec{k}} \times \frac{d\vec{k}}{dt}$$

$$= \nabla_k E_n(\vec{k})\cdot\frac{d\vec{k}}{dt}$$

so that $$\frac{\hbar\,d\vec{k}}{dt} = -e\,\vec{E} \equiv \vec{F} \text{ (the applied force)} \quad ...(9)$$

i.e., the response to an external force varying slowly in space and time is equal to the derivative of the crystal momentum and not the derivative of the electron momentum, Eq. (9) describes the behaviour of a packet of Bloch waves localized to about $\Delta r$ in real space and to about $\Delta k \sim 1/(\hbar\nabla r)$ in the reciprocal space.

## 12.22.3 Motion in Fields

If the force, $$\vec{F} = -e\,(\vec{\varepsilon} + \vec{v}\times\vec{B}) \quad ...(10)$$

varies in time at the scale of the transition frequencies between bands, the evolution of the system can no longer be described by the motion of the point $\vec{k}$ within a given band, but by transitions between bands. (This is true of the effect of light on a semiconductor.)

Equation (9) is an important relation: in a crystal $\hbar\,d\vec{k}/dt$ is equal to the external force on the electron. In free space $d(m\vec{v})/dt$ is equal to the force, the force term in this equation also includes the Lorentz force on an electron in a magnetic field, under low magnetic fields which are not so strong that it breaks down the band structure. Thus, for an electron of group velocity $\vec{v}$ in a constant magnetic field $\vec{B}$:

$$\hbar \frac{d\vec{k}}{dt} = -\frac{e}{c}\,\vec{v} \times \vec{B} \quad \text{(CGS units)}$$

or

$$\hbar \frac{d\vec{k}}{dt} = -e\,\vec{v} \times \vec{B} \quad \text{(SI units)} \qquad ...(11)$$

(where R.H.S. is the Lorentz force).

The rate of change of the wave vector is

$$\frac{d\vec{k}}{dt} = -\frac{e}{\hbar^2}(\nabla_k E) \times \vec{B} \quad \text{(using eq. (7) for } \vec{v}\text{)}. \qquad ...(12)$$

Here both sides of the equation refer to $\vec{k}$-space.

We see from the vector cross product in the above equation that in a magnetic field, an electron moves in $\vec{k}$-space in a direction normal to the direction of the gradient of energy $E$ so that *the electron moves on a surface of constant energy.*

The value of the projection $k_B$ of $\vec{k}$ on $\vec{B}$ is constant during the motion. The motion is $\vec{k}$-space is on a plane normal to the direction of $\vec{B}$, and the orbit is defined by the intersection of this plane with a surface of constant energy, i.e., electrons in a static magnetic field move on a curve of constant energy on a plane normal to $\vec{B}$. (Note that a surface of constant energy in $\vec{k}$-space is called a Fermi surface (see chapter 10) An electron on the FS will move in a curve on the F.S. There are three types of orbits (Fig. 12.38)

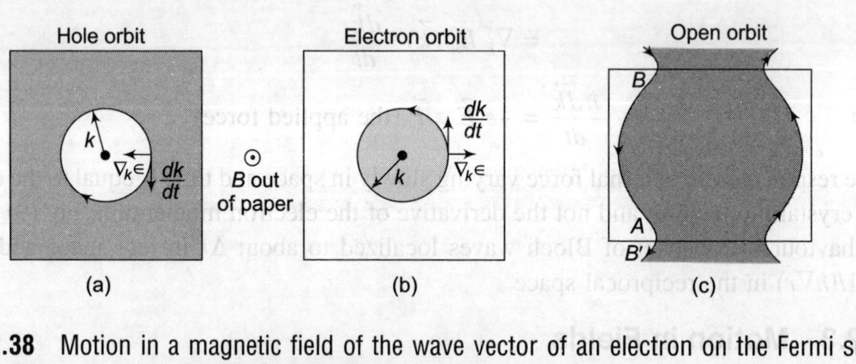


**Figure 12.38** Motion in a magnetic field of the wave vector of an electron on the Fermi surface, in (a) and (b) for Fermi surface topologically equivalent to those of Fig. 10. In (a) the wave vector moves around the orbit in a clockwise direction; in (b) the wave vector moves around the orbit in a counterclockwise direction. The direction in (b) is what we expect for a free electron of change $-e$: the smaller **k** value have the lower energy, so that the filled electron states lie inside the Fermi surface. We call the orbit in (b) **electronlike**. The sense of the motion in a magnetic field is opposite in (a) that in (b), so that we refer to the orbit in (a) as **homelike**. A hole move as a particle of charge $e$. In (c) for a rectangular zone we show the motion on an open orbit in the periodic see scheme. This is topologically intermediate between hole orbit and an electron orbit

In the open orbits, the particle on reaching the zone boundary at $A$ is instantly umklapped back to $B$, where $B$ is equivalent to $B'$ because they are connected by a reciprocal lattice vector. (The open orbits have an important effect on magnetoresistance.)

Vacant orbitals near the top of an otherwise filled band give rise to hole-like orbits as in Figs. (12.39) and (12.40). A view of a possible energy surface in three dimensions is given in Fig. (12.41).

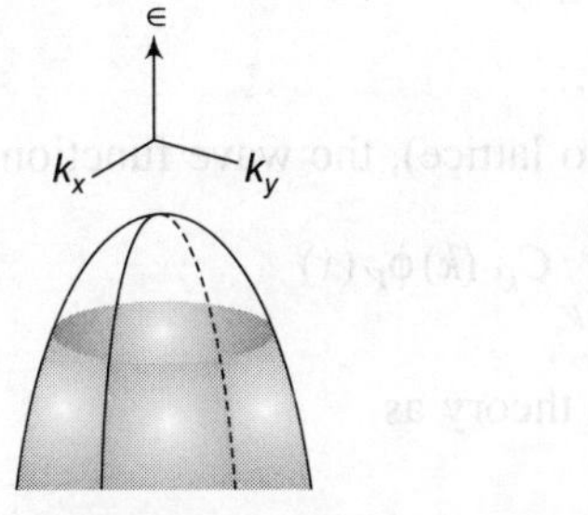


**Figure 12.40** Vacant states near the top of an almost filled band is a two-dimensional crystal. This figure is equivalent to Fig. 12.38 (a)

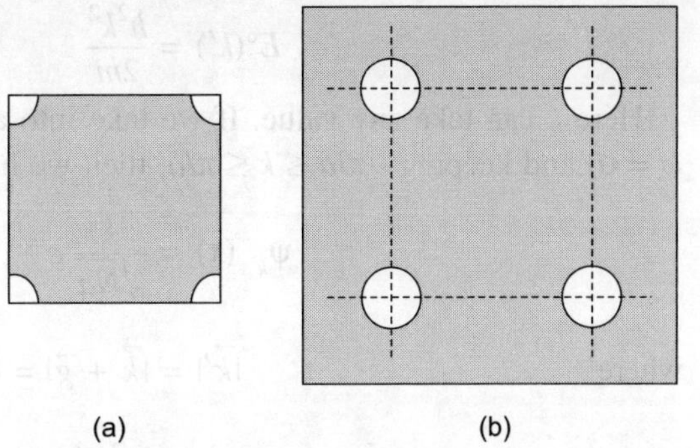


**Figure 12.39** (a) Vacant states at the corners of an almost-filled band, drawn in the reduced zone scheme. (b) In the periodic zone scheme the various parts of the Fermi surface are connected. Each circles forms a hole-like orbit. The deferent circles are entirely equivalent to each other, and the density of states is that of a single circle. (The orbits need not be true circles: for the lattice shown it is only required that the orbits have fourfold symmetry.)

## 12.23 NEARLY FREE ELECTRON CASE

For convenience, we may close the zero of potential energy such that

$$\int_0^a U(x)\, dx = 0 \qquad ...(1)$$

where $a$ is the periodicity of the one-dimensional lattice ($U = 0$ corresponds to the unperturbed wave functions). The unperturbed wave functions are the plane waves

$$\phi_k(x) = \frac{1}{\sqrt{Na}} e^{ikx} \qquad ...(2)$$

where the wave functions are normalised over $N$ atoms. The unperturbed electron energies are

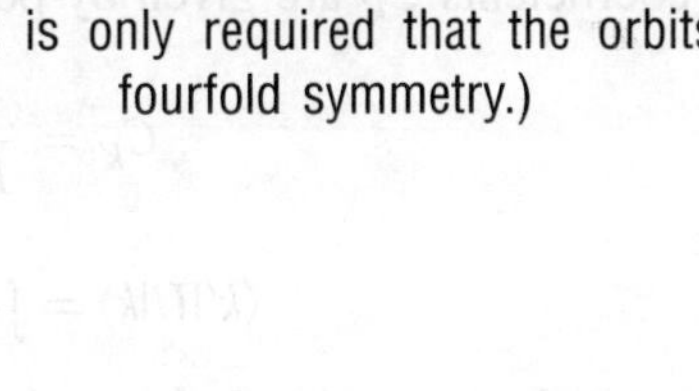


**Figure 12.41** Constant energy surface in the Brillouin zone of a simple cubic lattice for the assumed energy band $k = -a - 2\gamma$ (cos $k_xa$ + cos $k_ya$ + cos $k_za$) (a) Constant energy surface $e = -a$. The filled volume contains one electron per primitive cell

$$E^\circ(k') = \frac{\hbar^2 k^2}{2m} \qquad \text{...(3)}$$

Here $k$ can take any value. If we take into account the periodicity of the lattice (keeping $U = 0$) and keeping $-\pi/a \le k \le \pi/a$, then we have

$$\psi_{k'}(x) = \frac{1}{\sqrt{Na}} e^{ik'x} \qquad \text{...(4)}$$

where

$$|\vec{k}'| = |\vec{k} + \vec{g}| = k + n\frac{2\pi}{a} \qquad \text{...(5)}$$

and

$$E^\circ(k') = \frac{h^2}{2m}\left(k + n\frac{2\pi}{a}\right)^2 \qquad \text{...(6)}$$

where $n = 0, \pm 1, \pm 2, \ldots$

As we introduce the periodic potential (due to lattice), the wave functions are given by

$$\psi_k(x) = \phi_k(x) + \sum_{k' \neq k} C_{k'}(k)\phi_{k'}(x) \qquad \text{...(7)}$$

where coefficients $C_{k'}$ are given by perturbation theory as

$$C_{k'} = \frac{\langle k'|U|k\rangle}{E^\circ(k) - E^\circ(k')} \qquad \text{...(8)}$$

with

$$\langle k'|U|k\rangle = \int \phi_k^* U \phi_k \, dx$$

The perturbed energies (corrected to second order) are

$$E(k) = E^\circ(k) + \sum_{k'} \frac{\langle k'|U|k\rangle \langle k|U|k'\rangle}{E^\circ(k) - E^\circ(k')} \qquad \text{...(9)}$$

Because of Eq. (1), the first-order term is zero.

Now, the periodic potential due to linear lattice can be expressed as a Fourier series expansion

$$U(x) = \sum_{n \neq 0} V_n \exp(igx) \qquad \text{...(10)}$$

where $g$ is a reciprocal lattice vector and $V_n^* = V_{-n}$ (because of inversion symmetry). Then

$$\langle k'|U|k\rangle = \begin{cases} V_n & \text{for } k' = k + n\dfrac{2\pi}{a} \, (\equiv k + g) \\ 0 & \text{otherwise} \end{cases} \qquad \text{...(11)}$$

As a result, the wave function becomes

$$\psi_k(x) = \left[\frac{1}{\sqrt{Na}} \sum_n C_n(k) \exp(igx)\right] e^{ikx} \qquad \text{...(12)}$$

where

$$C_0 = 1 \quad \text{and} \quad C_n = \frac{V_n}{E^\circ(k) - E^\circ\left(k + n\dfrac{2\pi}{a}\right)}; (n \neq 0) \qquad ...(13)$$

The perturbed energy is

$$E(k) = E^\circ(k) + \sum_{n\neq 0} \frac{V_n^* V_n}{E^\circ(k) - E^\circ\left(k + \dfrac{2\pi n}{a}\right)} \qquad ...(14)$$

Near the zone boundaries, the energy is either increased or decreased with respect to $E^\circ(k)$ depending on

$$E^\circ(k) - E^\circ(k') = \frac{\hbar^2}{2m}[k^2 - (k+g)^2]$$

$$= \frac{h^2}{8\pi^2 m}\left[k^2 - \left(k + \frac{2\pi n}{a}\right)^2\right]$$

$$= -\frac{h^2 n}{2\pi m a}\left(k + \frac{2\pi}{a}\right) \qquad ...(15)$$

At the zone boundaries

$$k = -\frac{n\pi}{a} = -k' = \frac{g}{2} \qquad ...(16)$$

Thus, the wave is reflected $(k - k')$ at the zone boundary. Near a zone boundary, we must have

$$\psi_k(x) = C_0\, e^{ikx} + C_n e^{i(k+g)x} \qquad ...(17)$$

where $C_0$ and $C_n$ are constants. If we substitute the above equation in Schrödinger's equation, we get

$$[E^\circ(k) - E(k) + V_n(x)]\, C_0 \exp \exp (ikx)$$

$$+ [E^\circ(k+g) - E(k) + V_n(x)]\, C_n \exp[i(k+g)x] = 0 \qquad ...(18)$$

If we multiply this equation, either by $\exp(ikx)$ or by $\exp[-i(k+g)x]$ and integrate, we get the following simultaneous equations

$$[E^\circ(k) - E(k)]\, C_0 + V_n\, C_n = 0 \qquad ...(19)$$

$$V_n^*\, C_n\, [E^\circ(k+g) - E(k)] = 0 \qquad ...(20)$$

These two equations have a non-trivial solution only if the determinant of coefficients vanishes, i.e., when

$$\begin{vmatrix} E^\circ(k) - E(k) & V_n^* \\ V_n^* & E^\circ(k+g) - E(k) \end{vmatrix} = 0 \qquad ...(21)$$

This leads to a quadratic equation in $E(k)$

$$[E^\circ(k) - E(k)]\ [E^\circ(k + g) - E(k)] = |V_n|^2 \qquad ...(22)$$

This has two solutions

$$2E(k) = [E^\circ(k) + E^\circ(k + g)] \quad \{[E^\circ(k) - E^\circ(k + g)]^2 + 4|V_n|^2\}^{1/2} \qquad ...(23)$$

At the zone boundary, $k = n\pi/a$

and $$E^\circ(k) = E^\circ(k + g).$$

Therefore, $$E^+\left(\frac{n\pi}{a}\right) = E^\circ\left(\frac{n\pi}{a}\right) + |V_n| \qquad ...(24a)$$

and $$E^-\left(\frac{n\pi}{a}\right) = E^\circ\left(\frac{n\pi}{a}\right) - |V_n| \qquad ...(24b)$$

Hence, the energy gap at the first zone boundary ($n = 1$) is

$$E^+ - E^- = 2V_1\ (\equiv 2|V_g|) \qquad ...(25)$$

The two roots of Eq. (23) give the dominant effect of the periodic potential on the energies of the two free electron levels $E^\circ$ $(k)$ and $E^\circ$ $(k + g)$ when $k$ is close to the Bragg plane determined by $\vec{g}$. These are plotted in Fig. 12.42.

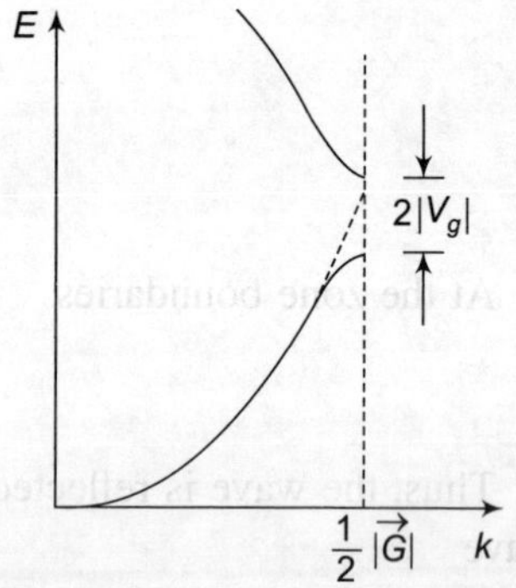


**Figure 12.42** Plot of the energy bands given by Eq. (13) for $\vec{k}$ parallel to $\vec{g}$. The lower band corresponds to the choice of minus sign in Eq. (13) and the upper band to a plus sign. When $\vec{k} = \frac{1}{2}\vec{g}$, the two bands are separated by a band gap of magnitude $2|V_g|$

## 12.24 ENERGY BANDS IN ONE DIMENSION GIVING RISE TO ZONE SCHEMES

Figure 12.43 (a) shows the free electron $E$ vs. $k$ parabola in one dimension. Figure (12.43) (b) illustrates first step in the construction of determining the distortion in the free electron parabola in the neighbourhood of a Bragg plane due to a weak periodic potential. If the Bragg plane is determined by $\vec{g}$, a second free electron parabola is drawn, centred on $g$. Figure 12.43 (c) shows step 2 in the construction to determine the distortion. The degeneracy of the two parabolas at $g/2$ is split. In Fig. 12.43 (d) those portions of Fig. 12.43 (c) corresponding to the original free electron parabola of Fig. 12.43 (a) are shown. Figure 12.43 (e) shows the effect of all additional Bragg planes on the free electron parabola. This way of displaying the electronic levels in a periodic potential is known as the *extended zone scheme.* In this scheme, different energy bands $[E(k)]$ are drawn in different zones in $k$-space. In Fig. 12.43 (f) the levels of Fig. 12.43 (e) reduced to first Brillouin zone are displayed. This corresponds to the way known as *reduced zone scheme.* (All wave vectors are transformed to the first Brillouin zone by addition of appropriate reciprocal lattice vectors so that all energy bands $[E(k)]$ are shown in the first

Brillouin zone.) Figure 12.43 (g) depicts electron levels of Fig. 12.43 (e) or Fig. 12.43 (f) in a *repeated zone scheme*. Every band [$E(k)$] is drawn in every zone in this scheme.

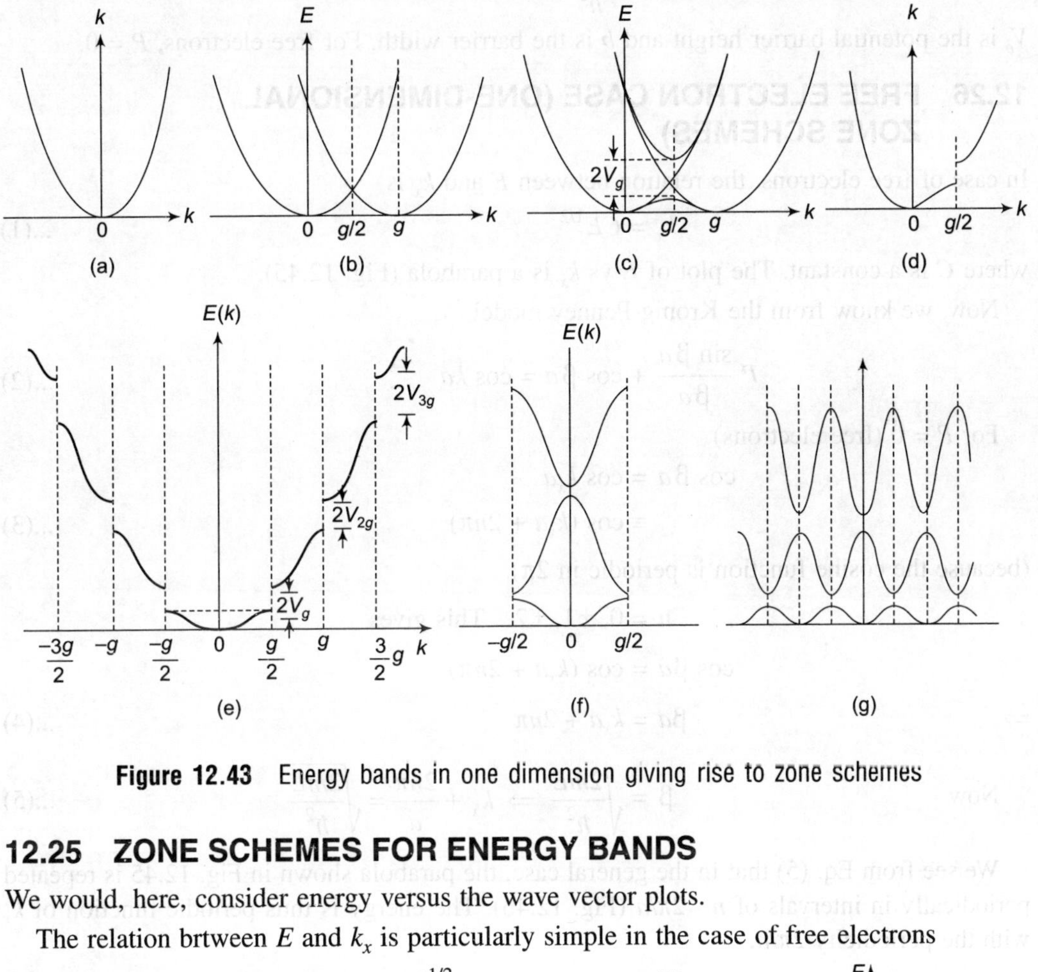


**Figure 12.43** Energy bands in one dimension giving rise to zone schemes

## 12.25 ZONE SCHEMES FOR ENERGY BANDS

We would, here, consider energy versus the wave vector plots.

The relation brtween $E$ and $k_x$ is particularly simple in the case of free electrons

$$k_x = \text{const} \times E^{1/2}. \qquad ...(1)$$

The plot $E(k_x)$ is a parabola (Fig. 12.44).

For one-dimensional case, we have Bloch functions

$$\psi(x) = u(x) \cdot e^{ikx}$$

where $u(x)$ possesses the periodicity of the lattice and, from Kronig-Penney model:

$$P\frac{\sin(\alpha a)}{\alpha a} + \cos(\alpha a) = \cos ka \qquad ...(2)$$

where $$\alpha = \left(\frac{2mE}{\hbar^2}\right)^{1/2} \quad (E = \text{energy}) \qquad ...(3)$$

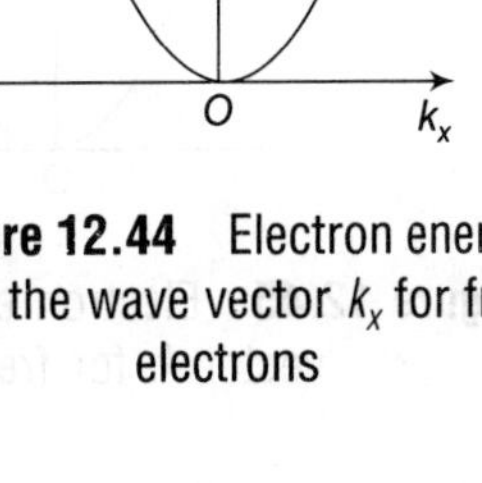


**Figure 12.44** Electron energy $E$ vs the wave vector $k_x$ for free electrons

$a$ is the lattice spacing.

$$P = \frac{maV_o}{\hbar^2} b$$

$V_o$ is the potential barrier height and $b$ is the barrier width. For free electrons, $P = 0$.

## 12.26 FREE ELECTRON CASE (ONE-DIMENSIONAL ZONE SCHEMES)

In case of free electrons, the relation between $E$ and $k_x$ is

$$k_x = CE^{1/2} \qquad \text{...(1)}$$

where $C$ is a constant. The plot of $E$ vs $k_x$ is a parabola (Fig. 12.45).

Now, we know from the Kronig-Penney model

$$P \frac{\sin \beta a}{\beta a} + \cos \beta a = \cos ka \qquad \text{...(2)}$$

For $P = 0$ (free electrons)

$$\cos \beta a = \cos k_x a$$
$$\equiv \cos (k_x a + 2n\pi) \qquad \text{...(3)}$$

(because the cosine function is periodic in $2\pi$).

$n = 0, \pm 1, \pm 2$... This gives

$$\cos \beta a = \cos (k_x a + 2n\pi)$$

$$\Rightarrow \qquad \beta a = k_x a + 2n\pi \qquad \text{...(4)}$$

Now $$\beta = \sqrt{\frac{2mE}{\hbar^2}} \Rightarrow k_x + \frac{2n\pi}{a} = \sqrt{\frac{2mE}{\hbar^2}} \qquad \text{...(5)}$$

We see from Eq. (5) that in the general case, the parabola shown in Fig. 12.45 is repeated periodically in intervals of $n \cdot 2\pi/a$ (Fig. 12.46). The energy is thus periodic function of $k_x$ with the periodicity $2\pi/a$.

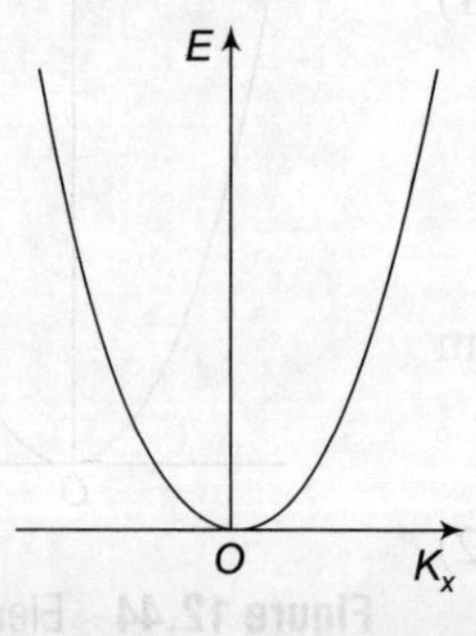


**Figure 12.45** Electron energy $E$ vs the wave vector $k_x$ for free electrons

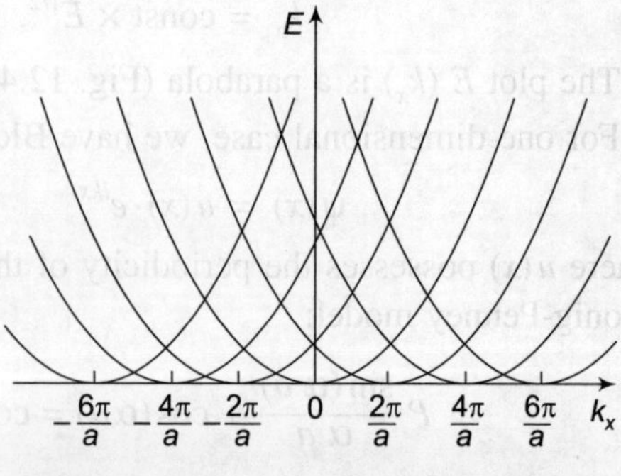


**Figure 12.46** Periodic repetition of Fig. 12.45 at the points $k_x = n \cdot 2\pi/a$

Now, we know that if an electron propagates in a periodic potential, we always observe discontinuities of the energies when $\cos k_x a = \pm 1$.

or $$k_x a = n\pi, \quad n = \pm 1, \pm 2, \pm 3....$$

or $$k_x = \frac{n\pi}{a} \qquad ...(6)$$

At these singularities, a deviation from the parabola $E$ vs $k_x$ curve occurs and the branches of the individual parabolas merge into the neighbouring ones as shown in Fig. 12.47.

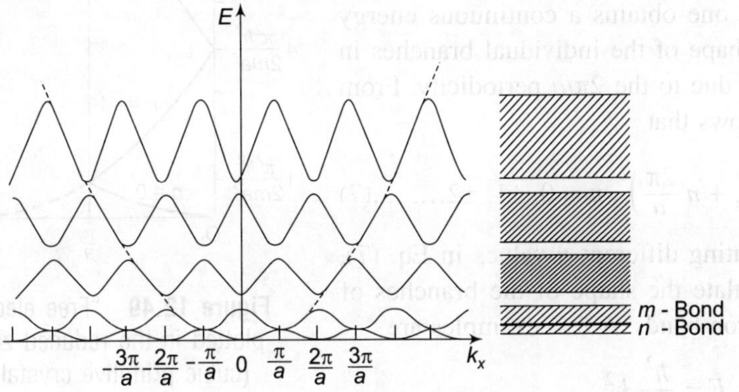


**Figure 12.47** Periodic zone scheme

The above considerations lead to an important result. The electrons in a crystal behave, for most $k_x$ values, like free electrons except when $k_x$ approaches the value $n \cdot \pi/a$.

Besides the "periodic zone scheme" of Fig. 12.47, two further zone schemes are common. We use mostly the "reduced zone scheme" (Fig. 12.49) the 'extended zone scheme', in which the deviations from the free electron at the critical points $k_x = n \cdot \pi/a$ are particularly easy to identify (Fig. 12.48).

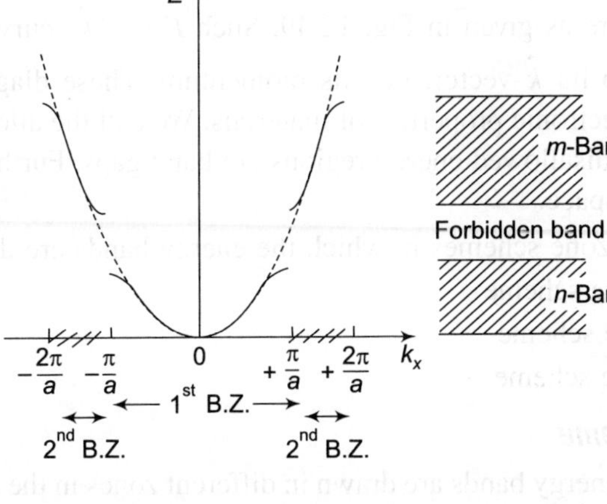


**Figure 12.48** Extended zone scheme. The first and second Brillouin zones (BZ) are shown

Sometimes, it is useful to plot free electrons in a reduced zone scheme. In doing so, one considers the width of the forbidden bands to be reduced until the energy gap between the individual branches disappears completely. This leads to the "free electron bands" which are shown in Fig. 12.49 for a special case. The well-known band character disappears for free electrons, however, and one obtains a continuous energy region. The shape of the individual branches in Fig. 12.48. is due to the $2\pi/a$ periodicity. From Eq. (5) it follows that

$$E = \frac{\hbar^2}{2m}\left(k_x + n\,\frac{2\pi}{a}\right)^2,\ n = 0, \pm 1, \pm 2\ldots \quad \ldots(7)$$

By substituting different $n$-values in Eq. (7), one can calculate the shape of the branches of the free electron bands. A few examples are

$n = 0$ gives $E = \dfrac{\hbar^2}{2m}\,k_x^2$

(parabola with $O$ as origin)

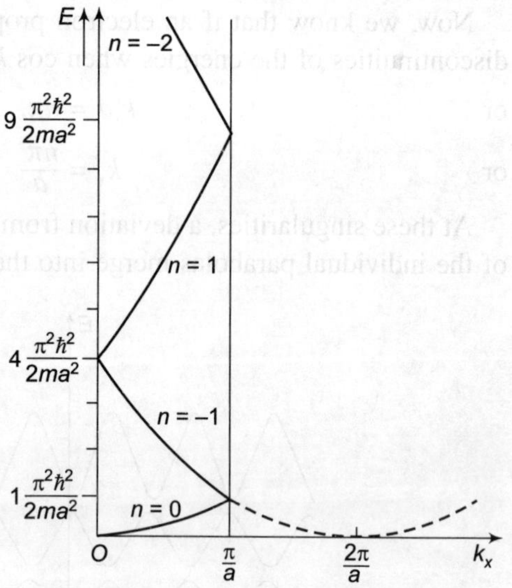


**Figure 12.49** "Free electron bands" plotted in the reduced zone scheme (cubic primitive crystal structure)

$n = -1$ gives $E = \dfrac{\hbar^2}{2m}\left(k_x - \dfrac{2\pi}{a}\right)^2$ $\left(\text{parabola with } \dfrac{2\pi}{a} \text{ as origin}\right)$

for $k_x = 0$ follows $E = 4\,\dfrac{\pi^2\hbar^2}{2ma^2}$

for $k_x = \dfrac{\pi}{a}$ follows $E = 1\,\dfrac{\pi^2\hbar^2}{2ma^2}$

The calculated data are as given in Fig. 12.49. Such $E$ vs. $|\vec{k}|$ curves actually relate the energy of an electron to its $\vec{k}$-vector, i.e., its momentum. These diagrams give important conclusions about the electronic properties of materials. We call the allowed bands as $n$-band or the $m$-band, etc., and disallowed energy regions are band gaps. Further, the wave vector $\vec{k}$ is defined in reciprocal space.

Thus, there are three zone schemes in which the energy bands are drawn:

(i) The extended zone scheme

(ii) The reduced zone scheme

(iii) The periodic zone scheme

### *The extended zone scheme*

In this scheme different energy bands are drawn in different zones in the $k$-space. Figure 12.48 gives the plot of dispersion curves in the extended zone scheme. The first band is shown in the first zone ($-\pi/a \le k \le \pi/a$) and the next higher in the second zone ($\pi/a \le k \le 2\pi/a$ and $-2\pi/a \le k \le -\pi/a$), and so on.

***The reduced zone scheme***

All energy bands are shown in the first Brillouin zone in this scheme. As an example, the free electron parabola is shown in this by Fig. 12.50. The curves in the two segments of the second zone are translated to the first zone by reciprocal vectors $2\pi/a$ and $-2\pi/a$, separately. Similarly, the energy band pictures in other zones are translated to the first zone by appropriate reciprocal lattice vectors.

***The periodic zone scheme***

Every band is drawn in every zone in this scheme.

The first three energy bands of a linear crystal as drawn in the three schemes are shown in Fig. 12.51 for the purpose of comparison.

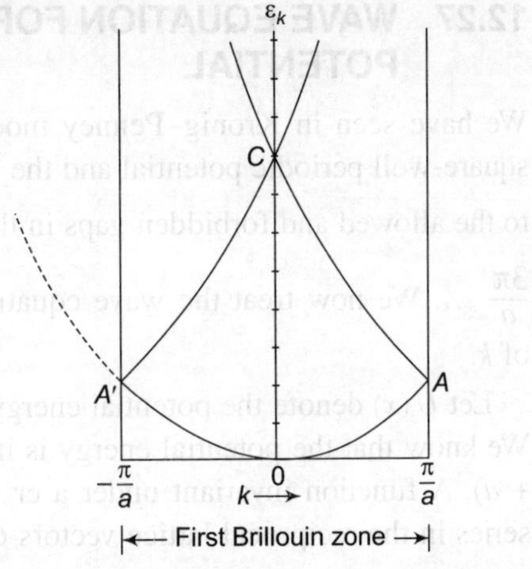


**Figure 12.50** The free electron energy parabola plotted in the reduced zone scheme for a one-dimensional lattice. The curve *AC*, when translated by $-2\pi/a$, reproduces the usual free electron curve for the negative *k*-values represented by the dashed curve. Similarly, the translation of A′C by $2\pi/a$ will give the curve for the positive *k*-values. This often gives a useful description of the band structure of a crystal

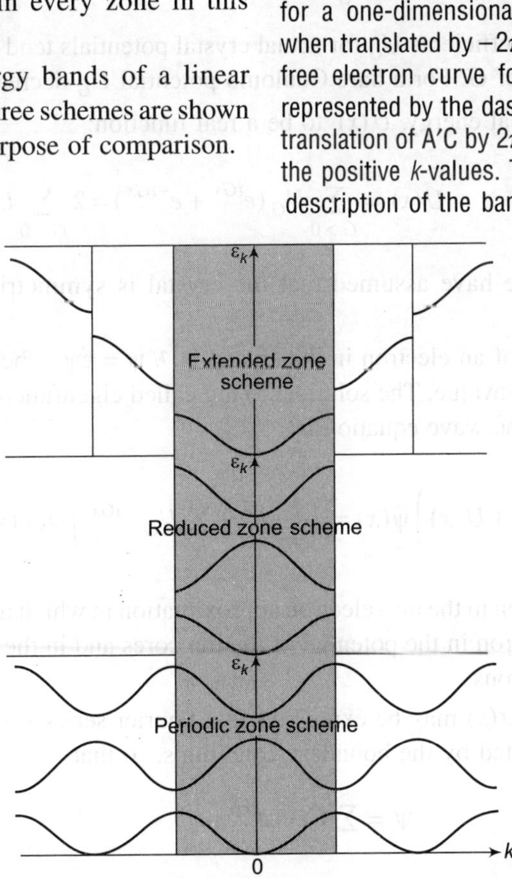


**Figure 12.51** First three energy bands of a one-dimensional crystal in the extended, reduced and periodic zone schemes

## 12.27 WAVE EQUATION FOR AN ELECTRON IN A PERIODIC POTENTIAL

We have seen in Kronig–Penney model for a one-dimensional array of atoms, there is a square-well periodic potential and the solution of the Schrödinger's wave equation gives rise to the allowed and forbidden gaps in the energy spectrum, with gaps occurring at $k = \frac{\pi}{a}, \frac{2\pi}{a}, \frac{3\pi}{a}$ .... We now treat the wave equation in detail for a general potential at general values of $k$.

Let $U(x)$ denote the potential energy of an electron in a linear lattice of lattice constant $a$. We know that the potential energy is invariant under a crystal lattice translation: $U(x) = U(x + a)$. A function invariant under a crystal lattice translation may be expanded as a Fourier series in the reciprocal lattice vectors $G$. We write the Fourier series for the potential energy as

$$U(x) = \sum_G U_G \, e^{iGx}. \qquad ...(1)$$

The values of the coefficients $U_G$ for actual crystal potentials tend to decrease rapidly with increasing magnitude of $G$. For a bare Coulomb potential $U_G$ decreases as $I/G^2$.

We want the potential energy, $U(x)$, to be a real function:

$$U(x) = \sum_{G>0} U_G (e^{iGx} + e^{-iGx}) = 2 \sum_{G>0} U_G \cos Gx. \qquad ...(2)$$

For convenience we have assumed that the crystal is symmetric about $x = 0$ and that $U_0 = 0$.

The wave equation of an electron in the crystal is $\mathcal{H}\psi = \varepsilon\psi$, where $\mathcal{H}$ is the Hamiltonian and $\varepsilon$ is the energy eigenvalue. The solutions $\psi$ are called eigenfunctions or orbitals or Bloch functions. Explicitly, the wave equation is

$$\left(\frac{1}{2m} p^2 + U(x)\right) \psi(x) = \left(\frac{1}{2m} p^2 + \sum_G U_G \, e^{iGx}\right) \psi(x) = \varepsilon \, \psi(x). \qquad ...(3)$$

Equation (3) is written in the one-electron approximation in which the orbital $\psi(x)$ describes the motion of one electron in the potential of the ion cores and in the average potential of the other conduction electrons.

The wave function $\psi(x)$ may be expressed as a Fourier series summed over all values of the wave vector permitted by the boundary conditions, so that

$$\psi = \sum_k C(k) \, e^{ikx} \qquad ...(4)$$

where $k$ is real. (We could equally well write the index $k$ as a subscript on $C$, as in $C_k$.)

The set of values of $k$ has the form $2\pi n/L$, because these values satisfy periodic boundary conditions over length $L$. Here $n$ is any integer, positive or negative. We do not assume, nor is it generally true, that $\psi(x)$ itself is periodic in the fundamental lattice translation, $a$. The translation properties of $\psi(x)$ are determined by the Bloch theorem.

Not all wave vectors of the set $2\pi n/L$ enter the Fourier expansion of any one Bloch function. If one particular wave vector $k$ is contained in a $\psi$, then all other wave vectors in the Fourier expansion of this $\psi$ will have the form $k + G$, where $G$ is any reciprocal lattice vector. We prove this result in Eq. (8).

We can label a wave function $\psi$ that contains a component $k$ as $\psi_k$ or, equally well, as $\psi_{k+G}$, because if $k$ enters the Fourier expansion then $k + G$ amy enter. The wave vectors $k + G$ running over $G$ are a restricted subset of the set $2\pi n/L$, as shown in Fig. 12.52.

We shall usually choose as a label for the Bloch function that $k$ say which lies within the first Brillouin zone. When other conventions are used, we shall say so. This situation differs from the phonon problem, where there are no components of the ion motion outside the first zone. The electron problem is like the X-ray diffraction problem because the electromagnetic field exists everywhere within the crystal and not only at the ions.

To solve the wave equation, substitute Eq. (4) in Eq. (3) to obtain a set of linear algebraic equations for the Fourier coefficients. The kinetic energy term is

$$\frac{1}{2m}p^2\psi(x) = \frac{1}{2m}\left(-i\hbar\frac{d}{dx}\right)^2\psi(x) = -\frac{\hbar^2}{2m}\frac{d^2\psi}{dx^2} = \frac{\hbar^2}{2m}\sum_k k^2 C(k)\,e^{ikx};$$

and the potential energy term is

$$\left(\sum_G U_G\,e^{iGx}\right)\psi(x) = \sum_G\sum_k U_G\,e^{iGx}\,C(k)\,e^{ikx}.$$

The wave equation is obtained as the sum

$$\sum_k \frac{\hbar^2}{2m}k^2 C(k)\,e^{ikx} + \sum_G\sum_k U_G C(k)\,e^{i(k+G)x} = \varepsilon\sum_k C(k)\,e^{ikx}. \qquad ...(5)$$

Each Fourier component must have the same coefficient on both sides of the equation. Thus,

$$\boxed{(\lambda_k - \varepsilon)\,C(k) + \sum_G U_G C(k-G) = 0.} \qquad ...(6)$$

with the notation

$$\lambda_k = \hbar^2k^2/2m. \qquad ...(7)$$

Equation (6) is a useful form of the wave equation in a periodic lattice, although unfamiliar because a set of algebraic equations has taken the place of the usual differential equation. The set appears unpleasant and formidable because there are, in principle, an infinite number of $C(k - G)$ to be determined. In practice a small number will often suffice perhaps two or four. Once we determine the C's from Eq. (6), the wave function (4), is given as

$$\psi_k(x) = \sum_G C(k - G)\, e^{i(k - G)x} \qquad ...(8)$$

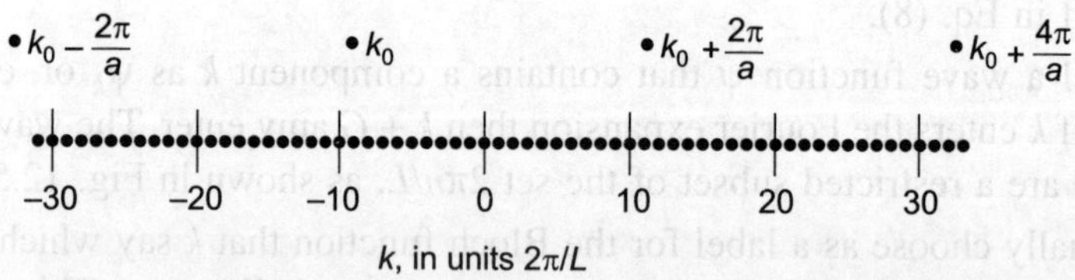

**Figure 12.52** The lower points represent values of the wave vector $k = 2\pi n/L$ allowed by the periodic boundary condition on the wave function over a ring of circumference $L$ composed of 20 primitive cells. The allowed values continue to $\pm\infty$. The upper points represent the first few wave vectors which may enter into the Fourier expansion of a wave function $\psi(x)$, starting from a particular wave vector $k = k_0 = -8(2\pi/L)$. The shortest reciprocal lattice vector is $2\pi/a = 20(2\pi/L)$

## 12.28 REDUCED ZONE SCHEME

It is always possible to select the wave vector index $k$ of any Bloch function to lie within the first Brillouin zone. The procedure is known as mapping the band in the reduced zone scheme.

If we encounter a Bloch function written as $\psi_{\mathbf{k}'}(\mathbf{r}) = e^{i\mathbf{k}'\cdot\mathbf{r}} u_{\mathbf{k}'}(\mathbf{r})$, with $\mathbf{k}'$ outside the first zone, as in Fig. 12.53, we may always find a suitable reciprocal lattice vector $\mathbf{G}$ such that $\mathbf{k} = \mathbf{k}' + \mathbf{G}$ lies within the first Brillioun zone. Then

$$\psi_{\mathbf{k}'}(\mathbf{r}) = e^{i\mathbf{k}'\cdot\mathbf{r}}\, u_{\mathbf{k}'}(\mathbf{r}) = e^{i\mathbf{k}\cdot\mathbf{r}}\, (e^{-i\mathbf{G}\cdot\mathbf{r}} u_{\mathbf{k}'}(\mathbf{r}))$$

$$\equiv e^{i\mathbf{k}\cdot\mathbf{r}} u_{\mathbf{k}'}(\mathbf{r}) = \psi_{\mathbf{k}}(\mathbf{r}), \qquad ...(1)$$

where $u_{\mathbf{k}}(\mathbf{r}) \equiv e^{-i\mathbf{G}\cdot\mathbf{r}}\, u_{\mathbf{k}'}(\mathbf{r})$. Both $e^{-i\mathbf{G}\cdot\mathbf{r}}$ and $u_{\mathbf{k}'}(\mathbf{r})$ are periodic in the crystal lattice, so $u_{\mathbf{k}}(\mathbf{r})$ is also, whence $\psi_k(\mathbf{r})$ is of the Bloch form.

Even with free electrons it is useful to work in the reduced zone scheme, as in Fig. 12.54. Any energy $\varepsilon_{\mathbf{k}'}$ for $\mathbf{k}'$ outside the first zone is equal to an $\varepsilon_{\mathbf{k}}$ in the first zone, where $\mathbf{k} = \mathbf{k}' + \mathbf{G}$. Thus, we need to solve for the energy only in the first Brillouin zone, for each band. An energy band is a single branch of the $\varepsilon_{\mathbf{k}}$ versus $\mathbf{k}$ surface.

In the reduced zone scheme we should not be surprised to find different energies at the same value of the wave vector. Each different energy characterizes a different band.

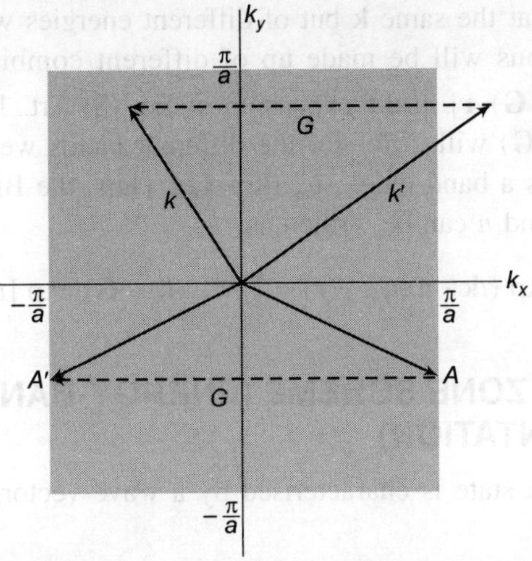


**Figure 12.53** First Brillouin's zone of a square lattice of side *a*. The wave vectror *k'* can be carried into the first zone by forming **k'** + **G**. The wave vector at a point *A* on the boundary is carried by **G** to the point *A'* on the opposite boundary of the same zone. Shall we count both *A* and *A'* as lying in the first zone? Because they can be connected by a reciprocal lattice vector, we count them as *one* identical point in the zone

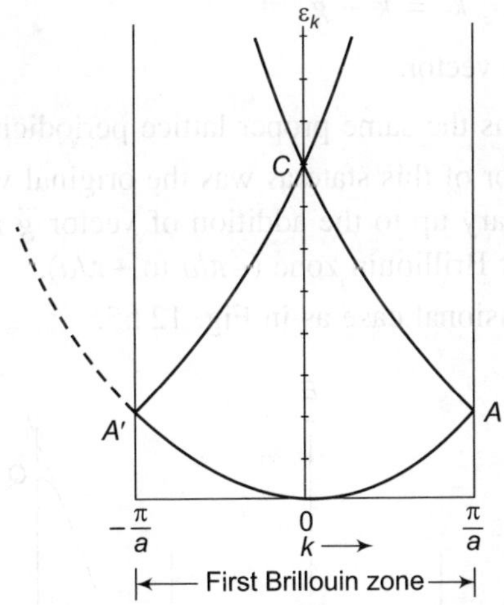


**Figure 12.54** Energy-wave vector relation $\varepsilon_k = \hbar^2k^2/2m$ for free electrons as drawn in the reduced zone scheme. This construction often gives a useful idea of the overall appearance of the band structure of a crystal. The branch *AC* if displaced by $-2\pi/a$ gives the usual free electron curve for negative *k*, as suggested by the dashed curve. The branch *A'C* if displaced by $2\pi/a$ gives the usual curve for positive *k*. A crystal potential $U(x)$ will introduce band gaps at the edges of the zone (as at *A* and *A'*) and at the centre of the zone (as at *C*). The point *C* when viewed in the extended zone scheme falls at the edges of the second zone. The overall width and gross features of the band structure are often indicated properly by such free electron bands in the reduced zone scheme

Two wave functions at the same **k** but of different energies will be independent of each other: the wave functions will be made up of different combinations of the plane wave components exp $[i(\mathbf{k} + \mathbf{G})\cdot\mathbf{r}]$ in the expansion of Eq. (8) Art. 12.2. Because the values of the coefficients $C(\mathbf{k} + \mathbf{G})$ will differ for the different bands we should add a symbol, say $n$, to the $C$'s to serve as a band index: $C_n(\mathbf{k} + \mathbf{G})$. Thus, the Bloch function for a state of wave vector **k** in the band $n$ can be written as

$$\psi_{n,k} = \exp(i\mathbf{k}\cdot\mathbf{r})\, u_{n,k}(\mathbf{r}) = \sum_G C_n(\mathbf{k} + \mathbf{G}) \exp[i(\mathbf{k} + \mathbf{G})\cdot\mathbf{r}].$$

## 12.29 REDUCED ZONE SCHEME (ENERGY-BAND REPRESENTATION)

Although every electron state is characterised by a wave vector, this is not unique. Since, we may write

$$\psi_{\vec{k}}(\vec{r}) = e^{i\vec{k}\cdot\vec{r}} u_{\vec{k}}(\vec{r})$$

$$= e^{i(\vec{k}+\vec{g})\cdot\vec{r}}\{e^{-i\vec{g}\cdot\vec{r}} u_{\vec{k}}(\vec{r})\}$$

$$\equiv e^{i\vec{k}'\cdot\vec{r}} u'(\vec{r})$$

where $$\vec{k}' = \vec{k} + \vec{g}$$

$\vec{g}$ being the reciprocal lattice vector.

The new function $u'(\vec{r})$ has the same proper lattice periodicity, so that $\vec{k}'$ is just as good a definition of the wave vector of this state as was the original vector $\vec{k}$. Thus, wave vectors of electronic states are arbitrary up to the addition of vector $\vec{g}$ and this arbitrariness allows us to bring all $\vec{k}$ into the first Brillioin zone ($-\pi/a$ to $+\pi/a$).

Let us consider one-dimensional case as in Fig. 12.55.

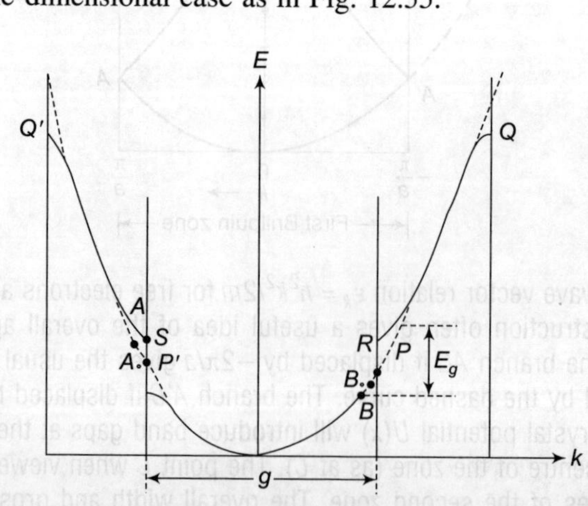


**Figure 12.55** Splitting of energy levels in one dimension

In Fig. 12.56 this has been redrawn, transforming all points into the central section of the $\vec{k}$-axis. It may be seen that $A$ and $B$, which represent linear combinations of free electron wave functions at $A_0$ and $B_0$, now lie at the same value of $\vec{k}$, split apart only by the perturbing potential (due to lattice). The curves $RQ$ and $SQ'$ of Fig. 12.55 are respectively represented by $S'Q$ and $R'Q'$ of Fig. 12.56 in the transformation. $Q$ and $Q'$ differing by $g$ in Fig. 12.55 trace the same point in Fig. 12.56. The upper curve is now continuous in $\vec{k}$ (like the lower curve and the two halves join precisely at $k = 0$).

Figure 12.56 represents what is known as a *reduced zone scheme*. It is characteristic of this scheme that the energy function $E(\vec{k})$ is necessarily multivalued and separates into branches or bands.

Finally, we could get Fig. 12.57 (from Fig. 12.55) depicting four bands in the reduced zone representation. We know that the wave vectors for electronic states, are arbitrary up to the addition of a vector of the reciprocal lattice. This arbitrariness allows us to bring all wave vectors into the first Brillouin zone. Consequently, we have the reduced zone scheme (Fig. 12.56). It is its characteristic that the energy function $E_k$ is continuous right across the zone, but is multivalued and separated into branches or *bands*, separated by forbidden bands. Figure 12.56 is for one-dimensional case. In three dimensions, the same reduction carries the Brillouin zone regions separated by possible energy discontinuities into the central zone. One can show that these regions fall into sets such that when transformed into the central cell, the various elements of the set jigsaw together to cover exactly the first zone. Each such set in its original positions makes up a single B.Z. We can pile the various zones on top of each other in the central zone and allot to each a branch of the energy function, which is a continuous function of $\vec{k}$ just as in one-dimensional case. Although, in our original picture (Fig. 12.58),

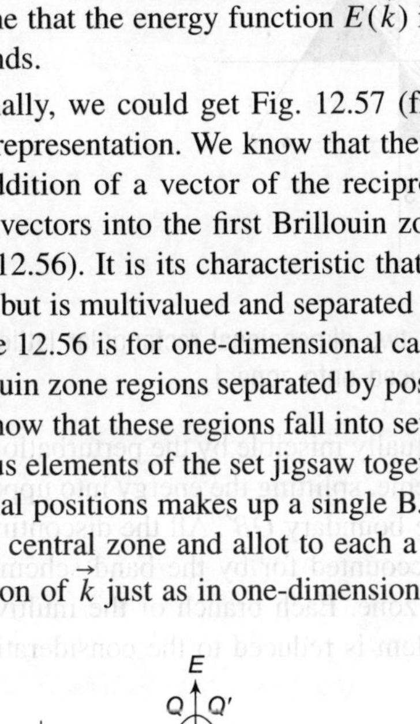


**Figure 12.56** Reduced zone scheme in one dimension

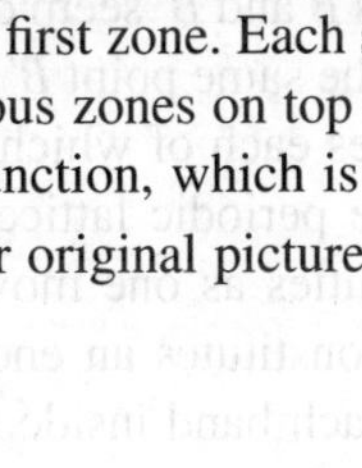


**Figure 12.57** Energy vs. reduced $\vec{k}$.

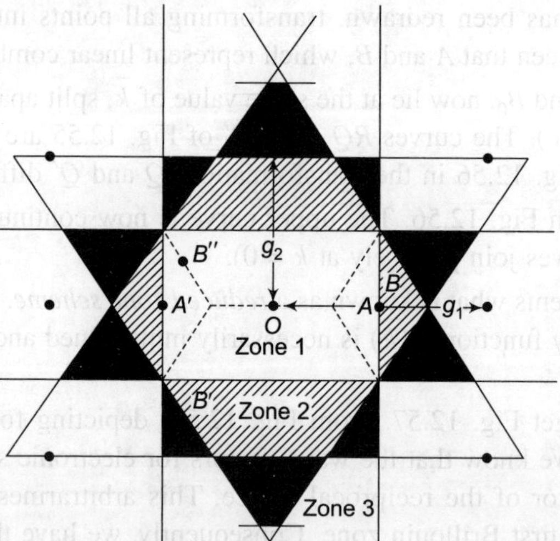


**Fig. 12.58** Brillouin zones in two dimensions for a two-dimensional rectangular lattice, showing how zone 2 may be reduced onto zone 1

points such as $B$ and $B'$ seem quite distinct, they are actually miscible by the perturbation, and both go into the same point $B'$ in the reduced zone scheme, splitting the energy into upper and lower branches each of which is continuous across the boundary $OB'$. All the discontinuities caused by the periodic lattice potential are exactly accounted for by the band scheme, but no discontinuities as one moves over the whole first zone. Each branch of the multivalued function $E_k$ constitutes an energy band, and the problem is reduced to the consideration of the form of each band inside the reduced zone.

## 12.30 FROM ISOLATED ATOMS TO THE FREE-ELECTRON LIMIT

Just how much influence the periodic lattice potential has on the electrons depends both on the magnitude of this potential and on the opportunities for atoms to interact, which varies with the interatomic spacing.

When atoms are sufficiently far apart, there is no interaction, and electrons are restricted to a discrete set of atomic states [Fig. 12.59 (a)]. For an interatomic spacing that permits some (not very much) overlap between atoms, the bands can be simulated by the tight binding model. (This was introduced by Bloch in 1928 in his initial discussion of energy bands.) This approach is also called the LCAO (linear combination of atomic orbitals) method, since in this case the wave functions for electrons in the relatively narrow bands can be approximated by an LCAO. (This approach is applicable, provided the band broadening is small compared with the energy gap; so this method is suitable for describing the broadening of $d$ bands in Cu or the deep $s$ and $p$ levels like $2s$ and $2p$ levels in Na but is not adequate for the ionized $s$ and $p$ states such as the $4s$ level in Cu or $3s$ and $3p$ levels in Na.)

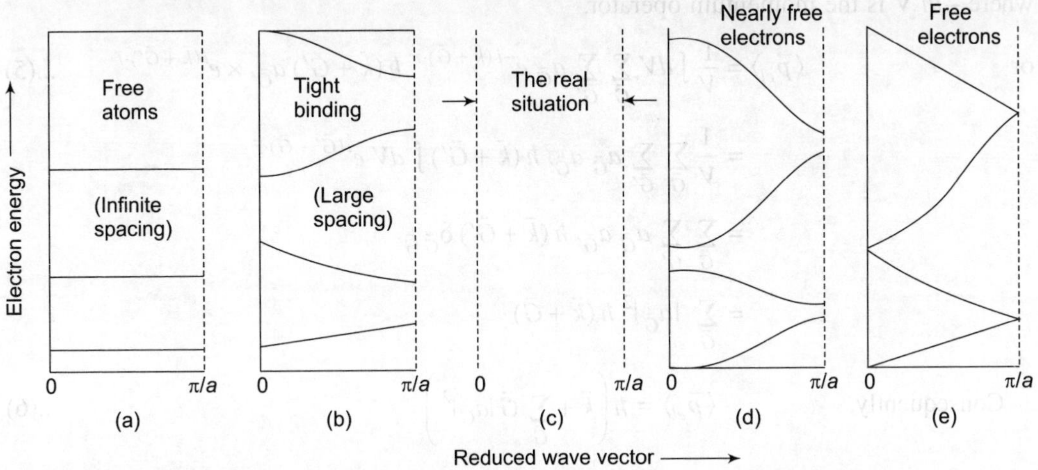


**Figure 12.59** A one-dimensional dependence of energy on wave vector for electron states as the coupling between adjacent atoms varies from zero (infinite separation) to the total absence of interatomic barriers

The nearly free electron model is appropriate when the periodic potential is rather small (and/or the interatomic separation is small enough to enforce considerable overlap between neighbouring atoms. Departures from the free electron dispersion relation are then apparent only in the regions of $\vec{k}$-space near the Brillouin zone boundaries.

## 12.31 CHANGE IN MOMENTUM AND CRYSTAL MOMENTUM

We have acceleration expression for electrons in the conduction band

$$m^* \frac{d\vec{v}}{dt} = e\vec{E} \qquad ...(1)$$

where $\vec{E}$ is the externally applied electric field. Since $\vec{p} = \hbar\vec{k}$ for the electron momentum for mobile electrons we have the acceleration equation

$$\hbar \frac{d\vec{k}}{dt} = e\,\vec{E} \qquad ...(2)$$

because the electron group velocities are $\hbar\vec{k}/m^*$.

We now start by evaluating the momentum $\vec{p}_{el}$ for an electron with a Bloch wave function, which can be written as

$$\psi(r) = V^{-1/2} \sum_{\vec{G}} a_{\vec{G}}\,(\vec{k})\, e^{i(\vec{k}+\vec{G})\cdot\vec{r}} \qquad ...(3)$$

This wave function is not a momentum eigenfunction, still we can evaluate the expectation value of the electron momentum as

$$\langle \vec{p}_{el} \rangle = \int dv\, \psi^*(\vec{r})\,(-i\hbar\nabla)\,\psi(\vec{r}) \qquad ...(4)$$

where $-i\hbar\nabla$ is the momentum operator.

or
$$\langle \vec{p}_{el}\rangle = \frac{1}{V}\int dV \sum_{\vec{G}}\sum_{\vec{G}'} a^*_{\vec{G}}\, e^{-i(\vec{k}+\vec{G})\cdot\vec{r}}\, \hbar(\vec{k}+\vec{G}')\, a_{\vec{G}'} \times e^{i(\vec{k}+\vec{G}')\cdot\vec{r}} \qquad ...(5)$$

$$= \frac{1}{V}\sum_{\vec{G}}\sum_{\vec{G}'} a^*_{\vec{G}}\, a_{\vec{G}'}\, \hbar(\vec{k}+\vec{G}')\int dV\, e^{i(\vec{G}'-\vec{G})\cdot\vec{r}}$$

$$= \sum_{\vec{G}}\sum_{\vec{G}'} a^*_{\vec{G}}\, a_{\vec{G}'}\, \hbar(\vec{k}+\vec{G}')\,\delta_{\vec{G}\vec{G}'}$$

$$= \sum_{\vec{G}} |a_{\vec{G}}|^2\, \hbar(\vec{k}+\vec{G})$$

Consequently,
$$\langle \vec{p}_{el}\rangle = \hbar\left(\vec{k} + \sum_{\vec{G}} \vec{G}\, |a_G|^2\right) \qquad ...(6)$$

where $\delta_{\vec{G}\vec{G}'}$ is the

$$\text{Kronecker delta} = I \quad \text{if} \quad \vec{G} = \vec{G}'$$
$$= 0 \quad \text{if} \quad \vec{G} \neq \vec{G}',$$

In Eq. (6), we have utilized the normalization condition

$$\sum_{\vec{G}} |a_G|^2 = 1.$$

The resulting Eq. (6) is just what we should expect from the fact that the plane wave $e^{i(\vec{k}+\vec{G})\cdot\vec{r}}$ is a momentum eigenfunction of eigenvalue $\hbar(\vec{k}+\vec{G})$, and $|a_{\vec{G}}|^2$ is the probability of the electron being found in the momentum eigenstate $\hbar\vec{k}$.

If the electron is now accelerated so that its wave vector $\vec{k}$ changes to $\vec{k} + \delta\vec{k}$, the expectation value of its momentum changes by

$$\vec{\delta}p_{el} = \hbar\left(\delta\vec{k} + \sum_{\vec{G}} \vec{G}\,\delta\vec{k}\cdot\frac{\partial}{\partial\vec{k}}|a_{\vec{G}}(\vec{k})|^2\right) \qquad ...(7)$$

A physical interpretation of the second term in Eq. (7) may be made by considering what happens when a free electron, incident on the crystal as if from outside is Bragg reflected. The electron state changes from $e^{i\vec{k}\cdot\vec{r}}$ (the free electron state) to $e^{i(\vec{k}+\vec{G})\cdot\vec{r}}$ (the free electron state) to $e^{i(\vec{k}+\vec{G})\cdot\vec{r}}$ (the Bragg state) and as a result, the lattice receives a recoil momentum

$$\delta\vec{p}_{latt} = -\hbar\vec{G} \qquad ...(8)$$

The *total change of momentum*

### 12.31.1 Crystal Momentum

In absence of the crystal potential (or in presence of a uniform potential $V(r)$, we have plane-waves for an electron (since, the electron is then free)

$$\psi_k = C \exp (i \vec{k} \cdot \vec{r}) \quad ...(1)$$

Then, the mean value of the operator $\vec{p}$ for an electron would be $\hbar k$

$$\vec{p}\, \psi_k = -i\hbar \nabla (C \exp i \vec{k} \cdot \vec{r})$$

$$= -i\hbar (i k) (C \exp i \vec{k} \cdot \vec{r})$$

$$= \hbar \vec{k} \psi_k \quad ...(2)$$

where $\hbar \vec{k}$ is the free electron momentum on the other hand for a Bloch electron

$$\left. \begin{aligned} \psi_k &= e^{i \vec{k} \cdot \vec{r}}\, u_k(\vec{r}) \\ \psi_k^* &= e^{-i \vec{k} \cdot \vec{r}}\, u_k^*(\vec{r}) \end{aligned} \right\} \quad ...(3)$$

and expectation value of momentum is

$$\langle \vec{p} \rangle_k = \int \psi_k^* (\vec{r}) (-i\hbar \nabla) \psi_k (\vec{r})\, d^3\vec{r}$$

$$= \int e^{-i\vec{k}\cdot\vec{r}}\, u_k^* (\vec{r}) (-i\hbar\nabla) (e^{i\vec{k}\cdot\vec{r}} u_k(\vec{r}))\, d^3\vec{r}$$

$$= \int e^{-i\vec{k}\cdot\vec{r}}\, u_k^* (\vec{r}) (-i\hbar) [e^{i\vec{k}\cdot\vec{r}}\, \nabla u_k(\vec{r}) + i\vec{k}\, e^{i\vec{k}\cdot\vec{r}}\, u_k (\vec{r})]\, d^3\vec{r}$$

$$= \int u_k^* (\vec{r}) [-i\hbar \nabla + \hbar k]\, u_k (\vec{r})\, d^3\vec{r} \quad ...(4)$$

i.e., momentum $\vec{p}$ is to be replaced by $\hbar \vec{k} - i\hbar \nabla$ where $\hbar \vec{k}$ is the crystal momentum.

## 12.32 ELECTRON TUNNELLING AND ENERGY BANDS

### 12.32.1 Transmission and Reflection of Electrons through an Arbitrary Potential

In this selection, we consider the electron propagation in one-dimensional crystals from the point of view to the *electron optics in solid state,* via the determination of the reflected and

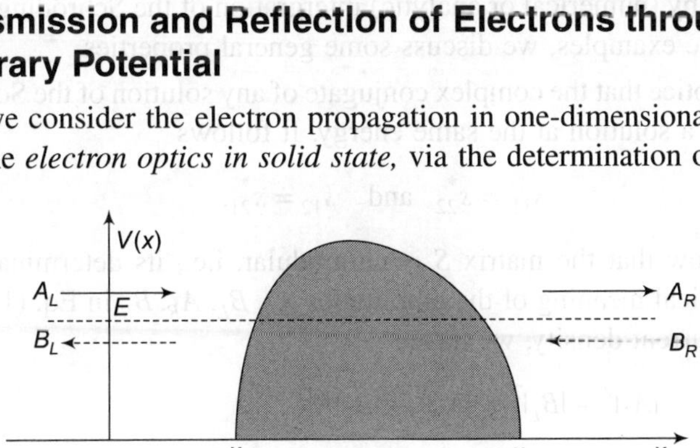


**Figure 12.60** Schematic representation of a potential (of finite range and arbitrary shape) connecting two leads at zero potential. Arrows indicate the wave vectors of the propagating plane waves (of energy $E$), and $A_L$, $B_L$, $A_R$, $B_R$ the amplitudes

transmitted components of a wave impinging on a given potential. We present first some general features of the elastic tunnelling through a potential of arbitrary shape; then, in the case of periodic potentials, we illustrate the formation of the energy bands from the point of view of quantum tunnelling.

Consider a potential of arbitrary profile, extending in the region $x_L \le x \le x_R$, and connecting two semi-infinite regions (called "leads") at the same constant potential, taken to be zero for simplicity (see Fig. 12.60.) The general solution of the Schrödinger equation for a positive energy $E$, in the left and right leads, can be written as

$$\left.\begin{aligned}\psi_L(x) &= A_L\, e^{iqx} + B_L\, e^{-iqx} \quad & x \le x_L \\ \psi_R(x) &= A_R e^{iqx} + B_R\, e^{-iqx} \quad & x \le x_R\end{aligned}\right\} \qquad ...(1)$$

where $q(E) = \sqrt{2mE/\hbar^2}$. Since the Schrödinger equation is linear and of second order, the amplitudes $(A_R, B_R)$ on the right-side lead depend linearly from the amplitudes $(A_L, B_L)$ on the left-side lead; we can thus write

$$\begin{pmatrix} A_R \\ B_R \end{pmatrix} = S(E)\begin{pmatrix} A_L \\ B_L \end{pmatrix} = \begin{pmatrix} s_{11}(E) & s_{12}(E) \\ s_{21}(E) & s_{22}(E) \end{pmatrix}\begin{pmatrix} A_L \\ B_L \end{pmatrix} \qquad ...(2)$$

where $S(E)$ is called *transfer matrix.*

From Eq. (2) we can write

$$\left.\begin{aligned}A_R &= S_{11}A_L + S_{12}B_L \\ B_R &= S_{21}A_L + S_{22}B_L\end{aligned}\right\} \qquad ...(3)$$

(for simplicity of notations, at times the energy dependence of variables is not indicated explicitly). The transfer matrix $S$, corresponding to a given potential $V(x)$ and chosen energy $E$, can be obtained by (numerical or analytic) intergration of the Schrödinger equation; before considering specific examples, we discuss some general properties.

First of all we notice that the complex conjugate of any solution of the Schrödinger equation is on its own right a solution at the same energy. It follows

$$s_{11} = s_{22}^* \quad \text{and} \quad s_{12} = s_{21}^*. \qquad ...(4)$$

We can also show that the matrix $S$ is unimodular, i.e., its determinant equals unity. In fact, from the physical meaning of the amplitudes $A_L$, $B_L$, $A_R$, $B_R$ in Eq. (1), and conservation of the electronic current density, we have

$$|A_L|^2 - |B_L|^2 \equiv |A_R|^2 - |B_R|^2. \qquad ...(5)$$

Using Eq. (3) and Eq. (4), we can rewrite Eq. (5) in the form

$$\begin{aligned}|A_L|^2 - |B_L|^2 &= |s_{11}A_L + s_{12}B_L|^2 - |s_{21}A_L + s_{22}B_L|^2 \\ &= [|s_{11}|^2 - |s_{22}|^2]\,[|A_L|^2 - |B_L|^2].\end{aligned}$$

We have thus $|s_{11}|^2 - |s_{22}|^2 = \det S = 1$: *the transfer matrix between two leads at the same potential is thus unimodular.*

### *Reflection and Transmission Amplitudes and Coefficients*

The transfer matrix provides an intuitive description of the electron tunnelling through the given potential region. Consider the stationary solution of the Schrödinger equation in the form of an impinging wave partially reflected and partially transmitted through the

$$\begin{cases} A_R = s_{11}A_L + s_{12}B_L; \\ 0 = s_{21}A_L + s_{22}B_L \end{cases} \quad \text{...(6)}$$

potential region. In this specific case we have $B_R \equiv 0$, and Eq. (3) read from these two equations, we can easily obtain the reflection and transmission amplitudes and coefficients as follows.

The *reflection amplitude r*, defined as $r = B_L/A_L$, is given by

$$\boxed{r = -\frac{s_{21}}{s_{22}}}; \quad \text{...(7a)}$$

the *reflection coefficient* is

$$R = r\,r^* = |r|^2 = \left|\frac{s_{21}}{s_{22}}\right|^2 \quad \text{...(7b)}$$

The *transmission amplitude* $t = A_R/A_L$ is given (remember the unimodularity of the transfer matrix $S$)

$$\boxed{t = \frac{1}{s_{22}}}; \quad \text{...(8a)}$$

the *transmission coefficient* is

$$T = t\,t^* = |t|^2 \left|\frac{1}{s_{22}}\right|^2. \quad \text{...(8b)}$$

We can easily verify the general property $R + T = 1$. The transfer matrix can be conveniently written in the form

$$S = \begin{pmatrix} 1/t^* & -r^*/t^* \\ -r/t & 1/t \end{pmatrix},$$

with a clear physical meaning of its matrix elements.

In the following we need to connect the transfer matrix $S$, corresponding to a given potential $V(x)$, with the transfer $S(d)$ corresponding to the rigidly displaced potential $V(x - d)$. It is seen by simple elaboration of the defining Eq. (1) that

$$S(d) = \begin{pmatrix} s_{11} & s_{12}\,e^{-2iqd} \\ s_{21}\,e^{2iqd} & s_{22} \end{pmatrix}; \quad \text{...(9)}$$

in shifting the potential along the $x$-axis by the displacement $d$, the diagonal matrix elements of the transfer matrix do not change, while the off-diagonal elements acquire the phase exp $(\pm 2iqd)$.

### *Transfer Matrix for a Rectangular Barrier and for a Piecewise Potential*

As an application of the concepts explained so far, we determine explicitly the transfer matrix for a rectangular barrier; such a matrix also constitutes the basic ingredient to build up the transfer matrix of any piecewise potential.

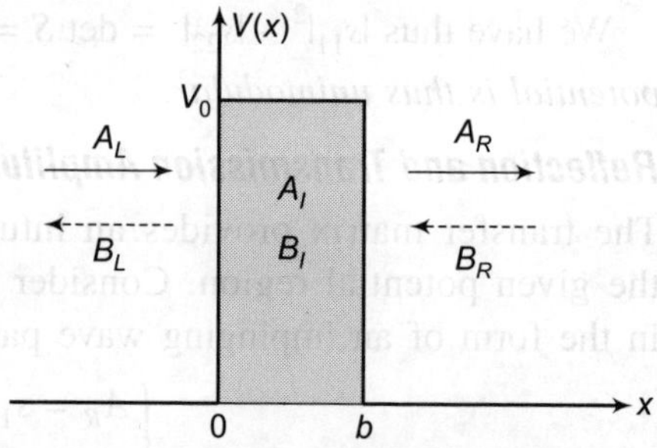


**Figure 12.61** Potential barrier of height $V_0$ and width $b$ connecting two leads at zero potential

Consider the potential barrier of height $V_0$ in the interval $0 \le x \le b$, shown in Fig. 12.61. In the whole $x$-axis, we can distinguish three regions: the left lead, the intermediate barrier region, and the right lead. In the three regions, the general solution of the Schrödinger equation for energies $0 < E < V_0$ has the form

$$\psi_L(x) = A_L e^{iqx} + B_L e^{-iqx} \qquad x < 0$$

$$\psi_I(x) = A_I e^{\beta x} + B_I e^{-\beta x} \qquad 0 < x < b$$

$$\psi_R(x) = A_R e^{iqx} + B_R e^{-iqx} \qquad x < b$$

where $q(E) = \sqrt{2mE/\hbar^2}$ and $\beta(E) = \sqrt{2m(V_0 - E)/\hbar^2}$.

The standard boundary conditions of continuity of wave function and its derivative at $x = 0$ give

$$\left.\begin{aligned} A_L + B_L &= A_I + B_I \\ A_L\, iq - B_L\, iq &= A_I \beta - B_I \beta \end{aligned}\right\} \qquad \text{...(10)}$$

From Eq. (10) we can express $(A_I, B_I)$ in terms of $(A_L, B_L)$ in the matrix form

$$\begin{pmatrix} A_I \\ B_I \end{pmatrix} = \frac{1}{2\beta}\begin{pmatrix} (iq+\beta) & (-iq+\beta) \\ (-iq+\beta) & (iq+\beta) \end{pmatrix}\begin{pmatrix} A_L \\ B_L \end{pmatrix}. \qquad \text{...(11a)}$$

Similarly, we can consider the boundary conditions at $x = b$ and we can express $(A_R, B_R)$ in terms of $(A_I, B_I)$ in the form

$$\begin{pmatrix} A_R \\ B_R \end{pmatrix} = \frac{1}{2iq}\begin{pmatrix} (iq+\beta)\, e^{(-iq+\beta)b} & (iq-\beta)\, e^{(-iq-\beta)b} \\ (iq-\beta)\, e^{(iq+\beta)b} & (iq+\beta)\, e^{(iq-\beta)b} \end{pmatrix}\begin{pmatrix} A_I \\ B_I \end{pmatrix}. \qquad \text{...(11b)}$$

The direct multiplication of the transfer matrices in Eq. (11a) and Eq. (11b) provides the transfer matrix $S$ for the rectangular barrier; with straightforward calculations, we obtain

$$\boxed{\begin{array}{ll} s_{11} = e^{-iqb}\left[\cosh \beta b + i\,\dfrac{q^2 - b^2}{2q\beta}\sinh \beta b\right] & s_{22} = s_{11}^* \\ s_{12} = e^{-iqb}\,(-i)\,\dfrac{q^2 + \beta^2}{2q\beta}\sinh \beta b & s_{21} = s_{12}^* \end{array}} \quad ...(12)$$

Expressions (12) provide the transfer matrix of a barrier of height $V_0$ in the interval [0, *b*]; for barriers shifted by a displacement *d*, appropriate phase factors (see Eq. 9) are to be included. The transfer matrix for any arbitrary piecewise potential is obtained simply multiplying in the appropriate order the matrices corresponding to each component barrier.

With the transfer matrix given in Eq. (12), we can analyse explicitly the electron tunnelling through a rectangular potential barrier; from the expression of the matrix element $s_{11}$ we obtain

$$|s_{11}|^2 = 1 + \frac{(q^2 + \beta^2)^2}{4q^2\beta^2}\sinh^2 \beta b.$$

The transmission coefficient of the barrier $T(E) = 1/|s_{11}|^2$ is thus

$$T(E) = \frac{1}{1 + \dfrac{V_0^2}{4E(V_0 - E)}\sinh^2\sqrt{\dfrac{2m(V_0 - E)\,b^2}{\hbar^2}}} \qquad 0 \le E \le V_0. \quad ...(13a)$$

For $E \ge V_0$ similar calculations can be performed and the transmission coefficient becomes

$$T(E) = \frac{1}{1 + \dfrac{V_0^2}{4E(E - V_0)}\sin^2\sqrt{\dfrac{2m(E - V_0)\,b^2}{\hbar^2}}} \qquad E \ge V_0. \quad ...(13b)$$

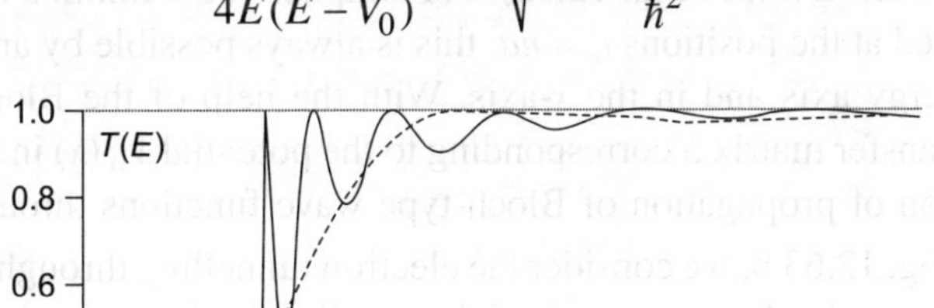


**Figure 12.62** Transmission coefficient through a barrier of height $V_0$ = 1 eV and *b* = 5 Å (dashed line), and through a barrier of height $V_0$ = 1 eV and length *b* = 20 Å (solid line)

For $E = V_0$ we have

$$T(V_0) = \frac{1}{1 + \frac{1}{4}\beta_0^2 b^2} \quad \text{with} \quad \hbar^2\beta_0^2/2m = V_0. \qquad ...(13c)$$

The typical behaviour of the transmission coefficient (13) of a rectangular barrier is reported in Fig. 12.62. For $0 \le E \le V_0$, $T(E)$ varies monotonically from zero to $T(V_0)$; for a reasonable thick barrier characterized by $\beta_0 b \gg 1$, the transmission is near to zero. For $E > V_0$ the transmission coefficient has an oscillatory behaviour and approaches asymptotically to 1. In particular $T(E)$ is exactly equal to 1 at energies such that the argument of the trigonometric function in Eq. (13b) equals an integer multiple of $\pi$.

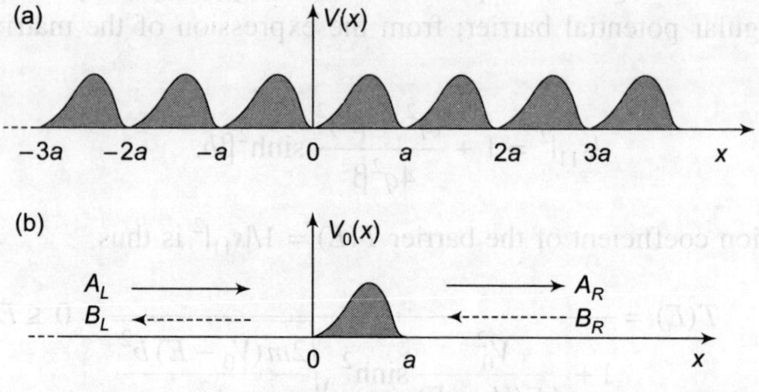


**Figure 12.63** Schematic representation of the periodic potential $V(x)$ of a one-dimensional crystal (a), and of the potential $V_0(x)$ within the unit cell (b)

## 12.32.2 Electron Tunnelling through a Periodic Potential

Consider a one-dimensional potential $V(x)$, of periodicity $a$, and let $V_0(x)$ denote the potential within the unit cell $0 \le x \le a$ (see Fig. 12.63). For simplicity, the minima of $V(x)$ are assumed to be zero and located at the positions $t_n = na$; this is always possible by an appropriate choice of origin in the energy axis and in the $x$-axis. With the help of the Bloch theorem and the knowledge of the transfer matrix $S$ corresponding to the potential $V_0(x)$ in the unit cell, we can express the condition of propagation of Bloch-type wave functions throughout the system.

As indicated in Fig. 12.63 b, we consider the electron tunnelling through the potential $V_0(x)$, which separates the two leads at zero potential extending in the regions $x \le 0$ and $x \ge a$. We specify Eq. (1) to the present geometry and we have

$$\left.\begin{aligned} \psi_L(x) &= A_L e^{iqx} + B_L e^{-iqx} & x \le 0 \\ \psi_R(x) &= (s_{11}A_L + s_{12}B_L)e^{iqx} + (s_{21}A_L + s_{22}B_L)e^{-iqx} & x \ge a \end{aligned}\right\} \qquad ...(14)$$

where $q(E) = \sqrt{2mE/\hbar^2}$ is the propagation wave number in the leads, the energy $E$ is positive, and $s_{ij}(E)$ are the matrix elements of the transfer matrix through the potential $V_0(x)$.

We look for solutions of Eq. (14) which satisfy, at the points $x = 0$ and $x = a$, the boundary conditions required by the Bloch theorem

$$\psi_R(a) = e^{ika}\, \psi_L(0) \quad \text{and} \quad \left(\frac{d\psi_R}{dx}\right)_{x=a} = e^{ika}\left(\frac{d\psi_L}{dx}\right)_{x=0} . \qquad ...(15)$$

Inserting Eq. (14) into Eq. (15), we obtain two linear homogeneous equations for the arbitrary coefficients $A_L$ and $B_L$; the two quations can be written in the compact matrix form

$$\begin{pmatrix} s_{11}\, e^{iqa} & s_{12}\, e^{iqa} \\ s_{21}\, e^{-iqa} & s_{22}\, e^{-iqa} \end{pmatrix}\begin{pmatrix} A_L \\ B_L \end{pmatrix} = e^{ika}\begin{pmatrix} A_L \\ B_L \end{pmatrix} \qquad ...(16)$$

The above equation shows that the $2 \times 2$ matrix in the first member of Eq. (16), has eigenvalue exp $(ika)$; since the matrix is unimodular, it has also the eigenvalue $\exp(-ika)$ and trace 2 cos $ka$. The condition of Bloch propagation and existence of allowed energy levels is thus,

$$\boxed{s_{11} e^{iqa} + s_{22}\, e^{-iqa} = 2 \cos ka} \qquad ...(17)$$

It is convenient to express the above equation in terms of the transmission amplitude $t = |t| \exp(i\phi)$; we have

$$s_{11}(E) = \frac{1}{t^*(E)} = \frac{1}{|t(E)|}\, e^{i\phi(E)}$$

The compatibility condition (17) for the tunnelling of Bloch-type wave functions takes the compact and significant form

$$\boxed{\frac{1}{|t(E)|} \cos\,[\phi(E) + q(E)\, a] = \cos ka} \qquad ...(18)$$

where $\phi(E)$ and $|t(E)|$ are the phase and the modulus of the transmission amplitude $t(E)$ through the potential $V_0(x)$ of the unit cell. Notice that Eq. (18) can be satisfied only for those values of $E$ for which the function $(1/|t|) \cos(\phi + qa)$ is in magnitude less than (or equal to) unity; since $|t(E)| \leq 1$, we can infer that the compatibility Eq. (18) leads quite generally to a sequence of allowed and forbidden energy regions.

It is instructive to analyse the band structure, produced by the periodic array of barriers of Fig. 12.60, from the point of view of the electron tunnelling framework. The transfer matrix $S$ for a single barrier is given by Eq. (12). The energy bands are determined by Eq. (32), which reads in the present case

$$\text{Re}\left\{\left[\cosh \beta b + i\frac{q^2 - \beta^2}{2q\beta} \sinh \beta b\right] e^{iq(\dot{a} - b)}\right\} = \cos ka$$

where Re denotes the real part of its argument. With $a - b = w$ (see Fig. 12.63), the above equation becomes

$$\cosh \beta b \cos qw - \frac{q^2 - \beta^2}{2q\beta} \sinh \beta b \sin qw = \cos ka. \qquad ...(19)$$

We, thus, see that the compatibility Eq. (19) for the tunnelling of Bloch-type wave functions through periodically repeated barriers, exactly coincides with equation for the ordinary boundary conditions.

## 12.33 METHODS OF CALCULATING BAND STRUCTURE

We can simplify the problem of calculating realistic energy-band structures by exploiting the translational symmetry of the lattice. The form of the Bloch functions implies that it is necessary to construct the solution within only the primitive unit cell of the reciprocal lattice or the first Brillouin zone. The solution for all other $k$ vectors outside of the Brillouin zone differs by only a phase factor $e^{ikR}$. In other words, every $k$ vector can be reduced to an equivalent $k$ vector within the first B.Z., i.e., the first B.Z. contains all the unique $k$ vectors in the system.

To solve the Schrödinger equation with a periodic potential, we can distinguish two main points of view. A first line of approach consists in expanding the crystal states in a convenient set of *energy independent Bloch functions*: Bloch sums made with atomic orbitals in the tight binding method, plane waves orthogonalized to core states in the orthogonalized plane-wave method, plane waves in the pseudopotential method.

**Table 12.3** Schematic Summary of Some Features of Methods of Band Structure Calculations

| *Point of view* | *Methods* | *Energy independent basis functions* |
|---|---|---|
| Whole crystal formulation | Tight binding method | Bloch sums |
| | Orthogonalized plane wave method | Bloch sums and plane waves |
| | Pseudopotential method | Plane waves |
| Single cell formulation | Cellular method | Spherical waves |
| | Augmented plane wave method | Spherical and plane waves |
| | Green's function method | Spherical waves |

A second line of approach (which includes the cellular method, the augmented plane wave method and the Green's function method) is based on the single cell formulation and adopts *energy dependent basis functions*. In the cellular formulation one considers the Wigner–Seitz cell (or subcells), an appropriate spherically symmetric potential within it, and the energy dependent solutions of the radial Schrödinger equation regular at the origin; these constitute the basis set for the expansion of the crystal wave functions, which must satisfy appropriate boundary conditions required by the Bloch's theorem; the modality of imposing the boundary conditions characterizes the different cellular methods.

In Table 12.3 we summarize in a schematic form the basic methods of band structure calculations and the basis functions used to describe the crystal states.

## 12.34 TIGHT BINDING APPROXIMATION

In the nearly free electron case, we observed that the electron wave function (outside the ion core) looked like nearly as plane waves. However, near the ion cores, they look like atomic orbitals. From this, Bloch conjectured in 1928 an entirely different scheme for the construction of crystal wave function. This has been referred to as the tight binding approximation. The tight binding approximation deals with the case in which the overlap of atomic wave functions is enough to require corrections to the pictures of isolated atoms, but not so much as to render the atomic description completely irrelevant. In this method (also known as the LCAO method), one starts from the wave function for an electron in a free atom and then constructs a crystal orbital which just equals the LCAOs and describes the electron in a periodic field of crystal as a whole.

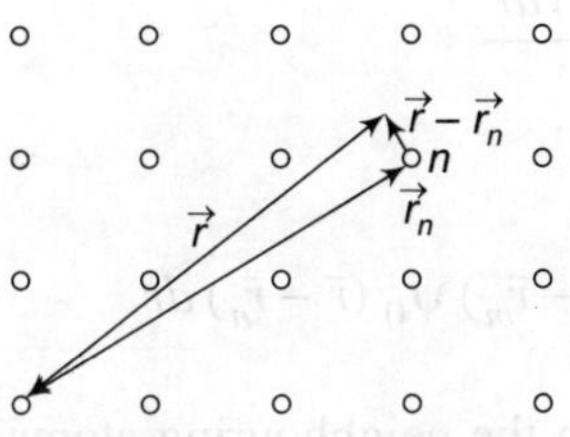


**Figure 12.64** A two-dimensional square-array of atoms

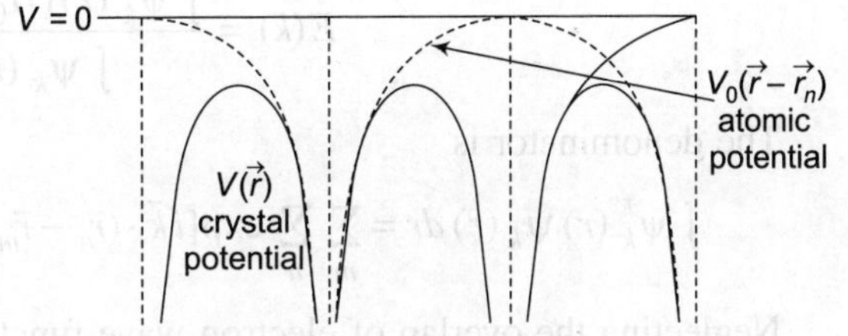


**Figure 12.65** Atomic and crystal potential for a one-dimensional lattice

Suppose we have an electron in a free atom and then a number of such atoms are brought close together to form a crystal. For instance, we have a two-dimensional square-array of atoms as shown in Fig. 12.64.

Let $\vec{r}$ and $\vec{r}_n$ be the position vector of the electron and $n$th atom in the crystal, respectively. The potential of an electron in a free atom is denoted by $V_0\,(r - r_n)$ and that in a crystal by $V(r)$. These are represented in Fig. 12.65.

In this approximation, it is assumed that the electron wave function is influenced by the nearest atom and is unaffected by other atoms in the lattice. The wave function of an electron for a free atom is then $\psi_k\,(\vec{r} - \vec{r}_n)$ and the crystal wave function may be written as a linear combination as follows

$$\psi_k\,(\vec{r}) = \sum_n \exp\,(i\vec{k}\cdot\vec{r}_n)\,\psi_0\,(\vec{r} - \vec{r}_n) \qquad \text{...(1)}$$

The summation is over all atoms of the crystal. The above wave function is a Bloch function since

$$\begin{aligned}\psi_k\,(\vec{r} + \vec{r}_m) &= \sum_n \exp\,(i\vec{k}\cdot\vec{r}_n)\,\psi_0\,(\vec{r} + \vec{r}_m - \vec{r}_n)\\ &= \sum_n \exp\,(i\vec{k}\cdot\vec{r}_n)\,\psi_0\,(\vec{r} - \vec{r}_n + \vec{r}_m)\\ &= \exp\,(i\,\vec{k}\cdot\vec{r}_m)\sum_n \exp\,[i\,\vec{k}\cdot(\vec{r}_n - \vec{r}_m)] \times \psi_0\,[(\vec{r} - (\vec{r}_n - \vec{r}_m))]\end{aligned}$$

$$= \exp(i\vec{k}\cdot\vec{r}_m)\,\psi_k(\vec{r}) \qquad ...(2)$$

The crystal Hamiltonian

$$H = H_0 + H_1 \qquad ...(3)$$

where

$$\left.\begin{aligned} H_0 &= -\frac{\hbar^2}{2m}\nabla^2 + V_0(\vec{r}-\vec{r}_n) \\ H_1 &= V(\vec{r}) - V_0(\vec{r}-\vec{r}_n) \end{aligned}\right\} \qquad ...(4)$$

where $V_0(\vec{r}-\vec{r}_n)$ is the potential due to $n$th ground state atom at the electron situated at $\vec{r}$.

The energy of the electron

$$E(k) = \frac{\int \psi_k^*(r)[H_0 + H_1]\,\psi_k(r)\,dr}{\int \psi_k^*(r)\,\psi_k(r)\,dr} \qquad ...(5)$$

The denominator is

$$\int \psi_k^*(r)\,\psi_k(r)\,dr = \sum_m \sum_n \exp[i\vec{k}\cdot(\vec{r}_n-\vec{r}_m)] \times \int \psi_0^*(\vec{r}-\vec{r}_m)\,\psi_0(\vec{r}-\vec{r}_n)\,dr \qquad ...(6)$$

Neglecting the overlap of electron wave functions between the neighbouring atoms,

$$\int \psi_0^*(\vec{r}-\vec{r}_m)\,\psi_0(\vec{r}-\vec{r}_n)\,dr = \delta_{nm} \qquad ...(7)$$

and

$$\int \psi_k^*(r)\,\psi_k(r)\,dr = N \qquad ...(8)$$

where $N$ is the total number of atoms in the crystal.

Hence,

$$E(k) = \frac{1}{N}\int \psi_k^*(r)\,(H_0 + H_1)\,\psi_k(r)\,dr \qquad ...(9)$$

$$= \frac{1}{N}\int \psi_k^*(r)\left[-\frac{\hbar^2}{2m}\nabla^2 + V_0(r-r_n)\right]\psi_k(r)\,dr$$

$$+ \frac{1}{N}\int \psi_k^*(r)[V(r) - V_0(r-r_n)]\,\psi_k(r)\,dr$$

$$= \frac{E_0}{N}\int \psi_k^*(r)\,\psi_k(r)\,dr$$

$$+ \frac{1}{N}\sum_m \sum_n \exp[i\vec{k}\cdot(\vec{r}_n-\vec{r}_m)]\int \psi_0^*(\vec{r}-\vec{r}_m)$$

$$[V(r) - V_0(\vec{r}-\vec{r}_n)]\,\psi_0(\vec{r}-\vec{r}_n)\,dr \qquad ...(10)$$

Here, we expect that every term in the summation from $n = 0$ to $n = N - 1$ to contain $N$ identical terms, which may be evaluated by substituting $r_n = 0$, then

$$E(k) = E_0 + \sum_m \exp(-i\vec{k}\cdot\vec{r}_m)\int \psi_0^*(\vec{r}-\vec{r}_m)[V(r) - V_0(r)]\,\psi_0(r)\,dr \qquad ...(11)$$

Furthermore, if we make an approximation that $\psi_0$ is spherically symmetric (e.g., ground state of alkali metal), then the contributions due to all nearest neighbours may be assumed to be identical. For $r_m = 0$, the above integral gives

$$\int \psi_0^* (\vec{r})\, [V(\vec{r}) - V_0(\vec{r})]\, \psi_0(\vec{r})\, d\,\vec{r} = -\alpha \quad ...(12)$$

and for the nearest neighbour atoms

$$\int \psi_0^* (\vec{r} - \vec{r}_m)\, [V(\vec{r}) - V_0(\vec{r})]\, \psi_0(\vec{r})\, d\,\vec{r} = -\gamma \quad ...(13)$$

where $r_m$ is a vector connecting the atom at the origin with a nearest neighbour, $\alpha$ and $\gamma$ are constants known as overlap integrals. Thus, equation for energy is simplified to the following form

$$E(\vec{k}) = E_0 - \alpha - \gamma \sum_m \exp\,(-i\,\vec{k} \cdot \vec{r}_m) \quad ...(14)$$

We observe that the energy of the electron in the crystal differs from that of the free atom by a constant $\alpha$ plus a term which depends on the wave vector $\vec{k}$. It is the last part which transforms the discrete atomic levels into an energy band in the solid.

The sum in Eq. (14) is only over the nearest neighbours. It is observed that since the free ion potential $V_0$ is more binding (i.e., more negative) than the lattice potential $V(r)$, $V - V_0$ is always negative (and $\alpha$ and $\gamma$ are always positive). Thus, $E < E_0$.

Let us now use these results to calculate the energy bands for the three cubic Bravais lattices.

(i) *The simple-cubic lattice:* Here any given ion has six neighbours at

$$\vec{r}_m = (\pm a, 0, 0), (0, \pm a, 0), (0, 0, \pm a) \quad ...(15)$$

where $a$ is the lattice constant. We then have

$$E = (E_0 - \alpha) - 2\gamma(\cos k_x a + \cos k_y a + \cos k_z a) \quad ...(16)$$

(ii) *The body-centred-cubic lattice*: The eight nearest neighbours are at

$$\vec{r}_m = \left(\pm\frac{a}{2}, \pm\frac{a}{2}, \pm\frac{a}{2}\right) \quad ...(17)$$

so that

$$E = (E_0 - a) - 8\gamma \cos\left(\frac{k_x a}{2}\right) \cos\left(\frac{k_y a}{2}\right) \cos\left(\frac{k_z a}{2}\right) \quad ...(18)$$

(iii) *The face-centred-cubic lattice:* Twelve nearest neighbours are at

$$\vec{r}_m = (0, \pm a, \pm a), (\pm a, 0, \pm a), (\pm a, \pm a, 0) \quad ...(19)$$

and

$$E = (E_0 - \alpha) - 4\gamma \left[\cos\left(\frac{k_y a}{2}\right) \cos\left(\frac{k_z a}{2}\right) + \cos\left(\frac{k_z a}{2}\right) \cos\left(\frac{k_x a}{2}\right) + \cos\left(\frac{k_x a}{2}\right) \cos\left(\frac{k_y a}{2}\right)\right] \quad ...(20)$$

We see the $s$-band is described by a constant energy term $(E_0 - \alpha)$ and a term which depends on the electron wave vector $\vec{k}$. Thus, we have an energy band of width 12 $\gamma$ eV centred about the energy $(E_0 - \alpha)$ eV for the simple cubic lattice. From our definition of $\gamma$ (Eq. 13), it is noted that the greater the overlap of the electron wave functions of neighbouring ions, the greater is $\gamma$ and the wider is the energy band. The tight binding approximation, however, is based on the assumption that the charge clouds of neighbouring atoms overlap very little. Thus, it applies only when $\gamma$ is small, so that (in this approximation) the energy band is narrow.

In the other two cases, the energy bands have widths $16\gamma$ and $24\gamma$, respectively. Thus, in the tight binding approximation, each discrete free atom energy level has an associated energy band (Fig. 12.66).

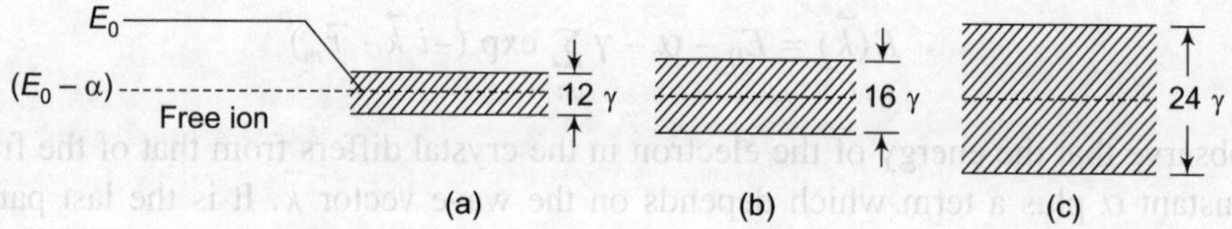


**Figure 12.66** $s$ Bands for: (a) Simple cubic, (b) bcc, (c) fcc lattices in the tight binding approximation

### 12.34.1 Energy Surfaces

The tight binding approximation implies that the potential energy of the valence electrons in the field of their parent and neighbouring ions is large compared with their kinetic energy. This implies that the separation between ions is small compared with the electron in de Broglie wavelength or

$$|\vec{k}|\, a \ll 1$$

Using this approximation

$$\cos ka \simeq 1 - \frac{k^2a^2}{2} \qquad \text{...(21)}$$

i.e., for small values of $k$, we may expand the cosines and retain up to the second term. Then,

$$E(k) \simeq (E_0 - \alpha) - 2\gamma\left[\left(1 - \frac{k_x^2\,a^2}{2}\right) + \left(1 - \frac{k_y^2\,a^2}{2}\right) + \left(1 - \frac{k_z^2\,a^2}{2}\right)\right]$$

$$= (E_0 - \alpha) - 6\gamma + \gamma a^2(k_x^2 + k_y^2 + k_z^2)$$

or

$$E(k) = (E_0 - \alpha) - 6\gamma + \gamma k^2a^2 + \ldots \quad \text{(for simple cubic)} \qquad \text{...(22)}$$

Similarly,

$$E(k) = (E_0 - \alpha) - 8\gamma + \gamma k^2a^2 + \ldots \quad \text{(for bcc)} \qquad \text{...(23)}$$

and

$$E(k) = (E_0 - \alpha) - 12\gamma + \gamma k^2a^2 + \ldots \quad \text{(for fcc)} \qquad \text{...(24)}$$

In view of our general expression for effective mass, we see that in each of these cases

$$m^* = \frac{\hbar^2}{\partial^2 E/\partial k^2} = \frac{\hbar^2}{2\gamma a^2} \left( \because \frac{\partial E}{\partial k_1} = 2\gamma a \sin k_1 a \right) \quad ...(25)$$

This indicates that for a narrow band (i.e., small γ), the electron behaves as through it was extremely heavy. For simple-cubic lattice,

$$E_{min} = E(k)_{bottom} = E_0 - \alpha - 6\gamma \quad ...(26a)$$

$$(\text{at } k_x = k_y = k_x = 0)$$

$$E_{max} = E(k)_{top} = E_0 - a + 6\gamma \quad ...(26b)$$

$$(\text{at } k_x = k_y = k_x = \pm\pi/a).$$

Therefore, width of the energy band,

$$E_{top} = E_{bottom} = 12\gamma \quad ...(27)$$

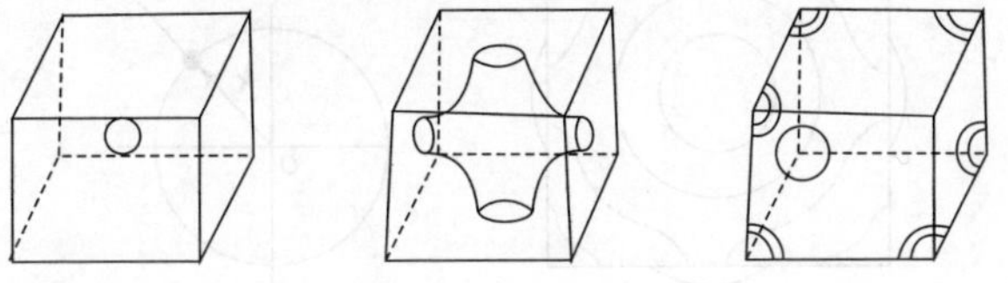

(a) (b) (c)

**Figure 12.67** Energy surfaces for the simple-cubic lattice (tight binding approximation) for energy: (a) near the bottom of the band, (b) about half way up the band, (c) near the top of the band

The values given by Eq. (27) refer to the bottom region of the band and conform to the constant spherical energy surfaces [Fig. 12.67 (a)] (i.e., near $k = 0$). That is, the state of lowest energy is at the centre of the zone and the states of highest energy are at the corners [Fig. 12.67 (c)]. As the energy becomes larger ($k > 0$), the energy surfaces bulge more and more towards the centres of the zone faces, until eventually they touch and intersect the zone boundary. The energy surfaces that intersect the zone boundary do so orthogonally (since the normal derivative of $E(k)$ vanishes on the bounding planes). A cross section through the energy surfaces at equal energy intervals in any one of the coordinate planes is shown in Fig. 12.68.

In Fig. 12.68 (a), the energy surfaces are spherical around the zone centre for only small $k$ values. The surfaces are also spherical with centres at the corner of the zone for small $k$ values measured from the corners. In Fig. 12.68 (b) (i.e., nearly free electron approximation), energy surfaces are spherical up to fairly large value of $k$.

## 12.35 VELOCITY OF THE BLOCH ELECTRON (QUALITATIVE IDEA)

Inside a solid, the electron – motion is not free, but is modified due to periodic potential of the lattice (refer to Kronig-Penney model and Bloch theorem).

Now, we would consider the motion of the Bloch electrons in solids. An electron in a state $\psi_{\vec{k}}$ moves through the crystal with a velocity directly related to energy of that state. Consider first the case of a free particle. Its velocity $\vec{v} = \vec{p}/m_0$ since $\vec{p} = \hbar\,\vec{k}$, it follows that for a free electron, velocity

$$\vec{v} = \frac{\hbar\,\vec{k}}{m_0} \qquad ...(28(i))$$

i.e., the velocity is proportional to and parallel to the wave vector $\vec{k}$ (Fig. 12.69 (a)).

For a Bloch electron, the velocity is also a function of $\vec{k}$ but the functional relationship is not as simple as Eq. 28 (i). To derive this relationship, we use a well-known formula in wave propagation, i.e., the group velocity of a wave-packet is given by

$$\vec{v} = \nabla_{\vec{k}}\ \omega(\vec{k}) \qquad ...(28(ii))$$

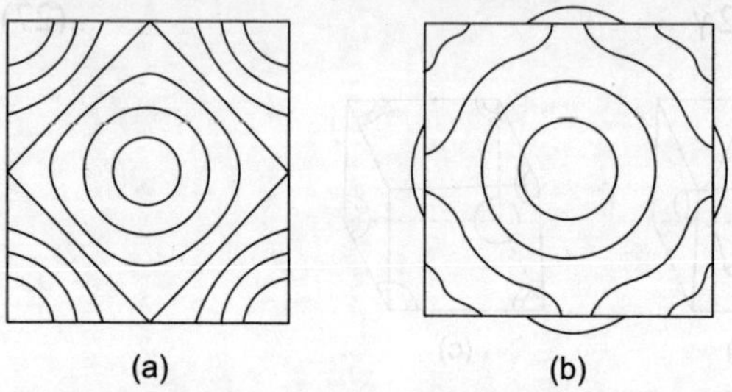


**Figure 12.68** Electron energy contours for simple-cubic lattice in (a) Tight binding approximation: (b) Nearly free electron approximation

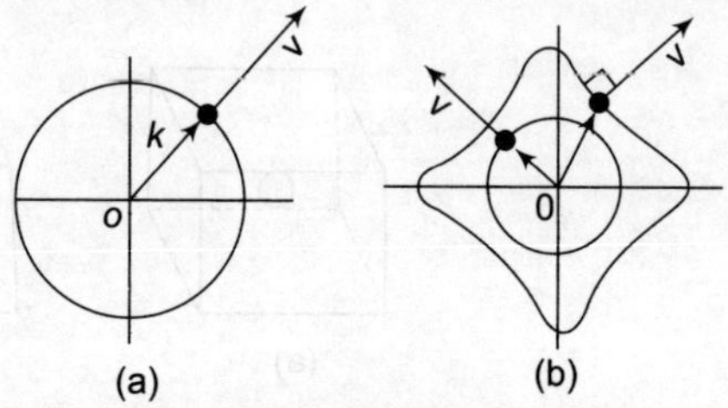


**Figure 12.69** Velocity of (a) a free electron, (b) a Bloch electron

where $\omega$ is the frequency and $\vec{k}$, the wave vector of the wave packet. Applying this to the electron-wave in the crystal, and using $\omega = E/\hbar$, we may write for the velocity of the Bloch electron.

$$\vec{v} = \frac{1}{\hbar}\nabla_{\vec{k}}\ E(\vec{k}) \qquad ...(28(iii))$$

which states that the velocity of an electron in state $\vec{k}$ is proportional to the gradient of energy in $\vec{k}$-space. Since the gradient vector is perpendicular to the contour-lines (according to vector-analysis), it follows therefore that the velocity $\vec{v}$ at every point in $\vec{k}$-space is normal to the energy contour passing though that point, (Fig. 12.69(b)). Because these contours are in general non-spherical, therefore velocity is not necessarily parallel to the wave vector $\vec{k}$, unlike the case of a free particle.

It is to be noted that near the centre of the zone, where the standard dispersion relation

$$E = \hbar^2 k^2/(2m^*) \qquad ...(29)$$

holds true, the relation (28 (i)) leads to

$$\vec{v} = \frac{\hbar \vec{k}}{m^*} \quad ...(30)$$

which is of the same form as the relation for a free particle (28(i)) except that $m_0$ has been replaced by $m^*$, the effective mass. Thus, near the centre of the zone, $\vec{v}$ is parallel to $\vec{k}$ and points radially outwards (as shown in Fig. 12.69 (b)). It is near the zone boundaries at which the energy contours are so distorted that this simple relationship between $\vec{v}$ and $\vec{k}$ is destroyed, then one must resort to the more general result, viz., Eq. (28 iii).

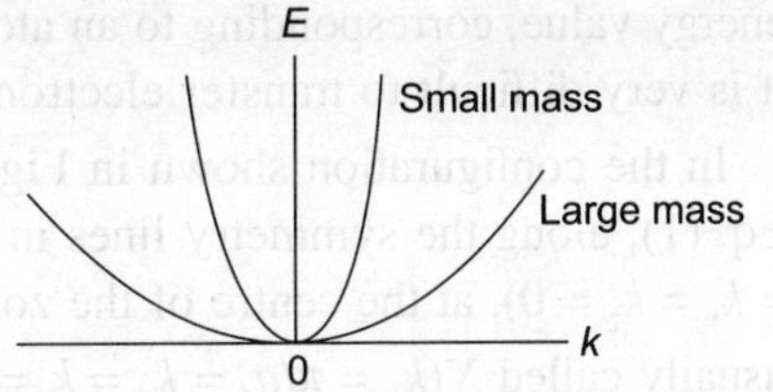


**Figure 12.70** The inverse relationship between effective mass and curvature of energy band

It should also be noted here that $m^*$ is inversely proportional to the curvature of the band: where the curvature is large (i.e., $\partial E/\partial k^2$), the mass is small and a small curvature implies a large mass (Fig. 12.70).

## 12.36 ENERGY BANDS IN A SIMPLE CUBIC LATTICE

We have seen that each atom in a simple cubic structure, has six nearest neighbours, at positions ($\pm a$, 0, 0), (0, $\pm a$, 0) and (0, 0, $\pm a$) so that, for this particular case, becomes

$$\varepsilon = \varepsilon_0 - a - 2\gamma\,(\cos k_x a + \cos k_y a + \cos k_z a) \quad ...(1)$$

Each electron state of energy $\varepsilon_0$ in the free atom is now confined to an energy band of width $12\gamma$ which directly depends upon the overlap $\gamma$ between the functions of adjacent atoms in the solid. The energy is a minimum at $k_x = k_y = k_z = 0$ and its maximum values occur at the corners of the Brillouin zone, with coordinates ($\pm\pi/a$, $\pm\pi/a$, $\pm\pi/a$). If there are $N$ atoms in the solid, each band which corresponds to a non-degenerate state $\psi_0(\vec{r})$ contains $2N$ states (including spin). For small values of $k$ we may expand the cosine up to the second order, which gives

$$\varepsilon \cong \varepsilon_0 - \alpha - 2\gamma\left[\left(1 - \frac{k_x^2 a^2}{2}\right) + \left(1 - \frac{k_y^2 a^2}{2}\right) + \left(1 - \frac{k_z^2 a^2}{2}\right)\right]$$

$$= \varepsilon_0 - \alpha - 6\gamma + \gamma k^2 a^2 \quad ...(2)$$

Thus, near the bottom of the band, the electron energy depends upon the square of $k$ in agreement with the results for both free and nearly-free electrons. Since the effective mass, depends inversely upon the overlap of adjacent orbitals

$$m_e^* = \frac{\hbar^2}{2\gamma a^2} \quad ...(3)$$

it follows that narrow bands are characterized by large effective masses, as we would expect from the definition. In other words, in the limit where the atomic distances become large (and the overlap $\gamma$ becomes very small, such that the energy band is approaching a single discrete energy value, corresponding to an atomic energy level, the large effective mass indicates that it is very difficult to transfer electrons from one atom to another.

In the configuration shown in Fig. 12.71, it is convenient to plot the dispersion relation, Eq. (1), along the symmetry lines in the simple cubic Brillouin zone, joining the point $\Gamma(k_x = k_y = k_z = 0)$, at the centre of the zone, and points of high symmetry on the zone boundary, usually called $X(k_x = \pi/a, = k_y = k_z = 0)$, $M(k_x = k_y = \pi/a, k_z = 0)$ and $W(k_x = k_y = k_z = \pi/a)$. The dispersion relation, Eq. (1), successively reduces to

$$\varepsilon_{\Gamma X} = \varepsilon_0 - \alpha - 4\gamma - 2\gamma \cos k_x a \qquad (k_y = k_z = 0)$$

$$\varepsilon_{XW} = \varepsilon_0 - \alpha + 2\gamma - 2\gamma(\cos k_y a + \cos k_z a) \qquad (k_x = \pi/a) \qquad \text{...(4)}$$

$$\varepsilon_{\Gamma M} = \varepsilon_0 - \alpha - 2\gamma - 4\gamma \cos \frac{ka}{\sqrt{2}} \qquad \left(k_x = k_y = \frac{k}{\sqrt{2}}, k_z = 0\right)$$

$$\varepsilon_{MW} = \varepsilon_0 - \alpha - 2\gamma + 4\gamma \cos kza \qquad (k_x = k_y = \pi a)$$

and is represented in Fig. 12.71. In the isotropic effective mass approximation, the dispersion relation for nearly-free electrons, is given by

$$\varepsilon = \varepsilon_\Gamma + \frac{\hbar^2 k^2}{2m_e^*} \qquad \text{...(5)}$$

and this becomes, along the specified symmetry lines

$$\varepsilon_{\Gamma X} = \varepsilon_\Gamma + \frac{\hbar^2 k_x^2}{2m_e^*}, \quad \varepsilon_{XW} = \varepsilon_\Gamma + \frac{\hbar^2}{2m_e^*}\left(\frac{\pi^2}{a^2} + k_y^2 + k_z^2\right) \qquad \text{...(6)}$$

$$\varepsilon_{\Gamma M} = \varepsilon_\Gamma + \frac{\hbar^2}{2m_e^*}(k_x^2 + k_y^2), \quad \varepsilon_{MW} = \varepsilon_\Gamma + \frac{\hbar^2}{2m_e^*}\left(2\frac{\pi^2}{a^2} + k_z^2\right)$$

These equations are also represented, as dashed lines, in Fig. 12.71.

The density of electronic states, as given by[6] may also assume the form[7] for nearly-free electrons, provided $m_e$ is replaced by the effective mass

$$N(\varepsilon) = \frac{V}{2\pi^2}\left(\frac{2m_e^*}{\hbar^2}\right)^{\frac{3}{2}} (\varepsilon - \varepsilon_\Gamma)^{\frac{1}{2}} \qquad \text{...(7)}$$

and $\varepsilon_\Gamma$ denotes the bottom of the energy band. For tightly bound electrons, near the bottom of the energy band, where $\varepsilon$ has the form (2):

---

6 $N(\varepsilon) = \dfrac{2V}{(2\pi)^3} \displaystyle\int_{S_\varepsilon} \dfrac{dS_\varepsilon}{|\nabla_{\vec{k}} \varepsilon|}$; a factor of 2 comes on the R.H.S. because each $\vec{k}$-state corresponds to one of the two possible spin states (or eigenvalues $\pm\hbar/2$). $S_\varepsilon$ is the surface of constant energy $\varepsilon$.

7 $N(\varepsilon) = \dfrac{V}{2\pi^2}\left(\dfrac{2m_c}{\hbar^2}\right)^{3/2} \sqrt{\varepsilon}$

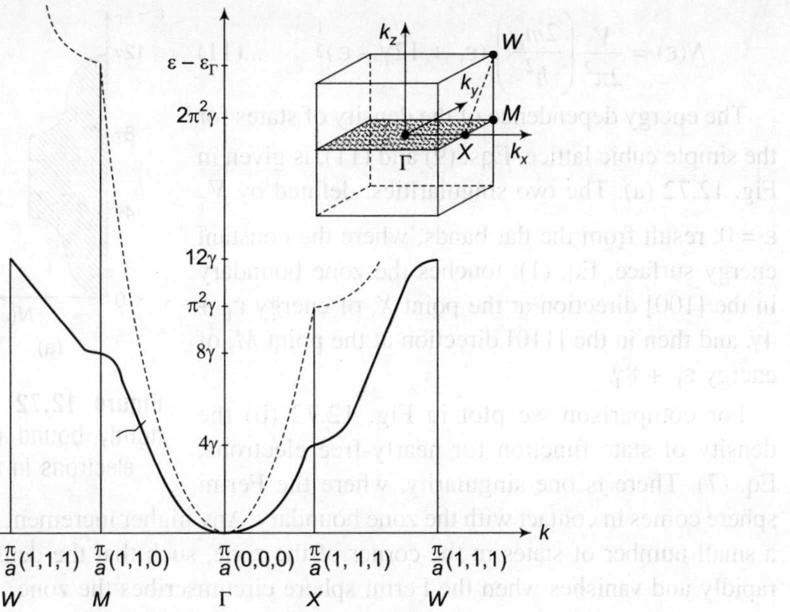


**Figure 12.71** Lowest energy bands for tightly bound and nearly-free (dashed lines) electrons for a simple cubic structure (inset)

$$\varepsilon = \varepsilon_\Gamma + \gamma k^2 a^2 \qquad \text{...(8)}$$

one has $\partial\varepsilon/\partial k = 2\gamma a^2 k$, and hence, substituting into[8] yields:

$$N(\varepsilon) = \frac{2V}{(2\pi)^3}\frac{4\pi k^2}{2\gamma a^2 k} = \frac{V}{2\pi^2}\frac{(\varepsilon - \varepsilon_\Gamma)^{\frac{1}{2}}}{(\gamma a^2)^{\frac{3}{2}}} = \frac{V}{2\pi^2}\left(\frac{2m_e^*}{\hbar^2}\right)^{\frac{3}{2}}(\varepsilon - \varepsilon_\Gamma)^{\frac{1}{2}} \qquad \text{...(9)}$$

This is the same as the parabolic density of states of nearly-free electrons, Eq. (7). In a similar manner, about the corners of the Brillouin zone we may expand

$$\cos\left(\frac{\pi}{a} - \delta k_i\right)a = \cos(\pi - \delta k_i a) = -\cos(\delta k_i a) = -1 + \frac{(\delta k_i)^2 a^2}{2} \qquad (i = x, y, z)$$

such that the energies near the top of the band are given by Eq. (1) in the form

$$E = E_0 - \alpha + 6\gamma - \gamma(\delta k)^2 a^2 = \varepsilon_\Gamma + 12\gamma - \gamma(\delta k)^2 a^2 \qquad \text{...(10)}$$

This immediately leads to a density of state function which is vanishing at the top of the band, where the energy is $\varepsilon_\Gamma + 12\gamma$, and has a parabolic increase for decreasing energy

[8] $N(\varepsilon)\ \frac{2V}{(2\pi)^3}\int \frac{dS_\varepsilon}{|\nabla_{\vec{k}}\,\varepsilon|}$; a factor of 2 comes on the R.H.S. because each $\vec{k}$-state corresponds to one of the two possible spin states for eignvalues $\pm\hbar/2$. $S_\varepsilon$ is the surface of constant energy ε.

$$N(\varepsilon) = \frac{V}{2\pi^2}\left(\frac{2m_e^*}{\hbar^2}\right)(\varepsilon_\Gamma + 12\gamma - \varepsilon)^{\frac{1}{2}} \quad ...(11)$$

The energy dependence of the density of states for the simple cubic lattice, Eqs. (9) and (11), is given in Fig. 12.72 (a). The two singularities, defined by $\nabla_{\vec{k}}\,\varepsilon = 0$, result from the flat bands, where the constant energy surface, Eq. (1), touches the zone boundary in the [100] direction at the point $X$, of energy $\varepsilon_\Gamma + 4\gamma$, and then in the [110] direction at the point $M$, of energy $\varepsilon_\Gamma + 8\gamma$.

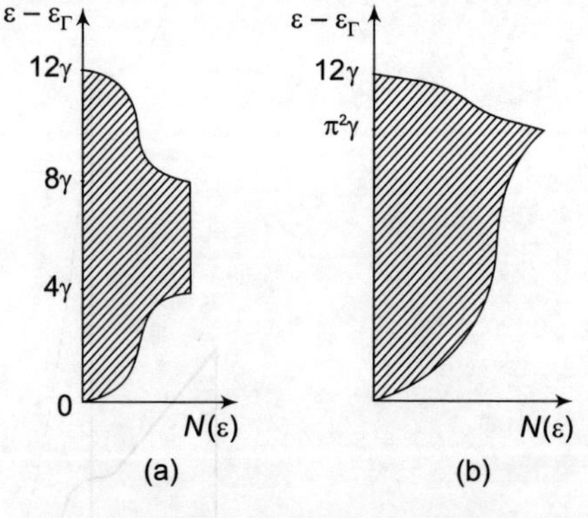


**Figure 12.72** Density of states for tightly bound (a) and nearly-free (b) electrons in a simple cubic lattice

For comparison we plot in Fig. 12.72 (b) the density of state function for nearly-free electrons, Eq. (7). There is one singularity, where the Fermi sphere comes in contact with the zone boundary. Any higher increment of energy encompasses a small number of states in the corner of the zone, such that the density of states decreases rapidly and vanishes when the Fermi sphere circumscribes the zone.

## 12.37 bcc LATTICE (TIGHTLY BOUND ELECTRONS)

We will find the dispersion relation for tightly bound electrons in a body-centred cubic lattice, with the primitive cell of side $a$.

Vectors $\vec{R}_n$ that connect the origin with its nearest neighbours. For a body-centred cubic crystal there are 8 nearest neighbours at

$$\vec{R}_n = \frac{a}{2}(\pm 1, \pm 1, 1), \frac{a}{2}(\pm 1, \pm 1, -1)$$

We may choose $\vec{k} = (k_x, k_y, k_z)$, such that

$$\begin{aligned}\sum_n e^{-i\vec{k}\cdot\vec{R}_n} &= e^{-ia(k_x+k_y+k_z)/2} + e^{-ia(k_x+k_y+k_z)/2} \\ &+ e^{-ia(k_x-k_y+k_z)/2} + e^{-ia(k_x-k_y+k_z)/2} \\ &+ e^{-ia(k_x+k_y-k_z)/2} + e^{ia(k_x+k_y-k_z)/2} \\ &+ e^{-ia(-k_x+k_y+k_z)/2} + e^{ia(-k_x+k_y+k_z)/2} \\ &= 2\left(\cos\frac{k_x+k_y+k_z}{2}a + \cos\frac{k_x-k_y+k_z}{2}a\right. \\ &\left. + \cos\frac{k_x+k_y-k_z}{2}a + \cos\frac{-k_x+k_y+k_z}{2}a\right)\end{aligned}$$

This yields

$$\sum_n e^{-i\vec{k}\cdot\vec{R}_n} = 4\cos\frac{k_y a}{2}\left(\cos\frac{k_x+k_y}{2}a + \cos\frac{k_x-k_y}{2}a\right)$$

$$= 8 \cos \frac{k_x a}{2} \cos \frac{k_y a}{2} \cos \frac{k_z a}{2}$$

which means that the dispersion relation reads

$$\varepsilon = \varepsilon_0 - \alpha - 8\gamma \cos \frac{k_x a}{2} \cos \frac{k_y a}{2} \cos \frac{k_z a}{2}$$

## 12.38 TWO-DIMENSIONAL SQUARE LATTICE

We will find the constant energy lines for a tightly bound electron in a two-dimensional square lattice with interatomic separation, $a$.

The equation of the lines of constant energy is derived from Eq. (1), by putting $k_z = 0$ and regarding $\varepsilon$ as a parametric constant

$$\cos k_x a + \cos k_y a = \frac{\varepsilon_0 - \alpha - \varepsilon}{2\gamma} = \text{const.}$$

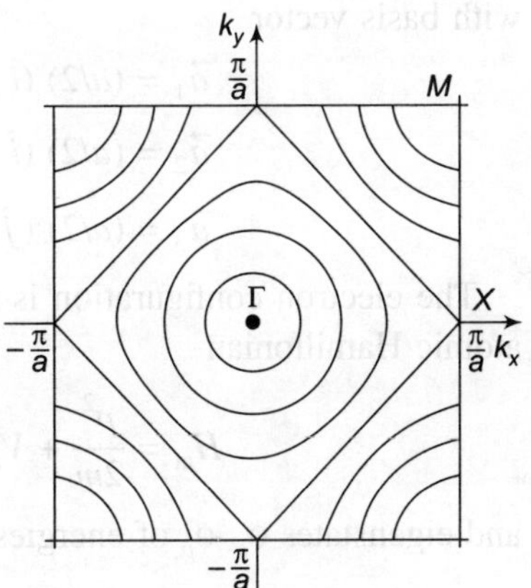


**Figure 12.73**

Near the bottom of band, we may use the expansion Eq. (2), which reduces to the equation of a circle in the $(k_x, k_y)$ plane

$$k_x^2 + k_y^2 = \frac{\varepsilon}{\gamma a^2} - \frac{\varepsilon_0 - \alpha - 4\gamma}{\gamma a^2}$$

In other words, near the bottom of the band, electrons behave like free electrons. At the origin, where $k_x = k_y = 0$, we have $\varepsilon_\Gamma = \varepsilon_0 - \alpha - 4\gamma$. As the value of $\varepsilon$ increases, the approximation $ka \ll 1$ is no longer valid, and the constant energy lines become distorted circles, until $\varepsilon = \varepsilon_0 - \alpha$. For this value of energy one obtains

$$\cos k_x a + \cos k_y a = 0 \quad \text{or} \quad k_x a = \pm k_y a \pm \pi$$

This defines a square having the $X$ points, of coordinates $(\pm\pi/a, 0)$ and $(0, \pm\pi/a)$, as corners. The area $2\pi^2/a^2$ of this square is half that of the first Brillouin zone, and hence it represents the Fermi surface if the two-dimensional lattice is made of monovalent atoms. The constant energy can be written as $\varepsilon_X = E_0 - \alpha = \varepsilon_\Gamma + 4\gamma$. Higher energies correspond to distorted circles approaching the circular lines about the corners of the Brillouin zone, described by Eq. (10), which reduces to

$$(\delta k_x)^2 + (\delta k_y)^2 = \frac{\varepsilon_0 - \alpha + 4\gamma}{\gamma a^2} - \frac{\varepsilon}{\gamma a^2}$$

This is the equation of a circle about the point $M$, of energy $\varepsilon_M = \varepsilon_0 - \alpha + 4\gamma = \varepsilon_\Gamma + 8\gamma$.

## 12.39 BAND STRUCTURE OF COLUMN IV ELEMENTS: (CALCULATED BY TIGHT BINDING METHOD)

Here, we give an idea of the band structure of column IV elements (calculated by the tight bonding method) for energies close to the band gap. This calculation formally holds for diamond, silicon, germanium and grey tin. The crystal is fcc with two atoms per cell (one at the origin, the other translated in the direction [111] of the quarter $\vec{b}$ of the principal diagonal of the cube (Fig. 12.74).

Each atom is at the centre of a regular tetrahedron. The primitive cell is rhombohedral with basis vector

$$\left.\begin{aligned}\vec{a}_1 &= (a/2)\,(\hat{i} + \hat{j})\\ \vec{a}_2 &= (a/2)\,(\hat{i} + \hat{k})\\ \vec{a}_3 &= (a/2)\,(\hat{j} + \hat{k})\end{aligned}\right| \quad ...(1)$$

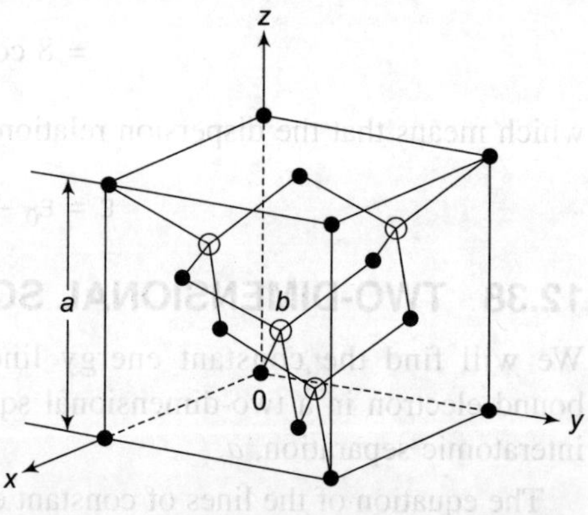


**Figure 12.74** Crystalline structure of silicon. The basis consist of two atoms, that at the origin and atom *b* with coordinates (*a*/4) (1, 1, 1). The Bravais lattice is FCC

The electron configuration is $ns^2\ np^2$ with the atomic Hamiltonian

$$H_{\text{at}} = \frac{p^2}{2m} + V(\vec{r} - \vec{b}) \quad ...(2)$$

and eigenstates $\phi_s$, $\phi_p$ of energies $E_s$, $E_p$. We consider

$$\left.\begin{aligned}\phi_0 &= (\phi_s + \phi_x + \phi_y + \phi_z)/2\\ \phi_1 &= (\phi_s - \phi_x - \phi_y + \phi_z)/2\\ \phi_2 &= (\phi_s - \phi_x + \phi_y - \phi_z)/2\\ \phi_3 &= (\phi_s + \phi_x - \phi_y - \phi_z)/2\end{aligned}\right| \quad ...(3)$$

These functions, called hybrid orbitals $sp_3$, are not atomic eigenfunctions but form a basis for the tensor product $s \otimes p$. They have the essential property of "pointing" respectively in the directions [1, 1, 1], [–1, –1, 1] [–, 1, –1] and [1,–1,–1], i.e., towards the vertices of a tetrahedron in the observed direction of the covalent bond. The matrix elements of $H_{\text{at}}$ in this basis are

$$\left.\begin{aligned}\langle\phi_i|H_{\text{at}}|\,\phi_i\rangle &= \frac{E_s + 3\,E_p}{4}\\ \langle\phi_i|H_{\text{at}}|\,\phi_i\rangle &= \frac{E_s - E_p}{4}\end{aligned}\right| \quad ...(4)$$

where $l$ denotes an orbital other than $i$ localized at the same atom.

Similarly, we consider hybrid orbitals pointing in the opposite directions

$\phi'_0 = (\phi_s - \phi_x - \phi_y - \phi_z)/2$ pointing towards [– 1, –1, – 1]

$\phi'_1 = (\phi_s + \phi_x + \phi_y - \phi_z)/2$ pointing towards [+1, +1, –1]

$\phi'_2 = (\phi_s + \phi_x - \phi_y + \phi_z)/2$ pointing towards [+1, –1, +1]

$$\phi'_3 = (\phi_s - \phi_x + \phi_y + \phi_z)/2 \quad \text{pointing towards } [-1, +1, +1] \qquad ...(5)$$

We seek a Bloch function solution for the crystal of the form

$$\psi_{n,k} = C \sum_j \exp(i\vec{k}\cdot\vec{R}_j) \sum_{i=0,1,2,3} [A_i\, \phi_i(\vec{r}-\vec{R}_j) + A'_i\, \phi'_i(\vec{r}-\vec{R}_j-\vec{b})] \qquad ...(6)$$

where the index $j$ denotes the particular site in the lattice, the index $i$ one of the four orbitals (Eq. 3) and $C$ is a normalization coefficient.

The crystal Hamiltonian is

$$H = -\frac{\hbar^2}{2m}\Delta + \sum_j \{V(\vec{r}-\vec{R}_j) + V(\vec{r}-\vec{R}_j-\vec{b})\} \qquad ...(7)$$

and therefore the eigenstate problem

$$H\,\psi_{n,\vec{k}}(\vec{r}) = E_{n,\vec{k}}\,\psi_{n,\vec{k}}(\vec{r}) \qquad ...(8)$$

Consider the functions $\phi_i$ and $\phi'_i$ centred respectively at the origin and at (1/4, 1/4, 1/4) and let us take their product with Eq. (8). We get

$$\left.\begin{aligned} \langle\phi_i\,|H|\,n,\vec{k}\rangle &= E_{n,\vec{k}}\,\langle\phi_i\,|n,\vec{k}\rangle \quad \text{(a)} \\ \langle\phi_i\,|H|\,n,\vec{k}\rangle &= E_{n,\vec{k}}\,\langle\phi_i\,|n,\vec{k}\rangle \quad \text{(b)} \end{aligned}\right| \qquad ...(9)$$

We can rewrite Eq. (7) in the form

$$H = H_{at} + \sum_{j\neq 0} V(\vec{r}-\vec{R}_j) + \sum_j V(\vec{r}-\vec{R}_j-\vec{b}) \qquad ...(10)$$

where $H_{at}$ is the electron Hamiltonian for the atom at the origin. One can thus see that the only contributions on the left in Eqs. (9a) and (9b) come either from the atomic terms or from the interaction term coupling $\phi_i$ or $\phi'_i$ to the only orbital from the nearest neighbour which points toward it (neglecting terms of the form $\langle\phi_i\,|V_{i\neq j}|\,\phi_i\rangle$ and interactions between orbitals centred on two neighbouring atoms but not pointing to each other). The interaction or "transfer" integrals are all equal and negative since the potential is attractive and the functions $\phi$ and $\phi'$ have the same sign in the region where they overlap. We set

$$\langle\phi\,|V|\,\phi'\rangle = -\lambda. \qquad ...(11)$$

Let us consider the site at the origin: $\phi_0$ points towards the site $\vec{b} = (a/4)\,(1, 1, 1)$ and is thus only coupled to the orbital $\phi'_0$ centred at $\vec{b}$, with the coupling coefficient $\exp i\,\vec{k}\cdot\vec{a}_0 = 1$. $\phi_1$ being directed along [–1, –1, 1] is coupled to the orbital centred at the site $(a/4)\,(-1, -1, 1) = \vec{b} - \vec{a}_1$, or $\phi'_1\,[\vec{r} - (-a_1)]$ with the coefficient $\exp i\,\vec{k}\cdot\vec{R}_j = \exp(-i\,\vec{k}\cdot\vec{a}_1)$, etc. Finally Eqs. (9a) and (9b) can be written as

$$\left[\frac{E_s + 3E_p}{4} - E_{n,\vec{k}}\right] A_i + \frac{E_s - E_p}{4} \sum_{l\neq i} A_l - \lambda \exp(-i\vec{k}\cdot\vec{a}_i)\, A'_i = 0 \qquad ...(12)$$

$$-\lambda \exp i\vec{k}\cdot\vec{a}_i\, A_i + \left[\frac{E_s + 3E_p}{4} - E_{n,\vec{k}}\right] A'_i + \frac{E_s - E_p}{4} \sum_{l \neq i} A'_l = 0 \quad ...(13)$$

This system of homogeneous equations gives the eight coefficients $A_i$ and $A'_i$ which specify the wave function (Eq. 6), provided the determinant of the coefficients vanishes. One thus obtains the following secular equation, with

$$X = E_{n,\vec{k}} - E_p, \quad \delta = \frac{E_p - E_s}{4}, \quad \alpha_n = \exp(-i\,\vec{k}\cdot\vec{a}_n) \quad ...(14)$$

$$\begin{vmatrix} x+\delta & \delta & \delta & \delta & \lambda\alpha_0 & 0 & 0 & 0 \\ \delta & x+\delta & \delta & \delta & 0 & \lambda\alpha_1 & 0 & 0 \\ \delta & \delta & x+\delta & \delta & 0 & 0 & \lambda\alpha_2 & 0 \\ \delta & \delta & \delta & x+\delta & 0 & 0 & 0 & \lambda\alpha_3 \\ \lambda\alpha_0^* & 0 & 0 & 0 & x+\delta & \delta & \delta & \delta \\ 0 & \lambda\alpha_1^* & 0 & 0 & \delta & x+\delta & \delta & \delta \\ 0 & 0 & \lambda\alpha_2^* & 0 & \delta & \delta & x+\delta & \delta \\ 0 & 0 & 0 & \lambda\alpha_3^* & \delta & \delta & \delta & x+\delta \end{vmatrix} = 0. \quad ...(15)$$

Setting $\Phi = \left|\frac{1}{4}\sum_n \alpha\ \alpha_n\right|$, a careful calculation shows that the above determinant can be written as

$$(x^2 - \lambda^2)^2\,(x^2 + 4\delta x - \lambda^2 + 4\,\delta\,\lambda\,\Phi),$$

$$(x^2 + 4\,\delta x - \lambda^2 - 4\,\delta\,\lambda\,\Phi) = 0 \quad ...(16)$$

with

$$\Phi(\vec{k}) = \frac{1}{2}\left[1 + \cos\frac{k_x a}{2}\cos\frac{k_y a}{2} + \cos\frac{k_x a}{2}\cdot\cos\frac{k_z a}{2} + \cos\frac{k_y a}{2}\cdot\cos\frac{k_z a}{2}\right]^{1/2} \quad ...(17)$$

Equation (16) gives $x(\Phi)$, i.e., the dispersion relation $E_n(\vec{k})$.

### Resulting Band Structure

We obtain four flat bands (for which $E_n$ does not depend on $\vec{k}$) which correspond to the doubly degenerate solutions $x_1 = \lambda$ and to the similarly doubly degenerate solutions $x_2 = -\lambda$.

The broad bands are associated with the other solutions of Eq. (16), i.e.,

$$x_3 = -2\delta + [4\delta^2 + \lambda^2 + 4\delta\lambda\Phi]^{1/2}$$

$$x_4 = -2\delta - [4\delta^2 + \lambda^2 + 4\delta\lambda\Phi]^{1/2}$$

$$x_5 = -2\delta + [4\delta^2 + \lambda^2 - 4\delta\lambda\Phi]^{1/2}$$

$$x_6 = -2\delta - [4\delta^2 + \lambda^2 - 4\delta\lambda\Phi]^{1/2} \quad ...(18)$$

The band structure has a different shape for $\lambda > 2\delta$ or $\lambda < 2\delta$. It is shown in Fig. 12.75 (a) for two directions of the vector $\vec{k}$ in the zone and in the special case $\lambda = \delta$. There are $8N$ electrons to be placed in these levels since the primitive cell contains two atoms each possessing four valence electrons. Taking account of the spin degeneracy of each atomic level, we see that at zero temperature, the bands $x_4$, $x_6$ and $x_2$ are filled. We are thus dealing here with a metal ($\lambda < 2\delta$) as the band $x_5$ is empty and very near in energy to the filled levels.

By contrast, if $x > 2\delta$, we have a semiconductor. This is shown in Fig. 12.75 (b), for the special case $\lambda = 3\delta$. The $8N$ electrons fill the $x_4$, $x_6$ and $x_2$ bands and the material is an insulator at zero temperature. The width of the band gap is

$$E_g = x_{5,\,k=0} - x_2 = 2\lambda - 4\delta. \quad ...(19)$$

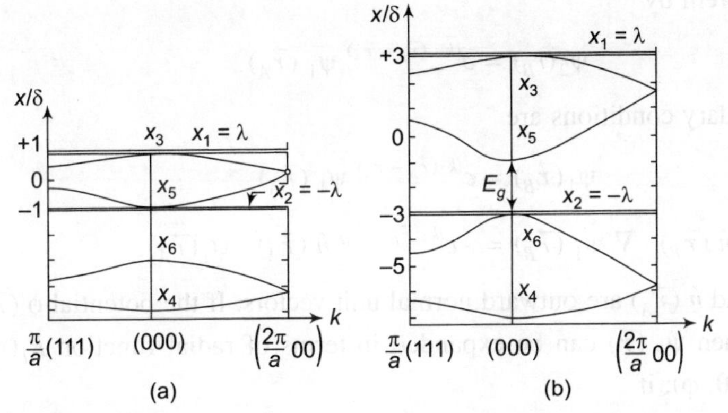


**Figure 12.75** Band structure of an element of column IV of the periodic table, calculated by the tight binding (LCAO) method. The transfer integral is $-\lambda$, and the distance between the atomic $s$ and $p$ levels is $\left(\frac{E_p - E}{4}\right) = 4\delta$. The bands $x_4$, $x_6$ and $x_2$ are full. (a) $\lambda = \delta$, the material is a conductor (b) $\lambda = 3\delta$, the material is an insulator. (After B, Sapoval and C. Hermann, *Physics of Semiconductors*, Springer (2006))

## 12.40 CELLULAR METHOD

Slater, in 1934, introduced the cellular method in the band structure calculations which has its origin in the method of Wigner and Seitz, who assumed that the electrons in a metal move independently within a regular array of fixed positive ions. Since the electron wave functions obey Bloch's theorem, it is only necessary to find out the wave functions in one Wigner Seitz cell of the structure such as shown in Fig. 12.76. For simplifying the theory, the Wigner Seitz cell was replaced by a sphere with the same volume as of cell and the potential $\phi\,(\vec{r})$

was then assumed to be spherically symmetrical. The one electron wave function $\psi_k(\vec{r})$ is then subject to the boundary conditions that $\psi_k(\vec{r})$ and its derivative are continuous at the surface of the Wigner Seitz sphere. This method, when applied by Wigner and Seitz, gave satisfactory results for energy bands at $\vec{k} = 0$, but not at other points of Brillouin zone.

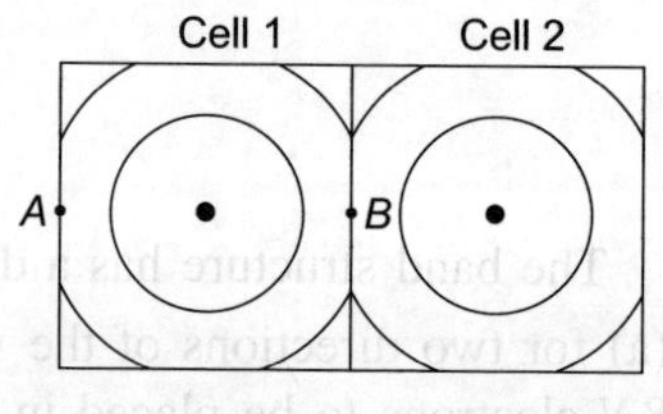


**Figure 12.76** Cellular method potential for monatomic square lattice

Slater, in cellular method, accounted the shape and the symmetry of the Wigner Seitz unit cell instead of approximating it by a sphere. From Fig. 12.76, the boundary condition on wave function at the interface between two cells is

$$\psi_1(\vec{r}_B) = \psi_2(\vec{r}_B) \quad ...(1)$$

The wave function $\psi_2(\vec{r}_B)$ can be related to the wave function in cell one at $A$ as a result of Bloch's theorem by

$$\psi_2(\vec{r}_B) = e^{i\vec{k}\cdot(\vec{r}_B - \vec{r}_A)}\,\psi_1(\vec{r}_A) \quad ...(2)$$

So the boundary conditions are

$$\psi_1(\vec{r}_B) = e^{ik\cdot(\vec{r}_B - \vec{r}_A)}\,\psi_1(\vec{r}_A)$$

$$\hat{n}(\vec{r}_B)\cdot\nabla\psi_1(\vec{r}_B) = -e^{ik\cdot(\vec{r}_B - \vec{r}_A)}\,\hat{n}(\vec{r}_A)\cdot\psi_1(\vec{r}_A) \quad ...(3)$$

where $\hat{n}(\vec{r}_B)$ and $\hat{n}(\vec{r}_A)$ are outward normal unit vectors. If the potential $\phi(\vec{r})$ is spherically symmetrical, then $\psi_k(\vec{r})$ can be expanded in terms of radial function $R_l(r)$ and spherical harmonics $Y_l^m(\theta, \varphi)$; if

$$\psi_k(\vec{r}) = u_k(\vec{r})\,e^{i\vec{k}\cdot\vec{r}} \quad ...(4)$$

We may write

$$u_k(\vec{r}) = \sum_l \sum_m A_{klm}\,R_l(r)\,Y_l^m(\theta, \varphi) \quad ...(5)$$

When such a function is substituted in boundary conditions Eq. (3), then these equations become

$$\sum_l \sum_m A_{klm}\,R_l(r_B)\,Y_l^m(\theta_B, \varphi_B) = e^{i\vec{k}\cdot(\vec{r}_B - \vec{r}_A)}$$

$$\sum_l \sum_m A_{klm}\,R_l(r_A)\,Y_l^m(\theta_A, \varphi_A)\,\hat{n}(\vec{r}_B)\cdot\nabla$$

$$\sum_l \sum_m A_{klm}\,R_l(r_B)\,Y_l^m(\theta_B, \varphi_B)$$

$$= e^{-i\vec{k}\cdot(\vec{r}_B - \vec{r}_A)}\,\hat{n}(\vec{r}_A)\cdot\nabla\sum_l \sum_m A_{klm}\,R_l(r_A)\,Y_l^m(\theta_A, \varphi_A) \quad ...(6)$$

By using a number of different pairs of boundary points $\vec{r}_A$ and $\vec{r}_B$ in Eq. (6), a set of linear simultaneous algebraic equations is obtained for the unknown coefficients $A_{klm}$ which on solution then gives first $A_{klm}$ then the wave functions and finally the energy eigenvalues, $E(\vec{k})$.

There are, basically two major difficulties with the cellular method:

(1) The potential used has a discontinuous derivative whenever the boundary between two cells is crossed, whereas in fact, the potential is quite flat in such regions.

(2) There are computational difficulties involved in numerically satisfying a boundary condition over the surface of the Wigner–Seitz primitive cell, a fairly complex polyhedral structure, (i.e., it involves a lot of numerical efforts to match two wave functions at the cell boundary, particularly for states $\vec{k} = 0$).

These problems are taken care of in the augmented plane wave (APW) method by using a *muffin-tin potential.*

## 12.41 MUFFIN-TIN POTENTIAL

Figure 12.77 shows the cellular method potential and Fig. 12.78, the muffin-tin potential.

Formally, the muffin-tin potential can be defined (for all $\vec{R}$) by

$$U(\vec{r}) = V(|\vec{r} - \vec{R}|) \text{ when } |\vec{r} - \vec{R}| < r_0 \text{ (the } \textit{core} \text{ or } \textit{atomic region}) \quad \text{...(1a)}$$

$$= V(r_0) = 0 \text{ when } |r - R| > r_0 \text{ (the } \textit{interstitial region}) \quad \text{...(1b)}$$

where $r_0$ is less than half the nearest-neighbour distance.

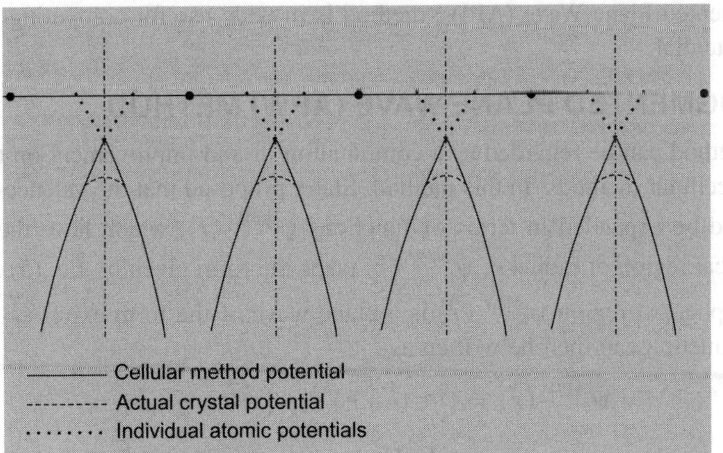


**Figure 12.77** The Cellular method potential has a discontinuous derivative midway between lattice points, but the actual potential is quite flat there

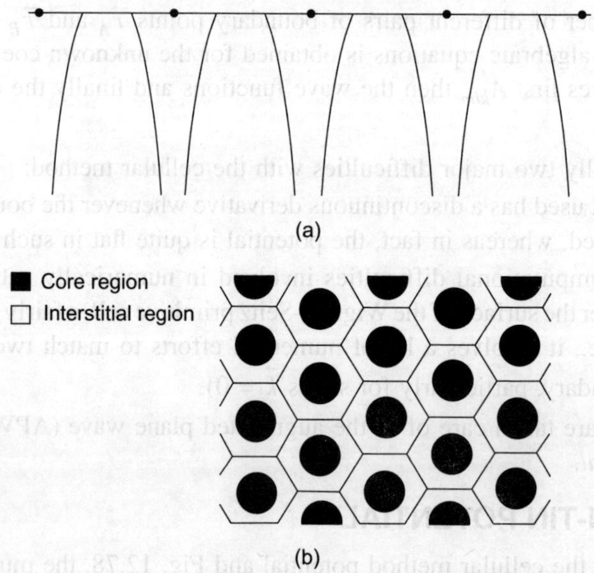


**Figure 12.78** (a) The muffin-tin potential, plotted along a line of ions (b) The muffin-tin potential is constant (zero) in the interstitial regions and represents an isolated ion in each core region

If we agree that the function $V(r)$ is zero when its argument exceeds $r_0$, then we can write $U(\mathbf{r})$ very simply as

$$U(\mathbf{r}) = \sum_{\mathbf{R}} V(|\mathbf{r} - \mathbf{R}|). \quad ...(2)$$

The Augmented Plane Wave (APW) method is in wide use for computing the bands in a muffin-tin potential.

## 12.42 AUGMENTED PLANE WAVE (APW) METHOD

The APW method can be regarded as a combination of and improvement on the nearly free electron and cellular methods. In this method, Slater proposed that the valence-electron wave function could be expanded in terms of functions $\psi_k^{APW}(\vec{r})$, which have the property that within spherical region of radius $r_i$, $\psi_k^{APW}(\vec{r})$ takes the form given by Eq. (5) Art. 12.40 and outside this spherical region $\psi_k^{APW}(\vec{r})$ is a plane wave of the form $\exp\{i(\vec{k} + \vec{g})\cdot\vec{r}\}$. The APW wave function can then be written as

$$\psi_k^{APW}(\vec{r}) = A^0 \in (r - r_i) \exp\{i(\vec{k} + \vec{g})\cdot\vec{r}\}.$$

$$+ \sum_l \sum_m A_{klm} \in (r_i - r)\, R_l(r)\, Y_l^m(\theta, \varphi) \quad ...(1)$$

where $\in(x)$ is a step function defined by

$$\in(x) = 1 \quad \text{if} \quad x \geq 0$$

$$= 0 \quad \text{if} \quad x < 0 \quad ...(2)$$

The use of a wave function of the form corresponds to using a muffin-tin potential of the form as shown in Fig. 12.79. The coefficients $A_{klm}$ are determined in such a way that $\psi_k^{APW}(\vec{r})$ is continuous across the spherical surface of radius $r_i$. In spherical polar coordinates, we may expand the plane wave as

$$e^{i\vec{k}\cdot\vec{r}} = 4\pi \sum_l \sum_m (i)^l j_l(kr)\, Y_l^m(\theta, \varphi)\, Y_l^{m*}(\theta_k, \varphi_k) \quad ...(3)$$

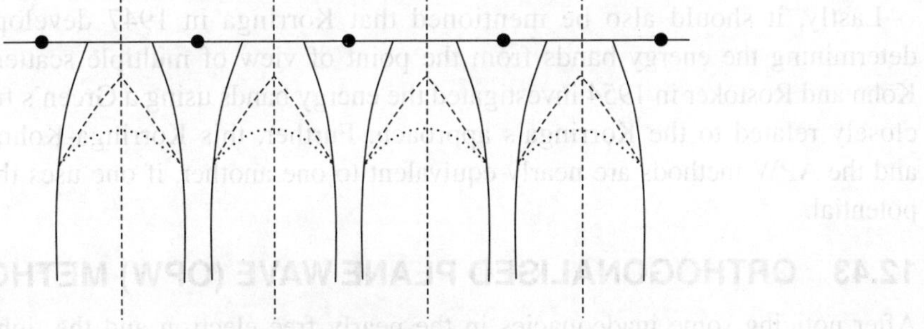


**Figure 12.79**

where $j_l(kr)$ is a spherical Bessel function and $\vec{k}$ is expressed in spherical polar coordinates as $(k, \theta_k, \varphi_k)$. The continuity of $\psi_k^{APW}(\vec{r})$ can be achieved if

$$A_{klm} = 4\pi(i)^l\, Y_l^{m*}(\theta_{k+g}, \varphi_{k+g})\, [j_l(|\vec{k}+\vec{g}|\, r_i)\, A^0]/R_l(r_i) \quad ...(4)$$

It should be mentioned that because $R_l(r_l)$ depends on $E(\vec{k})$, the value of $A_{klm}$ also depends on $E(\vec{k})$. The wave function $\psi_k(\vec{r})$ of an electron can then be expanded in terms of these APW's as

$$\psi_k(\vec{r}) = \sum_g a_g\, \psi_k^{APW}(\vec{r}) \quad ...(5)$$

where the summation is over all the vectors $(\vec{k}+\vec{g})$ that are related to $\vec{k}$ by a translation of the reciprocal lattice. The secular determinant we obtain finally is similar to that which we obtained in the OPW method

$$\det \left\| \left[ \frac{\hbar^2}{2m}(\vec{k}+\vec{g})^2 - E(\vec{k}) \right] \delta_{gg'} + \phi_{gg'}^{APW} \right\| = 0 \quad ...(6)$$

where

$$\phi_{gg'}^{APW} = 4\pi r_i^2 \left\{ -[(\vec{k}+\vec{g}')\cdot(\vec{k}+\vec{g}) - E(\vec{k})] \frac{j_l(|\vec{g}'-\vec{g}|\, r_i)}{|\vec{g}'-\vec{g}|} + \sum_{l=0}^{\infty} (2l+1)\, P_l(\cos\theta_{gg'})\, j_l(|\vec{k}+\vec{g}'\, r_i)\, j_l(|\vec{k}+\vec{g}|\, r_i) \frac{R_l'(r_i, E)}{Rl(r_i, E)} \right\} \quad ...(7)$$

Here $\theta_{gg'}$ is the angle between $\vec{g}$ and $\vec{g}'$ and $R'_l$ is the derivative of $R_l$. The variational principle is used because of the dependence of $A_{klm}$ on $E(\vec{k})$ as mentioned earlier. It is therefore not possible to solve Eq. (6) directly and one has to choose some trial value of the energy to evaluate the determinant, repeat the process number of times until a particular value of energy is obtained finally that makes the determinant vanish. With modern computing techniques, it is possible to achieve excellent convergence by including enough APW's and hence making the APW scheme of calculating energy bands quite successful.

Lastly, it should also be mentioned that Korringa in 1947 developed a method for determining the energy bands from the point of view of multiple scattering of waves and Kohn and Rostoker in 1954 investigated the energy bands using a Green's functions technique closely related to the Korringa's approach. Further, this Korringa-Kohn-Rostoker (KKR) and the APW methods are nearly equivalent to one another, if one uses the same muffin-tin potential.

## 12.43 ORTHOGONALISED PLANE WAVE (OPW) METHOD

After noticing some inadequacies in the nearly free electron and the tight binding models. Herring, in 1940, introduced the orthogonalised plane wave (OPW) method for energy band calculation.

In this method, it is necessary to make a clear distinction between the core electrons and the valence electrons. The core wave functions are localised about the lattice sites and the

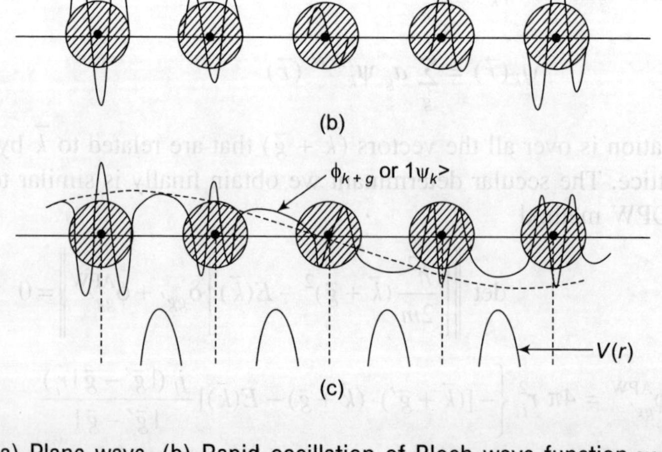


**Figure 12.80** (a) Plane wave, (b) Rapid oscillation of Bloch wave function $\psi_k$ near the atomic sites, (c) Synthesis of orthogonalised plane wave (OPW) = Plane wave-core function)

valence wave functions have a rapid oscillatory behaviour, with a tail. If Bloch functions are formed from ion-core orbitals with small overlap and are expanded in terms of plane waves, the series converges slowly and contains contribution from a large number of Brillouin zones. Furthermore, if the valence wave fuction is approximated by a few plane waves, then it fails to create the rapid oscillatory behaviour in core region. Herring, therefore, proposed the construction of valence wave functions from plane wave using the fact that these wave functions must be orthogonal to the ion-core electron wave functions.

Accordingly, Herring defined the OPW function by

$$\phi_{k+g} = \exp\,[i(\vec{k} + \vec{g}) \cdot \vec{r}] + \sum_m A_c\, \psi_k^c\,(\vec{r} - \vec{r}_m) \qquad ...(1)$$

where $\psi_k^c\,(\vec{r} - \vec{r}_m)$ is a core function centred at the lattice site $\vec{r}_m$. The summation is over all core states with Bloch wave vector $\vec{k}$. A logical relation is

$$\text{OPW} = (\text{plane wave}) - (\text{core function}) \qquad ...(2)$$

The constant $A_c$ can be determined from orthogonality condition

$$\int \psi_{\vec{k}}^{c*}\,(\vec{r} - \vec{r}_m)\, \phi_{k+g}\,(\vec{r})\, d\vec{r} = 0 \qquad ...(3)$$

This gives the coefficient $A_c$ as

$$A_c = -\int \psi_{\vec{k}}^{c*}\,(\vec{r} - \vec{r}_m) \exp\,[i(\vec{k} + \vec{g}) \cdot \vec{r}\,]\, d\vec{r} \qquad ...(4)$$

Since both the plane wave $\exp\,[i(\vec{k} + \vec{g}) \cdot \vec{r}\,]$ and the core wave function $\psi_k^c$ separately satisfy the Bloch condition. Therefore, OPW function $\phi_{k+g}$ will also satisfy it. We may, therefore, expand the Bloch function $\psi_k(r)$ of the valence electrons (i.e., the actual eigenstate of the Schrödinger wave equation) in terms of the basis function $\phi_{k+g}$ as

$$\psi_{\vec{k}}(\vec{r}) = \sum_g a_g\, \phi_{k+g}(r)$$

or

$$|\psi_k\rangle = \sum_g a_g\, |\phi_{k+g}\rangle \qquad ...(5)$$

Substituting this into the Schrödinger equation

$$H\psi_k(r) = E\psi_k(r)$$

or

$$(T + V - E)\, |\psi_k\rangle = 0 \qquad ...(6)$$

But

$$|\phi_{k+g}\rangle = e^{i(\vec{k} + \vec{g}) \cdot \vec{r}} - \sum_m \psi_k^c\,(\vec{r} - \vec{r}_m) \times \int e^{i(\vec{k} + \vec{g}) \cdot \vec{r}}\, \psi_k^{c*}\,(\vec{r} - \vec{r}_m)\, dr$$

$$= |\vec{k} + \vec{g}\rangle - \sum_m |\psi_{cm}\rangle \langle\psi_{cm}|\, \vec{k} + \vec{g}\rangle \qquad ...(7)$$

From Eqs. (5) – (7), we get a secular equation

$$\sum_g a_g\, (T + V - E)|\vec{k} + \vec{g}\rangle - (T + V - E) \sum_m |\psi_{cm}\rangle \langle\psi_{cm}|\vec{k} + \vec{g}\rangle \qquad ...(8)$$

This is simplified further by taking into account

$$(T + V)\,|\psi_{cm}\rangle = E_c|\psi_{cm}\rangle \qquad ...(9)$$

where $E_c$ is an atomic energy eigenvalue of the appropriate core level and $|\psi_{cm}\rangle$ is the core wave function centred at the lattice site $m$. Now operating from left $\langle k + g'|$ and rearranging the terms, we obtain

$$\sum_g a_g \left\{ \left[ \frac{\hbar^2}{2m}(k+g)^2 - E \right] \langle k+g'|k+g\rangle + \langle k+g'|V_{\text{eff}}(r)|k+g\rangle \right\} = 0 \qquad ...(10)$$

where

$$\langle k + g'|V_{\text{eff}}(r)|k + g\rangle = \langle k + g|V|k + g\rangle + \sum_m (E - E_c)\,\langle k + g'|\psi_{cm}\rangle\langle\psi_{cm}|k + g\rangle \qquad ...(11)$$

defines a new effective weak potential known as OPW pseudopotential.

## 12.44 PSEUDOPOTENTIALS (EMPIRICAL) METHOD

The electronic states in a crystal are obtained by solving the following non-relativistic Schrödinger equation in the one-electron approximation

$$\left[ -\frac{\hbar^2}{2m} + V(\vec{r}) \right] \psi_n(\vec{r}) = E_n(\vec{k})\psi_n(\vec{r}) \qquad ...(1)$$

However, it is possible only when we know the crystal potential $V(\vec{r})$. In the following we will show how the Schrödinger equation is solved to a good approximation by using empirical parameters, known as *pseudopotentials*, and the orthogonality of the wave functions.

First, we assume the electron wave functions of the core states and their energies are given by $\varphi_j$ and $E_j$ respectively. We then have

$$H\,|\varphi_j\rangle = [H_0 + V(\vec{r})]\,|\varphi_j\rangle = E_j|\varphi_j\rangle \qquad ...(2)$$

where

$$H_0 = -\frac{\hbar^2}{2m}\nabla^2 \qquad ...(3)$$

The Bloch functions $|\psi\rangle$ of the electrons in the valence and conduction bands are orthogonal with the above wave functions $|\varphi_j\rangle$; in other words

$$\langle\varphi_j|\psi\rangle = 0 \qquad ...(4)$$

The above orthogonality is satisfied when we choose the wave functions given by

$$|\psi(\vec{k}, \vec{r})\rangle = |\chi_n(\vec{k}, \vec{r})\rangle - \sum_j \langle\varphi_j|\chi_n\rangle\,\varphi_j\rangle \qquad ...(5)$$

This is evident from the relation

$$\langle\varphi_{j'}|\psi\rangle = \langle\varphi_{j'}|\chi_n\rangle - \sum_j \langle\varphi_j|\chi_n\rangle\langle\varphi_{j'}|\varphi_j\rangle$$

$$= \langle\varphi_{j'}|\chi_n\rangle - \sum_j \langle\varphi_j|\chi_n\rangle\,\delta_{j'j} \equiv 0. \qquad ...(6)$$

$$(\because \quad \delta_{j'j} \equiv 0 \text{ for } j = j')$$

We have to note that we require no restriction to $|\chi_n\rangle$. Therefore, we use plane waves for $|\chi_n(\vec{k}, \vec{r})\rangle$ and in this scheme (Eq. (5)) is called the OPW (orthogonalized plane wave). Substituting Eq. (5) into Eq. (1), we find

$$H\,|\chi_n\rangle - \sum_j \langle\varphi_j\,|\chi_n\rangle\, H\,|\varphi_j\rangle = E_n(\vec{k}) \times \left\{\left|x_n\right\rangle - \sum_j \left\langle\varphi_j\,|\chi_n\right\rangle \left|\varphi_j\right\rangle\right\} \quad ...(7)$$

and then we obtain the following relation

$$H\,|\chi_n\rangle + \sum_j\,[E_n(\vec{k}) - E_j]\,|\varphi_j\rangle\,\langle\varphi_j|\chi_n|\rangle = E_n(\vec{k})\,|\chi_n\rangle \quad ...(8)$$

$$(\because H|\varphi_j\rangle = E_j\,|\varphi_j\rangle)$$

Introduction of a new parameter

$$V_p \equiv \sum_j\,[E_n(\vec{k}) - E_j]\,|\varphi_j\rangle\,\langle\varphi_j| \quad ...(9)$$

will lead to

$$[H + V_p]\,|\chi_n\rangle = E_n(\vec{k})\,|\chi_n\rangle \quad ...(10)$$

or

$$[H_0 + V(\vec{r}) + V_p(\vec{r})]\,|\chi_n\rangle = E_n(\vec{k})\,|\chi_n\rangle \quad ...(11)$$

where $E_n(\vec{k})$ is the energy of the band $n$ we are interested in. There exists the following inequality between the energies of the core states, $E_j$, and the energies of the valence and conduction bands $E_n(\vec{k})$

$$E_n(\vec{k}) > E_j \quad ...(12)$$

and thus we find that

$$V_p > 0 \quad ...(13)$$

We may rewrite Eq. (11) as

$$[H_o + V_{ps}(\vec{r})]\,|\chi_n\rangle = E_n(\vec{k})\,|\chi_n\rangle \quad ...(14)$$

$$V_{ps} = V(\vec{r}) + V_p(\vec{r}) \quad ...(15)$$

and it may be possible to make $V_{ps}$ small enough, since $V(\vec{r}) < 0$. The new potential $V_{ps}(\vec{r})$, called the *pseudopotential*, is also periodic, and we can expand it as the Fourier series

$$V_{ps}(\vec{r}) = \sum_j V_{ps}(\vec{G}_j)\,e^{-i\vec{G}_j\cdot\vec{r}} \quad ...(16)$$

where the Fourier coefficients $V_{ps}(\vec{G}_j)$ are given by

$$V_{ps}(\vec{G}_j) = \frac{1}{\sqrt{\Omega}} \int_\Omega V_{ps}(\vec{r})\,e^{i\vec{G}_j\cdot\vec{r}}\,d^3\vec{r} \quad ...(17)$$

For the reason stated above, the potential $V_{ps}(\vec{r})$ may be chosen as small as possible, so that the potential $V_{ps}(\vec{G}_j)$ is expressed with a small number of the Fourier coefficient $V_{ps}$

$(\vec{G}_j)$, in other words, we may choose several values of $V_{ps}(\vec{G}_j)$ and neglect the other values because of their smallness.

We should note that $|V_{ps}(\vec{r})|$ is smaller than $|V(\vec{r})|$, but it does not mean that $V(\vec{r})$ converges with only a small number of its Fourier coefficients. The *empirical pseudopotential method* is based on the approximation that the Fourier coefficients $V_{ps}(\vec{r})$ are empirically chosen so that the shape of the critical points and their energies show good agreement with experimental observation.

Energy band calculations based on the empirical pseudopotential method take into account as few the pseudopotentials $V_{ps}(\vec{G}_j)$ as possible and uses the Bloch functions of the free-electron bands for the wave functions $|\chi_n\rangle$. The energy bands are obtained by solving

$$\left[-\frac{\hbar^2}{2m}\nabla^2 + V_{ps}(\vec{r})\right]|\chi_n(\vec{r})\rangle = E_n\,|\chi_n(\vec{r})\rangle. \qquad ...(18)$$

$$|\chi_n(\vec{r})\rangle = e^{i(\vec{k}+\vec{G}_j)\cdot\vec{r}} \qquad ...(19)$$

$$V_{ps}(\vec{r}) = \sum_{j'} V_{ps}(\vec{G}_{j'})\,e^{-i\vec{G}_j^n\cdot\vec{r}} \qquad ...(20)$$

Then the energy band structures are given by the following eigenvalue equation

$$\left[-\frac{\hbar^2}{2m}\nabla^2 + \sum_{j'} V_{ps}(\vec{G}_{j'})\,e^{-i\vec{G}_{j'}\cdot\vec{r}}\right]e^{i(\vec{k}+\vec{G}_j)\cdot\vec{r}}$$

$$= E_n(\vec{k})\,e^{i(k+\vec{G}_j)\cdot\vec{r}} \qquad ...(21)$$

The eigenvalues and eigenfunctions of the above equation are obtained by solving the following matrix equation. First we introduce

$$H_p = -\frac{\hbar^2}{2m}\nabla^2 + V_{ps}(\vec{r}) \qquad ...(22)$$

and rewrite Eq. (21) as

$$H_p\,|\vec{k}+\vec{G}_j\rangle = E_n(\vec{k})\,|\vec{k}+\vec{G}_j\rangle \qquad ...(23)$$

$$|\vec{k}+\vec{G}_j\rangle = e^{i(\vec{k}+\vec{G}_j)\cdot\vec{r}} \qquad ...(24)$$

Then the solutions are equivalently obtained by solving the determinant

$$\|\langle\vec{k}+\vec{G}_i|\,H_p\,|\vec{k}+\vec{G}_j\rangle - E(\vec{k})\,\delta_{ij}\| = 0 \qquad ...(25)$$

Once we know the Fourier coefficients $V_{ps}(\vec{G}_j)$, the solutions are easily calculated with a personal computer. The Fourier coefficients of the pseudopotentials reported by Cohen and Bergstresser are shown in Table 12.4.

**Table 12.4** Pseudopotentials for Several Semiconductors in Units of Rydberg

| | $V_3^S$ | $V_8^S$ | $V_{11}^S$ | $V_3^A$ | $V_4^A$ | $V_{11}^A$ |
|---|---|---|---|---|---|---|
| Si | −0.21 | +0.04 | +0.08 | 0 | 0 | 0 |
| Ge | −0.23 | +0.01 | +0.06 | 0 | 0 | 0 |
| Sn | −0.20 | 0.00 | +0.04 | 0 | 0 | 0 |
| GaP | −0.22 | +0.03 | +0.07 | +0.12 | +0.07 | +0.02 |
| GaAs | −0.23 | +0.01 | +0.06 | +0.07 | +0.05 | +0.01 |
| AlSb | −0.21 | +0.02 | +0.06 | +0.06 | +0.04 | +0.02 |
| InP | −0.23 | +0.01 | +0.06 | +0.07 | +0.05 | +0.01 |
| GaSb | −0.22 | 0.00 | +0.05 | +0.06 | +0.05 | +0.01 |
| InAs | −0.22 | 0.00 | +0.05 | +0.08 | +0.05 | +0.03 |
| InSb | −0.20 | 0.00 | +0.04 | +0.06 | +0.05 | +0.01 |
| ZnS | −0.22 | +0.03 | +0.07 | +0.24 | +0.14 | +0.04 |
| ZnSe | −0.23 | +0.03 | +0.06 | +0.18 | +0.12 | +0.03 |
| ZnTe | −0.22 | 0.00 | +0.05 | +0.13 | +0.10 | +0.01 |
| CdTe | −0.20 | 0.00 | +0.04 | +0.15 | +0.09 | +0.04 |

The diamond-type crystal structure such as Ge and Si contain two atoms $A$ and $B$ ($A = B$) in the unit cell, and the zinc-blende-type crystal structure has two different atoms $A$ and $B$ ($A \neq B$). When a new vector $\tau = \left(\frac{a}{8}\right)(111)$ is defined, the atomic positions of $A$ and $B$ are given by $\vec{R}^A = -\vec{\tau}$ and $\vec{R}^B = \vec{\tau}$, respectively. Therefore, we obtain

$$V_{ps}(\vec{G}_j) = V^s(\vec{G}_j)\cos(\vec{G}_j\cdot\vec{\tau}) + i\,V^A(\vec{G}_j)\sin(\vec{G}_j\cdot\vec{\tau}) \qquad ...(26)$$

$$V^s(\vec{G}_j) = [V_A(\vec{G}_j) + V_B(\vec{G}_j)]/2. \qquad ...(27)$$

$$V^A(\vec{G}_j) = [V_A(\vec{G}_j) - V_B(\vec{G}_j)]/2. \qquad ...(28)$$

From the definition in a diamond type crystal, we have $V^A(\vec{G}_j) = 0$. In Table 12.4, the pseudopotentials are defined by using the relations $V^s(\vec{G}_j) = V_j^A$ and $V^A(\vec{G}_j) = V_j^A$. The energy band calculations carried out by Cohen and Bergstresser reveal that the choice of appropriate values for the pseudopotentials $V_3^S$, $V_8^S$, $V_{11}^S$, $V_3^A$, $V_8^A$, $V_{11}^A$ and the neglect of higher-order values give band structures in good agreement with experimental results. The pseudopotential values determined by Cohen and Bergstresser for typical semiconductors are given in Table 12.4. Energy band structures calculated by Eq. (21) with about 130 plane waves are shown in Fig. 12.81. In the calculations, the free-electron states for $7E_o < E < 21E_o$, where $E_o = (\hbar^2/2m)(2\pi/a)^2$, are approximated by using Löwdin's perturbation method, and the matrix for the free-electron states for $E \leq 7E_o$ are diagonalized. This is equivalent to solving a matrix of $\approx 20 \times 20$ components. The aforementioned method is sometimes also called the local pseudopotential method.

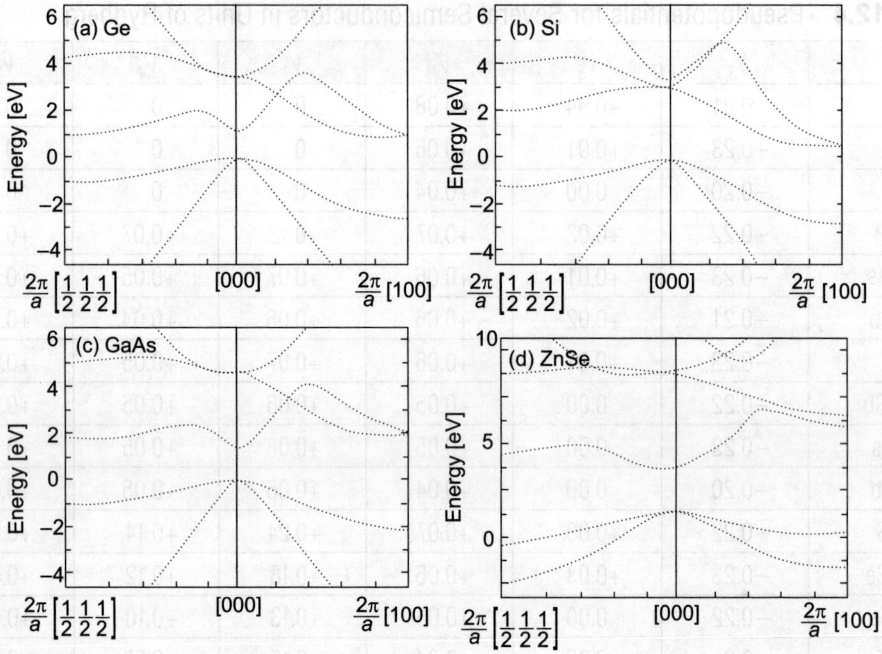


**Figure 12.81** Energy band structures calculated by the empirical pseudopotential method, (a) Ge, (b) Si, (c) GaAs, (d) ZnSe. The spinorbit interaction is not Included

## 12.45 THREE-DIMENSIONAL ENERGY BAND PICTURES

We know that because of $E(\vec{k})$ periodicity, all information pertaining to the electronic properties is contained in the first Brillouin zone. Equivalently, the energy $E_{\vec{k}}$ for $\vec{k}$, outside the first zone is identical to the energy $E_{\vec{k}}$ within the first zone if a suitable vector $\vec{G}$ can be found such that

$$\vec{k}' = \vec{k} + \vec{G}.$$

We have already seen this in the previous section in Fig. 12.81 plotted in the reduced zone scheme. We will now produce three-dimensional zone pictures for bcc and fcc structures. It would be expected that the energy bands would not be alike in different directions in $\vec{k}$-space. We would now consider this in detail using the bcc structure first.

$$E_{\vec{k}'} = \frac{\hbar^2 \vec{k}'^2}{2m}$$

$$\Rightarrow \qquad E_{\vec{k}} = \frac{\hbar^2}{2m}(\vec{k} + \vec{G})^2$$

Figure 12.82 shows three important directions in $\vec{k}$ space inside the BZ of a bcc lattice.

In three dimensions the equation analogous to reads

$$E_k = \frac{\hbar^2}{2m}(\mathbf{k} + \mathbf{G})^2 \qquad ...(1)$$

In Fig. 12.82 three important directions in **k**-space are inserted into the first Brillouin zone of a bcc lattice. They are the [100] direction from the origin (Γ) to point $H$, the [110] direction from Γ to $N$, and the [111] direction from Γ to $P$. These directions are commonly labelled by the symbols Δ, Σ, and Λ, respectively. Figure 12.83 depicts the bands, calculated by using Eq. (1) for these distinct directions in **k**-space. The sequence of the individual subgraphs is established by convention and can be followed using Fig. 12.82.

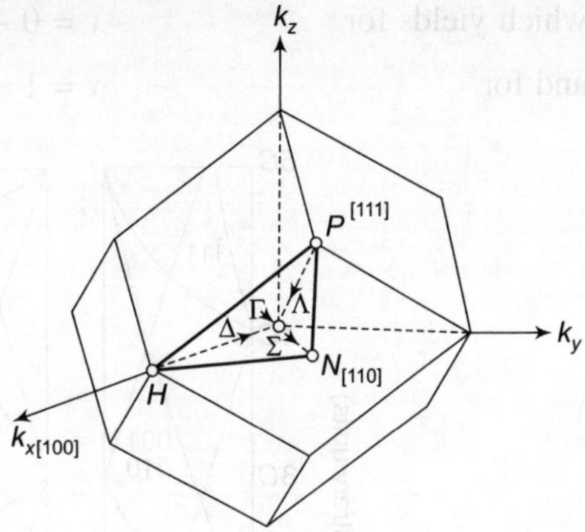


**Figure 12.82** First Brillouin zone of the bcc crystal structure

We now show how some of these bands are calculated for a simple case. We select the Γ – $H$ direction as an example. We vary the modulus of the vector $\mathbf{k}_{\Gamma H} = \mathbf{k}_x$ between 0 and $2\pi/a$, the latter being the boundary of the Brillouin zone (see Fig. 2.14). For this direction, Eq. (1) becomes

$$E = \frac{\hbar^2}{2m}\left(\frac{2\pi}{a}x\mathbf{i} + \mathbf{G}\right)^2, \qquad ...(2)$$

whereas $x$ may take values between 0 and 1. To start with, let **G** be **0**. Then Eq. (2) reads

$$E = \frac{\hbar^2}{2m}\left(\frac{2\pi}{a}\right)^2 (x\mathbf{i})^2 = Cx^2 \qquad ...(3)$$

This yields the well-known parabolic $E(\mathbf{k})$-dependence. The curve which represents (3) is labelled (000) in Fig. 2.22.

Now we let $h_1 = 0$, $h_2 = -1$, and $h_3 = 0$. Then we obtain by using

$$\vec{G} = 2\pi\,(h_1 a^* + h_2 b^* + h_3 c^*)$$

and

$$\begin{cases} a^* = \frac{1}{a}(0\,1\,1) \\ b^* = \frac{1}{a}(1\,0\,1) \\ c^* = \frac{1}{a}(1\,1\,0) \end{cases}$$

$$\Rightarrow \qquad \vec{\mathbf{G}} = -\frac{2\pi}{a}(\hat{\mathbf{i}} + \hat{\mathbf{k}}) \qquad ...(4)$$

Combining Eq. (2) with Eq. (4) provides

$$E = \frac{\hbar^2}{2m}\left[\frac{2\pi x}{a}\hat{\mathbf{i}} - \frac{2\pi}{a}(\hat{\mathbf{i}} + \hat{\mathbf{k}})\right]^2 = C[\hat{\mathbf{i}}\,(x-1) - \hat{\mathbf{k}}]^2 \qquad ...(5)$$

$$= C[(x-1)^2 + 1] = C(x^2 - 2x + 2),$$

which yields for $x = 0 \rightarrow E = 2C$

and for $x = 1 \rightarrow E = 1C.$

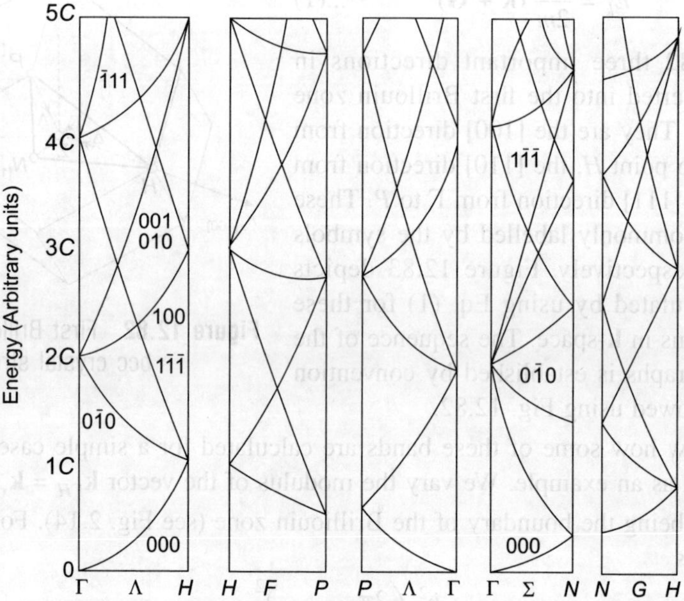


**Figure 12.83** Energy bands of the free electrons for the bcc structure. The numbers given on the branches are the respective $h_1$ Values (see the calculation in the text). Compare to Fig. 2.11 $C = \hbar^2 2\pi^2/ma^2$, see (3)

We obtain the band labelled $(0\bar{1}0)$ in Fig. 12.83. Similarly, all bands in Fig. 12.83 can be calculated by variation of the $h$ values, **k**-directions and by using Eq. (1).

The free electron bands are very useful for the following reason: by comparing them with the band structures of actual materials, an assessment is possible if and to what degree the electrons in that material can be considered to be free.

In Figs. 12.84 and 12.85 the first Brillouin zone and the free electron bands of an fcc structure are shown.

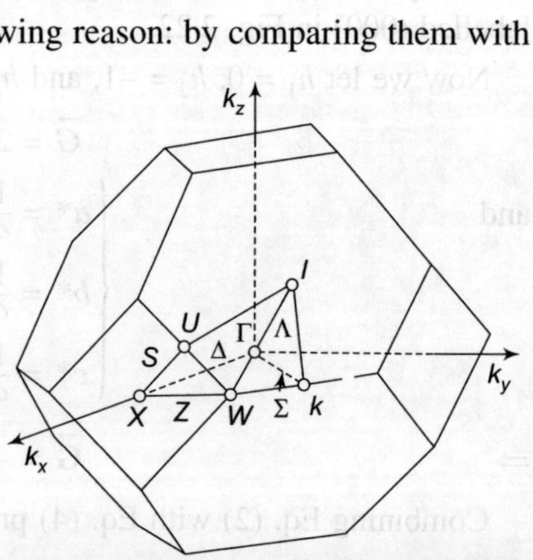


**Figure 12.84** First Brillouin zone of the fcc structure

## 12.46 FREE-ELECTRON BANDS (EMPTY-LATTICE BANDS)

In order to calculate energy band structures of a semiconductor the following procedures are required to carry out the calculations. These are

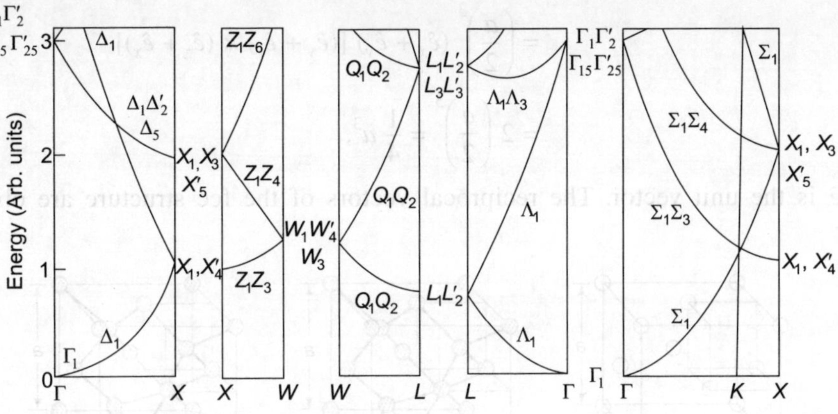


**Figure 12.85** Free electron bands of the fcc structure

1. Calculate the first Billouin zone.
2. Calculate the energy band structures in the limit of zero potential energy. This procedure is to obtain the free-electron bands or empty-lattice bands and plot the energy as a function of wave vector **k** in the reduced zone scheme.
3. Then calculate the energy bands using an appropriate method.

In the energy band calculation the most important procedure is to obtain the empty-lattice bands, which are calculated by assuming zero lattice potential $V(\mathbf{r}) = 0$ and keeping the lattice periodicity. In other words we assume the wave functions are given by the free-electron model with the wave vectors of the electrons in the periodic potential. Such an energy band structure is called empty-lattice bands or free-electron bands and thus the band structure exhibits the characteristics of the lattice periodicity.

Here we will show an example of empty-lattice bands in the case of the face-centred cubic (fcc) lattice. First, we calculate the Brillouin zone of the fcc lattice. Figure 12.86a shows the fcc structure. The diamond structure is obtained by displacing the lattice atom by the amount ($a$/4, $a$/4, $a$/4), which is shown in Fig. 12.86b. Therefore, the diamond structure belongs to the fcc structure. Diamond (C), Si and Ge have this diamond structure. On the other hand, the displaced lattice atoms are different from the original atoms, and the structure is called the zinc-blende structure, which is shown in Fig. 12.86c. Crystals such as GaAs, GaP, AlAs, InAs, InSb belong to the zinc-blende structure. The fundamental vectors and volume $v$ of a fcc lattice are defined by

$$\vec{a} = \frac{a}{2}(\hat{e}_x + \hat{e}_y),$$

$$\vec{b} = \frac{a}{2}(\hat{e}_y + \hat{e}_z),$$

$$\vec{c} = \frac{a}{2}(\hat{e}_z + \hat{e}_x),$$

$$v = \vec{a} \cdot (\vec{b} \times \vec{c}),$$

$$= \left(\frac{a}{2}\right)^3 (\hat{e}_x + \hat{e}_y) [(\hat{e}_y + \hat{e}_z) \times (\hat{e}_z + \hat{e}_x)]$$

$$= 2\left(\frac{a}{2}\right)^3 = \frac{1}{4} a^3, \qquad \text{...(1)}$$

where $\hat{e}$ is the unit vector. The reciprocal vectors of the fcc structure are obtained as follows:

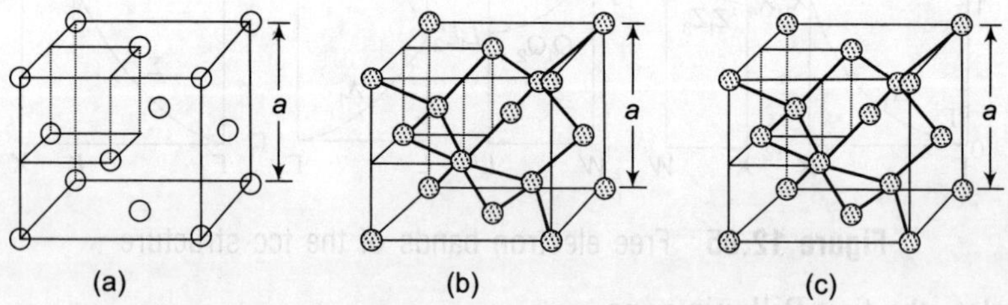


**Figure 12.86** (a) Face-centred cubic lattice, (b) The diamond crystal lattice is obtained by displacing the lattice atoms of (a) by (*a*/4, *a*/4, *a*/4), when the displaced lattice atoms are different from the original lattice atoms, the crystal structure is called the zinc-blende crystal structure

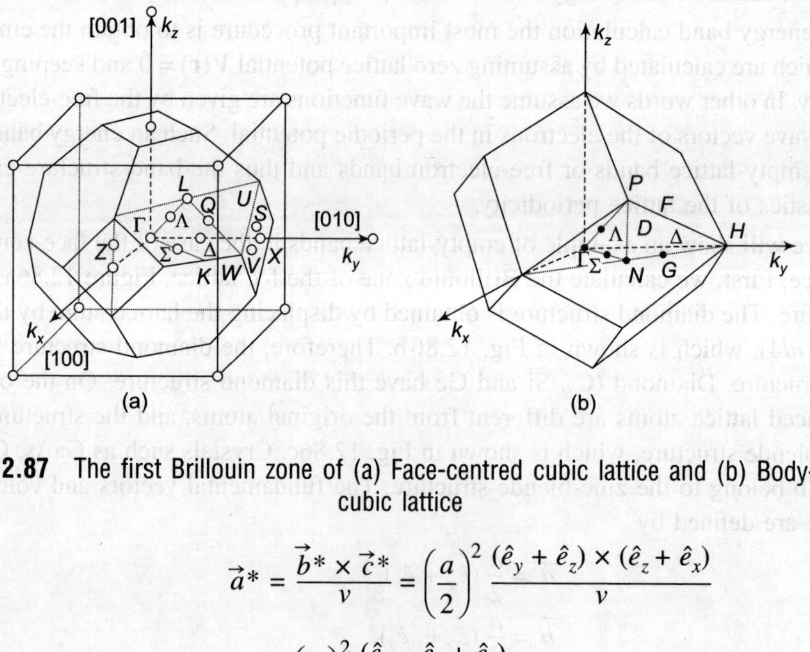


**Figure 12.87** The first Brillouin zone of (a) Face-centred cubic lattice and (b) Body-centred cubic lattice

$$\vec{a}^* = \frac{\vec{b}^* \times \vec{c}^*}{v} = \left(\frac{a}{2}\right)^2 \frac{(\hat{e}_y + \hat{e}_z) \times (\hat{e}_z + \hat{e}_x)}{v}$$

$$= \left(\frac{a}{2}\right)^2 \frac{(\hat{e}_x - \hat{e}_z + \hat{e}_y)}{a^3/4}$$

$$= \frac{1}{a} (\hat{e}_x + \hat{e}_y - \hat{e}_z), \qquad \text{...(2)}$$

$$\vec{b}^* = \frac{1}{a}(-\hat{e}_x + \hat{e}_y + \hat{e}_z), \qquad \text{...(3)}$$

$$\vec{c}^* = \frac{1}{a}(\hat{e}_x - \hat{e}_y + \hat{e}_z). \qquad \text{...(4)}$$

From these results we find that the reciprocal lattices of the fcc lattice form body-centred cubic lattices. Therefore, the reciprocal lattice vectors $G$ of the fcc lattice are given by

$$\vec{G} = 2\pi(n_1\vec{a}^* + n_2\vec{b}^* + n_3\vec{c}^*). \qquad \text{...(5)}$$

The Brillouin zone of the fcc lattice is defined by

$$k^2 = (\vec{k} - \vec{G}_l)^2 \qquad \text{...(6)}$$

or

$$2\vec{k} \cdot \vec{G}_l = G_l^2. \qquad \text{...(7)}$$

Using the above equation the first Brillouin zone is easily calculated, as shown in Fig. 12.87a. For comparison the first Brillouin zone of body-centred cubic lattices, which form face-centred cubic lattices, is shown in Fig. 12.87b.

Since the lattice potential is 0 in the empty-lattice model, the energy of an electron is given by the free-electron model

$$\varepsilon(\vec{k}) = \frac{\hbar^2}{2m}k^2. \qquad \text{...(8)}$$

We plot the energy $\varepsilon(\vec{k})$ versus wave vector $\vec{k}$ curves in the reduced zone scheme, using the relation

$$\vec{k}' = \vec{k} - \vec{G}, \qquad \text{...(9)}$$

and choose $\vec{k}'$ in the first Brillouin zone. Then the empty lattice bands are given by

$$\varepsilon(\vec{k}') = \frac{\hbar^2}{2m}(\vec{k}' + \vec{G})^2. \qquad \text{...(10)}$$

For better understanding we list several reciprocal lattice vectors from the lowest orders, which are given by

$$\vec{G}_0 = \frac{2\pi}{a}(0, 0, 0), \qquad \text{...(11a)}$$

$$\vec{G}_3 = \frac{2\pi}{a}(\pm 1, \pm 1, \pm 1), \qquad \text{...(11b)}$$

$$\vec{G}_4 = \frac{2\pi}{a}(\pm 2, 0, 0), \qquad \text{...(11c)}$$

$$\vec{G}_8 = \frac{2\pi}{a}(\pm 2, \pm 2, 0), \qquad \text{...(11d)}$$

$$\vec{G}_{11} = \frac{2\pi}{a}(\pm 3, \pm 1, \pm 1). \qquad \text{...(11e)}$$

Putting these values in Eq. (10), the empty-lattice bands (free-electron bands) are easily calculated. In the following we use $\vec{k}$ instead of $\vec{k}'$ and take account of $\vec{k}$ in the first Brillouin zone. As an example we calculate the energy bands along the direction (100) in the $\vec{k}$-space shown in Fig. 12.89. In other words, we calculate the energy band structures ε versus $\vec{k}$ from the Γ point to the $X$ point. Since $k_y = k_z = 0$ in this direction, putting the reciprocal lattice vectors into Eq. (8), the energy is given by the following equations:

$$\vec{G}_0 : \varepsilon = k_x^2, \quad \text{...(12a)}$$

$$\vec{G}_3 : \varepsilon = (k_x \pm 1)^2 + (\pm 1)^2 + (\pm 1)^2,$$

$$= \begin{cases} (k_x - 1)^2 + 2 & \text{(4-fold degeneracy)} \\ (k_x + 1)^2 + 2 & \text{(4-fold degeneracy)} \end{cases}, \quad \text{...(12b)}$$

$$\vec{G}_4 : \varepsilon = \begin{cases} k_x^2 + 4 & \text{(4-fold degeneracy)} \\ (k_x - 2)^2 & \text{(single state)} \\ (k_x + 2)^2 & \text{(single state)} \end{cases} \quad \text{...(12c)}$$

where the energy is measured in the units $\hbar^2 (2\pi/a)^2/2m$ and the wave vector $\vec{k}$ in the units $2\pi/a$. When we plot these relations, we obtain the curves shown in Fig. 12.88.

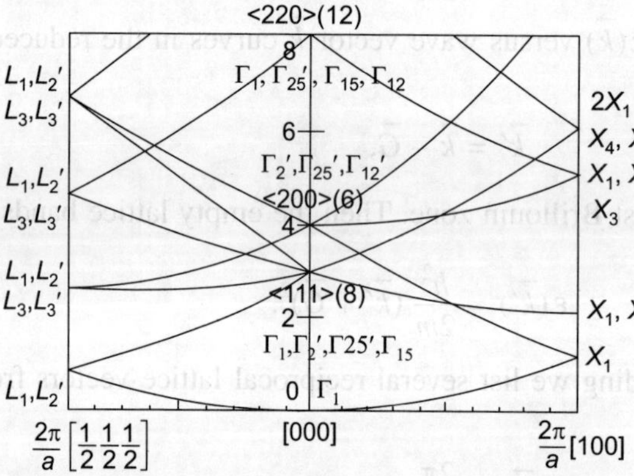


**Figure 12.88** Empty-lattice bands (free-electron bands) of a face-centred cubic lattice. <111>, <200>, and <220> represent the reciprocal lattice vectors $\vec{G}_3$, $\vec{G}_4$, and $\vec{G}_8$, respectively, and the numbers in ( ) show the degeneracy of the wave functions

Next we calculate the ε versus $\vec{k}$ curves in the (111) direction of $\vec{k}$-space, or along the direction from the Γ point to the $L$ point. The results are

$$\vec{G}_0 : \varepsilon = k_x^2 + k_y^2 + k_z^2 \equiv k_{111}^2,$$

$$\vec{G}_3 : \varepsilon = (k_x \pm 1)^2 + (k_y \pm 1)^2 + (k_z \pm 1)^2,$$

$$\vec{G}_4 : \varepsilon = \begin{cases} (k_x \pm 2)^2 + k_y^2 + k_z^2 & \text{(2-fold degeneracy)} \\ k_x^2 + (k_y \pm 2)^2 + k_z^2 & \text{(2-fold degeneracy)}, \\ k_x^2 + k_y^2 + (k_z \pm 2)^2 & \text{(2-fold degeneracy)} \end{cases} \quad ...(13)$$

where $k_x^2 + k_y^2 + k_z^2 = k_{111}^2$, and $k_x + k_y + k_z = k_{111}/\sqrt{3}$. Using these results we obtain the energy bands, $\varepsilon - \vec{k}$ curves in the direction ⟨111⟩, which are shown in the left half of Fig. 12.88. In Fig. 12.88, the notation of the point group for $O_h$ is used to represent the symmetry properties of the Brillouin zone edge.

## 12.47 ENERGY BAND FORMATION IN IDEAL CRYSTALS

In a three-dimensional crystal, the periodicity differs in different directions, depending on the crystal symmetry. The solution is obtained in the reciprocal lattice i.e., *k*-space.

Consider an fcc unit cell (Si, Ge and GaAs all have this lattice structure). In this case, the reciprocal lattice has bcc structure, which forms the Brillouin zone. The specific directions of importance here correspond to the [100], [110] and [111] directions in the real crystal space. In reciprocal space, [000] is denoted by Γ, [100] by *X* and [111] by *L*.

Consider first the valence band. For the fcc lattice, the valence band maximum occurs at Γ; actually, two maxima coincide at Γ at the valence band edge. The effective masses of holes in the two cases will be different. The effective mass is higher for smaller curvature of the *E*-**k** at the maxima or minima. Holes occupying states in the *E*-**k** diagram of lower curvature will be called *heavy holes* and those occupying states of higher curvature will be called *light holes.* As we shall discuss in greater detail later, the mobility, which is a measure of the response to an applied electric field, varies inversely with effective mass. So, light holes will respond more strongly to applied forces than heavy holes. Note that both types occupy similar energy states in the same energy band.

The conduction band minimum and the valence band maxima $E_C$ and $E_V$ may not occur at the same **k** values. If the **k** values are identical, it is a direct *band-gap* semiconductor. Otherwise, it is an *indirect band-gap* semiconductor. In the conduction band, for the fcc lattice, minimas occur at Γ and in the *L* and *X* directions, corresponding to the (000), (111) and (100) directions; we denote the energy difference of the minimas from the valence band edge in the three cases by $E_\Gamma$, $E_L$ and $E_X$ respectively. Obviously, the lowest of the three energies will correspond to the band gap. The values of $E_\Gamma$, $E_L$, and $E_X$ differ for the three semiconductors Ge, Si and GaAs. For GaAs, $E_\Gamma$ is the lowest, and hence, it is a direct band-gap semiconductor. For Ge, $E_L$ is the lowest. For Si, $E_X$ is the lowest. Both are therefore indirect band-gap semiconductors. A sketch of an *E*-**k** diagram of a material like silicon along the [100] and [111] directions is shown in Fig. 12.89.

In the conduction band, the minimas are called *valleys*. The minimum of the minimas in the band forms the conduction band edge $E_C$ and lies in the *Central Valley*. The other valleys are called *satellite valleys*. If the difference between the minimas in the central valley and one of the satellite valleys is very small, the *Gunn effect* may be observed. The Gunn effect[9] causes certain bulk semiconductors to behave as a negative differential conductivity device.

The band gap in a semiconductor is not actually a constant but varies a little with temperature. An empirical relationship for this change is given by

$$E_g(T) = E_{g0} - (\alpha_T T^2)/(T + \beta_T) \quad ...(1)$$

In Eq. (1), $E_{g0}$ is the band gap at 0 K and $\alpha_T$ and $\beta_T$ are empirical constants.

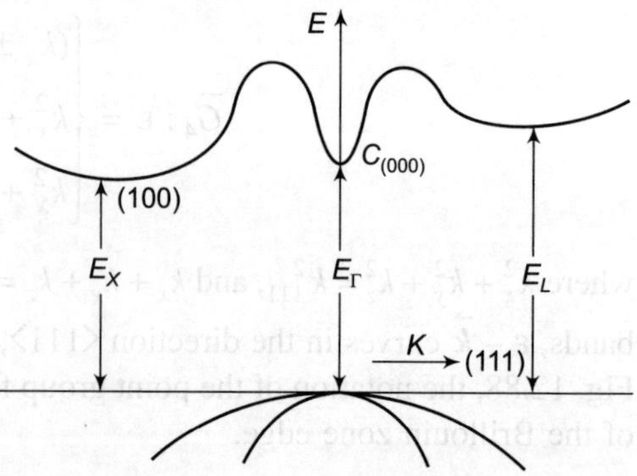


**Figure 12.89** Sketch of *E*-**k** diagram of indirect band gap material e.g. silicon

So far, we have looked at the band formations in an ideal crystal. However, in a real crystal, impurities and defects can be present, whether introduced deliberately or present inadvertently. This distorts the regular crystalline order of the periodic potential. If the number density of such imperfections is small, the problem can be looked upon as a superposition of a perturbation of the regular periodic potential present in the array. In such a case, in addition to the solutions for Bloch functions of the form exp ($j\mathbf{k}x$), we obtain additional solutions of the form exp ($-\mathbf{k}x$). These are evanescent solutions; allowed solutions for a value of $E$ exists only for specific values of **k**, i.e., for specific locations in the crystal. Such allowed electron states are called localized states or localized energy levels. We will generally refer to such localized energy levels occurring within the band gap as traps. Electrons occupying such energy levels are not free; on application of an applied force, they cannot acquire momentum, i.e., an adjacent value of **k**.

## 12.48 BAND STRUCTURES AND TYPICAL EXAMPLES

### 12.48.1 Basic Assumptions

Band theory is an approximate theory of the perfect lattice in a form which emphasizes the translational symmetry of the system. It uses the approximations, namely the static lattice, one-electron, and Koopmans' approximations. Corrections to these are needed before comparison with real crystals, especially because of the neglect of correlation and the fact that Koopmans' result is only valid when the effective one-electron potential is not sensitive to the actual occupancies of the one-electron orbitals.

### 12.48.2 Band Structures: General Features

The one-electron eigenstates of band theory are classified by a band index $n$ and a wave vector **k**. We now consider the eigenvalues $E_n(\mathbf{k})$ for typical insulators and semiconductors. The eigenstates which are occupied at absolute zero temperature constitute the valence bands; the unoccupied states belong to the various conduction bands. Collective modes, like the plasmon modes, are usually ignored in band calculations.

[9] We have considered Gunn effect in Chapter 11. (Section 11.47).

Here we consider the qualitative features of $E_n(\mathbf{k})$. Some of these concern the extrema of the bands—the lowest states of the conduction band and the highest valence band states. Thus, we are interested in the values of **k** for which $E_n(\mathbf{k})$ is extremal, whether the energy surface is degenerate near the extremum, and the energy surface near the extremum. Other properties, such as the density of states, are also clearly important.

The band extrema occur at points $\mathbf{k}_{n,\alpha}$ where $\nabla_k E_n(\mathbf{k})$ vanishes. As a rule, the lowest conduction band has just one minimum at the centre of the zone, $\mathbf{k} = 0$. This is so for most alkali halides and II-VI and III-V compounds. In order important cases the conduction band has several equivalent extrema away from $\mathbf{k} = 0$. Examples are given in Table 12.5. The maxima of the valence bands are almost invariably at the centre of the zone, $\mathbf{k} = 0$. For direct-band-gap materials the extrema of the lowest conduction and highest valence bands lie at the same point of the zone; others, such as diamond, silicon and germanium, have an indirect band gap (Fig. 12.90).

**Table 12.5** Positions of Conduction Band Minima in the Brillouin Zone

| *Crystal* | *Number of minima* | *Position in zone* |
|---|---|---|
| Diamond Silicon | 6 | (*K*00) and equivalent |
| Germanium | 4 | (*LLL*) and equivalent |
| GaP<br>AlSb | 3 | (*K*00) and equivalent |
| Most others | 1 | (000) |

For the materials of interest to us, the conduction bands are non-degenerate near their extrema. There is just a single energy surface $E_c(\mathbf{k})$, although it may exhibit a different type of degeneracy through having several equivalent minima. The lowest conduction band is often overlapped by higher conduction band, but this complication will rarely be important. In most cases, however, the valence bands are degenerate. This occurs when the valence band is formed primarily from a full shell of atomic *p*-electrons, for example, the valence band degeneracy can be lifted by spin-orbit coupling (Elliott 1954) or by low-symmetry crystal fields. Such splittings are shown for the Wurtzite and zinc blende structure in Fig. 12.91. The pin-orbit splitting $\Delta_{so}$ is largest when the corresponding atomic spin-orbit coupling is large, a for anions of high atomic number. Thus, $\Delta_{so}$ is larger for germanium than for silicon or diamond, and larger for iodides than for the other alkali halides.

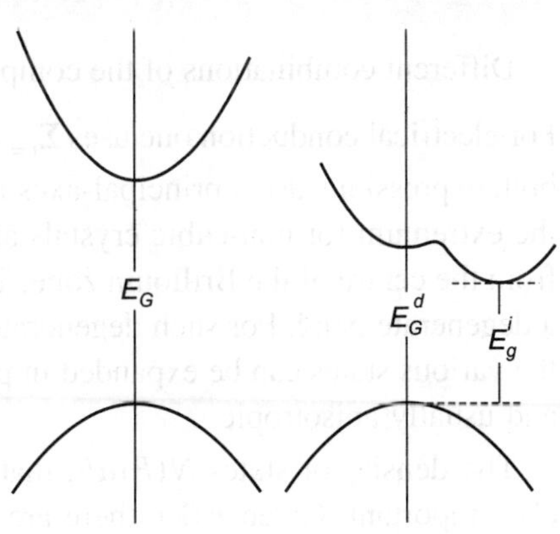


**Figure 12.90** Band structures with direct (a) and Indirect (b) gaps

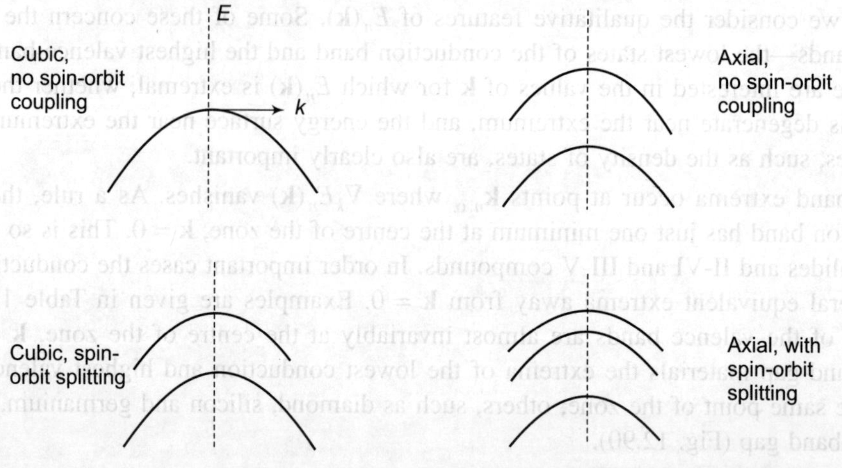


**Figure 12.91** Effects of spin-orbit coupling and low-symmetry fields on band structure

The energy surface near the extrema of a non-degenerate band can be obtained by expansion

$$E_n(\mathbf{k}) \simeq E_n(\mathbf{k}_n) + \frac{1}{2}\sum_{i,j}(k-k_n)\left.\frac{d^2E_n(\mathbf{k})}{dk_i\,dk_j}\right|_{\mathbf{k}=\mathbf{k}_n}(k-k_n)_j + \ldots \qquad \ldots(1)$$

If the expansion is truncated at this point we have the effective-mass approximation, with an effective-mass tensor

$$\left(\frac{1}{m}\right)_{ij} = \frac{1}{\hbar^2}\left.\frac{d^2E_n(\mathbf{k})}{dk_i\,dk_j}\right|_{\mathbf{k}=\mathbf{k}_n}. \qquad \ldots(2)$$

Different combinations of the components of the mass tensor enter the various predictions. For electrical conduction one uses $\Sigma_{i=1,3}\, m_i^{-1}$, and for the density of states $(\Pi_{i=1,3}\, m_i)^{\frac{1}{3}}$, where both expressions use a principal-axes representation. The energy surface are anisotropic near the extremum for non-cubic crystals and also for cubic crystals where the extrema lies away from the centre of the Brillouin zone. The expansion (1) cannot be used near the extremum of a degenerate band. For such degenerate cases the matrix elements of the Hamiltonian among the various states can be expanded in powers of $(\mathbf{k} - \mathbf{k}_n)$; the energy surfaces are complicated and usually anisotropic.

The density of states $N(E)\,dE$, that is, the number of states between $E$ and $(E + dE)$, is also important. Given $E(\mathbf{k})$, there are general numerical methods for evaluating the density of states. The most important feature is that $N(E)$ shows singular behaviour. We write the density of states for electrons of one spin as

$$N_n(E) = \frac{1}{\Omega_z}\int_z d^3\mathbf{k}\,\delta\{E - E_n(\mathbf{k})\}$$

$$= \frac{1}{\Omega_z} \int_E dS / |\nabla_{\mathbf{k}} E_n(\mathbf{k})|, \qquad ...(3)$$

where $\Omega_z$ is the volume of the Brillouin zone and the second integral is over an energy surface of constant energy, $E$. Clearly, there is a singularity at the critical points (van Hove 1953; Phillips 1956) where $\nabla_{\mathbf{k}} E_n(\mathbf{k})$ is zero.

Critical points are involved in defect problems in several ways. Thus, the corresponding band wave functions are important in constructing wave functions of weakly bound defects. Secondly, transitions such as photoionization have a probability which depends on the density of states and which may show structure corresponding to the critical points. Thirdly, the band-structure data used in defect calculations is often best obtained by modulation methods which exploit the singularities. Exactly analogous critical points occur in phonon densities-of-states.

### 12.48.3 Band Structures: Wave Functions

A. *Band functions and Bloch functions.* The one-electron wave functions of band theory are found from the equation

$$\mathcal{H}_0 \psi_n(\mathbf{k}, \mathbf{r}) = E_n(\mathbf{k}) \psi_n(\mathbf{k}, \mathbf{r}) \qquad ...(4)$$

The band functions $\psi_n(\mathbf{k}, \mathbf{r})$ can be written to make their translational symmetry explicit

$$\psi_n(\mathbf{k}, \mathbf{r}) = e^{i\mathbf{k}\cdot\mathbf{r}}\, u_n(\mathbf{k}, \mathbf{r}) \qquad ...(5)$$

We adopt as normalizations of the Bloch functions and band functions

$$\int_\infty d^3\mathbf{r}\, \psi_n^*(\mathbf{k}, \mathbf{r})\, \psi_{n'}(\mathbf{k}', \mathbf{r}) = \delta_{nn'}\, \delta(\mathbf{k} - \mathbf{k}'), \qquad ...(6)$$

$$\int_{\text{cell}} d^3\mathbf{r}\, u_n^*(\mathbf{k}, \mathbf{r})\, u_{n'}(\mathbf{k}', \mathbf{r}) = \delta_{nn'}/\Omega_z$$

$$= \int_{\text{cell}} d^3\mathbf{r}\, \psi_n^*(\mathbf{k}, \mathbf{r})\, \psi_{n'}(\mathbf{k}', \mathbf{r}) \qquad ...(7)$$

The Bloch functions are periodic in the lattice, that is, $u_n(\mathbf{k}, \mathbf{r}) = u_n(\mathbf{k}, \mathbf{r} + \tau)$ for any translation $\tau$ which leaves the lattice invariant. These functions are not orthogonal for different values of $\mathbf{k}$; indeed

$$\int_{\text{cell}} d^3\mathbf{r}\, u_n^*(\mathbf{k}, \mathbf{r})\, u_{n'}(\mathbf{k}', \mathbf{r})$$

$$= \delta_{nn'}/\Omega_z + \text{terms linear and higher in } (\mathbf{k} - \mathbf{k}') \qquad ...(8)$$

But the $u_n(\mathbf{k}, \mathbf{r})$ for a given $\mathbf{k}$ and for all bands $n$ form a complete orthonormal set of states satisfying the equation

$$\left( \mathcal{H}_0 + \frac{\hbar}{m_0} \mathbf{k} \cdot \pi + \frac{\hbar^2 k^2}{2m_0} \right) u_n(\mathbf{k}, \mathbf{r}) = E_n(\mathbf{k})\, u_n(\mathbf{k}, \mathbf{r}) \qquad ...(9)$$

In this equation $\pi$ is the momentum, $\mathbf{p}$. Plus a correction term $(\sigma \Lambda \nabla V)/4 m_0 c^2$ which appears when there is spin-orbit coupling.

B. *Wannier functions.* The theory of localized defects can be developed in terms of the band functions or Bloch functions of the host lattice. But it is convenient to adopt different basis functions, related to the Bloch functions, which are localized in space. One set of functions commonly used are the Wannier functions (Wannier 1937).

$$a_n(\mathbf{r}-\mathbf{R}_0) = \frac{1}{\sqrt{N}} \sum_{\mathbf{k}} \exp(-i\mathbf{k}\cdot\mathbf{R}_0)\, \psi_n(\mathbf{k},\mathbf{r})$$

$$= \frac{1}{\sqrt{\Omega_z}} \int_z d^3\mathbf{k} \exp(-i\mathbf{k}\cdot\mathbf{R}_0)\, \psi_n(\mathbf{k},\mathbf{r}). \qquad ...(10)$$

For the Wannier functions $\langle a_n(\mathbf{r}-\mathbf{R}_0)|\mathcal{H}|a_n(\mathbf{r}-\mathbf{R}_0)\rangle$ is just the mean band energy $\Sigma_{\mathbf{k}}$ $E_n(R)/\Sigma_{\mathbf{k}}\, l$. They are defined so as to be orthogonal for different bands and for different lattice sites $\mathbf{R}_0$. Other choices are discussed by Ferreira & Parada (1970).

Formal properties of Wannier functions are discussed by many authors. The localization properties and those of related functions are discussed by Blount (1962, p. 327) and Lix (1971). Extremal properties of the expectation value of the Hamiltonian are analysed by Koster (1953), Parzen (1953), and Blount (1962, p. 324). One property important in defect calculations concerns the choice of phase. Whilst any phase can be used if chosen consistently, certain choices have advantages. One might choose the phases to localize the Wannier functions as much as possible, by minimizing the expectation value of $r^2$, for example, (Blount 1962, p. 327). Another possibility is to ensure continuity of symmetry even when bands cross at special points in the Brillouin zone (e.g., Callaway & Hughes 1967; Ferreira & Parada 1970). Or the Wannier functions may be defined by imposing the condition $E_{n+1}(\mathbf{k}) > E_n(\mathbf{k})$ throughout the zone. Special problems are discussed by Turner & Goodings (1965) and Goodings & Harris (1969).

The Wannier functions reduce to atomic functions in the limits of extreme tight binding. If there is one atom in the unit cell, and if the overlaps of the atomic functions $\phi(\mathbf{r})$ centred on sites separated by $R_\mu$ are $S_\mu$, then (Landshoff 1936; Wannier 1937; Löwdin 1950)

$$a(\mathbf{r}-\mathbf{R}_0) = \phi(\mathbf{r}-\mathbf{R}_0) - \frac{1}{2} \sum_{\mu\neq 0} S_\mu \phi(\mathbf{r}-\mathbf{R}_0-\mathbf{R}_\mu) + ... \qquad ...(11)$$

The small admixtures of wave functions at other sites ensure the orthogonality properties of the Wannier functions, and enhance the values of $a(\mathbf{r})$ at these sites over the unorthogonalized values. When there is more than one atom per cell, special precautions must be taken if Wannier functions are to be associated with individual crystal atoms (Hughes 1968).

## 12.49 BAND STRUCTURES: EXAMPLES

1. *Relation to constituent atoms.* A prototype band structure can be constructed to show features typical of insulators and semiconductors. The constituent atoms are first arranged on a lattice of the correct symmetry and large spacing. Secondly, the electronic configurations of the atoms are changed to those of the final structure. Thus, the electron transfers needed to create ionic crystals, or the hybridization for covalent crystals, are made at this stage. For our

model system we assume that the highest-energy orbitals which are occupied are *p*-orbitals and that the lowest unoccupied orbitals (not necessarily on the same species of ion) are *s*-orbitals. Thirdly, the lattice parameter is reduced to the correct crystal volume. The discrete levels of the ions are broadened out by the interactions between the ions to give the energy bands. The widths of the bands are a measure of the interaction between ions. Clearly, the ratio of the widths to the splitting of the ionic levels depend on the nature of the band. The ratio is small in the tight-binding limit, where intra-ionic terms are dominant, but is large in the nearly-free-electron approximation.

2. *Typical band structure.* The highest valence band in the model system is almost entirely derived from the *p*-orbitals, and the lowest conduction band *s*-orbitals. The band structure near the zone centre is shown in Fig. 12.92. The effective mass can be calculated using Eq. (9) and assuming a matrix element $P \equiv \langle s|p_z|z\rangle/m$. Perturbation theory gives the band energies near the extrema as

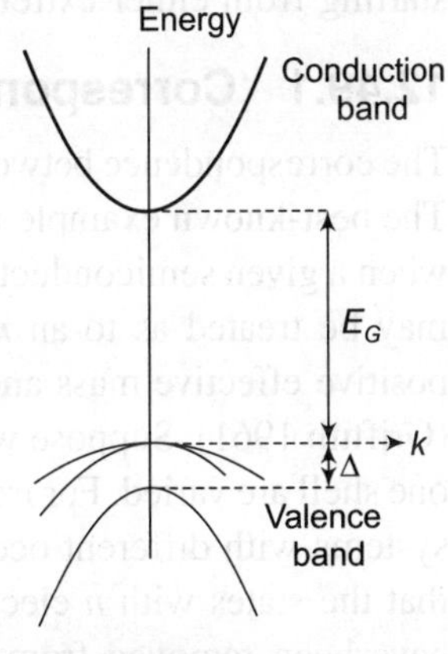


**Figure 12.92** 'Typical' band structure

Conduction band: (electron energies)

$$E = E_{c0} + \frac{1}{2}\frac{\hbar^2 k^2}{m}\left\{1 - \frac{2}{3} mP^2 \left(\frac{2}{E_G} + \frac{1}{E_G + \Delta}\right)\right\},$$

Valence band: (hole energies)

$$E_1 = E_{v0} + \frac{1}{2}\frac{\hbar^2 k^2}{m},$$

$$E_2 = E_{v0} + \frac{1}{2}\frac{\hbar^2 k^2}{m}\left\{1 + \frac{2}{3} mP^2 \left(\frac{2}{E_G}\right)\right\},$$

$$E_3 = E_{v0} + \frac{1}{2}\frac{\hbar^2 k^2}{m}\left\{1 + \frac{2}{3} mP^2 \left(\frac{1}{E_G + \Delta}\right)\right\} - \Delta \qquad \text{...(1)}$$

For small **k** the valence band is split by the spin-orbit coupling, which separates $E_3$ from $E_1$ and $E_2$ by $\Delta$, and by the terms in *P*. $E_1$, and $E_2$ form a 'heavy-hole band' $E_1$ and a 'light-hole band' $E_2$.

We now compare this model with valence crystals.

*Valence crystals: diamond, silicon, germanium.* Both the conduction and valence bands are complicated by the closeness of *s*- and *p*-orbitals, which leads to hybridization. For diamond $sp^3$ hybrid orbitals are formed, and a good description of the valence band can be obtained as linear combinations of bonding orbitals formed from the hybrids. Clearly the promotion energy (which is $E(2s\ 2p^3) - E(2s^2\ 2p^2)$ for carbon) is an important parameter. For silicon and germanium admixtures of *d*-orbitals are important.

For all these crystals the conduction band has minima away from the centre of the Brillouin zone. Further, the widths of the conduction and valence bands are large compared with the band gaps. This is typical of systems where the nearly-free-electron model, rather than the tight-binding model, is appropriate. One can get reasonable descriptions of the band structures starting from either extreme.

### 12.49.1 Correspondence Between Electrons and Holes

The correspondence between electrons and holes is used in several forms in solid state physics. The best-known example is of transport in semiconductors. Suppose there are no free carriers when a given semiconductor contains $N$ electrons. Then the same sample with $(N-n)$ electrons may be treated as to an $n$-particle problem in which the $n$ carriers are supposed to have a positive effective mass and a positive charge. Another example occurs in crystal-field theory (Griffith 1961). Suppose we are interested in the states of systems in which the occupancies of one shell are varied. For iron-group transition-metal ions in cubic symmetry we might consider systems with different occupancies of the $t_2$ levels derived from the $3d$ shell. Griffith shows that the states with $n$ electrons in this shell can be related to the states in which $n$ electrons have been removed from a full shell. Thus, in our example, the states of $t_2^n$ are related to those of $t_2^{6-n}$. In this case the phases defining the various states must be chosen with care for the correspondence to be useful.

In this section we have a very limited aim. We write down the matrix elements of a Hamiltonian between various Slater determinants, and rewrite them to correspond to contributions from a full band of $N$ electrons, together with corrections from the $n$ holes or electrons additional to the band. Thus, the situation is closer to the example of transport theory rather than the crystal-field case.

The total Hamiltonian is the sum of $\mathcal{H}_1$, which consists of one-electron operators, and $\mathcal{H}_2$, the electron–electron interaction,

$$\mathcal{H} = \mathcal{H}_1 + \mathcal{H}_2 = \sum_i \mathcal{H}_1 + \frac{1}{2}\sum_{i,j} V_{ij}. \qquad ...(2)$$

We consider the matrix elements of $\mathcal{H}$ among determinantal wave functions $|I\rangle$ and $|J\rangle$ describing a system with $(N-n)$ electrons.

First we consider the expectation value

$$\langle I|\mathcal{H}|I\rangle = \sum_{i=1}^{N-n} \langle i|\mathcal{H}_1|i\rangle + \frac{1}{2}\sum_{i=1}^{N-n}\sum_{j=1}^{N-n} (\langle ijij\rangle - \langle ij\, ji\rangle)$$

$$= E_N - \sum_{\substack{i=1 \\ \text{(empty)}}}^{N-n} \langle i|\mathcal{H}_1|i\rangle - \sum_{\substack{i=1 \\ \text{(all)}}}^{N-n}\sum_{\substack{j=1 \\ \text{(empty)}}}^{N-n} (\langle ijij\rangle - \langle ij\, ji\rangle)$$

$$+\frac{1}{2}\sum_{\substack{i=1 \\ \text{(empty)}}}^{n}\sum_{j=1}^{n} (\langle ijij\rangle - \langle ij\, ji\rangle). \qquad ...(3)$$

where the electron-electron interaction integrals are defined by

$$\langle a\, b\, c\, d\rangle = \int d^3\mathbf{r}_1 \int d^3\mathbf{r}_2 a^*(\mathbf{r}_1)\, b^*(r_2) \frac{e^2}{|r_1 - r_2|} c(r_1) d(r_2). \qquad ...(3a)$$

This equation contains four terms. The first $E_N$ is the energy of the $N$-electron system with the same Hamiltonian

$$E_N = \sum_{i=1}^{N} \langle i|\mathcal{H}|i\rangle + \frac{1}{2}\sum_{i=1}^{N}\sum_{j=1}^{N} (\langle ijij\rangle - \langle ijji\rangle). \qquad ...(4)$$

The second term gives the 'one-hole' elements. They are exactly like one-electron terms but with reversed signs. For these terms a system of $n$ holes can be treated like $n$ electrons if the signs of the various parts of the Hamiltonian are reversed. Some of the changes are obvious: holes are attracted by the Coulomb potentials of negative ions, whereas electrons are attracted by those of positive ions. Others are less obvious: the spin-orbit coupling and electric-field gradients should be reversed in sign. The third term, the interaction of holes with the full shell of electrons, also has an extra negative sign over the form for the interaction of $n$ extra electrons with the full shell. Finally, the fourth term in (3) is the interaction of the holes with each other; this is positive, as expected.

These results are adequate for Hartree–Fock theory, where only one determinantal wave function is involved. When we consider configuration interaction we need matrix elements between two different functions $|I\rangle$ and $|J\rangle$. Suppose $\tau$ is the order of the permutation needed to get the identical one-electron functions in $|I\rangle$ and $|J\rangle$ in the same order. Then, if $|I\rangle$ and $|J\rangle$ differ by just a single one-electron wave function $|k\rangle$

$$\langle I|\mathcal{H}|J\rangle = \langle k^I |\mathcal{H}_1| k^J\rangle + (-1)^\tau \sum_{i=1}^{N} (\langle k^I ik^J i\rangle - \langle ik^I k^J i\rangle)$$

$$+ (-1)^{\tau+1} \sum_{\substack{i=1 \\ \neq k}}^{n} (\langle k^I ik^J i\rangle - \langle ik^I k^J i\rangle), \qquad ...(5)$$

and what is essentially the electron-hole interaction changes sign. If $|I\rangle$ and $|J\rangle$ differ by two functions $|k\rangle$ and $|l\rangle$ we find

$$\langle I|\mathcal{H}|J\rangle = (-1)^\tau (\langle k^I l^I k^J l^J\rangle - \langle k^I l^I k^J l^J\rangle). \qquad ...(6)$$

There is no one-electron term here, and the results are essentially the same for $n$ extra electrons and $n$ holes.

### 12.49.2 Calculated Band Structure of GaAs and InP

The band structure of bulk GaAs and InP, calculated from the pseudopotential method of Cohen and Bergstresser (1966) are shown in Figs. 12.93 and 12.94 respectively. The points marked by a Γ represent the (000) points in $k$ space. The points marked by an $L$ are at the (111) point in the Brillouin zone while the points marked by an $X$ are at the (100) point in the first Brillouin zone. In all these calculations, the hole energy minimum corresponds to 0.0 eV in the figures and occurs at the Γ point the Brillouin zone. The figures show the energy-band structure calculated along the three principal directions in the first Brillouin zone. Recall that for a cubic symmetry crystal the three principal directions are along [100], [110], and [111]. In the figures shown, these directions correspond to moving along $k$ as $\Gamma - X$, $\Gamma - K$, and $\Gamma - L$, respectively.

Note that three valence bands are present in the plots for GaAs and InP. These bands are the heavy, light, and split-off hole bands, respectively. The light hole band is the band splitting from the $\Gamma$ point in each figure with the greater curvature and hence lighter mass. The split-off valence band arises from the spin-orbit effect. This band is termed the *split-off band*

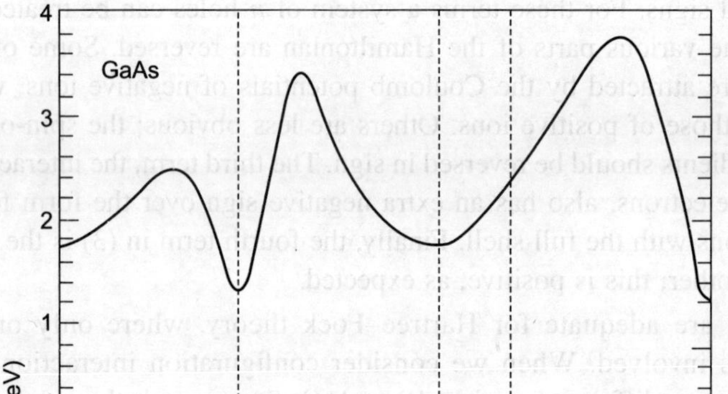


**Figure 12.93** Calculated energy-band structure of bulk GaAs based on the pseudopotential approach

since it is typically no longer degenerate with the heavy and the light hole bands at the $\Gamma$ point, as can be seen from Figs. 12.93 and 12.94.

It is interesting to examine the first conduction bands of the calculated band structures of GaAs and InP, shown in Figs. 12.93 and 12.94. As can be seen from the figures, the minimum of the first conduction band in GaAs and InP the conduction-band minimum

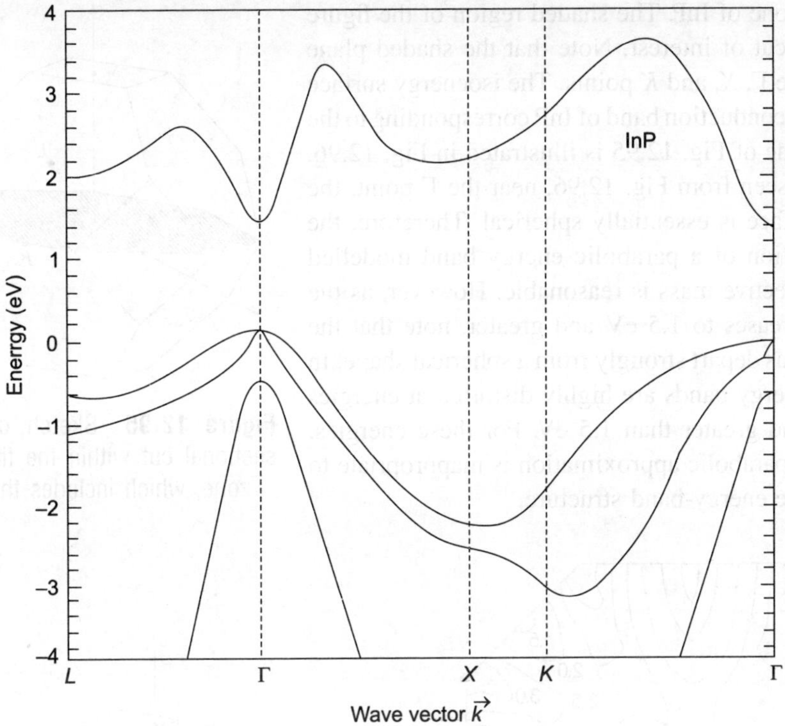


**Figure 12.94** Calculated energy-band structure of bulk InP based on the pseudopotential approach

occurs at the Γ point. Since the minimum hole energy in all the materials occurs at the Γ point, the difference in location in $k$ space of the minimum energy of the conduction band has important ramifications in the electrical and the optical properties of these materials. A material such as GaAs or InP, in which the conduction-band minimum and the valence-band minimum (minimum hole energy) occur at the same point in $k$ space, is called a direct-gap semiconductor. A material such as silicon, in which the conduction-band minimum appears at a different point to the valence-band minimum, is called an indirect-gap semiconductor.

Since the minima appear at the same point in $k$ space in direct-gap semiconductors, an electron can be promoted from the valence band across the energy gap into the conduction band without any change in momentum. As a result, optical transitions can occur to first order in direct-gap semiconductors. In indirect-gap semiconductors, optical transitions require an intermediary step before they can occur. In indirect-gap semiconductor a two-step process must occur in order for an optical transition to occur.

It is of further interest to examine the isoenergy surfaces in the conduction bands of these materials. For purposes of illustration, isoenergy surfaces of the first conduction band of InP are presented for two different cuts in the first Brillioun zone. Figure 12.95 shows the first

Brillouin zone of InP. The shaded region of the figure shows the cut of interest. Note that the shaded plane includes the Γ, *X*, and *K* points. The isoenergy surface for the first conduction band of InP corresponding to the shaded plane of Fig. 12.95 is illustrated in Fig. 12.96. As can be seen from Fig. 12.96, near the Γ point, the energy surface is essentially spherical. Therefore, the approximation of a parabolic energy band modelled with an effective mass is reasonable. However, as the energy increases to 1.5 eV and greater, note that the energy bands depart strongly from a spherical shape. In fact, the energy bands are highly distorted at energies equal to and greater than 1.5 eV. For these energies, usage of a parabolic approximation is inappropriate to describe the energy-band structure.

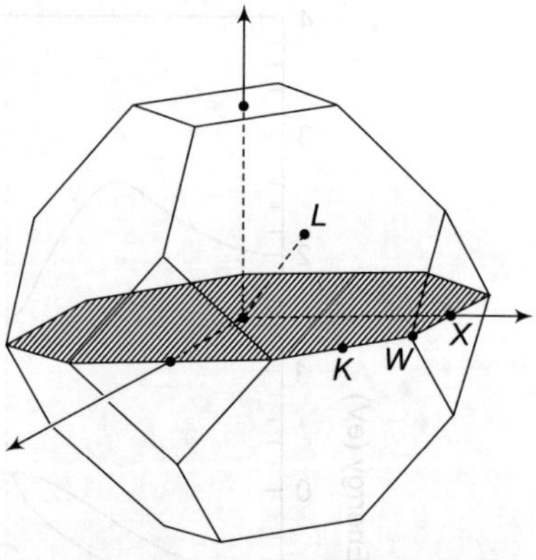


**Figure 12.95** Sketch of a cross-sectional cut within the first Brillouin zone, which includes the *X* point

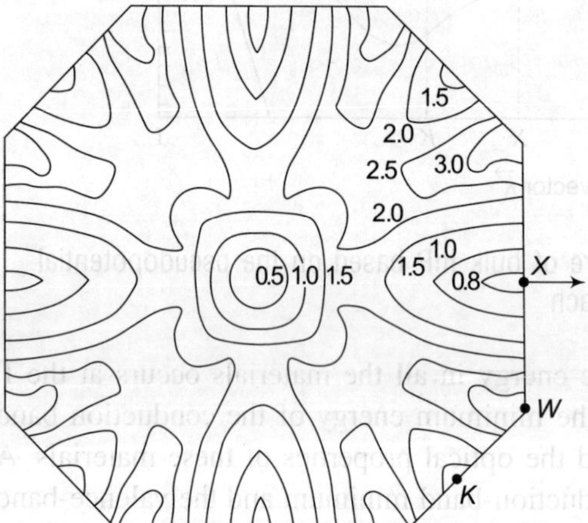


**Figure 12.96** Sketch of the isoenergy surfaces of the first conduction band of bulk InP within the cross-sectional cut of the first Brillouin zone shown in Fig. 12.95

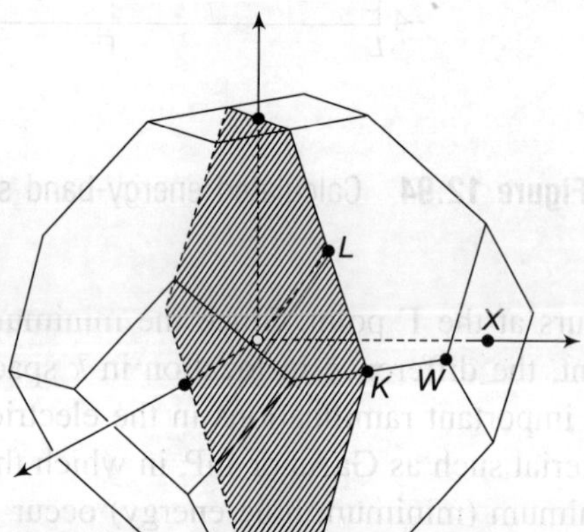


**Figure 12.97** Sketch of a cross-sectional cut within the first Brillouin zone, which includes the *L* point

Finally, Figs. 12.97 and 12.98 present a different cut in the first Brillouin zone and the corresponding isoenergy surface of the first conduction band of InP. In this case, the *L* and the *K* points are included. Inspection of Fig. 12.98 shows that the energy bands are again highly distorted at high energies. Again, there is a strong departure from parabolic energy bands.

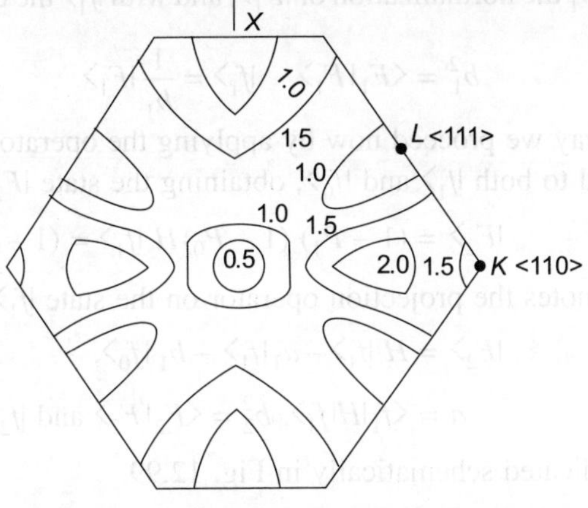


**Figure 12.98** Sketch of the isoenergy surfaces of the first conduction band of bulk InP within the cross-sectional cut first Brillouin zone shown in Fig. 12.97

## 12.50 THE LANCZOS OR RECURSION METHOD

The Lanczos method (1950), or the closely related recursion method (Haydock, Heine and Kelly, 1972), is a very convenient approach for the determination of the eigen-solutions of matrices, especially those of very large rank and sparse character (i.e., with many matrix elements equal to zero). In the field of electronic state calculations, the recursion method has been originally introduced in connection with a local basis representation of the electronic states in solids; successively it has been used also in connection with other basis sets (i.e., plane waves, muffin-tin orbitals). The method is very useful for periodic systems; it becomes really eassential in aperiodic materials, when a very large number of orbitals must be taken into account and when translational symmetry is (partially or completely) absent, so that the concepts and simplifications implied by the Bloch thermo are not at work.

The essential principle of the Lanczos recursion method is very simple and very general at the same time. Consider a quantum system, an operator $H$, and a number $N$(arbitrary large) of orthonormal basis states $\{|\phi_i\rangle\}$ $(I = 1, 2, \ldots, N)$. Starting from any given states $|f_0\rangle$ belonging to the space spanned by $\{|\phi_i\rangle\}$, and operating with $H$, *the recursion method provides a one-dimensional chain representation of the original quantum system.*

To illustrate the procedure, consider an operator $H$ (usually, the Hamiltonian of the system) and an initial normalized state $|f_0\rangle$, arbitrarily chosen; the starting state $|f_0\rangle$ is often addressed as the *seed state* of the procedure. We apply the operator $H$ to the initial state and subtract from $H\,|f_0\rangle$ its projection on the initial state, so to obtain a new state $|F_1\rangle$, orthogonal to $|f_0\rangle$. The operation of orthogonalization can be conveniently performed by means of the projection operator $P_0 = |f_0\rangle\,\langle f_0|$, and the state $|F_1\rangle$ can be expressed as

$$|F_1\rangle = (1 - P_0)\,H\,|f_0\rangle = H|f_0\rangle - a_0|f_0\rangle, \quad \text{...(1a)}$$

where $$a_0 = \langle f_0|H|f_0\rangle. \quad \text{...(1b)}$$

Let us indicate with $b_1$ the normalization of $|F_1\rangle$, and with $|f_1\rangle$ the corresponding normalized state, namely

$$b_1^2 = \langle F_1 | F_1 \rangle, \quad |f_1\rangle = \frac{1}{b_1} |F_1\rangle.$$

In a rather similar way we proceed now by applying the operator $H$ to the state $|f_1\rangle$; then $H|f_1\rangle$ is orthogonalized to both $|f_1\rangle$ and $|f_0\rangle$, obtaining the state $|F_2\rangle$. We have

$$|F_2\rangle = (1 - P_1)(1 - P_0) H |f_1\rangle = (1 - P_1 - P_0) H |f_1\rangle,$$

where $P_1 = |f_1\rangle\langle f_1|$ denotes the projection operator on the state $|f_1\rangle$. We thus obtain

$$|F_2\rangle = H|f_1\rangle - a_1|f_1\rangle - b_1|f_0\rangle, \quad \text{...(2a)}$$

where
$$a = \langle f_1 | H | f_1\rangle, \; b_2^2 = \langle F_2 | F_2\rangle \text{ and } |f_2\rangle = \frac{1}{b_2} |F_2\rangle. \quad \text{...(2b)}$$

The procedure is indicated schematically in Fig. 12.99.

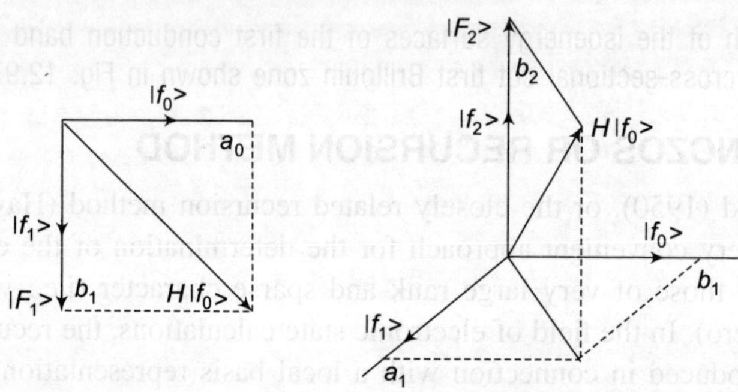


**Figure 12.99** Schematic representation of the lanczos recursion procedure

Let us now proceed to the next logical step of the sequence, and here we arrive at the key point of the ingenious, and yet so simple, Lanczos procedure. Consider, in fact, the state $|F_3\rangle$ obtained by orthogonalization of $H|f_2\rangle$ to the previous state $f_0, f_1, f_2$; we have

$$|F_3 = (1 - P_2)(1 - P_1)(1 - P_0) H|f_2\rangle = (1 - P_2 - P_1 - P_0) H|f_2\rangle.$$

We notice that $P_0 H |f_2\rangle \equiv 0$ since $\langle f_0 | H | f_2\rangle \equiv 0$; *thus the iteration procedure leads to a three-term relation*. The most remarkable feature of the method is that the interaction never includes more than three terms.

Iterating $n$ times the procedure, orthogonalizing the new generated state $H|f_n\rangle$ only to the two predecessors, $|f_n\rangle$ and $|f_{n-1}\rangle$ we obtain

$$|F_{n+1}\rangle = H|f_n\rangle - a_n|f_n\rangle - b_n|f_{n-1}\rangle. \quad \text{...(3a)}$$

The next pair of coefficients are given by

$$b_{n+1}^2 = \langle F_{n+1} | F_{n+1}\rangle \text{ and } a_{n+1} = \langle f_{n+1} | f_{n+1}\rangle. \quad \text{...(3b)}$$

The three-term relation (3a) is all we need to generate the orthonormal basis set $\{|f_n\rangle\}$ and the set of coefficients $\{a_n\}$ and $\{b_n\}$. This is a remarkable advantage, both conceptual and technical, in the specific calculations carried out by a computer code. On the basis set $\{|f_n\rangle\}$, the operator $H$ is represented by the tridiagonal matrix

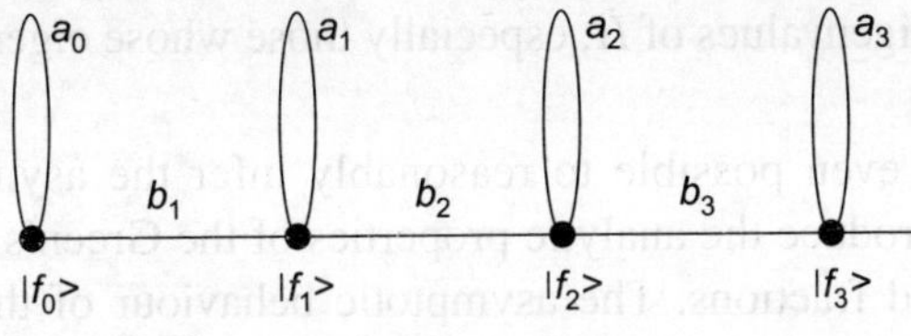


**Figure 12.100** Schematic representation of the linear chain generated by the Lanczos procedure

$$H = \begin{pmatrix} a_0 & b_1 & & & \\ b_1 & a_1 & b_2 & & \\ & b_2 & a_2 & b_3 & \\ & & \cdot & \cdot & \cdot \\ & & & \cdot & \cdot \end{pmatrix}. \quad \text{...(4a)}$$

The operator (4a) can also be written in the form

$$H = \sum_{n=0}^{\infty} a_n |f_n\rangle\langle f_n| + \sum_{n=0}^{\infty} b_{n+1} [|f_n\rangle \langle f_{n+1}| + |f_{n+1}\rangle \langle f_n|] \quad \text{...(4b)}$$

and can be represented diagrammatically as in Fig. 12.100.

The most important feature of the tridiagonal matrix (4a) is the possibility to obtain very easily the Green's function; in particular the matrix element $G_{00}$ $(E)$ of the Green's function is given by the continued fraction expansion

$$G_{00}(E) = \langle f_0| \frac{1}{E - H} |f_0\rangle = \cfrac{1}{E - a_0 - \cfrac{b_1^2}{E - a_1 - \cfrac{b_2^2}{E - a_2 - \dots}}} \quad \text{...(5a)}$$

From $G_{00}$ $(E)$ the local density-of-states projected on $|f_0\rangle$ of the Hamiltonian operator $H$ can be obtained; and the local density-of-states is given by

$$n_0(E) = -\frac{1}{\pi} \lim_{\varepsilon \to 0^+} \text{Im } G_{00}(E + i\varepsilon) \quad \text{...(5b)}$$

where $\varepsilon$ is an infinitesimal positive quantity.

The recursion method allows (at least in principle) the one-dimensional chain representation of any quantum mechanical system, with an arbitrary number $N$ of states. No explicit diagonalization of the original matrix $H$ is required for performing such a transformation. The one-dimensional chain representation allows to describe economically the system through the parameters $a_n$ and $b_n$ (i.e., $2N - 1$ parameters rather than $N(N + 1)/2$ matrix elements $\langle\phi_i|H|\phi_j\rangle$; for $N = 1000$ for instance, we have to compare approximately two thousand with one million).

A peculiar feature of the Lanczos procedure is that the number of steps to be carried out in specific problems may be reasonably small (say ten up to hundred or so). Once the original matrix $H$ is put into tridiagonal form, successive diagonalizations of small order tridiagonal matrices allow to infer the eigenvalues of $H$, especially those whose eigenfunctions have large overlap with the seed state.

In some problems, it is even possible to reasonably infer the asymptotic behaviour of expansion, Eq. (5a) and reproduce the analytic properties of the Green's function through the analytic theory of continued fractions. The asymptotic behaviour of the continued fraction coefficients $a_n$ and $b_n$ is connected with the compactness of the spectrum and with the presence of critical points. We cannot dwell on this and other interesting aspects of the recursion method, and on its interplay with other formalisms, such as the method of moments, the memory function approach, and the equation of motion method [see for instance the review articles by G. Grosso and G. Pastori Parravicini, *Adv. Chem. Phys.* **63**, 81 (1985); **63**, 133 (1985) and references quoted therein]; we rather prefer to clarify the procedure with an example.

## 12.50.1 Illustrative Example of the Recursion Method

To illustrate the recursion method, we consider for instance the case of a "quadratum", which is a simple square lattice with a single orbital per site and nearest neighbour interactions. The Hamiltonian is given by

$$H = E_0 \sum_{ij} |\phi_{ij}\rangle \langle\phi_{ij}|$$
$$+ t \sum_{ij} [|\phi_{ij}\rangle\langle\phi_{i+1,j}| + |\phi_{ij}\rangle\langle\phi_{i-1,j}| + |\phi_{i+1,j}\rangle\langle\phi_{ij}| + |\phi_{i-1,j}\rangle\langle\phi_{ij}|]$$
$$+ t \sum_{ij} [|\phi_{ij}\rangle\langle\phi_{i,j+1}| + |\phi_{ij}\rangle\langle\phi_{i,j-1}| + |\phi_{i,j+1}\rangle\langle\phi_{ij}| + |\phi_{i,j-1}\rangle\langle\phi_{ij}|]. \quad ...(6)$$

We take for convenience $E_0 = 0$ and $t = 1$. The Hamiltonian operator, Eq. (6) is represented by a sparse matrix (non-vanishing interactions are confined to nearest neighbour orbitals), and is indicated graphically in Fig. 12.101.

Before stating the Lanczos procedure, we need to establish explicitly the effect of the operator $H$ on a generic state of the form

$$|u\rangle = \sum_{ij} c_{ij} |\phi_{ij}\rangle \quad ...(7a)$$

From Eq. (6), we see that

$$H\,|u\rangle = \sum_{ij} \gamma_{ij} |\phi_{ij}\rangle \quad ...(7b)$$

with $\gamma_{ij} = c_{i-i,j} + c_{i+1,j} + c_{i,j-1} + c_{i,j+1}$. ...(7c)

With the help of Eq. (7), one can easily carry out the Lanczos procedure; choosing as seed state the orbital at the origin, i.e., $|f_0\rangle = |\phi_{00}\rangle$, one obtains

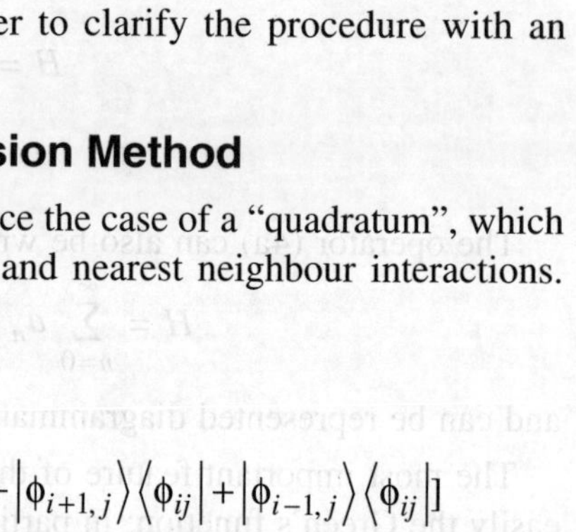


**Figure 12.101** Schematic graphical representation of a simple square lattice, with a single orbital per site, and nearest neighbour interactions; $\phi_{ij}$ indicates the orbital localized in the position $(i, j)$ (the edge of the quadratum is taken as unit)

$$|F_1\rangle = |\phi_{0,1}\rangle + |\phi_{0,-1}\rangle + |\phi_{1,0}\rangle + |\phi_{-1,0}\rangle;$$

the normalization of $|F_1\rangle$ gives $b_1^2 = \langle F_1|F_1\rangle = 4$. At the second step, one obtains

$$|F_2\rangle = |\phi_{1,1}\rangle + |\phi_{-1,1}\rangle + |\phi_{-1,-1}\rangle + |\phi_{1,-1}\rangle + \frac{1}{2}[|\phi_{2,0}\rangle + |\phi_{0,2}\rangle + |\phi_{-2,0}\rangle + |\phi_{0,-2}\rangle].$$

The normalization of $|F_2\rangle$ gives $b_2^2 = \langle F_2|F_2\rangle = 5$. The procedure can be continued so to obtain all the desired continued fraction parameters (with the help of a straight-forward computer code). The density-of-states can then be computed from Eq. (5b).

## 12.51 THE GREEN'S FUNCTION METHOD (KKR METHOD)

The Green's function method was proposed by Korringa (1947), Kohn and Rostoker (1954) (and therefore called also KKR method) and Morse (1956) in different though equivalent forms. In the KKR method the Schrödinger equation is transformed into an equivalent intergral equation and, once a muffin-tin potential is assumed, the problem of boundary conditions is elegantly solved through the fulfilment of appropriate surface intergrals, related to the Green's theorem. A nice feature of the method is the sharp separation of the procedure into two parts: (i) structural aspects of the lattice; (ii) phase shifts of the spherical muffin-tin potential.

From the point of view of electronic state calculation for periodic crystals, it is generally agreed that KKR is even more rapidly convergent than the APW method; but this advantage is to some extent counterbalanced by the somewhat more complicated expressions of the KKR structural coefficients. However, the real peculiar novelty of the KKR methodology (i.e., sharp separation of *geometry* from *potential*) has made the method generalizable to the treatment of impurities, clusters, homogeneously disordered alloys, and also photon energy bands in metallic ceramics. Probably these are the most important modern aspects of the KKR method; however, we keep in line with the purpose of this chapter and we limit ourselves to the study of electrons in fully periodic materials.

### 12.51.1 Scattering Integral Equation for a Generic Potential

In this section we briefly summarize the Green's function technique to transform into integral form the Schrödinger equation, here rewritten for convenience as

$$(E + \nabla^2)\, \psi(\mathbf{r}) = V(\mathbf{r})\, \psi(\mathbf{r}). \qquad ...(1)$$

The free-particle Green's function $g(\mathbf{r}, \mathbf{r}_0, E)$ for a δ-like source at $\mathbf{r}_0$ is defined by means of the equation

$$(E + \nabla^2)\, g(\mathbf{r}, \mathbf{r}_0, E) = \delta(\mathbf{r} - \mathbf{r}_0) \qquad ...(2)$$

and the requirement that $g(\mathbf{r}, \mathbf{r}_0, E)$ obeys the same boundary conditions as $\psi(\mathbf{r})$. It is straightforward to verify that the differential eigenvalue Eq. (1) is equivalent to the integral eigenvalue equation

$$\psi(\mathbf{r}) = \phi(\mathbf{r}) + \int g(\mathbf{r}, \mathbf{r}_0, E)\, V(\mathbf{r}_0)\, \psi(\mathbf{r}_0)\, d\mathbf{r}_0 \qquad ...(3)$$

where $\phi(\mathbf{r})$ denotes a solution at energy $E$ (if any) of the homogeneous differential equation $(E + \nabla^2)\,\phi(r) = 0$. The equivalence of Eq. (3) with Eq. (1) follows by direct application of the operator $(E + \nabla^2)$ to both members of Eq. (3).

It is convenient the express the Green's function $g(\mathbf{r}, \mathbf{r}_0, E)$ of the free-particle Hamiltonian $H_0 = -\nabla^2$ without sticking to the **r**-representation. The free-particle retarded Green's function $g(E)$ can be expressed as

$$g(E) = \frac{1}{E + iE - H_0} = \frac{1}{E + i\varepsilon + \nabla^2} \quad \varepsilon \to 0^+ \qquad \text{...(4)}$$

it is understood that the real energy $E$ is always accompanied by an infinitesimal positive imaginary part (sometimes, for brevity of notations, the imaginary part is not explicitly indicated; however, the imaginary part $i\varepsilon$ must never be overlooked, and must be reinstated whenever necessary). The integral eigenvalue Eq. (3) can now be written in the form

$$|\psi\rangle = |\phi\rangle + \frac{1}{E + i\varepsilon - H_0} V\,|\psi\rangle \qquad \text{...(5)}$$

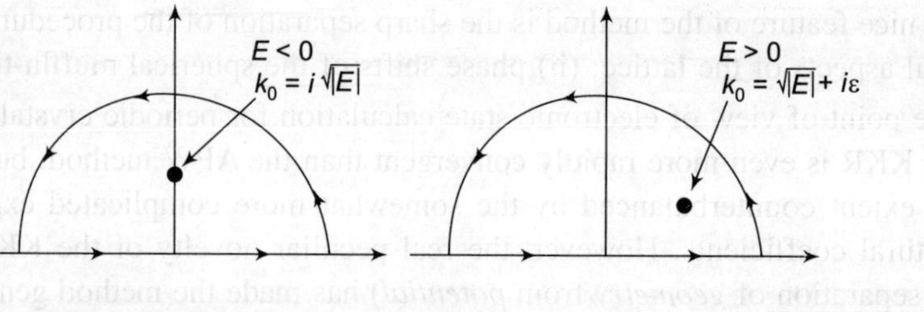


**Figure 12.102** Integration contour for evaluating the Green's function of the free-electron problem

It is convenient to obtain a spectral representation of the Green's function $g(E)$ in terms of the eigenvalues and eigenfunctions of the operator $H_0$ itself; these are plane waves $W(\mathbf{k}, \mathbf{r}) = (1/\sqrt{V})\exp(i\mathbf{k}\cdot\mathbf{r})$ of energy $k^2$. The Green's function, Eq. (4) can be evidently written as

$$g(E) = \frac{1}{E + i\varepsilon - H_0} \sum_{\mathbf{k}} |W_{\mathbf{k}}\rangle\langle W_{\mathbf{k}}| = \sum_{\mathbf{k}} \frac{1}{E + i\varepsilon - k^2} |W_{\mathbf{k}}\rangle\langle W_{\mathbf{k}}| \qquad \text{...(6a)}$$

In the **r**-representation we have

$$g(\mathbf{r}, \mathbf{r}_0, E) = \frac{1}{V} \sum_{\mathbf{k}} \frac{e^{i\mathbf{k}\cdot\mathbf{r}}\; e^{-i\mathbf{k}\cdot\mathbf{r}_0}}{E + i\varepsilon - k^2}. \qquad \text{...(6b)}$$

Notice that the free-particular Green's function depends on the difference $\mathbf{r} - \mathbf{r}_0$, i.e., $g(\mathbf{r}, \mathbf{r}_0, E)$ has the form $g(\mathbf{r} - \mathbf{r}_0, E)$.

The discrete sum over **k** in Eq. (6b) can be replaced (as usual) with $V/(2\pi)^3$ times the integral on the **k** variable; we have

$$g(\mathbf{r}-\mathbf{r}_0, E) = \frac{1}{(2\pi)^3}\int \frac{1}{E+i\varepsilon-k^2}\, e^{i\mathbf{k}\cdot(\mathbf{r}-\mathbf{r}_0)}\, d\mathbf{k}, \qquad \text{...(7a)}$$

where $d\mathbf{k}$ denotes the volume element in the reciprocal space. Performing the integration we obtain

$$g(\mathbf{r}-\mathbf{r}_0, E) = -\frac{1}{4\pi}\frac{e^{i\alpha|\mathbf{r}-\mathbf{r}_0|}}{|\mathbf{r}-\mathbf{r}_0|}, \qquad \text{...(7b)}$$

where $\alpha^2 = E$ or, more specifically,

$$\alpha(E) = \begin{cases} \sqrt{E} & \text{for } \; E>0 \\ i\sqrt{|E|} & \text{for } \; E<0 \end{cases}. \qquad \text{...(7c)}$$

The integration in Eq. (7a) has been carried out with the following procedure. We introduce polar coordinates in **k** space with $\mathbf{r}-\mathbf{r}_0$ as $k_z$ direction, and obtain

$$g(\mathbf{r}-\mathbf{r}_0, E) = \frac{1}{(2\pi)^3}\, 2\pi \int_0^\infty \frac{k^2}{E+i\varepsilon-k^2}\, dk \int_0^\pi e^{ik|\mathbf{r}-\mathbf{r}_0|\cos\theta}\, \sin\theta\, d\theta.$$

The integral in the angular variable θ can be performed and, with straightforward manipulations, we have

$$g(\mathbf{r}-\mathbf{r}_0, E) = \frac{1}{4\pi^2 i\,|\mathbf{r}-\mathbf{r}_0|}\int_{-\infty}^{\infty} \frac{k}{E+i\varepsilon-k^2}\, e^{ik|\mathbf{r}-\mathbf{r}_0|}\, dk. \qquad \text{...(8)}$$

Consider first the integration in Eq. (8) in the case $E < 0$. The integral can be easily evaluated closing the integration contour in the upper part of the complex plane, as indicated in Fig. 12.102. We have to consider one simple pole at $k = k_0 \equiv i\sqrt{|E|}$ and the residue of the integrand function is $(-1/2)\exp(ik_0|\mathbf{r}-\mathbf{r}_0|)$. We thus obtain Eq. (7b) for negative energies. A similar procedure proves that Eq. (7b) holds also for positive.

## 12.51.2 Scattering Integral Equation for a Periodic Muffin-tin Potential

In the case of a periodic crystal potential, we can specify the integral Eq. (3) taking into account that the wave functions must be of Bloch type; we have

$$\psi(\mathbf{k}, \mathbf{r}) = \int_{\text{crystal}} g(\mathbf{r}-\mathbf{r}_0, E)\, V(\mathbf{r}_0)\, \psi(\mathbf{k}, \mathbf{r}_0)\, d\mathbf{r}_0, \qquad \text{...(9a)}$$

where the integration extends on the crystal volume; the solution $\phi(\mathbf{r})$ of the homogeneous equation is set equal to zero, since in general there is no incident plane wave in the infinite crystal.

The wave function $\psi(\mathbf{k}, \mathbf{r})$, at any two points **r** and $\mathbf{r} + \mathbf{t}_n$ related by a translation vector, takes values related by the Bloch phase factor $\exp(i\mathbf{k}\cdot\mathbf{t}_n)$. This allows to restrict the integration in Eq. (9a) to a single unit cell (for convenience the Wigner–Seitz cell centred at $\mathbf{t}_n = 0$). We obtain the following integral equation

$$\boxed{\psi(\mathbf{k},\mathbf{r}) = \int_{\Omega\text{ws}} g(\mathbf{k},\mathbf{r}-\mathbf{r}_0,E)\,V(\mathbf{r}_0)\,\psi(\mathbf{k},\mathbf{r}_0)\,d\mathbf{r}_0}, \qquad \text{...(9b)}$$

where the kernel of the integral equation has the expression

$$\boxed{g(\mathbf{k},\mathbf{r}-\mathbf{r}_0,E) = \sum_{\mathbf{t}_n} e^{i\mathbf{k}\cdot\mathbf{t}_n}\, g(\mathbf{t}-\mathbf{r}_0-\mathbf{t}_n,E)} \qquad \text{...(9c)}$$

and is called *the Greenian of vector* **k.** Using Eq. (7b), we can express the Greenian of vector **k** in the form

$$g(\mathbf{k},\,\mathbf{r}-\mathbf{r}_0,\,E) = -\frac{1}{4\pi}\sum_{\mathbf{t}_n} e^{i\mathbf{k}.\mathbf{t}_n}\,\frac{e^{i\alpha|\mathbf{r}-\mathbf{r}_0-\mathbf{t}_n|}}{|\mathbf{r}-\mathbf{r}_0-\mathbf{t}_n|}, \qquad \text{...(10)}$$

where the sum is over the translation vectors of the direct lattice.

We now assume the muffin-tin form of the potential; the integration region in Eq. (9b) can then be reduced to the sphere $\Omega_s$ of radius $r_s$ ($\geq r_{\text{MT}}$) inscribed in the Wigner–Seitz polyhedron (we limit our attention to simple crystals, since the extension to composite crystals only requires some straightforward generalizations). We have thus that the crystal eigenfunctions of vector **k** satisfy the integral equation

$$\psi(\mathbf{k},\,\mathbf{r}) = \int_{\Omega_s} g(\mathbf{k},\mathbf{r}-\mathbf{r}_0-E)\,V(\mathbf{r}_0)\,\psi(\mathbf{k},\mathbf{r}_0)\,d\mathbf{r}_0.$$

Because of Eq. (1), we have also

$$\int_{\Omega_s} g(\mathbf{k},\mathbf{r}-\mathbf{r}_0,E)\,(E+\nabla_0^2)\,\psi(\mathbf{k},\mathbf{r}_0)\,d\mathbf{r}_0 = \psi(\mathbf{k},\mathbf{r}). \qquad \text{...(11)}$$

We use the identity

$$g\nabla^2\psi = (\nabla^2 g)\,\psi + \nabla\cdot[g\nabla\psi - \psi\nabla g],$$

where $g$ and $\psi$ are any two continuous and differentiable functions. We insert this identity into Eq. (11) and use the Green's theorem to transform the volume integral of div $(g\nabla\psi - \psi\nabla g)$ into a more convenient surface integral; we arrive at the basic "boundary condition" equation

$$\boxed{\int_{S_0}\left[g(\mathbf{k},\mathbf{r}-\mathbf{r}_0,E)\frac{\partial\psi(\mathbf{k},\mathbf{r}_0)}{\partial r_0} - \psi(\mathbf{k},\mathbf{r}_0)\frac{\partial g(\mathbf{k},\mathbf{r}-\mathbf{r}_0,E)}{\partial r_0}\right]_{r_0=r_s} dS_0 = 0}, \qquad \text{...(12)}$$

where $S_0$ is the surface of the sphere $\Omega_s$ and **r** is any point within it. In a crystal, with muffin-tin form of the potential, any crystal wave function of vector **k** and energy $E$ is such that the surface integral, defined in the left-hand side of Eq. (12), identically vanishes for any $\mathbf{r} \in \Omega_s$. In a crystal, with no muffin-tin form of the potential, a similar result holds, but the great complication occurs that the integral involves the surface of the primitive cell rather than the quite convenient spherical surface, $S_0$.

In order to perform explicitly the surface integrals in Eq. (12) need an expression of the Greenian in terms of appropriate (structure) coefficients, multiplied by factorized functions of **r** and $\mathbf{r}_0$ separately. The desired expression of the Greenian is

$$g(\mathbf{k}, \mathbf{r}-\mathbf{r}_0, E) = \alpha \sum_{lm} j_l(\alpha r)\, Y_{lm}(\mathbf{r})\, n_l(\alpha r_0)\, Y^*_{lm}(\mathbf{r}_0) + \sum_{lml'm'} j_l(\alpha r)\, Y_{lm}(\mathbf{r})\, \Gamma_{lm,l'm'}(\mathbf{k}, E)\, j_{l'}(\alpha r_0)\, Y^*_{l'm'}(\mathbf{r}_0) \quad ...(13)$$

where $\alpha = \sqrt{E}$, **r** and $\mathbf{r}_0$ are within the Wigner–Seitz cell, and $r < r_0$; in the case $r > r_0$ we have to exchange $r$ and $r_0$ in the first term in the right-hand side of Eq. (13) the coefficients $\Gamma_{lm,l'm'}(\mathbf{k}, E)$ are called *structure coefficients*.

As in the cellular method, also in the Green's function method the crystal wave functions are expanded in spherical waves

$$\psi(\mathbf{k}, \mathbf{r}_0) = \sum_{l''m''} c_{l''m''}(\mathbf{k})\, R_{l''}(E, r_0)\, Y_{l''m''}(\mathbf{r}_0) \quad ...(14)$$

but now the unknown coefficients $c_{lm}$ and the crystal eigenvalues are determined in a natural way through the fulfilment of the boundary conditions summarized by Eq. (12). As usual $R_l(E, r)$ indicates the solution, regular at the origin, of the radial Schrödinger equation. For convenience, the normalization of $R_l(E, r)$ is chosen so as to satisfy the following asymptotic behaviour

$$R_l(E, r) = \cos \eta_l(E) \cdot j_l(\alpha r) - \sin \eta_l(E) \cdot n_l(\alpha r) \quad \text{for} \quad r \geq r_{MT} \quad ...(15)$$

where $j_l$ and $n_l$ are the spherical Bessel and Neumann functions, and $n_l(E)$ are the *phase shifts* due to the muffin-tin potential at the energy $E$ and for the wave of angular momentum $l$. It is useful to recall the Wronskian relations for Bessel and Neumann functions

$$[j_l(x), n_l(x)] \equiv j_l(x)\, n'_l(x) - j'_l(x)\, \eta_l(x) = \frac{1}{x^2}$$

from it and Eq. (15), the following two Wronskian relations hold for $r = r_s \geq r_{MT}$

$$[j_l, R_l] = -\sin \eta_l(E) \frac{1}{\alpha r_s^2}, \quad [n_l, R_l] = -\cos \eta_l(E) \frac{1}{\alpha r_s^2}.$$

We now insert expression (13) for the Greenian and expansion (14) for the trial wave function into Eq. (12); we obtain

$$\alpha \sum_{lm} j_l(\alpha r)\, Y_{lm}(\mathbf{r})\, [n_l, R_l]\, c_{lm} + \sum_{lml'm'} j_l(\alpha r)\, Y_{lm}(\mathbf{r})\, \Gamma_{lm,l'm'}\, [j_{l'}, R_{l'}]\, c_{l'm'} = 0$$

for any $\mathbf{r} \in \Omega_s$. Because of this arbitrariness, it follows

$$\alpha \cos \eta_l(E)\, c_{lm} + \sum_{l'm'} \Gamma_{lm,l'm'}(\mathbf{k}, E) \sin \eta_{l'}(E)\, c_{l'm'} = 0$$

The above system of linear homogeneous equations for the arbitrary coefficients $c_{lm}$ leads to the compatibility determinantal equation

$$\boxed{\left\| \Gamma_{lm,l'm'}(\mathbf{k}, E) + \sqrt{E} \cot g\, \eta_l(E)\, \delta_{ll'}\, \delta_{mm'} \right\| = 0} \quad ...(16)$$

The determinant in Eq. (16) is plotted versus the energy $E$ and its zeroes provide the crystal eigenvalues at the chosen **k** vector.

A remarkable feature of the basic secular Eq. (16) is the net separation of the information concerning the crystal structure, completely embodied in the structure coefficients $\Gamma_{lm,l'm'}$ (**k**, $E$) and the effect of the potential summarized by the phase shifts $\eta_l(E)$. Furthermore, the dimension of the secular equation needed in actual calculations is expected to be reasonably small; for instance, to reproduce the $s$, $p$ and $d$ character of the bands of interest, it is sufficient to consider the angular momentum values $l = 0, 1, 2$ and the secular equation becomes a $9 \times 9$ matrix. From a formal point of view the KKR method is the most satisfactory tool to solve the problem of boundary conditions, when the muffin-tin form for the potential is assumed.

## 12.52 THE MOST IMPORTANT DETAILS OF THE BAND STRUCTURE OF A SEMICONDUCTOR

If the effective mass is a scalar, i.e., independent of direction, the theoretical treatment becomes particularly simple. The electrons then behave like free electrons but with a different value of mass. The same is true for holes which are accorded a positive charge as well as a (scalar) positive effective mass. This very simple model is complicated by the following possibilities (see Fig. 12.103). In this figure, first a band structure has been illustrated where individual

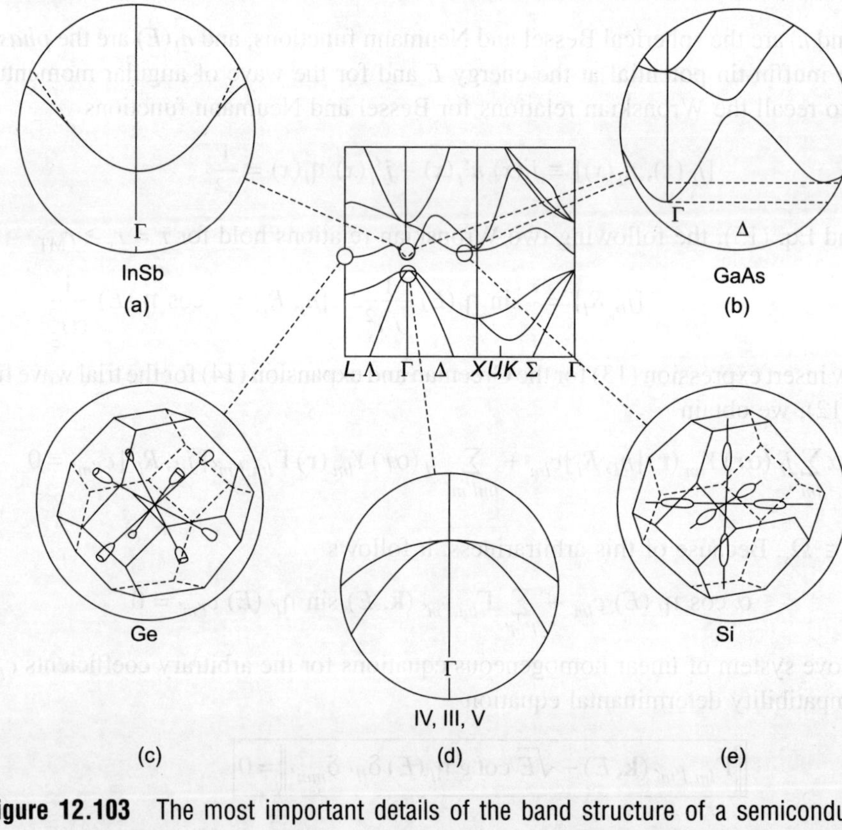


**Figure 12.103** The most important details of the band structure of a semiconductor

regions have been lifted out. Let us begin with the point Γ, the midpoint of the Brillouin zone. If the lower edge of the conduction band lies at this point (Fig. 12.103a), we have in general an isotropic parabolic and, i.e., states which can be described by a directionally independent constant effective mass. Only with increasing energy does the $E(k)$ dependence deviate from a parabola, the effective mass then becoming energy dependent. This case is realized for instance in the semiconducting compound InSb.

The upper edge of the valence band in many semiconductors lies at $k = 0$ (those of the group IV element—diamond, Si, Ge, α-Sn—and the III-V compounds, for example). Here, however, two different sub-bands have a common extremum (Fig. 12.103d). Holes can then appear near the extremum in both sub-bands. In this case the free charge carrier modal has to take account of two different sorts of holes existing together.

If the extrema of a band lie outside $k = 0$, then from symmetry considerations a number of equivalent extrema must be present. Surfaces of constant energy are often ellipsoidal in the vicinity of such extrema. This case, exemplified by Si and Ge, is shown in Figs. 12.103 c, e.

Only Fig. 12.103b remains to be discussed. It shows the overlapping of two sub-bands with extrema which are close together in energy but at different points in the Brillouin zone. Here too we have to consider two sorts of carriers (electrons) with differing effective masses. Since the minima lie at different energies, the ratio between the concentrations of the two sorts of electrons will become temperature dependent. An example of this case is the band structure of GaAs.

## 12.53 THE ENERGY BANDS FOR Si, Ge AND GaAs

The calculated energy bands for Ge, Si, and GaAs are shown in Fig. 12.104 (from Chelikowsky and Cohen 1976). The most important features of the band structures shown in Fig. 12.104 are represented by the minima of the lowest conduction band and by the maxima of the highest valence band because the states with energies much higher than the minima of the conduction band are empty and the states with energies much higher than the minima of the conduction band are empty and the states with energies well below the maxima of the valence band are completely filled. The model that summarizes the most important features of the band structure of the cubic semiconductors is shown in Fig. 12.105 (from Jacoboni and Reggiani 1979). In all important cubic semiconductors, the top of the highest filled (valence) band corresponds to the Γ point of the first Brillouin zone, that is, to point $k = (0, 0, 0)$. The bands shown in Fig. 12.105a are typical for Si, Ge, III-V, and most II-VI semiconductors where the top valence band and the lowest conduction band are separated by an energy gap. The tops of the valence bands are located at point Γ; two of these bands (light and heavy holes) have the same energy at this point (i.e., they are degenerate), and the third one I split (it is called a split-off band). The lowest minimum of the conduction band in different semiconductors corresponds to different points of the first Brillouin zone. For example, in germanium, the $L$ minimum (corresponding to the point (1,1,1) of the first Brillouin zone) is the lowest. The lowest minimum of the conduction band in silicon is located along the Δ axis (corresponding to the direction $(k, 0, 0)$ of the first Brillouin zone), close to the $X$-point (corresponding to the

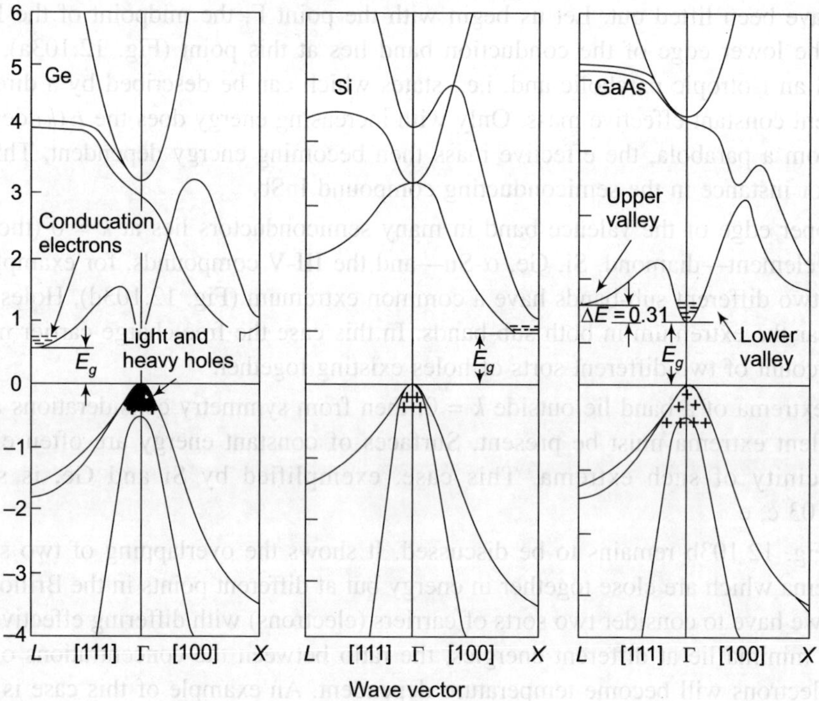


**Figure 12.104** Computed energy bands for silicon, germanium, and gallium arsenide (from J.R. Chelikowky and M.L. Cohen, "Nonlocal pseudopotential calculations for the electronic structure of eleven diamond and zinc-blend semiconductors," *Phys. Rev*, B14, p. 556 (1976) and S.M. Sze, *Physics of Semiconductor Devices*, John Wiley & Sons, (1981)

point (1, 0, 0) of the first Brillouin zone). In GaAs, the lowest minimum of the conduction band is at point Γ (0, 0, 0), i.e., at the same value of the wave vector **k** as the top of the valence band. Such a semiconductor is called a direct gap semiconductor. Silicon and germanium are indirect gap semiconductors.

The band structure of Fig. 12.105b occurs in gapless semiconductors such as grey tin and HgTe. There is no energy gap! This leads to very interesting and unusual properties of these materials. Using, for example, CdHgTe alloys, one may create materials with narrow energy gaps, to register interband transitions caused by infrared photons and thus make sensitive infrared detectors.

Near the minima of the conduction band the energy versus wave vector dependencies, such as shown in Fig. 12.104, can be approximated by one of the following functions

$$E(k) = \hbar^2 k^2/(2m^*) \text{ (spherical)} \qquad ...(1)$$

$$E(k) = \hbar^2/2\ (k_x^2/m_x + k_y^2/m_x + k_z^2/m_z) \text{ (ellipsoidal)} \qquad ...(2)$$

Wave vector $\vec{\mathbf{k}}$ and energy $E$ in Eqs. (1) to (2) are measured from the corresponding minima of the energy band. Equation (1) represents a parabolic dependence of energy on

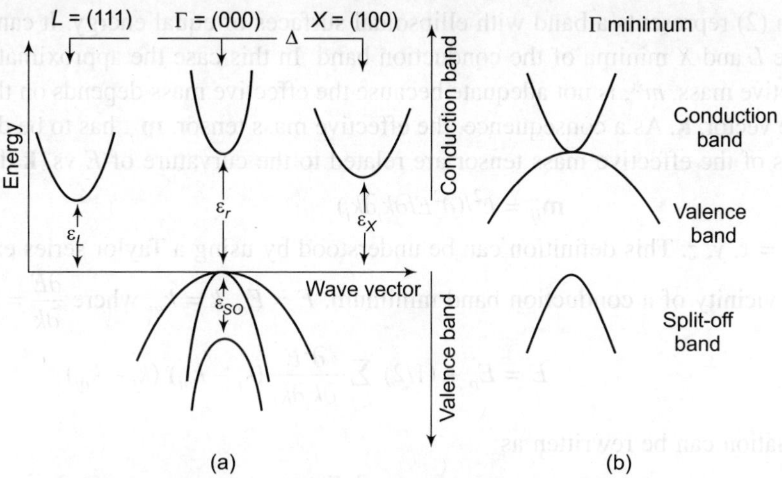


**Figure 12.105** Important minima of the conduction band and maxima of the valence band in cubic semiconductors

the wave vector, similar to that of an electron in free space, $E(k) = \hbar^2k^2/(2m_e)$. However, the value of $m^*$ (which is called the effective mass) can be very different from a free electron mass, $m_e$. For example, in gallium arsenide $m^* \approx 0.067\ m_e$. This difference is caused by the periodic crystal potential. Equation (1) corresponds to a band with a single scalar effective mass and spherical surfaces of equal energy. It is an appropriate model for the Γ minimum of the conduction band. A similar equation,

$$E(k) = E_1 - \hbar^2k^2\ (2m_p^*)\ \text{(spherical)} \qquad ...(3)$$

can be used for the split-off valence band. Equation (1) is also the simplest possible model of the band structure and is frequently used for crude estimates of transport properties. The similarity with the free electron motion is very useful for providing an insight into the physics of electronic motion. When Eq. (1) is applicable, one can write the Schrödinger equation for a semiconductor crystal[10] simply substituting $m_e$ by $m^*$, provided that potential $U$ varies slowly over distances on the order of the lattice constant. For example, we can find energy levels related to a positive charge introduced into the semiconductor solving the Schrödinger equation with the Coulombic potential (given by equation[11]) (we just have to substitute $\varepsilon_o$ by $\kappa\varepsilon_o$, where $\kappa$ is the dielectric constant of the semiconductor, and $m_e$ by $m^*$).

The importance of the effective-mass concept is related to the fact that in many cases electrons and holes occupy energy states in the vicinity of minima and maxima, shown in Fig. 12.105. On the other hand, even a casual look at Fig. 12.104 is enough to convince us that this concept should be used with caution, especially in high electric fields when electron and energies can be very high.

10 $-\dfrac{\hbar^2}{2m_e}\nabla^2\psi + U(\vec{r})\,\psi = E\,\psi$

11 $U(\vec{r}) = -\dfrac{q^2}{4\pi\varepsilon_0|\vec{r}|}$

Equation (2) represents a band with ellipsoidal surfaces of equal energy. It can be used to describe the $L$ and $X$ minima of the conduction band. In this case the approximation using a single effective mass, $m^*$, is not adequate because the effective mass depends on the direction of the wave vector, **k**. As a consequence, the effective mass tensor, $\mathbf{m}_{ij}$, has to be defined. The components of the effective mass tensor are related to the curvature of $E$ vs. **k** dependence

$$\mathbf{m}_{ij} = \hbar^2/(\partial^2 E/\partial k_i \partial k_j) \qquad ...(4)$$

Here $i, j = x, y, z$. This definition can be understood by using a Taylor series expansion of $E(k)$ in the vicinity of a conduction band minimum, $E = E_o$, $k = k_o$, where $\frac{\partial E}{\partial k} = 0$

$$E = E_o + (1/2) \sum_{i,j} \frac{\partial^2 E}{\partial k_i \partial k_j} (k_i - k_{oi})(k_j - k_{oj}) \qquad ...(5)$$

This equation can be rewritten as

$$E = E_o + (1/2)\hbar^2 \sum_{i,j} (k_i - k_{oi})(k_j - k_{oj})/m_{ij} \qquad ...(6)$$

where components of the effective mass tensor, $\mathbf{m}_{ij}$, are defined by Eq. (4).

As a result of the rotational symmetry of the ellipsoids of equal energy for the $L$ and $X$ minima of a conduction band (see Figs. 12.106 and 12.107). Equation 2 can be rewritten in a simpler form

$$E(k) = \frac{\hbar^2}{2}\left(\frac{k_l^2}{m_l} + \frac{k_t^2}{m_t}\right) \qquad ...(7)$$

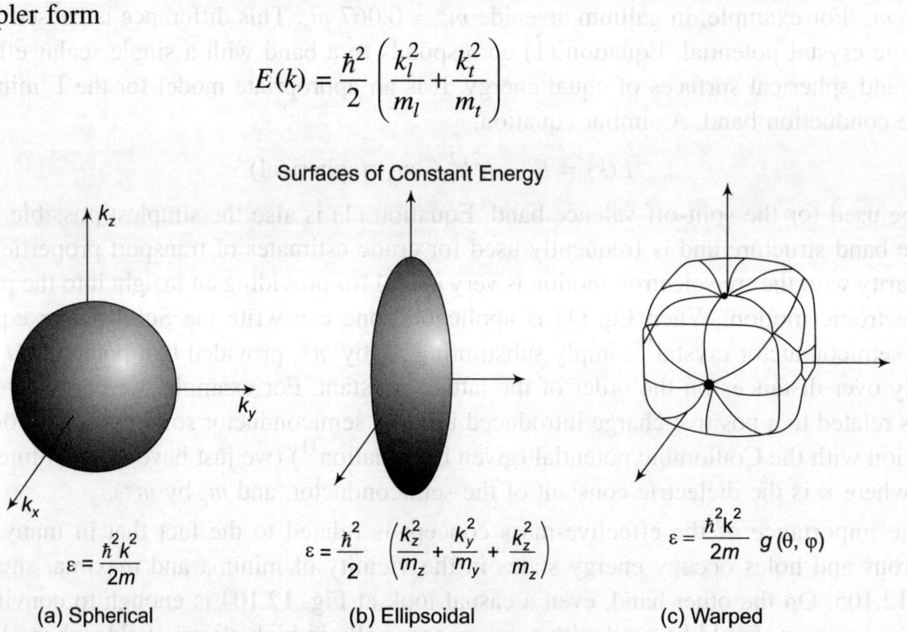


**Figure 12.106** Surfaces of equal energy in **k**-space (from C. Jacoboni and L Regriani, Bulk hot-electron properties of cubic semiconductors, *Advances in Physics*, Vol. 28, No. 4, pp 493–553, (1979), (a) Spherical equal energy surface, (b) Ellipsoidal equal energy surface, (c) "Warped" equal energy surfaces

where $1/m_l$ and $1/m_t$ are the longitudinal and transverse components of the inverse effective mass tensor, and $k_l$ and $k_t$ are the longitudinal and transverse components of the wave vector.

The dependence of energy on the wave vector near the top of the valence band was calculated by Dresselhaus et al. (1955)

$$E(k) = E_V - \hbar^2 k^2 g(k)/2m^* \qquad ...(8)$$

where

$$g = A_g \pm [B_g^2 + C_g^2 (k_x^2 k_y^2/k^4 + k_x^2 k_z^2/k^4 + k_y^2 k_z^2/k^4)]^{1/2} \qquad ...(9)$$

Here $A_g$, $B_g$, and $C_g$ are constants. Plus and minus signs in Eq. (6) correspond to the heavy and light holes, respectively. These bands are called warped bands because of the "warped" surface of equal energy (see Fig. 12.106c).

More realistic models should take into account the deviation of the $E$ vs **k** relationship from parabolicity. In the simplest way it can be done by substituting for $E$ in Eq. (1) with

$$\gamma(E) = E(1 + \alpha E) \qquad ...(10)$$

For example, for the minimum of the conduction band in the direct gap semiconductors such as GaAs we have

$$E(1 + \alpha E) = \frac{\hbar^2 k^2}{2m_n} \qquad ...(11)$$

Equation (11) takes into account the increase of the electron effective mass with energy, which can be noticed as the decrease in the curvature of $E$ vs **k** dependencies with an increase in $E$ in Fig. 12.104.

Equations describing the dispersion relationships in the vicinity of the symmetry points may be derived using the *k-p method*, first developed by Kane (1957). In this method, equation, $\psi_{\vec{k}}(\vec{r}) = e^{i\vec{k}\cdot\vec{r}}\, u_{\vec{k}}(\vec{r})$ is substituted into the Schrödinger equation. The resulting equation for the Bloch amplitude $u_{\mathbf{k}}(r)$ near the symmetry points of the Brillouin zone contains a term $\mathbf{k}\cdot\mathbf{p}$, where $\mathbf{p}$ is the momentum operator. This term is considered a perturbation. (For this reason the approach is called the *k-p* method.) The solution of the equation for the Bloch amplitude $u_{\mathbf{k}}(r)$ is then found based on the assumption that $u_{\mathbf{k}}(r)$ can be presented as a linear combination of the wave functions corresponding to the top valence bands and the bottom conduction band. In particular, this method yields the following approximate expression for the nonparabolicity constant

$$\alpha = \frac{1}{E_g}\left(1 - \frac{m_n}{m_e}\right)^2 \qquad ...(12)$$

Here $E_g$ is the energy gap, $m_n$ is the effective electron mass, and $m_e$ is the free electron mass (see, for example, Blakemore 1982). Band gaps and effective masses of some cubic semiconductors are given in Table 12.6 (after Jacoboni and Reggiani 1979).

All parameters characterizing the band structure are temperature dependent. According to Blakemore (1982), the temperature dependencies of energy gaps can be described by the following phenomenological equation

$$E_g(T) = E_{go} - \alpha_{temp}T^2/(T + \beta_{temp}) \quad ...(13)$$

where the values of the energy $E_{go}$ at $T = 0$ and coefficients $\alpha_{temp}$ and $\beta_{temp}$ are given for Si, Ge and GaAs in Table 12.7. For GaAs it is also useful to know that the energy gaps between the top of the valence band and the bottom of $L$ and $X$ minima, $E_L$ and $E_X$, vary with temperature. The phenomenological equations describing these dependencies for GaAs are

$$E_L(T) = 1.815 - 6.05 \times 10^{-4}\, T^2/(T + 204) \text{ (eV)} \quad ...(14)$$

and

$$E_X(T) = 1.981 - 4.60 \times 10^{-4}\, T^2/(T + 204) \text{ (eV)} \quad ...(15)$$

where temperature $T$ is in degree kelvin (from Blakemore 1982).

**Table 12.6** Band Gaps and Effective Masses of Some Cubic Semiconductors (After Jacoboni and Raggiani 1979)

| | $E_\Gamma$ (eV) | $E_L$ (eV) | $E_\Delta$ (eV) | $E_{so}$ (eV) | Electron effective masses | | | Hole effective masses | |
|---|---|---|---|---|---|---|---|---|---|
| | | | | | $m_l$ | $m$ | $m_t$ | $m_{hh}$ | $m_{hl}$ |
| Si | 4.08 | 1.87 | 1.13 | 0.04 | 0.98 | — | 0.19 | .53 | .16 |
| Ge | 0.89 | 0.76 | 0.96 | 0.29 | 1.64 | — | 0.082 | .35 | .043 |
| AlP | 3.3 | 3.0 | 2.1 | 0.05 | — | — | — | .63 | .2 |
| AlAs | 2.95 | 2.67 | 2.16 | 0.28 | 2.0 | — | — | .76 | .15 |
| AlSb | 2.5 | 2.39 | 1.6 | 0.75 | 1.64 | — | 0.23 | .94 | .14 |
| GaP | 2.24 | 2.75 | 2.38 | 0.08 | 1.12 | — | 0.22 | .79 | .14 |
| GaAs | 1.42 | 1.71 | 1.90 | 0.34 | — | 0.067 | — | .62 | .074 |
| GaSb | 0.67 | 1.07 | 1.30 | 0.77 | — | 0.045 | — | .49 | 0.46 |
| InP | 1.26 | 2.0 | 2.3 | 0.13 | — | 0.080 | — | .85 | .089 |
| InAs | 0.35 | 1.45 | 2.14 | 0.38 | — | 0.023 | — | .6 | .027 |
| InSb | 0.17 | 1.5 | 2.0 | 0.81 | — | 0.014 | — | .47 | 0.15 |
| ZnS | 3.8 | 5.3 | 5.2 | 0.07 | — | 0.28 | — | — | — |
| ZnSe | 2.9 | 4.5 | 4.5 | 0.43 | — | 0.14 | — | — | — |
| ZnTe | 2.56 | 3.64 | 4.26 | 0.92 | — | 0.18 | — | — | — |
| CdTe | 1.80 | 3.40 | 4.32 | 0.91 | — | 0.096 | — | — | — |

**Table 12.7** Temperature Dependence of the Energy Gaps (from Blakemore 1982)

| Semiconductor | $E_{go}$ (eV) | $\alpha_{temp}$ (eV/K$^2$) | $\beta_{temp}$ (K) |
|---|---|---|---|
| Si | 1.170 | $4.73 \times 10^{-4}$ | 636 |
| Ge | 0.7437 | $4.77 \times 10^{-4}$ | 235 |
| GaAs | 1.519 | $5.405 \times 10^{-4}$ | 204 |

## 12.54 BAND STRUCTURES OF SILICON AND GERMANIUM

Band structures are calculated by fitting the measured physical quantities, such as the band gap, the positions of points of high symmetry in the Brillouin zone (critical points), and the curvature of energy surfaces (the effective mass). The calculated band structure of Si and Ge are shown in Fig. 12.107. The features of the two band schemes appear quite different in contrast to the qualitative similarity. This is obviously the effect of the difference in electron wave functions associated with $3s^23p^2$ and $4s^24p^2$ configurations.

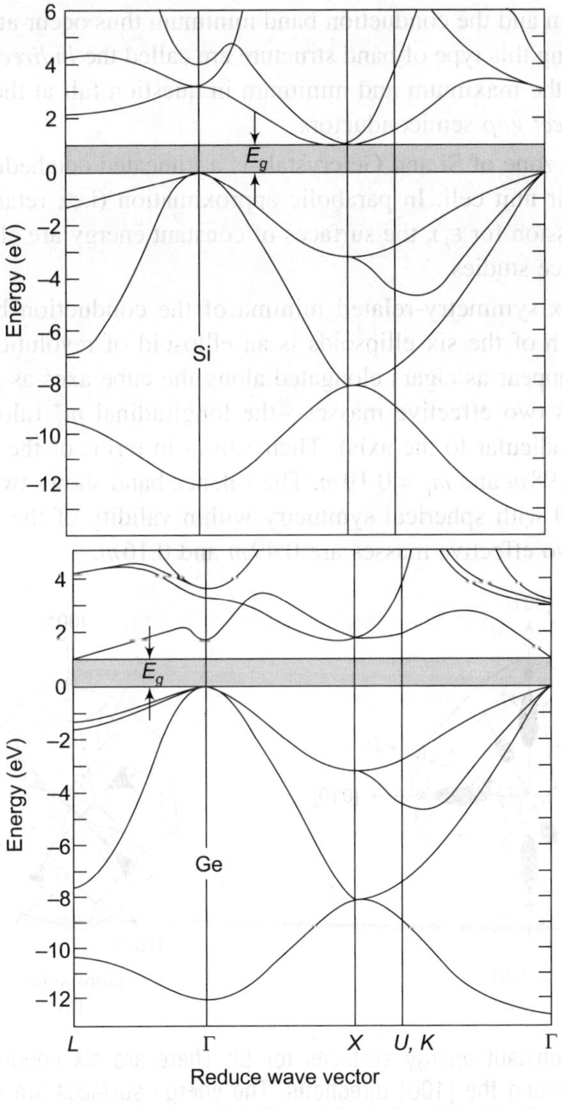


**Figure 12.107** Calculated band structures of Si and Ge [after J.R. Chelikowsky, M.L Cohen, *Phys. Rev.* **B14**, 556 (1976)]. The structures confirm the indirect gap nature for both Si and Ge. For Ge, the spin-orbit splitting is also considered

The symbols, Γ, $X$ and $L$ stand for positions of certain points of high symmetry in the Brillouin zone. They refer to the points at the zone centre (000), $\frac{2\pi}{a}$ (100) and $\frac{2\pi}{a}\left(\frac{1}{2}\frac{1}{2}\frac{1}{2}\right)$, respectively, where $a$ is the lattice constant. The valence band maximum occurs at the zone centre, i.e., $k = 0$ for both Si and Ge. But the conduction band minimum occurs for $k$ along the [100] direction in Si and for $k$ along the [111] direction in Ge. This amounts to saying that electrons of the lowest energy in the conduction band have their wave vectors oriented along the [100] direction in Si and along the [111] direction in Ge. In both materials the valence and maximum and the conduction band minimum thus occur at different values of $k$. Semiconductors having this type of band structure are called the *indirect gap* semiconductors and those for which the maximum and minimum in question fall at the same value of $k$ are referred to as the *direct gap* semiconductors.

The first Brillouin zone of Si and Ge crystals is a truncated octahedron appropriate to the fcc symmetry of their unit cell. In parabolic approximation (i.e., retaining terms up to the order $k^2$ in the expression for $\varepsilon_k$), the surfaces of constant energy are ellipsoids as confirmed by cyclotron resonance studies.

In Si there are six symmetry-related minima of the conduction band at points in the [100] directions. Each of the six ellipsoids is an ellipsoid of revolution about a cubic axis by symmetry. They appear as cigars elongated along the cube axes as shown in Fig. 12.108 (a). The electron has two effective masses—the longitudinal $m_l^*$ (along the axis) and the transverse $m_t^*$ (perpendicular to the axis). Their values in terms of the free electron mass $m$ are given as—$m_l^* = 0.98\,m$ and $m_t^* = 0.19\,m$. The valence band shows two degenerate maxima both located at $k = 0$ with spherical symmetry within validity of the ellipsoidal expansion (Fig. 12.107). The two effective masses are $0.49\,m$ and $0.16\,m$.

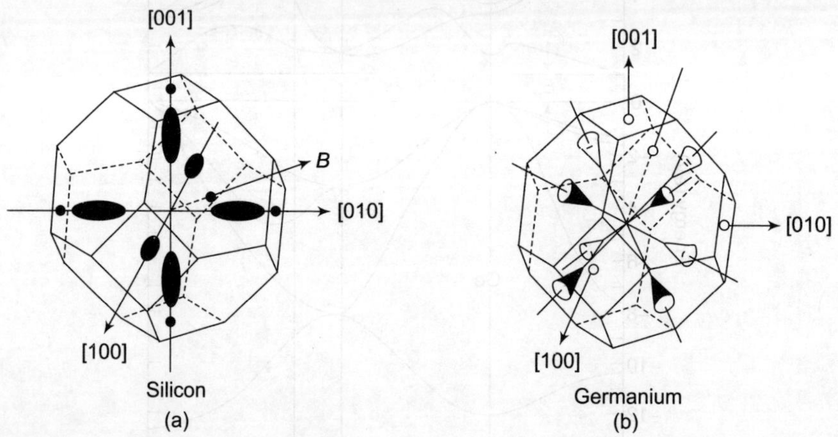


**Figure 12.108** (a) Constant energy surfaces for Si: There are six conduction band minima of cigar shape at points along the [100] directions. The energy surfaces are ellipsoids of revolution with elongation along the cube axes. (b) Constant energy surfaces for Ge: There are four symmetry related conduction band minima on the parallel hexagonal faces of the Brillouin zone. the energy surfaces are ellipsoids of revolution with elongation along the [111] directions

The conduction band minima in Ge occurs at the zone boundaries in the [111] directions. The minima on parallel hexagonal faces of the zone correspond to the same energy levels giving four symmetry-related conduction band minima [Fig. 12.108 (b)]. The constant energy surfaces are ellipsoids of revolution with elongation along the [111] directions and effective masses, $m_l^* = 1.57\,m$ and $m_t^* = 0.082\,m$. The two degenerate valence band maxima in this case give the characteristic effective masses of $0.28\,m$ and $0.44\,m$.

Properties of holes are equally important as those of electrons for the study of semiconductors. Details of the cyclotron resonance studies show that the structure of the valence band maximum near $\Gamma (k = 0)$ in Si and Ge is more complicated than what appears in Fig. 12.107. In addition to the two degenerate bands at $\Gamma$, there is third one that is split off slightly by energy band $\Delta$ towards a lower energy band (Fig. 12.109). The splitting is caused by the spin-orbit interaction. On the basis of the effective mass values, the two degenerate bands are attributed to the light and heavy holes.

The holes associated with the identity of the third band are named as 'split-off holes'. The split-off energy $\Delta$ has been estimated at 0.044 eV in Si and 0.29 eV in Ge.

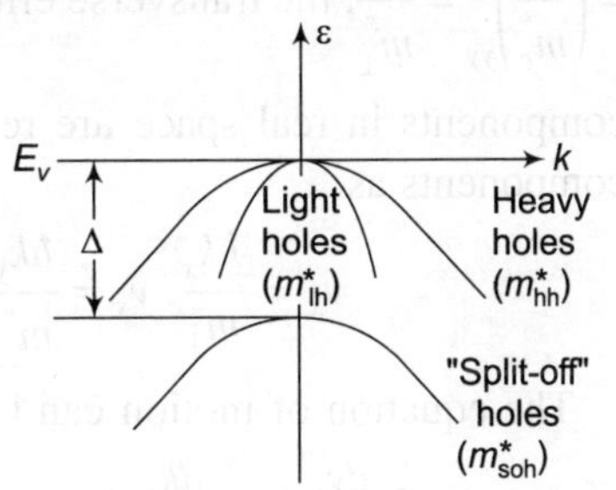


**Figure 12.109** Structure of the valence band (qualitative) in Si or Ge near the top showing the hole bands with the spin-orbit interaction taken into account. The split-off hole band is separated from the top at $k = 0$ by $\Delta$, contributed by the spin-orbit interaction

## 12.55 TYPICAL ENERGY BANDS IN SEMICONDUCTORS AND CYCLOTRON RESONANCE EXPERIMENT

The energy bands in semiconductors will completely depend on the number of electrons excited into the conduction band and the holes so created in the valence band. The electrons will lie almost in those energy levels which are near to the conduction band minima and similarly the holes will be found almost in levels close to the valence band maxima. Therefore, the energy band diagram can be simply depicted by the quadratic equations in the neighbourhood of such extrema. This will give the constant energy surface about the extrema as ellipsoidal in shape. Typical energy band diagram, without taking into account the spin-orbit coupling, in case of germanium is presented in Fig. 12.110. The valance band minima is at $\vec{k} = 0$ ($\Gamma$ point), where two degenerate bands with different curvatures meet, giving rise to pocket of holes with distinct effective masses. These effective masses are experimentally measured by the technique of cyclotron resonance.

Let us initially investigate for simplicity the case of a single ellipsoidal energy surface. If we measure the wave vector from the band edge, then the energy can be written in the form

$$E\,(\vec{k}) = \hbar^2 \left( \frac{k_x^2 + k_y^2}{2m_\perp^*} + \frac{k_z^2}{2m_\parallel^*} \right) \qquad ...(1)$$

Here the tensor component of the effective mass are $\left(\frac{1}{m_c^*}\right)_{zz} = \frac{1}{m_\parallel^*}$, the longitudinal effective mass and $\left(\frac{1}{m_c^*}\right)_{xx} = \left(\frac{1}{m_c^*}\right)_{yy} = \frac{1}{m_\perp^*}$, the transverse effective mass. The velocity components in real space are related to the wave vector components as

$$v_x = \frac{\hbar k_x}{m_\perp^*},\ v_y = \frac{\hbar k_y}{m_\perp^*},\ v_z = \frac{\hbar k_z}{m_\parallel^*} \qquad ...(2)$$

The equation of motion can be written as

$$m_\perp^* \frac{dv_x}{dt} = \hbar \frac{dk_x}{dt} = e\varepsilon_x + e(v_y B_z - v_z B_y) \qquad ...(3a)$$

$$m_\perp^* \frac{dv_y}{dt} = \hbar \frac{dk_y}{dt} = e\varepsilon_y + e(v_z B_x - v_x B_z) \qquad ...(3b)$$

$$m_\parallel^* \frac{dv_z}{dt} = \hbar \frac{dk_z}{dt} = e\varepsilon_z + e(v_x B_y - v_y B_x) \qquad ...(3c)$$

Now $B_z = B\cos\theta$

$B_x = B\sin\theta$

$B_y = B\cos 90° = 0$

(i.e., $\vec{B}$ lies in the $x$-$z$ plane). Electric field $\vec{\varepsilon}$ is along the $y$-axis: $E_y = E_0 e^{iwt}$, $\varepsilon_x = \varepsilon_z = 0$.

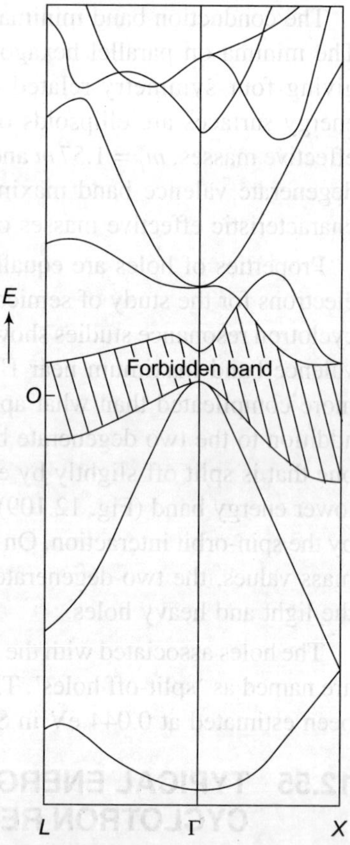


**Figure 12.110** Typical energy bands of germanium (spin orbit coupling is neglected)

Substituting the values for ε, $v$'s and $B$'s we get

$$m_\perp^* \frac{dv_x}{dt} = e\,v_y B\cos\theta - 0$$

or
$$\frac{dv_x}{dt} = \frac{eB}{m_\perp^*} v_y \cos\theta \equiv w_\perp v_y \cos\theta \qquad ...(4a)$$

$$\frac{dv_y}{dt} = \frac{e\varepsilon_0 e^{i\omega t}}{m_\perp^*} + \frac{eB}{m_\perp^*} v_z \sin\theta - \frac{eB}{m_\perp^*} v_x \cos\theta$$

$$\equiv \frac{eE_0 e^{i\omega t}}{m_\perp^*} + \omega\, v_z \sin\theta - \omega_\perp v_x \cos\theta \qquad ...(4b)$$

$$\frac{dv_z}{dt} = -\omega_{11} v_y \sin\theta \qquad ...(4c)$$

where
$$\omega_\perp = \frac{eB}{m_\perp^*} \text{ and } \omega_\parallel = \frac{eB}{m_\parallel^*} \qquad ...(5)$$

We assume oscillatory solutions of the type

$$\left.\begin{aligned} x &= x_o e^{i\omega t} \\ y &= y_o e^{i\omega t} \\ z &= z_o e^{i\omega t} \end{aligned}\right| \qquad ...(6)$$

Then, from Eqs. (4) and (6)

$$-\omega^2 x_o e^{i\omega t} = \omega \frac{dy}{dt} \cos\theta = \omega_\perp y_o (i\omega) e^{i\omega t} \cos\theta \qquad ...(7a)$$

$$-\omega^2 y_o e^{i\omega t} \left( = \frac{dv_y}{dt} \right) = \frac{e\varepsilon_0}{m_\perp^*} + \omega_\perp v_z \sin\theta - \omega_\perp v_x \cos\theta \qquad ...(7b)$$

$$-\omega^2 z_o e^{i\omega t} \left( = \frac{dv_z}{dt} \right) = -\omega_\parallel v_y \sin\theta = -\omega_\parallel (i\omega) y_o e^{i\omega t} \sin\theta$$

$$= -\omega\omega_\parallel y_o \sin\theta \, e^{i\omega t} \qquad ...(7c)$$

$$\Rightarrow \qquad z_o = i \frac{\omega_\parallel}{\omega} y_o \sin\theta \qquad ...(7d)$$

From Eq. (7a) $\qquad -\omega^2 x_o = i\omega\omega_\perp y_o \cos\theta$

or $$x_o = -i \frac{\omega_\perp}{\omega} y_o \cos\theta \qquad ....(8a)$$

From Eq. (7b)

$$-\omega^2 y_o e^{i\omega t} = \frac{eE_o}{m_\perp^*} + \omega_\perp (i\omega z_o) e^{i\omega t} \sin\theta$$

$$- \omega_\perp (i\omega x_o) e^{i\omega t} \cos\theta$$

or $$-\omega^2 y_o = \frac{e\varepsilon_o}{m_\perp^*} + i\omega\omega_\perp z_o \sin\theta - i\omega\omega_\perp x_o \cos\theta$$

$$= \frac{e\varepsilon_o}{m_\perp} + i^2 \omega\omega_\perp \frac{\omega_\parallel}{\omega} y_o \sin^2\theta + i^2 \frac{\omega\omega_\perp^2}{\omega} y_o \cos^2\theta$$

(Using Eq. 7d for $z_o$ and 8a for $x_o$),

$$= \frac{e\varepsilon_o}{m_\perp} - \omega_\parallel \omega_\perp y_o \sin^2\theta - \omega_\perp^2 y_o \cos^2\theta$$

$$= \frac{e\varepsilon_o}{m_\perp} - y_o (\omega_\perp \omega_\parallel \sin^2\theta - \omega_\perp^2 y_o \cos^2\theta)$$

Thus, $$y_o = -\frac{e\varepsilon_o}{m_\perp^*} \frac{1}{\omega^2 - \omega_\perp^2 \cos^2\theta - \omega_\perp^2 \omega_\parallel \sin^2\theta} \qquad ...(9a)$$

Substituting Eq. (9) for $y_o$ in Eq. (8a)

$$x_o = \frac{e\varepsilon_o}{m_\perp^*} \frac{i\omega_\perp \cos\theta}{\omega(\omega^2 - \omega_\perp^2 \cos^2\theta - \omega_\perp \omega_\parallel \sin^2\theta)} \quad \text{...(9b)}$$

and

$$z_o = \frac{e\varepsilon_o}{m_\perp^*} \frac{i\,\omega_\parallel \sin\theta}{\omega(\omega^2 - \omega^2 \cos^2\theta - \omega_\perp \omega_\parallel \sin^2\theta)} \quad \text{...(9c)}$$

The cyclotron resonance takes place when

$$\omega^2 = \omega_\perp^2 \cos^2\theta + \omega_\perp \omega_\parallel \sin^2\theta$$

or

$$\omega^2 = e^2B^2 \left[\frac{\cos^2\theta}{m^{*2}_\perp} + \frac{\sin^2\theta}{m_\perp^* m_\parallel^*}\right] \quad \text{...(10)}$$

Then, the cyclotron effective mass is given by

$$\left(\frac{1}{m_c^*}\right)^2 = \left(\frac{\cos^2\theta}{m_\perp^{*2}}\right) + \left(\frac{\sin^2\theta}{m_\perp^* m_\parallel^*}\right) \quad \text{...(11)}$$

Thus, due to magnetic field, charged particles move in circular orbits at an angular frequency $\omega = \frac{eB}{m_c^*}$ (independent of their energy) and thus, can absorb energy from a suitably phased radio frequency electric field $\varepsilon_o e^{i\omega t}$ at this frequency. Some classic cyclotron resonance results obtained at Berkeley (in 1955) by Dresselhaus, Kip and Kittel are given in Fig. 12.111.

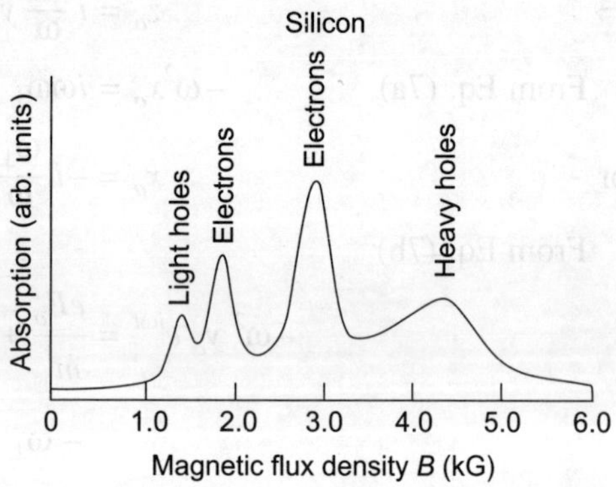


**Figure 12.111** Sketch of cyclotron resonance for silicon (near 24 × $10^3$ Mc/s and 4°K, $\vec{B}$ at 30° with [100] and in (110) plane)

In the experiments on Si, we impose a magnetic field and seek the natural frequencies of oscillatory motion. Cyclotron resonance absorption will occur when electric field with polarization in the plane of motion has a frequency ω equal to the frequency of motion due to the magnetie field (Eq.10). Thus, resonances appear (Fig. 12.111) whenever there is absorption of e.m. energy by the electrons (in the magnetic field multiples of cyclotron frequency.

At the resonance frequency, Eq. (9) indicates that the particle orbits are elliptical in nature lying in a plane perpendicular to the field $\vec{B}$. For ellipsoidal energy surfaces, it is found that the resonance frequency depends on the orientation of the sample relative to the magnetic field. For silicon, if the field $\vec{B}$ is along the $z$-direction's then θ = 0° for the two ellipsoids whose major axes are along the $k_z$ direction and θ = 90° for the other four ellipsoids as shown in Fig. 12.112. In terms of the free electron mass, the effective mass along the axis (i.e., the longitudinal effective mass) is $m_{\parallel}^* \cong 1.0\ m_e$ while the effective masses perpendicular to the axis (the transverse effective mass) are $m_{\perp}^* \simeq 0.2\ m_e$.

The constant energy surfaces $E(k)$ (Eq. (1)) near the conduction band minima for Si, Ge and GaAs are depicted in Fig. 12.112.

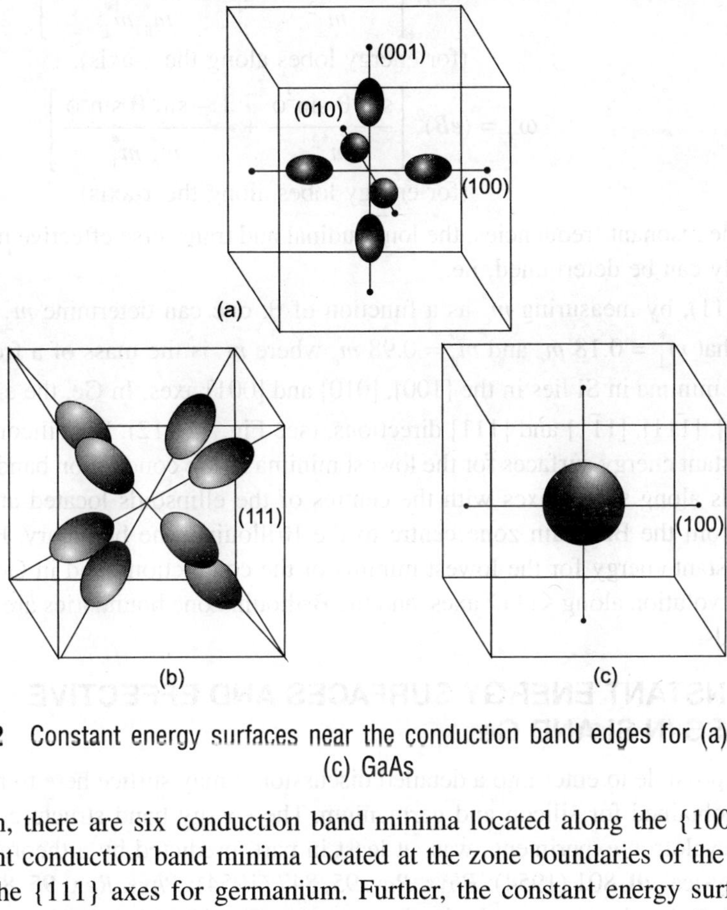


**Figure 12.112** Constant energy surfaces near the conduction band edges for (a) Si (b) Ge and (c) GaAs

For silicon, there are six conduction band minima located along the {100} axes, while there are eight conduction band minima located at the zone boundaries of the first Brillouin zone along the {111} axes for germanium. Further, the constant energy surfaces near the bottom of the conduction bands are ellipsoidal for Si and Ge and spherical for GaAs.

In the passing, it is to be mentioned here that if direction of $\vec{B}$ is as shown in Fig. 12.113 such that

$$B_x = B \sin\theta \cos\phi$$

$$B_y = B \sin\theta \sin\phi$$

$$B_z = B \cos\theta$$

then, there will be three resonant frequencies (instead of ω), viz., $\omega_x$, $\omega_y$ and $\omega_z$ which can be determined by

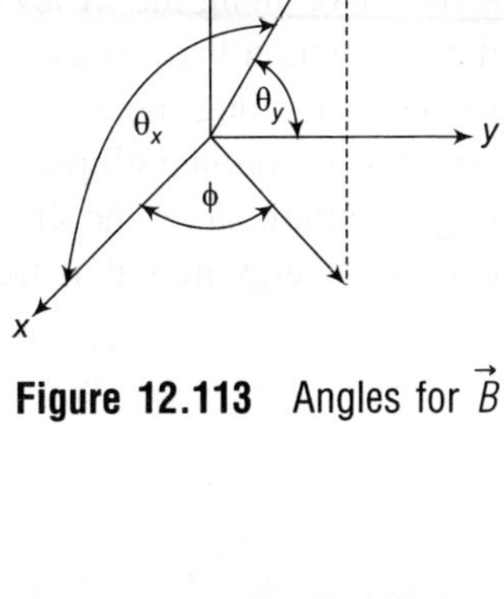


**Figure 12.113** Angles for $\vec{B}$

$$\omega_z^2 = (eB)^2 \left[\frac{\cos^2\theta}{m_\perp^{*2}} + \frac{\sin^2\theta}{m_\parallel^* m_\perp^*}\right]$$

(for energy lobes along the $z$-axis).

$$\omega_x^2 = (eB)^2 \left[\frac{\sin^2\theta\cos^2\phi}{m_\perp^{*2}} + \frac{1-\sin^2\theta\cos^2\phi}{m_\parallel^* m_\perp^*}\right]$$

(for energy lobes along the $x$-axis).

and
$$\omega_y^2 = (eB)^2 \left[\frac{\sin^2\theta\sin^2\phi}{m_\perp^{*2}} + \frac{1-\sin^2\theta\sin^2\phi}{m_\parallel^* m_\perp^*}\right]$$

(for energy lobes along the $x$-axis).

Knowing the resonant frequencies, the longitudinal and transverse effective masses $m_\parallel^*$ and $m_\perp^*$ respectively can be determined, i.e.,

From Eq. (11), by measuring $m_c^*$ as a function of θ, one can determine $m_\perp^*$ and $m_\parallel^*$. It is found for Si that $m_\perp^* = 0.18\ m_e$ and $m_\parallel^* = 0.98\ m_e$ where $m_e$ is the mass of a free electron.

The energy minima in Si lies in the [100], [010] and [001] axes. In Ge, the energy minima lie in the [111], $[\bar{1}11]$, $[1\bar{1}1]$ and $[11\bar{1}]$ directions. (see Fig. 12.112). For silicon, Fig. 12.112 (a) shows constant energy surfaces for the lowest minima of the conduction band for Si. There are 6 ellipsoids along ⟨100⟩ axes with the centres of the ellipsoids located at about 3/4 of the distance from the Brillouin zone centre to the Brillouin zone boundary. Figure 12.112 (b) shows constant energy for the lowest minima of the conduction band in Ge. There are 8 ellipsoids of revolution along ⟨111⟩ axes, and the Brillouin zone boundaries are in the middle of the ellipsoids.

## 12.56 CONSTANT ENERGY SURFACES AND EFFECTIVE MASS IN Si AND Ge

Since it is not possible to enter into a detailed discussion it may suffice here to mention some of the results obtained for silicon and germanium. The energy band structure of Si and Ge derived from cyclotron experiments was, at least in part, predicted by a theoretical study of F. Herman, Physica, **20**, 801 (1954); *Phys. Rev*, 95, 847 (1954); *Phys. Rev.*, **95**, 847 (1954). As an example of the band structures obtained, we give in Fig. 12.114 the energy as function of the wave vector along the ⟨100⟩ direction for silicon. It is observed that the minimum energy in the conduction band does not correspond to $k = 0$ but that there are in all six minima located somewhere along the ⟨100⟩ direction for silicon. It is observed that the minimum energy in the conduction band does not correspond to $k = 0$ but that there are all six minima located somewhere along the six ⟨100⟩ axes. In the vicinity of these minima, the constant energy surfaces are prolate ellipsoids of revolution. Similar minima occur for the conduction band in germanium along the ⟨111⟩ axes. Choosing one of these minima as origin, the surface of constant energy may thus be represented by an expression of the from

$$E(k) = \hbar^2\left(\frac{k_x^2 + k_y^2}{2m_t} + \frac{k_z^2}{2m_l}\right) \quad ...(1)$$

where $m_t$ and $m_l$ are called, respectively, the transverse and longitudinal electron mass. For Si and Ge the cyclotron resonance experiment lead to the following results at 4°K:

Silicon: $m_l = 0.19m;\ m_l = 0.98m$

Germanium: $m_t = 0.82m;\ m_l = 1.57m$

where $m$ is the free electron mass.

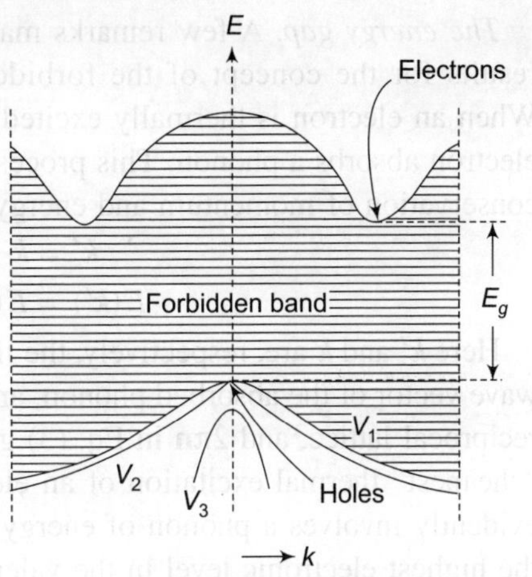


**Figure 12.114** Schematic representation of the energy band structure in Si along a <100> Axis. [After F. Herman, *Proc. IRE*, 43, 1703 (1955)]

The maximum energy for the valence band in both silicon and germanium occurs for $k = 0$, according to the results of cyclotron resonance experiments; furthermore, this maximum is common to two bands which meet at $k = 0$. The constant-energy surfaces near $k = 0$ for these two bands are warped and are given by the expression

$$E(k) = -\frac{\hbar^2}{2m}[Ak^2 \pm \sqrt{B^2k^4 + C^2(k_x^2k_y^2 + k_y^2k_z^2 + k_z^2k_x^2)}] \quad ...(2)$$

where $A$, $B$, and $C$ are constants. The negative and positive roots correspond, respectively, to the highest ($V_1$) and second highest ($V_2$) valence band. If one approximates the warped surface by sphere, one may determine the average hole mass in the two bands from the experimental value of $A$, $B$, and $C$. In this approximation, one obtains

Silicon: $m^*_{v_1} = 0.49m;\ m^*_{v_2} = 0.16m$

Germanium: $m^*_{v_1} = 0.28m;\ m^*_{v_2} = 0.044m$

We should note here that the form of expression (2) was indicated by the theory of spin-orbit splitting for these crystals.[12]

It is observed from Fig. 12.114 that there is a third valence band $V_3$ which is separated from the $V_1$ and $V_2$ bands as a result of spin-orbit interaction. The maximum of the $V_3$ band lie slightly below that of the two other bands. Near the maximum of the $V_3$ band, the constant energy surfaces are spherical; the effective masses are:

Silicon: $m^*_{v_3} = 0.24m$

Germanium: $m^*_{v_3} = 0.077m.$

The energy difference between the top of the $V_3$ band and the common maximum of the $V_1$ and $V_2$ bands has been estimated to be 0.035 eV for Si and 0.28 eV for Ge. It will be evident that the relative hole populations of the $V_3$ and $V_1$, $V_2$ bands is a function of temperature.

[12] G. Dresselhaus, A. F. Kip, and C. Kittel, *Phys. Rev.*, **95**, 568 (1954); **98**, 368 (1955); R. J. Elliott, *Phys. Rev.*, **96**, 266 (1954); **96**, 280 (1954).

*The energy gap.* A few remarks may be made here about the consequences of the above results for the concept of the forbidden energy gap and its experimental determination. When an electron is thermally excited from the valence band into the conduction band, the electron absorbs a phonon. This process is governed by the *s* election rules corresponding to conservation of momentum and energy

$$k' = k + q + 2\pi n$$

$$E(k') = E(k) + \hbar\omega_q \qquad ...(3)$$

Here $k'$ and $k$ are, respectively, the final and the initial wave vector of the electron; **q** is the wave vector of the absorbed phonon, and $\hbar\omega_q$ is the energy of the phonon; **n** is a vector in the reciprocal lattice, and $2\pi\mathbf{n}$ in Eq. (3) guarantees that $\mathbf{k}'$ is a vector in the reduced zone. The "cheapest" thermal excitation of an electron from the valence band to the conduction band evidently involves a phonon of energy $\hbar\omega_q = E_g$ where $E_g$ is the energy difference between the highest electronic level in the valence band and the lowest level in the conduction band (see Fig. 12.114). Thus, $E_g$ may be obtained from the variation of the carrier concentration with temperature. We should note here that $E_g$ itself is a function of temperature (resulting from the expansion of the lattice).

It has been observed that the size of the energy gap $E_g$ (or the splitting between the valence and conduction bands) decreases with increase in the interatomic separation, i.e., at higher temperatures, the interatomic separation becomes longer (due to thermal expansion) and the gap $E_g$ becomes smaller. This fact is confirmed by the measured values of the band gap $E_g$ at different temperatures (Table 12.8).

**Table 12.8** Data on the Band Gap of Some Semiconductors. (*D* = direct gap, *I* = indirect gap)

| *Crystal* | *Type of gap* | $E_g$ *(eV)* | |
|---|---|---|---|
| | | *0°K* | *300° K.* |
| Si | I | 1.17 | 1.11 |
| Ge | I | 0.75 | 0.67 |
| GaAs | D | 1.52 | 1.43 |
| InP | D | 1.42 | 1.27 |
| CdS | D | 2.58 | 2.42 |
| CdSe | D | 1.84 | 1.74 |
| CdTe | D | 1.61 | 1.44 |
| ZnS | — | 3.91 | 3.6 |
| PbS | D | 0.29 | 0.34–0.37 |
| GaP | I | 2.32 | 2.25 |

## 12.57 BAND STRUCTURE AND VAN HOVE SINGULARITIES

Analogous to the thermal properties of the phonon system, in the case of electronic states, a knowledge of the DOS is sufficient to describe the energy content of the electron system. For certain electronic excitation mechanisms, e.g., non-angle resolved photoemission spectroscopy, (in which the experiment effectively integrates over all $\vec{k}$-direction), one can often interpret the spectra in terms of the number of electron states per energy interval, $dE$.

Once the energy surfaces $E(\vec{k})$ of the band structure are known, then in analogy to phonon DOS, the DOS is obtained by integrating over an energy shell $\{E(\vec{k}), E(\vec{k}) + dE\}$, in $\vec{k}$-space:

$$dz = \frac{V}{(2\pi)^3} \int_{E}^{E+dE} d\vec{k}$$

where $V/(2\pi)^3$ is the density of states in $\vec{k}$-space. If the volume element $d\vec{k}$ is separated into an area element $d\vec{S}_E$ on the energy surface and a component $dk_\perp$ normal to this surface, i.e.,

$$d\vec{k} = d\vec{S}_E \, dk_\perp$$

then with $dE = |\nabla_k E| \, dk$, one has

$$D(E)\,dE = \frac{1}{(2\pi)^3}\left[\int_{E(k)=\text{const.}} \frac{dS_E}{|\nabla_k E(\vec{k})|}\right] dE \qquad ...(1)$$

(Note that $d\vec{S}_E$ is along the direction of $d\vec{k}_\perp$).

This DOS is given here in relation to the real volume $V$ of the crystal. It is also to be noted that due to spin degeneracy, each state can accommodate 2 electrons.

An important structure in the function $D(E)$ is introduced by those points in $\vec{k}$-space for which $|\nabla_k E|$ vanishes, i.e., where the energy surfaces are flat. These points are known as *van Hove singularities* or critical points. In three dimensions $D(E)$ does not become singular near to these critical points because an expansion of $E(\vec{k})$ about the extremum ($E \propto k^2$) implies that $|\nabla_k E|^{-1}$ has a $1/k$ singularity. Thus, the integration of the $E(k)$ surface (Eq. 1) yields a linear $k$ dependence. In three dimensions the DOS near to a critical point therefore has the form shown in Fig. 12.115.

DOS may be obtained by integrating in $k$-space over the first Brillouin zone for a calculated band-structure, and may then be compared with experimental data (e.g., from photoemission spectroscopy). These provide an important point of connection between calculated band structures and the experimental data.

Important contributions, i.e., maxima in the DOS, are clearly correlated with flat portions of the $E(\vec{k})$ curves along directions of high symmetry. Also evident is the absolute band gap between the fully occupied valence band states and (at low temperatures) unoccupied conduction band states. This forbidden region has a width of about 0.7 eV for Germanium.

In the integration over $k$-space the main contributions to the DOS are derived from the critical points. Since the critical points usually occur along lines of, or at points of high symmetry in $k$-space, this justifies for preferred representation of band structures along lines of high symmetry, e.g., $\Gamma X$, $\Gamma L$, etc. In the intermediate regions, one can expect only minor contributions.

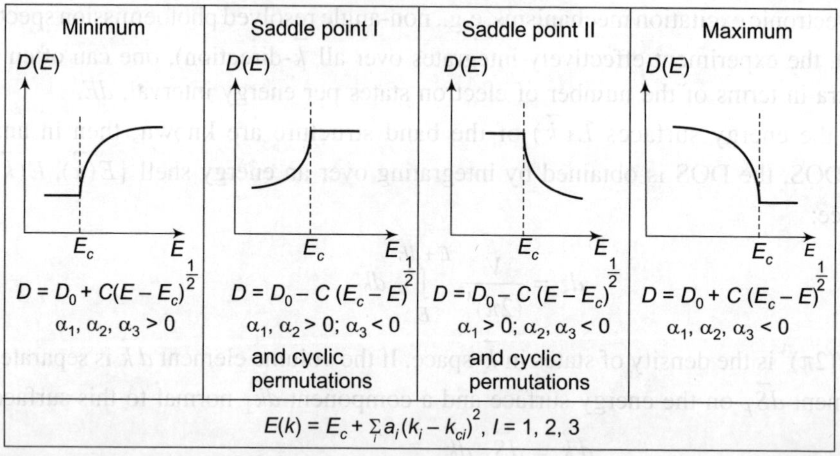


**Figure 12.115** Form of the density of states $D(E)$ in the vicinity of the four possible types of critical point in three dimensions. The energy of the critical points is denoted by $E_c$ and the corresponding $k$-space position by $k_{ci}$ ($i = 1, 2, 3$). In the parabolic approximation, the energy band has the form $E(k) = E_c + \Sigma_i a_i$ in the vicinity of a critical point, where $\alpha_i$ = const. The quantities $D_0$ and $C$ in the figure are also constants

The relationship between a calculated band structure and the corresponding DOS in case of semiconductor Germanium is shown is Fig. 12.116.

## 12.58 LIMIT OF THE BAND THEORY (MOTT TRANSITION)

Band models of solids are capable of explaining all the observed properties of metals. Yet, this model has a limitation which we are about to describe.

Consider the case of sodium (Na). This substance is a conductor because the 3's band is partially (or half) filled. Suppose Na is expanded somehow so that lattice constant $a$ can be increased arbitrarily. Would the material then remain conductor for an arbitrary value of $a$? The answer would be yes, because regardless of the value of $a$, $3s$ band would always be half full. Further, the model predicts that conductivity, $\sigma$, decreases as $a$ increases and the decrease is gradual.

In fact, however, this is not correct. As $a$ increases, a critical value $a_c$ is reached at which the conductivity drops to zero abruptly, rendering the solid as insulator and it remains so for all $a > a_c$. This is referred to as the metal-insulator of *Mott-transition.*

According to Band theory, Bloch electrons are assumed to be delocalized extending throughout the crystal. It is this delocalization which is responsible for metallic conductivity (The interaction between individual electrons is taken into account by Hartree–Fock approximation and it does not violate one-electron picture.)

However as $a$ increases, bandwidth decreases until it becomes quite small at sufficiently large $a$. Then band-model breaks down because it allows the presence of two (or more) electrons at the same lattice site.

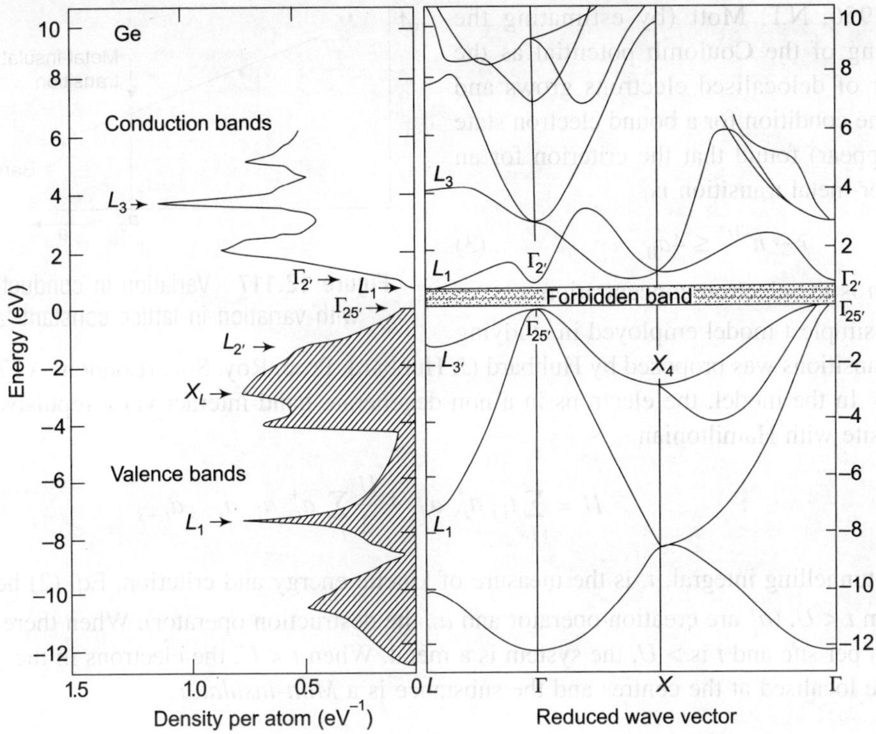


**Figure 12.116** Theoretically derived band structure $E(k)$ for germanium along directions of high symmetry (right), and the corresponding electronic density of states (left). A number of critical points, denoted according to their position in the Brillouin zone ($\Gamma$, $X$, $L$), can be seen to be associated with regions of the band structure where $E(k)$ has a horizontal tangent. The shaded region of the density of states corresponds to the states occupied by electrons [7.6]. [After: Harald Ibach and Hans Luth, *Solid State Physics* Springer Verlag (1996)]

In effect, for very large $a$, the proper electronic orbitals in a crystal are no longer of the Bloch-type. They are localised, centred around their respective sites which diminishes the effect of large Coulomb energy. As a result, the material becomes non-conducting.

Mean Coulomb interaction energy is $\overline{U} \sim e^2/\bar{r}$ and characteristic kinetic energy is given by

$$E_{\text{kin}} \sim \frac{\hbar^2}{m(\bar{r})^2} \quad ...(1)$$

At $E_{\text{kin}} < \overline{U}$, electrons become localised.

Then,
$$\bar{r} \geq \frac{\hbar^2}{me^2} \; (= a_0) \quad ...(2)$$

i.e., when the mean distance between the electrons is greater than the Bohr radius (in other words, when the overlapping of the wave functions of neighbouring centres must be small).

In 1956, N.F. Mott (by estimating the screening of the Coulomb potential as the number of delocalised electrons grows and using the condition for a bound electron state to disappear) found that the criterion for an insulator-metal transition is

$$\bar{r} \sim n^{-1/3} \leq 4a_0 \qquad ...(3)$$

where $n$ is the electron concentration.

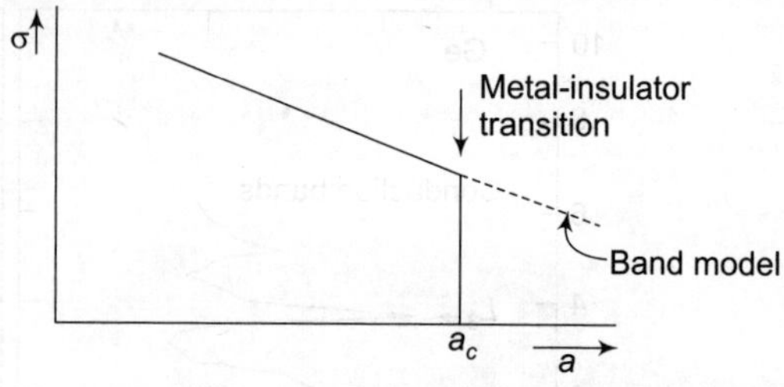


**Figure 12.117** Variation in conductivity with variation in lattice constant '*a*'

The simplest model employed in studying Mott transitions was proposed by Hubbard (J. Hubbard, Proc. Roy. Soc. (London) A 276, 238 (1963)). In the model, the electrons in a non-degenerate band interact via a repulsive force at one site with Hamiltonian

$$H = \sum_{ijs} t_{ij}\, a_{is}^{+}\, a_{js} + \frac{U}{2} \sum_{is} a_{is}^{+}\, a_{is}\, a_{i,-s}\, a_{i,-s} \qquad ...(4)$$

The tunnelling integral, $t$, is the measure of kinetic energy and criterion, Eq. (2) here has the form $t < U$. ($a_i^{+}$ are creation operator and $a_i$, the destruction operator). When there is one electron per site and $t$ is $> U$, the system is a metal. When $t < U$, the electrons in the ground state are localised at the centres and the substance is a *Mott-insulator*.

## EXERCISE

12.1 Explain formation of energy band in
   (a) Sodium atom
   (b) Sodium crystal.

12.2 Explain formation of energy bands in
   (a) Li atom
   (b) Li crystal

12.3 Explain energy bands in diamond.

12.4 Explain the concept of effective mass and derive an expression for the same. What is the significance of negative effective mass?

12.5 Explain the concept of holes and the hole band construction.

12.6 How does the existence of energy bands lead to the classification of solids into metals, semiconductors and insulators?

12.7 Give a comparison between semiconductors, metals and insulators.

12.8 Explain what are Bloch waves. What do you understand by crystal momentum?

12.9 State Bloch theorem. Give a proof of the Bloch theorem.

12.10 Discuss the Kronig–Penney model for linear lattice. How does it lead to the formation of energy bands in solids?

12.11 Discuss Kronig–Penney model for the energy band structure of solid. Show that each band can accommodate 2 $N$ electrons, where $N$ is the total number of atoms in the crystal.

12.12 Derive an expression for the velocity of the Bloch electron, or prove that $\vec{v} = \frac{1}{\hbar} \nabla_{\vec{k}} E_n(\vec{k})$.

12.13 Discuss electron motion in fields and explain: hole-orbit, electron-orbit and open orbits.

12.14 With reference to energy bands in one dimension, explain the formation of extended, reduced, and repeated zone schemes. Also give the sketches.

12.15 Describe the nearly free electron model for determining the electron energy bands and show that the model leads to finite discontinuities in energy at the zone-boundaries.

12.16 Describe the tight binding approximation for calculating the energy states of an electron in a solid.

12.17 Discuss the wave equation for an electron in a linear periodic potential

$$U(x) = \sum_{G} U_G \, e^{iGx}$$

($G$'s are the reciprocal lattice vectors)

12.18 Explain the reduced zone scheme in one dimension and give a plot of energy $E$ reduced wave-vector, $\vec{k}$.

12.19 Explain the concept of momentum and crystal momentum for a Bloch electron.

12.20 Find the constant energy lines for a tightly bound electron in a two-dimensional square lattice with interatomic separation $a$.

12.21 Explain the cellular method. What do you understand by muffin-tin potential?

12.22 Describe the augmented plane wave method.

12.23 Describe the orthogonalized plane wave (OPW) method for studying the bond structure of solids.

12.24 Write short notes on (a) van Hove singularitions and (b) Mott transmittion.

CHAPTER

# 13

# Magnetic Properties

## 13.1 INTRODUCTION

Matter is essentially electrical in nature. In nineteenth century, it was suggested by Ampere that atomic currents are responsible for the magnetic effects of matter. At that time, very little was known about the microscopic constituents of matter.

Main cause of magnetic properties of matter is a property of the electron known as its spin. In a sense, the word spin is a misnomer, because the electron does not really spin on an axis (being a fermion), but the term, in fact, describes the fact that an intrinsic angular momentum is associated with an electron, and a change having angular momentum also possesses a magnetic dipole moment in the behaviour of these moments in an external magnetic field contributes to macroscopic magnetic properties of matter.

As a consequence of the fact that magnetic phenomena originate from atomic currents, we should expect that when an external magnetic field is turned on, an emf will result on the atomic level which, by Lenz's laws, will be such that the magnetic field of the induced current will tend to cancel the external field. This is known as a *diamagnetic effect*, and always occurs when an external magnetic field is applied to matter.

In some substances there are no net atomic currents and effect of electron spin cancel out, so that normally there is no magnetic moment per atom. In such cases, the only magnetic effect is the diamagnetic one mentioned above.

In other substances, however, due either to currents or to uncancelled spins, there is a net magnetic moment per atom. In an external field, these will tend to line up in the direction of the field, and the external magnetic field is supplemented by the field of the aligned moments. This is known as a *paramagnetic effect*.

It usually exceeds the diamagnetic response which is also present. Paramagnetic substances increase $\vec{B}$, whereas diamagnetic ones decrease $\vec{B}$.

Diamagnetic effect is associated with neutral cores of the atoms whereas paramagnetism is associated with moments arising because of free electrons.

In some substances (notably iron), paramagnetic effects are greatly enhanced (by factors of $10^3$ to $10^6$) due to the fact that thousands of adjacent atoms are strongly coupled together and act cooperatively. These substances exhibit *ferromagnetic effect*. Since so many atoms are coupled together, the structure of ferromagnetic materials plays an important role in this effect. Ferromagnetic materials are of great technological importance.

The above three classes of materials can be described further as the following.

## 13.2 DIAMAGNETISM

*Diamagnetism* is the weakest type of magnetism. It occurs in all materials when placed in an external magnetic field, even those we generally think of as non-magnetic (such as wood, plastic, or water). Diamagnetism is explained by Faraday's law of induction: the application of an external magnetic field induces atomic magnetic dipole moments, which align in a direction to generate a magnetic field to oppose the change. Therefore, the magnetization produced in diamagnets is antiparallel to the applied field (Fig. 13.1a).

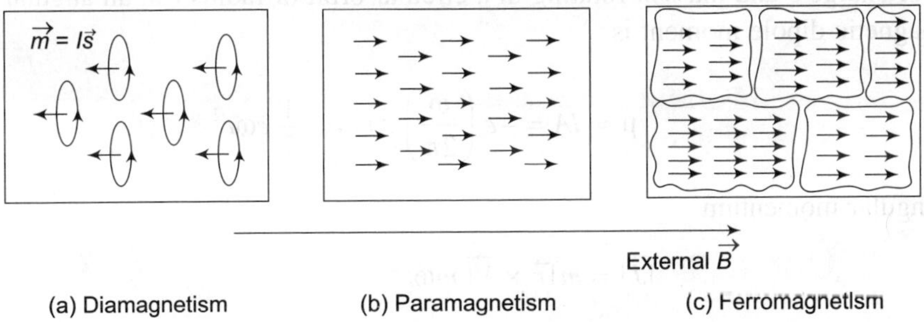


**Figure 13.1** (a) Diamagnetism: Alignment of induced dipole moments anti-parallel to the applied field, (b) Paramagnetism alignment of permanent dipole moments parallel to the applied field, and (c) Ferromagnetism: Alignment of permanent dipole moments parallel to applied field

## 13.3 PARAMAGNETISM

*Paramagnetism* occurs in materials that have unpaired electrons, such as Al or Na, and thus have permanent magnetic dipole moments due to the spin of the unpaired electron. In the absence of an external field, the permanent dipole moments point in random direction because of thermal motion, and so the material has no net magnetism. An external field aligns the dipole moments with the field, resulting in magnetization that is parallel to the external field (Fig. 13.1b). Paramagnetism overwhelms the much weaker diamagnetic effects.

## 13.4 FERROMAGNETISM

*Ferromagnetism* occurs only in some materials, such as iron, cobalt, nickel (Fe, Co, Ni) and produces the strongest magnetic fields like paramagnetic materials, ferromagnets have permanent atomic magnetic dipole moments from unpaired electron spins. The special property that sets them apart is the presence of so called *domains*, which are regions within the material where the magnetic dipole moments are aligned, even in the absence of an applied magnetic field. These regions (domains) are randomly oriented in the absence of an external field, thus the material has no net magnetism. The effect of an external field is to align the domains producing a strong magnetization ($\vec{M} = n\vec{m}$ [$Am^{-1}$] for $n$ dipoles per unit volume) (Fig. 13.1 (c)).

When the applied field is removed, the domains remain aligned. Therefore, ferromagnets retain their magnetization, thus producing permanent magnets. Heat, mechanical impacts, or reversed external fields can reduce the magnetization.

## 13.5 ORIGIN OF MAGNETIC PROPERTIES (IN AN ATOM)

We know that a small circular current-carrying loop has a magnetic moment associated with it. So motion of electrons in atoms is responsible for magnetism and quantised nature of electronic motion gives rise to the fundamental unit of magnetic dipole moment, the *Bohr magnetron*, $\mu_B$.

For a circular loop of area $A$ and carrying a current $I$, the dipole moment is $IA$. For an electron of charge $e$ and mass $m$ rotating in a circular orbit of radius $r$ at an angular velocity $\omega$, the magnetic dipole moment is

$$\mu = IA = -e\left(\frac{\omega}{2\pi}\right)\pi r^2 = -\frac{1}{2}\,e\omega r^2 \qquad \text{...(1)}$$

The angular momentum

$$|\vec{J}| = m|\vec{r} \times \vec{V}|\ m\omega r^2 \qquad \text{...(2)}$$

therefore,

$$\vec{\mu} = \left(\frac{e}{2m}\right)\vec{J} \qquad \text{...(3)}$$

The angular momentum is quantised in units of $h/(2\pi)$ where $h$ is Planck's constant. Therefore, the lowest non-zero value for $|\vec{\mu}|$ is $\mu_B = eh/4\pi m = 9.2741 \times 10^{-24}\ JT^{-1}$ (SI units) or $9.2741 \times 10^{-21}$ erg $Oe^{-1}$ (CGS units).

No electron can have a magnetic moment below $\mu_B$.

Thus, orbiting/spinning of electrons on the whole may impart a permanent magnetic moment to atoms.

In general,

$$\vec{\mu} = \gamma\vec{J} \qquad \text{...(4)}$$

where $\gamma$ is called the *magnetomechanical ratio*. For pure orbital motion of an electron, $\gamma = (-e/2m)$ and for pure electron spin, $\gamma = (-e/m)$.

Materials that lack permanent magnetic dipoles are called *diamagnetic*. If there are permanent magnetic dipoles associated with the atoms of a material, the material may be *paramagnetic, ferromagnetic, antiferromagnetic* or *ferromagnetic* depending on the interaction between the individual dipoles.

If the interaction between the atomic permanent dipole moments is zero or negligible, the material will be *paramagnetic*. A fundamental requirement for the existence of paramagnetism in solids is that the individual magnetic moments possess some degree of isolation. If they do not and if their wave functions overlap appreciably, then Pauli's exclusion principle will take them over and pair them up.

Magnetism is inseparable from quantum mechanics. *Ferromagnetism*, which is very strong, occurs when quantum mechanical exchange interactions align adjacent magnetic moments in the same direction. If the exchange interaction aligns the moments in opposite directions and only one type of moment is present, the material is called *antiferromagnetic*. If two (or more) types of moments are present (in opposite direction) so that there is a net moment equal to the difference, the material is called *ferrimagnetic* (Fig. 13.2).

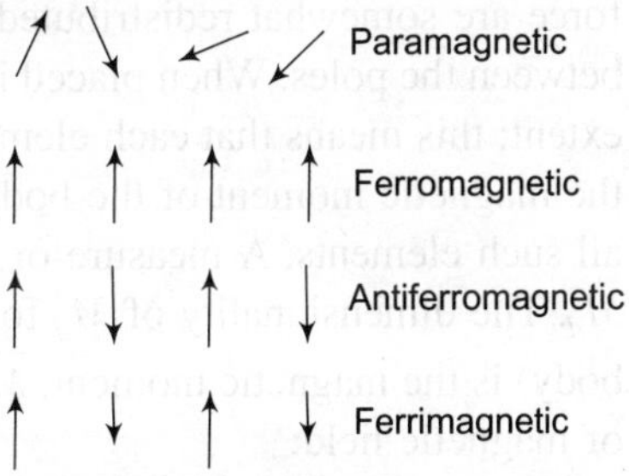


**Figure 13.2** Schematic of paramagnetic, ferromagnetic, antiferromagnetic and ferro-magnetic arrangement of spins

## 13.6 WHICH PROPERTY CHARACTERIZED THE MAGNETIC MATERIALS?

Magnetic materials are characterized by a non-zero value of the *time average microscopic current density* $\vec{j}_{\mu}(\vec{r})$ or equivalently, by a nonzero *time averaged microscopic magnetization* $M_{\mu}(\vec{r})$ at least in some region(s) within the primitive cell (*pc*).

Magnetic materials are classified as *antiferromagnetic* (if $\vec{m}_{pc} = 0$), or *ferromagnetic* (if $\vec{m}_{pc} \neq 0$), according to the value of the magnetic moment $\vec{m}_{pc}$, of the magnetic primitive cell,

where
$$\vec{m}_{pc} \equiv \int_{V_{pc}} \vec{M}_{\mu}(\vec{r})\, d^3r \qquad ...(1)$$

Ferromagnetic materials can be further divided into *simple ferromagnetic,* where $M_{\mu}(\vec{r})$ does not change sign or direction within the primitive magnetic cell, and *ferromagnetic*, where $M_{\mu}(\vec{r})$ does change direction (or even sign) within the primitive magnetic cell so that $|\vec{m}_{pc}|$ turns out to be smaller than with it would be if $M_{\mu}(\vec{r})$ did not change direction. Quite often it is more convenient to introduce magnetic moment of each atom *a*, i.e., $\vec{m}_a$, where

$$\vec{m}_a \equiv \int_{V_a} \vec{M}_{\mu}(\vec{r})\, d^3r \qquad ...(2)$$

and the integration is over the volume $V_a$ corresponding to atom *a*. Obviously,

$$\vec{m}_{pc} = \sum_a \vec{m}_a \; (a = \text{number of atoms in } pc)$$

## 13.7 MAGNETIC SUSCEPTIBILITY

Diamagnetism is associated with neutral cores of the atoms, whereas paramagnetism is associated with moments arising because of free electrons. At high temperature, materials are either diamagnetic or paramagnetic.

If a piece of some material is placed between the poles of a magnet, the magnetic lines of force are somewhat redistributed as compared with their distribution when there is nothing between the poles. When placed in a magnetic field, all substances are magnetized up to some extent; this means that each element in the volume of a body behaves as a small magnet, and the magnetic moment of the body (as a whole) is a vector sum of the magnetic moments of all such elements. A measure of magnetization of the body is the magnetic moment density $M_d$. The dimensionality of $M_d$ follows from the fact that $M_dV$ (where $V$ is the volume of the body) is the magnetic moment, $\vec{M}$. It follows that the dimensionality of $M_d$ is identical to that of magnetic field:

$$[M_d] = [H] \quad \text{...(1)}$$

A non-zero magnetization ($M$) appears only in response to a magnetic field, i.e.,

$$M = \chi\, H \quad \text{...(2)}$$

where $\chi$ is termed the magnetic susceptibility of the material in general, it is a tensor. For a homogeneous and isotropic material, it is a scalar (when the field is not too strong). It measures the magnetic moment $\vec{M}$ in an applied field $\vec{H}$.

## 13.8 MAGNETIC QUANTITIES

The magnetic induction or magnetic flux density ($B$) due to a magnetic field applied in vacuum is related to field strength $\vec{H}$ by

$$\vec{B} = \mu_0\, \vec{H} \quad \text{...(1)}$$

where $\vec{H}$ is in A/m. $\mu_0$ is called the permeability of free space.

$$\mu_0 = 1 \text{ (in e.m.u)}$$

$$= 4\pi \times 10^{-7} \text{ H/M (in MKS)} \quad \text{...(2)}$$

The unit of $\vec{B}$ is Gauss (in emu) or tesla (in MKS). The two systems are related as

$$1\text{T} = \frac{1\text{ Wb}}{\text{m}^2} = 10^4 \text{ Gauss} \quad \text{...(3)}$$

where Wb (weber) is the unit of magnetic flux $\phi$ defined by

$$\phi = \iint \vec{B}\cdot d\vec{a} \quad \text{...(4)}$$

If the field is applied in a solid, the magnetic induction is given by

$$B = \mu H \quad \text{...(5)}$$

where $\mu$ is the permeability of the solid material. The ratio $\mu/\mu_0$ called the relative permeability, $\mu_r$. The vectors $\vec{B}$, $\vec{H}$ and $\vec{M}$ are related as

$$\vec{B} = \mu_0\, \vec{H} + 4\,\pi M \text{ (in emu)} \quad \text{...(6)}$$

$$= \mu_0 (\vec{H} + \vec{M}) = \mu H$$

$$= \mu_0 (H + \chi H) = \mu H$$

or $\mu_0 H (1 + \chi) = \mu H, (\mu = \mu_0 \mu_r)$

$\Rightarrow$ $1 + \chi = \mu_r$ or $x = \mu_r - 1.$ ...(7)

## 13.9 MAGNETIC MOMENT OF A SINGLE FREE ION

The magnetic properties of solid materials are determined fundamentally by the characteristics of the individual atoms present and by the electrons associated with these atoms. While trying to understand the properties that are observed experimentally, we are concerned with the way in which all the atoms and electrons in the solid interact to give the whole effect.

The four quantum numbers $n$, $l$, $m_p$, and $m_s$ describe the state of an electron in an atom. Transitions between different $n$ levels do not concern us in considering the magnetic properties as the energy differences involved are too great and give rise to optical transitions (spectra).

$l$ can take the values 0, 2, ... $(n - 1)$.

The magnetic quantum numbers $m_l$ has $(2l + 1)$ possible values: $-l, (-i + 1), \ldots -1, 0, 1 \ldots (l - 1), l$. The spin quantum number $s = 1/2$ and $m_s = \pm 1/2$, For any allowed value of $l$, there are $2(2l + 1)$ levels.

In the first transition series of elements of the Periodic Table, the magnetism arises from $d$ electrons ($l = 2$) and in rare earths, it comes from $f$ electrons ($l = 3$).

How the electrons combine in an atom to produce a stable state of lowest energy (i.e., the ground state) is determined partly by the relative magnitudes of the energies of interactions of the following types:

- $u_{ik} \vec{l}_i \cdot \vec{s}_k$ is the spin-orbit interaction either for the interaction between the orbital angular momentum of one electron and its own spin ($i = k$) or between different electrons.
- $b_{ik} \vec{l}_i \cdot \vec{l}_k$ is the orbit-orbit interaction between different electrons.
- $c_{ik} \vec{s}_i \cdot \vec{s}_k$ is the spin-spin interaction (between electrons $i$ and $k$).

In fact, the orbital angular momentum of a given electron interacts more strongly with its own spin than with the spins of other electrons, i.e., $|a_{ii}| \gg |a_{ik}|$. This spin-orbit interaction is due to coupling between $\vec{l}_i$ and $\vec{s}_i$, through the interaction of both with the nuclear charge. $a_{ii}$ is largest for heavy atoms (i.e., atoms having high atomic number). Further, it has been experimentally found that in most atoms $a_{ik}$ or $a_{ii} < b_{ik}$ and $c_{ik}$.

The spins add vectorially to form a resultant $\vec{S}$ and orbital moments form a resultant $\vec{L}$, The spin-orbit interaction can then be regarded as a small perturbation (due to coupling between $\vec{L}$ and $\vec{S}$). When external magnetic fields are not too strong, $\vec{L}$ and $\vec{S}$ combine into a resultant $\vec{J}$ such that $|\vec{J}|$ can take the values

$$J = |L - S|, |L - S + 1| \ldots |L + S - 1i, |L + S|$$

The projection $M_J$ of $J$ in the direction of quantization is also quantized and has the values

$$M_J = -J_1, (-J + 1), \ldots, (J - 1), J$$

In a given atom, the maximum values of $L$ and $S$ are given by $\sum l_i$ and $\sum s_i$, respectively, but other values can occur between the limits of $\sum \pm l_i$ and $\sum \pm s_i$ give by $\sum m_l$ and $\sum m_s$. The choice of $L$ and $S$ is dictated by Hund's rules which first arose in interpreting atomic spectra.

## 13.10 HUND'S RULES

(i) The combination of $s_i$ that gives the lowest energy (i.e., the most stable) is that with the highest value of $(2S + 1)$.

(ii) If the first rule has been satisfied, there are several possible $L$ values, all having the same value of $(2S + 1)$, that with the largest $L$ will be the most stable.

The value of the total angular momentum $J$ is equal to $|L - S|$ when the shell is less than half full and to $[L + S]$ when the shell is more than half full. When the shell is just full, the application of the first rule gives $L = 0$, so that $J = S$.

The energies associated with rule (i) are of a larger order of magnitude than those .given by the second rule.

The following examples illustrate the application of the rules.

(i) $Co^{2+}$ ion in a $3d^7$ state $(l_i = 2)$

$$m_s \rightarrow \frac{1}{2} \quad \frac{1}{2} \quad \frac{1}{2} \quad \frac{1}{2} \quad \frac{1}{2} \quad -\frac{1}{2} \quad -\frac{1}{2} \quad -\frac{1}{2} \quad -\frac{1}{2} \quad -\frac{1}{2}$$

$$m_l \rightarrow 2 \quad 1 \quad 0 \quad -1 \quad -2 \quad 2 \quad 1 \quad 0 \quad -1 \quad -2$$

Counting the seven electrons from the left, we have

$S = 3/2;\ 2S + 1 = 4;\ L = 3$

(ii) $Dy^{3+}$ ion in a $4f^9$ state $(l_i = 3)$

$$m_s \rightarrow \frac{1}{2} \quad \frac{1}{2} \quad \frac{1}{2} \quad \frac{1}{2} \quad \frac{1}{2} \quad \frac{1}{2} \quad \frac{1}{2} \quad -\frac{1}{2} \quad -\frac{1}{2} \quad -\frac{1}{2} \quad -\frac{1}{2} \quad -\frac{1}{2} \quad -\frac{1}{2} \quad -\frac{1}{2}$$

$$m_D \rightarrow 3 \quad 2 \quad 1 \quad 0 \quad -1 \quad -2 \quad -3 \quad 3 \quad 2 \quad 1 \quad 0 \quad -1 \quad -2 \quad -3$$

Counting the nine electrons from the left, we have

$S = 5/2,\ 2S + 1 = 6;\ L = 5.$

## 13.11 MAGNETIC MOMENT OF A SINGLE ATOM, COMBINATION OF *L* AND *S* TO FORM *J* (AND LANDE'S SPLITTING FACTOR, *g*)

$$L = |\vec{L}| = \sqrt{l(l+1)}\,\hbar, \quad \text{where } \hbar = \frac{h}{2\pi} \qquad \ldots(1)$$

$$\mu_l = \left(\frac{e}{2m}\right) L \qquad \ldots(2)$$

$$S = |\vec{S}| = \sqrt{S(S+1)}\ \hbar; \quad \mu_s = 2\left(\frac{e}{2m}\right)S \qquad ...(3)$$

Since the charge on the electron is negative, the direction of $\vec{\mu}_l$ is opposite to that of $\vec{L}$. Similarly, $\vec{\mu}_s$ is directed opposite to $\vec{S}$. Vectors $\vec{L}$ and $\vec{S}$ combine to yield the resultant $\vec{J}$ (Fig. 13.3):

$$\vec{J} = \vec{L} + \vec{S}$$

When an external magnetic field is applied, $\vec{L}$ will tend to precess around $\vec{J}$ at the Larmor frequency. $\vec{S}$ will precess but at twice the Larmor frequency. If the external field is small, the coupling between $\vec{L}$ and $\vec{S}$ will be strong enough to maintain a resultant $\vec{J}\cdot\vec{J}$ will precess around the field direction with a compromise frequency. Since $\vec{L}$, $\vec{S}$ and $\mu_l\,\mu_s$, all precess around $\vec{J}$, their components, normal to $\vec{J}$, will average to zero. The resultant moment $\mu_j$ will be the sum of components of $\mu_l$ and $\mu_s$ parallel to $\vec{J}$.

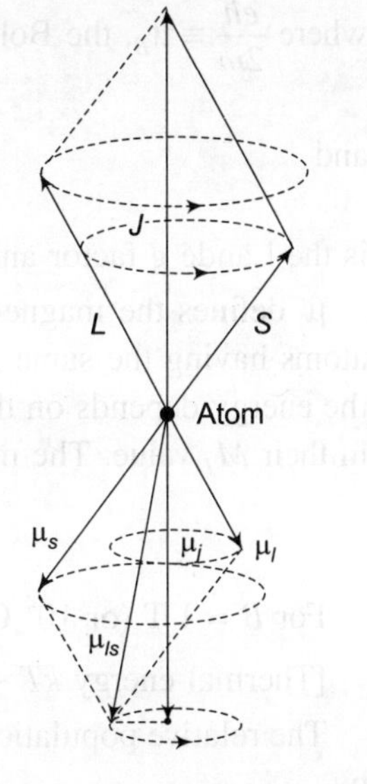


**Figure 13.3** Vector model of the atom

Component of $\mu_s$, along $\vec{J}$ is $\mu_l \cos(\vec{L}, \vec{J}) = L\dfrac{e}{2m}\cos(\vec{L}, \vec{J})$ ....(4)

Component of $\vec{\mu}_s$, along $\vec{J}$ is

$$S \times \frac{2e}{2m}\cos(\vec{S}, \vec{J}) \qquad ...(5)$$

$$|\vec{\mu}_j| = [L\cos(\vec{L}, \vec{J}) + 2S\cos(\vec{S}, \vec{J})]\,\frac{e}{2m} \qquad ...(6)$$

Applying the law of cosines,

$$S^2 = L^2 + J^2 - 2LJ\cos(\vec{L}, \vec{J}) \qquad ...(7)$$

$$\Rightarrow \qquad L\cos(\vec{L}, \vec{J}) = \frac{J^2 + L^2 - S^2}{2J} \qquad ...(8)$$

Similarly,

$$S\cos(\vec{S}, \vec{J}) = \frac{J^2 - L^2 + S^2}{2J} \qquad ...(9)$$

Substituting these in Eq. (6), we get

$$\mu_j = \left\{1 + \frac{J^2 - L^2 + S^2}{2J^2}\right\} J\,\frac{e}{2m}$$

or

$$= \left[1 + \frac{J(J+1) + S(S+1) - L(L+1)}{2J(J+1)}\right] \times \sqrt{J(J+1)}\,\frac{e\hbar}{2m}$$

or

$$= g\mu_B\sqrt{J(J+1)} \qquad ...(10)$$

where $\frac{e\hbar}{2m} \equiv \mu_B$, the Bohr magneton

and
$$g = 1 + \frac{J(J+1)\,S(S+1) - L(L+1)}{2J(J+1)} \qquad ...(11)$$

is the Landé $g$ factor and has the value 2 when $L = 0$ and the value $j$ when $S = 0$.

$\mu_j$ defines the magnetic moment of a single atom. In the absence of an applied field, all atoms having the same moment have the same energy. When a magnetic field $B$ is applied, the energy depends on the statistical occupation of the $(2J + 1)$ sublevels, all of which differ in their $M_J$ value. The magnetic energy of a given atom is

$$E_M = -\vec{\mu}_j \cdot \vec{B} = -g\mu_B\, M_J B \qquad ...(12)$$

For $B \simeq 1$ T (or $10^4$ Oe); $E_M \simeq 0.1 - 10\ \text{cm}^{-1}$

[Thermal energy $kT \sim 210\ \text{cm}^{-1}$ at $T = 300$ kJ]

The relative population $P(M_J)$ of a sublevel $M_J$ is given, according to statistical mechanics by

$$P(M_J) = \frac{\exp(-E_M/kT)}{\sum \exp(-E_M/kT)} \qquad ...(13)$$

## 13.12 CLASSIFICATION OF MAGNETIC MATERIALS

Materials can be classified into the following five categories:

If $\chi = 0$, then the material is non-magnetic, $\mu_r = 1$ for such material.

**(i)** ***Diamagnetic.*** Susceptibility $\chi < 0$ is magnetization opposes the applied field, $\mu_r < 1$ for such materials. The magnetic flux density $|\vec{B}|$ is less inside than outside. Susceptibility is independent of temperature.

***Examples.*** Organic solids like benzone, naphthalene, metals like copper, gold and silver, atoms with rare-gas configuration like He, Ne, etc.

**(ii)** ***Paramagnetic.*** Susceptibility $\chi$ is positive and has a small value $\mu_r > 1$. The flux density inside the material is slightly greater than outside.

***Examples.*** (a) All atoms and molecules with unfilled outer shells like Ti, etc.

(b) Salts of the transition elements, rare-earths and actinide series containing elements like $Cr^{3+}$, $Dy^{3+}$, etc.

(c) Metals like aluminium.

**(iii)** ***Ferromagnetic.*** $\chi$ is positive and very large. These materials possess spontaneous magnetisation. The atomic magnetic dipoles are arranged parallel to one another below a certain temperature known as the *ferromagnetic Curie temperature* ($T_f$). Some of the examples are Fe, Co, Ni, Gd. Above $T_f$, the behaviour of a ferromagnetic material is somewhat similar to that of a paramagnetic material.

**(iv)** ***Ferrimagnetic.*** The atomic dipoles are arranged antiparallel to one another but the moments in one direction have a larger magnitude so that the net magnetization is non-vanishing. The examples are $Fe_3O_4$, nickel ferrite and yttrium iron garnet. These materials become paramagnetic above the Curie temperature.

(v) ***Antiferromagnetic.*** In these materials, the net magnetisation in the absence of a field vanishes. The atomic dipoles are arranged antiparallel to one another so that the net moment is zero. Above the Neel temperature, $T_N$, the dipoles become randomly oriented as in a paramagnet. The examples are MnO, NiO, MnS, MnTe etc. $T_N$ plays a similar role in antiferromagnetic materials as does the ferromagnetic Curie temperature $T_f$ in ferromagnetic materials.

### 13.12.1 Classification on the Basis of Susceptibility

Materials may be broadly classified into three categories: (i) diamagnets, (ii) paramagnets and (iii) ferromagnets. If an atom or ion contains only paired electrons, the total magnetic moment is practically zero as a result of nullifying effect of the added orbital and spin angular moments. Such materials for which $\mu_r < 1$ and $\chi$ is negative, are called diamagnets. If the atom or ion contains one or more unpaired electrons and thus possesses a permanent magnetic moment, the material can be either paramagnetic or ferromagnetic. If there is no interaction between the adjacent magnetic moments and, thus, they can independently interact with the external magnetic field, the materials are called ideal paramagnets; for these, $\mu_r > 1$, $\chi$ is small, positive and varies inversely with the absolute temperature (Curie law). If a strong interaction between adjacent magnetic moment tends to align them parallel to one another, the materials are called ferromagnetic, for which $\chi$ values are very large, positive and also $B \simeq M$. Above a certain critical temperature, called the Curie temperature or Curie point ($T_c$), ferromagnets become paramagnetic and follow the so-called Curie–Weiss law, $\chi = \dfrac{C}{T - \theta}$, where $C$ and $\theta$ are the Curie and Weiss constants, respectively.

Under certain circumstances, the interaction between neighbouring magnetic moments may lead to an antiparallel alignment resulting in a vanishing resultant moment at 0 K; such materials are called antiferromagnetic Above a certain temperature, called the Néel temperature or Néel point ($T_N$), antiferromagnets become paramagnetic and follow the Curie–Weiss type relation $\chi = \dfrac{C}{T + \theta_N}$, where $\theta_N$ is a constant.

## 13.13 PARAMAGNETISM

We already know that paramagnetism arises because the individual atoms in the paramagnetic substance possess a net magnetic moment. The magnetic moment of the individual atoms in the bulk are randomly oriented due to thermal vibrations, consequently, there is no magnetic moment of the bulk. When subjected to an external magnetic field, the atoms experience a torque because of their magnetic moments. The effect of the torque is to simply align the magnetic moments of the atoms in the direction of the external field, resulting thereby in a net magnetization in the direction of the field.

Paramagnetism is a weak magnetism where intensity of magnetization is proportional to the applied magnetic field. As a result, the susceptibility $\chi$ is positive and is of the order of $10^{-3}$ to $10^{-5}$.

The $\chi$ of a paramagnetic substance obeys *Curie's law*:

$$\chi = \frac{C}{T}$$

where $C$ is known as Curie's constant.

### 13.13.1 Magnetic Susceptibility and Curie's Law (Statistical Approach)

The relationship between $\vec{H}$ and $\vec{M}$ is given by

$$\vec{B} = \vec{H} + 4\pi \vec{M}$$

($\vec{B}$ = magnetic induction, $\vec{H}$ = magnetic field)

is a unique property of a given material. This relationship is characterized by the *magnetic susceptibility* ($\chi$) defined as

$$\vec{M} = \chi \vec{H} \quad ...(1)$$

$\chi$ in general is a tensor which measures the ease with which a body acquires magnetization $\vec{M}$ (or $I$) in the presence of an applied field $\vec{H}$, the magnetic susceptibility is a scalar.

Materials for which $\chi < 0$ are called *diamagnets* and those for which $\chi > 0$ are called *paramagnets*.

$\chi$ is generally independent of temperature in diamagnets but varies inversely with the absolute temperature in paramagnets

$$\chi = \frac{C}{T} \quad ...(2)$$

This is known as *Curie's law*.

It can be derived using Maxwell–Boltzmann statistics as applied to a distribution of permanent magnetic moments, as follows.

### 13.13.2 Classical Theory of Paramagnetism

According to *Langevin*, a paramagnetic substance consists of a paramagnetic gas, whose each atom has a permanent magnetic moment, and the molecules do not exert any force on one another.

Let there be $N$ molecular magnets, each of magnetic moment $\mu$ per unit volume. In the absence of the external magnetic field. The atomic magnets are randomly oriented and there is no net magnetization. When a magnetic field is applied, the resulting torque on the atoms tend to align them in the direction of the applied field. This tendency is counteracted by the thermal agitation and a statistical equilibrium is established. Subsequently, we will find out the number ($dN$) of the molecular magnets that make an angle lying between $\theta$ and $\theta + d\theta$ with the field direction.

From Maxwell–Boltzmann statistics, $dN$ is proportional to

(i) $\sin\theta\, d\theta$

(ii) $e^{-\omega/kT}$

where $W = -\vec{\mu}\cdot\vec{B} = -\mu B\cos\theta$ is the magnetic potential energy of the molecular magnet and $\theta$ is the angle between field direction and the axis of the molecular magnet at absolute temperature, $T$.

Hence,
$$dN = Ae^{\mu B\cos\theta/kT}\sin\theta\, d\theta \quad ...(1)$$
where $A$ is a constant.

Therefore, the total number of molecular magnets inclined with the field direction at all possible angles is

$$N = \int dN = \int_0^{\pi} Ae^{\mu B\cos\theta/kT}\sin\theta\, d\theta$$

Let us put

$$\left(\frac{\mu B}{kT}\right) \equiv a. \text{ Then}$$

$$N = A\int_0^{\pi} e^{a\cos\theta}\sin\theta\, d\theta \quad ...(2)$$

To evaluate this integral, put

$$e^{a\cos\theta} = x$$

so that, differentiating

$$e^{a\cos\theta}\cdot(-a\sin\theta)\, d\theta = dx$$

or
$$e^{a\cos\theta}\cdot\sin\theta\, d\theta = -\frac{dx}{a}$$

When $\theta = 0$, $x = e^{a}$ and when $\theta = \pi$, $x = e^{-a}$

Substituting these values in Eq. (2), we get

$$N = -\frac{A}{a}\int_{e^{a}}^{e^{-q}} dx = \frac{A}{a}(e^{a} - e^{-a})$$

$$\therefore \quad A = \frac{aN}{e^{a} - e^{-a}} \quad ...(3)$$

The intensity of magnetization is defined as the magnetic moment per unit volume in the direction of the applied field.

The component of magnetic field of $dN$ number of molecular magnets, which are inclined to the field direction between the angles $\theta$ and $\theta + d\theta$, is $\mu\cos\theta\cdot dN$ (where $\mu$ are the individual dipole moments). Hence, the intensity of magnetization is

$$I = \int \mu \cos\theta \cdot dN$$

$$= A\mu \int_0^{\theta} e^{a\cos\theta} \cdot \sin\theta \cos\theta \, d\theta \quad \text{(using Eq. 1)} \qquad ...(4)$$

In order to evaluate this integral, we put

$$\cos\theta \equiv z, \quad \therefore \sin\theta \, d\theta = -dz$$

when $\theta = 0$, $z = 1$ and when $\theta = \pi$, $z = -1$. Putting these values in Eq. (4), we get

$$I = -A\mu \int_1^{-1} z e^{az} \, dz = A\mu \int_{-1}^{+1} z e^{az} \cdot dz$$

$$= -A\mu \left[ \frac{ze^{az}}{a} - \int 1 \cdot \int e^{az} \, dz \right]_{-1}^{+1}$$

$$= -A\mu \left[ \frac{ze^{az}}{a} - \frac{e^{az}}{a^2} \right]_{-1}^{+1} = A\mu \left[ \frac{e^a}{a} - \frac{e^a}{a^2} - \left( -\frac{e^{-a}}{a} - \frac{e^{-a}}{a^2} \right) \right]$$

$$= -A\mu \left[ \frac{e^a + e^{-a}}{a} - \frac{e^a - e^{-a}}{a^2} \right]$$

$$= \frac{\mu a N}{e^a - e^{-a}} \left[ \frac{e^a + e^{-a}}{a} - \frac{e^a - e^{-a}}{a^2} \right] \qquad \text{(Using Eq. 3)}$$

$$= \mu N \left[ \frac{e^a + e^{-a}}{e^a - e^{-a}} - \frac{1}{a} \right]$$

$$= \mu N \left[ \cosh a - \frac{1}{a} \right] \qquad ...(5)$$

The term within the bracket on the R.H.S. of Eq. (5) is known as *Langevin's function* and is denoted by $L(a)$. Hence,

$$I = \mu N L(a) \qquad ...(6)$$

It follows from Eq. (6) that the intensity of magnetization depends both on the intensity of magnetizing field $H$ and the temperature $T$, as $a$ is a function of both $B$ and $T$.

Two cases may arise:

(1) *When the temperature is low and the field is high:*

In such a case, $a = \frac{\mu B}{kT}$ is very large and therefore, we must consider the limiting value of Langevin's function $L(a)$ when $a$ tends to infinity, Hence,

$$\lim_{a\to\infty} L(a) = \lim_{a\to\infty}\left[\frac{e^a + e^{-a}}{e^a - e^{-q}} - \frac{1}{a}\right] = 1$$

Thus, $$I = \mu N \equiv I_S$$

the saturation value of the intensity of magnetization. Thus, at low temperature and very high magnetic field, the paramagnetic substances acquire saturation.

Hence, from Eq. (6),

$$I = I_S\, L\,(a) \qquad ...(7)$$

(2) *When the temperature is high and the field is low*

In such a case, $a = \dfrac{\mu B}{kT}$ is very small and, therefore, we must consider the limiting value of Langevin's function $L(a)$ when $a$ tends to zero. Hence,

$$\lim_{a\to 0} L(a) = \lim_{a\to 0}\left[\frac{e^a + e^{-a}}{e^a - e^{-a}} - \frac{1}{a}\right]$$

or

$$\lim_{a\to\infty} L(a) = \lim_{a\to 0}\left[\frac{\left(1+a+\frac{a^2}{2!}+\frac{a^3}{3!}+...\right)+\left(1-a+\frac{a^2}{2!}-\frac{a^3}{3!}+...\right)}{\left(1+a+\frac{a^2}{2!}+\frac{a^3}{3!}+...\right)-\left(1-a+\frac{a^2}{2!}-\frac{a^3}{3!}+...\right)} - \frac{1}{a}\right]$$

$$= \lim_{a\to 0}\left[\frac{2\left(1+\frac{a^2}{2!}\right)}{2\left(a+\frac{a^3}{3!}\right)} - \frac{1}{a}\right]$$

$$= \lim_{a\to 0}\left[\frac{\left(1+\frac{a^2}{2}\right)}{a\left(1+\frac{a^2}{6}\right)} - \frac{1}{a}\right]$$

$$= \lim_{a\to 0}\frac{1}{a}\left[\left(1+\frac{a^2}{2}\right)\left(1+\frac{a^2}{6}\right)^{-1} - 1\right]$$

$$\approx \lim_{a\to 0}\frac{1}{a}\left[\left(1+\frac{a^2}{2}\right)\left(1-\frac{a^2}{6}\right) - 1\right]$$

$$= \lim_{a\to 0} \frac{1}{a}\left[1+\frac{a^2}{2}-\frac{a^2}{6}-\frac{a^4}{12}-1\right]$$

or
$$\lim_{a\to 0} L(a) = \lim_{a\to 0} \frac{1}{a}\left[\frac{3a^2-a^2}{6}\right] = \frac{a}{3}$$

and so, from Eq. (6), we have

$$I = \mu N \cdot \frac{a}{3} = \mu N \cdot \frac{\mu B}{3kT}$$

or
$$I = \frac{\mu^2 NB}{3\,kT} \qquad \text{...(8)}$$

Therefore, the susceptibility is

$$\chi = \frac{I}{H} = \frac{\mu_0 \mu^2 N}{3\,kT} \quad C \because B = \mu H \qquad \text{...(9)}$$

$\Rightarrow$
$$\chi = \frac{C}{T} \quad \text{where} \quad C = \frac{\mu_0 \mu^2 N}{3\,k} \qquad \text{...(10)}$$

*C* is Curie's constant and Eq. (10) is *Curie's law* obeyed by a large number of paramagnetic substances.

Here, we have assumed that

(i) the magnetic moments of all atoms are the same, and

(ii) all orientations are possible from $\theta = 0$ to $\theta = \pi$.

## 13.14 WEISS MOLECULAR FIELD THEORY OF PARAMAGNETISM AND CURIE–WEISS LAW

In the Langevin's theory, we have ignored the dependence of susceptibility ($\chi$) on temperature by assuming that molecular magnets do not exert force on one another depending upon their orientations. Actually, paramagnetic solids have susceptibility variation not according to Curie's law but according to the relation

$$\chi = \frac{C}{T-\theta} \qquad \text{...(1)}$$

where $\theta$ is a constant known as *paramagnetic Curie temperature.*

To explain the dependence given by Eq. (1), Weiss in 1907 extended Langevin's theory by introducing the concept *of internal molecular field.* He assumed that the effective field ($B_e$) acting on a molecule is the resultant of the externally applied field *H* and an internal field $H_i$ which is proportional to, and acts in the same direction as the intensity of magnetization so that

$$B_e = \mu_0 (H + H_i) = \mu_0 H + \mu_0 H_i$$
$$= B + \mu_0 \beta I \qquad \text{...(2)}$$

where $\beta$ is a constant, called *molecular field constants* or *Weiss factor.*

According to Weiss, the internal field $H_i$ (also called *molecular field*), comes into play because the molecules are mutually affected by their individual moments. Thus, we must replace the field $B$ in Langevin's theory by the effective field $B_e$, i.e.,

$$I = \frac{\mu^2 N B_e}{3\,kT} = \frac{\mu^2 N\,(B + \mu_0\,\beta I)}{3\,kT} \qquad ...(3)$$

or

$$I \times \left(1 - \frac{\mu_0 \mu^2\, N\beta}{3\,kT}\right) = \frac{\mu^2 NB}{3\,kT}$$

or

$$I = \frac{\mu^2 NB}{3kT\left(1 - \dfrac{\mu_0\mu^2\,N\beta}{3kT}\right)} = \frac{\mu^2 NB}{3k\left(T - \dfrac{\mu_0\mu^2\,N\beta}{3k}\right)}$$

Hence, the magnetic susceptibility is

$$\chi = \frac{1}{H} = \frac{\mu_0\mu^2 N}{3k\left(T - \dfrac{\mu_0\mu^2 N\beta}{3k}\right)}$$

$$\Rightarrow \qquad \chi = \frac{C}{T - \theta} \qquad ...(4)$$

where

$$C = \frac{\mu_0\mu^2 N}{3\,k} \qquad ...(5)$$

is the Curie constant,

and

$$\theta = \frac{\mu_0\mu^2\,N\beta}{3\,k} \qquad ...(6)$$

is the paramagnetic Curie temperature. Equation (4) is known as *Curie–Weiss law.*

It states that the susceptibility of a magnetic substance (in which there is a molecular field) varies inversely as the excess of temperature, over a certain critical temperature, $\theta$.

It follows from Eq. (4) that for $T < \theta$, the susceptibility of an otherwise paramagnetic material would become negative (i.e., a paramagnetic would become diamagnetic at a temperature $T < \theta$). Moreover, the Curie temperature ($\theta$) for most paramagnetics is so low that even the Curie's law holds even at room temperature.

## 13.15 PARAMAGNETISM AND CURIE'S LAW (ANALYTIC APPROACH)

The system that we want to discuss is the **two-state paramagnet**, we start by reviewing the basic microscopic physics.

### 13.15.1 Notations and Microscope Physics

The system consists of $N$ spin-1/2 particles immersed in a constant magnetic field $\vec{B}$ pointing in the $+z$ direction (see Fig. 13.4). Each particle behaves like a little compass needle feeling a torque that tries to align its magnetic dipole moment with the field. Because of this behaviour I'll refer to the particles as **dipoles**. For simplicity I'll assume that there are *no* interaction *between* dipoles—each dipole feels only the torque from the external field. In this case we say that the system is an **ideal** paramagnet.

**Figure 13.4** A two-state paramagnet, consisting of $N$ microscopic magnetic dipoles, each of which is either "up" or "down" at any moment. The dipoles respond only to the influence of the external magnets field $B$; they do not interact with their neighbours (except to exchange energy)

According to quantum mechanics, the component of a particle's dipole moment along a given axis cannot take on just any value—instead it is **quantized**, that is, limited to certain discrete values. For a spin-1/2 particle only *two* values are allowed, which I'll call simply "up" and "down" (along the $z$-axis). The magnetic field, pointing in the $+z$ direction, gives each dipole a preference for the *up* state. To flip a single dipole from up to down we would have to add some energy; the amount of energy required is $2\mu B$, where $\mu$ is a constant related to the particles of magnetic moment (essentially the "strength" of the effective compass needle). For the sake of symmetry, I'll say that the energy of a dipole that points up is $-\mu B$, so that the energy of a dipole that the points down is $+\mu B$, (see Fig. 13.5).

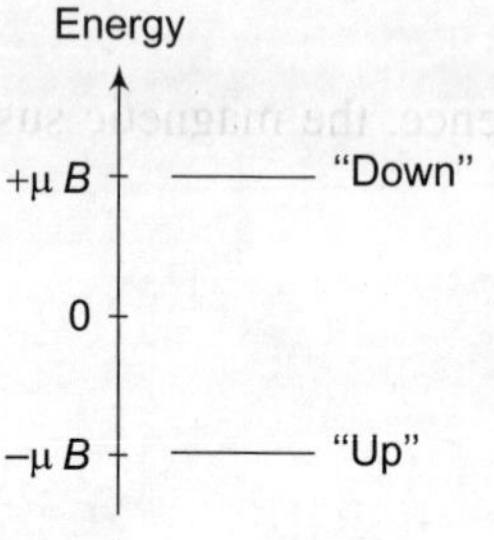


**Figure 13.5** The energy levels of a single dipole in an ideal two-state paramagnet are $-\mu B$ (for the "up" state) and $+\mu B$ (for the "down" state)

The *total* energy of the system is

$$U = \mu B(N_\downarrow - N_\uparrow) = \mu B\,(N - 2N_\uparrow) \quad ...(1)$$

where $N_\uparrow$ and $N_\downarrow$ are the numbers of up and down dipoles, respectively, and $N = N_\uparrow + N_\downarrow$. I'll define the **magnetization**, $M$, to be the total magnetic moment of the whole system. Each "up" dipole has magnetic moment $+\mu$ and each "down" dipole has magnetic moment $-\mu$, so the magnetization can be written

$$M = \mu\,(N_\uparrow - N_\downarrow) = -\frac{U}{B} \quad ...(2)$$

We would like to know how $U$ and $M$ depend on temperature.

Our first task is to write down a formula for the multiplicity. We will keep $N$ fixed, and consider each different value of $N_\uparrow$ (and hence $U$ and $M$) to define a different microstate. Then this system is mathematically equivalent to a collection of $N$ coins with $N_\uparrow$ heads, and the *multiplicity* is simply

$$\Omega\,(N_\uparrow) = \binom{N}{N_\uparrow} = \frac{N!}{N_\uparrow!N_\downarrow!} \quad ...(3)$$

### 13.15.2 Analytic Solution

Now that we have studied most of the physics of this system through numerical calculations, let us go back and use analytic methods to derive some more general formulas to describe these phenomena.

I will assume that the number of elementary dipoles is *large*, and also that at any given time the numbers of up and down dipoles are separately large. Then we can simplify the multiplicity function (3) using Stirling's approximation. Actually, it's easiest to just calculate the entropies:

$$\begin{aligned} S/k &= \ln N! - \ln N_{\uparrow}! - \ln (N - N_{\uparrow})! \\ &\approx N \ln N - N - N_{\uparrow} \ln N_{\uparrow} + N_{\uparrow} - (N - N_{\uparrow}) \ln (N - N_{\uparrow}) + (N - N_{\uparrow}) \qquad ...(4) \\ &= N \ln N - N_{\uparrow} \ln N_{\uparrow} - (N - N_{\uparrow}) \ln (N - N_{\uparrow}) \end{aligned}$$

From here on the calculations are fairly straightforward but somewhat tedious.

To find the temperature, we must differentiate $S$ with respect to $U$. It is simplest to first use the chain rule and Eq. (1) to express the derivative in terms of $N_{\uparrow}$:

$$\frac{1}{T} = \left(\frac{\partial S}{\partial U}\right)_{N,B} = \frac{\partial N_{\uparrow}}{\partial U}\frac{\partial S}{\partial N_{\uparrow}} = -\frac{1}{2\mu B}\frac{\partial S}{\partial N_{\uparrow}}. \qquad ...(5)$$

Now just differentiate the last line of Eq. (4) to obtain

$$\frac{1}{T} = \frac{k}{2\mu B}\ln\left(\frac{N - U/\mu B}{N + U/\mu B}\right) \qquad ...(6)$$

Notice from this formula that $T$ and $U$ always have opposite signs.

Equation (6) can be solved for $U$ to obtain

$$U = N\mu B\left(\frac{1 - e^{2\mu B/kT}}{1 + e^{2\mu B/kT}}\right) = -N\mu B \tanh\left(\frac{\mu B}{kT}\right), \qquad ...(7)$$

where tanh is the hyperbolic tangent function.[1] The *magnetization* is therefore

$$M = N\mu \tanh\left(\frac{\mu B}{kT}\right) \qquad ...(8)$$

The hyperbolic tangent function is plotted in Fig. 13.6; it rises from the origin with a slope of 1, then flattens to an asymptotic value of 1 as its argument goes to infinity. So at very small positive temperatures the system is completely magnetized (as we saw before), while as $T \to \infty$, the magnetization goes to zero. To obtain negative magnetization, as described above.

---

[1] The definitions of the basic hyperbolic functions are $\sinh x = \frac{1}{2}(e^x - e^{-x})$, $\cosh x = \frac{1}{2}(e^x + e^{-x})$, and $\tanh x = (\sinh x)/(\cosh x)$. From these definitions you can easily show that $\frac{d}{dx}\sinh x = \cosh x$ and $\frac{d}{dx}\cos x = \sinh x$ (with no minus sign).

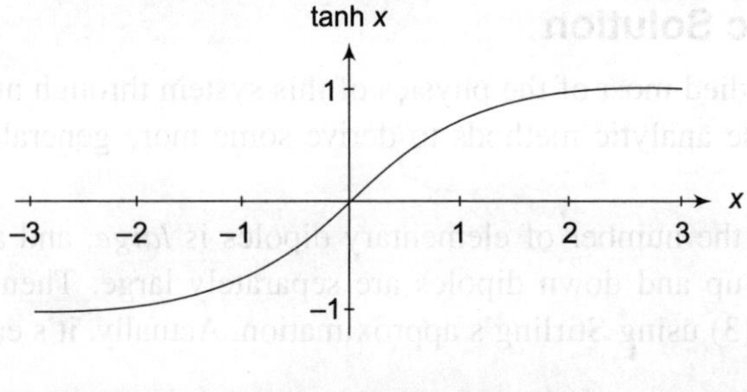


**Figure 13.6** The hyperbolic tangent function. In the formulas for the energy and magnetization of two-state paramagnet, the argument ($x$) of the hyperbolic tangent is $\mu B/kT$

To calculate the heat capacity of the paramagnet, just differentiate Eq. (7) with respect to the $T$:

$$C_B = \left(\frac{\partial U}{\partial T}\right)_{N,B} = Nk \cdot \frac{(\mu B / kT)^2}{\cosh^2 (\mu B / kT)}. \quad ...(9)$$

This function approaches zero at both low and high $T$, as we also saw in the numerical solution.

In a real-world paramagnet, the individual dipoles can be either electrons or atomic nuclei. Electronic paramagnetics occurs when there are electrons with angular momentum (orbital or spin) that is not compensated by other electrons; the circular currents then give rise to magnetic dipole moments. The number of possible states for each dipole is always some small integer, depending on the total angular momentum of all the electrons in an atom or molecule. The simple case considered here, with just two states, occurs when there is just one electron per atom whose spin is uncompensated. Ordinarily, this electron would also have orbital angular momentum, but in some environments the orbital motion is "quenched" by the neighbouring atoms, leaving only the spin angular momentum.

For a two-state paramagnet the value of the constant μ is the **Bohr magneton**,

$$\mu_B \equiv \frac{eh}{4\pi m_e} = 9.274 \times 10^{-24} \text{ J/T} = 5.788 \times 10^{-5} \text{ eV/T} \quad ...(10)$$

(Here $e$ is the electron's charge and $m_e$ is its mass.) If we take $B = 1$ T (a pretty strong magnet), then $\mu B = 5.8 \times 10^{-5}$ eV. But at room temperature, $kT \approx 1/40$ eV. So at ordinary temperature (more than a few kelvin), we can assume $\mu B/kT \ll 1$. In this limit, $\tanh x \approx x$, so the magnetization becomes

$$M \approx \frac{N\mu^2 B}{kT} \quad \text{(when } \mu B \ll kT\text{)}. \quad ...(11)$$

The fact that $M \propto 1/T$ was discovered experimentally by Pierre Curie and is known as Curie's law; it holds in the high-temperature limit for all paramagnets, even those with more than two angular momentum states. In this limit the heat capacity falls off in proportion to $1/T^2$.

## 13.16 QUANTUM THEORY OF PARAMAGNETISM

In the classical theory, it was assumed that all orientation of the molecular magnets with respect to the field direction are possible. Quantum Mechanics, however, contradicts this assumption and suggests that only certain orientation with respect to the field direction given by $\cos\theta = \frac{MJ}{J}$, are possible where $J$ is the resultant quantum number ($= L + S$) and $M_J$ is the magnetic quantum number having $(2J + 1)$ values from $-J$ to $+J$.

Let us consider a paramagnetic substance containing $N$ atoms per unit volume and let $J$ be the total angular momentum quantum number associated with each atom. Then component of atomic magnetic moment in the field direction is

$$\mu_j = g\mu_B M_J \qquad \text{...(1)}$$

where $g$ is the Lande's splitting factor and $\mu_B$ is the Bohr magneton.

The magnetic energy of an atom is

$$E_\mu = -\vec{\mu}_j \cdot \vec{B} = -g\,\mu_B M_J B \qquad \text{...(2)}$$

The relative population $P(M_J)$ of a sublevel $M_J$ is

$$P(M_J) = \frac{\exp(-E_M/kT)}{\Sigma \exp(-E_M/kT)} = \frac{\exp(g\mu_B M_J B/kT)}{\sum\limits_{M_J} \exp(g\mu_B M_J B/kT)} \qquad \text{...(3)}$$

In expression (2), as $M_J$ depends upon $\theta$ (see Fig. 13.7), the potential energy $E_M$ also depends upon $\theta$.

The number of atomic magnets having a particular orientation is proportional to $e^{-E_M/kT}$. The total atomic magnetic moment is obtained by summing over all the sublevels $M_J = -J$ to $M_J = +J$.

$$\langle\mu_j\rangle = \sum_{\mu_J} g\mu_B M_J P(M_J) \qquad \text{...(4)}$$

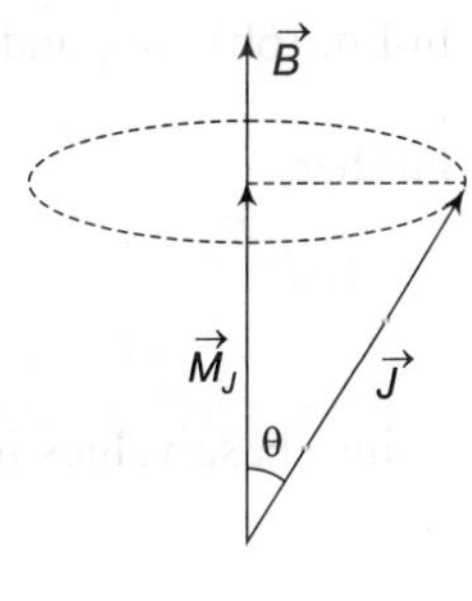


**Figure 13.7** Precession of $\vec{J}$ around $M_J$

And the total intensty of magnetization ($M$) is

$$M = N\langle\mu_j\rangle$$

$$= N\left[\frac{\sum\limits_{-J}^{J} g\mu_B M_J \exp(g\mu_B M_J B/kT)}{\sum\limits_{-J}^{J} \exp(g\mu_B M_J B/kT)}\right] \qquad \text{...(5)}$$

The term within the bracket in Eq. (5) gives the statistical average of the atomic magnetic moments.

Two cases arise

(1) The temperature is high and the field is low.

(2) The temperature is low and the field is high.

**Case 1:** In such a case, $\dfrac{g\mu_B M_J B}{kT}$ is small and Eq. (5) takes the form

$$M = N\left[\frac{\sum_{-J}^{J} M_J\, g\mu_B\left(1+\dfrac{M_J g\mu_B B}{kT}\right)}{\sum_{-J}^{J}\left(1+\dfrac{M_J g\mu_B B}{kT}\right)}\right] \qquad \text{...(6a)}$$

or

$$M = N\left[\frac{\sum_{-J}^{J} M_J\, g\mu_B + \sum_{-J}^{J} \dfrac{M_J^2\, g^2\, \mu_B^2}{kT}\, B}{\sum_{-J}^{J} 1 + \sum_{-J}^{J} \dfrac{M_J g\mu_B}{kT}\, B}\right] \qquad \text{...(6b)}$$

Now

$$\sum_{-J}^{J} M_J\, g\mu_B = g\mu_B \sum_{-J}^{J} M_J$$

$$= g\mu_B\{-J + (-J+1) + ... + (-1) + 0 + (1) + ... + (J+1) + J\}$$

$$= g\mu_B \times 0 = 0 \qquad \text{...(7)}$$

and

$$\sum_{-J}^{J} 1 = (2J+1) \qquad \text{...(8)}$$

In Eq. (6b), $g\mu_B$ and $kT$ are constants and can be taken outside the summation sign.

Further,

$$\sum_{-J}^{J} M_J^2 = J^2 + (J-1)^2 + ... + (J-1)^2 + J^2$$

$$= \frac{J(J+1)\,(2J+1)}{3} \qquad \text{...(9)}$$

Using these values in Eq. (6b), we get

$$M = N\left[0 + \frac{g^2\, \mu_B^2\, B}{kT}\cdot\frac{J(J+1)\,(2J+1)}{J}\right] / (2J+1)$$

$$= N\,\frac{g^2\mu_B^2\, B}{3\, kT}\cdot J\,(J+1)$$

The susceptibility is therefore

$$\chi = \frac{M}{H} = \frac{M}{(\mu_0 H)}\,\mu_0$$

$$= \frac{\mu_0 N g^2\, \mu_B^2\, J(J+1)}{3\, kT} \qquad (\because\ \ \mu_0 H = B)$$

or

$$\chi = \frac{\mu_0 N\, \mu_J^2}{3\, kT} = \frac{C}{T} \qquad \text{...(10)}$$

where $$C = \frac{\mu_0 N \mu_J^2}{3k} \qquad ...(11)$$

and $$\mu_J = g\mu_B \sqrt{J(J+1)} \qquad ...(12)$$

Denoting $$g\sqrt{J(J+1)} \equiv P_{\text{eff}}$$

We have $$\mu_J = P_{\text{eff}}\, \mu_B$$

i.e., $P_{\text{eff}}$ is the effective number of Bohr magneton. The dependence of $\chi$ on temperature $T$, given by Eq. (10) is identical with the classical result except that $\mu$ is replaced by $\mu_J$.

Table 13.1 shows calculated and experimental values of $P_{\text{eff}}$ for rare earth ions and they are found to be in good agreement.

**Table 13.1** Effective Number of Bohr Magneton for Some Trivalent Rare Earth Ions

| *Ion* | *Configuration* | $P_{eff}$ *(calculated)* | $P_{eff}$ *(experimental)* |
|---|---|---|---|
| $Ce^{3+}$ | $4f^1\,5s^2p^6$ | 2.54 | 2.4 |
| $Pr^{3+}$ | $4f^2\,5s^2p^6$ | 3.58 | 3.5 |
| $NCl^{3+}$ | $4f^3\,5s^2p^6$ | 3.62 | 3.5 |
| $Sm^{3+}$ | $4f^5\,5s^2p^6$ | 0.84 | 1.5 |
| $Gd^{3+}$ | $4f^7\,5s^2p^6$ | 7.94 | 8.0 |
| $Tb^{3+}$ | $4f^8\,5s^2p^6$ | 9.72 | 9.5 |

Still, the theory fails to explain the paramagnetism of transition group elements. As assumed by van Vleck, that every particle is in the same $L$, $S$ and $J$ state, is not true and it depends on the energy of transition (i.e., the energy $h\nu$ required to excite the particle to the higher energy states.

The following two situations may arise:

**(a) When the energy of transition is very large compared to kT, (i.e., hν ≫ kT).** As sufficiently high energy is required to raise the atoms to higher energy state (*wide multiplets*) most of the atoms remain in their low energy states and probability of occupation of higher levels is small. In this case

$$\chi = \frac{\mu_0\, N g^2 \mu_B^2\, J(J+1)}{3\,kT} \qquad ...(13)$$

Equation (13) is strictly accurate and Curie's law is followed. The effective number of Bohr magneton is

$$P_{\text{eff}} = g \cdot \sqrt{J(J+1)} \qquad ...(14)$$

There is a good agreement between calculated and experimental values of $P_{\text{eff}}$.

**(b) When the energy of transition is very small compared to kT, (hν ≪ kT).** In this case, probability of occupation of higher energy states is large (there are *narrow multiplets*). van Vleck argued that $LS$ coupling is weak compared to the coupling of $L$ or $S$ with $\vec{B}$. Thus, the ions behave as if they are in a strong magnetic field, consequently the orbital and spin moments are separately quantized with

$$M_L = L, L_{-1}, ..., -L, \text{ with } g_L = 1$$

$$M_S = S, S_{-1}, ..., -S, \text{ with } g_S = 2$$

The total magnetic moment will, therefore, be $M_L + 2M_S$ ($\because\ g_L = 1$ and $g_S = 2$). The equation for the intensity of magnetization becomes

$$M\,(\text{or } I) = N\,\mu_B \left[\frac{\sum_{-L}^{L} M_L\, e^{M_L \mu_B B/kT}}{\sum_{-L}^{L} e^{M_c \mu_B B/kT}} + \frac{\sum_{-S}^{S} 2M_S\, e^{2M_S \mu_B B/kT}}{\sum_{-S}^{S} e^{2M_S \mu_B B/kT}}\right]$$

$$= N\,\mu_B \left[\frac{\sum_{-L}^{L} M_L \left(1 + \frac{M_L \mu_B B}{kT}\right)}{\sum_{-L}^{L}\left(1 + \frac{M_L \mu_B B}{kT}\right)} + \frac{\sum_{-S}^{S} 2M_S \left(1 + \frac{2M_S \mu_B B}{kT}\right)}{\sum_{-S}^{S}\left(1 + \frac{2M_S \mu_B B}{kT}\right)}\right] \quad ...(15)$$

Now, $$\sum_{-L}^{L} M_L = 0; \sum_{-L}^{L} M_L^2 = \frac{L(L+1)(2L+1)}{3}$$

$$\sum_{-L}^{L} 1 = (2L+1); \sum_{-S}^{S} M_S = 0 \quad \left(\because\ M_s = \pm\frac{1}{2}\right)$$

$$\sum_{-S}^{S} M_s^2 = \frac{S(S+1)(2S+1)}{3}; \sum_{-S}^{S} 1 = (2s+1).$$

Substituting these values in Eq. (15)

$$M = N\,\mu_B \left[\frac{\mu_B B\, L(L+1)}{3\,kT} + \frac{4\mu_B\, BS\,(S+1)}{3\,kT}\right] \quad ...(16a)$$

$$= \frac{N\,\mu_B^2\, B}{3\,kT}\,[L(L+1) + 4s\,(S+1)] \quad ...(16b)$$

Therefore, the magnetic susceptibility is

$$\chi = \frac{M}{H} = \frac{N\mu_0\mu_B^2}{3\,kT}\,[L(L+1) + 4S\,(S+1)] \quad ...(17)$$

and the effective number of Bohr magneton is

$$P_{\text{eff}} = \sqrt{L(L+1) + 4S(S+1)} \quad ...(18)$$

Thus, Curie's law is again obeyed, but the effective number of Bohr magneton is different from that in the case of wide multiplets.

### 15.16.1 Comparison between Theory and Experiment

Quantum theory of paramagnetism explains very well the measured values of $P_{\text{eff}}$ for most of the rare earth ions. The experimental value of the effective number of Bohr magnetons, $P_{\text{eff}}$, for iron transition group do not agree with the theoretical values calculated from Eqs. (14) and (18). To explain this, Bose, in 1927, suggested that the paramagnetism of ions arises entirely due to the spin motion and 3*d* orbital electron plays no role in magnetic moment. Hence, we may write

$$P_{\text{eff}} = \sqrt{4S(S+1)} \qquad ...(19)$$

Table 13.2 shows that the value of $P_{\text{eff}}$ obtained from the above expression agrees quite satisfactorily with experimental value.

**Table 13.2** Effective Number of Bohr Magneton for Iron Group Ions

| *Ion* | *Configuration* | $P_{eff} = g\sqrt{J(J+1)}$ | $P_{eff} = \sqrt{4S(S+1)}$ | *Experimental value* |
|---|---|---|---|---|
| $Ti^{3+}$, $V^{4+}$ | $3d^1$ | 1.55 | 1.73 | 1.8 |
| $V^{3+}$ | $3d^2$ | 1.63 | 2.83 | 2.8 |
| $Gr^{2+}$, $V^{2+}$ | $3d^3$ | 0.77 | 3.87 | 3.8 |
| $Mn^{2+}$ | $3d^4$ | 0.00 | 4.90 | 4.9 |
| $Fe^{3+}$ | $3d^5$ | 5.92 | 5.92 | 5.9 |
| $Fe^{2+}$ | $3d^6$ | 6.70 | 4.90 | 5.4 |
| $Co^{2+}$ | $3d^7$ | 6.54 | 3.87 | 4.8 |
| $Ni^{2+}$ | $3d^8$ | 5.59 | 2.83 | 5.2 |

Thus, the iron group ions behave as if the orbital magnetic momentum does not contribute at all to the total angular momentum. In such a case, we speak of *quenching of orbital momentum*. The quenching may not necessarily be complete, it may be partial.

In the rare earth group, the 4*f* electrons responsible for paramagnetism lie deep inside the ions and thus, are partly shielded from the electrostatic field of the neighbours by the outer 5*s* and 5*p* shells. Because of the shielding of 4*f* electrons, their orbits are practically the same as in the case of free ions.

Thus, in the case of iron group, coupling between $\vec{L}$ and $\vec{S}$ vectors is broken and the orbital momenta are quenched whereas in the case of rare earth ions. Coupling between $\vec{L}$ and $\vec{S}$ vectors is quite strong and one can specify the states by their $J$ values.

**Case 2: When the temperature is low and the field is high**

Then, the spacing of the multiplet levels is very large compared with the splitting of the magnetic sublevels. Assuming this, the total magnetic moment in the field direction is obtained by summing over all the sublevel, i.e.,

$$\langle \mu_J \rangle = \sum_{M_J} g\mu_B M_J P(M_J) \qquad ...(1)$$

or

$$\frac{\langle \mu_J \rangle}{g\mu_B} = \frac{\sum M_J \exp(g\mu_B M_J B / kT)}{\sum_{M_J} \exp(g\mu_B M_J B / kT)} \qquad ...(2)$$

To evaluate this expression, we put

$$u = \ln\left[\sum \exp\left(\frac{M_J y}{J}\right)\right] \equiv \ln v \quad ...(3)$$

where $$y = \frac{J\, g\, \mu_B B}{kT} \quad \text{and} \quad v = \sum \exp\left(\frac{M_J y}{J}\right) \quad ...(4)$$

Then $$\frac{du}{dy} = \frac{du}{du}\cdot\frac{dv}{dy} = \frac{\sum (M_J/J)\exp(M_J y/J)}{\sum \exp(M_J y/J)} \quad ...(5)$$

$$= \frac{\langle \mu_J \rangle}{J g \mu_B}$$

Thus, we have to evaluate the sum

$$v = \sum_{-J}^{+J} \exp\left(\frac{M_J y}{J}\right) \quad ...(6)$$

This may seem to be the sum of a simple geometric progression as follows. Let

$$z = \exp\left(\frac{y}{J}\right); \Rightarrow \quad \exp\left(\frac{M_J y}{J}\right) = Z^{M_J} \quad ...(7)$$

Then $$v = \sum_{-J}^{+J} Z^{M_J} = Z^{-J}(1 + Z + Z^2 + ... + Z^{2J})$$

$$= \frac{Z^{-J}(Z^{(2J+1)} - 1)}{Z - 1} = \frac{Z^{(J+1/2)} - Z^{-(J+1/2)}}{Z^{1/2} - Z^{-1/2}}$$

$$= \frac{\exp[1 + 1/2\, J)y] - \exp[-(1 + 1/2\, J)y]}{\exp(y/2J) - \exp(-y/2J)}$$

$$= \frac{\sinh(1 + 1/2J)y}{\sinh(y/2\, J)} \quad ...(8)$$

$$\Rightarrow \frac{dv}{dy} = \frac{(1+1/2J)\left[\sinh(y/2J)\cdot\cosh\left(1+\frac{1}{2J}\right)y\right] - \left(\frac{1}{2J}\right)\left[\sinh\left(1+\frac{1}{2J}\right)y\cdot\cosh\left(\frac{y}{2J}\right)\right]}{\left[\sinh\left(\frac{y}{2J}\right)\right]^2} \quad ...(9)$$

Also $$\frac{du}{dv} = \frac{1}{v} = \frac{\sinh(y/2J)}{\sinh(1+1/2J)y} \quad ...(10)$$

Hence, $$\frac{du}{dy} = \left(\frac{du}{dv}\right)\left(\frac{dv}{dy}\right)$$

or $$\frac{\langle \mu_J \rangle}{J g \mu_B} = \left(1 + \frac{1}{2J}\right)\coth\left[\left(1 + \frac{1}{2J}\right)y\right] - \frac{1}{2J}\coth\left(\frac{y}{2J}\right) \quad ...(11)$$

(Using Eqs. (9) and (10))

where $y = Jg\mu_B B/kT$. The R.H.S. of Eq. (11) is known as the Brillouin function

$$B(J, y) = \left(1+\frac{1}{2J}\right)\coth\left[\left(1+\frac{1}{2J}\right)y\right] - \frac{1}{2J}\coth\left(\frac{y}{2J}\right) \qquad ...(12)$$

Thus, we have magnetic moment per but volume

$$M \equiv \langle\mu_j\rangle N = N_g J\mu_B B_J(y) \qquad ...(13)$$

where $y = Jg\mu_B JB/kT$; $N$ is the number density of atoms and $B$ being the magnetic field.

Figure 13.8 shows the Brillouin functions for $J = 1/2$, 3/2, 5/2 and $\infty$. The curve for $J = \infty$ will be corresponding to the Langevin function

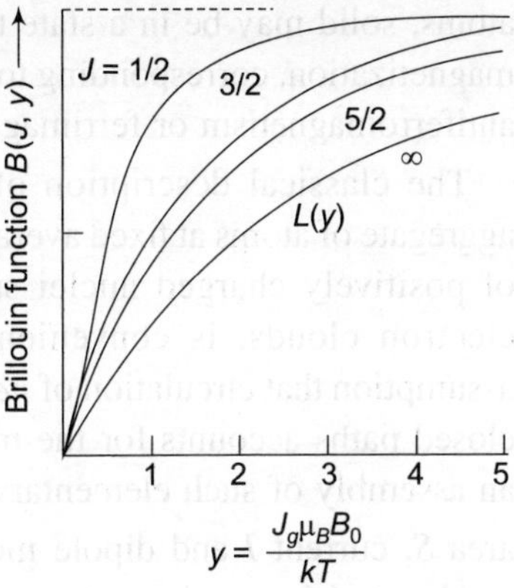


**Figure 13.8** Brillouin function for different values of $J$

$$L(y) \equiv \coth y - \frac{1}{y}.$$

$$M = Ng\mu_B JB_J(y) \qquad ...(13a)$$

The maximum value of magnetisation, represented by $M_S$, will be $Ng\mu_B J$, because $M_J$ will then have its maximum magnitude $J$, the magnetic axes of the atoms being along the magnetic field lines. Hence, Eq. (13) can be written as

$$M = M_S B_J(y) \qquad ...(14)$$

The following two cases are of special interest

(i) When $J = 1/2$, then

$$\frac{M}{M_S} = \tanh y = y - \frac{y^3}{3} + \frac{2}{15}y^5 + ...$$

(ii) When $J = \infty$, then in the limit, we get

$$\frac{M}{M_S} = \coth y - \frac{1}{y} = L(y)$$

which is the classical Langevin function.

Equation (14) is in good agreement with the experimental result obtained in the case of $GdI_2(SO_4).8H_2O$, at liquid helium temperature down to 1.3 K and magnetic fields up to 20,000 oersted. Thus, at low temperature and high field, we get the phenomenon of saturation.

## 13.17 DIAMAGNETISM AND MAGNETIZATION

Experiments have shown, that solids where the electrons are localized on isolated atoms as described by the tight-binding approximation display larger magnetic effects than those with nearly-free electrons, described by the weak-binding model. This suggests that magnetism is mainly of atomic origin and can be understood by treating the solid as a system of isolated atoms. If any interaction between the atoms is negligible, solids exhibit just diamagnetic or paramagnetic behaviour. If there is a cooperative interaction between the atoms, solid may be in a state that shows spontaneous magnetization, corresponding to either ferromagnetism, antiferromagnetism or ferrimagnetism.

The classical description of neutral matter, as an aggregate of atoms at fixed average positions, consisting of positively charged nuclei surrounded by negative electron clouds, is conveniently completed by the assumption that circulation of *negative electrons* around closed paths accounts for the magnetic properties. For an assembly of such elementary current loops, each of area $S$, current $I$ and dipole moments $\vec{\mu} = I\vec{S}$, which are located in an elementary volume $V$ smaller than the macroscopic dimensions but larger than the atomic dimensions, the magnetic moment per unit is defined by:

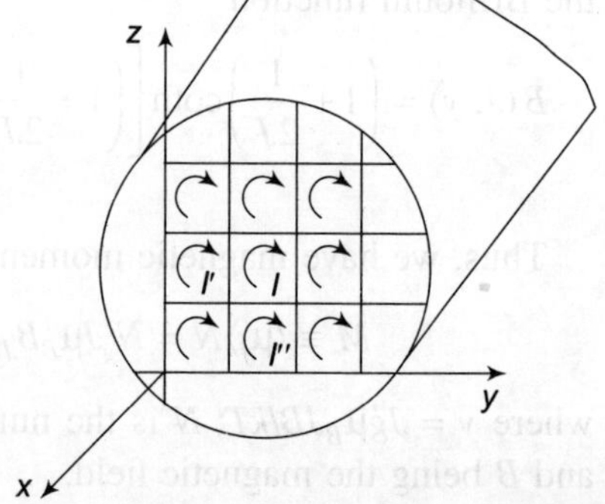


**Figure 13.9** Magnetization current density in a volume element

$$\vec{M} = \frac{1}{V}\sum\vec{\mu} \qquad ...(1)$$

This is called *magnetization*, and is the analogue of polarization $\vec{P}$ for dielectric solids. A physical picture of a volume $V$ of magnetic material is shown in Fig. 13.9 relative to a Cartesian coordinate system.

Assuming a rate of change of magnetization, three neighbouring volume elements, each of dimensions $dV = dx\,dy\,dz$, will have magnetic moments $d\vec{\mu}$, $d\vec{\mu}'$, $d\vec{\mu}'$ which must be associated with different current $I$, $I'$, $I''$. We can write the $x$-components of these magnetic moments as

$$d\mu_x = I\,dydz = M_x dV$$

$$d\mu'_x = I'\,dydz = \left(M_x + \frac{\partial M_x}{\partial y}dy\right)dV$$

$$d\mu''_x = I''\,dydz = \left(M_x + \frac{\partial M_x}{\partial z}dz\right)dV$$

so that rate of change of $M_x$ is related to a net current flow in both $z$- and $y$-directions

$$(I - I')\,dy\,dz = \left(-\frac{\partial M_x}{\partial y}dy\right)dx\,dy\,dz \quad \text{or} \quad I_{(x)z} = I - I' = -\frac{\partial M_x}{\partial y}dx\,dy$$

$$(I'' - I)\,dy\,dz = \left(\frac{\partial M_x}{\partial z}dz\right)dx\,dy\,dz \quad \text{or} \quad I_{(x)y} = I'' - I = -\frac{\partial M_x}{\partial z}dx\,dz$$

By similar arguments the rate of change of $M_y$ must be related to a net current flow in both $x$- and $z$-direction

$$I_{(y)x} = -\frac{\partial M_y}{\partial z}dy\,dz, \quad I_{(y)z} = \frac{\partial M_y}{\partial x}dx\,dy$$

and a net current flow along the $x$- and $y$-axis is associated with the change in $M_z$

$$I_{(z)x} = \frac{\partial M_z}{\partial y}\, dy\, dz, \quad I_{(y)z} = \frac{\partial M_z}{\partial x}\, dx\, dz$$

It follows that we can introduce a magnetization current density $\vec{j}_M$, of components

$$j_{M_x} = \frac{I_{(z)x} + I_{(y)x}}{dydz} = \frac{\partial M_z}{\partial y} - \frac{\partial M_y}{\partial z}, j_M = \frac{\partial M_x}{\partial z} - \frac{\partial M_z}{\partial x}, j_{M_z} = \frac{\partial M_x}{\partial y} - \frac{\partial M_y}{\partial x}$$

or, in a compact form

$$\vec{j}_M = \nabla \times \vec{M} \qquad \text{...(2)}$$

A *magnetization current density* ($\vec{j}_M$) is associated with the atomic current loops wherever there is a spatial rate of change of magnetization, a situation analogous to that for a dielectric, where the polarization charge requires that the divergence of polarization should not be zero. Hence, in a uniformly magnetized solid, the magnetization current density is zero. When using Ampere's law to investigate the magnetization currents of density $\vec{j}_M$, and hence, reserving the symbol $\vec{j}$ for the free current density only, one obtains

$$\nabla \times \vec{B} = \mu_0\, (\vec{j} \times \vec{j}_M) = \mu_0\, (\vec{j} + \nabla \times \vec{M}) \qquad \text{...(3)}$$

For convenience this relation is rearranged so that only free current density appears on the right-hand side

$$\nabla \times \left( \frac{\vec{B}}{\mu_0} - \vec{M} \right) = \vec{j} \qquad \text{...(4)}$$

and Ampére's law in a magnetic material takes the differential form

$$\nabla \times \vec{H} = \vec{j} \qquad \text{...(5)}$$

## 13.18 GENERAL FORMULATION IN CASE OF A UNIFORM FIELD

In the presence of a uniform magnetic field, the Hamiltonian of an ion (or atom) is modified in the following major ways:

(1) In the total electronic kinetic energy $T_0 = \sum p_i^2/2m$, the momentum of each electron of charge $-e$ is replaced by

$$\vec{p}_i \rightarrow \vec{p}_i + \frac{e}{c} \vec{A}\, (\vec{r}_i) \qquad \text{...(1)}$$

where $\vec{A}$ is the vector potential. $\vec{A}$ in a uniform field $\vec{H}$ has the form

$$\vec{A} = -\frac{1}{2} \vec{r} \times \vec{H} \qquad \text{...(2)}$$

so that the conditions

$$\vec{H} = \nabla \times \vec{A} \quad \text{and} \quad \nabla \cdot \vec{A} = 0 \qquad \text{...(3)}$$

are both satisfied.

(2) The interaction energy of the field with each electron spin $\vec{s}_i = \frac{1}{2} \vec{\sigma}_i$ must be added to the Hamiltonian

$$\Delta\mathcal{H} = g_0\,\mu_B\,H\,S_z, \left(\vec{S}_z = \sum_i \vec{s}_z^{\,i}\right) \qquad ...(4)$$

Here $\mu_B$, the Bohr magneton is given by

$$\mu_B = \frac{e\hbar}{2mc} = 0.927 \times 10^{-20} \text{ erg/G.}$$

$$= 0.579 \times 10^{-8} \text{ eV/G.}$$

## 13.19 THE INTERACTION OF ORBITAL MOTION WITH MAGNETIC FIELD (ORIGIN OF DIAMAGNETIC SUSCEPTIBILITY AND VAN VLECK PARAMAGNETISM)

The simplest atomic system to start our analysis with is the *H* atom. We consider therefore a single electron moving round the nucleus of an atom with period $\tau$. This is equivalent to a current loop carrying current $j = -e/\tau$. The magnetic moment of the loop is $-eS/\tau c$, where *S* is the area of the loop. But $S/\tau = 1/2\ m$, where *l* is the orbital angular momentum of the electron. Therefore the magnetic moments of the loop is

$$\mu_L = -\frac{e}{2mc}\,\mathbf{l}. \qquad ...(5)$$

Now, in the presence of an applied magnetic field **B**, the classical equation of motion of the electron is

$$m\ddot{\mathbf{r}} = \frac{e}{c}\,\dot{\mathbf{r}} \times \mathbf{B} - \frac{\partial V}{\partial \mathbf{r}} \qquad ...(6)$$

where $V(r) = -e^2/r$ is the potential energy function for the electron in the *H* atom. If we transform to a set of axes rotating with angular velocity $\omega = (e/2mc)\mathbf{B}$, then ignoring terms of order $\omega^2$ the new equation of motion becomes

$$\ddot{\mathbf{r}} = 2\omega \times \mathbf{r} + \frac{e}{mc}\,\mathbf{B} \times \mathbf{r} - \frac{1}{m}\frac{\partial V}{\partial \mathbf{r}}$$

$$= -\frac{1}{m}\frac{\partial V}{\partial \mathbf{r}}. \qquad ...(7)$$

Therefore the net effect of the magnetic field is to impress upon the electron a (Lamor) precession with angular velocity $\omega$.

The angular momentum of the electron in the atom is

$$\mathbf{l} = m\mathbf{r} \times (\omega \times \mathbf{r}) = \mathbf{m}[\omega r^2 - (\omega \cdot \mathbf{r})\mathbf{r}]. \qquad ...(8)$$

Thus, if **B** is put along the *z* axis, then we have

$$l_x = l_y = 0 \quad \text{and} \quad l_z = m\omega\,(r^2 - r^2\cos^2\theta) = m\omega(x^2 + y^2)$$

where $\theta$ is the angle between **r** and **B**. The mean value of $l_z$ is therefore given by

$$\frac{\bar{l}_z}{m\omega} = (\bar{x}^2 + \bar{y}^2) = 2\,(\bar{x}^2) = \frac{2}{3}\,\bar{r}^2$$

and we may write the induced magnetic moment for a system of *N* such atoms as

$$M = -\frac{Ne}{2mc}\,\bar{l}_z = -\frac{Ne^2}{6mc^2}\,\bar{r}^2\,B \equiv \chi^B. \qquad ...(9)$$

It follows that the susceptibility is negative,

$$\chi = -\frac{Ne^2}{6mc^2}\,\bar{r}^2; \qquad ...(10)$$

i.e., the induced magnetic moment is diamagnetic. This is the classical Langevin formula for orbital diamagnetism.

This same result can be derived more straightforwardly from quantum mechanics. We consider the simplest example of magnetic interaction Hamiltonian we can construct for an atomic system; namely,

$$H = \frac{1}{2m}\left(\mathbf{p} + \frac{e}{c}\mathbf{A}\right)^2 + V(\mathbf{r}) \qquad ...(11)$$

where **A** is the vector potential defining the applied magnetic field $\mathbf{B} \equiv \text{curl } \mathbf{A}$; and the momentum of the electron relative to the nucleus is $\mathbf{p} = -i\hbar\nabla$. This Hamiltonian may be written as a sum of two terms, namely, an unperturbed Hamiltonian $p^2/2m + V(\mathbf{r})$, and a perturbation

$$H' = \frac{ie\hbar}{2mc}(\nabla\cdot\mathbf{A} + \mathbf{A}\cdot\nabla) + \frac{e^2}{2mc^2}\mathbf{A}^2 \qquad ...(12)$$

The (uniform) magnetic field determines the vector potential in the Landau gauge as

$$\mathbf{A} = \frac{1}{2}\mathbf{B}\times\mathbf{r} \qquad ...(13)$$

Thus, we may eliminate **A** from $H'$ to find

$$H' = \frac{e}{2mc}\mathbf{r}\times\mathbf{p}\cdot\mathbf{B} + \frac{e^2}{8mc^2}(\mathbf{B}\times\mathbf{r})^2 \qquad ...(14)$$

This shows that the electron is coupled to **B** via its orbital angular momentum

$$\mathbf{l} = \mathbf{r}\times\mathbf{p} \qquad ...(15)$$

so that the magnetic field induces a dipole moment, $\mu_l \equiv -(e/2mc)\mathbf{l}$, as before For **B** along the $z$ axis, we have explicitly

$$H' = -\frac{ie\hbar B}{2mc}\left(x\frac{\partial}{\partial y} - y\frac{\partial}{\partial x}\right) + \frac{e^2B^2}{8mc^2}(x^2+y^2). \qquad ...(16)$$

The second of the two terms in this expression leads to the Langevin formula (10) for the *diamagnetic susceptibility* for an atom in its ground state. The first term has zero expectation value in this (ground state), but it is responsible for the *Van Vleck paramagnetism* of an atom whose excited states have non-zero orbital moment and can therefore be mixed with the ground state by the term in $H'$ proportional to $B$.

## 13.20 QUANTUM THEORY OF DIAMAGNETISM

We know that for a charged particle moving in an electromagnetic field, the Hamiltonian ($H'$) is

$$H' = \frac{1}{2m}\left(\vec{p} + \frac{e}{c}\vec{A}\right)^2 - e\phi \qquad ...(1)$$

where $p$ is the momentum of the particle, $\phi$ is scalar potential and $\vec{A}$ is magnetic vector potential.

Since
$$p = -i\hbar\,\nabla$$

$$H = \frac{1}{2m}\left(-i\hbar\nabla + \frac{e}{c}\vec{A}\right)^2 - e\phi \qquad \text{...(2)}$$

We have
$$\nabla = \vec{i}\frac{\partial}{\partial x} + \vec{j}\frac{\partial}{\partial y} + \vec{k}\frac{\partial}{\partial z}$$

$$\vec{A} = \vec{i}A_x + \vec{j}A_y + \vec{k}A_z$$

Therefore,
$$\left(-i\hbar\nabla + \frac{e}{c}\vec{A}\right)_x^2 = \left(-i\hbar\frac{\partial}{\partial x} + \frac{e}{c}A_x\right)\left(-i\hbar\frac{\partial}{\partial x} + \frac{e}{c}A_x\right)$$

$$= -\hbar^2\frac{\partial^2}{\partial x^2} - i\hbar\frac{\partial}{\partial x}\left(\frac{e}{c}A_x\right) - \frac{e}{c}A_x i\hbar\frac{\partial}{\partial x} + \frac{e^2}{c^2}A_x^2$$

$$= -\hbar^2\frac{\partial^2}{\partial x^2} - i\hbar\left[A_x\frac{\partial}{\partial x}\left(\frac{e}{c}\right) + \frac{e}{c}\frac{\partial A_x}{\partial x}\right] - \frac{e}{c}A_x i\hbar\frac{\partial}{\partial x} + \frac{e^2}{c^2}A_x^2$$

$$= -\hbar^2\frac{\partial^2}{\partial x^2} - i\hbar\frac{e}{c}\left(\frac{\partial A_x}{\partial x} + 2A_x\frac{\partial}{\partial x}\right) + \frac{e^2}{c^2}A_x^2$$

Here, we have implicitly operated over some function, say $\psi$.

Writing all the three components, $x$, $y$, $z$, we have

$$\left(i\hbar\nabla + \frac{e}{c}\vec{A}\right)^2 = -\hbar^2\nabla^2 - i\hbar\frac{e}{c}\,\text{div}\,\vec{A} - 2i\hbar\frac{e}{c}\left(A_x\frac{\partial}{\partial x} + A_y\frac{\partial}{\partial y} + A_z\frac{\partial}{\partial z}\right) + \frac{e^2}{c^2}A^2$$

Thus, the Hamiltonian becomes

$$H' = -\frac{\hbar^2}{2m}\nabla^2 - \frac{i\hbar e}{mc}\,\text{div}\,\vec{A} + \frac{i\hbar e}{mc}\left(A_x\frac{\partial}{\partial x} + A_y\frac{\partial}{\partial y} + A_z\frac{\partial}{\partial z}\right) + \frac{e^2}{2mc^2}A^2 - e\phi$$

Putting $\text{div}\,\vec{A} = \frac{1}{c}\frac{\partial\phi}{\partial t}$, we get

$$H' = -\frac{\hbar^2}{2m}\nabla^2 - \frac{i\hbar e}{mc}\frac{\phi}{c} - \frac{i\hbar e}{mc}\left(A_x\frac{\partial}{\partial x} + A_y\frac{\partial}{\partial y} + A_z\frac{\partial}{\partial z}\right) + \frac{e^2}{2mc^2}A^2 - e\phi \qquad \text{...(3)}$$

The uniform external field $H$ gives the vector potential in the Landau gauge as

$$\vec{A} = \frac{1}{2}\vec{H}\times\vec{r} \qquad \text{...(4)}$$

For $\vec{H}$ along the $z$-axis, $H_x = H_y = 0$ and $H = H_z$ so that

$$A_x = -\frac{1}{2}yH_z = -\frac{1}{2}yH$$

$$A_y = -\frac{1}{2} xH_z = \frac{1}{2} xH$$

$$A_z = 0$$

Therefore, $$A_x \frac{\partial}{\partial x} + A_y \frac{\partial}{\partial y} + A_z \frac{\partial}{\partial z} = \frac{1}{2} H \left( x\frac{\partial}{\partial y} - y\frac{\partial}{\partial x} \right)$$

and $$\dot{\phi} \equiv \frac{\partial \phi}{\partial t} = \frac{\partial}{\partial t}\left( -\frac{e}{r} \right) = 0$$

Also $$A^2 = \vec{A} \cdot \vec{A}$$

$$= \left[ \vec{i}\left( -\frac{1}{2} yH \right) + \vec{j}\left( \frac{1}{2} xH \right) \right] \cdot \left[ \vec{i}\left( -\frac{1}{2} yH \right) + \vec{j}\left( \frac{1}{2} xH \right) \right]$$

$$= \frac{1}{4} y^2 H^2 + \frac{1}{4} x^2 H^2$$

$$= \frac{1}{4} H^2 (x^2 + y^2)$$

Putting these values, we get for $H'$

$$H' = -\frac{\hbar^2}{2m}\nabla^2 - \frac{i\hbar e}{mc}\frac{H}{2}\left( x\frac{\partial}{\partial y} - y\frac{\partial}{\partial x} \right) + \frac{e^2}{8mc^2} H^2 (x^2 + y^2)$$

Thus, the additional term in the Hamiltonian due to the magnetic field is

$$H' = \frac{i\hbar e}{mc} H \left( x\frac{\partial}{\partial y} - y\frac{\partial}{\partial x} \right) + \frac{e^2 H^2}{8mc^2} (x^2 + y^2) \qquad ...(5)$$

Now, we note that

(i) First term can be written as

$$H' = \frac{eH}{2\ mc}\left[ x\left( -i\hbar\frac{\partial}{\partial y} \right) - y\left( -i\hbar\frac{\partial}{\partial x} \right) \right]$$

$$= \frac{eH}{2\ mc} (xp_y - yp_x)$$

$$= \frac{eH}{2\ mc} (\vec{r} \times \vec{p})_z = \frac{eH}{2\ mc} L_z$$

where $L_z$ is the $Z$ component of angular momentum. Therefore magnetic field induces a dipole moment

$$\vec{\mu}_l = -\left( \frac{e}{2mc} \right) \vec{L}_z \qquad ...(6)$$

Therefore, first term $= -(-\vec{\mu}_l \cdot \vec{H})$

This term is proportional to the orbital angular momentum component, $L_z$; therefore, in mononuclear system this gives rise only to paramagnetism.

(ii) The second term, if written for a unit volume of a substance containing $N$ atoms, containing $Z$ electrons, would become

$$ZN\frac{e^2H^2}{8mc^2}\sum_i (x_i^2 + y_i^2) = NZ\frac{e^2H^2}{8mc^2}\overline{\rho}^2 \qquad ...(7)$$

where $\overline{\rho}^2$ represents the mean squares of the radii of the projections of orbits on a plane perpendicular to $H$. Now, if the magnetic field produces induced dipole moments in the substance, then the corresponding energy term would be quadratic in $H$. Thus, the second term is the energy term associated with the diamagnetism of the solid. On comparing it with $-\frac{1}{2}\chi_{\text{dia}}H^2$, we get

$$\chi_{\text{dia}} = -NZ\left(\frac{e^2}{4mc^2}\right)\overline{e}^2 \qquad ...(8)$$

Since $$\overline{\rho}^2 = \frac{2}{3}\overline{r}^2$$

Therefore, $$\chi_{\text{dia}} = -NZ\left(\frac{e^2}{6mc^2}\right)\overline{r}^2 \qquad ...(9)$$

This expression for $\chi_{\text{dia}}$ is to be corrected by substituting the values of $\overline{r}^2$ using quantum orbital theory. For hydrogen-like atoms, $\overline{r}^2$ from the Bohr's theory is given by

$$\overline{r}^2 = a_0^2\frac{n_r^2}{Z^2}\left(\frac{5}{2}n_r^2 - \frac{3}{2}k^2\right) \qquad ...(10)$$

where $n_r$ is radial quantum number and $k$ is azimuthal number, $Z$ is the effective nuclear charge and $a_0$ is the Bohr radius (normal states orbital radius of hydrogen atom).

$$a_0 = 0.528 \times 10^{-8} \text{ cm}$$

Therefore, atomic diamagnetic susceptibility is given by

$$\chi_a = -0.79 \times 10^{-6}\frac{n_r^2}{Z^2}\left(\frac{5}{2}n_r^2 - \frac{3}{2}k^2\right)$$

Theoretical value of susceptibility was found to be much smaller than the experimental value. To make the expression more accurate, Van Vleck gave modified expression as

$$\chi_a = -0.79 \times 10^{-6}\frac{n_r^2}{Z^2}\left(\frac{5}{2}n_r^2 - \frac{3l(l+1)}{2}\right)$$

where $l$, the orbital quantum number $= k - 1$

For molecular hydrogen, the susceptibility will be

$$\chi_m = 2\chi_a$$

$$= 2 \times (-0.79 \times 10^{-6})\frac{n_r^2}{Z^2}\left(\frac{5}{2}n_r^2 - \frac{3l(l+1)-1}{2}\right)$$

$$= 2 \times (-0.79 \times 10^{-6}) \times \left(\frac{5}{2} + 0 + \frac{1}{2}\right)$$

$$= -4.74 \times 10^{-6} \qquad ...(11)$$

The experimental value of $\chi_m$ for hydrogen is found to be $-4.00 \times 10^{-6}$, which is in good agreement with the above calculated value.

**Example 13.1** Derive the classical Langevin formula for orbital diamagnetism.

***Solution*** The electron moving about the nucleus represents a current circulating in a closed loop, and this produces a magnetic dipole moment:

$$\vec{\mu} = I\vec{S} = \int \frac{dq}{dt}\, d\vec{S}$$

where the integration is over the orbital area. If $\vec{r}$ is the position vector of the electron and $\vec{v}$ its velocity, $d\vec{S}$ is a vector of magnitude equal to the area of a triangle of which $\vec{v}\,dt$ is the base. The direction of $d\vec{S}$ is normal to the base. Hence,

$$d\vec{S} = \frac{1}{2}(\vec{r} \times \vec{v}\,dt)$$

On substitution, one obtains

$$\vec{\mu} = \frac{1}{2}\int dq\,(\vec{r} \times \vec{v}) = \frac{1}{2m_e}(\vec{r} \times m_e\vec{v})\int dq = -\frac{e}{2m_e}\vec{L}.$$

In the presence of an applied field, the equation of motion of the electron is:

$$m_e \frac{d\vec{v}}{dt} = -e(\vec{E} + \vec{v} \times \vec{B}_0)$$

Assuming, for simplicity, that $\vec{B}_0 = (0, 0, B_0)$, this can be written as:

$$\begin{pmatrix} \ddot{x} \\ \ddot{y} \\ \ddot{z} \end{pmatrix} = -\frac{e}{m_e}\begin{pmatrix} E_x \\ E_y \\ E_z \end{pmatrix} - \frac{eB_0}{m_e}\begin{pmatrix} \dot{y} \\ -\dot{x} \\ 0 \end{pmatrix}$$

It is convenient to consider the magnetic field removed, and transform the equation of motion to a set axes rotating with constant angular velocity (ω) about the $z$-axis, by using the formal rule

$$\frac{d}{dt} \to \frac{d}{dt} + \vec{\omega}\times \quad \text{or} \quad m_e\left(\frac{d\vec{v}}{dt} + \vec{\omega} \times \vec{v}\right) = -e\vec{E}$$

where, according to the same rule, $\vec{v} = d\vec{r}/dt + \omega \times \vec{r}$. This yields

$$m_e\left(\frac{d^2\vec{r}}{dt^2} + 2\vec{\omega} \times \frac{d\vec{r}}{dt} + \vec{\omega} \times \vec{\omega} \times \vec{r}\right) = -e\vec{E}$$

or:
$$\begin{pmatrix} \ddot{x} \\ \ddot{y} \\ \ddot{z} \end{pmatrix} = -\frac{e}{m_e}\begin{pmatrix} E_x \\ E_y \\ E_z \end{pmatrix} - 2\omega\begin{pmatrix} \dot{y} \\ -\dot{x} \\ 0 \end{pmatrix} + \omega^2\begin{pmatrix} x \\ y \\ 0 \end{pmatrix}$$

Ignoring terms of order $\omega^2$, it is apparent that the equations of motion become identical if

$$\omega = \omega_L = \frac{eB_0}{2m_e}$$

which is called the Larmor frequency. This means that the net effect of the magnetic field is to impress upon the electron a precession with angular velocity, $\omega_L$. In other words, the angular momentum $\vec{L}$, which is normal to the plane of the orbit, precesses about the field direction. Each electron acquires, as a result of precession, an additional angular momentum

$$\Delta\vec{L} = m_e\vec{r} \times (\vec{\omega}_L \times \vec{r}) = m_e[\omega_L r^2 - (\vec{\omega}_L \cdot \vec{r})\vec{r}]$$

such that $\Delta L_x = \Delta L_y = 0$ and the component along the $z$-axis is

$$\Delta L_z = m\omega_L(r^2 - r^2\cos^2\theta) = \frac{eB_0}{2}(r^2 - z^2) = \frac{eB_0}{3}\langle r^2\rangle$$

The induced magnetic moment reads

$$\Delta\mu = \frac{e^2B_0}{6m_e}\langle r^2\rangle$$

and this leads to the diamagnetic susceptibility,

$$\chi_d = \frac{N\mu_0}{B_0}\Delta\mu = -\frac{Ne^2\mu_0}{6m_e}\langle r^2\rangle$$

which is the classical Langevin formula.

**Example 13.2** Find the quantum representation of the magnetic moment induced by the presence of an external magnetic field $\vec{B}_0 = \nabla \times \vec{A}$.

***Solution*** The classical current density $\vec{j}_l = \rho\,\vec{v}$, which corresponds to the orbital motion of electron charge $-e$ around the nucleus

$$-e = \int \rho dV$$

gives rise to an orbital magnetic moment $\vec{\mu}_l$, assigened to each electron orbit of angular momentum $\vec{L}$ in the form

$$\vec{\mu}_l = -\frac{e}{2m_e}\vec{L} = \frac{1}{2m_e}\int_V \rho\,\vec{L}dV = \frac{1}{2m_e}\int_V \rho(\vec{r}\times\vec{p})\,dV$$

$$= \frac{1}{2}\int_V (\vec{r}\times\rho\vec{v})dV = \frac{1}{2}\int_V (\vec{r}\times\vec{j}_l)\,dV$$

In the presence of an applied magnetic field $\vec{B}_0 = \nabla \times \vec{A}$, the total current density reads

$$\vec{j} = \rho \vec{v} = \frac{\rho}{m_e} (\vec{p} + e\vec{A})$$

and its quantum mechanical counterpart can be written as

$$\hat{\vec{j}} = -\frac{e}{2m_e} [\langle\psi| -i\hbar\nabla + e\vec{A}\,|\psi\rangle + \langle\psi|\, i\hbar\nabla + e\vec{A}\,|\psi\rangle]$$

$$= \frac{ie\hbar}{2m_e} (\psi^*\nabla\psi - \psi\nabla\psi^*) - \frac{e^2}{m_e} \vec{A}\psi^*\psi = \hat{\vec{j}}_l + \hat{\vec{j}}_d$$

where the charge density has been taken as $\rho = -e\psi^*\psi$, where $\psi^*\psi$ is the probability density, and the symmetrisation rule has been associated with the substitution $\vec{p} \rightarrow -i\hbar\nabla$. The first term

$$\hat{\vec{j}}_l = \frac{ie\hbar}{2m_e} (\psi^*\nabla\psi - \psi\nabla\psi^*)$$

is the quantum representation of the orbital current density, as it yields

$$\vec{\mu}_l = \frac{1}{2} \int_V (\vec{r} \times \hat{\vec{j}}_l)\, dV = \frac{ie\hbar}{4m_e} \int_V [\psi^* (\vec{r} \times \nabla)\, \psi - \psi\, (\vec{r} \times \nabla)\, \psi^*]\, dV$$

$$= -\frac{e}{4m_e} \left[ \int_V \psi^*\, \hat{L}\psi\, dV + \int_V (\hat{L}\psi)^*\, \psi\, dV \right]$$

which is similar to the classical result. The second term gives rise, in the presence of $B_0 = \nabla \times \vec{A}$, to an additional magnetic moment, of magnitude

$$\vec{\mu}_d = \frac{1}{2} \int_V (\vec{r} \times \vec{j}_d)\, dV = -\frac{e^2}{2m_e} \int \psi^*\, (\vec{r} \times \vec{A})\psi\, dV = -\frac{e^2}{2m_e} \left\langle \psi | \vec{r} \times \vec{A}\, | \psi \right\rangle$$

This is the expectation value of the operator

$$\hat{\vec{\mu}}_d = -\frac{e^2}{2m_e} (\vec{r} \times \vec{A})$$

which is associated with the diamagnetic moment, induced by the presence of an external magnetic filed.

**Example 13.3** Calculate the magnetic polarizability of the atom in the presence of an applied magnetic field, $\vec{B}_0$.

***Solution*** The magnetic polarizability is defined as the coefficient of proportionality between the diamagnetic moment $\vec{\mu}_d$, derived in Example 13.2 and the applied field, $\vec{B}_0$

$$\vec{\mu}_d = -\frac{e^2}{2m_e} \langle\psi|\, \vec{r} \times \vec{A}\, |\psi\rangle, \qquad \vec{\mu}_d = -\alpha\, \vec{B}_0$$

The induced magnetic moment always has an orientation antiparallel to $\vec{B}_0$. If the applied field is uniform, it is common practice to choose $\vec{A} = (\vec{B}_0 \times \vec{r})/2$ such that

$$\nabla \times \vec{A} = \nabla \times \left(\frac{1}{2}\vec{B}_0 \times \vec{r}\right) = \frac{1}{2}[(\nabla \cdot \vec{r})B_0 - (B_0 \cdot \nabla)\vec{r}] = \frac{1}{2}(3\vec{B}_0 - \vec{B}_0) = \vec{B}_0$$

On substitution, this yields

$$\vec{\mu}_d = -\frac{e^2}{2m_e}\left\langle \psi | \vec{r} \times \left(\frac{1}{2}\vec{B}_0 \times \vec{r}\right) | \psi \right\rangle = -\frac{e^2}{4m_e}\langle \psi | (\vec{r} \cdot \vec{B}_0)\vec{r} - r^2 \vec{B}_0 | \psi \rangle$$

which is linear in $\vec{B}_0$. The diamagnetic moment is significant for *s*-electrons only, which are described by wave functions $\psi_{n00}$ $(l = 0)$, because the contribution of the orbital magnetic moment vanishes in these states. The wave functions $(\psi_{n00})$ are spherically symmetric, which implies

$$\iint x_i x_j \sin\theta \, d\theta \, d\varphi = 0 \quad (i \neq j)$$

$$\iint x_i^2 \sin\theta \, d\theta \, d\varphi = \frac{1}{3}\iint r^2 \sin\theta \, d\theta \, d\varphi$$

and hence, one obtains

$$\vec{\mu}_d = -\frac{e^2 \vec{B}_0}{4m_e}\iiint (r^2 - z^2)|\psi_{n00}|^2 \, dV = -\frac{e^2 \vec{B}_0}{6m_e}\iiint r^2 |\psi_{n00}|^2 \, dV = -\alpha \vec{B}_0$$

We may evaluate the magnetic polarizability, for simplicity, in the ground state of hydrogen

$$\psi_{100} = \frac{1}{\sqrt{\pi a_0^3}} e^{-r/a_0}$$

where $a_0$ is the Bohr radius. This yields

$$\alpha_H = \frac{e^2}{6m_e}\iiint r^2 |\psi_{100}|^2 \, dV = \frac{e^2}{6m_e}\frac{1}{\pi a_0^3}\iiint r^4 e^{-2r/a_0} \sin\theta \, d\theta \, d\varphi \, dr$$

$$= \frac{e^2}{6m_e}\frac{4}{a_0^3}\int_0^\infty r^4 e^{-2r/a_0} \, dr = \frac{e^2 a_0^2}{2m_e}$$

## 13.21 WEISS CLASSICAL THEORY OF FERROMAGNETISM

To explain the phenomenon of ferromagnetism, Weiss assumed that a ferromagnetic material, at any temperature below the Curie point, is saturated magnetically to a value appropriate to that temperature even though the applied field is zero. This magnetization is known as *spontaneous magnetization*. To explain this, he put forward the idea of *domains*. According to Weiss, even though a ferromagnetic substance is everywhere magnetically self saturated, it is divided into small regions, called *domains*. In each domain the magnetization vectors are all parallel so that each domain is magnetized to saturation but, for different domains, the magnetization vectors are oriented in such different directions that the resultant intensity of magnetization is zero (Refer to Fig. 13.10). Therefore, for the specimen to exhibit bulk magnetization, the alignment of some of the magnetization vectors parallel to the direction of applied field is necessary.

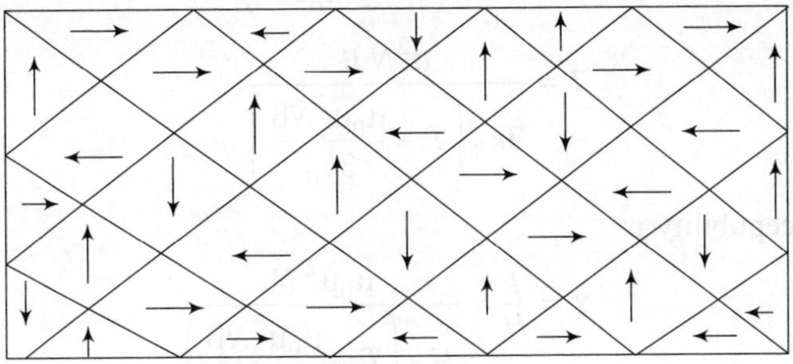

**Figure 13.10** Alignment of magnetization vectors in different domains in the absence of external field

As in the paramagnetic case, he represented the total effective field by

$$B_e = \mu_0(H + H_1) = \mu_0(H + \beta I) \qquad \text{...(1)}$$

Under this assumption, Weiss theory of ferromagnetism is reduced to Langevin's theory of paramagnetism with $H_e$ replacing $H$. According to Langevin's theory, the intensity of magnetization is given by

$$I = I_S\, L(a)$$

where $I_S =$

$\mu N$ is the saturation value of magnetization and $L(a)$ is the Langevin function.[2] In a ferromagnetic substance, therefore, we have

$$I = I_S\, L(a)$$

with $$a = \frac{\mu B_e}{k_B T} = \frac{\mu_0 \mu (H + \beta I)}{k_B T}$$

$\Rightarrow$ $$a = \frac{\mu(\mu_0 H + \mu_0 \beta I)}{k_B T}\ \frac{\mu(B + \mu_0 \beta I)}{k_B T} \qquad \text{...(2)}$$

As we have seen, at high temperature and low field, $a$ is very small and $L(a) = \frac{a}{3}$. Therefore, at high temperature and low field,

$$I = I_S \cdot \frac{a}{3} = \mu N \cdot \frac{\mu (B + \mu_0 \beta I)}{3 k_B T} = \frac{\mu^2 N (B + \mu_0 \beta I)}{3\, k_B T}$$

$\Rightarrow$ $$I \left(1 - \frac{\mu_0 \mu^2 N \beta}{3\, k_B T}\right) = \frac{\mu^2 N B}{3\, k_B T}$$

---

2 Also see Art. 13.13

$$\Rightarrow \qquad I = \frac{\mu^2 N B}{3k_B T\left(1 - \dfrac{\mu_0 \mu^2 N\beta}{3k_B T}\right)}$$

$$\Rightarrow \qquad I = \frac{\mu^2 N B}{3k_B\left(T - \dfrac{\mu_0 \mu^2 N\beta}{3k_B}\right)}$$

Hence, the susceptibility is

$$\chi = \frac{I}{H} = \frac{\mu_0 \mu^2 N}{3k_B\left(T - \dfrac{\mu_0 \mu^2 N\beta}{3k_B}\right)}$$

$$\Rightarrow \qquad \chi = \frac{C}{T - T_f} \qquad ...(3)$$

where $$C = \frac{\mu_0 \mu^2 N}{3\,k_B}$$

is the Curie constant, and

$$T_f = \frac{\mu_0 \mu^2 N\beta}{3\,k_B} \qquad ...(4)$$

is another constant, known as the *ferromagnetic Curie temperature.*

Equation (3) is the *Curie–Weiss law* which is obeyed by ferromagnetic substances above the ferromagnetic Curie temperature, $T_f$.

To examine the condition for spontaneous magnetization, we consider Eq. (2) which gives

$$\mu_0 \beta I = \frac{a k_B T}{\mu} - B \qquad ...(5)$$

In the absence of external field, $B = 0$ and Eq. (5) reduces to

$$\mu_0 \beta I = \frac{a k_B T}{\mu}$$

$$\Rightarrow \qquad I = \frac{a k_B T}{\mu_0 \mu \beta} \qquad ...(6)$$

Equation (6) gives the expression for spontaneous magnetization of different domains in a ferromagnetic material. Now,

$$\frac{I}{I_S} = \frac{a k_B T}{\mu_0 \mu \beta} \cdot \frac{1}{\mu N} = \frac{a k_B T}{\mu_0 \mu^2 N\beta} = \left[\frac{T}{3\left(\dfrac{\mu_0 \mu^2 N\beta}{3kB}\right)}\right] a$$

$$\Rightarrow \qquad \frac{I}{I_S} = \left(\frac{T}{3T_f}\right) a \qquad ...(7)$$

where we have used Eq. (4).

In general, whether the field is zero or not, the intensity of magnetization is given by

$$I = I_S L(a)$$

$$\Rightarrow \qquad \frac{I}{I_S} = L(a) \qquad ...(8)$$

Therefore, for a ferromagnetic substance to exhibit spontaneous magnetization, both Eqs. (7) and (8) must be simultaneously satisfied.

The required for a ferromagnetic substance to exhibit spontaneous magnetization is given by the points of intersection of the plots of Eqs. (7) and (8) as shown in Fig. 13.12.

It is obvious from Fig. 13.11 that the straight line, for which $T < T_f$, intersects the curve at non-zero point which implies there is some value of $\frac{I}{I_S}$, even though the externally applied field is zero. Hence, spontaneous magnetization occurs for $T < T_f$. The line $T = T_f$ represents the tangent to the curve at the origin at which $\frac{I}{I_S} = 0$ and, therefore, there cannot be any possibility of spontaneous magnetization for $T = T_f$. Although the point $O$ also represents a possible solution, the magnetization corresponding to this point is unstable. All the lines for which $T > T_f$, no intersection with the curve is possible and hence, spontaneous magnetization cannot arise.

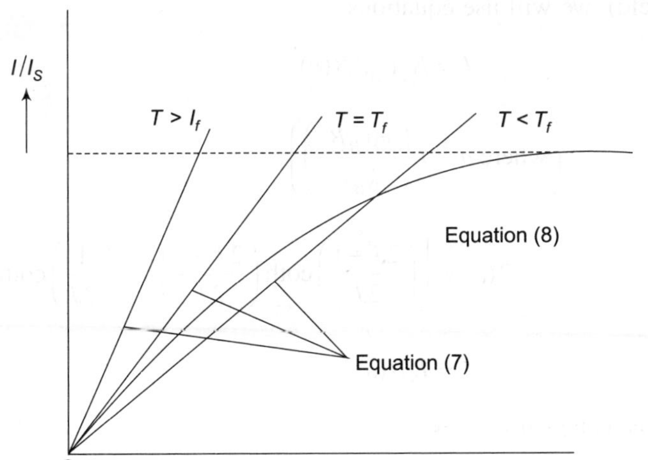


**Figure 13.11** Plots of Eqs. (7) and (8)

Now, the straight line represented by Eq. (7) will intersect the Langevin's curve if its slope is less than that of the Langevin's curve at the origin. The slope of the straight line is $\left(\frac{T}{3T_f}\right)$ and that of the Langevin's curve at the origin is $\frac{1}{3}$. Hence, the straight line will intersect the Langevin's curve if,

$$\frac{T}{3T_f} < \frac{1}{3}$$

$$\Rightarrow \qquad T < T_f$$

Thus, $T_f$ is the temperature below which a ferromagnetic substance exhibits spontaneous magnetization and the domains are formed.

An inspection of Eq. (6) (Section 13.14) and Eq. (4) (Section 13.21) shows that the value obtained for ferromagnetic Curie temperature ($T_f$) is identical with that of the paramagnetic Curie temperature, θ. Hence, the Weiss theory does not distinguish between the paramagnetic and ferromagnetic Curie temperatures.

## 13.22 QUANTUM THEORY OF FERROMAGNETISM

Let us consider a unit volume of a ferromagnetic substance containing $N$ atoms per unit volume placed in a magnetic field $B$ is given by

$$B_e = \mu_0(H + H_i) = \mu_0(H + \beta I) \qquad ...(1)$$

Because we are considering the ferromagnetic substance as the paramagnetic substances with an internal molecules field $\beta I$, which is in addition to the externally applied field $B$ ($B_e$ is the effective field), we will use equations.

$$I = N_g \mu_B J B(a) \qquad ...(2)$$

$$\left(\text{where } a = J\left(\frac{g\mu_B B_e}{k_B T}\right)\right) \qquad ...(3)$$

$$B(a) = \left[\left(\frac{2J+1}{2J}\right)\coth\left(\frac{2J+1}{2J}\right)a - \left(\frac{1}{2J}\right)\coth\left(\frac{1}{2J}\right)a\right] \qquad ...(4)$$

and

$$I = I_S B(a)$$

derived for paramagnetic substances.

$$I = I_S B(a) \qquad ...(5)$$

where

$$I_S = Ng\mu_B J \qquad ...(6)$$

and $$a = Jx = \frac{Jg\mu_B B_e}{k_B T} = \frac{Jg\mu_B \mu_0 (H + \beta I)}{k_B T} = \frac{Jg\mu_B (B + \mu_0 \beta I)}{k_B T} \quad ...(7)$$

For spontaneous magnetization we put $B = 0$ and hence, the expression for $a$ becomes

$$a = \frac{Jg\mu_B \mu_0 \beta I}{k_B T} \quad ...(8)$$

$$\Rightarrow \quad I = \frac{ak_B T}{Jg\mu_B \mu_0 \beta} \quad ...(9)$$

Hence, $$\frac{I}{I_S} = \frac{ak_B T}{Jg\mu_B \mu_0 \beta} \cdot \frac{1}{Ng\mu_B J}$$

$$\Rightarrow \quad \frac{I}{I_S} = \frac{ak_B T}{\mu_0 \beta N J^2 g^2 \mu_B^2} \quad ...(10)$$

We have

$$\frac{I}{I_S} = B(a) \quad ...(11)$$

Since $I$ must satisfy both Eqs. (10) and (11), its value at any given temperature may be determined by the graphical solution of these two equations. For the intersection of the straight line represented by Eq. (10) with the Brillouin function curve represented by Eq. (11), it is necessary that the slope of the straight line at the origin be less than that of the curve.

Now, the Brillouin function can be written as

$$B(a) = \left(\frac{J+1}{3J}\right) a - \left(\frac{J+1}{3J}\right) \cdot \left(\frac{2J^2 + 2J + 1}{30J^2}\right) a^3 - ... \quad ...(12)$$

Since near origin, $a \ll 1$, Eq. (12) can be approximated as

$$B(a) = \left(\frac{J+1}{3J}\right) a \quad ...(13)$$

The slope of the curve at the origin is $\left(\frac{J+1}{3J}\right)$ and that of the straight line represented by Eq. (10) is $\frac{k_B T}{\mu_0 \beta N J^2 g^2 \mu_B^2}$. Therefore, for the spontaneous magnetization to occur,

$$\frac{k_B T}{\mu_0 \beta N J^2 g^2 \mu_B^2} < \frac{J+1}{3J} \quad ...(14)$$

Equation (14) gives the condition for spontaneous magnetization which occurs for $T < T_f$. At temperatures greater than $T_f$, the substance behaves like a paramagnetic substance. At temperatures greater than $T_f$, the thermal agitation opposes the tendency of Weiss field to align the molecular magnets, whereas at temperatures below $T_f$, the Weiss field energy exceeds the thermal energy and consequently, most of the molecular magnets are aligned in the same direction, resulting in the magnetization of the material even in the absence of the externally applied field, the so-called phenomenon of spontaneous magnetization.

Using Eq. (14), the ferromagnetic Curie temperature is obtained by replacing $T$ by $T_f$ and the less than sign (<) by the sing of equality (=). Hence,

$$\frac{k_B T_f}{\mu_0 \beta N J^2 g^2 \mu_B^2} = \frac{J+1}{3J}$$

$$T_f = \frac{\mu_0 \beta N J^2 g^2 \mu_B^2}{kB} \cdot \left(\frac{J+1}{3J}\right) \quad ...(15)$$

It follows from Eq. (15) that the ferromagnetic Curie temperature ($T_f$) is directly proportional to the molecular field constant β, *a* result one expects, as the tendency for parallel alignment increases as β increases and consequently, the transition temperature from ferromagnetic to paramagnetic substance also increases.

From Eqs. (10) and (15), we have

$$\frac{I}{I_S} \cdot T_f = \left(\frac{a k_B T}{\mu_0 \beta N J^2 g^2 \mu_B^2}\right) \cdot \left(\frac{\mu_0 \beta N J^2 g^2 \mu_B^2}{k_B}\right) \cdot \left(\frac{J+1}{J}\right)$$

$$\frac{I}{I_S} = \left(\frac{J+1}{J}\right)\left(\frac{T}{T_f}\right) a \quad ...(16)$$

Figure 13.12 shows plots of $I/I_S$ versus $T/T_f$ for iron, cobalt and nickel for three different values of $J$, vis. $J = \infty$, 1 and $\frac{1}{2}$. The dots in the figure represent the experimental points for Fe, Co and Ni. It is observed that the experimental values are close to the curve for $J = \frac{1}{2}$.

This close agreement between the theoretical and experimental values reveals that the magnetization of a ferromagnetic material is due to the electron spin rather than due to the orbital motion of the electron.

Maurain's researches on thin films of iron give evidence to the existence if very strong intrinsic molecular fields. The magnitude of these molecular fields can be estimated with a knowledge of the Weiss factor, β. Table 13.3 gives values of Weiss factor β and the internal molecular field $H_1$ for different ferromagnetic materials, taking $I$ at normal external field strength of a few hundred oersted.

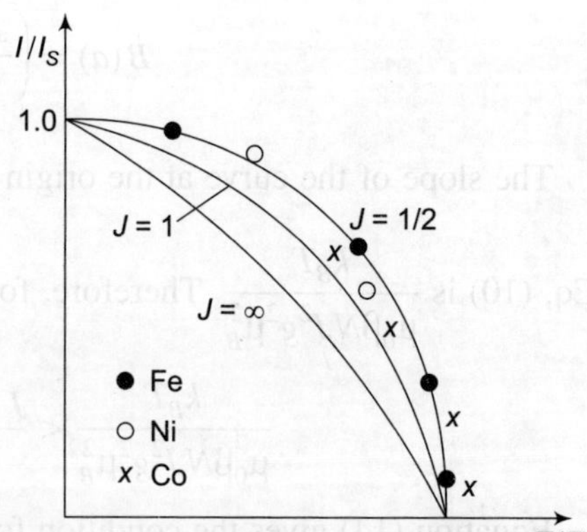

**Figure 13.12** Variation of $I/IS$ with $T/T_f$

**Table 13.3** Values of Weiss Factor and the Internal Molecular Field

| *Ferromagnetics* | $\beta$ | $H_i = \beta I$ (ampere/metre) |
|---|---|---|
| Iron | 3850 | $8.2 \times 10^4$ |
| Nickel | 12700 | $7.94 \times 10^4$ |
| Cobalt | 33200 | $17.9 \times 10^4$ |
| Magnetite | 6180 | $11.08 \times 10^4$ |

## 13.23 ORIGIN OF DOMAINS IN A FERROMAGNET

According to Landau and Lifshitz, the domain structure is a natural consequence of the various contributions to the energy – exchange, anisotropy and magnetic, of a ferromagnetic body.

We may understand the origin of domains with a reference to Fig. 13.13.

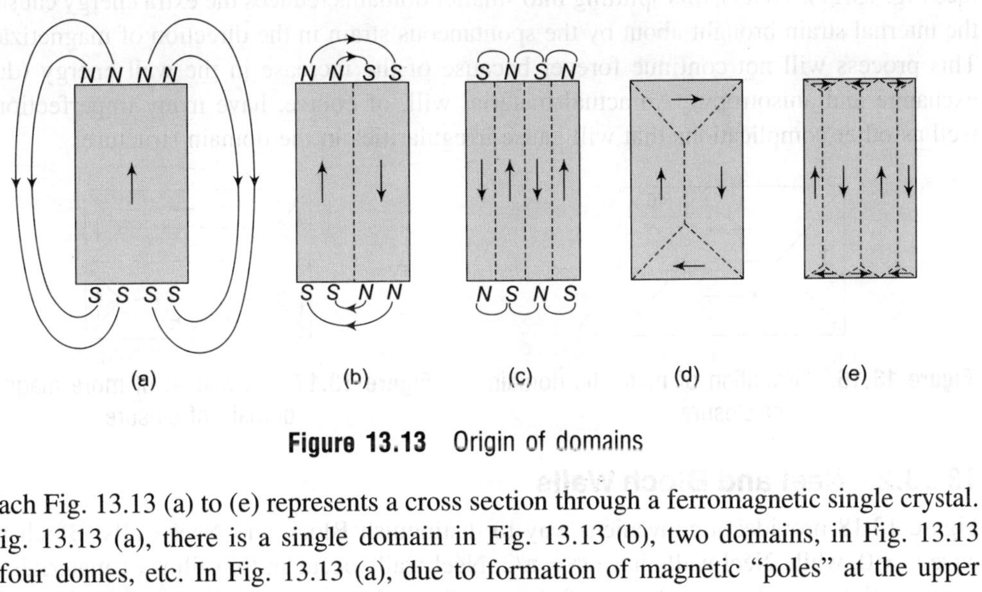


**Figure 13.13** Origin of domains

Each Fig. 13.13 (a) to (e) represents a cross section through a ferromagnetic single crystal. In Fig. 13.13 (a), there is a single domain in Fig. 13.13 (b), two domains, in Fig. 13.13 (c), four domes, etc. In Fig. 13.13 (a), due to formation of magnetic "poles" at the upper and lower edge respectively. This configuration will have a high value of magnetic energy $(1/8\pi) \int B^2 \, dV$. The magnetic energy density for this configuration will be of the order of $M_s^2 \approx 10^6$ erg/cm$^3$, where $M_S$ is the saturation magnetization (in CGS units).

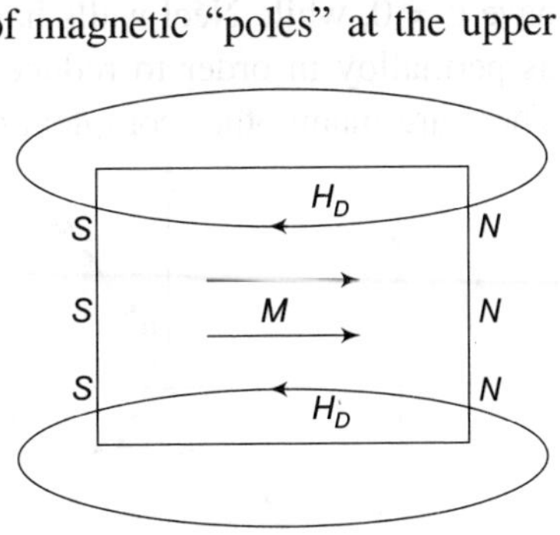


**Figure 13.14** Magnetic domain formation within a material

### 13.23.1 Formation of Magnetic Domains

We now give a qualitative account of the formation of domains. Consider a cubic material, originally magnetized along an easy direction as shown in Fig. 13.14. Because the magnetization ($M$) and demagnetizing fields have opposite directions this configuration has large magnetostatic energy. The magnetostatic energy can be reduced if the material splits into domains as shown in Fig. 13.15.

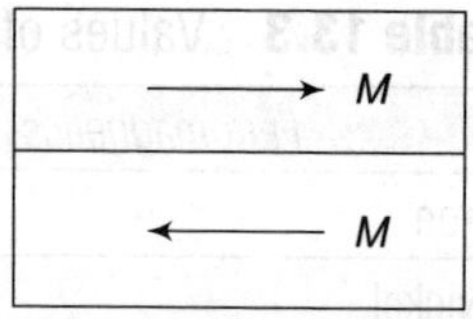


**Figure 13.15** Magnetic-domain splitting within a material

Since the density of surface poles is $+\boldsymbol{M}\cdot\boldsymbol{n}$ where $\boldsymbol{n}_M$ is the outward normal, at an interface the net *magnetic charge* per unit area is

$$(\boldsymbol{M}_2 - \boldsymbol{M}_1)\cdot\boldsymbol{n}_{M_2},$$

where $\boldsymbol{n}_{M_2}$ is a unit pointing from region 1 to region 2. Thus, when $\boldsymbol{M}\cdot\boldsymbol{n}$ is continuous, there are no demagnetizing fields (assuming also $M$ is uniform in the interior). Thus, (for typical magnetic materials with cubic symmetry) the magnetostatic energy can be further reduced by forming domains of closure, as shown in Fig. 13.16. The overall magnetostrictive and strain energy can be reduced by the formation of more domains of closure (see Fig. 13.17). That is, this splitting into smaller domains reduces the extra energy caused by the internal strain brought about by the spontaneous strain in the direction of magnetization. This process will not continue forever because of the increase in the wall energy (due to exchange and anisotropy). An actual material will, of course, have many imperfections as well as other complications that will cause irregularities in the domain structure.

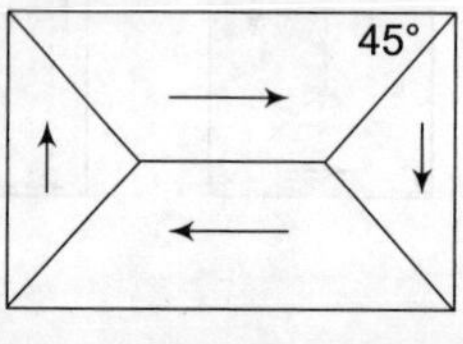


**Figure 13.16** Formation of magnetic domain of closure

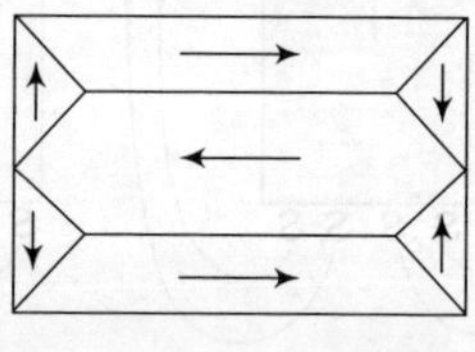

**Figure 13.17** Formation of more magnetic domain of closure

## 13.23.2 Néel and Bloch Walls

Figure 13.18 provides a convenient way to distinguish Bloch and Néel walls. Bloch walls have $\varphi = 0$, while Néel walls have $\varphi = \pi/2$. Néel walls occur in thin films of materials such as permalloy in order to reduce surface magnetostatic energy as suggested by Fig. 13.19. There are many other complexities involved in domain wall structures.

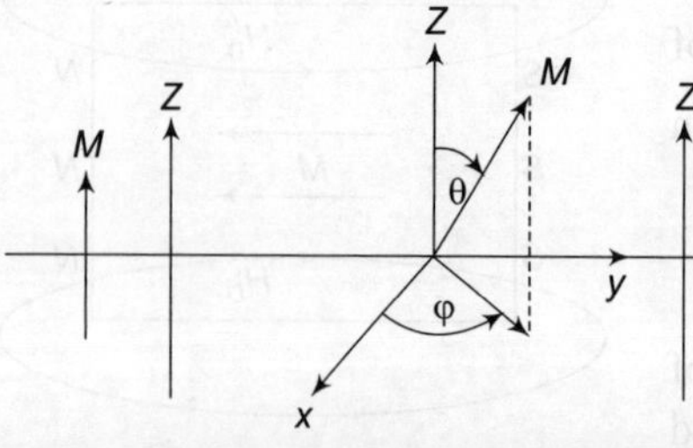


**Figure 13.18** Bloch wall: $\varphi = 0$; Néel wall: $\varphi = \pi/2$

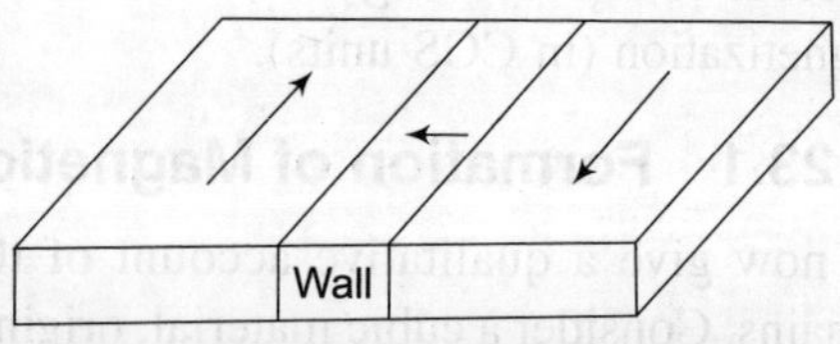


**Figure 13.19** Néel wall in thin film

### 13.23.3 Methods of Observing Domains

We briefly summarize five methods.

1. Bitter patterns—a colloidal suspension of particles of magnetite is placed on a polished surface of the magnetic material to be examined. The particles are attracted to regions of nonuniform magnetization (the walls) and hence the walls are readily seen by a microscope.
2. Faraday and Kerr effects—these involve rotation of the plane of polarization on transmission and reflection (respectively) from magnetic substances.
3. Neutrons—since neutrons have magnetic moments they experience interaction with the internal magnetization and its direction, see Bacon GE, "Netron Diffraction," Oxford 1962 (p. 355ff).
4. Transmission electron microscopy (TEM)—moving electrons are influenced by forces due to internal magnetic fields.
5. Scanning electron microscopy (SEM)—moving secondary electrons sample internal magnetic fields.

## 13.24 HYSTERESIS, REMANENCE, AND COERCIVE FORCE

Consider an unmagnetized ferromagnetic well below its Curie temperature. We can understand the material being unmagnetized if it consists of a large number of domains, each of which is spontaneously magnetized, but that have different direction of magnetization so the net magnetization averages to zero.

The magnetization changes from one domain to another through thin but finite-width domain walls. Typically, domain walls are of thickness of about $10^{-7}$ metres or some hundred of atomic spacings, while the sides of the domains are a few micrometres and larger.

The hysteresis loop can be visualized by plotting $M$ vs. $H$ or $B = \mu_0(H + M)$ (in SI units) $= H + 4\pi M$ (in Gaussian units) (see Fig. 13.20). The *virgin curve* is obtained by starting in an ideal demagnetized state in which one is at the absolute minimum of energy.

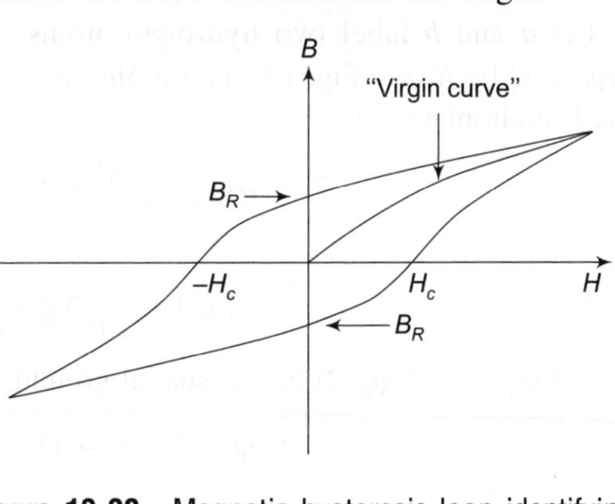


**Figure 13.20** Magnetic hysteresis loop identifying the virgin curve.
$H_c \equiv$ coercive "force"
$B_R$ = remanence.
$M_s = [(B - H)/4\pi]_{H \to \infty}$ = saturation magnetization.
$M_R = B_R/4\pi$ = remnant magnetization.

When an external field is turned on, "favourable" domains have lower energy than "unfavourable" ones, and thus the favourable ones grow at the expense of the unfavourable ones.

Imperfections determine the properties of the hysteresis loop. Moving a domain wall generally increases the energy of a ferromagnetic material due to a complex combination of interactions of the domain wall with dislocations, grain boundaries, or other kinds of defects. Generally, the first part of the virgin curve is reversible, but as the walls sweep past defects one enters an irreversible region, then in the final approach to saturation, one generally has some rotation of domains. As $H$ is reduced to zero, one is left with a remanent magnetization (in a metastable state with a "local" rather than absolute minimum of energy) at $H = 0$ and $B$ only goes to zero at $-H_c$ the coercive "force".[3] For permanent magnetic materials, $M_R$ and $H_c$, should be as large as possible. On the other hand, soft magnets will have very low coercivity. The hysteresis and domain properties of magnetic materials are of vast technological importance, but a detailed discussion would take us too far afield. See Cullity.[4]

## 13.25 HEISENBERG HAMILTONIAN (THE HEITLER–LONDON METHOD)

In this section, we develop the Heisenberg Hamiltonian and then relate our results to various aspects of the magnetic state. The first method that will be discussed is the Heitler–London method. This discussion will have at least two applications. First, it helps us to understand the covalent bond, and so relates to our previous discussion of valence crystal. Second, the discussion gives us a qualitative understanding of the Heisenberg Hamiltonian. This Hamiltonian is often used to explain the properties of coupled spin systems. The Heisenberg Hamiltonian will be used in the discussion of magnons. Finally, as we will show, the Heisenberg Hamiltonian is useful in showing how an electrostatic exchange interaction approximately predicts the existence of a molecular field and hence gives a fundamental qualitative explanation of the existence of ferromagnetism.

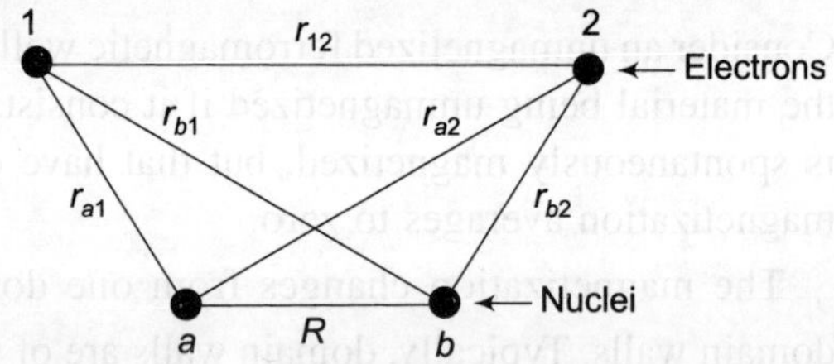


**Figure 13.21** Model for two hydrogen atoms

Let $a$ and $b$ label two hydrogen atoms separated by $R$ (see Fig. 13.21). Let the separated ($R \to \infty$) hydrogen atoms be described by the Hamiltonians.

$$\mathcal{H}_0^a(1) = \frac{\hbar^2}{2m}\nabla_1^2 - \frac{e^2}{4\pi\varepsilon_0 r_{a1}}, \quad ...(1)$$

and

$$\mathcal{H}_0^b(2) = \frac{\hbar^2}{2m}\nabla_2^2 - \frac{e^2}{4\pi\varepsilon_0 r_{b2}}. \quad ...(2)$$

Let $\psi_a(1)$ and $\psi_b(2)$ be the spatial ground-state wave function, that is

$$\mathcal{H}_0^a \psi_a(1) = E_0 \psi_a(1), \quad ...(3)$$

or

$$\mathcal{H}_0^b \psi_b(2) = E_0 \psi_b(2),$$

where $E_0$ is the ground-state energy of the hydrogen atom. The zeroth-order hydrogen molecular wave functions may be written

[3] Some authors define $H_C$ as the field that reduced $M$ to zero.

[4] BD Cullity, *Introduction to Magnetic Materials*, Addison-Wesley (1972).

$$\psi_{\pm} = \psi_a(1)\,\psi_b(2) \pm \psi_a(2)\,\psi_b(1). \qquad ...(4)$$

In the Heitler–London approximation for un-normalized wave functions

$$E_{\pm} \cong \frac{\int \psi_{\pm}\mathcal{H}\psi_{\pm}d\tau_1 d\tau_2}{\int \psi_{\pm}^2 d\tau_1 d\tau_2}, \qquad ...(5)$$

where $d\tau_i = dx_i dy_i dz_i$ and we have used that wave functions for stationary states can be chosen to be real. In Eq. (5),

$$\mathcal{H} = \mathcal{H}_0^a(1) + \mathcal{H}_0^b(2) - \frac{e^2}{4\pi\varepsilon_0}\left(\frac{1}{r_{a2}} + \frac{1}{r_{b1}} - \frac{1}{r_{12}} - \frac{1}{R}\right). \qquad ...(6)$$

Working out the details when Eq. (4) is put into Eq. (5) and assuming $\psi_a(1)$ and $\psi_b(2)$ are normalized we find

$$E_{\pm} = 2E_0 + \frac{e^2}{4\pi\varepsilon_0 R} + \frac{K \pm J_E}{1 \pm S}, \qquad ...(7)$$

where $$S = \int \psi_a(1)\,\psi_b(1)\,\psi_a(2)\,\psi_b(2)\,d\tau_1 d\tau_2 \qquad ...(8)$$

is the overlap integral,

$$K = \frac{e^2}{4\pi\varepsilon_0}\int \psi_a^2(1)\,\psi_b^2(2)\,V(1,2)\,d\tau_1 d\tau_2 \qquad ...(9)$$

is the Coulomb energy of interaction, and

$$J_E = \frac{e^2}{4\pi\varepsilon_0}\int \psi_a(1)\,\psi_a(2)\,\psi_b(1)\,\psi_b(2)V(1,2)\,d\tau_1 d\tau_2 \qquad ...(10)$$

is the exchange energy. In Eqs. (9) and (10),

$$V(1,2) = \frac{e^2}{4\pi\varepsilon_0}\left(\frac{1}{r_{12}} - \frac{1}{r_{a2}} - \frac{1}{r_{b1}}\right). \qquad ...(11)$$

The corresponding normalized eigenvectors are

$$\psi^{\pm}(1,2) = \frac{1}{\sqrt{2(1 \pm S)}}\,[\psi_1(1,2) \pm \psi_2(1,2)], \qquad ...(12)$$

where $$\psi_1(1,2) = \psi_a(1)\,\psi_b(2), \qquad ...(13)$$

$$\psi_2(1,2) = \psi_a(2)\,\psi_b(1). \qquad ...(14)$$

So far there has been no need to discuss spin, as the Hamiltonian did not explicitly involve it. However, it is easy to see how spin enters. $\psi^+$ is a symmetric function in the interchange of coordinates 1 and 2, and $\psi^-$ is an antisymmetric function in the interchange of coordinates 1 and 2. The total wave function that includes both space and spin coordinates must be antisymmetric in the interchange of all coordinates. Thus in the total wave function, an antisymmetric function of spin must multiply $\psi^+$, and a symmetric function of spin must multiply $\psi^-$. If we denote $\alpha(i)$ as the "spin-up" wave function of electron $i$ and $\beta(j)$ as the "spin-down" wave function of electron $j$, then the total wave functions can be written as

$$\psi_T^+ = \frac{1}{\sqrt{2(1+S)}}(\psi_1 + \psi_2)\frac{1}{\sqrt{2}}[\alpha(1)\beta(2) - \alpha(2)\beta(1)], \quad ...(15)$$

$$\psi_T^- = \frac{1}{\sqrt{2(1-S)}}(\psi_1 - \psi_2)\begin{cases} \alpha(1)\alpha(2), \\ \frac{1}{\sqrt{2}}[\alpha(1)\beta(2) + \alpha(2)\beta(1)], \\ \beta(1)\beta(2). \end{cases} \quad ...(16)$$

Equation (15) has total spin equal to zero, and is said to be a *singlet* state. It corresponds to antiparallel spins. Equation (16) has total spin equal to one (with three projections of +1, 0, –1) and is said to describe a *triplet* state. This corresponds to parallel spins. For hydrogen atoms, $J$ in Eq. (10) is called the *exchange integral* and is negative. Thus, $E_+$ (corresponding to $\psi_T^+$) is lower in energy than $E_-$ (corresponding to $\psi_T^-$), and hence the singlet states is lowest in energy. A calculation of $E_\pm - E_0$ for $E_0$ labelling the ground states of hydrogen is sketched in Fig. 13.22. Let us now pursue this two-spin case in order to write an effective spin Hamiltonian that describes the situation. Let $S_1$ and $S_2$ be the spin operators for particles 1 and 2. Then

$$(S_1 + S_2)^2 = S_1^2 + S_2^2 + 2S_1 \cdot S_2 \quad ...(17)$$

Since the eigenvalues of $S_2^1$ and $S_2^2$ are $3\hbar^2/4$ we can write for appropriate $\phi$ in the space of interest

$$S_1 \cdot S_2\phi = \frac{1}{2}\left[(S_1 + S_2)^2 - \frac{3}{2}\hbar^2\right]\phi \quad ...(18)$$

In the triplet (or parallel spin) state, the eigenvalue of $(S_1 + S_2)^2$ is $2\hbar^2$, so

$$S_1 \cdot S_2\phi_{\text{triplet}} = \frac{1}{4}\hbar^2\phi_{\text{triplet}} \quad ...(19)$$

In the singlet (or antiparallel spin) state, the eigenvalue of $(S_1 + S_2)^2$ is 0, so

$$S_1 \cdot S_2\phi_{\text{singlet}} = -\frac{3}{4}\hbar^2\phi_{\text{singlet}} \quad ...(20)$$

Comparing these result to Fig. 13.22, we see can formally write an effective spin Hamiltonian for the two electrons on the two different atoms

$$\mathcal{H} = -2JS_1 \cdot S_2 \quad ...(21)$$

where $J$ is often simply called the *exchange constant* and $J = J(R)$, i.e., it depends on the separation $R$ between atoms. By suitable choice of $J(R)$, the eigenvalues of $\mathcal{H} - 2E_0$ can reproduce the curves of Fig. 13.22. Note that $J > 0$ gives the parallel spin

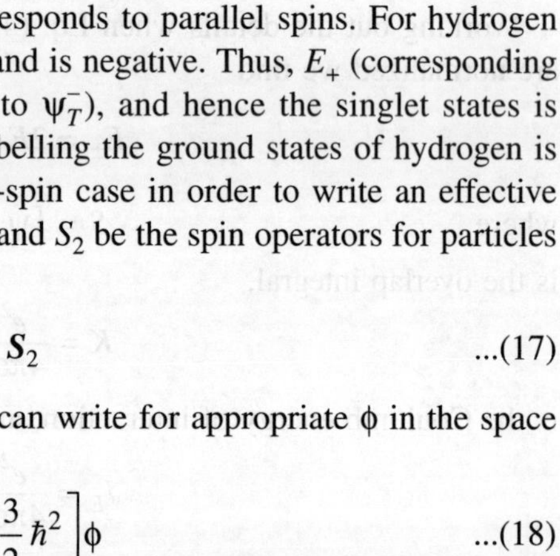


**Figure 13.22** Sketch of result of the Heitler–London theory applied to two hydrogen atoms ($R/R_0$ is the distance between the two atoms in Bohr radii)

case the lowest energy (ferromagnetism) and $J < 0$ (the two-hydrogen-atom case – this does not *always* happen, especially in a solid) gives the antiparallel spin case the lowest energy (antiferromagnetism). If we have many atoms on a lattice, and if there is an exchange coupling between the spin of the atoms, we assume that we can write a Hamiltonian

$$\mathcal{H} = -\sum'_{\substack{\alpha,\beta \\ \text{(electrons)}}} J_{\alpha,\beta}\, \mathbf{S}_\alpha \cdot \mathbf{S}_\beta . \qquad ...(22)$$

If there are several electrons on the same atom and if $J$ is constant for all electrons on the same atom, then we assume we can write

$$\sum J_{\alpha,\beta}\, \mathbf{S}_\alpha \cdot \mathbf{S}_\beta \cong \sum_{\substack{k,l \\ \text{(atoms)}}} J_{k,l} \sum_{\substack{i,j \\ \text{(electrons on } k,l \text{ atoms)}}} \mathbf{S}_{ki} \cdot \mathbf{S}_{lj}$$

$$= \sum_{k,l} J_{k,l} \left(\sum_i \mathbf{S}_{ki}\right)\left(\sum_j \mathbf{S}_{lj}\right) \qquad ...(23)$$

$$= \sum_{k,l} J_{k,l}\, \mathbf{S}_k^T \cdot \mathbf{S}_l^T ,$$

where $\mathbf{S}_k^T$ and $\mathbf{S}_l^T$ refer to the spin operators associated with atoms $k$ and $l$. Since $\sum' \mathbf{S}_a \cdot \mathbf{S}_p J_{\alpha\beta}$ differs from $\sum \mathbf{S}_a \cdot \mathbf{S}_\beta J_{\alpha\beta}$ by only a constant and $\sum'_{k,l} J_{k,l} \mathbf{S}_k^T \cdot \mathbf{S}_l^T$ differs from $\sum_{k,l} J_{k,l} \mathbf{S}_k^T$ by a constant, we can write the effective spin Hamiltonian as

$$\mathcal{H} = \sum'_{k,l} J_{k,l}\, \mathbf{S}_k^T \cdot \mathbf{S}_l^T \qquad ...(24)$$

here unimportant constants have not been retained. This last expression is called the *Heisenberg Hamiltonian* for a system of interacting spins in the absence of external field.

This form of the Heisenberg Hamiltonian already tells us two important things:

1. It is applicable to atoms with arbitrary spin.
2. Closed shells contribute nothing to the Heisenberg Hamiltonian because the spin is zero for a closed shell.

Our development of the Heisenberg Hamiltonian has glossed over the approximations that were made. Let us now return to them. The first obvious approximation was made in going from the two-spin case to the $N$-spin case. The presence of a third atom can and does affect the interaction between the original pair. In addition, we assumed that the exchange interaction between all electrons on the same atom was a constant.

Another difficulty with the extension of the Heitler–London method to the $n$-electron problem is the so-called "overlap catastrophe." This will not be discussed here.

## 13.26 THE BLOCH WALL ENERGY

The boundary surface between two domains is known as the Bloch wall. The wall has a certain thickness because the spin orientation changes only gradually across the domain boundary. If the adjacent domain spin are antiparallel then the Bloch wall is called the 180° wall. Similarly, one have 90° wall (Fig. 13.23).

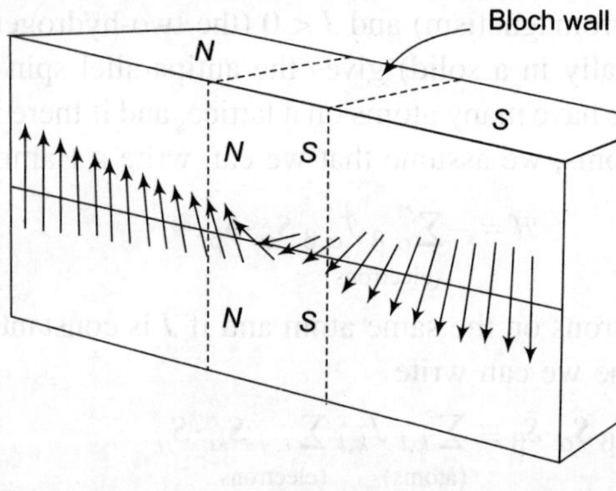


**Figure 13.23** The structure of a 180° Bloch wall separating two domains

In a Bloch wall, the entire change in spin direction between domains magnetized in different directions does not occur in one sudden jump across a single atomic plane, instead, the change takes place in a gradual way extending over many atomic planes. The reason for this is the fact for a given total change of spin direction, the exchange energy is lower when the change is distributed over many spins.

Now, we know that according to Heisenberg's model, the exchange energy is

$$E_{ex} = -2\, J_e\, \vec{S}_i \cdot \vec{S}_j \qquad \text{...(1a)}$$

or

$$E_{ex}(\phi) = -2\, J_e\, S^2 \cos \phi \qquad \text{...(1b)}$$

where $\phi$ is the angle between two spin directions assuming $|\vec{S}_i| = |\vec{S}_j| = S$

$$\text{since } \cos \phi = 1 - \frac{\phi^2}{2} + \ldots$$

Therefore,

$$E_{ex}(\phi) = -2\, J_e S^2 \left(1 - \frac{\phi^2}{2} + \ldots\right)$$

For small angle $\phi$,

$$\Delta E_{ex} = E_{ex}(\phi) - E_{ex}(0) \approx J_e\, S^2\, \phi^2 \qquad \text{...(2)}$$

This gives the increase in exchange energy when two spins are rotated from exact parallelism to make some small angle $\phi$ with each other.

Now, let $\phi_0$ denote the total change of angle between two domains and the change occurs in $N$ equal steps, so that the change of angle between neighbouring spins is $\phi_0/N$. Then, the exchange energy between each pair of neighbouring atoms is

$$(\Delta E_{ex})_{\text{pair}} = J_e S^2 \left(\frac{\phi_0}{N}\right)^2 \qquad \text{...(3)}$$

and the total energy of the array of atoms

$$(\Delta E_{ex})_{\text{total}} = \frac{J_e S^2 \phi_0^2}{N} \qquad \text{...(4)}$$

This shows that exchange energy decrease when $N$ increases. Now, one may argue that why does not the wall become infinitely thick (to increase $N$)? This is explained by the concept of *anisotropy energy*.

Since the spins within the wall are nearly all directed away from the easy direction, an anisotropy energy is associated with the wall which is roughly proportional to the thickness of the wall. This energy also needs to be minimum. Thus, the actual thickness and the energy of the wall is, therefore, the result of compromise between the energies, viz., the exchange energy and the *anisotropy energy*.

To make a rough estimate of the thickness and the energy of a Bloch wall, we consider a wall parallel to the cube face of a simple cubic lattice, separating two domains magnetised in opposite direction. Its energy per unit area may be written as

$$E = E_{ex} + E_{\text{anis}} \qquad \text{...(5)}$$

where

$$E_{ex} = \frac{1}{a^2} \times \frac{J_e S^2 \pi^2}{N} \quad (\because \phi_0 = \pi) \qquad \text{...(6)}$$

The anisotropy energy $E_{\text{anis}}$ is roughly equal to the anisotropy constant $K$ times the volume $Na$ of the wall. Hence,

$$E = \frac{J_e S^2 \pi^2}{N a^2} + KNa \qquad \text{...(7)}$$

where $a$ is the lattice constant.

Putting $dE/dN = 0$, we obtain the equilibrium value of $N$ as

$$N = \left( \frac{J_e S^2 \pi^2}{K a^2} \right)^{1/2} \qquad \text{...(8)}$$

The total energy per unit area of the Bloch wall is obtained by substituting the value of $N$ from Eq. (8) into Eq. (7). This gives

$$E = 2S\pi \left( \frac{J_e K}{a} \right)^{1/2} \qquad \text{...(9)}$$

It is to be noted that the above treatment is rather crude, more accurate calculations for a 180° wall in a (100) plane gives

$$E = 2 \left( \frac{2KJ_e S^2}{a} \right)^{1/2} \qquad \text{...(10)}$$

The thickness of the Bloch wall is estimated to be of the order of 1000 Å involving a couple of hundred spins. The energy of the wall is about 1 erg/cm$^2$.

## 13.27 A TWO ELECTRON SYSTEM (SINGLET AND TRIPLET) STATES

We will subsequently see that the Pauli's exclusion principle can lead to magnetic effects even when there are no spin-dependent terms in the Hamiltonian (due to inherent endowment of spins to the electrons).

Let us consider a spin-independent Hamiltonian $H$ for a two-electron system. Since $H$ is independent of spin, (the general state function ($\psi$) will be the product of a purely (orbital) state function $\psi_0$ and spins dependent part) so wave function $\psi(\vec{r}_1, \vec{r}_2)$ satisfies the Schrödinger equation

$$H\,\psi(\vec{r}_1, \vec{r}_2) = \frac{\hbar^2}{2m}(\nabla_1^2 + \nabla_2^2)\,\psi(\vec{r}_1, \vec{r}_2) + V(\vec{r}_1, \vec{r}_2)\,\psi(\vec{r}_1, \vec{r}_2)$$

$$= E\psi(\vec{r}_1, \vec{r}_2) \qquad \text{...(1)}$$

with any linear combination of the four-spin states

$$|\uparrow\uparrow\rangle, |\downarrow\uparrow\rangle, |\uparrow\downarrow\rangle, |\downarrow\downarrow\rangle$$

There are four possible combinations so as to have definite values of the total spin $S$ and its component $S_z$ (along $z$-axis). These are given in Table 13.4.

**Table 13.4** Four Possible Spin States for a Electron System

| *Spin states ($\psi$'s)* | $S$ | $S_z$ |
|---|---|---|
| $\frac{1}{\sqrt{2}}(\lvert\uparrow\downarrow\rangle - \lvert\downarrow\uparrow\rangle)$ | 0 | 0 |
| $\lvert\uparrow\uparrow\rangle$ | 1 | 1 |
| $\frac{1}{\sqrt{2}}(\lvert\uparrow\downarrow\rangle + \lvert\downarrow\uparrow\rangle)$ | 1 | 0 |
| $\lvert\downarrow\downarrow\rangle$ | 1 | –1 |

The state with $S = 0$ is known as the singlet state and it changes sign under exchange of spins of two electrons. The other three states (with $S = 1$) are known as the triplet states and these do not change sign (if the spins of two electrons are interchanged).

Now, the Pauli's exclusion principle dictates that the total wave function $\psi(\vec{r}_1, \vec{r}_2)$ ($\equiv \psi_0\,\psi_s$) must change sign under the simultaneous interchange of both space and spin coordinates. It therefore follows that solutions to the orbital Schrödinger equation (1) that do not change sign under interchange of $\vec{r}_1$ and $\vec{r}_2$ (i.e., symmetric solutions) must describe states with $S = 0$ (singlet spin states) and those who change sign (antisymmetric orbital solutions) must have $S = 1$ (i.e., require triplet spin states).

If $E_s$ and $E_t$ are the lowest eigenvalues of Eq. (1) associated with the singlet (symmetric) and triplet (antisymmetric) solutions respectively, then the ground states will have spin 0 or 1 depending on whether $E_s$ is less than or greater than $E_t$ (for energy to be minimum). But, for two-electron system, there is a theorem which states that the ground state wave function for Eq. (1) must be symmetric so that $E_s$ must be less than $E_t$ (i.e., ground state must have zero total spin).

## 13.28 CALCULATION OF THE SINGLET TRIPLET SPLITTING (EXCHANGE-SPLITTING)

The singlet–triplet energy splitting measures the extent to which the $S = 0$ spin alignment of two electrons is more favourable than the $S = 1$ alignment.

We first consider the independent electron approximation (i.e., ignoring the electron–electron Coulomb interaction in $V(\vec{r}_1, \vec{r}_2)$ retaining only the interaction of each electron with the two ions (taken to be at $\vec{R}_1$ and $\vec{R}_2$). The two electron Schrödinger equation is then

$$(H_1 + H_2)\,\psi(\vec{r}_1, \vec{r}_2) = E\psi(\vec{r}_1, \vec{r}_2) \qquad ...(1)$$

where
$$H_i = \frac{\hbar^2}{2m}\nabla_i^2 - \frac{e^2}{(\vec{r}_i - \vec{R}_1)} - \frac{e^2}{(\vec{r}_i - \vec{R}_2)} \qquad ...(2)$$

Since the Hamiltonian in Eq. (1) is a sum of one-electron Hamiltonians, we can consider one-electron Schrödinger equation as follows

$$H\psi\,(\vec{r}) = E\psi(\vec{r}) \qquad ...(3)$$

If $\psi_0(\vec{r})$ and $\psi_1(\vec{r})$ are two solutions to Eq. (3) of lowest energy (with $\varepsilon_0 < \varepsilon_1$), then the symmetric solution of lowest energy to the approximate two-electron Schrödiger Eq. (2) is

$$\psi_s\,(\vec{r}_1, \vec{r}_2) = \psi_0(\vec{r}_1)\,\psi_0(\vec{r}_2)$$

$$E_s = 2\varepsilon_0 \qquad ...(4)$$

and the lowest antisymmetric solution is

$$\psi_T\,(\vec{r}_1, \vec{r}_2) = \psi_0(\vec{r}_1)\,\psi_1(\vec{r}_2)\,\psi_0(\vec{r}_2)\,\psi_1(\vec{r}_1)$$

$$E_t = \varepsilon_0 + \varepsilon_1 \qquad ...(5)$$

Then, splitting $\qquad E_s - E_t = \varepsilon_0 - \varepsilon_1$

which is consistant with the requirement, $E_s < E_t$.

Here, the wave function Eq. (4) does not hold good exact Schrödinger Eq. (1) section 13.27, when the protons are very far apart and it fails to account for electron–electron coulomb interaction.

For a two-atom 'solid' (i.e., $N = 2$), the tight binding method permits

$$\left.\begin{aligned}\psi_0(\vec{r}) &= \phi_1(\vec{r}) + \phi_2(\vec{r})\\ \psi_1(\vec{r}) &= \phi_1(\vec{r}) - \phi_2(\vec{r})\end{aligned}\right\} \qquad ...(6)$$

where $\phi_i(\vec{r})$ is the ground states electron wave function for a single hydrogen atom. In this case (i.e., for well-separated protons), the two-electron wave functions [Eqs. (4) and (5)] (for independent electron approximation) become

$$\psi_s\,(\vec{r}_1, \vec{r}_2) = \phi_1(\vec{r}_1)\,\phi_2(\vec{r}_2) + \phi_2(\vec{r}_1)\,\phi_1(\vec{r}_2) \qquad ...(7)$$

and
$$\psi_T\,(\vec{r}_1, \vec{r}_2) = 2\,[\phi_2(\vec{r}_1)\,\phi_1(\vec{r}_2) - \phi_1(\vec{r}_1)\,\phi_2(\vec{r}_2)] \qquad ...(8)$$

(This corresponds to Heitler–London case).

In this case, the singlet-triplet splitting is

$$E_S - E_T = \frac{\int(\psi_s^* H \psi_s)\, d\tau}{\int(\psi_s^* \psi_s)\, d\tau} - \frac{\int(\psi_T^* H \psi_T)\, dr}{\int(\psi_T^* \psi_T)\, dr} \quad ...(9a)$$

where $H$ is the full Hamiltonian of Eq. (1), Art. 13.27.

In the limit of large spatial separation, this splitting can be shown to be given by

$$\frac{1}{2}(E_S - E_T) = \int d\vec{r}_1 d\vec{r}_2\, [\phi_1(\vec{r}_1)\phi_2(\vec{r}_2)]$$

$$\left(\frac{e^2}{|\vec{r}_1 - \vec{r}_2|} + \frac{e^2}{|\vec{R}_1 - \vec{R}_2|} - \frac{e^2}{|\vec{r}_1 - \vec{R}_1|} - \frac{e^2}{|\vec{r}_2 - \vec{R}_2|}\right)[\phi_2(\vec{r}_1)\phi_1(\vec{r}_2)] \quad ...(9b)$$

Because this is a matrix element between two states that differ solely through the exchange of coordinates of two electrons, this singlet-triplet energy difference is referred to as an *exchange splitting* or (when viewed as a source of magnetic interaction) as an *exchange interaction*.

## 13.29 HEISENBERG SPIN HAMILTONIAN

There is a way to express the dependence of the spin of a two-electron system on the singlet-triplet energy splitting, which is of fundamental importance in analysing the energetics of the spin configuration of real insulating solids.

To construct the spin Hamiltonian for a two-electron system, we note each individual electron spin operator satisfies $\vec{S}_i^2 = \frac{1}{2}\left(\frac{1}{2}+1\right) = \frac{3}{4}$ so that the total spin $\vec{S}$ satisfies

$$\vec{S}^2 = (\vec{S}_1 + \vec{S}_2)^2 = \frac{3}{2} + 2\,\vec{S}_1 \cdot \vec{S}_2 \quad ...(1)$$

Since $\vec{S}^2$ has the eigenvalue $S(S+1)$ in states of spin $S$, it follows from Eq. (1) that the operator $\vec{S}_1 \cdot \vec{S}_2$ has eigenvalue $-\frac{3}{4}$ in the singlet ($S = 0$) state and $+\frac{1}{4}$ in the triplet ($S = 1$) states. Consequently, the operator

$$H^{\text{spin}} = \frac{1}{4}(E_s + 3E_T) - (E_s - E_T)\,\vec{S}_1 \cdot \vec{S}_2 \quad ...(2)$$

has eigenvalue $E_s$ in the single state and $E_T$ in each of the three triplet states. By redefining the zero of the energy, we omit the first (constant) term which is common to all four states, and write

$$H^{\text{spin}} = -J\,\vec{S}_1 \cdot \vec{S}_2, \text{ where } J = E_s - E_T \quad ...(3)$$

It will favour parallel spins if $J$ is positive and antiparallel if $J$ is negative. The above expression is for $N = 2$. When $N$ is large, the form of the spin Hamiltonian is simply that for the two-spin case, summed over all pairs of ions

$$H^{\text{spin}} = -\sum J_{ij}\,\vec{S}_i \cdot \vec{S}_j \quad ...(4)$$

**Example 13.4** *Comparison of Magnetic Interaction Energy with $k_BT$.* A typical magnetic field achievable with an electromagnet with iron core is about $10^4$ $G = 1$ $T$. Let us compare the magnetic interaction energy $\mu H$ of an electron spin magnetic dipole moment with $k_BT$ at room temperature.

Since the electron spin magnetic dipole moment is the Bohr magneton

$$\mu = \frac{e\hbar}{2m} = 9.3 \times 10^{-24} \text{ J/T},$$

we find

$$\mu H = 9.3 \times 10^{-24} J = 5.8 \times 10^{-5} \text{ eV}.$$

At room temperature, $k_BT \simeq \frac{1}{40}$ eV. Therefore

$$\frac{\mu H}{k_BT} \simeq 40 \times 5.8 \times 10^{-5} = 2.3 \times 10^{-3}.$$

For this reason, the approximation $\mu H/k_BT \ll 1$ is quite good at ordinary temperatures.

### 13.29.1 Magnetic Order

Among the paramagnetic materials, it is possible especially at low temperatures to distinguish three different subclasses of magnetic materials called ferromagnets, antiferromagnets, or ferrimagnets. The basis for the distinction is provided by *magnetic order* with spontaneous magnetization, i.e., the alignment of spontaneous atomic or molecular moments with respect to one another and with respect to the crystallographic axes. Simple (schematic) cases are illustrated in Fig.13.24 in which each arrow indicates a magnetic moment and its orientation.

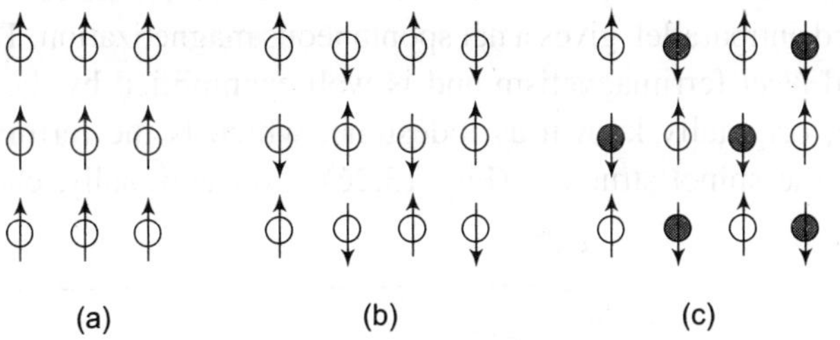

**Figure 13.24** Different types of magnetic order. (a) ferromagnetic; (b) antiferromagnetic; (c) ferrimagnetic

Magnetic ordering introduces a *magnetic unit cell* that may be equal to or a small multiple of the crystallographic unit cell. It may be simple or complex. The simplest type of order consists of the parallel alignment of all the atomic moments in a crystallographic unit cell, as shown in Fig. 13.24(a), which corresponds to ferromagnetic order and occurs in EuO. The simplest type of antiferromagnetic order consists of two like ferromagnetic subarrays of atomic moments directed opposite (antiparallel) to one another, as shown in Fig. 13.25(b): Real antiferromagnetic crystals, $MnF_2$ and MnO, are shown in Fig. 13.25.

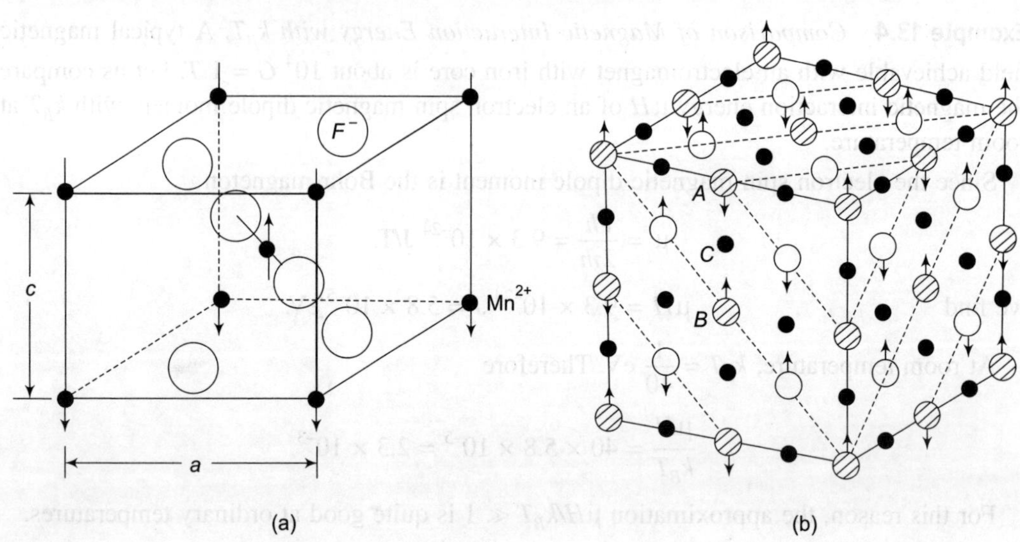


**Figure 13.25** (a) Antiferromagnetic ordering in $MnF_2$. [*After* R.A. Erickson, *Phys. Rev.* **90**, 779 (1953)]; (b) Antiferromagnetic ordering of MnO below the Curie temperature, showing one magnetic unit cell, which comprises eight chemical unit cells

⇕ $Mn^{+2}$ ions, with cross-hatching thus ⇕ to indicate an individual sub-lattice. O oxygen ions.

There is antiferromagnetic coupling between second nearest manganese neighbours, such as *A* and *B*, via the intervening oxygen, *C*.

The simplest type of ferromagnetic order consists of two unlike ferromagnetic subarrays that, though coupled antiparallel, gives a net spontaneous magnetization. This type of magnetic order is also called *Neel* ferrimagnetism and is well exemplified by the naturally occurring mineral magnetite, originally known as lodestone, which is the ferromagnetic iron oxide $Fe[Fe_2]O_4$ having the spinel structure (Fig. 13.26). Conventionally, chemical formulas for

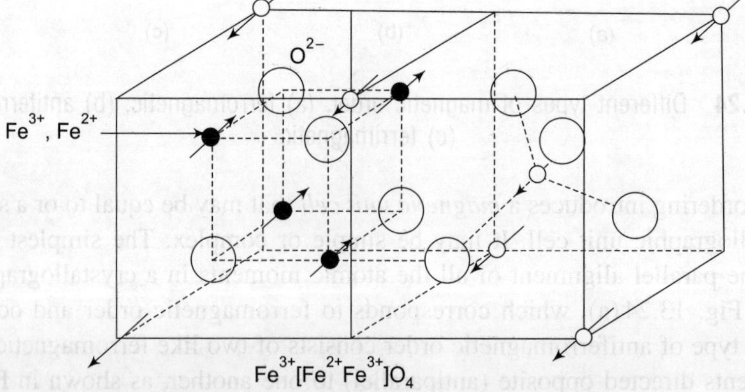


**Figure 13.26** Ferrimagnetic ordering in magnetite

the spinel structure are written with the tetrahedral site ($A$) cations outside and the octahedral site ($B$) cations inside square brackets to distinguish the two cation subarrays.

Other types of configurations are known. For example, in antiferromagnetic manganese dioxide ($MnO_2$) (shows in Fig. 13.27), which has the rutile structure, the magnetic configuration is simple spiral or *helical*. On passing down the $c$ axis, the turn angle between the direction of magnetic moments in successive basal planes in not 180° as in $MnF_2$ but $2\pi/7$ or 51.6°.

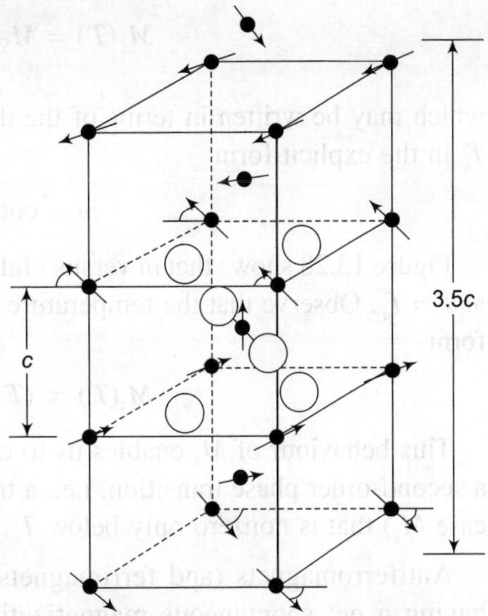


**Figure 13.27** Antiferromagnetic ordering in $MnO_2$. [*After* R.A. Erickson. *Phys. Rev.* **90**, 779 (1953)]

## 13.29.2 The Curie Point and the Neel Temperature

As already mentioned, magnetic order exists only below certain critical temperatures ($T_c$) called the Curie point; $T_c$ = 774, 1131, and 372°C for Fe, Co, and Ni, respectively. Above $T_c$, all ferromagnets show normal paramagnetic behaviour, with the difference that, in Eq. (2) (section 13.13.1), the absolute temperature $T$ has to be replaced by the "distance" $T - T_c$ from the Currie point, so that

$$\chi = \frac{C}{T - T_c}. \quad \text{...(1a)}$$

This is known as the *Curie–Weiss law*. In order to explain this bchaviour, one postulates the existence of an internal magnetic field, called the Weiss molecular field or exchange field, $H_w$, which is proportional to the existing magnetization

$$\mathbf{H}_W = \frac{T_c}{C}\mathbf{M}. \quad \text{...(1b)}$$

The resultant field $\mathbf{H} + \mathbf{H}_W$ now yields the total magnetization according to Curie's law, as

$$\mathbf{M} = \frac{C}{T}(\mathbf{H} + \mathbf{H}_W) \quad \text{...(1c)}$$

so that on eliminating $H_W$ between the two equations the Curie–Weiss law follows.

Moreover, the magnetization is given by Langevin formula

$$M(T) = N\mu L(y), \quad y \equiv \frac{\mu H + \mu H_W}{k_B T} \quad \text{...(2a)}$$

Therefore, when $T$ = 0°K, $L(y) = L(\infty) = 1$ and we obtain a spontaneous magnetization (i.e., magnetization at zero applied field, $H$ = 0), $M_s(0) = N\mu$; at finite temperature, the spontaneous magnetization is determined by the equation

$$M_s(T) = M_S(0)\, L(y_s), \quad y_s = \frac{H_W}{k_B T} \equiv \frac{3 M_s(T)\, T_c}{M_s(0)\, T}, \qquad \text{...(2b)}$$

which may be written in terms of the dimensionless variables, $m = M_1(T)/M_s(0)$ and $t = T/T_c$ in the explicit form

$$m = \coth(3mt) - (3mt)^{-1}.$$

Figure 13.28 shows that $m$ versus $t$ falls off continuously as $t$ increases and vanishes ($m = 0$) at $T = T_c$. Observe that the temperature dependence of the saturation magnetization is of the form

$$M_s(T) \propto (T - T_c)^{-1/2} \quad (\text{for } T < T_c). \qquad \text{...(2c)}$$

This behaviour of $M_s$ enables us to classify the paramagnetic–ferromagnetic transition as a second-order phase transition, i.e., a transition characterized by an order parameter (in this case $M_s$) that is nonzero only below $T_c$.

Antiferromagnets (and ferrimagnets, since these differ from antiferromagnets only in having a net spontaneous magnetization) also obey the Curie–Weiss law with negative value of $T_c$ (which is usually denoted by θ). For ferromagnets, $T_c$ is the temperature below which the ferromagnetic ordering is effective (in the Weiss molecular field approximation), whereas θ, being a negative temperature, cannot represent the temperature below which the antiferromagnetic ordering is effective; this latter temperature is called the *Neel point*, $T_N$. These differences are exhibited by the curve of $1/\chi$ versus $T$ shown in Fig. 13.29.

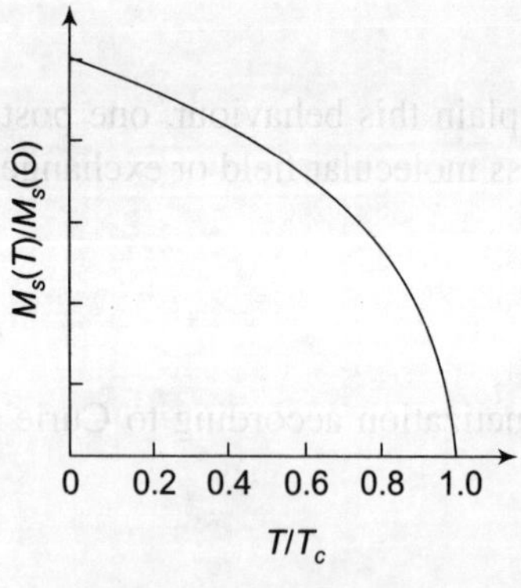


**Figure 13.28** The spontaneous magnetization versus the temperature

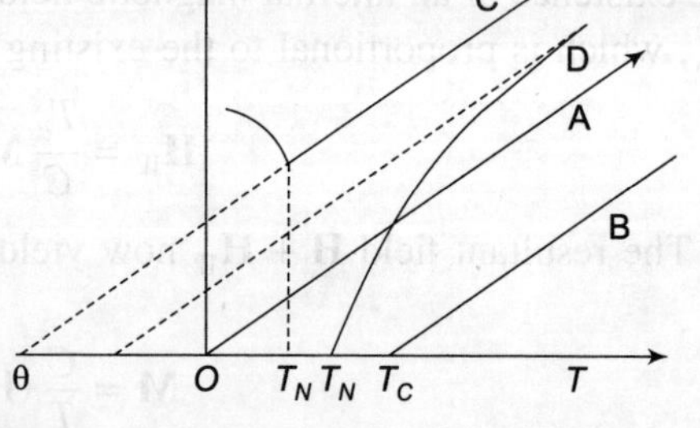


**Figure 13.29** The reciprocal of susceptibility versus temperature. (A) Simple paramagnet obeying the Curie law. (B) Ferromagnetic obeying Curie–Weiss law with positive $T_c$. (C) Antiferromagnet obey Curic-Weiss law with negative θ and positive ordering temperature (Neel temperature) $T_N$. (D) Ferromagnetic (special case of antiferromagnet).

A simple model (due to Neel) leads to a relation between $\theta$ and $T_N$. In Neel's model, the Curie law for the two (type $A$ and type $B$) sublattices of the antiferromagnetic arrays is expressed in the forms

$$M_A = \frac{C}{T}(H - \beta M_B - \varepsilon M_A) \quad \text{...(3a)}$$

$$M_B = \frac{C}{T}(H - \beta M_A - \varepsilon M_B) \quad \text{...(3b)}$$

where $\beta$ and $\varepsilon$ are coupling constants.

Three equations apply provided $T > T_N$. To have $M_A$ and $M_B$ finite at $H = 0$, we must require

$$\begin{vmatrix} T + \varepsilon C & \beta C \\ \beta C & T + \varepsilon C \end{vmatrix} = 0;$$

i.e., $(T + \varepsilon C)^2 = (\beta C)^2$, giving the Neel temperature

$$T_N = (\beta - \varepsilon)\, C.$$

There exists solution of Eq. (3a) and (3b); namely, $M_A = M_B$, for which

$$\frac{M_A}{H} = \frac{C}{T + C(\beta + \varepsilon)} = \frac{M_B}{H}$$

so that the susceptibility is

$$\chi = \frac{M_A + M_B}{H} = \frac{2C}{T + \theta} \quad \text{...(3c)}$$

where $\theta = C(\beta + \varepsilon)$. Thus, we find a relation between $T_N$ and $\theta$ of the form

$$\frac{\theta}{T_N} = \frac{\beta + \varepsilon}{\beta - \varepsilon}. \quad \text{...(4)}$$

This ratio is prevented from becoming too large through the appearance of different types of magnetic ordering. In system where a large ratio has been found experimentally, the conditions for spontaneous atomic moments appear to be only barely fulfilled, and the assumption that the Weiss constant ($T_c/C$) and the atomic moments are independent of temperature begin to breakdown.

Manganese and chromium and many compounds (for example, oxides) containing members of the $3d$ transition metal series exhibit antiferromagnetism, For MnO, $\theta$ = 610 °K, $T_N$ – 122 °K, and $\theta/T_N$ = 5.0.

## 13.30 RELATION BETWEEN EXCHANGE INTEGRAL, WEISS-CONSTANT (β) AND FERROMAGNETIC CURIE TEMPERATURE

Quantum mechanically, the probability that an electron 1 in atom a will exchange place with an electron 2 in atom $b$, or the exchange integral, is given by

$$J_e = \int \psi_b(1)\psi_a(2)\left(\frac{1}{r_{ab}} - \frac{1}{r_{a2}} - \frac{1}{r_{b1}} + \frac{1}{r_{12}}\right)\psi_b(1)\,\psi_a(2)\,dr_1\,dr_2 \quad \text{...(1)}$$

where $\psi_a(1)$ is the characteristic function of electron 1 when it is in atom $a$, $\psi_b(2)$ of electron 2 in atom $b$, $\psi_a(2)$ that when electron 2 is in atom $a$, and $\psi_b(1)$ that when electron 1 is in atom $b$; $r_{12}$ is the distance between the electrons, $r_{a2}$ that between the atom $a$ and electron 2, $r_{b1}$ that between the atom $b$ and electron 1 and $r_{ab}$ that between the nuclei of atoms $a$ and $b$. In general, this integral is negative because the attraction between the nuclei and the electrons is greater than the repulsion between the two nuclei and between the two electrons. However, it is positive when there exists a certain ratio of distance $D$ between the neighbouring atoms in the metal crystal to the radius $r$ of the energy shell in which the electron spins are uncompensated—a state of affairs found in rare earths and in metals of iron group. According to Slater, this ratio $D/r$ must be greater than 3.0, but not much greater. For iron, cobalt and nickel this ratio is respectively 3.26, 3.64 and 3.94.

Because of the importance of exchange integral $J_e$, we would like it to relate to the Weiss constant ($\beta$) and to the ferromagnetic Curie temperature, $T_f$. To simplify the procedure, we will assume, following Stoner, that exchange intergral is negligible except for the nearest neighbours and that its value is $J_e$ for all neighbouring pairs.

The exchange energy of a given atom $i$ with its neighbours can be written as

$$E_e = -2J_e \sum_j \vec{S}_i \cdot \vec{S}_j \qquad ...(2)$$

where the summation extends over all the nearest neighbours of $i$. Stoner assumed that the instantaneous value of the neighbouring spins may be replaced by their time-average values.

Therefore, if there are $Z$ nearest neighbours then, we can write Eq. (2) as

$$E_e = -2ZJ_e[S_{xi} \langle S_{xj}\rangle + S_{yi} \langle S_{yj}\rangle + S_{zi} \langle S_{zj}\rangle] \qquad ...(3)$$

where $S_{xi}$, $S_{yi}$, $S_{zi}$ are the components of spin of the $i$th atom along three mutually perpendicular axes $x$, $y$ and $z$ respectively, and $\langle S_{xj}\rangle$, $\langle S_{yi}\rangle$ and $\langle S_{zy}\rangle$ are the average values of the spins of the nearest neighbour, the $j$th atom, along three nutually perpendicular axes $x$, $y$ and $z$ respectively.

If the magnetization is along the $z$ direction then

$$\langle S_{xy}\rangle = \langle S_{yi}\rangle = 0$$

and
$$\langle S_{zj}\rangle = \frac{I}{g\mu_B N} \qquad ...(4)$$

Therefore, Eq. (3) reduces to

$$E_e = -2Z\, J_e\, S_{zi}\left[\frac{I}{g\mu_B N}\right] \qquad ...(5)$$

Now, the maximum component of an atomic dipole moment associated with a quantum number $S$ is $g\mu_B N$. Hence, the potential energy of the $i$th atom having spin $S_{zi}$ in the Weiss field $\beta I$ will be

$$E = -g\,\mu_B S_{zi}\beta I \qquad ...(6)$$

and this must be equal to the exchange energy as given by Eq. (5). That is,

$$-g\mu_B S_{zi}\beta I = -2ZJ_e S_{zi}\left[\frac{I}{g\mu_B N}\right]$$

$$\beta = \frac{2ZJ_e}{Ng^2 m_B^2} \quad ...(7)$$

Equation (7) gives the relation between the molecular field constant (β) and the exchange integral, $J_e$.

Substituting this value of β, given by Eq. (7), in Eq. (15, section 13.22), we get

$$T_f = \left(\frac{\mu_0 2ZJ_e}{Ng^2\mu_B^2}\right)\left(\frac{NJ^2g^2\mu_B^2}{k_B}\right)\left(\frac{J+1}{3J}\right)$$

$$T_f = \frac{2\mu_0 Z J_e \cdot J(J+1)}{3k_B} \quad ...(8)$$

Since the magnetization of ferromagnetic materials arise due to the spin of the electrons and the exchange force appears in the form of spin-spin interaction, we replace $J$ by $S$ in Eq. (8) and get

$$T_f = \frac{2\mu_0 Z J_e \cdot S(S+1)}{3k_B} \quad ...(9)$$

## 13.31 ORIGIN OF FERROMAGNETIC STATES

The magnetic moment of an atom is given by the sum of the magnetic moments of all its electrons, including both the orbital and spin moments. Each electron contributes one Bohr magneton to the total magnetic moment of the atom. Since all filled shells have zero total angular momentum, they also have zero total magnetic moment. Therefore, atoms or ions, which possess only filled shells, have no permanent moments and hence, they cannot be paramagnetic or ferromagnetic. No exception to this rule has yet been reported. The inert gases, e.g., helium, argon, krypton, etc., and the ions such as $Na^+$ and $Cl^-$ are all diamagnetic.

Three well-known ferromagnetic substances are iron (Fe), cobalt (Co) and nickel (Ni). These three elements have closed shells for 1*s*, 2*s*, 2*p*, 3*s* and 3*p* levels. Since a closed shell possesses no magnetic moment, the 18 inner electrons in the above closed shells do not participate in the magnetic process. Attention must, therefore, be focused on the 3*d* and 4*s* levels. Now, the 4*s* electrons are mainly responsible for the binding and electrical conductivity on the metals while the 3*d* electrons are responsible for ferromagnetism. The theoretical value of the atomic magnetic moments can, therefore, be obtained merely by finding the number of impaired 3*d* electrons. In Fe, Co and Ni, their distribution is as given below:

| | | | | | | |
|---|---|---|---|---|---|---|
| Fe: | $3d^6$ | ↑↓ | ↑ | ↑ | ↑ | ↑ |
| Co: | $3d^7$ | ↑↓ | ↑↓ | ↑ | ↑ | ↑ |
| Ni: | $3d^8$ | ↑↓ | ↑↓ | ↑↓ | ↑ | ↑ |

| | | |
|---|---|---|
| $4s^2$ | ↑ | ↓ |
| $4s^2$ | ↑ | ↓ |
| $4s^2$ | ↑ | ↓ |

The above spin arrangements have been obtained by applying *Hund's rule* which states that the spin of the electron of a shell always adds together in such a way as to give the maximum angular momentum and magnetic moment. For example, in a *d* shell, there are 10 states, 5 with spin up and 5 with spin down. If this shell contains 5 electrons, all of them will have their spins in the same direction, say up, which is the maximum angular momentum a *d* shell can have, since a sixth electron must have its spin down, cancelling one of the first five. Again, when 10 electrons are present, the total spin is zero, with 5 up and 5 down.

Accordingly, the atomic magnetic moments of Fe, Co, and Ni should be 4, 3 and 2 Bohr magnetons respectively. The actual values, however, are found to be 2.22, 1.70 and 0.61 Bohr magneton for Fe, Co and Ni, respectively.

We, therefore, see that even though each atom has an integral number of electrons, the effective number of magnetic moment carriers come out to be non-integral. This non-integral number of electron spin per atom is explained on the basis of zone theory as follows.

The 3*d* and 4*s* zones of these metals overlap, as shown in Fig. 13.30 for nickel. In general, there is, on an average a fractional number of electrons in each band. In case of nickel, the 10 electrons of *d* and *s* bands may be distributed among them and we find that on an average, there are 9.4 electrons in the 3*d* band and 0.6 in the 4*s* band. In case of cobalt, out of 9 valence electrons, 8.3 go to the 3*d* band and 0.7 to the 4*s* band. Iron with 8 valence electrons supplies 7.8 electron to the 3*d* band 0.2 electron to the 4*s* band, where the division in all cases is only approximate.

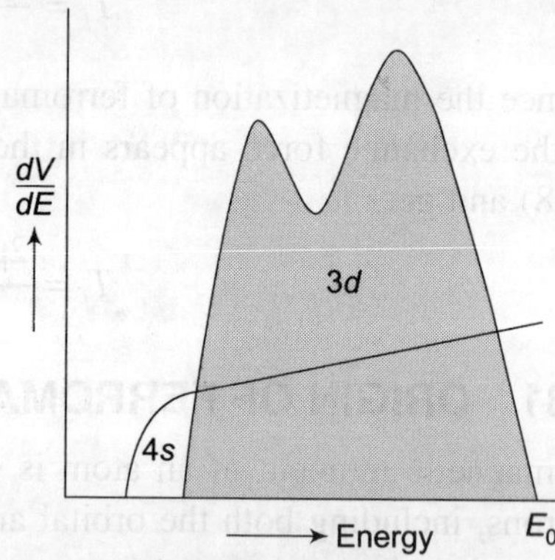


**Figure 13.30** Overlapping of 3*d* and 4*s* zones in case of nickel

If the 3*d* band contains 10 electrons, as allowed by Bohr–Bury scheme, then 5 of them will have their spins parallel and 5 anti-parallel to an externally applied field. In a partially filled band, we may still have 5 spins set parallel to the field when saturation is attained, but the remaining electrons must set themselves anti-parallel to the field. Thus, in the case of nickel, with 9.4 electrons in the 3*d* band, 5 are parallel to the field and remaining 4.4 are anti-parallel to the field so that the nickel atom gives a net parallel spin of $(5 - 4.4)\ \mu_B = 0.6\ \mu_B$. Similarly, cobalt gives a net parallel spin of $(5 - 3.3)\ \mu_B = 1.7\ \mu_B$ and iron $(5 - 2.8)\ \mu_B = 2.2\ \mu_B$.

The accuracy of this model can easily be checked by alloying nickel with copper or zinc. Since the atomic numbers of nickel, copper and zinc are 28, 29 and 30 respectively, the substitution of a copper atom for a nickel atom introduces one additional electron in the crystal, whereas zinc introduces 2 electrons. It is only reasonable to expect that these electrons will prefer the lower-lying 3*d* states of nickel and, therefore, as more copper or zinc added, the net parallel spin of the nickel atom decreases. Consequently, the intensity of magnetization and hence, the magnetic susceptibility of nickel should decrease. It has been found from saturation magnetization experiments that it is indeed the case and the magnetization of nickel becomes virtually zero when 60% of copper or 30% of zinc is added to it, that is, when enough electrons are added to completely fill the $3d$ zone nickel.

### 13.31.1 Exchange Integral in Ferromagnets

According to Heisenberg, if two neighbouring magnetic atoms have spins $S_i$ and $S_j$, then their potential energy can be expressed as

$$U = \text{constant} - 2\, J_e\, \vec{S}_i \cdot \vec{S}_j \qquad ...(1)$$

where the second term on the R.H.S is the *exchange energy* and is a consequence of the *exchange force*.

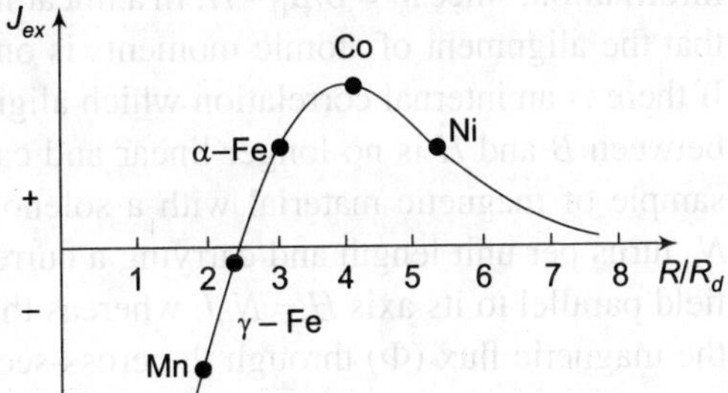


**Figure 13.31** The Néel-Slater curve for the transition metal crystals

It may appear from Eq. (1) that the spin on one atom is directly coupled to the spin on its neighbour, which is not the actual case. Actually, quantum mechanically, the exchange integral $J_e$ is a negative quantity for a small distance $r_{AB}$ ($\equiv R_d$) between neighbouring magnetic atoms ($A$ and $B$), which (for small $r_{AB}$) makes parallel spins energetically unfavourable (similar to $H_2$ molecule case). However, when $r_{AB}$ is larger than a critical distance, the exchange integral can be positive (see Fig. 13.31), $R_d$ is the radium of the $3d$ orbit and parallel spins are then energetically favoured. Ferromagnetism thus places requirement on the minimum spacing of magnetic neighbours and the spacing of magnetic neighbours should not be too large, since then, the exchange integral declines to zero when $r_{AB}$ is very large. This accounts for the ferromagnetic behaviour of Fe, Co and Ni. In these atoms, states are occupied in accordance with the Pauli's principle to give a minimum clectron energy for an appropriate value of $r_{AB}$. However, the element Mn (manganese) although has a large unpaired spin, is not ferromagnetic in the elemental form because $r_{AB}$ is too small. It can be madc ferromagnetic by incorporating it in a compound (such as MnAs and MnSb) which increases the Mn-Mn distance far enough to make the exchange integral positive (the lattice constants of MnAs and MnSb are respectively 2.85 Å and 2.89 Å, whereas that of pure Mn is 2.58 Å).

The crystallographic alignment of the magnetic moments in the α-Fe, Co and Ni ferromagnetic solids is illustrated in Fig. 13.32.

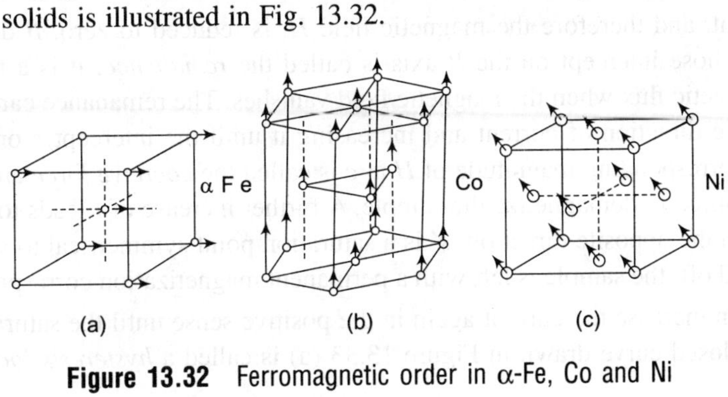


**Figure 13.32** Ferromagnetic order in α-Fe, Co and Ni

## 13.32 FERROMAGNETIC MATERIALS

The most common experimental approach to the magnetic behaviour of a ferromagnetic material is to represent the magnetic induction ($\vec{B}$) for various field strengths, $\vec{H} = \vec{B}_0/\mu_0$. Alternatively, plots of magnetization ($\vec{M}$) in terms of $\vec{H}$ are used, which contain the same information, since $\vec{M} = \vec{B}/\mu_0 - \vec{H}$. In a linear magnetic material, this constitutive equation shows that the alignment of atomic moments is only maintained by an applied magnetic induction. If there is an internal correlation which aligns the atomic magnetic moments, the relationship between $\vec{B}$ and $\vec{H}$ is no longer linear and can be experimentally obtained using a cylindrical sample of magnetic material with a solenoid of length $l$ close-wound over it, consisting of $N_0$ turns per unit length and carrying a current $I$. The solenoid provides an uniform magnetic field parallel to its axis $H = N_0 I$, whereas the magnetic induction ($B$) is measured in terms of the magnetic flux ($\Phi$) through the cross-sectional area $A$ of the sample.

As the current through the solenoid increases we obtain the magnetization curve $Os$, as in Fig. 13.33 (a), characteristic of the unmagnetized material, which corresponds to the alignment of magnetic dipoles, until the saturation point $s$ is reached. A *differential permeability* ($dB/dH$) is defined from the curve $Os$ rather than the constant permeability of a linear material.

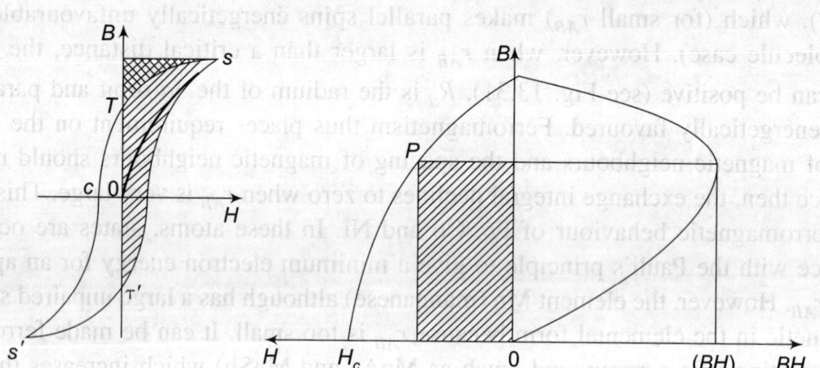


**Figure 13.33** $B(H)$ hysteresis loop (a) and the energy product $(BH)_{max}$ (b)

If the current, and therefore the magnetic field $H$, is reduced to zero, $B$ decreases along the curve $sr$, whose intercept on the $B$ axis is called the *remanence*. It is a measure of the remaining magnetic flux when the magnetic field vanishes. The remanance can be eliminated by reversing the direction of current and increasing it until the intercept $c$ on the $H$ axis is reached. The corresponding magnitude of $H$ at $c$ is called the *coercive force* and is a measure of the field required to demagnetize the sample. A further increase of $I$ leads to the alignment of all dipoles in the opposite direction: $s'$ is a saturation point symmetrical to $s$. If the current is now switched off, the sample is left with a permanent magnetization corresponding to point $r'$. We may then increase the current again in the positive sense until the saturation point $s$ is reached. The closed curve drawn in Figure 13.33 (a) is called a *hysteresis loop*. The energy spent in taking the magnetic material around a hysteresis cycle is proportional to the area of the loop.

Let us consider a cycle which starts with the conditions corresponding to point $r'$, in Fig. 13.33 (a). When the current and the field are increasing from $r'$ to $s$, the magnetic induction will also be increased, generating a back e.m.f. which opposes the increase in current

$$\text{e.m.f.} = -\frac{d}{dt}\int BdS = -(N_0 l)\, A\frac{dB}{dt} = -N_0 V\frac{dB}{dt}$$

where $N_0 l$ is the number of turns of the solenoid $V = lA$ is the volume of the sample. The spent energy is measured by the work done against the back e.m.f.:

$$dU = -(\text{e.m.f.})\, I\,dt = N_0 VI\,dB = VH\,dB$$

and hence, the energy required in going $r'$ to $s$, which represented in Fig. 13.33 (a) by the shaded area, is

$$U_{r's} = V\int_{r'}^{s} HdB$$

When the current is decreasing in the same direction $s$ to $r$, the polarity of the back e.m.f. is reversed and the energy

$$U_{sr} = -V\int_{s}^{r} HdB$$

is returned from the sample. The corresponding area must be subtracted from the shaded one and the result is equal to $V$ times half the ares contained within the hysteresis loop. In a similar manner the net energy supplied to execute the second half of the cycle is found to be equal to $V$ times the second half of the loop, so that the energy required for one cycle is

$$U_h = V\oint HdB \qquad ...(1)$$

where the line integral is evaluated around the hysteresis loop.

The suitability of ferromagnetic materials for applications is determined principally from the characteristics shown by their hysteresis loops. For a permanent magnet, the energy Eq. (1), i.e., the area of the loop, should be high in order to retain the magnetization as much as possible, by high remanence and coercivity (*hard magnetic material*). Closely related to the area enclosed by the hysteresis loop is the maximum energy product, which is the maximum $BH$ value, shown in Fig. 13.33 (b). It is apparent the magnets operate in the second quadrant of their hysteresis loop, where $(BH)_{max}$ is measured by the shaded area corresponding to the operating point $P$. For a material subjected to an alternating current, such as a transformer core material, the power dissipated is proportional to the frequency $\nu$, which gives the number of hysteresis cycles per second, and these hysteresis losses are minimized in case of a narrow hysteresis loop which exhibits low coercivity and high saturation (*soft magnetic material*).

The process of magnetization of ferromagnets can be explained assuming that long-range magnetic order exists on the interatomic scale, but at some larger scale the magnetizations of completely aligned volumes, or *domains*, are randomly oriented in the demagnetized state. The magnetization ($M_s$) within each domain is almost saturated, due to large number of atomic dipoles, typically $10^{12}$ to $10^{15}$, which are aligned parallel to each other. The magnetization of a specimen will consist of converting its domain configuration to a single domain, with magnetization parallel to the applied field. Consider, for simplicity, the rotation of the magnetization, caused by the action of an applied field on an array of isolated particles, which are small enough to be single domain. Experiment shows that, in crystals, magnetization exhibits a preferred direction, called the easy axis, which is energetically favourable. In other words, if the magnetization is rotated out of this direction by the application of a field, there must be an increase in energy, known as the *anisotropy energy*. The most obvious is the uniaxial anisotropy, found in single crystal cobalt, where there is a single easy axis corresponding to the $c$-axis of the hexagonal unit cell shown in Fig. 13.32(b).

**Example 13.5** Find the spin wave contribution to the low-temperature heat capacity of a ferromagnet, assuming the spin wave dispersion relation to be $\omega = Ak^2$.

***Solution*** The internal energy of a magnon system has a form similar to that used for a phonon system, where the mean energy per magnon mode can also be written as

$$E(\omega) = \frac{\hbar\omega}{e^{\hbar\omega/k_BT} - 1}$$

The density of states is given by

$$N(\omega) = \frac{V}{(2\pi)^3}\frac{4\pi k^2}{d\omega/dk} = \frac{V}{4\pi^2}A^{\frac{3}{2}}\omega^{\frac{1}{2}}$$

On substitution into Eq. $U = \int N(\omega)\,E(\omega)\,d\omega$ yields:

$$U = \frac{V\hbar}{4\pi^2 A^{\frac{3}{2}}}\int_0^{\omega_{\max}}\frac{\omega^{3/2}d\omega}{e^{\hbar\omega/k_BT} - 1} = \frac{V}{4\pi^2}\frac{(k_BT)^{\frac{5}{2}}}{(A\hbar)^{\frac{3}{2}}}\int_0^{\omega_{\max}}\frac{x^{\frac{3}{2}}\,dx}{e^x - 1}$$

For low temperatures the upper limit is large and can be replaced by $\infty$, making the integral is the independence of $T$, such that

$$C_V = \frac{V}{4\pi^2}\frac{5\,k_B}{2}\left(\frac{k_BT}{A\hbar}\right)^{\frac{3}{2}}\int_0^{\infty}\frac{x^{\frac{3}{2}}\,dx}{e^x - 1}$$

This is known as the $T^{\frac{3}{2}}$ law for the low-temperature behaviour of the heat capacity of ferromagnets, which is observed experimentally.

## 13.33 ANTIFERROMAGNETISM

We know that the behaviour of a ferromagnet may be explained quantum mechanically in terms of the exchange interaction. It is, therefore, assumed that the exchange integral ($J$) is positive.

When the exchange integral ($J$) is negative, favouring an ordered antiparallel orientation of electron spins, one has an antiferromagnetic substance (Fig. 13.34). The susceptibility versus temperature variation for antiferromagnetic material is shown in Fig. 13.35 where, unlike ferromagnetism, the antiferromagnetic susceptibility exhibits a maximum at the Neel temperature, $T_N$. Above $T_P$ it is given by

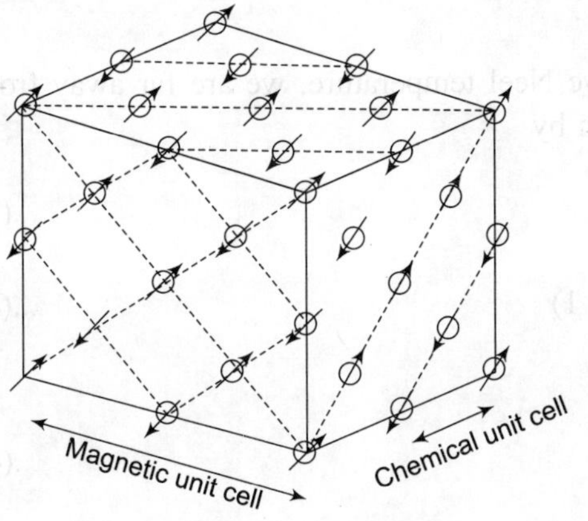


**Figure 13.34** Antiferromagnetic coupling in MnO. Only the magnetic ions ($Mn^{2+}$) are shown and the diamagnetic ions ($O^{2-}$) are omitted

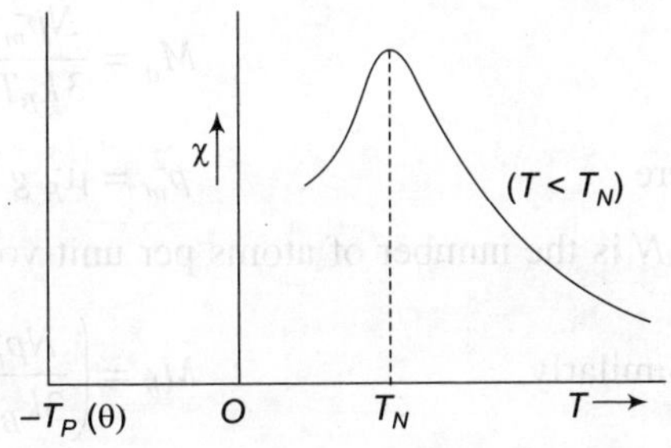


**Figure 13.35** Antiferromagnetic susceptibility temperature variation. Notice the peak at $T_N$. ($T_N$ is Neel temperature)

$$\chi = \frac{C}{T + T_p}$$

where $T_P$ is Weiss constant and below $T_P$ there are two values depending on whether the applied field is parallel or perpendicular to the sublattice magnetization.

This characteristic feature may be explained on the following model.

### 13.33.1 The Two-Sublattice Model

Consider a crystal containing two types of atoms $A$ and $B$ distributed over two interpenetrating lattice $s$, for example, let the $A$ atoms occupy the corner points of an elementary cube and the $B$ atoms being located at the centres of these cubes. Furthermore, let the interaction between the atoms be such that the $A$ spins tend to line up antiparallel to the $B$ spins. At low temperature, this interaction is very effective and in an external field the resulting magnetisation will be small. As the temperature is raised, the efficiency of the interaction decreases and the susceptibility increases. Finally, a critical temperature, $T_N$ (known as Neel temperature) is reached above which the spins are free and the antiferromagnetic material becomes paramagnetic, i.e., $\chi$ decreases with further increase in temperature.

Let there be two sublattice arrays $A$ and $B$. We assume that all nearest neighbours of an $A$ atom are $B$ atoms and vice versa. Further assume that besides an antiferomagnetic $AB$ interaction, there are also antiferomagnetic interactions $AA$ and $BB$. Let the molecular field at $A$ and $B$ be $H_a$ and $H_b$. Then

$$H_a = H \quad \alpha M_a \quad \beta M_b \qquad \text{...(1a)}$$

$$H_b = H - \beta M_a - \alpha M_b \qquad \text{...(1b)}$$

where $H$ is the applied field, $M_a$ and $M_b$ represent the magnetisations of $A$ and $B$, $\alpha$ is the interaction parameter for two atoms having like spins ($AA$ or $BB$) and $\beta$ is the interaction parameter for two unlike atoms.

Now we consider two temperature regions.

**Case I:** $T > T_N$, i.e., when temperature is above Neel temperature, we are far away from saturation and magnetisation of lattice $A$ is given by

$$M_a = \frac{Np_m^2}{3k_BT} H_a \qquad \text{...(2)}$$

where $$p_m^2 = \mu_B^2 \, g^2 \, J(J+1) \qquad \text{...(3)}$$

and $N$ is the number of atoms per unit volume.

Similarly, $$M_b = \left(\frac{Np_m^2}{3k_BT}\right) H_b \qquad \text{...(4)}$$

$$M = M_a + M_b = \left(\frac{Np_m^2}{3k_BT}\right) [2H - (\alpha + \beta)\, M]$$

$$M\left[1 + \frac{Np_m^2}{3k_BT}(\alpha + \beta)\right] = \frac{2Np_m^2}{3k_BT} H$$

or $$\chi = \frac{M}{H} = \frac{2Np_m^2/3k_B}{T + [Np_m^2\,(\alpha + \beta)]/3k_B}$$

or $$\chi = \frac{C'}{T + T_p} \text{ (where } C' = 2Np_m^2/3k_B) \qquad \text{...(5)}$$

Comparing this with expression

$$\chi = \frac{M}{H} = \frac{C}{T + T_p}$$

for the susceptibility of a ferromagnet above Curie temperature, it is seen that an antiferromagnet contains $T + T_p$ rather than $T - T_p$. Further, the constant $C'$ is twice the Curie constant ($C$) for individual $A$ or $B$ lattices (i.e., $C' = 2C$).

**Case II:** $T < T_N$, at the Neel temperature $T_N$ itself, we are still sufficiently far away from saturation. Thus, in the absence of applied magnetic field, we may write $T = T_N$ and from Eq. (2), we have

$$M_a = \frac{Np_m^2}{3k_BT_N}(\alpha M_a + \beta M_b)$$

or $$\left[1+\frac{Np_m^2}{3k_BT}\alpha\right]M_a+\frac{Np_m^2}{ek_BT}\beta M_b=0 \qquad ...(6)$$

Similarly, $$\frac{Np_m^2}{3k_BT_N}\beta M_a+\left[1+\frac{Np_m^2}{3k_BT_N}\alpha\right]M_b=0 \qquad ...(7)$$

These equations have a non-vanishing solution for $M_a$ and $M_b$ only if the determinant of their coefficient vanishes.

Since $$C'=\frac{2Np_m^2}{3k_B} \quad \text{or} \quad \begin{vmatrix} \left(1+\frac{C'}{2T_N}\alpha\right) & \frac{C'}{2T_N}\beta \\ \frac{C'}{2T_N}\beta & \left(1+\frac{C'}{2T_N}\alpha\right) \end{vmatrix}=0$$

or $$\left[1+\frac{C'}{2T_N}\alpha\right]^2-\frac{C'^2}{4T_N^2}\beta^2=0$$

or $$1+\frac{C'^2}{4T_N^2}\alpha^2+\frac{C'}{T_N}\alpha-\frac{C'^2}{4T_N^2}\beta^2=0$$

or $$T_N^2+\frac{C'^2}{4}\alpha^2+C\alpha T_N-\frac{C'^2}{4}\beta^2=0$$

or $$T_N^2+C\alpha T_N+\frac{C'^2}{4}(\alpha^2-\beta^2)=0$$

or $$T_N=\frac{-C'\alpha\pm\sqrt{C'^2\alpha^2-C'^2(\alpha^2-\beta^2)}}{2}$$

or $$T_N=\frac{C'(\beta-\alpha)}{2} \qquad ...(8)$$

From this expression it is quite obvious that Neel temperature increases as $AB$ interactions ($\beta$) are stronger, whereas it decreases with $AA$ or $BB$ interaction ($\alpha$ increasing).

In this model the Neel temperature is not identical with $T_p$. Setting up a relation between $T_N$ and $T_P$ with the help of Eq. (5), we get

$$T_p=\frac{Np_m^2}{3k_B}(\alpha+\beta)=\frac{C'(\alpha+\beta)}{2}$$

Therefore, $$\frac{T_N}{T_p}=\frac{\beta-\alpha}{\beta+\alpha} \qquad ...(9)$$

### 13.33.2 Susceptibility Below Neel's Temperature

For simplicity, we assume only *AB* interaction, i.e., assume $\alpha = 0$. We consider two-sublattice model (a) with the applied magnetic field perpendicular to the axis of spins (b) with the field parallel to the axis of spins.

**Case I:** For the calculation of susceptibility, an equilibrium is determined by the balance of an external force and a restoring force. In this case, field tends to align the dipoles along the field, but as a result of tendency for *A* and *B* to remain antiparallel, a compromise is obtained in which the dipoles make a certain ($\phi$) with original spin direction.

Consider one of the dipoles *B* as made up if two unit poles. The forces on the positive pole are *H* and $-\beta\, M_a$; the forces on the negative pole are equal but of opposite sign. In equilibrium, the resultant force should lie along the line joining the two poles, so that for small angles, we have

$$2\beta\, M_a \phi = H \qquad ...(10)$$

which is opposite in direction of *H* [Fig. 13.36(a)]

Since $M_a = M_b$, the total magnetization alone the external field direction is

$$M = (M_a + M_b)\,\phi = \frac{H}{\beta} \qquad ...(11)$$

the susceptibility $\chi_\perp = \frac{M}{H} = \frac{1}{\beta}$ is independent of temperature.

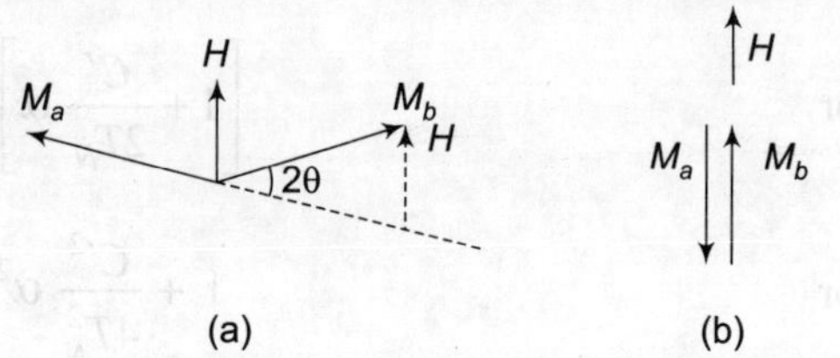


**Figure 13.36** Calculation of; (a) Perpendicular, (b) Parallel susceptibilities at OK

**Case II:** Calculation of $\chi_{\parallel}$ corresponding to an applied along the axis of spins is much complicated, Van Vleck, on the basis of his calculations, has shown that susceptibility rises smoothly from zero to $\chi\,(T_N)$ as the temperature increases. At absolute zero, all the spins are parallel or antiparallel to the field so that no torque is exerted on them. Consequently, there is no net magnetization, i.e., at $T = 0$ K, in the parallel orientation, the magnetic energy is not changed if the spin systems *A* and *B* make equal angles with the field. Consequently, $\chi_{\parallel}$ at $T = 0$ K is zero. Figure 13.37 shows measurements on $MnF_2$.

For the high-temperature region, we recollect that

| *para* | *ferro* | *antiferro* | |
|---|---|---|---|
| $\chi = C/T$ | $\chi = C/(T - T_p)$ | $\chi = C/(T + T_p)$ | ...(12) |
| | for $T > T_f (= T_p)$ | for $T > T_N$ | |

## 13.34 FERRIMAGNETISM

If the antiferromagnetic interaction in a solid is such that the spins of neighbouring atoms are antiparallel but unequal in magnitude, the solid exhibits a net spontaneous magnetization and is known as a ferrimagnetic material. The net magnetic moment of these materials

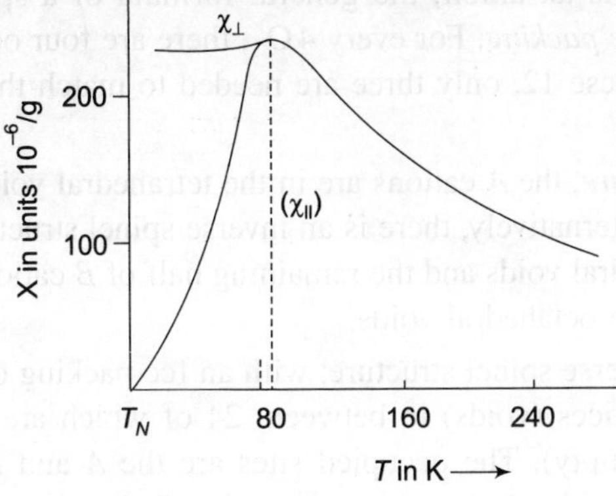


**Figure 13.37** Magnetic susceptibility of $MnF_2$

disappears above a Curie temperature (analogous to Neel temperature) at which thermal energy randomizes the individual magnetic moments and the material becomes paramagnetic.

## 13.34.1 Ferrites and their Properties

Ferrites form a class of ferromagnetic materials which may be prepared from electrically non-conducting ceramic materials. A common ferrite material is magnetite [$Fe^{2+}O^{2} \cdot Fe_2^{3+} O_3^{2-}$] ferrites form a class of mixed oxides with chemical formula (MO, $Fe_2O_3$), where *M* is a divalent metal such as Mn, Co, Fe, Ni, Cu or Zn.

The general electric and magnetic characteristics of ferrites (in addition to their ferromagnetic property) arc:

(i) They are insulators with very high resistivity (~$10^{12}$ Ω-cm). As a consequence, there will be no eddy current loss as usually notcd with iron, at high frequencies.

(ii) High microwave dielectric constant (~10 to 12) (low dielectric loss).

(iii) A high magnetic permeability (500–1000) for the mixed Ni–Zn ferrites ($Ni_{0.36}Zn_{0.64}Fe_2O_4$) at low frequencies and ~10 at high frequencies (≥ 300 MHz).

(iv) Ferrites show the same type of hysteresis curve, domain structure and motion similar to ferromagnets; but the coercive force and saturation magnetization (though appreciable) is smaller than that of ferromagnetic materials.

(v) They have a Curie temperature that varies from 100°C to several hundred °C.

(vi) They are mechanically hard, brittle and difficult to machine.

(vii) They are to be prepared using ceramic method, and it is not possible to cast them from the molten state because oxygen is then driven off from the material.

(viii) The resistivity depends on the preparation technique (and can be anywhere between 1 and $10^{12}$ Ω-cm).

Ferrites are similar in structure to the mineral *spinel*. In general, the crystal class spinel consists of ionic crystal with more than two types of ions, there are two different cations

$A^{2+}$ and $B^{3+}$ and oxygen is an anion; the general formula of a spinel being $AB_2O_4$. *The oxygen anions form the fcc packing*. For every $4O^-$, there are four octahedral sites and eight tetrahedral sites. Out of these 12, only three are needed to match the cations of the general formula $AB_2O_4$.

In *normal spinel structure*, the *A* cations are in the tetrahedral voids and the *B* cations are in the octahedral voids. Alternatively, there is an inverse spinel structure, where half of the *B* cations occupy the tetrahedral voids and the remaining half of *B* cations plus all the *A* cations are distributed randomly in octahedral voids.

A ferrite consists of inverse spinel structure, with an fcc packing of unit cells each having 32 $O^-$ ions, and 96 interstices (voids) in between, 24 of which are filled with cations (the remaining 72 voids are empty). The occupied sites are the *A* and *B* sites. There are eight occupied *A* sites (surrounded by 4 $O^-$) and 16 occupied *B* sites (surrounded by 6 $O^-$). For *B* = Fe (i.e. $B_2O_3 = Fe_2O_3$), we have the usual ferrite structure where different *A* cations can be present in varying proportions.

### 13.34.2 Spin Arrangement in the Ferrite $FeO \cdot Fe_2O_3$

The ferric ions ($Fe^{3+}$) are in a state with spin $S = 5/2$ and zero orbital moment. Therefore, each ion contributes $5\,\mu_B$ to the saturation moment. The ferrous ions ($Fe^{2+}$) have a spin of 2 and should contribute $4\,\mu_B$ except. For a residual moment which may be negligible. Thus, the effective number of magnetrons per $Fe_3O_4$ formula unit = $2 \times 5 + 4 = 14$, if all spins were parallel. However, the observed value is 4.1. The discrepancy is matched if the moments of the $Fe^{3+}$ ions are antiparallel to each other; then the observed moments arise only from the $Fe^{2+}$ ions.

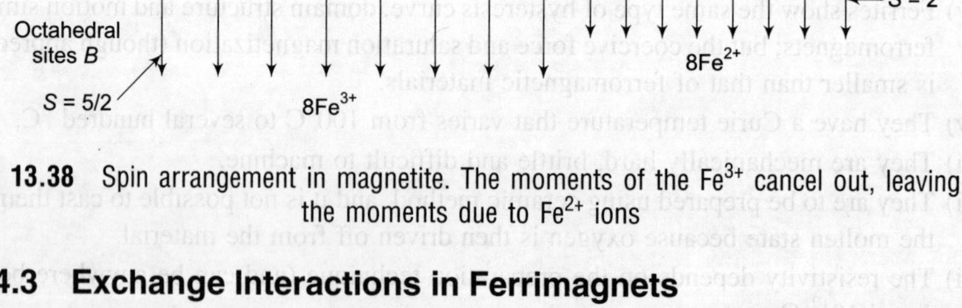


**Figure 13.38** Spin arrangement in magnetite. The moments of the $Fe^{3+}$ cancel out, leaving only the moments due to $Fe^{2+}$ ions

### 13.34.3 Exchange Interactions in Ferrimagnets

A remarkable feature (which is common to all spinels) is that all exchange interactions (*AA*, *AB*, *BB*) favour antiparallel alignment of the spins connected by the interaction. Yet, the *AB* interaction is the strongest, because the *A* spins are also parallel to each other and the *B* spins are also parallel to each other, such that $A$ spins may be *antiparallel* to the $B$ spins. All exchange integrals $J_{AA}$, $J_{AB}$ and $J_{BB}$ are negative. If $J$ in $U = -2J\,\vec{S}_i \cdot \vec{S}_j$ is positive, then the exchange integral would yield ferromagnetism, otherwise, if $\vec{J}$ is negative, the exchange integral would be antiferromagnetic.

We now proceed to see that three antiferromagnetic interactions could result in ferrimagnetism. The mean exchange fields for the $A$ and $B$ (spin) sublattices may be written as

$$B_A = -\lambda M_A - \mu M_B$$

$$B_B = -\mu M_A - \nu M_B \quad ...(1)$$

taking all constants $\lambda$, $\mu$, $\nu$ to be positive. The minus sign then corresponds to an antiparallel interaction.

The interaction energy density is

$$U = \frac{1}{2}(B_A \cdot M_A + B_B \cdot M_B)$$

$$= \frac{1}{2}\lambda M_A^2 + \mu M_A \cdot M_B + \frac{1}{2}\nu M_B^2 \quad ...(2)$$

This is lower when $M_A$ is antiparallel to $M_B$ than when $M_A$ is parallel to $M_B$. The energy when antiparallel to each other should be compared with zero, because, a positive solution is

$$M_A = M_B = 0 \quad ...(3)$$

Thus, when $$\mu M_A \cdot M_B > \frac{1}{2}(\lambda M_A^2 + \nu M_B^2) \quad ...(4)$$

the ground state will have $M_A$ directed oppositely to $M_B$. Under certain conditions, there may still be (antiparallel) spin arrays of further lower energy.

### 13.34.4 Applications of Ferrimagnetic Materials

Table 13.5 gives the properties of some ferrites.

**Table 13.5** Some Ferrite Materials

| *Material* | *Curie temperature (°C)* | *Saturation magnetization (Wb/m²)* | *Initial permeability* |
|---|---|---|---|
| $MnFe_2O_4$ | 300 | 0.49 | 250 |
| $Fe_3O_4$ | 585 | 0.60 | 70 |
| $NiFe_2O_4$ | 585 | 0.34 | 10 |
| $MgFe_2O_4$ | 440 | 0.15 | 10 |

In certain ferrimagnetic materials, the domains can be altered in their shape by the application of an external field. These domains can be used to store information. The materials of interest are the rare earth orthoferrites and the rare earth iron garnets.

The applications of (ceramic) ferrites can be broadly put into the following four categories:

(i) *Hard ferrites* (Ferrites for permanent magnets):

*Examples*: barium ferrite ($BaFe_{12}O_{19}$), strontium ferrite ($SrFe_{12}O_{19}$).

These have high value of uniaxial anisotropy field, high coercive force and high resistivity. These are used as focusing magnets for television tubes.

(ii) *Soft ferrite*: (Ferrites for transformers and inductors):

*Examples:* manganese-zinc ferrite, nickel-zinc ferrite.

These have high permeability, low coercive force, low eddy current losses thus ability to operate upto high frequencies (~10 to 1000 MHz).

These are used as inductor cores, transformer cores, television transformers and aerial rods.

(iii) *Rectangular loop ferrites* (For data storage):

*Examples:* Mn—magnesium ferrite, Mn—copper ferrite and Li—Ni ferrite

These ferrites have hysteresis loops which are almost rectangular in shape. This property makes them suitable for use in a magnetic memory core in computers.

(iv) *Microwave ferrites:*

*Examples:* Commercial ceramic ferrites (e.g., nickel ferrite, magnesium ferrite) commercial garnets [YI (Al) GYI(Al, Gd)G].

These are used in the frequency range of 1 – 100 GHz. In this frequency range, electromagnetic waves interact with the spin-magnetic moments in the ferrite. The phenomenon is known as *Faraday rotation.* (i.e., rotation of the polarization plane of an electromagnetic wave as it travels through a ferrite in the direction of an applied magnetic field). The chief applications of microwave ferrites are as *isolators* (to reject specific plane polarized waves) *gyrators* (rotation of plane of polarized wave) gyrators (rotation of plane of polarization), *phase shifters* and circulators. These find application in radar communication.

## 13.35 DIRECT EXCHANGE AND SUPEREXCHANGE

The magnetic interaction which arises from the direct Coulomb interaction among electrons from the two ions is known as direct exchange. It often happens that two magnetic ions are separated by a non-magnetic ion (i.e., one with all electronic shells closed) (e.g., in MnO, two Mn atoms are separated by oxygen). It is then possible for the magnetic ions to have magnetic interaction mediated by the electrons in their common non-magnetic neighbour (which is more important than their direct exchange interaction). This type of magnetic interaction is called *superexchange* (Fig. 13.39).

## 13.36 SUPEREXCHANGE IN ANTIFERROMAGNETS

We now consider microscopic origin of antiferromagnetism exhibited by a great majority of ionic crystals (e.g., $MnO_3$ $MnF_2$ and sulfides or chalcogenides). (A few other ionic compounds, e.g., EuO and $CrO_2$ are ferromagnetic).

In an antiferromagnet such as MnO (refer to Fig. 13.34), which has NaCl structure, the spins in a single [111] plane are parallel. While those in adjacent [111] planes are antiparallel.

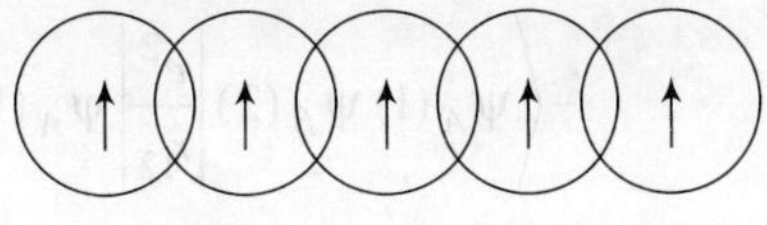

(a)

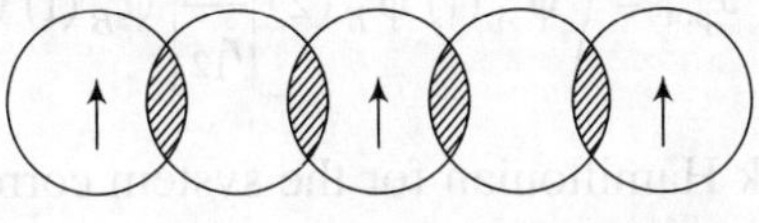

(b)

**Figure 13.39** Schematic representation of: (a) Direct exchange in which the magnetic ions interact because their charge distributions overlap. (b) Superexchange in which magnetic ions with non-overlapping charge distribution interact because both have overlap with the same non-magnetic ion

The antiferromagnetic coupling is between second nearest neighbour cations such as $A \equiv Mn^{2+}$ and $B \equiv Mn^{2+}$, via the intervening (nearest neighbour) anion $C \equiv O^{2-}$. The calculation of the exchange integral in this situation is to be based on bonding and antibonding symmetrised linear combination of atomic orbitals, $\psi_A$ and $\psi_B$. If the configuration were perfectly ionic, each Mn ion would have a single unpaired electron in *d*-state, and the oxygen ion would have in its (outermost occupied) *p*-states, two electrons with antiparallel spins (Fig. 13.40). This may be understood as follows. If atom *A* has + spin, there is room for an electron of – spin to come in from the oxygen *p*-state. If atom *B* has – spin, then it can take in the other electron of the pair from the oxygen. Thus, the interaction between the spins of the *p*-electrons (on the oxygen) is transformed into an interaction between the spins of the *d*-shells in the spins of the *d*-shells in the magnetic ions.

The primary coupling between the *d*-electrons is an indirect one, effected via the *p*-states of the intervening $O^{2-}$. This type of coupling was first demonstrated by P.W. Anderson[5] and is known as superexchange coupling (also known as kinetic exchange). (It is different from direct and antibonding symmetrised linear combination of atomic orbitals (LACOs) $\psi_A$ and $\psi_B$, formed from the four orbitals

$$\psi^A_{d\uparrow},\quad \psi^C_p\downarrow,\quad \psi^C_p\uparrow,\quad \text{and}\quad \psi^B_d\downarrow,$$

by the expressions

$$J_{\text{kin}} = \frac{-2b^2}{V} \qquad \text{...(1)}$$

$$b = \langle \psi_A | H | \psi_B \rangle \qquad \text{...(2)}$$

$$U = \left\langle \psi_A(1)\,\psi_A(2) \left| \frac{e^2}{r_{12}} \right| \psi_A(1)\,\psi_A(2) \right\rangle$$

[5] [*Phys. Rev.* **79**, 350 (1950)].

$$- \left\langle \psi_A(1)\,\psi_B(2) \left| \frac{e^2}{r_{12}} \right| \psi_A(1)\,\psi_B(2) \right\rangle \qquad ...(3)$$

$$J_{\text{pot}} = \left\langle \psi_A(1)\,\psi_B(2) \left| \frac{e^2}{r_{12}} \right| \psi_B(1)\,\psi_A(2) \right\rangle \qquad ...(4)$$

Here $H$ is the Hartree–Fock Hamiltonian for the system corresponding to antiparallel spin arrangement $a$ and $b$ is called the transfer integral. In a periodic structure, generalized atomic orbitals must be used and $b$ becomes

$$b_{ij} = \langle \phi(\vec{r} - \vec{R}_i)/H(R_j)/\phi(\vec{r} - \vec{R}_j) \rangle \qquad ...(5)$$

$\phi(\vec{r} - \vec{R}_i)$ being the localised wave function for an electron on an atom at $\vec{R}_i$. Exchange parameter $J_{\text{kin}}$ become $J_{ij}^{\text{kin}}$.

According to Goodenough and Kanamori[6] interactions between half filled orbitals are antiferromagnetic.

$$J_{ij}^{\text{kin}} = \frac{-2b_{ij}^2}{4S^2U} \qquad ...(6)$$

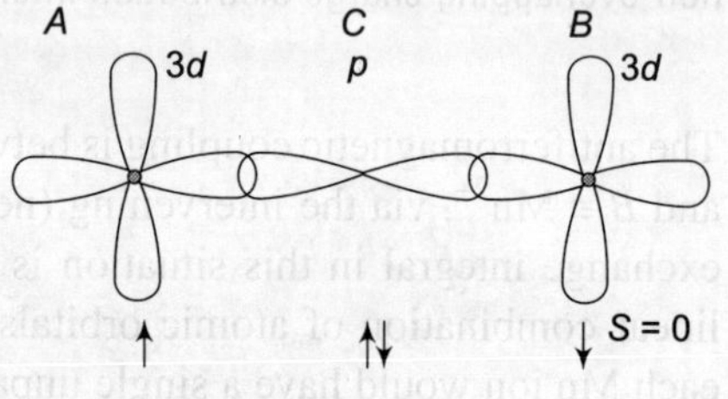


**Figure 13.40** Spin orientations of four electrons in the 3$d$ and $p$-orbitals of Mn-O-Mn configuration (resultant spin S = 0)

The antiferromagnetic ordering temperature ($T_N$) is proportional to the exchange energy parameter

$$kT_N \sim Z\,J_{ij}\,S(S+1) \approx \frac{Zb^2}{V} \qquad ...(7)$$

where $Z$ is the number of nearest neighbours of the magnetic atom.

Further, electrons of up- or down-spin may be transferred from a full orbital to an empty orbital [as in Fig. 13.41(b)], but the interatomic exchange energy $\Delta_{ex}$ given by $\Delta_{ex}/U^2$ favours ferromagnetic coupling. As a result, third-order perturbation theory gives the second rule for superexchange, viz., interactions between full and half-filled orbitals or between half-filled and empty orbitals are ferromagnetic with

$$J_{ij}^{\text{kin}} = \frac{+2b_{ij}^2\,\Delta_{ex}}{4S^2U} \qquad ...(8)$$

The above two rules give antiferromagnetism for MnO, CoO, NiO etc., and ferromagnetism for $CrO_2$. ($Mn^{2+}$ is $3d^5$, $Cr^{2+}$ is $3d^4$).

6 [*Phys. Rev.* **100**, 564 (1955); *J. Phys. Chem. Solids* **10**, 87 (1959)].

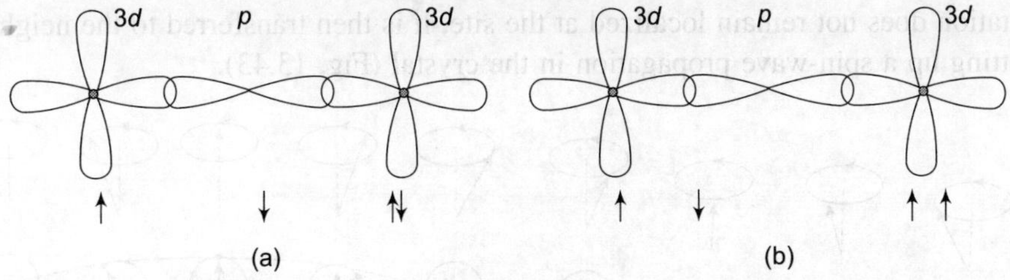


**Figure 13.41** (a) Cation $d$ shell ≥ half filled (antiferromagnetic), (b) Cation $d$ shell < half filled (ferromagnetic)

### 13.36.1 Antiferromagnetism in MnO

Qualitatively, antiferromagnetism in MnO is explained as follows.

In MnO, oxygen ion $O^{2-}$ has the inert gas configuration $1s^2\ 2s^2\ 2p^6$. But the lowest energy state is an admixture of states $O^{2-}$, $O^-$ and the neutral O atoms. That is, an electron on the $O^{2-}$ ion spends a part of the time near the metal ion from which it came originally.

In MnO, imagine one of the electrons of $O^{2-}$ is transferred to $Mn^{2+}$ on the left which has five unpaired spins in the $3d$ subshell. The spin that is transferred to $Mn^{2+}$ (on the left, as in Fig. 13.42) should have orientation antiparallel to the five $3d$ spins. The remaining spin $O^{2-}$, which is pointing down, can now lower its energy through exchange with the spins in the $3d$ subshell on $Mn^{2+}$ on the right. If this exchange has a negative sign, the spin on this $Mn^{2+}$ (on the right) should be pointing up so that spins on $Mn^{2+}$ (on the left) and $Mn^{2+}$ (on the right) get coupled antiferromagnetically, mediated by oxygen atom. This is the basic principle of superexchange mechanism.

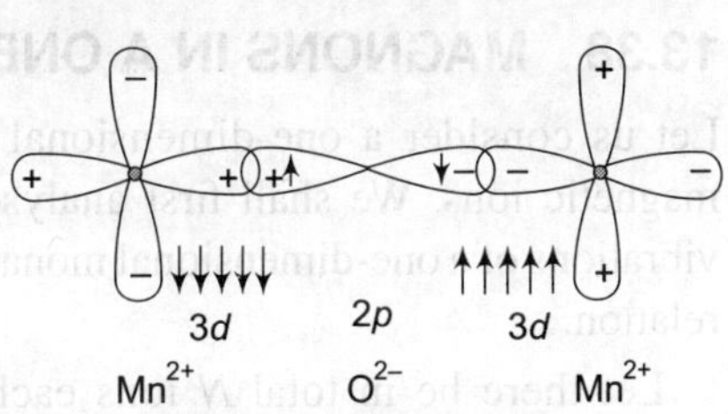


**Figure 13.42** Superexchange in MnO

## 13.37 SPIN WAVES AND MAGNONS

Individual elementary magnetic moments in a magnetic crystal are essentially completely ordered at $T = 0$ K. We know that on increasing the temperature, the magnetization decreases towards zero at critical temperature ($T_c/T_N/T_f$). However, a deviation from the ordered state may well be produced at $T = 0$ K by providing the spin system with an excitation from an external source (presuming $T = 0$ K as the ground state of the system). A state in which one or more spins are reversed relative to their orientation at absolute zero, represents an excited state of the system.

Let us consider the ground state of a ferromagnet at absolute zero (with all spins aligned). It is to be noted here that the mechanism of spin alignment is not magnetic in origin rather it is an attribute of Pauli's exclusion principle with energy given by exchange interaction and energy of a spin depending on the magnetisation in its immediate neighbourhood.

On the Weiss model, the lowest excited state is obtained by reversing a single spin; but since each spin is coupled to its neighbours, this reversal of a single spin is not a normal mode and the motions of all spins are coupled together through exchange interaction. As a result, the excitation does not remain localized at the site, it is then transferred to the neighbouring spins setting up a spin-wave propagation in the crystal (Fig. 13.43).

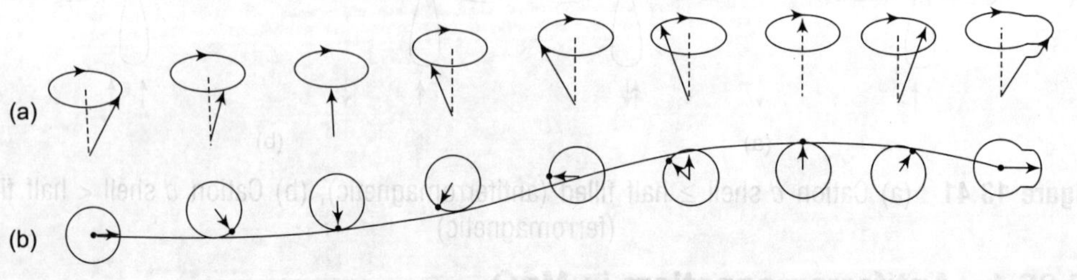


**Figure 13.43** Formation of a spin wave over a line of spins: (a) a perspective view of charging spin orientations, (b) spin vectors as viewed from top

The quantized unit of spin wave energy is called a *magnon*. (The total magnetization decreases by one unit of spin with the creation of one magnon). It is a quantum mechanical concept and magnons are elementary excitations of a spin system.

## 13.38 MAGNONS IN A ONE-DIMENSIONAL FERROMAGNET

Let us consider a one-dimensional ferromagnet consisting of a linear chain of identical magnetic ions. We shall first analyse the problem classically, similar to analysis of lattice vibrations of a one-dimensional monatomic chain. We will finally obtain the magnon dispersion relation.

Let there be in total $N$ ions each of spin $S$ coupled by the exchange interaction with neighbouring spins and exchange energy being given by

$$E_{ex} = -\sum_i \sum_{j \neq i} J_{ij} \vec{S}_i \cdot \vec{S}_j \qquad ...(1)$$

The system is initially in ground state with spins treated as classical vector magnetic moments

$$\vec{\mu}_i = \gamma \hbar \vec{S}_i \qquad ...(2)$$

where $\gamma$ is the gyromagnetic ratio (or magnetomechanical ratio). We can express Eq. (1) as

$$E_{ex} = -\frac{1}{2} \sum_i \vec{\mu}_i \cdot \vec{B}_i \qquad ...(3)$$

where

$$\vec{B}_i = \frac{2J}{\gamma \hbar} \sum_{j \neq i} J_{ij} \vec{S}_j \qquad ...(4)$$

For simplicity, we consider only interactions between nearest neighbours, so that the effective magnetic field $\vec{B}_n$ at the $n$th atom is given by

$$\vec{B}_n = \frac{2J}{\gamma \hbar} (\vec{S}_{n-1} + \vec{S}_{n+1}) \qquad ...(5)$$

[because $E_{ex} = -2J \vec{S}_n (\vec{S}_{n-1} + \vec{S}_{n+1})$ or using Eq. (4)].

We now equate torque to rate of change of (spin) angular momentum to obtain the gyroscope equation of motion of a spin

$$\hbar \frac{d\vec{S}_n}{dt} = \vec{\mu}_n \times \vec{B}_n$$

therefore $$\hbar \frac{d\vec{S}_n}{dt} = 2J\,\vec{S}_n \cdot (\vec{S}_{n-1} + \vec{S}_{n+1})$$

[Using Eqs. (2) and (5)

$$= 2J(\vec{S}_n \times \vec{S}_{n+1} + \vec{S}_n \times \vec{S}_{n-1}) \qquad ...(6)$$

This equation of motion is non-linear. It can be linearised by expressing $\vec{S}_B$ as

$$\vec{S}_n = \vec{S}_z + \vec{\sigma}_n \qquad ...(7)$$

where $\vec{S}_z$ is a constant vector parallel to the mean magnetization and $\vec{\sigma}_n$ is a small spin vector in the $x$–$y$ plane. This is illustrated in Fig. 13.44.

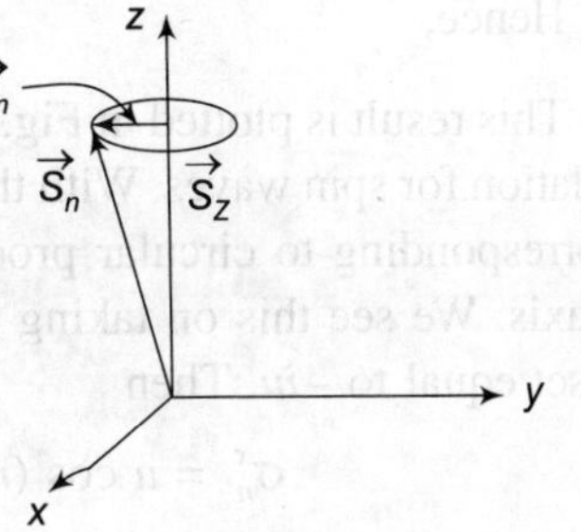


**Figure 13.44** Procession of single spin in a classical spin wave

Note that $\vec{S}_z$ is a constant vector. $d\vec{S}_z/dt = 0$. Substituting Eq. (7) in Eq. (6) and retaining only terms of the first order in $\sigma_n$, we get

$$\hbar \frac{d\vec{\sigma}_n}{dt} = 2J\vec{S}_z \times (\vec{\sigma}_{n+1} + \vec{\sigma}_{n-1}) + 2J\vec{\sigma}_n \times 2\vec{S}_z$$

or $$\hbar \frac{d\vec{\sigma}_n}{dt} = 2J\vec{S}_z \times (\vec{\sigma}_{n-1} - 2\vec{\sigma}_n + \vec{\sigma}_{n+1}) \qquad ...(8)$$

From this equation, the components of spin motion along the $x$- and $y$-direction may be written as

$$\hbar\left(\frac{d\sigma_n}{dt}\right)_y = 2JS_z\,(\sigma_{n-1} - 2\sigma_n + \sigma_{n+1})_x \qquad ...(9a)$$

$$\hbar\left(\frac{d\sigma_n}{dt}\right)_x = -2JS_z\,(\sigma_{n-1} - 2\sigma_n + \sigma_{n+1})_y \qquad ...(9b)$$

By analogy with photon-problem, we look for travelling wave solution of the form

$$\sigma_n^x = u \exp\,[i(nka - \omega t)] \qquad ...(10a)$$

$$\sigma_n^y = v \exp\,[i(nka - \omega t)] \qquad ...(10b)$$

where $u$ and $v$ are constants and $a$ is the lattice constant ($n$ is an integer)

$$-i\omega u = \frac{2JS_z}{\hbar}(2 - e^{-ika} - e^{ika})v$$

$$-i\omega u = \frac{4JS_z}{\hbar}(1 - \cos ka)v \qquad ...(11)$$

and
$$-i\omega v = \frac{4JS_z}{\hbar}(1 - \cos ka)u \quad ...(12)$$

These two equations have a solution for $u$ and $v$ if the determinant of the coefficients vanishes, i.e.,

$$\begin{vmatrix} i\omega & \frac{4JS_z}{\hbar}(1-\cos ka) \\ -\left(\frac{4JS_z}{\hbar}\right)(1-\cos ka) & i\omega \end{vmatrix} = 0 \quad ...(13)$$

Hence,
$$\hbar\omega = 4JS_z(1 - \cos ka) \quad ...(14)$$

This result is plotted in Fig. 13.45 and is the dispersion relation for spin waves. With this solution, we find $v = -iu$ corresponding to circular procession of each spin about $z$-axis. We see this on taking real parts of Eq. (10) with $v$ set equal to $-iu$. Then

$$\sigma_n^x = u\cos(n\,ka - \omega t) \quad ...(15a)$$

$$\sigma_n^y = u\sin(n\,ka - \omega t) \quad ...(15b)$$

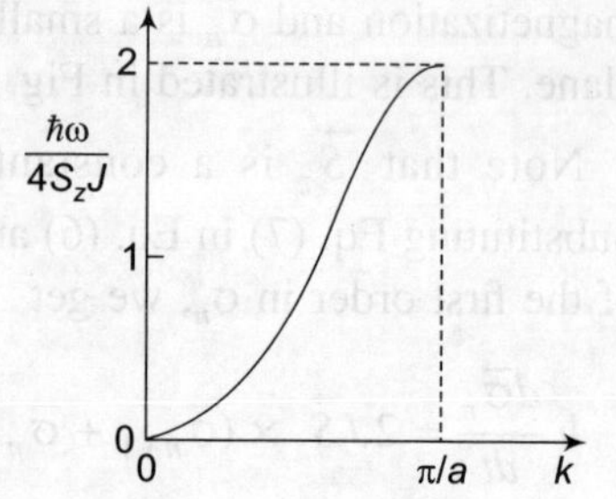


**Figure 13.45** Dispersion relation for magnons in a one-dimensional ferromagnet considering only nearest neighbour interactions

In the long wavelength limit, $ka \ll 1$ so that $(1 - \cos ka) \simeq \frac{1}{2}(ka)^2$ and the frequency

$$\omega = \left(\frac{2JS_z a^2}{\hbar}\right)k^2 \quad ...(16)$$

Near $\vec{k} = 0$, the energy of spin wave can be written as

$$E = \hbar\omega = 2JS_z a^2 k^2 \equiv \frac{\hbar^2 k^2}{2m^*}$$

with
$$m^* = \left(\frac{\hbar^2}{4JS_z a^2}\right)$$

The energy is thus of just of the same form as for electron, but we have no exclusion principle. We can have as many spin wave quanta (magnons) as we like with a given frequency, and like phonons, magnons are bosons.

## 13.39 THE BLOCH $T^{3/2}$ LAW

The concept of spin wave was introduced by Felix Bloch (1930) who correlated the thermal excitation of magnons with the decrease in spontaneous in magnetization from its saturated value at absolute zero. He went on to set up a law that describes the temperature dependence of the decrease in magnetization.

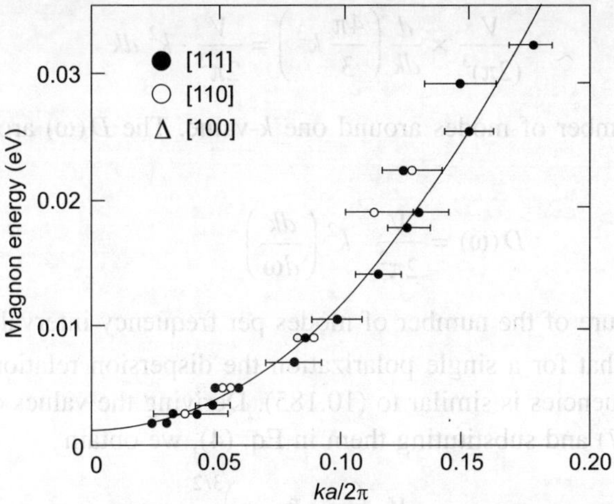


**Figure 13.46** Magnon dispersion relation in a ferromagnetic alloy (92% cobalt and 8% iron) determined from an inelastic neutron scattering experiment at room temperature. The energy is plotted against the dimensionless equivalent of wave vector. [After R.N. Sinclair and B.N. Brockhouse, *Phys. Rev.*, **120**, 1638 (1960).]

To derive the Bloch's law we essentially need to determine the average number of thermally excited magnons at thermal equilibrium. This number gives a measure of the decrease in magnetization, since with the excitation of each magnon there is a decrease in magnetization by one unit of spin moment.

It is evident that the structure of energy levels of magnons is identical to those of a harmonic oscillator. This allows us to treat magnons in a manner similar to that in practice for the radiation oscillators of Planck. Thus, the average number of magnons in the mode **k** at thermal equilibrium is given by the Planck distribution function

$$\langle n_{\mathbf{k}} \rangle = \frac{1}{\exp(\hbar\omega/k_B T) - 1} \qquad ...(1)$$

Then, the total number of thermally excited magnons at temperature $T$ is estimated to be equal to

$$\sum_{\mathbf{k}} n_{\mathbf{k}} = \int D(\omega) \langle n(\omega) \rangle \, d\omega \qquad ...(2)$$

where $D(\omega)$ denotes the density of the magnon modes.

There is a single polarization for magnons for each value of **k** and all useful values of **k** (the unique values) lie in the first Brillouin zone, as is the case with phonons. The integral in Eq. (2) is evaluated by emphasizing the analogy with phonons.

The periodic boundary condition gives the number of modes per unit volume in the $k$-space as $V/(2\pi)^3$, where $V$ is the volume of the crystal. Therefore, the number of modes of the wave vector measuring between $k$ and $(k + dk)$ for a spherical distribution of momentum in the $k$-space is equal to

$$\frac{V}{(2\pi)^3} \times \frac{d}{dk}\left(\frac{4\pi}{3}k^3\right) = \frac{V}{2\pi^2} \cdot k^2\, dk \qquad ...(3)$$

This gives the number of modes around one $k$-value. The $D(\omega)$ around this $k$-value can then be given as

$$D(\omega) = \frac{V}{2\pi^2} \cdot k^2 \left(\frac{dk}{d\omega}\right) \qquad ...(4)$$

Since it is a measure of the number of modes per frequency interval.

It can be shown that for a single polarization the dispersion relation for cubic lattice in the limit of low frequencies is similar to (10.185). Deriving the values of $k^2$ and $dk/d\omega$ from Eq. 16, section 13.37) and substituting them in Eq. (4), we obtain

$$D(\omega) = \frac{V}{4\pi^2}\left(\frac{\hbar}{2J_{ex}\, Sa^2}\right)^{3/2} \omega^{1/2} \qquad ...(5)$$

At very low temperatures the limits of the integration may be placed at 0 and $\infty$ because $\omega \to \infty$, $\langle n(\omega)\rangle \to 0$ exponentially.

We rewrite Eq. (2) to express the total number of magnons as

$$\sum_{\mathbf{k}} n_{\mathbf{k}} = \frac{V}{4\pi^2}\left(\frac{k_B T}{2J_{ex}\, Sa^2}\right)^{3/2} \int_0^{\infty} \frac{x^{1/2}}{\exp(x)-1} \text{ with } x = \frac{\hbar\omega}{k_B T}$$

$$= (0.0587)\,\frac{V}{a^3}\left(\frac{k_B T}{2J_{ex}\, S}\right)^{3/2} \qquad ...(6)$$

For a cube crystal with $N$ magnetic ions per unit cell, the contribution of a single spin to the saturation magnetization $M(0)$ is

$$\frac{M(0)}{(N/a^3)\, S} = \frac{M(0)\, a^3}{NS} \qquad ...(7)$$

Since the total spin is reduced from its saturation value $NS$ by one unit per magnon, the change in magnetization is given by

$$\Delta M = \frac{\Sigma n_k}{V} \times \frac{M(0)a^3}{NS} = \frac{0.0587}{NS}\left(\frac{k_B T}{2J_{ex}\, S}\right)^{3/2} \cdot M(0) \qquad ...(8)$$

Therefore, the fractional change of magnetization is written as

$$\frac{\Delta M}{M(0)} = \frac{0.0587}{NS}\left(\frac{k_B T}{2J_{ex}\, S}\right)^{3/2} \qquad ...(9)$$

The above relation is famous as the $T^{3/2}$ law and has been verified by experiments. The heat capacity contribution by thermal excitation of magnons at low temperatures also varies as $T^{3/2}$.

## 13.40 FERROMAGNETIC ANISOTROPY AND MAGNETOSTRICTION

The $B$ versus $H$ curve of a single crystal of ferromagnetic material depends on the crystal orientation relation to the field. Figure 13.47 shows $B$ versus $H$ curves for single crystal of iron and nickel, with the applied field along <111>, <110>, and <100> directions. In each case, one direction is more easily magnetized than the others. Saturation occurs at a lower field in this direction. For bcc iron, the direction of *easy magnetization* is <100>, and *hard magnetization* is <111>. In fcc nickel, the reverse is true. On this account the different grains in a polycrystalline material approach saturation differently. The grains whose directions of easy magnetization lie near the field direction saturate at low applied fields; those whose directions of hard magnetization lie near the applied field direction rotate their moment into the field direction only at higher fields. The work required to rotate all the domains against the anisotropy is called *magnetocrystalline energy*.

When the magnetic electron spin dipole moments of the atoms in a solid are rotated into alignment, the length of the bonds between the atoms changes. The fields of the dipoles themselves affect the atoms changes. The fields of the dipoles themselves affect the atomic spacing, as they may attract or repel each other. Therefore, the shape and volume of a ferromagnetic solid changes as it is magnetized. The principal change, called the *magnetostriction*, is a reversible strain along the axis of magnetization. Depending upon the material, the solid may expand or contract.

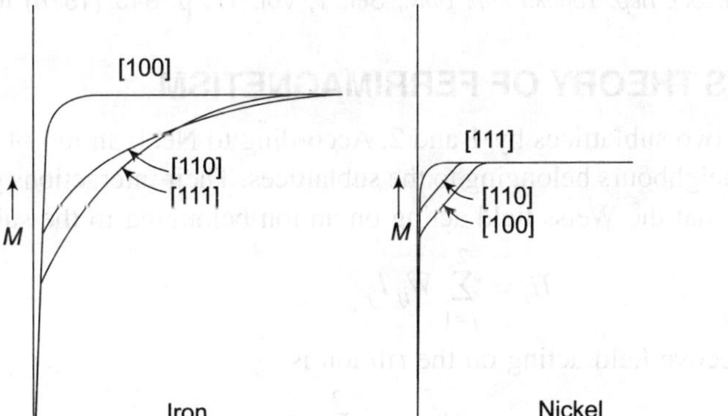


**Figure 13.47** Magnetization versus applied field for field direction along the [100], [110], and [111] directions of iron and nickel single crystals, [After K. Honds and S. Kaya, *Sci, Rep. Tôhoku Imp. Univ.,* Ser. I, Vol. **15**, p. 721 (1926) for the Iron, and S. Kaya, *Sci, Rep. Tôhoku Imp. Univ.,* Ser. I, Vol. 17, pp. 639, 1157 (1928) for the nickel]

The magnetostriction is anisotropic, not only because magnetization curve is anisotropic, but also because the elastic properties of the crystal are anisotropic. For any given crystal direction, the magnetostriction approaches a constant value at high magnetic fields. Magnetostriction and magnetization usually saturate at the same time. Figure 13.48 shows magnetostriction versus applied field curves for nickel and iron single crystals at <100>, <110>, and <111> field orientations. The strain $\delta l/l$ along the field axis is plotted as a function of the field intensity in Fig. 13.48.

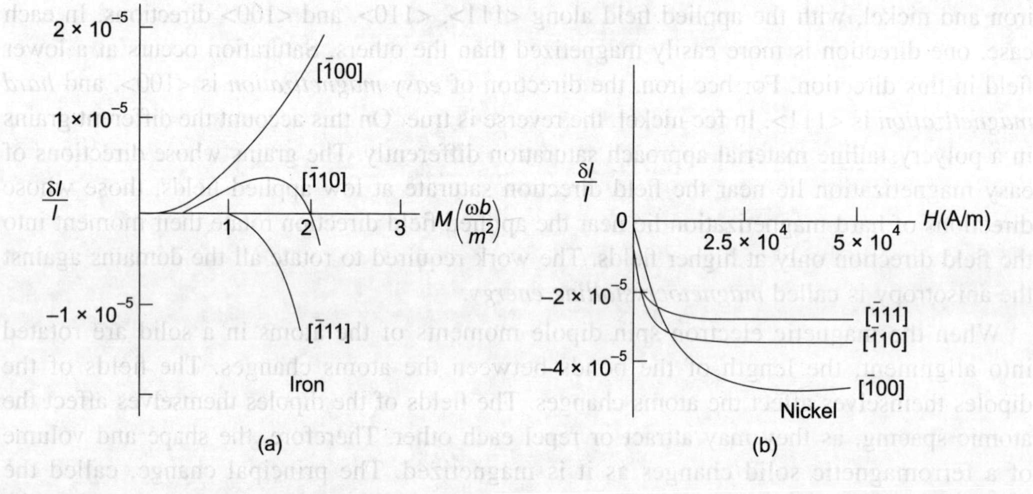


**Figure 13.48** Magnetostriction data for iron and nickel single crystals showing the anisotropy [After W.L. Webster, *Proc. Roy. Soc.* (*London*), Ser. A, Vol. 109, p. 570 (1925) for iron, and Y. Masiyama, *Sci. Rep. Tôhoku Imp. Univ.*, Ser. 1, Vol. 17, p. 945 (1928) for nickel.]

## 13.41 NEEL'S THEORY OF FERRIMAGNETISM

Let us denote the two sublattices by 1 and 2. According to Neel, an ion of a given sublattice interacts with its neighbours belonging to the sublattices. Their interactions give rise to distant internal fields, so that the Weiss field acting on an ion belonging to the sublattice is

$$H_i = \sum_{j=1}^{2} W_{ij} I_j \qquad ...(1)$$

So that the effective field acting on the $i$th ion is

$$H_e = H_i + H = \sum_{j=1}^{2} W_{ij} I_j + H \qquad ...(2)$$

Since action and reaction are equal and opposite, $W_{ij} = -W_{ij}$. For antiferromagnetic coupling between the two sublattices, $W_{12}$ is negative. It is customary to use the following definitions

$$n = -W_{12},\ n\alpha = W_{11},\ n\beta = W_{22}$$

Two different cases may arise:

1. Above the Curie temperature, where the arrangement of spines is disturbed, the intensity of magnetization $I_1$ and $I_2$ of the sublattices 1 and 2 respectively may be assumed to follow the Curie law $I = \left(\frac{C}{T}\right) H$, where $C$ is the Curie constant. Hence,

$$I_iT - C_iH_e = 0$$

$$\Rightarrow \quad I_iT - C_i\left[H + \sum_{j=1}^{2} W_{ij} I_j\right] = 0 \quad ...(3)$$

Therefore,

$$I_1T - C_1\,[H + W_{11}I_1 + W_{12}I_2] = 0$$

$$I_1T - C_1\,[H + n\,\alpha\, I_1 - nI_2] = 0$$

$$\Rightarrow \quad I_1 = \left[1 - \frac{naC_1}{T}\right] + I_2\left[\frac{nC_1}{T}\right] - \frac{C_1H}{T} = 0 \quad ...(4)$$

and

$$I_2T - C_2\,[H + W_{21}I_1 + W_{22}I_2] = 0$$

$$\Rightarrow \quad I_2T = C_2\,[H - nI_1 + n\beta I_2] = 0$$

$$\Rightarrow \quad I_1\left[\frac{nC_2}{T}\right] + I_2\left[1 - \frac{n\beta C_2}{T}\right] - \frac{C_2H}{T} = 0 \quad ...(5)$$

The Curie constant $C_1$ and $C_2$ for the two sublattices depend upon the properties of the ions in the two sublattices.

Solving simultaneously Eqs. (4) and (5) by Cramer's rule, we get

$$I_1 = \frac{\dfrac{H}{T}\left[C_1 - \dfrac{nC_1C_2}{T}(1+\beta)\right]}{1 - \dfrac{n}{T}(\alpha C_1 + \beta C_2) + \dfrac{C_1C_2n^2}{T^2}(\alpha\beta - 1)} \quad ...(6)$$

and

$$I_2 = \frac{\dfrac{H}{T}\left[C_2 - \dfrac{nC_1C_2}{T}(1+\alpha)\right]}{1 - \dfrac{n}{T}(\alpha C_1 + \beta C_2) + \dfrac{C_1C_2n^2}{T^2}(\alpha\beta - 1)} \quad ...(7)$$

The total intensity of magnetization will be

$$I = I_1 + I_2$$

$$\Rightarrow \quad I = \frac{\dfrac{H}{T}\left[C_1 - \dfrac{nC_1C_2}{T}(1+\beta) + C_2 - \dfrac{nC_1C_2}{T}(1+\alpha)\right]}{1 - \dfrac{n}{T}(\alpha C_1 + \beta C_2) + \dfrac{C_1C_2n^2}{T^2}(\alpha\beta - 1)}$$

$$\Rightarrow \quad I = \frac{\dfrac{H}{T}\left[C - \dfrac{nC_1C_2}{T}(2+\alpha+\beta)\right]}{1 - \dfrac{n}{T}(\alpha C_1 + \beta C_2) + \dfrac{C_1C_2n^2}{T^2}(\alpha\beta - 1)}, \text{ where } C_1 + C_2 = C$$

$$\Rightarrow \qquad \frac{H}{I} = \frac{1}{\chi} = \frac{1 - \frac{n}{T}(\alpha C_1 + \beta C_2) + \frac{C_1 C_2 n^2}{T^2}(\alpha\beta - 1)}{\frac{1}{T}\left[C - \frac{nC_1C_2}{T}(2 + \alpha + \beta)\right]}$$

$$\Rightarrow \qquad \frac{1}{\chi} = \frac{T^2 - nT(\alpha C_1 + \beta C_2) + C_1 C_2 n^2 (\alpha\beta - 1)}{C\left[T - \frac{nC_1C_2}{C}(2 + \alpha + \beta)\right]} \qquad ...(8)$$

By actual division, Eq. (8) gives

$$\frac{1}{\chi} = \frac{1}{C}\left[T - \theta_a - \frac{\theta_b^2}{T - \theta}\right] \qquad ...(9)$$

where

$$\theta_a = \frac{n(\alpha C_1^2 + \beta C_2^2 - 2C_1C_2)}{C} \qquad ...(10)$$

$$\theta = \frac{nC_1C_2}{C}(2 + \alpha + \beta) \qquad ...(11)$$

$$\theta_b = \frac{n\sqrt{C_1C_2}}{C}[C_1(1 + \alpha) - C_2(1 + \beta)] \qquad ...(12)$$

Figure 13.49 gives a plot of $\frac{1}{\chi}$ versus $T$. As is evident from Eq. (9), the term involving $\theta_b$ provides the curvature to the curve. From the shape of the experimental curve, one can find the value of $\chi_a \left(= \frac{C}{\theta_a}\right)$, $\theta_b$ and $\theta$. For an antiferromagnetic material composed of two identical sublattices, $C_1 = C_2$ and $\alpha = \beta$ and therefore, from Eq. (12), $\theta_b = \theta$.

Hence, the expression for $\frac{1}{\chi}$, given by Eq. (9), reduces to Curie–Weiss law.

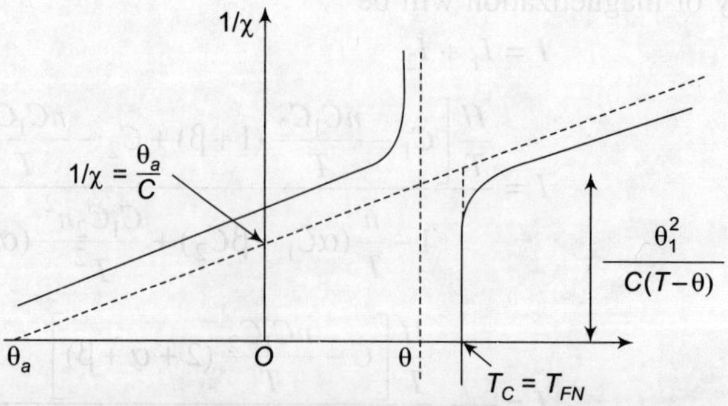


**Figure 13.49** Plot of $1/\chi$ against $T$ for a ferromagnetic material

2. Below the Curie temperature, there is a spontaneous magnetization, just as we have in the case of ferromagnets. The magnetization in this case is equal to the vector sum of the magnetizations of the two anti-parallel sublattices and hence, has the magnitude of

$$I_S = |I_1 - I_2| \quad ...(13)$$

To investigate the temperature dependence of $I_s$, we first consider the temperature dependence of $I_1$ and $I_2$ separately. Figure 13.50 and Fig. 13.51 show respectively the temperature dependence of $I_1$ and $I_2$ when $\beta > \alpha$ and $\alpha > \beta$.

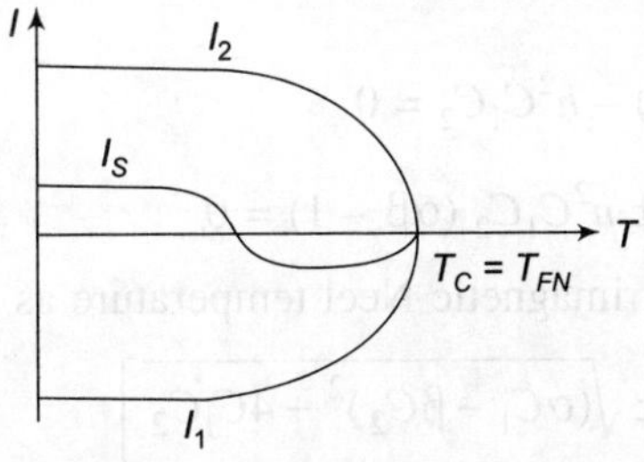


**Figure 13.50** Temperature dependence of $I_1$ and $I_2$ for ferrimagnetic material when $\beta > \alpha$

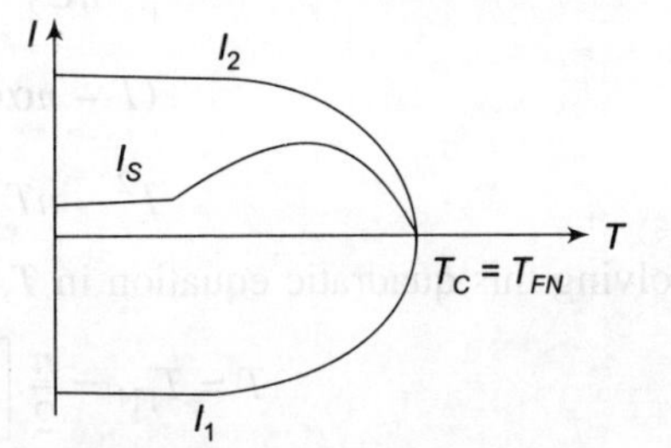


**Figure 13.51** Temperature of $I_1$ and $I_2$ for ferrimagnetic material when $\alpha > \beta$

Because of the saturation effect, we cannot employ Curie–Weiss law. We must, therefore, replace the earlier expressions for $I_1$ and $I_2$, as given by Eqs. (6) and (7) respectively, by

$$I_1 = N\, g\mu_B S \cdot B\left(\frac{g\mu_B S H_1}{k_B T}\right) \quad ...(14)$$

and

$$I_2 = N\, g\mu_B S \cdot B\left(\frac{g\mu_B S H_2}{k_B T}\right) \quad ...(15)$$

where $H_1$ and $H_2$ are the effective fields acting on the sublattices 1 and 2 respectively. Neel's theory is found to describe the observations quite well.

## 13.42 FERRIMAGNETIC NEEL TEMPERATURE

As pointed out earlier, ferrimagnetic Neel temperature is the temperature at which transition from ferrimagnetism to paramagnetism takes place. It can be determined by using Eqs. (4) and (5) with $H = 0$, since it is the temperature at which $I_1$ and $I_2$ have a finite value even in the absence of an externally applied field. Putting $H = 0$ in these equations, we get

$$I_1\left[1 - \frac{n\alpha C_1}{T}\right] + I_2\left[\frac{nC_1}{T}\right] = 0 \quad ...(16)$$

and

$$I_1\left[\frac{nC_2}{T}\right] + I_2\left[1 - \frac{n\beta C_2}{T}\right] = 0 \quad ...(17)$$

The above equations can be solved simultaneously. They will have non-zero values for $I_1$ and $I_2$ only if the determinant of their coefficient vanishes. That is,

$$\begin{vmatrix} \left(1-\dfrac{n\alpha C_1}{T}\right) & \left(\dfrac{nC_1}{T}\right) \\ \left(\dfrac{nC_2}{T}\right) & \left(1\dfrac{n\beta C_2}{T}\right) \end{vmatrix} = 0$$

$$\Rightarrow \qquad \begin{vmatrix} T - n\alpha C_1 & nC_1 \\ nC_2 & T - n\beta C_2 \end{vmatrix} = 0$$

$$\Rightarrow \qquad (T - n\alpha C_1)\,(T - n\beta C_2) - n^2 C_1 C_2 = 0$$

$$\Rightarrow \qquad T^2 - nT\,(\alpha C_1 + \beta C_2) \pm n^2 C_1 C_2\,(\alpha\beta - 1) = 0$$

Solving this quadratic equation in $T$, we get the ferrimagnetic Neel temperature as

$$T = T_{\text{FN}} = \frac{n}{2}\left[(\alpha C_1 + \beta C_2) \pm \sqrt{(\alpha C_1 - \beta C_2)^2 + 4C_1C_2}\right] \qquad \text{...(18)}$$

Equation (18) shows that there are two transition temperatures which is not possible. As both the sublattices are tied together, only one transition temperature can be physically realized and this is the one that corresponds to the largest real root. The other solution corresponds to less stable phase.

Hence, the ferrimagnetic Neel temperature is given by

$$T_{\text{FN}} = \frac{n}{2}\left[(\alpha C_1 + \beta C_2) + \sqrt{(\alpha C_1 - \beta C_2)^2 + 4C_1C_2}\right] \qquad \text{...(19)}$$

## 13.43 ANTIFERROMAGNETISM, FERRIMAGNETISM, AND OTHER TYPES OF MAGNETIC ORDER AND SUSCEPTIBILITY

Antiferromagnetism is similar to ferromagnetism except that the lowest-energy state involves adjacent spins that are antiparallel rather than parallel (but see the end of this section). As we will see, the reason for this is a change in sign (compared to ferromagnetism) for the coupling parameter or *exchange integral.*

Ferrimagnetism is similar to antiferromagnetism except that the paired spins do not cancel and thus the lowest-energy state has a net spin.

Examples of antiferromagnetic substances are FeO and MnO. The temperature at which an antiferromagnetic substance becomes paramagnetic is known as the *Néel* temperature.

Examples of ferrimagnetism are $MnFe_2O_4$ and $NiFe_2O_2$.

We now discuss these in more detail by use of mean field theory.[7] We assume near-neighbour and next-nearest-neighbour coupling as shown schematically in Fig. 13.5. The figure is drawn for an assumed ferrimagnetic order below the transition temperature. *A* and *B*

[7] See also. e.g., Kittel [7.39 p 458ff].

represent two sublattices with spins $S_A$ and $S_B$. The coupling is represented by the exchange integrals $J$ (we assume that $J_{BA} = J_{AB} < 0$ and these $J$ dominate $J_{AA}$ and $J_{BB} > 0$). Thus, we assume the effective field between $A$ and $B$ has a negative sign. For the effective field we write

$$B_A = -\omega\mu_0 M_B + \alpha_A\mu_0 M_A + B, \quad ...(1)$$

$$B_B = -\omega\mu_0 M_A + \beta_B\mu_0 M_B + B, \quad ...(2)$$

where $\omega > 0$ is a constant proportional to $|J_{AB}| = |J_{BA}|$, while $\alpha_A$ and $\beta_B$ are constants proportional to $J_{AA}$ and $J_{BB}$. The $M$ represent *magnetization* and $B$ is the external field (that is the magnetic induction $B = \mu_0 H_{\text{external}}$).

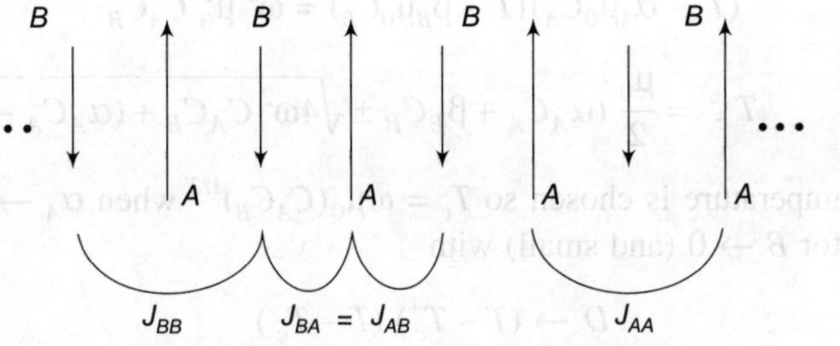


**Figure 13.52** Schematic to represent ferrimagnets

By the mean field approximation with $B_{SA}$ and $B_{SB}$ being the approximate Brillouin functions.[8]

$$M_A = N_A g_A S_A \mu_B B_{S_A} \ (\beta g_A \mu_B S_A B_A) \quad ...(3)$$

$$M_B = N_B g_B S_B \mu_B B_{S_B} \ (\beta g_B \mu_B S_B B_B) \quad ...(4)$$

The $S_A$, $S_B$ are quantum numbers (e.g., 1, 3/2, etc., labelling the spin). In the above, $N_i$ is the number of ions of type $i$ per unit volume, $g_A$ and $g_B$ are the *Lande g-factors* (note we are using $B$ not $\mu_0 H$), $\mu_B$ is the Bohr magneton and $\beta = 1/(k_B T)$.

Defining the Curie constants

$$C_A = \frac{N_A S_A (S_A + 1)\, g_A^2\, \mu_B^2}{3k}, \quad ...(5)$$

$$C_B = \frac{N_B S_B (S_B + 1)\, g_B^2\, \mu_B^2}{3k}, \quad ...(6)$$

we have if $B_A/T$ and $B_B/T$ are small

$$M_A = \frac{C_A B_A}{T}, \quad ...(7)$$

---

8 Defined by $B_J(y) = \frac{2J+1}{2J} \coth\left(\frac{2J+1}{2J} y\right) - \frac{1}{2J} \coth\left(\frac{y}{2J}\right)$.

$$M_B = \frac{C_B B_B}{T} \quad ...(8)$$

This holds above the ordering temperature when $B \to 0$ and even just below the ordering temperature provided $B \to 0$ and $M_A$, $M_B$ are very small. Thus, the equations determining the magnetization become

$$(T - \alpha_A \mu_0 C_A) M_A + \omega \mu_0 C_A M_B = C_A B \quad ...(9)$$

$$\omega \mu_0 C_B M_A + (T - \beta_B \mu_0 C_B) M_B + = C_B B \quad ...(10)$$

If the external field $B \to 0$, we can have nonzero (but very small) solutions for $M_A$, $M_B$ provided

$$(T - \alpha_A \mu_0 C_A)(T - \beta_B \mu_0 C_B) = \omega^2 \mu_0^2 C_A C_B \quad ...(11)$$

So $$T_c^{\pm} = \frac{\mu_0}{2}\left(\alpha_A C_A + \beta_B C_B \pm \sqrt{4\omega^2 C_A C_B + (\alpha_A C_A - \beta_B C_B)^2}\right) \quad ...(12)$$

The critical temperature is chosen so $T_c = \omega \mu_0 (C_A C_B)^{1/2}$ when $\alpha_A \to \beta_B \to 0$, and so $T_c = T_c^+$. Above $T_c$ for $B \to 0$ (and small) with

$$D \to (T - T_c^+)(T - T_c^-)$$

$$M_A = D^{-1}[T - \beta_B \mu_0 C_B) C_A - \omega \mu_0 C_A C_B] B$$

$$M_B = D^{-1}[T - \alpha_A \mu_0 C_A) C_B - \omega \mu_0 C_A C_B] B$$

The reciprocal magnetic susceptibility is then given by

$$\frac{1}{\chi} = \frac{B}{\mu_0 (M_A + M_B)} = \frac{D}{\mu_0 \{T(C_A + C_B) - [(\alpha_A + \beta_B) + 2\omega]\, \mu_0 C_A C_B\}} \quad ...(13)$$

Since $D$ is quadratic in $T$, $1/\chi$ is liner in $T$ only at high temperatures (*ferrimagnetism*). Also note

$$\frac{1}{\chi} = 0 \quad \text{at} \quad T = T_c^+ = T_c.$$

In the special case where two sublattices are identical (and $\omega > 0$), since $C_A = C_B \equiv C_1$ and $\alpha_A = \beta_B \equiv \alpha_1$,

$$T_c^+ = (\alpha_1 + \omega)\, C_1 \mu_0, \quad ...(14)$$

and after cancelling, $$\chi - 1 = \frac{[T - C_1 \mu_0 (\alpha_1 - \omega)]}{2 C_1 \mu_0} \quad ...(15)$$

which is linear in $T$ (*antiferromagnetism*).

This equation is valid for $T > T_c^+ = \mu_0(\alpha_1 + \omega) C_1 \equiv T_N$, the Néel temperature.

Thus, if we define

$$\theta \equiv C_1(\omega - \alpha_1)\mu_0$$

$$\chi_{AF} = \frac{2\mu_0 C_1}{T + \theta} \quad \text{...(16)}$$

Note: $$\frac{\theta}{T_N} = \frac{\omega - \alpha_1}{\omega + \alpha_1}$$

We can also easily derive result for the ferromagnetic case. We choose to drop out one sublattice and in effect double the effect of the other to be consistent with previous work.

$$C_A = C_A^F \equiv 2C_1, \beta_B = 0, C_B = 0,$$

so $$T_c = \mu_0 \alpha_A^F C_A^F = 2C_1\mu_0\alpha_1 \quad (\text{if } \alpha_1 \equiv \alpha_A^F).$$

Then, $$\chi = \frac{\mu_0 M_A}{B} = \frac{\mu_0 T(2C_1)}{T(T - 2C_1\mu_0\alpha_1)} = \frac{2C_1\mu_0}{T - 2C_1\mu_0\alpha_1} \quad \text{...(17)}$$

The paramagnetic case is obtained from neglecting the coupling so

$$\chi = \frac{2C_1\mu_0}{T} \quad \text{...(18)}$$

The reality of antiferromagnetic has been absolutely determined by neutron $d$ if fraction that shows the appearance of magnetic order below the critical temperature.

## 13.44 MAGNETISM SHOWN BY QUASI-FREE ELECTRONS

The quasi-free electron in metals shows both a paramagnetic and diamagnetic effect. Paramagnetic is a fairly weak induced magnetic in the direction of the applied field. Diamagnetism is a very weak induced magnetization opposite the direction of the applied field.

### 13.44.1 Pauli Spin Paramagnetism

The paramagnetism of quasi-free electrons is called Pauli spin paramagnetism. It is a simple application of Fermi–Dirac statistics to electrons.

Electronic paramagnetism (positive contribution to $\chi$) is found in the following:

(i) atoms, molecules and lattice defects possessing an odd number of electron (since the total spin of the system cannot be zero), e.g., sodium atoms, gaseous nitric oxide, $F$ centres in alkali halides;

(ii) free atoms/ions with a partly filled inner shell, e.g., transition electron; and

(iii) metals.

If classical theory is used to calculate paramagnetic susceptibility of the free electrons (assuming that an electron has associated with it and magnetic moment of one Bohr magneton $\mu_B$), a Curie-type result ($M = N\,\mu_B^2\,B/k_BT$) would be obtained. This, however, gives an unsatisfactory account of the paramagnetic non-ferromagnetic metals is independent of temperature.

Pauli showed that the application of the Fermi-Dirac distribution corrects the theory as anticipated according to experimental results.

Pauli, in 1927, noted that the exclusion principle permits two electrons of opposite spin to occupy the same region of $\vec{k}$-space (i.e., to have identical spatial wave functions). So, we can draw the total density of states versus energy as the sum of two components, each governed by half the $g(E)$, where

$$g(E) = \frac{V}{2\pi^2}\left(\frac{2m}{\hbar^2}\right)^{3/2} E^{1/2} \qquad ...(1)$$

In the absence of a magnetic field, states in one-half in Fig. 13.53. (a) are as likely to be occupied as those in the other half (representing an opposite spin direction)

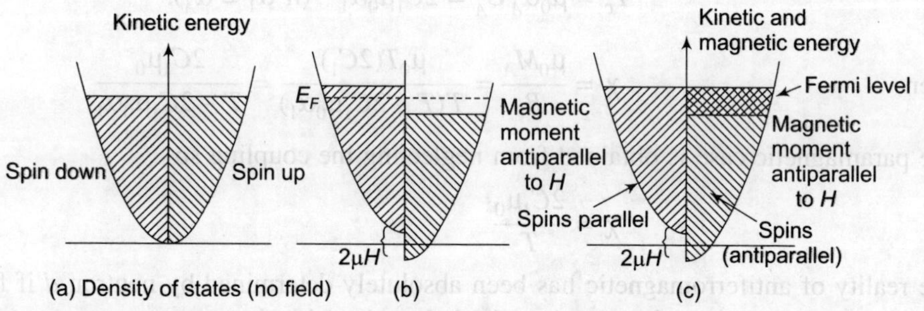


**Figure 13.53** The free electron model to describe paramagnetism in presence of magnetic field $\vec{H}$. (a) Equal occupancy of density of states in absence of magnetic field, (b) with magnetic field applied, and (c) the number of electrons in the "up" and "down" bands adjust to make the energies equal at the Fermi level. The cross-hatched area shows excess of up-spin electrons

Suppose now that Fermi gas of electron is placed in a magnetic field $\vec{H}$. Then, the energy of a particle with spin parallel to $\vec{H}$ is

$$E = \frac{p^2}{2m} - \mu H \qquad ...(2)$$

and with spin antiparallel to $\vec{H}$ is

$$E = \frac{p^2}{2m} + \mu H \qquad ...(3)$$

where $\mu$ is the intrinsic magnetic moment of the particle.

At $T = 0°$K, all levels up to Fermi level will be filled up. Therefore, the kinetic energy of the first group will range from 0 to $(E_F + \mu H)$ while that of the second group from 0 to $(E_F + \mu H)$.

Now, in order to preserve a common electrochemical potential, some of the electron of the lower energy (i.e., spin anitparallel for magnetic moments parallel to $\vec{H}$) flip over and become parallel to the field as shown in Fig. 13.53. This result in a positive magnetic moment or paramagnetism.

We now derive the expression for the paramagnetic susceptibility of this free electron gas.

Let us denote the particles in the two groups by $N^+$ and $N^-$. Then

$$N^{+} = \frac{V}{4\pi^2\hbar^3}(2m)^{3/2} \int_0^{E_F+\mu H} E^{1/2}\, dE \qquad ...(4)$$

$$= \frac{V}{4\pi^2\hbar^3}(2m)^{3/2}\,(E_F+\mu H)^{3/2}$$

$$= \frac{V}{6\pi^2\hbar^3}(2m)^{3/2}\,(E_F+\mu H)^{3/2} \qquad ...(5)$$

Similarly,

$$N^{-} = \frac{V}{6\pi^2\hbar^3}(2m)^{3/2}\,(E_F+\mu H)^{3/2} \qquad ...(6)$$

So, the net magnetic moment

$$M = \mu(N^{+} - N^{-}) = \frac{\mu V(2m)^{3/2}}{6\pi^2\hbar^3}[(E_F+\mu H)^{3/2} - (E_F - \mu H)^{3/2}] \qquad ...(7)$$

If the field $H$ is low,

$$M = \frac{\mu V(2m)^{3/2}}{6\pi^2\hbar^3}\left(E_F^{3/2} + \frac{3}{2}E_F^{1/2}\,\mu H - E_F^{3/2} + \frac{3}{2}E_F^{1/2}\,\mu H\right)$$

$$= \frac{\mu V}{6\pi^2\hbar^3} \times 3\,E_F^{1/2}\mu H \qquad ...(8)$$

Hence, the susceptibility

$$\chi_0 = \frac{M}{VH} = \frac{\mu^2(2m)^{3/2}\sqrt{E_F}}{2\pi^2\hbar^2} \qquad ...(9)$$

Substituting for $(2m)^{3/2}$ from

$$E_F = \frac{\hbar^2}{2m}\left(\frac{3\pi^2 N}{V}\right)^{2/3} \qquad ...(10)$$

we get

$$\chi_0 = \frac{3\mu^2 N}{2\,E_F V} = \frac{3}{2}\frac{\mu^2 N}{k_B T_F V} \qquad ...(11)$$

where $T_F$ is Fermi temperature. Thus, $\chi_0$ is independent of temperature ($T_F$ is a constant for a given metal). This is Pauli paramagnetism. The experimental measurements agree with the above theoretical predictions. In some cases, deviations between theory and experiment are due to the contribution from the diamagnetic susceptibility of the ion cores in the metal.

## 13.45 LANDAU DIAMAGNETISM

It has been mentioned above that quasi-free electrons show a diamagnetic effect. Also, in 1930, Landau showed theoretically that conduction electrons in a metal should have a diamagnetic susceptibility, due to the orbital motion of the electrons, although classically, it is predicted that a gas of free electrons should have zero susceptibility. This non-zero diamagnetism appears because the energy levels of the electrons (which have a quasi-continuous character in the absence of a magnetic field) become grouped into highly degenerate discrete levels in the presence of a field. There are known as *Landau levels*.

The main part of this discussion will be devoted to solving exactly the quantum mechanical problem of a free electron moving in a region in which there is a constant magnetic field. We will find that we obtain a particularly simple set of energy levels. Then standard statistical-mechanical calculations can be made, from which we could make a prediction of the magnetic susceptibility of the electron gas. Since the statistical mechanical analysis is rather complicated, we will give only outline.

In SI units, neglecting spin effects, the Hamiltonian of an electron in a constant magnetic field described by a vector potential $A$ is (here $e > 0$)

$$\mathcal{H} = \frac{1}{2m}(\boldsymbol{p} + e\boldsymbol{A})^2 = \frac{\hbar^2}{2m}\nabla^2 + \frac{e\hbar}{2mi}\nabla\cdot\boldsymbol{A} + \frac{e\hbar}{2mi}\boldsymbol{A}\cdot\nabla + \frac{e^2}{2m}\boldsymbol{A}^2. \qquad ...(1)$$

Using $\nabla\cdot(\boldsymbol{A}\psi) = \boldsymbol{A}\cdot\nabla\psi + \psi\nabla\cdot\boldsymbol{A}$, we can formally write the Hamiltonian as

$$\mathcal{H} = -\frac{\hbar^2}{2m}\nabla^2 + \frac{e\hbar}{2mi}\nabla\cdot\boldsymbol{A} + \frac{e\hbar}{mi}\boldsymbol{A}\cdot\nabla + \frac{e^2}{2m}\boldsymbol{A}^2. \qquad ...(2)$$

A constant magnetic field in the $z$ direction is described by the non-unique vector potential

$$\boldsymbol{A} = -\frac{\mu_0 Hy}{2}\hat{\boldsymbol{i}} + \frac{\mu_0 Hx}{2}\hat{\boldsymbol{j}}. \qquad ...(3)$$

To cheek this result we use the defining relation

$$\mu_0\boldsymbol{H} = \nabla\times\boldsymbol{A}. \qquad ...(4)$$

and after a little manipulation it is clear that Eq. (3) and Eq. (4) imply $H = H\hat{\boldsymbol{k}}$. It is also easy to see that $A$ defined by Eq. (3) implies

$$\nabla\cdot\boldsymbol{A} = 0, \qquad ...(5)$$

Combining Eqs. (2), (3) and (5), we find that the Hamiltonian for an electron in a constant magnetic field is given by

$$\mathcal{H} = -\frac{\hbar^2}{2m}\nabla^2 + \frac{e\hbar\mu_0 H}{2mi}\left(x\frac{\partial}{\partial y} - y\frac{\partial}{\partial x}\right) + \frac{e^2\mu_0^2 H^2}{8m}(x^2 + y^2). \qquad ...(6)$$

It is perhaps worth pointing out that Eq. (6) plus a central potential is a Hamiltonian often used for atoms. In the atomic case, the term $(x\,\partial/\partial y - y\,\partial/\partial x)$ gives rise to paramagnetism (orbital), while the term $(x^2 + y^2)$ gives rise to diamagnetism. For free electron, however, we will retain both terms as it is possible to obtain an exact energy eigenvalue spectrum of Eq. (6).

The exact energy eigenvalue spectrum of Eq. (6) can readily be found by making three transformations. The first transformation that it is convenient to make is

$$\psi(x, y, z) = \phi(x, y, z)\exp\left(\frac{ie\mu_0 H}{2}\frac{xy}{\hbar}\right) \qquad ...(7)$$

Substituting Eq. (7) into $\mathcal{H}\psi = E\psi$ with $\mathcal{H}$ given by Eq. (6), we see that $\phi$ satisfies the differential equation

$$-\frac{\hbar^2}{2m}\nabla^2\phi - \frac{e\hbar\mu_0 H}{im}x\frac{\partial\phi}{\partial y} + \frac{H^2\mu_0^2 e^2}{2m}x^2\phi = E\phi. \quad ...(8)$$

A further transformation is suggested by the fact that the effective Hamiltonian of Eq. (8) does not involve $y$ or $z$ so $p_y$ and $p_z$ are conserved

$$\phi(x, y, z) = F(x)\exp[-i(k_y y + k_z z)] \quad ...(9)$$

This transformation reduces the differential equation to

$$\frac{d^2F}{dx^2} + (A + Bx)^2 F = CF \quad ...(10)$$

or more explicitly

$$-\frac{\hbar^2}{2m}\frac{d^2F}{dx^2} + \frac{1}{2m}[\hbar k_y - (H\mu_0)(ex)]^2 F = \left(E - \frac{\hbar^2 k_z^2}{2m}\right)F \quad ...(11)$$

Finally, if we make a transformation of the dependent variable $x$,

$$x^1 = x - \frac{\hbar k_y}{eH\mu_0} \quad ...(12)$$

then we find
$$-\frac{\hbar^2}{2m}\frac{d^2F}{d(x^1)^2} + \frac{e^2H^2\mu_0^2}{2m}(x^1)^2 F = \left(E - \frac{\hbar^2 k_z^2}{2m}\right)F \quad ...(13)$$

Equation (13) is the equation of a harmonic oscillator. Thus, the allowed energy eigenvalues are

$$E_{n,k_z} = \frac{\hbar^2 k_z^2}{2m} + \hbar\omega_c\left(n + \frac{1}{2}\right) \quad ...(14)$$

where $n$ is an integer and

$$\omega_c \equiv \left|\frac{eH\mu_0}{m}\right| \quad ...(15)$$

is just the *cyclotron* frequency.

This quantum mechanical result can be given a simple classical meaning. We think of the electron as describing a helix about the magnetic field. The helical motion comes from the fact that, in general, the electron may have a velocity parallel to the magnetic field (which velocity is unaffected by the magnetic field) in addition to the component of velocity that is perpendicular to the magnetic field. The linear motion has the kinetic energy $p^2/2m = h^2k_z^2/2m$, while the circular motion is quantized and is mathematically described by harmonic oscillator wave function.

It is at this stage that the rather complex statistical-mechanical analysis must be made. Landau diamagnetism for electron in a periodic lattice requires a still more complicated analysis. The general method is to compute the free energy and concentrate on the terms that are monotonic in $H$. Then thermodynamics tells us how to relate the free energy to the magnetic susceptibility.

A beginning is made by calculating the partition function for a canonical ensemble,

$$Z = \Sigma_i \exp(-E_i/kT), \qquad ...(16)$$

where $E_i$ is the energy of the whole system in state $i$, and $i$ may represent several quantum numbers. (Proper account of the Pauli principal must be taken in calculating $E_i$ from Eq. (14).) The Helmholtz free energy $F$ is then obtained from

$$F = -kT \ln Z \qquad ...(17)$$

and from this the magnetization is determined

$$M = \frac{\partial F}{\mu_0 \partial H}. \qquad ...(18)$$

Finally the magnetic susceptibility is determined from

$$\chi = \left(\frac{\partial M}{\partial H}\right)_{H=0} \qquad ...(19)$$

The approximate result obtained for free electron is

$$\chi_{\text{Landau}} = -\frac{1}{3}\chi_{\text{Pauli}} = -N\mu_0\mu^2/2\,kT_F. \qquad ...(20)$$

Physically, Landau diamagnetism (negative $\chi$) arises because the coalescing of energy levels (described by Eq. (14)) increases the total energy of the system. Fermi–Dirac statistics plays an essential role in measuring the average energy increase.

## 13.46 SPIN GLASSES

Magnetic ions in dilute magnetic alloys such as Mn in Cu, Au and Pt have their spin magnetic moments appearing as 'frozen' in random directions due to some structural disorder. This means that the interaction between two spins alternates between ferromagnetic and antiferromagnetic with distance, and a given spin will receive contradictory signals for the direction it should align from different spins in the sample. The first characteristic of a spin glass is the appearance of a sharp 'cusp' in the a.c. susceptibility *vs* temperature curve, as shown in Fig. 13.54. The second characteristic of the spin glass is the non-existence of a long range magnetic order. Considering the alternating sign of the exchange as the essential characteristic of a spin glass, a model was proposed with short range exchange ($J^{ex}$) with probability distribution $P(J^{ex})$

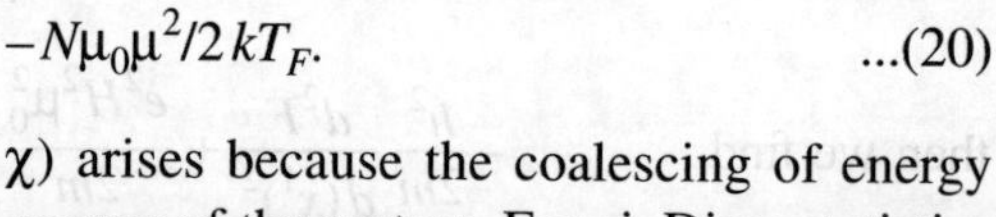


**Figure 13.54** Variation of a.c. susceptibility with temperature in Cu-Mn (Cu 0.06 atomic %, Mn 0.94 atomic %) cusp is observed at $T_f$ = 10 K

giving non-zero weight to both positive and negative $J^{ex}$. In Fig. 13.55, a phase diagram of spin glass is presented in which bonds with strength $+J^{ex}$ and $-J^{ex}$ are present with probability $P$ and $(1 - P)$ respectively. For $P = \frac{1}{2}$ there is a paramagnetic (PM) phase at high temperature and a spin glass (SG) phase at low temperature. As $P$ is increased at $T = 0$, there is a transition from the spin glase phase to a ferromagnetic (FM) phase. The spin glass phenomenon has also been observed in non-metallic materials like $Eu_xSr_{1-x}S$, amorphous materials like $Al_{0.63}$ $Gd_{0.37}$, insulators like $MnO \cdot Al_2O_3\text{-}SiO_2$, metallic glasses like $(Fe_xNi_{1-x})_{75}P_{16}B_5Al_3$.

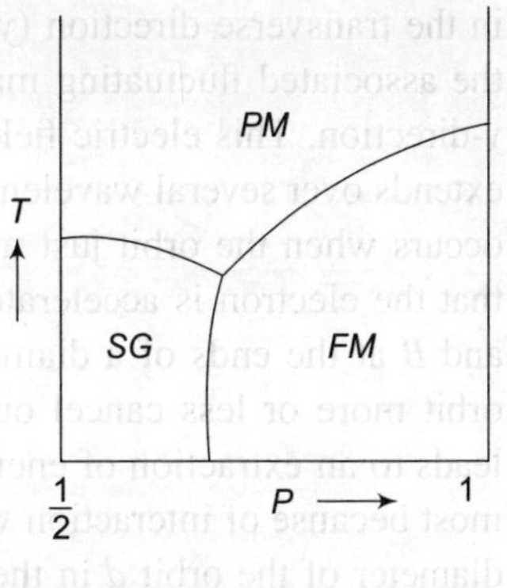


**Figure 13.55** Phase diagram for a spin glass

## 13.47 THE MAGNETO-ACOUSTIC EFFECT

When ultrasonic waves of frequencies less than 200 MHz are sent through a single crystal of very pure metal kept at liquid helium temperatures under a transverse magnetic field $H$, it is found that the measured attenuation of the waves propagating along a known crystal direction shows an oscillatory behaviour with period $1/H$ (Fig. 13.56). This phenomenon is known as the magneto-acoustic effect. It is necessary that the mean free path of the electron be large compared to the wavelength of the ultrasonic waves in the metal.

This condition is expressed by stating that if $q$ is the wave vector of the sound wave and 1 is the mean free path of the electron, than: $l \gg \frac{2\pi}{q}$; this means effectively that: $ql \gg 1$.

Pippard has given a simple explanation for the origin of these magneto-acoustic oscillations under this condition. The velocity of the conduction electrons in the metal, the Fermi velocity, is of the order of $10^8$ cm/sec, and is very large compared to the velocity of sound in the metal ($10^5$ cm/scc). One can then consider that the electron in its orbit sees a stationary wave pattern in the metal. The effect of the ultrasonic wave is to alter the lattice ion positions locally. If the wave propagating along the $x$-direction is transverse, a movement of the ions

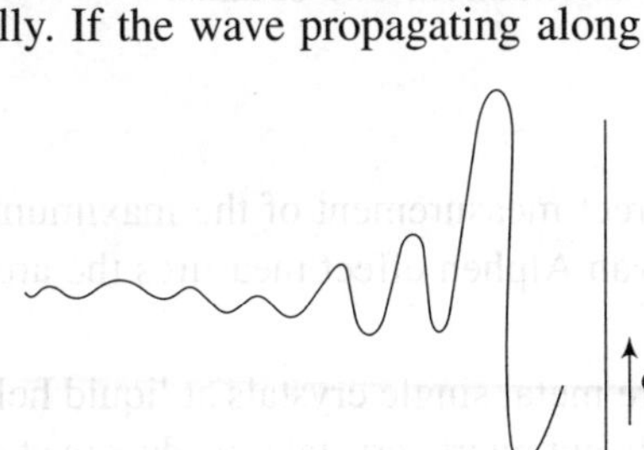


**Figure 13.56** Ultrasonic attenuation (∝) as a function wavelength and $H\lambda$, where $\lambda$ is the applied a magnetic field

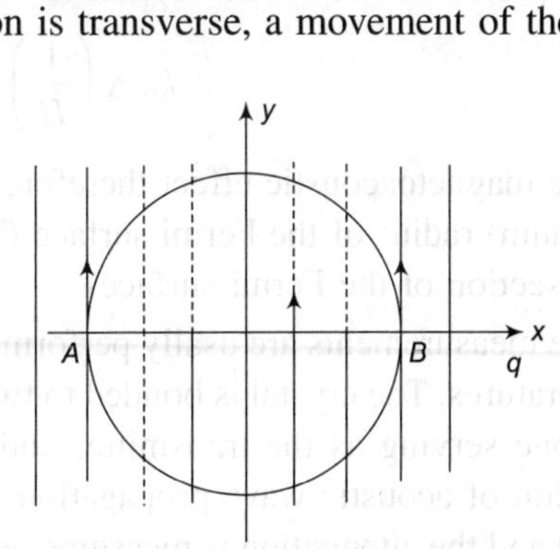


**Figure 13.57** Matching of electron orbit with an integral number of half wavelengths of the incident sound wave in the solid

in the transverse direction ($y$-direction) constitutes an alternating current which by virtue of the associated fluctuating magnetic field produces an induced tranverse electric field in the $y$-direction. This electric field acts on the electron. If the helical electron orbit in real space extends over several wavelengths of the sound wave, then the maximum effect on the electron occurs when the orbit just matches an integral number of half wavelengths (Fig. 13.57), so that the electron is accelerated in the same direction as its rotation at the extremal points $A$ and $B$ at the ends of a diameter of the orbit. (The effects on other on other portions of the orbit more or less cancel out because the field acting on the electron is alternating). This leads to an extraction of energy from the ultrasonic wave and thereby the wave is attenuated most because of interaction with the electron. It follows that a resonance will occur when the diameter of the orbit $d$ in the $x$-direction is equal to

$$\left(n+\frac{1}{2}\right)\lambda,$$

where $n$ is an integer. Since $d = \dfrac{\hbar c}{(-e)H_z}\, 2\, k_0$,

when $k_0$ is the extremal radius of the Fermi surface in the $y$-direction the magnetoacoustic resonance condition is obtained as

$$\frac{k_0}{H_z} = -\frac{e}{2\,hc}\left(n+\frac{1}{2}\right)\lambda \qquad \text{...(1)}$$

A series of maxima are obtained in the attenuation coefficient as the magnetic field is varied for a given $q$. If $H_1$ and $H_2$ are the two values of the field at which consecutive resonances are obtained which corresponds to a change in the orbit diameter by $\lambda$, then writing

$$\Delta\left(\frac{1}{H}\right) = \frac{1}{H_2} - \frac{1}{H_1}$$

as the period in $1/H$ of the oscillations in the acoustic attenuation, one obtains:

$$k_0\,\Delta\left(\frac{1}{H}\right) = -\frac{e}{2\hbar c}\lambda \qquad \text{...(2)}$$

The magnetoacoustic effect therefore provides a direct measurement of the maximum (or minimum) radius of the Fermi surface (The de Haas-Van Alphen effect measures the area of cross section of the Fermi surface).

The measurements are usally performed on very pure metal single crystals at liquid helium temperatures. The crystal is bonded to two appropriately cut quartz crystal transducers at each end, one serving as the transmitter and the other as the receiver of ultrasonic waves. The direction of acoustic wave propagation is aligned parallel to an axis of the crystal (e.g., $q \parallel [100]$) and the attenuation is measured as a function of $1/H_z$, where $H_z$ is the applied magnetic field. The magneto-acoustic effect shows up as a set of oscillations of decreasing amplitudes and the period in $\frac{1}{H}$, $\Delta\left(\frac{1}{H}\right)$, gives the Fermi wave vector in the direction of propagation of the acoustic wave. It has been said that the magneto-acoustic effect acts as a caliper for the Fermi surface.

## 13.48 GIANT QUANTUM OSCILLATIONS

In 1961 Gurevich, Skobov and Firsov showed that it is possible to have sharp spike-like oscillations in ultrasonic attenuation in metals placed in a high magnetic field. The period of oscillations is very nearly equal to the de Hass-Van Alphen oscillations. They have been called *giant quantum oscillations* and the study of these oscillations can yield the area of cross section of the Fermi surface normal to the applied magnetic field, the cyclotron effective mass and the $g$-factor of the electron. The oscillations are called 'giant' because they are comprised of sharp maxima separated by wide minima.

The origin of the giant quantum oscillations can be understood in the following way: We recall that in a free electron gas placed in a steady magnetic field $H$, the energy of an electron (of effective mass $m^*$) is given by Landau's formula

$$E(n, k_z) = \left(n + \frac{1}{2}\right)\hbar\omega_c + \frac{\hbar^2 k_z^2}{2m^*} \qquad ...(3)$$

where $n$ is the Landau level quantum number, $\omega_c$ is the cyclotron frequency and $k_z$ is the component of the wave vector of the electron in the field direction.

Suppose now we send an acoustic wave of wave vector $q$ through a metal containing quasi-free electron of mass ($m^*$) in the presence of the field. We can picture the propagation of ultrasonic waves in the metal as the propagation of a stream of phonons of energy ($\hbar\omega_q$) and momenta, $\hbar q$. A phonon is absorbed when it collides with an electron and the electron makes a transition from the initial state ($n$, $k_z$) to a final state ($n'$, $k_z$), energy and the momentum component parallel to the field being conserved in the process. Then the following conservation equation must be satisfied

$$\hbar k'_z = \hbar k_z + \hbar q_z \qquad ...(4)$$

(momentum conservation)

$$\left(n' + \frac{1}{2}\right)\hbar\omega_c + \frac{\hbar^2 k_z'^2}{2m^*} = \left(n + \frac{1}{2}\right)\hbar\omega_c + \frac{\hbar^2 k_z{}^2}{2m^*} + \hbar\omega_q \qquad ...(5)$$

(energy conservation from Eq. (3))

Let $v_s$ be the sound velocity so that: $\omega_q = v_s q$. If we combine Eqs. (4) and (5) and consider $q_z$ to be small compared to $k_z$, $k_z'$, then

$$(k_z'^2 - k_z^2) = (k_z' - k_z)\,(k_2' - k_z) \cong 2k_z q_z$$

and the equation to be satisfied for phonon absorption is

$$\omega_c(n' - n) + \frac{\hbar k_z q_z}{m^*} = q v_s \qquad ...(6)$$

In high magnetic fields, where $\omega_c$ is large, the Landau level separation ($\hbar\omega_c$) is large. The phonon energy ($\hbar\omega_q$) will then not be sufficient to excite the electron from a Landau level $n$

to higher levels $n'$. Absorption of sound can then only take place if there is no change in the Landau level quantum number $n$. Hence, from Eq. (6) the condition for ultrasonic absorption is:

$$\frac{\hbar k_z q_z}{m^*} = q\,v_s \qquad ...(7)$$

or

$$k_z = \frac{m^*\,v_s}{h}\frac{q}{q_z} = \frac{m^* v_s}{\hbar\cos\theta} \qquad ...(8)$$

where $\cos\theta = q_z/q$ which means that $\theta$ is the angle between the magnetic field direction (z-direction) and the propagation direction of the sound wave. If $q$ is parallel to **H**, $\theta = 0°$, and Eq. (8) reads

$$\frac{\hbar k_z}{m^*} = v_s \qquad ...(9)$$

Hence, ultrasonic absorption can take place only for electrons whose velocity component in the field direction, $v_z$, is equal to the sound velocity. This absorption condition $v_z = v_s$, which we write as

$$k_z = k_0 = m^*\,v_s/\hbar,$$

is called a selection rule. Note that since the sound velocity ($v_s$) is very small in comparison to the Fermi velocity

$$\left(\frac{v_s}{v_F} \sim 10^{-3}\right),$$

$k_0$ is very small compared to the Fermi wave vector $\mathbf{k}_F$ unless $q$ is nearly perpendicular to **H**.

The transition of an electron in a Landau on the absorption of a phonon must take place from an occupied state to an unoccupied states because of the Pauli principle. Unoccupied states are available only near the Fermi energy and hence it follows that ultrasonic absorption can only occur if the electron energy $kT$ of the Fermi level, $E_F$ (Fig. 13.58). (It is necessary, of course, that thermal broadening of levels be small, i.e., the condition: $\hbar\omega_c \gg kT$ must be fulfilled).

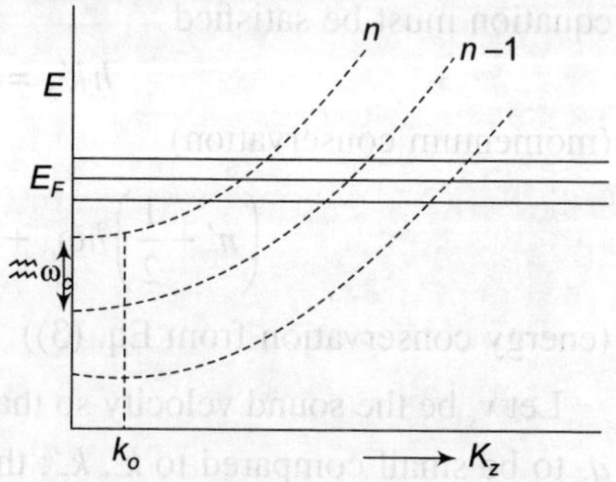


**Figure 13.58** Transition of an electron between Landau levels

This means that the ultrasonic frequency be such that

$$\hbar\omega_q \ll kT, \text{ where } T \sim 1\text{ K}.$$

In order to see how the giant quantum arise, let us plot the Landau level energies as a function of $k_z$. For a given quantum number $n$, each level is a parabola, as follows from

Eq. (3). Thus, we get series of parabolae with vertices separated by $\hbar\omega_c$, as indicated in Fig. 13.58. We consider the Landau levels near the Fermi energy, $E_F$.

Let $k_0$ be the wave vector $\frac{m^* v_s}{\hbar \cos \theta}$. Now, as the applied magnetic field is varied gradually, the parabolas move bodily upwards and it may happen that the point representing $k_0$ comes within the shaded area of width $kT$. Then sound absorption will take place sharply. If $kT$ is small compared to $\hbar\omega_c$, an absorption peak of very small width (a spike) will be produced. A procession of points cross the Fermi level one after another and we get a series of sharp absorption spikes separated by wide minima. These are the giant quantum oscillations (Fig. 13.59).

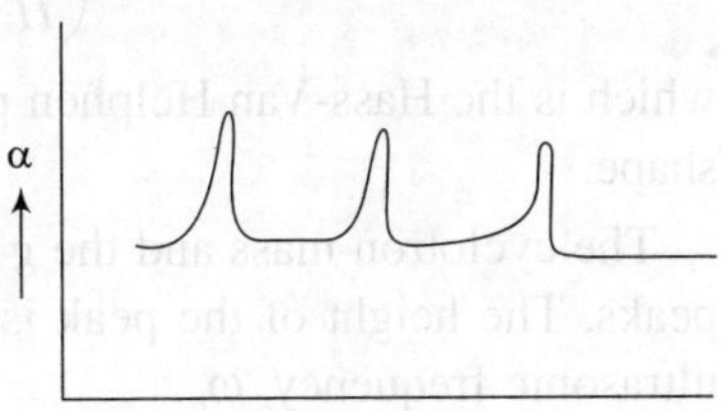


**Figure 13.59** Giant quantum oscillation spikes

Suppose the absorption peak due to the $n$th Landau level occurs at a field $H_1$. The next peak occurs when the $(n - 1)$th Landau levels cross the Fermi level. We then have

$$\frac{1}{H_1} = \frac{\left(n + \frac{1}{2}\right)}{E_F - \frac{\hbar^2 k_0^2}{2m^*}} \left(\frac{-e\hbar}{m^* c}\right)$$

$$\approx \frac{\left(n + \frac{1}{2}\right)}{E_F} \cdot \left(\frac{-e\hbar}{m^* c}\right)$$

since $k_0 \lll k_F$ if 0 is not close to 90°.

Similarly,
$$\frac{1}{H_2} = \frac{\left(n - \frac{1}{2}\right)}{E_F - \frac{\hbar^2 k_0^2}{2m^*}} \left(\frac{-e\hbar}{m^* c}\right)$$

$$\approx \frac{\left(n - \frac{1}{2}\right)}{E_F} \left(\frac{-e\hbar}{m^* c}\right)$$

so that the period in $1/H$, $\Delta(1/H)$, of the oscillations is given by

$$\Delta\left(\frac{1}{H}\right) = \frac{1}{H_1} - \frac{1}{H_2} \approx \frac{-e\hbar}{m^* c\, E_F}$$

$$= -\frac{2e}{\hbar c\, kF^2} \quad ...(10)$$

For quasi-free electrons, the area of cross section of the Fermi surface at the plane $k_z = k_0$, normal to field direction, is given by.

$$A(E_F) = \pi k_F^2$$

Then, $$\Delta\left(\frac{1}{H}\right) = \frac{2\pi(-e)\hbar}{c.A(E_F)} \quad ...(11)$$

which is the Hass-Van Helphen period. Relation (11) is valid also for a Fermi surface of any shape.

The cyclotron mass and the $g$-factor of the electron can be obtained from the width of the peaks. The height of the peak is found to be proportional to the ratio $H_z/T$, and also to the ultrasonic frequency, $\omega_q$.

It should be pointed out that the effect of electron spin on the Landau levels is to split each Landau level into a doublet, obtained by adding to the expression (3).

$$\pm\frac{1}{2}g\mu_B H$$

The effect of this splitting would be to give rise to two sets of spikes in the giant quantum oscillation spectrum, which can be revealed under stringent experimental conditions.

The giant quantum oscillations have been observed in semi-metals like bismuth, gallium, zinc and other materials (for a review of the work see the article by Y. Shapira in *Physical Acoustics*, Vol. 5 (Ed) Mason W.P. (1968) Academic Press). Ultrasonic frequencies of the order of 300 MHz, magnetic field strengths of the order of one hundred kilogauss and temperatures of the order of 1K have been used.

The condition: $$\hbar\omega_c > kT$$

is rather strict in this case. The purity of the samples affects the peak widths and need to be high.

The giant quantum oscillations promise to become a very accurate method of determining the croos-sectional areas of the Fermi surface like the de Hass-Van Alphen effect. They have also been called Landau level oscillations.

## EXERCISE

13.1 How do permanent magnetic moments originate in atoms and ions? Explain their various types of arrangements in solids and mention the magnetism associated with each.

13.2 Give quantum theory of paramagnetism and explain how it removes the shortcomings of the Langevin's theory.

13.3 What are the characteristics of ferromagnets? Discuss the nature of their ferromagnetic carriers. Derive the Curie–Weiss law for their susceptibility.

13.4 What is a Bloch wall? Calculate its thickness and energy per unit area. What are units of the thickness of this wall?

13.5 What is hysteresis? Give its domain theory. Discuss the origin of domains.

13.6 Distinguish among diamagnetism, paramagnetism and ferromagnetic susceptibility on the basis of classical theory.

13.7 Derive an expression for paramagnetic susceptibility of free electron on the basis of classical laws. Discuss its theory of free electron paramagnetism.

13.8 Discuss the Pauli's theory of free electron paramagnetism.

13.9 Draw temperature dependence of the susceptibility of all types of magnetic materials. Explain Heisenberg's exchange interaction in ferromagnetism.

13.10 Explain the origin of diamagnetic susceptibility. What is origin is van Vleck paramagnetism.

13.11 Give quantum theory of diamagnetism.

13.12 Define magnetic susceptibility and derive Curie's law. What is Langevin function.

13.13 Derive the classical Langevin formula for orbital diamagnetism.

13.14 Discuses Weiss classical theory of ferromagnetic.

13.15 Explain origin of domains in ferromagnets.

13.16 Give a qualitative account of the formation of domains.

13.17 What do you understand by Neel wall and Bloch wall?

13.18 Explain the concepts of hysteresis, remanance and coercive force in case of ferromagnets.

13.19 Derive an expression for Heisenberg Hamiltonian using the Heitler London method.

13.20 Explain the concepts of Heisenberg spin Hamiltonian.

13.21 Explain the origin of ferromagnetic states in Fe, Co and Ni.

13.22 Discuss the two-sublattice model for antiferromagnetism.

13.23 What is ferrimagnetism? Explain this for ferrite [FeO, $Fe_2O_3$].

13.24 Explain exchange interaction in ferromagnets.

13.25 Explain superexchange in antiferromagnets.

13.26 Explain magnetoacoustic effect in brief.

13.27 What are magnons. Discuss magnons in a one-dimensional ferromagnet.

13.28 Discuss Neel's theory of Ferrimagnetism.

13.29 Explain Landau diamagnetism.

CHAPTER

# 14

# Optical Properties

## 14.1 INTRODUCTION

At the beginning of this century the study of the interactions of light with matter (black body radiation, etc.) laid the foundations for quantum theory. Today, optical methods are among the most important tools for elucidating the electron structure of matter. Most recently, a number of optical devices such as lasers, photodetectrors, waveguides, etc., have gained considerable technological importance. They are used in communication, fiber optics, medical diagnostics, night viewing, solar applications, optical computing, or for other optoelectronic purposes. Traditional utilizations of optical materials for windows, antireflection coatings, lenses, mirrors, etc, should be likewise mentioned. All taken, it is well justified to spend a chapter of this book on the optical properties of materials.

Before we start our discourse we need to define the optical constants. We make use of some elements of physics.

## 14.2 INDEX OF REFRACTION, *n*

When light passes from an optically "thin" into an optically dense medium, one observes that in the dense medium, the angle of refraction $\beta$ (i.e., the angle between the refracted light beam and a line perpendicular to the surface) is smaller than the angle of incidence, $\alpha$. This well-known phenomenon is used for the definition of the refractive power of a material and is called Snell's law

$$\frac{\sin \alpha}{\sin \beta} = \frac{n_{med}}{n_{vac}} = n. \qquad ...(1)$$

Commonly, the index of refraction for vacuum $n_{vac}$ is arbitrarily set to be unity. The refraction is caused by the different velocities, $c$, of the light in the two media

$$\frac{\sin \alpha}{\sin \beta} = \frac{c_{vac}}{c_{med}}. \qquad ...(2)$$

Thus, if light passes from vacuum into a medium, we find

$$n = \frac{c_{\text{vac}}}{c_{\text{med}}} = \frac{c}{v}. \qquad ...(3)$$

The magnitude of the refractive index depends on the wavelength of the incident light. This property is called dispersion. In metals, the index of refraction varies in addition with the angle of incidence. This is particularly true when $n$ is small.

## 14.3 DAMPING CONSTANT, *k*

Metals damp the intensity of light in a relatively short distance. Thus, to characterize the optical properties of metals, an additional material constant is needed.

We make use of the electromagnetic wave equation which mathematically describes the propagation of light in a medium. The derivation of this wave equation from the well-known Maxwell equations does not further our understanding of the optical properties.

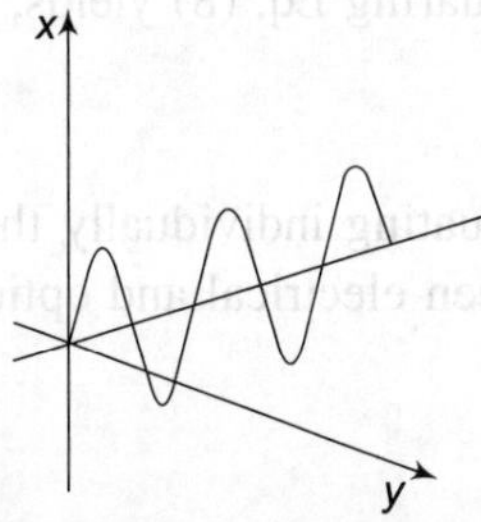


**Figure 14.1** Plane polarized wave which propagates in the positive *z*-direction and vibrates in the *x*-direction

For simplification, we consider a plane polarized wave which propagates along the positive $z$-axis and which vibrates in the $x$-direction (Fig. 14.1). We neglect possible magnetic properties. For this special case, the electromagnetic wave equation reads

$$c^2 \frac{\partial^2 \mathscr{E}_x}{\partial z^2} = \varepsilon \frac{\partial^2 \mathscr{E}_x}{\partial t^2} + 4\pi\sigma \frac{\partial \mathscr{E}_x}{\partial t}, \quad \text{(CGS units)} \qquad ...(4)$$

where $\mathscr{E}_x$ is the $x$-component of the electric field strength,[1] $\varepsilon$ is the dielectric constant, and $\sigma$ is the (a.c.) conductivity. The solution to Eq. (4) is commonly achieved by using the following trial solution

$$\mathscr{E}_x = \mathscr{E}_0 \exp\left[i\omega\left(t - \frac{zn}{c}\right)\right] \qquad ...(5)$$

where $\mathscr{E}_0$ is the maximal value of the electric field strength (amplitude) and $\omega = 2\pi v$ is the angular frequency. Differentiating Eq. (5) once with respect to time, and twice with respect to time and $z$, and inserting these values into Eq. (4) yields

$$\hat{n}^2 = \varepsilon - \frac{4\pi\sigma}{\omega} i = \varepsilon - \frac{2\sigma}{v} i. \qquad ...(6)$$

Equation (6) leads to an important result: The index of refraction is generally a complex quantity as inspection of the right-hand side of Eq. (6) indicates. We denote for clarity the complex index of refraction by $\hat{n}$. As is true for all complex quantities, the complex index of refraction consists of a real and an imaginary part

$$\hat{n} = n_1 - in_2 \qquad ...(7)$$

[1] We use, for the electric field strength, $\mathscr{E}$ to distinguish it from the energy.

In the literature, the imaginary part of the index of refraction $n_2$ is often denoted by "$k$" and Eq. (7) is then written as

$$\hat{n} = n - ik \qquad \text{...(8)}$$

We will call $n_2$ or $k$ the *damping constant*. (In some books $n_2$ or $k$ is named the *absorption constant*. We will not follow this practice because the latter term is extremely misleading. Other authors call $k$ the *attenuation index* or the *extinction coefficient* which we will not use either in this context.)

Squaring Eq. (8) yields, together with Eq. (6),

$$\hat{n} = n^2 - k^2 - 2nik = \varepsilon - \frac{2\sigma}{v} i \qquad \text{...(9)}$$

Equating individually the real and imaginary parts of (9) yields two important relations between electrical and optical constants

$$\varepsilon = n^2 - k^2 \qquad \text{...(10)}$$

$$\sigma = nkv \qquad \text{...(11)}$$

Let us return to Eq. (9). The right-hand side is the difference between two dielectric constants (a real and an imaginary one). Thus, the left side must be a dielectric constant too and Eq. (9) may be rewritten as

$$\hat{n}^2 = n^2 - k^2 - 2nik = \hat{\varepsilon} = \varepsilon_1 - i\varepsilon_2 \qquad \text{...(12)}$$

Equating individually the real and imaginary parts in Eq. (12) yields

$$\varepsilon_1 = n^2 - k^2 \qquad \text{...(13)}$$

and (with Eq. (11))

$$\varepsilon_2 = 2nk = \frac{2\sigma}{v} \qquad \text{...(14)}$$

Similarly as above, $\varepsilon_1$ and $\varepsilon_2$ are called the real and the imaginary parts of the complex dielectric constant $\hat{\varepsilon}$, respectively. ($\varepsilon_1$ in Eq. (13) is identical to $\varepsilon$ in Eq. (10)). $\varepsilon_2$ is often called the *absorption product* or, briefly, the *absorption*.

We consider a special case: For insulators ($\sigma \approx 0$) it follows from Eq. (11) that $k \approx 0$ (see also Table 14.1). Then Eq. (10) reduces to $E = n^2$ (*Maxwell relation*).

From Eqs. (10), (11), (13), and (14) one obtains

$$n^2 = \frac{1}{2}\left(\sqrt{\varepsilon^2 + \left(\frac{2\sigma}{v}\right)^2} + \varepsilon\right) + \frac{1}{2}\left(\sqrt{\varepsilon_1^2 + \varepsilon_2^2} + \varepsilon_1\right) \qquad \text{...(15)}$$

$$k^2 = \frac{1}{2}\left(\sqrt{\varepsilon^2 + \left(\frac{2\sigma}{v}\right)^2} - \varepsilon\right) + \frac{1}{2}\left(\sqrt{\varepsilon_1^2 + \varepsilon_2^2} - \varepsilon_1\right) \qquad \text{...(16)}$$

**Table 14.1** Characteristic Penetration Depth, *W*, and Damping Constant, *k*, for Some Materials ($\lambda$ = 589.3 nm)

| *Material* | *Water* | *Flint glass* | *Graphite* | *Gold* |
|---|---|---|---|---|
| *W* (cm) | 32 | 29 | $6 \times 10^{-6}$ | $1.5 \times 10^{-6}$ |
| *k* | $1.4 \times 10^{-7}$ | $1.5 \times 10^{-7}$ | 0.8 | 3.2 |

It should be emphasized that Eqs. (10) – (16) are only valid if $\varepsilon$, $\sigma$, *n*, and *k* are measured at the same wavelength, because these "constants" are wavelength dependent. For small frequencies, however, the d.c. values for $\varepsilon$ and $\sigma$ can be used with good approximation, as will be shown later. Finally, it should be noted that the above equations are only valid for optically isotropic media; otherwise $\varepsilon$ becomes a tensor.

We return now to Eq. (5) in which we replace the index of refraction by the general expression Eq. (8). This yields

$$\mathscr{E}_x = \mathscr{E}_0 \exp\left[i\omega\left(t - \frac{z(n - ik)}{c}\right)\right] \qquad \text{...(17)}$$

which may be rewritten to read

$$\boxed{\mathscr{E}_x = \mathscr{E}_0 \underbrace{\exp\left[-\frac{\omega k}{c} z\right]}_{\text{Damped amplitude}} \cdot \underbrace{\exp\left[i\omega\left(t - \frac{zn}{c}\right)\right]}_{\text{Undamped wave}}.} \qquad \text{...(18)}$$

Equation (18) is now the complete solution of the wave equation (4). It represents a damped wave and expresses that in matter the amplitude decreases exponentially with increasing *z* (Fig. 14.2). The constant *k* determines how much the amplitude decreases, i.e., *k* expresses the degree of damping of the light wave. We understand now why *k* is termed the *damping constant*.

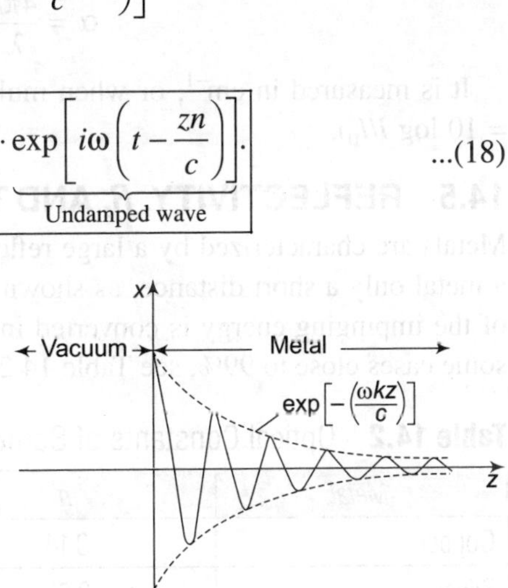

**Figure 14.2** Modulated light wave. The amplitude decreases exponentially

The result which we just obtained is well known to electrical engineers. They observe that at high frequencies the electromagnetic waves are conducted only on the outer surface of a wire. They call this phenomenon the (*normal*) *skin effect*.

## 14.4 CHARACTERISTIC PENETRATION DEPTH, *W*, AND ABSORBANCE, $\alpha$

The field strength ($\mathscr{E}$) is hard to measure. Thus, the intensity, *I*, which can be measured effortlessly with light-sensitive devices (such as a photodetector) is commonly used. The intensity equals the square of the field strength. Thus, the damping term in Eq. (18) may be written as

$$I = \mathscr{E}^2 = I_0 \exp\left(-\frac{2\omega k}{c} z\right) \qquad ...(19)$$

We define a characteristic penetration depth, $W$, as that distance at which the intensity of the light wave, which travels through a material, has decreased to $1/e$ or 37% of its original value, i.e., when

$$\frac{I}{I_0} = \frac{1}{e} = e^{-1}. \qquad ...(20)$$

This definition yields in conjunction with Eq. (19)

$$z = W = \frac{c}{2\omega k} = \frac{c}{4\pi \nu k} = \frac{\lambda}{4\pi k} \qquad ...(21)$$

Table 14.1 presents values for $k$ and $W$ for some materials obtained by using sodium vapour light ($\lambda = 589.3$ nm).

The inverse of $W$ is sometimes called the (exponential) *attenuation* or the *absorbance* which is, by making use of Eqs. (21), (14), and (11),

$$\alpha = \frac{4\pi k}{\lambda} = \frac{2\pi \varepsilon_2}{\lambda n} = \frac{4\pi\sigma}{nc} = \frac{2\omega k}{c} \qquad ...(21a)$$

It is measured in $cm^{-1}$, or when multiplied by 4.3 in decibels (dB) per centimeter (1 dB = 10 log $I/I_0$).

## 14.5 REFLECTIVITY, *R*, AND TRANSMISSIVITY, *T*

Metals are characterized by a large reflectivity. This stems from the fact that light penetrates a metal only a short distance, as shown in Fig. 14.2 and Table 14.1. Thus, only a small part of the impinging energy is converted into heat. The major part of the energy is reflected (in some cases close to 99%, see Table 14.2. In contrast to this, visible light penetrates into glass

**Table 14.2** Optical Constants of Some Metals ($\lambda = 600$ nm)

| *Metal* | *n* | *k* | *R[%]* |
|---|---|---|---|
| Copper | 0.14 | 3.35 | 95.6 |
| Silver | 0.05 | 4.09 | 98.9 |
| Gold | 0.21 | 3.24 | 92.9 |
| Aluminum | 0.97 | 6.0 | 90.3 |

much farther than into metals, i.e., approximately seven orders of magnitude more, see Table 14.1. As a consequence, very little light is reflected by glass. Nevertheless, a piece of glass about one or two metres thick eventually dissipates a substantial part of the impinging light into heat. (In practical applications, one does not observe this large reduction in light intensity because windows are, as a rule, only a few millimeters thick.) It should be noted that typical windowpanes reflect the light on the front as well as on the back.

The ratio between reflected intensity ($I_R$) and incoming intensity ($I_0$) of the light serves as a definition for the *reflectivity*,

$$R = \frac{I_R}{I_0} \qquad ...(22)$$

Quite similarly, one defines the ratio between the transmitted intensity, $I_T$, and the impinging light intensity as the *transmissivity*,

$$T = \frac{I_T}{I_0}. \qquad ...(22a)$$

Experiments have shown that for insulators, $R$ depends solely on the index of refraction. For perpendicular incidence one finds

$$R = \frac{(n-1)^2}{(n+1)^2}. \qquad ...(23)$$

This equation can be derived from the Maxwell equations.

We know already that $n$ is generally a complex quantity. By definition, however, $R$ has to remain real. Thus, the modulus of $R$ becomes

$$R = \left|\frac{\hat{n}-1}{\hat{n}+1}\right|^2 \qquad ...(24)$$

which yields

$$R = \frac{(n-ik-1)}{(n-ik+1)} \cdot \frac{(n+ik-1)}{(n+ik+1)} = \frac{(n-1)^2+k^2}{(n+1)^2+k^2} \qquad ...(25)$$

(*Beer's equation*). The reflectivity is a unitless material constant and is often given in per cent of the incoming light (see Table 14.2). $R$ is, like the index of refraction, a function of the wavelength of the light.

The reflectivity is a function of $\varepsilon_1$ and $\varepsilon_2$. We shall derive this relationship by performing a few transformations. Equation (25) is rewritten as

$$R = \frac{n^2+k^2+1-2n}{n^2+k^2+1+2n} \qquad ...(26)$$

(1) $n^2 + k^2 = \sqrt{(n^2+k^2)^2} = \sqrt{n^4+2n^2k^2+k^4}$

$$= \sqrt{n^4-2n^2k^2+k^4+4n^2k^2} = \sqrt{(n^2-k^2)^2+4n^2k^2}$$

$$= \sqrt{\varepsilon_1^2+\varepsilon_2^2}, \qquad ...(27)$$

(2) $2n = \sqrt{4n^2} = \sqrt{2(n^2+k^2+n^2-k^2)} = \sqrt{2(\sqrt{\varepsilon_1^2+\varepsilon_2^2}+\varepsilon_1)}$. ...(28)

Inserting Eqs. (27) and (28) into Eq. (26) provides

$$R = \frac{\sqrt{\varepsilon_1^2 + \varepsilon_2^2} + 1 - \sqrt{2(\varepsilon_1^2 + \varepsilon_2^2 + \varepsilon_1)}}{\sqrt{\varepsilon_1^2 + \varepsilon_2^2} + 1 + \sqrt{2(\sqrt{\varepsilon_1^2 + \varepsilon_2^2} + \varepsilon_1)}}. \quad ...(29)$$

## 14.6 HAGEN–RUBENS RELATION

Our next task is to find a relationship between reflectivity and conductivity. For small frequencies (i.e., $\nu < 10^{13}\ \mathrm{s}^{-1}$) the ratio $\sigma/\nu$ for metals is very large (d.c. conductivity[2] $\sigma_0$ $10^{17}\ \mathrm{s}^{-1}$, $\varepsilon \approx 10$). Thus, we obtain

$$\frac{\sigma}{\nu} \approx \frac{10^{17}}{10^{13}} \gg \varepsilon \quad ...(30)$$

Then Eqs. (15) and (16) reduce to

$$n^2 \approx \frac{\sigma}{\nu} \approx k^2 \quad ...(31)$$

The reflectivity may now be rewritten by combining Eq. (25) with Eq. (31) to read

$$R = \frac{n^2 + 2n + 1 + k^2 - 4n}{n^2 + 2n + 1 + k^2} = 1 - \frac{4n}{2n^2 + 2n + 1}. \quad ...(32)$$

If $2n + 1$ is neglected as small compared to $2n^2$ (which can be done only for small frequencies for which $n$ is much larger than 1), then Eq. (32) reduces by using Eq. (31) to

$$R = 1 - \frac{2}{n} = 1 - 2\sqrt{\frac{\nu}{\sigma}} \quad ...(33)$$

Finally, we set $\sigma = \sigma_o$ which is again only permissible for small frequencies, i.e., in the infrared region of the spectrum. This yields the Hagen–Rubens equation

$$\boxed{R = 1 - 2\sqrt{\frac{\nu}{\sigma_o}}} \quad ...(34)$$

which states that in the infrared (IR) region metals with large electrical conductivity are good reflections. This equation was found empirically by Hagen and Rubens from reflectivity measurements in the IR and was derived theoretically by Drude. As stated above, the Hagen–Rubens relation is only valid at frequencies below $10^{13}\ \mathrm{s}^{-1}$ or, equivalently, at wavelengths larger than about 30 μm.

### 14.6.1 Frequency Dependence of Reflectivity

The calculation of the frequency dependence of the optical constants is accomplished by using the well-known equations for vibrations, whereby the interactions of electrons with atoms are taken into account by a damping term which is assumed to be proportional to the velocity of the electrons. The free electron theory describes, to a certain degree, the dispersion of the optical constants quite well. This is schematically shown in Fig. 14.3, in which the spectral

[2] The conductivity in cgs units is expressed in $\mathrm{s}^{-1}$.

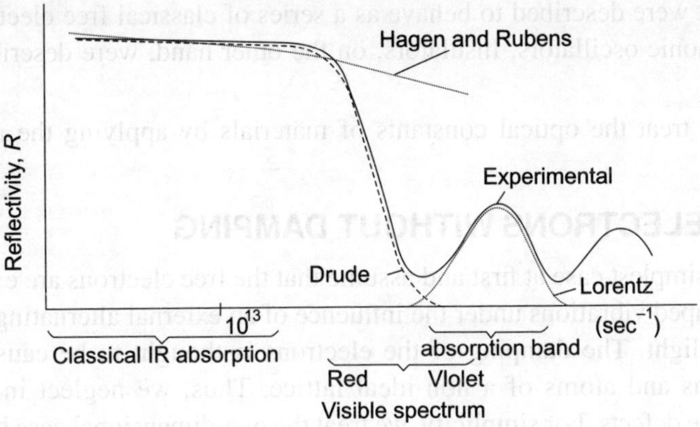


**Figure 14.3** Schematic frequency dependence of the reflectivity of metals, experimentally (solid line) and according to three models

dependence of the reflectivity is plotted for a specific case. The Hagen–Rubens relation reproduces the experimental findings only up to $10^{13}$ s$^{-1}$. In contrast to this, the Drude theory correctly reproduces the spectral dependence of $R$ even in the visible spectrum. Proceeding to yet higher frequencies, however, the experimentally found reflectivity eventually rises and then decreases again. Such an *absorption band* cannot be explained by the Drube theory. For its interpretation, a new concept needs to be applied.

Lorentz postulated that the electrons should be considered to be bound to their nuclei and that an external electric field displaces the positive charge of an atomic nucleus against the negative charges of its electron cloud. In other words, he represented each atom as an electric dipole. Retracting forces were thought to occur which try to eliminate the displacement of charges. Lorentz postulated further that the centres of gravity of the electric charges are identical if no external forces are present. However, if one shines light onto a solid, i.e., if one applies an alternating electric field to the atoms, then the dipoles are thought to perform forced vibrations. Thus, a dipole is considered to behave similarly as a mass which is suspended on a spring, i.e., the equations for a harmonic oscillator may be applied. An oscillator is known to absorb a maximal amount of energy when excited near its resonance frequency (Fig. 14.4). The absorbed energy is thought to be dissipated mainly by diffuse radiation. Figure 14.4 resembles an absorption band as shown in Fig. 14.3.

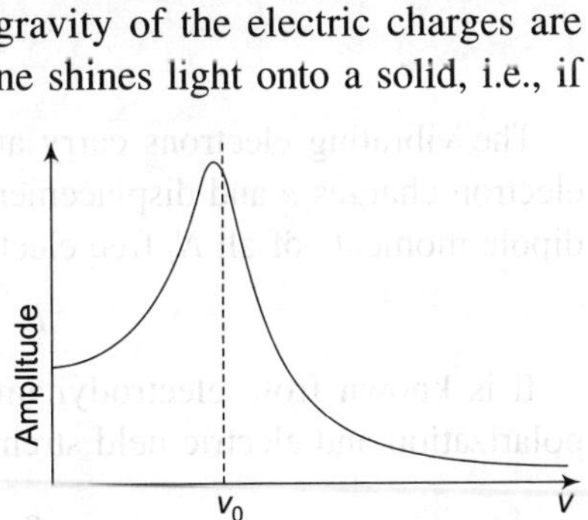


**Figure 14.4** Frequency dependence of the amplitude of a harmonic oscillator which was excited to perform forced vibrations, assuming weak damping, $\nu_0$ is the resonance frequency

Thirty or forty years ago, many scientists considered the electrons in metals to behave at low frequencies as if they were free and at higher frequencies as if they were bound. In other words, electrons in a metal under the influence of light were described to behave as a series of classical free electrons and a series of classical harmonic oscillators. Insulators, on the other hand, were described by harmonic oscillators only.

We shall now treat the optical constants of materials by applying the above-mentioned theories.

## 14.7 FREE ELECTRONS WITHOUT DAMPING

We consider the simplest case at first and assume that the free electrons are excited to perform forced but undamped vibrations under the influence of an external alternating field, i.e., under the influence of light. The damping of the electrons is thought to be caused by collisions between electrons and atoms of a non-ideal lattice. Thus, we neglect in this section the influence of lattice defects. For simplicity, we treat the one-dimensional case because the result obtained this way does not differ from the general case. Thus, we consider the interaction of plane polarized light with the electrons. The momentary value of the field strength of a plane polarized light wave is given by

$$\mathscr{E} = \mathscr{E}_0 \exp(i\omega t) \qquad ...(1)$$

where $\omega = 2\pi\nu$ is the angular frequency, $t$ is the time, and $\mathscr{E}_0$ is the maximal value of the field strength. The equation describing the motion of an electron which is excited to perform forced, harmonic vibrations under the influence of light is

$$m\,\frac{d^2x}{dt^2} = e\mathscr{E} = e\mathscr{E}_0 \exp(i\omega t) \qquad ...(2)$$

where $e$ is the electron charges, $m$ is the electron mass, and $e\,\mathscr{E}$ is the modulus of the excitation force. The stationary solution of this vibrational equation is obtained by forming the second derivative of the trial solution, $x = x_0 \exp(i\omega t)$ and inserting it into Eq. (2). This yields

$$x = -\frac{e\mathscr{E}}{m4\pi^2\nu^2} \qquad ...(3)$$

The vibrating electrons carry an electric dipole momentum which is the product of the electron charges $e$ and displacement, $x$. The polarization ($P$) is defined to be the sum of the dipole moments of all $N_f$ free electrons per cubic centimetre

$$P = exN_f \qquad ...(4)$$

It is known from electrodynamics that the dielectric constant may be calculated from polarization and electric field strength as follows:[3]

$$\varepsilon = 1 + 4\pi\,\frac{P}{\mathscr{E}}. \qquad ...(5)$$

Inserting Eqs. (3) and (4) into Eq. (5) yields

$$\hat{\varepsilon} = 1 - \frac{e^2N_f}{\pi m\nu^2}. \qquad ...(6)$$

[3] Equation (5) reads in the SI system: $\varepsilon = 1 + P/\varepsilon_0\mathscr{E}$.

(It is appropriate to use in the present case the *complex* dielectric constant, see below.) The dielectric constant equals the square of the index of refraction $n$. Equation (6) thus becomes

$$\hat{n}^2 = 1 - \frac{e^2 N_f}{\pi m v^2}. \quad ...(7)$$

We consider two special cases:

(a) For small frequencies, the term, $e^2N_f/\pi m v^2$, is larger than one. Then $\hat{n}^2$ is negative and $\hat{n}$ imaginary. An imaginary $\hat{n}$ means that the real part of $\hat{n}$ disappears. Equation (25) (section 14.5) becomes, for $n = 0$,

$$R = \frac{(n-1)^2 + k^2}{(n+1)^2 + k^2} = \frac{1 + k^2}{1 + k^2} = 1,$$

i.e., the reflectivity is 100% (see Fig. 14.5).

(b) For large frequency (UV light), the term $e^2N_f/\pi m v^2$ becomes smaller than one. Thus, $\hat{n}^2$ is positive and $\hat{n} \equiv n$ real (but smaller than one). The reflectivity of real values of $\hat{n}$, i.e., for $k = 0$, becomes

$$R = \frac{(n-1)^2}{(n+1)^2}.$$

i.e., the material is essentially transparent for these wavelengths (and perpendicular incidence) and therefore behaves optically like an insulator, see Fig. 14.5.

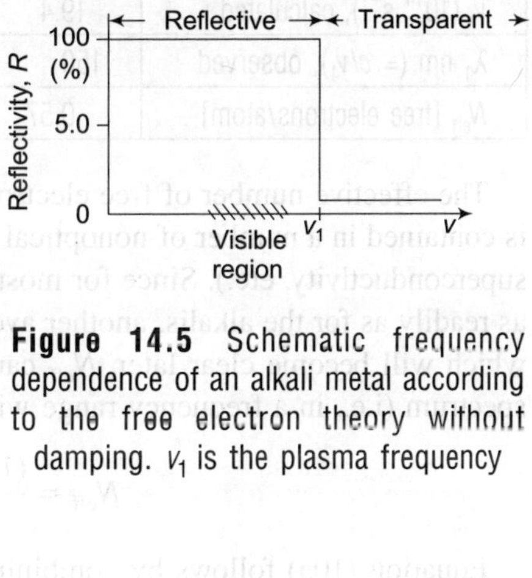


**Figure 14.5** Schematic frequency dependence of an alkali metal according to the free electron theory without damping. $\nu_1$ is the plasma frequency

We define a characteristic frequency $\nu_1$, often called the *plasma frequency*, which separates the reflective from the transparent region (Fig. 14.5). The plasma frequency is defined by setting $e^2N_f/\pi m \nu_1^2$ equal to unity which yields

$$\nu_1^2 = \frac{e^2 N_f}{\pi m}. \quad ...(8)$$

We conclude from (6) that the dielectric constant becomes zero at the plasma frequency because of $e^2N_f/\pi m \nu_1^2 = 0$. $\hat{\varepsilon} = 0$ is the condition for a *plasma oscillation*, i.e., a fluid-like oscillation of the entire electron gas.

The alkali metals behave essentially as shown in Fig. 14.5. They are transparent in the near UV and reflect the light in the visible region. This result indicates that the $s$-electrons of the outer shell of the alkali metals can be considered to be free.

Table 14.3 contains some measured, as well as some calculated, plasma frequencies. For the calculations, applying Eq. (8), *one* free electron per atom was assumed. This means that $N_f$ was set equal to the number of atoms per volume, $N_a$. (The latter quantity is obtained by using

$$N_a = \frac{N_0 \cdot \delta}{M}, \quad \text{...(9)}$$

where $N_0$ is the Avogadro constant, $\delta$ = density, and $M$ = atomic mass.)

We note in Table 14.3 that the calculated and the observed values for $\nu_1$ are only identical for sodium. This may be interpreted to mean that only in sodium exactly *one* free electron per atom is contributing to the electron gas. For other metals an "effective number of free electrons" is commonly introduced which is defined to be the ratio between the observed and calculated $\nu_1^2$ value

$$\frac{\nu^2_{1(\text{observed})}}{\nu^2_{1(\text{calculated})}} = N_{\text{eff}} \quad \text{...(10)}$$

**Table 14.3** Plasma Frequencies and Effective Numbers of Free Electrons for Some Alkali Metals

| *Metal* | *Li* | *Na* | *K* | *Rb* | *Cs* |
|---|---|---|---|---|---|
| $\nu_1$ ($10^{14}$ $s^{-1}$), observed | 14.6 | 14.3 | 9.52 | 8.33 | 6.81 |
| $\nu_1$ ($10^{14}$ $s^{-1}$), calculated | 19.4 | 14.3 | 10.34 | 9.37 | 8.33 |
| $\lambda_1$ nm (= $c/\nu_1$), observed | 150 | 210 | 290 | 320 | 360 |
| $N_{\text{eff}}$ [free electrons/atom] | 0.57 | 1.0 | 0.8 | 0.79 | 0.67 |

The effective number of free electrons is a parameter which is of great interest because it is contained in a number of nonoptical equations (such as the Hall constant, electrotransport, superconductivity, etc.). Since for most metals the plasma frequency, $\nu_1$, cannot be measured as readily as for the alkalis, another avenue for determining $N_{\text{eff}}$ has to be found. For reasons which will become clear later, $N_{\text{eff}}$ can be obtained by measuring $n$ and $k$ in the red or IR spectrum (i.e., in a frequency range without absorption bands, Fig. 14.3) and by applying

$$N_{\text{eff}} = \frac{(1 - n^2 + k^2)\,\nu^2\,\pi m}{e^2}. \quad \text{...(10a)}$$

Equation (10a) follows by combining Eq. (6) with Eq. (10) (section 14.3) and replacing $N_f$ by $N_{\text{eff}}$.

## 14.8 FREE ELECTRONS WITH DAMPING (CLASSICAL FREE ELECTRON THEORY OF METALS)

The simple reflectivity spectrum as depicted in Fig. 14.5 is seldom found for metals. We need to refine our model. We postulate that the motion of electrons in metals is damped. More specifically, we postulate that the velocity is reduced by collisions of the electrons with atoms of a nonideal lattice. Lattice defects may be introduced into a solid by interstitial atoms, vacancies, impurity atoms, dislocations, grain boundaries, or thermal motion of the atoms.

To take account of the damping, we add to the vibration equation (2, section 14.7) damping term $y\,(dx/dt)$ which is proportional to the velocity

$$M\frac{d^2x}{dt^2}+\gamma\frac{dx}{dt}=e\mathscr{E}=e\mathscr{E}_0\exp(i\omega t) \quad ...(11)$$

We determine first the damping factor, γ. For this we write a particular solution of Eq. (11) which is obtained by assuming that the electrons drift under the influence of a steady or slowly varying electric field with a velocity, $v'$ = const. through the crystal. (The drift velocity of the electrons, which is caused by an external field, is superimposed on the random motion of the electrons). The damping is depicted to be a friction force which counteracts the electron motion, $v'$ = const. yields

$$\frac{d^2x}{dt^2}=0. \quad ...(12)$$

By using Eq. (12), Eq. (11) becomes

$$\frac{e\mathscr{E}}{\gamma}=\frac{dx}{dt}=v'. \quad ...(13)$$

The drift velocity is

$$v'=\frac{j}{eN_f} \quad ...(14)$$

where $j$ is the current density (i.e., that current which passes in one second through an area of one square centimetre). $N_f$ is the number of free electrons per cubic centimetre. The current density is connected with the d.c. conductivity ($\sigma_0$) and the field strength ($\mathscr{E}$) by Ohm's law

$$j=\sigma_0\mathscr{E}. \quad ...(15)$$

Inserting Eqs. (14) and (15) into Eq. (13) yields

$$\gamma=\frac{N_f e^2}{\sigma_0}. \quad ...(16)$$

Thus, Eq. (11) becomes

$$m\frac{d^2x}{dt^2}+\frac{N_f e^2}{\sigma_0}\frac{dx}{dt}=e\mathscr{E}=e\mathscr{E}_0\exp(i\omega t) \quad ...(17)$$

We note that the damping term in Eq. (17) is inversely proportional to the conductivity, i.e., proportional to the resistivity. This result makes sense.

The stationary solution of Eq. (17) is obtained, similarly as in section 14.7 by differentiating the trial solution $x = x_o\exp(i\omega t)$ by the time, and inserting first and second derivatives into Eq. (17) which yields

$$-mw^2x+\frac{N_f e^2}{\sigma_o}x\omega i=\mathscr{E}e. \quad ...(18)$$

Rearranging Eq. (18) provides

$$x=\frac{\mathscr{E}}{\dfrac{N_f e\omega}{\sigma_0}i-\dfrac{m\omega^2}{e}}. \quad ...(19)$$

Inserting Eq. (19) into Eq. (4) yields the polarization

$$P = \frac{eN_f\mathscr{E}}{\frac{N_f e\omega}{\sigma_0} i - \frac{m\omega^2}{e}} \qquad ...(20)$$

With Eqs. (20) and (5) the complex dielectric constant becomes

$$\hat{\varepsilon} = 1 + 4\pi \frac{P}{\mathscr{E}} = 1 + \frac{1}{\frac{\nu}{2\sigma_0} i - \frac{m\pi}{N_f e^2}\nu^2} \qquad ...(21)$$

The term $N_f e^2/m\pi$ is set, as in Eq. (8), equal to $\nu_1^2$, which reduces Eq. (21) to

$$\hat{\varepsilon} = 1 + \frac{1}{\frac{\nu}{2\sigma_0} i - \frac{\nu^2}{\nu_1^2}} = 1 + \frac{\nu_1^2}{i\nu\, \nu_1^2/2\sigma_0 - \nu^2} \qquad ...(22)$$

The term $\nu_1^2/2\sigma_0$ in Eq. (22) has the unit of a frequency. Thus, for abbreviation, we define a *damping frequency*

$$\nu_2 = \frac{\nu_1^2}{2\sigma_0} \qquad ...(23)$$

(Table 14.4 lists values for $\nu_2$ which were calculated using experimental $\sigma_0$ and $\nu_1$ values). Now Eq. (22) becomes

$$\hat{\varepsilon} = 1 + \frac{\nu_1^2}{i\nu\nu_2 - \nu^2} \qquad ...(24)$$

where $\hat{\varepsilon}$ is, as usual, identical to $\hat{n}^2$

$$(\hat{n})^2 = n^2 - 2nki - k^2 = 1 - \frac{\nu_1^2}{\nu^2 - \nu\nu_2 i}. \qquad ...(25)$$

Multiplying the numerator and denominator of the fraction in Eq. (25) by the complex conjugate to the denominator ($\nu^2 + \nu\nu_2 i$) allows us to equate individually real and imaginary parts. This provides the Drude equations for the optical constants.

$$\boxed{n^2 - k^2 = \varepsilon_1 = 1 - \frac{\nu_1^2}{\nu^2 + \nu_2^2}} \qquad ...(26)$$

and

$$\boxed{2nk = \varepsilon_2 = \frac{2\sigma}{\nu} = \frac{\nu_2}{\nu}\frac{\nu_1^2}{\nu^2 + \nu_2^2}} \qquad ...(27)$$

**Table 14.4** Resistivities, Conductivities, and Damping Frequencies for Some Metals

| *Metal* | *Li* | *Na* | *K* | *Rb* | *Cs* | *Cu* | *Ag* | *Au* |
|---|---|---|---|---|---|---|---|---|
| $\rho_0$ ($\mu\Omega$ cm)[a] | 8.55 | 4.2 | 6.15 | 12.5 | 20 | 1.67 | 1.59 | 2.35 |
| $\sigma_0$ ($10^{17}$ s$^{-1}$) | 1.05 | 2.14 | 1.46 | 0.72 | 0.45 | 5.37 | 5.66 | 3.83 |
| $\nu_2$ ($10^{12}$ s$^{-1}$) | 10.1 | 4.8 | 3.1 | 4.82 | 5.15 | 4.7 | 4.35 | 5.9 |

[a]*Handbook of Chemistry and Physics*, 1977; room temperature values.

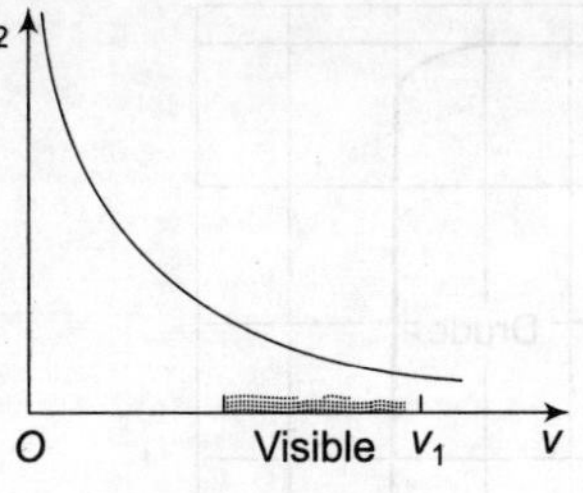


**Figure 14.6** The absorption $E_2 = 2nk$ versus frequency ($\nu$) according to the free electron theory (schematic)

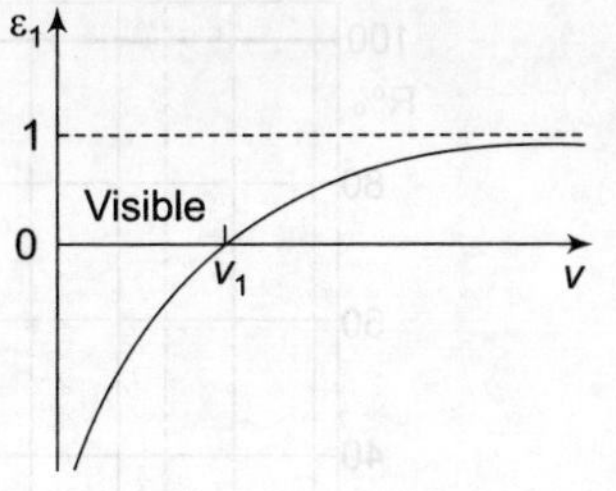


**Figure 14.7** The polarization $E = n^2 - k^2$ as a function of frequency according to the Drude theory (schematic)

with the characteristic frequencies

$$\nu_1 = \sqrt{\frac{e^2 N_f}{\pi m}} \qquad ...(28)$$

and

$$\nu_2 = \frac{\nu_1^2}{2\sigma_0}. \qquad ...(29)$$

The absorption ($E_2$) and the polarization ($\varepsilon_1$) are plotted in Figs. 14.6 and 14.7 as a function of frequency, making use of Eqs. (27) and (26).

## 14.9 SPECIAL CASES

For the *UV*, *visible*, and *near IR* regions, the frequency varies between $10^{14}$ and $10^{15}$ s$^{-1}$. The average damping frequency $\nu_2$ is $5 \times 10^{12}$ s$^{-1}$ (Table 14.4). Thus, $\nu^2 \gg \nu_2^2$. Equation (27) then reduces to

$$\varepsilon_2 = \frac{\nu_2}{\nu}\frac{\nu_1^2}{\nu^2}. \qquad ...(30)$$

With $\nu \approx \nu_1$ (Table 14.3), we obtain

$$\varepsilon_2 = \frac{\nu_2}{\nu} \qquad ...(31)$$

Equation (31) confirms that $\varepsilon_2$ plotted versus the frequency yields a hyperbola with $\nu_2$ as parameter (Fig. 14.6). For *very small frequencies* ($\nu^2 \ll \nu_2^2$), we may neglect $\nu^2$ in the denominator of Eq. (27). This yields with Eq. (23)

$$nk\nu = \sigma = \frac{1}{2}\frac{\nu_1^2}{\nu_2} = \sigma_0. \qquad ...(32)$$

Thus, in the far IR the a.c. conductivity $\sigma$ and the d.c. conductivity $\sigma_0$ may be considered to be identical.

## 14.10 REFLECTIVITY

The reflectivity of metals is calculated using Eq. (31) in conjunction with Eqs. (26) and (27) section 14.9, see Fig. 14.8. We notice that the experimental behaviour for not too high frequencies (Fig. 14.3) is essentially reproduced.

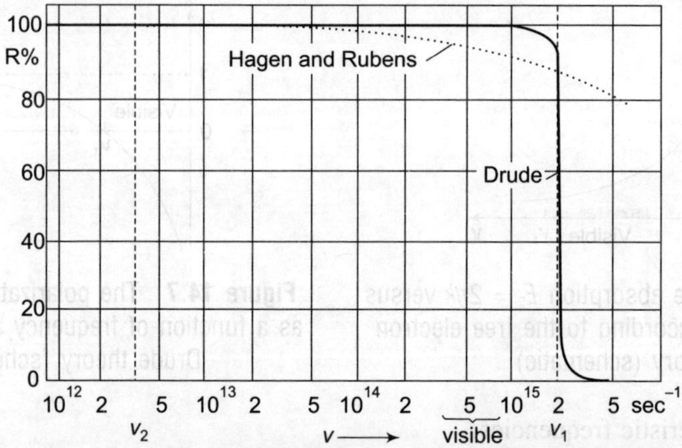


**Figure 14.8** Calculated spectral reflectivity for a metal using the exact Drude equation (solid line), and the Hagen–Rubens equation (10.34) using $\nu_1 = 2 \times 10^{15}$ s$^{-1}$ and $\nu_2 = 3.5 \times 10^{12}$ s$^{-1}$

## 14.11 BOUND ELECTRONS (CLASSICAL ELECTRON THEORY OF DIELECTRIC MATERIALS)

The preceding sections have shown that the optical properties of metals can be described and calculated quite well in low-frequency range by applying the free electron theory. We mentioned already that this theory has its limits at higher frequencies at which we observe that light is absorbed by metals as well as by nonmetals in a narrow frequency band. To interpret these absorption bands, Lorentz postulated that the electrons are bound to their respective nuclei. He assumed that under the influence of an external electric field, the positively charged nucleus and the negatively charged electron cloud are displaced with respect to each other (Fig. 14.9). An electrostatic force tries to counteract this displacement. For simplicity, we describe the negative charge of the electrons to be united in one point. Thus, we describe the atom in an electric field as consisting of a positively charged core which is bound quasi-elastically to *one* electron (electric dipole, Fig. 14.10). A bound electron, thus, may be compared to a mass which is suspended from a spring. Under the influence of an alternating electric field (i.e., by light), the electron is thought to perform forced vibrations. For the description of these vibrations, the well-known equations of mechanics dealing with a harmonic oscillator may be applied. This will be done now.

We first consider an isolated atom, i.e., we neglect the influence of the surrounding atoms upon the electron. An external electric field with force

$$e\mathscr{E} = e\mathscr{E}_0 \exp(i\omega t) \quad ...(1)$$

periodically displaces an electron from its rest position by a distance, $x$. This displacement is counteracted by a restoring force $\kappa \cdot x$ which is proportional to the displacement, $x$. Then, the vibration equation becomes

$$m\frac{d^2x}{dt^2} + \gamma'\frac{dx}{dt} + \kappa x = e\mathscr{E}_0 \exp(i\omega t). \qquad ...(2)$$

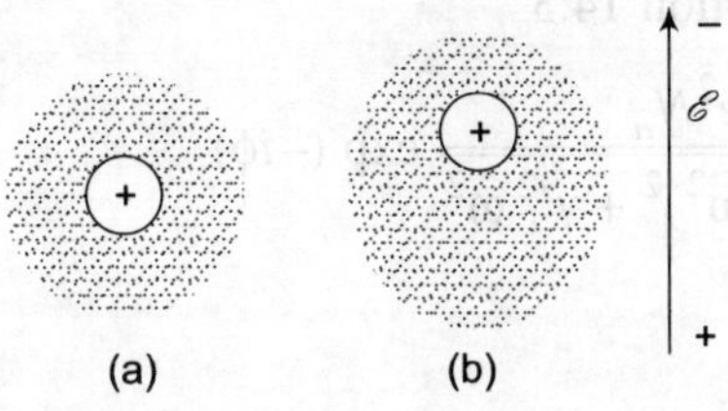


**Figure 14.9** An atom is represented by a positively charged core and a surrounding negatively charged electron cloud (a) in equilibrium and (b) in an external electric field

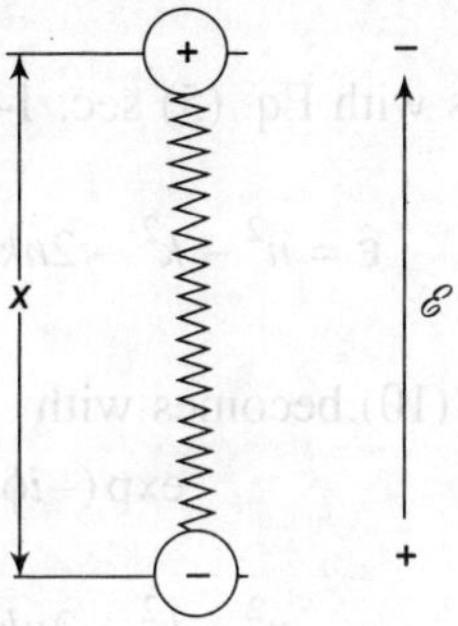

**Figure 14.10** Quasi-elastic bound electron in an external electric field (harmonic oscillator)

The factor κ is the *spring constant* which determines the binding strength between the atom and electron. Each vibrating dipole (e.g., an antenna) loses energy by radiation. Thus, $\gamma'$ $(dx/dt)$ represents the damping of the oscillator by radiation ($\gamma'$ = damping constant). The stationary solution of Eq. (2) for weak damping is (see Appendix 1)

$$x = \frac{e\mathscr{E}_0}{\sqrt{m^2(\omega_0^2 - \omega^2)^2 + \gamma'^2\omega^2}} \exp[i(\omega t - \phi)] \qquad ...(3)$$

where
$$\omega_0 = 2\pi\nu_0 = \sqrt{\frac{\kappa}{m}} \qquad ...(4)$$

is called the resonance frequency of the oscillator, i.e., that frequency at which the electron vibrates freely without an external force. (ϕ) is the phase difference between forced vibration and the excitation force of the light wave. It is defined to be

$$\tan\phi = \frac{\gamma'\omega}{m(\omega_0^2 - \omega^2)} = \frac{\gamma'\nu}{2\pi m(\nu_0^2 - \nu^2)}. \qquad ...(5)$$

As in the previous sections, we calculate the optical constants starting with the polarization $P$ which is the product of the dipole moment $e \cdot x$ of *one* dipole times the number of all dipoles (oscillators) $N_a$. As before, we assumed *one* oscillator per atom. Thus, $N_a$ is identical to the number of atoms per unit volume. We obtain

$$P = exN_a. \qquad ...(6)$$

Inserting Eq. (3) yields

$$P = \frac{e^2 N_a \mathscr{E}_0 \exp[i(\omega t - \phi)]}{\sqrt{m^2(\omega_0^2 - \omega^2)^2 + \gamma'^2\omega^2}}. \qquad ...(7)$$

With
$$\exp[i(\omega t - \phi)] = \exp(i\omega t) \cdot \exp(-i\phi) \qquad ...(8)$$

we obtain
$$P = \frac{e^2 N_a \mathscr{E}}{\sqrt{m^2(\omega_0^2 - \omega^2)^2 + \gamma'^2 \omega^2}} \exp(-i\phi), \quad ...(9)$$

which yields with Eq. (5) sec. 14.7 and Eq. (12), section 14.3

$$\hat{\varepsilon} = n^2 - k^2 - 2nki = 1 + \frac{4\pi e^2 N_a}{\sqrt{m^2(\omega_0^2 - \omega^2)^2 + \gamma'^2 \omega^2}} \exp(-i\phi) \quad ...(10)$$

Equation (10) becomes with

$$\exp(-i\phi) = \cos\phi - i\sin\phi, \quad ...(11)$$

$$n^2 - k^2 - 2nki = 1 + \frac{4\pi e^2 N_a}{\sqrt{m^2(\omega_0^2 - \omega^2)^2 + \gamma'^2 \omega^2}} \cos\phi - i\frac{4\pi e^2 N_a}{\sqrt{m^2(\omega_0^2 - \omega^2)^2 + \gamma'^2 \omega^2}} \sin\phi \quad ...(12)$$

The trigonometric terms in Eq. (12) are replaced, using Eq. (5), as follows:

$$\cos\phi = \frac{1}{\sqrt{1+\tan^2\phi}} = \frac{m(\omega_0^2 - \omega^2)}{\sqrt{m^2(\omega_0^2 - \omega^2) + \gamma'^2 \omega^2}} \quad ...(13)$$

Separating the real and imaginary parts in Eq. (12) finally provides the optical constants

$$\varepsilon_1 = n^2 - k^2 = 1 + \frac{4\pi e^2 m N_a(\omega_0^2 - \omega^2)}{m^2(\omega_0^2 - \omega^2)^2 + \gamma'^2 \omega^2} \quad ...(14)$$

that is,

$$\boxed{\varepsilon_1 = 1 + \frac{4\pi e^2 m N_a(\nu_0^2 - \nu^2)}{4\pi^2 m^2(\nu_0^2 - \nu^2)^2 + \gamma'^2 \nu^2}} \quad ...(15)$$

and
$$\varepsilon_2 = 2nk = \frac{4\pi e^2 m N_a \gamma' \omega}{m^2(\omega_0^2 - \omega^2)^2 + \gamma'^2 \omega^2}$$

or
$$\boxed{\varepsilon_2 = \frac{2e^2 N_a \gamma' \nu}{4\pi^2 m^2(\nu_0^2 - \nu^2)^2 + \gamma'^2 \nu^2}} \quad ...(16)$$

The frequency dependencies of $\varepsilon_1$ and $\varepsilon_2$ are plotted in Figs. 14.11 and 14.12. Figure 11.9 resembles the dispersion curve for the index of refraction as it is experimentally obtained for dielectrics. Figure 14.12 depicts the absorption product $\varepsilon_2$ in the vicinity of the resonance frequency, $\nu_0$ (absorption band) as experimentally observed for dielectrics. Equations (15) and (16) reduce to the Drude equations $\nu_0 \rightarrow 0$ (no oscillators).

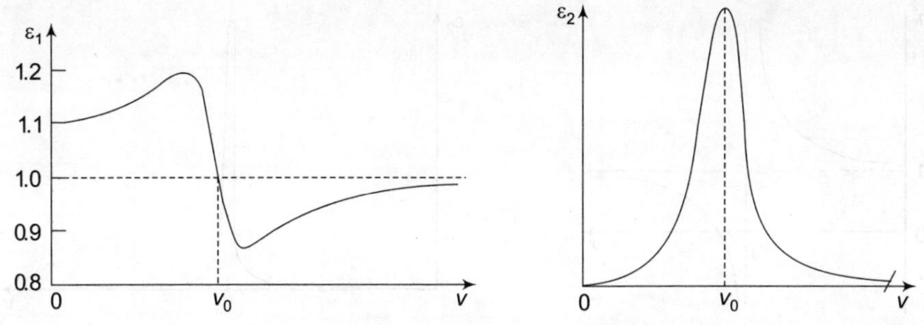


**Figure 14.11 and 14.12** Frequency dependence of polarization $E_1 = n^2 - k^2$ and absorption $E_2 = 2nk$, as calculated with Eqs. (15) and (16), respectively, using characteristic values for $N_a$ and $\gamma'$

## 14.12 DISCUSSION OF THE LORENTZ EQUATIONS FOR SPECIAL CASES

### 14.12.1 High Frequencies

We observe in Fig. 14.12 that $\varepsilon_2$ approaches zero at high frequencies and far away from any resonances (absorption bands). In the same frequency region, $\varepsilon_1 = n^2 - k^2$ and thus essentially $n$ assumes the constant value 1 (Fig.14.11). This is consistent with experimental observations that X-rays are not refracted and are not absorbed by many materials. (Note, however, that high energetic X-rays interact with the *inner* electrons, i.e., they may be absorbed by the $K$, $L$, …, etc., electrons. Metals are, therefore, opaque for high energetic X-rays).

### 14.12.2 Small Damping

We consider the case for which the radiation-induced energy loss of the oscillator is very small. Then, $\gamma'$ is small. With $\gamma'^2\nu^2 \ll 4\pi^2 m^2(\nu_0^2 - \nu^2)^2$ (which is only valid for $\nu \neq \nu_0$), equation (15) reduces to

$$\varepsilon_1 = n^2 - k^2 = 1 + \frac{e^2 N_a}{\pi m(\nu_0^2 - \nu^2)} \qquad ...(17)$$

Figure 14.13 depicts a sketch of (17). We observe that for small damping $\varepsilon_1$ (and thus essentially $n^2$) approaches infinity near the resonance frequency. A dispersion curve, such a Fig. 14.13 is indeed observed for many dielectrics (glass, etc.).

### 14.12.3 Absorption Near $\nu_0$

Electrons absorb most energy from light at the resonance frequency, i.e., $\varepsilon_2$ has a maximum near $\nu_0$. For small damping, the absorption band becomes an absorption line (see Fig. 14.14). Inserting $\nu = \nu_0$ into Eq. (16) yields

$$\varepsilon_2 = \frac{2e^2 N_a}{\gamma' \nu_0}, \qquad ...(18)$$

which shows that the absorption becomes large for small damping ($\gamma'$).

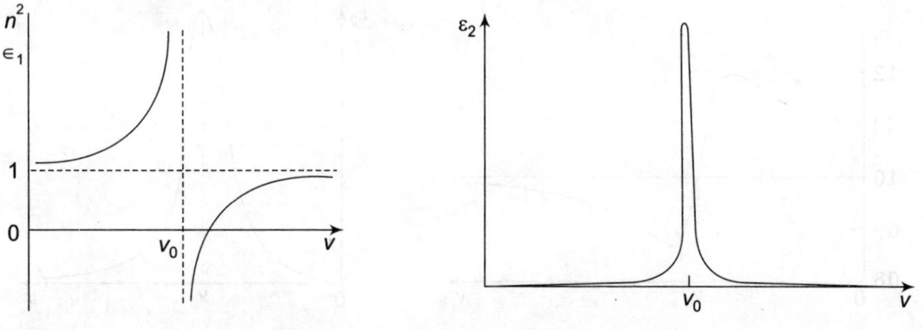


**Figure 14.13 and 14.14** The functions $E_1$ ($n^2$) and $E_2$ respectively, versus frequency according to the bound electron theory for the special case of small damping

### 14.12.4 More Than One Oscillator

At the beginning of Section 14.11 we assumed that *one* electron is quasi-elastically bound to a given nucleus; in other words, we assumed *one* oscillator per atom. This assumption is certainly a gross simplification, as one can deduce from the occurrence of multiple absorption bands in experimental optical spectra. Thus, each atom has to be associated with a number of $i$ oscillators each having an oscillator strength, $f_i$. The $i$th oscillator vibrates with its resonance frequency, $v_{0i}$. The related damping constant is $\gamma'_i$. (This description has its equivalent in the mechanics of a system of mass points having one basic frequency and higher harmonics.) If all oscillators are taken into account, Eqs. (15) and (16) become

$$\varepsilon_1 = n^2 - k^2 = 1 + 4\pi e^2 m N_a \sum_i \frac{f_i(v_{0i}^2 - v^2)}{4\pi^2 m^2 (v_{0i}^2 - v^2)^2 + \gamma_i'^2 v^2}, \quad ...(19)$$

$$\varepsilon_2 = 2nk = 2e^2 N_a \sum_i \frac{f_i v \gamma_i'}{4\pi^2 m^2 (v_{0i}^2 - v^2)^2 + \gamma_i'^2 v^2}. \quad ...(20)$$

Equations (19) and (20) reduce for weak damping (see above) to

$$\varepsilon_1 = n^2 - k^2 \approx n^2 = 1 + \frac{e^2 N_a}{\pi m} \sum_i \frac{f_i}{v_{0i}^2 - v^2}, \quad ...(21)$$

$$\varepsilon_2 = 2nk = \frac{e^2 N_a}{2\pi^2 m^2} \sum_i \frac{f_i v \gamma_i}{(v_{0i}^2 - v^2)^2}. \quad ...(22)$$

## 14.13 CONTRIBUTIONS OF FREE ELECTRONS AND HARMONIC OSCILLATORS TO THE OPTICAL CONSTANTS

In the previous section, we ascribed two different properties to the electrons of a solid. In Section 14.9 we postulated that $N_f$ electrons move freely in metals under the influence of an electric field and that this motion is damped by collisions of the electrons with vibrating lattice atoms and lattice defects. In Section 14.11 we postulated that a certain number of electrons are quasi-elastically bound to $N_a$ atoms which are excited by light to perform forced vibrations. The energy loss was thought to be by radiation.

The optical properties of metals may be described by postulating a certain number of free electrons and a certain number of harmonic oscillators. Both the free electrons and the oscillators contribute to the polarization. Thus, the equations for the optical constants may be rewritten, by combining Eq. (26), (Sec. 14.8) Eq. (27) (Sec. 14.8), Eqs. (19) and (20)

$$\varepsilon_1 = 1 - \frac{v_1^2}{v^2 + v_2^2} + 4\pi e^2 m N_a \sum_i \frac{f_i(v_{0i}^2 - v^2)}{4\pi^2 m^2 (v_{0i}^2 - v^2)^2 + \gamma_i'^2 v^2}, \quad \text{...(23)}$$

$$\varepsilon_2 = 2nk = \frac{v_2}{v}\frac{v_1^2}{v^2 + v_2^2} + 2e^2 N_a \sum_i \frac{f_i v \gamma'_i}{4\pi^2 m^2 (v_{0i}^2 - v^2)^2 + \gamma_i'^2 v^2}. \quad \text{...(24)}$$

Figures 14.15 and 14.16 depict schematically the frequency dependence of

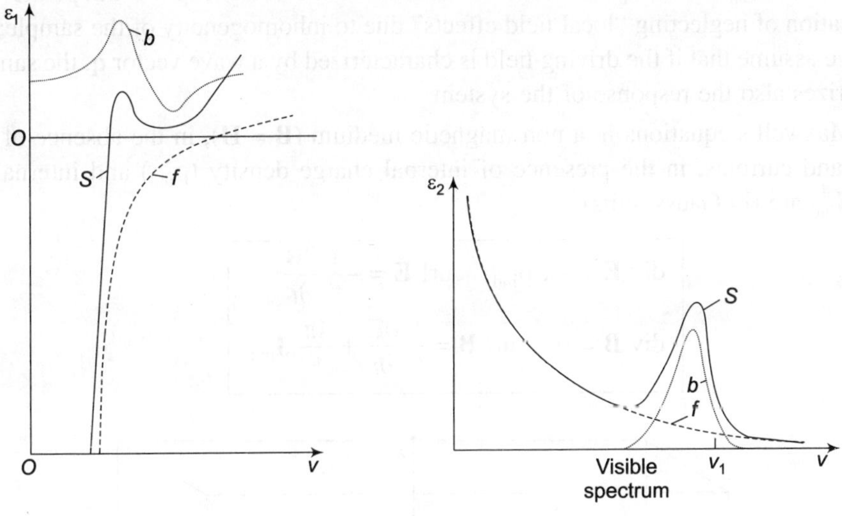


**Figure 14.15 and 14.16** Frequency dependence of $\varepsilon_1$ and $\varepsilon_2$ according to Eqs. (23) and (24). ($i$ = 1). $f$ = free electron theory; $b$ = bound electron theory; $S$ = summary curve (schematic)

$\varepsilon_1$ and $\varepsilon_2$ as obtained by using Eqs. (23) and (24). These figures also show the contributions of free and bound electrons on the optical constants. The experimentally found frequency dependence of $\varepsilon_1$ and $\varepsilon_2$ resembles these calculated spectra quite well.

## 14.14 MACROSCOPIC THEORY OF OPTICAL CONSTANTS IN HOMOGENEOUS MATERIALS

The propagation of electromagnetic waves in materials is described by the Maxwell's equations, supplemented by appropriate constitutive equations of the specific media. In this section we summarize the phenomenological constants, more commonly investigated in experiments. The considerations of the present section are rather general and may apply to homogeneous materials of different nature (metals, semiconductors and insulators), while the following sections refer almost exclusively to simple metals.

We consider for simplicity non-magnetic media ($\mathbf{B} = \mathbf{H}$; $\mu = 1$) in the absence of external charges and currents ($\rho_{ext} = 0$; $\mathbf{J}_{ext} = 0$). We focus our attention on (macroscopically) *homogeneous materials in the linear response regime*; this means that changes of the internal charge density $\rho_{ind}$ and of the internal current density $\mathbf{J}_{ind}$ induced by a driving electromagnetic field (of small intensity) are proportional to the field; furthermore, if the driving field is periodic in space and time, the same space and time dependence holds for the response of the system. (We notice that even an ideally perfect crystal is not rigorously homogeneous on a microscopic scale; for instance, the internal charge density has the crystal periodicity, and varies sharply in the unit cell. In general, the response of a crystal to a driving field of wave vector ($\mathbf{q}$) includes not only a plane wave of vector $\mathbf{q}$, but also plane waves of vectors $\mathbf{q} + \mathbf{g}_m$, where $\mathbf{g}_m$ are reciprocal lattice vectors. In the following we adopt the standard simplification of neglecting "local field effects" due to inhomogeneity of the sample; in other words, we assume that if the driving field is characterized by a wave vector $\mathbf{q}$, the same vector characterizes also the response of the system.

The Maxwell's equations in a non-magnetic medium ($\mathbf{B} = \mathbf{H}$), in the absence of external charges and currents, in the presence of internal charge density ($\rho_{ind}$) and internal current density $\mathbf{J}_{ind}$ are (in Gauss units)

$$\boxed{\begin{array}{ll} \text{div}\,\mathbf{E} = 4\pi\,\rho_{ind} & \text{curl}\,\mathbf{E} = -\dfrac{1}{c}\dfrac{\partial \mathbf{B}}{\partial t} \\ \text{div}\,\mathbf{B} = 0 & \text{curl}\,\mathbf{B} = \dfrac{1}{c}\dfrac{\partial \mathbf{E}}{\partial t} + \dfrac{4\pi}{c}\mathbf{J}_{ind} \end{array}} \quad ...(1)$$

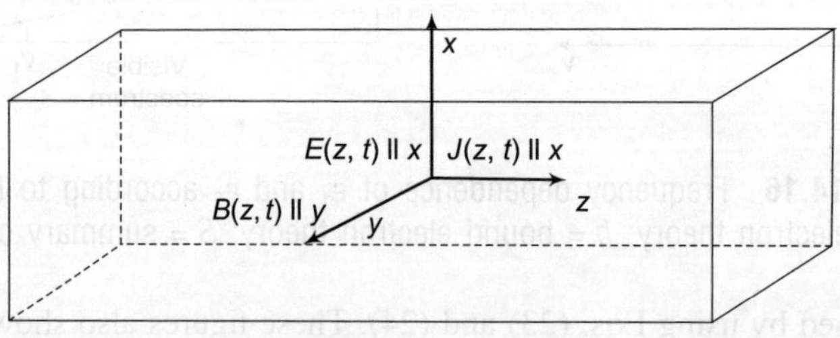


**Figure 14.17** Geometry chosen for the description of transverse electromagnetic field in isotropic materials. The electric field and the internal density current are in the *x*-direction, the magnetic field is in the *y*-direction, the wave propagation is along the *z*-direction

We consider the propagation of *transverse electromagnetic wave of given angular frequency* ($\omega$) *in homogeneous materials*, and specify the Maxwell's equations in the geometry of Fig.14.17. We take the electric field ($\mathbf{E}$) in the $x$-direction, the magnetic field ($\mathbf{B}$) in the

$y$-direction, and the current density ($\mathbf{J}_{ind}$) parallel to the electric field in the isotropic medium; we can write

$$\mathbf{E}(\mathbf{r}, t) = E(z)e^{-i\omega t}\ (1, 0, 0)\ \mathbf{E} \parallel x \qquad ...(2a)$$

$$\mathbf{B}(\mathbf{r}, t) = B(z)e^{-i\omega t}\ (0, 1, 0)\ \mathbf{B} \parallel y \qquad ...(2b)$$

$$\mathbf{J}_{ind}(\mathbf{r}, t) = J(z)e^{-i\omega t}\ (1, 0, 0)\ \mathbf{J}_{ind} \parallel x. \qquad ...(2c)$$

From Eq. (2a) we have div $\mathbf{E} = 0$; combined with the first Maxwell's equation, it entails $\rho_{ind} = 0$. From Eq. (2b) we have div $\mathbf{B} = 0$, consistently with the second Maxwell's equation. The remaining two Maxwell's equations give

$$\frac{dE(z)}{dz} = \frac{i\omega}{c} B(z) \qquad ...(3a)$$

$$-\frac{dB(z)}{dz} = -\frac{i\omega}{c} E(z) + \frac{4\pi}{c} J(z). \qquad ...(3b)$$

By eliminating $B(z)$ we obtain

$$\boxed{\frac{d^2E(z)}{dz^2} = -\frac{\omega^2}{c^2} E(z) - \frac{4\pi i\omega}{c^2} J(z)} \qquad ...(4)$$

This is the basic information, provided by the Maxwell's equations, on the electric field and current density; to proceed further we must establish between them a phenomenological or microscopic constitutive relationship.

### *Local and non-local constitutive relationship for the conductivity*

The simplest and often adopted constitutive equation, linking the current density to the electric field, is a local relationship of the type

$$J(z) = \sigma(\omega)\, E(z) \qquad ...(5)$$

where the frequency-dependent proportionality constant ($\sigma(\omega)$) is called the complex *conductivity constant,* or *complex conductivity function*, of the medium (the time dependence of the fields J and E is assumed to be characterized by a given angular frequency $\omega$). According to Eq. (5), the current density at a given point in the material is proportional to the value of the electric field at the same point: this is the so-called *local-response regime*. A local relationship between current and field can be justified when the average distance travelled by the carriers is small with respect to the length of spatial variation of the electric field; otherwise, one should follow the trajectory of the carriers to find the effect of the spatially varying electric field: this is the so-called *non-local-response regime.*

In the non-local-response regime, the relationship between current density and electric field must be generalized in the form

$$J(z) = \int \sigma(z - z', \omega)\, E(z')\, dz' \qquad ...(6a)$$

where $\sigma(z, z', \omega) = \sigma(z - z', \omega)$ represents the generalized complex conductivity function of the *homogeneous* medium, at a given frequency, $\omega$. If we multiply both members of Eq. (6a) by exp $(iqz)$ and integrate in $dz$, we obtain

$$J(q) = \sigma(q, \omega)\, E(q), \qquad \text{...(6b)}$$

where

$$\sigma(q, \omega) = \int \sigma(z - z', \omega)\, e^{iq(z-z')}\, dz = \int \sigma(z, \omega)\, e^{iqz}\, dz \qquad \text{...(6c)}$$

and, similarly, $J(q)$ and $E(q)$ are the Fourier transforms of $J(z)$ and $E(z)$. Notice that the local-response regime is recovered when the generalized conductivity becomes $\delta$-like function of the form $\sigma(z - z', \omega) = \sigma(\omega)\, \delta(z - z')$; in this case the conductivity $\sigma(q, \omega)$ given by Eq. (6c) is independent of $q$, and one has $\sigma(q, \omega) = \sigma(q = 0, \omega) = \sigma(\omega)$. Thus, the *local-response corresponds to neglecting the wave vector dependence of the conductivity.*

### Local relationship for the conductivity and expression of the optical constants

In the rest of this section, we assume that a local constitutive equation holds between the current density and the electric field. Once the local approximation, Eq. (5) between $J$ and $E$ is assumed, we can easily obtain the other frequency-dependent optical constants (or optical functions), which are commonly used for the description of the optical properties of matter. The basic Eq. (4), together with the assumption of Eq. (5), gives

$$\frac{d^2E(z)}{dz^2} = \frac{\omega^2}{c^2}\left[\frac{1 + 4\pi i\omega\, \sigma(\omega)}{\omega}\right] E(z). \qquad \text{...(7)}$$

The solution of this equation inside the material is a damped (or undamped) wave, which can be written in the form

$$E(z) = E_0 e^{i(\omega/c)N z}, \qquad \text{...(8a)}$$

where the *complex refractive index* $(N(\omega))$ is given by

$$\boxed{N^2 = 1 + \frac{4\pi i\sigma(\omega)}{\omega}} \qquad \text{...(8b)}$$

The magnetic field associated to the electric field, Eq. (8a) can be obtained from Eq. (3a), and is given by

$$B(z) = N\, E_0 e^{i(\omega/c)Nz} = NE(z) \qquad \text{...(8c)}$$

It is customary to write the complex refractive index in the form $N = n + ik$, where $n$ is the ordinary *refractive index* and $k$ the *extinction coefficient*; Eq. (8a) takes the form

$$E(z) = E_0 e^{i(\omega/c)nZ}\, e^{-(\omega/c)kz} \qquad \text{...(9)}$$

From Eq. (9), we see that the velocity of the electromagnetic wave in the medium is $c/n$; the penetration depth (*classical skin depth*, defined as the distance at which the field amplitude drops of $1/e$) is

$$\boxed{\delta(\omega) = \frac{c}{\omega k(\omega)}} \qquad \text{...(10a)}$$

The damping of the electromagnetic wave is related to the absorption of electromagnetic energy. Since the intensity of an electromagnetic field is proportional to $|E(z)|^2$, from Eq. (9) we have $I(z) = I_0 \exp(-2\omega kz/c)$. The absorption coefficient is, thus,

$$\boxed{\alpha(\omega) = \frac{2\omega k(\omega)}{c} \equiv 2/\delta(\omega)} \quad \text{...(10b)}$$

Equation (8b) can also be written in the more effective form, $N^2 = \varepsilon$, where $\varepsilon$ is the *complex dielectric function*. Indicating $\varepsilon = \varepsilon_1 + i\varepsilon_2$, we have

$$\varepsilon_1 = n^2 - k^2 \text{ and } \varepsilon_2 = 2nk;$$

and inversely

$$n^2 = \frac{1}{2}(\varepsilon_1 + \sqrt{\varepsilon_1^2 + \varepsilon_2^2}) \quad \text{and} \quad k^2 = \frac{1}{2}(-\varepsilon_1 + \sqrt{\varepsilon_1^2 + \varepsilon_2^2}).$$

From Eq. (8b), and $N^2 = \varepsilon$, it is seen that the conductivity and the dielectric function satisfy the relation

$$\boxed{\varepsilon(\omega) = 1 + \frac{4\pi i\sigma(\omega)}{\omega}}. \quad \text{...(11)}$$

Separation of the real and imaginary part gives

$$\varepsilon_1(\omega) = 1 - \frac{4\pi\sigma_2(\omega)}{\omega} \text{ and } \varepsilon_2(\omega) = \frac{4\pi\sigma_1(\omega)}{\omega};$$

these equations link the real and imaginary part of the dielectric function to the imaginary and real part of the conductivity.

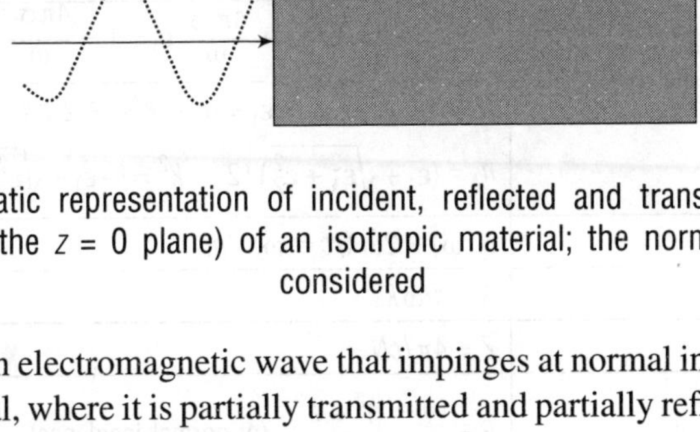


**Figure 14.18** Schematic representation of incident, reflected and transmitted electromagnetic waves at the surface (the $z = 0$ plane) of an isotropic material; the normal incident situation is considered

We consider now an electromagnetic wave that impinges at normal incidence on the surface of an isotropic material, where it is partially transmitted and partially reflected. In the geometry of Fig. 14.18 we have for the spatial part

$$E(z) = E_t e^{i(\omega/c)N x} \quad z > 0 \qquad \text{...(12a)}$$

$$E(z) = E_i e^{i(\omega/c)z} + E_r\, e^{-i(\omega/c)z} \quad z < 0. \qquad \text{...(12b)}$$

From Eq. (12), the continuity condition at $z = 0$ of the electric field $E(z)$ parallel to the surface gives $E_t = E_i + E_r$. From Eq. (3a) and Eqs.(12), the continuity condition of the $y$-components of the magnetic field gives $N\,E_t = E_i - E_r$. The ratio $E_r/E_i$ equals $(1 - N)/(1 + N)$; thus *the reflectivity R at normal incidence* is

$$R = \left|\frac{E_r}{E_i}\right|^2 = \left|\frac{1-N}{1+N}\right|^2 = \frac{(n-1)^2 + k^2}{(n+1)^2 + k^2}. \qquad \text{...(13)}$$

Another convenient measurable quantity for describing a surface is the *surface impedance* Z defined as

$$Z = \frac{4\pi}{c}\,\frac{E(0)}{B(0)}. \qquad \text{...(14a)}$$

From Eq. (3a) and Eq. (8a) we have also

$$Z = \frac{4\pi i\omega}{c^2}\left(\frac{E(z)}{dE(z)/dz}\right)_{z=0} = \frac{4\pi}{cN}. \qquad \text{...(14b)}$$

The relationships between optical constants are summarized in Table 14.5.

Before closing, it is very instructive to consider the implications of the occurrence $\varepsilon_1(\omega) > 0$ and $\varepsilon_2(\omega) = 0$ for some given $\omega$ region; then $n(\omega) = \sqrt{\varepsilon_1(\omega)}$ and $k(\omega) = 0$. From Eqs. (2) and Eq. (9), it is seen that the medium can sustain *undamped transverse waves* (constituted by an electromagnetic field and an induced current density) of frequency ($\omega$) and propagation vector, **q**, with dispersion relation

$$q = \frac{\omega}{c}\,n, \quad \text{or equivalently} \quad \omega = \frac{c}{n}\,q = \frac{c}{\sqrt{\varepsilon_1(\omega)}}\,q;$$

**Table 14.5** Relationships between Optical Constants

| | |
|---|---|
| Conductivity | $\sigma = \sigma_1 + i\,\sigma_2$ |
| Dielectric constant | $\varepsilon = \varepsilon_1 + i\sigma_2$ $\varepsilon = 1 + \frac{4\pi i\sigma}{\omega}$ $\varepsilon_1 = 1 - \frac{4\pi\sigma_2}{\omega}$ $\varepsilon_2 = \frac{4\pi\sigma_1}{\omega}$ |
| Refractive index | $N = n + ik \cdot \varepsilon = N^2 \varepsilon_1 = n^2 - k^2 \varepsilon_2 = 2\,nk$<br>$n_2 = (\varepsilon_1 + \sqrt{\varepsilon_1^2 + \varepsilon_2^2})/2 \quad k^2 = (-\varepsilon_1 + \sqrt{\varepsilon_1^2 + \varepsilon_2^2})/2$ |
| Absorption coefficient | $\alpha(\omega) = 2\omega k/c \equiv \omega\varepsilon_2/nc$ |
| Classical skin depth | $\delta = c/\omega k$ |
| Surface impedance | $Z = 4\pi/cN$ |
| Reflectivity | $R = \frac{(n-1)^2 + k^2}{(n+1)^2 + k^2}$ (at normal incidence) |

notice that in the local response regime we are considering, it holds $\varepsilon_1(q, \omega) = \varepsilon_1(0, \omega) = \varepsilon_1(\omega)$. The undamped transverse waves of the system can be interpreted as elementary excitations of energy ($\hbar\omega$) and momentum, $\hbar\mathbf{q}$. We should also remember that in the case of longitudinal waves and longitudinal dielectric function, $\phi$ *Plasmon* excitations are inferred from the vanishing of both $\varepsilon_1$ and $\varepsilon_2$.

## 14.15 THE DRUDE THEORY OF THE OPTICAL PROPERTIES OF FREE CARRIERS

Consider a free-electron gas with $n$ carriers per unit volume, each with effective mass ($m$) and charges ($-e$); the carriers are embedded in a uniform background of neutralizing positive charges. The classical equation for the motion of an electron in a dissipative medium in the presence of an electric field $\mathbf{E}(\mathbf{r}, t) = \mathbf{E}_0 \exp(i\mathbf{q}\cdot\mathbf{r} - i\omega t)$ of frequency ($\omega$) and wave vector ($\mathbf{q}$) is

$$m\,\ddot{\mathbf{r}} = -\frac{m}{r}\,\dot{\mathbf{r}} + (-e)\,\mathbf{E}_0 e^{i(\mathbf{q}\cdot\mathbf{r} - \omega t)} \qquad ...(1)$$

where $\mathbf{r}(t)$ is the coordinate of the particle, and $\tau$ is a phenomenological relaxation time. The viscous damping term $-(m/\tau)\dot{\mathbf{r}}$ is introduced without specifying the dissipative mechanisms; in a wide sense it is due to the random collisions between the electron and whatever kind of impurities, phonons and imperfections in the crystal. [Strictly speaking, the second term in the right-hand side of Eq. (1) should be accompanied by its complex conjugate, so that the electric field acting on the electron is real; for simplicity, we omit the complex conjugate term, since nothing would change either in the formal treatment or in the final results].

The term *spatial dispersion* has been coined to refer to the wave vector dependence of the electric field and optical constants. Suppose the spatial excursions of $\mathbf{r}(t)$, in the whereabouts of some point $\mathbf{r}_0$, are much smaller than the wavelength of the driving field; in this case, in the exponential in the right-hand side of Eq. (1) we can replace $\mathbf{r}(t)$ with $\mathbf{r}_0$ (without loss of generality we can take $\mathbf{r}_0 \equiv 0$), and we obtain

$$m\,\ddot{\mathbf{r}} = -\frac{m}{\tau}\,\dot{\mathbf{r}} - e\,\mathbf{E}_0 e^{-i\omega t} \qquad ...(2)$$

*We give now a criterion to establish when the electron dynamics can be described by the simple Eq. (2), which neglects the spatial dispersion of the oscillating electric field.* Consider first the case $\omega\tau \ll 1$, corresponding to a relaxation time ($\tau$) much smaller than the period, $T = 2\pi/\omega$ of the electromagnetic field. The displacement of an electron with typical Fermi velocity ($u_F$) between two collisions is about the mean free path, $\Lambda_F = v_F\tau$; the spatial dispersion of the electromagnetic field of wavelength, $\lambda = 2\pi/q$, is irrelevant if $\Lambda_F \ll \lambda$, or equivalently

$$v_F\tau \ll \frac{1}{q}\ (\omega\tau \ll 1). \qquad ...(3a)$$

In the case $\omega\tau \gg 1$, the displacement of an electron with velocity ($v_F$) during the period $T = 2\pi/\omega$ of the electromagnetic field is $\approx v_F/\omega$; when $\omega\tau \gg 1$ the spatial dispersion of the electromagnetic field becomes irrelevant if

$$v_F\,\frac{1}{\omega} \ll \frac{1}{q}\ (\omega\tau \gg 1). \qquad ...(3b)$$

We can thus conclude that Eq. (2) is justified when condition (3a), or condition (3b), is verified.

The conditions (3), here obtained on the basis of simple qualitative considerations, are confirmed by the study of the optical properties of metals with the Boltzmann equation. Such a study shows that the spatial dispersion of the electromagnetic field can be neglected provided the dimensionless parameter

$$s(q, \omega) = \frac{iqv_F\tau}{1 - i\omega t} \qquad ...(4)$$

satisfies the condition

$$|s(q, \omega)| \ll 1 \qquad ...(5)$$

the equivalence of Eq. (5) with Eqs. (3a) and (3b) can be seen by inspection.

Having established the criterion Eq. (5) for the validity of Eq. (2), we insert $\mathbf{r}(t)$ in the form $\mathbf{r}(t) = \mathbf{A}_0 \exp(-i\omega t)$ into Eq. (2) and obtain

$$\mathbf{A}_0 = \frac{e\tau}{m} \frac{1}{\omega(i + \omega\tau)} \mathbf{E}_0.$$

The free-carrier contribution to the current density is

$$\mathbf{J} = n(-e)\, \dot{\mathbf{r}} = n(-e)(-i\omega)\, A_0 e^{-i\omega t} = \frac{ne^2\tau}{m} \frac{1}{1 - i\omega\tau} \mathbf{E}_0 e^{-i\omega t}$$

The frequency dependent complex conductivity is thus

$$\sigma(\omega) = \frac{ne^2\tau}{m} \frac{1}{1 - i\omega\tau} = \sigma_0 \frac{1}{1 - i\omega t} \qquad ...(6)$$

where $\sigma_0 = ne^2 r/m$ is the static conductivity. Notice that in general the conductivity $\sigma(q, \omega)$ should depend both on the frequency ($\omega$) and on the wave vector $q$ of the driving electric field; the Drude theory neglects spatial dispersion and thus provides only $\sigma(q \to 0, \omega)$, also denoted $\sigma(0, w)$ or simply $\sigma(\omega)$.

From Eq. (6) and Eq. (11) (section 14.14), the complex dielectric constant becomes

$$\boxed{\varepsilon(w) = 1 - \frac{\omega_p^2}{\omega(\omega + i/\tau)}} \qquad ...(7)$$

where $\omega_p$ denotes the free-electron plasma frequency

$$\omega_p^2 = \frac{4\pi ne^2}{m} \qquad ...(8)$$

and $n$ is the electron density. From Eq. (7) the real and imaginary parts of the dielectric function are

$$\boxed{\varepsilon_1(\omega) = 1 - \frac{\omega_p^2\tau^2}{1 + \omega^2\tau^2}} \quad \text{and} \quad \boxed{\varepsilon_2(\omega) = \frac{\omega_p^2\tau}{\omega(1 + \omega^2 r^2)}} \qquad ...(9)$$

Inserting into Eq. (8) the dimensionless parameter, $r_s$, related to the electron density by the expression $(4/3)\,\pi r_s^3\, a_B^3 = 1/n$, we obtain

$$\hbar^2\omega_p^2 = \frac{12}{r_s^3}\,\frac{\hbar^2}{2ma_B^2}\,\frac{e^2}{2a_B} \quad \text{and} \quad \hbar\omega_p = \sqrt{\frac{12}{r_s^3}}\ \text{Rydberg.}$$

For ordinary metals $2 < r_s < 6$, typical values of $\hbar\omega_p$ are in the range 3 – 17 eV.

For relaxation times we have typically $\tau \approx 10^{-14}$ sec, and the quantity $\gamma = \hbar/\tau$ is of the order of 0.1 eV.

In the study of the optical properties of free carriers in metals, we can roughly distinguish three frequency regions, called non-relaxation region ($\omega \ll 1/\tau$), relaxation region ($1/\tau \ll \omega \ll \omega_p$) and ultraviolet region ($\omega \approx \omega_p$ and $\omega > \omega_p$).

### Non-relaxation region ($\omega\tau \ll 1 \ll \omega_p\tau$ or equivalently $\hbar\omega \ll \gamma \ll \hbar\omega_p$)

In the non-relaxation region ($\omega\tau$) is negligible with respect to 1, and the dielectric function (9) becomes

$$\varepsilon_1(\omega) \approx -\omega_p^2\tau^2 \quad \text{and} \quad \varepsilon_2(\omega) \approx \frac{\omega_p^2\tau}{\omega}.$$

We see that $\varepsilon_1(\omega)$ is negative and constant, while $\varepsilon_2(\omega)$ is singular for $\omega \approx 0$. Notice that $|\varepsilon_2| \gg |\varepsilon_1|$.

From the relation $N^2 = E = iE_2$, it follows $N \approx \varepsilon_2(1 + i)/\sqrt{2}$. The refractive index $n$ and the extinction coefficient $k$ are almost equal and given by

$$n(\omega) \approx \kappa(\omega) \approx \sqrt{\frac{\varepsilon_2(\omega)}{2}} = \sqrt{\frac{\omega_p^2\,\tau}{2\omega}}.$$

Thus, also $n(\omega)$ and $k(\omega)$ are singular for $\omega \approx 0$. Since $n(\omega) \gg 1$ the reflectivity is almost 1 and metal is strongly reflecting.

The penetration depth is

$$\delta(\omega) = \frac{c}{\omega k(\omega)} \approx \frac{c}{\omega_p}\sqrt{\frac{2}{\omega\tau}} \gg \frac{c}{\omega_p}.$$

To estimate the penetration depth in ordinary metals, assume as typical values $\tau \approx 10^{-14}$ sec, $\hbar\omega_p \approx 10$ eV and then $\omega_p \approx 10^{16}$ rad/sec ($\hbar = 6.852\cdot 10^{16}$ eV·Sec). The Debye length, defined as $c/\omega_p$, is of the order of $10^{-6}$ cm; for $\omega \approx 100$ rad/sec the penetration depth is of the order of one cm; the penetration depth is of the order of the Debye length at the high frequency limit of the non-relaxation region.

### Relaxation Region ($1 \ll \omega\tau \ll \omega_p\tau$ or equivalently $\gamma \ll \hbar\omega \ll \hbar\omega_p$)

When $\omega\tau \gg 1$, we can perform a series expansion of $(1 + i/\omega\tau)^{-1}$ in Eq. (7) and the dielectric function becomes

$$\varepsilon(\omega) \approx 1 - \frac{\omega_p^2}{\omega^2} + i\frac{\omega_p^2}{\omega^3\tau} \qquad ...(10)$$

Thus, $\varepsilon_1(\omega) \approx -\omega_p^2/\omega^2$ is still negative, and the imaginary part of the dielectric function is much smaller than the real part. From $N^2 = (n + ik)^2 = \varepsilon_1 + i\varepsilon_2$, and taking into account that $\varepsilon_1 < 0$ and $|\varepsilon_2| \ll |\varepsilon_1|$, we obtain

$$\kappa(\omega) \approx \sqrt{-\varepsilon_1} = \frac{\omega_p}{\omega} \text{ and } n(\omega) \approx \frac{1}{2}\frac{\varepsilon_2}{\sqrt{-\varepsilon_1}} = \frac{1}{2}\frac{\omega_p}{\omega^2\tau}.$$

The above relation shows that $k(\omega) \gg n(\omega) \gg 1$; since the refractive index $n(\omega)$ is still much larger than 1, the reflectivity is almost 1 and the metal is strongly reflecting also in the relaxation region. The penetration depth is now approximately a constant, given by the Debye length.

***Ultraviolet region ($\omega \approx \omega_p$ and $\omega > \omega_p$)***

In this region, Eq. (9) for the dielectric function read

$$\varepsilon_1(\omega) \approx 1 - \frac{\omega_P^2}{\omega^2} \text{ and } \varepsilon_2(\omega) = \frac{\omega_p^2}{\omega^3\tau} \approx 0.$$

We see that $\varepsilon_1(\omega)$ is positive for $\omega \geq \omega_p$; the reflectivity changes from (almost) one to (almost) zero, when $\omega$ increases above the plasma frequency, and the metal becomes transparent for $\omega > \omega_p$. The schematic behaviour of the dielectric function of a metal in the Drude model is given in Fig. 14.19.

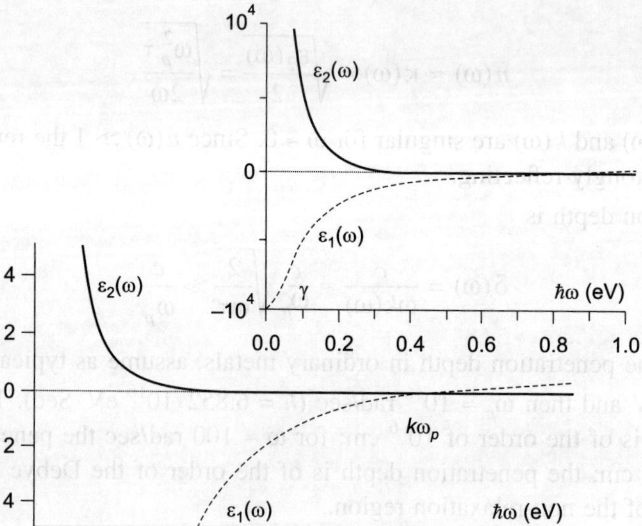


**Figure 14.19** Behaviour of real and imaginary part of the dielectric function of a free-electron gas with the Drude model; we have taken $\hbar\omega_p$ = 10 eV and $\gamma = \hbar/\tau$ = 0.1 eV

By gathering data about any optical process, we can gain information about the inner workings of the solid.

## 14.16 MACROSCOPIC PROPERTIES

We start here by relating the dielectric properties to optical properties, particularly those involving absorption and reflection. The complex dielectric constant and the relation of its two components by the Kronig–Kramers relation, is particularly important. The imaginary part relates to the absorption coefficient. We assume the total charge density $P_{\text{total}} = 0$, $\vec{j} = \sigma\vec{E}$ and $\mu = \mu_0$ (no internal magnetic effects). We assume a wavelength large compared with atomic dimensions but small compared with the dimensions of the sample. We start with Maxwell's equations (in SI units):

$$\nabla \cdot \vec{E} = 0,\ \nabla \times \vec{E} = -\mu_0 \frac{\partial \vec{H}}{\partial t} = -\frac{\partial \vec{B}}{\partial t} \quad ...(1)$$

$$\nabla \cdot \vec{B} = 0,\ \nabla \times \vec{H} = \vec{j} + \frac{\partial \vec{D}}{\partial t}$$

$$\vec{D} = \epsilon_0 \vec{E} + \vec{P} = \varepsilon \vec{E} \quad ...(2)$$

$$\vec{B} = \mu_0 (\vec{H} + \vec{M}) = \mu \vec{H}$$

From Eq. (1)

$$\nabla \times \frac{\partial \vec{H}}{\partial t} = \frac{\partial \vec{j}}{\partial t} + \frac{\partial^2 \vec{D}}{\partial t^2} \quad ...(3)$$

$$\nabla \times (\nabla \times \vec{E}) = \mu_0 \frac{\partial}{\partial t}(\nabla \times \vec{H})$$

$$= -\mu_0 \frac{\partial}{\partial t}\left(\vec{j} + \frac{\partial \vec{D}}{\partial t}\right)$$

or

$$\nabla(\nabla \cdot \vec{E}) - \nabla^2 \vec{E} = -\mu_0 \frac{\partial \vec{j}}{\partial t} - \mu_0 \frac{\partial^2 \vec{D}}{\partial t^2}$$

or

$$\nabla^2 \vec{E} = \mu_0 \sigma \frac{\partial \vec{E}}{\partial t} + \mu_0 \frac{\partial^2}{\partial t^2}(\varepsilon \vec{E}) \quad ...(4)$$

$$(\because\ \vec{j} = \sigma \vec{E}).$$

We now look for solutions of Eq. (4) for each Fourier component

$$\vec{E}(k, \omega) = \vec{E}_0 \exp[i(\vec{k} \cdot \vec{r} - \omega t)] \quad ...(5)$$

keeping in mind $\varepsilon \equiv \varepsilon(\vec{k}, \omega)$.

Substituting Eq. (5) in Eq. (4), we get

$$i^2 k^2 = -\mu_0 \sigma (i\omega) + (-i\omega\mu_0)(-i\omega\mu_0)\varepsilon$$

or $$k^2 = \mu_0 (\varepsilon\omega^2 + i\,\sigma\,\omega) \qquad \text{...(6a)}$$

or, since $c = 1/(\mu_0\varepsilon_0)^{1/2}$, we have

$$k = \frac{\omega}{c}\left(\frac{\varepsilon}{\varepsilon_0} + i\frac{\sigma}{\varepsilon_0\,\omega}\right)^{1/2} \qquad \text{...(6b)}$$

For an insulator, $\sigma = 0$, so

$$k = \frac{\omega}{v} = \frac{\omega n}{c} = \frac{\omega}{c}\sqrt{\frac{\varepsilon}{\varepsilon_0}} \qquad \text{...(6c)}$$

where $n$ is the index of refraction. It is then natural to define a complex dielectric constant ($\varepsilon_c$) and a complex index of refraction ($n_c$) so,

$$k = \frac{\omega}{c}\,n_c \qquad \text{...(7)}$$

where $$n_c = \sqrt{\frac{\varepsilon}{\varepsilon_0} + i\frac{\sigma}{\varepsilon_0\omega}} = n + in_i \qquad \text{...(8)}$$

Letting $$\varepsilon_c = \varepsilon_r + i\varepsilon_i = \frac{\varepsilon(k,\omega)}{\varepsilon_0} + i\,\frac{\sigma}{\varepsilon_0\omega} \qquad \text{...(9)}$$

Squaring both sides and equating real and imaginary parts, we find

$$\varepsilon_r(k,\omega) = n^2 - n_i^2 \qquad \text{...(10)}$$

and $$\varepsilon_i(k,\omega) = 2nn_i \qquad \text{...(11)}$$

Now, assuming the wave propagates in the $z$ direction, if we substitute

$$k = \frac{\omega}{c}(n + in_i) \qquad \text{...(12)}$$

we have $$E = E_0 \exp\left[i\omega\left(\frac{nz}{c} - t\right)\right]\exp\left(-\frac{\omega}{c}n_i z\right) \qquad \text{...(13)}$$

So, since energy in the wave is proportional to $|E|^2$, we have that the absorption coefficient is given by

$$\alpha = \frac{2n_i\omega}{c} \qquad \text{...(14)}$$

Another readily measured quantity can be related to $n$ and $n_i$. If we apply appropriate boundary conditions to a solid surface, we can show as noted below that the reflection coefficient for normal incidence is given by

$$R = \frac{(n-1)^2 + n_i^2}{(n+1)^2 + n_i^2} \qquad \text{...(15)}$$

This relation follows directly from the Maxwell relations. From Faraday's law, we can show that the tangential component of $E$ is continuous, and from Ampere's law we can show the tangential component of $H$ is continuous. Further manipulation leads to the desired relation. Let us work this out. For normal incidence from the vacuum on a surface at $z = 0$, the incident and reflected waves can be written as

$$E_{i+r} = E_1 \exp\left[i\omega\left(\frac{z}{c} - t\right)\right] + E_2 \exp\left[-i\omega\left(\frac{z}{c} + t\right)\right] \quad ...(16)$$

and the refracted wave is given by

$$E_{rf} = E_0 \exp\left[i\omega\left(n_c \frac{z}{c} - t\right)\right] \quad ...(17)$$

where $n_c$ is the complex index of refraction. Since

$$\nabla \times E + \frac{\partial \boldsymbol{B}}{\partial t} = 0 \quad ...(18)$$

we can use the loop of Fig. 14.20 to write

$$\int_A (\nabla \times E) \cdot ds + \int \frac{\partial \boldsymbol{B}}{\partial t} \cdot ds = 0 \quad ...(19)$$

or $$\int_C \boldsymbol{E} \cdot d\boldsymbol{r} = -\frac{d}{dt} \int B \cdot ds \quad ...(20)$$

as $$(E_T^1 - E_T^2) l + O(\varepsilon) = -\frac{d}{dt}(B_\perp 2l\varepsilon), \quad ...(21)$$

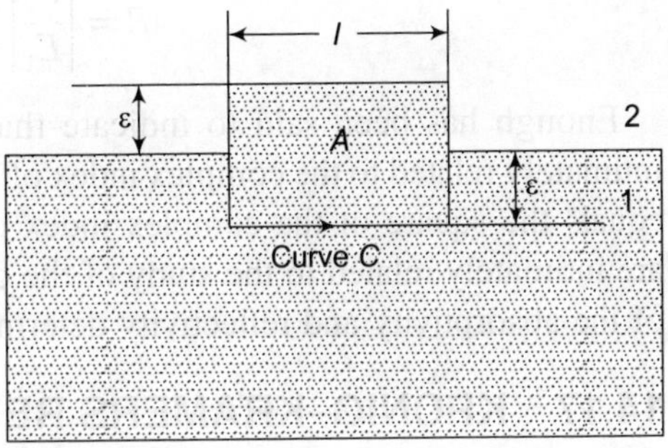


**Figure 14.20** Loop used for deriving field boundary conditions (notice this ε is a distance)

where the subscript $T$ means the tangential component of the electric field, and the subscript $\perp$ means perpendicular to the page of the paper. Taking the limit as $\varepsilon \to 0$, we obtain

$$E_T^1 - E_T^2 = 0 \quad ...(22)$$

or the tangential component of $E$ is continuous. In a similar way we can use

$$\nabla \times \boldsymbol{H} = \boldsymbol{j} + \frac{\partial \boldsymbol{D}}{\partial t} \quad ...(23)$$

to show that

$$(H_T^1 - H_T^2) l + O(\varepsilon) = \int_A \boldsymbol{j} \cdot ds + \frac{d}{dt}(D_\perp 2l\varepsilon) = j_\perp (2l\varepsilon) + \frac{d}{dt}(D_\perp 2l\varepsilon). \quad ...(24)$$

Again taking the limit as $\varepsilon \to 0$, we find

$$(H_T^1 - H_T^2) = 0 \quad ...(25)$$

or that the tangential component of $H$ is also continuous. Continuity of the tangential component of $H$ requires (using $\nabla \times \boldsymbol{E} = -\partial \boldsymbol{B}/\partial t$, proper constitutive relations, and (16) and (17))

$$n_c E_0 = E_1 - E_2. \qquad \text{...(26)}$$

Continuity of the tangential component of $E$ requires Eqs. (16) and (17)

$$E_0 = E_1 + E_2. \qquad \text{...(27)}$$

Adding these two equations give

$$E_1 = \frac{E_0\,(n_c + 1)}{2}. \qquad \text{...(28)}$$

Subtracting these equations give

$$E_2 = \frac{E_0\,(-n_c + 1)}{2}. \qquad \text{...(29)}$$

Thus, the reflection coefficient is given by

$$R = \left|\frac{E_2}{E_1}\right|^2 = \left|\frac{1 - n_c}{1 + n_c}\right|^2 = \left|\frac{(n-1)^2 + n_i^2}{(n+1)^2 + n_i^2}\right| \qquad \text{...(30)}$$

Enough has been said to indicate that the theory of the optical properties of solids is intimately related to the complex index of refraction of solids. The complex dielectric constant equals the square of the complex index of refraction. Thus, the optical properties of solids are intimately related to the study of the dielectric properties of solids, and the measurement of the absorptivity and reflectivity determine $n$ and $n_i$, and hence, $\varepsilon_r$ and $\varepsilon_i$.

## 14.17 KRONIG–KRAMERS RELATIONS (A)

We will give a quantum description of the absorption of radiation, but first it is helpful to derive the Kronig–Kramers equations, which give a relation between the real and imaginary parts of the dielectric constant. Let $\varepsilon$ be a complex function of $\omega$ that coverages in the upper half-plane. We need to define the Cauchy principal value $P$ with $a$ real for the following equations and diagrams

$$P\int_{-\infty}^{\infty} \frac{\varepsilon(\omega)\,d\omega}{\omega - a} = \frac{1}{2}\left[\int_{C'} \frac{\varepsilon(\omega)\,d\omega}{\omega - a} + \int_{C''} \frac{\varepsilon(\omega)\,d\omega}{\omega - a}\right] \qquad \text{...(31)}$$

as shown by Fig. 14.21. It is assumed that the integral over the large semicircles is zero. From complex variables, we know that if $C$ encloses $a$ and if $f$ has no singularity in $C$, then

$$\oint_C \frac{f(Z)\,dZ}{Z - a} = 2\pi_i f(a). \qquad \text{...(32)}$$

Using the definition of Cauchy principal value, since we have the integral

$$\int_{C''} = 0 \quad P \to \frac{1}{2}\left(\int_{C'} - \int_{C''}\right). \qquad \text{...(33)}$$

Thus,

$$P\int \frac{\varepsilon(\omega)\,d\omega}{\omega - a} = \frac{1}{2} \int_{\substack{\text{small circle} \\ \rho \to 0}} \frac{\varepsilon(\omega)\,d\omega}{\omega - a} = i\pi\varepsilon(a) \qquad \text{...(34)}$$

and we have used that $\varepsilon(\omega)$ on the big circle is zero (actually, to achieve this we should use that $\varepsilon(\omega) = [\varepsilon_r(\omega) - 1] + i\varepsilon_i(\omega)$, which we will put in explicitly at the end). Taking real and imaginary parts we then have,

$$P\int_{-\infty}^{\infty} \frac{\mathrm{Re}[\varepsilon(\omega)]\,d\omega}{\omega - a} = -\pi\,\mathrm{Im}[\varepsilon(a)] \qquad ...(35)$$

and

$$P\int_{-\infty}^{\infty} \frac{\mathrm{Im}[\varepsilon(\omega)]\,d\omega}{\omega - a} = +\pi\,\mathrm{Re}[\varepsilon(a)]. \qquad ...(36)$$

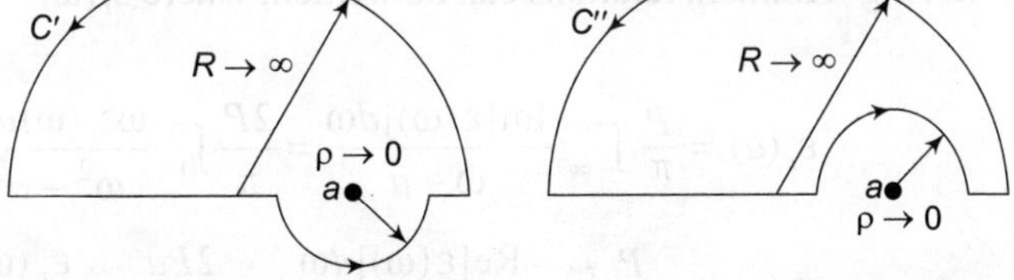


**Figure 14.21** Contours used for Cauchy principal value

There are some other ways to write these relationships,

$$\varepsilon_r(\omega) = \frac{1}{\pi} P\int_{-\infty}^{\infty} \frac{\varepsilon_i(\omega)}{\omega - a}\,d\omega = \frac{1}{\pi} P\left[\int_0^{\infty} \frac{\varepsilon_i(\omega)}{\omega - a}\,d\omega + \int_{-\infty}^{0} \frac{\varepsilon_i(\omega)}{\omega - a}\,d\omega\right] \qquad ...(37)$$

But, the second term can be written

$$-\int_0^{-\infty} \frac{\varepsilon_i(\omega)}{\omega - a}\,d\omega = \int_0^{-\infty} \frac{[-\varepsilon_i(\omega)]}{-\omega + a}\,d(-\omega) = \int_0^{\infty} \frac{[-\varepsilon_i(-\omega)]}{\omega + a}\,d\omega \qquad ...(38)$$

and $\varepsilon^*(\boldsymbol{r}, t) = \varepsilon(\boldsymbol{r}, t)$, so $\varepsilon(-q, -\omega) = \varepsilon^*(q, \omega)$. Therefore,

$$\varepsilon^*(-\omega) = \varepsilon(\omega) \qquad ...(39)$$

or

$$\varepsilon_r(-\omega) = \varepsilon_r(\omega) \qquad ...(40)$$

and

$$\varepsilon_i(-\omega) = -\varepsilon_i(\omega) \qquad ...(41)$$

We get

$$\frac{1}{\pi} P\int_{-\infty}^{0} \frac{\varepsilon_i(\omega)}{\omega - a}\,d\omega = \int_0^{\infty} \frac{\varepsilon_i(\omega)}{\omega + a}\,d\omega. \qquad ...(42)$$

We can thus write the real component of the dielectric constant as

$$\varepsilon_r(a) = \frac{P}{\pi}\left[\int_{-\infty}^{0} \frac{\varepsilon_i(\omega)\,d\omega}{\omega - a} + \int_0^{\infty} \frac{\varepsilon_i(\omega)\,d\omega}{\omega + a}\right] = \frac{2P}{\pi}\int_0^{\infty} \frac{\omega\varepsilon_i(\omega)\,d\omega}{\omega^2 - a^2} \qquad ...(43)$$

and similarly the imaginary component can be written

$$\varepsilon_i(a) = -\frac{P}{\pi}\int_{-\infty}^{\infty} \frac{\varepsilon_r\,d\omega}{\omega - a} = -\frac{P}{\pi}\left[\int_0^{\infty} \frac{\varepsilon_r\,d\omega}{\omega - a} + \int_{-\infty}^{0} \frac{\varepsilon_r\,d\omega}{\omega - a}\right]$$

$$= \frac{P}{\pi}\left[\int_0^{\infty}\frac{\varepsilon_r d\omega}{\omega - a} + \int_{-\infty}^{0}\frac{\varepsilon_r(-\omega)\,d(-\omega)}{-\omega + a}\right] \qquad ...(44)$$

$$= -\frac{P}{\pi}\left[\int_0^{\infty}\frac{\varepsilon_r(\omega)d\omega}{\omega - a} + \int_{\infty}^{0}\frac{\varepsilon_r(\omega)\,d\omega}{\omega + a}\right]$$

$$= -\frac{P}{\pi}\left[\int_0^{\infty}\frac{\varepsilon_r(\omega)d\omega}{\omega - a} - \int_{\infty}^{0}\frac{\varepsilon_r(\omega)\,d\omega}{\omega + a}\right] = -\frac{2aP}{\pi}\int_0^{\infty}\frac{\varepsilon_r(\omega)d\omega}{\omega^2 - a^2}.$$

In summary, the Kronig–Kramers relations can be written, where $\varepsilon_r(\omega) \to \varepsilon_r(\omega) - 1$ should be substituted

$$\varepsilon_r(a) = \frac{P}{\pi}\int_{-\infty}^{\infty}\frac{\text{Im}[\varepsilon(\omega)]d\omega}{\omega - a} = \frac{2P}{\pi}\int_0^{\infty}\frac{\omega\varepsilon_i(\omega)d\omega}{\omega^2 - a^2} \qquad ...(45)$$

$$\varepsilon_i(a) = -\frac{P}{\pi}\int_{-\infty}^{\infty}\frac{\text{Re}[\varepsilon(\omega)]d\omega}{\omega - a} = -\frac{2Pa}{\pi}\int_0^{\infty}\frac{\varepsilon_r(\omega)d\omega}{\omega^2 - a^2} \qquad ...(46)$$

## 14.18 ABSORPTION OF ELECTROMAGNETIC RADIATION (QUANTUM APPROACH)

We now give a fairly general discussion of the absorption process by quantum mechanics. Although much of the discussion is more general, we have in mind the absorption due to transitions between the valence and conduction bands of semiconductors. If $-e$ is the electronic charge, and if we assume the electromagnetic field is described by a vector potential $A$ and a scalar potential $\phi$, the Hamiltonian describing the electron in the field is in SI

$$\mathcal{H} = \frac{1}{2m}[\mathbf{p} + e\mathbf{A}]^2 - e(\phi + V) \qquad ...(1)$$

where $V$ is the potential in the absence of an electromagnetic field; $V$ would be a periodic potential if the electron were in a solid. We will use the Coulomb gauge to describe the electromagnetic field so $\phi = 0$, $\nabla\cdot\boldsymbol{A} = 0$ and the fields are given by

$$\boldsymbol{E} = -\frac{\partial \boldsymbol{A}}{\partial t},\ \boldsymbol{B} = \nabla\times A \qquad ...(2)$$

The Hamiltonian can then be written

$$\mathcal{H} = \frac{1}{2m}[p^2 + e\boldsymbol{A}\cdot\boldsymbol{p} + e\boldsymbol{p}\cdot\boldsymbol{A} + e^2A^2] - \text{eV} \qquad ...(3)$$

The terms quadratic in $A$ will be ignored as they are normally small compared to the terms linear in $A$. Further in the Coulomb gauge, we can write

$$\boldsymbol{p}\cdot\boldsymbol{A}\,\psi \propto \nabla\cdot(\boldsymbol{A}\psi) = (\nabla\cdot\boldsymbol{A})\psi + (\boldsymbol{A}\cdot\nabla)\psi = \boldsymbol{A}\cdot\nabla\psi, \qquad ...(4)$$

so that the Hamiltonian can be written

$$\mathcal{H} = \mathcal{H}_0 + \mathcal{H}',\ \mathcal{H}_0 = \frac{p^2}{2m} - \text{eV};$$

where the perturbation is

$$\mathcal{H}' = \frac{e}{m}\boldsymbol{A}\cdot\boldsymbol{P}. \qquad ...(5)$$

We assume the matrix element responsible for electronic transitions will be in the form $\langle f|H|i\rangle$, where $i$ and $f$ refer to the initial and final electron states and $\mathcal{H}'$ is the perturbing Hamiltonian. We assume the vector potential is given by

$$\boldsymbol{A}(\boldsymbol{r}, t) = a\boldsymbol{e}\{\exp[i(\boldsymbol{k}\cdot\boldsymbol{r} - \omega t)] + \exp[-i(\boldsymbol{k}\cdot\boldsymbol{r} - \omega t)]\} \qquad ...(6)$$

where $\boldsymbol{e}\cdot\boldsymbol{k} = 0$ and $a^2$ is given by

$$a = \sqrt{\overline{E^2}/2\omega^2} \qquad ...(7)$$

where $\overline{E^2}$ is the averaged squared electric field. Then,

$$P_{i\to f}^{\text{absorption}} = \frac{2\pi}{\hbar}a^2\frac{e^2}{m^2}\left|\langle f|\exp(i\boldsymbol{k}\cdot\boldsymbol{r})\boldsymbol{e}\cdot\boldsymbol{p}|i\rangle\right|^2 \qquad ...(8)$$

and for emission

$$P_{i\to f}^{\text{absorption}} = \frac{2\pi}{\hbar}a^2\frac{e^2}{m^2}\left|\langle f|\exp(-i\boldsymbol{k}\cdot\boldsymbol{r})\boldsymbol{e}\cdot\boldsymbol{p}|i\rangle\right|^2 \qquad ...(9)$$

## 14.19 DIRECT AND INDIRECT ABSORPTION COEFFICIENTS (B)

Let us now look at the absorption coefficient. Using Bloch wave functions ($\psi_k = e^{i\boldsymbol{k}\cdot\boldsymbol{r}}u(\boldsymbol{r})$), we have

$$\langle f|\exp(i\boldsymbol{k}\cdot\boldsymbol{r})\boldsymbol{e}\cdot\boldsymbol{p}|i\rangle = \int u_f^* \exp[i(\boldsymbol{k} - \boldsymbol{k}_f + \boldsymbol{k}_i)\cdot\boldsymbol{r}]\hbar\boldsymbol{e}\cdot\boldsymbol{k}_i u_i\, d\Omega$$

$$+ \int u_f^* \exp[i(\boldsymbol{k} - \boldsymbol{k}_f + \boldsymbol{k}_i)\cdot\boldsymbol{r}]\hbar\boldsymbol{e}\cdot\boldsymbol{p}u_i\, d\Omega. \qquad ...(10)$$

The first integral can be written as proportional to

$$\int \psi_f^*\psi_i \exp(i\boldsymbol{k}\cdot\boldsymbol{r})d\Omega = \sum_j \exp[i(\boldsymbol{k} - \boldsymbol{k}_f + \boldsymbol{k}_i)\cdot\boldsymbol{R}_j]$$

$$\times\int_{\Omega_c} u_f^* \exp[i(\boldsymbol{k} - \boldsymbol{k}_f + \boldsymbol{k}_i)\cdot\boldsymbol{r}]\,u_i d\Omega \qquad ...(11)$$

$$\cong N\int_{\Omega_c} \psi_f\psi_i d\Omega \cong 0$$

by orthogonality and assuming $k$ is approximately zero, where we have also used

$$\sum_j \exp[i(\boldsymbol{k} - \boldsymbol{k}_f + \boldsymbol{k}_i)\cdot\boldsymbol{R}_j] = \delta_{\boldsymbol{k}}^{\boldsymbol{k}_f - \boldsymbol{k}_i}(N) \qquad ...(12)$$

and $\Omega_c$ is the volume of a unit cell. The neglect of all terms but the $k = 0$ terms (called the electric dipole approximation) allows a similar description of the emission term. Following a similar procedure for the second term in Eq. (10), we obtain for absorption,

$$\langle f \,|\, \exp(i\boldsymbol{k}\cdot\boldsymbol{r})\boldsymbol{e}\cdot\boldsymbol{p} \,|\, i\rangle = N\int_{\Omega_c} u_f^* \,\boldsymbol{e}\cdot\boldsymbol{p}u_i d\Omega \qquad ...(13)$$

with $k = 0$ and $\boldsymbol{k}_i = \boldsymbol{k}_f$.

Notice in the electric dipole approximation since $k_i = k_f$, we have what are called *direct optical transitions.* If something else such as phonons is involved, direct transitions are not required but the whole discussion must be modified to include this new physical ingredient. The electric dipole transition probability for photon absorption per unit time is

$$P_{i\to f}^{\text{abs}} = \frac{2\pi}{\hbar}\sum_{k}\frac{\overline{E^2}}{2\omega^2}\frac{e^2N^2}{m^2}\left|\int_{\Omega_c} u_f^* \,\boldsymbol{e}\cdot\boldsymbol{p}\, u_i d\Omega\right|^2 \delta[E_c(\boldsymbol{k}) - E_v(\mathbf{k}) - \hbar\omega] \qquad ...(14)$$

The powder (per unit volume) lost by the field due to absorption in the medium is the transition probability per unit volume $P$ multiplied by the energy of each photon (where in carrying out the sum over $\boldsymbol{k}$ in Eq. (14), we will assume we are summing over $\boldsymbol{k}$ states per unit volume). Carrying out the manipulations below, we finally find an expression for the coefficient and, hence, the imaginary part of the dielectric constant. The power lost equals

$$P\hbar\omega = -\frac{dI}{dt}, \qquad ...(15)$$

where $I$ is the energy/volume. But,

$$\frac{dI}{dt} = \frac{dI}{dx}\frac{dx}{dt} = \alpha I\left(\frac{c}{n}\right) \qquad ...(16)$$

where $\alpha = 2n_i\omega/c$, and $n_i = \varepsilon_i/2n$. Thus,

$$-\frac{dI}{dt} = \frac{\varepsilon_i\omega I}{n^2} = P\hbar\omega \qquad ...(17)$$

Using
$$I = \frac{1}{2}n^2\,\varepsilon_0\,\overline{E^2}\times 2 \qquad ...(18)$$

where $n = (\varepsilon/\varepsilon_0)^{1/2}$ if $\mu = \mu_0$ and the factor of 2 comes from both magnetic and electric fields carrying current, we find

$$\varepsilon_i(\omega) = \frac{P\hbar}{\varepsilon_0}\frac{1}{\overline{E^2}} \qquad ...(19)$$

Using the Kronig–Kramers relations, we can also derive an expression for the real part of the dielectric constant. Defining

$$|M_{vc}| = \left|\int_{\Omega} u_f^* \,\boldsymbol{e}\cdot\boldsymbol{p}u_i d\Omega\right| \qquad ...(20)$$

we have (using Eqs. (19)), (20), and (14)

$$\varepsilon_i = \frac{\pi}{\varepsilon_0}\left(\frac{e}{m\omega}\right)^2 \Sigma_k |M_{vc}|^2\, \delta(E_c - E_v - \hbar\omega) \qquad ...(21)$$

and by Eq. (45), section (14.17),

$$\varepsilon_r = 1 + \frac{e^2}{m\varepsilon_0}\Sigma_k\left(\frac{2}{m\hbar\omega_{cv}}\frac{|M_{vc}|^2}{\omega_{cv}^2 - \omega^2}\right) \qquad ...(22)$$

(where $E_c - E_v \equiv \hbar\omega_{cv}$ and $\delta(ax) = \delta(x)/a$ has been used). Recall that the $\Sigma_k$ has to be per unit volume and the oscillator strength is defined by

$$f_{vc} = \frac{2|M_{vc}|^2}{m\hbar\omega_{cv}}. \qquad ...(23)$$

Classically, the oscillator strength is the number of oscillators per unit volume with frequency, $\omega_{cv}$. Thus, the real part of the dielectric constant can be written

$$\varepsilon_r = 1 + \frac{e^2}{m\varepsilon_0}\Sigma_k\left(\frac{f_{vc}}{\omega_{cv}^2 - \omega^2}\right). \qquad ...(24)$$

We want to work this out in a little more detail for direct absorption edges. For direct transitions between parabolic valence and conduction bands, effective mass concepts enter because one has to deal with both the valence band and conduction band. For parabolic bands we write

$$E_{vc} = E_g + \frac{\hbar^2 k^2}{2\mu} \qquad ...(25)$$

where

$$\frac{1}{\mu} = \frac{1}{m_c} + \frac{1}{m_v}. \qquad ...(26)$$

The joint density of states per unit volume is then given by

$$D_j = \left[\frac{\sqrt{2}\,\mu^{3/2}}{\pi^2\hbar^3}\right]\sqrt{E_{vc} - E_g} \quad \text{where} \quad E_{vc} > E_g, \qquad ...(27)$$

and

$$D_j = 0 \quad \text{where} \quad E_{vc} < E_g. \qquad ...(28)$$

Thus, we obtain that the imaginary part of the dielectric constant is given by

$$\varepsilon_i(\omega) = \begin{cases} K\sqrt{\Omega - 1}, & \Omega > 1 \\ 0 \quad , & \Omega > 1 \end{cases}, \qquad ...(29)$$

where

$$K = \frac{2e^2(2\mu)^{3/2}|M_{vc}|^2\sqrt{E_g}}{m^2\omega^2\hbar^3}, \qquad ...(30)$$

and

$$\Omega = \frac{\hbar\omega}{E_g}. \qquad ...(31)$$

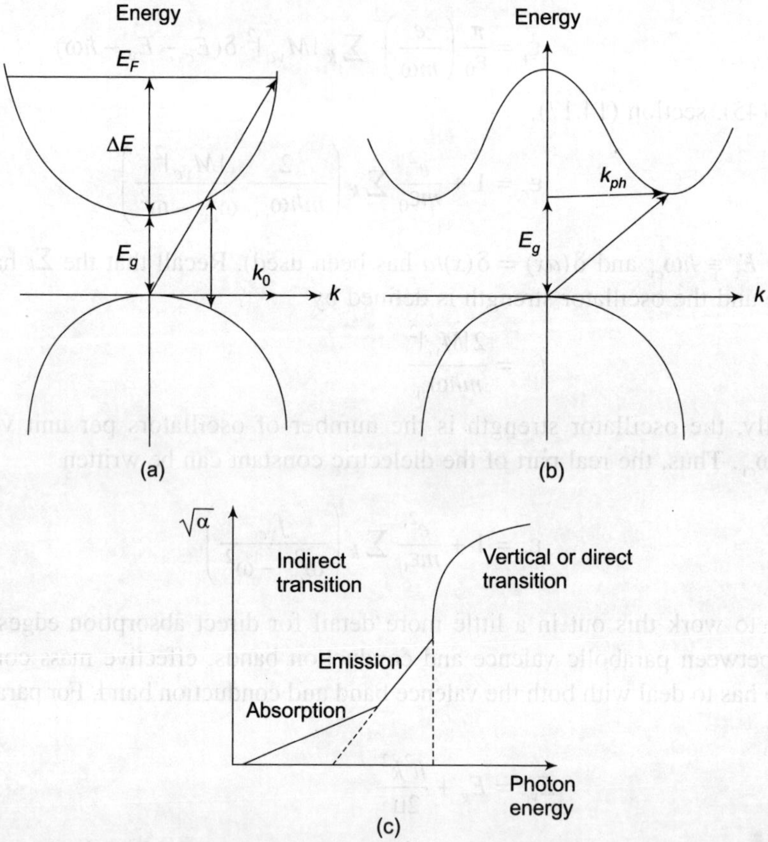


**Figure 14.22** (a) Direct transitions and indirect transitions due to band filling; (b) Indirect transitions, where $k_{ph}$ is the phonon wave vector; (c) Vertical transitions dominate indirect transitions when energy is sufficient to cause them emission and absorption. Refer to phonons in all sketches

From this, one then has an expression for the absorption coefficient (since $\alpha = \omega\varepsilon_i/nc$). Thus, for direct transitions and parabolic bands, a plot of the square of the absorption coefficient as a function of the photon energy should be a straight line, at least over a limited frequency. Figure 14.22 illustrates direct and indirect transitions and absorption. Indirect transitions are discussed below.

The fundamental absorption edge due to the band gap determines the apparent colour of semiconductors as seen by transmission.

We now want to discuss indirect transitions. So far, our analysis has assumed a direct band gap. This means that the $k$ of the initial and final electronic states defining the absorption edge are almost the same (as has been mentioned, the $k$ of the photon causing the absorption is negligible, compared to the Brillouin zone width, for visible wavelengths). This is not true for the two most common semiconductors, Si and Ge. For these semiconductors, the maximum energy of the valence band and the minimum energy of the conduction band do not occur at the same $k$ vectors, one has what is called an indirect band gap, semiconductor. For a minimum energy transition across the band gap, something else, typically a phonon, must be involved in order to conserve wave vector. The requirement of having, for example, a phonon being involved reduces the probability of the event; see Fig. 14.22 b, c, Fig. 14.23, Eq. (36), and consider also Fermi's Golden Rule.

Even in a direct band gap semiconductor, processes can cause the fundamental absorption edge to shift from direct to indirect, see Fig. 14.22 (a). For degenerate semiconductors, the optical absorption edge may be a function of the carrier density. In simple models, the location of the Fermi energy in the conduction band can be estimated on the free-electron model. When the Fermi energy is above the bottom of the conduction band, the $k$ vector of the minimum energy that can cause a transition has also shifted from the $k$ of the conduction band minimum. Now direct transitions will originate from deeper sates in the valence band, they will be stronger than the threshold energy transitions, but of higher energy.

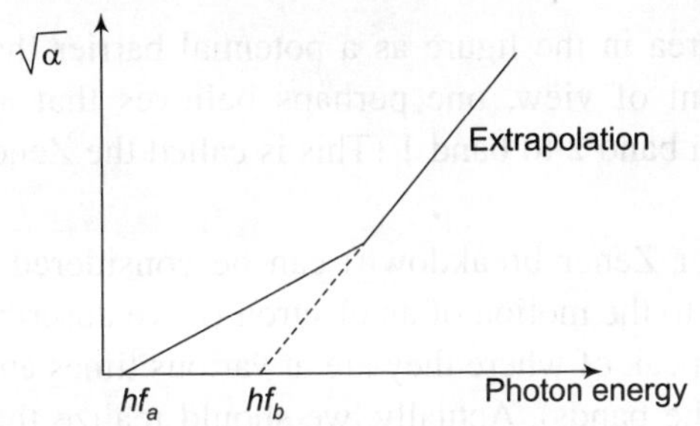


**Figure 14.23** Indirect transitions: $hf_a = E_g - E_{phonon}$, $hf_b = E_g + E_{phonon} = E_g = (hf_a + hf_b)/2$, sketch

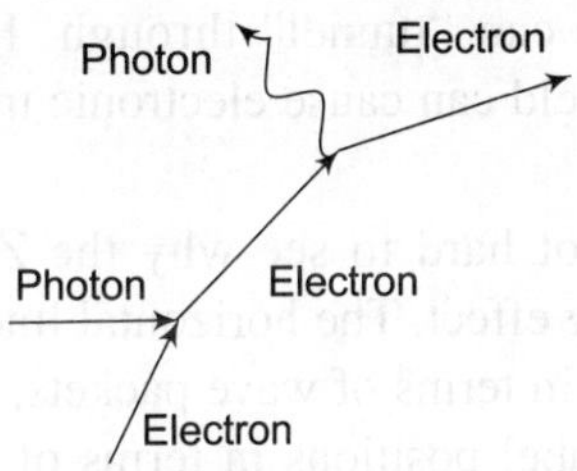


**Figure 10.24** An indirect process viewed in two steps

For indirect transitions, we can write the energy and momentum conservation conditions as follows

$$\boldsymbol{k}' = \boldsymbol{k} + \boldsymbol{K} \pm \boldsymbol{q} \qquad ...(32)$$

where $\boldsymbol{K}$ = photon $\cong 0$ and, $\boldsymbol{q}$ = photon (= $k_{ph}$ in Fig. 14.22 b). Also

$$E(\boldsymbol{k}') = E(\boldsymbol{k}) + \hbar\omega \pm \hbar\omega_{\boldsymbol{q}} \qquad ...(33)$$

where $\hbar\omega$ = photon, and $\hbar\omega_q$ = phonon.

**Note:** Although the photon makes the main contribution to the transition energy, the phonon carries the burden of insuring that momentum is conserved. Now the Hamiltonian for the process would look like

$$\mathcal{H}' = \mathcal{H}'_{\text{photon}} + \mathcal{H}'_{\text{phonon}} \qquad ...(34)$$

where $$\mathcal{H}'_{\text{phonon}} = \frac{e}{m}\boldsymbol{p}\cdot\boldsymbol{A} \qquad ...(35)$$

and $$\mathcal{H}'_{\text{phonon}} = \sum_{\sigma kq} M_{kq}\,(a^{+}_{-\boldsymbol{q}} + a_{\boldsymbol{q}})\,c^{+}_{\boldsymbol{k}+\boldsymbol{q},\sigma}\,c_{\boldsymbol{k},\sigma} \qquad ...(36)$$

One can sketch the indirect process as a two-step process in which the electron absorbs a photon and changes state then absorbs or emits a phonon. See Fig. 14.24.

We mention as an aside another topic of considerable interest. We discuss briefly optical absorption in an electric field. The interesting feature of this phenomenon is that in an electric field, optical absorption can occur for photon energies lower than the normal bandgap energies. The increased optical absorption due to an electric field can be qualitatively understood by thinking about pictures such as in Fig. 14.25. This figure does not present a rigorous concept, but it is helpful.

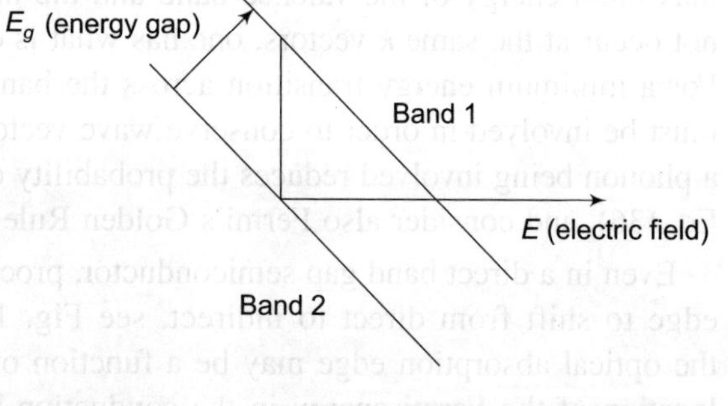


**Figure 14.25** Qualitative effect of an electric field on the energy bands in a solid

Very simply, we can think of the triangular area in the figure as a potential barrier that electrons can "tunnel" through. From this point of view, one perhaps believes that an electric field can cause electronic transitions from band 2 to band 1 (This is called the Zener effect).

It is not hard to see why the Zener effect (or Zener breakdown) can be considered a tunnelling effect. The horizontal line corresponds to the motion of an electron (if we describe electrons in terms of wave packets, then we can speak of where they are at various times and we can label positions in terms of distances in the bands). Actually, we should realize that this horizontal line corresponds to the electric field causing the electron to make transitions to higher and higher stationary states in the crystal. When the electron reaches the top of the lower band, we normally think of the electron as being Bragg reflected. However, we should remember what we mean by the energy gap.

The energy gap, $E_g$, does not represent an absolutely forbidden gap. It simply represents energies corresponding to attenuated, nonpropagating wave functions. The attenuation will be of the form $e^{-Kx}$, where $x$ represents the distance travelled ($K$ is real and greater than zero) and $K$ is actually a function of $x$, but this will be ignored here. The electron gains energy from the electric field $E$ as $|eEx|$. When the electron has travelled $x = |E_g/eE|$, it has gained sufficient energy to get into the bottom of the upper band if it started at the top of the lower band. In order for the process to occur, we must require that the electron's wave function not be too strongly attenuated, i.e., Zener breakdown will occur if $1/K \gg |E_g/eE|$. To see the analogy to tunnelling, we observe that the electron's wave function in the triangular region also behaves as $e^{-Kx}$ from a tunnelling viewpoint (also with $K$ a function of $x$), and that the larger we make the electric field, the thinner the area we have to tunnel across, so the greater a band-to-band transition. A more quantitative discussion of this effect is obtained by evaluating $K$ not from the picture, but directly from the Schrödinger equation. The $x$ dependence on $K$ turns out to be fairly easy to handle in the WKB approximation.

Finally, we can summarize the results for many cases in Table 14.6. Absorption coefficients α for various cases (parabolic bands) can be written

$$\alpha = \left(\frac{A}{hf}\right)(hf + \beta - E_g)^{\gamma} \qquad ...(37)$$

where γ, β depend on the process as shown in Table 14.6. When phonons are involved we need to add both the absorption and emission (±) possibilities to get the total absorption coefficient. A very clean example of optical absorption is given in Fig. 14.26. Good optical absorption experiments on InSb were done in the early days by Gobeli and Fan. In general, one also needs to take into account the effect of temperature. For example, the indirect allowed terms should be written

$$\alpha = \left(\frac{A'}{hf}\right)\left[\frac{(hf + \beta - E_g)^2}{\exp(\beta/kT) - 1} + \frac{(hf - b - E_g)^2}{\exp(\beta/kT) - 1}\exp(\beta/kT)\right], \qquad ...(38)$$

where $A'$ is a constant independent of the temperature.

**Table 14.6** Absorption Coefficients

| | γ | β |
|---|---|---|
| Direct, allowed | 1/2 | 0 |
| Direct, forbidden | 3/2 | 0 |
| Indirect, allowed | 2 | $\pm hf_q$ (phonons) |
| Indirect, forbidden | 3 | $\pm hf_q$ (phonons) |

γ and β are defined by (37).

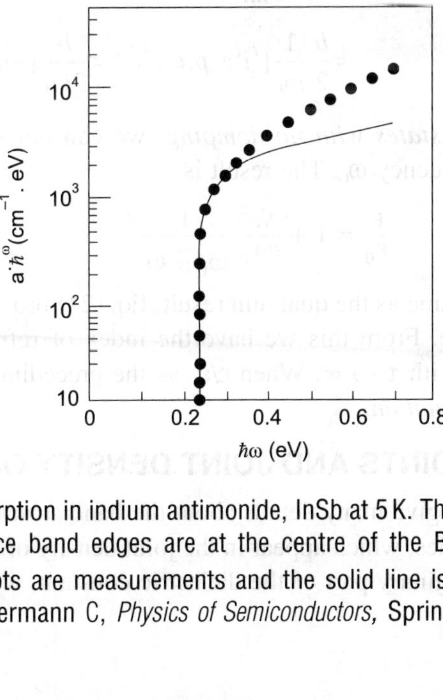


**Figure 14.26** Optical absorption in indium antimonide, InSb at 5 K. The transition is direct because both conduction and valence band edges are at the centre of the Brillouin zone, $k = 0$. Notice the sharp threshold. The dots are measurements and the solid line is $(\hbar\omega - E_g)^{1/2}$. Adapted from Sapoval B and Hermann C, *Physics of Semiconductors*, Springer Verlag, 1988)

## 14.20 OSCILLATOR STRENGTHS AND SUM RULES (A) (DIELECTRIC CONSTANT AND SELLMEIRR'S RELATION)

Let us define the oscillator strength by

$$f_{ij} = b\omega_{ij}\, |\langle i|e\cdot r|j\rangle|^2$$

We will show this is equivalent to the previous definition with the proper choice of $b$ by using commutation relations to cast it in another form. From $[x, p_x] = i\hbar$ we can show

$$[\mathcal{H}, x] = -\frac{i\hbar}{m}p_x \qquad ...(1)$$

Also,
$$[\mathcal{H}, e\cdot r] = -\left(\frac{i\hbar}{m}\right) e\cdot p, \qquad ...(2)$$

therefore
$$\langle i|e\cdot p|j\rangle = im\omega_{ij}\,\langle i|e\cdot r|j\rangle. \qquad ...(3)$$

Thus, we can write the oscillator strength as,

$$f_{ij} = b\,\frac{|\langle i|e\cdot p|j\rangle|^2}{m^2\omega_{ij}} \qquad ...(4)$$

which is consistent with how we wrote it before, if $b = -2m/\hbar$ see Eqs. (23), (20). It is also interesting to show that the oscillator strength obeys a sum rule. If $e = i$, then

$$\sum_j f_{ij} = b\sum_j \omega_{ij}\left|\langle i|\,e\cdot r\,|j\rangle\right|^2$$

$$= \frac{b}{2}\sum_j \frac{1}{im}\{\langle i|e\cdot p|j\rangle\langle j|e\cdot r|i\rangle - \langle i|e\cdot r|j\rangle\langle j|e\cdot p|i\rangle\} \qquad ...(5)$$

$$= \frac{b}{2}\frac{1}{im}[\langle i|e\cdot p, e\cdot r|i\rangle] = \frac{b}{2im}[-i\hbar] = 1.$$

Classically, *for bound states with no damping*, we can derive the dielectric constant. Assume $N$ states with frequency $\omega_0$. The result is

$$\frac{\varepsilon}{\varepsilon_0} = 1 + \frac{Ne^2}{m\varepsilon_0}\frac{1}{\omega_0^2 - \omega^2} \qquad ...(6)$$

Note that it is just the same as the quantum result, Eq. (24) provided the oscillator strength from one oscillator is one. From this we have the index of refraction, and it is given by $n^2 = \varepsilon/\varepsilon_0$, since $\varepsilon$ is real with $\tau \to \infty$. When $\varepsilon/\varepsilon_0$ as the preceding, the resulting equation is often called *Sellmeier's equation*.

## 14.21 CRITICAL POINTS AND JOINT DENSITY OF STATES (A)

Optical absorption spectra give many details about the band structure. This can be explained by the Van Hove singularities, which appear in the joint density of states as mentioned below. In the integral for the imaginary part of the dielectric constant, we had an expression of the form, Eq. (21),

$$\varepsilon_i \propto \frac{2}{\omega^2} \int \frac{d^3\boldsymbol{k}}{(2\pi)^3} |M_{vc}|^2\, \delta(E_c - E_v). \qquad ...(7)$$

A property of delta functions can be written as

$$\int_a^b g(x)\,\delta[f(x)]\,dx = \sum_{x_p} g(x_p) \frac{1}{\left.\frac{\partial f}{\partial x}\right|_{x=x_p}}, \qquad ...(8)$$

where $x_p$ are the zeros of $f(x)$. From which we conclude that the imaginary part of the dielectric constant can be written as

$$\varepsilon_i \propto \frac{2}{\omega^2} \frac{1}{(2\pi)^3} \int_S \frac{dS|M_{vc}|^2}{|\nabla_{\boldsymbol{k}}(E_c - E_v)|_{E_c - E_v = \hbar\omega}} \qquad ...(9)$$

where $dS$ is a surface of constant $\hbar\omega = E_c - E_v$. The joint density of states is defined as

$$J_{vc} = \int \frac{2}{(2\pi)^3} \frac{dS}{|\nabla_{\boldsymbol{k}}(E_c - E_v)|_{E_c - E_v = \hbar\omega}} \qquad ...(10)$$

and typically the matrix element $M_{vc}$ is a slowly varying function compared with the joint density of states. Now the joint density of states is a strongly varying function of $k$ where the denominator is zero, i.e., where

$$\nabla_{\boldsymbol{k}}(E_c - E_v) = 0. \qquad ...(11)$$

Both valence and conduction band energies must be periodic functions in reciprocal space and so must their difference and from this it follows that there must be a point for which the denominator vanishes (smooth periodic functions have analytic maxima and minima). These critical points lead to singularities in the density of states, the Van Hove singularities. At very highly symmetrical points in the Brillouin zone, we can have critical points due to the gradient of both conduction and valence energies vanishing, at other critical points only the gradient of the difference vanishes. *Critical points are defined by the band structure, and in turn, they help determine the absorption coefficient. Reversing the process, studying the absorption coefficient gives information on the band structure.*

## EXERCISE

14.1 Define index of refraction ($n$) and explain damping constant, $k$.

14.2 Define characteristic penetration depth $W$, absorbance $\alpha$, reflectivity $R$ and transmissivity, $T$.

14.3 Obtain Hagen–Rubens relation between reflectivity ($R$) and conductivity, $\sigma$.

14.4 Discuss free electrons with damping, i.e., classical free electron theory of metals and derive Drude equations for the optical constants.

14.5 Discuss classical electron theory of dielectric materials (bound electrons) and derive expressions for optical constants.

14.6 Derive Kronig–Kramers relations.

CHAPTER

# 15

# Superconductivity (Conventional and High $T_C$)

## 15.1 CONCEPT OF SUPERCONDUCTIVITY (CRITICAL TEMPERATURE)

The discovery of superconductivity is associated with the milestone work of Kammerlingh Onnes in 1911 on the electrical resistivity of mercury. He was investigating the electrical properties of frozen mercury and at low temperatures, it was observed that the electrical resistance of Hg completely disappeared suddenly on approaching 4.2 K.

Thus, Hg at 4.2 K enters a new state, which owing to its particular electrical properties is called the state of superconductivity (Fig. 15.1).

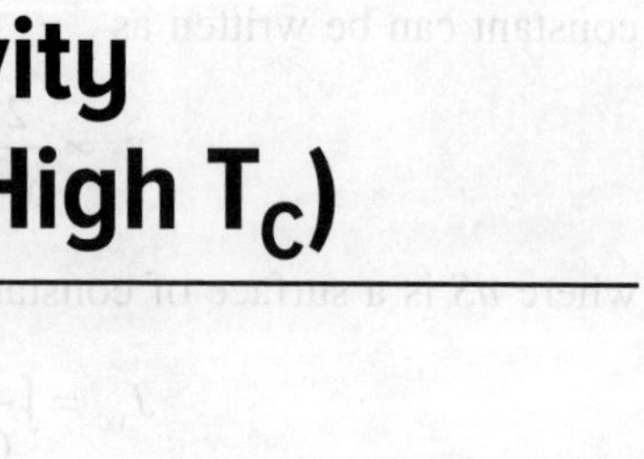


**Figure 15.1** Temperature dependence of a normal metal and superconductor

Superconductivity is exhibited by many elements, alloys, binary and ternary compounds, organic superconductors and superconductors lately discovered in 1986 by Bednorz and Muller called high temperature superconductors.

The temperature at which there is total vanishing of the electrical resistance is called the **critical temperature ($T_c$)** of the material. Some examples of the superconductors are as follows.

## 15.2 OCCURRENCE ELEMENTAL SUPERCONDUCTORS

Be (0.0.5 K) As (1.10 K), Pb (7 K), Nb (9.2 K) Mn (0.02 K), Zn (0.9 K). Fe, Co and Ni do not display superconductivity.

*Alloys.* $Nb_3Al_{0.75}\ Ge_{0.25}$ (20.7 K).

*Binary compounds* (viz., $A_3B$ compounds having A-15 crystal structure) shown in Fig. 15.2.

In $A_3B$, $A$ is one of the transition metals such as Nb, V, Ta or Zr and the $B$ atom comes from the III-A or IV-A column of the Periodic Table and is a metal or a semiconductor, e.g.,

$V_3$ Ga (16.5K, $V_3Si$) (17K)

$Nb_3$ Sn (18K), $Nb_3Al$ (17.5K)

$Nb_3$ Ge (23.2 K)

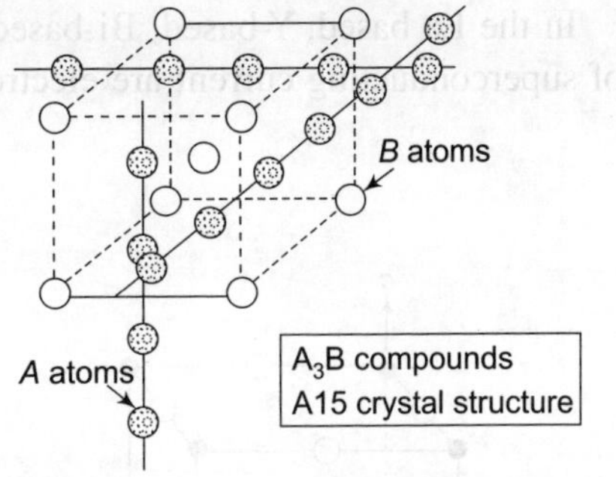


**Figure 15.2** Crystal structure of $Nb_3Sn$

Prior to 1986 (viz., the year of discovery of the high $T_c$ super conductors, the highest $T_c$ known was that of $Nb_3$ Ge (23.2 K).

The discovery of Bednorz and Muller in 1986 of the superconductivity in barium doped lanthanum cuprate ($La_{2-x}$ $Ba_x$ $CuO_4$, the first high $T_c$ superconductor ($T_c$ = 35 K) gave a fresh novel impetus to the subject of superconductivity [J.G. Bednorz and K.A. Müller, *Z. Phys* B**64**, 189 (1986)]. Soon other families of superconducting cuprate oxides were discovered, and the achievements of materials with critical temperatures above 77 K (the boiling point of liquid nitrogen) passed suddenly from dream to reality.

## 15.3 HIGH TEMPERATURE SUPERCONDUCTORS

These oxide superconductors are defect perovskite like cuprate materials, which were discovered lately in 1986 by Bednorz and Müller (they were awarded the Nobel prize for their remarkable discovery). This discovery waited for nearly 70 years after the discovery of superconductivity. These materials have $T_c$s as high as 90–125 K (prior to this, $T_c$ achieved was below 23.2K, corresponding to that of $Nb_3Ge$). Bednorz and Müller discovered high temperature superconductivity ($T_c$ ~ 35 K) in defect perovskite like oxide material $La_{2-x}Ba_xCuO_4$. These materials are layered structures in which sheets of copper and oxygen atoms alternate with sheets of rare-earth (and oxygen) atoms. Soon after, Paul Chu and his coworkers discovered the so called 123 oxides of the general formula $LnBa_2Cu_3O_{7-\delta}$ (Ln = Y, Nd, Sm, Eu, Gd, Dy, Ho, Er, Tm or Yb) with $T_c$ values in the 90 K region. The discovery of materials with superconductivity above the liquid-$N_2$ temperature raised much hope and prompted intensive search for new classes of oxides with still higher $T_c$s. Two series of compounds belonging to the Bi–Sr–Ca–Cu–O and Tl–Ba–Ca–Cu–O systems have been found to exhibit superconductivity between 60 and 125 K.

It is noteworthy that all the high $T_c$ cuprates possess "defect perovskite layers" and all except 123 compounds contain rock salt type oxide layers. The crystal structure of perovskite and rock salt type layers are shown in Fig. 15.3.

The structure in bulk corresponds to $K_2NiF_4$ structure, depicted in Fig. 15.4 ($La_2CuO_4$). In this Ba/Sr is substituted on Lanthanum sites. The perovskite layers consist of corner sharing $CuO_6$ octahedra.

Some important superconductors, their $T_c$s and year of discovery are given in Table 15.2.

In the La based, Y-based, Bi-based, TI-based and Hg-based cuprate families, the carriers of superconducting current are electron-vacancies or holes (pairs).

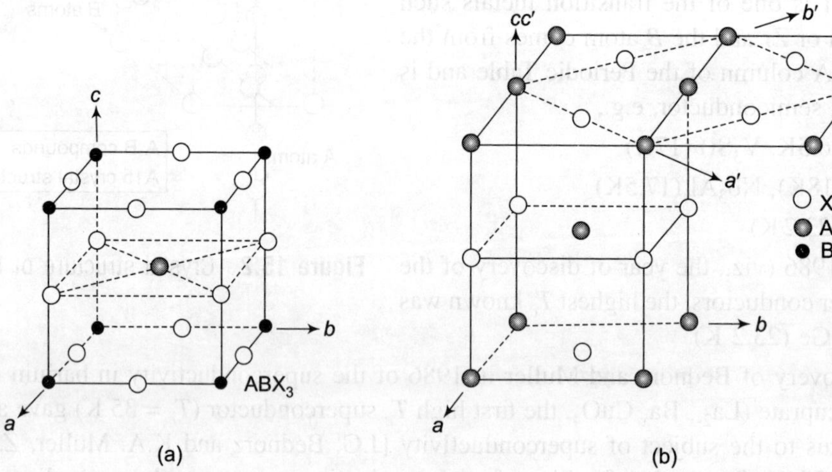


**Figure 15.3** Crystal structure of (a) perovskite and (b) rock salt. The conventional unit cell is face centred and is defined by axes *a*′, *b*′ and *c*′

By contrast, in the HTSC discovered in 1989 ($Nd_{2-x}Ce_xCuO_4$), the carriers are electron-pairs as verified by Hall-coefficient measurement.

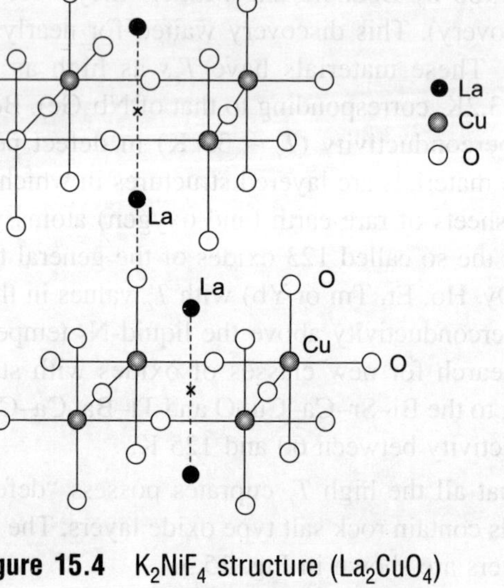


**Figure 15.4** $K_2NiF_4$ structure ($La_2CuO_4$)

## 15.3.1 Variation of $T_c$ with Oxygen Stoichiometry

The critical temperature $T_c$ of $Y_1Ba_2Cu_3O_{7-x}$ also varies with $x$-value and the material becomes non-superconducting at $x \cong 0.6$ (Fig. 15.6).

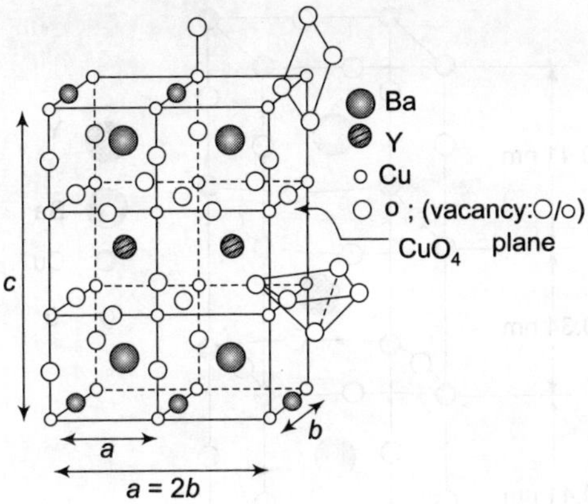


**Figure 15.5** Crystal structure with oxygen disappearing from alternate chains in $YBa_2Cu_3O_{6.5}$ and complete disappearance oxygen from $YBa_2Cu_3O_{7-x}$ to give $YBa_2Cu_3O_6$. (It has been shown using neutron diffraction techniques and high resolution electron microscopy that the oxygen deficit is situated in the $CuO_4$ square plane layers and not in the $CuO_5$ pyramids.) Every 1/*x*th chain oxygen is missing for $YBa_2Cu_3O_{7-x}$

The $T_c$ remains around 90 K from $x = 0$ to $x \approx 0.2$ (for $x = 0.25$, there are alternately "full" and half oxygen chains) and then shows a plateau at 60 K when $x = 0.3–0.4$. $T_c$ is 45 K, when $x = 0.5$. Structural studies have revealed that oxygen vacancies may be ordered when $x = 0.5$ and 0.75 and no clear oxygen vacancy ordering has been found in the $x = 0.3 – 0.4$ range, (i.e., in the plateau region). For $x = 0.5$ (i.e., $YBa_2Cu_3O_{6.5}$), fully oxygenated Cu–O chains (as in $YBa_2Cu_3O_7$) are present along *b*-axis, alternating with fully vacant $O_4$ sites (as in $YBa_2Cu_3O_6$) (see Fig. 15.5).

Thus, for $x = 1.0$ there are no chain oxygen atoms and $T_e$ is zero. The value of *x* in a particular sample depends on the processing conditions, especially temperature and oxygen partial pressure. Even though *x* itself is temperature and atmosphere dependent, under equilibrium conditions at high temperature, the transition appears always to occur at $x = 6.5$.

Figure 15.7 shows the cell-parameters for the orthorhombic structure of Y–Ba–Cu–O. The double $CuO_2$ layers are a feature common to all the HTSCs and are believed to be the key to the occurrence of high temperature superconductivity in these oxides.

The anisotropic structure of these compounds is reflected in the anisotropy of the relevant intrinsic electronic properties of these materials, as shown in Table 15.1.

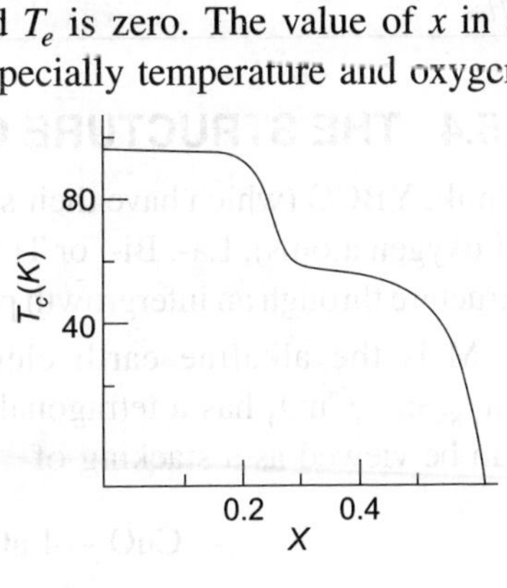


**Figure 15.6** Variation of the superconducting transition temperature $T_c$ of $YBa_2Cu_3O_{7-x}$ with the oxygen stoichiometry *x*

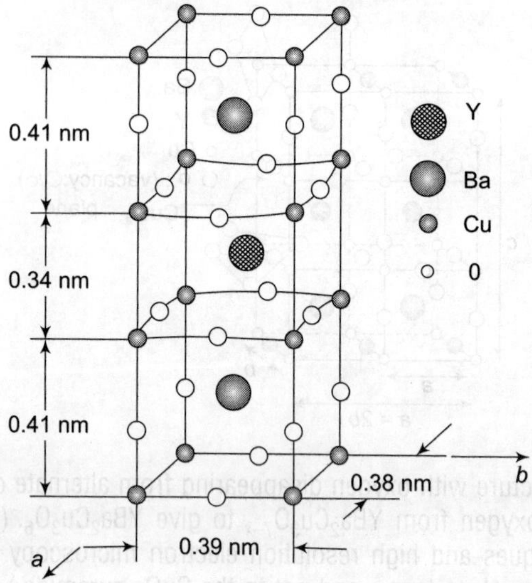


**Figure 15.7** The orthorhombic (distorted) crystal structure of YBCO superconductor

**Table 15.1** Relevant Intrinsic Properties of YBa2Cu3O7–*x*

| | *Parallel to c-axis* | *Parallel to ab-plane* |
|---|---|---|
| ξ (Å) | ~2–3 | ~15–20 |
| λ (Å) | 1,500 | 7,500 |
| $J_C$ (Acm$^{-2}$ at 77 K) | $10^4$ | $10^6$ |
| $H_{c2}$ | 150 T | 30 T |

## 15.4 THE STRUCTURE OF $La_{2-x}M_xCuO_4$

Unlike YBCO (which have their structure derived from ideal perovskite by an ordered removal of oxygen atoms), La-, Bi-, or Ti-based cuprates have structures derived from ideal perovskite structure through an intergrowth phenomenon. Structure of $La_{2-x}M_xCuO_4$ is shown in Fig. 15.8.

M is the alkaline earth element (Sr or Ba). It can be substituted on the La-site. $La_{1.85}Sr_{0.15}CuO_4$ has a tetragonal $K_2NiF_4$ type of layered perovskite structure. The structure can be viewed as a stacking of

$$—CuO_2—LaO—LaO—CuO_2—LaO—LaO—CuO_2—$$

$La_{2-x}M_xCuO_4$ shows the highest $T_c$ around a critical value of $x$ (0.15 for Sr and 0.2 for Ba with $T_c s$ of 30 and 35 K respectively).

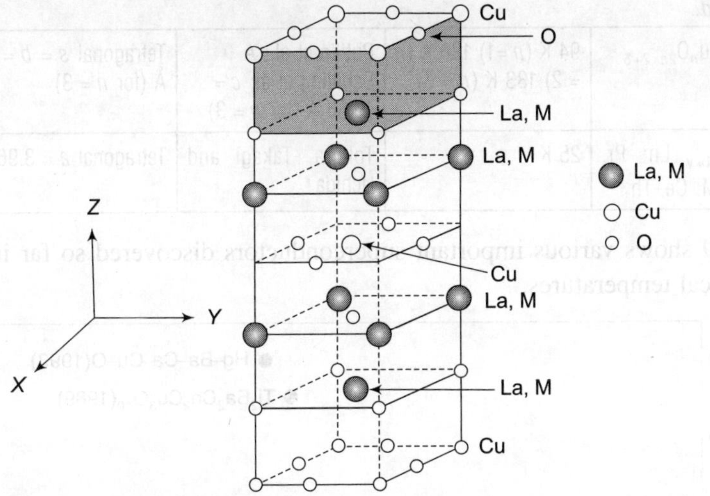


**Figure 15.8** The crystal structure of $La_{2-x}M_xCuO_4$ (M = Ba or Sr). For $x = 0$, $a = b = 3.77$ Å, $c = 13.25$ Å. M atoms occupy La site in $La_2CuO_4$

## 15.5 SOME IMPORTANT HIGH-TEMPERATURE SUPERCONDUCTORS

These oxide superconductors are defect perovskite-like cuprate materials, which were discovred lately in 1986, by Bednorz and Müller. This discovery had to wait for nearly 70 years after the discovery of superconductivity. These materials have $T_c$'s as high as 90-125 K, and prior to this discovery the $T_c$ achieved was below 23.2 K (corresponding to that of $Nb_3Ge$). In 1986, and Müller discovered the first superconductor of the high-$T_c$ family, viz., $La_{2-x}Ba_xCuO_4$ with a $T_c$ of ~35 K. (They were awarded thc Nobel prize for this remarkable discovery.) In subsequent 3 or 4 years, other 'families' of high-$T_c$ cuprate oxide materials were discovered, the important ones are surmised in the Table 15.2.

**Table 15.2** Some Important High-$T_c$ Superconductors

| *High $T_c$ superconductor (family)* | *$T_c$* | *Discovered by* | *Structure of the materials* |
|---|---|---|---|
| 1. $La_{2-x}Ba_xCuO_4$ | 35 K | Bednorz & Müller (1986) | Tetragonal ($K_2NiF_4$ structure) ($a = 3.79$ Å, $c = 13.25$ Å) |
| 2. $Y_1Ba_2Cu_3O_{7-x}$ | 90 K | M.K. Wu and C.W. Chu | Orthorhombic (becomes tetragonal for $x = 1$) $a = 3.82$ Å, $b = 3.89$ Å, $c = 11.70$ Å |
| 3. $Bi_2Sr_2Ca_{n-1}Cu_nO_{2n+4}$ ($n$ = 1, 2, 3) | (7–20) K ($n$ = 1) 85 K($n$ = 2) 110 K ($n$ = 3); | Michel et al. Maeda et al.; Tarascan et al. | Pseudo-tetragonal $a = b = 5.41$ Å $c = 2(9.0 + 3.2n)$ Å |
| 4. $Tl_mBa_2Ca_{n-1}Cu_nO_{2n+2+m}$ (two families) (i) $m$ = 1 (single TlO layer) (ii) $m$ = 2 (double TlO layer) | Highest for $n$ = 4(122 K) $n$ = 3(125 K) | Sheng and Hermann | Tetragonal (orthorhombic only for $m$ = 2) and $n$ = 1) [$a = 3.80$ Å, $c = (6.3 + 3.2n)$ Å for $m$ = 1] [$a = 3.80$ Å, $c = 2(8.2 + 3.2n)$ Å for $m$ = 2] |

*Contd...*

**Table 15.2** *Contd...*

| | | | |
|---|---|---|---|
| 5. $HgBa_2Ca_{n-1}Cu_nO_{2n+2+\delta}$ | 94 K ($n$ =1) 120 K ($n$ = 2) 133 K ($n$ = 3) | Putilin et al.; A Schilling et al; $c$ = 15.86 Å (for $n$ = 3) | Tetragonal $s$ = $b$ = 3.86 Å, $c$ = 15.86 Å (for $n$ = 3) |
| 6. $Ln_{2-x}M_xCuO_{4+y}$ Ln; Pr, Nd, Sm, Eu M: Ce, Th | 25 K | Tokura, Takagi and Uchida | Tetragonal $a$ = 3.961 Å $c$ = 12.130 Å |

Figure 15.9 shows various important superconductors discovered so far in various years and their critical temperatures.

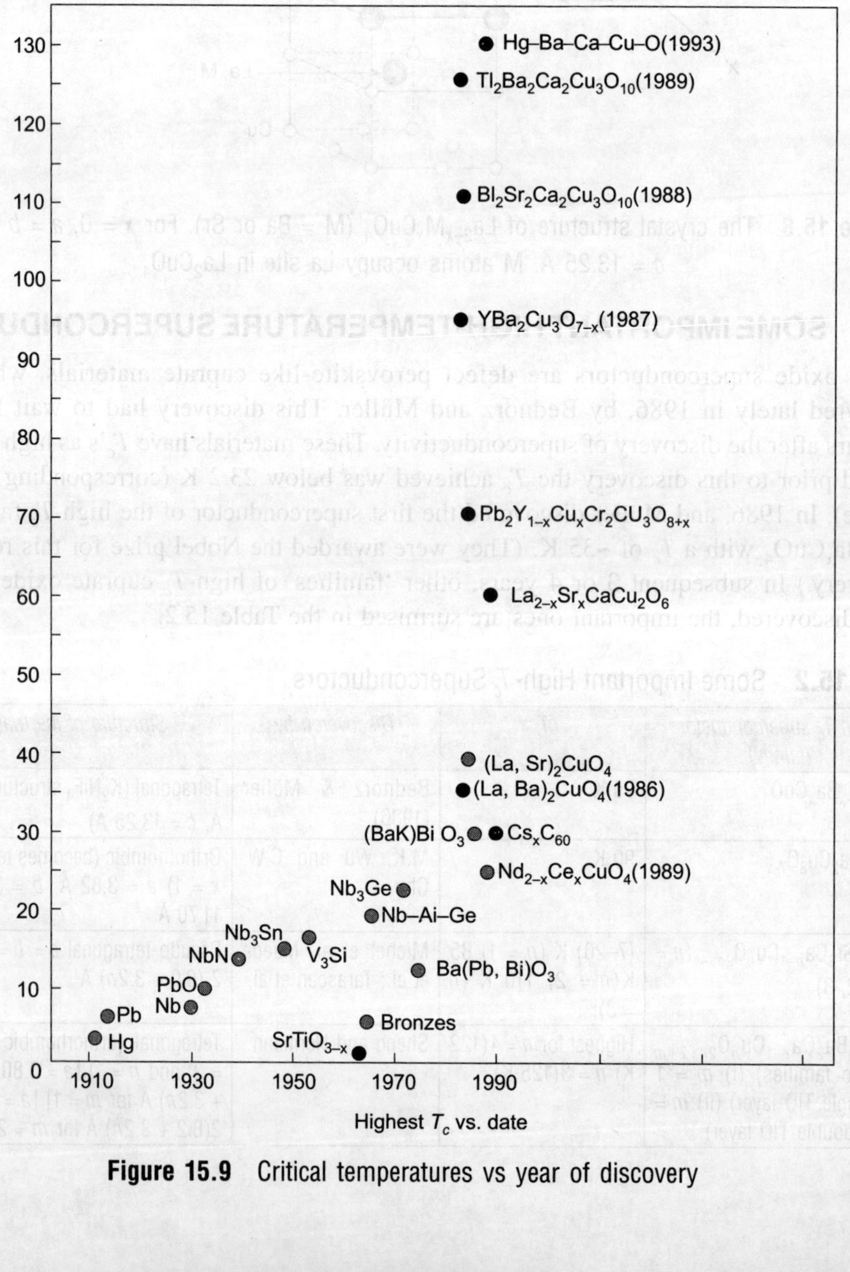


**Figure 15.9** Critical temperatures vs year of discovery

The discovery of high-$T_c$ materials with superconductivity well above boiling point of liquid nitrogen ($LN_2$) was notable from the viewpoint of technological applications. Also, the availability of these materials raised the hopes for reducing the cost of cryogenics involved in operating at $LN_2$ temperatures compared to that operating at liquid helium temperature ($LN_2$ costs only ₹10/litre whereas liquid helium costs about 1000 times higher). Further, thermal insulation required for cryostats was also reduced.

## 15.6 ISOTOPE EFFECT

It has been observed experimentally that the superconducting transition temperature ($T_c$) varies with the isotopic mass $M$ of the material as

$$T_c \propto M^{-\beta} \quad ...(1)$$

where $\beta$ is called the isotope effect coefficient and is fiven by

$$\beta = 0.5\,[1 - 0.01\,\{N(0)\,V\}^{-2}] \quad ...(2)$$

where $N(0)$ is the density of single particle states for one spin at the Fermi level and $V$ is the potential between the electrons. $\beta$ has the value 0.45 to 0.50 for many superconductors. There are notable exceptions where the value of $\beta$ is quite small like $Nb_3Sn$, Mo etc.

In has been found that in mercury, $T_c$ varies from 4.185 to 4.146 K as isotopic mass varies from 1995 amu to 203.4 amu. Since Debye temperature $\theta_D$ is proportional to the velocity of sound, which varies as $M^{-1/2}$, therefore $T_c$ can be related to $\theta_D$ as

$$T_c \propto \theta_D \quad \text{or} \quad \frac{T_c}{\theta_D} = \text{a constant} \quad ...(3)$$

Since Debye temperature depends on lattice vibrations, the above relation implies that lattice vibrations play an important role in superconductivity as also suggested by Fröhlich[1] prior to the BCS theory.

## 15.7 CRITICAL MAGNETIC FIELD

It was discovered by K. Onnes that application of magnetic fields destroys superconductivity. The minimum magnetic field necessary to destroy superconductivity is called the critical magnetic field. It is found to be dependent on temperature and the variation is given by

$$H_c(T) = H_c(0)\left[1 - \left(\frac{T}{T_c}\right)^2\right] \quad ...(4)$$

This is known as Tuyn's law.

The temperature dependence of $H_c$ is shown in Fig. 15.10.

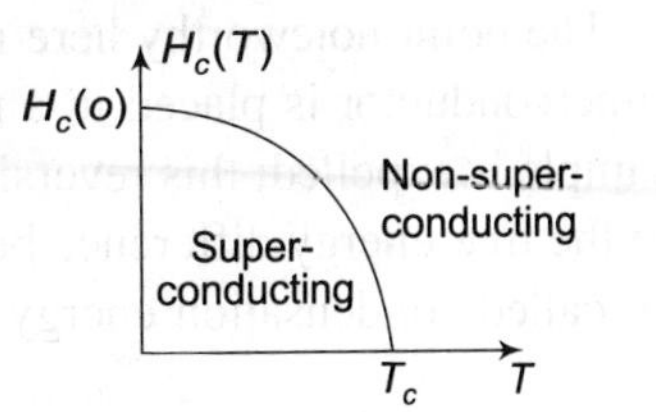


**Figure 15.10** Critical magnetic field as a function of temperature

[1] H. Fröhtich., *Phys. Rev.* **79**, 845 (1950).

$H_c(0)$ is the critical field at absolute zero, $H_c(0)$ and $T_c$ are constants and characteristic of the material. Equation (1) represents the equation of phase boundary between the superconducting and the normal states.

## 15.8 MEISSNER EFFECT

In metals other than ferromagnetics, the magnetic field created by the elementary atomic currents are oriented chaotically in the metal and cancel out, therefore magnetic induction $B = 0$ in the absence of an external field. In presence of an external field $H$, there appears a finite induction $\vec{B}$ given by $\vec{B} = \mu \vec{H}$ where $\mu$ is the permeability.

$\mu > 1$ (for paramagnetics)

$\mu < 1$ (for diamagnetics)

$\mu = 0$ (for superconductors)

i.e., in the superconducting state, a superconductor exhibits perfect diamagnetism. Meissner effect is the phenomenon which relates diamagnetism with superconductors. This is as follows.

Meissner and Ochsenfeld discovered in 1933 that below the critical temperature ($T < T_c$) if a superconductor is placed in a magnetic field, the magnetic field is expelled from the interior of a superconductor This is known as Meissner effect (Fig. 15.11). Further, if $H > H_c$ (or $T > T_c$), the flux penetrates the superconductor, because then the material is non-super conducting and behaves like a normal conductor.

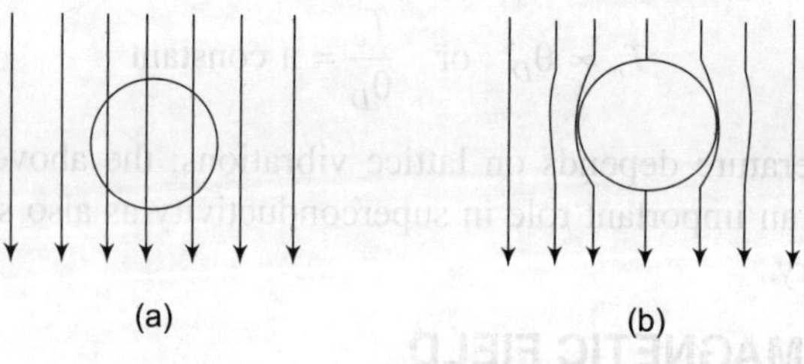

**Figure 15.11** The Meissner effect (a) for $T > T_c$ or $H > H_c$, normal conductor (b) $T < T_c$ and $H < H_c$, material is superconductor

The point noteworthy here is that not only a magnetic field is excluded (for $T < T_c$) as the superconductor is placed in a magnetic field, but also a field present in an originally normal sample is expelled; this reversible nature of the Meissner effect is related thermodynamically to the free-energy difference between the normal and superconducting states in zero field, the so called condensation energy of the superconducting state, i.e.,

$$f_n(T) - f_s(T) = \frac{H_c^2(T)}{8\pi} \quad \text{(C.G.S. units)}$$

where $f_n$ and $f_s$ are the Helmholtz free energies/volume in the respective phases in zero field.

## 15.9 COMPARISON BETWEEN A SUPERCONDUCTOR AND A VERY GOOD (OR IDEAL) CONDUCTOR

The electron-lattice interaction is the basis mechanism of electrical resistance in an ordinary metal. Metals, such as Au, Ag and Cu are excellent conductors but do not display superconductivity, because very good conductivity indicates that electrical resistance is very low. Therefore, the interaction between electrons and lattice is very weak, which at low temperatures does not create sufficient inter-electron attraction to overcome the Coulomb repulsion. Therefore, there is no transition to superconducting state.

*A perfect conductor does not exhibit Meissner effect, i.e., no flux expulsion on colling.*

In fact, the magnetic behaviour of an ideal conductor depends on whether the sample is first cooled to below $T_c$ before applying the field (1) or the field is applied and then cooled (2).

For a superconductor, the final states are identical, regardless of whether $B_{ext}$ is switched on before, or after cooling.

For any closed path enclosing an area $a$ in must be true that

$$IR = V = \iint \text{curl}\, \vec{E} \cdot \mathbf{d}\, \vec{a} = -\frac{\partial \vec{B}}{\partial T} \cdot \vec{A} \qquad \text{...(1)}$$

(using Maxwell's equation) vanishing resistance implies magnetic flux $\vec{B} \cdot \vec{A}$ through the loop may not alter, that is, magnetic field inside must be maintained (what was initially there) both after cooling and after switching off the external field, $B_{ext}$.

On switching off $B_{ext}$ in the cooled state, this requirement is satisfied (in both $A$ and $B$, Fig. 15.12), because the process of switching off $B_{ext}$ induces persistent currents inside the material surface, which maintain the value of magnetic field in the interior Now, in sense of (Eq. 1), an ideal conductor (for $B_{ext} = 0$ and $T < T_c$) may adopt two different states ((c.) or (f)) depending on the order of events leading to this state.

Thus, we have two different states (d) and (g) for an ideal conductor. However, if a superconductor was merely such an "ideal conductor", the superconducting state would not be a stste in the thermodynamic sense. In fact, in the superconductor, not only $\frac{\partial \vec{B}}{\partial T} = 0$, but also $\vec{B} = 0$ independent of the path by which the state (c) or (f) is reached.

This Meissner effect, which represents the property of ideal diamagnetism (for a superconductor) follows independent of vanishing of electrical resistance. However, an ideal conductor does not show ideal diamagnetism (as evident from state (g)), i.e., external field is not expelled, if cooling is done after the application of external field.

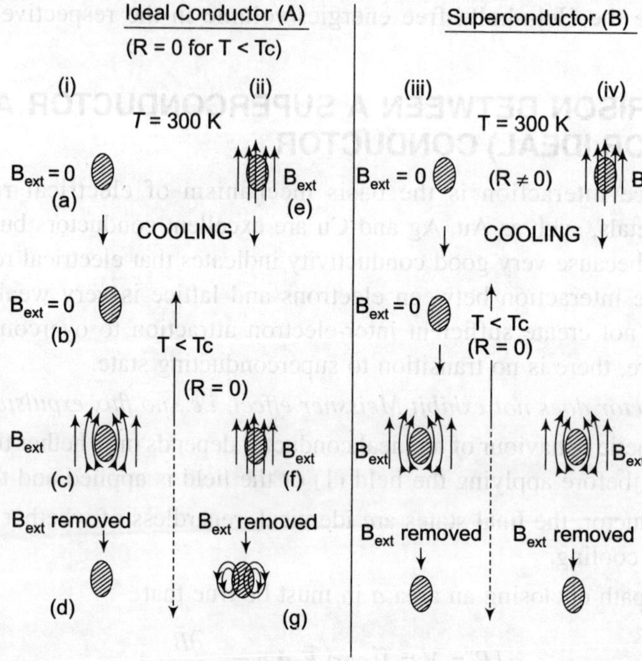


**Figure 15.12** Comparison of magnetic behaviours of an conductor and a superconductor. Both have $R = 0$ for $T < T_c$

## 15.10 SPECIFIC HEAT

The specific heat $C_v$ in a normal conductor at low temperatures consists of two contributions, $C_v^e$ from the electrons in the conduction band $C_v^l$ frob the lattice. Thus,

$$C_v = C_v^e + C_v^l = \gamma T + \beta T^3$$

$C_v^e$ is linearly proportional to $T$ while $C_v^l$ is proportional to $T^3$ and is explained by Debye theory. The contribution to the specific heat by the electrons ($C_v^e$) in the super-conducting state is plotted in Fig. 15.13 and is compared with their contribution in the normal state.

The two curves are obtained by measuring the specific heat $C_v$ with and without a magnetic field whose strength exceeds the critical field $H_c(T)$. From this, the electronic contribution $C_v^e$ is obtained by subtracting the lattice contribution $C_v^l$. When the field is applied the metal is in the normal state, and when it is not applied it is in the superconducting state. The presence of the magnetic field in the normal state does not significantly modify the contribution to the specific heat by the electrons in the conduction band.

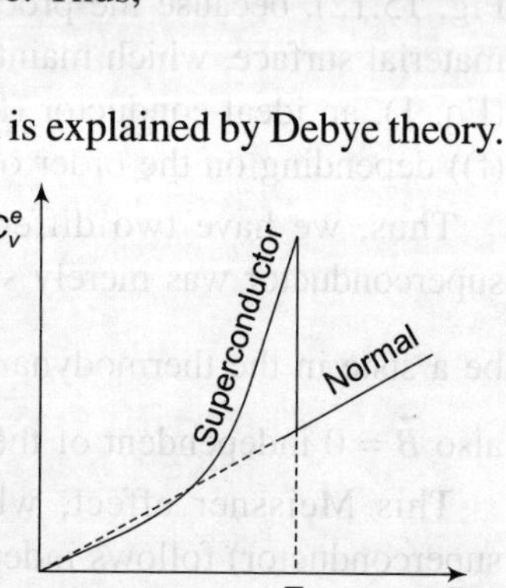


**Figure 15.13** The temperature dependence of electronic specific heat of a conductor in the normal and superconducting states

## 15.11 THE DENSITY OF STATES AND ENERGY GAP

The behaviour of the specific heat in the superconducting state is a strong indication of the existence of an energy gap in the excitation spectrum of the electrons in the conduction band. Figure 15.14 illustrates this by comparing the difference between the occupancy of states in a normal metal and a superconductor. Figure 15.14 (a) gives density of state in the absence of superconductivity (which can be obtained by applying $H > H_c$ (0)). The superconducting ground state for zero temperature is illustrated in Fig. 15.14 (b). This shows a zero density of states for energies within $\pm\Delta(0)$ or $2\Delta_0$. At $T = 0$, no electrons are excited to higher states.

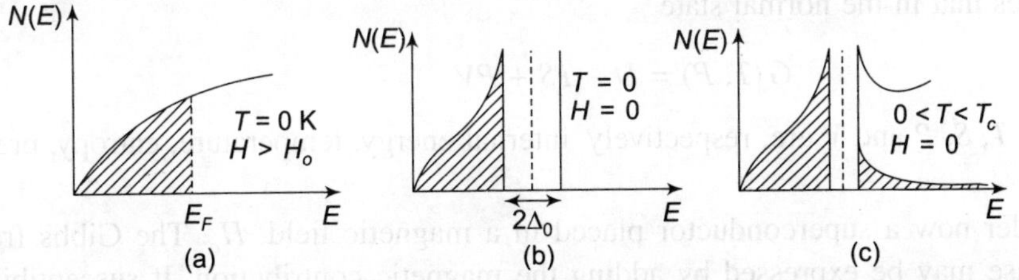


**Figure 15.14** Density and occupancy of states (a) in absence of superconductivity. (b) in superconducting ground state, (c) superconducting state at finite temperature

Figure 15.14 (c) shows consequence of a finite temperature ($T$) less than $T_c$. The superconducting energy gap $\Delta(T)$ is now smaller than $\Delta_0$. A fraction of electron number is now in states above $E_F(0) + \Delta(T)$, leaving behind some unoccupied states below $E_F(0) - \Delta(T)$.

As the temperature is increased, the energy gap $\Delta(T)$ decreases in a manner as shown Fig. 15.15.

This dependence of $\Delta$ on $T$ is explained by the BCS theory. As the temperature is increased more electrons get excited to the available states above the energy gap and $C_v^e$ increases exponentially till $T = T_c$.

The value of the energy gap at 0 K (i.e., $2\Delta_0$) of some of the elements is shown in Table 15.3.

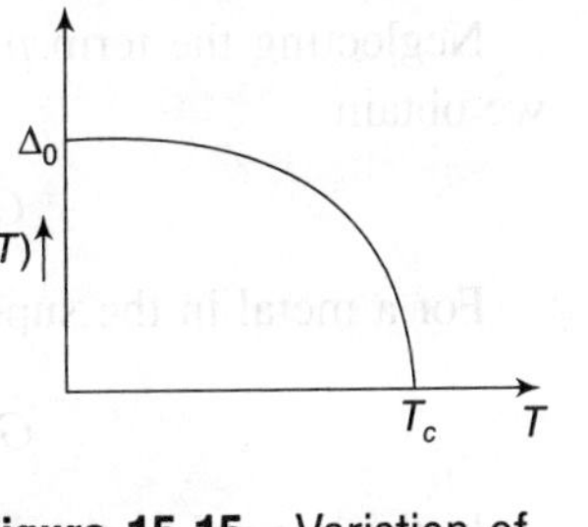


**Figure 15.15** Variation of energy gap as a function temperature

**Table 15.3** Energy Gap $2\Delta_0$ at 0 K for Some Elements

| *Element* | $2\Delta_0$ (meV) | $T_c$(K) | $2\Delta_0/k_BT_c$ |
|---|---|---|---|
| Nb | 3.05 | 9.50 | 3.8 |
| Ta | 1.40 | 4.48 | 3.6 |
| Al | 0.34 | 1.20 | 3.3 |
| Sn | 1.15 | 3.72 | 3.5 |
| Pb | 1.90 | 7.19 | 4.3 |
| Hg | 1.65 | 4.15 | 4.6 |

The ratio of $2\Delta_0$ to $k_BT_c$, according to the BCS theory is 3.52, the same for all superconductors. The ratios (given in last column of the Table 15.3) show good agreement with the theory.

## 15.12 THERMODYNAMICS OF A SUPERCONDUCTOR

The change from the normal to the superconducting state is a thermodynamically reversible transition. We now show that this change occurs because the Gibbs free energy in the superconducting state ($G_S$) is lower than its value ($G_N$) in the normal state. We will also show that at the transition temperature $T_c$, the specific heat in the superconducting state, is three times that in the normal state

$$G(T, P) = U - TS + PV \qquad ...(1)$$

where $U$, $T$, $S$, $P$ and $V$ are, respectively, internal energy, temperature, entropy, pressure and volume.

Consider now a superconductor placed in a magnetic field, $H_a$. The Gibbs free energy in this case may be expressed by adding the magnetic contribution. If susceptibility ($\chi$) is independent of $H$, the magnetic energy is

$$U_M = \int_0^{H_i} \mu_0\, HdM = \int_0^{H_i} \mu_0\, H\chi\, dH$$

$$= -\frac{1}{2}\mu_0\chi H_a^2 \qquad ...(2)$$

Neglecting the term $pV$ for a magnetic solid and introducting the magnetic energy term, we obtain

$$G\,(T, H_a) = U - TS - 1/2\,\mu_0\chi H_a^2 \qquad ...(3)$$

For a metal in the superconducting state, $\chi = -1$. So,

$$G_S(T, H_a) = G_S(T, 0) + 1/2\,\mu_0 H_a^2 \qquad ...(4)$$

In the normal state, $\chi \cong 10^{-6}$ emu, which is six orders of magnitude smaller, so that we may assume $\chi \cong 0$, i.e.,

$$G_N(T, H_a) = G_N(T, 0) \qquad ...(5)$$

At $H_a = H_c$, the normal and superconducting phases are in equilibrium at the same temperature and field. This requires that

$$G_S(T, H_c) = G_N(T, H_c) = G_N(T, 0)$$

$$G_S(T, 0) + 1/2\,\mu_0 H_c^2\,(0) \qquad ...(6)$$

or $$G_N(T, 0) = G_S(T, 0) = 1/2\,\mu_0 H_c^2$$

From Eqs. (4) – (6) it is seen that for $H_a > H_c$, $G_s > G_N$, while for $H_a < H_c$, $G_s < G_N$ and the superconducting state is stable for $H_a < H_c$. The Gibbs free energies ($G_S$ and $G_N$) from Eqs. (4) and (5) are plotted in Fig. 15.16. The transition from the superconducting to the normal state for $H_a > H_c$ occurs because $G_s$ exceeds $G_N$ at this field.

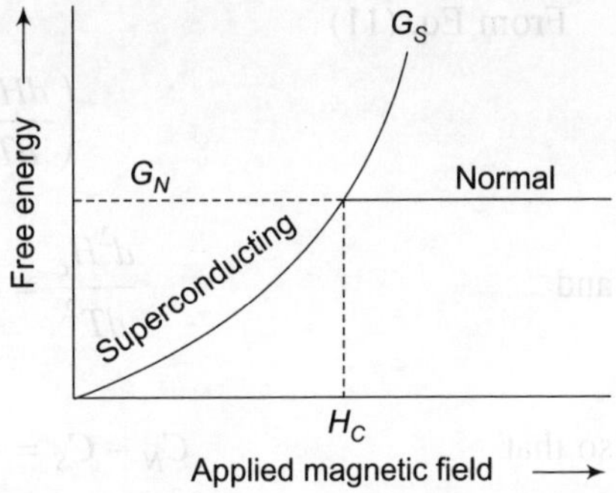


**Figure 15.16** Free energy in the normal and superconducting states vs. applied magnetic field

The normal and superconducting phases are in equilibrium along the critical field curve $H_c(T)$ given in Fig. 15.16. For all the points of the curve, Eq. (6) is satisfied and hence for such points

$$dG_N(T, H_c) = dG_S(T, H_c)$$

Therefore, $$\left(\frac{\partial G_N}{\partial T}\right)_{H_c} dT + \left(\frac{\partial G_N}{\partial H_c}\right)_T dH_c = \left(\frac{\partial G_S}{\partial T}\right)_{H_c} dT + \left(\frac{\partial G_s}{\partial H_c}\right)_T dH_c \quad ...(7)$$

Now, $G = U - TS + 1/2\, \mu_0 H_a^2$. For a constant field $H_a$, $dG = dU - T\, ds - S\, dT$ and $dU = T\, dS = C_v dT$, so that

$$S = -\left(\frac{\partial G}{\partial T}\right)H_c \quad ...(8)$$

and $$C_v^e = T\left(\frac{\partial S}{\partial T}\right)H_c \quad ...(9)$$

From Eqs. (6) – (8), we obtain the relation between the entropy in the normal state ($S_N$) and that in the superconducting state ($S_S$) as

$$S_N = S_S - 1/2\, \mu_0 \frac{d}{dT}(H_c^2)$$

$$= S_S - \mu_0 \left[H_c \frac{dH_c}{dT}\right] \quad ...(10)$$

We have $$H_c(T) = H_c(0)\, [1 - (T/T_c)^2]. \text{ So,}$$

$$\frac{dH_c(T)}{dt} = -2\, H_c(0)\, T/T_c^2 < 0 \quad ...(11)$$

The entropy in the superconducting state is lower than that in the normal state because $H_2(dH_c/dT)$ is always negative. In other words, the superconducting state is more ordered than the normal state.

Using Eqs. (9) and (10), we obtain the difference in the specific heats, $C_N$ and $C_S$, in normal and superconducting states as

$$C_N - C_S = -T\, \mu_0 \left[\left(\frac{dH_c}{dT}\right)^2 + H_c \frac{d^2 Hc}{dT^2}\right] \quad ...(12)$$

From Eq. (11)

$$\left(\frac{dH_c}{dT}\right)^2 = 4H_c^2(0)\frac{T^2}{T_c^4}$$

and
$$\frac{d^2H_c}{dT^2} = d/dT\left[\frac{dH_c(T)}{dT}\right] = -2\frac{H_c(0)}{T_c^2}$$

so that
$$C_N - C_S = -\mu_0 T\left[-H_c\frac{2H_c(0)}{T_c^2} + \frac{4T^2}{T_c^4}H_c^2(0)\right]$$

$$= -\mu_0\frac{2H_c(0)}{T_c}\left[\frac{2T^3}{T_c^3}H_c(0) - \frac{T}{T_c}H_c\right]$$

$$= -\mu_0\frac{2H_c^2(0)}{T_c}\left[\frac{2T^3}{T_c^3} - \frac{T}{T_c}\frac{H_c}{H_c(0)}\right]$$

$$= -\mu_0\frac{2H_c^2(0)}{T_c}\left[\frac{2T^3}{T_c^3} - \frac{T}{T_c} + \frac{T^3}{T_c^3}\right]$$

$$= -\mu_0\frac{2H_c^2(0)}{T_c}\left[\frac{3T^3}{T_c^3} - \frac{T}{T_c}\right] \quad ...(13)$$

Experimentally, it is observed that specific heat in the superheat in the superconducting state has no term linear in $T$. Therefore, it is natural to identify the linear term in Eq. (13) with the contribution from the normal state.

Then,
$$C_N = \gamma T = \frac{2H_c^2(0)}{T_c^2}\mu_0 T \quad ...(14)$$

or
$$\gamma = \frac{2H_c^2(0)}{T_c^2}\mu_0 \quad ...(15)$$

The value of $H_c(0)$ and $T_c$ for cadmium are 30 Gauss and 0.52 K. So $\gamma$ from Eq. (15) is 52.97 J m$^{-3}$deg$^{-2}$. The experimental value of $\gamma$ from the electronic specific heat is 53.13 Jm$^{-3}$deg$^{-2}$. Thus, there is good agreement with theory.

The specific heat for the superconducting state is given by

$$C_S = \frac{6H_c^2(0)\mu_o}{T_c^2}\left(\frac{T}{T_c}\right)^3$$

$$= 3\,\gamma\,T_c \left[\frac{T}{T_c}\right]^3$$

The value of $C_s$ at $T = T_c$ is three times the value of $C_N$ at the same temperature.

## 15.13 THE LONDON THEORY AND PENETRATION DEPTH

F. London and H. London in 1935 showed that the magnetic flux lines are not completely expelled from a superconductor, but remain confined in a thin surface layer. This induces a flow of supercurrent in a closed circuit within the thin surface layer. They postulated a two-fluid model for electron with super fluid and normal densities $n_s$, $n_n$ and velocities $v_s$ and $v_n$ respectively. If $n_o$ = number of electrons per unit volume then

$$n_o = n_s + n_n \quad \text{...(1)}$$

The equations of motions for the superfluid electrons is

$$m\frac{dv_s}{dt} = -eE \quad \text{...(2)}$$

The current density of the superfluid electrons is

$$j_s = -en_s v_s \quad \text{...(3)}$$

Equations (2) and (3) yield

$$\frac{d\vec{j}_s}{dt} = \frac{n_s e^2}{m}E \quad \text{...(4)}$$

This is called the *first London equation*. Taking curl of Eq. (4)

$$\nabla \times \frac{d\vec{j}_n}{dt} = \frac{n_s e^2}{m}\,\text{curl}\,E$$

$$= \frac{n_s e^2}{m}\frac{\partial \vec{B}}{\partial t} \quad \text{...(5)}$$

(using Maxwell's equation curl $\vec{E} = -\partial \vec{B}/\partial t$). Integrating Eq. (5) with respect to time and choosing the constant of integration to be zero to be consistent with the Meissener effect, we have

$$\nabla \times \vec{j}_s = -\frac{n_s e^2}{m}\vec{B} \quad \text{...(6)}$$

This is called the *second London equation*. We may derive the Meissner effect from second London equation by using Maxwell's equation

$$\nabla \times \vec{B} = \mu_0\,\vec{j}_s \quad \text{...(7)}$$

Taking the curl of this equation

$$\nabla \times \nabla \times \vec{B} = \mu_0\,\nabla \times \vec{j}_s \quad \text{...(8)}$$

or $$\text{grad}(\text{div}\,\vec{B}) - \nabla^2\vec{B} = \nabla \times \vec{j}_s$$

or $$-\nabla^2\vec{B} = \mu_0\,\nabla \times \vec{j}_s \qquad ...(9)$$

Now from Eqs. (6) and (9)

$$\nabla^2\vec{B} = \frac{1}{\lambda^2}\vec{B} \qquad ...(10)$$

where $\lambda = (m/\mu_0 n_s e^2)^{1/2}$ is called *London penetration depth.* ...(11)

For a superconductor to the right of the plane, $x = 0$, Eq. (10) has the solution

$$\vec{B}(x) = \vec{B}(0)\exp(-x/\lambda) \qquad ...(12)$$

This indicates that $\vec{B}$ does not penetrate very deeply into superconductor, but penetrates only a distance $\lambda$ within the surface (it implies the Meissner effect). The penetration depth $\lambda$ range from 300 to 500 Å depending on the material. The penetration depth also depends on temperature and becomes quite large as temperature approaches the transition temperature. This is given by (see Fig. 15.17).

$$\frac{\lambda(T)}{\lambda(0)} = \left[1 - \left(\frac{T}{T_c}\right)^4\right]^{-1/2} \qquad ...(13)$$

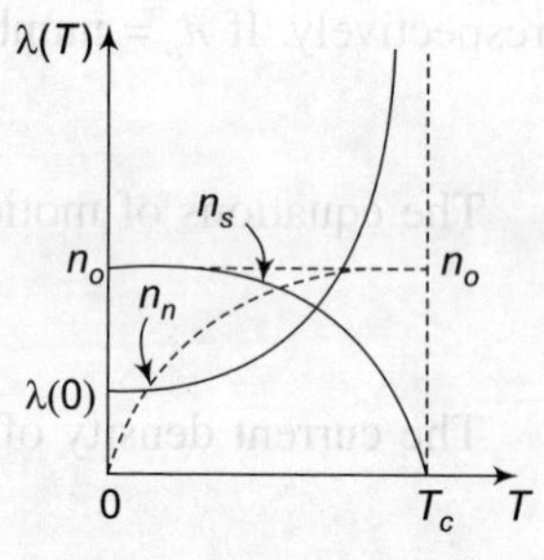


**Figure 15.17** Variation of penetration depth with temperature

The London penetration depth and the number of super electrons $n_s$ are inversely related to each other and $n_s$ is given by

$$n_s = n_0\left[1 - \left(\frac{T}{T_c}\right)^4\right] \qquad ...(14)$$

or $$\psi = \frac{n_s}{n_0}\left(1 - \frac{T^4}{T_c^4}\right) \qquad ...(15)$$

where $\psi$ is called the order parameter.

### 15.13.1 Diamagnetism of a Different Kind

For normal magnetic materials

$$\vec{B} = \mu_0(\vec{H} + \vec{M})$$

$$\vec{B}_0 \equiv \mu_0\vec{H}$$

Mostly there is a linear relationship between $\vec{B}_0$ and $\vec{M}$ given by

$$\mu_0\vec{M} = \chi\vec{B}_0$$

giving $$\chi = \frac{\mu_0\vec{M}}{\vec{B}_0}$$

or the magnetic susceptibility.

$$\therefore \qquad \boxed{\chi = \frac{\vec{M}}{\vec{H}}}$$

$$\text{or} \qquad \vec{M} = \chi \vec{H}$$

$$\vec{M} + \chi \vec{M} = \chi \vec{H} + \chi \vec{M}$$

$$\text{or} \qquad \vec{M} = \frac{\chi \vec{H} + \chi \vec{M}}{(1+\chi)} = \frac{\chi (H + M)}{(1+\chi)}$$

$$\text{or} \qquad \vec{M} = \frac{\chi \mu_0 (\vec{H} + \vec{M})}{\mu_0 (1+\chi)}$$

Thus, we obtain

$$\vec{M} = K \vec{B} \qquad ...(16)$$

where $K \equiv \dfrac{\chi}{\mu_0 (1+\chi)}$, a material constant.

The persistent current $\vec{j}_s$ in a superconductor, placed in a magnetic field, is thought non-uniform magnetization $\vec{M}(r)$ over the penetration depth. The current density $\vec{j}_s$ is related to effective magnetization $\vec{M}$ by the following equation

$$\vec{j}_s = \text{curl } \vec{M} \equiv \nabla \times \vec{M} \qquad ...(17)$$

$$\Rightarrow \qquad \nabla \times \vec{j}_s = \nabla \times \nabla \times \vec{M} \qquad ...(18)$$

Using Eq. (18) in the second London equation

$$\text{curl } \vec{j}_s = -\left(\frac{n_s e^2}{m}\right)\vec{B}$$

we obtain

$$\nabla \times \nabla \times \vec{M} = -\left(\frac{n_s e^2}{m}\right)\vec{B} \qquad ...(19)$$

$$\text{or} \qquad \text{curl curl } \vec{M} = -\left(\frac{1}{\mu_0 \lambda_L^2}\right)\vec{B} \qquad ...(20)$$

$$\text{where} \qquad \lambda_L \equiv \left(\frac{m}{\mu_0 n_s e^2}\right)^{1/2} \qquad ...(21)$$

is the London penetration depth. The Eq. (20) gives a different dependence of $\vec{M}$ on $\vec{B}$ than that represented by Eq. (16), viz., for a normal magnetic material. This stresses the fact that superconductivity is not merely a perfect diamagnetism rather a diamagnetism of a quite new kind, through involvement of the London penetration depth.

## 15.14 PIPPARD'S EQUATION AND COHERENCE LENGTH

If one looks at the expression of the penetration depth as given by

$$\lambda = \left(\frac{m}{\mu_0 n_s e^2}\right)^{1/2} \qquad ...(1)$$

then impurities in a material should not change appreciably the penetration depth. Pippard did an experiment by measuring the microwave surface impedance of Sn (tin) diluted with small amounts of indium at about 3-cm wavelength and estimated the penetration depth from the surface impedance values. According to London's theory, there should not be an appreciable change in the penetration depth of tin, but Pippard observed that penetration depth changes roughly by a factor of 2 at low temperatures, and this led to the non-local modification of the London theory and the concept length $\xi$, which depends on the mean free path.

We have
$$\vec{B} = -\frac{m}{n_s e^2}\,\text{curl}\,\vec{j}_s$$

or
$$\vec{H} = -\frac{m}{\mu_0 n_s e^2}\,\text{curl}\,\vec{j}_s$$

Also,
$$\vec{H} = \text{curl}\,\vec{A}$$

$\therefore$
$$\vec{A} = -\frac{m}{\mu_0 n_s e^2}\,\vec{j}_s \qquad ...(2)$$

Pippard postulated that current density at any point in the superconductor depends on the vector potential $\vec{A}$ not at that point, but integrated over a domain of radius $\xi$, known as *coherence length*. Further, $\xi$ must depend on the mean free path of the electrons in order to account for the variation of penetration depth with impurity content. The current density is given by the generalised expession as

$$\vec{j}_s(\vec{r}) = \int K(r, r')\,\vec{A}(r')\,dr' \qquad ...(3)$$

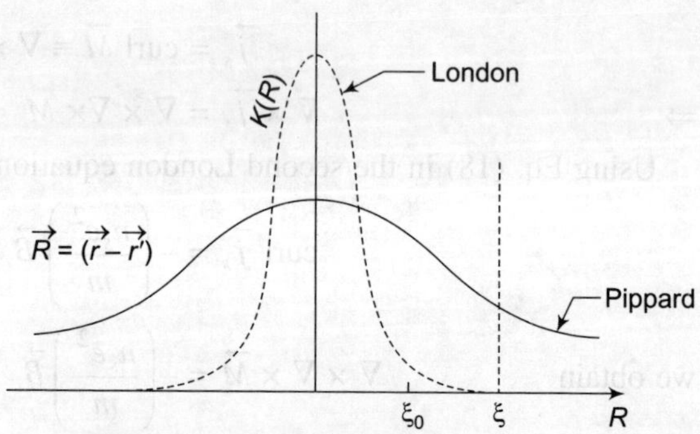


**Figure 15.18** Kernel function $K(R)$ vs. $R$ (representation of coherence length)

The width of the kernel function $K(\vec{r} - \vec{r}')$ defines the *Pippard's coherence length,* $\xi$. Pippard suggested the form of kernel as (see Fig. 15.18)

$$K(r, r') = \exp\,[-(r - r')\,(1/\xi)] \qquad ...(4)$$

with
$$\frac{1}{\xi} = \frac{1}{\xi_0} + \frac{1}{\alpha} \qquad ...(5)$$

where $\xi$ is the electromagnetic coherence length (which is of the order of $10^{-4}$ cm), $\alpha$ is an empirical constant of over unity and $\xi_0$ is intrinsic coherence length.

The apearance of the electron mean free path shows that the coherence length is reduced by the presence of impurities in a superconductor and, therefore, the ratio

$$k_L = \frac{\lambda}{\xi} \qquad ...(6)$$

is a number which is controlled by alloying or by doping a superconductor, $k_L$ is known as *Ginzberg–Landau parameter*.

## 15.15 VARIATION OF ORDER PARAMETER ACROSS A 'SUPERCONDUCTOR-NORMAL INTERFACE

The variation of order parameter ($\psi$) through an interface between normal and superconducting regions is depicted in Fig. 15.19.

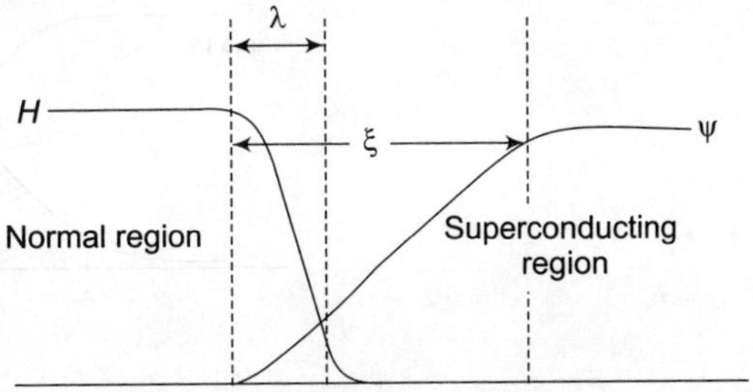


**Figure 15.19** Variation of magnetic field H and the order parameter $\psi$ across a normal and superconduction interface in a type I superconductor

For a pure semiconductor, an expression for intrinsic coherence length ($\xi_0$) is obtained as follows. Since the electron states responsible for superconductivity lie within $kT_c$ of the Fermi surface, from uncertainty principle their lifetime ($\tau$) is given by

$$(kT_c)\tau \simeq \hbar \qquad ...(1)$$

If $v_F$ is the electron velocity at the Fermi surface, the wave function must extend over a distance $\xi_0 = v_F \tau$, so

$$\xi_0 = \frac{\hbar v_F}{kT_c} \simeq \frac{\hbar v_F}{2\Delta} \qquad ...(2)$$

A more refined form of this relation is

$$\xi_0 = \frac{2\,\hbar\, v_F}{\pi \Delta} \qquad ...(3)$$

## 15.16 TYPE-I AND TYPE-II SUPERCONDUCTORS

Both coherence length and penetration depth diverge in the same way near $T = T_c$. So, a constant is $\kappa$ (kappa) is defined as

$$\kappa = \frac{\lambda}{\xi} \qquad ...(1)$$

which will remain finite as $T \to T_c$. This is known as Grzberg – Landau parameter and can be considered a characteristic of a material. The magnitude of $\kappa$ defines two kinds of superconductors. For $K \gg 1$ (i.e., $\lambda \ll \xi$) we have type-1 superconductors and superconductors having $k \gg 1$ (i.e., $\lambda \gg \xi$) are type II supercondnctors. Figures 15.20 (a) and 15.20 (b) depict spatial dependence of the order parameter $\psi(x)$ and the magnetic field $H(x)$ at the superconductor – normal interface for respectively, type-1 and type-II superconductors.

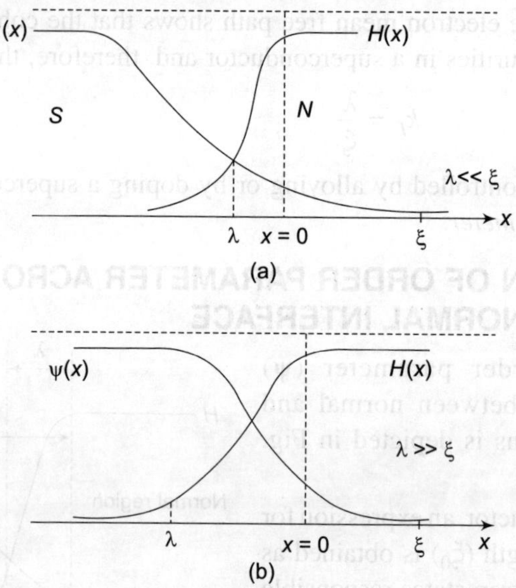


**Figure 15.20** (a) Type-I and (b) Type-II superconductors

Consider a plane boundary between the superconducting ($S$) and normal ($N$) regions in the same material. At the boundary, The magnetic flux density must rise from zero in the superconductor to $H_c$ in the normal region. Over a distance in penetration depth, $\lambda$ (as required by the condition for thermodynamic equilibrium). Likewise, the superconducting wave function $\psi$, (and hence the density of paired electrons) must fall to zero over a distance ~ $\xi$ (the coherence length). Thus, the formation of a S–N boundary involves an increase in energy (or loss of condensation energy $\simeq \xi_0 \left(\frac{\mu_0 \, Hc^2}{2}\right)$ per unit area and as reduction in the magnetic $\simeq \lambda \left(\frac{\mu_0 Hc^2}{2}\right)$ per unit area.

The sign of the surface energy

$$\sigma_s = (\xi - \lambda) \frac{\mu_0 Hc^2}{2} \qquad \text{...(2)}$$

is then determined by the relative magnitudes of the coherence and penetration length. Thus, a type-I superconductor has a positive surface energy and type-II superconductor has a negative surface energy in an applied magnetic field.

The main characteristics type-1 *superconductor* are as follows.

(i) They are soft in nature and behave roughly as ideal superconductors.

(ii) These show complete Meissner effect. The magnetisation versus applied magnetic filed behaviour is shown in Fig. 15.21. In this case the magnetisation drops suddenly at critical field ($H = H_c$). The values of $H_c$ for this class of superconductors are always too low. In coils, they are used as superconducting magnets.

(iii) They may be called as Pippard superconductors. Their behaviour may not be described by the London theory, while it is possible by Pippard's non-local theory. Also Bardeen–Cooper–Schrieffer (BCS) theory can be utilised in the explanation of their various properties like specific heat, tunnelling effects and infrared absorption, etc.

(iv) The London's penetration depth ($\lambda$) and Pippard's coherence length ($\xi$) are used in differentiating both types of superconductors. For type-I superconductors

$$\frac{\lambda}{\xi} = k < 1/\sqrt{2} \quad \text{...(3)}$$

Generally, all simple metallic superconductrs such as Al, In, Sn and Pb are classified as type-I. For alloys of low-concentration effective mass $m^*$, the Fermi velocity is rather high.

(v) Generally, type-I superconductors can be transformed into type-II superconductors by adding impurities or by creating certain defect in the crystal lattice.

The main characteristics of type-II *superconductors* are as follows.

(i) The transition metals of low electrical conductivity and low electronic mean free path in the normal state and alloys like $Nb_3Sn$ and $V_3Ga$ are classified in this group.

(ii) The magnetisation versus applied magnetic field behaviour of type-II superconductors is shown in Fig. 15.21. In this type the magnetisation drops earlier before reaching the critical field ($H = H_c$). and the electrical properties in the superconducting state are also seen up to field $H_{c2}$, which is greater than $H_c$. The state between field values

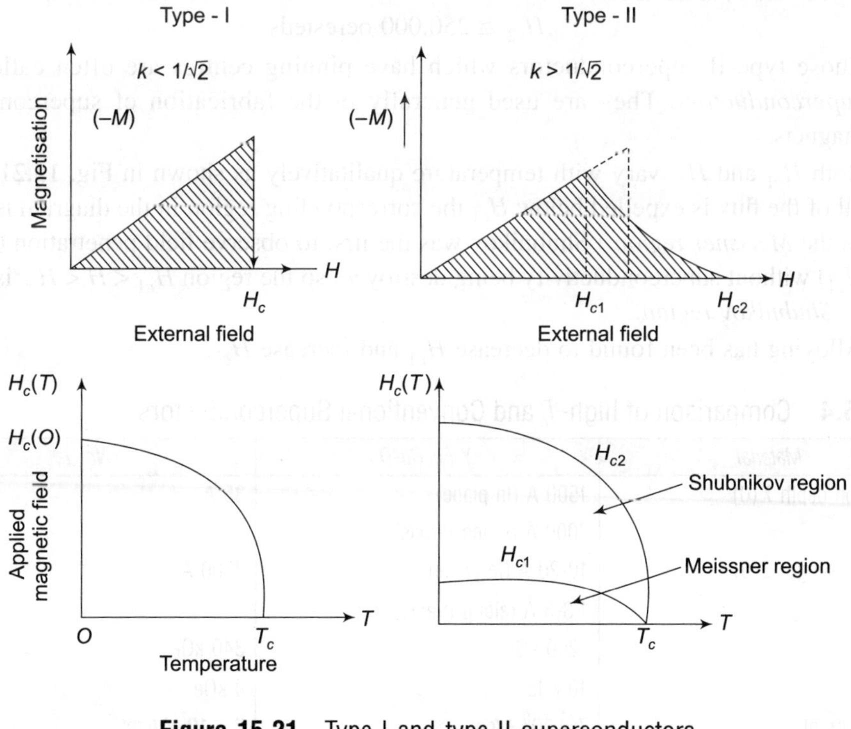


**Figure 15.21** Type-I and type-II superconductors

$H_{c1}$ and $H_{c2}$ is known as the *vortex state* or the mixed state. The transitions at these points are of second order. The of $H_{c1}$ is related to $H_c$ by

$$H_{c1} = \frac{\lambda}{\xi} H_c \qquad ...(4)$$

$$B = 0 \text{ for } H < H_{c1}$$

Flux enters the superconducting specimen as the field $H$ exceeds the critical field, $H_{c1}$. According to Abrikosov, it does so in the form of flux vortices, which assume a triangular lattice structure to minimise their interaction energy. Abrikosov modified Eq. (4) in the manner

$$H_{c1} = \frac{\xi}{\lambda} H_c \, [\log(\lambda/\xi) - 0.27] \qquad ...(5)$$

and

$$H_{c2} = \sqrt{2}\,(\lambda/\xi)\, H_c \qquad ...(6)$$

(iii) For type-II superconductors

$$\frac{\lambda}{\xi} \equiv K > 1/\sqrt{2} \qquad ...(7)$$

(iv) Type-II superconductors are called 'London superconductors'. Their behaviour may be described by London theory.

(v) Type-II superconductors in large fields exhibit no electrical resistance. Near $T = 0$, $H_{c1} \cong (10 - 100)$ Oeresteds

$$H_{c2} \cong 250{,}000 \text{ oeresteds}$$

(vi) Those type-II superconductors which have pinning centres are often called *hard superconductors*. They are used generally in the fabrication of superconducting magnets.

Both $H_{c1}$ and $H_{c2}$ vary with temperature qualitatively as shown in Fig. 15.21. (Since all of the flux is expelled below $H_{c1}$ the corresponding region in the diagram is known as the *Meissner region.*) Shubnikov was the first to observe field penetration (for $H > H_{c1}$) without superconductivity being destroyed, so the region $H_{c1} < H < H_{c2}$ is known as *Shubnikov region*.

(vii) Alloying has been found to decrease $H_{c1}$ and increase $H_{c2}$.

**Table 15.4** Comparison of high-$T_c$ and Conventional Superconductors

| *Material* | *Y-Ba-Cu-O* | *$Nb_3Ge$* |
|---|---|---|
| Penetration depth $\lambda(0)$ | 1500 Å (In plane)<br>7000 Å (Along $c$-axis) | 35 Å |
| Coherence length $\xi(0)$ | 10-20 Å (in plane)<br>1.3-3 Å (along $c$-axis) | 3500 Å |
| $Hc_2(0)$ | 1200 kOe | 340 kOe |
| $Hc(0)$ | 10 kOe | 4 kOe |
| Critical current $J_c$ (4 K, 0 H) | $7 \times 10^6$ A/cm$^2$ | $2 \times 10^7$ A/cm$^2$ (Thin film) |

*Critical current* ($J_c$) is the maximum lossless current that a superconductor can sustain before the voltage develops. Table 15.4 expresses a comparison of parameters of typical high $T_c$ material (Y–Ba–Cu–O) and a conventional superconductor ($Nb_3Ge$).

## 15.17 THE MIXED STATE (IN TYPE II SUPERCONDUCTOR)

In a type-II superconductor, in the mixed state, there is a fine subdivision into normal and superconducting phases. There are normal cores (cylinders of normal regions parallel to the applied magnetic field) embedded in the superconductor material. The microscopic normal region is surrounded by circulating supercurrents. Thus, the field penetration in mixed state occurs in the form of a lattice of vortex lines (normal regions) see Fig. 15.22. Each vortex line Fig. 15.23 has a normal force of radius, $\xi$. At the centre of the core the field is maximum, and it drops to $1/e$ of its maximum value over a distance $\lambda t$. A flux vortex (also known as fluxoid) carries a quantum of magnetic flux

$$\phi_0 = 2.07 \times 10^{-15} \text{ Wabers} \quad ...(1)$$

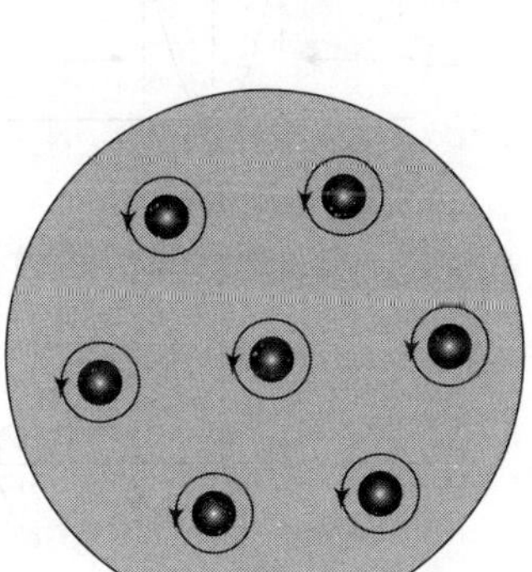

**Fig. 15.22** The triangular lattices of vortices in mixed state

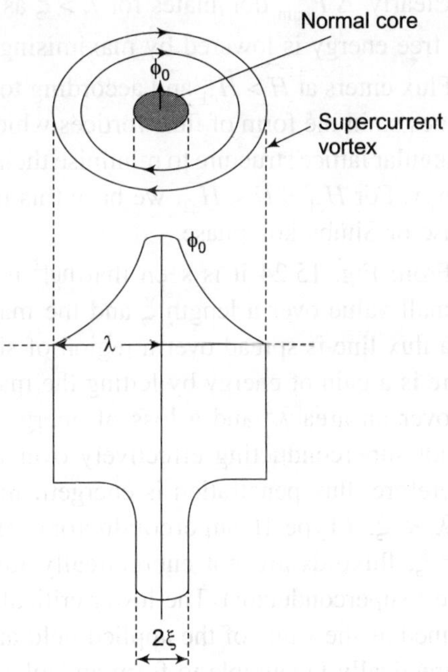


**Figure 15.23** Structure of a vortex line

At $H_{c1}$, the first flux vortex is nucleated, and with increasing field, their equilibrium separation is reduced such that at $H_{c2}$ the normal cores begin to overlap and then bulk of the normal cores begin to overlap and then bulk of the material turns normal (i.e., non-superconducting).

The state between $H_{c1}$ and $H_{c2}$ as known as the mixed state or vortex state (because there is partial flux penetration and there are both superconducting and normal regions in the form of vortices). At $H \geq H_{c2}$, the superconductivity is destroyed. The phase change at $H_c$ represents a first-order transition. The transition at $H_{c1}$ and $H_{c2}$ are of second order.

The free energy $$F_n > F_s \quad ...(2)$$

The free energy cost per unit surface area $(F_n - F_s)$ results from sacrificing the condensation energy in a large of width, $\xi$ (the coherence length), i.e.,

$$\Delta F_{\text{cost}} \text{ varies as } \left(\frac{Hc^2}{2\mu_0}\right)\xi \quad ...(3)$$

($H_c$ is known as the thermodynamic critical field). The free energy of a superconductor increases in an applied field. The free energy-gain derives from allowing the field to invade a distance equal to the penetration depth $\lambda$, i.e.,

$$\Delta F_{\text{gain}} \sim -\left(\frac{Hc_{pp}^2}{2\mu_0}\right)\lambda. \quad ...(4)$$

Clearly, $\Delta F_{\text{gain}}$ dominates for $\lambda > \xi$ as in a type-II superconductor. When $H$ exceeds $H_{c1}$, the free energy is lowered by maximising the $S/N$ interface area. $(H > H_{c1} \Rightarrow B \neq 0)$.

Flux enters at $H > H_{c1}$ and according to Mirikosov, it does so in the form of flux vortices which assume a triangular lattice structure to minimise their interaction energy. For $H_{c1} < H < H_{c2}$, we have this intermediate phase or Shubnikov phase.

From Fig. 15.24 it is seen that $|\psi|^2$ is reduced to a small value over a length $\xi$ and the magnetic field of a flux line is spread over a region of size $\lambda$. Thus, there is a gain of energy by letting the magnetic field in over an area $\lambda^2$ and a loss of energy, because it is not superconducting effectively over an area, $\xi^2$. Therefore, flux penetration is energetically favoured, if $\lambda < \xi$, (Type-II superconductor). However, if $\lambda < \xi$, fluxoids are not energetically favoured (for type-I superconductor). The lower critical field $H_{c1}$ is defined as the value of the applied field at which it is energetically favourable to form an isolated fluxoid.

$$H_{c1} = \frac{H_c}{k\sqrt{2}} \text{ where } k \text{ is GL parameter.} \quad ...(5)$$

For the formation of an isolated fluxoid, the necessary condition is $\lambda > \xi$. A detailed treatment modifies this to

$$k > \frac{1}{\sqrt{2}}. \quad ...(6)$$

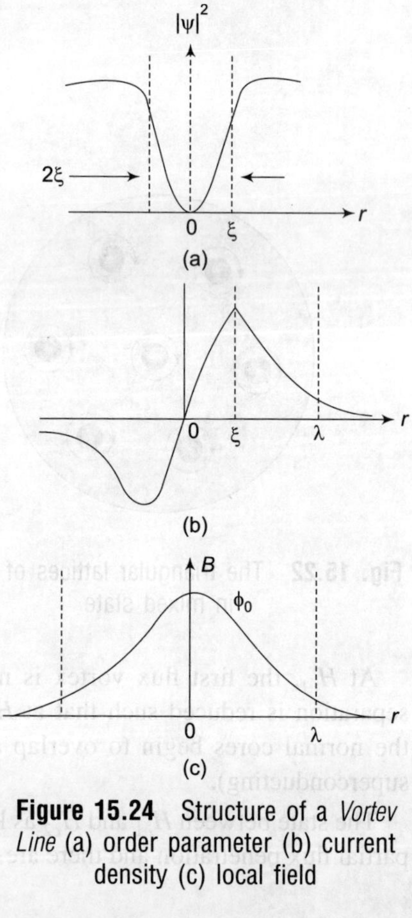


**Figure 15.24** Structure of a *Vortev Line* (a) order parameter (b) current density (c) local field

## 15.18 CRITICAL CURRENT AND SILSBEE'S RULE

By critical current, $J_c$ is meant the maximum lossless current, a superconductor can sustain before the voltage develops.

In general, there can be two contributions to the current flowing on the surface of superconductor. Consider, for example, a superconducting wire along which we are passing a current from some external source. We call this, the transport current (because it transports charge into and out of the wire). If the wire is in an applied magnetic field, screening currents circulate so as to cancel the flux inside. These screening currents are superimposed on the transport current, and at any point, the current density is

$$\vec{J}_t = \vec{J}_i + \vec{J}_H \qquad ...(1)$$

where $\vec{J}_t$ is due to transport current and $\vec{J}_H$ arises from screening current.

The superconductivity breaks (i.e., a resistance appears), if the magnitude of total current density $\vec{J}_t$ at any point exceeds the critical current density $J_c$.

If the total current flowing is sufficiently large, the current density at the surface will reach the critical value ($J_c$) and the associated magnetic field strength at the surface will have a value, $H_c$. Conversely, a magnetic field of strength ($H_c$) at the surface is always associated with a surface super-current density, $J_c$. This leads to the hypothesis:

"A super-current loses its zero resistance at any point on the surface, when the total magnetic field strength, due to transport current and applied magnetic field, exceeds the critical field $H_c$."

Clearly the stronger the applied magnetic field, the smaller is the critical current. If there is no applied magnetic field, the only magnetic field will be that generated by any transport current. In that case, the critical current will be that generated by any transport current. In the case, the critical current will be that which generates the critical magnetic field ($H_c$) at the surface. This special case (of above) is known as *Silsbee's rule*. $H_c$ is known as bulk or thermodynamic critical field.

## 15.19 CRITICAL CURRENT OF A WIRE

Consider a long superconducting wire of circular cross section with radius $a \gg \lambda$, carrying a current $I$. This current produces a circumferential self field (at the surface) of magnitude

$$H = \frac{2I}{ca}. \qquad ...(2)$$

When this field reaches the critical field $H_c$, it will destroy the superconductivity, which is known as the *Silsbee-criterion*. Thus, the critical current will be

$$I_c = \frac{caH_c}{2} \qquad ...(3)$$

which scales with the perimeter, and not the cross-sectional area of the wire. This current flows in a surface layer of a constant thickness, $\lambda$. Since the cross-sectional area of this surface layer will be $2\pi a\lambda$, the critical current density will be

$$J_c = \frac{I_c}{2\pi\lambda a} \quad \text{...(4)}$$

$$= \frac{cH_c}{4\pi\lambda} \equiv J_d \quad \text{(The depairing current density.)} \quad \text{...(5)}$$

(The *depairing current density* sets an upper limit to the $J_c$ and is defined by the energy at which super-electrons are excited over the superconducting gap.)

The thermodynamic critical field ($H_c$) for pure element is small (of the order of a few hundred oersteds). Niobium, lead or tin are not useful for high field magnets, although zero field $J_d$ is very large. For Pb–Bi alloy, it was found (during 1928) that a field of 25 kOe (2.5 Tesla) was necessary to quench superconductivity. But, a solenoid of this alloy proved disappointing as it did not conduct as expected from Silsbee's rule.

From Eq. (4) it is seen the critical current ($J_c$) is inversely proportional to the radius of the sample and represents the current at which vortices first enter the sample. The critical current occurs, when the surface field is equal to $H_c$. This is called *Silsbee's rule.*

Some typical figures for high performance conventional type-II superconductors are given in Table 15.5.

The figures in Table 15.5 are to be borne in mind for use of high temperature superconductors in cryogenic technology.

**Table 15.5** Critical Current and Upper Critical Field

| *Alloy* | $J_c$ | $H_{c2}$ |
|---|---|---|
| NbTi | $3 \times 10^5$ A cm$^{-2}$ at 5T | 11 T (4.2 K) |
| | | 15 T (0 K) |
| $Nb_3Sn$ | $10^6$ A cm$^{-2}$ at 5T | 28 T (0 K) |
| $V_3Ga$ | $2 \times 10^6$ A cm$^{-2}$ at 5T (highest $J_c$) | – |
| $PbGd_{0.2}Mo_6S_8$ | – | 54 T (42) K |
| | | 60 T (0 K) (highest) |

## 15.20 CRITICAL CURRENT IN MIXED STATE

The resistance-less current in a homogeneous type-II superconductor is limited to a value, which just produces field $H_{c1}$ at its surface, as given by Silsbee's rule. Above $H_{c1}$ the sample contains magnetic vortex line threading through bulk of the same in the form of a triangular lattice called flux line lattice (FLL).

Now, if we pass a current (called transport current) $\vec{J}$, which is perpendicular to the vortex-cores, a Lorentz driving force equal to $\frac{1}{c}(\vec{B} \times \vec{J})$ acts on the vortices, tending them to move. If the vortex lines move in response to the Lorentz force, they will dissipate the energy in the flowing current, specifically the dissipation induces a voltage, and thus, resistance in a sample known as flux-flow resistance.

In a homogeneous wire, there is no counteracting force preventing this. Clearly to carry a large current with resistance, the vortices must be pinned, so that their motion is inhibited there.

## 15.21 FLUX QUANTISATION[2]

Consider a superconducting ring. There may be magnetic field $\vec{B}$ threading the ring, but in the superconducting state, field will be zero within the body of the ring. Since the fields are zero, the current density ($\mu_0 \vec{j}_s = \nabla \times \vec{B}$) must also be zero. The currents are restricted to depth, $l$ from the surface. We may therefore define a path of integration around the ring within the interior of the superconductor such that the current density is everywhere zero along the path $l$, i.e.,

$$\int \vec{j} \cdot \vec{dl} = 0 \qquad ...(1)$$

The general form of the London equation is

$$\vec{J}_s = \frac{n_s}{2}(e/m_e)(\hbar \Delta S - 2es\vec{A}) \qquad ...(2)$$

where $S$ is the phase of the wavefunction. In a solid block of supereconductor, the wavefunction must be single valued, i.e., $\Delta S = 0$. But, when the superconductor has a hole passing through it, the change in phase around a closed path $l$ (such as shown in Fig. 15.25) must be zero, i.e., phase $S$ can increase or decrease by multiples of $2\pi$ without altering $\psi = \psi_0 e^{iS(r)}$. (The dimensions of the superconductor should be much greater than $\lambda$ so that we can define a path like $l$.)

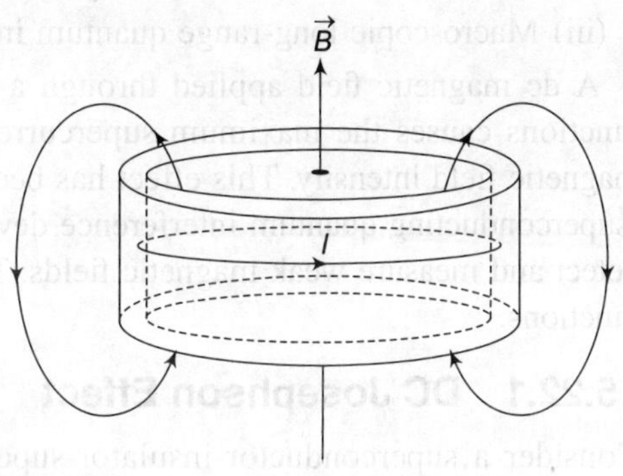


**Figure 15.25** A small superconducting ring

$$\oint \vec{J} \cdot \vec{dl} = \frac{n_s}{2}(e/m_e)\left[\oint \hbar \nabla S \cdot dl - \int 2e\vec{A} \cdot \vec{dl}\right] = 0 \qquad ...(3)$$

or $$2\pi n\hbar = 2e \int \vec{A} \cdot \vec{dl}$$

or $$2\pi n = \frac{2e}{\hbar} \iint (\nabla \times \vec{A}) \cdot \vec{da} \text{ (from Stokes theorem)} \qquad ...(4)$$

$$= \frac{2e}{\hbar} \iint \vec{B} \cdot \vec{da}$$

or $$2\pi n = \frac{2e}{\hbar}\phi = \frac{\phi}{\phi_0} \qquad ...(5)$$

where $$\phi_0 = \frac{\hbar}{2e} \qquad ...(6)$$

where $\phi$ is the magnetic flux which passes through the hole.

---

[2] Flux quantisation was discovered by Deaver and Fairbank and Doll and Zabuer in 1961.

Although flux is excluded from the bulk of the superconductor, the equation above tells us that flux can pass through a hole in it and in units of ($\hbar/2e$). This flux quantum is extremely small and equal to $2 \times 10^{-15}$ Webers.

## 15.22 JOSEPHSON JUNCTION AND JOSEPHSON EFFECT

When one superconductor is separated from another by a thin layer of insulator, the superconductor-insulator-superconductor junction is called a *weak link* or *Josephson*[3] *junction* and Cooper pairs can tunnel asross the thin insulator from one superconductor to the other superconductor.

The effects of pair tunnelling include the following:

(i) *DC Josephson Effect:* A dc current flows across the junction in the absence of any electric or magnetic field.

(ii) *AC Josephson Effect:* A dc voltage applied across the junction causes r.f. current oscillations across the junction. Further, an r.f. voltage applied with the dc voltage can cause a dc current across the junction.

(iii) Macroscopic long-range quantum inference.

A dc magnetic field applied through a superconducting ring containing two Josephson junctions causes the maximum supercurrent to show interference effects as a function of magnetic field intensity. This effect has been utilised in sensitive devices known as SQUIDs (superconducting quantum interference devices), which are sensitive magnetometers used to detect and measure weak magnetic fields. This forms an important application of Josephson junctions.

### 15.22.1 DC Josephson Effect

Consider a superconductor-insulator-superconductor junction (see Fig. 15.26).

Let $\psi_1$ be probability amplitude of electron pairs on one side of junction and $\psi_2$ be the amplitude on the other side. Let both superconductors be identical and both be at zero potential.

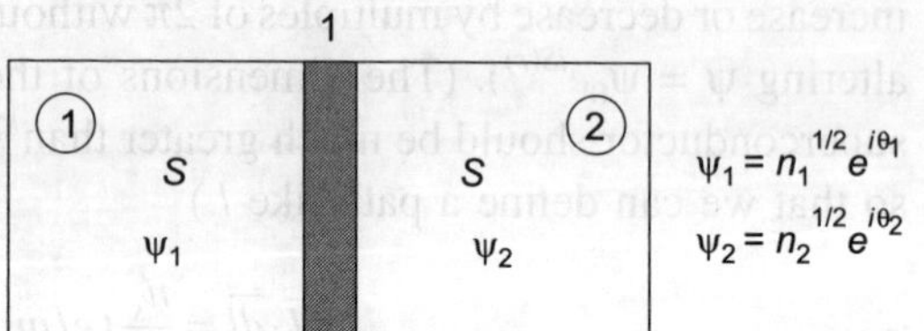


**Figure 15.26** An S-I-S junction

The time-dependent Schrödinger equation, $i\hbar\, \partial\psi/\partial t = H\psi$, applied to the two amplitudes gives

$$i\hbar \frac{\partial \psi_1}{dt} = \hbar\, T\psi_2$$

$$i\hbar \frac{\partial \psi_2}{\partial t} = \hbar\, T\psi_1 \qquad ...(1)$$

Here $\hbar T$ represents the effect of electron pair coupling or transfer interaction across the insulator. $T$ has the dimensions of $(\text{time})^{-1}$. It is a measure of leakage of $\psi_1$ into the region 2 and of $\psi_2$ into region 1. If insulator is thick, $T = 0$ (No pair tunnelling).

---

[3] Giever and Josephson were awarded the Nobel prize in physics for the discovery of supercurrent tunnelling in 1973.

$$\frac{\partial \psi_1}{\partial t} = \frac{1}{2} n_1^{-1/2} e^{i\theta_1} \frac{\partial n_1}{\partial t} + i\psi_1 \frac{\partial \theta_1}{\partial t} = -iT\psi_2 \quad ...(2)$$

$$\frac{\partial \psi_2}{\partial t} = \frac{1}{2} n_2^{-1/2} e^{i\theta_2} \frac{\partial n_2}{dt} + i\psi_2 \frac{\partial \theta_2}{\partial t} - iT\psi_1 \quad ...(3)$$

We multiply Eq. (2) by $n_1^{-1/2} e^{i\theta_1}$ to obtain

$$\frac{1}{2} \frac{\partial n_1}{\partial t} + i n_1 \frac{\partial \theta_1}{\partial t} = -iT(n_1 n_2)^{1/2} e^{-i\delta} \quad ...(4)$$

where $\delta = (\theta_2 - \theta_1)$.

Similarly, we multiply Eq. (3) by $n_2^{-1/2} e^{-i\theta_2}$ to obtain

$$\frac{1}{2} \frac{\partial n_2}{\partial t} + i n_2 \frac{\partial \theta_2}{\partial t} = -iT(n_1 n_2)^{1/2} e^{-i\delta} \quad ...(5)$$

Now,

$$e^{i\delta} = \cos\delta + i \sin\delta$$

$$e^{i\delta} = \cos\delta - i \sin\delta$$

Equating the real and imaginary parts Eq. (4) gives

$$\frac{\partial n_1}{\partial t} = 2T(n_1 n_2)^{1/2} \sin\delta \quad ...(6a)$$

$$\frac{\partial \theta_1}{\partial t} = -T(n_1 n_2)^{1/2} \cos\delta \quad ...(6b)$$

Similarly, from Eq. (5)

$$\frac{\partial n_2}{\partial t} = -2T(n_1 n_2)^{1/2} \sin\delta \quad ...(7a)$$

$$\frac{\partial \theta_2}{\partial t} = -T(n_1 / n_2)^{1/2} \cos\delta \quad ...(7b)$$

If $n_1 = n_2$ as identical superconductors 1 and 2, we have from Eqs. [6(a)] and [7(b)]

$$\frac{\partial \theta_1}{\partial t} = \frac{\partial \theta_2}{\partial t} \text{ or } \frac{\partial}{\partial t}(\theta_2 - \theta_1) = 0 \quad ...(8)$$

From Eqs. [6(a)] and [7(a)], we have

$$\frac{\partial n_1}{\partial t} = \frac{\partial n_2}{\partial t} \quad ...(9)$$

that is, density of pairs changes at equal and opposite rates when there is a supercurrent flow across the junction. The current density per unit area $J_s$ is given as

$$\vec{J_s} = e^* \frac{\partial}{\partial t} |\psi_2|^2 = e^* \frac{\partial}{\partial t} [\psi_2 \psi_2^*] \quad ...(10)$$

$$= e^* \frac{\partial}{\partial t} [n_2 e^{i\theta_2} e^{-i\theta_2}]$$

or $$\vec{J}_s = \left(e^* \frac{\partial n_2}{\partial t}\right) = -e^* \frac{\partial n_1}{\partial t} \qquad ...(11)$$

$$\vec{J}_s = -2e^* T (n_1 n_2)^{1/2} \sin(\theta_2 - \theta_1)$$

$$= -2e^* T (n_1 n_2)^{1/2} \sin \delta$$

$$= J_0 \sin \delta \qquad ...(12(a))$$

where $J_0$ is the maximum zero-voltage current that can be passed by the junction. $J_0$ is proportional to the transfer interaction ($T$) and depends upon the details of overlap of the two wavefunctions.

With no applied voltage, a dc current $J_s$ flows across the junction with a value between $J_0$ and $-J_0$ according to the value of $\delta = (\theta_2 - \theta_1)$. This is dc Josephson effect. The current voltage characteristic is depicted in Fig. 15.27.

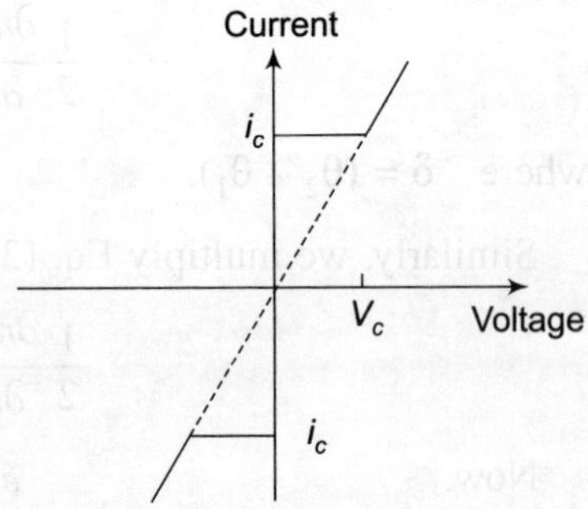


**Figure 15.27** Current-voltage characteristics of a Josephson junction at voltages above $V_c$, the junction has a finite resistance

The tunnelling supercurrent (a macroscopic quantity) is determined directly by the phase of the wave functions. Ambegaokar and Baratoff obtained the following expression for $J_0$:

$$J_0(t) = \left(\frac{\pi\Delta}{2eR_N}\right) \tanh\left(\frac{\Delta}{2t}\right) \qquad ...(12(b))$$

where $\Delta$ is the energy gap, $R_N$, the normal state resistance and, $t$, the temperature. The maximum zero voltage current $J_0$, is therefore temperature dependent and maximum at 0°K. It is thus

$$J_0(o) = \frac{\pi\Delta(0)}{2eR_N} \qquad ...(13)$$

### 15.22.2 AC Josephson Effect

Let $V$ be the voltage applied across the S-I-S junction (it can be applied because the junction is an insulator (Fig. 15.28). An electron pair of charge $q = -2E$ will experience a potential energy difference of ($qV$) on passing across the junction, i.e., a pair on one side is at potential energy (–eV) and a pair in the other side at (+eV). The equations of motion are

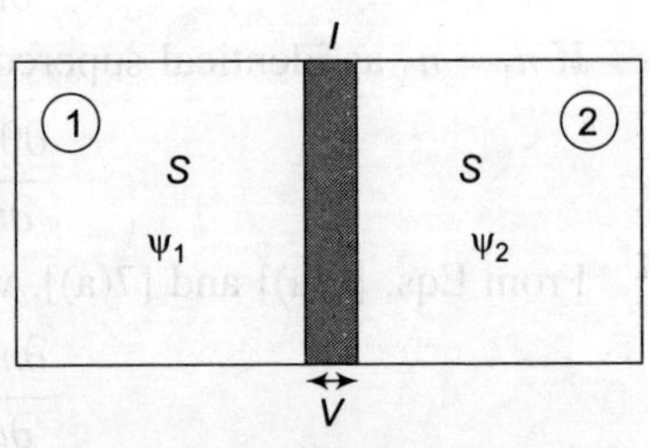


**Figure 15.28** An S-I-S junction with a voltage applied

$$i\hbar \frac{\partial \psi_1}{\partial t} = \hbar T \psi_2 - eV\psi_1 \qquad ...(1)$$

$$i\hbar \frac{\partial \psi_2}{dt} = \hbar T \psi_1 + eV\psi_2 \qquad ...(2)$$

$$\psi_1 = n_1^{1/2}\, e^{i\theta_1}$$

$$\psi_2 = n_2^{1/2}\, e^{i\theta_2} \qquad ...(3)$$

$$\frac{\partial \psi^1}{\partial t} = \frac{1}{2}\, n_1^{-1/2}\, e^{i\theta_1} \frac{\partial n_1}{\partial t} + i\psi_1 \frac{\partial \theta_1}{\partial t} \qquad ...(4)$$

$$= -iT\,\psi_2 + i\frac{\mathrm{eV}}{\hbar}\,\psi_1 \qquad ...(5)$$

$$\frac{\partial \psi_2}{\partial t} = \frac{1}{2}\, n_2^{-1/2}\, e^{i\theta_2} \frac{\partial n_2}{\partial t} + i\psi_2 \frac{\partial \theta_2}{\partial t}$$

$$= -iT\,\psi_1 + i\frac{\mathrm{eV}}{\hbar}\,\psi_2 \qquad ...(6)$$

We multiply Eq. (5) by $n_1^{1/2}\, e^{-i\theta_1}$ to obtain

$$\frac{1}{2}\frac{\partial n_1}{\partial t} + in_1 \frac{\partial \theta_1}{\partial t} = ie\frac{Vn_1}{\hbar} - iT(n_1\, n_2)^{1/2}\, e^{i\delta}$$

where $\delta = (\theta_2 - \theta_1)$. ...(7)

This equation breaks up into the real part (which is without voltage)

$$\frac{\partial n_1}{\partial t} = 2\,T(n_1\, n_2)^{1/2}\,\sin\delta \qquad ...(8)$$

and the imaginary part

$$\frac{\partial \theta_1}{\partial t} = \left(\frac{eV}{\hbar}\right) - T\left(\frac{n_2}{n_1}\right)^{1/2} \cos\delta \qquad ...(9)$$

Further, from Eq. (6), we get

$$\frac{1}{2}\frac{\partial n_2}{\partial t} + in_2 \frac{\partial \theta_2}{\partial t} = ie\frac{Vn_2}{\hbar} - iT(n_1\, n_2)^{1/2}\, e^{-i\delta} \qquad ...(10)$$

Therefore,

$$\frac{\partial n_2}{\partial t} = -2\,T(n_1\, n_2)^{1/2}\,\sin\delta \qquad ...(11)$$

and

$$\frac{\partial \theta_2}{\partial t} = -\left(\frac{\mathrm{eV}}{\hbar}\right) - T\left(\frac{n_1}{n_2}\right)^{1/2} \cos\delta \qquad ...(12)$$

From Eqs. (9) and (12), both with $n_1 = n_2$, we have

$$\frac{\partial(\theta_2 - \theta_1)}{\partial t} = \frac{\partial \delta}{\partial t} = \frac{2\mathrm{eV}}{\hbar} \qquad ...(13)$$

By integration of Eq. (13) we see that with a dc voltage across the junction, the relative phase of the probability varies as

$$\delta(t) = \delta(0) - \left(\frac{2eVt}{\hbar}\right) \qquad ...(14)$$

The current is now given by

$$J = J_0 \sin \delta(t)$$

$$J = J_0 \sin\left[\partial(0) - \left(\frac{2eVt}{\hbar}\right)\right] \qquad ...(15)$$

that is, the current oscillates with a frequency

$$\omega = \left(\frac{2eV}{\hbar}\right) \qquad ...(16)$$

This relation says that a photon of energy $\hbar\omega$ = 2 eV is emitted or absorbed when an electron pair crosses the barrier. A dc voltage of 1 μV produces a frequency of ~484 MHz.

## 15.23 TYPES OF JOSEPHSON JUNCTIONS

Figure 15.29 a–c shows possible arrangements that yield a weakly coupled contact (Josephson junction) between two superconductors.

Figure (15.29a) shows a junction, which consists of two superconducting thin films separated by an insulating oxide film. This insulating layer is an oxide of one of the superconductors and may be intentionally created or simply be the natural oxide film, which exists on the surface of most of the metals exposed to the atmosphere. In this type of junction, the two superconductors are coupled by tunnelling process (tunnelling of Cooper pairs).

Figure (15.29b) is essentially two pieces of superconductors connected by a very small bridge with dimensions of the order of a coherence length. It is usually an evaporated thin film, with the bridge formed by evaporating through an appropriate mask or by photo-etch techniques. The mechanism by which two superconductors are coupled is not simply Cooper-pair tunnelling.

Figure (15.29c) shows a point contact junction. It is simply a small area contact between a micro-size sharp point on one piece of superconductor and another piece of superconductor. It may be of particular interest, because it is extremely simple to fabricate. It can behave like a weak link or tunnel junction, depending on the contact pressure and the nature of oxide films on the superconductors at the contact (critical current can be adjusted by varying applied pressure). See also Fig. 15.30.

### 15.23.1 Typical Current Voltage Characteristics for the above Types of Junctions

The *I-V* characteristic for a thin film tunnel Josephson junction displays a zero voltage current (dc Josephon current), followed by switching along the circuit load line to the single particle tunnelling curve. If the two superconductors are identical, this consists of a very small current at voltages below $2\Delta/e$. At the gap voltage, there is a sharp rise in current followed by an asymptotic approach to the linear normal state *I–V* curve (Fig. 15.31).

The *I–V* characteristic for a weak link junction displays a zero voltage current followed by a smooth transition to the normal state current.

It is worthwhile to mention here that the Josephson tunnelling was observed for the first time by Meissner and Holm in 1932. They observed that the contact resistance between two metals vanishes as soon as both become superconductors, a manifestation of Josephson tunnelling. Figure 15.32 shows equivalent circuit of a Josephson junction.

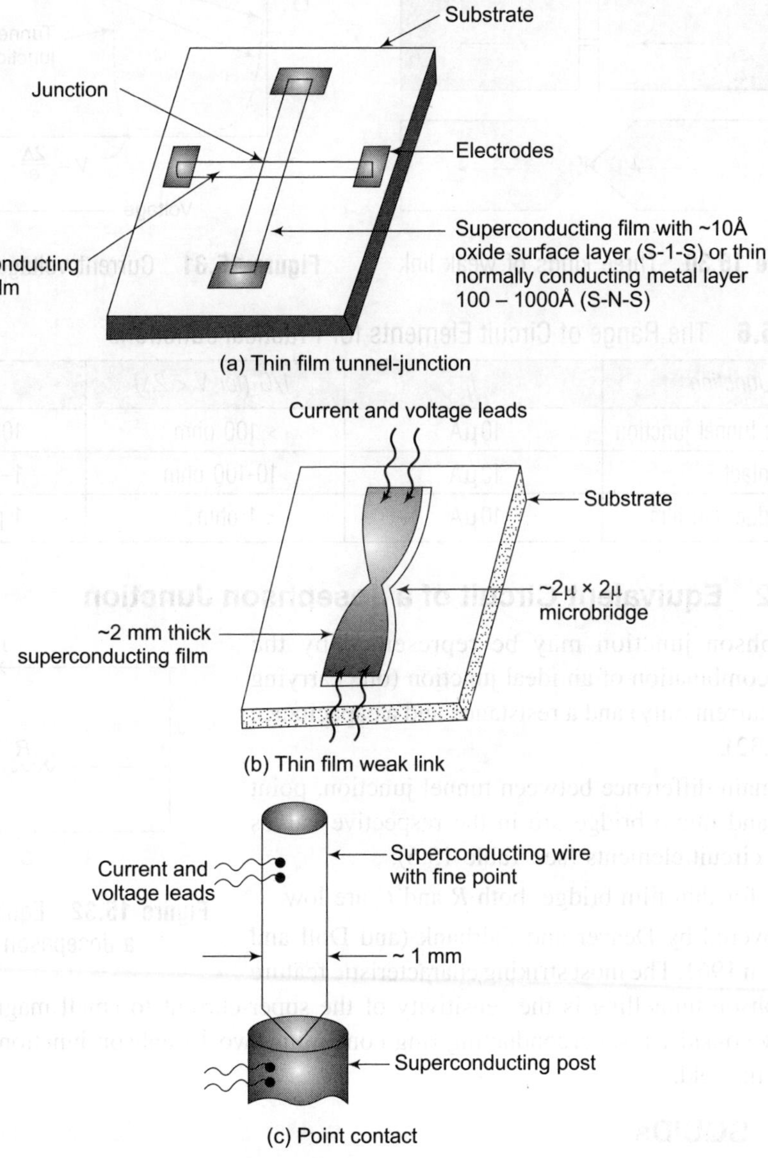


**Figure 15.29** Some Josephson junctions (a) thin film tunnel junction, (b) thin film weak-link, (c) a point contact

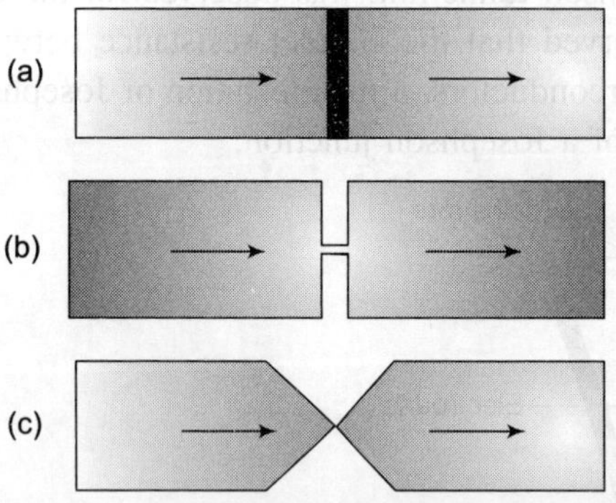


**Figure 15.30** Three kinds of weak link

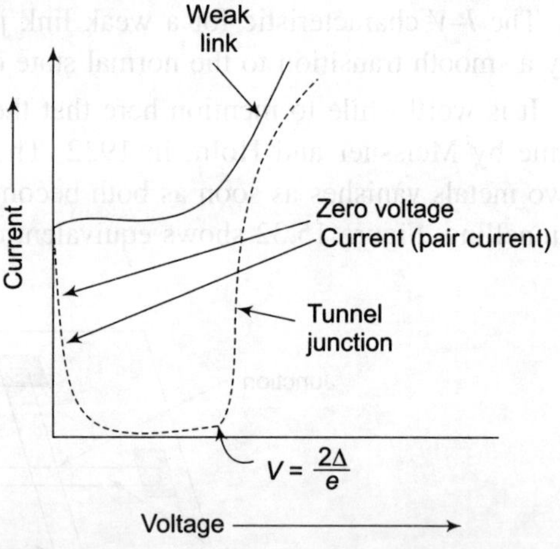


**Figure 15.31** Current voltage characteristics

**Table 15.6** The Range of Circuit Elements for Practical Junctions

| *Junction* | $I_j$ | *1/G (for V < 2Δ)* | *C* |
|---|---|---|---|
| Thin film tunnel junction | 10 μA | > 100 ohm | 100–1000 pF |
| Point contact | 10 μA | 10-100 ohm | 1–1000 pF |
| Microbridge thin film | 10 μA | < 1 ohm | 1 pF |

## 15.23.2 Equivalent Circuit of a Josephson Junction

A Josephson junction may be represented by the parallel combination of an ideal junction (one carrying a super-current only) and a resistance and a capacitance (Fig. 15.32).

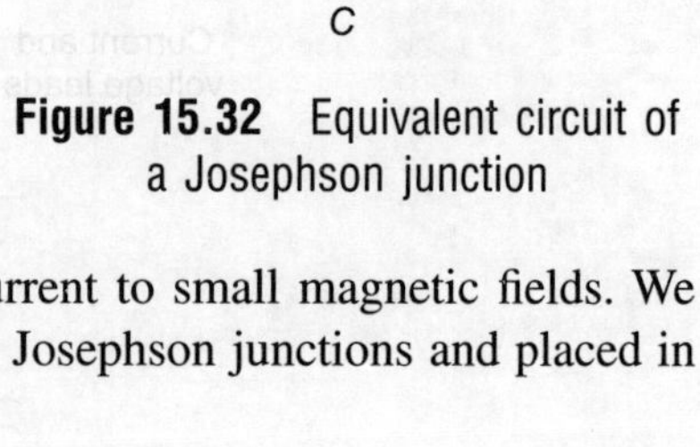


**Figure 15.32** Equivalent circuit of a Josephson junction

The main difference between tunnel junction, point contact and micro-bridge are in the respective values of these circuit elements (see Table 15.6).

Thus, for thin film bridge, both $R$ and $C$ are low.

Discovered by Deaver and Fairbank (and Doll and Nabuer) in 1961. The most striking characteristic feature of Josephson tunnelling is the sensitivity of the super-current to small magnetic fields. We will now consider a superconducting ring containing two Josephson junctions and placed in a magnetic field.

## 15.24 SQUIDs

A superconducting ring containing either one or two Josephson junctions can be used as a very sensitive flux meter. Such a device is known as a superconducting quantum interference device or (SQUID). There are two types of SQUIDs (Fig. 15.33).

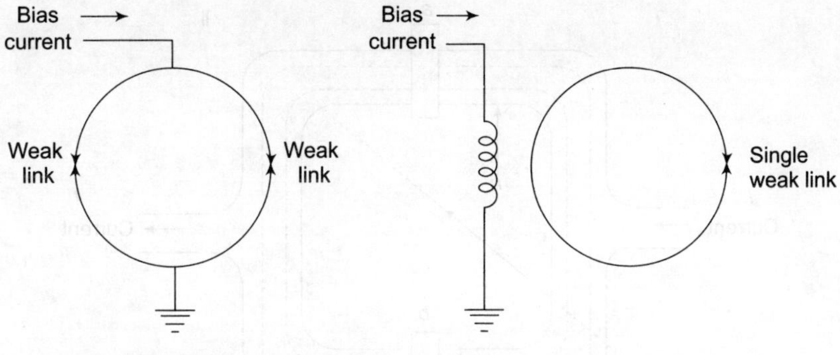


**Figure 15.33** Schematic represent of dc and rf SQUIDs

1. A dc SQUID, contains two Josephson junctions in a superconducting ring and is operated by direct current. The critical current of the SQUID oscillates as a function of the externally applied flux ($\phi$) with a period of one flux quantum ($\phi_0$). If the SQUID is biased with constant current, the voltage across the SQUID is also periodic in $\phi$. Hence, essentially, it is a flux-to-voltage transducer.
2. An r.f. SQUID, contains only one Josephson junction in the superconducting ring, which is coupled to a tank circuit and is operated by r.f. current. The r.f. voltage across the tank circuit is also periodic in $\phi$.

## 15.24.1 DC SQUID (A Superconducting Loop with Two Josephson Junction)

This problem was investigated by Mercereau and coworkers in detail. Since the Josephson effect arises because of the phase coherence between the two superconductors, it is expected that a two-junction device, consisting of a pair of Josephson tunnel junctions connected by superconducting links, would exhibit interference behaviour similar to an optical interference in so far as the critical current for the device is concerned.

*Case I.* The junctions are identical and there is no applied magnetic field (Fig. 15.34).

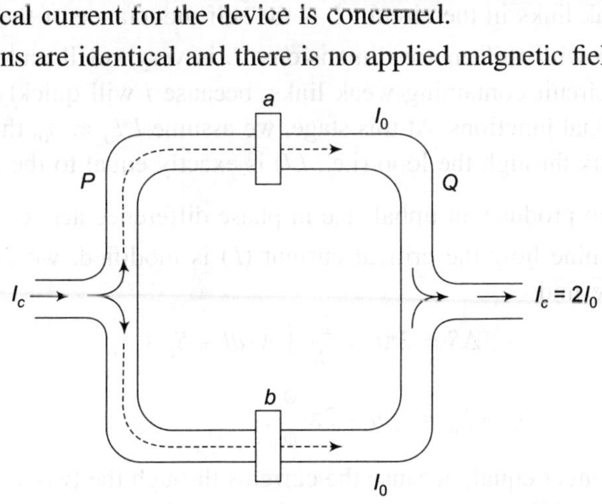


**Figure 15.34** Superconducting ring with two weak links. There is no applied field

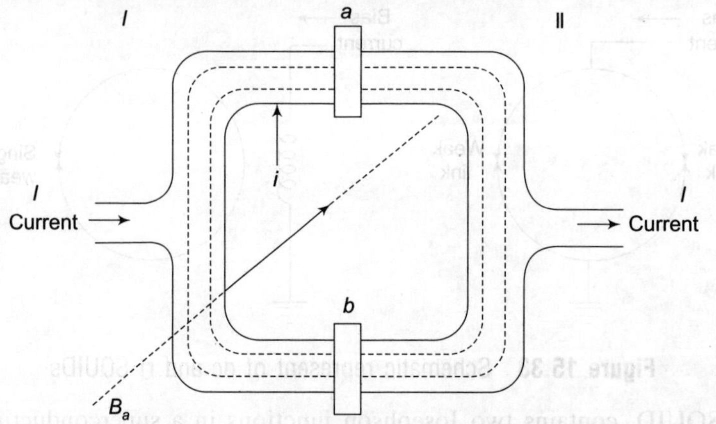


**Figure 15.35** Superconductors I and II with two weak links (a) and (b). A magnetic field $B$ is applied

**Case I.** The junctions are identical and there is no applied magnetic field (Fig. 15.34) (Bias) current $f_c$ from an external source divides equally between the two possible paths so that the phase difference across the junctions is the same ($S_a = S_b$), so that no voltage appears between $P$ and $Q$

and $$l_c = 2I_0. \quad ...(1)$$

**Case II.** There is an externally applied magnetic field. An applied magnetic field perturbs the situation in two ways (Fig. 15.35).

1. A circulating current ($i$) flows along the inner surface of the loop. This current would prevent any flux from passing through the loop and we would find that

$$Li = \phi_a = (B_a) \times (\text{area of the loop}) \quad ...(2)$$

if there were no weak links in the circuit. $L$ is the self inductance of the loop.

For a single loop with millimetre dimensions, $L$ is very small and flux cancellation is not possible in the circuit containing weak links, because $i$ will quickly exceed the critical currents $I_0$ of individual junctions. At this stage, we assume $LI_0 \ll \phi_0$, the flux quantum, i.e., $L \ll \phi_0/I_0$, so that flux through the loop (i.e., $LI$) is exactly equal to the applied flux $\phi_a$.

2. Second effect is to produce an imbalance in phase difference across the junctions.

In order to determine how the critical current ($I$) is modified, we need to calculate the phases $S_a$ and $S_b$. We find

$$\Delta S = 2\pi n = \frac{2e}{\hbar} \int A \cdot dl + S_a + S_b$$

or $$S_a + S_b = 2pn - 2\pi \frac{\phi_a}{\phi_0}. \quad ...(3)$$

$S_a$ and $S_b$ are no longer equal, because the currents through the two junctions are different, being $(i + 1/2)$ and $(i - 1/2)$ when measured in the same clockwise direction. This imbalance in phase can be expressed as

$$S_a = \left(\pi n - \frac{\pi\phi_a}{\pi\phi_0}\right) - S_0 \quad ...(4)$$

$$S_b = \left(\pi n - \frac{\pi\phi_a}{\pi\phi_0}\right) + S_0 \quad ...(5)$$

where $(S_b - S_a) = 2S_0$, is the total phase shift. Here $S_a$ and $S_b$ are phases at (**a**) and (**b**) in presence of flux, $\phi_a$. Equations (3 – 5) show that the phases $S_a$ and $S_b$ and, therefore, the phase difference can be controlled by varying the applied flux, $\phi_a$. The situation is identical to that of two coherent light beams, which are made to interfere (over paths *a* and *b*). The phenomenon of super-current interference becomes evident, if we consider the total current ($I$) due to the currents $I_a$ and $I_b$ in presence of magnetic field.

So in Eqs. (4) and (5), there is a constant phase shift which depends on the nature of the barriers (*a*) and (*b*) and the applied voltage.

$$I_a \cong I_0 \sin S_a \quad ...(6)$$

$$I_b \cong I_0 \sin S_b \quad ...(7)$$

$$I = (I_a - I_b) = I_0\,[\sin S_a - \sin S_b]$$

$$= I_0 \left\{ \sin\left[\pi\left(n - \frac{\phi_a}{\phi_0}\right) - S_0\right] - \sin\left[\pi\left(n - \frac{\phi_a}{\phi_0}\right) + S_0\right]\right\}$$

$$= I_0 \left[2\cos\left(\pi\left(n - \frac{\phi_a}{\phi_0}\right)\right)\sin S_0\right] = 2I_0 \cos\pi\left(n - \frac{\phi_a}{\phi_0}\right)\sin S_0 \quad ...(8)$$

or

$$I = \left(2I_0 \cos\frac{\phi_a}{\phi_0}\right)\sin S_0 \quad ...(9)$$

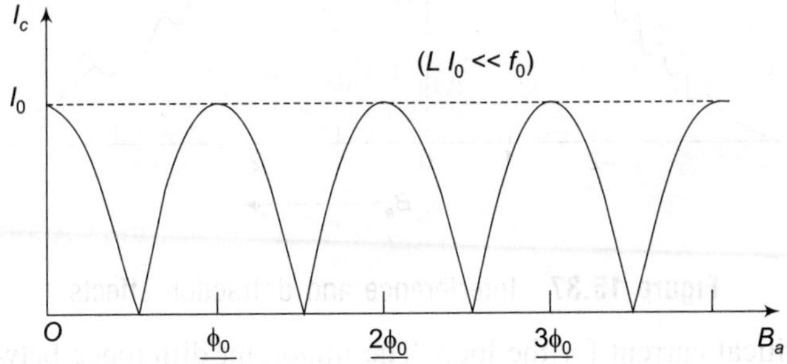


**Figure 15.36** The variation of critical current is a dc SQUID as a function of applied magnetic field. "*a*" is the area of the loop hole [3]

that is, critical current is now given by

$$I_c = 2I_0\left(\cos\frac{\phi_a}{\phi_0}\right) \quad ...(10)$$

Hence, the super-current through two Josephson junctions in a superconducting ring varies as the cosine of the magnetic flux through the ring. The maximum values of current ($I_c$) are obtained, whenever the loop contains an integral number of flux units. Figure 15.36 shows how the maximum current that flows (without a voltage appearing across the circuit) is modulated by applied flux threading the loop. However, this pattern is not observed in practice, since the circulating current does provide some screening of applied flux and coupled flux is $\phi' = \phi - Li$. Then, the interference pattern is unchanged, but the depth of modulation is reduced because the current minima are no longer zero. The modulation of interference pattern is optimum when

$$2(I_c)_{max} \propto \phi_0/L \quad \text{(where } L \text{ is ring inductance)}$$

or

$$(2LIc/\phi_0) \equiv \beta_c.$$

$\beta_c$ is known as the *modulation parameter* of a SQUID.

Figure 15.37 shows experimentally measured current through a pair of Josephson contacts as a function of the magnetic field through the loop. We may write

$$I(\phi_a) = I_c(0)\cos(\phi_a/\phi_0), \quad ...(11)$$

where

$$I_c(0) = 2I_0 \sin S_0 \quad ...(12)$$

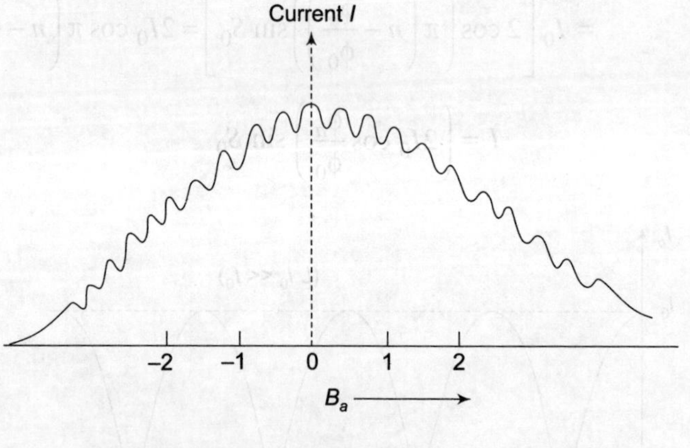


**Figure 15.37** Interference and diffraction effects

is the zero-flux critical current for the loop. The important difference between this case and that of a single junction is that, in this case, the appropriate area is the loop area and not area of a single junction, so that the magnetic field sensitivity is increased manifold.

In Fig. 15.37 the oscillations each correspond to a flux quantum and result from the cosine interference term. Thus, the interference between the Cooper pair tunnelling currents of two parallel Josephson junctions can be exploited to produce extremely sensitive magnetometers (SQUIDs).

### 15.24.2 The Characteristics of an Ideal DC SQUID

The SQUID, in essence, is a flux to voltage transducer; a tiny charge in magnetic flux is converted into a voltage.

The dc SQUID consists of two Josephson junctions arranged on superconducting ring. A current applied to the SQUID, called bias current, divides between the junctions, and if greater than the critical current, produces a voltage across the SQUID. Plotting this current against the voltage yields the characteristic curve as in Fig. 15.38.

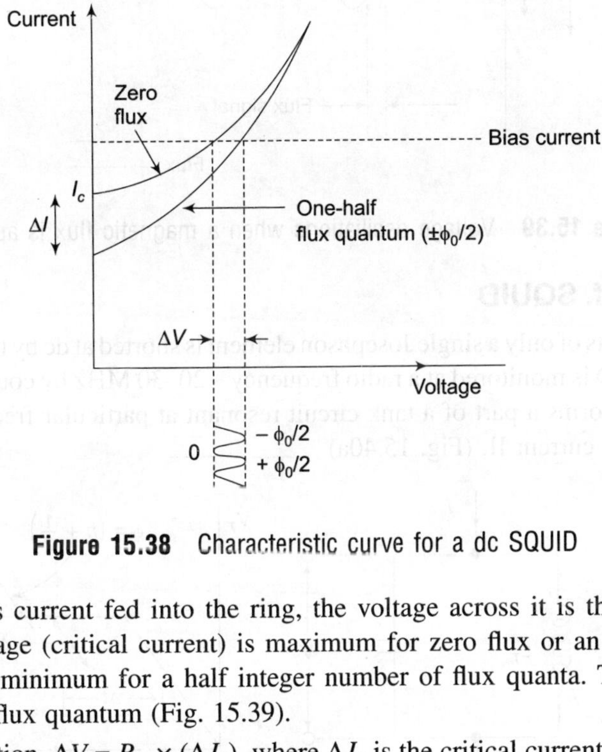


**Figure 15.38** Characteristic curve for a dc SQUID

For a fixed bias current fed into the ring, the voltage across it is the magnetic field is changed. The voltage (critical current) is maximum for zero flux or an integral number of flux quanta and a minimum for a half integer number of flux quanta. The period of these oscillations is the flux quantum (Fig. 15.39).

Voltage modulation, $\Delta V = R_D \times (\Delta I_0)$, where $\Delta I_0$ is the critical current modulation and $R_D$ is the dynamic resistance of the SQUID ($I_0$ is the critical current for individual junction). The depth of critical current modulations $\Delta I$, depends on the product $LI_0$. $L$ must be kept small to critical current modulations $\Delta I$, depends on the product $LI_0$. $L$ must be kept small to reduce the effect of thermal noise, but not so small as to limit the flux-input coupling. $I_0$ must also be small to avoid hysteresis in the weak links, again, with a lower limit determined by the size of $\Delta I$ and the dynamic resistance $R_D$, which falls with decreasing $I_0$.

For these reasons, a value of

$$\frac{LI_0}{\phi_0} \approx \frac{1}{2} \qquad ...(13)$$

is chosen and the modulation depth is then about 50% or $f_0/2L$. The device is usually operated in the finite voltage regime with $I > I_c$.

Then $$\Delta V = R_D \times \Delta I_0 \quad ...(14)$$

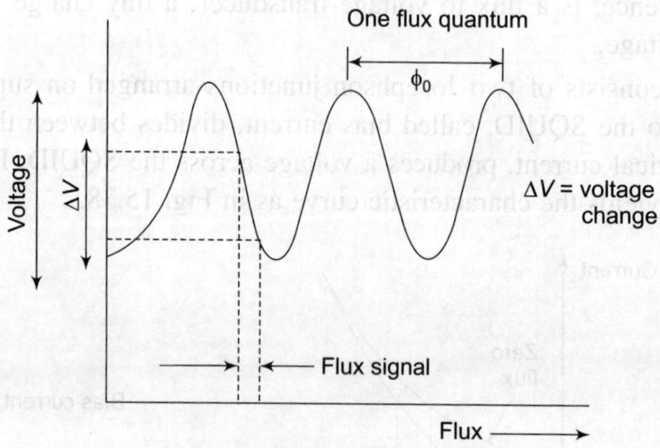


**Figure 15.39** Voltage oscillations when a magnetic flux is applied

## 15.24.3 The r.f. SQUID

The r.f. squid consists of only a single Josephson element is shorted at dc by the superconducting loop. Such a SQUID is monitored at a radio frequency ~ 20–30 MHz by coupling an r.f. current from a coil which forms a part of a tank circuit resonant at particular frequency supplied by a constant r.f.-drive current II. (Fig. 15.40a)

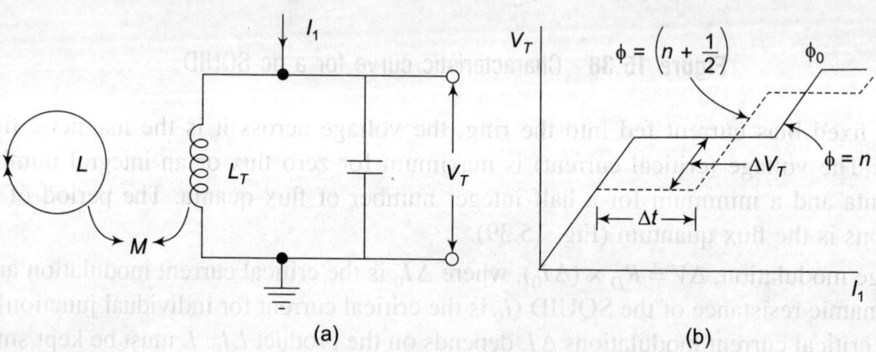


**Figure 15.40** (a) r.f. SQUID, (b) relation between tank voltage $V_T$ and $I_f$ for the cases of integral and half integral flux quanta (For intermediate values of flux, the voltage steps occur at intermediate values of $V_T$)

The magnitude of the r.f. loss from the resonant circuit due to its coupling to the loop depends periodically on the dc flux enclosed by the loop, with period, $\phi_0$. This flux modulated r.f. loss modulates the $Q$ of the tank circuit and hence the magnitude of the r.f. voltage $V_T$

across it (Fig. 15.40 b). This r.f. voltage is detected to provide the flux-dependent readout from the device. With the SQUIDs, one can measure even the extremely small magnetic fields of human brain.

### 15.24.4 Applications of SQUIDs

A SQUID is essentially a flux to voltage transducer, so, it can be used to measure magnetic flux, change in flux, current or voltage. These find the following applications:

1. Medical diagnosis (e.g., activities of brain).
2. Sensing geological changes deep within the earth-crust (for ore deposits).
3. Prospecting to submarine hunting.
4. Earthquake scientists, may, one day, use to predict likely earthquake periods by magnetically monitoring strain build up deep beneath the earth.
5. Measurement of diamagnetism of new HTSCs.

## 15.25 QUANTUM INTERFEROMETER FOR MEASUREMENT OF MAGNETIC FIELD STRENGTH

The system shown in Fig. 15.41 is the quantum mechanical equivalent of Young's slits in optics. The circuit is made up of superconducting material and *X* and *Y* are weak links. The super-current flowing at *W* divides, part going via *X* and part going via *Y* and then recombines at *Z*. The route via *X* is left unchanged while *Y* is subjected to varying magnetic fields and thus to a varying phase angle. The quantum waves are coherent and the recombined wave at *Z* reflects the differing phases between the *X* and *Y* routes. This controls the magnitude of the super-current at *Z*. Measuring this current as a function of applied field at *Y* gives a graph such as Fig. 15.42. Counting the 'fringes' and also following the structure of the current within the fringes gives a sensitive measure of the magnetic field strength at *Y*.

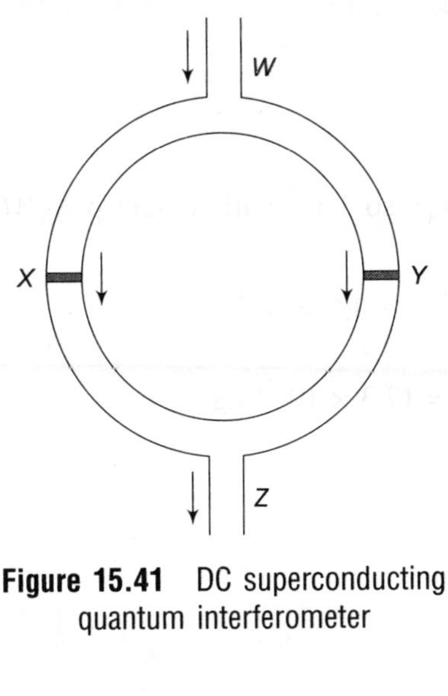


**Figure 15.41** DC superconducting quantum interferometer

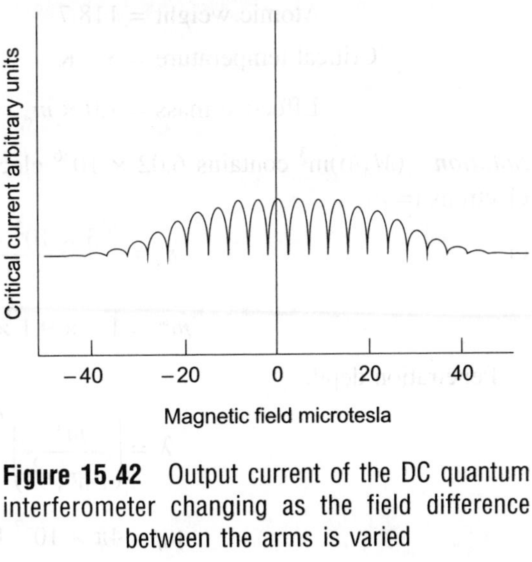


**Figure 15.42** Output current of the DC quantum interferometer changing as the field difference between the arms is varied

This leads to a very sensitive magnetometer, called the SQUID magnetometer. In practice there are refinements and developments that make it reliable and convenient to use but these are not covered here. A typical working arrangement for measuring susceptibility at low temperature is illustrated schematically in Fig. 15.43.

Limiting sensitive for the method are indicated in Table 15.7. It is easily the most sensitive instrument for measuring field strengths, magnetizations and susceptibilities. Its disadvantages are that the sensor must be maintained at a constant, very low temperature and the problem of screening against the unwanted influence of stray external magnetic fields is not trivial.

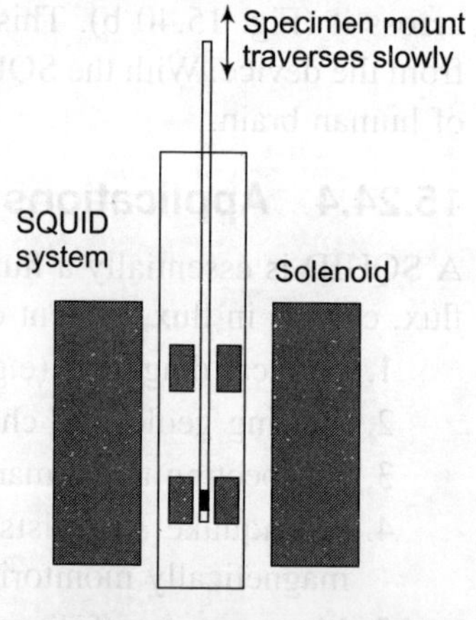

**Figure 15.43** Schematic SQUID magnetometer

**Table 15.7** Sensitivity of SQUID Magnetometer for 1 g Samples

| | *cgs/emu* | *SI* |
|---|---|---|
| Specific magnetization | $10^{-8}$ | $10^{-8}$ |
| Mass susceptibility | $10^{-12}$ | $10^{-8}$ |
| Field | $10^{-8}$ | $10^{-12}$ |
| Induction | $10^{-15}$ | $10^{-19}$ |

**Example 15.1** Assuming that each atom provides one conduction electron, calculate the London penetration depth from the following data

$$\text{Density} = 7.3 \times 10^3 \text{ kg/m}^3$$

$$\text{Atomic weight} = 118.7$$

$$\text{Critical temperature} = 3.7 \text{ K}$$

$$\text{Effective mass} = 1.0 \times m_e$$

***Solution*** $(M_A/\rho)\text{m}^3$ contains $6.02 \times 10^{26}$ electrons ($\equiv N_A$), so 1 $\text{m}^3$ will contain $\rho\, N_A/M_A$ electrons (= $n$)

or
$$n_s = \frac{7.3 \times 10^3 \times 6.02 \times 10^{26}}{118.7} = 3.7 \times 10^{28}$$

$$m^* = 1.9 \times 9.1 \times 10^{-31} \text{ kg} = 17.3 \times 10^{-31} \text{ kg}$$

Penetration depth

$$\lambda = \left(\frac{m^*}{\mu_0 n_s e^2}\right)^{1/2}$$

$$\mu_0 = 4\pi \times 10^{-7} \text{ Henry/m}$$

$$\therefore \quad \lambda = \left[\frac{17.3\times10^{-31}}{4\pi\times10^{-7}\times3.7\times10^{28}\,(1.6\times10^{-19})^2}\right]^{1/2}$$

$$= 15 \times 10^{-13}\text{ m} = 0.0015 \times 10^{-9}\text{ m.}$$

## 15.26 GINZBERG–LANDAU THEORY

There are basically two theoretical approaches to describe various superconducting phenomena:

(i) a phenomenological theory effective through the London equations, and

(ii) the microscopic (BCS) theory.

However, the comprehensive theoretical picture remains incomplete without considering the impact and consequences of yet another theory known as the Ginzberg–Landau theory (GL) (proposed 7 years earlier than BCS theory).

According to Landau, an ordered thermodynamic phase is characterised by a non-zero order parameter, and accordingly, the superconducting state is described in terms of a complex order parameter ($\psi$) that vanishes at $T \le T_c$. This theory is accorded a semiphenomenological status. Near the critical temperature $|T - T_c| \le T_c$, the theory is very much simplified and the behaviour of superconductors is described by GL equations.

According to London, the wave function of the superconducting electrons is rigid, i.e., it does not change when a magnetic field is applied. In GL theory, the absolute rigidity is modified. Parameter ($\psi$) is defined such that $|\psi|^2 = n_s$, the density of superconducting electrons ($\psi$) is a kind of effective wave-function of the superconducting electrons). The phenomenon of superconductivity is considered to be a second-order phase transition, i.e., there is a critical temperature ($T_c$) and a positive parameter $n_s$, which is zero when $T > T_c$ and is finite when $T < T_c$. The Gibb's free energy $F(T, V, n_s)$ therefore, can be expanded as

$$F(T, V, n_s) = F(T, V, O) + \alpha\, n_s + \frac{1}{2}\,\beta\, n_s^2 \qquad ...(1)$$

The equilibrium condition $\partial F/\partial n_s = 0$ gives

$$\alpha + \beta n_s = 0 \quad \text{or} \quad n_s = -\alpha/\beta \qquad ...(2)$$

Sine $F$ is to be minimum, for stability $\partial^2 F/\partial n_s^2 > 0$ and therefore $\beta$ is positive and from Eq. (2), therefore $\alpha$ should be negative. So, to obtain a phase transition of second kind, we assume

$$\alpha\,(T_c) = 0$$

$$\alpha\,(T) = (T - T_c)\left(\frac{\partial \alpha}{\partial T}\right)_{T=T_c} \qquad ...(3)$$

Substituting for $n_s$ from Eq. (2) into Eq. (1) we get

$$F_s = F_n - \frac{\alpha^2}{2\beta} \qquad ...(4)$$

But
$$F_s - F_n = -\frac{H_c^2}{8\pi} \quad ...(5)$$

So we have
$$\frac{\alpha^2}{2\beta} = \frac{H_c^2}{8\pi} \quad ...(6)$$

Here, $n_s$ is always constant and the magnetic field enters as a term $H^2/(8\pi)$ in the free energy of the superconducting phase.

Since $\psi$ is a king of effective wave function, under the presence of an external magnetic field $H$, the free energy will not only increase by $H^2/(8\pi)$ per unit volume, but also by an extra term connected with the appearance of a gradient of $\psi$ (because $\psi$ is not rigid in the presence of the magnetic field). To preserve gauge invariance, GL, assumed the extra term to be

$$\frac{1}{2m}\left|-i\hbar\,\nabla\,\psi + \frac{e^*}{c}\vec{A}\,(\vec{r})\,\psi\right|^2 \quad ...(7)$$

where $\vec{A}\,(\vec{r})$ is the vector potential of the applied field and $e^*$ the effective charge.

Thus,

$$F_s = F_n + \int\left\{\alpha|\psi|^2 + \frac{1}{2}\beta|\psi|^4 + \frac{\mu^2}{8\pi} + \frac{1}{2m}\left|\left(i\hbar\,\nabla\,\psi - \frac{e^*}{c}\vec{A}\psi\right)\right|^2\right\}dv \quad ...(8)$$

This expression is gauge invariant, i.e., invariant under the simultaneous transformation

$$\left.\begin{aligned}\psi' &= \psi\exp\left(\frac{ie^*}{\hbar c}\right)\phi\\ \vec{A}' &= \vec{A} + \nabla\phi\end{aligned}\right\} \quad ...(9)$$

(In fact $e^* = 2e$). We now minimise the expression (8) with respect to $\psi^*$ and $\vec{A}$. Variation with respect to $\psi^*$ gives

$$\delta F = \int\left\{\alpha\psi + \beta|\psi|^2\,\psi + \frac{1}{2m}\left[\left(-i\hbar\,\nabla\,\psi - \frac{e^*}{c}\psi\right)\left(i\hbar\,\nabla - \frac{e^*}{c}\vec{A}\right)\right]\right\}\delta y^*\,dv \quad ...(10)$$

Integrating the last term by parts, we get

$$\frac{1}{2m}\left(\int\delta\psi^*\left(-i\hbar\nabla - \frac{e^*}{c}\vec{A}\right)^2\,\psi\,dv + \iint\hat{n}\cdot\left(-i\hbar\nabla - \frac{e^*}{c}\vec{A}\right)\psi\,\delta\psi^*\,da\right)$$

The surface integral vanishes because $\delta\psi^*$ is arbitrary, hence we get

$$\frac{1}{2m}\left(-i\hbar\nabla - \frac{e^*}{c}\vec{A}\right)^2\,\psi + \alpha\psi + \beta\,|\psi|^2\,\psi = 0 \quad ...(11)$$

and the boundary condition

$$\hat{n}\cdot\left(-i\hbar\nabla - \frac{e^*}{c}\vec{A}\right) = 0$$

Variation with respect to $\vec{A}$ yields the equation

$$\nabla \times \nabla \times \vec{A} = -\nabla^2 \vec{A} = \frac{4\pi}{c}\vec{J} \quad ...(12)$$

where $$\vec{J} = -\frac{ie^*\hbar}{2m}(\psi^* \nabla\psi - \psi\nabla\psi^*) - \frac{e^{*2}}{mc}|\psi|^2 \vec{A} \quad ...(13)$$

(using the gauge $\nabla \cdot \vec{A} = 0$), so, the fundamental equations of the GL theory are

$$\frac{1}{2m}\left(-i\hbar\nabla - \frac{e^*}{c}\vec{A}\right)^2 \psi + \alpha\psi + \beta|\psi|^2 \psi = 0 \quad ...(14)$$

and $$\vec{J} = -\frac{ie^*\hbar}{2m}(\psi^* \nabla\psi - \psi\nabla^*) - \frac{e^{*2}}{mc}|\psi|^2 \vec{A} \quad ...(15)$$

## 15.27 THE BRADEEN-COOPER-SCHRIEFFER (BCS) THEORY (QUALITATIVE IDEAS)

Theoretical understanding of the phenomenon of superconductivity came more than 50 years after its discovery. The microscopic theory was developed by J. Bardeen, L.N. Cooper and J.R. Schrieffer in 1957 [J. Barden, L.N. Cooper and J.R. Schrieffer, *Phys. Rev.* 108, **1175** (1957)]. Prior to this theory, it was predicted by H. Frohlich that the lattice vibrations could yield an attractive interaction between electrons. J. Bhatt and M.R. Schafroth showed that a gas of bosons, each with an electric charge, undergoing Bose–Einstein condensation would show the Meissner effect. It was followed by suggestion of Leon Cooper (in 1956) that electrons in superconducting state form bound pairs. The presence of electrons does not prevent the formation of pairs as the long-range Columb repulsion is largely suppressed by screening due to strong correlation effects of the other electrons.

The microscopic theory of superconductivity was put forward by Bardeen, Cooper and Schrieffer in 1957. Its main points are as follows:

(i) Two electrons interact via lattice vibrations. The electron-lattice interaction leads to an electron–electron interaction. As a result, electrons form bound pairs in the superconducting state.

(ii) The electron–electron interaction is mediated by the exchange of virtual phonons between two electrons of a pair.

(iii) The pairing is such that an electron with spin-up and momentum $\vec{k}$ is paired with spin-down and momentum $-\vec{k}$, and that anugular momentum of the pair is zero.

(iv) The coherence length is the same as the size of the pair, and energy gap observed corresponds to binding energy of the pair.

Consider a metal with conduction electrons inside the Fermi sphere. Suppose two electrons (labelled 1 and 2) lie just inside the Fermi surface. If they form a bound pair, their motion is correlated, provided they are close to the Fermi surface. The pair formation leads to the condensation *Bose-Eienstein condensation* of all electrons in the vicinity of Fermi surface into a sea of *Cooper pairs*.

Let us now consider how the attraction is brought about in the first place. Since the electron 1 is near the Fermi surface, its speed is great (= Fermi velocity). As it passes by the positive ions, ions being heavy, respond rather slowly. By the time the ions feel and completely respond to the electron 1, it leaves its initial region, at least partially, stimulating an overcompensation (Fig. 15.44). In the language of field theory, due to lattice distortion (caused by electron scattering), a virtual *y* phonon is created, i.e., each electron is surrounded by a phonon cloud and the electrons 1 and 2 interact by the exchange of a phonon (Fig. 15.44).

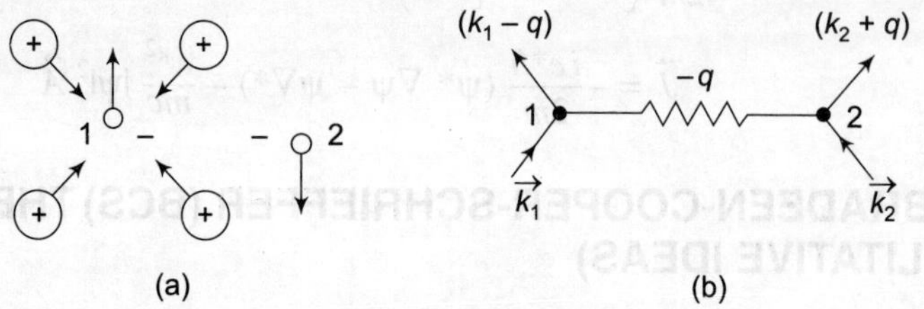


**Figure 15.44** (a) The electron–electron interaction, (b) The phonon-mediated pairing

The phonon of momentum *q*, emitted by electron 1, is absorbed by the electron 2. As a result, the electrons 1 and 2 acquire momenta $(k_1 - q)$, and $(k_1 + q)$, respectively. This interaction is strongest when the two electrons 1 and 2 have opposite momenta and opposite spins

$$\vec{k}_1 = \vec{k}_\uparrow$$

$$\vec{k}_2 = -\vec{k}_\downarrow \quad ...(1)$$

The availability of phonons is restricted to a small energy range $\hbar w_D = k_B\,\Theta_D$ (Fig. 15.46), where $w_D$ is the Debye frequency and $\Theta_D$, the Debye temperature. The interaction is attractive if the energy difference

$$e_{k_1} - e_{k_2} = \hbar w_D \quad ...(2)$$

The pair of two electrons is chosen so that there is maximum possibility of transition from any one pair state to all other pair states without any change in the total momentum

$$(\vec{k}_1 + \vec{k}_2) = (\vec{k'}_1 + \vec{k'}_2) = ... = ... = \vec{k} \quad ...(3)$$

that is, constant momentum for each pair. In the ground state, maximum overlap of phase with largest number of possible transitions is achieved when $\vec{k} = 0$, i.e., when electrons forming a pair have equal and opposite momenta ($\vec{k}$ and $-\vec{k}$) see Fig. 15.45. A detailed consideration shows that energetically, it is most favourable to restrict the pairing of electrons of opposite spins, i.e., $(\vec{k}_\uparrow - \vec{k}_\downarrow)$.

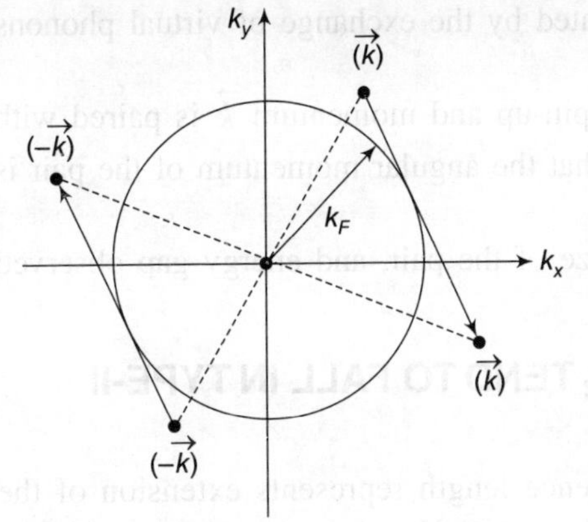


**Figure 15.45** *k*-space representation of the scattering of an electron pair from $(k, -k)$ to the state

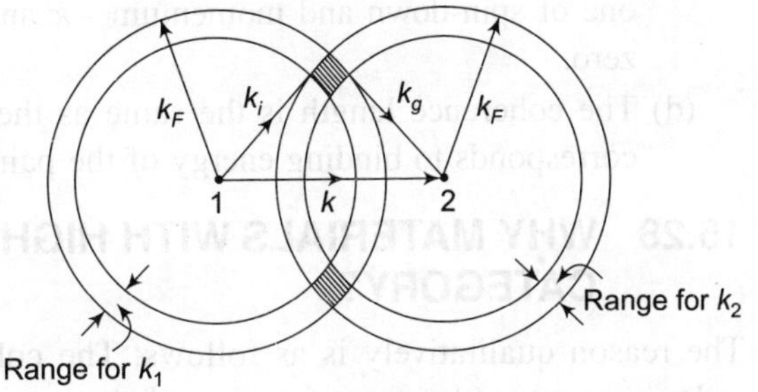


**Figure 15.46** Two shells of radius $k_F$ and thickness $\Delta k$. All pairs of momenta $k_i$, $k_j$ satisfy $k_i + k_j = k$

The energy of pairs is lower than the total energy of free electrons. As a result, a 'gap' is formed near the Fermi energy, which is approximately given by

$$2\Delta = 3.52\ k_B T_c \qquad \text{...(4)}$$

where $k_B$ is Boltzmann constant and $T_c$, the critical temperature.

Below $T_c$, the energy gap parameters ($2\Delta$) rises steeply from zero, accordingly

$$2\Delta(T) \cong 2.52\ k_B\ T_c\ (1 - T/T_c)^{1/2} \qquad \text{...(5)}$$

For $0 < T < T_c$, the pairs are continually being scattered by lattice vibrations. But, since the pair assembly involves a high degree of correlation in their motion, the net momentum is always conserved. Therefore, when an electric field is applied, no momentum is transferred from the moving pairs to the lattice and lattice periodicity has no influence on the electron pairs. Therefore, the wave function is not localised, but has infinite extent. The wave function is given by

$$\psi_G = \psi(r_1, r_2)\ \psi(r_3, r_4) \ldots \psi(r_{n_{s-1}}, r_{n_{s-1}}) \qquad \text{...(6)}$$

where $n_s$ = number of super electrons and number of pairs = $n_s/2$. All the pairs have the same wave function

$$\psi(r', r'') = \sqrt{n_s}\ e^{iS(r)} \qquad \text{...(7)}$$

where the phase function $S(r)$ characterises the coherent state.

The superelectrons (i.e., pairs) provide lossless current, and the superconducting state offers no electrical resistance as the normal electrons, though present, cannot carry any current as they are shorted out by the super electrons.

The *main points* are as follows:

(a) The electron-lattice interaction leads to an electron-electron interaction and electrons form bond pairs in the superconducting state ($T \le T_c$).

(b) The electron – electron interaction is mediated by the exchange of virtual phonons between two electrons of a pair.

(c) The pairing is such that an electron with spin-up and momentum $\vec{k}$ is paired with one of spin-down and momentum $-\vec{k}$ and that the angular momentum of the pair is zero.

(d) The coherence length is the same as the size of the pair, and energy gap observed corresponds to binding energy of the pair.

## 15.28 WHY MATERIALS WITH HIGH $T_C$ TEND TO FALL IN TYPE-II CATEGORY?

The reason qualitatively is as follows. The coherence length represents extension of the order parameter (the wave function of the super-electrons). Using the position momentum uncertainty relation, we write

$$\xi = \frac{\hbar}{\Delta p} \qquad ...(1)$$

where $\Delta p$ is the uncertainty in momentum. But a super-electron lies within an energy interval $k_B T_c$ from the Fermi surface and hence, the uncertainty of its energy is

$$\Delta E \cong k_B T_c. \qquad ...(2)$$

Since $E = p2/(2m)$, it follows that

$$\Delta E = p\frac{\Delta p}{m} \qquad ...(3)$$

or $\Delta p \sim \Delta E = k_B T_c$ which substituted into (1) gives

$$\xi \propto \frac{1}{T_c}. \qquad ...(4)$$

Therefore, the greater the $T_c$, the shorter the coherence length.

Transition metals and alloys usually fall in the type-II category. Their coherence length is shortened by the relatively large amount of scattering present.

Nearly 25 years ago, all known superconductors were having $T_c < 20$ K.

## 15.29 WHY IT IS EXTREMELY DIFFICULT TO OBTAIN HIGHER $T_C$?

From the microscopic (BCS) theory, the $T_c$ is given by

$$k_B T_c = (\hbar\omega)\, e^{-1/\lambda_{\text{eff}}} \qquad ...(5)$$

or

$$k_B T_c \cong (k_B \theta)\, e^{-1/\lambda_{\text{eff}}} \qquad ...(6)$$

$k_B\theta$ denotes the energy range near the Fermi level, in which conduction electrons attract to form Cooper pairs.

(1) If "attraction" is due to interaction with phonons, then $\theta = \theta_D$, the Debye temperature of the material.

(2) $k_B\theta_D$ is of the order of $\hbar\omega_D$, i.e., energy of highest frequency of phonons that can propagate in the substance.

(3) $\lambda_{eff}$ characterises the attractive force within the framework of the BCS theory (i.e., in the weak coupling limit).

$$\lambda_{eff} = N(0)\, V \ll 1, \quad ...(7)$$

where $N(0)$ is density of states in normal phase at the Fermi level, $V$ is density of states in normal phase at the Fermi level, $V$ is average matrix element of the interaction energy ($U$) between the electrons. $\lambda_{eff}$ is a measure of the attractive force (Fig. 15.47).

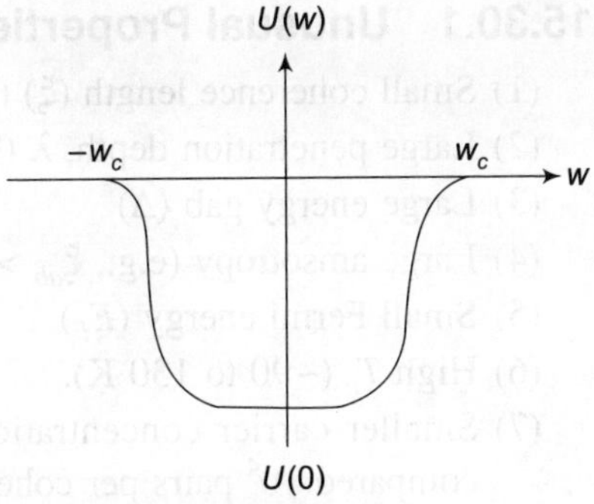


**Figure 15.47** Interaction energy between the electrons

The BCS model (the weak coupling case) gives

$$\frac{2\Delta(0)}{k_B T_c} = 3.53 \quad ...(8)$$

and $$\Delta C = 1.43\, T_c, \quad ...(9)$$

where $\Delta C$ is specific heat jump at $T_c$ and $\gamma$ is coefficient in the law

$$C_n = \gamma T$$

for electronic specific heat in normal state.

If the coupling is weak, then, within the phonon-mechanism

$$T_c \ll \theta_D. \quad ...(10)$$

Specifically, for $\lambda_{eff} = 1/2$ and $\theta = \theta_D$, the $T_c$ is 0.135 $\theta_D$, i.e.,

$$T_c < 40 \text{ K} \quad \text{for} \quad \theta_D < 300 \text{ K}. \quad ...(11a)$$

With an increase in $\lambda_{eff}$, phonon frequencies become lower, $\theta_D$ falls

$$\therefore \quad T_c \sim 20 - 40 \text{ K}. \quad ...(11b)$$

## 15.30 RESEMBLANCE BETWEEN HTSC AND CONVENTIONAL SUPERCONDUCTORS

The following are the points of resemblances:

(1) Superconducting state is made up of paired carriers

(2) Presence of a true energy gap ($\Delta(k) \neq 0$ for all values of $k$)

(3) Jump in specific heat at $T = T_c$

(4) The pair wave function has $\delta$-wave symmetry

(5) Presence of a Fermi-surface.

### 15.30.1 Unusual Properties of HTSCs

(1) Small coherence length ($\xi$) (This reduces the pinning energy implying weak pinning)
(2) Large penetration depth, $\lambda$ (This softens the flux line lattice (FLL)),
(3) Large energy gab ($\Delta$)
(4) Large amsotropy (e.g., $\xi_{ab} > \xi_c$) (This is due to layered structure.)
(5) Small Fermi energy ($E_F$)
(6) High $T_c$ (~90 to 130 K).
(7) Smaller carrier concentration (There are only $\leq 5$ pairs per coherence volume as compared $10^4$ pairs per coherence volume for BCS superconductor)
(8) High $H_{c2}\left(=\dfrac{\phi_0}{2\pi\xi^2}\right)$
(which is anisotropic: much larger when applied parallel to the planes than when applied perpendicular to the planes) ~ 1M gauss (for YBCO) which is 3 to 5 times the fields available presently.
(9) Heat transport dominated by lattice. ($k_{el} \ll k_{lat.}$)
(10) The jump in heat capacity is not sharp.
(11) (a) Sound velocity (elastic constants) is dramatically affected at $T_c$ (unaffected for conventional superconductors).
(b) Presence of low-lying optical phonon modes.
(12) The ratio $\dfrac{\Delta(0)}{E_F}$, which indicates fraction of electrons paired, is much larger than in conventional superconductors.

It may be mentioned here that in HTSCs, the existence of pairs is reflected experimentally by

(i) ac Josephson effect
(ii) direct measurement of $\phi_o$ (flux quantum) and
(iii) Little-Parks oscillations.

## 15.31 THE SUPERCONDUCTING STATE

The fact that apparently there is no scattering of electrons by the atoms of the crystal lattice leads to the conclusion that the wave function describing the electrons in the superconducting state must be different substantially from those in the normal state. Since (in super-conducting state) lattice periodicity has no influence on the electrons, therefore the wave function is not localized but has infinite extent. In view of the uncertainty relation, this implies a precisely definite momentum. Ground state (i.e., superconducting state) is described by

$$\psi_G = \psi(r_1, r_2)\,\psi(r_3, r_4)\,\ldots\,\psi(r_{ns-1}, r_{ns})$$

where $n_s$ = number of super-electrons
(there are $n_s/2$ pairs)
i.e., the wave function is the product of the pair wave functions. All the pairs have the same wave function given by

$$\psi (r', r'') = \sqrt{n_s}\ e^{is(\vec{r})}$$

where the phase function $S(\vec{r})$ characterises the coherent state.

As the temperature is lowered, the electrons having energy close to Fermi energy get a chance to interact with ion-lattice via photons due to reduced scattering at low temperatures. As a result, Cooper pairs get formed, the length scales over which their motion is correlated is the *coherence-length*:

$$\xi(T) = \frac{0.74\,\xi_0}{\sqrt{(1-T)}};\ \xi_0 = \frac{\hbar\, v_F}{\pi\Delta_0}$$

$V_F$ is the Fermi velocity and $\Delta_o$ is the energy gap (at 0°K). (Gap between superconducting pairs and normal electrons.) *The paired electrons move without resistance whereas single electrons do not.*

Before pairing, electrons are separate entities (fermions) having random momenta (due to scattering). When pairs form, two electrons of a pair have zero net momentum and opposite spins, so that is current carrying state they do not undergo inherent scattering, i.e., a pair is like a singlet and all pairs acquire same momentum in the same direction. Same momenta implies the same value of phase gradient for all pairs (momentum $\vec{P} = \hbar\nabla S \equiv 2me\ v$). Therefore, no change in current flow. Their motions are correlated (*phase coherence*). The only scattering event which will reduced the current will be when a pair is imparted with an energy is greater than binding energy of the pair (i.e., $\geq \Delta_o$). For $I < I_c$, there is no way to impart this energy. The long range Coulomb repulsion between two electrons of a pair has been largely suppressed by screening due to strong correlation effects of the other electrons.

Penetration depth

$$\lambda_{L_0} = \sqrt{\frac{m^* c^2}{4\pi n_S q^2}} \quad \text{...(1a)}$$

## 15.32 THE COHERENCE LENGTH

Consider the Ginzburg–Landau equation in the absence of magnetic fields (**A** = 0). Then, in one dimension from first Ginzburg–Landau equation,

$$\left(-\frac{\hbar^2}{2m^*}\frac{d^2}{dx^2} + \alpha + \beta|\psi|^2\right)\psi = 0. \quad \text{...(1)}$$

We define $$f = \frac{\psi}{\psi_0} \text{ where } \psi_0 = \sqrt{-\frac{\alpha}{\beta}}. \quad \text{...(2)}$$

Then $$\left(\frac{-\hbar^2}{2m^*\alpha}\frac{d^2}{dx^2} + 1 - |f|^2\right) f = 0. \quad \text{...(3)}$$

When $f$ has no gradients $|f| = 1$, which would correspond to being well inside the superconductor. We assume a boundary at $x = 0$ between a normal and a superconductor so $f = 1$, $\psi \to \psi_0$ as $x \to \infty$.

Defining $$\xi^2(T) = \frac{-\hbar^2}{2m^*\alpha} \qquad ...(4)$$

and letting $f = 1 + g$, where $g$ is small, then

$$\xi^2(T)\,\frac{d^2g}{dx^2} + 1 + g - (1+g)(1+g)^2 = 0 \qquad ...(5)$$

Keeping only first order in $g$ since it is small,

$$\xi^2(T)\,\frac{d^2g}{dx^2} - 2g \cong 0 \qquad ...(6)$$

$$g(x) \cong \exp[-\sqrt{2}x/\xi(T)]. \qquad ...(7)$$

Thus, the wave function attains its characteristic value of $\psi_0$ in a distance $\xi(T)$. $\xi(T)$ is called coherence length. The coherence length measures the "range" or

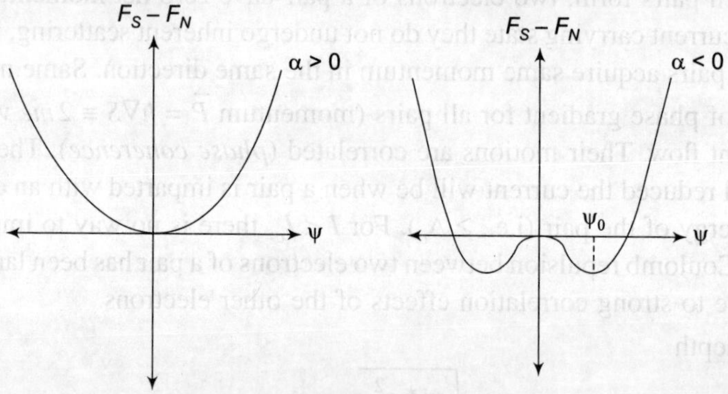


**Figure 15.48** Free energy change at the transition temperature. See (8) for how $\alpha$ enter the free energy with no fields or gradients

"size" of Cooper pairs or the distance necessary for the superconducting wave function to change much.

Let us discuss the coherence length further. First, let us review a little about the free energy. The plots for $F_S - F_N$ are shown in Fig. 15.48. The superconducting transition clearly appears at $T = T_c$ or $\alpha = 0$. The free energy for no fields or gradients is

$$F_S - F_N = \alpha\psi^2 + \frac{\beta}{2}\psi^4 \qquad ...(8)$$

so $$\frac{\partial}{\partial\psi}(F_S - F_N) = 0 \qquad ...(9)$$

gives $$\psi^2 = \psi_0^2 = -\frac{\alpha}{\beta} \qquad ...(10)$$

at the minimum. Thus, at the minimum

$$F_S - F_N = -\frac{\alpha^2}{2\beta} \quad ...(11)$$

which would also be the stabilization energy. From thermodynamics, if $F = U - TS$, and $dU = TdS = -MdH$, then $dF = -MdH$ at constant $T$. For perfect diamagnetism,

$$M = \frac{-1}{4\pi} H \quad ...(12)$$

So,

$$dF_S = \frac{1}{4\pi} HdH \quad ...(13)$$

and

$$F_S(H) - F_S(0) = \frac{H^2}{8\pi}. \quad ...(14)$$

At $H = H_c$, the critical field that destroys superconductivity,

$$F_N(H_c) = F_S(H_c). \quad ...(15)$$

$F_N$ is almost independent of $H$ so,

$$F_N(H_c) = F_N(0) \quad ...(16)$$

We show the idea schematically in Fig. 15.49. Therefore,

$$F_S(0) - F_N(0) = -\frac{H_c^2}{8\pi} = -\Delta F \quad ...(17)$$

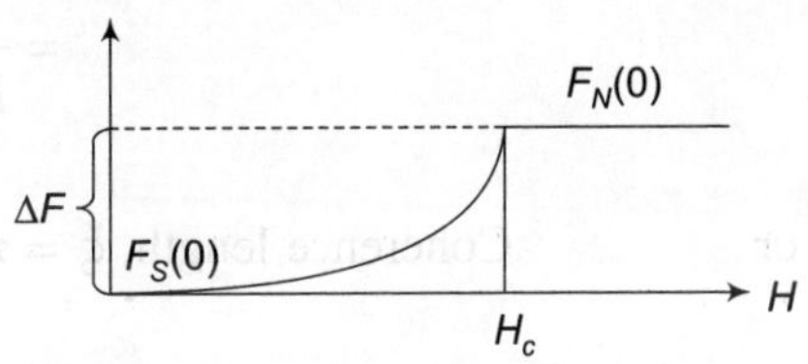

**Figure 15.49** Free energy as a function of field

would also be the negative of the stabilization energy, or using Eq. (11)

$$\frac{H_c^2}{8\pi} = \frac{\alpha^2}{2\beta} \quad ...(18)$$

$$\alpha = \sqrt{\beta \frac{H_c^2}{4\pi}} \quad ...(19)$$

Notice deep inside in the superconductor

$$\psi_S^2 = -\frac{\alpha}{\beta} = n_S = \frac{n}{2} \quad ...(20)$$

and

$$\lambda^2 = \frac{m^* c^2}{4\pi q^2 n_S} \quad ...(21)$$

So,

$$\alpha = -\beta\left(\frac{n}{2}\right) \quad ...(22)$$

by Eq. (20) and thus by Eq. (21) and Eq. (19)

$$\alpha = -\frac{2e^2\lambda^2 H_c^2}{m^* c^2} \quad ...(23)$$

Combining Eq. (4) and Eq. (23), an expression for the coherence length can be derived as follows

From Eq. (4) $$\xi^2(T) = -\frac{\hbar^2}{2m^*\alpha}$$

From Eq. (23) $$\therefore \quad \alpha = -\frac{2e^2\lambda^2 H_c^2}{m^*c^2}$$

So $$\xi^2(T) = \frac{-\hbar^2}{2m^*}\frac{m^*c^2}{(-2e^2\lambda^2 H_c^2)}$$

$$= \frac{\hbar^2}{4e^2\lambda^2 H_c^2}$$

Substituting for $\lambda^2$ from Eq. (1a), we get

$$\xi^2 = \frac{\hbar^2}{4e^2 H_c^2} \times \frac{4\pi n_s e^2}{m^* c^2}$$

$$= \frac{\pi n_s \hbar^2}{H_c^2 m^* c^2}$$

or Coherence length, $$\xi = \pi\left(\frac{\hbar^2}{m^* c^2}\right)\frac{n_s}{H_c^2} \quad ...(24)$$

## 15.33 A CRUDE ESTIMATE OF SPATIAL EXTENSION OF $\Psi_{BCS}$ ($\xi_0$)

Coherence of the BCS state follows from the fact that a large member of Cooper pairs are found to be interlocked over the spatial extent of the BCS wavefunction. It is only in the energy range from ($E_f - 2\Delta_0$) to ($E_f + 2\Delta_0$) that one particle occupation in the superconductor is modified compared to that of a normal conductor. The one particle wave functions that constitutes the two identical particles (i.e., pair) wave functions, therefore originate from this energy range. This concept helps us in making a crude estimate of the spatial extension of $\psi_{BCS}$ in the following way.

If we take the energy uncertainty for electrons in Cooper pairs as $2(2\Delta_0)$:

$$2(2\Delta_0) \sim \delta\left(\frac{p^2}{2m}\right) \approx \frac{p_F}{m}(\delta p)$$

where $\delta p$ denotes the momentum uncertainty. Using the uncertainty relation, the spatial extent ($\psi_{BCS}$) is obtained as

$$\xi_0 \sim \delta x \sim \frac{\hbar}{\delta p} \approx \hbar \times \frac{p_F}{m \cdot 2(2\Delta_0)}$$

$$\approx \frac{\hbar^2 k_F}{2m(2\Delta_0)}$$

$$\approx \frac{1}{k_F}\left(\frac{E_F}{2\Delta_0}\right)$$

Since $E_F/(2\Delta_0)$ is typically of the order of $10^3$ or $10^4$ and $k_F \sim 10^8$ cm$^{-1}$. Therefore, $\xi_0 \approx 10^3$ to $10^4$ Å.

This magnitude turns out to be of the order of the Cooper pair size, interpreted as the intrinsic coherence length. With $E_F/2\Delta_0) \approx 10^4$, we can assume that out of $10^{23}$ electrons per cm$^3$, about $10^9$ form Cooper pairs. Therefore, within the space occupied by a single Cooper pair (~ $10^{-12}$ cm$^3$), there lie centres of gravity of about $10^6$ to $10^7$ other Cooper pairs, forming a Bose–Einstein condensate and it is thus highly misleading to view the pairs as independent bosons, and they are, in fact, spatially anchored to one another. This leads to high spatial coherence and stability of the many body BCS state, analogical to the photons coherence in a Laser beam. This makes it possible for some of the quantum mechanical effects (e.g., flux-quantization) to become macroscopically observable.

## 15.34 COOPER PAIRS[4]

Let us discuss Cooper pairs a little more. A large clue as to the nature of the unusual character of the superconducting state was obtained by L. Cooper in 1956. He showed that the Fermi sea was unstable if electrons interacted by an attractive mechanism—no matter how weak.

Consider the normal Fermi sea of electron with a well-defined Fermi energy, $E_F$. Now added two more electrons interacting with an attractive interaction $V(1, 2)$ and suppose the only interaction with the other electrons is via the Pauli principle.

We write the Schrödinger wave equation for the two electrons as

$$\left[-\frac{\hbar^2}{2m}\nabla_1^2 - \frac{\hbar^2}{2m}\nabla_2^2 + V(1,2)\right]\psi(1,2) = E\psi(1,2) \qquad ...(1)$$

We seek a solution of the form

$$\psi(1, 2) = \frac{V}{(2\pi)^3} A(1, 2)\, \frac{1}{V} \int e^{i\boldsymbol{k}\cdot(r_1 - r_2)} f(\boldsymbol{k})d\boldsymbol{k}, \qquad ...(2)$$

where $A(1, 2)$ is the antisymmetric spin zero spin wave function

$$A(1, 2) = \frac{1}{\sqrt{2}}\,[\alpha(1)\,\beta(2) - \alpha(2)\,\beta(1)],$$

with $\alpha$, $\beta$ being the usual spin-up and down-wave function (note $A^\dagger A = 1$) and $f(\mathbf{k}) = + f(-\boldsymbol{k})$ so that the spatial wave function is symmetric (it can be shown that * with spin 1 and antisymmetric wave function yields no energy shift, at least in our approximation, and in any case such wave functions correspond to $p$-state pairs that we are not considering). Note that the spatial wave function pairs off the electrons into $(\boldsymbol{k}, -\boldsymbol{k})$ states.

---

4 See Cooper L., "Bound Electrons Pairs in a Degenerate Fermi Gas" *Phys Rev* **104**, 1187–1190 (1956).

Inserting Eq. (2) into Eq. (1), we have

$$\frac{1}{(2\pi)^3} A(1, 2) \int \left[ \left( \frac{\hbar^2}{2m} k^2 + \frac{\hbar^2}{2m} k^2 + V \right) f(\boldsymbol{k}) e^{i\boldsymbol{k}\cdot(\boldsymbol{r}_1 - \boldsymbol{r}_2)} \right] d\boldsymbol{k}$$

$$= \frac{A(1, 2)}{(2\pi)^3} \int E f(\boldsymbol{k})\, e^{i\boldsymbol{k}\cdot(\boldsymbol{r}_1 - \boldsymbol{r}_2)}\, d\boldsymbol{k}. \quad ...(3)$$

Now multiplying by

$$A^\dagger(1, 2) \frac{1}{V} e^{-i\boldsymbol{k}'\cdot(\boldsymbol{r}_1 - \boldsymbol{r}_2)},$$

and integrate over $\boldsymbol{r}_1$ and $\boldsymbol{r}_2$ and we obtained ($\boldsymbol{r} = \boldsymbol{r}_1 - \boldsymbol{r}_2$, $V(r_1, r_2) = V(\boldsymbol{r}_1 - \boldsymbol{r}_2) = V(\boldsymbol{r})$, and $E_k = \hbar^2 k^2/2m$)

$$\iint e^{-i\boldsymbol{k}'\cdot\boldsymbol{r}} \, [2E_k + V(\boldsymbol{r})] f(\boldsymbol{k}) e^{i\boldsymbol{k}\cdot\boldsymbol{r}} \, d\boldsymbol{r}d\boldsymbol{k} = \int E f(\boldsymbol{k})\, e^{i(\boldsymbol{k} - \boldsymbol{k}')\cdot\boldsymbol{r}} \, d\boldsymbol{k}. \quad ...(4)$$

Using $$\frac{1}{(2\pi)^3} \int e^{i\boldsymbol{k}\cdot\boldsymbol{r}} \, d\boldsymbol{k} = \delta(\boldsymbol{k}),$$

and $$V_{\boldsymbol{k}',\boldsymbol{k}} = \frac{1}{V} \int e^{i\boldsymbol{k}\cdot\boldsymbol{r}} \, V(\boldsymbol{r}) e^{i\boldsymbol{k}\cdot\boldsymbol{r}} \, d\boldsymbol{r},$$

we obtain $$[2E_{k'} - E] f(\boldsymbol{k}') + \frac{V}{(2\pi)^3} \int f(\boldsymbol{k}') V_{\boldsymbol{k},\boldsymbol{k}'} \, d\boldsymbol{k} = 0 \quad ...(5)$$

We suppose $$V_{\boldsymbol{k}',\boldsymbol{k}} = -V_0 < 0 \quad \text{for} \quad E_F < E_k, \quad E_{k'} < E_F + \hbar\omega_D.$$
$$= 0 \quad \text{otherwise.}$$

We divide by $2E_{k'} - E$ and integrate over $\boldsymbol{k}'$ and obtain (after cancelling)

$$1 = V_0 \frac{V}{(2\pi)^3} \int \frac{dk'}{2E_{k'} - E} \quad ...(6)$$

Note that in the limit of large volumes

$$\frac{V/N}{(2\pi)^3} \int d\boldsymbol{k}' () \leftrightarrow \frac{1}{N} \sum_{k'} \leftrightarrow \int N(E')()dE',$$

where $N(E)$ is the density of state for one spin per unit cell ($N$ unit cells). Thus, with $E_{\text{pair}} = E$.

$$1 = V_0 \int_{E_F}^{E_F + \hbar\omega_D} \frac{N(E')}{2E' - E_{\text{pair}}} dE'. \quad ...(7)$$

Note we can replace $N(E') \cong N(E_F)$ because $\hbar\omega_D \ll E_F$ so we obtain

$$1 = \frac{V_0 N(E_F)}{2} \ln \frac{2E_F + 2\hbar\omega_D - E_{\text{pair}}}{2E_F - E_{\text{pair}}} \quad ...(8)$$

Let $\delta = 2E_F - E_{\text{pair}}$ so

$$\delta = \hbar\omega_D \frac{\exp\left(\dfrac{-1}{V_0 N(E_F)}\right)}{\sinh \dfrac{1}{V_0 N(E_F)}} \qquad ...(9)$$

and in the week coupling limit

$$\delta = 2\hbar\omega_D \exp\left(\frac{-2}{V_0 N(E_F)}\right) \qquad ...(10)$$

We note in particular, the following points:

1. A pair electron wave function that is independent of the electron of $\boldsymbol{r}_1 - \boldsymbol{r}_2$ is said to be an *s* wave function, which is consistent with an antisymmetric spin wave function.
2. $\delta$ is not an analytic function of $V_0$ so ordinary perturbation theory would not work.
3. In the BCS theory one considers pairing of all electrons.
4. For $\delta > 0$ then the Fermi sea is unstable with respect to the formation of Cooper pairs.

## 15.35 PHASE COHERENCE

The superconducting ground state is represented by many electron, wave functions

$$\psi_G(\vec{r}_1, \vec{r}_2, ... \vec{r}_{ns}) = \phi(\vec{r}_1, \vec{r}_2)\,\phi(\vec{r}_3, \vec{r}_4) ... \phi(\vec{r}_{n_{s-1}}, r_{ns}) \qquad ...(1)$$

where $n_s$ = is the total number of superelectrons and so, these are $n_s/2$ pairs $\vec{r}_n$ is the position coordinate of $n^{\text{th}}$ electron and the $\phi$'s are the same for all pairs. The local superfluid density is given by

$$n_s(\vec{r}) = |\psi_G(\vec{r})|^2 \qquad ...(2)$$

and the super-current density is given by

$$\vec{J}(\vec{r}) = -i\frac{e\hbar}{2m_e}[\psi_G^* \nabla \psi_G - \psi_G \nabla \psi_G^*] - \frac{e^2}{m_e}\vec{A}(\vec{r})\,\psi_G^*\psi_G \qquad ...(3)$$

where $\vec{A}(\vec{r})$ is the vector potential defined by

$$\vec{B} = \nabla \times \vec{A}$$

For steady state conditions,

$$\psi_G(\vec{r}) \equiv \psi(\vec{r}) = \psi_0\, e^{is(\vec{r})} = \sqrt{n_s}\, e^{is(\vec{r})} \qquad ...(4)$$

where the phase $S(\vec{r})$ is a real function of position $\vec{r}$. Then, Eq. (3) becomes

$$\vec{J}_s = \frac{n_s}{2}\frac{e}{m_e}(\hbar\nabla s - 2e\vec{A}) \qquad ...(5)$$

$(\because \quad \psi^*\psi = 1)$

The superconductor in superconducting state is a Bose–Einstein condensate made up of electron-pairs ($n_s/2$) mass 2 $m_e$ and charge $2e$.

The superconducting state may be thought of as a Bose–Einstein condensate, as the electron-pairs are bosons and not fermions.

Equations (4) and (5) imply we cannot describe a supercurrent in homogeneous materials unless $\psi_G$ $(r)$ is complex. The physical situation depends on the variation of the phase and not simply on the magnitude of $\psi$.

From Eq. (5), since $\vec{J}_s = \frac{n_s}{2} e \, \vec{v}_s$

$$m_e^* \vec{v}_s (\vec{r}) = \hbar \nabla s (\vec{r}) - 2e^* \vec{A} (\vec{r}). \quad ...(6)$$

In the absence of a magnetic field, the phase function $S(\vec{r})$ plays the part of a velocity potential for the superfluid velocity, $\vec{v}_s$. Therefore, if $s$ did not vary in space, there would be no supercurrent. The phase function $S(\vec{r})$ characterises the electron–ordinary, represents the same phase of all pairs (termed as the phase–coherence).

In the absence of $\vec{A}$, the local value of $S$ varies with position at a rate proportional to $v_s$, i.e., $\nabla S \propto \vec{v}_s$ along the direction of flow.

## 15.36 THE PHONON-INDUCED ELECTRON-ELECTRON INTERACTION

The origin of the phenomenon related to the superconductivity is a small attractive interaction between electrons; in the conventional superconductivity it is generally accepted that this indirect interaction occurs via the phonon field (Fröhlich attractive mechanism). While the quantitative analysis of effective interactions constitute a formidable problem, from a qualitative point of view simple models can be considered to evidentiate some basic aspects; in particular it is not difficult to show that one should expect an attractive interactions for electrons lying near the Fermi surface in an energy shell of the order of $\hbar\omega_D$, $\omega_D$ being the Debye frequency. To make this results plausible, we exploit a well-known standard canonical transformation, and then we apply it to the indirect interaction via phonon field.

### *A Canonical Transformation*

Let us consider a Hamiltonian $H$; we can always perform a canonical transformation of the type

$$\tilde{H} = e^{-S} H e^{S} = \left(1 - S + \frac{1}{2!}S^2 - ...\right) H \left(1 + S + \frac{1}{2!}S^2 + ...\right)$$

$$= H + [H, S] + \frac{1}{2!} [[H, S], S] + \frac{1}{3!} [[[H, S], S] S] + ... \quad ...(1)$$

with $S$ any arbitrary operator. It is evident that the eigenvalues of $\tilde{H}$ and $H$ coincide.

Suppose that the operator $H$ is split in the form

$$H = H_0 + H_1 \quad ...(2)$$

Then Eq. (1) reads

$$\tilde{H} = H_0 + H_1 + [H_0, S] + \frac{1}{2}[[H_0, S], S] + \ldots$$

We can exploit the arbitrariness of $S$ to satisfy the relation

$$H_1 + [H_0, S] = 0. \qquad \ldots(3)$$

With such a choice, $\tilde{H}$ becomes

$$\tilde{H} = H_0 + \frac{1}{2}[H_1, S] + \ldots$$

If we imagine to replace $H_1$ with $\lambda H_1$ (with $\lambda$ eventually put equal to 1) we see that the terms neglected in $\tilde{H}$ are at least of order $\lambda^3$. Thus, the eigenvalues of $H$ and the eigenvalues of $\tilde{H}$, where

$$\tilde{H} = H_0 + H_{\text{indirect}}$$

$$H_{\text{indirect}} = \frac{1}{2}[H_1, S]$$

are equal to order $\lambda^2$ in the coupling parameters $\lambda$.

From Eq. (3), we see that the operator $S$ in a representation in which $H_0$ is diagonal is

$$\langle n|S|m\rangle = \frac{\langle n|H_1|m\rangle}{E_m - E_n}$$

(to avoid singularities we assume that the diagonal matrix elements of $H_1$ in a representation in which $H_0$ is diagonal are zero; if not so, these terms could be included in $H_0$). The expression of the matrix elements of $H_{\text{indirect}}$ on the eigenstates of $H_0$ is

$$\langle f|H_{\text{indirect}}|i\rangle = \frac{1}{2}\langle f|H_1 S - S H_1|i\rangle = \frac{1}{2}\sum_{\alpha}[\langle f|H_1|\alpha\rangle\langle\alpha|S|i\rangle - \langle f|S|\alpha\rangle\langle\alpha|H_1|i\rangle]$$

$$= \frac{1}{2}\sum_{\alpha}\langle f|H_1|\alpha\rangle\langle\alpha|H_1 S|i\rangle\left[\frac{1}{E_i - E_\alpha} + \frac{1}{E_f - E_\alpha}\right]. \qquad \ldots(4)$$

The meaning of $H_{\text{indirect}}$ can be further clarified if we note that $\langle i|H_{\text{indirect}}|i\rangle$ is the standard second order correction given by perturbation theory.

### *Indirect electron—electron interaction*

The mechanism leading to the indirect electron-electron interaction via phonons is schematically represented in Fig. 15.50 one electron of wave vector **k** "emits" a phonon of wave vector **q** and is scattered into the state **k** – **q**; the phonon of wave vector **q** is immediately "absorbed" by another electron of wave vector **k′** that scatters into the state **k′** + **q**. The quantum mechanical analysis of this process provides a net attractive interaction between pairs of electrons, whose energy difference is smaller than the phonon energy, $\hbar\omega_{\mathbf{q}}$. To show this, we can apply the canonical transformation considered before to obtain the effective electron-electron interaction.

We consider final and initial electron states, coupled with the vacuum phonon state (the indirect interaction we obtain is, however, independent from the phonon occupation numbers).

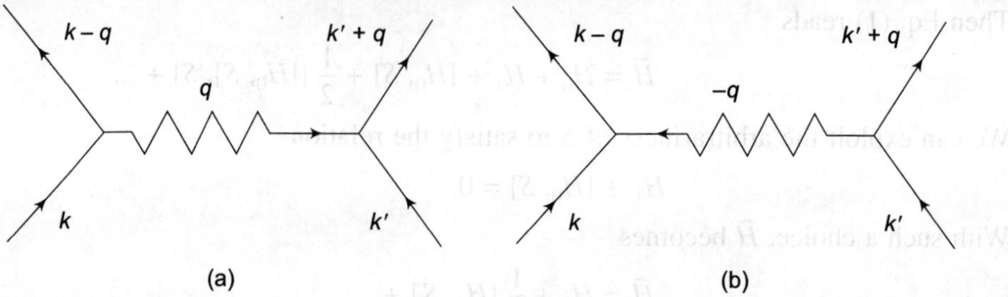


**Figure 15.50** Origin of the electron-electron interaction via phonon

With reference to Fig. 15.50, the states of interest and their unperturbed energy (neglecting electron-phonon interaction) are:

$$|i\rangle = |\mathbf{k}, \mathbf{k}'; 0\rangle \qquad E_i = E_{\mathbf{k}} + E_{\mathbf{k}'}$$
$$|\alpha\rangle = |\mathbf{k}-\mathbf{q}, \mathbf{k}'; 1\rangle \qquad E_\alpha = E_{\mathbf{k-q}} + E_{\mathbf{k}'} + \hbar\omega_{\mathbf{q}} \qquad \text{...(5)}$$
$$|f\rangle = |\mathbf{k}-\mathbf{q}, \mathbf{k}'+\mathbf{q}; 0\rangle \qquad E_f = E_{\mathbf{k-q}} + E_{\mathbf{k'+q}}.$$

Let us consider the electron-phonon interaction term of the type

$$H_1 = M_{\mathbf{q}}\, e^{-i\mathbf{q}\cdot\mathbf{r}}\, a_{\mathbf{q}}^{\dagger} + M_{\mathbf{q}}^{*}\, e^{+i\mathbf{q}\cdot\mathbf{r}}\, a_{\mathbf{q}}. \qquad \text{...(6)}$$

The matrix element for the scattering of an electron from the state **k** into the state **k** – **q** with creation of a phonon **q** on the vacuum phonon state is $M_{\mathbf{q}}\langle 1|a_{\mathbf{q}}^{\dagger}|0\rangle = M_{\mathbf{q}}$; similarly, the scattering amplitude from the state **k′** + **q** with absorption of a phonon **q** is $M_{\mathbf{q}}^{*}$. Applying Eq. (4) with the states (5) we obtain

$$\langle f|H_{\text{indirect}}|i\rangle = |M_{\mathbf{q}}|^2 \left[\frac{\hbar\omega_{\mathbf{q}}}{(E_{\mathbf{k}} - E_{\mathbf{k-q}})^2 - \hbar^2\omega_{\mathbf{q}}^2} + \frac{\hbar\omega_{\mathbf{q}}}{(E_{\mathbf{k'+q}} - E_{\mathbf{k}'})^2 - \hbar^2\omega_{\mathbf{q}}^2}\right]. \qquad \text{...(7)}$$

By assuming $\hbar\omega_{\mathbf{q}} \approx \hbar\omega_D$, it is seen that in the energy shell of the order of $\hbar\omega_D$ around the Fermi energy this virtual process leads to an attractive interaction. It is customary to add to the above indirect matrix element, Eq. (7) the term calculated from Fig. 15.50b; the sum gives

$$\langle f|H_{\text{indirect}}|i\rangle = |M_{\mathbf{q}}|^2 \left[\frac{\hbar\omega_{\mathbf{q}}}{(E_{\mathbf{k}} - E_{\mathbf{k-q}})^2 - \hbar^2\omega_{\mathbf{q}}^2} + \frac{\hbar\omega_{\mathbf{q}}}{(E_{\mathbf{k'+q}} - E_{\mathbf{k}'})^2 - \hbar^2\omega_{\mathbf{q}}^2}\right]. \qquad \text{...(8)}$$

The indirect interaction, Eqs. (7) or (8) have been calculated assuming initial and final electron states coupled to the vacuum phonon state. It is easily seen that the indirect interaction is independent from the phonon state considered; the key point is just the *non-commutation* between the phonon creation and annihilation operators appearing in Eq. (6). [In the hypothetical case of “ordinary scattering” with commutating operators, no indirect interaction would occur].

## 15.37 BARDEEN, COOPER SCHRIEFFER'S MICROSCOPIC THEORY

It is based on electron-phonon interaction.

The total Hamiltonian for interacting electrons and phonons (with $\hbar = 1$) and neglecting the zero point energy of lattice vibrations, is

$$H = H_0 + H_{ep} = \left( \sum_q \omega_q \, a_q^+ \, a_q + \sum_k \varepsilon_k \, c_k^+ \, c_k \right)$$

$$= i \sum_{q,k} B_q \, C_{k+q}^+ \, C_k \, (a_q - a_{-q}^+)$$

The first two terms on the R.H.S. within the brackets are the unperturbed Hamiltonians, $H$. The third term is the electron-phonon interaction term, the first term is the Hamiltonian for phonons only and the second term is the Hamiltonian for electrons only $a$'s are the phonon creation and annihilation operators, and $C$'s are the electron creation and annihilation operators.

It can be shown[5] that the interaction Hamiltonian can be written as

$$H = H_0 + H_1$$

with

$$H_1 \equiv \frac{1}{2} |B_q|^2 \sum_{k,k'} C_{k'+q}^+ \, C_{k'} \, C_{k-q}^+ \, C_k$$

$$\times \left( \frac{1}{\varepsilon_k - \varepsilon_{k-q} - \omega_q} - \frac{1}{\varepsilon_{k'} - \varepsilon_{k'+q} + \omega_q} \right) \qquad ...(1)$$

The above Hamiltonian is in the second quantization notation and the $C$'s satisfy the commutation relations

$$C_k \, C_{k'}^+ + C_{k'}^+ \, C_k = \delta_k^{k'} \qquad ...(2)$$

$$\left. \begin{aligned} C_k^+ \, C_{k'}^+ + C_{k'}^+ \, C_k^+ = 0 \\ C_k \, C_{k'} + C_{k'} \, C_k = 0 \end{aligned} \right\} \qquad ...(3)$$

The Hamiltonian of Eq. (1) describes a process called a *virtual exchange of a phonon* between two electrons. It has the diagrammatic representation shown in Fig. 15.51.

Note that Eq. (1) is independent of the number of phonons in mode $q$, and it is the effective electrons Hamiltonian with phonons in the single mode $q$. To get the effective Hamiltonian with phonons in all modes, we merely have to sum over the modes of $q$. Thus, the total effective interaction Hamiltonian is given by

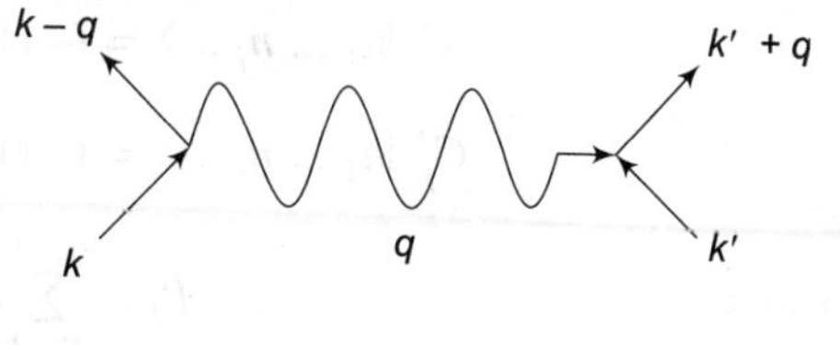


**Figure 15.51** The virtual exchange of a phonon of wave vectors ***q***. The ***k*** are the wave vectors of the electrons. This is the fundamental process of superconductivity

[5] James D. Patterson and Bernard C. Bailey.

$$\mathcal{H}_1 = \frac{1}{2} \sum_{q} \sum_{k, k'} |B_q|^2 C^\dagger_{k'+q} C_{k'} C^\dagger_{k-q} C_k$$

$$\times \left( \frac{1}{\varepsilon_k - \varepsilon_{k-q} - \omega_q} - \frac{1}{\varepsilon_{k'} - \varepsilon_{k'+q} + \omega_q} \right) \qquad ...(4)$$

By dropping further terms that do not involve the interaction of electrons (terms not involving both $\boldsymbol{k}$ and $\boldsymbol{k'}$) and by making variable changes, we can reduce this Hamiltonian to

$$\mathcal{H}_1 = \sum_{q} \sum_{k, k'} |B_q|^2 \frac{\omega_q}{(\varepsilon_k - \varepsilon_{k-q})^2 - \omega_q^2} C^\dagger_{k'+q} C^\dagger_{k+q} C_k C_{k'} \qquad ...(5)$$

From the above equation, we see that there is an attractive electron-electron interaction $|\varepsilon_k - \varepsilon_{k-q}| < |\omega_q|$. We will assume, appropriate excitation energies, that the main interaction is attractive. In this connection, most of the electron energies of interest are near the Fermi energy, $\varepsilon_F$. A typical phonon energy is the Debye energy, $\hbar\omega_D$ (or cutoff frequency with $\hbar = 1$). Many approximations have already been made, and so a very simple criterion for the dominance of the attractive interaction will be assumed. It will be assumed that the interaction is attractive when the electronic energies are in the rage of

$$\varepsilon_F - \hbar\omega_D < \varepsilon_k < \varepsilon_F + \hbar\omega_D \quad (\hbar \neq 1 \text{ here}) \qquad ...(6)$$

The state that do not satisfy this criterion are not directly involved in the superconducting transition, so their properties are of no particular interest. Hence, the effective Hamiltonian can be written in the following form (*fourth major approximation*):

$$\mathcal{H}_1 \equiv -\sum_{q} \sum_{k, k'} V_q C^\dagger_{k'+q} C^\dagger_{k+q} C_k C_{k'}. \qquad ...(7)$$

For simplicity, we will assume the $V_q$ is positive and found from experiment, that $V_q = V_{-q}$ and $V_q = 0$, unless $\boldsymbol{q}$ is such that Eq. (6) is satisfied

The Fermion operators (i.e., $C$'s) satisfy

$$C_j \, |n_1 \ldots n_j \ldots\rangle = (-1)^{P_j} n_j \, |n_1 \ldots (1 - n_j) \ldots\rangle \qquad ...(8)$$

$$C_j^+ \, |n_1 \ldots n_j \ldots\rangle = (-1)^{P_j} (1 - n_j) \, |n_1 \ldots (1 + n_j) \ldots\rangle \qquad ...(9)$$

where $$P_j = \sum_{P=1}^{j-1} n_p \qquad ...(10)$$

Notice the attraction in sign defined by Eq. (9). This attraction is very important for discovering the nature of the lowest energy state. When we begin to guess a trial wave function (if we pay no attention to the alteration of sign), the presence of the interaction will result in little lowering of the energy. What we need is a way of selecting the trial wave function so that most of the matrix elements of individual terms in the second sum in Eq. (6) are negative. The way to do this for the ground state is by grouping the electrons into *Cooper pairs*.

These are some assumptions necessary to construct a minimum energy wave function, for the ground state wave function, it will be assumed that the Bloch states are occupied only in pairs. In fact, the superconducting ground state is a coherent superposition of Cooper pairs. The Hamiltonian concerns the wave vector, and only pairs with equal total momentum will be considered, i.e., $\vec{k} + \vec{k}' = \vec{K}$, where $\vec{K}$ is the same for each pair. Reasonably $\vec{K} = 0$ for the ground (non-current carrying) state of the pairs.

## 15.38 VARIATIONAL DETERMINATION OF GROUND STATE (i.e., AT $T$= 0°K) AND $\Delta_0$ FOR A SUPERCONDUCTOR IN THE BCS THEORY

In the Cooper treatment, we come to know that the normal electron gas in the presence of a (net) attractive electron-electron interaction is unstable, even with respect to the formation of a single Cooper pair, A realistic theory must be able to describe a *cooperative condensation process* in which many pairs of electrons of the normal Fermi sea are formed so to minimize the total energy of the system at $T = 0$.

These demanding physical and formed requirements can be adequately described by the following BCS variational form of the ground state wave function for superconductors

$$|\psi_S\rangle = \prod_{\mathbf{k}} (u_{\vec{k}} + v_{\vec{k}}\, c^{+}_{\vec{k}\uparrow}\, c^{+}_{-\vec{k}\downarrow})\, |0\rangle \quad ...(1)$$

The normal ground-state of a free-electron metal is represented as a slater determinant and in second quantization formalism

$$|\psi_N\rangle = \prod_{\mathbf{k}}^{\mathbf{k}<\mathbf{k_F}} (c^{+}_{\vec{k}\uparrow}\, c^{+}_{-\vec{k}\downarrow})\, |0\rangle \quad ...(2)$$

where $|0\rangle$ is the vacuum state, $c^{+}_{\vec{k}\uparrow}$ creates an electron with moment $\hbar\vec{k}$ and spin up (and obvious meaning for $c^{+}_{-\vec{k}\downarrow}$). $u_{\vec{k}}$ and $v_{\vec{k}}$ are real and even functions of $\vec{k}$, chosen in such a way to minimize the ground state energy.

Under the constraint $u_{\mathbf{k}}^2 + v_{\mathbf{k}}^2 = 1$ (which assures the normalization to unity of $\Psi_S$), $v_{\mathbf{k}}$ represent the probability amplitude that the pair ($\mathbf{k}\uparrow$, $-\mathbf{k}\downarrow$) is occupied and $u_{\mathbf{k}}$ represents the probability amplitude that the pair ($\mathbf{k}\uparrow$, $-\mathbf{k}\downarrow$) is unoccupied. In the superconducting ground state described by $\Psi_S$, electrons are involved only as pairs; in fact, by carrying out explicitly the products in Eq. (2) we see that there is a term without Cooper pairs and terms with one, two, and so on Cooper pairs. Notice that the wave function (2) of the superconductor is reduced to the wave function of the normal metal in the particular case in which $u_{\mathbf{k}}$ and $v_{\mathbf{k}}$ are given by

$$\begin{cases} u_{\mathbf{k}} = 0 \quad v_{\mathbf{k}} = 1 \quad \text{for} \quad k < k_F \\ u_{\mathbf{k}} = 1 \quad v_{\mathbf{k}} = 0 \quad \text{for} \quad k < k_F \end{cases} \quad ...(3)$$

An aspect of the trial wave function (2) is that the total number of electrons is not well defined. Actually, it is not convenient to be tied up with a fixed number of electrons; in a manner which is usual in such cases, we proceed to minimize the expectation value $\langle\Psi_S|H|\Psi_S\rangle$ under the constraint $\langle\Psi_S|N_{op}|\Psi_S\rangle = N$, where

$$N_{op} = \sum_{\mathbf{k}} (c^{\dagger}_{k\uparrow}\, c_{k\uparrow} + c^{\dagger}_{-k\downarrow}\, c_{-k\downarrow})$$

is the particle number operator and $N$ is the number of electrons in the actual metal. The standard method to treat minimization plus one (or more) appropriate constraint consists in introducing one (or more) Lagrange multiplier $\mu$, minimizing without restraints the quantity $\langle\Psi_S|H - \mu|\,\Psi_S\rangle$ and finally determining through the condition $(\Psi_S|N_{op}|\,\Psi_S) = N$; it turns out to be the Fermi energy, and it is same for the normal or superconducting state.

In essence we have thus arrived at minimizing of the quantity

$$\boxed{W_S = \langle\Psi_S|H_{BCS}|\,\Psi_S\rangle} \qquad \text{...(4a)}$$

where $H_{BCS} \equiv H - \mu N_{op}$ is the so called Bardeen–Cooper–Schrieffer Hamiltonian given by

$$\boxed{H_{BCS} = \sum_{\mathbf{k}} \varepsilon_k\, (c^{\dagger}_{\mathbf{k}\uparrow} c_{\mathbf{k}\uparrow} + c^{\dagger}_{-\mathbf{k}\downarrow} c_{-\mathbf{k}\downarrow}) + \sum_{\mathbf{k}\mathbf{k}'} U_{\mathbf{k}\mathbf{k}'}\, c^{\dagger}_{\mathbf{k}\uparrow} c^{\dagger}_{-\mathbf{k}\downarrow} c_{-\mathbf{k}\downarrow} c_{-\mathbf{k}\uparrow}} \qquad \text{...(4b)}$$

and $\varepsilon_{\mathbf{k}} = E_{\mathbf{k}} - \mu = \hbar^2\mathbf{k}^2/2m - \mu$ denotes the single particle energy measured with respect to the Fermi energy.

The expectation value (4a) can be easily evaluated keeping in mind the practical recipe of using the anticommutation rules for systematically shifting the fermion annihilation operators to the right (or creation operators to the left) until they arrive to operate on the vacuum state (giving zero). Here are some matrix elements of interest. We first note that

$$\langle 0|\,(u_{\mathbf{k}} + v_{\mathbf{k}} c_{-\mathbf{k}\downarrow}\, c_{\mathbf{k}\uparrow})\,(u_{\mathbf{k}} + v_{\mathbf{k}}\, c^{\dagger}_{\mathbf{k}\uparrow}\, c^{\dagger}_{-\mathbf{k}\uparrow})\,|0\rangle = u^2_{\mathbf{k}}\, v^2_{\mathbf{k}}. \qquad \text{...(5a)}$$

the normalization of the wave function $|\Psi_S|$ is given by

$$\langle\Psi_S|\Psi_S\rangle = \prod_{\mathbf{k}} (u^2_{\mathbf{k}} + v^2_{\mathbf{k}}) = 1 \quad \text{if} \quad u^2_{\mathbf{k}} + v^2_{\mathbf{k}} = 1 \quad \text{for every } \mathbf{k}. \qquad \text{...(5b)}$$

Similarly, for the number operator we have

$$\langle\Psi_S|c^{\dagger}_{\mathbf{k}\uparrow}\, c_{-\mathbf{k}\uparrow}|\Psi_S\rangle = \langle 0|\,(u_{\mathbf{k}} + v_{\mathbf{k}} c_{-\mathbf{k}\downarrow}\, c_{\mathbf{k}\uparrow})\, c^{\dagger}_{\mathbf{k}\uparrow}\, c_{-\mathbf{k}\uparrow}\,(u_{\mathbf{k}} + v_{\mathbf{k}}\, c^{\dagger}_{\mathbf{k}\uparrow}\, c^{\dagger}_{-\mathbf{k}\uparrow})\,|0\rangle = v^2_{\mathbf{k}}. \qquad \text{...(5c)}$$

We have also

$$\langle\Psi_S|c^{\dagger}_{\mathbf{k}\uparrow}\, c^{\dagger}_{-\mathbf{k}\uparrow}|\Psi_S\rangle = \langle\Psi_S|c_{-\mathbf{k}\downarrow}\, c_{\mathbf{k}\uparrow}|\Psi_S\rangle = u_{\mathbf{k}}\, v_{\mathbf{k}}. \qquad \text{...(5d)}$$

Using the matrix element, Eq. (5) we obtain for the superconductor ground-state energy the following expression

$$W_S = \langle\Psi_S|H_{BCS}|\,\Psi_S\rangle = 2\sum_{\mathbf{k}} \varepsilon_{\mathbf{k}}\, v^2_{\mathbf{k}} + \sum_{\mathbf{k}\mathbf{k}'} U_{\mathbf{k}\mathbf{k}'}\, u_{\mathbf{k}}\, v_{\mathbf{k}}\, u_{\mathbf{k}'}\, u_{\mathbf{k}'} \qquad \text{...(6)}$$

at this stage our problem has become a standard algebraic minimization problem of $W_S$ as a function of the $u_{\mathbf{k}}$ and $v_{\mathbf{k}}$ (with **k** in the reciprocal space).

In order to minimize $W_S$ under the constraint $u_{\mathbf{k}}^2 + v_{\mathbf{k}}^2 = 1$, let us represent $u_{\mathbf{k}}$ and $v_{\mathbf{k}}$ in polar form

$$\begin{cases} u_{\mathbf{k}} = \cos\theta_{\mathbf{k}} \\ v_{\mathbf{k}} = \sin\theta_{\mathbf{k}} \end{cases}$$

The superconductor ground-state energy, Eq. (6) becomes

$$W_S = 2\sum_{\mathbf{k}} \varepsilon_{\mathbf{k}} \sin^2\theta_{\mathbf{k}} + \frac{1}{4}\sum_{\mathbf{k}\mathbf{k}'} U_{\mathbf{k}\mathbf{k}'} \sin 2\theta_{\mathbf{k}} \sin 2\theta_{\mathbf{k}'}.$$

The condition $\partial W_S/\partial\theta_{\mathbf{k}} = 0$ gives

$$2\varepsilon_{\mathbf{k}} \sin 2\theta_{\mathbf{k}} + \sum_{\mathbf{k}'} U_{\mathbf{k}\mathbf{k}'} \cos 2\theta_{\mathbf{k}} \sin 2\theta_{\mathbf{k}'} = 0;$$

we re-write this minimization condition in terms of $u_{\mathbf{k}}$ and $v_{\mathbf{k}}$ and obtain

$$2\varepsilon_{\mathbf{k}} u_{\mathbf{k}} v_{\mathbf{k}} + \sum_{\mathbf{k}'} U_{\mathbf{k}\mathbf{k}'} (u_{\mathbf{k}}^2 - v_{\mathbf{k}}^2) u_{\mathbf{k}'} v_{\mathbf{k}'} = 0 \qquad \text{...(7)}$$

for every **k**.

To solve this set of self-consistent equations for $u_{\mathbf{k}}$ and $v_{\mathbf{k}}$, let us define the *energy gap parameters*, $\Delta_{\mathbf{k}}$ (the origin of the name will be cleared below)

$$\Delta_{\mathbf{k}} = -\sum_{\mathbf{k}'} U_{\mathbf{k}\mathbf{k}'} u_{\mathbf{k}'} v_{\mathbf{k}'}. \qquad \text{...(8)}$$

The variational condition, Eq. (7) thus becomes

$$\boxed{2\,\varepsilon_{\mathbf{k}} u_{\mathbf{k}} v_{\mathbf{k}} - \Delta_{\mathbf{k}} (u_{\mathbf{k}}^2 - v_{\mathbf{k}}^2) = 0} \qquad \text{...(9)}$$

The solution of Eq. (9), together with the normalization condition $u_{\mathbf{k}}^2 + v_{\mathbf{k}}^2 = 1$ gives

$$u_{\mathbf{k}}^2 = \frac{1}{2}\left[1 + \frac{\varepsilon_{\mathbf{k}}}{\sqrt{\varepsilon_{\mathbf{k}}^2 + \Delta_{\mathbf{k}}^2}}\right] \quad \text{and} \quad u_{\mathbf{k}}^2 = \frac{1}{2}\left[1 - \frac{\varepsilon_{\mathbf{k}}}{\sqrt{\varepsilon_{\mathbf{k}}^2 + \Delta_{\mathbf{k}}^2}}\right] \qquad \text{...(10)}$$

The choice of sign Eq. (10) has been determined by the requirement that, for vanishing $U_{\mathbf{k}\mathbf{k}'}$ interaction and thus for vanishing $\Delta_{\mathbf{k}}$, the standard normal ground state described by Eq. (3) is obtained. If we use Eq. (10) to calculate $u_{\mathbf{k}'} v_{\mathbf{k}'}$, and insert the result into Eq. (8) we see that the energy gap parameters are determined by the self-consistent equations

$$\boxed{\Delta_{\mathbf{k}} = -\frac{1}{2}\sum_{\mathbf{k}'} U_{\mathbf{k}\mathbf{k}'} \frac{\Delta_{\mathbf{k}'}}{\sqrt{\varepsilon_{\mathbf{k}'}^2 + \Delta_{\mathbf{k}'}^2}}} \qquad \text{...(11)}$$

for every **k**.

A trivial solution of Eq. (11) is $\Delta_{\mathbf{k}} = 0$ for every $\mathbf{k}$; this corresponds to the normal metal state with electrons filling the Fermi sphere up to $k_F$. In general, the integral equation (11) is not easy to be solved. For simplicity, and in analogy with the procedure adopted in the treatment of a single *s*-wave Cooper pair, we assume that the matrix elements $U_{\mathbf{kk}'}$ can be considered constant in an appropriate energy shell around the Fermi level; in the so-called *average potential approximation*, we assume

$$U_{\mathbf{kk}'} = \begin{cases} -U_0/N, & \text{if } |\varepsilon_{\mathbf{k}}|, |\varepsilon_{\mathbf{k}'}| < \hbar\omega_D \\ 0, & \text{otherwise} \end{cases} \qquad \text{...(12)}$$

where $U_0 > 0$ is a *positive constant*, independent of the state $\mathbf{k}$ and $\mathbf{k}'$ in the shell $|\varepsilon_{\mathbf{k}}|$, $|\varepsilon_{\mathbf{k}'}| < \hbar\omega_D$ and also independent of the number $N$ of unit cells of the crystal. From the average potential approximation of Eq. (12), and from (8), it follows that

$$\Delta_{\mathbf{k}} = \begin{cases} \Delta_0, & \text{if } |\varepsilon_{\mathbf{k}}| < \hbar\omega_D \\ 0, & \text{otherwise} \end{cases}$$

The integral Eq. (11) simplifies as

$$1 = \frac{1}{2} U_0 \frac{1}{N} \sum_{\mathbf{k}'} \frac{1}{\sqrt{\varepsilon_{\mathbf{k}'}^2 + \Delta_0^2}} \quad \text{with} \quad -\hbar\omega_D < \varepsilon_{\mathbf{k}'} < \hbar\omega_D.$$

As usual the sum over $\mathbf{k}'$ can be converted into an integral; assuming the electron density-of-state as constant in the small energy shell of interest around the Fermi level, one obtains

$$1 = \frac{1}{2} U_0 n_0(E_F) \int_{-\hbar\omega_D}^{\hbar\omega_D} \frac{d\varepsilon}{\sqrt{\varepsilon^2 + \Delta_0^2}} \qquad \text{...(13a)}$$

as before, $n_0(E_F) = D_0(E_F)/N$ denotes the density-of-states for one spin direction per unit cell at the Fermi energy.

The integral in Eq. (13a) can be carried out considering that the primitive of $1/\sqrt{x^2 + a^2}$ is arc sinh $(x/a)$; Eq. (13a) becomes

$$1 = U_0 n_0(E_F) \text{ arc sinh } \frac{\hbar\omega_D}{\Delta_0}$$

It follows $\Delta_0 = \dfrac{\hbar\omega_D}{\sinh[1/U_0 n_0(E_F)]}$.

In the week coupling limit, when $U_0 n_0(E_F)$ is rather small with respect to unity, and $\exp(-1/U_0 n_0) \ll 1$, one obtains

$$\boxed{\Delta_0 = 2\hbar\omega_D \exp[-1/U_0 n_0(E_F)]} \qquad \text{...(13b)}$$

Until now, we have confined our discussion of the BCS theory of superconductivity within the "weak coupling limit" (i.e., $\Delta_0 \ll \hbar\omega_D$), and "*s-wave symmetry of the pairing state*". We wish to mention that the BCS theory has been generalized by various authors to describe strong-coupling situations [see W. L. MacMillan. *Phys. Rev.* **167**, 331 (1978) and in particular the works of Eliashberg and other authors, quoted therein].

## 15.39 BCS HAMILTONIAN, $T_C$ AND 2Δ (O)

Returning to the mainstream of the BCS argument, the above reasoning can be used to pick out the best wave function to use as a trial wave function for evaluating the ground-state energy by variational principle. For mathematical convenience, it is easier to place these assumptions directly in the Hamiltonian also, due to exchange, the spins in the Cooper pairs are usually opposite. Thus, the interaction part of the Hamiltonian is now written with $K$[6] = 0,

$$\mathcal{H}_1 = \sum_{k,q} V_q\, C^{\dagger}_{k+q\uparrow}\, C^{\dagger}_{-k+q\downarrow}\, C_{-k\downarrow}\, C_{-k\uparrow}. \quad ...(1)$$

Next, assume a "BCS Hamiltonian" for interacting pairs consistent with Eq. (1), with $\boldsymbol{k} + \boldsymbol{q} \to \boldsymbol{k}$, $\boldsymbol{k} \to \boldsymbol{k}'$, $V_q = V_{k-k'} = -V_{k,k'}$

$$\mathcal{H} = \sum_{k\sigma} \varepsilon_k\, C^{\dagger}_{k\sigma} C_{k\sigma} + \sum_{k,k'} V_{k,k'}\, C^{\dagger}_{k\uparrow}\, C^{\dagger}_{-k\downarrow}\, C_{-k'\downarrow}\, C_{k'\uparrow} \quad ...(2)$$

where $$\varepsilon_k = \frac{\hbar^2 k^2}{2m} - \mu \quad ...(3)$$

and where μ is the chemical potential. Also

$$\mathcal{H} \equiv \mathcal{H}_0 + \mathcal{H}_1 \quad ...(4)$$

and note $$V_{k,k'} = V_{k',k} = V^*_{k,k'} \quad ...(5)$$

As before $C$ are Fermion (electron) annihilation operators, and $C^{\dagger}$ are Fermion (electron) creation operators. Defining the pair creation and annihilation operators

$$b^{\dagger}_k = C^{\dagger}_{k\uparrow}\, C^{\dagger}_{-k\downarrow} \quad ...(6)$$

$$b_k = C_{-k\downarrow}\, C_{k\uparrow} \quad ...(7)$$

and defining

$$\overline{b}_k = \frac{Tr(e^{-\beta H} b_k)}{Tr(e^{-\beta H})} \quad ...(8)$$

we can show $\overline{b}^{\dagger}_k = \overline{b}^*_k$ using $Tr(AB) = Tr(BA)$. We can also show in the representation we use that $\overline{b}^*_k = \overline{b}_k$. We define

$$\Delta_k = -\sum_{k'} V_{k,k'}\, \overline{b}_{k'} = \Delta^*_k \quad ...(9)$$

As we will demonstrate later, this will turn out to be the gap parameter. We can write the interaction term as

$$\mathcal{H}_1 = \sum_{k,k'} V_{k,k'}\, b^{\dagger}_k\, b_{k'}. \quad ...(10)$$

[6] $\vec{K} = \vec{k} + \vec{k}'$ only pairs with equal total momentum are comidered. $\vec{K}$ is the same for each pair.

Note $$b_{k'} = \overline{b}_{k'} + \delta b_{k'} = \overline{b}_{k'} + (b_{k'} - \overline{b}_{k'}) \qquad ...(11)$$

and $$b^\dagger_{k'} = \overline{b}^*_{k'} + \delta b^\dagger_{k'} = \overline{b}^*_{k'} + (b^\dagger_{k'} - \overline{b}^*_{k'}) \qquad ...(12)$$

$$b^\dagger_k = \overline{b}^*_k + \delta b^\dagger_k = \overline{b}_k + \delta b^\dagger_k \qquad ...(13)$$

and we will neglect $(\delta b_{k'}) \times (\delta b^\dagger_k)$ terms. (This is sort of a mean-field-like approximation for pairs) Thus, using Eqs. (12) and (13) and neglecting $O(\delta b^2)$ terms, we can write

$$\begin{aligned} b^\dagger_k\, b_{k'} &= (\overline{b}_k + \delta b^\dagger_k)(\overline{b}_{k'} + \delta b_{k'}) \\ &= \overline{b}_k\, \overline{b}_{k'} + \overline{b}_{k'}\, \delta b^\dagger_k + \overline{b}_k\, \delta b_{k'} \qquad ...(14) \\ &= \overline{b}_k\, \overline{b}_{k'} + \overline{b}_{k'}\, b^\dagger_k - \overline{b}_k\, \overline{b}_{k'}, \end{aligned}$$

assuming $\overline{b}_k$ is real. Also,

$$\mathcal{H}_1 = \sum_{k,k'} V_{k,k'}(\overline{b}_{k'}\, b^\dagger_k + \overline{b}_{k'}\, \overline{b}_k,\ \vec{b}_k,\ \vec{b}_k). \qquad ...(15)$$

Thus, $$\mathcal{H} = \sum_{k\sigma} \varepsilon_k C^\dagger_{k\sigma} C_{k\sigma} - \sum_k (\Delta_k b^\dagger_k + \Delta_k b_k - \Delta_k \overline{b}_k). \qquad ...(16)$$

We now diagonalize by a Bogoliubov–Valatin transformation

$$C_{k'} = u_k \alpha_k + v_k \beta^\dagger_k \qquad ...(17)$$

$$C_{-k'} = u_k \beta_k - v_k \alpha^\dagger_k \qquad ...(18)$$

where $u_k^2 + v_k^2 = 1$ (to preserve anticommutation relations) $u_k$ and $v_k$ are real, and the α and β given by

$$\alpha^\dagger_k = u_k C^\dagger_{k\uparrow} - v_k C_{-k\downarrow} \qquad ...(19)$$

$$\beta^\dagger_k = u_k C^\dagger_{k\downarrow} + v_k C_{k\uparrow} \qquad ...(20)$$

are Fermion operators obeying the usual anticommutation relations. The $\alpha^\dagger_k$, and $\beta^\dagger_k$ create "bogolons". The algebra gets a bit detailed here and one can *skip* along unless curious,

$$b_k = C_{-k\downarrow}\, C_{k\uparrow} = (-v_k \alpha^\dagger_k + u_k \beta_k)(u_k \alpha_k + v_k \beta^\dagger_k) \qquad ...(21)$$

$$b^\dagger_k = C^\dagger_{k\uparrow}\, C^\dagger_{-k\downarrow} = (u_k \alpha^\dagger_k + v_k \beta_k)(-v_k \alpha_k + u_k \beta^\dagger_k) \qquad ...(22)$$

$$C^\dagger_{k\uparrow}\, C_{-k\downarrow} = (u_k \alpha^\dagger_k + v_k \beta_k)(u_k \alpha_k + v_k \beta^\dagger_k) \qquad ...(23)$$

$$C^\dagger_{-k\downarrow}\, C_{-k\downarrow} = (-v_k \alpha_k + u_k \beta^\dagger_k)(-v_k \alpha^\dagger_k + u_k \beta_k) \qquad ...(24)$$

$$b^\dagger_k = -u_k v_k\, \alpha^\dagger_k \alpha_k + u_k v_k\, \beta_k \beta^\dagger_k - v_k^2\, \beta_k \alpha_k + u_k^2\, \alpha^\dagger_k \beta^\dagger_k \qquad ...(25)$$

$$b_k = -v_k u_k \alpha_k^\dagger \alpha_k + u_k v_k \beta_k \beta_k^\dagger - v_k^2 \alpha_k^\dagger \beta_k^\dagger + u_k^2 \beta_k \alpha_k \quad \text{...(26)}$$

$$C_{k\uparrow}^\dagger C_{-k\uparrow} = u_k^2 \alpha_k^\dagger \alpha_k + v_k^2 \beta_k \beta_k^\dagger + u_k v_k \alpha_k^\dagger \beta_k^\dagger + u_k v_k \beta_k \alpha_k \quad \text{...(27)}$$

$$C_{-k\downarrow}^\dagger C_{-k\downarrow} = u_k^2 \beta_k^\dagger \beta_k + v_k^2 \alpha_k \alpha_k^\dagger - u_k v_k \beta_k^\dagger \alpha_k^\dagger - v_k u_k \alpha_k \beta_k \quad \text{...(28)}$$

$$\left.\begin{aligned} \mathcal{H} = \sum_k [&2\varepsilon_k u_k v_k \alpha_k^\dagger \beta_k^\dagger + 2\varepsilon_k \beta_k \alpha_k u_k v_k \\ &+ \varepsilon_k (u_k^2 - v_k^2)(\alpha_k^\dagger \alpha_k + \beta_k^\dagger \beta_k) + 2\varepsilon_k v_k^2 \\ &+ \Delta_k u_k v_k (\alpha_k^\dagger \alpha_k + \beta_k^\dagger \beta_k) \\ &- \Delta_k u_k v_k + \Delta_k v_k^2 \beta_k \alpha_k - \Delta_k u_k^2 \alpha_k^\dagger \beta_k^\dagger \\ &+ \Delta_k^\dagger u_k v_k (\alpha_k^\dagger \alpha_k^\dagger \beta_k^\dagger) - \Delta_k^* u_k v_k \\ &+ \Delta_k v_k^2 \alpha_k^\dagger \beta_k^\dagger - \Delta_k^* u_k^2 \beta_k \alpha_k] + \textstyle\sum_k \Delta_k \bar{b}_k. \end{aligned}\right\} \quad \text{...(29)}$$

Rewriting this we get

$$\left.\begin{aligned} \mathcal{H} = \textstyle\sum_k \{&(2\Delta_k u_k v_k - \varepsilon_k (v_k^2 - u_k^2))(\alpha_k^\dagger \alpha_k + \beta_k^\dagger \beta_k) \\ &+ (2\varepsilon_k u_k v_k - \Delta_k (v_k^2 - u_k^2))(\alpha_k^\dagger \beta_k^\dagger + \beta_k \alpha_k)\} + G \end{aligned}\right\} \quad \text{...(30)}$$

where $G = \sum (2\varepsilon_k v_k^2 - 2\Delta_k u_k v_k + \Delta_k \bar{b}_k).$

Next, choose

$$2\varepsilon_k u_k v_k + \Delta_k (v_k^2 - u_k^2) = 0 \quad \text{...(31)}$$

so as to diagonalize the Hamiltonian. Also, using $u_k^2 + v_k^2 = 1$ let

$$v_k^2 = \frac{1}{2} - a \quad \text{...(32)}$$

$$v_k^2 - u_k^2 = -2a \quad \text{...(33)}$$

$$u_k v_k = \sqrt{\frac{1}{4} - a^2}. \quad \text{...(34)}$$

$$2\varepsilon_k \sqrt{\frac{1}{4} - a^2} = \Delta_k 2a \quad \text{...(35)}$$

$$\varepsilon_k^2 \left(\frac{1}{4} - a^2\right) = \Delta_k^2 a^2 \quad \text{...(36)}$$

Thus,

$$a = \frac{\varepsilon_k}{2\sqrt{\varepsilon_k^2 + \Delta_k^2}}. \quad \text{...(37)}$$

Rewriting

$$\mathcal{H} = [2\Delta_k u_k v_k - \varepsilon_k (v_k^2 - u_k^2) \times (\alpha_k^\dagger \alpha_k + \beta_k^\dagger \beta_k)] + G. \quad \text{...(38)}$$

But, define
$$E_k = \sqrt{\varepsilon_k^2 + \Delta_k^2} \quad ...(39)$$

$$a = \frac{\varepsilon_k}{2E_k} \quad ...(40)$$

and thus
$$2u_k v_k = \frac{\Delta_k}{E_k} \quad ...(41)$$

Thus, after a bit of algebra,

$$2\Delta_k u_k v_k - \varepsilon_k(v_k^2 - u_k^2) = E_k \quad ...(42)$$

So
$$\mathcal{H} = \sum_k E_k(\alpha_k^\dagger \alpha_k + \beta_k^\dagger \beta_k)] + G \quad ...(43)$$

and $G$ can be put in the form

$$G = \sum_k (\varepsilon_k - E_k + \Delta_k \bar{b}_k). \quad ...(44)$$

Note by Fig. 15.52 and Eq. (39) how $E_k$ predicts a gap, for clearly $E_k \geq \Delta_0$. Continuing

$$b_k = v_k u_k \alpha_k^\dagger \alpha_k + u_k v_k \beta_k \beta_k^\dagger - v_k^2 \alpha_k^\dagger \beta_k^\dagger + u_k^2 \beta_k \alpha_k \quad ...(45)$$

But $\bar{b}_k$ involves only diagonal terms, so ussing an appropriate anticommutation relation

$$\bar{b}_k = u_k v_k (1 - \overline{\alpha_k^\dagger \alpha_k} - \overline{\beta_k^\dagger \beta_k}) \quad ...(46)$$

so
$$\bar{b}_k = u_k v_k (1 - 2n_k) \quad ...(47)$$

where
$$n_k = \frac{1}{e^{\beta E_k} + 1} = f(E_k) \quad ...(48)$$

$f(E_k)$ is of course the Fermi function but it looks strange without the chemical potential, This is because $\alpha^\dagger$, $\beta^\dagger$ do not change the particle number. Therefore,

$$\begin{aligned} \Delta_k &= -\sum V_{k,k'} \bar{b}_{k'} \\ &= -\sum V_{k,k'} u_{k'} v_{k'} [1 - 2f(E_{k'})] \\ &= -\sum_{k'} V_{k,k'} \frac{\Delta_{k'}}{2E_{k'}} [1 - 2f(E_{k'})]. \end{aligned} \quad ...(49)$$

Now assume that (not using $\hbar = 1$)

$$\Delta_{k'} = \Delta \quad \text{when} \quad |\varepsilon_k| < \hbar\omega_D \quad ...(50)$$

$$V_{k,k'} = -V \quad \text{when} \quad |\varepsilon_k| < \hbar\omega_D \quad ...(51)$$

$$\Delta_k = 0 \quad \text{when} \quad |\varepsilon_k| < \hbar\omega_D \quad ...(52)$$

and
$$V_{k,k'} = 0 \quad \text{when} \quad |\varepsilon_k| > \hbar\omega_D \quad ...(53)$$

where $\omega_D$ is the Debye frequency (see Eq. (6), Art. 15.37)). So,

$$\Delta = V \sum_{|\varepsilon_{k'}| < \hbar\omega_D} \frac{\Delta[1 - 2f(E_{k'})]}{2E_{k'}}. \qquad ...(54)$$

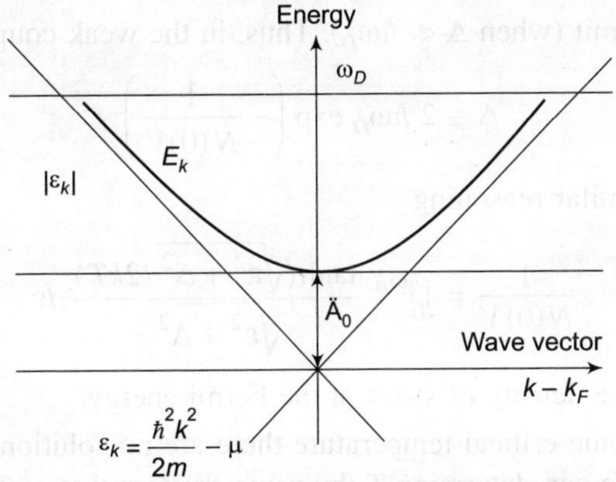


**Figure 15.52** Gap in single-particle excitations near the Fermi energy

For $T = 0$, then

$$\Delta = \sum_{k'} V \frac{\Delta}{2\sqrt{\varepsilon_{k'}^2 + \Delta^2}}, \qquad ...(55)$$

and for $T \neq 0$, then

$$\Delta = \sum_{k'} \frac{V\Delta}{2\sqrt{\varepsilon_{k'}^2 + \Delta^2}} \tanh\left(\frac{E_{k'}}{2kT}\right). \qquad ...(56)$$

We can then write

$$\Delta \cong \int_0^{\hbar\omega_D} N(E)V \frac{\Delta}{\sqrt{E^2 + \Delta^2}} dE \qquad ...(57)$$

If we further suppose that $N(E) \cong$ constant $\cong N(0) \equiv$ the density of states at the Fermi level, then Eq. (57) becomes

$$\frac{1}{N(0)V} = \int_0^{\hbar\omega_D} \frac{dE}{\sqrt{E^2 + \Delta^2}} = \ln(E + \sqrt{E^2 + \Delta^2})\Big|_0^{\hbar\omega_D}$$

$$= \ln\left(\frac{\hbar\omega_D + \sqrt{(\hbar\omega_D)^2 + \Delta^2}}{\Delta}\right). \qquad ...(58)$$

This equation can be written as

$$\exp\left(-\frac{1}{N(0)V}\right) = \frac{\Delta}{\hbar\omega_D + \sqrt{(\hbar\omega_D)^2 + \Delta^2}}$$

$$= \frac{\Delta}{\hbar\omega_D\,(\sqrt{1+(\Delta/\hbar\omega_D)^2}+1)} = \frac{\Delta}{2\hbar\omega_D} \quad ...(59)$$

in the weak coupling limit (when $\Delta \ll \hbar\omega_D$). Thus, in the weak coupling limit, we obtain

$$\Delta \cong 2\,\hbar\omega_D \exp\left(-\frac{1}{N(0)V}\right) \quad ...(60)$$

From Eq. (56) by similar reasoning

$$\frac{1}{N(0)\,V} = \int_0^{\hbar\omega_D} \frac{\tanh(\sqrt{\varepsilon^2+\Delta^2}/2kT)}{\sqrt{\varepsilon^2+\Delta^2}}\,d\varepsilon \quad ...(61)$$

where, again, $N(0)$ is the density of ststes at the Fermi energy.

For $T$ greater than some critical temperature there are no solutions for $\Delta$, i.e., the energy gap no longer exists. We can determine $T_c$ by using the fact that at $T = T_c$, $\Delta = 0$. This says that

$$\frac{1}{N(0)\,V} = \int_0^{\hbar\omega_D} \frac{\tanh(\varepsilon/2kT_c)}{\varepsilon}\,d\varepsilon \quad ...(62)$$

In the weak coupling approximation, when $N(0)V \ll 1$, we obtain from Eq. (62) that

$$kT_c = 1.14\;\hbar\omega_D \exp(-1/N(0)V). \quad ...(63)$$

Equation Eq. (63) is a very important equation. It depends on three material properties:

(a) The Debye frequency $\omega_D$.

(b) $V$ that measures the strength of the electron-phonon coupling.

(c) $N(0)$ that measures the number of electrons available at the Fermi energy.

Note that typically $\omega_D \propto (m)^{-1/2}$, where $m$ is the mass of atoms. This leads directly to the isotope effect. Note also the energy gap $E_g = 2\Delta(0)$ at absolute zero.

We can combine this result with our result for the energy gap parameter in the ground state to derive a relation between the energy gap at absolute zero and the critical superconducting transition temperature with no magnetic field. By Eq. (63) and (60), we have that

$$\Delta(0) = 2\hbar\omega_D \exp\,(-1/N(0)\,V) = \frac{2}{1.14}\,kT_c, \quad ...(64)$$

or

$$2\Delta(0) = 3.52\;kT_c. \quad ...(65)$$

Note that our expression for $\Delta(0)$ and $T_c$ both involve the factor $\exp(-1/N(0)\,V)$; that is, a power series (in $V$) expansion for both $\Delta(0)$ and $T_c$ have an essential singularity in $V$. We could not have obtained reasonable results if we had tried ordinary perturbation theory because with ordinary perturbation theory, we cannot reproduce the effect of an essential singularity in the perturbation.

The above discussion is valid for weakly coupled superconductors (i.e., $(T_c/\theta_D)^2 \geq (500)^{-2}$).

## 15.40 PREDICTIONS OF THE BCS THEORY

The BCS theory correctly predicts the observed thermal and electromagnetic properties of many superconductors, especially those for which $T_c \ll \Theta_D$. This is the *weak-coupling limit* and is appropriate for superconductors in which the electron-phonon interaction is not too strong [i.e., when $\rho(E_F)U \ll 1$]. Since the BCS theory involves complicated many-body quantum-mechanical effects, a detailed discussion is not presented here. Instead, only the important predictions of the standard weak-coupling BCS theory are listed in Table 15.8.

Note that the superconducting energy gap $2\varepsilon(0)$, the condensation energy $G_n - G_s$, and the transition temperature $T_c$ are all predicted to be exponential functions of the product $\rho(E_F)U$ of the electron density of states and the net interaction strength, $U$. Typical values of the net interaction strength in weak-coupling superconductors correspond to $\rho(E_F)U \approx 0.3$, which, corresponds to $T_c \approx \Theta_D/30$. The weak-coupling limit is not appropriate for superconductors such as Pb and Hg, where $T_c/\Theta_D \approx 0.1$. Pb and Hg are thus said to be *strong-coupling superconductors*. The high-$T_c$ superconductor $YBa_2Cu_3O_7$ with $T_c = 92$ K and $\Theta_D \approx 400$ K is also a strongcoupling superconductor since $T_c/\Theta_D \approx 0.23$.

**Table 15.8** Important Predictions of the BCS Theory in the Weak-Coupling Limit, $\rho(E_F)U \ll 1^a$

| |
|---|
| 1. Superconducting energy gap at $T = 0$ K: $$2\varepsilon(0) = \frac{2\hbar\omega(q)}{\sinh[1/\rho(E_F)U]} \approx 4\hbar\omega(q)\exp\left[-\frac{1}{\rho(E_F)U}\right] = 3.52\,k_BT_c$$ |
| 2. Ground-state condensation energy of a superconductor at $T = 0$ K and $H = 0$: $$G_n - G_s = \frac{\mu_0 H_{c0}^2}{2} = \frac{2\rho(E_F)[\hbar\omega(q)]^2}{\exp[2/\rho(E_F)U] - 1} \approx 2\rho(E_F)[\hbar\omega(q)]^2\exp\left[-\frac{2}{\rho(E_F)U}\right] = \frac{\rho(E_F)\varepsilon(0)^2}{2}$$ |
| 3. Superconducting transition temperature: $$k_BT_c = 1.14\hbar\omega(q)\exp\left[-\frac{1}{\rho(E_F)U}\right] = \frac{2\varepsilon(0)}{3.52}$$ |
| 4. Specific heat jump at $T_c$: $$C_{ec} - C_{en} = 1.43\gamma T_c$$ |
| 5. Isotope effect for $T_c$: $$T_c \propto \hbar\omega(q) \propto M^{-1/2}$$ |

[a] $H_{c0}$, thermodynamic critical magnetic field at $T = 0$ K. $\rho(E_F)$, density of electronic states per unit energy per spin at the Fermi energy $E_F$ in the normal state, $\hbar\omega(q)$, average phonon energy and range of attractive interaction [in the BCS theory $\hbar\omega(q)$ is often approximated by $k_B\Theta_D$, where $\Theta_D$ is the Debye temperature]: $U$, strength of the net attractive interaction between pairs of electrons; $Y$, coefficient of linear term in electronic specific heat in the normal state, equal to $2\pi^2k_B^2\,\rho(E_F)/3$ in the free electron model $M$, average ionic mass.

## 15.41 MOTT INSULATOR

An insulator with a magnetic moment at each lattice site (i.e., a magnetic insulator) is known as Mott insulator. The unusuality of a Mott insulator is that strong correlations suppress the charge fluctuations, which are required for a finite conductivity. However, doping produces holes in copper-oxygen planes and it becomes susceptible to charge density wave (CDW): charge on a site changes, as the excitation moves, i.e., there is a gap for charge excitations, but no gap for excitations. Further, due to doping, long range antiferromagnetic order is replaced by short range order, and superconductivity emerges as the ground state.

## 15.42 THE FIRST CUPRATE $La_{2-x}M_xCuO_4$ (THE PHASE-DIAGRAM)

A crystalline solid with an odd number of electrons or holes per unit cell is, in general, a Pauli paramagnetic metal (i.e., electrons in a metal constitute a Fermion gas and exhibit a temperature independent paramagnetism). However, the compound $La_2CuO_4$ has one hole per unit cell is insulator and anti-ferromagnetic with an ordering or Neel temperature ~ 210 K. It has an orthorhombically distorted $K_2NiF_4$ structure and transforms to regular $K_2NiF_4$ (tetragonal) structure above 530 K.

Let us consider first, $La_2CuO_4$ as a chemical compound. The valence requirements are satisfied for Cu in the 2+ state, and hence, can be written as $[(La^{3+})_2Cu^{2+}](O^{2-})_4$. Here, $Cu^{2+}$ has an unfilled shell configuration, namely, $d^9$ or one hole in the filled $d^{10}$ shell. So, $La_2CuO_4$ has one "hole" per unit cell. $La_2CuO_4$, instead of being a Pauli paramagnetic metal, is an anti-ferromagnetic insulator (at temperatures $T \leq T_N$ (~ 210 K for $La_2CuO_4$), the Neel temperature, there is anti-ferromagnetic ordering of magnetic moments).

Now suppose La in $La_2CuO_4$ is doped by a fraction of Sr (i.e., it becomes $La_{2-x}Sr_xCuO_4$). Since Sr is divalent, therefore, one should have either $Cu^{3+}$ or $O^{1-}$ for $Sr^{2+}$. Spectroscopic evidence shows that the latter is favoured. Therefore, the addition of a Sr atom introduces a hole in the filed oxygen *p*-shell and the concentration of holes is proportional to *x*. If these holes move, one has a metal. This is what happens at temperatures from 0 to 30 K. There is a rapid disappearance of anti-ferromagnetism with increase in *x*, which is absent for $x > 0.02$. Figure 15.53 shows the phase diagram of this compound in the $T - x$ plane. There is an insulator to metal transition for $x \geq 0.02$. Then, there is a phase at $T < 20$ K and for $0.02 < x < 0.06$, there are frozen disordered magnetic moments (spin glass state). This is followed by the superconducting phase, which persists for $0.06 < x < 0.30$. Hall effect measurements confirm that carriers are positively charged (i.e., holes) with their density, approximately equal to *x* per unit cell. Thus, $La_{2-x}M_xCuO_4$ shows the highest $T_c$ around a critical value of *x* (0.15 and 0.2, respectively in case of Ba and Sr with $T_c$s of 30 and 35 K).

Electron doped superconductors can be produced by substituting the trivalent rare earth with a tetravalent ion, e.g., $Nd_{2-x}Ce_xCuO_4$.

## 15.43 THE CHARGE-TRANSFER MODEL OF A HIGH $T_C$ CUPRATE SUPERCONDUCTOR

All cuprate superconductors can be described by the general compositional formula:

$$(AO)_m\,(BO)_2\,(M)_{n-1}\,(CuO_2)_n,$$

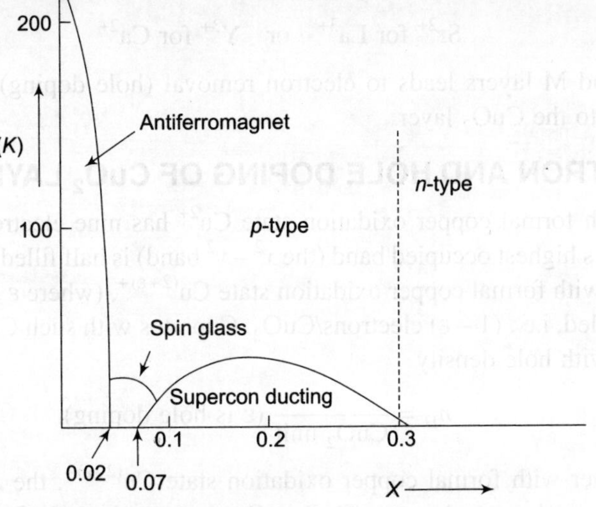


**Figure 15.53** The phases diagram of $La_{2-x}Sr_xCuO_4$ as a function of *T* and *x*

where *A* is Bi or Ti (or Hg) and B is La/Sr/Ba and M is Ca/Y.

Thus, $(AO)_m$ forms the rock salt block and $(M)_{n-1}(CuO_2)_n$ forms the perovskite block. The layered structure of a cuprate HTSC may be viewed, as consisting of

(1) Conduction layer ($CuO_2$)

(2) Charge reservoir layers: $(AO)_m(BO)_2(M)_{n-1}$

as shown in Fig. 15.54.

This is known as charge-transfer model of a high $T_c$ cuprate superconductor. (Superconducting cuprates are generically doped charge transfer insulators in the strong (local) correlation or Mott-limit).

Hole carriers are created in the conduction layer, when electrons are transferred to the charge reservoir layers. All superconducting cuprates have a fraction 0.05–0.30 of mobile holes per $CuO_2$ unit cell in the $CuO_2$ (i.e., *ab*) planes.

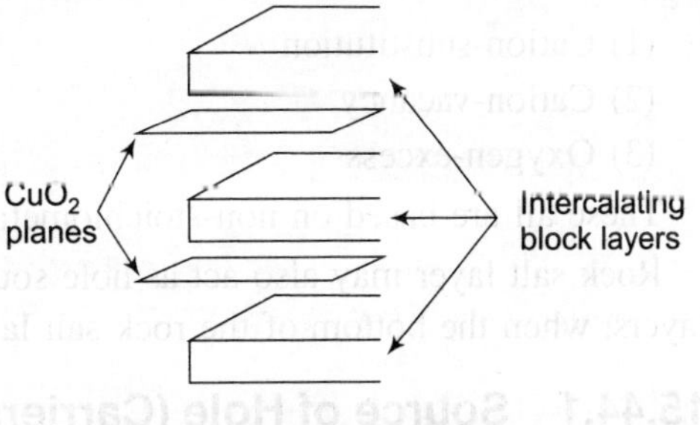


**Figure 15.54** Charge-transfer model of a high $T_c$ cuprate superconductor. $CuO_2$ planes form the conduction layer whereas intercalating block layers act as charge reservoir layers

The carrier number density is controlled by (a) the overall chemistry and (b) amount of charge transferred. The amount of charge transfer depends on

(1) Structure

(2) Available oxidation states of atoms

(3) Competition between charge-transfer and oxidation/reduction of metal atoms in the charge reservoir layer

Structural oxidation-reduction chemistry, e.g., cation substitution, such as

$$Sr^{2+} \text{ for } La^{3+} \quad \text{or} \quad Y^{3+} \text{ for } Ca^{2+}$$

in the AO, BO and M layers leads to electron removal (hole doping) or electron addition (electron doping) to the $CuO_2$ layer.

## 15.44 ELECTRON AND HOLE DOPING OF $CuO_2$ LAYERS

A $CuO_2$ layer with formal copper oxidation state $Cu^{2+}$ has nine electrons to fill its *d*-block bands, therefore, its highest occupied band (the $x^2 - y^2$ band) is half filled (one electron/$CuO_2$). For a $CuO_2$ layer with formal copper oxidation state $Cu^{(2+\varepsilon)+}$, (where $\varepsilon > 0$), the $x^2 - y^2$ band is less than half filled, i.e., $(1 - \varepsilon)$ electrons/$CuO_2$. Cuprates with such $CuO_2$ layers are *p*-type superconductors with hole density

$$n_H = \frac{\varepsilon}{CuO_2 \text{ unit}} \quad (\varepsilon \text{ is hole doping}).$$

For a $CuO_2$ layer with formal copper oxidation state $Cu^{(2-\varepsilon)+}$, the $x^2 - y^2$ band is more than half filled (i.e., $(1 + \varepsilon)$ electrons/$CuO_2$). Cuprates with such $CuO_2$ layers are *n*-type superconductors with electron density

$$n_E = \frac{\varepsilon}{CuO_2 \text{ unit}} \quad (\varepsilon \text{ is electron doping}).$$

The holes or electrons of the cuprates are associated with the in-plane $x^2 - y^2$ band. The charge transfer $\varepsilon$ is the linear combination of same site $p_x$, $p_y$ oxygen (hole) orbitals with local $d_{x^2-y^2}$ symmetry. Hole doping of the cuprate $CuO_2$ layers is achieved by several means, e.g.,

(1) Cation-substitution

(2) Cation-vacancy

(3) Oxygen-excess

These all are based on non-stoichiometry of chemical composition.

Rock salt layer may also act as hole source. Holes may also be introduced into the $CuO_2$ layers, when the bottom of the rock salt layer bands lies below the Fermi level.

### 15.44.1 Source of Hole (Carriers) in Various Cuprate Families

It has been speculated that are produced

(1) In YBCO, due to mobile carriers in the Cu–O chains

(2) In BSCCO, Bi in BiO layers are insulating. Holes are created by excess O-atoms in the BiO layer, and by Sr-deficiency. For lead-doped BSCCO (i.e., BPSCCO),

$$Pb^{4+} + 2Cu^{3+} \rightarrow 2Cu^{3+} + Pb^{2+}$$

(3) TIO double rock salt layers TI:2212, 2223, etc., are important in creating holes in the $CuO_2$ planes

$$TI^{3+} + 2Cu^{2+} \rightarrow 2Cu^{3+} + TI^{1+}$$

(4) TIO single layers are insulating in Thallium mono-layer cuprates. Holes are produced, because copper atoms are already of high enough oxidation state:

$$Cu^{2+1/n} \text{ where } n = 1, 2 \text{ or } 3 \text{ and}$$

$$2 + 1/n = +3.00, +2.50 \text{ or } + 2.33$$

(5) In mercury cuprates, holes are produced due to non-stoichiometry of the HgO layers. There is partial oxygen-occupancy, represented by $HgO_\delta\,(0 < \delta < 1)$:

$$Hg_1Ba_2CuO_{4+\delta}$$

or $$(HgO_\delta)2\,(BaO)\,CuO_2$$

$$Hg^{+2}O^{-\delta'}\;(2BaO)\;Cu^{+2+x}O^{-2-2}$$

Due to $1 > \delta'$ (> 0), a change in copper valence of $+x$ is to be mentioned here that Hg: 1201 is overdoped. Further, since the number of Cu atoms per formula unit is more in Hg: 1212 and Hg: 1223 as compared to Hg: 1201, these are, therefore, underdoped like Bi- and Ti-cuprates.

## 15.45 THE CONDUCTION PLANE IN CUPRATES

Neutral atom configurations for copper and oxygen are

$$Cu^0 = (Ar)\,3d^{10}4s^1$$

$$O^0 = (Be)\,2p^4$$

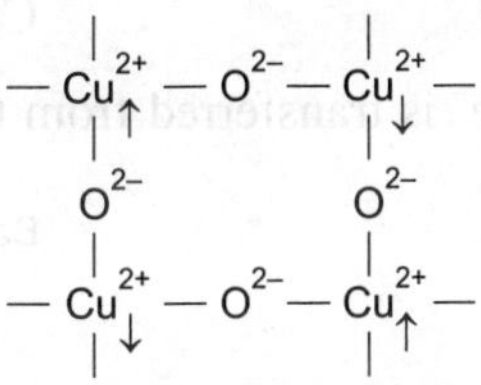


**Figure 15.55** Conduction plane in cuprates ($CuO_2$ plane)

since $O^{2-}$ = $(Be)\,2p^6$ is a closed shell configuration (i.e., stable configuration corresponding to neon), therefore, oxygen favours the valency $2^-$. This implies copper should have valency $2^+$. Hence, the configuration in the *ab*- conductor plane is as shown in Fig. 15.55, i.e., the configuration is $CuO_2$ per each O-taking $2e^{-1}$ (one each) from two closest Cu atoms).

$$Cu^{2+} = (Ar)\,3d^9\ \boxed{\uparrow\downarrow}\boxed{\uparrow\downarrow}\boxed{\uparrow\downarrow}\boxed{\uparrow\downarrow}\boxed{\uparrow}$$

Thus, copper – 3*d* and oxygen –2*p* orbitals are the relevant ones for electronic properties of $CuO_2$ planes.

The copper 3*d* orbitals combine with each other, such that there result five degenerate orbital states shown in Fig. 15.56.

The $d_{3z^2-r^2}$ along *z*-axis mixed well with *c*-axis oxygen and has lower energy than $d_{x^2-y^2}$. The energy splitting is due to

(1) electrostatic field of oxygen-ions at corners of octahedra

(2) covalent overlap (i.e., admixture) between copper-*d* and oxygen-*p* wave functions.

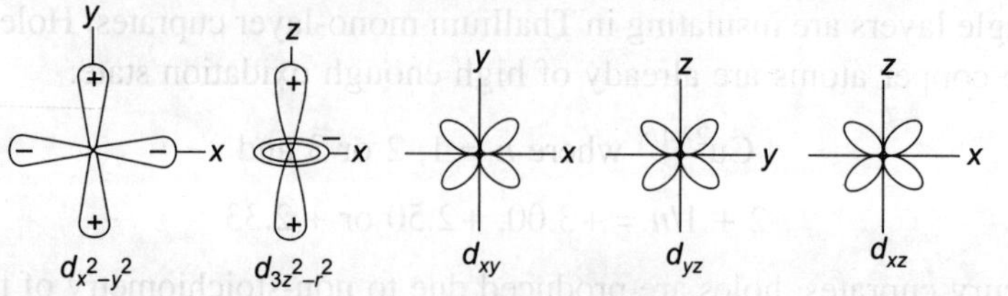


**Figure 15.56** *d*-orbital of copper (*x* and *y* coordinates are in the *ab* plane)

## 15.46 INTER-LAYER SUPERCONDUCTION IN Y, $Ba_2$ $Cu_3$ $O_{7-X}$

For the $d^9$ situation, if the degeneracy between $d_{z2}$ and $d_{x_2-y_2}$ levels as lifted (as for square planar Cu), there is charge disproportionation of the type

$$2Cu^{2+} \rightarrow Cu^{+} + Cu^{3+}$$

in the ionic approximation due to O – O overlap in the Cu – $O_2$ planes.

$$2\,O^{-2} \rightarrow 2O^{-} + Ce^{-}$$

$$Cu^{+3} + 1e^{-} \rightarrow Ce^{+2}$$

$$Cu^{+3} + 2O^{-} \rightarrow Cu^{+2}\,O^{-2} + 1\text{ hole}^{+}$$

$1e^{-}$ is transferred from the conduction layer to the BaO layer, Via O atom of BaO sheet.

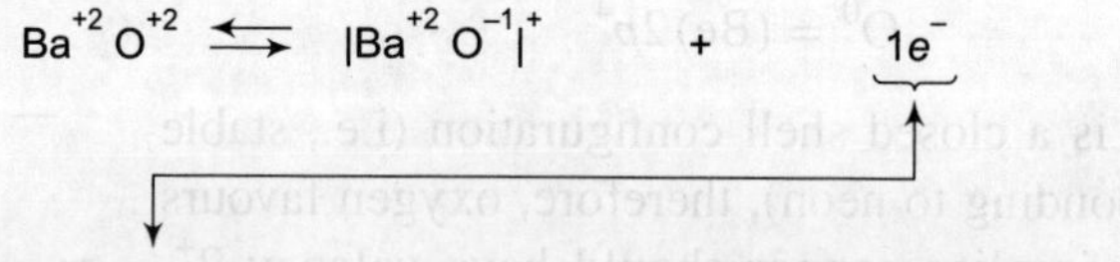


$Y^{3+}$
/ $Cu^{2+}(O^{2-})_2/^{-2}$ ($CuO_2$ plane)
/ $Ba^{2+}(O^{-})/^{+}$
/ $Cu^{+}(O^{2-})_2/^{-2}$ (Cu–O chain)
/ $Ba^{2+}(O^{-})/^{+}$
/ $Cu^{2+}(O^{2-})_2/^{-2}$ ($CuO_2$ plane)
$Y^{3+}$

(see Fig. 15.57) *charges in* $Y_1Ba_2Cu_3O_7$.

There is one missing electron (i.e., 1 hole$^{+}$) per $Cu^{2+}$ ion in the $CuO_2$ (i.e., conduction) plane, and conduction is by hole-pairs:

$$Cu^{2+}_{\sigma} + 1\text{ hole}_{-\sigma} \rightarrow Cu^{3+}$$

SPINON ($d^9$ spin) Holon (Spinless Boson).

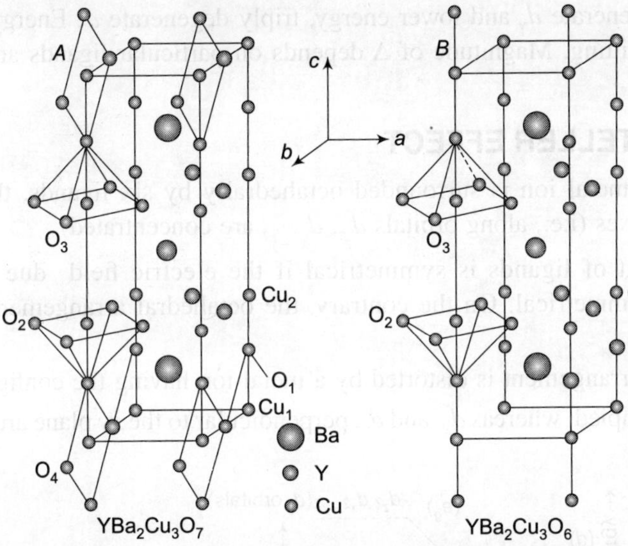


**Figure 15.57** The structure of Y—Ba—Cu—0. for $Y_1Ba_2Cu_3O_7$ (i.e., orthorhombic structure) $a_0$ = 3.823 Å. $b_0$ = 3.885 Å. For $Y_1Ba_2Cu_3O_6$ (i.e., tetragonal structure) $a_t = b_t$ = 3.86 Å

Figure 15.57 shows the structure of $Y_1Ba_2Cu_3O_{6+x}$ ($x$ = 0 for tetragonal and $x$ = 1 for orthorhombic structures respectively). $Cu_2$ atoms lie in $CuO_2$ (i.e., conduction planes and Cu 1 atoms are in Cu – O chains.

## 15.47 OCTAHEDRAL LIGAND FIELD

The crystal field theory considers the effect of ligand fields, i.e., electrostatic field due to ligands (the surrounding atoms/ions) on the *d*-orbitals of the central metal atom.

When negatively charged ligands approach the central metal atom, the electrons of the central atom are subjected to repulsive forces, which are obviously strongest in the "approach directions", i.e., to say, the energy of the orbitals, which point in these directions is, therefore, increased. So, the *d*-electrons will tend to occupy those orbitals that are farthest away from approach directions of ligands. The most common approach directions or rather the arrangement of ligands around a central atom are octahedral (or tetrahedral).

When a transition metal ion is surrounded octahedrally by six ligands (Fig. 15.58), their directions of approch coincide with *x*-, *y*-, *z*-axes (which are the directions along which $d_{z^2}$ and $d_{x^2-y^2}$ orbitals are concentrated). Therefore, these orbitals are raised in energy. However, the $d_{xy}$, $d_{yz}$ and $d_{xz}$, which points away from the ligands are lowered in energy. Consequently, there are two sets of higher

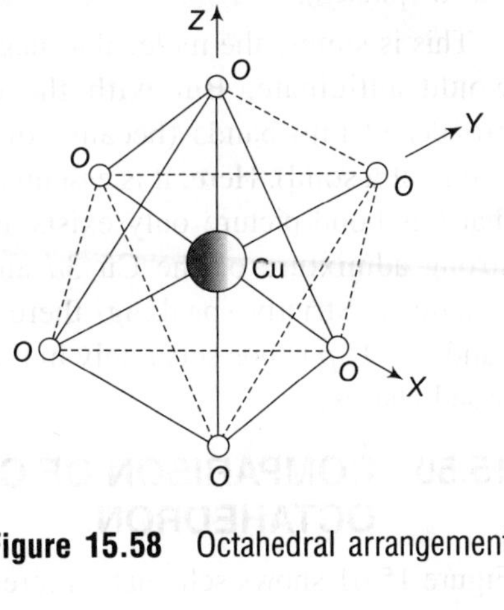


**Figure 15.58** Octahedral arrangement

energy, doubly degenerate $d_\gamma$ and lower energy, triply degenerate $d$. Energy difference $\Delta$ is the crystal field splitting. Magnitude of $\Delta$ depends on particular ligands and on the central metal ion.

## 15.48 JAHN–TELLER EFFECT

When a transition metal ion is surrounded octahedrally by six ligands, their direction of approach $x$-, $y$-, $z$-axes (i.e., along orbitals $d_{z^2}$, $d_{x^2-y^2}$ are concentrated).

The arrangement of ligands is symmetrical if the electric field due to non-bonding ($d$ –) electron is symmetrical. On the contrary, the octahedral arrangement is likely to be distorted.

The octahedral arrangement is distorted by a metal-ion having the configuration $d^9$.

$d_{xy}$ is doubly occupied, whereas $d_{xz}$ and $d_{yz}$ perpendicular to the $xy$ plane are singly occupied (Fig. 15.59).

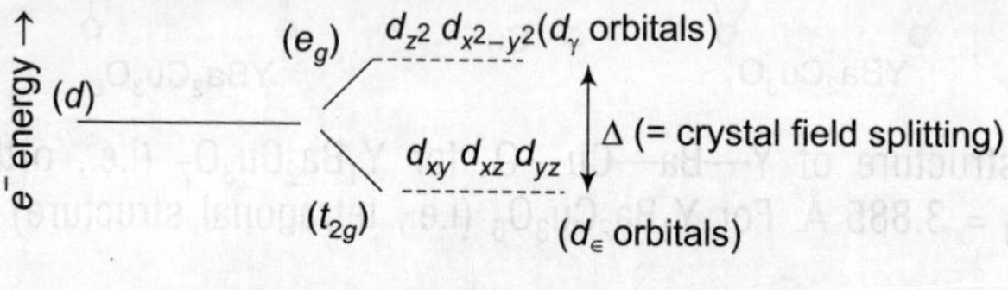


**Figure 15.59** Splitting of $d$-levels

Thus, when the ligands approach the central metal ion $Cu^{2+}$ octahedrally, they can do so more closely along the two $z$-directions. As a result, the octahedron is elongated, making $d_{x^2-y^2}$ more anti-bonding and, therefore, last to be occupied. The one $d_{x^2-y^2}$ "hole" accounts qualitatively for the distortion of octahedron.

## 15.49 ENERGY LEVELS OF COPPER

A schematic energy level diagram for a square planar $Cu^{II}O_4$ unit is given in the Fig. 15.60 (for $Cu–O^2$ planes).

This is simply the molecular diagram a chemist would anticipate. But with the energy levels broadened into bands (because of the "infinite" nature of a solid). Here, it is essential to emphasize that this bond-picture only exists as a result of a strong admixture of the Cu $3d$ and O $2p$ wave functions (Strictly speaking, there are no Cu $3d$ bands or O $2p$ bands separately in the copper oxide based phases).

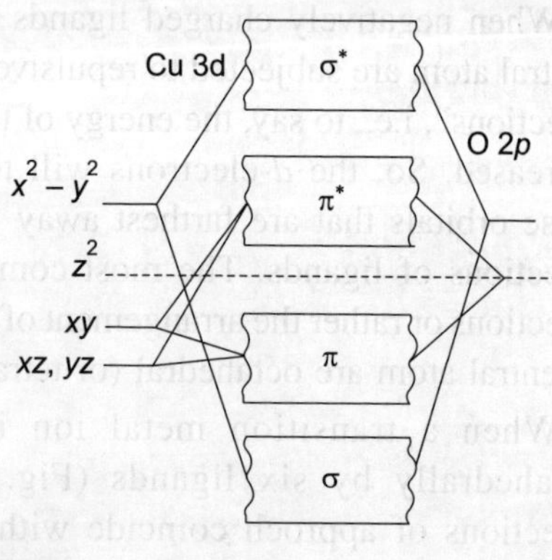


**Figure 15.60** Schematic energy bands for copper in square planar coordination

## 15.50 COMPARISON OF $Cu^{3+}$ AND $Cu^{2+}$ IONS IN THE OXIDE OCTAHEDRON

Figure 15.61 shows schematic representation of orbitals for octahedrally coordinated copper ions in oxides for $Cu^{3+}$ with $3d^8$ configuration. The orbitals transforming as base function

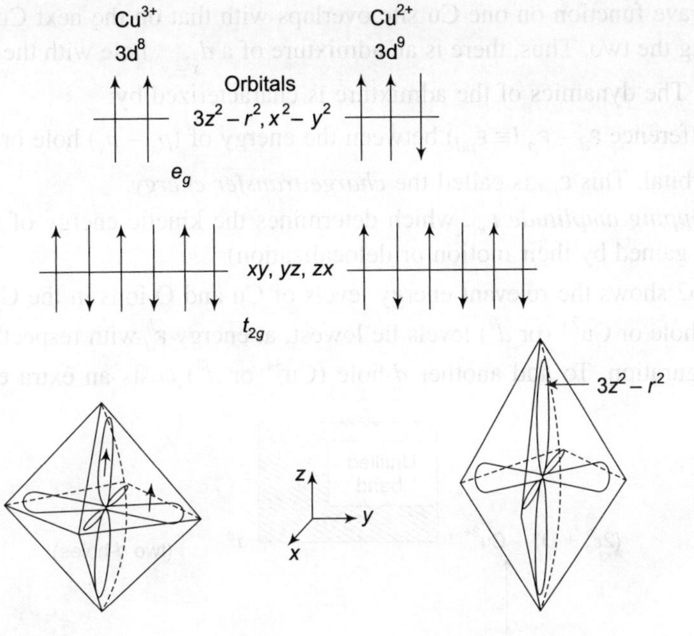


**Figure 15.61** Elongation of the octahedron (in case of $Cu^{2+}$)

of the cubic $\varepsilon_g$ group are half-filled, thus a singlet ground state is formed. In the presence of $Cu^{2+}$ with $3d^9$ configuration, the ground state is degenerate, and a spontaneous distortion of the octahedron occure to remove this degeneracy (the *Jahn-Teller effect*).

### 15.50.1 The Hamiltonian and the Relevant Energy Levels in the Conduction Plane

All the five $3d$ orbitals are fully occupied, except $d_{x^2-y^2}$. There is one missing $d_{x^2-y^2}$ electron (i.e., a hole) per $Cu^{2+}$ ion. Thus, we have

$$Cu^{2+} = (\text{Ar})\ 3d^9 \text{ (one hole), (energy } \varepsilon_d),$$

$$Cu^{3+} = (\text{Ar})\ 3d^8 \text{ (two hole), (energy } 2\varepsilon_d + U)$$

and $Cu^+ = (\text{Ar})\,3d^{10}$ (no hole), (energy zero), i.e., the single $d$-hole is in an orbital with local $d_{x^2-y^2}$ symmetry. All the above three states can be described by the equation

$$H_0 = \sum \varepsilon_d\, n_{d\sigma} + U n_{d\uparrow}\, n_{d\downarrow}$$

where $H_0$ is the energy operator (Hamiltonian) $n_{d\sigma}$ is the number of $d$ holes with spin $\sigma$. This number has two values namely zero and one. The energy $U$ is large (~ 8 eV) and positive and correlates two $d$ holes on the same site away from each other (making $d^8$, unlikely) (see Fig. 15.62).

The hole wave function on one Cu site overlaps with that on the next Cu site via oxygen site intervening the two. Thus, there is an admixture of a $d_{x^2-y^2}$ hole with the surrounding ($p_x$, $p_y$) hole state. The dynamics of the admixture is characterized by:

(1) The difference $\varepsilon_d - \varepsilon_p$ ($\equiv \varepsilon_{pd}$) between the energy of ($p_x - p_y$) hole orbital and a $d_{x^2-y^2}$ hole orbital. This $\varepsilon_{pd}$ is called the *charge transfer energy*.

(2) The *hopping amplitude* $t_{pd}$, which determines the kinetic energy of the *d*-holes (i.e., energy gained by their motion or delocalization).

Figure 15.62 shows the relevant energy levels of Cu and O ions in the $CuO_2$ planes.

The one-*d* hole or $Cu^{2+}$ (or $d^9$) levels lie lowest, at energy $\varepsilon_d^h$ with respect to the no *d*-hole (or $d^{10}$) configuration. To add another *d*-hole ($Cu^{3+}$ or $d^8$) costs an extra energy ($\varepsilon_d^h + U$),

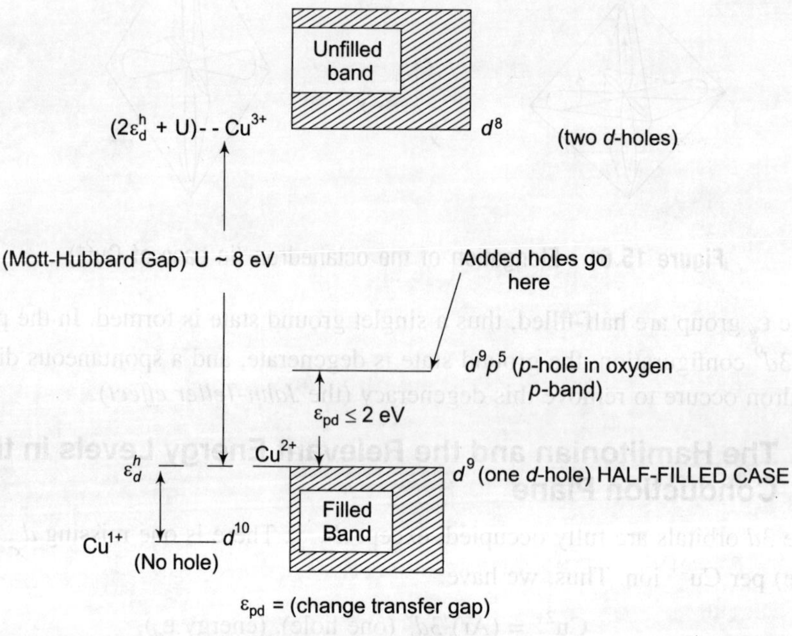


**Figure 15.62** The relevant energy levels of Cu and O ions in the Cu–$O_2$ plane [1]

where $U$ is the large (~ 8 eV) single site Mott–Hubbard repulsion energy. The large value of $U$ leads to the stoichiometric one *d*-hole per site (material $La_2CuO_4$) and makes it an insulator. The energy $U$ is due to Coulomb repulsion between two $d$ holes on the same site (i.e., a spin-up and a spin-down hole on the same site). The feature that distinguishes cuprates from model one band ($d$ band) Mott insulators is the presence of oxygen *p*-hole levels $\varepsilon_p^h$ between $d^8$ and $d^9$ energies, i.e. in the correlation or Mott Hubbard gap. Thus, the minimum energy required to transfer a *d*-hole to an unoccupied site is

$$(\varepsilon_p^h - \varepsilon_p^d) \approx 2 \text{ eV (from neutron scattering experiments)}$$

the charge transfer being to the same site linear combination of $p_x$, $p_y$ oxygen hole orbitals with local $d_{x^2-y^2}$ symmetry.

### *Addition of holes*

When holes are added to $La_2CuO_4$, they go into the oxygen *p*-hole band, and hop from site to site, both directly (with hopping amplitube $t_{pp}$ = 0.7 eV) and via admixture with *d* states ($t_{pd}$ = 1 eV). The ratio $|\varepsilon_{pd}|/t_{pd} = (\varepsilon_p - \varepsilon_d)/t_{pd}$ (i.e., the ratio of charge transfer energy to the hopping amplitude) determines the hole concentration. If the ratio increases, the hole concentration decreases.

All the superconducting cuprates have a fraction *x* (0.05 to 0.30) of mobile holes in the *ab* plane per $CuO_2$ unit cell. Thus, the superconducting cuprates, generically, are doped charge transfer insulators in the strong (local) correlation (or Mott) limit. Suppose $La_2CuO_4$ is doped with Sr (in place of La). One should have either $Cu^{3+}$ or $O^{1-}$ for each $Sr^{2+}$, $O^{1-}$ is favoured. Therefore, addition of a Sr atom introduction a hole in the otherwise filled oxygen *p*-shell. The concentration of holes is proportional to *x* in $La_{2-x}Sr_xCuO_4$. $T_c$ reaches a maximum around an optimal hole-concentration.

## 15.51 HOLE SUPERCONDUCTIVITY IN CUPRATE OXIDES

The essential feature of oxide superconductors is the existence of a network of anions, through which conduction occurs. Figure 15.63 shows schematically the relevant oxygen and copper orbitals in the $CuO_2$ planes. Conductivity in these materials occurs dominantly in the planes.

Optical experiments have shown that the O2*p* holes have at least 95% $p_{x,y}$ character but have not determined whether it is *p*σ or *p*π orbitals. However, NMR results on $^{17}O$ and $^{63}Cu$ have unequivocally shown that BCS like superconductivity is due to oxygen holes and Cu is decoupled as far as superconductivity is concerned. Since thr *p*σ orbitals are strongly

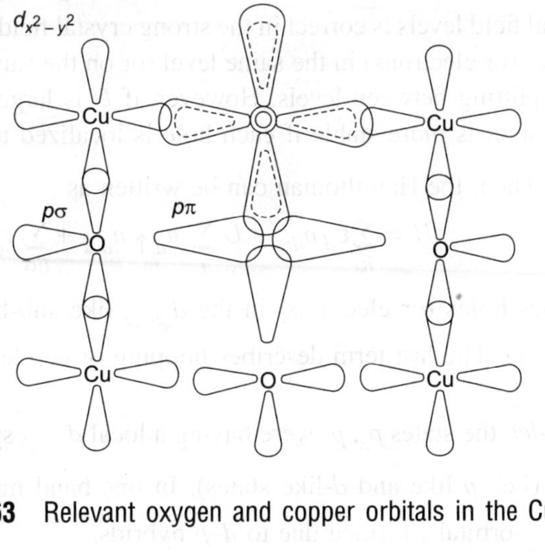


**Figure 15.63** Relevant oxygen and copper orbitals in the $CuO_2$ planes

hybridized with the Cu $d_{x^2-y^2}$ orbitals, the combination of these observations leaves no doubt that the conduction oxygen holes that are responsible for superconductivity reside in the plane $p\pi$ orbitals. There is a direct overlap between $p\pi$ orbitals that gives rise to conduction without involving the cation.

When an electron hops from the oxygen $p\pi$ orbital of the $p\pi$ orbital of the neighbouring O, it will drastically alter the states of the other $p$-electrons on these anions. This is due to the large Coulomb repulsion between that electron and the other electrons in the $p$-orbitals of the anions involved. The wave functions of all the remaining $p$-electrons on the anion, where the electron was, will contract (as they experience less repulsion), when a $p$-electron is removed, and the ones, where the electron hops, will expand (as shown schematically in the Fig. 15.63 This contraction of the charge density of the $O^-$ anion relative to that of the $O^-$ anion will lead to two effects, when the holes conduct through the $p\pi$ network: first, a dynamical band narrowing, as it occurs with small polarons (which is responsible for the anomalous normal state properties of these materials) and second, a pairing interaction that gives rise to a high $T_c$ superconductivity.

## 15.52 TWO-BAND AND ONE-BAND HUBBARD MODELS

For the $CuO_2$ lattice, the energies of different (admixture) state are described by a two-band Hubbard model with Hamiltonian:

$$H = \sum_{i\sigma} \varepsilon_d n_{d_i}\sigma + \sum_i U n_{d_i\downarrow} n_{d_i\uparrow} + \sum_{i\sigma} \varepsilon_p n_{pi\sigma} + \sum t_{pd} (C^+_{pi\sigma} C_{di\sigma} + C^+_{di\sigma} C_{pi\sigma}).$$

The last term describes the removal of a $d$ hole (spin $\sigma$) at site $i$ (destruction operator $C_{di\sigma}$) and its appearance as a $p$-hole with the same local $d_{x^2-y^2}$ symmetry around site $i$ (creation operator $C^+_{pi\sigma}$). This process and its reverse $C^+_{pi\sigma}\, c_{pi\sigma}$, both have the same hopping amplitude $t_{pd}$ (kinetic energy of $d$-hole due to delocalization). The energy ($U$) tends to correlate two $d$-holes on the same site away from each other (it is large and positive).

The picture of crystal field levels is correct in the strong crystal field limit, i.e., when energy ($U$) of putting two holes (or electrons) in the same level (or on the same site) is much smaller than the crystal field splitting between levels. However, if $U$ is larger than the width of the $d_{x^2-y^2}$ sub-band, the system is more stable if each hole is localized to a lattice site (i.e., the Mott insulating state). Then, the Hamiltonian can be written as

$$H = \sum_{i\sigma} \varepsilon_d n_{di\sigma} + U \sum_i n_{d_i\uparrow} n_{d_i\downarrow} + \sum_{ij\sigma} t_{ij}\, C^+_{i\sigma} C_{j\sigma}$$

This model describes holes (or electrons) in the $d_{x^2-y^2}$ like sub-band. This is known as *one band Hubbard model*. The last term describes hopping of an electron from site $i$ to site $j$; with amplitude, $t_{ij}$.

In the *two-band model*, the states $p_x$, $p_y$ were having a local $d_{x^2-y^2}$ symmetry and they were considered separately (i.e., $p$-like and $d$-like states). In one-band model, the $d_{x^2-y^2}$ derived band is a result of $d_{x^2-y^2}$ orbitals, formed due to $d$–$p$ hybrids.

In the *two-band model*, the *p* hole has a tendency to bind anti-ferromagnetically to the nearest *d*-hole and to form a singlet. This spinless entity can be thought of as a missing electron (hole) in half filling (in the one-band Hubbard model). Thus, the low energy dynamics of the two models are very similar.

## 15.53 RESONATING VALENCE BOND MODEL OF HTSCs (CUPRATE SUPERCONDUCTORS): ANDERSOSON's THEORY OF SUPERCONDUCTIVITY

**Introduction.** The high Tc materials have their crystal structure built up from unit cells which are elongated along the *c*-axis and have an essential quadratic base in the *a-b* plane. The first class is $La_{2-x}M_xCuO_4$ with impurity content $x \leq 0.15$ and M = Ba, Sr (the second class is $YBa_2Cu_3O_{6+x}$ with variable oxygen content $x \leq 1$). The superconductivity in these materials is basically due to copper oxygen subsystem out of the four atomic species. There are copper-oxygen planes in which each Cu is the centre of an octahedron of *O*'s which is elongated in the *c*-direction. In the Y-123 compound, there are additional copper and oxygen atoms which alternate in chains along the *b*-axis and are basically responsible for providing careers for superconductivity. In particular, in both ($La_{2-x}M_xCuO_4$ and $YBa_2Cu_3O_{6+x}$) the $x = 0$ version is insulating and antiferromagnetic. It is only with increasing *x* that the metallic state and superconductivity sets in as evident from the phase digram (Fig. 15.64) for the lanthanum compounds. In the antiferromagnetic state, these both materials are insulators (Mott insulators).

A surprising feature is found in the temperature dependence of the electrical resistivity (ρ) in the normal (non-superconducting state and low temperatures). One finds that in the *c*-direction,

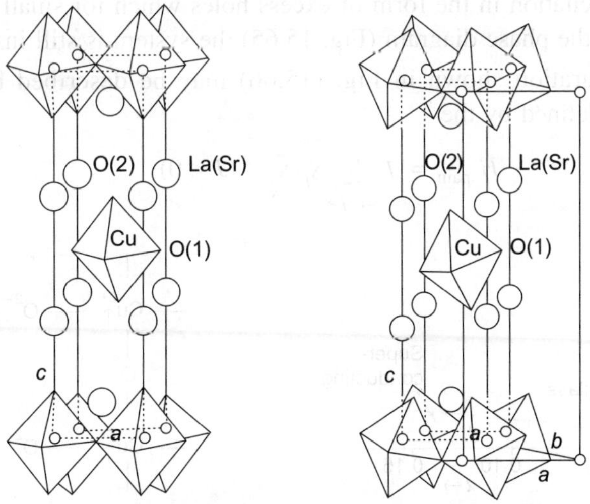


**Figure 15.64** Unit cell of tetragonal (left) and orthorhombic (right) $La_{1.85}Sr_{0.15}CuO_4$

the material is semiconducting, $\rho \propto \frac{1}{T}$, while in the *ab* plane $\rho \propto T$. (The latter *T*-dependence is different from the usual law $\rho \propto T^2$ found for metals at low temperatures.)

### 15.53.1 The Ground State

The essential physics happens in the Cu-O planes. Consequently, the valencies which those atoms assume in the crystal are of primary importance. The electronic configuration of neutral atoms are:

$$Cu^0 = (Ar)\ 3d^{10}4s^1$$

$$O^0 = (Be)\ 2p^4 \qquad ...(1)$$

where (Ar) and (Be) are respectively the configuration of argon and beryllium.

$$Cu^+ = (Ar)\ 3p^{10}$$

and $$O^{2-} = (Be)\ 2p^6 \qquad ...(2)$$

are closed shell configurations, i.e., oxygen favours the valency $2^-$.

In the pure ($x = 0$) compound $La_2CuO_4$, La acts as a donor of one electron. Further, since the configuration in the *a-b* plane is $CuO_2$, copper enters with the valency $^{2+}$. Neutral O has $2e^-$ less (than $O^{2-}$) and neutral Cu has one $e^-$ less, since copper enters as $d^9$ in the structure. So, copper will have valency $2^+$ in the structure. As a result, $Cu^{2+}$ contains (i.e., $3d^9$) one hole (missing electron) and hence has a spin 1/2. In the non-superconducting state, there is an antiferromagnetic order; the spins of the copper atoms in the square lattice of the *a-b* planes alternate and there is also an alteration in the consecutive *a-b* planes. From the theoretical point of view this antiferromagnetic spin arrangement of the "pure" compounds is considered to be the ground state of the electronic system. Then, the *M* impurities in $La_{2-x}M_xCuO_4$ measured by *x*, then create excitation in the form of excess holes which for small *x* must be localized (since according to the phase diagram (Fig. 15.65) the system is still insulating.

The spin configuration shown in Fig. (15.66) may be described by a 2-dimensional Heisenberg model defined by the

$$H_{\text{spain}} = J \sum_{<i,j>} \vec{S}_i \cdot \vec{S}_j \quad (J < 0) \qquad ...(3)$$

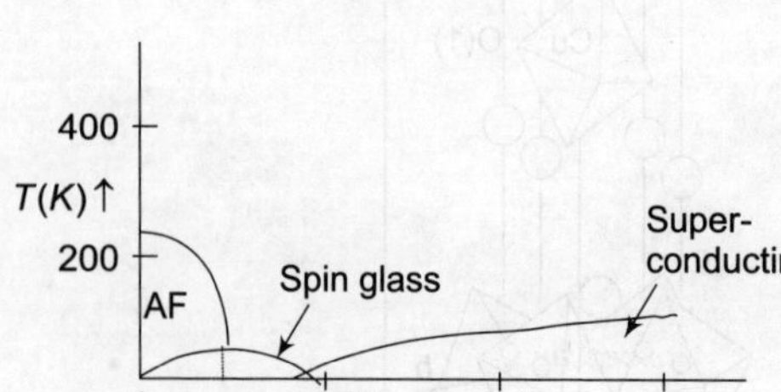


**Figure 15.65** Phase diagram of $La_{2-x}M_xCuO_4$ M = Sr, Ba

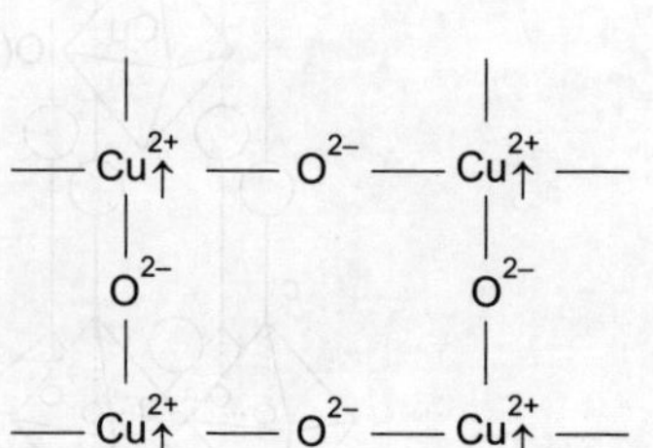


**Figure 15.66** Antiferromagnetic spin arrangement of Cu-O lattice in the *a-b* planes

where $\vec{S}$ is the vector operator for spin 1/2 and the sum runs over nearest neighbour sites $i, j$ of the Cu-lattice. The AF arrangement which has long range order, is very close in energy to a spin-configuration without long range order in which neighbouring spins $i, j$ are paired off into a singlet state. The latter is obtained by operating on the vacuum state with

$$b^{+}_{i,j} = (1/\sqrt{2})\,(C^{+}_{i\uparrow}\,C^{+}_{j\downarrow} - C^{+}_{i\uparrow}\,C^{+}_{j\downarrow}) \qquad ...(4)$$

where $C^{+}_{i\sigma}$ is the creation operator for a hole with spin $\sigma$ (up or down) on a copper site.

These singlet "dimers" (two-component objects) created by the operators of the above equation are called "valence bonds" by Anderson.[7] The Heisenberg model (Eq. 3) is not an accurate description of the localized charge on the copper sites. It is better described by the planar Hubbard model

$$H_{\text{charge}} = -t \sum_{<i,j>\sigma} C^{+}_{i\sigma}\,C_{j\sigma} + U\sum_{i} n_{i\uparrow}\,n_{i\downarrow} \qquad ...(5)$$

Here the first term decribes hopping of the holes betweem neighbouring sites $i$ and $j$ while the second term is a simplified form of the Coulomb repulsion between a spin-up and a spin-down hole on the same site $i$, $n_{i\sigma}$ being the number operator of a hole at $i$ with spin $\sigma$. (Holes with parallel spins cannot occupy the same site because of the Pauli's principle.)

Since an AF square lattice may be divided into sublattices with up and down spins only, there is an enormous number of configurations in which all up and down spins (not only neighbours) are paired into singlets. For the ground state, linear combinations of all these configurations may be considered. This leads to Anderson's notion of the "resonating valence bond" (RVB)[8] which expresses the fact that a valence bond cycle in one configuration may switch, leading to another configuration and back again (Fig. 15.67).

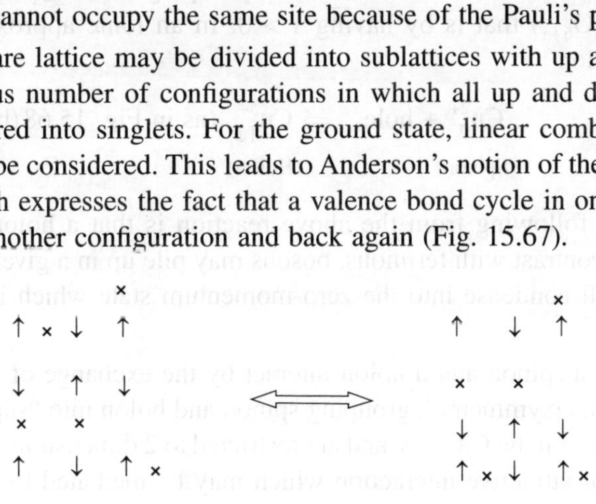

**Figure 15.67** Nearest-neighbour valence bonds (marked by crosses)

## 15.53.2 Theoretical Model for Excitations

According to Anderson, two types of excitations may be generated on top of the RVB ground state. The simplest one which does not require the introduction of additional charges (holes), is just an isolated unpaired spin $\sigma$ (up or down) as shown in Fig. 15.68. By definition, this excitation called "spinon" does not carry a charge. It is essentially (in a purely ionic

[7] P.W. Anderson, *Science*, **235**, 1196 (1987).

[8] *Ibid.*

approximation) just a $Cu_{\sigma}^{2+}$. Since this single spin may break up a neighbouring valence bond and pair with one of the former partners, it is obvious that a spinon can move through the Cu lattice very freely. It has a spin 1/2 and therefore it is fermion.[9]

The second type of excitation of the RVB model is a single charge on a Cusite called "holon" A hole represents an unoccupixd non-bonded[10]

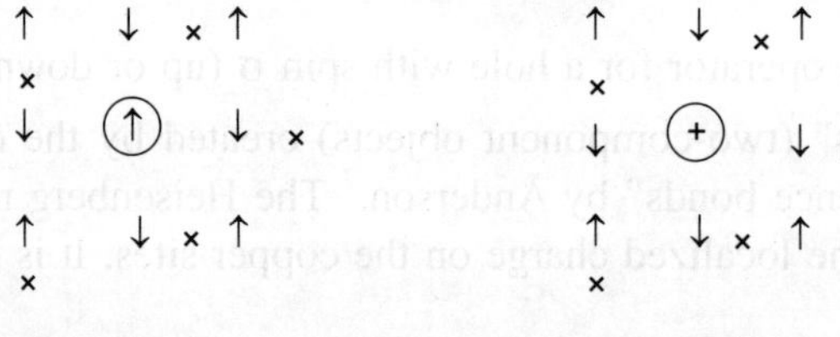


**Figure 15.68** Unpaired spin or "spinon" $Cu_{\uparrow}^{2+}$ (a) and excess hole or "holon" $Cu_{\uparrow}^{3+}$ (b) in a valence bond configuration

Excess charges are introduced into the lattice by doping (in $La_{2-x}M_xCuO_4$) or by excess oxygen (in $YBa_2Cu_3O_{6+x}$) that is by having $x > 0$. In an ionic approximation, a holon is obtained by the reaction

$$\underset{\text{(Spinon)}}{Cu_{\sigma}^{2+}} + \text{hole}_{-\sigma} \Rightarrow \underset{\text{(Holon)}}{Cu_{\uparrow\downarrow}^{3+}} \quad \text{(as in Fig. 15.68 (b)).} \qquad ...(6)$$

The key property following from the above reaction is that a holon has spin zero and hence is a boson. In contrast with fermions, bosons may pile up in a given state. In particular, at $T = 0$ they may all condense into the zero-momentum state which is the Bose–Einstein condensation.

As Eq. (6) shows, a spinon and a holon interact by the exchange of a hole. This relation may be viewed as "supersymmetry", grouping spinon and holon into "superpartners". In fact, the holons are localized on the Cu-ions, and are restricted to 2 dimensions. Holon condensation requires some kind of attractive interaction which may be mediated by spinon exchange or by interlayer coupling.

### 15.53.3 Pairing Mechanism

In resonating valence bond theory, two different condensation mechanisms have been proposed. The first is the holon condensation discussed above. The second mechanism in the

9 A broken singlet bond excitation gives rise to two spin 1/2 objects, i.e., zero charge spin half fermions called spinons (Fig. 15.68 a).

10 Site of charge $+e$ and zero spin (Fig. 15.68 b). Holons are bosom and are created by extraneous $p$-type doping. Holons give Bose condensate at $T = T_c$.

framework of RVB theory is condensation of the valence bonds of Eq. (3) meaning that the order parameter is now defined as:

$$\Delta = \sqrt{2}\ \langle b_{i,j} \rangle$$

This Δ is degenerate with respect to the "up" and sub-lattices as well as to the *a*- and *b*-directions.

Since there are also "up" and "down" holon condensates the question may be asked whether the two condensation mechanisms discussed here are equivalent in a supersymmetric sense.

In the models based on the assumption of hopping *Cu*-holes, two different pairing mechanisms have been proposed. In the first, the AF background acts like a "spin bag" which preferentially attracts two holes with opposite spin on nearest-neighbour sites and thus leads to pairing. Superconductivity then is obtained by a conventional BCS formalism.

The second type of model makes use of the fluctuating valence reaction

$$Cu^{3+}_{\uparrow\downarrow} - O^{2-} - Cu^{3+}_{\uparrow\downarrow} \Rightarrow Cu^{2+}_{\uparrow} - O^{0}_{\uparrow\downarrow} - Cu^{2+}_{\downarrow}$$

which also explains the oxygen loss mechanism because of the neutral oxygen. This reaction leads to a coupling between the two holes on the $Cu^{3+}$ and the two holes on $O^0$. Approximating the latter by a doubly charged spin-zero boson, the zero-momentum projection of this interaction has the form

$$H_{\text{int}} = W \sum_k a^+_{k\uparrow}\, a^+_{-k\downarrow}\, b + herm.conj.$$

Here $a^+_{k\sigma}$ is the creation operator for a hole of momentum *k* and *b* the annihilation operator for the doubly charged boson on the *O*-site. The coupling constant (*W*) may be expressed in terms of an extended Hubbard model in which the *O*-ions are explicitly taken into account.

## 15.54 ARGUMENTS AGAINST THE APPLICABILITY OF THE ORIGINAL BCS THEORY TO THE HTSCs

(a) Observed values of $T_c$ are probably too high (i.e., some interaction much stronger than electron-phonon interaction is at work).

(b) A linear term in the specific neat versus tamperature curve has been observed in La–, Y– and Tl-compounds at low termperatures. The BCS theory predicts an exponentially small electronic specific heat if $T \ll T_c$.

(The linear term is present in tetragonal Y123, and absent in non-superconducting $La_{2-x}\ Ba_x\ CuO_4$ and Bi based BSCCO.)

(c) Raman scattering and infra-red absorption measurements imply energy gap (Δ) might vary along certain lines or at certain points in momentum space, i.e., for electrons with momentum along certain directions, like heavy fermion superconductons e.g. $UBe_{13}$.

(d) Tunnelling experiments provide evidence against existence of nodes in energy gab (i.e., a reasonably well defined energy gap with very few states, if any, in it.)

(e) The spectral features associated with Δ barely shift in energy when $T_c$ is approached, as if the gap were nearly constant upto $T_c$ (i.e., no temperature dependence of Δ).

## 15.55 IDENTIFICATION OF NON-PHONONIC MECHANISMS

The isotope effect does not give an unequivocal answer as to whether a non-phononic mechanism is present in high $T_c$ superconductivity. If there is a relationship between $T_c$ and the ion mass $M$, then there is definitely a lattice contribution to pairing, but it would be incorrect to try to correlate the value of $\alpha$ (or its deviation from 0.5) with the degree of phonon-participation. High $T_c$ by itself does not imply the presence of non-phononic mechanism. In principle, it is quite realistic that both the mechanisms (phononic and non-phononic) may coexist.

## 15.56 SCENARIO OF THEORIES FOR SUPERCONDUCTIVITY IN HTSCs

Figure 15.69 depicts the scenario of various theories given for high $T_c$ superconductors. The L.H.S. corresponds to Fermi liquid like theories with electron-phonon (top) and electron-electron (bottom) interactions. The R.H.S. corresponds to non-Fermi liquid theories with electron-phonon (top) and electron-electron (bottom) correlations.

The various proposals for pairing mechanism range from conventional models (i.e., pairing mediated by various type of Bosonic excitations) to exotic models (based on quasi-particles in two dimensions that obey anyonic statistics).

The Bosonic excitations may include (i) ordinary phonons or unusual lattice vibrations associated with the (presumed) polarizability of oxygen, (ii) electronic excitations and (iii) magnetic excitations. Both (ii) and (iii) could be at energies higher than the phonons and therefore, could lead to higher $T_c$'s.

In regard to the exotic models, the experimental results are inconclusive so far.

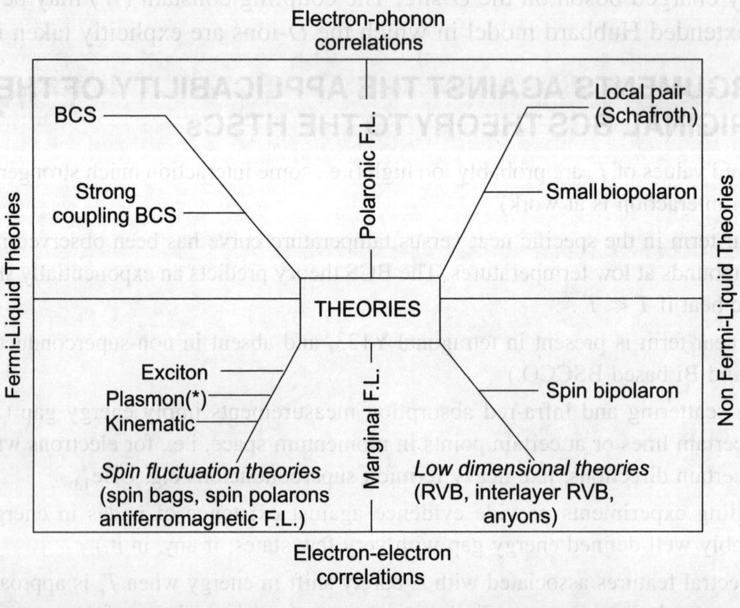


**Figure 15.69** Various theories for high $T_c$ superconductivity. ((*) electron excitons are called plasmons)

It is frequently assumed that the strong coupling variant of the BCS theory can be employed to explain the high temperature superconductivity, because it successfully accounts for the properties of the classic superconductors also. Evidences for this fact stem from a number of observations;

(a) Electron pairing seen in Shapiro steps displayed by Josephson tunnel junctions biased with both ac and dc voltages. The steps have voltage spacing $h\nu/(2e)$ which implies that the electrons are bound into Cooper pairs.

(b) Electron tunnelling and e.m. absorption measurements indicate presence of an energy gap in the distribution of energy levels available to the system. (It is of the same order of magnitude at temperatures $T \ll T_c$ as predicted by BCS theory.) Values somewhat larger than 3.53 $k_B T_c$ are known to be associated with either (i) strong electron phonon coupling or (ii) strong electron-other virtual excitation coupling.

## 15.57 CHARACTERISTIC PROPERTIES OF HIGH-TEMPERATURE SUPERCONDUCTORS (HTSCs)

Some of the common structural and physical properties of oxide high-$T_c$ cuprate superconductors are as follows.

(i) They have their structure derived from the ideal perovskite structure (therefore termed a defect perovskite structure), either through an intergrowth phenomenon or by an ordered removal of oxygen atoms.

(ii) They have a layered crystal structure consisting of one or more $CuO_2$ layers. Each copper atom in such a plane is strongly covalently bonded, in a nearly square-planar arrangement, to four oxygen atoms at a distance of $\cong 1.9$ Å (the Cu—O bond length).

(iii) Copper is present in the mixed valence state involving a partial oxidation of $Cu^{2+}$ to $Cu^{3+}$.

(iv) There is a charge transfer to, and from, the $CuO_2$ layers, which is induced by doping near the metal insulator phase boundary existing in all oxide high-temperature superconductors.

(v) The sign of Hall coefficient, $R_H$, is positive, indicating that the charge carriers are holes in all oxide high $T_c$ superconductors (except $(Ln_{1-x}M_x)_2$ $CuO_{4-y}$ which is an electron superconductor). $R_H$ for a magnetic field $H$ parallel to the $ab$ plane is much smaller than for $H$ parallel to the $c$-axis.

(vi) When rare earth (RE) ions replace Y in $YBa_2Cu_3O_7$, $T_c$ remains virtually unchanged, and most of the transport properties are unaffected. However, the specific heat data of (RE) $Ba_2Cu_3O_7$ compounds, which are within 2% of the $YBa_2Cu_3O_7$ for $T > T_c$, show drastic differences at low temperatures. This arises due to the magnetic interactions between the paramagnetic ions as well as between these ions and the crystal field. This establishes that superconductivity and magnetism can coexist in the RE compounds and the RE ions do not participate in superconductivity.

(vii) The normal state transport properties are highly anisotropic. In Y-Ba-Cu-O, there is a linear temperature dependence of the $ab$ plane resistivity ($\rho_{ab}$). But, the $c$-axis resistivity ($\rho_c$) shows a semiconducting ($\propto T^{-1}$) dependence for less oxygenated samples and ($\rho_c \propto T$) for well-oxygenated samples $\rho_c/\rho_{ab} \sim 30$ at room temperature. Similar anisotropic behaviour is observed for other HTSCs.

(viii) They are all type-II superconductors. The lower critical field $H_{c1}$ and the upper critical field $H_{c2}$ are highly anisotropic ($H_{c2}\|(0) = 120\ T$ and $H_{c2}^{\perp}(0) = 510\ T$ for $YBa_2Cu_3O_7$).

(ix) The two fundamental superconducting parameters, viz., the coherence length ($\xi$) and the penetration depth ($l$), are also highly anisotropic (e.g., $\xi_{ab}$ = 20–30 Å and $\xi_c$ = 2–3 Å for $YBa_2Cu_3O_7$). Further, the magnitude of coherence length is quite small as compared to that for conventional superconductors.

(x) There is near absence of the isotope effect in high-$T_c$ cuprate superconductors.

(xi) The value of $2\Delta(0)/k_BT_c$ is smaller for tunnelling of pairs perpendicular to the *ab*-plane (in the range 3.8–4), while its value parallel to the plane is in the range 5.8 – 6.0.

(xii) The low-temperature specific heat of HTSCs in the superconducting phase is due to three contributions

$$C = \alpha T^{-2} + \gamma T + \beta T^3$$

The first term gives a Schottky-like anomaly and arises from magnetic impurities. The second is a linear contribution similar to that for free electrons in a metal. The third is Debye contribution from the lattice. The absence of the exponential dependence of the electronic specific heat (expected on the basis of the BCS theory) is a major departure from the properties of the conventional superconductors. The linear term is now believed to arise from the presence of impurities.

(xiii) The phononic mechanisms of superconductive (the BCS theory) has certain limitations in theoretically explaining the mechanism of superconductivity in these cuprate high $T_c$ superconductors. Various other mechanisms have been proposed (e.g., resonating valence bond model, spin-bag model, bipolar on model, etc.), but till date no suitable mechanism is able to explain satisfactorily various normal state and superconducting properties of these materials.

## 15.58 APPLICATIONS OF SUPERCONDUCTIVITY

Many possible applications of superconductivity have been explored since the discovery of this phenomenon. Table 15.9 lists major present as well as potential applications under various bases of application.

**Table 15.9** Applications of Superconductivity

| *Bases for application* | *Application* |
|---|---|
| $R = 0$ and High $J_c$. | 1. Magnets for variety of applications |
| | 2. Passive microwave devices |
| | 3. Interconnects in microelectronics |
| | 4. Electrical energy transport by cables |

*Contd...*

**Table 15.9** *Contd...*

| | |
|---|---|
| Josephson tunnelling | 1. Microwave detectors and mixers |
| | 2. In physical measurements (SQUIDS, etc.) |
| | 3. Computers (fast logic and memory circuits) |
| | 4. Plasma and space |
| | 5. In medicine (magnetic encephalography) |
| High $J_c$ at high $H_{c2}$ | 1. Electrical power industry |
| | 2. Plasma confinement (in high energy physics) |
| | 3. In transport (levitation trains, NHD-propelled ships) |
| | 4. Medicine (Nuclear Magnetic Resonance Tomography) |

### 15.58.1 Superconducting Magnets

Since $R = 0$, there is no energy dissipation ($I^2R$ loss), a current setup in a closed loop of a superconductor persists forever without decay. Many superconductors carry (without energy loss) 100–1000 times larger current densities compared to normal conductors like copper. These facts have made possible the utilisation of superconductors for building lightweight, high-intensity compact magnets, useful in various applications.

First operational superconductor magnet (developed in 1957) was made with Nb wire, which could produce 0.7 T. In 1962, a superconductor was prepared using NbSn, which could develop 7.0 T. Presently, superconductor magnets are able to produce 17 of magnetic field.

The superconducting magnets offer the advantage of getting high homogeneity and stability of the field ensures reliability as compared to resistive (permanent) magnets. Their use in magnetic resonance imaging ensures a superior image quality. In high energy physics, particles can be accelerated, deflected and focussed by the fields of superconducting magnets. In bubble chambers, high magnetic fields are used to defect particles, forcing them to follow a curved path; the curvature and direction of their track determines the momentum and charge of the particle.

### 15.58.2 Applications of High–$T_c$ Oxide Superconductors

The SQUIDs are the basis of the most sensitive magnetometers available today. These currently use the conventional superconductors and operate at liquid helium temperature, i.e., 4.2 K. It is expected that SQUIDs made from oxide superconductors and operating at liquid nitrogen temperature (77 K) will be soon available commercially. This will improve the ruggedness and utility of the device and extend its use to major areas, specially military and medical fields.

In VLSI (very large scale integration) technology, the use of HTSC interconnections working at 77 K offers several advantages and it is expected that in high-speed computers and telecommunications this will be used extensively in near future.

The use of HTSCs for generation, storage, transmission and transformation of electrical power in electrical industry is being assessed by major industries all over the world. The three parameters, viz., critical temperature $T_c$, critical current, $J_c$ and critical field, $H_c$ are continually being improved to develop the power industry. The current density and magnetic field requirement for different applications are listed in Table 15.10.

**Table 15.10** $J_c$ and $H_c$ Requirements for Various Applications in Electrical Power Industry

| *Application* | *$J_c$ (A/cm$^2$)* | *$H_c$ (tesla)* |
|---|---|---|
| Power transmission | $2 \times 10^4$ | 1 |
| Power generation | $2 \times 10^4$ | 5–7 |
| Transformer | $10^4$ | 1 |
| Power storage | $10^3 - 10^4$ | 5–10 |
| MRI scanners | $10^4$ | 2 |
| Mineral separation | $10^4$ | 2–5 |

One application of superconductivity that has gained popularity in medical field is magnetic resonance imaging. In this fairly intense magnetic fields are created which simulate the resonance of hydrogen atoms in the human body. This helps in the building up of a two-dimensional picture of human organs.

The possibility of applications of HTSCs in magnetically elevated trains and in magnetic mineral separation are also being pursued, although to make cost-effective devices for use in these fields appear to be somewhat remote.

In Table 15.11 we give a subjective "top ten" of superconductivity research.

**Table 15.11** Top 10 of Superconductivity (Subjective)

| *Person* | *Achievement* | *Date/comments* |
|---|---|---|
| 1. H. Kammerlingh Onnes | Liquefied He<br>Found resistance of Hg → 0 at 4.19 K | 1908 – Started low-*T* physics<br>1911 – Discovered superconducting state<br>1911 – Nobel Prize |
| 2. W. Meissner and R. Ochsenfeld | Perfect diamagnetism | 1933 – Flux exclusion |
| 3. F. and H. London | London equations and flux expulsion | 1935 – B proportional to curl of *J* |
| 4. V. L. Ginzburg and L.D. Landau<br>A. A. Abrikosov<br>L. P, Gor'kov | Phenomenological equations<br>Improvement to GL equations, Type II<br>GL limit of BCS and order parameter | 1950 – Eventually GLAG equations<br>1962 – Nobel Prize, Landau<br>2003 – Nobel Prize, Ginzburg<br>1959 – Order parameter proportional to gap parameter |
| 5. A. B. pippard | Nonlocal electrodynamics | 1953 – *x* and *l* dependent on mean free path in alloys |
| 6. J. Bardeen, L. Cooper and J. Schrieffer | Theory of superconductivity | 1957 – e.g., see (15.27)<br>1972 – Nobel Prize (all three) |
| 7. I. Giaver | Single-particle tunnelling | 1960 – Get gap energy<br>1973 – Nobel Prize |

*Contd...*

**Table 15.11** *Contd...*

| | | |
|---|---|---|
| 8. B. D. Josephson | Pair tunnelling | 1962 – SQUIDS and metrology<br>1973 – Nobel Prize |
| 9. Z. Fisk, et al. | Heavy fermion "exotic" superconductors | 1985 – Pairing different than BCS, probably |
| 10. J.G. Bednorz and K. A. Müller | High-temperature superconductivity | 1986 – Now, $T_c$s are over 100 K<br>1987 – Nobel Prize (both) |

**Example 15.2** Calculate the critical which can flow through a long thin superconducting wire of aluminium of diameter $10^{-3}$m. Critical magnetic field for Al is $7.9 \times 10^3$ A/m.

***Solution*** The critical current flowing through a wire is given by $I_c = 2\pi r H_c$.

Here $$2r = 10^{-3}\text{m}.$$

$$H_c = 7.9 \times 10^3 \text{ A/m}.$$

$$\therefore \quad I_c = \pi \times 2r \times H_c$$

$$= \pi \times 10^{-3} \times 7.9 \times 10^3$$

$$= 24.806 \text{ amperes.}$$

**Example 15.3** Critical temperature of a sample with isotopic mass of 204.87 is 19.2 K. Find $T_c$ when isotopic mass changes to 218.87.

***Solution*** $$T_c M^{1/2} = \text{constant}$$

Here $$T_{c1} = 19.2\text{K for } M_1 = 204.87$$

$$= 18.58 \text{ K.}$$

**Example 15.4** Calculate the energy gap $E_g$ for Nb ($T_c = 9.25$K) at $T = 0°$K and find the minimum photon wavelength needed to break the Cooper pair.

***Solution*** At 0°K, energy gap $E_g = 3.54\, k_B T_c$

or $$E_g = 3.54\,(1.38 \times 10^{-23}\text{J/K})\,(9.25 \text{ K})$$

$$= 4.25 \times 10^{-22} \text{ J}$$

$$= \frac{4.25 \times 10^{-22}}{1.6 \times 10^{-19} \times 10^{-6}} \text{ MeV} = 2.82 \text{ MeV},$$

Also $$E_g = \frac{hc}{\lambda_{\min}} \therefore \lambda_{\min} = \frac{hc}{E_g} = 4.3 \times 10^{-4}\text{m}.$$

**Example 15.5** Experiments on superconducting sample show that at 12 K, a critical field of 15 T exists and at 10 K, the critical field is 18 T, calculate the critical temperature.

***Solution*** We have $$H_c = H_0\left[1 - \left(\frac{T}{T_c}\right)^2\right]$$

$$\text{At } 12\,\text{K}, H_c = 15T$$

$$\text{At } 10\,\text{K}, H_c = 18T$$

$$\therefore \quad H_c(12) = H_0\left[1-\left(\frac{12}{T_c}\right)^2\right]$$

$$H_c(10) = H_0\left[1-\left(\frac{10}{T_c}\right)^2\right]$$

$$\frac{H_c(12)}{H_c(10)} = \frac{\left[1-\left(\frac{12}{T_c}\right)^2\right]}{\left[1-\left(\frac{10}{T_c}\right)^2\right]} = \frac{15}{18}.$$

$$\text{or} \quad \frac{T_c^2-(12)^2}{T_c^2-(12)^2} = \frac{15}{18} \text{ or } \frac{T_c^2-144}{T_c^2-100} = \frac{5}{6}$$

$$6T_c^2 - 864 = 5T_c^2 - 500$$

$$\text{or} \quad T_c^2 = 864 - 500 = 364$$

$$\therefore \quad T_c = 19.08\text{ K}$$

## EXERCISE

15.1 Enumerate the properties of type-I and type-II superconductors.

15.2 Derive London equations and discuss how do they help in explaining the superconducting state.

15.3 What is Meissner effect? Obtain an expression for the London penetration depth of magnetic field in a superconductor.

15.4 What are Coopper pairs? Explain their formation in a superconductor.

15.5 Give a qualitative treatment of BCS theory.

15.6 Give a quantitative treatment of BCS ground state. Show that energy gap (at 0°K) is given by

$$\Delta_0 = 2\hbar\omega_D \exp\left[-\frac{1}{V_0 Z(E_S^0)}\right]$$

(where $V_0$ is restricted to $\pm\,\hbar\omega_D$ and $Z(E_F^0)$ is pair density of states).

15.7 Using the parabolic temperature dependence of the critical magnetic field curve, show that

$$C_n - C_s = \frac{2\mu_0^2}{\mu_0 T_c}\left[\frac{T}{T_C} - 3\left(\frac{T}{T_C}\right)^3\right]$$

15.8 The critical temperature for mercury with isotopic mass 199.5 is 4.185 K. Calculate its critical temperature when its isotopic mass changes to 203.4. (*Ans.* 4.139 K)

15.9 Estimate the London penetration depth from the following data:

$$T_c = 3.7 \text{ K}$$

$$\text{Density} = 7.3 \text{ g/cm}^3$$

$$\text{Atomic weight} = 118.7$$

$$m^* \text{ (effective mass)} = 1.9 \times m_e$$

where $m_e$ is mass of a free electron.

15.10 The penetration depths for Pb are 396 Å and 1730 Å at 3 K and 7 K, respectively. Calculate the critical temperature. (*Ans.* 7.19 K)

15.11 Write short notes on

(i) Occurrence of superconductivity

(ii) Meissner effect

(iii) Ginzberg–Landau theory

15.12 Enumerate the properties of high-$T_c$ superconductors.

15.13 Briefly describe the Resonating Valence Bond Theory of High $T_c$ superconductivity.

15.14 Derive London's equations and define penetration depth. What is perfect diamagnetism?

15.15 Explain phase-coherence in the superconducting state.

15.16 Explain the concept of coherence length and derive an expression for the same.

15.17 Explain the mixed statc in Type II superconductivity.

15.18 Explain the variation of $T_c$ with oxygen stoichiometry in $\gamma_1$ $Ba_2Cu_3O_{7-x}$ superconductor.

CHAPTER

# 16

# Thermoelectric, Photovoltaic and Magnetoelectric Effects

In a common thermocouple, a voltage is induced between the hot and cold junctions of two dissimilar metals which increases monotonically with increasing temperature difference. The temperature difference produces changes in the distribution of electrons in the energy states at the hot and cold ends, leading to a flow of electricity. Three experimental observations of thermoelectricity, the Seebeck (1822), Thomson, and Peltier (1834) effects, are related thermodynamically by the Kelvin (Thomson) relations. In semiconductors, much larger thermoelectric effects occur than in metals. Thermoelectric temperature measurement, refrigeration, and heating, and generation of electrical power are all in limited use today. The efficiency of a thermoelectric device depends on the Seebeck coefficient, thermal and electrical conductivity, and the operation temperatures.

## 16.1 INTRODUCTION

Consider the conducting rod shown in Fig. 16.1 with one end maintained hot, and the other cold. At the hot end, electrons will, on an average, be excited to higher energies; the Fermi-Dirac distribution will have more electrons above $E_F$ and fewer below. The higher-energy electrons at the hot end are able to lower their energies by diffusing to the cold end. Thus, the cold end becomes negatively charged, the hot end positively charged, and a voltage is induced along the rod. The induced voltage causes an electrical current to flow, which is equal to the voltage divided by electrical resistance of the rod. Equilibrium

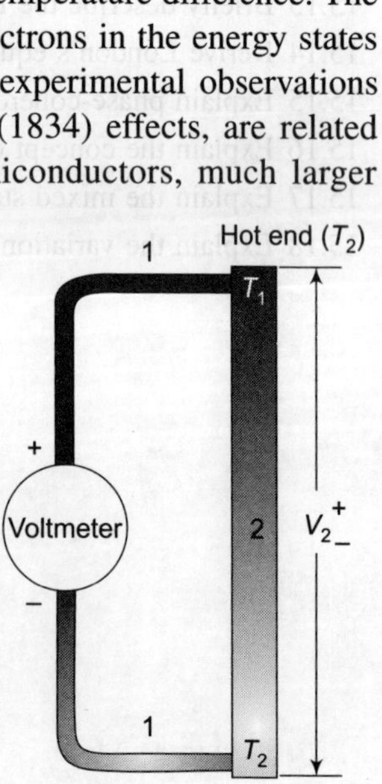


**Figure 16.1** Schematic of circuit used to measure thermal electromotive force

is reached when the current is equal to the flow of electrons due to the temperature difference, and the induced voltage is constant. It should be noted that the induced voltage between the ends of the rod depends *only* on the temperatures of the ends. Whatever the distribution of temperatures, the same voltage will appear between the ends.

To measure the induced voltage, a voltmeter is connected to the ends of the rod as shown in Fig. 16.1. If the connecting wires, *1*, are of the same material as the rod, 2, the temperature difference will induce the same voltage in the connecting wires as in the rod, and there will thus be no voltage across the meter. On the other hand, if the connectors are of another material, a different voltage will be induced in *1*, and the net voltage, $V_1 - V_2$, will be observed at the meter. If we raise the temperature difference $T_1 - T_2$ a small amount, $\Delta T$, the induced voltages will rise a small amount. The net voltage, designated as $V_{12}$, will change. The rate of change of voltage with temperature is defined as the *thermoelectric power*, $S_{12}$ of the junction *1-2*:

$$S_{12} = \frac{dV_{12}}{dT} = \frac{dV_1}{dT} - \frac{dV_2}{dT} \quad ...(1)$$

Although $S_{12}$ is called the thermoelectric power of the junction, it is not at all a property of the junction. It depends only on the bulk properties, $S_1$ and $S_2$, of the two materials. The *term power*, though inappropriate, is nevertheless widely used.

The thermally induced voltage, $V_{12}$, called the *Seebeck potential*, is often used in measuring temperature. One junction of the thermocouple is held at a known temperature. The voltage, $V_{12}$, is then a function of the temperature of the other junction. Thermoelectric powers have been compiled for a large number of materials and pairs, and calibrated thermocouple wire is a commercial item. In common use are copper (up to 315°C), Constantan (60 Cu-40 Ni) and iron (both up to 950°C), Chromel (90 Ni-10 Cr) and Alumel (94 Ni-2 A1-3 Mn-1 Si), both up to about 1200°C. For higher temperatures up to 1500°C, couples of platinum and various platinum-rhodium alloys are used, and 1500°C, tungsten-rhenium alloys.

At room temperature and above, thermoelectric potentials in the millivolt range are common; at low temperatures, the potentials are normally in the microvolt range. The thermoelectric power is very sensitive to structural defects so that thermocouple materials must be very carefully processed. In general, anything which increases the resistivity, such as cold work, will increase the thermoelectric power.

In view of the way that the voltages are defined in Fig. 16.1 and in the first paragraph of this section, it is expected that the thermoelectric power of a conductor or of any material in which conduction electrons are the majority carrier will be negative. Since only differences in thermoelectric potentials are measurable, it is difficult to determine directly the absolute thermoelectric power. Fortunately, the power is zero for a superconducting material, which provides a standard. The absolute thermoelectric power can also be calculated from the Thomson heat as discussed below.

## 16.2 THOMSON EFFECT

Referring again to the rod of Fig. 16.2 with one end hot and the other cold, the ends will be charged, and there will be a voltage along the length of the rod. If conventional electrical current passes down the rod from the hot end to the cold end, electrons flow from cold to hot; as they flow from the negative to the positive end of the rod. This increases their potential energies by absorbing energy in the form of heat. If the current flow is reversed, the electrons will decrease their potential energies by emitting heat. This absorption or evolution of heat must be added to the $I^2R$ power loss or Joule heat together with the heat flow due to the thermal gradient in the rod, to get the total heat production and flow. The absorption or evolution of heat is called the *Thomson effect*, and the relation

$$\frac{dQ(\text{Thomson})}{dt} = -\mu_T J_x \frac{dT}{dx} \quad ...(2)$$

defines the *Thomson coefficient*, $\mu_T$. $J_x$ and $dT/dx$ are the current density and thermal gradient along the rod of Fig. 16.2, and $dQ/dt$ is the heat evolved per unit volume per second. (Note that the $Q$ in this definition has a sign opposite to that common in thermodynamics. Here, it is positive when the heat is emitted.)

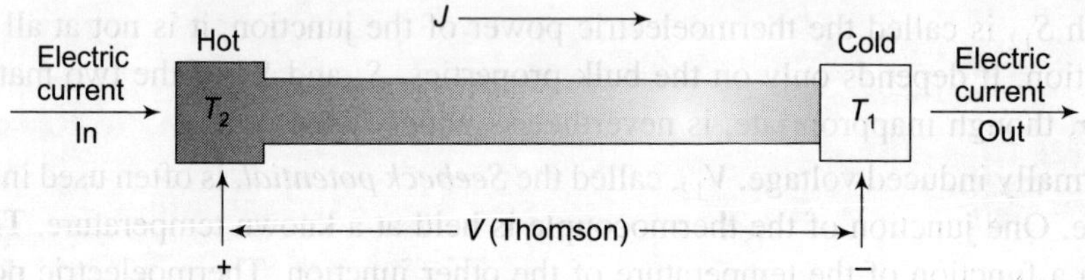


**Figure 16.2** Schematic of Thomson effect. In addition to the Ohm's law voltage, a second voltage is observed, due to the temperature difference. When the current is passed, additional heat is absorbed or emitted, depending on the directions of the current and thermal gradient

## 16.3 PELTIER EFFECT

A third important thermoelectric effect illustrated in Fig. 16.3 is the *Peltier effect*. When current is passed through the thermocouple junction, heat is absorbed or evolved, depending on the direction of the current. The *Peltier heat*, $\pi_{12}$, is defined as the reversible heat evolved at the junction per unit time, per unit electric current flowing from 1 to 2, or

$$\frac{dQ(\text{junction})}{dT} = \pi_{12} \quad ...(3)$$

From this definition, $\pi_{12} = -\pi_{21}$.

In the discussion of the Thomson effect, it was stated that electrons can convert heat to potential energy and vice versa. In general, electrons can transport heat in the form of thermal energy and potential energy. Precisely, how much heat the electrons carry depends on temperature, electrical fields, and the material through which the electrons flow. The heat transported per mole of electrons is different for different materials, and therefore their thermoelectric powers and Thomson coefficient differ. When electrons flow from material 1 to material 2 in Fig. 16.3, they change the amount of heat that they carry, and the difference must be absorbed or given off, depending on the direction of current flow at the junction. Consequently, the Peltier effect is proportional to the difference in the heat carried by the conduction electrons of the two materials. The absolute Peltier heat may be defined in the same manner as the absolute thermoelectric power

$$\pi_{12} = \pi_1 - \pi_2 \quad ...(4)$$

Neither the Peltier nor Seebeck effect depends on the *nature* of the junction between the two materials. The junction may be soldered, brazed, spot-welded, or fused. The effects are bulk material and not junction properties.

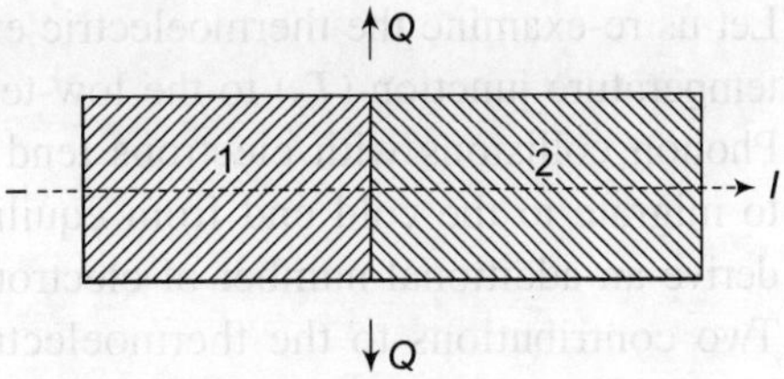


**Figure 16.3** Schematic of the Peltier effect. If current is passed through the 1-2 junction in one direction, heat is evolved. If the current direction is reversed, heat is absorbed

## 16.4 KELVIN (THOMSON) RELATIONS

The three parameters which describe thermoelectric behaviour are the thermoelectric power $S$, the Thomson coefficient $\mu_T$, and the Peltier heat, $\pi$. These quantities are not independent. Thomson applied the first and second laws of thermodynamics to show that only one of the three is needed to specify to the other two. Although Thomson's derivation was not rigorous, since classical thermodynamics is not directly applicable to this problem,[1] his answers are correct.

$$\mu_T = T\frac{dS}{dT} \quad ...(5)$$

$$\pi = TS \quad ...(6)$$

(Remember that $S$ here refers to the thermoelectric power, not the entropy.) If $S$ is known as a function of temperature, Eqs. 5 and 6 give $\pi$ and $\mu_T$. If $\mu_T$ rather than $S$ is known Eq. 5 can be integrated to obtain $S$

$$S(T)–S(0) = S(T) = \int_0^T \frac{\mu_T}{T}\,dT\,. \quad ...(7)$$

$S(0)$ can be dropped from Eq. (7) because the third law of thermodynamics implies that all thermoelectric effects vanish at 0°K. It is then possible, from Eq. (7) to calculate the absolute thermoelectric power of a conductor at any temperature.

## 16.5 PHONON DRAG

In the previous description of thermoelectricity it was assumed that the energy of conduction electrons may be converted freely to heat or lattice vibrations. In other words, electrons colliding with phonons give up their energy. The reverse was also assumed to hold. If energy is exchanged in electron-phonon collisions, momentum must be exchanged also.

---

1 The techniques of *irreversible thermodynamics* are necessary. See, for instance, Part III of the *Thermodynamics* by H.B. Callen, Wiley, New York, 1960.

Let us re-examine the thermoelectric experiment of Fig. 16.1. Heat will flow from the high-temperature junction ($T_2$) to the low-temperature junction ($T_1$) providing a net phonon flow. Phonon collisions with electrons tend to drive the electrons along as well. Electrons tend to migrate to the cold end from equilibrium considerations alone. The heat flow will also derive an additional number of electrons to the cold end. This effect is called *phonon drag*. Two contributions to the thermoelectric power are therefore observed. The phonon-drag contribution is only considerable at low temperatures. It has a maximum below about 100° K and falls off rapidly at higher temperatures, and vanishes at 0° K. Phonon-drag components must also be included in the Peltier and Thomson heats of Eqs. (5) and (6). The flow of electrons will *push* heat along the rod, and from junction to junction, by collisions with phonons as well as by the equilibrium effects.

## 16.6 THERMOELECTRICITY IN SEMICONDUCTORS

In semiconductors, thermoelectric effects are many times larger than in metals. The three coefficients, $S$, $\mu_T$, and $\pi$, may be positive or negative, whereas in the ideal simple conductor, they are negative. Such behaviour results from arrangement of the energy bands in semiconductors and the details of the occupation of the states by electrons. To account for the algebraic sign of the three coefficients in semiconductors, let us recall section 16.1. If electrons are the majority carrier, $S$ is negative. According to Eqs. (5) and (6), $\pi$ and $\mu_T$ should, also be negative. If holes are the majority carrier, more electrons are excited into the acceptor levels at the hot end, and thus more holes are available. Electrons at the cold end near the top of the valence band can lower their energies by moving into the holes at the hot end. The hot end becomes negatively charged, and the cold end positively charged. This behaviour is opposite to that discussed in Section 16.1. Therefore, in $p$-type semiconductors, $S$, $\pi$, and $\mu_T$ are all positive. A similar consequence prevails in the Hall effect. Indeed, calculations show that the Hall coefficient can be determined from the thermoelectric coefficients and vice versa.

The thermoelectric powers of semiconductors are two orders of magnitude larger than for nonferromagnetic metals (refer to Table 16.1). One reason is that the carrier density of semiconductors is sensitive to temperature, and the hot end will have more conduction electrons per unit volume than the cold end, or more holes, depending on whether we are dealing with a $p$-type or an $n$-type material. The more important reason arises from the

**Table 16.1** Thermoelectric Power, Thermal Conductivity, and Electrical Resistivity of Various Metals and Semiconductors at Various Temperatures

| *Meterial* | *Temperature (°C)* | *S (Microvolts/Deg K)* | $\rho$ *(ohm-m)* | $\sigma_T$ *(Watts/m-Deg K)* |
|---|---|---|---|---|
| | | *Metals (Nonferromagnetic)* | | |
| Al | 100 | −0.20 | $2.85 \times 10^{-8}$ | 210 |
| Cu | 100 | +3.98 | $1.7 \times 10^{-8}$ | 390 |
| Ag | 100 | +3.68 | $1.59 \times 10^{-8}$ | 420 |
| W | 100 | +5.0 | $0.55 \times 10^{-8}$ | 167 |

*Contd...*

**Table 16.1** *Contd..*

| | | | | |
|---|---|---|---|---|
| *Semiconductors* | | | | |
| $(Bi, Sb_2)Te_3$ | 25 | +195 | $9 \times 10^{-6}$ | 1.47 |
| $Bi_2(Te, Se)_3$ | 100 | −210 | $10.5 \times 10^{-6}$ | 1.6 |
| ZnSb | 200 | +220 | $23 \times 10^{-6}$ | 1.6 |
| InSb | 500 | −130 | $4.5 \times 10^{-6}$ | 7.5 |
| Ge | 700 | −210 | $14.5 \times 10^{-6}$ | 20 |
| $TiO_2$ | 727 | −200 | $148 \times 10^{-6}$ | 3.4 |
| $(Li_{0.05}\ Ni_{0.95}O)$ | 1100 | +225 | $80 \times 10^{-6}$ | 3.5 |

forbidden energy band. In an ideal conductor, most of the carriers are near the Fermi level. In an *n*-type extrinsic semiconductor, the majority carriers occupy states in the conduction band, high above the Fermi level. Consequently, the heat transported by an electrical current in an *n*-type semiconductor exceeds the heat transported by the same current in an ideal conductor. The application of this reasoning to the junction between an *n*-type semiconductor and a metal is shown in Fig. 16.4. The average energy of the conduction electrons decreases considerably as the junction is crossed and energy is lost in the form of heat. If the flow of current were reversed, the conduction electrons would increase their average energy by absorbing heat.

In the case of *p*-type semiconductors, the carriers lie well below the Fermi level in the valence band, and the heat transported is far less than in the ideal conductor. The large difference, once more, gives rise to a large Peltier effect, but of the opposite sign. A better

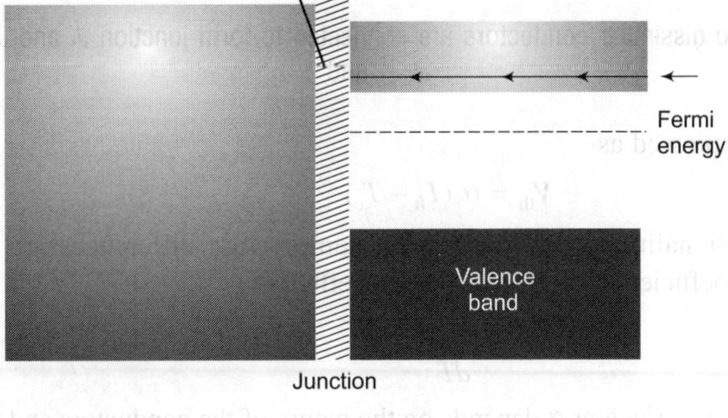


**Figure 16.4** The Peltier effect at a junction between a metal and an *n*-type semiconductor. The average energy of the conduction electron changes as the junction is crossed. For flow of electrons to the left, heat is given off, equivalent to the reduction in energy. If the flow were to the right, the average energy would increase by the absorption of heat at the junction

heat- or cold-producing junction than that of Fig. 16.4 is therefore a *p-n* semiconductor junction.

The rapid diffusion of the carriers, together with the high electrical resistivities further increases the thermoelectric effects in semiconductors. Since the thermoelectric voltage is determined by the equilibrium between the diffusion of electrons from hot to cold, and the reverse flow of electrons due to the induced voltage, high diffusivity builds up large voltages, and high resistivity reduces the return flow.

## 16.7 THERMOELECTRIC EFFECTS

### 16.7.1 Seebeck Effect

When two dissimilar metals or semiconductors are connected as a circuit and the junctions are held at different temperatures, $T_h$ and $T_c$, respectively, thermoelectric phenomenon takes place. In 1823, T. Seebeck discovered that an electromotive force ($V_{th}$) appears in a circuit composed of two dissimilar conductors, *A* and *B*, if the junctions, $J_1$ and $J_2$, between the conductors are held at temperatures, $T_h(T_{hot})$ and $T_c$ $(T_{cold})$ respectively as shown in Fig. 16.5. The emf developed is known thermal emf. This emf is measured breaking into two terminals either conductor *A* or *B* Fig. 16.5.

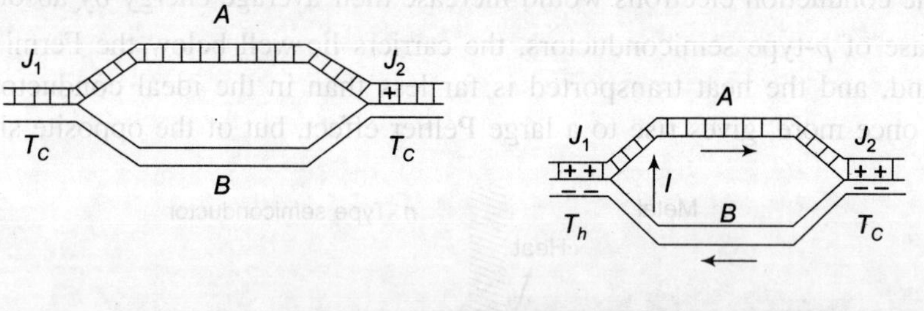


**Figure 16.5** Two dissimilar conductors are connected to form junction $J_1$ and $J_2$ and maintained at $T_h$ and $T_c$

The emf is expressed as

$$V_{th} = \alpha\,(T_h - T_c) \qquad \text{...(1a)}$$

α is the proportionality constant, when the temperature differences are small, then the thermoelectric coefficient α, called Seebeck coefficient is

$$\alpha = \frac{dV_{th}}{dT}$$

the thermoelectric coefficient α depends on the nature of the conductors and the temperature. There are three sources of the origin of thermal emf. They are:

(a) a change in the contact potential difference $\Delta V_c$ with difference in temperature ($\Delta T$),

(b) a directional carrier flow that arises in the conductor under the action of a temperature gradient (volume component, $\Delta V_v$)

(c) a change that is brought about due to change in mobility or diffusion of charge carriers ($\Delta V_d$). We shall examine the nature of these sources. The observed specific thermal emf is the resultant of all the above components.

#### 16.7.1.1 Contact Potential Difference

Let us consider two dissimilar semiconductors *A* and *B* brought into contact and kept at $T_c$ to start with. Observe Fig. 16.6 when the temperature is at $T_c$, the two semiconductors are in equilibrium state. Let the conductor *A* be associated with Fermi level $Ef_1$ and the conductor *B* with Fermi level $Ef_2$. Let $\phi_1$ and $\phi_2$ be the work functions of these semiconductors *A* and *B* respectively. The difference in work functions $(\phi_2 - \phi_1)$ should be equal to the contact potential difference that may arise between *A* and *B*. As the temperature of both junctions is same, the contact potential differences at the junctions $J_1$ and $J_2$ are equal in magnitude and opposite in direction. Therefore, the resultant potential difference $V_{th}$ is zero.

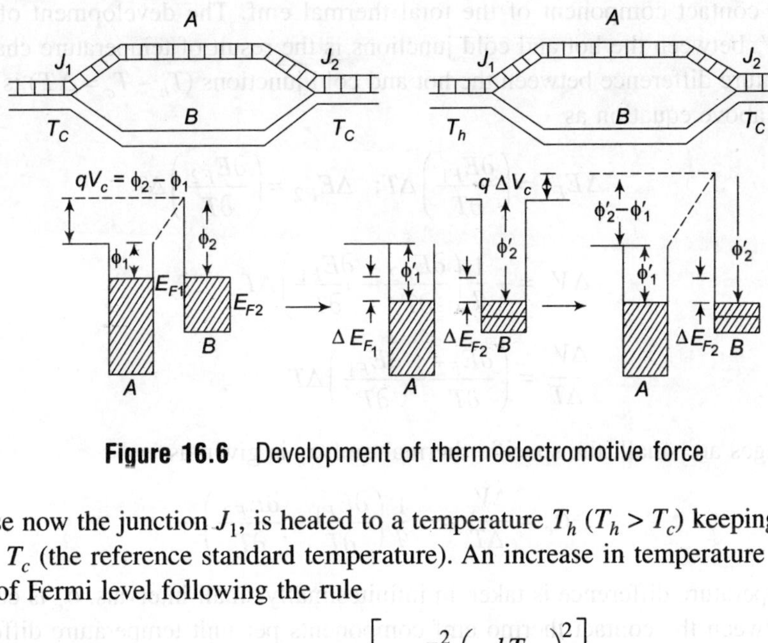


**Figure 16.6** Development of thermoelectromotive force

Suppose now the junction $J_1$, is heated to a temperature $T_h$ ($T_h > T_c$) keeping the junction $J_2$ cold at $T_c$ (the reference standard temperature). An increase in temperature results in the lowering of Fermi level following the rule

$$E_F(T) = F_F(0)\left[1 - \frac{\pi^2}{12}\left(\frac{kT}{E_F(0)}\right)^2\right] \quad ...(2)$$

where $E_F(0)$ is the Fermi energy at $T = 0°K$, which is given as

$$E_F(0) = \frac{\eta^2}{2m}(3\pi^2 n)^{2/3} \quad ...(3)$$

here *n* refers to the concentration of change carriers. The above relations are true for *n*-type semiconductors and metals.

Let the decrease in Fermi level energies be $\Delta E_{F1}$ and $\Delta E_{F2}$ in the semiconductors, *A* and *B* respectively. Obviously, a changes in Fermi levels should result in a change in work functions.

Let the semiconductor $A$ possess the work function, $\phi'_1 = \phi_1 - \Delta E_{F1}$, and the semiconductor ($B$) possess the work function, $\phi'_2 = \phi_2 - \Delta E_{F2}$. They are not equal as the semiconductors $A$ and $B$ are dissimilar. The equilibrium is therefore disturbed from what was existing at $T_c$. A new equilibrium can be set up only by flow of electrons from the semiconductor $A$ to semiconductor $B$ until the two Fermi levels come to the same height. There is an electric field, and therefore, a contact potential difference, $V_c + \Delta V_c$, at the junction $J_1$. This potential difference is higher than that at $J_2$ at $T_c$.

$$V_c + \Delta V_c = \left(\frac{1}{q}\right)(\phi'_2 - \phi'_1) = \left(\frac{1}{q}\right)(\phi_2 - \phi_1) - \left(\frac{1}{q}\right)(\Delta E_{F2} - \Delta E_{F1})$$

or

$$\Delta V_c = -\left(\frac{1}{q}\right)(\Delta E_{F2} - \Delta E_{F1}) \quad ...(4)$$

$\Delta V_c$ is the contact component of the total thermal emf. The development of potential difference $\Delta V_c$ between the hot and cold junctions is the result of temperature change at $J_1$. If the temperature difference between the hot and cold junctions ($T_h - T_c = \Delta T$) is small, we can write the above equation as

$$\Delta E_{F1} = \left(\frac{\partial E_{F1}}{\partial T}\right)\Delta T; \quad \Delta E_{F2} = \left(\frac{\partial E_{F2}}{\partial T}\right)\Delta T$$

$$\therefore \quad \Delta V_c = -\frac{1}{q}\left(\frac{\partial E_{F2}}{\partial T} - \frac{\partial E_{F1}}{\partial T}\right)\Delta T \quad ...(5a)$$

$$\therefore \quad \frac{\Delta V_c}{\Delta T} = \left(\frac{\partial E_{F2}}{\partial T} - \frac{\partial E_{F1}}{\partial T}\right)\Delta T \quad ...(5b)$$

If the changes are small, the specific thermal emf $\alpha_c$ is given as

$$\alpha_c = \frac{\Delta V_c}{\Delta T} = -\frac{1}{q}\left(\frac{\partial E_{F1}}{\partial T} - \frac{\partial E_{F2}}{\partial T}\right) \quad ...(6)$$

when the temperature difference is taken in infinitesimally small intervals. $\alpha_c$ is equal to the difference between the contact thermo emf components per unit temperature difference for the individual semiconductors making up the circuit.

$$\alpha_c = \frac{1}{q}\frac{dE_F}{\partial T} \quad ...(7)$$

Differentiating Eq. (2) and substituting for $\frac{dE_F}{dT}$ in Eq. (7), we have

$$\alpha_c = -\frac{\pi^2 k}{6qE_F(0)}T \quad ...(8)$$

It is evident that $\alpha_c$ depends on the temperature. For a nondegenerate $n$-type semiconductor, the Fermi energy is given as

$$E_F = kT \ln \left( \frac{n}{N_c} \right) \quad ...(9)$$

where $n$ is the number of charge carriers in $n$-type semiconductor, $N_c$ is the effective number of states given as

$$N_c = 2 \left( \frac{2\pi m\, kT}{h^2} \right)^{3/2} \quad ...(10)$$

where $m$ is the mass of the electron.

$$E_F = kT \ln \frac{nh^3}{2\,(2\pi m\, kT)^{3/2}} \quad ...(11)$$

Differentiating Eq. (11) with reference to temperature $T$, we get

$$\frac{dE_F}{dT} = \left[ k \ln \frac{nh^3}{(2\pi m kT)^{3/2}} + k \frac{d}{dT} (\ln n) - \frac{3}{2} k \right]$$

$$\alpha_c = \frac{k}{q} \left[ \frac{E_F}{kT} + T \frac{d}{dT} (\ln n) - \frac{3}{2} \right] \quad ...(12)$$

### 16.7.1.2 Volume Component of Thermo emf

In semiconductors, the charge carrier concentration is strongly dependent on temperature and increases very strongly with rise of temperature. Therefore, the charge carrier concentration at the hot end junction builds up rapidly and it is much higher than that at the cold end. This gives rise to a concentration gradient from hot end to the cold end. The charge carriers tend to diffuse along the semiconductor to the cold end. There will be a negative space charge build up at the cold end and a positive space charge at the hot end giving rise to an electric field. This results in a potential difference ($\Delta V_v$) which is the volume component of the thermo emf.

Under steady state conditions, the charge transport by diffusion is balanced by the drift flow caused by the electric field which is directed from hot end to the cold end. The drift flow is given as $n\langle v_d \rangle = +\, n\, \mu_n\, E$, where $\langle v_d \rangle$ is the drift velocity and $\mu_n$ is the mobility of electrons in the $n$-type semiconductor, $E$ is the electric field that is built up. The current density due to the diffusion of the charge carriers is $-D_n \left( \frac{dn}{dx} \right)$, where $D_n$ is the diffusion coefficient of electron and $\left( \frac{dn}{dx} \right)$ is the concentration gradient of the electrons along the semiconductor at every point hot end to cold end. Then in equilibrium conditions,

$$n\mu_n\, E = -D_n \left( \frac{dn}{dx} \right) \quad ...(13)$$

The relation between the diffusion coefficient $D_n$ and the mobility $\mu_n$ is

$$D_n = \frac{\mu_n}{q}\, kT \quad ...(14)$$

The diffusion takes place in the semiconductor from hot junction to cold junction regardless of the sign of charge.

$$n\mu_n(E_x) = \frac{\mu_n\, kT}{q}\left(\frac{dn}{dx}\right)$$

$$\therefore \quad E_x = -\frac{k}{q}\frac{T}{n}\left(\frac{dn}{dT}\right)\left(\frac{dT}{dx}\right)$$

$$\therefore \quad \frac{E_x}{\left(\dfrac{dT}{dx}\right)} = -\frac{kT}{q}\left(\frac{d}{dT}\ln n\right)$$

$$\therefore \quad \left(\frac{dV_v}{dT}\right) = -\frac{kT}{q}\frac{d}{dT}(\ln n) \qquad \text{...(15)}$$

### 16.7.1.3 Temperature Dependence of Mobility and Diffusion Coefficient

As the temperature is increased at the hot junction, there is a rise in the rate of thermal motion of the charge carriers. The energy of the charge carriers is increased resulting in the changes in the mean free path ($\lambda$) of the charge carriers. The mean free path is proportional to the (energy)$^\lambda$. The constant $\lambda$ depends on the mechanism of scattering of charge carriers. They can be scattered by the vibrations of the lattice or by the impurities. The mobility variation with temperature varies as $T^{\gamma^{-1/2}}$.

Let the electric field generated be $E_x$ along the $x$ direction. Then, for $n$-type semiconductor

$$qn\mu_n E'_x = -qn\frac{dD_n}{dx}$$

i.e.,

$$E'_x = -\frac{1}{\mu_n}\frac{dD_n}{dx} \qquad \text{...(16)}$$

substituting for $D_n$ in terms of the mobility $\mu_n$, we get

$$E'_x = -\frac{1}{\mu_n}\frac{d}{dx}\left(\frac{\mu_n kT}{q}\right) \qquad \text{...(17)}$$

$$\alpha'_v = \frac{dV'_v}{dT} = \frac{k}{q\mu_n}\left[T\frac{d\mu_n}{dT} + \mu_n\frac{d}{dT}T\right]$$

$$= \frac{k}{q}\left[T\frac{d}{dT}\ln\mu_n + 1\right] \qquad \text{...(18)}$$

$\mu_n$ is found to vary with temperature as

$$\mu_n = CT^{\gamma - 1/2} \qquad \text{...(19)}$$

where $C$ is a constant, and $\gamma = 0$ for phonon scattering and $\gamma = 2$ for impurity scattering, then

$$\therefore \quad \frac{dV_c}{dT} = \frac{k}{q}\left[T\frac{d}{dT}\ln CT^{\gamma-1/2} + 1\right]$$

$$= \frac{k}{q}\left[T\frac{d}{dT}(\ln C) + T\frac{d}{dT}\ln T^{\gamma-1/2} + 1\right] = \frac{k}{q}\left[\gamma + \frac{1}{2}\right] \quad ...(20)$$

adding all the components of Seebeck coefficients, the resultant specific thermal emf for $n$-type semiconductor is

$$\alpha_n = -\left(\frac{k}{q}\right)\left(\gamma + 2 - \frac{E_{fn}}{kT}\right) \quad ...(21)$$

similarly, for a $p$-type semiconductor, Seebeck coefficient is

$$\alpha_p = \left(\frac{k}{q}\right)\left(\gamma + 2 - \frac{E_{fp}}{kT}\right) \quad ...(22)$$

where $E_{fp}$ is the Fermi level distance from valence band in terms of energy.

Semiconductors can have mixed type of conduction, i.e., both electrons and holes contribute to the conductivity. Then, we can have Seebeck coefficient as

$$\alpha_{np} = \left(\frac{\alpha_n \mu_n n + \alpha_p \mu_p p}{n\mu_n + p\mu_p}\right) \quad ...(23)$$

for a good intrinsic semiconductor, $n = p \approx n_1$

$$\therefore \quad \alpha_i = \left(\frac{\alpha_n \mu_n + \alpha_p \mu_p}{\mu_n + \mu_p}\right) \quad ...(24)$$

in the case of an intrinsic semiconductor, $E_{Fi} = -Eg/2$, and $(\gamma + 2)$ is far less than $E_F/kT$.

$$\therefore \quad \alpha_i \approx -\frac{Eg}{2qT}\left(\frac{\mu_n - \mu_p}{\mu_n + \mu_p}\right) \quad ...(25)$$

but $\mu_n > \mu_p$, therefore $\alpha_i$ is of negative sign in intrinsic semiconductor.

At higher temperatures all semiconductors whether $n$-type or $p$-type, are intrinsic in nature and they all exhibit negative Seebeck coefficient. The $p$-type semiconductor enters into intrinsic region at a particular temperature. The Seebeck coefficient changes sign from positive to negative at that temperature.

At low temperatures $E_{FP}/kT$ is proportional to temperature and $\alpha_p$ also varies in a similar way. In the region of impurity depletion, $-E_{FP}/kT = \ln \dfrac{N_v}{N_a}$.

But $N_v \propto T^{3/2}$, hence $\alpha_p$ slightly increases with the temperature in this region. The variation of $\alpha_p$ and $\alpha_n$ with temperature is shown in Fig. 16.7.

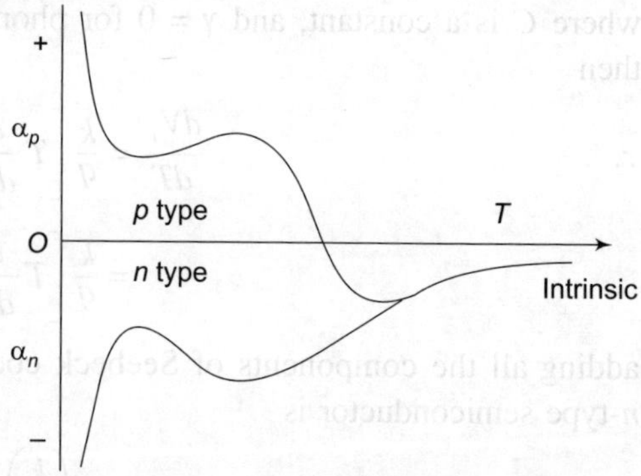


**Figure 16.7** $\alpha_n$, $\alpha_p$ as functions of temperature

### 16.7.2 Peltier Effect

When current is set up through the junction of two dissimilar materials, there is a production or absorption of heat at the junction. This absorption or production of heat is different from Joule heat that may be generated due to passage of charge. This phenomenon is known as Peltier effect. The quantity of heat evolved or absorbed ($Q_p$) is proportional to the charge. It traverses across the junction where $I$ is the current in the conductor.

$$Q_p = \pm \pi It \qquad ...(26)$$

where $\pi$ is called Peltier coefficient. Peltier effect is the inverse of the Seebeck effect. In Peltier effect, the current exits through the dissimilar semiconductor circuit forming two junctions. It produces the temperature difference between the junctions. In Seebeck effect, the temperature difference between the junctions gives rise to the thermo emf, thereby giving rise to a current in the circuit. The two coefficients at a temperature $T$ are correlated as

$$\pi = \alpha T \qquad ...(27)$$

consider a neutral junction, which is a junction between a metal and a semiconductor. At the contact (Ohmic contact) the two materials have the same work function. There is no depletion layer at the junction. Let the semiconductor be a $n$-type semiconductor. The Fig. 16.8 depicts the situation. The Fermi level of semiconductor and that of metal are at the same height. The electrons moving from metal to semiconductor have to face a potential barrier of height $-E_{Fn}$.

Now, apply a voltage to the contacts as shown in Fig. 16.8 (b). Then the Fermi level in the semiconductor is raised along with the bottom of the conduction band and the top of the valence band. The height is given by $qV$. There is a gradation of the levels from left to right. As electrons flow from right to left, there is a current existing in the system from left to right. The electrons going from semiconductor to metal

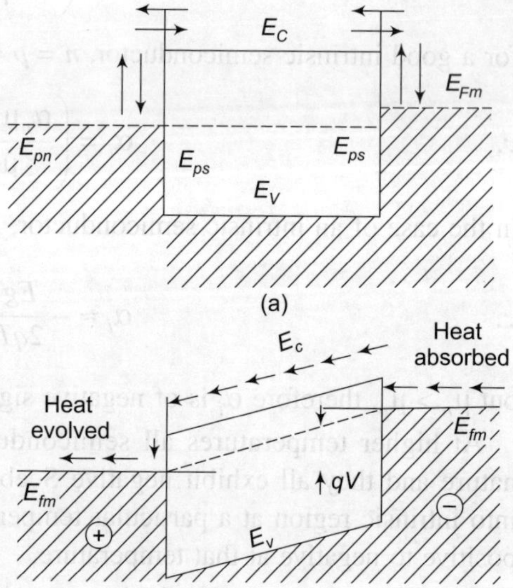


**Figure 16.8** (a) Energy diagram for metal-semiconductor metal system at thermal equilibrium (b) A current flows through it

on the left junction have more energy than in the metal. The mean energy of the electrons in the semiconductor is $E_n = (\gamma + 2)kT$. Thus, each electron traversing from semiconductor to metal gives up energy to the junction by an amount equal to $[E_n = (\gamma + 2)kT]$. This results in Peltier heat at the junction apart from Joule heating.

The Peltier, coefficient is given by

$$\pi = \frac{k}{q}\left[\gamma + 2 - \frac{E_{fn}}{kT}\right]T \qquad ...(28)$$

at the junction on the right side there is a cooling effect as the electrons have to move from metal to semiconductor resulting in absorption of energy from the material.

### 16.7.3 Thomson Effect

The conductors *A* to *B* have conduction of heat through them. If there is no current, a uniform temperature distribution exists in each semiconductor. If a current exists, the temperature distribution in each conductor is altered. This is not due to Joule effect. The additional charge in the temperature distribution is called the Thomson effect. Not withstanding the Joule effect, the heat must be supplied or extracted laterally at all places along the conductors to restore the initial temperature distribution. This effect is called the Thomson effect. The rate at which Thomson heat is transferred into a small region of a conductor with a current ($I$) and supporting a temperature difference ($dT$) is equal to $\tau\, I\, dT$, where $\tau$ is called the Thomson coefficient.

### 16.7.4 Applications

The Peltier effect and Seebeck are widely used in applications in various fields of technology. The thermoelectric effect enables us to convert thermal energy to electrical energy or vice versa. Thermal convectors or thermoelectric generators are possible. Heat flow in solids is primarily due to electrons, which carry energy. Electrons flow down a temperature gradient because they have more energy and high velocities at the hot end than at the cold end. The electrical conductivity ($\sigma$) and the thermal conductivity ($K_t$) are related as $\frac{\sigma T}{K_t} \approx \frac{e^2}{2k^2}$.

The thermal conductivity affects the thermoelectric effects. The materials useful for thermal conversion into electricity should have a lower thermal conductivity so that $\sigma/K_t$ ratio is high. Proper doping of donors or acceptors can control this parameter in the semiconductors. The efficiency of a thermal converter is defined by $\alpha^2\, \sigma/K_t$.

The thermal converters were in use even before second world war. In the war, they were useful in radio receivers.

**Thermoelectric generator and refrigerator.** Seebeck effect is concerned about the net conversion of thermal energy into electrical energy with the appearance of electric current. The Peltier effect is involved in the reversible evolution or absorption of heat that occurs when an electric current exits through the junction between two dissimilar conductors. Thomson effect is concerned about reversible evolution or absorption of heat that occurs whenever charge carriers traverse a region where a temperature gradient exits along a single homogeneous conductor. The three coefficients are related as

$$\alpha_{AB}\,T = \pi_{AB} \quad \text{and} \quad T\,\frac{d\alpha_{AB}}{dT} = \tau_A - \tau_B \qquad \text{...(29)}$$

the thermoelectric effects in a circuit having two dissimilar conductors can provide useful heating, cooling or power generation. The efficiency of a thermoelectric generator and the coefficient of performance of a thermoelectric heat pump are independent of the capacity of the unit.

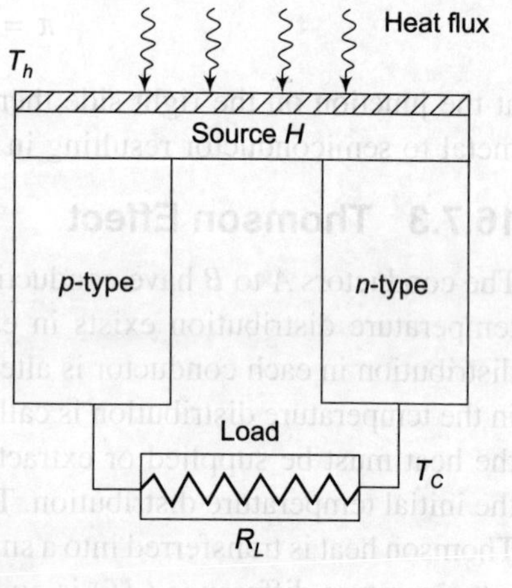


**Figure 16.9** Thermoelectric generator

**Thermoelectric generator.** Consider a thermoelectric circuit as shown in Fig. 16.9. The circuit consists of two dissimilar thermoelectric elements ($n$ and $p$) in contact with a heat source ($H$) at temperature ($T_h$) and a heat sink at temperature, $T_c$. Under steady state conditions, a temperature difference of $T_h - T_c$ is established across the elements. This causes a current flow through these branches and this current flows through load, $R_L$. Therefore, directly heat is converted into the electrical energy. The load ($R_L$) can be any device. The conversion efficiency $n_g$ is given as ($n_g$) = power supplied to Load ($W_2$)/heat absorbed at hot junction per unit time ($H_2$).

The power in $R_L$ is simply $I^2 R_L$. If Thomson heating is neglected, heat absorbed at $T_h$ is made up of two terms: (1) The Peltier heat ($H_1$) and (2) The heat withdrawn from junction, $H_2$.

We have,
$$H_1 = \pi_{np}\,I = \alpha_{np}\,IT_h \qquad \text{...(30)}$$

and
$$I = \frac{(T_h - T_c)}{R_t + R_L} \qquad \text{...(31)}$$

$$H_2 = K_t\,(T_h - T_c)^{-1/2}\,I^2 R_L \qquad \text{...(32)}$$

$$\therefore \quad n_g = \frac{I^2 R_L}{\alpha_{np}\,IT + k_t(T_h - T_c) - \frac{1}{2}I^2 R_L} \qquad \text{...(33)}$$

The main parameter that decreases the efficiency is the conduction of heat along the conductors and heat dissipation in the conductor. The efficiency to the fullest level can never be realized as conduction and heat dissipation can never be made zero. The efficiency in these cases will be of the order of 3–7%. This is very small as compared to the ordinary power generators whose efficiencies lie between 30 and 40%. Still they are used in specific applications. The heat evolved in the radioactive decay of radioisotope is used in radioisotope thermoelectric generators (RTG). A RTG working with radioactive isotope of $Ce^{144}$ gives a power of 150–200 watts. The efficiency nowadays has reached 15%–20%. For example,

Voyager spacecraft used the RTG systems. RTGs also find application in cardiac pacemakers, navigational buoys, remote navigational lights, cathodic-protection in oil pipes like the protection of the east-west pipeline across Saudi Arabia. The advantage of these generators lies in their high reliability, no noise, no pollution, and simple circuitry.

Many materials are available for possible utility in thermoelectric generators. Semiconductors possess large Seebeck coefficients, good electrical conductivity and poor thermal conductivity. Thus, a thermocouple of *p*-type $Bi_2$ $Te_3$ joined to *n*-type $Bi_2$ $Te_3$, has a value of Seebeck coefficient 425 μV/deg. This is much more than the Seebeck coefficients of metals.

**Thermoelectric Refrigerator.** The metal thermocouples are not suited to extract heat by the Peltier effect, because the difference in Seebeck coefficients between two thermocouple metals is very small. To produce ice, $T$ should be about 270 K. If we take metals, a current of 20 amps and about 10 couples in series are needed to obtain Peltier heat current which is given as $Q = 270\ C(\alpha_{AB}) \times 20 \times 10$. If the junctions are of Cu and Fe, $\alpha_{AB} = 13.7\ \mu V/$ deg and $Q$ would be only 0.74 W. But in metals, large Joule effect arises and the efficiency of performance would be very small. Semiconductors are useful. Consider *p*- and *n*-type semiconductors joined as shown in Fig. 16.10.

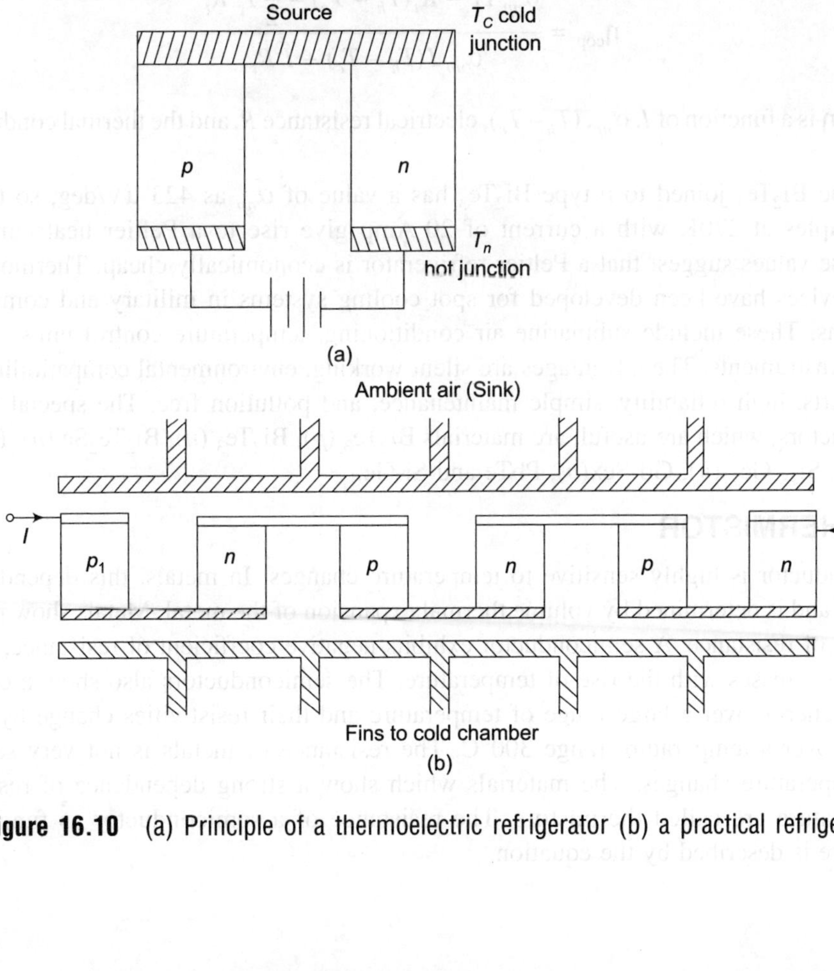


**Figure 16.10** (a) Principle of a thermoelectric refrigerator (b) a practical refrigerator

A current ($I$) is passed through a thermoelectric junction (Fig. 16.10a) with one of its junctions in thermal contact with a heat source and the other in contact with a heat sink. Under these conditions, the device pumps heat from the heat source to the heat sink. The device is maintained under steady state conditions and the temperatures of hot junction (normally kept at ambient air) and cold junction (a cold chamber) are $T_h$ and $T_c$ respectively. The coefficient of performance is

$$\eta_{cop} = \frac{\text{Cooling obtained } (Q_L)}{\text{Power input } (V_I)} \quad ...(34)$$

where $V_I$, $I$ is the input dc power supplied. At cold junction, the Peltier heat is removed. This is opposed by the thermal conduction of heat, along the semiconductor from heat sink to cold chamber. One half of the Joule heat produced in the thermoelectric circuit also opposes it. The cooling obtained is

$$Q_c = \alpha_{ap}\, It - K_t\,(T_h - T_c)^{-1/2}\, I^2 R_t \quad ...(35)$$

The power input is

$$V_I\, I = \alpha_{np}\, I(T_h - T_c) + I^2\, R_t \quad ...(36)$$

$$\eta_{cop} = \frac{\alpha_{np}\ \text{IT} - K_t(T_h - T_c) - \frac{1}{2} I^2\, R_t}{\alpha_{np}\ I(T_h - T_c) + I^2 R_t} \quad ...(37)$$

Hence, $\eta$ is a function of $I$, $\alpha_{np}$, $(T_h - T_c)$, electrical resistance $R_t$ and the thermal conductivity $K_t$.

A $p$-type $Bi_2Te_3$ joined to $n$-type $Bi_2Te_3$ has a value of $\alpha_{np}$ as 423 μV/deg, so that ten thermocouples at 270K with a current of 20 Amp give rise to a Peltier heat current of 23W. These values suggest that a Peltier refrigerator is economically cheap. Thermoelectric cooling devices have been developed for spot cooling systems in military and commercial applications. These include submarine air conditioning, temperature control units, cooling of optical instruments. The advantages are silent working, environmental compatibility, non-moving parts, high reliability, simple maintenance, and pollution free. The special type of semiconductors, which are useful, are materials $Bi_2Te_3$ ($p$), $Bi_2Te_3$ ($n$), $Bi_2Te_2Se$ ($n$), $PbS_n$ Te +MnTe($n$), $Si_{78}$ $Ge_{22}$($n$), Gd Sex($n$), PbTe and Si-Ge.

## 16.8 THERMISTOR

A semiconductor is highly sensitive to temperature changes. In metals, this dependence is very weak and is determined by volume thermal expansion of the metal. Metals show positive coefficient of resistance. A semiconductor exhibits negative coefficient of resistance, i.e., its resistance decreases with the rise of temperature. The semiconductors also show a constant activation energy over a large range of temperature and their resistivities change by nearly 1% – 6% over a temperature range 300°C. The resistance of metals is not very sensitive to the temperature changes. The materials which show a strong dependence of resistance on temperature are called thermistors. The resistance of a semiconductor as function of temperature is described by the equation.

$$R_T = A \exp\left(\frac{B}{T}\right) \quad ...(38)$$

where $A$ is a constant and $B = E/k$. $E$ is called the activation energy for the conduction process. Over a limited temperature range the above equation can be expanded as

$$R_T = R_t\left(\frac{B}{T} + \frac{C}{T^2} + \frac{D}{T^3} +\right) \quad ...(39)$$

where $B$, $C$, $D$ are constants. The semiconductors that are useful as thermistor are the mixed oxides of nickel, cobalt, manganese, copper, tin, uranium, ion, magnesium titanium, etc. The actual materials are Li-doped (Mn, Ni, Co) O, $(NiMn)_3 O_4$, $(NiMnFeCo)_3 O_4$, etc. The compounds are the oxides. MnO has five $3d$ electrons resulting in partially filled band. NiO, CoO have $3d^5$ and $3d^7$ configurations. All the materials are basically insulators. By creating some fraction of trivalent ions in the lattice of the crystal they can be made semiconducting. For example, incorporation of lithium in nickel oxide produces some trivalent nickel ions. Then the electrons can be exchanged between the ions of nickel, i.e., between $Ni^{2+}$ and $Ni^{3+}$ ions. The electron is transferred from divalent nickel ion to trivalent nickel ion. Therefore, the mobility of the charge carriers is localised to the ions in the lattice of the crystal. The localised electron in the ionic polarization field is known as small polaron. This polaron is responsible for the semiconducting nature of the oxides. The localised energy levels will be in the forbidden gap. This method of creating mixed valence states and thereby conduction by aliovalent doping of the crystal is known as controlled valency technique. The mobility of the polarons is extremely small compared to the mobility of the charge carriers in normal conduction process. The mixtures of metal oxides are sintered in required shapes at 980–1000°C. They exhibit high temperature coefficient and the resistance falls off very rapidly with temperature as shown in Fig. 16.11.

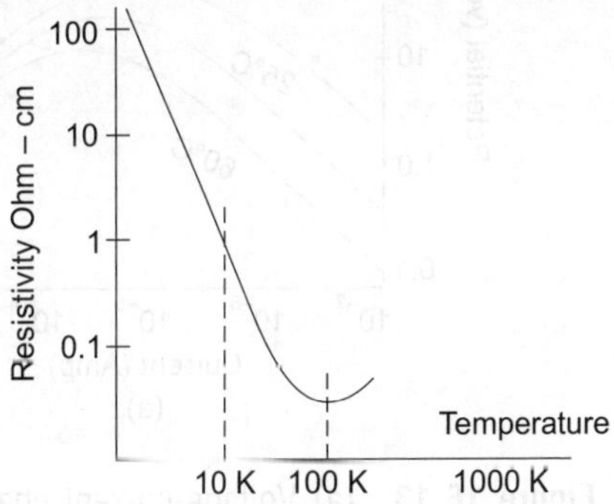


**Figure 16.11** Thermistors are sensitive to temperature

The resistivity for normal thermistor varies in the range, 10–10°Ω-m over a temperature of 600°C. The resistance of a metal like platinum changes by a factor 10 in this temperature range. A number of configurations exist for thermistors. NTC thermistors are normally very small in size, but are complex in construction and material composition. They come in a variety of sizes and shapes to satisfy the requirements of the device. The major and very common types are bead type, washers, dies, or rods. The Fig. 16.12 depicts them.

A thermistor is very sensitive to temperature changes. Therefore, it can be used as a temperature sensor or transducer. Thermistors exhibit three important characteristics.

(1) Temperature sensitive resistance (Fig. 16.11) and a negative temperature coefficient.

(2) Voltage – current characteristic (Fig. 16.13 (a))

(3) Current – time characteristic (Fig. 16.13 (b))

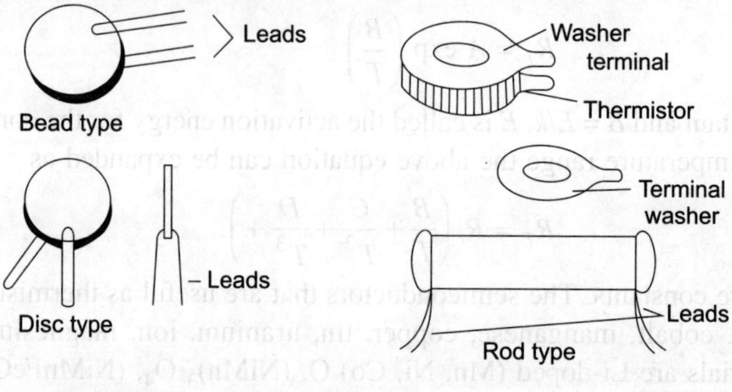


**Figure 16.12** The various configurations of thermistor

The temperature resistance characteristic is useful for the temperature control and measurement.

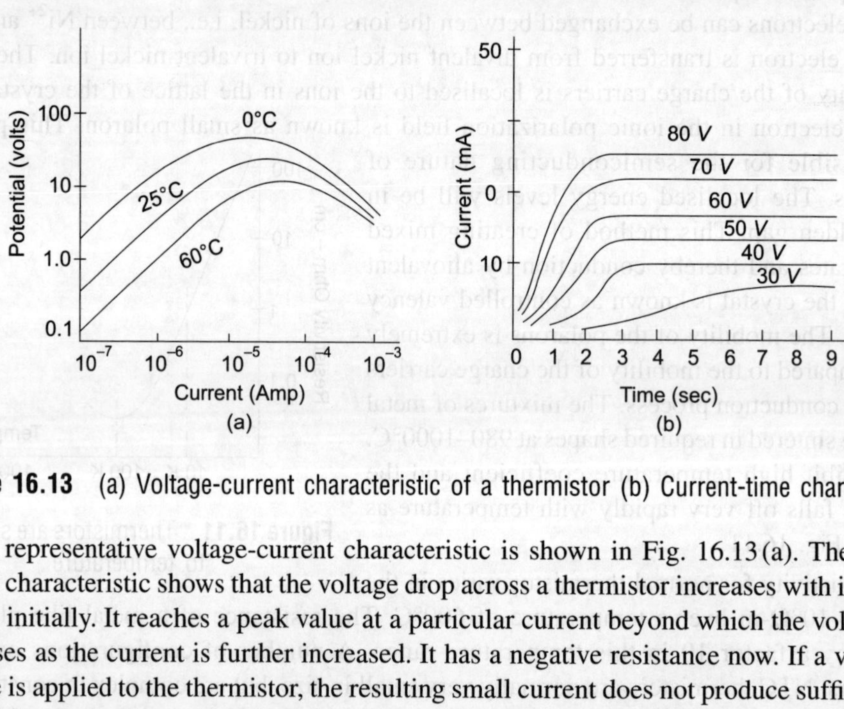


**Figure 16.13** (a) Voltage-current characteristic of a thermistor (b) Current-time characteristic

The representative voltage-current characteristic is shown in Fig. 16.13(a). The voltage-current characteristic shows that the voltage drop across a thermistor increases with increasing current initially. It reaches a peak value at a particular current beyond which the voltage drop decreases as the current is further increased. It has a negative resistance now. If a very small voltage is applied to the thermistor, the resulting small current does not produce sufficient heat to raise the temperature of the thermistor above the temperature of the surroundings. Under this condition, Ohm's law is obeyed. Larger currents at larger applied voltages produce sufficient heat to raise the temperature of the thermistor thereby reducing its resistance. Then, more current is drawn by the thermistor and its resistance decreases further. The current continues to increase until heat dissipation of the thermistor equals the power input. Therefore, at certain temperature the resistance of a thermistor is a function of the power dissipated in itself. At this point the thermistor temperature is normally well above the ambient temperature and the resistance would have dropped to a value approximately $10^{-3}$ times its original value. This self-heating can be maintained low by maintaining low currents in the thermistor. The size of the thermistor also can be reduced. In such a case, the response time of the thermistor will be good.

There is a time delay to reach the maximum current as a function of applied voltage. The current build-up takes some time to reach the steady state. This helps in introducing a time delay in an electronic circuit when it is inserted in the circuit.

Thermistors are useful as temperature sensors, temperature controllers, flow meters, pressure meters, etc. In all the cases, the change in resistance is the parameter, which is measured. Wheaston bridge is helpful in the measurement of resistance to a great accuracy. Resistance changes can easily be measured with the bridge. The bridge is shown in Fig. 16.14.

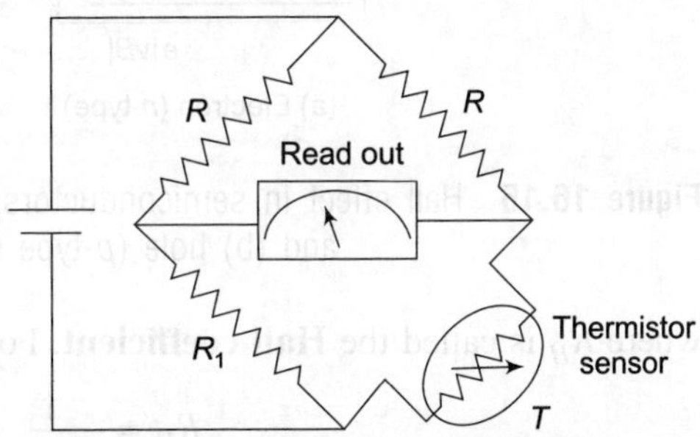


**Figure 16.14** The resistance bridge

## 16.9 PHENOMENOLOGICAL THEORY OF THE HALL EFFECT

The application of a magnetic field perpendicular to the current through a conducting material results in the generation of an electric field perpendicular to both the current and the magnetic field. This phenomenon was discovered in 1879 by E.H. Hall and is called the Hall effect. This phenomenon has been successfully used to investigate the electronic properties of semiconductors and metals. In this section, first we deal with the Hall effect in a simple way in which all the electrons have the same velocity and later we will discuss a more accurate treatment using the Boltzmann transport equation.

First, let us consider an infinitely long sample, as shown in Fig. 16.15, and assume that a uniform current density **J** flows along the sample. An electron with drift velocity ($v$) experience the Lorentz force, $\mathbf{F} = -ev \times \mathbf{B}$, where $-e$ is the charge of the electron. Let the electron density be $n$; the current density is then given by $\mathbf{J} = n(-e)v$. When the current flows in the $+x$ direction, the velocity component is $v_x < 0$ and thus the electron drifts in the $x$-direction. In a magnetic field applied in the $z$ direction, the Lorentz force for the electron is given by $F_y = -e(|v_x| B_z) = -e|v_x| B_z < 0$, and thus the electron is forced to move in the $y$-direction. Noting that the Hall effect is normally measured as the electric field induced in the $y$-direction without any flow of current in this direction, negative charges due to the electrons will accumulate on the front surface as shown in Fig. 16.15 and immobile positive charges (donors) will appear on the back surface, resulting in an electric field in the $y$-direction. The induced electric field (Hall field) $E_y < 0$ will balance the Lorentz force and a steady state will be achieved so that $F_y + (-e)E_y = 0$. This electric field $E_y$ is called the **Hall Field**, and we find the following relation

$$E_y = v_x B_z = -\frac{B_z J_x}{ne} \equiv R_H J_x B_z \qquad ...(1)$$

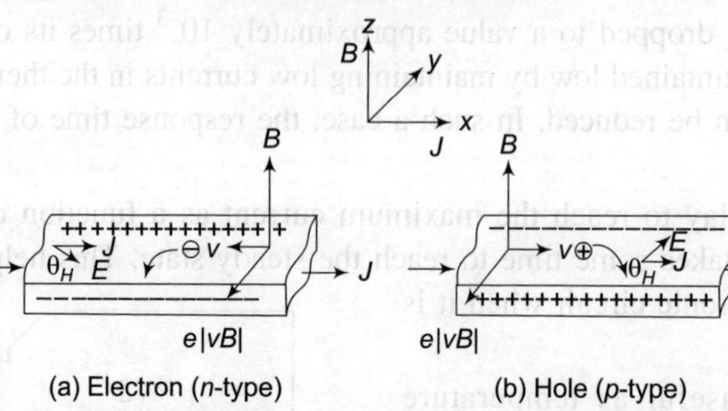


**Figure 16.15** Hall effect in semiconductors. Hall fields for (a) electron (*n*-type semiconductor) and (b) hole (*p*-type semiconductor) are reversed

where $R_H$ is called the **Hall coefficient**. For electrons the Hall coefficient is given by

$$R_H = -\frac{1}{ne} \qquad ...(2)$$

The angle $\theta_H$ between $E_x$ and $E_y$ is called the **Hall angle**, which is given by

$$\tan\theta_H = \frac{E_y}{E_x} = -\mu_e B_z = \omega_c \tau. \qquad ...(3)$$

In Eq. (3) $J_x = ne\mu_e E_x, v_x = -\mu E_x$ and

$$\omega_c = \frac{eB_z}{m^*},\ \mu_e = \frac{eT}{m^*} \qquad ...(4)$$

are used and $\omega_c$ is the cyclotron angular frequency. The Hall effect is schematically shown in Fig. 16.15 (a) for negative charge carriers (electrons in *n*-type semiconductors). On the other hand, Fig. 16.15 (b) shows the Hall effect for holes (*p*-type semiconductors). A hole current in the +*x* direction is produced by hole drift in the +*y* direction ($v_x > 0$), and the Lorentz force for a hole in the presence of a magnetic field in the *z*-direction directs it along the $-y$ direction, resulting in positive and negative charge accumulation on the $+y$ surface and $-y$ surface, respectively. This charge accumulation gives rise to a Hall field in the $+y$ direction, which is opposite to the case for electrons. The Hall coefficient for holes with density *p* is therefore given by

$$R_H = \frac{1}{pe}. \qquad ...(5)$$

If we let the width in the *y* direction of a semiconductor sample be *w* and the thickness in the *z*-direction be $t_1$ then we obtain the Hall voltage in the *y* direction $V_H = wE_y$. From current $I_x = J_x wt$ and from Eq. (1) the Hall voltage is expressed as

$$V_H = \frac{R_H}{t} I_x B_z. \qquad ...(6)$$

Measurement of the Hall effect gives the Hall coefficient ($R_H$) from Eq. (6) and the electron density (*n*) or hole density (*p*) are determined from Eq. (2) or Eq. (5) respectively.

It should be noted here that the Hall coefficients given by Eqs. (2) and (5) are derived under the assumption that the carrier relaxation time ($T$) is independent of the carrier energy. When the relaxation time ($\tau$) depends on the carrier energy and the effective mass is not scalar, the Hall coefficient is given by the following relation instead of Eq. (2).

$$R_H = -\frac{r_H}{ne}, \qquad ...(7)$$

where $r_H$ is called the scattering factor of the **Hall coefficient**, and is determined from the scattering mechanisms and distribution function of the carriers. A detailed treatment will be given in next section.

Using (42) of section 5.4, for the conductivity $\sigma$ the Hall coefficient, Eq. (2) is rewritten as

$$|R_H|\,\sigma = \mu. \qquad ...(8)$$

This relation is valid only for $r_H = 1$ as seen from Eq. (7), and in general use the relation

$$|R_H|\,\sigma = r_H\mu. \qquad ...(9)$$

As discussed later, it is very difficult to determine the scattering factor ($r_H$) in semiconductors. We define the Hall mobility by

$$|R_H|\,\sigma = \mu_H \qquad ...(10)$$

where $\mu_H$ is called the **Hall mobility**. The Hall mobility defined above has the same dimensions as the drift mobility but differs by a factor, $r_H$, or in other words, the following relation holds

$$\frac{\mu_H}{\mu} = r_H \qquad ...(11)$$

Although the treatment stated above is based on a simplified model, the results give good information to interpret the Hall effect. In the following we will derive relations useful for understanding magnetotransport in semiconductors, which may be extended to the more detailed general case. The equation of motion for an electron is written as

$$m^*\frac{dv}{dt} + \frac{m^*v}{\tau} = -e(\boldsymbol{E} + \boldsymbol{v} \times \boldsymbol{B}). \qquad ...(12)$$

when a static electric field is applied to a semiconductor, the steady state condition $dv/dt = 0$ gives rise to a steady-state drift velocity $v$ given by

$$\boldsymbol{v} = -\frac{e\tau}{m^*}\boldsymbol{E} \equiv -\mu\boldsymbol{E}. \qquad ...(13)$$

The velocity is the drift velocity. In the presence of a magnetic field $B$, the velocity is given by

$$v = -\frac{e\tau}{m^*}(\boldsymbol{E} + \boldsymbol{v} \times \boldsymbol{B}). \qquad ...(14)$$

If the magnetic field is along the $z$ direction, the above equation leads to

$$v_x = -\frac{eT}{m^*}(E_x + v_y \times B_z), \qquad ...(15a)$$

$$v_y = -\frac{e\tau}{m^*}(E_y - v_x B_z), \qquad \text{...(15b)}$$

$$v_z = -\frac{e\tau}{m^*} E_z. \qquad \text{...(15c)}$$

These relations are rewritten[2] as

$$v_x = -\frac{e}{m^*}\left[\frac{\tau}{1+\omega_c^2\tau^2} E_x - \frac{\omega_c\tau^2}{1+\omega_c^2\tau^2} E_y\right], \qquad \text{...(16a)}$$

$$v_y = -\frac{e}{m^*}\left[\frac{\omega_c\tau^2}{1+\omega_c^2\tau^2} E_x + \frac{\tau}{1+\omega_c^2\tau^2} E_y\right], \qquad \text{...(16b)}$$

$$v_z = -\frac{e}{m^*} \tau E_z. \qquad \text{...(16c)}$$

Since the current density is given by $\mathbf{J} = n(-e)\boldsymbol{v}$, we obtain

$$J_x = \frac{ne^2}{m^*}\left[\frac{\tau}{1+\omega_c^2\tau^2} E_x - \frac{\omega_c\tau^2}{1+\omega_c^2\tau^2} E_y\right], \qquad \text{...(17a)}$$

$$J_y = \frac{ne^2}{m^*}\left[\frac{\omega_c\tau^2}{1+\omega_c^2\tau^2} E_x + \frac{\tau}{1+\omega_c^2\tau^2} E_y\right], \qquad \text{...(17b)}$$

$$J_z = \frac{ne^2}{m^*} \tau E_z. \qquad \text{...(17c)}$$

Therefore, writing the $i$ component of the current as

$$J_i = \sigma_{ij} E_j, \quad |J| = |\sigma|\,|E| \qquad \text{...(18)}$$

we find

$$|\sigma| = \begin{bmatrix} \sigma_{xx} & \sigma_{xy} & 0 \\ \sigma_{yx} & \sigma_{yy} & 0 \\ 0 & 0 & \sigma_{zz} \end{bmatrix}, \qquad \text{...(19)}$$

---

[2] $v_x = -\frac{e\tau}{m^*} E_x - v_y \frac{eB_z}{m^*}\tau; \quad \left(\frac{eBz}{m^*} \equiv \omega_c\right)$

$$v_x = -\frac{e\tau}{m^*} E_x - v_y\,\omega_c\tau \qquad \text{...(i)}$$

$$v_y = -\frac{e\tau}{m^*} E_y - v_x\,\omega_c\tau \qquad \text{...(ii)}$$

Substitute for $v_y$ from Eq. (ii) into Eq. (i)

$$v_x = -\frac{e\tau}{m^*} E_x + \frac{e}{m^*}\omega_c\tau^2 E_y - v_x\,\omega_c^2\tau^2$$

or $v_x(1+\omega_c^2\tau^2) = -\frac{e\tau}{m^*} E_x + \frac{e}{m^*}\omega_c t^2 E_y$ ...(iii)

$$\sigma_{xx} = \sigma_{yy} = \frac{ne^2}{m^*} \cdot \frac{\tau}{1 + \omega_c^2 r^2}, \qquad \text{...(20a)}$$

$$\sigma_{xy} = -\sigma_{yx} = -\frac{ne^2}{m^*} \cdot \frac{\omega_c r^2}{1 + \omega_c^2 \tau^2}, \qquad \text{...(20b)}$$

$$\sigma_{zz} = \frac{ne^2}{m^*} \tau \equiv \sigma_0. \qquad \text{...(20c)}$$

Next, we will discuss the Hall effect, where we assume the current is in the $x$ direction and the magnetic field is applied in the $z$ direction. The current density in the $x$ and $y$ directions are then given by

$$J_x = \sigma_{xx} E_x + \sigma_{xy} E_y, \qquad \text{...(21a)}$$

$$J_y = \sigma_{yx} E_x + \sigma_{yy} E_y, \qquad \text{...(21b)}$$

Since the Hall effect is measured without current flow in the $y$ direction, i.e., $J_y = 0$, this condition gives rise to the following relation

$$E_y = -\frac{\sigma_{yx}}{\sigma_{yy}} E_x = \frac{\sigma_{xy}}{\sigma_{xx}} E_x = -\omega_c \tau E_x. \qquad \text{...(22)}$$

Inserting $E_y$ into $J_x$ of (21a), we get

$$J_x = \frac{\sigma_{xx}^2 + \sigma_{xy}^2}{\sigma_{xx}} E_x. \qquad \text{...(23)}$$

It is very interesting to point out that Eqs. (20a) and (20b) result in

$$J_x = \frac{ne^2}{m^*} \tau E_x = ne\mu E_x = \sigma_0 E_x. \qquad \text{...(24)}$$

This result means that the current density ($J_x$) is independent of the magnetic field $B_z$ under the assumption we have used above, and that the resistivity is not changed by the application of a magnetic field, resulting in no magnetoresistance effect. We know that this conclusion conflicts with the experimental observations and that all semiconductors exhibit magnetoresistance. This conflicting conclusion arises from the following incorrect assumptions: (1) isotropic effective mass (spherical energy surface), (2) electron relaxation time ($r$) independent of electron energy and all electrons drifting with the same velocity $v$, (3) shape of the sample not considered ( the length is infinite). Although these assumptions will lead to wrong conclusions, the relations derived here give the correct expressions when we introduce the averaging procedures later. For this reason we shall proceed with the theoretical analysis of magnetotransport with the above simplified assumptions.

From Eqs. (21b) and (23) we have

$$E_y = \frac{\sigma_{xy}}{\sigma_{xx}} E_x = \frac{\sigma_{xy}}{\sigma_{xx}^2 + \sigma_{xy}^2} J_x \equiv R_H B_x J_x, \qquad \text{...(25)}$$

and the Hall coefficient is generally expressed as

$$R_H = \frac{\sigma_{xy}}{\sigma_{xx}^2 + \sigma_{xy}^2} \frac{1}{B_z}. \quad ...(26)$$

Under the above assumption, substituting Eqs. (20a), (20b) or Eq. (24) into Eq. (25), we obtain

$$E_y = -\omega_c r E_x = -\mu B_z \frac{J_x}{\sigma_0} = -\frac{1}{ne} B_x J_x \quad ...(27)$$

and thus $R_H = -1/ne$, which is equivalent to Eq. (2).

Next, we will consider the case where two types of carriers are present. Let us identify two types of carriers by using subscripts 1 and 2. We, therefore, obtain

$$J_x = [\sigma_{xx}^{(1)} + \sigma_{xx}^{(2)}] E_x + [\sigma_{xy}^{(1)} + \sigma_{xy}^{(2)}] E_y, \quad ...(28a)$$

$$J_y = [\sigma_{xy}^{(1)} + \sigma_{xy}^{(2)}] E_x + [\sigma_{xx}^{(1)} + \sigma_{xx}^{(2)}] E_y. \quad ...(28b)$$

Let us define the electrical conductivity for each carrier by $\sigma_1$ and $\sigma_2$, and the Hall coefficient by $R_1$ and $R_2$. Then, from Eqs. (7.20a) and (7.20b) we have for carrier 1

$$\sigma_{xx}^{(1)} = \frac{\sigma_1}{1 + \sigma_1^2 R_1^2 B_z^2}, \quad \sigma_{xy}^{(1)} = -\frac{\sigma_1^2 R_1 B_z}{1 + \sigma_1^2 R_1^2 B_z^2} \quad ...(29)$$

and similar relations for carrier 2. From the conduction for achieving the Hall effect we put $J_y = 0$ and in the same way as we derived Eq. (26) we obtain

$$R_H B_z = \frac{\sigma_{xy}^{(1)} + \sigma_{xy}^{(2)}}{[\sigma_{xx}^{(1)} + \sigma_{xx}^{(2)}]^2 + [\sigma_{xy}^{(1)} + \sigma_{xy}^{(2)}]^2}. \quad ...(30)$$

Inserting Eq. (29) and the same expression for carrier 2 in this equation we find

$$R_H = \frac{\sigma_1^2 R_1 (1 + \sigma_2^2 R_2^2 B_z^2) + \sigma_2^2 R_2^2 (1 + \sigma_1^2 R_1^2 B_z^2)}{(\sigma_1 + \sigma_2)^2 + \sigma_1^2 \sigma_2^2 (R_1 + R_2)^2 B_z^2}. \quad ...(31)$$

It is very important to point out here that Eq. (31) is valid even if the relaxation time $\tau$ ($\tau_1$ the $\tau_2$) depends on the carrier energy, and thus it gives the general expression for the Hall coefficient with two types of carriers.

Since $\sigma_1 |R_1| = \mu_1$ and $\sigma_2 |R_2| = \mu_2$, in the weak magnetic field case such as $\mu_1 B_z \ll 1$ and $\mu_2 B_z \ll 1$, Eq. (31) may be approximated as

$$R_H(0) = \frac{\sigma_1^2 R_1 + \sigma_2^2 R_2}{(\sigma_1 + \sigma_2)^2}. \quad ...(32)$$

On the other hand, in the high magnetic field case such that $\sigma_1 |R_1| B_z = \mu_1 B_x \gg 1$ and $\sigma_2 |R_2| B_x = \mu_2 B_x \gg 1$ are satisfied, we obtain

$$R_H(\infty) = \left( \frac{1}{R_1} + \frac{1}{R_2} \right)^{-1} \quad ...(33)$$

As an example, we consider the case of two carriers, electrons and holes, with respective mobilities, $\mu_e$ and $\mu_h$, densities, $n$ and $p$, and assume $r_H = 1$ for simplicity. Inserting $\sigma_1 = neu_e$, $\sigma_2 = peu_h$, $R_1 = -1/ne$ and $R_2 = 1/pe$ into Eq. (31) we obtain

$$R_H = \frac{(p - nb^2) + b^2\mu_h^2 B_x^2 (p - n)}{(bn + p)^2 + b^2 u_h^2 B_z^2 (p - n)^2} \cdot \frac{1}{e}, \quad ...(34)$$

where $b = \mu_e/\mu_h$ is the mobility ratio. Due to the assumptions described above, the above equation is valid for $r_H = 1$ and thus for the case where the relaxation times are independent of energy and are constant. From this equation we find that the Hall coefficient becomes zero or $R_H = 0$, when the following relation is satisfied:

$$p = \frac{nb^2 (1 + \mu_h^2 B_x^2)}{1 + b^2 \mu_h^2 B_z^2}. \quad ...(35)$$

The Hall coefficients at low magnetic fields, $R_H(0)$, and at high magnetic fields, $R_H(\infty)$, are given as follows from Eqs. (32) and (33), respectively,

$$R_H(0) = \frac{pu_h^2 - nu_e^2}{e(pu_h + nu_e)^2} = \frac{p - b^2 n}{e(nb + p)^2}, \quad ...(36)$$

$$R_H(\infty) = \frac{1}{e(p - n)}. \quad ...(37)$$

## 16.10 THE PHOTOVOLTAIC (DEMBER) EFFECT AND DEMBER VOLTAGE

Figure 16.16 shows a schematic diagram of an intrinsic photoconductor under illumination and bias conditions. In the intrinsic photo conduction process, electron hole pairs are generated when photons with energies exceeding the band gap energy of the semiconductor are absorbed.

Let $R$ denote the reflection coefficient of the semiconductor, $\alpha$, the absorption coefficient, $\phi_0$ the photon flux density ($\phi_0 = I_o/h\nu$) and $I_o$, the incident light intensity per unit area (W/cm$^2$). Figure 16.16 shows the geometry, where $d$ is the sample thickness ($d \sim 2$ μm). Let $s$ be the surface recombination velocity.

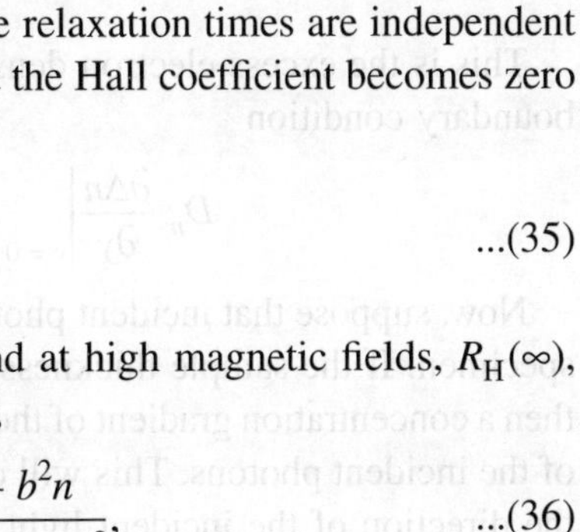


**Figure 16.16**

As shown in Fig. 16.16. the excess carrier densities as a function of distance along the $y$-direction can be obtained by solving the continuity equation for excess electron density:

$$D_n \frac{\partial^2 \Delta n}{\partial y^2} - \frac{\Delta n}{\tau_n} = -g_E = -\alpha\, \phi_0 (1 - R)\, e^{-\alpha y} \quad ...(1)$$

where $g_E$ is the generation rate for electron-hole pairs (for $\alpha d \gg 1$). $\tau_n$ is the excess carrier (electron) lifetime. The surface recombination lifetime is given by

$$\tau_s = \frac{d}{2s}.$$

If the electron diffusion length

$$L_n = (D_n \tau_n)^{1/2}$$

is much smaller than the sample thickness $d$, then Eq. (1) has the solution

$$\Delta n_0 = \frac{\alpha L_n I_0 (1 - R)\tau_n}{h\nu (\alpha L_n + 1)(L_n + s\tau_n)} \qquad ...(2)$$

This is the excess electron density at $y = 0$. Note that Eq. (2) has been obtained using the boundary condition

$$D_n \left.\frac{\partial \Delta n}{\partial y}\right|_{y=0} = s\, \Delta n|_{y=0}.$$

Now, suppose that incident photons with $h\nu \geq Eg$ are impinging on a $p$-type semiconductor specimen. If the sample thickness ($d$) is much larger than the inverse absorption coefficient, then a concentration gradient of the photo injected excess carriers is established in the direction of the incident photons. This will cause electron, and hole-currents to flow by diffusion along the direction of the incident light. The net diffusion current would be zero if the mobilities of electrons and holes are equal in the semiconductor. In general, the electron and hole mobilities are not equal, as a result, an unbalanced electron and hole diffusion current will create an internal electric field along the direction of the incident light. The polarity of this internal electric field tends to assist hole diffusion and retard electron diffusion. This is *internal photovoltaic (Dember) effect*.

As a result, an internal electric field is established known as the *Dember- field*. To derive an expression for the Dember voltage, $V_d$, the components of electron and hole current densities (i.e., $J_{ny}$ and $J_{py}$) along the direction of incident photons can be written as

$$J_{ny} = qn\, \mu_n\, \varepsilon_y + qD_n \frac{\partial n}{\partial y} \qquad ...(3)$$

$$J_{py} = qp\, \mu_p\, \varepsilon_y - qD_p \frac{\partial p}{\partial y} \qquad ...(4)$$

where

$$n = n_0 + \Delta n$$

$$p = p_0 + \Delta p$$

$$\Delta n = \Delta p$$

$(\mu_n/D_n)$ and $(\mu_p/D_p)$ are related by the Einstein relations

$$D_n = \left(\frac{kT}{q}\right)\mu_n; \quad D_p = \left(\frac{kT}{q}\right)\mu_p \qquad ...(5)$$

The total current density in the $y$-direction

$$J_y = J_{ny} + J_{py} = q(bn + p)\, \mu_p\, \varepsilon_y + (b - 1)\, q\, D_p \frac{\Delta n}{\partial y} \qquad ...(6)$$

where $b = \frac{\mu_n}{\mu_p}$, the ratio of the mobilities.

Under open circuit conductions, $J_y = 0$ and the Dember field

$$\varepsilon_y = -\left(\frac{kT}{q}\right)\frac{(b-1)}{(bn+p)}\frac{\partial \Delta n}{\partial y} \qquad ...(7)$$

Note that Dember field vanishes if $b = 1$.

This Dember field developed inside will enhance the diffusion of holes and retard the diffusion of electrons. The resulting Dember voltage between the front side (i.e., illuminated side) and the back side of the sample is obtained by integrating Eq. (7) from $y = 0$ to $y = d$. For the small injection case (i.e., $\Delta n \ll n_0$),

$$V_d = \int_0^d -\varepsilon_y \, dy$$

$$= \left(\frac{kT}{q}\right)\int_{\Delta n_0}^{\Delta n_d} \frac{-(b-1)}{(bn_0+p_0)} d\Delta n \qquad ...(8)$$

where $\Delta n_0$ is the excess electron density at $y = 0$ and $\Delta n_d$ is the excess electron density at $y = d$. For a thick sample $\alpha d \gg 1$, and $\Delta n_d \ll \Delta n_0$, Eq. (8) becomes

$$V_d = \left(\frac{kT}{q}\right)\frac{(b-1)}{(bn_0+p_0)}\Delta n_0. \qquad ...(9)$$

Now, the excess electron density ($\Delta n_0$) at $y = 0$ is related to the incident light intensity ($I_0$) by Eq. (2). Substituting Eq. (2) into Eq. (9) yields

$$V_d = \left(\frac{kT}{q}\right)\frac{\alpha L n\, I_0\,(1-R)\,\tau_n\,(b-1)}{h\nu\,(\alpha L_n+1)\,(L_n+\Delta\tau_n)\,(bn_0+p_0)} \qquad ...(10)$$

Equation (10) shows that under low injection, the Dember voltage varies linearly with the light intensity, $I_0$. In contrast to the photochemical effect, the PV effect requires no external applied voltage, and hence can be used to generate electrical power using the PV effect in a semiconductor. (Details of the PV devices (i.e., solar-cells) have been discussed in section 11.51).

### 16.10.1 The Photomagnetoelectric Effect

The photomagnetoelectric (PME) effect refers to the voltage (or current) developed in a semiconductor specimen as a result of the interaction of an applied magnetic field with the diffusion current produced by the photogenerated excess carriers. A PME open-circuit voltage (or short-circuit current) is developed in the *x*-direction when a magnetic field is applied in the *z*-direction and the incident light is in the *y*-direction of the specimen, as illustrated in Fig. 16.17.

The development of a PME field in a semiconductor under the influence of a magnetic field and incident light may be explained as follows: Electron-hole pairs generated near the surface of a semiconductor specimen by the incident photons are diffused into the specimen in the direction of incident light. The time-invariant magnetic flux density ($B$) along the $z$-direction deflecting holes in the positive $x$-direction and electrons in the negative $x$-direction results in a net PME short-circuit current flowing in the positive $x$-direction. Under open-circuit conditions, a PME voltage is developed in the $x$-direction.

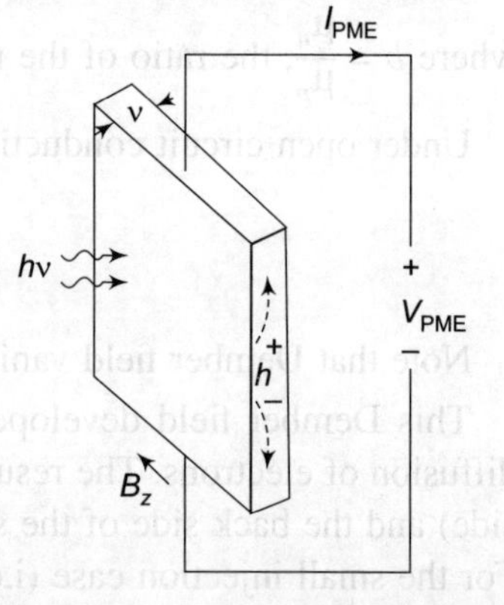


**Figure 16.17** Schematic diagram showing the photomagnetoelectric (PME) effect in a semiconductor

In order to derive the expression for the PME open-circuit voltage of PME short-circuit current, consider a semi-infinite semiconductor slab. The following equations hold for the rectangular slab shown in Fig. 16.17. For a small magnetic field (i.e., $\mu B \ll 1$) and the small-injection case, the Hall angles are given by

$$\tan \theta_p \approx \theta_p = \mu_p B \quad \text{...(1)}$$

$$\tan \theta_n \approx \theta_n = \mu_n B \quad \text{...(2)}$$

$$\theta = \theta_p - \theta_n = \mu_p (1 + b) B, \quad \text{...(3)}$$

where $\Delta n = \Delta p \ll n_0$ or $p_0$ and $b = \mu_n/\mu_p$ is the electron and hole mobility ratio.

The hole and electron current density can best be described by vector equations with $\hat{i}, \hat{j}, \hat{k}$ denoting the unit vectors along the $x$-, $y$-, and $z$-axes, respectively

$$\mathbf{J}_p \cong J_{py}\hat{\mathbf{J}} + \theta_p J_{py} \hat{\mathbf{J}} \times \hat{\mathbf{k}}, \quad \text{...(4)}$$

$$\mathbf{J}_n \cong J_{ny}\hat{\mathbf{J}} + \theta_n J_{ny} \hat{\mathbf{J}} \times \hat{\mathbf{k}}. \quad \text{...(5)}$$

Here $J_{py}$ and $J_{ny}$ are given respectively by

$$J_{py} = q\left(\mu_p p\varepsilon_y - D_p \frac{\partial \Delta p}{\partial y}\right), \quad \text{...(6)}$$

$$J_{ny} = q\left(\mu_n n\varepsilon_y + D_n \frac{\partial \Delta n}{\partial y}\right), \quad \text{...(7)}$$

Thus, the total current density is given by

$$\mathbf{J} = \mathbf{J}_p + \mathbf{J}_n = (J_{px} + J_{nx})\,\hat{\mathbf{i}} + (J_{py} + J_{ny})\,\hat{\mathbf{J}}. \quad \text{...(8)}$$

Since the total current density in the $y$-direction is zero, Eq. (8) reduces to

$$J = J_x = (J_{px} + J_{nx}) = [q\,(p\mu_p + n\mu_n)\,\varepsilon_x + \theta_p J_{py} + \theta_n J_{ny}]. \quad \text{...(9)}$$

Solving Eq. (8) through Eq. (9) yields

$$J_{py} = -J_{ny} = -qD\left(\frac{\partial \Delta n}{\partial y}\right). \quad ...(10)$$

where $D = D_n\,(n+p)/(bn+p)$ is the effective diffusion coefficient. Substituting Eq. (10) into Eq. (9), one obtains

$$J_x = J_{nx} + J_{px} = q\mu_p\,(p + nb)\varepsilon_x - q(b+1)\mu_p BD\left(\frac{\partial \Delta n}{\partial y}\right). \quad ...(11)$$

To derive the PME electric field or PME open-circuit voltage along the $x$-direction, it is usually not sufficient to assume that $J_x = J_{nx} + J_{px} = 0$. Such a solution would lead to $J_x$ being a function of $z$, which is incorrect, because for a constant magnetic field the electric field must be irrotational. With $\partial\varepsilon_z/\partial x = 0$, it follows that $\partial E_x/\partial_z = 0$. Thus, the correct boundary condition is given by

$$\int_0^d (J_{nx} + J_{px})\, dy = 0. \quad ...(12)$$

For the small-injection case, with $\Delta n = \Delta p \ll n_0$ or $p_0$, the PME electric field in the $x$-direction becomes

$$\varepsilon_x = \frac{(b+1)BD}{d(n_0 b + p_0)} \int_0^d \frac{\partial \Delta_n}{\partial y}\, dy = -\frac{BD(b+1)}{d(n_0 b + p_0)} = (\Delta n_0 - \Delta n_d) \quad ...(13)$$

The PME short-circuit current can be obtained by setting $\varepsilon_x = 0$ in Eq. (11) and then integrating the equation from $y = 0$ to $y = d$. The result is

$$I_{\text{PME}} = \int_0^d q(b+1)\mu_p BD\left(\frac{\partial \Delta n}{\partial y}\right) dy$$

$$= qW\,(b+1)\,\mu_p\, BD\,(\Delta n_0 - \Delta n_d). \quad ...(14)$$

For $ad \gg 1$ and $d \gg L_n$, one may assume that $\Delta n_d \approx 0$. Substituting Eq. (2) section 16.10 into Eq. (14), one obtains

$$I_{\text{FME}} = \frac{qW(b+1)\mu_p I_0 (1-R) B L_n^2}{h\nu\,(L_n + s\tau_n)} \cdot \frac{\alpha L_n}{(1+\alpha L_n)} \quad ...(15)$$

In Eq. (15), the quantum yield is assumed equal to 1. Thus, at low magnetic fields, the PME short-circuit current is directly proportional to the magnetic flux density $B$. For $\alpha L_n \gg 1$, $I_{\text{PME}}$ becomes independent of the wavelength.

The ratio of the PME short-circuit given by Eq. (15) to the photocurrent[3] is given by

---

[3] $$I_{\text{ph}} = \frac{qI_0 W L_n\, \mu_p\,(1+b)\,\tau_n (1-R)V}{\ell(L_n + s\tau_n)\, h\nu}\left[1 + \frac{s\tau_n}{L_n(1+\alpha L_n)}\right]$$

After Sheng S. Li, *Semiconductor Physical Electronics*, Springer (2007)

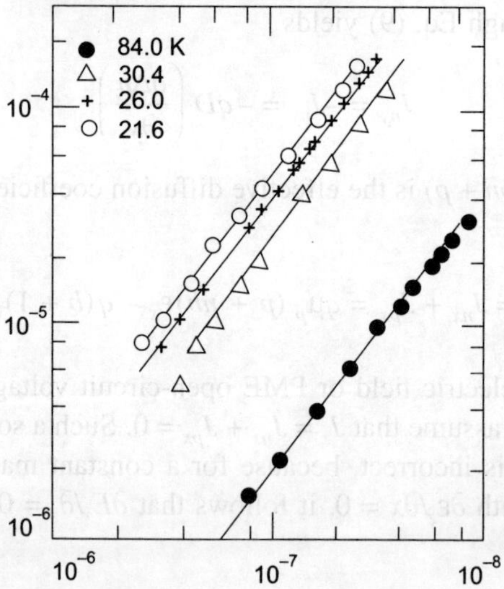


**Figure 16.18** PME short-circuit current versus photoconductance for an Au-doped silicon specimen with $N_{Au} = 5 \times 10^{16}$ cm$^{-3}$. After Sheng S. Li, *Semiconductor Physical Electronics*, Springer (2007)

$$\frac{(I_{\text{PME}}/B)}{(I_{\text{ph}}/\varepsilon_x)} = \frac{\alpha L_n^2}{\left(1 + \alpha L_n + \dfrac{s\tau_n}{L_n}\right)\tau_n} \qquad \text{...(16)}$$

If $\alpha L_n$ is much larger than 1 and $s\tau_n/L_n$, then Eq. (16) reduces to

$$\frac{(I_{\text{PME}}/B)}{(I_{\text{ph}}/\varepsilon_x)} = \frac{L_n}{\tau_n} = \left(\frac{D_n}{\tau_n}\right)^{1/2} \qquad \text{...(17)}$$

where $L_n = (D_n\tau_n)^{1/2}$ is the electron diffusion length. Equation (17) provides a direct means for determining the minority carrier lifetime from the PMC and PC effect measurements. The electron diffusion constant can be determined from the electron mobility data using the Einstein relation.[4]

The PME effect has been used in determining the minority carrier lifetimes in a semiconductor. This method is particularly attractive for semiconductors with very short carrier lifetimes for which the transient photoconductivity decay method fails. Studies of the PME effect have been reported in various semiconductors such as Ge, Si, InSb, and GaAs. Figure 16.18 shows the PME short-circuit current versus photoconductance for an Au-doped silicon sample measured at different temperatures. The nonlinear relationship between $I_{\text{PME}}$ and $\Delta G$ is due to the effect of minority carrier trapping in the Au-doped silicon sample.

---

[4] $D_n = \left(\frac{kT}{q}\right)\mu_n,\ D_p = \left(\frac{kT}{q}\right)\mu_p$

## EXERCISE

16.1 Explain the concept of thermal electromotive force. Define thermoelectric power.

16.2 What is Thomson effect? Define Thomson coefficient.

16.3 What is Peltier effect? Explain the Peltier effect between a metal and a semiconductor.

16.4 What is Seebeck effect? Explain it.

16.5 What are thermistors?

16.6 Give phenomenological theory of the Hall effect.

16.7 Give theory of photo magnetoelectric effect.

16.8 Discuss the Dember-effect in a semiconductor and derive an expression for the Dember voltage for a thick sample.

CHAPTER

# 17

# Dielectrics, Ferroelectrics and Piezoelectrics

## 17.1 INTRODUCTION

Dielectric or electrical insulating materials are understood as the materials in which electrostatic fields can persist for a long time. These materials offer a very high resistance to the passage of electrical current under the action of the applied dc voltage and hence sharply differ in their basic properties from conductive materials.

A *dielectric* is a substance whose basic electrical property is the ability to be polarized and in which an electrostatic field can exist. A dielectric is used to prevent the leakage of electrical charges in electrical engineering devices. In this chapter, we will mainly consider basically two important properties, viz., piezoelectric and ferroelectric properties. Before this, we will briefly discuss electrostriction and electrets.

The mechanical forces arising in a derivative under the action of an electric field tend to deform the dielectric. Theory shows that the nature of deformation is determined by the sign of the dielectric of permittivity of matter with respect to its density. If permittivity increases, in other words $\partial\varepsilon/\partial D > 0$, with the compression of matter, i.e., with growth of its density $D$, the body tends to reduce its volume; on the other hand, if $\partial\varepsilon/\partial D < 0$, the body tends to increase its volume. Since, as a rule, $\partial\varepsilon/\partial D > 0$ in real dielectrics, **dielectrics are compressed in an electric field**. The deformation of a dielectric in an electric field is known as **electrostriction**.

As a rule, the quantitative effect of electrostriction manifests itself negligibly and is often concealed by other phenomena. However, in a number of cases, this effect becomes quite obvious: for example, high voltage porcelain insulators secured with cement mortar are apt to crack due to electrostriction processes.

The deformation of a body under electrostriction is proportional to the square of the field intensity ($E^2$) and is not altered when the direction of the field is reversed; **electrostriction is an even effect**. The electrostriction observed in all substances should not be linked to the inverse piezoelectric effect which manifests itself only in some crystalline substances – piezoelectrics and a more pronounced in a number of cases from a quantitative aspect than electrostriction. The sign of deformation in a piezoelectric effect depends upon the direction of the electric field, and is reversed when the direction of the field changes; the **piezoelectric effect** is **an odd effect**.

In insulating materials (dielectrics), electrons are very tightly bound to the atoms (and the forbidden gap in the energy band picture is comparatively larger). Consequently, electrons cannot be made free, however they can only be displaced a bit within the molecule under the application of an external electric field and their cummulative effect accounts for the characteristic behaviour of dielectric materials. Thus, under the action of a strong electric field, the centres of the positive charges are displaced slightly in one direction (i.e., the direction of the field) and that of negative charges in the opposite direction. This gives rise to a local electric dipole moment throughout the crystal relative to lattice deformation.

## 17.2 MACROSCOPIC DESCRIPTION

The problem of interaction of electric or magnetic energy with matter has continued to inspire scientific activity over the last two centuries and has been a problem of perennial interest to the scientists, in general, and the physicists, in particular.

The whole electromagnetic frequency spectrum is governed by Maxwell's field equations which describe qualitatively how a time varying electric field is accompanied by a megnetic field and vice versa,

$$\text{div } \vec{D} = 4\pi\rho, \quad \text{div } \vec{B} = 0$$

$$\text{curl } \vec{E} = -\frac{1}{C}\frac{d\vec{B}}{dt}; \quad \text{curl } \vec{H} = \frac{1}{C}\left[4\pi i \quad \frac{d\vec{D}}{dt}\right]$$

Since both fields cause energy storage and dissipation in matter, a pair of parameters are required to characterise material as a carrier of electromagnetic waves.

Various such parameters have been introduced depending on the specific interest of the observer. The scientist and the engineer concerned with the low frequency range refer, in general, to complex permittivity and complex permeability; the communication engineer to a complex propagation vector; in optics, the complex index of refraction is used by physicists and electrical engineer use complex impedance and the chemists refer to dielectric constant and power factor.

Although conventionally dielectrics are thought to be insulators, Maxwell's equation treat matter as a dielectric medium. Moreover, the atomic structure of matter does not confine itself to such a rigid motion. Indeed, an attempt to explain the bulk properties of matter such as dielectric, magnetic, thermal, mechanical, etc., in terms of the distribution of atoms or ions as we find then in the bulk material, has yielded valuable information.

For a conventional dielectric, the property of practical importance and theoretical significance is its ability to increase the amount of charge a parallel plate condenser can hold at a given potential when it is inserted in between the condenser plates. The molecular interpretation of this property is that under the influence of an applied field, the dielectric is polarized by the formation of dipole chains which bind with their free ends, counter changes on the condenser plates, thereby reducing the effective potential between the plates without reducing the charge on them. It may be seen that the higher the polarization, the greater is the reduction in potential.

## 17.3 DIELECTRIC POLARIZATION

An atom consists of a central nucleus, containing positive charges with tiny, negatively charged particles revolving around it in different orbits. In an atom or molecule, the centres of gravity of positive charges and that of the negative charges may or may not coincide. If the two centres of gravity coincide, the atom or molecule is said to be *non-polar*. A few examples of non-polar molecules are $H_2$, $N_2$ $CCL_4$, $CH_4$, etc. If the two centres of gravity do not coincide, the atom or molecule is said to be *polar*. In such a case, the molecule or atom may be replaced by two equal and opposite point charges separated by a small distance. It, thus, becomes an electric dipole and has a permanent dipole moment. In the absence of an externally applied field, the molecules of a dielectric orient at random in all possible directions and hence, the dipole moment per unit volume of a dielectric is zero, whether it consists of polar molecules or non-polar molecules. But, when an electric field is applied across a dielectric, there is a relative displacement between the centres of gravity of positive and negative charges. This phenomenon of relative displacement between the centres of the positive and negative charges is known as *polarization* and the dielectric is said to be *polarized*.

In the absence of an applied electric field, even the polar molecules do not show any polarity, because they are oriented at random. When an electric field is applied across the dielectric specimen, all of them (polar molecules as well as induced polar molecules) align themselves along the direction of the applied electric field, and hence, show a net polarity.

Consider a dielectric slab placed between two parallel metallic plates, as shown in Fig. 17.1, between which an electric field is applied. Even though the dielectric fills the distance between the plates, there is always some gap between them, although of molecular dimension. The applied electric field aligns the polar as well as non-polar molecules in its own direction such that the positive charges move towards the negatively charged plate and the negative charges move towards the positively charged plate. Assuming that all the molecules have equal dipole moment, the net charge within the dielectric is zero and the induced charges appear

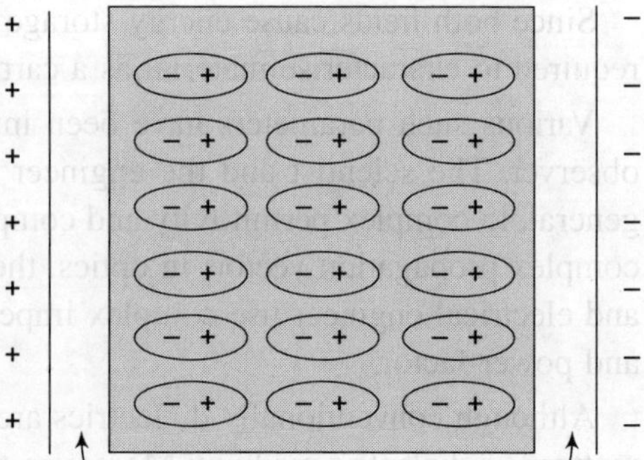


**Figure 17.1** A dielectric slab between two parallel metallic plates

only at the boundary surfaces of the dielectric, parallel to the charged plates. These charges are known as *fictitious*, or *bound* charges and the charges on the metallic plates are called *real charges*. Evidently, the electric field due to the polarized dielectric is in a direction opposite to that of the applied electric field. Hence, the resultant field within the dielectric is always less than the field outside.

If $\vec{E}$ is the average electric field acting on a molecule and $\vec{p}$ be the dipole moment induced, then obviously

$$\vec{p} \propto \vec{E}$$

$$\Rightarrow \quad \vec{p} = \varepsilon_0 \alpha \vec{E} \quad \text{...(1)}$$

where $\alpha$ is a constant, known as *molecular polarizability*. Equation (1) is valid only for small values of the electric field, $\vec{E}$. For large values of $\vec{E}$, that is, fields exceeding $10^7$ Vm$^{-1}$, $\vec{p}$ decreases with increasing $\vec{E}$. This phenomenon is known as *saturation*.

If $n$ is the number of molecules per unit volume of the specimen, then the total induced electric dipole moment, known as *polarization*, will be

$$\vec{P} = n\vec{p} = \varepsilon_0 n \alpha \vec{E} \quad \text{...(2)}$$

Consider a dielectric slab placed in an electric field so that fictitious charges appear on its opposite faces, as shown in Fig. 17.2. If $\sigma_p$ is the surface density of charge on the end faces of the slab, $A$ be the cross-sectional area of the slab shown shaded in Fig. 17.2, and $l$ be the length of the slab, then total charge induced on each of the end faces of the slab $= \sigma_p A$

Total electric dipole moment $= \sigma_p A l$ ...(3)

If $|\vec{P}|$ is the polarization, then the total dipole moment is $|\vec{P}|\, Al$. Hence,

$$|\vec{P}|\, Al = \sigma_p Al$$

$$\Rightarrow \quad |\vec{P}| = \sigma_p \quad \text{...(4)}$$

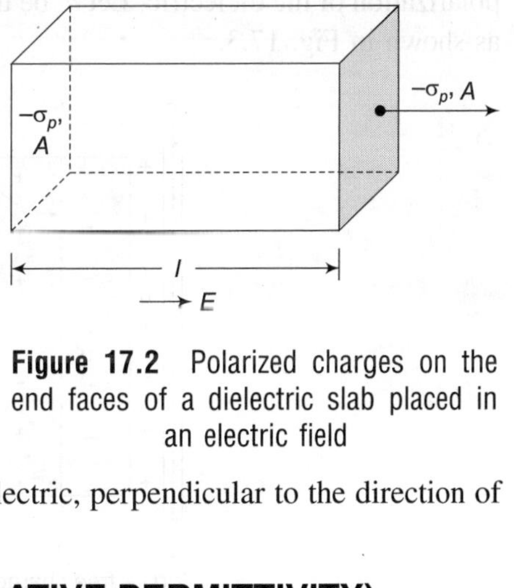


**Figure 17.2** Polarized charges on the end faces of a dielectric slab placed in an electric field

Hence, polarization may be defined as the surface density of charge induced in the end faces of the dielectric, perpendicular to the direction of the applied electric field.

## 17.4 DIELECTRIC CONSTANT (OR RELATIVE PERMITTIVITY)

When a dielectric slab is placed between two plates having charges $+Q$ and $-Q$, fictitious charges appear on the surfaces of the dielectric which are perpendicular to the direction of the applied field. The direction of the depolarizing field that arises due to the fictitious charges is in a direction opposite to that of the applied electric field. Hence, the effect of the dielectric is simply to reduce the field within the dielectric.

The dielectric constant is defined as the ratio of the field ($\vec{E}_0$) in free space to that of the field $\vec{E}$ in the dielectric and is represented by $\varepsilon_r$. Hence,

$$\varepsilon_r = \frac{\vec{E}_0}{\vec{E}} = \frac{\dfrac{1}{4\pi\varepsilon_0}\left(\dfrac{Q}{r^3}\right)\vec{r}}{\dfrac{1}{4\pi\varepsilon}\left(\dfrac{Q}{r^3}\right)\vec{r}}$$

$$\Rightarrow \qquad \varepsilon_r = \frac{\varepsilon}{\varepsilon_0} \qquad ...(5)$$

where $\varepsilon_0$ and $\varepsilon$ are the absolute permittivities of the free space and the dielectric.

Hence, the relative permittivity of a medium is defined as the ratio of the absolute permittivity of the medium to the absolute permittivity of the free space.

## 17.5 ELECTRIC DISPLACEMENT

The electric field $(\vec{E})$ at any point depends upon the medium in which the point is located. We define a quantity $\vec{D}$, known as *electric displacement* or the *electric induction*, which is independent of the medium but depends only on the distribution of charges and their magnitudes. To introduce this vector, we consider a pair of metallic plates across which a potential difference exists. Let $q$ and $-q$ be the charges on the plates. These charges are called real charges. When a dielectric slab is placed in between the metallic plates, charges appear on the end faces of the dielectric which are parallel to the metallic plates due to the polarization of the dielectric. Let $q'$ be the polarized charges on the end faces of the dielectric, as shown in Fig. 17.3.

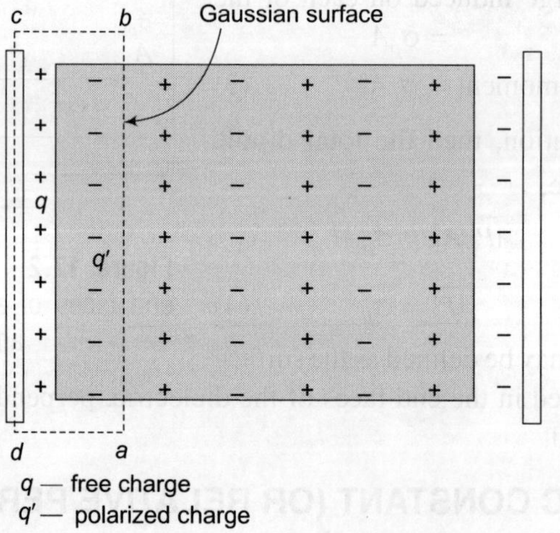


**Figure 17.3** Introducing the displacement vector

Now construct a Gaussian surface as shown by the dotted lines in Fig. 17.3. If $S$ is the area of either plate, then the electric flux over the surface $ab$ is

$$\sum \vec{E} \cdot \Delta\vec{S} = \sum E\Delta S \cos 0° = ES$$

The flux over the area *ad* or *bc* is

$$\sum \vec{E}\cdot\Delta\vec{S} = \sum E\Delta S \cos 90° = 0$$

The flux over the area *cd* is

$$\sum \vec{E}\cdot\Delta\vec{S} = 0$$

as the electric field inside a conductor is zero. The total electric flux over the Gaussian surface is, therefore, *ES*. The total charge enclosed by the Gaussian surface is

$$Q = q - q'$$

According to Gauss' theorem, the total electric flux over an enclosed surface is equal to $1/\varepsilon_0$ times total charge enclosed by the surface. That is,

$$ES = \frac{1}{\varepsilon_0}(q - q')$$

$$\Rightarrow \quad E = \frac{q}{\varepsilon_0 S} - \frac{q'}{\varepsilon_0 S}$$

$$\Rightarrow \quad \varepsilon_0 E = \sigma - \sigma_p$$

$$\Rightarrow \quad \varepsilon_0 E + \sigma_p - \sigma$$

$$\Rightarrow \quad \varepsilon_0 E + P = \sigma$$

$$(\varepsilon_0 E + P)S = \sigma S = q \qquad ...(6)$$

where $\sigma$ and $\sigma_p$ are the surface densities of real and polarized charges respectively.

Equation (6) suggests that an auxiliary vector

$$D = \varepsilon_0 E + P \qquad ...(7)$$

may be introduced whose flux over any closed surface equals the total charge enclosed by that surface. This auxiliary vector (*D*) is called electric displacement vector. It, therefore, follows from Eqs. (6) and (7) that

$$DS = q$$

$$\Rightarrow \quad D = \frac{q}{S} = \sigma, \text{ surface density of free charge} \qquad ...(8)$$

Equation (7) can also be expressed in vector form as

$$\vec{D} = \varepsilon_0 \vec{E} + \vec{P} \qquad ...(9)$$

## 17.6 ELECTRIC SUSCEPTIBILITY, $\chi$

When an electric field is applied to a non-polar substance, its molecules are polarized and aligned along the direction of the applied field. The magnitude of the polarization depends upon the magnitude of the applied electric field so that

$$\vec{P} \propto \vec{E}$$

$$\Rightarrow \quad \vec{P} = \varepsilon_0 \chi_e \vec{E} \qquad ...(10)$$

where $\chi_e$ is a constant, known as *electric susceptibility* of the dielectric and has no unit.

Using Eq. (10), Eq. (9) can now be expressed as

$$\vec{D} = \varepsilon_0 \vec{E} + \varepsilon_0 \chi_e \vec{E}$$

$$\Rightarrow \quad \vec{D} = \varepsilon_0 (1 + \chi_e) \vec{E} = \varepsilon_0 \varepsilon_r \vec{E} = \varepsilon \vec{E} \quad ...(11)$$

where $\varepsilon_r$ is a dimensionless constant, known either as *dielectric coefficient* or *dielectric constant* or *relative permittivity* of the dielectric material. The quantity ($\varepsilon$) is known as the *permittivity of the dielectric*.

In free space, $\vec{P} = 0$, $\vec{E} = \vec{E}_0$ and hence, Eq. (11) takes the form

$$\vec{D} = \varepsilon_0 (1 + \chi_e) \vec{E} = \varepsilon_0 \varepsilon_r \vec{E} = \varepsilon_0 \vec{E}_0 \quad ...(12)$$

$$\Rightarrow \quad \varepsilon_r \vec{E} = \vec{E}_0$$

$$\Rightarrow \quad \varepsilon_r = \frac{\vec{E}_0}{\vec{E}} \quad ...(13)$$

Thus, the dielectric constant may be defined as the ratio of electric field in free space and that in the dielectric. It also shows how many times the electric field in the dielectric is reduced in comparison to the field in free space.

From Eq. (12), we get

$$\varepsilon_r = (1 + \chi_e) \quad ...(14)$$

Equation (14) is valid so long as $\varepsilon_r$ and $\chi_e$ are independent of the applied electric field. For high fields, it may be possible for the electrons to be knocked out of the molecules and the material no longer remains an insulator, but starts conducting. This situation is referred to as the *breaking down* of the dielectric.

The maximum electric field which a dielectric can withstand without breaking down is known as its *dielectric strength*.

## 17.7 POLARIZATION AND POLARIZATION DENSITY

The dielectric properties of solids arise from the polarization produced by an electric field, and are described in terms of a specific function which must be added to both the field and the electric potential when the laws of electrostatics are applied to matter in the condensed state. As the solid is a large collection of ions and valence electrons, we differentiate between *conductors*, with a nearly-free flow of electric charge and *dielectrics*, having a negligible content of free electrons. In conductors the effect of applying an electric field is to redistribute the free charge until the internal field is cancelled, and hence, the net charge inside a conductor is zero under static conditions. When an electric field is applied to a dielectric, each nucleus of charge $q = +Ze$ will tend to be displaced in the direction of the field, and the corresponding cloud of tightly bound electrons will move against the field, as illustrated in Fig. 17.4 (a). The relative displacement, given by the position vector ($\vec{r}$) of the centre of the negative cloud with respect to the nucleus, allows the restoring force due to the electrons to balance the force on the nucleus due to the applied field. As a result, an electric dipole moment $\vec{p} = q\vec{r}$ is produced in each atom. As a consequence of the induced dipole moments, layers of negative and positive charges develop on the two surfaces of an element of dielectric, normal to the direction of the applied field $\vec{E}$, as in Fig. 17.4 (b). The elementary charge induced in a layer of thickness $r = |\vec{r}|$ and area $dS$ can be expressed in terms of the *polarized charge density* $\sigma_P = dq/dS$ as follows:

$$\sigma_P dS = \left(\frac{1}{V}\sum q\right) r dS = \left(\frac{1}{V}\sum qr\right) dS = \left(\frac{1}{V}\sum p\right) dS = P dS = \vec{P}\cdot d\vec{S}$$

which can be interpreted using $P \cdot d\vec{S} = \vec{P}\cdot\vec{e}_n dS$.

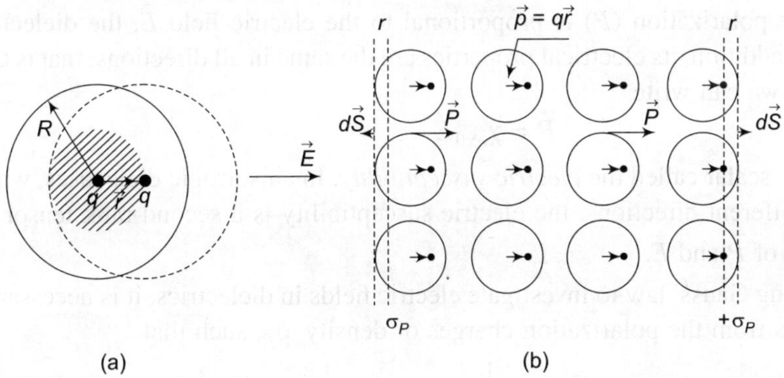


**Figure 17.4** Polarization of an atom (a) and of an element of dielectric (b)

In other words, the surface charge density ($\sigma_P$) induced on a dielectric equals the component along the outward normal $\vec{P}\cdot\vec{e}_n$ of a vector defined by

$$\vec{P} = \frac{1}{V}\sum \vec{p} \qquad ...(1)$$

and called the *polarization* or the dipole moment per unit volume of the element of dielectric. We may integrate the surface charge density over the closed surface of a given volume $V$ to obtain

$$\oint_S \sigma_P \, dS = \oint_S \vec{P}\cdot d\vec{S} = \int_V (\nabla\cdot\vec{P}) dV$$

where we have used Gauss' divergence theorem. Since each element of dielectric remains neutral under an applied electric field, we must introduce a volume polarization charge, of density, $\rho_P = dq/dV$, such that the total charge is zero

$$\oint_S \sigma_P \, dS + \int_V \rho_P \, dV = 0$$

Comparison between the foregoing equations gives

$$\int_V \rho_P \, dV = -\oint_S \sigma_P \, dS = -\int_V (\nabla\cdot\vec{P}) dV$$

which is true however small the volume, $V$. It follows that the charge density resulting from polarization of the bulk is

$$\rho_P = -\text{div}\vec{P} = -\nabla\cdot\vec{P} \quad ...(2)$$

Therefore, there is a volume polarization charge density distributed through the bulk of dielectric wherever the divergence of polaration is not zero, a result similar to that stated by Gauss' law, which relates the variation of $\vec{E}$ at a point in space to the density of charges at the same point. However, we must bear in mind that both $\sigma_P = \vec{P}\cdot\vec{e}_n$ and $\rho_P = -\nabla\cdot\vec{P}$ are bound charge densities arising from the bound electrons and nuclei in the dielectric solid. As the polarization is determined by the charge separation in the individual atoms, a functional relationship between $\vec{P}$ and the applied field $\vec{E}$ may always be defined.

When the polarization $(\vec{P})$ is proportional to the electric field $\vec{E}$, the dielectric is called *linear*. If, in addition, its electrical properties are the same in all directions, that is the dielectric is *isotropic*, we can write

$$\vec{P} = \chi_e\varepsilon_0\vec{E} \quad ...(3)$$

where $\chi_e$ is a scalar called the *electric susceptibility*. In anisotropic dielectrics, where $\vec{P}$ and $\vec{E}$ may have different directions, the electric susceptibility is a second-rank tensor relating the components of $\vec{P}$ and $\vec{E}$.

When using Gauss' law to investigate electric fields in dielectrics, it is necessary to include contributions from the polarization charges of density, $\rho_P$, such that

$$\nabla\cdot\vec{E} = \frac{1}{\varepsilon_0}(\rho + \rho_P) = \frac{1}{\varepsilon_0}(\rho - \nabla\cdot\vec{P})$$

where $\rho$ stands for the free-charge density only. It is convenient to separate the distribution of free charges from that due to polarization, in the form

$$\nabla\cdot(\varepsilon_0\vec{E} + \vec{P}) = \rho \quad \text{or} \quad \nabla\cdot\vec{D} = \rho \quad ...(4)$$

$\vec{D}$ is called the *electric displacement* and must always be associated with a particular dielectric

$$\vec{D} = \varepsilon_0\vec{E} + \vec{P} \quad ...(5)$$

whereas the electric field $\vec{E}$ is defined irrespective of the environment. In a linear isotropic dielectric we may use Eq. (3) and the displacement $\tilde{D}$ becomes

$$\vec{D} = \varepsilon_0\vec{E} + \chi_e\varepsilon_0\vec{E} = \varepsilon_0(1 + \chi_e)\vec{E} = \varepsilon_0\varepsilon_r\vec{E} \quad ...(6)$$

where: $\varepsilon_r = 1 + \chi_e$ ...(7)

is known as the *relative permittivity*. $\varepsilon_r$, is a constant for isotropic dielectrics, however for anisotropic materials $\chi_e$ (and so $\varepsilon_r$) is represented by a second-rank tensor,

## 17.8 MECHANISM OF POLARISATION

In molecules as well as atoms, an external field will displace the electrons with respect to the corresponding nuclei (*electronic polarisation*) (see Fig. 17.5). Over and above this, however, a displacement of atoms or ions within the molecule may be caused by an external field. For example, in an HCl molecule, an external field will change the interionic distance to some extent, leading to a change in the dipole moment (*atomic* or *ionic polarisation*).

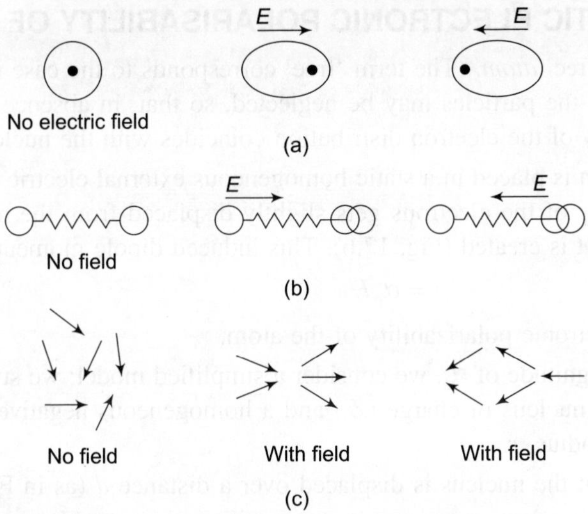


**Figure 17.5** Mechanism of dielectric polarisation: (a) electronic polarisation, (b) ionic polarisation, (c) orientational or dipolar polarisation

Apart from these two causes (electronic and ionic), there may be *orientational* or *dipolar polarisation*; this arises in a substance built up of polar molecules possessing a permanent electric dipole moment. Under an applied electric field, polar molecules tend to rotate and orient themselves in the direction of the field. This alignment tendency is opposed to the thermal agitation of the molecules. The higher the temperature, the greater the tendency towards randomness and the smaller the degree of orientation.

Thus, dipolar polarisation is strongly temperature dependent. (It occurs mainly in gases and liquids where molecules are fairly free to rotate.)

We can write total polarisation,

$$P = P_e + P_i + P_o \qquad ...(1)$$

where subscripts $e$, $i$ and $o$ correspond to electronic, ionic and orientational polarisation, respectively.

Further,

$$\left.\begin{aligned} P_e &= \alpha_e NE \\ P_i &= \alpha_i NE \\ P_o &= \alpha_o NE \end{aligned}\right\} \qquad ...(2)$$

where $\alpha_e$ is the electronic polarisability, $\alpha_i$ is the ionic polarisability and $\alpha_o$ is the orientation polarisability

So, we get $$\alpha = \alpha_e + \alpha_i + \alpha_o \qquad ...(3)$$

where $\alpha$ is the total polarisability.

## 17.9 THE STATIC ELECTRONIC POLARISABILITY OF ATOMS

Let us consider a 'free' *atom.* (The term 'free' corresponds to the case in which the neutral interaction between the particles may be neglected, so that, in absence of an electric field, the centre of gravity of the electron distribution coincides with the nucleus.

Suppose the atom is placed in a static homogeneous external electric field, $E$. As a result, the centre of gravity of the electrons gets slightly displaced from the nucleus and a small, finite dipole moment is created (Fig. 17.6). This induced dipole moment is given by

$$p = \alpha_e E \qquad ...(4)$$

where $\alpha_e$ is the electronic polarizability of the atom.

To obtain the magnitude of $\alpha_e$, we consider a simplified model: we suppose that the atom is represented by a nucleus of charge ($Z_e$) and a homogeneous negative charge distribution inside a sphere of radius, $r$.

Now suppose that the nucleus is displaced over a distance $d$ (as in Fig. 17.7 by external field, $E$.

The restoring force is equal to the force exerted on the nucleus by a negative charge, i.e., in equilibrium

$$E = \frac{(Ze)d}{4\pi\varepsilon_0 r^3} \quad \text{or} \quad ZeE = \frac{(Ze)^2 d}{4\pi\varepsilon_0 r^3} \qquad ...(5)$$

$$\therefore \qquad d = \frac{4\pi\varepsilon_0 r^3 E}{Ze} \qquad ...(6)$$

This gives the induced dipole moment

$$p = Z_{ed} = 4\pi\varepsilon_0 r^3 E \qquad ...(7)$$

and the electronic polarisability

$$\alpha_e = \frac{p}{E} = 2\pi\varepsilon_0 r^3 \qquad ...(8)$$

Note that $\alpha_e$ has dimensions of volume. For $r = 10^{-10}$ m, $\alpha_e$ is of the order of $10^{-40}$ m$^3$. It has been found that $\alpha_e$ is independent of temperature at moderate temperatures.

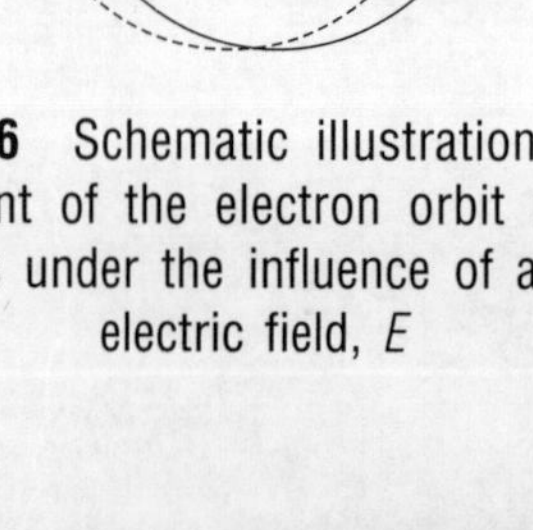


**Figure 17.6** Schematic illustration of slight displacement of the electron orbit relative to the nucleus under the influence of an external electric field, $E$

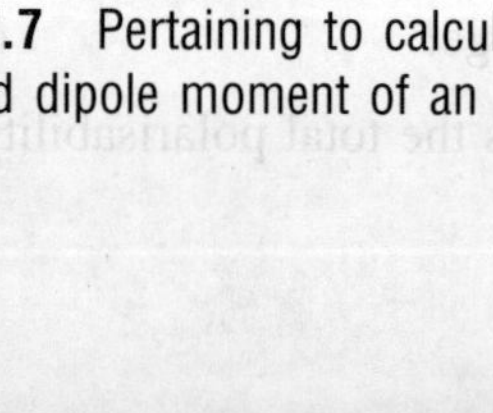


**Figure 17.7** Pertaining to calculation of induced dipole moment of an atom

## 17.10 IONIC POLARISABILITY

It is found only in ionic substances. In an electric field, the resultant torque lines up the dipole parallel to the field at the absolute zero of temperature. The field produces forces on the two charges of opposite sign so that distance between them is changed by some amount. The balance between the electrostatic force and the interatomic force due to stretching or compressing gives the value of the charge and distance between two neighbouring ions of opposite charge.

This enables us to calculate the ionic dipole moment and hence the ionic polarizability, $\alpha_i$. It is also independent of temperature at moderate temperatures.

## 17.11 DIPOLAR POLARISABILITY

Dipolar or orientation polarisation occurs in liquids or solids that have asymmetric molecules whose permanent dipole moments can be aligned by an external electric field.

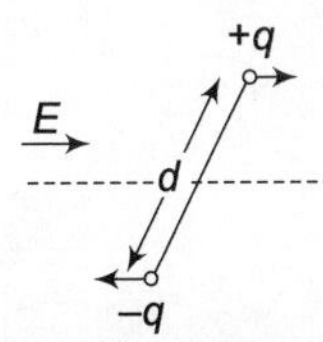


**Figure 17.8** The couple acting on the dipole due to an applied (external) field, $\vec{E}$

Let us consider a system of $n$ permanent dipoles, with dipole moment $\vec{p}$ subjected to an external field $\vec{E}$ parallel to $x$-axis. The work required to bring one of the dipole molecules into a position where $\vec{p}$ makes an angle $\theta$ with $\vec{E}$ is given by (see Fig. 17.8)

$$U = -\vec{p} \cdot \vec{E} = -pE \cos \theta \qquad ...(9)$$

According to Boltzmann distribution law, the probability of finding a dipole within the solid angle $d\Omega$ (i.e., angle between $\theta$ and $\theta + d\theta$) is proportional to the Boltzmann factor (see Fig. 17.9)

$$f(\theta) = \exp\left(-\frac{U}{kT}\right) = \exp\left(\frac{pE\cos\theta}{kT}\right) \qquad ...(10)$$

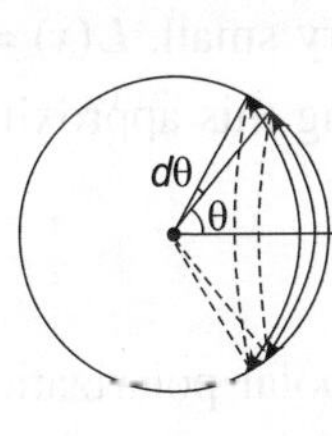


**Figure 17.9** Regarding calculation of dipolar polarisation

Consequently, the average moment contribution per dipole $p$ laying within the solid angle $d\Omega$ and be determined by integrating over all angles from parallel alignment ($\theta = 0$) to antiparallel alignment ($\theta = 180°$)

$$\vec{p} = \frac{\int_0^\pi p_x\, f(\theta)\, d\Omega}{\int_0^\pi f(\theta)\, d\Omega} \qquad ...(11)$$

where $p_x = p \cos \theta,\ d\Omega = 2\pi \sin \theta\, d\theta.$ ...(12)

Hence,

$$\vec{p} = \frac{\int_0^\pi p \cos\theta\, 2\pi \sin\theta \exp(pE\cos\theta / kT)\, d\theta}{\int_0^\pi 2\pi \sin\theta \exp(p\varepsilon\cos\theta / kT)\, d\theta} \qquad ...(13)$$

Let us substitute $x = \frac{pE}{kT}$ and $y = \cos\theta \times \frac{pE}{kT}\, dy = -\frac{pE}{kT}\sin\theta\, d\theta$. Then

$$\vec{p} = \frac{p\int_{-x}^{+x} y\exp(y)\,dy}{x\int_{-1}^{+1}\exp(y)\,dy} = \frac{p}{x}\frac{[ye^y - e^y]_{-x}^{x}}{[e^y]_{-x}^{x}}$$

$$= \frac{p}{x}\left\{\frac{xe^x - e^x + xe^{-x} + e^{-x}}{e^x - e^{-x}}\right\}$$

$$= p\left\{\frac{e^x - e^{-x}}{e^x - e^{-x}} - \frac{1}{x}\right\} = p\left[\cosh x - \frac{1}{x}\right] \quad ...(14)$$

$$\equiv pL(x) \quad ...(15)$$

where $L(x) = \coth(x) - \frac{1}{x}$ is known as the Langevin function since this formula was first derived by Langevin in 1905. Later, in 1912 Debye applied this formula to an electric dipole gas. Figure 17.10 shows a plot of this function. If $x$ is very small, $L(x) \simeq x/3$.

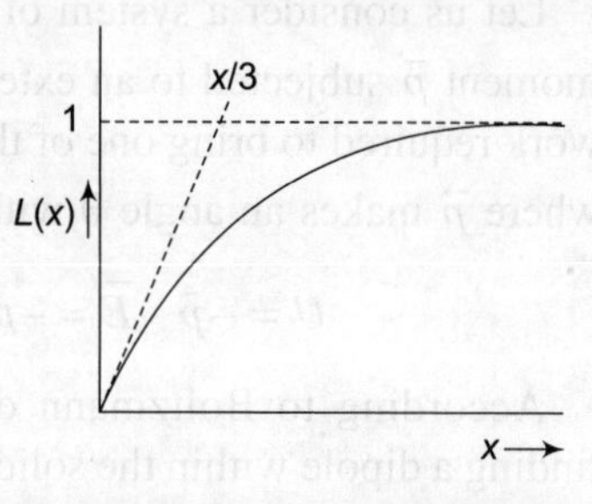


**Figure 17.10** Langevin function

Using this approximation, the dipolar polarization

$$\vec{p} \simeq \frac{p^2E}{3kT} \quad ...(16)$$

and dipolar polarization ($\alpha_d$) is given by

$$\alpha_d = \frac{p^2E}{3kT} \quad ...(17)$$

If $\alpha_0$ denotes the sum of the electronic and ionic polarizabilites, then the total polarizability

$$\alpha = \alpha_0 + p^2/(3kT) \quad ...(18)$$

This is known as Langevin–Debye equation.

Figure 17.11 shows the plot of molar polarizabilities as a function of $1/T$ for $CH_4$, $CH_3Cl$, $CH_2Cl_2$, $CHCl_3$ and $CCl_4$. This plot enables us to determine $p$ and through the determination of dipole moment, it is possible to know the molecular structure. Typical values of dipole moments are

(HCl) = $1.03 \times 10^{-18}$ e.s.u.

(HBr) = $0.79 \times 10^{-18}$ e.s.u.

($H_2O$) = $1.87 \times 10^{-18}$ e.s.u.

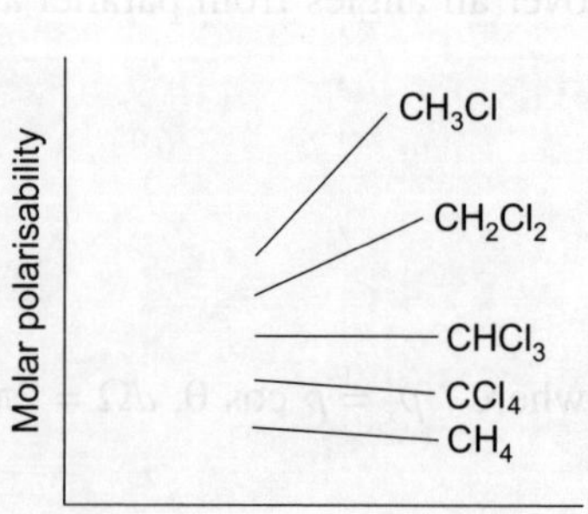


**Figure 17.11** Molar polarisability as function of $1/T$

### 17.11.1 Depolarisation Field

One contribution to the electric field inside a body is that of the applied electric field $\vec{E}_0$ defined as the field produced by fixed charges external to the body. However, this causes a uniform polarization within the body.

Then, the only contributions to the macroscopic field $E$ are $E_0$ and $\vec{E}_d$, i.e.,

$$\vec{E} = \vec{E}_0 + \vec{E}_d \qquad ...(19)$$

where $\vec{E}_d$ is the field due to the uniform polarization and is called the *depolarisation field* because within the body it tends to oppose the applied field, $\vec{E}_0$.

## 17.12 MOLECULAR FIELD IN A DIELECTRIC

The electric field that is responsible for polarising a molecule of the dielectric is called the molecular field $\vec{E}_m$. This is the electric field at a molecular position in the dielectric; it is produced by all external sources and by all polarised molecules in the dielectric except the molecule situated at the point under consideration. It is to be noted here that $\vec{E}_m$ need not be the same as the macroscopic electric field. $\vec{E}_m$ may be calculated as follows.

Let us consider that a thin dielectric sample is polarized by placing it between two (uniformly), oppositely charged parallel plates [Fig. 17.12 (a)]. Now, we cut out a small piece of the dielectric leaving a spherical cavity surrounding the point where molecular field is to be computed. The dielectric that is left is treated as a continuum from the macroscopic point of view. Now, we put the dielectric back into the cavity (molecule by molecule) except for the molecule at its centre (where $\vec{E}_m$ is to be computed). The molecule replaced inside are to be treated not as continuum, but as individual dipoles.

Within the bulk of the dielectric (outside cavity), polarization is uniform and $\vec{p}$ is parallel to the field producing it. The part of the dielectric outside the cavity may be replaced by a

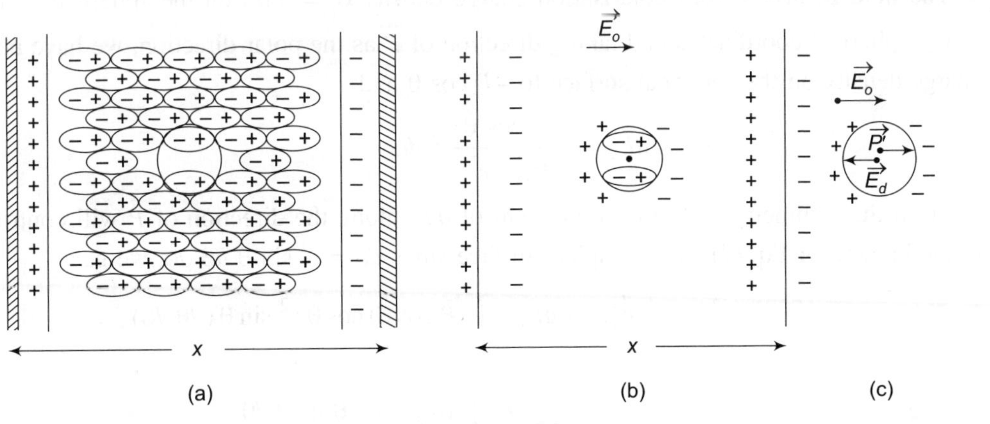


**Figure 17.12** (a) A dielectric slab between two parallel oppositely charged plates, (b) replacement of the dielectric outside the cavity by a system of polarised charges, (c) The depolarisation field, $\vec{E}_d$ is opposite to $\vec{p}$

system of polarisation charges [Figs. 17.12(b) and 17.13]. Consequently, the electric field at the centre of the cavity is

$$\vec{E}_m = \vec{E}_o + \vec{E}_d + \vec{E}_s + \vec{E}' \quad ...(1)$$

where $\vec{E}_o$ is the field due to charged parallel plates, $\vec{E}_d$ is the depolarised field, $\vec{E}_s$ is the field due to polarisation charge on the cavity surface and $\vec{E}'$ is that due to all of the dipoles inside the cavity.

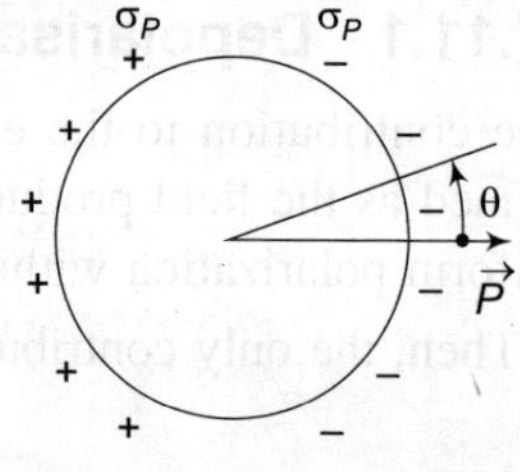

**Figure 17.13** Pertaining to calculation of $\vec{E}_s$ just outside the cavity

$$\vec{E}_o = \frac{\sigma}{\varepsilon_0} \quad ...(2)$$

where σ is the surface charge density on the plates. The depolarising field is also produced by two parallel planes of charge with density $\sigma_p = \pm p$, so

$$\vec{E}_d = -\frac{\vec{P}}{\varepsilon_0} \quad ...(3)$$

Just outside the electric slab,

$$\vec{D} = \varepsilon_0 \vec{E} = \varepsilon_0 \vec{E}_0 - \vec{P} \quad ...(4)$$

where $\vec{E}$ is the macroscopic electric in the dielectric. From Eq. (4)

$$\vec{E} = \vec{E}_o - \frac{\vec{P}}{\varepsilon_o} = \vec{E}_o + \vec{E}_d \quad ...(5)$$

Therefore, from Eqs. (1) and (4)

$$\vec{E}_m = \vec{E} + \vec{E}_s + \vec{E}' \quad ...(6)$$

This is a general result independent of the geometry of Fig. 17.12.

The field $\vec{E}_s$ arises from polarisation charge density $\sigma_p = \pm|\vec{P}|$ on the spherical surface. Using spherical coordinates and taking direction of $\vec{P}$ as the polar direction, we have bound charge density on the spherical surface to $-P\cos\theta$ and

$$d\vec{E}_s = \frac{(-P\cos\theta)}{4\pi\varepsilon_0 r^2}\hat{r}\, da \quad ...(7)$$

From the symmetry, only the component of $d\vec{E}_s$ along the direction of $\vec{P}$ will contribute to the integral of Eq. (7) over complete surface since $da = r^2 \sin\theta\, d\theta\, d\phi$, we get

$$\vec{E}_s = \int_s d\vec{E}_s = \int_s (\vec{P}\cos\theta)\cos\theta\, r^2 \sin\theta\,(d\theta\, d\phi)$$

$$= \frac{1}{4\pi\varepsilon_0}\vec{P}\int_0^{2\pi} d\phi \int_0^{\pi} \cos^2\theta \sin\theta\, d\theta$$

$$= \frac{2\pi\vec{P}}{4\pi\varepsilon_0}\int_0^{\pi}\cos^2\theta\sin\theta\, d\theta$$

$$= \frac{\vec{P}}{3\varepsilon_0} \qquad ...(8)$$

Now, in most of the cases (e.g., gas or liquids) homogeneous isotopic dielectrics, $\vec{E}' = 0$ (in general, it is not zero, however, we restrict ourselves to the case $\vec{E}' = 0$).

Thus, from Eqs. (2), (3), (5), (6) and (8), we found the molecular field

$$E_m = \left(\frac{\sigma}{\varepsilon_0} - \frac{P}{\varepsilon_0}\right) + \frac{P}{3\varepsilon_0}$$

or
$$\vec{E}_m = \vec{E} + \frac{\vec{P}}{3\varepsilon_0} \qquad ...(9)$$

where $\vec{E}_m$ is the macroscopic field within the dielectric.

## 17.13 CLAUSIUS-MOSSOTTI EQUATION

Equation (9) gives the electric field ($E_m$) that acts upon a single molecule of the dielectric. Under the action of this field, the molecule displays an electric moment ($\vec{p}$) in the direction of the field, i.e.,

$$\vec{p} = \alpha \vec{E}_m \qquad ...(10)$$

where $\alpha$ is molecular polarsability. If there are $N$ molecules per unit volume, then the polarisation

$$\vec{P} = N\vec{p} = N\alpha \vec{E}_m$$

$$= N\alpha\left(\vec{E} + \frac{\vec{p}}{3\epsilon_0}\right)$$

or
$$\vec{E} = \vec{P}\left(\frac{1}{N\alpha} - \frac{1}{3\varepsilon_0}\right) \qquad ...(11)$$

But
$$\vec{P} = \varepsilon_0 \chi_e \vec{E}$$

$$= \varepsilon_0 (\varepsilon_r - 1)\vec{E} \qquad (\because \ \varepsilon_r = 1 + \chi_e)$$

or
$$\vec{E} = \frac{\vec{P}}{\varepsilon - \varepsilon_0} \qquad ...(12)$$

(where $\varepsilon = \varepsilon_r \varepsilon_0$)

($\vec{E}_m$ is the field at the centre of the gravity)

$\vec{E}$ is the macroscopic field inside the dielectric)

Comparing Eqs. (11) and (12)

$$\vec{P}\left[\frac{1}{N\alpha} - \frac{1}{3\varepsilon_0}\right] = \frac{\vec{P}}{\varepsilon - \varepsilon_0}$$

or $$\frac{1}{N\alpha} = \frac{1}{\varepsilon - \varepsilon_0} + \frac{1}{3\varepsilon_0}$$

$$= \frac{3\varepsilon_0 + \varepsilon - \varepsilon_0}{3\varepsilon_0\,(\varepsilon - \varepsilon_0)}$$

or $$\frac{1}{N\alpha} = \frac{\varepsilon + 2\varepsilon_0}{3\varepsilon_0 - (\varepsilon - \varepsilon_0)}$$

or $$\frac{N\alpha}{3\,\varepsilon_0} = \frac{\varepsilon - \varepsilon_0}{\varepsilon + 2\varepsilon_0}$$

or $$\frac{N\alpha}{3\varepsilon_0} = \frac{\varepsilon_r - 1}{\varepsilon_r + 2} \quad ...(13)$$

Equation (13) is known as the *Clausius–Mossotti relation.*

The number of molecules per unit volume ($N$) is proportional to the density of the gas. Taking the molecular polarizability $\alpha$ to be a constant, Eq. (13) shows that $N$ is directly proportional to $\left(\frac{\varepsilon_r - 1}{\varepsilon_r + 2}\right)$. Now $\varepsilon_r$ can be experimentally determined and $N$ can be calculated from the density of the gas at a given temperature and pressure. Thus, the molecular polarizability ($\alpha$) can then be calculated. As the molecular polarizability ($\alpha$) is directly proportional to the cube of the radius of the molecule, the radius of the molecule, hence, can be determined. The value of the molecular radius so obtained agrees fairly well with the values obtained by other methods and this proves the validity of the Claussius-Mossotti relation. This holds only in the case of monoatomic gases and weak solutions.

## 17.14 CLASSICAL THEORY OF ELECRONIC POLARISABILILTY

The total polarisability of a dielectric has three parts: electronic, ionic and dipolar. The electronic contribution arises from the displacement of the electron shell relative to a nucleus.

The dielectric constant at optical frequencies (of the applied field) arises entirely from the electronic polarisability. In the optical range, Eq. (13) (Clausius–Mossotti relation) reduces to

$$\frac{n^2 - 1}{n^2 + 2} = \frac{1}{3\varepsilon_0}\Sigma N_j\,\alpha_j \quad \text{(elecronic)} \quad ...(14)$$

where $N_j$ and $\alpha_j$ are, respectively, the concentration and polarisability of $j$th atoms and $n^2 = \varepsilon_r$, where $n$ is the refractive index.

An electron bound harmonically to an atom in a local field $E_{\text{loc}}$ sin $\omega t$ will show resonance absorption at a frequency $\omega = \omega_0 \equiv \sqrt{K/m}$, where $K$ is the force constant. The displacement of the electron is given by

$$-eE_{\text{loc}} = Kx = m\,\omega_0^2\,x \quad ...(15)$$

so that the static electronic polarisability is

$$\alpha_e = \frac{p}{E_{\text{loc}}} = \frac{ex}{E_{\text{loc}}} = \frac{e}{E_{\text{loc}}}\left(\frac{-eE_{\text{loc}}}{m\omega_0^2}\right)$$

or
$$\alpha_e = \left(\frac{e^2}{m\omega_0^2}\right) \qquad ...(16)$$

In general, the electronic polarisability will depend on frequency and is obtained by solving the equation of motion

$$m\frac{d^2x}{dt^2} + m\omega_0^2 x = -eE_{\text{loc}} \sin \omega t \qquad ...(17)$$

For
$$x = x_0 \sin \omega t;\ \dot{x} = x_0\, \omega \cos \omega t$$

and
$$\ddot{x} = -x_0\, \omega^2 \sin \omega t$$

$\therefore$
$$m(-\omega^2 + \omega_0^2)\, x_0 = -eE_{\text{loc}} \qquad ...(18)$$

or
$$x_0 = \frac{-e\,E_{\text{loc}}}{m(\omega_0^2 - \omega^2)} \qquad ...(19)$$

The dipole moment has the amplitude

$$p_0 = -ex_0 = \frac{e^2\, E_{\text{loc}}}{m\,(\omega_0^2 - \omega^2)} \qquad ...(20)$$

so that electronic polarizability

$$\alpha_e = \frac{p_0}{E_{\text{loc}}} = \frac{e^2/m}{m\,(\omega_0^2 - \omega^2)} \qquad ...(21)$$

In quantum theory, the corresponding expression is

$$\alpha_e = \frac{e^2}{m}\sum_j \frac{f_{ij}}{(\omega_{ij}^2 - \omega^2)} \qquad ...(22)$$

where $f_{ij}$ is the oscillator strength of the electric dipole transition between the atomic states, $i$ and $j$.

## 17.15 FREQUENCY DEPENDENCE OF DIFFERENT POLARISABILITIES

We know that polarisability $\alpha$ is given by

$$\alpha = \alpha_e + \alpha_i + \alpha_d$$

The response of each type of polarisation (electronic, ionic or dipolar) to the impressed frequency of the field is different. For example, while electronic polarisation can be quickly built up in $10^{-14}$ to $10^{-15}$ sec, ionic polarisation may require a time period ~$10^{-11}$ – $10^{-13}$ sec. As a consequence, while both ionic and electronic polarisabilities respond in the regions from microwave to infrared, only the electronic polarisability contributes to the polarisation in the optical and ultraviolet regions. Figure 17.14 depicts the frequency dependence of the different types of polarisabilities. The dipolar and ionic contributions are small at high frequencies because of the inertia of the molecules and ions.

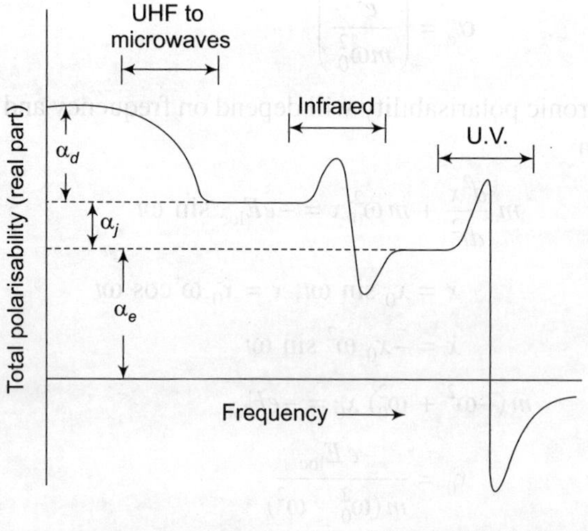


**Figure 17.14** Frequency dependence of the various contributions to the total polarisability

## 17.16 DIELECTRIC CONSTANT AND ALTERNATING FIELDS

Assuming small displacements, Hooke's law is obeyed and the treatment of the material behaviour in an external alternating electric field is similar to the problem of a classical harmonic oscillator undergoing forced oscillations. Let the external field be given by

$$E^* = E_0 \exp(i\omega t) \qquad \text{...(1)}$$

The equation of motion of a particle is

$$m\left(\frac{d^2x}{dt^2} + \gamma\frac{dx}{dt} + \omega_0^2 x\right) = eE_0 \exp(i\omega t) \qquad \text{...(2)}$$

Here, $x$ and $\omega_0$ are the displacement and the natural angular frequency of the particle, respectively. The second term on the L.H.S is due to damping and accounts for emission of radiation from the accelerating particle, is damping constant. The solution of Eq. (2) is

$$x = \frac{eE^*}{m(\omega_0^2 - \omega^2 + i\gamma\omega)} \qquad \text{...(3)}$$

Consequently, the complex susceptibility of a dipolar gas is given by

$$\chi^* = \frac{Ne^2}{m\varepsilon_0(\omega_0^2 - \omega^2 + i\gamma\omega)} \qquad \text{...(4)}$$

$$= \frac{Ne^2}{m\varepsilon_0}\left[\frac{\omega_0^2 - \omega^2 - i\gamma\omega}{\omega_0^2 - \omega^2 + i\gamma\omega(\omega_0^2 - \omega^2 - i\gamma\omega)}\right]$$

$$= \frac{Ne^2}{m\varepsilon_0}\left[\frac{\omega_0^2 - \omega^2 - i\gamma\omega}{(\omega_0^2 - \omega^2)^2 + \gamma^2\omega^2)}\right]$$

$$= \frac{Ne^2(\omega_0^2 - \omega^2)}{m\varepsilon_0[(\omega_0^2 - \omega^2)^2 + \gamma^2\omega^2]} - \frac{Ne^2 i\gamma\omega}{m\varepsilon_0[(\omega_0^2 - \omega^2)^2 + \gamma^2\omega^2]} \quad ...(5a)$$

$$\equiv \chi' - i\chi'' \quad ...(5b)$$

where $\chi'$ and $\chi''$ are the real and imaginary parts of the complex susceptibility. If the damping were absent, $\chi''$ would be zero (no absorption), since energy absorbed is proportional to $\chi''$. The real and imaginary parts of the complex dielectric constant $\varepsilon^*$ are respectively,

$$\varepsilon'(\omega) = 1 + \frac{Ne^2}{m\varepsilon_0}\left[\frac{(\omega_0^2 - \omega^2)}{(\omega_0^2 - \omega^2)^2 + \gamma^2\omega^2)}\right] \quad ...(6)$$

$$\varepsilon'(\omega) = \frac{Ne^2}{m\varepsilon_0}\left[\frac{\gamma\omega}{(\omega_0^2 - \omega^2)^2 + \gamma^2\omega^2)}\right] \quad ...(7)$$

Further, $$\varepsilon^* = (\varepsilon' - i\varepsilon'') \quad ...(8)$$

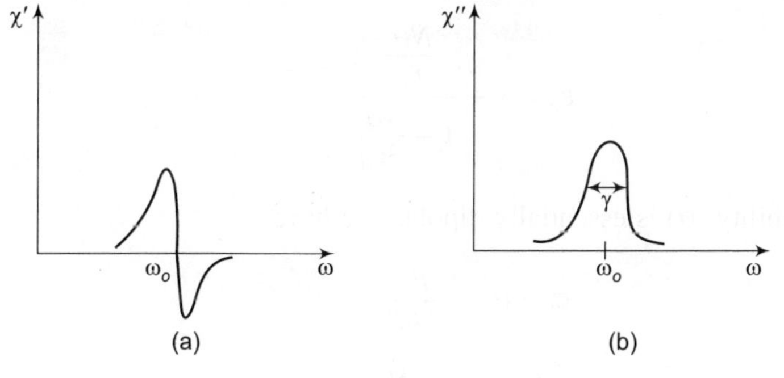


**Figure 17.15** Frequency dependence of the complex susceptibility in the neighbourhood of natural angular frequency $\omega_0$: (a) Real part, (b) Imaginary part

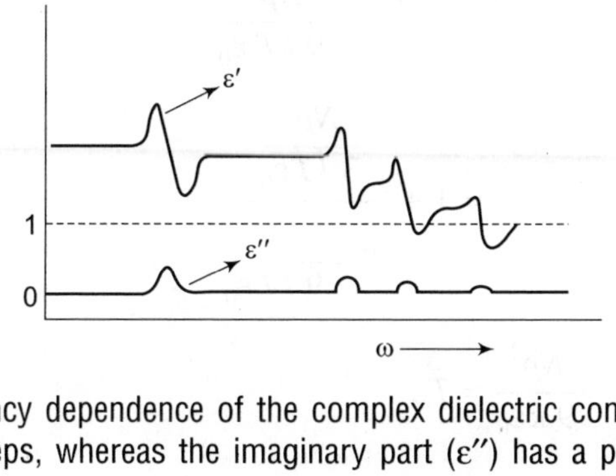


**Figure 17.16** Frequency dependence of the complex dielectric constant. The real part ($\varepsilon'$) decreases in steps, whereas the imaginary part ($\varepsilon''$) has a peak at each step

## 17.17 CLAUSIUS–MOSSOTTI CATASTROPHE

Polar (i.e., orientational) polarisability leads to an interesting consequence, we have

$$P = N\alpha\, E_m = N\alpha \left( E + \frac{P}{3\varepsilon_0} \right) \qquad \text{...(1)}$$

or

$$P\left(1 - \frac{N\alpha}{3\varepsilon_0}\right) = N\alpha E$$

or

$$P = \frac{N\alpha E}{1 - \dfrac{N\alpha}{3\varepsilon_0}} \qquad \text{...(2)}$$

Further,

$$P = \varepsilon_0\, \chi_e E = \varepsilon_0 (\varepsilon_r - 1) E$$

$$= \varepsilon_0\, \varepsilon_r E - \varepsilon_0 E \qquad \text{...(3)}$$

From Eqs. (2) and (3)

$$\varepsilon_0\, \varepsilon_r E = \varepsilon_0 E + \frac{N\alpha E}{1 - \dfrac{N\alpha}{3\varepsilon_0}}$$

$$\varepsilon_r = 1 + \frac{\dfrac{N\alpha}{\varepsilon_0}}{1 - \dfrac{N\alpha}{3\varepsilon_0}}$$

The polarisability ($\alpha$) is essentially dipolar, we have

$$\alpha \simeq \alpha_d = \frac{p^2}{3\,kT} \qquad \text{...(4)}$$

and

$$\varepsilon_r = 1 + \frac{\dfrac{Np^2}{3\,kT\,\varepsilon_0}}{1 - \dfrac{Np^2}{9\,kT\,\varepsilon_0}}$$

$$= 1 + \frac{\dfrac{Np^2}{3\,kT\varepsilon_0}}{T - \dfrac{Np^2}{9\,kT\varepsilon_0}}$$

If we call

$$\frac{Np^2}{9\,k\varepsilon_0} = T_c, \qquad \text{...(5)}$$

we get $$\varepsilon_r = 1 + \left(\frac{3T_c}{T - T_c}\right) \qquad ...(6)$$

Thus, as we approach $T_c$ from the high-temperature side, the dielectric constant will follow an increase according to Curie–Weiss law. At $T = T_c$, the (static) dielectric constant will be infinite. These are the characteristics exhibited by ferroelectrics. Thus, ferroelectricity in crystals can be thought as arising out of the catastrophe appeared in the above discussion. This catastrophe is referred to as the Clausius–Mossotti catastrophe. In case of water, this is expected to occur at 1200°C.

## 17.18 PERMANENT POLARISATION, FERROELECTRICITY

We know that it is the molecular field ($\vec{E}_m$) that is responsible for polarizing the individual molecules. The relationship between $\vec{E}_m$ and the macroscopic electric field ($\vec{E}$) is given by

$$\vec{E}_m = \vec{E} + \frac{\vec{P}}{3\varepsilon_0} \qquad ...(7)$$

In most cases, the polarisation is proportional to $\vec{E}$, then $\vec{E}_m$ vanishes when $\vec{E}$ goes to zero. However, under certain conditions, for some materials. Equation (7) is also compatible with a permanent (or spontaneous) polarisation. When $\vec{E}$ is set equal to zero,

$$\vec{E}_m = \frac{\vec{P}_0}{3\varepsilon_0} \qquad ...(8)$$

i.e., if a polarisation $\vec{P}_0$ exists, it will create an electric field at the molecule which tends to polarise the molecule. Let there be $N$ molecules per unit volume, so

$$\vec{P}_0 = N\alpha\vec{E}_m = \frac{N\alpha}{3\varepsilon_0}\vec{P}_0 \qquad ...(9)$$

or $$P_0\left(1 - \frac{Na}{3\varepsilon_0}\right) = 0$$

This is satisfied when

$$\frac{N\alpha}{3\varepsilon_0} = 1 \quad (\text{for } P_0 \neq 0) \qquad ...(10)$$

This is the condition for a permanent polarisation.

For most materials, $N\alpha/(3\varepsilon_0) < 1$ and ordinary dielectric behaviour results. For a few crystalline materials, however, the condition (10) is met. Such materials are called *ferroelectric* because their electrical properties are analogous to magnetic properties of ferromagnetic materials.

## 17.19 FERROELECTRICS AND PIEZOELECTRICS

Ferroelectrics are crystal dielectrics that possess, in a certain temperature range, a number of unusual properties.

These properties were discovered for the first time in Seignette salt, and hence ferroelectrics are sometimes called Seignette-electrics. Crystals of Seignette salt have clearly pronounced *anisotropy*, i.e., their properties depend on the orientation of the crystals with respect to the direction of the electric field. The unusual properties can only be observed when the crystal is in a special position with respect to the condenser plates. In any other position the crystals will have the same properties as ordinary dielectrics.

Ferroelectrics include such dielectrics as potassium phosphate, potassium niobate, and barium metatitanate. Over a hundred ferroelectrics are known at present.

The primary feature of a ferroelectric is the anomalous dependence of its dielectric constant on temperature. We know that the dielectric constant (ε) of common dielectrics changes very little with heating. The dielectric constant temperature diagrams for a ferroelectrics has one or more very sharp maxima where ε can reach several thousands. The temperatures at which these maxima occur are called Curie temperatures, or Curie points. For instance, Seignette salt has two Curie temperatures (+22.5°C and –15°C), while barium metatitanate has only one Curie temperature, (+80°C). If a ferroelectric has just one Curie temperature, its anomalous properties show up at temperature below it, whereas for one with two Curie temperatures, its properties are anomalous in the temperature in the temperatures between the Curie points.

The second feature of a ferroelectric is a nonlinear dependence of its polarization on the strength ($E_0$) of the external electric field. It follows that the polarization ($P$) is proportional to $E_0$ in ordinary dielectrics. The relationship in ferroelectrics is more complex in nature, being different in different ferroelectrics. The dielectrics parameters (ε) is then clearly not a constant but depends on $E_0$.

The third and the most remarkable feature of ferroelectrics is that the dependence of the polarization on $E_0$ has the appearance of a hysteresis loop, like the magnetization curve of a ferromagnet. Suppose a ferroelectric that is not polarised is put into the electric field between two capacitor plates and the strength of the field is increased. The polarization ($P$) will rise nonlinearly and reach saturation at a certain value of $P_s$. The polarization will not change even if $E_0$ is increased further. If the field in the capacitor then decreases, The $P$ versus $E_0$ graph will not be the same as the curve obtained for rising $E_0$; the new curve will lie above the old one. When $E_0$ is down to zero, the polarization will be nonzero and equal to a certain value $P_0$ that is called the *residual polarization*. The residual polarization persists in the ferroelectrics even in the absence of an external electric field. To reduce the polarization to zero, a field ($E_0$) must be applied in the opposite direction; this field is called an electric *coercive force*.

There is a clear similarity between the electric properties of ferroelectrics and the magnetic properties of ferromagnets, which is exactly why the dielectrics are called ferroelectrics.

The behaviour of ferroelectrics can be explained as follows. Strong interaction between the particles making up the crystals results in the apperance of separate *domains* within the ferroelectric, domains being regions with nonzero polarization. Theses domains were first discovered in barium metatitanate, and they can be seen through a polarizing microscope.

The polarization vector has different directions in different domains, so that the electric moments of the whole specimen is close to zero. This corresponds to the lowest energy state, since otherwise, polarization charges would appear on the surface of the specimen, and a field would be induced around it containing additional energy.

The polarization directions of individual domain change in the external electric field to the direction of the field. In addition, the most advantageously oriented domains can grow by pushing their boundaries into neighbouring domains. The resulting electric moment of the ferroelectric becomes nonzero. The presence of polarized domains in ferroelectrics has an essential effect on their electric properties.

The domains only exist within a certain temperature range where the ferroelectric properties appear.

So far we have only considered cases when the polarization of a dielectric was brought about by an asymmetry in the localization of positive and negative charges and was caused by an external electric field. This asymmetry can be produced in some crystals by deforming them mechanically. If such a crystal is compressed or stretched in a certain direction, it becomes polarized, and polarization charges appear on its surface. They can then be registered, for instance, by an electrometer. This phenomanon is called the *piezoelectric effect.* The electric moment that appears in the crystals due to mechanical deformations depends on the crystal lattice direction along which the deformation occurs and on the crystal lattice type. The piezoelectric effect only occurs in crystals whose structure does not possess a centre of symmetry.

Crystals that exhibit the piezoelectric effect are called *piezoelectrics*. They include Seignette salt, ammonium phosphate, potassium phosphate, ethylenediamine tartrate, quartz, and tourmaline.

Let us discuss the piezoelectric effect in some detail, taking quartz as an example. Quartz crystals have four symmetry axes. An axis is a symmetry axis when a rotation of the crystal around it through an angle $\alpha = 360°/n$ brings the crystal into a position indistinguishable from its original position. If $n = 2$, the symmetry axis is said to be a twofold axis, if $n = 3$, it is a threefold axis, etc. Quartz crystals have one threefold symmetry axis called the *optic axis* and three twofold axes called the *electrical axes*. The latter are at right angles to the optic axis and at 120° to each other.

Suppose we have a specimen in the form of a parallelepiped cut from a quartz crystal so that its $z$-axis is the optic axis of the crystal, and its $x$-axis is one of the electric axes.

If we apply a force $f_x$ to the specimen compressing it in the direction of the $x$-axis, polarization charges $+Q$ and $-Q$ will appear on the faces perpendicular to this axis. The value of $Q$ is proportional to the compressing force $f_x$ and does not depend on the specimen's dimensions, i.e.,

$$Q = kf_x, \qquad ...(1)$$

where $k$ is the *piezoelectric constant* of the material. This is called *longitudinal piezoelectric effect.*

However, if we apply a compressive (or tensile) force ($\mathbf{f}_z$) to the crystal in the direction of the $z$-axis no polarization will occur.

If a compressive force ($\mathbf{f}_y$) is applied in the direction of the $y$-axis, polarization charges will appear like those in the first case and on the faces perpendicular to the $x$-axis. However, the signs of the charges will be opposite to the ones in the first case. The charges will also be proportional to the value of compressive force, $f_y$, but now they also depend on the dimensions $a$ and $b$ of the specimen along the $x$- and $y$-axes respectively, i.e.,

$$Q = k\,\frac{b}{a}\,f_y,$$

where $k$ is the same constant as in Eq. (1)

This is called *transverse piezoelectric effect*. It appears because the force compressing the specimen along the $y$-axis causes a secondary expansion of the specimen along the $x$-axis.

If the force applied to the crystal is not compressive but tensile, the signs of the polarization charges change.

The *reverse piezoelectric effect* occurs as well. If an electric field is applied to metal electrodes fixed onto opposite faces of a crystal, the dimensions of the crystal change a little, the crystal becoming longer or shorter depending on the direction of the field.

The inevitability of the reverse piezoelectric effect can easily be shown to follow from the law of the conservation of energy. Suppose a compressive force (**f**) is applied to a dielectric specimen. When there is no piezoelectric effect, the work of the external forces is equal to the potential energy of the specimen that is elastically compressed. If there is a piezoelectric effect, polarization charges appears on the specimen faces, and an electric field appears that contains an additional energy. According to the law of the conservation of energy, the work performed when the specimen is compressed will be greater, and therefore an additional force counteracting the compression will appear. This is the force of the reverse piezoelectric effect. It acts in a direction opposite to that of the force of the direct piezoelectric effect, i.e., if the charges $+Q$ and $-Q$ appear on the specimen faces *in compression*, the specimen will *expand* if the same charges appear due to an applied electric field, and vice versa.

## 17.20 FERROELECTRICS

Some dielectrics are called *ferroelectrics*, because they exhibit a polarization versus electric field curve very similar to the $B$–$H$ curve of a ferromagnetic material. Such a *ferroelectric hysteresis loop* is show in Fig. 17.17 with the *remanent polarization* ($P_r$) and the *coercive field*, $\varepsilon_c$.

(Domain structure can be observed in ferroelectric barium titanate ($BaTiO_3$) using polarized light).

The spontaneous polarization in ferroelectrics vanishes at some critical temperature called the *ferroelectric Curie temperature* Below the Curie temperature, $D$ and $P$ are not linear functions of $\varepsilon$, hence in describing $\varepsilon_r$, it is more convenient to consider the initial relative

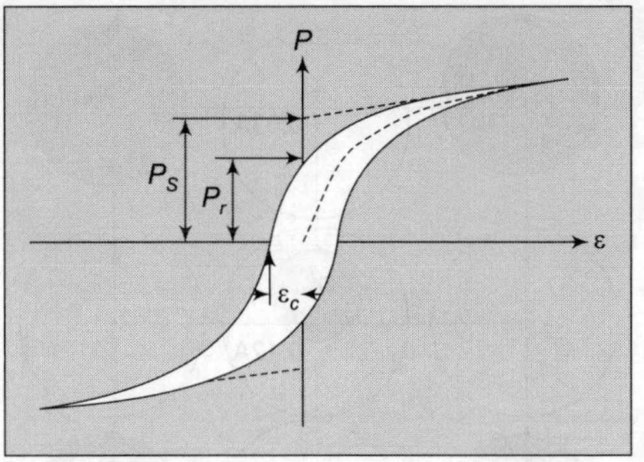


**Figure 17.17** A ferroeleclectric hysteresis curve, with coercive field, $\varepsilon_c$, the saturation, polarization $P_s$, and the remanent polarization, $P_r$ indicated

permittivity. Above the Curie temperature, the relative permittivity varies according to the Curie–Weiss law. Above the Curie temperature barium titanate ($BaTiO_3$) has the Perovskite structure, which is shown in Fig. 17.18, in two different ways. Figure 17.18 shows the $Ti^{4+}$ ion at the centre of a cubic unit cell, with oxygen ions at the faces and Ba ions at the corners. Figure 17.18 puts the Ba ion at the cube centre, and Ti ions, octahedrally coordinated by oxygen ions, at the corners. *When a field is applied below the Curie temperature, all the anions move in one direction, and the cations in the other, destroying the cubic symmetry, and leaving the unit cell with a net dipole moment*. When the field is reduced to zero the ferroelectric is left with a remnant polarization. This can only be removed by applying a negative coercive field, $\varepsilon_c$.

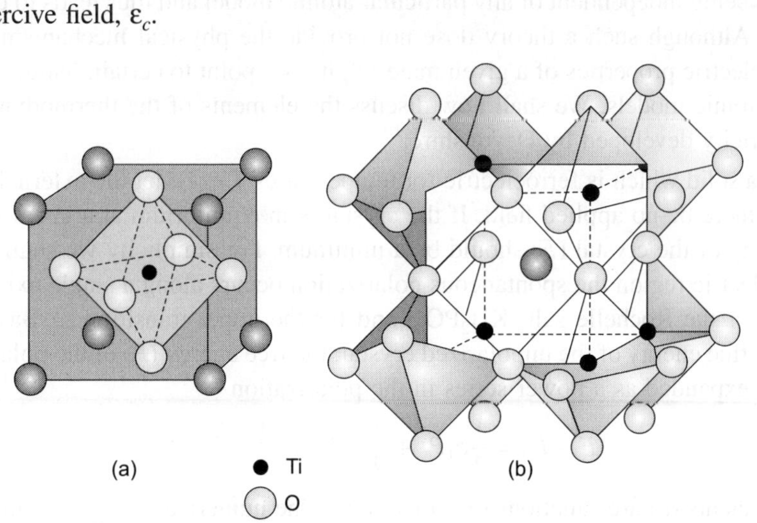


Ion positions in ideal Perovskite structure, i.e., $BaTiO_3$ above the Curie temperature

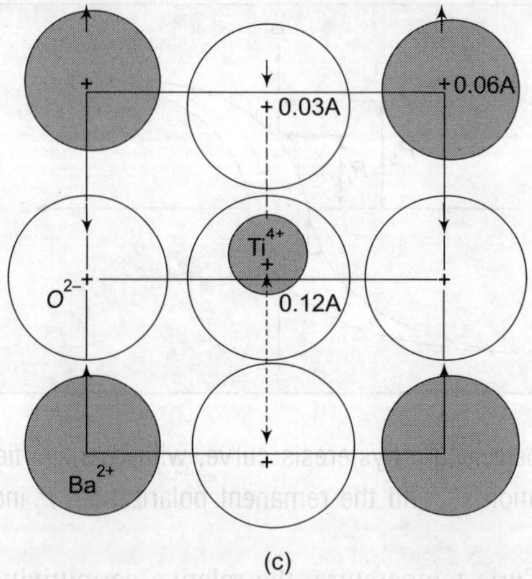


(c)

Top view of unit cell (a) showing the shifting of the ions below the Curie temperature

**Figure 17.18** Crystal structure and ferroelectricity in barium titanate

## 17.21 THERMODYNAMICS OF FERROELECTRIC TRANSITIONS

It is of interest to investigate the behaviour of a ferroelectric in the vicinity of its transition temperature ($T_c$) on the basis of thermodynamic arguments. A thermodynamic theory has the advantage of being independent of any particular atomic model and thus leads to quite general conclusions. Although such a theory dose not provide the physical mechanism responsible for the ferroelectric properties of a given material, it does point to certain features one should look for in atomic models. We shall now discuss the elements of the thermodynamic theory of ferroelectricity developed by Devonshire.[1]

Consider a solid which is ferroelectric for temperatures $T < T_c$; let the external pressure be zero and let there be no applied field. If the crystal is in equilibrium at a given temperature, the free energy of the crystal ($F$) should be a minimum. For simplicity we shall assume that in the ferroelectric region the spontaneous polarization occurs along a single axis; this would be the case for the Rochelle salt, $KH_2PO_4$; and for the upper transition of $BaTiO_3$. Let $F_0$ represent the free energy of the unpolarized crystal; the free energy ($F$) of the polarized crystal may then be expanded as a power series in the polarization

$$F - F_0 = \frac{1}{2}c_1P^2 + \frac{1}{4}c_2P^4 + \frac{1}{6}c_3P^6 + \ldots \qquad \ldots(1)$$

The coefficients ($c$) are functions of temperature; the numerical factors are introduced for later convienence. Note that since we want the free energy to be the same for "positive" and

[1] For a review of this work see A. F. Devonshire. "Theory of Ferroelectrics." *Advances in Physics* (quarterly suppl. of *Phil. Mag.*), **3**, April 1954, p. 85.

"negative" polarization along the polar axis, only even powers of $P$ are included. In thermal equilibrium $(\partial F/\partial P)_T = 0$ so that spontaneous polarization satisfies the equation

$$0 = c_1 P_s + c_2 P_s^3 + c_3 P_s^5 + \ldots \qquad \ldots(2)$$

It is observed that $P_s = 0$ is always a root of this equation and that this will correspond to a minimum of the free energy if $c_1$ is positive. If $c_1$, $c_2$ and $c_3$ are all positive, the root $P_s = 0$ will correspond to the only minimum of the free energy and thus spontaneous polarization would not occur. However, if as a result of the temperature dependence the coefficient ($c_1$) would become negative, $F$ would have a maximum for $P_s = 0$ and there would be at least one nonvanishing value for $P_s$ for which $F$ would be a minimum, i.e., spontaneous polarization would occur.

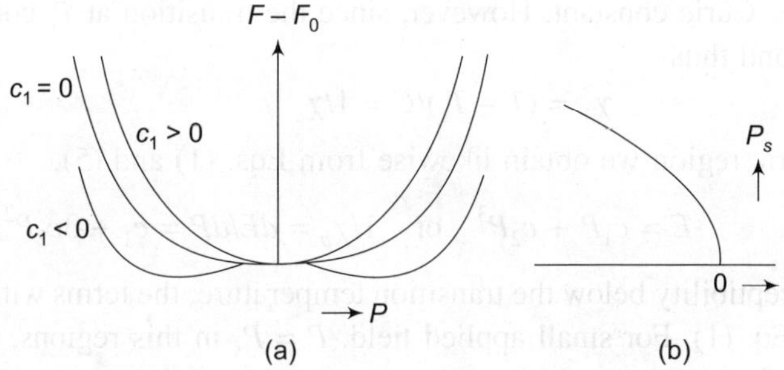


**Figure 17.19** Second-order transition. In (a) the free energy is given schematically as function of polarization for various values of $c_1$; in (b) the spontaneous polarization as function of $c_1$. The critical temperature corresponds to $c_1 = 0$

Consequently, if $c_1$ changes continuously with temperature from a positive to a negative value, the equilibrium of the crystal changes from an unpolarized to a spontaneously polarized state. In order to discuss the properties in the vicinity of the vicinity of the transition temperature, we shall consider two cases of particular interest.

### 17.21.1 Second-order Transition

If the coefficients $c_2$, $c_3$, ... are all positive and the value of $c_1$ varies from positive to negative as the temperature is lowered, one obtains free–energy curves as illustrated in Fig. 17.19(a). The corresponding spontaneous polarization as function of temperature is indicated in Fig. 17.19(b). The transition temperature corresponds to $c_1 = 0$. Assuming in Eq. (2) that the term with $c_3$ is negligible, one obtains for the spontaneous polarization,

$$P_s^2 = -c_1/c_2 \qquad \ldots(3)$$

Note that $P_s$ is a continuous function of temperature; a transitive of this type is not associated with a latent heat but with a discontinuity in the specific heat and is called a *second-order transition*. We shall return to this point below.

Let us now consider the susceptibility of the crystal above and below the transition temperature. For this purpose it is necessary to apply a small electric field to the crystal.

Now, for a crystal under zero pressure in an applied field $E$, we may write according to thermodynamics,

$$dF = -SdT + EdP \quad ...(4)$$

Hence the applied field may be written as $E = (\partial F/\partial P)_T$. Above the transition temperature the polarization will be small for small applied fields, and in this region we may neglect all terms on the right–hand side of Eq. (1) except the first. We thus obtain for $T > T_c$,

$$E = \partial F/\partial P = c_1 P \quad \text{and} \quad 1/\chi_n = dE/dP = c_1 \quad ...(5)$$

where $\chi_a$ is the susceptibility above the critical temperature; the coefficient ($c_1$) is evidently equal to the reciprocal of the susceptibility, $\chi_n$. However, we know that in this temperature range the susceptibility is given by the Curie–Weiss law, $\chi_n = C/(T = \theta)$, so that $c_1 = (T - \theta)/C$, where $C$ is the Curie constant. However, since the transition at $T_c$ corresponds to $C_1 = 0$, we have $\theta = T_c$ and thus

$$\chi_1 = (T - T_c)/C = 1/\chi_a \quad ...(6)$$

In the ferroelectric region we obtain likewise from Eqs. (1) and (5),

$$E = c_1 P + c_2 P^3 \quad \text{or} \quad 1/\chi_b = dE/dP = c_1 + 3c_2 P^2 \quad ...(7)$$

where $\chi_b$ is the susceptibility below the transition temperature; the terms with powers $\geq 6$ have been neglected in Eq. (1). For small applied field, $P \simeq P_s$ in this regions, so that according to Eq. (7) and Eq. (3), we have

$$1/\chi_b = -2c_1 \quad ...(8)$$

If we assume that the temperature dependence of $c_1$ on the ferroelectric side of $T_c$ is still given by Eq. (6) we obtain

$$1/\chi_b = 2(T_c - T)/C \quad ...(9)$$

The temperature dependence of the reciprocal of the susceptibility on both sides of the transition temperature as given by Eq. (6) and Eq. (9) is illustrated in Fig. 17.20(a). Note that the slope in the ferroelectric region is twice that above the transition temperature.

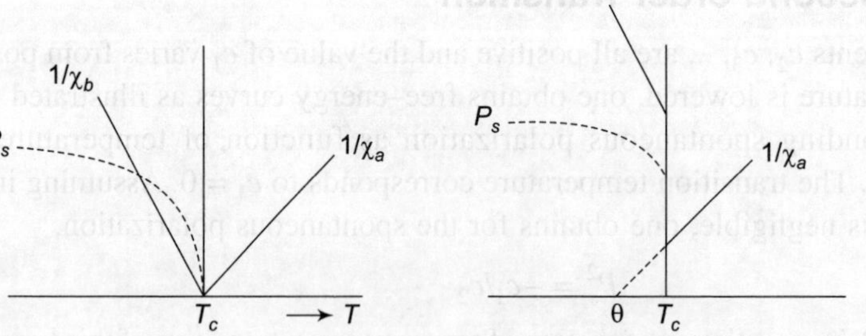


**Figure 17.20** Reciprocal susceptibility near the critical temperature. (a) For a second-order transition; (b) for a first-order transition. The corresponding spontaneous polarization are indicated by the dashed curves; in (a) $P_s$ is continuous, in (d) discontinuous at $T_c$

In connection with the remark made above that the transition under discussion is of the second order, let us consider the entropy associated with the spontaneous polarization. According to Eq. (4) and Eq. (1), the entropy is given by

$$S = -(\partial F/\partial T)_P = S_0 - \frac{1}{2}P^2(\partial c_1/\partial T) - \frac{1}{4}P^4(\partial c_2/\partial T) + \cdots$$

where $S_0$ is the entropy of the unpolarized crystal. To the first approximation we may then write

$$S - S_0 \simeq -\frac{1}{2}P^2(\partial c_1/\partial T) \qquad \text{...(10)}$$

Since $P$ is a continuous function of temperature for the case under consideration and since the slope of $P^2$ has a discontinuity at $T = T_c$, there should be a discontinuity in the specific heat, but no latent heat, i.e., the transition is of the second order. This type of transition is observed in Rochelle salt and in $KH_2PO_4$.

### 17.21.2 First-order Transitions

We have seen that spontaneous polarization requires the coefficient ($c_1$) to be negative. Furthermore, we have seen that if at the same time ($c_2$) is positive, a second order transition results. We shall now consider the case for which $c_2$ is negative and $c_3$ is positive. Under these circumstances it is possible for the free energy curves to have a minimum value for a nonzero value of the polarization to coexist with a minimum for $P_s = 0$. Assuming that $c_1$ varies from positive to negative values as the temperature is lowered, one obtains free energy curves of the type indicated in Fig. 17.21. A transition from the nonpolarized state to a spontaneously polarized state will now occur when the minimum of the free energy corresponding to $P_s = 0$ become equal to the minimum associated with a nonzero value for $P_s$. It will be evident that in this case the polarization jumps at the critical temperature from zero to some non-zero value, i.e., the polarization as function of temperature exhibits

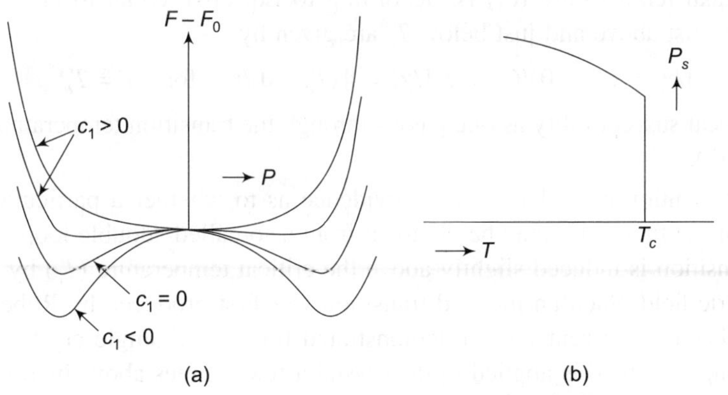


**Figure 17.21** First-order transition. In (a) the free energy is represented as function of $P$ for different value of $c_1$. In (b) the spontaneous polarization is given as function of $T$; note the discontinuity at $T_c$. (After Devonshire, ref. 33)

a discontinuity at $T = T_c$ as shown in Fig. 17.21 (b). According to Eq. (10) the entropy will also be discontinuous at $T = T_c$ and there will be a latent heat, i.e., the transition is of the first order.

In the absence of an external field we obtain from the equilibrium condition $(\partial F/\partial P)_T = 0$ and from Eq. (1) for the nonvanishing value of the spontaneous polarization the equation

$$0 = c_1 + c_2 P_s^2 + c_3 P_s^4 + \ldots \qquad \ldots(11)$$

At the critical temperature ($T_c$) the quantity, $P_s\,(T_c)$ should satisfy Eq. (1) as well as the condition mentioned above that $F(T_c) = F_0(T)$. According to Eq. (1) we thus have also

$$0 = \frac{1}{2}\, c_1 P_s^2(T_c) + \frac{1}{4} c_2 P_s^4(T_c) + \frac{1}{6} c_3 P_s^6(T_c) + \ldots \qquad \ldots(12)$$

From this equation and Eq. (11) as applied to the critical temperature we then find the relations

$$P_s^2(T_c) = -\frac{3}{4}\,(c_2/c_3);\;\; c_1 = \frac{3}{16}\,(c_2^2/c_3);\; P_s^4(T_c) = 3\,c_1/c_3 \qquad \ldots(13)$$

The first of these results shows that the polarization is discontinuous at the critical temperature (Fig. 17.21).

We shall now consider the susceptibility on both sides of the critical temperature. As in case (i), the coefficient ($c_1$) in the region above the temperature ($T_c$) is again equal to $1/\chi_a$. In this region the susceptibility follows the Curie–Weiss law, so that

$$\chi_0 = C/(T - \theta) \text{ and } c_1 = (T - \theta)/C \qquad \ldots(14)$$

where $\theta$ is somewhat smaller then $T_c$. We leave it to the reader to show that by similar arguments as used under (i) and by making use of the relation (13) we find for the susceptibility below the critical temperature,

$$1/\chi_b = 4\,c_1 \qquad \ldots(15)$$

At the critical temperature ($c_1$) is, according to Eq. (14), equal to $(T_c - \theta)/C$ and the susceptibilities just above and just below $T_c$ are given by

$$1/\chi_a = (T_c - \theta)/C \text{ and } 1/\chi_b = 4\,(T_c - \theta)/C \quad \text{for} \quad T = T_c \qquad \ldots(16)$$

The reciprocal susceptibility as one passes through the transition temperature is illustrated in Fig. 17.20 (b).

We should mention here that decisive evidence as to whether a particular ferroelectric transition is of the first order may be obtained from a so-called "double loop" experiment in which the transition is induced slightly above the critical temperature ($T_c$) by application of a strong electric field. Such an induced transition was first produced by Roberts in ceramic material and has more recently been demonstrated for a good single crystal of $BaTiO_3$ by Merz.[2] A strong $a-c$ field is applied to the crystal a few degrees above its normal transition temperature. At zero applied field the crystal is nonferroelectric but at a critical value of the applied field the polarization increases rapidly and upon reversal of the field hysteresis

[2] S. Roberts, *Phys, Rev.*, **85**, 925 (1952); W. J. Merz, *Phys. Rev.*, **91**, 513 (1953).

is observed. The hysteresis loop is not complete, however, and for low applied fields the behaviour is normal again (see Fig. 17.22). A double hysteresis loop obtained in this manner can only occur if the transition is of the first order, as may be understood in the following manner. In the absence of the free energy for $P_s = 0$ is equal to the minimum associated with nonvanishing value of the spontaneous polarization. For a crystal subjected to a field $E$, however, the induced transition occurs when $F - EP$ rather then $F$ has the same value as the minimum at the origin. Such induced transition can evidently occur only if the free energy curves are of the type illustrated in Fig. 17.21 and not if they are of the type corresponding to Fig. 17.19. Hence, the double loop experiment distinguishes between first- and second-order transitions. Since a double loop has been observed for $BaTiO_3$ the upper transition of this material is evidently of the first order. We should note that it is usually not possible to obtain a clear-cut distinction between a first- or second-order transition from measurement of the spontaneous polarization as function of temperature, since $P_s$ rises rapidly just below $T_C$ even for a second-order transition. For further details on the thermodynamic theory of ferroelectricity and for a treatment of antiferroelectric transitions we refer the reader to A.F. Devonshire, *op. cit.*

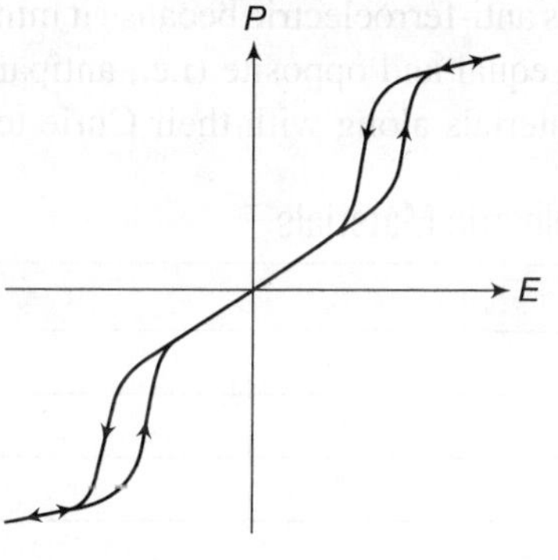


**Figure 17.22** Schematic representation of a double hysteresis loop, of the type observed for $BaTiO_3$, slightly above the transition temperature

## 17.22 APPLICATIONS OF FERROELECTRIC MATERIALS

The most important ferroelectric materials are Rochelle salt (tetrahydrate of potassium tartrate), $BaTiO_3$, $SrTiO_3$, $KH_2PO_4$, $LiNbO_3$ and $KNbO_3$.

Ferroelectric materials are of particular importance when a high dielectric constant is required, e.g., when a ferroelectric ceramic (lead zirconate titanate) is used as a dielectric for capacitor, it is possible to make it smaller by an order of magnitude than when paper or mica is used. Further, the high value of the electromechanical coupling constant of ferroelectrics makes them highly desirable for piezoelectric transducers (with the expectation of frequency-determining devices that are made of quartz). All ferroelectrics are piezoelectric, but not all piezoelectrics are ferroelectrics (quartz is piezoelectric but not ferroelectric). Ferroelectric $BaTiO_3$ is piezoelectric.

Ferroelectrics are also used as electrooptic materials whose refractive indices are controlled by an electric field, e.g., lithium insolates ($LiNbO_3$).

When a piezoelectric crystal is heated, the interatomic distance is increased in an asymmetric way and causes changes in the polarisation causing a potential difference within the crystal. This is called the *piezoelectric effect* and can be produced by heating and cooling the crystal.

## 17.23 ANTIFERROELECTRICITY

We have seen that a ferroelectric material analogous to a ferromagnets, induces ordered dipoles below its transition temperature in order to minimize its energy. However (analogous to antiferromagnets), there is a group of antiferroelectric solids that also have induce ordered dipoles below a characteristic temperature but do not demonstrate spontaneous bulk polarisation because each dipole is antiparallel to its neighbouring dipoles.

Thus, sodium mobate ($NaNbO_3$) and lead zirconate ($PbZrO_3$) (both having perovskite structure) are antiferroelectric, whereas $BaTiO_3$ and $KNbO_3$ are ferroelectric. Similarly, we find potassium dihydrogen phosphate (KDP) is ferroelectric, but the isomorphism ammonium dihydrogen phosphate (ADP) is anti-ferroelectric because it minimizes its energy by having two interpenetrating sublattices of equal and opposite (i.e., antiparallel) polarizations. Table 17.1 lists some antiferroelectric materials along with their Curie temperatures.

**Table 17.1** Some Antiferroelectric Materials

| *Material* | $T_c(K)$ |
|---|---|
| $WO_3$ | 1010 |
| $PbZrO_3$ | 506 |
| $PbHfO_3$ | 488 |
| $NH_4H_2PO_4$ | 148 |
| $NH_4H_2AsO_4$ | 216 |

Antiferroelectrics do not exhibit dielectric hysteresis. However, they may undergo a transition to ferroelectric state under the influence of a strong external electric field or upon cooling to low enough temperatures.

## 17.24 PIEZOELECTRICITY

Some crystalline solids become electrically polarized when they are elastically strained. As a converse phenomenon, these same solids became strained when an external electric field is applied in certain directions. This phenomenon is called *piezoelectricity*. This is a property which is only possible for ionic solids that crystallize in structures lacking a centre of inversion. This phenomenon is of great technological importance and was discovered in 1881 by P. Curie and J. Curie. Of the 32-point groups that can be envisaged in three dimensions, 21 lack a centre-of-inversion symmetry. One of these have so high a degree of symmetry in other respects that is does not show piezoelectricity. Thus, remaining 20 types of point group symmetry are compatible with piezoelectric behaviour. However, only a relatively minor fraction of the solids utilizing these 20-point groups show a piezoelectricity effect large enough to be measurable. A few of them are listed in Table 17.2.

**Table 17.2** Some Piezoelectricity Solids[3]

| *Material* | *Chemical formula* | *Relative value of piezoelectric constant* |
|---|---|---|
| Rochelle salt | $NaKC_4H_4O_6 \cdot 4H_2O$ | Very large |
| Lead zirconium titanate (PZT) | $Pb_xZr_{1-x}TiO_3$ | Large |
| Ammonium dihydrogen Phosphate (ADP) | $NH_4H_2O_4$ | Large |
| Potassium dihydrogen Phosphate (KDP) | $KH_2PO_4$ | Moderate |
| Tourmaline | $(FeCrLiNaK)_4\,Mg_{12}B_6Al_{16}H_8Si_{12}O_{63}$ | Small |
| α-Quartz | $SiO_2$ | Small |

It is to be noted here that both ferroelectricity and piezoelectric occur in non-centro symmetrical crystals, which are 20 in number. Piezoelectricity occurs in all of these 20 but ferroelectricity occurs only in 10, viz., those providing a favourable axis of polarity. Thus, all ferroelectrics are piezoelectric, but all piezoelectrics are not ferroelectric. For example, tourmaline and quartz are piezoelectric, but not ferroelectric.

Because of the lack of the centre of inversion (and complex structure) electrical behaviour of piezoelectric (under in applied field) is not isotropic in nature. However, a simple picture of the phenomena can be presented in a schematic one-dimensional notation by the following equations

$$P - Zd + E\chi; \quad e = Zs + Ed \qquad ...(1)$$

These are the known as piezoelectric equations. Here $P$ is the polarisation, $Z$ the stress $d$ the piezoelectricity strain constant, $E$ the electric field, $\chi$ the dielectric susceptibility, $e$ the elastic strain and $s$ the elastic compliance constant. The value of $d$ for quartz is $d = 10^{-7}$ cm/stat volt, and for $BaTiO_3$ it is $d = 10^{-5}$ cm/stat volt.

In real crystals, the tensile, compressional or shear strains produced by an external electric field may be anisotropic (depending on the crystal orientation). In view of this fact, the piezoelectric strain constants are defined as

$$d_{ik} = \left(\frac{\partial e_k}{\partial E_i}\right)_z \qquad ...(2)$$

where $i \equiv x, y, z$ and $k \equiv xx, yy, zz, yz, zx, xy$.

These $d_{ik}$ form a third rank tensor.

3 After *Solid State Physics*, J.S. Blakemore, Cambridge University Press (1985).

Depending on the application and the desired behaviour, a crystal is cut so as to have the parallel faces of the crystal in a specific orientation, e.g., in quartz, mechanical stress directed along its optic axis (i.e., perpendicular to the hexagonal faces) does not produce piezoelectricity and maximum effect is produced when the crystal is X-cut, i.e., has faces perpendicular to the sides of the hexagonal faces.

### 17.24.1 Applications of Piezoelectricity

(i) The piezoelectric constants of quartz are quite small, but the great mechanical and thermal strengths of this solid have made it pre-eminent for highly stable piezoelectric oscillators. These devices control the frequencies of radio transmitters and quartz clocks, particularly quartz wristwatches.

(ii) The highest possible efficiency is a very desirable attribute. In view of this, Rcohelle salt has been used as a sensitive transducer in phonographs. A shaped ceramic block of BaTiO or lead zirconium titanate (PZT) should be preferred, since these are quite efficient and additionally, and are resistant to heat and moisture.

(iii) Piezoelectric oscillators are used to convert mechanical pulses into electrical ones and *vice versa*. The crystal in these devices work as a transducer. The acoustic pulses are used in underwater search (*sonars*) and other applications,

(iv) The piezoelectric effect in synthetic polyvinylidenfluoride ($PVF_2$) is about five times stronger than that in quartz. The $PVF_2$ films are recently used in applications such as monitoring blood pressure and respiration.

(v) The piezoelectric materials are used as delay lines in communication devices.

## 17.25 APPLICATIONS TO PLASMA

### 17.25.1 What is a Plasma?

The term plasma first appeared in Physics in the 1902. *Plasma* is a conducting gas, i.e., a gas which contains a noticeable proportion of charged particles (electrons and ions). Plasma, as a whole is electrically neutral.

A metal may be described as an assembly of positive ion cores (atomic nuclei) and conduction electrons that are almost free to move over the hole of the crystal. The condition of charge neutrality is maintained because of the balance between the negative charge on conduction electrons and an equal concentration of positive charge on ions. Thus, a metal is a good example of plasma.

### 17.25.2 Metal Plasma

Consider a metal in which a negative charge is introduced at a point taken as origin. Let the volume charge density at the point be ρ. If the charge imbalance because of a momentary fluctuation in charge density lowers the electron concentration at the point by $\delta N$, then the resulting charge density at that point would be $(\rho - e\delta N)$. From Poisson equation

$$\nabla^2 V = \frac{1}{\varepsilon_0}(\rho - e\delta N) \quad ...(1)$$

Assuming the excitation energy eV to be << $kT$ and the electron concentration $N$ to be too low to permit the use of classical statistics, we have

$$\delta N = N_0\left[\exp\left(\frac{eV}{kT}-1\right)\right] \simeq N_0\left(\frac{eV}{kT}\right) \quad ...(2)$$

where $N_0$ is the equilibrium electron concentration in the absence of fluctuation in equilibrium). Substituting the value of $N$ from Eq. (1) in Eq. (2), we have

$$\left(\nabla^2 + \frac{N_0 e^2}{\varepsilon_0 kT}\right)V = \frac{\rho}{\varepsilon_0} \quad ...(3)$$

It has the solution

$$V = \frac{\rho}{r}\, e^{-r/\lambda_D} \quad ...(4)$$

where

$$\lambda_D = \left(\frac{\varepsilon_0 kT}{N_0\, \varepsilon^2}\right)^{1/2} \quad ...(5)$$

$\lambda_D$ is known as the Debye length. Relation (4) expresses the fact that a given electron is screened from other electrons because of a fluctuation of the electron charge density. Screening is such that when other electrons are at a distance $\lambda_D$ from the given electron, the normal Coulomb potential is lowered by a factor $1/e$.

### 17.25.3 Plasma Oscillations and Plasmons

A plasma oscillation in a metal is a collective longitudinal excitation of the conduction electron gas. A *plasmon* is a quantum of a plasma oscillation (or plasma waves). While analysing plasma waves, we assume the ions to be at rest and their charge to be uniformly distributed over the 'gas' volume.

We now follow a classical approach to calculate the dielectric constant of a plasma in the absence of an external perturbation. We consider the two-dimensional case as shown in Fig. 17.23.

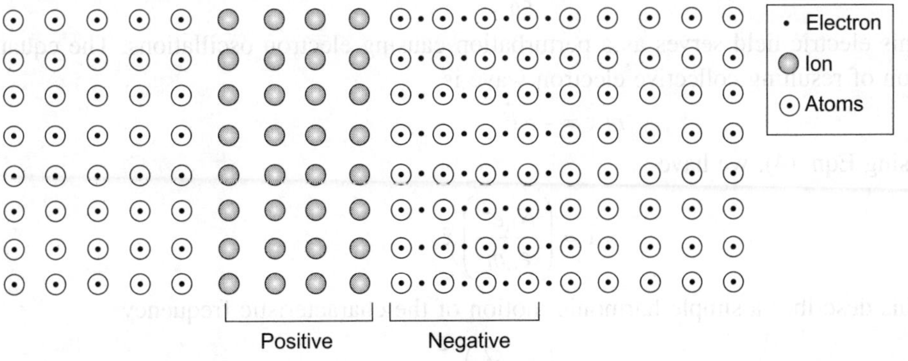


**Figure 17.23** Displacement of electrons in a two-dimensional plasma

On account of a random fluctuation in the equilibrium positions of electrons, two nearby regions of positive and negative charge densities are created, which means that in these two adjoining regions, the charge neutrality condition is destroyed. We now try to calculate the frequency of a given plasma electron.

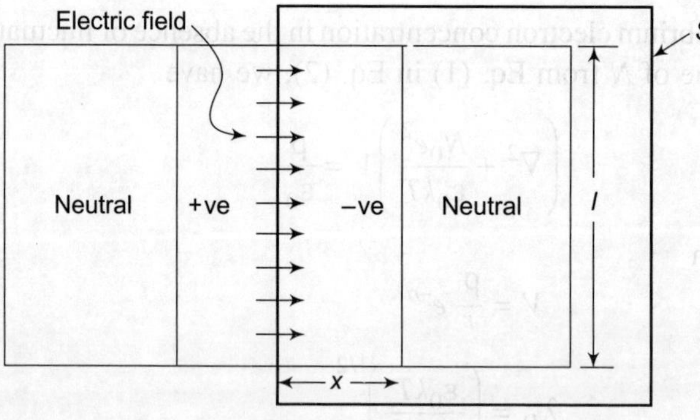


**Figure 17.24** Electric field in two-dimensional plasma. The heavy line shows the boundary of surface over which $\int \vec{E}\cdot d\vec{a}$ is evaluated

Let a small volume element enclosed by surface $S$ in the negatively charged region have a charge. Then, from Gauss law, the electric field in this small volume satisfies

$$\int_s \vec{E}\cdot d\vec{a} = \frac{q}{\varepsilon_0} \qquad ...(6)$$

From Fig. 17.24 for two-dimensional case

$$\int_s \vec{E}\cdot d\vec{a} = -lE \qquad ...(7)$$

Further, $$q = -lxN_0e \qquad ...(8)$$

Thus, $$\vec{E} = \frac{N_0 e\vec{x}}{\varepsilon_0} \qquad ...(9)$$

This electric field serves as a perturbation causing electron oscillations. The equation of motion of resulting collective electron wave is

$$m\ddot{\vec{x}} = -e\vec{E} \qquad ...(10)$$

Using Eqn. (4), we have

$$\ddot{\vec{x}} = \left(\frac{N_0e^2}{\varepsilon_0 m}\right)\vec{x} \qquad ...(11)$$

This describes a simple harmonic motion of the characteristic frequency

$$\omega_p = \left(\frac{N_0e^2}{\varepsilon_0 m}\right)^{1/2} \qquad ...(12)$$

This, is known as the *plasma frequency*.

In the presence of an external field, both $\vec{E}$ and $\vec{x}$ exhibit oscillatory character of the form $\sim\exp(-i\omega t)$, where ω is the angular frequency of the external perturbing field. Assuming the positive ion to be at rest, under the influence of $\vec{E}$, an electron acquires a dipole moment, and bulk polarisation of the plasma is given by

$$P(\omega) = -N_0 ex \qquad ...(13)$$

$$m\ddot{x} = -m\omega^2 x = -eE(\omega) \qquad ...(14)$$

But
$$\varepsilon(\omega) = \frac{D(\omega)}{\varepsilon_0 E(\omega)} = 1 + \frac{P(\omega)}{\varepsilon_0 E(\omega)} \qquad ...(15)$$

Using Eqs. (13) and (14), we get

$$\varepsilon(\omega) = 1 - \frac{N_0 e^2}{\varepsilon_0 m\omega^2} \qquad ...(16)$$

or
$$\varepsilon(\omega) = 1 - \omega_p^2/\omega^2 \qquad ...(17)$$

This expresses the dielectric constant of the free electron gas, when the frequency (ω) matches $\omega_p$, the dielectric constant $\varepsilon(\omega) = 0$. This condition refers to the longitudinal plasma oscillations and the energy quanta ($\hbar\omega_p$) are the *plasmons*. During such oscillations, the electron gas is moved as a whole relative to the positive ion background.

### 17.25.4 Experimental Setup for Plasmon Excitation

We may excite a plasmon by passing an electron through a thin metallic film (Fig. 17.25) or by reflecting an electron and a photon from a film. The charge of the electron couples with the electrostatic field fluctuations of the plasma oscillations (Fig. 17.26). The reflected or transmitted electron will show an energy loss equal to integral of the plasmon energy.

Table 17.3 shows a comparison of observed and calculated values of plasmon energies.

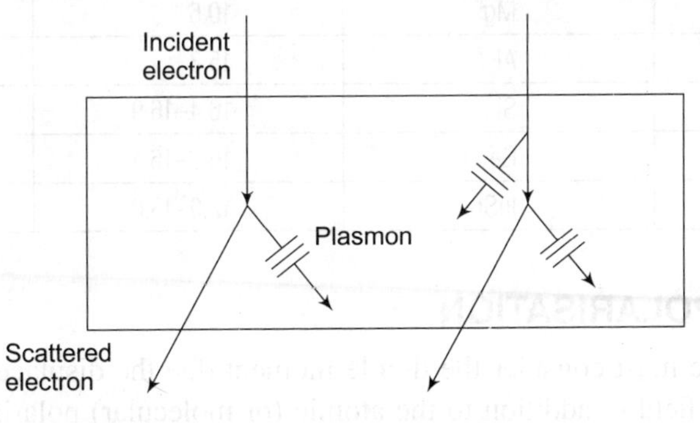


**Figure 17.25** Creation of the plasmon in a metal film by inelastic scattering of an electron. The incident electron typically has an energy 1-10 keV (plasmon energy ~10 eV). An event is also shown in which two plasmons are created

Note that it is equally possible to excite collective plasma oscillations in dielectric films. The calculated plasma energies of Si, Ge and InSb are based on four valence electrons per atom. The plasma oscillation in a dielectric is physically the same as in metals: the entire valence electron sea oscillates back and forth with respect to the ion cores.

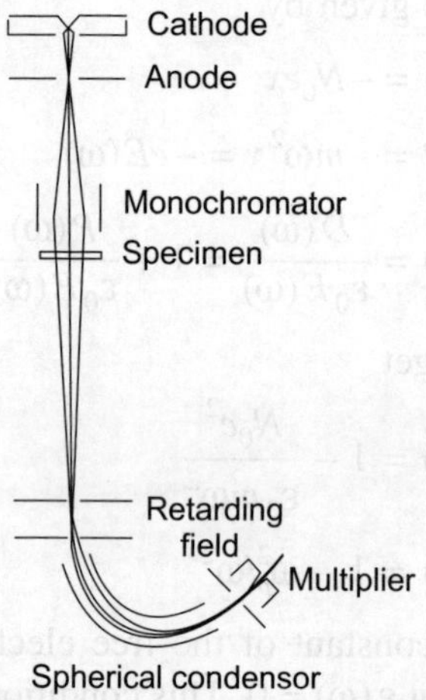


**Figure 17.26** A spectrometer with electrostatic analyzer for the study plasmon excitation by electrons

**Table 17.3** Volume Plasmon Energies

| | *Material* | *Observed (eV)* | *Calculated $\hbar\omega_p$ (eV)* |
|---|---|---|---|
| Metals | Li | 7.12 | 8.02 |
| | Na | 5.71 | 5.95 |
| | K | 3.72 | 4.29 |
| | Mg | 10.6 | 10.9 |
| | Al | 15.3 | 15.8 |
| Dielectrics | Si | 16.4–16.9 | 16.0 |
| | Ge | 16.0–16.4 | 16.0 |
| | InSb | 12.0–13.0 | 12.0 |

## 17.26 IONIC POLARISATION

In ionic crystals, we must consider the dipole moment due the displacement of the charged ions by the electric field in addition to the atomic (or molecular) polarisation.

The displacement polarisation is given by

$$\alpha_{dis} = \frac{e^2}{M(\overline{\omega}^2 - \overline{\omega}^2)} \quad ...(1)$$

where $M$ is the mass of the ion and the resonant frequency ($\overline{\omega}$) is the characteristic of the lattice vibrational frequencies and $\hbar\overline{\omega} \approx \hbar\omega_D$.

Equation (1) does not take into account the atomic (or molecular) polarisability. The most naive way to incorporate this is to simply add the two types of contribution as

$$\alpha = \alpha_A + \alpha_{\text{dis}} \qquad ...(2)$$

In conjunction with the Clausius–Mossotti relation, the approximation expressed by Eq. (2) leads to a dielectric constant, $\varepsilon(\omega)$, for an ionic crystal given by

$$\frac{\varepsilon(\omega) - 1}{\varepsilon(\omega) + 2} = \frac{1}{3\varepsilon_0}\left(\alpha_A + \frac{N_i e^2}{M(\overline{\omega}^2 - \omega^2)}\right) \qquad ...(3)$$

where $\alpha_A$ is the atomic polarisability and $N_i$ the density of pairs. We now attempt to derive a general relation for the dielectric constant by considering two limiting cases of frequency.

(i) When $\omega \to 0$ (or $\omega \ll \overline{\omega}$) the static dielectric constant is given by

$$\left(\frac{\varepsilon_s - 1}{\varepsilon_s + 2}\right) = \frac{1}{3\varepsilon_0}\left(\alpha_A + \frac{N_i e^2}{M\overline{\omega}^2}\right) \qquad ...(4)$$

(ii) When $\omega \gg \overline{\omega}$ (i.e., $\omega$ approaching optical frequencies), we have

$$\frac{\varepsilon_\infty - 1}{\varepsilon_\infty + 1} = \frac{\alpha_A}{3\varepsilon_0} \quad (\overline{\omega} \ll \overline{\omega} \ll \text{atomic frequency}) \qquad ...(5)$$

Now, it is convenient to write $\varepsilon(\omega)$ in terms of $\varepsilon_0$ and $\varepsilon_\infty$ (i.e., static dielectric constant of crystal and the dielectric constant at optical frequencies, respectively). From Eqs. (4) and (5)

$$\frac{N_i e^2}{M\overline{\omega}^2} = 3\varepsilon_0\left[\left(\frac{\varepsilon_s - 1}{\varepsilon_s + 2}\right) - \frac{\varepsilon_\infty - 1}{\varepsilon_\infty + 2}\right] \qquad ...(6)$$

Using Eqs. (5) and (6), in Eq. (3) becomes

$$\frac{\varepsilon(\omega) - 1}{\varepsilon(\omega) + 2} = \frac{\varepsilon_\infty - 1}{\varepsilon_\infty + 2} + \frac{1}{3\varepsilon_0}\left[\frac{N_i e^2}{M\overline{\omega}^2\left(-\frac{\omega^2}{\overline{\omega}^2} + 1\right)}\right]$$

which, in solving for $\varepsilon(\omega)$ gives

$$\varepsilon(\omega) = \varepsilon_\infty + \frac{(\varepsilon_\infty - \varepsilon_s)}{(\omega^2/\omega_T^2) - 1} \qquad ...(7)$$

where

$$\omega_T^2 = \overline{\omega}^2\left[\frac{\varepsilon_\infty + 2}{\varepsilon_s + 2}\right] = \overline{\omega}^2\left[1 - \frac{\varepsilon_s - \varepsilon_\infty}{\varepsilon_s + 2}\right] \qquad ...(8)$$

Here $\omega_T$ is the long-wave limiting frequency of transverse optical modes of the crystal.

### 17.26.1 Application to Long-Wavelength Optical Modes of Ionic Crystals

In a long-wavelength ($\vec{k} \approx 0$) optical mode, the oppositely charged ions in each primitive cell undergo opposite displacements, giving rise to a non-vanishing polarization $P$ such that

$$\vec{D} = \varepsilon \vec{E} = \varepsilon_0 \vec{E} + \vec{P} \quad ...(9)$$

In the absence of free charge,

$$\nabla \cdot \vec{D} = 0 \quad ...(10)$$

Further, the microscopic field, $E_{\text{micro}}$, is given by

$$E_{\text{micro}} = -\nabla \phi_{\text{micro}} \quad ...(11)$$

where $\phi_{\text{micro}}$ is the corresponding potential. Ignoring the effect of the $-\partial B/\partial t$ term (of Maxwell's equation), we have

$$\nabla \times E_{\text{micro}} = \nabla \times (-\nabla \phi_{\text{micro}}) = 0 \quad ...(12)$$

In cubic ionic crystal $\vec{D}$ is parallel to $\vec{E}$ (i.e., is not a tensor), and therefore both $\vec{D}$ and $\vec{E}$ are parallel to $\vec{P}$. The spatial variations of these three vectors is given by

$$\left.\begin{matrix} \vec{D} \\ \vec{E} \\ \vec{P} \end{matrix}\right\} = \text{Re}\left\{\begin{matrix} \vec{D} \\ \vec{E} \\ \vec{P} \end{matrix}\right\} \exp(i\vec{k}\cdot\vec{r}) \quad ...(13)$$

Then $\nabla \cdot \vec{D} = 0 \;\Rightarrow\; \vec{k} \cdot \vec{D}_0 = 0$, which requires that

$$\vec{D} = 0 \quad \text{or} \quad \vec{D}, \vec{E} \quad \text{and} \quad \vec{P} \perp \vec{k} \quad ...(14)$$

while $\nabla \times \vec{E} = 0 \;\Rightarrow\; \vec{k} \times \vec{E}_0 = 0$, which requires that

$$\vec{E} = 0 \quad \text{or} \quad \vec{E}, \vec{D} \quad \text{and} \quad \vec{P} \parallel \vec{k}. \quad ...(15)$$

In a longitudinal optical mode $\vec{P}$ is parallel to $\vec{k}$ and Eq. (14) therefore requires and $\vec{D}$ must vanish, i.e.,

$$\vec{E} = -\frac{\vec{P}}{\varepsilon_0} \quad \text{and} \quad \varepsilon = 0 \quad \text{(longitudinal mode)} \quad ...(16i)$$

On the other hand, in a transverse optical mode, $\vec{P}$ is perpendicular to $\vec{k}$, which is consistent with Eq. (15) only if $\vec{E}$ vanishes, i.e.,

$$\vec{E} = 0, \quad \varepsilon = \frac{\vec{D}}{\varepsilon_0 \vec{E}} = \infty \quad \text{(transverse mode)} \quad ...(16ii)$$

According to Eq. (14) $\varepsilon = \infty$ when $\omega^2 = \omega_T^2$. Therefore, the above result identifies $\omega_T$ as the frequency of long-wavelength transverse optical mode.

### 17.26.2 Condition for Transverse Optical Mode

For a transverse optical mode $\vec{P} \perp \vec{k}$ and $\vec{E} = 0$, $\varepsilon = \infty$. ...(17)

This condition has been derived in the electrostatic approximation (neglecting $-\partial\vec{B}/dt$). But actually, for a non-magnetic isotropic medium, the wave equation is

$$\mu_0\varepsilon_0\varepsilon(\omega, \vec{k}) \frac{\partial^2 E}{\partial t^2} = \nabla^2 E \quad ...(18)$$

and for $\vec{E} \sim \exp[i(\vec{k}\cdot\vec{r} - \omega t)]$, we have

$$\varepsilon(\omega, \vec{k})\, \mu_0\varepsilon_0\,\omega^2 = k^2$$

or

$$\varepsilon(\omega, \vec{k}) = \frac{k^2c^2}{\omega^2} \quad ...(19)$$

So, the condition, $\varepsilon = \infty$ correspond to $kc \gg \omega$.

Since the magnitude of the optical phonon wave vector matches with the order of the Debye wave vector, $kc \gg k_D s$ where $s$ is the speed of sound in the crystal

$$\Rightarrow \qquad \frac{k}{k_D} \gg \left(\frac{s}{c}\right) \quad ...(20)$$

Since $s/c = 10^{-4}$ to $10^{-5}$ and $k_D$ compare with the size of the Brillouin zone, therefore the electrostatic approximation is fully justified except for optical modes having wave vectors only a small fraction of a percentage of dimension of the zone from $k = 0$.

## 17.27 THE INTERACTION OF ELECTROMAGNETIC WAVES WITH OPTICAL MODES (POLARIZATION)

We discussed longitudinal optical phonons and transverse optical phonons in chapter 5, but we differed the treatment of interaction of transverse optical phonons with transverse e.m. waves. At resonance (i.e., $\omega = \omega_T$), the phonon-photon coupling entirely changes the character of the propagation and a forbidden band is stabilized that has nothing to do with the periodically of the lattice. The coupling is implicit in Maxwell's equations and expressed by the dielectric function.

$$\varepsilon(\omega) = \varepsilon_\infty + \frac{(\varepsilon_\infty - \varepsilon_s)}{\left(\dfrac{\omega^2}{\omega_T^2}\right) - 1} \quad ...(1)$$

Figure 17.27 shows a plot of $\varepsilon$ versus $\omega$ using the above relation. It is to be noted that $\varepsilon$ is negative between $\omega_T$ and $\omega_L$, making $k$ imaginary. This means that no radiation can propagate over this range of $\omega$, known as reststrahlem band. There occurs a resonance at frequency ($\omega_T$) due to coupling between phonon and photon. The quantum of coupled phonon-photon transverse wave field (in the resonance region, also known as crossover region) is termed *polariton*. (It is a new elementary excitation, whose interaction with other elementary excitation in the solids represents the real absorption process.)

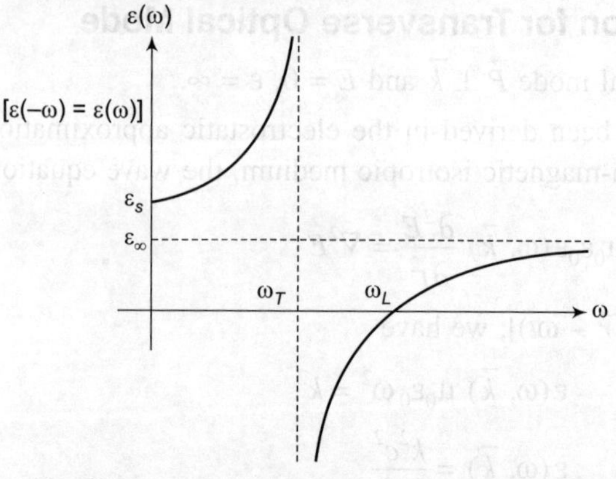


**Figure 17.27** Frequency dependents dielectric constant for a diatomic ionic crystal

Outside the region of crossover, ω is plotted versus *k* in Fig. 17.28.

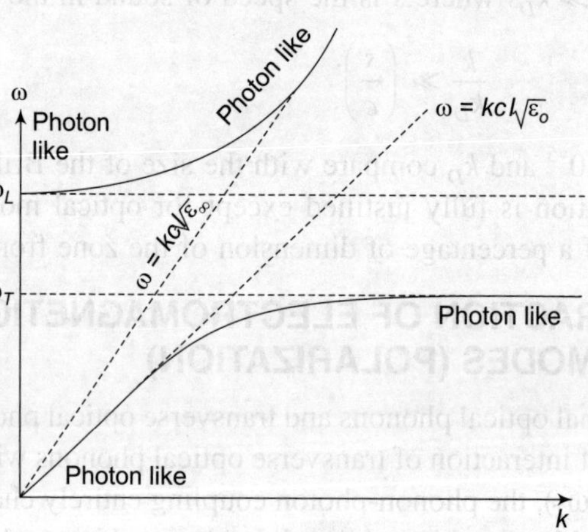


**Figure 17.28** Solution to the dispersion relation ($\omega = kc\sqrt{\varepsilon(\omega)}$) for transverse electromagnetic modes propagating in a diatomic crystal. In the linear regions, one mode is clearly photon like and one clearly optical phonon like. In the curve regions, there exist polariton. (After *Solid State Physics* by N.W. Ashcroft and N.D. Mermin, Saunders College Publishing (1976))

We now derive an expression for polariton frequency

Since $$\varepsilon(\omega)\,\omega^2 = k^2c^2 \qquad \text{...(2i)}$$

∴ $$\varepsilon(\omega)\,\omega^2 = \frac{k^2}{\mu_0\varepsilon_0} \qquad \text{...(2ii)}$$

or $$\left[\frac{\varepsilon_0\vec{E}+\vec{P}}{\varepsilon_0 E}\right]\omega^2 = \frac{k^2}{\mu_0\varepsilon_0}$$

or $$\vec{P}\omega^2 + \left[\varepsilon_0\,\omega^2 - \frac{k^2}{\mu_0}\right]\vec{E} = 0 \qquad ...(3)$$

Neglecting electronic polarisation, we have

$$\vec{P} = \frac{N_i e^2}{M(\overline{\omega}^2 - \omega^2)}\vec{E}$$

or $$M(\overline{\omega}^2 - \omega^2)\,\vec{P} - N_i e^2\,\vec{E} = 0 \qquad ...(4)$$

A solution of Eqs (3) and (4) exists if the determinant of the coefficients of $\vec{E}$ and $\vec{P}$ is zero

$$\begin{vmatrix} \omega^2 & (\varepsilon_0\,\omega^2 - k^2/\mu_0) \\ M(\overline{\omega}^2 - \omega^2) & -N_i e^2 \end{vmatrix} = 0 \qquad ...(5)$$

For $k = 0$, there are two roots, $\omega = 0$ (for the photon)

and $$\omega^2 = \overline{\omega}^2 + \frac{N_i e^2}{\varepsilon_0 M} \qquad ...(6)$$

Using above equation, this expression becomes

$$\omega^2 = \omega_T^2\left(\frac{\varepsilon_s + 2}{\varepsilon_\infty + 2}\right) + \frac{N_i e^2}{M\varepsilon_0} \qquad ...(7)$$

This gives polariton frequency at $k = 0$.

Solutions to the dispersion relation for the transverse optical modes in an ionic diatomic lattice are shown in Fig. 17.28. There are two branches, one lying completely below $\omega_T$ and the order completely above $\omega_L$. In the linear regions of each of the continuous curves, one mode is clearly phonon like. The curved regions are mixed in character and correspond to the region of resonance or the crossover region. These describe the dispersion of polariton.

### 17.27.1 LST Relation

The theory discussed above (for polariton) also provides the relationship between $\omega_L$ and $\omega_T$. For $\omega = \omega_L$, $\varepsilon = 0$. This condition imposed on Eq. (7) gives

$$\frac{\omega_L^2}{\omega_T^2} = \frac{\varepsilon_s}{\varepsilon_\infty} \qquad ...(8)$$

This relation is known as the *Lydiane-Sachs-Teller relation.* Here $\varepsilon_D$ is the static dielectric constant and $\varepsilon_\infty$ is the high frequency limit of the dielectric function (defined to include the core electron contribution). For soft modes with $\omega_T \to 0$, we see that $\varepsilon_s \to \infty$, a characteristic of ferroelectricity.

## 17.28 DEBYE EQUATIONS

Consider a solid having electronic, ionic and dipolar polarizations. The net polarization is

$$P = P_e + P_i + P_d \qquad ...(1)$$

If static electric field is switched on, it takes a definite time for the polarization to attain its final value. We assume that $P_e$ and $P_i$ attain their final values instantaneously while $P_d$ is time dependent, i.e., $P_d = P_d(t)$. Let $P_d(t)$ be represented by equation

$$P_d(t) = P_{ds}(1 - e^{-t/\tau}) \qquad ...(2)$$

where $P_{ds}$ is the saturated value of $P_d$ which is attained after a long time $t$ and $\tau$ is the relaxation time. The rate of change of polarization is

$$\frac{dP_d}{dt} = \frac{1}{\tau} P_{ds} e^{-t/\tau} = \frac{1}{\tau}\{P_{ds} - P_{ds}(1 - e^{-t/\tau})\}$$

$$\equiv \frac{1}{\tau}[P_{ds} - P_d(t)] \qquad ...(3)$$

This equation may also be used for alternating electric field provided $P_{ds}$ is replaced by a time dependent polarization $P_{ds}(t)$; $P_{ds}(t)$ being the value of $P_{ds}$ attained instantaneously when the field is $E(t)$. Let $E(t) = E_0 e^{i\omega t}$. Then, in this case

$$\frac{dP_d}{dt} = \frac{1}{\tau}[P_{ds}(t) - P_d] \qquad ...(4)$$

Now, we define an instantaneous permittivity ($\varepsilon_\infty$) such that

$$P_e + P_i = (\varepsilon_\infty - 1)\varepsilon_0 E \qquad ...(5)$$

$$P_{ds}(t) = P_s - (P_e + P_i)$$

$$\equiv \varepsilon_r \varepsilon_0 E - (\varepsilon_\infty - 1)\varepsilon_0 E - \varepsilon_0 E$$

$$= (\varepsilon_r - \varepsilon_\infty)\varepsilon_0 E \qquad ...(6)$$

$\varepsilon_r$ being the static permittivity corresponding to the static polarization ($P_s$) substituting Eq. (6) in Eq. (4), we get

$$\frac{dP_d}{dt} = \frac{1}{\tau}[(\varepsilon_r - \varepsilon_\infty)\varepsilon_0 E_0 e^{i\omega t} - P_d]$$

$$\frac{dP_d}{dt} + \frac{P_d}{\tau} = \frac{1}{\tau}(\varepsilon_r - \varepsilon_\infty)\varepsilon_0 E_0 e^{i\omega t} \qquad ...(7)$$

This differential equation is of the form

$$\frac{dy}{dx} + P(x)y = Q(x) \qquad ...(8)$$

(with $y = P_d$ and $x = t$)

The solution is

$$y = e^{-\int P(x)dx} \int Q(x) e^{\int P(x)dx}\, dx + Ce^{-\int P(x)dx} \qquad ...(9)$$

Letting $\frac{1}{t} = P(x),\ Q(x) = \left(\frac{\varepsilon_r - \varepsilon_\infty}{\tau}\right)\varepsilon_0 E_0 e^{i\omega t}$, we get

$$P_d(t) = e^{-\int \frac{dt}{\tau}} \int (\varepsilon_r - \varepsilon_\infty)\varepsilon_0 E_0\, e^{i\omega t} e^{\int \frac{dt}{\tau}}\, dt + Ce^{-\int \frac{dt}{\tau}}$$

$$= e^{-\frac{t}{\tau}} \left(\frac{\varepsilon_r - \varepsilon_\infty}{\tau}\right)\varepsilon_0 E_0 \int \exp\left(i\omega + \frac{1}{\tau}\right) t dt + C \exp(-t/\tau)$$

$$= \left(\frac{\varepsilon_r - \varepsilon_\infty}{i\omega\tau + 1}\right)\varepsilon_0 E_0 \exp(i\omega t) + C \exp(-t/\tau) \qquad ...(10)$$

The 2nd term on the RHS represents transient and need not be considered. The total polarization is given by

$$P(t) = P_e + P_i + P_d(t) \qquad ...(11)$$

The electric displacement is

$$D(t) = \varepsilon_0 E + P$$

$$= \varepsilon_0 E + [P_e + P_i + P_d(t)]$$

$$= \varepsilon_0 E + (\varepsilon_\infty - 1)\varepsilon_0 E + \left(\frac{\varepsilon_r - \varepsilon_\infty}{i\omega\tau + 1}\right)\varepsilon_0 E$$

$$\equiv \varepsilon_0 \varepsilon_r^* E \qquad ...(12)$$

Using Eqs. (11) and (12) and neglecting the transient term, we get

$$\varepsilon_r^* \equiv \varepsilon'_r - i\varepsilon'_r = \varepsilon_\infty + \left(\frac{\varepsilon_r - \varepsilon_\infty}{1 + i\omega\tau}\right) \qquad ...(13)$$

Separating the real and imaginary parts,

$$\varepsilon'_r = \varepsilon_\infty + \frac{\varepsilon_r - \varepsilon_\infty}{1 + \omega^2\tau^2} \qquad ...(14)$$

$$\varepsilon''_r = \left(\frac{\varepsilon_r - \varepsilon_\infty}{1 + \omega^2\tau^2}\right)\omega\tau \qquad ...(15)$$

and
$$\tan\delta = \frac{(\varepsilon_r - \varepsilon_\infty)\,\omega\tau}{\varepsilon_r + \varepsilon_\infty \omega^2\tau^2} \qquad ...(16)$$

The above equations for $\varepsilon'_r(\omega)$, $\varepsilon''_r(\omega)$ and $\tan\delta$ are called the *Debye equations.* Figure 17.29 shows plots of $\varepsilon'_r$, $\varepsilon''_r$ and $\tan\delta$ as a function of $\omega\tau$.

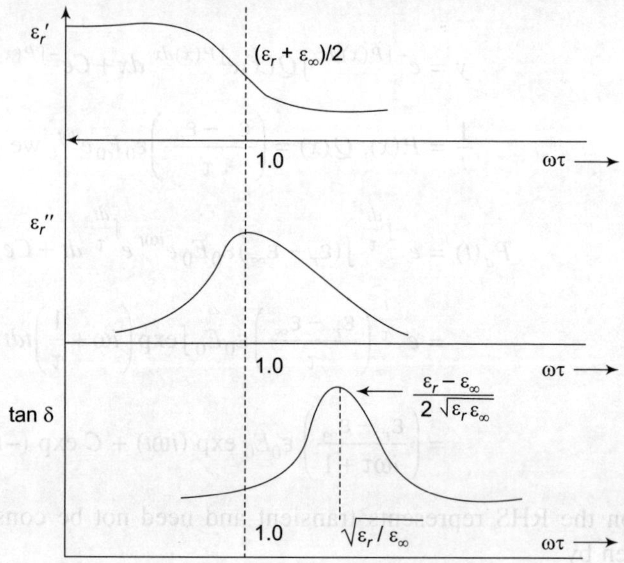


**Figure 17.29** Variations of $\varepsilon'_r$, $\varepsilon''_r$ and tan δ as a function of ωτ

it may be seen that the dielectric loss, which is proportional to $\varepsilon_r$ exhibits a maximum when $\omega\tau = 1$ and the real part ($\varepsilon'_r$) becomes equal to the static value ($\varepsilon_r$) when the frequency is very low. At such a low frequency, there is no dielectric loss and the dipoles contribute fully towards polarization. If however $\omega\tau > 1$, the dipoles are unable to follow the alternating field and then $\varepsilon'_r \to \varepsilon_\infty$, the instantaneous dielectric permittivity.

Thus for $\omega\tau = 1$, $\varepsilon'_r$ has an inflection point whereas $\varepsilon_r$ has a peak value and tan δ has a maximum at $\omega t = \sqrt{\varepsilon_r / \varepsilon_\alpha}$. Quite often, the polarization process in real materials cannot be characterized by a single relaxation time but a spectrum of relaxation times. In that case, the plots of the Debye equation deviate from the ideal. Another way of treating dielectric data is then to plot $\varepsilon_r$ is a function of $\varepsilon'_r$ at different frequencies as was suggested by Cole and Cole. Such a Cole-Cole plot for an ideal Debye dipolar liquid is a semicircle with its centre on the $\varepsilon'_r$ axis and the radius of $(\varepsilon_r - \varepsilon_\infty)/2$ (Fig. 17.30).

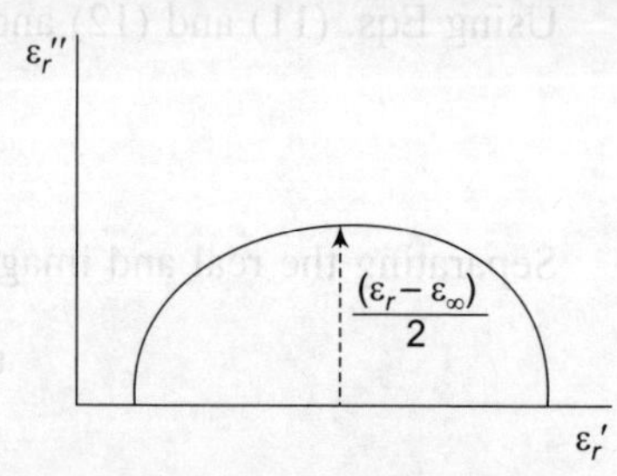


**Figure 17.30** Cole-Cole plot

## 17.29 IONIC SOLIDS

Most of the dielectric solids are ionic crystals, and we consider, for simplicity, those of cubic symmetry, with alkali halide structure, in which the unit cell consists of an anion of mass $M_-$ and a cation of mass $M_+$. Cubic crystals are isotropic with respect to elementary excitations which are spread out over many unit cells, because their wavelength is large compared to the unit cell size ($ka \ll 1$). As a result, the dielectric properties can almost always be described in terms of the macroscopic constants that are defined for continuous media. The origin of the dielectric response of ionic solids, through the mechanism of ionic polarization, is revealed more clearly in the simple case of a linear ionic crystal, with results that are easily generalized to three dimensions.

The unit cell, of length $a$, of an ionic linear chain, contains one cation $M_+$ and one anion $M_-$, with equilibrium positions $pa$ and $\left(p + \frac{1}{2}\right)a$ respectively, as shown in Fig. 17.31.

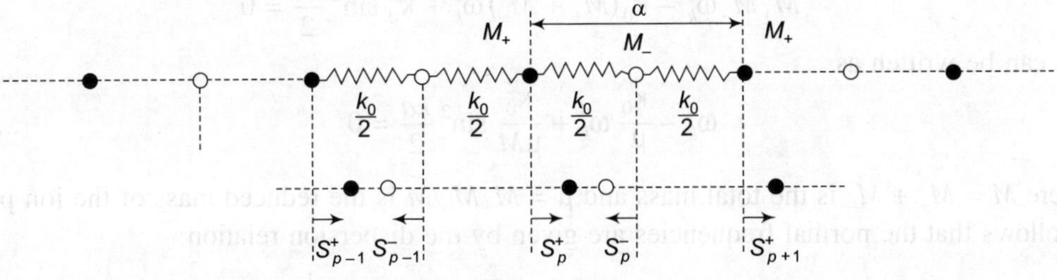


**Figure 17.31** The linear diatomic chain model

In the harmonic approximation, the equation of motion for the $p$th ion, derived for the monatomic lattice, is replaced by two equations of motion associated with the $p$th unit cell:

$$M_+ \frac{d^2 s_p^+}{dt^2} = \frac{\kappa_0}{2}(s_p^- + s_{p-1}^- - 2s_p^+),$$

$$M_- \frac{d^2 s_p^-}{dt^2} = \frac{\kappa_0}{2}(s_p^+ + s_{p+1}^+ - 2s_p^-) \qquad ...(1)$$

It was assumed that only nearest neighbours interact and the coupling constant is the same for all the ion pairs. The displacements of the $M_+$ (or $M_-$) ions, which are all equivalent by translational symmetry, are related to each other in the same way as in the case of the monatomic chain, and hence, we assume solutions of the form:

$$s_p^+ = A_+ e^{i(kpa - \omega_k t)}$$

$$s_p^- = A_- e^{i[k(p + 1/2)a - \omega_k t]} \qquad ...(2)$$

Substitution into Eq. (1) yields two linear, homogenous equations in $A_+$ and $A_-$:

$$\left.\begin{aligned} -\omega_k^2 M_+ A_+ &= \kappa_0\left(A_- \cos\frac{ka}{2} - A_+\right) \\ -\omega_k^2 M_- A_- &= \kappa_0\left(A_+ \cos\frac{ka}{2} - A_-\right) \end{aligned}\right\} \qquad ...(3)$$

The vibrational modes are determined by the characteristic equation:

$$\begin{vmatrix} \kappa_0 - M_+\omega_k^2 & -\kappa_0 \cos\dfrac{ka}{2} \\ -\kappa_0 \cos\dfrac{ka}{2} & \kappa_0 - M_-\omega_k^2 \end{vmatrix} = 0 \qquad ...(4)$$

which is a quadratic equation in $\omega_k^2$:

$$M_+M_-\omega_k^4 - \kappa_0(M_+ + M_-)\omega_k^2 + \kappa_0^2 \sin^2\frac{ka}{2} = 0$$

and can be written as

$$\omega_k^4 - \frac{\kappa_0}{\mu}\omega_k^4 + \frac{\kappa_0^2}{\mu M}\sin^2\frac{ka}{2} = 0 \qquad ...(5)$$

where $M = M_+ + M_-$ is the total mass and $\mu = M_+M_-/M$ is the reduced mass of the ion pair. It follows that the normal frequencies are given by the dispersion relation:

$$\omega_k^2 = \frac{\kappa_0}{2\mu} \pm \frac{\kappa_0}{2\mu}\left[1 - \frac{4\mu}{M}\sin^2\frac{ka}{2}\right]^{\frac{1}{2}} \qquad ...(6)$$

which is represented in Fig. 17.32 in the reduced zone scheme, assuming an almost continuous distribution of the $\omega_k$ values. It is apparent that two normal modes correspond to each $k$ value, and hence there are $2N$ normal modes in all, if $N$ is the number of unit cells.

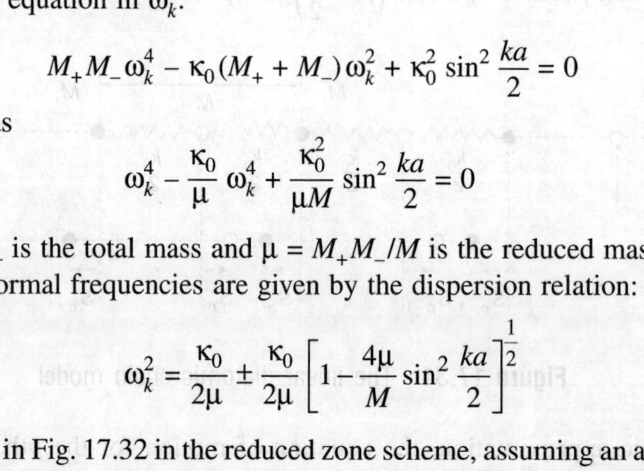


**Figure 17.32** The dispersion relation for a linear ionic crystal

The two branches of the dispersion relation define one *acoustic* (low frequency) and one *optical* (high frequency) mode per unit cell. For small $k$ values, the solutions (6) take the form

$$\omega_{O\max}^2 = \frac{\kappa_0}{\mu}, \quad \omega_A^2 = \frac{\kappa_0}{M}\left(\frac{ka}{2}\right)^2 \qquad ...(7)$$

whereas, at the Brillouin zone boundary ($k = \pi/a$), one obtains

$$\omega_{O\min}^2 = \frac{\kappa_0}{M_+},\ \omega_{A\max}^2 = \frac{\kappa_0}{M_-} \qquad ...(8)$$

The two branches are separated by a frequency gap $\Delta\omega = \omega_{O\min} - \omega_{A\max}$, which narrows as the difference $M_- - M_+$ decreases. There will be no frequency gap if $M_- = M_+$, when the ionic lattice can be treated as a linear monatomic chain with the unit cell length of $a/2$. The amplitude ratio of the two types of atoms, as given by Eq. (3) is a function of frequency of the form

$$\frac{A_+}{A_-} = \frac{\kappa_0 - M_-\omega_k^2}{\kappa_0 \cos(ka/2)} = \frac{1 - \dfrac{M_-}{2\mu} \mp \dfrac{M_-}{2\mu}\left[1 - \dfrac{4\mu}{M}\sin^2\dfrac{ka}{2}\right]^{\frac{1}{2}}}{\cos(ka/2)} \qquad ...(9)$$

The lower sign corresponds to the acoustic branch, where in the low frequency or long wavelength limit ($ka \ll 1$) the amplitude ratio is approximately unity. It is positive for any $k$ value in the first Brillouin zone, and hence all the atoms move in the same way. The optical branch is described by the upper sign in Eq. (9) which shows that long wavelength optical modes ($ka \ll 1$) correspond to $A_+/A_- = -M_-/M_+$. In other words, for the long wavelength vibrations, neighbouring ions move in opposite directions, such that the centre of mass of each ionic cell is fixed. The optical modes give rise to an electric polarization and can therefore be excited by electromagnetic waves.

Consider now an ionic crystal containing $N$ ion pairs per unit volume, with ionic masses $M_+$ and $M_-$ and charges $\pm ze$, $z$ being the valence. At any given instant all the positive ions have a displacement $\vec{s}_+$ and all the negative ions a displacement $\vec{s}_-$ from their equilibrium positions, which leads to polarization

$$\vec{P} = \frac{1}{V}\sum \vec{p} = Nze(\vec{s}_+ - \vec{s}_-) \qquad ...(10)$$

In the long wavelength limit, we may assume that at any instant in time all the ions of the same sign experience the same displacement from equilibrium. Hence, we may drop the unit cell label in Eq. (1) and, in the presence of an effective field $\vec{E}_{\text{eff}}$ at each ion site, one obtains

$$M_+\frac{d^2\vec{s}_+}{dt^2} = ze\vec{E}_{\text{eff}} - \kappa_0(\vec{s}_+ - \vec{s}_-)$$

$$M_-\frac{d^2\vec{s}_-}{dt^2} = -ze\vec{E}_{\text{eff}} - \kappa_0(\vec{s}_- - \vec{s}_+) \qquad ...(11)$$

Subtracting the two equations and substituting $\vec{P}$ from Eq. (10), yields:

$$\frac{d_2\vec{P}}{dt^2} = \frac{Nz^2e^2}{\mu}\vec{E}_{\text{eff}} - \frac{\kappa_0\vec{P}}{\mu} \qquad ...(12)$$

where μ is the reduced mass. It is convenient to introduce the so-called free-ion plasma frequency

$$\Omega_p^2 = \frac{Nz^2e^2}{\mu\varepsilon_0} \qquad ...(13)$$

such that Eq. (12) becomes

$$\frac{d_2\vec{P}}{dt^2} = \Omega_p^2\varepsilon_0\vec{E}_{\text{eff}} - \frac{\kappa_0}{\mu}\vec{P} \qquad ...(14)$$

Both $\vec{E}_{\text{eff}}$ and $\vec{P}$ are functions of the macroscopic field $\vec{E}$ within the dielectric medium, which satisfies the axioms of electrostatics. In the absence of free carriers (ρ = 0) Gauss' law reads

$$\nabla \cdot (\varepsilon_0\vec{E} + \vec{P}) = 0 \qquad ...(15)$$

whereas the circulation law is

$$\nabla \times \vec{E} = 0 \qquad ...(16)$$

In cubic crystals, $\vec{P}$ is always parallel to $\vec{E}$, and assuming that

$$\vec{E} = \vec{E}_0 e^{i(\vec{k}\cdot\vec{r} - \omega t)}, \quad \vec{P} = \vec{P}_0 e^{i(\vec{k}\cdot\vec{r} - \omega t)} \qquad ...(17)$$

it follows from Eqs. (15) and (16) that the two vectors must simultaneously satisfy

$$\vec{k}\cdot(\varepsilon_0\vec{E} + \vec{P}) = 0 \qquad ...(18)$$

$$\vec{k} \times \vec{E} = 0 \qquad ...(19)$$

Since $E_{\text{eff}}$, has the same time dependence as $\vec{E}$ and $\vec{P}$, Eq. (14) becomes

$$-\omega^2\vec{P} = \Omega_p^2\varepsilon_0\vec{E}_{\text{eff}} - \frac{\kappa_0}{\mu}\vec{P} \qquad ...(20)$$

where $\vec{E}_{\text{eff}}$ takes different values for the *longitudinal* optical modes, where $\vec{P}$ is parallel to $k$, and for the *transverse* optical modes, with $\vec{P}$ perpendicular to $\vec{k}$.

In a longitudinal mode, Eq. (19) is always satisfied, since $\vec{E}$ is parallel to $\vec{P}$, and hence to $\vec{k}$, while Eq. (18) requires that

$$\vec{E} = -\frac{\vec{P}}{\varepsilon_0} \quad \text{or} \quad \vec{E}_{\text{eff}} = \vec{E} + \frac{\vec{P}}{3\varepsilon_0} = -\frac{2\vec{P}}{3\varepsilon_0} \qquad ...(21)$$

where use has been made of Eq. (9), (section 17.12 ). Substituting into Eq. (20), yields:

$$-\omega^2\vec{P} = \Omega_p^2\,\varepsilon_0\left(-\frac{2\vec{P}}{3\varepsilon_0}\right) - \frac{\kappa_0\vec{P}}{\mu}$$

and this defines $\omega = \omega_L$, which is the frequency of the long-wavelength ($\vec{k} \to 0$) longitudinal optical mode

$$\omega_L^2 = \frac{\kappa_0}{\mu} + \frac{2}{3}\Omega_p^2 \quad ...(22)$$

In a transverse optical mode, where $\vec{E}$ and $\vec{P}$ are both perpendicular to $\vec{k}$, such that Eq. (18) is always satisfied, the condition (19) implies that

$$\vec{E} = 0 \quad \text{or} \quad \vec{E}_{\text{eff}} = \vec{E} + \frac{\vec{P}}{3\varepsilon_0} = \frac{\vec{P}}{3\varepsilon_0} \quad ...(23)$$

Substituting into Eq. (20) yields the allowed frequencies

$$-\omega^2\vec{P} = \Omega_p^2\varepsilon_0\left(\frac{\vec{P}}{3\varepsilon_0}\right) - \frac{\kappa_0\vec{P}}{\mu}$$

which are denoted by $\omega_T$ and read

$$\omega_T^2 = \frac{\kappa_0}{\mu} - \frac{1}{3}\Omega_p^2 \quad ...(24)$$

For an arbitrary field $\vec{E}$, we may substitute $\vec{E}_{\text{eff}}$, into Eq. (20), which allows us to derive the susceptibility

$$\chi(\omega) = \frac{\vec{P}}{\varepsilon_0\vec{E}} = \frac{\Omega_p^2}{\kappa_0/\mu - \frac{1}{3}\Omega_p^2 - \omega^2} = \frac{\Omega_p^2}{\omega_T^2 - \omega^2} \quad ...(25)$$

and then the relative permittivity:

$$\varepsilon_r(\omega) = 1 + \chi(\omega) = \frac{\omega_L^2 - \omega^2}{\omega_T^2 - \omega^2} = \varepsilon_r(\infty)\frac{\omega_L^2 - \omega^2}{\omega_T^2 - \omega^2} \quad ...(26)$$

where it is apparent that $\varepsilon_r(\omega) \to 1$ if $\omega \to \infty$. This result is illustrated in Fig. 17.33 (a). For $\omega = 0$, Eq. (26) reduces to the so-called *Lyddane-Sachs-Teller relation*

$$\frac{\varepsilon_r(0)}{\varepsilon_r(\infty)} = \frac{\omega_L^2}{\omega_T^2} \quad ...(27)$$

between the static dielectric constants $\varepsilon_r(0)$ and $\varepsilon_r(\infty)$ which are indicated in Fig. 8.7 (a), and the longitudinal and transverse optical mode frequencies. It is convenient to substitute the electric field, Eq. (17), into the wave equation for dielectric matter, which reads

$$\nabla^2\vec{E} - \varepsilon\mu\frac{\partial^2\vec{E}}{\partial t^2} = 0 \quad \text{or} \quad k^2 = \frac{\omega^2}{c^2}\varepsilon_r(\omega) \quad ...(28)$$

where it is assumed that $\mu_r \approx 1$. It follows from Eq. (28) that in the interval $\omega_T < \omega < \omega_L$ the wave vector is wholly imaginary because $\varepsilon_r(\omega) < 0$. The field amplitude is exponentially decreasing in this frequency range, because substitution of $k = (\omega/c)\sqrt{-\varepsilon_r(\omega)} = ik_i$ into Eq. (17) yields

$$\vec{E}(\vec{r}, t) = \vec{E}_0 e^{-k_i\xi} e^{i\omega t}$$

where it is assumed that $\vec{k} = k\vec{e}_n$ and $\xi = \vec{e}_n \cdot \vec{r}$. This behaviour corresponds to the dashed line of $\varepsilon_r(\omega)$ shown in Fig. 17.33 (a), which is confirmed by experiments on dielectric materials.

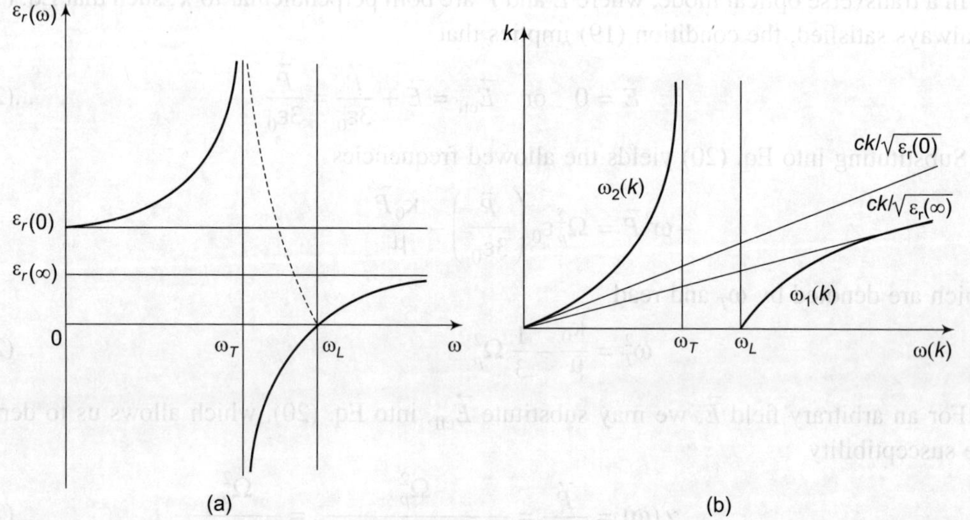


**Figure 17.33** Relative permittivity $\varepsilon_r(\omega)$ as a function of frequency (a) and the dispersion relation $\omega(k)$ for polaritons (b) in ionic solids

When Eq. (26) is substituted into Eq. (28), to give

$$\frac{c^2k^2}{\omega^2} = \varepsilon_r(\infty)\frac{\omega_L^2 - \omega^2}{\omega_T^2 - \omega^2} = \frac{\varepsilon_r(0)\,\omega_T^2 - \varepsilon_r(\infty)\,\omega^2}{\omega_T^2 - \omega^2}$$

one obtains the dispersion relation for electromagnetic waves in ionic crystals, in the form

$$\varepsilon_r(\infty)\,\omega^4 - [\varepsilon_r(0)\,\omega_T^2 + c^2k^2]\,\omega^2 + c^2k^2\omega_T^2 = 0 \qquad \text{...(29)}$$

The two solutions

$$\omega^2 = \frac{1}{2\,\varepsilon_r(\infty)}\left\{\varepsilon_r(0)\,\omega_T^2 + c^2k^2 \pm [(\varepsilon_r(0)\omega_T^2 + c^2k^2)^2 - 4c^2k^2\,\omega_T^2\,\varepsilon_r(\infty)]^{\frac{1}{2}}\right\} \qquad \text{...(30)}$$

reduce, in the long wavelength limit ($k \to 0$), to

$$\omega_1(k) = \sqrt{\frac{\varepsilon_r(0)\,\omega_T^2 + c^2k^2}{\varepsilon_r(\infty)}} = \sqrt{\omega_L^2 + \frac{c^2k^2}{\varepsilon_r(\infty)}}, \quad \omega_2(k) = \frac{ck}{\sqrt{\varepsilon_r(0)}} \qquad \text{...(31)}$$

and exhibit a distinct asymptotic behaviour

$$\omega_1(k) = \frac{ck}{\sqrt{\varepsilon_r(\infty)}}, \quad \omega_2(k) = \omega_T \qquad ...(32)$$

The dispersion curve, shown in Fig. 17.33 (b), consists of two branches. The high frequency branch $\omega_1(k)$ has a linear region, which is characteristic of photon dispersion, Eq. (32), while the low frequency branch behaves like the phonon transverse optical mode in the constant frequency region, Eq. (32). The two branches are separated by a frequency gap, $\omega_T < \omega < \omega_L$. The elementary excitations of the ionic lattice vibrations by an electromagnetic field are often referred to as *polaritons*, which are associated with standing electromagnetic waves in ionic crystals, that carry the polarization energy of ion vibration modes.

**Example 17.1** Find the force constant $\kappa_0$ and $\kappa$ for the nearest-neighbour and next-nearest-neighbour interactions in a linear ionic chain, with lattice constant *a*.

***Solution*** Assuming the potential energy per ion pair of the form

$$E(R) = \frac{A_n}{R_n} - \frac{Az^2e_0^2}{R}$$

where the first term is the Born–Mayer repulsive potential, $A = 2 \ln 2$ is the Madelung constant for the linear chain of ions, and $z$ is the valence, the condition of static equilibrium at $R_0 = a/2$ yields

$$\left(\frac{\partial U}{\partial R}\right)_{R=a/2} = N\left(\frac{-nA_n}{R^{n+1}} + \frac{Az^2e_0^2}{R^2}\right)_{R=a/2} = 0 \text{ or } A_n = \frac{(2\ln 2)z^2e_0^2}{n}\left(\frac{a}{2}\right)^{n-1}$$

In the configuration shown in Fig. 17.31, the nearest-neighbour interaction consists of both a repulsive force

$$F_+ = \frac{nA_n}{(a/2 + s_p^+ - s_{p-1}^-)^{n+1}} - \frac{nA_n}{(a/2 + s_p^- - s_p^+)^{n+1}}$$

$$\approx \frac{n(n+1)A_n}{(a/2)^{n+2}}(s_{p-1}^- + s_p^- - 2s_p^+)$$

and an attractive force

$$F_- = -\frac{z^2e_0^2}{(a/2 + s_p^+ - s_{p-1}^-)^2} + \frac{z^2e_0^2}{(a/2 + s_p^- - s_p^+)^2} \approx -\frac{2z^2e_0^2}{(a/2)^3}(s_{p-1}^- + s_p^- - 2s_p^+)$$

By taking, as in the first Eq. (1)

$$F_{\kappa_0} = F_+ + F_- = \frac{\kappa_0}{2}(s_{p-1}^- + s_p^- - 2s_p^+)$$

it follows that the nearest neighbours interact with a force constant

$$\kappa_0 = 2\left[\frac{n(n+1)A_n}{(a/2)^{n+2}} - \frac{2z^2a_0^2}{(a/2)^3}\right] = \frac{4z^2e_0^2}{(a/2)^3}[(n+1)\ln 2 - 1]$$

As $n$ is of the order of 9, it is apparent that the main contribution to the nearest-neighbour force constant comes from the repulsive interaction.

The next-nearest-neighbour interaction, is purely electrostatic, will be written in the form

$$F_k = \frac{z^2 e_0^2}{(a + s_p^+ - s_{p-1}^+)^2} + \frac{z^2 e_0^2}{(a + s_{p+1}^+ - s_p^+)^2}$$

$$= \frac{2z^2 e_0^2}{a^3}(s_{p-1}^+ + s_{p+1}^+ - 2s_p^+) = \frac{\kappa}{2}(s_{p-1}^+ + s_{p+1}^+ - 2s_p^+)$$

One obtains $\kappa = 4z^2 e_0^2/a^3$, which is one of magnitude less than $\kappa_0$.

**Example 17.2** Find the attenuation constant of the linear ionic lattice vibrations in the frequency gap, $\kappa_0/M_- < \omega^2 < \kappa_0/M_+$.

***Solution*** The ion displacements have the form (Eq. 2), where the $k$ values associated with a given frequency range are provided by the characteristic Eq. (4) in the form

$$\cos\frac{ka}{2} = \sqrt{\left(1 - \frac{M_+}{\kappa_0}\omega^2\right)\left(1 - \frac{M_-}{\kappa_0}\omega^2\right)} = i\delta$$

where $\delta$ is a real number if $\omega$ is in the frequency gap

$$\delta = \sqrt{\left(1 - \frac{M_+}{\kappa_0}\omega^2\right)\left(\frac{M_-}{\kappa_0}\omega^2 - 1\right)}$$

It follows that the wave number $k$ can be written in a complex representation, which on substitution yields

$$\cos\frac{(k + ik_i)a}{2} = \cos\frac{ka}{2}\cosh\frac{k_i a}{2} - 1\sin\frac{ka}{2}\sinh\frac{k_i a}{2} = i\delta$$

and this implies that

$$\cos\frac{ka}{2} = 0 \quad \text{or} \quad \frac{ka}{2} = \pm\frac{\pi}{2}$$

$$\sin\frac{ka}{2}\sinh\frac{k_i a}{2} = \pm\sinh\frac{k_i a}{2} = \delta \quad \text{or} \quad \frac{k_i a}{2} = \mp\sinh^{-1}\delta$$

Substituting into Eq. (1) one obtains

$$s_p^+ = A_+ e^{2ip(k + ik_i)a/2}\, e^{-i\omega_k t} = \pm A_+ e^{-pak_i}\, e^{-i\omega_k t}$$

These vibrations are damped by the lattice, with the attenuation constant

$$k_i = \frac{2}{a}\sinh^{-1}\delta$$

In other words, the lattice cannot support vibrational modes in the frequency gap, because the associated displacements are rapidly attenuated.

**Example 17.3** Find the frequencies of normal modes in a bidimensional ionic solid with a square lattice of edge $a$, in which ions of masses $M_+$ and $M_-$ are alternately placed, assuming that only nearest neighbours interact.

***Solution*** In the sequence of neighbouring ions depicted in Fig. 17.34, the repeat distance is $2a$. The equations of motion of the two kinds of ion read

$$M_+ \frac{d^2 s^+_{2p,2q}}{dt^2} = \frac{\kappa_0}{2}(s^-_{2p+1,2q} + s^-_{2p-1,2q} - 2s^+_{2p,2q}) + \frac{\kappa_0}{2}(s^-_{2p,\,2q+1} + s^-_{2p,2q-1} - 2s^+_{2p,2q})$$

$$M_- \frac{d^2 s^+_{2p+2,2q}}{dt^2} = \frac{\kappa_0}{2}(s^+_{2p+2,2q} + s^+_{2p,2q} - 2s^-_{2p+1,2q}) + \frac{\kappa_0}{2}(s^+_{2p+1,2q+1} + s^+_{2p+1,2q-1} - 2s^-_{2p+1,2q})$$

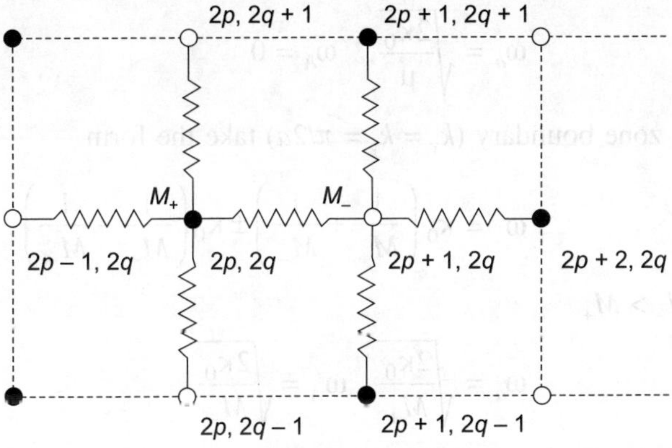


**Figure 17.34**

The longitudinal displacements of the two kinds of ion will usually have different amplitudes

$$s^+_{2p,2q} = A_+ e^{i(2pk_xa + 2qk_ya - \omega t)}$$

$$s^+_{2p+1,2q} = A_- e^{i[(2p+1)k_xa + 2qk_ya - \omega t]}$$

Substituting into the equations of motion yields

$$-\omega^2 M_+ A_+ = \frac{\kappa_0}{2}[A_-(e^{ik_xa} + e^{-ik_xa}) - 2A_+] + \frac{\kappa_0}{2}[A_-(e^{ik_ya} + e^{-ik_ya}) - 2A_+]$$

$$= \kappa_0 (A_- \cos k_xa + A_- \cos k_ya - 2A_+)$$

and similarly

$$-\omega^2 M_- A_- = \kappa_0 (A_+ \cos k_x a + A_+ \cos k_y a - 2A_-)$$

which can be written as

$$(M_+\omega^2 - 2\kappa_0)A_+ + \kappa_0(\cos k_x a + \cos k_y a)\, A_- = 0$$

$$\kappa_0(\cos k_x a + \cos k_y a)A_+ + (M_-\omega^2 - 2\kappa_0)A_- = 0$$

The characteristic equation provides the dispersion relation

$$\omega^4 - \frac{2\kappa_0}{\mu}\omega^2 + \frac{\kappa_0^2}{M\mu}[4 - (\cos k_x a + \cos k_y a)^2] = 0$$

where $M = M_+ + M_-$, $\mu^{-1} = M_+^{-1} + M_-^{-1}$, and $M_+ M_-$, $\mu M$. There are two branches

$$\omega^2 = \frac{\kappa_0}{\mu} \pm \frac{\kappa_0}{\mu}\sqrt{1 - \frac{\mu}{M}[4 - (\cos k_x a + \cos k_y a)^2]}$$

which, in the long wavelength limit ($k_x = k_y = 0$), reduce to

$$\omega_o = \sqrt{\frac{2\kappa_0}{\mu}}, \quad \omega_A = 0$$

and at the Brillouin zone boundary ($k_x = k_y = \pi/2a$) take the form

$$\omega^2 = \kappa_0\left(\frac{1}{M_+} + \frac{1}{M_-}\right) \pm \kappa_0\left(\frac{1}{M_+} - \frac{1}{M_-}\right)$$

or, assuming that $M_- > M_+$

$$\omega_o = \sqrt{\frac{2\kappa_0}{M_+}}, \quad \omega_A = \sqrt{\frac{2\kappa_0}{M_-}}$$

The results are similar to those obtained for a linear ionic lattice, except that the Brillouin zone boundary is now situated at $\pm\pi/2a$, compared with $\pm\pi/a$ for the linear chain. This allows us to stress that the range covered by a Brillouin zone is determined by the lattice repeat distance $2a$ rather than the nearest-neighbour separation, $a$.

## 17.30 FREQUENCY-DEPENDENT POLARIZABILITY (QUANTUM TREATMENT) AND SELLIMEYER'S EQUATION

The dielectric response of ionic solids for high frequency applied electric fields can be discussed in two ways. First, in terms of the ionic polarization process through relative movement of positive and negative ions, under the influence of the applied electric field. Secondly, in terms of the electronic polarization mechanism, which occurs within every dielectric solid and consists of the displacement of electrons in the atom relative to the nucleus. The effect of frequency on polarization of a dielectric is shown schematically in Fig. 17.35. When an electromagnetic wave of relatively low frequency passes through an ionic solid, both the valence electrons and the displacements of the ion cores contribute to the relative permittivity, $\varepsilon_r(0)$. A resonance effect occurs in the infrared region, at angular frequencies of the order or $10^{13}$ rad/s. For higher frequency waves, where only bound electrons can respond, the relative permittivity has a rather smaller value $\varepsilon_r(\infty)$, as is shown in Fig. 17.33 (a). The electronic polarization remains up to an angular frequency of about $10^{14}$ red/s, which is in the optical range. Another resonance effect usually occurs in this region.

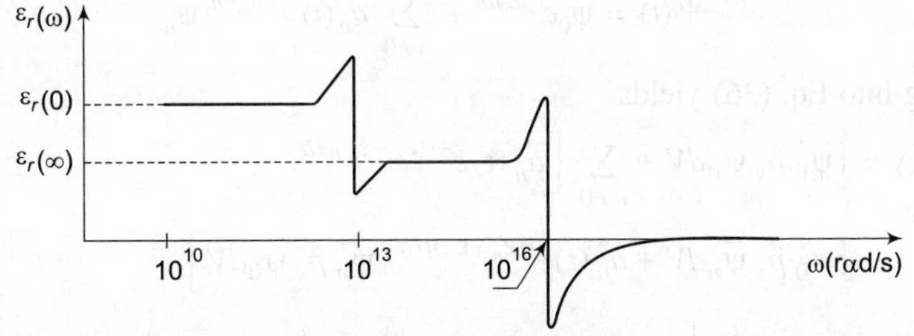


**Figure 17.35** Frequency dependence of the ionic and electronic contributions to the relative permittivity of an ionic solid

The resonant behaviour of the ionic polarizability can be derived by substituting $\vec{P} = N\alpha_i(\omega)\varepsilon_0 E_{\text{eff}}$, where $N$ is the number of ion pairs and $\alpha_i(\omega)$ is the ionic polarizability, into Eq. (20), which yields

$$\alpha_i(\omega) = \frac{\Omega_p^2/N}{\kappa_0/\mu - \omega^2} = \frac{z^2e^2}{\mu\varepsilon_0(\omega_i^2 - \omega^2)} \qquad ...(33)$$

The singularity in the ionic polarizability, predicted by Eq. (33) at the natural ionic resonant frequency, $\omega_i = \sqrt{\kappa_0/\mu}$, is actually avoided because of dielectric loss, which is associated with a frictional term that has been neglected in the equations of ionic motion, Eqs. (11). Furthermore, comparison of Eq. (33) with Eqs. (26) shows that the presence of a large density of polarizable ions has the effect of shifting the resonant frequency of the relative permittivity to a lower value

$$\omega_T = \left(\omega_i^2 - \frac{1}{3}\omega_p^2\right)^{\frac{1}{2}} \qquad ...(34)$$

where use has been made of Eq. (24)

The polarizability of bound electrons induced by an oscillating electric field is associated with a time-dependent perturbation of the atomic system, similar to that introduced for static field,

$$V(t) = -\vec{p}\cdot\vec{E}(t) = -p_x E(e^{i\omega t} + e^{-i\omega t}) \qquad ...(35)$$

It is assumed, for simplicity, that the electric field is applied in the $x$-direction. As the perturbing potential contains the time explicitly, the expectation value of the dipole moment

$$\langle p_x \rangle = \int \Psi^*(t)\hat{p}_x \Psi(t)\, dV \quad ...(36)$$

should be calculated in terms of the wave functions given by the time-dependent perturbation theory in the form

$$\Psi(t) = \psi_0 e^{-i\varepsilon_0 t/\hbar} + \sum_{n\neq 0} a_n(t)\, e^{-i\varepsilon_n t/\hbar}\, \psi_n \quad ...(37)$$

Substituting into Eq. (36) yields

$$\langle p_x \rangle = \int \psi_0^* \hat{p}_x \psi_0 dV + \sum_{n\neq 0} \Big[ a_n(t) e^{-i(\varepsilon_n-\varepsilon_0)t/\hbar}$$

$$\int \psi_0^* \hat{p}_x \psi_n dV + a_n^*(t) e^{i(\varepsilon_n-\varepsilon_0)t/\hbar} \int \psi_n^* \hat{p}_x \psi_0 dV \Big]$$

$$= \langle p_{x0} \rangle + \sum_{n\neq 0} \Big[ \langle \psi_0 | \hat{p}_x | \psi_n \rangle a_n(t)\, e^{i\omega_{n0}t} + \langle \psi_n | \hat{p}_x | \psi_0 \rangle\, a_n^*(t) e^{i\omega_{n0}t} \Big] \quad ...(38)$$

where only the terms linear in the electric field have been retained, and the angular frequencies $\hbar\omega_{n0} = \varepsilon_n - \varepsilon_0$ are associated with natural oscillations, which are expected to disappear in the presence of damping. The coefficients $a_n(t)$ are given, to a first approximation, by

$$a_n(t) = -\frac{i}{\hbar}\int_0^t \langle \psi_n | \hat{V}(t) | \psi_0 \rangle\, e^{i(\varepsilon_n-\varepsilon_0)t'/h}\, dt'$$

$$= -\frac{i}{\hbar}\langle \psi_n | \hat{p}_x | \psi_0 \rangle E \int_0^t [e^{i(\omega_{n0}+\omega)t'} + e^{i(\omega_{n0}-\omega)t'}]dt'$$

$$= \frac{\langle \psi_n | \hat{p}_x | \psi_0 \rangle E}{\hbar}\left[\frac{e^{i(\omega_{n0}+\omega)t}}{\omega_{n0}+\omega} + \frac{e^{i(\omega_{n0}-\omega)t}}{\omega_{n0}-\omega}\right]$$

$$= e^{i\omega_{n0}t}\frac{\langle \psi_n | \hat{p}_x | \psi_0 \rangle E}{\hbar(\omega_{n0}^2-\omega^2)}(2\omega_{n0}\cos\omega t - 2i\omega\sin\omega t)$$

where we have considered the large-$t$ limit and a slowly rising rather than abrupt perturbation at $t = 0$. Substituting this result and its complex conjugate into Eq. (38) yields

$$\langle p_x \rangle = \langle p_{x0} \rangle + \frac{2}{\hbar}(2E\cos\omega t)\sum_{n\neq 0}\frac{\omega_{n0}\left|\langle \psi_n | \hat{p}_x | \psi_0 \rangle\right|^2}{\omega_{n0}^2-\omega^2} \quad ...(39)$$

Since in Eq. (39), $2E\cos\omega t = E(t)$, it follows that the frequency-dependent electronic polarizability reads

$$\alpha_e(\omega) = \frac{2}{\hbar}\sum_{n\neq 0}\frac{\omega_{n0}\left|\langle \psi_n | \hat{p}_x | \psi_0 \rangle\right|^2}{\omega_{n0}^2-\omega^2} = \frac{e^2}{m_e}\sum_{n\neq 0}\frac{f_{n0}}{\omega_{n0}^2-\omega^2} \quad ...(40)$$

The real dimensionless constant, characteristic of the $\psi_0 \rightarrow \psi_n$ transition

$$f_{n0} = \frac{2m_e\omega_{n0}}{\hbar} |\langle\psi_n|\hat{x}|\psi_0\rangle|^2 \qquad ...(41)$$

is called the *oscillator strength*. In the low frequency limit, $\alpha_e(\omega)$ given by Eq. (40) reduces to the static polarizability, which is easily obtained for $\omega \rightarrow 0$. The electronic polarization, Eq. (40), is written as a sum over the contributions of many electron resonances, associated with atomic transitions. When the electromagnetic energy $\hbar\omega$ matches the energy difference of two electron levels, the electron undergoes a transition to a higher state. In the absence of damping, it returns to the initial state by emission of a photon at ultraviolet wavelengths and even shorter. Therefore, it is apparent that the schematic representation given in Fig. (17.35) is only indicative for the frequency range in which electronic polarizability is significant.

The quantum mechanical substantiates the Lorentz model of description electronic polarizability, in which the electrons are considered to be bound to fixed sites by quasi-elastic forces. The classical Lorentz model provides a simple approach to the optical constants of lossy dielectrics, which are the most susceptible to simple experimental determination. The equation of motion for a bound electron in a harmonic local field, involving a restoring force and a phenomenological representation of the electron momentum loss as a result of collisions, assumes the form

$$m_e \frac{d^2\vec{r}}{dt^2} + 2m_e\beta\frac{d\vec{r}}{dt} + \kappa_0\vec{r} = -e\vec{E}_{\text{eff}}\, e^{-i\omega t} \qquad ...(42)$$

where $\kappa_0$ and $2\beta$ denote the force and the damping constants, respectively. Equation (42) describes the forced damped harmonic motion, and it is satisfied by substituting $\vec{r} = \vec{r}_0 e^{-i\omega t}$, which gives

$$\vec{r} = \frac{-e/m_e}{\kappa_0/m_e - \omega^2 - 2i\beta\omega}\vec{E}_{\text{eff}} \qquad ...(43)$$

This solution is associated with the *steady-state* oscillation of electrons at the forcing frequency of the harmonic field. If $N$ is the number of bound electrons per unit volume, each of them displaced a distance $\vec{r}$ from their equilibrium position, the electronic polarization of the medium has the form $\vec{P} = -Ne\vec{r}$ which, upon substitution of Eq. (43), reads:

$$\vec{P} = \frac{Ne^2/m}{\kappa_0/m_e - \omega^2 - 2i\beta\omega}\vec{E}_{\text{eff}} = \frac{\varepsilon_0\omega_p^2}{\kappa_0/m_e - \omega^2 - 2i\beta\omega}E_{\text{eff}} \qquad ...(44)$$

where $\omega_p$ is the *plasma frequency*

$$\omega_p = \frac{Ne^2}{m_e\varepsilon_0} \qquad ...(45)$$

It is apparent that $\Omega_P$, Eq. (13), reduces to $\omega_p$ for $z = 1$ and $\mu \equiv m_e$. If we consider the steady-state motion of bound electrons, the local field $\vec{E}_{\text{eff}}$ can be given by $E_{\text{eff}} = \left(\vec{E} + \frac{\vec{P}}{3\varepsilon_0}\right)$ and the appropriate form of Eq. (44) is

$$\vec{P} = \frac{\varepsilon_0 \omega_p^2}{(\kappa_0/m_e - \omega^2) - 2i\beta\omega}\left(\vec{E} + \frac{\vec{P}}{3\varepsilon_0}\right) \quad ...(46)$$

which can be solved for $\vec{P}$ to give

$$\vec{P} = \frac{\varepsilon_0 \omega_p^2}{(\kappa_0/m_e - \omega_p^2/3) - \omega^2 - 2i\beta\omega}\vec{E} = \frac{\varepsilon_0 \omega_p^2}{\omega_0^2 - \omega^2 - 2i\beta\omega}\vec{E} \quad ...(47)$$

where the resonant frequency ($\omega_0$) is taken as

$$\omega_0 = \left(\frac{\kappa_0}{m_e} - \frac{1}{3}\omega_p^2\right)^{\frac{1}{2}} \quad ...(48)$$

Equation (44) defines a complex electron susceptibility

$$\chi_e = \frac{\omega_p^2}{\omega_0^2 - \omega^2 - 2i\beta\omega} \quad ...(49)$$

which describes a lossy dielectric, even if the free charges are absent ($\sigma = 0$). The complex permittivity can be obtained by substituting Eq. (49) in

$$\varepsilon_c = \varepsilon_0 (1 + \chi_e)$$

$$\Rightarrow \quad \varepsilon_c = \varepsilon_0 \left(1 + \frac{\omega_p^2}{\omega_0^2 - \omega^2 - 2i\beta\omega}\right) \quad ...(50)$$

One may introduce optical constants in a form similar to that suitable for transparent solids, $\varepsilon = \varepsilon_0\varepsilon_r = \varepsilon_0 n^2$, provided the refractive index is taken to be a complex quantity. If $\omega_p$ is small enough for the absolute value of the complex number on the right-hand side to be small when compared with unity, for all frequencies, we may approximate

$$n + in_i = \left(1 + \frac{\omega_p^2}{\omega_0^2 - \omega^2 - 2i\beta\omega}\right)^{\frac{1}{2}} \cong 1 + \frac{1}{2}\,\frac{\omega_p^2}{\omega_0^2 - \omega^2 - 2i\beta\omega} \quad ...(51)$$

which gives the optical constants $n$ and $n_i$ as

$$n \cong 1 + \frac{1}{2}\,\frac{\omega_p^2\,(\omega_0^2 - \omega^2)}{(\omega_0^2 - \omega^2) + 4\beta^2\omega^2} \quad ...(52)$$

$$n_i \cong \frac{\omega_p^2\,\beta\omega}{(\omega_0^2 - \omega^2)^2 + 4\beta^2\omega^2}$$

In the so-called transparent region which occurs for frequencies below the resonance frequency $\omega_0$, where $(\omega_0^2 - \omega^2)^2 \gg 4\beta^2\omega^2$, Eq. (52) show that $n_i \cong 0$ and we obtain the dispersion *relation* gives the frequency dependence of $n$ as

$$n \cong 1 + \frac{1}{2}\frac{\omega_p^2}{\omega_0^2 - \omega^2} \qquad ...(53)$$

This refractive index is greater than unity and increases with increasing frequency $dn/d\omega > 0$. Such a behaviour, which is characteristic of most of the ionic and molecular crystal in the visible region of the spectrum, is called *normal dispersion*.

In the vicinity of the resonance frequency ($\omega_0$) we may set $\omega \cong \omega_0$ so that $\omega_0^2 - \omega^2 \cong 2\omega_0 (\omega_0 - \omega)$ and Eq. (52) reduce to

$$n \cong 1 + \frac{\omega_p^2}{4\omega_0}\frac{(\omega_0 - \omega)}{(\omega_0 - \omega)^2 + \beta^2} \qquad ...(54)$$

$$n_i \cong \frac{\omega_p^2}{4\omega_0}\frac{\beta}{(\omega_0 - \omega)^2 + \beta^2}$$

Both $n$ and $n_i$ depend on frequency as plotted in Fig. 17.36. The frequency range $\omega = \omega_0 \pm \beta$ over which $n_i$ is significantly different from zero is called the *absorption region*. In this region $n$ reaches a maximum for $\omega = \omega_0 - \beta$ and then falls to a minimum at $\omega = \omega_0 + \beta$. This behaviour is referred to as *anomalous dispersion*, where $dn/d\omega < 0$.

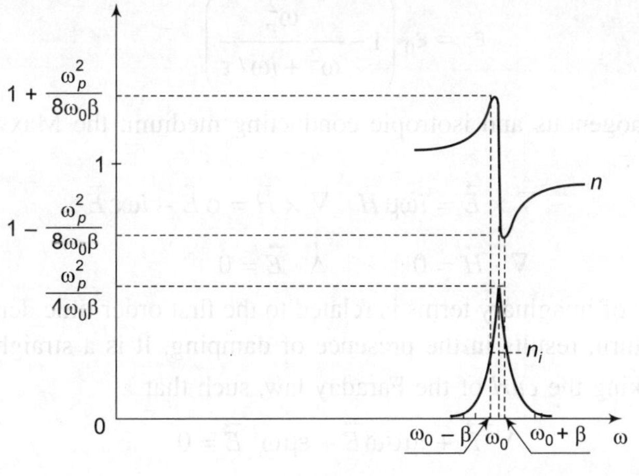


**Figure 17.36** Dependence of $n$ and $n_i$ on frequency in the vicinity of the resonance frequency, $\omega_0$

If there are several kinds of elastically bound electrons, we may think of certain resonance frequencies ($\omega_j$) and the dispersion relation (51) becomes

$$n + in_i \cong 1 + \frac{\omega_p^2}{2}\sum_j \frac{f_j}{\omega_j^2 - \omega^2 - 2i\beta\omega} \qquad ...(55)$$

where $f_j$ is interpreted as the oscillator strength introduced by Eq. (41). In the transparent region Eq. (55) reduces to

$$n \cong 1 + \frac{\omega_p^2}{2} \sum_j \frac{f_j}{\omega_j^2 - \omega^2} \qquad ...(56)$$

which is known as *Sellmeyer's equation*, when given in terms of wavelength instead of frequency, and is commonly used for experimental fitting of measured values of *n*. At very low frequencies, Eq. (56) takes the form of the *Maxwell relation* for the static refractive index of transparent media

$$n \cong 1 + \frac{\omega_p^2}{2} \sum \frac{f_j}{\omega_j^2} \qquad ...(57)$$

## 17.31 OPTICAL CONSTANTS OF METALS

The Lorentz model approach can be applied to the classical *free-electron model*, which is appropriate for good conductors, by setting $\omega_0 = 0$, which in effect neglects the binding force. However, the constant $(2\beta)$ now has a different meaning from that in Eq. (42) and represents the damping due to the resistance of the conductor, characterized by a relaxation time $\tau = 1/2\beta$ associated with electron scattering. By taking $\omega_0 = 0$, the complex permittivity (50) is obtained in terms of as

$$\varepsilon_c = \varepsilon_0 \left( 1 - \frac{\omega_p^2}{\omega^2 + i\omega/\tau} \right) \qquad ...(58)$$

In a linear, homogenous and isotropic conducting medium, the Maxwell equations take the form

$$\nabla \times \vec{E} = i\omega\mu\vec{H} \quad \nabla \times \vec{H} = \sigma\vec{E} - i\omega\varepsilon\overleftarrow{E} \qquad ...(59)$$

$$\nabla \cdot \vec{H} = 0 \qquad \Delta \cdot \vec{E} = 0$$

where the presence of imaginary terms is related to the first order time derivatives of the field vectors, which in turn, results in the presence of damping. It is a straightforward matter to eliminate $\vec{H}$, by taking the *curl* of the Faraday law, such that

$$\nabla^2 \vec{E} + i\mu\sigma\omega\vec{E} + \varepsilon\mu\omega^2 \vec{E} = 0 \qquad ...(60)$$

This reduces to the usual Helmholtz form

$$\nabla^2 \vec{E} + \varepsilon_c \mu\omega^2 \vec{E} = 0 \qquad ...(61)$$

if the real permittivity is replaced by the *complex permittivity*, $\varepsilon_c$

$$\varepsilon_c = \varepsilon + i\frac{\sigma}{\omega} = \varepsilon_0(\varepsilon_r + i\varepsilon_i) = \varepsilon_0 \left( \varepsilon_r + i\frac{\sigma}{\varepsilon_0\omega} \right) = \varepsilon_0 \left( 1 + \chi_e + i\frac{\sigma}{\varepsilon_0\omega} \right) \qquad ...(62)$$

Since in a free-electron gas there is no polarization charge, we must assume that $\chi_e = 0$, so that a frequency-dependent or optical *conductivity* $(\sigma)$ is obtained from Eqs. (58) and (62) as

$$\sigma(\omega) = \frac{\omega_p^2 \varepsilon_0 \tau}{1 - i\omega\tau} = \frac{\sigma_0}{1 - i\omega\tau} \qquad ...(63)$$

where $\sigma_0$ is the low frequency limit, identified as the static conductivity. In view of the relation $\varepsilon_c = \varepsilon_0 n_c^2 = \varepsilon_0 (n + in_i)^2$, when separated into real and imaginary parts, Eq. (62) gives

$$\varepsilon_r = n^2 - n_i^2 = 1 - \frac{\omega_p^2}{\omega^2 + 1 + \tau^2}$$

$$\varepsilon_i = 2nn_i = \frac{\omega_p^2}{\omega^2 + 1/\tau^2}\left(\frac{1}{\omega\tau}\right) \qquad ...(64)$$

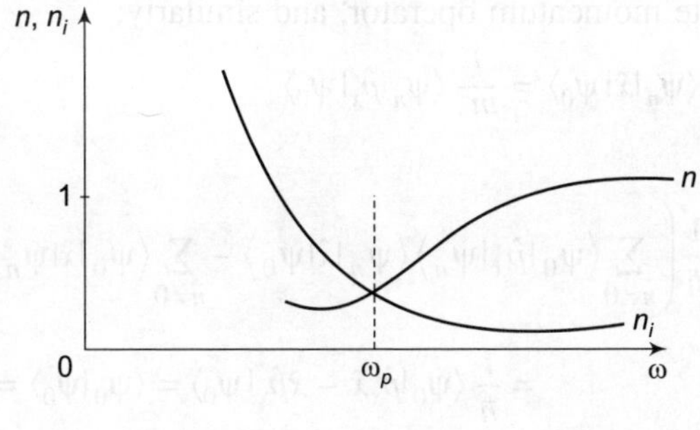


**Figure 17.37** Dependence of $n$ and $n_i$ on frequency for a metallic solid

A plot of $n$ and $n_i$ as function of frequency is given in Fig. 17.37. It shows that the extinction index ($n_i$) assumes large values at low frequencies, when the refractive index is less than unity plasma frequencies for metals correspond to the visible region of the spectrum. At high frequencies good conductors are transparent, since $n_i$ becomes negligible of $\omega > \omega_p$. The onset of transparency is defined, using Eq. (64), by the condition that $\varepsilon_r$ is zero, and hence

$$\omega^2 = \omega_p^2 - 1/\tau^2 \qquad ...(65)$$

In other words, metals are nearly perfect reflectors for incident radiation with a frequency below the plasma frequency.

**Example 17.4** Find the high frequency limit of the electronic polarizability, $\alpha_e(\omega)$.

***Solution*** Assuming that $\omega \to \infty$, we may neglect $\omega_{n0}^2 \ll \omega^2$ in the denominator of Eq. (40), which yields:

$$\alpha_e(\omega) = -\frac{e^2}{m_e\omega^2} \sum_{n \neq 0} f_{n0}$$

It is convenient to put $f_{n0}$, Eq. (8.68), in the form

$$f_{n0} = \frac{m_e\omega_{n0}}{\hbar} \langle \psi_0 | \hat{x} | \psi_n \rangle \langle \psi_n | \hat{x} | \psi_0 \rangle - \frac{m_e\omega_{0n}}{\hbar} \langle \psi_0 | \hat{x} | \psi_n \rangle \langle \psi_n | \hat{x} | \psi_0 \rangle$$

and substitute the matrix elements of the position operator by using the evolution equation for operators that do not involve time explicitly

$$i\hbar \frac{d\hat{x}}{dt} = [\hat{x}, \hat{H}_0] \quad \text{or} \quad i\hbar \frac{\hat{p}_x}{m_e} = [\hat{x}, \hat{H}_0]$$

One immediately obtains;

$$\begin{aligned} \langle \psi_0 | [\hat{x}, \hat{H}_0] | \psi_n \rangle &= \langle \psi_0 | \hat{x}\hat{H}_0 - \hat{H}_0 \hat{x} | \psi_n \rangle \\ &= (\varepsilon_n - \varepsilon_0) \langle \psi_0 | \hat{x} | \psi_n \rangle = \hbar\omega_{n0} \langle \psi_0 | \hat{x} | \psi_n \rangle \end{aligned}$$

This yields $\qquad \omega_{n0} \langle \psi_0 | \hat{x} | \psi_n \rangle = \dfrac{i}{m_e} \langle \psi_0 | \hat{p}_x | \psi_n \rangle$

where $\hat{p}_x$ is the conjugate momentum operator, and similarly;

$$\omega_{0n} \langle \psi_n | \hat{x} | \psi_0 \rangle = \frac{i}{m_e} \langle \psi_n | \hat{p}_x | \psi_0 \rangle$$

It follows that

$$\begin{aligned} \sum_{n\neq 0} f_{n0} &= \frac{1}{\hbar} \left( \sum_{n\neq 0} \langle \psi_0 | \hat{p}_x | \psi_n \rangle \langle \psi_n | \hat{x} | \psi_0 \rangle - \sum_{n\neq 0} \langle \psi_0 | \hat{x} | \psi_n \rangle \langle \psi_n | \hat{p}_x | \psi_0 \rangle \right) \\ &= \frac{i}{\hbar} \langle \psi_0 | \hat{p}_x \hat{x} - \hat{x}\hat{p}_x | \psi_0 \rangle = \langle \psi_0 | \psi_0 \rangle = 1 \end{aligned}$$

where use has been made of the basic commutation relation $[\hat{p}_x, \hat{x}] = -i\hbar$ for the position and momentum conjugate operators. This result is known as the *Thomas–Kuhn sum rule*. By substitution into $\alpha_e(\omega)$ one obtains the electronic polarizability at very high frequencies as

$$\alpha_e(\omega) = -\frac{e^2}{m_e \omega^2}$$

**Example 17.5** Derive the optical conductivity of metals from the Boltzmann transport equation in the presence of an oscillatory electric field.

***Solution*** The linearized form of the Boltzmann equation reads

$$\frac{\partial f_1}{\partial t} = -\vec{v} \cdot \nabla f - \frac{1}{\hbar} \vec{F} \cdot \nabla_{\vec{k}} f - \frac{1}{\tau} f_1$$

where $\vec{F} = -e\vec{E} = -eE_0 e^{i(\vec{k}\cdot\vec{r} - \omega t)}$, and hence, it is reasonable to assume that $f_1 = f - f_0 = f_1(0)$ $e^{i(\vec{k}\cdot\vec{r} - \omega t)}$. This yields

$$f_1 \left[ 1 - i\omega\tau \left( 1 - \frac{\vec{v}\cdot\vec{k}}{\omega} \right) \right] = \frac{\tau e}{\hbar} \vec{E} \cdot \nabla_{\vec{k}} f = \tau e (\vec{E} \cdot \vec{v}) \frac{\partial f_0}{\partial \varepsilon}$$

because $\nabla_{\vec{k}} f = (\partial f_0 / \partial \varepsilon) \nabla_{\vec{k}} \varepsilon = (\partial f_0 / \partial \varepsilon) \hbar \vec{v}$. At high frequencies where $\vec{v} \cdot \vec{k} \ll \omega$, the distribution function is approximated by

$$f_1 = \frac{\tau e(\vec{E}\cdot\vec{v})}{1 - i\omega\tau}\frac{\partial f_0}{\partial \varepsilon}$$

The current density can be calculated as a sum over the contributions from individual electrons of the form

$$\vec{j} = -\frac{e}{V}\sum_{\vec{k}}\vec{v}(\vec{k})f_1(\vec{r},\vec{k}) = -\frac{e^2\vec{E}}{V}\sum_{\vec{k}}\frac{\tau\vec{v}\cdot\vec{v}}{1-i\omega\tau}\frac{\partial f_0}{\partial \varepsilon}$$

where the relaxation time is assumed to be independent of $\vec{k}$, such that it is convenient to factor $(1 - i\omega\tau)^{-1}$ from the sum

$$\vec{j} = \frac{1}{1 - i\omega\tau}\left(-\frac{e^2}{V}\sum_{\vec{k}}\tau\vec{v}\cdot\vec{v}\frac{\partial f_0}{\partial \varepsilon}\right)\vec{E} = \frac{\sigma_0}{1 - i\omega\tau}\vec{E}$$

It is apparent that $\sigma_0$ stands for static conductivity $\sigma_0 = ne\mu_e = Ne\mu_e/V$. Thus, as a result, one obtains

$$\sigma(\omega) = \frac{\sigma_0}{1 - i\omega\tau}$$

which is the same as the frequency dependent conductivity derived in Eq. (63).

$$\chi = \left(\frac{\partial P}{\partial E}\right)_T = \frac{1}{2\alpha_0(T - T_C)} \approx \varepsilon_r \quad (T > T_C)$$

This is often called the Curie-Weiss law for ferroelectric.

**Example 17.6** Show that the electric field created by a cubic array of atomic dipoles, with identical orientations, averages to zero inside a spherical region.

***Solution*** The expression for the field due to an atomic dipole simplifies considerably if detarmined at a distance from the charges which is large compared with $d$, as in the arrangement shown in the Fig. 17.38. If we take the origin of the coordinate system at the dipole centre, the electric potential due to the two charges at the centre ($C$) of the spherical dielectric rigon, with position vector $\vec{r}$, may be expressed as

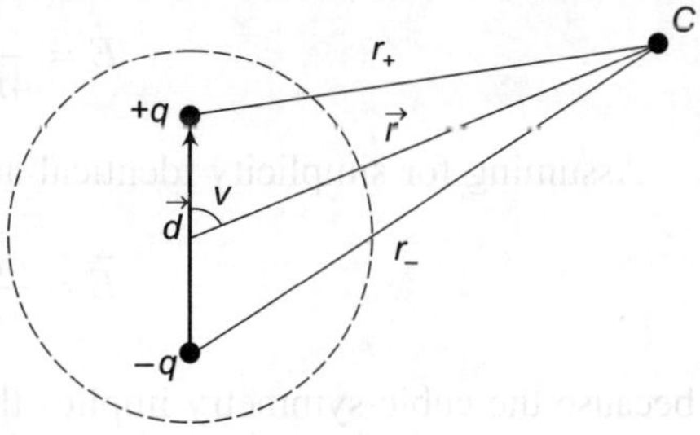


**Figure 17.38**

$$V(r) = -\frac{q}{4\pi\varepsilon_0}\int_0^{r_+}\frac{dr_+}{r_+^2} + \frac{q}{4\pi\varepsilon_0}\int_0^{r_-}\frac{dr_-}{r_-^2} = \frac{q}{4\pi\varepsilon_0}\left(\frac{1}{r_+} - \frac{1}{r_-}\right)$$

It is convenient to expand $r_+$ and $r_-$ into power series of the ratio $d/2r$ in the form

$$\frac{1}{r_+} = \frac{1}{\sqrt{r^2 + d^2/4 \mp rd\cos\theta}} = \frac{1}{r}\left(1 + \frac{d^2}{4r^2} + \frac{d}{r}\cos\theta\right)^{-\frac{1}{2}}$$

$$= \frac{1}{r}\left[1 \pm \frac{d}{2r}\cos\theta + \frac{1}{2}(3\cos^2\theta - 1)\left(\frac{d}{2r}\right)^2 \pm ...\right]$$

so that the potential is given in terms of the dipole moment $\vec{p}$ by

$$V(r) = \frac{qd\cos\theta}{4\pi\varepsilon_0 r^2} = \frac{p\cos\theta}{4\pi\varepsilon_0 r^2} = \frac{\vec{p}\cdot\vec{r}}{4\pi\varepsilon_0 r^3}$$

The electic field due to the atomic electric dipole is obtained by further differentiation

$$\vec{E} = -\nabla V(r) = -\nabla\left(\frac{\vec{p}\cdot\vec{r}}{4\pi\varepsilon_0 r^3}\right) = -\frac{1}{4\pi\varepsilon_0}\left[\frac{1}{r^3}\nabla(\vec{p}\cdot\vec{r}) + (\vec{p}\cdot\vec{r})\nabla\left(\frac{1}{r^3}\right)\right]$$

We can now write $\nabla(\vec{p}\cdot\vec{r}) = \vec{p}$, and since

$$\frac{\partial r}{\partial x} = \frac{\partial}{\partial x}(x^2 + y^2 + z^2)^{\frac{1}{2}} = \frac{x}{r}, \quad \frac{\partial r}{\partial y} = \frac{y}{r}, \quad \frac{\partial r}{\partial z} = \frac{z}{r}$$

it follows that

$$\nabla\left(\frac{1}{r^3}\right) = -\frac{3}{r^4}\left(\frac{x\vec{e}_x}{r} + \frac{y\vec{e}_y}{r} + \frac{z\vec{e}_z}{r}\right) = -\frac{3\vec{r}}{r^5}$$

Hence,
$$\vec{E} = \frac{1}{4\pi\varepsilon_0}\left[3\frac{(\vec{p}\cdot\vec{r})\vec{r}}{r^5} - \frac{\vec{p}}{r^3}\right]$$

For a group of atomic dipoles $\vec{p}_i$, located at distances $r_i$ from the point $C$, the total field at the centre of the sphere is

$$\vec{E} = \frac{1}{4\pi\varepsilon_0}\sum_i \frac{3(\vec{p}_i\cdot\vec{r}_i)\vec{r}_i - r_i^2\vec{p}_i}{r_i^5}$$

Assuming for simplicity identical atomic dipoles parallel to the $z$-axis, one obtains

$$\vec{E} = \frac{\vec{p}}{4\pi\varepsilon_0}\sum_i \frac{3z_i^2 - r_i^2}{r_i^5} = \frac{\vec{p}}{4\pi\varepsilon_0}\sum_i \frac{2z_i^2 - x_i^2 - y_i^2}{r_i^5} = 0$$

because the cubic symmetry implies that the $x$-, $y$- and $z$-directions are equivalent

$$\sum_i \frac{x_i^2}{r_i^5} = \sum_i \frac{y_i^2}{r_i^5} = \sum_i \frac{z_i^2}{r_i^5}$$

and hence, the electric field must vanish.

**Example 17.7** Show that the induced atomic dipole moment is not affected by thermal oscillation of charges.

***Solution*** If the thermal motion of charges is neglected, and the effective field, $E_{\text{eff}} \approx E$, is taken to be along the $x$-axis, the equilibrium condition on each atom yields

$$qE_{\text{eff}} = \kappa_0 \langle x \rangle \quad \text{or} \quad \langle p \rangle = q\langle x \rangle = \frac{q^2 E_{\text{eff}}}{\kappa_0} \approx \frac{q^2 E}{\kappa_0}$$

where $\kappa_0$ stands for the force constant. In the presence of thermal oscillations, the potential energy becomes

$$\varepsilon = \frac{1}{2}\kappa_0 x^2 - qxE$$

and the mean dipole moment is given by

$$\langle p \rangle = \frac{\int_{-\infty}^{\infty} p e^{-\varepsilon/k_B T} dx}{\int_{-\infty}^{\infty} e^{-\varepsilon/k_B T} dx} = \frac{\int_{-\infty}^{\infty} qx e^{-\kappa_0 x^2/2k_B T} e^{qxE/k_B T} dx}{\int_{-\infty}^{\infty} e^{-\kappa_0 x^2/2k_B T} e^{qxE/k_B T} dx}$$

For weak applied fields, we have $e^{qxE}/k_B T \approx 1 + qxE/k_B T$, and hence, evaluating the Gaussian integrals, one obtains

$$\langle p \rangle = \frac{\dfrac{q^2 E}{k_B T}\dfrac{\sqrt{\pi}}{2}\left(\dfrac{2k_B T}{\kappa_0}\right)^{\frac{3}{2}}}{\dfrac{\sqrt{\pi}}{2}\left(\dfrac{2k_B T}{\kappa_0}\right)^{\frac{1}{2}}} = \frac{q^2 E}{\kappa_0}$$

which is the same as the value derived neglecting thermal oscillations.

## 17.32 ELECTRETS

An electret is understood as a body composed of dielectric, which preserves its polarization for a long time after the external electric field that has produced this polarization is removed, and which sets up an electrostatic field in the surrounding space.

The word "electret" is formed by an analogy with word "magnet". An electret establishes an electric field for a long time in the same way as a permanent magnet sets up a continuous magnetic field. The existence of electrets was surmised by the Russian academician F. Epinus, M. Lomonosov's contemporary, and later by the English physicist O. Heaviside which introduced in 1886 the term "Electret" and studied some theoretical problems related to electrets. But it was only in 1921 that the Japanese researcher Mototaro Eguchi actually received and investigate the electrets. After this, the electrets have become the object of numerous experimental and theoretical studies and found an increasingly wider practical application.

Figure 17.39 illustrates the installation designed by Eguchi in which the first electrets have been obtained. A round metal bath $A$ with an inlet $B$ placed on insulators $F$ and support $D$ is adjusted strictly horizontal with the help of screws $E$. The bath is filled with a molten dielectric (a mixture of wax and resin) after which electrode $C$ suspended from a thread with counterweight $G$ is let down until it touches the surface of the dielectric, and a direct-current voltage is applied between electrodes *A* and *C*. This voltage is maintained until the molten dielectric cools down and solidifies in the electric field. Such a dielectric becomes an electret; the opposite surfaces of the disc have unlike charges which can persist for years, especially if the electret is shorted (similar to the magnet with the poles short circutied through a ferromagnet). If the outer layer of the dielectric are cut off the charge of the electret will not be changed, while two electrets are obtained when the original electret is cut along the neutral plane (in the same way as when a magnet is cut in two parts). If an electret is moistened or intensely irradiated by X-rays, most of the charge is lost, but after drying or cessation of irradiation the charge is restored. A completely molten electret will be fully discharged.

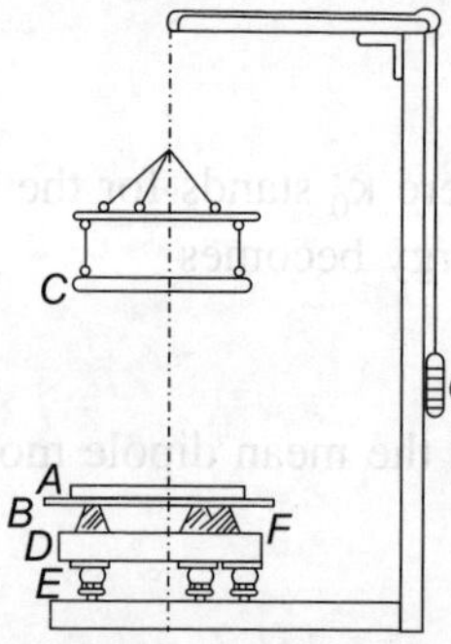


**Figure 17.39** Eguchi's installation for obtaining thermoelectrets

It has been found later that when a heated dielectric is cooled in an electric field, electrets (so-called thermoelectrets) can be obtained from various dielectrics—both organic (polymethylmethacrylate, polytetrafluoroethylene, ebonite, ceresin, naphthalene, beewax, etc.) and inorganic (titanates of various metals – barium, magnesium, zinc, calcium, etc., steatite, sulphur, etc). The intensity of the polarizing electric field ($E_p$) appreciably affects properties of an electret. When $E_p$ is small, (approx. up to I Mv/m, the obtained electret possesses charges opposite to the polarizing of the polarizing voltage, Fig. (17.40a). If $E_p$ is > I Mv/m the electret receives charges of the same sing as the electrodes with which the surfaces of the electret were in contact during polarization, Fig. (17.40b).

The charges on the first type are known as **heterocharges** and of the second type as **homocharges**. The stability of these charges can be different. An example is illustrated in Fig. (17.41) which shows that in this case after a comparatively rapid drop of the heterocharge, a homocharge is set up which is reduced rather slowly.

If an electret is properly manufactured the charges produced on its side may have a surface density up to $1.3 \times 10^{-5}$ c/m$^2$. This density of charges corresponds to the intensity of the field near the surface of the electret up to 1–4 Mv/m, which approaches the magnitude of the

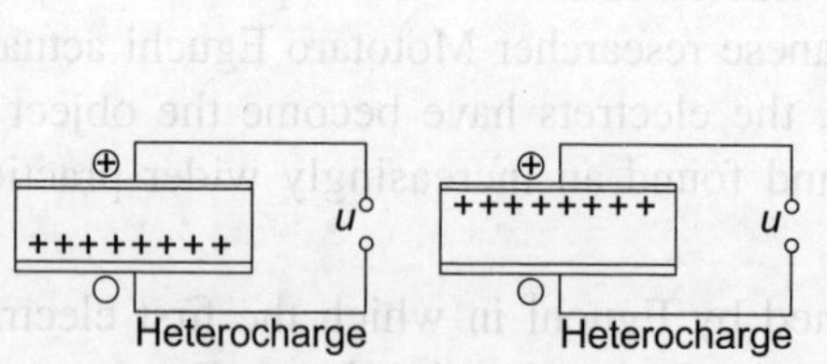


**Figure 17.40** Heterocharge and homocharge of electrets

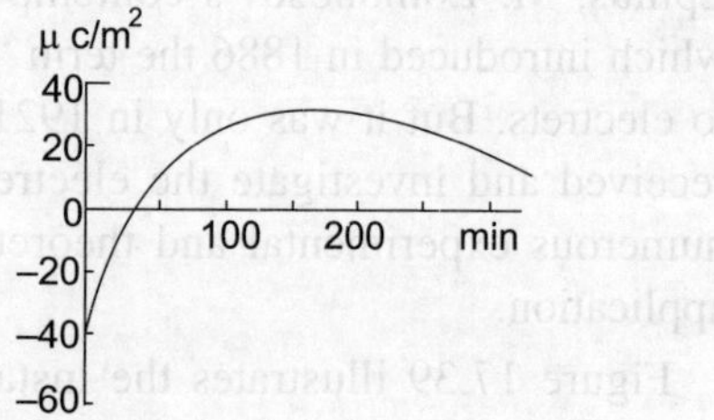


**Figure 17.41** Charge density of an electret made of Carnauba wax vs. time elapsed after polarization of the electret

electric strength of the air. An electret obtained in an atmosphere of increased pressure may have a surface density of the charge up to $5 \times 10^{-5}$ c/m$^2$.

The method of manufacture of electrets (thermoelectrets) described above and employed under an increased temperature is not only one. In 1937, G. Nadjakov of Bulgaria devised a new type of electret, **photoelectret**. Such electrets are made of photoconductive (i.e., acquiring a higher electrical conductivity with illumination) dielectrics. To make a photoelectret, a dielectric placed into a strong electric field is illuminated after which the electric field is taken off. Photoelectrets possess a charge (usually a heterocharge) which can be preserved for a long time if the photoelectret is kept in darkness (when exposed to light the charge disappears).

**Electroelectrets** are obtained under the active of an electric field without heating a dielectric **radioelectrets** are produced by radioactive radiation and **mechanoelectrets**, by a mechnical effect, i.e., deformation or friction of a dielectric. Under an ionizing radiation, γ-rays, for example, some dielectrics become **pseudoelectrets** obtained without the application of votlage. **Magnetoelectrets** first discovered by C. Bhatnagar of India from when a heated metal (polymethyl metha crylate, for example) is acted upon by a strong permanent magnetic field with a subsequent cooling also in a magnetic field.

Electrets are employed for a continuous generation of an electric field without an external source of voltage—in electret microphones, vibrotransmitters, electrometers, electrostatic generators, for forcussing an electronic beam in electron—ray tubes, etc. Photoelectrets are utilized to make rapid photographic prints without liquid solutions.

## 17.33 DESIGN CRITERIA FOR DIELECTRICS

Designers of capacitors require material with large dielectric constant. According to Clausius – Mosotti equation these meterials should be ionic compounds with large atomic polarizabilities and large ionic polarizability. Titanium oxide which has a dielectric constant of about 100 is a good example of such a material. The use of molecules or atom group with permanent dipole moment is not desirable because they involve large dielectric losses at high frequencies.

In microwave application one requires materials with low dielectric constant. According to the Clausius-Mosotti equation this is achieved by designing non-ionic compounds of atoms with small atomic polarizabilities and no free dipoles. Some organic compounds such as polystrene and polyethylene are examples of this group.

It is often important to design oscillator tank circuits with temperature independent tuning. This can be done by using a capacitor with a negative temperature coefficient as part of the tuned circuit capacitance. $TiO_2$ capacitors are very useful in this respect.

The following design criteria hold for low-loss dielectrics.

1. There should be no dipoles in the dielectric either of the bound or free type.
2. There should be no ions that can move more or less freely. Free ions in the cleavage planes of mica for instance give extra losses, the dielectric losses of mica are reduced by boiling in strong $H_2SO_4$.
3. Ions that fit tightly in their proper position and that can move to another position only upon overcoming a large activation energy give small losses.

4. The material should not show ionic conductivity since that gives rise to larger loss factor at low frequencies. Besides this ionic conductivity may ultimately destroy the dielectric by electrolysis.

## 17.34 PIEZOELECTRIC EFFECT

The word Piezo in Greek means pressure and therefore Piezoelectricity means "pressure electricity". In certain crystals which lack certain symmetrical properties, a mechanical stress produces an electric polarization and reciprocally an applied field produces a mechanical strain. These effects are called the direct and the inverse piezoelectric effects.

The physical cause of this effect could be understood by looking at the crystal structure of the compound $MgSO_4$ $6H_2O$. The crystal consists of two types of atom groups: the group $Mg^{++}$, $6H_2O$ which is an $Mg^{2+}$ ion surrounded symmetrically by six water molecules and $S^{4+}(O^{2-})_3$ group which has a pyramid structure with the S atom on top. All $(SO_3)^{2-}$ pyramids point with their top in the same direction, it is thus understandable that a polarization may occur if these pyramids are compressed in the direction of their axes. The crystal should therefore be piezoelectric. On the other hand consider the compound Zn $(BrO_3)_2$ $6H_2O$. The $Zn^{++}$. $6H_2O$ groups are arranged in the same way as the $Mg^{++}$. $6H_2O$ group in $MgSO_4$. $6H_2O$ and the $(BrO_3)^-$ groups have again a pyramidal structure with the Br atom on top. But there are here twice as many $(BrO_3)^-$ groups as there are $(SO_3)^{2-}$ groups in $MgSO_4 \cdot 6H_2O$ and the tops of the pyramidal groups $(BrO_3)^-$ point alternately in opposite direction, hence a compression along their axis cannot result in a net polarisation. Consequently, Zn $(BrO_3)_2$ $6H_2O$ is not piezoelectric.

The best known piezoelectric material is quartz. Lord Kelvin in his explanation of piezoelectric effect considered each silicon atom to be positively charged and each oxygen double atom to be represented by a negative charge (Fig. 17.42). Under a compressional stress the angle θ increases but the distance between *A* and *B* remains unchanged. On comparing the new arrangement with the old positive charges have a net displacement towards the left and the negative charges towards the right resulting in a net dipole moment pointing in the direction *BA*, Fig. (17.42 b). For a shearing stress the moment of atom *D* towards the right is compensated by the moment of *F* towards the left. The downward movement of atoms *A*, *D*

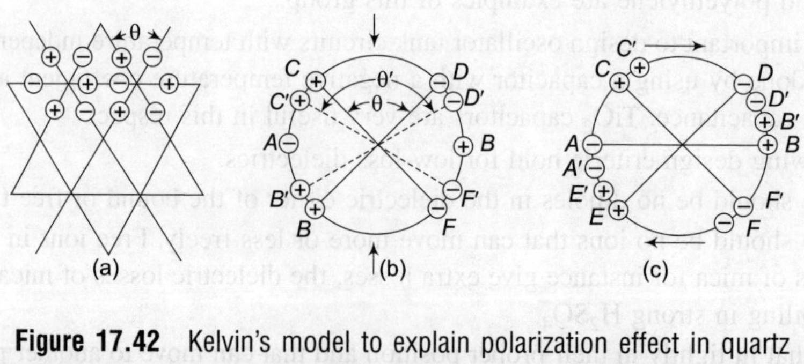


**Figure 17.42** Kelvin's model to explain polarization effect in quartz

and $F$ are additive. Therefore the cluster of atoms has a net dipole moment pointing upwards Fig. (17.42 c).

Considering the stress on a elementary cube, the stress exerted on the area $ABCD$ can be resolved into three components, $T_{xx}$, $T_{yx}$, $T_{zx}$ (Fig. 17.43). The first subscript denotes the direction of the stress and the second subscript indicates the direction of the normal to the plane, $ABCD$. $T_{xx}$ denotes the tensile stress. $T_{yx}$ and $T_{zx}$ denote the shear stress. For torque to balance around the $z$-axis, $T_{xy} = T_{yx}$. Therefore, there are altogether three tensile stresses

$$T_1 = T_{xx},\ T_2 = T_{yy},\ T_3 = T_{zz} \qquad ...(1)$$

and three shear stresses

$$T_4 = T_{yz} = T_{xy},\ T_5 = T_{xz} = T_{zx},\ T_6 = T_{xy} = T_{yx} \qquad ...(2)$$

The net force in the $x$-direction is given by

$$F_x = \left(\frac{\partial T_1}{\partial x}\,dx\right) dy\, dz + \left(\frac{\partial T_6}{\partial y}\,dy\right) dx\, dz + \left(\frac{\partial T_5}{\partial z}\,dz\right) dx\, dy \qquad ...(3)$$

$$= \left(\frac{\partial T_1}{\partial x} + \frac{\partial T_6}{\partial y} + \frac{\partial T_5}{\partial z}\right) dx\, dy\, dz \qquad ...(4)$$

Similar expressions can be obtained for $F_y$ and $F_z$.

For *strain components* consider two points $P(0, 0, 0)$ and $Q(x, y, z)$ under stress, Fig. (17.43 b). The two points are shifted to $P'(\xi_1, \eta_1, \zeta_1)$ and $Q'(\xi_1 + \xi, \eta_2 + y, \zeta_2 + z)$. Then

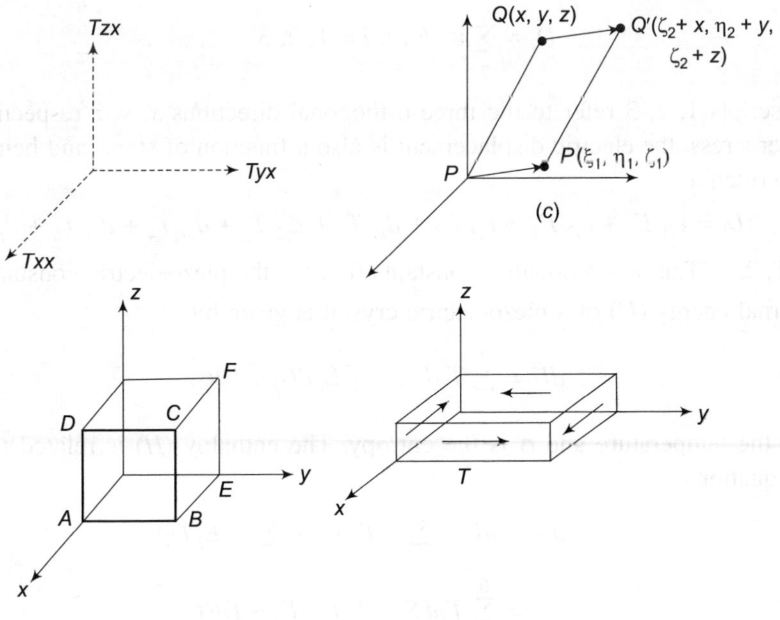


**Figure 17.43** Diagrams defining the mechanical stresses and strains in crystals

$$\xi^2 = \xi_1 + \frac{\partial \xi}{\partial x} x + \frac{\partial \xi}{\partial y} y + \frac{\partial \xi}{\partial z} z$$

$$\eta_2 = \eta_1 + \frac{\partial \eta}{\partial x} x + \frac{\partial \eta}{\partial y} y + \frac{\partial \eta}{\partial z} z \qquad \text{...(5)}$$

$$\zeta_2 = \zeta_1 + \frac{\partial \zeta}{\partial x} x + \frac{\partial \zeta}{\partial y} y + \frac{\partial \zeta}{\partial z} z$$

The linear strains in the $x$, $y$ and $z$-directions are given by

$$S_1 = \frac{\partial \xi}{\partial x},\ S_2 = \frac{\partial \eta}{\partial y},\ S_3 \frac{\partial \zeta}{\partial z} \qquad \text{...(6)}$$

and the strains around the $x$, $y$ and $z$ axes are defined as

$$S_4 = \frac{\partial \xi}{\partial y} + \frac{\partial \eta}{\partial z},\ S_5 = \frac{\partial \xi}{\partial z} + \frac{\partial \xi}{\partial x},\ S_6 = \frac{\partial \xi}{\partial y} + \frac{\partial \eta}{\partial x} \qquad \text{...(7)}$$

According to Hooke's law, stresses are proportional to the strains

$$T_i = \sum_j C_{ij}\, S_j,\ i, j = 1 \text{ to } 6. \qquad \text{...(8)}$$

or the inverse relation $$S_i = \sum_j S_{ij}\, T_i,\ i, j = 1 \text{ to } 6. \qquad \text{...(9)}$$

Since most piezoelectric crystals are anisotropic, the relationship between $D$ and $E$ is given by

$$D_i = \sum_j \varepsilon_{ij}\, E_j,\ i, j = 1, 2, 3. \qquad \text{...(10)}$$

The subscripts 1, 2, 3 refer to the three orthogonal directions $x$, $y$, $z$ respectively. For a crystal under stress, the electric displacement is also a function of stress and hence Eq. (10) should be written as

$$D_i = \varepsilon_{i1} E_1 + \varepsilon_{i2} E_2 + \varepsilon_{i3} E_3 + d_{i1} T_1 + d_{i2} T_3 + d_{i4} T_4 + d_{i5} T_5 + d_{i6} T_6 \qquad \text{...(11)}$$

where $i = 1, 2, 3$. The proportionality constants ($d_{ij}$) are the piezoelectric constants.

The internal energy ($U$) of a piezoelectric crystal is given by

$$dU = \sum_j^6 T_j dS_j + \sum_i^3 E_i\, dD_i + Td\sigma \qquad \text{...(12)}$$

where $T$ is the temperature and $\sigma$ is the entropy. The enthalpy ($H$) is related to $U$ by the following equation

$$dH = dU - \sum d(T_j\, S_j) - \sum d(E_i\, D_i) \qquad \text{...(13)}$$

$$= \sum_j^6 T_j dS_j + \sum_i^3 E_i\, dD_i + Td\sigma \qquad \text{...(14)}$$

For $dH$ to be total differential we have

$$S_j = \frac{\partial H}{\partial T_j} \quad \text{and} \quad D_i = \frac{\partial H}{\partial E_i}$$

i.e.,

$$\frac{\partial S_i}{\partial E_i} = \frac{\partial D_i}{\partial T_j} \qquad ...(15)$$

According to Eq. (15), the inverse relationship between $S$ and $E$ can be written as

$$S_j = d_{j1}\, E_i + d_{j2}\, E_2 + d_{j3}\, E_3 + S_{j1}\, T_i + s_{j2}\, T_2 + s_{j3}\, T_3 + s_{j4}\, T_4 + s_{j5}\, T_5 + s_{j6}\, T_6 \qquad ...(16)$$

### 17.34.1 A Simple Example of the Application of General Relations to Piezoelectric Effect

Consider a thin dielectric slab (Fig. 17.44) whose thickness and width are much smaller than its length. Electrodes are plated to the surfaces normal to the $z$-axis and an alternating voltage is applied across the electrodes. Our task is to find an equivalent circuit for the dielectric slab taking the piezoelectric effect into consideration. At present we restrict the discussion to longitudinal vibrating mode of the slab, i.e., along the $x$-direction. Under such circumstances, the surfases normal to the $y$ and $z$ axes can be considered free of stresses. Thus, $T_2 = T_3 = 0$ and $T_{yz} = T_{xy} = T_{xz} = 0$ or $T_4 = T_5 = T_6 = 0$. Since $E_1 = E_2 = 0$. Equation (16) reduces to

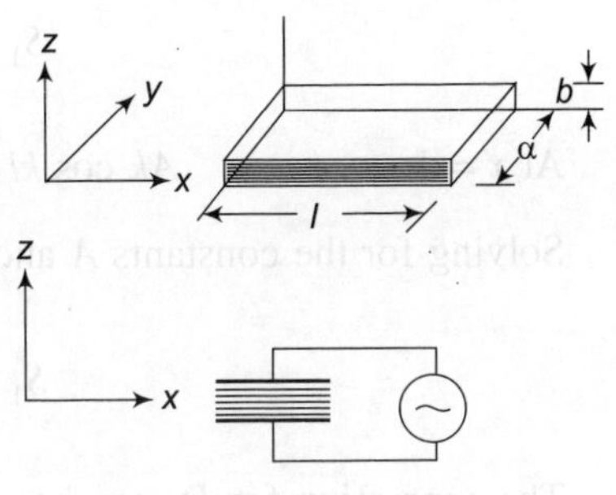


**Figure 17.44** A thin piezoelectric crystal under an alternating electric field

$$S_1 = S_{11}\, T_1 + d_{13} E_3 \qquad ...(1)$$

The displacements of the tube in the $x$, $y$ and $z$ directions are given by $\xi$, $\eta$, and $\zeta$, respectively. Hence, according to Newton's law of motion the net force acting on the cube in the $x$-direction is given by

$$F_x = \rho\, dx\, dy\, dz\, \frac{d^2\xi}{dt^2} \qquad ...(2)$$

where $\rho$ is the density. Equating Eq. (2) with Eq. (4) and remembering $T_5 = T_6 = 0$ we have

$$\rho\, \frac{d^2\xi}{dt^2} = \frac{\partial T_1}{\partial x} \qquad ...(3)$$

By substituting Eq. (1) into Eq. (3) and letting $S_1 = \left(\frac{\partial \xi}{\partial x}\right)$ Eq. (3) is further transformed into

$$\rho\, \frac{d^2\xi}{dt^2} = \frac{1}{s_{11}} \cdot \frac{d^2\xi}{dx^2} - \frac{d_{13}\, \partial\, E_3}{s_{11}\, \partial x} \qquad ...(4)$$

Since the applied field is independent of $x$, the last term in Eq. (4) vanishes. Thus, Eq. (4) is a simple wave equation, the solution of which is given by

$$\xi = A \sin kx + B \cos kx \qquad ...(5)$$

where $k = \omega/v$, $\omega$ is the angular frequency of oscillation

and $$v = \frac{1}{\sqrt{(\rho s_{11})}}$$

is the velocity of sound propagation in the plated crystal.

The boundary conditions are those for a free crystal at $x = 0$, and $x = 1$, $T_1 = 0$. Thus, from Eqs. (1) and (5) we have, at $x = 0$

$$S_1 = \frac{\partial \xi}{\partial x} = Ak = d_{13} E_3. \qquad ...(6)$$

At $x = 1$, $$Ak \cos kl - Bk \sin kl = d_{13} E_3. \qquad ...(7)$$

Solving for the constants $A$ and $B$ and substituting values obtained in Eq. (5), we get

$$S_1 = \frac{d_{13} E_3 \cos [k(1 - 2x)/2]}{\cos (kl/2)} \qquad ...(8)$$

The expression for $D_3$ can be obtained from Eqs. (11) and (1) (section 17.33) as follows

$$D_3 = \varepsilon_{33} E_3 + d_{13} T_1 = \left(\varepsilon_{33} - \frac{d_{13}^2}{S_{11}}\right) E_3 + \frac{d_{13}}{S_{11}} S_1 \qquad ...(9)$$

Using Eq. (9) and integrating Eq. (9) over the surface normal to the $z$-axis, the current is given by

$$I_3 = i\,\omega \int_0^1 D_3\, a\, dx = i\omega \left[\left(\varepsilon_{33} - \frac{d_{13}^2}{S_{11}}\right) al + \frac{d_{13}^2}{s_{11}} \frac{2a}{k} \tan\left(\frac{k\,1}{2}\right)\right] E_3 \qquad ...(10)$$

and the equivalent admittance,

$$Y = \frac{I_3}{E_3 b} = i\,\omega \frac{a\,1}{b}\left[\left(\varepsilon_{33} - \frac{d_{13}^2}{S_{11}}\right) + \frac{d_{13}^2}{s_{11}} \frac{2}{k1} \tan\left(\frac{k\,1}{2}\right)\right] \qquad ...(11)$$

A quick look at Eq. (11), shows that $Y$ possesses resonant and antiresonant frequencies. From equivalent circuit (Fig. 17.45), let the electromechanical coupling coefficient ($K$) be defined as

$$K = \frac{d_{13}}{\sqrt{(\varepsilon_{33} s_{11})}} \qquad ...(12)$$

At very low frequencies

$$Y = i\,\omega\,\varepsilon_{33}\frac{a1}{b} = i\,\omega(C_0 + C_1) \quad ...(13)$$

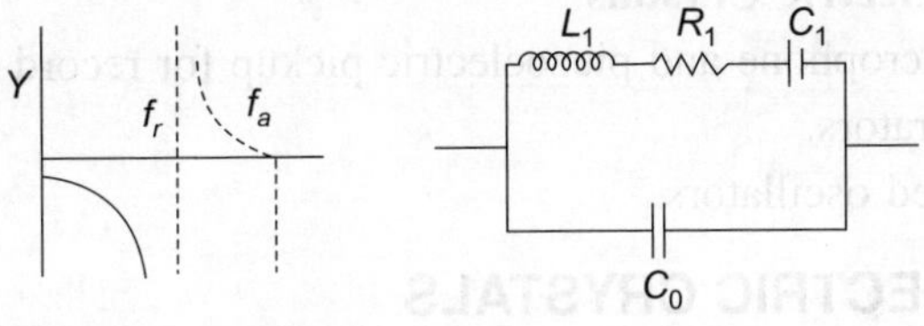


**Figure 17.45**

and at resonance frequency $f$, $Y = \infty$. $kl/2 = \pi/2$

or $$f_r = \frac{1}{21\sqrt{(\rho s_{11})}} = \frac{1}{2\pi\sqrt{(L_1 C_1)}} \quad ...(14)$$

At antiresonance frequency $f_a$, $Y = 0$

or $$\frac{1-k^2}{k^2} = \frac{2v}{\omega_a 1}\tan\left(\frac{\omega_a 1}{2v}\right) \quad ...(15)$$

If $f_a = f_r + \Delta f$ and $\Delta f/f_r \ll 1$, then tan $(\omega_a/2v)$ is approximately equal to $2v/(2\pi\Delta f1)$. But $\omega_r 1/2v = \pi/2$.

Then from Eq. (15), we have

$$\frac{\Delta f}{f_r} = \frac{4}{\pi^2}\cdot\frac{K^2}{1-K^2} \quad ...(16)$$

From the equivalent circuit

$$\omega_a^2 L_1\left(\frac{C_0 C_1}{C_0 + C_1}\right) = 1 \quad \text{and} \quad \omega_r^2 L_1 C_1 = 1 \quad ...(17)$$

and the ratio $\Delta f/f_r \approx (\omega_a^2 - \omega_r^2)/2\omega_r^2$ becomes

$$\frac{\Delta f}{f_r} = \frac{C_1}{2C_0} = \frac{4}{\pi^2}\frac{K^2}{1-K^2} \quad ...(18)$$

(i) Parameters in the equivalent circuit ($L_1$, $C_1$ and $C_0$) can be completely determined from measurement of $f_r$, $f_a$ and $Y$ at low frequencies.

(ii) The small resistance in equivalent circuit represents the internal losses in the crystal which includes the loss of energy at the clamping points and the loss of energy due to setting up of air waves by the crystal motion.

(iii) Quality factor ($Q$) is defined as $\omega\,L_i/R_1$ and is as high as $3 \times 10^6$.

(iv) Parameters $\varepsilon_{33}$, $s_{11}$ and $d_{13}$ can be determined from the same set of measurements.
(v) There are other more complicated modes of motion besides the simple longitudinal mode involving different elastic and electromechanical constants.

**Applications of Piezoelectric Crystals**

1. Piezoelectric microphone and piezoelectric pickup for record players.
2. Ultrasonic generators.
3. Crystal-controlled oscillators.

## 17.35 FERROELECTRIC CRYSTALS

A crystal is said to be ferroelectric if it exhibits a spontaneous polarization in the absence of an applied field. Substances like Rochelle salt ($NaKC_4H_4O_6 \cdot 4H_2O$), potassium dihydrogen phosphate ($KH_2PO_4$), barium titanate (Ba $TiO_3$) are typical examples. The existence of spontaneous polarization is necessary but is not a sufficient condition for ferroelectricity. Ferroelectric crystals also lack an inversion centre, i.e., the centre of positive charge of the crystal does not coincide with the centre of negative charge, This can be illustrated as such.

For example, a crystal has a centre of inversion *O* Fig. (17.46). Atoms *A* and *B* possessing negative charges are symmetrically located with respect to *O* resulting in a cancellation of dipole moments. The same applies to atoms *C* and *D* possessing positive charges. Under isotropic stress, atoms will be symmetrically displaced with respect to the centre of symmetry hence such a crystal cannot be ferroelectric. Diamond structure in the cubic system is such an example because for every atom in the diamond structure there is always another atom oppositely situated. Therefore a crystal having a diamond structure cannot be ferroelectric. Among the 32 crystal classes, 12 crystal classes have centre of inversion and therefore are not ferroelectric. Notably among those are $CuSO_4$ (triclinic), $CaSO_4 \cdot 2H_2O$ (monoclinic), $BaSO_4$ (orthorhombic), $ZnSiO_4$ (tetragonal), $CaCO_3$ (trigonal), $Be_3Al_2(SiO_3)_6$ (hexagonal) and NaCl (cubic).

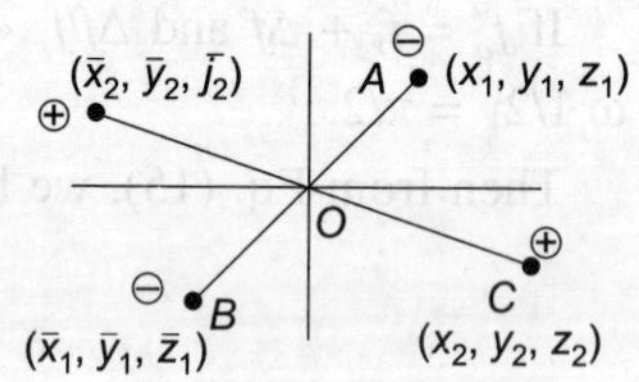


**Figure 17.46**

Besides the centre of inversion, there are other elements of symmetry such as planes of symmetry and axes of symmetry. A combination of such symmetry elements can also eliminate the possible existence of axes of symmetry. A combination of such symmetry elements can also eliminate the possible existence of a polar axis and hence the possibility of ferroelectricity. This can also be understood as under.

Suppose a crystal has a twofold symmetry about the *x*- and *y*-axes. By *n*-fold symmetry, we mean that the crystal duplicates itself if it is rotated about an axis by $2\pi/n$ radians. For a dipole moment with vector components $(u, v, w)$ there must be another dipole moment $(u, -v, -w)$ by virtue of twofold symmetry about the *x*-axis. The symmetry requirments about the *y*-axis and further two vectors with components $(u, v, -w)$ and $(-u, -v, w)$. The resultant dipole moment is obviously zero, therefore a crystal possessing the said symmetry cannot be ferroelectric. Beside 12 classes among the 32 crystal classes possessing a centre of inversion additional 10 classes have symmetry elements which exclude the possible existence of a polar axis and hence ferroelectricity. However, structural changes which destroy the symmetry of a crystal can bring about ferroelectricity particularly with a change of temperature.

## 17.36 CHANGE OCCURRING IN CRYSTAL STRUCTURE DURING POLARIZATION

Mostly known ferroelectric materials are classified into three main groups typically characterized by (1) Rochelle salt (2) Potassium dihydrogen phosphate and (3) Barium Titanate.

The crystal first discovered to be ferroelectric is Rochelle salt of chemical composition $NaKC_4H_4O_6 \cdot 4H_2O$, a double tartarate of sodium and potassium with four molecules of water. Above 24°C the crystal is orthorhombic Fig. (17.47). In orthorhombic region the three crystal axes are axes of twofold symmetry and therefore the crystal is not ferroelectric.

Below the transition temperature the crystal is slightly distorted and transformed into monoclinic structure. In the ferroelectric region the spontaneous polarization is along the *x*-axis. The relative dielectric constant of Rochelle salt along different axes *y* and *z* are shown as function of temperature Fig. (17.48) In the temperature range from 24°C to 17°C a hysteresis

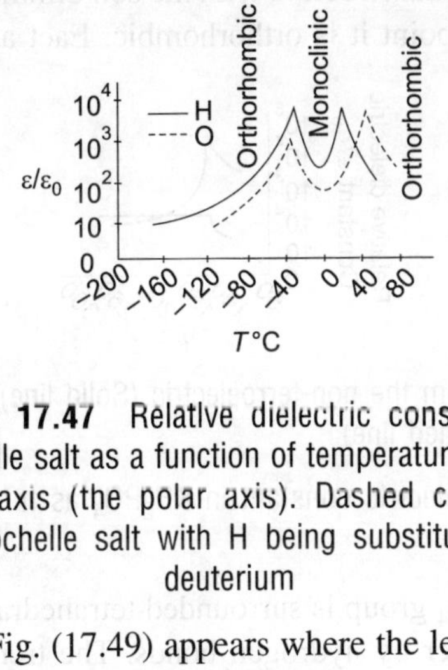


**Figure 17.47** Relative dielectric constant of Rochelle salt as a function of temperature along the x-axis (the polar axis). Dashed curve is for Rochelle salt with H being substituted by deuterium

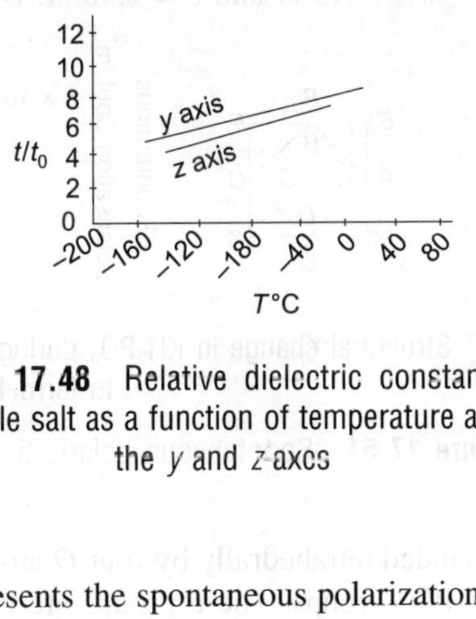


**Figure 17.48** Relative dielectric constant of Rochelle salt as a function of temperature along the *y* and *z* axcs

loop Fig. (17.49) appears where the length (*OC*) represents the spontaneous polarization (*P*) and *OA* the coercive field, $E_c$.

The critical temperature at which a crystal becomes ferroelectric is called the *Curie temperature*. For Rochelle salt, there exists an upper and lower *Curie temperature*. The substitution of hydrogen by deuterium changes the properties of the crystal from solid to dotted curves Fig. 17.50 (a) and Fig. 17.50 (b). The deuterons for protons nearly doubles the Curie temperature although the fractional change in the molecular weight is less than two per cent. This extraordinary isotope shift is believed to be a quantum effect involving the mass dependence of the de Broglie wavelength. Deuterium

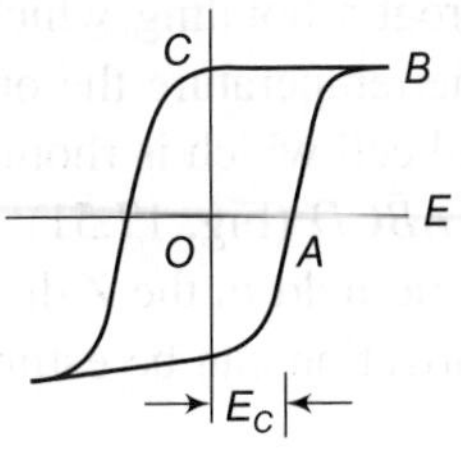


**Figure 17.49** Hysteresis loop for Rochelle salt between 30°C and –20°C

substitution also changes the spontaneous polarization and coercive force. There are other tartarates isomorphous with Rochelle salt with the replacement of potassium by Rb, TI or $NH_4$ but none of these is ferroelectric.

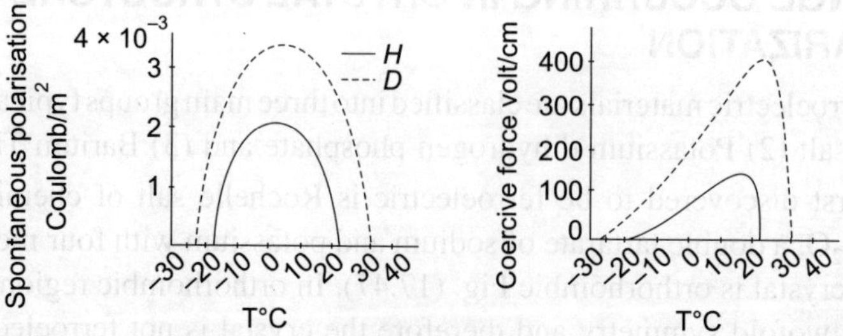


**Figure 17.50**

The most important member of the tetragonal phosphate group of ferroelectric substances is $KH_2PO_4$ (potassium dihydrogen phosphate). It has a Curie point –150°C.

Above the Curie temperature the crystal has a tetragonal structure with the cell dimmension as $a = b = 7.45$ Å and $e = 6.96$ Å. Below the Curie point it is orthorhombic. Eact atom is

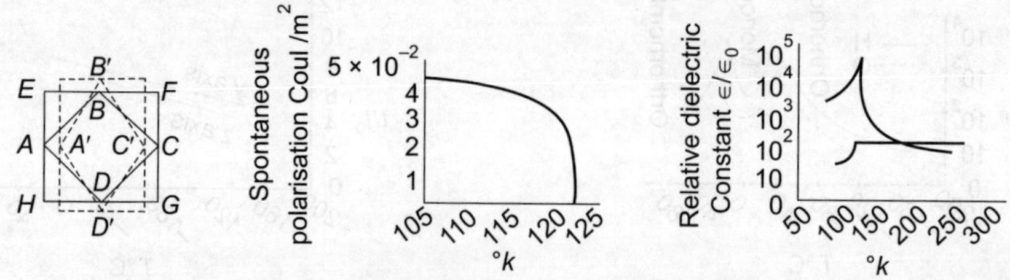


Structural change in $KH_2PO_4$ during the transition from the non-ferroelectric (Solid line) to ferroelectric phase (dashed line)

**Figure 17.51** Spontaneous polarization and relative dielectric constant in $KH_2PO_4$ as function of temperature

surrounded tetrahedrally by four *O* atoms and each $PO_4$ group is surrounded tetrahedrally by other $PO_4$ groups. These groups are connected together by hydrogen bonds. The trensition to the ferroelectric phase brought about by an ordering of the protons involved in the hydrogen bonding which is verified by X-ray and neutron diffraction studies. Below the Curie temperature the original aquare base becomes slightly distorted into a rectangle. The small cell which is rhombic may be considered being the result of shear stress applied to the cell *ABCD* (Fig. 17.51). Because of the piezoelectric effiect the shear strain ($S_6$) induces an electric field in the *Z* direction. Based on this model de Quervain proposes that spontaneous polarization can be estimated as

$$P_s = \varepsilon_c E_z = \varepsilon_e \frac{S_6}{d_{36}} \qquad \text{...(1)}$$

where $\varepsilon_c$ is the dielectric constant along the *c* axis or *Z* axis and $d_{36}$ is the piezoelectic constant. These two quantities can be measured first above the Curie temperature and the ratio $\varepsilon_e/d_{36}$ can be compared with the value of $P_s/S_6$ measured just below the Curie temperature. A good agreement between these two ratios found experimentally indicates that piezoelectricity and spontaneous polarization are caused by the same deformation of the crystal cell. Other ferroelectric crystals belonging to the tetragonal phosphate family include:

| | |
|---|---|
| $KD_2PO_4$ | 213°K |
| $KH_2ASO_4$ | 92°K |
| $RbH_2PO_4$ | 146°K |
| $CsH_2AsO_4$ | 143°K |
| $CsD_2AsO_4$ | 212°K |

Among ferroelectric material barium titanate $BaTiO_3$ has been most throughly investigated and best understood. Above 120°C (Curie point) $BaTiO_3$ has the ideal perovskite structure with a Ba atom sitting at the corners of a cube and *O* atoms at the mid points of the cube edges (Fig. 17.52).

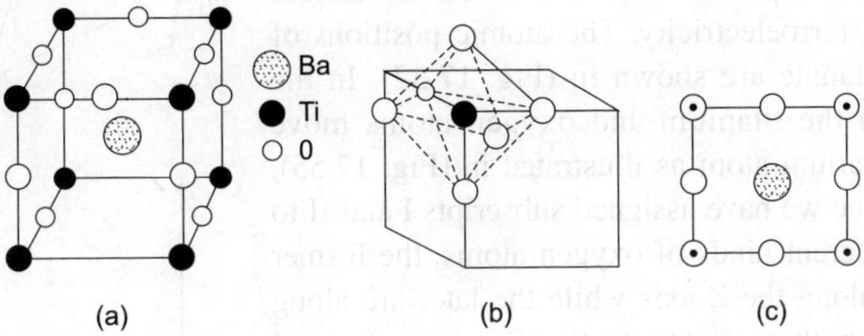


**Figure 17.52** Projection upon (010) Plane Ba (1/2, 1/2, 1/2), Ti (0, 0, 0), (0, 0, 1/2), (0, 1/2, 0), (1/2, 0, 0)

At the Curie point the structure becomes slightly stretched parallel to one cubic edge. The new structure is tetragonal and the stretched edge becomes the polar axis. In comparison to Rochelle salt and potassium dihydrogen phosphate in which one ferroelectric phase exists. $BaTiO_3$ exhibits three different ferroelectric phases. The temperature dependence of the relative dielectric constant and that of the spontaneous polarization of $BaTiO_3$ are shown in Figs. (17.53) and (17.54).

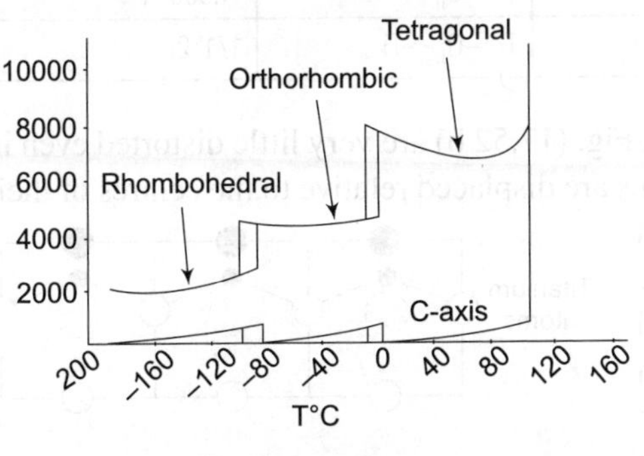


**Figure 17.53**

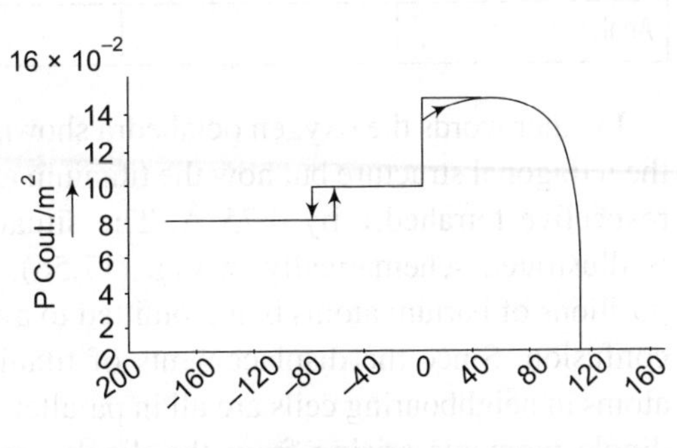


**Figure 17.54**

Between 5°C and –90°C the structure becomes orthorhombic by slightly stretching the original cubic structure along one face diagonal and compressing it along the other. These two face diagonals are now the new orthorhombic axes and one of the two face diagonals becomes the new polar axis. Below –90°C the structure changes again, this time into a rhombohedron formed by slightly stretching the original cube along its body diagonal which now becomes the new polar axis. The three-phase tetragonal, orthorhombic and rhombohedral are all ferroelectric. The quantity (*P*) measured experimentally is the component of $P_s$ normal to the surface of the specimen which coincides with the face of the cube. As the polar axis changes from cubic edge to face diagonal and then to the body diagonal, it is found that the value of *P* changes from $P_s$ to $\phi$ to $P_s/\sqrt{2}$ and then to $P_s/\sqrt{3}$ indicating that $P_s$ is essentially constant in the three ferroelectric phases. It may be instructive to see qualitatively how structural changes can bring about ferroelectricity. The atomic positions of cubic barium titanate are shown in (Fig. 17.52). In the tetragonal form the titanium and oxygen atoms move relative to the barium atom as illustrated in (Fig. 17.55). For easy reference we have assigned subscripts I and II to indicate two different kinds of oxygen atoms, the former being oriented along the *Z* axis while the later are along the *X* and *Y* axes with respect to the titanium atom. Several important points may be observed about this tetragonal structure. First the actual bond lengths between the barium and the surrounding oxygen atoms are practically the same while those between the titanium and oxygen atoms are quite different depending on the orientation of the oxygen atoms. The various bond lengths and angles in the tetragonal $BaTiO_3$ are given as follows:

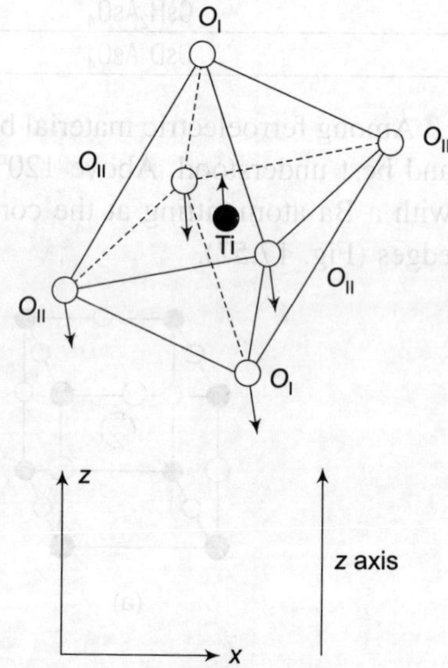


**Figure 17.55**

| Length | Ba—$O_I$ | 2.824 Å<br>2.800 Å | Ti—$O_I$ | 1.869 Å<br>2.167 Å |
|---|---|---|---|---|
| | Ba—$O_{II}$ | 2.886 Å | Ti—$O_{II}$ | 1.999 Å |
| Angle | | | Ti—$O_{II}$—Ti | 171°28 |

In other words the oxygen octahedra shown in Fig. (17.52 b) are very little distorted even in the tetragonal structure but now the titanium atoms are displaced relative to the centres of their respective tetrahedra by 0.75 Å. The situation is illustrated schematically in Fig. (17.56), the positions of barium atoms being omitted to avoid confusion. Since the displacements of titanium atoms in neighbouring cells are all in parallel, the dipole moments arising from the displacement

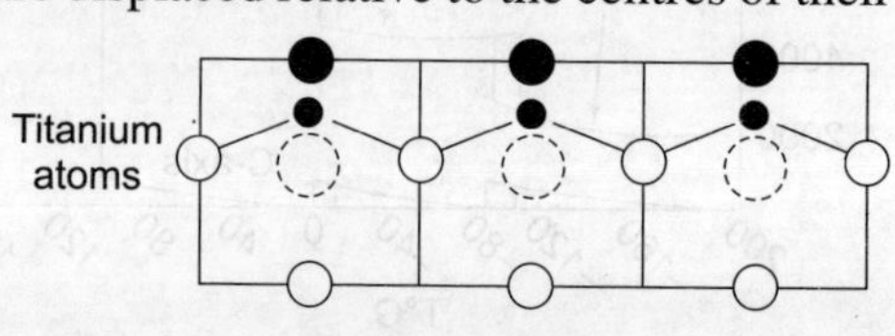


**Figure 17.56**

are additive, This corresponds to the state of ferroelectricity. In the perovskite family besides $BaTiO_3$, following are the other notable ferroelectric materials with their respective transition temperatures.

| | |
|---|---|
| $NaTiO_3$ | 750°K. |
| $RbTiO_3$ | 520°K. |
| $LiTiO_3$ | 473°K. |
| $PbTiO_3$ | 763°K. |
| $KNbO_3$ | 698°K. |

Ferroelectric crystals are therefore order-disorder or displacive according to whether the transition is associated with the ordering of ions or is associated with the displacement of a whole sublattice of ions of one type relative to another sublattice. The order-disorder class of ferroelectrics includes crystals with hydrogen bonds in which the motion of the protons is related to the ferroelectric prooerties as in potassium dihydrogen phosphate and isomorphous salts. The displacive class of ferroelectrics includes ionic crystal structures closely related to the perovskite like $BaTiO_3$ and ilemenite structures.

The occurrence of spontaneous polarization is well understood from X-ray and neutron diffraction studies.

## 17.37 FERROELECTRIC DOMAINS

A *domain* is defined as a region within which the direction of spontaneous polarization is constant and the region which separates two domains is called the *domain wall*. When the volumes of domains in opposite senses are equal the crystal as a whole will appear to be unpolarised. The total dipole moment of a crystal is therefore changed by the movement of the walls between domains or by nucleation of new domains.

Charges are produced on the surfaces perpendicular to the direction of the polarization. The electric field is therefore not zero and there is a depolarizing field due to the surface charges. The electrostatic energy due to the depolarizing field may be written as

$$U_E = \frac{1}{3} \Gamma P_S^2 V_S \qquad ...(1)$$

where $V_s$ is the volume of the sample.

$\Gamma$ is the depolarising factor which depends upon the shape of the sample.

If a sample is divided in a number of "domains" each showing uniform saturation polarization ($P_E$) and being surrounded by domains of opposite polarity, the depolarizing field will be reduced, a reason for the formation of domains.

Adjacent domains are separated by walls in which the polarization changes gradually from $+P_s$ to $-P_s$ whereas the strain changes gradually from $+e_s$ to $-e_s$. The electrostatic energy $U_E$ decreases strongly with an increase in the number of domains because of the decrease in depolarizing field, thus the splitting up into domains continues until the total energy is minimum.

The energy per square metre stored in a wall consists of two parts:

(1) The *dipole part* due to change in polarization from $+P_s$ to $-P_s$ in the wall.

(2) The *elastic part* due to change in stress from $+e_s$ to $-e_s$.

It is elastically favourable if the change in polarization comes about gradually and therefore dipole part is inversely proportional to the width of the wall.

It is elastically favourable if the change in strains comes about abruptly and thus the elastic part is directly proportional to the width of the wall. The wall thickness will thus be such that stored energy per square metre is minimum. In ferroelectric substances the wall thickness is of the order of a few lattice spacings and the wall energy is about $10^{-2}$ joule/m$^2$ or 10 erg/m$^2$.

If the directions of polarization in the two sides of the wall differ by 180°, they are called 180° walls (Fig. 17.57).

180° walls can occur in ferroelectric substances having only one possible direction of spontaneous polarization like Rochelle salt or potassium dihydrogen phosphate.

If the directions of polarization in the two domains separated by the wall are 90° to each other than that domain wall is known as 90° wall. In $BaTiO_3$ there are three possible directions

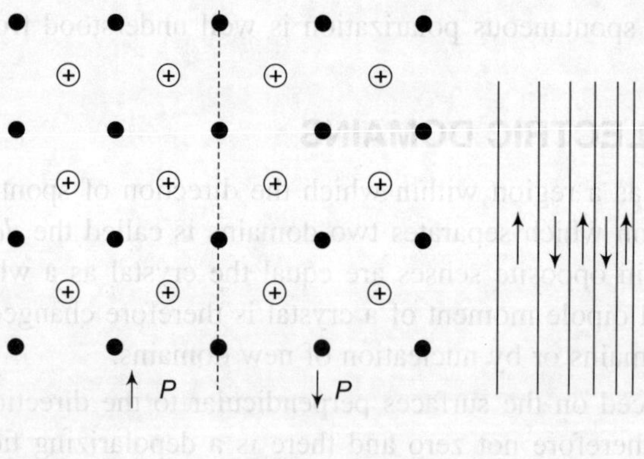


**Figure 17.57** (a) Schematic drawing of atomic displacements between domains polarized in opposite directions (b) 180° boundaries between domains

of spontaneous polarization and therefore it may happen that direction of polarization on both sides of the wall may differ by 90°. In tetragonal $BaTiO_3$ (experiments by Hooton and Merz) both 90° and 180° walls are present (Fig. 17.58).

### 17.37.1 Difference between Ferroelectric and Ferromagnetic Domains

1. Ferromagnetism arises due to a near perfect alignment of permanent magnetic dipoles while ferroelectricity is due to enhanced ionic polarization.
2. Ferroelectric domain wall is extremely thin perhaps several lattice constants as compared to several hundred lattice constants for a ferromagnetic domain wall.

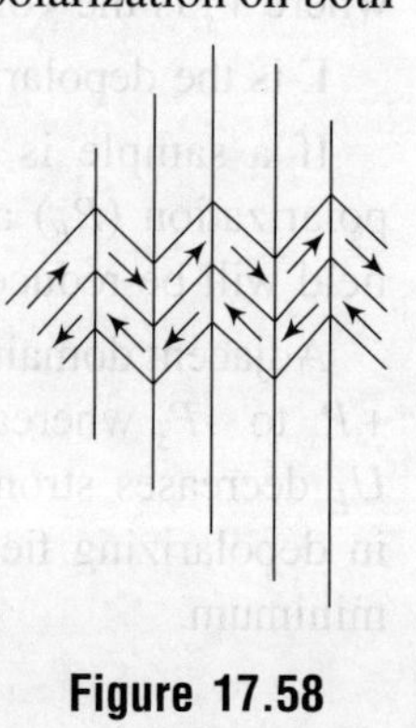

**Figure 17.58**

3. Since the energies associated with the parallel and antiparallel phase are close so that a ferroelectric state can be induced in some antiferroelectric crystals by the application of a strong electrostatic field which is not possible with antiferroelectric crystals by the application of a strong electrostatic field which is not possible with antiferromagnetic materials.
4. The motion of domain wall in ferroelectrics is not simple. The wall motion appears to result from the repeated nucleation of steps by thermal fluctuations along the parent wall. The nucleation rate is the controlling factor in the propagation of the wall, while in the case of ferromagnetic domain wall the spin direction gradually changes from the preferred direction of the one domain to the preferred direction of the next domain.

## 17.38 EXPERIMENTAL EVIDENCE OF THE EXISTENCE OF DOMAIN STRUCTURE

1. Domains can be made visible by optical means. The refractive index depends upon direction. $BaTiO_3$ has two different indices of refraction parallel and perpendicular to the tetragonal axis.
2. Experimental evidence of domains has also been obtained by photography under polarzed light (A. Von Hippel *Rev. Moden Physics* 1950).
3. Chemical etching rates are vastly different for different domains. The surface having positive polarization being more rapidly attacked by etchant HCL.
4. There is experimental evidence from X-ray analysis.

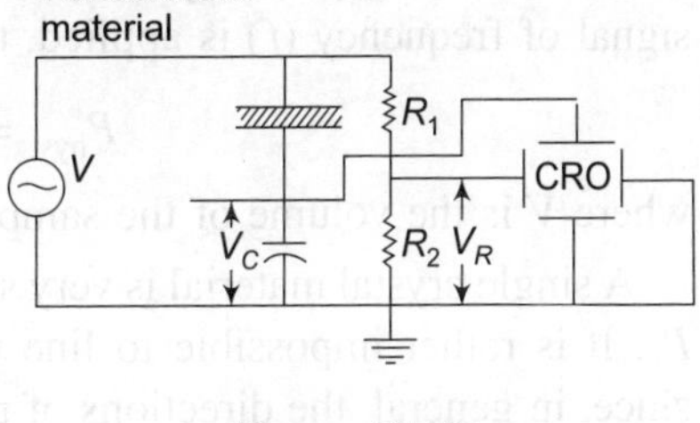

**Figure 17.59**

### 17.38.1 Hysteresis

To measure the displacement-field characteristic a thin sheet of ferroelectric material of electrode area $A$ and thickness $d$ is taken to which a voltage $V$ is applied. Thus, if $Q$ is the charge on the capacitor electrodes, then

$$Q = DA \quad \text{and} \quad F = \frac{A}{d}$$

($D$, $F$) characteristic corresponds to the ($Q$, $V$) characteristic. Hence $C$ is a suitable capacity in series with the sample. The voltage ($V_R$) which is a measure of the applied voltage is connected to the horizontal plates of an oscillograph and the voltage ($V_C$) is a measure of the charge ($Q$) on the electrodes of the ferroelectric sample. The oscillograph will show a hysteresis loop.

When the field ($F$) is reduced to zero a fine value of polarization called *remanent polarisation* ($P_r$) can be measured. In order to eliminate the remanent polarization ($P_r$), an electric field of opposite direction must be applied. This is called the coercive force, $F_c$. If the linear part of the hysteresis loop occurring at sufficiently large applied field strengths is extrapolated back to the polarization axis, one obtains the value of the *spontaneous polarization,* $P_S$.

The occurrence of hysteresis loop is understandable on the basis of the domain structure. In unpolarized materials a domain has equal chance of being polarized positively and negatively. If a field is applied, new positive polarized domains will grow at the expense of negatively polarized domains; the number of new domains increases at first slowly, then more rapidly, until finally all domains are lined up in the direction of the applied field. If the applied field is now reduced to zero many domains will remain aligned thus causing a remanent polarization, $P_r$. An opposing electric field $-F_C$ has to be applied in opposite direction to remove this alignment. If the opposing field is large the material will gradually become polarized in opposite direction until saturation polarization sets in. If the field is now reduced to zero many domains remain aligned thus causing the remanent polarization, $-P_r$: an opposing field ($+F_C$) has to be applied to remove this alignment.

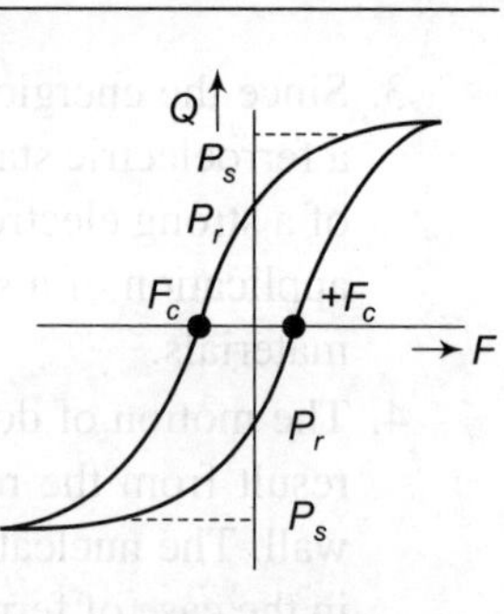

**Figure 17.60**

Energy is lost if the polarization is run through a complete cycle; the loss of energy per unit volume per cycle is $\oint E\, dP$. This is important for $a - c$ operation; if a sufficiently large signal of frequency ($f$) is applied, the power loss ($P_{\text{hyst}}$) of the sample is

$$P_{\text{hyst}} = \int V_S \oint F \cdot dP \quad \text{...(1)}$$

where $V$ is the volume of the sample.

A single crystal material is very suitable for the determination of the saturation polarization, $P_s$. It is rather impossible to line up all the domains in the directions of the applied field since, in general, the directions of preferred polarization of a crystallite do not coincide with the applied field; a polycrystalline material is not very suitable for a determination of $P_s$.

## 17.39 ANTIFERROELECTRIC CRYSTALS

There are certain substances in which it is more favourable for adjacent rows of atoms to move in *opposite* direction. Such substances are called *antiferroelectric*. For example, in tungsten trioxide $WO_3$ the bond angle W—O—W remains at 180° as in cubic phase resulting in antiparallel displacements of tungsten atoms Fig. 17.61. See Fig. 17.56 for comparison.

In such circumstances the dipole moments from neighbouring cells are antiparallel to each other and consequently a complete cancellation of polarization results. Other antiferroelectric crystals include $NH_4H_2PO_4$ (148°K) and $NH_4H_2ASO_4$ (216°K) in the tetragonal phosphate group solid solutions of (K· $NH_4$) $NaC_4H_4O_6$ $4H_2O$ in the Rochelle salt group $Pb_2Nb_2O_7$ $(NH_4)_2$ $H_2IO_6$ and $Ag_2$ $H_3$ $IO_3$ in the miscellaneous classification. The essential difference between the electrical behaviour of an antiferroelectric and a ferroelectric crystal is that the former dose not show a hysteresis loop. Above the transition temperature the dielectric constant of an antiferroelectric crystal increases with decreasing temperature as in ferroelectric crystal. Antiferroelectric compounds might be useful as capacitor materials in view of the large dielectric constant.

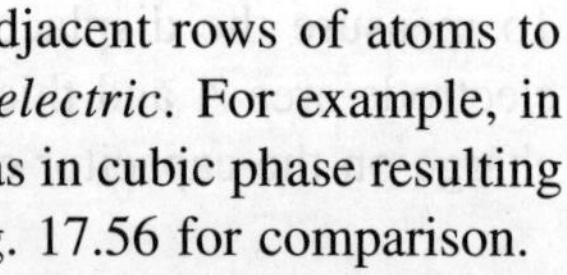

**Figure 17.61** $WO_3$

## 17.40 PIEZOELECTRIC MATERIALS[4]

A material such as $BaTiO_3$ changes its dimensions in an electric field because the negative charges are pulled towards the positive electrode, and the positive charge are pulled towards the negative electrode, thus increasing the dipole length, *d*. This displacement also increases the dipole moment *Qd*, and the polarization, $\mathcal{P}$, since the latter is the total of the dipole moments $\Sigma Qd$ per unit volume, *V*.

This sequence of effects provides a means of changing mechanical energy into electrical energy and vice versa. To understand this possibility, refer to Fig. 17.62 (a). The cooperative alignment of the dipole moments of the many unit cells gives a polarization that collects positive charges at one end of the crystal and negative charges at the other end. Now consider two alternatives, as shown in Fig. 17.62 (b and d). (1) Compress (or pull) the crystal with a stress, *s*. There is a strain *e*, which is dictated by the elastic modulus. This strain changes the dipole length (*d*) and directly affects the polarization (= $\Sigma$ *Qd*/*V*) since *Q* and *V* remain essentially constant. With a smaller polarization (from compression), there is excess of charge density on the two ends of the crystal. If the two ends are isolated, a voltage differential develops (Fig. 17.62b). If they are in electrical contact, electrons will shift from one end to the other (Fig. 17.62c). No pressure is applied in Fig. 17.62(d); rather, a voltage is applied, which increases the charge density at the two ends. The negative charges within the $BaTiO_3$ are pulled on way, and the positive charges are pulled in the opposite direction, thus changing not only the dipole length, *d*, but also that dimension of the crystal.

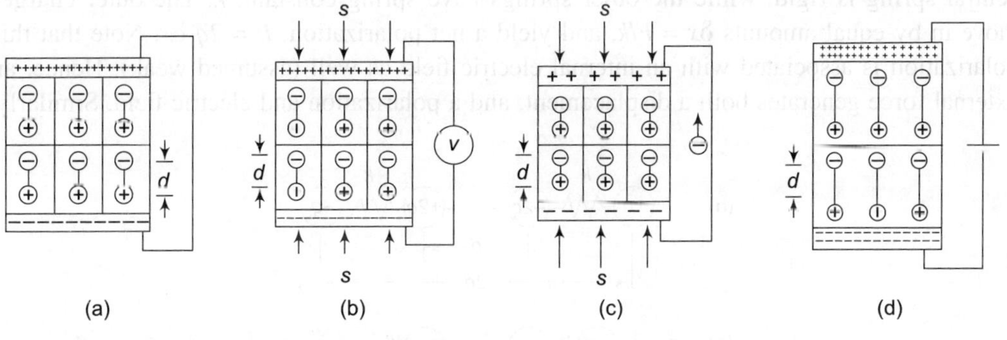


**Figure 17.62 Piezoelectric Material (Schematic).** (a) No external field: Centres of positive and negative charges are not coincident. (b) Pressure applied: Compression leads to voltage differential. (c) Pressure applied and electrical path present: Compression leads to charge transfer. (d) External potential applied: Dipole lengthening produces dimensional change

These two situations indicate how mechanical forces and dimensions can be interchanged with electrical charges and voltages. Devices with these capabilities are called *transducers*. Materials with these characteristics are called *piezoelectric* (i.e., pressure-electric). Crystals

---

[4] The most widely used piezoelectric ceramics are $PbZrO_3$–$PbTiO_3$ solid solutions called PZTs. They have the structure of $BaTiO_3$. However, their ferroelectric Curie points are higher than that of $BaTiO_3$, and they have a greater range of design applications.

of the $BaTiO_3$ type are used for pressure gauges, for tactile sensors of robotic manipulators, for phonograph cartridges, and for high frequency sound generators.

Quartz crystals ($SiO_2$) also are piezoelectric. They are produced for special applications in circuits requiring frequency control. Once they are cut to a selected geometry, the elastic vibrational frequency is constant to one part in $10^8$! Thus, in resonance, the crystal can control the frequency of an electronic circuit to that same precision. Fine watch circuits and control circuits for radio broadcasting are but two applications the use the piezoelectric characteristics of quartz crystals.

## 17.41 PIEZOELECTRIC INTERACTION

**Piezoelectric Materials.** Materials that are not inversion-symmetric, that is, ones where inverting all the atomic positions from $\boldsymbol{R}_i$ to $-\boldsymbol{R}_i$ does not return the original crystal, and at the same time include ions with different equilibrium charges, will often display an effect known as *piezoelectricity*. This effect can be understood using Fig. 17.63.

At equilibrium, the four-atom chain in Fig. 17.63 has charges $q_i = \pm 2q$ at $x_i = \pm d/2$ and $q_i = \mp q$ at $x_i = \pm d$. The net polarization, defined as

$$P = \sum_{i=1}^{4} q_i x_{ij} \qquad \text{...(1)}$$

is equal to zero. When an compressional force ($F$) is applied to both ends of the chain, the central spring is rigid, while the outer springs have spring constant, $k$. The outer charges move in by equal amounts $\delta x = F/k$, and yield a net polarization, $P = 2q\ \delta x$. Note that this polarization is associated with an internal electric field as well (assumed weak). Hence, an external force generates both a displacement, and a polarization and electric field. Similarly,

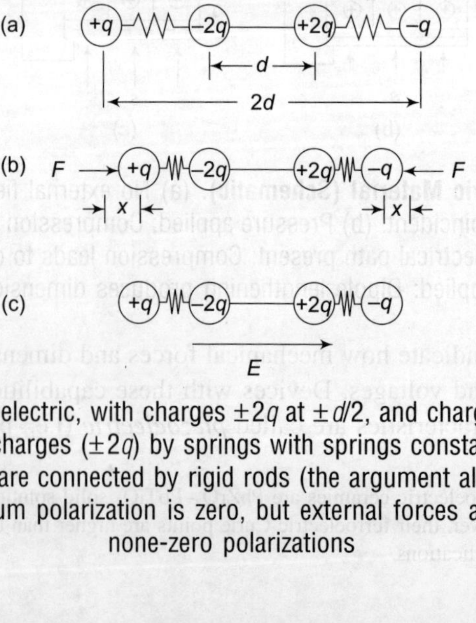


**Figure 17.63** Model piezoelectric, with charges $\pm 2q$ at $\pm d/2$, and charges $\pm q$ at $\pm d$. The charges ($\pm q$) are connected to the charges ($\pm 2q$) by springs with springs constant $k$, and for simplicity we assume the charges ($\pm 2q$) are connected by rigid rods (the argument also works if one takes very stiff springs). The equilibrium polarization is zero, but external forces and electric fields generate none-zero polarizations

when an external electric field ($E$) is applied, the outside charges move in by equal amounts $\delta x' = qE/k$, yielding a polarization $P = 2(q^2/k)E$; in addition to the polarization, an internal force, or stress, is generated.

We therefore find that system has its displacement, polarization, force and electric field forming a coupled set, each generating the other in turn. If we take the force and the electric field as the independent variables, we can write these relations in the form

$$\left.\begin{aligned} \delta x &= (1/k)\,F + (q/k)\,E, \\ P &= (2q/k)\,F + (2q^2/k)\,E. \end{aligned}\right\} \qquad ...(2)$$

Note that the choice of dependent and independent variables is arbitrary; we could equally well have chosen the displacement ($\delta x$) and the polarization, $P$.

In a real piezoelectric solid, the same type of behaviour is found: The lack of inversion symmetry in the charge distribution, along with differing displacements of one charge type compared to the other, generates displacements (strains) and polarizations from externally applied forces (stresses) and electric fields. The constitutive relations corresponding to (3.133) for a three-dimensional solid are usually written in terms of the *electric displacement*, $\boldsymbol{D}$, instead of the polarization, $\boldsymbol{P}$. The electric displacement is defined through

$$\boldsymbol{D} = \varepsilon_0 \boldsymbol{E} + \boldsymbol{P}, \qquad ...(3)$$

where $\varepsilon_0 = 8.85 \times 10^{-12}$ F/m is the permittivity of vacuum, and $P$ the polarization density. In a linear, isotropic dielectric solid, with relative permittivity $\varepsilon_r$ and dielectric constant, $\varepsilon = \varepsilon_r \varepsilon_0$, the polarization density is given by $\boldsymbol{P} = (\varepsilon_r - 1)\,\varepsilon_0 \boldsymbol{E}$, and the electric displacement is simply $\boldsymbol{D} = \varepsilon \boldsymbol{E}$.

In a crystalline non-piezoelectric solid, the corresponding linear relation between $\boldsymbol{D}$ and $\boldsymbol{E}$ involves the 3 × 3 dielectric tensor, $\varepsilon_{\alpha\beta}$, as now the polarization ($\boldsymbol{P}$) is not necessarily co-linear with the field $\boldsymbol{E}$. The relation is then, in component form,

$$D_\alpha = \sum_{\beta=1}^{3} \varepsilon_{\alpha\beta} E_\beta, \quad (\alpha = 1 \text{ to } 3) \qquad ...(4)$$

or in tensor form,

$$\boldsymbol{D} = \varepsilon \cdot \boldsymbol{E}. \qquad ...(5)$$

The variables used to describe a continuum piezoelectric solid are the six-vector strain $\varepsilon$, the six-vector stress $\tau$, the electric field $\boldsymbol{E}$, and the electric displacement, $\boldsymbol{D}$.

In a piezoelectric solid, the relation between $\boldsymbol{D}$ and $\boldsymbol{E}$ must include the effect of the strain $\varepsilon$. The constitutive relation for $\boldsymbol{D}$ has the form

$$D_\alpha = \sum_{\beta=1}^{3} \varepsilon^{\varepsilon}_{\alpha\beta} E_\beta + \sum_{\mu=1}^{6} e_{\alpha\mu}\, \varepsilon_\mu, \quad (\alpha = 1 \text{ to } 3) \qquad ...(6)$$

where $\varepsilon^{\varepsilon}$ is the form for the dielectric tensor when the strain is chosen as the independent variable. We have introduced a new matrix $e$, which has 3 rows and 6 columns, and is known as the piezoelectric stress matrix. This matrix gives the relation between the electric displacement ($\boldsymbol{D}$) and the strain, ε.

The corresponding relation for the stress (τ) is

$$\tau_\mu = -\sum_{\beta=1}^{3} \varepsilon_{\beta\mu} E_\beta + \sum_{\nu=1}^{6} c^E_{\mu\nu}\,\varepsilon_\nu, \quad (\mu = 1 \text{ to } 6) \qquad ...(7)$$

with the elastic constant $c^E$ relating the stress and strain when the electric field is treated as an independent variable. Note the interchange of indices for the piezoelectric stress constant matrix ($e$) that appears in (Eq. 7). We will indicate the transpose by using an underline, so that $\underline{e}_{\beta\mu} = e_{\mu\beta}$.

These relations can be written in matrix form as

$$\left.\begin{aligned} \boldsymbol{D} &= \varepsilon^{\varepsilon} \cdot \boldsymbol{E} + e\cdot\varepsilon, \\ \tau &= -\underline{e} \cdot \boldsymbol{E} + c^E \cdot \varepsilon, \end{aligned}\right\} \qquad ...(8)$$

where $\underline{e}$ the transpose of the 3 × 6 matrix $e$, yielding a 6 × 3 matrix.

There are a number of other forms for the piezoelectric relations (8); for example, if we choose the electric field $\boldsymbol{E}$ and the stress τ as the independent parameters, the relations are

$$\left.\begin{aligned} \boldsymbol{D} &= \varepsilon^{\tau} \cdot \boldsymbol{E} + d\cdot\varepsilon, \\ \varepsilon &= -\underline{d} \cdot \boldsymbol{E} + s^E \cdot \tau, \end{aligned}\right\} \qquad ...(9)$$

where $\varepsilon^\tau$ is the dielectric tensor with the stress τ as the independent parameter. These relations involve a different piezoelectric strain matrix, $d$. The terms ($d_{\alpha\nu}$) form a 3 × 6 tensor, related to the piezoelectric constants ($e_{\alpha\nu}$) through relation

$$e_{\alpha\nu} = \sum_{\mu=1}^{6} d_{\alpha\mu}\, c^E_{\mu\nu}. \quad (\alpha = 1 \text{ to } 3,\ \nu = 1 \text{ to } 6) \qquad ...(10)$$

The dielectric tensors are related by

$$\varepsilon^{\varepsilon}_{\alpha\beta} = \varepsilon^{\tau}_{\alpha\beta} - \sum_{\mu\nu=1}^{6} d_{\alpha\mu}\, c^E_{\mu\nu}\, d_{\nu\beta} \quad (a, b = 1 \text{ to } 3) \qquad ...(11)$$

Yet another form is

$$\left.\begin{aligned} \boldsymbol{E} &= \beta^{\varepsilon} \cdot \boldsymbol{D} - h\cdot\varepsilon, \\ \tau &= -\underline{h} \cdot \boldsymbol{D} + c^D \cdot \varepsilon, \end{aligned}\right\} \qquad ...(12)$$

where $\beta^\varepsilon$ is the inverse of the permittivity tensor ε, $h$ is another piezoelectric matrix, defined by

$$h = \beta^{\varepsilon} \cdot e, \qquad ...(13)$$

and the elastic tensor $c^D$ is given by

$$c^D = \underline{h}\cdot e + c^E. \qquad ...(14)$$

## 17.42 COMMON DIELECTRIC MATERIALS AND THEIR APPLICATIONS

### (Solid Dielectrics)

| *Material* | *At 1 Mhz k or $\varepsilon_r$/tan $\delta$* | | *Dielectric strength × $10^6$ V/m.* | *Applications* |
|---|---|---|---|---|
| 1. Glass ($SiO_2$) Quartz/ Pyrex | 7 | 0.0005 | 10 | (a) Envelope of electric bulbs<br>(b) x-rays tubes<br>(c) capacitors |
| 2. Porcelain (Ceramines) | 6 | 0.02 | 5 | (a) Switches<br>(b) Plug holders<br>(c) Telephones<br>(d) Thermocouples<br>(e) Transmission lines |
| 3. Poly-ethylene | 2.3 | 0.0004 | 4 | (a) Sleeves over wire<br>(b) Radio frequency work |
| 4. Polyvinlychloride (PVC) | 3.4 | 0.5 | 2 | (a) Flexible wire<br>(b) Tubings<br>(c) Gaskets<br>(d) wire covering |
| 5. Nylon | 3.5 | 0.02 | 15 | (a) Insulation for windings<br>(b) Capacitors |
| 6. Bakelite | 4.4 | 0.028 | 15 | (a) Switch bases and covers<br>(b) Electrical appliance handles<br>(c) Lamps holders |
| 7. Rubber | 2.7 | 0.003 | 25 | (a) Insulation for electric wires<br>(b) Cables<br>(c) Ropes<br>(d) Coatings<br>(e) Transformer and meter windings<br>(f) For fabricating impegraneted paper |
| 8. Mica | 5 | 0.0005 | 100 | (a) Best insulating material<br>(b) Switchgears<br>(c) Commutator segment separations<br>(d) Electric iron<br>(e) Hot plates<br>(f) High frequency applications |

## EXERCISE

17.1 Assuming that an atom consists of a uniform sphere of negative charge with radius $r$ surrounding a point positive charge, show that the polarizability is equal to $4\pi\varepsilon_0 r^3$.

17.2 What are the various polarizabilities? Discuss the classical theory of electronic polarizability and obtain corresponding dispersion relation.

17.3 Derive Clausius-Mossotti relation. Show that the dielectric constant of a ferroelectric follows the Curie-Weiss law above the Curie temperature.

17.4 Discuss and explain Clausius-Mossotti catastrophe.

17.5 Discuss orientational polarizability and obtain an expression known as Langevin-Debye equation.

17.6 Discuss classical theory of electronic polarizability. Show that it is given by

$$\alpha_e = \frac{e^2/m}{(\omega_0^2 - \omega^2)}.$$

17.7 Derive expression for complex susceptibility and complex dielectric constant of a dielectric placed in an alternating electric field

$$E^* = E_0 \exp(i\omega t)$$

17.8 Derive the condition for a permanent polarization in case of ferroelectrics. What are the properties of ferroelectric substances.

17.9 Explain ferroelectric domains.

17.10 Explain the phenomenon of piezoelectricity. Discuss its applications.

17.11 What is ferroelectricity? Name any two ferroelectric materials and their applications.

17.12 Derive Debye equation. What is a Cole-Cole plot?

17.13 What is a plasma? Discuss plasma oscillations in a metal and show that its dielectric constant is given by

$$\varepsilon(\omega) = 1 - \omega_p^2/\omega^2$$

where

$$\omega_p = \sqrt{\left(\frac{N_0 e^2}{\varepsilon_0 m}\right)}$$

17.14 Discuss frequency dependence of complex dielectric constant in case of alternating fields.

CHAPTER

# 18

# Lasers and Mössbauer Effect

## 18.1 INTRODUCTION: (LASERS)

The word LASER is an acronym for Light Amplification by Stimulated Emission of Radiation. The basic mechanism of light emission from a LASER is the same as that from a LED. However, in contrast to the incoherent light of a light emitting diode, LASER light has spatial and temporal coherence. Spatial coherence means that all the points on the wavefront have the same phase, and this results in a highly directional beam of light. Temporal coherence means that, at any point in space, the wave amplitude varies sinusoidally with time, hence, the emitted radiation is monochromatic.

A laser is the result of three phenomena which occur when an electromagnetic wave interacts with a material, viz.,

(i) spontaneous emission,

(ii) stimulated emission, and

(iii) absorption.

## 18.2 SPONTANEOUS EMISSION

Let us consider two energy levels 1 and 2 with energies $E_1$ and $E_2$ respectively ($E_1 < E_2$). Let the level 1 be the ground level. Now suppose that an atom of the material is initially in level 2. Since $E_1 < E_2$, the atom will tend to decay to level 1, during this decay, the atom releases energy ($E_2 - E_1$). When this energy released is in the form of electromagnetic wave, the process is called spontaneous emission. The frequency of the radiated wave is

$$\nu = \frac{E_2 - E_1}{h}$$

Thus, the spontaneous emission is characterised by the emission of photon of energy ($h\nu = E_2 - E_1$) when the atom decays from level 2 to level 1. The decay can also occur in the non-radiative way. In that case, the energy $(E_2 - E_1)$ is released in some form other than electro-magnetic radiation. Larger the number of atoms in level 2; more is the probability for it to decay to level 1.

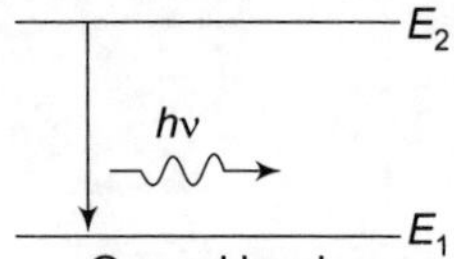


**Figure 18.1**

## 18.3 STIMULATED EMISSION

Suppose that the atom is initially in the level 2. Now, suppose when an electromagnetic waves is incident on the material, this wave has the same frequency as the atomic frequency, so there is a finite probability that this wave will force the atoms to undergo transition from level 2 to level 1. In this case, the energy $(E_2 - E_1)$ is released in the form of electromagnetic wave which adds up to the incident wave. This process is called stimulated emission and will result in the light amplification.

As we know that in the spontaneous emission, the electromagnetic waves emitted by the atom have no fixed phase relation with that emitted by another atom and so, the emitted wave can be emitted in any direction. However, in stimulated emission, the process is forced by the incident electromagnetic wave, the emission of any photon adds in phase to that of the incoming wave. There is a phase relationship between the incoming and the emitted wave, so the induced emission adds the photons travelling in the same direction as the induced photons. Hence, the intensity of the wave increases as it induces emission in a medium containing excited electrons. This is the principle behind laser operation. (The incoming wave also determines the direction of the emitted wave.) It is to be noted here that the idea of stimulated emission was first propounded theoretically by Einstein in 1917.

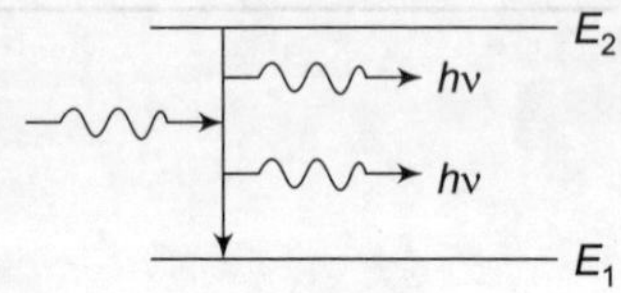


**Figure 18.2**

## 18.4 ABSORPTION

In this case, we assume the atom to lie in the level 1 initially. If it is ground level, then the atom remains in this level unless some external energy is given to it. Suppose an electromagnetic wave of frequency ($v$) be incident upon the material. The energy difference $(E_2 - E_1)$ required by the atom to undergo the transition is obtained from the energy of the incident electromagnetic wave. This process is called absorption.

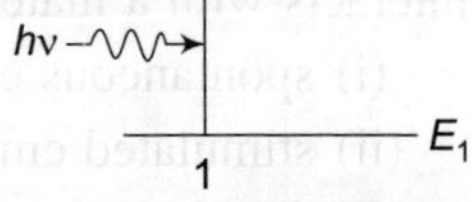


**Figure 18.3**

In terms of photons, the above three processes can be described as follows:

In spontaneous emission, the atom decays from level 2 to level 1 through the emission of a photon. In stimulated process, the incident photon stimulates the $2 \rightarrow 1$ transition and then we have two photons—the stimulated photon and the stimulating photon. In the absorption process, the incident photon is simply absorbed to give the $1 \rightarrow 2$ transition.

## 18.5 POPULATION INVERSION (DEFINED)

We know that in general, under ordinary conditions, the population of atoms in the ground level will be more than in the higher level (i.e., $N_1 > N_2$). However, we would subsequently see that for laser action, we necessarily require a non-equilibrium condition $N_2 > N_1$, for the laser material to act as an amplifier, i.e., population of the higher level greater than the population of the ground level. This condition is called *population inversion*.

The population inversion means that the population difference $(N_2 - N_1) < 0$ is opposite in sign to that which exists under ordinary conditions $(N_2^e - N_1^e < 0)$. The material which exhibits the phenomenon of population inversion is known as *active material*.

If the transition frequency $v = (E_2 - E_1)/h$ falls in the microwave region, this type of amplifier is called a MASER-amplifier. But if the transition frequency falls in the optical region, the amplifier is called LASER-amplifier. The word laser is used for frequency of visible light, in far or infrared, in the ultraviolet or even in the X-ray region.

As we know that an amplifier can be made an oscillator by applying a suitable positive feedback, in the microwave range, this is done by placing the active material in a resonant cavity, having a resonance at the frequency, $\nu_0$. But, in the case of lasers, the feedback is generally obtained by placing the active material between two highly reflecting mirrors, which acts like a resonant cavity.

## 18.6 SCHEMATIC OF A TYPICAL LASER

### 18.6.1 Laser Cavity

The lasing material is contained in a long narrow container called the cavity. A simple laser cavity consists of two absolutely parallel, optically flat plane mirrors. However, in general, the mirrors can be curved having radii of curvature, $R_1$ and $R_2$.

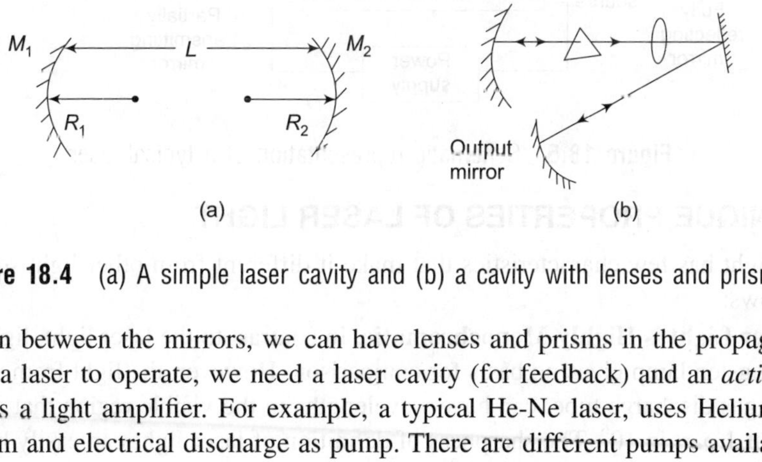


**Figure 18.4** (a) A simple laser cavity and (b) a cavity with lenses and prisms

Actually in between the mirrors, we can have lenses and prisms in the propagation path of light. For a laser to operate, we need a laser cavity (for feedback) and an *active medium* which acts as a light amplifier. For example, a typical He-Ne laser, uses Helium-Neon as active medium and electrical discharge as pump. There are different pumps available:

(i) electrical (DC or r.f.)
(ii) chemical
(iii) optical
(iv) thermal
(v) nuclear

Most of the common lasers use electrical or optical pumping.

Figure 18.5 shows schematic representation of a typical laser.

In this case, a plane electromagnetic wave travelling in a direction orthogonal to the mirrors bounces back and forth between the two mirrors and gets amplified on each passage through the active material. One of the mirrors is partially transparent, so that a useful output beam can be obtained.

Thus, a laser device amplifies and produces a highly directional, high intensity beam that most often has a very pure frequency or wavelength. It comes in sizes ranging from approximately one tenth the diameter of a human hair to the size of a very large building, in powers ranging from $10^{-9}$ to $10^{20}$ $W$ and in wavelengths ranging from the microwave to the soft X-ray spectral regions with corresponding frequencies from $10^{11}$ to $10^{17}$ Hz respectively. Lasers have pulse energies as high as $10^4$ J and pulse duration as short as $6 \times 10^{-5}$ s. A laser is a specialised light source that can be used when its certain unique properties are required.

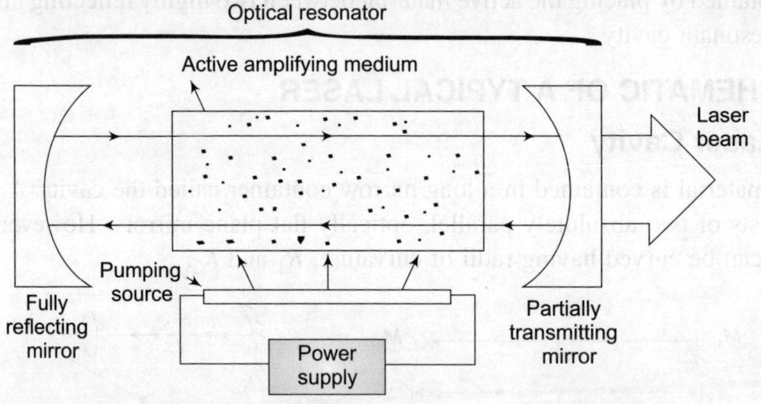


**Figure 18.5** Schematic representation of a typical laser

## 18.7 UNIQUE PROPERTIES OF LASER LIGHT

The laser light has few characteristics that make it different from other light sources. These are as follows:

1. **Laser Light is Highly Monochromatic.** In contrast to the laser light, light emitted by a tungsten lamp gives no basis for comparison. However, the light from selected lines in a gas discharge tube, can have wavelengths in the visible region that are precise to about 1 part in $10^6$. The sharpness of definition of laser light can easily be a thousand times greater, i.e., 1 part in $10^9$. This property is due to the following two facts: (i) only an electromagnetic wave of frequency $\nu_{12}$ can be amplified. (ii) Since the mirror arrangement forms a resonant cavity, oscillations can occur only at resonant frequencies of this cavity.
2. **Laser Light is Highly Directional.** As the active material is placed between plane-parallel reflecting surfaces, only electromagnetic wave which propagates along cavity

direction or very near to it, is sustained in the cavity and others leak out. Thus, high directionality is achieved.

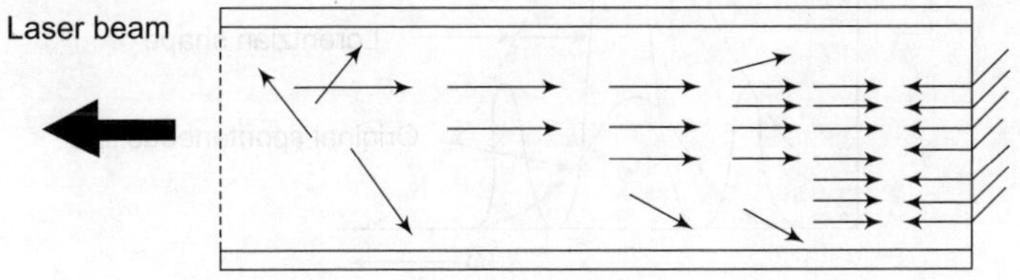


**Figure 18.6** Schematic representation of a laser cavity and the build-up of (uni) directional beam

The beam divergence is given by

$$\alpha_d = \frac{1.22\lambda}{D}$$

where $\lambda$ is the wavelength of the beam and $D$ is its diameter.

3. **Laser Light is Highly Coherent.** The laser light should be both space and time coherent. It may be noted here that a source may have little time coherence but perfect space coherence. Example of such space coherent source is the light from a distant star.
   - **(a)** ***Spatial Coherence:*** Suppose the two points $P_1$ and $P_2$ at time $t_0$ lie in the same wavefront of some given electromagnetic wave and $E_1(t), E_2(t)$ be the corresponding electric fields at those points. The two are in same phase at $t_0$. Now if this phase difference remains zero at time *t,* the coherence between the two points is said to be perfect. However, if it occurs for any other two points on the wavefront then we say that the wave has perfect spatial coherence. The directional property of a laser beam is due to space-coherency. The spatial coherence of a laser beam, for example, a He-Ne laser can be easily demonstrated with the help of a diffraction grating and Fraunhofer interference of a plane wave is obtained in the far field pattern on a screen.
   - **(b)** ***Temporal Coherence:*** Consider the electric field of the electromagnetic wave at a point *P* at time *t* and $1 + \tau$. If the time delay ($\tau$) remains same, the phase difference between the two fields will remain same. But if the phases agree for all time intervals $\tau$, the light is said to be perfectly time-coherent. Extreme monochromaticity is the manifestation of this property of temporal coherence.
4. **Lorentzian Shape.** Laser oscillation is marked by a dramatic narrowing of the spectral and angular distribution of the spontaneous emission radiation at a particular level of excitation called the threshold. A qualitative representation of the line narrowing is shown in Fig. 18.7.

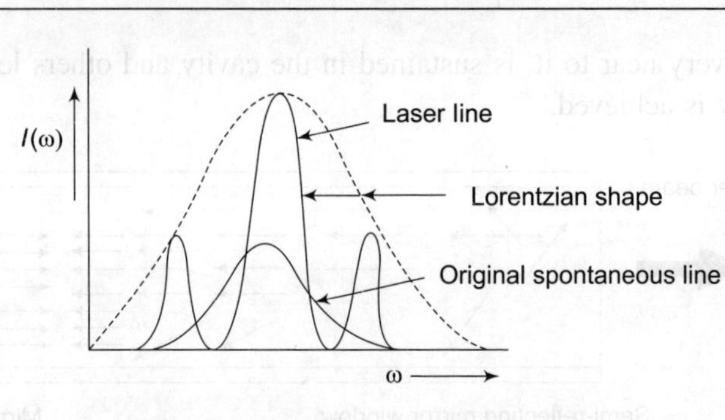


**Figure 18.7** Laser line narrowing

Light from a natural source, consists of a large number of statistically independent wave tracks each of about $10^{-8}$ second duration. The distribution is Gaussain. In case of laser, the axial mode lying closest to the atomic resonance has the highest gain and thus the laser light concentrates its line width around the resonance resulting in line narrowing. Laser light can thus achieve an enormous spectral purity.

5. **Laser Light can be Sharply Focused.** This property is related to the parallelism of the laser beam. As for light from a star, the size of the focused spot for a laser beam is limited only by diffraction effects and not by the size of the source. Flux densities for focused laser light of $10^{15}$ W/cm$^2$ are readily achieved. On the other hand, an oxyacetylene flame has a flux density only of $10^3$ W/cm$^2$.

Apart from the above characteristics, a laser light also has the following important peculiarities which differentiate it from thermal light resources.

Laser light is both spatially and temporally coherent, whereas light from other sources is mostly incoherent. The small time bandwidth $\Delta f$, and the spatial bandwidth $\Delta f_x$, achievable with significant energy by the laser light, are nearly impossible to reproduce by thermal sources. The differences are so dramatic that although both laser light and light from other sources are the same electromagnetic waves, it would be almost correct to say that there is a fundamental difference between them.

## 18.8 THE EINSTEIN COEFFICIENTS IN RELATION TO SPONTANEOUS AND STIMULATED EMISSION

Let there be two energy levels ($E_1$ and $E_2$) of a material (out of an innumerable levels possessed by the material). Let us suppose $E_1 < E_2$ and $E_2$ be the excited level and $E_1$, the ground level. If the atom is in level 2 then it will have a tendency to decay to level 1. The atom may emit electromagnetic radiation of frequency.

$$\nu_{12} = \frac{E_2 - E_1}{h} \quad ...(1)$$

This process is *spontaneous emission*. The decay can also occur in another way (non-radiative way). In this latter process the energy difference is imparted to the surrounding atoms in the form of kinetic energy.

If $N_2$ denotes the population density of level 2, the spontaneous decay rate will be proportional to 2 and given by

$$\left(\frac{dN_2}{dt}\right)_{sp} = -AN_2 \quad ...(2)$$

The spontaneous emission coefficient $A$ is termed the Einstein $A$ coefficient. The quantity

$$(\tau)_{sp} \equiv \frac{1}{A} \quad ...(3)$$

is called *spontaneous emission lifetime*

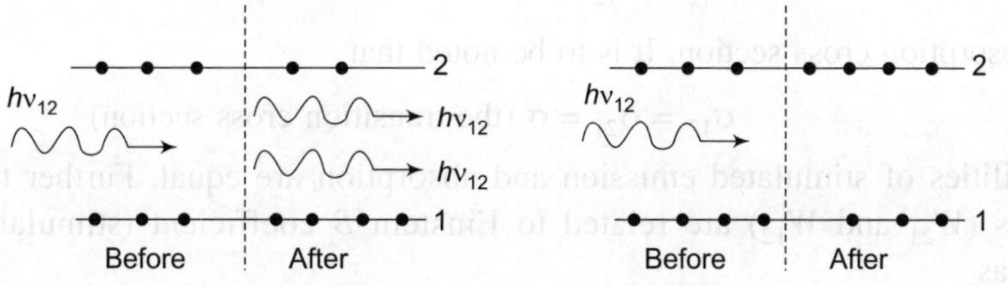


**Figure 18.8** Energy transitions in an atom (a) stimulated emission (b) absorption

Suppose an electromagnetic wave of frequency equal to $v_{12} = \frac{E_2 - E_1}{h}$ is made to strike the material. This wave has a finite probability of inducing atom to undergo transition from initial level 2 to level 1. In this process, photon of the frequency ($v_{12}$) is released in addition to the incident one. This phenomenon is known as stimulated emission. This is different from spontaneous emission. In case of spontaneous emission, the atom emits electromagnetic wave which has no definite phase relationship with that emitted by another atom. In stimulated emission, the emitted radiation is in phase with the incident radiation. The stimulated emission rate depends on the population density $N_2$, and is given by

$$\left(\frac{dN_2}{dt}\right) = -W_{21}N_2 \quad ...(4)$$

$W_{21}$ is called the stimulated transition probability. (Its reciprocal is called stimulated emission lifetime.) In addition to the particular transition, $W_{21}$ depends on the intensity of the incident wave

$$W_{21} = \sigma_{21}F \quad ...(5)$$

where $F$ is the incident photon flux and $\sigma_{21}$ is a quantity having dimensions of area. $\sigma_{21}$ is known as *stimulated emission cross section* and depends only upon the characteristics of the given transition.

***Absorption***

Figure 18.8 (b) depicts the inverse process, viz., absorption. If an electromagnetic wave of frequency $\nu_{12} = \frac{E_2 - E_1}{h}$ is incident on the atom, then there is a finite probability that atom makes a transition from ground level (1) to the excited state (2). The rate of absorption is given by

$$\frac{dN_1}{dt} = -W_{12}N_1 \qquad ...(6)$$

where $N_1$ is the population density of level 1 and the absorption probability is

$$W_{12} = \sigma_{12}F$$

$\sigma_{12}$ is the absorption cross section. It is to be noted that

$$\sigma_{12} = \sigma_{21} = \sigma \text{ (the transition cross section)}$$

i.e., probabilities of stimulated emission and absorption are equal. Further the transition probabilities ($W_{21}$ and $W_{12}$) are related to Einstein $B$ coefficient (stimulated emission coefficient) as

$$W_{12} = B_{12}\rho(\nu)$$

$$W_{21} = B_{21}\rho(\nu) \qquad ...(7i)$$

Here $\rho$ is electromagnetic energy density and $W_{12} = W_{21}$.

## 18.8.1 Interrelation between Einstein Coefficients

If absorption and spontaneous emission were the only processes operative, the principle of detailed balance would give

$$\rho(\omega_{21})B_{12}N_1 = A_{21}N_2 \qquad ...(7ii)$$

where $\rho$ is the radiation density.

$$\therefore \qquad \rho(\omega_{21}) = \frac{A_{21}}{B_{12}}\frac{N_2}{N_1} \qquad ...(7iii)$$

In thermal equilibrium, the total number of atoms $N$ will be distributed over the available levels according to the Boltzmann distribution

$$N_i = \frac{Ng_i \exp(-E_i / k_B T)}{\sum_i \exp(-E_i / k_B T)}$$

where $g_i$ is the statistical weight of the level $i$, $T$, the absolute temperature, $E_i$, the energy of $i$th level and $k_B$ is Boltzmann constant

$$\frac{N_2}{N_1} = \frac{g_2 \exp(-E_2/k_B T)}{g_1 \exp(-E_1/k_B T)}$$

It may be assumed that

$$g_2 = g_1 = 1$$

$$\therefore \quad \frac{N_2}{N_1} = \frac{\exp(-E_2/k_BT)}{\exp(-E_1/k_BT)}$$

$$= \exp[-(E_2 - E_1)/k_BT] = \exp(-\hbar\omega_{21}/kT) \qquad ...(8)$$

Hence Eq. (7iii) changes to

$$\rho(\omega_{21}) = \frac{A_{21}}{B_{12}} \exp(-\hbar\omega_{21}/k_BT) \qquad ...(9)$$

Now, it is an experimental fact, that the black body radiation is given by

$$\rho(v) = \frac{8\pi h v^3}{c^3} \cdot \frac{1}{e^{hv}/k_BT - 1} \qquad ...(10)$$

where $\rho(v)$ is the energy density of photons having frequency, $v$.

Let us consider a medium which has two levels $E_2$ and $E_1$. If this medium is placed in a black body enclosure, in equilibrium, the number transmitted from $E_2$ to $E_1$ would be equal to that from $E_1$ to $E_2$. The number of photons generated per unit time per unit volume would be given by

$$N_2(W_{21} + A_{21}) \qquad ...(11)$$

where $W_{21}$ is the stimulated emission rate. The term $A_{21}$ is due to spontaneous emission rate.

The photons lost per unit volume will be given by

$$N_1 W_{21} \qquad ...(12)$$

where $W_{12}$ is the transition rate of spontaneous emission rate from $E_1$ to $E_2$.

From Eqs. (11) and (12) and Eq. (7*i*), we obtain

$$\frac{N_2}{N_1} = \frac{\rho(v)B_{12}}{\rho(v)B_{21} + A_{21}} \qquad ...(13)$$

Also from Eq. (8) $$\frac{N_2}{N_1} = e^{-hv/k_BT} \qquad ...(14)$$

Thus, $$\frac{N_2}{N_1} = \frac{\rho(v)B_{12}}{\rho(v)B_{21} + A_{21}} \qquad ...(15)$$

or $$\frac{B_{12}e^{hv/k_BT} - B_{21}}{e^{hv/k_BT} - 1} = \frac{c^3}{8\pi h v^3} A_{21} \text{ (using Eq. 14)} \qquad ...(16)$$

The left-hand side of the above equation is dependent on the temperature $T$, whereas the right hand side is independent of temperature. In order that this be valid at all temperatures, we must have

$$B_{12} = B_{21} = \frac{c^3}{8\pi h v^3} A_{21} \qquad ...(17)$$

This is the famous Einstein relationship between $A$ and $B$ coefficients. Thus, obtain for stimulated emission rate

$$W_1 = W_{12} \equiv W_{21} = \frac{c^3}{8\pi h v^3} A_{21}\, \rho(v) \qquad ...(18)$$

or Rate,

$$W_i = \frac{c^3}{8\pi h v^3} \frac{1}{t_{\text{spon}}} \rho(v) \qquad ...(19)$$

($\because$ $A_{21}$ is inverse to the spontaneous emission rate $t_{\text{spon}}$).

## 18.9 POPULATION INVERSION

Under the conditions of thermal equilibrium, we have

$$\frac{N_2}{N_1} = \frac{g_2}{g_1} \exp(-E_2/kT)/\exp(-E_1/kT)$$

and the lower energy level ($E_1$) of the two levels atomic system contains more atoms than the upper energy level, $E_2$. This situation prevails at room temperatures and is depicted in Fig. 18.9 (a). However, to achieve stimulated emission and optical amplification it is required to create a non-equilibrium distribution of atoms such that the population of the upper energy level is greater than that of the lower energy level (i.e., $N_2 > N_1$). This condition is known as population inversion and is depicted in Fig. 18.9 (b).

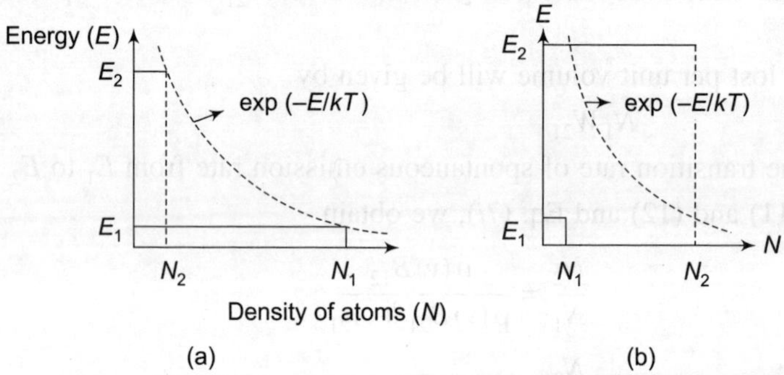


**Figure 18.9** Populations in a two-level system (a) Boltzmann distribution for a system in thermal equilibrium (b) a non-equilibrium situation depicting population inversion

### 18.9.1 Light Amplification by Population Inversion

Population inversion in a laser is brought about by pumping. Figure 18.10 shows a comparison of the effect of passage of radiation through a non-inverted and an inverted population. In each case, the population is shown as being made up of nine molecules of the lasing medium. In non-inverted population, three molecules are in the excited state and six are in the ground state. Three of the incoming photons (Fig. 18.10 (a)) are absorbed by the medium (producing 3 additional excited molecules). However, the radiation stimulates emission of two photons from excited molecules. Thus, there is a net deficit of one photon.

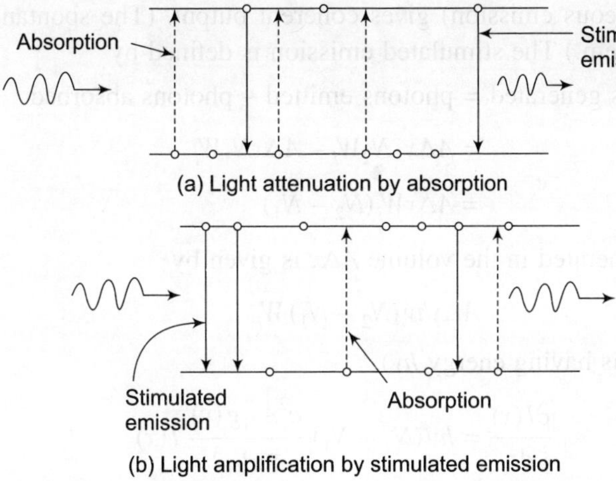


**Figure 18.10** Passage of radiation through (a) a non-inverted population (b) an inverted population

In Fig. 18.10 (b) there is a net gain in photons in the inverted system because the extent of stimulated emission is more than the absorption.

## 18.10 LASING GAIN CONSTANT AND INVERSION TEMPERATURE

To obtain an expression for the lasing gain constant, we require to relate the photon's density to the intensity. For this we proceed as follows. We recollect that photons travel with a velocity $c$ (the velocity of light in the particular medium). Let the intensity of the light beam be $I(x)$ at $x = x$ and $I(x + \Delta x)$ at $x = x + \Delta x$ as shown in Fig. 18.11.

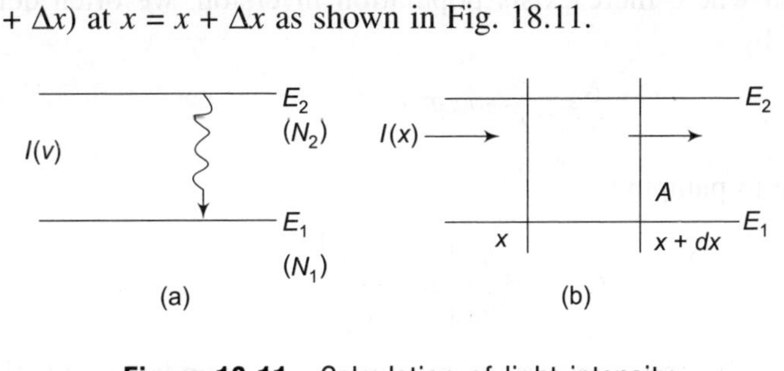


**Figure 18.11** Calculation of light intensity

Let us consider the volume $\Delta x A$ as shown in Fig. 18.11 (b) where $A$ is the cross-sectional area. We can write

$$I(x + \Delta x) = I(x) + \Delta x \frac{\partial I(x)}{\partial x} \qquad ...(1)$$

The second term on the right hand side arises because in the volume $(A\Delta x)$, some photons are generated. However, the number of photons generated per unit time by the stimulation process (which unlike spontaneous emission) gives coherent output. (The spontaneous emission is like a noise to the system.) The stimulated emission is defined by

$$\text{photons generated} = \text{photons emitted} - \text{photons absorbed}$$

$$= A\Delta x \cdot N_2 W_i - A\Delta x N_1 W_i$$

$$= A\Delta x W_i (N_2 - N_1)$$

Thus, the power generated in the volume $A\Delta x$ is given by

$$A\Delta x h\nu (N_2 - N_1) W_i$$

(since each photon is having energy $h\nu$).

Hence
$$\frac{\partial I(x)}{\partial x} = h\nu (N_2 - N_1) \frac{c^2 A_{21} g(\nu)}{8\pi h\nu^3} I(x) \qquad ...(2)$$

or
$$I(x) = I_0 e^{2\nu x} \qquad ...(3)$$

where
$$\nu = (N_2 - N_1) \frac{c^2 g(\nu)}{16\pi\nu^2 t_{\text{spon}}} \qquad ...(4)$$

Thus, the amplification factor is proportional to $(N_2 - N_1)$. Ordinarily it is given by

$$\frac{N_2}{N_1} = e^{-h\nu/k_B T} \text{ or } (N_2 - N_1) = (e^{-h\nu/k_B T} - 1) N_1 \qquad ...(5)$$

In general, we do not have amplification unless and until we pump the medium, i.e., energize the medium so that $N_2 > N_1$. Then we can have amplification.

In the situation where there exists population inversion, we often define an effective temperature ($T_{\text{eff}}$) by

$$\frac{N_2}{N_1} = e^{-h\nu/(k_B T_{\text{eff}})} \qquad ...(6)$$

If $N_2 > N_1$, due to pumping.

$$T_{\text{eff}} = -\frac{h\nu}{k_B}\left[\ln \frac{N_2}{N_1}\right]^{-1} \qquad ...(7)$$

Thus, theoretically, the population inversion corresponds effectively to a negative temperature.

## 18.11 THRESHOLD CONDITION FOR LASER OSCILLATION AND LASER RATE EQUATIONS

Although population inversion between the energy levels providing the laser transition is necessary for oscillation to be established, it is not alone sufficient for lasing to occur. In addition, a minimum threshold gain within the amplifying medium must be attained to initiate and sustain laser oscillations. This threshold gain may be evaluated by considering the change in energy of a light beam as it passes through the amplifying medium. For convenience, all the losses (except those due to transmission through the mirrors) may be included in a single loss coefficient per unit length $\alpha$ $cm^{-1}$. It is assumed that the amplifying medium has a length $L$ and completely fills the region between the two mirrors which have reflectivities $r_1$ and $r_2$. On each round trip, the beam passes through the medium twice. Hence, the fractional loss of the light beam is

$$\text{Fractional loss} = r_1 r_2 \exp(-2\alpha L) \qquad \text{...(1)}$$

Further, it is found that the increase in beam intensity resulting from stimulated emission is exponential. So, if the gain coefficient per unit length produced by stimulated emission is $g$ $cm^{-1}$, the fractional round trip gain is given by

Fractional gain = $\exp(2gL)$

Hence, $\exp(2gL) \times r_1 r_2 \exp(-2gL) = 1$ ...(2)

and $r_1 r_2 \exp 2(g - \alpha) L = 1$

Rearranging this expression, we get for the threshold gain per unit length,

$$g_{th} = \alpha + \frac{1}{2L} \ln\left(\frac{1}{r_1 r_2}\right) \qquad \text{...(3)}$$

The second term on the right hand side represents the transmission loss through the mirrors. If we denote by $N_T$, the threshold population inversion density, (i.e., the minimum population inversion density needed to start lasing action and then sustain it), then it can be shown to be given by

$$N_T = (N_2 - N_1)_{th} = \frac{16\pi v^2 t_{spon}}{c^2 g(v)}\left[\alpha - \frac{1}{2L}\ln(r_1 r_2)\right] \qquad \text{...(4)}$$

The cavity losses are characterised by defining the cavity decay constant, $t_c$. This is associated with decay of an electric field as it bounces back and forth between the mirrors and the lasing medium is not pumped, i.e.,

$$E(t) = E_0 e^{-t/tc} \qquad \text{...(5)}$$

If $T$ denotes the time taken by the electric field to make a round trip through the cavity, then

$$T = \frac{2L}{V} \qquad \text{...(6)}$$

Thus, if we observe the output at any mirror, then we will find the decaying pulses of light coming out of the mirror at every $T$ seconds.

Equation (4) shows that threshold is reached when the population inversion $(N_2 - N_1)$ reaches a threshold value known as *critical inversion.* Once this is reached, the oscillation will build up from the spontaneous emission. This is the basis of the laser.

### 18.11.1 Non-radiative Decay

An atom can also undergo a non-radiative decay from level 2 to level 1. In this case, the energy difference ($E_2 - E_1$) is given to the surrounding molecules in the form of translational, vibrational or rotational energy.

The phenomenon can be described through a characteristic lifetime known as the non-radiative lifetime, $\tau_{nr}$.

As a result of simultaneous occurrence of both radiative and non-radiative decay, the time variation of the upper state population $N_2$ can be written as

$$\frac{dN_2}{dt} = -\left(\frac{N_2}{\tau_{sp}} + \frac{N_2}{\tau_{nr}}\right) \equiv -\frac{N_2}{\tau} \quad \text{...(7)}$$

The overall lifetime $\tau$ is given by

$$\frac{1}{\tau} = \frac{1}{\tau_{sp}} + \frac{1}{\tau_{nr}} \quad \text{...(8)}$$

### 18.11.2 Saturation

Let us study the time evolution of the population of a two-level system in the presence of an incident monochromatic electromagnetic wave of intensity ($I$) and frequency, $\omega$. Let us consider a monochromatic line. We suppose that the total population ($N_t$) does not vary with time while the populations ($N_1$ and $N_2$) of levels 1 and 2 vary with time.

When the atoms are transferred from one level to another, there will always be a band of energy levels. In the stimulated transition, the photons will be emitted. Even though we have supplied energy, still there may not be additional photons. This means that though the atoms have been transferred from level 2 to level 1, still it has not contributed any photons to the pre-exciting photons. This has been known as *non-radiative decay*. Let $\tau_{nr}$ denote the lifetime of the non-radiative decay and $\tau_{st}$, the lifetime of the stimulated emission by the radiation. Hence, the atom transfer from one level to another can take place in two ways. The rate of change of population will be consisting of radiative and non-radiative processes and so will be given by

$$\frac{dN^2}{dt} = \left(\frac{N_2}{\tau_{nr}} + \frac{N_2}{\tau_{st}}\right) \equiv \frac{N_2}{\tau}$$

or

$$\frac{dN_2}{dt} = -\frac{N_2}{\tau}$$

Negative sign has been placed to take into account decrease of population. If $N_1$ and $N_2$ denote the populations of levels 1 and 2 respectively, then total population

$$N_t = N_1 + N_2 \quad \text{...(9)}$$

If $N_1 > N_2$, it leads to absorption

If $N_1 < N_2$, it leads to population inversion

and if $N_2 = N_1$, it leads to saturation.

In fact
$$\frac{dN_2}{dt} = -W(N_2 - N_1) - \frac{N_2}{\tau} \qquad ...(10)$$

where $WN_1$ = Probability of absorption

and $WN_2$ = Probability of transition, per unit time.

Equation (10) gives the rate of change of $N_2$ with time as the difference between the rate ($WN_1$) of stimulated transitions upward to the level 2 and the rate of downward transitions to level 1, and ($N_2/\tau$) accounts for lifetimes of radiative and non-radiative decays.

If we put

$$N_1 - N_2 = \Delta N$$

then
$$\frac{d}{dt}(\Delta N) = \frac{dN_1}{dt} - \frac{dN_2}{dt} \qquad ...(11)$$

In Eq. (9), $N_1$ and $N_2$ change with time but $N_t$ is constant in time. Therefore, $\frac{d}{dt}(N_1 + N_2)$ will remain constant in time. So,

$$\frac{d}{dt}(N_1 + N_2) = 0 = \frac{dN_1}{dt} + \frac{dN_2}{dt}$$

or
$$\frac{dN_1}{dt} = -\frac{dN_2}{dt} \qquad ...(12)$$

From Eqs. (11) and (12), we get

$$\frac{d}{dt}(\Delta N) = -\frac{2dN^2}{dt} \qquad ...(13)$$

Equation (10) can be written as,

$$\frac{dN^2}{dt} = -W(\Delta N) - \frac{N_2}{\tau}$$

$$= -\frac{1}{2}\left(-2W\,\Delta N + \frac{2N_2}{\tau}\right)$$

Adding and subtracting $N_1/\tau$, we get

$$\frac{dN_2}{dt} = -\frac{1}{2}\left[-2W\,\Delta N + \frac{2N_2}{\tau} + \frac{N_1}{\tau} - \frac{N_1}{\tau}\right]$$

$$= -\frac{1}{2}\left[-2W\,\Delta N + \frac{N_1 + N_2}{\tau} + \frac{N_2 - N_1}{\tau}\right]$$

$$= -\frac{1}{2}\left[-2W\,\Delta N + \frac{N_t}{\tau} - \frac{\Delta N}{\tau}\right]$$

or
$$-2\frac{dN_2}{dt} = -\left[2W(\Delta N) + \frac{\Delta N}{\tau}\right] + \frac{N_t}{\tau} \qquad ...(14)$$

From Eqs. (13) and (14), we get

$$\frac{d(\Delta N)}{dt} = -\Delta N\left(2W + \frac{1}{\tau}\right) + \frac{N_t}{\tau} \qquad ...(15)$$

Hence, we see that the rate of change of population depends upon

(i) $W$, i.e., depends upon the intensity ($I$) of incident light or strength of pump.

(ii) $\tau$, i.e., depends upon the lifetime of the pump.

(iii) $N_t$, i.e., depends upon the total population of the levels.

When $\frac{d}{dt}(\Delta N) = 0$, i.e., at steady state, we have

$$\Delta N = \frac{N_t}{1 + 2W\tau} \qquad ...(16)$$

When the intensity $I$ increases, $W$ increases, hence $\Delta N$ will decrease and in the limit $I \to \infty$. $\Delta N \to 0$, i.e., $N_1 \to N_2 \approx \frac{N_t}{2}$. In this case, the population of the two levels becomes equal. This phenomenon is called as *saturation.*

Now, in order to maintain the population difference, the system must absorb the energy from the pump. The power absorbed per unit volume by the system to maintain the population difference is given by

$$\frac{dP}{dV} = \Delta N \cdot W(\omega\hbar) \qquad ...(17)$$

(= (population difference/volume) × probability × energy)

$$= W\,\omega\hbar\,\frac{N_t}{1 + 2W\tau} \quad \text{(using Eq. 16)} \qquad ...(18)$$

Hence, at saturation, when $W\tau \gg 1$,

$$\left(\frac{dP}{dV}\right)_s = \frac{(\hbar\omega)N_t}{2\tau} \qquad ...(19)$$

This equation shows that the power $(dP/dV)_s$ which must be absorbed to keep the system in saturation is equal to the power lost by the material due to the decay of the upper states.

## 18.12 METHODS OF ACHIEVING POPULATION INVERSION

There are primarily four types of methods for achieving population inversion:

(i) Optical pumping

(ii) Electrical pumping

(iii) Chemical method

(iv) Semiconductor laser.

**(i) Optical Pumping.** When the atoms are exposed to electromagnetic radiation (photons) of frequency $\nu_{12}$, then there is selective absorption due to which the atoms are raised to excited state, e.g., ruby or neodymium laser and liquid state (dye) lasers. It is particularly suitable for solid state lasers.

**(ii) Electrical Pumping.**

(a) ***Excitation by Electrons.*** This method is used in gas lasers. Electrons are released from the atoms due to intense electric discharge through gas. The electrons are then accelerated to high velocities due to the strong electric field inside the discharge tube. When these electrons collide with the neutral gas atoms, a fraction of atoms are raised to excited state.

(b) ***Inelastic Collision between Atoms.*** If a gas has two different kinds of atoms (say $P$ and $Q$) then during an electric discharge, some of these atoms are raised to excited states ($P\ Q^*$ or $P^*Q^*$). Suppose $Q^*$ is in a metastable state having a mean life longer than the mean life of an excited atom ($\approx 10^{-8}$s).

If the excitation energy of $P^*$ and $Q^*$ are nearly equal, then some of $Q$ atoms may be raised to excited states due to inelastic collisions with $P^*$. $P^* + Y \rightarrow P + Y^*$. As a result, the number of $Q^*$ increases. This type of collision is known as collision of second kind. This method is used in He-Ne laser.

**(iii) Excitation by Chemical Method.** Sometimes an atom or a molecule which is a product of a chemical reaction is produced in an excited state, e.g., $H$ and $F$ combine to produce $HF$ molecule in the excited state and the number of such (excited) molecules may be considerably larger than those in the ground state (separate atoms), causing thereby population inversion. This method is used in excimer laser.

**(iv) Semiconductor Diode Laser.** The lasing material in these lasers consists of a $p$-$n$ homojunction (i.e., of a single crystal) made of heavily doped (~ $10^{18}$ $cm^{-3}$) $n$ and $p$ type semiconductor material such as GaAs. With a forward bias, the electrons concentrate on the $p$-side and the holes, on the $n$-side. The recombination of the electrons and holes in the junction region gives rise to laser action. If the current density at the junction is high enough, there may be population inversion between the electrons and hole levels. Thus, stimulated emission may take place if the optical gain in semiconductor cavity is higher than the junction layer (recombination) loss.

## 18.13 RUBY LASER

It is a three-level laser. Theodore Maiman, in 1960, demonstrated the first successful laser action by using a crystal of ruby as a laser material. Ruby is crystalline $Al_2O_3$ containing a small (0.05%) percentage of $Cr^{3+}$ ions (density $1.62 \times 10^{19}/cm^3$). This doping is done by adding small amounts of $Cr_2O_3$ in the melt of highly purified $Al_2O_3$. The normal green of $Cr^{3+}$ is modified to red by the distortion of the local crystal field stemming from the replacement of an $Al^{3+}$ ion by a slightly larger $Cr^{3+}$ ion. The bulk of the material $Al_2O_3$ does not participate in the laser action, it rather acts as a host, but the $Cr^{3+}$ ions from the active material. Ruby is a three-level laser. Maiman's first laser was in the form shown in Fig. 18.12.

The ruby was in the form of cylinder ~ 5 cm in length and ~ 0.3 cm in diameter. The ends of the ruby rod are grounded and polished plane and parallel (the ends are flat to better than

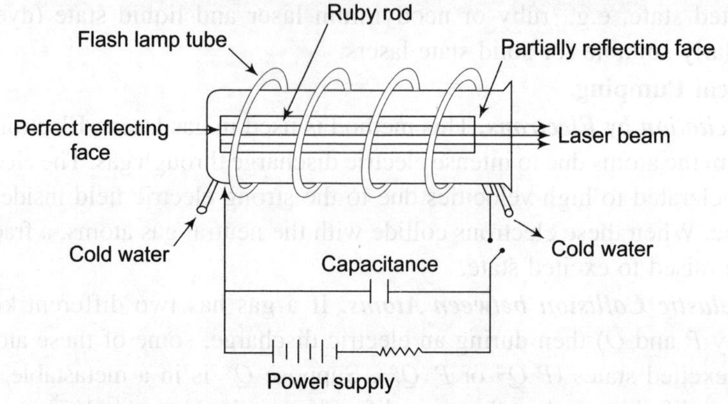


**Figure 18.12** A typical flash lamp pumped pulsed ruby laser

λ/10 and parallel to within 2 seconds of an arc). One end is silvered to give a completely reflecting surface and the other end is translucent. The rod is kept inside a glass tube which is surrounded by a tube in the form of a coil which is filled with xenon gas. The ends of this flash tube are connected to a power supply to produce flashes of light. Since a larger proportion of energy used in producing flashes goes away as heat energy. Cold water is allowed to circulate continuously in the glass tube.

Instead of flash lamp, pumping scheme, one may use other pumping schemes such as that shown in Fig. 18.13 in which the pump lamp and the laser rod are placed along the two foci of an elliptical cylindrical reflector. It is well known that the elliptical reflector focuses the light emerging from one focus into the other focus of the ellipse, leading to an efficient focusing of pump light on the laser rod.

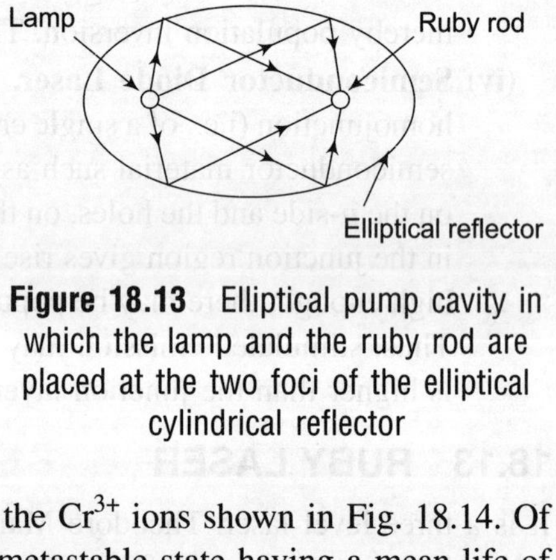


**Figure 18.13** Elliptical pump cavity in which the lamp and the ruby rod are placed at the two foci of the elliptical cylindrical reflector

### 18.13.1 Operation

The ruby laser is based on the three energy levels in the $Cr^{3+}$ ions shown in Fig. 18.14. Of the three levels shown, the first excited state $E_1$ is a metastable state having a mean life of about 3 milliseconds, which is about $10^5$ times longer than the normal mean life of an excited atomic state.

The higher excited state ($E_2$) has a mean life of $\sim 10^{-8}$s. Thus, there are two main pump bands $4F_1$ and $4F_2$ centred around 0.42 μm (blue) and 0.55 μm (green) respectively. Each pump band is about 0.1 μm wide.

When the xenon flash lamp is switched on, it provides light flashes of wavelength 5500 Å. This excites the $Cr^{3+}$ ions to the excited state ($E_2$) which subsequently make transitions to the lower excited state, $E_1$ (within $\sim 10^{-8}$ s) through phonon assisted (non-radiative) transition via

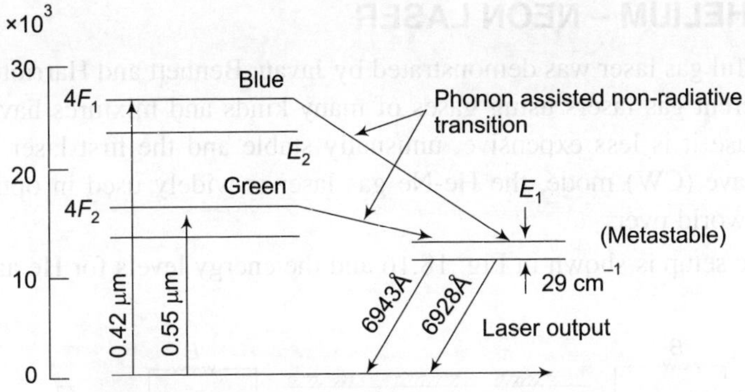


**Figure 18.14** Energy levels of $Cr^{3+}$ in ruby laser

losing some energy in collision with the lattice. Since the ions can remain in the metastable state, for comparatively longer periods ($10^{-3}$ s), more and more ions become trapped in this state making their number very large as compared to the ground state (provided pumping excitation from Xe flash lamp is sufficiently intense). In order that induced emission be greater than induced absorption, more than half the $Cr^{3+}$ ions in the ruby rod must be in the metastable state. The metastable state ($E_1$) slowly decays in ~$10^{-3}$ by emitting a sharp doublet with components 6943 Å and 6928 Å (the laser output beam). The rod's length must be precisely an integral number of half wavelength so that radiation trapped in it forms an optical standing wave.

The population inversion is very difficult to sustain continuously and in practice the ruby laser is pulsed. Typical pulses might consist of 2 J pulses persisting for 10 ns, corresponding to a power of 0.2 GW. Even during the short period of a few tens of microseconds in which the laser is oscillating, the output is a highly irregular function of time with the intensity having random amplitude fluctuations of varying duration (Fig. 18.15). This is called laser spiking.

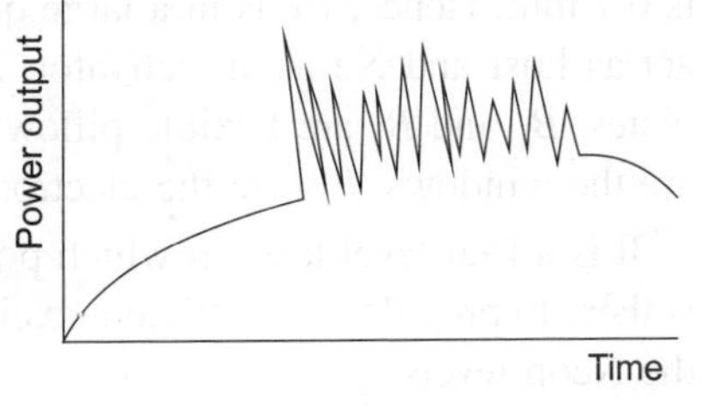


**Figure 18.15** Laser spiking in ruby laser beam

Ruby laser is one of the important practical lasers. The absorption bands of ruby are very well matched with the emission spectra of practically available flash lamps so that an efficient use of the pump can be made. It also has a favourable combination of a long lifetime and a narrow line width. It is also attractive from an application point of view since its output lies in the visible region where photographic emulsions and photodetectors are much more sensitive than in the infrared region. Thus, ruby laser finds application in holography and laser ranging. They are also used in measuring such plasma properties as electron density and temperature. Ruby lasers are also used to remove tattoos and also skin lesions resulting from excess melanin.

## 18.14 THE HELIUM – NEON LASER

The first successful gas laser was demonstrated by Javan, Bennett and Harriott in 1961. Since then, many different gas lasers using gases of many kinds and mixtures have been put into operation. Because it is less expensive, unusually stable and the first laser to be operative in continuous wave (CW) mode, the He-Ne gas laser is widely used in optics and physics laboratories the world over.

The schematic setup is shown in Fig. 18.16 and the energy levels for He and Ne are given in Table 18.1.

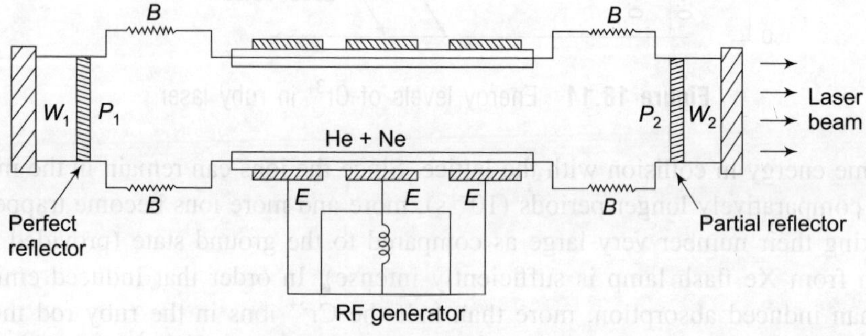


**Figure 18.16** The He–Ne gas laser arrangement

A discharge tube is filled with Ne and He gas. The He pressure is 1 mm and Ne pressure is 0.1 mm. Hence, He is in a large quantity (9 parts of He and 1 part of Ne) and so, He will act as host and Ne as an activator. $P_1$ and $P_2$ are the optically plane and parallel reflector plates. $B_1$ and $B_2$ are flexible pillows to adjust the distance between $P_1$ and $P_2$. $W_1$ and $W_2$ are the windows. *E*'s are the electrodes. Here we use an *r.f.* generator.

It is a four level laser in which population inversion is achieved by electric discharge. He is there to provide more efficient excitation only and actual laser emissions take place through the Neon levels.

**Table 18.1** The Energy Levels and their Designations for He and Ne

| *Element* | *Electron Configuration* | *Level Designation* | *Element* | *Electron Configuration* | *Level Designation* |
|---|---|---|---|---|---|
| He | $1s^2$ | $1\,S_0$ | | | 6 |
| He | $1s\,2s$ | $3S_1$ | Ne | $2p^5 3p$ | 7 |
| | | $1\,S_0$ | | | 8 |
| Ne | $2p^6$ | $1\,S_0$ | | | 9 |
| | | | | | 10 |
| | | $^3P_2$ | | | $^3P_2$ |
| Ne | $2p^5 3s$ | $^3P_1$ | Ne | $2p^5 4s$ | $^3P_1$ |
| | | $^3P_0$ | | | $^3P_0$ |
| | | $^1P_1$ | | | $^1P_1$ |

*Contd...*

**Table 18.1** *contd...*

| | | | | | |
|---|---|---|---|---|---|
| Ne | $2p^5 3p$ | 1<br>2<br>3<br>4<br>5 | Ne | $2p^5 5s$ | $^3P_2$<br>$^3P_1$<br>$^3P_0$<br>$^1P_1$ |

The spacing of the mirrors is equal to an integral number of half wavelength of the laser light. An electric discharge is produced in the gas by electrodes connected to a high frequency electric source. As a result, collisions between ionised atoms and electrons carrying the discharge current excite the atoms to various energy states and at a sufficiently high voltage, an orange-red glow discharge is produced. The energy level diagram is shown in Fig. 18.17.

The normal state of helium is a $^1S_0$ level arising from two valence electrons in 1*s* orbit. The excitation of either one of these electrons to the 2*s* orbit finds the atom in a $^1S_0$ or a $^3S_1$ state, both quite metastable since transitions to the normal state are forbidden by selection

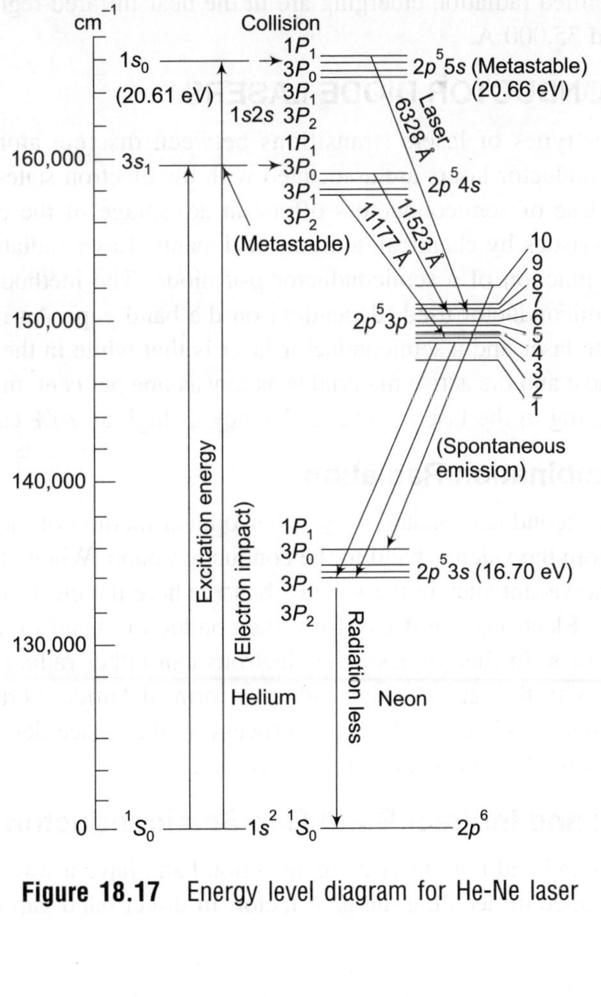


**Figure 18.17** Energy level diagram for He-Ne laser

rules. (He atom excited to the 1$s$ 2$s$ state cannot return to the ground state by emitting a 20.61 eV photon because both the states have zero total angular momentum and a photon must carry away at least one $\hbar$ of angular momentum). However, the He atom can lose energy by energy exchange collisions with Ne-atoms initially in the ground state. A 1$s$ 2$s$ He atom, with its initial energy 20.61 eV and a little (0.05 eV) additional (kinetic) energy (provided by kinetic energy of atoms) can collide with a Ne atom in the ground state, exciting it to the 5$s$ excited state at 20.66 eV and leaving the He atom in the $^1s_2$ ground state. Consequently, the population in the 5$s$ state is substantially enhanced. Thus, He atoms help to achieve a population inversion in the Ne atoms. Stimulated emission from 20.66 eV metastable state then results in the emission of highly coherent laser light at 6328 Å, due to transition to a lower excited state at 18.70 eV. This state ($2p^5 3p$) decays by spontaneous emission. The remaining excitation energy is lost in collisions with the tube walls through radiation less transitions.

Since the electron impacts exciting the He and Ne atoms occur all the time, it is a continuous wave (CW) laser. Further since the laser transition does not terminate at the ground state, the power needed for excitation is less than that in a 3 level laser.

Most of the amplified radiation emerging are in the near infrared region of the spectrum, between 10,000 and 35,000 Å.

## 18.15 SEMICONDUCTOR DIODE LASERS

In contrast to other types of lasers (transitions between discrete atomic energy levels), transitions in semiconductor lasers are associated with the electron states in the valence and conduction bands. Use of semiconductors offers an advantage of the comparative ease to adjust their characteristics by chemical addition of dopants. Laser radiation can be obtained by pumping at the junction of a semiconductor *p-n* diode. The method of pumping varies with the type of semiconductor used (depending on the band gap). An important difference between a solid state laser and a semiconductor laser is that while in the former, the bulk of the material is the host and the active material is as low as one per cent, in the latter, the entire material is participating in the laser action. Efficiency as high as 70% can be achieved.

### 18.15.1 Recombination Radiation

Excitation of the semiconductor material by either optical method or electrical method, can transfer electrons from the valence band to the conduction band. When electrons move to the conduction band, the vacant sites in the valence band (where the electrons were located) are referred to as holes. Electrons can decay from the conduction band to the valence band by recombining with holes. In this process, the electrons can either radiate the energy or give it up via collisions with the lattice which induce a form of lattice vibration called phonon relaxation in the material. Which of these two processes takes place depends on whether the semiconductor is a direct or indirect band gap material.

### 18.15.2 Direct and Indirect Band Gap Semiconductors

The top of valence band and bottom of the conduction band have a wave vector dependence which has a significance in deciding the gain factor. In direct band gap material, the lowest conduction band minimum and the highest valence band maximum correspond to the same crystal momentum or wave vector in the Brillouin zone but in indirect band gap semiconductor, the above two are at different wave vectors. The optical transition in indirect band gap semiconductors is a first order process, since momentum is automatically conserved and the optical gain is high. In the case of indirect semiconductor, the transition involves phonons also and collisional phonon relaxation in the material produces acoustic waves and unwanted heating of the material. In this process the quantum yield of radiation is low, i.e., optical transitions are weaker and the available gain is smaller.

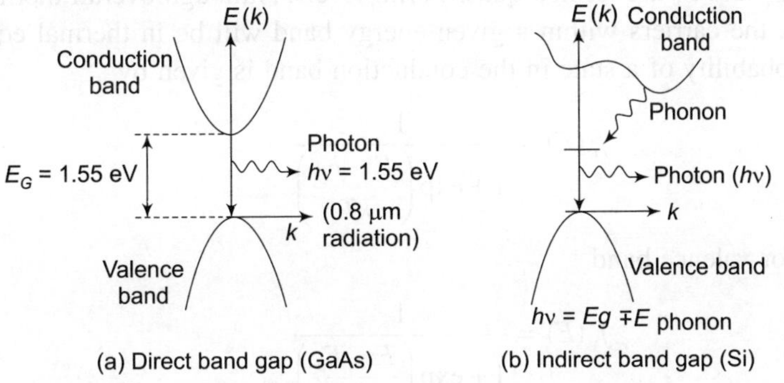


**Figure 18.18** (a) Direct band gap and (b) Indirect band gap semiconductor materials

### 18.15.3 Condition for Laser Action in a Semiconductor

In case of metals/semiconductor solids, the probability of occupation, $F(E)$, of any energy state $(E)$ is given by Fermi Dirac statistics

$$(E) = \frac{1}{1+\exp\left(\dfrac{E-F_0}{kT}\right)} \qquad \text{...(1)}$$

where $F_0$ is known as Fermi level for the system. $F_0$ represents the boundary between fully occupied and completely empty levels at $T = 0$ K. When there is a perfect thermal equilibrium between the conduction and valence band, single Fermi level gives the distribution of electrons over the whole range of energies of both bands (Fig. 18.19(a)).

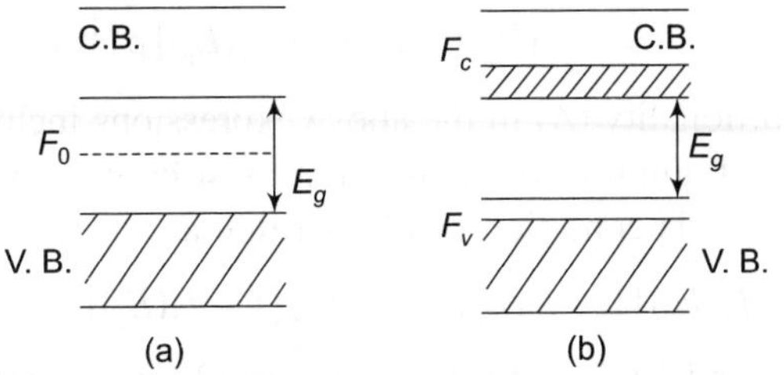


**Figure 18.19** Energy levels of a semiconductor (a) $T$ = 0 K (b) ambient temperature. Some electrons excited to the conduction band

The electrons may be raised from the valence band to the conduction band by (say) optical excitation with energy $E > E_g$. The semiconductor in this condition is not in equilibrium. Nevertheless, the equilibrium is reached within each band in a very short time (~ $10^{-13}$ s) in which the electrons in conduction band will have dropped to the lowest level in that band; and any electrons near the top of the valence band will also have dropped to the lowest unoccupied level, leaving the top of valence band full of holes (Fig. 18.19 (b)). Thus, there is population inversion between bottom of conduction band and top of valence band which may give rise to laser action. (Valence band is empty of electrons down to an energy ($F_v$) and the CB is full upto $F_c$). $F_c$ and $F_v$ are termed quasi-Fermi levels. Although overall thermal equilibrium does not exist, the carriers within a given energy band will be in thermal equilibrium. The occupation probability of a state in the conduction band is given by

$$f_c(E) = \frac{1}{1+\exp\left(\dfrac{E-F_c}{kT}\right)} \quad ...(2)$$

Similarly, for valence band

$$f_v(E) = \frac{1}{1+\exp\left(\dfrac{E-F_v}{kT}\right)} \quad ...(3)$$

At equilibrium $\quad F_c = F_v = F_o$

When a beam of light is incident on such a semiconductor, the number ($N_a$) of quanta absorbed per unit time will be proportional to:

(i) the probability per unit time of direct transition from valence band to conduction band, $B_{vc}$,

(ii) the density of incident radiation $\rho(W)$,

(iii) the probability of the concerned state in the valence band being occupied: $f_v(E_v)$, and

(iv) the probability that the upper state in conduction band is empty $[1 - f_c(E_c)]$

That is, $\quad N_a = AB_{vc} f_v(E_v)[1 - f_c(E_c)]\rho(W) \quad ...(4)$

Similarly, the number of quanta ($N_e$) emitted per unit time by stimulated emission is

$$N_e = AB_{cv} f_c(E_c)\,[1 - f_v(E_v)]\,\rho(W) \quad ...(5)$$

The constant of proportionality ($A$) in the above expressions includes the density of states of the two bands. For a net amplification, emission must be greater than absorption, i.e., $N_e > N_a$. Assuming $B_{vc} = B_{cv}$, this condition can be written as

$$f_c(E_c)[1 - f_v(E_v)] > f_v(E_v)[1 - f_c(E_c)] \quad ...(6)$$

Substituting for $f_c(E_c)$ and $f_v(E_v)$ from Eqs. (2) and (3) and simplifying, we get

$$F_c - F_v > E_c - E_v = hv$$

This is the necessary condition for stimulated emission to be dominant over absorption in an intrinsic semiconductor. If the semiconductor contains impurities and the impurity energy states are either the initial or final states, it is necessary only to use the quasi-fermi levels for the impurities with the proper degeneracy for the states.

### 18.15.4 *p-n* Junction Laser

A simple way of achieving population inversion is to use a semiconductor in the form of a *p-n* junction diode heavily doped with donors (on *n*-side) and acceptors (on *p*-side). Figure 18.20 depicts the energy level diagram.

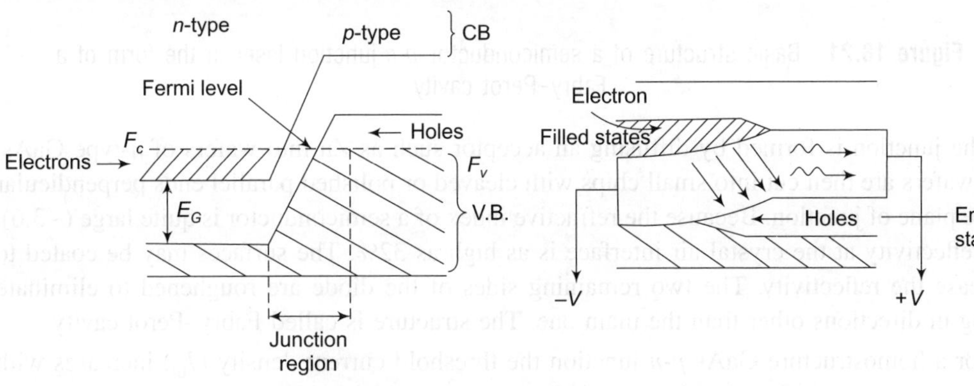


(a) No voltage applied (b) With a forward voltage

**Figure 18.20** Energy levels of a *p-n* junction, (a) equilibrium with no voltage applied, (b) with forward voltage, there results recombination emission

When no voltage is applied, electrons from the *n* side flow to the *p*-side until an electrical potential barrier ($V_B$) is built up due to presence of immobile ions (electrons and holes have recombined in the junction region—the depletion layer). This prevents further flow of charges (Fig. 18.20 (a)). Due to a forward voltage (Fig. 18.20 (b)) we have injected electrons on *n*-side and injected holes on *p*-side, both in high concentrations. A high hole concentration means a large number of empty sites into which electrons can fall, i.e., a population inversion in the depletion region. Hence, laser action will take place in this region when the current flow through the diode exceeds a certain threshold value. Similarly, population inversion has also occurred on the *n*-side due to filled levels near the bottom of the conduction band. The recombination of electrons and holes can only take place in the junction region due to lowering of energy gap in presence of forward voltage resulting in an ease for carrier flow and consequent recombination. This results in generation of coherent (laser) radiation. On the bulk *p*-side and *n*-side on both sides, the recombination energy is dissipated through non-radiative processes and so, the laser action is confined to the very thin planar junction region.

## 18.16 SEMICONDUCTOR HOMOJUNCTION LASER STRUCTURE

Semiconductor lasers were first fabricated in 1962 using GaAs in the form of a diffused *p-n* homojunction. Figure 18.21 shows the basic structure of a *p-n* junction laser.

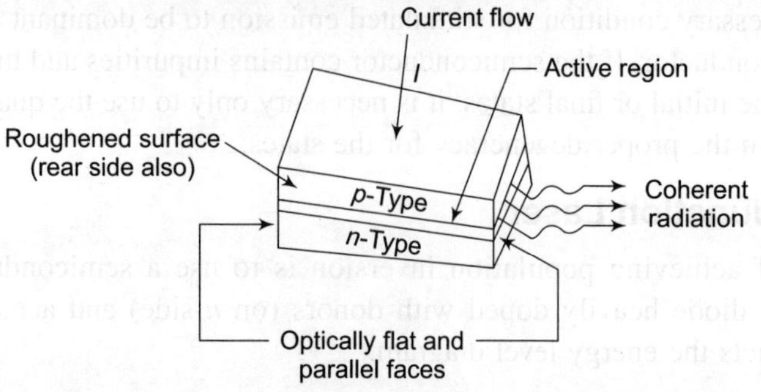


**Figure 18.21** Basic structure of a semiconductor *p-n* junction laser in the form of a Fabry–Perot cavity

The junction is formed by diffusing an acceptor such as Zn into wafers of *n*-type GaAs. The wafers are then cut into small chips with cleaved or polished parallel ends perpendicular to the plane of junction. Because the refractive index of a semiconductor is quite large (~ 3.6), the reflectivity at the crystal air interface is as high as 32%. The surfaces may be coated to increase the reflectivity. The two remaining sides of the diode are roughened to eliminate lasing in directions other than the main one. The structure is called Fabry–Perot cavity.

For a homostructure GaAs *p-n* junction the threshold current density ($J_{th}$) increases with temperature. A typical value of $J_{th}$ is ~ 500 A/(mm)$^2$ at room temperature. Such large current density imposes serious difficulties in continuous operation of these lasers. Only pulsed operation is possible. The threshold current density can be reduced by using heterojunctions instead of homojunction. A typical single heterojunction laser consisted of a *p-n* junction between GaAs and GaA1As. At the heterojunction there is carrier confinement as well as optical confinement. The room temperature threshold current density is reduced to ~ 100 A/(mm)$^2$.

## 18.16.1 Advantages of Semiconductor Laser over Other Lasers

Semiconductor lasers differ from other lasers in the following important aspects:

1. In conventional lasers, the quantum transitions occur between discrete energy levels, whereas in semiconductor lasers, the transitions are associated with the band properties of the material.
2. A semiconductor laser is very compact in size (≃ $10^{-4}$ cm$^3$). Because the active region is very narrow (~ or < 1 μm) the divergence of the beam is considerably larger than in a conventional laser.
3. The spatial and spectral characteristics of a semiconductor laser are strongly influenced by the properties of the junction medium (such as band gap and refractive index variation).
4. For the *p-n* junction laser, the laser action is produced by simply passing a forward current through the diode itself. The result is a very efficient overall system that can

be modulated easily by modulating the current. Since semiconductor lasers have very short photon lifetimes, modulation at high frequencies can be achieved.

### 18.16.2 Disadvantages of Semiconductor Lasers

In spite of several advantages, there are unfortunately some drawbacks. The lifetime of these lasers is rather short unless they are operated at low temperatures and in a pulsed mode. The optical transitions in the semiconductors are rather broad of the order of 2 nm yielding line shape function with a width of the order of about 30 $cm^{-1}$ for a wavelength of 840 nm.

## 18.17 SEMICONDUCTOR HETEROJUNCTION INJECTION LASER

In the preceding section, we have considered the *p-n* junction laser fabricated from a single crystal semiconductor material. This is known as a homojunction. However, we may have improved radiative properties by the use of heterojunctions, i.e., *p-n* junction fabricated from two different semiconductor materials having different energy band gaps. Devices fabricated with heterojunctions are said to have heterostructure. Homojunction lasers can operate only at very low temperatures. In order to make room temperature lasers, it became necessary to develop heterostructures.

Heterojunctions may be either isotype (*p-p* or *n-n*) or anisotype (*p-n*). The isotype heterojunction provides a potential barrier within the structure which is useful for the confinement of minority carriers to a small active region. This is known as carrier confinement. It effectively reduces the carrier diffusion length and thus the volume within the structure where radiative recombination may take place. This technique is used for the fabrication of injection lasers. Use of anisotype heterojunctions having sufficiently large band gap differences improves the injection efficiency of either electrons or holes. Both types of heterojunction (i.e., isotype/anisotype) provide a dielectric step due to the different refractive indices at either side of the junction. This is used to provide radiation confinement in active region (optical waveguide effect) in injection laser. The confinement efficiency depends on the magnitude of the step, which in turn, depends on difference in band gap energies and the wavelength of radiation.

Heterostructures are used to provide potential barriers in injection lasers. When a double heterojunction structure (DH), e.g., *p-p-n* is implemented, the resulting carrier and optical confinement reduce the threshold currents necessary for lasing by a factor of about 100. The layer structure and an energy band diagram for a DH injection laser are illustrated in Fig. 18.22.

There is a heterojunction on either side of the active layer. When a forward voltage corresponding to the band gap energy of the active layer is applied, a large number of electrons are injected from *n*-side to *p*-side (active layer) and holes to the active layer from the left *p*-end. Consequently, laser oscillations commence. The injected carriers are confined to the active layer by the energy barriers provided by the heterojunctions which are made within the diffusion length of the injected carriers. Further, a refractive index step at the heterojunction provides a radiation confinement to the active layer (because, in effect, the

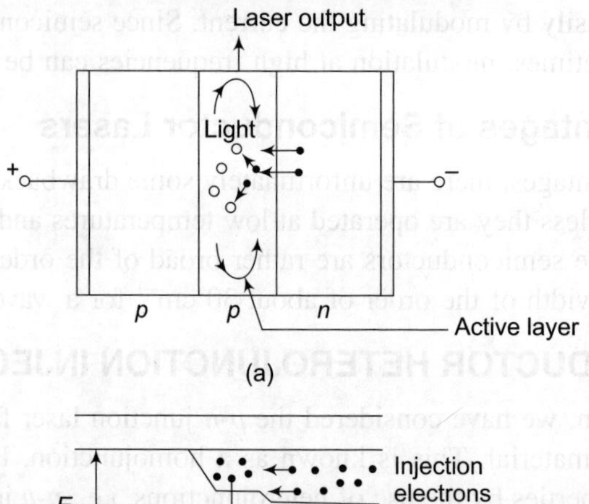


**Figure 18.22** The double heterojunction injection laser (a) the layer structure with an applied forward bias (b) energy band diagram with a *p-p* heterojunction on the left and a *p-n* heterojunction on the right

active layer behaves as a dielectric waveguide to confine the electroluminescence within the active region). Figure 18.23 illustrates a broad area GaAs/A1GaAs double heterojunction injection laser.

Lasing takes place across the whole width of the device and so two sides of the device are roughened to reduce unwanted emission in these directions. Further, to provide optical confinement in the horizontal plane, stripe geometry is used. The stripe is formed by the creation of high resistance areas on either side of the active layer by using the proton bombardment technique. The forward voltage is applied to the $p^+$–GaAs and $n^+$-GaAs which are placed at the two sides of the confinement layers.

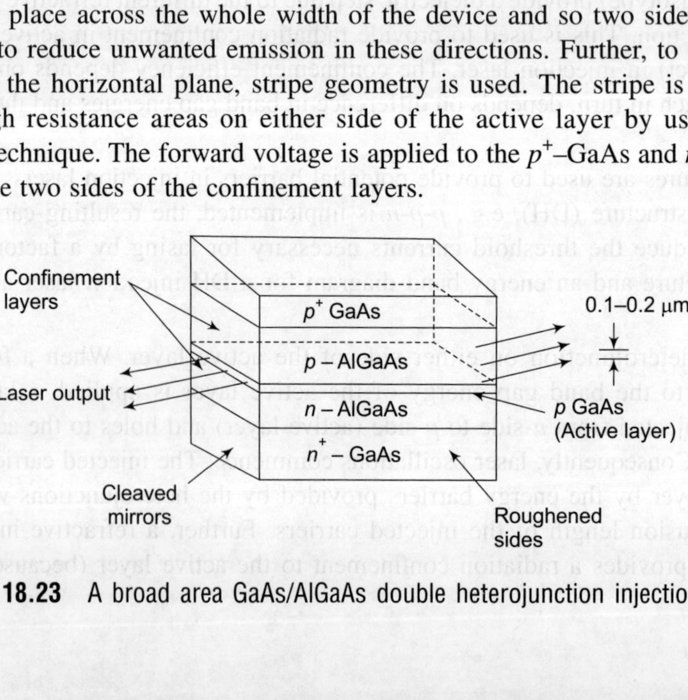


**Figure 18.23** A broad area GaAs/AlGaAs double heterojunction injection laser

Because the charge carriers in heterojunction lasers are confined to a much smaller region than in homojunction lasers, the heat deposition is much lower while the threshold current density (and thus the recombination radiation) are still at a sufficiently high level. Double heterojunction structures provide additional index of refraction variations providing better guiding of optical wave when operated as laser.

## 18.18 POPULATION INVERSION IN A SEMICONDUCTOR LASER AND COMPARISON OF EMISSION WITH A LED

In a laser diode, the doping levels on the two sides of the junction are kept lower than those in a tunnel diode. Thus, the *I-V* characteristics of a laser diode does not exhibit a negative resistance. Figure 18.24 shows the energy band diagram of a *p-n* junction at thermal equilibrium, and Fig. 18.24 shows the same structure with a forward bias that is larger than the band gap. Figure 18.24 (b) shows that the region near the junction is not depleted of mobile carriers but rather contains a high concentration of electrons in the conduction band and vacant states in the valence band. Thus, a condition of population inversion occurs in a thin region about the junction. When photons of appropriate energy impinge on this region, they will induce electron transitions from the conduction to the valence band and cause stimulated emission.

The meaning of population inversion in Fig. 18.24 (b) can be understood by referring to quasi-Fermi levels $E_{Fn}$ and $E_{Fp}$, which are shown by the dashed lines in the figure. Note that in the neutral *n*- and *p*-regions, far away from the junctions, $E_{Fn}$ and $E_{Fp}$ coincide. However, it is evident from Fig. 18.24 (b) that, in the population inversion region near the junction, $E_{Fn}$ and $E_{Fp}$ are separated from each other by an energy equal to $E_g + q(\phi_n + \phi_p)$. Population inversion has occurred in this region in the sense that, in the energy range between $E_{Fn}$ and $E_{Fp}$, more states are occupied in the conduction band than in the valence band. It is clear that the incident photon that can induce stimulated emission should have an energy $E_g < hv < [E_g + q(\phi_n + \phi_p)]$.

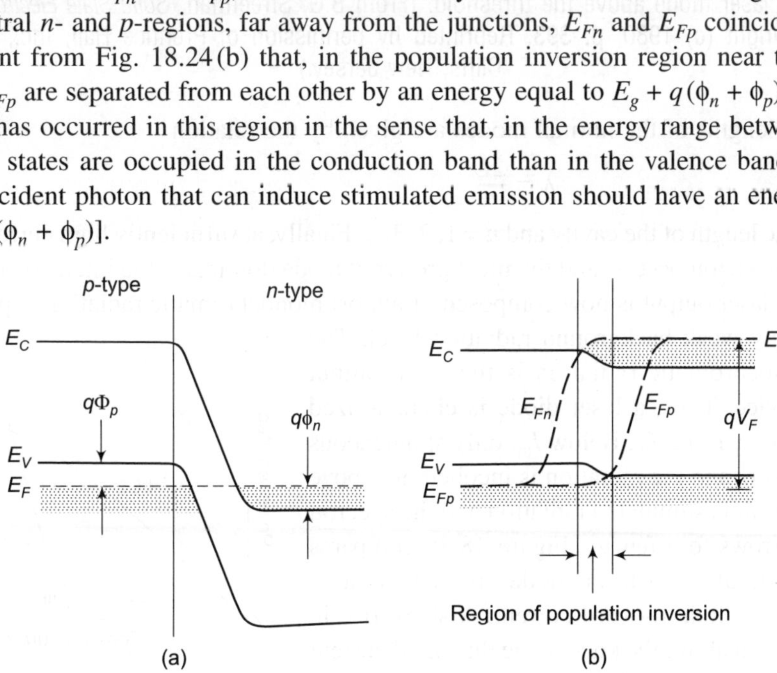


**Figure 18.24** Energy band diagram of a *p-n* junction laser diode (a) under thermal equilibrium and (b) at a large forward bias in excess of the band gap voltage

In a forward-biased *p-n* junction, the injected electron-hole pairs can also recombine spontaneously and cause spontaneous emission of photons in energy ranging from $E_g$ to $E_g + q(\phi_n + \phi_p)$. Spontaneous radiation dominates before the occurrence of strong population inversion at a large forward bias. Figure 18.25 shows the typical plots of the intensity of the emitted radiation as a function of $h\nu$. At low bias currents, a spontaneous emission spectrum with energies between $E_g$ and $E_g + q(\phi_n + \phi_p)$ is obtained (Fig. 18.25a). As the diode current is gradually increased, a point is reached where significant population inversion exists near the junction region. Consequently, stimulated emission occurs at frequencies corresponding to the normal modes of the Fabry–Perot cavity as shown in plot (b).

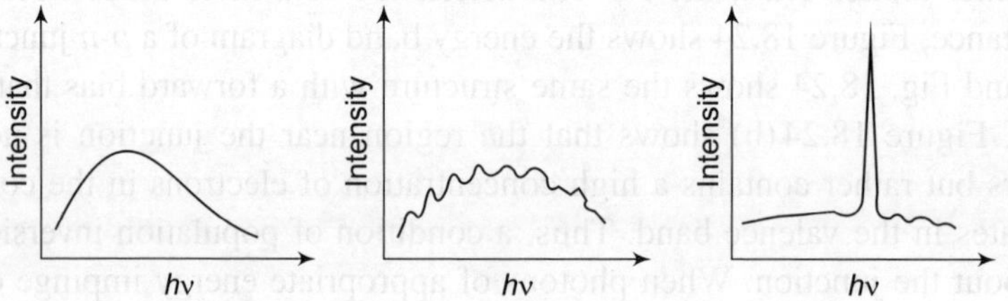


**Figure 18.25** Light intensity as a function of photon energy for a *p-n* junction laser diode: (a) incoherent emission below the laser threshold, (b) laser modes at the onset of laser threshold, and (c) dominant laser mode above the threshold. (From B.G. Streetman, *Solid State Electronic Devices*, 2nd ed., copyright (c) 1980, p. 393. Reprinted by permission of Prentice Hall, Inc., Englewood Cliffs, New Jersey.)

The wavelengths of the normal modes are given by the relation

$$\lambda = \frac{2L}{n} \qquad ...(1)$$

where $L$ is the length of the cavity and $n = 1, 2, 3\ldots$. Finally, at sufficiently high currents, strong population inversion occurs, and the most preferred mode dominates the intensity as shown in plot (c). The laser output is now composed of almost monochromatic radiation superimposed on a relatively weak background radiation level. The intense monochromatic radiation is the main output from the device. Thus, a laser diode is characterized by a threshold current, $I_{th}$. Below $I_{th}$, only spontaneous emission occurs and the radiation is incoherent. Above $I_{th}$, the diode emits stimulated radiation and the spectral emission narrows to a few Å. Figure 18.26 compares the power outputs of a laser diode and a LED as a function of the diode forward current. A sharp rise in the power output above the knee of the threshold current is observed in the case of the laser diode. However, the power from an LED increases continuously with the bias current, and no threshold is observed.

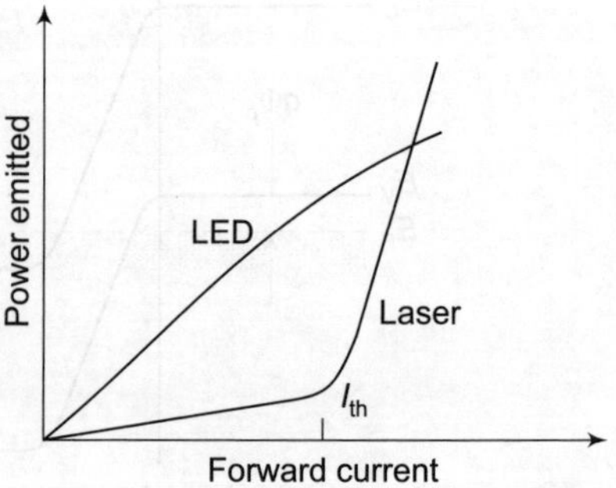


**Figure 18.26** Power emitted from a laser diode and from a LED as a function of the bias current

## 18.19 MÖSSBAUER EFFECT

Increasing the quantum energy beyond the X-rays leads to the spectral range of $\gamma$ radiation. This radiation originates from a relaxation of excited and quantized states of the nuclei. The idea of spectroscopy using $\gamma$-ray quanta was suggested by Kuhn in 1929. A number of attempts to observe resonance scattering with $\gamma$ radiation were made but were all without success. The problems of nuclear resonance fluorescence and absorption were complicated due to the large amounts of recoil energies given to the nucleus by the emitted $\gamma$-radiations. However, R. Mössbauer reported for the first time recoil free resonance absorption for $\gamma$-rays in 1958. This technique is now known as *Mössbauer spectroscopy*.

Mössbauer spectroscopy is the study of $\gamma$-ray absorption (or emission) spectra for transitions between nuclear states. For the discovery of this spectroscopic method, Mossbauer was awarded the Nobel Prize in 1961.

## 18.20 PRINCIPLE OF MÖSSBAUER SPECTROSCOPY (NRF)

Nuclear resonance absorption is in principle expected to occur when gamma radiation emitted in a transition is reabsorbed by another nucleus of the same kind as depicted in Fig. 18.27.

Imagine, now an isolated atom of mass ($M$) in an excited nuclear state ($E_i$) above the ground state, $E_f$. Let $E_0$ be the energy difference, $E_i - E_f$. If $p$ is the recoil momentum, then kinetic energy of the recoil nucleus

$$E_{\text{re}} = \frac{p^2}{2M} \qquad ...(1)$$

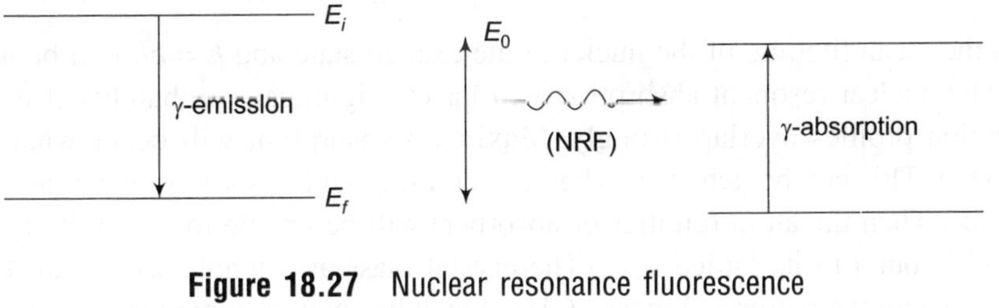


**Figure 18.27** Nuclear resonance fluorescence

The emitted gamma ray will have an energy

$$E_\gamma = E_0 - \frac{p^2}{2M} = E_0 - E_{\text{re}} \qquad ...(2)$$

Therefore, the emitted spectral line will be shifted from the expected position, $E_0$ (as shown in Fig. 18.28).

By the law of conservation of momentum the recoil momentum of the atom is equal to the momentum of the emitted $\gamma$-ray photon (given by the de Broglie relationship), i.e.,

$$p = \frac{h}{\lambda} = \frac{E_\gamma}{c} \quad (\because \; c = \nu\lambda) \qquad ...(3)$$

where $c$ is the velocity of light.

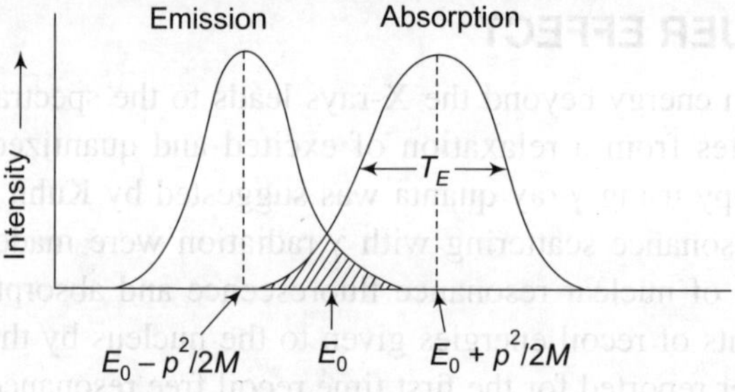


**Figure 18.28** Overlap of emission and absorption lines

Hence, $$E_{re} = \frac{p^2}{2M} = \frac{E_\gamma^2}{2Mc^2} = \frac{(E_0 - E_{re})^2}{2Mc^2} \simeq \frac{E_0^2}{2Mc^2} \quad (\because E_{re} \ll E_0) \qquad ...(4)$$

When the same atom absorbs the γ-ray, its nucleus will have an energy

$$E_\gamma = E_0 + \frac{p^2}{2M} = E_0 + E_{re} \qquad ...(5)$$

Thus, the nuclear resonant absorption can only be observed as a successive γ-ray emission and absorption, and the effect of the recoil is to introduce a difference, $E_\gamma - E_\gamma = 2E_{re}\ E_0^2/Mc^2$, between the energies of the emitted and absorbed γ-rays in the resonant process. The line width ($\delta v/v$) is governed by the energy-time uncertainty relation.

$$\Gamma_E \times \tau = \hbar \text{ or } \Gamma_E = \frac{\hbar}{\tau}; \quad (\Gamma_E \text{ is energy uncertainty})$$

where $\tau$ is the mean lifetime of the nuclei in the excited state and $\hbar = h/2\pi$, $h$ being Planck's constant. The nuclear resonant absorption will have a significant probability if the emission and absorption profiles overlap strongly. Maximum absorption will occur when the recoil energy is zero. This can be achieved when the emitting and absorbing nuclei are found in a crystal lattice. Then the atom (emitter or absorber) will be unable to recoil freely because it is chemically bound to its lattice site. (The crystal mass as a whole will recoil.) Equations (2) and (5) contain the reciprocal mass *(1/M),* which has now become the mass of the crystal (containing ~ $10^{15}$ atoms).

Thus, the recoil energy of the atom becomes very small, it cannot rupture a chemical bond, rather it becomes the property of the lattice and the lattice as a whole vibrates to dissipate the recoil energy. In other words, there is a finite probability in a solid matrix that a γ-ray photon be emitted or absorbed without recoil or thermal broadening.

## 18.21 EXPERIMENTAL METHOD

The experimental method is based on using a solid matrix containing the excited nuclei of a given isotope called the *source* next to which is placed a second matrix containing the same *isotope* in ground state, called the absorber. There are a large number of radioactive isotopes produced artificially which can be suitable for the study of Mössbauer effect, e.g., $Fe^{57}$, $Zn^{67}$

and $Sn^{119}$ have been commonly used. We have seen that the amount of overlap of the energy profiles for the source and absorber decides the extent of resonant absorption, so the basic experimental technique relies on creating *Doppler velocity shift* by some device.

A movement of the source and absorber relative to each other with velocity ($v$) produces a Doppler shift in frequency

$$\Delta\nu = \pm\left(\frac{v}{c}\right)\nu$$

or energy shift, $\Delta E = \pm\dfrac{vE}{c}$

where $c$ is the velocity of light and $\nu$, the frequency of γ-ray.

The nuclear transition energy involved is usually 10–100 keV $mol^{-1}$ and is far more energetic than chemical bond ($10^{-6}$ keV $mol^{-1}$) or molecular vibration energy ($10^{-6}$ keV $mol^{-1}$). The energy time uncertainty relation and overlap condition puts a limit on the value of excited state lifetime of the source material, because natural width of the resonance line depends on the lifetime of the excited state. Consequently, it is desirable that the excited state lifetime of the source should be in the range $10^{-6}$ s $< \tau < 10^{-10}$ s.

Consider the source $^{57}Co$ ($I$ = 7/2, $e\ Q$ < 0) which belongs to $^{57}Fe^*$ through electron capture. This excited nucleus comes to stable $^{57}Fe$ through emission of γ-rays. The complete energy diagram is shown in Fig. 18.29.

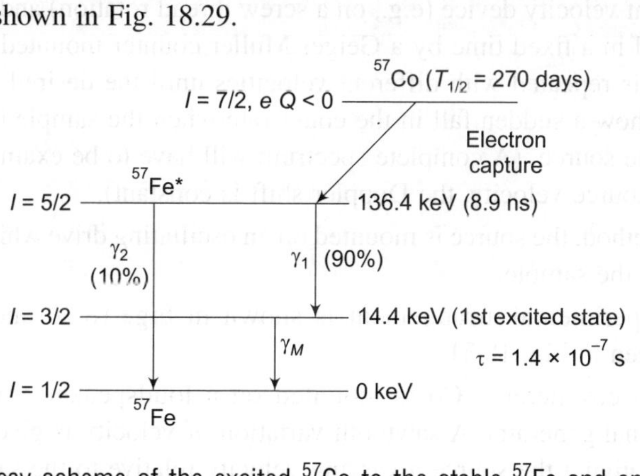


**Figure 18.29** Decay scheme of the excited $^{57}Co$ to the stable $^{57}Fe$ and emission of delayed γ-rays

The absorbers are usually stable isotopes of elements with which the Mössbauer effect can be observed. As emitter, one selects the excited state of the same isotope, e.g., $^{57}Co$ serves as an emitter and $^{57}Fe$ stable isotope (which absorbs the 14.4 keV γ radiation) serves as good absorber.

A transition from $I$ = 3/2 → 1/2 involving an energy change of $E_0$ = 23.07 × $10^{-16}$ kJ/nuclei is of interest. The frequency of the emitted γ-ray is $\nu$ = 34.88 × $10^{17}$ Hz. The absorber placed in the path of the emitted γ-ray leads to the transition, $I = 1/2 \rightarrow 3/2$. The transitions in any stationary system are not sharp due to the uncertainty relation:

$$\Gamma_E \times \tau = \hbar = 1.054 \times 10^{-34} \text{ Js}$$

where $\Gamma_E$ is the energy uncertainty and $\tau$ is the mean lifetime of the nuclei in the excited state. Since $\tau = 1.4 \times 10^{-7}$ s in the present case, therefore

$$\Gamma_E = \frac{\hbar}{\tau} = 7.5 \times 10^{-28} \text{ J}$$

and, therefore, frequency uncertainty

$$\delta\nu = 1.14 \times 10^6 \text{ Hz},$$

and the corresponding line width

$$\delta\nu/\nu = 3.27 \times 10^{-13}$$

This is much smaller spectroscopic widths found in other spectroscopic techniques (viz., for NMR ≈ $10^{-8}$; for *IR* ≈ $10^{-4}$). So, a very sharp absorption line is observed.

## 18.22 APPARATUS FOR MÖSSBAUER SPECTROSCOPY

Experimentally, two methods are possible. In one method, the source ($^{57}$Co) is mounted on a mechanical constant velocity device (e.g., on a screw thread rotation) and the total number of counts is registered in a fixed time by a Geiger Muller counter mounted behind the sample. The same process is repeated with different velocities until the desired velocity is covered. GM counter will show a sudden fall in the count rate when the sample begins to absorb the γ-ray emitted by the source. (A complete spectrum will have to be examined point by point, since for any one source velocity, the Doppler shift is constant).

In the second method, the source is mounted on an oscillating drive which gives it a varying velocity relative to the sample.

The general experimental arrangement is shown in Fig. 18.30 and that of a modern spectrometer is given in Fig. 18.31.

In a modern spectrometer, $^{57}$Co is mounted on a loudspeaker coil whose motion is controlled by a signal generator. A sawtooth variation of velocity is given. This provides an oscillatory motion giving the source a varying velocity relative to the sample (zero velocity at the extremities and maximum at the centre).

The absorbing sample is refrigerated. A scintillation counter, or a semiconductor detector may be used for detection. The signal is then fed to a multichannel analyser which collects the results and sums it over each cycle. The discriminator rejects the background (non-resonant) radiation. The final Mössbauer spectrum is displayed (as counts/s) as a function of relative velocity between the source and absorber. The absorption by the sample is shown by a fall in counts/s. A relative velocity of 0.1 mm/s is sufficient to produce a marked reduction of the 14.4 keV $^{57}$Fe transition.

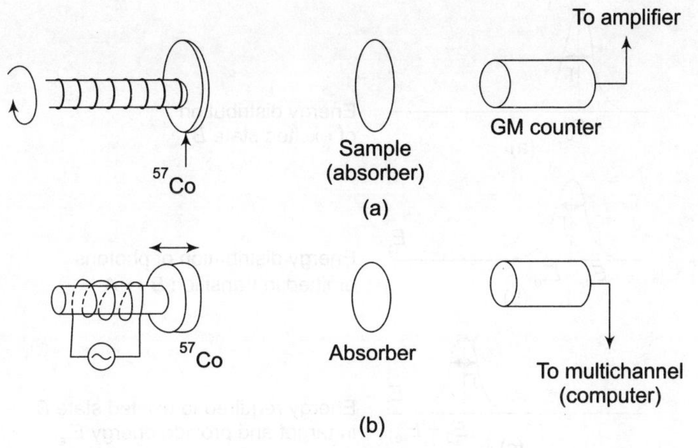


**Figure 18.30** Two experimental arrangements for Mössbauer spectroscopy (a) employing a screw-thread drive (b) using an oscillating drive for the source

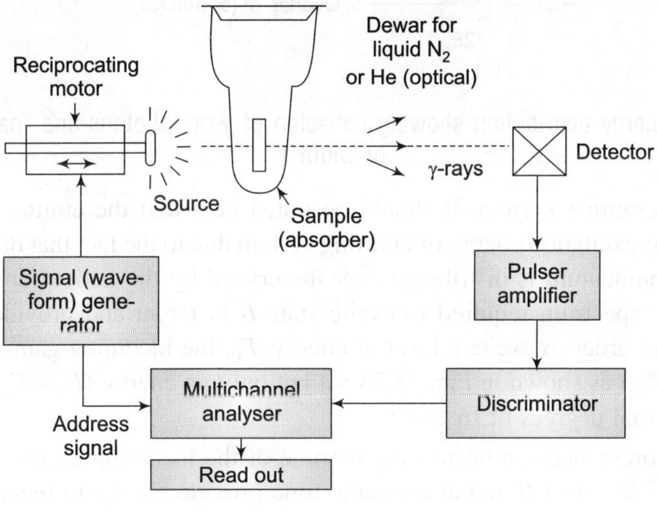


**Figure 18.31** Schematic arrangement of a modern Mössbauer spectrometer

## 18.23 RESONANCE FLUORESCENCE

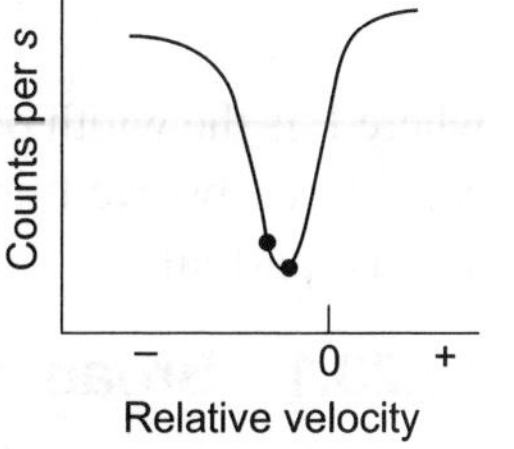


**Figure 18.32** Typical Mössbauer spectrum

The Mössbauer effect is known as recoilless gamma resonance fluorescence in which no part of energy is expended in recoil of the nucleus emitting or absorbing the gamma quanta. In resonance fluorescence, a photon of energy ($E_r$) and momentum ($p$) strikes a target of mass $M$ (initially at rest) and the entire photon-momentum is transferred to the target. The target thus recoils and the energy of recoil (i.e., energy distribution of excited state is as shown in Fig. 18.33 (a). Figure 18.33 (b) shows energy distribution of photons emitted in transition $B \rightarrow A$. This energy is obtained from

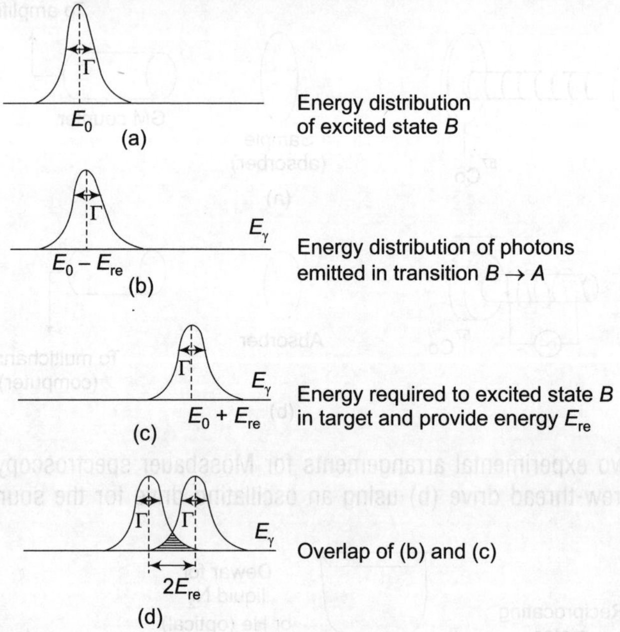


**Figure 18.33** Energy distribution showing emission of $\gamma$-ray photons and "natural line shape" of width $\Gamma$

de-excitation of emitting system. It should be noted here that the emitted photon has less energy than the de-excitation energy of emitting system due to the fact that the excited nucleus recoils and the momentum is distributed over the crystal by the lattice forces. Figure 18.33 (c) shows energy spectrum required to excite state $B$ in target and provide centre of mass energy, $E_{re}$, i.e., in order to excite a level of energy, $E_0$, the incoming gamma ray must have an energy $(E_0 + E_{re})$ as shown in Fig. 18.33 (c) but only an energy $(E_0 - E_{re})$ is available for excitation of internal degrees of freedom.

Resonance fluorescence can occur only if some of the incoming photons possess enough energy to "reach" the state $B$ and at the same time provide energy to recoiling system, i.e., for resonance fluorescence, the absorption line would overlap the emission line. But in actual case only the overlapping part of the spectra, (Figs. 18.33(b) and 18.33 (c)) is responsible for resonance fluorescence as shown in Fig. 18.7 (d). The condition for overlap is

$$2\,E_{re} < \Gamma$$

where $\Gamma$ is the width of the "natural line shape". It is thus clear that the resonance fluorescence cannot be observed unless it is possible for the emission line and absorption line to overlap at least partially.

### 18.23.1 Broadening of Spectral Lines

One of the factors which makes the emission line and absorption line to overlap is the natural broadening of the spectral lines. Apart from natural broadening some other factors are as follows:

(a) The collision of an excited state atom with neighbouring atoms induces a transition to the ground state. Consequently, the lifetime would be shortened, broadening the spectral line.

(b) In the γ-ray emission process, the de-excitation of the system takes place through several mechanisms such as radioactive transitions, internal conversion, change in the valence state of the atom, etc. All these tend to decrease the lifetime consequently broadening the γ-ray line. When the excited state decays in several different ways, the total width is equal to the sum of all partial widths related by the uncertainty principle correspondingly,

$$\Gamma = \sum \Gamma_i = h\left(\sum \frac{1}{T_i}\right)$$

(c) Besides above-mentioned broadening processes, there is yet another process called Doppler broadening which arises out of movement of the emitting and absorbing system as a result of temperature. The Doppler broadening has no effect on the value of the absorption integral over the whole spectrum and there is no displacement of the centre of the resonance region. The intensity however decreases rapidly as the distance from the centre of the resonance region is increased.

### 18.23.2 Cross Section of Resonance Process

Resonance experiments with γ-rays are usually performed by either measuring the scattered intensity (resonance fluorescence or resonance scattering) or by determining the attenuation of a beam due to resonance absorption. The cross section for these two processes, for an incident γ-ray of energy ($E$) and wavelength ($\lambda$) can be calculated for thin absorber using the expressions

$$\sigma_{\text{scatt.}}(E) = (2\pi\lambda^2)\left(\frac{2I_B+1}{2I_A+1}\right)\left[\frac{\Gamma_r^2}{4(E-E_{\text{re}})^2+\Gamma^2}\right]$$

$$\sigma_{\text{abs}}(E) = (2\pi\lambda^2)\left(\frac{2I_B+1}{2I_A+1}\right)\left[\frac{\Gamma\Gamma_r^2}{4(E-E_{\text{re}})^2+\Gamma^2}\right]$$

In these experiments $\Gamma$ is the total width of the absorption line, $\Gamma_r$ its γ-ray width and the maximum resonance cross section $2\pi\lambda^2\left(\frac{2I_B+1}{2I_A+1}\right)$ can be designated as $\sigma_0$. $I_A$ is the spin of the ground state $A$ and $I_B$, the spin of the excited state $B$.

Both $\sigma_{\text{scatt.}}$ and $\sigma_{\text{abs}}$ vary as the square of the wavelength of the incoming radiation, therefore the maximum resonance cross section for γ-rays is

$$\sigma_0 = 2\pi\lambda^2\left(\frac{2I_B+1}{2I_A+1}\right)$$

For γ-rays with energies ranging from a few keV to a few MeV, $\sigma_0$ is $10^6$ to $10^{12}$ times smaller than the corresponding cross-section for optical radiation. For example, for the sodium $D_1$ line which corresponds to a transition $2p_{3/2} \rightarrow 2s_{1/2}$, the scattering cross section is $4.6 \times 10^6$ times the scattering cross section for 14.4 keV γ rays from $Fe^{37}$ arising out of a transition from the spin state 3/2 to the spin state 1/2. It should however be noted that in the γ ray region, the total absorption line width (Γ) is different from the γ ray line width due to branching as well as internal conversion.

If from a certain level, we have several transitions giving rise to γ-rays of various energies and if $\Gamma_i$ is the width of $i^{th}$ γ ray and $\alpha_i$, the corresponding internal conversion coefficient, then, the total width

$$\Gamma = \sum \Gamma_i + \sum \Gamma_i \alpha_i$$

If however, the level totally de-excites by emission of one single γ-ray then

$$\Gamma = \Gamma_i + \Gamma_i \alpha = \Gamma_i (1 + \alpha)$$

or
$$\Gamma_i = \frac{\Gamma}{1+\alpha}$$

Consequently, the scattering cross section and the absorption cross section are modified to

$$\sigma_{scatt}(E) = 2\pi\lambda^2 \left(\frac{2I_B+1}{2I_A+1}\right) \frac{\Gamma^2}{4(E-E_{re})^2+\Gamma^2} \frac{1}{(1+\alpha)^2}$$

$$\sigma_{abs}(E) = 2\pi\lambda^2 \left(\frac{2I_B+1}{2I_A+1}\right) \frac{\Gamma^2}{4(E-E_{re})+\Gamma^2} \frac{1}{(1+\alpha)}$$

It is clear that the scattering cross section is reduced by $\left(\frac{1}{1+\alpha}\right)^2$ and the absorption cross-section is reduced by $\left(\frac{1}{1+\alpha}\right)$. For example, in case of isotope $Fe^{37}$ emitting 14.4 keV γ-ray with a corresponding internal conversion coefficient of 9, the scattering cross section is reduced by 100 while in case of $Ge^{73}$ emitting 13.5 keV γ-ray, the internal conversion coefficient is 3600 and the scattering cross section is reduced by $7 \times 10^8$. The latter decrease is far greater because of large conversion coefficient. Here, it has been assumed that the emission line is very sharp, in other words, the incident radiation has been assumed to be delta function.

If the incident radiation has an intensity distribution $I(E)$, then, effective scattering and absorption cross sections are given by

$$\sigma_{\text{eff. (scatt.)}} = \int 2\pi\lambda^2 \left(\frac{2I_B+1}{2I_A+1}\right) \frac{\Gamma^2}{4(E-E_{\text{re}})^2+\Gamma^2} \frac{1}{(1+\alpha)^2} I(E)\,dE$$

$$\sigma_{\text{eff. (scatt.)}} = \int 2\pi\lambda^2 \left(\frac{2I_B+1}{2I_A+1}\right) \frac{\Gamma^2}{4(E-E_{\text{re}})^2+\Gamma^2} \frac{1}{(1+\alpha)} I(E)dE$$

The actual lines are much wider because of Doppler broadening and it can be seen by averaging over all possible values of $E$ of the incoming $\gamma$-ray energy that effective cross section is reduced tremendously.

## 18.24 APPLICATIONS OF MÖSSBAUER SPECTROSCOPY

The most useful feature of the Mössbauer effect is the sharpness of the $\gamma$-ray emission (line width $\approx 10^{-13}$ as compared to $10^{-8}$ for NMR and $10^{-4}$ for IR). The width is purely due to nuclear levels and no phonon-excitation is involved. A fractional line width ($\delta\nu/\nu$) of $10^{-13}$ implies that the energy of the $\gamma$-ray emitted is measurable to an accuracy of 1 part in $10^{13}$. Such an accuracy in measurement is not achievable even with lasers. Thus, $\gamma$-rays provide us the most stable and accurately defined electromagnetic radiation source for use in experiments. Some of the important applications of the Mössbauer effect are as follows:

1. Isomer shift or chemical shift in the Mössbauer spectra reflects differences in the *s*-electron density around the emitter and absorber nuclei.
2. When the absorber atom is in a reasonably symmetrical environment a single line is observed in the Mössbauer spectrum. When the environment is asymmetrical a feeble internal electric field operates giving rise to a quadrupole splitting. This is used in structure determination.
3. Biochemical applications: Since the Mössbauer effect produces a powerful probe of the chemical state and the environment of iron atoms, it can be applied to the study of proteins and enzymes.

In order to utilize the Mössbauer effect to obtain information about different nuclear orientations or the interaction of nuclei with surrounding electrons, some means must be found to modify the energy of the gamma-ray source systematically by a few megahertz, because if the emitting and absorbing nuclei were in different nuclear orientations, e.g., if the nuclei were in different chemical compounds, an exact energy match would not be achieved and resonance absorption would not occur. The most convenient way of making these energy adjustments is through the *doppler effect*.

## 18.25 USE OF THE DOPPLER EFFECT TO VARY THE GAMMA-RAY ENERGY

Mössbauer was the first to utilize the doppler effect to modify the energy of the gamma ray source to match the difference in energy between the ground state and the excited state of the absorber, and this principle has found almost universal acceptance, although the mechanisms now employed bear little resemblance to those originally used. Other methods of varying the energy, e.g., utilizing the thermal red shift or ultrasonically produced sidebands, have been proposed for special applications but have not been adopted for general use.

In the doppler effect the gamma-ray source is moved with a velocity ($v$) relative to the absorber. The modified frequency of the gamma-ray ($v$) will be related to its original frequency ($v_0$) by

$$v = v_0\left(1+\frac{vc}{c^2}\right) = v_0\left(1+\frac{v}{c}\cos\theta\right) \quad ...(1)$$

where θ is the angle between the axis of symmetry and the optical axis. Since energy ($E$) is directly proportional to the frequency ($E = hv$), the significance of Eq. (1) is best visualized in terms of the energy. If an absorber and a gamma ray source are travelling directly towards each other with a velocity $v$, the energy will seem to be $E_0$ $(1 + v/c)$, where $E_0$ is the inherent energy of the emitted gamma ray and θ = 1. On the other hand, if the absorber and the gamma ray source are moving directly away from each other (cos θ = –1), the energy will seem to be $E_0(1 - v/c)$. Clearly then, the doppler effect affords a means of varying the energy of the gamma ray. To estimate the magnitude of the doppler shift ($\Delta v$) which is possible with ordinary velocity changes, the following equation can be used

$$\Delta v = v_0\frac{v}{c} \quad ...(2)$$

For example, with the 14.4 keV gamma ray of $^{57}$Fe, a velocity of 1 cm/s would give a frequency shift of 116 MHz [Eq. (2)]. Since the line width of $^{57}$Fe is about 1.6 MHz (calculated earlier from the uncertainty principle), a frequency scan of 116 MHz is more than adequate to scan an entire Mössbauer spectrum. A relative velocity of 1 cm/s is easily produced in the laboratory, and in fact, velocities considerably less than that are sufficient for most studies. Mechanical motion devices based on gears and cams have met with only limited success because wear and bearing vibrations tend to cause problems. Most of the devices now used to move the source relative to the absorber are electromechanical feedback systems, e.g., a microphone coil fed by some type of waveform generator.

## 18.26 INTERPRETATION OF MÖSSBAUER SPECTRA

The raw data of a Mossbauer spectrum are the relative number of gamma-rays per second (counts per second) being transmitted through the absorbing sample as a function of the relative velocity of the source and sample. A typical Mössbauer spectrum is shown in Fig. 18.34. Note that the ordinate has been transcribed to per cent absorption units, where the maximum count rate is taken to be 100 per cent transmittance or 0 per cent absorption, and a zero (or background) count rate would be taken as 0 per cent transmittance or 100 per cent absorption. The standard convention in presenting the abscissa is to represent motion of the source towards the absorber by positive velocities and motion of the source away from the absorber by negative velocities. The parameters of a Mössbauer spectrum of greatest interest to a chemist are the size of the absorption peak ε, the line width $w$, the isomer shift (IS), quadrupole splitting (QS), magnetic hyperfine structure (MHFS), and the temperature coefficients of these parameters.

**Size of the Absorption Peak (ε):** This obviously depends on the number of absorber nuclei in the optical path, but at present it is very difficult to make quantitative use of this fact because a variety of interference effects and background-scattering contributions from the environment near the system contribute to the counting rate. In most experiments the nuclear transition of interest is accompanied by other radiations which may constitute an interference. With $^{57}$Fe, for example, the 14.4 keV gamma ray is of primary interest, but for every 14.4 keV gamma ray there is an equal number of 123 keV precursor events, as well as 9 per cent of 137 keV radiation and numerous low-energy *x*-rays both from Auger effects and scattering from the environment. Despite careful energy resolution, some of this radiation will reach the detector and contribute to the overall counting rate, thus rendering the absolute value of the Mössbauer peak height variable and unpredictable.

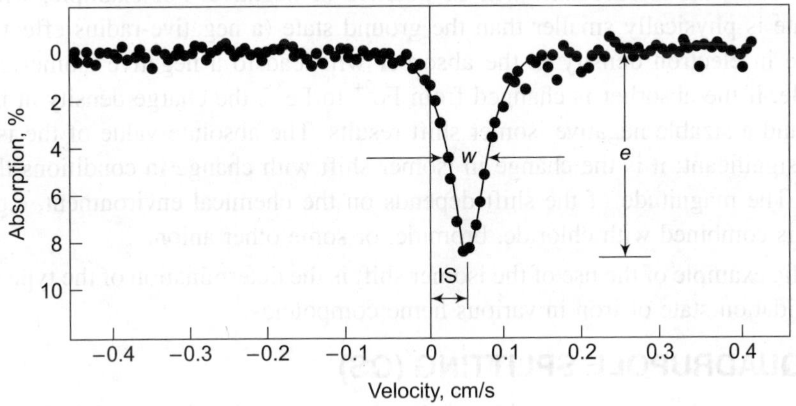


**Figure 18.34** Typical Mössbauer spectrum (ferricenium bromide at 20 K using a $^{57}$Fe source). (*After* G K Wertheim, *"Mossbauer Effect Principles and Applications,"* p. 51, Academic Press., New York, 1964)

A further complication is that the Mössbauer resonance is with a particular isotope among many isotopes that may be present, and the problem of a low or variable percentage abundance of that isotope can lead to large relative errors in attempts to reproduce the value of the peak height.

**Line Width (*w*):** This depends theoretically on the stability of the excited state and the Heisenberg uncertainty principle, as already noted. In practice, however, the observed line width is always wider than the theoretical value because of experimental complications. Experimental imperfections in the velocity drive, the presence of solid-state defects and impurities, thermal effects, and the finite thickness of both the source and the absorber all tend to broaden the resonance line. The most important use of the resonance line width is as a measure of the quality of a given Mössbauer source. When a standard absorber is used, the magnitude of *w* for two or more sources can be compared, and the source giving the smallest *w* should be selected.

**Isomer Shift (IS):** This is the Mössbauer parameter which yields the greatest amount of chemical information. A few authors refer to it as the *chemical shift* but isomer shift is the term most widely used, since the effect depends on the difference in the nuclear radius of the ground and excited (isomeric) states. Experimentally, the isomer shift is defined as the displacement of the resonance maximum from zero velocity, as shown in Fig. 18.34.

The isomer shift arises from the fact that the nucleus of an atom occupies a finite volume and interacts as a region of charge with its electronic environment. There are two fundamental requirements for the creation of an isomer shift: (1) the radii of the nuclear excited state and the ground state must differ, in order that each state interacts differently with the electronic environment; (2) the absorber and source must be in chemically different environments, e.g., in two different oxidation states, so that the electron density at the nucleus will be different for the absorber and source. According as the excited state radius is larger or smaller than the ground state, the isomer shift will be positive or negative. For example, with $^{57}Fe$ the excited state is physically smaller than the ground state (a negative-radius effect), and thus an increase in electron density in the absorber will lead to a negative isomer shift. Thus, for example, if the absorber is changed from $Fe^{2+}$ to $Fe^{3+}$, the charge density at the nucleus increases and a sizable negative isomer shift results. The absolute value of the isomer shift is not too significant; it is the change in isomer shift with change in conditions that is most important. The magnitude of the shift depends on the chemical environment, e.g., whether the cation is combined with chloride, bromide, or some other anion.

A specific example of the use of the isomer shift is the determination of the type of bonding and the oxidation state of iron in various heme compounds.[1]

## 18.27 QUADRUPOLE SPLITTING (QS)

This phenomenon is often more sensitive to the nature of the chemical bond, its ionic character, for example, than the isomer shift. Quadrupole splitting results from a non-symmetrical electric field gradient around the nucleus. This might be caused by a nonspherical (oblate, or flattened) nucleus or (of more interest in Mossbauer spectroscopy) by a distortion of the normally spherical electron distribution of the filled electronic shells by the valence-electron configuration. This nonsymmetrical field has the effect of removing the degeneracy of the spin states in the excited state of nuclei as illustrated in Fig. 18.35. In Fig. 18.35 (a) the energy levels for a free nuclei are shown on the left. Transitions between these states represent resonance absorption at zero relative velocity. In the centre of Fig. 18.35 (a) the displacement due to an isomer shift is shown. The displacement for the excited state is different from that for the ground state because the chemical environment is affecting the two states differently. At the right of Fig. 18.35 (a) is shown the effect of an asymmetric electrostatic field gradient at the nucleus. The quadrupole coupling between the electric field and the spin of the nucleus causes a splitting of the upper spin state (one is raised by an energy $EQ$, and one is lowered by the same amount). Selection rules indicate that transitions to the two new states should be equally probable. Hence, a Mössbauer spectrum of an element which is subject to such quadrupole splitting will show two lines of equal intensity. This is illustrated with the spectrum of biferrocenyl shown in Fig. 18.35 (b).

[1] A. J. Bearden, T. H. Moss, W. S. Caughey, and C. A. Beaudreau, *Proc. Natl. Acad. Sci. U.S.*, **53:** 1246 (1965).

Note from Fig. 18.35 (b) that the isomer shift is measured from the zero velocity point to the exact midpoint between the two peaks, since in the absence of the quadrupole field, the singlet peak would appear at that point.

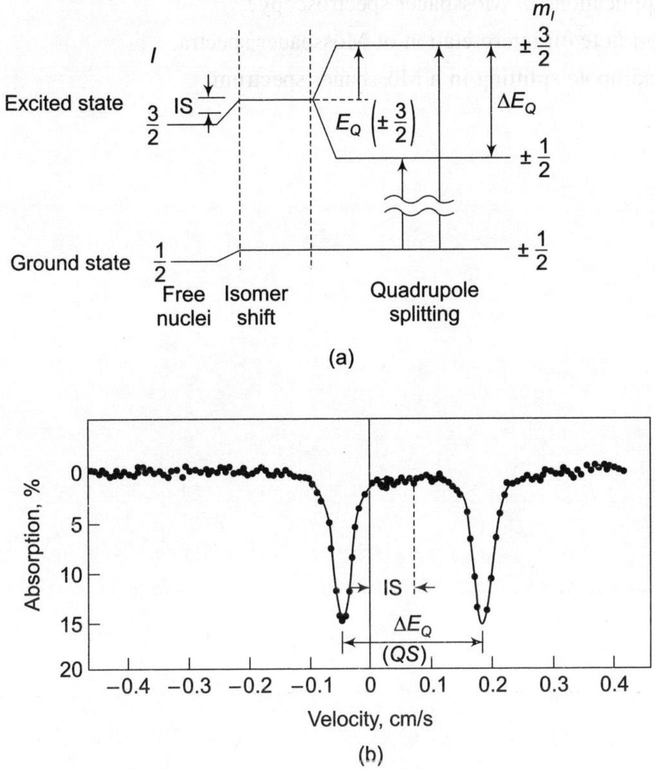


**Figure 18.35** The origin of quadrupole splitting interaction

## EXERCISE

1. What do you understand by 'LASER'? What are the unique properties of laser light?
2. Define Einstein A and B coefficients. Derive the relation between them.
3. Explain the terms 'spontaneous emission' and 'stimulated emission'.
4. Explain 'population inversion' and light amplification by population inversion.
5. Explain critical inversion for laser.
6. Describe the methods of achieving the population inversion.
7. Describe the ruby laser and explain its working.
8. What is a semiconductor diode laser? Derive the necessary condition for stimulated emission to be dominant over absorption in this case.
9. Explain the working of a *p-n* junction laser. What are the advantages of semiconductor laser over other lasers?

10. Explain nuclear resonance fluorescence.
11. Describe schematically an arrangement for Mössbauer spectroscopy.
12. Explain resonance fluorescence and the condition to observe resonance fluorescence.
13. What are applications of Mössbauer spectroscopy?
14. Write a short note on interpretation of Mössbauer spectra.
15. Explain quadrupole splitting in a Mössbauer spectrum.

CHAPTER

# 19

# Born Oppenheimer, Hartree and Hartree – Fock Approximations (Reduction to One Electron Problem)

## 19.1 INTRODUCTION

The proper choice of the potential $U(\vec{k})$ appearing in the one-electron Schrödinger equation

$$-\frac{\hbar^2}{2m}\nabla^2 \psi(\vec{r}) + U(\vec{r})\,\psi(\vec{r}) = E\psi(\vec{r})$$

is a subtle problem. Underlying this problem is the question of how best to represent the effects of electron–electron interactions, a fact we ignore by working in the independent electron approximation.

The universe of traditional solid state physics is described by the crystalline lattice. The principal components are the elementary excitations within the lattice as for instance the *phonons* are quanta of lattice vibrations. Another is the *electron*, basic component in all of the physics of crystalline solids. By an electron in a solid we basically mean something a little different from a free electron, i.e., we mean a dressed electron or an electron plus certain of its interactions, e.g., an electron in a solid with an effective mass.

Before describing the method of "reduction" to a one electron problem, we have to discuss the Born–Oppenheimer approximation. After making the Born–Oppenheimer approximation, we still have to deal with many electron problems. A way to reduce the many electron problem approximately to an equivalent one electron problem is given by the Hartree and Hartree–Fock methods. After these the Thomas Fermi method and then the density functional method allow (at least in principle), the exact evaluation of some ground state properties are also important and can be regarded as an extension of the Hartree–Fock method and have been much used in the recent years.

## 19.2 THE BASIC HAMILTONIAN

The Schrödinger equation is the starting point for all quantitative calculations of solid state properties. We begin by setting up the Hamiltonian for the entire problems. It is made up of the kinetic energy of all particles in the solid and of their interaction energies. The solid is comprised of two groups of electrons—*valence electrons* which contribute to chemical bonding and *core electrons* which are tightly bound in the closed shells of the lattice ions and which scarcely influence the properties of the solid. Consequently, we usually consider the *valence electrons* and the *lattice ions* as independent constituents of the solid. It is not always possible to make such a clear distinction, yet, it is the first of our approximations.

The Hamiltonian then, consists of the kinetic energy of all valence electrons (we can, subsequently drop the prefix "valence") and all ions, the energy associated with all the interactions between these particles and eventually, the energy associated with interactions with external fields.

$$H = H_{\text{el}} + H_{\text{ion}} + H_{\text{el-ion}} + H_{\text{ex}} \qquad \text{...(1)}$$

For the time being, we neglect the last term. For the electron part, we write

$$H_{\text{el}} = H_{\text{el}},\, k_{\text{in}} + H_{\text{el-el}} = \sum_k \frac{p_k^2}{2m_e} + \frac{1}{8\pi\varepsilon_0} \sum_{k,k'}{}' \frac{e^2}{|\vec{r}_k - \vec{r}_{k'}|} \qquad \text{...(2)}$$

where the last term is the Coulomb interaction term.

The sums are over all the electron indices, excluding $k = k'$ for the interaction term. $\vec{p}_k$, $\vec{r}_k$ and we are the momentum, position and mass of an electron of index, $k$.

For the ion-part of the Hamiltonian, we correspondingly write

$$H_{\text{ion}} = H_{\text{ion}},\, k_{\text{in}} + H_{\text{ion-ion}}$$

$$= \sum_i \frac{Pi^2}{2Mi} + \frac{1}{2} \sum_{i,i'}{}' V_{\text{ion}}\,(\vec{R}_i - \vec{R}_{i'}) \qquad \text{...(3)}$$

where we have identified ion parameters with capital letters. Here, we have assumed that ion–ion interaction can be described as a sum over two–particle interactions, each dependent only on the difference in the ion co-ordinates, $\vec{R}_i$.

For the electron–ion interaction, we put

$$H_{\text{el-ion}} = \sum_{k,i} V_{\text{el-ion}}\,(\vec{r}_k - \vec{R}_i) \qquad \text{...(4)}$$

A further division is required at this stage. A feature of the crystal is their symmetry, resulting from the periodic arrangements of the ions in the lattice, however, it is the equilibrium position about which the ions vibrate which show strong periodicity, rather than their actual positions at any instant. Thus, we divide the ion–ion interaction into two components, one describing the interaction when the ions are in their equilibrium position, and the other, a correction to account for the vibration of the lattice.

$$H_{\text{ion-ion}} = H_{\text{ion-ion}} + H_{\text{ph}} \qquad ...(5)$$

$$H_{\text{el-ion}} = H_{\text{el-ion}} + H_{\text{el-ph}} \qquad ...(6)$$

The index 'ph' stands for phonons, the Eqs. (1) to (6) provide the foundation for the quantum mechanical treatment of most solid–state properties. The next step is the transition from the Hamiltonian *function* to the Hamiltonian *operator*. In the co-ordinate representation, the Hamiltonian operator then depends on the co-ordinates of all electrons and ions. Correspondingly, the wave function on which H operates becomes a function of all these co-ordinates.

This form of the Hamiltonian allows only limited account to be taken of the electron spin. However, for most of the problems, the non-relativistic Schrödinger equation without spin-orbit coupling terms is adequate. It is not possible to rigorously solve the quantum mechanical problem. Approximations have to be made. In solid-state theory, two simplifications are commonly made. In solving a given problem, individual terms of the Hamiltonian may be neglected, or only partially considered, or handled subsequently as perturbations. This simplified problem is then further simplified using the symmetry of the lattice. The particular approximations made will depend on the questions we are seeking to answer and on the nature of the solid under examination.

There is a problem in completely neglecting individual terms of the Hamiltonian, since all interactions, being always Coulomb interactions, are equally strong. For this reason, the *electron component* ($H_{\text{el}}$) in Eq. (2) for example, cannot be considered alone, because it describes an electron gas whose charge is not compensated by ions as in the solid. As a first approximation, Eq. (2) has to be extended at least to include a constant space charge $P_+$, representing the average charge of the ions, and the interaction of the electrons with this space charge. If we include these two additions in a single term $H_+$, we can write

$$H_{\text{el}} = -\sum_k \frac{\hbar^2}{2m}\nabla_k^2 + \frac{1}{8\pi\varepsilon_0}\sum_{k,k'}{}' \frac{e^2}{|\vec{r}_k - \vec{r}_{k'}|} + H_+ \qquad ...(7)$$

The electron gas is considered here to be embedded in a constant positive background. This model is often described as "*jellium*" *model*. Lattice symmetry retreats completely into the background whilst the properties of the electron gas, particularly the electron-electron interaction, are to the fore. Many of the properties of metals can be described by means of this approximation.

The model can be improved by replacing the uniform distribution of space charge by a distribution in which the ions are taken to be fixed in their equilibrium positions, $\vec{R}_i^0$. The electron-ion interaction is then described by the term $H^0_{\text{el-ion}}$ from Eq. (6). In this case, lattice symmetry is taken into account. The Hamiltonian, however, then becomes too complicated that it cannot be dealt with without making further approximations.

Starting from Eqs. (1) and (5), we can handle the *movement of the ions* in a similar way, ending up with an equation which corresponds to Eq. (7). In this case we introduce a constant negative space charge (ρ-) to take account of the neglected electrons and a term to represent their interaction with the ions. Combining both additions into a single term $H_{-}$, we have

$$H_{ion} = -\sum_i \frac{\hbar^2}{2M_i}\nabla_i^2 + \frac{1}{2}\sum_{ii'}{}' V_{ion}(R_i - R_{i'}) + H_{-} \quad \text{...(8)}$$

The second term can be subdivided according to (5). This Hamiltonian forms the basis for the study of lattice dynamics which is treated in detail in Section 3.3.

The two terms ($H_+$ and $H_-$) in Eqs. (7) and (8) exactly cancel each other. In Eq. (1) we are then (apart from $H_{ex}$) left with only $H_{el\text{-}ion}$, the term which couples electron and ion motion. If, following Eq. (6), we separate the interaction of the electrons with the static ion lattice from $H_{el\text{-}ion}$ and add this component to $H_{el}$, the only remaining coupling between electrons and ions is the *electron-phonon interaction,* $H_{el\text{-}ph}$. This coupling can be dealt with in most cases by perturbation theory, following the solution of Eqs. (7) and (8).

We have been able, through the above, to divide the total solid-state problem into two components—the movement of the electrons in a stationary lattice and the movement of the ions in a uniform space charge of electrons. This decoupling of the total system really requires rigorous justification. The so-called *adiabatic approximation* (Born–Oppenheimer method) is often used to provide such justification. It is based on the following argument: Electrons and ions have very different masses. The ions can respond only slowly to a change in the electron configuration, while the electrons respond adiabatically to a change in the positions of the ions. As far as the motion of the electrons is concerned, it is only the instantaneous configuration of the ions which is of interest. As a first approximation we can therefore adopt for the electrons a Schrödinger equation of the form

$$(H_{el} + H_{el\text{-}ion})\psi = E_{el}\psi \quad \text{...(9)}$$

in which the coordinates of the ions are kept fixed. The wave function depends only on the coordinates of the electrons. The coordinates of the ions appear in the equation as parameters. As the starting point for a solution of the total problem we now use the product

$$\Psi = \psi(r_1 \dots r_N; R_1 \dots R_{N'})\,\varphi(R_1 \dots R_{N'}) \quad \text{...(10)}$$

where the ψ are solutions of (9) and $N, N'$ denote the number of electrons and ions, respectively. Putting this into the Schrödinger equation with (1) as the Hamiltonian we have

$$H\Psi = (H_{el} + H_{ion} + H_{el\text{-}ion})\psi\varphi$$
$$= \psi(H_{ion} + E_{el})\varphi - \sum_i \frac{\hbar^2}{2M_i}(\varphi\nabla_i^2\psi + 2\nabla_i\varphi\nabla_i\psi). \quad \text{...(11)}$$

If the last term here were absent, Eq. (10) would be a separation ansatz which effectively decouples electron and ion motion. For the movement of the ions an equation of the form

$$(H_{\text{ion}} + E_{\text{el}})\varphi = E\varphi \qquad ...(12)$$

would follow, in which $E_{\text{el}}$ still depends on the ion positions and therefore provides a contribution from the electrons to the potential energy of the ions. Equation (12) is a Schrödinger equation which involves only the coordinates of the ions. It therefore describes the *motion of the ions*. To describe the *motion of the electrons*, we replace in Eq. (9) the instantaneous positions of the ions by their mean positions, i.e., we replace $H_{\text{el–ion}}$ by $H^0_{\text{el–ion}}$.

The last term in Eq. (11) couples the electron and ion systems. One can show that it provides only a small contribution to the total energy of the system in the state, $\Psi$. This cannot, however, be taken as a proof that this term only describes a weak interaction which can subsequently be treated by perturbation theory.

## 19.3 THE FOUR BASIC APPROXIMATIONS

The general formulation of the study of solids starts with the many-body Schrödinger equation

$$\hat{H}\psi = E\psi \qquad ...(1)$$

for all the ions and electrons that through their mutual binding make up the condensed matter. The Hamiltonian $\hat{H}$ is of the form

$$\hat{H} = \hat{H}_e + \hat{H}_{ei} + \hat{H}_i \qquad ...(2)$$

where $\hat{H}_e$ is the electronic part, $\hat{H}_i$ is the ionic part and $\hat{H}_{ei}$ is the electron–ion interaction. Since a typical number of particles in a solid is of the order of the Avogadro number, i.e., $\sim 10^{24}$, the explicit solution to Eq. (1) is out of question. Some drastic approximations have to be introduced. At a first level of accuracy, the following four approximations are widely used, because, in combination, they make Eqs. (1) and (2) tractable, without missing (in most cases) much of the physics.

### A1: Born – Oppenheimer or Adiabatic Approximation

For the purpose of studying the motion of electrons, the ions are considered as immobile. This approximation is justified physically because of the huge difference in mass between ions and electrons. Indeed the ratio $\frac{M_a}{m_e}$ is typically of the order of 10,000 or larger. As a result, the motion of ions is very slow in comparison with that of the electrons and the ions hardly have time to move during the characteristic electronic time of motion.

**A1** is important, because it allows us to study the combined motion of ions and electrons in two stages: first, to examine the motion of electrons, the ions are considered immobile (at this first stage only); second, to examine the motion of ions, the electrons are assumed to adjust instantaneously to each temporary configuration of the ions.

### A2: One-Electron Approximation

The electronic properties can be obtained from the solutions to the one-electron Schrödinger equation in the presence of a self-consistent effective potential.

This is indeed a very drastic approximation. It took several decades of research to provide a partial justification for **A2** and to obtain a reasonable form for the self-consistent potential $V(\vec{r})$. In view of **A2**, the electronic properties are determined by the solutions to the one-electrons Schrödinger equation

$$\left[-\frac{\hbar^2}{2me}+V(\vec{r})\right]\psi_i(\vec{r})=a_i\,\psi_i\,(\vec{r}) \qquad ...(3)$$

$$V(\vec{r})=\int(\{\psi_i(\vec{r})\}) \qquad ...(4)$$

where Eq. (4) is essentially an algorithm for obtaining the effective potential $V(\vec{r})$ in terms of the solutions $\psi_i\,(\vec{r})$ to Eq. (3).

## A3: Harmonic Approximation for Ions in Solids

*Ions in solids undergo small coupled harmonic oscillations around their equilibrium positions.*

Because of the very definition of a solid, ions have fixed equilibrium positions in space and undergo oscillations around these positions. Since the amplitude, $\Delta x$, of these oscillations is typically very small in comparison with the nearest neighbour ionic separation, $d$ ($\Delta x/d$ is no more than a few per cent), their nature to a very good approximation is harmonic, which means that the force ($\boldsymbol{F}_\ell$) on the $\ell$th ion is proportional to a linear combination of the ionic displacements, $\{\boldsymbol{u}_n\}$, from the equilibrium positions

$$F_{\ell,i}=\sum_n\sum_{j=1}^{3}K_{\ell n,ij}\,(u_{nj}-u_{\ell j}) \qquad ...(5)$$

where $\{K_{\ell n,ij}\}$ are the effective "spring constants" connecting ion $\ell$ to ion *n; i, j* denote the Cartesian components of the vector ($\boldsymbol{u}$) and the tensor, $\boldsymbol{k}$. Taking into account that (a) in the eigenoscillations, by definition, all ions move with the same angular frequency, so that their time dependence is of the form exp $(-i\omega t)$; and (b) that $F_{\ell,i}=m_\ell \ddot{u}_{\ell i}=-\omega^2 m_\ell u_{\ell i}$, we have

$$\sum_n\sum_{j=1}^{3}A_{\ell n,ij}\,u_{n,j}=0, \qquad ...(6)$$

where $A_{\ell\ell,ij}=\omega^2 m_\ell\delta_{ij}+K_{\ell\ell,ij}-\sum_{n=1}^{N_a}K_{\ell n,ij}$ and $A_{\ell n,ij}=K_{\ell n,ij}$ for $\ell\neq n$;

$\{A_{\ell n,ij}\}$ is a $3N_a\times 3N_a$ matrix multiplying the $3N_a$ component column matrix $\{u_{n,j}\}$, $\ell$, $n$ = 1, ..., $N_a$; $i, j$ = 1, 2, 3. $N_a$ is the number of ions (or atoms) in the solid. If we manage to diagonalize the matrix $\{A_{\ell n,ij}\}$, the problem has been reduced to that of $3N_a-6$ one-dimensional independent harmonic oscillator (the remaining six degrees of freedom are associated with the rigid translation and the rigid rotation of the solid as a whole).

It must be stressed that the exact diagonalization of such a huge matrix as the $\{A_{\ell n,ij}\}$, as well as the exact solution of Eq. (3) over the *whole extent of the solid*, is still an impossible task. It is the fourth approximation that overcomes these huge obstacles to a solution.

## A4: Periodicity

*Crystalline solids are structurally and electronically periodic.*

If solids were *perfect* crystals without defects, or foreign atoms, or phononic displacements, then **A4** would be an exact statement that, in combination with **A1**, would imply that $v(r)$ in Eq. (3) and $A_{\ell n, ij}$ in Eq. (6) are periodic. However, even the best and purest monocrystals have some point defects and contain some traces of foreign atoms; furthermore, they are not free of some line defects called dislocations; and of course, phononic displacements are always present. Nevertheless, **A4** is a very good approximation not only for monocrystals but also for polycrystalline solids (because, usually, the size of each crystallite is much larger than the lattice constant). However, for some transport quantities, such as the electrical conductivity in metals at very low temperatures or in semiconductors, structural and chemical defects are the main factors determining the values of these quantities.

For crystalline solids, we stress the fact that **A4** implies the validity of Bloch's theorem according to which the following relations hold:

$$\psi(r + R_n) = e^{ik \cdot Rn}\psi(r) \qquad ...(7)$$

$$u_{\ell+n} = e^{iq \cdot Rn}u_\ell \qquad ...(8)$$

where $\hbar k$ and $\hbar q$ are the crystal momenta for electrons and phonons respectively and $R_n$ is any *lattice* vector. Equation (7) means that there is no need to solve Eq. (3) over the whole volume of the solid; it is enough to solve it in a single primitive cell and use Eq. (7) to extend it over the whole volume. Similarly, Eq. (8) implies that all the $u_{\ell+n}s$, $n \neq 0$, can be expressed in terms of the $u_\ell s$ of the atom(s) in a single primitive cell; thus, the size of the linear homogeneous system is dramatically reduced from $3N_a$ equation to $3N_{a,pc}$, where $N_{a,pc}$ is the number of ions (or atoms) in a single primitive cell; $N_{a,pc}$ can be as low as one. This is indeed a tremendous simplification, which is used in the LCAO electronic case.

The combination of Eqs. (3) and (7), allows the determination[1] of the electronic band structure, $E_n(k)$, and the corresponding eigenfunctions, which are of the form $\psi_{k,n}(r) = w_{k,n}(r)$ exp $(ik \cdot r)$.

Similarly, the combination of Eqs. (6) and (8) allows the determination of the relation between crystal wave vector ($q$) and the ionic eigenfrequencies, $\omega_s = \omega_s(\boldsymbol{q})$, where $s = 1, ..., 3N_{a,pc}$.

## 19.4 THE BORN–OPPENHEIMER APPROXIMATION

The energy imparted to a solid can be stored in two ways, i.e., there are two main reservoirs, viz., the phonons (lattice vibrations) and the electrons. Generally, we discuss these two independently, but this is jus an approximation. This approximation is clear-cut in insulators (electrons playing a negligible role), but in metals, it is much harder to justify. In this latter case, the intellectual framework goes by the name Born–Oppenheimer approximation. (This approximation, actually, paves the way for a systematic study of electron-phonon interactions in solids by employing the perturbation theory).

[1] For details, see *The Physics of Solids* by Eleftherios. N. Economou, Springer, Heidelberg (2010).

### 19.4.1 The Approximation

In case of solids, the crystal Hamiltonian is rather complex due to electronic and nuclear motions and their interactions and the most fundamental problem is to solve the many-particle Schrödinger wave equation

$$H_c \psi = i\hbar \frac{\partial \psi}{\partial t} \quad ...(1)$$

where $H_c$ is the crystal Hamiltonian. In general, the wave function ($\psi$) is a function of all electronic and nuclear co-ordinates of the time $t$. i.e.,

$$\psi = \psi(\vec{r}_i, \vec{R}_\ell, t) \quad ...(2)$$

where $\vec{r}_i$ are the electronic co-ordinates and $\vec{R}_\ell$, the nuclear co-ordinates.

The Hamiltonian $H_c$ can be represented as

$$H_c = -\sum_i \frac{\hbar^2}{2m}\nabla_i^2 - \sum_l \frac{\hbar^2}{2M_\ell}\nabla_\ell^2 + \frac{1}{2}\sum_{i,j}{}' \frac{e^2}{4\pi\varepsilon_0 |\vec{r}_i - \vec{r}_j|} - \sum_{i,\ell} \frac{e^2 Z_\ell}{4\pi\varepsilon_0 |\vec{r}_i - \vec{R}_\ell|} + \frac{1}{2}\sum_{\ell,\ell'}{}' \frac{e^2 Z_\ell Z_{\ell'}}{4\pi\varepsilon_0 |\vec{R}_\ell - \vec{R}_{\ell'}|} \quad ...(3)$$

In this expression, $m$ is the mass of the electron, $M_e$ is the mass of the nucleus located at $\vec{R}_\ell$, $Z_e$ is the nuclear charge (the atomic number) of the nucleus located at $R_e$, $e$ is the electronic charge. The sums over $i$ and $j$ run over all electrons and the primes over the summation means that $i = j$ terms and $\ell = \ell'$ terms are to be omitted. (The sums over $\ell$ and $\ell'$ run over all nuclear co-ordinates). The various terms have the following interpretations.

The first term is the operator representing the kinetic energy of the electrons. The second term is the operator representing the kinetic energy of the nuclei. The third term is the Coulomb potential energy of interaction between the electrons. The fourth term is the Coulomb potential energy of interaction between the electrons and the nuclei, and the fifth term is the Coulomb potential energy of interaction between the nuclei.

The magnetic interactions are negligibly small, thus relativistic effects have been ignored.

It is to be noted here that the wave function of Eq. (2) is a function of the spin degrees of freedom as well as the space degrees of freedom. Thus, it is assumed here that $\vec{r}_i$ gives both, the position and the spin. However, $\nabla_i^2$ is just the ordinary (spatial) Laplacian.

For the sake of brevity, it is convenient here to let $T_E$ be the kinetic energy of the electrons, $T_N$ be the kinetic energy of the nuclei and $U$ be the total Coulomb energy of interaction of the nuclei and the electrons. Then Eq. (3) becomes

$$H_C = T_E + U + T_N = H_0 + T_N \quad ...(4)$$

where
$$H_0 = T_E + U \quad ...(5)$$

Now, nuclei have large masses and hence, in general, have small kinetic energies. Thus, in the expression

$$H_C = H_0 + T_N,$$

We can regard $T_N$ as a small perturbation on $H_0$ and we can make a perturbation expansion.

Let $M_0$ denote a mean nuclear mass and we define

$$K \equiv \left(\frac{m}{M_0}\right)^{1/4}.$$

and, if we define

$$H_L = -\sum_{\ell} \frac{M_0}{M_\ell} \frac{\hbar^2}{2m} \nabla_\ell^2 \quad ...(6)$$

then
$$T_N = K^4 H_L \quad ...(7)$$

and the total Hamiltonian then has the form

$$H_c = H_0 + K^4 H_L \quad ...(8)$$

The time-independent Schrödinger equation that we wish to solve is

$$H_c \, \psi(\vec{r}_i, \vec{R}_\ell) = E \, \psi(\vec{r}_i, \vec{R}_\ell) \quad ...(9)$$

Assuming nuclei as fixed at positions $\vec{R}_\ell$, we have for electrons

$$H_0 \, \phi(\vec{r}_i, \vec{R}_\ell) = E^\circ \, \psi(\vec{r}_i, \vec{R}_\ell) \quad ...(10)$$

Born and Huang have made a perturbation expansion of the solution of Eqn. (9) in powers of $K$. They have shown that if the wave function is evaluated to second order in $K$, then a product separation of the form

$$\psi_n(\vec{r}_i, \vec{R}_\ell) = \phi_n(\vec{r}_i) \times (\vec{R}_\ell)$$

(where $n$ labels an electronic state) is possible. The assertion that the total wave function can be written as a product of the electronic wave function (depending only on electronic coordinates with nuclei at fixed positions) times the nuclear wave function (depending only on nuclear co-ordinates with the electrons in some fixed state) is the physical content of the *Born–Oppenheimer approximation*. In this approximation, the electrons provide a potential energy for the motion of the nuclei while the moving nuclei continuously deform the wave function of the electrons (rather than causing any sudden changes). Thus, this is also called the adiabatic approximation.

It turns out when the wave function is evaluated to second order in $K$ that the effective potential energy of the nuclei involves nuclear displacements to fourth order and lower. Expanding the nuclear potential energy to second order in the nuclear displacements yields the harmonic approximation. Terms higher than second order are called *anharmonic terms.*

Apparently, in practice, the adiabatic approximation does not break down quite so quickly. Thus, a somewhat simpler development of the B–O approximation is sometimes useful. In this development, we attempt to find a solution for $\psi$ in Eq. (9) of the form

$$\psi(\vec{r}_i, \mathrm{R}_\ell) = \sum_n \psi_n(\vec{R}_\ell)\,\phi_n(\vec{r}_i, \vec{R}_\ell) \quad ...(11)$$

The $\phi_n$ are eigenfunctions of Eq. (10).

Substituting into Eq. (9), gives

$$\sum_n H_c\,\psi_n\,\phi_n = E\sum_n \psi_n\,\phi_n$$

or

$$\sum_n H_0\,\psi_n\,\phi_n = \sum_n T_N\,\psi_n\,\phi_n = E\sum_n \psi_n\phi_n$$

or

$$\sum_n E_n^0\,\psi_n\,\phi_n + \sum_n T_N\,(\psi_N\,\phi_n) = E\sum_n \psi_n\phi_n \text{ (using Eq. 10).} \quad ...(12)$$

Noting that

$$T_N(\psi_N\phi_n) = (T_N\psi_n)\,\phi_n + \psi_n(T_N\phi_n) + 2\sum_\ell (\hat{P}_\ell\,\phi_n)\cdot(\hat{P}_\ell\,\psi_n)\frac{1}{2M_\ell}$$

where $T_N = \sum_\ell \frac{1}{2M_\ell}\hat{P}_\ell^2, \left(= -\hbar^2\sum_\ell \frac{1}{2M_\ell}\nabla^2_{\vec{R}_\ell}\right)$

We can write Eqn. (12) as

$$\sum_n E_n^0\,\psi_n\,\phi_n + \sum_n (T_N\,\psi_n)\,\phi_n + \sum_n \psi_n\,(T_N\,\phi_n)$$

$$+ \sum_\ell\sum_n \frac{1}{M_\ell}(\hat{P}_\ell\,\phi_n)\cdot(\hat{P}_\ell\,\psi_n)$$

$$- E\sum_n \psi_n\,\phi_n = 0.$$

or

$$\sum_{n!} \phi_{n^1}\,(T_N + E_n^0 - E)\,\psi_n + \sum_{n'_1} \psi_{n'}\,T_N\,\phi_{n'}$$

$$+ \sum_{n'}\sum_\ell \frac{1}{M_\ell}(\hat{P}_\ell\,\phi_{n'})\cdot(\hat{P}_\ell\,\psi_{n'}) = 0 \quad ...(13)$$

Pre-multiplying Eq. (13) by $\phi_n^*$

$$\sum_{n'} \phi_n^*\,\phi_{n'}\,(T_N + E_n^0 - E)\,\psi_n + \sum_{n'} \phi_n^*\,\psi_{n'}\,T_N\,\phi_{n'}$$

$$+ \sum_{n'} \sum_{\ell} \frac{1}{M_\ell} \phi_n^* P_\ell \phi_{n'} - P_\ell \psi_{n'} = 0. \quad ...(14)$$

Integrating Eq. (14) over the electronic co-ordinates, we get

$$(T_N + E_n^0 - E)\psi_n + \sum_{n'} C_{nn'} \psi_{n'} = 0 \quad ...(15)$$

$$\left( \because \sum_{n'} \int \phi_n^* \phi_{n'} \, dz = \sum_{n'} \delta_{nn'} = 1 \right)$$

where $$C_{nn'} = \sum_{\ell i} \frac{1}{M_\ell} (Q_{nn'}^{\ell i} P_{\ell i} + R_{nn'}^{\ell i}) \quad ...(16)$$

and $$Q_{nn'}^{\ell i} = \int \phi_n^* P_{\ell i} \phi_{n'} \, d\tau \quad ...(17)$$

$$R_{nn'}^{\ell i} = \frac{1}{2} \int \phi_n^* P_{\ell i}^2 \phi_{n'} \, d\tau \quad ...(18)$$

(the sum over $i$ goes from 1 to 3, labelling the $x$, $y$ and $z$ components). The integral in Eq. (17) and (18) is over electronic co-ordinates.

For stationary states, the $\phi$'s can be chosen to be real and so it can be seen that the diagonal elements of $Q$ vanish

$$Q_{nn'}^{\ell i} = \int \phi_n P_{\ell i} \phi_n \, dz$$

$$= \frac{\hbar}{\partial i} \frac{\partial}{\partial X_{\ell i}} \int \phi_n^2 \, d\tau = 0 \quad ...(19)$$

$$(\because \int \phi_n^2 \, d\tau - \text{constant})$$

Thus, we see that the effect of the diagonal elements of $c$ is a multiplication effect and not an operator effect. Therefore, the diagonal elements of $c$ can be added to $E_n^0$ to give an effective potential energy $U_{\text{eff}}$. Then Eq. (15) can be written as

$$(T_N + U_{\text{eff}} - E)\psi_n + \sum_{\substack{n^1 \\ (\neq n)}} C_{nn'} \psi_{n'} = 0 \quad ...(20)$$

If the $C_{nn'}$ vanish, then we can split the discussion of the electronic and nuclear motions apart as in the adiabatic approximation. Otherwise, we cannot. For metals, there appears to be no reason to suppose that the effect of the $c$ is negligible. This is because the excited states are continuous in energy with the ground state, and so the sum in Eq. (20) goes over into an integral.

In general, perturbation theory indicates that if there is a large energy gap between the ground and excited electronic states, then an adiabatic approximation may be valid. Now, can we speak of lattice vibrations in metals without explicitly also discussing the electrons? The answer is no. However, for completely free electrons (whose wave functions do not depend at all on the $\vec{R}_\ell$), it is clear that all the $c$ vanish and the adiabatic or B – O approximation is valid, i.e., in metals, when the electrons can be thought of as almost free, the effect of the $c$ is small enough to be neglected in zeroth – order approximation. Then B – O approximation can be applied to conduction electrons in metals. Further, strange effects may appear in metals due to the fact that the coupling between electrons and lattice vibrations is not negligible. The phenomenon of superconductivity is also due to this coupling, since, we can always write the Hamiltonian as

$$H = H\text{ (electrons)} + H\text{ (lattice vibrations)} + H\text{ (coupling)} \qquad ...(21)$$

In basic terms, the B – O approximation discussed above allows the wave function of a molecule to be broken into its electronic and nuclear (vibrational, rotational) components:

$$\psi_{\text{tot}} = \psi_{\text{el}} \times \psi_{\text{nucl}}.$$

In the first step of the B – O approximation the electronic Schrödinger equation is solved yielding the wave function $\psi_{\text{el}}$ depending on electrons only. During this solution, the nuclei are fixed in a certain configuration, very often the equilibrium configuration. If the effects of the quantum mechanical nuclear motion are to be studied, for instance, a vibrational spectrum is required, this electronic computation must be repeated for many different nuclear configurations. The set of electronic energies thus computed becomes a function of the nuclear co-ordinates. In the second step of B – O approximation, this function serves as a potential in a Schrödinger equation *containing only the nuclei.*

Thus, the electronic Schrödinger equation

$$H_e(\vec{r}, \vec{R})\, \psi_{\text{el}}(\vec{r}, \vec{R}) = E_e\, \psi_{\text{el}}(\vec{r}, \vec{R})$$

is solved. Here $\vec{r}$ stands for all electronic co-ordinates and $\vec{R}$ for all nuclear co-ordinates. Obviously, the electronic energy eigenvalue ($E_e$) depends on the chosen position ($\vec{R}$) of the nuclei. Varying these positions ($\vec{R}$) in small steps and repeatedly solving the electronic Schrödinger equation, one obtains $E_e$ as a function of $\vec{R}$. This is the potential energy surface (PES): $E_e(\vec{R})$. Because this procedure of recomputing the electronic wavefunctions as a function of an infinitesimally changing nuclear geometry is reminiscent of the conditions for the adiabatic theorem, this manner of obtaining a PES is often referred to as the *adiabatic approximation* and the PES itself is called an *adiabatic surface*.

In the second step of the B – O approximation, the nuclear kinetic energy, $T_n$ (containing partial derivatives with respect to the components of $\vec{R}$) is reintroduced and the Schrödinger equation for the nuclear motion

$$[T_n + E_e(\vec{R})]\, \phi(\vec{R}) = E\, \phi(\vec{R})$$

is solved. This second step of the B – O approximation involves separation of vibrational, translational and rotational motions. (This can be achieved by applications of the Eckart conditions). The eigenvalue ($E$) is the total energy of the molecule, including contributions from electrons, nuclear vibrations and overall rotation and translation of the molecule.

## 19.5 THE VARIATIONAL PRINCIPLE

The variational principle derived in this section is often called the *Rayleigh – Ritz variational principle*. It is of great practical importance. It gives us a way of constructing energies that have a value greater than or equal to the ground state energy of the system, i.e., a way of constructing upper bounds for the energy. This will be used to derive both the Hartree and Hartree – Fock equations subsequently in the later sections.

Let $H$ be a positive definite Hermitian operator with eigenvalues $(E_\mu)$ and eigenkets, $|\mu>$. Since $H$ is positive definite and Hermitian, it has a lowest $E_\mu$ and the $E_\mu$ are real.

Let the $E_\mu$ be labelled so that $E_0$ is the lowest. Let $|\psi>$ be an arbitrary ket (not necessarily normalised) in the space of interest and define a quantity $Q(\psi)$ such that

$$Q(\psi) = \frac{<\psi|H|\psi>}{<\psi|\psi>} \qquad \text{...(1)}$$

The eigenkets $|\mu>$ are assumed to form a complete set so that

$$|\psi| = \sum_\mu a_\mu \, |\mu> \qquad \text{...(2)}$$

Since $H$ is Hermitian, we can assume that the $|\mu>$ are orthonormal, and we find

$$<\psi|\psi> = \sum_{\mu',\mu} a_{\mu'}^{*} a_\mu <\mu'|\mu>$$

$$= \sum_{\mu',\mu} a_{\mu'}^{*} a_\mu \, \delta_{\mu'\mu} = \sum_\mu |a_\mu|^2 \qquad \text{...(3)}$$

and

$$<\psi|H|\psi> = \sum_{\mu',\mu} a_{\mu'}^{*} a_\mu \langle \mu'|H|\mu \rangle$$

$$= \sum_\mu |a_\mu|^2 E_\mu \qquad \text{...(4)}$$

$Q$ can then be written as

$$Q(\psi) = \frac{\sum_\mu E_\mu |a_\mu|^2}{\sum_\mu |a_\mu|^2} = \frac{\sum_\mu E_0 |a_\mu|^2}{\sum_\mu |a_\mu|^2} + \frac{\sum_\mu (E_\mu - E_0)|a_\mu|^2}{\sum_\mu |a_\mu|^2}$$

or

$$Q(\psi) = E_0 + \frac{\sum_\mu (E_\mu - E_0)|a_\mu|^2}{\sum_\mu |a_\mu|^2} \qquad \text{...(5)}$$

Since $E_\mu > E_0$ and $|a_\mu|^2 \geq 0$, we can immediately conclude from Eq. (5) that

$$Q(\psi) \geq E_0 \qquad \text{...(6)}$$

Summarising, we have

$$\frac{<\psi|H|\psi>}{\psi|\psi>} \geq E_0 \qquad ...(7)$$

Equation (7) is the basic equation of the variational principle. Suppose $\psi$ is a trial wave function with a variable parameter, $\eta$. Then the $\eta$ that are the best if $Q(\psi)$ is to be as close to the lowest eigenvalue as possible (or as close to the ground – state energy if $H$ is the Hamiltonian) are among the $\eta$ for which

$$\frac{\partial Q}{\partial \eta} = 0 \qquad ...(8)$$

For the $\eta = \eta_b$ that solves Eq. (8) and minimizes $Q(\psi)$, $Q(\psi(\eta_b))$ is an approximation to $E_0$. By using successively more sophisticated trial wavefunctions with more and more variable parameters (this is where the hard work comes in), we can get as close to $E_0$ as desired.

$Q(\psi) = E_0$ exactly only if $\psi$ is an exact wave function corresponding to $E_0$.

Thus, in the applications of this method, it is customary to begin with a family of states, known as the *trial functions* or *states*, depending on $n$ parameters $\eta_0$, $\eta_1$, .... $\eta_n$. With them we compute the expectation value of $H$ and we obtain an upper bound on $E_0$ depending on these parameters. The result obtained is then minimized with respect to the $n$'s and in this way, we obtain the best bound of the whole family.

## 19.6 MULTIELECTRON ATOMS AND HARTREE EQUATIONS

To calculate the energy levels of a many-electron system, we consider the Hamiltonian for the motion of $Z$ electrons about a nucleus of mass $M$ and charge, $Ze$:

$$\hat{H} = \sum_{i=1}^{z}\left(-\frac{\hbar^2}{2m_e}\nabla_i^2 - \frac{Ze^2}{r_i}\right) + \sum_{i=1}^{z}\sum_{j>1}\frac{e^2}{|\vec{r}_i - \vec{r}_j|} \qquad ...(1)$$

$$= \sum_{i=1}^{z}\hat{H}_i + \hat{V} \qquad ...(2)$$

The 1st term on the R.H.S. of Eq. (1) is the kinetic energy of the electrons, the 2nd term is the electrostatic interaction between the nucleus and each of the $Z$ electrons and the last (i.e., the third term) is the mutual interaction between all pairs of electrons, $V$. The single particle operators $\hat{H}_i$ for the individual electrons are assumed to satisfy the eigenvalue equations

$$\hat{H}_i\,\psi_{n_i \ell_i}\,m_i\,m_{si} = \left(-\frac{\hbar^2}{2m_e}\nabla_i^2 - \frac{Ze^2}{r_i}\right)\psi_{n_i \ell_i}\,m_i\,m_{si}$$

$$= E_{ni}\,\psi_{n_i \ell_i}\,m_i\,m_{si} \qquad ...(3)$$

where $a = 1, 2 ... z$.

Here, it is no longer a good approximation to treat the term $\hat{V}$ as a perturbation. A better approximation is provided by the variation method, which accounts for the motion of each electron in a central potential energy, consisting of the nuclear attraction $-\frac{Ze^2}{r_i}$ (reduced by the screening action of the other electrons).

It is convenient to choose first, for simplicity, a trial wave function which is the product of single electron wave functions

$$\Phi = \phi_{E_1}(\vec{r}_1, m_{si}) \ldots \phi_{E_i}(\vec{r}_i, m_{si}) \ldots \phi_{E_Z}(\vec{r}_2, m_{sz}) \qquad \ldots(4)$$

where it is assumed that all the $\phi_{E_i}$ are normalized and orthogonal, either in spatial or spin coordinates, such that there is no more than one electron in the same state $(\vec{r}_i, m_{si})$. Because of the spin-independence of the Hamiltonian, Eq. (4), we obtain the expectation value of energy in the form

$$E = \int_1 \ldots \int_i \ldots \int_z \Phi^* \hat{H} \Phi \, dv_1 \ldots dv_i \ldots dv_z \qquad \ldots(5)$$

$$= \int_1 \ldots \int_i \ldots \int_z \phi^*_{E_1}(\vec{r}_1) \ldots \phi^*_{E_i}(\vec{r}_i) \ldots \phi^*_{E_z}(\vec{r}_z) \left( \sum_{i=1}^{z} \hat{H}_i + \hat{V} \right)$$

$$\times \phi_{E_1}(\vec{r}_1) \ldots \phi_{E_i}(\vec{r}_i) \ldots \phi_{Ez}(\vec{r}_z) \, dv_1 \ldots dv_i \ldots dv_z$$

$$= \sum_{i=1}^{z} \left[ \int \phi^*_{E_i}(\vec{r}_i) \hat{H}_i \, \phi_{Ei}(\vec{r}_i) \, dv_i \right.$$

$$\left. + \sum_{j>i} \int \phi^*_{Ei}(\vec{r}_i) \left( e^2 \int \frac{|\phi_{Ej}(\vec{r}_j)|^2}{|\vec{r}_i - \vec{r}_j|} dv_j \right) \phi_{Ei}(\vec{r}_i) \, dv_i \right]$$

or

$$E = \sum_{i=1}^{z} \int \phi^*_{Ei}(\vec{r}_i) \left( \hat{H}_i + \sum_{j>i} \hat{V}_{ij} \right) \phi_{Ei}(\vec{r}_i) \, dv_i \qquad \ldots(6)$$

The operator $\hat{V}_{ij}$ represents the effect of the electron density, $P_j = e\,|\phi_{Ej}(\vec{r}_j)|^2$ on the electron at $\vec{r}_i$ such that

$$\hat{V}_{ij}\, \phi_{Ei}(\vec{r}_i) = e^2 \int \frac{|\phi_{E_0}(\vec{r}_0)|^2}{|\vec{r}_i - \vec{r}_j|} \phi_{Ei}(\vec{r}_i) \, dv_j \qquad \ldots(7)$$

$$\left( = \int \frac{e\,P_j}{|\vec{r}_i - \vec{r}_j|} \phi_{Ei}(\vec{r}_i) \, dv_j \right)$$

The operator $(\hat{V}_{ij})$ is introduced in order to apply the variational method to the single electron problem. Assuming that an arbitrary variation $(\delta\Phi)$ is exclusively due to the variation of $\phi_{ei}(\vec{r}_i)$, namely

$$\delta\Phi = \phi_{E_1}(\vec{r}_1) \ldots \delta\phi_{Ei}(\vec{r}_i) \ldots \phi_{Ez}(\vec{r}_z)$$

we must find the minimum of the energy $(E_i)$ given by

$$E_i = \int \phi^*_{Ei}(\vec{r}_i)\left(H_i + \sum_{j \neq i} \hat{V}_{ij}\right)\phi_{Ei}(\vec{r}_i)\, dv_i \qquad ...(8)$$

which includes all the terms of Eq. (6) involving the single electron function $\phi_{Ei}(\vec{r}_i)$. The variational equation which reads

$$\delta E_i = \left\langle \delta\phi_{Ei} \left| \hat{H}_i + \sum_{j \neq i} \hat{V}_{ij} \right| \phi_{Ei} \right\rangle + \left\langle \phi_{Ei} \left| \hat{H}_i + \sum_{j \neq i} \hat{V}_{ij} \right| \delta\phi_{Ei} \right\rangle = 0 \,. \qquad ...(9)$$

is restricted to the variations $\delta\phi_{Ei}$ compatible with the normalization condition on $\delta\phi_{Ei}$ viz.,

$$\delta \langle \phi_{Ei} | \phi_{Ei} \rangle = \langle \delta\phi_{Ei} | \phi_{Ei} \rangle + \langle \phi_{Ei} | \delta\phi_{Ei} \rangle = 0 \qquad ...(10)$$

By using the Lagrangian multiplier ($\varepsilon$) such that

$$\delta_{Ei} - \varepsilon\delta \langle \phi_{Ei} | \phi_{Ei} \rangle = 0$$

we obtain, using Eqs. (9) and (10)

$$\delta_{Ei} - \varepsilon\, [\langle \delta\phi_{Ei} | \phi_{Ei} \rangle + \langle \phi_{Ei} \mid \delta\phi_{Ei} \rangle\} = 0$$

or

$$\delta_{Ei} - \varepsilon\, [\langle \delta\phi_{Ei} | \phi_{Ei} \rangle - \varepsilon \langle \phi_{Ei} \mid \delta\phi_{Ei} \rangle = 0$$

or

$$\left\langle \delta\phi_{Ei} \left\{ \left| \hat{H}_i + \sum_{j \neq i} \hat{V}_{ij} \right| \phi_{Ei} \right\rangle - \varepsilon \left| \phi_{Ei} \right\rangle \right\} + \left\langle \phi_{Ei} \left| \hat{H}_i + \sum_{j \neq i} \hat{V}_{ij}\, \delta\phi_{Ei} \right\rangle - \varepsilon \left\langle \phi_{Ei} \left| \delta\phi_{Ei} \right\rangle = 0$$

or

$$\left\langle \delta\phi_{Ei} \left| \left( \hat{H}_i + \sum_{j \neq i} \hat{V}_{ij} \right) - \varepsilon \right| \phi_{Ei} \right\rangle + \left\langle \phi_{Ei} \left| \left( \hat{H}_i + \sum_{j \neq i} \hat{V}_{ij} \right) - \varepsilon \right| \delta\phi_{Ei} \right\rangle$$

This means that $\phi_{Ei}$ satisfies the eigenvalue equation

$$\left( \hat{H}_i + \sum_{j \neq i} \hat{V}_{ij} \right) | \phi_{Ei} \rangle = \varepsilon | \phi_{Ei} \rangle \qquad ...(11)$$

Multiplying by $|\phi_{Ei}\rangle$ and integrating over $\vec{r}_i$ (where $|\phi_{Ei}\rangle = \phi_{Ei}(\vec{r}_i)$), we obtain from Eq. (8) that $\varepsilon = E_i$ so that the energy eigenfunction and eigenvalue of the $i$th electron are determined by solving the single particle equation

$$\left( \hat{H}_i + \sum_{j \neq i} \hat{V}_{ij} \right) \phi_{Ei}(\vec{r}_i) = E_i\, \phi_{Ei}(\vec{r}_i) \qquad ...(12)$$

The set of $z$ simultaneous equations of the form (12) written for each electron, of the atom, are called the *Hartree equations*.

It is clear that the terms $\hat{V}_{ij}\,\phi_{Ei}(\vec{r}_i)$ having the explicit form Eq. (7), can only be calculated if $\phi_{Ej}(\vec{r}_j)$ are already known, which implies that the Hartree equations cannot be simultaneously solved in closed form. However, an iterative procedure can be applied, assuming an initial set of functions $\phi_{Ei}^{(k)}(\vec{r}_i)$ to obtain the potential energies

$$V_{ij}^{(0)} = \langle \phi_{Ei}^{(0)} | \hat{V}_{ij} | \phi_{Ei}^{(0)} \rangle$$

and then solving numerically Eq. (12) for a new set of single electron functions, $\phi_{Ei}^{(1)}(\vec{r}_i)$. This procedure is separated until the wave functions $\phi_{Ei}^{(k)}(\vec{r}_i)$ and the eigenvalues $E_i^{(k)}$ have values which are no longer different from those used in the equations solved for them.

The Hartree equation for the *i*th electron Eq. (12) can be written again in terms of the average potential arising from all the other electrons and the nucleus

$$V_{\text{eff}}(\vec{r}_i) = \sum_{j\neq i} \hat{V}_{ij} - \frac{Ze^2}{r_i} = \sum_{j\neq i} e^2 \int \frac{|\phi_{Ej}(\vec{r}_j)|^2}{|\vec{r}_i - \vec{r}_j|}\, dv_j - \frac{Ze^2}{r_i} \qquad ...(13)$$

A convenient approximation consists of replacing $V_{\text{eff}}(\vec{r}_i)$ by its average over the angle which is a central potential

$$V_{\text{eff}}(r_i) = \int V_{\text{eff}}(\vec{r}_i)\, d\Omega_i / 4\pi.$$

Substituting into Eq. (12) gives the so called *central field approximation* to the Hartree equations

$$\left( -\frac{\hbar^2}{2m_e} \nabla_i^2 + V_{\text{eff}}(r_i) \right) \phi_{Ei}(\vec{r}_i) = E_i\, \phi_{Ei}(\vec{r}_i) \qquad ...(14)$$

Because the Hamiltonian contains only central forces and does not involve the spin, it commutes with both the orbital and spin angular momentum operators, such that the solutions $\phi_{Ei}(\vec{r}_i)$ are eigenstates of the angular momentum, of the form

$$\phi_{ni\,\ell i\,mi\,msi}(\vec{r}_i) - R_{ni\,\ell i}(r)\, Y_{\ell i\,mi}(0, \phi)\, \chi_{msi} \qquad ...(15)$$

called orbitals.

The expectation value of energy, in a state of the multi-electron atom which is written as a simple product of orbitals, Eq. (4) is obtained from Eq. (6) as

$$E = \sum_{i=1}^{z} E_{ni} + \sum_{i=1}^{z} \sum_{j>i} \int \phi_{Ei}^*(\vec{r}_i) \hat{V}_{ij}\, \phi_{Ei}(\vec{r}_i)\, dv_i$$

$$= \sum_{i=1}^{z} E_{ni} + \frac{1}{2} \sum_{i=1}^{z} \sum_{j>i} \int \phi_{Ei}^*(\vec{r}_i) \hat{V}_{ij}\, \phi_{Ei}(\vec{r}_i)\, dv_i$$

$$= \sum_{i=1}^{z} E_{ni} + \frac{1}{2} \sum_{i=1}^{z} \sum_{j\neq i} \frac{|\phi_{Ei}(\vec{r}_j)|^2 |\phi_{Ei}(\vec{r}_i)|^2}{|\vec{r}_i - \vec{r}_j|}\, dv_i\, dv_j \qquad ...(16)$$

where each electron-pair is counted only once, although the summation is unrestricted except for the requirement, $i \neq j$.

## 19.7 THE HARTREE APPROXIMATION AND HARTREE – FOCK APPROXIMATION

When applied to electrons, the Hartree method neglects the effects of antisymmetry of many electron wave functions. It also neglects *correlations*. Despite these deficiencies, the Hartree approximation can be very useful, e.g., when applied to many – electron atoms. The Hartree approximation is also useful for gaining a crude understanding of the quasi free-electron picture of metals. Finally, it is easier to understand the Hartree-Fock method as well as the density functional method by slowly building up the requisite ideas. The Hartree approximation is a first step.

In the Hartree approximation, the electrons are independent, moving in the average field of the other electrons in the system. In terms of the variational approach, we may write the wave function as a product of single-particle functions

$$\psi(\vec{r}_1, \vec{r}_2 ...) = \psi_1(\vec{r}_1) ... \psi_n(\vec{r}_n) \qquad ...(1)$$

For this form of wave function, the variational principle requires that each of the functions $\psi_i(\mathbf{r}_i)$, satisfies a one-electron Schrödinger equation of the form

$$\left(-\frac{\hbar^2}{2m}\nabla^2 + V_{ext} + \Phi_i\right)\psi_i(\mathbf{r}) = \varepsilon_i \psi_i(\mathbf{r}), \qquad ...(2)$$

where $V_{\text{ext}}$ is the potential due to the nuclei, and the Coulomb potential ($\Phi_i$) is given by Poissons's equation and arises from the average field of the other electrons. The state of the system is then defined by the single particle functions $\psi_i(\mathbf{r})$, the eigenvalues ($\varepsilon_i$) and the occupancy of the "orbitals" so defined. Fermi statistics can be incorporated into this picture by replacing the product wave function by a single (Slater) determinantal function, often called a "configuration". The resultant (Hartree–Fock, HF) approximation leads to an additional "exchange" term $V_x^{\text{HF}}$ in the Schrödinger equation that is non-local in the sense that

$$V_x^{\text{HF}}\,\psi(\mathbf{r}) \equiv \int d\mathbf{r}'\, \mathbf{V}_x^{HF}(\mathbf{r},\mathbf{r}')\, y(\mathbf{r}') \qquad ...(3)$$

Nevertheless, the single-particle picture, with the wave function described in terms of orbitals with particular spins and occupation numbers, is unchanged. The inclusion of Fermi statistics improves the total energy calculation, and the HF approximation has been an indispensable benchmark in molecular physics since its inception over 60 years ago.

Hartree–Fock (often referred to in the chemical literature as "self-consistent field") calculations, and the picture that results, are very familiar, but it has been known for many years that the total energies so found are not accurate enough for many purposes. The lowest-lying configuration is generally only one of many with comparable energies, and an improved energy results if we take a linear combination. This procedure, where effects beyond the Hartree–Fock approximation ("correlation effects') are included by improving the many-particle wave function, is known as "configuration interaction" (CI). It leads, in principle, to the exact wave function, from which many properties of interest can be calculated, but the numerical effort required increases dramatically with increasing electron number. In practice, numerically exact total energies can be found using CI only for systems with well below 100 electrons.

### 19.7.1 The Hartree Equations and Hartree – Approximation

To represent many-electron wave functions, it is convenient to consider preliminarily a complete set of orthonormal one-electron orbitals $\{\phi_i(\mathbf{r})\}$. From this, we can form a complete set of orthonormal *spin-orbitals* $\{\psi_i(\mathbf{r}\sigma)\}$, where $\psi_i(\mathbf{r}\sigma) = \phi_i(\mathbf{r})\chi_i(\sigma)$ denotes the product of the *spatial orbital,* $\phi_i(\mathbf{r})$ and the *spin function,* $\chi_i(\sigma)$ (either spin-up function ($\alpha$) or spin-down function, $\beta$). In short-hand notations, space and spin variables are (frequently) omitted. In order to avoid possible source of confusion, throughout this chapter, we use the notation $\psi_i$ exclusively to denote spin-orbitals; similarly, we use the notation $\phi_i$ and $\chi_i$ exclusively to denote the orbital part and the spin part of the spin-orbital, $\psi_i$. Thus, the arguments of $\psi_i$, $\phi_i$ and $\chi_i$ are understood to be $\psi_i(\mathbf{r}\sigma)$, $\phi_i(\mathbf{r})$ and $\chi_i(\sigma)$, respectively.

In the Hartree theory the ground-state wavefunction of the many-body system is expressed as simple product of orthonormalized one-electron spin-orbitals in the form

$$\Psi_0(\mathbf{r}_1\sigma_1, \mathbf{r}_2\sigma_2, ..., \mathbf{r}_N\sigma_N) = \psi_1(\mathbf{r}_1\sigma_1)\psi_2(\mathbf{r}_2\sigma_2) \,...\, \psi_N(\mathbf{r}_N\sigma_N). \qquad ...(1)$$

In Eq. (1), the many-body wavefunction ($\Psi_0$) is constructed by assigning any given electron to some given spin-orbital (for instance the electron of coordinates ($\mathbf{r}_1\sigma_1$) is assigned to the spin-orbital $\Psi_1$, the electron $\mathbf{r}_2\sigma_2$ is assigned to the spin-orbital $\psi_2$, and so on). It is apparent that the simple Hartree product (1) does not have the correct antisymmetry character for interchange of space and spin coordinates of any two electrons; the Pauli principle is taken into account ad hoc, avoiding multiple occupancy of any given spin-orbital. Furthermore, a simple product of type (1) completely neglects any correlation in the position of the electrons. Because of these serious drawbacks, we mention the Hartree theory mostly for its historical role, and confine our discussion to some heuristic considerations [later, for the conceptually more important Hartree – Fock equations and Kohn – Sham equations, we use rigorous variational principles].

The electronic charge density, $\rho(\mathbf{r})$, corresponding to the Hartree wavefunction (1) is given by

$$\rho(\mathbf{r}) = (-e) \sum_j^{(\text{occ})} \phi_j^*(\mathbf{r})\,\phi_j(\mathbf{r}), \qquad ...(2)$$

where the sum runs over all occupied spin-orbitals, entering the ground state, $\psi_0$. The Coulomb potential (also called the *Hartree potential*) corresponding to the electronic charge density (2) is

$$V_{\text{coul}}(\mathbf{r}) = \sum_j^{(\text{occ})} \int \phi_j^*(\mathbf{r}')\frac{e^2}{|\mathbf{r}-\mathbf{r}'|}\phi_j(\mathbf{r}')\,d\mathbf{r}'. \qquad ...(3)$$

In the Hartree self-consistent approximation, it is assumed that each electron moves in the effective field corresponding to the Coulomb potential generated by the charge distribution of all the other $N - 1$ electrons; for simplicity, to avoid the determination of as many (slightly) different effective fields as the number of orbitals, it is assumed that the effective field for any electron is given by the Hartree potential, Eq. (3). We thus obtain that the spin-orbitals entering in the product wave function, Eq. (1) satisfy the Hartree equations

$$\left[\frac{\mathbf{p}^2}{2m} + V_{\text{nucl}}(\mathbf{r}) + V_{\text{coul}}(\mathbf{r})\right]\psi_i = \varepsilon_i \psi_i. \quad ...(4)$$

The Hartree potential is defined in terms of the occupied orbitals $\psi_i$ and must thus be determined in a self-consistent way (usually with appropriate iterations). From an initial reasonable guess of the functions, $\psi_1, \psi_2, ..., \psi_N$, one evaluates the space charge distribution, Eq. (2) and the corresponding Hartree potential, Eq. (3). A new set of improved wave functions is then obtained from the solution of the Hartree Eq. (4). The corresponding Hartree potential (or, more often, an appropriate weighted average of input and output ones) is used to start a new cycle. Cycles are repeated up to self-consistency of input and output functions and potentials.

## 19.8 IDENTICAL PARTICLES AND DETERMINANTAL WAVE FUNCTIONS

In this section we summarize some basic results on the antisymmetry principle for fermions, and in particular the possibility to write the basis functions of a many-body system in the form of determinantal states. It is well known that a many-electron wave function must be antisymmetric for interchange of the coordinates of any two electrons. Consider an $N$-electron system and a set $\{\psi_i\}$ $(i = 1, 2, ..., N)$ of *orthonormal one-particle spin-orbitals*. A proper antisymmetric $N$-electron wave function takes the form

$$\Psi_0(\mathbf{r}_1\sigma_1, \mathbf{r}_2\sigma_2, ..., \mathbf{r}_N\sigma_N) = A\{\psi_1(\mathbf{r}_1\sigma_1)\psi_2(\mathbf{r}_2\sigma_2) ... \psi_N(\mathbf{r}_N\sigma_N)\} \quad ...(5a)$$

where $A$ denotes the antisymmetrization operator

$$A = \frac{1}{\sqrt{N!}}\sum_{i=1}^{N!}(-1)^{p_i} P_i \quad ...(5b)$$

(the sum extends to all the $N!$ permutations ($P_i$) of the electronic coordinates, and $(-1)^{p_i}$ equals +1 or –1 for permutations of even or odd class with respect to the fundamental one). Notice that $\Psi_0$ is normalized to one, if the composing spin-orbitals ($\psi_i$) are orthonormal.

Expression (5) can be conveniently written in the determinantal form suggested by Slater

$$\Psi_0(\mathbf{r}_1\sigma_1, \mathbf{r}_2\sigma_2, ... \mathbf{r}_N\sigma_N) = \frac{1}{\sqrt{N!}}\begin{vmatrix} \psi_1(\mathbf{r}_1\sigma_1) & \psi_1(\mathbf{r}_2\sigma_2) & ... & \psi_1(\mathbf{r}_N\sigma_N) \\ \psi_2(\mathbf{r}_1\sigma_1) & \psi_2(\mathbf{r}_2\sigma_2) & ... & \psi_2(\mathbf{r}_N\sigma_N) \\ ... & ... & ... & ... \\ \psi_N(\mathbf{r}_1\sigma_1) & \psi_N(\mathbf{r}_2\sigma_2) & ... & \psi_N(\mathbf{r}_N\sigma_N) \end{vmatrix}. \quad ...(6)$$

It is evident that the interchange of two columns changes the sign of the determinant consistently with the antisymmetry property of the wave function; moreover, occupancy of the same spin-orbital by two electrons gives two equal rows and thus the determinant equals zero.

An important property automatically embodied in determinantal wave functions is that electrons with parallel spin are (correctly) kept apart. To better realize this point, consider for simplicity the determinantal state, Eq. (6) in the particular case in which the spin-orbitals have *all spin parallel* (for instance spin up). The determinantal state, Eq. (6) then keeps the form

$$\Psi_0(\mathbf{r}_1\sigma_1, \mathbf{r}_2\sigma_2, \ldots \mathbf{r}_N\sigma_N) = \frac{1}{\sqrt{N!}}\begin{vmatrix} \phi_1(\mathbf{r}_1) & \phi_1(\mathbf{r}_2) & \ldots & \phi_1(\mathbf{r}_N) \\ \phi_2(\mathbf{r}_1) & \phi_2(\mathbf{r}_2) & \ldots & \phi_2(\mathbf{r}_N) \\ \ldots & \ldots & \ldots & \ldots \\ \phi_N(\mathbf{r}_1) & \phi_N(\mathbf{r}_2) & \ldots & \phi_N(\mathbf{r}_N) \end{vmatrix} \alpha(1)\alpha(2)\ldots\alpha(N). \quad \ldots(7)$$

It is evident that *nodes* of $\Psi_0$ occur whenever $\mathbf{r}_i \equiv \mathbf{r}_j$; thus any two electrons cannot be in the same spatial position (with the same spin).

The many-body wave functions can be written in a more compact form, leaving implicit the space and spin coordinates of the electrons. Two frequently used short-hand notations for the wavefunction ($\Psi_0$) are

$$\Psi_0 = A\{\psi_1\psi_2\ldots\psi_N\} \quad \text{or} \quad \Psi_0 = \frac{1}{\sqrt{N!}}\det(\psi_1\psi_2\ldots\psi_N); \quad \ldots(8)$$

both notations clearly list the spin-orbitals entering the antisymmetrization operator or forming the Slater determinant.

## 19.9 MATRIX ELEMENTS BETWEEN DETERMINANTAL STATES

The purpose of this section is to provide the explicit expression of the matrix elements of one-electron and two-electron operators between determinantal states. These expressions constitute a technical (but nevertheless essential) tool for handling the Hartree–Fock theory, as well as the second quantization formalism of the many-body problem. [The readers more interested in the structure and physical meaning of the Hartree–Fock equations can skip this section, using their own knowledge when meeting matrix elements between determinantal states].

### *Expectation Values of One-particle and Two-particle Operators on Determinantal States*

The many-body electron Hamiltonian contains two types of operators. One is the sum of one-particle operators of the form

$$H_e = \sum_{i=1}^{N} h(\vec{r}_i) + \frac{1}{2}\sum_{i\neq j}^{N}\frac{e^2}{r_{ij}} \quad \text{where } h(\vec{r}) = \frac{\vec{p}^{\,2}}{2m} + V_{\text{nucl}}(\vec{r}) \quad \ldots(9a)$$

$$G_1 = \sum_{i}^{N} h(\mathbf{r}_i); \quad \ldots(9b)$$

the other type is the sum of two-particle operators of the form

$$G_2 = \frac{1}{2}\sum_{i \neq j}^{N} \frac{e^2}{|\mathbf{r}_i - \mathbf{r}_j|} \qquad \text{...(9c)}$$

which describes electron-electron interactions. We wish now to express the matrix elements of one- and two-particle operators between Slater determinants in terms of monoelectronic and bi-electronic integrals, respectively. The elaboration is based on the elementary properties of determinants and the invariance of $G_1$ and $G_2$ under permutations of the electronic coordinates.

We begin with the evaluation of the expectation value of $G_1$ on a given determinantal state $\Psi_0$; we have

$$\langle\Psi_0|G_1|\Psi_0\rangle = \frac{1}{N!}\langle\det\{\psi_1\psi_2 \ldots \psi_N\}|G_1|\det\{\psi_1\psi_2 \ldots \psi_N\}\rangle \qquad \text{...(10a)}$$

In the above expression, we can just select the identical permutation in the determinant in the left part; in fact, all the other $N! - 1$ terms contributing to the determinant in the left side are equal to the selected one. We can thus cancel $N!$ at the denominator of Eq. (10a), and write

$$\langle\Psi_0|G_1|\Psi_0\rangle = \langle\psi_1\psi_2 \ldots \psi_N|G_1|\det\{\psi_1\psi_2 \ldots \psi_N\}\rangle. \qquad \text{...(10b)}$$

Because of the orthonormality of spin-orbitals it is also evident that only the identical permutation survives in the determinant on the right part of the above matrix element. We have thus

$$\langle\Psi_0|G_1|\Psi_0\rangle = \langle\psi_1\psi_2 \ldots \psi_N|G_1|\psi_1\psi_2 \ldots \psi_N\rangle. \qquad \text{...(10c)}$$

This relation shows that, for what concerns the matrix elements of the operator $G_1$, it is irrelevant to use antisymmetrized products of spin-orbitals or simple (Hartree) products of the same spin-orbitals.

Inserting the expression (9b) of $G_1$ into expression (10c), we obtain

$$\boxed{\langle\Psi_0|G_1|\Psi_0\rangle = \sum_i \langle\psi_i|h|\psi_i\rangle}. \qquad \text{...(11)}$$

Thus, we see that the expectation value on a determinantal state of a sum of one-electron operators equals the sum of the expectation values of the one-electron operator on the spin-orbitals, which enter the determinantal state.

We evaluate now the expectation value of $G_2$ on a given determinantal state; we have

$$\langle\Psi_0|G_2|\Psi_0\rangle = \frac{1}{N!}\langle\det\{\psi_1\psi_2 \ldots \psi_N\}|G_2|\det\{\psi_1\psi_2 \ldots \psi_N\}\rangle$$

$$= \{\psi_1\psi_2 \ldots \psi_N\}|G_2|\det\{\psi_1\psi_2 \ldots \psi_N\}\rangle. \qquad \text{...(12a)}$$

Consider for instance the contribution of the term $e^2/r_{12}$. We restore for clarity the space and spin electronic coordinates, and we easily verify that

$$\langle \psi_1(\mathbf{r}_1\sigma_1)\psi_2(\mathbf{r}_2\sigma_2)\ \dots\ \psi_N(\mathbf{r}_N\sigma_N) \left| \frac{e^2}{r_{12}} \right| \det\{\psi_1(\mathbf{r}_1 s_1)\psi_2(\mathbf{r}_2\sigma_2)\dots\psi_N(\mathbf{r}_N\sigma_N)\}\rangle.$$

$$= \langle \psi_1\psi_2 \left| \frac{e^2}{r_{12}} \right| \psi_1\psi_2\rangle - \langle \psi_1\psi_2 \left| \frac{e^2}{r_{12}} \right| \psi_1\psi_2\rangle; \qquad \text{...(12b)}$$

In fact only the identical permutation and the permutation (of odd class) that interchanges coordinates $\mathbf{r}_2\sigma_2$ with coordinates $\mathbf{r}_1\sigma_1$ survive. The definition and elementary properties of the Coulomb and exchange bi-electronic integrals appearing in the right-hand side of Eq. (12b) are discussed in Appendix A; in particular we remind the reader that bi-electronic exchange integrals are definite positive quantities and are different from zero only for parallel spins of the spin-orbitals entering in their definition.

The expectation value of $G_2$ on a given determinantal state thus becomes

$$\boxed{\langle \Psi_0 | G_2 | \Psi_0 \rangle = \frac{1}{2}\sum_{ij}\left[\left\langle \psi_i\psi_j \left| \frac{e^2}{r_{12}} \right| \psi_i\psi_j \right\rangle - \left\langle \psi_i\psi_j \left| \frac{e^2}{r_{12}} \right| \psi_j\psi_i \right\rangle\right]}. \qquad \text{...(13)}$$

In Eq. (13) we can write indifferently $\sum_{ij}$ or $\sum_{i\neq j}$, since Coulomb and exchange contributions exactly cancel each other for $i \neq j$. We can thus recast Eq. (13) in the form

$$\langle \Psi_0 | G_2 | \Psi_0 \rangle = \frac{1}{2}\sum_{i\neq j}\left\langle \psi_i\psi_j \left| \frac{e^2}{r_{12}} \right| \psi_i\psi_j \right\rangle - \frac{1}{2}\sum_{i\neq j}\left\langle \psi_i\psi_j \left| \frac{e^2}{r_{12}} \right| \psi_i\psi_j \right\rangle \qquad \text{...(14)}$$

It is important to notice explicitly that the *electron-electron repulsive energy* $\langle \Psi_0 | G_2 | \Psi_0 \rangle$ *on the antisymmetrized product of spin-orbitals, is always lower than the electron-electron repulsive energy on the simple Hartree product of the same spin-orbitals*; the difference is just represented by the second term in the right-hand side of Eq. (14). Since exchange bi-electronic integrals are positive quantities, different from zero only for spin-orbitals with parallel spins, we link the decrease in energy with the physical fact that electrons with parallel spin are kept apart in real space in Slater determinantal states (as already discussed in Eq. (7)).

In the following we report other matrix elements of interest of one-electron and two-electron operators among determinantal states; however we omit from now on any proof, as this could be easily worked out following "*mutatis mutandis*" the reasoning leading to Eq. (11) and Eq. (13).

### *Determinantal States with Replacement of One Spin-orbital*

Consider two $N$-electron determinantal states which differ by one of the one-particle functions, say

$$\Psi_0 = A\{\psi_1\psi_2\ \cdots\ \psi_m\ \cdots\ \psi_N\}$$

and

$$\Psi_{\mu,m} = A\{\psi_1\psi_2\ \cdots\ \psi_m\ \cdots\ \psi_N\}.$$

The notation $(\Psi_{\mu,m})$ helps to memorize that the spin-orbital $(\psi_m)$ has been replaced by the spin-orbital, $\psi_\mu$, all other spin-orbitals being equal and in the same sequence.

It can be easily verified that

$$\boxed{\langle \Psi_{\mu,m} | G_1 | \Psi_0 \rangle = \langle \psi_\mu | h | \psi_m \rangle} \qquad ...(14a)$$

Thus, we see that the matrix elements of $G_1$ connecting two determinantal states, which differ by one of the one-particle functions, equal the matrix elements of the one-particle operator between the two (replacing and replaced) spin-orbitals. For the operator $G_2$ we have

$$\boxed{\langle \Psi_{\mu,m} | G_2 | \Psi_0 \rangle = \sum_j \left[ \left\langle \psi_\mu \psi_j \left| \frac{e^2}{r_{12}} \right| \psi_m \psi_j \right\rangle - \left\langle \psi_\mu \psi_j \left| \frac{e^2}{r_{12}} \right| \psi_j \psi_m \right\rangle \right]} \qquad ...(15)$$

### Determinantal States with Replacement of Two Spin-orbitals

Consider two determinantal states which differ by two of the one-particle functions, say

$$\Psi_0 = A\{\psi_1 \psi_2 \ldots \psi_m \ldots \psi_n \ldots \psi_N\}$$

and
$$\Psi_{\mu\nu,mn} = A\{\psi_1 \psi_2 \ldots \psi_\mu \ldots \psi_\nu \ldots \psi_N\}.$$

A sum of one-electron operators, as $G_1$, has vanishing matrix elements between determinantal states which differ by two spin-orbitals. The matrix elements of $G_2$ are given by

$$\boxed{\langle \Psi_{\mu\nu,mn} | G_2 | \Psi_0 \rangle = \left\langle \psi_\mu \psi_\nu \left| \frac{e^2}{r_{12}} \right| \psi_m \psi_n \right\rangle - \left\langle \psi_\mu \psi_\nu \left| \frac{e^2}{r_{12}} \right| \psi_n \psi_m \right\rangle.} \qquad ...(16)$$

Finally, the matrix elements of the operators $G_1$ and $G_2$ between determinantal states which differ by *more than two spin-orbitals are zero.*

In this section we have provided useful expressions of the matrix elements of the operator $G_1$ and $G_2$ between determinantal states. With respect to the matrix elements involving simple product wave functions, it can be observed that the antisymmetrization does not alter the matrix elements of the one-body operator, $G_1$, while, for the two-body operator $G_2$, exchange bi-electronic integrals are to be appropriately included. A more convenient formal and conceptual point of view to treat Slater determinants and related matrix elements is the use of the second quantization formalism, which is outlined in Appendix *E*.

## 19.10 THE HARTREE–FOCK EQUATIONS

### 19.10.1 Variational Approach and Hartree – Fock Equations

Let us consider a system of $N$ interacting electrons described by the Hamiltonian $H_e$ given in Eq. (9a). We attempt to describe *the ground state of the many-electron system with a single determinantal state*; the best possible choice of spin-orbitals entering this determinantal state is obtained minimizing the electron energy, using the variational principle. We find that the "optimized" orbitals are solutions of the self-consistent integro-differential equations, known as Hartree–Fock equations.

We approximate the ground state of the system with a single determinantal state in the form

$$\Psi_0 = A\{\psi_1 \psi_2 \dots \psi_N\}.$$

A determinantal state $\Psi_0$, constituted by doubly occupied spatial orbitals, is called *spin-restricted Slater determinant;* this spin configuration is the appropriate choice for *closed-shell systems*, where electrons with spin-up and spin-down balance for each spatial orbital. However, to discuss also *open-shell systems*, we do not put at this stage any restriction on the spin part of the spin-orbitals entering the ground statem, $\Psi_0$, which is thus, in general, a *spin-unrestricted Slater determinant.* The variational procedure concerns the optimized determination of the orbital parts of the spin-orbitals (for a given chosen spin configuration).

The expectation value $E_0 = \langle\Psi_0|H_e|\Psi_0\rangle$ of the electronic Hamiltonian $H_e$ on the (normalized) state $\Psi_0$ can be calculated by means of Eq. (11) and Eq. (13); it is given by

$$E_0 = \sum_i \langle\psi_i|h|\psi_i\rangle + \frac{1}{2}\sum_{ij}\left[\left\langle \psi_i\psi_j\left|\frac{e^2}{r_{12}}\right|\psi_i\psi_j\right\rangle - \left\langle \psi_i\psi_j\left|\frac{e^2}{r_{12}}\right|\psi_j\psi_i\right\rangle\right] \quad ...(17)$$

According to the variational principle we vary the $N$ contributing spin-orbitals $\{\psi_i\}$ $(i = 1, 2, ..., N)$ until the energy $E_0 \equiv E_0(\{\psi_i\})$ achieves its minimum value. This minimum value is higher (or at most equal) than the correct ground-state energy. The variation of the functional $E_0(\{\psi_i\})$ must be done under the constraint that $\{\psi_i\}$ are orthonormal

$$\langle\psi_i|\psi_j\rangle = \delta_{ij} \ (i, j = 1, 2, ..., N).$$

With the standard technique of undetermined Lagrange multipliers, we minimize without constraints the functional

$$G(\{\psi_i\}) = \sum_i \langle\psi_i|h|\psi_i\rangle + \frac{1}{2}\sum_{ij}\left[\left\langle \psi_i\psi_j\left|\frac{e^2}{r_{12}}\right|\psi_i\psi_j\right\rangle - \left\langle \psi_i\psi_j\left|\frac{e^2}{r_{12}}\right|\psi_j\psi_i\right\rangle\right] - \sum_{ij}\varepsilon_{ij}\langle\psi_i|\psi_j\rangle$$

where $\varepsilon_{ij}$ are at this stage $N^2$ undermined Lagrange multipliers.

We can now calculate the variation $\delta G$ when the orbital part of the spin-orbitals $(\psi_i)$ are changed by an infinitesimal amount, $\psi_i + \delta\psi_i$. Consider, for instance, the variation of the term $\langle\psi_i|h|\psi_i\rangle$; we have to first order in the variations.

$$\delta\langle\psi_i|h|\psi_i\rangle = \langle\delta\psi_i|h|\psi_i\rangle + \langle\psi_i|h|\delta\psi_i\rangle.$$

Similar considerations can be performed for any term appearing in the expression of $G(\{\psi_i\})$, and the variation $\delta G$ can be calculated.

We can consider the variations $\delta\psi_i$ and $\delta\psi_i^*$ completely independent; for instance, taking the variations $\delta\psi_i^*$ arbitrary and the variations $\delta\psi_i \equiv 0$ we obtain for $\delta G$

$$\delta G = \sum_i \langle\delta\psi_i|h|\psi_i\rangle + \sum_{ij}\left[\left\langle \delta\psi_i\psi_j\left|\frac{e^2}{r_{12}}\right|\psi_i\psi_j\right\rangle - \left\langle \delta\psi_i\psi_j\left|\frac{e^2}{r_{12}}\right|\psi_j\psi_i\right\rangle\right] - \sum_{ij}\varepsilon_{ij}\langle\delta\psi_i|\psi_j\rangle \quad ...(18)$$

In stationary conditions $\delta G = 0$; since the variations $\delta\psi_i^*$ are arbitrary, we have that the quantity multiplying any chosen $\delta\psi_i^*$ in Eq. (18) must be zero. We obtain that the best spin-orbitals composing the ground determinantal state satisfy the set of non-linear integro-differential equations

$$\left[\frac{\mathbf{p}^2}{2m}+V_{\text{nucl}}(\mathbf{r})+V_{\text{coul}}(\mathbf{r})+V_{\text{exch}}\right]\psi_i=\sum_j^{(\text{occ})}\varepsilon_{ij}\,\psi_j, \qquad \text{...(19a)}$$

with the Coulomb and exchange operators defined as

$$V_{\text{coul}}\,\psi_i(\mathbf{r}\sigma)=\sum_j^{(\text{occ})}\psi_i(\mathbf{r}\sigma)\int\psi_j^*(\mathbf{r}_2\sigma_2)\frac{e^2}{|\mathbf{r}-\mathbf{r}_2|}\psi_j(\mathbf{r}_2\sigma_2)\,d(\mathbf{r}_2\sigma_2) \qquad \text{...(19b)}$$

$$V_{\text{exch}}\,\psi_i(\mathbf{r}\sigma)=-\sum_j^{(\text{occ})}\psi_j(\mathbf{r}\sigma)\int\psi_j^*(\mathbf{r}_2\sigma_2)\frac{e^2}{|\mathbf{r}-\mathbf{r}_2|}\psi_i(\mathbf{r}_2\sigma_2)\,d(\mathbf{r}_2\sigma_2) \qquad \text{...(19c)}$$

(the sum over $j = 1, 2, ..., N$ has been relabelled as sum over the occupied spin-orbitals).

If we perform explicitly the integration over the spin variables in Eq. (19b), we see that the Coulomb operator is the standard classical Hartree potential $V_{\text{coul}}(\mathbf{r})$, defined by Eq. (3). The exchange operator of Eq. (19c) (also improperly called "exchange potential") is an integral operator, since the wave function on which it operates enters under integral sign. If the integration over the spin variables is performed in Eq. (19c), it is seen that the occupied spin-orbitals contributing to $V_{\text{exch}}\,\psi_i(\mathbf{r}\sigma)$ are only the *spin-orbitals with spin parallel to* $\psi_i(\mathbf{r}\sigma)$.

We can always perform a unitary transformation within the basis of occupied spin-orbitals $\psi_i$ to put the matrix $\varepsilon_{ij}$ in diagonal form, $\varepsilon_{ij} = \varepsilon_i\delta_{ij}$; this is possible because both the Coulomb and exchange operators, Eq. (19) keep the same form under any unitary transformation. We arrive thus at the *canonical Hartree–Fock equations*

$$\boxed{\left[\frac{\mathbf{p}^2}{2m}+V_{\text{nucl}}(\mathbf{r})+V_{\text{coul}}(\mathbf{r})+V_{\text{exch}}\right]\psi_i=\varepsilon_i\psi_i}\;; \qquad \text{...(20a)}$$

the electronic Coulomb and exchange potentials, again expressed by Eq. (19b) and Eq. (19c), can be written in the form

$$\boxed{V_{\text{coul}}(\mathbf{r})=\sum_j^{(\text{occ})}\left\langle\psi_j(\mathbf{r}_2\sigma)\left|\frac{e^2}{|\mathbf{r}-\mathbf{r}_2|}\right|\psi_j(\mathbf{r}_2\sigma)\right\rangle} \qquad \text{...(20b)}$$

and

$$\boxed{V_{\text{exch}}\,\psi_i(\mathbf{r}\sigma)=-\sum_j^{(\text{occ})}\left\langle\psi_j(\mathbf{r}_2\sigma)\left|\frac{e^2}{|\mathbf{r}-\mathbf{r}_2|}\right|\psi_i(\mathbf{r}_2\sigma)\right\rangle\psi_j(\mathbf{r}\sigma)} \qquad \text{...(20c)}$$

The Hartree–Fock equations can be written in a more compact form by introducing the Fock operator

$$F = h(\mathbf{r}) + V_{\text{coul}}(\mathbf{r}) + V_{\text{exch}}, \qquad \text{...(21a)}$$

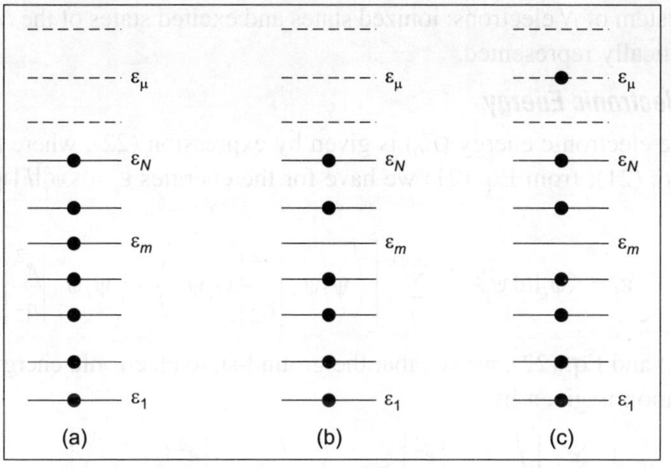


**Figure 19.1** Schematic representation (a) of occupied spin-orbitals (solid lines), in increasing order of energy, for the ground state of an *N*-electron system in the Hartree–Fock approximation; virtual spin-orbitals are represented with broken lines. (b) represents an ionized state of the *N*-electron system, with an electron removed from the occupied spin-orbital $\psi_m$ (and transferred to infinity). (c) represents an excited state of the *N*-electron system, with an electron removed from the occupied spin-orbital $\psi_m$ and transferred to the virtual spin-orbital, $\psi_\mu$.

where $h(\mathbf{r}) = \mathbf{p}^2/2m + V_{\text{nucl}}(\mathbf{r})$; we have

$$F\psi_i = \varepsilon_i \psi_i \qquad \text{...(21b)}$$

The Fock operator is defined in terms of the occupied spin-orbitals ($\psi_i$) and must thus be determined in a self-consistent way (usually with appropriate iterations). The Fock operator reduces the many-electron problem to a one-particle problem (within the stated approximations); although this is a major formal and practical achievement, the highest caution must be applied for a correct interpretation of the meaning of eigenfunctions and eigenvalues of the Fock operator in connection with the original many-electron problems under investigation; we thus analyse further some general properties of the Hartree – Fock equations.

## 19.10.2 Ground-state Energy, Ionization Energies and Transition Energies

Consider the Hartree – Fock approximation for a system of $N$ interacting electrons and suppose we know (at least in principle) the exact self-consistent Fock operator of the system, its eigenfunctions ($\psi_i$) and the spin-orbital energies, $\varepsilon_i$. The $N$ eigenfunctions of lower energy of the Fock operator represent the occupied spin-orbitals, used to construct the Fock operator itself; all the other eigenfunctions of the operator ($F$) are unoccupied, do not enter the construction of $F$ itself, and are called *virtual spin-orbitals*. In Fig. 19.1 we represent schematically occupied and virtual spin-orbitals, in increasing order of energy, for an interacting system of $N$ electrons; ionized states and excited states of the $N$-electron system are also schematically represented.

### Ground-state Electronic Energy

The ground-state electronic energy ($E_0$) is given by expression (22), where the $\psi_i$ satisfy the Hartree–Fock Eq. (21); from Eq. (21) we have for the energies $\varepsilon_i = \langle\psi_i|F|\psi_i\rangle$ the following expression

$$\varepsilon_i = \langle\psi_i|h|\psi_i\rangle + \sum_j^{(\text{occ})}\left[\left\langle\psi_i\psi_j\left|\frac{e^2}{r_{12}}\right|\psi_i\psi_j\right\rangle - \left\langle\psi_i\psi_j\left|\frac{e^2}{r_{12}}\right|\psi_j\psi_i\right\rangle\right]. \quad ...(22)$$

From Eq. (17) and Eq. (22), we see that the ground-state electronic energy in the Hartree–Fock approximation is given by

$$E_0^{(\text{HF})} = \sum_i^{(\text{occ})}\varepsilon_i - \frac{1}{2}\sum_{ij}^{(\text{occ})}\left[\left\langle\psi_i\psi_j\left|\frac{e^2}{r_{12}}\right|\psi_i\psi_j\right\rangle - \left\langle\psi_i\psi_j\left|\frac{e^2}{r_{12}}\right|\psi_j\psi_i\right\rangle\right]. \quad ...(23)$$

Equation (23) shows that the ground-state electronic energy *is not equal to the sum of the Hartree–Fock spin-orbital energies of the occupied states*; this can be understood intuitively from the fact that otherwise the Coulomb and exchange interactions would be counted twice. The total energy of the electronic-nuclear system in the Hartree–Fock approximation is obtained adding to $E_0^{(\text{HF})}$ the nucleus-nucleus Coulomb repulsion (1/2) $\sum_{I\neq J}Z_IZ_Je^2/R_{IJ}$ which constitutes the last term of the Hamiltonian of Eq. (1).

It is to be noticed that, in agreement with the variational principle, the Hartree–Fock energy, $E_0^{(\text{HF})}$, is *higher* than the exact ground-state energy, $E_0^{(\text{exact})}$, of the many-body system (the difference, $E_0^{(\text{HF})} - E_0^{(\text{exact})}$ is called *correlation energy*)). In spite of the importance and achievements of the Hartree–Fock approximation, corrections beyond it are often to be considered; this is due to the fact that a single determinantal state, even when built up with the best possible orbitals, remains in general a rather poor representation of the complicated ground-state wavefunction of a many-particle system.

### Ionization Energy and Koopmans' Theorem

Consider an (ideal) ionization process in which an electron in the occupied state ($\psi_m$) is removed from the system and transferred to infinity (i.e., far enough from the electronic system) with zero kinetic energy; it is assumed that no modification (or "relaxation") of the other electronic orbitals occurs in the ionization process.

In the Hartree–Fock approximation, the energy $E_0^{(\text{HF})}(N)$ of the ground state is given by Eq. (17) the energy $E_0^{(\text{HF})}(N-1)$ of the ionized states given by the same expression (17), provided the spin-orbital ($\psi_m$) is excluded from the sums. We have for the energy difference

$$E_0^{(\text{HF})}(N) - E_0^{(\text{HF})}(N-1)$$

$$= \langle\psi_m|h|\psi_m\rangle + \sum_j^{(\text{occ})}\left[\left\langle\psi_m\psi_j\left|\frac{e^2}{r_{12}}\right|\psi_m\psi_j\right\rangle - \left\langle\psi_m\psi_j\left|\frac{e^2}{r_{12}}\right|\psi_j\psi_i\right\rangle\right] \equiv \varepsilon_m.$$

We thus obtain the following important result, known as *Koopmans' theorem: the energy required to remove (without relaxation) an electron from the spin-orbital* ($\psi_m$) *is simply the opposite of the Hartree–Fock eigenvalue,* $\varepsilon_m$. The parameters ($\varepsilon_m$) appearing in the Hartree–Fock Eq. (20) as purely formal Lagrange multipliers are thus rescued to physical interpretation by the Koopmans' theorem.

### Excited States and Transition Energies

We consider now the ground state, $\Psi_0 = A\{\psi_1\psi_2 \cdots \psi_m \cdots \psi_N\}$, and the excited state, $\Psi_{\mu,m} = A\{\psi_1\psi_2 \cdots \psi_\mu \cdots \psi_N\}$, where an electron has been removed from the occupied spin-orbital ($\psi_m$) and transferred in the previously unoccupied spin-orbital ($\psi_\mu$) (we assume again that all the other occupied orbitals are unchanged); we want to determine the transition energy

$$\Delta E = \langle \Psi_{\mu,m} | H_e | \Psi_{\mu,m} \rangle - \langle \Psi_0 | H_e | \Psi_0 \rangle.$$

The quantity $\langle \Psi_0 | H_e | \Psi_0 \rangle$ is given by expression (17); with straightforward manipulations we can obtain also $\langle \Psi_{\mu,m} | H_e | \Psi_{\mu,m} \rangle$; the transition energy ($\Delta E$) then becomes

$$\begin{aligned} \Delta E = {} & \langle \psi_\mu | h | \psi_\mu \rangle - \langle \psi_m | h | \psi_m \rangle \\ & + \sum_j^{(\text{occ})} \left[ \left\langle \psi_\mu \psi_j \left| \frac{e^2}{r_{12}} \right| \psi_\mu \psi_j \right\rangle - \left\langle \psi_\mu \psi_j \left| \frac{e^2}{r_{12}} \right| \psi_j \psi_\mu \right\rangle \right] \\ & - \left[ \left\langle \psi_\mu \psi_m \left| \frac{e^2}{r_{12}} \right| \psi_\mu \psi_m \right\rangle - \left\langle \psi_\mu \psi_m \left| \frac{e^2}{r_{12}} \right| \psi_m \psi_\mu \right\rangle \right] \\ & - \sum_j^{(\text{occ})} \left[ \left\langle \psi_j \psi_m \left| \frac{e^2}{r_{12}} \right| \psi_j \psi_m \right\rangle - \left\langle \psi_j \psi_m \left| \frac{e^2}{r_{12}} \right| \psi_m \psi_j \right\rangle \right] \end{aligned}$$

Using Eq. (22) for the spin-orbital energies, we obtain

$$\boxed{\Delta E = \varepsilon_\mu - \varepsilon_m - \left[ \left\langle \psi_\mu \psi_m \left| \frac{e^2}{r_{12}} \right| \psi_\mu \psi_m \right\rangle - \left\langle \psi_\mu \psi_m \left| \frac{e^2}{r_{12}} \right| \psi_m \psi_\mu \right\rangle \right]}. \qquad \text{...(24)}$$

The physical meaning of Eq. (24) is transparent indeed: to promote an electron from the occupied spin-orbital $\psi_m$ to the unoccupied spin-orbital ($\psi_\mu$) we need an energy equal to the difference ($\varepsilon_\mu - \varepsilon_m$) of the spin-orbital energies, corrected by the *Coulomb* and *exchange* energy between the "extra electron" in the spin-orbital ($\psi_\mu$) and the "missing electron" (hole) left in the initial spin-orbital, $\psi_m$.

In crystals, $\psi_\mu$ and $\psi_m$ are normalized Bloch functions, and thus the bi-electronic integrals are of the order of $O(1/N)$ (where $N$ is the number of unit cells of the crystal). This appears to provide a good justification for neglecting the "electron-hole interaction", expressed by the bielectronic integrals in square brackets in Eq. (24), thus when the wave functions are itinerant, we expect $\Delta E \approx \varepsilon_\mu - e_m$ for the transition energies. Although for Bloch functions, bielectronic integrals are of the order of $O(1/N)$, we must remember that also the allowed **k** vectors (and hence wave functions for each band) in the Brillouin zone equal the number $N$ of unit cells of the crystal; thus it is not *a priori* justified to neglect $N$ contributions of order $O(1/N)$, and the exciton theory, how to include electron-hole corrections, whenever necessary.

The major obstacle to the solution of the Hartree–Fock equations is the integro-differential nature of the Fock operator. Various physical and technical elaborations have been devised for it in the literature; it would be impossible to describe all of them, but we wish here to mention, in their general lines, the ideas of the Slater approximation and the Roothaan approximation, because of their basic role in solid state physics and in atomic and molecular physics, respectively.

## 19.11 HARTREE–FOCK–SLATER APPROXIMATION

The Slater approximation starts from the fact that the exchange integral operator in the homogeneous free-electron gas can be well approximated with a local potential. Then Slater proposes to replace, also for non-homogeneous systems, the exchange integral operator with the ordinary local potential, corresponding to the free-electron gas of the same local density $n(\mathbf{r})$; in this way the exchange operator takes the form

$$V_{\text{exch}}^{(\text{Slater})}(\mathbf{r}) = -\frac{3}{2}\frac{e^2}{\pi}\,[3\pi^2 n(\mathbf{r})]^{1/3}.$$

With the above approximation, the Fock operator, Eq. (21) becomes an ordinary differential operator with a local potential to be determined self-consistently. The method (with appropriate implementations) has been of wide use mainly in solid state physics.

The Hartree–Fock–Slater approximation becomes particularly agile when applied to atoms, because of the spherical symmetry of the (average) atomic electron density, $n(r)$. The effective atomic potential in the Hartree–Fock–Slater approximation is related to $n(r)$ in the form

$$V^{(\text{HFS})}(r) = -\frac{Ze^2}{r} + \frac{e^2}{r}\int_0^r 4\pi r'^2 n(r')\,dr' + e^2\int_r^\infty 4\pi r' n(r')\,dr' - \frac{3}{2}\frac{e^2}{\pi}\,[3\pi^2 n(r)]^{1/3}. \quad \text{...(25)}$$

The first term in the right-hand side of Eq. (25) is the nuclear potential, the second and the third are the Hartree potential of the spherically symmetric charge distribution $(-e)\,n(r)$; the last is the Slater local exchange potential (to be precise the potential $V^{(\text{HFS})}(r)$ should be appropriately corrected to preserve the asymptotic behaviour $-e^2/r$, when acting on occupied wave functions). Usually, the Numerov method is adopted to solve the radial Schrödinger equations; self-consistency of wave functions and potential can be achieved with reasonable modest computational labour. Complete calculations (programme code included) for all atoms is reported, for instance, by F. Herman and S. Skillman, *Atomic Structure Calculations* (Prentice Hall, Englewood Cliffs, New Jersey 1963).

## 19.12 DERIVATION OF HARTREE–FOCK EQUATIONS IN SECOND QUANTIZATION NOTATION

There really aren't many new ideas introduced in this section. Its purpose is to gain some familiarity with the second quantization notation for fermions. Of course, the idea of the variational principle will still have to be used.

According to Appendix G, if the Hamiltonian is of the form

$$\mathcal{H}(x_1, \dots x_n) = \sum_{i=1}^{N} \mathcal{H}(i) + \frac{1}{2}\sum_{i,j}{}' V(ij)^2$$

Then we can write

$$\mathcal{H} = \sum_{i,j} \mathcal{H}_{ij} a_i^\dagger a_j + \frac{1}{2} \sum_{i,j,l} V_{ij,kl} a_j^\dagger a_i^\dagger a_k a_l, \qquad \dots(1)$$

where the $\mathcal{H}_{ij}$ and the $V_{ij,kl}$ are matrix elements of the one- and two-body operators,

$$V_{ij,kl} = V_{ji,kl} \quad \text{and} \quad a_i a_j^\dagger + a_j^\dagger a_i = \delta_{ij}. \qquad \dots(2)$$

The rest of the anticommutators of the $a$ are zero.

We shall assume that the occupied states for the normalized ground state, $\Phi$ (which is a Slater determinant) that minimizes $\langle\Phi|\mathcal{H}|\Phi\rangle$ are labelled from 1 to $N$. For $\Phi$ giving a true extremum, as we saw in the Section on the Hartree approximation, we require only that

$$\langle\delta\Phi|\mathcal{H}|\Phi\rangle = 0. \qquad \dots(3)$$

It is easy to see that if $\langle\Phi|\Phi\rangle = 1$, then $|\Phi\rangle + |\delta\Phi\rangle$ is still normalized to first order in the variation. For example, let us assume that

$$|\delta\Phi\rangle = (\delta s)\, a_{k^1}^\dagger a_i 1\, |\Phi\rangle \quad \text{for} \quad k^1 > N,\ i^1 \le N, \qquad \dots(4)$$

where $\delta s$ is a small number and where all one-electron states up to the $N$th are occupied in the ground state of the electron system. That is, $|\delta\Phi\rangle$ differs from $|\Phi\rangle$ by having the electron in state $\Phi_i^1$ go to state $\Phi_k^1$. Then

$$\begin{aligned}(\langle\Phi| + \langle\delta\Phi|)\,(|\Phi\rangle + |\delta\Phi\rangle)&\\ &= (\langle\Phi| + \langle\Phi|\, a_{i^1}^\dagger a_{k^1}\, \delta s^*)\,(|\Phi\rangle + a_{k^1}^\dagger a_{i^1}\, \delta s|\Phi\rangle) \qquad \dots(5)\\ &= 1 + (\delta s)^* \langle\Phi|a_{i^1}^\dagger a_{k^1}|\Phi\rangle + \delta s\langle\Phi| + a_{k^1}^\dagger a_{i^1}|\Phi\rangle + O(\delta s)^2\\ &= 1 + O(\delta s)^2.\end{aligned}$$

According to the variational principle, we have as a basic condition.

$$0 = \langle\delta\Phi|\mathcal{H}|\Phi\rangle = (\delta s)^* \langle\Phi|\mathcal{H}a_{i^1}^\dagger a_{k^1}|\Phi\rangle. \qquad \dots(6)$$

Combining Eqs. (1) and (6) yields

---

[2] $\mathcal{H}(i)$ is one particle operator and $V(ij)$ is a two particle operator.

$$0 = \sum_{i,j} \mathcal{H}_{i,j} \langle\Phi| a^\dagger_{i^1} a_{k^1} a^\dagger_i a_j |\Phi\rangle$$
$$+ \frac{1}{2} \sum_{i,j,k,l} V_{ij,kl} \langle\Phi| a^\dagger_{i^1} a_{k^1}\, a^\dagger_j a^\dagger_i\, a_k a_l |\Phi\rangle \qquad ...(7)$$

where the summation is over all values of *i, j, k, l* (both occupied and unoccupied).

There are two basically different matrix elements to consider. To evaluate them we can make use of the anticommutation relations. Let us do the simplest one first. $\Phi$ has been assumed to be the Slater determinant approximation to the ground state, so

$$\langle\Phi| a^\dagger_{i^1} a_{k^1} a^\dagger_i a_j |\Phi\rangle = \langle\Phi| a^\dagger_{i^1} (\delta^i_{k^1} - a^\dagger_i a_{k^1}) a_j |\Phi\rangle$$
$$= \langle\Phi| a^\dagger_{i^1} a_j |\Phi\rangle \delta^i_{k^1} - \langle\Phi| a^\dagger_{i^1} a^\dagger_i a_{k^1} a_j |\Phi\rangle.$$

In the second term $a_{k^1}$ operating to the right gives zero (the only possible result of annihilating a state that isn't there). Since $a_j|\Phi\rangle$ is orthogonal to $a_{i^1}|\Phi\rangle$ unless $i^1 = j$, the first term is just $\delta^j_{i^1}$. Thus, we obtain

$$\langle\Phi| a^\dagger_{i^1} a_{k^1} a^\dagger_i a_j |\Phi\rangle = \delta^j_{i^1}\, \delta^i_{k^1}. \qquad ...(8)$$

The second matrix element in Eq. (7) requires a little more manipulation to evaluate

$$\langle\Phi| a^\dagger_{i^1} a_{k^1}\, a^\dagger_j a^\dagger_i a_k a_l |\Phi\rangle$$
$$= \langle\Phi| a^\dagger_{i^1} (\delta^j_{k^1} - a^\dagger_j a_{k^1}) a^\dagger_i a_k a_l |\Phi\rangle$$
$$= \delta^j_{k^1} \langle\Phi| a^\dagger_{i^1} a^\dagger_i a_k a_l |\Phi\rangle - \langle\Phi| a^\dagger_{i^1} a^\dagger_j a_{k^1} a^\dagger_i a_k a_l |\Phi\rangle$$
$$= \delta^j_{k^1} \langle\Phi| a^\dagger_{i^1} a^\dagger_i a_k a_l |\Phi\rangle - \langle\Phi| a^\dagger_{i^1} a^\dagger_j (\delta^i_{k^1} - a^\dagger_i a_{k^1}) a_k a_l |\Phi\rangle$$
$$= \delta^j_{k^1} \langle\Phi| a^\dagger_{i^1} a^\dagger_i a_k a_l |\Phi\rangle - \delta^i_{k^1} \langle\Phi| a^\dagger_{i^1} a^\dagger_j a_k a_l |\Phi\rangle$$
$$+ \langle\Phi| a^\dagger_{i^1} a^\dagger_j a^\dagger_i a_{k^1} a_k a_l |\Phi\rangle.$$

Since $a_{k^1}|\Phi\rangle = 0$, the last matrix element is zero. The first two matrix elements are both of the same form, so we need to evaluate only one of them

$$\langle\Phi| a^\dagger_{i^1} a^\dagger_i a_k a_l |\Phi\rangle = -\langle\Phi| a^\dagger_i a^\dagger_{i^1} a_k a_l |\Phi\rangle$$
$$= -\langle\Phi| a^\dagger_i (\delta^k_{i^1} - a_k a^\dagger_{i^1}) a_l |\Phi\rangle$$
$$= -\langle\Phi| a^\dagger_i a_l |\Phi\rangle \delta^k_{i^1} + \langle\Phi| a^\dagger_i\, a_k a^\dagger_{i^1} a_l |\Phi\rangle$$
$$= -\delta_i^{l \le N}\, \delta^k_{i^1} - \langle\Phi| a^\dagger_i a_k (\delta^l_{i^1} - a_l a^\dagger_{i^1}) |\Phi\rangle.$$

$a^\dagger_{i^1}|\Phi\rangle$ is zero since this tries to create a fermion in an already occupied state. So

$$\langle\Phi| a^\dagger_{i^1} a^\dagger_i a_k a_l |\Phi\rangle = -\delta_i^{l \le N}\, \delta^k_{i^1} + \delta^l_{i^1}\, \delta_i^{k \le N}.$$

Combining with previous results, we finally find

$$\langle\Phi| a^{\dagger}_{i^1}\, a_{k^1} a^{\dagger}_{j}\, a^{\dagger}_{i}\, a_k a_l |\Phi\rangle = \delta^{j}_{k^1}\, \delta^{l}_{i^1}\, \delta^{k\le N}_{i} - \delta^{j}_{k^1}\, \delta^{l\le N}_{i}\, \delta^{k}_{i^1}$$

$$-\delta^{i}_{k^1}\, \delta^{l}_{i^1}\, \delta^{k\le N}_{j} - \delta^{j}_{k^1}\, \delta^{l\le N}_{j}\, \delta^{k}_{i^1}. \qquad ...(9)$$

Combining Eqs. (7), (8) and (9), we have

$$0 = \sum\nolimits_{i,j} \mathcal{H}_{i,j}\, \delta^{j}_{i^1}\, \delta^{i}_{k^1}$$

$$+ \frac{1}{2} \sum\nolimits^{N}_{ijkl} V_{ij,kl}\, (\delta^{j}_{k^1}\, \delta^{l}_{i^1}\, \delta^{k}_{i} + \delta^{i}_{k^1}\, \delta^{l}_{j}\, \delta^{k}_{i^1} - \delta^{j}_{k^1}\, \delta^{l}_{i}\, \delta^{k}_{i^1} - \delta^{i}_{k^1}\, \delta^{l}_{i^1}\, \delta^{k}_{j}),$$

or $$0 = \mathcal{H}_{k^1 i^1}$$

$$+ \frac{1}{2} \left(\sum\nolimits^{N}_{i=1} V_{ik^1,\, ii^1} + \sum\nolimits^{N}_{j=1} V_{k^1 j,\, i^1 j} - \sum\nolimits^{N}_{i=1} V_{ik^1,\, i^1 i} - \sum\nolimits^{N}_{j=1} V_{k^1 j,\, j i^1}\right).$$

By using the symmetry in the $V$ and making dummy changes in summation variables this can be written as

$$0 = \mathcal{H}_{k^1 i^1} + \sum\nolimits^{N}_{j=1} (V_{k^1 j, i^1 j} - V_{k^1 j, j i^1}). \qquad ...(10)$$

Equation (10) suggests a definition of a one-particle operator called the self-consistent one-particle Hamiltonian

$$\mathcal{H}_C = \sum\nolimits_{ki} [\mathcal{H}_{ki} + \sum\nolimits^{N}_{j=1}(V_{kj,ij} - V_{kj,ji})]\, a^{\dagger}_{k} a_i. \qquad ...(11)$$

At first glance we might think that this operator is identically zero by comparing it to Eq. (10). But in Eq. (11) $k^1 > N$ and $i^1 < N$, whereas in Eqs. (10), (11) there is no such restriction.

An important property of $\mathcal{H}_C$ is that it has no matrix elements between occupied ($i^1$) and normally unoccupied ($k^1$) levels. Letting $\mathcal{H}_C = \sum_{ki} f_{ki}\, a^{\dagger}_{k} a_i$, we have

$$\langle k^1 | \mathcal{H}_C | i^1\rangle = \sum\nolimits_{ki} f_{ki} \langle k^1 | a^{\dagger}_{k} a_i | i^1\rangle$$

$$= \sum\nolimits_{ki} f_{ki} \langle 0 | a_{k^1} a^{\dagger}_{k} a^{\dagger}_{i^1} | 0\rangle$$

$$= \sum\nolimits_{ki} f_{ki} \langle 0 | a^{\dagger}_{k} a_{k^1} - \delta^{k}_{k^1})(a^{\dagger}_{i^1} a_{i^1} - \delta^{i}_{i^1}) | 0\rangle.$$

Since $a_i\, |0\rangle = 0$, we have

$$\langle k^1 | \mathcal{H}_C | i^1\rangle = +f_{k^1 i^1} = 0$$

by the definition of $f_{ki}$ and (10).

We have shown that $\langle\delta\Phi|\mathcal{H}|\Phi\rangle = 0$ (for $\Phi$ constructed by Slater determinants) if, and only if, Eq. (10) is satisfied, which is true if, and only if, $\mathcal{H}_C$ has no matrix elements between occupied ($i^1$) and unoccupied ($k^1$) levels. Thus, in a matrix representation $\mathcal{H}_C$ is in block diagonal form since all $\langle i^1 | \mathcal{H} | k^1\rangle = \langle k^1 | \mathcal{H} | i^1\rangle = 0$. $H_C$ is Hermitian, so that it can be diagonalized. Since it is already in block diagonal form, each block can be separately diagonalized. This means that the new occupied levels are linear combinations of the old occupied levels only and the new occupied levels are linear combinations of the old unoccupied levels only. By new levels we mean those levels that have wave functions $\langle i|$, $\langle j|$ such that $\langle i|\mathcal{H}_C|j\rangle$ vanishes unless $i = j$.

Using this new set of levels, we can say

$$\mathcal{H}_C = \sum_i \varepsilon_i a_i^\dagger a_i. \qquad \text{...(12)}$$

In order that Eqs. (12) and (11) are equivalent, we have

$$\mathcal{H}_{ki} + \sum_{j=1}^{N} (V_{kj,\, ij} - V_{kj,\, ji}) = \varepsilon_i \delta_{ki}. \qquad \text{...(13)}$$

These equations are the Hartree–Fock equations. (We have established that $\langle\delta\Phi|\mathcal{H}|\Phi\rangle = 0$ (for $\Phi$ a Slater determinant) implies Eq. (13). It is also true that the set of one-electron wave functions for which Eq. (13) is true minimizes $\langle\Phi|\mathcal{H}|\Phi\rangle$, where $\Phi$ is restricted to be a Slater determinant of the one-electron functions.

## 19.12.1 Hermitian Nature of the Exchange Operator

In this section, the Hartree–Fock "Hamiltonian" will be proved to be Hermitian. If the Hartree–Fock Hamiltonian, in addition, has nondegenerate eigenfunctions, then we are guaranteed that the eigenfunctions will be orthogonal. Regardless of degeneracy, the orthogonality of the eigenfunctions was built into the Hartree–Fock equations from the very beginning. More importantly, perhaps, the Hermitian nature of the Hartree–Fock Hamiltonian guarantees that its eigenvalues are real. They have to be real. Otherwise, Koopmans' theorem would not make sense.

The Hartree–Fock Hamiltonian is defined as that operator $\mathcal{H}^F$ for which

$$\mathcal{H}^F u_k = \varepsilon_k u_k. \qquad \text{...(14)}$$

$\mathcal{H}^F$ is then defined by (taking care of the spin summations) as

$$\mathcal{H}^F = \mathcal{H}_1 + \sum_j \int \psi_j^*(r_2) V(1,2) \psi_j(r_2)\, d\tau_2 + A_1, \qquad \text{...(15)}$$

where
$$A_1 \psi_k(r_1) = -\sum_{j(\|k)} \int \psi_j^*(r_2) V(1,2) \psi_k(r_2)\, d\tau_2\, \psi_j(r_1),$$

and $A_1$ is called the *exchange operator.*

For the Hartree–Fock Hamiltonian to be Hermitian we have to prove that

$$\langle i|\mathcal{H}^F|j\rangle = \langle j|\mathcal{H}^F|i\rangle^*. \qquad \text{...(16)}$$

This property is obvious for the first two terms on the right-hand side of Eq. (15) and so needs only to be proved for $A_1$:

$$\begin{aligned}
\langle l|A_l|m\rangle^* &= -\left(\sum_{j(\|m)} \int \psi_l^*(r_1) \int \psi_j^*(r_2) V(1,2) \psi_m(r_2) \psi_j(r_1)\, d\tau_2\, d\tau_1\right)^* \\
&= -\left(\sum_{j(\|m)} \int \psi_l^*(r_1) \psi_j(r_1) \int \psi_j^*(r_2) V(1,2) \psi_m(r_2)\, d\tau_2\, d\tau_1\right)^* \\
&= -\left(\sum_{j(\|m)} \int \psi_m(r_1) \psi_j^*(r_1) \int \psi_j(r_2) V(1,2) \psi_l^*(r_2)\, d\tau_2\, d\tau_1\right)^* \\
&= \langle m|A_l|l\rangle.
\end{aligned}$$

In the proof,[3] use has been made of changes of dummy integration variable and of the relation, $V(1, 2) = V(2, 1)$.

### 19.12.2 The Fermi Hole

The exchange term (when the interaction is the Coulomb interaction energy and $e$ is the magnitude of the charge on the electron) is

$$A_1\psi_i(r_1) \equiv -\sum_{j(\|i)} \int \frac{e^2}{4\pi\varepsilon_0 r_{12}} \psi_j^*(r_2)\psi_i(r_2)\,d\tau_2 \cdot \psi_j(r_1)$$

$$= -\sum_{j(\|i)} \int \frac{e}{4\pi\varepsilon_0 r_{12}} \left(\frac{e\psi_j^*(r_2)\psi_i(r_2)\psi_j(r_1)}{\psi_i(r_1)}\right)\psi_i(r_1)\,d\tau_2$$

$$A_1\psi_i(r_1) = \int \frac{(-e)}{4\pi\varepsilon_0 r_{12}} \rho(r_1, r_2)\,\psi_i(r_1)\,d\tau_2, \qquad ...(17)$$

where $$\rho(r_1, r_2) = \frac{e\sum_{j(\|i)} \psi_j^*(r_2)\psi_i(r_2)\psi_j(r_1)}{\psi_i(r_1)}.$$

From Eq. (17) we see that exchange can be interpreted as the potential energy of interaction of an electron at $r_1$ with a charge distribution with charge density, $\rho(r_1, r_2)$. This charge distribution is a mathematical rather than a physical charge distribution.

Several comments can be made about the exchange charge density, $\rho(r_1, r_2)$:

1. $$\int \rho(r_1, r_2)\,d\tau_2 = +e\int \sum_{j(\|i)} \psi_j^*(r_2)\,d\tau_2 \cdot \frac{\psi_j(r_1)}{\psi_i(r_1)}$$

$$= e\sum_{j(\|i)} \delta_i^j \cdot \frac{\psi_j(r_1)}{\psi_i(r_1)} = +e. \qquad ...(18)$$

Thus, we can think of the total exchange charge as being of magnitude, $+e$.

2. $\rho(r_1, r_2) = e\sum_{J(\|i)} |\psi_j(r_1)|^2$, which has the same magnitude and opposite sign of the charge density of parallel spin electrons.
3. From Eqs. (1) and (2), we can conclude that $|\rho|$ must decrease as $r_{12}$ increases.
4. It is convenient to think of the Fermi hole and exchange charge density in the following way: in $\mathcal{H}^F$, neglecting for the moment $A_1$, the potential energy of the electron is the potential energy due to the ion cores and all the electrons. Thus, the electron interacts with itself in the sense that it interacts with a charge density constructed from its own wave function. The exchange term cancels this unwanted interaction in a sense, but it cancels it out locally. That is, the exchange term ($A_1$) cancels the potential energy of

[3] A matrix elements in Eq. (16) would vanish of $i$ and $j$ did not refer to spin states which were parallel.

interaction of electrons with parallel spin in the neighbourhood of the electron with given spin. Pictorially, we say that the electron with given spin is surrounded by an exchange charge hole (or Fermi hole of charge $+e$).

The idea of the Fermi hole still does not include the description of the *Coulomb correlations* between electrons due to their mutual repulsion. In this respect the Hartree–Fock method is no better than the Hartree method. In the Hartree method, the electrons move in a field that depends only on the average charge distribution of all other electrons. In the Hartree–Fock method, the only correlations included are those that arise because of the Fermi hole, and these are simply due to the fact that the Pauli principle does not allow two electrons with parallel spin to have the same spatial co-ordinates. We could call these kinematic correlations (due to constraints) rather than dynamic correlations (due to forces).

## 19.13 THE THOMAS – FERMI METHOD

Using the fact that electrons obey the Fermi–Dirac statistics, Thomas and Fermi (in 1926 and 1928) developed a simple method to compute the approximate form of the average potential, $V(r)$, acting on the atomic electrons. This model is also useful as the starting point of more sophisticated calculations.

The mathematical difficulties of a numerical solution of the set of integro-differential equations of Hartree method increase enormously when the number of electrons in the atom is larger. One can then use the statistical method suggested by Thomas and Fermi to describe the basic features of the electron distribution and the field in complex atoms.

The Thomas – Fermi statistical model assumes that the potential function, $V(r)$, produced by the nucleus and all the other electrons, (except the one whose motion is under consideration) is spherically symmetric and varies slowly enough in an electron wavelength so that many electrons can be localized within a volume over which the potential changes only by a small fraction of itself the electrons can, then be treated by statistical mechanics, obeying Fermi-Dirac statistics.

### 19.13.1 Thomas – Fermi Model

The Thomas Fermi model is appropriate to atoms with sufficient number of electrons which permits the systems to be treated in a statistical sense. The model views an atom as a spherically symmetric gas of $Z$ electrons which surrounds a nucleus of charge Ze. It is further assumed that the potential in the medium $\Phi(r)$ varies sufficiently slowly so that it may be taken as constant in small volumes of the electron gas. Then the concept of Fermi energy is applicable. We can equate the Fermi energy to the maximum kinetic energy of an electron in an aggregate of electrons at zero kelvin. This description holds in the present model also because at normal temperature $k_B T \ll \Phi(r)$.

The relation between $E_F$ and electron density $n(r)$ is given by

$$E_{\max}^{\text{kin}} \equiv E_F = \frac{\hbar^2}{2m}[3\pi^2 n(r)]^{2/3} \qquad \text{...(1)}$$

The energy of an electron in the medium at a distance $r$ from the nucleus is

$$E(r) = \frac{p^2}{2m} - e\Phi(r) \qquad ...(2)$$

where we have used $e = |e|$.

Equation (2) indicates that for $\frac{p^2}{2m} > e\phi\,(r)$ $E > 0$ which implies that the election escapes the atom. As this does not occur in the present model, we may conclude that maximum kinetic energy may also be set equal to $e\phi$. With Eq. (1) we may write

$$e\phi\,(r) = \frac{\hbar^2}{2m}\,[3\pi^2 n(r)]^{2/3} \qquad ...(3)$$

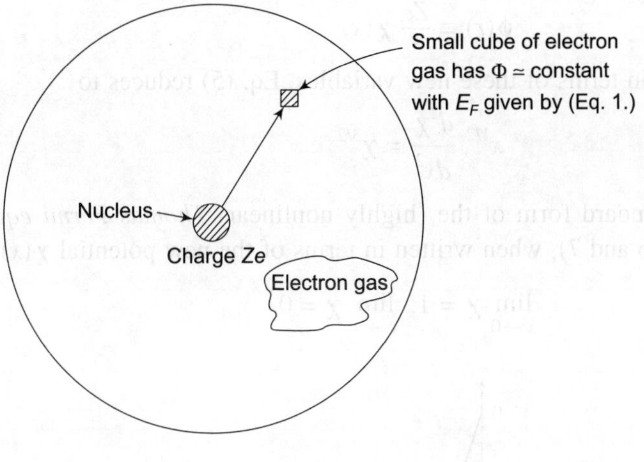


**Figure 19.2** The Thomas–Fermi statistical atom

An additional relation between the potential $\phi(r)$ and density $n(r)$ is given by Poisson's equation

$$\nabla^2\phi = 4\pi en \quad \text{(CGS units)} \qquad ...(4)$$

Working in the given spherically symmetric geometry, and eliminating $n(r)$ from the latter two equations gives

$$\nabla^2\phi = \frac{1}{r}\frac{d^2}{dr^2}(r\phi) = \frac{4\pi e}{3\pi^2}\frac{(2me)^{3/2}}{\hbar^3}\,\phi^{3/2} \qquad ...(5)$$

This is the desired single equation for $\phi(r)$.

The following boundary conditions apply to the potential.

Near the origin $\phi(r)$ is due, predominantly to the nucleus, and we write

$$\phi \simeq \frac{Ze}{r}$$

or equivalently

$$\lim_{r \to 0} r\phi(r) = Ze \qquad ...(6)$$

At large $r$, we obtain

$$\lim_{r \to \infty} r\phi(r) = 0. \qquad ...(7)$$

The resulting Eq. (5) may be written in a more concise form in terms of new dimensionless variables $(x, \chi)$, defined by

$$r = ax \qquad ...(8)$$

$$a \equiv \frac{1}{2}\left(\frac{3\pi}{4}\right)^{2/3} \frac{\hbar^2 {}^* m e^2}{Z^{1/3}} = \frac{0.885}{Z^{1/3}} a_0. \qquad ...(9)$$

$$\phi(r) \equiv \frac{Ze}{r} \chi(x) \qquad ...(10)$$

When written in terms of these new variables, Eq. (5) reduces to

$$x^{1/2} \frac{d^2\chi}{dx^2} = \chi^{3/2} \qquad ...(11)$$

This is the standard form of the (highly nonlinear) *Thomas-Fermi equation*. Boundary conditions (Eqs. 6 and 7), when written in terms of the new potential $\chi(x)$ appear as

$$\lim_{x \to 0} \chi = 1, \lim_{x \to \infty} \chi = 0 \qquad ...(12)$$

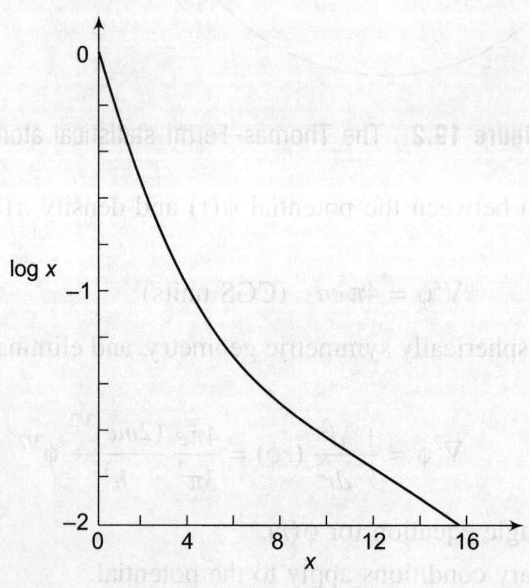


**Figure 19.3** Semilogarithmic plot of solution to the Thomas–Fermi eq. (11) subject to the boundary conditions (12)

The semilogarithmic plot shown in Fig. 19.3 reveals the rapid decay of the potential $\chi$ with increase in $x$. Thus, for example, for $x > 4$, $\chi < 0.1$.[4]

## 19.14 DENSITY FUNCTIONAL THEORY (DFT)

In contrast to the previous approaches, DFT is a *ground-state* theory in which the emphasis is on the *charge density* as the relevant physical quantity. DFT in the local density approximation (LDA) has proved to be highly successful in describing structural and electronic properties in a vast class of materials. Furthermore, LDA is computationally (relatively) simple. For these reasons LDA has become a common tool in first principle calculations aimed at describing or even predicting properties of complex condensed matter systems and reliable predictions of structures and energy differences in molecules or clusters can be made.

As in the Thomas–Fermi approach, the focus of the DF formalism is on the electron density $n(\vec{r})$, rather than the wave function of the system.

The basic theorems of the DF formalism were derived by Hohenberg and Kohn (1964). They showed that

(1) Ground-state (GS) properties of a system of electrons and ions in an external field $V_{\text{ext}}$ can be determined from the electron density $n(\vec{r})$ alone.

(2) The total energy ($E$) is such a functional of the density, and $E[n]$ satisfies the variational principle.

$$E[N] \geq E_{GS}.$$

The density for which the equality holds is the ground state density, $n_{GS}$.

### 19.14.1 (DF) Formalism

The usual implementation of the formalism results from the observation by Kohn and Sham (1965) that the minimization of $E[n]$ is simplified if we write (using atomic units $c = \hbar = m = 1$) as

$$E[n] = T_0[n] + \int d\vec{r}\, n(\vec{r})\left(V_{\text{ext}}(\vec{r}) + \frac{1}{2}\phi(\vec{r})\right) + E_{xc}[n] \quad ...(1)$$

where $T_0$ is the kinetic energy that a system with density ($n$) would have in the absence of electron-electron interactions, $\phi(\vec{r})$ is the Coulomb potential, and $E_{xc}$ defines the exchange-correlation energy.

The variational principle yields

$$\frac{\delta E[n]}{\delta n(\vec{r})} = \frac{\delta T_0}{\delta n(\vec{r})} + V_{\text{ext}}(\vec{r}) + \Phi(\vec{r}) + \frac{\delta E_{xc}[n]}{\delta n(\vec{r})} = \mu \quad ...(2)$$

where $\mu$ is the Lagrange multiplier associated with the requirement of constant particle number. If we compare this with the corresponding equation for a system with an effective potential $V(\vec{r})$ but without electron-electron interactions,

[4] Numerical integration of the Thomas–Fermi Eq. (11) was originally performed by V. Bush and S. H. Caldwell, *Phys. Rev.* **38**, 1898 (1931).

$$\frac{\delta E[n]}{\delta n(\vec{r})} = \frac{\delta T_0}{\delta n(\vec{r})} + V(\vec{r}) = \mu \qquad ...(3)$$

We see that the mathematical problems (2) and (3) are identical, provided that

$$V(\vec{r}) = V_{\text{ext}} + \Phi(\vec{r}) + \frac{\delta E_{\text{xc}}[n]}{\delta n(\vec{r})} \qquad ...(4)$$

The solution of Eq. (3) can be found by solving the Schrödinger equation for non-interacting particles,

$$\left(-\frac{1}{2}\nabla^2 + V(\vec{r})\right) \psi_i(\vec{r}) = \varepsilon_i \psi_i(\vec{r}) \qquad ...(5)$$

yielding $n(\vec{r}) = \sum_{i=1}^{N} |\psi_i(\vec{r})|^2$ ...(6)

It is necessary to satisfy the condition, Eq. (3), and this can be achieved in a self-consistent procedure. From the solutions $\psi_i$ we determine the density and the related potential, and then we solve Eq. (5) again to determine new values of $\psi_i$. This cycle is repeated until the input and output solutions are identical.

From the solution of a single-particle equation we can determine the total energy and density of the lowest state of a system of electrons and ions, and all quantities derivable from them. In contrast to the Hartree–Fock potential, the effective potential $V(\vec{r})$ is *local*.

## 19.15 THE DENSITY FUNCTIONAL THEORY AND THE KOHN–SHAM EQUATIONS

In the previous sections, the many-electron problem has been attacked by approximating as better as possible the exact ground-state many-electron wavefunction, $\Psi_G(\mathbf{r}_1\sigma_1, ..., \mathbf{r}_N\sigma_N)$.In the density functional theory the emphasis shifts from the ground-state wave function to the much more manageable ground-state *one-body electron density,* $n(\mathbf{r})$. The density functional theory shows that the ground-state energy of a many-particle system can be expressed as a functional of the one-body density; minimization of this functional allows, in principle, the determination of the actual ground-state density. The success of the theory is also to provide a reasonable approximation of the functional to be minimized. The peculiarity of the density functional approach to the many-body theory, is to attain rigorously a one-electron Schrödinger equation with a local effective potential in the study of the ground-state properties of the many-electron systems.

### 19.15.1 The Hohenberg–Kohn Theorem

Consider a system of $N$ electrons, described by the standard many-electron Hamiltonian, $H_e$. For our reasoning, it is convenient to decompose $H_e$ into the sum of an "internal" part (kinetic energy of the electrons plus electron-electron Coulomb interactions) and "external part (in our specific case the external part is given by the electronic-nuclear interactions). With obvious rearrangement of notations, we have

$$H_e = H_{\text{int}} + V_{\text{ext}} \quad \text{...(1a)}$$

$$H_{\text{int}} = T + V_{\text{ee}} = \sum_i \frac{\mathbf{p}_i^2}{2m} + \frac{1}{2} \sum_{i \neq j} \frac{e^2}{|\mathbf{r}_i - \mathbf{r}_j|} \quad \text{...(1b)}$$

$$V_{\text{ext}} = \sum_i v_{\text{ext}}(\mathbf{r}_i) \text{ with } v_{\text{ext}}(\mathbf{r}) \equiv V_{\text{nucl}}(\mathbf{r}) = -\sum_I \frac{z_I e^2}{|\mathbf{r} - \mathbf{R}_I|}. \quad \text{...(1c)}$$

For simplicity, we suppose that the ground-state $|\Psi_G\rangle$ is non-degenerate (in principle, any degeneracy can be removed by an arbitrary small perturbation that appropriately lowers the symmetry of the system).

Let us consider as only variable of the many-electron problem the external potential, $v_{\text{ext}}(\mathbf{r})$; the mass of the electrons, their charge, their number $N$, the form of the internal interactions are vice versa supposed to be fixed. The Hohenberg–Hohn theorem states that *there is a one-to-one correspondence between the ground-state density of an N electron system and the external potential acting on it*; in this sense the ground-state electron density becomes the variable of interest.

The first part of the theorem is almost trivial. Suppose, in fact, that we know $v_{\text{ext}}(\mathbf{r})$ and hence the total Hamiltonian ($H_e$) of the system. Thus, we know exactly (at least in principle) all its eigenfunctions and eigenvalues; in particular we know the ground-state, $\Psi_G(\mathbf{r}_1, \mathbf{r}_2, ..., \mathbf{r}_N)$. [In the following, for simplicity, we confine our attention to systems without spin polarization, and do not indicate explicitly the spin-variables, since this is not essential to the "abstract" reasoning we are going to present]. The ground-state ($\Psi_G$) depends on the chosen external potential, $v_{\text{ext}}(\mathbf{r})$, and is also denoted with the short-hand functional notation, $\Psi_G[v_{\text{ext}}]$. From the knowledge of $\Psi_G$ we obtain the one-body ground-state density, $n(\mathbf{r})$, defined as

$$n(\mathbf{r}) = \langle \Psi_G(\mathbf{r}_1, \mathbf{r}_2, ..., \mathbf{r}_N)| \sum_i \delta(\mathbf{r} - \mathbf{r}_i) |\Psi_G(\mathbf{r}_1, \mathbf{r}_2, ..., \mathbf{r}_N)\rangle. \quad \text{...(2)}$$

Incidentally, we notice the property

$$\langle \Psi_G | V_{\text{ext}} | \Psi_G \rangle = \int n(\mathbf{r}) v_{\text{ext}}(\mathbf{r})\, d\mathbf{r}, \quad \text{...(3)}$$

which follows from definition, Eq. (50) and from the fact that the potential ($V_{\text{ext}}$) can be written as

$$V_{\text{ext}} = \sum_i \int v_{\text{ext}}(\mathbf{r}) \delta(\mathbf{r} - \mathbf{r}_i)\, d\mathbf{r}.$$

We can summarize the contents of the previous reasoning by the following scheme:

$$v_{\text{ext}}(\mathbf{r}) \Rightarrow \Psi_G\, [v_{\text{ext}}] \Rightarrow n(\mathbf{r}).$$

Thus, the knowledge of $v_{\text{ext}}(\mathbf{r})$ entails the knowledge of $\Psi_G\, [v_{\text{ext}}]$ and hence the knowledge of $n(\mathbf{r})$. In other words there exists a functional that links $n(\mathbf{r})$ and $v_{\text{ext}}(\mathbf{r})$, and we write

$$n(\mathbf{r}) = F[v_{\text{ext}}]. \quad \text{...(4)}$$

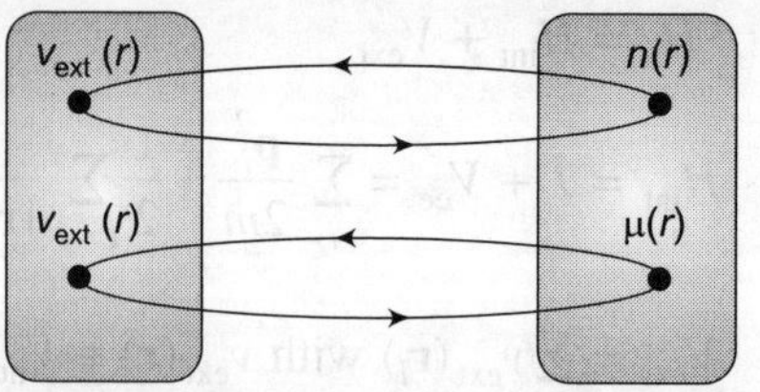


**Figure 19.4** Schematic representation of the Hohenberg–Kohn theorem. Two different ground-state density functions must correspond to two different potentials, and vice versa

It is also evident that two external potentials, which differ by a constant in the whole space, lead to the same $n(\mathbf{r})$ (and can be considered essentially the same, upon appropriate shift by an additive constant). The conceptual scheme of this reasoning is indicated in Fig. 19.4, focusing on $v_{ext}(\mathbf{r})$ and following the arrows towards right. The real novelty of the theorem is that the functional relation, Eq. (4) can be inverted in the form

$$\boxed{v_{ext}(\mathbf{r}) = G[n(\mathbf{r})]}, \quad ...(5)$$

which means that from the knowledge of the ground-state density $n(\mathbf{r})$ we can determine uniquely the external potential (to within a non-essential additive constant) and thus the Hamiltonian of the system.

In order to prove Eq. (5), we should establish that for any given pair of external potentials, $v_{ext}(\mathbf{r})$ and $\bar{v}_{ext}(\mathbf{r})$, such that $v_{ext}(\mathbf{r}) \neq \bar{v}_{ext}(\mathbf{r})$ we have $n(\mathbf{r}) \neq \bar{n}(\mathbf{r})$. This follows from the minimum property of the ground-state energy. Consider, in fact, the two Hamiltonians

$$H = H_{int} + V_{ext} \quad ...(6a)$$

$$\overline{H} = H_{int} + \overline{V}_{ext}, \quad ...(6b)$$

and their ground-states, $|\Psi_G\rangle$ and $|\overline{\Psi}_G\rangle$, with eigenvalues $E_G$ and $\overline{E}_G$ respectively. We have

$$\langle\overline{\Psi}_G|H|\overline{\Psi}_G\rangle = \langle\overline{\Psi}_G|H_{int} + \overline{V}_{ext} + V_{ext} - \overline{V}_{ext}|\overline{\Psi}_G\rangle$$

$$= \overline{E}_G + \langle\overline{\Psi}_G|V_{ext} - \overline{V}_{ext}|\overline{\Psi}_G\rangle$$

$$= \overline{E}_G + \int \bar{n}(\mathbf{r})\,[v_{ext}(\mathbf{r}) - \bar{v}_{ext}(\mathbf{r})]\,d\mathbf{r}$$

where the last passage follows from Eq. (3). Since the ground eigenvalue of $H$ is strictly lower than the mean value of the Hamiltonian on any other state, we have

$$E_G < \overline{E}_G + \int \bar{n}(\mathbf{r})\,[v_{ext}(\mathbf{r}) - \bar{v}_{ext}(\mathbf{r})]\,d\mathbf{r}. \quad ...(7a)$$

From a similar reasoning using $\langle\Psi_G|\overline{H}|\Psi_G\rangle$, one obtains the above expression with barred and unbarred quantities exchanged

$$\overline{E}_G < E_G + \int n(\mathbf{r})\,[\bar{v}_{ext}(\mathbf{r}) - v_{ext}(\mathbf{r})]\,d\mathbf{r}. \quad ...(7b)$$

Equation (7a) and Eq. (7b) are obviously incompatible in the hypothesis $n(\mathbf{r}) = \bar{n}(\mathbf{r})$. Therefore for $v_{ext}(\mathbf{r}) \neq \bar{v}_{ext}(\mathbf{r})$ it must be $n(\mathbf{r}) \neq \bar{n}(\mathbf{r})$, which completes the proof of the Hohenberg–Kohn theorem (schematically represented in Fig. 19.4)

The Hohenberg–Kohn theorem has far reaching consequences, that require simple (although subtle) elaborations. First of all, it is convenient to define other functionals of the ground-state density of an electron system. We start from the following (obvious) functionals of the external potential $v_{ext}(\mathbf{r})$

$$v_{ext}(\mathbf{r}) \Rightarrow \Psi_G[v_{ext}] \begin{cases} \Rightarrow & E[v_{ext}] \\ \Rightarrow & T[v_{ext}] \\ \Rightarrow & V_{ee}[v_{ext}] \end{cases} \quad \text{...(8)}$$

The above logical sequence allows one to state that the ground-state energy is a functional $E[v_{ext}]$ of the external potential $v_{ext}(\mathbf{r})$; the expectation value of the electron kinetic energy on the ground-state wave function is a functional $T[v_{ext}]$, and the expectation value of the electron-electron interaction on the ground-state wave function is a functional, $V_{ee}[v_{ext}]$.

Because of the Hohenberg–Kohn theorem, $v_{ext}(\mathbf{r})$ and $n(\mathbf{r})$ are in one-to-one correspondence. We thus have that the ground-state energy is a functional $E[n]$ of the ground-state density, that the expectation value of the electronic kinetic energy is a functional, $T[n]$, of the ground-state density and so is also the expectation value, $V_{ee}[n]$, of the electron-electron interaction.

The most important consequence of the Hohenberg–Kohn theorem is the formulation of a variational principle concerning the ground-state density of a system. Consider a system of interacting electrons in a given nuclear external potential, denoted by $v_{ext}(\mathbf{r})$. Following Hohenberg and Kohn, we consider the following functional

$$E^{(HK)}[n(\mathbf{r}); v_{ext}(\mathbf{r})] \equiv \langle \Psi_G[n] | T + V_{ee} + V_{ext} | \Psi_G[n] \rangle, \quad \text{...(9)}$$

where $v_{ext}(\mathbf{r})$ is taken fixed, $n(\mathbf{r})$ is allowed to vary, and $\Psi_G[n]$ is the ground-state of the system with ground-state density, $n(\mathbf{r})$. It is evident that the absolute minimum of the energy functional, Eq. (9) occurs when $\Psi_G[n]$ is the ground-state of the operator $H_e = T + V_{ee} + V_{ext}$, i.e., when $n(\mathbf{r})$ is the exact electron density of the system.

The properties of the Hohenberg–Kohn energy functional, Eq. (9), here rewritten in the form

$$\boxed{E^{(HK)}[n(\mathbf{r}); v_{ext}(\mathbf{r})] = T[n(\mathbf{r})] + V_{ee}[n(\mathbf{r})] + \int v_{ext}(\mathbf{r}) n(\mathbf{r}) d\mathbf{r}} \quad \text{...(10)}$$

can be summarized as follows. The energy functional, $E^{(HK)}[n(\mathbf{r}); v_{ext}(\mathbf{r})]$, which exists and is unique, is *minimal at the exact ground-state density,* and *its minimum gives the exact ground-state energy* of the many-body electron system. We also notice that the functional $F[n] = T[n] + V_{ee}[n]$ is *universal*, i.e., it does not depend on $v_{ext}$; this good news, however, is no compensation for the fact that the functional $F[n]$ is not known explicitly, and must be appropriately approximated.

### The Kohn–Sham Equations

The Kohn–Sham equations are obtained by minimizing the functional, Eq. (10) with respect to $n(\mathbf{r})$. To carry out explicitly the variational procedure, we decompose the ground-state density $n(\mathbf{r})$ of an interacting electron system into the sum of $N$ independent orbital contributions of the form

$$n(\mathbf{r}) = \sum_i \phi_i^*(\mathbf{r})\phi_i(\mathbf{r}), \qquad \text{...(11)}$$

where $\{\phi_i(\mathbf{r})\}$ (1, 2, ..., $N$) are orthonormal orbitals.

We can argue that the decomposition, Eq. (11) is exact and unique for any chosen (well behaved) density, $n(\mathbf{r})$. In fact, we can always imagine a fictitious *system of non-interacting electrons, whose ground-state density* $n_0(\mathbf{r})$ *equals* $n(\mathbf{r})$. The Hohenberg–Kohn theorem applies to any electron system, whether interacting or not. For the system of non-interacting electrons of density $n_0(\mathbf{r})$, the Hohenberg–Kohn theorem guarantees the existence and unicity of an external potential $v^0_{\text{ext}}(\mathbf{r})$, producing the ground-state density, $n_0(\mathbf{r})$. If $\{\phi_i(\mathbf{r})\}$ ($i$ = 1, 2, ..., $N$) denote the $N$ orbitals of lowest energy of the system of non-interacting electrons, the ground-state wave function is just the Slater determinant $A\{\phi_1 \phi_2 ... \phi_N\}$ and the electron density is given by $n_0(\mathbf{r}) = \sum_i \phi_i^*(\mathbf{r})\phi_i(\mathbf{r})$. Since $n_0(\mathbf{r})$ and $n(\mathbf{r})$ have been taken equal, the decomposition, Eq. (11) is fully justified also for the interacting electrons, regardless of the fact that *the exact ground-state of an interacting electron system is not the Slater determinant formed with the orbitals contributing to* $n(\mathbf{r})$.

Before adopting the variational procedure on the functional, Eq. (10), it is convenient to extract from it the inter-electron Coulomb interaction, $V_{\text{H}}[n]$ (called Hartree potential)

$$V_{\text{H}}[n] = \frac{1}{2}\int n(\mathbf{r})\frac{e^2}{\mathbf{r}\;\mathbf{r}'|}n(\mathbf{r}')\,d\mathbf{r}\,d\mathbf{r}' \equiv \frac{1}{2}\sum_{ij}\left\langle \phi_i\phi_j\left|\frac{e^2}{r_{12}}\right|\phi_i\phi_j\right\rangle,$$

and the kinetic energy, $T_0[n]$ (of a system of non-interacting electrons with the same density) defined as

$$T_0[n] = \sum_i \left\langle \phi_i(\mathbf{r})\left|-\frac{\hbar^2\nabla^2}{2m}\right|\phi_i(\mathbf{r})\right\rangle.$$

The Hohenberg–Kohn functional, Eq. (10) can be written as

$$E^{(\text{HK})}[n(\mathbf{r}); v_{\text{ext}}(\mathbf{r})] = T_0[n] + V_{\text{H}}[n] + \int v_{\text{ext}}(\mathbf{r})n(\mathbf{r})d\mathbf{r} + E_{\text{xc}}[n], \qquad \text{...(12a)}$$

where the *exchange-correlation functional*, $E_{\text{xc}}[n]$, is defined as

$$E_{\text{xc}}[n] = T[n] - T_0[n] + V_{\text{ee}}[n] - V_{\text{H}}[n]. \qquad \text{...(12b)}$$

The functional, Eq. (12a) can be recast in the form

$$\boxed{E^{(\text{HK})}[n(\mathbf{r}); v_{\text{ext}}(\mathbf{r})] = \sum_i \left\langle \phi_i\left|\frac{\hbar^2\nabla^2}{2m} + v_{\text{ext}}\right|\phi_i\right\rangle + \frac{1}{2}\sum_{ij}\left\langle \phi_i\phi_j\left|\frac{e^2}{r_{12}}\right|\phi_i\phi_j\right\rangle + E_{\text{xc}}[n]} \qquad \text{...(13)}$$

According to the standard variational procedure, we vary the $N$ contributing orbitals $\phi_1$, $\phi_2$, ..., $\phi_N$ so to make stationary the energy functional, Eq. (13), under the constraint of orthonormalization of the wave functions $\{\phi_i\}$. The variational expression for the first two terms in the right-hand side of Eq. (13) is trivial and need not to be repeated here. The only novelty from a technical point of view is thus calculation of the variation, $\delta E_{xc}[n]$; the variation of the functional is defined as

$$\delta E_{\text{xc}}[n] = \int V_{\text{xc}}(\mathbf{r})\, \delta n(\mathbf{r})\, d\mathbf{r} = \int V_{\text{xc}}(\mathbf{r})\, \delta \sum_i \phi_i^*(\mathbf{r})\phi_i(\mathbf{r})\, d\mathbf{r} \quad ...(14)$$

where $$V_{\text{xc}}(\mathbf{r}) \equiv \frac{\delta E_{\text{xc}}[n]}{\delta n(\mathbf{r})} \quad ...(15)$$

is the functional derivative of $E_{\text{xc}}[n]$.

A straightforward variational calculation of the functional, Eq. (13), keeping in mind Eq. (14) and Eq. (15), leads to the Kohn–Sham equations

$$\boxed{\left[-\frac{\hbar^2\nabla^2}{2m} + V_{\text{nucl}}(\mathbf{r}) + V_{\text{coul}}(\mathbf{r}) + V_{\text{xc}}(\mathbf{r})\right]\phi_i(\mathbf{r}) = \varepsilon_i \phi_i(\mathbf{r}),} \quad ...(16)$$

where $V_{\text{coul}}(\mathbf{r})$ denotes the Hartree potential, $V_{\text{xc}}(\mathbf{r})$ is the functional derivative of $E_{\text{xc}}[n]$, and $V_{\text{nucl}}(\mathbf{r})$ is the external potential in consideration. Once the Kohn–Sham orbitals and energies have been determined, the exact total ground-state energy, Eq. (13), of the electronic system can be expressed as:

$$E_0 = \sum_i \varepsilon_i - \frac{1}{2}\sum_{ij}\left\langle \phi_i\phi_j \left| \frac{e^2}{r_{12}} \right| \phi_i\phi_i \right\rangle + E_{xc}[n] - \int V_{xc}(\mathbf{r})n(\mathbf{r})d\mathbf{r}. \quad ...(17)$$

We notice that the Kohn–Sham equations are standard differential equations with a rigorously local effective potential $V_{\text{eff}}(\mathbf{r}) = V_{\text{nucl}}(\mathbf{r}) + V_{\text{coul}}(\mathbf{r}) + V_{\text{xc}}(\mathbf{r})$; any difficulty in the procedure has been confined to a reasonable guess of the exchange-correlation functional $E_{xc}[n]$ (which is known only in principle). Conceptually, the Kohn–Sham equations determine exactly the electron density and the electronic energy of the ground-state; however, the orbital energies ($\varepsilon_i$) appearing in Eq. (16) are and remain purely formal Lagrange multipliers (not rescued for physical interpretation by any Koopmans' theorem); any identification of $\varepsilon_i$ with (occupied or non-occupied) one-particle energies is to be justified (and often heuristically corrected) situation by situation. Experience shows that the calculations, performed within one version or another of the density functional theory, tend to underestimate the energy band gap in semiconductor and insulators; however, the general trend of the dispersion curves of the valence and the conduction bands is often represented to reasonably accuracy.

### *Local Density Approximation for the Exchange-correlation Functional*

The formal definition of the exchange-correlation functional $E_{\text{xc}}[n]$, provided by Eq. (12b), would be of no practical use, unless workable approximations can be given. Among them, the most popular is the *local density approximation,* which is particularly justified in systems with reasonably slowly varying spatial density, $n(\mathbf{r})$. In the local density approximation (LDA), the exchange-correlation functional, Eq. (12b) is approximated in the form

$$E_{xc}^{(LDA)}[n(\mathbf{r})] = \int \varepsilon_{xc}(n(\mathbf{r}))\, n(\mathbf{r})\, d\mathbf{r}, \quad \text{...(18a)}$$

where $\varepsilon_{xc}(n(\mathbf{r}))$ is the many-body exchange-correlation energy per electron *of a uniform gas of interacting electrons of density*, $n(\mathbf{r})$. The exchange-correlation potential produced by the functional, Eq. (18a) becomes

$$V_{xc}^{(LDA)}(\mathbf{r}) \equiv \frac{\delta E_{xc}^{(LDA)}[n]}{\delta n(\mathbf{r})} = \varepsilon_{xc}(n(\mathbf{r})) + n(\mathbf{r})\frac{d\varepsilon_{xc}(n(\mathbf{r}))}{dn(\mathbf{r})}. \quad \text{...(18b)}$$

In the local density approximation, the total ground-state energy, $E_0^{(LDA)}$, takes the expression

$$E_0^{(LDA)} = \sum_i \varepsilon_i - \frac{1}{2}\int n(\mathbf{r})\frac{e^2}{|\mathbf{r}-\mathbf{r}'|}n(\mathbf{r}')\,d\mathbf{r}\,d\mathbf{r}' - \int n(\mathbf{r})\frac{d\varepsilon_{xc}(n(\mathbf{r}))}{dn(\mathbf{r})}n(\mathbf{r})\,d\mathbf{r}, \quad \text{...(18c)}$$

where $d\varepsilon_{xc}(n(\mathbf{r}))/dn(\mathbf{r})$ denotes $d\varepsilon_{xc}/dn$ calculated for $n = n(\mathbf{r})$.

Among the various forms, adopted for $\varepsilon_{xc}$, we report here the expression given by J. P. Perdew and A. Zunger, *Phys. Rev.* B**23**, 5048 (1981), who have parametrized the numerical results obtained by Ceperley and Alder with Monte Carlo calculations. For the unpolarized homogeneous electron gas, we have

$$\varepsilon_{xc}(r_s) = \varepsilon_x(r_s) + \varepsilon_c(r_s)$$

$$\varepsilon_x(r_s) = -\frac{0.4582}{r_s}$$

$$\varepsilon_c(r_s) = \begin{cases} -0.1423/(1+1.0529\sqrt{r_s}+0.3334\,r_s) & \text{for } r_s \geq 1 \\ -0.0480+0.0311\ln r_s - 0.0116\,r_s + 0.0020\,r_s & \text{for } r_s \leq 1 \end{cases}$$

In the above expressions, energies are in Hartrees (1 Hartree = 2 Rydberg); $r_s$ is the usual dimensionless parameter defined by $(4\pi/3)(r_s a_B)^3 = 1/n$; the distinction between the exchange term and the correlation term is just for comparison with the Hartree–Fock approach. The density functional theory, especially in conjunction with the local density approximation, has become the most successful tool for the investigation of the chemical and physical ground state properties of electronic systems.

## EXERCISE

19.1 Write short notes on the following:

(a) Born–Oppenheimer or adiabatic approximation

(b) One–electron approximation

(c) Harmonic–approximation for ions in solids.

19.2 Discuss and derive the Rayleigh–Ritz variational principle.

19.3 Derive the Hartree equations for a multi-electron atom.

19.4 What is central field approximation?

19.5 Derive the Hartree–Fock equations and define Fock–Operator.

19.6 What is Koopmans' theorem?

19.7 What is Hartree–Fock–Slater approximation?

19.8 Give a basic idea of Density Functional Theory.

CHAPTER

# 20

# Nanostructures and Nanophysics

## 20.1 INTRODUCTION

All materials are composed of grains, which in turn, comprise many atoms. These grains are usually invisible to the naked eye, depending on their size. Conventional materials have grains varying in size anywhere from 100's of microns (μm) to millimetres (mm). A micron (μm) is a micrometre or a millionth ($10^{-6}$) of a metre. An average human hair is about 100 μm in diametre. A nanometre (nm) is even smaller in dimension than a μm and is a billionth ($10^{-9}$) of a metre. A nanocrystalline material has grains of the order of 1–100 nm. The average size of an atom is of the order of 1–2 Å in radius. A nanometre comprises 10 Å, and hence in one nm, there may be 3–5 atoms, depending on the atomic radii.

*Nanomaterials* (nano-crystalline materials) are materials possessing grain sizes of the order of a billionth of a metre and they manifest extremely fascinating and useful properties. There are often advantages of making things smaller. Especially in modern semiconductor industry, miniaturization of devices often delivers more efficient performance. But the first and foremost question is what should be the limits of miniaturization. Theoretically, devices can be miniaturized to atomic level, whose sizes can be varied from fraction of a nanometre to few hundreds of nanometres.

*Nanotechnology* deals with these kinds of human-designed working devices produced in a controlled and reproducible manner. The idea of miniaturization of the size-scale is fundamentally interesting due to the fact that as size approaches the atomic scale, the relevant physics associated with the systems changes from classical domain to quantum mechanical regime. These changes in the behaviours from classical to atomic scale via "mesoscopic" domain can be explained by modern quantum-mechanical treatments, but, in several specific cases, more complex treatments and new methodologies are needed to explain the behaviour of the system. It should be noted that these changes from classical physics to quantum-mechanical nanophysics may mean that some existing devices will fail, while the same changes can open up possibilities for new devices.

Nanotechnology is the creation and utilization of materials, devices and systems through the control of matter synthesis on the nanometre length-scale; i.e., at the level of atoms, molecules and supramolecular structures. The essence of nanotechnology is the ability to work at these levels to generate larger structures with fundamentally new molecular organization. Such materials can be rationally designed to exhibit novel and significantly improved physical, chemical and biological properties, phenomena and processes because of their size. When characteristic structural features are intermediate to some extent between isolated atoms and bulk materials, in the range of about $10^{-9}$ to $10^{-7}$ m, the objects often display physical attributes, substantially different from those displayed by either atoms or bulk materials. Structures in the nanometre dimensions are fundamentally different because it is in the quantum length scale. Thus, nanostructures are natural home for engineered quantum effects.

Further nanomaterials can be classified according to their dimension:

- Materials with two *dimensions* in the nanoscale are layers, such as a ***thin films*** or ***surface coatings***.
- Materials that are nanoscale in one *dimension* are *nanowires* and ***nanotubes***.
- Materials that are nanoscale in *three dimensions* are particles ***quantum dots*** (tiny particles of semiconductor materials). Nanocrystalline materials, made up of nanometre sized grains, also fall into this category.

Two principal factors cause the properties of nanomaterials to differ significantly from other materials: increased relative surface area and quantum effects. These factors can change or enhance properties such as reactivity, strength and electrical properties, optical characteristics.

## 20.2 NANOSCIENCE AND NANOTECHNOLOGY

Currently, **nanotechnology is described as a revolutionary discipline** in terms of its possible impact on industrial applications. Nanotechnology offers potential solutions to many problems using emerging nanotechniques.

Depending on the strong interdisciplinary character of nanotechnology **there are many research fields and several potential applications** that involve nanotechnology.

*Nanoscience and nanotechnology* are an extension of the field of materials science, and materials science departments at universities around the world in conjunction with physics, mechanical engineering, bioengineering, and chemical engineering departments are leading the breakthroughs in nanotechnology.

The term ***nanotechnology*** means the technology which exploits phenomena and structures that can only occur at the nanometer scale, which is the scale of several atoms and small molecules (A nanometre is 1 billionth of a metre). Such phenomena include:

- ***Quantum Confinement***—this can result in different electromagnetic and optical properties of a material between nanoparticles and the bulk material.
- ***Gibbs–Thomson Effect***—this is the lowering of the melting point of a material when it is nanometres in size; and such structures including carbon nanotubes (CNTs).

Nanotechnology is based on the fact that the properties of materials become markedly different when their size approaches that of a few hundreds or tens of atoms. More broadly, nanotechnology includes the many techniques used to create structures at a size scale below 100 nm, including those used for fabrication of *nanowires*, those used in *semiconductor fabrication* such as deep ultraviolet lithography, electron beam lithography, focused ion beam machining, *nanoimprint lithography* atomic layer deposition, and *molecular vapour deposition,* and further including molecular self-assembly techniques such as those employing di-block *copolymers.*

Other forms of nanotechnology being developed include tiny sensors called nano-units, of which some simple types are available: "smart materials" that change in response to light or heat; "nano-bots" – tiny mobile robots that have yet to be developed but are theoretically possible; and self-assembling nanomaterials that can be assembled into larger equipment.

The related term ***nanoscience*** is used to describe the interdisciplinary fields of science devoted to the study of nanoscale phenomena employed in nanotechnology. This is the world of atoms, molecules, macromolecules, quantum dots, and macromolecular assemblies, and is dominated by surface effects such as van der Waals force attraction, hydrogen bonding, electronic charge, ionic bonding, covalent bonding, hydrophobicity, hydrophilicity, and quantum mechanical tunnelling.

## 20.3 QUANTUM BEHAVIOUR OF NANOMETRIC WORLD

On atomic scale, the failure of classical concepts leads to the quantum mechanical approach to understand the behaviour of matters in such low dimensions. A clear understanding of these nanophysical rules is the key to understand the deviations of the devices that will appear as the sizes are reduced to atomic scale.

### 20.3.1 Applications of Schrödinger Equation

**(a) Potential Box: Particle Trapped in 3D (Nanodot)**

If we consider a particle inside a 3D infinite trap (Fig. 20.1) of volume $L^3$ with impenetrable walls, then we need to consider the generalized Schrödinger equation in three dimensions as

$$-\left(\frac{\hbar^2}{2m}\right)\nabla^2\psi(\vec{r}) + V(\vec{r})\,\psi(\vec{r})\,\psi(\vec{r}) = E\psi(\vec{r}) \qquad ...(1)$$

where $\nabla^2 = \dfrac{\partial^2}{\partial x^2} + \dfrac{\partial^2}{\partial y^2} + \dfrac{\partial^2}{\partial z^2}$ ...(2)

and $\vec{r}$ is the position vector. And the corresponding solution will be

$$\psi_n(x, y, z) = \left(\frac{2}{L}\right)^{3/2} \sin\left(\frac{n_x \pi x}{L}\right) \sin\left(\frac{n_y \pi y}{L}\right) \sin\left(\frac{n_z \pi z}{L}\right) \qquad ...(3)$$

and

$$E_n = \left(\frac{h^2}{8mL^2}\right)(n_x^2 + n_y^2 + n_z^2) \qquad ...(4)$$

where $n_x = 1, 2, 3, ...,$ etc. These results can also be used for boxes of unequal dimensions, $L_x$, $L_y$ and $L_z$, as

$$E_{n_x n_y n_z} = \left(\frac{h^2}{8m}\right)\left(\frac{n_x^2}{L_x^2} + \frac{n_y^2}{L_y^2} + \frac{n_z^2}{L_z^2}\right) \quad ...(5)$$

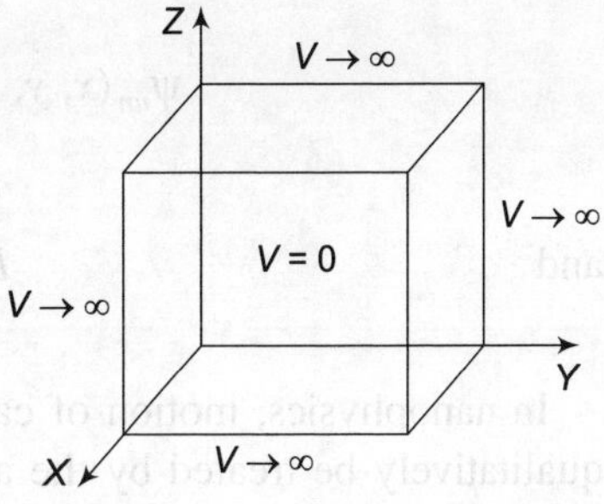


**Figure 20.1** Three-dimensional infinite potential well

It is apparent from the above equation that the energy is dependent on three quantum numbers, $n_x$, $n_y$, and $n_z$. The lowest energy is $E_{111}$, and non-zero (for $n_x = 1$, $n_y = 1$ and $n_z = 1$). The next energy level corresponds to $E_{211}$ and is equivalent to $E_{121}$ and $E_{112}$ (for either one of the $n$-values as 2 and the other two as 1). Therefore, the second energy level has three states for same energy as $\psi_{211}$, $\psi_{121}$ and $\psi_{112}$. The number of states that have the same energy is called degeneracy of that energy level. Hence, the second energy level is thus threefold degenerate.

Nanoparticles with a fraction of a nanometre to a few tens of nanometres can be treated as the classic examples of three-dimensional confinement of carriers in a small 3D space. Carriers are confined within the 3D space of this nanoparticle and discrete energy levels will be formed with a little modification due to quantum confinement effect.

**(b) Electron trapped in 2D plane: Nanosheet**

The above-mentioned results can also be used for 2D spaces (Fig. 20.2) where carriers are confined in one-dimension, say along Z-direction to a very small thickness $t$, and free to move along X- and Y-directions. In nanophysics, this is the case of nanosheets where carriers are allowed to move along a two-dimensional plane. In this case, the modification of wave function and energy will respectively be

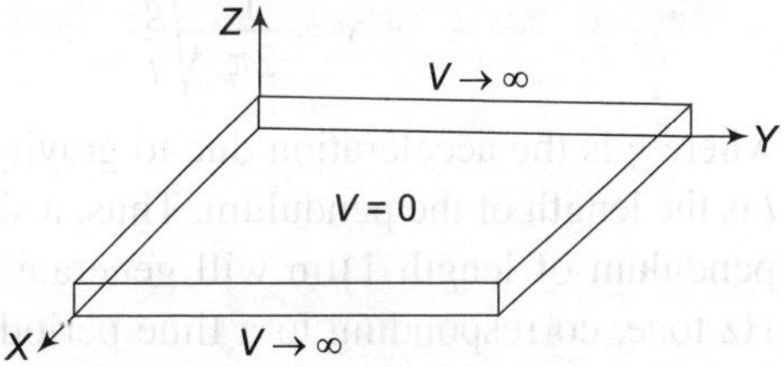


**Figure 20.2** Two-dimensional space. Particles can move along X- and Y-directions freely but confined along Z-direction by a small distance, $t$

$$\text{or } \psi_n(x, y, z) = \left(\frac{2}{t}\right)^{1/2} \sin\left(\frac{n_z \pi z}{t}\right) e^{ik_x x} e^{ik_y y} \quad ...(6)$$

and
$$E_n = \left(\frac{h^2}{8mt^2}\right) n_z^2 + \frac{\hbar^2 k_x^2}{2m} + \frac{\hbar^2 k_y^2}{2m} \quad ...(7)$$

where the quantum number ($n_z$) is called the sub-band index and for $n = 1$ the carrier is assumed to be in the first sub-band. In general, the electron motion in a semiconductor band is described by introducing the effective mass of electrons (and holes).

**(c) Electrons moving in one-dimension: Nanowire/rod/belt**

As shown in Fig. 20.3 in nanowire, carriers are confined in both directions (say, $Y$ and $Z$) to small dimension, $t$ (assuming square cross-section, $t^2$) and allowed to move freely along the length of $x$-axis. In this case, the wave function and the energy will respectively be

$$\psi_{nn}(x, y, z) = \left(\frac{2}{t}\right) \sin\left(\frac{n_y \pi_y}{t}\right) \sin\left(\frac{n_z \pi z}{t}\right) e^{ik_x x} \quad ...(8)$$

and

$$E_n = \left(\frac{h^2}{8mt^2}\right)(n_y^2 + n_z^2) + \frac{\hbar^2 k_y^2}{2m} \quad ...(9)$$

In nanophysics, motion of carriers in carbon nanotubes or similar kind of nanowires can qualitatively be treated by the above equation, although the actual bound states and energy expressions are more involved and complicated.

## 20.4 SIZE EFFECTS IN SMALLER SYSTEMS (PRE-QUANTUM)

### 20.4.1 Increase in Response Time

Many artificial systems show interesting behaviour when scaled down to smaller sizes. For example, the fundamental frequency of a simple harmonic oscillator increases for smaller systems. The resonance frequency of a simple pendulum is given by

$$\nu = \frac{1}{2\pi}\sqrt{\frac{g}{l}} \quad ...(1)$$

where $g$ is the acceleration due to gravity and $l$ is the length of the pendulum. Thus, a simple pendulum of length 1 μm will generate 1000 Hz tone, corresponding to a time period of 1 ms. Therefore, the miniature version of the simple pendulum has a much faster response time with respect to a normal pendulum which has a resolution of 1 $s$ for a pendulum length of 1 m.

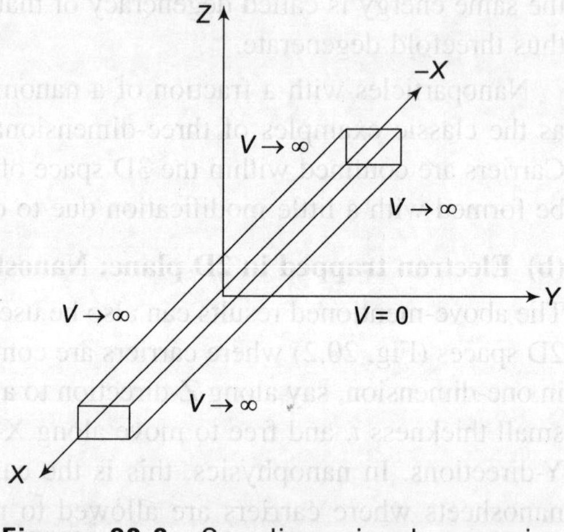


**Figure 20.3** One-dimensional nanowire. Particles are only allowed to move freely along X-axis and confined along Y- and Z-directions by a small distance, $t$ (assuming square cross-section)

### 20.4.2 Decrease in Thermal Time Constant in Smaller Systems

If a hot body ($T > 0$°K) of heat capacity per unit volume C, is connected to a cold body ($T = 0$°K) through a heat channel of length $l$, cross-section ($a$) and thermal heat capacity ($k_{\text{Th}}$) (Fig. 20.4), then the rate of flow of heat $d\Theta/dT$, from hot to cold body can be written as

$$\frac{d\Theta}{dT} = \frac{k_{\text{Th}}\, aT}{l} \quad ...(2)$$

$$\frac{d\Theta}{dt} = -CV\frac{dT}{dt} \quad ...(3)$$

where $V$ is the volume of the hot body and negative sign indicates the loss of heat from the hot body. Hence, from Eqs. (2) and (3), we get

$$-CV\frac{dT}{dt} = \frac{k_T aT}{l} \quad ...(4)$$

and

$$\frac{dT}{T} = -\left(\frac{k_T a}{ICV}\right)dt \quad ...(5)$$

which leads to the solution

$$T = T_0 e^{-t/t_{th}} \quad ...(6)$$

where $t_{th} = lCV/k_T a$ is the thermal time constant, which under isotropic environment has the dimension $[L]^2$. Therefore for thermal time constant strongly decreases with the size of the system. This phenomenon is used for data storage system where the data bits are recorded with a resistively heated AFM tip and are detected by the integrated piezoelectric (PZT) sensor.

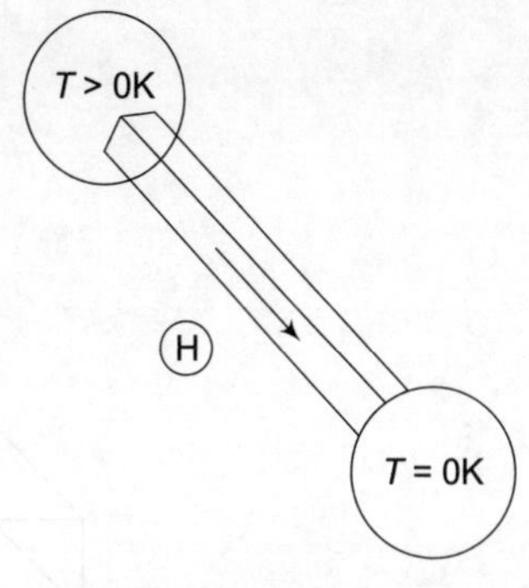


**Figure 20.4** Heat transfer from a hot body to a cold body through a heat channel

### 20.4.3 Disappearance of Friction in Highly Symmetric Molecular System

In molecular-scale systems, the viscous and frictional forces are assumed to be zero (Drexler, K.E., *Nanosystems*, Wiley, NY, USA (1992)). Theoretical modelling shows that the self-alignment of the nanometre scale wheels and axles in vacuum leads to absolutely zero space for any fluid in the interface between two surfaces. Therefore, these moving parts will no longer experience any fluid viscous force between them, but may encounter some frictional force (if any) at a much lower level than ordinary mechanical frictional forces observed in macroscopic moving parts. For example, in nested carbon nanotubes, which are essentially rolled sheets of graphite with no dangling bonds perpendicular to their surfaces, there are no molecules between the surfaces of two adjacent tubes, and thus, the medium between the very closely spaced moving elements is vacuum. As a result, the motion between two neighbouring tubes is almost frictionless.

## 20.5 EXCITONS

The excitons are bound states of photo-excited and coulomb-correlated electron-hole pairs, produced momentarily by light illumination of energy greater than the band gap of the material concerned. The electron-hole pairs, thus produced, will orbit around each other, having equal and opposite charges (cf. Fig. 20.5). These electron-hole pairs are elementary excitation, or a quasi-particle of a solid and combine in a short time giving off a flash of light. As the exciton results from the binding of the electron with its hole, the exciton has slightly less energy than the unbound electron and hole. The wave function of the bound is hydrogenic. However, the binding energy is much smaller and the size is much bigger than a hydrogen atom because of the effects of screening and the effective mass of the constituents in the material. Also unlike in the hydrogen atom, where the nucleus is assumed to be stationary and electrons circle around the nucleus, in the case of excitons, the electron-hole pairs circle

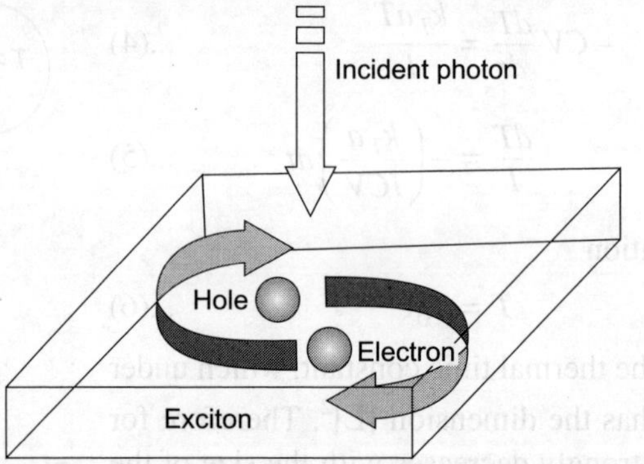


**Figure 20.5** Schematic diagram of the formation of an exciton inside a material

around each other with respect to their centre of mass. Therefore, the Schrödinger equation should contain terms involving both electron and hole. Hence, the equation will look like (in Cartesian coordinate).

$$-\left(\frac{h^2}{2m_h^*}\nabla_h^2+\frac{\hbar^2}{2m_e^*}\nabla_e^2\right)\psi-k\frac{e^2}{|\vec{r}_e-\vec{r}_h|}=E\psi \quad ...(1)$$

where $m_h^*$ and $m_e^*$ are the effective masses of the hole and the electron respectively, $\vec{r}_e$ and $\vec{r}_h$ are the position vectors for the electron and the hole with respect to the centre of mass of the system and $k$ is the Coulomb constant, which involves the dielectric constant of the material involved.

Depending on the materials in question, there are two types of excitons that can exist. In semiconductors, the dielectric constant is generally large, and as a result, screening tends to reduce the Coulomb interaction between electrons and holes. This results in the formation of **Mott-Wannier exciton**, which has a radius much larger than the lattice spacing. As a result, the effect of the lattice potential can be incorporated into the effective masses of the electron and the hole, and because of the lower masses and the screened Coulomb interaction, the binding energy is usually much less than a hydrogen atom, typically of the order 0.1 eV. On the other hand, when the dielectric constant of the material is very small, the Coulomb interaction between the electron and the hole becomes very strong and the excitons tend to be much smaller, of the same order as the unit cell, so the electron and the hole sit on the same cell. This type of excitons are called **Frenkel excitons,** whose binding energy is typically of the order of 1.0 eV.

## 20.6 QUANTUM CONFINEMENT EFFECT IN NANOMATERIALS

In terms of hydrogen atom analogy, the band gap is the ionization limit of the hydrogenic electron-hole bound states. Absorption of photon at higher energies creates free electrons and holes with excess kinetic energies inside the semiconductor. Also the dielectric constants of most of the inorganic semiconductors range from 5 to 12, which implies that the Coulomb interaction is almost screened. Also, the small effective masses of electrons and the holes inside the semiconducting material imply that the localization energies for the electrons and the holes are large. Thus, the excitation wave function extends over a large region within the material. For example, the lowest 1*s* exciton of CdS has an approximate diameter of 6 nm. Therefore, when the crystallite size is comparable to this dimension, 'bulk' energy level scheme will not be valid and electron-hole interactions will dominate the dynamics. So in the nanometric dimensions of the material, the energy level scheme depends upon the size, shape and nature of the material. Proper formation of the potential energy terms along with the Schrödinger equation and wave functions show that the band gap of the nanocrystalline material increases with respect to the bulk value, i.e., there will be a blue-shift ($\Delta E$) of the band gap as the crystalline size approaches from 'bulk' size to lower dimensions of nanometric range as shown below

$$\Delta E = E_{g\,(\text{nano})} - E_{g\,(\text{bulk})} = \frac{h^2 n^2}{8\mu^* r^2} - \frac{1.8e^2}{\varepsilon r} - 0.248\,E_R^* \qquad ...(2)$$

where $r$ is the radius of the nanoparticle, $n$ is an integer, $\mu^* = [(m_h^*)^{-1} + (m_e^*)^{-1}]^{-1}$ is the reduced mass of the effective masses of the electron and the hole respectively, $\varepsilon$ is the dielectric constant of the semiconducting material, $E_R^*$ is the effective Rydberg energy. The first term on the right-hand-side expression in the equation represents the particle-in-a-box quantum localization energy and has a $1/r^2$ dependence for both the electron and the hole. The second term represents the Coulomb energy with an $1/r$ dependence and the third term is responsible for the spatial correlation. The $n$ values larger than 1 in Eq. (2) correspond to the exciton transitions to higher excited states. In the limit of large $r$, the value of $E_{g\,(\text{nano})}$ approaches that of $E_{g\,(\text{bulk}\,r)}$. This band gap enhancement in nanomaterials is experimentally observed in many semiconducting materials as shown in Fig. 20.6 and described in Table 20.1.

**Table 20.1** Increase in the Optical Band Gap of p-$CuAlO_2$ nanoparticles with Decrease in the Average Particle Size[1], Showing the Quantum Confinement Effect of the Band Gap Enhancement in Semiconductor Nanocrystals

| *Particle size (nm)* | *Band gap (eV)* |
|---|---|
| > 90 (bulk) | 3.34 |
| 60 | 3.60 |
| 30 | 3.72 |
| 20 | 3.84 |
| 10 | 3.94 |

[1] Tyler, T., V.V. Zhirnov, A.V. Kvit, D. Kang and J.J. Ilren, Electron emission from diamond nanoparticles on metal tips, *Appl. Phys. Lett.*, **82** (2003), p. 2904.

The optical and luminescent properties of nanoparticles are greatly dependent on this quantum size effect. The CdSe quantum dots are found to emit different colours as their sizes change. As the size decreases, the emitted light wavelength moves towards the blue region. Thus, by changing the size of the quantum dots, one can generate different wavelengths

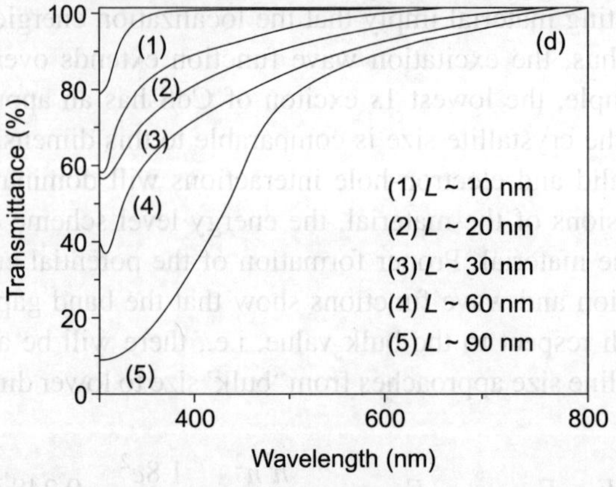


**Figure 20.6** Optical transmittance spectra of the p-$CuAlO_2$ nanoparticles (*L* is the average particle size) showing the blue-shift of the band-gap edges with decrease in the particle size, indicating the size effects of band-gap enhancement in semiconductor nanocrystals

of light through them under irradiation. Similarly, Fig. 20.7 shows the size effect in the photoluminescence spectra of p-$CuAlO_2$ nanoparticles. As the size decreases, the emission peak shifts towards low wavelength (blue-shift) region.

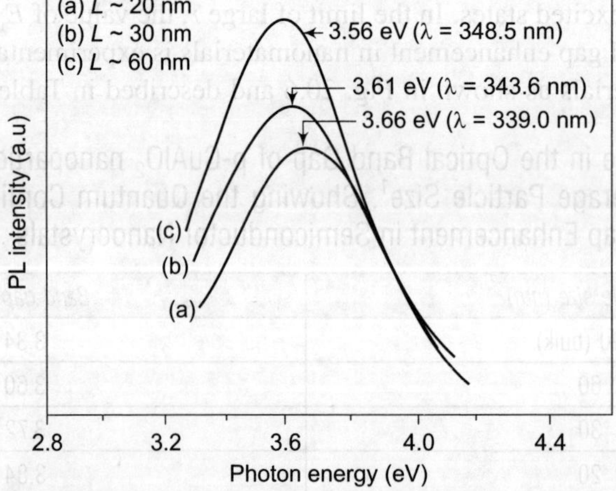


**Figure 20.7** Photoluminescence spectra of p-$CuAlO_2$ nanoparticles showing a blue-shift in the emission peak with a decrease in the particle-size [*]

## 20.7 GRAPHITE STRUCTURE

Carbon atoms in graphite arrange themselves in planar structures, only one atom thick, called *graphenes*. Graphenes are stocked on top of each other, being held together by weak van der Waals forces. Within each graphene the carbon atom from a 2–D honeycomb lattice as shown in Fig. 20.8.

Graphite is an allotrope of carbon in which $sp^2$ hybridization binds each carbon atom in a layer to three nearest neighbours so that strong covalent bonds exist in the layers. The different layers themselves are bonded by van der Waals forces. The C – C bond lengths in the layers are 1.42 Å and between the layers are 3.35 Å due to presence of weak van der Waals forces.

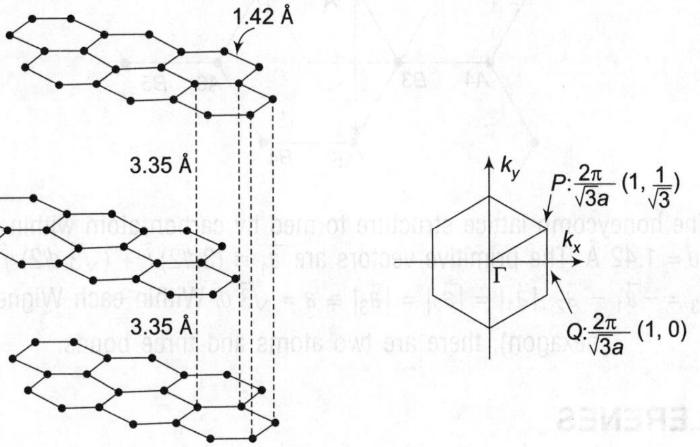


**Figure 20.8** (a) Graphite structure (b) The first BZ of the honeycomb direct lattice; P: $(2\pi/3d)$ $(1, 1/\sqrt{3})$ and Q: $(2\pi/3d)$ (1, 0) in cartesian coordinates. The bond length $d$ is 1.42 Å (i.e. C–C bond length)

It is obvious from the structure of the graphene that the relevant hybridization is $sp^2$ with the $p_z$ orbital of each carbon atom perpendicular to the plane of the graphene. Note that the coupling of the $p_z$ orbitals is maximized when these orbitals are parallel to each other.

The honeycoub lattice structure in a graphene is shown in Fig. 20.9.

The successive planes are stacked parallel to each other with a displacement of the hexagons by one bond length, so that every alternate C-atom in the layer lies vertically above a C-atom in the layer below is vertically below an atom in the layer above. Due to strong covalent bonds, conductivity within a layer is low. However, due to weak nature of interlayer van der Waals forces and parallel $p_z$ orbitals, the interlaminar bonds are almost metallic.

Within a layer, the three $sp^2$ bonds are coplanar with an interbond angle of 120°. The fourth bonding electron of carbon atom is delocalised and resonates between the three $sp^2$ bonds. Its mobility accounts for a hundredfold increase in electrical and thermal conductivity in a direction parallel to the sheets as compared to perpendicular to the sheets (high anisotropy). The weak intersheet van der Waals bonding explains the softness of graphite as compared to diamond which is another crystalline form of carbon, but the hardest known mineral.

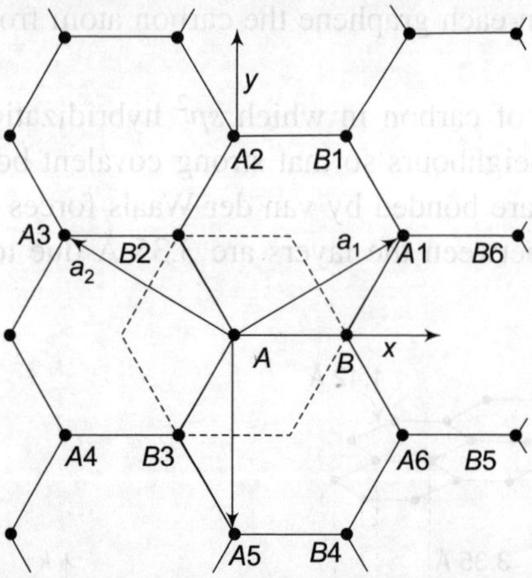


**Figure 20.9** The honeycomb lattice structure formed by carbon atom within each graphene. The bond length is $d$ = 1.42 Å. The primitive vectors are $\vec{a}_1 \equiv (3d/2)\,\hat{i} + (\sqrt{3}\,d/2)\,\hat{j}$ and $\vec{a}_2 = -(3d/2)\,\hat{i} + (\sqrt{3}\,d/2)\,\hat{j}$. $\vec{a}_3 = -\vec{a}_1 - \vec{a}_2$. $|\vec{a}_1| = |\vec{a}_2| = |\vec{a}_3| \equiv a = \sqrt{3}\,d$. Within each Wigner–Seitz cell (dotted hexagon), there are two atoms and three bonds.

## 20.8 FULLERENES

The material composed of $C_{60}$ molecules that contain sixty carbon atoms in a network of $sp^2$ bonding which forms a spherical structure, is known as *buckminster fullerene.* In this material, graphite acquires curvature due to the presence of pentagonal rings in the hexagonal graphite sheet. The prototypical fullerene (i.e., $C_{60}$ molecule) was discovered by Kroto, Heath, O'Brien, Curl and Smalley in 1985. It is approximately 1 nm in diameter. The structure is illustrated in Fig. 20.10. The starting three-dimensional framework is the icosahedron which has 20 (i.e., twenty) equal and equilateral triangular faces. By shipping off each of the twelve vertices of this (Fig. 20.10(a)), we obtain the truncated icosahedron (having the symmetry of a soccer-ball).

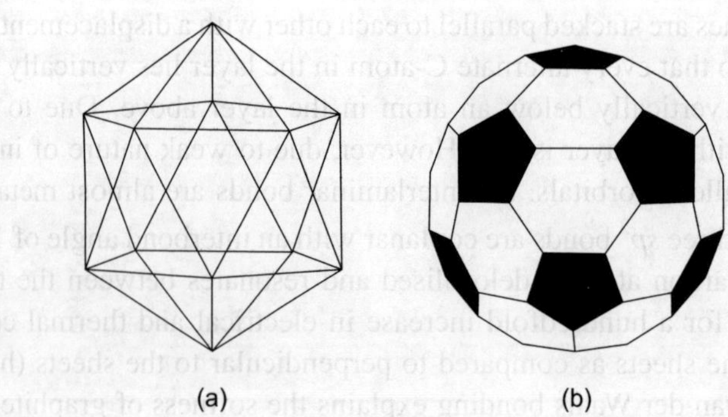


**Figure 20.10** Structure of $C_{60}$ molecule; **(a)** the icosahedron, **(b)** the truncated icosahedron ($C_{60}$)

In fact, a soccer ball has the same geometric configuration as a fullerene. These ball-like molecules bind with each other in the solid state to form a crystal lattice having a fcc structure shown in Fig. 20.10 (c). In the lattice, each $C_{60}$ molecule is separated from its nearest neighbour by 1 nm (the distance between their centres is 1 nm), and they are held together by weak van der Waals forces. Because $C_{60}$ is soluble in benzene, single crystals of it can be grown by slow evaporation from benzene solutions.

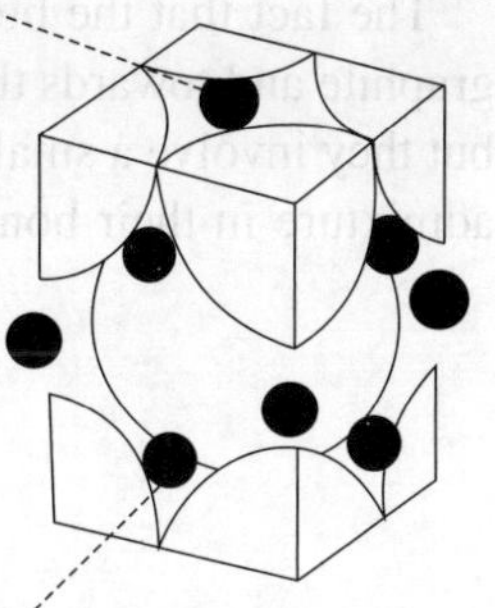


**Figure 20.10 (c)** Crystal lattice unit cell of $C_{60}$ molecules (large spheres) doped with alkali atoms (dark circles). Adapted from F.J. Owens and C.P. Poole, *The New Superconductors*, Plenum (1998)

### 20.8.1 Alkali-Doped $C_{60}$

In the face-centred cubic fullerene structure, 26% of the volume of the unit cell is empty, so alkali atoms can easily fit into the empty spaces between the molecular balls of the material. When $C_{60}$ crystals and potassium metal are placed in evacuated tubes and heated to 400°C, potassium vapour diffuses into these empty spaces to form the compound $K_3C_{60}$. The $C_{60}$ crystal is an insulator, but when doped with an alkali atom it becomes electrically conducting Figure 20.10 (c) shows the location of the alkali atoms in the lattice where they occupy the two vacant tetrahedral sites and a larger octahedral site per $C_{60}$ molecule. In the tetrahedral site the alkali atom has four surrounding $C_{60}$ balls, and in the octahedral site there are six surrounding $C_{60}$ molecules. When $C_{60}$ is doped with potassium to form $C_{60}$, the potassium atoms become ionized to form $K^+$ and their electrons become associated with the $C_{60}$, which becomes a $C_{60}^{3-}$ triply negative ion, i.e., each $C_{60}$ has three extra electrons that are loosely bonded to the $C_{60}$, and can move through the lattice making $C_{60}$ electrically conducting. In this case the $C_{60}$ to be electron doped. Thus, $K_3C_{60}$ becomes superconducting at low temperatures. Unlike high temperature superconductors which are hole-superconductors, $K_3C_{60}$ is an electron superconductor. Later, it was found that many alkali atoms could be doped into the $C_{60}$ lattice, and the superconducting transition temperature increased to as high as 33 kelvin in $Cs_2RbC_{60}$.

In this molecule the nuclei of the 60 carbon atoms are located in a very symmetrical way on the surface of a sphere of diameter[2] 6.825 Å; each atom is connected to three neighbouring atoms; thus the number of bonds $n_b$ is 3 × 60/2 = 90. These bonds define the edges of a polyhedron, while the nuclei of the carbon atoms are at its vertices. According to Euler's theory the number ($nf$) of the faces is equal to, $n_e - n_v + 2$, where $n_e = n_b$ is the number of edges and $n_v$ is the number of vertices. Thus, in the present case, there are 32 faces, 12 of which are regular pentagon and 20 almost[3] regular hexagons as shown in Fig. 20.11. Each pentagon is surrounded only by hexagons as shown in Fig. 2011.

[2] The value 1034 A often quoted in the literature for C60's diameter refers to the outer diameter which includes the finite size of each carbon atom.

[3] The bond lengths in each hexagon alternate between 1.395 and 1.455 A.

The fact that the bond lengths around each pentagon are 1.455 Å, i.e., larger than that of graphite and towards the value in diamond, suggest that the pentagon bonds are not purely $sp^2$ but they involve a small admixture of $sp^3$; anyway, both the presence of pentagons and the $sp^3$ admixture in their bonds are necessary for allowing a flat structure such as graphene to fold

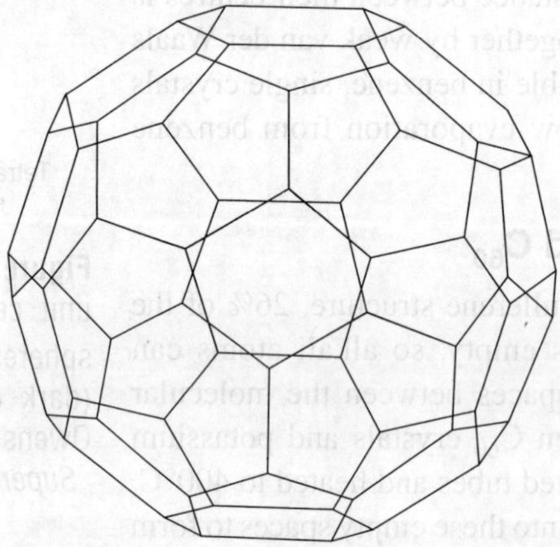

**Figure 20.11** The hollow almost spherical molecule $C_{60}$ consists of 60 carbon atoms at the vertices of a truncated regular icosahedron; it has 90 edges, 60 vertices and 32 faces (12 regular pentagons and 20 almost regular hexagons). The length of the bonds shared by two hexagons is 1.395 Å, while the length of the bonds shared by a pentagon and a hexagon is 1.455 Å. (In graphite the bond length is 1.421 Å, and in benzene it is 1.395 Å, while in diamond it is 1.545 Å)

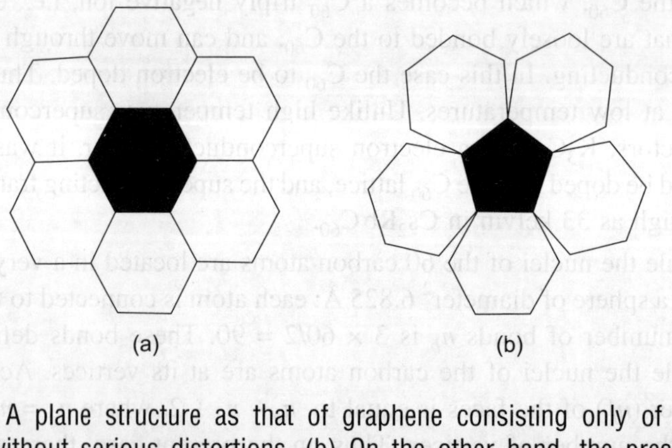


**Figure 20.12** (a) A plane structure as that of graphene consisting only of regular hexagons cannot be folded without serious distortions. (b) On the other hand, the presence of a regular pentagon surrounded by hexagons allows some space for the hexagons to undergo a small rotation around each pentagon side, as to be connected smoothly and to from a convex structure without distortions. This small rotation makes the solid angle at each pentagon side less than 180°; as a result the pentagon bonds acquire a small $sp^3$ component

into a polyhedron without serious distortions as shown in Fig. 20.12. The procedure described in the caption of Fig. 20.12 implies that three sides of each hexagon (the ones shared by pentagons) are rotation axes (hence, they acquire an $sp^3$ component), while the other three (between the rotation axes) remain almost intact, and consequently, have an $sp^2$–$sp^2$ bond and a $p_\perp$–$p_\perp$ bond[3] as in benzene. This is shown in Fig. 20.13, where the edges with the $sp^3$ admixture that acted as rotation axes are single lines, while the almost-intact hexagon edges are shown as double lines indicating the double bond. This picture is strongly supported by the fact that the double bond length is 1.395 Å the same as in benzene.

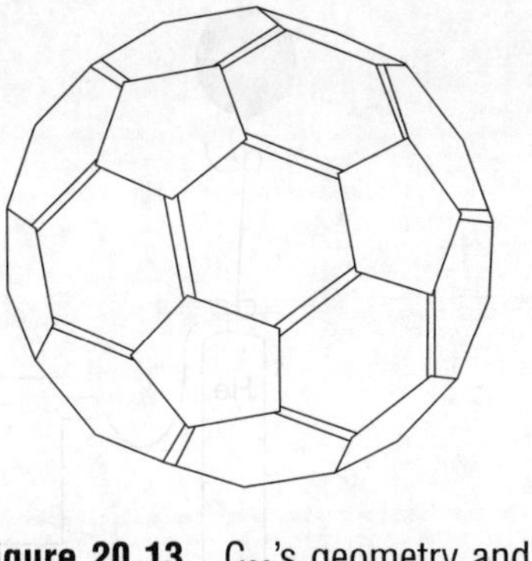

**Figure 20.13** $C_{60}$'s geometry and bond nature. The double bonds (30 in total) are the same as those in benzene and have the same length of 1.395 Å. The single lines, (60) in total, indicate a bond intermediate between $sp^2$ and $sp^3$ and have a length of 1.455 Å.

## 20.9 CREATION OF FULLERENES

### 20.9.1 Discovery

Richard Smalley at Rice University in Houston, Texas was studying strange chains of carbon atoms found in space through microwave spectroscopy (a science that studies the absorption spectra of stellar particles billions of kilometers away) to identify what compounds are found in space. Smalley had been studying clusters of metal atoms with the help of Robert Curl, using an apparatus Smalley had in his laboratory. This laser-supersonic cluster beam apparatus had the ability to vaporize nearly any known material into plasma using a laser, which is a highly concentrated beam of light with extremely high energy.

Smalley's apparatus, shown in Fig. 20.14, fires a high energy laser beam at a rotation disk of graphite in a helium-filled vacuum chamber. Helium is used because it is an inert gas and therefore does not react with the gaseous carbon. The intense heating of the surface of the graphite breaks the C—C bonds because of the intense energy. Once vaporized, the carbon atoms cool and condense in the high-pressure helium gas, colliding and forming new bond arrangement. Immediately upon cooling several degrees above absolute zero in a chamber, the carbon leads to a mass spectrometer for further analysis.

A mass spectrometer uses an atom or molecule's weight and electric charges to separate it from other molecules. This is done by ionizing the molecules, which is done by bombarding the molecules with high energy electrons which then knocks off electrons. If an electron is removed from an otherwise neutral molecule, then the molecule becomes a positively charged ion or cation. The charged particles are then accelerated by passing through electric plates and then filtered through a slit. A stream of charged particles exits the slit and is then deflected by a magnetic field into a curved path. Because all the particles have a charge +1, the magnetic field exerts the same amount of force on them, however, the more massive ions are deflected less, and thus a separation occurs. By adjusting the strength of the accelerating electric plates or the deflecting magnetic field, a specific mass can be selected to enter the

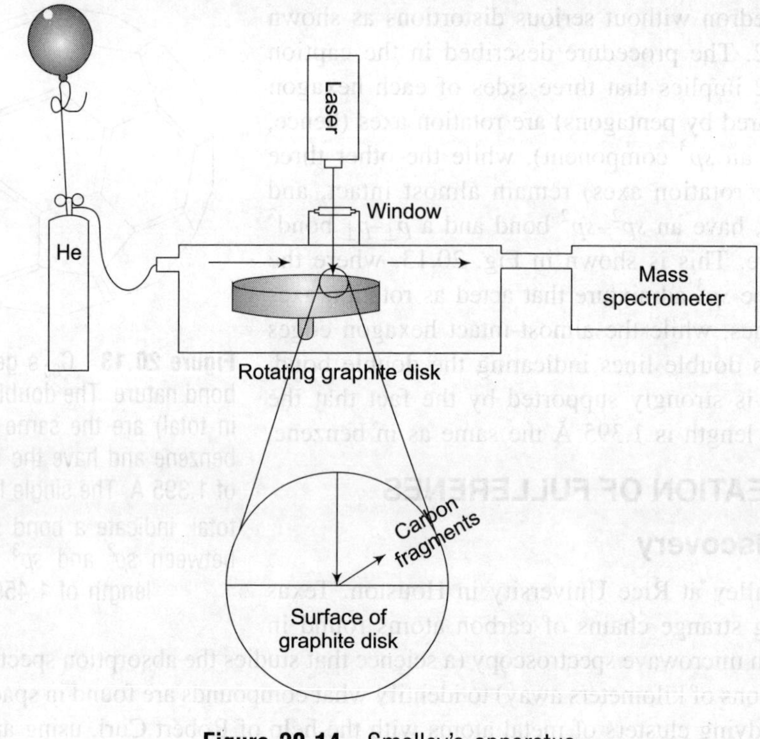


**Figure 20.14** Smalley's apparatus

receptor on the end. After adjusting the experiment, it became greatly evident that the most dominant molecule measured was 720 amu (atomic mass units). By dividing this number by the mass of a single carbon atom (12 amu), it was deduced that the molecule was comprised of 60 carbon atoms (720/12 = 60).

The next task was to develop a model for the structure of C60, this new allotrope of carbon. Because it was overwhelmingly dominant, Smalley reasoned the molecule had to be very stable. The preferred geometry for stable molecule would reasonably be spherical, because this would mean that all bonding capabilities for carbon would be satisfied. If it were a chain or sheet like graphite, the carbon atoms could still bond at the ends, but if it were circular all ends would meet. Another hint as to the arrangement of the molecule was that there must be a high degree of symmetry for a molecule as stable as C60. Constructing a model that satisfied these requirements was fairly difficult and the group of scientists experimented with several models before coming to a conclusion. As a last resort, Smalley made a paper model by cutting out paper pentagons and hexagons in which he tried to stick them together so that the figure had 60 vertices. Smalley found that he created a sphere made out of 12 pentagons interlocking 20 hexagons to make a ball. The ball even bounced. To ensure that the shape fulfilled the bonding capabilities of carbon, Kroto and Curl added sticky labels to represent double bonds. The resulting shape is that of a truncated icosahedron, the same as that of a soccer ball. Smalley, Curl, and Kroto named the molecule buckminsterfullerene after the American architect and engineer Richard Buckminster Fuller who used hexagons and pentagons for the basic design of his geodesic domes.

Eleven days after they had begun, the scientist submitted their discovery to the prestigious journal *Nature* in a manuscript title "C60 Buckminsterfullerene." The journal received it on the 13th of September and published it on the 14th of November 1985. The controversial discovery sparked approval and criticism for a molecule that was remarkably symmetrical and stable.

### 20.9.2 How Buckyballs are Made?

Experimentally, Smalley, Kroto, and Curl, first created the buckyballs using Smalley's laser-supersonic cluster beam apparatus to knock carbons off of a plate and into a high pressure stream of helium atoms. They would be carried off and immediately be cooled to only a few degrees above absolute zero, where they would aggregate and form these buckyballs. This method however, resulted in low yields of buckyballs, and it took nearly five years until in 1990 newer methods developed by American and German scientists could manufacture buckyballs in large quantities.

The common method today involves transmitting a large current between two graphite electrodes in an inert atmosphere, such as Helium. This gives rise to a carbon plasma arc bridging the two electors, which cools instantaneously and leaves behind a sooty residue from which the buckyballs can be extracted.

### 20.9.3 Properties of Buckyballs

Since buckyballs are still relatively new, their properties are still being heavily studied. Buckyballs' unique shape and electron bonding give them interesting properties on the physical level, and on the chemical level.

1. It was **extermely stable** (physically as well as chemically).
2. Also, **fullerenes are resilient to impact and deformation**. This means, that squeezing a buckyball and then releasing it would result in its popping back in shape. Or perhaps, if it was thrown against an object it would bounce back; ironically just like the very soccer ball it resembles.
3. Buckyballs, though sparingly soluble in many solvents, are in fact, the only known carbon allotropes to be soluble.
4. Fullerene's hollow structure allows them to hold other atoms inside them.

## 20.10 USES OF BUCKYBALLS

The discovery of buckyballs has given the information about allotropes of carbon never before conceived. More importantly, these buckyballs might allow engineers and doctors do what was never before possible. Following are some of the applications/uses for buckyballs:

1. **Drug Treatments.** Buckyballs are now being considered for use in the field of medicine, both as diagnostic tools and drug candidates. C Sixty, a Toronto based company that

specializes in medical uses of fullerenes, plans to test on humans two new fullerene-based drugs for Lou Gehrig's disease and HIV in the near future.

2. **Gadolinium Carriers.** Another medical use for buckyballs is taking place in the field of diagnostics. Buckyballs unique cage-like structure might allow it to take the place of other molecules in shuttling toxic metal substances through the human body during MRI scans. Usually, the metal gadolinium is attached to another molecule and sent into the body to provide contrast on the MRI scans, but unfortunately these molecules are excreted from the system quickly to reduce the chance of toxic poisoning in the subject. Lon Wilson of Rice University and researchers at TDA Research have encased gadolinium inside buckyballs, where they cannot do harm to the patient, allowing them to remain inside the body longer, but still appear in MRIs.
3. **Nano STM.** The scanning tunnelling microscope (STM) is one of the foremost tools in microscopy today; boasting the ability to map out the topology of material surfaces at atomic resolution (i.e., on the order of 0.2 nanometres). The STM achieves this feat by bringing a needle point, functioning as a probe, within just several nanometres of a sample's surface. At these minute scales, even small disturbances can cause the tip to crach into the sample and deform it. A possible solution to this problem would be the replacement of the standard needle point with a buckyball. As fullerenes bear amazing resilience due to their spherical geometry, and would resist distortions from such collisions.
4. **Buckyballs in Curcuits.** European scientists are aiming to use buckyballs in circuit. So far, they have been able to attach a single fullerene to a copper surface, and then, through a process called ***shrink wrapping***, fitted its centre with a metal ion and made it smaller to increases electric conductivity by a hundred times.
5. **Lubricants.** Because of their shaped, they could be used equivalently to ball bearings, and thus allow surface to roll over each other, making the fullerenes equivalently lubricants.
6. **Superconductors.** It has been shown that fitting a potassium ion in the buckyball causes it to become superconductive.
7. **Catalysts.** Attaching metals onto the surface of fullerenes offers the possibility for buckyballs to become catalysts.

## 20.11 CARBON CLUSTERS

Clusters are small aggregates of atoms. C60 molecule has a spherical cage structure, but C20, a cluster of carbon, has three isomers, viz., cage structure, half spherical cauldron shape (with a pentagon and five hexagons) and a ring structure. (Fig. 20.15).

The ring structure is predicted to be the generator of a nanotube.

In clusters, the number of atoms lying on the surface is higher than the number of atoms inside, i.e., surface-to-volume ratio is quite large compared to the bulk. So, the clusters are ideal for improved and enhanced catalytic actions.

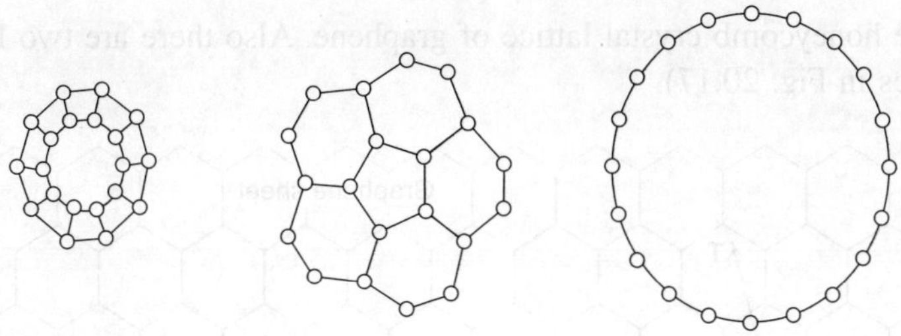


**Figure 20.15** Isomers of C20 cluster; (a) cage structure (b) cauldron shape and (c) ring structure

## 20.12 CARBON NANOTUBES

Ijiyama of NEC Japan in 1991 first announced the synthesis of carbon nanotubes. In his experiments on the arc discharge between graphite electrodes, alongwith the collected shoot particles, there were some fine tube-like stuctures. On careful analysis it was found, that they are carbon nanotubes. Its structure consists of a single sheet of graphite, rolled into a tube, both ends of which are capped with C60 fullerene hemispheres. The tubes can be either open at their ends or capped at one or both ends with half a spheroidal fullerene.

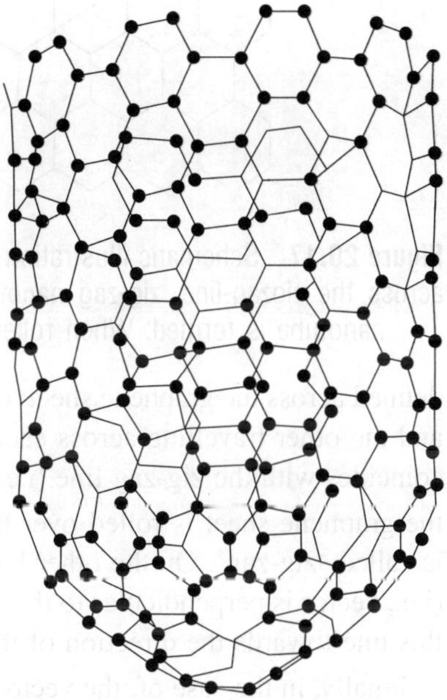

**Figure 20.16** Fullerene nanotube

### 20.12.1 Types and Structures of CNTs

Carbon nanotubes (CNTs) are allotropes of carbon. Λ single walled carbon nanotube (SWNT) is a one-atom thick sheet of graphite (called graphene) rolled up into a seamless cylinder with diameter of the order of a nanometer. The results in a nanostructure where the length-to-diameter ratio exceeds 1,000,000. On the other hand, multi-walled carbon nanotubes (MWNTs) are basically made up of a number of concentric tubes. Such cylindrical carbon molecules have novel properties that make them potentially useful in many applications in nanotechnology, electronics, optics and other fields of materials science. They exhibit extraordinary strength and unique electrical properties, and are efficient conductors of heat.

The choice of rolling axis relative to the hexagonal network of graphene sheet (called pitch) and the radius of the closed cylinder allows for different types of SWNT's, which vary from insulating to conducting. The way the graphene sheet is wrapped is represented by a pair of indices ($n_1$, $n_2$) called the chiral vector Fig. 20.17 (represented by the unit vectors ($\mathbf{r}_1 \cdot \mathbf{r}_2$) as shown in Fig. 20.17). The integers ($n_1$ and $n_2$) denote the number of unit vectors along two directions in the honeycomb crystal lattice of graphene. Also there are two lines (shown by thick dotted lines in Fig. 20.17)

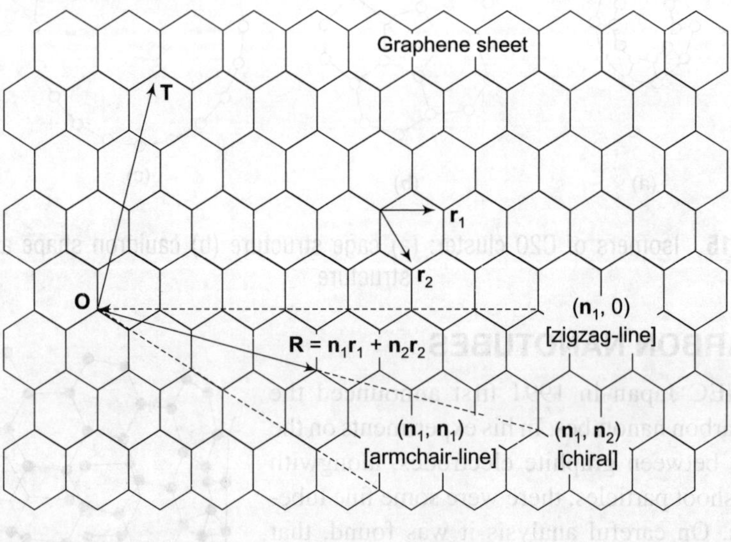


**Figure 20.17** Schematic illustration of the formation of different types of nanotubes. When rolled across the zigzag-line, zig-zag nanotube is formed. When rolled across armchair line, armchair nanotube is formed. When rolled across any arbitrary line *R*, chiral nanotube is formed

defined across the graphene sheet: one along the horizontal zig-zag bonds (called zig-zag line) and the other travelling across each hexagon (called armchair line). If $n_2 = 0$ then vector **R** coincides with the zig-zag line (i.e., **T** will be perpendicular to the zig-zag line). And now if the graphene sheet is rolled over this zig-zag line towards the direction of $\vec{T}$, the nanotube is called 'zig-zag'. On the other hand, when $n_1 = n_2$, the vector $\vec{R}$ falls on the armchair line (i.e., vector is perpendicular to the armchair line) and if the graphene sheet is now rolled over this line towards the direction of the nanotubes are called 'armchair'.

Finally, in the case of, the vector will be directed in any arbitrary direction, defined by the magnitudes of $n_1$ and $n_2$, and when the sheet is rolled along that axis, the nanotube is called 'chiral'. Figure 20.18 (a), (b), (c) are the schematic structures of armchair, zig-zag and chiral nanotubes.

The properties of nanotubes are determined by their diameter and chiral angle, both of which depend on $n_1$ and $n_2$. The diameter ($d_t$) of a nanotube is given by

$$d_t = \frac{\sqrt{3}}{\pi} a_{cc} \sqrt{n_1^2 = n_1 n_2 + n_2^2}$$

where length of chiral vector is

$$L = \left| c_h^* \right| = \sqrt{c_h^* \, c_h} = a_{cc} \sqrt{n_1^2 + n_1 n_2 + n_2^2}$$

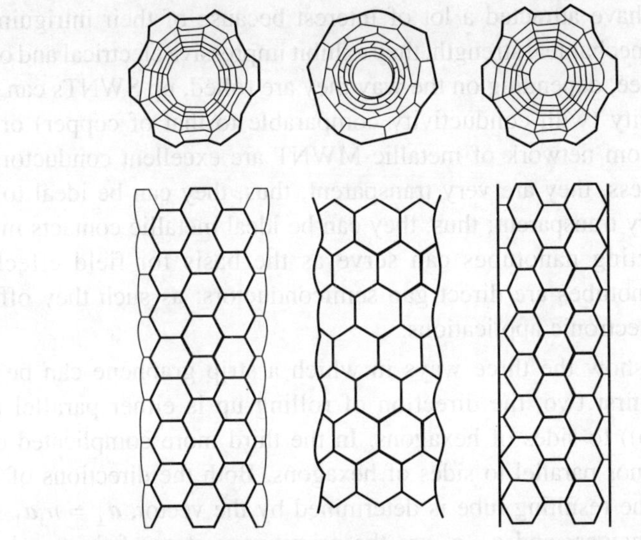


**Figure 20.18** Armchair, zig-zag and chiral nanotubes

The constant, $a_{cc}$ (= 0.142 nm) represents the nearest neighbour C–C distance.

$$d_t = \frac{1.732}{3.14} \times 0.142 \sqrt{n_1^2 + n_1 n_2 + n_2^2}$$

or $$d_t = 0.0783 \sqrt{n_1^2 + n_1 n_2 + n_2^2} \text{ nm.}$$

The chiral angle θ is defined as the angle between $\vec{c}_h$ and $\vec{r}_1$, which can be expressed as

$$\theta = \tan^{-1}\left[\frac{\sqrt{3}n_1}{2n_2 + n_1}\right]$$

with values of θ in the range $0 < \theta < 30°$.

Measurements of the nanotube diameter and the chiral angle have been made using STM (scanning tunnelling microscopy) and TEM (transmission electron microscopy). However, it remains a major challenge to determine $d_t$ and θ at the same time as measuring a physical property such as resistivity. This is partly because the nanotubes are very small and also the carbon atoms are in constant thermal motion. Further, the nanotubes can be damaged by the electron beam in the microscope.

## 20.13 ROLLING OF GRAPHENE STRIPS

There are single-wall nanotubes (SWNT) and multiple wall nanotubes (MWNT) in which SWNT are nested one within another. There are also random networks of carbon nanotubes. Typical size of a SWNT is about 10–20 Å in diameter, while the length can be as short as several nanometres and as long as many micrometres.

Carbon nanotubes have attracted a lot of interest because of their intriguing properties. Besides their unusual mechanical strength, they exhibit impressive electrical and optoelectronic features. As we shall see, depending on the way they are rolled, up SWNTs can be very good conductors of electricity (with conductivity comparable to that of copper) or they can be semiconductors. Random network of metallic MWNT are excellent conductors, while, due to their minute thickness, they are very transparent; thus, they can be ideal to their minute thickness, they are very transparent; thus, they can be ideal metallic contacts in photovoltaic systems. Semiconducting nanotubes can serve as the basis for field effect transistors; furthermore, these nanotubes are direct gap semiconductors; as such they offer intriguing possibilities for optoelectronic applications.

In Fig. 20.19, we show the three ways in which a strip graphene can be rolled up to create a tube. In the first two, the direction of rolling up is either parallel (case (a)) or perpendicular (case (b)) to sides of hexagons. In the third more complicated case (c), it is neither perpendicular nor parallel to sides of hexagons. Both the directions of rolling up as well as perimeter of the resulting tube is determined by the vector, $\boldsymbol{a}_\perp = n_1\boldsymbol{a}_1 + n_2\boldsymbol{a}_2 \equiv (n_1, n_2)$ where $n_1$, $n_2$ are integers and $\boldsymbol{a}_1$, $\boldsymbol{a}_2$ are the primitive vectors of the graphene structure shown in Fig. 20.8. (In the present case $\boldsymbol{a}_2$ has been chosen as $-\boldsymbol{a}_2$ of Fig. 20.9).

Another characteristic vector of the SWNT is the $\boldsymbol{a}_\parallel$ (defined in Fig. 20.19) which determines the periodicity in the direction parallel to the axis of the nanotube. The magnitude of the vector $\boldsymbol{a}_\perp$ and $\boldsymbol{a}_\parallel$ are

$$|\boldsymbol{a}_\perp| = \sqrt{3}\,d\,(n_1^2 + n_2^2 + n_1 n_2)^{1/2}, \qquad ...(1)$$

$$|\boldsymbol{a}_\parallel| = \sqrt{3}\,d\,(m_1^2 + m_2^2 + m_1 m_2)^{1/2}, \qquad ...(2)$$

where $d = 1.42$ A is the bond length in graphene. According to (1), the diameter of 10 Å in the zig-zag case implies that $n \simeq 13$. From the orthogonality of $\boldsymbol{a}_\parallel$ and $\boldsymbol{a}_\perp$ follows that $\boldsymbol{a}_\parallel \cdot \boldsymbol{a}_\perp = 0$ or $n_1 m_1 + n_2 m_2 + \frac{1}{2}(n_1 m_2 + n_2 m_1) = 0$ or

$$1 + qq' + \frac{1}{2}(q' + q) = 0 \qquad ...(3)$$

where $q \equiv n_2/n_1$ and $q' \equiv m_2/m_1$. Knowing $q$, we determine from (3) $q' = -\left(1 + \frac{1}{2}q\right)/\left(q + \frac{1}{2}\right)$ and then $\boldsymbol{a}_\parallel = (m_1, q'm_1)$ with $|m_1|$, $|q'm_1|$ the smallest integers.

**Example 20.1** Show that the vector all for the three cases shown in Fig. 20.19 are the following: (a) $\boldsymbol{a}_\parallel = \boldsymbol{a}_1 - \boldsymbol{a}_2$; (b) $\boldsymbol{a}_\parallel = \boldsymbol{a}_1 - 2\boldsymbol{a}_2$; (c) $\boldsymbol{a}_\parallel = 4\boldsymbol{a}_1 - 5\boldsymbol{a}_2$.

In the armchair case, the period in the direction normal to the tube axis is define by $\boldsymbol{a}_3 \equiv \boldsymbol{a}_1 + \boldsymbol{a}_2 = (1, 1)$, while the period in the parallel direction is $\boldsymbol{a}_4 \equiv \boldsymbol{a}_1 - \boldsymbol{a}_2 = (1, \bar{1})$. It is very easy to show that the dot product of $\boldsymbol{a}_3$ and $\boldsymbol{a}_4$ is zero, as expected, while the cross product of $\boldsymbol{a}_3$ and $\boldsymbol{a}_4$ is twice as large in magnitude as the quantity $|\boldsymbol{a}_1 \times \boldsymbol{a}_2|$. These relations mean that the vectors $\boldsymbol{a}_3$ and $\boldsymbol{a}_4$ define a rectangular primitive cell of twice the area of the primitive cell of the graphene lattice. Hence, the first Brillouin zone corresponding to $\boldsymbol{a}_3$ and $\boldsymbol{a}_4$ is a

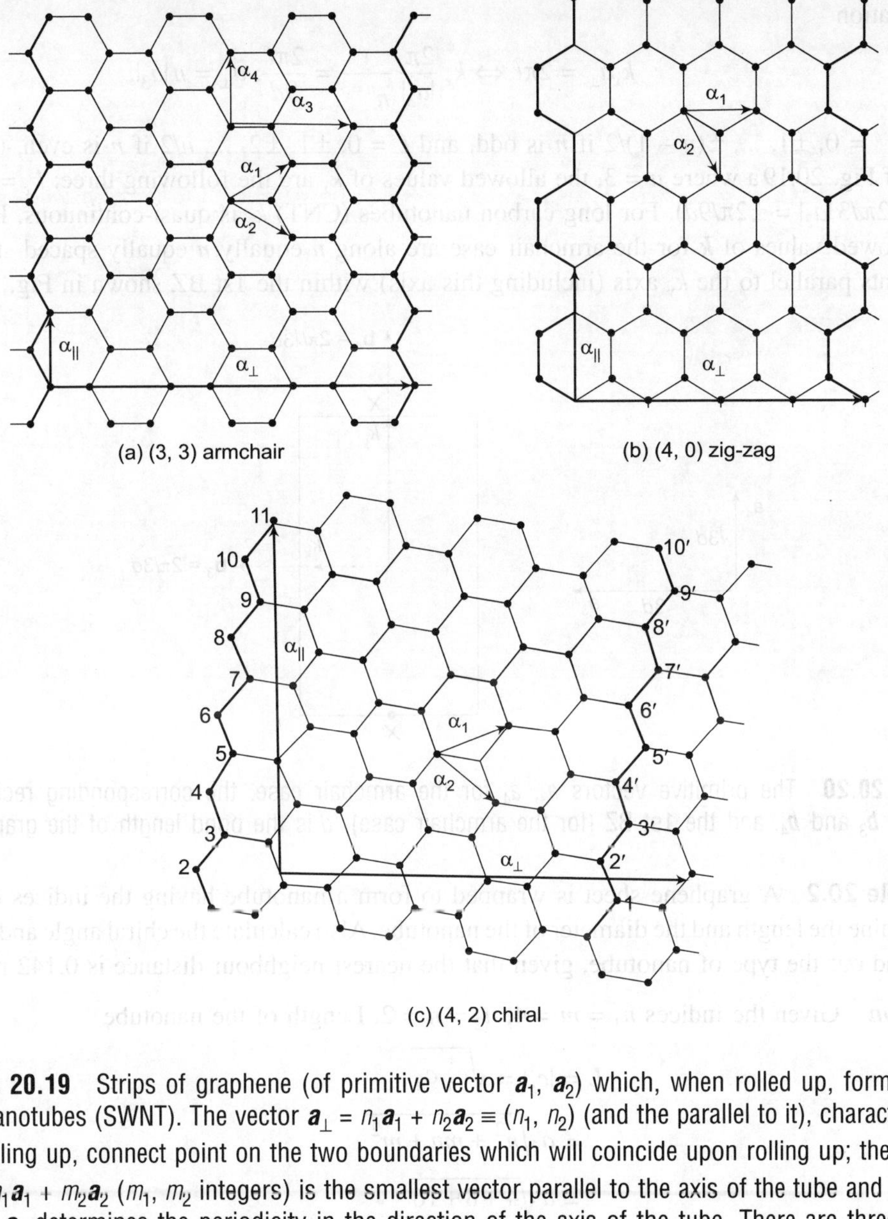


**Figure 20.19** Strips of graphene (of primitive vector $\boldsymbol{a}_1$, $\boldsymbol{a}_2$) which, when rolled up, form single wall nanotubes (SWNT). The vector $\boldsymbol{a}_\perp = n_1\boldsymbol{a}_1 + n_2\boldsymbol{a}_2 \equiv (n_1, n_2)$ (and the parallel to it), characterizing the rolling up, connect point on the two boundaries which will coincide upon rolling up; the vector $\boldsymbol{a}_{\|} = m_1\boldsymbol{a}_1 + m_2\boldsymbol{a}_2$ ($m_1$, $m_2$ integers) is the smallest vector parallel to the axis of the tube and normal to $\boldsymbol{a}_\perp$; $\boldsymbol{a}_{\|}$ determines the periodicity in the direction of the axis of the tube. There are three types of SWNT: (a) the $\boldsymbol{a}_\perp = (n, n)$ which is called armchair, (b) the $\boldsymbol{a}_\perp = (n, 0)$ which is called zigzag, and (c) the $\boldsymbol{a}_\perp = (n_1, n_2)$ with $n_1 > n_2 > 0$ which produces a chiral tube

rectangular of half the area of the 1st BZ of graphene as shown in Fig. 20.20. However, not all points in this 1st BZ are allowed. Because of the rolling up in the $a3$ direction, we have periodic boundary conditions which restrict the values of $k_x$ (see Fig. 20.16) to those satisfying the relation

$$k_x a_\perp = 2\pi\ell \Leftrightarrow k_x \frac{2\pi}{|a_3|} \frac{\ell}{n^{n=3}} = \frac{2\pi\ell}{9d};\ a_\perp = n|a_3|, \qquad ...(4)$$

where $\ell = 0, \pm 1, ..., \pm(n-1)/2$ if $n$ is odd, and $\ell = 0, \pm 1, \pm 2, ..., n/2$ if $n$ is even. (In the case of Fig. 20.19a where $n = 3$, the allowed values of $k_x$ are the following three: $k_x = 0$ and $k_x = \pm 2\pi/3|a_3| = \pm 2\pi/9d$). For long carbon nanotubes (CNT) $k_y$ is quasi-continuous. Hence, the allowed values of $\boldsymbol{k}$ for the armchair case are along $n$ equally $n$ equally spaced straight segments parallel to the $k_y$ axis (including this axis) within the 1st BZ shown in Fig. 20.20.

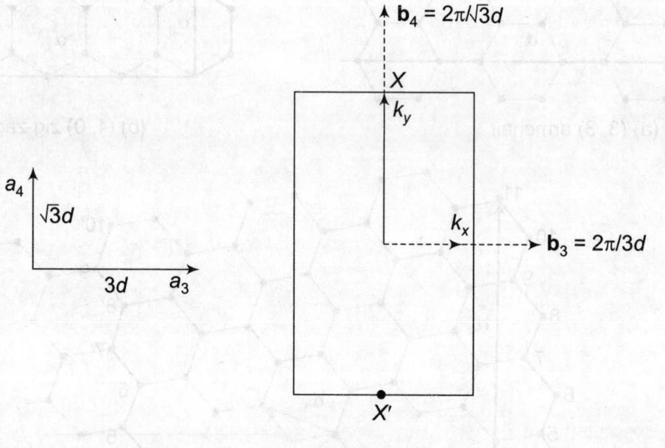


**Figure 20.20** The primitive vectors $a_3$, $a_4$ for the armchair case, the corresponding reciprocal vector $\boldsymbol{b}_3$ and $\boldsymbol{b}_4$, and the 1st BZ (for the armchair case); $d$ is the bond length of the graphene

**Example 20.2** A graphene sheet is wrapped to form a nanotube having the indices (4, 2). Determine the length and the diameter of the nanotube. Also calculate the chiral angle and using this find out the type of nanotube, given that the nearest neighbour distance is 0.142 nm.

***Solution*** Given the indices $n_1 = m = 4$, $n_2 = n = 2$. Length of the nanotube

$$L = |c_h| = \sqrt{c_h^* c_h}$$

$$= a\sqrt{n^2 + mn + m^2}$$

$$= a\sqrt{4+8+16}$$

$$= 5.29\ a \text{ units}$$

Diameter of the nanotube

$$d_t = \frac{\sqrt{3}}{\pi} a_{cc} \sqrt{m^2 + mn + n^2}$$

$$= \frac{1.732}{3.14} \times (0.142 \text{ nm}) \times \sqrt{28}$$

$$= 0.0783 \times 5.29 = 0.4142 \text{ nm}$$

(where $a_{cc}$ is the nearest neighbour C–C distance, chiral angle

$$\theta = \tan^{-1}\left[\frac{\sqrt{3}\times 2}{2\times 4+2}\right]; \quad \begin{pmatrix} n_1 \equiv n = 2 \\ n_2 \equiv m = 4 \end{pmatrix}.$$

$$= \tan^{-1}[0.3464] = 19.1°$$

Types of nanotube:

As $m \neq n \neq 0$, also $0 < \theta < 30°$

Hence, this is a chiral nanotube.

## 20.14 FABRICATION OF CARBON NANOTUBES

CNTs are placed in two broad categories:

(i) Single-walled carbon nanotubes (SWCNTs)

(ii) Multi-walled carbon nanotubes (MWCNTs)

Conceptually, SWCNTs can be considered to be formed by the rolling of a single layer of graphite (called a graphene layer) into a seamless cylinder. A MWCNT can similarly be considered to be a coaxial assembly of cylinders of SWCNTs, one within another, the separation between tubes is about equal to that between the layers in natural graphite.

A single-walled nanotube can have a diameter of 2 nm and a length of 100 μm, making it effectively a one-dimensional structure called a *nanowire.*

CNT synthesis requires three things: *carbon feed*, a *metal catalyst* and *heat*. In the case of SWCNT production requires a metal catalyst, whereas producing MWCNT (which is through carbon arc discharge), it does not require metal catalyst. Prominent methods for CNT synthesis are:

1. Laser-vaporization technique
2. Carbon arc discharge technique
3. Chemical vapour deposition method
4. Electrolysis

**1. Laser Evaporation Technique.** Figure 20.21 illustrates the apparatus for making CNTs by laser evaporation. A quartz tube containing argon gas and a graphite target are heated to 1200°C. Contained in the tube, but somewhat outside the furnace, is a water-cooled copper collector. The graphite target contains small amounts of Co and Ni that act as catalytic nucleation sites for the formation of the tubes. An intense pulsed laser beam is incident on the target, evaporating carbon from the graphite. The argon then sweeps the carbon atoms from the high temperature zone to the colder copper collector on which they condense into nanotubes. Tubes 10–20 nm in diameter and 100 μm long can be made by this method.

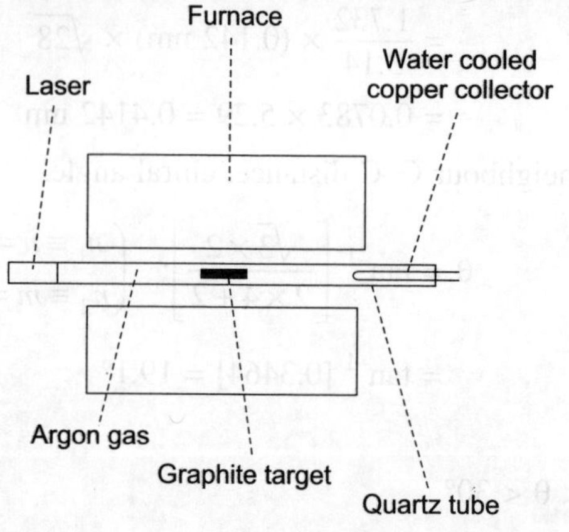


**Figure 20.21** Experimental arrangement for synthesizing carbon nanotubes by laser evaporation

**2. Carbon-Arc Discharge Technique.** Nanotubes can also be synthesized using a carbon arc. A potential of 20–25 V is applied across carbon electrodes of 5–20 μm diameter and separated by 1 mm at 500 torr pressure of flowing helium. Carbon atoms are ejected from the positive electrode and form nanotubes on the negative electrode. As the tubes form, the length of the positive electrode decreases, and a carbon deposit forms on the negative electrode. To produce single walled nanotubes, a small amount of cobalt, nickel or iron is incorporated as a catalyst in the central region of the positive electrode. If no catalysts are used, the tubes are nested or multi-walled types (MWNT), which are nanotubes within nanotubes, as illustrated in Fig. 20.22. The carbon arc method can produce single walled nanotubes of diameters 1–5 nm with a length of 1 μm.

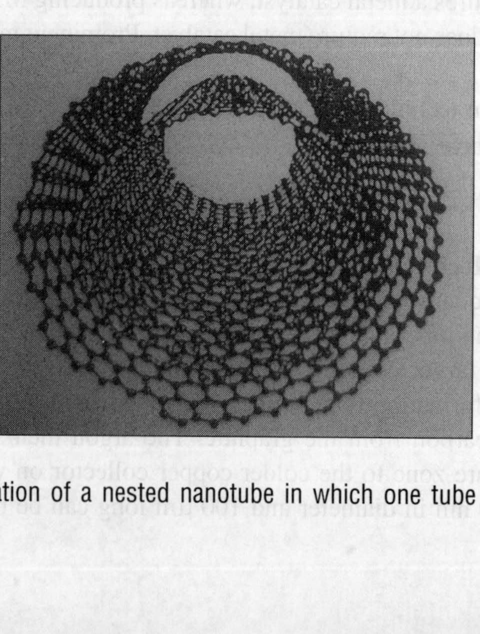

**Figure 20.22** Illustration of a nested nanotube in which one tube is inside the another

**3. The Chemical Vapour Deposition Method.** The chemical vapour deposition method involves decomposing a hydrocarbon gas such as methane ($CH_4$) at 1100°C. As the gas decomposes, carbon atoms are produced that then condense on a cooler substrate that may contain various catalysts such as iron. This method produced tubes with open ends, which does not occur when other methods are used. This procedure allows continuous fabrication and may be the most favourable method for scale-up and production.

**4. Electrolysis.** It is a less common method for CNT production. MWCNTs are formed when a current is passed through two graphite electrodes immersed in molten ionic salts, such as LiC1 at 600°C. After electrolysis, the carbonaceous material is extracted by dissolving the ionic salt in distilled water and separating the dispersion by filtration. The material produced contains MWCNTs, carbon-encapsulated particles, amorphous carbon and carbon filaments.

The mechanism of nanotube growth is not understood. Since the metal catalyst is necessary for the growth of SWNTs, the mechanism must involve the role of the Co or Ni atoms. One proposal referred to as the "scooter mechanism" suggests that atoms of the metal catalyst attach to the dangling bonds at the open end of the tubes, and that these atoms scoot around the rim of the tube, absorbing carbon atoms as they arrive.

Generally, when nanotubes are synthesized, the result is a mix of different kinds, some metallic and some semiconducting. A group at IBM has developed a method to separate the semiconducting from the metallic nanotubes. The separation was accomplished by depositing bundles of nanotubes, some of which are metallic and some semiconducting, on a silicon wafer. Metal electrodes were then deposited over the bundle. Using the silicon wafer as an electrode, a small bias voltage was applied that prevents the semiconducting tubes from conducting, effectively making them insulators. A high voltage is then applied across the metal electrodes, thereby sending a high current through the metallic tubes but not the insulating tubes. This causes the metallic tubes to vaporize, leaving behind only the semiconducting tubes.

## 20.15 PROPERTIES OF CARBON NANOTUBES

**1. Electrical Properties.** CNT properties are dependent on their configuration. For example, one can say that for typical diameters, all armchair SWCNTs (single wall carbon nanotubes) and 1/3$^{rd}$ of all zig-zag nanotubes are *metallic*, the rest are semiconducting, i.e., the metallic conduction occurs when

$$n_1 - n_2 = 3l$$

where $l$ is an integer.

Otherwise, the nanotube is a semiconductor. Thus, all armchair ($n_1 = n_2$) nanotubes are metallic and nanotubes (5, 0), (6, 4), (9, 1), etc., are semiconducting. Theoretically, metallic nanotubes can have an electrical current density more than 1000 times greater than metals such as silver and copper.

**2. Thermal Properties.** All nanotubes are expected to be very good thermal conductors along the tube, exhibiting a property called *ballistic conduction;* but good insulators laterally to the tube axis. It is predicted that CNTs can transmit upto 3000 W/mK at room temperature, compared to copper (a metal well known for its good thermal conductivity) which only transmits 385 W/mK.

**3. Mechanical Properties.** CNTs are the strongest and stiffest materials on earth in terms of tensile strength and elastic modules respectively. This strength results from the covalent $sp^2$ bonds between the individual carbon atoms, CNTs have a tensile strength of 150 GPa as compared to steel which has tensile strength of 0.4 GPa only. Since CNTs have a low density for a solid of 1.3 to 1.4 g/cm$^3$, its specific strength of up to 48000 kN.m. kg$^{-1}$ is the best of known materials compared to high carbon steels (154 kN.m kg$^{-1}$).

**Table 20.2** Comparison of Mechanical Properties of CNTs and structural Materials

| *Materials* | *Young's modulus (GPa)* | *Tensile strength (GPa)* | *Density g/cm$^3$* |
|---|---|---|---|
| Single Walled CNTs | 1054 | ~150 | 1.3–1.4 |
| Multiwalled CNTs | 1200 | ~150 | 2.6 |
| Steel | 208 | 0.04 | 7.8 |
| Epoxy | 3.5 | 0.05 | 1.25 |
| Wood | 16 | 0.08 | 0.6 |

## 20.16 OTHER PROPERTIES OF CNTs

These can be summarized as follows:

1. CNTs have high electrical conductivity.
2. CNTs have very high tensile strength.
3. CNTs are highly flexible: can be bent considerably without damage.
4. CNTs are very elastic ~ 18% elongation to failure.
5. CNTs have high thermal conductivity (~ 3000 W/m K) in the axial direction with small values in the radial direction.
6. CNTs have a low thermal expansion coefficient
7. CNTs are good electron field emitters.
8. CNTs have a high aspect ratio (length ≈ 1000 × diameter)
9. CNTs have high Young's modulus of over 1 TPa Vs 70 GPa for aluminium, 700 GPa for C-fibre.
10. High Young's modulus of CNTs confirms that CNTs are indeed, rather soft in the radial direction.

## 20.17 EFFECT OF DEFECTS ON CNT PROPERTIES

As with any material, the existence of defects affects the material properties. Defects can occur in the form of ***atomic vacancies.*** High levels of such defects can lower the tensile strength by up to 85%. Another form of defect that may occur in carbon nanotubes is known as the ***Stone Wales Defect,*** which creates a pentagon and heptagon pair by rearrangement of the bonds.

Because of the very small structure of CNTs, the tensile strength of the tube depends on its weakest segment in a similar manner to a chain, where a defect in a single link diminishes the strength of the entire chain.

The tube's electrical properties are also affected by the presence of defects. A common result is the lowered conductivity through the defective region of the tube. Some defect formation in armchair-type tubes (which can conduct electricity) can cause the region surrounding that defect to become semiconducting. Furthermore single monoatomic vacancies induce magnetic properties.

The tube's thermal properties are heavily affected by defects. Such defects lead to *phonon scattering,* which in turn, increases the relaxation rate of the phonons. This reduces the mean free path, and reduces the thermal conductivity of nanotube structures. Phonon transport simulations indicate that ***substitutional defects*** such as nitrogen or boron will primarily lead to scattering of high frequency optical phonons. However, larger scale defects such as Stone Wales defects cause phonon scattering over a wide range of frequencies, leading to a greater reduction in thermal conductivity.

**Example 20.3** Prove that the diameter ($d_t$) of ($m$, $n$) carbon nanotube is related to chiral angle ($\theta$) by the relation

$$d_t = \frac{3a_{cc}}{2\pi}\frac{n}{\sin\theta}$$

where $a_{cc}$ is the constant that represents the nearest neighbour C–C distance.

*Solution* We know that diameter ($d_t$) of a ($m$, $n$) CNT is given as

$$d_t = \frac{\sqrt{3}\,a_{cc}}{\pi}\sqrt{m^2 + mn + n^2} \qquad \text{...(i)}$$

and chiral angle is given as

$$\theta = \tan^{-1}\left(\frac{\sqrt{3}\,n}{2m+n}\right) \qquad \text{...(ii)}$$

Rearranging Eq. (ii), we have

$$\sin\theta = \frac{\sqrt{3}\,n}{\sqrt{(2m+n)^2 + 3n^2}}$$

$$= \frac{\sqrt{3}\,n}{\sqrt{4m^2 + n^2 + 4mn + 3n^2}}$$

$$= \frac{\sqrt{3}\,n}{\sqrt{4m^2 + 4mn + 4n^2}} = \frac{\sqrt{3}\,n}{2\sqrt{m^2 + mn + n^2}}$$

or $$\sqrt{m^2+mn+n^2} = \frac{\sqrt{3}n}{2\sin\theta} \quad \text{...(iii)}$$

Substitute this value in Eq. (i), we get

$$d_t = \frac{\sqrt{3}a_{cc}}{\pi}\frac{\sqrt{3}n}{2\sin\theta}$$

$$d_t = \frac{3a_{cc}}{\pi}\frac{n}{\sin\theta}$$

Hence, **proved.**

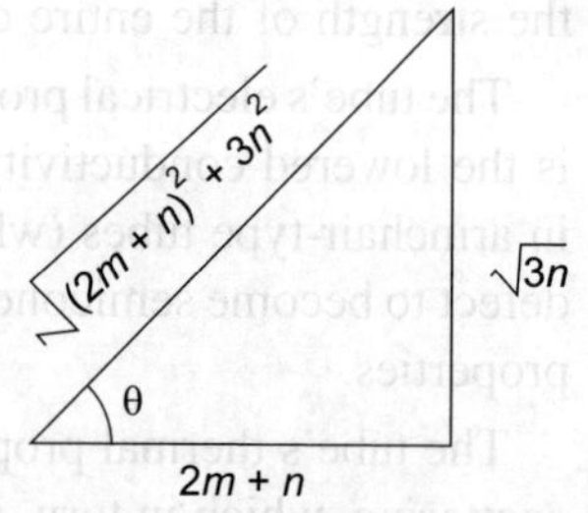


**Figure 20.23**

## 20.18 USES OF CARBON NANOTUBES

The strength and flexibility of carbon nanotubes make them of potential use in controlling other nanoscale structures, which suggests they will have an important role in nanotechnology engineering. The uses/applications of CNTs can be studied under the following sub-headings:

**1. Structural.** As CNTs possesses good mechanical properties, a variety of structures have been proposed ranging from everyday items like clothes and sports gear to combat jackets and space elevators.

**2. Composites**

(i) The high aspect ratio (length-to-radius ratio) and high conductivity of CNTs makes them excellent for conducting composites. The first use of such composites was by Hyperion for electrostatically applying paint onto car components.

(ii) Another use of CNT composites is as antistatic shielding on airplane wings and fuse lages.

(iii) Further use of CNT is for shielding of electromagnetic interference.

(iv) A fourth application of conducting composites is as a transparent conductor.

**3. Field Emission.** The high aspect ratio also makes CNTs ideal field-emission (FE) materials. FE is the emission of electrons from a solid under an intense electric field.

(i) Application of FE from CNTs is as electron guns for next generation scanning electron microscopes (SEMs) and transmission electron microscopes (TEMs).

(ii) A second use of CNTs is as FE cathodes in high power microwave amplifiers such as Klystrons for base stations.

(iii) A third possible application is as electron sources for miniature X-ray sources. Here an electron beam from a CNT cathode is accelerated to strike a metal target to create X-rays.

(iv) A fourth application is in FE displays (FEDs). The FED is a flat panel display in which the electron beams from many nanotube electron guns are controlled to create an image on a screen of phosphor pixels.

The advantages of FEDs over LCDs are a *higher video rate, higher power efficiency* because it is an emissive display, and wider operating temperature range. Beside this FEDs possess excellent picture quality, low cost, long life span and short response time.

**4. Electrochemistry.** Another area of potential application is electrochemistry. The large surface area and low resistivity of CNTs makes them of great interest in electrochemistry.

(i) As we know that the capacitance is given as

$$C = \frac{\varepsilon A}{d}$$

where ε is the dielectric constant, $A$ is the electrode surface area (in this case) and $d$ is the electrochemical double layer (ECDL) thickness (~ 0.4 nm).

This capacitance value for CNT is very large because of large $A$ and small $d$ values. This has led to the use of CNT for the electrodes in so-called ***supercapacitors.*** Supercapacitors consist of two carbon-based electrodes on a mylar backing, an ion-permeable membrane, and an electrolyte.

The performance of supercapacitors is judged in terms of their energy density (per unit volume) or $\frac{1}{2}CV^2$ and power density or $\frac{V^2}{R}$, where $R$ is their series discharge resistance.

(ii) CNTs are used to increase the conductivity of porous carbon electrodes and thus improve the power density of Li ion batteries.

(iii) Tabular structure of CNTs suggest that capillarity effects are important. Thus, CNTs could be very useful for the storage of hydrogen.

**5. Electronics.** CNTs can carry the highest current density of any metal, $10^9 - 10^{10}$ A/cm$^2$, over 1000 times that for Cu, before failing as a result of electromigration. It is well known that feature sizes in Si integrated circuits are continually reducing. This forces the interconnects between each transistor to carry over larger current densities.

CNTs have also been made into field-effect transistors (FETs). In comparing nanotube FETs with Si FETs, the performance is 20 times better. However, a nanotube FET occupies a huge wafer area compared to its diameter because of its connections, so the 'economic' figure of merit is actually quite poor.

CNTs have also been shown to be good sensing elements for pressure, fluid-flow and alcohol and are potential candidates for system-on-a chip (SOC) sensors.

## 20.19 OTHER APPLICATIONS OF CNTs

(i) Carbon nanotubes have also been implemented in nanoelectromechanical systems (NEMS), including *mechanical memory elements* (NRAM) and *nanoscale electric motors.*

(ii) Carbon nanotubes have also been proposed as a possible *gene delivery vehicle* and for use in combination with radio frequency (RF) fields to destroy cancer cells.

(iii) Eikos Inc of Franklin, Massachusetts and Unidym Inc. of Silicon Valley, California are developing transparent, electrically conductive film carbon nanotubes to replace indium tin oxide (ITO). Carbon nanotube films are substantially more mechanically robust than ITO films, making them ideal for *high reliability touch screens and flexible displays.* Printable water-based inks of carbon nanotubes are desired to enable the production of these films to replace ITO. Nanotube films show promise for use in displays for computers, cell phones, PDAs and ATMs.

(iv) Carbon nanotubes are said to have the strength of diamond, and research is being made into weaving them into clothes to create *stab-proof* and *bullet-proof clothing.* The nanotubes would effectively stop the bullet penetrating the body but the force and velocity of the bullet would be likely to cause broken bones and internal bleeding.

(v) A flywheel made of carbon nanotubes could be spun at extremely high velocity on a floating magnetic axis, and potentially store energy at a density approaching that of conventional fossil fuels. Since energy can be added to and removed from flywheels very efficiently in the form of electricity, this might offer a way of storing electricity, making the electrical grid more efficient and variable power suppliers (like wind turbines) more useful in meeting energy needs. The practicality of this depends heavily upon the cost of making massive, unbroken nanotube structures and their failure rate under stress.

(vi) Many researchers and corporations have already developed *CNT based air* and *water filtration* devices. It has been reported that these filters cannot only block the smallest particles but also kill most bacteria.

**Conclusion and Summary**

From the above-mentioned uses and applications of carbon nanotube in the area of nanoscience and nanotechnology, we can say that some applications are commercialized, some are near commercialized and others are meant for research and development (R&D). Hence these applications can be summarized as follows:

**Commercialized**

- Composites
- Battery materials

**Near Commercialized**

- Field emission

**R&D**

- Electronic
- Sensor
- $H_2$ storage
- Catalysts
- Adsorbent

## 20.20 GROWTH TECHNIQUES OF NANOMATERIALS

During the past 30 years research on nanoscale, it is becoming increasingly clear that no single fabrication technique is universally applicable for the production of nanometre scale semiconductor devices. In spite of the fact that lithographic techniques have been preferred methods for fabrication of some devices, yet nonlithographic techniques are becoming promising for many other nanodevices, which we will consider in the subsequent sections.

### 20.20.1 Top-down vs Bottom-up Techniques

There are two different approaches of material fabrication. One is the top-down approach and the other is the bottom-up approach. In top-down technique, generally a bulk material is taken and machined to modify into the desired shape and product. Examples of this type of technique are the manufacturing of integrated circuits using a sequence of steps such as crystal growth, lithography, etching, ion implantation, etc. For nanomaterial synthesis, ball-milling is an important top-down approach, where macrocrystalline structures are broken down to nanocrystalline structures, but original integrity of the material is retained. Sometimes this method is used to prepare nanostructured metal oxides by chemical reaction between two constituents during crushing. The crystallites are allowed to react with each other by the supply of kinetic energy during milling process to form the required nanostructured oxide.

On the other hand, bottom-up technique is used to build something from basic materials, for example, assembling materials from the atoms/molecules up, and, therefore very important for nano-fabrication. Unlike lithographic technique of top-down approach, which is extensively used in silicon industry, this bottom-up non-lithographic approach of nanomaterial synthesis is not completely proven in manufacturing yet but has great potential to become important alternative to lithographic process. Examples of bottom-up technique are self-assembly of nanomaterials, sol-gel technology electrodeposition, physical and chemical vapour deposition (PVD, CVD), epitaxial growth, laser ablation, etc. Details of these procedures are discussed in the subsequent sections.

### 20.20.2 e-beam Evaporation

This technique is based on the heat produced by high energy electron beam bombardment on the material to be deposited. The electron beam is generated by an electron gun, which uses the thermionic emission of electrons produced by an incandescent filament (cathode). Emitted electrons are accelerated towards an anode by a high difference of potential (kV). The material itself or the crucible or a near perforated disc can act as the anode. A magnetic field is often applied to bend the electron trajectory, allowing the electron gun to be positioned below the evaporation line (as shown in Fig. 20.24). The energetic impact of the electron beam raises the temperature of the evaporant, creating a hot zone on the material or a heated 'oven' in the case of a crucible. At a sufficiently elevated temperature, the material will begin to evaporate. As electrons can be localized, it is possible to obtain a much localized heating on the material to evaporate, with a high density of evaporation power (several kW). This allows controlling the evaporation rate, from low to very high values, and best of all, the chance of depositing materials with high melting point (W, Ta, C, etc.). Cooling the crucible avoids contamination problems from heating and degasification. An intrinsic part of the evaporation process results in a small fraction of the vapour being in an ionized state. This phenomenon can be harnessed to determine the deposition rate without recourse to film thickness monitors

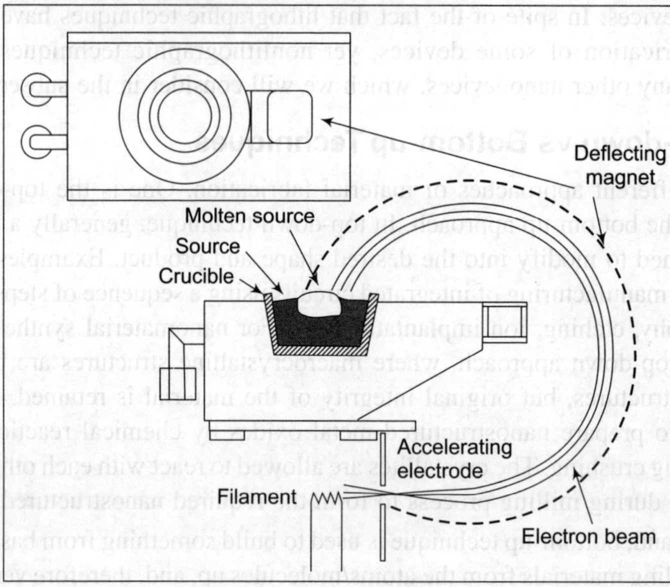


**Figure 20.24** Schematic diagram of an e-beam evaporator system. Electron beams are accelerated from a heated filament towards the source by accelerating anodes and magnetic field. Due to the bombardment of highly energetic electron beam, local heat is generated at the source, which evaporates the material

during evaporation. All evaporators are equipped with flux monitoring plates which not only measure the ion current in the beam, but also act to extract the vast majority of ions, leaving the beam predominantly neutral.

### 20.20.3 Chemical Vapour Deposition

Chemical vapour deposition (CVD) is a chemical process used to produce high-purity, high-performance solid materials. The process is often used in the semiconductor industry to produce thin films. It is a technique for synthesizing materials in which chemical components in vapour phase react to form a solid film at some surface (substrate). The occurrence of chemical reaction is central to this means of thin film growth, as is the requirement that the materials must start out in the vapour phase. Ability to control the components and physical conditions of the gas phase, the solid surface and the envelope that surrounds them determine the capability to control the properties of the thin film deposited by the CVD process. Micro- and nano-fabrication processes widely use CVD to deposit materials in various forms, including monocrystalline, polycrystalline, amorphous, and epitaxial. These materials include: silicon, carbon fibre, carbon nano-fibres/nano-filaments/nanotubes/nanorods, $SiO_2$, silicon-germanium, tungsten, silicon carbide nanostructures, silicon nitride nanomaterials, silicon oxynitride, titanium nitride, and various high-*k* dielectrics. The CVD process is also used to produce synthetic diamonds and nano-diamonds. A schematic diagram of a CVD chamber is shown in Fig. 20.25.

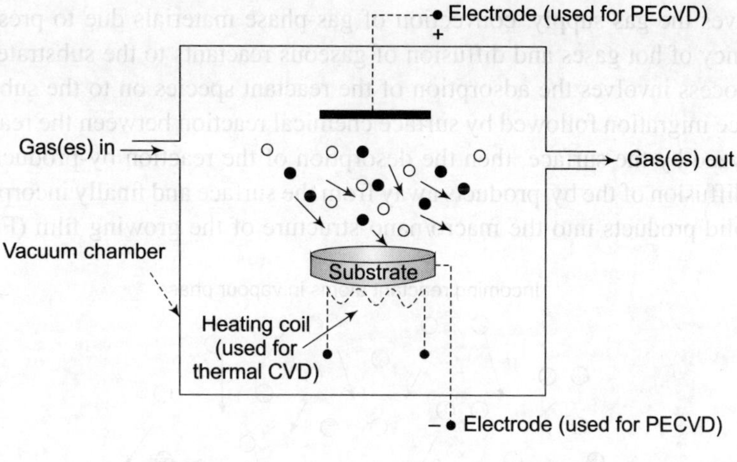


**Figure 20.25** Schematic diagram of a CVD system. A heating coil is used for thermal CVD and electrodes are used for plasma enhanced (PE) CVD

CVD is a sequential process which starts from the initial vapour phase, progresses through a series of quasi-steady sub-processeses, and culminates in the formation of a solid thin film of its final micro- or nano-structure. The sequence is illustrated by the flowchart in Fig. 20.26

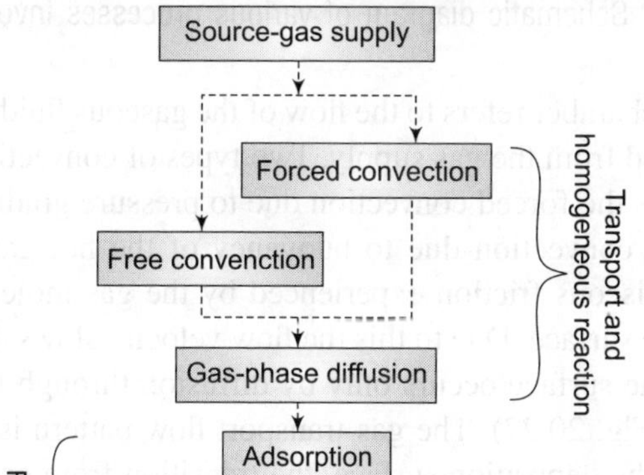


**Figure 20.26** Flowchart of the various steps involved in a CVD process

and schematically represented in Fig. 20.27. In general, the CVD process can be divided into two broad steps, one is transport of gas phase materials to the reaction zone and the reaction of materials and the other is the deposition of the film on the substrate. The transport process involves the gas supply, convection of gas-phase materials due to pressure gradient and/or buoyancy of hot gases and diffusion of gaseous reactants to the substrate surface. The deposition process involves the adsorption of the reactant species on to the substrate surface sites by surface migration followed by surface chemical reaction between the reactant species, usually catalyzed by the surface, then the desorption of the reaction by-products takes place followed by diffusion of the by-products away from the surface and finally incorporation of the condensed solid products into the macro/nano-structure of the growing film (Fig. 20.27).

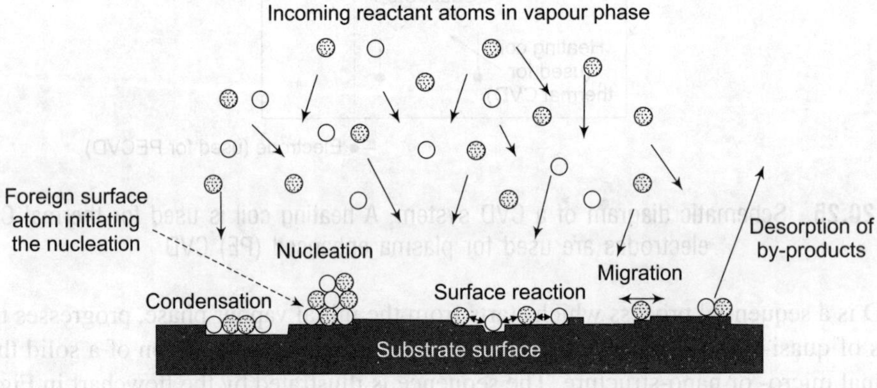


**Figure 20.27** Schematic diagram of various processes involved in CVD

Convection in a CVD chamber refers to the flow of the gaseous fluid as it moves through the reactor after being injected from the gas supply. Two types of convection generally take place inside the chamber, one is the forced convection due to pressure gradient across the chamber and the other is the free convection due to buoyancy of the hot gases. Another important process involved is the viscous friction experienced by the gas molecules, when they come very close to the substrate surface. Due to this the flow velocity slows down and the remaining transport of reactant to the surface occurs only by diffusion through the relatively stationary boundary layer of fluid (Fig. 20.27). The gas-transport flow pattern is an important aspect to determine how far from the deposition surface, the transition from convection to diffusion is occurring and which of these two transport processes is limiting the reactant arrival rate at the substrate surface. Once the transition from convection to the diffusion takes place, the source gas molecules are absorbed to the surface followed by the surface reaction to produce the required material. The important factors that influence the homogeneous reaction are the gas residence time near the surface and gas heating, both of which are functions of the gas flow rate and flow pattern. Depending on these factors, the reaction often begins in the gas phase rather than occurring entirely at the substrate surface. Generally, the gas phase reactions are homogeneous and produce powdery material and surface reactions are heterogeneous and produce thin films.

### 20.20.4 Pulsed Laser Deposition

Pulsed laser deposition (PLD) is another deposition technique used for thin film deposition as well as nanomaterial synthesis. In this deposition technique, a high power pulsed laser beam is focused inside a vacuum chamber to strike a target of the desired material and composition to vaporize the target and deposit materials on a substrate. The deposition generally takes place in ultra high vacuum or in the presence of some background gas of certain pressure. Some neutral gases such as Ar are used at some desired pressure to control the deposition rate as well as the film composition whereas reactive gases like oxygen or nitrogen are sometimes used to deposit oxide or nitride materials. PLD has been utilized to fabricated high quality crystalline films. It has gained a great deal of attention in the past few years for its ease of use and success in depositing materials of complex stoichiometry. PLD was the first technique used to successfully deposit a superconducting $YBa_2Cu_3O_7$ thin film. Since that time, many materials that are normally difficult to deposit by other methods, especially multi-element oxides, nitrides, metallic multilayers and various superlattices have been successfully deposited by PLD. In the 1990s the development of new laser technology, such as lasers with high repetition rate and short pulse durations, made PLD a very competitive tool for the growth of thin, well-defined films with complex stoichiometry.

The basic setup of PLD is shown in Fig. 20.28, which is very simple relative to many other deposition techniques, but the physical phenomena of laser-target interaction and film growth are quite complex. There are many different arrangements to build a deposition chamber for PLD. The target material which is evaporated by the laser is normally found as a rotating disc attached to a support. However, it can also be sintered into a cylindrical rod with rotational

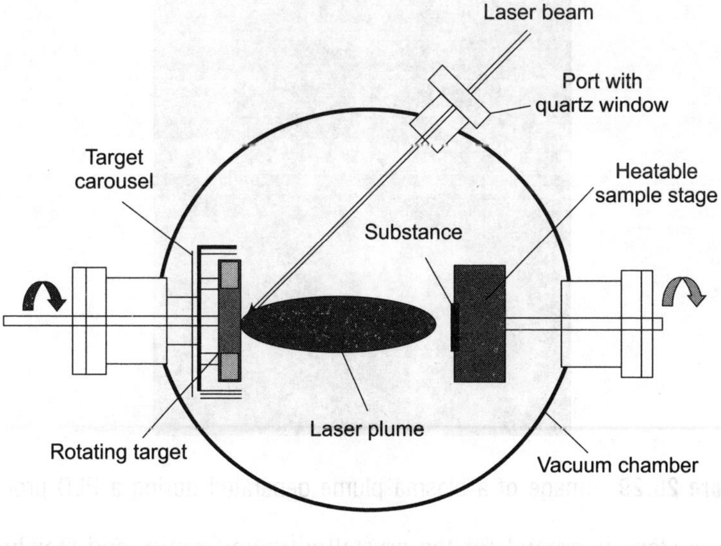


**Figure 20.28** Basic setup of a PLD system. Laser beam is collimated through lenses to the target for ablation. Target can be rotated. A laser plume is generated between target and the substrate. The substrate can be rotated as well as heated (if needed) for uniform film deposition

motion and a translational up and down movement along its axis. This special configuration allows not only the utilization of a synchronized reactive gas pulse but also of a multicomponent target rod with which films of different multilayers can be created. The main advantage of PLD derives from the laser material removal mechanism; PLD relies on a photon interaction to create an ejected plume of material from any target. The vapour (plume) is collected on a substrate placed at short distance from the target. Though the actual physical processes of material removal are quite complex, one can consider the ejection of material to occur due to rapid explosion of the target surface due to superheating. Unlike thermal evaporation, which produces a vapour composition dependent on the vapour pressures of elements in the target material, the laser induced expulsion produces a plume of material with stoichiometry similar to the target. It is generally easier to obtain the desired film stoichiometry for multi-element materials using PLD than with other deposition technologies. When the laser pulse is absorbed by the target, energy is first converted to electronic excitation and then into thermal, chemical and mechanical energy resulting in evaporation ablation, plasma formation and even exfoliation. The ejected spices expand into the surrounding vacuum in the form of a plume (Fig. 20.29) containing many energetic species including atoms, molecules, electrons, ions, clusters, particulates and molten globules, before depositing on the typically hot substrate. The process of PLD can generally be divided into four stages:

(i) Laser ablation of the target material and creation of a plasma
(ii) Dynamics of the plasma
(iii) Deposition of the ablation material on the substrate
(iv) Nucleation and growth of the film on the substrate surface

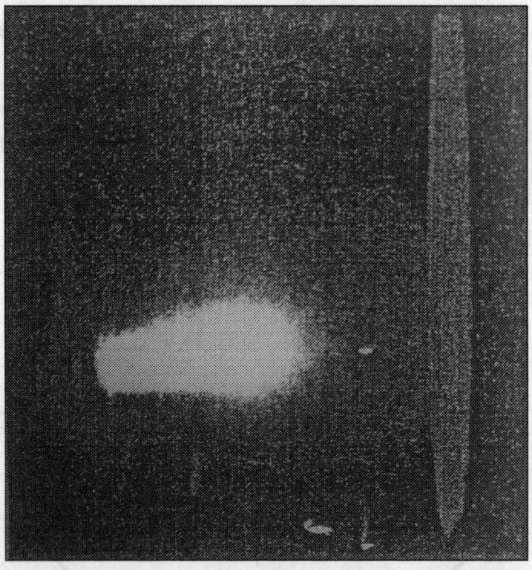

**Figure 20.29** Image of a plasma plume generated during a PLD process

Each of these steps is crucial for the crystallinity, uniformity and stoichiometry of the resulting film. The first step of laser ablation of the target material upon laser irradiation and the creation of plasma are very complex processes. Laser ablation is the process of removing material from a solid (or occasionally liquid) surface by irradiating it with a laser beam. At low laser flux, the material is heated by the absorbed laser energy and evaporates or sublimes. At high laser flux, the material is typically converted to a plasma.

Usually, laser ablation refers to removing material with a pulsed laser, but it is possible to ablate material with a continuous wave laser beam if the laser intensity is high enough. The removal of atoms from the bulk material is done by vaporization of the bulk at the surface region in a state of non-equilibrium and is caused by a Coulomb explosion. In this process, the incident laser pulse penetrates into the surface of the material within the penetration depth. The dimension of the penetration depth is dependent on two main factors, e.g., the laser wavelength and the index of refraction of the target material at the applied laser wavelength and is typically in the region of 10 nm for most materials. The strong electrical field generated by the laser light is sufficiently strong to remove the electrons from the bulk material of the penetrated volume. This process occurs within 10 pico-seconds of a nano-seconds laser pulse and is caused by non-linear processes such as multiphoton ionization, which are enhanced by microscopic cracks at the surface, voids, and nodules (which enhances the electric field). The free electrons oscillate within the electromagnetic field of the laser light and can collide with the atoms of the bulk material thus transferring some of their energy to the lattice of the target material within the surface region. The surface of the target is then heated up and the material is vaporized from the target surface and ablated out with stoichiometry as in the target. The instantaneous ablation rate is highly dependent on the fluences of the laser shining on the target. The temperature of the generated plasma plume is typically $10^4$ K and the laser pulses can vary over a very wide range of duration (milliseconds to femtoseconds) and can be precisely controlled. This makes laser ablation very valuable for both research and industrial applications.

In the second stage of PLD deposition, the material expands in a plasma parallel to the normal vector of the target surface towards the substrate due to Coulomb repulsion and recoil from the target surface. The emitted materials tend to move towards the substrate according to the laws of gas dynamics and show the forward peaking phenomenon. The film uniformity is significantly dependent on the spot size of the laser and the plasma temperature. The target to substrate distance is another parameter that governs the angular spread of the ablated materials. It has also been observed that a mask placed close to the substrate could reduce the spreading. The spatial distribution of the plasma plume is dependent on the background pressure inside the PLD chamber. The dependency of the plume shape on the pressure can be described in three regions:

1. The vacuum region, where the plume is very narrow and forward directed and almost no scattering of ablated material clusters with the background gas molecules occurs in this region.
2. The intermediate region, where a separation of the high energetic ions from the less energetic species is observed. The time-of-flight (ToF) data can be fitted to a shock wave model; however, other possible models can also be fitted.
3. High pressure region, where a more diffusion-like expansion of the ablated material is observed. Naturally, this scattering is also dependent on the mass of the background gas and can influence the stoichiometry of the deposited film. The most important

consequence of increasing the background pressure is the slowing down of the high energetic species in the expanding plasma plume. It has been shown that particles with kinetic energies around 50 eV can resputter the film already deposited on the substrate. This results in a lower deposition rate and can furthermore result in a change in the stoichiometry of the film.

The third stage of the PLD process is the most important stage to determine the quality of the deposited thin film. The ejected high-energy species can impinge onto the substrate surface and may induce various types of damages to the substrate. These energetic species sputter some of the surface atoms and a collision region is formed between the incident flow and the sputtered atoms. Film grows after a thermalized region is formed. The region serves as a source for condensation of particles. When the condensation rate is higher than the rate of particles supplied by the sputtering, thermal equilibrium condition can be reached quickly and film grows on the substrate surface at the expense of the direct flow of the ablation particles and the thermal equilibrium obtained.

The final nucleation-and-growth stage of crystalline films depends on many factors such as the density, energy, ionization degree, and the type of the condensing material, as well as the temperature and the physico-chemical properties of the substrate. The two main thermodynamic parameters for the growth mechanism are the substrate temperature ($T_S$) and the supersaturation, $S_m$. They can be related by the following equation

$$S_m = kT_S \ln\left(\frac{\eta}{\eta_e}\right)$$

where $k$ is the Boltzmann constant, $\eta$ is the actual deposition rate, and $\eta_e$ is the equilibrium value at the temperature, $T_S$. The nucleation process depends on the interfacial energies between the three phases present – the substrate, the condensing material and the vapour. The minimum-energy shape of a nucleus is like a cap. The critical size of the nucleus depending on the driving force, i.e., the deposition rate and the substrate temperature. For the large nuclei, as a characteristic of small supersaturation, they create isolated patches (islands) of the film on the substrate which subsequently grow and coalesce together. As the supersaturation increases, the critical nucleus shrinks until its height reaches one atomic diameter and its shape is that of a two-dimensional layer. For large supersaturation, the layer-by-layer nucleation will happen for incompletely wetted foreign substrates. The crystalline film growth depends on the surface mobility of the adatom (vapour atoms). Normally, the adatom will diffuse through several atomic distances before sticking to a stable position within the newly formed film. The surface temperature of the substrate determines the adatom's surface diffusion ability. High temperature favours rapid and defect free crystal growth, whereas low temperature or large supersaturation crystal growth may be overwhelmed by energetic particle impingement, resulting in disordered or even amorphous structures. In the PLD process, due to the short laser pulsed duration (~10 ns) and hence the small temporal spread ( 10 ms) of the ablated materials, the deposition rate can be enormous (~10 mm/s). Consequently, a layer-by-layer nucleation is favoured and ultra-thin and smooth film can be produced. In addition, the rapid deposition of the energetic ablation species helps to raise the substrate surface temperature. In this respect PLD tends to demand a lower substrate temperature for crystalline film growth. Although PLD has a much lower average deposition rate than other deposition techniques such as molecular beam epitaxy (MBE) and sputtering, depending on the repetition rate of the laser it can be demonstrated that the fraction of stable nucleation sites is orders of magnitudes higher for PLD than for sputtering and MBE. This has several implications on the growth mechanism in PLD. The critical nucleus radius is smaller and is in the range of 1 or two atoms. Since the nucleation rate is proportional to the nucleation site density and the rate of impingement, it is much higher in the case of PLD. Furthermore, the high density of nucleation sites also increases the smoothness of the deposited film. This is why PLD is such an outstanding method for the growth of thin films.

The main advantages of pulsed laser deposition are:

- Conceptually simple: a laser beam vaporizes a target surface, producing a film with the same composition as the target.
- Versatile: many materials can be deposited in a wide variety of gases over a broad range of gas pressures. Especially for nanomaterial growth, PLD is very useful in the sense that by adjusting the laser pulses, one can control the deposition rate, so also the nanostructure of the grown film.
- Fast: high quality samples can be grown reliably in 10 or 15 minutes.
- Scalable: as complex oxides move towards volume production.

## 20.21 LUMINESCENCE

There are many ways of production of light for spectroscopic investigations. The important ones are temperature radiation and all kinds of *luminescence* (i.e., the phenomenon of emission of optical radiation, e.g., fluorescence, electroluminescence, etc.). In temperature radiation of gases, the atoms/molecules are excited to light emission by collision among themselves, the necessary energy as emission is derived from the kinetic energy of the colliding particles, (i.e., a high temperature is necessitated).

Some bodies generally emit light when heated. Some materials emit light after they have been irradiated with visible or ultraviolet light, with X-rays, γ-rays, electrons or other particles or when placed in an electric field, etc. The emitted light may be in the visible part of the spectrum although the temperature of the emitting body is low (room temperature and below). Such cold emission of light is termed *luminescence* and the bodies exhibiting it are termed *luminophors.* The luminescene excited by light is termed *photoluminescence*. In contrast to thermal radiation, luminescence is a non-equilibrium process. An important feature of luminescence is its long duration in comparison with the period of optic oscillation equal to $10^{-13}$ to $10^{-15}$ s. The emission of light in the process of luminescene continues at least $10^{-10}$ s after the excitation has ceased. In some instances, the emission of light may continue for seconds, minutes, hours and even months after the excitation has ceased. In accordance with the duration of light emission, photoluminescence is divided into *phosphorescence* and *fluorescence*.

The energy gained by an atom or a molecule by the absorption of a photon does not remain in that molecule but is lost by any of several mechanisms, e.g., conversion of a part into heat, lowering the net energy of the molecule to the lowest vibrational and rotational level within the same electronic (singlet) level. The rest of the energy is then released as radiation and the molecule then returns to the ground state. This is known as *fluorescence*. The emitted radiation has less energy per photon than the exciting radiation and hence a longer wavelength, e.g., a number of organic compounds fluoresce in the visible region when irradiated with ultraviolet light.

In some molecules, it is possible for a non-radiative transition to occur from an excited singlet state to the corresponding triplet state from which the remaining energy is radiated when the molecule comes to the ground state. The probability of transition from the triplet to the ground state is however very low. Hence, radiation may last for some time after exciting radiation is cut off. This persistent radiation is called *phosphorescence*. Phosphorescence takes place over periods longer than $10^{-5}$ seconds and may continue longer than a few minutes or hours after irradiation has ceased.

## 20.22 NANOSTRUCTURES: QUANTUM WELLS, WIRES AND DOTS

When the size or dimension of a material is continuously reduced from a large or macroscopic size, such as a metre or a centimetre, to a very small size, the properties remain the same at first, then small changes begin to occur, until finally when the size drops below 100 nm, dramatic changes in properties can occur. If one dimension is reduced to the nanorange while the other two dimensions remain large, then we obtain a structure known as a *quantum well*. If two dimensions are so reduced and one remains large, the resulting structure is referred to as a *quantum wire*. The extreme case of this process of size reduction in which all three dimensions reach the low nanometre range is called a *quantum dot*. The word *quantum* is associated with these three types of nanostructures because the changes in properties arise from the quantum-mechanical nature of physics in the domain of the ultra-small. Figure 20.30 illustrates these processes of diminishing the size for the case of rectilinear geometry, and Fig. 20.31 presents the corresponding reductions in curvilinear geometry. In this chapter we are interested in all three of these ways to decrease the size. In other words, we probe the dimensionality effects that occur when one, two, or all three dimensions become small. Of particular interest is how the electronic properties are altered by these changes. Jacak et al. (1998) have surveyed the field of quantum nanostructures.

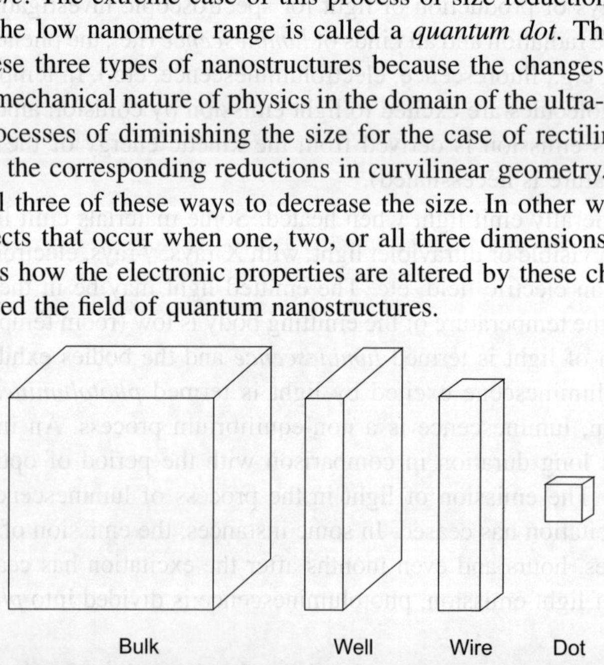


**Figure 20.30** Progressive generation of rectangular nanostructures

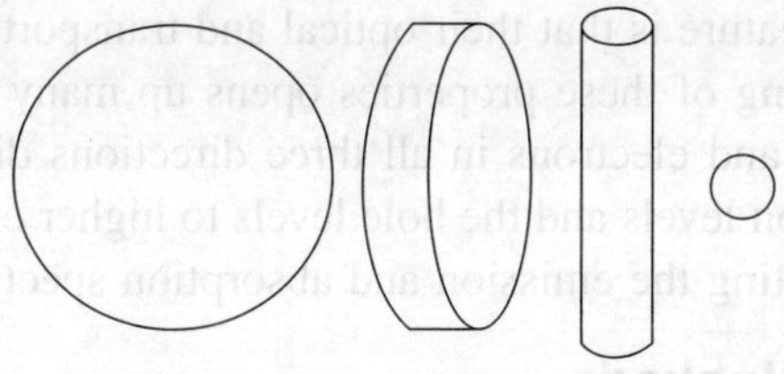


**Figure 20.31** Progressive generation of curvilinear nanostructures

## 20.23 QUANTUM WIRE

A quantum wire, shown in Fig. 20.32, is an electrically conducting wire, in which quantum effects affect transport properties. Due to the confinement of conduction electrons in the transverse direction of the wire, their transverse energy is quantized into a series of discrete value, $E_0$ ('ground state energy', with lower value), $E_1$, $E_2$, ... $E_n$. One consequence of this quantization is that the classical formula for calculating the electrical resistivity of a wire, viz.,

$$R = \rho \frac{l}{A}$$

is not valid for quantum wires. Instead, an exact calculation of the transverse energies of the confined electrons has to be performed to calculate the resistance of the wire. Following the quantization of electron energy, the resistance of the wire is also found to be quantized.

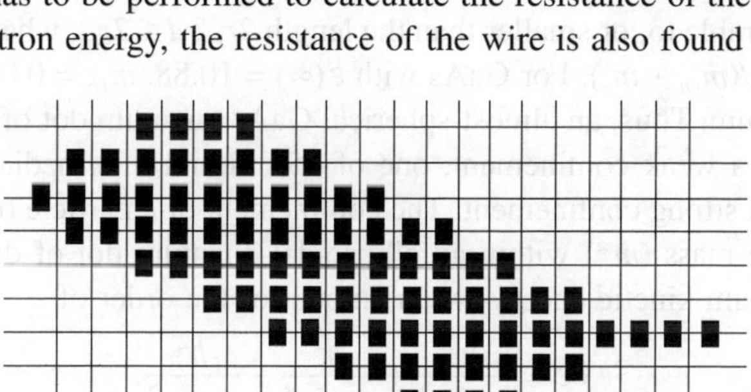

**Figure 20.32** A typical quantum wire

A quantum wire confines the motion of electrons or holes in two spatial dimensions and allow free propagation in the third. Semiconductor nanowires are mainly cylindrical nano structures with a diameter somewhere between 20 nm and 100 nm.

## 20.24 QUANTUM DOTS

### 20.24.1 Introduction

Quantum dots are semiconductor nanoparticles (usually Si or III-V semiconductors) grown or deposited on a substrate. Their sizes can be as small as 2 nm and as large as several hundred nm. Their main feature is that their optical and transport properties depend on their size and shape. The tailoring of these properties opens up many technological possibilities. The confinement of holes and electrons in all three directions discretizes the energy levels and pushes both the electron levels and the hole levels to higher energies, increasing thus the effective gap and blue shifting the emission and absorption spectrum.

### 20.24.2 Coulomb Blockade

Quantum dots with a proper setup can act somehow as a single electron field effect transistor; at certain values of the gate voltage, $v_p = (e/C_d)\ (N_0 \pm 1/2)$, the energy cost of injecting an extra electron in the QD can disappear; at these values of $V_p$ the conductance exhibits sharp peaks. The energy cost of injecting an extra electron in the QD is due to the electrostatic repulsion of other extra electrons already residing in the QD; this is the *Coulomb blockade effect*; (named after Charles–Augustin de Coulomb). It is the increased resistance at small bias voltages of an electronic device comprising at least one low-capacitance tunnel junction).

### 20.24.3 An Overview

A quantum dot (QD) is usually a semiconductor nanoparticle deposited, or grown, or etched in a substrate; the diameter D of a QD is in the range from 2 nm to about 100 nm, or larger, depending on the method of fabrication.

The main characteristic of a quantum dot is the confinement of the electrons and the holes in all of its three dimensions.[4,5] The term confinement means that the available length for movement is comparable to, or smaller than the length $2r_{ex}$, $d \le 2r_{ex}$, where $2r_{ex} = (2\varepsilon(\infty)/m^*)\, a_B$ with $m^* = m_{\ell h} m_c/(m_{\ell h} + m_c)$. For GaAs with $\varepsilon(\infty) = 10.88$, $m_{\ell h} = 0.082 m_e$, $m_c = 0.067\, m_e$ we obtain $2r_{ex} \simeq 31$ nm. Thus, an almost-spherical GaAs quantum dot of diameter, let us say, of 50 nm, provides a weak confinement, one of 25 nm and intermediate confinement, and one, e.g., of 10 nm a strong confinement. The confinement of a particle of a particle (electron or hole) of effective mass ($m^*$) within an ellipsoidal quantum dot of dimensions $d_x$, $d_y$, $d_z$, increases the minimum kinetic energy by an amount of the order of

$$\delta\varepsilon \simeq \frac{\hbar^2}{2m_x^* d_x^2} + \frac{\hbar^2}{2m_y^* d_y^*} + \frac{\hbar^2}{2m_z^* d_z^2}.$$

As a result, the energy levels in a quantum dot made of a direct-gap semi-conductor (e.g., CdSe with $E_g = 1.83$ eV) are different from those of the same material in the bulk. By making the quantum dot smaller and smaller one blue-shifts the absorption and emission spectrum so that the fluorescence emission can change from red, yellow, green, blue, to violet and beyond. The possibility of tailoring the emission and absorption spectrum and controlling the quantum yield by changing the size and the shape of quantum dots is one reason for the

---

[4] If the confinement in the vertical direction is much stronger than in the planar directions, the QDs can be treated as 2-D disks of finite radius. Such 2-D QDs appear in a 2-D electron gas in graphene, etc; they are important for controlling transport properties and, hence, they are considered as elements in various devices (sec. 20.25).

[5] In a **quantum well** the motion of electrons (or holes) in one direction is restricted, leaving them free to move in the other two dimensions. In the **quantum wires** there is confinement in two directions and unrestricted possibility of motion in the third direction.

increased interest in these nanoparticles. The superior optical properties of quantum dots open up many technological possibilities such as increasing the yield of silicon photovoltaic cells, of developing white light emitting diodes and making displays, of creating artificial fluorophores for molecular imaging, etc. Quantum dots are studied also as a possible vehicle for quantum computers, since they offer the all-important advantage of solid state, namely the integrability with the existing semiconductor technology. Finally, quantum dots (QDs) can be connected by tunnel barriers to external conducting leads acting this way like single-electron transistors and exhibiting *Coulomb blockade* effects; these electronic aspects can be used for improved flash memory devices.

The simplest way for the fabrication of QDs is by colloidal synthesis according to which the QD material is supplied in a liquid suspension or dispersed in plastic host and then processed onto the substrate; QDs of this type can be as small as 2–10 nm. Self-assembled QDs nucleate spontaneously under certain conditions during molecular beam epitaxy or during metallorganic vapour phase epitaxy when there is a lattice mismatch between the material of the QD and that of the substrate; this way of growing QDs, which produces sizes in the range 10–50 nm, is common in studies aiming at quantum cryptography and quantum computing. Quantum dots, defined by lithographic patterning or by etching the heterojunction at which 2D electron (or hole) gas is trapped, can exceed in lateral size the 100 nm and their main use is in transport properties. QDs appear sometimes accidentally in quantum wells due to fluctuations in the thickness of the well.

## 20.25 QDs AND COULOMB BLOCKADE

In Fig. 20.33, we show schematically the experimental setup that can inject electron(s) to the QD, while its state can be influenced by the voltage ($V_p$) of the electrode, *P*. To start with, we shall ignore the kinetic energy of the injected electron(s) (and the external current) because the main contribution to the energy of the electrons is the electrostatic energy. The charge on the QD is $q_d = -Ne$, where *N* is the number of the injected electrons; there is also induced charge ($q_p$) on the plunger.

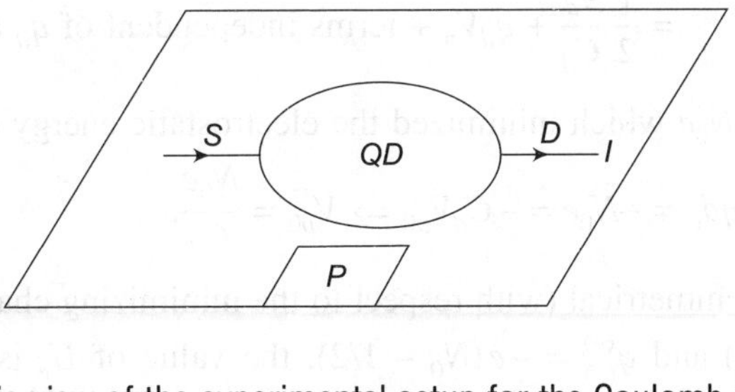


**Figure 20.33** Schematic view of the experimental setup for the Coulomb blockade measurements. The current (*I*) through the QD from source (S) to drain (D) is measured, as well as the voltage $V_{SD}$ between the contact points, *S*, *D*. There is also a metallic contact, *P*, called plunger, connected to an external reservoir which controls the voltage VP of the plunger

The electrostatic energy ($U_e$) is

$$U_e = \frac{1}{2}[q_d V_d + q_p V_p] + [Q_r - q_p] V_p \quad ...(1)$$

where $Q_r$ is the charge on the (huge) reservoir (when $q_p = 0$) which controls the voltage of the plunger. Now the charges on the QD, $q_d$, and the plunger, $q_p$, can be expressed in terms of the corresponding voltages ($V_d$ and $V_p$) through the diagonal and off-diagonal matrix elements of the capacitance matrix

$$q_d = C_d V_d + C_{dp} V_p \quad ...(2)$$

$$q_p = C_{pd} V_d + C_p V_p \quad ...(3)$$

where the off-diagonal matrix elements are symmetric and negative[6] ($C_{dp} = C_{pd}$). If no current flows through the dot and $V_S = V_D = V_d$, then the charge on the dot would depend on the difference, $V_d - V_p$, since there is no other quantity on which $q_d$ may depend. Hence,

$$C_d = -C_{pd} = -C_{dp} \quad ...(4)$$

From Eqs. (2) and (4), we have

$$V_d = \frac{q_d}{C_d} + V_p \quad ...(5)$$

and from Eqs. (3), (4) and (5)

$$q_p = -C_d V_d + C_p V_p = -C_d \frac{q_d}{V_p} - C_d V_p + C_p V_p \quad ...(6)$$

We substitute Eqs. (5) and (6) in Eq. (1) and obtain

$$U_e = \frac{1}{2}\frac{q_d^2}{C_d} + \frac{1}{2} q_d V_p - \frac{1}{2} q_p V_p + Q_r V_p$$

$$= \frac{1}{2}\frac{q_d^2}{C_d} + \frac{1}{2} q_d V_p + \frac{1}{2} q_d V_p + \text{terms independent of } q_d \text{ and } V_d$$

$$= \frac{1}{2}\frac{q_d^2}{C_d} + q_d V_p + \text{terms independent of } q_d \text{ and } V_d \quad ...(7)$$

The value of $q_d = -N_0 e$ which minimized the electrostatic energy ($U_e$) is given by

$$qd_0 = -N_0 e = -C_d V_{p0} \Rightarrow V_{p0} = \frac{N_0 e}{C_d}. \quad ...(8)$$

Notice that, at the symmetrical (with respect to the minimizing charge ($qd_0 = -eN_0$) value of $q_d^{(+)} = -e(N_0 + 1/2)$ and $q_{(d)}^{(-)} = -e(N_0 - 1/2)$, the value of $U_e$ is the same. Hence, the injection of an extra electron (such as their number changes from $N_0 - \frac{1}{2}$ to $N_0 = \frac{1}{2}$ does not cost any energy. This zero energy cost for an electron injection implies easy transfer of electron from the sources $S$ to the drain $D$, and consequently, large current, $I$, even with a

[6] See Landau and Lifshitz' book on *Electrodynamics of Continuous Media*. Eqs. (2.8) and (2.11).

minute voltage difference, $V_{SD}$. Hence, if we plot the current ($I$) vs. the plunger voltage ($V_p$) or the conductance, $G \equiv I/V_{SD}$ vs. $V_p$, we expect equally spaced sharp peaks when $V_p = (e/C_d)(N_0 \pm 1/2)$; two consecutive such peaks are at a distance $\delta V_p = e/C_d$ apart. The experimental data confirm this impressive theoretical prediction associated with the disappearence of the Coulomb blockade effect at the values of $V_p = (e/C_d)\left(N_0 \pm \frac{1}{2}\right)$. If we assume a voltage $V_{SD}$ = 0.5 mV and ballistic propagation of the electron within the QD we obtain a conductance of about $0.5\,e^2/h$ which clearly is an ideal upper limit.

## 20.26 PREPARATION OF QUANTUM NANOSTRUCTURES

One approach to the preparation of a nanostructure, called the *bottom-up* approach, is to collect, consolidate, and fashion individual atoms and molecules into the structure. This is carried out by a sequence of chemical reactions controlled by catalysts. It is a process that is widespread in biology where, for example, catalysts called *enzymes* assemble amino acids to construct living tissue that forms and supports the organs of the body.

The opposite approach to the preparation of nanostructures is called the *top-down method*, which starts with a large-scale object or pattern and gradually reduces its dimension or dimensions. This can be accomplished by a technique called *lithography* which shines radiation through a template on to a surface coated with a radiation-sensitive resist; the resist is then removed and the surface is chemically treated to produce the nanostructure. A typical resist material is the polymer polymethyl methacrylate $[C_5O_2H_8]_n$ with a molecular weight in the

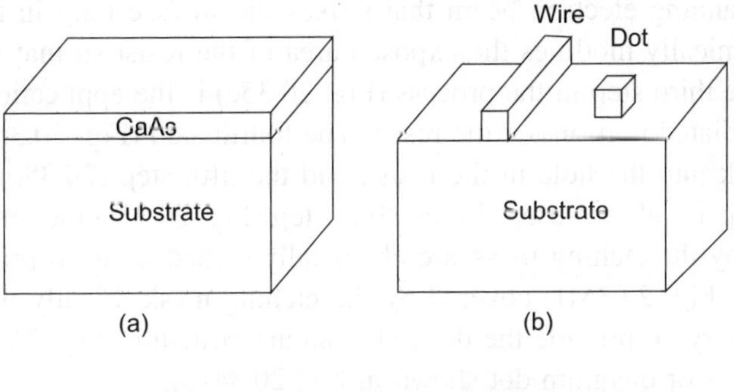


**Figure 20.34** (a) Gallium arsenide quantum well on a substrate; (b) quantum wire and quantum dot formed by lithography

range from $10^5$ to $10^6$ Da. The lithographic process is illustrated by starting with a square quantum well located on a substrate, as shown in Fig. 20.34. The final product to be produced from the material (e.g., GaAs) of the quantum well is either a quantum wire or a quantum dot, as shown in Fig. 20.34. The steps to be followed in this process are outlined in Fig. 20.31.

The first step of the lithographic procedure is to place a radiation-sensitive resist on the surface of the sample substrate, as shown in Fig. 20.35 (a). The sample is then irradiated by an electron beam in the region where the nanostructure will be located, as shown in Fig. 20.35 (b).

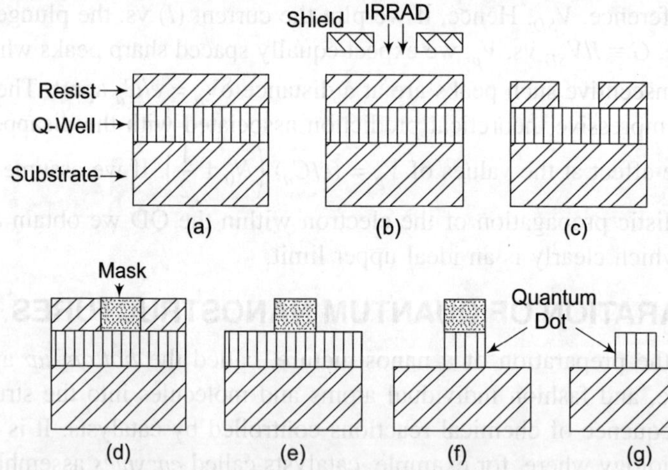


**Figure 20.35** Steps in the formation of a quantum wire or quantum dot by electron-beam lithography; (a) initial quantum well on a substrate, and covered by a resist; (b) radiation with sample shielded by template; (c) configuration after dissolving irradiated portion of resist by developer; (d) disposition after addition of etching mask; (e) arrangement after removal of remainder of resist; (f) configuration after etching away the unwanted quantum-well material; (g) final nanostructure on substrate after removal of etching mask

This can be done by using either a radiation mask that contains the nanostructure pattern, as shown, or a scanning electron beam that strikes the surface only in the desired region. The radiation chemically modifies the exposed area of the resist so that it becomes soluble in a developer. The third step in the process (Fig. 20.35c) is the application of the developer to remove the irradiated portions of the resist. The fourth step (Fig. 20.35d) is the insertion of an etching mask into the hole in the resist, and the fifth step (20.35e) consists in lifting off the remaining parts of the resist. In the sixth step (Fig. 20.35f) the areas of the quantum well not covered by the etching mask are chemically etched away to produce the quantum structure shown in Fig. 20.35(f) covered by the etching mask. Finally the etching mask is removed, if necessary, to provide the desired quantum structure (Fig. 20.35g), which might be the quantum wire or quantum dot shown in Fig. 20.34(b).

The most common process used is *electron-beam lithography,* which makes use of an electron beam for the radiation. Other types of lithography employ neutral atom beams (e.g., Li, Na, K, Rb, Cs), charged ion beams (e.g., $Ga^+$), or electromagnetic radiation such as visible light, ultraviolet light, or X-ray. When laser beams are utilized, frequency doublers and quadruplers can bring the wavelength into a range (e.g., $\lambda \sim 150$ nm) that is convenient for quantum-dot fabrication. Photochemical etching can be applied to a surface activated by laser light.

The lithographic technique can be used to make more complex quantum structures than the quantum wire and quantum dot shown in Fig. 20.34(b). For example, one might start with a multiple quantum-well structure of the type illustrated in Fig. 20.36, place the resist on top of it, and make use of a mask film or template with six circles cut out of it, as portrayed at the top of Fig. 20.36. Following the lithographic procedure outlined in Fig. 20.35, one can produce the 24-quantum-dot array consisting of six columns, each containing four stacked

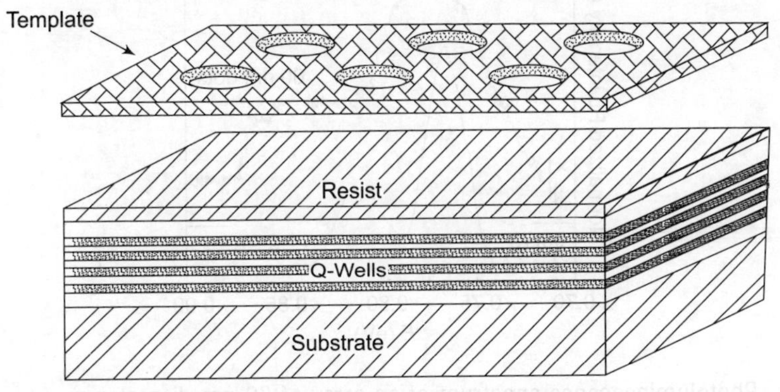


**Figure 20.36** Four-cycle multiple-quantum-well arrangement mounted on a substrate and covered by a resist. A radiation shielding template for lithography is shown at the top

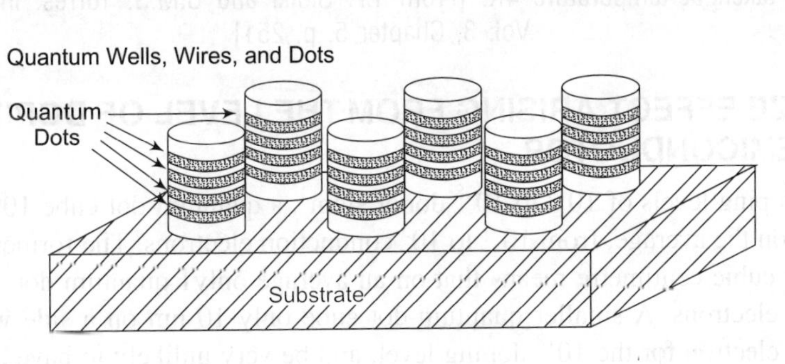


**Figure 20.37** Quantum-dot array formed by lithography from the initial configuration of Fig. 20.36. This 24-fold quantum-dot array consists of six columns of four stacked quantum dots

quantum dots, that is sketched in Fig. 20.37. As an example of the advantages of fabricating quantum dot arrays, it has been found experimentally that the arrays produce a greatly enhanced photoluminescent output of light. Figure 20.38 shows a photoluminescence (PL) spectrum from a quantum-dot array that is much more than 100 times stronger than the spectrum obtained from the initial multiple quantum wells.

The main peak of the spectrum of Fig. 20.38 was attributed to a localized exciton (LE). This magnitude of enhancement shown in the figure has been obtained from initial samples containing, typically, a 15-period superlattice (SL) of alternating 3 nm-thick layers of Si and $Si_{0.7}Ge_{0.3}$ patterned into quantum-dot arrays consisting of 300 nm high columns with 60 nm diameters and 200 nm separations.

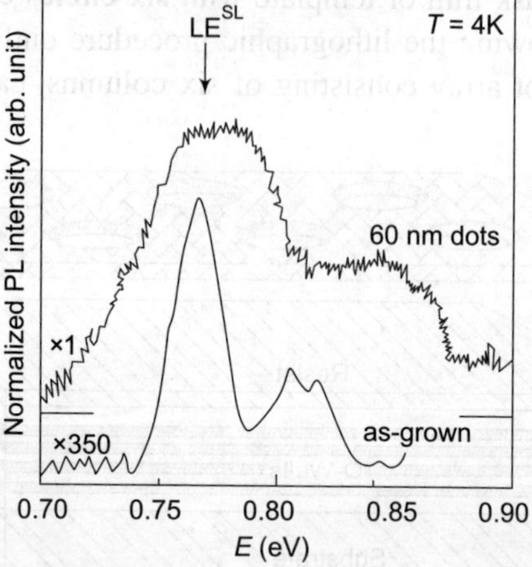


**Figure 20.38** Photoluminescence spectrum of an array of 60-nm-diameter quantum dots formed by lithography, compared with the spectrum of the initial as grown multiple quantum well. The intense peak at 0.7654 eV is attributed to localized excitons (LE) in the superlattice (SL). The spectra were taken at temperature 4K. [From T.P. Sidiki and C.M.S. Torres, in Nalwa (2000), Vol. 3, Chapter 5, p. 251]

## 20.27 SIZE EFFECT ARISING FROM THE LEVEL OF DOPING OF A SEMICONDUCTOR

For typical doping levels of $10^{14}$ to $10^{18}$ donor s/cm$^3$, a quantum-dot cube 100 nm on a side would have, on the average, from $10^{-1}$ to $10^3$ conduction electrons. The former figure of $10^{-1}$ electrons per cubic centimetre means that on an average only1 quantum dot in 10 will have one of these electrons. A smaller quantum-dot cube only 10 nm on a side would have, on an average 1 electron for the $10^{18}$ doping level, and be very unlikely to have any conduction electrons for the $10^{14}$ doping level. A similar analysis can be made for quantum wires and quantum wells and the results shown in Table 20.3 demonstrate that these quantum structures are typically characterized by very small numbers or concentrations of electrons that can carry current. This results in the phenomena of single-electron tunnelling and the Coulomb blockade which we will discuss later.

**Table 20.3** Conduction Electron Content of Smaller Size (on left) and Larger Size (on right) Quantum Structures Containing Donor Concentrations of $10^{14}$–$10^{18}$ cm$^{-3}$

| *Quantum Structure* | *Size* | *Electron Content* | *Size* | *Electron Content* |
|---|---|---|---|---|
| Bulk material | — | $10^{14}$–$10^{18}$ cm$^{-3}$ | | $10^{14}$–$10^{18}$ cm$^{-3}$ |
| Quantum well | 10 nm thick | 1–$10^4$ $\mu$m$^{-2}$ | 100 nm thick | 10–$10^5$ $\mu$m$^{-2}$ |
| Quantum well | 10 × 10-nm cross section | $10^{-2}$–$10^2$ $\mu$m$^{-1}$ | 100 nm × 100 nm cross section | 1–$10^4$ $\mu$m$^{-1}$ |
| Quantum dot | 10 nm on a side | $10^{-4}$–1 | 100 nm on a side | $10^{-1}$–$10^3$ |

## 20.28 CONDUCTION ELECTRONS AND DIMENSIONALITY

We are accustomed to studying electronic systems that exist in three dimensions and are large or macroscopic in size. In this case the conduction electrons are delocalized, and move freely throughout the entire conducting medium such as a copper wire. It is clear that all the wire dimensions are very large compared to the distances between atoms. The situation changes when one or more dimensions of the copper becomes so small that it approaches several times the spacings between the atoms in the lattice. When this occurs, the delocalization is impeded, and the electrons experience confinement. For example, consider a flat plate of copper that is 10 cm long, 10 cm wide, and only 3.6 nm thick. This thickness corresponds to the length of only 10 unit cells, which means that 20% of the atoms are in unit cells at the surface of the copper. The conduction electrons would be delocalized in the plane of the plate, but confined in the narrow dimension, a configuration referred to as a *quantum well*. A *quantum wire* is a structure such as a copper wire that is long in one dimension, but has a nanometer size as its diameter.

**Table 20.4** Delocalization and Confinement Dimensionalities of Quantum Nanostructures

| *Quantum structure* | *Delocalization dimensions* | *Confinement dimensions* |
|---|---|---|
| Bulk conductor | 3 (*x*, *y*, *z*) | 0 |
| Quantum well | 2 (*x*, *y*) | 1 (*z*) |
| Quantum well | 1 (*z*) | 2 (*x*, *y*) |
| Quantum dot | 0 | 3 (*x*, *y*, *z*) |

The electrons are delocalized and move freely along the wire, but are confined in the transverse directions. Finally, a *quantum dot*, which might have the shape of a tiny cube, a short cylinder, or a sphere with low nanometer dimensions, exhibits confinement in all three spatial dimensions, so there is no delocalization. Figures 20.30 and 20.31, as well as Table 20.4, summarize these cases.

## 20.29 DELOCALIZATION AND FORMULAS FOR DIMENSIONALITY

The dimensionality of the system (e.g., quantum well, quantum wire or quantum dots), the dimensionality of the system has a pronounced effect on its properties. This is one of the important new factors that makes nanoscience so interesting. In what follows, we would give some important formulas that would summarize the role played by the dimensionality.

### 20.29.1 Delocalization

The region occupied by a system of conduction electrons delocalised in one-, two-, and three-dimensional co-ordinate space are the length, area and volume given in the first column of Table 20.5. Column 2 of the table gives the size of the unit cell in reciprocal or $k$ space, and column 3 gives the size of the Fermi region that is occupied by the delocalized electrons, where the Fermi energy ($E_1$) has the value, $E_F = \hbar^2 k_F^2/2m$, and in this region $E < E_F$. Column 4 gives expressions for $k^2$ in the three systems. The numbers of electrons ($N$) in the occupied

**Table 20.5** Properties of Coordinate and *K* Space in One, Two and Three Dimensions

| *Coordinate Region* | *k-space Unit Cell* | *Fermi Region* | *Value of $k^2$* | *Dimensions* |
|---|---|---|---|---|
| Length $L$ | $2\pi/L$ | $2k_F$ | $k_X^2$ | One |
| Area $A = L^2$ | $(2\pi/L)^2$ | $\pi k_F^2$ | $k_X^2 + k_Y^2$ | Two |
| Volume $V = L^3$ | $(2\pi/L)^3$ | $4\pi k_F^3/3$ | $k_X^2 + k_Y^2 + k_Z^2$ | Three |

regions of column 3 at the temperature of absolute zero, as well as the density of states, $D(E)$, defined by the expression $D(E) = dN(E)/dE$, are given in Table 20.6. we see from this table that the density of states decreases with the energy for one dimension, is constant for two dimensions, and increases with increasing energy for three dimensions. Thus, the number of electrons and the density of states as functions of the energy have quite different behaviours for the three cases, as indicated by the plots of Figs. 20.39, 20.40 and 20.43.

### 20.29.2 Partial Confinement

The conduction electrons in nanostructures can be partially confined and partially delocalized, depending on the shape and the dimensions of the structure. One limiting case is a quantum dot in which they are totally confined, and the other limiting case is a bulk material, in which they are all delocalized. The intermediate cases are a quantum wire, which is long in one dimension and very small in its transverse directions; and a quantum well, which is a flat plate nanosized in thickness and much larger in length and width. The quantum wire exhibits electron confinement in two dimensions and delocalization in one dimension, and the quantum well reverses these characteristics.

**Table 20.6** Number of Electrons $N(E)$ and Density of States $D(E) = dN(E)/dE$ as Function of Energy ($E$) for Electrons Delocalized in One, Two, and Three Spatial Dimensions, where $A = L^2$ and $V = L^3$

| *Number of Electrons N* | *Density of States D(E)* | *Delocalization Dimensions* |
|---|---|---|
| $N(E) = \dfrac{4k_F}{2\pi/L} = \dfrac{2L}{\pi}\left(\dfrac{2m}{\hbar^2}\right)^{1/2} E^{1/2}$ | $D(E) = \dfrac{L}{\pi}\left(\dfrac{2m}{\hbar^2}\right)^{1/2} E^{-1/2}$ | 1 |
| $N(E) = \dfrac{2\pi k_F^2}{(2\pi/L)^2} = \dfrac{A}{2\pi}\left(\dfrac{2m}{\hbar^2}\right) E$ | $D(E) = \dfrac{A}{2\pi}\left(\dfrac{2m}{\hbar^2}\right)$ | 2 |
| $N(E) = \dfrac{2(4\pi k_F^3/3)}{(2\pi/L)^3} = \dfrac{V}{3\pi^2}\left(\dfrac{2m}{\hbar^2}\right)^{3/2} E^{3/2}$ | $D(E) = \dfrac{V}{2\pi^2}\left(\dfrac{2m}{\hbar^2}\right)^{3/2} E^{1/2}$ | 3 |

## 20.30 THE DENSITY OF STATES

The number of conduction electrons with a particular energy depends on the value of the energy and also on the dimensionality of the space. This is because in one dimension the size of the Fermi region containing electrons has the length 2 $k_F$ in two dimensions it has the area of the Fermi circle $\pi k_F^2$, and in three dimensions it has the volume of the Fermi sphere, $4\pi k_F^3/3$. These expressions are listed in column 3 of Table 20.5. If we divide each of these Fermi regions by the size of the corresponding $k$-space unit cell listed in column 2 of this table, and make use of Eq. ($E_F = \hbar^2 k_F^2/2m$) to eliminate $k_F$, we obtain the dependence of the number of electrons ($N$) on the energy ($E$) given on the left side of Table 20.7 and shown plotted in Fig. 20.39. The slopes of the lines, $N(E)$, shown in Fig. 20.39 provide the density

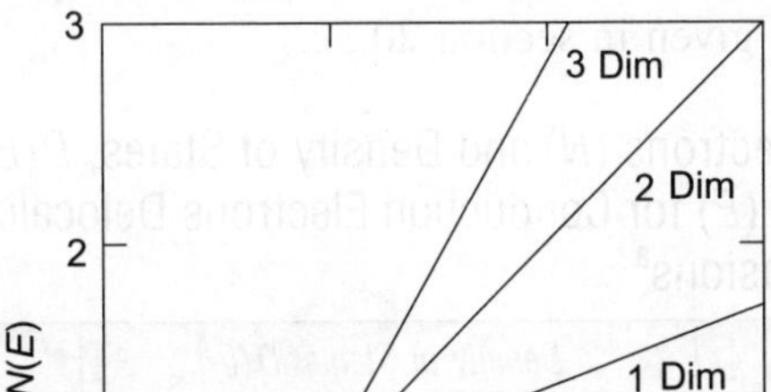


**Figure 20.39** Number of electrons $N(E)$ plotted as a function of the energy ($E$) for conduction electrons delocalized in one (quantum wire), two (quantum well), and three dimensions (bulk material)

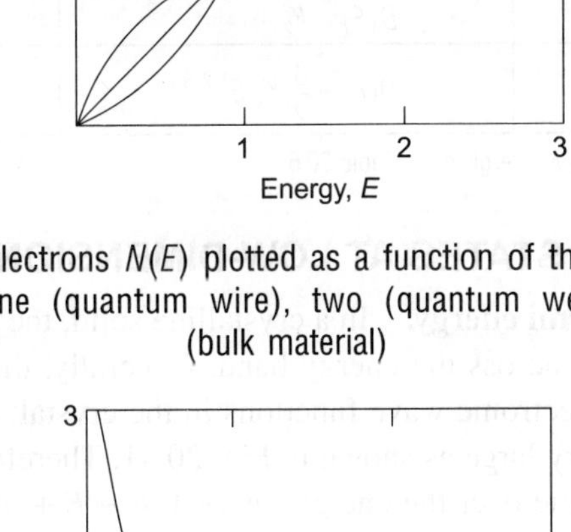


**Figure 20.40** Density of states, $D(E) = dN(E)/dE$ plotted as a function of the energy ($E$) for conduction electrons delocalized in one (Q-wire), two (Q-Well), and three (bulk) dimensions.

of states $D(E)$, which is defined more precisely by the mathematical derivative, $D(E) = dN/dE$, corresponding to the expression, $dN = D(E)\,dE$. This means that the number of electrons ($dN$) with an energy ($E$) within the narrow range of energy, $dE = E_2 - E_1$, is proportional to the density of states at that value of energy. The resulting formulas for $D(E)$ for the various dimensions are listed in the middle column of Table 20.7 and are shown plotted in Fig. 20.40. We see that the density of states decreases with increasing energy for one dimension, is constant for two dimensions and increases with increasing energy for three dimensions. Thus, the density of states has quite a different behaviour for the three cases. These equations and plots of the density of states are very important in determining the electrical, thermal, and other properties of metals and semiconductors, and make it clear why these features can be so dependent on the dimensionality. Examples of how various properties of materials depend on the density of states are given in section 20.32.

**Table 20.7** Number of Electrons ($N$) and Density of States, $D(E) = dN(E)/dE$ as a Function of the Energy ($E$) for Conduction Electrons Delocalized in One, Two, and Three Spatial Dimensions[a]

| *Number of Electrons, N* | *Density of Status, D(E)* | *Delocalization Dimension* |
|---|---|---|
| $N = K_1 E^{1/2}$ | $D(E) = \frac{1}{2} K_1 E^{-1/2}$ | 1 |
| $N = K_2 E$ | $D(E) = K_2$ | 2 |
| $N = K_3 E^{3/2}$ | $D(E) = \frac{3}{2} K_3\, E^{1/2}$ | 3 |

[a] The value of the constant $K_1$, $K_2$, and $K_3$ are giving in Table 20.6.

## 20.31 DENSITY OF STATES AT LOW-DIMENSIONAL STRUCTURES

**Fermi-Dirac statistics, fermi energy.** In a crystalline solid, the energy and volume density of electronic states change across the energy band. Generally, the number of energy levels, which correspond to the electronic wave functions in the crystal, in the central region of the band are expected to be very large as shown in Fig. 20.41. Therefore, the number of electron energy states per unit volume over the energy interval, $E \sim E + dE$, are called the density of states (DOS) and defined as $\rho(E)\,dE$, where $\rho(E)$ is the DOS. Thus, the number of states per unit volume up to some energy ($\acute{E}$) will be

$$N_v(\acute{E}) = \int_0^{\acute{E}} \rho(E)\,dE \qquad ...(1)$$

The expression for the density of states for a 3D crystal is

$$\rho(E)_{3D} = \frac{1}{2\pi^2}\left(\frac{2m^*}{\hbar^2}\right)^{3/2} (E \sim E_{pot})^{1/2} \qquad ...(2)$$

where $m^*$ is the electron effective mass in a band, $E_{pot}$ is a potential energy, which may be the conduction band edge or valence band edge, $E_c$ or $E_v$, respectively ($\hbar$ is the reduced form of Planck's constant $h$, where $\hbar = h/2\pi = 1.0546 \times 10^{-34}$ Js).

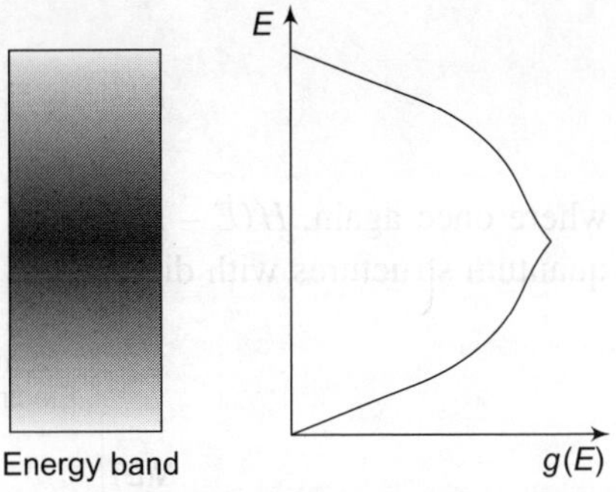


**Figure 20.41** Density of states across an energy band

For 2D structures, such as a quantum well, the confinement of carriers in one spatial direction results in the formation of quantum states for motion in this direction. Therefore, the density of states for 2D structures is modified accordingly and can be expressed as

$$\rho(E)_{2D} = \frac{m^*}{\pi\hbar^2} \sum_i H(E - E_i) \quad ...(3)$$

where $H(E - E_i)$ is the Heaviside function. It takes the value of zero when $E$ is less than $E_i$ and 1, when $E$ is equal to or greater than $E_i$. $E_i$ is the $i$th energy level within the 2D quantum well.

For 1D structures, like quantum wires, motion of charge carriers are allowed only in one direction and other two directions are quantized. The density of states in this case will then be expressed as

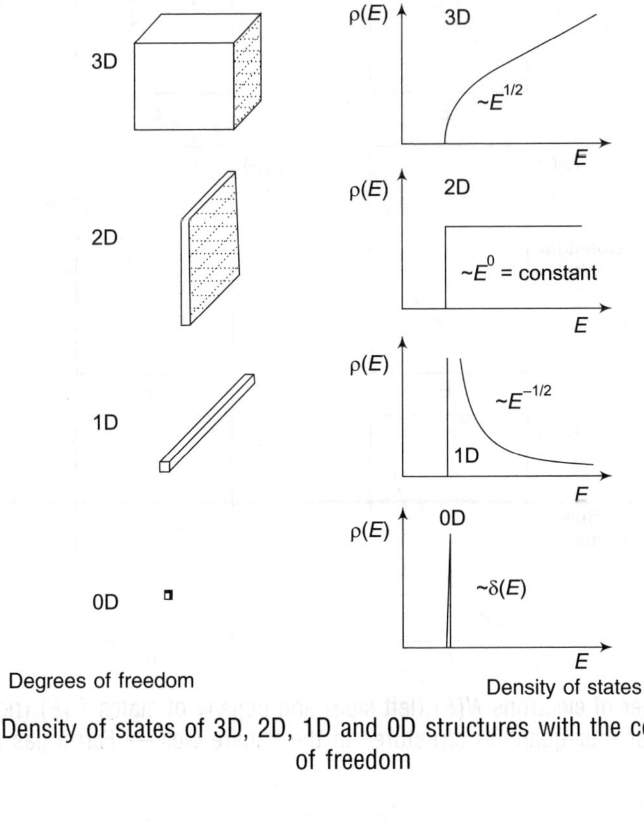


**Figure 20.42** Density of states of 3D, 2D, 1D and 0D structures with the corresponding degrees of freedom

$$\rho(E)_{1D} = \frac{1}{\pi}\left(\frac{2m^*}{\hbar^2}\right)^{1/2} \sum_i \left(\frac{n_i H(E - E_i)}{(E - E_i)^{1/2}}\right) \qquad \text{...(4)}$$

where once again, $H(E - E_i)$ is the Heaviside function and $n_i$ is the degeneracy factor. For quantum structures with dimensions lower than 2, it is possible for the same energy level to

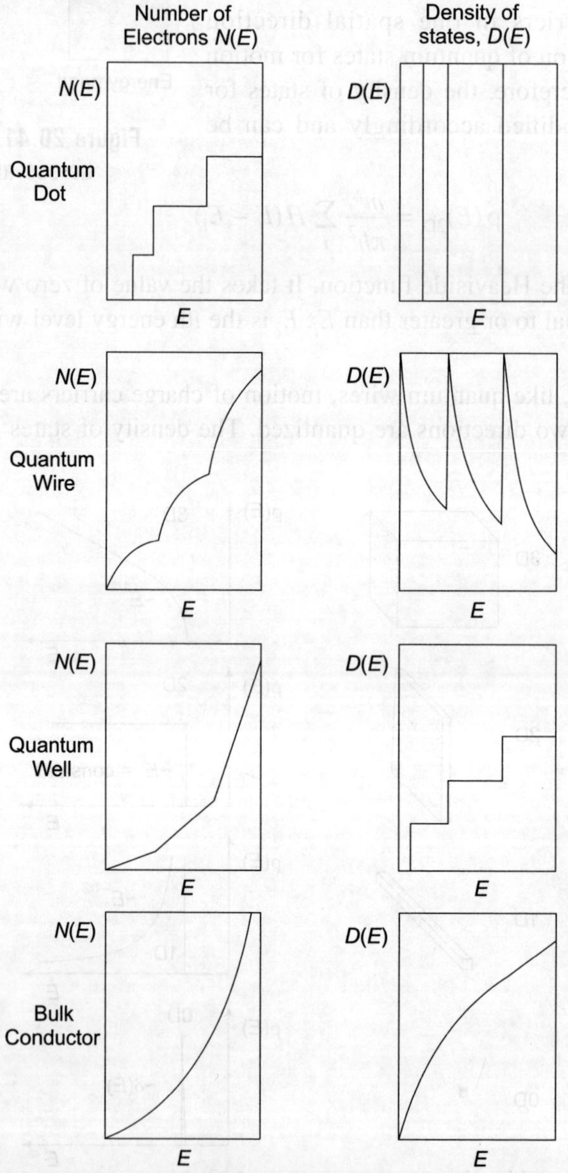


**Figure 20.43** Number of electrons $N(E)$ (left side) and density of states $D(E)$ (right side) plotted against the energy for four quantum structures in the square well — Fermi gas approximations

occur for more than one arrangement of confined states. To account for this, a second factor $n_i(E)$, is introduced.

For 0D structures, like quantum dots, all three spatial directions are quantized and the expression for the density of states follows a δ-function as

$$\rho(E)_{0D} = \sum_i 2\delta(E - E_i) \quad ...(5)$$

where the summation is over all of the $i$ quantum states. The densities of states for 3D, 2D, 1D and 0D electron systems for one quantized level are schematically shown in Fig. 20.42.

## 20.32 PROPERTIES DEPENDENT ON DENSITY OF STATES

We have discussed the density of states, $D(E)$, of conduction electrons, and have shown that it is strongly affected by the dimensionality of a material. Phonons or quantized lattice vibrations also have a density of states $D_{PH}(E)$ that depends on the dimensionality, and like its electronic counterpart, it influences some properties of solids, but our principal interest is in the density of states, $D(E)$, of the electrons. In this section we mention some of the properties of solids that depend on the density of states, and we describe some experiments for measuring it.

The specific heat of a solid ($C$) is the amount of heat that must be added to it to raise its temperature by one degree Celsius (centigrade). The main contribution to this heat is the amount that excites lattice vibrations, and this depends on the phonon density of states, $D_{PH}(E)$. At low temperatures there is also a contribution to the specific heat ($C_{el}$) of a conductor arising from the conduction electrons, and this depends on the electronic density of states at the Fermi level: $C_{el} = \pi^2 D(E_F) k_R^2 T/3$, where $k_B$ is the Boltzmann constant.

The susceptibility ($\chi = M/H$) of a magnetic material is a measure of the magnetization ($M$) or magnetic moment per unit volume that is induced in the material by the application of an applied magnetic field, $H$. The component of the susceptibility arising from the conduction electrons, called the *Pauli susceptibility*; is given by the expression, $\chi_{el} = \mu_B^2 D(F_F)$, where $\mu_B$ is the unit magnetic moment called the *Bohr magneton,* and is hence $\chi_{el}$ characterized by its proportionality to the electronic density of states, $D(E)$, at the Fermi level, and its lack of dependence on the temperature.

When a good conductor such as aluminium is bombarded by fast electrons with just enough energy to remove an electron from a particular Al inner-core energy level, the vacant level left behind constitutes a hole in the inner-core behind. An electron from the conduction band of the aluminium can fall into the vacant inner core level to occupy it, with the simultaneous emission of an X-ray in the process. The intensity of the emitted X-radiation is proportional to the density of states of the conduction electrons because the number of electrons with each particular energy that jumps down to fill the hole is proportional to $D(E)$. Therefore, a plot of the emitted X-ray intensity versus the X-ray energy ($E$) has a shape very similar to a plot of $D(E)$ versus $E$. These emitted X-rays for aluminium are in the energy range from 56 to 77 eV.

Some other properties and experiments that depend on the density of states and can provide information on it are photoemission spectroscopy, Seebeck effect (thermopower) measurements, the concentrations of electrons and holes in semiconductors, optical absorption determination of the dielectric constant, the Fermi contact term in nuclear magnetic resonance (NMR), the de Haas-van Alphen effect, the superconducting energy gap, and Josephson junction tunnelling in superconductors. It would take us too far a field to discuss any of these topics. Experimental measurements of these various properties permit us to determine the form of the density of states, $D(E)$, both at the Fermi level ($E_F$) and over a broad range of temperature.

## 20.33 SINGLE-ELECTRON TUNNELLING

We have been discussing quantum dots, wires, and wells in isolation, such as the ones depicted in Figs. 20.30–20.34. To make them useful, they need coupling to their surroundings, to each other, or to electrodes that can add or subtract electrons from them. Figure 20.40 shows an isolated quantum dot or island coupled through tunnelling to two leads, a source lead that supplies electrons, and a drain lead that removes electrons for use in the external circuit.

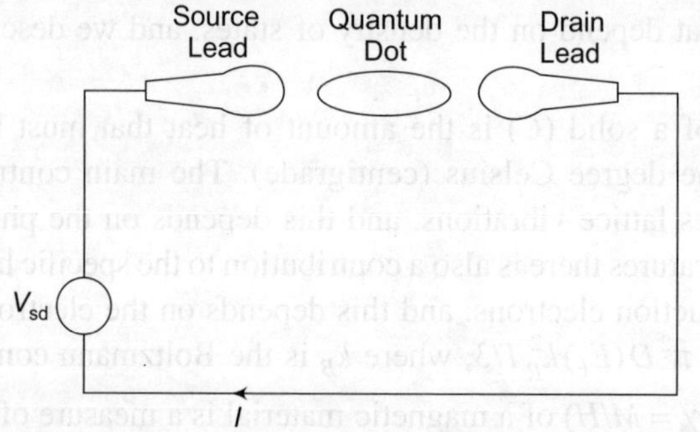


**Figure 20.44** Quantum dot coupled to an external circuit through source and a drain leads

The applied voltage ($V_{sd}$) causes direct current ($I$) to flow, with electrons tunnelling into and out of the quantum dot. In accordance with Ohm's law, $V = IR$, the current flow ($I$) through the circuit of Fig. 20.44 equals the applied source-drain voltage ($V_{sd}$) divided by the resistance ($R$), and the main contribution to the value of $R$ arises from the process of electron tunnelling from source to quantum dot, and from quantum dot to drain. Figure 20.41 shows the addition to the circuit of a capacitor-coupled gate terminal. The applied gate voltage ($V_g$) provides a controlling electrode or gate that regulates the resistance ($R$) of the active region of the quantum dot, and consequently regulates the current flow ($I$) between the source and drain terminals. This device, as described, functions as a voltage-controlled or field-effect-controlled transistor, commonly referred to as an FET. For large or macroscopic dimensions the current flow is continuous, and the discreteness of the individual electrons passing through the device manifests itself by the presence of current fluctuations or shot noise. Our present interest is in the passage of electrons, one by one, through nanostructures based on circuitry of the type sketched in Fig. 20.45.

For an FET-type nanostructure the dimensions of the quantum dot are in the low nanometer range, and the attached electrodes can have cross sections comparable in size. For disk and spherical shaped dots of radius $r$, the capacitance is given by

$$C = 8\varepsilon_0 \left(\frac{\varepsilon}{\varepsilon_0}\right) r \quad \text{disk} \qquad ...(1)$$

$$C = 4\pi\varepsilon_0 \left(\frac{\varepsilon}{\varepsilon_0}\right) r \quad \text{sphere} \qquad ...(2)$$

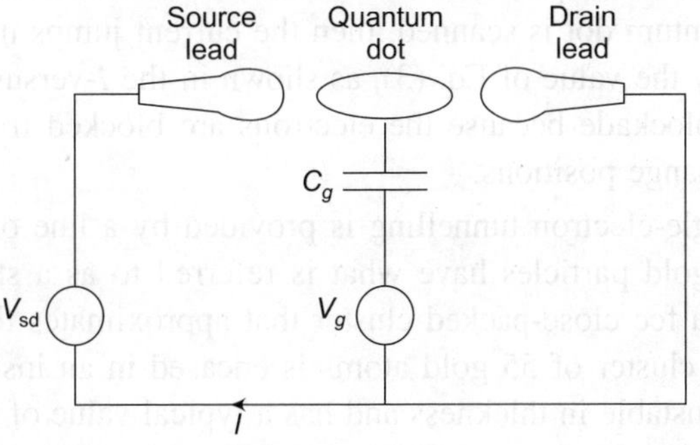


**Figure 20.45** Quantum dot coupled through source and a drain leads to an external circuit containing an applied bias voltage $V_{sd}$, with an additional capacitor-coupled terminal through which the gate voltage $V_g$ controls the resistance of the electrically active region

where $\varepsilon/\varepsilon_0$ is the dimensionless dielectric constant of the semiconducting material that forms the dot, and $\varepsilon_0 = 8.8542 \times 10^{-12}$ F/m is the dielectric constant of free space. For the typical quantum dot material GaAs we have $\varepsilon/\varepsilon_0 = 13.2$, which gives the very small value $C = 1.47 \times 10^{-18}\, r$ farad for a spherical shape, where the radius ($r$) is in nanometres. The electrostatic energy ($E$) of a spherical capacitor of charge ($Q$) is changed by the amount $\Delta E \sim eQ/C$ when a single electron is added or subtracted, corresponding to the change in potential, $\Delta V = \Delta E/Q$

$$\Delta V = e/C \cong 0.109/r \text{ volts} \qquad ...(3)$$

where $r$ is in nanometres. For a nanostructure of radius, $r = 10$ nm, this gives a change in potential of 11 mV, which is easily measurable. It is large enough to impede the tunnelling of the next electron.

Two quantum conditions must be satisfied for observation of the discrete nature of the single-electron charge transfer to a quantum dot. One is that the capacitor single-electron charging energy $e^2/2C$ must exceed the thermal energy $k_BT$ arising from the random vibrations of the atoms in the solid, and the other is that the Heisenberg uncertainty principle be satisfied by the product of the capacitor energy ($e^2/2C$) and the time, $\tau = R_TC$ required for charging the capacitor.

$$\Delta E\ \Delta t = \left(\frac{e^2}{2C}\right)(R_T C) > h \qquad ...(4)$$

where $R_T$ is the tunnelling resistance of the potential barrier. These two tunnelling conditions correspond to

$$\frac{e^2}{2C} \gg k_B T \qquad ...(5a)$$

$$R_T \gg \frac{h}{e^2} \qquad ...(5b)$$

where $h/e^2 = 25.813\ \mathrm{k\Omega}$ is the quantum of resistance. When these conditions are met and the voltage across the quantum dot is scanned, then the current jumps in increments every time the voltage changes by the value of Eq. (3), as shown in the *I*-versus-*V* characteristics. This is called a Coulomb blockade because the electrons are blocked from tunnelling except at the discrete voltage change positions.

An example of single-electron tunnelling is provided by a line of ligand-stabilized $Au_{55}$ nanoparticles. These gold particles have what is referred to as a structural magic number of atoms arranged in a fcc close-packed cluster that approximates the shape of a sphere of diameter 1.4 nm. The cluster of 55 gold atoms is encased in an insulating coating called a ligand shell that is adjustable in thickness and has a typical value of 0.7 nm. Single-electron tunnelling can take place between two of these $Au_{55}$ ligand-stabilized clusters when they are in contact, with the shell acting as the barrier for the tunnelling.

## 20.34 SINGLE ELECTRON TRANSISTOR

The invention of the transistor by John Bardeen and William Shockley in 1948 triggered a new era in electronics. Originally designed simply to emulate the vacuum tube, scientists soon found that this solid-state device could offer much more. The great potential of the transistor for speed, miniaturization and reliability has been fully exploited since well controlled materials such as pure single-crystal silicon became available. According to the latest "road-map" for the microelectronics industry, microchips containing one billion transistors and operating with a clock cycle of a billionth of a second will be on the market just a few years into the new millennium. As transistors continue to shrink, researchers have devised the so-called single-electron tunnelling transistor–a device that exploits the quantum effect of tunnelling to control and measure the movement of single electrons. Experiments have shown that charge does not flow continuously in these devices but in a quantized way. Indeed, single-electron transistors are so sensitive to charge that they can be used as extremely precise electrometers. The single electron transistor or SET is a type of switching device that uses controlled electron tunnelling to amplify current. A SET is made from two tunnel junctions that share a common electrode. A tunnel junction consists of two pieces of metal separated by a very thin (~ 1 nm) insulator (Fig. 20.46). The only way for electrons in one of the metal electrodes to travel to the other electrode is to tunnel through the insulator. Since tunnelling is a discrete process, the electric charge that flows through the tunnel junction flows in multiples of *e*, the charge of a single electron.

The electrical behaviour of the tunnel junction depends on how effectively the barrier transmits electron-waves, which decreases exponentially with its thickness, and on the number of electron-wave modes that impinge on the barrier, which is given by the area of the tunnel junction divided by the square of the electron wavelength. A single-electron transistor exploits the fact that the transfer of charge through the barrier becomes quantized when the junction is made sufficiently resistive.

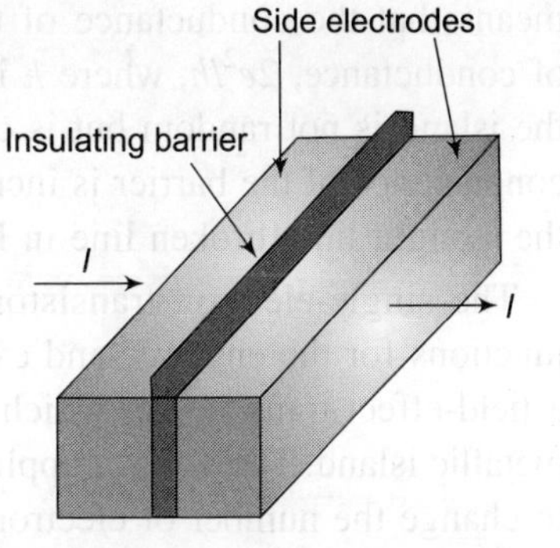


**Figure 20.46** Schematic structure of a tunnel junction consists of a very thin insulating barrier sandwiched between two metallic electrodes

This quantization process is shown particularly clearly in a simple system known as a single-electron box [Figs. 20.47 (a) and (b)]. If a voltage source charges a capacitor, *C*, through an ordinary resistor, the charge on the capacitor is strictly proportional to the voltage and shows no sign of charge quantization. But if the resistance is replaced by a tunnel junction, the metallic area between the capacitor plate and one side of the junction forms a conducting "island" surrounded by insulating materials. In this case, the transfer of charge onto the island becomes quantized as the voltage increases, leading to the so-called Coulomb staircase [Fig. 20.47 (c)].

This Coulomb staircase is only seen under certain conditions. First, the energy of the electrons due to thermal fluctuations must be significantly smaller than the Coulomb energy, which is the energy needed to transfer a single electron onto the island when the applied voltage is zero.

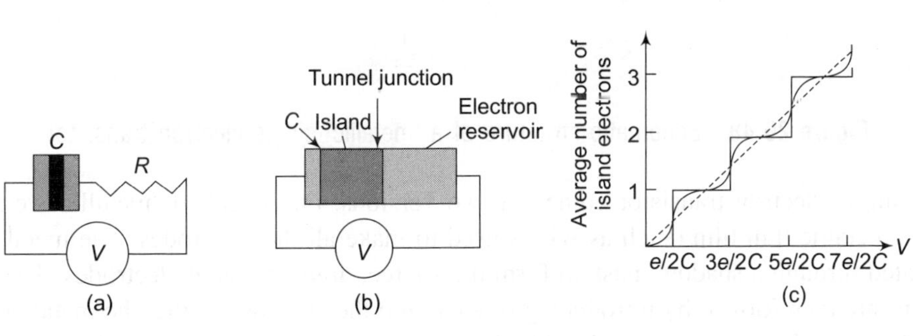


**Figure 20.47** (a) Charging of a capacitor through a resistor is linear, (b) 'electron in a box'-replacing *R* by a tunnel junction shows quantization of charge, and (c) staircase voltage showing charge quantization. The steps are sharper for more resistive barriers and lower temperature

This Coulomb energy is given by $e^2/2C$, where $e$ is the charge of an electron and $C$ is the total capacitance of the capacitor and the tunnel junctions. Secondly, the tunnel effect itself should be weak enough to prevent the charge of the tunnelling electrons from becoming delocalized over the two electrodes of the junction, as happens in chemical bonds. This means that the conductance of the tunnel junction should be much less than the quantum of conductance, $2e^2/h$, where $h$ is Planck's constant. In a single-electron box, the charge on the island is not random but is controlled by the applied voltage. As the temperature or the conductance of the barrier is increased, the steps become rounded and eventually merge into the straight line [broken line in Fig. 20.47 (c)] typical of an ordinary resistor.

The single-electron transistor can be viewed as an electron box that has two separate junctions for the entrance and exit of single electrons (Fig. 20.48). It can also be viewed as a field-effect transistor in which the channel is replaced by two tunnel junctions forming a metallic island. The voltage applied to the gate electrode affects the amount of energy needed to change the number of electrons on the island.

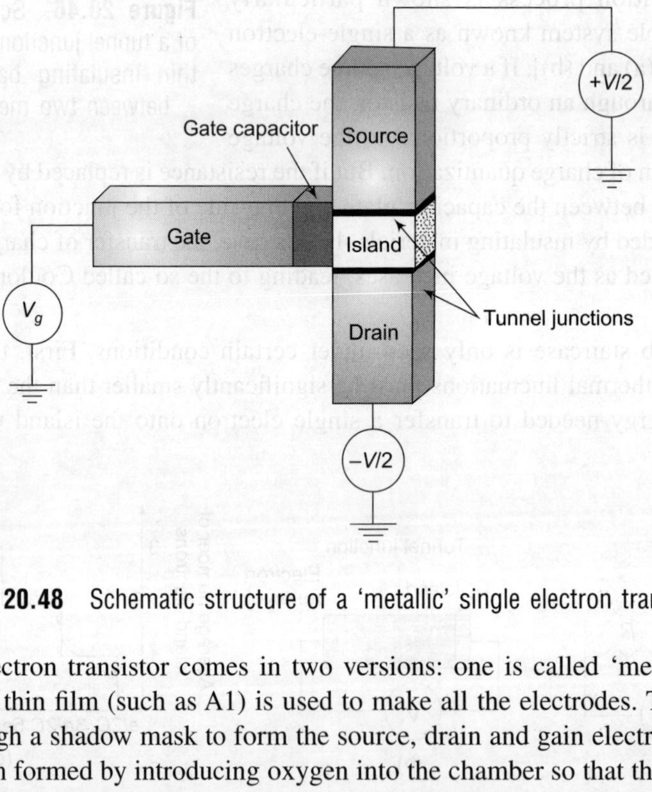


**Figure 20.48** Schematic structure of a 'metallic' single electron transistor

The single-electron transistor comes in two versions: one is called 'metallic' version in which a metallic thin film (such as A1) is used to make all the electrodes. The metal is first evaporated through a shadow mask to form the source, drain and gain electrodes. The tunnel junctions are then formed by introducing oxygen into the chamber so that the metal becomes coated by a thin layer of its natural oxide. Finally, a second layer of the metal – shifted from the first by rotating the sample – is evaporated to form the island. In the other version, called 'semiconducting', the source, drain and island are usually obtained by "cutting" regions in a two-dimensional electron gas formed at the interface between two layers of semiconductors such as A1GaAs and GaAs [Figs. 20.49 (a)] and (b)]. In this case, the conducting regions are defined by metallic electrodes patterned on the top semiconducting layer. Negative voltages applied to these electrodes deplete the electron gas just beneath them, and the depleted regions can be made sufficiently narrow to allow tunnelling between the source, island and drain.

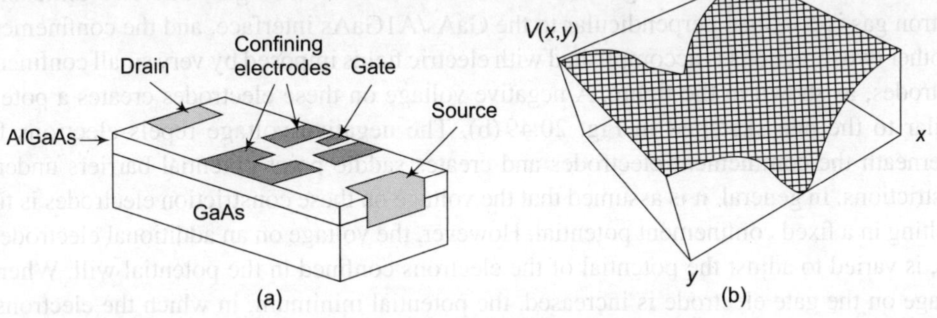


**Figure 20.49** (a) Schematic structure of a 'semiconducting' single electron transistor, and (b) sketch of the electrostatic potential energy distribution of a moving electron at the GaAs/AlGaAs interface

Moreover, the electrode that shapes the island can be used as the gate electrode. In this case of semiconducting version of the single-electron transistor, the island is often referred to as a quantum dot, since the electrons in the dot are confined in all three directions. Recently, various research groups showed that quantum dots can behave like artificial atoms. Indeed, it has been possible to construct a new periodic table that describes dots containing different numbers of electrons.[7]

For the operation of a single-electron transistor, an electron needs to hop onto the island, and for that, its energy must be equal to the Coulomb energy, $e^2/2C$. When both the gate voltages and the bias voltage are zero, electrons do not have enough energy to enter into the island and current does not flow. As the bias voltage between the source and drain is increased, an electron can pass through the island when the energy in the system reaches the Coulomb energy. This effect is known as the Coulomb blockade, and the critical voltage needed to transfer an electron onto the island, equal to $e/C$, is called the Coulomb gap voltage. In the case of no electrons in the island, if the bias voltage is kept below the Coulomb gap voltage and the gate voltage is increased, the energy of the system gradually increases. On the other hand, in the case of one excess electron in the island, the energy of the system gradually decreases. At the gate voltage corresponding to the point of maximum slope on the Coulomb staircase, both of these cases equally qualify as the lowest energy states of the system. This lifts the Coulomb blockade, allowing electrons to tunnel into and out of the island. Thus, gate voltage controls the 'open' or 'close' of the single-electron transistor.

In the case of 'semiconducting' single electron transistor, as shown in Fig. 20.49 (a), a semiconducting material (in this case GaAs) is separated from metal electrodes by an insulator (here A1GaAs), doped with Si, which donates electrons. These electrons fall into the GaAs, because their energy is lower in the latter material. The resulting positive charge on the Si atoms creates a potential that holds the electrons at the GaAs/A1GaAs interface, creating a two-dimensional electron gas. The source and drain contacts allow for the movement of electrons

[7] Leo Kouwenhoven and Charles Marcus, Quantum dots, *Physics World*, pp. 35-39, (1998).

from an external circuit through the two-dimensional electron gas. The two-dimensional electron gas is confined perpendicular to the GaAs/A1GaAs interface, and the confinement in the other two directions is accomplished with electric fields imposed by very small confinement electrodes, as shown in the figure. A negative voltage on these electrodes creates a potential similar to the one sketched in Fig. 20.49 (b). The negative voltage repels electrons from underneath the confinement electrodes and creates saddle point potential barriers under the constrictions. In general, it is assumed that the voltage on these constriction electrodes is fixed, resulting in a fixed confinement potential. However, the voltage on an additional electrode, the gate, is varied to adjust the potential of the electrons confined in the potential will. When the voltage on the gate electrode is increased, the potential minimum, in which the electrons are trapped, becomes deeper. This causes the number of trapped electrons to increase. However, unlike a conventional transistor, in which the charge increases continuously, the charge in the trap increases in discrete steps, and this is reflected in the conductance between source and drain, which shows a periodical variation with respect to gate voltage. A calculation of the capacitance between the gate-electrode and the droplet of confined electrons shows that the voltage between two peaks or two valleys is just that necessary to add one electron to the droplet. The name "single electron transistor" comes from the observation that the transistor turns on and off again every time a single electron is added to it.

The equivalent electrical circuit of the SET structure shown in Fig. 20.48 is shown in Fig. 20.50 (a). It is a double tunnel junction in series and biased with an ideal voltage source $V_b$, with a gate voltage ($V_g$) capacitively coupled to the island, which controls the current flow. The charges on junction 1, junction 2 and the whole island can be written as

$$q_1 = C_1 V_1 \qquad ...(1)$$

$$q_2 = C_2 V_2 \qquad ...(2)$$

with

$$V_b = V_1 + V_2 \qquad ...(3)$$

$$C_\Sigma = C_1 + C_2 + C_g \qquad ...(4)$$

and

$$q = q_2 - q_1 + q_0 = -ne + q_0 + C_g(V_g - V_2) \qquad ...(5)$$

with

$$n = n_1 - n_2 \qquad ...(6)$$

where $n_1$ and $n_2$ are the numbers of electrons tunnelled through the first junction (and entering the island) and second junctions (after leaving the island) respectively. $q_0$ is the background charge, which is induced by stray capacitances and impurities near the island, arises from various circuit connections and elements and are always present (not shown in the figure). The importance of $V_g$ is that $q_0$ can be changed at will because the gate additionally polarizes the island and the voltages across the junctions will be

$$V_1 = \frac{(C_2 + C_g) V_b - C_g V_g + ne - q_0}{C_\Sigma} \qquad ...(7)$$

and

$$V_2 = \frac{C_1 V_b + C_g V_g - ne + q_0}{C_\Sigma} \qquad ...(8)$$

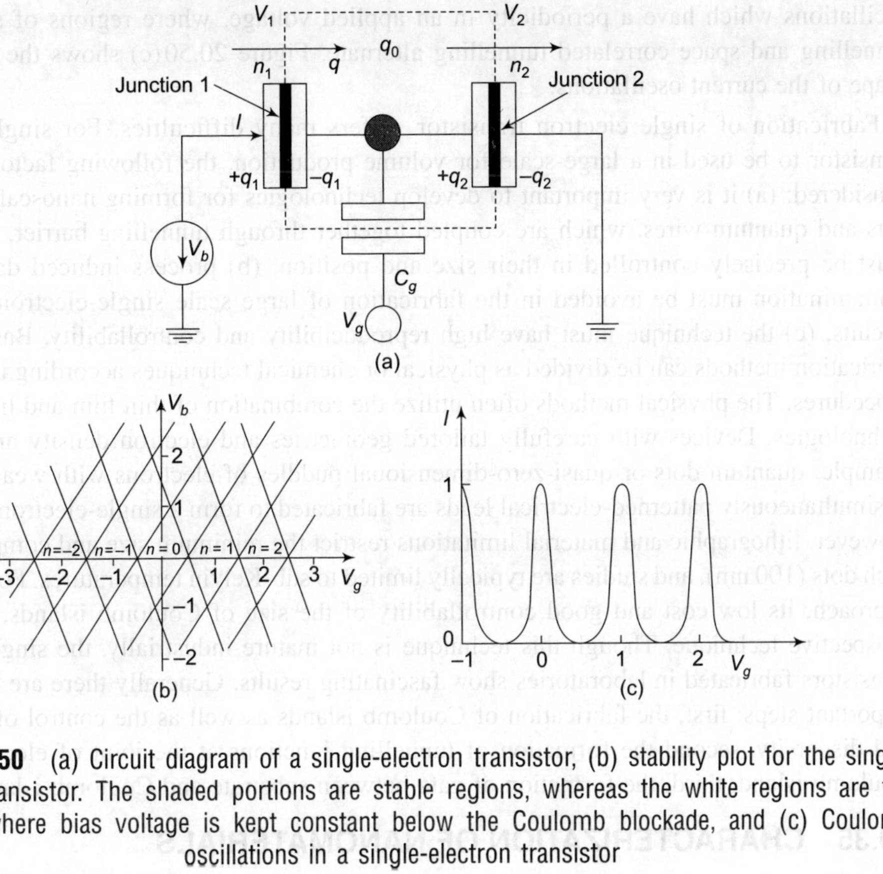

**Figure 20.50** (a) Circuit diagram of a single-electron transistor, (b) stability plot for the single-electron transistor. The shaded portions are stable regions, whereas the white regions are the portions where bias voltage is kept constant below the Coulomb blockade, and (c) Coulomb oscillations in a single-electron transistor

The calculations of electrostatic energy and free energy include the energy stored in $C_1$, $C_2$ and $C_g$ and the work done by the gate voltage. The change in free energy after a tunnel event in junctions one and two becomes

$$\Delta F_1^{\pm} = \frac{e}{C_\Sigma}\left(\frac{e}{2} \pm ((C_2 + C_g)V_b - C_g V_g + ne - q_0)\right) \quad ...(9)$$

$$\Delta F_2^{\pm} = \frac{e}{C_\Sigma}\left(\frac{e}{2} \pm (V_b C_1 + C_g V_g - ne + q_0)\right) \quad ...(10)$$

At zero temperature only transitions with a negative change in free energy ($\Delta F_1 < 0$ or $F_2 < 0$) are allowed. These conditions may be used to generate a stability plot in the $V_b - V_g$ plane, as shown in Fig. 20.50(b). The shaded regions correspond to stable regions with an integer number of excess electrons on the island, neglecting any non-zero background charge. If the gate voltage is increased, and the bias voltage is kept constant below the Coulomb blockade ($V_b < e/C_\Sigma$), which is equivalent to a cut through the stable regions in the stability plot, parallel to the $x$-axis, the current will oscillate with a period of $e/C_g$. These are Coulomb oscillations which have a periodicity in an applied voltage, where regions of suppressed tunnelling and space correlated tunnelling alternate. Figure 20.50(c) shows the qualitative shape of the current oscillations.

Fabrication of single electron transistor suffers many difficulties. For single electron transistor to be used in a large-scale for volume production, the following factors must be considered: (a) it is very important to develop technologies for forming nanoscale quantum dots and quantum wires, which are coupled together through tunnelling barrier, and which must be precisely controlled in their size and position, (b) process induced damage and contamination must be avoided in the fabrication of large scale single-electron transistor circuits, (c) the technique must have high reproducibility and controllability. Basically, the fabrication methods can be divided as physical or chemical techniques according to the main procedures. The physical methods often utilize the combination of thin film and lithographic technologies. Devices with carefully tailored geometries and electron density are got. For example, quantum dots or quasi-zero-dimensional puddles of electrons with weak coupling to simultaneously patterned-electrical leads are fabricated to form a single-electron transistor. However, lithographic and material limitations restrict the minimum size and composition of such dots (100 nm), and studies are typically limited to sub-Kelvin temperatures. For chemical approach, its low cost and good controllability of the size of Coulomb islands, made it a prospective technique. Though this technique is not mature industrially, the single-electron transistors fabricated in laboratories show fascinating results. Generally there are three most important steps: first, the fabrication of Coulomb islands as well as the control of their size and dispersity, second the formation of tunnelling junctions at the joint of electrodes and Coulomb island, third, the formation of gate between substrate and Coulomb islands.

## 20.35 CHARACTERIZATION OF NANOMATERIALS

### 20.35.1 Scanning Probe Microscopy (SPM)

**(a) Introduction.** The recent tremendous advancement of materials science in general and nanostructured materials in particular is due to the discovery and development of modern characterization equipment. The most elegant and useful of these equipment is the scanning tunnelling microscope (STM) and atomic force microscope (AFM). In the early 1980s two IBM scientists, Binning & Rohrer [1], developed a new technique for studying surface structure—scanning tunnelling microscopy (STM). The invention was quickly followed by the development of a whole family of related techniques which, together with STM, may be classified in the general category of scanning probe microscopy (SPM) techniques. Of these later techniques, the most important is atomic force microscopy (AFM). The development of these techniques has without doubt been the most important event in the surface science field in recent times, and opened up many new areas of science and engineering at the atomic and molecular level.

**(b) Basic Principles of SPM Techniques.** All the techniques are based upon scanning a probe (typically called the *tip* in STM, since it literally is sharp metallic tip) just above a surface whilst monitoring some interaction between the probe and the surface (Fig. 20.51).

- **STM** is the tunnelling current between a metallic tip and a conducting substrate which ate in very close proximity but not actually in physical contact.
- **AFM** is the van der Waals forces between the tip and the surface; this may be either the short range repulsive force (in contact-mode) or the longer range attractive force (in non-contact mode).

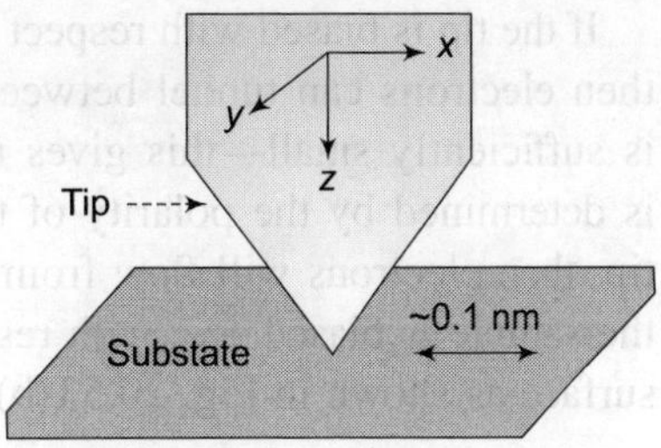


**Figure 20.51** Tip–substrate geometry of SPM

For the techniques to provide information on the surface structure at the atomic level (which is what they are capable of doing!):

1. The position of the tip with respect to the surface must be very accurately controlled (to within about 0.1 Å) by moving either the surface or the tip.
2. The tip must be very sharp—ideally terminating in just a single atom at its closest point of approach to the surface.

The attention paid to the first problem and the engineering solution to it is the difference between a good microscope and a not so good microscope—it need not worry us here, sufficient to say that it is possible to accurately control the relative positions of the tip and the surface by ensuring good vibrational isolation of the microscope and using sensitive piezoelectric positioning devices.

Tip preparation is a science in itself—having said that, it is largely serendipity which ensures that one atom on the tip is closer to the surface then all others. Let us look at the region where the tip approaches the surface in greater detail—the end of the tip will almost invariably show a certain amount of structure, with a variety of crystal facets exposed—and if we now go down to the atomic scale, there is a reasonable probability of ending up with a truly atomic tip as explained Figs. 20.52 (a), (b) and (c).

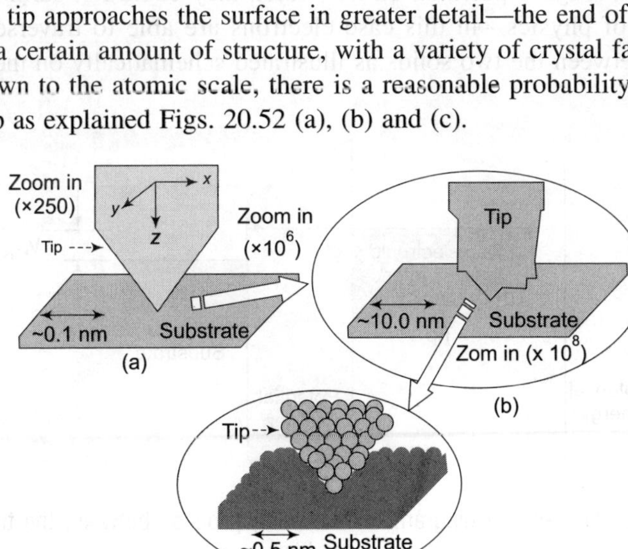


**Figure 20.52** (a) Schematic representation of the tip—sample geometry, (b) magnified versions of tip, and (c) at higher magnification

If the tip is biased with respect to the surface by the application of a voltage between them then electrons can tunnel between the two, provided the separation of the tip and surface is sufficiently small—this gives rise to a tunnelling current. The direction of current flow is determined by the polarity of the bias. If the sample is a biased – ve with respect to the tip, then electrons will flow from the surface to the tip as shown in Fig. 20.53 (a), whilst if the sample is biased + ve with respect to the tip, then electrons will flow from the tip to the surface as shown in Fig. 20.53 (b).

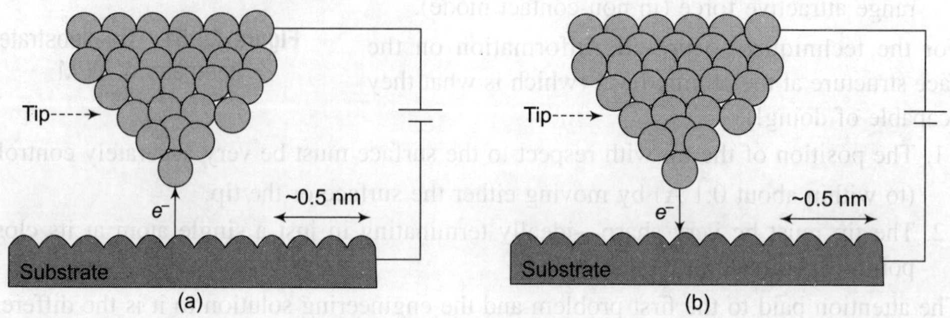


**Figure 20.53** Electron tunnelling due to sample biasing under (a) negative bias, and (b) positive bias

The name of the technique arises from the quantum mechanical tunnelling-type mechanism by which the electron can move the tip and substrate. Quantum mechanical tunnelling permits particles to tunnel through a potential barrier which they could not surmount according to the classical laws of physics—in this case electrons are able to traverse the classically-forbidden region between the two solids as illustrated schematically on the energy diagram in Fig. 20.54.

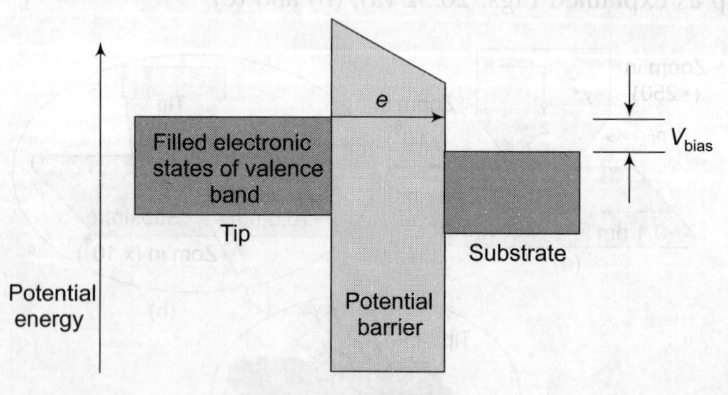


**Figure 20.54** Schematic energy diagram of tunnelling process between the tip and the sample under bias

This is an over-simplistic model of the tunnelling that occurs in STM but it is a useful starting point for understanding how the technique works. In the model, the probability of tunnelling is exponentially dependent upon the distance of separation between the tip and the surface: the tunnelling current is therefore a very sensitive function of this separation. Imaging of the surface topology may then be carried out in one of the two ways:

1. In constant height mode [in which the tunnelling current is monitored as the tip is scanned parallel to the surface, Fig. 20.55 (a)].
2. In constant current mode [in which the tunnelling current is maintained constant as the tip is scanned across the surface, Fig. 20.55 (b)].

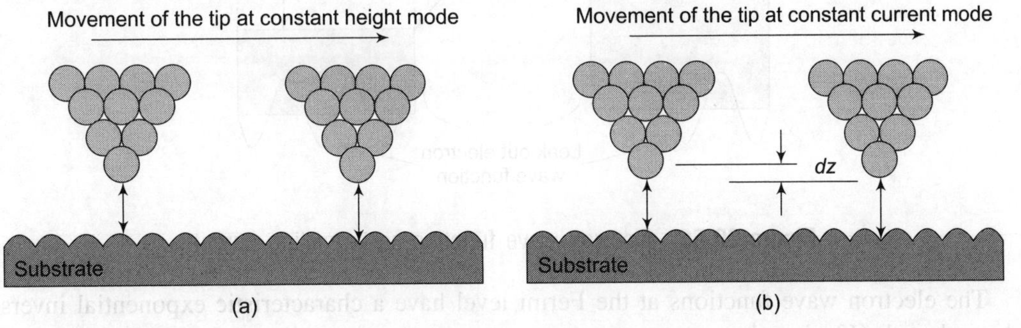


**Figure 20.55** (a) Constant height mode operation of the STM tip, (b) constant current mode operation of the STM tip

If the tip is scanned at what is nominally a constant height above the surface, then there is actually a periodic variation in the separation distance between the tip and the surface atoms. At one point the tip will be directly above the hollow sites on the surface and the tunnelling current will be much smaller [Fig. 20.55 (a)]. A plot of the tunnelling current vs. tip position therefore shows a periodic variation which matches that of the surface structure—hence it provides a direct 'image' of the surface (and by the time the data has been processed it may even look like a real picture of the surface!).

In practice, however, the normal way of imaging the surface is to maintain the tunnelling current constant whilst the tip is scanned across the surface. This is achieved by adjusting the tip's height above the surface so that the tunnelling current does not vary with the lateral tip position. In this mode, the tip will move slightly upwards as it passes over a surface atom, and conversely, slightly in towards the surface as it passes over a hollow, as shown in Fig. 20.55 (b). The image is then formed by plotting the tip height (strictly, the voltage applied to the $z$-piezo) vs. the lateral tip position.

## 20.36 THE DETAILS OF STM

The understand the theory of how STM works, it is vital to know what is tunnelling current, and how it relates to all the experimental observations.

### 20.36.1 Tunnelling Current

Tunnelling current originate from the wavelike properties of particles (electrons, in the case) in quantum mechanics. When an electron is incident upon a vacuum barrier with potential energy larger than the kinetic energy of the electron, there is still a non-zero probability that it may traverse the forbidden region and reappear on the other side of the barrier. It is shown by the leak out electron wave function in Fig. 20.56.

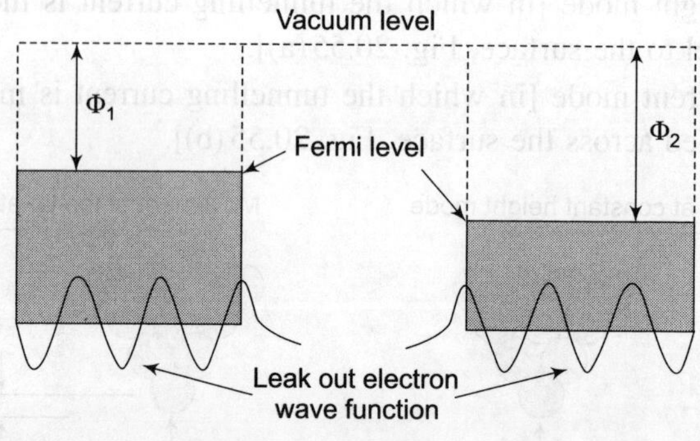


**Figure 20.56** Electron wave function as tunnelling curent

The electron wave functions at the Fermi level have a characteristic exponential inverse decay length ($K$) given by

$$K = \frac{\sqrt{2m\Phi}}{\hbar} \quad \text{...(1)}$$

where, $m$ is mass of electron, $\Phi$ is the tunnelling barrier height or the average work function of the tip and the sample. If two conductors are so close that their leak out electron wave functions overlap, as shown in Fig. 20.57, then a small voltage, $V$, applied between the tip and the sample will allow the overlapped electron wave function to permit quantum mechanical tunnelling and a current, $I$ will flow across the vacuum gap. At low voltage and temperature

$$I \propto e^{-2Kd} \quad \text{...(2)}$$

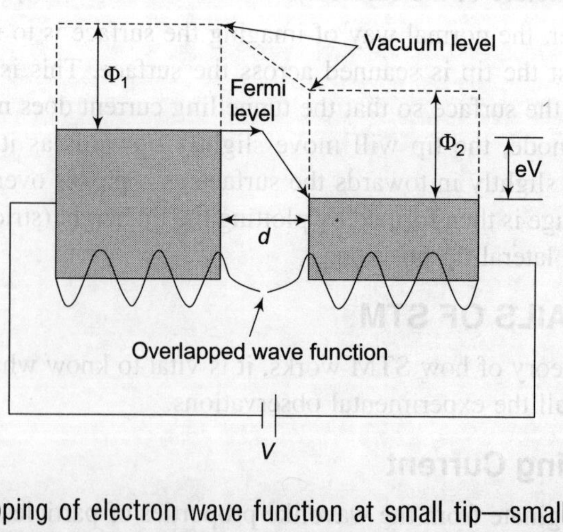


**Figure 20.57** Overlapping of electron wave function at small tip—small distance, under bias

where $d$ is the distance between the tip and the sample. If the distance increases by 1 Å, the current flow decreases by an order of magnitude, so that sensitivity to vertical distance is terribly high. As the tip scans across the surface, it gives atomic resolution image as shown in Fig. 20.58. This is the STM image of Si (111) = 7 × 7 surface, and the white spots represent the position of the atoms.

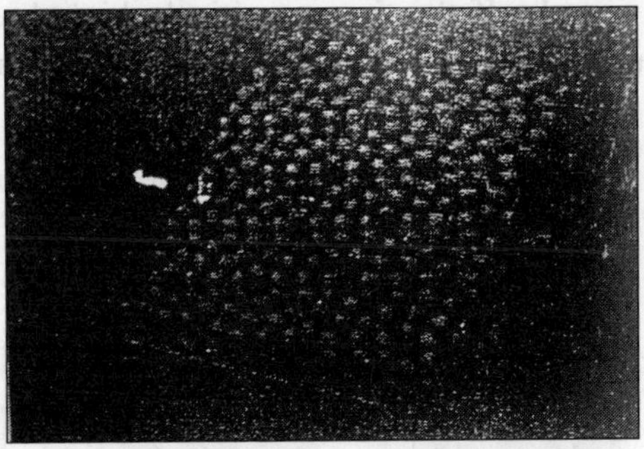

**Figure 20.58** STM image of a Si (111) surface, the white spots represent the position of the atoms

It is to be noted that the STM does not probe the nuclear position directly, but rather it is a probe of the electron density, so STM images do not always show the position of the atoms, and it depends on the nature of the surface and the sign and magnitude of the tunnelling current.

## 20.36.2 Local Barrier Height

Equation (2) obviously shows that the current exponentially depends on both the gap distance and the local barrier height. Change of current might be due to corrugation of the surface or to the locally varying local barrier height. The two effects can be separated by the relationship

$$\Phi \propto \left(\frac{\partial \ln I}{\partial d}\right) \qquad ...(3)$$

The tip is vibrating vertically and the current is measured, in theory, the local barrier height can be calculated. Again, if the tip rasters the surface, map of local barrier height can be obtained. The local barrier image also contains topographic features, but some questions related to the local barrier image also contains topographic features, but some questions related to the local barrier have so far remained unexplained. So, extra care needs to be taken in performing such experiments.

## 20.36.3 Local Density of States (LDOS)

Density of states (DOS) represents the amount of electrons exist at specific values of energy. The tunnelling conductance, σ (or $I/V$) is proportional to the LDOS.

$$\sigma \propto \rho(r_0, E) \quad ...(4)$$

where $\rho(r, E)$ is the local density of states of the sample. Keeping the gap distance constant, measure the current change with respect to the bias voltage can probe the LDOS of the sample. Moreover, changing the polarity of bias voltage can get local occupied and unoccupied states. When the tip is negatively biased electrons tunnel from the occupied states of the tip to the unoccupied states of the sample. If the tip is positively biased, electrons tunnel from the occupied states of the sample to the unoccupied states of the tip, as shown in Fig. 20.59.

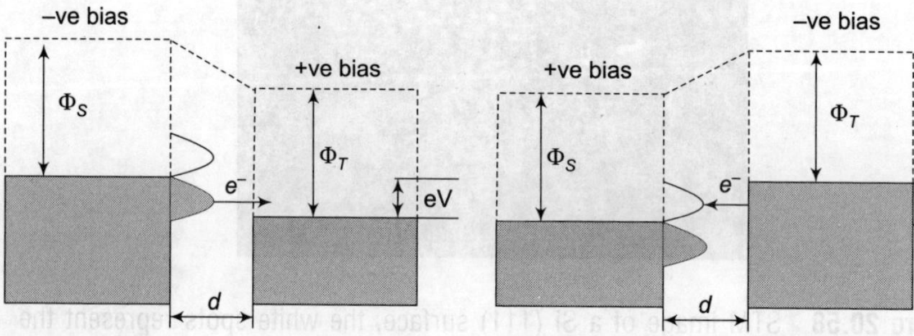


**Figure 20.59** Schematic diagram of the tip–sample biasing indicating the electrin transfer due to LDOS

Figure 20.60 shows the spectra for Si (111)–7 × 7 surface. The bottom spectrum [Fig. 20.60 (a)] is the area averaged tunnelling conductance measured by STM, and the top spectrum [Fig. 20.60 (b)] is the surface states spectrum measures photoemission and inverse photoemission. Both spectra show similar features.

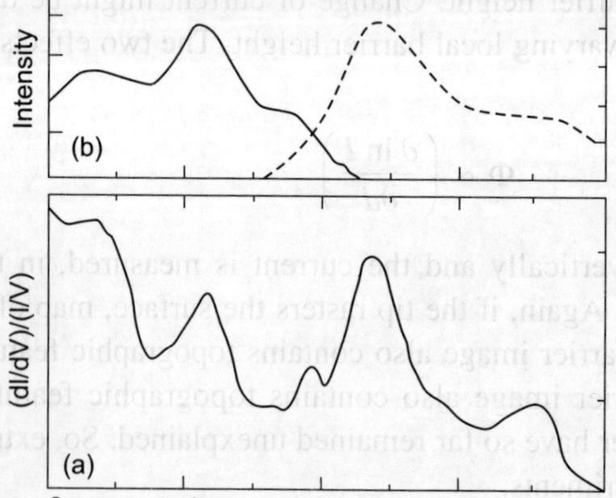


**Figure 20.60** Spectra for Si (111) surface for (a) tunnelling conductance measured by STM and (b) photoemission spectrum

### 20.36.4 The Detailed Calculations of Tunnelling Current

Based on the Bardeen's tunnelling current formalism

$$I = \frac{2e\pi}{\hbar} \sum_{TS} f(E_T)[1 - f(E_S + eV)] |M_{TS}|^2 \delta(E_T - E_S) \quad ...(5)$$

where $$f(E) = \exp\left(\left[\frac{(E-\mu)}{k_B T} + 1\right]^{-1}\right) \quad ...(6)$$

where $f(E)$ is the Fermi function, $V$ is the applied voltage, $M_{TS}$ the tunnelling matrix element between the tip wave function, $\psi_S$, $\delta(E)$ is the delta function, $E_i$ the energy of a state in the absence of tunnelling ($i = T, S$), and $\mu$ is the chemical potential, $k_B$ is Boltzmann constant, and $T$ is temperature.

The matrix element ($M_{TS}$) is obtained from

$$M_{TS} = -\frac{\hbar^2}{2m} \int dS\, (\psi_T^* \nabla \psi_S - \psi_S \nabla \psi_T^*) \quad ...(7)$$

where $m$ is the free electon mass, the integral is taken over a surface lying entirely within the vacuum barrier region between the tip and the sample,* denote the complex conjugate, and $\nabla$ is given by

$$\nabla = i\frac{\partial}{\partial x} + j\frac{\partial}{\partial y} + k\frac{\partial}{\partial z} \quad ...(8)$$

For small tunnelling voltage and low temperature, $f(E)$ can be expanded to obtain

$$I = \frac{2\pi}{\hbar} e^2 V \sum_{TS} [M_{TS}]^2 \delta(E_T - E_F)\delta(E_S - E_F) \quad ...(9)$$

Carrying out the integral in one dimension (plane wave) by taking $\psi_T^*$ and $\psi_S^*$ as the wave function of the tip and the sample, respectively, gives the explicit exponential dependence of the tunnelling current as a function of tip–sample separation. In particular, if the tip and sample are identical materials, the wave functions in the vacuum gap can be written as (according to the tip–sample geometry shown in Fig. 20.61)

$$\psi_S(r_0) = \psi_S^* e^{-Kr_0} \quad ...(10)$$

and $$\psi_T(r_0) = \psi_T^* e^{-K(d-r_0)} \quad ...(11)$$

where $K$ is the inverse decay length

$$K = \frac{\sqrt{2m\Phi}}{\hbar} \quad ...(12)$$

with $\Phi$, the local barrier height or average work function. Therefore, the tunnelling current exhibits an exponential dependence on the separation ($d$) given by

$$I \propto |\psi_S^*|^2 |\psi_T^*|^2 e^{-2Kd} \quad ...(13)$$

With the spherical tip configuration shown in Fig. 20.61, $M_{TS}$ may be evaluated to assess the potential spatial resolution STM. The surface wave functions are expanded in plane waves parallel to the surface with decaying amplitude into the vacuum

$$\psi_S^* = \Omega_S^{-1/2} \sum_G a_G \exp [-(K^2 + |k + G|^2)^{1/2} r_0] \exp [i(k + G)x] \quad ...(14)$$

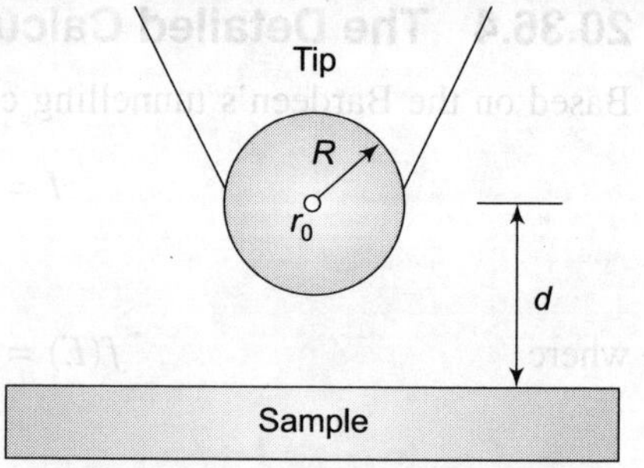


**Figure 20.61** Tip–sample geometry. The tip–sample distance is $d$ with respect to the centre of the tip atom (of radius, $R$) situated at a position $r_0$ with respect to the sample surface

where $\Omega_S$ is the sample volume, $K$ is inverse decay length, $k$ is the surface wave vector and $G$ is a two-dimensional reciprocal lattice vector of the surface. The first few factors of $a_G$ are typically of the order unity. The spherical tip ($R \gg k^{-1}$) wave function are expanded in similar form

$$\psi_T^* = \Omega_T^{-1/2} C_T K \, \mathrm{Re}^{KR} (K|r - r_0|)^{-1} e^{-K|r - r_0|} \quad ...(15)$$

where $\Omega_T$ is the probe volume and $R$ is the radius of curvature of the tip. The resulting matrix element is

$$M_{TS} = -\frac{\hbar^2}{2m} 4\pi \frac{(K \, \mathrm{Re}^{KR})}{(K\Omega^{1/2})} \psi_S(r_0) \quad ...(16)$$

where $r_0$ is the position of the centre of curvature of the tip. The tunnelling current is

$$I = \frac{32\pi^3 e^2 V}{\hbar} \Phi^2 D_T(E_F) \, R^2 K^{-4} \, e^{2Kr} \sum_S |\psi_S(r_0)|^2 \, \delta(E_S - E_F) \quad ...(17)$$

where $D_T(E_F)$ is the density of states per unit volume. Diving the current by bias voltage, the tunnelling conductance, $\sigma$ can be obtained as

$$\sigma \propto \rho(r_0, E) \quad ...(18)$$

where
$$\rho(r_0, E) = \sum_S |\psi_S(r_0)|^2 \, \delta(E_S - E_F) \quad ...(19)$$

Differentiating the logarithmic of the tunnelling gives the local barrier height, $\Phi$ as

$$F \infty \left(\frac{\partial \ln I}{\partial d}\right)^2 \quad ...(20)$$

## 20.37 APPLICATIONS OF NANOTECHNOLOGY

Since nanomaterials possess unique, beneficial chemical, physical, and mechanical properties, they can be used for a wide variety of applications. These applications include, but are not limited to the following:

• ***Next – Generation Computer Chips***

The microelectronics industry has been emphasizing miniaturization, whereby the circuits, such as transistors, resistors, and capacitors, are reduced in size. By achieving a significant reduction in their size, the microprocessors, which contain these components, can run much faster, thereby enabling computations at far greater speeds. However, there are several technological impediments to these advancements, including lack of the ultra-fine precursors to manufacture these components; poor dissipation of tremendous amount of heat generated by these microprocessors due to faster speeds; short mean time to failures (poor reliability), etc. Nanomaterials help the industry break these barriers down by providing the manufacturers with nanocrystalline starting materials, ultra-high purity materials, materials with better thermal conductivity, and longer-lasting, durable interconnections (connections between various components in the microprocessors).

• ***Better Insulation Materials***

Nanomaterials synthesized by the sol-gel technique results in foam-like structure called an "*aerogel*". These aerogels are porous and extremely lightweight; yet, they can withstand 100 times their weight. Aerogels are composed of three-dimensional, continuous networks of particles with air (or any other fluid, such as a gas) trapped at their interstices. Since they are porous and air is trapped at the interstices, aerogels are currently being used for insulation in offices, homes, etc. They are also being used as materials for "smart" windows, which darken when the sun is too bright and they lighten themselves, when the sun is not shining too brightly.

• ***Phosphors for High-Definition TV***

The resolution of a television, or a monitor, depends greatly on the size of the pixel. These pixels are essentially made of materials called "phosphors", which glow when struck by a stream of electrons inside the cathode ray tube (CRT). The resolution improves with a reduction in the size of the pixel, or the phosphors. Nanocrystalline zinc selenide, zinc sulphide, cadmium sulphide, and lead telluride synthesized by the sol-gel technique are candidates for improving the resolution of monitors.

• ***Low-Cost Flat-Panel Displays***

By synthesizing nanocrystalline phosphors, the resolution of display devices can be greatly enhanced, and the manufacturing costs can be significantly reduced. Also, the flat-panel displays constructed out of nanomaterials possess much higher brightness and contrast than the conventional ones owing to their enhanced electrical and magnetic properties.

• ***Tougher and Harder Cutting Tools***

Cutting tools made of nanocrystalline materials, such as tungsten carbide, tantalum carbide, tantalum carbide, and titanium carbide, are much harder, much more wear-resistant, erosion-resistant, and last longer than their conventional (large-grained) counterparts.

• ***Elimination of Pollutants***

Nanocrystalline materials possess extremely large grain boundaries relative to their grain size. Hence, nanomaterials are very active in terms of their chemical, physical, and mechanical properties. Due to their enhanced chemical activity, nanomaterials can be used as catalysts to react with such noxious and toxic gases as carbon monoxide and nitrogen oxide in automobile catalytic converters and power generation equipment to prevent environmental pollution arising from burning gasoline and coal.

• ***High Energy Density Batteries***

Nickel-metal hydride (Ni-MH) batteries made of nanocrystalline nickel and metal hydrides are envisioned to require far less frequent recharging and to last much longer because of their large grain boundary (surface) area and enhanced physical, chemical, and mechanical properties.

• ***High-Power Magnets***

The magnets made of nanocrystalline yttrium-samarium-cobalt grains possess very unusual magnetic properties due to their extremely large surface area. Typical applications for these high-power rare-earth magnets include quieter submarines, automobile alternators, land-based power generators, motors for ships, ultra-sensitive analytical instruments, and magnetic resonance imaging (MRI) in medical diagnostics.

• ***High-Sensitivity Sensors***

A carbon monoxide sensor made of zirconium oxide (zirconia) uses its chemical stability to detect the presence of carbon monoxide. In the event of carbon monoxide's presence, the oxygen atoms in zirconium oxide react with the carbon in carbon monoxide to partially reduce zirconium oxide. This reaction triggers a change in the sensor's characteristics, such as conductivity (or resistivity) and capacitance. The rate and the extent of this reaction are greatly increased by a decrease in the grain size. Hence, sensors made nanocrystalline materials are extremely sensitive to the change in their environment. Typical applications for sensors made out of nanocrystalline materials are smoke detectors, ice detectors on aircraft wings, automobile engine performance sensor, etc.

• ***Automobiles with Greater Fuel Efficiency***

The engine cylinders (liners) are currently being envisioned to be coated with nanocrystalline ceramics, such as zirconia and alumina, so that they retain heat much more efficiently and result in complete and efficient combustion of the fuel.

• ***Aerospace Components with Enhanced Performance Characteristics***

One of the key properties required of the aircraft components is the fatigue strength, which decreases with the component's age. By making the components out of stronger materials, the life of the aircraft is greatly increased. The fatigue strength increases with a reduction in the grain size of the material. Nanomaterials provide such a significant reduction in the grain size over conventional materials that the fatigue life is increased by an average of 200–300%. Furthermore, components made of nanomaterials are stronger and can operate at higher temperatures; aircraft can fly faster and more efficiently (for the same amount of aviation fuel). In spacecraft, elevated-temperature strength of the material is crucial because the components (such as rocket engines, thrusters, and vectoring nozzles) operate at much higher temperatures than aircraft and higher speeds. Nanomaterials are perfect candidates for spacecraft applications, as well.

- ***Longer-Lasting Satellites***

Nanomaterials, such as nanocrystalline tungsten-titanium diboride-copper composite, are potential candidates for enhancing ignitor's life and performance characteristics in satellites.

- ***Longer-Lasting Medical Implants***

Nanocrystalline silicon carbide (SiC) is a candidate material for artificial heart valves primarily due to its low weight, high strength, extreme hardness, wear resistance, inertness (SiC does not react with biological fluids), and corrosion resistance.

- ***Ductile, Machinable Ceramics***

Nanocrystalline ceramics, such as silicon nitride ($Si_3N_4$) and silicon carbide (SiC), have been used in automotive applications as high-strength springs, ball bearings, and valve lifters, because they possess good formability and machinability combined with excellent physical, chemical, and mechanical properties. They are also used as components in high-temperature furnaces. Nanocrystalline ceramics can be pressed and sintered into various shapes at significantly lower temperatures, whereas it would be very difficult, if not impossible, to press and sinter conventional ceramics even at high temperatures.

- ***Large Electrochromic* Display *Devices***

Nanocrystalline materials, such as tungstic oxide ($WO_3 . xH_2O$) gel are used in very large electrochromic display devices. The reaction governing electrochromism (a reversible coloration process under the influence of an electric field) is the double-injection of ions (or protons, H+) and electrons, which combine with the nanocrystalline tungstic acid to form tungsten bronze. These devices are primarily used in public billboards and ticker boards to convey information. Electrochromic devices are similar to liquid-crystal displays (LCD) commonly used in calculators and watches. However, electrochromic devices display information by changing colour when a voltage is applied. When the polarity is reversed, the colour is bleached. The resolution, brightness, and contrast of these devices greatly depend on the tungstic acid gel's grain size. Hence, nanomaterials are being explored for this purpose.

## 20.38 POTENTIAL RISKS OF NANOMATERIALS

The potential risks associated with the use of nanomaterials are as follows:

(i) First, nanomaterials massively damage the lungs. Ultra-fine particles from diesel machines, power plants and incinerators can cause considerable damage to human lungs. This is both because of their size (as they can get deep into the lungs) and also because they carry other chemicals including metals and hydrocarbons in with them.

(ii) Second, nanoparticles can get into the body through the skin, lungs and digestive system. This may help create free radicals that can cause cell damage. There is also concern that once nanoparticles are in the bloodstream, they will be able to cross the blood-brain barrier.

(iii) Third, the human body has developed a tolerance to most naturally occurring elements and molecules that it has contact with. It has no natural immunity to new substances and is more likely to find them toxic.

(iv) Fourth, the most dangerous nano-application use for military purposes is the nano-bomb that contains engineered self-multiplying deadly viruses that can continue to wipe out a community, country or even a civilization.

## EXERCISE

20.1 What are nanomaterials? What is meant by nanoscience and nanotechnology?

20.2 What is nanotechnology? What are its applications?

20.3 What are carbon nanotubes? What are their different types?

20.4 Explain the structure of carbon nanotubes. On the basis of structure, how CNTs can be classified?

20.5 What are the properties possessed by CNTs?

20.6 What is meant by CNT? Explain the commercialized application of CNTs in the area of nanoscience and nanotechnology.

20.7 Discuss the application of Schrödinger equation to a nanodot, nanosheet and nanowire.

20.8 A cubic quantum dot (QD) of GaAs has a = 7.0 nm. Calculate the lowest three energy levels. Calculate the same if the dimensions of the QD are $a$ = 10 nm, $b$ = 8 nm and $c$ = 12 nm, and $m^* = 0.067\ m_0$.

20.9 What are excitons? Why are excitonic effects more important in nanostructures than in bulk material?

20.10 "Bottom-up technique is more convenient for nano fabrication". Explain.

20.11 Describe the basic working principles of a "metallic" SET (single electron transistor).

20.12 Describe briefly the pulsed laser deposition method for nanomaterial thin film synthesis.

APPENDIX

# Physical Constants and Their SI Values

| *Constant* | *Symbol* | *Value* |
|---|---|---|
| 1. Velocity of Light | $c = 1/\sqrt{\mu_0 \varepsilon_0}$ | $2.9979 \times 10^8$ $ms^{-1}$ |
| 2. Permeability of Free Space | $\mu_0$ | $4\pi \times 10^{-7} = 1.2566 \times 10^{-6}$ $Hm^{-1}$ |
| 3. Permittivity of Free Space | $\varepsilon_0 \left(= \frac{1}{\mu_0 c^2}\right)$ | $8.8542 \times 10^{-12}$ FC |
| 4. Charge on Proton or Electron | $e$ | $\pm 1.6022 \times 10^{-19}$ C |
| 5. Rest mass of Electron | $m_e$ | $9.1096 \times 10^{-31}$ kg |
| 6. Rest mass of Proton | $m_p$ | $1.6725 \times 10^{-27}$ kg |
| 7. Rest mass of Neutron | $m_n$ | $1.6749 \times 10^{-27}$ kg |
| 8. Rest energy of Electron | $m_e c^2$ | 0.5110 MeV |
| 9. Rest energy of Proton | $m_p c^2$ | 938.25 MeV |
| 10. Rest energy of Neutron | $m_n c^2$ | 939.55 MeV |
| 11. Classical radius of Electron | $r_e = \frac{e^2}{4\pi\varepsilon_0 m_e c^2}$ | $2.81794 \times 10^{-15}$ m |
| 12. Proton-Electron mass ratio | $m_p/m_e$ | 1836.1090 |
| 13. Planck constant | $h \equiv 2\pi \hbar$ | $6.6262 \times 10^{-34}$ Js<br>($\hbar = 1.055 \times 10^{-34}$ Js) |
| 14. Boltzmann constant | $k_B \equiv \frac{R}{N_A}$ | $1.38.06 \times 10^{-23}$ $JK^{-1}$ |
| 15. Avogadro constant | $N_A$ | $6.0222 \times 10^{23}$ $mol^{-1}$ |
| 16. Universal gas constant | $R$ | 8.3144 $JK^{-1}$ $mol^{-1}$ |
| 17. Stefan Boltzmann constant | $\sigma = \frac{2\pi^5 k_B^4}{15h^3 c^3}$ | $5.6697 \times 10^{-8}$ |
| 18. Sommerfeld fine structure constant | $\alpha = \frac{e^2}{2\varepsilon_0 hc}$ | $7.2974 \times 10^{-3} = \frac{1}{137.036}$ |

| *Constant* | *Symbol* | *Value* |
|---|---|---|
| 19. Rydberg constant (Ground state energy of hydrogen) | $R_\infty = \dfrac{m_0 e^4}{8\epsilon_0^2 h^3 c}$ | $2.1799 \times 10^{-18}$ J |
| 20. Gravitational constant | $G$ | $6.6732 \times 10^{-11}$ Nm$^2$ kg$^{-2}$ |
| 21. Acceleration of free fall | $g$ | $9.8067$ ms$^{-2}$ |
| 22. Loschmidt number (number of molecules/m$^3$ of gas at STP) | $N_L$ | $2.6872 \times 10^{25}$ m$^{-3}$ |
| 23. Faraday constant | $F$ | $9.6487 \times 10^{-4}$ C mol$^{-1}$ |
| 24. Bohr radius of hydrogen ground state | $a_0 = \dfrac{4\pi\varepsilon_0 \hbar^2}{e^2 m_e}$ | 0.5292 Å (= $5.2918 \times 10^{-11}$ m) |
| 25. Compton wavelength | $\dfrac{h}{m_e c}$<br>$\hbar / m_e c$ | $2.4263 \times 10^{-12}$ m<br>$3.8616 \times 10^{-13}$ m |
| 26. Bohr magneton | $\mu_B = \dfrac{eh}{4\pi m}$ | $0.928 \times 10^{-23}$ A m$^2$<br>or $\dfrac{\text{J}}{\text{Wb.m}^2}$ |
| 27. Electron magnetic moment | $\mu_e = \dfrac{2.79\, e\hbar}{2\, m_p}$ | $1.41 \times 10^{-26}$ J/T. |
| 28. Nuclear magneton | $\mu_N = \dfrac{e\hbar}{2 m_p}$ | $5.0509 \times 10^{-27}$ J/T |
| 29. Proton magnetic moment | $\mu_p$ (= 2.7928 $\mu_N$) | $1.4106 \times 10^{-26}$ J/T |
| 30. Neutron magnetic moment | $\mu_n = \dfrac{1.91\, e\hbar}{2\, m_n}$ | $-9.66 \times 10^{-27}$ J/T |
| 31. Flux quantum | $\phi_0 = \dfrac{h}{2e}$ | $2.07 \times 10^{-15}$ Tm$^2$ |
| 32. Planck length | $(G\hbar/c^3)^{1/2}$ | $1.6 \times 10^{-35}$ m |
| 33. Planck mass | $(c\hbar/G)^{1/2}$ | $2.2 \times 10^{-8}$ kg |
| 34. Planck time | $(G\hbar^3/c^5)^{1/2}$ | $5.4 \times 10^{-44}$ s |

*Note:* N = newton (unit of force) K = kelvin (unit of temperature)
J = joule (unit of energy) T = tesla (unit of flux density)
C = Coulomb (unit of electric charge)
F = farad (unit of electric capacity) kg = kilogram
H = henry (unit of inductance) s = second
mol = mole
cal = calorie (unit of heat)

# Index

## D

## F

## I



## J

## N



## O



## P

## Q



## R

## S